CONSTRUCTION
Principles, Materials & Methods

CONSTRUCTION
Principles, Materials & Methods

HAROLD B. OLIN, A.I.A.
DIRECTOR OF ARCHITECTURAL AND CONSTRUCTION RESEARCH
UNITED STATES LEAGUE OF SAVINGS ASSOCIATIONS

JOHN L. SCHMIDT, A.I.A.
Vice President, The Institute of Financial Education

WALTER H. LEWIS, A.I.A.
Professor of Architecture, University of Illinois

THE INSTITUTE OF FINANCIAL EDUCATION
Chicago, Illinois
and
INTERSTATE PRINTERS AND PUBLISHERS, INC.
Danville , Illinois

FOURTH EDITION

CONSTRUCTION
Principles, Materials & Methods

Published jointly by The Institute of Financial Education, 111 East Wacker Drive, Chicago, Illinois 60601, and by the Interstate Printers and Publishers, Inc. Fourth Edition, first printing, June 1980.

Library of Congress Catalogue Card Number 80-82180. Printed in the United States of America.

INTRODUCTION

This edition of *Construction: Principles, Materials & Methods,* is the result of 16 years of research costing several million dollars. It is the end-result of an ongoing membership education program initiated by the United States League of Savings Associations in 1964 in the form of a looseleaf publication entitled the *Construction Lending Guide.* With the publication of this edition, the *Guide* program will be discontinued as an annual membership service. However, the research program aimed at continually expanding and updating the construction information in this book will continue.

The *Construction Lending Guide* was developed to provide members with up-to-date information on land planning, design, construction, appraising and construction lending for residential properties. The five-volume Guide was distributed to League members in split-ring binders to facilitate inserting new and revised pages. Because of the widespread interest in the *Construction Lending Guide* by groups outside the savings business, the *Guide* in 1966 also was published in hardbound form. Since then the contents have been continually expanded and updated.

This book consists of all of the published material in Volume 3 of the Guide as of this date. The fourth edition contains approximately 1250 pages and 45 topical sections. Each of the original sections has been reviewed by one or more experts in the field and revised accordingly. The new book thus contains not only 200 pages of new material, but 600 revised pages as well. Like the previous three editions, it is published at the initiative of the educational affiliate of the League, The Institute of Financial Education, in a joint venture with Interstate Printers and Publishers, Inc.

The United States League and The Institute gratefully acknowledge the painstaking work of the co-authors, the League's Architectural Department staff, and the many individuals and trade organizations who have cooperated in the preparation of this edition of *Construction: Principles Materials & Methods.*

On behalf of the United States League and The Institute we are pleased to make this book available to the savings business and the homebuilding industry.

William B. O'Connell
Executive Vice President
United States League of Savings Associations

Dale C. Bottom
Executive Vice President
The Institute of Financial Education

PREFACE

Prefaces to previous editions of this book spoke of pending technological breakthroughs, occasioned by the spiraling increases in the costs of labor, materials and land. If these conditions were true in the 1960's and 1970's, they are doubly true in the 1980's; but the wistful hope of sudden miraculous breakthroughs has been replaced by the realization that progress in homebuilding will not come in great leaps which will immediately revolutionize the industry, but as a steady evolutionary process.

The oil embargo of 1973 created another kind of awareness—that our fossil fuels and other natural resources are exhaustible, and that the cost of housing in the future will have to reflect both the higher price of energy and of scarcer resources. Given the expectation of continuing escalation in land and building costs, the imperative of the 1980's becomes to build housing more economically, more modestly and more durably.

Construction: Principles, Materials & Methods has anticipated the shift to a more long-term, life cycle, orientation. From the very beginning, this book has stressed *standards of quality* and *levels of suitability,* rather than the traditional building prescriptions in terms of uniformly applied minimum standards. By providing a basis of understanding, this book enables designers, builders and lenders to apply its recommendations intelligently. This flexibility encourages a more economical use of traditional materials and acceptance of new materials. Also, by focusing on the visible evidence of compliance with quality standards—such as grademarks, seals and labels—the book foreshadows the emerging trend toward greater consumer protection from short-lived, hazardous, or energy-wasteful products.

Special thanks are due Research Associate Christina Farnsworth, for her perseverence in obtaining expert reviews of old chapters and in revising these chapters accordingly; Ms. Farnsworth also prepared the new chapter on metrication. The contributions of Richard Laya, who prepared the new chapters on heat control, passive solar heating and pole foundations, deserve special mention. The list of valuable contributions would be incomplete without acknowledgement of the careful typing and preparation of graphics by Production Editor Betsy Pavichevich.

We would like to gratefully acknowledge also the many trade associations, professional societies, government agencies, manufacturers, and individuals who have provided valuable research material and illustrations, as well as text reviews and comments. Without their help this book would not have been possible; their names are listed at the beginning or end of each section.

Harold B. Olin, AIA
John L. Schmidt, AIA
Walter H. Lewis, AIA

THIS BOOK HAS A TWO-PART FORMAT:

MAIN TEXT

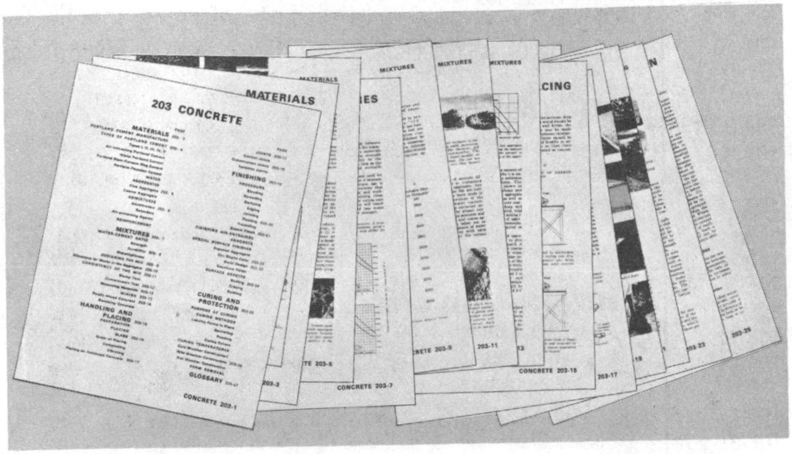

BASIC TERMINOLOGY AND PRINCIPLES, GENERAL DISCUSSIONS AND REFERENCE INFORMATION

WORK FILE

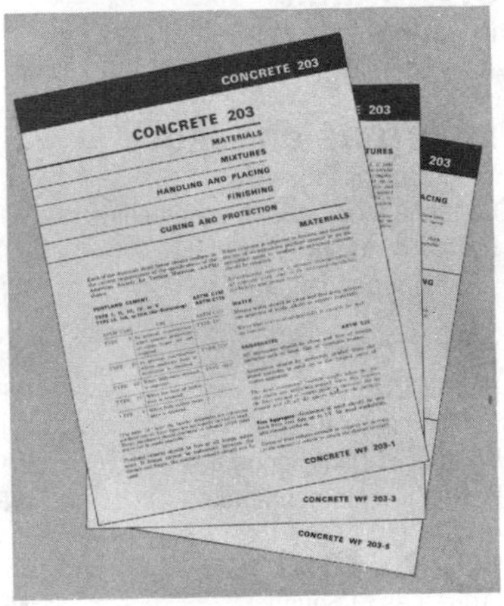

RECOMMENDATIONS AND INDUSTRY STANDARDS

Subjects are treated as SECTIONS and designated by three-digit numbers.

Where appropriate, a Main Text section is followed by a Work File which summarizes recommendations from the Main Text discussion and assembles industry specifications and standards.

Page numbers in both the Main Text headings and Work File consist of section numbers followed by consecutive numbers.

Generally, the Work File and Main Text headings are in identical order to permit easy reference from one to the other. Each page of the Work File is identifiable by a black title strip heading and the prefix WF before its page number.

TABLE OF CONTENTS

OF COMPLETED SECTIONS

101 CONSTRUCTION STANDARDS

CONSTRUCTION STANDARDS

INTRODUCTION

The homebuilding industry is made up of so many diverse interest groups, that it has not been possible to develop a single comprehensive set of criteria or standards acceptable to all concerned.

In selecting materials or determining the suitability of materials and methods, the specifier and the builder make reference to a variety of industry standards.

The preparation of standards may be the effort of a group of manufacturers, professionals or tradesmen seeking simplification and efficiency or the assurance of a minimum level of quality. Standards also may be the work of a governmental agency or other group interested in establishing minimum levels of safety and performance. Hence, standards take a variety of forms depending on source and purpose.

Some standards specify the constituents of a material, its physical properties and performance under stress and varying climatic conditions. Some may define terms, classify materials, state thicknesses, lengths and widths. Standards may spell out the methods of joining separate materials, of fabricating various parts of construction, or of assembling systems.

In general, construction standards have two basic objectives: (1) to establish levels of quality which may be recognized by the user, specifier, approver or buyer of a material, product or system; and (2) to standardize or simplify such variables as dimensions, varieties or other characteristics of specific products so as to minimize the variation in manufacture and use.

Construction standards are used by manufacturers, specifiers, consumers, communities and others. These standards may be used either separately or within *collections* of standards such as building codes, *HUD Minimum Property Standards,* architectural specifications and this *Construction Lending Guide.* These larger works have broader objectives than standards alone. Codes are concerned basically with minimum acceptable standards of public health, safety and welfare; *HUD Minimum Property Standards* establish minimum requirements of design and construction for the insurance of mortgage loans; architectural specifications establish requirements of materials, equipment, finish and workmanship as desired by an architect for a particular job. This *Construction* volume of the *Construction Lending Guide* has the purpose of providing a better understanding of the principles of construction, as well as gathering into a single reference source many recognized industry standards and specifications.

Many industry groups provide programs of certification in which materials and products are marked or labeled to indicate that they have been produced to meet particular standards of manufacture and/or performance. Certification, when based on independent inspection and testing according to industry specifications, can provide assurance that certain standards have been met.

This section discusses the types and purposes of standards used in residential construction and some of the many organizations active in establishing standards. Building codes are discussed in Section 602 Codes. Throughout the *Construction Lending Guide* reference is made to recognized industry specifications and standards, as well as to various certification programs.

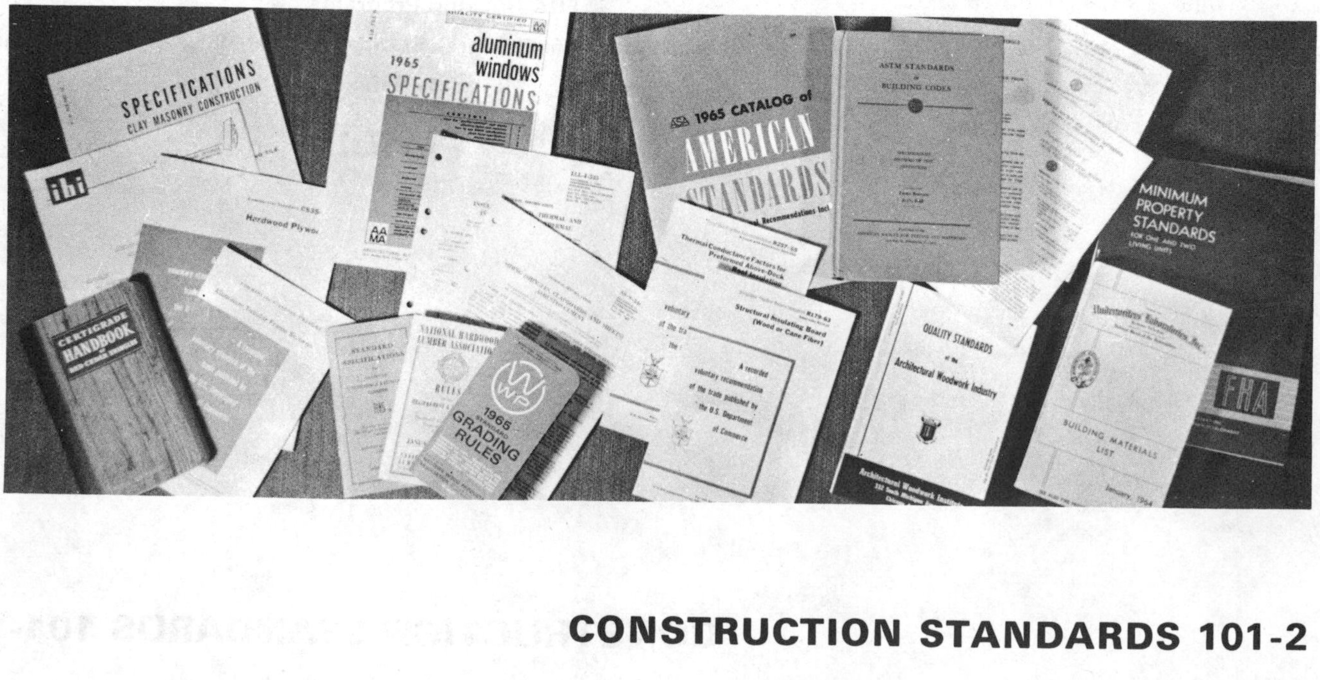

INDUSTRY STANDARDS

Manufacturers, tradesmen and suppliers, working through trade associations and industry societies, provide most of the active basic preparation of materials specifications and performance criteria. The Appendix to this section, page 101-10, lists some of the industry groups active in developing technical information and promulgating construction standards.

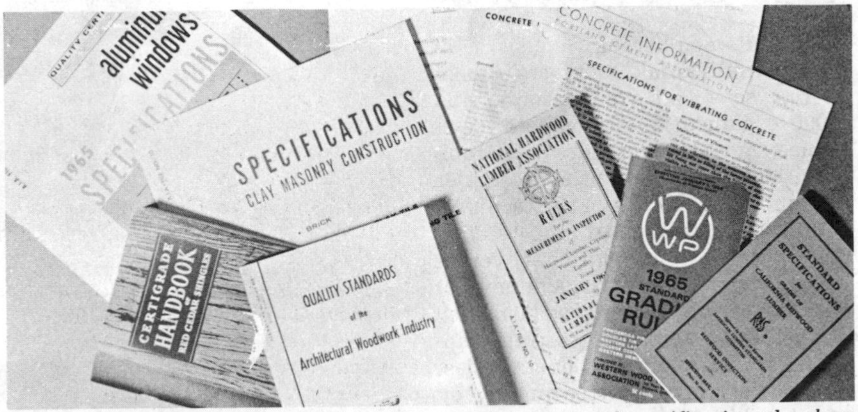

FIG. 1 The homebuilding industry uses many standards and specifications developed by trade associations for various materials, products and methods.

TRADE ASSOCIATIONS

A trade association is an organization of individual manufacturers or businesses engaged in the production or supply of materials and/or services of a similar nature. A basic function of a trade association is to promote the interests of its membership. Those interests generally are best served by the proper use of the groups' materials, products and services. The proper use of homebuilding materials and methods is guided largely by the development of suitable levels of quality for the manufacture, use and installation during construction. Some of the most important activities of trade associations are directed toward research into the use and improvement of materials and methods, and toward formulating specifications and performance *standards.* In many instances, trade associations also sponsor programs of *certification* in which labels, seals or other identifiable marks are placed on materials or products manufactured to particular standards or specifications.

Specifications—Standards

Specifications are clear and accurate descriptions of technical requirements of materials, products or services. They may state requirements for quality and the use of materials and methods to produce a desired product, system, application or finish. *Specifications become standards when they are adopted for use by a broad group of manufacturers, users or specifiers.* A few of the recognized industry standards are shown in Figure 1. Many more are referenced wherever appropriate throughout the *Construction Lending Guide.*

The American Plywood Association is a good example of a trade association active in developing standards. This association is supported by a large membership of manufacturers producing softwood plywood. The production of various types and grades of plywood requires the selection and classification of wood veneers according to strength and appearance. Producing plywood from these veneers demands careful manufacturing control of such factors as moisture content and adhesive type. The completed product, properly assembled, bonded and finished, must conform to a body of material specifications, manufacturing procedures and performance testing standards.

Plywood manufacturers, through their own research and combined efforts within the American Plywood Association, financially and technically support the research and development of criteria which the association formulates into industry specifications. The written specifications become the standards to which members of the association agree to produce plywood.

The standards developed by trade associations often are adopted by or used as a base from which other groups, such as the American National Standards Institute, may develop standards. For example, industry standards may be promulgated as ANSI Standards which are often incorporated into the *HUD Minimum Property Standards* or building codes.

Certification

Trade associations and industry groups also may provide assurance that established standards have been met in manufacture. Certification of quality may take the form of grademarks, labels or seals. For example, the American Plywood Association also maintains a continuing program of product testing during and after manufacture with grade *markings* applied directly to the plywood. The grademark is a visible statement that the specifications and standards of the American Plywood Association and appropriate Product Standard has been met.

The *Construction Lending Guide* will reference both the recognized industry specifications or *standards* appropriate to any material or product, as well as any broad industry-supported program of *certification.*

AMERICAN SOCIETY FOR TESTING AND MATERIALS (ASTM)

The American Society for Testing and Materials (ASTM) is an international, privately financed, nonprofit, technical, scientific and educational society. The objectives of the Society are "the promotion of knowledge of the materials of engineering, and the standardization of specifications and methods of testing."

ASTM membership consists of 11,000 individual members (engineers, scientists, educators) and 4,000 organizational members (companies, government agencies, universities) interested in the properties of materials. Some 110 technical committees formulate and recommend ASTM standards covering many types of materials; a number of administrative committees deal with publications, research, testing, consumer standards and other activities. The Society is supported mainly by membership dues, with some income from the sale of publications.

The Society has grown from the American Section of the International Society for Testing Materials, an organization formed in 1898 and incorporated in 1902 as the American Society for Testing Materials. In 1961, the name of the Society was changed to the American Society for Testing *and* Materials to emphasize its interest in basic information about materials.

Two general categories of information and publications are available from ASTM: (1) ASTM Standards which include definitions of terms, specifications and methods of test for materials and construction used throughout the industry (Fig. 2); and (2) data dealing with research and testing of materials—"Materials Research & Standards" published monthly, "Journal of Materials" published quarterly, and special technical publications (STP's and DP's) which cover symposia and collections of data.

ASTM Specifications are designated by the initials ASTM, followed by a code number and the year of last revision (Fig. 3). For example, ASTM C55-66T is the ASTM tentative specification for "Concrete Building Brick" as last revised in 1966.

Publications are sold directly by the American Society for Testing and Materials (1916 Race St., Philadelphia, Pa. 19103) or may be obtained through bookstores handling technical publications. Lists of prices and publications may be obtained by writing to the ASTM at the above address.

FIG. 2 ASTM Standards define terms and establish specifications and test methods for materials and their use.

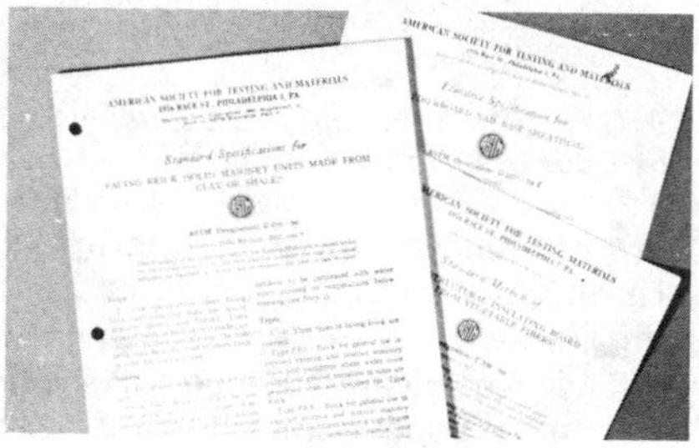

FIG. 3 Specifications are identified by name, followed by ASTM designation letter and number.

THE AMERICAN NATIONAL STANDARDS INSTITUTE (ANSI)

The American National Standards Institute (ANSI) is the name adopted by the United States of America Standards Institute (USASI) in October 1969. USASI was created in 1966 by the complete reorganization of the earlier standards organization, the American Standards Association (ASA). This restructuring provided for a Member Body Council (the former Standards Council of ASA), a Company Member Council and a Consumer Council.

The *Member Body Council*, comprised of approximately 140 national trade, professional and scientific associations, is responsible for establishing and maintaining procedures for the approval of standards as American National Standards.

The *Company Member Council* consists of several hundred industrial firms and representatives of labor and government. The Company Member Council and the *Consumer Council* provide membership on the Board of Directors and work closely with the Member Body Council in recommending areas of standardization deemed essential, and in reviewing standards.

ANSI is privately financed, its financial support being derived from voluntary membership dues and from the sale of the published American National Standards. These are national standards arrived at by common consent, are available for voluntary use, and often are incorporated in regulations and codes. The purposes of the American National Standards Institute are:

(1) To act as the national coordinating institute for voluntary standardization through which organizations may cooperate in reorganizing, establishing and improving standards.

(2) To further the voluntary standards movement as a means of advancing the national economy, benefiting public health, safety and welfare, and facilitating domestic and international trade and communications and understanding.

(3) To assure that the interests of the public (consumer, labor, industry and government) have appropriate protection and representation in standardization activity.

(4) To provide the means for determining the need for new standards, to assure activity by existing organizations competent to resolve the need, and to work toward establishing suitable groups for this purpose where such do not already exist—*but not to formulate standards itself.*

(5) To promote knowledge and voluntary use of standards.

(6) To stimulate the work of competent existing committees and organizations to formulate standards for recognition as American National Standards.

(7) To cooperate with federal, state and local governments in achieving optimum compatibility between government codes and standards and the voluntary standards of industry and commerce, and maximum common usage of American National Standards.

(8) To represent the United States in international standardization ac-

tivities concerned with civilian safety, trade and commerce, except where otherwise provided by treaty.

(9) To serve as a clearinghouse for information on standards and standardization work in the United States and of foreign countries.

ASA, the forerunner of ANSI and USASI, was founded during World War I to avoid duplication and waste in war production. In 1918, five leading American engineering societies —including The American Society of Mechanical Engineers, The American Society of Civil Engineers, and The American Society for Testing Materials—and three departments of the federal government—Commerce, War and Navy—formed the American Engineering Standards Committee to coordinate the development of national standards. This committee was reorganized in 1928 into the American Standards Association, which was later incorporated into USASI, and subsequently renamed ANSI.

Approximately 3,000 American National Standards have been developed and approved under the procedures of the Institute. These standards apply in the fields of engineering, industry, safety and consumer goods.

An American National Standard is designated by code number and date of last revision. For example, ANSI A12.1-1967 covers "Safety Requirements for Floor and Wall Openings, Railings and Toe Boards," which was originally adopted by ASA in 1932, later revised in 1967, and became an American National Standard upon adoption of the new name in 1969 (Fig. 4).

An American National Standards Institute catalog, price list and literature on ANSI and standards may be obtained by writing to the American National Standards Institute, 1430 Broadway, New York, N.Y. 10018.

UNDERWRITERS' LABORATORIES, INC. (UL)

Underwriters' Laboratories, Inc. (UL) is chartered as a nonprofit organization to establish, maintain and operate laboratories for the examination and testing of devices, systems and materials.

The main objectives of UL are: (1) "to determine the relation of various materials, devices, constructions, and methods to life, fire, and casualty hazards"; (2) "to ascertain, define, and publish Standards, Classifications, and Specifications for materials, devices, construction, and methods affecting such hazards"; and (3) to provide "other information tending to reduce and prevent loss of life and property from fire, crime, and casualty."

UL also contracts with manufacturers and other parties "for the examination, classification, testing, and inspection of buildings, materials, devices, and methods with reference to life, fire, and casualty hazards." Findings are circulated by various publications to insurance organizations, other interested parties and to the public.

The UL also contracts to provide product certification by attaching certificates or labels to examined,

FIG. 4 American National Standards formerly were known also as USA Standards and American Standards.

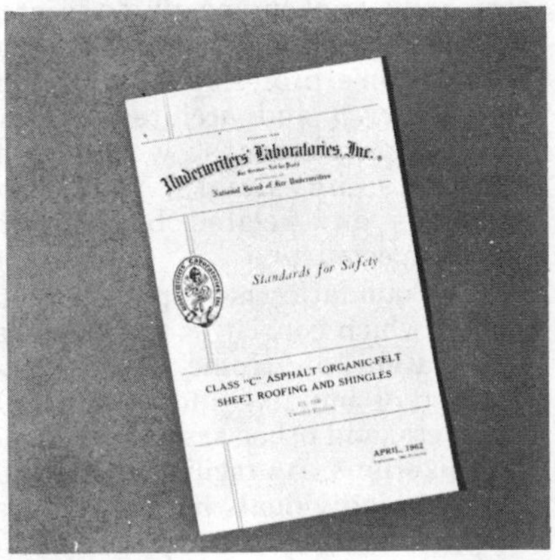

FIG. 5 UL Standards are concerned with the safety as well as the performance of products they certify.

tested or inspected materials, devices and products.

The Underwriters' Laboratories, Inc. originally was formed in 1894 and was subsidized by the stock insurance companies. Before the turn of the century, as new electrical devices and products came rapidly into the market, it became necessary to test and inspect them to insure public safety. The National Board of Fire Underwriters (now the American Insurance Association) organized and sponsored Underwriters' Laboratories, Inc. to meet this demand.

UL became self-supporting in about 1916. To sustain the testing program, UL contracts with the product submittor for testing, report and listing of devices, systems or materials on a time and material basis. The cost of the inspection service is provided for either by an annual fee or by service charges for labels, depending on the type of service. Materials and products carrying UL labels and certificates must meet published standards of performance and manufacture and are subjected to UL inspection during manufacture.

Although primarily interested in public safety, UL's policy is to list and label only products which *perform* their intended function. If a product does not perform with reasonable efficiency, even though it may be perfectly safe, it does not qualify for the UL label.

UL has a staff of approximately 1,300 (including 400 engineers and 300 inspectors) with four testing stations (listed below), and representatives in more than 200 cities throughout the country.

"Published Standards of Underwriters' Laboratories, Inc." is obtainable from the Chicago, Melville, N.Y. and Santa Clara, Cal. offices.

Underwriters' Laboratories, Inc. Standards are designated by the initials UL, followed by a code number (Fig. 5). No date of last revision is indicated by the number. For example, UL55B covers "Class 'C' Asphalt Organic-Felt Sheet Roofing and Shingles" and was last revised in 1962.

Lists of submittors whose products are listed and labeled are published annually (Fig. 6). These lists and other publications are available from the Underwriters' Laboratories, Inc., 207 E. Ohio St., Chicago, Ill. 60611.

NAHB RESEARCH FOUNDATION, INC.

The Research Foundation is a wholly owned subsidiary of the National Association of Home Builders. The objectives of the Foundation are:

(1) "To conduct and disseminate the results of research and development with respect to homes, apartments and light commercial structures for the purpose of lowering the cost and improving the quality of buildings constructed by the American homebuilding industry."

(2) "To conduct, for itself or by contract for others, tests and investigations into and on materials, products, systems and any and all other matters related to the design, construction, or occupancy of homes and other buildings."

(3) "To the extent permitted by law, to do any and all things necessary and proper to encourage lower construction costs and improved quality in the design and construction of residential and related structures."

(4) "To the extent permitted by law, to provide a system for labeling materials and other products and to grant a seal or certificate of quality or similar device."

The Research Foundation was created in 1965 and is the expansion of the NAHB Research Institute founded in 1952. The Foundation will continue the research activities initiated by the Institute. Those activities have included the design and construction of a number of "research houses" which incorporate many of the newest methods and materials available.

The Research Institute also developed the successful TAMAP system, which marked the first time that industrial engineering techniques were used to improve productivity in the design and construction of new homes.

As it became more evident that the work of the Research Institute was having far-reaching effects throughout the homebuilding industry, a number of manufacturers and related industry associations asked the Institute to perform research, development and testing services on a sponsored basis. Since then the Foundation has carried out product, standards and system research, development and evaluation studies for approximately 100 different clients. These included many leading companies in the building product and chemical industries, as well as trade associations.

The Research Foundation's activities are directed by a 17-man Board of Directors, 11 representing NAHB and six representing the fields of: engineering and science, education and labor, finance, building codes and government, architecture, and public interest. The organization is divided into four major operating portions—laboratory, industrial engineering, research houses and special studies. The Foundation's professional staff consists of civil and architectural engineers, architects, wood technologists, physicists, industrial engineers and related building industry specialists.

The Foundation is supported by its clients which consist of the National Association of Home Builders, a number of building industry manufacturers, and other associations and organizations. Its facilities are available to individual manufacturers, builders or trade associations, or to any combination of these on a group basis. The Research Foundation laboratory in Rockville, Md. is equipped with a broad range of facilities including equipment for ASTM standard tests, vibration study, acoustical measuring and temperature-humidity control testing.

Information is either published by the Foundation through NAHB, or is the sole property of the client. Published reports of research and development of standards are available from the NAHB Research Foundation, Inc., 627 Southlawn Lane, Rockville, Md. 20850.

THE AMERICAN SOCIETY OF HEATING, REFRIGERATING AND AIR CONDITIONING ENGINEERS (ASHRAE).

Since 1894, ASHRAE and its predecessor societies have pursued their objective of advancing the arts and sciences of heating, refrigeration, air conditioning and ventilation.

The Society's membership numbers over 30,000 members, in 22 international associate societies, conducts an extensive research program, publishes meeting transactions and establishes standards.

ASHRAE standards are established to assist industry and the public by offering a uniform method of testing for rating purposes, by suggesting safe practices in designing and installing heating, air conditioning and ventilating equipment, by providing proper definitions of the equipment, and providing other information which may serve to guide the industry. The creation of ASHRAE standards is determined by the need for them and conformance to them is voluntary.

ASHRAE standards are updated on a five-year cycle; each title is preceded by a number: the digits before the hyphen refer to the standard designation, the digits after the hyphen refer to the year of approval, revision or update. Thus, "ASHRAE Standard 90-75—Energy Conservation in New Building Design" describes a Standard of Designation 90, approved in 1975.

ASHRAE has developed standards not only in the traditional areas of heating, ventilating, and air conditioning equipment, but on such diverse subjects as fire safety in buildings, energy conservation, solar energy, pollution control, and ozone

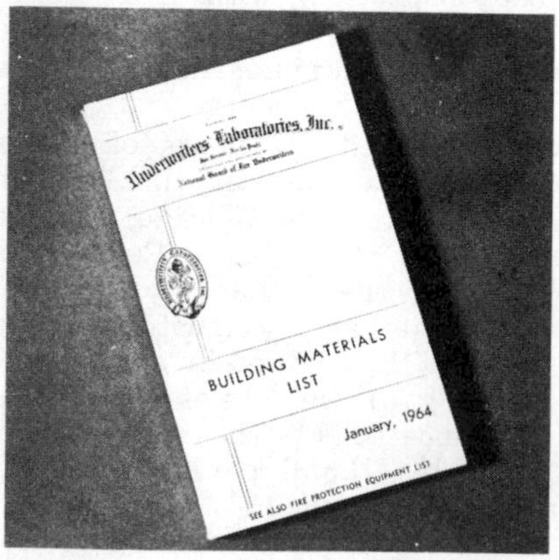

FIG. 6 The Underwriters' Laboratories, Inc., annually publishes a list of building materials which they certify.

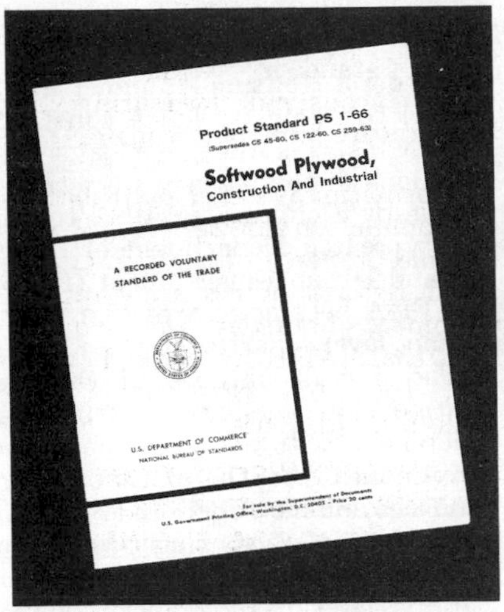

FIG. 7 PS1-66 is the first Product Standard published by the Department of Commerce.

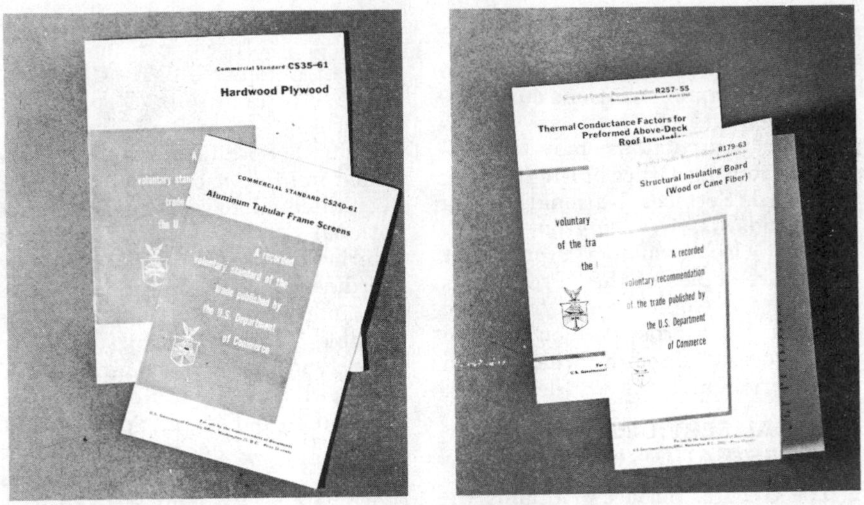

FIG. 8 Commercial Standards and Simplified Practice Recommendations developed under the direction of the Department of Commerce gradually are being replaced by Product Standards.

depletion.

Published reports of research, transactions, the handbook of fundamentals, and ASHRAE's standards are available from ASHRAE, 345 East 47th Street, New York, New York 10017.

DEPARTMENT OF COMMERCE

Manufacturers seek to encourage product acceptance and improve their own efficiency by establishing basic levels of quality for materials and products and by coordinating dimensions, terminology and other variables such as type and style.

By law, manufacturers cannot agree to establish unreasonable specifications which might rule out the success of individual competitors, nor can they engage in price fixing agreements; but the federal government recognizes the desirability of certain types of industry-supported standardization. The U. S. Department of Commerce provides for the development of Product Standards.

Product Standards (PS)

Product Standards are developed by manufacturers, distributors and users in cooperation with the Office of Engineering Standards Services of the National Bureau of Standards. The purpose of a Product Standard may be either (1) to establish standards of practice for sizes, dimensions, varieties or other characteristics of a specific product; or (2) to establish quality criteria, including standard methods of testing, rating, certifying and labeling of the manufactured products.

The adoption and use of a Product Standard is voluntary. However, when reference to a Product Standard is made in contracts, labels, invoices or advertising literature, the provisions of the standard are enforcible through usual legal channels as a part of the sales contract.

A Product Standard usually originates with the manufacturing segment of the industry. The sponsors may be manufacturers, distributors, or users of the specific product. One of these three elements of industry (the proponent) submits to the Office of Engineering Standards Services the necessary data to be used as the basis for developing a Product Standard. The Office by means of assembled conferences or letter referenda, or both, assists the sponsor group in arriving at a tentative standard of practice and thereafter refers it to the other elements of the same industry for approval or for constructive criticism which will be helpful in making necessary adjustments. The regular procedure of the Office assures continuous servicing of each Product Standard through review and revision whenever, in the opinion of the industry, changing conditions warrant such action.

A Product Standard is designated by the letters "PS," followed by an identification number and the last two digits of the year of issuance or last revision. For example, PS1-66 is the Product Standard for "Softwood Plywood, Construction and Industrial," issued in 1966, as the first Product Standard (Fig. 7). Product Standards will gradually replace the existing Commercial Standards (CS) and Simplified Practice Recommendations (SPR), previously published by the Depart-

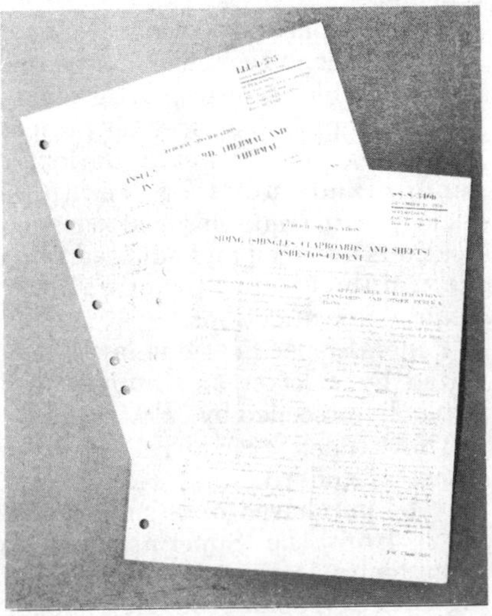

FIG. 9 Federal Specifications are developed for governmental procurement, but are used widely by industry.

ment of Commerce (Fig. 8).

A list of Product Standards, Commercial Standards, and Simplified Practice Recommendations currently available, and the price as well as ordering instructions may be obtained from the Office of Engineering Standards Services, National Bureau of Standards, Washington, D. C. 20234. Also available from that Office are copies of the "Procedures for the Development of Voluntary Product Standards," which explain the process through which such voluntary standards are developed.

GENERAL SERVICES ADMINISTRATION (GSA)

The General Services Administration (GSA) of the federal government develops Federal and Interim Specifications through the cooperation of federal agencies and industry groups. The purpose of these specifications is to standardize the variations and quality of materials and products being purchased by government agencies. Approximately 4,000 Federal and Interim Federal Specifications have been developed by the General Services Administration (Fig. 9).

Federal Specifications

Federal Specification projects are started when a government procurement need arises, when present specifications have become obsolete, or when revision is required for other reasons.

When a particular federal agency has specialized technical competence and available facilities, GSA assigns the development of specific projects to that agency. However, nationally recognized industry, technical society and trade association specifications are used and adopted to the maximum extent practicable in Federal Specifications.

Federal specifications are designated by a letter and number code typically preceded by "FS."

The "Index of Federal Specifications, Standards and Handbooks" may be purchased on a subscription basis from the Superintendent of Documents, U. S. Government Printing Office, Washington, D. C. 20402. The Index is available for reference at all GSA Business Service Centers; field offices of the Small Business Administration; most government purchasing offices; certain Department of Commerce field offices, and some Government Depository Libraries.

Federal Specifications often are issued in *interim* or temporary form by individual federal agencies with optional use by all agencies. These Interim Federal Specifications are coordinated with federal agencies and industry groups as much as possible. If justified by usage (in more than one agency) an Interim Federal Specification may be changed to a Federal Specification. Before this is done, however, the proposed specification is further coordinated with industry and with all federal agencies having an interest.

An Interim Federal Specification may be recognized by the double zero in front of the serial number and by the addition of the issuing agency's identification letters in parentheses after the serial number. For example, SS-S-0072a (GSA-FSS) designates an Interim Federal Specification for "Architectural Stone," as issued by the Federal Supply Service of GSA.

MILITARY AGENCIES

The federal government is the world's largest buyer of equipment, materials and supplies with annual purchases in the billions of dollars. Various departments within the Department of Defense have developed specifications covering materials, products or services used predominately by military activities.

Military Specifications

Military Specifications may be used by any interested civilian organization or specifier. Military Specifications are indexed by the title and code letter prefix MIL.

The Department of Defense maintains a three-part "Index of Specifications and Standards" issued annually on July 30, with cumulative monthly supplements. The Index lists Specifications alphabetically by subject, numerically and by federal supply classification.

DEPARTMENT OF HOUSING AND URBAN DEVELOPMENT (HUD)

The National Housing Act, enacted by Congress in 1934 and amended from time to time, created the Federal Housing Administration (FHA) to stimulate home construction by insuring mortgage loans. The functions of this agency were transferred by Congress in 1965 to the newly created Department of Housing and Urban Development (HUD) and FHA became part of this larger cabinet-level department.

The overall purpose of the Department of Housing and Urban Development is to assist in the sound development of the nation's communities and metropolitan areas. Encouragement of housing production through mortgage insurance and various subsidies has been one of HUD's chief objectives. Improvement in housing quality and in land planning standards has been another HUD objective mandated by Congress.

FHA Housing Programs

The FHA makes no loans, nor does it plan or build housing. It functions mainly as an insuring agency for mortgage loans made by private lenders, such as savings associations and commercial banks. For instance, through the Section 203(b) program FHA encourages lenders to make loans with low down payments and long maturities on one to four-family dwellings. The borrower pays an annual insurance premium of 1/2 of 1% of the average principle outstanding over the premium year. The Secretary of HUD sets the interest rate ceilings on FHA loans at a level required to meet market conditions. Another frequently used section, 221(d)(4), provides for mortgage insurance of new or rehabilitated low or middle-income rental housing.

The traditional role of the FHA was transformed in recent years as the agency became the administrator of interest-rate subsidy and rent-supplement programs authorized by Congress since 1965.

The section 235 program combines insurance with interest assistance payments for owner-occupied homes; in addition to insuring the loan, FHA pays part of the interest the borrower owes the mortgage lender. Section 236 offers insurance and interest assistance for rental projects. Section 237 provides insurance on loans to borrowers with poor credit histories. Section 238 authorizes insurance for mortgage loans in high-risk situations, such as transitional urban areas, not covered by other programs.

HUD Minimum Property Standards

Not all housing programs authorized by Congress involve mortgage insurance and hence not all are administered by FHA. For instance, Section 8 of the Housing Act of 1974 authorizes rental subsidies for leased low-income housing. The housing may be existing or new, financed conventionally or with FHA mortgage insurance.

Prior to 1973, FHA-insured private housing had to conform to the FHA Minimum Property Standards, while subsidized public housing was regulated by a different set of standards. With the adoption in that year of the HUD Minimum Property Standards (MPS), uniform standards became applicable to all HUD housing programs (Fig. 10).

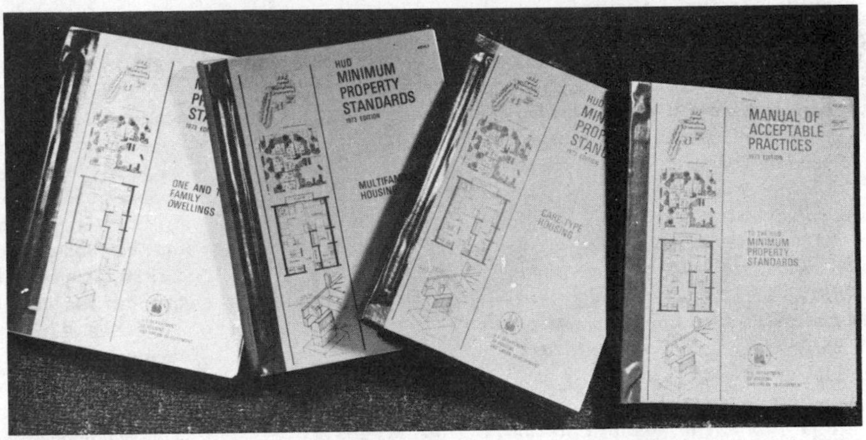

FIG. 10 The first three volumes of the HUD Minimum Property Standards contain mandatory requirements for different housing types. The fourth volume includes explanatory information, suggested design criteria and sound construction practices.

The MPS are intended to provide a sound technical basis for the planning and design of housing under the numerous programs of the Department of Housing and Urban Development. The standards describe those characteristics in a property which will provide present and continuing utility, durability, desirability, economy of maintenance, and a safe and healthful environment.

Environmental quality is considered throughout the MPS. As a general policy, development of all properties must be consistent with the national program for conservation of energy and other natural resources. Care must be exercised to avoid air, water, land and noise pollution and other environmental hazards.

The Minimum Property Standards consist of three volumes of *mandatory standards:* 1) MPS for One and Two Family Dwellings, HUD 4900.1; 2) MPS for Multifamily Housing, HUD 4910.1; and 3) MPS for Care-Type Housing, HUD 4920.1. Variations and exceptions for seasonal homes intended for other than year-round occupancy are listed in HUD 4900.1. Exceptions for elderly housing are listed in HUD 4900.1 and 4910.1. A fourth volume, the MPS Manual of Acceptable Practices, HUD 4930.1, contains *advisory* and illustrative material for the three volumes of mandatory standards.

Where no specific level of performance is stated in the standards, the Manual of Acceptable Practices may be used to determine acceptance or equivalence. This manual is not an additional standard, but is intended to provide information and data representing good current practice in residential design and construction technology.

Materials Releases

The Architectural Standards Division of the Federal Housing Administration issues "Material Releases" for specific proprietary products for which no recognized standards exist. Materials Releases relate only to technical elements of a product. Each release describes a product and its use and is issued to FHA field offices for guidance in determining the acceptance of the product.

The absence of a release for a particular product does not preclude its use. Materials Releases are not intended to indicate approval, endorsement or acceptance by FHA. Manufacturers of materials and products for which Materials Releases have been issued are not authorized to use them in any manner for sales promotion. Copies of Materials Releases are on file in FHA field offices but are not available for general distribution.

The FHA has additional provisions for the review of special materials, products or construction methods which it may be asked to insure. Design, materials, equipment and construction methods other than those described in the Minimum Property Standards are considered for use provided complete substantiating data satisfactory to the FHA are submitted. The local FHA field offices are authorized to accept variations from the standards for specific cases, subject to conditions outlined in the Standards.

Variations on an area or regional basis, or variations involving a substantial number of properties on a repetitive basis are authorized only after consideration of recommendations by the FHA field office and approval by the Architectural Standards Division. Under certain conditions some variations are established and published as Local Acceptable Standards (LAS) for the specific area.

APPENDIX

Acoustical Society of America
335 East 45th St., New York, NY 10017
Adhesive and Sealant Council, Inc.
1500 North Wilson Blvd., Arlington, VA 22209
Adhesives Manufacturers Association of America
111 E. Wacker Drive, Chicago, IL 60601
Air-Conditioning and Refrigeration Institute
1815 N. Fort Myer Dr., Arlington, VA 22209
Aluminum Association
818 Connecticut Ave., NW, Washington, DC 20006
American Association of Nurserymen
Suite 230, Southern Bldg., Washington, DC 20005
American Building Contractors Association
2476 Overland Ave., Suite 205,
Los Angeles, CA 90064
American Concrete Institute
P.O. Box 19150, Redford Station, Detroit, MI 48219
American Concrete Pipe Association
8320 Old Courthouse Road, Vienna, VA 22180
American Consulting Engineers Council
1015 15th St. NW, Washington, DC 20005
American Gas Association, Inc.
1515 Wilson Boulevard, Arlington, VA 22209
American Hardboard Association
205 West Touhy Ave., Park Ridge, IL 60068
American Home Lighting Institute
230 N. Michigan Ave., Chicago, IL 60601
American Institute for Design & Drafting
3119 Price Road, Bartlesville, OK 74003
American Institute of Architects
1735 New York Ave., NW, Washington, DC 20006
American Institute of Building Design
2730 Arden Way, Suite 138, Sacramento, CA 95825
American Institute of Constructors
1140 NW 63rd St., Oklahoma City, OK 73116
American Institute of Industrial Engineers
25 Technology Park/Atlanta, Norcross, GA 30092
American Institute of Landscape Architects
6810 North 2nd Place, Phoenix, AZ 85012
American Institute of Real Estate Appraisers
430 N. Michigan Ave., Chicago, IL 60611
American Institute of Steel Construction, Inc.
The Wrigley Building, 400 N. Michigan Ave.,
Chicago, IL 60611
American Institute of Timber Construction
333 W. Hampden Ave., Englewood, CO 80110
American Insurance Association
85 John St., New York, NY 10038
American Iron and Steel Institute
1000 16th St., NW, Washington, DC 20036
American Land Development Association
1000 16th St., NW, Washington, DC 20036
American Land Title Association
1828 L Street, Suite 705, Washington, DC 20036
American Lumber Standards Committee
20010 Century Boulevard, Germantown, MD 20767
American National Standards Institute, Inc.
1430 Broadway, New York, NY 10018
American Paper Institute
260 Madison Ave., New York, NY 10016
American Parquet Association, Inc.
1650 Union National Place, Little Rock, AR 72201
American Planning Association
1313 E. 60th St., Chicago, IL 60637
American Plywood Association
P.O. Box 11700, Tacoma, WA 98411
American Section of the International Solar Society, Inc.
American Technological University,
P.O. Box 1416, Killeen, TX 76541
American Society for Concrete Construction
329 Interstate Road, Addison, IL 60101
American Society for Testing and Materials
1916 Race Street, Philadelphia, PA 19103
American Society of Civil Engineers
345 East 47th St., New York, NY 10017
American Society of Consulting Planners
1717 N Street, NW, Washington, DC 20036
American Society of Heating, Refrigerating & Air-Conditioning Engineers, Inc.
345 E. 47th St., New York, NY 10017
American Society of Home Inspectors, Inc.
1629 K Street, NW, Suite 520, Washington, DC 20006
American Society of Landscape Architects
1900 M. Street, NW, Washington, DC 20036
American Society of Mechanical Engineers
345 E. 47th St., New York, NY 10017
American Society of Plumbing Engineers
15233 Ventura Boulevard, Suite 616,
Sherman Oaks, CA 91403
American Society of Professional Estimators
Box 3517, Los Angeles, CA 90051
American Society of Safety Engineers, Inc.
850 Busse Highway, Park Ridge, IL 60068
American Society of Sanitary Engineering
960 Illuminating Bldg., Cleveland, OH 44113
American Water Works Association, Inc.
6666 W. Quincy Ave., Denver, CO 80235
American Welding Society
2501 NW 7th St., Miami, FL 33125
American Wood Preservers Association
Suite 4444, 7735 Old Georgetown Rd.,
Bethesda, MD 20014
American Wood Preservers Bureau
Box 6085, 2772 S. Randolph St.,
Arlington, VA 22206
American Wood Preservers Institute
1651 Old Meadow Road, McLean, VA 22102
Architectural Aluminum Manufacturers Association
35 E. Wacker Drive, Chicago, IL 60601
Architectural Woodwork Institute
2310 South Walter Reed Drive,
Arlington, VA 22206
Asphalt Institute
Asphalt Institute Bldg., College Park, MD 20740
Asphalt Roofing Manufacturers Association
1800 Massachusetts Ave., NW #702,
Washington, DC 20036
Associated General Contractors of America
1957 E Street, NW, Washington, DC 20006
Associated Sheet Metal Contractors, Inc.
3000 W. Hallandale Beach Boulevard,
Hallandale, FL 33009
Association of Bituminous Contractors
2020 K St., NW, Suite 800, Washington, DC 20006

Association of Home Appliance Manufacturers
20 N. Wacker Drive, Chicago, IL 60606
Association of Soil & Foundation Engineers
8811 Colesville Road, Suite 225,
Silver Spring, MD 20910
Association of the Wall & Ceiling Industries
1711 Connecticut Ave., NW, Washington, DC 20009
Battelle Memorial Institute
Columbus Labs., 505 King Ave.,
Columbus, OH 43201
Brick Institute of America
1750 Old Meadow Road, McLean, VA 22102
Builders Hardware Manufacturers Association
60 East 42nd St., New York, NY 10017
Building Officials and Code Administrators International, Inc.
17926 S. Halsted St., Homewood, IL 60430
Building Research Advisory Board—
National Research Council
2101 Constitution Avenue, NW,
Washington, DC 20418
California Redwood Association
1 Lombard St., San Francisco, CA 94111
Carpet & Rug Institute
310 Holiday Drive, Box 2048, Dalton, GA 30720
Cast Iron Soil Pipe Foundation
P.O. Box 452, Hollywood Station,
Los Angeles, CA 90028
Cast Iron Soil Pipe Institute
1499 Chain Bridge Road, Suite 203,
McLean, VA 22101
Cellulose Insulation Association
P.O. Box 2533, Salt Lake City, UT 84110
Center for Building Technology
National Bureau of Standards,
Washington, DC 20234
Ceramic Tile Institute
700 N. Virgil Street, Los Angeles, CA 90029
Chamber of Commerce of the United States
Community and Regional Development Section,
1615 H St., NW, Washington, DC 20062
Chemical Manufacturers Association
1825 Connecticut Ave., NW,
Washington, DC 20009
Community Associations Institute
1832 M St., NW, Suite 101,
Washington, DC 20036
Concrete Reinforcing Steel Institute
Room 2110, 180 N. LaSalle St., Chicago, IL 60601
Construction Specifications Institute
1150 17th St., NW, Suite 300,
Washington, DC 20036
Contractors Mutual Association
1101 15th St., NW, Suite 207,
Washington, DC 20005
Copper Development Association, Inc.
405 Lexington Ave., New York, NY 10017
Council of American Building Officials
2233 Wisconsin Ave., NW,
Washington, DC 20007
Division of Building Research
National Research Council of Canada,
Ottawa, Canada, KIA OR6
Door and Hardware Institute
1815 N. Fort Myer Drive, Suite 412,
Arlington, VA 22209
Door Operator and Remote Control Manufacturers Association
410 N. Michigan Ave., Suite 960,
Chicago, IL 60611
Ductile Iron Pipe Research Association
1301 West 22nd St., Suite 509,
Oak Brook, IL 60521
Ecological Society of America
University of Oklahoma, Norman, OK 73019
Edison Electric Institute
1111 19th St., NW, Washington, DC 20036
Engineers Joint Council
345 East 47th St., New York, NY 10017
Expanded Shale, Clay, and Slate Institute
7401 Wisconsin Ave., Suite 414,
Bethesda, MD 20014
Fine Hardwoods—American Walnut Association
5603 West Raymond St., Suite "O",
Indianapolis, IN 46241
Flexicore Manufacturers Association
P.O. Box 1807, Dayton, OH 45401
Forest Insect and Disease Research
USDA—Forest Service RPE605, P.O. Box 2417,
Washington, DC 20013
Forest Products Laboratory
U.S. Dept. of Agriculture, P.O. Box 5130,
Madison, WI 53705
Gas Appliance Manufacturers Association
1901 Fort Myer Dr., Arlington, VA 22209
Glass Tempering Association—
Laminators Safety Glass Association
White Lakes Professional Building,
3310 Harrison, Topeka, KS 66611
Gypsum Association
1603 Orrington Ave., Evanston, IL 60201
Hardwood Plywood Manufacturers Association
P.O. Box 2789, Reston, VA 22090
Home Ventilating Institute
4300 - L Lincoln Ave.,
Rolling Meadows, IL 60008
IIT Research Institute
10 W. 35th St., Chicago, IL 60616
Illuminating Engineering Society of North America
345 East 47th St., New York, NY 10017
Indiana Limestone Institute of America, Inc.
Suite 400, Stone City Bank Bldg.,
Bedford, IN 47421
Institute of Electrical & Electronics Engineers, Inc.
345 E. 4th St., New York, NY 10017
Institute of Environmental Sciences
940 E. Northwest Highway,
Mt. Prospect, IL 60056
Institute of Real Estate Management
430 N. Michigan Ave., Chicago, IL 60611
Institute of Roofing and Waterproofing Consultants
1800 North Argyle Avenue, Suite 301,
Los Angeles, CA 90028
Insulating Glass Certification Council
1640 West 32 Place, Hialeah, FL 33012

International Conference of Building Officials
5360 W. Workman Mill Road,
Whittier, CA 90601

International Masonry Institute
823 15th Street, NW, Suite 1001,
Washington, DC 20005

Interprofessional Council on Environmental Design
c/o National Society of Professional Engineers,
2029 K St., NW, Washington, DC 20006

Iron League of Chicago, Inc.
131 S. Marion St., Oak Park, IL 60302

Laser Institute of America
4100 Executive Park Drive,
Cincinnati, OH 45241

Lightning Protection Institute
48 N. Ayer Street, Harvard, IL 60033

Manufactured Housing Institute
1745 Jefferson Davis Highway, Suite 511,
Arlington, VA 22202

Maple Flooring Manufacturers Association, Inc.
1800 Pickwick Ave., Glenview, IL 60025

Marble Institute of America
c/o Drake Marble Co., 60 Plato Avenue,
St. Paul, MN 55107

Mason Contractors Association of America
17 West 601 - 14th St.,
Oak Brook Terrace, IL 60181

Mechanical Contractors Association of America, Inc.
5530 Wisconsin Ave., NW, Suite 750,
Washington, DC 20015

Metal Lath/Steel Framing Association
221 N. LaSalle St., Chicago, IL 60601

Mineral Insulation Manufacturers Association, Inc.
382 Springfield Ave., Suite 312,
Summit, NJ 07901

Model Code Standardization Council
c/o International Conference of Building Officials,
5360 S. Workman Mill Rd., Whittier, CA 90601

NAHB Research Foundation, Inc.
P.O. Box 1627, 627 Southlawn Ln.,
Rockville, MD 20850

National Academy of Code Administration
Suite 502, 1970 Chain Bridge, McLean, VA 22102

National Academy of Engineering
2101 Constitution Ave., NW,
Washington, DC 20418

National Academy of Sciences
2101 Constitution Ave., NW,
Washington, DC 20418

National Association of Architectural Metal Manufacturers
221 N. LaSalle, Suite 2026, Chicago, IL 60601

National Association of Counties
1735 New York Ave., Washington, DC 20006

National Association of Floor Covering Distributors
13 - 186 Merchandise Mart, Chicago, IL 60654

National Association of Home Builders of the U.S.
15th and M Streets, NW, Washington, DC 20005

National Association of Home Manufacturers
6521 Arlington Blvd., Falls Church, VA 22042

National Association of Housing and Redevelopment Officials
2600 Virginia Avenue, NW, Washington, DC 20037

National Association of Housing Cooperatives
1012 14th St., Suite 805,
Washington, DC 20005

National Association of Minority Contractors, Inc.
1835 K Street, NW, Suite 600,
Washington, DC 20006

National Association of Mirror Manufacturers
5101 Wisconsin Ave., Washington, DC 20016

National Association of Mutual Savings Banks
200 Park Ave., New York, NY 10017

National Association of Plumbing-Heating-Cooling Contractors
1016 20th St., NW, Washington, DC 20037

National Association of Realtors
430 N. Michigan Ave., Chicago, IL 60601

National Association of Women in Construction
2800 W. Lancaster Ave., Fort Worth, TX 76107

National Building Code
American Insurance Association, 85 John St.,
New York, NY 10038

National Building Materials Distributors Association
55 E. Monroe St., Chicago, IL 60603

National Bureau of Standards
Center for Building Technology,
Washington, DC 20234

National Coal Association
1130 17th St., NW, Washington, DC 20036

National Concrete Masonry Association
2302 Horsepen Road, P.O. Box 781,
Herndon, VA 22070

National Conference of States on Building Codes and Standards
1970 Chain Ridge Road, McLean, VA 22102

National Council of Acoustical Consultants, Inc.
66 Morris Ave., Springfield, New Jersey 07081

National Council of Architectural Registration Boards
1735 New York Ave., NW, Suite 700,
Washington, DC 20006

National Electrical Contractors Association, Inc.
7315 Wisconsin Ave., Washington, DC 20014

National Electrical Manufacturers Association
2101 L St., NW, Washington, DC 20037

National Environmental Development Association
No. 3 National Press Building,
Washington, DC 20045

National Fire Protection Association
Suite 570 - South, 1800 M St., NW,
Washington, DC 20036

National Forest Products Association
1619 Massachusetts Ave., NW,
Washington, DC 20036

National Housing Conference
1126 16th St., NW, Washington, DC 20036

National Institute of Building Sciences
1015 15th St., NW, Suite 700,
Washington, DC 20005

National Lime Association
5010 Wisconsin Ave., NW, Washington, DC 20016

National Kitchen Cabinet Association
136 St. Matthews Ave., Louisville, KY 40207

National Landscape Association, Inc.
230 Southern Building, Washington, DC 20005

National League of Cities
1620 Eye Street, NW, Washington, DC 20006
National Mineral Wool Insulation Association, Inc.
382 Springfield Ave., Suite 312,
Summit, NJ 07901
National Oak Flooring Manufacturers Association
804 Sterick Building, Memphis, TN 38103
National Oil Jobbers Council
1707 H St., NW, Washington, DC 20006
National Paint and Coatings Association
1500 Rhode Island Ave., NW,
Washington, DC 20005
National Particleboard Association
2306 Perkins Place, Silver Spring, MD 20910
National Pest Control Association, Inc.
Suite 1100, 8150 Leesburg Pike,
Vienna, VA 22180
National Precast Concrete Association
825 East 64th St., Indianapolis, IN 46220
National Ready Mixed Concrete Association
900 Spring St., Silver Spring, MD 20910
National Remodelers Association
50 East 42nd St., New York, NY 10017
National Research Council
2101 Constitution Ave., NW,
Washington, DC 20418
National Retail Hardware Association
770 N. High School Rd., Indianapolis, IN 46224
National Roofing Contractors Association
1515 North Harlem Ave., Oak Park, IL 60302
National Safety Council
444 N. Michigan Ave., Chicago, IL 60611
National Sand & Gravel Association
900 Spring St., Silver Spring, MD 20910
National Sash and Door Jobbers Association
20 N. Wacker Drive, Chicago, IL 60606
National Slate Association
445 W. 23rd St., New York, NY 10011
National Society of Professional Engineers
2020 K Street, NW, Washington, DC 20006
National Swimming Pool Institute
2000 K Street, NW, Washington, DC 20006
National Terrazzo and Mosaic Association, Inc.
3166 Des Plaines Ave., Des Plaines, IL 60018
National Wholesale Hardware Association
1900 Arch Street, Philadelphia, PA 19103
National Woodwork Manufacturers Association
205 West Touhy, Park Ridge, IL 60068
Northeastern Lumber Manufacturers Association, Inc.
4 Fundy Rd., Falmouth, ME 04105
Northern Hardwood and Pine Manufacturers Association, Inc.
501 Northern Building, Green Bay, WI 54301
Office of Standards Information, Analysis, and Development
National Bureau of Standards,
U.S. Dept. of Commerce, Washington, DC 20234
Painting and Decorating Contractors of America
7223 Lee Highway, Falls Church, VA 22046
Perlite Institute, Inc.
45 West 45th St., New York, NY 10036
Pittsburgh Testing Laboratory
850 Poplar St., Pittsburgh, PA 15220
Plastic Pipe Institute
355 Lexington Ave., New York, NY 10017
Plastics in Construction Council
355 Lexington Ave., New York, NY 10017
Plumbing, Heating, Cooling Information Bureau
35 E. Wacker Drive, Chicago, IL 60601
Portland Cement Association
5420 Old Orchard Road, Skokie, IL 60077
Prestressed Concrete Institute
20 N. Wacker Drive, Chicago, IL 60606
Producers Council
1717 Massachusetts Avenue, NW,
Washington, DC 20036
Product Fabricating Service, Inc.
2402 Daniels St., Madison, WI 53704
Property Management Association of America
8811 Colesville Rd., Suite 225,
Silver Spring, MD 20910
Red Cedar Shingle and Handsplit Shake Bureau
Suite 275, 515 - 116th Ave., NE,
Bellvue, WA 98004
Resilient Floor Covering Institute
1030 15th St., NW, Suite 350,
Washington, DC 20005
Sealed Insulating Glass Manufacturers Association
111 E. Wacker Drive, Suite 600,
Chicago, IL 60601
Small Homes Council—Building Research Council
1 E. St. Mary's Rd., Champaign, IL 61820
Society of American Registered Architects
2011 West Pershing Road, Chicago, IL 60609
Society of American Value Engineers
P.O. Box 210887, Dallas, TX 75211
Society of American Wood Preservers, Inc.
1401 Wilson Boulevard, Suite 205,
Arlington, VA 22209
Society of Fire Protection Engineers
60 Batterymarch Street, Boston, MA 02110
Society of Plastics Engineers, Inc.
656 W. Putnam Ave., Greenwich, CT 06830
Society of Real Estate Appraisers
2600 Virginia Ave., NW, Washington, DC 20037
Society of the Plastics Industries, Inc.
355 Lexington Ave., New York, NY 10017
Society of Women Engineers
345 East 47th St., Rm. 305, New York, NY 10017
Solar Energy Industries Association, Inc.
1001 Connecticut Ave., NW, Suite 800,
Washington, DC 20036
Southern Building Code Congress International
900 Montclair Road, Birmingham, AL 35213
Southern Cypress Manufacturers Association
805 Sterick Building, Memphis, TN 38103
Southern Forest Products Association
P.O. Box 52468, New Orleans, LA 70152
Southern Hardwood Lumber Manufacturers Association
805 Sterick Building, Memphis, TN 38103
Southern Pine Inspection Bureau
4709 Scenic Highway, Pensacola, FL 32504
Southwest Research Institute
6220 Culebra Road, P.O. Drawer 28510,
San Antonio, TX 78284

Sprayed Mineral Fiber Manufacturers Association Inc.
One Wall Street, Suite 2400,
New York, NY 10005

SRI International
333 Ravenswood, Menlo Park, CA 94025

Steel Deck Institute
P.O. Box 3812, St. Louis, MO 63122

Steel Door Institute
712 Lakewood Center North,
Cleveland, OH 44107

Steel Joist Institute
1703 Parham Rd., Richmond, VA 23229

Steel Window Institute
1230 Keith Building, Cleveland, OH 44115

Stucco Manufacturers Association, Inc.
14006 Ventura Blvd., Sherman Oaks, CA 91423

Superintendent of Documents
U.S. Government Printing Office,
Washington, DC 20402

Test Boring Association, Inc.
481 Main Street, New Rochelle, NY 10801

Thermal Insulation Manufacturers Association, Inc.
7 Kirby Plaza, Mt. Kisco, NY 10549

Tile Contractors' Association of America
112 N. Alfred, Alexandria, VA 22314

Tile Council of America
P.O. Box 326, Princeton, NJ 08540

Underwriters Laboratories, Inc.
333 Pfingsten Road, Northbrook, IL 60062

Urban Land Institute
1200 18th Street, NW, Washington, DC 20036

Urethane Foam Contractors Association
1406 Third National Building, Dayton, OH 45402

Urethane Institute
355 Lexington Ave., New York, NY 10017

U.S. Dept. of Housing and Urban Development
451 Seventh St., SW, Washington, DC 20411

U.S. Forest Products Laboratory
P.O. Box 5130, Madison, WI 53705

Vermiculite Association
52 Executive Park South, Atlanta, GA 30329

Water Quality Association
477 E. Butterfield Rd., Lombard, IL 60148

West Coast Lumber Inspection Bureau
P.O. Box 23145, Portland, OR 97223

Western Red Cedar Lumber Association
Yeon Building, Portland, OR 97204

Western Red & Northern White Cedar Association
P.O. Box 2786, New Brighton, MN 55112

Western Wood Products Association
1500 Yeon Building, Portland, OR 97204

Window Shade Manufacturers Association
Executive Plaza, 1211 West 22nd St.,
Oak Brook, IL 60521

Wire Reinforcement Institute
7900 Westpark Drive, McLean, VA 22102

Zinc Institute, Inc.
292 Madison Ave., New York, NY 10017

We gratefully acknowledge the following for their cooperation in the preparation of this section: American Society for Testing and Materials; American National Standards Institute; Federal Housing Administration; NAHB Research Foundation, Inc.; Underwriters' Laboratories, Inc.; U. S. Department of Commerce, National Bureau of Standards.

104 MOISTURE CONTROL

MOISTURE CONTROL

INTRODUCTION

Moisture in the *visible liquid state, water,* is relatively simple to understand and to cope with in construction. Roofs are built to shed it, earth is graded to divert it, pipes are used to conduct it and, generally, its control is fairly well considered in building design.

But moisture also is present as an *invisible gas, water vapor,* in the air all around us and in cavities of construction assemblies: in stud and joist spaces; in cores and furred spaces of masonry. As absorbed water it is present in most construction materials and affects their properties, especially dimensional stability. Moisture may be present also *as a solid, ice,* within the range of temperatures encountered in most parts of the country.

Moisture is not damaging to construction in its vapor form. However, it is very changeable and it becomes much more dangerous as it condenses (liquefies) or as it freezes (solidifies). Hidden condensation over a period of time can cause decay of organic materials, corrosion of metal products, blistering of paint coatings and—if it freezes—cracking or spalling of concrete and masonry.

Condensation and freezing may occur either as a result of a general temperature drop in the area, or the migration of water vapor to areas of lower temperature. Therefore moisture problems are most likely in colder climates, in heated occupied buildings, which usually generate substantial amounts of water vapor.

Houses built 40 or more years ago were not so well built and sealed against the weather as they are today. This permitted considerable air leakage, dissipating a lot of interior moisture to the outside harmlessly through the structure. Considerations of comfort and economy require that modern houses be built smaller, tighter, warmer in winter and cooler in summer.

Also, just as dehumidification is an important aspect of summer comfort conditioning, so is deliberate humidification an integral part of typical winter heating systems. All of these factors create conditions conducive to moisture problems in modern residential construction, unless suitable provisions to cope with them are included in the original design and construction.

The properties and behavior of water in its various forms are an important consideration in many aspects of building construction, ranging from soil mechanics to roofing. This section limits discussion to the principles of water vapor movement, problems associated with its condensation to visible water, and methods of controlling condensation in buildings.

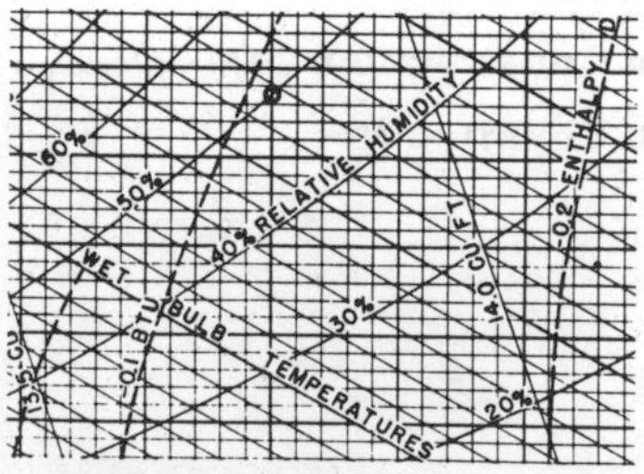

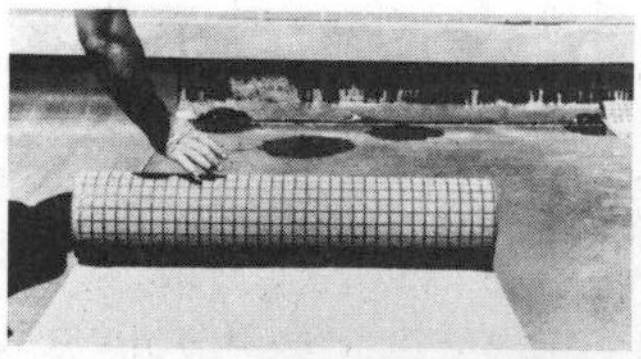

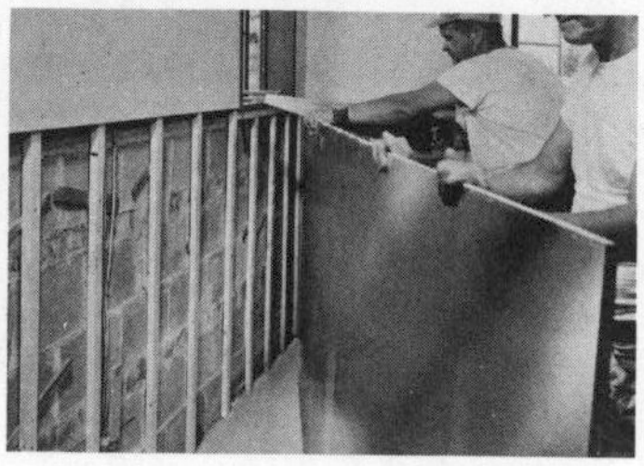

WATER VAPOR PROPERTIES

Moisture can exist in any one of several states: as liquid, water; as solid, ice; and in gaseous form, as water vapor. The reverse of *condensation*, the change of water from a liquid to a gas, is called *evaporation*. The transition from ice directly into water vapor is termed *sublimation*.

The change of state from water or ice to vapor requires considerable heat input, called *latent heat*, measured in British Thermal Units, (BTU's). Hence, when water evaporates from any surface, that surface is cooled; the perspiring skin, for example, is an effective cooling system of the human body. Like any warm body, the skin also gives off heat to the surrounding atmosphere. The heat absorbed by a substance without changing its state is known as *sensible heat* and also is measured in BTU's.

When water turns to vapor, it mixes with the air and occupies all the available space. In many ways water vapor acts independently of the air because its general properties do not depend on the presence of air. However, when the air either is moved suddenly, is heated or is cooled, the water vapor is similarly affected. It is therefore convenient to consider moist air as a mixture of both dry air and water vapor.

FIG. 1 TEMPERATURE/HUMIDITY RELATIONSHIPS*

Temperature	Actual M.C. at given temp. (grains/lb. dry air)	Maximum M.C. at dew point (grains/lb. dry air)	Relative Humidity (% RH)	Condensed Moisture** (grains/lb. dry air)
70°F	54	108	54/108 = 50	none
60°F	54	77	54/77 = 70	none
50°F	54	54	54/54 = 100	none
40°F	36	36	36/36 = 100	54 − 36 = 18
30°F	24	24	24/24 = 100	54 − 24 = 30

* At 29.92 in. Hg pressure.
** As air at 50% RH and 70°F drops in temperature (read down in first column).

RELATIVE HUMIDITY

The measure of the amount of water vapor in air is known as *relative humidity* (RH) and is stated as a percentage. The percentage is a ratio of the *actual* amount of water vapor in the air, as compared to the *maximum* amount the air can contain, at a specific temperature and atmospheric pressure. For instance, air of 50% RH contains half the amount of vapor it could hold, at that air temperature and barometric pressure; at 100% RH it contains the maximum, and such air is said to be *saturated*.

An observable law of physics is that the warmer a mass of air, the more water vapor it can hold. Therefore if the temperature of saturated air is lowered, some of the vapor must be given up by condensation. Up to the point of condensation the relative humidity increases as the temperature of the air is lowered, while maintaining the amount of moisture.

As demonstrated in Figure 1, air at 70°F can hold twice as much moisture as the same volume of air at 50°F, when the air is completely saturated; hence at 70°F it is said to be half-saturated or to have a relative humidity of 50% (50% RH). If the temperature were lowered below the saturation point, say to 40°F, the relative humidity would remain at 100% but approximately 18 grains of moisture per pound of dry air would condense into visible water on cool surfaces. Further cooling to 30°F would result in the removal of a total of 30 grains per pound of air (Fig. 1).

Human Comfort

Air movement, relative humidity and temperature—all affect the rate at which perspiration evaporates from the skin and the body loses heat. Excessively low air humidity encourages more evaporation and increases the sensation of coolness, undesirable in the winter. On the other hand, too high a humidity reduces skin evaporation, retards body cooling and makes people feel warmer in the summer. Either too little or too much humidity can be uncomfortable, or even unhealthy.

High humidities at higher temperatures can contribute to fatigue and reduced working efficiency. Air that is too dry can irritate nasal and sinus passages, can contribute to chapped and dry skin, and can encourage the generation of annoying static electricity.

However, the specific "comfort" level of humidity or temperature is largely an individual, subjective, judgment and depends on many factors such as prevailing climate, seasonal variations and the individual's age, health, living habits, clothing, and activities.

As mentioned above, cold air cannot hold nearly as much water vapor as warmer air. Hence, when cold outside air, which has a low moisture content (even though its relative humidity may be high), is brought inside and heated, extremely low relative humidity results. For instance, as shown in Figure 1, saturated air at 30°F can hold 24 grains of moisture per lb. of air, hence half-saturated air (50% RH) at 30°F has a moisture content of 12 grains/lb. of air. If such air were to be heated to 70°F, it would be capable of holding 108 grains, and its relative humidity would be 100% x 12/108, or 11.1% RH. The resulting humidity level is too low for comfort; 25% to 40% RH generally is considered conducive to good health, but hard to maintain in cold weather.

Effective Temperature

This is an arbitrary index which combines into a single value the effects of temperature, humidity and air movement, on the sensation of warmth or cold felt by the human body. It has been developed by the American Society of Heating, Refrigerating and Air-Conditioning Engineers (ASHRAE) in cooperation with the U.S. Public Health Service and is based on extensive polls of trained individuals subjected to various temperature/humidity conditions.

Effective temperature (ET) is the temperature of saturated air (100% RH) which gives the same physical sensation of warmth as various other combinations of aiı temperature and relative humidity, with relatively still rates of air movement ranging from 15 to 25 feet per minute (fpm).

Winter Comfort It has been found that 98% of the people tested were comfortable in the winter season at 68 ET (100% RH and 68°F). Combinations of humidity and temperature which gave this same feeling of warmth are tabulated in Figure 2. This table demonstrates that in winter, as the relative humidity is reduced, higher temperatures are necessary to maintain the same sensation of warmth; conversely a higher relative humidity produces warmth with lower temperatures. However the thermal and vapor permeance characteristics of exterior construction impose practical limits on interior humidity, which vary with the severity of winters.

In areas of low winter design temperatures (0°F or lower), not many houses can tolerate 40% RH without developing condensation and moisture problems. At this interior humidity and exterior temperature combination, even sealed insulating glass will develop condensation (Fig. 3). Single glazing

FIG. 2 **TEMPERATURE-HUMIDITY EQUIVALENTS OF OPTIMUM SUMMER & WINTER ET'S***

WINTER OPTIMUM—ET 68° **		SUMMER OPTIMUM—ET 71° **	
Relative Humidity	Dry Bulb Temperature	Relative Humidity	Dry Bulb Temperature
100%	68°F	100%	71°F
80%	69°F	80%	73.1°F
70%	70°F	70%	74.3°F
60%	71.1°F	60%	75.6°F
50%	72.2°F	50%	76.9°F
40%	73.5°F	40%	78.3°F
30%	74.8°F	30%	79.7°F
20%	76.1°F	20%	81.2°F
10%	77.5°F	10%	83.0°F

* Air movement of 15 to 25 fpm.
** Considered comfortable by 98% of people tested.

FIG. 3 **RECOMMENDED MAXIMUM INTERIOR RH TO PREVENT CONDENSATION DAMAGE***

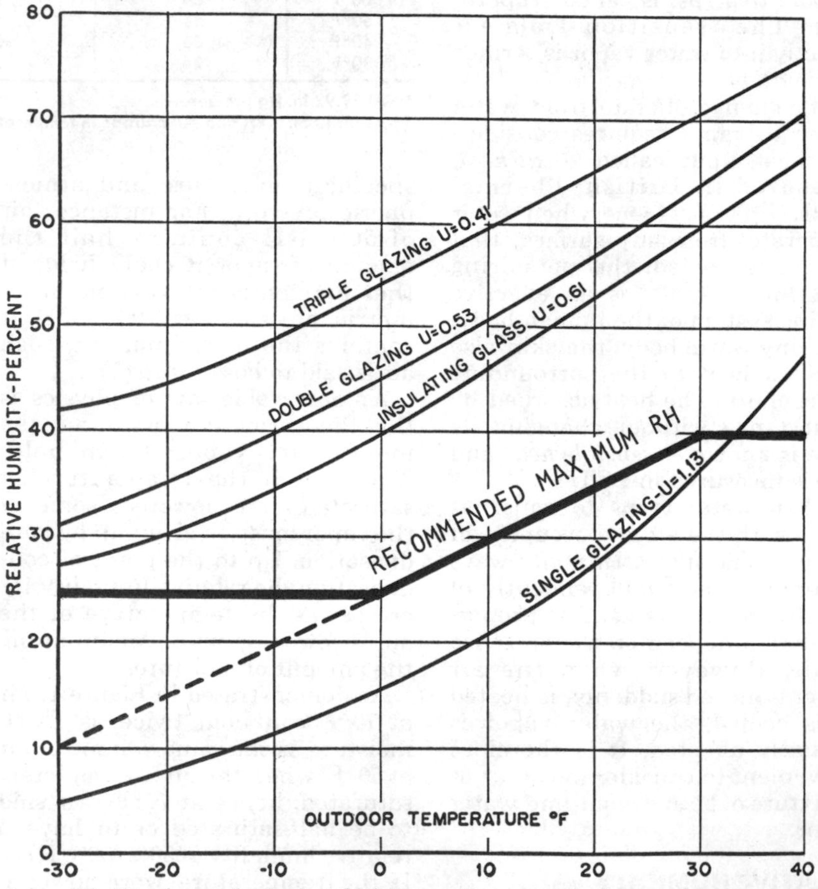

* Below 0°F, RH values maintained at 25% to provide acceptable degree of occupant comfort (solid line); however, to minimize condensation on single glazing, use RH values represented by dashed line.

and highly conductive surfaces such as metal door and window frames without thermal break characteristics will show condensation even at 15% RH. To minimize condensation and yet provide for human comfort, it is recommended that approximately 25% RH be maintained when the outside temperature is 0°F or lower; the humidity can be maintained at 30% when the outside temperature is 10°F, at 35% when 20°F and at 40% when 30°F outside (Fig. 3). Humidities above 40% RH generally are not recommended during the winter heating season.

Summer Comfort In summer, 98% of those tested were comfortable at 71 ET (100% RH and 71°F). Other combinations of relative humidity and temperature that give the same physical sensation in summer are tabulated in Figure 2.

The dashed lines in the ASHRAE comfort chart (Fig. 4) constitute the effective temperatures. Each line represents graphically the equivalent combinations of temperature and humidity. By following the 68 ET and 71 ET lines to their intersections with the solid diagonal RH lines, the corresponding dry bulb temperatures (horizontal coordinate) can be selected. The temperature-humidity equivalents shown in Figure 2 can thus be derived from this chart.

FIG. 4 **ASHRAE COMFORT CHART SHOWING ET'S & SUBJECTIVE RESPONSES TO TEMPERATURE-HUMIDITY COMBINATIONS**

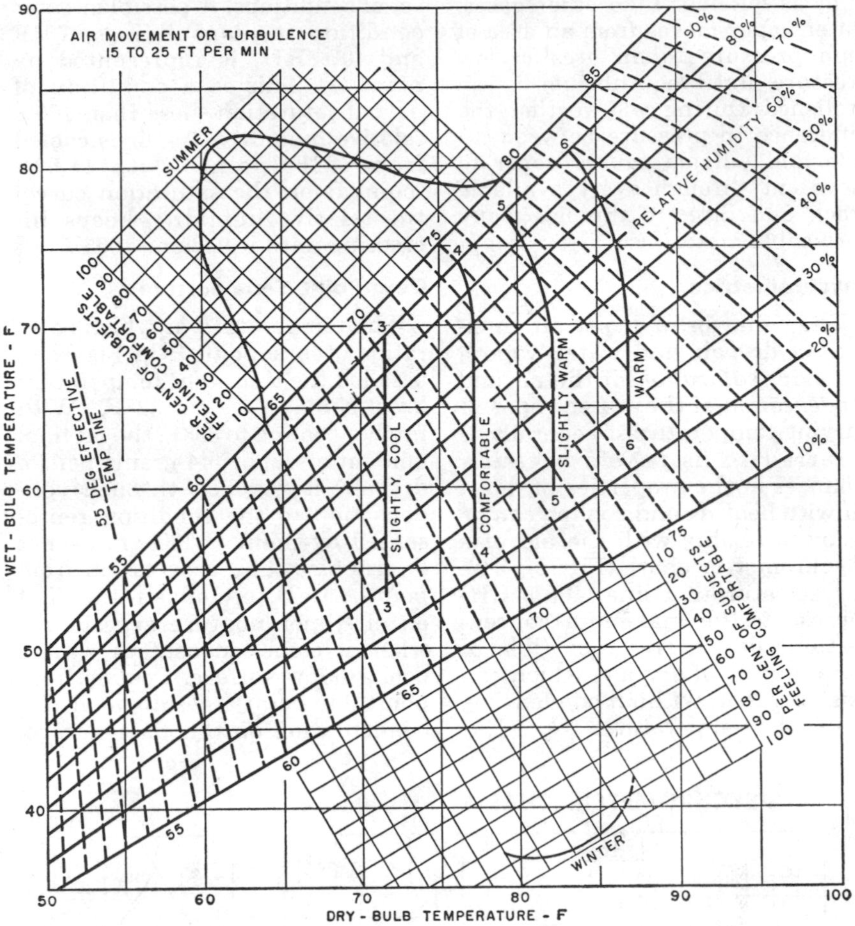

It will be noted that the optimum summer ET 71 line strikes near the top of the summer curve representing the distribution of responses by the subjects tested. From this curve it can be established that nearly all subjects in the summer were comfortable at 71 ET, while 50% were comfortable within a range of 68 ET to 75 ET. Similarly, by reading the winter comfort curve, it can be seen that nearly all subjects were comfortable at 68 ET, while about 85% of subjects were comfortable between 65 ET and 70 ET.

The effective temperature index however is useful only in establishing human response to temperature-humidity conditions *immediately* upon entering a room. Where subjects were tested after a *three hour stay,* it was found that humidity played a much lesser role in determining comfort. These findings are represented by solid lines 3, 4, 5 and 6 in Figure 4. For both summer and winter, it was found that a dry bulb temperature of 77.5°F is equally comfortable for humidities from 25% RH up to 60% RH. Only above 60% RH (curved portion of line 4) must the temperature be reduced to maintain the same degree of comfort. The lines represent averaged responses for both summer and winter conditions; actually there is a spread of about 1.2°F between winter and summer responses. The maximum comfort line therefore would be at approximately 77°F in winter and 78°F in summer.

VAPOR MOVEMENT

The movement of water vapor is largely independent of air movement. Water vapor establishes a pressure proportional to the amount of water vapor present within the air mix. Air with more vapor has a higher vapor pressure. Vapor moves through air by *diffusion* from regions of high vapor pressure to regions of lower pressure, without relying on air circulation to carry it.

Vapor pressure is commonly expressed in inches of mercury. At 70°F and 100% relative humidity, the vapor pressure is 0.739″ of mercury (Hg), while at 0°F and 100% RH the vapor pressure is only 0.0377″ Hg, or about 1/20 of the saturated condition.

With a winter condition of 0°F and 75% RH, an outside vapor pressure of 0.027″ Hg would exist, while inside a building heated to 70°F and with 35% relative humidity, vapor pressure would equal 0.259″ Hg. The vapor pressure in-

0.259″ Hg. The vapor pressure inside would be nearly 10 times as high as outside. Like other gases, water vapor moves from an area of high pressure to an area of low pressure until equilibrium is established. During cold weather, the difference in pressure between inside and outside causes vapor to move out through every available crack and directly through many permeable materials.

Psychrometrics

The behavior of moist air (mixture of dry air and water vapor), involving *saturation* of the air and *condensation* of the water vapor at varying temperatures, generally is referred to as *psychrometrics*. Changes in the properties of moist air with heating and cooling can be followed readily with the aid of a psychrometric chart (Fig. 5). The curved saturation line (100% RH) represents the maximum concentrations of water vapor (lbs. of water per lb. of dry air) which can exist as vapor, without condensing, at various temperatures.

Figure 6 represents a simplified psychrometric chart for a specific set of conditions. A possible winter condition inside buildings, 70°F and 40% RH, is represented by point A. This is a condition of partial saturation—less than 100% relative humidity. As air is cooled from 70°F (point A) to 44.6°F (point B on the saturation curve) the relative humidity keeps increasing until it become 100%.

Dew Point Temperature

The temperature at which saturation occurs (point B in Figure 6) is called the *dew point* temperature (*frost point,* if below 32°F). With further cooling to 35°F, the original amount of vapor (44 grains per lb. dry air) is reduced through condensation to the condition represented by point C (30 grains per lb. dry air). The progression from point A to B to C illustrates what an air-vapor mixture experiences when it comes in contact with a cool window surface. Cooling from point B to point C results in visible condensation on the glass surface.

FIG. 5 **PSYCHROMETRIC CHART***

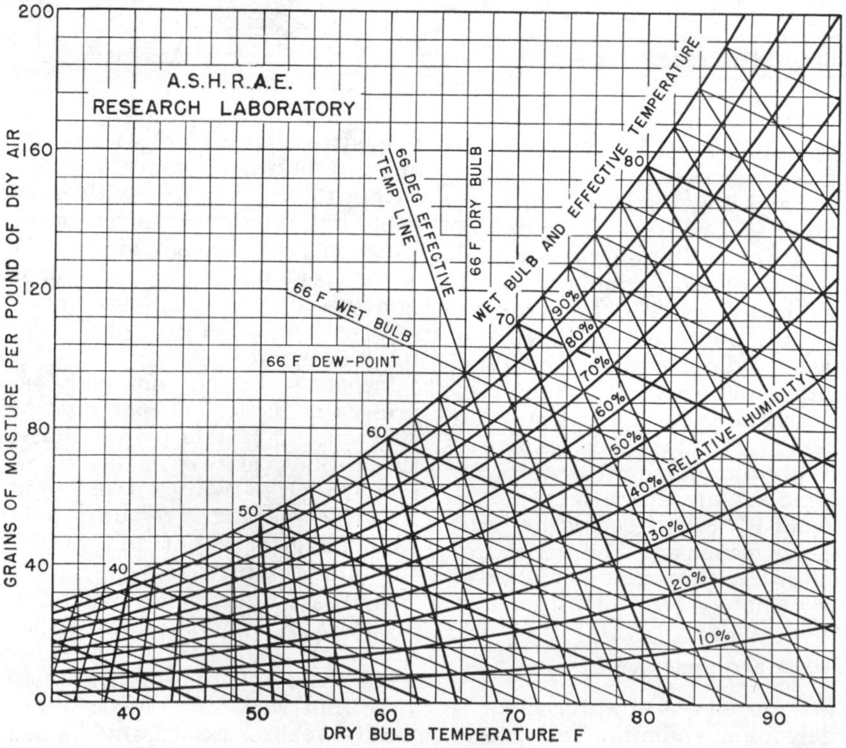

* With ET lines, for persons at rest, normally clothed, in still air.

FIG. 6 **USE OF PSYCHROMETRIC CHART ILLUSTRATED**

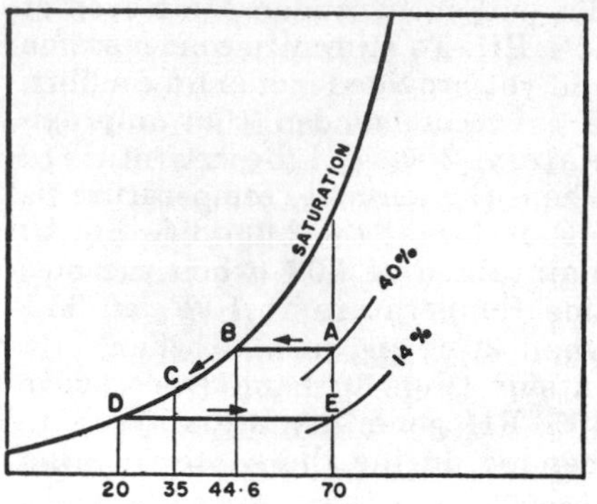

If point C were below 32°F, the condensation would be in the form of frost.

Once the temperature drops below the dew point (or frost point) the vapor pressure at the window surface is also reduced. This reduction in vapor pressure causes diffusion currents within the room, moving water vapor continuously to the window surface to be condensed as long as the concentration of vapor is maintained in the room.

By moving from point D to point E, (Fig. 6) a common winter phenomenon is illustrated, where outside air at 20°F and 100% RH is heated to 70°F, with a resulting decrease in humidity to 14% RH. This example explains why the relative humidity within buildings is greatly reduced in extremely cold weather, when cold outside air enters the house and is heated to room temperature.

High dew point temperatures are common in the summertime. From the psychrometric chart it can be seen that for a dry bulb temperature of 80°F, the dew point temperature would be 60°F if the RH is 50% (Fig. 5). However if the RH were to be raised to 90%, the dew point would rise to 78°F. As the humidity rises the dew point gets closer to the dry bulb temperature but cannot exceed it. If on the other hand, the temperature is raised to 90°F and the RH is maintained at 90% (by adding moisture to the air), the dew point would rise to 87°F.

ADSORPTION AND ABSORPTION

The surfaces of most materials

have an affinity for water and a film of water molecules often is held to the surface by molecular forces of attraction. This phenomenon is known as *adsorption.* Surface films at low humidities may be only one molecule thick; at moderate humidities, multi-molecular films may be established. The film thickness, and therefore the amount of water held in equilibrium with the surrounding atmosphere, is roughly proportional to the relative humidity. At humidities close to 100%, the films become so relatively thick, that small surface pores may become completely filled and larger capillaries may be partially filled. At saturation conditions, all voids in the material are completely filled.

Fibrous materials such as wood, plywood and fiberboards present very large effective surfaces to water molecules, so that the amount of water held on the surface, and subsequently *absorbed into* these materials through surface pores, is relatively large even at moderate humidities. These materials are said to be *hygroscopic.* Materials which present relatively small surfaces, such as most metals, are not penetrated by the water molecules and take on only minute quantities of water, as surface films.

Dimensional Instability

Changes of moisture content can effect significant dimensional changes in many construction materials. The dimensional changes occurring in wood are probably the best known. With a change from an air-dry condition of 12% or 15% moisture content (m.c.) to oven-dry condition (0% m.c.), the dimensional changes (shrinkage) which occur are approximately 0.1% longitudinally, 2% radially and 4% tangentially.

The continuous movement of dimensionally unstable materials must be carefully considered in building design and detailing. For example, millwork elements lap one another to permit movement without developing unsightly cracks; air gaps around wood strip floors allow expansion without buckling in humid weather.

Deterioration of Materials

Moisture can cause or contribute to the breakdown of materials by: (1) chemical changes such as the rusting of steel, (2) physical changes such as the spalling of masonry by frost action, or (3) biological processes such as the decaying of wood. Measures to control liquid water (other than water vapor) such as flashings, waterproofing and storm drainage, are discussed in appropriate sections of Divisions 300 and 400.

Since air saturation is most likely to take place at low temperatures when the risk of freezing also is enhanced, condensation and freezing both are major design con-

FIG. 7 WATER VAPOR PERMEANCE OF BUILDING MATERIALS

Material	Permeance (Perm)
Materials Used in Construction	
1″ Concrete (1:2:4 mix)	3.2
Brick masonry (4″ thick)	0.8
Concrete block (8″ cored, limestone aggregate)	2.4
Asbestos-cement board (0.2″ thick)	0.54
Plaster on metal lath (3/4″)	15.0
Plaster on plain gypsum lath (with studs)	20.0
Gypsum wallboard (3/8″ plain)	50.0
Gypsum sheathing (1/2″ asphalt impregnated)	10.0
1″ Structural insulating board (sheathing quality)	20—50
Structural insulating board (interior, uncoated, 1/2″)	50—90
Hardboard (1/8″ standard)	11.0
Hardboard (1/8″ tempered)	5.0
Built-up roofing (hot-mopped)	0.0
1″ Wood, sugar pine	0.4—5.4
Plywood (douglas-fir, interior, glue, 1/4″ thick)	1.9
Thermal Insulations, 1″ thick	
Air (still)	120.0
Cellular glass	0.0
Corkboard	2.1—9.5
Mineral wool (unprotected)	116.0
Expanded polyurethane (R-11 blown)	0.4—1.6
Expanded polystyrene—extruded	1.2
Expanded polystyrene—bead	2.0—5.8
Unicellular synthetic flexible rubber foam	0.02—0.15
Plastic and Metal Foils and Films	
Aluminum foil (1 mil)	0.0
Aluminum foil (0.35 mil)	0.05
Polyethylene (2 mil)	0.16
Polyethylene (4 mil)	0.08
Polyethylene (6 mil)	0.06
Building Papers, Felts, Roofing Papers*	
Duplex sheet, asphalt laminated, aluminum foil one side (43)	0.002
Saturated and coated roll roofing (326)	0.05
Kraft paper and asphalt laminated, reinforced 30-120-30 (34)	0.3
Blanket thermal insulation back-up paper, asphalt coated (31)	0.4
Asphalt-saturated and coated vapor-barrier paper (43)	0.2—0.3
Asphalt-saturated but not coated sheathing paper (22)	3.3
15 lb. asphalt felt (70)	1.0
15 lb. tar felt (70)	4.0
Single-kraft, double infused (16)	31.0
Protective Coatings	
Paint—2 coats	
Aluminum varnish on wood	0.3—0.5
Enamel on smooth plaster	0.5—1.5
Various primers plus 1 coat flat oil paint on plaster	1.6—3.0
Water emulsion on interior insulating board	30.0—85.0
Paint—3 coats	
Exterior paint, white lead and oil on wood siding	0.3—1.0
Exterior paint, white lead-zinc oxide and oil on wood	0.9
Styrene-butadiene latex coating, 2 oz./sq. ft.	11.0
Polyvinyl acetate latex coating, 4 oz./sq. ft.	5.5
Asphalt cut-back mastic, 1/16″ dry	0.14
Hot melt asphalt, 2 oz./sq. ft.	0.5

* Numbers in parentheses are weights in lbs. per 500 sq. ft.

siderations. Also, condensation may occur over a period of time in concealed locations within wall, floor/ceiling or roof assemblies and not become apparent until a conspicuous failure occurs.

Effect on Heat Flow

Moisture can have a marked effect on the transmission of heat through building materials. When present in liquid form, it increases the conductivity of materials by increasing the available paths of heat flow.

Hygroscopic insulations such as vegetable fiber products which readily absorb and hold moisture, do not dry out as quickly as non-absorbent fiber insulations and lose their thermal effectiveness for longer periods. The non-absorptive products however have limited moisture storage capacity, and begin "dripping" sooner, possibly damaging interior finishes.

WATER VAPOR TRANSMISSION

The amount and rate of water vapor transmitted through building materials is dependent on: (1) the *gradient* (differential) in vapor pressure from one side of the material to the other, (2) the *area* of the material, and (3) its *permeance* (ability to permit vapor passage).

Permeance and Permeability

Water vapor *permeance* is a measure of water vapor flow through a material *of specific thickness,* or an assembly of several materials. The unit of permeance, *perm,* states the amount of vapor flow in grains per hour, per sq. ft. of surface, per 1″ Hg (mercury) vapor pressure gradient.

Water vapor *permeability* is the permeance of a *1″ thickness* of a homogeneous substance. Permeability is designated in *perm-inches* and states the amount of vapor flow through 1″ of material, in grains per hour, per sq. ft. of surface, per 1″ Hg vapor pressure gradient.

Most building materials are permeable to a degree. Some materials such as metals and glass are completely *impermeable* (vapor-resistant) and permit no passage of water vapor. Such materials have a perm rating (permeance) of 0.00. Materials which either are impermeable or have a permeance of 1 perm or less are referred to as vapor barriers. However, modern construction methods demand better vapor barriers than those approaching 1 perm; suitable barriers frequently must have a rating of 0.5 perm or less

The overall permeance (M_t) of an assembly of materials can be calculated from the permeances of the individual components. The calculation is similar to that used in determining overall coefficient of heat transmittance (U) from individual material conductances (C):

$$M_t = \frac{1}{\frac{1}{M_1} + \frac{1}{M_2} + \frac{1}{M_3} \cdots + \frac{1}{M_n}} ;$$

where M_t = overall, total, permeance and M_1, M_2, M_3, through M_n = individual permeances of materials in an assembly. Permeances for representative construction materials are given in Figure 7.

Many of the permeance numbers in Figure 7 were obtained by different test methods. Since different test methods can produce somewhat varying permeance ratings, it is not valid to compare the relative performance of those materials with small variations in their permeances. For instance, it is not nearly as meaningful to compare "hot melt asphalt, 2 oz./sq. ft." (perm—0.5) with "asphalt-saturated paper" (perm—0.2-0.3). Either of these two or any one of the vapor barrier materials (under 1 perm) can be more readily compared to products with permeances substantially higher than 1 perm.

Water-Vapor Transmission

Sometimes it is more significant to report the vapor resistance of materials at a *specific pressure difference,* rather than at 1″ Hg, as in permeance and permeability. *Water-vapor transmission* (WVT) states the weight of vapor (grains) transmitted per unit of time (typically one hour) per unit of area (one sq. ft.); values may be high or low depending on the vapor pressure gradient chosen for the test. WVT data can be converted to perms by the following formula:

$$\text{Permeance} = \text{WVT rating}/\Delta p,$$

where WVT rating equals weight of vapor transmitted in grains per hour per sq. ft. and Δp equals vapor-pressure gradient in the test, in inches of mercury.

CONDENSATION
CAUSES AND REMEDIES

Condensation problems are most likely to occur in occupied, heated buildings. An average family normally generates about 22.5 lbs. of vapor in a 24 hour period, but in some cases moisture production can be more than 50 lbs. (Fig. 8). Condensation occurs when interior vapor-laden air contacts surfaces cooled by heat loss to cold exterior air. In winter, condensation may collect in the form of water or frost on the glass and other cold surfaces such as the metal frames of windows and sliding glass doors.

Although condensation forming on a surface may enter surface pores as fast as it forms and thus become invisible, any condensation on a visible surface is referred to as *surface condensation* to distinguish it from *concealed condensation,* occurring within a material or assembly.

INTERIOR VENTILATION

Summer cooling does not usually create serious concealed condensation problems. The cooled interior air seldom produces conditions within the wall conducive to condensation. However, during periods of high humidity, condensation may form on cool basement walls and floors.

Likelihood of both surface and concealed condensation can be greatly minimized by introducing dry night air, or by exhausting excessive water vapor directly at its source, such as in the kitchen, bath and laundry. In most houses, a sufficient amount of fresh air generally is available through infiltration to make up the air lost by exhaust.

During cold weather the interior relative humidity often becomes too low for human comfort, and mechanical humidification of living spaces is required. Even then, ventilation in kitchen, laundry and bath areas is desirable for odor control and to reduce excessively high local humidities which could easily result in surface condensation.

FIG. 8 SOURCES OF MOISTURE IN TYPICAL HOME*

Source or Function	Moisture Generated (lbs.)
Breathing and perspiring	13
Cooking	5
Bathing	1
Dishwashing	3.5**

* For family of 4, in 24 hours.
** As much as 30 lbs. or more of moisture can be added by clothes washing and inside drying.

The effectiveness of ventilation in reducing humidity is shown in Figure 9. It illustrates the small amount of water vapor escaping into the well-insulated walls and ceilings of a typical 2,000 sq. ft. residence provided with vapor barriers (the floor is not considered). Ventilation at the rate of 2,000 cfh is near the minimum for odor control when cooking. With indoor temperature at 70°F and RH of 40%, ventilation at the rate of 2,000 cfh removes approximately 21 lbs. of moisture per day; at the same time, less than 1.5 lbs. escapes into the walls and ceilings of the structure. It is not suggested that continual mechanical ventilation is required to get rid of excess moisture generated in the home. In most homes, the action of kitchen and bath exhaust fans is supplemented by substantial exfiltration (air leakage) through walls, windows and doors to keep moisture in check. However, unusually tight houses located in sheltered areas may have to rely on more extended ventilation than indicated by odor removal efforts alone. This is particularly true in periods of cold weather when RH must be limited to prevent condensation (Fig. 3).

SURFACE CONDENSATION

Condensation occurs when a surface is colder than the dew point of the nearby air. The temperature of interior surfaces such as walls, ceilings or floors is dependent on:

FIG. 9 WATER VAPOR BALANCE MAINTAINED BY VENTILATION

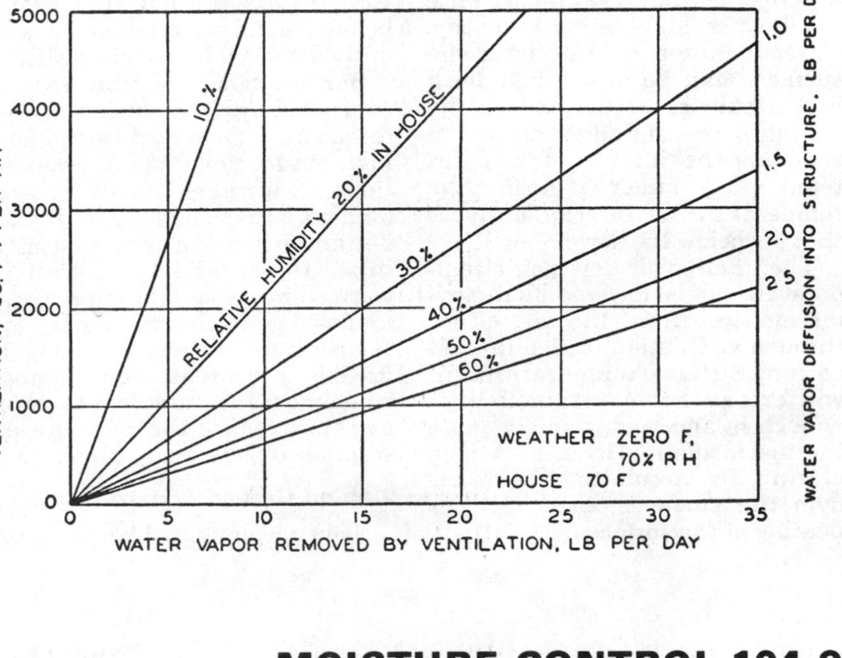

(1) the air temperatures both inside and outside the building, and (2) the heat transmittance coefficient (U value) of the wall, ceiling or floor assembly. Ratings of *overall* heat transmittance for an assembly can be derived by formula from individual transmittances of the materials in the assembly.

Figure 10 shows the relative humidities at which condensation will appear on interior surfaces with various U values in a room with air temperature of 70°F. Since condensation can appear at any sufficiently cold spot, values given in Figure 10 should be used with caution. U values constitute an average for large surface areas, within which there may be spots where the heat transmittance is higher—such as near structural supports or metal outlet boxes. Also, the inside surface temperature of a wood stud wall generally will be lower at the base due to air stratification inside the stud cavity. As a result, the maximum limit of relative humidity for a non-homogeneous wall is lower than might be inferred from its average U value.

Glass areas normally are the most likely surfaces in heated buildings on which condensation might appear; however, an uninsulated wall surface shielded from radiation and/or convection by furniture may also be quite cold. Surfaces which are close to the dew point temperature of the air, over a period of time may experience swelling, mold, staining or decay.

Condensation on interior room surfaces can be controlled both by suitable construction and by operating precautions such as: (1) reducing the interior dew point temperature and/or (2) raising the temperatures of interior surfaces that are below the dew point.

The interior air dew point temperature can be lowered by removing moisture from the air, either through ventilation or dehumidification. Surface temperatures in winter can be maintained high enough by incorporating adequate thermal insulation, by using double glazing, by circulating warm air over the surfaces, or by direct heating of the surface.

FIG. 10 RELATIVE HUMIDITY AND CONDENSATION

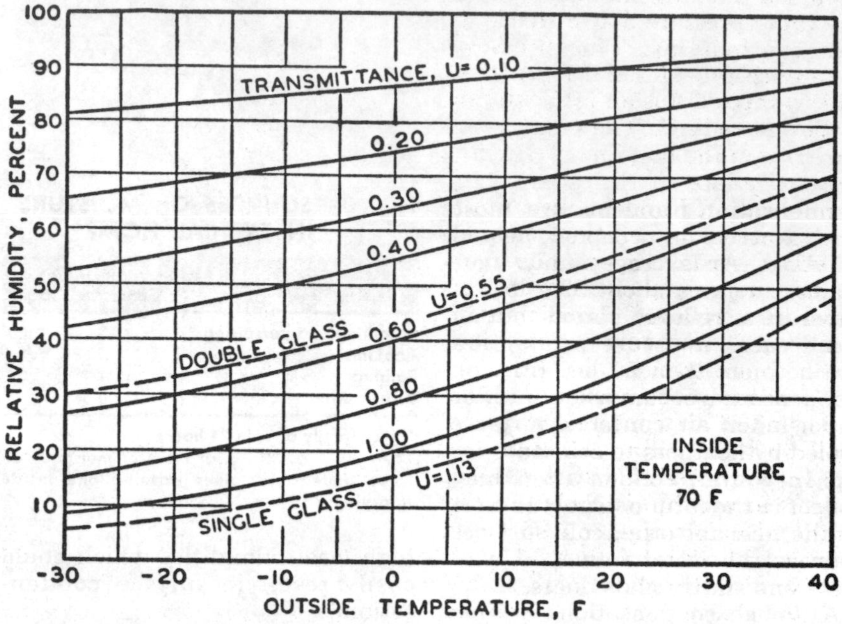

Below Grade Construction

Visible surface condensation may occur in summer on concrete basement walls and floors and on exposed cold water pipes. Because of their large mass and the insulating effect of the adjacent soil, basement floors and walls tend to maintain a relatively cool and constant temperature. If the dew point temperature within the space rises above the temperature of walls, floors and pipes, condensation results. Finish surfaces in below-ground spaces can be seriously damaged by condensation.

Surface condensation can be controlled by: (1) reducing the air intake from the outside at times of high outdoor dew point temperatures, (2) warming the surfaces, or (3) most successfully, by deliberate dehumidification. For instance, pipes to water closets can be warmed by using lukewarm water for flushing; in climates having low temperatures at night, it may be feasible to ventilate only at night when the air contains less moisture and thus reduce the moisture absorption of hygroscopic materials.

Slab-on-Ground Construction

Slabs on ground, like basement walls, tend to maintain a relatively constant temperature. Surface condensation can develop when the slab is sufficiently cold in relation to the warm interior atmosphere having a high dew point. In climates where high dew point temperatures frequently occur in summer, condensation may damage the finish flooring laid on the slab. Thick rugs or carpets aggravate the problem because they insulate the slab from interior warmer air and retain the slab at a low temperature. Because woven, knitted or tufted floor coverings usually are vapor permeable, moisture may condense under the covering and may promote decay, mold and musty odor.

Methods of controlling surface condensation are similar to those for above-grade construction: deliberate dehumidification and increasing surface temperature. The slab can be kept warm by insulating the slab from cooler ground temperatures with rigid insulation at the perimeter and with a granular well-drained base course under the entire slab.

CONCEALED CONDENSATION

In the winter, the continual production of vapor inside an

occupied building normally raises the vapor pressure above that outside the building. This pressure difference provides the force that causes vapor diffusion into exterior walls and ceilings. The vapor pressure rise in the building is directly proportionate to the amount of vapor produced and inversely proportionate to its ability to escape. In other words, the pressure rises as more vapor is produced, but the pressure is relieved if it is permitted to diffuse freely to areas of lower concentration.

As water vapor moves through a portion of a structure, such as a wall or floor/ceiling assembly, it may reach a point within the assembly at which it is cooled to the point of condensation. In older houses without insulation, so much heat was lost through the assembly that the entire joist or stud cavity often was warm enough to be above the dew point (Fig. 11a). The dew point temperature was reached somewhere in the sheathing or siding, but condensation was rarely a problem because moisture was carried off by air leakage through joints in the sheathing and/or siding which were usually of the board-type.

Although the theoretical dew point line in an insulated frame wall generally lies well within the cavity or even within the insulation (Fig. 11b), the vapor does not condense at this plane but moves on through the vapor-permeable fibrous insulation until it reaches the sheathing. Modern sheathing and siding products frequently are made in broad panels rather than narrow boards, thus reducing the opportunity of moisture diffusion by air leakage through joints; also many siding and sheathing materials such as plywood and hardboard are inherently vapor resistant so that only small amounts of vapor can move directly through the material itself. All of these factors, coupled with the generally higher interior humidity levels maintained by modern heating systems, account for the greater incidence of condensation problems in modern construction, where an adequate vapor barrier is not provided.

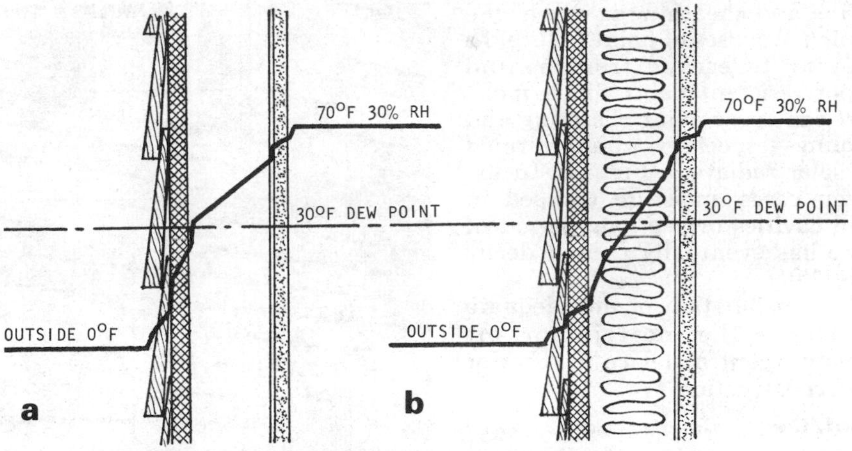

FIG. 11 When no insulation is used in the wall cavity, so much heat is lost that temperature in the cavity frequently is above dew point (a); with insulation, dew point is reached within the cavity (b).

FIG. 12 Sheathing reversed to show accumulation of frost on inside surface; cavity was insulated but was not protected with vapor barrier.

Wall Construction

When water vapor is allowed to enter a wall assembly and condensation occurs within its outer cold elements, frost or water may develop, depending on the outside temperature (Fig. 12). In weather that is continuously cold for a long period, the frost may build back into the wall cavity or fibrous insulation forming a mass of ice. If the outside temperature rises frequently, the frost or ice may melt and be absorbed by hygroscopic materials such as wood, or it may run down non-absorbing or already-saturated surfaces (Fig. 13).

In masonry walls, water seepage to the outside may occur harmlessly when the weather is above freezing, but water seepage into the building must be prevented. The inclusion of weep holes and base flashings encourages runoff to the outside and bars seepage to the interior (see Masonry Walls, page 343-11).

Accumulation of moisture within the insulation lowers the efficiency of the insulation, causing greater heat loss and other condensation problems. Trapped moisture in siding and sheathing is one of the principal causes of paint failure. In seeking to escape from behind vapor resistant paint films, moisture can cause paint blistering and peeling—especially when warmed by solar radiation (Fig. 14). In extreme cases, moisture trapped in wall cavities over a long period of time has eventually caused decay (Fig. 15).

The installation of an adequate barrier is the most important moisture control device in exterior wall construction.

Roof/Ceiling Construction

The principles of vapor movement and control within roof/ceiling construction are much the same as in walls. Water vapor from interior spaces moves toward the colder outside and may condense in concealed spaces of the roof/ceiling assembly. Some roofing materials, such as wood shingles can "breathe" and are affected little by trapped vapor; others such as built-up roofs are likely to be ruptured by excessive moisture. When a built-up roof blisters and cracks, permitting water to enter the roofing, continual and rapid deterioration is imminent.

In buildings or in spaces with high relative humidities, such as laundry, sauna or pool enclosures, special consideration must be given to properly placed effective barriers and adequate insulation. In typical roof/ceiling construction with insulation between joists, adequate ventilation and a barrier on the warm side of the insulation, no problem exists. But plank and beam roof construction with rigid insulation above the wood deck may require a vapor barrier under the insulation and/or details which permit *venting* (facilitating vapor diffusion) of the insulation.

Condensation within unventilated roof/ceiling spaces may collect in the form of ice, frost or water on the rafters, roof boards and on projecting nails (Fig. 16). Water also may be retained between the roof sheathing and impermeable exterior roof covering such as built-up roofs. On sunny days, even at low temperatures, frost or ice may melt and seep to

FIG. 13 Condensation formed frost and ice behind siding during cold weather, then weeped through laps in siding as the weather moderated, staining surface.

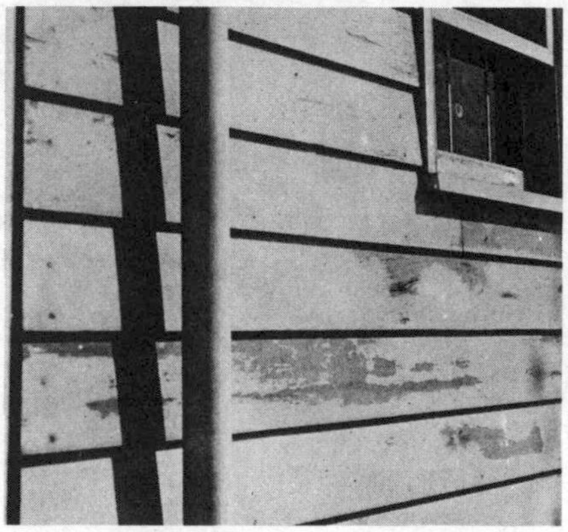

FIG. 14 Blistering and deterioration of paint surfaces caused by vapor pressure behind paint film.

FIG. 15 Decay of studs and sheathing due to frequent condensation in wall cavity.

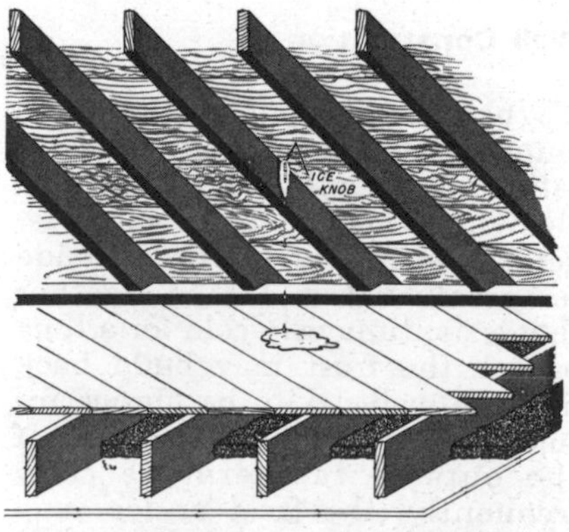

FIG. 16 Dripping ceilings frequently have been traced to condensation on protruding nails in unventilated attics.

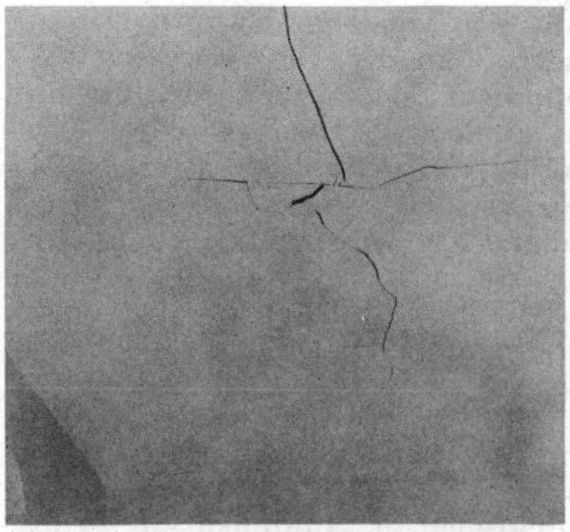

FIG. 17 In this house, repeated wetting of ceiling caused plaster cracking and staining.

the ceiling below causing staining and, possibly, damage to gypsum board and plaster ceilings (Fig. 17). Rapid deterioration and decay of roof members also is possible in extreme cases.

Adequate *vapor barriers* and the *ventilation* of roof/ceiling spaces above the insulation are essential.

Crawl Space Construction

In crawl spaces, water vapor movement from the ground below generally is a greater problem than vapor movement from inhabited spaces above. Moisture in the soil can rise as much as 11 ft. above the water table. Depending on ground water tables, character of the soil and drainage around the foundation, large amounts of water may be introduced into the crawl space by evaporation from the ground (Fig. 18).

Excessive vapor build-up can cause condensation on the outer ends of the floor joists and other structural members which are cooled by heat loss to the outside (Fig. 19). Water vapor from the crawl space may also enter walls, move upward within stud spaces and even reach roof construction when the wall structure permits.

Controlling evaporation from the ground with vapor barrier and providing adequate ventilation are simple preventive measures which will avoid these problems.

FIG. 18 DAILY EVAPORATION FROM CRAWL SPACE FLOOR

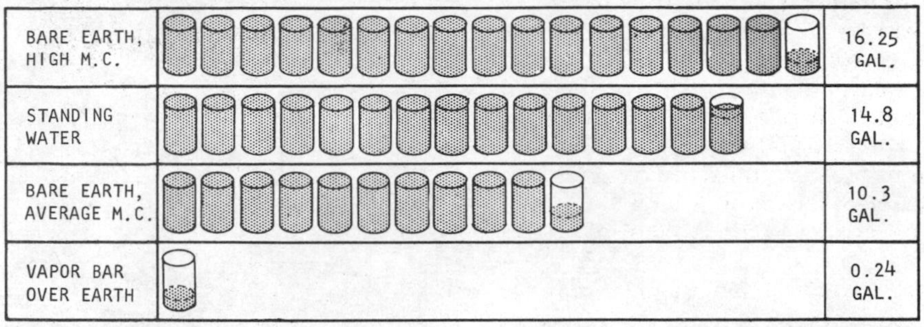

BARE EARTH, HIGH M.C.
16.25 GAL.
STANDING WATER
14.8 GAL.
BARE EARTH, AVERAGE M.C.
10.3 GAL.
VAPOR BAR OVER EARTH
0.24 GAL.

FIG. 19 Beads of condensation formed on joists of a crawl space not provided with foundation vents or ground cover.

This page deliberately left blank.

DESIGN AND CONSTRUCTION RECOMMENDATIONS

Moisture condensation in wall, floor and roof/ceiling assemblies can be controlled (1) by providing a vapor barrier on the interior side of the assembly to limit vapor entrance into the construction and (2) by ventilating structural spaces (wall cavities, attic areas, crawl spaces) to dissipate vapor to the outside. This double line of defense is desirable in all areas where *severe* or even *moderate* condensation hazard exists. The continental United States can be divided into three condensation zones, based mainly on the isotherms of the 0°F and 20°F winter design temperatures (Fig. 20).

When construction limitations make it impractical to provide a vapor barrier, more reliance is placed on ventilation; conversely, when adequate ventilation is hard to achieve, more vapor-resistant interior construction is indicated.

In the following discussion the term "ventilation" describes typical methods of dissipating moisture from structural spaces through continuous air passages linked to the general atmosphere. "Venting" describes the deliberate inclusion of air leaks or gaps in an otherwise vapor-resistant finish or membrane. In venting, air spaces are not necessarily interconnected and air movement is non-existent or sluggish; ventilation infers much more positive air movement and more rapid moisture dissipation.

VAPOR BARRIER SELECTION

Although a vapor barrier by definition is any material with a permeance of 1 perm or less, not all materials are equally suitable for use in a particular assembly. Some materials are inherently better vapor barriers than others; however, the effectiveness of the barrier depends on the method of installation, as well as its physical properties. For instance, 1 mil thick aluminum foil has a perme-

FIG. 20 **CONDENSATION HAZARD ZONES OF THE U.S.**

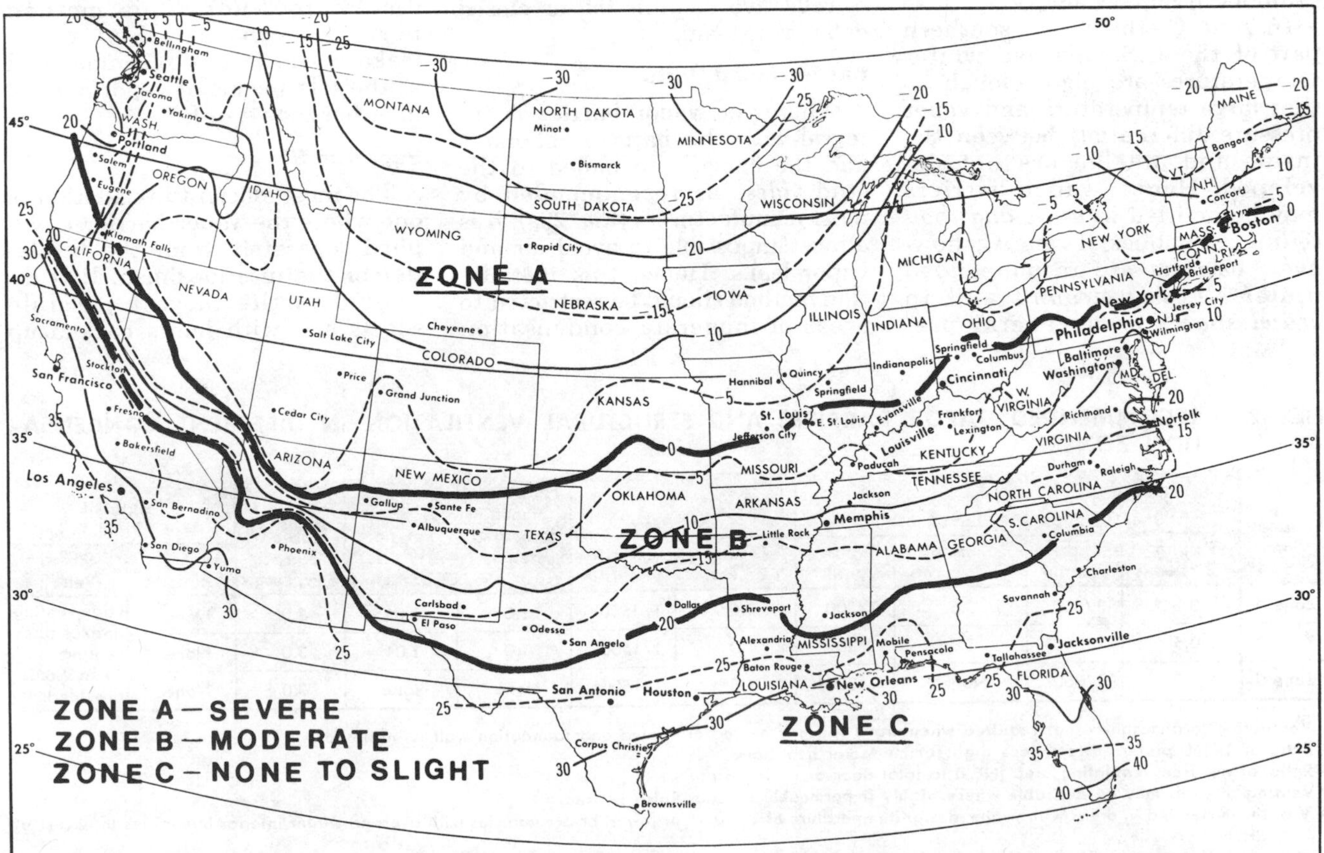

ance of 0.00; the permeance of the typical 0.35 mil paper-backed product is likely to be approximately 0.05, when installed in large sheets across the face of the studs. The practice of stapling aluminum-faced batts to the sides of wood studs may increase the installed permeance approximately 20-fold, to as high as 1 perm.

Selection of a suitable barrier depends on a variety of factors such as prevailing interior and exterior humidities, length and severity of the winter season, materials and methods of construction. In general, the severity of the winter season, as expressed by the winter design temperature is a prime factor in determining the degree of vapor protection required (Fig. 20).

Figure 21 summarizes vapor barrier and ventilation recommendations for roof, wall and floor assemblies in the three zones. Structural ventilation is recommended for joist roofs and crawl spaces in all zones. Vapor barriers are recommended for most assemblies in zones A and B.

In zone C, the warm southern part of the U.S., average winter temperatures are high enough so that large temperature and vapor pressure differentials between the inside and outside are not developed. Hence, vapor barriers may be omitted in *walls and roof/ceiling* assemblies of zone C; however, barriers are recommended under *slabs-on-ground* and in *crawl spaces* as a general precaution against moisture in this, as in all other zones.

Figure 21 recommends three levels of barrier efficiency: (a) permeance of 0.1 perms or less—usually aluminum foils or plastic films; (b) 0.5 perms or less—asphalt-coated felts or asphalt-laminated papers; (c) 1.0 perms or less—same as above, but with less asphalt per sq. ft.

Depending on the type, vapor barriers may be incorporated with, or installed separately from, other materials in the construction. Typical vapor barriers in residential construction include: (1) a separate, plastic or metallic membrane, job-installed over framing members and under the interior finish material; (2) a metallic foil, factory-applied on the back of an interior finish material such as gypsum wallboard; (3) aluminum-faced paper or asphalt-laminated paper, factory-applied over batt or blanket insulation; or (4) insulations made of inherently vapor-resistant materials, such as layers of reflective metallic foil or closed cell plastic foam.

Barrier-faced Batts

Where the vapor barrier is integral with the batts or blankets and the flanges are nailed to the stud *sides*, as recommended by most manufacturers (Fig. 22a), it is almost impossible to avoid air and vapor leaks. Hence this installation method should be restricted to areas of moderate condensation hazard, as in zone B. In zone A, the vapor barrier flanges should be secured across the stud *faces*. or a separate, continuous, vapor barrier should be used (Fig. 22c).

Foil-backed Boards

Insulating gypsum lath and gypsum wallboard have an aluminum foil barrier laminated to the back of each board (Fig. 22b). However, breaks exist at board joints—typically at 16″ x 48″ intervals for lath and at 48″ x 96″ intervals for wallboard. These joints permit some vapor leakage (especially where the joint is not backed up by a structural member) and the efficiency of the barrier is reduced.

Foil-backed Wallboard Use of this product results in effective vapor isolation if the boards are installed vertically, with all joints over studs. It is suitable for vapor control in exterior walls in all zones.

Foil-backed Gypsum Lath Lath boards generally are installed horizontally with joints perpendicular to studs. This method permits a great deal of vapor leakage and is not recommended as the sole means of vapor control in either zone A or B.

Separate Sheets

The most efficient installation is one where the vapor barrier is applied separately from the insulation or the interior finish. It is installed in the largest possible sheets and with joints backed up

FIG. 21 **RECOMMENDED VAPOR BARRIERS AND STRUCTURAL VENTILATION IN DIFFERENT CONDENSATION ZONES**

Condens. Zone	Attic-less Joist Roof/Ceiling		Attic-type Joist Roof/Ceiling		Crawl Space		Floor Over Open Unheated Area —	Exterior Walls	Slab-on-Ground	Insulated Wood Plank Roof	
	Ceiling V.B., Perms	Struct. Ventil.[3]	Ceiling V.B., Perms	Struct. Ventil.[3]	Ground V.B., Perms	Struct. Ventil.[1,3]	V.B., Perms	V.B., Perms	V.B., Perms	V.B., Perms	Struct. Ventil.
Zone A	0.1	1/300	0.5	1/300	1.0	1/1500	1.0	1.0[4]	3.0	None[5]	Edge venting and/or stack venting when V.B. is provided.
Zone B	0.5	1/300[2]	1.0	1/300[2]	1.0	1/1500	1.0	1.0[4]	3.0	None	
Zone C	1.0	1/300[2]	None	1/300[2]	1.0[6]	1/1500[6]	None	None	3.0	None	

[1] Ventilation to atmosphere not required when crawl space is heated or cooled and foundation wall is insulated.
[2] Ratio of 1/150 preferred to relieve high summer solar heat gain.
[3] Ratio of net free ventilating area (NFA) to total floor or ceiling area.
[4] Venting of walls may be desirable where highly impermeable exterior finish is used.
[5] V.b. recommended in areas with winter design temperature of −10° F or lower, or occupancies with average winter interior humidities in excess of 40% RH.
[6] When crawl space is not heated or cooled, v.b. may be omitted provided vent area is increased to 1/150.

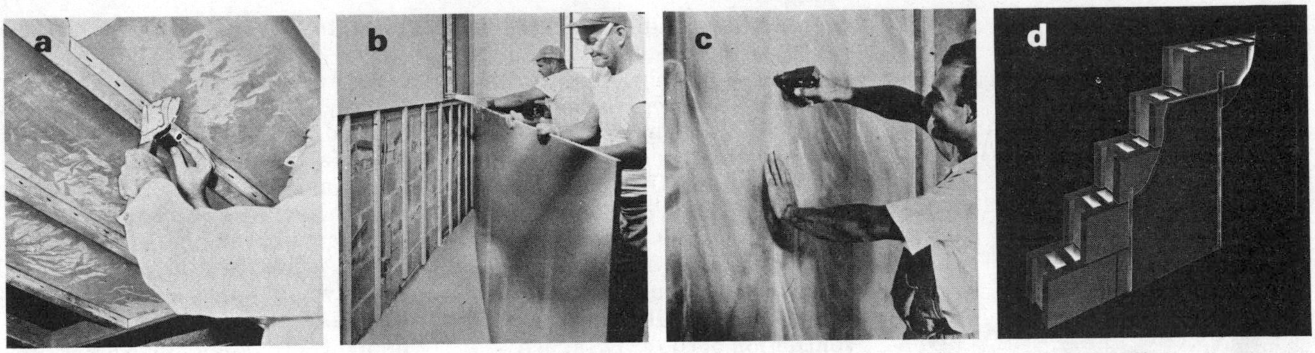

FIG. 22 Vapor control can be provided by using (a) barrier-faced insulating batts, (b) foil-backed gypsum board, (c) separate polyethylene sheet and (d) insulating boards with inherent vapor resistance.

by framing members (Fig. 22c). Insulation can be provided as pressure-fit blankets or rigid boards.

Polyethylene film and Kraft-paper supported aluminum foil are typical products available in roll widths from 3′ to 20′. Best results are obtained when joints are made over a structural member. This "vaporproofing" method is highly recommended for zone A.

Insulating Boards

A number of synthetic insulating boards—such as foamed glass, polystyrene and polyurethane—are being marketed; many of these materials have a closed cell construction and are inherently good barriers (Fig. 22d). However, their permeability varies with manufacturing process and should be checked in each case to determine that the required permeance can be achieved in the thickness contemplated. Also boards should be butted tightly together and joints should be sealed with a vapor-resistant mastic.

Paints and Other Coatings

Although many types of paints have proven to be good vapor barriers under laboratory conditions, the uncertainties of field application make reliance on paint coatings somewhat questionable. In new construction, the more reliable factory-produced vapor barriers are recommended.

However, paint coatings can be used correctively in older structures, and as an adjunct to other vapor control methods. Several manufacturers now market products for specific use as vapor barriers. Alkyd, aluminum and bituminous coatings generally have low permeability (high resistance) to vapor. Vapor properties of specific products depend on formulation, hence the manufacturer should be consulted. Most coatings are more vapor-resistant when their finish appearance is shiny or glossy.

BARRIER AND VENTILATION DESIGN

Ventilating openings in exposed locations often are provided with louvers to keep out wind and rain; screening usually is included to bar insects and vermin. Ventilating requirements generally are given as *net free area*, which is the space actually available to the moving air, after deducting obstructions such as the screening and louvers. Figure 23 shows how to calculate the *required gross area* of various types of ventilating openings.

Walls

Ventilation of exterior wall cavities may provide adequate vapor control in a moderate climate. However, wall openings designed to move substantial amounts of air and vapor may waste considerable heat in colder climates. *Hence, in walls, as in most other assemblies, vapor barriers are the primary means of vapor control; wall ventilation (more typically venting), when provided, is an adjunct to vapor control with suitable barriers.*

FIG. 23 REQUIRED GROSS AREA OF VENT OPENINGS

VENT COVERING	R.G.A.* / N.F.A.
Hardware cloth, 1/4″ mesh	1
Screening, 1/8″ mesh	1 1/4
Insect screen, 1/16″ mesh	2
Louvers & hardware cloth, 1/4″ mesh	2
Louvers & screening, 1/8″ mesh	2 1/4
Louvers & insect screen, 1/16″ mesh	3

* R.G.A. means required gross area; N.F.A. means net free area. Net free area can be determined from Fig. 21.

Recognizing the high vapor resistance of plastic and metal sidings, manufacturers of these products usually include ventholes. No special venting efforts are required with wood and plywood horizontal *board* sidings; these are self-venting at the overlap. However, plywood and hardboard *panel* sidings require greater attention to effective interior vapor barriers and may require special venting details.

In older homes without vapor barriers, painted siding finishes may resist vapor flow sufficiently to trap moisture within stud spaces. In zones A and B, the resulting vapor pressure under the paint film may cause blistering and peeling. In these cases, ventilating stud spaces with individual vents at top and bottom (Fig. 24) will help dry out the stud cavities. However, as vents become clogged with cobwebs and dust, ultimately it still is necessary to provide an interior vapor barrier.

It is essential to form a continuous barrier envelope on the inside of all walls, using a material with a permeance of 1.0 perm or less. Relatively permeable materials should be used on the

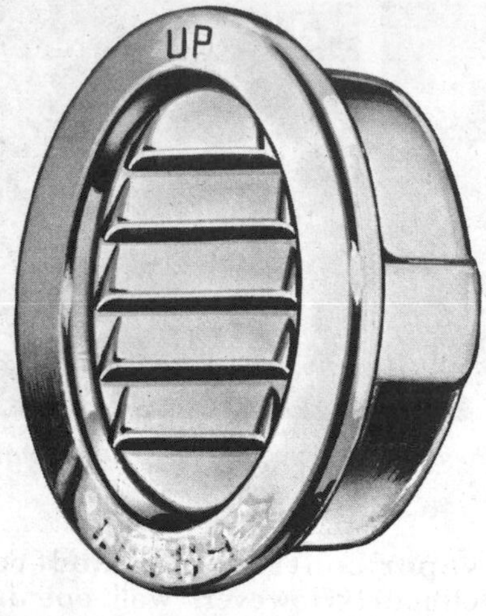

FIG. 24 Prefabricated cup vents can relieve vapor pressure in walls of older homes not equipped with barriers.

outside of the wall; the sheathing paper should be a "breathing" type, having a permeance of 5 perms or more.

(Special emphasis is placed on the sheathing paper because the 15 lb. asphalt-saturated felt typically used for this purpose also is used for built-up roofing and often is too vapor-resistant for use in the wall. Figure 7 shows 15 lb. asphalt felt and tar felt as having a permeance of 1.0 and 4.0 perms respectively; these are average figures—permeance of a particular product can be higher or lower depending on the degree of saturation with bitumens.)

A good rule of thumb to insure safe outward diffusion of vapor is that the combination of exterior materials should have a total installed permeance of at least 5 times the total permeance of the interior vapor barrier plus interior finish materials. Thus, if the *interior* barrier-and-finish permeance is 1 perm, the total *exterior* permeance should be 5 perms or more. Conversely, when it is known that vapor-resistant finishes such as plywood or hard board panels will be used on the exterior, a more effective barrier and installation method should be used.

Crawl Spaces

When a crawl space underlies a heated or air-conditioned area, it is generally more economical to limit heat loss (or gain) by insulating the foundation wall rather than the floor above. Under these conditions, ventilation to the *outside* is eliminated and the crawl space is likewise heated or cooled. Introducing nominal amounts of air from the supply side (or conducting air to the return side) of an air handling unit will insure adequate ventilation of the crawl space.

In closed crawl spaces not equipped with a means of ventilation, moisture rising by capillary action from the ground can cause condensation problems. To stop the moisture at its source, houses with either heated or cooled crawl spaces in all zones should be provided with a vapor barrier laid directly over the soil as a ground cover.

A typical vapor barrier consists of 4 or 6 mil polyethylene film, a coated 30 lb. per sq. ft. or heavier roofing felt or roll roofing. The barrier should be installed in largest possible widths, with edges lapped 6″ and taped or sealed (unless ballast is used), and the sides turned up approximately 6″ against the foundation walls. A ballast of pea gravel or sand over the barrier is desirable to keep it in place and to prevent mechanical damage.

Ventilation to the atmosphere is recommended in all zones when the foundation wall is *not* insulated. When a vapor-resistant ground cover is included, at least 2 vents should be provided, with a total net free area not less than 1/1500 of the crawl space area (Fig. 25). Vents should be placed as high as possible, in opposite walls. Vents are not required if the crawl space is completely open on one side to a basement; however, if the crawl space area is greater than the basement, it should be separately ventilated by supplying or returning air.

In older buildings or where the crawl space is not cooled or heated, the ground cover can be omitted, provided the net free area of vents is increased 10 times, to 1/150 of the crawl space area. In the absence of a ground cover, more positive ventilation is required and at least four vents should be provided—one in each wall.

In zone A, crawl spaces are usually heated and ventilation is limited. If the crawl space will be ventilated in the winter, it is necessary to insulate the floor to reduce heat loss and improve occupant comfort. During periods of cold weather, the air in the crawl space would have a low temperature and low moisture content, hence there would be vapor pressure *outward* from the dwelling. To prevent condensation on the joists, a vapor barrier would be required on the *upper*, warm side of the insulation. Normal barrier-faced blankets have nailing flanges on the same side as the barrier. If the insulation is installed from below the joists, special reverse flange blankets or barrier-wrapped blankets should be used (Figs. 36 and 37). Wide polyethylene sheets laid over the subfloor or joists are most effective, but barrier-faced or barrier-wrapped blankets are acceptable.

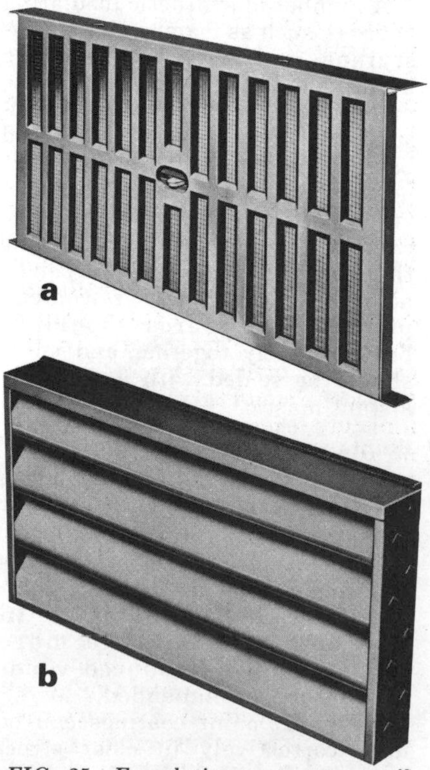

FIG. 25 Foundation vents are available in (a) grill type with closing damper and (b) louver type.

Roof/ceiling Construction

Joist-type roof/ceiling assemblies always should be provided with ventilation. Generally, a net free area not less than 1/300 of the ceiling is recommended, in addition to a suitable vapor barrier. In older buildings not provided with vapor barriers, the vent openings should have a net free area not less than 1/150 of the ceiling area.

This larger volume of ventilation may be desirable for attic heat control in the moderate and southern zones, regardless of the presence of barriers. The effectiveness of the ventilation system in removing both heat and moisture depends on the volume of air moved, and the extent to which the air flow reaches into all parts of the attic or joist space.

Ventilation of roof/ceiling spaces is particularly important because this area is a natural depository of moisture from inside the house. Convection currents can carry moisture-laden air up the wall cavities and into the roof space through openings inadvertently left in the top plates. Light fixtures and other equipment installed in ceilings create air and vapor leaks.

Ideally, all such vapor leaks into roof/ceiling spaces should be eliminated; in practice this is not possible and a fair amount of moisture will penetrate past the ceiling. Fortunately, suitable ventilation details can dissipate this moisture safely to the atmosphere. In the case of an attic-type roof, ventilation alone could take care of all the moisture (except in the coldest part of zone A, with winter design temperature of —10° F or lower).

The recommendation of a vapor barrier in attic-type roofs is based perhaps as much on the desire to maintain the interior relative humidity, as it is to minimize moisture hazards in attics. On the other hand, in attic-less joist roofs, ventilation does not involve as much air movement, moisture cannot be dissipated as readily, and reliance on the quality and continuity of the vapor barrier is much greater.

Attic-type Joist Roofs Extensive ventilation tests have determined that *wind direction* and *velocity*, as well as the *vent type*, greatly affect the distribution and volume of air movement within attic spaces. "Dead" spots of minimum air flow appeared to be pockets of potential condensation. Hence, ventilation recommendations based only on the ratio of vent opening to ceiling area may be inadequate. Vents should be carefully selected and located for maximum air flow and uniform air distribution; generally, it is recommended that at least 1/2 of the required vent area be provided at the ridge or upper half of the roof, the balance at the eaves. The important characteristics of typical residential attic vents are described below.

Roof Louvers (Fig. 26a) when used alone contribute little to effective air flow through an attic space. However, when placed in the upper half of the attic space and combined with *soffit vents*, (Fig. 26b) in equal amounts, the air distribution is improved and air flow is almost doubled. The combined system also becomes more responsive to wind, increasing air volume in direct proportion to wind velocity.

Continuous Soffit Vents (Fig. 26b) produce substantial air flow regardless of wind direction, and the flow is directly proportionate to wind velocity. However, most of the air flow remains near the attic floor, indicating a need for vent openings in the upper roof or ridge for good circulation.

Gable End Louvers (Fig. 26c) are most effective for wind directions perpendicular or diagonal to the louver face. However, air flow is 1/2 to 1/3 this amount when the wind direction is parallel to the louver face, perpendicular to the ridge. Although the addition of soffit vents does not improve air flow for the ideal condition, under adverse wind conditions, the addition of continuous soffit vents increases air flow by 50%.

Continuous Ridge Vents (Fig. 26d) are unique in that they provide air flow due to temperature differences with roof slopes as low as 4/12. They provide about the same air flow at 0 wind velocity as at 10 mph; however, air flow volume is several times larger with

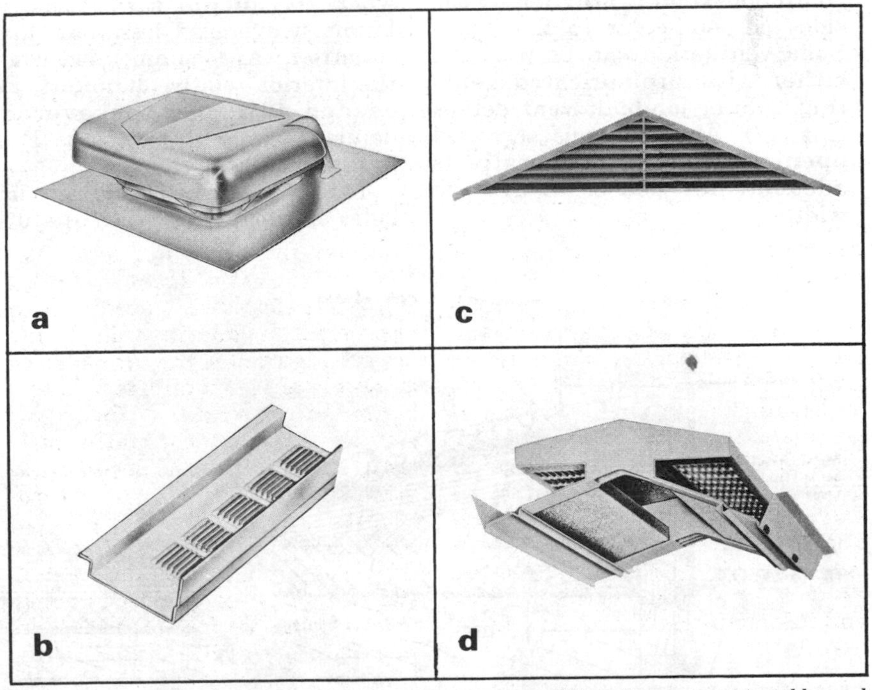

FIG. 26 Typical roof ventilators: (a) roof louvers, (b) soffit vents, (c) gable end louvers, (d) continuous ridge vent.

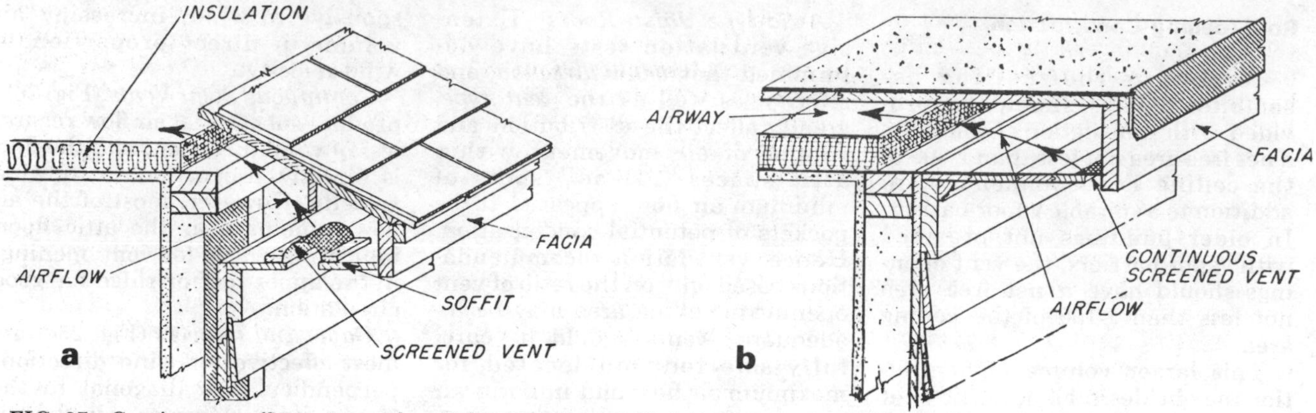

FIG. 27 Continuous soffit vents can be used to ventilate both (a) attic-type and (b) attic-less joist roofs.

wind parallel to the ridge as it is with the wind perpendicular. The combination of continuous ridge vents with continuous soffit vents produces the greatest volume and most uniform air flow under most wind conditions.

Attic-less Joist Roofs Flat or shed roofs of single joist construction should be provided with effective ventilation, as well as an efficient vapor barrier under the insulation. Ventilation of single joist roofs should include outlets from every joist space, either with individual cup vents (Fig. 24) or with continuous soffit ventilation on the sides perpendicular to the joists. Soffit ventilation can be provided either with prefabricated vents (Fig. 26b) or job-built vent details (Fig. 27). A continuous screened opening 3/4″ wide generally is adequate for houses of average width.

Wood Plank Roof/ceiling Because built-up roofing, typically used with insulated wood decks, is a nearly perfect vapor barrier (0.00 perms), it is impossible for vapor to be diffused through it. A recent nationwide study of roofing problems demonstrated that a vapor barrier on the underside of the insulation actually can cause more problems than it solves, by trapping moisture within the insulation. The study concluded that a vapor barrier generally was not needed in residential construction, nor in other occupancies where the interior winter relative humidity averaged less than 40%. A barrier was recommended when the interior relative humidity exceeded 40% *and* the average January temperature was 35°F or lower.

The 35°F January isotherm coincides approximately with the 0°F isotherm shown in Figure 20; humidities over 40% are encountered in public shower rooms, kitchens, laundries and pool areas. In single family structures, only pool areas merit special consideration; other high moisture areas, such as baths and kitchens, usually are equipped with exhaust fans to draw off the moisture.

Unless a vapor barrier is used on the underside of the insulation, it is not necessary to dissipate moisture with structural ventilation or venting. The insulation generally is absorptive enough to hold the vapor for short periods of time without condensation and the wood plank deck is permeable enough to permit harmless dissipation of the vapor back into the interior, when the vapor pressure is reversed.

However, *if a vapor barrier is provided* (in exceptionally cold

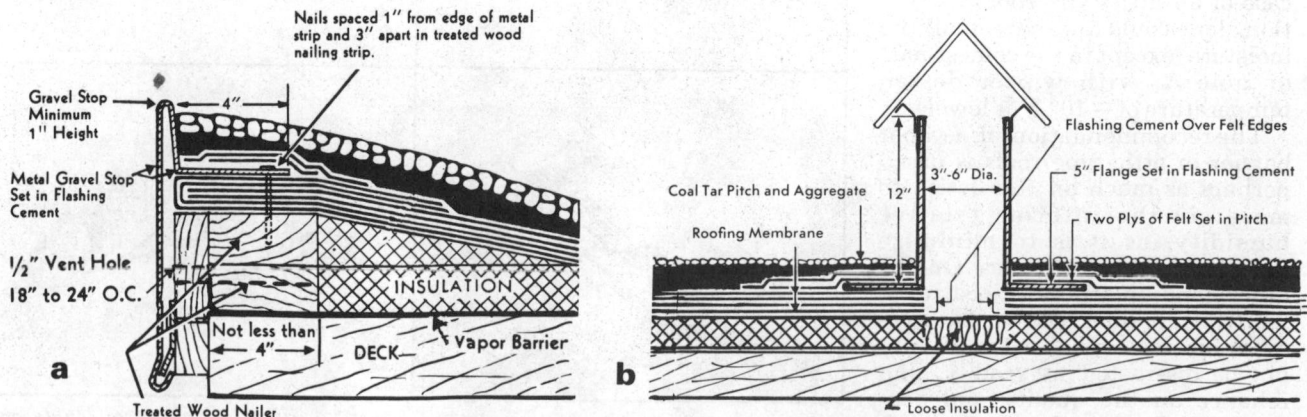

FIG. 28 Insulated wood plank decks may require (a) edge venting or (b) stack venting.

climates and with high interior relative humidities), the possibility of trapping moisture between the barrier and the roofing is greatly increased. The moisture may be present during construction, or may enter the insulation later through leaks from above or below. In any case, moisture cannot readily escape through the roofing or the vapor barrier by diffusion, hence peripheral air leaks (venting) must be provided to permit dissipation of moisture.

Edge Venting is a method of connecting the insulation with the exterior atmosphere through built-in vent openings at the perimeter of the roof (Fig. 28a). Edge venting is adequate for roofs not exceeding 40′ in width; for wider roofs, stack venting as well as eave venting is required.

Stack Venting consists of simple sheet metal stacks about 3″ in diameter, reaching through the roofing membrane into the insulation (Fig. 28b). One stack should be provided for each 900 sq. ft. of roof area.

Slabs-on-Ground Whether at grade or below grade (basements), such slabs are in direct contact with the soil, which often contains large amounts of vapor or free water. Although 4″ of concrete has good resistance to vapor (perm rating of less than 1.0), the vapor pressure from the moist slab underside to the interior may be sufficiently high and prolonged enough to cause large amounts of vapor to be *diffused through* the slab. Additional moisture, in vapor or liquid form, can seep through cracks and structural joints in the slab. Hence, a separate vapor barrier under the slab generally is recommended.

Where the slab is underlain by well-drained granular fill, and in arid regions where irrigation and heavy sprinkling is not practiced, the vapor barrier *may* not be necessary. However, because a barrier can be installed at nominal cost during construction and because it is costly to correct moisture problems later, a barrier is recommended under slabs of habitable rooms in all zones. The vapor barrier should have a permeance of less than 0.30 perms, should be installed with 6″ laps and should be fitted carefully around utility and other service openings.

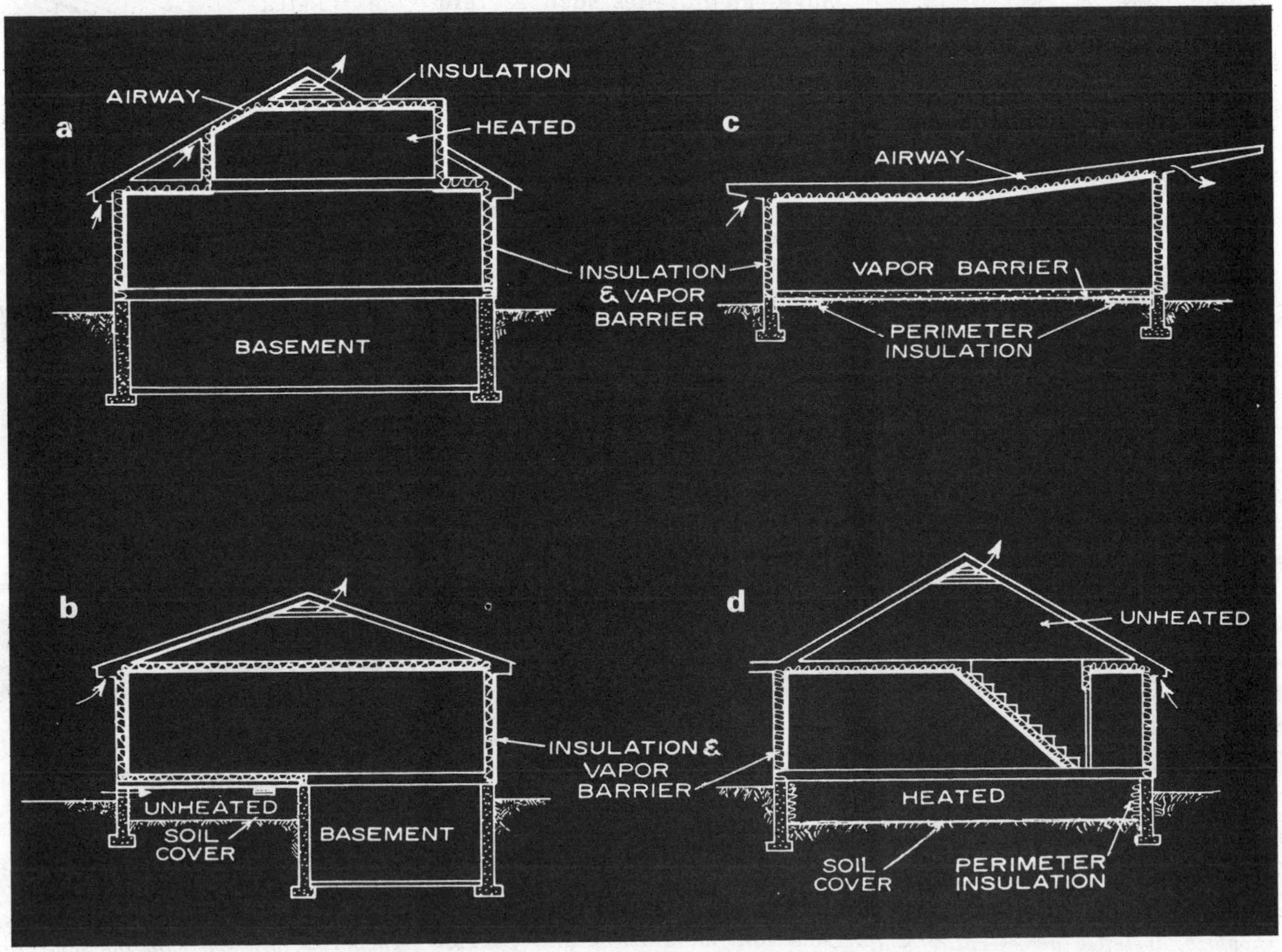

FIG. 29 Location of insulation and vapor barrier in (a) 1-1/2 story basement-type house; (b) one-story house with unheated crawl space; (c) slab-on-grade house; (d) one-story house with heated crawl space.

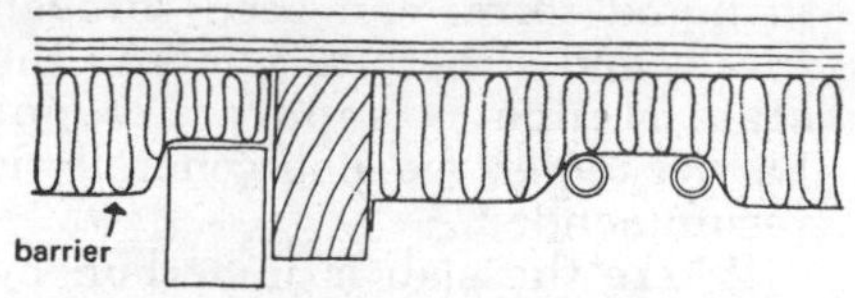

FIG. 30 Insulation should be packed behind pipes, ducts or electrical boxes, filling the space between them and the exterior finish.

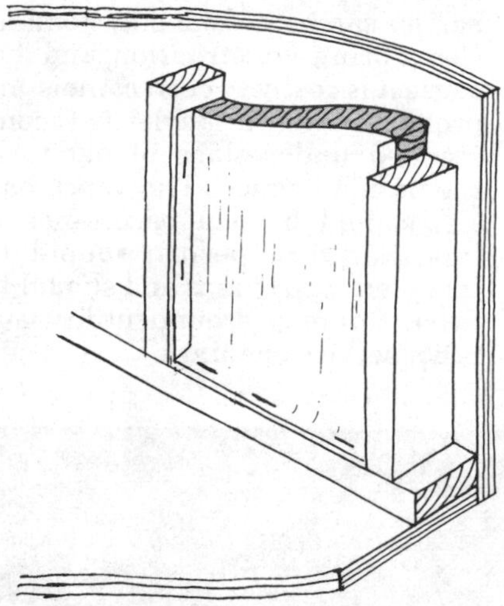

FIG. 31 When barrier-faced blankets are inset-stapled, they should be fit snugly to top and bottom plates.

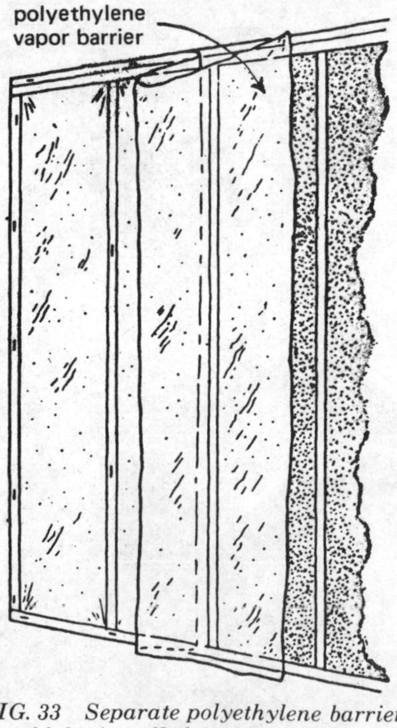

FIG. 33 Separate polyethylene barrier should be installed across entire wall surface, cutting out for door and window openings.

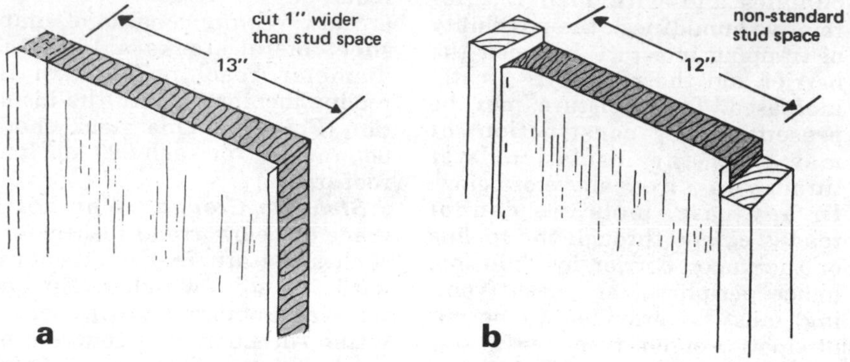

FIG. 32 For stud spaces narrower than 14-1/2", standard barrier-faced blankets should be cut down to form a stapling flange (a); then flanges should be smoothly stapled to stud sides (b).

Vapor barrier materials suitable for slab-on-ground construction include single-layer membranes such as polyethylene film, or multiple-layer membranes such as reinforced waterproof paper with a polyethylene film extrusion-coated onto both sides. When polyethylene film is used over gravel or crushed stone, or under any structurally reinforced slab, it should be at least of 6 mil nominal thickness; thinner, 4 mil film can be used over sand or firmly compacted soil.

INSTALLATION PRECAUTIONS

Three fundamental principles should be observed in the application of vapor barriers: (1) the barrier should be as near as possible to the warm (heated in winter) side of the insulation, or interior face of the assembly; (2) the barrier should be as continuous as possible, with a minimum of joints, breaks, openings or air leaks; and (3) the barrier should be installed by a method suitable for the specific condensation hazard (Figs. 20 and 21). The parts of a residence typically requiring a vapor barrier (and often insulation as well) are illustrated in Figure 29.

Since the vapor barrier frequently is an integral part of the insulation, effective vapor control generally depends on the proper selection and attachment of the insulation. Insulation and barrier should be fitted neatly behind piping and equipment (Fig. 30). In

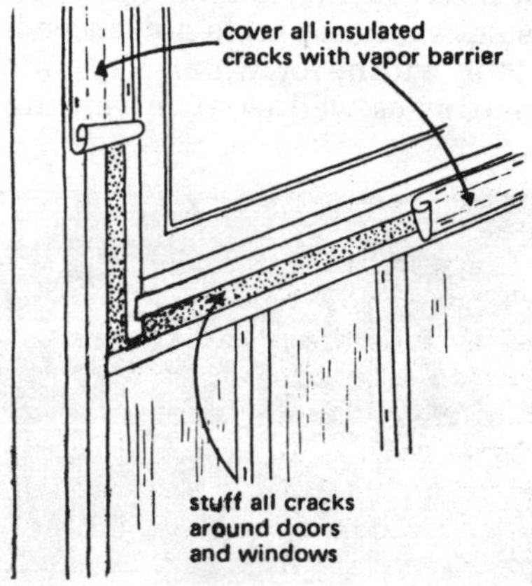

FIG. 34 Spaces between window or door frame and the rough framing should be packed with insulation and covered with vapor barrier.

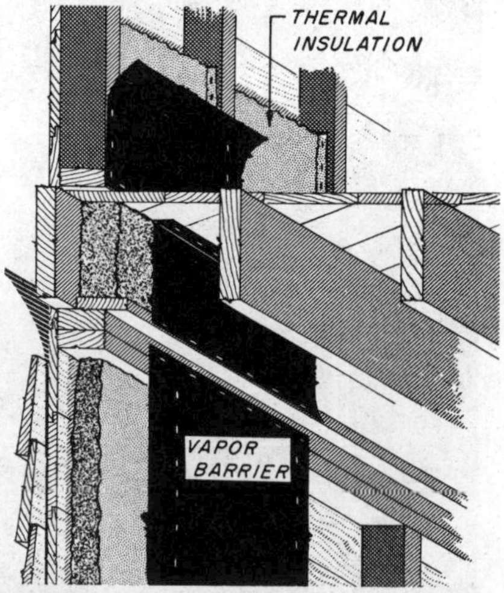

FIG. 35 Outside edge of floor cavities should be provided with the same insulation and barrier protection as adjacent walls.

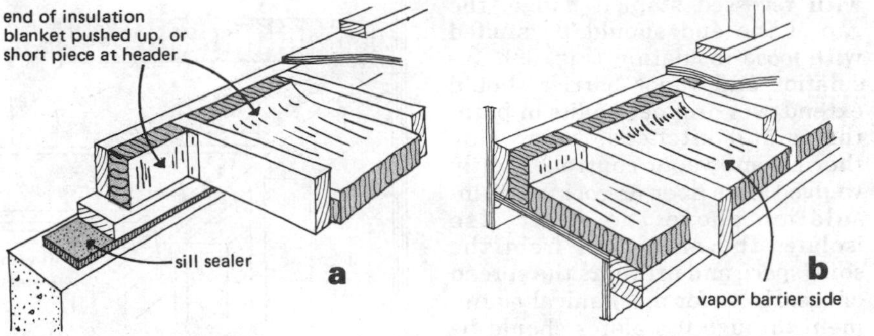

FIG. 36 *Insulating blankets are laid with barrier up in (a) floors over unheated crawl spaces and (b) floor overhangs. Reverse flange blankets can be installed from below.*

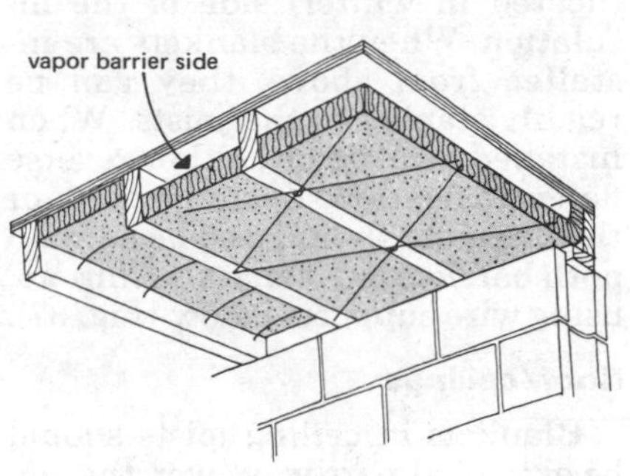

FIG. 37 *Wire supports can be used in place of stapling, when blankets are installed from below.*

using barrier-faced insulating blankets, the flanges should be snugly stapled approximately 6″ o.c. to framing members, so as to avoid air leaks (Fig. 31).

Walls

Gaps in the insulation near the bottom and/or the top of the wall cavity encourage convection currents, which carry heat and vapor from the inside to the outside face of the cavity. It is recommended therefore that the insulation be pushed tight against the back of the sheathing and that it fill the cavity from top to bottom (Fig. 31).

Batts and blankets are made to fit a standard 14-1/2″ wide cavity resulting from 16″ o.c. stud or joist spacing. Non-standard spaces less than 14-1/2″ wide should be fitted with blankets cut 1″ oversize to form a stapling flange on the cut size (Fig. 32). Non-standard spaces wider than 14-1/2″ can best be protected by a separate barrier.

If the standard spacing of members in a structure is 12″ or 24″ o.c., pressure-fit insulation with a separate barrier should be used (Fig. 33). Cracks around window and door frames should be stuffed with loose insulation and covered with a separate barrier (Fig. 34).

Floors

The perimeter of the floor assembly just behind the band joist is frequently overlooked when insulation and vapor barrier is installed. This is particularly true when the floor is between two habitable spaces and no insulation is used in the joist cavities (Fig. 35).

Where the floor is over an open-air area or an unheated crawl space, insulation is likely to be required in the floor. Here too, care should be taken to provide thermal and vapor control near the perimeter (Fig. 36). The vapor barrier should be on the upper

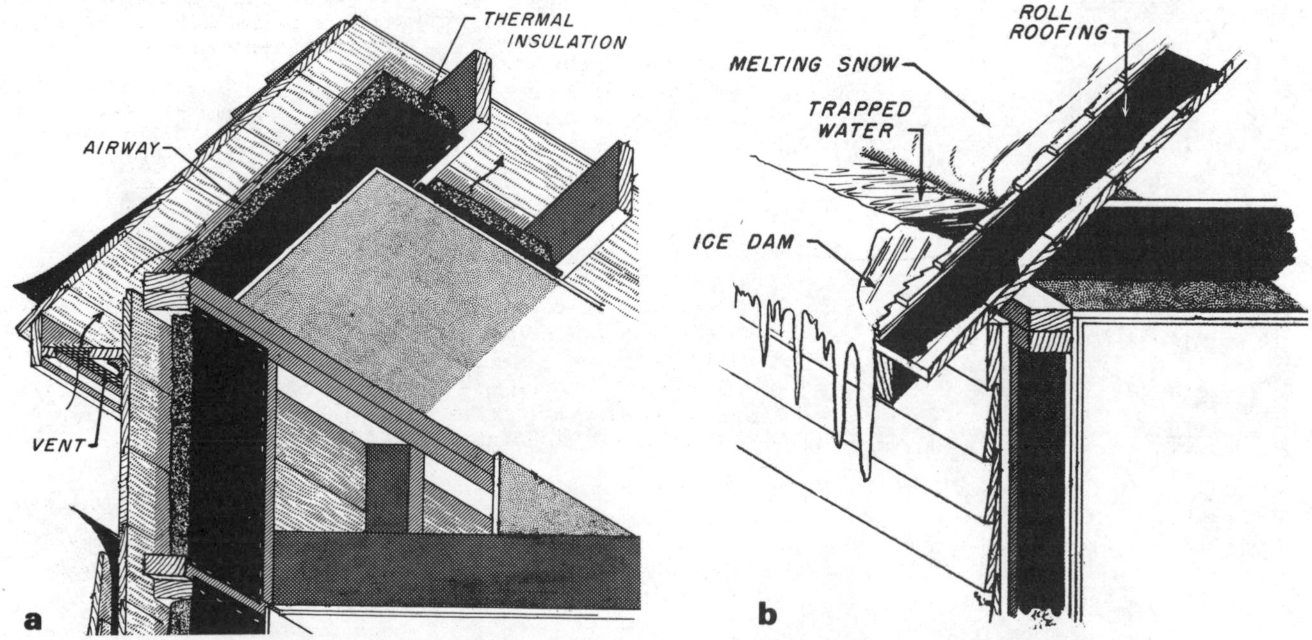

FIG. 38 *(a) Keeping an airway open over the insulation promotes good ventilation and equalizes roof temperatures; (b) if the airway is blocked, the eave is colder than the rest of the roof, causing ice dams and possible leakage.*

(heated in winter) side of the insulation. When the blankets are installed from above, they can be readily stapled to the joists. When installed from below, either reverse flange insulation should be used, or the blankets should be placed with both barrier and flanges facing up, using wire supports below (Fig. 37).

Roof/ceilings

Blankets in ceiling joists should be extended part way over the top plate but not enough to block eave ventilation (Fig. 38). When blankets are installed from below with recessed stapled flange, the gap at the end should be stuffed with loose insulation (Fig. 39). Insulation and vapor barrier should extend over dropped soffits in bathrooms and kitchens, to provide thermal and vapor control. Tightly wedged glass fiber or rock wool insulation above the soffit also isolates the attic area from the soffit space and prevents the spread of fire. Holes for mechanical equipment through top plates should be filled with loose insulation to minimize air and vapor leaks into the roof space.

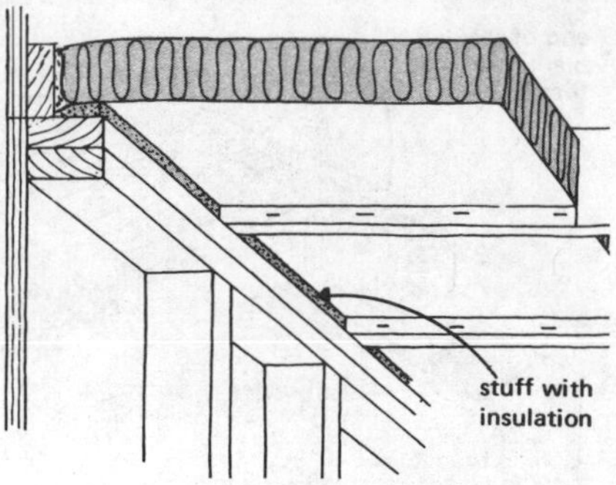

FIG. 39 Space between top plates and inset blankets should be filled with loose insulation.

We gratefully acknowledge the assistance of the following organizations with the preparation of the text and/or permission to use illustrations: American Society of Heating, Refrigerating and Air-Conditioning Engineers; Forest Products Laboratory; National Mineral Wool Insulation Association; Koppers Company, Inc.; Owens Corning Fiberglas.

MOISTURE CONTROL 104

CONTENTS

WATER VAPOR PROPERTIES

RELATIVE HUMIDITY

Relative humidity in residential structures should be controlled during the heating season within the limits indicated in Fig. WF1.

Although a relative humidity (RH) of 25% or higher is still considered desirable during the winter season for health reasons, the effect of relative humidity on occupant comfort has been overestimated in the past.

Fig. WF2a shows both the results of older tests measuring the effect of temperature and humidity on human comfort immediately upon entering a room (curved lines top and bottom) and recent tests on occupants after a three hour stay (solid vertical lines).

Unlike the optimum summer and winter ET lines derived from earlier studies (Fig. WF2b), the newer data suggest: (1) that a DB temperature of 77.5°F is most comfortable and (2) that humidity has no perceptible effect on the sensation of warmth within a range of 25% to 60% RH.

VAPOR MOVEMENT

Wall, floor and ceiling assemblies should be designed and constructed so as to minimize vapor migration into structural cavities.

Vapor moves independently of the air, from areas of higher vapor concentration and higher temperatures to areas of lower concentration and lower temperatures.

FIG. WF1 **RECOMMENDED MAXIMUM INTERIOR RH TO PREVENT CONDENSATION DAMAGE***

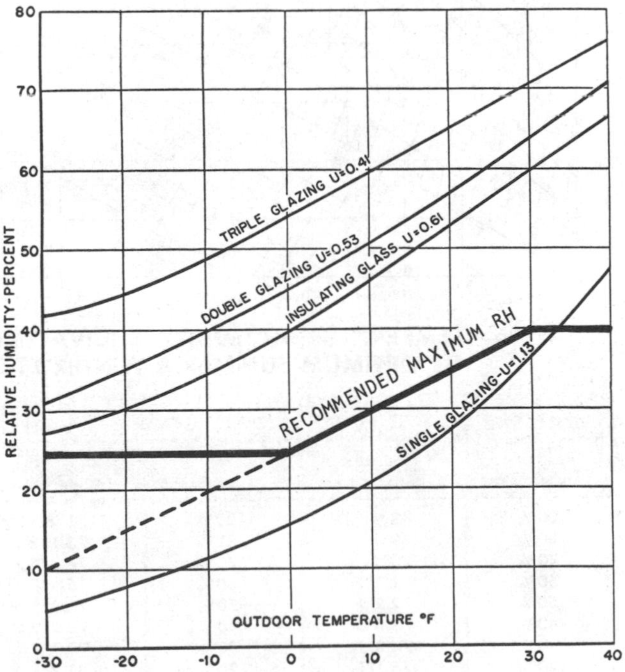

* Below 0°F, RH values are maintained at 25% to provide acceptable degree of occupant comfort (solid line); however, to minimize condensation on single glazing, use RH values represented by dashed line.

(Continued) WATER VAPOR PROPERTIES

Water vapor becomes a hazard when it condenses (liquifies) inside the structural cavity of a floor, wall or roof assembly. Here it can cause reduced effectiveness of insulating materials, decay of organic materials and deterioration of protective coatings.

FIG. WF2a ASHRAE COMFORT CHART SHOWING ET'S & SUBJECTIVE RESPONSES TO TEMPERATURE-HUMIDITY COMBINATIONS

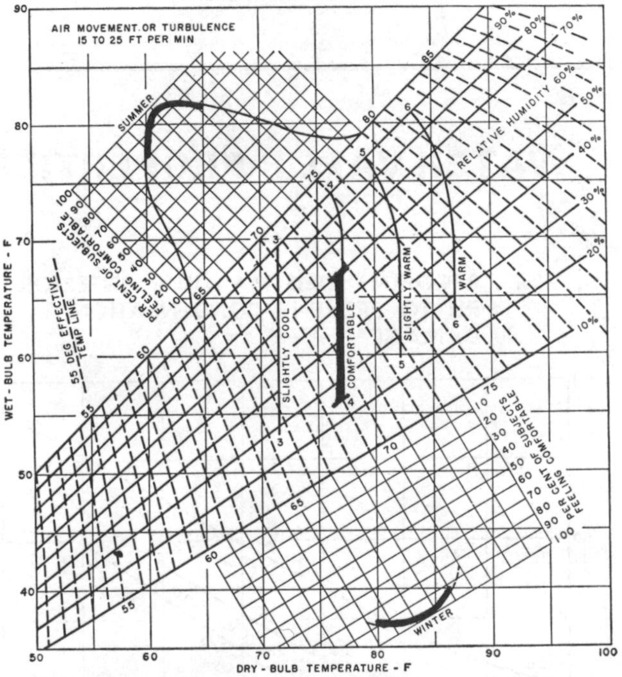

FIG. WF2b TEMPERATURE-HUMIDITY EQUIVALENTS OF OPTIMUM SUMMER & WINTER ET'S*

WINTER OPTIMUM—ET 60° **		SUMMER OPTIMUM—ET 71° **	
Relative Humidity	Dry Bulb Temperature	Relative Humidity	Dry Bulb Temperature
100%	68°F	100%	71°F
80%	69°F	80%	73.1°F
70%	70°F	70%	74.3°F
60%	71.1°F	60%	75.6°F
50%	72.2°F	50%	76.9°F
40%	73.5°F	40%	78.3°F
30%	74.8°F	30%	79.7°F
20%	76.1°F	20%	81.2°F
10%	77.5°F	10%	83.0°F

* Air movement of 15 to 25 fpm.
** Considered comfortable by 98% of people tested.

FIG. WF3 WATER VAPOR PERMEANCE OF BUILDING MATERIALS

Material	Permeance (Perm)
Materials Used in Construction	
1" Concrete (1:2:4 mix)	3.2
Brick masonry (4" thick)	0.8
Concrete block (8" cored, limestone aggregate)	2.4
Asbestos-cement board (0.2" thick)	0.54
Plaster on metal lath (3/4")	15.0
Plaster on plain gypsum lath (with studs)	20.0
Gypsum wallboard (3/8" plain)	50.0
Gypsum sheathing (1/2" asphalt impregnated)	10.0
1" Structural insulating board (sheathing quality)	20—50
Structural insulating board (interior, uncoated, 1/2")	50—90
Hardboard (1/8" standard)	11.0
Hardboard (1/8" tempered)	5.0
Built-up roofing (hot-mopped)	0.0
1" Wood, sugar pine	0.4—5.4
Plywood (douglas-fir, interior, glue, 1/4" thick)	1.9
Thermal Insulations, 1" thick	
Air (still)	120.0
Cellular glass	0.0
Corkboard	2.1—9.5
Mineral wool (unprotected)	116.0
Expanded polyurethane (R-11 blown)	0.4—1.6
Expanded polystyrene—extruded	1.2
Expanded polystyrene—bead	2.0—5.8
Unicellular synthetic flexible rubber foam	0.02—0.15
Plastic and Metal Foils and Films	
Aluminum foil (1 mil)	0.0
Aluminum foil (0.35 mil)	0.05
Polyethylene (2 mil)	0.16
Polyethylene (4 mil)	0.08
Polyethylene (6 mil)	0.06
Building Papers, Felts, Roofing Papers*	
Duplex sheet, asphalt laminated, aluminum foil one side (43)	0.002
Saturated and coated roll roofing (326)	0.05
Kraft paper and asphalt laminated, reinforced 30-120-30 (34)	0.3
Blanket thermal insulation back-up paper, asphalt coated (31)	0.4
Asphalt-saturated and coated vapor-barrier paper (43)	0.2—0.3
Asphalt-saturated but not coated sheathing paper (22)	3.3
15 lb. asphalt felt (70)	1.0
15 lb. tar felt (70)	4.0
Single-kraft, double infused (16)	31.0
Protective Coatings	
Paint—2 coats	
Aluminum varnish on wood	0.3—0.5
Enamel on smooth plaster	0.5—1.5
Various primers plus 1 coat flat oil paint on plaster	1.6—3.0
Water emulsion on interior insulating board	30.0—85.0
Paint—3 coats	
Exterior paint, white lead and oil on wood siding	0.3—1.0
Exterior paint, white lead-zinc oxide and oil on wood	0.9
Styrene-butadiene latex coating, 2 oz./sq. ft.	11.0
Polyvinyl acetate latex coating, 4 oz./sq. ft.	5.5
Asphalt cut-back mastic, 1/16" dry	0.14
Hot melt asphalt, 2 oz./sq. ft.	0.5

* Numbers in parentheses are weights in lbs. per 500 sq. ft.

(Continued) WATER VAPOR PROPERTIES

ADSORPTION AND ABSORPTION

Materials such as wood, plywood and fiberboards have a large internal area of pores, hence they not only adsorb *moisture (hold it on the surface), but also* absorb *moisture (into the pores), changing their moisture content. High moisture content over a period of time can cause decay, swelling and paint blistering.*

WATER VAPOR TRANSMISSION

The water vapor permeance of component materials should be considered in designing exterior walls, floors and roofs of habitable spaces.

Most building materials are vapor permeable to a degree. Some materials, such as metals and glass, are completely impermeable to vapor; such materials have a perm rating (permeance) of 0.00.

Materials with a perm rating of 0.00 to 1.0 are considered vapor barriers. Fig. WF3 shows the permeance of typical building materials.

CONDENSATION CAUSES AND REMEDIES

INTERIOR VENTILATION

Exhaust fans should be provided at the main sources of moisture generation such as in kitchens, baths and laundries.

Fig. WF4 illustrates typical sources of household moisture, totaling 22.5 lbs. Fig. WF5 demonstrates that ventilation at the rate of 2000 cubic ft./hr. would exhaust 21 lbs. of moisture, allowing only 1.5 lbs. of moisture to migrate through the walls on a cold winter day when the vapor pressure out is high.

Exhaust fans make it possible to limit interior relative humidity during cold weather when the hazard of vapor migration and condensation is greatest, as well as during warm weather when a lower humidity contributes to occupant comfort.

SURFACE CONDENSATION

Exterior walls, roof/ceilings and floors should be adequately insulated to prevent surface condensation.

Surface condensation occurs when an exterior surface loses heat fast enough to reach the dew point (condensation temperature). Adequate insulation reduces heat loss and increases surface temperature thereby reducing condensation hazard. Where improved insulation is not feasible, warm air currents will increase surface temperature and dissipate moisture. Keeping interior RH within the range recommended in Fig. WF1 also will minimize surface condensation.

FIG. WF4 **SOURCES OF MOISTURE IN TYPICAL HOME***

Source or Function	Moisture Generated (lbs.)
Breathing and perspiring	13
Cooking	5
Bathing	1
Dishwashing	3.5**

* For family of 4, in 24 hours.
** As much as 30 lbs. or more of moisture can be added by clothes washing and inside drying.

FIG. WF5 **WATER VAPOR BALANCE MAINTAINED BY VENTILATION**

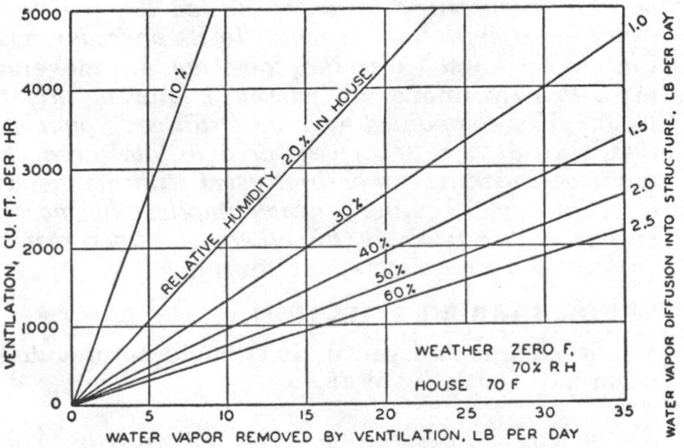

(Continued) CONDENSATION CAUSES AND REMEDIES

CONCEALED CONDENSATION

Exterior walls, roof/ceilings and floors should be protected with suitable vapor barriers and/or ventilation provisions.

The sources of moisture typically are within the house and the vapor barrier is placed near the interior finish to prevent moisture migration into the concealed structural spaces. However, in heated crawl spaces the suspended floor is uninsulated and the ground constitutes the chief source of moisture (Fig. WF6), hence the vapor barrier should be used as a ground cover.

FIG. WF6 DAILY EVAPORATION FROM CRAWL SPACE FLOOR

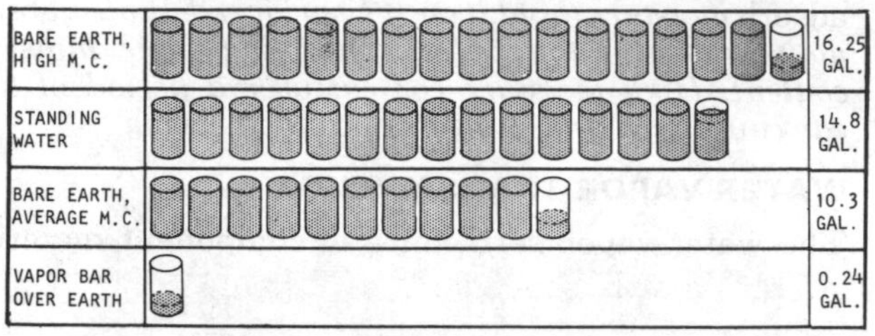

DESIGN AND CONSTRUCTION RECOMMENDATIONS

The continental United States can be divided into three condensation zones based mainly on the isotherms of the 0°F and 20°F winter design temperatures (Fig. WF7).

A double line of defense against moisture condensation—both a vapor barrier and ventilation—is recommended in all areas where severe or moderate condensation hazard exists (Fig. WF7). However, in wall assemblies more reliance is placed on the barrier, and air movement through the cavity is discouraged to minimize heat transfer.

In this discussion, the term "ventilation" is used both in its broad, all-inclusive sense—any attempt at dissipating moisture by using air as the vehicle—as well as in a more specific sense. In its narrower meaning, "ventilation" describes positive air movement through continuous air passages, and in this text usually is accompanied by a quantitative recommendation, such as a ratio of vent area to total area. This narrower meaning can be contrasted with another type of ventilation, "venting", which implies the mere inclusion of air leaks in an otherwise vapor resistant surface, as a means of vapor diffusion.

VAPOR BARRIER SELECTION

Vapor barriers and ventilation should be provided in accordance with Fig. WF8.

Vapor barriers are recommended for most assemblies in zones A and B. In zone C, the warm southern part of the U.S., large temperature and vapor pressure differentials between the inside and outside are not developed. Hence, vapor barriers may be omitted in walls and roof/ceiling assemblies of zone C.

Figure WF8 recommends three levels of barrier efficiency: (a) permeance of 0.3 perms or less—usually aluminum foils or plastic films; (b) 0.5 perms or less—asphalt-coated felts or asphalt-laminated papers; (c) 1.0 perms or less—same as above, but with less asphalt per sq. ft.

Barrier-faced Batts

In zone A, the vapor barrier flanges should be secured across the stud *faces*, or a separate, continuous, vapor barrier should be used.

Where flanges are nailed to the stud sides, as recommended by most manufacturers, it is almost impossible to avoid air and vapor leaks. Hence, this installation method should be restricted to areas of moderate condensation hazard, as in zone B.

Foil-backed Boards

Breaks in gypsum board joints—typically at 16" x 48" intervals for lath and at 48" x 96" intervals for wallboard—permit vapor leakage unless the joint is backed up by a structural member.

Foil-backed wallboard intended to act as a vapor

(Continued) DESIGN AND CONSTRUCTION RECOMMENDATIONS

barrier in zones A and B should be installed vertically, with all joints over studs.

Foil-backed gypsum lath, normally installed horizontally with joints perpendicular to studs, is not recommended as the sole means of vapor control in either zone A or B.

Separate Sheets

Separate vapor barriers should be installed in the largest possible sheets and with joints backed up by framing members.

When combined with suitable insulations they are particularly suitable for severe condensation conditions, as in zone A.

Polyethylene film and Kraft-paper supported aluminum foil are typical barrier products available in rolls of varying widths from 3 to 20 ft. Rigid boards and pressure-fit mineral blankets are typical insulations.

Insulating Boards

The permeability of insulating boards should be checked in each case to determine that the required permeance can be achieved in the thickness contemplated. Boards should be butted tightly together and joints should be sealed with a vapor-resistant mastic.

A number of synthetic insulating boards—such as foamed glass, polystyrene and polyurethane—are being marketed; many of these materials have a closed cell

FIG. WF7 CONDENSATION HAZARD ZONES OF THE UNITED STATES

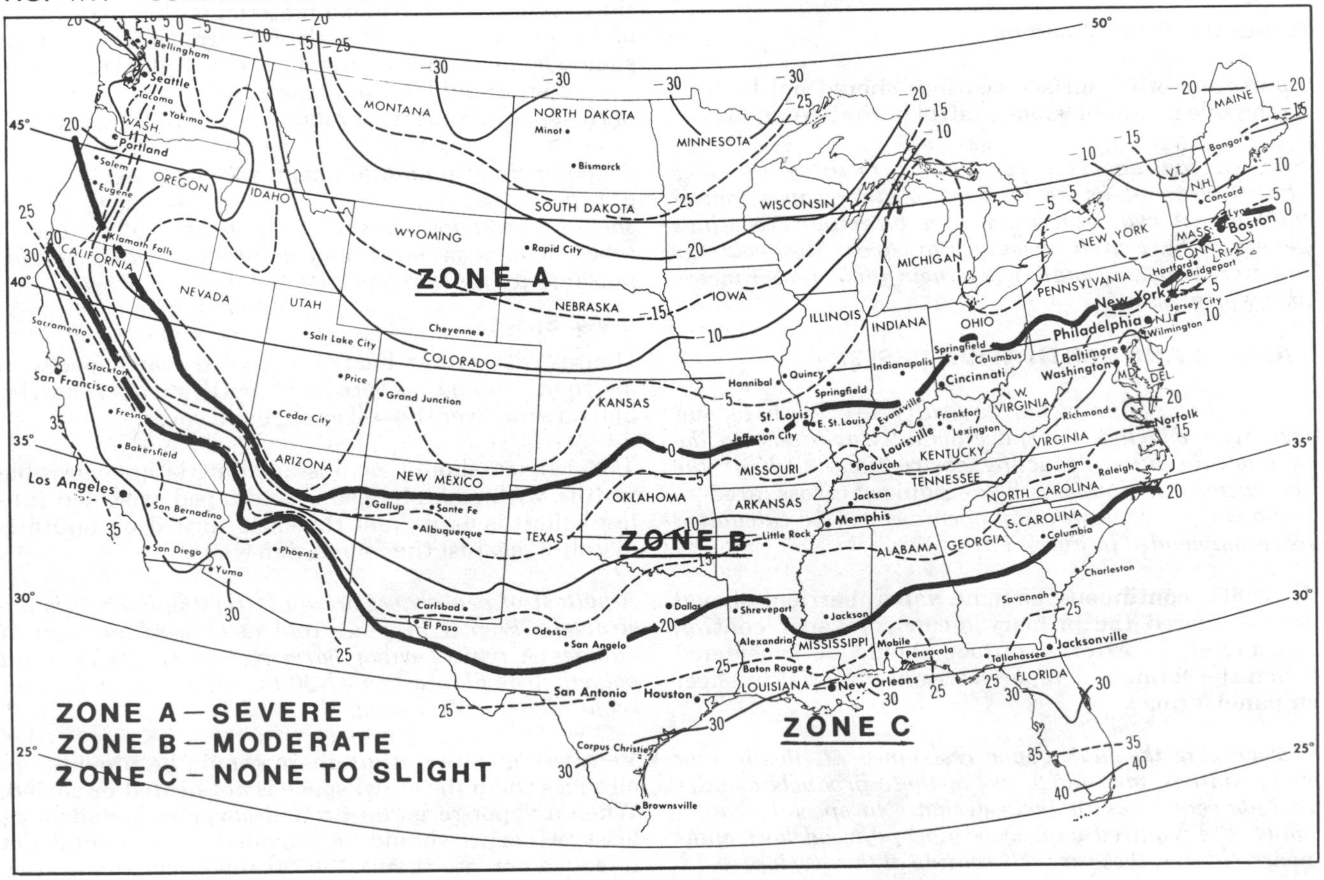

(Continued) DESIGN AND CONSTRUCTION RECOMMENDATIONS

FIG. WF8 RECOMMENDED VAPOR BARRIERS AND STRUCTURAL VENTILATION IN DIFFERENT CONDENSATION ZONES

Condens. Zone	Attic-less Joist Roof/Ceiling		Attic-type Joist Roof/Ceiling		Crawl Space		Floor Over Open Unheated Area —	Exterior Walls	Slab-on-Ground	Insulated Wood Plank Roof	
	Ceiling V.B., Perms	Struct. Ventil.[3]	Ceiling V.B., Perms	Struct. Ventil.[3]	Ground V.B., Perms	Struct. Ventil.[1,3]	V.B., Perms	V.B., Perms	V.B., Perms	V.B., Perms	Struct. Ventil.
Zone A	0.1	1/300	0.5	1/300	1.0	1/1500	1.0	1.0[4]	3.0	None[5]	Edge venting and/or stack venting when V.B. is provided.
Zone B	0.5	1/300[2]	1.0	1/300[2]	1.0	1/1500	1.0	1.0[4]	3.0	None	
Zone C	1.0	1/300[2]	None	1/300[2]	1.0[6]	1/1500[6]	None	None	3.0	None	

[1] Ventilation to atmosphere not required when crawl space is heated or cooled and foundation wall is insulated.
[2] Ratio of 1/150 preferred to relieve high summer solar heat gain.
[3] Ratio of net free ventilating area (NFA) to total floor or ceiling area.
[4] Venting of walls may be desirable where highly impermeable exterior finish is used.
[5] V.b. recommended in areas with winter design temperature of −10° F or lower, or occupancies with average winter interior humidities in excess of 40% RH.
[6] When crawl space is not heated or cooled, v.b. may be omitted provided vent area is increased to 1/150.

construction and are inherently good barriers.

Paints and Other Coatings

Paints and other surface coatings should not be used as the sole means of vapor control in zones A and B.

Surface coatings can be used correctively in older structures, and as an adjunct to other vapor control methods. Alkyd, aluminum and bituminous coatings generally have high resistance to vapor. Most coatings are more vapor-resistant when their finish appearance is shiny or glossy.

BARRIER AND VENTILATION DESIGN

Ventilating requirements generally are given as net free area, *which is the space actually available to the moving air, after deducting obstructions such as the screening and louvers. The* required gross area *of various types of ventilating openings can be calculated as recommended in Fig. WF9.*

In walls, continuous airtight vapor barriers should be considered the primary means of vapor control; venting of the exterior surface should be considered when the siding is a vapor-resistant material in sheet or panel form.

Recognizing the high vapor resistance of plastic and metal sidings, manufacturers of these products usually include vent-holes in their design. No special venting efforts are required with wood and plywood horizontal board *sidings; these are self-venting at the overlap.*

It is essential to form a continuous barrier envelope on the inside of walls, using a material with a permeance of 1.0 perm or less. Relatively permeable materials should be used on the outside of the wall; the sheathing paper should be a "breathing" type, having a permeance of 5 perms or more.

A good rule of thumb to insure safe outward diffusion of vapor is that the combination of exterior materials should have a total installed permeance of at least 5 times the total permeance of the interior vapor barrier plus interior finish materials.

Crawl Spaces

Houses with either heated or cooled crawl spaces in all zones should be provided with a vapor barrier laid directly over the soil as a ground cover.

The barrier should be installed in largest possible widths, with edges lapped 6″ and taped or sealed (unless ballast is used), and the sides turned up approximately 6″ against the foundation walls.

A ballast of pea gravel or sand over the barrier is desirable to keep it in place and to prevent mechanical damage. A typical vapor barrier consists of 4 or 6 mil polyethylene film, a coated 30 lb. per sq. ft. or heavier roofing felt or roll roofing.

Ventilation to the atmosphere should be provided in all zones when the crawl space is *not* heated or cooled. When a vapor-resistant ground cover is included, at least two vents should be provided, with a total net free area not less than 1/1500 of the crawl space area.

(Continued) DESIGN AND CONSTRUCTION RECOMMENDATIONS

FIG. WF9 **REQUIRED GROSS AREA OF VENT OPENINGS**

Vent Covering	R.G.A.* / N.F.A.
Hardware cloth, 1/4″ mesh	1
Screening, 1/8″ mesh	1¼
Insect screen, 1/16″ mesh	2
Louvers & hardware cloth, 1/4″ mesh	2
Louvers & screening, 1/8″ mesh	2¼
Louvers & insect screen, 1/16″ mesh	3

*R.G.A. means required gross area; N.F.A. means net free area. Net free area can be determined from Fig. WF8.

Vents should be placed as high as possible, in opposite walls. Vents are not required if the crawl space is completely open on one side to a basement; however, if the crawl space area is greater than the basement, it should be separately ventilated by supplying or returning air.

In older buildings or where the crawl space is not cooled or heated, the ground cover can be omitted, provided the net free area of vents is increased ten times, to 1/150 of the crawl space area. In the absence of a ground cover, more positive ventilation is required and at least four vents should be provided—one in each wall.

When the crawl space is not heated or cooled and the space is ventilated to the atmosphere, the overlying floor should be protected with thermal insulation and a vapor barrier on the upper side of the insulation.

Normal barrier-faced blankets have nailing flanges on the same side as the barrier. If the insulation is installed from below the joists, special reverse flange blankets or barrier-wrapped blankets should be used.

Roof/ceiling Construction

Joist-type Roof/ceilings These assemblies should be provided with a *net free vent area* of not less than 1/300 of the ceiling area, in addition to a suitable vapor barrier. In older buildings, not provided with vapor barriers, the vent openings should have a net free area not less than 1/150 of the ceiling area.

This larger volume of ventilation may be desirable for attic heat control in the moderate and southern zones, regardless of the presence of barriers. The effectiveness of the ventilation system in removing both heat and moisture depends on the volume of air moved, and the extent to which the air flow reaches into all parts of the attic or joist space.

Wood Plank Roof/ceiling A vapor barrier should not be used in insulated roofs of structures where the interior winter relative humidity averages less than 40%. A barrier is recommended between the wood deck and rigid insulation when the interior relative humidity exceeds 40% *and* the average January temperature is 35° or lower.

The 35°F January isotherm coincides approximately with the 0°F isotherm shown in Figure WF7; humidities over 40% are encountered in public shower rooms, kitchens, laundries and pool areas. In single family structures, only pool areas merit special consideration; other high moisture areas, such as baths and kitchens, should be equipped with exhaust fans to draw off the moisture.

If a vapor barrier is provided, venting should be considered to permit dissipation of moisture from the insulation. Edge venting can be used for roofs under 40 ft. in width. For wider roofs, additional *stack* venting should be provided, in the amount of one 3″ stack for each 900 sq. ft. of roof area.

Slabs-on-ground A barrier is recommended under slabs of habitable rooms in all zones. The vapor barrier should have a permeance of less than 0.30 perms, should be installed with 6″ laps and should be fitted carefully around utility and other service openings.

Vapor barrier materials suitable for slab-on-ground construction include single-layer membranes such as polyethylene film, or multiple-layer membranes such as reinforced waterproof paper with a polyethylene film extrusion-coated onto both sides.

When polyethylene film is used over gravel or crushed stone, or under any structurally reinforced slab, it should be at least of 6 mil nominal thickness; 4 mil film can be used over sand or firmly compacted soil.

INSTALLATION PRECAUTIONS

Three fundamental principles should be observed in the application of vapor barriers: (1) the barrier should be as near as possible to the warm (heated in winter) side of the insulation, or interior of the assembly; (2) the barrier should be as continuous as possible, with a minimum of joints, breaks, openings or air leaks; and (3) the barrier should be installed by a method suitable for the specific condensation hazard (Figs. WF7 and WF8).

Since the vapor barrier frequently is an integral part of

(Continued) DESIGN AND CONSTRUCTION RECOMMENDATIONS

the insulation, effective vapor control frequently depends on the proper selection and attachment of the insulation.

Insulation and barrier should be fitted neatly behind piping and equipment. In using barrier-faced insulating blankets, the flanges should be snugly stapled approximately 6″ o.c. to framing members, so as to avoid air leaks around the flanges.

Walls

Wall insulation should be pushed tight against the back of the sheathing and should fill the cavity from top to bottom.

Gaps in the insulation near the bottom and/or top of the wall cavity encourage convection currents, which carry heat and vapor from the inside to the outside face of the cavity.

Cracks around window and door frames should be stuffed with loose insulation and covered with a separate barrier.

Floors

When the exterior walls are insulated and vapor protected, similar thermal and vapor protection should be provided behind the band-joist at the perimeter of the floor cavity.

The perimeter of the floor assembly just behind the band-joist is frequently overlooked when insulation and vapor barrier are installed. This is particularly true when the floor is between two habitable spaces and no insulation is used in the joist cavities.

Where the floor is over an open-air or an unheated crawl space, thermal insulation and vapor protection should be considered. The vapor barrier should be on the upper (heated in winter) side of the insulation.

When the blankets are installed from above, they can be readily stapled to the joists. When installed from below, either reverse flange insulation can be used, or the blankets can be placed with both barrier and flanges facing up, using wire supports below.

Roof/ceilings

Blankets in ceiling joists should be extended part way over the top plate, but should not block eave ventilation. Insulation and vapor barrier should extend over dropped soffits in bathrooms and kitchens, to provide thermal and vapor control.

Tightly wedged glass fiber or rock wool insulation above the soffit also isolates the attic area from the soffit space and prevents the spread of fire.

Holes provided for mechanical equipment through top plates should be filled with loose insulation to minimize air and vapor leaks into the roof or attic space. A separate vapor barrier should be provided when poured insulation is used between ceiling joists.

105 HEAT CONTROL

INTRODUCTION

Early theories defined heat as a material substance called *caloric*—a weightless fluid which had the power of penetrating, expanding and transforming other materials into vapor. It was thought that a substance was hot or cold, depending on the proportion of caloric contained in its pores.

The theory of heat was placed on a sound scientific basis in the early 19th century by James P. Joule, an English physicist. Joule's experiments led to the principle of conservation of energy, which stated that heat energy was not lost as it was used, but was transformed into an equivalent amount of another kind of energy. For instance, heat energy can be converted to kinetic (motion) energy by a steam engine, and from kinetic energy to electric energy by a generator.

The human organism sustains life by converting organic food substances into heat energy. Normally, excess heat energy is produced and the body dissipates this excess to the surrounding atmosphere. Each organism has an optimum rate of body heat loss depending on physical exertion—whether at rest or engaged in physical activity. When the body loses heat at the ideal rate it feels comfortable; when too fast—cold; when too slow—hot. *The objective of interior heat control is to create conditions at which the body loses heat at the optimum rate.*

Heat energy generally flows from a warmer region to one that is cooler—to the outside of a dwelling in winter and to the inside in summer. This heat flow affects interior surface and air temperatures.

Cold is the absence of heat. A warm window radiates heat, but a cold window does not radiate cold. A person's sensation of coldness is caused by a rapid loss of body heat to the cold window surface. Likewise, the movement of air over the body produces the sensation of coldness because it carries off heat, not because it brings "coolth."

Modern man has come to expect a relatively narrow range of uniform temperatures within his dwelling, and the ability to select that range at will. In a century of cheap energy, it was relatively easy to strongarm the interior environment into a narrow band of comfort. As energy costs increase, both architects and homeowners are beginning to examine the benefits of permitting a broader range of interior temperatures and adjusting the mode of dress to the temperature.

For instance, willingness to put on a sweater in winter might make a 65° F interior air temperature acceptable, while removing one's tie would probably make 80° F quite tolerable in summer. The potential gain from accepting a 5° F variation up or down from traditional thermostat settings is a 15-30% energy savings in average heating/cooling costs.

Until the energy crunch of the '70s, traditional approaches in interior heat control emphasized reliance on energy-intensive mechanical systems. These systems pumped heat in or out of the dwelling and relied on the interior shell only to minimize heat gain in summer and heat loss in winter. More recently, architects have begun to design the building envelope so as to permit heat flow through the shell when advantageous. For instance, buildings can be designed to make use of sunshine for heating in the winter and of natural breezes for cooling in the summer.

The following subsections discuss: 1) the principles of heat flow and human comfort; 2) the use of insulations and other suitable materials to minimize heat flow through the building envelope; 3) appropriate building design which selectively either insulates the interior *from*, or integrates the interior *with*, the outdoor environment. This flexibility makes the modern home a more efficient machine for living, with reduced energy consumption and less reliance on mechanical systems.

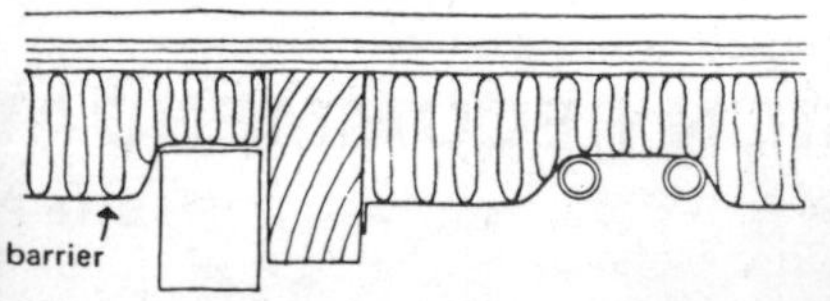

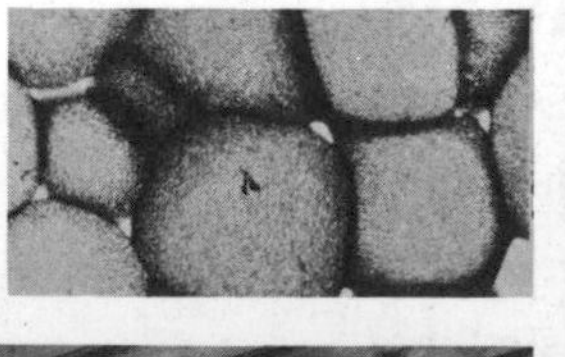

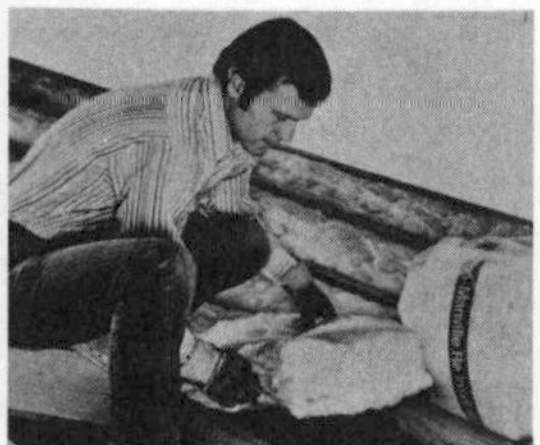

HEAT & COMFORT

Heat is a form of energy. It is produced in any substance by the movement of its molecules—the tiny particles of matter which make up that substance. When heat is added, the molecular movement inside that substance is increased. The intensity of movement determines the physical state of the substance, whether solid, liquid or gas. *Temperature*, recorded in degrees Fahrenheit (°F) or degrees Celsius (°C) is a measure of that intensity (Fig. 1). The *British Thermal Unit* (BTU) is a measure of heat quantity and is equal to the amount of heat required to raise 1 lb. of water 1° F measured at sea level.

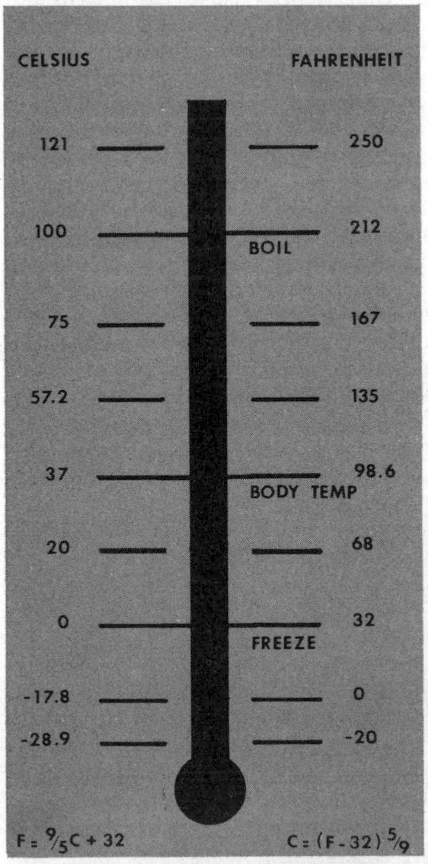

FIG. 1 Temperature equivalents on the Celsius (metric) and Fahrenheit (English) scales. Thermometers can be calibrated to a temperature of —273.1° C (—459.6° F), known as absolute zero*—the point at which all molecular movement stops.*

SENSIBLE AND LATENT HEAT

That energy which causes a rise in temperature but does not alter the physical state of a substance is *sensible heat.* Heat can be added to a substance without altering its physical state only up to a certain temperature. For example, if a teapot (Fig. 2) with 1 lb. of water is placed on a burner, each BTU of heat added raises the temperature 1° F. If the water started at 62° F, it would take 150 BTUs to bring the water to a boil, at 212° F (Fig. 2).

That energy which would be used to convert the water to steam if the pot were left on the burner is the *latent heat of vaporization.* To change the entire 1 lb. of water to steam would require 970 BTUs.

To change 1 lb. of ice at 32° F to water would require 144 BTUs—*the latent heat of fusion* of water. Only after all of the first 144 BTUs were used and all of the ice melted, would the additional heat be used to raise the water temperature above 32° F.

METHODS OF HEAT TRANSFER

Heat energy moves from a warmer to a colder area by *conduction, radiation* or *convection.* The body gives up heat by all of these methods, and also by *evaporation* (perspiration).

Conduction

The movement of heat through a *solid or liquid substance* as heated particles transfer their thermal energy to adjacent particles is called *conduction* (Fig. 3). Generally, dense materials such as metals can transfer more heat faster and are termed *conductors.* Less dense substances such as wood or plastic retard heat flow and are termed *insulators*.

Radiation

The transmission of heat by invisible light waves independent of any medium is termed *radiation* (Fig. 4). The radiant heat from the sun, for example, reaches the earth by traveling through the vacuum of space and the atmosphere. Radiant energy can travel through any non-opaque substance.

An opaque body absorbs radiant

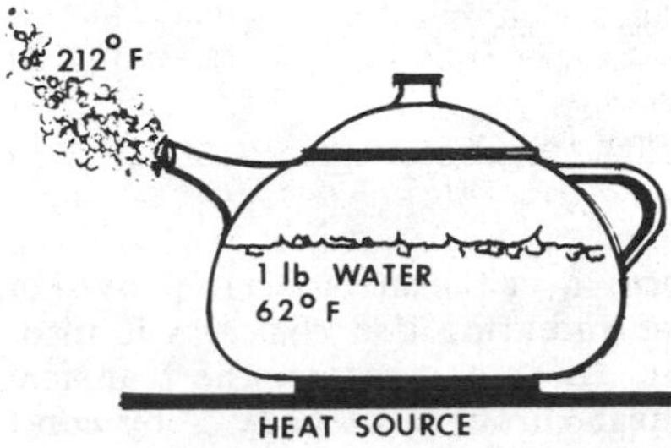

FIG. 2 Heat added beyond the first 150 BTUs will be used to change water to vapor but will not raise the temperature of the water or the vapor above 212° F.

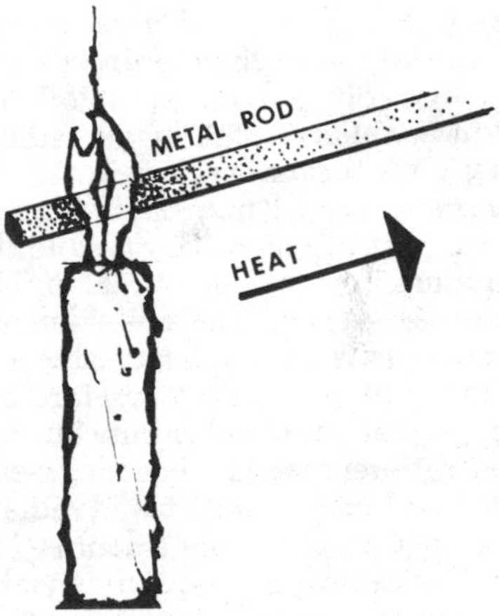

FIG. 3 Heat from the flame is conducted from one end of the rod to the other.

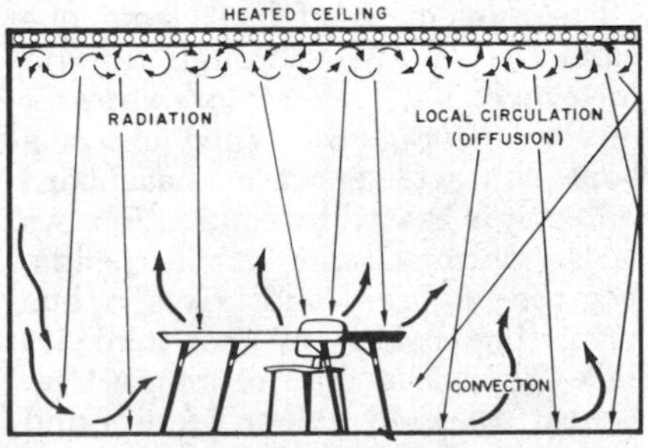

FIG. 4 Heated ceiling radiates energy to interior objects warming them; these pass on their heat to the air by conduction and convection.

FIG. 5 REFLECTIVITY/EMMISSIVITY CHART

Material	Reflectivity, %	Emmissivity, %
Built up roof, gravel aggregate	7	93
Slate, dark soil, other dark textures	15	85
Grass, dry	30	70
Copper foil:		
tarnished	36	64
new	75	25
Paint:		
light grey	25	75
red	26	74
aluminum	46	54
light green	50	50
light cream	65	35
white	75	25
Fresh snow	80	20
Aluminum foil	95	5

energy over the surface struck by the solar radiation and converts it into heat. After absorption, the transfer of heat through the body is by conduction; some heat is also reradiated from the warm body into the air.

Reflectivity Materials vary in their acceptance and rejection of the radiant energy which strikes them. Those which reject much of the radiant energy are called reflectors, because of their chief property, reflectivity. Shiny aluminum foil, for instance, reflects 95% of all radiant energy; white paint reflects 75%.

Emmissivity Emmissivity is the counterpart of reflectivity. While the aluminum foil reflects 95% of the radiant heat, it absorbs the remainder as sensible heat; the emmissivity of the foil is therefore 5% (Fig. 5). The practical significance of such differences is that high-emmissivity surfaces absorb and radiate more energy. Air temperatures immediately above asphalt pavements, for example, can be as much as 20° F higher than nearby shaded areas.

Convection

The movement of heat from one region to another through a *fluid substance*, such as air or water is convection (Fig. 6). When air in a room contacts a warm baseboard heater, it is heated by conduction. As it is warmed, the air expands, becomes lighter and rises, thus permitting colder air to take its place. The warmed air often gives up its heat to a cold wall or window and settles to the floor to be reheated. Thus a *convection loop* is established, which distributes the heat throughout the space.

Evaporation

The human body at rest loses heat mainly by convection and radiation. When engaged in strenuous activity or when the air is too warm, those methods are insufficient to carry off all the excess heat. The body then

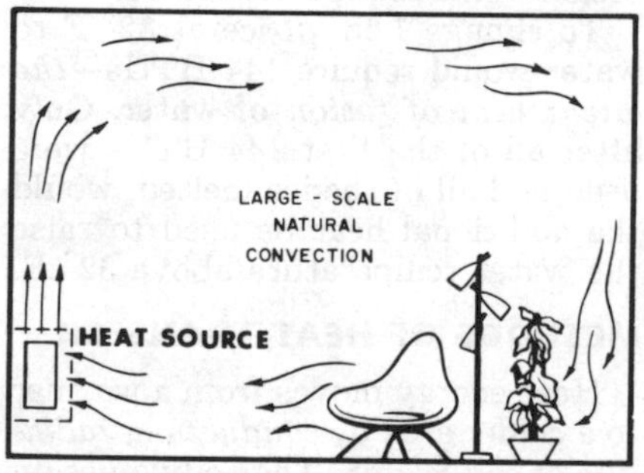

FIG. 6 As air is warmed by heater it rises; after it gives up heat to walls and ceilings, it drops to the floor.

utilizes perspiration, which cools the body by withdrawing from it the latent heat of vaporization. Low humidity aids cooling by perspiration; high humidity inhibits it.

RELATIVE HUMIDITY

Most air contains invisible moisture in the form of water vapor. When the moisture becomes visible, it is no longer vapor, but a liquid. A cloud or puff of steam is water in an intermediate stage—vapor in the process of becoming liquid, or vice versa. The conversion from vapor to a liquid is called *condensation.*

Relative humidity (RH) is the ratio of moisture actually present in the air, compared to the maximum that can exist at a given temperature without condensation. An RH of 50% means the air holds half the total water vapor it could possibly contain at that temperature. The maximum relative humidity is dependent upon the *ambient* (general air) *temperature,* because warm air can hold more water vapor than cooler air (Fig. 7). When all possible moisture is held by the air (100% RH), its *saturation point* or dew point has been reached. For additional discussion of relative humidity and saturation see Section 104 Moisture Control.

Relative humidity is measured by a *sling psychrometer* (Fig. 8). The device consists of two thermometers mounted side by side on a frame. A wetted cloth sack is attached to one of the thermometers and both are whirled through the air. The tempera-

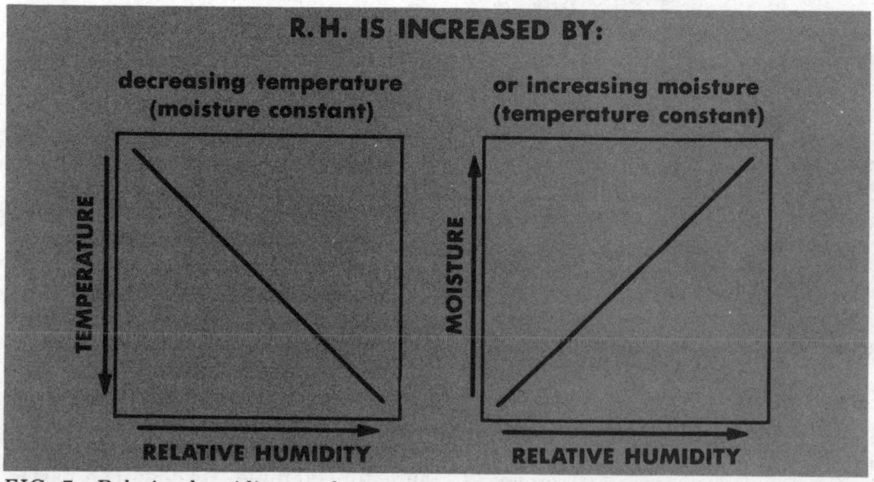

FIG. 7 Relative humidity can be increased by decreasing temperature or increasing amount of moisture.

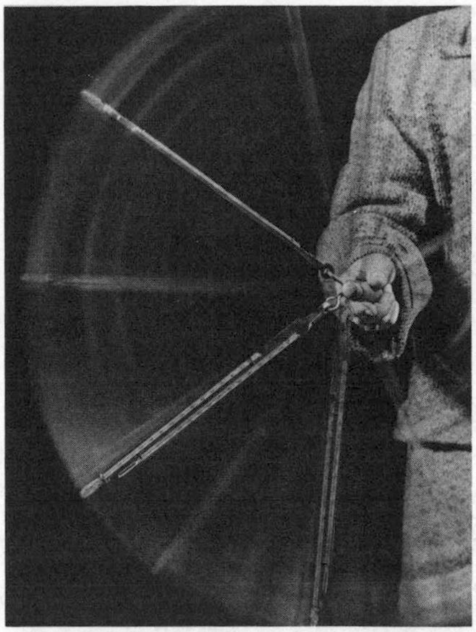

FIG. 8 Sling psychrometer is whirled to obtain both dry-bulb and wet-bulb temperatures; with these, RH can be determined from psychrometric chart.

FIG. 9 **PSYCHROMETRIC CHART**

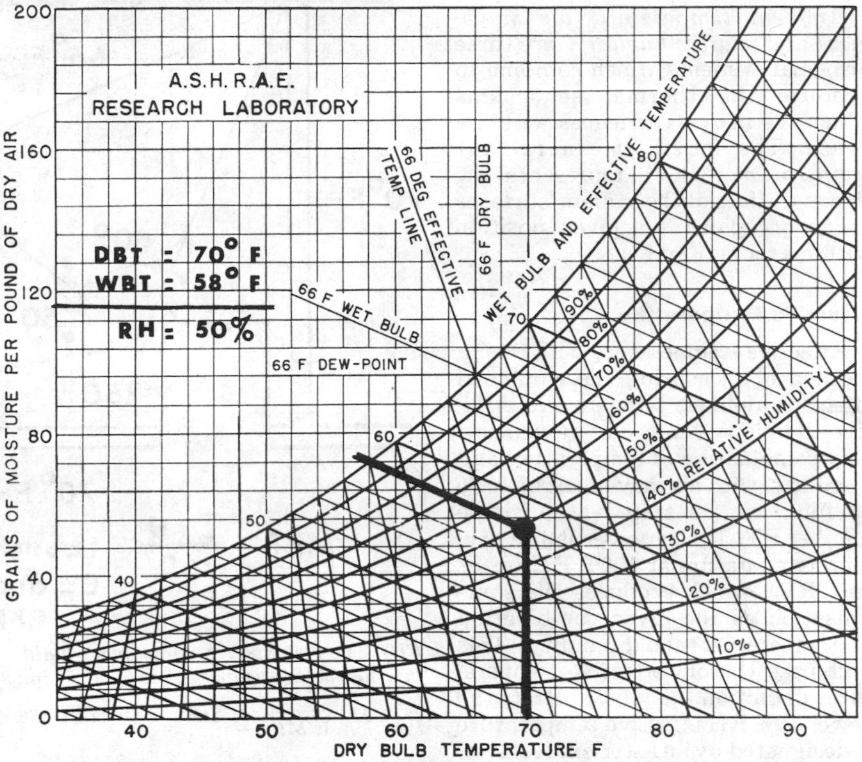

* **With ET lines, for persons at rest, normally clothed, in still air.**

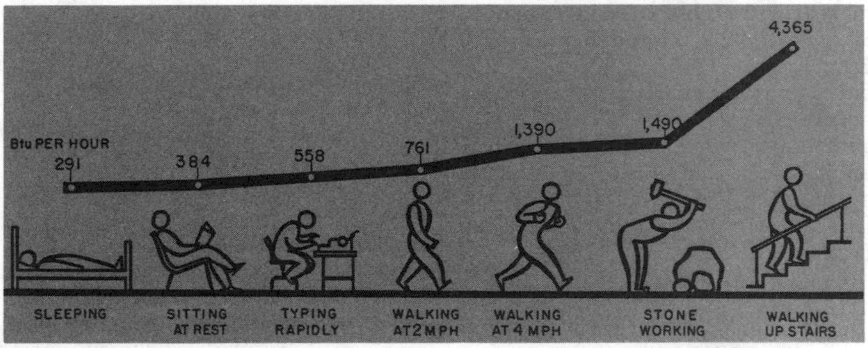

FIG. 10 The body's production of heat depends on how strenuous the activity is.

ture reading given by the uncovered thermometer is the *dry-bulb temperature* (DBT)—the common way of measuring and stating temperature. Ambient temperatures are usually stated in DBT. The reading shown by the wetted thermometer is the *wet-bulb temperature* (WBT)—the reading more commonly used in the design of mechanical systems.

The rapid movement of the two thermometers through the air causes some of the water to evaporate from the wetted thermometer. The process of evaporation uses up latent heat energy, resulting in a lower temperature reading. The greater the difference between wet and dry-bulb temperatures, the lower the relative humidity. When the dry-bulb temperature and wet-bulb temperature are known, the relative humidity can be determined from a psychrometric chart (Fig. 9). For instance, if the DBT is 70° F and WBT is 58° F, then the RH is 50%.

INTERIOR COMFORT REQUIREMENTS

The body is a heat producing mechanism, generating heat energy from oxidized food. In most cases, it produces more heat than is needed for thermal stability. The amount of extra heat which must be dissipated depends upon the activity being performed (Fig. 10) and ranges between 200 BTUs and 4,500 BTUs per hour.

Environmental Control

Control of the interior environment helps the body regulate its temperature and achieve the ideal comfort condition—a sensation that is "neither too warm nor too cool."

During the past 75 years, periodic studies to determine ideal comfort conditions have been conducted by the American Society of Heating, Refrigerating and Air-Conditioning Engineers, Inc. (ASHRAE). They have shown that the ideal winter temperature range has risen steadily—from 65-70° F in the 1900s to 72-78° F in the 1970s.

It seems that the desire for warmer temperatures is a result of cultural conditioning rather than biological necessity. With reduced energy availability and higher cost, this trend may be halted and even reversed. A reversal could be effected with minor alterations in dress habits, such as the use of warmer indoor clothing in winter and lighter, looser fitting apparel in summer.

Comfort Factors

Ambient temperature, air movement and *relative humidity* are three principal elements which combine to produce the thermal sensations which the body experiences within a space. The fourth, *mean radiant temperature* (MRT), represents the thermal effect of the radiant surfaces on an occupant at a given position within a room (Fig. 11).

Effective Temperature

Effective temperature (ET) is a method of measuring the combined effect of ambient temperature, relative humidity and air movement. Any combination of temperature and humidity, with air movement of 15 to 25 fpm, can be stated as a single number ET. That number is the dry-bulb temperature at 100% RH (or at 50% RH), which produces the same sensation as the given combination of temperature and humidity. This text uses the old effective temperature definition at 100% RH; the newer 50% RH effective temperature is designated by an asterisk, ET*.

Early ASHRAE tests suggested that the majority of people were most comfortable in winter at ET 68° F (ET* 72.2° F) and in summer at ET 71° F (ET* 76.9° F). From the psychrometric chart it is possible to obtain DBT/RH equivalents at more common seasonal humidities; according to this chart (Fig. 9) summer optimum is near 77° F DBT at 50% RH and winter optimum near 75° F DBT at 30% RH.

One of the defects in the use of optimum ETs as a comfort index is that the effect of relative humidity is exaggerated in importance. Later tests concluded that within a temperature range of 60° to 80° F and a relative humidity range of 25% to 60%, evaporative body cooling is nearly constant. That is, the body at rest loses roughly the same amount of heat by perspiration at 25% as it does at 60% RH, so long as the temperature does not exceed about 80° F. However, different temperatures within the 25-60% RH range do produce varying sensations of warmth.

Another problem with using optimum ETs as a comfort index is that they ignore the effect of radiant surfaces on the occupants (Fig. 12).

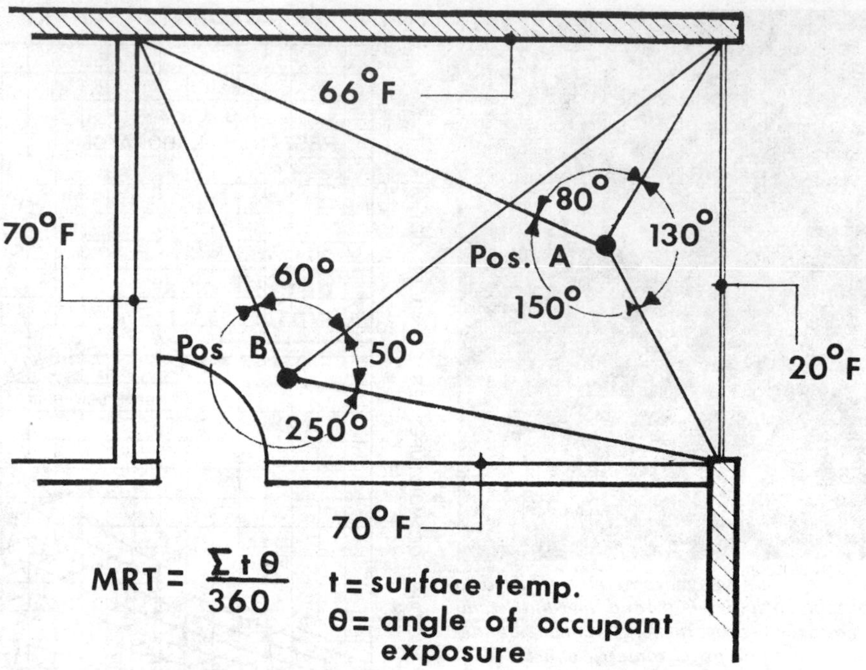

FIG. 11 Example: An apartment building has masonry and glass exterior walls with interior surface temperatures of 66° F and 20° F. Interior partitions are at 70° F.

$$\text{Position A: } MRT = \frac{\Sigma t\Phi}{360} = \frac{(20 \times 50) + (66 \times 60) + (70 \times 150)}{360} = \frac{18{,}680}{360} = 51^\circ F$$

$$\text{Position B: } MRT = \frac{\Sigma t\Phi}{360} = \frac{(20 \times 50) + (66 \times 60) + (70 \times 250)}{360} = \frac{22{,}460}{360} = 62^\circ F$$

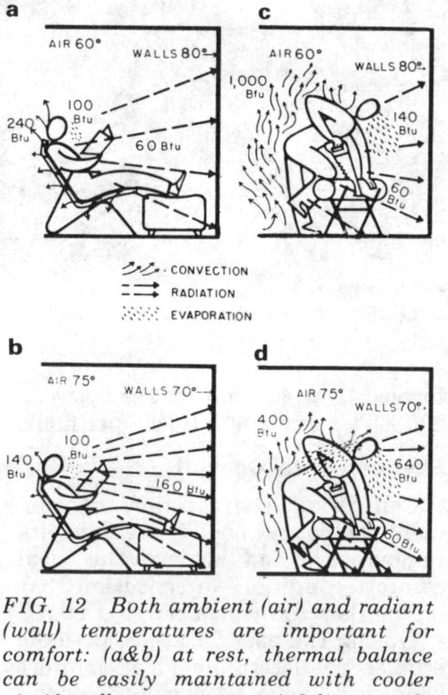

FIG. 12 Both ambient (air) and radiant (wall) temperatures are important for comfort: (a&b) at rest, thermal balance can be easily maintained with cooler air if walls are warmer; (c&d) at work, cool air removes more body heat, requiring less perspiration to maintain balance.

In summer, cool walls aid body heat dissipation because they absorb radiant body heat which cannot be removed by convection due to high air temperature. The body can therefore lose by radiation the heat it would otherwise have to eliminate at greater stress, by perspiration. In winter, warmer wall temperatures compensate for low air temperature by preventing radiant body heat loss.

ASHRAE THERMAL COMFORT ENVELOPE

ASHRAE Standard 55-74, Thermal Environmental Conditions for Human Occupancy, specifies generally acceptable comfort conditions for sedentary and slightly active, healthy, normally clothed people at altitudes of sea level to 7,000 ft.

The *comfort envelope* developed in this standard (Fig. 13) outlines acceptable thermal conditions for 80% of tested Americans and Canadians. It integrates the effect of: 1) dry-bulb temperature, 2) mean radiant temperature, 3) relative humidity and 4) air movement. In this chart the effects of DBT and MRT are combined in the *adjusted dry-bulb temperature* (ADBT) scale and the effect of relative humidity is shown by the water vapor pressure scale. Air movement is assumed at less than 70 fpm.

THERMAL DESIGN CRITERIA

There are several sets of comfort design criteria currently in use and proposed for use. For maximum energy conservation the Conservation Guideline Temperatures of the Federal Energy Administration should be used as interior design temperatures.

ASHRAE Energy Conservation Standard

ASHRAE Standard 90-75, Energy Conservation in New Building Design, provides performance oriented criteria for the design of the building envelope and of the mechanical and electrical systems. This standard refers to the comfort envelope parameters established in ASHRAE Standard 55-74 for winter and summer design criteria (Fig. 13).

Considering that low relative humidities generally accompany the heating season and high humidities are characteristic of the cooling season in most of the United States, it can be assumed that most designers would be operating near the bottom of the comfort zone in winter and the top in summer. If energy conservation is a design objective, it can be further assumed that the design ADBT would be near 73° F in winter and 78° F in summer.

ASHRAE Standard 90-75 does not require humidification as part of the heating function, but if humidification is provided it should be limited to 30% RH. Summer humidity control should be provided, but the designer is free to select a design humidity which will minimize energy usage.

In conventionally heated buildings with thermally efficient walls and roofs, the radiant effect can be ignored and DBT rather than ADBT values may be used. Thermal design generally can be based on the ambient, DBT values, unless large areas of glass or radiant panel heating systems are intended. Several solar heating designs are based on the capability of materials such as concrete and masonry to store and radiate heat; the radiant effect is certainly an important consideration in such designs, and either MRT or ADBT values should be used.

FIG. 13 ASHRAE COMFORT ENVELOPE

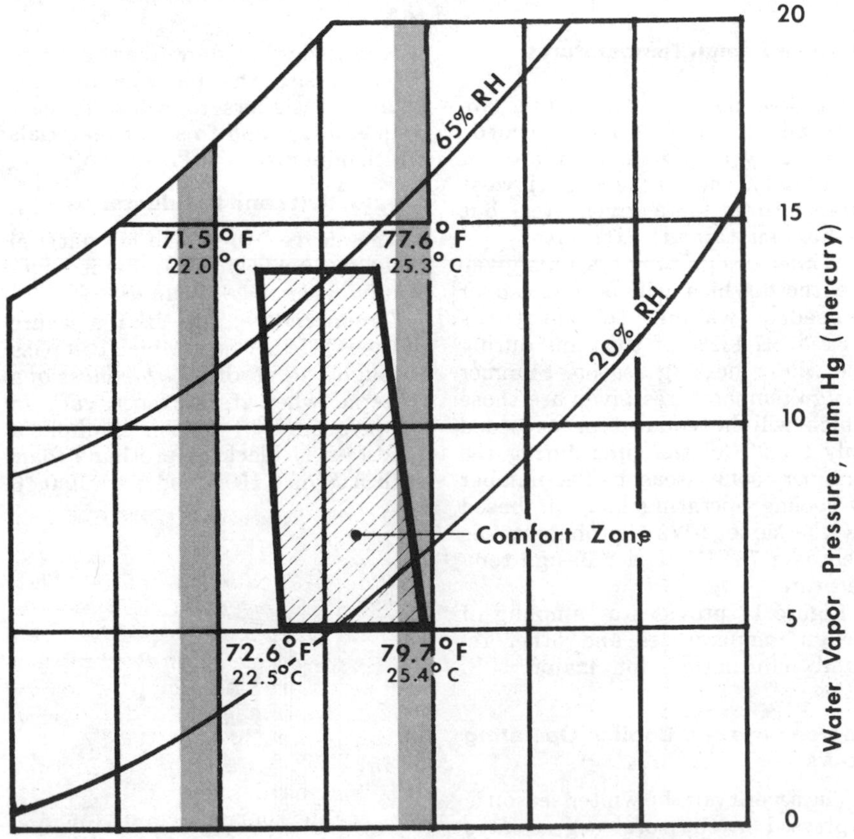

✱ Air velocity assumed to be less than 70 fpm

HUD Standards

The HUD Minimum Property Standards for One and Two Family Dwellings adheres to the historic interior comfort criteria—70° F DBT for winter and 75° F DBT for summer ambient air. This standard does not make any specific requirements for relative humidity or radiant temperature.

FEA/DOE Recommendations

In an effort to conserve energy resources, in 1974 the Federal Energy Administration (FEA) established Conservation Guideline Temperatures. These are stated as ranges, rather than single numbers. *The guideline winter temperatures are 68°-70° F (20°-21.1° C) DBT; the guideline summer temperatures are 78°-80° F (25.6°-26.7° C) DBT.* The guideline winter range falls entirely outside the ASHRAE comfort envelope, whereas the summer range straddles the boundary. This prompted the FEA to authorize a study to determine whether the proposed temperatures would be acceptable to the general public.

Based on a series of surveys conducted by the John B. Pierce Foundation in 1975 and 1976 it was concluded that *the FEA Conservation Guideline Temperatures could be made acceptable to 80% of the population* by several simple strate-

FIG. 14 REPRESENTATIVE OUTDOOR DESIGN TEMPERATURES, DEGREE DAYS AND COOLING OPERATING HOURS[1]

City	Winter Design Temperature (97½%) (°F)	Degree Days Heating (DDH)	Summer Design Temperature (2½%) (°F[2])	Annual Cooling Operating Hours
Anchorage, AK	−18	10,860	68M	60
Phoenix, AZ	34	1,680	107H	1,930
Los Angeles, CA	43	1,960	80M	460
San Francisco, CA	38	3,040	77M	170
Denver, CO	1	6,150	91H	700
Washington, DC	17	4,240	91M	960
Miami, FL	47	200	90M	2,940
Chicago, IL	− 4	6,640	89M	650
Wichita, KS	7	4,640	98M	1,030
Minneapolis, MN	−12	8,250	89M	590
New York, NY	15	4,880	87M	770
Cincinnati, OH	6	4,830	90M	870
Tulsa, OK	13	3,730	98M	1,230
Rapid City, SD	− 7	7,370	92H	610
Houston, TX	32	1,410	94M	1,880
Salt Lake City, UT	8	5,990	95H	770
Richmond, VA	17	3,910	92M	1,010
Spokane, WA	2	6,770	90H	460
Green Bay, WI	− 9	8,100	85M	430

[1]Data from Insulation Manual: Homes Apartments; Second Edition, 1979, by NAHB Research Foundation.
[2]These numbers represent temperatures that will be equaled or exceeded 2½% of the time during the summer cooling season. Daily temperature fluctuation is (a) low (L) if it is less than 15° F, (b) medium (M) if it is 15° F to 25° F, and (c) high (H), if it exceeds 25° F.

gies. For the summer, lighter clothing and more air movement is recommended; for the winter, somewhat heavier clothing and greater care in the covering of legs and feet is suggested.

The Federal Energy Administration in 1977 has been absorbed and superseded by the U.S. Department of Energy (DOE). Much of the consumer oriented material originating from the DOE recommends a household thermostat setting of 65° F in winter and 80° F in summer.

Neither the DOE recommendations nor the FEA Guidelines should be applied to rooms occupied by the elderly or sick. Therefore, the more temperate sides of the FEA Guideline Temperatures—68° F and 78° F—appear to be more acceptable and are recommended as interior design temperatures.

CLIMATE AND WEATHER CONDITIONS

In order to design a building envelope and mechanical system capable of maintaining interior comfort conditions, it is necessary to predict likely heat gain or loss. To do this the designer must know not only what indoor temperature is desired, but also the outdoor temperatures that are likely to be encountered.

Outdoor Design Temperatures

Outdoor design temperatures are determined from weather records over a 25 year period. They are not absolutely the highest or lowest temperatures for a given area, but rather sustained extremes.

Winter design temperatures given are those which will be equaled or exceeded (warmer temperatures logged) 97-1/2% of the time during the winter heating season. Summer design temperatures given are those which will be equaled or exceeded only 2-1/2% of the time during the summer cooling season. The number of cooling operating hours is based on the same 2-1/2% probability as well as a 78° F interior design temperature.

Figure 14 provides a sampling of design temperatures and other related information for major U.S. cities.

Degree Days and Cooling Operating Hours

The severity of the winter season is expressed in the total *degree days heating* (DDH), which is the sum of daily degree days when the temperature dropped below 65° F. Daily degree days is the numerical difference between 65 and the average recorded temperature of that day. For instance, if there were 10 days in the season when the temperature dropped to 45° F, then the DDH for each day is 20 and the total DDH for the season is 200.

Cooling operating hours measure the intensity of the annual cooling season. They are the hours an air conditioning unit must operate in order to maintain an indoor temperature of 78° F, 97-1/2% of the time.

The higher the number of total degree days or annual cooling operating hours for the season, the greater the heating or cooling demand (Fig. 14). Minneapolis, Minnesota, with 8,250 DDH and 590 cooling operating hours, has a severe heating season but a mild cooling season. On the other hand, Miami, Florida, with 200 DDH and 2,940 cooling operating hours, has a minimal heating season but a severe cooling season.

HEAT TRANSFER THROUGH MATERIALS

Knowledge of how various materials resist the passage of heat enables designers to calculate heat gain and loss and to select materials which minimize heat flow.

Conductivity and Conductance

The degree to which a material transfers heat is rated by its *conductivity (k)* or *conductance (C).*

Conductivity This is a measure of the BTUs per hour which can pass through 1 sq. ft. of a *1" thickness* of a material when its surfaces vary in temperature by 1° F. For example, a 1' x 1' x 1" block of urethane foam with a *k* of 0.16 would permit 0.16

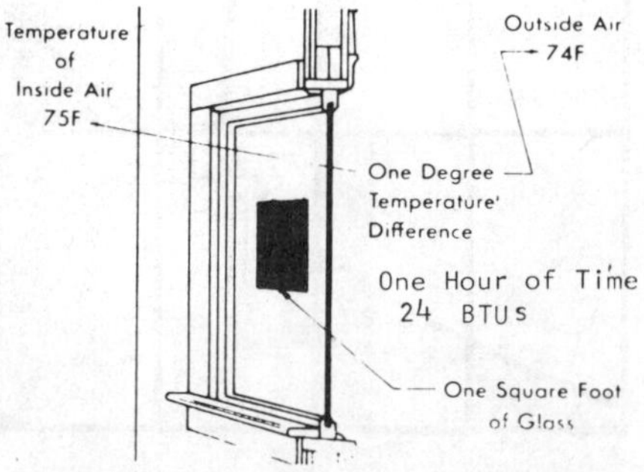

FIG. 15 If the C-value of a single glazed window is 24, then a 10 sq. ft. window with 10 degree temperature differential will conduct 2,400 BTUs per hr.

BTUs to pass in one hour, if one surface of the block were at 74° F and the other at 75° F.

Conductance For products which are not homogeneous or have air cavities (such as hollow concrete block), it is not meaningful to determine heat transfer per 1" of thickness. Such products are given a conductance (*C*) rating based on the number of BTUs transmitted in one hour, through 1 sq. ft. of stated thickness, when the surfaces vary in temperature 1° F. For instance, a single thickness of 1/4" glass has a *C*-value of 24. Thus, 24 BTUs are transferred in 1 hour if one face of the glass is at 74° F and the other 75° F (Fig. 15).

Resistance

From Figure 16 it can be seen that materials which are normally considered good insulators are those with low *k* and *C*-values. However, because these values are stated as multi-digit decimals smaller than 1.0, they are difficult to understand and compare. As a result, the use of *resistance (R)* numbers has increased as a method of rating building products and thermal insulations. The *R*-value is computed by taking the reciprocal of *C*, *(1/C)* or the reciprocal of *k*, *(1/k)*. In the first case the *R*-value must be qualified as being for 1" thickness; in the second, it is simply the *R* value of the prroduct, whatever its thickness.

Resistance (*R*) represent, the ability of a material to retard heat flow. *R*-values also are convenient because they are directly proportional to insulating value and can therefore be added or subtracted easily in order to determine the thermal performance of a product or assembly. For example, an 8" concrete block with a *C*-value of 0.90 would have an *R*-value of 1.11 (Fig. 16). If a 1" slab of polystyrene with an *R*-value of 3.57 were added, the assembly would have a total *R*-value of 4.68 not counting surface films. The molecules of air clinging to the surfaces contribute some insulating value, so *R*-0.68 should be added for still inside air, and *R*-0.17 for outside air at 15 mph. The final total for the assembly is *R*-5.53.

Many insulating products are labeled with their *R*-value on the wrapper. The number given is usually the value of the product only, without the air spaces and films, which make up the total installed resistance. Some insulations, such as multiple-layer foil, however, depend on air spaces and films for performance and can only be described in terms of installed resistance.

FIG. 16 **HEAT TRANSFER COEFFICIENTS OF REPRESENTATIVE BUILDING PRODUCTS**[1]

Material or Product	Conductivity k	Conductance C	Resistance per 1" thickness (1/k)	Resistance per thickness shown (1/C)
Concrete	12.0		0.08	
Face Brick	9.0		0.11	
Hollow concrete block, 8"		0.90		1.11
Stucco	5.0		0.20	
Metal lath & plaster, 3/4"		7.70		0.13
Gypsum board, 1/2"		2.22		0.45
Plywood, 1/2"		1.60		0.62
Pine, fir, other softwoods	0.80		1.25	
Oak, maple, other hardwoods	1.10		0.91	
Asphalt shingles		2.27		0.44
Built-up roofing, 3/8"		3.00		0.33
Wood shingles		1.06		0.94
Structural insulation board, 1/2"		0.76		1.32
Mineral wool batts, 3"—4"		0.09		11.00
5"—6"		0.05		19.00
6-1/2"—7"		0.05		22.00
8-1/2"—9"		0.03		30.00
Expanded polystyrene, 1"		0.28		3.57
Inside surface air film[2]				0.68
Outside surface air film[3]				0.17
Air space 3/4", nonreflective[4]		1.0		1.01
Air space 3/4", reflective[4]				3.48

[1]At 75°F mean temperature, from ASHRAE 1977 Fundamentals Handbook.
[2]Heat flow horizontal, still air.
[3]Heat flow any direction, 15 mph wind.
[4]Heat flow horizontal.

FIG. 17 **U VALUE COMPUTATION**

FRAME WALL CONSTRUCTION — INSULATED U AND R_T VALUES

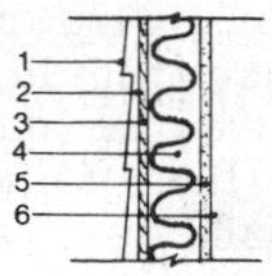

CONSTRUCTION	R
1. Outside Air Film	0.25
2. Wood Siding	0.81
3. ½" Insul. Sheathing	1.32
4. Insulation	11.00
5. ⅝" Gyp. Bd.	0.56
6. Inside Air Film	0.68
	R_T = 14.62

$1/R_T = U = 0.07$

MASONRY WALL CONSTRUCTION — UNINSULATED U AND R_T VALUES

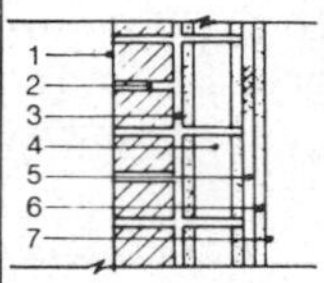

CONSTRUCTION	R
1. Outside Air Film	0.25
2. 4" Common Brick	0.80
3. ½" Cement Mortar	0.10
4. 4" Lt. Wt. Conc. Block	1.50
5. ¾" Air Space*	—
6. ⅝" Gyp. Bd.	0.56
7. Inside Air Film	0.68
	R_T = 3.89

$1/R_T = U = 0.26$

Transmittance

Transmittance (*U*) is the number of BTUs transferred in 1 hr. per 1 sq. ft. of a building assembly, for each 1° F temperature difference between inside and outside surfaces. The *U*-value is calculated by adding the *R*s for all elements of the building assembly (R_T) and taking the reciprocal of the total ($1/R_T$). If the R_T for the assembly equals 20, then the *U*-value would be 0.05 (1/20). Sample *U*-value calculations are given in Figure 17.

Performance of Masonry and Concrete

Traditional heat flow calculations are based on the theory that design conditions are sustained for long periods and heat flows continually from the warmer to the cooler side at a constant rate. This *steady-state* concept of thermodynamics is useful in designing heating or cooling systems where the building shell is not made of massive materials and the day-to-night outdoor temperature variation is not great.

It has been found in practice that heavy masonry and concrete buildings behave *dynamically* and do not generally lose or gain heat as rapidly as would be indicated by the steady-state theory. This is especially true in arid sunny climates where there are relatively warm days in winter and cool nights in summer. Under these conditions, massive buildings respond dynamically, acting as storers of heat—rather than as flow resistors —the way insulated lightweight buildings do.

In summer, massive structures not only store heat but are able to slow down its movement into the building until late afternoon, when the air and structure start cooling off. Likewise in winter, heat movement out of the building during the night is slowed by the mass of the structure until the sun warms the walls, and by late afternoon heat actually moves inward. Designers can use this *thermal lag* to help maintain interior comfort conditions with a minimum of man-made equipment.

The principle of using heavy, dense materials as *thermal mass* was well known to the Indian builders of 1000 year-old Arizona cliff communities, as well as to modern adobe homeowners throughout the sunbelt states. Despite limited insulating value earth sheltered and adobe homes remain surprisingly cool in summer and warm in winter. The effectiveness of a material as thermal mass depends on its *specific heat*—that is, its ability to store heat. (See 539 Passive Heating & Cooling).

The dynamic performance of concrete masonry and other massive shells can be accurately calculated by computers, programmed with detailed winter and summer weather information. To perform such calculations manually, utilizing the steady-state approach, is difficult, but a reasonable approximation can be made using a dimensionless correction factor called the *M Factor.* The Brick Institute of America can provide more information about the thermal performance of masonry walls.

When the assembly of structural and finish materials in the building envelope permits excessive heat flow, insulating products must be included in roof, walls and floors.

Insulation products are classified according to form as: 1) *loose fill;* 2) *flexible;* 3) *rigid;* 4) *reflective;* and 5) *foamed-in-place.* Thermal properties are given in Figure 18.

LOOSE FILL INSULATIONS

Fibers, granules or chips are the usual components of loose fill insulations. They are either poured or blown into attics or wall cavities. Most loose fill insulation in use today is manufactured from several inert materials collectively termed *mineral wool.* Organic fill materials, *cellulosic fibers,* are being increasingly used.

Pouring and Blowing Wool

Three types of mineral matter are included under this heading—*rock, slag* and *glass wool.* All are moisture and vermin resistant. They are less often used in walls than flexible insulation, because of a tendency for the large clumps of fiber to catch on nails or other obstructions, leaving unfilled voids. The most frequent application is between ceiling joists in attics where the wool can be poured from bags or blown by machine (Fig. 19).

Vermiculite

Vermiculite also is used in areas where it can be readily poured or blown, such as masonry cavities (Fig. 20). It is manufactured by expanding aluminum magnesium silicate—a form of mica. Vermiculite is non-combustible, vermin and moisture-resistant and does not compress in the cavity.

Cellulosic Fiber

Cellulosic fiber insulations are composed of recycled newsprint, wood chips and other organic fibers. They can be installed by either blow-

FIG. 18 **RESISTANCE (R) VALUES OF INSULATING PRODUCTS***

Product	R per 1" thickness
LOOSE FILL	
Mineral fiber (rock, slag or glass)	2.20-3.00
Cellulose	3.70
Perlite	2.70
Vermiculite	2.13
FLEXIBLE	
Mineral wool batts	3.10-3.70
RIGID	
Cellular glass	2.63
Expanded polystyrene (extruded)	5.00
Expanded polystyrene (molded)	3.57
Expanded polyurethane	6.25
Mineral fiberboard	3.45
Polyisocyanurate	7.20
REFLECTIVE	
Aluminum foil**	3.48
FOAMED-IN-PLACE	
Urea formaldehyde	4.20
Polyurethane	6.25

*All values are for 75° F mean temperature.
**Thickness of foil not a factor; 3/4" air space on room side.

FIG. 19 Large areas can be quickly insulated with pressurized hose.

ing or pouring into cavities, and can be combined with adhesives for sprayed application.

The fibers are chemically treated to resist fire and to minimize smoke contribution. However, certain clearances adjacent to heat producing fixtures must be maintained. They have a neutral pH to prevent reaction with metal and will not decay after extended exposure to moderate dampness. Cellulosic insulation should be UL listed and labeled to insure that the material is fire resistant.

FLEXIBLE INSULATIONS

These products are composed of either mineral wool or cellulosic fibers and are produced in *batt* and *blanket* form. The two are similar in appearance and composition, but differ in length—batts are 48″ long, blankets are longer and are packaged in large rolls (Fig. 21).

Because of the inorganic nature of mineral wool, products made from it are naturally resistant to fire, moisture and vermin. The cellulosic fiber products are chemically treated to achieve those properties.

Flexible insulations are produced in several types: 1) *wrapped*—with kraft paper on the edges and a vapor barrier on one or both sides; 2) *faced*—with a vapor barrier on one side only; 3) *friction-fit*—without any covering, the interlaced fibers having sufficient resilience to remain upright in the cavity.

Batts and blankets are manufactured to fit standard 16″ and 24″ o.c. framing patterns. Faced and wrapped products are provided with a stapling flange for secure installation. Product thickness ranges from 3″ to 7″ or more. Thickness, however, is not a reliable measure of insulating effectiveness, and should be accompanied by an *R*-number stamped on the wrapper or facing. For instance, typical batts for 2″ x 4″ walls are rated *R*-11, but some are made 3″ thick, others are 3-1/2″. In some areas, extra dense insulation rated at *R*-13 is available in 3-5/8″ thickness.

FIG. 22 Insulating sheathing provides thermal resistance and wind bracing.

FIG. 20 Vermiculite can be poured directly from the sack into masonry cavity.

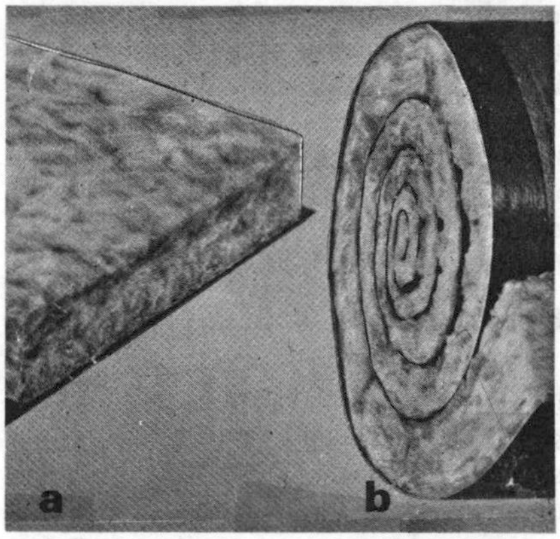

FIG. 21 Batt insulation is produced in panels (a); blanket insulation in rolls which can be cut to size (b).

RIGID INSULATIONS

Rigid insulations are composed of either organic, inorganic or synthetic materials. A variety of products which can be used in all parts of the building envelope fall within this classification. Those included are: 1) *structural wall insulation;* 2) *fiberboard;* 3) *structural deck insulation;* and 4) *rigid board insulation.*

Structural Wall Insulation

Wood or cane fiber panels are used primarily for sheathing wood frame walls, and are of three types: 1) *regular density;* 2) *intermediate density;* and 3) *nail base.* Panels are coated or impregnated with asphalt, or given some other treatment to provide water resistance. Occasional wetting and drying during construction will not damage the material.

Regular density sheathing is manufactured in 1/2″ and 25/32″ thicknesses and in 2′ x 8′, and 4′ x 9′ sizes. *Intermediate* density sheathing is a denser product than regular sheathing. It is available only in 1/2″ thickness and in 4′ x 8′ and 4′ x 9′ sizes. The 2′ x 8′ panels with tongue-and-grooved edges are used only for horizontal applications, while 4′ x 8′ and longer panels are installed with the long dimension vertical. If the 4′ x 8′ sheets are installed vertically, they should be nailed at 6″ o.c. at the intermediate support and 3″ o.c. at the edges (Fig. 22).

Nail base sheathing is dense enough to support exterior wood or asbestos shingles without intermediate nailing strips. It is manufactured in the same sizes as the intermediate density product.

Fiberboard

This general purpose, low-cost insulation board sometimes is used as an interior wall finish in cottages and temporary buildings. However, because of its softness, fiberboard wall panels are seldom used in "permanent" homes.

Fiberboard for wall use is available with flame resistant predecorated factory finish. It is manufactured 1/2″ and 3/4″ thick in tongue-and-grooved plank, or panel form. The 1/2″ thickness can be used when the supporting members are spaced 16″ o.c., 3/4″ thickness is recommended when 24″ spacing is used.

Fiberboard ceiling tile can be used in both temporary and "permanent" buildings. It is typically stapled to furring strips or applied with mastic to any plane smooth surface (Fig. 23). It can also be installed with wood or metal hangers to form a suspended ceiling.

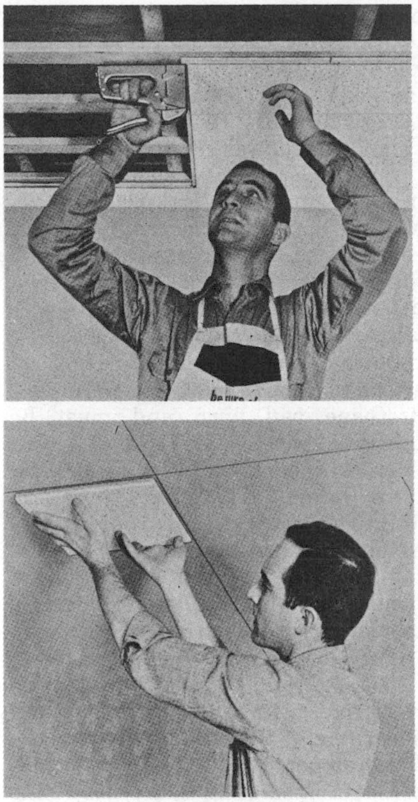

FIG. 23 Insulation board tiles can be stapled to furring strips (a), or mastic applied to most smooth surfaces (b).

Structural Insulating Roof Deck

This board product is designed to provide the structural deck as well as thermal insulation and interior ceiling finish. It is used primarily on flat or sloping roofs with exposed beams. The underside of the decking is factory-finished in various designs or textures—some with acoustical value (Fig. 24).

Structural insulating roof decks are of two basic types: 1) *lightweight aggregate boards* and 2) *wood fiber boards.* The boards are available in a variety of thicknesses and sizes, in either butt-joint or tongue-and-groove edge profiles (Fig. 25).

Because the underside of the deck is also the interior finish, care must be taken not to damage the surface during erection. In areas with high interior humidities, such as bathrooms or laundries, adequate roof insulation and ventilation should be provided to prevent condensation.

Rigid Board Insulation

This is a preformed *non-structural* board for use over structural roof decks, at the foundation perimeter and in masonry cavities (Fig. 26). It not only provides thermal resistance, but also serves as a base for the roofing membrane. Products in this category are manufactured from 1) *urethane foam;* 2) *polystyrene foam;* 3) *polyisocyanurate foam;* 4) *expanded perlite;* 5) *cellular glass;* and 5) *glass fibers.*

All of the plastic foams are good vapor barriers, thus preventing wall moisture from escaping unless certain precautions are taken: A continuous polyethylene vapor barrier

FIG. 24 Insulating deck is nailed directly to beams and is exposed to the interior.

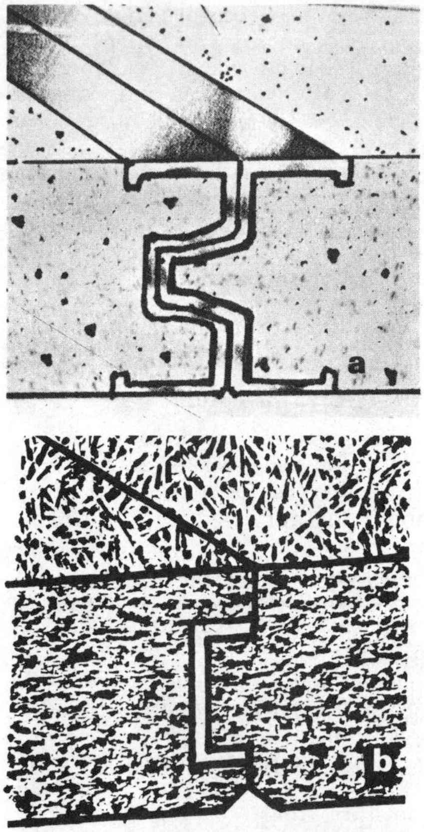

FIG. 25 Insulating deck of lightweight perlite aggregate (a), or wood fibers (b).

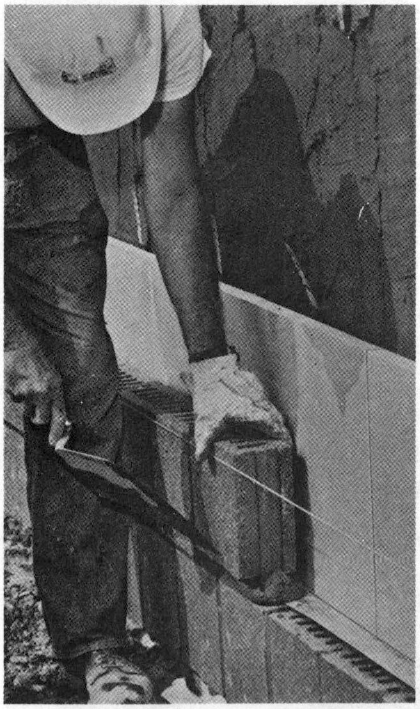

FIG. 26 Insulation is inserted in cavity as the wall is constructed.

on the room side of the studs, and venting of the wall cavity, are highly recommended. Specific venting methods vary with the manufacturer, but his recommendations should be followed (Fig. 27).

Urethane foam is a cellular plastic product noted for its excellent resistance to heat transmission and deterioration from chemicals. When formed into rigid boards, the low-density closed-cell structure produces a lightweight and thermally efficient product. Urethane foam has a resistance per 1″ of 6.25. However, urethane will burn and must be protected by some type of finish material if used for interior applications. It may also expand after prolonged exposure to moisture. Its uses extend to every element of the building envelope; roof and foundation insulation, and as non-structural insulating wall sheathing.

Polyisocyanurate foam is manufactured in rigid boards faced with reflective aluminum foil for wall sheathing. It has the highest resistance per 1″ (7.2) of any of the common insulating materials. Like urethane, it must be protected from flame in interior applications. In cold climates special venting strips should be used to prevent condensation in the wall.

Polystyrene foam, also a plastic cellular material, is produced in both *molded* and *extruded* types. The molded type is commonly termed "beadboard" because it is made up of compressed pellets or beads. The extruded type has a denser structure of non-connecting cells. These insulations have similar properties to the urethanes, but with lower resistance values per 1″ —*R*-3.57 truded smooth; *R*-4.0 for the extruded cut cell type.

Extruded polystyrene insulation is resistant to the transmission of vapor and absorption of moisture. Molded polystyrene has an open-cell structure which makes it less resistant to moisture. All polystyrene is sensitive to daylight and will deteriorate after prolonged exposure. Both types, like all cellular plastics, will give off carbon monoxide gas while burning. They do not smolder or add a significant amount of fuel to the fire, however. Polystyrene boards (molded and extruded) can be used as

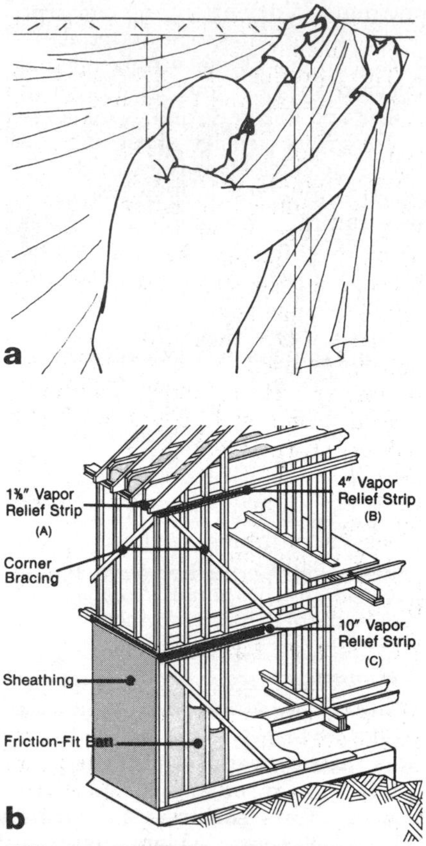

FIG. 27 To prevent trapping moisture within wall when using plastic sheathing, provide (a) continuous vapor barrier on the inside and (b) vents to the outside.

wall sheathing if wind bracing is provided by other means.

When 1″ of polyisocyanurate foam board is combined with *R*-13 mineral wool batts, a thermal resistance of *R*-20 can be achieved for insulation only even with a conventional 2″ x 4″ stud wall.

Expanded perlite and mineral binders are combined to make boards with weight characteristics similar to the plastic insulations. Unlike the cellular plastics, perlite boards are inorganic and will not burn or smolder (Fig. 28). However, the *R*-value per 1″ is approximately 2.63 —considerably lower than for the plastics. Perlite should be kept dry because its tendency to absorb water

Cellular glass is also used to produce a non-combustible and lightweight board insulation. The closed cell structure of the glass makes it impervious to water vapor and moisture absorption. *R*-value per 1″ is about 2.5, comparable to expanded perlite insulation.

This product is also used on exterior decks and promenades, and as perimeter insulation for slabs and foundations. It is available in thicknesses ranging from 1-1/2″ to 4″ and can be produced in tapered form on special order. When used under the waterproof membrane, the tapered form permits positive drainage of the roof or promenade.

Glass fiber insulation board is generally used at the foundation perimeter. It is most frequently placed along slab edges or inside reduces thermal resistance. heated crawl spaces, but can also be used as board insulation under built-up roofing.

This insulation product is available in 2′ x 4′ x 1′ boards which can be cut to size in the field.

REFLECTIVE INSULATION

This type of insulation is composed of one or more layers of metallic foil. Reflective insulation is different from other insulations in that its insulating effectiveness depends on how shiny the reflective surface remains over time, whether a 3/4″ air space is provided on the warm side and whether it is intended to reduce heat flow in or out of the structure.

Reflective insulations include sheets of paper-backed aluminum foil rolled into blankets of single or multiple layers and foil-backed gypsum wallboard.

FOAMED-IN-PLACE INSULATIONS

These plastic-based materials are available as liquid components or expandable pellets, which may be poured or sprayed in place to form semi-rigid masses. Fibrous materials mixed with liquid binders may also be sprayed in place.

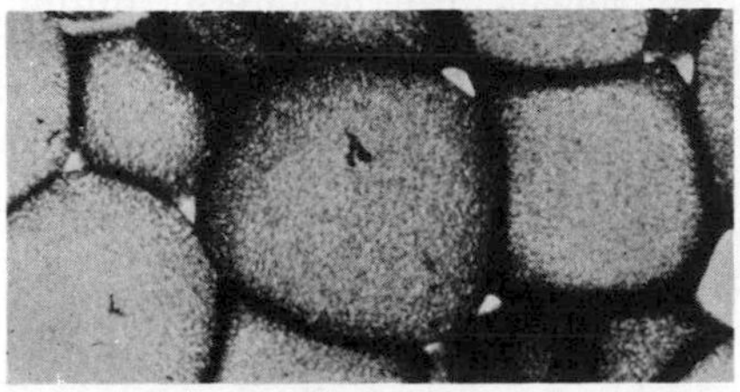

FIG. 28 Each bead of perlite is a honeycomb of dead air cells sealed by glass-like membranes.

The foaming process is based on a chemical reaction between two or more components, which expands the mixture by as much as 30 times. After expansion, the foam solidifies into a closed-cell plastic that fully cures in about 24 hours.

Urethane Foam

This foam has the highest resistance per 1″ of any insulation (*R*-6.25), is resistant to moisture and most chemicals and has fire-retardant capabilities if chemically treated. It has a wide variety of applications, including floor and wall cavities (Fig. 29), roofs and structural sandwich panels. The *R*-value per 1″ may vary with field conditions at the time of placement.

Urethane foam is particularly effective where a high *R*-value must be achieved in limited space or on irregularly shaped construction. Care must be taken not to overfill a cavity, because it expands to its ultimate volume *after injection.*

Urea Formaldehyde

This insulation is particularly suitable for use in cavities of existing buildings because it foams to full dimension *before injection* into the cavity. The danger of expanding foam bursting the cavity is therefore eliminated. It can be spread in an open horizontal space or injected into a closed cavity in a manner similar to loose fills. The most common use is for sidewalls where thickness is limited.

Urea formaldehyde foam (UFF) has a lower resistance per 1″ (*R*-4.2) than urethane but has a higher value than blown-in materials. It will not settle, is not affected by moisture and is not flammable. Shrinkage of 1-3% is common however, and may cause the solidified foam to pull away from framing members. Care should also be taken to vent the noxious odor which is released during the curing process.

GENERAL RECOMMENDATIONS

There are several ways to control heat flow through walls, floors and ceilings: 1) the assembly can be composed of materials which are inherently good insulators; 2) mass insulating materials such as batts, blankets or poured insulations can be added to the assembly; 3) reflective insulations can be added to one or both surfaces of an air space created by a joist, rafter or stud; and 4) solid masonry or other dense material which has the capacity to store heat can control heat flow. Whatever method is employed, it must be correctly installed to achieve the full benefit.

Conventional Installation

The building envelope should be designed to minimize heat flow through floors, walls and ceilings (Fig. 30). In conventionally heated and cooled homes, the insulation is generally placed near the *inside* of the building envelope. In some solar heated houses, the insulation is placed near the *outside* of the wall.

Ceilings Flexible insulation with vapor barriers should be installed so that the barrier faces the living space (Fig. 30). Loose fill or flexible insulation without the barrier should have a separate polyethylene film placed below the insulation facing the living space.

For maximum effectiveness against vapor transmission the kraft paper facing should overlap the face of the framing member. If such a

FIG. 29 Polyurethane can be foamed-in-place before interior finish is applied.

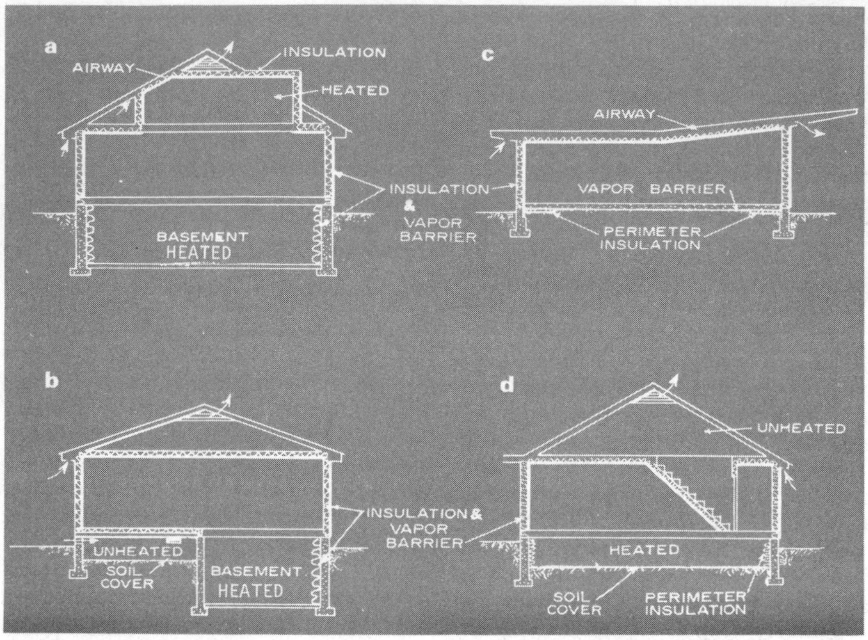

FIG. 30 Insulation and vapor barrier should wrap around all habitable areas.

procedure is not appropriate because of the application of finish materials such as drywall, a separate vapor barrier should be installed.

The insulation should extend entirely across the top plate to prevent heat loss at the joint. If necessary, the junction between the insulation and the plate should be stuffed with loose insulation (Fig. 31). When eave vents are used, the insulation should not block the air movement between the vent and the space above the insulation (Fig. 32).

If loose fill insulation is used, baffles should be installed to prevent the insulation from covering the vents. Cellulosic insulation should also be kept from contact with potential sources of heat such as electrical fixtures by providing at least 3″ of clearance along each face of the fixture.

Walls Flexible insulation in stud spaces should touch the sheathing or siding and fit tightly against the top and bottom plates. If it will not interfere with installation of finish materials, the flanges should be attached to the stud face to make a continuous vapor barrier.

If the stud face must be kept clear or if insulation without a vapor barrier is used, the inside face of the wall should be completely covered with polyethylene (Fig. 33). Any necessary openings can be cut out later.

Narrow cracks around window and door openings should be packed with loose insulation and covered with polyethylene (Fig. 34). Insulation should also be packed around pipes, ducts and electrical boxes to prevent "cold spots" in the wall (Fig. 35).

Masonry walls can be insulated by installing flexible insulation between furring strips (Fig. 36) or by applying board insulation directly to the wall. Board insulation which is not inherently fire resistant should be protected with gypsum wallboard.

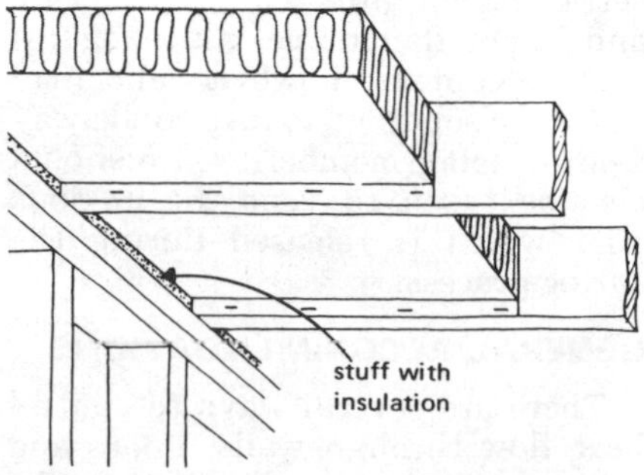

FIG. 31 Inset stapling, which keeps joist faces clear, is preferred for gypsum wallboard application; as an alternate, insulation can be laid from the top after wallboard is installed.

Closed cell plastic boards have a high degree of resistance to moisture and vermin and can also be installed on the outer faces of masonry and frame walls. However, polystyrene board insulation must be painted or otherwise protected from exposure to daylight. It should also have some type of protection to prevent dents and chips if used in an exposed area.

Floors and Crawl Spaces Floors over crawl spaces can be protected from heat loss by installing the insulating material *against the foundation walls* if the crawl space is heated (Fig. 37). If the space is unheated, the insulation should be placed *between the floor joists* (Fig. 38). In either case, the vapor barrier should face the living space.

In basements and heated crawl spaces, sill sealer insulation should be used to prevent air infiltration between sill and foundation (Fig. 39). In addition, the insulating sheathing should extend to the base of the plate.

The space behind the header joist should be covered with short pieces of flexible insulation. If the wall is insulated as in Figure 37 the end of the blanket can be extended upward and pushed against the header. Cantilevered floors should be insulated as shown in Figure 40.

When perimeter insulation is needed for slab-on-grade construction, rigid board insulation should be placed as shown in Figure 41.

Unconventional Systems

In some passive solar houses, the insulation is placed on the outside face of the wall, rather than the inside, as in conventional houses. The basic objective of such solar houses is to absorb solar heat in masonry or concrete walls during sunshine hours, and to release (radiate) this heat slowly into the interior during sunless periods. To prevent rapid cooling of the warm masonry during sunless periods, insulation is placed on its *outside face.*

STANDARDS OF QUALITY

Insulating products are manufactured to various standards, including those set by the General Services Administration and known as *Fed-*

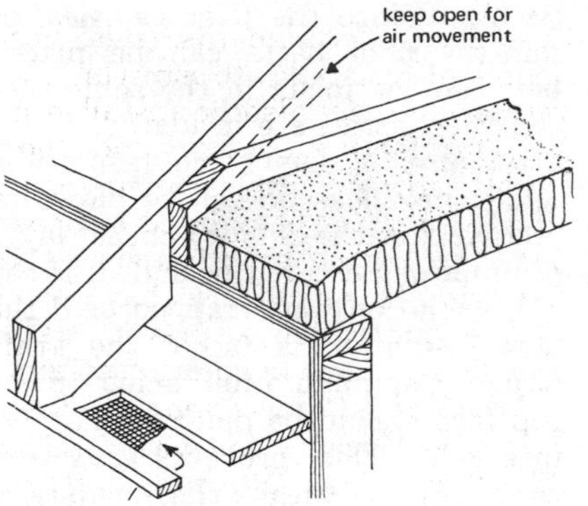

FIG. 32 Eave vents kept clear for air circulation.

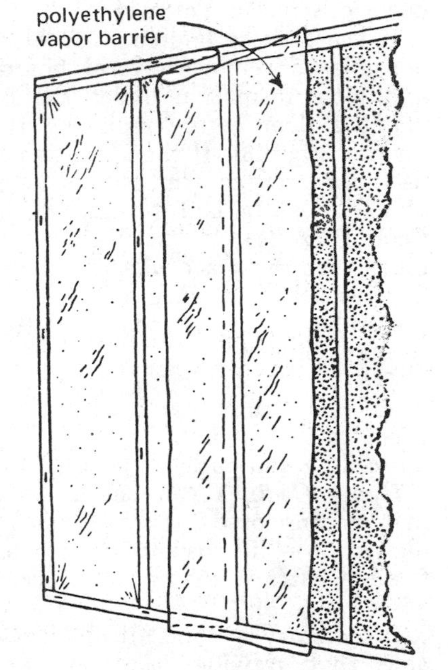

FIG. 33 Continuous polyethylene film over inside stud faces reduces vapor transmission and air infiltration.

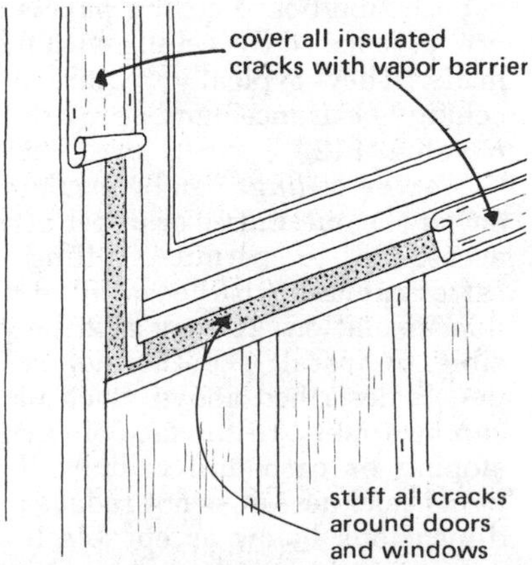

FIG. 34 Cracks around windows and doors should be packed with insulation and covered with polyethylene.

eral Specifications. Loose fill mineral wool, for example, should conform to Federal Specification HH-I-1030A, which establishes the thermal conductance and resistance requirements for such products.

Methods of testing formulated by the American Society for Testing and Materials (ASTM) measure a product's conformance to prescribed standards. Figure 42 summarizes current standards of quality.

REINSULATION OF EXISTING BUILDINGS

Although the concepts of heat control discussed above also apply to existing construction, it is not always possible to insulate an older structure in the same way and to the same level as a new building. Frequently it is not economical to add insulation to less accessible parts of a completed building such as walls and floors.

In general, attics with no insulation should be insulated to the same levels as new construction (Fig. 82). If the attic already has 6″ of batt insulation or 8″ of loose fill insulation, the benefits of additional insulation should be examined on a

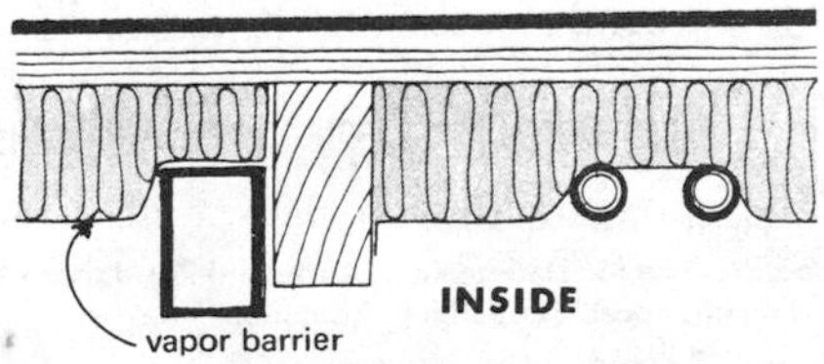

FIG. 35 Insulation should be pushed behind pipes, ducts and electrical boxes.

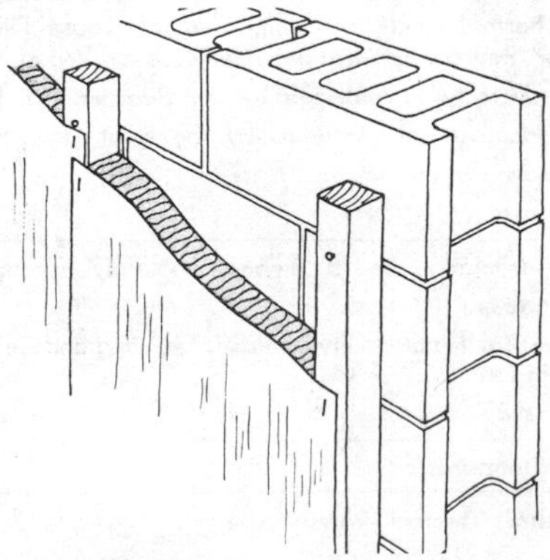

FIG. 36 Flexible insulation can be applied between furring strips, behind the wall finish.

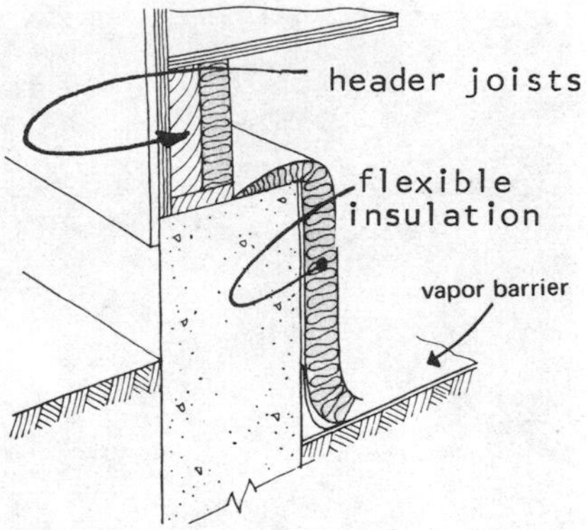

FIG. 37 The inside walls of heated crawl spaces should be insulated; a vapor barrier should cover the ground.

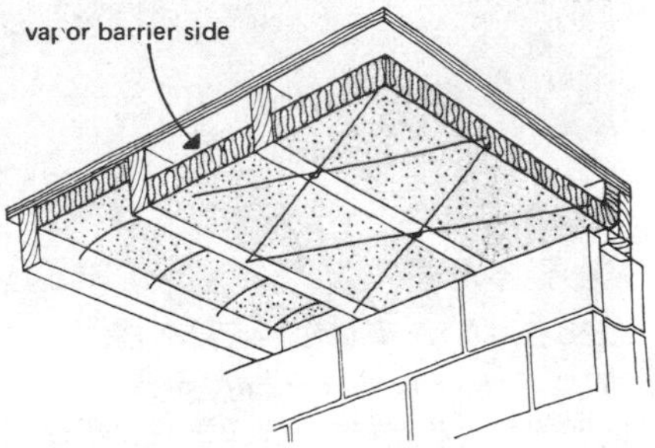

FIG. 38 If the crawl space is unheated, insulation should be placed between joists and supported with wires.

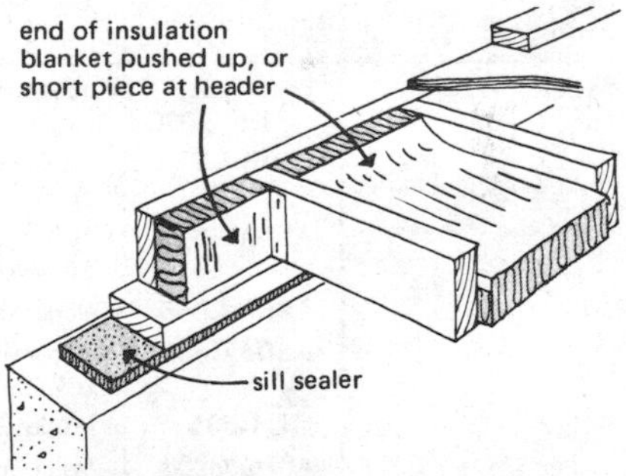

FIG. 39 Sill sealer is placed on top of foundation wall; insulation should be fitted snugly against band joist.

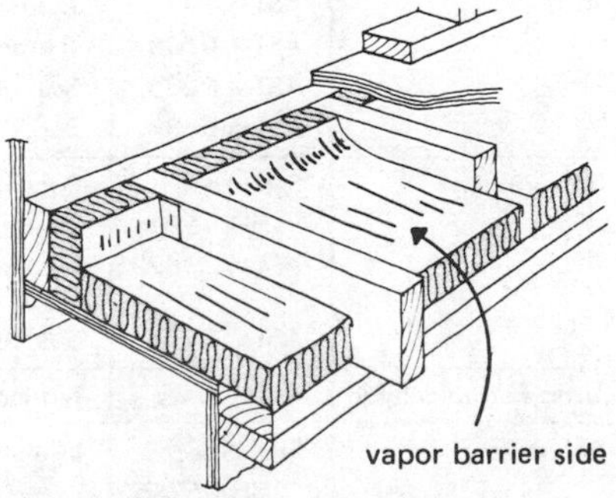

FIG. 40 Small areas such as soffits should not be overlooked and be carefully insulated.

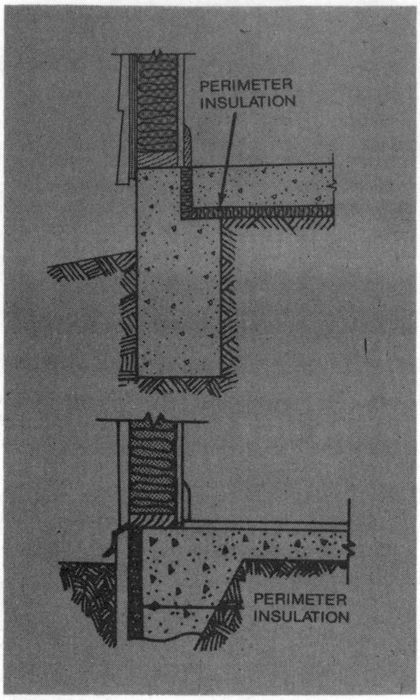

FIG. 41 Placement of perimeter insulation for slab on grade construction.

case by case basis. Other parts of the building envelope, such as floors and walls, should be reinsulated to those levels when it is economically practical.

Whenever insulation is added to an existing structure, proper vapor control is essential. This involves installation of a vapor barrier facing the living space, provision for adequate cavity ventilation or both. Section 104 Moisture Control provides a more detailed description of proper techniques.

Attics

Attics are often the most economical area to reinsulate because they are normally accessible and can account for a substantial part of the building heat loss or gain. Where the insulation is placed usually depends on whether the attic has been left unfinished or whether finish materials have been added to floors, walls and ceilings.

Unfinished Attic When the attic is unfloored, loose fill insulation can be blown into the joist cavities, or mineral wool batts can be placed between the joists in the same way that new attics are insulated. If the attic already has some insulation, additional loose fill can be blown or poured over the old, or another layer of unfaced batts can be added. If the old insulation was wrapped or if the new insulation is faced, the kraft paper or aluminum foil facing on the top face should be punctured at 12″ intervals. This prevents moisture condensation within the insulation. Condensed moisture could reduce insulating effectiveness and cause decay of structural members.

If the attic is floored, it is often possible to remove several boards and blow insulation under the remaining floor before replacing the boards (Fig. 43). If the space is not used for storage, unfaced batts or blankets can be laid directly on the floor. (Fig. 44). If there is a possibility that the space will be finished and heated in the future, it may be preferable to attach the insulation to the sloping rafters.

Finished Attic This space usually has three parts which may need insulation; upper ceiling, sloping ceiling and outer attic (Fig. 45).

Upper Ceilings can be insulated like an unfloored attic with either blown loose fill insulation, or with flexible batts if the space is accessible.

Proprietary systems are also available that provide insulation and ceiling finish in one integral panel. In such systems, mineral wool batts are attached to the back side of typical fiberboard ceiling panels and are fastened to the ceiling with metal grids. They typically require 3″ of ceiling clearance and provide an R-value of 12.

Sloping ceilings can be insulated if either the outer attic or upper attic is accessible to permit stuffing the rafter cavities with loose fill or flexible insulation. It may also be possible to install the thermal ceiling panels described above. Such panels can be applied to the face of existing sloping or cathedral ceilings if the panel thickness does not reduce room dimensions below acceptable levels.

Outer attics can be insulated either at the knee wall and outer attic floor or in the sloping rafters. It is gen-

FIG. 42 SUMMARY OF INSULATING PRODUCT STANDARDS

Material	Standard	Description
Mineral Wool	Fed. Spec. HH-I-1030a	Insulation, Thermal (Mineral fiber for Pneumatic or Poured Application)
	HH-I-521e	Insulation Blankets, Thermal (Mineral Fiber for Ambient Temperature)
	ASTM C236	Test for Thermal Conductance and Transmittance of Built-Up Sections . . .
	ASTM C665	Mineral Fiber Blanket Thermal Insulation
	ASTM C687	Recommended Practice for Determining of Thermal Resistance of Low Density Fibrous Loose Fill Building Insulation
	LLL-1-535	Insulation Board, Thermal
	ASTM C208	Insulating Board (Cellulosic Fiber)
	ASTM C209	Method of Test for Insulating Board
Cellulosic Fiber	Fed. Spec. HH-I-515c	Insulation, Thermal; and Insulation, Thermal (Loose Fill for Pneumatic or Poured Application): Cellulose or Wood Fiber
	ASTM C177	Thermal Conductivity of Materials by the Guarded Hot Plate
	ASTM C518	Thermal Conductivity of Materials by the Heat Flow Meter
	ASTM C236	See Titles Above
	ASTM C687	See Titles Above
Polyurethane	Fed. Spec.	Polyurethane Insulation Board Thermal and Polyisocyanurate
	HH-I-530a	Interim Amendment 3
	ASTM D1692	Method of Test for Flammability of Plastic Sheeting and Cellular Plastics Burning Rate
	ASTM C177	See Titles Above
Urea Formaldehyde		No Industry Standards
Polystyrene	Fed. Spec. HH-I-524b	Insulation Board, Thermal (Polystyrene)
	ASTM C177	See Titles Above
	ASTM D696	See Titles Above

erally preferable to insulate the knee wall and outer attic floor when there are eave vents, or eave vents can be installed, and if the outer attic is sealed from the living space and is not to be used for storage.

Walls

The basic types of walls—*conventional stud, brick veneer* and *masonry cavity*—are most economically insulated at the time of construction. They can be effectively insulated later, only if there was no insulation installed initially, and if there are no major mechanical or electrical obstructions in the wall.

Stud Walls Thermal efficiency of these walls can be improved by adding insulation to the stud cavity or to the exterior wall face before adding new siding.

Stud cavities are typically reinsulated with either blown mineral wool, cellulose, or foamed-in-place UFF foam.

Before the insulation is added, the wall should be checked for existing insulation by removing the face plate of a wall receptacle. If no insulation is present, the cavity should be probed with a plumb-bob or other weight to detect any obstruction which might cause blockages. The insulation is blown or squirted into the wall through holes drilled between the studs (Fig. 46). Exterior installation is preferable, because clapboards and shingles can more readily be removed and replaced than can parts of the interior finish.

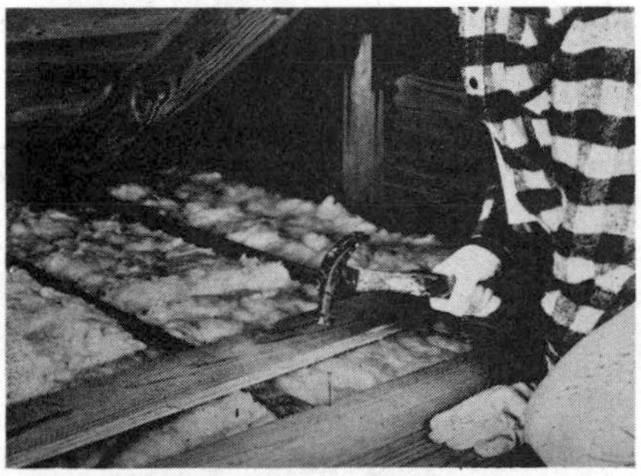

FIG. 43 Cavities should not be overfilled so that the insulation is compacted when the floor boards are replaced.

FIG. 44 Insulation over an existing floor should be fitted snugly together to prevent thermal leaks.

Exterior re-siding can provide an opportunity for adding plastic foam sheathing board under the new finish. The simple procedure requires only that the insulation be attached to the existing siding before new siding is applied (Fig. 47). Special metal channels accommodate the extra wall thickness around door and window frames. Although this procedure is not as thermally beneficial as fully insulating the stud cavity, the addition of 1″ rigid polystyrene board for example, can add 3.57 to the total R-value of the wall.

Brick Veneer Insulation can only be added to this type of wall by blowing or foaming. It is often difficult, however, to gain access to the stud cavity except in cases where the facia board near the eaves can be removed to expose an opening to the cavity (Fig. 48).

Masonry Cavity Such walls are rarely reinsulated because of the difficulty of gaining access to the cavity. However, it may be possible to remove several bricks or a facia board and thus squirt the insulation into the cavity. In addition to UFF or mineral wool, vermiculite and perlite can be used for such applications.

Foundations

Existing foundations of concrete or masonry can be insulated from either inside or outside the wall. Outside application is most frequently used when the interior already has a wall finish which would be covered by the new insulation.

Inside the Wall Mineral wool batts or blankets or rigid plastic foam boards are the typical materials used for such applications.

Batts or blankets are installed by furring out the wall with wood strips,

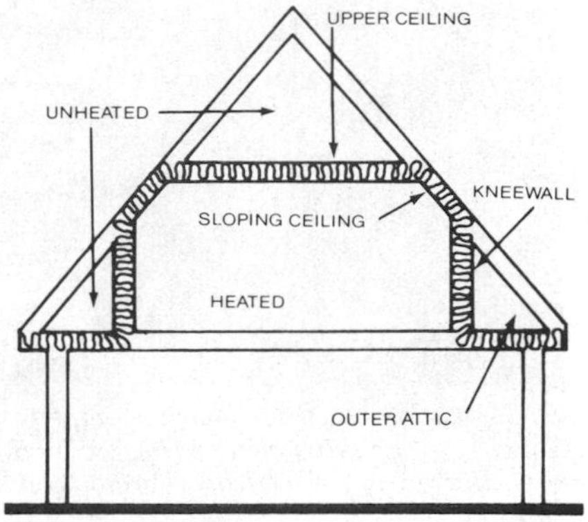

FIG. 45 A layer of insulation should be provided between any heated or conditioned space and any unheated or unconditioned space.

FIG. 46 Wall reinsulation is a three step process: open the wall cavity (a); add insulation (b); close the cavity (c).

FIG. 47 Insulating board completely covers the existing walls, thus reducing air infiltration as well as thermal transmission.

FIG. 48 Masonry cavity walls can be insulated by removing several bricks and blowing in insulation.

FIG. 49 If 2" x 4" studs are used, the wall can be insulated to R-11; 2" x 3" can accommodate R-7 insulation.

FIG. 50 A wall covering should be applied over paper-faced insulation.

and stapling the insulation to the strips (Fig. 49). The added wall thickness may not be acceptable for inhabited above-ground rooms, but can easily be accommodated in basements that are not used as primary living space.

Batts and blankets are the best materials for do-it-yourselfers and other non-professionals because they require less exacting installation. Mineral wool has the additional advantage of being fireproof if a paper backing is not attached. If the batts are paper-faced, they should be protected from flame by some type of wall covering (Fig. 50).

Rigid plastic boards can be installed on the interior wall between traditional 1" x 2" and 2" x 2" wood furring strips with the wall finish attached to the strips. There are also various proprietary systems that use mastics, adhesives, or metal channels in place of furring. Adhesives or mastics are thermally more efficient than the furring system, and can be installed at a lower cost by a contractor familiar with their use. However, these systems can only be used if the wall is straight and true, which basement walls often are not. In addition, mastics and adhesives can only be used on clean, dry, unpainted surfaces.

Plastic boards offer the advantage that they require less space to achieve a particular *R*-value. Extruded polystyrene 1-1/2" thick, for example, has an *R*-value of over 7.0, roughly equivalent to 2-1/2" of flexible mineral wool. A major disadvantage of foamed plastic board is that it must be covered with drywall to be safe from fire.

Outside the Wall Only the plastic foam boards—urethane and extruded polystyrene—are suitable for this installation because of their resistance to moisture. The mastic or adhesive normally used to secure the board to the masonry requires a sound, unpainted masonry or stucco surface for a good bond. To be most effective, the insulation should extend below the ground at least 2' (Fig. 51).

Polystyrene and urethane board are soft materials that can be dented and chipped if not protected by some harder material such as cement asbestos board. Several commercial systems available involve bonding a layer of glass fiber reinforcement to the insulation and applying a stucco-like finish. Polystyrene is also sensitive to daylight and must be protected to prevent deterioration.

Outside-the-wall reinsulation is often less economical than inside-the-wall, because of higher installation and material costs but it has the advantage that installation can proceed without disrupting interior spaces or covering up interior finises.

Crawl Space

If the space is accessible, it can be insulated in the same way as for new construction. Special attention should be given to the area between the floor joists and behind the header joist—an often ignored or under-insulated spot (Fig. 52). If the crawl space is not accessible it can be insulated on the outside of the wall as described for basements.

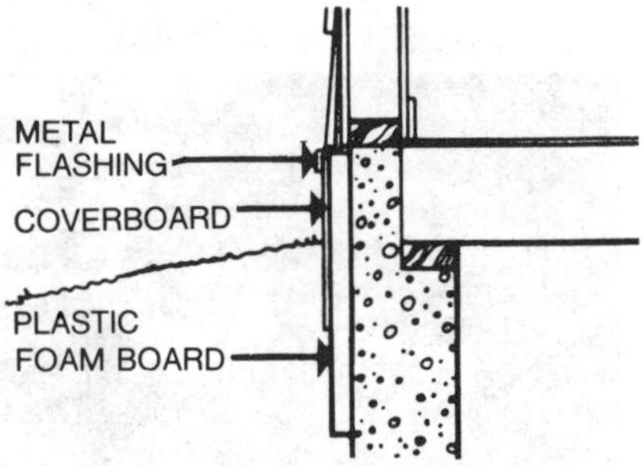

FIG. 51 Plastic foam board should be protected at the top edge with metal flashing and on its face with a hard waterproof coverboard.

FIG. 52 Strips of insulation should be stuffed behind header joists.

WINDOW & DOOR RECOMMENDATIONS

In a typical building, windows and doors lose and gain more heat than any other part of the structure. In some climates, such heat loss or gain can account for as much as 70% of the total heating load and 50% of the cooling load according to a study by the Department of Housing and Urban Development.

Most heat is lost through *infiltration*—air leakage—through cracks around the frame, and *conduction*—warm air coming in contact with the cold window and door surfaces. Most heat is gained through *radiation*—transmission of heat by invisible electromagnetic waves through the glass. The necessary window and door openings should be constructed so that unwanted heat loss and heat gain is minimized. The following are recommended strategies that will make such heat gain and heat loss reductions possible. Many have been suggested by the National Bureau of Standards' publication *Window Design Strategies to Conserve Energy.*

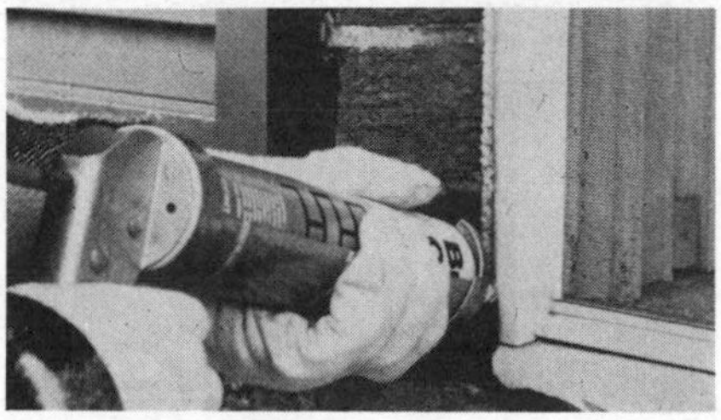

FIG. 53 Caulking bead should overlap the frame and the wall to insure a tight seal.

HEAT LOSS

The best way to prevent heat loss through windows is with: tightly sealed openings; multiple glazing (storm sash or insulating glass); and insulating shutters or tight fitting roll shades.

In most parts of the country, these methods can reduce heating costs, and improve interior comfort by reducing drafts and radiant body heat loss. In zones IV and V of Figure 82a, multiple glazing may not be justified and therefore should be decided on a case-by-case basis, taking into account fuel and material costs.

Tightly Sealed Windows and Doors

Tightly sealed windows and doors are highly recommended in all climates, regardless of whether the need for cooling or heating is predominant.

To achieve a tight seal around windows and doors, they should be: 1) caulked at the joint between the wall and window; 2) weatherstripped where practical; and in the case of doors, 3) closely fitted at the bottom.

Caulking Caulking should be used wherever the window or door frame meets the wall (Fig. 53). It should be applied to a clean, dry surface on a warm day. All caulking has a limited life and should be renewed periodically as necessary.

Caulking is usually sold in cartridge tubes and is applied with a gun. Bigger cracks require a cord or rope type which is applied by hand as a backup for the final coat of tube type caulk (Fig. 54).

Caulking performance varies according to chemical composition. Durability and other desirable properties range from the cheapest oil-based compounds, to the relatively expensive synthetics, latexes and silicones. Figure 55 lists the principal uses and characteristics of several common caulking compounds.

Weatherstripping This product is applied to window and door frames to prevent air leakage. It should be installed so the resilient part seals out air by pressing against the door or window sash.

The following are basic weatherstripping types: (1) foam rubber stripping; (2) felt stripping; (3) rolled vinyl; (4) casement stripping; (5) spring metal and (6) interlocking metal.

Foam rubber stripping with either adhesive or wood backing is easy to

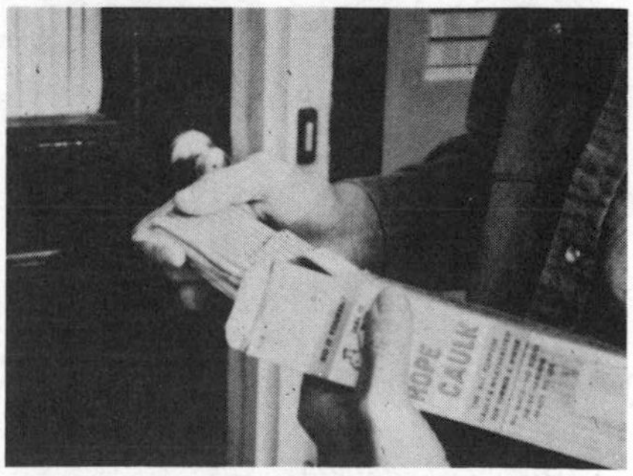

FIG. 54 Rope caulk can be forced into large cracks with a putty knife or large screwdriver.

FIG. 55 **CAULKING USES AND CHARACTERISTICS**

Performance Level	Caulking Type	Uses	Durability (Years)	Adhesion	Shrinkage Resistance	Cure (Days)
Basic Performance	oil and resin caulks	glazing	1 to 7	fair	poor	up to 1 year
	polybutane cord or rope	wide gaps	1 to 2	none	excellent	Remains moist
Intermediate Performance	nonacrylic latex; PVA	indoor and protected surfaces	2 to 10	good, except to metal	fair	3
	acrylic latex	indoor and protected surfaces	2 to 10	excellent except to metal	fair	3
	butyl rubber	metal-to-masonry	7 to 10	excellent	fair	7
	chlorosulfonated polyethylene	all surfaces	15 to 20	good	good	60 to 90
	neoprene	concrete	15 to 20	excellent	excellent	30 to 60
High Performance	polysulfide	all surfaces	20	excellent	excellent	7
	polyurethane	all surfaces	20	excellent	excellent	4 to 14
	silicone	all surfaces	20 or more	good, excellent with primer	excellent	2 to 5

install, but lacks durability. The adhesive type can be used wherever there is no friction such as at the top and sides of frames, and at the top and bottom of double hung windows (Fig. 56a).

Wood-backed foam rubber is more conspicuous than the adhesive type and can be used effectively only on doors (Fig. 56b).

Felt stripping has the same advantages and disadvantages as foam rubber; however, it also collects dirt more readily. Metal-backed felt, which provides more durability, is also available.

Rolled vinyl is available with or without a metal backing. Some types are manufactured with tubular gaskets that can either be hollow or filled with foam (Fig. 56c). Rolled vinyl is very durable and can be installed on the exterior of double hung, sliding and casement windows, as well as on doors.

Casement window stripping can be one of several kinds depending upon whether the window is wood or metal. Wood casements can be sealed with felt, adhesive-backed foam or spring metal. For metal casements, a special vinyl gasket with a deep groove can be used to slip onto all edges of the frame. A vinyl-to-metal adhesive must be used to bond the gasket to the frame so that the window closes flat against the vinyl strips. Most such weatherstripping is factory applied.

Spring metal stripping is available in two types—*flat stripping,* which is most commonly used for double hung windows, and *V-strip,* which is used for doors. The flat stripping is inserted into the narrow side channels of the window sash on the bottom rails of the upper and lower sash and on the top rail of the upper sash.

The V-strip is a double-over strip of springy metal that fits between door edge and jamb (Fig. 56d). When the door is open, the metal springs apart; when the door is closed, it is forced together to fit tightly between door and jamb. Both types are very durable and inconspicuous after installation, and they each provide an excellent seal. They will last indefinitely but are more costly to install than other types of weatherstripping.

Interlocking metal stripping is produced as *interlocking channels* and *interlocking J-strips.* They are difficult to install because alignments must be exact for a secure interlock. Both provide excellent, durable seals. The channels are subject to damage,

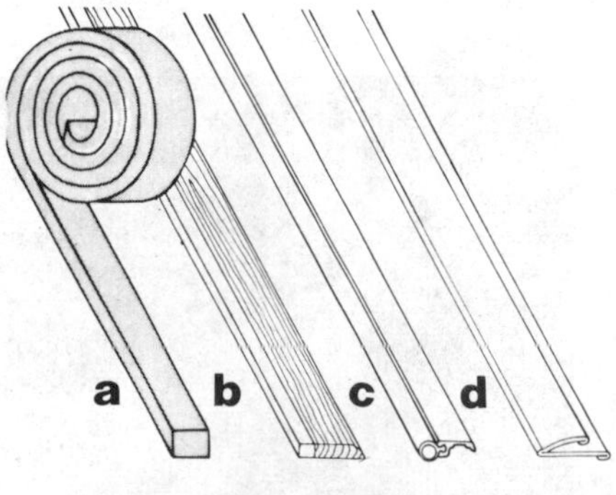

FIG. 56 Common types of weatherstripping: (a) adhesive-backed foam rubber; (b) wood-backed foam rubber; (c) rolled vinyl; (d) v-strip.

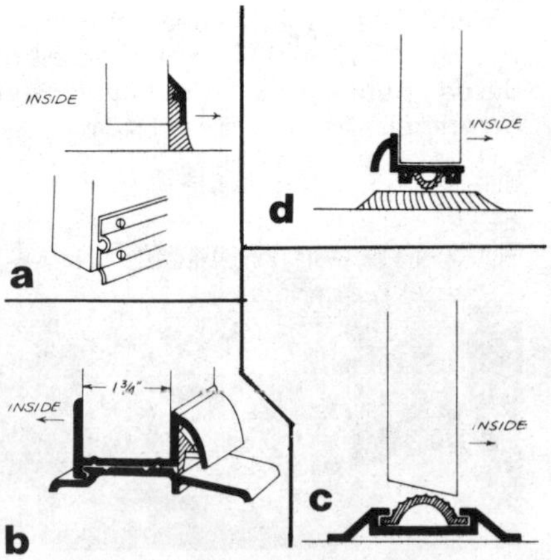

FIG. 57 Door bottoms: (a) sweeps; (b) door shoe; (c) vinyl bulb threshold; (d) interlocking threshold.

however, because they are exposed. The J-strips are not visible when installed; they are not exposed and therefore, not subject to damage.

Door Bottoms Air leakage through cracks at door bottoms can be stopped with: 1) sweeps attached to the door bottom; 2) door shoes; or 3) weatherproof thresholds.

Sweeps are easy to install because they do not require removal of the door. The most common types consists of a felt or rubber flap mounted in a metal channel (Fig. 57a). They seal most effectively on flat thresholds. They drag on carpets, however, and will eventually wear down from friction. To avoid carpet drag, automatic door sweeps can be used. These are spring-loaded and flip up when the door is opened, down when it is closed.

Door shoes are metal frames with attached vinyl bulbs that press against the threshold to form a tight seal (Fig. 57b). They are very durable because the bulbs can be replaced when worn. Door shoes are best suited for flat, unworn thresholds. They are difficult to install, however, because the door must be removed and trimmed to fit.

Weatherproof thresholds can be either the *vinyl bulb* or *interlocking type.* These thresholds are difficult to install because a good fit requires that the door bottom be trimmed at an 1/8″ bevel. The vinyl-bulb threshold contains a flexible vinyl strip which is compressed when the door is closed (Fig. 57c). The threshold itself is metal and very durable, and, although the vinyl bulb will wear out, it can be replaced.

The interlocking threshold consists of a metal sill that is shaped to interlock tightly with the metal channel attached to the door bottom (Fig. 57d). It is durable and provides an excellent weather seal but is difficult to install because proper alignment is critical.

Multiple Glazing

Glass is a good conductor of heat. One sq. ft. of 1/4″ clear glass can conduct 6 to 10 times more BTUs per hour than a square foot of a typical frame wall. The rate of heat transfer through the glass is so rapid that added thickness is of almost no value. However, if layers of glass are separated by an air space, the path of conduction is interrupted, and the rate of heat flow is reduced. Heat can then pass through the air space primarily by *radiation* and *convection* and only minimally by conduction.

The width of the air space will affect thermal performance if it is at least 3/16″ but no wider than 5/8″. That is, from 3/16″ up to 5/8″, the heat flow is reduced as the width is increased. Beyond 5/8″, increasing the air space does not reduce heat flow because the wider space allows the air to circulate more freely and develop *convection* currents within the space that transport heat from the warm glass to the colder glass. The *increased convection heat loss* more than offsets any slight reduction in *conductive heat loss* through the air. A space of less than 3/16″ is of little value because across such a short distance, heat is readily conducted by the air.

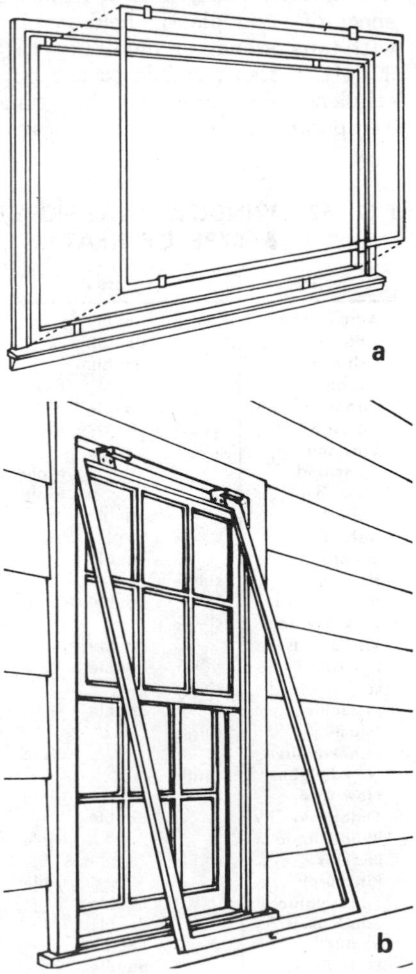

FIG. 58 Clip-on metal rimmed storm panels are typically used with awning or casement windows (a); storm panels for double hung windows are attached to the frame (b).

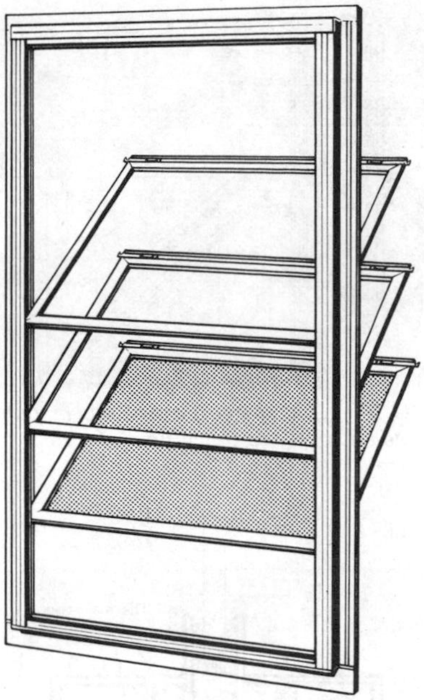

FIG. 59 These windows are also called "triple track" because of the three panels which move in separate tracks.

There are two ways to establish the necessary thermal air space: with *storm sash* or with *insulating glass.*

Storm sash These are commonly termed "storm windows." They control heat flow by making the window more airtight to reduce infiltration, and by providing an insulating layer of air between the window sash and the exterior to reduce conductive transfer.

They are recommended in Zones I, II, and III of the map in Figure 82. In these areas, storm windows can significantly reduce heating costs and minimize condensation on interior windows.

In some air conditioned buildings where natural ventilation is not possible, it may be desirable to leave storm windows in place during the summer to reduce heat gain through the glass and frames.

Storm windows can be of the *conventional, combination* or *fixed-in-*

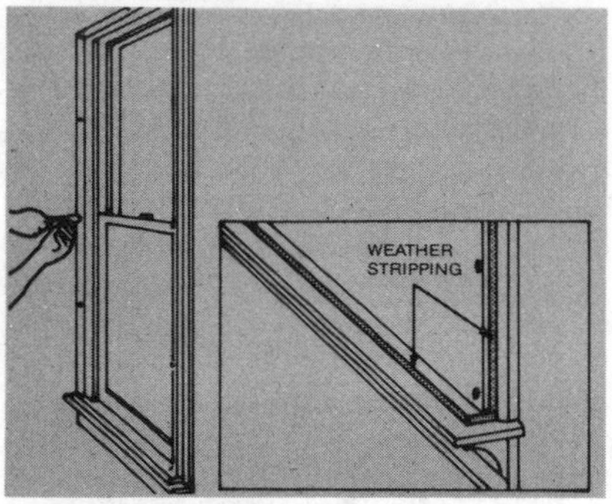

FIG. 60 These temporary storm panels attach directly to the window frame.

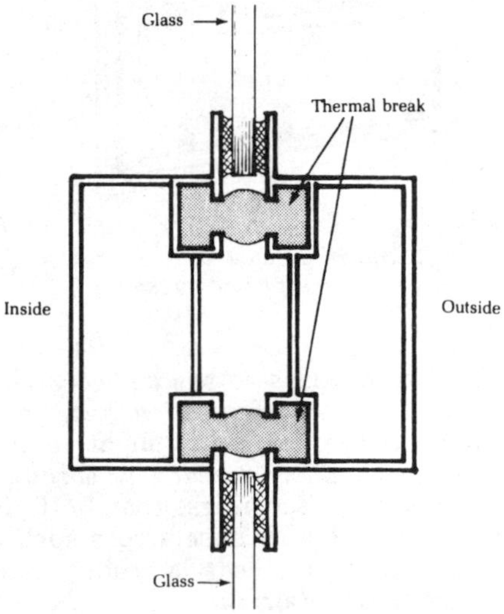

FIG. 61 The shaded thermal break reduces conductive heat transmission through the metal frame.

place interior types. They can be manufactured with frames of wood, metal, or rigid plastic. Storm doors are most often made of aluminum, although wood is still sometimes used.

The following is a general discussion of the various types of storm sash. Sections 202 and 214 provide a more detailed description.

Conventional storm panels are relatively inexpensive sash, usually made of wood or metal frames with single glazing. They are usually hung in place with special clips attached to each sash or to the frame. (Fig. 58). They are completely independent of the screens, so they typically are installed in the fall and taken down in spring. If the building is air conditioned, and not properly orientated for natural ventilation, they can be left up year-round to reduce cooling loads.

Combination storm windows are more expensive units, are permanently installed, and include not only storm sash but screens as well (Fig. 59). Frames are usually aluminum and should have an anodized or baked enamel finish. The window and screen sash should move smoothly in the frame tracks and seal tightly when closed. They should have tight joints and suitable weatherstripping to insure a tight seal.

Fixed-in-place interior storm windows are inexpensive *temporary* installations intended for the handy homeowner. This *interior* storm window usually consists of a frameless sheet of rigid plastic held in place with tape or snap-on mouldings (Fig. 60). It is also possible to use polyethylene film taped to the frame. Temporary plastic windows provide an amount of insulating value that is comparable to ordinary storm windows because it is the *air films* along the glazing surfaces and the air space between that reduce the heat loss, not the material thickness.

Insulating Glass Sealed insulating glass units, like storm sash, improve interior comfort and reduce conductive heat flow. Insulating glass eliminates condensation and has the convenience of being part of the prime window or door. It does not, however, have the added protection against infiltration that a separate storm panel provides.

If a metal frame is used to hold the glass, it should have a *thermal break* to prevent a thermal "short circuit," that is, heat flow being resisted by the insulating glass, but flowing freely through the aluminum frame. This is important because metal conducts heat many times faster than wood. The thermal break is usually polyurethane, vinyl, or some

FIG. 62 WINDOW GLAZING GUIDELINES BY LOCATION & TYPE OF HEAT

City	Oil Heat	Elec. Resistence Heat	Heat Pump
Albuquerque	double	triple	double
Atlanta	double	double	single
Baltimore	double	triple	double
Boston	triple	triple	triple
Charlotte	double	double	single
Chicago	triple	triple	double
Cincinnati	double	triple	double
Cleveland	triple	triple	triple
Columbus	triple	triple	double
Dallas	double	double	single
Detroit	triple	triple	triple
El Paso	double	double	single
Houston	single	single	single
Indianapolis	double	double	single
Jacksonville	single	double	single
Kansas City	double	double	double
Las Vegas	double	double	single
Los Angeles	single	double	single
Memphis	double	double	single
Miami	single	single	single triple
Minneapolis	triple	triple	single
New Orleans	single	single	triple
New York	triple	triple	single
Oklahoma City	double	double	double
Philadelphia	triple	triple	single
Phoenix	single	double	double
Pittsburgh	triple	triple	single
San Antonio	double	double	double
San Francisco	double	triple	single
Seattle	double	double	single
St. Louis	double	double	single
San Diego	single	single	triple
Washington	double	triple	

*Guidelines developed in October 1977 and based on the following assumptions: (1) 7 year period to recoup investment; (2) annual increase in energy cost of 10% in current dollars; and (3) finance interest rate of 9%.

FIG. 63 **DOOR *R* VALUES**

Door Type	R-Value
1¾″ exterior door	2.04
1¾″ exterior door and aluminum storm door (50% glass)	3.03
1¾″ exterior door and wood storm door (50% glass)	3.70
Metal clad insulated door	8.00-15.00

other low conductive material that provides a separation between the outside and inside frames (Fig. 61).

An aluminum frame with a good thermal break has an *R*-value of 1.66, which is approximately equal to that of double-pane insulating glass. A comparable frame with no thermal break would have an *R*-value of 1.18.

In some areas it may be economically practical to use three layers of glass separated by air spaces. Three layers of sealed glass with two 1/4″ air spaces, for example, has an *R*-value of 2.13 compared to 1.72 for typical insulating glass. This technique of *triple glazing* can also be accomplished by installing storm sash over double-glazed windows.

According to the NAHB Thermal Performance Guidelines, current fuel prices warrant the use of triple glazing in many cities including those listed in Figure 62.

Exterior Door Protection

Front and rear entry doors are often the second greatest source of heat loss after windows. The two most effective techniques to reduce such heat loss are: *storm doors* and *insulating doors.*

Storm Doors Like storm windows, storm doors reduce heat loss by reducing both *conduction* and *infiltration.* For example, a typical 1-3/4″ solid exterior door has an *R*-value of about 2. When a wood storm door is added, the *R*-value is increased to approximately 3.70 (Fig. 63).

Insulating Doors These have an inner core of plastic foam insulation, have high *R*-values (*R*-8 to *R*-15), and are even more effective than the combination of prime and storm doors in reducing heat flow. These doors frequently have metal skins and magnetic weatherstrips that are more airtight and minimize heat loss by infiltration (Fig. 64).

Window Insulation

For maximum reduction of heat flow through windows, more thermal protection than insulating glass or storm panels is desirable. Even with double glazing, a window (including glass and frame) loses heat approximately 6 to 10 times faster than a typical insulated wood-framed wall. Windows can be protected against heat loss by thermal shutters or thermal drapes and shades.

Thermal Shutters Covering a window with a rigid insulating material such as polyurethane or polystyrene is one of the most effective ways to control heat loss. The table in Figure 65 lists several common insulating materials and the resulting *R*-values when they are combined with 1/4″ plate glass. They are usually installed on the interior, but they can be placed on the exterior if the insulation is protected from the weather.

A common way to achieve protection from the elements is by facing the insulation with aluminum, galvanized steel or exterior plywood. If foam plastic is used, it must also be protected from fire. Plywood or metal cladding will also serve this function.

The shutters can be hinged, tracked, or mounted with clips. Removable shutters can be taken off and stored during sunny periods or warm weather. Hinged or tracked shutters can be folded or pushed back out of the way when they are not needed.

If the insulating material is placed in direct contact with the glass, the heat loss is only reduced in proportion to the *R*-value of that material. If the shutter fits so that a gap exists between it and the sash, conductive heat loss is more effectively reduced because of the insulating dead air space. Infiltration can also be reduced if cracks around the frame are covered. For such a reduction, however, a tight fit is crucial. If the panel fits loosely, air can circulate between the glass and the panel. Such air movement can carry heat away when the air passes over the glass. The result is a dramatic reduction in overall insulating effectiveness.

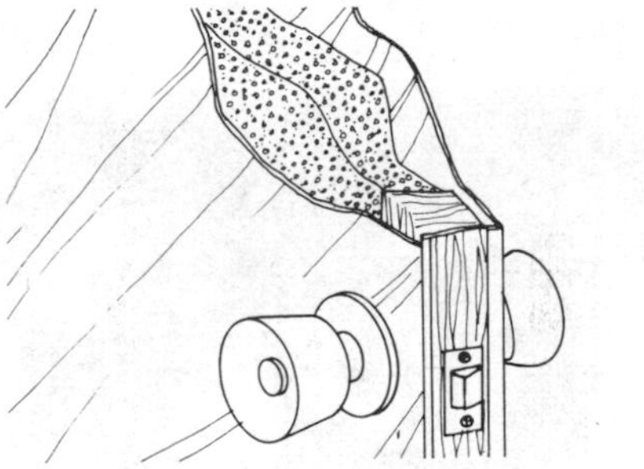

FIG. 64 Plastic foam insulation fills the inner core of insulating doors.

Thermal Drapes Drapes offer a more traditional way of controlling window heat flow. The surface temperature of an ordinary drape, for example, is usually much higher than that of the adjacent window glass. Such a situation reduces radiant heat loss from occupants and therefore improves thermal comfort.

For maximum benefit from window covers, the room air must be prevented from passing between the glass and insulating material. When the air is allowed to pass over the glass, the improvement from a loose fitting drape is nominal—an increase to *R*-0.94 from *R*-0.85 for an undraped window. If the heated air

FIG. 65 **WINDOW INSULATIONS**

Material	R-Value Per 1″	Window R-Value	
		with 1″ insulation	with 2″ insulation
Expanded polystyrene (extruded)	4.00	5.00	9.00
Expanded polystyrene (molded)	3.57	4.54	8.11
Polyurethane	6.25	7.14	13.39

from a wall or floor register passes between the covering and the glass, heat loss is actually more rapid than if no covering were provided.

By contrast, when the warm room air is prevented from circulating behind the drape, the R-value of the window and shade can increase to 1.13. The curtains must be tightly sealed at the bottom and sides.

Thermal Drapes can be fastened to the window frame with hooks, snaps, or "velcro"—a textured fabric that bonds securely to itself or to soft or wooly fabrics. If long drapes are used, they should touch the floor.

FIG. 66 Common shutter types: (a) rolling shutters move in horizontally mounted tracks; (b) Bahama shutters cover entire window opening; (c) Sarasota shutters extend over half the window opening (top) and are typically combined with side-hinged shutters.

Thermally efficient drapes are usually made of tightly woven fabric. The drape can be further improved by adding a liner or reflective material, foam backing, or additional layers of fabric.

HEAT GAIN

The best ways to prevent unwanted heat gain through glass areas is with shading devices and special glazing techniques. Tightly sealed openings, multiple glazing, and insulating shutters and curtains (discussed above) will also prevent heat gain in summer.

Shading Devices

In general, summertime sun controls are more effective if installed on the *exterior* because their shade prevents the heat producing sunlight from entering the building. This is critical because only a portion of the solar radiation that enters can be reflected back to the exterior. The remainder is absorbed and then radiated and conducted to the interior.

Although not as effective, interior shading devices have the advantage of greater accessibility and ease of management. The most effective shading devices are exterior shutters, exterior blinds and shades, awnings and architectural projections.

Less efficient but still beneficial are interior blinds and opaque roll shades.

Exterior Shutters A shutter that effectively blocks out all direct sunlight can reduce solar heat gain through a window by 80%. The overall shading performance, however, depends upon how well any heat absorbed into the shutter itself is dissipated to the outside air. Those that absorb large amounts of heat radiate a significant percentage of that heat into the building. Light-colored shutters, which reflect most of the sunlight before it can be absorbed, are therefore superior to dark shutters.

The most desirable shutters are those that have operable-tilt slats that can be adjusted to block the sun and also let air circulate through them. Non-operable louvered shutters will also permit natural air flow but cannot be adjusted to efficiently track the sun's angle. For shutters to most effectively prevent heat gain, air should be allowed to circulate between the window and the shutter. Such air circulation dissipates any heat absorbed by the shutters. Most shutters are manufactured of wood or aluminum. Vinyl is most frequently used to decorative shutters. Some shutters are not suitable for use as operable shutters because they are molded with an attractive surface on one side only. Others have simulated louvers with no open spaces between slats. Several common shutter types are illustrated in Figure 66.

FIG. 67 Soffits above the windows house the motorized roller which stores the blinds (a). Even large openings can be protected, such as this screened porch (b).

Tight-fitting shutters will provide resistance to heat loss even if air can circulate behind them. If the fit is not tight, or the louvers cannot be completely closed, shutters will still reduce heat loss by reducing infiltration and preventing the scouring action of the wind from disturbing the insulating film of air at the outer surface of glass.

Exterior Roll Blinds and Shades These devices can range from a

FIG. 68 **SOLAR TRANSMITTANCES OF AWNING MATERIALS**

Material	Direct Transmittance (%)	Diffuse Transmittance (%)
Canvas	0.00	0.00
Plastic	0.25	0.15
Aluminum	0.00	0.20

FIG. 69 **HEAT GAIN THROUGH SINGLE GLAZED WINDOWS WITH AWNINGS**

Window Orientation	Awning Type	Heat Gain Per 100 Sq. Ft. Glass Surface (BTU per day)	Heat Excluded	
			Total (BTU per day)	% Reduction
South	No awning	62,200	—	—
	White canvas	22,500	39,700	64
	Dark green canvas	27,700	34,500	55
	Dark green plastic	35,600	26,600	43
West	No awning	84,200	—	—
	White canvas	19,500	64,700	77
	Dark green canvas	23,900	60,300	72
	Dark green plastic	34,800	49,400	59

FIG. 71 Venetian blinds can be made to cover large glazed areas such as sliding glass doors.

simple bamboo or canvas shade to a motorized roll blind that can be operated from the interior. Like shutters, such devices are highly effective because they stop the sunlight from entering a building. Because they provide an opaque covering of the window itself, they are effective in stopping low-angle sun on west elevations. A light color is most effective because it reflects the heat producing sunlight.

Motorized Roll Blinds They are usually made of aluminum, vinyl or wood and therefore offer the added benefit of year-round use and durability (Fig. 67). They can also reduce heat loss in winter by as much as 50% by acting like a tight-fitting storm window. Independent shade control of each window and ease of operation are other advantages. Their prime disadvantage is that exterior view and natural ventilation is reduced. One manufacturer, however, provides a blind which can be projected away from the window like an awning, to provide view and ventilation.

Awnings How well an awning shades a window depends on how opaque the material is to both direct sunlight and diffuse light from the sky. The more opaque a material, the more complete is the shade provided. The table in Figure 68 gives the light transmittance values of several common awning materials. Transmittance values are computed as a ratio of light penetration to total light present.

The exposed surfaces of the awning should be a light color to minimize the amount of sunlight absorbed. When sunlight is absorbed by the awning, heat is produced and its temperature is raised. The result is heat transfer through the window.

A clean, white canvas awning or a slatted, light aluminum awning, for example, reflects between 80 and 91% of the sunlight that strikes its surface. A dirty awning is less reflective and therefore absorbs more sunlight. By comparison, a dark green canvas awning reflects only 21%, and a dark green plastic awning reflects 27% of the sunlight.

A typical light-colored awning as described above can reduce heat gain by 64% on south-facing windows and 77% on west-facing windows. Figure 69 lists the net effectiveness of several types of awnings.

Heat from sunlight absorbed by a dark-colored fabric will build up under the awning and be transferred to the window unless air is permitted to circulate under the awning. Fabric awnings are commonly installed with a continuous gap between the top of the awning and the wall to prevent hot air from being trapped under the awning. Slatted aluminum awnings inherently provide air circulation between the horizontal slats.

Awnings should provide adequate coverage of the window area for the specific orientation of the window. A south-facing window requires only a minimal horizontal projection to be completely shaded throughout the day. An east- or west-facing window needs an awning that extends down a substantial distance to block out early morning or late afternoon sun. In addition, the sides of the awning

FIG. 70 **ROOF OVERHANG FOR COMPLETE SUMMER WINDOW SHADING***

Latitude • North, Degrees	Overhang Width	
	South Exposure	Southeast/Southwest Exposure
28	1′-5″	3′-10″
32	1′-11″	4′-2″
36	2′-5″	4′-8″
40	3′-0″	5′-4″
44	3′-6″	6′-0″

*Assume window sill 2′ from floor.

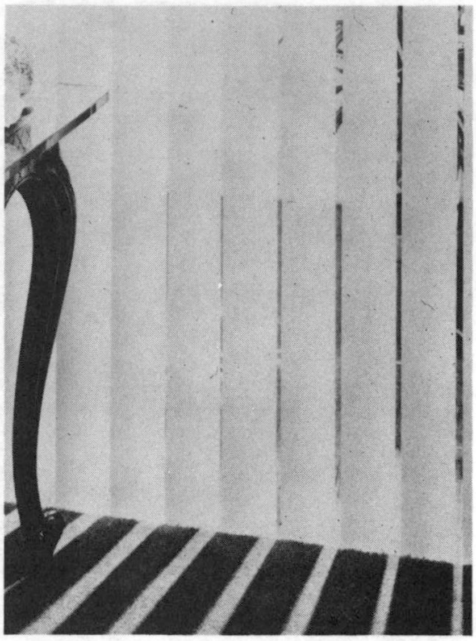

FIG. 72 Vertical blinds extend to the floor and therefore reduce heat-robbing convection currents.

should be closed to prevent sunlight from angling in behind the awning on south-facing windows.

Architectural Projections Windows can be shaded by horizontal or vertical planes that extend beyond the face of the glass to intercept the summer sun. If properly designed, these projections will not interfere with the penetration of the beneficial winter sun, or block exterior views.

A projecting roof, termed an *overhang,* is one of the most common ways to shield residential windows from sunlight. The table in Figure 70 lists the horizontal projection necessary to provide complete summer window shading with an overhang.

A deeply recessed window, or one with closely spaced horizontal or vertical fins, can also be effective. In general, east- and west-facing windows are more effectively shaded by vertical projecting planes. South-facing windows are best shaded by horizontal planes.

The underside of an overhang ideally should be dark because it will transmit less ground reflected sunlight through the glass. The net result of such shading is that heat gain through a completely protected window can be reduced by 80%.

Interior Blinds The most common of such devices are *venetian blinds* (Fig. 71). They are slatted horizontally and can be tilted to reflect a maximum amount of sunlight out the window or to direct daylight toward the ceiling for glare free uniform daylighting. Blinds, however, provide only minimal reduction in heat loss in winter because they do not effectively entrap a layer of air.

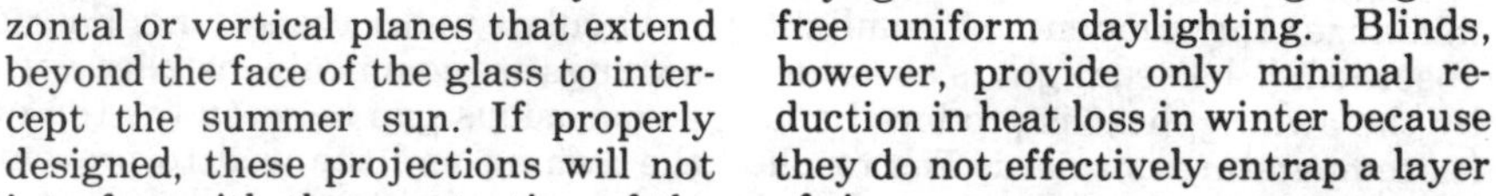

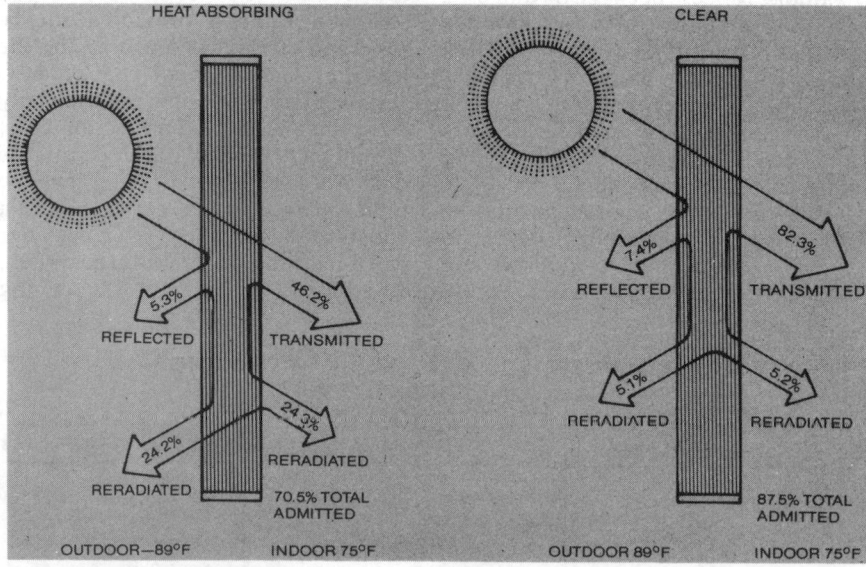

FIG. 73 Heat-absorbing glass reduces window heat gain because it significantly reduces direct solar transmission.

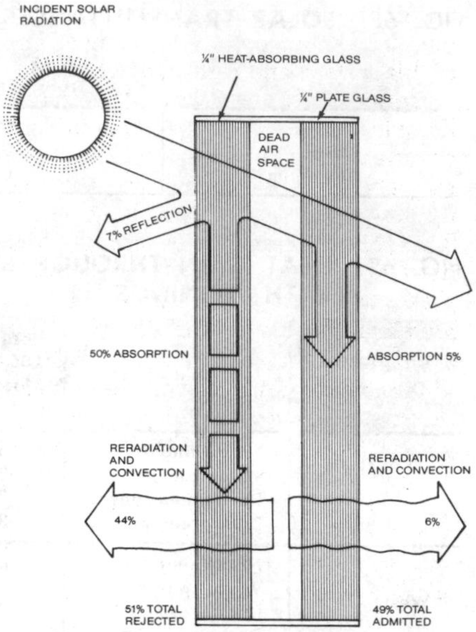

FIG. 74 When heat-absorbing and clear plate are combined as sealed insulating glass, only 49% of radiation striking the window is transmitted inside.

Vertical blinds offer a similar degree of protection against heat gain. They can be significantly more efficient in reducing heat loss in winter, however, because they more effectively trap an insulating layer of air between themselves and the glass (Fig. 72).

Interior Roll Shades The most effective type of interior protection against window heat gain is an *opaque* or *translucent* roll shade. They can be easily installed with simple brackets on the window frame and when not in use, roll up for unobtrusive storage.

A completely opaque shade with a white outer face can reflect 80% of the sunlight back through the glass.

A roll shade can have a dark color on one side which effectively absorbs sunlight and a white surface on the reverse side which effectively reflect sunlight. By simply reversing the shade, from dark side facing out in winter, to reflective side out in summer, the shade can perform both as a solar collector and shading device.

Special Glazing

When interior or exterior shading is not practical, it is possible to use special types of glass to keep the sun out. The most common are heat-absorbing glass and reflecting glass.

Heat-absorbing Glass If a metallic oxide is added to the glass during the manufacturing process, the glass can absorb significantly more sunlight—particularly ultraviolet light. This characteristic distinguishes heat-absorbing glass from *tinted* glass, which may merely exclude the visible light useful for illumination.

The sunlight absorbed by the glass is reradiated and convected as heat to the inside or outside, depending upon which side is cooler. The result is that in summer, more heat is dissipated to an air conditioned interior, because of its cooler temperature. In winter, more heat is dissipated to the outside, because it is then cooler.

However, the net effect is still beneficial because 17% less heat-producing sunlight is admitted through heat-absorbing glass than through clear glass (Fig. 73).

The performance of heat-absorbing glass is substantially improved when it is used as the outer sheet of an insulating unit (Fig. 74). In an insulating unit, the absorbed heat must first bridge the air space to the inner sheet of glass before it can be radiated and convected into the building interior. Because heat transfer is slowed by the insulating air space, more heat is dissipated to the outside air, which is in direct contact with the heat-absorbing glass. Furthermore, the outward rate of heat dissipation is greatly accelerated if there is any wind.

Some designers have proposed a double-glazed reversible window sash consisting of clear glass on one side and heat-absorbing glass on the other. The heat-absorbing side can face the outside in summer to dissipate heat outward. It can be reversed in winter so that the heat-absorbing glass faces inside and dissipate its heat into the building.

Reflective Glass When sunlight strikes a window, it is either transmitted, reflected, or absorbed and radiated as heat. By increasing the amount of reflected sunlight, the amount that can be absorbed and transmitted or radiated is reduced.

Reflective glass is more effective as the outer sheet of insulating glass than as single glazing (Fig. 75). This is because, despite its name, reflective glass absorbs more sunlight than clear glass that transmits most of the sunlight which strikes its surface. The result is a heat build-up in the reflective glass. When that heat is concentrated in the outer sheet of insulating glass, it is more readily dissipated to the outside air because the air space between glass layers impedes the inward flow of heat.

Some of the disadvantages of reflective glass are that in winter,

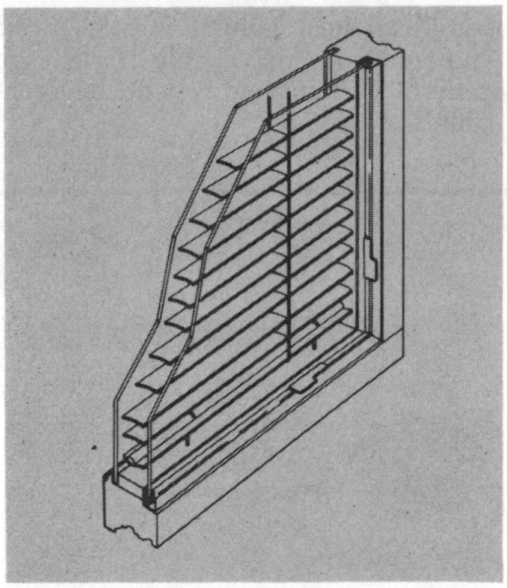

FIG. 76 With blinds fully closed, solar heat gain is reduced by approximately 82%.

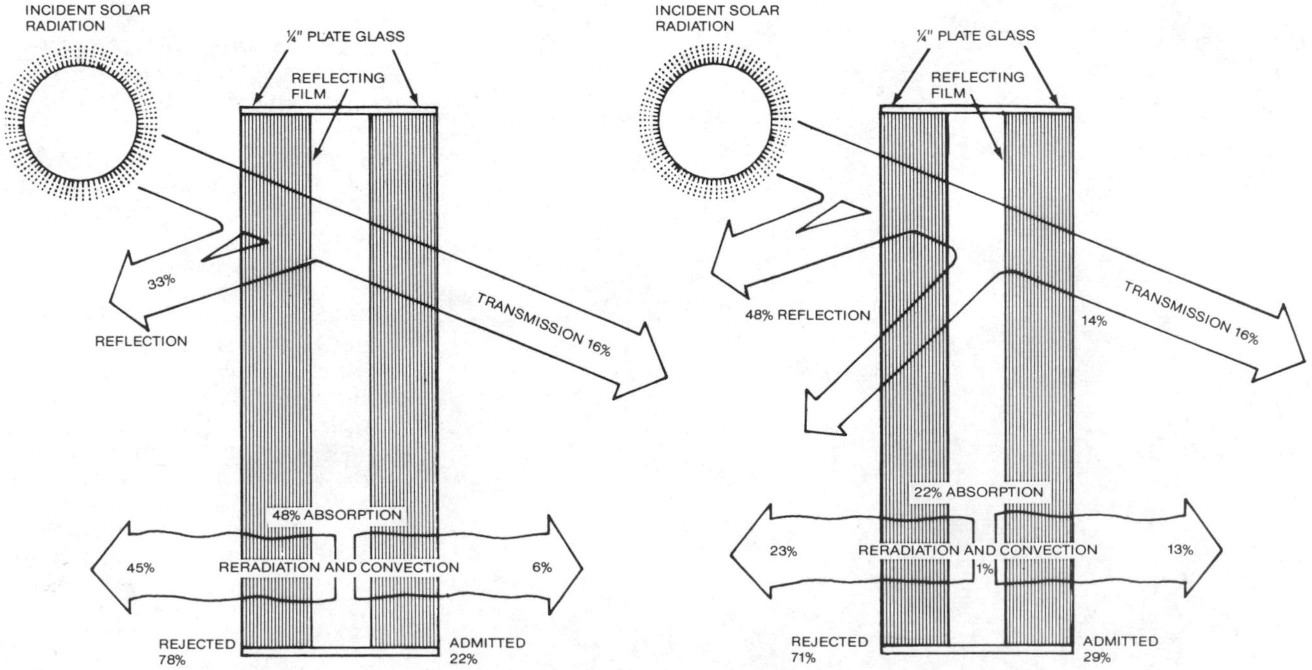

FIG. 75 Insulating glass with a reflective film on the outer sheet, admits 7% less radiation than similar glass with an inner sheet with reflective film. The first bars 78%, the second bars 71% of the incident solar energy.

beneficial heat gain is reduced, and transmission of visible light for general illumination is reduced year-round.

Integral Blinds At least one proprietary system includes venetian blinds encased between layers of glass to reduce heat gain and heat loss (Fig. 76). When fully closed this system can provide an *R*-value of 2.50 compared to an *R*-value of 2.04 for a similar window with no blinds.

DESIGN RECOMMENDATIONS

Buildings should regulate heat flow in a way which will 1) minimize dependence on mechanical equipment and 2) take advantage of exterior environmental conditions. Energy saving features can be incorporated economically in buildings if considered early in the planning and design phases.

SITE PLANNING

In locating the building on the site, the following should be considered: 1) general climate and microclimate, 2) terrain and landscape elements, 3) orientation to the sun.

Climate

Climatic analysis can provide valuable data on which to base energy efficient design. Such an analysis should include data on the amount and angle of sunshine which strikes the site, prevailing winds, temperature ranges and precipitation.

U.S. Weather Service and private meteorologists maintain regional data on these factors. However, natural terrain and proximity to bodies of water may substantially modify the weather information recorded for a particular region. To avoid siting a building in an unsuitable area, the climate variations of each site (*microclimate*) should be examined.

For example, sites in warm climates which act as natural dead-air basins should be avoided. In such areas, temperatures may exceed those of surrounding areas by as much as 35° F. In other areas prevailing winds and natural terrain can combine to produce unusual wind, snow massing or rain conditions not typical to a particular climate. Air quality also should be studied to determine whether natural ventilation can be used for cooling.

WINTER: LEAFLESS TREES LET SUN THROUGH.
SUMMER: LEAVES SHADE HOUSE.
WINTER: EVERGREENS PROTECT HOUSE.
SUMMER: LEAVES DIRECT BREEZE INTO HOUSE.

FIG. 78 Proper trees in the right place can reduce energy needs.

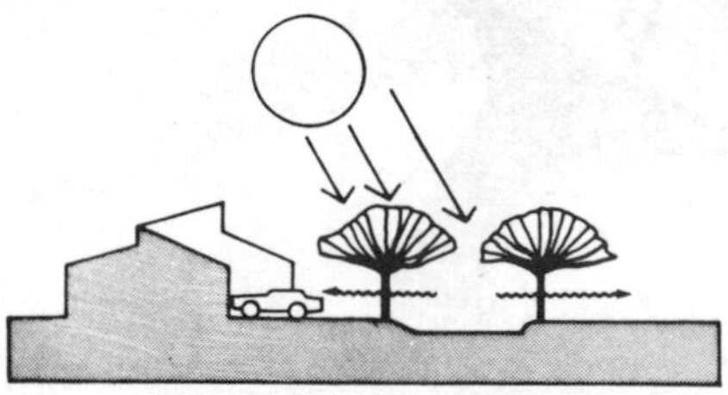

FIG. 77 Shading paved areas prevents heat build up and channels breezes.

Terrain & Landscape Elements

Terrain configurations and plant material can be used to control heat flow into and out of buildings.

Plantings can prevent reflection or reradiation of solar energy from paved areas and adjacent buildings (Fig. 77). Exterior walls can reduce heat transmission and earth berms can also be used to shield walls and outdoor living spaces.

Coniferous (evergreen) trees can be located to block prevailing winter winds. Deciduous (leaf dropping) trees on south and west exposures can shade walls and windows in summer but drop their leaves to allow desirable winter sun penetration (Fig. 78). Ponds and fountains fed by collected rainwater act as natural coolants to reduce air temperature around building exteriors (Fig. 79).

Orientation

In cold regions, the area of north facing windows should be limited to reduce heat loss in winter. The area of north facing walls can be reduced by berming and sloping roofs (Fig. 80). In hot climate areas, west wall and glass exposures should be minimized to reduce cooling loads from intense afternoon summer sun.

Buildings in cold regions should be sited so that most windows face

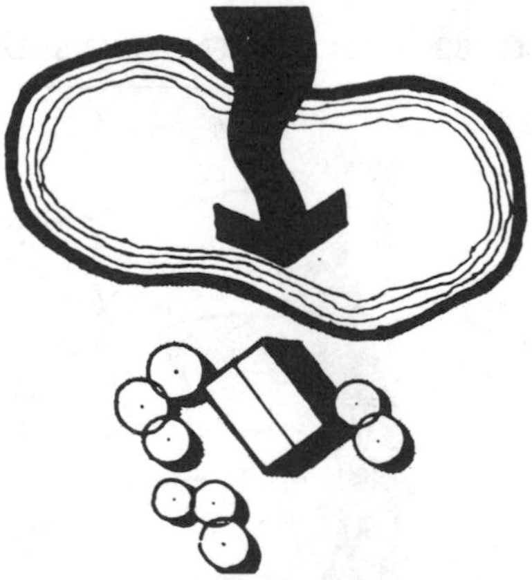

FIG. 79 Air flowing over cool water acts as natural air conditioner.

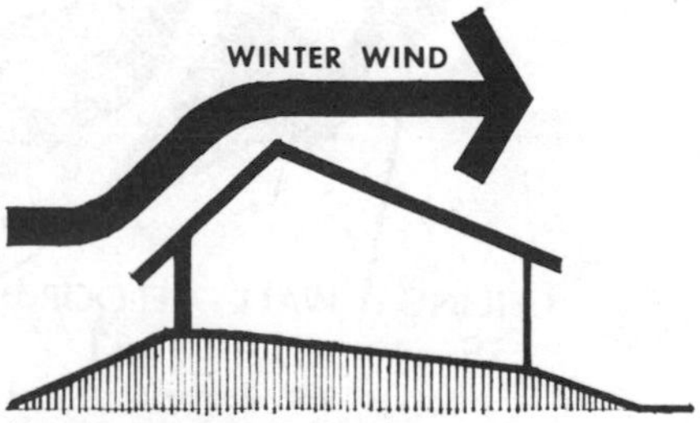
WINTER WIND

FIG. 80 Sloping roofs also act as a wind deflector.

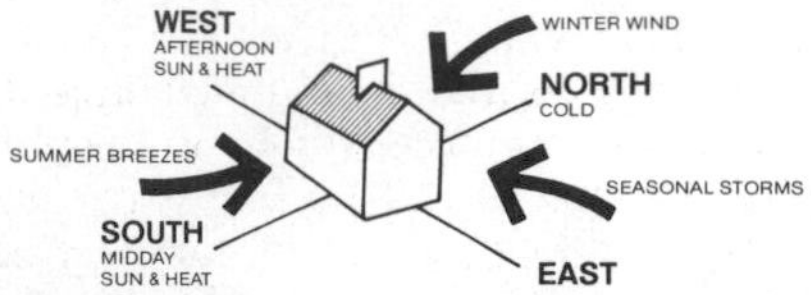

FIG. 81 Careful orientation takes advantage of natural elements.

south, southeast or southwest to take advantage of the warming sun. Orientation that will buffer against hot breezes, is desirable in hot climates. Prevailing summer breezes which can provide ventilation and natural cooling should be considered when siting a building or locating windows (Fig. 81).

BUILDING DESIGN

Thermally efficient buildings can be achieved through 1) careful room organization and 2) effective building envelope design.

Room Organization

In areas subject to either extreme heat or cold, compact design reduces weather-exposed areas and reduces heat flow. Garages, closets, corridors and other service spaces can be located against exterior walls to act as insulating buffers.

Inactive living spaces such as kitchens and laundry areas, which contain heat producing appliances should be located on north and east sides. This location permits wintertime use of their excess heat for warming and permits heat dissipation in summer.

Active living spaces should face south or southwest to take advantage of warming afternoon sun in winter. Such spaces should be protected against overheating in summer by external shading devices such as louvers, overhangs, awnings and trees. Interior shading with blinds and drapes is less effective but acceptable. Rooms likely to remain vacant for long periods should be located where they can be closed off readily. Lower ceiling heights can also reduce the volume of air to be heated or cooled. Built-in furnishings should be located away from supply and return air grilles to prevent obstruction of internal air flows. The omission of unnecessary partitions will also facilitate natural ventilation and reduce cooling loads.

Building Envelope

Adequate insulation of the building envelope is essential to controlling heat flow. In some climates, materials with high thermal capacities are effective in slowing heat gain or loss. Walls, floors and ceilings of one- to four-family frame dwellings should have insulation meeting the R-values recommended in Figure 82a.

Floors Where solar radiation penetrates interior spaces, floors can act as *heat sinks.* Heat sinks are building components consisting of dense heavy materials such as concrete or masonry. They are desirable in cold climates because the material can absorb and store heat for long periods before releasing it to the interior by radiation and convection.

The floors of living areas above unheated crawl spaces or basements

FIG. 82a **RECOMMENDED INSULATION *R*-VALUES**

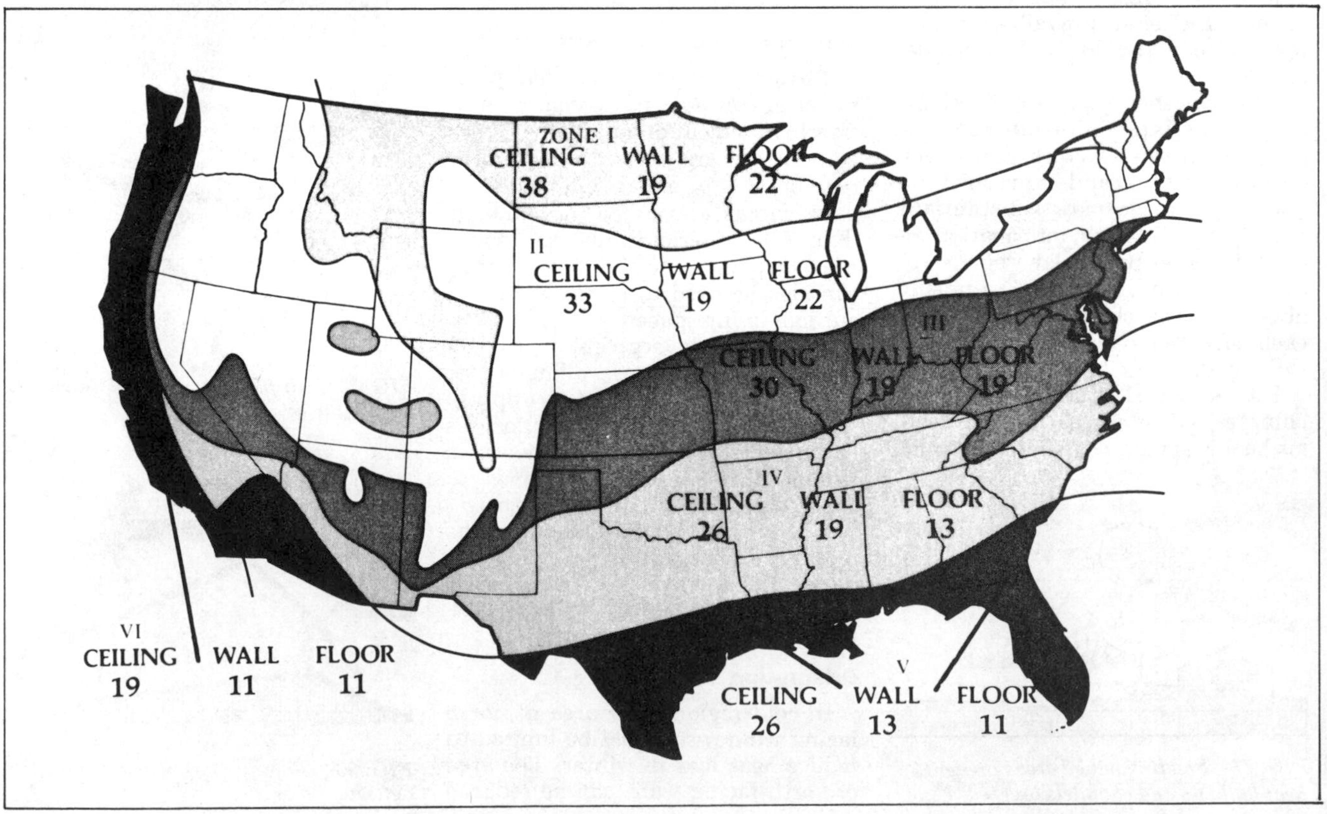

should be insulated as recommended in Figure 82a. If the crawl space is heated, the exterior walls should be insulated to an R-value of 19.

Slab-on-ground floors should be insulated with 2″ of moisture resistant insulation under the outer edge of the slab or adhered to the foundation wall (Fig. 41). Depending on the types of insulation selected, the R-value could range from 5.26 to 12.5 per 2″.

Walls Heat transmission through wall assemblies can be reduced by selecting materials which are inherently good insulators or by adding a suitable amount of insulation (Fig. 82a). To accommodate even more insulation, 2″ x 6″ wall studs are sometimes used. The University of Illinois "Low-Cal House" is an example of a more elaborate system. It features double-framed walls with staggered studs spaced a distance of 8-1/2″ from face to face to permit installation of R-33 insulation.

Wind deflecting devices such as *berms, walls* or *fences* can also be employed to reduce infiltration along exterior walls exposed to cold winds by steering the wind away from the wall.

Roofs Heat transmission through the roof can be reduced by: 1) use of optimum amounts of insulation (Fig. 82a); 2) use of shiny or light colored roof materials to reflect heat; 3) ventilation of attic spaces; 4) increasing the thermal mass to achieve a time lag in the heat flow; 5) use of louvered rather than solid overhangs to free heated air trapped beneath eaves; and 6) use of steeply pitched roofs on the windward side of the building to deflect cold winds and reduce the amount of roof exposure (Fig. 80).

EXISTING AND NEW HOUSE INSULATION RECOMMENDATIONS

A number of voluntary and mandatory standards are currently in use nationally for new housing (Fig. 82b). All of these will be superseded by the National Building Energy Performance Standards (BEPS) for new buildings, if and when the U.S. Department of Energy (DOE) issues

FIG. 82b INSULATION RECOMMENDATIONS FOR NEW AND EXISTING HOUSES IN EIGHT REPRESENTATIVE CITIES

CITY	FUEL	INSULATION LEVEL RECOMMENDATIONS								
		EXISTING HOUSES				NEW HOUSES				
		DOE-RCS[2]	FmHA[3]	EEI[2]	"ENERGY $"[2]	HUD-MPS[3]	FmHA[3]	EEI[2]	NAHB-TPG[2]	OCF ZONE[2]
		C W F	C W F	C W F	C W F	C W F	C W F	C W F	C W F	C W F
Tampa, FL (700 DDH)	Gas	19 x 0	19 11 11		30 x 0	19 11 0	19 11 11		19 12 0	
	Oil	19 x 0	19 11 11		30 x 0	19 11 0	19 11 11		19 12 0	
	Elec.	19 x 0	19 11 11	30 11 19	33 x 0	19 11 0	19 11 11	30 18 19	22 12 11	26 13 11
	Heat Pump	19 x 0	19 11 11	30 11 19	30 x 0	19 11 0	19 11 11	30 18 19		
Jackson, MS (2260 DDH)	Gas	19 x 0	22 12 11		30 x 11	19 11 0	22 12 11		19 12 11	
	Oil	19 x 0	22 12 11		33 x 19	19 11 0	22 12 11		19 12 11	
	Elec.	22 x 0	22 12 11	30 11 19	33 x 19	22 12 0	22 12 11	30 18 19	19 12 19	26 19 11
	Heat Pump	19 x 0	22 12 11	30 11 19	30 x 11	19 11 0	22 12 11	30 18 19		
Atlanta, GA (2990 DDH)	Gas	22 x 0	30 17 19		30 x 11	22 11 0	30 17 19		19 12 11	
	Oil	22 x 0	30 17 19		38 x 19	22 11 0	30 17 19		22 12 11	
	Elec.	30 x 11	30 17 19	30 11 19	38 x 19	30 17 11	30 17 19	30 18 19	30 14 19	26 19 11
	Heat Pump	22 x 0	30 17 19	30 11 19	30 x 11	22 11 0	30 17 19	30 18 19		
Washington, DC (4240 DDH)	Gas	30 x 11	30 17 19		33 x 19	30 12 11	30 17 19		19 12 19	
	Oil	30 x 11	30 17 19		44 x 22	30 12 11	30 17 19		30 14 19	
	Elec.	30 x 19	30 17 19	30 11 19	49 x 22	30 17 19	30 17 19	30 18 19	30 18 19	30 19 19
	Heat Pump	30 x 11	30 17 19	30 11 19	33 x 19	30 12 11	30 17 19	30 18 19		
Portland, OR (4700 DDH)	Gas	30 x 11	30 17 19		33 x 19	30 12 11	30 17 19		19 12 19	
	Oil	30 x 11	30 17 19		38 x 22	30 12 11	30 17 19		30 14 19	
	Elec.	30 x 19	30 17 19	30 11 19	38 x 22	30 17 19	30 17 19	30 18 19	30 14 19	19 11 11
	Heat Pump	30 x 11	30 17 19	30 11 19	30 x 19	30 12 11	30 17 19	30 18 19		
Columbus, OH (5670 DDH)	Gas	30 x 11	30 17 19		30 x 19	30 12 11	30 17 19		19 12 19	
	Oil	30 x 11	30 17 19		44 x 22	30 12 11	30 17 19		30 14 19	
	Elec.	30 x 19	30 17 19	30 11 19	44 x 22	30 17 19	30 17 19	30 18 19	30 18 19	30 19 19
	Heat Pump	30 x 19	30 17 19	30 11 19	30 x 19	30 17 19	30 17 19	30 18 19		
Chicago, IL (6640 DDH)	Gas	30 x 11	38 17 19		38 x 22	30 12 11	38 17 19		30 14 19	
	Oil	30 x 11	38 17 19		49 x 22	30 12 11	38 17 19		30 16 19	
	Elec.	38 x 19	38 17 19	30 11 19	49 x 22	38 17 19	38 17 19	30 18 19	30 18 19	33 19 22
	Heat Pump	38 x 19	38 17 19	30 11 19	33 x 19	38 17 19	38 17 19	30 18 19		
Minneapolis, MN (8250 DDH)	Gas	38 x 19	38 17 19		38 x 22	38 17 19	38 17 19		30 14 19	
	Oil	38 x 19	38 17 19		49 x 22	38 17 19	38 17 19		30 18 19	
	Elec.	38 x 19	38 17 19	30 11 19	60 x 22	38 17 19	38 17 19	30 18 19	38 18 22	38 19 22
	Heat Pump	38 x 19	38 17 19	30 11 19	44 x 22	38 17 19	38 17 19	30 18 19		

[1]Data as of November 1979; numbers given are R-values, for roof/ceiling (C), walls (W) and floor (F).

[2]Voluntary Guidelines: (a) DOE-RCS DOE Residential Conservation Program; (b) EEI = Edison Electric Institute; (c) "Energy $" = Make the Most of Your Energy Dollars, National Bureau of Standards; (d) NAHB-TPG = Thermal Performance Guidelines; (e) OCF Zone = Recommendations of Figure 82a.

[3]Mandatory Standards: (a) FmHA = Farmers' Home Administration; (b) HUD MPS = Minimum Property Standards for One and Two Family Dwellings.

BEPS as a final rule (expected in November 1980).

In 1978, the U.S. Congress authorized DOE and HUD to study the impact on consumers, builders, lenders and others of mandatory standards for *existing* houses. It is not likely that such requirements will be adopted as mandatory national standards in the near future, but a variety of insulation guidelines are currently in use. Figure 82b summarizes current mandatory and voluntary guidelines for new and existing housing.

The differences among the various guidelines and the column marked *OCF Zone* are relatively small. The OCF Zone recommendations are taken from the six-zone map shown in Figure 82a and developed by Owens-Corning Fiberglas Corporation in 1975 for electrically heated and cooled new homes.

Although the OCF Zone recommendations are not at all sensitive to fuel type, and are not as sensitive to climatic variations as some others, they are recommended for use by conventional lenders until more demanding standards (such as BEPS) are mandated by federal, state or local governments. On properties insured by an agency of the federal government (HUD, FmHA and VA), appropriate federal mandatory standards should be employed.

We gratefully acknowledge the assistance of the following for the use of their publications as references and for permission to use photographs: American Board Products Assn.; Brick Institute of America; Celotex Corp.; Dow Chemical Corp.; Johns Manville Sales Corp.; Levelor Lorentzen Inc.; National Association of Home Builders Research Foundation; National Bureau of Standards; Mineral Insulation Manufacturers Association; Owens Corning Fiberglas Corp.; The Pease Co.; Willard Shutter Co.; W. R. Grace Co.

HEAT CONTROL 105

CONTENTS

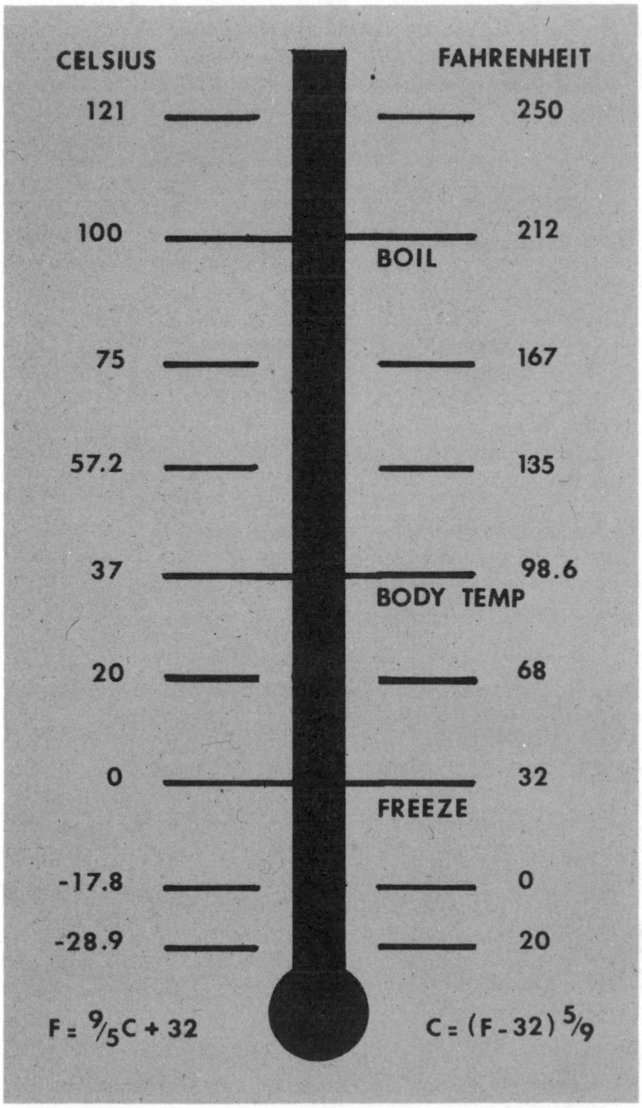

FIG. WF1 Temperature equivalents on the Celsius (metric) and Fahrenheit (English) scales. Thermometers can be calibrated to a temperature of —273.1°C (—459.6°F), known as absolute zero—the point at which all molecular movement stops.

HEAT & COMFORT

Heat is the result of the molecular movement of a substance. The intensity of that movement determines its physical state—whether solid, liquid or gaseous—and is measured in either degrees Fahrenheit (F) or Celsius (C) (Fig. WF1). The British Thermal Unit (BTU) measures heat quantity and equals the amount of heat required to raise 1 lb. of water 1° F. In the metric system, heat quantity is measured by the calorie (cal) and the joule. One calorie equals the amount of heat required to raise 1 gram of water 1° C. One joule equals 4.184 cal.

METHODS OF HEAT TRANSFER

Heat can move into or out of a building by conduction (through a solid or liquid), convection (through a fluid medium), or by radiation (by infrared light waves, independent of any medium).

Materials can either accept or reject radiant energy. Materials that reject most radiant energy, such as shiny aluminum, have a high degree of reflectivity. Dark materials, such as asphalt pavement, which absorb more radiant energy, have a high degree of emmissivity (Fig. WF2).

FIG. WF2 **REFLECTIVITY/EMMISSIVITY CHART**

Material	Reflectivity, %	Emmissivity, %
Built up roof, gravel aggregate	7	93
Slate, dark soil, other dark textures	15	85
Grass, dry	30	70
Copper foil:		
tarnished	36	64
new	75	25
Paint:		
light grey	25	75
red	26	74
aluminum	46	54
light green	50	50
light cream	65	35
white	75	25
Fresh snow	80	20
Aluminum foil	95	5

(Continued) **HEAT & COMFORT**

FIG. WF3 ASHRAE COMFORT ENVELOPE

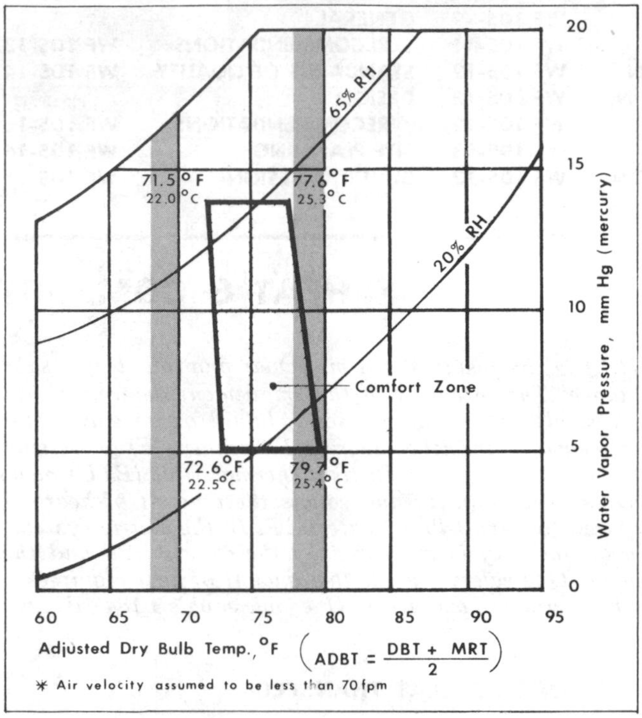

RELATIVE HUMIDITY

Air contains moisture in the form of water vapor. The ratio of moisture actually present in the air compared to the amount that can exist without condensation is percent relative humidity (% RH). When all possible moisture is held by air at a given temperature its dew point is reached.

Relative humidity should be kept at a level consistent with the comfort recommendations listed below.

COMFORT FACTORS

The following are factors which should be considered in any heating/cooling system: (1) ambient temperature; (2) air movement; (3) relative humidity; and (4) temperature of room surfaces.

Ambient air and radiant surface temperatures were combined in a single measure, adjusted dry bulb temperature (ADBT), in the American Society of Heating, Refrigerating and Air Conditioning Engineers' (ASHRAE) comfort envelope. The comfort envelope outlines acceptable thermal conditions for 80% of tested Americans and Canadians (Fig. WF3).

Adjusted dry bulb temperatures and relative humidity levels should be controlled within the ranges shown in Figure WF3. Interior surface temperatures in winter should not be more than 5° F below ambient air temperature, to prevent radiant heat loss from body and interior spaces. In summer, walls should be kept slightly cooler than air temperature to aid in body heat dissipation.

In 1974 the Federal Energy Administration (FEA) recommended more energy conserving interior temperatures (Guideline Temperatures) ranging from 68° to 70° F for winter and 78° to 80° F for summer. Even though FEA temperatures are lower in winter and higher in summer than those recommended by ASHRAE, FEA concluded after a series of surveys that the Guideline Temperatures could be made acceptable to 80% of Americans if heavier clothing were worn in winter and lighter clothing in summer.

Interior summer design temperature should be 78° F, interior winter design temperature should be 68° F.

HEAT TRANSFER THROUGH MATERIALS

Wall, floor and roof assemblies should be constructed with materials that minimize heat flow out of the building in winter and into the building in summer.

Heat flow can be resisted by insulating materials (such as plastic foams or mineral fibers) or can be slowed by massive dense materials (such as thick masonry or concrete).

All building materials transfer heat to some degree. Heat transfer can be measured by any of the following: (1) conductivity (k); (2) conductance (c); (3) transmittance (U); or (4) resistance (R) (Fig. WF4).

The ability of dense materials to retard heat flow depends on their ability to store heat and is measured by specific heat and heat capacity (Fig. WF5).

HEAT LOSS AND HEAT GAIN ANALYSIS

Heating and air conditioning equipment is sized on the basis of heat loss and heat gain calculations for the particular area in which the building is situated. These calculations also provide a measure of the thermal efficiency of the building.

The following terms should be understood to perform a heat loss or heat gain analysis.

R-Value *A material's resistance to the flow of heat; the larger the R-value, the greater the resistance and the better the insulating value (Fig. WF6).*

Heat Loss *The amount of heat flow in BTUs per hour or per season, from inside a structure to outside, at given design conditions.*

(Continued) **HEAT & COMFORT**

FIG. WF4 **HEAT TRANSFER COEFFICIENTS OF REPRESENTATIVE BUILDING PRODUCTS**[1]

Material or Product	Conductivity k	Conductance C	Resistance — R per 1" thickness (1/k)	Resistance — R per thickness shown (1/C)
Glass 1/4"		1.13		0.88
Concrete	12.00		0.08	
Face Brick	9.00		0.11	
Hollow concrete block, 8"		0.90		1.11
Stucco	5.00		0.20	
Metal lath & plaster, 3/4"		7.70		0.13
Gypsum board, 1/2"		2.22		0.45
Plywood, 1/2"		1.60		0.62
Pine, fir, other softwoods	0.80		1.25	
Oak, maple, other hardwoods	1.10		0.91	
Asphalt shingles		2.27		0.44
Built-up roofing, 3/8"		3.00		0.33
Wood shingles		1.06		0.94
Structural insulation board, 1/2"		0.76		1.32
Mineral wool batts, 3"—4"		0.09		11.00
5"—6"		0.05		19.00
6-1/2"—7"		0.05		22.00
8-1/2"—9"		0.03		30.00
Expanded polystyrene, 1"		0.28		3.57
Inside surface air film[2]				0.68
Outside surface air film[3]				0.17
Air space 3/4", nonreflective[4]		1.00		1.01
Air space 3/4", reflective[4]				3.48
Wood Door 1-3/4", hollow		.49		2.04

[1]At 75°F mean temperature, from ASHRAE 1977 Fundamentals Handbook.
[2]Heat flow horizontal, still air.
[3]Heat flow any direction, 15 mph wind.
[4]Heat flow horizontal.

FIG. WF5 **HEAT STORAGE CAPABILITIES OF BUILDING MATERIALS**

Material	Specific Heat BTU per lb. per °F	Density lb. per cu. ft.	Heat Capacity BTU per cu. ft. per °F
Water (40°F)	1.000	62.5	62.5
Steel	0.120	489.0	58.7
Cast Iron	0.120	450.0	54.0
Copper	0.092	556.0	51.2
Aluminum	0.214	171.0	36.6
Basalt	0.200	180.0	36.0
Marble	0.210	162.0	34.0
Concrete	0.220	144.0	31.7
Asphalt	0.220	132.0	29.0
Ice (32°F)	0.487	57.5	28.0
Glass	0.180	154.0	27.7
White Oak	0.570	47.0	26.8
Brick	0.200	123.0	24.6
Limestone	0.217	103.0	22.4
Gypsum	0.260	78.0	20.3
Sand	0.191	94.6	18.1
White Pine	0.670	27.0	18.1
White Fir	0.650	27.0	17.6

FIG. WF6 **RESISTANCE (R) VALUES OF INSULATING PRODUCTS**

Product	R per 1" thickness
LOOSE FILL	
Mineral fiber (rock, slag or glass)	2.20-3.00
Cellulose	3.70
Perlite	2.70
Vermiculite	2.13
FLEXIBLE	
Mineral wool batts	3.10-3.70
RIGID	
Cellular glass	2.63
Expanded polystyrene (extruded)	5.00
Expanded polystyrene (molded)	3.57
Expanded polyurethane	6.25
Mineral fiberboard	3.45
Polyisocyanurate	7.20
REFLECTIVE	
Aluminum foil**	3.48
FOAMED-IN-PLACE	
Urea formaldehyde	4.20
Polyurethane	6.25

*All values are for 75°F mean temperature.
**Thickness of foil not a factor; 3/4" air space on room side.

Sensible Heat Gain *The amount of heat gain in BTUs per hour that can be recorded as a rise in temperature.*

Latent Heat Gain *Heat introduced into a structure by moisture in the air and which must be removed by the cooling system. The latent heat gain is commonly calculated as 30% of the sensible heat gain.*

Indoor Design Temperature *The ideal comfort temperatures that are to be maintained inside the structure in winter and summer. They should correspond with either the ASHRAE comfort envelope or preferrably the*

FIG. WF7 **REPRESENTATIVE OUTDOOR DESIGN TEMPERATURES, DEGREE DAYS, COOLING HOURS AND SUMMER OUTDOOR DAILY RANGE***

City	Winter Design Temperature (97½%) (°F)	Degree Days Heating (DDH)	Summer Design Temperature (2½%) (°F)	Annual Cooling Operating Hours	Summer Outdoor Daily Range
Anchorage, AK	−18	10,860	68	, 60	15
Phoenix, AZ	34	1,680	107	1,930	27
Los Angeles, CA	43	1,960	80	460	20
San Francisco, CA	38	3,040	77	170	14
Denver, CO	1	6,150	91	700	28
Washington, DC	17	4,240	91	960	18
Miami, FL	47	200	90	2,940	15
Chicago, IL	− 4	6,640	89	650	20
Wichita, KS	7	4,640	98	1,030	23
Minneapolis, MN	−12	8,250	89	590	22
New York, NY	15	4,880	87	770	16
Cincinnati, OH	6	4,830	90	870	21
Tulsa, OK	13	3,730	98	1,230	24
Rapid City, SD	− 7	7,370	92	610	28
Houston, TX	32	1,410	94	1,880	18
Salt Lake City, UT	8	5,990	95	770	32
Richmond, VA	17	3,910	92	1,010	21
Spokane, WA	2	6,770	90	460	28
Green Bay, WI	− 9	8,100	85	430	23

* Design temperatures and summer outdoor daily range DDH from *ASHRAE Handbook of Fundamentals*
Cooling hours over 80° F from *Engineering Weather Data Manual*, rounded to neatest 50.

FEA Guideline Temperatures.

Outdoor Design Temperatures *High and low temperatures that can be expected to occur regularly in a particular area (Fig. WF7).*

Design Temperature Difference *The difference between the outside design temperature (t_o) and the inside design temperature (t_i).*

Infiltration *The leakage of heat into or out of a building through cracks in the building structure, especially around windows and doors (Fig. WF8).*

Transmittance (U) *The number of BTUs per hour transferred through 1 sq. ft. of a building assembly (such as a wall) for each 1° F temperature difference between inside and outside surfaces. It is calculated by adding the Rs for all elements of the building assembly (R_T) and taking the reciprocal ($1/R_T$) of the total (Fig. WF9).*

Heat Loss Formulas

Heat flows into or out of a building at the following locations: (1) through exterior walls, roofs, ceilings and glass areas; (2) at the perimeter of a building built on a slab; (3) through floors over unheated spaces; and (4) by infiltration around windows and doors. Figures WF10 and WF11 illustrate a method of calculating the contribution to overall heat loss or heat gain of each of the areas listed above.

FIG. WF8 **WINDOW AND DOOR INFILTRATION VALUES**

Window Type	Infiltration Airflow * cu. ft. per ft. of crack		
	10 mph	20 mph	30 mph
DOUBLE-HUNG WOOD FRAME:			
Non-weatherstripped	21	59	104
Weatherstripped	13	36	63
DOUBLE-HUNG METAL FRAME			
Non-weatherstripped	47	104	170
Weatherstripped	19	46	76
RESIDENTIAL METAL CASEMENT	18	47	74
WOOD OR METAL DOOR:			
Non-weatherstripped	69	154	249
Weatherstripped	19	51	92

* Reduce values above by 1/3 when storm window (or door) is added to weatherstripped construction, by 1/2 when added to non-weatherstripped construction.

Heat Loss Calculation

The following procedure should be used to determine building heat loss (Fig. WF10).

(1) *Establish Design Conditions by determining: (1) the winter outdoor design temperature (t_o) from Figure WF7; and (2) the desired indoor design temperature (t_i) from the ASHRAE Comfort Envelope or 68° F, the FEA Guideline Temperature. The design temperature difference is equal to the difference between the established indoor and outdoor design temperatures ($t_i - t_o$).*

FIG. WF9 U-VALUE CALCULATION PROCEDURE

Building Part	Construction Materials	R-value
Roof/Ceiling	Outside Air Film	0.17
	Shingles	0.44
	Building Paper	0.06
	Plywool ½"	0.62
	Attic Air Film	0.61
	Insulation	19.00
	Gypsum Board, ½"	0.45
	Inside Air Film	0.61
	Total R-value (R_T)	21.96
	U-value ($1/R_T$)	**0.045**
Wall	Outside Air Film	0.17
	Siding, Wood ½" x 8" Lapped	0.81
	Sheathing, Plywood ½"	0.62
	Insulation	11.00
	Interior Finish Gyp. Bd. ½"	0.45
	Inside Air Film	0.68
	Total R-value (R_T)	13.73
	U-value ($1/R_T$)	**0.073**
Header Joist	Outside Air Film	0.17
	Siding	0.81
	Sheathing	0.62
	Header, Wood 1½"	1.88
	Insulation	11.00
	Inside Air Film	0.68
	Total R-value (R_T)	15.16
	U-value ($1/R_T$)	**0.066**
Sill	Outside Air Film	0.17
	Siding	0.81
	Sheathing	0.62
	Sill — Wood 5½"	6.88
	Inside Air Film	6.88
	Total R-value (R_T)	9.16
	U-value ($1/R_T$)	**0.109**
Foundation	Outside Air Film	0.17
	Conc. Blk 8"	1.11
	Insulation	5.00
	Interior Finish Gyp. Bd. ⅜"	0.32
	Inside Air Film	0.68
	Total R-value (R_T)	7.28
	U-value ($1/R_T$)	**0.137**

(2) *Calculate the U-values for roof, wall and glass areas by summing the Rs for all parts of such an assembly, including air films, (R_T) and taking the reciprocal ($1/R_T$) of that figure (Fig. WF9).*

(3) *Calculate the Surface Area (A) of glass and all similarly constructed roof, wall and floor areas.*

(4) *Calculate building heat loss (H_c) through exterior roof and wall construction and through exterior glass by multiplying the area (A) by its computed U-value, by the design temperature difference ($t_i - t_o$).*

(5) *Calculate the floor heat loss (H_f) by the same method used for walls, roof and glass, with the following qualifications:*

- *If heating unit is in the basement and ducts or pipes are uninsulated, no floor loss calculation is necessary.*
- *For floors over unheated basements, use 1/3 of the design temperature difference.*
- *For floors over unvented crawl spaces with insulated crawl space walls, use 1/2 of the design temperature difference.*
- *For floors over vented crawl spaces or other open spaces, use the full design temperature difference.*

(6) *Calculate the building perimeter (P) in feet if the floor is of slab construction.*

(7) *Calculate the floor slab edge heat loss (H_e) by multiplying a factor of 0.55 by the slab perimeter (P) in feet if the slab edge is insulated or a factor of 0.8 by (P) in feet if the slab edge is uninsulated, by the design temperature difference ($t_i - t_o$).*

(8) *Calculate the length of crack (L) in feet around window and door openings.*

(9) *Select infiltration airflow volume in cu ft per hr (cfh) per ft of crack from Figure WF8.*

(10) *Calculate the infiltration heat loss (H_i) for each opening by multiplying the length of crackage (L) by 0.018 times the per ft infiltration volume in cu ft. per hr (cfh), by the design temperature difference ($t_i - t_o$).*

(11) *Calculate the total building heat loss (H_T) by summing H_c, H_i, and H_e or H_f if appropriate.*

FIG. WF10 **HEAT LOSS CALCULATION EXAMPLE***

Example: A single family house, 30′ x 40′, one story over a heated basement with wood frame walls, insulated concrete block foundation and weatherstripped wood sash; located in Washington, D.C.

Heat Loss: Transmission

Construction	Area (sq. ft.)		Design Temp. Difference (°F)		U-Value (See Fig. WF9)		Heat Loss H_c (BTUs per hr.)
Glass 1/4″	180.0	×	52	×	1.130†	=	10,577
Wood Doors 1 3/4″ hollow	40.0	×	52	×	0.490†	=	1,019
Walls	1184.0	×	52	×	0.073	=	4,494
Header Joist 7 1/2″	92.5	×	52	×	0.066	=	317
Sill 1 1/2″	18.5	×	52	×	0.109	=	105
Foundation 9″	111.0	×	52	×	0.137	=	791
Roof/Ceiling	1320.0	×	52	×	0.045	=	3,089
							Total H_c 20,392

Heat Loss: Infiltration

Opening	Crack Length (ft.)		Design Temp. Difference (°F)		Infiltration Airflow (Cu. ft. per hr. per ft.) (See Fig. WF8)		Reduction Constant		Heat Loss H_i (BTUs per hr.)
Windows	125	×	52	×	36	×	0.018	=	2,412
Doors	40	×	52	×	51	×	0.018	=	1,908
									Total H_i 4,330

Heat Loss: Total

Heat Loss Factor	Heat Loss in BTUs per hr.
Transmission H_c	20,392
Floor H_f	—
Edge H_e	—
Infiltration H_i	4,330
	Total H_T 24,722

*Follow 11 step procedure under "Heat Loss Calculation."
†See Fig. WF4 for C-value which here equals U-value.

Heat Gain Calculation

The following procedure should be used to determine building heat gain (Fig. WF11).

(1) *Establish the design conditions by determining the following: (1) the winter outdoor design temperature (t_o) from Figure WF7; and (2) the desired indoor design temperature (t_i) from the ASHRAE Comfort Envelope or 78° F, the FEA Guideline Temperature. The design temperature difference is equal to the difference between the established indoor and outdoor design temperatures ($t_i - t_o$).*

(2) *Calculate the U-value for roof and walls by summing the Rs for all parts of each assembly (R_T) and taking the reciprocal ($1/R_T$) of that value (Fig. WF9).*

(3) *Calculate the surface area (A) of all similarly constructed roof, walls and floors.*

(4) *Calculate heat gain through unglazed (opaque) roof and wall sections (H_o) by multiplying the calculated area (A), by the calculated U-value, by the Design Equivalent Temperature Difference (DETD) from Figure WF12.*

(5) *Calculate heat gain through glass areas (H_g) by the following procedure:*

(a) *Locate the glass according to its building face*

FIG. WF11 HEAT GAIN CALCULATION EXAMPLE*

Example: A three bedroom single family house, 30′ x 40′, one story over a heated basement with wood frame walls, insulated concrete block foundation and single glazed weatherstripped wood sash with venetian blinds. Windows are equally distributed on each building face (N, S, E, and W); roof is dark color; located in Washington, D.C.

Heat Gain: Unglazed Roof and Wall Sections

Construction**	Area (sq. ft.)		U-Value (see Fig. WF9)		DETD (see Fig. WF12)		Heat Gain H_o (BTUs per hr.)
Doors	40.0	×	0.490†	×	22.6	=	443
Walls	1184.0	×	0.073	×	22.6	=	1953
Header	92.5	×	0.066	×	22.6	=	138
Sill	18.5	×	0.019	×	22.6	=	46
Roof/Ceiling	1320.0	×	0.045	×	43.0	=	2554
							Total H_o 5134

Heat Gain: Glazed Areas

Glass Type	Outdoor Design Temp. °F	Orientation	Area (sq. ft.)		Solar Heat Gain Factor (see Fig. WF13)		Heat Gain H_g (BTUs per hr.)
Single ¼″	94	N	45	×	22.2	=	999
Single ¼″	94	S	45	×	30.2	=	1359
Single ¼″	94	E	45	×	55.2	=	2484
Single ¼″	94	W	45	×	55.2	=	2484
							Total H_g 7326

Heat Gain: People

Occupants: 2 × 3 bedrooms = 6 people (estimated)
6 people × 225 BTUs per hr. per person = 1350

Total Heat Gain, H_p = 1350 BTUs per hr.

Heat Gain: Appliances

Total Heat Gain H_a = 1200 BTUs per hr. (estimated)

Heat Gain: Total Sensible

Heat Gain Factor	Heat Gain in BTUs per hr.
Unglazed Roof and Wall Sections H_o	5134
Glazed Areas H_g	7326
People H_p	1350
Appliances H_a	1200
	Total H_{TS} 15010

Heat Gain: Total Latent

H_{TL} = ⅓ H_{TS} = 15,010 BTUs per hr.
H_{TL} = 5,003 BTUs per hr.

Heat Gain: Total Sensible + Total Latent

$H_T = H_{TS} + H_{TL}$
H_T = 15,010 + 5,003 = 20,013 BTUs per hr.

*Follow 11 step procedure under "Heat Gain Calculation."
**Only above grade construction considered.
†See Fig. WF4 for C-value which here equals U-value.

FIG. WF12 DESIGN EQUIVALENT TEMPERATURE DIFFERENCES

Design Temperature, ° F *	85		90			95			100		105	110
Daily Temperature Range **	L	M	L	M	H	L	M	H	M	H	H	H
WALLS AND DOORS												
Frame and veneer-on-frame	17.6	13.6	22.6	18.6	13.6	27.6	23.6	18.6	28.6	23.6	28.6	33.6
Masonry walls, 8-in. block or brick	10.3	6.3	15.3	11.3	6.3	20.3	16.3	11.3	21.3	16.3	21.3	26.3
Wood doors	17.6	13.6	22.6	18.6	13.6	27.6	23.6	18.6	28.6	23.6	28.6	33.6
CEILINGS AND ROOFS												
Ceilings under naturally vented attic or vented flat roof—dark	38.0	34.0	43.0	39.0	34.0	48.0	44.0	39.0	49.0	44.0	49.0	54.0
—light	30.0	26.0	35.0	31.0	26.0	40.0	36.0	31.0	41.0	36.0	41.0	46.0
Built-up roof, no ceiling—dark	38.0	34.0	43.0	39.0	34.0	48.0	44.0	39.0	49.0	44.0	49.0	54.0
—light	30.0	26.0	35.0	31.0	26.0	40.0	36.0	31.0	41.0	36.0	41.0	46.0
Ceilings under unconditioned rooms	9.0	5.0	14.0	10.0	5.0	19.0	15.0	10.0	20.0	15.0	20.0	25.0
FLOORS												
Over unconditioned rooms	9.0	5.0	14.0	10.0	5.0	19.0	15.0	10.0	20.0	15.0	20.0	25.0
Over basement, heated crawl space or concrete slab on ground	0	0	0	0	0	0	0	0	0	0	0	0
Over unheated crawl space	9.0	5.0	14.0	10.0	5.0	19.0	15.0	10.0	20.0	15.0	20.0	25.0

*Interpolate for values between those listed.
**Daily Temperature Range (see Fig. WF 7)

L (Low) Calculation Value: 12
Range: Less than 15 deg.

M (Medium) Calculation Value: 20
Range: 15 to 25 deg.

H (High) Calculation Value: 30
Range: More than 25 deg.

FIG. WF13 SOLAR HEAT GAIN FACTORS FOR RESIDENTIAL STRUCTURES*

Outdoor Design Temp., °F	*Regular Single Glass*						*Regular Double Glass*						*Heat Absorbing Double Glass*					
	85	90	95	100	105	110	85	90	95	100	105	110	85	90	95	100	105	110
							No Awnings or Inside Shading											
North	23	27	31	35	38	44	19	21	24	26	28	30	12	14	17	19	21	23
NE and NW	56	60	64	68	71	77	46	48	51	53	55	57	27	29	32	34	36	38
East and West	81	85	89	93	96	102	68	70	73	75	77	79	42	44	47	49	51	53
SE and SW	70	74	78	82	85	91	59	61	64	66	68	70	35	37	40	42	44	46
South	40	44	48	52	55	61	33	35	38	40	42	44	19	21	24	26	28	30
							Draperies or Venetian Blinds											
North	15	19	23	27	30	36	12	14	17	19	21	23	9	11	14	16	18	20
NE and NW	32	36	40	44	47	53	27	29	32	34	36	38	20	22	25	27	29	31
East and West	48	52	56	60	63	69	42	44	47	49	51	53	30	32	35	37	39	41
SE and SW	40	44	48	52	55	61	35	37	40	42	44	46	24	26	29	31	33	35
South	23	27	31	35	38	44	20	22	25	27	29	31	15	17	20	22	24	26
							Roller Shades Half-Drawn											
North	18	22	26	30	33	39	15	17	20	22	24	26	10	12	15	17	19	21
NE and NW	40	44	48	52	55	61	38	40	43	45	47	49	24	26	29	31	33	35
East and West	61	65	69	73	76	82	54	56	59	61	63	65	35	37	40	42	44	46
SE and SW	52	56	60	64	67	73	46	48	51	53	55	57	30	32	35	37	39	41
South	29	33	37	41	44	50	27	29	32	34	36	38	18	20	23	25	27	29
							Awnings											
North	20	24	28	32	35	41	13	15	18	20	22	24	10	12	15	17	19	21
NE and NW	21	25	29	33	36	42	14	16	19	21	23	25	11	13	16	18	20	22
East and West	22	26	30	34	37	43	14	16	19	21	23	25	12	14	17	19	21	23
SE and SW	21	25	29	33	36	42	14	16	19	21	23	25	11	13	16	18	20	22
South	21	24	28	32	35	41	13	15	18	20	22	24	11	13	16	18	20	22

*Shown in BTUs per sq. ft. per hr. for the hours of 5:30 a.m. to 6:30 p.m., and for 30° and 40° N latitude. Tabulated factors are applicable for an indoor design temperature of 75°F and for outdoor design temperatues as indicated. Interpolate for values between those listed.

N, S, E, W.

(b) Calculate the surface area of each particular type of glass (single, double or heat absorbing).

(c) Select the appropriate solar heat gain factor from the chart in Fig. WF13.

(d) Multiply the glass area (A) by the heat gain factor for the total window heat gain in BTUs per hour.

(e) If glass is partially or completely shaded by a permanent shading device such as an overhang, the shaded portion is treated as north facing glass. Any unshaded portion is considered separately. The shade line is computed by multiplying overhang width by shade line factor from Figure WF14. Glass above the line is in shade, and glass below the line is in sun.

(6) Heat gain from infiltration (H_i) is normally insignificant in summer and can be ignored.

(7) Estimate the sensible heat gain from people (H_p) at 225 BTUs per hour per person. For residences, estimate the number of occupants at twice the number of bedrooms.

(8) Estimate appliance loads (H_a) at 1,200 BTUs per hour.

(9) Calculate total sensible heat gain (H_{TS}) by summing H_o, H_g, H_p and H_a.

(10) Estimate total latent heat gain (H_{TL}) as 1/3 H_{TS}.

(11) Calculate total heat gain as the sum of total latent heat gain (H_{TL}) and total sensible heat gain (H_{TS}).

FIG. WF14 SHADE LINE FACTORS

Window Orientation	Latitude						
	25°	30°	35°	40°	45°	50°	55°
E	0.8	0.8	0.8	0.8	0.8	0.8	0.8
SE	1.9	1.6	1.4	1.3	1.1	1.0	0.9
S	10.1	5.4	3.6	2.6	2.0	1.7	1.4
SW	1.9	1.6	1.4	1.3	1.1	1.0	0.9
W	0.8	0.8	0.8	0.8	0.8	0.8	0.8

Note: Distance shadow line falls below the edge of the overhang equals shade line factor multiplied by width of overhang. Values are average for five hours of greatest solar intensity on August 1.

Example: Assume a window height of 7′ from floor and a 2′ overhang at the same height; south face, 40° N latitude. Shade line factor from table is 2.6, multiplied by 2′ equals 5.2′. The top 5.2′ of the window is in shade.

WINDOW AND DOOR RECOMMENDATIONS

Windows and doors are often responsible for more than 50% of the heat loss and heat gain of a building. Most heat is lost by infiltration and conduction. Most heat is gained by radiation.

Window and door openings should be constructed so that heat flow through them can be controlled and used to supplement mechanical cooling or heating systems.

HEAT LOSS

Heat loss through windows and doors can be reduced with tightly sealed openings. Windows can also be protected with multiple glazing (storm sash and/or insulating glass) and insulating shutters or curtains. Doors are generally protected by storm doors or vestibules. These techniques will reduce heating costs, minimize condensation on interior surfaces, and improve interior comfort by reducing drafts and radiant body heat loss.

Tightly Sealed Openings

Windows and doors in all climatic regions should be

(Continued) **WINDOW AND DOOR RECOMMENDATIONS**

FIG. WF15 **CAULKING USES AND CHARACTERISTICS**

Performance Level	Caulking Type	Uses	Durability (Years)	Adhesion	Shrinkage Resistance	Cure (Days)
Basic Performance	oil and resin caulks	glazing	1 to 7	fair	poor	up to 1 year
	polybutane cord or rope	wide gaps	1 to 2	none	excellent	Remains moist
Intermediate Performance	nonacrylic latex; PVA	indoor and protected surfaces	2 to 10	good, except to metal	fair	3
	acrylic latex	indoor and protected surfaces	2 to 10	excellent except to metal	fair	3
	butyl rubber	metal-to-masonry	7 to 10	excellent	fair	7
	chlorosulfonated polyethylene	all surfaces	15 to 20	good	good	60 to 90
	neoprene	concrete	15 to 20	excellent	excellent	30 to 60
High Performance	polysulfide	all surfaces	20	excellent	excellent	7
	polyurethane	all surfaces	20	excellent	excellent	4 to 14
	silicone	all surfaces	20 or more	good, excellent with primer	excellent	2 to 5

tightly sealed with caulking around the frames, weatherstripping inside the frames, and in the case of doors, a tight fitting threshold or door sweep at the bottom.

Caulking Wherever window or door frames meet the wall, caulking should be applied. Caulking should be applied to a clean, dry surface on a warm day and should be periodically renewed as necessary.

Caulking performance varies according to chemical composition, proper preparation and application. Durability ranges from oil-based compounds to butyl rubber, latexes and silicones (Fig. WF15).

Weatherstripping Window and door frames should be weatherstripped to prevent air leakage. The stripping should be installed so the resilient part seals out air by pressing against the door or window sash.

The following are the most common types of weatherstripping: (1) foam rubber stripping; (2) felt stripping; (3) rolled vinyl; (4) casement stripping; (5) spring metal; and (6) interlocking metal. Each is particularly suited for certain applications as described in the Main Text.

Multiple Glazing *Single glass conducts 6 to 10 times more BTUs per hr. than a typical frame wall. However, that rate of heat transfer can be considerably reduced if two or more layers of glass are separated by dead air spaces. Although the air spaces interrupt flow by conduction, heat can still pass through the air space by radiation and convection.*

The optimum air space thickness between layers of multiple glazing is about 5/8", with only minor loss in effectiveness as the space is reduced to 3/16". Insulating effectiveness decreases significantly as the air space is increased or decreased beyond these dimensions. However, a typical storm window with up to 6" of separation, produces worthwhile results, both in decreased conduction and infiltration.

To establish this necessary air space, storm sash or insulating glass can be used.

Storm Sash *These attachments to the primary window are commonly termed "storm windows" and are classified as conventional, combination or fixed-in-place interior types. They control heat flow by reducing direct heat transfer and infiltration.*

Storm windows should be installed if the building is located in Zones I, II, or III of the map in Figure WF18. If combination storm windows with sliding panels are installed, they should have tight joints and suitable weatherstripping for a tight seal.

Insulating Glass *Sealed insulating glass units, like storm panels, improve interior comfort and reduce conductive heat flow. In northern areas, insulating glass eliminates condensation and has the convenience of being part of the prime window. It does not, however, have the added protection against infiltration which a separate storm sash provides.*

(Continued) WINDOW AND DOOR RECOMMENDATIONS

In some areas it may be economically practical to use three layers of glass (triple glazing) separated by air spaces (Fig. WF16). Triple insulating glass with two 1/4" air spaces, for example, has an R-value of 2.13 compared to 1.72 for double sealed insulating glass. This technique of triple glazing can also be accomplished by installing storm sash over double-glazed windows.

If a metal frame is used to hold the glass in the primary window, it should have a "thermal break" of a low conductive material to interrupt the flow of heat from the warm to the cold surface.

Window Insulation

For additional control of heat flow, windows can be protected by thermal shutters or thermal drapes and shades (see Main Text).

Exterior Door Protection

Exterior entry doors are often the second greatest area of heat loss after windows. Heat loss here can be reduced by storm doors and insulating doors.

Storm Doors *Like storm windows, they reduce heat loss by conduction and infiltration. A typical 1-3/4" solid wood door, for example, has an R-value of approximately 2.00. When a wood storm door is added, the R-value is increased to 3.70.*

Insulating Doors *These doors have an inner core of plastic foam, high R-values (R-8 to R-15) and are even more effective than the combination of prime and storm doors in reducing conductive heat flow. These doors frequently have metal skins and magnetic weatherstrips, which are highly effective in reducing heat loss by air leakage.*

HEAT GAIN

Excessive heat gain through glass areas can be prevented with one or more of the following techniques: (1) shading devices; (2) special glass; (3) multiple glazing; and (4)

FIG. WF16 **WINDOW GLAZING GUIDELINES BY LOCATION & TYPE OF HEAT**

City	Oil Heat	Elec. Resistance Heat	Heat Pump
Albuquerque	double	triple	double
Atlanta	double	double	single
Baltimore	double	triple	double
Boston	triple	triple	triple
Charlotte	double	double	single
Chicago	triple	triple	double
Cincinnati	double	triple	double
Cleveland	triple	triple	triple
Columbus	triple	triple	double
Dallas	double	double	single
Detroit	triple	triple	triple
El Paso	double	double	single
Houston	single	single	single
Indianapolis	double	double	single
Jacksonville	single	double	single
Kansas City	double	double	double
Las Vegas	double	double	single
Los Angeles	single	double	single
Memphis	double	double	single
Miami	single	single	single triple
Minneapolis	triple	triple	single
New Orleans	single	single	triple
New York	triple	triple	single
Oklahoma City	double	double	double
Philadelphia	triple	triple	single
Phoenix	single	double	double
Pittsburgh	triple	triple	single
San Antonio	double	double	double
San Francisco	double	triple	single
Seattle	double	double	single
St. Louis	double	double	single
San Diego	single	single	triple
Washington	double	triple	

*Guidelines developed in October 1977 and based on the following assumptions: (1) 7 year period to recoup investment; (2) annual increase in energy cost of 10% in current dollars; and (3) finance interest rate of 9%.

(Continued) WINDOW AND DOOR RECOMMENDATIONS

insulating shutters. Multiple glazing and insulating shutters were discussed above.

Shading Devices

Summertime shading devices preferably should be installed on the exterior of windows. Shades are more effective when they block solar radiation before it penetrates through the glass. This is true because only a portion of the solar radiation that enters a building can be reflected back to the exterior. The remainder is absorbed and radiated to the interior.

The most effective shading devices are the following: (1) exterior shutters; (2) exterior blinds and shades; (3) awnings; and (4) architectural projections. Less efficient but still beneficial are (1) interior blinds; and (2) opaque roll shades. See Main Text for more detailed discussion.

In general, light colors should be chosen for all window shading devices.

Light colors reflect most sunlight before it can be transmitted through the window and into the space. An exception is a roof overhang, which works more effectively when its underside is dark because it will reflect less sunlight through the glass.

Special Glazing

When interior or exterior shading is not practical, special types of glass should be used to keep the sun out.

Heat absorbing glass and reflective glass are most commonly used for this task. They are described on Main Text page 105-28.

BUILDING INSULATIONS

In most cases, the assembly of structural and finish materials will not adequately control the flow of heat into or out of the building. To make a building more energy efficient, insulating products are generally included in roofs, walls and floors.

The amount of insulation added to each part of the building envelope should be specified by *R*-value.

Insulations are classified according to form: (1) loose fill; (2) flexible; (3) rigid; (4) reflective; and (5) foamed-in-place. All types are rated according to their ability to resist heat flow (R-value).

LOOSE FILL INSULATIONS

Loose fill insulations are either poured or machine-blown into structural cavities and are manufactured from mineral wool (rock, slag and glass wool) or cellulosic fiber (recycled newsprint, wood chips and other organic fibers). R-values range from 2.13 for vermiculite to 3.70 for cellulose (Fig. WF6).

Cellulosic insulation should be treated to resist fire and vermin and have a neutral pH. To insure that the material is fire resistant, it should be U.L. listed and labeled.

Loose fill insulation which is not free flowing should not be used in wall or other closed cavities where the large fibers may catch on nails or other obstructions that could leave uninsulated pockets.

FLEXIBLE INSULATIONS

These products are manufactured in batt and blanket form from mineral wool or cellulosic fibers. Flexible insulations are made in the following forms: (1) wrapped—with kraft paper on the edges and a vapor barrier on one or both sides; (2) faced—with a vapor barrier on one side only; (3) friction-fit—without any covering, the interlaced fibers having sufficient resilience to remain upright in the cavity.

Flexible insulations are generally used in areas where it is not practical to install loose fill or where the attached foil or kraft paper facing is desired as a vapor barrier.

Flexible insulation composed of organic fibers should be treated to resist fire and vermin and have a neutral pH. To insure that the material is fire resistant, it should be U.L. listed and labeled.

Insulation with kraft paper or foil paper facing should be protected from flame.

RIGID INSULATIONS

Rigid insulations are useable in all parts of the building envelope and are manufactured in the following forms: (1) structural wall insulation; (2) fiberboard; (3) structural deck insulation; and (4) rigid board insulation. R-values range from 2.28 per 1″ for certain types of structural insulating deck, to 6.25 per 1″ for expanded polyurethane foam board (Fig. WF6).

Some rigid insulations are manufactured from flammable materials and should be protected by fire-resistant finishes such as gypsum wallboard.

When polyisocyanurate board is used as sheathing in cold climates, it should be installed with special venting strips to prevent condensation.

All polystyrene products should be protected from exposure to daylight to avoid ultraviolet deterioration. Molded polystyrene should be protected from contact with water to prevent absorption by its open cell structure. Expanded perlite board should be likewise protected from water.

REFLECTIVE INSULATION

Reflective insulation should face an air space of at least 3/4″ and should remain free of dust or other materials that could reduce its reflective qualities.

FOAMED-IN-PLACE INSULATIONS

In general, foamed-in-place insulations are of two basic types: urethane foam and urea formaldehyde foam (UFF). Each is created by a chemical reaction that expands a mixture of components by as much as 30 times. In approximately 24 hours, each type solidifies into a cellular plastic. R-values are 4.20 per 1″ for UFF and 6.25 per 1″ for urethane (Fig. WF6).

Urethane Foam

This type of insulating foam should be used in open cavities or where it is possible to accurately compute the amount of foam needed. It will burn and should be protected from flame by a fire-resistant finish.

Urethane foam expands to its ultimate volume after injection and can burst a closed cavity if the space is overfilled. It is most appropriate where a high R-value (6.25 per 1″) must be achieved in a limited space or on irregularly shaped construction.

Urea Formaldehyde Foam

This type of insulating foam should only be installed in well-vented areas so the noxious odors it produces can be carried away. It should also be protected from flame.

Unlike urethane, urea formaldehyde (UFF) foams to its ultimate dimension before injection into a cavity and is therefore suitable for closed or open spaces. It will, however, shrink, which may cause significant reduction in insulating effectiveness.

GENERAL RECOMMENDATIONS

The flow of heat through ceilings, walls and floors can be controlled by one of the following techniques: (1) by constructing the assembly with materials that are inherently good insulators; (2) by adding to the assembly products such as loose fill, flexible, rigid, reflective or foamed-in-place insulations; or (3) by using solid masonry or other dense material that has the capacity to store heat and slow down heat flow.

Conventional Installations

For conventionally heated and cooled buildings, the objective of heat control is to block the flow of heat through the ceilings, walls and floors that surround habitable areas.

To most effectively block the flow of heat, the insulation should be placed near the inside of the building envelope.

Ceilings Flexible insulation with vapor barriers should be installed so that the barrier faces the living space. Loose fill or flexible insulation without the barrier should have a separate polyethylene film placed below the insulation facing the living space.

For maximum effectiveness against vapor transmission, the kraft paper facing should overlap the face of the framing member. If such a procedure is not appropriate because of the application of finish materials such as drywall, a separate vapor barrier should be installed.

The insulation should extend entirely across the top plate to prevent heat loss at the joint. If necessary, the junction between the insulation and the plate should be stuffed with loose insulation. When eave vents are used, the insulation should not block the air movement between the vent and the space above the insulation.

If loose fill insulation is used, eave baffles or special cardboard ducts should be installed to insure air movement from eaves to general attic. All insulation should be kept at least 3″ away from sources of heat such as recessed electrical fixtures.

Walls Flexible insulation in stud spaces should touch the sheathing or siding and fit tightly against the top and bottom plates.

Gaps in the insulation at the top or bottom of the cavity may allow convection currents which carry heat and vapor from the inside to the outside of the cavity.

The flanges of the batts or blankets should be attached to the stud face to make a continuous vapor barrier if it will not interfere with installation of finish materials.

The inside face of the wall should be completely covered with a vapor barrier such as polyethylene film or aluminum foil; if the stud faces must be kept clear or if friction fit insulation is used. Openings should be cut out later. The barrier material should be installed in the largest possible sheets to minimize vapor leakage. Barrier joints should be made over framing member.

Cracks around window and door openings should be packed with loose insulation and covered with polyethylene to reduce air infiltration. Insulation should also be packed in openings around pipes, ducts and electrical boxes to prevent "cold spots" in the wall.

Floors/Crawl Spaces Floors over crawl spaces should be insulated by installing the insulating material against the foundation walls if the crawl space is heated. If the space is not heated the insulation should be placed between the floor joists. The vapor barrier should be on the warm-in-winter side facing the living space in both cases.

Sill sealer should be used in basements and in heated crawl spaces to prevent air infiltration between sill and foundation or between header joist and foundation. In addition, the insulating sheathing should extend to the base of the plate.

Short pieces of flexible insulation should be fitted against the header and between floor joists, or the end of the blanket should be pushed against the header. Cantilevered floors should be insulated along any face exposed to the weather.

When perimeter insulation is needed for slab-on-ground construction, the rigid board insulation should extend inward at least 24″ from the perimeter or downward to below the frost line.

STANDARDS OF QUALITY

Insulating products should conform to the industry standards listed in figure WF17.

FIG. WF27 **SUMMARY OF INSULATING PRODUCT STANDARDS**

Material	Standard	Description
Mineral Wool	Fed. Spec. HH-I-1030a	Insulation, Thermal (Mineral Fiber for Pneumatic or Poured Application)
	HH-I-521e	Insulation Blankets, Thermal (Mineral Fiber for Ambient Temperature)
	ASTM C236	Test for Thermal Conductance and Transmittance of Built-Up Sections . . .
	ASTM C665	Mineral Fiber Blanket Thermal Insulation
	ASTM C687	Recommended Practice for Determining of Thermal Resistance of Low Density Fibrous Loose Fill Building Insulation
	LLL-1-535	Insulation Board, Thermal
	ASTM C208	Insulating Board (Cellulosic Fiber)
	ASTM C209	Method of Test for Insulating Board
Cellulosic Fiber	Fed. Spec. HH-I-515C	Insulation, Thermal; and Insulation, Thermal (Loose Fill for Pneumatic or Poured Application): Cellulose or Wood Fiber
	ASTM C177	Thermal Conductivity of Materials by the Guarded Hot Plate
	ASTM C518	Thermal Conductivity of Materials by the Heat Flow Meter
	ASTM C236	See Titles Above
	ASTM C687	See Titles Above
Polyurethane	Fed. Spec. HH-I-530a	Polyurethane Insulation Board Thermal and Polyisocyanurate Interim Amendment 3
	ASTM D1692	Method of Test for Flammability of Plastic Sheeting and Cellular Plastics Burning Rate
	ASTM C177	See Titles Above
Urea Formaldehyde		No Industry Standards
Polystyrene	Fed. Spec. HH-I-524b	Insulation Board, Thermal (Polystyrene)
	ASTM C177	See Titles Above
	ASTM D696	See Titles Above

DESIGN RECOMMENDATIONS

Buildings should regulate heat flow in ways that will minimize dependence on mechanical equipment and take advantage of exterior environmental conditions.

SITE PLANNING

In locating the building on the site, the following should be considered: (1) general climate and microclimate, (2) terrain and landscape elements, and (3) orientation to the sun.

Climate

Climatic analysis should include data on the amount and angle of sunshine which strikes the site, prevailing winds, temperature ranges and precipitation.

The climatic variation of each site (microclimate) should be examined to avoid siting a building in an unsuitable area. Sites in warm climates that act as natural dead air basins should be avoided.

In some areas, prevailing winds and natural terrain can combine to produce unusual wind, snow massing, or rain conditions not typical to a particular climate. In dead air basins, temperatures may exceed those of surrounding areas by as much as 35° F.

Outdoor air quality should be studied to determine whether natural ventilation can be used for cooling.

Terrain & Landscape Elements

Terrain configurations and plant material should be used to control heat flow into and out of buildings.

Plantings can prevent reflection or reradiation of solar energy from paved areas and adjacent buildings and earth berms can be used to shield walls and outdoor living spaces.

Coniferous (evergreen) trees should be located to block prevailing winter winds. Deciduous (leaf dropping) trees should be used on south and west exposures to shade walls and windows in summer but allow desirable winter sun penetration.

Ponds and fountains fed by collected rainwater can be used as natural coolants to reduce air temperature around building exteriors.

Orientation

In cold regions, the area of north facing windows should be limited to reduce heat loss in winter. The area of north facing walls should, if possible, be reduced by berming and sloping roofs.

In hot climate areas, west wall and glass exposures should be minimized to reduce cooling loads from intense afternoon summer sun.

Buildings in cold regions should be sited so that most windows face south or southeast to take advantage of the warming sun.

Orientation that will buffer against hot breezes should be considered in hot climates. Prevailing summer breezes which can provide ventilation and natural cooling should be considered when siting a building or locating windows.

BUILDING DESIGN

Thermally efficient buildings should be achieved through careful room organization and effective building envelope design.

Room Organization

In areas subject to either extreme heat or cold, buildings should have compact design to reduce weather exposed areas and resultant heat flow.

Garages, closets, corridors and other service spaces should be located against those exterior walls where they can act as insulating buffers against prevailing winter winds or hot summer sun, whichever is more important locally.

Kitchens, laundry areas, and rooms that contain heat-producing appliances should be located on north and east sides. Such a location permits wintertime use of their excess heat for warming and permits heat dissipation in summer.

Active living spaces such as living and dining rooms should face south or southeast to take advantage of warming afternoon sun in winter. These spaces should be protected against overheating in summer by external shading devices such as louvers, overhangs, awnings or trees.

Interior shading with blinds and drapes is less effective than external shading of windows.

Rooms likely to remain vacant for long periods should be located where they can be closed off readily to reduce the volume of air which must be heated or cooled. Built-in furnishings should be located away from supply and return air grilles to prevent obstruction of internal air flows. Unnecessary partitions should be omitted to facilitate natural ventilation and reduce cooling loads.

Building Envelope

Walls, floors and ceilings of one- to four-family frame dwellings should have insulation meeting the R-values recommended in Figure WF18.

FIG. WF18 **RECOMMENDED INSULATION R-VALUES**

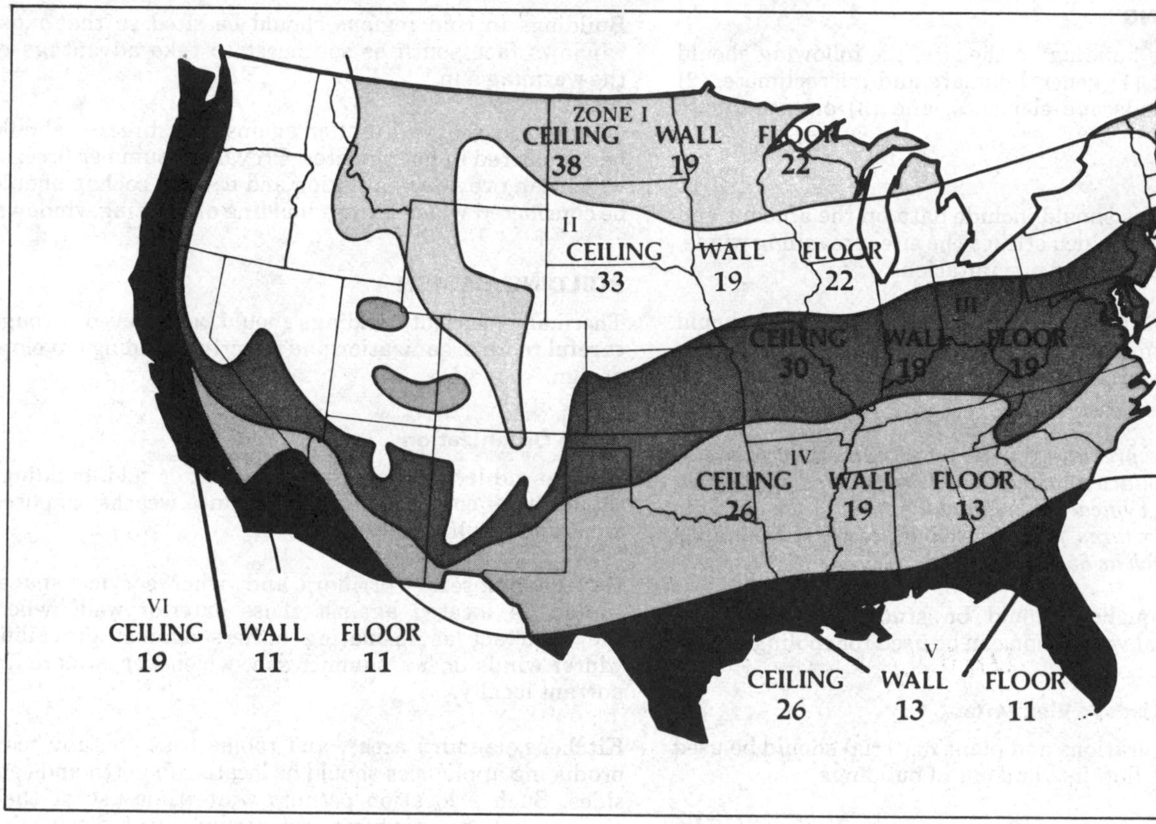

106 SOUND CONTROL

SOUND CONTROL

INTRODUCTION

Contemporary living is noisy. Today's electrical servants—washers, dryers, dishwashers, garbage disposers, refrigerators, vacuum cleaners, radios, TV and more—make life easier and more enjoyable but at the same time create noise problems. Central heating and cooling equipment produce noise and carry it into every room. Outdoors, noise levels have increased also, and sound created by high-speed roads, fast trains and jet airliners add to the problem. Today people live closer to their neighbors and engage in an increasing amount of outdoor leisure activity.

Many materials and construction methods can contribute to the problem. The use of hard, flat surfaces—glass, plaster, gypsum board, wood paneling, masonry, tile and metals—reflect up to 98% of the sound that strikes them. Noise levels within a room grow higher and more irritating as one reflected sound adds to another. The improper use of thin, lightweight materials, including some preassembled components, can create or heighten the problem of transmission of sound between rooms, between floors and between living units.

The practice of commercial sound conditioning began 30 years ago in churches, theaters and auditoriums; a little later it was used in classrooms, offices, hospitals, restaurants and other commercial buildings. Today, in homes and apartments, in fact in any building where noise or good hearing conditions are a problem, the building may be considered obsolete before it is built unless plans include considerations of sound control. When preplanned, sound control can usually be accomplished inexpensively. After a building is completed, corrections in the construction to control unwanted sound transmission are almost always costly.

In this section, two basic acoustical problems are considered: (1) *room acoustics,* where the goal is noise reduction for good hearing conditions within a space; and (2) *sound isolation,* where the goal is the control of sound transmission from one space to an adjacent space as a result of airborne sound or structural vibrations. Solutions to these problems require an understanding of the physics of sound and a knowledge of acoustical materials and the construction methods used in sound conditioning practices. This section discusses the properties of sound, identifies the basic acoustical problems and explains suitable solutions.

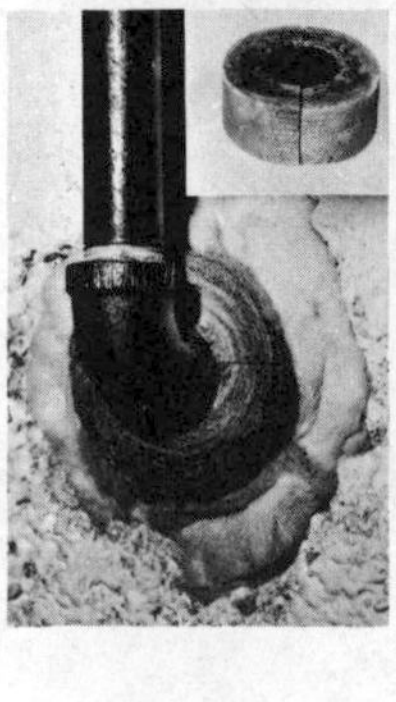

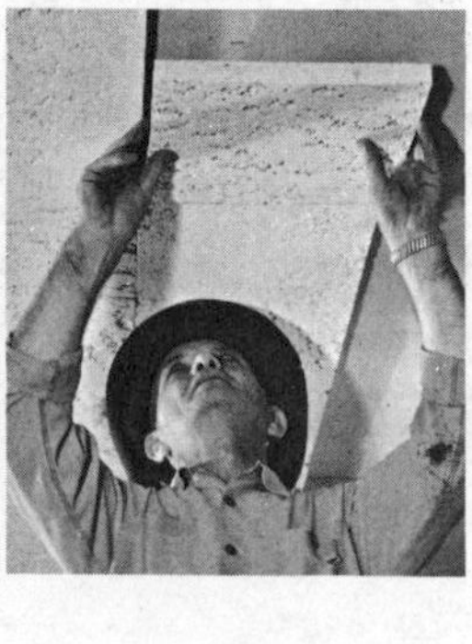

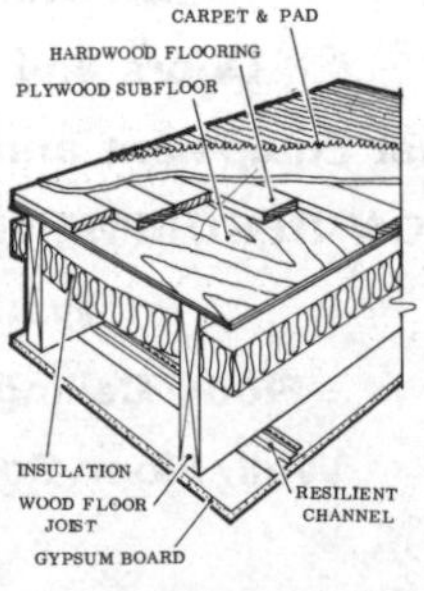

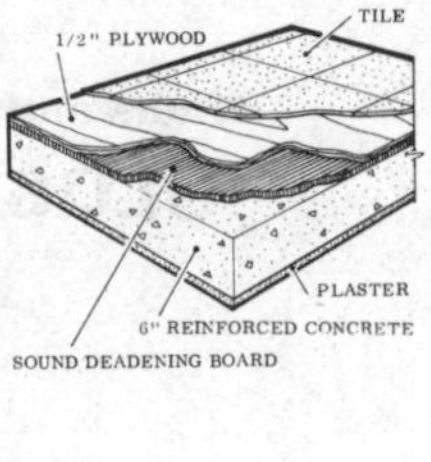

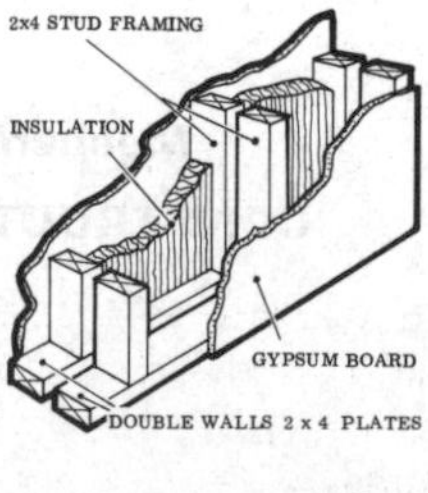

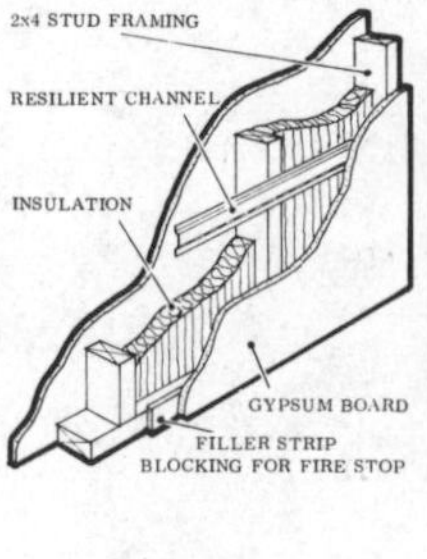

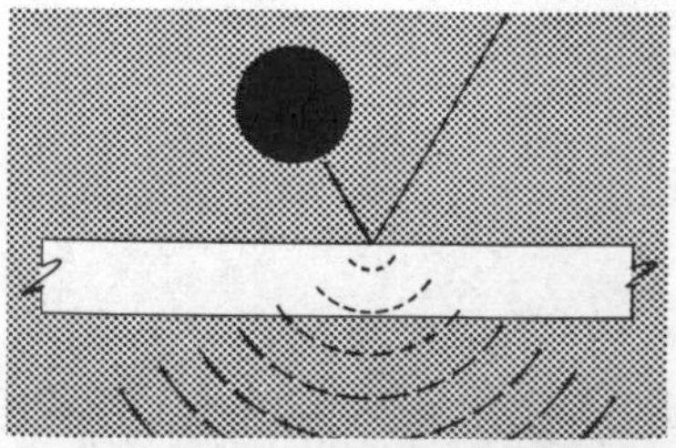

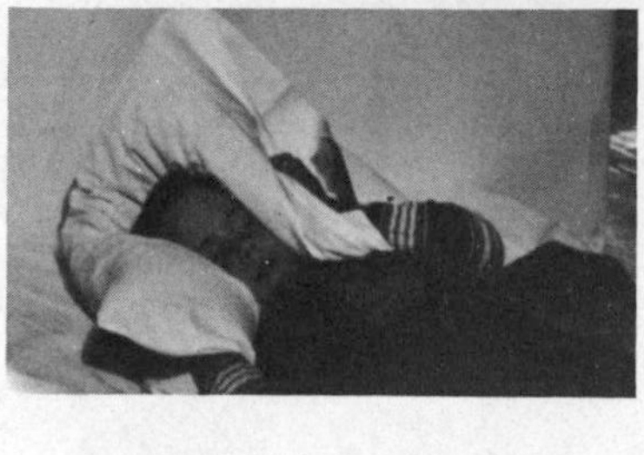

PROPERTIES OF SOUND

Sound is a vibration in any elastic medium. In air, sound may be defined as the movement of air molecules in a wave motion. Sound waves are set up by any vibrating body, they travel through the air at a definite speed, and when their frequency and intensity are within certain ranges, they produce the sensation of hearing.

A vibrating body causes the layer of air particles next to its surface to vibrate. These air particles in turn transmit their motion to the next air layer. Because air is compressible and has weight, a small interval of time is required for the first air layer to set the second layer in motion. The next air layers are similarly set in motion after time intervals that are directly proportional to their distance from the source of the sound. Such action constitutes a sound wave. The action produced when a stone is dropped into water is analogous (Fig. 1).

WAVE MOTION

Water waves travel only along the surface and are circular in shape. Sound waves, on the other hand, spread out in all three dimensions and are spherical in shape. Each air particle vibrates only in the direction in which the wave is traveling; that is, along a straight line drawn between the particles and the source of the sound wave. Sound travels in air at a speed of about 1130 feet per second (in water at about 4500 feet per second; in steel about 15,000 feet per second), its speed being the same regardless of the pitch or loudness of the sound. Outdoors, sound does not reach our ears until after a time interval of about 5 seconds for each mile of distance from its source (Fig. 2). The fact that sound does not travel instantaneously is the cause of echoes and reflection indoors.

FREQUENCY

One complete round trip of a vibrating body, starting from its neutral position, moving to one side then to the other side and back to neutral, is called a cycle (Fig. 3). The number of cycles performed in one second is termed the *frequency.* Frequency is measured in cycles per second (cps) and may be expressed in terms of Hertz (Hz). Each air particle in a sound wave must necessarily have the same frequency as the source. For example, if the sound from a loud-speaker has a frequency of 1000 cycles per second (1000 cps), it means the loud-speaker diaphram and every air particle through which the sound wave passes are vibrating back and forth 1000 times each second. The performance and efficiency of acoustical materials in absorbing sound and the ability of constructions to transmit sound, varies depending on the frequency of the sound.

Wave Length

The length of a sound wave is calculated by dividing the velocity of sound by the frequency:

$$\text{Wave length} = \frac{\text{Velocity}}{\text{Frequency}}$$

Wave length may also be defined as the distance the sound travels during a single vibration. Since the velocity of sound is constant, every frequency has a different wave length. A sound with a frequency of 1000 cps has a wave length equal to about one foot. To reflect a sound effectively, the reflecting surface must be equal to or greater in size than the wave length. A surface one foot square will reflect sound 1000 cps and higher, but sound below 1000 cps will bend around such a surface and pass beyond. A sound with a frequency of 100 cps has a wave length of 11.3 feet. This dimension usually exceeds the ceiling height and often the wall length of residential rooms, thus the sound wave does not fit properly in the room and the sound is not reproduced within the room with fidelity. This accounts for the poor reproduction in small rooms of low frequency sounds from hi-fi or other sources.

Octaves and Tones The frequency scale is laid out in *octaves* in the same manner as a piano keyboard. The term octave is taken from musical terminology and is defined as the interval between any two sounds having a frequency ratio of 2 to 1. For example, fre-

FIG. 1—Sound is a wave motion in air caused by a vibrating source. Wave motion in water is analogous.

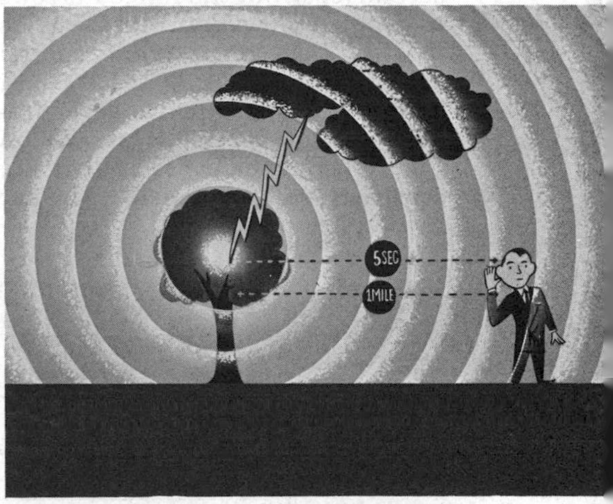

FIG. 2—Light travels almost instantly, but sound requires a time interval of about 5 sec. for each mile from the source.

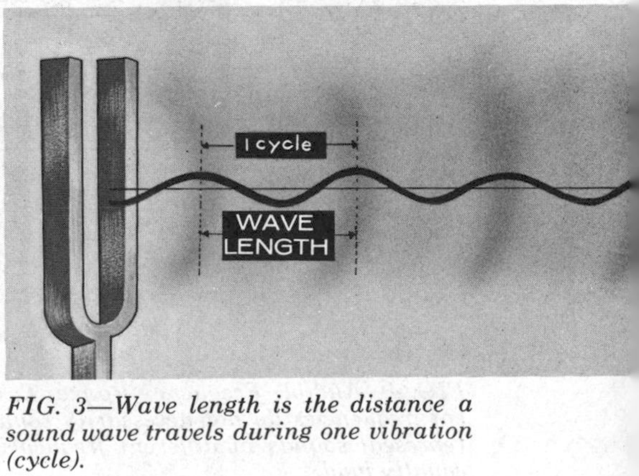

FIG. 3—Wave length is the distance a sound wave travels during one vibration (cycle).

quencies between 125 and 250 cps or between 250 and 500 cps compose an octave. A *pure tone* is a sound having only one frequency, while a *complex tone* is a sound composed of a number of frequencies simultaneously.

Pitch The hearing sensation produced by the frequency of a sound is called *pitch.* The greater number of vibrations per second, the higher the pitch. High frequency sounds such as squeaks and whistles are heard as high pitches and low frequency sounds such as rumbles and roars are heard as low pitches. Pitch is expressed in tones of the musical scale.

The audible frequency range is from about 20 to 20,000 cycles per second (cps), but the range narrows as people grow older. Music covers most of this range, while speech extends from about 100 to 5,000 cps. The most important range of frequencies for speech is shown on Figure 4. But the ear is not equally sensitive to all frequencies. The ear is most sensitive to sounds with frequencies between 1000 to 4000 cps, shown by the low dip on the Threshold of Hearing curve in Figure 4. Sounds between 250 and 1000 cps or between 4000 and 8000 cps must be more intense and below 250 or above 8000 cps must be considerably more intense in order to be heard. Most sounds consist of a mixture of numerous frequencies at different relative intensities.

Quality The quality of a tone as judged by the ear is determined by the distribution of its numerous frequencies. The higher frequency components are called harmonics or overtones when they are exact multiples of the fundamental frequency. The distribution and relative intensity of the overtones of a musical sound determine the quality (*timbre*) of the sound. The sounds illustrated in Figure 5 are complete tones composed of a combination of related frequencies. The vertical distance of the graph is proportional to the loudness.

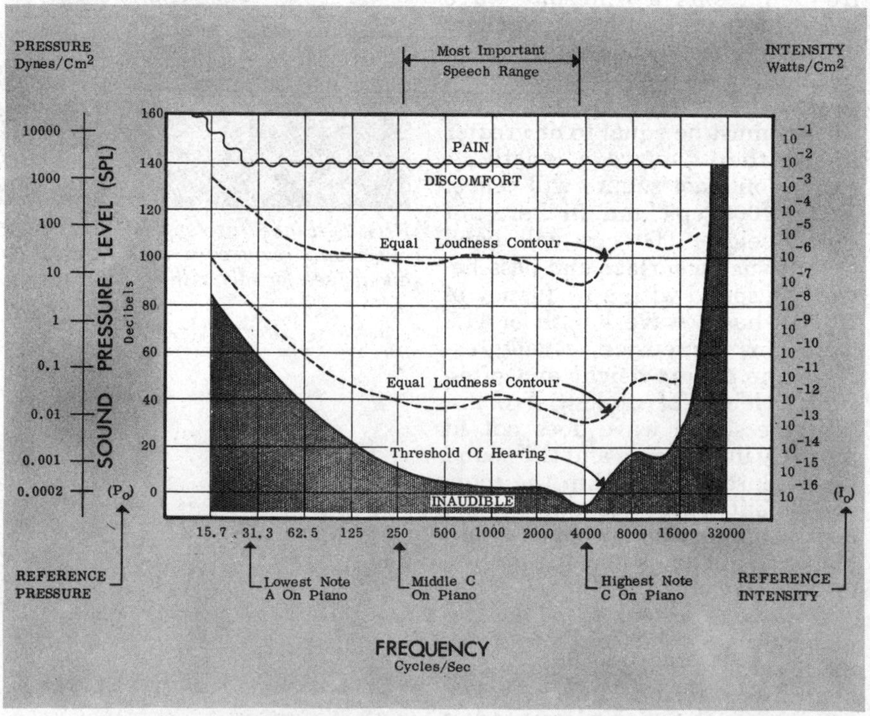

FIG. 4—Audible Frequency Range. Two sounds of the same intensity but of different frequencies do not necessarily sound equally loud. Equal Loudness Contours represent sounds of different frequencies that appear to the average ear to be equally loud.

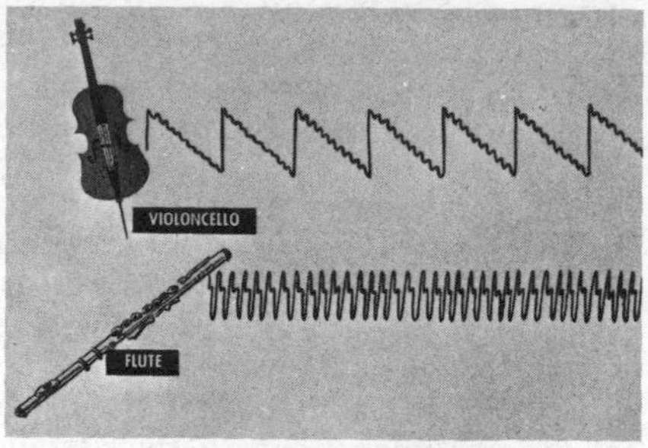

FIG. 5—Musical instruments produce sound of many related frequencies.

Sound Levels

A sound source may vibrate with small or great intensity, depending on the amount of force that sets it into vibration. Sound levels are expressed in *decibels* (db). The decibel scale extends from 0 db (threshold of hearing) to over 120 db (threshold of feeling). Zero decibels represents a fixed value of sound intensity that is slightly less than that of the faintest audible sound. Sound at 120 db is so loud as to cause the sensation of tickling in the ear. Sound as high as 180 db, generated by some rocket engines, can cause structural damage to buildings and can be fatal to humans.

A decibel is equivalent to the smallest change in sound intensity that can be detected by the average human ear. However, a decibel cannot be measured directly and has no fixed absolute value. It is a logarithmic unit that expresses the *ratio* between a given sound level being measured and a reference point. In other words, a decibel tells by what proportion one sound level is greater or less than another. The greater the sound level, the higher the rating on the decibel scale.

Sound levels in decibels are calculated from either a ratio of sound *pressures* (P) or sound *intensities* (I) using a reference point that corresponds to the faintest audible sound.

Pressure is a force per unit of area. Sound pressure at a given distance from a source is the force of moving air molecules spread over a spherical area and is expressed in dynes (units of force) per square centimeter.

Power is expended when a force is applied over a certain distance in a given unit of time. Sound power in watts (units of power) is a basic expression of the flow of sound energy. Sound power is spread over a sphere of increasing area as the sound wave radiates further from the source. Sound power per unit of area is called sound *intensity* (I) and is expressed in watts per square centimeter.

The reference pressure, marked (Po) in Figure 4 is 0.0002 dynes/cm^2 and is roughly equivalent to the smallest amount of pressure that will cause the eardrum to vibrate. The corresponding reference intensity, marked (Io) in Figure 4, is 10^{-16} watt/cm^2.

The ratio in decibels between any measured pressure (P) and the reference pressure (Po) is called the *Sound Pressure Level (SPL)* and is found by formula #1 given in Figure 6. The ratio in decibels between any intensity (I) and the reference intensity (Io) is called the *Sound Intensity Level* (IL) and is given by formula #2 in Figure 6.

SOUND PRESSURE LEVEL

$$(1)\ SPL = 20\ Log_{10}\frac{P}{P_o}$$

SOUND INTENSITY LEVEL

$$(2)\ IL = 10\ Log_{10}\frac{I}{I_o}$$

FIG. 6—Formulas for computing the ratio in decibels between (1) Pressure of two sounds, or (2) Intensity of two sounds.

The decibel scale makes values computed for either pressure or intensity directly comparable. Although intensity levels are convenient for discussion, intensity cannot be measured directly, hence in practice sound levels are determined by measuring *pressures* as read on a sound level meter and expressing the measurement as Sound Pressure Level on a decibel scale.

The simplest way to understand sound levels on the decibel scale is to compare them to common, easily recognized sounds (Fig. 7).

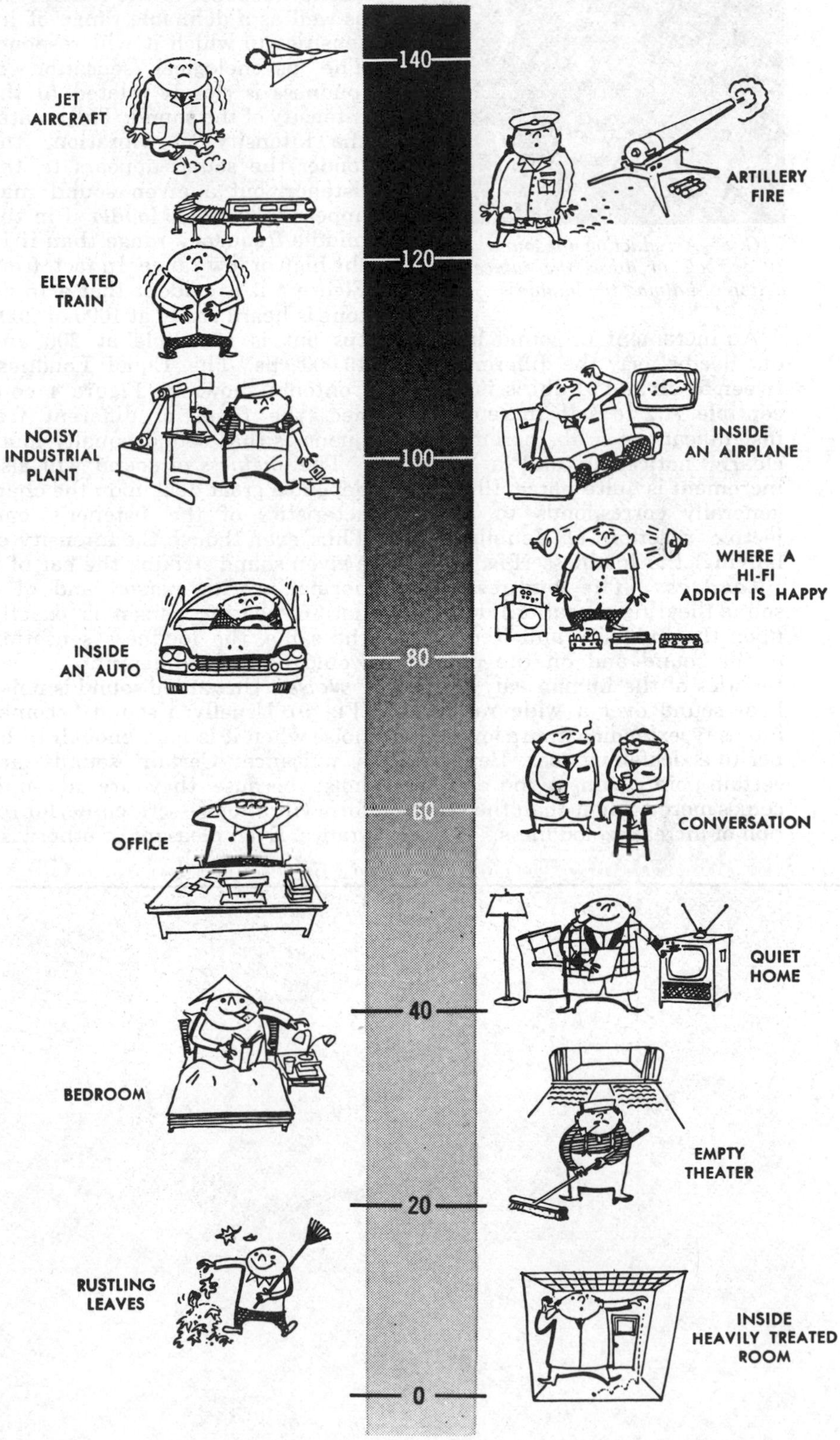

FIG. 7—Sound Pressure Level in decibels of certain sounds.

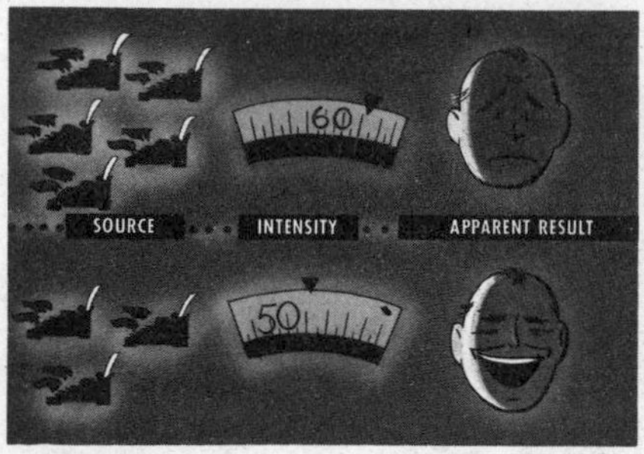

FIG. 8—A reduction in sound level of 10 decibels produces the subjective reaction of halving the loudness.

An increment in sound level of one decibel (say the difference between 50 db and 51 db) is just perceptible. A 2 to 3 db increment is insignificant, a 5 db increment is clearly noticeable and a 10 db increment is quite perceptible and generally corresponds to a subjective reaction of doubling (or halving) the loudness (Fig. 8).

Loudness The loudness of a sound (hearing sensation) depends upon the intensity and frequency of the sound and on the characteristics of the human ear. People hear sound over a wide range of intensity, extending from a low whisper to a deafening roar. Beyond a certain point, pain in the ears becomes more evident than the sensation of increasing loudness.

There are limitations in the frequency response of the human ear as well as a definable range of intensities to which it will respond. The psychological sensation of loudness is closely related to the intensity of the sound. The greater the intensity of vibration, the louder the sound appears to the listener, but a given sound may appear to be much louder if in the middle frequency range than if in the high or low range. In fact, from Figure 4 it is evident that a 15 db tone is heard easily at 1000 or 2000 cps but is inaudible at 200 and 10,000 cps. The Equal Loudness Contours shown in Figure 4 connect the tones of different frequencies that sound equally loud.

The *loudness* of sound will also depend a great deal upon the characteristics of the listener's ear. Thus, even though the intensity of a given sound striking the ear of a normal hearing person and of a hard-of-hearing person is exactly the same, the loudness sensation would be quite different.

Noise Unwanted sound is noise (Fig. 9). Usually a sound becomes noise when it is loud enough to be a nuisance. Certain sounds are noise because they are non-uniform vibrations—scratching, for instance. Even pleasant or otherwise acceptable sound becomes noise whenever it interferes with concentration, conversation or other tasks to be performed. It is difficult to say how loud a sound should be to make it objectionable, since it depends on the kind of sound and the person who hears it.

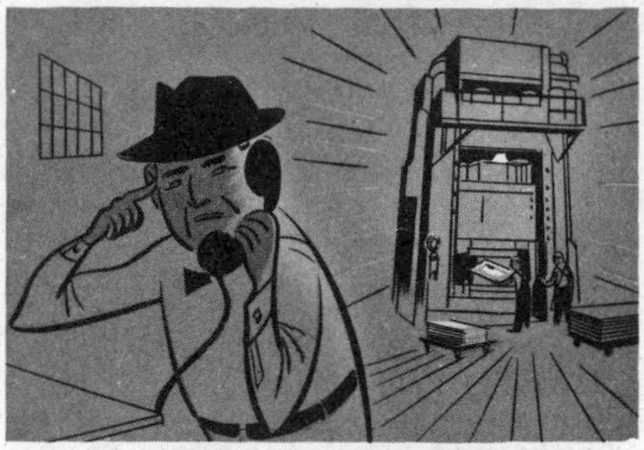

FIG. 9—Noise is unwanted sound.

In the following text, certain criteria are established to help determine the degree of sound control. *Room Acoustics,* page 106-7, explains sound conditioning practices to help *control noise within* a space; *Sound Isolation,* page 106-11, explains the practices to control airborne and structure-borne *sound transmission between* spaces; *Design and Construction,* page 106-19, explains planning principles and precautions to obtain and maintain suitable noise levels.

Room acoustics is a branch of architectural acoustics that deals with both *acoustical correction* and *noise reduction.*

ACOUSTICAL CORRECTION

Acoustical correction involves the detailed planning and shaping of a room to establish the best possible hearing conditions for *wanted* sound within the space. The primary problem is to provide an environment—such as in auditoriums, concert halls and similar spaces—that makes it possible to hear the sounds distinctly in ample loudness and with as faithful a replica of the source as is possible in all locations within the room. Acoustical correction, involving engineering calculations, the selection and location of acoustical and other finish materials, often in conjunction with sound amplification systems, requires the skill and judgment of an acoustical specialist and is therefore outside the scope of the *Construction Lending Guide.* In this section room acoustics deals only with noise reduction.

NOISE REDUCTION

Noise reduction involves the treatment of room surfaces with acoustical materials to alleviate the discomfort and distraction caused by the reflection of sound *unwanted* within the space. The primary problem is to reduce the reflection of sound and prevent a build up of the noise level due to continued reflection.

Reflection of Sound

When a sound source is operating, sound waves travel outward in all directions radially from the source. When the sound waves encounter an obstacle or surface, such as a wall, their direction of travel is changed, or reflected.

The reflection of sound follows the same laws as the reflection of light from a mirror: (1) the direction of travel of the reflected sound always makes the same angle with the surface as that of the incident sound, that is, the angle of incidence equals the angle of reflection; and (2) the reflected sound waves travel exactly in the same manner as they would if they had originated at an "image" of the sound source, located the same distance behind the wall as the real source is in front of the wall (Fig. 10).

Multiple Reflection When a sound source is started in a room having reflective surfaces, sound waves are reflected back and forth by these surfaces and almost immediately the room is filled with sound waves traveling in every direction. This is called multiple reflection (Fig. 11). Multiple reflection causes a build-up of intensity throughout the room, while the source is operating, to a level greater than there would be if no reflection occurred. It makes noises unnaturally and unnecessarily loud and makes all noise last longer than necessary.

Aside from annoyance and fatigue, excessive reflection interferes with telephoning, conversation, detection of alarms, sound signals and any other type of work or play dependent on accurate hearing.

Reverberation If the sound source is stopped, the reflected waves do not simply cease to exist at that instant, but continue to travel back and forth, reverberating or bouncing between the room surfaces. The total sound energy in the room gradually diminishes due to the partial absorption of the sound at each contact with a sound absorbing surface (Fig. 12).

If a listener is in the room, the reflected waves strike the ear in rapid succession so that he usually does not hear them as distinct repetitions of the original sound. Instead, he hears the original sound being drawn out or prolonged after the source is stopped, and steadily dying out until it becomes inaudible.

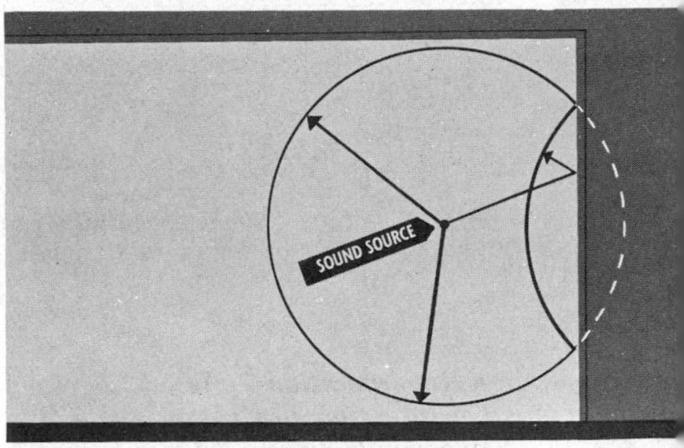

FIG. 10—Sound travels radially in all directions and is reflected by room surfaces.

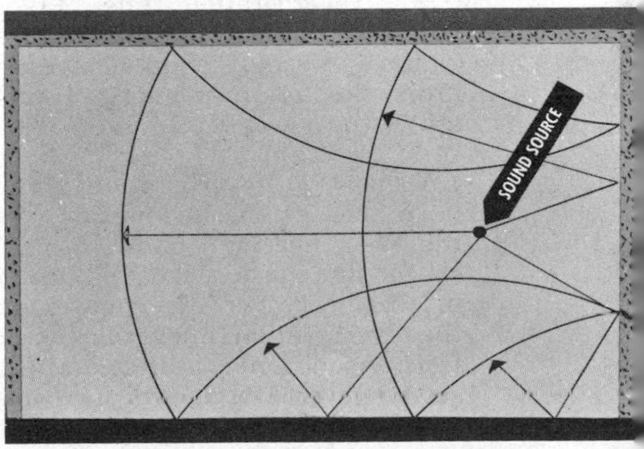

FIG. 11—Multiple reflection of sound waves causes (1) sound levels to build up, and (2) sound to reverberate longer.

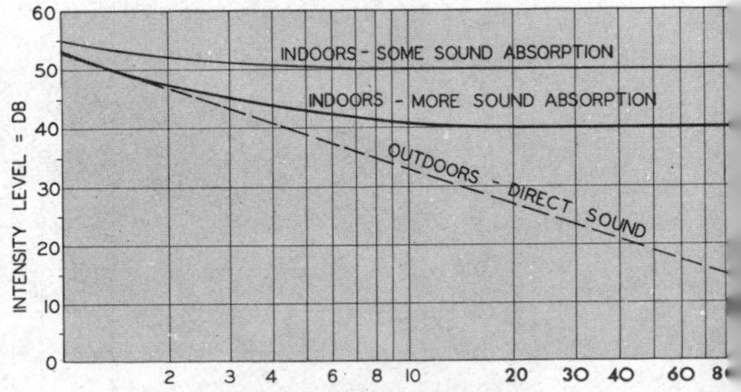

FIG. 12—Relation of distance from source to the resulting sound level in decibels. In open air, the level drops 6 db for each doubling of the distance from the source.

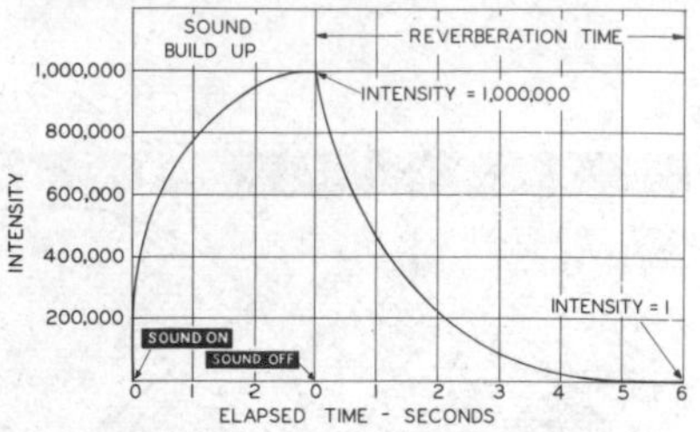

FIG. 13—Curve showing building up and dying out of sound in a room.

Reverberation Time The *reverberation time* of a room is the time in seconds required for a sound to die out to an intensity one millionth the intensity it had at the moment the source was stopped. On the decibel scale this is a drop of 60 db (Fig. 13).

The reverberation time (T) of any room depends only on its size (volume, V) and sound absorbing properties (absorption units, a) and not on the power of the source (Fig. 14).

Values of reverberation time have been established which give the best hearing conditions for speech and music (Fig. 15). Speech intelligibility becomes poorer as the reverberation time is increased. Both insufficient loudness and excessive reverberation will decrease speech intelligibility.

Echoes An echo is a single reflection which can be heard as a distinct repetition of the original sound. Echoes are likely to occur in large rooms where there is an appreciable time lag between the original sound and the reflected sound. The reflected sound must be heard at least 1/17th second later than the direct sound to be heard as an echo. In other words, the reflected sound must travel a path from source to listener at least 65′ longer than the path of the direct sound. Large rooms make possible the time lag for hearing an echo, and when reverberation time is short the echo is heard more easily above general reverberation.

Echoes produced by concave surfaces are louder than those produced by flat surfaces because a curved surface converges and focuses the reflected sound, increasing its intensity in the same manner as the curved reflector in a flashlight focuses light (Fig. 16). A ceiling having its center of curvature near the floor line is the most objectionable type of curved ceiling from the standpoint of focused echoes, since the focal point is brought closer to the audience than when the center of curvature is higher or lower.

Focusing is best corrected by breaking up the curve or changing its shape. Treatment with acoustical materials is rarely enough.

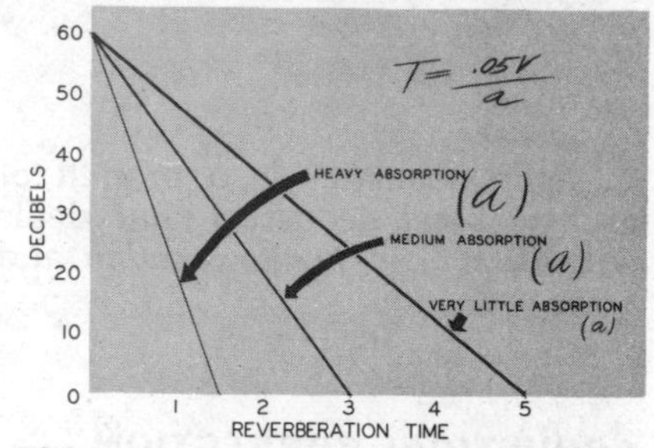

FIG. 14—Reverberation time is important for speech intelligibility and quality reproduction of music. T = Reverberation time in sec., V = room volume in ft.³, and a = room absorption in sabins.

Absorption of Sound

Ordinary smooth hard building surfaces are highly reflective and reflect up to 98% of the sound that strikes them. This reflection causes the sound to build up to a high level and to die out slowly.

Acoustical materials are designed to absorb most of the sound striking them and reflect less than 50% back into the room. These sound absorbing materials owe their absorption efficiency to the fact that they are highly porous.

The amount of absorption depends on the thickness of porous material, the size and number of

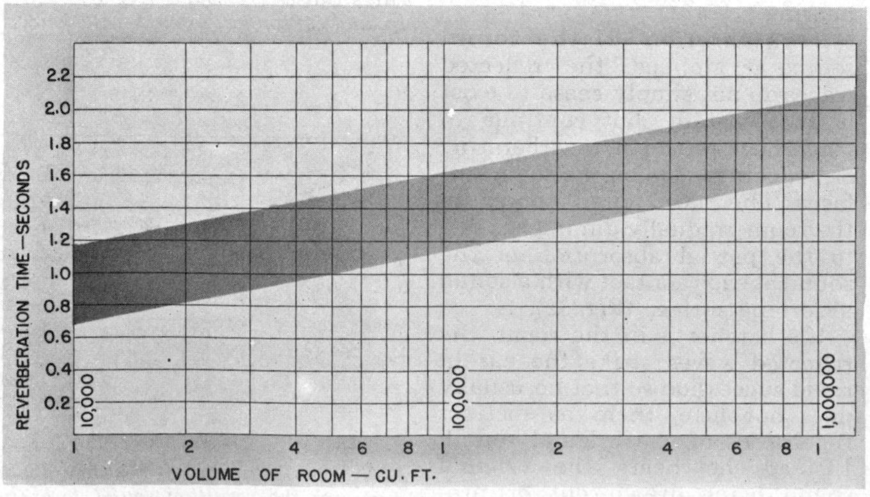

FIG. 15—Range of acceptable reverberation times at a frequency of 500 cps for rooms of various sizes.

FIG. 16—(A) Reflections from a flat surface are divergent. (B) Concave surfaces focus sound at points (f) that may be louder than the source (s).

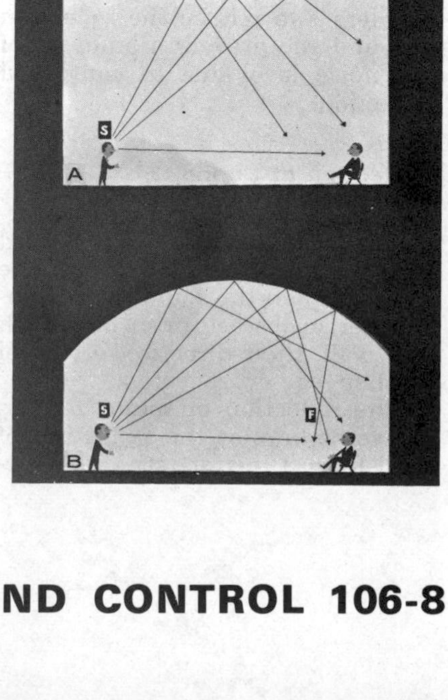

pores, and the frequency of the sound. Air particles moving in and out of the pores cause friction and sound energy is dissipated as heat (Fig. 17). Various materials and combinations of materials designed for this purpose are discussed in other sections where appropriate.

When a sound absorbing material is placed over a hard surface in a room, sound does not build up as much and will die out faster (Fig. 18). Although the sound level goes down, no matter how much sound absorbing material is introduced into the room, the level cannot drop below the level of the direct sound from the source to the listener. Sound absorbing materials cannot "suck up" the sound from the source; they can only absorb what reaches it. This is analogous to a light in a room—if the walls are white (reflective), the whole room is very bright; but if the walls are painted black (absorptive), the whole room becomes very dark, even though the same source at the same brightness level continues to emit the same amount of light. The brightness of the room is what changes. Acoustical materials act essentially the same with sound as does the black paint with light.

In practice, it is impractical and uneconomical to obtain sound level reductions of much more than 10 decibels by the use of sound absorbing materials. For example, between 5 to 7 decibels is the reduction usually accomplished when the entire ceiling of a family room in a home or a normal office is covered with acoustical tile.

FIG. 18 Hard surfaces can be treated with sound absorbing materials which reduce the build-up of sound and cause it to die out faster.

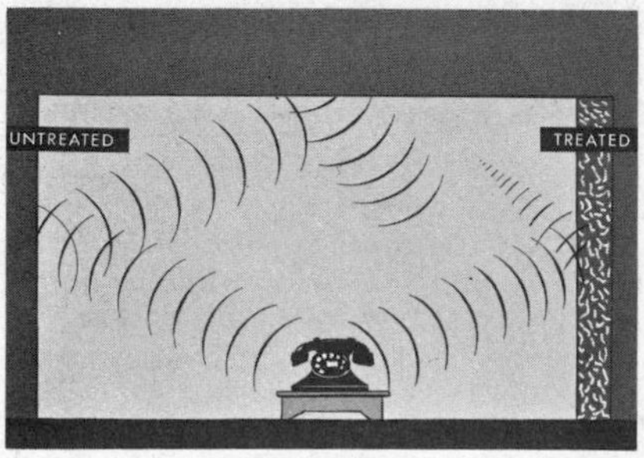

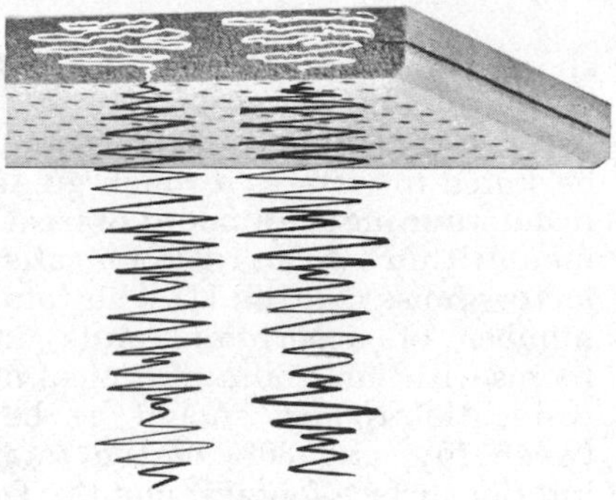

FIG. 17 Sound absorbing materials have interconnecting pore spaces. Friction of moving air within the pores transforms much of the sound energy into heat.

However, such a reduction is obvious to the user of the space and the improvement in communications and comfort is almost always conspicuous.

Measuring Sound Absorption One *unit of absorption* (called a *Sabin*) is equal to 1 sq. ft. of totally absorbent surface. An open window 1 sq. ft. in area is equivalent to one sound absorbing unit because theoretically all of the sound that strikes it passes on to the outdoors, and none is reflected back into the room. As far as a listener in the room is concerned, the open window has 100% absorption.

Absorption Coefficients (a) are assigned to materials and measure the percentage of incident sound energy absorbed by the material. The coefficients vary theoretically from 0 to 1—a perfectly reflective material having an $a = 0$, and a perfectly absorptive material having an $a = 1$. If a material has an absorption coefficient of $a = 0.50$, it means the material will absorb 50% of the sound that strikes it and reflects the remaining 50%. The absorption coefficient of every material varies with the frequency of the sound, thus it is necessary to test materials at more than one frequency in order to cover the complete range of high and low frequency sounds. Materials are normally tested at six different frequencies—125, 250, 500, 1000, 2000 and 4000 cps—and the information may be plotted as a curve (Fig. 19).

The numerical average of absorption coefficients at the four middle frequencies—250, 500, 1000 and 2000 cps—rounded off to the closest 5% is called the *Noise Reduction Coefficient.*

Noise Reduction Coefficient (NRC) is a single number rating used for calculating the required amount of sound absorbing material and permits comparison selection of products by representing average acoustical performance.

Because small differences in acoustical efficiencies cannot be detected by the human ear, and since tests which determine these efficiencies cannot be carried out to pinpoint accuracy, the selection of an acoustical material should normally be based on a 10 point range of NRC's (Fig. 20). In addition to the sound absorption rating of a material, such other properties as light reflectivity and flame resistance also generally are considered. Only in special cases where high absorption is required does NRC alone determine the selection. Sound absorption coefficients and noise reduction coefficients of acoustical materials are published by the manufacturers of acoustical materials.

Controlling Noise With Sound Absorbing Materials Acoustical treatment controls noise by reducing multiple reflection of sound waves, and as a result the loudness level of all reflected sound is reduced. Also, reverberation time is

FIG. 19 Absorption coefficients and other information are published by the manufacturers of acoustical materials.

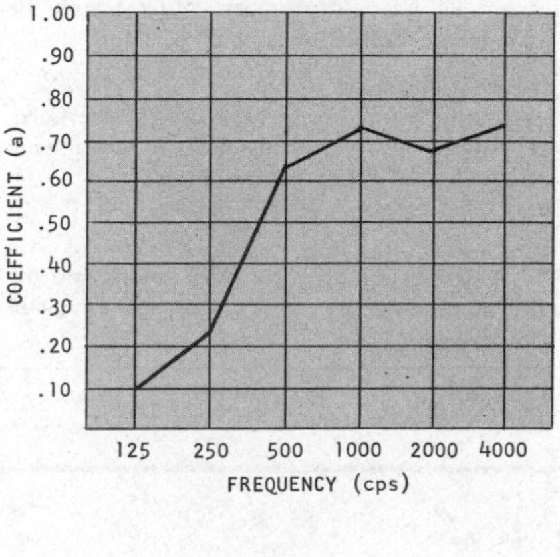

FIG. 21

NOISE REDUCTION EXAMPLE

Noise reduction with acoustical tile can be illustrated by assuming a family (or living) room 14′ x 24′ x 8′, finished with a wood floor and plaster walls and ceiling. Absorption *before* treatment (a_1) is found by multiplying the area of each material by its coefficient*.

Material	Area (sq. ft.)		Absorption Coefficient* (500 cps)		Units
Wood Floor	336	@	.01	=	3.36
Plaster Walls	608	@	.03	=	18.24
Plaster Ceiling	336	@	.03	=	10.08
Furniture & People	(estimated)			=	12.32
Total Absorption Before Treatment (a_1)				=	44.00

Assume the entire ceiling is treated with acoustical tile having a noise reduction coefficient (NRC) of .70. The net coefficient is .67 (after subtracting .03 for the plaster covered with tile) and will furnish 225.12 net added absorption units. The total absorption *after* treatment (a_2) will then be:

$$44 + 225.12 = 269.12 \text{ units}$$

Dividing a_2 by a_1 $\left(\frac{269.12}{44} = 6.12\right)$, the absorption will be increased 6.12 times, meaning the intensity of the reflected sound will be reduced to 1/6.12 or 16.3% of its value before treatment. Referring to Fig. 22, an absorption ratio of 6.12 corresponds to a reduction in loudness of 45%!

Checking these calculations with the noise reduction guidelines given in the text, the total number of units after treatment (269.12) is 21% of the total room surface (1280 sq. ft.) which is near the lower limit of the recommended range of 20% to 30%. The absorption ratio of 6.12 is well within the recommended range of 3 to 10, indicating there will be a satisfactory contrast with the conditions before treatment.

Using the formula T = .05V/a, the reverberation time before treatment is reduced from 3 sec. to .49 sec. after treatment.

* A partial list of absorption coefficients for building materials and furnishings is given in the Work File.

shortened so that the noise is not prolonged unnecessarily.

Noise reduction guidelines may be stated to serve as a rough guide in determining the amount of treatment within a given room for satisfactory noise control: (1) The total number of absorption units in rooms with low ceilings, typical of residential rooms, should be between 20% and 30% of the total interior surface footage; and (2) To produce a satisfactory improvement in an existing room, the total absorption after treatment should be between 3 and 10 times the absorption before treatment (see example in Fig. 21).

A procedure for calculating the percent reduction in loudness of reflected sound produced by acoustical treatment is outlined as follows: (1) Determine the number of absorption units in the room before treatment (a_1); (2) to this add the net number of units furnished by the treatment to give the total absorption in the room after treatment (a_2); (3) divide a_2 by a_1 to obtain the absorption ratio $\left(\frac{a_2}{a_1}\right)$; (4) determine the percent loudness reduction from Figure 22.

In average residential rooms it is desirable to reduce the reverberation time to about 1/2 second and in large rooms used for music as well as speech to not more than one second. In rooms such as the family room, kitchen, utility room, corridors and stairways, the problem is usually one of reducing reverberation enough to lower noise levels about 5 to 8 db. Unless the room shape presents special problems, covering the ceiling with a good sound absorbing material is usually adequate treatment.

Hallways generally need sound absorbing materials on the ceiling as well as the floor to reduce the "piping" of sound along the hallway to other rooms.

In small rooms, such as bedrooms full of upholstered furniture and heavily carpeted, it is likely that additional absorption in the

FIG. 20—NRC rates the efficiency of acoustical materials. A 10 point NRC range is used to specify or select materials, since a smaller difference is seldom detectable in a completed installation.

form of acoustical tile will be unnecessary.

Transmission Through Sound Absorbing Materials Acoustical materials are used to absorb sound. Although some materials may be effective in preventing sound transmission, most acoustical surfacing materials are not. Sound absorbing surfaces—whether tile, board, carpet and pad or other material—when used in a room where noise originates, act indirectly in reducing sound transmission to adjoining rooms by lowering the noise level in the room where the sound originates. Used in an adjoining room, they will similarly lower the level of the transmitted noise. These reductions, however, usually amount to only a few decibels. Acoustical surface treatment will supplement but will not take the place of sound isolation, which is discussed on the following pages.

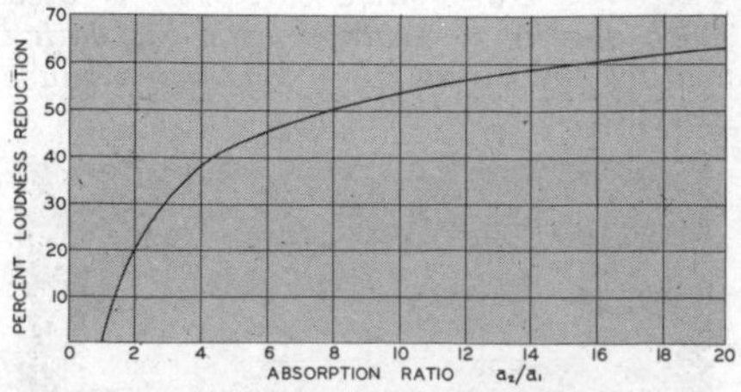

FIG. 22—Relation of reduction in loudness (in %) to the absorption ratio.

SOUND ISOLATION

Sound isolation deals with the design of building constructions to control noise transmission through walls, floors and ceilings which may occur as a result of diaphramatic vibration.

Sound is transmitted through the construction when the wall, floor and ceiling surfaces vibrate. The surface may be set into vibration by either (1) the alternating air pressure of incident sound waves, called *airborne* sound transmission, or (2) by direct mechanical contact or impact caused by vibrating equipment, a footstep or an object dropped on a floor. Transmission as a result of direct mechanical contact or impact is called *structure-borne* sound transmission. In either case, the vibrating surfaces generate new sound waves of reduced intensity in the room or space where the sound originates as well as in the room or space on the opposite side (Fig. 23).

AIRBORNE SOUND TRANSMISSION

Airborne sound is produced by indoor and outdoor sources that radiate sound directly into the air —people talking, a TV or phonograph playing, the hum of mechanical equipment, traffic noises. Airborne sound may be transmitted to adjoining rooms or space (1) directly through the intervening constructions, and (2) by way of flanking paths such as open doors or along inter-connecting structure (Fig. 24). The problem of isolating airborne sound is to provide wall and floor/ceiling constructions that will reduce the level of transmission to the point where it is either inaudible or not annoying or interfering.

Measuring Airborne Sound Transmission

In order to determine the performance of walls and floor/ceilings in isolating airborne sound transmission, the *transmission loss* of construction must be measured*.

Transmission loss is the number of decibels that a sound loses in being transmitted through a wall or floor/ceiling. In both laboratory and field tests of walls for airborne transmission, the procedure is to generate a steady sound on one side of the wall and measure the sound level in decibels on both sides. The transmission loss of the wall is then determined from the measured difference in sound level on the two sides. Thus, in Figure 25, a sound of a given frequency with an intensity level of 70 db on one side of a wall will have a reduction in intensity to 40 db on the other side when the wall has a transmission loss of 30 db at this frequency. The higher the transmission loss, the more efficient the

** Testing should be in accordance with current ASTM E-90 specifications.*

FIG. 23—Sound is transmitted by: (A) alternating air pressure of airborne sound, and (B) impact or direct mechanical contact.

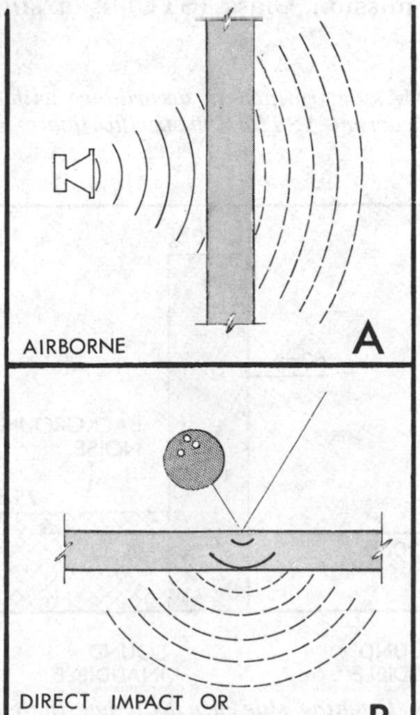

FIG. 24—Sound may be transmitted: (1) through the intervening construction; (2) through open air paths; or (3) along the interconnecting structure.

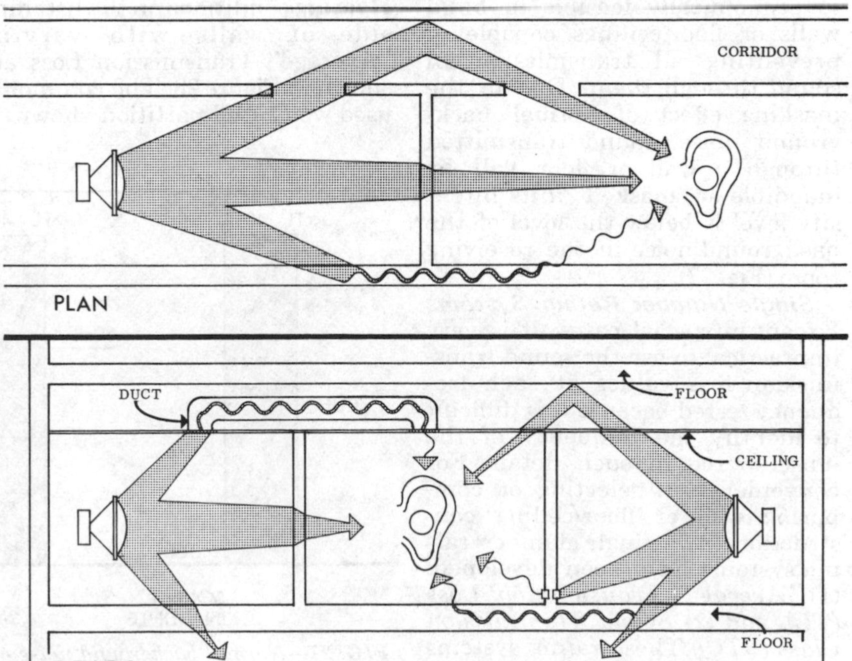

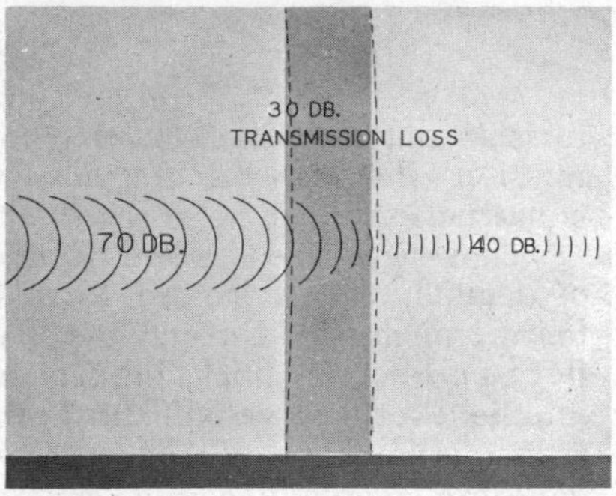

FIG. 25—Transmission loss is the difference in sound level on opposite sides of a construction. The wall here has a transmission loss of 30 db.

construction. Transmission loss is a physical property of the wall or floor/ceiling which depends only on the materials and methods used in the construction, and not on the loudness of the sound striking the surface. However, transmission loss varies considerably depending on the frequency of the sound, and therefore constructions are tested at numerous frequencies and plotted as a curve as shown in Figure 26. As shown in this example, constructions are generally much more efficient for high frequency sounds than for those of low frequency.

Actually, it is seldom necessary or economically feasible to build walls or floor/ceilings capable of preventing all transmission of sound through them. Due to the masking effect of normal background noise, sound transmitted through a wall or floor will be inaudible or "masked" if its intensity level is below the level of the background noise in the receiving room (Fig. 27).

Single Number Rating Systems

Except in special cases, it is very impractical to use the sound transmission loss values at each frequency tested because it is difficult to identify the frequency of the noise source in such detail. For convenience in selecting or comparing wall or floor/ceiling constructions, two single number rating systems have been developed: (1) *"Average" Transmission Loss (TL)*, and (2) *Sound Transmission Class (STC)*. These rating systems can be used to determine if a wall or floor construction will offer sufficient resistance to sound transmission, so that the transmitted level will be below the background noise level in the "listening" room.

"Average" Transmission Loss (TL) is the numerical average of the transmission loss values measured at nine frequencies—125, 175, 250, 350, 500, 700, 1000, 2000, 4000 cps. It has been used for many years by the National Bureau of Standards, Riverbank Acoustical Laboratories and other laboratories for rating airborne sound transmission through walls and floors and in the past has been the most commonly used basis for comparison of published data. "Average" Transmission Loss (TL) as measured by the average nine test frequencies is a fair indication of the sound isolation value of a wall or floor/ceiling construction if there are no sharp dips along the curve and if the values being compared are based on identical tests performed by the same laboratory under identical conditions.

To estimate the bare minimum "Average" Transmission Loss (TL) required of a construction to obtain privacy, subtract the level of the background noise on one side of the construction from the level of sound that strikes the other side. Hearing conditions on the listening side of walls with varying "Average" Transmission Loss are shown in Figure 28. The commonly used wood stud partition shown in

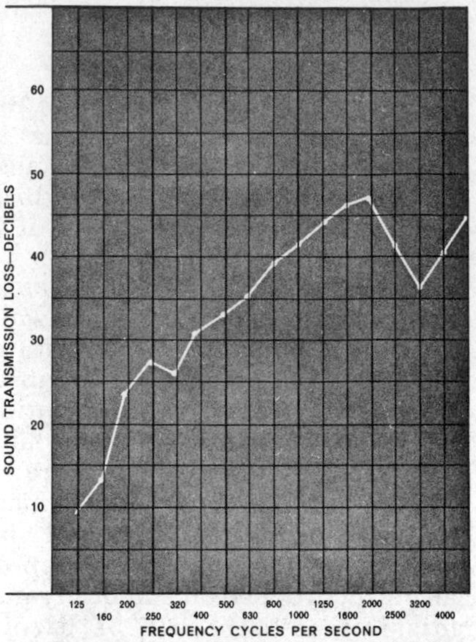

FIG. 26—This curve shows the transmission loss for a wood wall faced with 1/2" gypsum board nailed directly to studs 16" o.c.

Figure 26 has an "Average" Transmission Loss of 31 db, based on the curve shown, which offers little resistance to sound transmission.

*Sound Transmission Class (STC)** is the newer and recognized method of rating performance of sound isolating constructions. The Sound Transmission Class (STC) is a single

**As determined in accordance with the current ASTM E90 specification.*

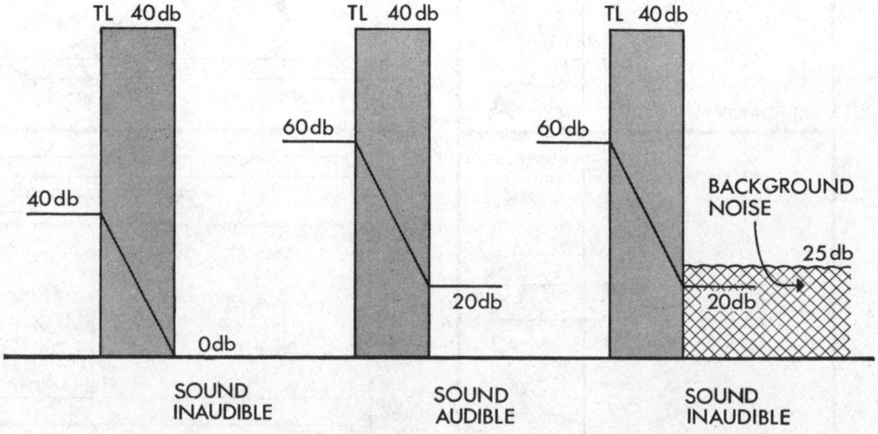

FIG. 27—Normal background noise on the listening side of a wall has the effect of masking transmitted sound that might otherwise be objectionable.

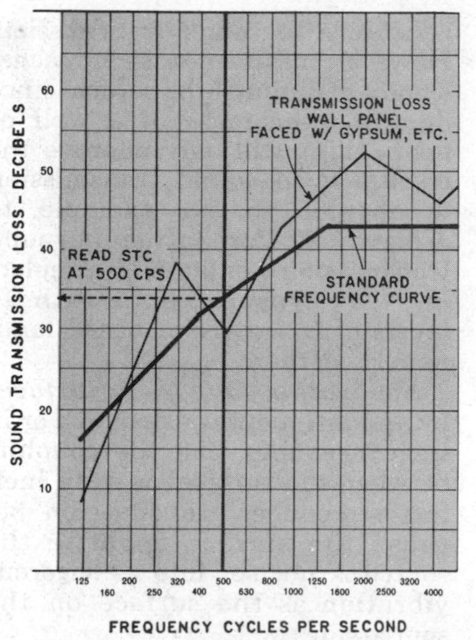

FIG. 29 An STC rating results from comparing a transmission loss test curve with a standard frequency curve (heavy line) and is read at 500 Hz (cps).

number rating for airborne sound which represents the transmission loss performance of a wall or floor at *all* test frequencies. The higher the STC rating, the more efficient the construction.

The STC rating is derived by plotting the curve of transmission loss for a given construction, tested at 16 frequencies; the transmission loss test curve is then compared to a standard frequency curve, also called a *reference contour* (Fig. 29). The standard frequency curve is drawn on a transparent overlay and is moved vertically to a position where the transmission loss curve remains above the standard curve at any frequency. However, certain tolerances (deviations) are allowed: (1) a single unfavorable deviation (where the sound transmission loss value falls *below* the standard curve) may not exceed 8 dB; (2) the sum of unfavorable deviations may not exceed 32 dB. The STC rating is the numerical value which corresponds to the sound transmission loss value at 500 Hz (Fig. 29).

The ASTM test procedure was first issued in 1950 and has been extensively revised since that time. Many of the currently published STC ratings for various constructions are based on tests performed prior to revision of ASTM E90 in 1966. STC values derived from earlier tests will vary somewhat from tests on similar constructions performed in accordance with ASTM E90-66T, depending on the conditions of testing such as number of test frequencies used, size of panel tested and application of the standard frequency curve.

The STC rating system is recommended in preference to the "Average" Transmission Loss because it gives a more reliable indication of speech or noise transmission by partitions whose curves have a sharp dip in a narrow frequency band.

In general, it has been found from experience that if the level of conversational speech is reduced by a wall just to the point where it cannot be understood in an adjoining room, the wall will provide acceptable overall privacy between rooms (Fig. 30).

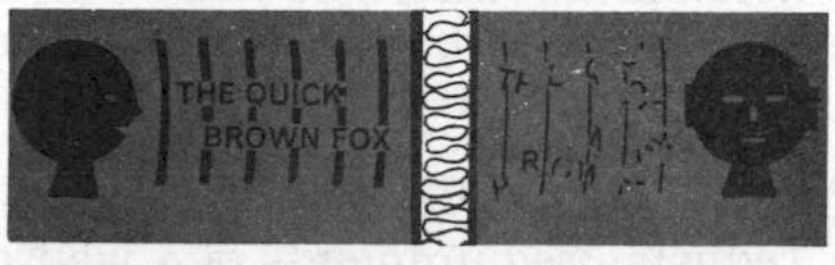

FIG. 30 A wall generally provides acceptable privacy between rooms when the level of conversational speech is reduced so that it cannot be understood.

To estimate the bare minimum Sound Transmission Class (STC) required of a construction to obtain privacy, subtract the level of the background noise on one side of the construction from the level of sound that strikes the other side. The examples in Figure 31 illustrate the approximate effectiveness of walls with varying STC numbers. Values of STC for several constructions are given in the Work File.

TRANSMISSION LOSS (TL)	HEARING CONDITIONS
30 dB or less	Normal speech can be understood quite easily and distinctly through the wall.
30 to 35 dB	Loud speech can be understood fairly well. Normal speech can be heard but not easily understood.
35 to 45 dB	Loud speech can be heard, but is not easily intelligible. Normal speech can be heard only faintly, if at all.
40 to 45 dB	Loud speech can be faintly heard but not understood. Normal speech is inaudible.
45 dB or greater	Very loud sounds, such as loud singing, brass musical instruments or a radio at full volume can be heard only faintly or not at all.

FIG. 28 Sound insulating properties of constructions according to Transmission Loss (TL); background noise level on the "listening" side assumed at 30 dB.

SOUND TRANSMISSION CLASS (STC)				SPEECH AUDIBILITY[2]
BACKGROUND NOISE[1]				
None	20dB	30dB	40dB	
40 To 50	25 To 35	15 To 25		Normal speech easily understood
50 To 60	35 To 45	25 To 35	15 To 25	Loud speech easily understood. Normal speech 50% understood
60 To 70	45 To 55	35 To 45	25 To 35	Loud speech 50% understood Normal speech faintly heard but not understood
70 To 80	55 To 65	45 To 55	35 To 45	Loud speech faintly heard but not understood Normal speech usually inaudible
	65 To 75	55 To 65	45 To 55	Loud speech usually inaudible

1 Background noise levels on "A" scale of sound level meter.
2 "Loud" speech taken as 10dB higher than "normal" speech; intelligibility taken as percent of sentences understood; normal source and receiving room absorption considered.

FIG. 31 Speech audibility through constructions according to their Sound Transmission Class (STC) rating as influenced by background noise levels.

Controlling Sound Transmission

Several variables affect the control of sound transmission to provide a suitable acoustical environment including: (1) the frequency and sound level of the sources; (2) the sound level that will be acceptable within the space; and (3) the sound-reducing characteristics of the intervening construction. The *principles* of controlling airborne sound transmission are discussed below; for detailed discussion of the *methods* to control airborne sound transmission, see *Design & Construction,* page 106-19.

Acoustical Planning The sound reducing requirements of a wall or floor/ceiling construction bear directly on its cost. Often many of the special techniques of construction used for sound control can be minimized or eliminated by preliminary acoustical planning and common sense. The principles are simple: locate rooms as far from noise sources as possible; segregate noisy rooms from quiet rooms; isolate and/or reduce noise sources; use space as a buffer between rooms and living units; avoid room-to-room or apartment-to-apartment open air paths for sound; and choose quiet appliances and mechanical equipment.

Airborne Transmission Through Walls, Floors and Ceilings The sound isolating efficiency of a construction depends on (1) its weight or mass; (2) effective separation of each face of a wall or floor/ceiling construction; and (3) the addition of sound absorbing materials between the effectively separated surfaces. Hence, the basic methods of controlling airborne sound transmitted through walls and floors/ceilings include: (1) increasing the weight of the construction; and/or (2) devising discontinuous construction methods that will permit each surface of the same wall or floor/ceiling to be built independent of one another; and (3) as an adjunct to discontinuous construction, adding sound absorbing materials in the cavity of frame wall and floor/ceiling constructions.

Weight has the effect of damping vibration. If a construction had an infinitely great mass it could not be set into vibration by airborne sound pressures, regardless of their intensity. If the mass were extremely small, even faint sound pressures would move it. The ability of a wall or floor/ceiling to transmit sound is inversely dependent upon its mass—the heavier or more massive it is, the greater its resistance to sound transmission. However, massiveness soon reaches a point of diminishing return, since doubling the mass of a wall or floor/ceiling will only improve the resistance to sound transmission by about 5 db. For example, to achieve a 20 db reduction in sound transmission would require replacing 1/2″ gypsum board with 16 layers of 1/2″ gypsum board laminated together.

Separating Opposite Surfaces by discontinuous construction or in such a manner that the coupling between the surfaces is very ineffective reduces transmission because the surface opposite the source is not set into as vigorous vibration as the surface on the source side.

One method is to use a construction in which the opposite faces are built independently of one another with no structural connections between them (Figs. 32 & 33). In wood frame constructions such as those with staggered studs or with two rows of studs, the framing provides vibration isolation between the two wall surfaces. The surfaces also can be isolated with resilient underlayment materials between outer wall surfaces and the structural framing (Fig. 34), thus reducing trans-

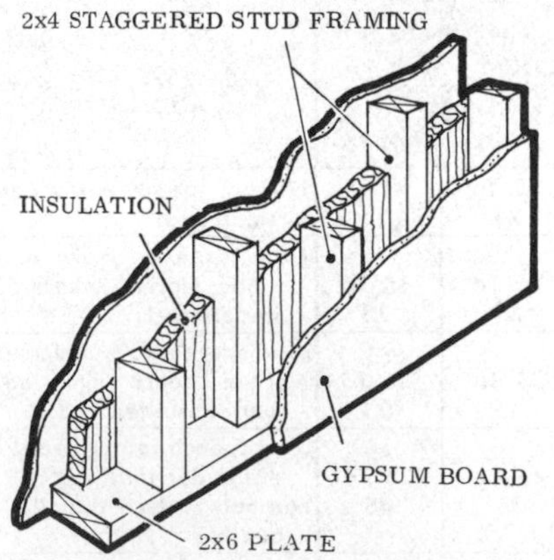

FIG. 32 Staggered studs provide discontinuous construction by effectively separating opposite surfaces. Addition of insulation improves performance.

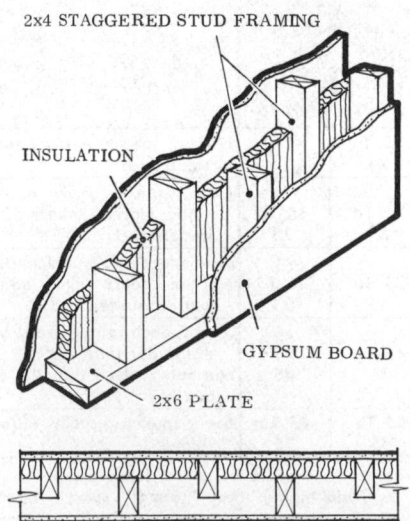

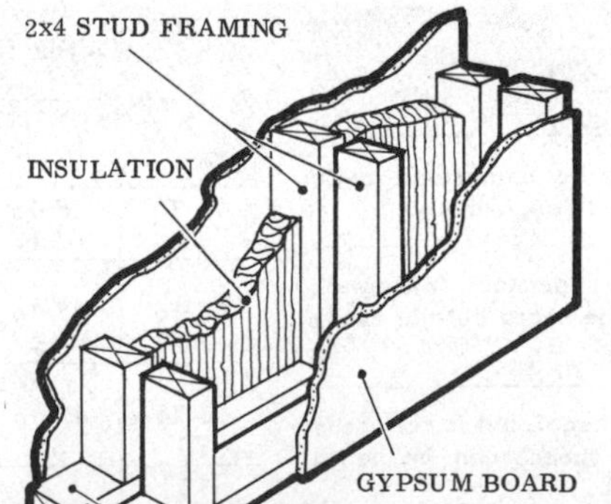

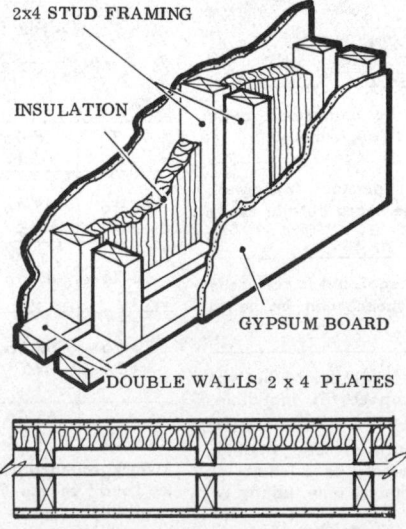

FIG. 33 Double row of studs, each row on separate plates, provides discontinuous construction; performance increases with increased thickness of insulation.

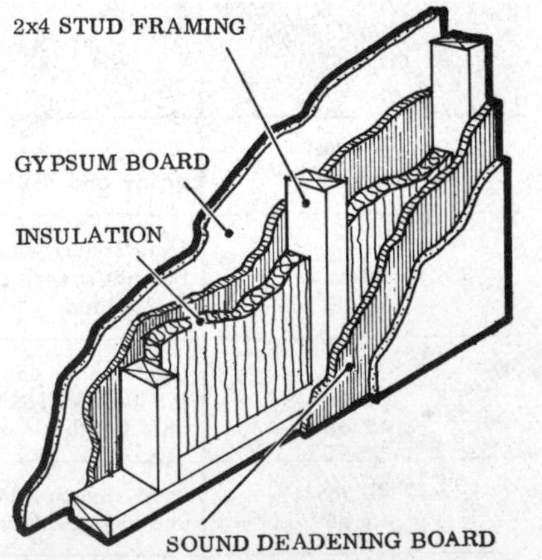

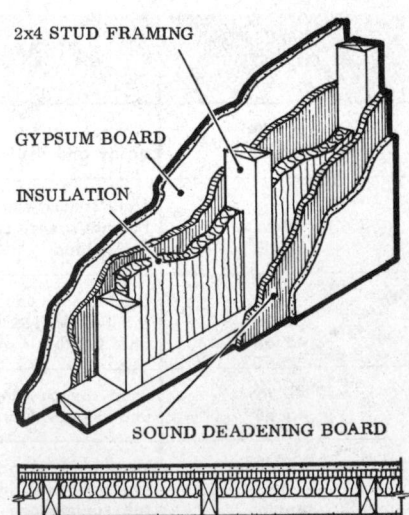

FIG. 34 Sound deadening board reduces transmission by resilient mounting of surfaces and can be used with single, double or staggered stud walls.

mission by isolating the surfaces, as an automobile spring isolates the body of the auto from the road. Sound in one room striking one side of the wall causes it to vibrate but, because of mechanical separation the vibration of the other side is greatly reduced. Double walls are much more effective than single walls of the same overall weight.

Another method is to attach each face to the structural support yet still have them mechanically separated using flexible connections or resilient mountings such that the opposite faces do not vibrate sympathetically (Fig. 35). Thus much of the sound energy is dissipated while vibrating one surface and the mountings, instead of being transmitted to the opposite surface. The performance improvement by discontinuous construction or resilient mountings can be increased significantly by placing sound absorbing insulation blankets within the wall or floor/ceiling to absorb sound energy in the cavity as shown in Figure 32.

Attempting to isolate airborne sound transmission by separating opposite surfaces alone has limitations. Doubling the free air space between the opposite surfaces of a wall or floor/ceiling construction

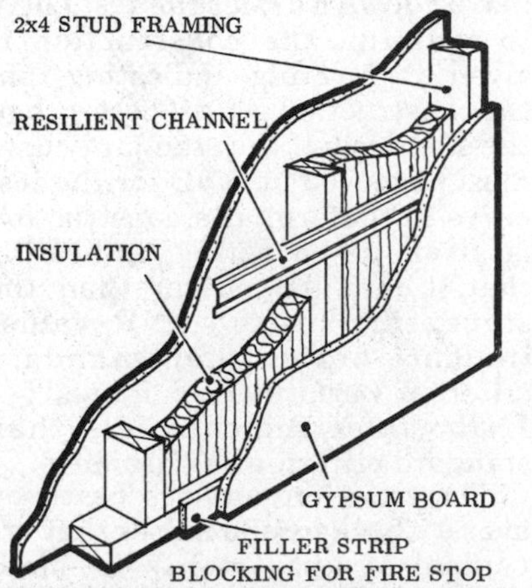

FIG. 35 Resilient channels (on one side of the wall) flexibly mount the surface; performance increases with increased thickness of insulation.

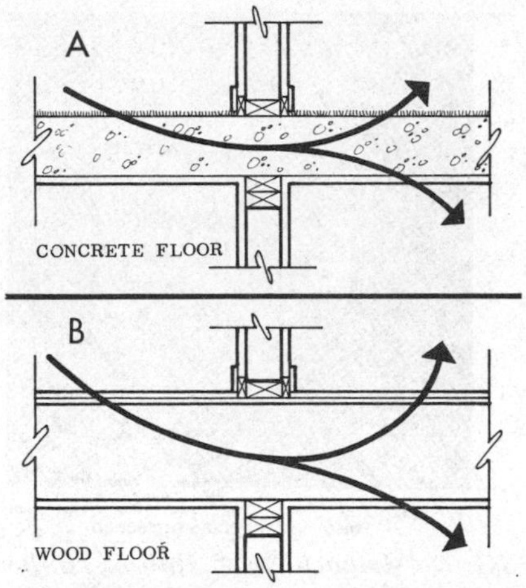

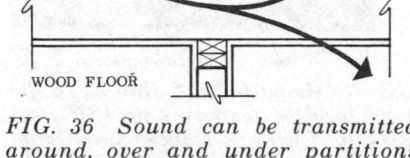

FIG. 36 Sound can be transmitted around, over and under partitions through (A) structure-borne flanking paths and (B) airborne flanking paths.

improves its resistance to sound transmission by only approximately 5 db, assuming a constant mass. Within practical limits of conventional building materials, *both* mass and discontinuous construction are required in order for a wall to reach a 50 to 55 db "Average" Transmission Loss. Transmission loss values greater than 50 to 55 db in a construction are not practical regardless of mass and isolation, without construction methods that will control transmission through flanking paths.

Airborne Transmission Through Flanking Paths It is normal to construct walls and floor/ceilings as continuous surfaces to divide living units into rooms. Windows, doors and other openings are required in them. These horizontal and vertical surfaces and their openings, as well as construction imperfections such as cracks and holes, create *flanking paths* that transmit sound. Just as snapping a rope will cause it to snake the entire length, sound impulses in a wall or floor/ceiling construction create vibrations that may transmit sound around, over or under a partition even though the partition itself has a high resistance to transmission (Fig. 36).

Open-Air Flanking Paths are created by openings through which sound travels virtually undiminished. A doorway or window opening is an enormous hole through a wall. It is an ideal sound leak, and can render the wall almost acoustically ineffective unless properly sealed.

To reduce the number of flanking paths, doors and windows should be airtight, heavy enough to minimize transmission through the construction itself, and be installed to maintain airtightness.

Flanking Paths Through Construction Imperfections are illustrated in Figure 24. Walls, ceilings and floors are usually pierced by innumerable pipes, ducts, conduits, lighting fixtures, electrical boxes, and similar equipment. Every one of these is a potential sound leak. Often they are inadvertently arranged to become ideal "speaking tubes;" air conditioning and ventilating ducts are frequently offenders of this type. Medicine cabinets or outlet boxes, placed back-to-back are common sound leaks. Such air leaks must be closed, openings caulked or sealed, fixtures staggered, and ducts lined with absorbent materials or equipped with baffles which interrupt the direct path of the sound and assist in absorbing it.

Acoustical Treatment Airborne sound transmission between two rooms usually cannot be controlled satisfactorily solely by the use of acoustical surfacing materials in one or both rooms. Covering the surfaces of the ceiling or wall with acoustical materials, or the floor with carpeting, adds little to their transmission loss. By reducing the level of sound reflected from the acoustically treated surfaces within the room, the amount of sound to be transmitted to adjoining rooms will be reduced. However, experience has shown that a considerably greater reduction in the transmitted sound usually is necessary for satisfactory results than is possible to obtain with acoustical surface materials.

STRUCTURE-BORNE SOUND TRANSMISSION

Structure-borne sound is produced when floor, wall or ceiling constructions are set into vibration by direct impact or direct mechanical contact. The vibrating surface then generates new sound waves on *both* sides of the construction. The floor is especially critical in this type of noise transmission. Typically, structure-borne noise is caused by objects that strike or slide on a floor or wall construction such as footsteps, dropped toys, cooking utensils or other objects, moving furniture, door slamming or by riding toys directly into the wall. It is also caused by mechanical equipment, fixtures and appliances which communicate their own vibrations to the intervening construction. These include garbage disposers, washers, dryers, motors and fans on heating, cooling and ventilating equipment, vacuum cleaners, toilets, bathtubs, showers.

The problem of isolating structure-borne sound is to provide floor/ceiling constructions that will reduce the level of transmission to the point where it is either inaudible or not annoying or interfering.

Measuring Impact Sound Transmission

To measure the relative performance of floor/ceiling constructions, a constant sound source of impact noises is created on top of the construction and the sound level then measured in the receiving room or adjoining areas (Fig. 37).

The isolation against impact noise provided by a given floor/ceiling construction is determined by means of a standard "tapping machine,"* which produces a series of uniform impacts at a uniform rate on the floor under test. The Impact Sound Pressure Level (ISPL) measured in decibels, produced by this standard tapping on the floor is measured in the receiving room below and electronically analyzed into different bands of

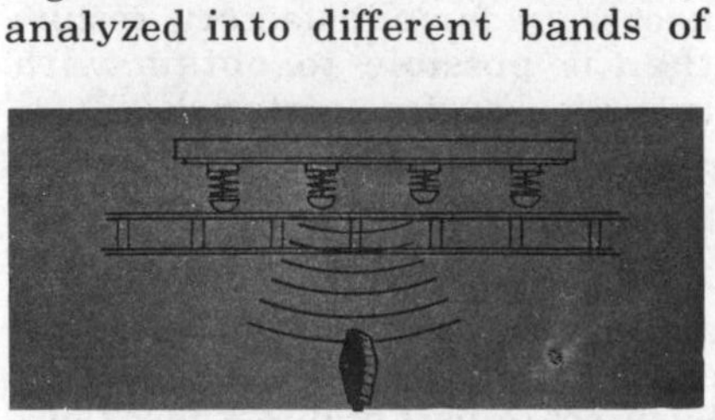

FIG. 37 Impact sound transmission through floor/ceiling systems is measured using a standard tapping machine.

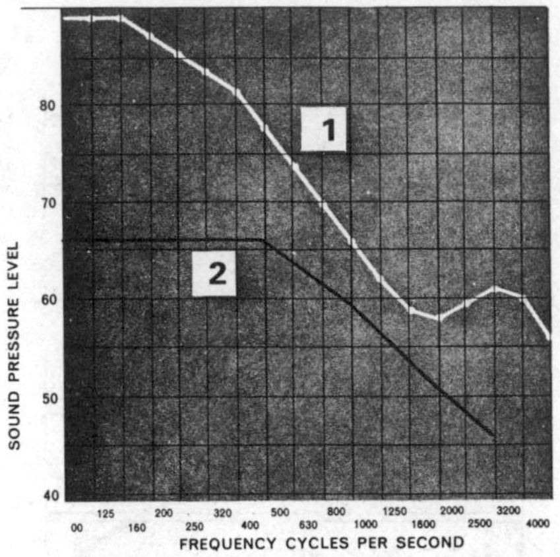

FIG. 38 Impact Noise Rating (INR) is obtained bv comparing the ISPL test curve (#1) with a standard criterion curve (#2).

frequency, so that a curve can be plotted showing how the sound energy in the receiving room is distributed over the audible frequency range (Curve #1 in Fig. 38). The standard tapping machine is placed in specified locations on the floor above, and the sound pressure levels are measured in the receiving room below. It is assumed that a construction that will transmit little sound with the standard tapping machine will perform similarly with other varied types of impact sound. However, because the noise produced by the standard tapping machine does not duplicate the sound produced by many types of impact, studies are currently being conducted to develop additional methods for determining impact sound performance.

In the case of airborne sound transmission, the difference between the sound levels in the sending and receiving rooms is measured and is expressed as sound transmission *loss.* Unlike the airborne sound transmission loss curves, Impact Sound Pressure Level (ISPL) curves measure the *actual* sound levels from a standard noise source *transmitted* through the construction. The lower the transmitted level, the better the acoustical performance of the construction.

Impact Noise Rating (INR) To evaluate the performance of a floor/ceiling construction, a single number system, Impact Noise Rating (INR), was developed. The ISPL test curve for the construction is plotted; a *standard criterion curve* (curve #2 in Fig. 38) then is laid over the ISPL test curve. The Impact Noise Rating (INR) can be determined by comparing the two curves and will indicate whether the performance of the construction is poorer than or is better than the standard. The impact noise rating is determined by moving the standard criterion curve up or down until the ISPL test curve meets the following tolerances: (1) the mean amount by which the ISPL test curve exceeds the standard criterion curve may be 2 dB or less, as averaged over the 16 one-third octave frequency bands between 100 and 3200 Hz; (2) the test curve may not exceed the criterion curve by more than 8 dB at any one frequency.

The INR value indicates how far below or above the standard criterion the ISPL test curve for a given construction falls. For example, if the test curve meets the standard curve within the allowable tolerances, the INR would be 0. If the standard curve must be moved down 5 dB for the test curve to conform, the construction is given a +5 rating, indicating that the construction is 5 dB better than the standard; if the standard curve must be moved up 5 dB for the test curve to conform, the construction is given a −5 rating, indicating that it is 5 dB poorer than the standard. Plus (+) INR values indicate better than standard criterion performance; minus (−) INR values indicate less than standard criterion performance.

The criterion curve is based on average background noises that are typical of apartments in moderately quiet suburban neighborhoods. For quieter areas, floor constructions with an INR of +5 to +10 should be used. For noisy urban districts a construction with a −5 INR may suffice. INR values for various floor/ceiling construc-

**International Organization For Standardization, ISO Recommendation #140-1960 "Field and Laboratory Measurements of Airborne and Impact Sound Transmission."*

tions are given in the Work File.

Impact Insulation Class (IIC) In an effort to reconcile and correlate the expression of various rating systems for airborne sound transmission and impact sound transmission, the Federal Housing Administration has devised a new rating system for impact sound transmission called the *Impact Insulation Class (IIC)*.

Using the same standard impact tester as is used for the INR system, the ISPL curve for a construction is plotted as shown in Figure 38. A reference contour on a transparent overlay is moved vertically until only the following allowable tolerances exist: (1) a single unfavorable deviation—where an ISPL value lies above the contour —may not exceed 8 dB, and (2) the sum of the unfavorable deviations may not be greater than 32 dB. The IIC rating is the value on the right scale corresponding to the ISPL value at 500 Hz (Fig. 39).

The IIC rating system parallels the STC rating system in that performance of the floor/ceiling system is expressed with positive values in ascending degrees of impact sound isolation. The larger the IIC rating, the better the construction.

On the average, an IIC rating for a given construction is approximately 51 dB higher than the INR rating. For example, an INR rating of +4 would be comparable to an IIC rating of approximately 55. However, exact values for IIC ratings should be determined on the basis of individual test data.

Controlling Structure-Borne Sound Transmission

Most building materials, especially the hard and dense materials, transmit vibrations readily. The steel frame of a building, for example, can telegraph a vibration throughout a large structure. Clicking of high heels on a concrete floor can be heard distinctly in the room below if no other construction intervenes. The vibration of the structure sets up sound waves which can often be heard distinctly at great distances from the source. Impact noise caused by a floor or wall being set into vibration by direct mechanical contact is then

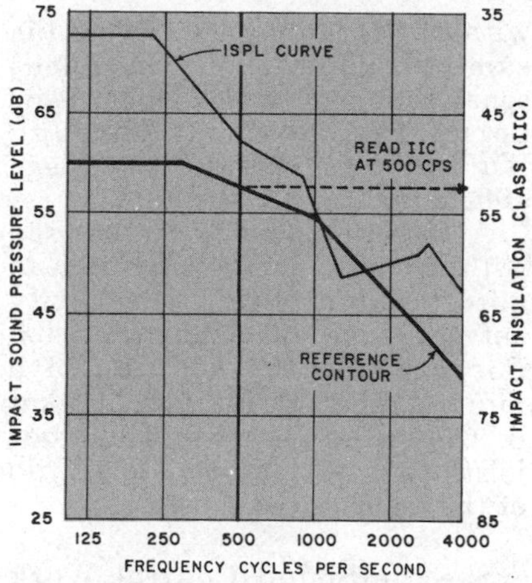

FIG. 39 The Impact Insulation Class (IIC) is obtained by comparing the ISPL test curve with a standard criterion curve and is read as a positive rating.

radiated from both sides. This vibration may also be transmitted throughout the structure to walls and re-radiated as sound to adjoining spaces. Since footsteps, children romping and playing, and moving furniture on floors constitute the major impact problem, control of impact noise is treated here as primarily a floor problem.

Impact isolation methods in general are more effective in reducing the high fequency com-

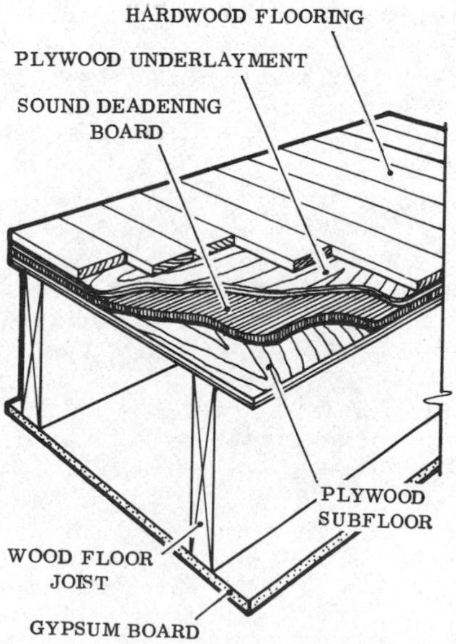

FIG. 40 Resilient underlayment (shaded) reduces impact sound transmission by "floating" the finished floor from the subfloor or structural floor.

ponents of impact noise than the low frequency "thumping" noises caused by heavy impacts. The limitation on low frequency isolation is more stringent for wood joist floors than for concrete floors because of the inherent lower weight and greater flexibility of wood floor joist construction. Methods to control structure-borne sound transmission are discussed below.

Cushioning The Impact Direct mechanical impact causes the structure to vibrate. Cushioning the impact diminishes structural vibration and is one of the most effective methods of isolating impact noise. In living rooms, bedrooms or other areas where they can be used, carpet with heavy padding is an effective surface treatment to cushion impact but is most effective in the upper frequency range. Carpet and carpet padding, however, are air permeable and, like most sound absorbing materials, do not contribute significantly to airborne sound isolation.

Floating The Floor Structure Finished floors which easily transmit impact noises to the structure below may be *floated* (separated) from the structure or subfloor below by the use of resilient underlayment materials. The degree of isolation and therefore increased resistance to impact noise transmission depends on using no nails or other mechanical fasteners to attach the finish floor to the structure, and on the resilient characteristics of the underlayment material (Fig. 40).

In general, the greater the resilience of the underlayment material the greater the isolation. Like carpeting, the resilient underlayment performs primarily as a cushion and only secondarily in resisting airborne sound transmission through the construction.

Separated Ceiling Construction Isolation of the ceiling from the structural joists or floor slab can be achieved by mounting the ceiling on resilient clips or channels (Fig. 41) or on independent ceiling joists. Built in conjunction with a floated floor, this type of construction can result in a system

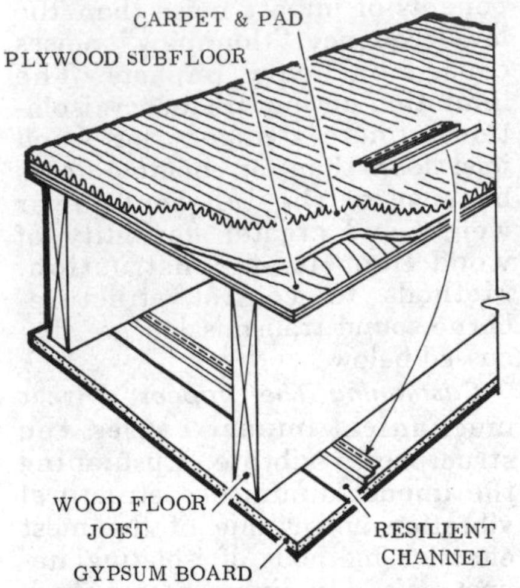

FIG. 41 Resilient channels resiliently separate the ceiling finish from the structural framing and control both airborne and impact sound transmission.

with an IIC rating of 57. This is the same principle of discontinuous construction discussed under Airborne Transmission Through Walls, Floors and Ceilings, page 106-14.

Absorbent materials can be used with discontinuous construction to absorb sound within the cavity between floor and ceiling (Fig. 42). For example, the addition of a mineral fiber insulating blanket in the space between the floor joists absorbs approximately 7 db of the transmitted noise.

See the Sound Control Work File for performance recommendations and ratings for walls and floors to isolate both airborne and structure-borne sound transmission.

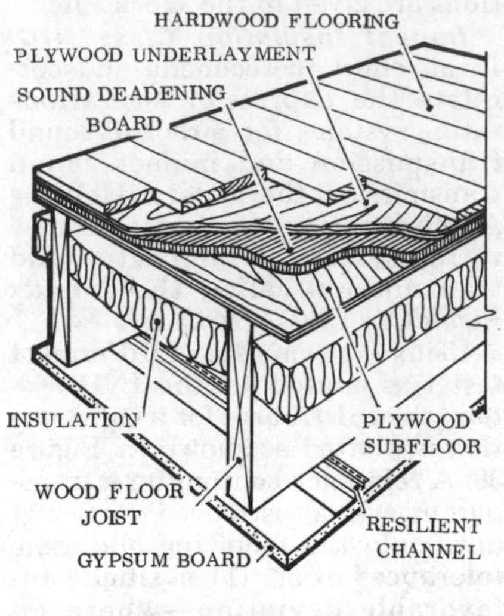

FIG. 42 A resiliently mounted ceiling with floated floor construction effectively resists impact sound transmission.

This subsection deals with: (1) *criteria and planning principles* to aid in the design and selection of wall, floor/ceiling, mechanical equipment and other systems and materials to control noise within buildings; and (2) *construction recommendations,* including precautions essential to obtaining and maintaining suitable sound isolation between rooms and between living units.

DESIGN AND PLANNING

Planning for sound control involves overcoming problems of loud exterior and interior noise at the sources; orienting quiet living spaces away from high-level noise sources; careful interior design and building layout, especially with respect to the shape of the buildings and the location of door and window openings and other sound leaks. Generally, the objectives in the planning for sound control are to reduce noise by (1) quieting the sound at its source; (2) separating the source from the listener by distance; and (3) buffering the transmission of noise between the source and the listener by using barriers such as walls and floor/ceilings that have improved resistance to sound transmission.

Background Noise Sources and Levels

The source and intensity of outdoor noise at the site affect the desired noise reducing effectiveness of exterior walls and roofs and also have an important effect on the selection of interior wall and floor-ceiling systems. Noise characteristics of the site should be studied to determine existing and potential noise levels (Fig. 43).

Theoretically, airborne sound travels from a *point noise source* in a spherical pattern (see Fig. 1, page 106-3). The sound level of a point noise source is reduced 6 dB each time the distance from the source is doubled (see Fig. 12, page 106-7). But fast moving traffic actually is a *line* noise source rather than a point source. Sound from a line noise source travels in a cylindrical rather than a spherical pattern. In order to reduce the noise level from a line source 6 dB, the distance between noise source and listener must be increased 4 times.

FIG. 43 **TYPICAL OUTDOOR BACKGROUND NOISE LEVELS**

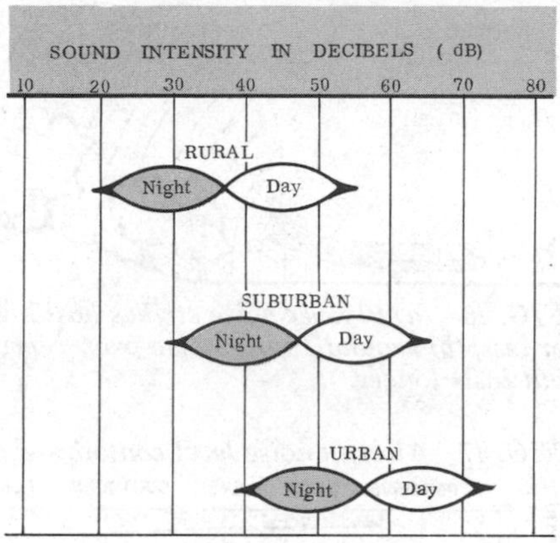

The direction and speed of air movement, air temperature and obstructions such as buildings, trees and terrain all affect the sound level. For example, where the wind moves unobstructed, sites upwind of major noise sources will be quieter than sites an equal distance away from the noise on the downwind side.

FIG. 44 Hard surfaces reflect sound; soft, porous surfaces such as landscaping will absorb some sound.

Hard exterior surfaces of buildings reflect a large proportion of the sound energy which strikes them (Fig. 44). Multiple reflection of high sound sources striking a large building can materially increase the noise problem at a site. Curved buildings or buildings designed to form courtyards having the concave side facing a noise source also will result in an especially difficult noise problem. Future development of nearby vacant land should be anticipated and the type and level of probable or possible noise sources should be investigated.

If the proposed site is near high noise sources—a business district, a highway, an expressway, railway traffic or an airport—both daytime and nighttime outdoor noise levels should be surveyed. Where the average nighttime background noise level is unusually high or if there are infrequent but intermittent unusually loud or annoying noises, expert acoustical advice should be sought.

Automobile Traffic Normally, automobile traffic is a primary source of outdoor background noise. Background noise levels are affected by such local variations as topography, landscaping and the masking or magnifying effect of other buildings. Typical outdoor noise levels are shown in Figure 45. High-speed highways and dense

FIG. 45 **APPROXIMATE SOUND LEVELS FOR AUTOMOBILE TRAFFIC AT GROUND LEVEL**

Cars Per Minute*	Distance From Traffic Lane		
	40′	100′	500′
1	50 dB	43 dB	29 dB
10	59 dB	52 dB	38 dB
100	63 dB	60 dB	47 dB

*The maximum capacity of one lane of cars moving freely at approximately 30 mph is 2000 cars per hour or 34 cars per lane per minute.

city traffic generally produce higher noise levels than moderate-speed suburban roads carrying an equivalent number of cars. Trucks increase highway noise about 15 dB, and trucks going uphill will produce the highest highway noise levels. Unless the highway is already "saturated" with traffic, a future increase in traffic generally can be expected with resultant increase in noise.

Aircraft Traffic The rapid expansion of jet transport operations throughout the country and the recent introduction of small commercial and business jet aircraft create noise problems in communities of any size. Methods for minimizing aircraft noise include techniques for quieting engines, reducing noise exposure by takeoff and landing operational procedures, and reducing the number of people affected by controlling land use around airports through zoning regulations.

Two types of aircraft noise are of particular concern: (1) noise generated by *revving engines* on the ground and (2) *flyover* noise. Figure 46a illustrates that the noise level during flyover changes very rapidly with time, with the maximum noise reached and sustained for only perhaps a second or less, and that sound waves during this short time interval strike a building from all directions.

Figure 46b illustrates the noise produced by revving engines while the aircraft is stationary on the ground. In contrast to flyover noise, because the relationship between aircraft and building is fixed, only one or two exterior surfaces of the building may be directly exposed to aircraft noise, while other surfaces may be partially shielded from the sound radiated. There also may be intervening obstacles such as hills, houses or other terrain features which will partially block the sound transmitted from the aircraft to the building.

Figure 47a illustrates typical noise level contours of jet aircraft during takeoffs; Figure 47b is an example of noise level contours for the revving of one engine of a jet aircraft.

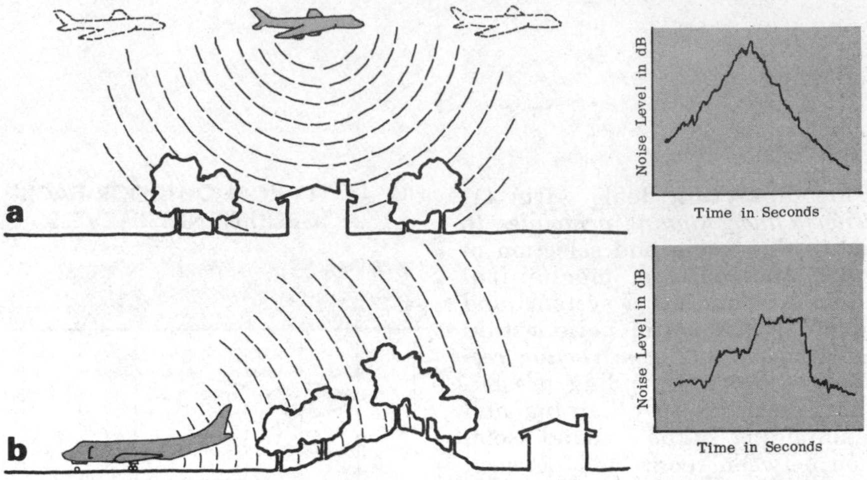

FIG. 46 (a) Flyover noise strikes buildings from all angles but lasts only a second or less; (b) ground runup noise from revving engines strikes fewer building surfaces but lasts longer.

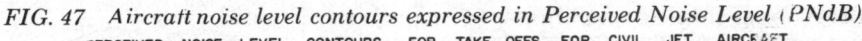

FIG. 47 Aircraft noise level contours expressed in Perceived Noise Level (PNdB).

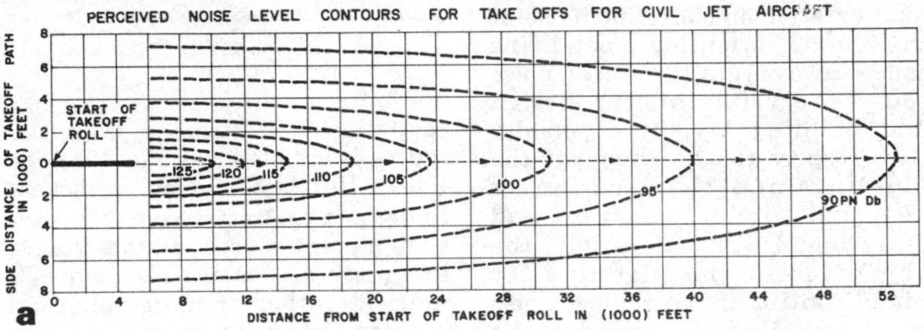

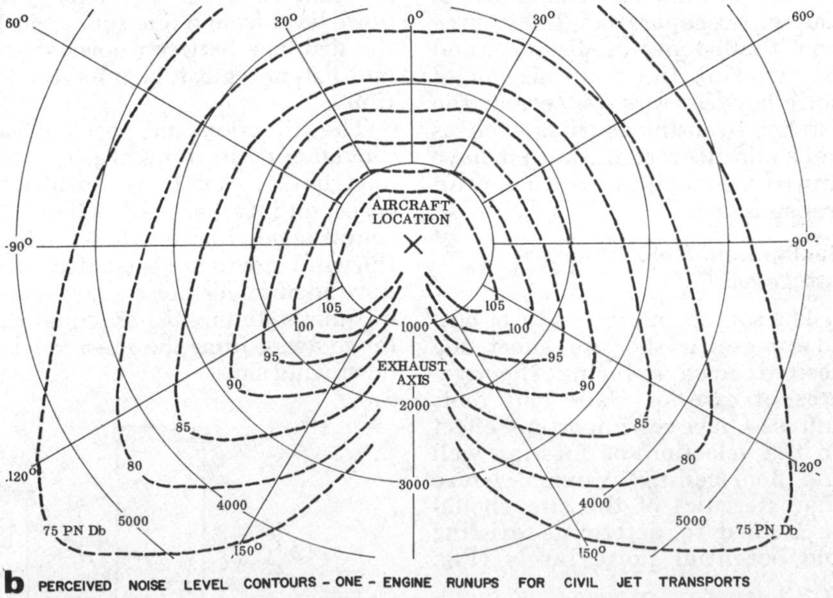

*PNdB, Perceived Noise Level, in decibels, is used to describe aircraft noise levels. The perceived noise level is a single number rating of aircraft noise used to rate the relative acceptability or noisiness of aircraft sounds. The perceived noise level is calculated from measured noise levels and correlates well with subjective responses to various kinds of aircraft noise

Of all exterior wall and roof surfaces, windows offer the least resistance to noise transmission. The difference in perceived noise level (PNdB, Fig. 47) between exterior and interior noise levels from aircraft is illustrated in Figure 48. Generally, the lower values in each category will be observed for houses having window areas comprising more than 20% of the total exposed wall area, with relatively lightweight wood frame walls and roofs (weighing less than 8 psf). The higher values generally will be obtained for houses with relatively small airtight windows and airtight storm windows of good quality. An example of aircraft noise reductions through various exterior wall and roof construction assemblies are summarized in Figure 49.

Site Planning

Site planning can reduce problems caused by exterior noise by: (1) increasing the distance between the noise source and the buildings; (2) siting the buildings to minimize the effect of direct sound transmission, reflection and/or reverberation from existing or possible future buildings; and (3) using natural topography, other buildings, barrier walls and in some cases landscaping to shield buildings from major noise sources.

FIG. 48 PERCEIVED NOISE LEVELS FROM AIRCRAFT

Window Condition	Difference between Interior and Exterior Noise Levels
Windows Open	10 to 20 PNdB
Windows Closed*	20 to 30 PNdB
Storm Sash Installed*	25 to 35 PNdB

*For maximum performance, glass should be sealed airtight within frames; storm sash should be separated by at least 2″ air space between panes.

Building Location Buildings reflect noise readily. Orientation which places rows of buildings parallel to each other results in build-up or focusing of the noise by multiple reflection of the sound. Acoustically, the best orientation for a building occurs when the long axis of the building is perpendicular to the noise source (Fig. 50a).

Courtyards created by a series of buildings located to form a U shape can be especially noisy when the open end of the U faces the major source of noise (Fig. 50b). If the courtyard is desired for other design considerations, having the open end of the U face away from the noise source is acoustically better. Landscaping in the courtyard will help to reduce reverberation of sound within the courtyard.

FIG. 49 SUMMARY OF NOISE REDUCTION THROUGH EXTERIOR CONSTRUCTIONS

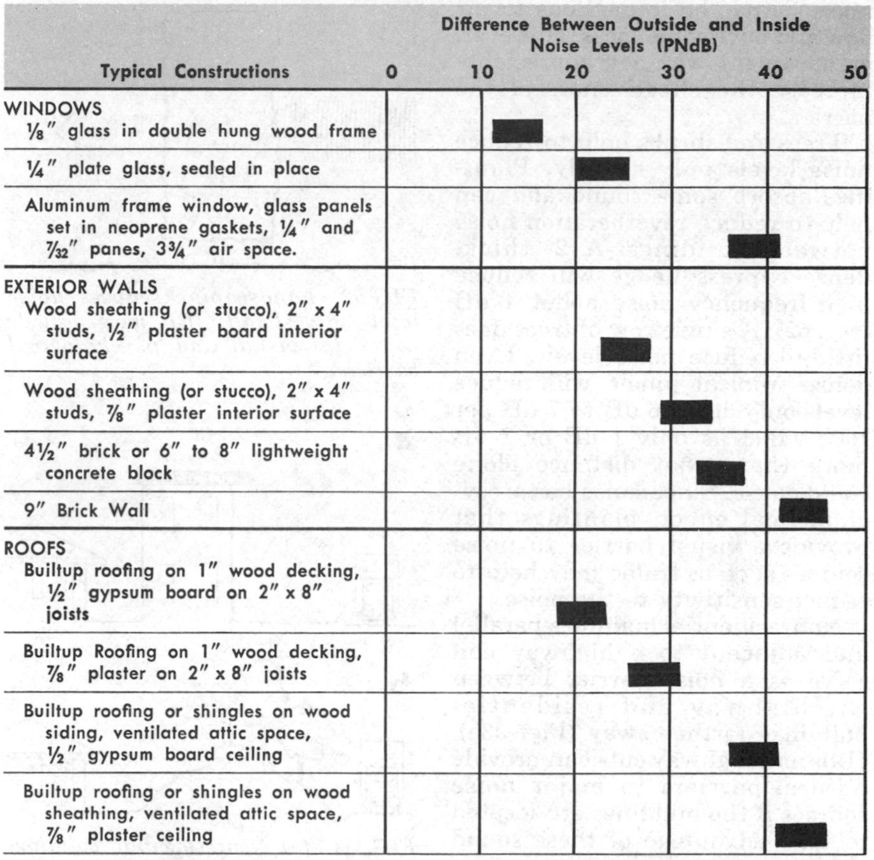

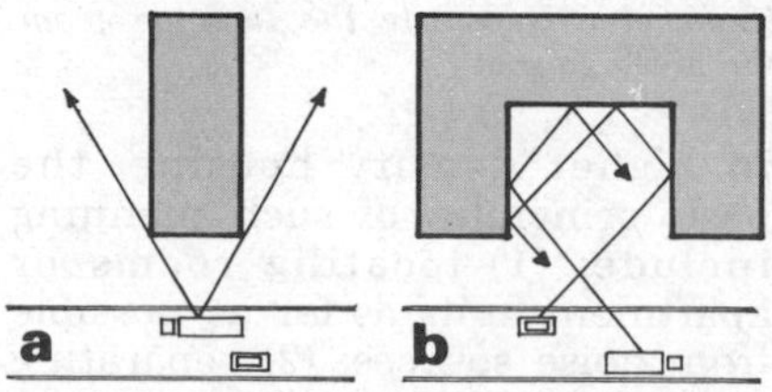

FIG. 50 (a) Perpendicular orientation exposes less wall area to noise source; (b) sound reverberates in U-shaped courtyard.

Noise Barriers Buildings, walls, landscaping and topography can be used as barriers between the noise source and the listener to help shield the exterior noise source. A barrier's effectiveness depends on the frequency of the sound, the angle of the *sound shadow* and the height of the barrier (Fig. 51). Higher frequency sounds are controlled more effectively by barriers than are low frequency sounds. For greatest effectiveness, barriers should be as high as possible, be

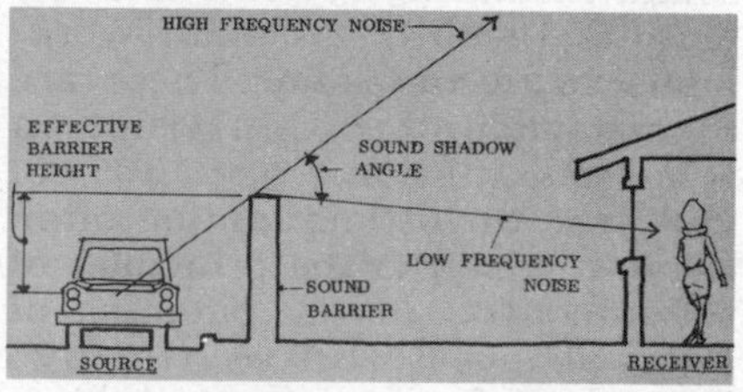

FIG. 51 Buildings and walls serving as barriers should be as high as possible and as close to the noise source as possible.

located as close as possible to the noise source, be impervious to air flow and have some mass. However, great weight will not appreciably increase the effectiveness of the barrier.

Trees and shrubs help to reduce noise levels only slightly. Plantings absorb some sound and can help to reduce reverberation noise between buildings. A 2′ thick, dense cypress hedge will reduce high frequency noise about 4 dB (Fig. 52). A single row of trees does little to reduce noise levels. Even dense tropical jungle will reduce levels only about 6 dB to 7 dB per 100′, which is only 1 dB or 2 dB more than what distance alone could cause. Since noise has a psychological effect, plantings that provide a visual barrier to noise sources such as traffic may help to reduce sensitivity to the noise.

Nonresidential buildings parallel and adjacent to a highway can serve as a noise barrier between the highway and residential buildings farther away (Fig. 53a). Hills and highway cuts can provide natural barriers to major noise sources if the buildings are located to take advantage of these sound shadow barriers (Fig. 53b).

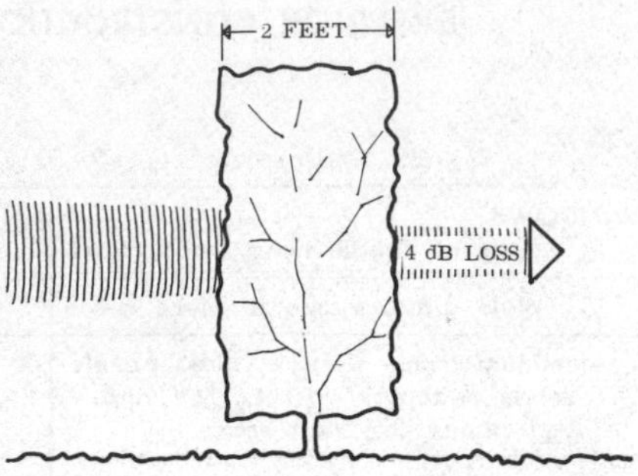

FIG.52 Landscaping reduces noise levels only slightly, but may provide beneficial visual and psychological barriers.

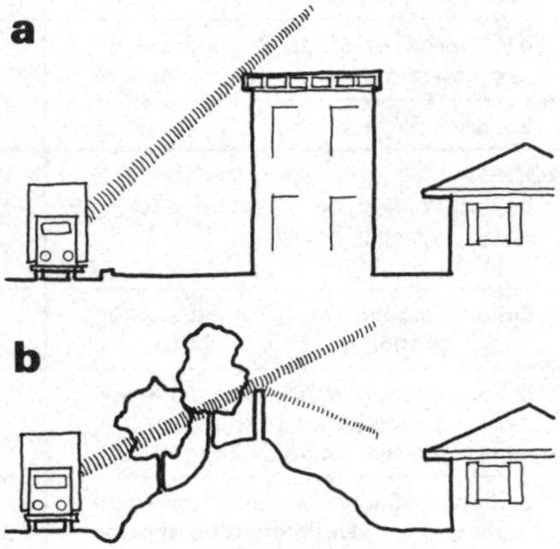

FIG. 53 (a) Nonresidential buildings and (b) natural terrain and man-made road cuts and fills can serve as noise barriers.

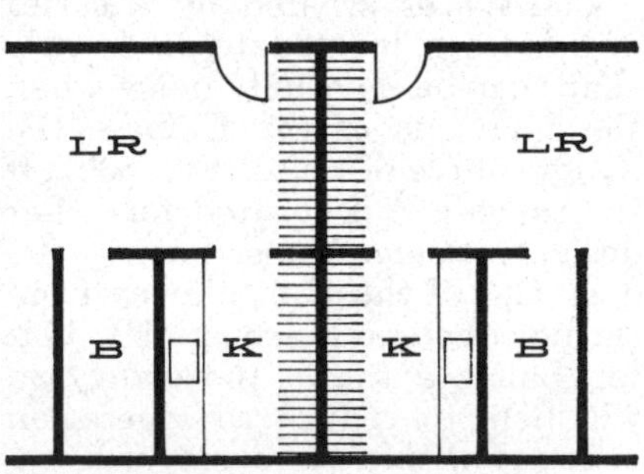

FIG. 54 Buffers such as closets or bookshelves separate the listener from the noise source.

Interior Planning

Careful planning is essential to reducing the effect of noise in both single-family and multifamily residences. Preplanning and design of interior spaces with sound privacy in mind often can minimize or eliminate the need for costly construction techniques to control noise. The problems associated with sound transmission through walls and floor/ceiling systems can be solved by selecting suitable sound insulating systems and demanding good-to-excellent workmanship in the field. These are discussed under Construction Recommendations, page 106-28. However, too little attention often has been given to the principles of sound control in the *planning* of today's houses and apartments. The aim of sound control is to create an environment that is acoustically comfortable, rather than to attempt to render rooms *soundproof*. In single-family and in higher density housing, the basic principles of such planning include: (1) locating rooms or apartment units as far as possible from noise sources; (2) separating noisy areas from quiet areas by isolating the noise source or using buffer spaces; (3) avoiding air paths for the travel of sound between rooms or apartments, especially in the design, size and placement of doors and windows; (4) actually reducing the noise level at the source by selecting mechanical equipment and appliances which operate quietly.

Normally the occupant can control the intensity of noise sources within his own apartment unit or house. Because one usually is more tolerant of noise created by his own family, the problem of sound transmitted between apartment units most often is emphasized in multifamily design. It is almost impossible to correct problems caused by faulty interior design after construction without great expense. Therefore, it is essential that acoustical principles be recognized in the *initial* design stages, whether in single-family or multifamily construction.

The following discussion is concerned primarily with design of apartment living units, but most of the principles are valid for single-family design as well.

Space Layout Residential living spaces within the living unit normally can be zoned acoustically into three general areas: (1) work and play areas that are relatively noisy such as kitchen and family room; (2) study and sleeping areas that should be quiet such as bedrooms; and (3) living areas that fall between the two extremes such as dining room, living room, foyer and hallway.

People and their activities establish acceptable noise levels for different areas. Bedrooms are expected to be quiet and therefore the sound reduction requirements for constructions between bedrooms are higher than the requirements for those between kitchens where noise is more tolerable.

Buffer Spaces such as corridors, stairways or rooms can help to provide acoustical buffers against airborne sound transmission between apartments. Closets and bookcases backed up to common walls help to reduce airborne sound transmission between apartments by forcing furniture to be placed away from the common walls and thus increasing the distance between noise sources in each apartment unit (Fig. 54).

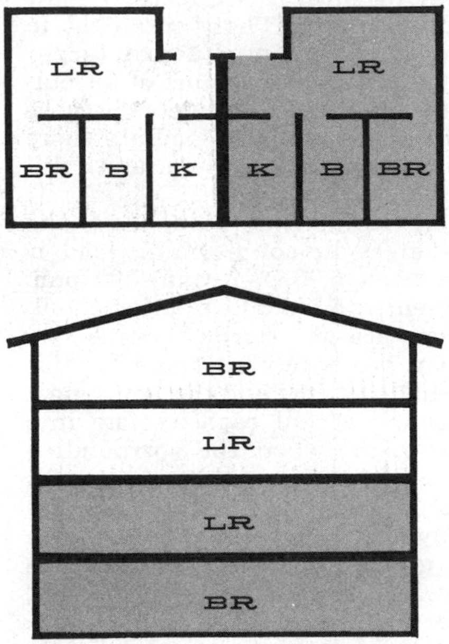

FIG. 55 *Mirroring the plan results in approximately equal noise level requirements on both sides of common walls and floor/ceiling constructions.*

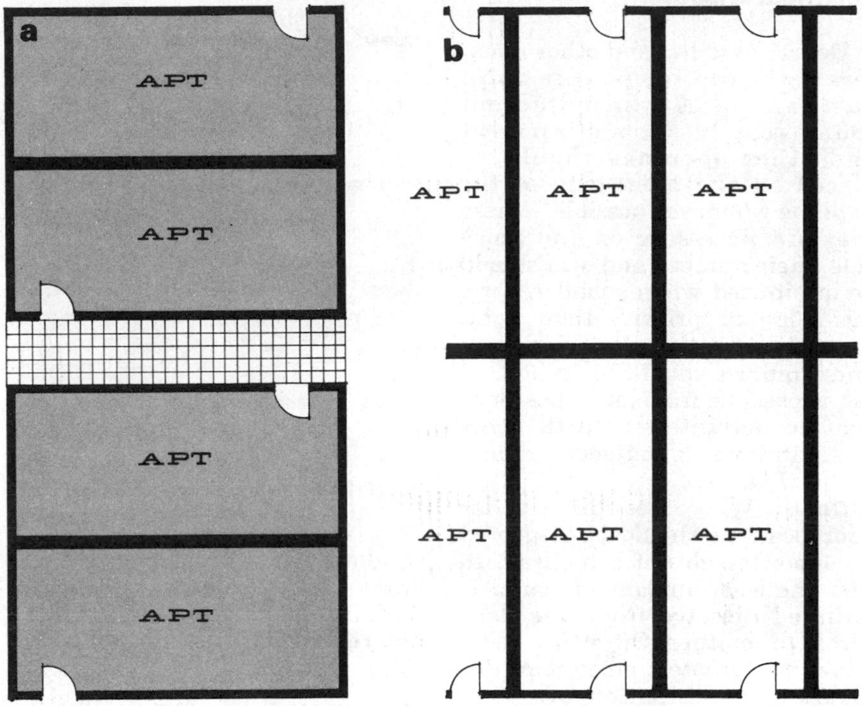

FIG. 57 *(a) Row planning reduces the number of common walls between units. (b) Block planning can result in as many as three common walls for each unit.*

Mirror Plans often are used in apartment design. It is acoustically preferable that the plan be mirrored horizontally and matched vertically where possible (Fig. 55). This type of planning places noisy areas adjacent to other noisy areas and quiet areas adjacent to other quiet areas so that the sound control requirements and therefore the cost of intervening walls and floor/ceilings can be as low as possible. For example, kitchens and bathrooms which have similar noise levels are located adjacent to and above and below kitchens and bathrooms in adjoining apartments. Similarly, quieter areas such as bedrooms and living rooms should be located adjacent to and above and below quiet areas in adjoining apartments. The noise level requirements on both sides of walls and floor/ceiling constructions will be the same, and the full effect and cost of improved sound control measures will benefit all apartments. Some additional precautions that should be observed in mirror planning are illustrated in Figure 56.

Row Plans where apartments are arranged in-line provide better acoustical performance than *block* plans where each apartment may have up to three common walls which must act as sound barriers (Fig. 57). In *row planning,* with halls between every other apartment, each apartment has only one common wall (Fig. 57).

Room Proportions can be planned to reduce sound transmission. The higher the ratio of common wall to floor area, the more sound a common wall will transmit (Fig. 58). And, when rooms have large absorptive surfaces, such as acoustical tile, carpeting or heavy drapes, there is a slight reduction in the amount of sound available that can be transmitted. In a room with large hard-surfaced areas, a large portion of the sound is reflected, whereas large soft-surfaced areas will absorb sound.

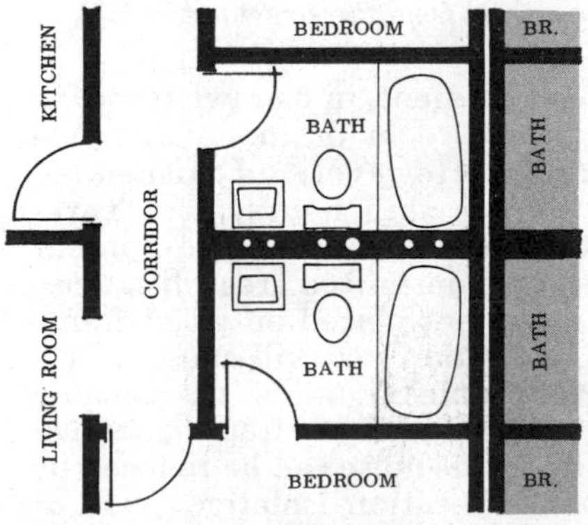

FIG. 56 *Plumbing should be located in walls within each apartment unit rather than in common walls between apartment units.*

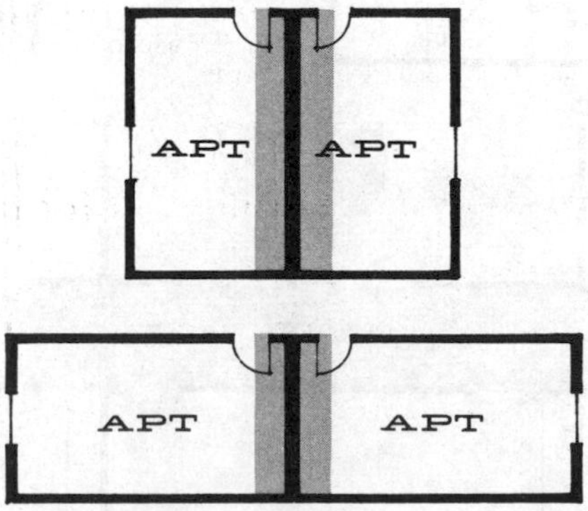

FIG. 58 *Room proportions can reduce the ratio of common wall to floor area. Hard-surfaced areas should be minimized, soft, absorbent areas increased.*

Doors and Windows

Doors, windows and other openings in the construction are direct air-to-air paths which transmit sound readily. Doors, windows and other openings should be placed on the quiet side of the building whenever possible. Where they are necessary on the noisy side, their number and size should be minimized when sound control has a higher priority than light, view or ventilation. Exterior doors and windows should be located as far as possible from, or in positions that do not interfere with doors and windows in adjacent apartments (Fig. 59a).

Doors Where possible, interior door locations should be staggered on opposite sides of a hallway, so that the least amount of sound is radiated directly from one door across to another. Offsetting doors allows the greatest amount of diffusion and dissipation of sound before it reaches adjoining units (Fig. 59b).

Windows Noise is transmitted in both directions from open windows. To minimize noise from one apartment to another, the windows should be placed as far as possible from the common wall. Casement

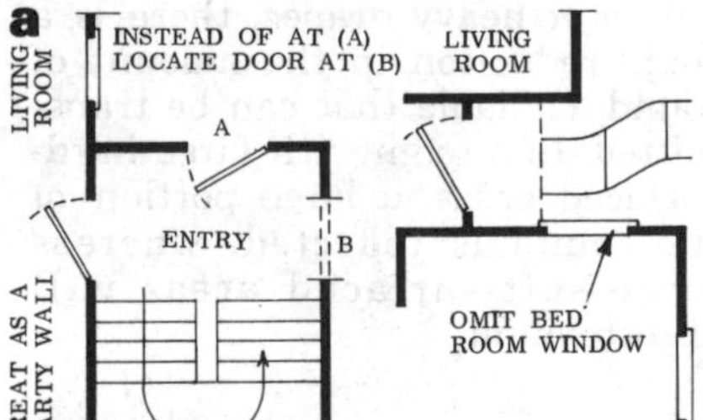

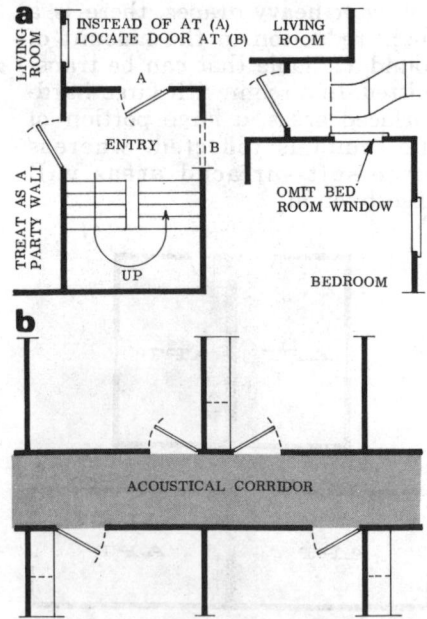

FIG. 59 Placement of doors and windows should not allow sound to travel easily from one unit to another.

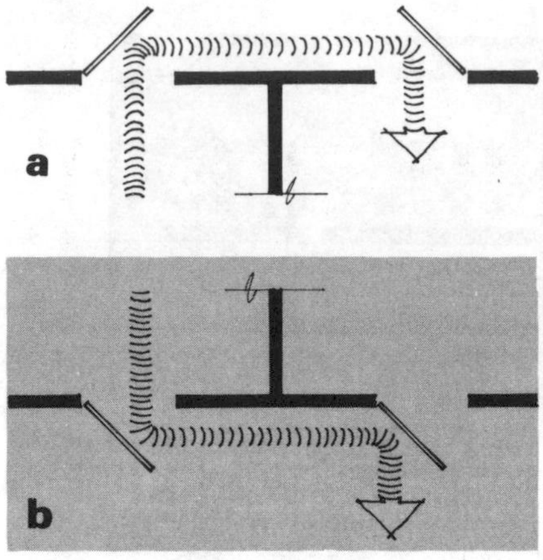

FIG. 60 Casement windows should not reflect sound from one unit to another; (a) improper, (b) proper arrangement.

windows should be arranged to operate in the same direction so that the sound from one room is not reflected into another (Fig. 60).

Mechanical Equipment and Systems

Mechanical equipment and systems produce airborne noise as well as structure-borne noise. While the noise at the source cannot be eliminated, it can be minimized by selection of equipment, design layout and construction precautions.

Heating and Cooling Equipment Motors and motor driven compressors, blowers, pumps and flowing air and liquids are sources of noise vibration—whether they are part of a small individual package unit or part of a large central system serving several apartments. Heating and cooling equipment can have a wide range of noise levels. It is generally more economical to install a quietly operating (even if more expensive) unit than to attempt to reduce the noise level of a less expensive unit by acoustical construction. Often, a better quality unit with more than adequate capacity at slight increase in cost will help to create a quieter installation and will perform more efficiently over the years.

Most types of furnaces for single-family houses or individual apartment units are manufactured for either heating only or for initial or later installation of air conditioning. Those intended to provide cooling have a larger blower for more volume of air output; the furnace is quieter because the larger fan can operate more slowly and still move a larger volume of air.

In residential occupancy, the greatest air conditioning load is between 5:00 p.m. and 7:00 p.m. when exterior and interior conditions are at extremes. Some systems are barely adequate for this peak load time and therefore must operate at full capacity late into the night when the surrounding noise level is greatly reduced and equipment noise is generally annoying. It is recommended that equipment be selected that has greater than adequate capacity, particularly for fans and pumps. Reducing fan speeds only slightly can produce noise level reductions that will more than justify the cost of additional power.

Vibration from mechanical equipment usually is not a significant problem when the equipment is mounted on a concrete slab placed on ground, located

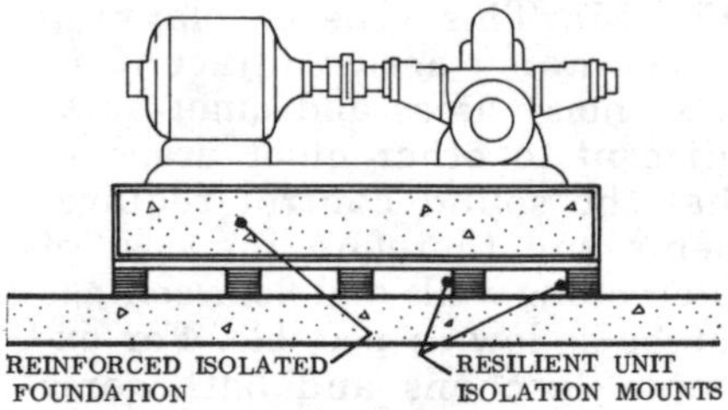

FIG. 61 Mounting mechanical equipment on resilient foundations isolates vibrations from the structure.

in a basement, in a crawl space, in a utility room or in an attached garage. However, on suspended concrete slab floors, structure-borne vibration from equipment can be transmitted great distances. Equipment placed on wood frame construction also will cause acoustical problems unless it is isolated from the structural framing. Structure-borne noise can be reduced by using vibration isolation pads or mounts between the equipment and the supporting floor structure (Fig. 61). Whenever possible, mechanical equipment should be mounted resiliently and all piping, refrigerant lines, condensation lines, electrical connections and duct systems should be flexibly mounted to the unit to prevent transfer of vibrations to walls and floor/ceilings. Flues should be isolated from walls by an air gap, especially where they pass through walls where the room is intended to be quiet.

Complex or heavy equipment installations should be designed by structural and mechanical engineering specialists based on acoustical analysis. Whole machinery rooms can be floated or isolated from the building, or special foundations can be constructed to isolate individual pieces of equipment. Noise levels for heating and cooling capacities of equipment should be obtained. Reliable acoustical data can be obtained for mechanical engineering purposes. Many manufacturers can furnish or develop sound pressure output data based on standard testing for most types of equipment, from which sound pressure levels can be computed for a specific space.

Location of Mechanical Equipment The mechanical equipment room should be buffered from the rest of the building by design. Areas containing furnaces, pumps, compressors, laundry and similar equipment, areas enclosing elevator shafts, incinerator chutes and other potential sources of noise should be located both horizontally and vertically as far as possible from quiet areas such as bedrooms and living rooms.

In single-family and low-rise apartments, an exterior wall location generally is best acoustically (Fig. 62a). Unfortunately, in many homes and apartments, unit forced air heating and cooling equipment often is placed in a closet, an alcove or a utility room near the center of the living unit to produce shorter and therefore more economical duct runs (Fig. 62b). A single air return is located near the center of the floor plan; often a grille in the wall or door of the furnace enclosure is all that is provided. This type of planning

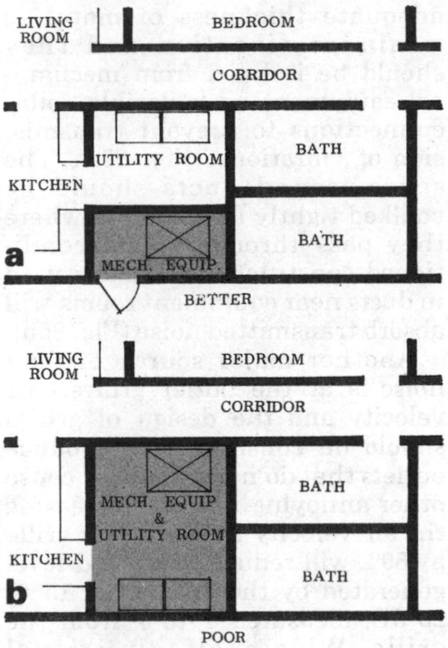

FIG. 62 Acoustically, the mechanical equipment room should be located on an exterior wall (a), rather than in a central location (b).

places major noise sources in the center of the living space.

One way to confine the noise of a mechanical equipment room is to provide three solid walls around the furnace room with the entry door to the outside for servicing (Fig. 62a). If the service door is installed in an interior location, it should be a solid core door and should be gasketed around the perimeter to render the door airtight (see page 106-39).

Mechanical systems also are a source of exterior noise. While a cooling coil is located inside at the furnace plenum, the location of the water or air-cooled condensing unit must be located outside. Because ordinances sometimes govern the allowable noise level at property lines, condensers often must be located adjacent to the residence in front or rear locations where the distance is greatest to the next building. This creates an exterior noise source at the exterior wall. An alternate location for the condensing unit would be on a vibration-isolated platform on the roof (Fig. 63). This would keep the unit away from windows, porches, patios or property lines.

Unit air conditioners installed in exterior walls in multifamily residences should be located as far as possible from adjacent apartments, as operation of one unit may disturb neighbors who may not have their units in operation at the same time.

Duct Systems Careful planning and installation of duct systems is critical in controlling sound transmission. Metal ducts act like speaking tubes in transmitting noises. Noise may be: (1) from equipment such as fans and motors; (2) from sound that is transmitted through thin walls into the duct and then conducted through the duct; or (3) created within the duct by air rushing through the system or by expansion and contraction of metal ducts. The sound does not necessarily have to enter or leave a metal duct through an opening or ventilator but can be transmitted through walls, particularly where the ducts pass vertically through living spaces (Fig. 64). Planning can minimize the sound transmitted from one room to another through the open air paths of the duct system.

Locating duct outlets back-to-back, as is practiced in return air systems, practically eliminates any benefit obtained from sound

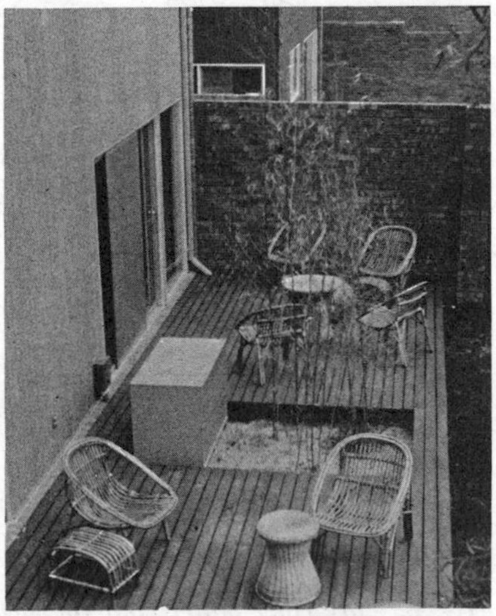

FIG. 63 An exterior condenser should not be placed so that loud noise is created near windows, doors or outdoor living areas, as shown above.

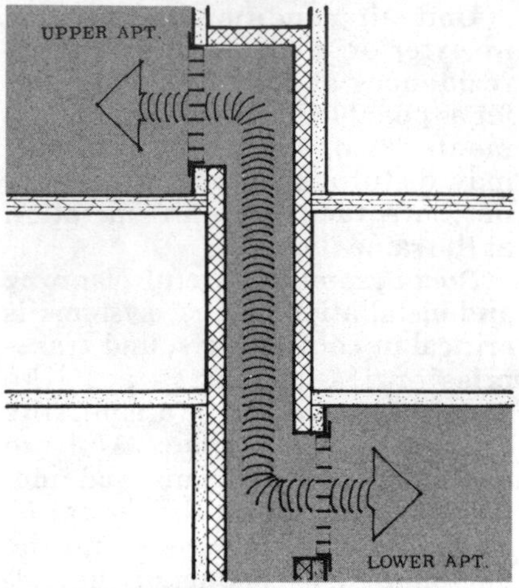

FIG. 64 Duct systems can conduct noise between units. Ducts should not be housed in common walls or floor/ceilings between apartment units.

conditioned walls (Fig. 65a). A 6′ length of fibrous duct material adjacent to the outlet generally will provide reduction of furnace noises and will prevent short circuiting sound through ductwork between rooms (Fig. 65b).

The ducts should be formed of adequate thickness of metal to minimize vibration and they should be isolated from mechanical equipment with flexible collar connections to prevent transmission of vibrations (Fig. 65c). The space around ducts should be caulked tightly for isolation where they pass through sound conditioned construction. Mufflers used in ducts near equipment rooms will absorb transmitted noise (Fig. 65d).

Another major source of duct noise is at the outlet grilles. Air velocity and the design of grilles should be considered to produce outlets that do not whistle or cause other annoying noises. Decreasing the air velocity at the outlet grille by 50% will reduce the sound level generated by the grille from 15 to 20 dB, measured 3′ to 5′ from the grille. Wide angle directional grilles generate the most noise, while wire mesh or perforated metal grilles with little air spread capacity appear to be the quietest.

If a door is to provide acoustical privacy, it cannot be used as a means of returning air to the furnace. Separate air return registers can be placed in each room, connected to a sound absorbing duct system. This eliminates blower and furnace noises and reduces cross-talk from room to room. Sound absorbing air inlets can be used to allow air intake from corridors without admitting noise.

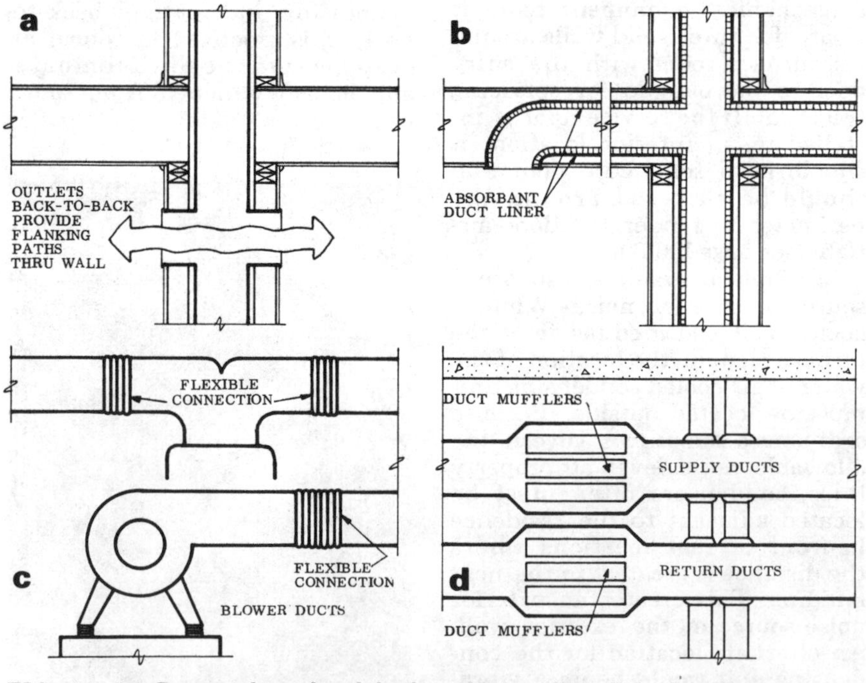

FIG. 65 (a) Duct outlets placed back to back will nullify the effect of sound isolating walls. (b) Fibrous insulation in the duct, particularly near outlets, helps to reduce the noise. (c) Flexible connections reduce the transmission of vibrations from equipment to the duct system. (d) Duct mufflers are effective in reducing noise transmitted from the furnace.

Piping High water pressure and velocity and quick closing valves in supply piping systems combine to produce plumbing noises. If the dwelling does not exceed three stories and the water pressure at the street exceeds 35 to 45 psi, a pressure reducing valve can be installed, preferably as far

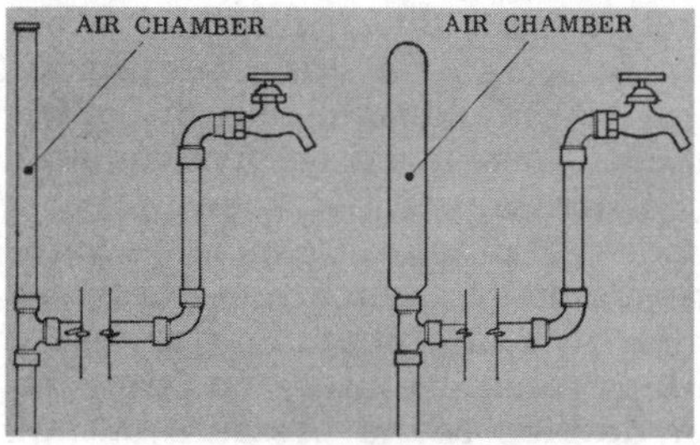

FIG. 66 Air in the chamber, acting as a shock absorber, dissipates pressure surges caused by quick closing valves, and thereby reduces water hammer.

from the building as feasible. In high-rise apartments, combinations of pressure reducing valves and pressure regulating systems may be necessary to provide suitable and uniform pressure levels at the various floor levels. Quick closing faucet valves shut off the water flow quickly causing the pipe to vibrate and may produce what is called *water hammer* (Fig. 66).

Noise resulting from water rushing through piping can be minimized by planning the system with a minimum of bends and by reducing the number of fittings and valves to a minimum. Valves and right angle fittings produce substantially more noise than straight runs of pipe due to water turbulence. The noise produced by 4 elbows is about 10 dB higher than the noise produced in a straight run of equivalent size pipe under conditions of turbulent flow.

Long runs of hot water supply piping produce creaking noises as the piping expands or contracts from hot or cooled water. Long,

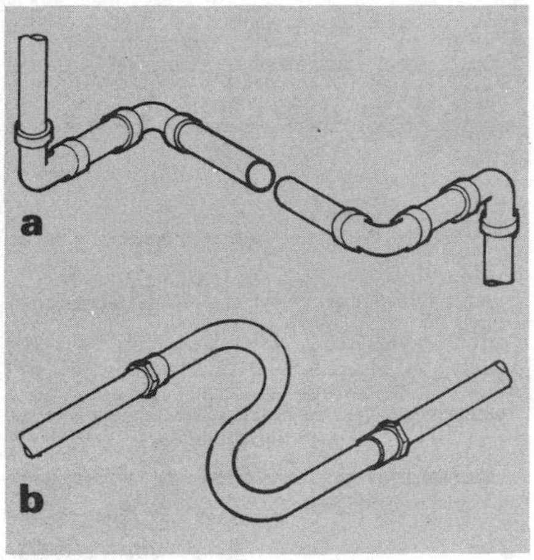

FIG. 67 Flexible connections, using (a) swing arms in piping or (b) S-curves in tubing, minimize noises caused by expansion and contraction in long runs of piping.

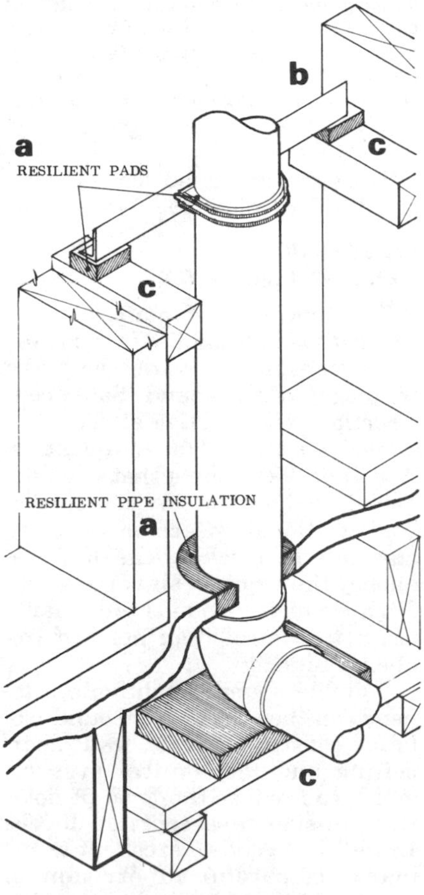

FIG. 68 Vibrations from a plumbing stack can be isolated from the structure by resiliently mounting vertical supports and resiliently supporting horizontal runs.

straight runs of iron or steel piping should be designed with flexibility with at least one end having a swing arm to permit movement (Fig. 67a). Flexible L or M piping bent into an S-shaped curve at one end will allow for movement in long runs of copper piping (Fig. 67b). Supports should be designed to permit the piping to expand and contract without binding.

Supply piping to units which use large amounts of water such as dishwashers and water closets should not be undersized.

When pipes are rigidly connected to structures with hangers, clamps or straps, sound resulting from pipe vibration can be transmitted through the structure and can be amplified from one apartment unit to another. *Resilient* pipe insulation between the pipe supports and the pipe should be used to help reduce the transmission of these noises (Fig. 68a). The number of pipe supports should be kept to the minimum required to structurally support the pipe, and supports should be attached to the most massive structural element available.

Running drain piping through bedroom or other quiet living area walls should be avoided. Waste dropping down vertical stacks causes piping to vibrate, especially when it hits horizontal runs, unless the pipe is *resiliently* mounted. When vertical stacks are hung on floor joists, vibrations are transmitted to the structure (Fig. 68b). Standard pipe clamps and a resilient vertical support or wood bracket should be installed with the end of the support resting on the vertical stack with a 1″ thick isolation pad resting on the support (Fig. 68c). A short section of pipe insulation around the pipe where it passes through wall sills will keep it in horizontal alignment and will isolate it from the structure.

It is essential that all holes and openings where piping passes through common walls be sealed with a resilient material or should be tightly packed with insulation, or both, to isolate vibration of the pipe from the structural framing

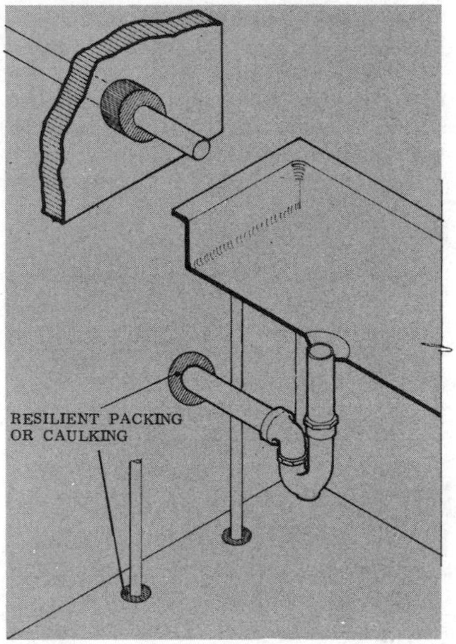

FIG. 69 Where pipes pass through walls, openings must be tightly packed or sealed airtight with resilient materials to isolate vibrations and to close paths for airborne sound transmission.

and to seal against air leaks (Fig. 69).

Fixtures and Appliances Plumbing fixtures and appliances often can be selected and installed to reduce the amount of noise. For example, a siphon jet water closet having an adjustable-rate water supply inlet float valve will operate more quietly. The water closet can be mounted on a floating floor and the noise transmission from fixtures can be reduced by using a neoprene or rubber gasket between the base of the water closet and the floor (Fig. 70), and/or by a re-

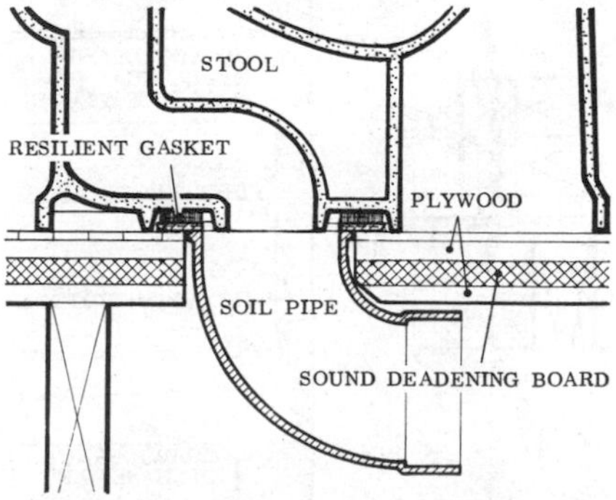

FIG. 70 Water closet vibrations can be minimized by resilient mounts and floated floor construction.

silient underlayment between two layers of floor sheathing. Flexible caulking or tightly packed insulation should be used to seal the space between the toilet bend and the floor sheathing.

Appliances should be selected

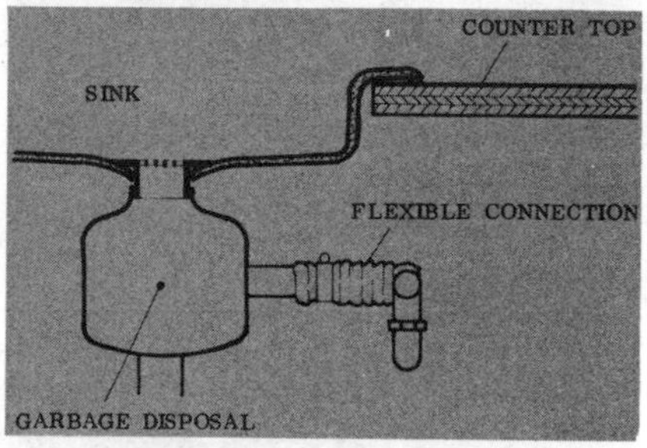

FIG. 71 Flexible piping connections prevent equipment vibrations from being transmitted to the piping.

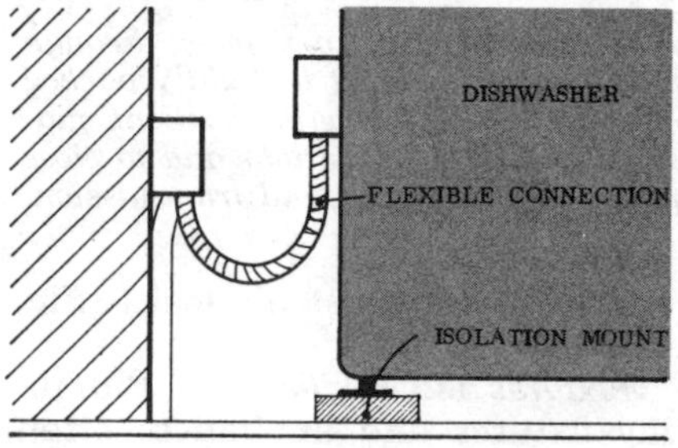

FIG. 72 Wiring in rigid conduit can be connected to vibrating equipment with flexible cable.

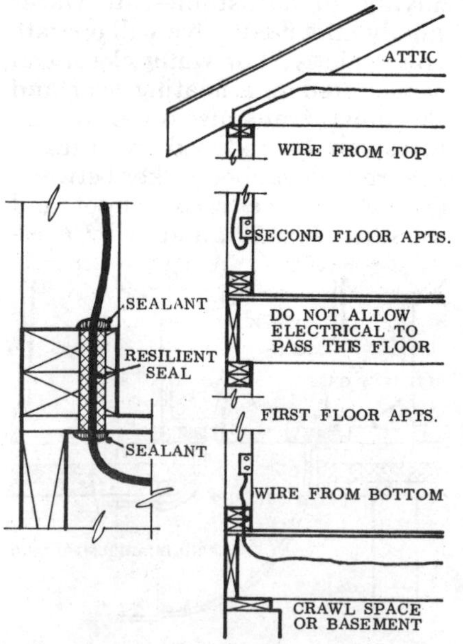

FIG. 73 Each apartment should be wired independently and all air passages should be sealed airtight.

for quiet operation. Vibrating appliances such as dishwashers, disposers, washers and dryers should not be rigidly connected to the structure. Resilient or flexible isolation gaskets should be provided between the appliance and the floor, wall or counter to isolate vibrations, unless a floated floor system is used. Piping connections to appliances should be flexible enough to prevent vibrations from being transmitted to the piping (Fig. 71). Electrical connections to appliances also should be flexible, whenever possible; a loop as shown in Figure 72 will reduce vibration transmission.

Electrical Systems Improper electrical installations can reduce the acoustical performance of floor, ceiling and wall constructions. Causes for reduced performance are: (1) open air transmission paths where wires pass through walls, floors and/or wall plates; (2) openings provided for outlets, switches or fixtures; and (3) noise vibration transmitted by wires or fixtures rigidly connecting two otherwise separated structural member, or connecting equipment to the structure.

In multifamily housing, each apartment should be wired as a unit, keeping wiring within the walls of the same unit. *All holes* should be *tightly caulked* with resilient, nonhardening compound (Fig. 73).

Convenience outlets, switches, telephone and other such installations as intercoms should not be placed back-to-back: (1) convenience outlets should be placed at least 36″ apart on the opposite sides of the wall. In staggered stud walls, installations should be offset horizontally by at least three complete stud spaces (Fig. 74a). (2) A horizontal distance of 24″ should be maintained between switches and outlets, and they should not be located in the same stud space (Fig. 74b). (3) Wall fixtures and equipment should be kept 24″ apart on opposite sides of the wall (Fig. 74c).

Lighting Fixtures and other electrical fixtures should be surface-mounted (never recessed) and they should not be connected to joists above resiliently mounted ceiling

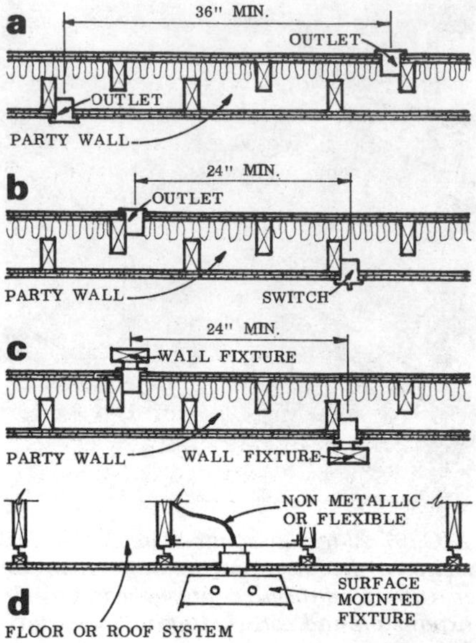

FIG. 74 (a, b & c) Electrical outlets, switches, fixtures and equipment should not be mounted back to back, should not be located in the same stud space and should not be spaced closer than shown above. (d) Ceiling fixtures should be surface-mounted, rather than recessed.

finishes or connected across separated ceiling joists (Fig. 74d).

CONSTRUCTION RECOMMENDATIONS

The efficiency of materials and construction assemblies in reducing noise transmission varies with the frequency of the sound. Some constructions are effective at high frequencies, others at low frequencies. A wall or floor/ceiling that will adequately reduce the transmission of conversational speech may be inadequate in blocking the low frequency thump of an air compressor. High frequency sounds are usually easier to control than are low frequency sounds.

Suitable sound conditioning depends on the selection of sound isolating constructions for wall, floor/ceiling and mechanical systems which can reduce the level of noise transmission to acceptable levels. In addition, good-to-excellent workmanship, careful supervision of construction and coordination between the various trades are *essential* to achieving effective sound control. Slight variations from installation instructions, which may seem insignificant, can nullify the effectiveness of costly sound control measures—nails which are too long that short circuit resilient mountings and failure to seal small cracks or holes are examples of common mistakes. Manufacturer's instructions specifying sound control installation procedures should be followed precisely by workmen. If possible, the workmen and other key personnel should be taught the reasons for care in applying sound control measures. Assuming that a partition selected on the basis of test data will perform adequately, obtaining suitable workmanship in the field is the greatest single problem in achieving satisfactory sound conditioned houses or apartments. Quality workmanship is absolutely essential in the successful application of sound control principles.

Wall Systems

Sound isolating wall systems may be constructed of wood or metal framing, concrete or unit masonry. Any wall system intended to control sound transmission should be built properly to provide maximum acoustical performance. Regardless of the system selected, it is essential that the surfacing materials be installed over the entire wall area—especially behind tubs, soffits, cabinets and duct enclosures where the structural framing might otherwise be left exposed. This precaution is particularly critical for common walls between living units to assure that sound will not be transmitted through an unfinished space (Fig. 75). Resilient, nonhardening caulking must be used at the floor and ceiling of all partitions and around all perforations such as electrical outlets, plumbing and other openings to make the wall airtight (Fig. 76).

The majority of recent sound transmission test data for wood and metal frame construction has been obtained with gypsum board as the surfacing material because of its extensive use and ease of installation in the laboratory, without the necessity of drying and curing. The following discussion is based primarily on evaluation of such testing, but the principles and conclusions evolved are applicable to lath and plaster and other surfacing materials of comparable weight applied to similar constructions. Representative STC ratings for lath and plaster, as well as other typical constructions, are given in the Work File.

Wood Framing Systems Resistance to sound transmission through wood frame walls can be improved by the following measures: (1) increasing the weight of the wall and the surfacing material, (2) including sound absorbing blankets in the free air space within the wall, (3) resilient mounting of the surfacing materials, (4) using staggered framing members to effectively separate the finished wall surfaces, and (5) using two separate, independent stud walls separated by an air space.

Conventional Single Stud Walls, finished with one layer of gypsum board on each side (Fig. 77), offer

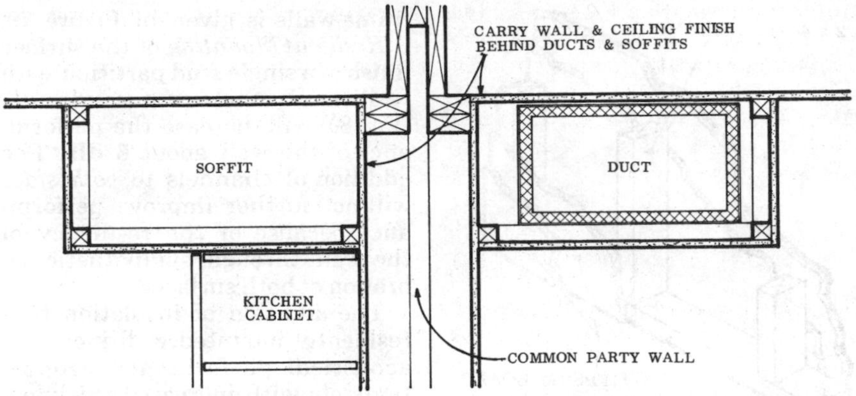

FIG. 75 Surface finishes should cover all of common wall and ceiling areas—behind ducts, soffits, stairways and above suspended ceilings—to prevent sound transmission through unfinished spaces.

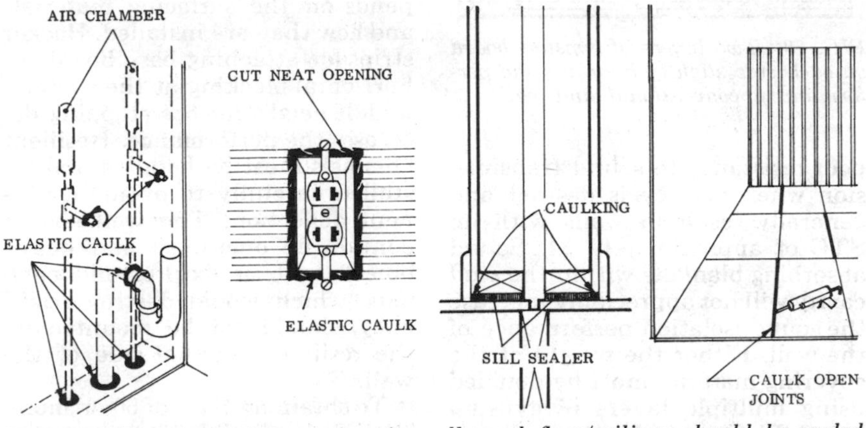

FIG. 76 All construction openings in walls and floor/ceilings should be sealed with nonhardening, resilient caulking or resilient material.

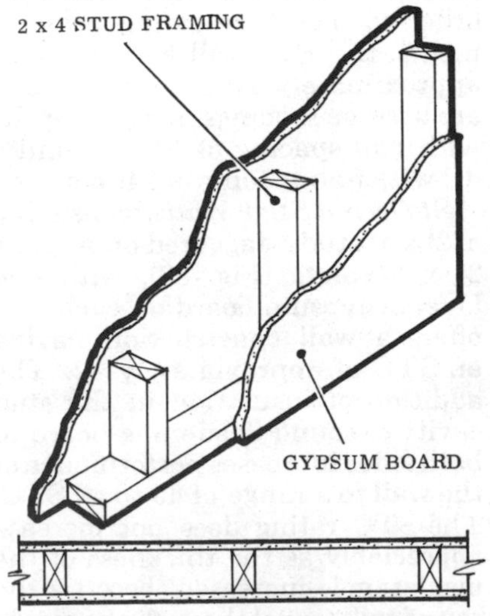

FIG. 77 Conventional stud walls permit loud speech and other noises to be transmitted from one unit to another.

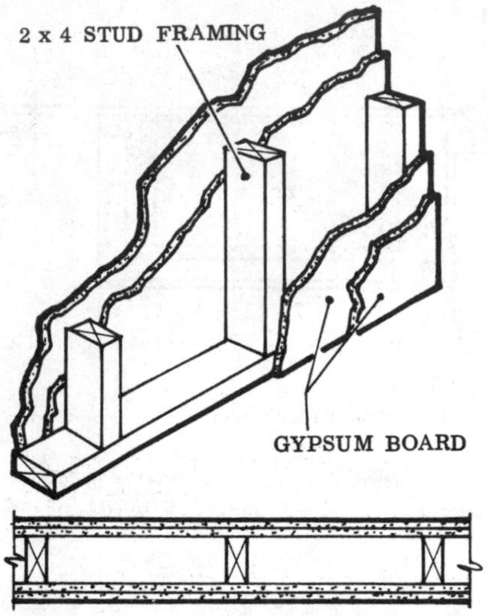

FIG. 78 Two layers of gypsum board on each side slightly improves the performance of conventional stud walls.

poor resistance to sound transmission when privacy is desired and generally result in walls with an STC of approximately 34. Sound absorbing blankets within the stud cavity will not appreciably increase the sound isolation performance of the wall. Either the weight of the covering material must be doubled using multiple layers of gypsum board (Fig. 78), or the two wall surfaces must be mechanically separated to achieve an STC rating of more than 40. A summary of the acoustical performance of wood frame walls is given in Figure 79.

Resilient Mounting of the surface finish of a single stud partition with resilient channels *on one side only* (Fig. 80) will increase the performance of the wall about 6 dB. The addition of channels to *both sides* will not further improve performance because of the resiliency of the wall through sympathetic vibration of both surfaces.

The addition of insulation to a resiliently mounted wall increases acoustical performance proportionately with increased thickness of the insulation and can result in as much as a 10 dB improvement. Noise isolation performance depends on the surfacing materials and how they are installed. Backer strips for attaching base boards or horizontal blocking at the top and middle height for fire stopping decreases the performance. Resilient channels must be handled and installed carefully to avoid permanent deflection. The wall should contain no openings, and heavy objects such as fixtures and cabinets (which would destroy resiliency) should not be mounted on the resilient channel side of the wall.

To obtain an STC of 50 or more, without regard for fire ratings, channels must be used throughout, in place of gypsum backing strips, to totally float the surfacing materials and resilient caulking used on all edges.

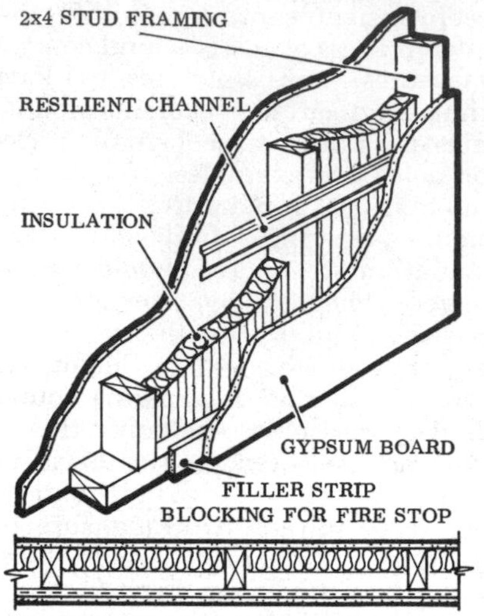

FIG. 80 Resilient mounting of gypsum board finish on one side improves transmission resistance of conventional wall.

A wall with resilient channels on one side generally results in the thinnest sound isolating system in wood frame construction and can be constructed with proper blocking to give a one hour fire rating.

Sound deadening board applied to both sides of a single stud wall will provide resilient mounting if the surfacing materials are *correctly laminated* to the sound deadening board (Fig. 81), and can result in a wall with an STC of approximately 46. However, if the surfacing materials are nailed in order to meet fire rating requirements, the STC will be reduced—approximately 10 dB when nails are used at spacings of 8″ o.c., 8 dB with nail spacing of 16″ o.c. and 4 dB with nail spacing of 24″ o.c.

Staggered Stud Walls, consisting of 2″ x 4″ studs staggered on a *single* 2″ x 6″ plate (Fig. 82), with one layer of gypsum board on each side offers a wall construction having an STC of approximately 42. The addition of insulation in the stud cavity or sound deadening board to both sides increases performance of the wall to a range of 45 to 49 STC. The STC rating does not increase appreciably as the thickness of the insulation is increased, because the sound will travel through the single plate. Therefore, a minimum thickness of 2″ or 3″ of mineral fiber

FIG. 79 ACOUSTICAL PERFORMANCE OF TYPICAL WOOD FRAME WALLS[1]

Wall Construction	Approximate STC Rating		
	Without Insulation	With Insulation[2]	Sound Deadening Board Both Sides[3]
Single Stud Wall Single layer surfaces	34	36	46
Double layer surfaces	40	—	—
Single Stud Wall with resilient channel mounting one side	40	50[4]	—
Staggered Stud Wall	41	49	50
Double Stud Wall	43	53	50

1. As illustrated in Figs. 77, 78, 80, 81 & 82.
2. Minimum thickness of insulation (2″ or 3″).
3. Gypsum board surfaces laminated, with exception of double stud wall which was laminated and nailed 8″ o.c. into studs and floor and ceiling plates.
4. Resilient channels installed throughout; fire-rated wall with fire stopping top and middle heights or backer strips for attaching wall base, STC 46.

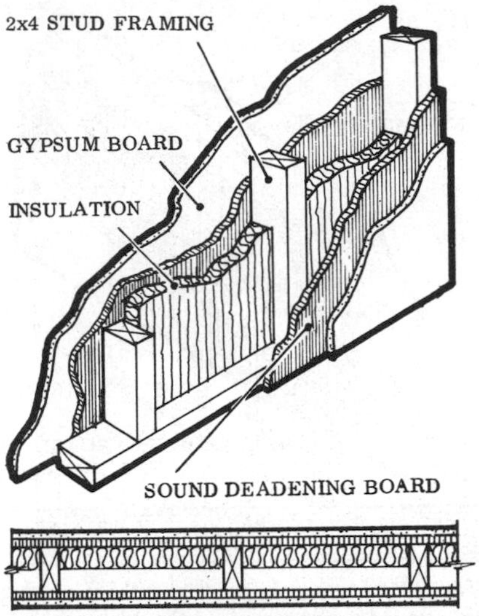

FIG. 81 Sound deadening board also provides resilient mounting and improves performance; surfaces should be laminated to provide best performance.

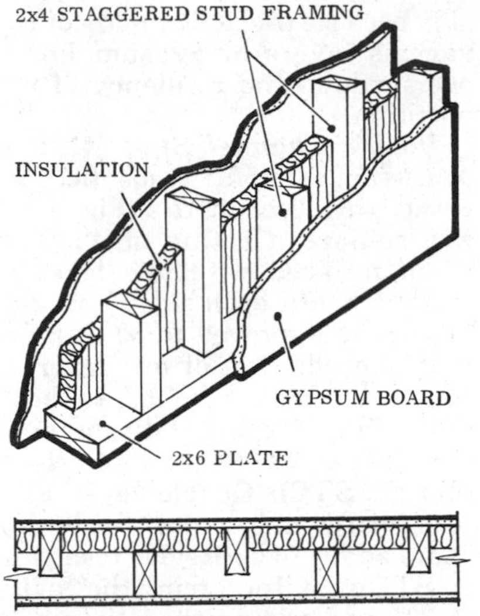

FIG. 82 Staggered studs on a single plate with insulation provides good sound isolation and cabinets can be mounted without reducing performance.

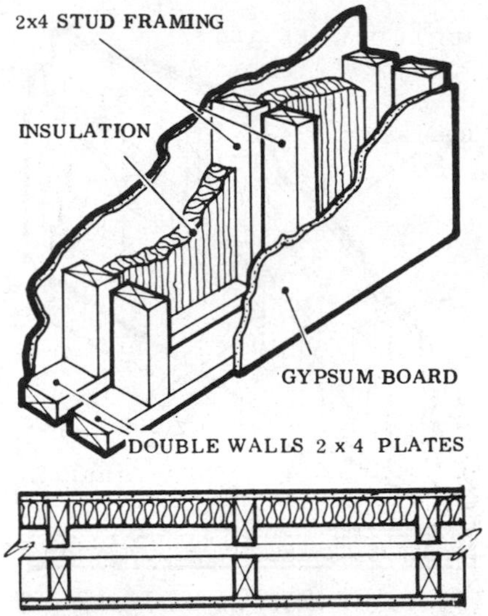

FIG. 83 Separate (double) stud walls offer most versatile sound isolating walls, less affected by variations in materials and workmanship than others.

blanket insulation within the cavity can be used effectively.

An advantage of a staggered stud construction is that cabinets can be surface-mounted on the wall framing without impairing its acoustical effectiveness, because they are mounted on independent framing members.

Double Stud Walls consist of two separate 2″ x 4″ stud walls on *two separate* 2″ x 4″ or 2″ x 3″ plates spaced 1/4″ to 1″ apart (Fig. 83). When compared with the conventional single stud wall, the double stud wall will exhibit an increased performance of about 9 dB (43 STC). The isolation performance of the double stud wall with separate plates improves with increased thickness of the insulation. Ratings as high as 59 STC can be achieved. The addition of sound deadening board to both sides of a double stud wall will result in an STC of approximately 50.

The double stud wall can accommodate surface mounting of cabinets without reduced effectiveness and less performance variations due to materials and workmanship result with this type of construction. However, to perform acoustically, the "separate walls" should not be tied together in any manner when plumbing or wiring is installed.

Metal Framing Systems Because of the resilient nature of metal channel studs themselves, the two surfaces of the wall vibrate independent of each other and may be considered to be resiliently mounted. Therefore, the acoustical performance of nonload-bearing metal stud walls differs from that of structural load-bearing stud walls. For example, the addition of resilient channels to mount surface finishes, used effectively for wood frame walls, does not appreciably improve the performance of metal stud walls.

The difference between the acoustical performance of walls constructed with 3-5/8″ metal studs and 2-1/2″ metal studs, or using 1/2″ and 5/8″ gypsum board, is considered negligible for practical purposes. A summary of the acoustical performance of metal channel stud walls is given in Figure 84.

Single Metal Channel Stud Walls with single layer surfaces achieve an STC of approximately 37. By

FIG. 84 **ACOUSTICAL PERFORMANCE OF TYPICAL METAL CHANNEL STUD WALLS**[1]

Gypsum Board Surfacing	Approximate STC Rating					
			Sound Deadening Board One Side		Sound Deadening Board Two Sides	
	Without Insulation	With Insulation[2]	Without Insulation	With Insulation[2]	Without Insulation	With Insulation[2]
Single layer each side	38	45	43	49	47	54
Double layer one side, single layer one side	44	50	46	51	52	54
Double layer each side	49	57	53	57	54	57

1. As illustrated in Figs. 85 and 86.
2. Minimum thickness of insulation (2″ or 3″).

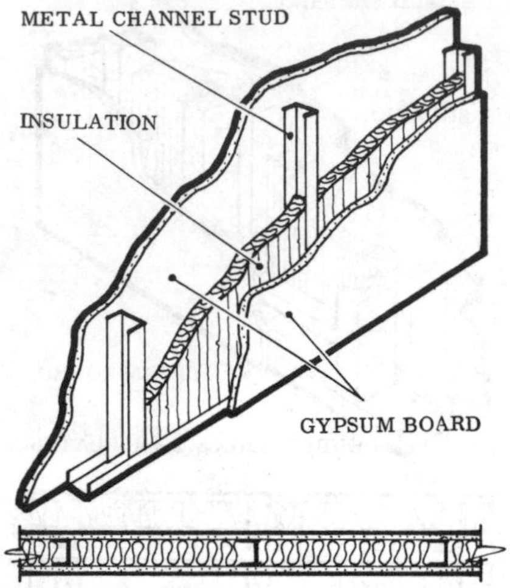

FIG. 85 A metal channel stud wall acts as a resilient system without the use of resilient mountings.

adding a minimum thickness of 2″ to 3″ of insulation within the cavity of a wall with single layer gypsum board surfaces each side (Fig. 85), the acoustical performance can be improved about 9 dB, bringing the wall's STC to about 46. The addition of one layer of mineral fiber sound deadening board to one side of the wall will improve the STC rating to 5 to 6 dB, bringing the STC to 43 without insulation within the cavity, and 49 with insulation. By adding sound deadening board to both sides, plus the addition of insulation in the cavity (Fig. 86), an STC of 52 to 54 can result.

Additional layers of surfacing material to either one or both sides will increase the STC rating of the wall 5 to 6 dB for each layer added. The addition of insulation to a wall with two layers of gypsum board on one side will improve performance an additional 6 to 8 dB. Sound deadening board will perform the same as adding a single layer of gypsum board with an increase of approximately 5 dB for each layer (one or both sides).

The effects of adding gypsum board, insulation and sound deadening board in metal stud construction are cumulative (Fig. 84). However, the improvement is not as great with the addition of sound deadening board to both sides when double layer gypsum board is added to both sides. This is probably because the added mass of the various layers of gypsum board overshadows the resiliency of the partition.

Double Channel Stud Walls—consisting of 1-5/8″ studs tied together with bracing ties (Fig. 87)—will achieve a rating of 42 STC when surfaced with one layer of gypsum board each side. This rating can be improved to 50 or more with the addition of one layer of sound deadening board to both sides (STC 50) or by adding a minimum thickness of 2″ to 3″ of insulation (52 STC). Double layer gypsum board and sound deadening board added to both sides results in an STC of 53. Increasing the cavity to 12″ and adding three layers of insulation (9″) to a wall with single layer gypsum board each side can result in an STC of 55; another layer of gypsum board added to one or both sides of this construction can result in a 59 or 60 STC.

Concrete and Unit Masonry The sound isolating performance of walls with sealed surface pores is closely correlated with the weight of the wall and performance of such walls parallels what is known as

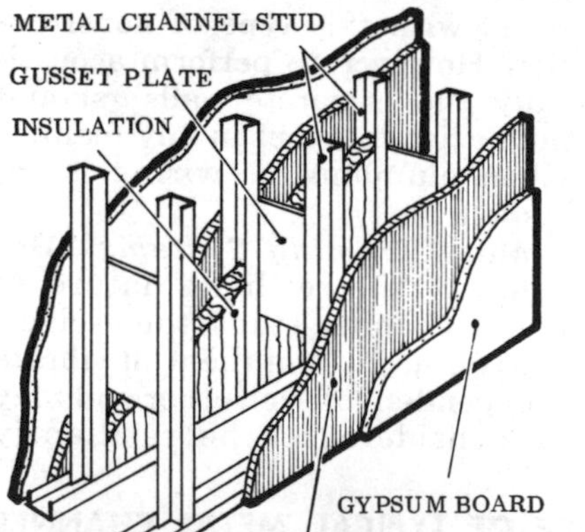

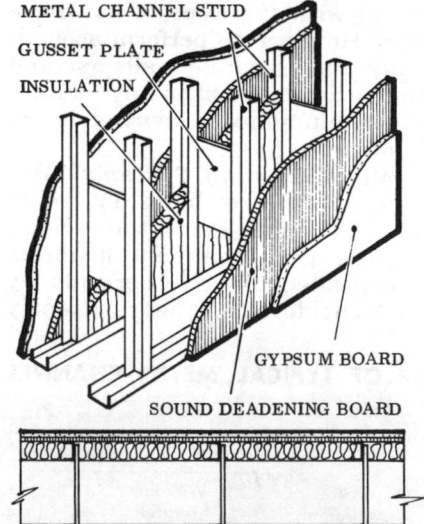

FIG. 87 With increased cavity depth, insulation thickness, and/or sound deadening board, double channel stud (chase) walls can achieve excellent resistance to sound transmission, and mechanical lines can be included within the wall cavity.

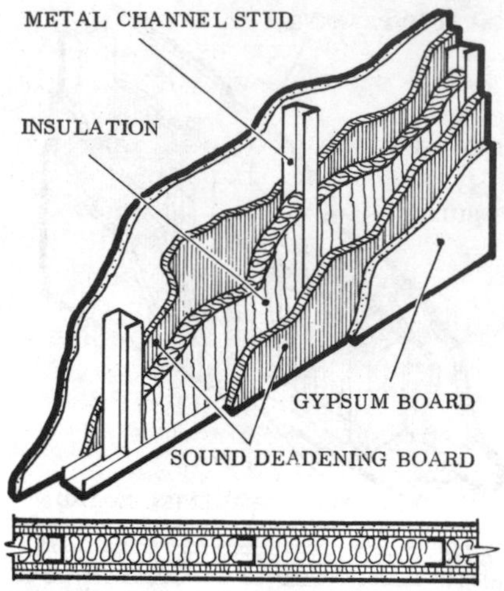

FIG. 86 Metal channel stud walls are measurably improved with addition of insulation, sound deadening board and added layers of gypsum board.

the *limp mass law relationship.* The mass law relationship holds true only for limp *homogeneous* partitions that are *nonporous* and have uniform physical properties throughout the entire wall panel. This relationship between transmission loss and mass can be expressed by the formula:

$$TL = 20 \log m + 20 \log f - 33$$

where m = the mass of the wall stated in lbs. per sq. ft. and f is the frequency in cycles per second.

Heavier walls tend to reduce the transmission of sound more than lightweight walls. However, doubling the mass of a wall increases the transmission loss only 6 dB. As can be seen in Figure 88, to achieve transmission loss of greater magnitude, the increase in mass must be comparatively great.

When concrete and masonry walls have a very tight pore structure (are nonporous) they tend to transmit sound by acting as a diaphragm, literally not permitting sound to pass through them. When the wall surface has a porous nature, the pore structure is the principal transmission path, particularly at low frequencies. If the porous surface of the wall is thoroughly sealed with paint or plaster, transmission loss does not depend on pore structure but is determined by the weight and stiffness of the wall. Therefore, acoustical performance of a painted or plastered porous wall will be essentially the same as a nonporous wall of the same weight and stiffness.

When concrete and masonry walls are nonporous so that sound transmission through them is due to vibration rather than the passage of airborne sound through them, their STC can be predicted from the weight of the wall, providing there are no openings in the wall and all joints and cracks are sealed. Figure 88 shows that STC test values fall approximately 5 to 6 dB below STC values calculated according to the mass law formula.

A representative number of tests on masonry walls have indicated that STC values are comparable to some TL (average transmission loss) values derived from earlier testing, because the transmission loss curve for such walls is relatively uniform without major deviations. STC values can be *estimated* from TL values of earlier tests (Fig. 89) with some adjustment made to compensate for variations in conditions under which the tests were performed. When compared to 1966 test results, ratings on tests performed prior to 1957 appear to be from 7 to 10 dB too high; ratings on tests performed from 1957 to 1966 appear to be from 2 to 4 dB too high.

Concrete Masonry is produced in hollow or solid units manufactured from either lightweight aggregate or normal weight aggregate (sand-gravel). The normal weight aggregate (dense) units have a relatively tight pore structure, are considerably heavier than lightweight aggregate units and will perform acoustically according to their mass.

Lightweight concrete block typically contain a larger volume of pores (air spaces between aggregate particles and cement matrix) than the dense units and for this reason may excel in sound absorption but are much less effective in resisting sound *transmission*. Lightweight block of various manufacturers are subject to a wide variation in transmission loss when not sealed, because of large differences in pore structure. Thorough sealing of the surface pores increases the performance of lightweight units about 7 dB when painted, and about 10 dB when plastered (Fig. 90). When dense units are either painted or plastered, the increase is only about 2 dB. However, because of their greater mass, dense units perform considerably better than lightweight units. Cavity wall construction and combinations of dense and porous units offer improved acoustical performance and can provide STCs of 50 or better (Fig. 90). Using a combination of two 4″ lightweight wythes with 3/8″ or more air space between and one wythe plastered on the cavity side, the faces of the wall need not be sealed and the sound absorbing porosity of the lightweight units can be retained.

Filling the core space of hollow units with sand or other aggregate does not appreciably increase the acoustical performance of the wall and will result in only a 1 or 2 dB improvement, when the wall is built properly. Filling the core spaces will improve the performance of a poorly built wall, but it will never be as good as one properly constructed.

Tests have indicated that hollow units may be better sound insulators than solid units of the same

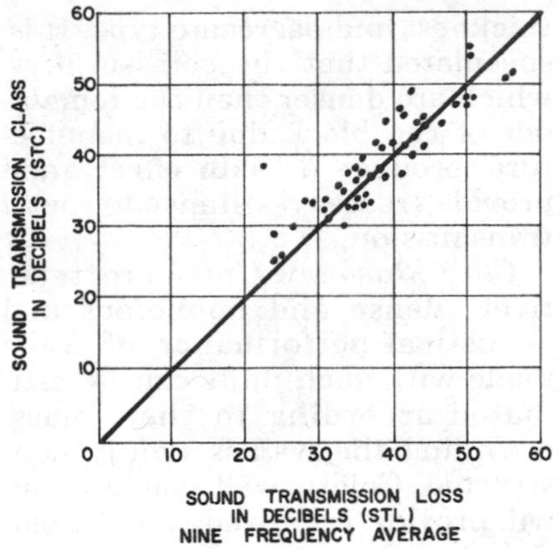

FIG. 89 STC values can be estimated from earlier Average Transmission Loss (TL) test values. Shown above, the relationship between STC and TL for 55 concrete masonry partitions.

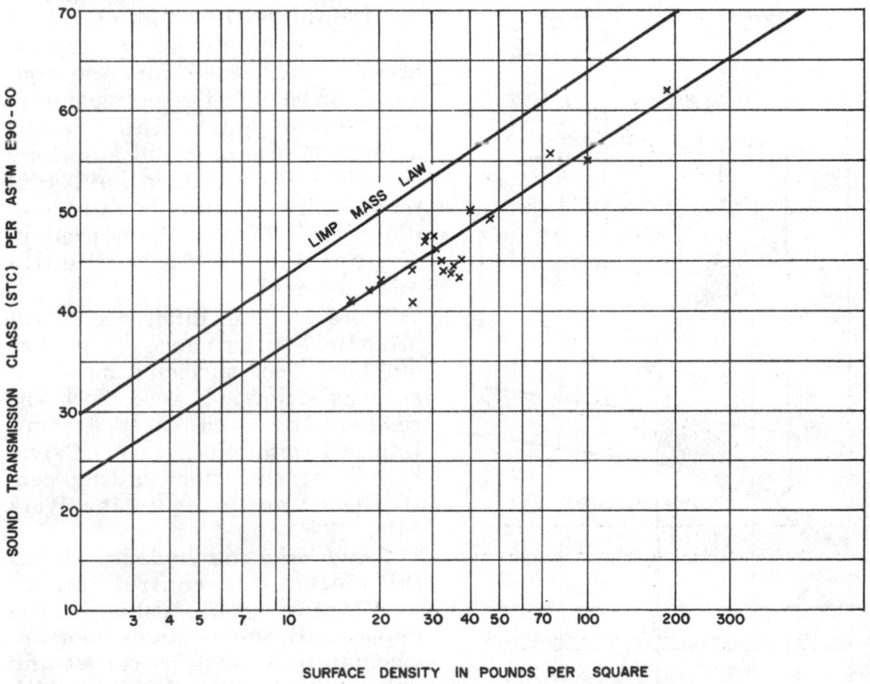

Based on tests performed by Riverbank Acoustical Laboratories.

FIG. 88 Performance of well built, nonporous masonry walls can be predicted from their weight according to a mass law formula. As shown above, performance of tested walls consistently falls approximately 5 to 6 dB below STC calculations.

thickness and aggregate type. It is speculated that the core surfaces, which are denser than the remainder of the block due to manufacture, produce a "skin effect" and provide greater resistance to sound transmission.

Clay Masonry Units are relatively dense and nonporous and acoustical performance of walls made with such units can be estimated according to their mass, providing the wall is tightly constructed. Cavity wall construction will provide improved sound isolation as with concrete masonry.

The sound isolating performance of *any* masonry wall will depend on the extent to which it is airtight. Even relatively fine cracks and open joints such as may occur through poor workmanship or at partition edges will destroy the acoustical effectiveness of the wall.

Because of the great number of mortar joints (and therefore possible acoustical leaks) it is recommended that all brick masonry walls be sealed with plaster where acoustical privacy is desired. Surfaces should be sealed over their entire area—behind cabinets and surfacing materials, in closets, etc., to eliminate flanking paths through these areas. Even if masonry walls are to be furred and surfaced with gypsum board as the interior finish, porous walls should be sealed by plastering or painting before being furred. (A sheet of polyethylene film over the surface does not seal the surface acoustically). There is generally sufficient leakage at the top and bottom of the wall to allow sound which passes through an untreated masonry wall to reach living units.

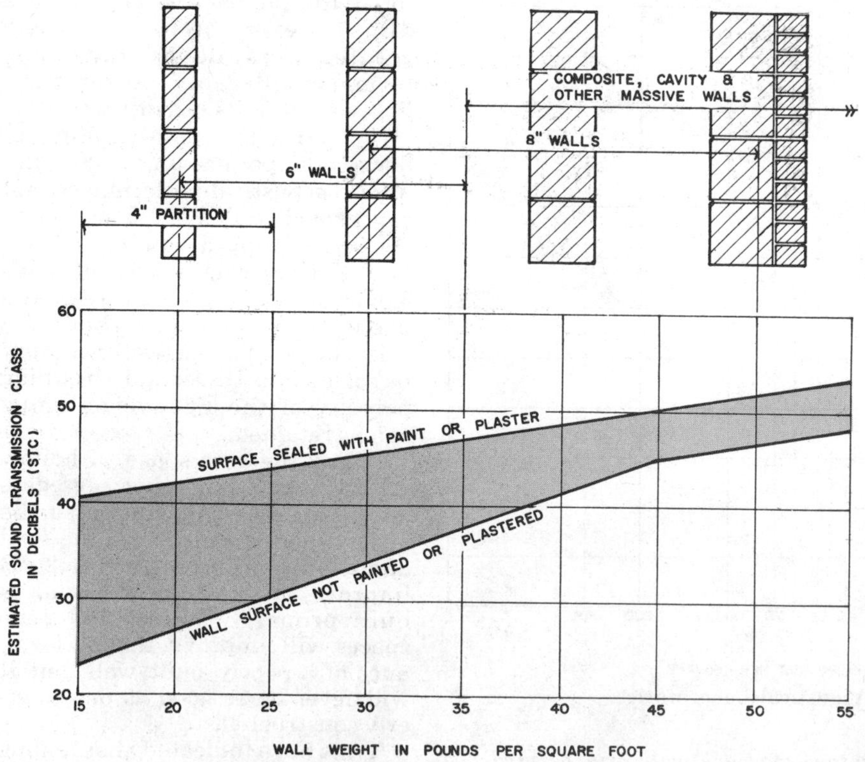

FIG. 90 Estimated STC values for sealed and porous concrete masonry walls according to weight of the wall in lbs. per sq. ft.

Floor/Ceiling Systems

Although airborne sound transmission is a consideration in floor/ceiling construction, the problem of structure-borne (impact) sound transmission is the prime problem. Generally, all of the conventional wood, concrete and metal floor/ceiling systems provide negligible resistance to impact sound transmission. Therefore, some sound isolating construction system must be provided where acoustical privacy is required between floor/ceiling systems. Most measures to control impact sound will also provide a good degree of resistance to airborne sound, providing the essential precautions of construction practices are followed (see page 106-37, Wall/Floor Connections).

The measures used to control structure-borne sound include: (1) cushioning the impact with soft floor surfacing materials, (2) providing discontinuous construction by floating the finished floor from the structural framing or subflooring, (3) providing discontinuous construction between the structural floor and the finished ceiling by resilient mounting of the ceiling or use of separate ceiling joists, and (4) including sound absorbing materials within the cavity of frame construction.

Cushioning the impact and floated floor systems can be used effectively with any type of structural floor system. Discontinuous ceiling construction generally is used only with wood or metal frame construction or for critical areas of reinforced concrete construction such as for ceilings under equipment rooms and service areas. The use of sound absorbing materials within the cavity of frame construction is only significantly effective when used in conjunction with a resiliently mounted ceiling.

There is little difference in the acoustical performance of the various hard floor surfacing materials such as stripwood, wood block and resilient tile and sheet flooring. Impact Insulation Classes (IICs) for representative floor/ceiling constructions are given in the Work File.

Cushioning the Impact Of all the methods to control impact sound transmission, cushioning the impact with soft resilient floor finish materials (such as carpet and pad) is the most effective—with any type of floor system. By dissipating a substantial amount of the impact energy as it strikes the surface, the amount of sound transmitted to the supporting structural system is substantially reduced. The Impact Insulation Class (IIC) of the carpet and pad combination vary with the thickness of the carpet, degree of matting of the pile and weight of both carpet and pad in oz. per sq. yd. (Fig. 91). However, no appreciable gain in airborne sound isolation is obtained by using carpet and pad or by adhering acoustical surfacing materials to the ceiling below. Therefore, when cushioning the impact is the only measure intended to control impact sound transmission, acoustical performance of the basic floor/ceiling system should be adequate to control airborne sound transmission as well.

Cushion-backed Vinyl Sheet, a comparatively recent development in resilient flooring products, may prove to be effective in controlling impact sound transmission on floor systems where carpeting is not desired, such as in kitchens, baths and hallways. Test data for a representative cushion-backed vinyl sheet product show that an IIC of 54 was achieved when this material was installed on a 6" precast hollow concrete slab. When the ceiling beneath the slab was resiliently mounted, an IIC of 59 was achieved. Addition of 2" of concrete topping to the slab increased these values by 4 dB.

Wood Frame Construction A conventional wood joist floor/ceiling system, finished with hard-surfaced flooring and gypsum board ceiling attached directly to the joists (Fig. 92) will receive an IIC of approximately 26. The addition of one or more layers of plywood or other subflooring will not appreciably affect this rating. All of the previously mentioned methods are used to control impact sound transmission in wood frame construction. As with other floor/ceiling systems, cushioning the impact with carpet and pad of sufficient thickness and density is the most effective. Carpet and/or pad used alone or with any of the following tested sound isolating systems will result in IICs ranging from 42 to 85.

Separated Ceiling construction between the structural floor and the ceiling can be accomplished either by resilient mounting of the ceiling finish with resilient channels or by attaching the ceiling finish to separate ceiling joists (Fig. 93). Either method of discontinuous construction is effective in controlling impact sound from above and airborne sound from either direction. However, the acoustical effectiveness of the construction is dependent on how well the construction techniques are employed to prevent sound from bypassing the ceiling through walls and other structural connections. As with resiliently mounted walls, no heavy objects such as electrical fixtures should be mounted on or recessed in the ceiling and airtight construction should be maintained.

FIG. 91 EFFECT OF CARPETING IN CONTROLLING IMPACT SOUND[1]

Carpet[2]	Pad	Floor/Ceiling Construction: Concrete Slab[3] IIC	Floor/Ceiling Construction: Wood Joist[4] IIC[5]
20 oz. Wool	(none)	53	—
40 oz. Wool	(none)	57	—
60 oz. Wool	(none)	65	—
44 oz. Wool	44 oz. sponge rubber bonded to carpet	68	54
20 oz. Wool	40 oz. felt	69	—
44 oz. Wool	40 oz. hair & jute covered with foam rubber face & back	70	59
40 oz. Wool	40 oz. felt	72	—
44 oz. Wool	40 oz. all hair	72	59
44 oz. Wool	66 oz. all hair	72	63
60 oz. Wool	40 oz. felt	73	—
44 oz. Wool	40 oz. hair & jute	73	61
44 oz. Wool	20 oz. combination hair & jute, foam rubber face, latex back	73	61
44 oz. Wool	Urethane foam	75	—
44 oz. Wool	40 oz. sponge rubber	76	65
44 oz. Wool	31 oz. foam rubber on burlap (3/8" thick)	79	67
44 oz. Wool	80 oz. sponge rubber	80	68

1. Based on tests performed by Kodaras Acoustical Laboratories, Inc. for the American Carpet Institute. IIC values approximated by adding 51 to original INR test values.
2. Carpet pile weight given in oz. (/sq yd.).
3. 5" concrete slab (sand gravel aggregate), no ceiling finish.
4. 2" x 8" wood joists, 16" o.c.; 5/8" T&G plywood nailed to joists as subfloor; 5/8" gypsum board nailed to joists as ceiling.
5. Values given only where test data is comparable to tests performed on concrete slab construction.

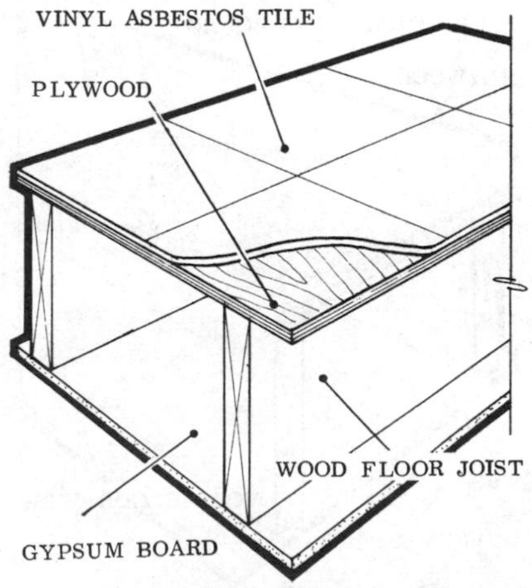

FIG. 92 Conventional wood joist floor construction provides poor resistance to impact sound transmission.

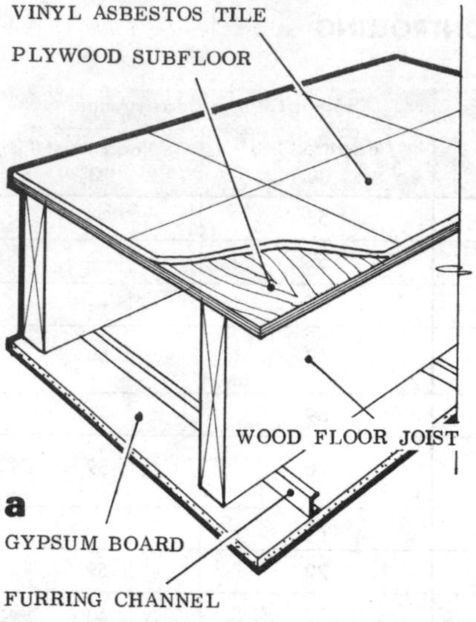

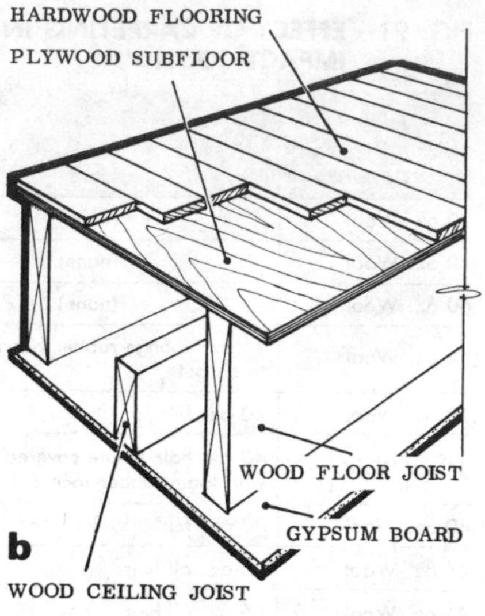

FIG. 93 Impact sound isolating performance can be improved by two methods of discontinuous ceiling construction: (a) resiliently mounting the ceiling finish; (b) ceiling finish carried on independent ceiling joists.

Resilient mounting of the ceiling or attaching the ceiling to separate joists can improve the IIC of the construction as much as 7 dB; the addition of sound absorbing material in the cavity of the resiliently mounted construction can provide an additional 7 dB increase. But separated construction alone will not achieve a floor/ceiling system with much more than 45 IIC with a hard-surfaced floor, even though an acceptable STC of 50 is obtained. This type of ceiling construction used in conjunction with a floated floor system with hard surface, however, can result in an IIC of 57 and an STC of 50 or more.

Floated Floor construction over wood joists and subfloor can be accomplished by using a *resilient* underlayment to separate the finished flooring and underlayment from the structural floor system (Fig. 94a), or by placing lightweight (cellular) concrete over the resilient underlayment (Fig. 94b). (See the Work File for variations of possible constructions). When resilient underlayment is used, the resilient material must be capable of resisting major deformation from superimposed loading. Such material is designed to carry only the relatively light loading of the finished floor, people and furniture.

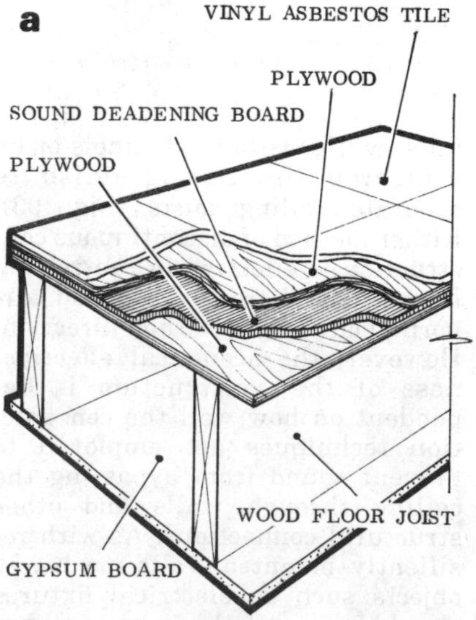

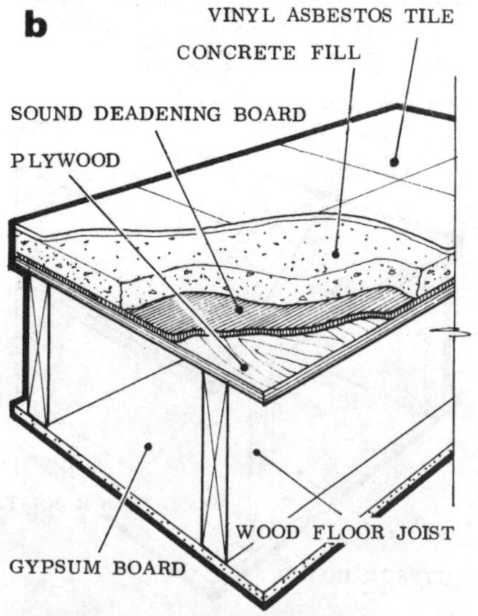

FIG. 94 Floated floor construction: (a) resilient underlayment separates the finished floor from the structural floor; (b) lightweight cellular concrete is placed over resilient underlayment to form a floating floor system.

The performance of the floated floor is dependent on the effective separation of the finished floor from the structural elements (see page 106-37, Wall/Floor Connections). The floating construction should not be rigidly connected with the surrounding elements. Adhesive generally is used to attach resilient underlayment but, in any case, the nailing or stapling of the finished floor or subfloor through the resilient underlayment into the supporting construction will nullify the sound isolating properties of the construction.

A basic wood joist construction can be improved approximately 7 dB with a floated floor system using resilient underlayment. To achieve a rating of more than 51 IIC, the floated floor can be used in conjunction with separated ceiling construction; and with the addition of sound absorbing material in the joist space, an IIC of 57 can be achieved.

Lightweight concrete placed directly over a plywood subfloor does not appreciably improve the impact sound resistance of a basic wood joist floor/ceiling construction. However, when the concrete is resiliently floated, the improvement is about 10 dB. This construction still only results in less than a 51 IIC and the addition of carpet and pad is necessary or the ceiling must be resiliently mounted to achieve a plus 51 IIC rating.

Concrete Slab Construction A plain concrete slab finished with resilient tile and plaster ceiling applied directly to the slab (Fig. 95) will receive an IIC of about 36. Weight and thickness of the slab has little effect in reducing the impact sound transmission to an acceptable level. Sound control measures to improve performance of concrete slab construction gen-

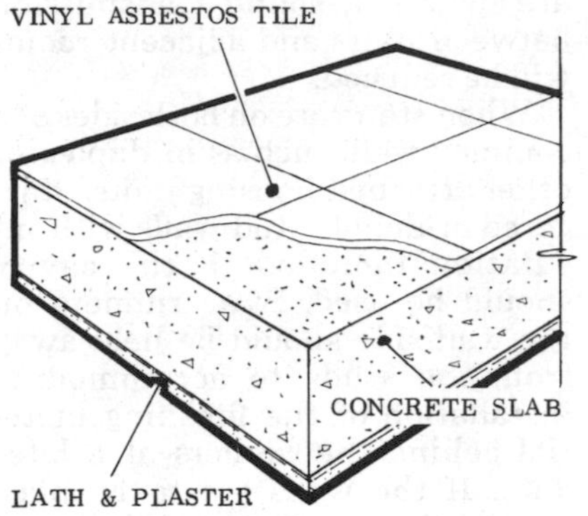

FIG. 95 Typical concrete slab construction with hard surface finish offers poor resistance to impact sound transmission.

erally are limited to the addition of a floated floor system and/or cushioning the impact with carpet and pad. Because of the considerations of practicality and cost, the ceiling under concrete slab construction is not often resiliently mounted and therefore test data for such a ceiling system is limited.

The addition of carpet and/or pad to a concrete slab provides somewhat better performance than does carpet and pad on wood frame construction, with IICs ranging from 53 to 80 or more, depending on the weight and density of the carpet and pad (Fig. 91).

Floated Floor construction to isolate the finished floor from the concrete slab is accomplished much in the same manner as for wood frame construction (Fig. 96). The provisions to keep the floated floor from rigid contact with structural elements should be observed (see page 106-37, Wall/Floor Connections). A hard-surfaced floor floated on a resilient underlayment over a 6″ concrete slab with plaster ceiling will result in an IIC of approximately 55.

The performance of precast concrete floor/ceiling systems is similar to that of placed concrete. Available test data on such systems is limited. Data for representative precast floor systems are given in the Work File.

Metal Frame Construction Steel bar joist and other similar metal floor systems can be improved by measures similar to those used for wood frame construction. Methods for floated floor and resiliently mounted ceiling systems closely parallel those used for wood frame construction. The acoustical effectiveness of steel bar joist construction, as well as metal deck, concrete floor joist and other such construction systems, depends primarily on sealing the wall/floor connections airtight. Walls must be brought above the suspended ceiling line and be sealed at the under side of the floor slab or deck. When walls run perpendicular to joists, proper sealing is difficult and expensive to achieve. Also, the relatively thin (2″) concrete slab normally placed over steel bar joist construction does not have sufficient mass to provide more than an STC in the high 40's. Therefore, when a higher STC is required for either floor/ceilings or walls, this type of construction has not proven effective. Even though walls can be constructed that will achieve higher STC's, the sound can bypass the wall through the flanking path provided by the relatively lightweight floor system.

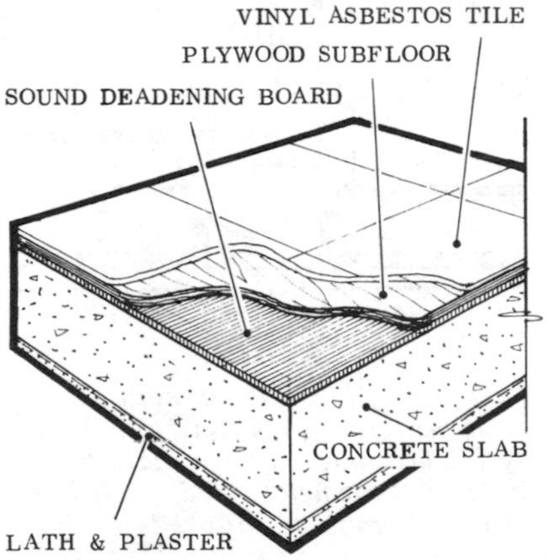

FIG. 96 Floating the finished floor over resilient underlayment improves performance of concrete slab considerably.

Current test data for evaluating metal frame floor/ceiling construction is not as extensive as data for other construction systems. Representative steel bar joist systems and available test data are given in the Work File.

Wall/Floor Connections

Where walls and floors are joined, noise problems from flanking paths are encountered. Both airborne and structure-borne sound may penetrate around and through the walls and floors—even though the wall and floor constructions are structurally correct—unless precautions are taken.

Airborne Sound Both the wall and floor systems selected should have comparable sound isolating abilities so that a weakness in one system will not diminish the effectiveness of the other.

Spaces between floor joists should be blocked at the juncture of walls and floors to reduce airborne sound transmission through such flanking paths (Fig. 97). Joints at top and bottom of the partition between the floor and ceiling should be sealed airtight with a re-

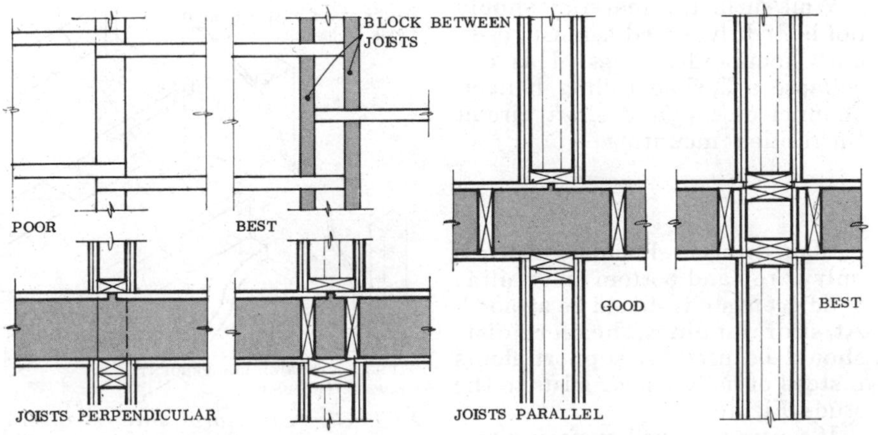

FIG. 97 Open air paths should be closed to minimize problems of airborne flanking noise. Spaces between joists should be blocked particularly when joists are framed perpendicular to walls.

silient, nonhardening acoustical sealant or resilient material (Fig. 98).

Structure-borne Sound Heavy loading of floated floors from structural walls will cause resilient underlayment to collapse, causing damage to floors and settling of the structure. Any underlayment material which is sufficiently rigid to support structural loading is considered too rigid to provide resilient isolation for a floated floor system. Therefore, bottom plates of a partition should *not* be supported on the top of a floating floor. The plates should be fastened instead to the subfloor or supporting structure.

Provision for a break in the subfloor under partitions will help to reduce impact noise transmission between adjacent rooms by breaking direct contact of the structural elements (Fig. 99a). If possible, floating floors should not touch partition edges (Fig. 99b). A cushion between the edge of the partition and the floor can be provided.

Floated concrete floors on wood joist construction should be confined to a single room to prevent transmission of impact noise along the floor (Fig. 100). Isolation joints around the perimeter of all rooms and across doorways are essential. Walls should be supported on a floor plate; polyethylene film should be installed over the resilient underlayment and isolation joints installed around the perimeter of the room. The concrete then is placed, making sure it is not directly in contact with the walls. Molding and quarter round should be used to cover the isolation joint.

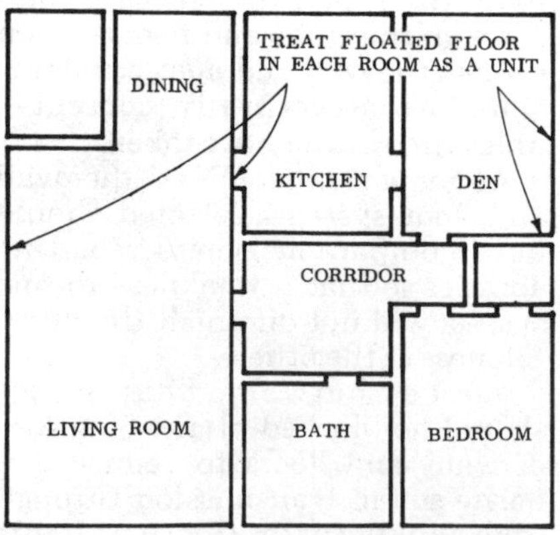

FIG. 100 A floated floor should be confined to a single room by isolating it at doorways and walls.

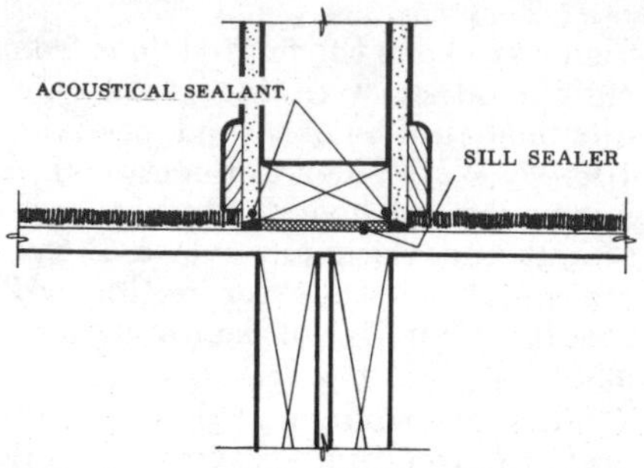

FIG. 98 Acoustical sealant should be used to seal all open air paths—at partition edges between floors and ceilings and all around edges.

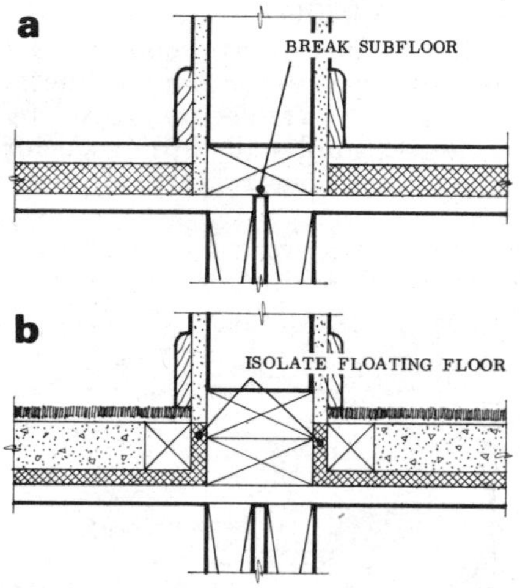

FIG. 99 Flanking structure-borne noise between rooms can be minimized by (a) breaking direct contact between structural elements, and (b) isolating floated floors from walls.

Wall surfacing material should not be tightly forced against resiliently mounted ceilings. This can collapse a flexible ceiling hanger, channel or clip and short circuit the resilient mounting.

Stairs

Stair stringers should be fastened only at top and bottom and nailing to adjacent studs should be avoided. At stair landings, header joists should be used to support floors instead of nailing floor joists to the studs (Fig. 101).

If continuous wall surfacing material is installed on the entire wall adjacent to stairwells before stairs are installed, sound transmission between stairs and adjacent rooms will be reduced.

When stairs are on both sides of a common wall, such as in duplex or other attached housing units, staggered or double stud walls with insulation included in the cavity should be used. Stair runners on the wall side should be held away from the studs to accommodate installation of the finishing material behind the runners at a later date. If the walls are to be plastered, it is recommended that temporary construction ladders, as used in fire-resistant construction, be used in order to permit complete plastering and to eliminate any breaks in the wall.

Finishing the stairs with carpet or resilient material such as rubber mats will substantially reduce impact noise transmission from stairs to adjacent apartment.

Hallways

When a heating plenum area is created by furring down the hallway ceiling as a hot air supply duct or an area for conditioned air ducts, the ducts often are installed before the interior ceiling is finished and no ceiling is installed above the ducts. This permits impact noise from the hallway to pass through the ceiling and down the open stud spaces in the walls of hallways and adjoining rooms. This problem can be solved by installing only register outlets and first duct section in the hallways, deferring the balance of

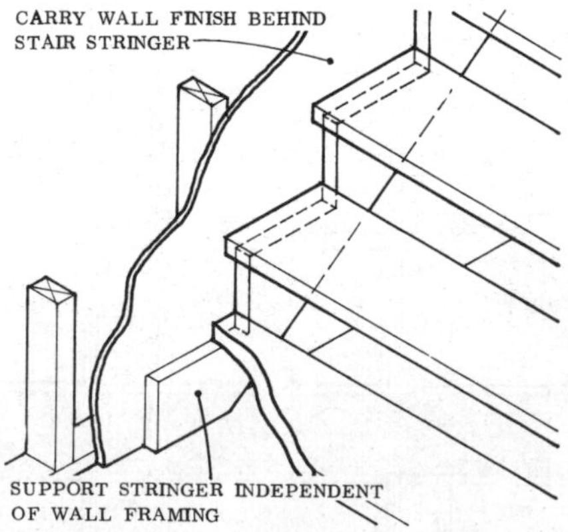

FIG. 101 Structural stair stringers should not be nailed to adjacent walls. Footfalls can be transmitted to the walls, if stair and walls are rigidly connected.

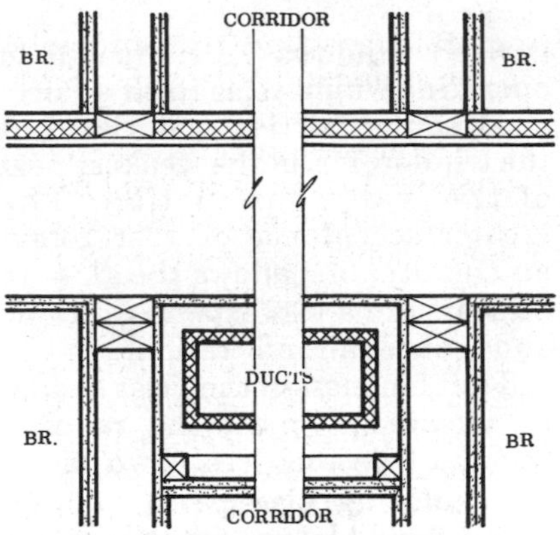

FIG. 102 Furred ceiling finishes often create sound transmission paths. Ducts can be installed after the walls and ceiling have been finished.

the installation until the interior finishes of the walls and ceilings are completed. Ducts then can be installed and a furred ceiling installed below (Fig. 102).

An alternate solution is to have the interior finishing material installed before the ducts are run, omitting the furred ceiling section. Then after air conditioning and heating installations are complete, ceiling panels can be laid on wall angles or moldings to form the hallway ceiling.

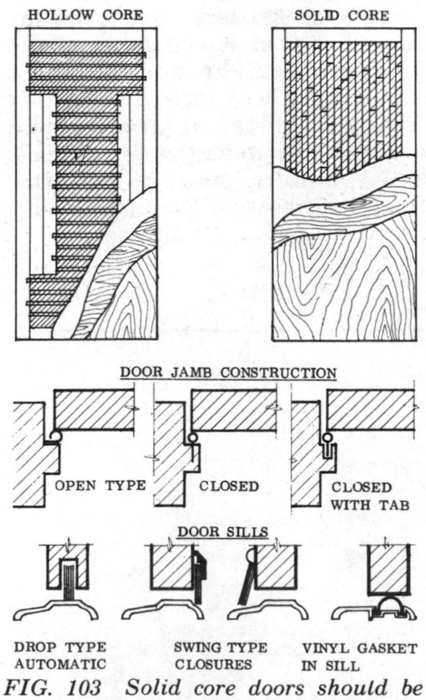

FIG. 103 Solid core doors should be used in sound conditioned walls, with the door sealed at jamb, head and sill.

Doors and Windows

Doors and windows are acoustically weak. Maximum acoustical performance requires that the edges and frames of doors and windows be sealed airtight. Packing and caulking around window and door frames is necessary to reduce sound transmission through cracks.

Doors Solid core or mineral filled doors are generally better sound isolators than hollow core doors (Fig. 103a). However, the amount of air space around the edges of the door is usually the controlling acoustical factor. Whenever possible, doors should be gasketed if sound privacy is required (Fig. 103b). The relative performance of solid and hollow core doors, gasketed and ungasketed is given in Figure 104. Soft weatherstripping provides a good seal at the top and sides of the door, but automatic threshold closers are needed to seal the bottom. While it may be costly to use gasketed doors, provision of a cushion such as sponge rubber between the door jam will reduce the impact energy of a door slam and make the door closure quieter. Door closure control also may be considered for apartments opening onto common hallways. Unfortunately, properly sealed doors, because they are tight-fitting, are harder to close and more prone to warping.

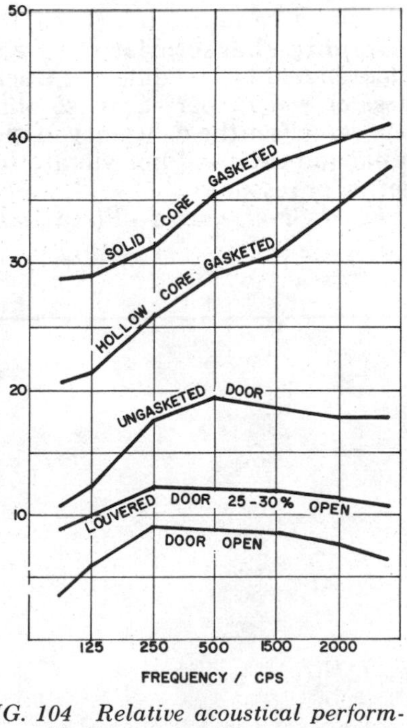

FIG. 104 Relative acoustical performance of hollow and solid core doors.

The use of undercut doors to allow for rugs or return air movement in forced air ventilation systems essentially destroys the acoustical effectiveness of the door (see page 106-26, Duct Systems).

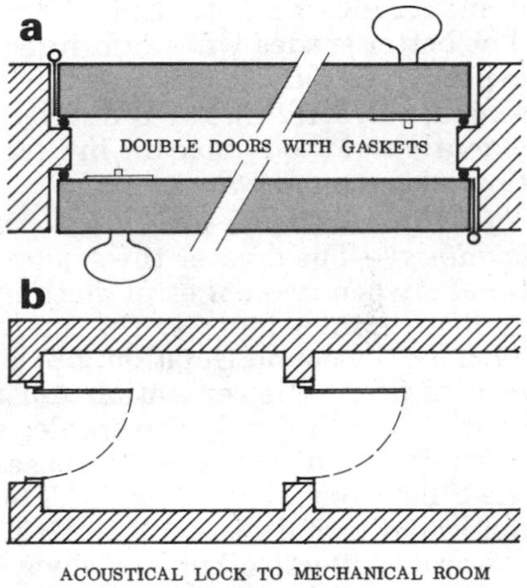

FIG. 105 "Acoustical locks" can be used to reduce sound transmission from high noise sources such as service areas and equipment rooms.

Sliding Doors generally provide little sound privacy because a relatively tight air seal cannot be maintained.

Equipment Room Doors usually are metal or metal clad over very heavy core materials, required for fire ratings. These doors will not provide good acoustical performance unless they are sealed with gaskets and door sills (Fig. 105a). Where equipment room doors lead to a hallway, a second door with better acoustical performance can be installed (Fig. 105b).

Windows Approximately 80% of the acoustical problems occurring with windows results from the quality of airtightness of construction and about 20% from the selection of the glass itself. Generally, sturdy, well weatherstripped and tightly closed windows are better sound insulators than loosely fitting windows; fixed sash performs better than operating sash. The relative acoustical performance of various types of windows is given in Figure 106.

Operating Single-glazed—Wood windows in grooved wooden tracks with weatherstripping, which are pressure sealed when closed, perform best. Wood windows in metallic or plastic slides are fair against air infiltration but are generally poor against resonant frequencies. Metal windows are available in many grades and quality levels. The better grades will perform acceptably if tight-fitting, properly sealed, correctly installed and sufficiently rigid to reduce air infiltration and resonant vibration.

Operating Double-glazed (storm windows)—The greater the separation between two panes of glass in double-glazed windows, the more efficient the sound isolation, but to be significant the separation must be relatively large. For example, a separation of from 2″ to 4″ increases the STC from 33 to only 34. The minimum useful distance is 4″; 8″ spacing is effective but not always practical, and 12″ is required to isolate severe noise. If only 2″ space is possible, it is considered more economical to double the weight of the single glass. Storm windows attached to casements are credited for added weight of glass only.

Each glass should be mounted in a heavy frame and securely fastened in place. Mounting the glass in the frame with flexible or pre-

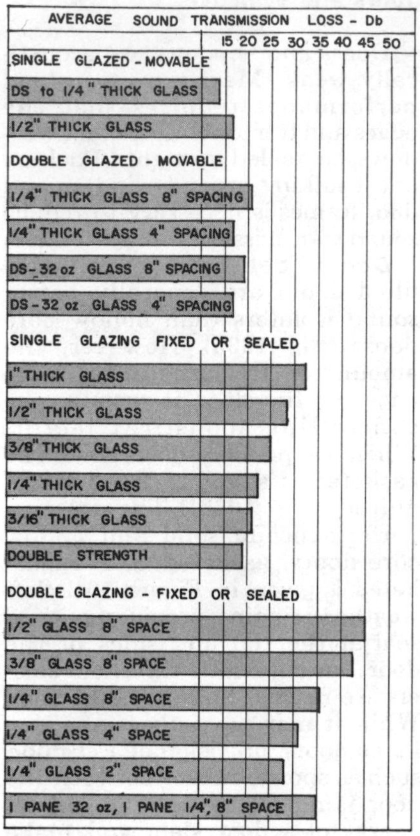

FIG. 106 Relative acoustical performance of various types of windows.

extruded caulking improves the damping characteristics. Each glass should be of a different thickness or weight per sq. ft. so that one can offset the deficiency of the other and they will not vibrate together or resonate.

Fixed Single-glazed—Plate glass in fixed windows, as compared to operating windows, is itself a fairly good sound barrier material; but the window is still the weakest part of the wall construction. The greater acceptance of year round air conditioning allows the glass to be sealed in a less expensive frame and reduces air infiltration.

The thickness of the glass should be selected to meet the requirements of decibel loss. Two layers of insulating glass with 1/2″ air space should be considered for fuel savings and prevention of condensation. Two layers of 1/4″ glass are approximately equal to one layer of 1/2″ glass, acoustically, if closely spaced.

The windows should be installed with a flexible mounting, making certain that the mounting is installed in a firm frame and is airtight to prevent flanking sound from bypassing the glass.

Fixed Double-glazed—For the most demanding noise problems, the most practical construction is two panes of glass separated by at least 8″ of enclosed air space. When built in an exterior wall, a jar of silica gel in the space will absorb vapor that might get through the interior seal. Glass should be arranged so that it can be removed for cleaning. The addition of a sound absorber, such as 1″ thick resilient material around the perimeter between the two glasses, will generally improve performance by absorbing some of the sound energy in the air space.

GLOSSARY

Absorption The ability of a material to absorb rather than reflect sound waves striking it by converting sound energy to heat energy within the material.

Absorption Unit See *Sabin.*

Absorption Coefficient (a) The ratio of sound absorbing effectiveness (at a specific frequency) of one sq. ft. of a material to one sq. ft. of perfectly absorptive material; usually expressed as a decimal value (such as .70) or in percent.

Acoustics The science of sound, including its production, transmission and effects.

Architectural Acoustics The acoustics of buildings and structures.

Room Acoustics The branch of architectural acoustics dealing with both *acoustical correction* and noise reduction.

Acoustical Correction Specially planning, shaping and equipping a space to establish the best possible hearing conditions for faithful reproduction of wanted sound within the space.

Airborne Sound Sound, produced by vibrating sources that radiate sound directly into the air, which is transmitted through air as a medium rather than through solids or the structure of the building (see *Structure-borne Sound Transmission).*

Ambient Sound The continuous existing sound level (background noise) in a room or space, which is a composite of sounds from both exterior and interior sources, none of which generally are identifiable individually by the listener (see *Background Noise).*

Attenuation, Sound Reduction of the energy or intensity of sound.

Average Transmission Loss (TL) The numerical average of the transmission loss values of a construction measured at nine frequencies. It is a single number rating for comparing the airborne sound transmission through walls and floors (see *Transmission Loss).*

Background Noise Normal sound always present in a space created either by outdoor sounds such as street traffic or indoor sounds such as ventilating noise, appliances, etc.

Common Wall A wall which separates adjacent dwelling units within an apartment building; also referred to as a *party wall.*

Decibel Logarithmic unit expressing the ratio between a given sound being measured and a reference point (see page 106-4 & 106-5).

Discontinuous Construction Any of several construction methods, such as the use of staggered studs, double walls, or resilient mounting of surfaces, used to break the continuous paths through which sound may be transmitted.

Dyne A unit of force, that if acting on a mass of one gram would accelerate it one centimeter per second per second.

Echo Reflected sound loud enough and received late enough to be heard as distinct from the source.

Flanking Paths A wall or floor/ceiling construction that permits sound to be transmitted along its surface; or any opening which permits the direct transmission of sound through the air.

Frequency The number of complete cycles of a vibration performed in one second, measured in cycles per second (cps) and expressed in Hertz (Hz).

Hertz (Hz) A unit of frequency of a periodic process equal to one cycle per second.

Impact Insulation Class (IIC) A single number rating developed by the Federal Housing Administration to estimate the impact sound isolation performance of floor/ceiling systems (see page 106-17).

Impact Noise Rating (INR) A single number rating used to compare and evaluate performance of floor/ceiling constructions in isolating impact noise transmissions (see page 106-16).

Impact Sound Pressure Level (ISPL) The sound level (in decibels) measured in the "receiving" room resulting from the transmission of sound through the floor/ceiling construction produced by a standard tapping machine.

Incident Sound Sound striking a surface, contrasted to reflected sound. The angle of incidence equals the angle of reflection.

Loudness The subjective response to sound, indicating the magnitude of the hearing sensation, dependent on the listener's ear.

Masking The effect produced by background noise which appears to diminish the loudness of transmitted noise.

Noise Unwanted sound.

Noise Reduction Reducing the level of unwanted sound by any of several means of acoustical treatment.

Noise Reduction Coefficient (NRC) A single number index of the noise reducing efficiency of acoustical materials. It is found by averaging the sound absorption coefficients at 250, 500, 1000 and 2000 cps.

Octave The interval between any two sounds having a frequency ratio of 2 to 1.

Perceived Noise Level (PNdB) A single number rating of aircraft noise in decibels, used to describe the acceptability or noisiness of aircraft sound. PNdB is calculated from measured interior and exterior noise levels and correlates well with subjective responses to various kinds of aircraft noise.

Reflection The return from surfaces of sound not absorbed on contact with the surfaces.

Resonance The sympathetic vibration, resounding or ringing of enclosures, room surfaces, panels, etc. when excited at their natural frequencies.

Reverberation Continuing travel of sound waves between reflective surfaces after the original source is stopped.

Reverberation Time (T) The time in seconds required for a sound to diminish 60 decibels after the source is stopped.

Sabin Measure of sound absorption of a surface, equivalent to 1 sq. ft. of a perfectly absorptive surface.

Separated Construction See *Discontinuous Construction.*

Short Circuit A bypassing connection or transmission path which tends to nullify or reduce the sound isolating performance of a building construction.

Sound A vibration in any elastic medium within the frequency range capable of producing the sensation of hearing.

Sound Isolation Materials or methods of construction designed to resist the transmission of airborne and structure-borne sound through walls, floors and ceilings.

Sound Pressure The instantaneous pressure at a point as a result of the sound vibration minus the static pressure at that point; the change in pressure resulting from sound vibration. It is measured in dynes per square centimeter. A sound pressure of 1 dyne per sq. cm. is about that of conversational speech at close range and is approximately equal to one millionth of atmospheric pressure.

Sound Pressure Level (SPL) The sound pressure when measured on the decibel scale; the ratio in decibels between any measured pressure and a reference pressure (see Fig. 6, page 106-5).

Sound Transmission The passage of sound through a material construction or other medium.

Airborne Sound Transmission Sound transmitted when a surface is set into vibration by the alternating air pressures of incident sound waves.

Structure-borne Sound Transmission Sound transmitted as a result of direct mechanical contact or impact caused by vibrating equipment, footsteps, object dropped, etc.

Sound Transmission Class (STC) A single number rating for evaluating efficiency of constructions in isolating airborne sound transmission. The higher the STC rating the more efficient the construction (see page 106-12).

Transmission Loss The decrease or attenuation in sound energy (expressed in decibels) of airborne sound as it passes through a building construction.

Watt A unit of power equal to 1 x 10^7 dyne per centimeter per second, which is the basic expression of the flow of sound energy.

We gratefully acknowledge the assistance of the following for use of their publications as references and for permission to use illustrations: Acoustical Materials Association; Celotex Corporation; Federal Housing Administration; Flexicore Company, Inc.; Gypsum Association; Insulation Board Institute; Kodaras Acoustical Laboratories, Inc.; NAHB Research Foundation, Inc.; National Concrete Masonry Association; National Gypsum Company; Owens-Corning Fiberglas Corporation; Portland Cement Association; Riverbank Acoustical Laboratories, IIT Research Institute; Structural Clay Products Institute; United States Gypsum Company.

SOUND CONTROL 106

CONTENTS

This Work File section: (1) identifies basic acoustical problems and summarizes methods of measuring and controlling noise; (2) discusses design and construction techniques essential to providing and maintaining acoustical performance of wall, floor/ceiling and other construction systems; (3) provides criteria for selecting wall and floor/ceiling systems to resist airborne and structure-borne (impact) sound transmission; and (4) provides an index to Sound Transmission Class (STC) ratings and Impact Noise Ratings (INR) for commonly used construction systems.

For a detailed discussion of the physics of sound see Main Text, page 106-3 through 106-18.

ROOM ACOUSTICS

Noise is unwanted sound. Sound waves are produced by vibrating bodies, and when their frequency and intensity are within certain ranges, the sensation of hearing is produced.

FREQUENCY

Frequency of sound is perceived by the ear as the pitch of a sound. High pitched sounds have high frequencies, low thumping sounds have low frequencies. Frequency is measured in cycles per second (cps) and is expressed in Hertz (Hz). Most sounds consist of a mixture of numerous frequencies at different relative intensities.

SOUND LEVEL

The intensity (level) of sound is expressed in decibels. A decibel (dB) is equivalent to the smallest change in sound intensity that can be detected by the human ear. The decibel scale extends from 0 dB (threshold of hearing) to over 120 dB (threshold of feeling). The simplest way to understand sound levels on the decibel scale is to compare them to commonly recognized sounds (Fig. WF1).

SOURCES OF SOUND

Exterior and interior noise can produce three types of sound problems: (1) multiple reflection of airborne sound *within* a room; (2) airborne sound transmitted through walls, floor/ceilings and openings, (3) structure-borne (impact) sound transmitted primarily through floor/ceilings and from the vibration of mechanical equipment systems.

Highly reflective, hard surfaces can cause airborne sound within a room to build up to annoying levels—higher even than that of the original source.

Noise levels produced by multiple reflection *within* rooms can be controlled by using acoustical materials such as ceiling tile, carpet and other materials which absorb sound energy and convert it into heat energy

(Continued) ROOM ACOUSTICS

FIG. WF1 SOUND LEVELS OF TYPICAL SOUNDS IN DECIBELS

Loudness	Sound Level in Decibels (dB)	
	160	Near jet engine.
	130	Threshold of painful sounds; limit of ear's endurance.
Deafening	120	Threshold of feeling (varies with frequency).
		18' from airplane propeller.
	110	
		Express train passing at high speed.
	100	Loud automobile horn 23' away.
		Noisy factory
Very Loud	90	
		Loud Hi-fi
	80	Subway train
		Motor trucks 15' to 50' away.
Loud	70	Stenographic room, average TV, loud conversation
	60	Average busy street.
		Noisy office or department store.
Moderate	50	Moderate restaurant clatter.
		Average office, noisy room, average conversation
	40	
		Soft radio music in apartment, average residence.
Faint	30	
	20	Average whisper 4' away.
Very Faint	10	Rustle of leaves in gentle breeze.
	0	Threshold of audibility.

through friction. Thus sound absorbing materials reduce multiple reflection by causing the sound to die out faster than if the sound strikes a hard, reflective surface. In addition, sound absorbing materials will slightly reduce airborne sound transmission by reducing the amount of sound that can be transmitted from one room to another.

It is impractical and uneconomical to obtain sound level reductions within a room of much more than 10 dB by treating reflective surfaces with sound absorbing materials. For example, a typical ceiling covered with acoustical tile provides a sound level reduction of only 5 to 7 dB, yet the improvement *within* the room is clearly noticeable.

MEASURING SOUND ABSORPTION

One unit of absorption (called a Sabin) is equal to 1 sq. ft. of totally absorbent surface. Absorption coefficients (a) measure the percentage of incident sound energy absorbed by a material. Coefficients vary theoretically from 0 to 1. A perfectly reflective material would have an $a = 0$. A material which is perfectly absorptive would have an $a = 1$. Materials normally are tested at 6 different frequencies (125, 250, 500, 1000 and 4000 Hz). Absorption coefficients for typical construction surfacing materials are given in Figure WF2.

Noise Reduction Coefficients (NRC)

To obtain a single number rating for a material to indicate its noise reducing efficiency, the absorption coefficients for the middle 4 test frequencies (250, 500, 1000 and 2000 Hz) are averaged arithmetically. This average is called the Noise Reduction Coefficient (NRC) and is expressed to the nearest multiple of .05. A difference of 10 points (.10) in NRC is seldom detectable in a completed installation, and therefore minor difference in NRC values are not usually significant.

Noise reduction guidelines may be stated to serve as a rough guide in determining the amount of treatment necessary within a room for satisfactory noise control:

(1) The total number of absorption units in rooms with low ceilings, typical of residential rooms, should be between 20% and 30% of the total interior surface square footage; and (2) To produce a satisfactory improvement in a room, the total absorption after treatment should be between 3 and 10 times the absorption before treatment.

(See Main Text, page 106-10 and Figure 21 for procedure for calculating noise reduction within rooms).

(Continued) ROOM ACOUSTICS

FIG. WF2 ABSORPTION COEFFICIENTS FOR TYPICAL CONSTRUCTION MATERIALS*

Material	Absorption Coefficients					
	125 Hz	250 Hz	500 Hz	1000 Hz	2000 Hz	4000 Hz
Brick						
Unglazed	.03	.03	.03	.04	.05	.07
Unglazed, painted	.01	.01	.02	.02	.02	.03
Carpet (heavy) on concrete slab	.02	.06	.14	.37	.60	.65
With 40 oz. hair-felt or foam rubber pad	.08	.24	.57	.69	.71	.73
With impermeable latex backing & 40 oz. hair-felt or foam pad	.08	.27	.39	.34	.48	.63
Concrete Block						
Porous	.36	.44	.31	.29	.39	.25
Painted	.10	.05	.06	.07	.09	.08
Fabrics						
Light velour (10 oz/sq. yd) hung straight in contact with wall	.03	.04	.11	.17	.24	.35
Medium velour (14 oz/sq. yd.) draped to half the area	.07	.31	.49	.75	.70	.60
Floors						
Concrete or Terrazzo	.01	.01	.015	.02	.02	.02
Resilient Flooring on concrete	.02	.03	.03	.03	.03	.02
Wood	.15	.11	.10	.07	.06	.07
Parquet in asphalt on concrete	.04	.04	.07	.06	.06	.07
Glass						
Large panes, heavy plate glass	.18	.06	.04	.03	.02	.02
Standard window glass	.35	.25	.18	.12	.07	.04
Gypsum Board, ½" nailed to 2" x 4" studs 16" o.c.	.29	.10	.05	.04	.07	.09
Marble or Glazed Tile	.01	.01	.01	.01	.02	.02
Plaster						
Smooth finish on brick or tile	.013	.015	.02	.03	.04	.05
Rough finish on lath	.02	.03	.04	.05	.04	.03
Smooth finish on lath	.02	.02	.03	.04	.04	.03
Plywood Paneling, ⅜" thick	.28	.22	.17	.09	.10	.11

*Absorption coefficients and other acoustical data for proprietary acoustical materials are published by the Acoustical Materials Association, Bulletin No. XXVIII.

SOUND ISOLATION

Sound isolation deals with the control of sound transmission through walls, floor/ceilings and openings. Sound is transmitted through constructions as a resulc of a vibration by either (1) *airborne sound* or (2) *structure-borne sound.*

AIRBORNE SOUND TRANSMISSION

Airborne sound travels through wall and floor/ceiling constructions or by way of open air flanking paths.

Methods to control airborne sound transmission include: (1) selecting constructions that reduce the degree of transmission by vibration, (2) closing paths of direct air transmission, and (3) using sound absorbing materials within the free air space of frame construction. Generally, a combination of these control measures is necessary to achieve suitable sound privacy between rooms (see Main Text, page 106-14). The primary objective, in any case, is to reduce the level of transmitted noise below the level of background noise on the listening side of the construction.

Due to the masking effect of normal background noise, sound transmitted through a wall or floor/ceiling con-

(Continued) **SOUND ISOLATION**

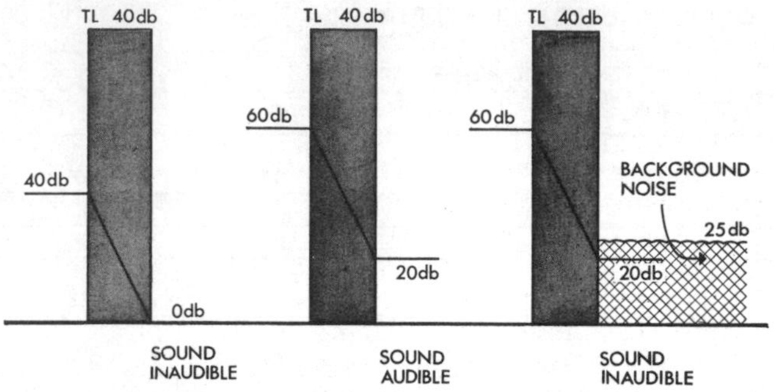

FIG. WF3 Normal background noise on the "listening" side of a wall has the effect of masking transmitted sound that might otherwise be objectionable.

struction will be inaudible or will be "masked" if its intensity is below the level of the background noise in the receiving room (Fig. WF3).

Direct Air Transmission

Airborne sound will travel wherever air passages exist. To achieve suitable isolation from airborne sound transmission, it is essential that all direct air passages be closed. (See page WF106-18 for specific design and construction precautions.)

Dampening Vibration

Constructions that have mass or that provide discontinuous construction (that effectively separate surface finishes) will reduce airborne sound transmission by reducing the degree to which the wall surfaces of the construction can vibrate.

A massive wall (such as concrete or masonry) will not vibrate as readily as a lightweight wall. However, since doubling the weight of the construction generally reduces sound transmission by about 5 dB, there is a practical and an economical limit to the sound isolating effectiveness achieved by increasing the mass of the construction.

Discontinuous construction refers to assemblies having opposite finished surfaces effectively separated or resiliently mounted so that the surface opposite the sound source is not set into vigorous vibration by the sound striking the surface on the source side of the wall. Discontinuous construction for wall systems can be provided by resilient mounting of surfaces, use of staggered stud or double stud walls (see Main Text page 106-14). Discontinuous construction for floor/ceiling systems can be provided by use of floated floor systems, resilient mounting of the ceiling finish or use of separate ceiling joists (see Main Text, page 106-17).

Absorbing Sound Within Constructions

Flexible insulation placed in the free air space of frame walls and floor/ceiling constructions improves performance by absorbing sound within the cavity, and can result in as much as a 10 dB improvement when the surfaces are effectively separated (see Main Text page 106-29 Wall Systems, and page 106-34 Floor/Ceiling Systems).

Measuring Airborne Sound Transmission

For convenience in selecting or comparing wall and floor/ceiling constructions, two single number rating systems have been developed: (1) "Average" Transmission Loss (TL), and (2) Sound Transmission Class (STC). These rating systems indicate the resistance to sound transmission provided by the construction.

"Average" Transmission Loss (TL) The "Average" Transmission Loss of a construction is the numerical average of transmission loss measured at nine or more frequencies (Fig. WF4). To estimate the bare minimum TL required of a construction to obtain privacy, the background noise level on one side of the construction

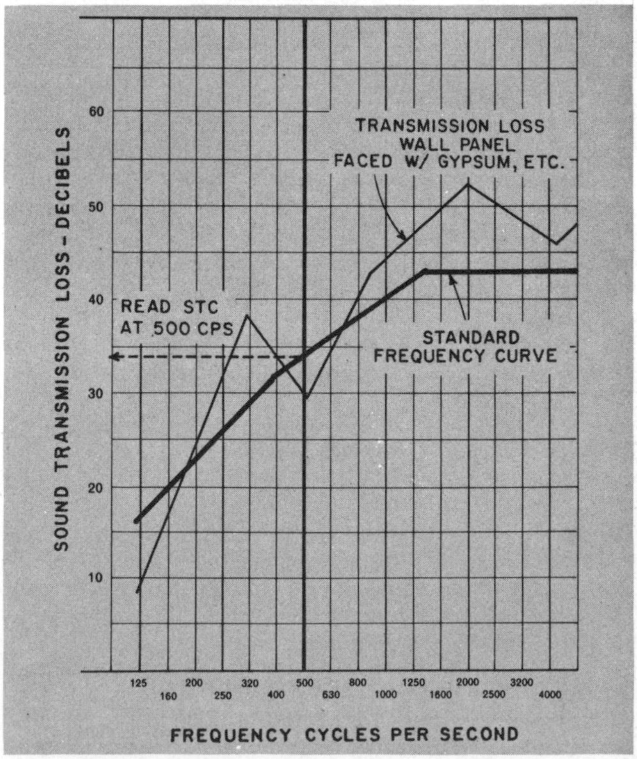

FIG. WF4 An STC rating results from comparing a transmission loss test curve with a standard frequency curve (heavy line) and is read at 500 Hz (cps).

(Continued) SOUND ISOLATION

FIG. WF5 SOUND ISOLATION OF WALLS IN TERMS OF TRANSMISSION LOSS

TRANSMISSION LOSS (TL)	HEARING CONDITIONS*
30 dB or less	Normal speech can be understood quite easily and distinctly through the wall.
30 to 35 dB	Loud speech can be understood fairly well. Normal speech can be heard but not easily understood.
35 to 45 dB	Loud speech can be heard, but is not easily intelligible. Normal speech can be heard only faintly, if at all.
40 to 45 dB	Loud speech can be faintly heard but not understood. Normal speech is inaudible.
45 dB or greater	Very loud sounds, such as loud singing, brass musical instruments or a radio at full volume can be heard only faintly or not at all.

*Background noise level assumed to be 30 dB.

is subtracted from the sound level that strikes the other side. The sound isolating properties of constructions according to their TL (assuming a background noise level of 30 dB) are given in Fig. WF5.

Sound Transmission Class (STC) A newer and now recognized method of rating performance of sound isolating constructions is according to Sound Transmission Class (STC). The STC of a construction represents the transmission loss performance of the construction at all test frequencies. The STC is derived by plotting the transmission loss for a given construction at 16 test frequencies*. The transmission loss test curve is then compared to a standard frequency curve (Fig. WF4). The standard curve is moved to a position where the TL test curve remains above it with the following tolerances (deviations) allowed: (1) a single unfavorable deviation (where a TL value falls below the standard curve) may not exceed 8 dB; (2) the sum of unfavorable deviations may not exceed 32 dB. The STC rating is the numerical value which corresponds to the TL value at 500 Hz (Fig. WF4).

The speech audibility through constructions according to their Sound Transmission Class (STC) rating as influenced by background noise levels is given in Figure WF26, page WF106-22.

**Testing should be in accordance with current ASTM E-90 specifications.*

STRUCTURE-BORNE SOUND

Structure-borne sound transmission through walls and floor/celings results when the surfaces of the construction are set into vibration by direct mechanical contact or mechanical impact. The structure supporting the surfaces is thus set into vibration, carrying the sound to other parts of the building. Structure-borne sound transmission can result from such sources as footsteps, dropped objects, doors slamming or vibration from mechanical systems and equipment. Floor/ceiling constructions are especially vulnerable to this type of sound transmission.

Methods to control impact sound transmission include: (1) cushioning the impact, (2) separated ceiling construction, (3) use of floated floor construction, (4) use of sound absorbing materials within the free air space of frame construction. Singly and in combination, these contruction measures reduce impact sound transmission by breaking the direct structural connections that could carry vibration into other rooms.

Cushioning the Impact

Carpet with padding is the most effective means of reducing impact sound transmission with all types of construction (see Main Text, page 106-35). By dissipating a substantial amount of impact energy as it strikes the surface, the amount of sound transmitted to the supporting structural system is substantially reduced. However, the use of carpet and pad is not significantly effective in reducing airborne sound transmission.

Separated Ceiling Construction

Discontinuous construction between the ceiling finish and the structural floor can be accomplished by use of resilient channels, clips or hangers or use of separate ceiling joists to mount the ceiling finish (see Main Text, page 106-36). Discontinuous ceiling construction will improve resistance to both airborne and structure-borne sound transmission.

Floated Floor Construction

The finished floor can be isolated from the structural floor or subfloor in most types of construction. This is achieved by using resilient underlayment to separate the finished floor from structural members (see Main Text, page 106-36). However, to be effective, care must be taken to assure that the wall/floor connections also provide discontinuous construction to prevent sound from traveling from the floor through the walls (see page WF106-18). Floated floor construction also provides a good degree of resistance to airborne sound transmission.

(Continued) **SOUND ISOLATION**

Absorbing Sound Within Constructions

The addition of sound absorbing materials within the free air space of frame floor/ceiling constructions is only significantly effective in controlling impact sound transmission when used with a resiliently mounted ceiling. However, it is effective in controlling airborne sound with all types of discontinous construction.

Measuring Structure-borne (Impact) Sound

To measure impact sound isolation performance of floor/ceiling constructions, the construction is tested with a standard tapping machine in accordance with ISO Recommendations #140-1960*. The Impact Sound Pressure Level (ISPL) is measured in decibels in the receiving room below and is electronically analyzed into different frequency bands so that a curve can be plotted, distributing the sound over the audible frequency range (Curve #1, Fig. WF6).

It is assumed that a construction that will transmit little sound with the standard tapping machine will perform similarly with other varied types of impact sound. However, because the noise produced by the standard tapping machine does not duplicate the sound produced by many types of impact, studies are currently being conducted to develop additional methods for determing impact sound performance.

Impact Sound Pressure Level curves measure the actual sound levels from a standard noise source transmitted through the construction. The lower the transmitted level, the better the acoustical performance of the construction.

Impact Noise Rating (INR) The Impact Noise Rating (INR) is a single number rating in decibels for evaluating the impact sound isolation performance of a floor/ceiling system. This rating is determined by test in accordance with ISO Recommendation #140-1960*. The ISPL curve is plotted for a construction and is compared with a standard criterion curve (Fig. WF6). The INR is determined by moving the standard curve up or down until the ISPL test curve meets the following tolerances: (1) the mean amount by which the test curve exceeds the standard curve may be 2 dB or less, as averaged over the 16 1/3 octave frequency bands between 100 and 3200 Hz; (2) the test curve may not exceed the criterion curve by more than 8 dB at any one frequency. The INR indicates how far below or above the standard curve the test curve for a construction falls. If the standard curve must be moved

*International Organization For Standardization, ISO Recommendation #140-1960 "Field and Laboratory Measurements of Airborne and Impact Sound Transmission."

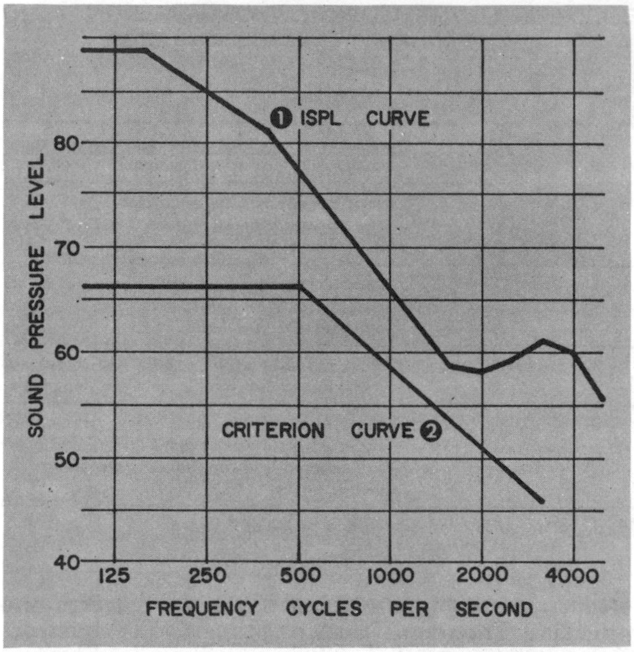

FIG. WF6 Impact Noise Rating (INR) is obtained by comparing the ISPL test curve with a standard criterion curve.

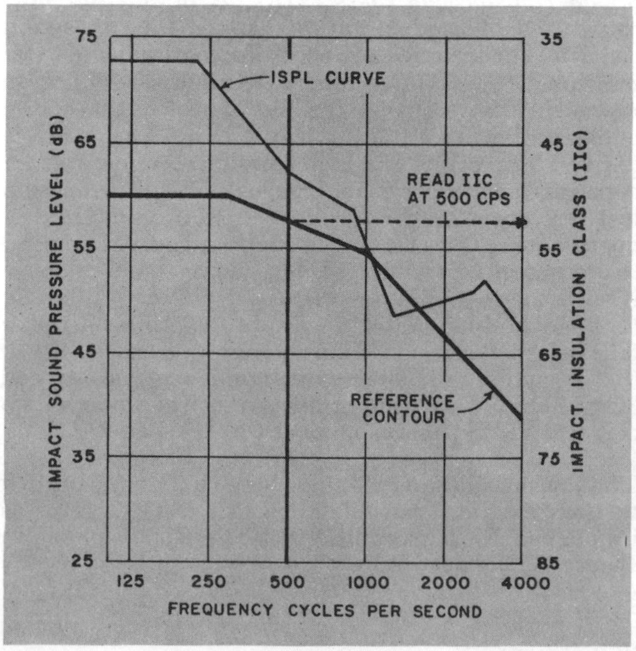

FIG. WF7 Impact Isolation Class (IIC) is obtained by comparing the ISPL test curve with a standard criterion curve and is read as a positive rating.

(Continued) SOUND ISOLATION

down 5 dB for the test curve to conform, the INR is +5, indicating the construction is 5 dB better than the standard; if the standard curve must be moved up 5 dB, the construction is given a −5 INR, indicating 5 dB poorer than the standard. The standard criterion curve is based on average background noise typical of apartments in moderately quiet suburban neighborhoods. Quieter or noiser areas may require higher or lower INR ratings, respectively.

Impact Insulation Class (IIC) The Impact Insulation Class (IIC) is a single number rating system devised by the Federal Housing Administration for rating impact sound isolation. The IIC results in a positive rating which more closely parallels the STC rating system for airborne sound transmission.

Using the same test procedure* as is used for the INR system, the ISPL test curve for a construction is plotted and is compared to a reference contour (Fig. WF7). The reference contour is moved vertically until the following allowable tolerances exist: (1) a single unfavorable deviation (where an ISPL value lies above the contour) may not exceed 8 dB; and (2) the sum of the unfavorable deviations may not be greater than 32 dB. The IIC rating is the value read on the right-hand scale corresponding to the ISPL value at 500 Hz (Fig. WF7).

On the average, an IIC rating for a given construction is approximately 51 dB higher than the INR rating. For example, an INR rating of +4 would be comparable to an IIC rating of approximately 55. However, exact values for IIC ratings should be determined on the basis of individual test data.

DESIGN AND CONSTRUCTION

The following recommendations deal with (1) criteria and planning principles to aid in the design and selection of wall, floor/ceiling, mechanical equipment and other systems and materials to control noise within buildings; (2) construction recommendations, including precautions essential to obtaining and maintaining suitable sound isolation between rooms and between living units; and (3) selection criteria for obtaining suitable sound isolation.

DESIGN AND PLANNING

The objectives in planning for sound control should be: (1) to quiet the noise at its source; (2) to separate the noise source from the listener by distance; and (3) to buffer the transmission of noise by using walls, floor/ceilings and other barriers having suitable resistance to sound transmission.

Background Noise Sources and Levels

Noise characteristics of the site, including the source and intensity of the sound, should be studied to determine existing and potential noise levels (Fig. WF8).

Future development of nearby vacant land should be anticipated and the type and level of probable or possible noise should be investigated.

FIG. WF8 **TYPICAL OUTDOOR BACKGROUND NOISE LEVELS**

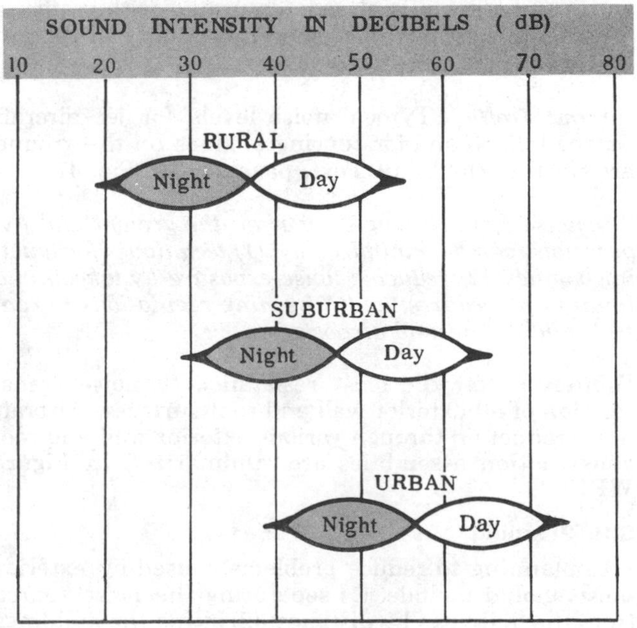

(Continued) DESIGN AND CONSTRUCTION

If the proposed site is near high noise sources—a business district, a highway, an expressway, railway traffic or an airport—both daytime and nighttime outdoor noise levels should be surveyed. Where the average nighttime background noise level is unusually high or if there are infrequent but intermittent unusually loud or annoying noises, expert acoustical advice should be sought.

Automobile Traffic Typical outdoor noise levels for automobile traffic are shown in Figure WF9.

FIG. WF9 APPROXIMATE SOUND LEVELS FOR AUTOMOBILE TRAFFIC AT GROUND LEVEL

Cars Per Minute*	Distance From Traffic Lane		
	40′	100′	500′
1	50 dB	43 dB	29 dB
10	59 dB	52 dB	38 dB
100	63 dB	60 dB	47 dB

*The maximum capacity of one lane of cars moving freely at approximately 30 mph is 2000 cars per hour or 34 cars per lane per minute.

High-speed highways and dense city traffic generally produce higher noise levels than moderate-speed suburban roads carrying an equivalent number of cars.

Unless a highway is already saturated with traffic, a future increase in traffic generally should be expected with resultant increase in noise.

Aircraft Traffic Typical noise levels for jet aircraft during takeoff and for revving engines on the ground are given in the Main Text, page 106-20, Fig. 47.

The noise from revving engines on the ground and flyover noise can be minimized by: (1) techniques for quieting engines, (2) reducing noise exposure by takeoff and landing procedures, and (3) zoning regulations to control land use around airports.

Windows offer the least resistance to noise transmission of all exterior wall and roof surfaces. Aircraft noise reduction through various exterior wall and roof construction assemblies are summarized in Figure WF10.

FIG. WF10 SUMMARY OF NOISE REDUCTION THROUGH EXTERIOR CONSTRUCTIONS

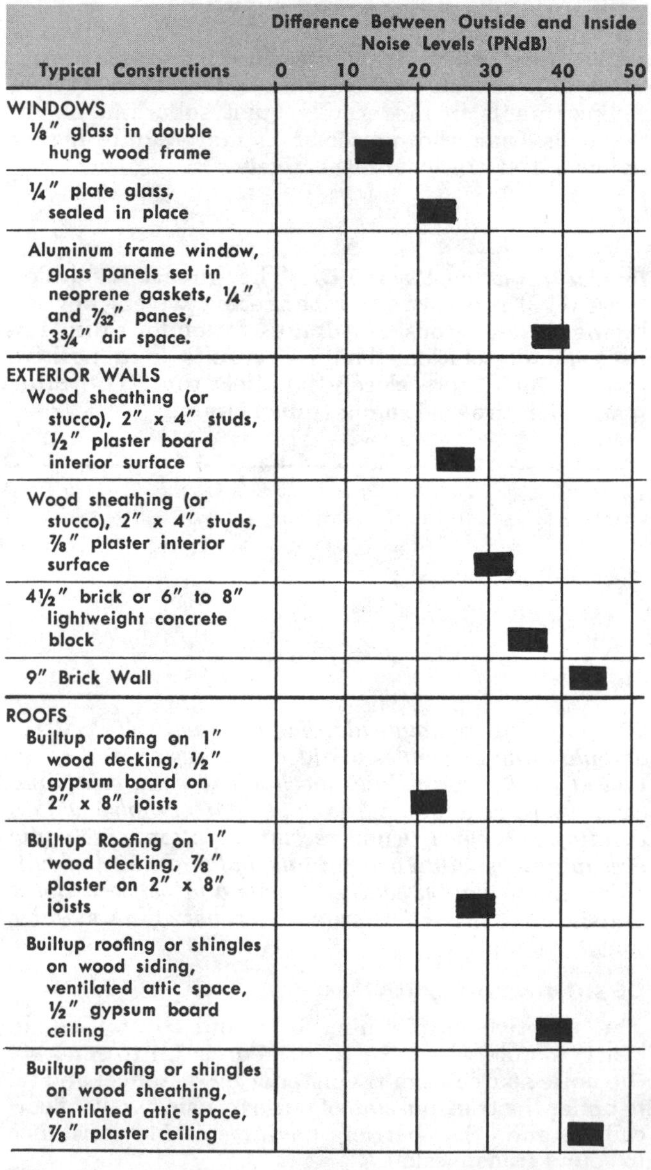
Difference Between Outside and Inside Noise Levels (PNdB)
Typical Constructions
0
10
20
30
40
50
WINDOWS
1/8″ glass in double hung wood frame
1/4″ plate glass, sealed in place
Aluminum frame window, glass panels set in neoprene gaskets, 1/4″ and 7/32″ panes, 3 3/4″ air space.
EXTERIOR WALLS
Wood sheathing (or stucco), 2″ x 4″ studs, 1/2″ plaster board interior surface
Wood sheathing (or stucco), 2″ x 4″ studs, 7/8″ plaster interior surface
4 1/2″ brick or 6″ to 8″ lightweight concrete block
9″ Brick Wall
ROOFS
Builtup roofing on 1″ wood decking, 1/2″ gypsum board on 2″ x 8″ joists
Builtup Roofing on 1″ wood decking, 7/8″ plaster on 2″ x 8″ joists
Builtup roofing or shingles on wood siding, ventilated attic space, 1/2″ gypsum board ceiling
Builtup roofing or shingles on wood sheathing, ventilated attic space, 7/8″ plaster ceiling

Site Planning

Site planning to reduce problems caused by exterior noise should include: (1) separating the noise source from the buildings by distance; (2) siting the buildings to minimize the effect of direct sound transmission, reflection or reverberation from existing or possible future buildings; and (3) using natural topography, other buildings, barrier walls and in some cases landscaping to shield buildings from noise sources.

Building Location Residential buildings should be located as far as possible from noise sources. To pre-

(Continued) **DESIGN AND CONSTRUCTION**

vent buildup of noise by multiple reflection, orientation should not place buildings parallel to each other. Acoustically, the best orientation for a building occurs when the long axis of the building is perpendicular to the noise source (Fig. WF11b).

If a courtyard is desired for other design considerations, the buildings should be oriented so that the end of the U faces away from the noise source. Landscaping in the courtyard will reduce some of the reflected sound.

Noise Barriers For greatest effectiveness, barriers should be as high as possible, be located as close as possible to the noise source, be impervious to air flow and have some mass (Fig. WF11a).

However, great mass will not appreciably increase the effectiveness of barriers, and landscaping will only slightly reduce noise levels. Plantings can help to reduce reverberation noise between buildings. Since noise has a psychological effect, plantings providing a visual barrier to noise sources may help to reduce sensitivity to the noise.

Nonresidential buildings parallel and adjacent to a highway can serve as a barrier between the highway and residential buildings. Hills and highway cuts also can provide natural barriers if the buildings are located to take advantage of these barriers (Fig. WF11c & d).

Interior Planning

Careful planning is essential to reducing the effect of noise in both single family and multifamily residences. Preplanning of interior spaces for sound privacy in the initial design stages can minimize or eliminate the need for costly construction techniques to control noise.

Basic acoustical planning principles should include: (1) locating rooms or apartments as far as possible from noise sources; (2) separating noisy areas from quiet areas by isolating the noise source or using buffer spaces; (3) avoiding air paths through which sound can travel between rooms or apartments, especially in the design, size and placement of doors and windows; and (4) reducing the noise level at the source by selecting quietly operating mechanical equipment and appliances.

Space Layout Buffer spaces such as corridors, stairways or rooms should be used to help provide acoustical buffers against airborne sound transmission between apartments (Fig. WF12a).

Also, room proportions should be planned to reduce sound transmission (Fig. WF12b).

Space layouts such as mirror plans and row plans (Fig. WF13) should be used so that the most effective use of sound isolating systems can be achieved. Such planning should reduce the number of common walls

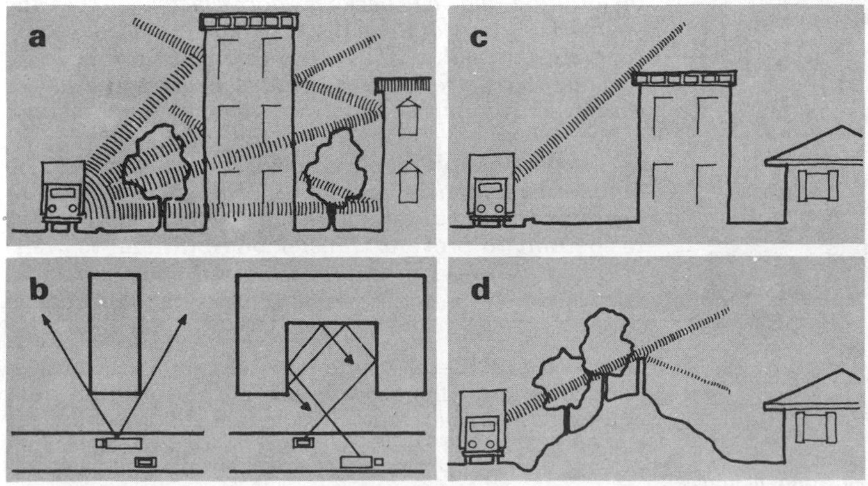

FIG. WF11 (a) Noise barriers should be as high as possible and located as close as possible to the noise source. (b) Perpendicular orientation exposes less wall area to noise source, sound reverberates in U-shaped courtyard. (c) Nonresidential buildings and (d) natural terrain and man-made road cuts serve as noise barriers.

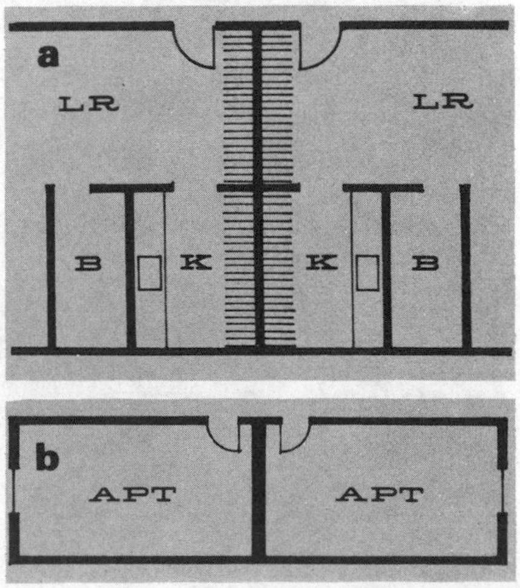

FIG. WF12 (a) Closets and bookshelves can serve as buffers to the noise source. (b) Room proportions can reduce ratio of common wall to floor area.

(Continued) DESIGN AND CONSTRUCTION

to a minimum and place areas with similar noise level requirements adjacent to, above and below each other.

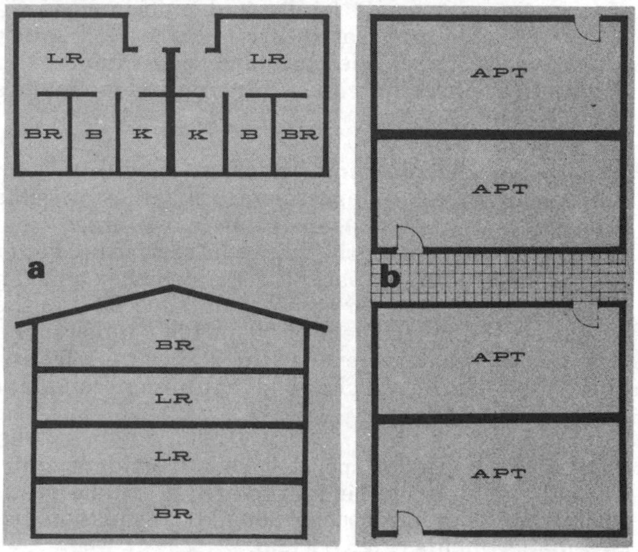

FIG. WF13 (a) Mirroring the plan results in approximately equal noise levels on both sides of common walls and floor/ceiling constructions. (b) Row planning reduces the number of common walls between units.

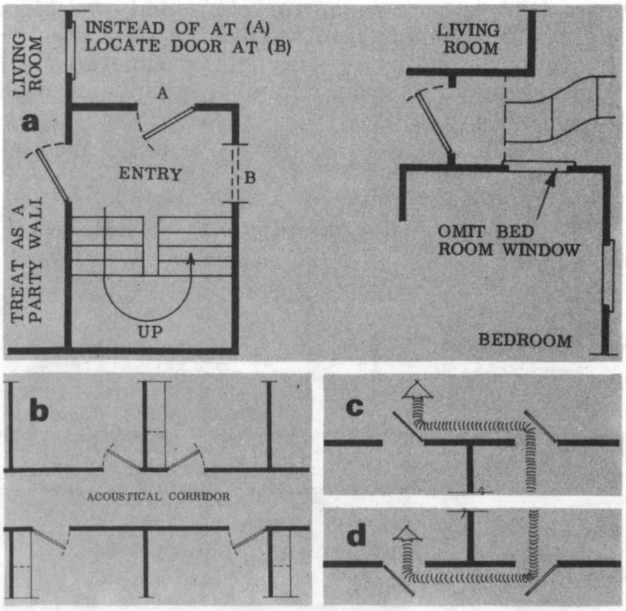

FIG. WF14 (a) Placement of exterior doors and windows should not allow sound to travel easily from one unit to another. (b) Interior doors should be staggered on opposite sides of hallway. Casement windows should not reflect sound from one unit to another—proper (c), improper arrangement (d).

Doors and Windows

Whenever possible, doors, windows and other openings should be placed on the quiet side of the building. If they are necessary on the noisy side and sound control has a higher priority than light, view and ventilation, the size and number of such openings should be minimized.

Exterior doors and windows should be located as far as possible from doors and windows in adjacent apartments. Interior doors should be staggered on opposite sides of a hallway to reduce sound radiated directly from one door to another (Fig. WF14).

Windows should be placed as far as possible from the common wall. Casement windows should be arranged to operate so that the sound from one room will not be reflected into another (Fig. WF14).

Mechanical Equipment Systems

The selection, design and installation of mechanical equipment and systems should minimize both structure-borne and airborne sound transmission.

Heating and Cooling Equipment It is recommended that equipment be selected that has greater than adequate capacity.

Whether a small package unit or a large central system serving several apartments, better quality heating and cooling equipment with more than adequate capacity at slight increase in cost often will help to create a quieter installation and will perform more efficiently over the years. Selecting equipment with greater than adequate capacity, particularly for fans and pumps, can produce considerable noise level reductions because the larger equipment runs more slowly and therefore more quietly.

Whenever possible, mechanical equipment should be mounted resiliently and all piping, electrical connections and duct systems should be flexibly mounted to the unit to prevent transfer of vibrations to walls and floor/ceilings. Vibration isolation pads should be used between the equipment and the supporting structure.

On wood and metal frame construction and on suspended concrete floors, structure-borne vibration from equipment can be transmitted great distances unless the equipment is isolated from the supporting structural construction.

Flues should be isolated from walls by an air gap, especially where they pass through walls of rooms which are intended to be quiet.

(Continued) DESIGN AND CONSTRUCTION

Complex or heavy equipment installations should be designed by engineering specialists, based on acoustical analysis.

Location of Mechanical Equipment Equipment rooms containing furnaces, pumps, compressors, laundry equipment, areas enclosing elevator shafts, incinerator chutes and other potential sources of noise should be located both horizontally and vertically as far as possible from living areas.

In single-family and low-rise apartments, unit forced air heating and cooling equipment preferably should be located on an exterior wall. These major noise sources should not be placed in a closet, an alcove or a utility room near the center of the living unit (Fig. WF15).

If the service door is installed in an interior location, it should be a solid core door and should be gasketed around the perimeter.

The exterior location of air-cooled condensing units should be as far as possible from windows, porches, patios and neighboring dwellings.

Unit air conditioners installed in exterior walls in multifamily residences should be located as far as possible from adjacent apartments so that operation of one unit will not disturb neighbors who may not have their units in operation at the same time.

Duct Systems Planning and installation of duct systems should minimize sound transmitted through

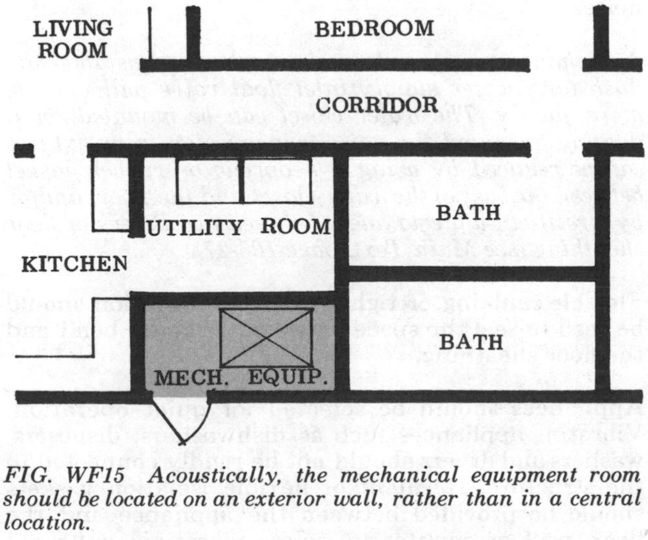

FIG. WF15 Acoustically, the mechanical equipment room should be located on an exterior wall, rather than in a central location.

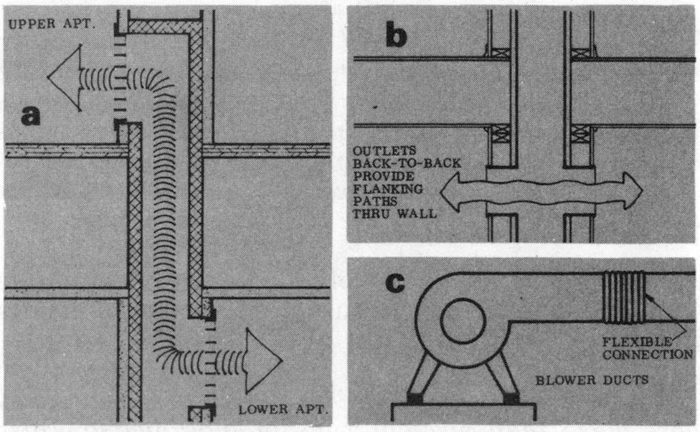

FIG. WF16 (a) Duct systems can conduct noise between units. (b) Duct outlets placed back to back will nullify effect of sound isolating walls. (c) Flexible connections reduce transmission of vibrations from equipment to duct system.

open air paths provided by openings and ventilators or transmitted through walls, particularly where the ducts pass vertically through living spaces (Fig. WF16).

Duct outlets should not be located back-to-back between rooms.

Ducts should be formed of adequate thickness of metal to minimize vibration, and they should be isolated from mechanical equipment with flexible collar connections to prevent transmission of vibrations. The space around ducts should be tightly caulked for isolation where ducts pass through sound conditioned construction.

Duct insulation and duct mufflers will help reduce furnace noises and will help to prevent short circuiting sound through ductwork between rooms (see Main Text, page 106-26).

Grilles should be designed to produce outlets that do not whistle or cause other annoying noises.

If a door is to provide acoustical privacy, it should not be used as a means of returning air to the furnace. Separate air return registers connected to a sound absorbing duct system should be placed in each room, or sound absorbing air inlets should be used to allow air intake from corridors without admitting noise.

Piping In planning of supply piping systems, precautions should be taken to reduce plumbing noises caused by high water pressure and velocity and quick closing valves.

(Continued) **DESIGN AND CONSTRUCTION**

If dwellings do not exceed 3 stories and the water pressure at the street exceeds 35 to 45 psi, a pressure reducing valve can be installed, preferably as far as feasible from the building. In high-rise buildings, combinations of pressure reducing valves and pressure regulating systems may be necessary to provide suitable and uniform pressure at the various floor levels.

Piping layout should be planned with a minimum number of bends and the minimum number of fittings and valves required.

Valves and right angle fittings produce substantially more noise than straight runs of pipe due to water turbulence. Noise produced by 4 elbows is about 10 dB higher than the noise produced in a straight run of equivalent size pipe under conditions of turbulent flow.

It is essential that all holes and openings where piping passes through common walls be sealed with resilient material or be tightly packed with insulation, or both, to isolate vibration of the pipe from the structural framing and to seal against air leaks (Fig. 17a).

Resilient pipe insulation should be used between pipe supports and piping to reduce transmission of noise resulting from pipe vibrations (Fig. WF17b). The number of supports should be kept to the minimum required to structurally support the pipe, and supports should be attached to the most massive structural element available.

Running drain piping through bedroom and living room walls should be avoided, unless a double sound isolating wall is provided.

Standard pipe clamps and resilient vertical support or wood bracket should be installed with the end of the support resting on the vertical stack with an isolation pad resting on the support (Fig. WF17b).

A short section of pipe insulation around the pipe where it passes through wall sills should be used to keep it in horizontal alignment and to isolate it from the structure (Fig. WF 17b).

Long, straight runs of hot water supply piping should be designed with flexibility to reduce noise produced by creaking as the piping expands or contracts from hot or cooled water (Fig. WF17c).

Supply piping to units which use large amounts of water (such as dishwashers and water closets) should not be undersized.

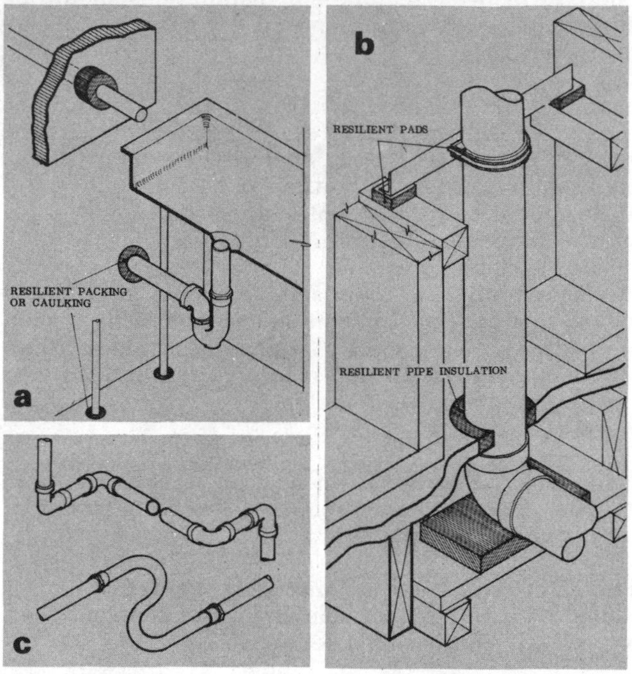

FIG. WF17 (a) Where pipes pass through walls, openings must be tightly packed or sealed airtight with resilient materials. (b) Vibrations from plumbing stack can be isolated from structure by resiliently mounting vertical supports and horizontal runs. (c) Flexible connections in piping and tubing minimize noise caused by expansion and contraction of long runs of piping.

Fixtures and Appliances Plumbing fixtures and appliances should be selected and installed to reduce noise.

A siphon jet water closet having an adjustable-rate flush tank water supply inlet float valve will operate more quietly. The water closet can be mounted on a floating floor and the noise transmission from fixtures can be reduced by using a neoprene or rubber gasket between the base of the water closet and the floor, and/or by a resilient underlayment between two layers of floor sheathing (see Main Text, page 106-27).

Flexible caulking or tightly packed insulation should be used to seal the space between the toilet bend and the floor sheathing.

Appliances should be selected for quiet operation. Vibrating appliances such as dishwashers, disposers, washers and dryers should not be rigidly connected to the structure. Resilient or flexible isolation gaskets should be provided between the appliance and the floor, wall or counter to isolate vibration; a floated

(Continued) DESIGN AND CONSTRUCTION

floor system also can be used to isolate vibration from the structural floor system. Piping connections to appliances should be flexible enough to prevent vibrations from being transmitted to piping. Electrical connections to appliances also should be flexible, whenever possible.

Electrical Systems Care should be taken in the installation of electrical systems so that the acoustical performance of floor/ceiling and wall constructions will not be reduced by: (1) open air paths where wires pass through walls, floor/ceilings and/or wall plates, or provided by openings for outlets, switches or fixtures; and (2) noise vibration transmitted by wires or fixtures rigidly connected to otherwise separated structural members or connecting equipment to the structure.

In multifamily housing, each apartment should be wired as a unit, keeping wires within the walls of the same unit. All holes should be tightly caulked with resilient, nonhardening compound (Fig. WF18a).

Convenience outlets, switches, telephone and other such installations should not be placed back to back: (1) Convenience outlets should be placed at least 36″ apart on opposite sides of the wall. In staggered stud walls, installations should be offset horizontally by at least 3 complete stud spaces (Fig. WF18b). (2) A horizontal distance of 24″ should be maintained between switches and outlets and they should not be located in the same stud space (Fig. WF18c). (3) Wall fixtures and equipment should be kept 24″ apart on opposite sides of the wall (Fig. WF18d).

Lighting fixtures and other electrical fixtures should be surface-mounted (not recessed), and they should not be connected to joists above resiliently mounted ceiling finishes or connected across separated ceiling joists (Fig. WF18e).

CONSTRUCTION RECOMMENDATIONS

Suitable sound isolating constructions for wall, floor/ceiling and mechanical systems should be selected to reduce noise transmission to acceptable levels (see page WF106-21 for Selection Criteria).

A wall or floor/ceiling construction that will adequately

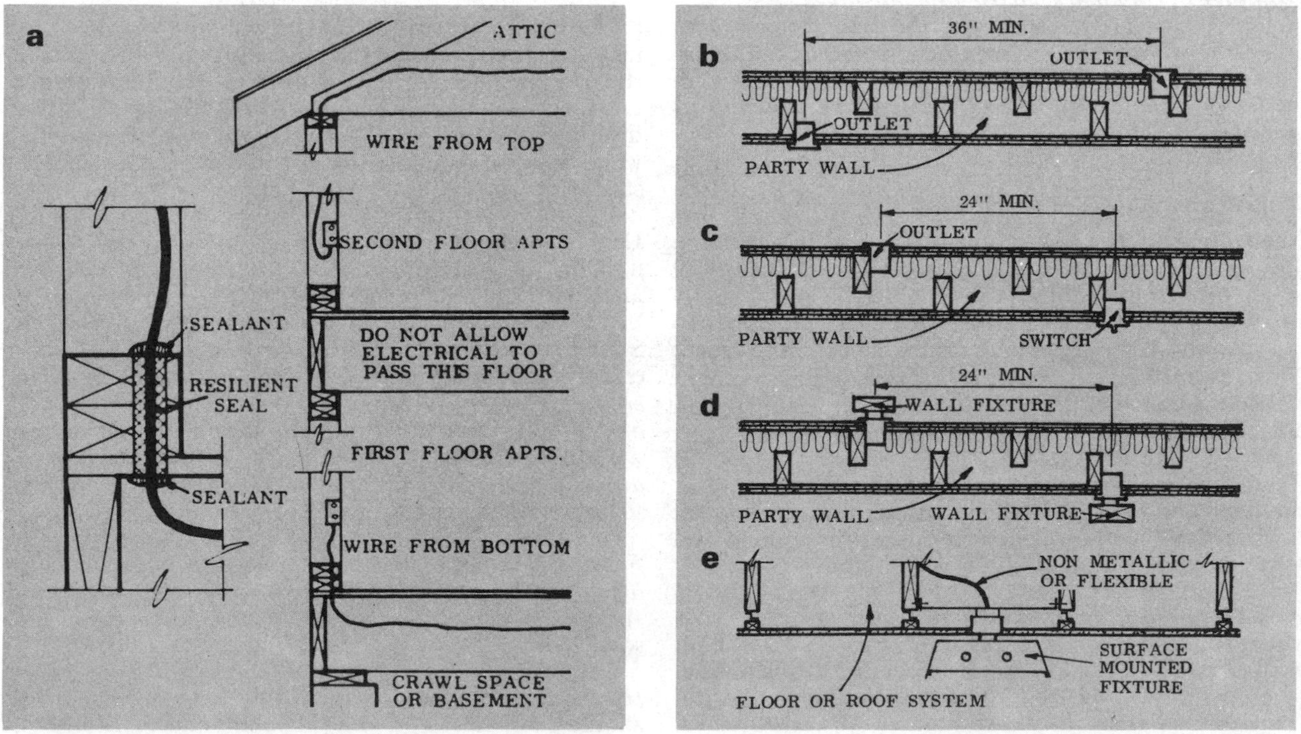

FIG. WF18 (a) Each apartment should be wired independently with air passages sealed airtight. (b, c & d) Electrical outlets, switches, fixtures and equipment should not be mounted back-to-back, should not be located in the same stud space and should not be spaced closer than shown above. (e) Ceiling fixtures should be surface mounted, rather than recessed.

(Continued) **DESIGN AND CONSTRUCTION**

reduce the transmission of conversational speech may be inadequate in blocking low frequency sounds of mechanical equipment. High frequency sounds are usually easier to control than low frequency sounds.

Good-to-excellent workmanship, careful supervision of construction and coordination of various trades should be maintained to achieve effective sound control. Workmen and other key personnel should be taught the reasons for care in applying sound control measures.

Quality workmanship is absolutely essential in the successful application of sound control principles.

Manufacturer's instructions specifying sound control installation procedures should be followed precisely by workmen. Slight variations from installation instructions, which may seem insignificant, can nullify the effectiveness of costly sound control measures.

Nails which are too long that short circuit resilient mounts and failure to seal small cracks or holes are examples of common construction mistakes. Assuming that a construction selected on the basis of test data will perform adequately, obtaining suitable workmanship in the field is the greatest single problem in achieving satisfactory sound conditioned apartments and houses.

FIG. WF19 ACOUSTICAL PERFORMANCE OF TYPICAL WOOD FRAME WALLS

Wall Construction	Approximate STC Rating			
	Without Insulation	With Insulation[1]	Sound Board	Deadening Both Sides[2]
Single Stud Wall Single layer surfaces	34	36		46
Double layer surfaces	40	—		—
Single Stud Wall with resilient channel mounting one side	40	50[3]		—
Staggered Stud Wall	41	49		50
Double Stud Wall	43	53		50

1. Minimum thickness of insulation (2" or 3").
2. Gypsum board surfaces laminated, with exception of double stud wall which was laminated and nailed 8" o.c. into studs and floor and ceiling plates.
3. Resilient channels installed throughout; fire-rated wall with fire stopping top and middle heights or backer strips for attaching wall base, STC 46.

Wall Systems

Regardless of the system selected, wall surfacing materials should be installed over the entire wall area—especially behind tubs, cabinets, duct enclosures, soffits and above suspended ceiling lines where the structural framing might otherwise be left exposed. This precaution is particularly critical for common walls between living units to assure that sound will not be transmitted through an unfinished area.

Resilient, nonhardening caulking should be used at the floor and ceiling of all partitions and around all perforations, such as electrical outlets, plumbing and other openings to make the wall airtight.

Wood Framing Systems Single conventional stud walls, finished with one layer of gypsum board on each side, offer poor resistance to sound transmission when privacy is desired. Where resistance to sound transmission through wood frame walls is required, one or more of the following measures should be selected, depending on the degree of sound control required:

(1) Increasing the weight of the wall and the surfacing materials; (2) Including sound absorbing blankets in the free air space within the wall; (3) Resilient mounting of the surfacing materials; (4) Using staggered framing members to effectively separate the finished wall surfaces; or (5) Using two separate independent stud walls separated by an air space.

Each system should be selected on the basis of test data for the system and should be installed in accordance with specified procedures.

Sound absorbing blankets within the stud cavity of a conventional single stud wall will not appreciably increase the sound isolating performance. However, they are effective when used in walls where the surfaces of the wall are effectively separated. A summary of the acoustical performance of wood frame walls is given in Figure WF19.

For a discussion of the various wood frame sound isolating walls, see Main Text, page 106-29. See page WF106-25 for an index to STC ratings for various constructions.

Resilient channels for mounting the surface finish should be installed on one side only; channels on both sides of the wall do not improve performance significantly because of the resiliency of the wall through sympathetic vibration of both surfaces.

(Continued) DESIGN AND CONSTRUCTION

For maximum performance, without regard for fire rating, resilient channels should be used throughout to mount the surface. Backer strips for attaching wall base or horizontal blocking at the top and middle height for fire stopping will decrease performance.

Resilient channels should be handled and installed carefully to avoid permanent deflection. The wall should contain no openings, and heavy objects such as fixtures and cabinets should not be mounted on the resilient channel side of the wall.

When sound deadening board is used for resilient mounting, surfacing materials should be correctly laminated to the sound deadening board for maximum performance. If surfacing materials are nailed to meet fire rating requirements, the STC of the wall is reduced: approximately 10 dB when nails are spaced 8″ o.c., approximately 8 dB with nail spacing of 16″ o.c., and approximately 4 dB with nail spacing 24″ o.c.

The addition of insulation to resiliently mounted walls increases acoustical performance proportionately with increased thickness of insulation.

Staggered stud walls, consisting of 2″ x 4″ studs staggered on a single 2″ x 6″ plate, offer effective sound isolating walls which can accommodate surface mounting of cabinets or equipment, because the surfaces of the wall are mounted on independent framing members.

For maximum performance, wall base should not be nailed into the bottom plate of a staggered stud wall.

The addition of insulation to the stud cavity or sound deadening board to both sides of a staggered stud wall will increase performance appreciably. However, the performance does not increase proportionately with increased thickness of insulation, because sound will travel through the single plate and a minimum of 2″ or 3″ thickness of insulation can be used effectively.

Double stud walls, consisting of two separate 2″ x 4″ stud walls on two separate 2″ x 4″ or 2″ x 3″ plates spaced 1/4″ to 1″ apart, can accommodate surface mounting of cabinets and equipment without reduced effectiveness; and less performance variations due to materials and workmanship result with this type of construction.

To perform acoustically, the "separated" double stud walls should not be tied together in any manner when plumbing, wiring or other systems are installed.

Performance of the double stud wall with separate plates will improve with increased thickness of insulation.

Metal Framing Systems Because of the resilient nature of metal channel studs themselves, the addition of resilient channels to mount surfaces does not appreciably improve performance of the wall. Where resistance to sound transmission through metal frame walls is required, one or more of the following measures should be selected, depending on the degree of sound control required:

(1) Increasing the weight of the surfacing material on one or both sides of the wall; (2) Adding sound deadening board to one or both sides of the wall; or (3) Adding sound absorbing insulation within the cavity of the wall.

Each system should be selected on the basis of individual test data for the system and should be carefully installed in accordance with specified procedures.

For a discussion of various metal frame sound isolating walls, see Main Text, page 106-31. See page WF106-25 for an index to STC ratings for various constructions.

FIG. WF20 ACOUSTICAL PERFORMANCE OF TYPICAL METAL CHANNEL STUD WALLS

Gypsum Board Surfacing	Approximate STC Rating					
			Sound Deadening Board One Side		Sound Deadening Board Two Sides	
	Without Insulation	With Insulation*	Without Insulation	With Insulation*	Without Insulation	With Insulation*
Single layer each side	38	45	43	49	47	54
Double layer one side, single layer one side	44	50	46	51	52	54
Double layer each side	49	57	53	57	54	57

*Minimum thickness of insulation (2″ or 3″).

(Continued) **DESIGN AND CONSTRUCTION**

Generally, the precautions recommended for resiliently mounted wood frame walls apply to metal frame walls. For maximum performance, the wall should contain no openings, and heavy objects such as cabinets or fixtures should not be mounted on either side of the wall.

A summary of the acoustical performance of metal channel stud walls is given in Figure WF20. The effects of adding gypsum board, insulation and sound deadening board in metal channel stud construction are cumulative.

The difference between the acoustical performance of walls constructed with 3-5/8″ metal studs and 2-1/2″ metal studs is considered negligible for practical purposes.

Double channel stud (chase) walls, consisting of two rows of 1-5/8″ studs tied together with bracing ties, provide effective sound isolating walls which can accommodate plumbing, wiring and other mechanical systems within the wall cavity. (See Main Text, page 106-32).

Concrete and Unit Masonry To perform acoustically, concrete and masonry walls should be sealed to make them airtight. Even relatively fine cracks and open joints, such as may occur through poor workmanship or at partition edges, will destroy the acoustical effectiveness of the wall.

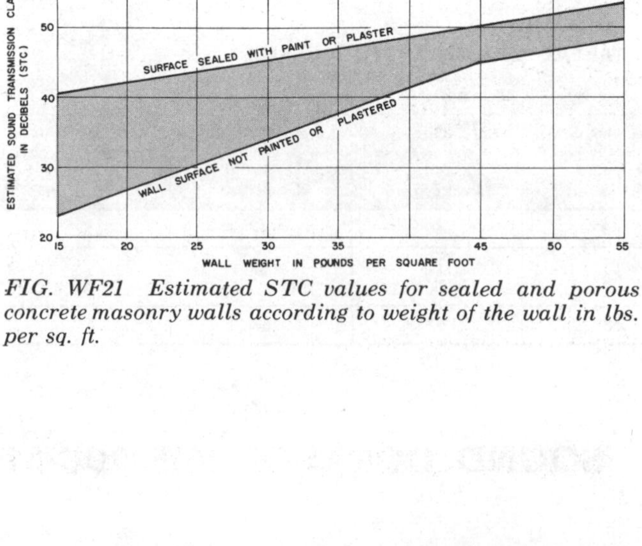

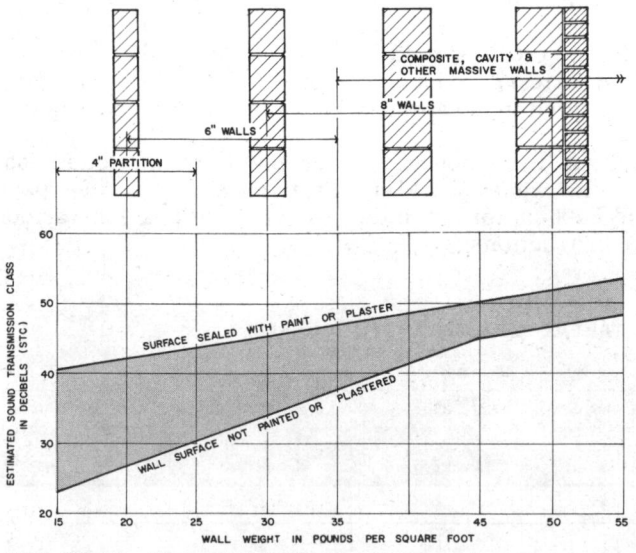

FIG. WF21 Estimated STC values for sealed and porous concrete masonry walls according to weight of the wall in lbs. per sq. ft.

Because of the great number of mortar joints (and therefore possible acoustical leaks), it is recommended that all brick masonry walls be sealed with plaster where acoustical privacy is desired.

Surfaces of concrete and masonry walls should be sealed over their entire area—behind cabinets and surfacing materials, in closets, etc.—to eliminate flanking paths through these areas. Even if walls are to be furred and surfaced with gypsum board as the interior finish, porous walls should be sealed by plastering or painting before being furred. Generally there is sufficient sound leakage at the top and bottom of the wall to allow sound which passes through an untreated masonry wall to reach living units, unless it is sealed.

Polyethylene film installed over the surface as a vapor barrier does not seal the surface acoustically.

Lightweight concrete block typically contain a greater volume of pores than dense units and for this reason may excel in sound absorption but are much less effctive in resisting sound transmission. Lightweight block of various manufacturers are subject to a wide variation of transmission loss when not sealed, because of large differences in pore structure. Thorough sealing of the surface pores increases the performance of lightweight units about 7 dB when painted and about 10dB when plastered (Fig. WF21). When dense units are either painted or plastered, the increase is only about 2 dB. However, because of their greater mass, dense units generally perform better than lightweight units.

Cavity wall construction and combinations of dense and porous units offer improved acoustical performance and can provide STCs of 50 or better (Fig. WF21). Using a combination of two 4″ lightweight wythes with a 3/8″ or more air space between and one wythe plastered on the cavity side, the faces of the wall need not be sealed and the sound absorbing porosity of the lightweight units can be retained.

For a discussion of concrete and unit masonry wall systems, see Main Text, page 106-32; see page WF106-25 for an index to STC ratings for various systems.

Floor/Ceiling Systems

Generally, all of the conventional wood, concrete and metal floor/ceiling systems provide negligible resistance to impact sound transmission. Where acoustical privacy is required, one or more of the following measures to control impact sound should be provided, depending on the degree of sound isolation required:

(Continued) DESIGN AND CONSTRUCTION

(1) Cushioning the impact with soft floor surfacing materials; (2) Providing discontinuous construction between the structural floor and the finished ceiling by resilient mounting of the ceiling or use of separate ceiling joists; (3) Providing floated floor construction to resiliently float the finished floor from the subfloor and/or structural floor; and (4) Including sound absorbing materials within the cavity of discontinuous frame construction.

Discontinuous ceiling construction and floated floor systems also will provide a good degree of resistance to airborne sound transmission, providing the essential precautions of construction practices are followed (see page WF106-18, Wall/Floor Connections).

FIG. WF22 EFFECT OF CARPETING IN CONTROLLING IMPACT SOUND ISOLATION[1]

		Floor/Ceiling Construction	
		Concrete Slab[3]	Wood Joist[4]
Carpet[2]	Pad	IIC	IIC[5]
20 oz. Wool	(none)	53	—
40 oz. Wool	(none)	57	—
60 oz. Wool	(none)	65	—
44 oz. Wool	44 oz. sponge rubber bonded to carpet	68	54
20 oz. Wool	40 oz. felt	69	—
44 oz. Wool	44 oz. hair & jute covered with foam rubber face and back	70	59
40 oz. Wool	40 oz. felt	72	—
44 oz. Wool	40 oz. all hair	72	59
44 oz. Wool	66 oz. all hair	72	63
60 oz. Wool	40 oz. felt	73	—
44 oz. Wool	40 oz. hair & jute	73	61
44 oz. Wool	20 oz. combination hair and jute, foam rubber face, latex back	73	61
44 oz. Wool	Urethane foam	75	—
44 oz. Wool	40 oz. sponge rubber	76	65
44 oz. Wool	31 oz. foam rubber on burlap (3/8" thick)	79	67
44 oz. Wool	80 oz. sponge rubber	80	68

1. Based on tests performed by Kodaras Acoustical Laboratories, Inc. for the American Carpet Institute. IIC values approximated by adding 51 to original INR test values.
2. Carpet pile weight given in oz. (/sq. yd.).
3. 5" concrete slab (sand gravel aggregate), no ceiling finish.
4. 2" x 8" wood joists, 16" o.c.; 5/8" T&G plywood nailed to joists as subfloor; 5/8" gypsum board nailed to joists as ceiling.
5. Values given only where test data is comparable to tests performed on concrete slab construction.

Using sound absorbing materials within the cavity of frame construction is significantly effective in controlling impact sound only when used in conjunction with a resiliently mounted ceiling.

There is little difference in the acoustical performance of the various hard floor surfacing materials such as stripwood, wood block and resilient tile and sheet flooring.

For a discussion of various wood, concrete, and metal floor/ceiling systems, see Main Text, page 106-34. See page WF106-25 for an index to INR ratings for various construction systems. For a discussion of FHA Impact Insulation Class (IIC) see page WF106-7 and page WF106-21, Selection Criteria.

Cushioning the Impact Where a particularly high degree of impact sound isolation is required, the use of carpet and padding will provide the greatest resistance to impact sound transmission for any type of floor/ceiling construction. Performance will vary depending on the thickness of the carpet, degree of matting of the pile and weight of both carpet and pad in oz. per sq. yd. (Fig. WF22).

When cushioning the impact is the only measure intended to control impact sound transmission, acoustical performance of the basic floor/ceiling system should be adequate to control airborne sound transmission as well.

No appreciable gain in airborne sound isolation is obtained by using carpet and pad or by adding acoustical surfacing materials to the ceiling below.

Wood Frame Construction Where an IIC of 51 or better is required, one of the following methods should be employed, depending on the degree of sound isolation required:

(1) Use of carpet and padding alone or in combination with a floated floor or separated ceiling construction; (2) Separated ceiling construction (either resiliently mounted ceiling or use of separate ceiling joists) in conjunction with a floated floor construction.

In separated ceiling construction, appropriate construction techniques should be employed to prevent impact sound from bypassing the ceiling through walls and other structural connections (see Page WF106-18, Wall/Floor Connections).

(Continued) DESIGN AND CONSTRUCTION

As with resiliently mounted walls, no heavy objects such as electrical fixtures should be mounted on or recessed in a resiliently mounted ceiling and airtight construction should be maintained.

In floated floor construction, resilient underlayment used to float the finished floor should be capable of resisting major deformation from loading.

The floating floor construction should not be rigidly connected with surrounding elements.

Nailing or stapling of the finished floor or subfloor through the underlayment into supporting construction should not be allowed.

Adhesive generally is used to attach resilient underlayment. Nailing or stapling of the finished floor or subfloor through the underlayment will nullify the sound isolating properties of the construction.

Lightweight concrete placed directly over a plywood subfloor does not appreciably improve the impact sound resistance of a basic wood joist floor/ceiling construction. However, when the concrete is resiliently floated, the improvement is about 10 dB, but the construction still will only result in less than 51 IIC, and the addition of carpet and pad is necessary or the ceiling must be resiliently mounted to achieve a plus 51 IIC rating.

Concrete Slab Construction To achieve an IIC of better than 51, one of the following measures should be used, depending on the degree of sound isolation required:

(1) Cushioning the impact with resilient material such as carpet and pad alone or in combination with a floated floor system; or (2) Use of a floated floor system.

Because of the considerations of practicality and cost, the ceiling under concrete slab construction is not often resiliently mounted. Weight and thickness of concrete slabs has little effect in reducing impact sound transmission to an acceptable level.

The provisions to keep the floated floor from rigid contact with structural elements should be observed (see page WF106-18, Wall/Floor Connections).

The performance of precast concrete floor/ceiling systems is similar to that of placed concrete. Available test data for such systems is limited; representative test data and precast floor systems are given in the Index, page WF106-25.

Metal Frame Construction The performance of steel bar joist and other similar floor/ceiling systems can be improved by measures similar to those used for wood frame construction. Methods for floated floor and resiliently mounted ceiling systems closely parallel those used for wood frame construction.

The acoustical effectiveness of steel bar joist construction, as well as other constructions such as metal deck and concrete joist systems, depends primarily on sealing the wall/floor connections airtight. Walls must be brought above the suspended ceiling line and be sealed at the under side of the floor slab or deck.

When walls run perpendicular to joists, proper sealing is difficult and expensive to achieve. Also, the relatively thin (2") concrete slab normally placed over steel bar joist construction does not have sufficient mass to provide more than an STC in the high 40's. Therefore, when a higher STC rating is necessary for either floors or walls this type of construction has not proven effective. Even though the wall system may provide suitable sound isolation, the sound can bypass the wall through the flanking path provided by the lightweight floor system.

For representative steel bar joist constructions and available IIC test data, see the Index, page WF106-25.

Wall/Floor Connections

Where walls and floors are joined, precautions should be taken to prevent both airborne and structure-borne sound from penetrating around and through walls and floors.

Even though the wall and floor constructions are structurally correct, sound can be transmitted through flanking paths unless precautions are taken.

Airborne Sound Both the wall and floor/ceiling systems selected should have comparable sound isolating abilities so that a weakness in one system will not diminish the effectiveness of the other.

Spaces between floor joists should be blocked at the juncture of walls and floors to reduce airborne sound through such flanking paths. Joints at the top and bottom of the partition between the floor and the ceiling should be sealed airtight with a resilient non-hardening acoustical sealant or resilient material (Fig. WF23).

Structure-borne Sound Floated floor construction should not be heavily loaded from structural elements. Heavy loading of a floated floor will cause the resilient

(Continued) DESIGN AND CONSTRUCTION

underlayment to collapse, causing damage to the floors and settling of the structure.

Bottom plates of a partition should not be supported on the top of the floating floor. The plates should be fastened instead to the subfloor or supporting structure.

The floating floor should not be rigidly connected to surrounding structural elements (Fig. WF23).

For maximum acoustical performance, provision for a break in the subfloor should be made under partitions to help reduce impact noise transmission between adjacent rooms by breaking direct contact of the structural elements.

Whenever possible, floating floors should not touch partition edges (Fig. WF23). A cushion between the edge of the partition and the floor can be provided.

Floated concrete floors on wood joist construction should be confined to a single room to prevent transmission of impact noise along the floor (Fig. WF23). Isolation joints around the perimeter of all rooms and across doorways should be provided. Walls should be supported on a floor plate, with polyethylene film installed over the resilient underlayment and isolation joints installed around the perimeter of the room before the concrete is placed. Molding should be used to cover the isolation joints.

Wall surfacing material should not be tightly forced against resiliently mounted ceilings. This can collapse a flexible ceiling hanger, channel or clip and short circuit the resilient mounting.

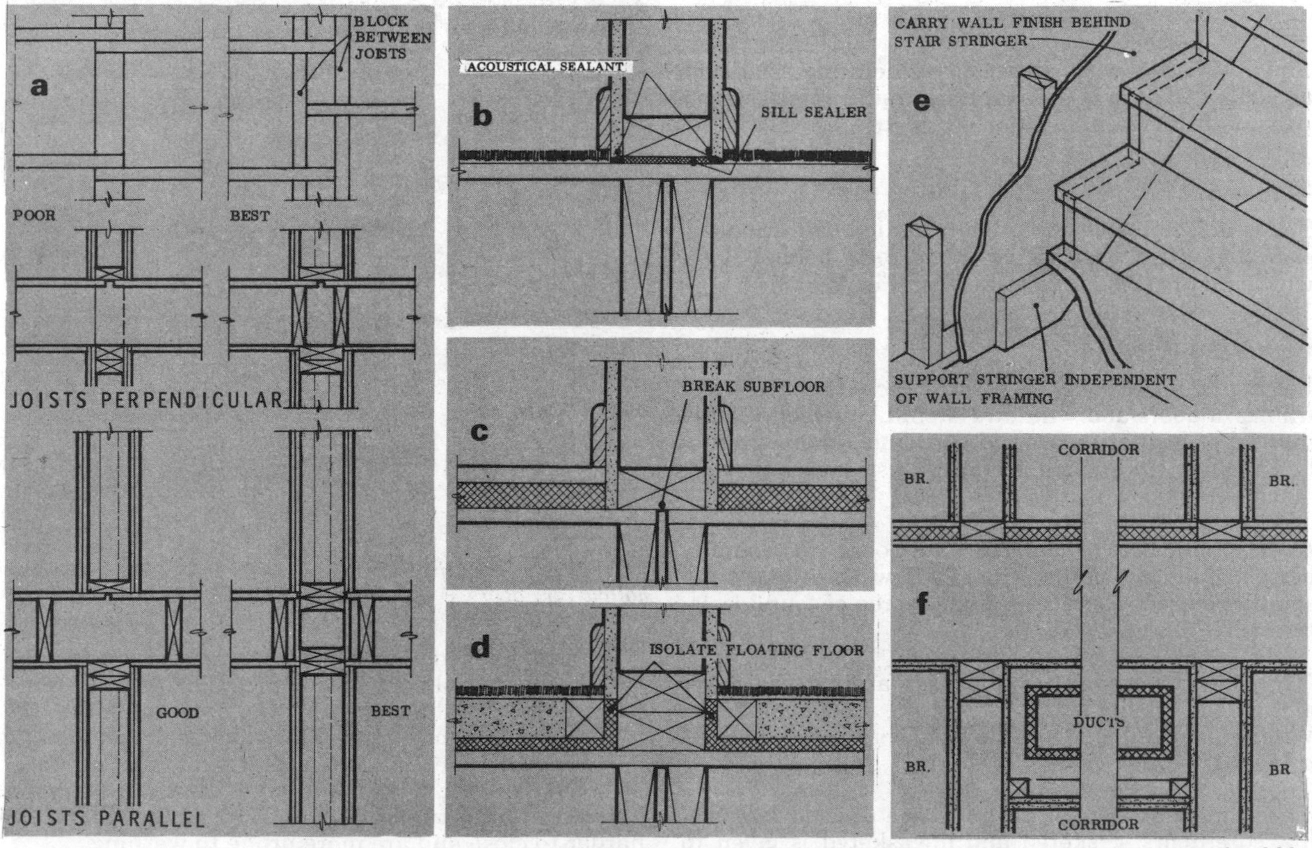

FIG. WF23 (a) Open air paths should be closed to minimize problems of airborne flanking noise. Spaces between joists should be blocked, particularly when joists are framed perpendicular to walls. (b) Acoustical sealant should be used to seal all open air paths. (c) Breaking direct contact between structural elements and (d) isolating floating floors from walls will minimize structure-borne noise. (e) Structural stair stringers should not be rigidly connected to adjacent walls. (f) Ducts can be installed after walls and ceiling are finished, with a furred ceiling added below.

(Continued) **DESIGN AND CONSTRUCTION**

Stairs

Stair stringers should be fastened only at top and bottom and nailing to adjacent studs should be avoided. At stair landings, a header joist should be used to support floors, instead of nailing floor joists to the studs (Fig. WF23).

Continuous wall surfacing material should be installed on the entire wall surface before stairs are installed to reduce sound transmission between stairs and adjacent rooms. If walls are to be plastered, it is recommended that temporary construction ladders be used in order to permit complete plastering and to eliminate any breaks in the wall.

When stairs are on both sides of a common wall, such as in duplex or other attached housing units, staggered or double stud walls with insulation are recommended in wood frame construction.

Finishing stairs with carpet or resilient material such as rubber mats will substantially reduce noise transmission from stairs to adjacent apartments.

Hallways

Surfacing material should be applied continuously over the entire wall and ceiling areas, behind ductwork and other such installations (Fig. WF23).

Doors and Windows

For maximum acoustical performance, the edges and frames of doors and windows should be sealed. Caulking and/or insulation around door and window frames is necessary to reduce sound transmission through cracks.

Doors Solid core or mineral filled doors are generally better sound insulators than hollow core doors and should be used wherever a high degree of sound isolation is required.

However, the amount of air space around the edges of the door usually is the controlling acoustical factor. Whenever possible, doors should be gasketed if sound privacy is required.

The relative acoustical performance of solid and hollow core doors, gasketed and ungasketed is given in Figure WF24.

Soft weatherstripping provides a good seal at the top and sides of the door, while automatic threshold closers are needed to seal the bottom. While it may be costly to use gasketed doors, provision of a cushion such as sponge rubber between the door and jamb will reduce the impact energy of a door slam and make door closure quieter. Door closure control may also be considered for apartments opening onto common hallways. Unfortunately, properly sealed doors are harder to close and are more prone to warping.

Use of undercut doors to allow for rugs or return air movement in forced air ventilation systems should not be allowed.

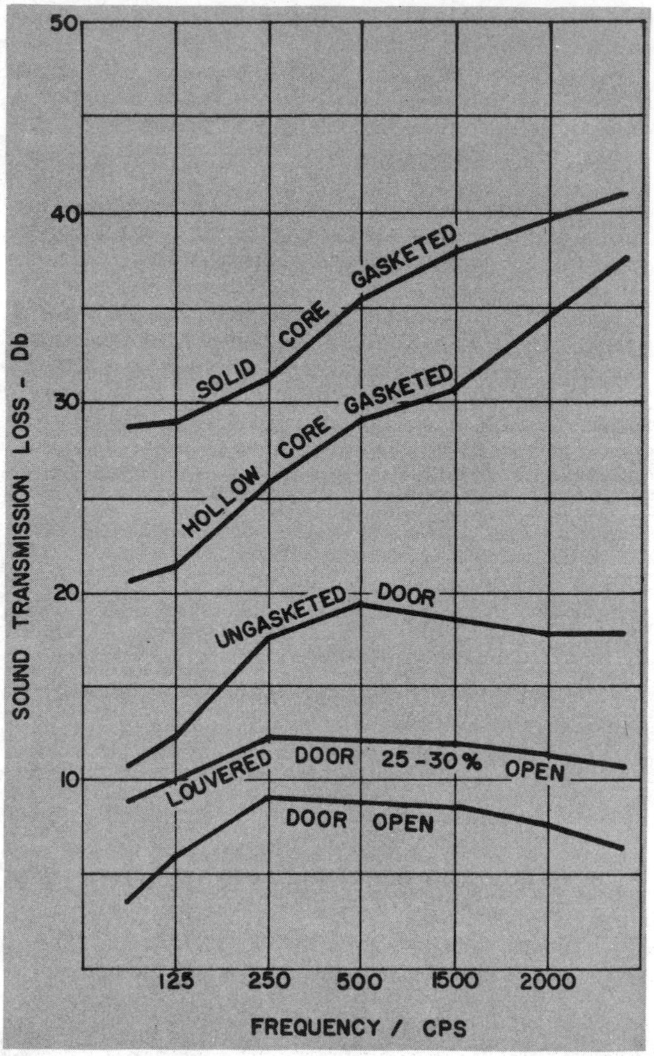

FIG. WF24 Relative acoustical performance of solid and hollow core doors.

(Continued) DESIGN AND CONSTRUCTION

Sliding doors should not be used where sound privacy is required because a relatively tight air seal cannot be maintained.

Equipment room doors should be sealed with gaskets and door sills for good acoustical performance. Where an equipment room door leads to a hallway, a second door with better acoustical performance is recommended.

Windows Windows should be sturdy, well weather-stripped and tightly closed. Approximately 80% of the acoustical problems occuring with windows results from the quality of airtightness of construction and about 20% from the selection of the glass itself. Fixed sash perform better than operating sash.

SELECTION CRITERIA

The following recommended criteria are suggested for selecting walls and floor/ceiling constructions to provide resistance to airborne and structure-borne sound transmission.

Generally, if the level of conversational speech is reduced by a wall or floor/ceiling to the point where it cannot be understood in an adjoining room, the construction will provide acceptable overall airborne sound privacy. To estimate the bare minimum STC requirement for a construction to obtain privacy, the level of background noise on one side of the construction is subtracted from the level of sound that strikes the other side. Figure WF26 illustrates the approximate effectiveness of walls or floor/ceilings with varying STC values at varying background noise levels.

General recommendations for floor/ceilings in terms of INR ratings are given in Fig. WF27. General recommendations for common walls and floor/ceilings separating living units in terms of STC and IIC ratings are given in Figs. WF28, WF29 and WF30.

FIG. WF26 SOUND ISOLATION OF WALLS IN TERMS OF STC VALUES

SOUND TRANSMISSION CLASS (STC)				SPEECH AUDIBILITY[2]
BACKGROUND NOISE[1]				
None	20dB	30dB	40dB	
40 To 50	25 To 35	15 To 25		Normal speech easily understood
50 To 60	35 To 45	25 To 35	15 To 25	Loud speech easily understood. Normal speech 50% understood
60 To 70	45 To 55	35 To 45	25 To 35	Loud speech 50% understood Normal speech faintly heard but not understood
70 To 80	55 To 65	45 To 55	35 To 45	Loud speech faintly heard but not understood Normal speech usually inaudible
	65 To 75	55 To 65	45 To 55	Loud speech usually inaudible

1. Background noise levels on "A" scale of sound level meter.
2. "Loud" speech taken as 10dB higher than "normal" speech; intelligibility taken as percent of sentences understood; normal source and receiving room absorption considered.

FIG. WF27 IMPACT NOISE RATING (INR) SELECTION CRITERIA[1]

Location of Floor/Ceiling[2]	Impact Noise Rating (INR)	
	Low Background Noise (Land-Use Intensity less than 6.0)[3]	High Background Noise (Land-Use Intensity 6.0 or higher)[3]
Separating Living Units	0	−2
Corridor above Living Units	+5	+2
Living Units above public Space or Service Area[4]	−5	−8
Public Space or Service Area above Living Units[5]	+5	+5

1. Adapted from FHA Minimum Property Standards for Multifamily Housing.
2. For living units above 8th floor in high-rise buildings use Low Background Noise column.
3. Refers to land use intensity rating determined by FHA based on land area, open space, livability space, recreation space and car storage capacity.
4. Not applicable to floors over storage rooms and similar spaces not containing mechanical equipment or where noise from living unit would not be objectionable.
5. Not recommended; where situation is unavoidable, special acoustical analysis, control and isolation of equipment may be required.

FIG. WF28 GENERAL CRITERIA FOR AIRBORNE AND IMPACT SOUND ISOLATION BETWEEN DWELLING UNITS*

Construction	Grade I	Grade II	Grade III
Wall Partitions	STC $\geq$ 55	STC $\geq$ 52	STC $\geq$ 48
Floor/Ceilings	STC $\geq$ 55 IIC $\geq$ 55	STC $\geq$ 52 IIC $\geq$ 52	STC $\geq$ 48 IIC $\geq$ 48

*Adapted from FHA "Guide to Airborne, Impact, and Structure Borne Noise Control in Multifamily Dwellings."

(Continued) DESIGN AND CONSTRUCTION

On the average, an IIC rating for a given construction is approximately 51 dB higher than the INR rating. For example, an INR rating of +4 would be comparable to an IIC rating of approximately 55. However, exact values for IIC and INR ratings should be determined on the basis of individual test data in accordance with prescribed procedures (see Main Text, pages 106-16 and 106-17).

The standard criterion curve used in determining INR ratings is based on average background noises that are typical of apartments in moderately quiet suburban neighborhoods. For quieter areas, constructions with an INR or +5 to +10 may be required. For noisy urban districts where the level of background noise is particularly high, an INR of −5 may suffice. In any case, criteria given generally should be considered as minimum recommendations. In the opinion of some acoustical experts, INR ratings in the order of +10 to +15 may be necessary to render a reasonable degree of tenant satisfaction.

FHA Recommended Criteria

The Federal Housing Administration (FHA) recommended criteria for noise control often are used as a basis in the design and construction of multifamily housing units. The following recommendations are included in the FHA *Guide to Airborne, Impact, Structure Borne Noise Control in Multifamily Dwellings.* The criteria are based on: (1) the level of ambient background noise represented by urban, suburban and rural locations; (2) minimum-, average- and high-income housing; and (3) the specific function of the wall or floor/ceiling construction within the building.

The FHA uses three grades of acoustic environment which allows the criteria to be applied to a wide range of urban developments, geographic locations, economic conditions and other factors. Constructions which meet the criteria will provide good sound isolation and will satisfy a majority of the occupants in a building which fits the conditions of the particular grade. Grade II is applicable to the largest percentage of multifamily construction and, therefore, can be considered as a basic guide.

Grade I This category includes primarily suburban and peripheral suburban residential areas, considered to have a "quiet" background noise level. The nighttime exterior background noise level might be between 35 and 40 dB or lower. In addition, the sound isolation criteria of this grade are applicable in certain special cases such as dwelling units above the eighth floor in high-rise buildings and "luxury" apartments, regardless of location.

Grade II This category is applicable primarily in residential urban and suburban areas considered to have the "average" noise environment. The nighttime exterior background noise levels might be between 40 and 45 dB.

Grade III Criteria included in this category should be considered minimal recommendations and are applicable in some urban areas which generally are considered as "noisy" locations. The nighttime exterior background noise levels might be about 55 dB or higher.

Wall and floor/ceiling constructions should have STC

FIG. WF29 SOUND ISOLATION CRITERIA FOR WALLS BETWEEN DWELLING UNITS[1]

Common Walls Separating Dwelling Units[2]			Sound Transmission Class (STC)		
			Grade I	Grade II	Grade III
Bedroom	and	Bedroom	55	52	48
Living Room	and	Bedroom	57	54	50
Kitchen[3]	and	Bedroom	58	55	52
Bathroom	and	Bedroom	59	56	52
Corridor	and	Bedroom[4]	55	52	48
Living Room	and	Living Room	55	52	48
Kitchen[3]	and	Living Room	55	52	48
Bathroom	and	Living Room	57	54	50
Corridor	and	Living Room[4,5]	55	52	48
Kitchen[3]	and	Kitchen[6]	52	50	46
Bathroom	and	Kitchen	55	52	48
Corridor	and	Kitchen[4,5]	55	52	48
Bathroom	and	Bathroom	52	50	46
Corridor	and	Bathroom[4]	50	48	46

1. Adapted from FHA "Guide to Airborne, Impact, and Structure Borne Noise Control in Multifamily Dwellings."
2. Acoustically, mirror planning is desirable where common walls separate rooms of similar function; when not possible, walls separating bedrooms or living rooms from dissimilar areas must have greater sound isolating properties.
3. Or dining, family or recreation room.
4. No entrance door leading from corridor to living unit assumed.
5. It is commonly and incorrectly assumed that walls containing entrance doors need be no better acoustically than the door. However, the basic corridor wall may separate sensitive living areas without entrance doors and therefore must have adequate sound isolating properties to assure privacy in these areas. (See page WF106-20 for recommendations for sound isolating construction of doors.)
6. Double wall construction is recommended to provide isolation from impact noise resulting from vibration of mechanical equipment and systems. See page WF106-14 for precautions concerning resiliently mounted walls.

(Continued) DESIGN AND CONSTRUCTION

FIG. WF30 SOUND ISOLATION CRITERIA FOR FLOOR/CEILINGS BETWEEN DWELLING UNITS[1]

Common Floor/Ceilings Separating Dwelling Units[2]			Grade I		Grade II		Grade III	
			STC	IIC	STC	IIC	STC	IIC
Bedroom	above	Bedroom	55	55	52	52	48	48
Living Room	above	Bedroom	57	60	54	57	50	53
Kitchen[3]	above	Bedroom	58	65	55	62	52	58
Family Room	above	Bedroom[4]	60	65	56	62	52	58
Corridor	above	Bedroom	55	65	52	62	48	58
Bedroom	above	Living Room[5]	57	55	54	52	50	48
Living Room	above	Living Room	55	55	52	52	48	48
Kitchen	above	Living Room	55	60	52	57	48	53
Family Room	above	Living Room[4]	58	62	54	60	52	56
Corridor	above	Living Room	55	60	52	57	48	53
Bedroom	above	Kitchen[5]	58	52	55	50	52	46
Living Room	above	Kitchen[5]	55	55	52	52	48	48
Kitchen	above	Kitchen	52	55	50	52	46	48
Bathroom	above	Kitchen	55	55	52	52	48	48
Family Room	above	Kitchen[4]	55	60	52	58	48	54
Corridor	above	Kitchen	50	55	48	52	46	48
Bedroom	above	Family Room[5]	60	50	56	48	52	46
Living Room	above	Family Room[5]	58	52	54	50	52	48
Kitchen	above	Family Room[5]	55	55	52	52	48	50
Bathroom	above	Bathroom	52	52	50	50	48	48
Corridor	above	Corridor	50	50	48	48	46	46

1. Adapted from FHA "Guide to Airborne, Impact, and Structure Borne Noise Control in Multifamily Dwellings."
2. Acoustically, mirror planning, where common floor/ceiling systems separate rooms of similar function, is desirable; when not possible, floor/ceilings separating bedroom or living rooms from dissimilar areas must have greater sound isolating properties. When bedrooms or living rooms are located below less sensitive living spaces, greater impact isolation is required.
3. Or dining, family or recreation room.
4. STC criteria also applies to vertical partitions between these two spaces as well.
5. Equivalent airborne sound isolation, but less impact sound isolation generally required for the converse situation.

and IIC ratings equal to or greater than the recommended criterion figures. A floor/ceiling which may provide adequate impact sound isolation may not provide suitable airborne sound isolation. The criteria for both airborne and impact sound isolation of floor ceiling constructions must be met.

Recommended basic criteria for airborne and impact sound isolation of common wall and floor/ceiling assemblies *which separate dwelling units of equivalent function* are given in Figure WF28. These criteria are based on STC and IIC ratings derived from *laboratory* measurements. Standard methods of test for field measurements as yet have not been formally adopted.

Wall Systems Figure WF29 gives the recommended criterion values for common walls, related to room or area function. Most of the typical conditions encountered for common walls separating dwelling units in multifamily construction are included. Some of these conditions are clearly undesirable and should be avoided by careful planning in the initial design of the building (see Interior Planning, page WF106-9). The purpose of this detail is to illustrate the importance of the acoustical separation between sensitive and nonsensitive areas. Where a partition between dwelling units is common to several functional spaces, the partition should meet the highest criterion value.

(Continued) **DESIGN AND CONSTRUCTION**

Floor/Ceiling Systems Figure WF30 includes criteria for most of the conditions encountered for common floor/ceiling constructions which separate units in multifamily dwellings. Some of these conditions are clearly undesirable and should be avoided by careful planning in the initial design of the building (see page WF106-9). Additional precautions and qualifying recommendations are given in the footnotes to Figure WF30.

Mechanical Equipment, Service and Business Areas When the placement of living units vertically or horizontally adjacent to mechanical equipment rooms cannot be avoided, special consideration should be given to provide suitable sound isolating construction. The recommended airborne sound criteria between mechanical equipment rooms and sensitive areas in dwelling units are:

Grade I STC ≥ 65
Grade II STC ≥ 62
Grade III STC ≥ 58

Mechanical equipment rooms include furnace-boiler rooms, elevator shafts, trash chutes, cooling towers, garages and similar areas.

The recommended criteria between mechanical equipment rooms and less sensitive areas, such as kitchens and family and recreation rooms, are:

Grade I STC ≥ 60
Grade II STC ≥ 58
Grade III STC ≥ 54

Double wall construction is usually necessary to achieve adequate acoustical privacy. Where living units are *above* noisy areas, the airborne sound isolation is important, but the impact sound isolation is not significant if structure-borne vibration is minimal. However, where mechanical equipment rooms are above living areas, the airborne sound isolation must be maintained, impact sound isolation becomes critical, and specially designed sound isolating constructions are required to assure quiet living areas.

Placing dwelling units vertically or horizontally adjacent to commercial establishments such as restaurants, bars, community laundries and similar businesses, should be avoided whenever possible. If such situations cannot be avoided, the recommended airborne isolation criteria between business areas and sensitive living areas are:

Grade I STC ≥ 60
Grade II STC ≥ 58
Grade III STC ≥ 56

If the living areas are situated above such areas, adequate impact isolation criteria should be:

Grade I IIC ≥ 60
Grade II IIC ≥ 58
Grade III IIC ≥ 56

However, if the business areas are above living areas, the impact isolation criteria values should be increased by at least 5 points.

If noise levels in mechanical equipment rooms or business areas exceed 100 dB, as measured using the "C" scale of a standard sound level meter, the airborne sound isolation criteria given above should be raised 5 points.

Walls Within Living Units Figure WF31 lists *suggested* criteria for airborne sound isolation for walls separating rooms *within* a given living unit.

Townhouses and row-houses where the living unit occupies more than one story should be separated by a double wall construction with a rating of 60 STC or greater. Suggested criteria between rooms in a given dwelling are the same as listed in Figure WF29 and, in addition, the floor/ceiling constructions should have IIC ratings which are at least numerically equivalent to or greater than the listed STC criteria.

Roof-top or indoor swimming pools, bowling alleys, ballrooms, tennis courts and gymnasiums require constructions specifically designed in accordance with acoustical analysis.

FIG. WF31 **SOUND ISOLATION OF WALLS WITHIN DWELLING UNITS[1]**

Walls Separating Rooms[2]			Sound Transmission Class (STC)		
			Grade I	Grade II	Grade III
Bedroom	and	Bedroom	48	44	40
Living Room	and	Bedroom	50	46	42
Bathroom	and	Bedroom	52	48	45
Kitchen	and	Bedroom	52	48	45
Bathroom	and	Living Room	52	48	45

1. Adapted from FHA "Guide to Airborne, Impact, and Structure Borne Noise Control in Multifamily Dwellings."
2. Doors leading to bedrooms and bathrooms preferably should be gasketed, solid-core doors to provide maximum privacy. See page WF106-11 and WF106-12 for recommendations concerning installation of piping, fixtures and appliances.

INDEX TO SOUND ISOLATING SYSTEMS

The following index includes sound performance data on wall and floor/ceiling systems used in residential construction. Data on systems which provide poor sound isolation are included to indicate the degree of improvement required to render such systems sufficiently resistant to sound transmission.

Unless otherwise noted, the data presented is based on tests performed by nationally recognized acoustical laboratories[1] in accordance with current ASTM[2] and ISO[3] test procedures.

The STC, INR, and IIC ratings shown in this section have been drawn from a variety of manufacturer's literature and testing laboratory reports. In some instances either only very old test information or little correlating test results can be found. Hence, some ratings have been estimated.

Whenever possible, a range of values is given to account for (1) slight variations in construction materials, (2) slight variation in application procedures, and (3) variation in testing facilities.

The reproducibility of laboratory tests on different samples of standard construction systems can range up to 5 dB at any one frequency. However, overall STC rating should be within 2 dB. The subjective response to decibel difference as judged by the ear can be summarized as follows: a difference of 1 to 3 dB is insignificant, a difference of 5 dB is clearly noticeable, and a difference of 10 dB is highly significant.

Strict adherence to good practices of design and construction (see Main Text, pages 106-19 through 106-40), excellent workmanship and installation in accordance with manufacturers' specific instructions are essential to the maximum sound isolating performance of any construction system.

WALL SYSTEMS

Improvement in acoustical performance by adding insulation blankets in the cavity of wood and metal stud wall varies depending on wall construction and thickness. Adding insulation to *single wood stud walls* improves performance slightly and is not considered economically justified. The addition of a minimum thickness of two inches to three inches of insulation improves the performance of *staggered wood stud walls* and *single channel metal stud walls,* but increasing the thickness improves performance slightly and is not considered economically justified. The acoustical performance increases proportionately with increased thickness of insulation in *double wood stud walls, double channel metal stud walls,* and *wood stud walls* having *resiliently mounted surface finishes.*

The STC ratings given in the clay masonry wall section are interpolated from Fig. WF21.

FLOOR/CEILING SYSTEMS

The use of plywood, fiberboard, hardboard or other types of underlayment is generally determined by the type of finish flooring or other practical construction techniques. With the exception of resilient underlayment materials in floated floor systems, these underlayment materials generally contribute insignificant impact sound isolating properties. Unless otherwise noted, the ranges given for impact sound isolation in terms of INR and Impact Insulation Class in terms of IIC, include constructions with and without underlayment.

Carpet and pad combinations are effective in improving the impact sound isolation performance of any type of flooring system. Improvement depends on carpet thickness, degree of pile matting and weight of carpet and pad in ounces per square yard. However, no appreciable gain in airborne sound isolation is obtained using carpet and pad combinations.

The addition of insulation in the cavity between the subfloor and ceiling finish in joist construction improves performance only slightly but is considered economically justified when the ceiling finish is resiliently mounted.

In the case of poured-in-place reinforced concrete slabs, impact sound isolation performance is not improved significantly by increasing the weight or thickness of the slab. The most economical control measures consist of cushioning the impact with carpet and pad and floating the finished floor.

Because basic construction systems to control impact sound isolation also will provide a good degree of resistance to airborne sound transmission, much of the test data for floor/ceiling construction are given in terms of INR and IIC, only. However, the STC values for a particular type of construction can be estimated, providing essential precautions of construction practices are followed. Where test data for STC ratings are not available, the STC values for the construction are given as estimates.

1. Data includes tests performed by: Cedar Knolls Acoustical Laboratories; Geiger and Hamme; Kodaras Acoustical Laboratories; National Bureau of Standards; National Gypsum Company; Ohio Research Corporation; Owens-Corning Fiberglass Sound Laboratory; Riverbank Acoustical Laboratories, IIT Research Institute.

2. ASTM E90-66T, Tentative Recommended Practice for Laboratory Measurement of Airborne Sound Transmission Loss of Building Partitions.

3. International Organization for Standardization (ISO) Recommendation R-140, "Field and Laboratory Measurements of Airborne and Impact Sound Transmission," January 1960.

WALL SYSTEMS

WOOD FRAME LOAD-BEARING WALLS—GYPSUM BOARD FINISH[1]

Single Stud Walls

Single Layer Finish[2]

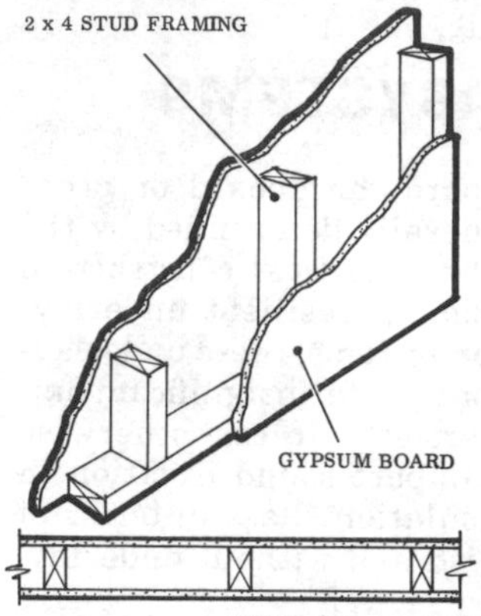

STC 32 to 36

Multilayer Finish[2]

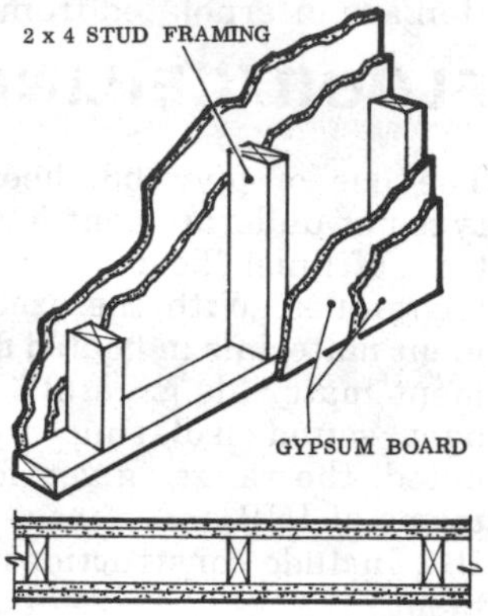

STC 38 to 41

Single Stud Walls

Single Layer Finish

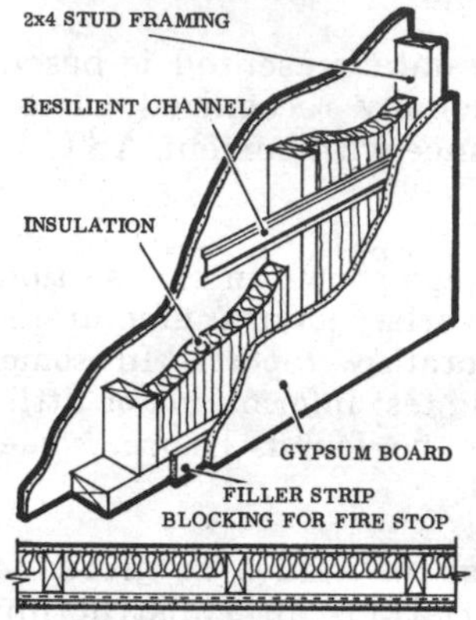

STC 40 to 42

With insulation[7]

STC 47 to 51[4]

Sound Deadening Board[5]

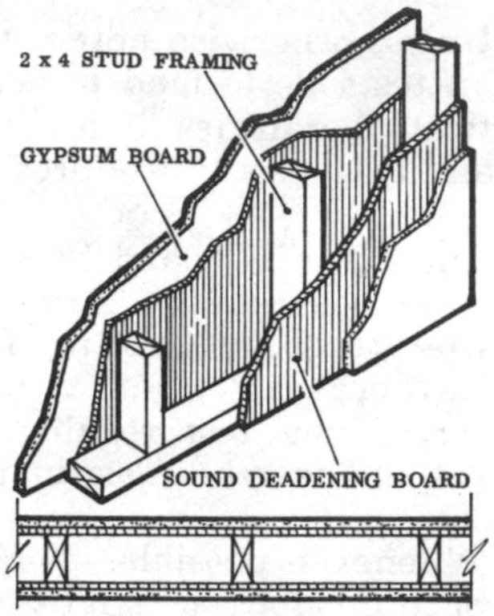

STC 47 to 51[4]

Staggered Stud Walls[6]

Resilient Finish[3]

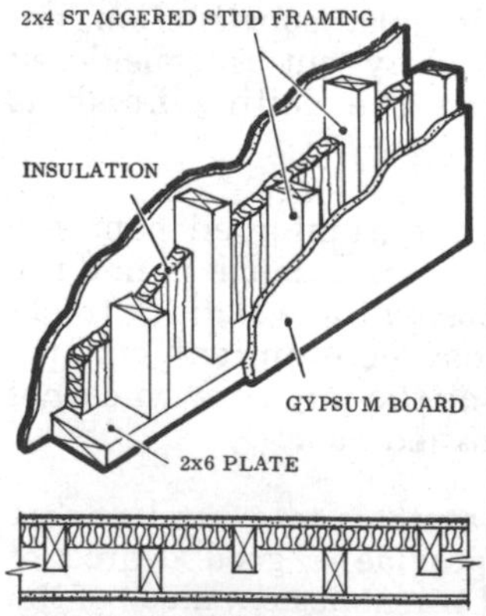

STC 38 to 42[4]

With insulation[8]

STC 46 to 48[4]

Sound Deadening Board[5]

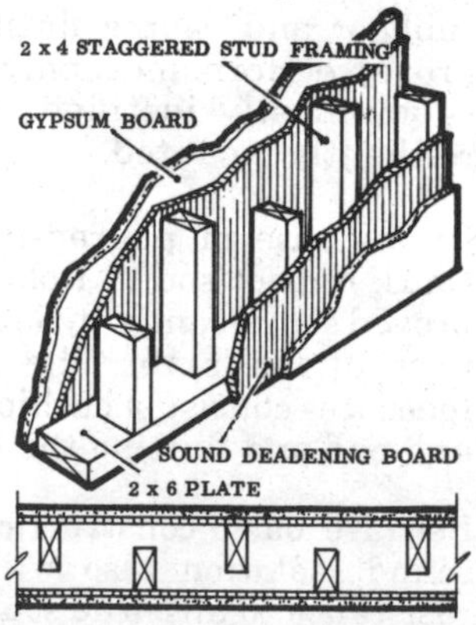

STC 46 to 50[4]

Double Stud Walls[6]

Single Layer Finish

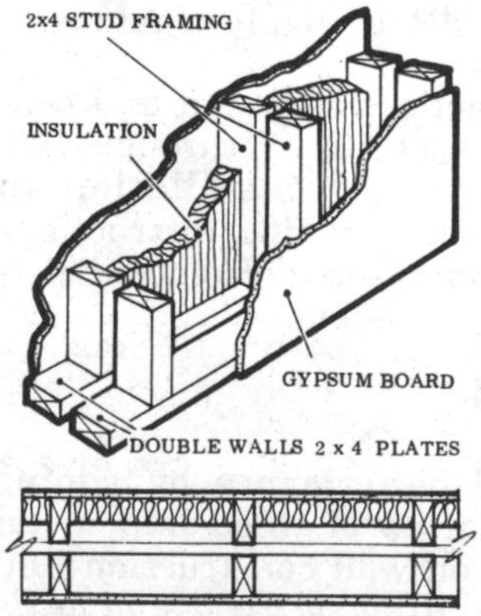

STC 42 to 44

With insulation[8]

STC 50 to 53[4]

Sound Deadening Board[5]

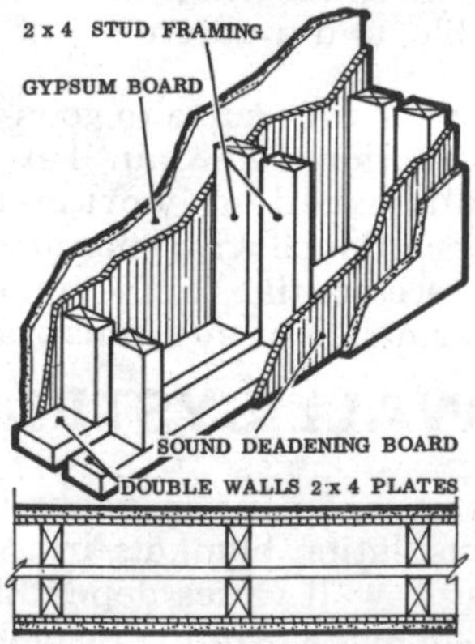

STC 50 to 53[9]

With insulation[8]

STC 54 to 58

NOTES:

1. Ranges include values for ⅜″, ½″ and ⅝″ thicknesses of gypsum board finish.
2. Addition of insulation does not appreciably improve this construction; an increase of approximately 2 to 4 dB can be expected regardless of thickness or density of insulation.
3. Resilient channels or clips may be used one side only to totally float the finish; use of blocking for fire stopping and use of backer strips to attach wall base reduces performance approximately 4 to 5 dB.
4. Range includes values derived from ASTM E90-61T, ASTM E90-66T and field test procedures.
5. Gypsum board finish laminated to sound deadening board; if gypsum board is nailed or screwed to meet fire ratings, performance decreases approximately 10 dB for fasteners spaced 8″ o.c., approximately 8 dB for fastener spacing 16″ o.c., and approximately 4 dB with fastener spacing 24″ o.c.
6. Ranges include values for 2x3 studs staggered on single 2x4 plate and 2x4 double stud walls on separate 2x3 plates.
7. A minimum thickness of 2″ to 3″ of insulation can be used effectively as performance does not increase proportionately with increased thickness of insulation.
8. Ranges include 1½″ to 3″ thickness of insulation; performance increases with increased thickness. On double stud walls, an STC of 58 to 60 can result with up to 9″ of insulation.

WALL SYSTEMS

WOOD FRAME LOAD-BEARING WALLS—PLASTER FINISH[11]

Single Stud Walls

Metal Lath & Plaster[2]

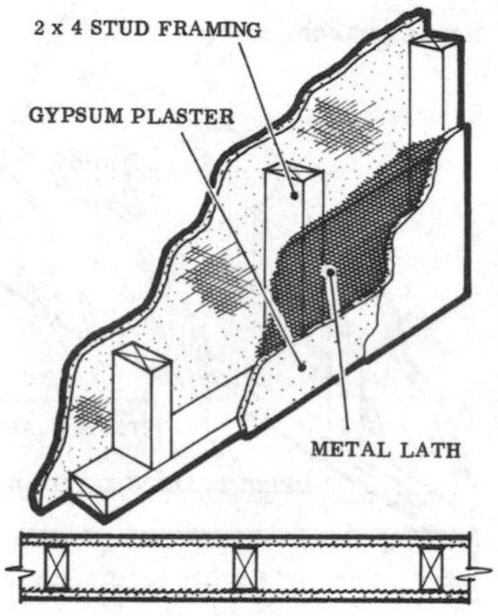

STC 41[10]

Resilient Gypsum Lath & Plaster[3]

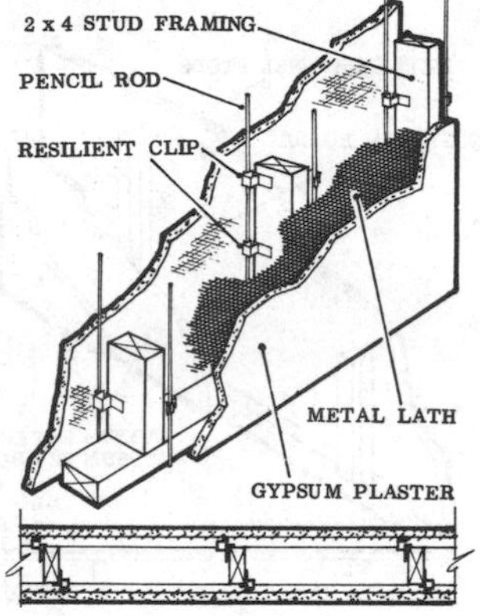

STC 43[9]

Single Stud Walls

Plaster Base, Veneer Plaster[2]

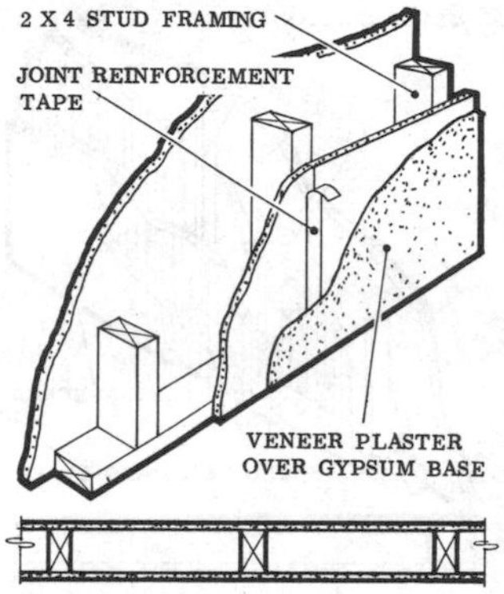

STC (est) 36

2 layer base 2 sides

STC (est) 42

Resilient Plaster Base, Veneer Plaster[3]

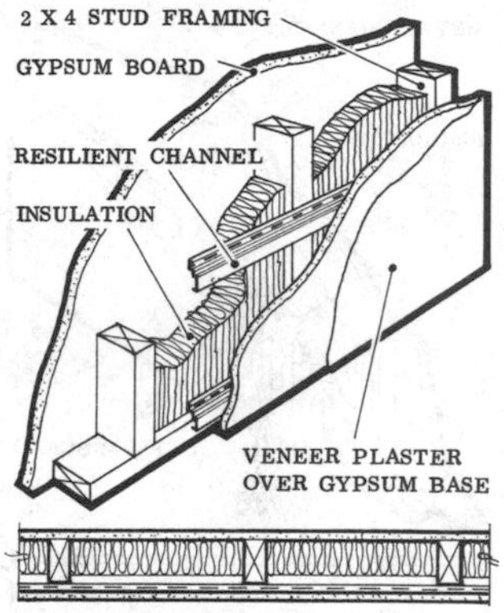

With insulation[8]

STC 49 to 50[4]

Single Stud Walls

Gypsum Lath & Plaster[2]

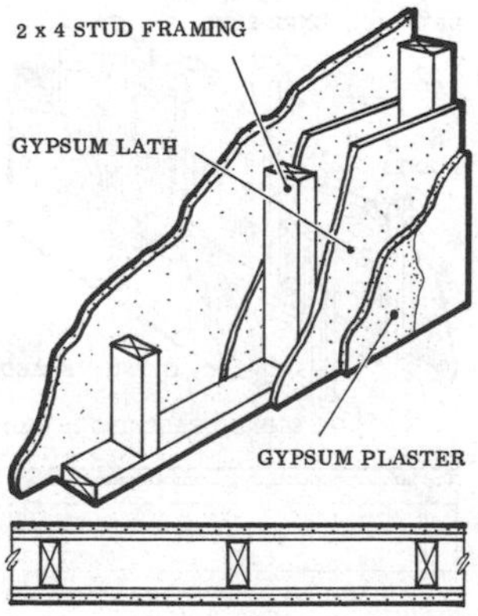

STC 41[10]

Resilient Metal Lath & Plaster[3]

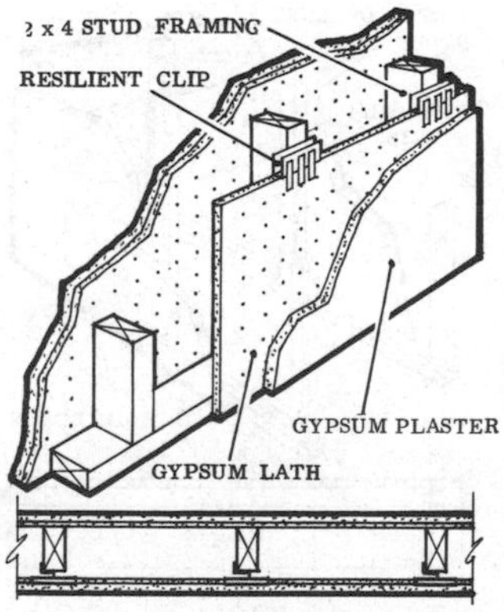

STC 45 to 47[10]

With insulation[8]

STC 50 to 54[4]

Single Stud Walls

Resilient Plaster Base & Multilayer Base, Veneer Plaster[3]

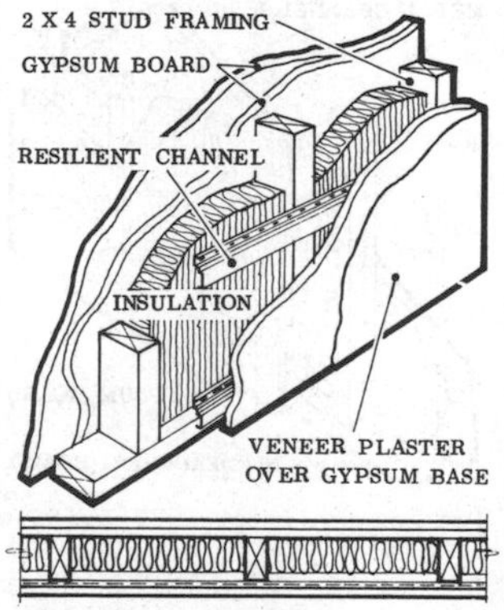

With insulation[8]

STC 53

Staggered Stud Walls[6]

Gypsum Lath & Plaster

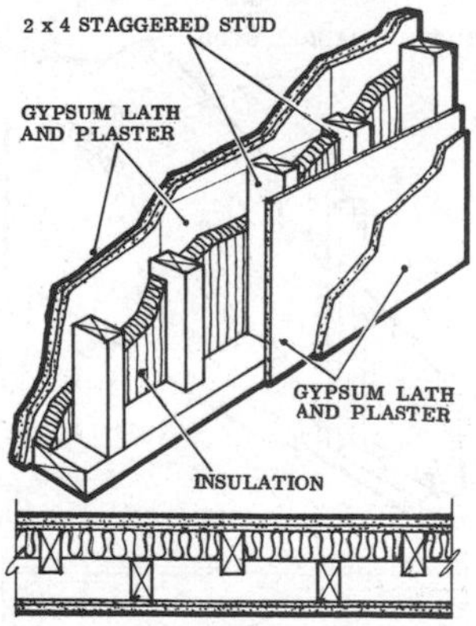

STC 42 to 43[9]

With insulation[7]

STC 50[10]

9. Based on ASTM E90-61T test procedures.
10. Based on tests performed prior to ASTM E90-61T.
11. Test data can vary widely depending on weight and thickness of plaster, and water content (due to length of curing) of plaster.
12. With exception of double stud walls, ranges include values for ½″ or ⅝″ gypsum board thickness on 2½″ or 3⅝″ wide studs.
13. Ranges include values for gypsum board screwed, or laminated and screwed, to sound deadening board or base layer.
14. Unbalanced Construction: single layer gypsum board one side, double layer gypsum board one side.
15. Ranges include values for ½″ and ⅝″ board thickness on 1⅝″ wide studs.
16. Up to 60 STC with 9″ of insulation; lower values result with lesser thicknesses of insulation.
17. Performance of concrete and unit masonry systems is highly dependent on minimizing the number of openings through the construction (to accommodate piping, wiring and other equipment) and on all openings being sealed airtight.
18. Sealed both sides unless otherwise indicated. Sealing with plaster is most effective; when surfaces are to be painted, use of block sealer and/or 2 coats of cement-based paint is more effective than 2 coats of resin-emulsion paint to seal against airborne sound transmission.

WALL SYSTEMS

METAL FRAME NONLOAD-BEARING WALLS—GYPSUM BOARD FINISH[12]

Single Stud Walls

Single Layer Finish

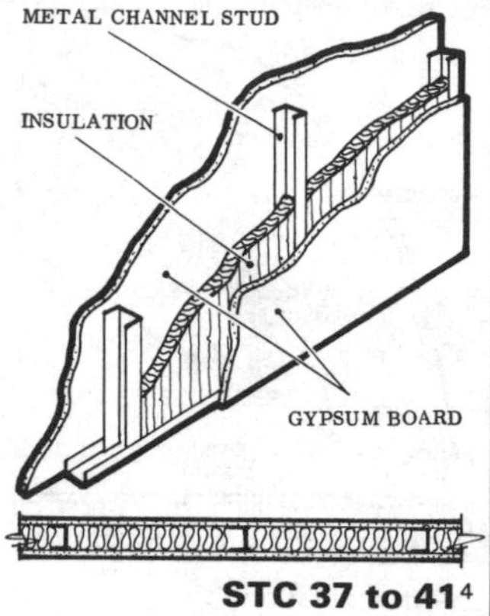

STC 37 to 41[4]
With insulation[8]
STC 44 to 48

Single Layer Finish Sound Deadening Board[13]

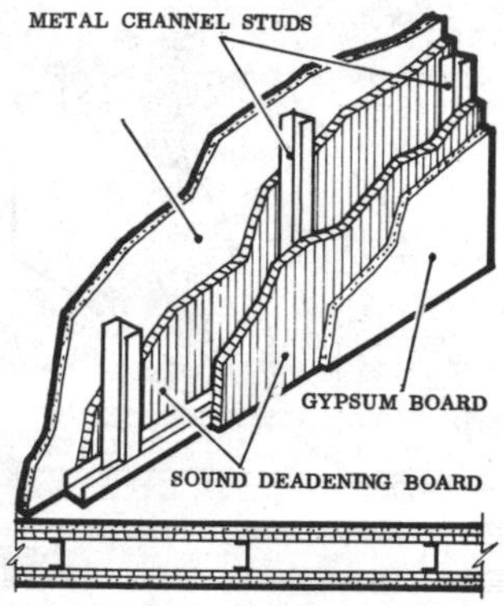

SDB 1 side	**STC**	**43**
SDB 2 sides	**STC 46 to 50**[4]	
With insulation[8]		
SDB 1 side	**STC 47 to 50**[4]	
SDB 2 sides	**STC 50 to 53**[4]	

Single Stud Walls

Unbalanced Finish[14]

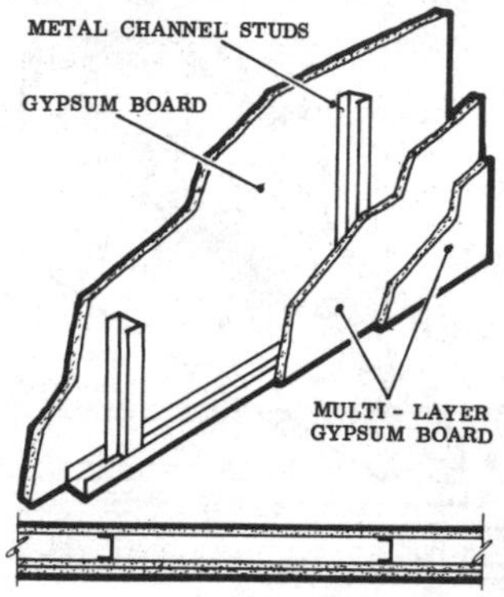

STC 44 to 47
With insulation[8]
STC 49 to 52

Unbalanced Finish[14] **Sound Deadening Board**[13]

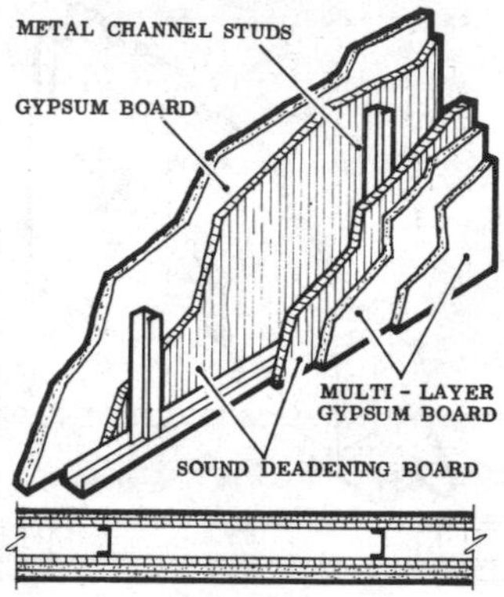

SDB 1 side	**STC**	**48**[9]
SDB 2 sides	**STC**	**52**[9]
With insulation[8]		
SDB 1 side	**STC**	**49**[9]
SDB 2 sides	**STC**	**54**[9]

Single Stud Walls

Multilayer Finish

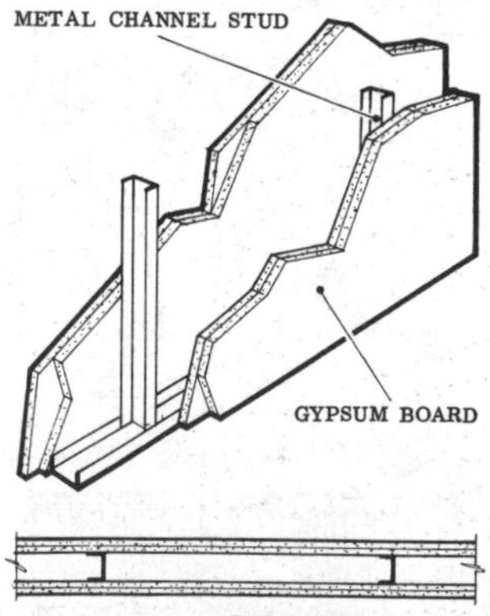

STC 48 to 50[4]
With insulation[8]
STC 53 to 58[4]

Multilayer Finish Sound Deadening Board[13]

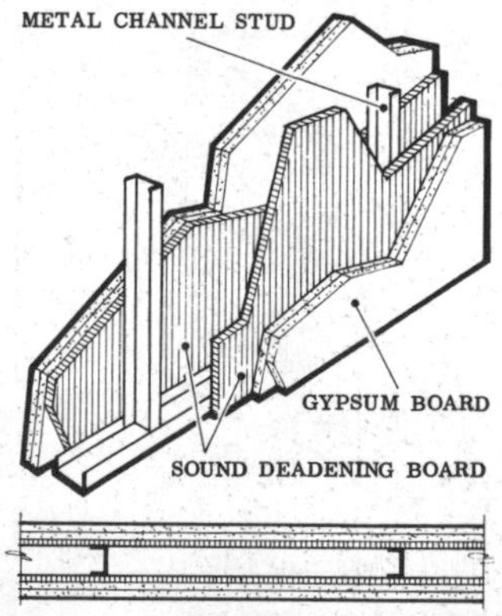

SDB 1 side	**STC**	**52**[9]
SDB 2 sides	**STC**	**54**[9]
With insulation[8]		
SDB 2 sides	**STC**	**57**[9]

Double Stud (Chase) Walls[15]

Single Layer Finish

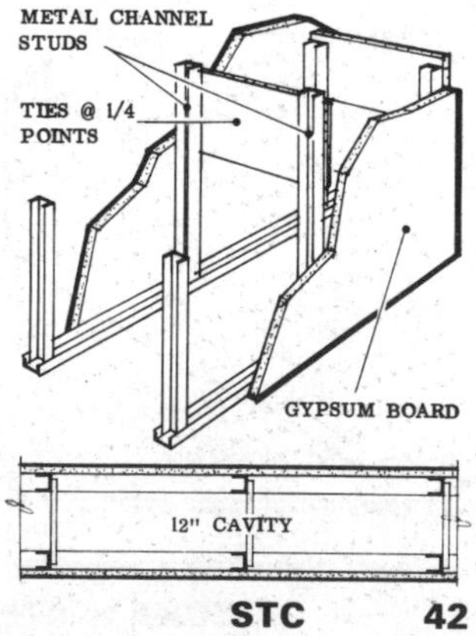

STC **42**
With insulation[8]
STC 52 to 55[16]

Single Layer Finish Sound Deadening Board[13]

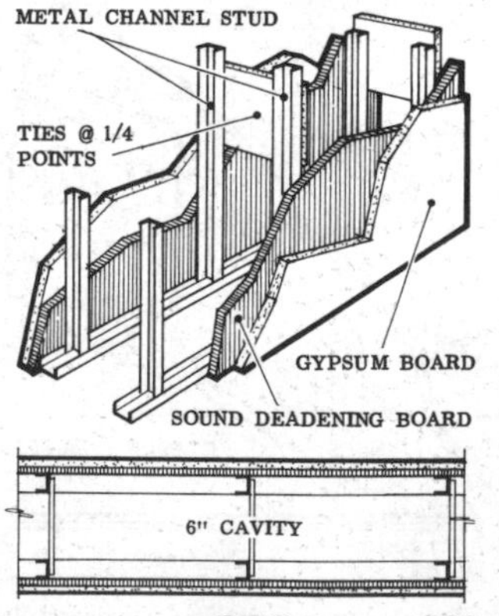

STC **50**[9]
With insulation[8]
STC **53**

NOTES:

1. Ranges include values for ⅜", ½" and ⅝" thicknesses of gypsum board finish.
2. Addition of insulation does not appreciably improve this construction; an increase of approximately 2 to 4 dB can be expected regardless of thickness or density of insulation.
3. Resilient channels or clips may be used one side only to totally float the finish; use of blocking for fire stopping and use of backer strips to attach wall base reduces performance approximately 4 to 5 dB.
4. Range includes values derived from ASTM E90-61T, ASTM E90-66T and field test procedures.
5. Gypsum board finish laminated to sound deadening board; if gypsum board is nailed or screwed to meet fire ratings, performance decreases approximately 10 dB for fasteners spaced 8" o.c., approximately 8 dB for fastener spacing 16" o.c., and approximately 4 dB with fastener spacing 24" o.c.
6. Ranges include values for 2x3 studs staggered on single 2x4 plate and 2x4 double stud walls on separate 2x3 plates.
7. A minimum thickness of 2" to 3" of insulation can be used effectively as performance does not increase proportionately with increased thickness of insulation.
8. Ranges include 1½" to 3" thickness of insulation; performance increases with increased thickness. On double stud walls, an STC of 58 to 60 can result with up to 9" of insulation.

WALL SYSTEMS

METAL FRAME NONLOAD-BEARING WALLS—GYPSUM BOARD FINISH[12] (CONT.)

Double Stud (Chase) Walls[15]

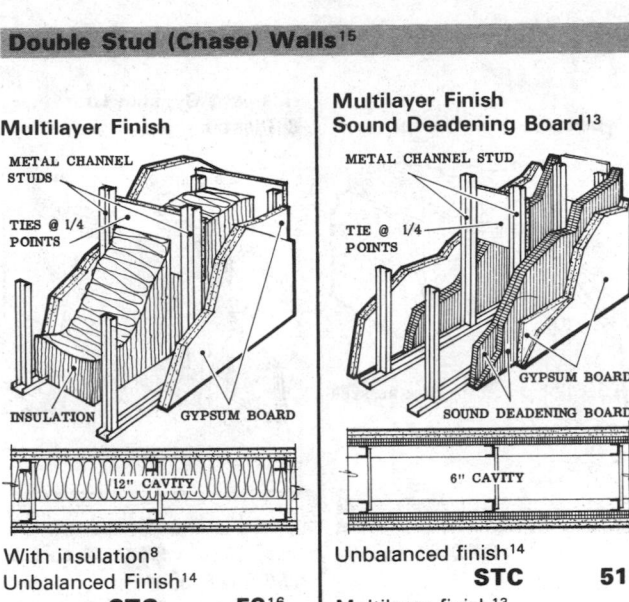

Multilayer Finish

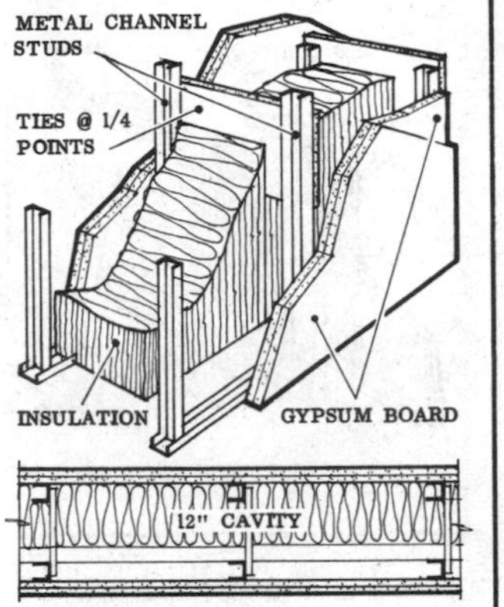

With insulation[8]
Unbalanced Finish[14]
STC 59[16]
Multilayer finish[13]
STC 55 to 60[4,16]

Multilayer Finish Sound Deadening Board[13]

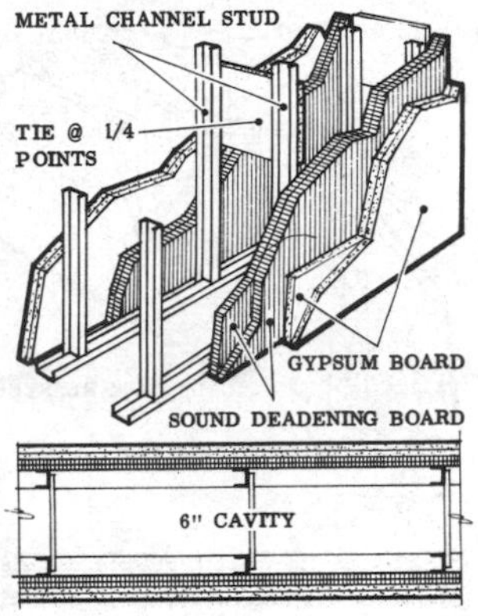

Unbalanced finish[14]
STC 51[9]
Multilayer finish[13]
STC 53[9]

METAL FRAME NONLOAD-BEARING WALLS—PLASTER FINISH[11]

Single Channel Stud Walls

Plaster Base & Veneer Plaster

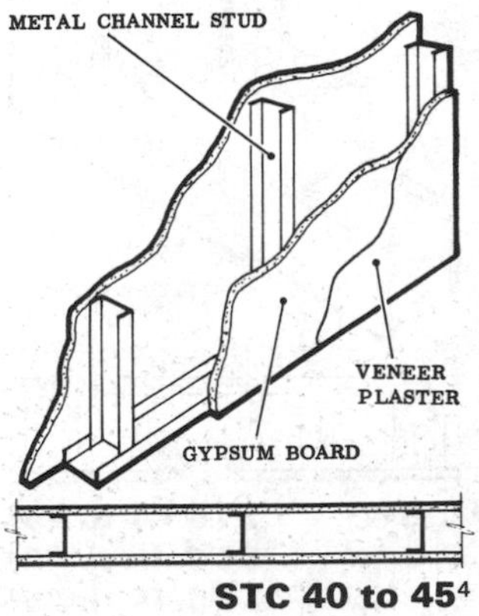

STC 40 to 45[4]

Multilayer Base[13] & Veneer Plaster

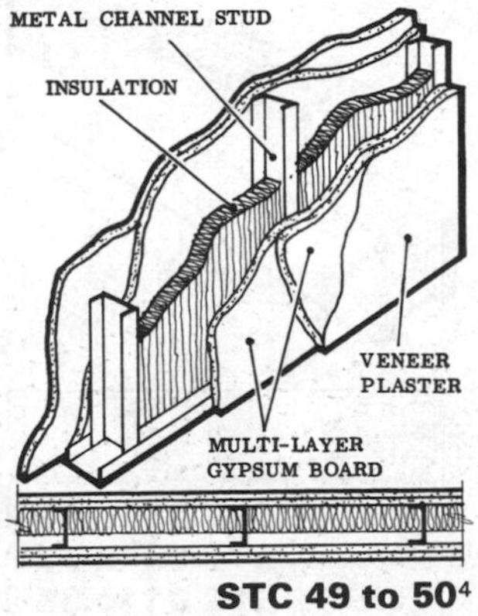

STC 49 to 50[4]
With insulation[8]
STC 52 to 53[4]

Single Channel Stud Walls

Gypsum Lath & Plaster

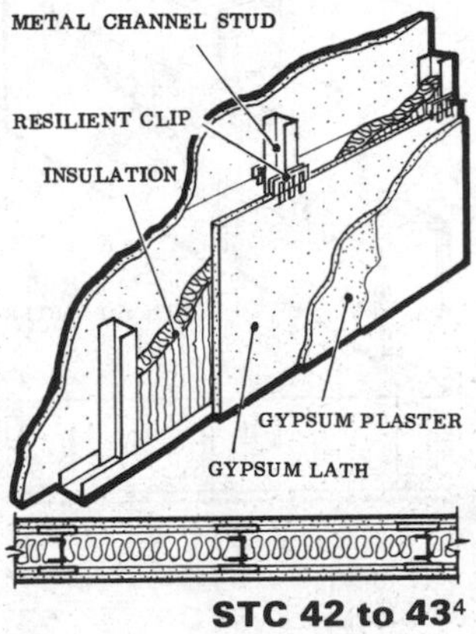

STC 42 to 43[4]
With insulation
STC 46

Resilient Gypsum Lath[3] & Plaster

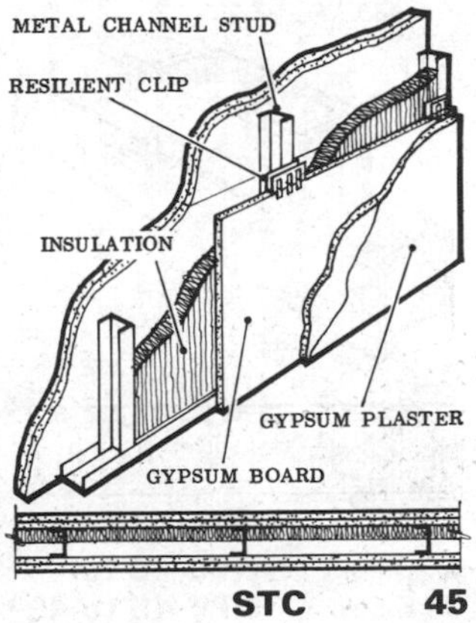

STC 45
With insulation[8]
STC 49

9. Based on ASTM E90-61T test procedures.
10. Based on tests performed prior to ASTM E90-61T.
11. Test data can vary widely depending on weight and thickness of plaster, and water content (due to length of curing) of plaster.
12. With exception of double stud walls, ranges include values for ½" or ⅝" gypsum board thickness on 2½" or 3⅝" wide studs.
13. Ranges include values for gypsum board screwed, or laminated and screwed, to sound deadening board or base layer.
14. Unbalanced Construction: single layer gypsum board one side, double layer gypsum board one side.
15. Ranges include values for ½" and ⅝" board thickness on 1⅝" wide studs.
16. Up to 60 STC with 9" of insulation; lower values result with lesser thicknesses of insulation.
17. Performance of concrete and unit masonry systems is highly dependent on minimizing the number of openings through the construction (to accommodate piping, wiring and other equipment) and on all openings being sealed airtight.
18. Sealed both sides unless otherwise indicated. Sealing with plaster is most effective; when surfaces are to be painted, use of block sealer and/or 2 coats of cement-based paint is more effective than 2 coats of resin-emulsion paint to seal against airborne sound transmission.

WALL SYSTEMS

METAL FRAME NONLOAD-BEARING WALLS—PLASTER FINISH[11] (CONT.)

Single Truss Stud Walls

Metal Lath & Plaster

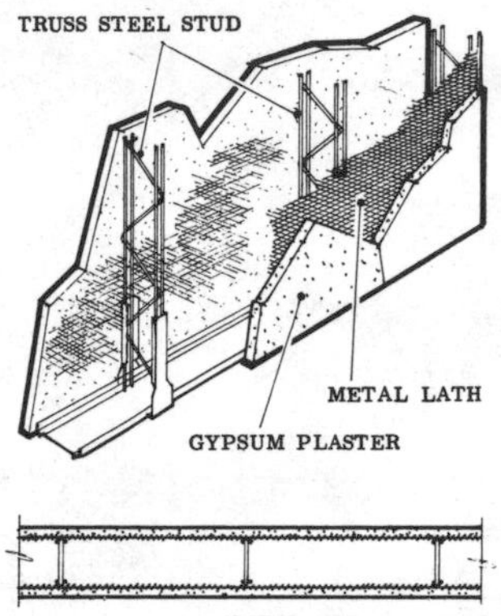

STC 38 to 41[9]

Resilient Metal Lath[3] & Plaster

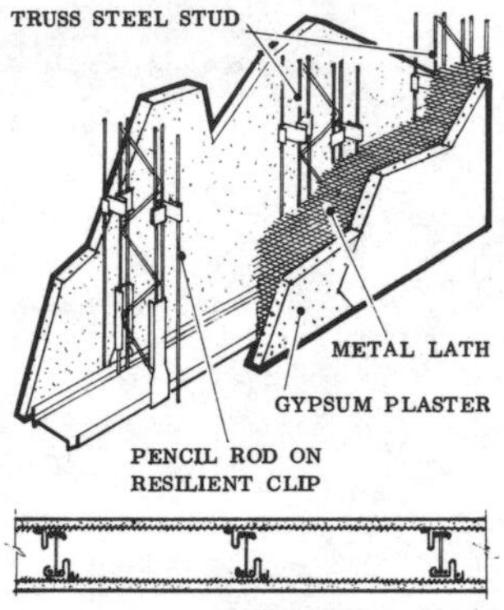

STC 46 to 50[4]

Single Truss Stud Walls

Gypsum Lath & Plaster

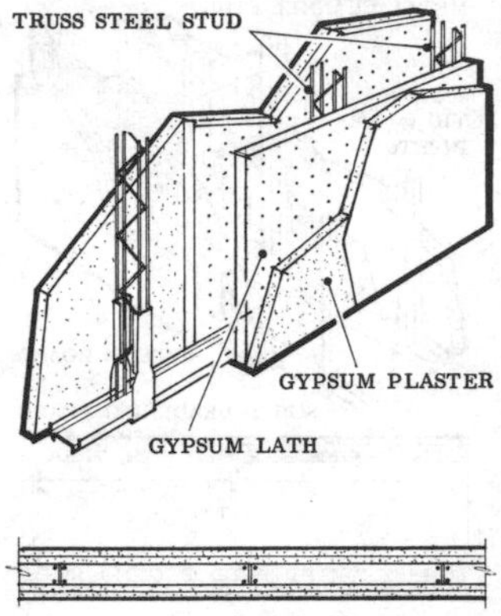

STC 41 to 45[4]

Resilient Gypsum Lath[3] & Plaster

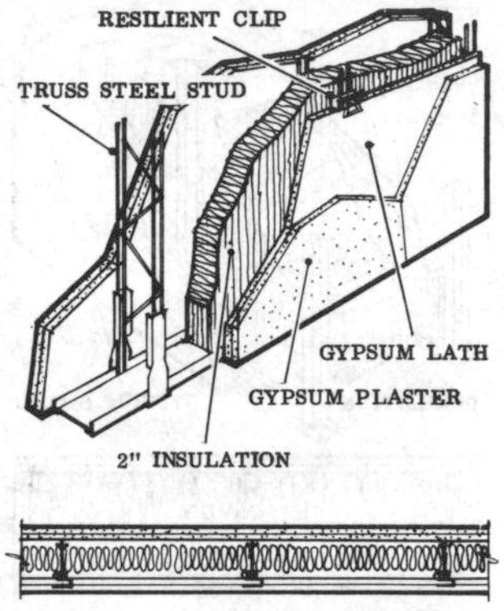

STC 46 to 50

With insulation[8]

STC 46 to 52[4]

CONCRETE AND CONCRETE MASONRY WALLS[17]

Nonporous (Normal Weight) Hollow Concrete Block Walls

Unsealed

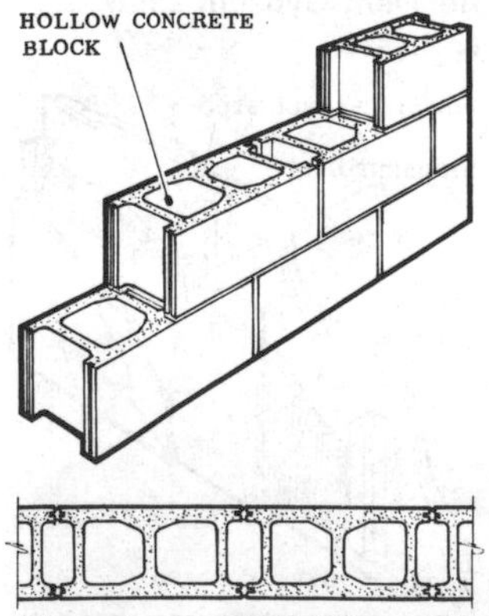

4″ Block	**STC 42 to 45[4]**
6″ Block	**STC 43 to 46[4]**
8″ Block	**STC 46 to 49[9]**

Sealed[18]

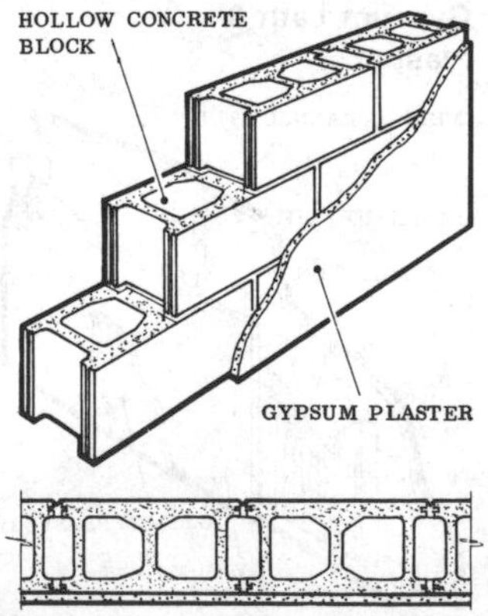

4″ Block	**STC 44 to 48[4]**
6″ Block	**STC 47 to 51[4]**
8″ Block	**STC 53[9]**

Porous (Lightweight) Hollow Concrete Block Walls

Unsealed

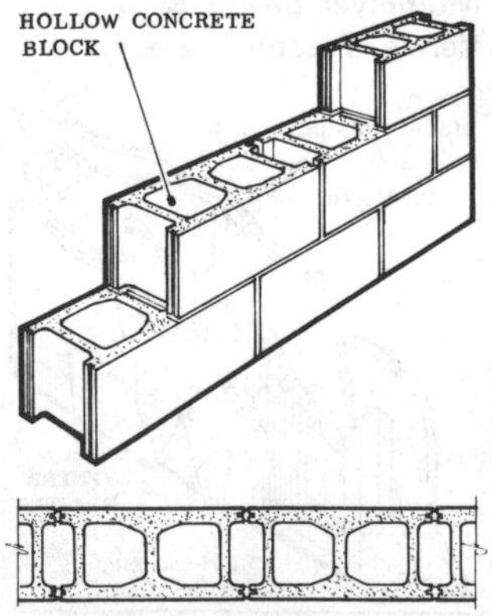

4″ Block	**STC 26 to 34[10]**
6″ Block	**STC 33 to 37[10]**
8″ Block	**STC 34 to 38[10]**

Sealed[18]

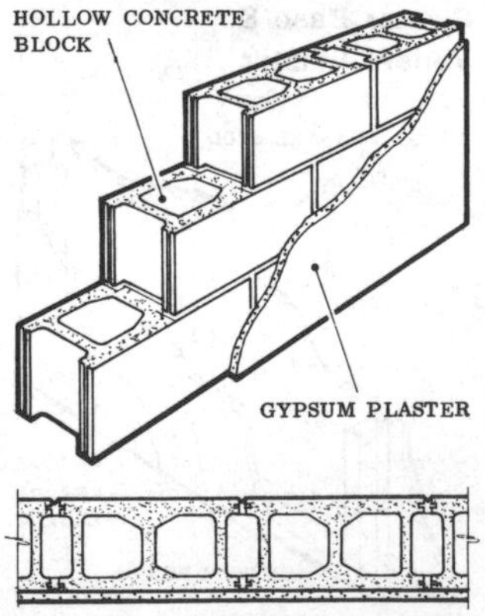

4″ Block	**STC 41 to 44[4]**
6″ Block	**STC 44 to 49[4]**
8″ Block	**STC 44 to 49[4]**

NOTES:

1. Ranges include values for ⅜″, ½″ and ⅝″ thicknesses of gypsum board finish.
2. Addition of insulation does not appreciably improve this construction; an increase of approximately 2 to 4 dB can be expected regardless of thickness or density of insulation.
3. Resilient channels or clips may be used one side only to totally float the finish; use of blocking for fire stopping and use of backer strips to attach wall base reduces performance approximately 4 to 5 dB.
4. Range includes values derived from ASTM E90-61T, ASTM E90-66T and field test procedures.
5. Gypsum board finish laminated to sound deadening board; if gypsum board is nailed or screwed to meet fire ratings, performance decreases approximately 10 dB for fasteners spaced 8″ o.c., approximately 8 dB for fastener spacing 16″ o.c., and approximately 4 dB with fastener spacing 24″ o.c.
6. Ranges include values for 2x3 studs staggered on single 2x4 plate and 2x4 double stud walls on separate 2x3 plates.
7. A minimum thickness of 2″ to 3″ of insulation can be used effectively as performance does not increase proportionately with increased thickness of insulation.
8. Ranges include 1½″ to 3″ thickness of insulation; performance increases with increased thickness. On double stud walls, an STC of 58 to 60 can result with up to 9″ of insulation.

WALL SYSTEMS

CONCRETE AND CONCRETE MASONRY WALLS[17] (CONT.)

Porous (Lightweight) Hollow Concrete Block Walls

Gypsum Board Laminated Both Sides

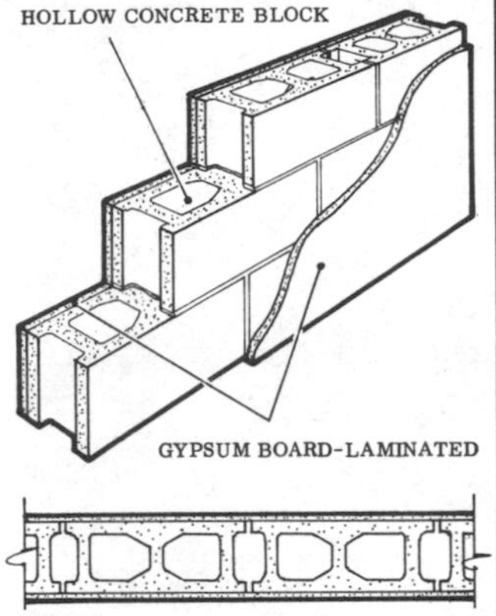

6″ Block **STC** **49**

Resilient Gypsum Board 1 Side, Sealed[18] 1 Side

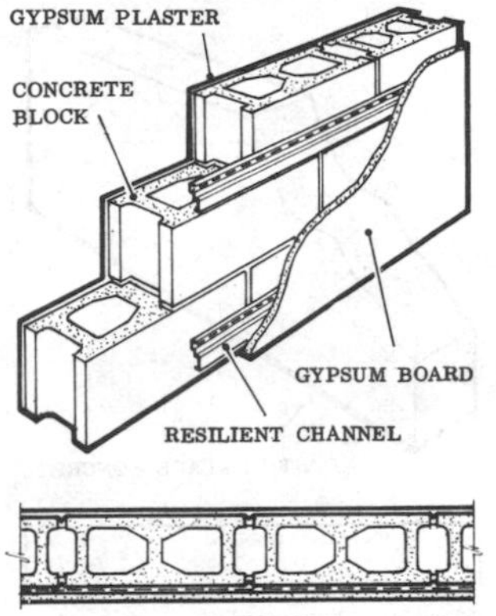

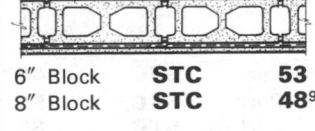

6″ Block **STC** **53**
8″ Block **STC** **48**[9]

Porous (Lightweight) Hollow Concrete Block Walls

Resilient Gypsum Board Both Sides

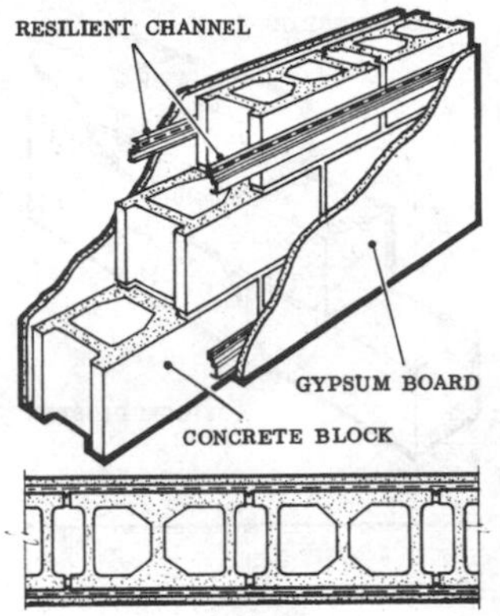

8″ Block **STC** **49**[9]

Cavity Wall Unsealed

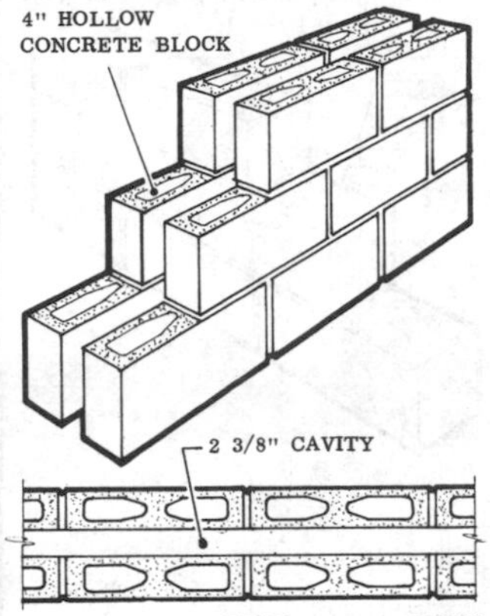

STC **38**[10]

Hollow Concrete Block Walls

Porous Cavity Wall Sealed[18]

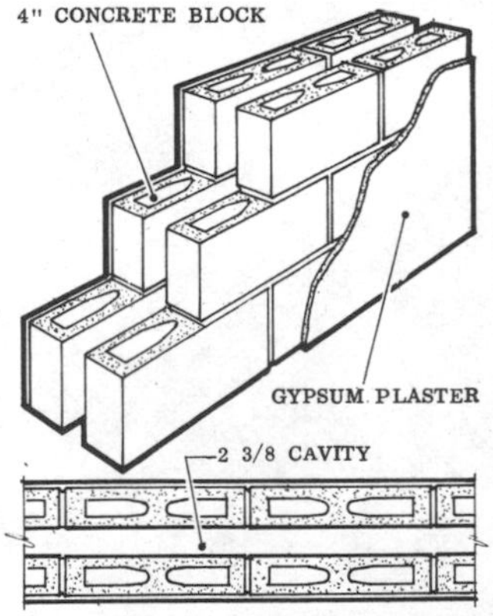

STC **49**[10]

Porous / Nonporous Cavity Wall Sealed[18]

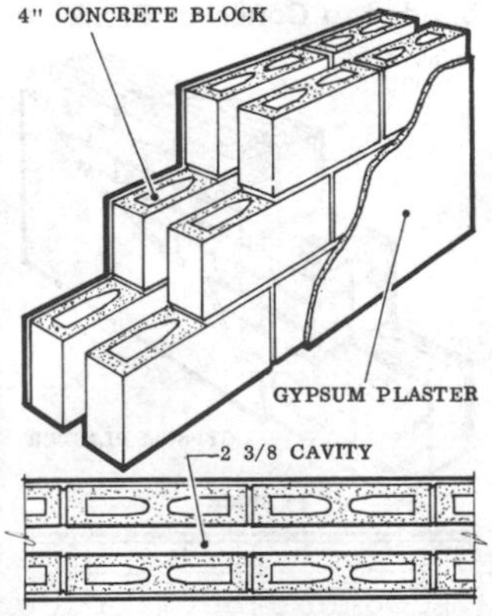

STC 51 to 52[10]

Nonporous (Normal Weight) Solid Concrete Masonry Walls

Unsealed

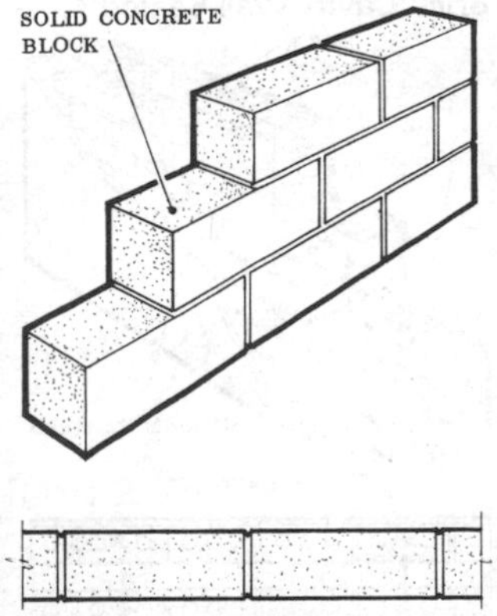

4″ Block **STC** **30**
6″ Block **STC** **38**
8″ Block **STC** **47**

Sealed[18]

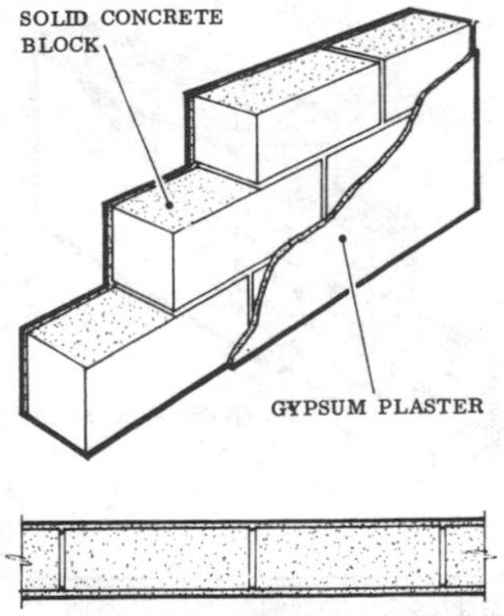

4″ Block **STC** **44**
6″ Block **STC** **48**
8″ Block **STC** **53**

9. Based on ASTM E90-61T test procedures.
10. Based on tests performed prior to ASTM E90-61T.
11. Test data can vary widely depending on weight and thickness of plaster, and water content (due to length of curing) of plaster.
12. With exception of double stud walls, ranges include values for ½″ or ⅝″ gypsum board thickness on 2½″ or 3⅝″ wide studs.
13. Ranges include values for gypsum board screwed, or laminated and screwed, to sound deadening board or base layer.
14. Unbalanced Construction: single layer gypsum board one side, double layer gypsum board one side.
15. Ranges include values for ½″ and ⅝″ board thickness on 1⅝″ wide studs.
16. Up to 60 STC with 9″ of insulation; lower values result with lesser thicknesses of insulation.
17. Performance of concrete and unit masonry systems is highly dependent on minimizing the number of openings through the construction (to accommodate piping, wiring and other equipment) and on all openings being sealed airtight.
18. Sealed both sides unless otherwise indicated. Sealing with plaster is most effective; when surfaces are to be painted, use of block sealer and/or 2 coats of cement-based paint is more effective than 2 coats of resin-emulsion paint to seal against airborne sound transmission.

WALL SYSTEMS

CONCRETE AND CONCRETE MASONRY WALLS[17] (CONT.)

Porous (Lightweight) Solid Concrete Masonry Walls

Unsealed

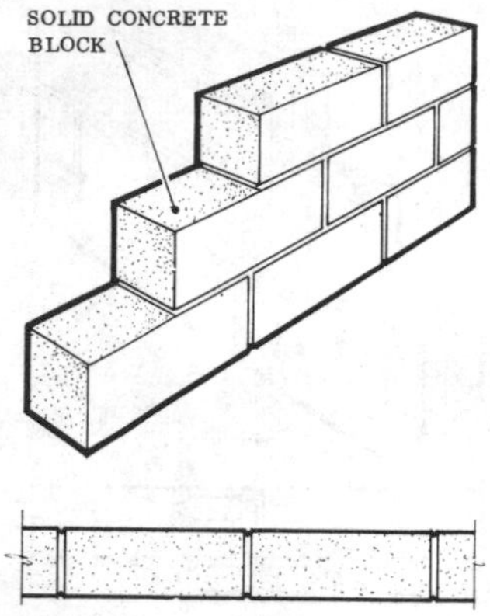

4″ Block	**STC**	**27**
6″ Block	**STC**	**36**
8″ Block	**STC**	**45**

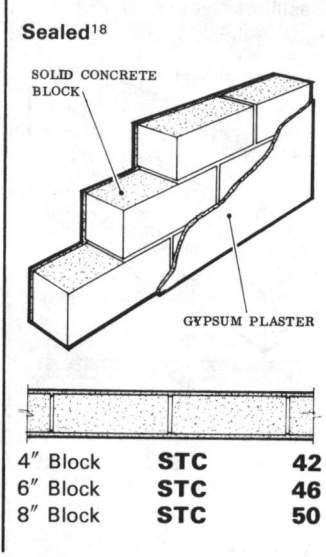

Sealed[18]

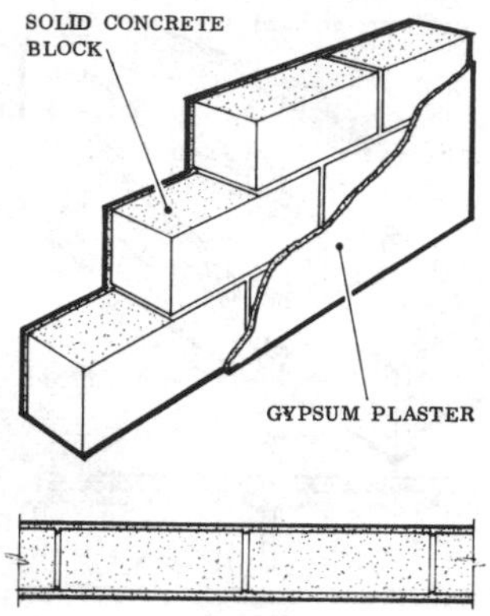

4″ Block	**STC**	**42**
6″ Block	**STC**	**46**
8″ Block	**STC**	**50**

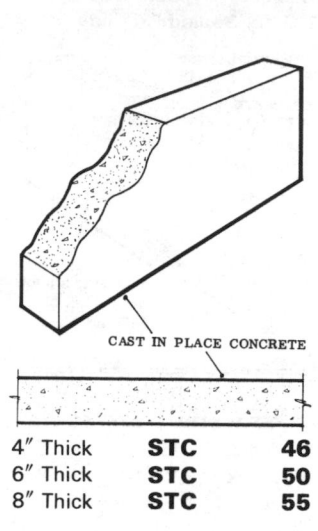

Poured-in-place Solid Concrete Walls

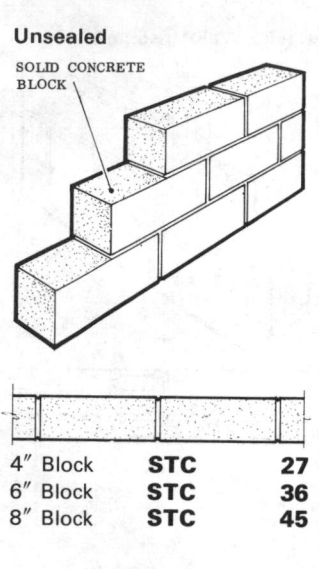

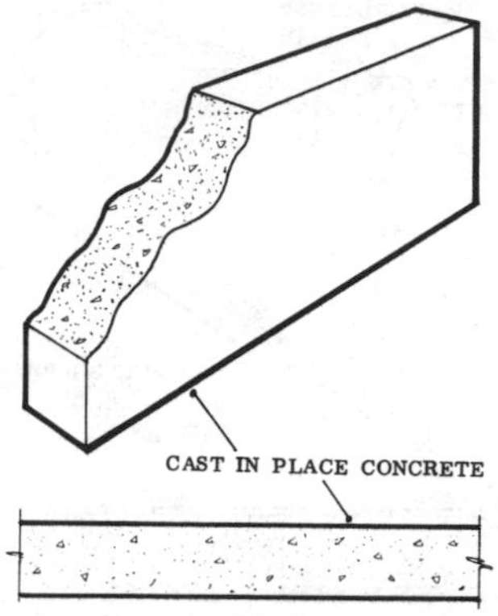

4″ Thick	**STC**	**46**
6″ Thick	**STC**	**50**
8″ Thick	**STC**	**55**

CLAY MASONRY WALLS[17]

Solid Wall

Brick Wall Sealed[18]

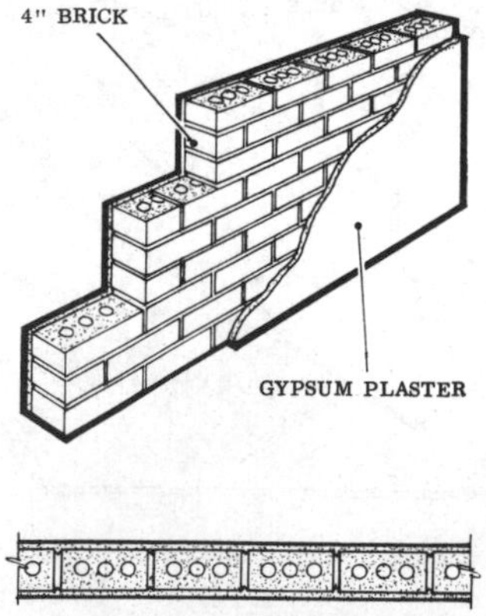

6″ Brick	**STC 45 to 50**
8″ Brick	**STC 50 to 52**
12″ Brick	**STC 55**

Cavity Wall

Brick Cavity Wall Sealed[18]

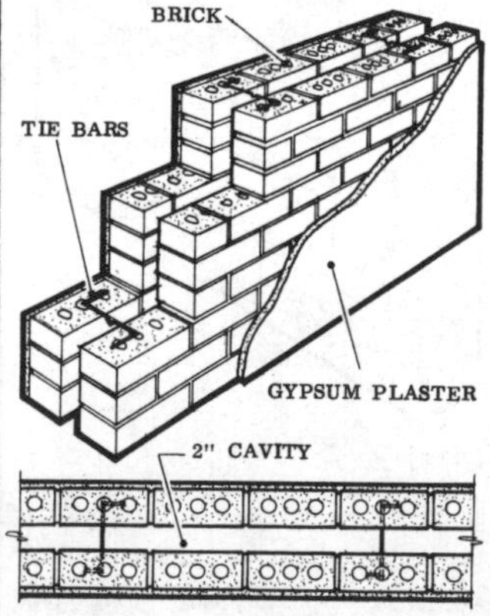

STC 45 to 50

Composite Wall

Composite Brick, bonded to Concrete Block

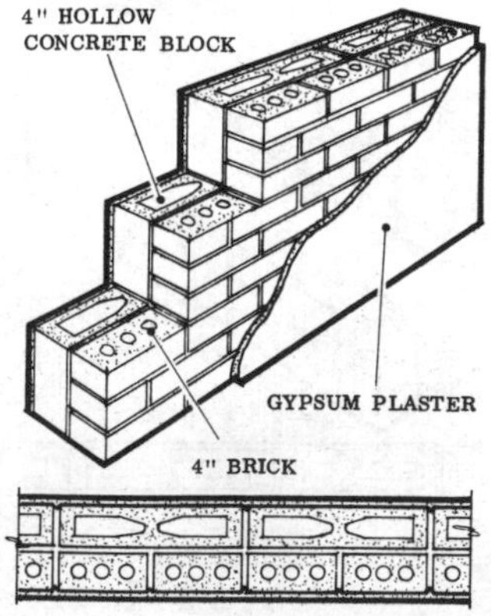

STC 45 to 50

9. Based on ASTM E90-61T test procedures.
10. Based on tests performed prior to ASTM E90-61T.
11. Test data can vary widely depending on weight and thickness of plaster, and water content (due to length of curing) of plaster.
12. With exception of double stud walls, ranges include values for ½″ or ⅝″ gypsum board thickness on 2½″ or 3⅝″ wide studs.
13. Ranges include values for gypsum board screwed, or laminated and screwed, to sound deadening board or base layer.
14. Unbalanced Construction: single layer gypsum board one side, double layer gypsum board one side.
15. Ranges include values for ½″ and ⅝″ board thickness on 1⅝″ wide studs.
16. Up to 60 STC with 9″ of insulation; lower values result with lesser thicknesses of insulation.
17. Performance of concrete and unit masonry systems is highly dependent on minimizing the number of openings through the construction (to accommodate piping, wiring and other equipment) and on all openings being sealed airtight.
18. Sealed both sides unless otherwise indicated. Sealing with plaster is most effective; when surfaces are to be painted, use of block sealer and/or 2 coats of cement-based paint is more effective than 2 coats of resin-emulsion paint to seal against airborne sound transmission.

FLOOR/CEILING SYSTEMS

WOOD JOIST FLOORS—GYPSUM BOARD CEILING FINISH[2]

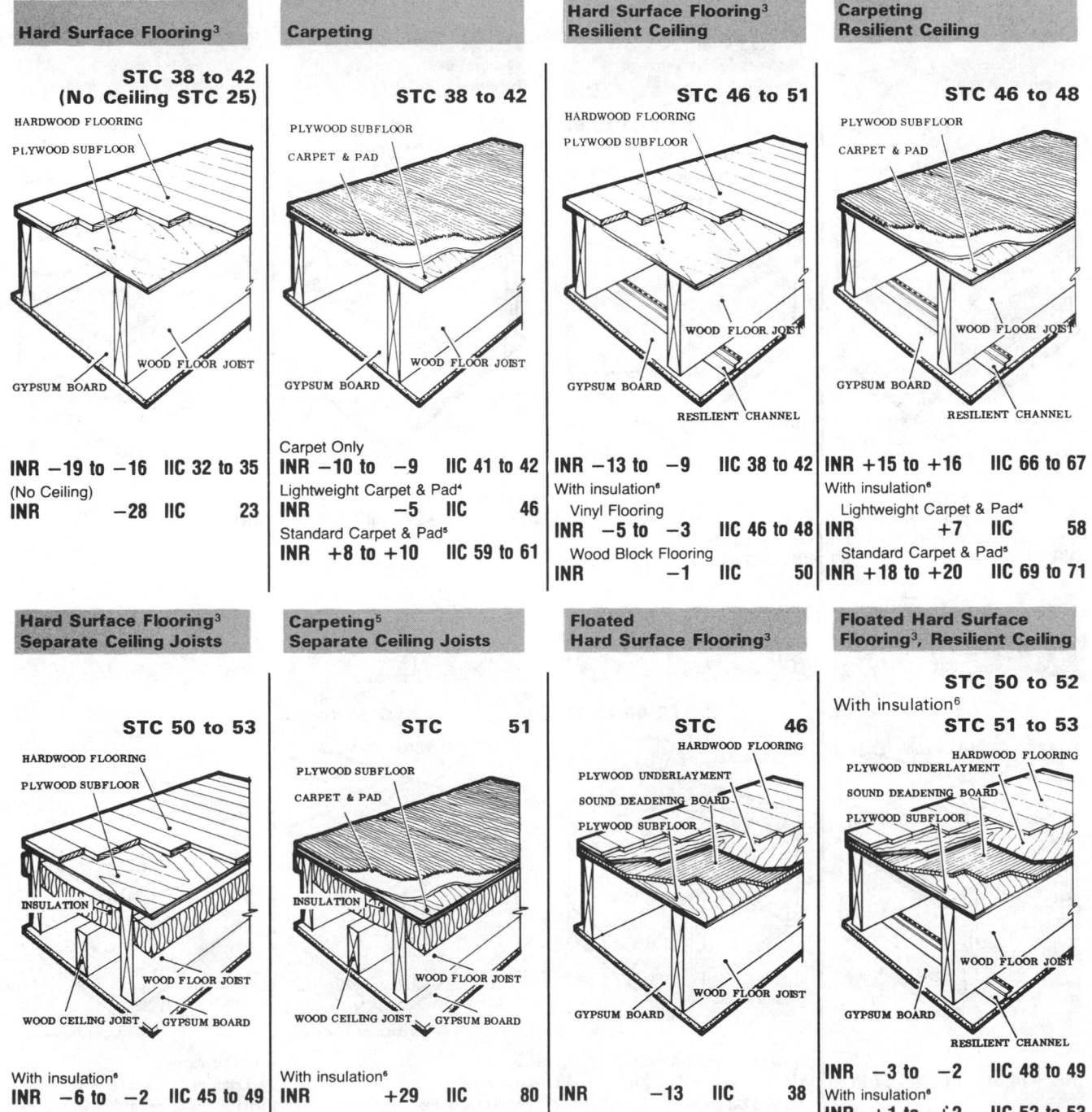

Hard Surface Flooring[3]

STC 38 to 42
(No Ceiling STC 25)

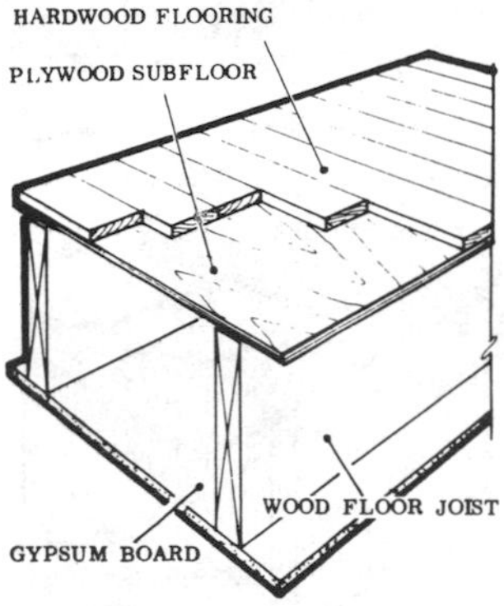

INR −19 to −16 IIC 32 to 35
(No Ceiling)
INR −28 IIC 23

Carpeting

STC 38 to 42

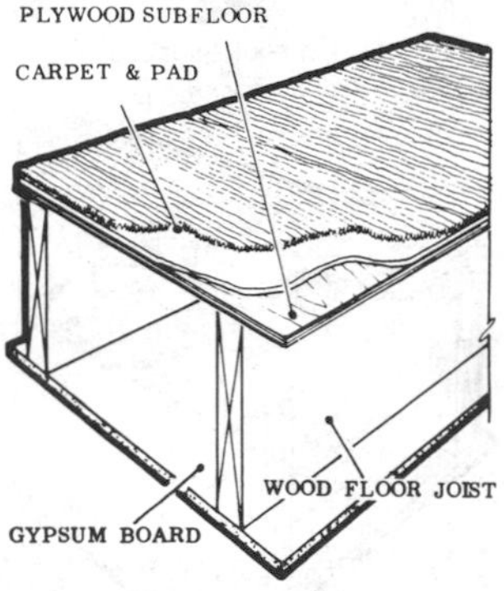

Carpet Only
INR −10 to −9 IIC 41 to 42
Lightweight Carpet & Pad[4]
INR −5 IIC 46
Standard Carpet & Pad[5]
INR +8 to +10 IIC 59 to 61

Hard Surface Flooring[3] Resilient Ceiling

STC 46 to 51

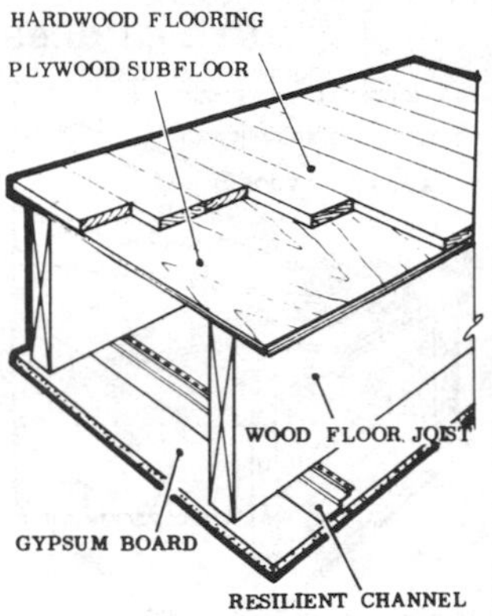

INR −13 to −9 IIC 38 to 42
With insulation[6]
Vinyl Flooring
INR −5 to −3 IIC 46 to 48
Wood Block Flooring
INR −1 IIC 50

Carpeting Resilient Ceiling

STC 46 to 48

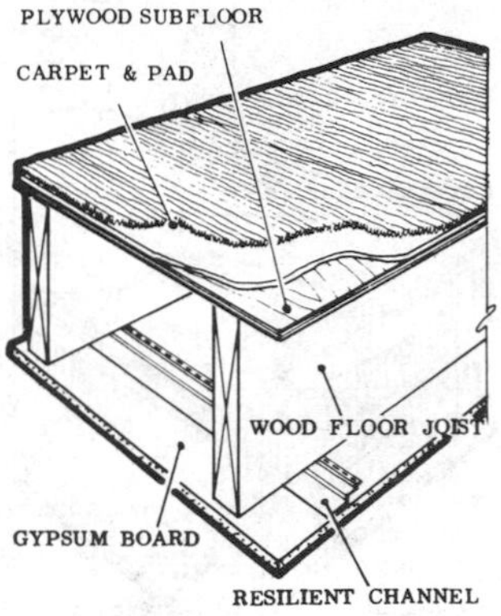

INR +15 to +16 IIC 66 to 67
With insulation[6]
Lightweight Carpet & Pad[4]
INR +7 IIC 58
Standard Carpet & Pad[5]
INR +18 to +20 IIC 69 to 71

Hard Surface Flooring[3] Separate Ceiling Joists

STC 50 to 53

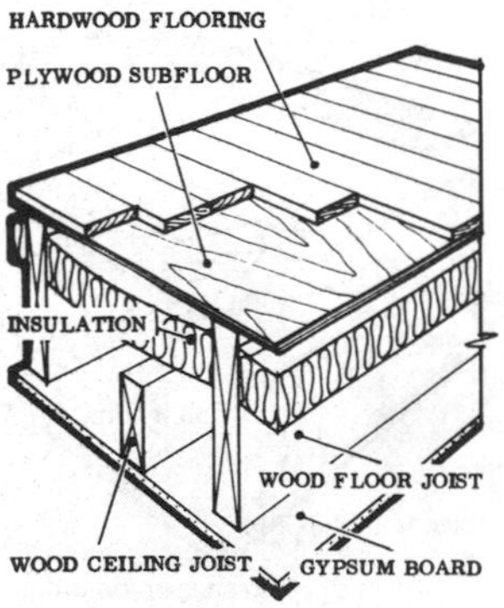

With insulation[6]
INR −6 to −2 IIC 45 to 49

Carpeting[5] Separate Ceiling Joists

STC 51

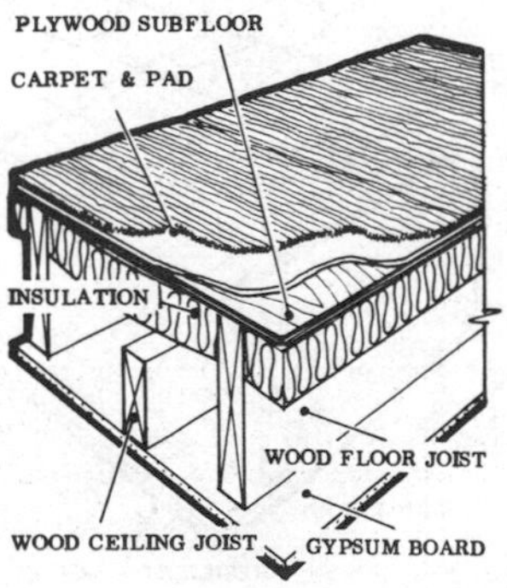

With insulation[6]
INR +29 IIC 80

Floated Hard Surface Flooring[3]

STC 46

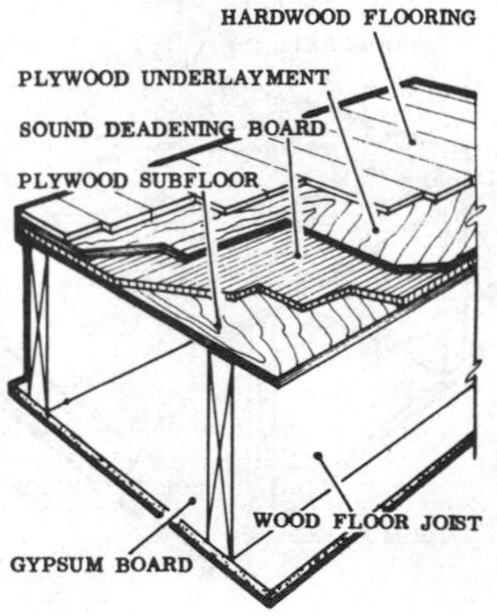

INR −13 IIC 38

Floated Hard Surface Flooring[3], Resilient Ceiling

STC 50 to 52
With insulation[6]
STC 51 to 53

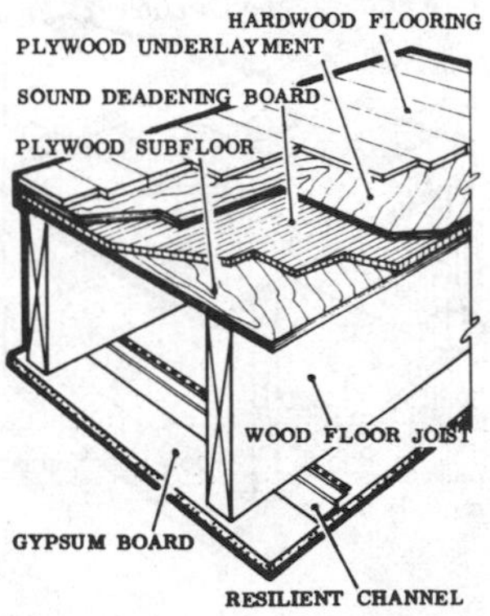

INR −3 to −2 IIC 48 to 49
With insulation[6]
INR +1 to +2 IIC 52 to 53

---

1. INR values are based on tests performed in accordance with ISO Recommendation R-140, "Field and Laboratory Measurements of Airborne and Impact Sound Transmission," January, 1960. IIC values approximated by adding 51 to original INR test values.
2. Ranges include value for ⅜", ½" and ⅝" thicknesses of gypsum board finish.
3. Hard surface finish flooring (i.e., stripwood, wood block, vinyl and vinyl asbestos) generally perform similarly in isolating impact sound. Unless otherwise indicated, ranges given include values for construction with any typical hard surface finish flooring.
4. Lightweight carpet and pad having pile weights of 20 oz. per sq. yd. and 28 oz. per sq. yd. respectively.
5. Standard weight carpet and pad having pile weights of 44 oz. per sq. yd. and 40 oz. per sq. yd. respectively. Ratings shown are based on standard weight combinations unless otherwise noted.
6. A minimum thickness of 2" to 3" of insulation can be used effectively as performance does not increase with increased thickness in floor/ceiling systems.
7. Ratings vary with thickness of concrete slabs. The low end of the range indicates value for 6" thickness; high end, 8" thickness.

FLOOR/CEILING SYSTEMS

WOOD JOIST FLOORS—GYPSUM BOARD CEILING FINISH[2] (CONT.)

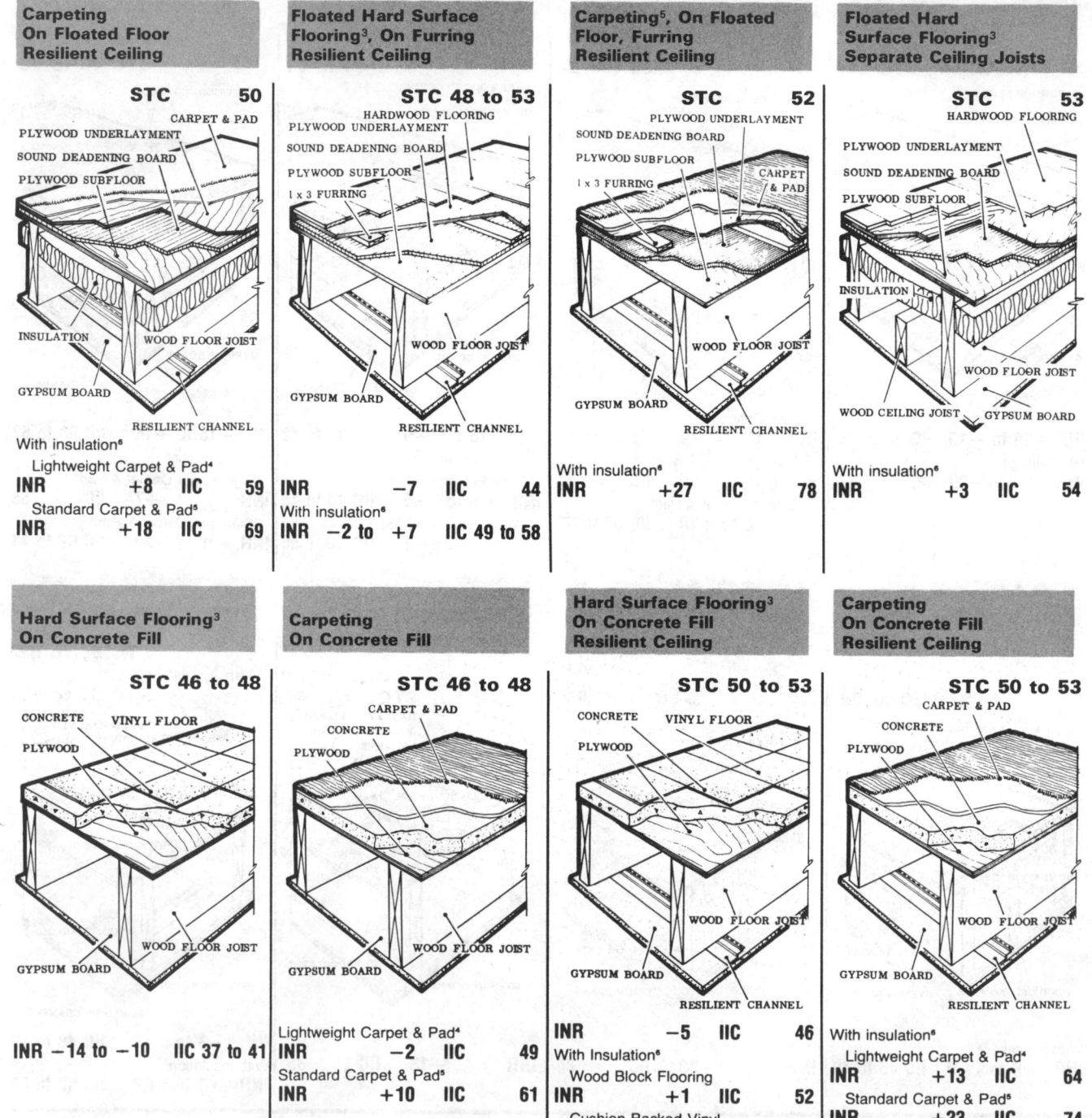

Carpeting On Floated Floor Resilient Ceiling

STC 50

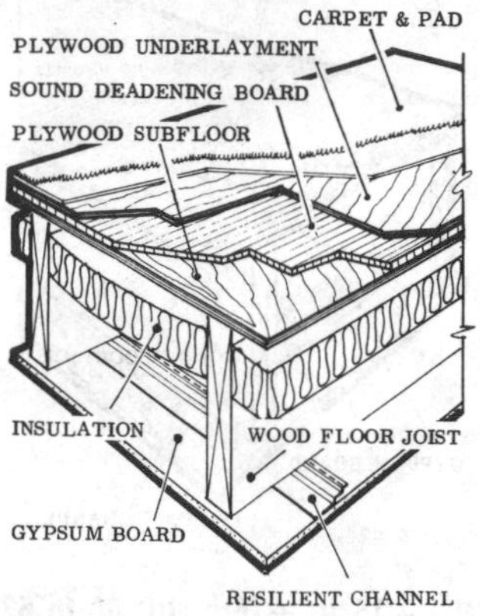

With insulation[6]

Lightweight Carpet & Pad[4]

INR +8 IIC 59

Standard Carpet & Pad[5]

INR +18 IIC 69

Floated Hard Surface Flooring[3], On Furring Resilient Ceiling

STC 48 to 53

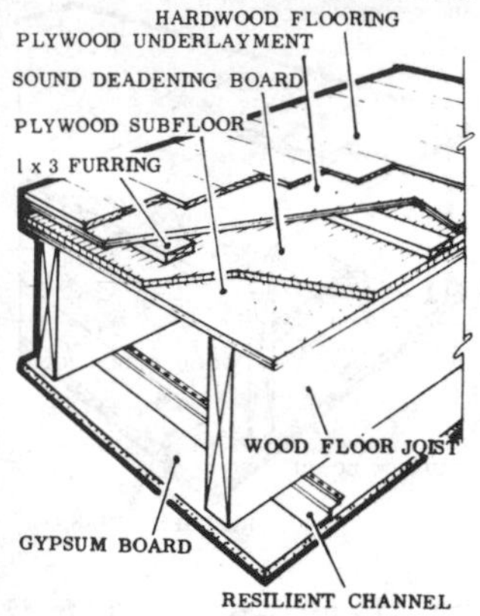

INR −7 IIC 44

With insulation[6]

INR −2 to +7 IIC 49 to 58

Carpeting[5], On Floated Floor, Furring Resilient Ceiling

STC 52

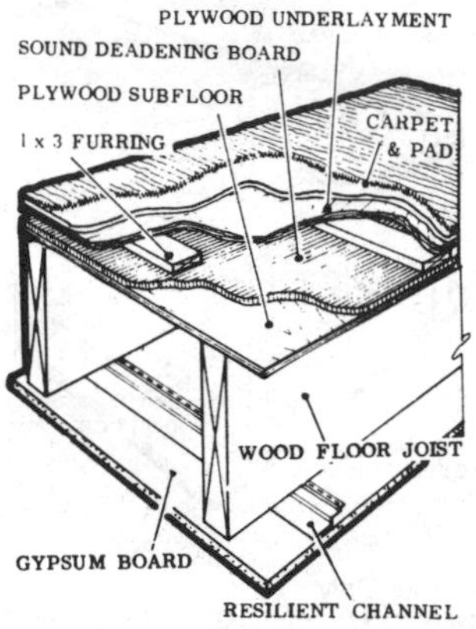

With insulation[6]

INR +27 IIC 78

Floated Hard Surface Flooring[3] Separate Ceiling Joists

STC 53

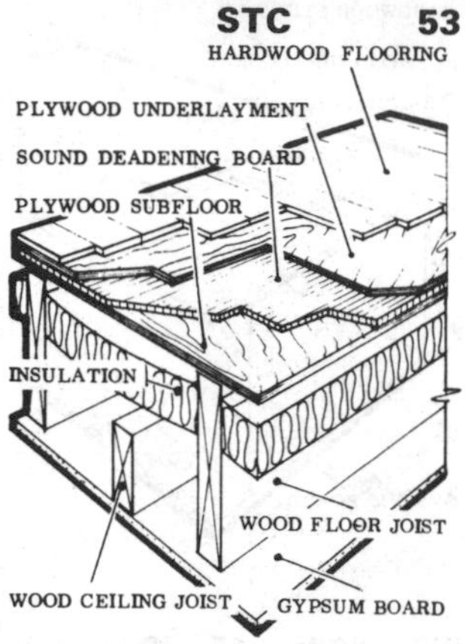

With insulation[6]

INR +3 IIC 54

Hard Surface Flooring[3] On Concrete Fill

STC 46 to 48

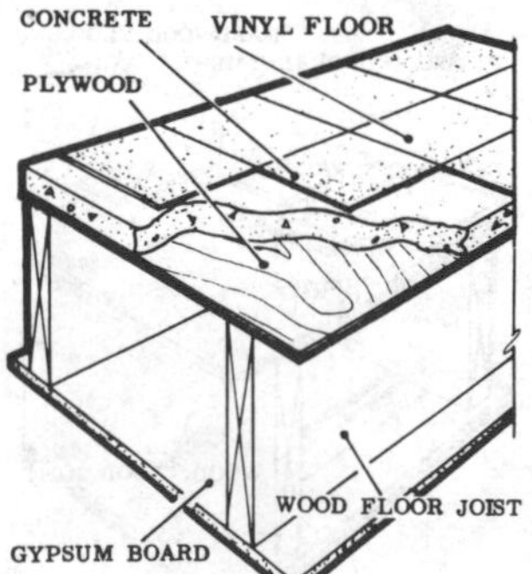

INR −14 to −10 IIC 37 to 41

Carpeting On Concrete Fill

STC 46 to 48

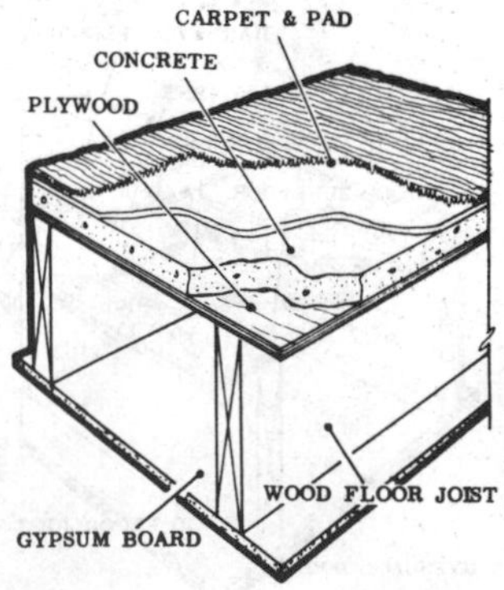

Lightweight Carpet & Pad[4]

INR −2 IIC 49

Standard Carpet & Pad[5]

INR +10 IIC 61

Hard Surface Flooring[3] On Concrete Fill Resilient Ceiling

STC 50 to 53

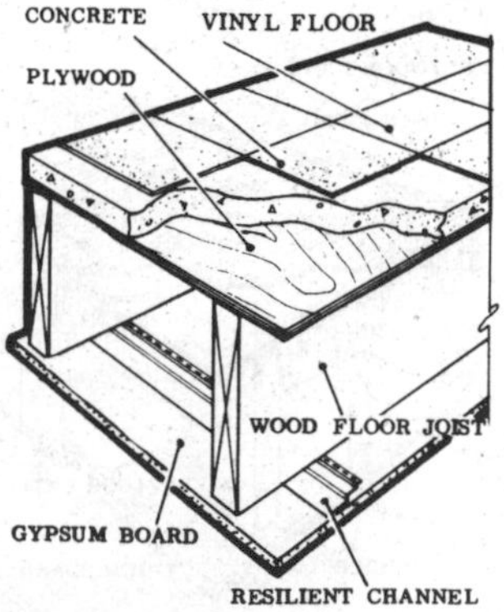

INR −5 IIC 46

With Insulation[6]

Wood Block Flooring

INR +1 IIC 52

Cushion Backed Vinyl

INR +5 IIC 56

Carpeting On Concrete Fill Resilient Ceiling

STC 50 to 53

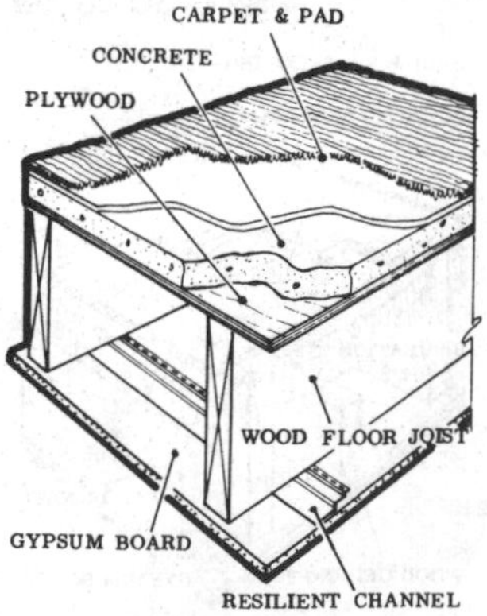

With insulation[6]

Lightweight Carpet & Pad[4]

INR +13 IIC 64

Standard Carpet & Pad[5]

INR +23 IIC 74

1. INR values are based on tests performed in accordance with ISO Recommendation R-140, "Field and Laboratory Measurements of Airborne and Impact Sound Transmission," January, 1960. IIC values approximated by adding 51 to original test values.
2. Ranges include value for ⅜", ½" and ⅝" thicknesses of gypsum board finish.
3. Hard surface finish flooring (i.e., stripwood, wood block, vinyl and vinyl asbestos) generally perform similarly in isolating impact sound. Unless otherwise indicated, ranges given include values for construction with any typical hard surface finish flooring.

FLOOR/CEILING SYSTEMS

WOOD/STEEL TRUSS FLOORS—GYPSUM BOARD CEILING FINISH[2]

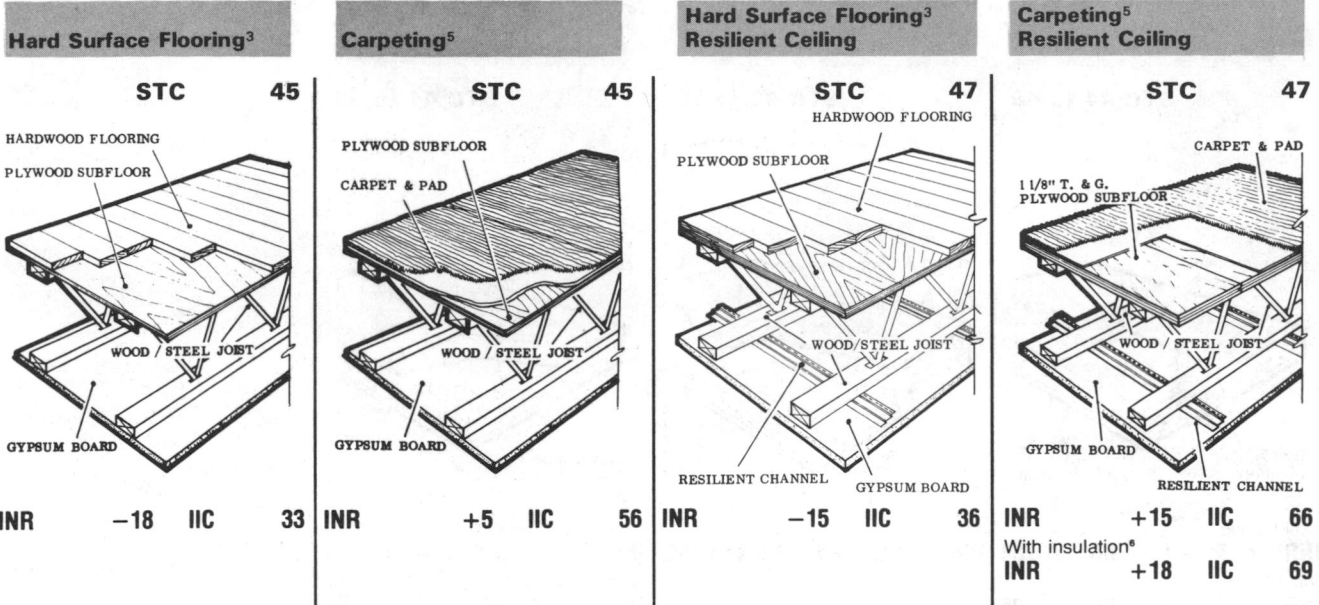

Hard Surface Flooring[3]

STC 45

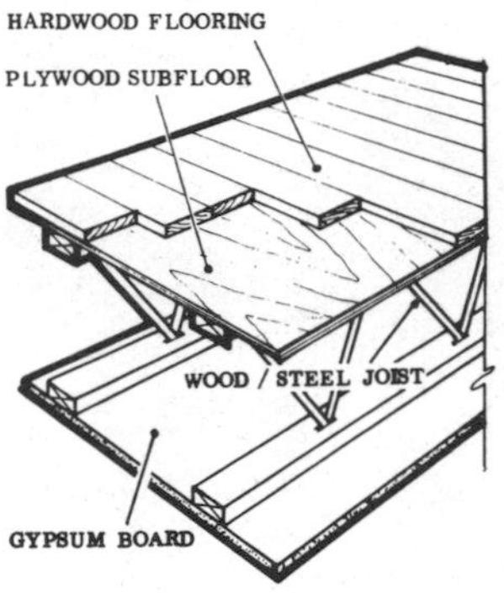

INR −18 IIC 33

Carpeting[5]

STC 45

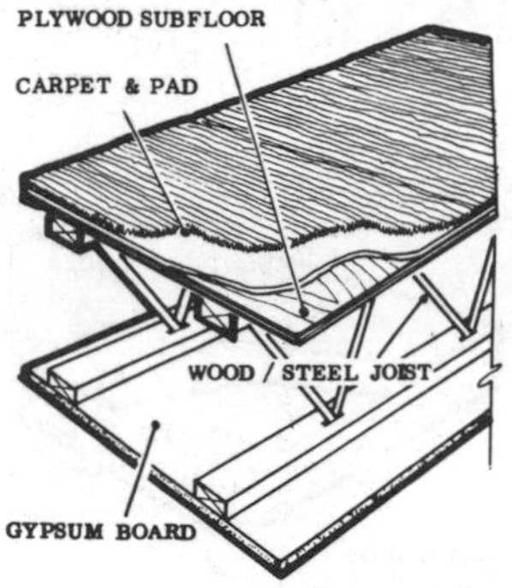

INR +5 IIC 56

Hard Surface Flooring[3]
Resilient Ceiling

STC 47

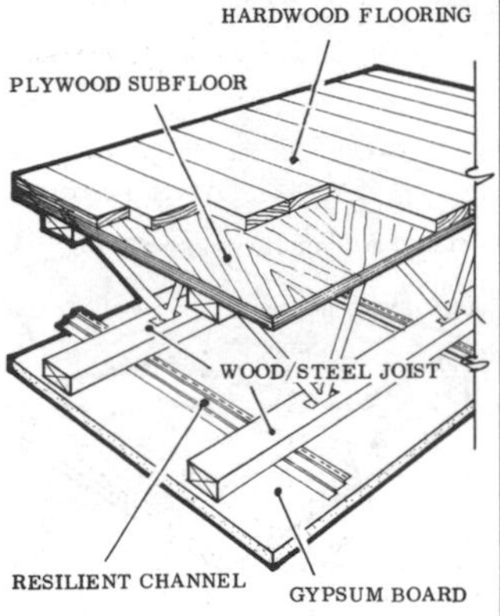

INR −15 IIC 36

Carpeting[5]
Resilient Ceiling

STC 47

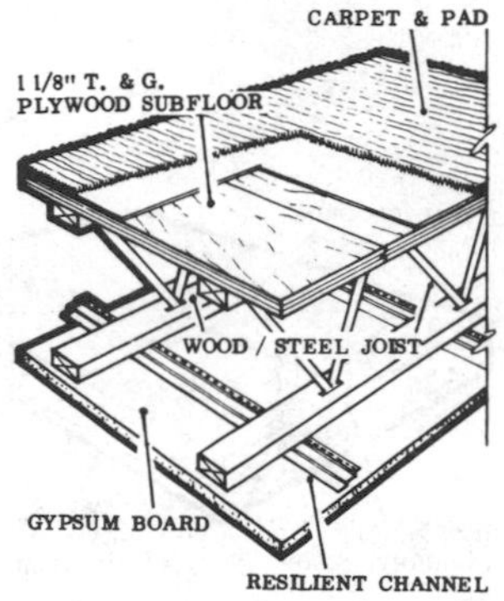

INR +15 IIC 66
With insulation[6]
INR +18 IIC 69

Hard Surface Flooring[3]
On Concrete Fill

STC 46

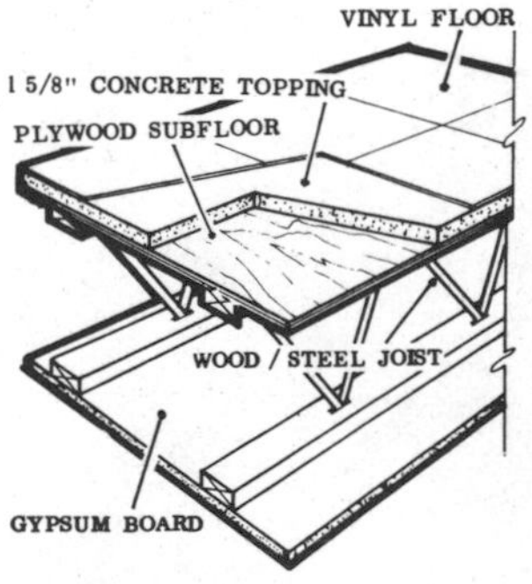

INR −14 IIC 37

Carpeting[5]
On Concrete Fill

STC 46

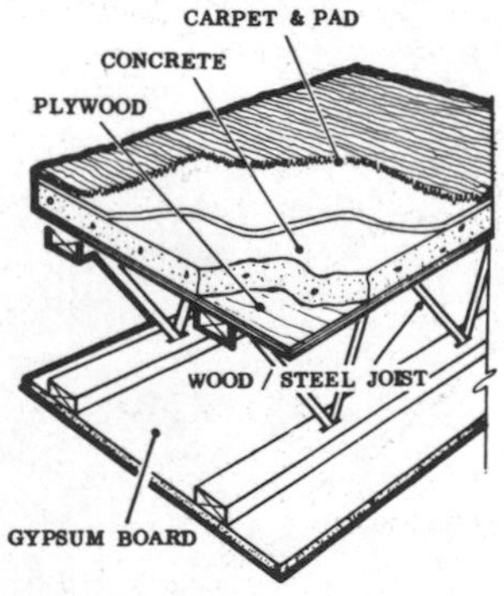

INR +11 IIC 62

4. Lightweight carpet and pad having pile weights of 20 oz. per sq. yd. and 28 oz. per sq. yd. respectively.
5. Standard weight carpet and pad having pile weights of 44 oz. per sq. yd. and 40 oz. per sq. yd. respectively. Ratings shown are based on standard weight combinations unless otherwise noted.
6. A minimum thickness of 2″ to 3″ of insulation can be used effectively as performance does not increase with increased thickness in floor/ceiling systems.
7. Ratings vary with thickness of concrete slabs. The low end of the range indicates value for 6″ thickness; high end, 8″ thickness.

FLOOR/CEILING SYSTEMS

POURED-IN-PLACE REINFORCED CONCRETE SLABS

Hard Surface Flooring[3] With or without Plaster Ceiling

STC 44 to 48

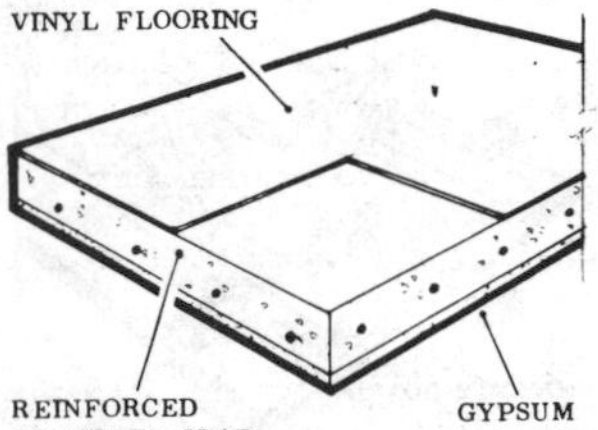

No Flooring
INR −17 IIC 34
Vinyl Flooring
INR −15 IIC 36
Wood Block Flooring
INR −9 to −6 IIC 42 to 45

Floated Hard Surface Flooring[3], With or Without Plaster Ceiling

STC 48 to 50

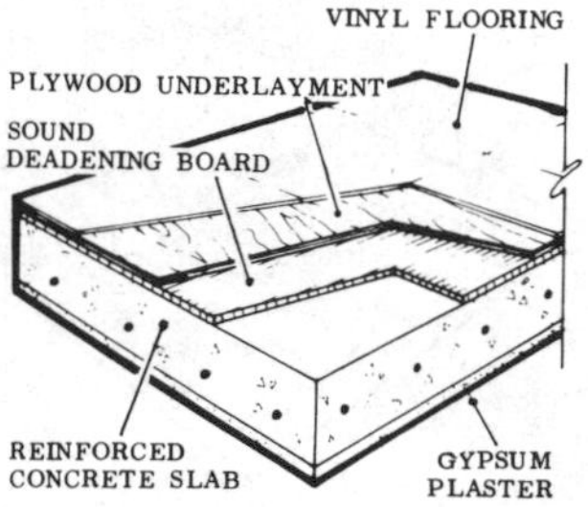

INR −7 to +4 IIC 44 to 55

Carpeting With or without Plaster Ceiling

STC 44 to 48

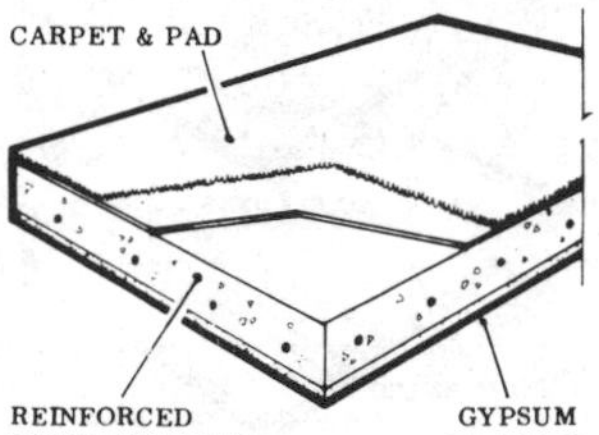

Lightweight Carpet Only
INR +2 IIC 53
Standard Carpet Only
INR +6 IIC 57
Standard Carpet & Pad[5]
INR +17 to +22 IIC 68 to 73

1. INR values are based on tests performed in accordance with ISO Recommendation R-140, "Field and Laboratory Measurements of Airborne and Impact Sound Transmission," January, 1960, IIC values approximated by adding 51 to original INR test values.
2. Ranges include value for ⅜", ½" and ⅝" thicknesses of gypsum board finish.
3. Hard surface finish flooring (i.e., stripwood, wood block, vinyl and vinyl asbestos) generally perform similarly in isolating impact sound. Unless otherwise indicated, ranges given include values for construction with any typical hard surface finish flooring.

FLOOR/CEILING SYSTEMS

PRECAST CONCRETE SLABS[7]

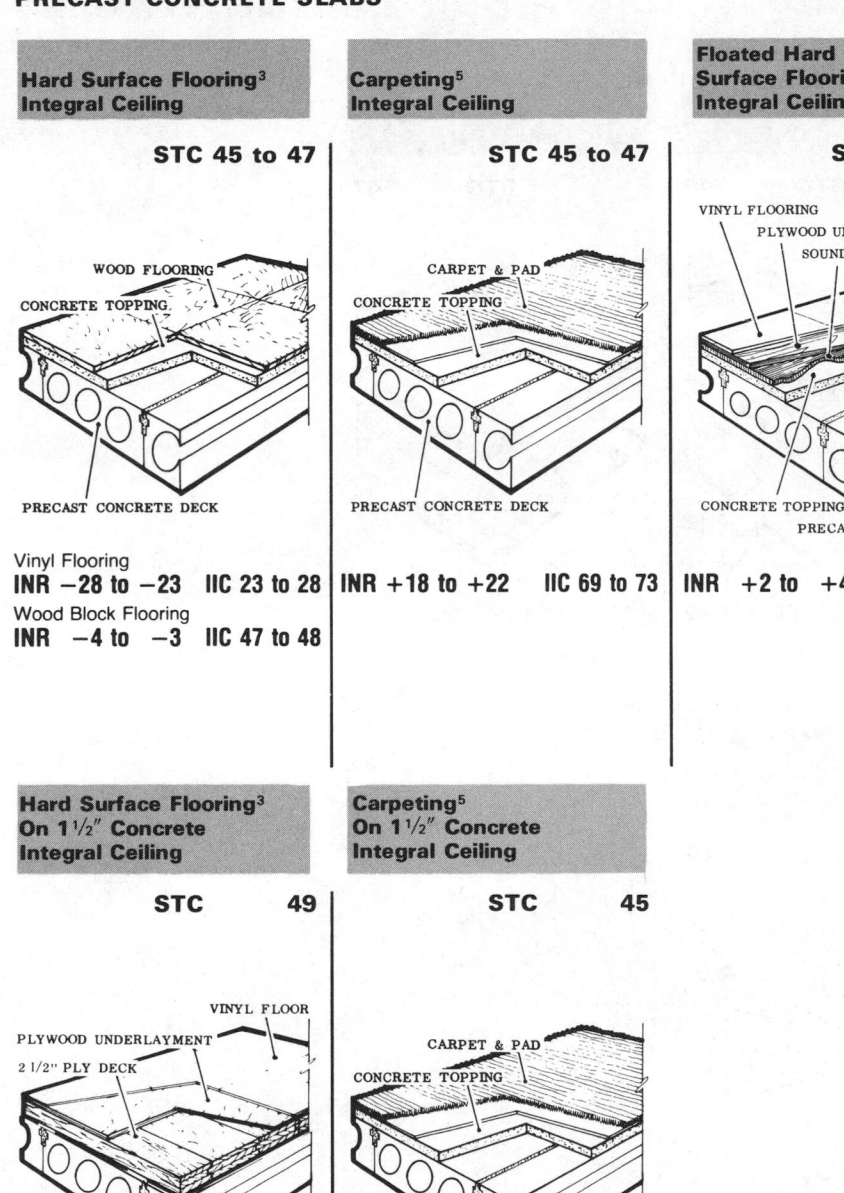

Hard Surface Flooring[3] Integral Ceiling

STC 45 to 47

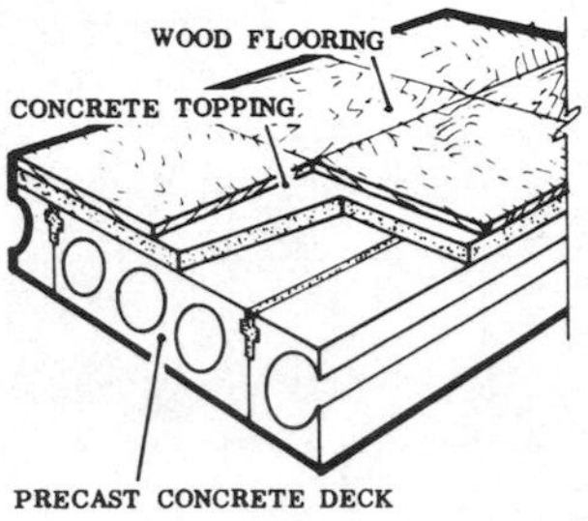

Vinyl Flooring
INR −28 to −23 IIC 23 to 28
Wood Block Flooring
INR −4 to −3 IIC 47 to 48

Carpeting[5] Integral Ceiling

STC 45 to 47

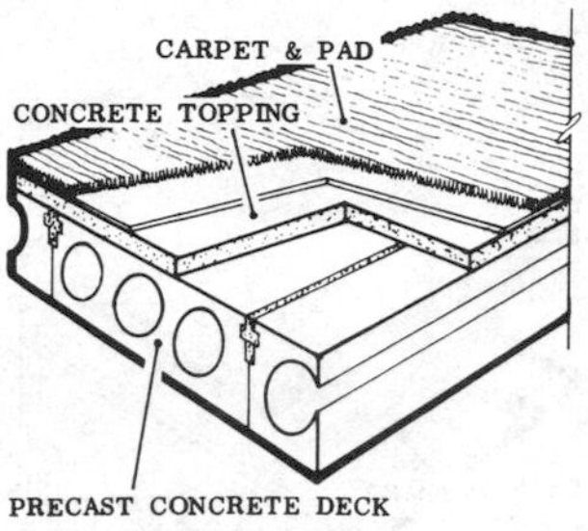

INR +18 to +22 IIC 69 to 73

Floated Hard Surface Flooring[3] Integral Ceiling

STC 47 to 51

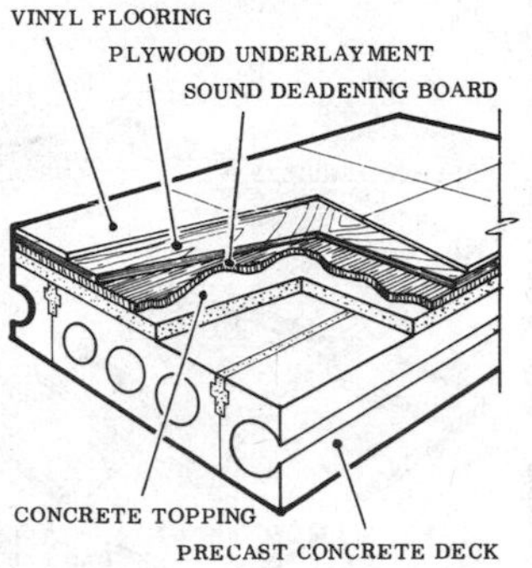

INR +2 to +4 IIC 53 to 55

Hard Surface Flooring[3] On 1½" Concrete Integral Ceiling

STC 49

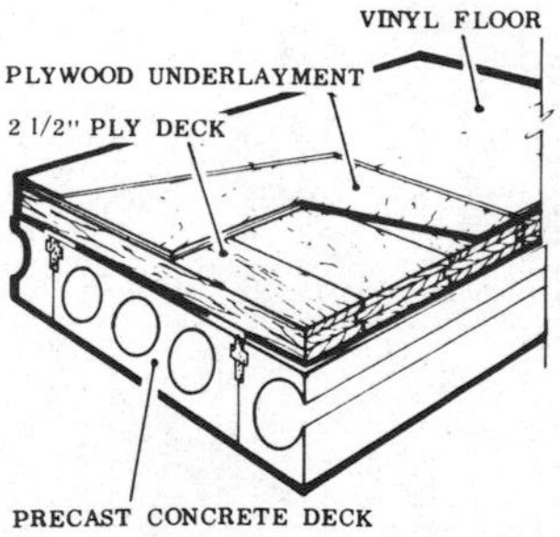

INR +2 to +4 IIC 53 to 55

Carpeting[5] On 1½" Concrete Integral Ceiling

STC 45

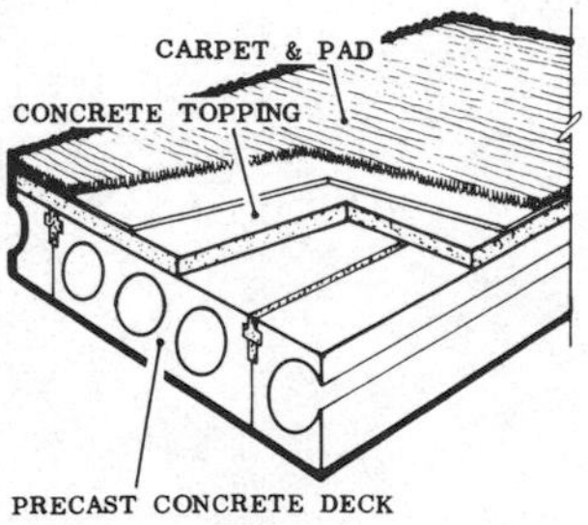

INR +25 IIC 76

4. Lightweight carpet and pad having pile weights of 20 oz. per sq. yd. and 28 oz. per sq. yd. respectively.
5. Standard weight carpet and pad having pile weights of 44 oz. per sq. yd. and 40 oz. per sq. yd. respectively. Ratings shown are based on standard weight combinations unless otherwise noted.
6. A minimum thickness of 2" to 3" of insulation can be used effectively as performance does not increase with increased thickness in floor/ceiling systems.
7. Ratings vary with thickness of concrete slabs. The low end of the range indicates value for 6" thickness; high end, 8" thickness.

FLOOR/CEILING SYSTEMS

STEEL BAR JOIST FLOORS

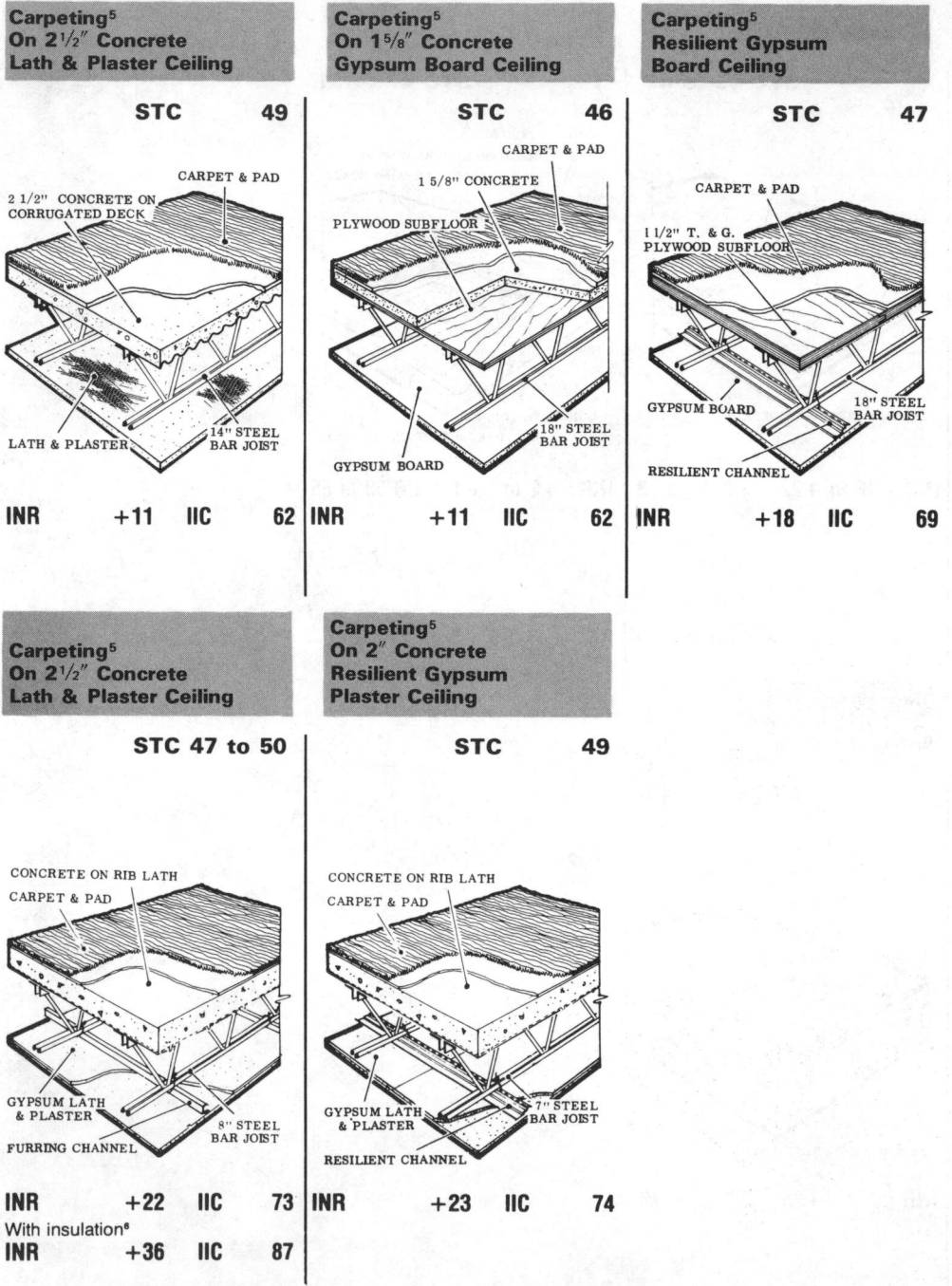

Carpeting[5]
On 2½″ Concrete
Lath & Plaster Ceiling

STC 49

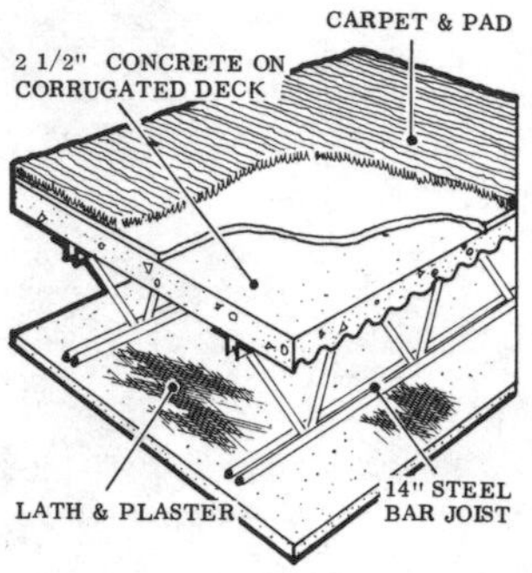

INR +11 IIC 62

Carpeting[5]
On 1⅝″ Concrete
Gypsum Board Ceiling

STC 46

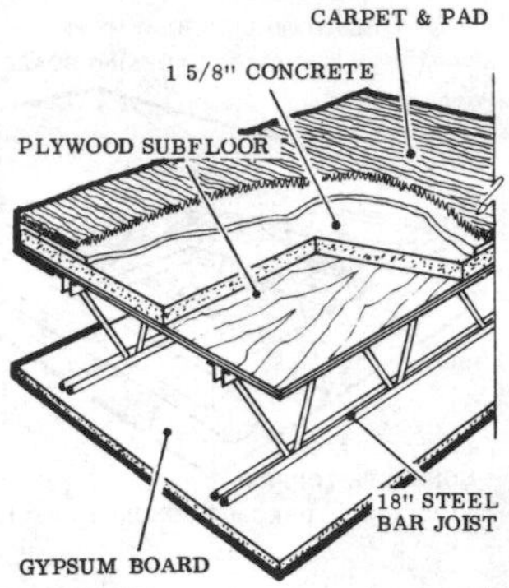

INR +11 IIC 62

Carpeting[5]
Resilient Gypsum
Board Ceiling

STC 47

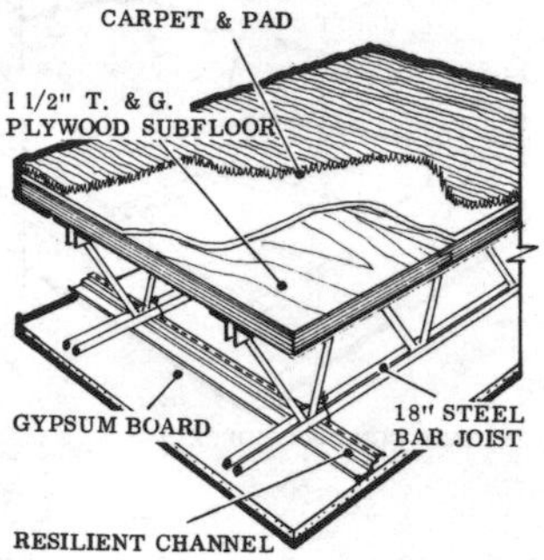

INR +18 IIC 69

Carpeting[5]
On 2½″ Concrete
Lath & Plaster Ceiling

STC 47 to 50

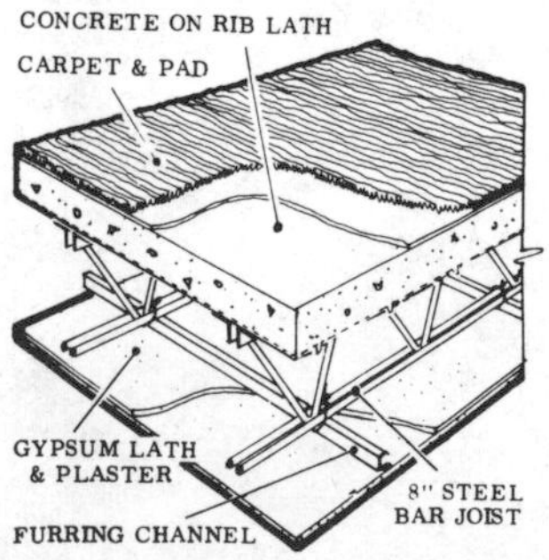

INR +22 IIC 73

With insulation[6]

INR +36 IIC 87

Carpeting[5]
On 2″ Concrete
Resilient Gypsum
Plaster Ceiling

STC 49

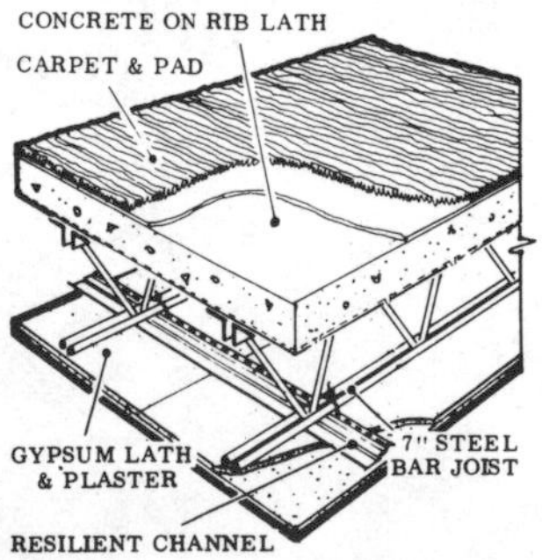

INR +23 IIC 74

1. INR values are based on tests performed in accordance with ISO Recommendation R-140, "Field and Laboratory Measurements of Airborne and Impact Sound Transmission," January, 1960. IIC values approximated by adding 51 to original INR test values.
2. Ranges include value for ⅜", ½" and ⅝" thicknesses of gypsum board finish.
3. Hard surface finish flooring (i.e., stripwood, wood block, vinyl and vinyl asbestos) generally perform similarly in isolating impact sound. Unless otherwise indicated, ranges given include values for construction with any typical hard surface finish flooring.

FLOOR/CEILING SYSTEMS

STEEL BAR JOIST FLOORS

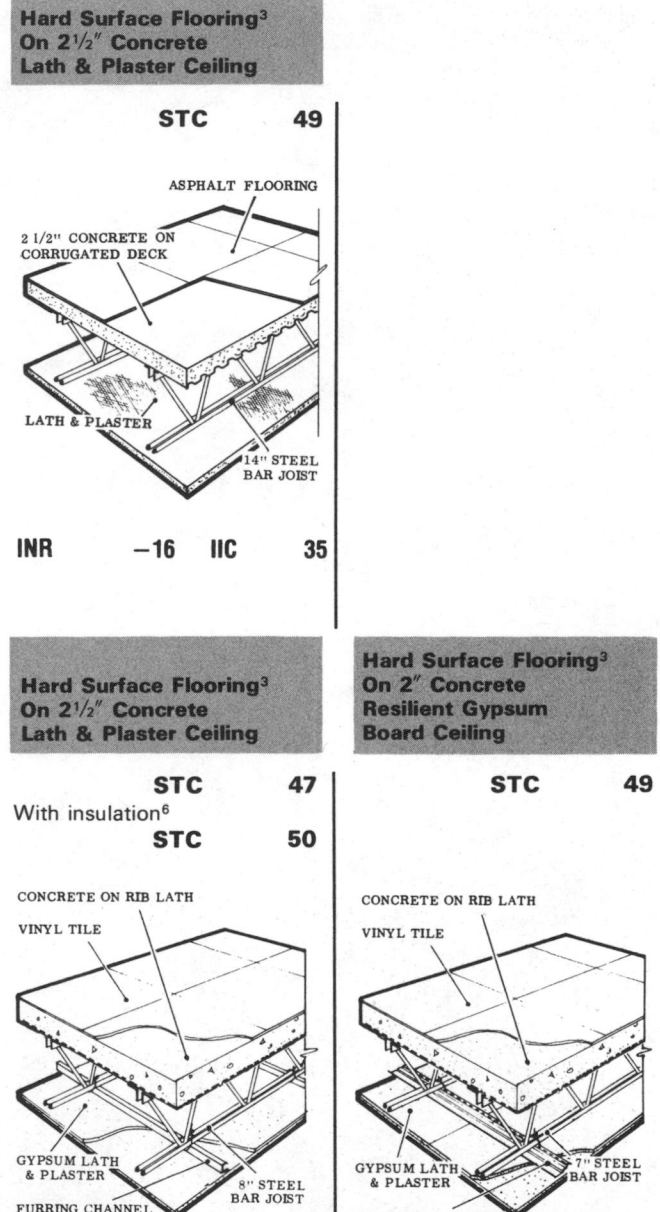

**Hard Surface Flooring[3]
On 2½″ Concrete
Lath & Plaster Ceiling**

STC 49

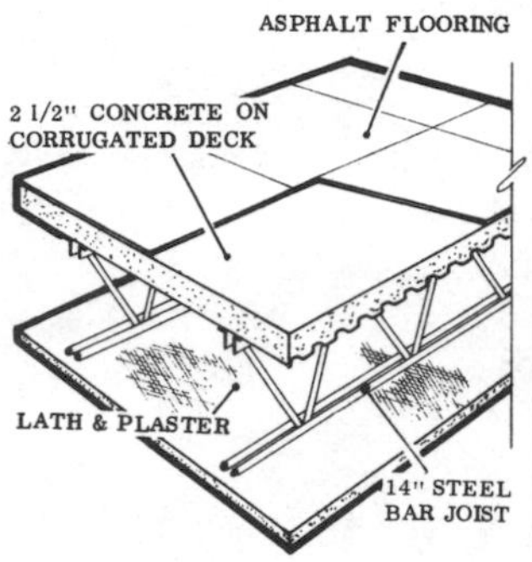

INR −16 IIC 35

**Hard Surface Flooring[3]
On 2½″ Concrete
Lath & Plaster Ceiling**

STC 47
With insulation[6]
STC 50

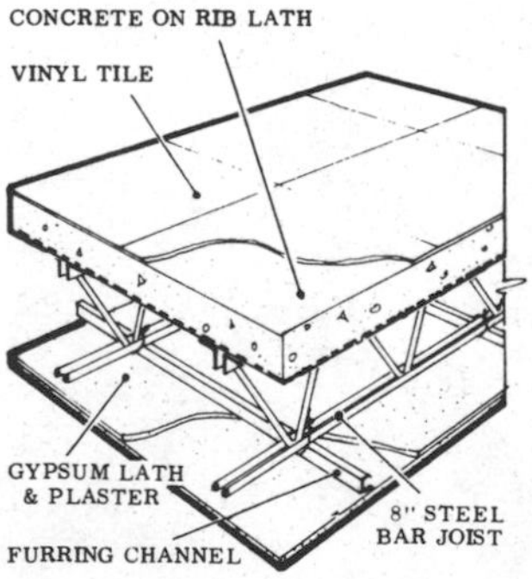

INR −25 IIC 26
With insulation[6]
INR −22 IIC 29

**Hard Surface Flooring[3]
On 2″ Concrete
Resilient Gypsum
Board Ceiling**

STC 49

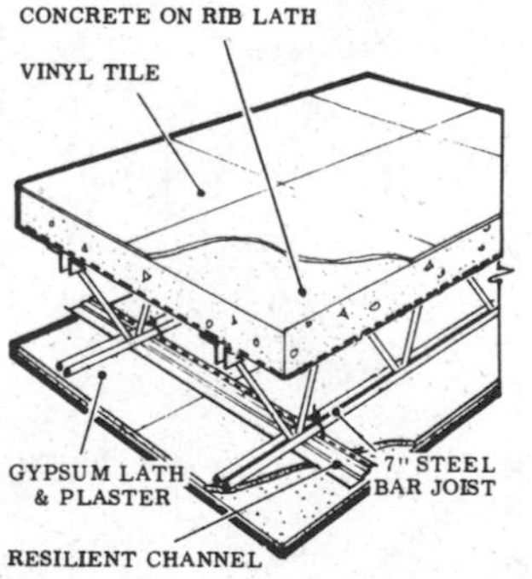

INR −10 IIC 41

4. Lightweight carpet and pad having pile weights of 20 oz. per sq. yd. and 28 oz. per sq. yd. respectively.
5. Standard weight carpet and pad having pile weights of 44 oz. per sq. yd. and 40 oz. per sq. yd. respectively. Ratings shown are based on standard weight combinations unless otherwise noted.
6. A minimum thickness of 2″ to 3″ of insulation can be used effectively as performance does not increase with increased thickness in floor/ceiling systems.
7. Ratings vary with thickness of concrete slabs. The low end of the range indicates value for 6″ thickness; high end, 8″ thickness.

108 PROPERTIES OF MATERIALS

PROPERTIES OF MATERIALS

INTRODUCTION

In the last few decades there has been a great increase in the number of materials available for use in construction. At least 300 new materials appear annually. In order to combine these materials effectively the designer, the builder, and even the lender—all need a basic understanding of the general properties of materials.

A material's suitability depends on its ability to perform in specific applications. The performance of a material in any situation is predictable, if the material's properties are known. There are four general groups of properties of interest to those in the construction industry: *mechanical, thermal, electrical* and *chemical.* These properties are dependent upon the structure of the material being examined.

The millions of materials which exist are all made up from just three basic building blocks of nature: atoms, ions and molecules. These are bonded together by one of several basic chemical bonds. All materials can be classified according to these basic structural elements and bonds into three major categories: (1) *ceramics and glasses,* which are hard, brittle and poor conductors of heat and electricity, (2) *metals,* which are more ductile than ceramics and are good conductors of heat and electricity, and (3) *molecular materials,* which have low melting temperatures, fair strength and are poor conductors of heat and electricity.

All construction materials fall into one of the above categories. Understanding materials is simply a matter of recognizing in which category a particular material belongs and remembering the properties of that category. For instance, glass wool and foamed polystyrene may have identical insulating capabilities, but the glass wool falls into the ceramic category and the foam into the molecular category. It is obvious that the lower melting temperature of the foam, which is characteristic of molecular materials, should be a source of concern to the designer.

This section is intended to expand and sharpen the reader's instinct for categorizing materials. First, the basic structure of matter is discussed, then the four sets of properties which determine a material's performance are considered in some detail. For more information on the properties of individual materials, see Division 200 sections such as 201 Wood, 203 Concrete and 209 Plastics.

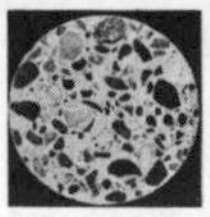

Matter is formed by the chemical bonding of atoms, ions and molecules. The particular combination of components and bonding methods determines final material properties.

BUILDING BLOCKS OF MATTER

There are about 100 chemical substances which cannot be subdivided into other substances. These basic materials, such as carbon, iron, hydrogen and oxygen, are called *chemical elements.* Just two chemical elements, oxygen and silicon, make up more than 3/4 of all the matter on earth; just eight chemical elements constitute more than 97% of all earthly substances (Fig. 1).

A given quantity of any element can be divided into smaller and smaller quantities until a minute particle is reached. This entity, the smallest particle which retains the properties of the original material, is called the *atom.*

Ordinarily, atoms are neutral particles, having no electrical charge. However, atoms may acquire a positive or negative charge by losing or attracting negatively charged particles; they are then termed *ions.* Other atoms may combine into simple units called *molecules.* All matter is formed from these three basic building blocks: atoms, ions and molecules.

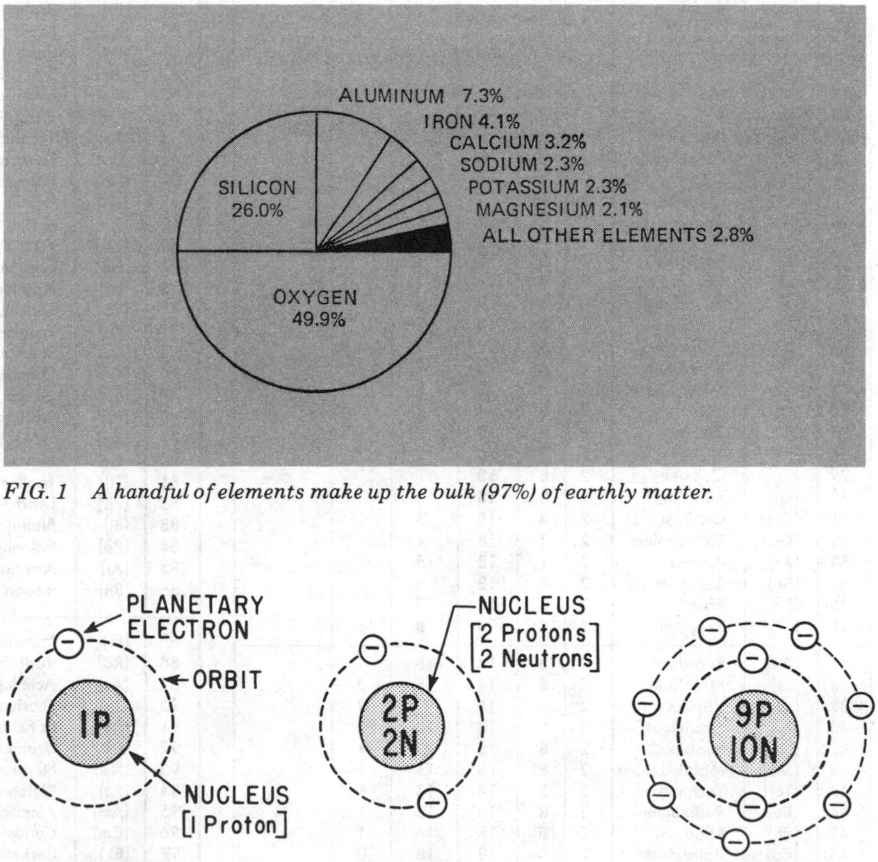

FIG. 1 A handful of elements make up the bulk (97%) of earthly matter.

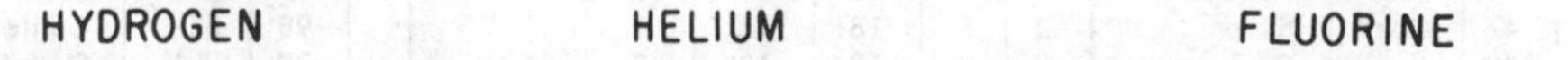

FIG. 2 Electrons move in orbits around the nucleus. Helium's two electrons move in the same orbit, but fluorine's 9 electrons move in several different orbits. Groups of electrons with identical or neighboring orbits comprise a shell; hydrogen and helium have one shell, fluorine two.

Atoms

The simplest picture of an atom is that of a miniature solar system (Fig. 2). The atom has at its core a *nucleus*, consisting of positively-charged *protons* and electrically-neutral *neutrons*. Orbiting about the nucleus are negatively charged *electrons*.

Protons and neutrons are much denser than electrons, so essentially all of the weight of an atom resides in the nucleus. The sum of the weights of the protons and neutrons of an atom is called the *atomic weight* of an element. Since the atom is electrically neutral, the positive charge of the protons in the nucleus is balanced by an equal negative charge from the electrons. The number of electrons, which is equal to the number of protons, is the *atomic number* of an element.

Groups of electrons moving in neighboring orbits, are said to belong to the same *shell* (Fig. 2). The first shell, closest to the nucleus, can only accommodate 2 electrons; the second a maximum of 8, the third 18 and fourth, 32. The distribution of

FIG. 3 **ELECTRON DISTRIBUTION IN ELEMENTS**

At. No.	Element (Symbol) & Name		Electrons in shells (#) 1	2	3	4	5
1	(H)	Hydrogen	1				
2	(He)	Helium	2				
3	(Li)	Lithium	2	1			
4	(Be)	Beryllium	2	2			
5	(B)	Boron	2	3			
6	(C)	Carbon	2	4			
7	(N)	Nitrogen	2	5			
8	(O)	Oxygen	2	6			
9	(F)	Fluorine	2	7			
10	(Ne)	Neon	2	8			
11	(Na)	Sodium	2	8	1		
12	(Mg)	Magnesium	2	8	2		
13	(Al)	Aluminum	2	8	3		
14	(Si)	Silicon	2	8	4		
15	(P)	Phosphorus	2	8	5		
16	(S)	Sulfur	2	8	6		
17	(Cl)	Chlorine	2	8	7		
18	(A)	Argon	2	8	8		
19	(K)	Potassium	2	8	8	1	
20	(Ca)	Calcium	2	8	8	2	
21	(Sc)	Scandium	2	8	9	2	
22	(Ti)	Titanium	2	8	10	2	
23	(V)	Vanadium	2	8	11	2	
24	(Cr)	Chromium	2	8	13	1	
25	(Mn)	Manganese	2	8	13	2	
26	(Fe)	Iron	2	8	14	2	
27	(Co)	Cobalt	2	8	15	2	
28	(Ni)	Nickel	2	8	16	2	
29	(Cu)	Copper	2	8	18	1	
30	(Zn)	Zinc	2	8	18	2	
31	(Ga)	Gallium	2	8	18	3	
32	(Ge)	Germanium	2	8	18	4	
33	(As)	Arsenic	2	8	18	5	
34	(Se)	Selenium	2	8	18	6	
35	(Br)	Bromine	2	8	18	7	
36	(Kr)	Krypton	2	8	18	8	
37	(Rb)	Rubidium	2	8	18	8	1
38	(Sr)	Strontium	2	8	18	8	2
39	(Y)	Yttrium	2	8	18	9	2
40	(Zr)	Zirconium	2	8	18	10	2
41	(Nb)	Niobium	2	8	18	12	1
42	(Mo)	Molybdenum	2	8	18	13	1
43	(Tc)	Technetium	2	8	18	14	1
44	(Ru)	Ruthenium	2	8	18	15	1
45	(Rh)	Rhodium	2	8	18	16	1
46	(Pd)	Palladium	2	8	18	18	0
47	(Ag)	Silver	2	8	18	18	1
48	(Cd)	Cadmium	2	8	18	18	2
49	(In)	Indium	2	8	18	18	3
50	(Sn)	Tin	2	8	18	18	4
51	(Sb)	Antimony	2	8	18	18	5

At. No.	Element (Symbol) & Name		Electrons in shells (#) 1	2	3	4	5	6	7
52	(Te)	Tellurium	2	8	18	18	6		
53	(I)	Iodine	2	8	18	18	7		
54	(Xe)	Xenon	2	8	18	18	8		
55	(Cs)	Cesium	2	8	18	18	8	1	
56	(Ba)	Barium	2	8	18	18	8	2	
57	(La)	Lanthanum	2	8	18	18	9	2	
58	(Ce)	Cerium	2	8	18	20	8	2	
59	(Pr)	Pra'mium	2	8	18	21	8	2	
60	(Nd)	Neodymium	2	8	18	22	8	2	
61	(Pm)	Promethium	2	8	18	23	8	2	
62	(Sm)	Samarium	2	8	18	24	8	2	
63	(Eu)	Europium	2	8	18	25	8	2	
64	(Gd)	Gadolinium	2	8	18	25	9	2	
65	(Tb)	Terbium	2	8	18	27	8	2	
66	(Dy)	Dysprosium	2	8	18	28	8	2	
67	(Ho)	Holmium	2	8	18	29	8	2	
68	(Er)	Erbium	2	8	18	30	8	2	
69	(Tm)	Thulium	2	8	18	31	8	2	
70	(Yb)	Ytterbium	2	8	18	32	8	2	
71	(Lu)	Lutetium	2	8	18	32	9	2	
72	(Hf)	Hafnium	2	8	18	32	10	2	
73	(Ta)	Tantalum	2	8	18	32	11	2	
74	(W)	Tungsten	2	8	18	32	12	2	
75	(Re)	Rhenium	2	8	18	32	13	2	
76	(Os)	Osmium	2	8	18	32	14	2	
77	(Ir)	Iridium	2	8	18	32	17	0	
78	(Pt)	Platinum	2	8	18	32	17	1	
79	(Au)	Gold	2	8	18	32	18	1	
80	(Hg)	Mercury	2	8	18	32	18	2	
81	(Tl)	Thallium	2	8	18	32	18	3	
82	(Pb)	Lead	2	8	18	32	18	4	
83	(Bi)	Bismuth	2	8	18	32	18	5	
84	(Po)	Polonium	2	8	18	32	18	6	
85	(At)	Astatine	2	8	18	32	18	7	
86	(Rn)	Radon	2	8	18	32	18	8	
87	(Fr)	Francium	2	8	18	32	18	8	1
88	(Ra)	Radium	2	8	18	32	18	8	2
89	(Ac)	Actinium	2	8	18	32	18	9	2
90	(Th)	Thorium	2	8	18	32	18	10	2
91	(Pa)	Pr'tinium	2	8	18	32	20	9	2
92	(U)	Uranium	2	8	18	32	21	9	2
93	(Np)	Neptunium	2	8	18	32	22	9	2
94	(Pu)	Plutonium	2	8	18	32	23	9	2
95	(Am)	Americium	2	8	18	32	24	9	2
96	(Cm)	Curium	2	8	18	32	25	9	2
97	(Bk)	Berkelium	2	8	18	32	26	9	2
98	(Cf)	Californium	2	8	18	32	27	9	2
99	(Es)	Einsteinium	2	8	18	32	28	9	2
100	(Fm)	Fermium	2	8	18	32	29	9	2
101	(Md)	Mendelvium	2	8	18	32	30	9	2
102	(No)	Nobelium	2	8	18	32	31	9	2

FIG. 4 **PERIODIC TABLE OF THE ELEMENTS**

Period	Group IA	Group IIA	TRANSITION ELEMENTS										Group IIIA	Group IVA	Group VA	Group VIA	Group VIIA	Group O
			Group IIIB	Group IVB	Group VB	Group VIB	Group VIIB	Group VIII			Group IB	Group IIB						
1	1 **H** 1.00797		Metals										Non-Metals				1 **H** 1.0079	2 **He** 4.0026
2	3 **Li** 6.939	4 **Be** 9.012											5 **B** 10.811	6 **C** 12.011	7 **N** 14.007	8 **O** 15.9994	9 **F** 18.998	10 **Ne** 20.183
3	11 **Na** 22.990	12 **Mg** 24.312											13 **Al** 26.98	14 **Si** 28.086	15 **P** 30.97	16 **S** 32.064	17 **Cl** 35.453	18 **Ar** 39.95
4	19 **K** 39.102	20 **Ca** 40.08	21 **Sc** 44.96	22 **Ti** 47.90	23 **V** 50.94	24 **Cr** 52.00	25 **Mn** 54.94	26 **Fe** 55.85	27 **Co** 58.93	28 **Ni** 58.71	29 **Cu** 63.54	30 **Zn** 65.37	31 **Ga** 69.72	32 **Ge** 72.59	33 **As** 74.92	34 **Se** 78.96	35 **Br** 79.91	36 **Kr** 83.80
5	37 **Rb** 85.47	38 **Sr** 87.62	39 **Y** 88.91	40 **Zr** 91.22	41 **Nb** 92.91	42 **Mo** 95.94	43 **Tc** 99	44 **Ru** 101.07	45 **Rh** 102.91	46 **Pd** 106.4	47 **Ag** 107.87	48 **Cd** 112.40	49 **In** 114.82	50 **Sn** 118.69	51 **Sb** 121.75	52 **Te** 127.60	53 **I** 126.90	54 **Xe** 131.30
6	55 **Cs** 132.90	56 **Ba** 137.34	57–71 **La** series*	72 **Hf** 178.49	73 **Ta** 180.95	74 **W** 183.85	75 **Re** 186.2	76 **Os** 190.2	77 **Ir** 192.2	78 **Pt** 195.1	79 **Au** 196.97	80 **Hg** 200.59	81 **Tl** 204.37	82 **Pb** 207.19	83 **Bi** 208.98	84 **Po** 210	85 **At** 210	86 **Rn** 222
7	87 **Fr** 223	88 **Ra** 226	89– **Ac** series†															

*Lanthanide Series
†Actinide Series

58 **Ce** 140.12	59 **Pr** 140.91	60 **Nd** 144.24	61 **Pm** 147	62 **Sm** 150.35	63 **Eu** 151.96	64 **Gd** 157.25	65 **Tb** 158.92	66 **Dy** 162.50	67 **Ho** 164.93	68 **Er** 167.26	69 **Tm** 168.93	70 **Yb** 173.04	71 **Lu** 174.97
90 **Th** 232.04	91 **Pa** 231	92 **U** 238.03	93 **Np** 237	94 **Pu** 239	95 **Am** 241	96 **Cm** 242	97 **Bk** 249	98 **Cf** 252	99 **Es** 254	100 **Fm** 253	101 **Md**	102 **No**	103 **Lw**

electrons around the nucleus gives elements some of their inherent chemical properties. Figure 3 shows the chemical symbols and common names of all 102 elements, as well as their atomic numbers (electron numbers) and shell distribution.

The elements also can be classified according to ascending atomic numbers and number of electron shells into a arrangement called the *periodic table* (Fig. 4). The period number represents the number of electron shells in each element; thus, all elements with one electron shell (H & He) are listed horizontally opposite Period 1; those with two shells, opposite Period 2, and so on up to 7. For each element, the chemical symbol is shown, with atomic number written above and the atomic weight below.

Each vertical column of the table is labeled with a Roman numeral, and elements within a column possess similar properties. For example, copper (Cu), silver (Ag) and gold (Au) in column IB are all dense, soft, and fairly chemically-inert metals. The elements on the right of the table exhibit non-metallic behavior, while those on the left exhibit metallic behavior to various degrees.

Ions

Under certain conditions electrons may leave an atom, resulting in a net positive charge. The atom then becomes a *positive ion.* This tendency is especially prevalent in elements of groups IA & IIA having just a few electrons in their outermost shells, since the attraction of the positive nucleus for these remote electrons is limited. Metallic elements exhibit this tendency for positive ionization most markedly.

Certain other elements show a strong tendency to acquire excess electrons in their outer orbits and to become *negative ions.* These elements, which usually have five or more electrons in their outermost shells to begin with, are almost invariably nonmetallic, and appear at the right of the periodic table (groups VIA and VIIA).

Molecules

A real material may consist of billions of one or more kinds of atoms (atomic substances), or of billions of several kinds of ions (ionic substances). Still other substances are formed by the joining together of units, each unit consisting of a number of atoms. These units are called *molecules,* and can only be broken into their constituent atoms with difficulty. A molecule of methane, for example, is composed of one carbon atom and four hydrogen atoms (Fig. 5).

Substances composed of molecules are called molecular materials. Water, for instance, is a molecular material. Each water molecule contains two hydrogen atoms and one oxygen atom, and has a triangular shape. When large numbers of molecules are bonded closely together, the water exists as ice. As the tightness of bonding of the triangular molecule is decreased, water assumes the liquid form. When bonding is very loose, water exists as water vapor. In all three forms of the material the triangular units remain intact, and the individual hydrogen and oxygen atoms are not free to leave their respective molecules.

Molecules can vary greatly in size, as shown in Figure 6. A molecule of water contains only 3 atoms, whereas a molecule of ethane contains 8 and butane 14. Some molecules contain thousands of individual atoms; molecules of this size are termed polymers and form the basis of materials such as plastics. (see Section 209 Plastics).

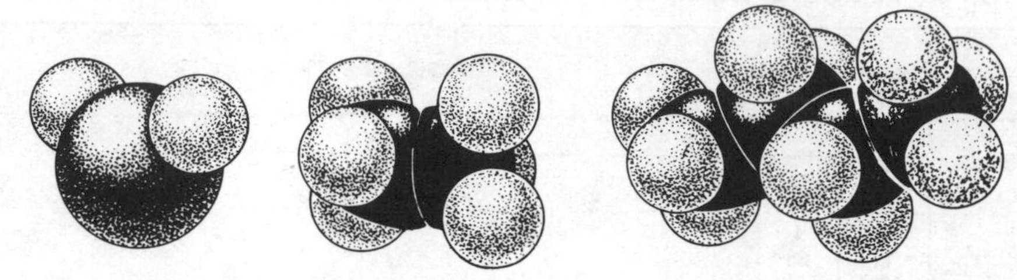

a **b** **c**

FIG. 6 Three molecules of varying sizes: (a) water, (b) ethane and (c) butane.

BONDING OF MATTER

One of the major reasons that atoms, ions and molecules bond together to form substances is the strong tendency in nature for outermost electron shells to contain as many pairs of electrons as possible. When a shell contains the maximum number of electron pairs it is said to be *filled.* The filled shell results in a low energy condition on the atomic scale, meaning the matter is relatively inert.

All systems in nature want to remain in the state of lowest energy; therefore, the filled outer shell condition produces materials which are stable and resist change. The elements which occur naturally with filled outer shells are the inert gases, such as helium (He), neon (Ne) and argon (Ar), found in group O at the far right of the periodic table. These elements refuse to combine with any other elements, including each other.

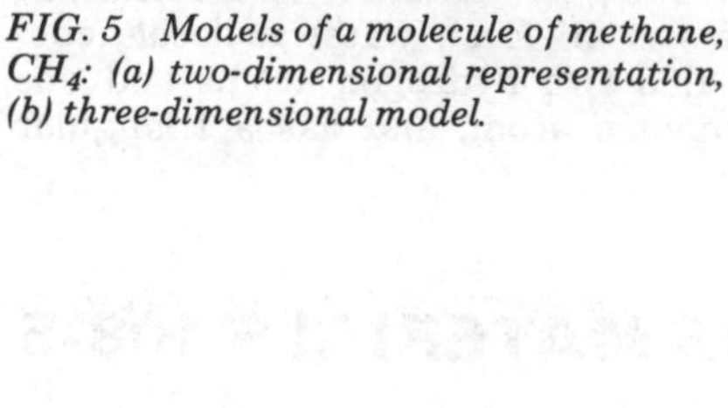

a **b**

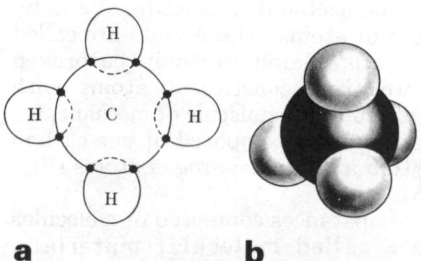

FIG. 5 Models of a molecule of methane, CH_4: (a) two-dimensional representation, (b) three-dimensional model.

Whenever the outer shell of an atom or ion of a non-inert gas element is filled, the atom or ion is said to have achieved the *inert gas configuration* (IGC), in which it is content to remain because of the stability of the situation. Nature moves atoms and ions so strongly toward the IGC that these particles can be thought of as employing many tricks, including capturing electrons, surrendering electrons, and sharing electrons with their neighbors in an effort to achieve chemical stability.

In exchanging or sharing electrons, atoms and ions interact with their neighbors, are attracted to them and bond into matter. There are four types of bonding: *ionic, metallic, covalent* and *secondary.*

Ionic Bonding

When atoms with almost empty outer electron shells exist in the vicinity of atoms with almost filled shells, the natural tendency of each set of atoms to achieve the IGC can be fulfilled. The atoms with few outer electrons surrender their "free" electrons and become positive ions, whereas atoms with many outer electrons accept these electrons to fill their outer orbit and become negative ions. The situation now exists in which positive ions and negative ions are in close proximity. The oppositely charged particles are attracted and bonded into matter by ionic bonding. Metals tend to have almost empty outer orbits, and nonmetals tend to have them almost filled; thus ionic substances usually consist of elements from the left and right hand groups of the periodic table, bonded together.

Probably the most familiar ionically-bonded substance is common table salt, NaCl. Atoms of sodium (Na) have just a single outer electron and therefore have a strong tendency to form positive Na^+ ions. Atoms of chlorine (Cl) have seven outer electrons and upon receiving one more electron, fill their outer orbits and achieve the IGC by becoming Cl^- ions. The two sets of ions are then bonded into NaCl (Fig. 7).

Ceramic materials, being combinations of metallic atoms and nonmetals, are bonded primarily through the ionic mechanism. Examples of this category of materials are brick, tile, portland cement, concrete and natural stone. Their properties follow from the quality of the ionic bond: they have high melting temperatures and are chemically inert because both ions have achieved the IGC and are tied by a strong ionic bond. Melting involves separating the bonds, while chemical attack requires separation and re-combination of ions. Both processes are difficult when the original substance is stable to begin with.

Ceramic materials tend to be brittle, because the regular and extremely rigid arrangement of positive and negative ions resists shape changes. They tend to shatter, rather than be forced from one shape into another. Brittleness generally is associated with good strength in compression but low strength in tension. These materials also are poor conductors of electricity and heat, because the electrons are tightly bound into the stable ions and are not free to conduct electrical or thermal energy in response to an applied voltage or temperature.

It is apparent that the typical

properties of ceramic materials follow logically from the nature of their constituent ions and of the ionic bond. Hundreds of new ceramic materials may be developed in the next few years. In general however, all of them will have high melting temperatures, high compressive strength, chemical inertness and good thermal and electrical resistance.

Metallic Bonding

Metallic atoms tend toward the IGC by attempting to surrender their few outer electrons, thus becoming positive ions. When large numbers of atoms are brought together, they each contribute these electrons to a mobile "sea" of electrons which circulates near the original parent atoms. The metal ions are now in a stable situation since they no longer possess their few outer electrons. They exist in a three dimensional shape, as a metal crystal, with an electron "sea" dispersed throughout the solid (Fig. 8).

This peculiar cooperative surrendering of electrons by metal ions results in the metallic bond. The characteristic properties of metals follow from the nature of this bond. They are strong and have fairly high melting temperatures because the metallic bond is strong. Unlike ceramics they are good conductors of heat and electricity because the electrons in the "sea" are extremely mobile and free to transport thermal and electrical energy.

Metals may or may not be chemically inert. Their ability to transport electrical current makes them susceptible to chemical degradation in the form of corrosion. The corrosion tendency sometimes outweighs the intrinsic strength of the metallic bond.

Covalent Bonding

Many elements found in groups IVA & VA of the periodic table (Fig. 4), such as carbon (C) and nitrogen (N), lack the strong tendency to form either positive ions or negative ions. These elements, with moderate numbers of electrons in their outer shells, reach the IGC by sharing electrons with similar elements. The process of mutual sharing of outer (valence) electrons by a cluster of atoms to create a stable entity, is known as *covalent bonding.*

Covalent bonding is the most common mechanism whereby small numbers of atoms are bound into molecules. These molecules are then joined together by weak secondary bonds, to produce molecular materials. The covalent bond is very strong, as is evidenced by the fact that it is very difficult to break up a molecule into its constituent atoms.

A few substances are produced entirely by covalent bonding—diamond is a familiar example. These substances are extremely hard and have very high melting temperatures, due to the strength of the covalent bonds. However, they seldom occur in nature, as suggested by the fact that diamonds are rarely found and then only in small amounts.

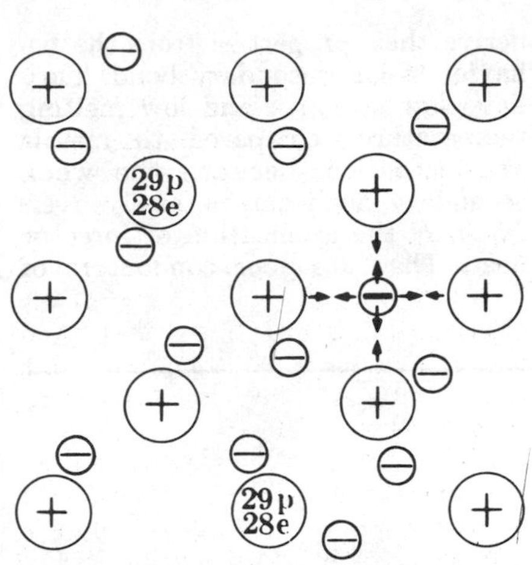

FIG. 8 Metallic bond results from the attraction of the positive cores of the atom (copper) to the negative electrons in the surrounding "sea".

Secondary Bonding

To become molecular materials, atoms are bound into molecules by covalent bonding, and the molecules then are joined by means of weak secondary bonds. These bonds arise because positive nuclei or negative electrons in one molecule feel the weak attraction of their opposites in neighboring molecules. The weak attractions bind adjacent molecules into the substance.

Molecular materials such as wood, plastics and bituminous products,

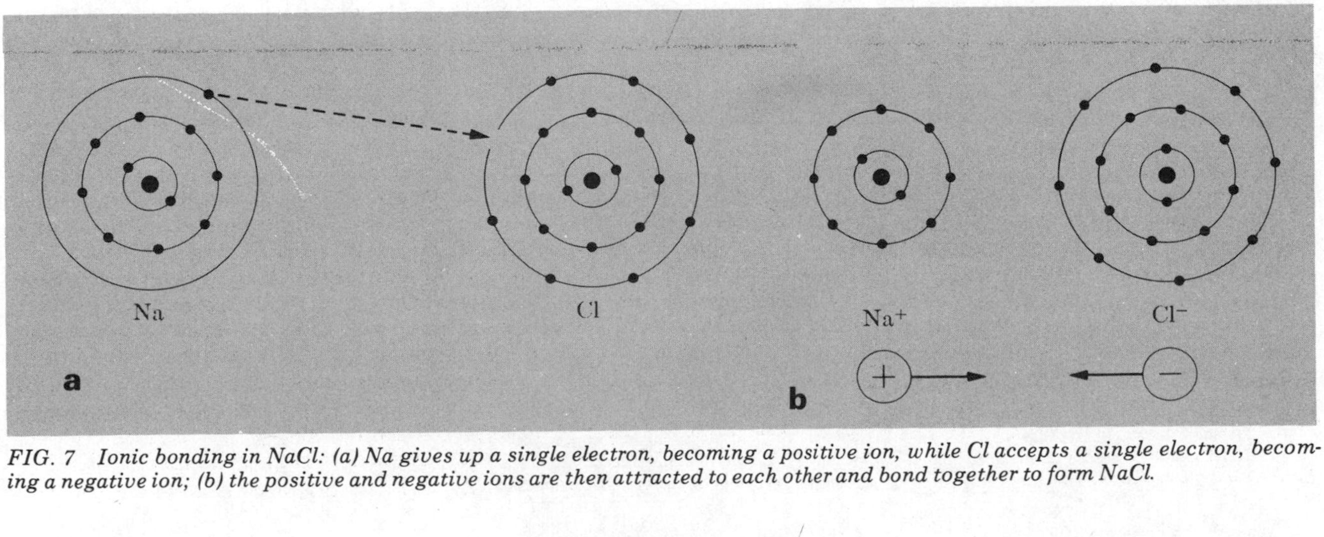

FIG. 7 Ionic bonding in NaCl: (a) Na gives up a single electron, becoming a positive ion, while Cl accepts a single electron, becoming a negative ion; (b) the positive and negative ions are then attracted to each other and bond together to form NaCl.

derive their properties from the behavior of the secondary bond. They have low strength and low melting temperatures compared to metals and ceramics, because the weak secondary bond can be easily overcome by the application of force or heat. They are poor conductors of heat and electricity, because individual electrons remain tightly constrained within covalently-bonded molecules.

Surprisingly, the weak secondary bond is not broken by many of the strong chemical compounds that attack metals and ceramics. Thus molecular materials are chemically inert in a large number of environments. They are attacked by molecular solvents such as acetone, but their resistance to attack by most salts, acids and industrial atmospheres makes them complimentary to metals and ceramics.

PROPERTIES OF MATTER

Anyone who has been alone in a building on a quiet evening knows that materials are "alive." The creaks and groans are in fact material responses to external stimuli such as heat, wind and gravity loads. These stimuli fall into four categories, which define the four major *properties* exhibited by materials: mechanical, thermal, electrical and chemical.

MECHANICAL PROPERTIES

Mechanical properties describe the response of a material to static (continuous) or dynamic (intermittent) loads.

Stress and Strain

When a force (load) of *x* lbs. is applied to an object, the object deflects (sags) or deforms (changes shape) *y* in. or fractions of an inch. However, knowing the load in pounds or the deflection in inches does not provide enough information to predict the likely response.

For instance, a load of 100 lbs. may induce a negligible reaction when distributed uniformly on the surface of a brick. The same load, *concentrated* at the point of a chisel, may shatter the brick. A deformation of 6″ may be negligible when averaged over the height of an entire building, but the same deformation may have catastrophic consequences if localized. It is necessary therefore to consider forces in relation to the area over which they act, and to view deformations with respect to the size of the region being deformed.

Stress (σ) is equal to the applied load (L) divided by the area (A) on which the load is acting:

$$\sigma = \frac{L}{A};$$

A 100 lb. load acting downward on the face of a typical 4″ × 8″ brick exerts a stress of only 3 lbs. per sq. in. (psi), well below the brick's inherent compressive strength. The same load, concentrated at the point of a chisel 0.5″ wide and sharpened to 0.02″ exerts a stress of 10,000 psi, well beyond the strength of the typical brick. In engineering texts, stress usually is designated by the Greek symbol σ when the applied load is at right angles to the area in question, as in tension (stretching) or compression (squeezing).

Stress takes into account the fact that an applied load may or may not be shared by a substantial area of a body. *Strain* (ε) correspondingly accounts for the fact that deformation may be distributed over a large area or localized in a small region. If the original dimension of a body along a certain line is designated d, and the deformation (change in that dimension) is designated Δ d, then:

$$\varepsilon = \frac{\Delta d}{d};$$

Strain is stated generally in % dimension change; thus, sometimes,

$$\varepsilon = \frac{\Delta d}{d} \times 100;$$

Elastic Deformation An applied load, no matter how small, always generates both a stress and a strain in a solid object. A feather floating downward and landing on an I-beam may cause the beam to deflect (sag), and the resulting strain can be measured with today's sophisticated instruments. When the feather is removed, the strain disappears. The object returns to its original dimensions.

In this case, deflection is present only so long as the load is present, and its effect can be reversed by simply removing the load. This reversible type of strain is called *elastic deformation.* Many forces acting on structures cause only elastic deformation. For instance, the swaying of a tall building under the action of wind is an example of elastic deformation. In general, elastic deformation is harmless since the structure is restored to its original configuration when the force is removed.

Most building criteria contain factors of safety which insure that a structure will not be stressed beyond its elastic range. The most common measure of a material's *stiffness,* or ability to resist elastic deformation, is its modulus of elasticity, E.

Plastic Deformation Permanent or irreversible deformation is called *plastic* deformation. This type of deformation is essential when aluminum ingot is formed into structural shapes. It would be most inconvenient if the shapes sprung back elastically into the shape of the original ingot. In the final structure however, load-bearing areas are made large enough so that applied loads will not generate large enough stresses to cause permanent, plastic, deformation.

The Stress-strain Test

Many of a material's mechanical properties can be accurately determined from its performance in a *stress-strain tensile* test. In this test a sample of material is stretched to fracture (breaking), and a continuous record is made of both the applied stress and resulting strain. The universal acceptance of certain parameters in the stress-strain test as illustrating mechanical behavior requires that such testing be conducted according to rigid standards. Description of test procedures for materials, ranging from structural steel to vinyl electrical insulation, are described by ASTM Standard Methods of Test. Tables of properties for various materials are presented in such reference works as the Metals Handbook and the Metals Progress Data book published by the American Society for Metals.

A typical ductile metal such as pure aluminum has a stress-strain

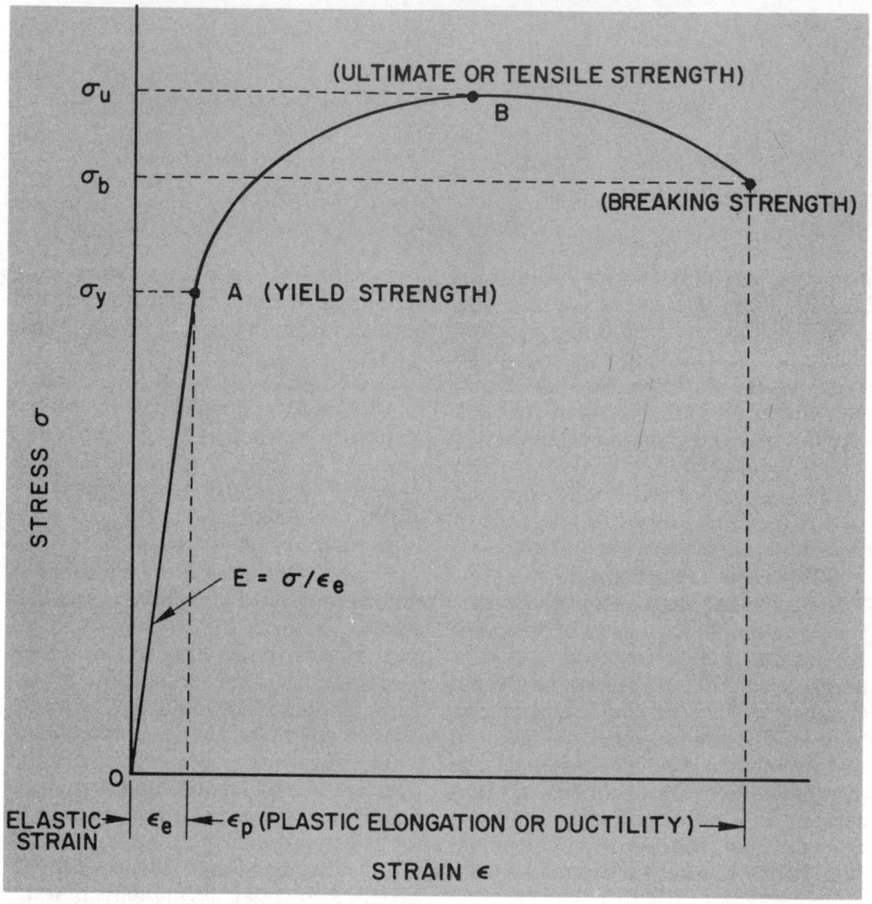

FIG. 9 **TYPICAL STRESS-STRAIN BEHAVIOR OF A DUCTILE METAL**

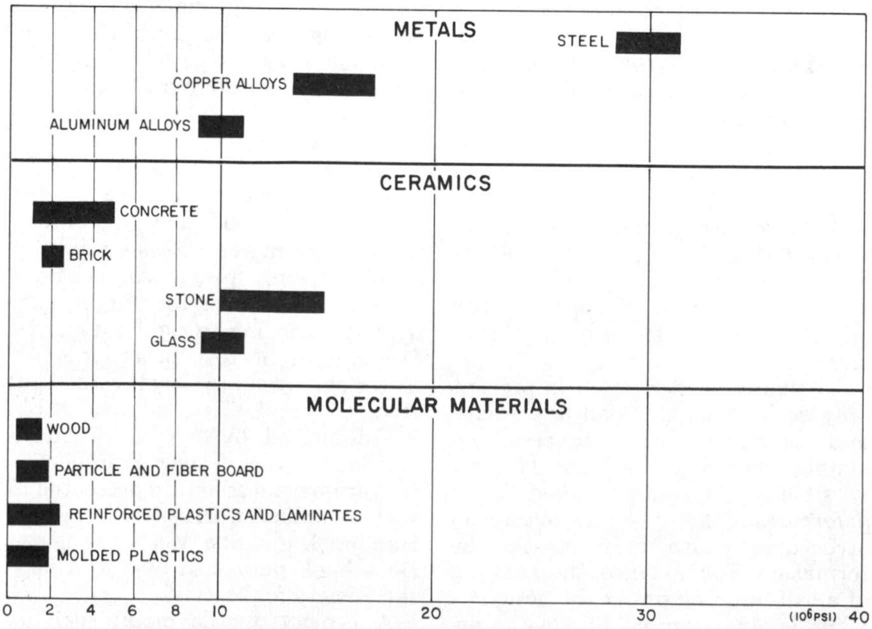

FIG. 10 **COMPARATIVE STIFFNESS OF CONSTRUCTION MATERIALS**

behavior as plotted in Figure 9. The behavior in region OA is elastic and linear. If any stress between O and A is applied and then removed, the strain in the body will return to 0%. The straight line shows that for each unit of increase in stress, the strain will increase by a constant amount—that is, strain is proportional to stress.

The proportionality constant between stress (σ) and strain (ε) is the modulus of elasticity (E) and is represented by the slope of line OA. It follows then that for stresses within the elastic range:

$$E = \frac{\sigma}{\varepsilon};$$

The modulus E is an important property because it enables designers to calculate the elastic strain accompanying a particular stress or loading condition. Figure 10 compares the E-values of construction materials belonging to the metals, ceramics and molecular categories. Metals generally are quite stiff; ceramics have slightly lower stiffness and plastics the lowest.

When the stress rises above A, into the non-linear portion of the curve, permanent deformation will appear in the body. If the stress then is reduced to O, the elastic strain (ε_e) in the body will disappear but some plastic (ε_p) will remain. The stress value on the curve marking the transition from elastic to plastic behavior is called the *yield point,* or *yield strength* (σ_y).

Most materials do not have stress-strain curves with clearly defined yield points. For these materials, a theoretical yield strength is calculated from the stress-strain curve. Because the yield strength marks the limit of usable strength for a ductile material, its value must be carefully determined. The precise definition of σ_y and detailed specifications for its determination are given in ASTM standards.

If stresses greater than σ_y are applied in the test, the material behaves as shown by the AB portion of the curve with the entire region AB representing plastic behavior. At the stress marked by point B, the material rapidly begins "necking down," as does a piece of chewing gum stretched to the breaking point. The quantity σ_u is the *ultimate*

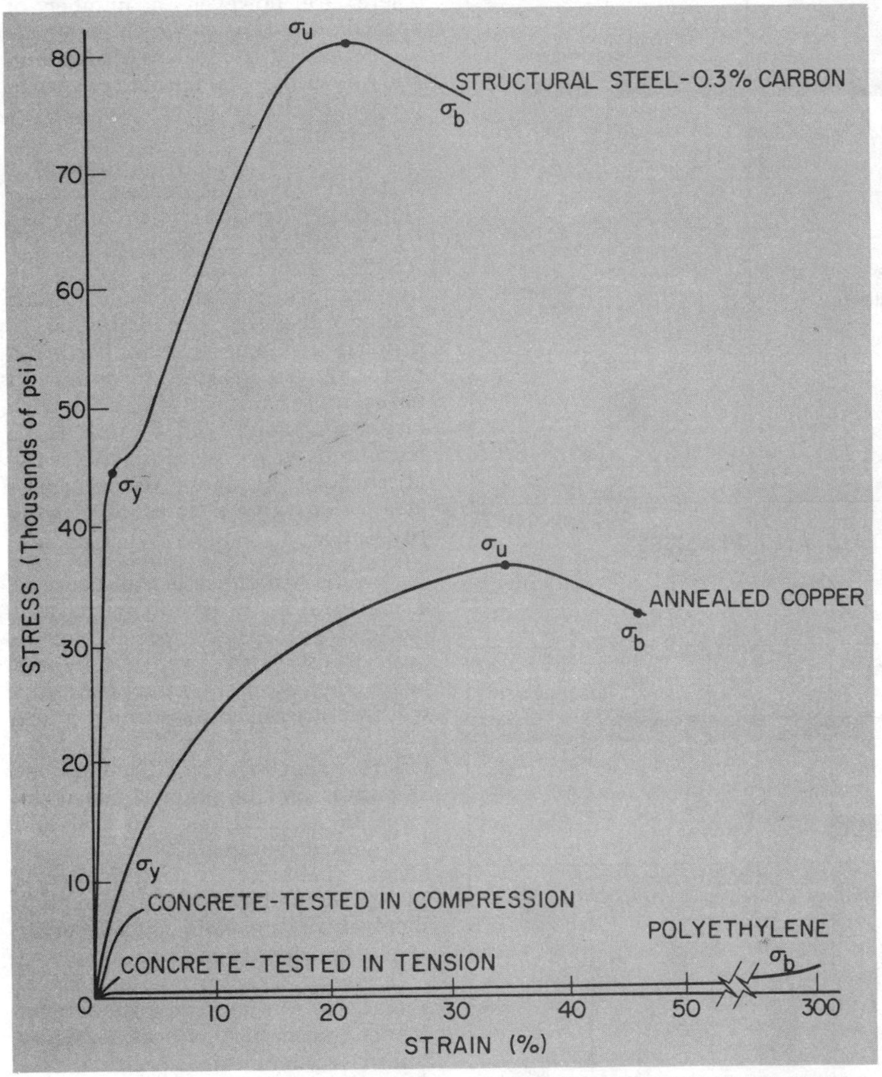

FIG. 11 STRESS-STRAIN BEHAVIOR OF CONSTRUCTION MATERIALS

(tensile) strength.

Beyond this point, deformation continues even with application of increasing smaller loads for a ductile material and thus leads to the eventual fracture of the material at stress σ_b—the fracture, or *breaking strength* of the material. The total plastic stretching of the material to failure is given by ε_p, the *elongation,* which is a measure of the material's ductility.

Figure 11 shows the stress-strain behavior of four typical construction materials: regular structural steel, annealed copper, portland cement concrete and polyethylene plastic.

It is evident that steel has a much higher E than copper, so that it deforms much less in the elastic range. The yield strength of steel is considerably higher than that of copper, indicating that it can be used at greater stresses without undergoing permanent deformation. On the other hand, steel stretches only about one half as much copper, and its greater strength is obtained at the price of higher brittleness (less ductility). The behavior of steel and copper is typical of the metallic category of materials, and is compared with that of plastics in Figure 12. Because of the strength of the metallic bond as compared with the secondary bond, metals resist permanent deformation more effectively than plastics. It is of interest to note the range of yield strengths that can be developed in metals by appropriate processing such as annealing, heat-treating or work-hardening. Because ceramics do not deform plastically before fracture, they generally are not characterized by a specific yield strength and were not included in Fig. 12.

Materials in the ceramic category are typified by the curves for concrete (Fig. 11). They have moderately high E values represented by slopes greater than for plastic but not as great as for steel. They respond rigidly to elastic stresses, showing small deformations for even fairly large stresses. At the same time they are very brittle, providing very little plastic deformation before fracture. The brittleness is accentuated when a ceramic material is stressed in tension, and this accounts for the common use of concrete, brick, tile and stone in bearing compressive loads, but not tensile loads. The absence of an elastic-plastic transition and a maximum in the stress-strain behavior means that yield strength, ultimate strength and breaking strength are the same for individual ceramic materials.

The polyethylene curve is typical of materials in the molecular category. The linear elastic region is limited and a small elastic stress produces a large deflection. This behavior may be a disadvantage when molecular materials are used as load-bearing elements. Molecular materials which show the greatest tendency to stretch elastically are the class of rubbers known as elastomers, used in rubber bands and sealants.

Under the most severe conditions metals and plastics can be stressed to ultimate strength, σ_u, before catastrophic fracture occurs, whereas ceramics can be stressed to breaking strength, σ_b, which for them is identical to yield strength, σ_y, and ultimate strength, σ_u. Figure 13 compares the extreme limits of strength for construction materials falling into each of the three categories.

Figures 11, 12 and 13 substantiate the generalizations about material properties derived from their structure. Metals are strong, elastically rigid and fairly ductile. Ceramics are

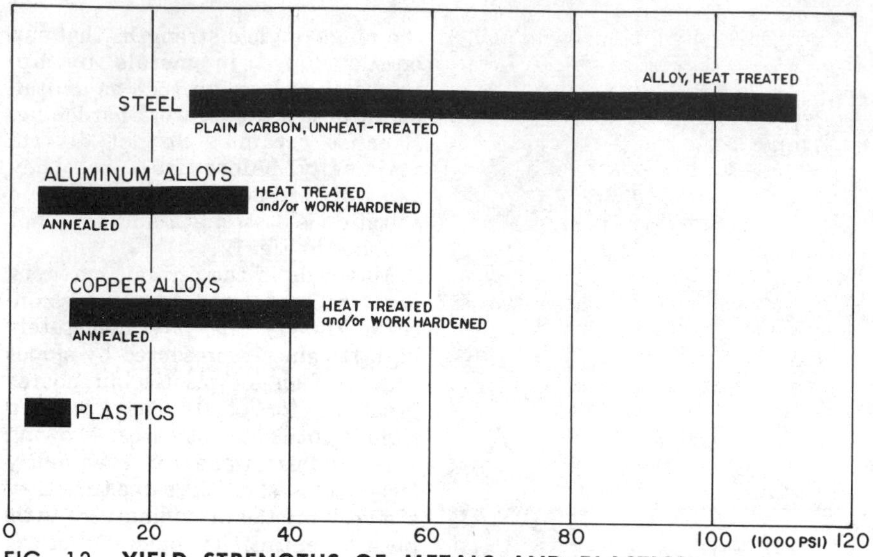

FIG. 12 **YIELD STRENGTHS OF METALS AND PLASTICS**

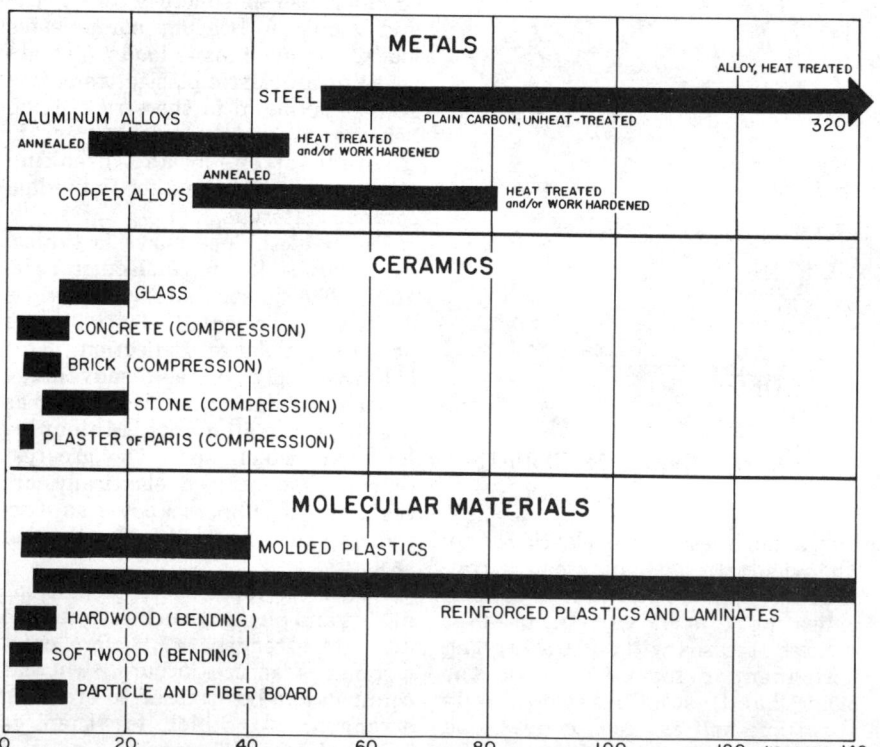

FIG. 13 **ULTIMATE STRENGTHS OF REPRESENTATIVE MATERIALS**

rigid, strong in compression and brittle, especially in tension. Molecular materials have adequate mechanical properties for many construction applications, but are limited by other properties.

Miscellaneous Mechanical Properties

Tabulated stress-strain data provide a quick means for gaging a material's mechanical performance. There are however, a number of special situations in which materials are subjected to mechanical loading, and knowledge of a number of special properties is necessary to predict performance.

Hardness is a measure of a material's ability to resist indentation or penetration. It is determined by various tests, described in the ASTM standards, in which indentors are forced into materials under carefully prescribed conditions. In general, the harder a material, the greater its wear and abrasion resistance. Materials with large values of E and σ_y tend to be hard and wear-resistant, so in the absence of hardness data, stress-strain data give a measure of wear resistance.

Fatigue resistance is a measure of a material's ability to withstand cyclic (repeated) stresses. When repeatedly stressed, even at stresses below yield strength, many materials will fracture without warning. Many airplane accidents are caused by failures due to fatigue. The lifetimes of pumps and other mechanical devices in a building's electrical and mechanical systems depend upon the fatigue behavior of the materials employed, and the use of adequate shock-absorbant mountings deserves careful consideration.

Damping capacity is a measure of a material's ability to dissipate or deaden mechanical vibration. Since sound is mechanical vibration, a material's ability to absorb sound is directly related to its damping capacity.

Impact strength or *toughness* marks a material's capacity to absorb impact without fracturing. It is defined as the total energy, from elastic deformation to fracture, which a material can absorb before breaking under impact. Two quantities, strength and ductility, affect this property: ceramics are strong, but do not deform significantly under static or dynamic load, so they lack toughness and shatter under impact. Plastics can stretch up to several hundred percent under load, but their strength is relatively low, hence they absorb little energy under impact. Metals, with good strength and ductility, are the toughest of the common construction materials.

THERMAL PROPERTIES

When subjected to temperature changes, a material may change its state (solidify, melt or vaporize), expand or contract, and conduct or reflect heat. These material responses to thermal energy are discussed below.

Melting Temperature

It has been noted earlier that certain mechanical properties, such as tensile strength, give an indirect measure of other properties, such as hardness. Melting temperature is a similar property. Naturally, materials with high melting points usually can be relied on to retain their mechanical properties over a greater temperature range. In addition, they tend to be stronger and more chemically inert than materials with lower melting points.

As a rule of thumb, materials with strong ionic bonding and high melting temperatures, such as ceramics, perform best at high temperatures. Metals perform moderately well, and molecular materials, with their weak secondary bonding, perform least well. Figure 14 compares graphically the temperature limits for continuous service of common construction materials and substantiates this rule of thumb. A more quantitative summary of the thermal properties of typical construction materials is given in Figure 15.

Thermal Conductivity

The main concern of the building designer is a material's response to heat while it is solid. Of special importance is the material's thermal conductivity, the ability to transfer heat from a region of high temperature to a region of lower temperature (Fig. 15). The free electrons in metals transport heat effectively, so metals typically exhibit the highest thermal conductivities; ceramics have much lower conductivities, and molecular materials the lowest.

Thermal conductivity is measured in BTU's per hour conducted, per sq. ft. of a material one inch thick, per one °F temperature differential (BTU /hr./sq.ft./in./°F). Conductivity is measured by placing a slab of the material one sq.ft. by one inch thick in a laboratory oven with its "hot" face one °F above its cold face. The heat energy in BTU's (British Thermal Units) conducted through the material in one hour is the material's thermal conductivity.

Thermal Expansion

One reason that buildings seem "alive" at times is that temperature

FIG. 14 SERVICE TEMPERATURES OF CONSTRUCTION MATERIALS

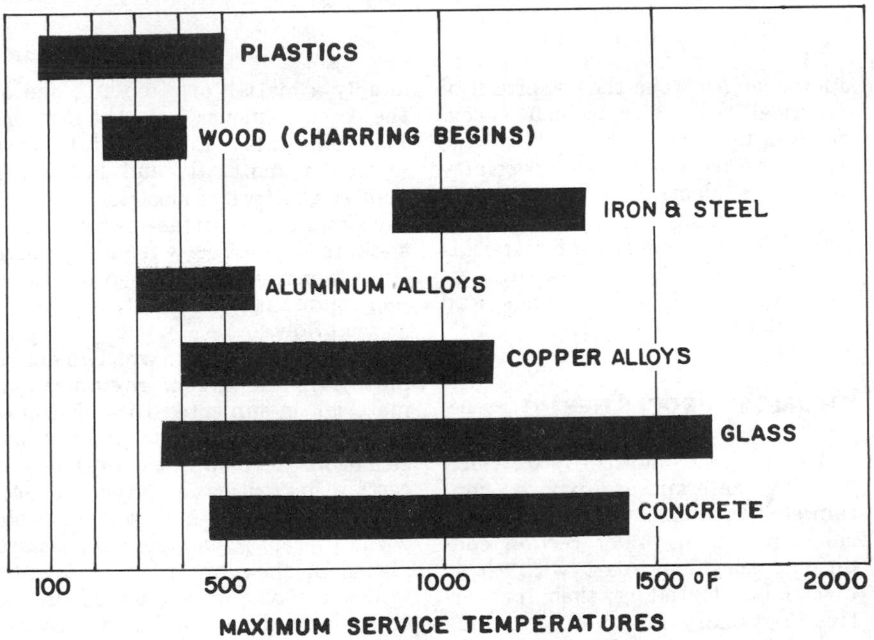

FIG. 15 THERMAL PROPERTIES OF TYPICAL CONSTRUCTION MATERIALS

Category	Material	Melting Temperature (°F)	Thermal Conductivity (BTU/ft²/°F/in/hr)	Thermal Expansion (10^{-6} in/in/°F)
Metals	Steels (Carbon or low alloy)	2700-2800	300-500	6.5
	Stainless steels	2700-2800	100-150	7
	Aluminum alloys	1100-1200	500-1500	9
	Copper alloys	1800-2000	500-2500	12.5
Ceramics	Brick	2000-5000	3-6	5
	Concrete	3000	5-10	7
	Plate glass	1500	5-6	1
Molecular Materials	Wood	400 (Chars)	1-2	10-30
	ABS Plastic	200-300	1-2	50
	Acrylics (PMMA)	200-300	1	30-50
	Vinyls (PV)	200-300	0.5-1	30-100
	Polyethylene (PE)	200	2-5	60
	Epoxies (EP)	200-500	1-2	30
	Silicones (SI)	400-600	1	1-100

changes are causing materials to expand and contract. The furnace ductwork which crackles as the furnace starts up is an example of thermal expansion in action. Thermal expansion is measured by the coefficient of thermal expansion, which is stated in terms of length change per unit of length per degree temperature change (in./in./°F).

Ceramics tend to have slightly lower coefficients of thermal expansion than metals, and plastics usually have much higher coefficients than metals (Fig. 15). It is apparent that this ranking is inverted compared to the melting temperature ranking; thus the lower the melting temperature, the higher the thermal expansion.

Structural steel has an expansion coefficient of 0.0000065 in/in/°F, which indicates that each one inch length of steel expands 6.5 millionths (10^{-6}) of an inch for every degree temperature rise. A steel bridge one mile long expands about 12″ as its temperature is raised from 70 to 100°F.

Thermal expansion is a critical property in applications where several materials are joined. For example, aluminum has twice the thermal expansion of steel. Thus, if aluminum trim is fixed to a steel base, the uneven expansion of the two materials will cause internal stresses and elastic deformation in the assembly. Noises, as well as unattractive buckling effects, may result. The need for properly designed expansion joints in such cases is evident.

ELECTRICAL PROPERTIES

The primary electrical property of interest to the building designer is electrical conductivity, a characteristic which is closely related to a material's thermal conductivity. The free electrons in metals allow heat and electrical energy to be conducted easily, hence conductivity is high. Ceramics have lower electrical conductivities and molecular materials have the lowest. This property accounts for the extensive use of molecular materials such as plastics in electrical insulation products.

Electrical conductivity usually is given in units of *mhos* per ft. of conductor length. It is the reciprocal of electrical resistivity, which is expressed in *ohms* per sq. ft. Thus, metals have *low* electrical *resistivity* and *high* electrical *conductivity*. Figure 15 summarizes the resistivities of typical construction materials and illustrates the excellent insulating capabilities of ceramics and molecular materials.

FIG. 16 **ELECTRICAL RESISTIVITY OF CONSTRUCTION MATERIALS**

Category	Material	Electrical Resistivity (ohm/ft)
Metals	Steel (Carbon or low alloy)	.000006
	Stainless steel	.00003
	Aluminum alloys	.000001
	Copper alloys	.0000005
Ceramics	Brick	4,000,000
	Concrete	4,000,000
	Plate Glass	10,000,000,000,000
Molecular Materials	ABS Plastic	1,000,000,000,000,000
		100,000,000,000,000
	Acrylics (PMMA)	
	Vinyls (PV)	100,000,000,000
	Polyethylene (PE)	1,000,000,000,000
	Epoxies (EP)	100,000,000,000,000
	Silicones (SI)	1,000,000,000,000

CHEMICAL PROPERTIES

The air and moisture to which building materials are exposed contain small amounts of active chemical compounds. Under certain conditions, they can react with these materials, degrading their properties. Typically however, ceramics resist chemical attack in practically all normal environments, and plastics resist attack from all except a few organic solvents.

The situation is more complicated in the case of metals. Metals degrade or corrode through the transport of minute amounts of electricity from certain regions, called *anodes,* to other regions, termed *cathodes.* The cathode accepts the electrons and remains intact, but the anode is degraded by the chemical reaction.

The essential elements in a corroding system are the anode, the cathode, and the current carrying medium (*electrolyte*) connecting them. The electrolyte is usually a solution of water and a gas, such as carbon dioxide or sulfur dioxide. Preventing corrosion, therefore, is simply a matter of removing one of the three elements from the system. For example, most paints are molecular materials and hence are poor conductors of electricity. Painting metallic surfaces not only shelters them directly from corrosive environments, but also provides a non-conductive barrier to electrical current flow.

Corrosion can occur when an entire material, or given region in a material, is subjected to a situation conducive to anodic behavior. Such situations develop in the presence of certain impurities in the metal (acting as cathodes) and when metals with different *galvanic potentials* are placed in close proximity. Galvanic tables ranking materials according to their tendency to become anodic can be used to predict potentially corrosive situation (Fig. 17).

Materials at the top of the list have a strong tendency to become anodic

FIG. 17 **GALVANIC SERIES OF COMMON METALS AND ALLOYS**

Electrolytic Tendency	Metal or Alloy
	Magnesium alloys
	Zinc
Anodic (read up)	Aluminum alloys
	Carbon steel
	Stainless steel (active)
	Lead
	Tin
	Brass
	Copper
Cathodic (read down)	Bronze
	Stainless steel (passive)
	Gold

and those toward the bottom cathodic. If a material from near the top of the list, such as carbon steel, is placed in contact with a material lower on the list, such as brass, corrosion is likely to result. The steel, acting as an anode, and the brass, acting as the cathode, will cause an electric current to flow. Corrosion of the steel is inevitable.

A number of points regarding galvanic reactions and corrosion problems should be remembered: (1) Intimate contact between materials widely separated in the galvanic table should be avoided. (2) Zinc can be used as a protective coating (galvanizing) on steel. Because it is higher than steel in the galvanic series, the zinc serves as a "sacrificial anode" and is intentionally permitted to corrode, thus saving the steel. (3) Tin is cathodic with respect to steel, so as a coating over steel (as in *terneplate* metal), it simply acts as a physical barrier to the corrosive environment. Should the tin coating be scratched or removed in spots, the exposed steel will corrode quickly. Care must be taken with terneplate, as with painted surfaces, to maintain a continuous film of the protective coating.

Surprisingly, aluminum and stainless steel are inherently "actively" anodic and susceptible to corrosion. The corrosion resistance which is normally associated with these metals develops when they are exposed to oxygen. The oxygen forms a thin, impervious, transparent oxide film which protects the underlying metal. The anodizing of aluminum alloys is simply an electro-chemical process which produces a thicker oxide coating than forms naturally.

It is essential therefore that, unless buried, these metals be exposed to the atmosphere if they are to exhibit their desired corrosion resistance.

We gratefully acknowledge the singular contribution of Professor Nicholas F. Fiore in the preparation of this section. We are further indebted to Professor Lawrence H. Van Vlack and the Addison-Wesley Publishing Co. for permission to reproduce copyrighted illustrations.

201 WOOD

WOOD

INTRODUCTION

Through continued lumber industry research, conservation and other technological advances, wood in its various forms remains a major resource in the construction of buildings.

Wood is a renewable resource. Wise forest management assures a permanent supply of timber to meet future foreseeable needs. Foresters can grow better trees faster than can nature unassisted by man, and now, as a result of reforestation and other conservation practices, the yearly growth of timber nationwide exceeds that harvested or lost to fire or disease.

Aside from its universal availability, wood possesses many properties that make it suitable to hundreds of end uses in construction. Wood has high strength relative to its weight in compression, tension, bending and resistance to impact. It can be worked easily to desired shapes with simple tools, has high durability when simple construction precautions are observed and has excellent insulating qualities. Wood is highly prized for its decorative character, which is derived from an infinite variety of grain, figure, texture, hue and other natural markings. These properties are important in lumber, plywood and laminated timber, as well as in other wood products—fiberboards, flooring, windows, doors and millwork (see Section 202—Wood Products).

Lumber is the product of the saw and planing mill. As cut from the log, it possesses a wide range of variable quality and thus is divided into commercial lumber grades. Each grade has a relatively narrow range in quality to provide a basis for selecting lumber to suit a particular use. Building construction is by far the biggest user of lumber. More than 80% of all new homes are built with a structural framework of lumber, and the remainder use wood in one or another of its various forms. An average house uses approximately 10,000 board feet of lumber in its various forms.

Plywood is a form of wood used both structurally and decoratively in construction. Structural benefits result from thin, crossbanded veneer sheets that are glued together into panels. This construction tends to equalize wood's strength properties, improve dimensional stability and increase resistance to both checking and splitting. Decoratively, face veneers may be arranged to form panels with an infinite variety of visual effects.

Laminated timber makes possible sizes and shapes in wood difficult or impossible to obtain in solid timber. Arches, beams, trusses, and other laminated members are custom-fabricated from stress-graded lumber, bonded together with structural glues stronger than the wood itself.

Lumber, plywood and laminated timber of *suitable* quality means that which will perform satisfactory service in its intended use. In this section, the general properties and characteristics of wood are described, and the manufacture, grading and uses of lumber, plywood and laminated timber are described and illustrated. This section provides a basis for understanding the general characteristics of wood and the proper selection and appropriate uses of lumber, plywood and laminated timber so that suitable quality in service may be achieved.

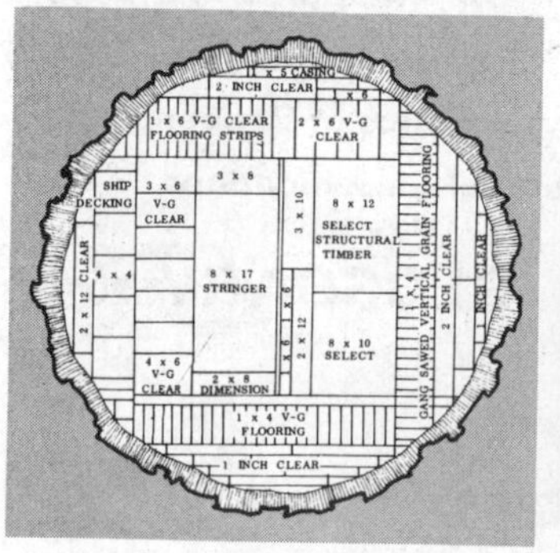

PROPERTIES OF WOOD

The wide variation in the characteristics of wood makes it suitable for use in a number of vastly different products—structural, non-structural and decorative. The general properties of wood that set it apart from other classes of materials, such as concrete or the metals and plastics, should be understood so that wood and wood products are used properly to perform satisfactorily in service.

IMPORTANT TREE SPECIES

The most important commercial species of wood and their areas of growth are shown in Figure W1. The division of species into two classes, *softwood* and *hardwood*, is a botanical difference and is not descriptive of the softness or hardness of the wood. Because of fundamental differences in microstructure and properties, softwoods and hardwoods differ in use, size standards and method of grading.

Softwoods

About 75% of the total lumber production is from softwood trees, providing structural and framing lumber, sheathing, roofing, subflooring, exterior siding, flooring, trim and interior paneling. Softwoods are called *conifers* or *coniferous* because most species bear cones. With few exceptions, softwoods have scalelike or needlelike leaves and are evergreen (Fig. W2).

Hardwoods

The remaining lumber production is from hardwood trees, providing flooring, interior paneling, and cabinet and furniture woods. Hardwoods are broadleaved trees and are called *deciduous* because they shed their leaves at the end of each growing season, except in the warmer climates (Fig. W2).

WOOD GROWTH

A tree grows by the production of new cells in the *cambium layer*, a zone of active cell growth directly under the bark (Fig. W3). Each

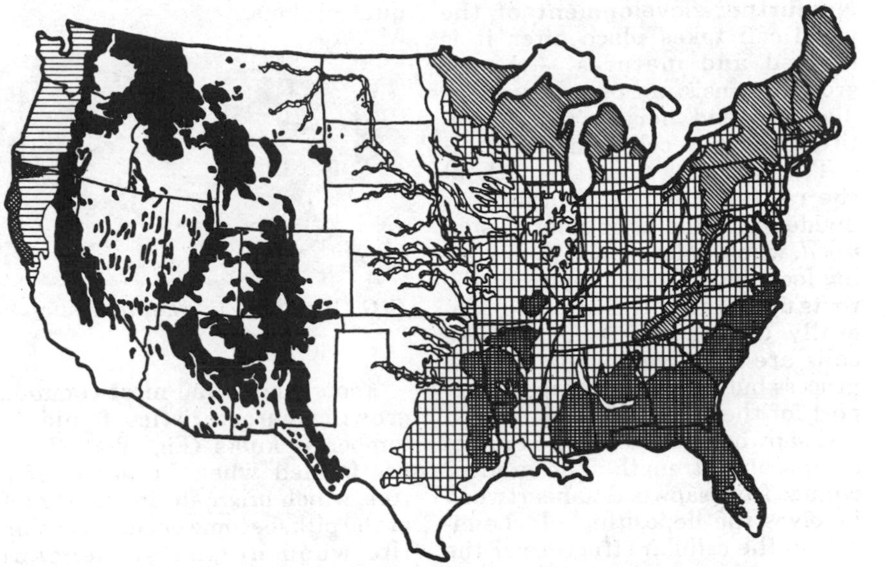

FIG. W1 AREAS OF TIMBER GROWTH IN THE U. S.

SOFTWOODS		HARDWOODS
WEST COAST REGION: Douglas fir, west coast hemlock, western red cedar, Sitka spruce	NORTHERN FORESTS: northern white pine, eastern spruce, jack pine, aspen, northern hemlock	NORTHERN FORESTS: maple, birch, beech, ash, black cherry
REDWOOD REGION: redwood, Douglas fir	SOUTHERN PINE REGION: loblolly pine, slash pine, shortleaf pine, longleaf pine, cypress	CENTRAL AND SOUTHERN HARDWOOD FORESTS: oaks, yellow poplar, gum, hickory, black walnut, basswood
WESTERN PINE REGION: ponderosa pine, Idaho white pine, Douglas fir, white fir, sugar pine, inland red cedar, western larch, Engelmann spruce, lodgepole pine, incense cedar		

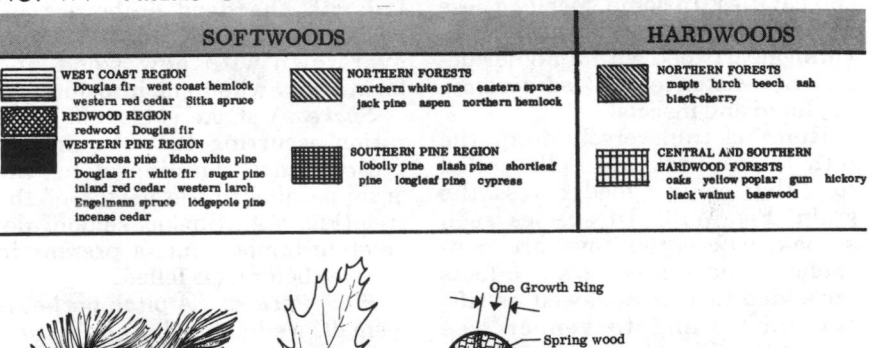

FIG. W2 *Trees can be divided broadly into two groups: (a) coniferous (softwoods) and (b) deciduous (hardwoods).*

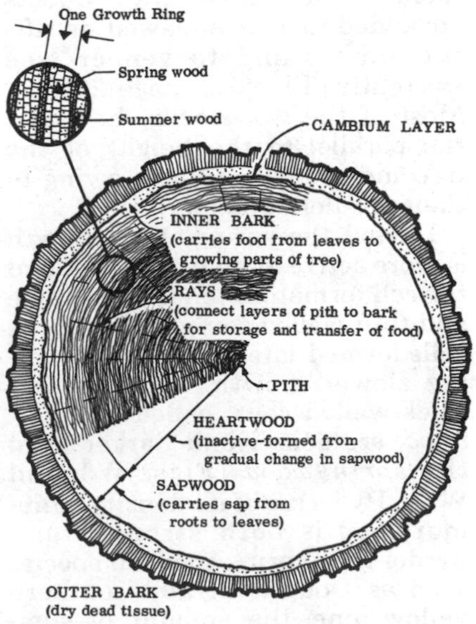

FIG. W3 *Cross-section of a living tree, showing active and inactive cell layers.*

year a completely new layer of wood cells envelops the preceding year's growth over the entire tree. Each year's growth can be seen visually as an annual growth ring. No further development of the wood cell takes place after it is formed and matures. All tree growth, consisting of increases in diameter and height, results entirely from new cell formation.

The wood between the pith at the center and the cambium is divided into the light-colored *sapwood*, whose cells are active in storing food and carrying sap from the roots up to the leaves and the generally darker *heartwood*, whose cells are inactive in the growth process but provide structural support for the tree. Generally speaking, sapwood and heartwood are of comparable strength. The gradual change from sapwood to heartwood involves the depositing of chemicals in the cellular structure of the heartwood, sometimes filling the cell cavities. In some species these chemicals are of a toxic nature, providing heartwood with a higher degree of resistance to wood-destroying fungi and insects.

Running transversely from the pith to the bark are cells called *rays* that convey food across the grain (Fig. W13). In species such as oak, where the rays are conspicuous, decorative grain effects are added to quarter-sawed or rift-cut lumber and to veneer (see Decorative Plywood, page 201-36). Most of the wood cells, however, run parallel to the height of the tree and are called fibers owing to their needlelike shape.

During the spring, tree growth is more active in many species, and the cell formation tends to be large and thin-walled compared to those cells formed later in the year during slower growth. The smaller, thick-walled cells, called *summerwood,* are denser and darker than the *springwood* (Figs. W3 and W13). By virtue of its density, summerwood is both stronger and harder than springwood. In species such as Douglas fir and southern yellow pine, the amount of summerwood can be used visually as an estimate of both its strength and its density.

Growth Characteristics

Certain growth characteristics affect the grade and use of wood and include knots, shakes and pitch pockets.

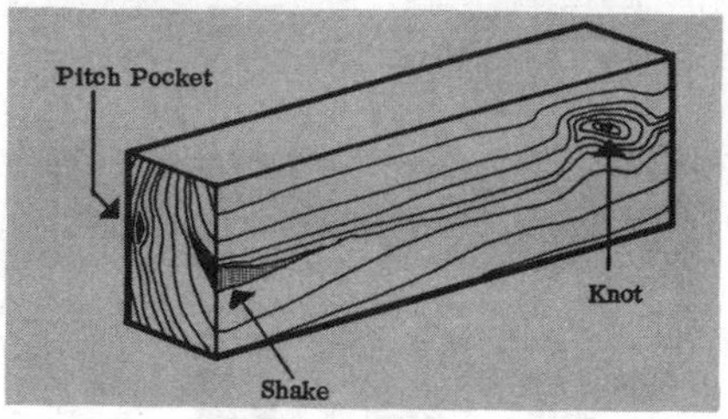

FIG. W4 Growth characteristics of wood.

Knots One of the most common growth characteristics found in lumber is knots (Fig. W4). They are formed when branches of a tree, which originate in the center at the pith, become enclosed within the wood during subsequent growth. If a branch dies and later falls off, the dead stubs are incorporated in the wood, become overgrown with new wood and form knots when cut into lumber.

Shakes A shake is a grain separation occurring between annual growth rings, running along the grain parallel to the height of the tree (Fig. W4). Shakes seldom develop in lumber unless present in the tree before it is felled.

Pitch Pockets A pitch pocket is a small, well-defined grain separation that may contain solid or liquid resin (Fig. W4). Pitch pockets may be found in Douglas fir, western larch, pine and spruce.

PHYSICAL PROPERTIES

The properties of wood are determined by its physical and chemical composition, and most characteristics of wood are related to its cellular structure.

Chemical Composition

The wood cells or fibers are primarily *cellulose*, cemented together with a material called *lignin.* The wood structure is approximately 70% cellulose, from 12% to 28% lignin and up to 1% ash-forming materials. These constituents give wood its hygroscopic properties, its susceptibility to decay and its strength. The bond between individual fibers is so strong that when tested in tension they commonly tear apart rather than separate. The remainder of wood, although not part of its structure, consists of extractives that give different species distinctive characteristics such as color, odor and natural resistance to decay.

Under chemical processes, the lignin in wood chips can be dissolved, freeing the cellulose fibers into pulp for use after further processing in paper and paperboard products. Cellulose also may be converted for use in textiles (such as rayon), plastics and into other products that depend on cellulose derivatives.

Hygroscopic Properties

Wood is *hygroscopic:* It expands when it absorbs moisture and shrinks when it dries or loses moisture. This property affects practically every single end-use of wood. Although the *wet* (green) condition is normal for wood throughout its life as a tree, most products made of wood require that it be used in a *dry* condition; hence, seasoning by drying to an acceptable *moisture content* is necessary.

Moisture Content *The moisture content of wood is the weight of the water it contains, expressed as a percentage of the weight of the wood when oven-dry.* The oven drying method, used to determine moisture content (m.c.) in the laboratory, is shown in Figure W5. For field use, the moisture meter is more practical: it gives an instantaneous reading by measuring the resistance to current flow between two pins driven into the wood.

In the living tree the amount of moisture varies widely: among species, among individual trees of the same species, among different parts of the tree and between sapwood and heartwood. Many softwoods have a large proportion of moisture in the sapwood with far less in the heartwood, while most hardwoods have about equal moisture content in both sapwood and heartwood. The extreme limits of moisture content in green soft-

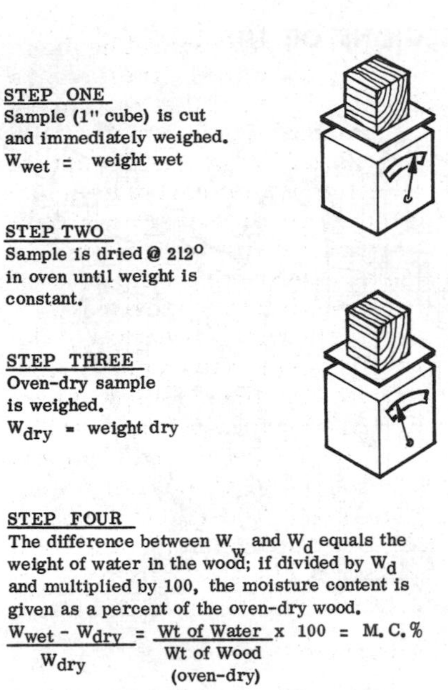
STEP ONE
Sample (1" cube) is cut and immediately weighed.
W_{wet} = weight wet
STEP TWO
Sample is dried @ 212°
in oven until weight is constant.
STEP THREE
Oven-dry sample is weighed.
W_{dry} = weight dry
STEP FOUR
The difference between W_w and W_d equals the weight of water in the wood; if divided by W_d and multiplied by 100, the moisture content is given as a percent of the oven-dry wood.
$\frac{W_{wet} - W_{dry}}{W_{dry}} = \frac{\text{Wt of Water}}{\text{Wt of Wood (oven-dry)}} \times 100 = \text{M.C.\%}$

FIG. W5 Oven-drying method of measuring moisture content of wood.

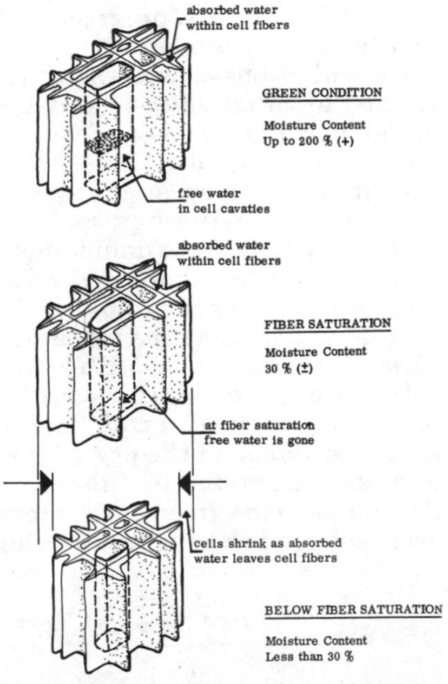
absorbed water within cell fibers
GREEN CONDITION
Moisture Content Up to 200 % (+)
free water in cell cavaties
absorbed water within cell fibers
FIBER SATURATION
Moisture Content 30 % (±)
at fiber saturation free water is gone
cells shrink as absorbed water leaves cell fibers
BELOW FIBER SATURATION
Moisture Content Less than 30 %

FIG. W6 Moisture content and wood cell shrinkage.

woods can be illustrated by comparing the moisture content in the heartwood of Douglas fir and southern pine, which may be as low as 30% to as much as 200% m.c. (twice the weight of the wood oven-dry) in the sapwood of various cedars and redwoods.

Fiber Saturation Point Moisture in green wood is present in two forms: *in the cell cavities* as free water and *within the cell fibers* as absorbed water (Fig. W6). As the wood dries, the cell fibers do not give off their absorbed water until all the free water is gone and the adjacent cell cavities are empty. The point at which the fibers are still fully saturated, but the cell cavities are empty, is called the *fiber saturation point,* occurring in most species at about 30% m.c. The significance of this condition is that it represents the point at which shrinkage *begins.* Even though green lumber may be cut with moisture contents as high as 200%, with subsequent drying to the fiber saturation point (30% m.c.) no shrinkage of the wood occurs until the cell cavities lose their free water. Only when the cell fibers begin to give off their absorbed water and start to constrict does the wood shrink.

Thus, the total shrinkage that the wood can experience takes place between its fiber saturation point and a theoretical moisture content of 0% (oven-dry condition). Within this range, shrinkage is practically proportional to moisture loss. For every 1% loss or gain in moisture content below 30%, the wood shrinks or swells, respectively, about 1/30 of the total. For example, at 15% m.c. the wood has experienced half of the total shrinkage possible. Wood in service practically never reaches a moisture content of 0% due to the influence of water vapor in the surrounding atmosphere; hence the total shrinkage possible is far less significant than the actual shrinkage probable.

Equilibrium Moisture Content After a tree is felled and cut into lumber, its moisture content begins to drop as moisture in the wood is lost to the surrounding air. Wood will give off or take on moisture from the surrounding air until the moisture within the wood has come to a point of equilibrium with the moisture in the air. The moisture content of the wood at this point is called the *equilibrium moisture content (e.m.c.).* The relation of atmospheric temperature and humidity to the equilibrium moisture content of wood is given in Figure W7 and shows, for example, that wood kept constantly in air at 70°F and 60% relative humidity eventually will balance at a moisture content of about 11%. Since temperature and humidity are not constant in service, wood is always subject to variations in moisture content as it seeks an equilibrium with the surrounding air. Variations tend to be gradual and seasonal, and the equilibrium moisture content of wood generally is less on the interior of heated buildings than in unheated buildings or exterior exposure.

The practical significance of the equilibrium moisture content is that it allows a prediction of the moisture content wood will attain in service. *To ensure that wood will experience only minor dimensional changes, ideally it should be fabricated and installed at a moisture content as close as possible to the equilibrium moisture content it will attain in service* (Fig. W8).

The above recommendation is most critical in the installation of strip flooring, wood doors and items of millwork, where a proper fit between adjoining wood elements is important. The moisture content of *interior* wood items in most of the country varies between 6% and 12% and these products therefore are fabricated in this range. However, since the wood may lose or absorb moisture during shipment, it may end up at the site with a moisture content substantially higher or lower than the local equilibrium moisture content. It is advisable to moisture-condition

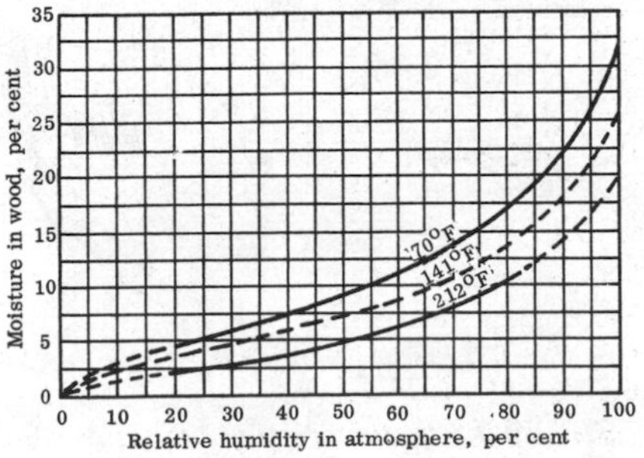

FIG. W7 Equilibrium moisture content for various temperatures and relative humidity.

such products by storing them within the space for several days before installation.

Exterior trim, siding and board sheathing are subject to greater variations in atmospheric humidity and temperature, hence also in equilibrium moisture content. The equilibrium moisture content outside is usually higher than indoors and may differ from season to season within the same area, as well as from region to region (Fig. W8). However, the conditions of installation are not so demanding as for interior elements and the normal fabricated moisture content of 12% to 15% is adequate for all areas.

Framing lumber is surfaced at over 19% m.c. (grademarked S-GRN; "S" for surfaced, "GRN" for green), at maximum 19% m.c. (grademarked S-DRY) and sometimes at maximum 15% m.c. (grademarked MC15).

To make sure that no individual piece exceeds the maximum moisture content required by the gradestamp, lumber is usually seasoned to an *average* moisture content several percentage points lower than the maximum. Thus, S-DRY lumber has an average moisture content of 15%; MC15 lumber is dried to an average 12%. Since the equilibrium moisture content of wood in most of the country is under 15%, lumber which has been milled in the green state may, after air drying in the lumber yard or on the site, actually have a much lower moisture content. *In the arid region, framing lumber should have a maximum moisture content of 15% at the time interior finishes are installed; in the damp region and remainder of the country, the lumber should have a moisture content of 19% or less* (Fig. W8).

In the arid region, either S-GRN or S-DRY lumber may achieve the desired maximum 15% m.c., if it is permitted to air-dry at the yard or site, because the lumber loses moisture quickly under these atmospheric conditions. Specifying MC15 merely provides the visual assurance of a grademark that this moisture content has been achieved at the mill. In other parts of the country, air drying is not so rapid and it is best to specify dry lumber (19% m.c. maximum) for all 2″ *thick* framing lumber.

An overwhelming quantity of lumber used in residential construction is 2″ or less in thickness. *Boards and dimension lumber 2″ or less thick should be specified as S-DRY, or MC15 and should be identified by the appropriate grademark.*

In most species boards and 2″ dimension lumber can be graded and sold either dry or green. All Southern pine lumber 2″ or less thick which has been graded according to SPIB rules is surfaced at 19% maximum moisture content and usually carries the S-DRY grademark. When specified, 2″ southern pine lumber may be surfaced at 15% maximum moisture content (grademarked KD) and heavy dimension lumber, 3″ and 4″ thick, may be obtained surfaced at 19% m.c. (S-DRY).

In most parts of the country lumber *3″ and 4″ thick* is not readily available in the dry grades, because most mills fabricate these sizes in the green state. However such lumber, although bearing

FIG. W8 **ATMOSPHERIC HUMIDITY REGIONS OF THE U. S.**

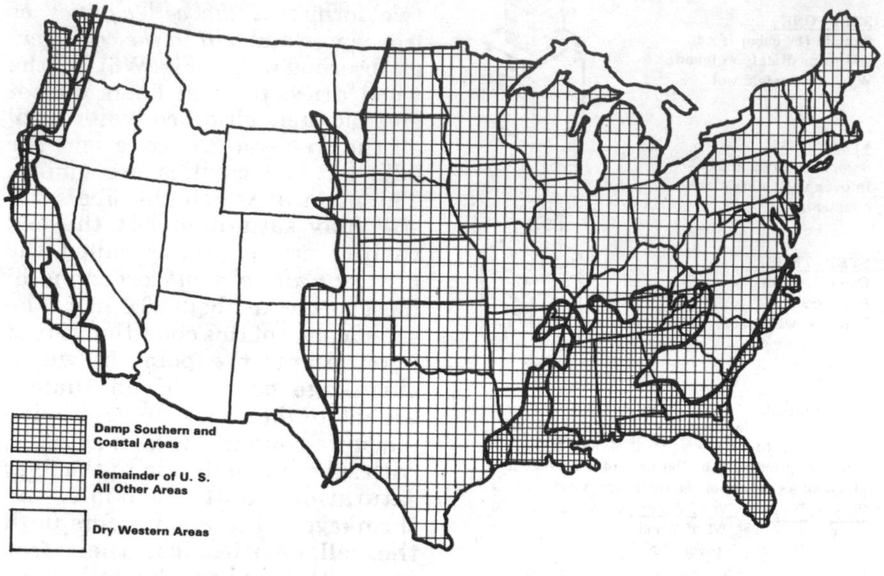

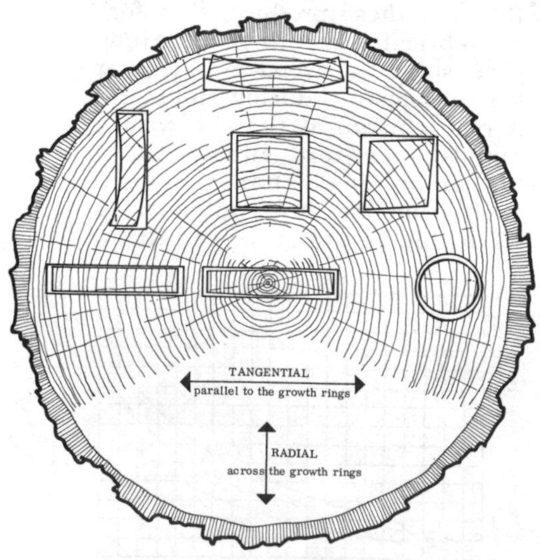

FIG. W9 Characteristic shrinkage and distortion of members as affected by the direction of the annual rings. Tangential shrinkage is about twice as great as radial.

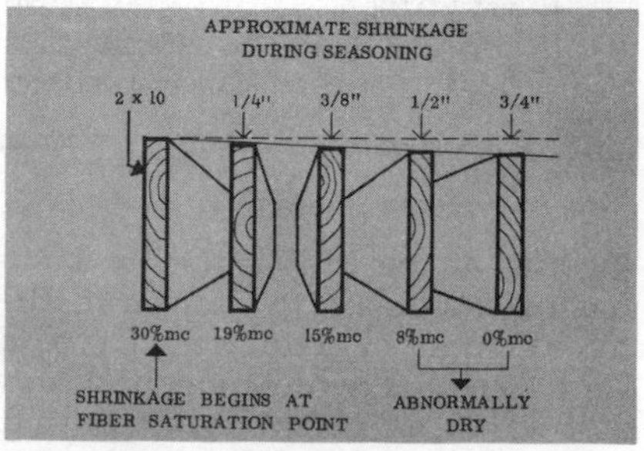

FIG. W10 Approximate shrinkage during seasoning of a 2 x 10 from green condition to theoretical 0% m.c.

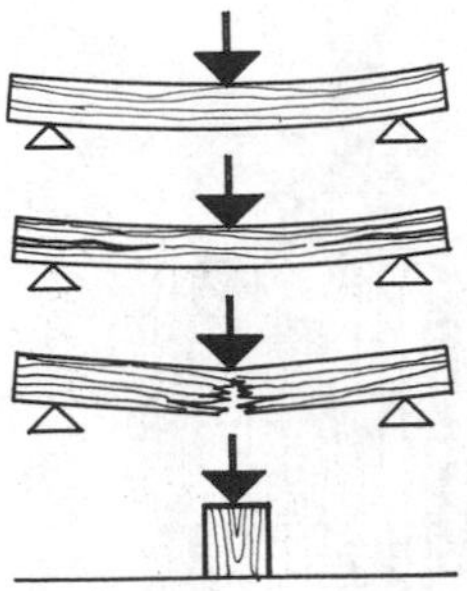

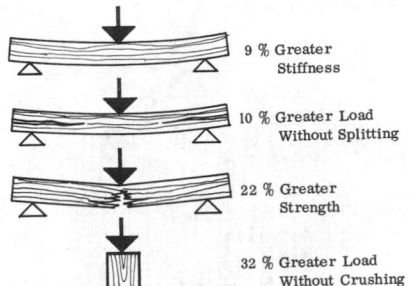

FIG. W11 Approximate increase in wood structural properties with a decrease in moisture content from 30% to 19%.

an S-GRN grademark, frequently has an acceptable moisture content due to air drying at the yard. *Larger pieces, 3″ and 4″ thick, should be specified to have 19% maximum moisture content and, in the absence of a grademark, should be checked to insure that the moisture content is as specified.*

Effects of Moisture Content on Properties The effects of moisture content on size and dimensional stability, strength and stiffness, decay resistance, paintability and other properties such as glue bonding are summarized below:

Size variations in wood occur only with changes in moisture content below the fiber saturation point. Shrinkage or swelling *parallel* to the grain (longitudinally with the height of the tree) is practically negligible, having little significance in construction applications. *Across* the grain, however, wood will shrink appreciably in both width and thickness. Shrinkage is greatest in the direction parallel to the annual growth rings (tangential), and is about 1/2 to 2/3 as much across these rings (radial). The combined effects of tangential and radial shrinkage in drying from the green condition are shown for various shapes in Figure W9.

In general, hardwoods shrink more than softwoods and heavier species shrink more than lighter ones. Members of a large cross section do not shrink as much proportionately because drying is not simultaneous in the inner and outer parts of the piece. Since the inner part dries more slowly than the outer portion, the wood near the surface is prevented from shrinking normally. The outer layers become set and tend to keep the inner portions from shrinking normally as they dry out. In softwood structural lumber, 6″ x 6″ or larger, a shrinkage of approximately 1/32″ per inch of face width may be expected in drying from a green condition to ordinary conditions in use.

Softwood structural lumber 2″ and 3″ thick is usually partially or fully seasoned when marketed. The average amount of shrinkage in service for material of these sizes usually does not exceed 1/32″ per inch of face width (Fig. W10).

Strength of wood increases as moisture content decreases. Increased strength of dry wood over green wood of the same dimensions is due to two causes: (1) strengthening and stiffening of the cell fibers as they dry out, and (2) increase in the compactness of the wood in a given volume. Wood dried from green to 5% m.c. may add from 2.5% to 20% to its density (weight of wood per unit of volume) and, in small pieces, end-crushing strength and bending strength may easily be doubled or, in some woods, tripled. The increase in strength in small, clear specimens is much greater than in large timbers containing defects because the increase in strength in large members is offset to a large extent by the influence of defects that develop during seasoning.

All strength properties are not equally affected by decreases in the moisture content. While crushing and bending strength increases greatly, others such as stiffness increase only moderately. Still others such as shock resistance may show a very slight decrease. Shock resistance or toughness depends on pliability as well as strength. Although drier wood is stronger, it does not bend as far as green wood before failure. The approximate increase in various strength properties due to a change from the fiber saturation point at 30% m.c. down to a moisture content of 19% is diagrammatically shown in Figure W11.

Decay Resistance is assured at a moisture content below 20%. In fact, the most effective and practical method to prevent decay is to control the moisture content of wood. Optimum conditions for decay occur when moisture content is about 25%. *Wood that is installed and maintained below a moisture content of 20% will not decay.* Most wood in use under protected situations is below this level. If the moisture content will exceed 20% due to use or exposure conditions, the wood should be treated with toxic chemicals to prevent decay (see page 201-13), or the naturally durable heartwood of certain species may be used (see page 201-11).

Paintability varies slightly with moisture content. When wood is wet it should not be painted. This refers to free water present on the surface and not to the moisture content of wood. Wood painted at 16% to 20% m.c. holds paint slightly longer than wood painted at 10% m.c.

Although the time required for paint to dry (harden) depends almost entirely on the paint, intensity of sunlight, temperature and relative humidity, drying is retarded when moisture content is excessive. Hence, in wood to be painted, moisture content should not exceed 20%.

Glue Bond is improved at reduced moisture content. Many wood products depend extensively on gluing, and the moisture content of the wood at the time of gluing affects both the strength of bond and later the *checking* and *warping* of glued members. Glues adhere well to wood with moisture contents up to 15% and higher, but when large changes in moisture content occur after gluing, shrinking or swelling stresses may weaken both the wood and joints. As with all wood products, the moisture content at the time of gluing (when increased by the moisture of the glue) ideally should approximate the average moisture content that the glued member will attain in service (Fig. W8).

Lumber used for *interior* millwork is satisfactorily glued at 5% to 6% m.c.; for *exterior* millwork, 10% to 12% m.c. Veneers used for plywood must be dried to 2% to 6% m.c. at the time of gluing.

FIG. W12 **AVERAGE WEIGHTS AND SPECIFIC GRAVITIES OF IMPORTANT WOOD SPECIES**

Species	Aver. Weight at 15% m.c. (lbs./cu. ft.)	Specific Gravity (oven dry)
Birch, sweet	45.4	.60
Birch, yellow	41.6	.55
Cedar, Western red	24.9	.33
Cypress, Southern	32.5	.43
Douglas fir	34.8	.46
Douglas fir, South	32.5	.43
Hem-fir	31.8	.42
Hickory	50.7	.67
Larch, Western	36.3	.48
Oak, red	46.1	.61
Oak, white	45.4	.60
Pine, Idaho white	27.9	.37
Pine, lodgepole	29.5	.39
Pine, Ponderosa	29.5	.39
Pine, Southern	40.8	.54
Redwood	29.5	.39
Spruce, Engelmann	24.6	.33
Spruce, Sitka	28.7	.38

Specific Gravity

Specific gravity is a ratio comparison of the weight of a material to water, and is obtained by dividing the weight of a material by the weight of an equal volume of water. Water has a specific gravity of 1. Wood fiber substance itself is heavier than water—its specific gravity being about 1.5. However, dry wood of most species floats in water because a proportion of its volume is occupied by air-filled cell cavities. The specific gravity of most species when oven-dry ranges from about 0.40 to 0.60 (Fig. W12). Variation among species in the size of cells and in the thickness of cell walls affects the amount of solid wood substance present and hence the specific gravity. Specific gravity is an approximate measure of solid wood substance and a general index of strength properties, although specific gravity values may be affected somewhat by gums, resins and extractives which contribute little to strength. The relationship of specific gravity to wood strength is evident in the practice of assigning higher basic stress values to lumber such as Douglas fir and southern yellow pine designated as *dense* (Fig. W14).

Weight of Wood Average weights per cu. ft. of various commercial woods at 15% m.c. are given in Figure W12. Weights, naturally, would differ at other moisture contents. Individual pieces will vary from these averages due to variations in density and sapwood thickness within each piece.

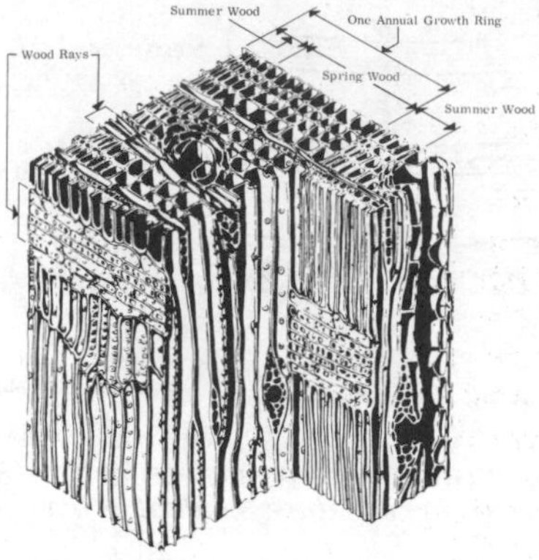

FIG. W13 Cell structure of wood. The top of the minute cube is parallel to the end surface of a log and represents a block about 1/32″ on one side.

Porosity of Wood Void space in wood occurs as: (1) coarse, microscopic capillaries within the basic cell system, and (2) submicroscopic capillaries within the cell walls which are at a maximum when the wood is saturated with moisture, as in the green or freshly cut condition. The magnitude of the porosity of wood is demonstrated by the fact that 1 cu. in. of green wood has about 15 sq. ft. of internal cell wall surface area and about 22,000 sq.

FIG. W14a **WORKING STRESSES (Allowable Unit Stresses in lbs./sq. in.) for several common species and grades of dimension lumber 2″ to 4″ thick 5″ and wider***

(1) SPECIES	(2) GRADES	(3a) Extreme fiber in bending, Fb** Single member uses	(3b) Repetitive member uses	(4) Tension parallel to gain, Ft	(5) Horizontal shear, Fv.	(6) Compression perpendicular to grain, Fc	(7) Compression parallel to grain, Fc	(8) Modulus of elasticity, E
Douglas Fir-Larch	Select Structural	1,800	2,050	1,200	95	385	1,400	1,800,000
	No. 1	1,500	1,750	1,000	95	385	1,250	1,800,000
	No. 2	1,250	1,450	650	95	385	1,050	1,700,000
	No. 3	725	850	375	95	385	675	1,500,000
	Appearance	1,500	1,750	1,000	95	385	1,500	1,800,000
Hem Fir (Western Hemlock & Western true Firs)	Select Structural	1,400	1,650	950	75	245	1,150	1,500,000
	No. 1	1,200	1,400	800	75	245	1,050	1,500,000
	No. 2	1,000	1,150	525	75	245	875	1,400,000
	No. 3	575	675	300	75	245	550	1,200,000
	Appearance	1,200	1,400	800	75	245	1,250	1,500,000
Southern Pine	Select Structural	1,750	2,000	1,150	90	405	1,350	1,700,000
	No. 1	1,450	1,700	975	90	405	1,250	1,700,000
	No. 2	1,200	1,400	625	90	405	1,000	1,600,000
	No. 3	700	800	350	90	405	625	1,400,000

* This table is for illustrative purposes mainly; for design values of other species and grades, see Work File. Values shown are for normal loading conditions at maximum 19% m.c., as in most covered structures.

**Values in bending for "repetitive member uses" are intended for design of members spaced not over 24″ o.c., in groups of not less than three members and when joined by floor or roof sheathing elements.

ft. (about half the surface area of a football field) of internal submicroscopic capillary surface area in the cell walls (Fig. W13).

The enormous internal surface area gives wood many of its unique properties: (1) ease of impregnation with toxic chemicals for decay and insect protection, and with moisture repellents for slowing down moisture exchange to minimize shrinkage; (2) excellent insulating properties; (3) shrinking and swelling in response to variations in moisture content; and (4) ease of adherence of paint films and adhesives, many new synthetic resins adhering with such tenacity that they resist virtually any exposure condition.

Structural Properties

Unlike crystalline materials such as steel or concrete, wood is fibrous. The length of fibers varies from about 1/25″ in hardwoods to from 1/8″ to 1/3″ in softwoods. The strength of wood, however, does not depend on the length of the fibers but on the thickness of their walls and on the direction of the fibers relative to applied loads. The fibers in wood are oriented with their length essentially parallel to the vertical axis of the tree, and the strength of wood parallel to these fibers (parallel to the grain) is quite different from its strength perpendicular to the fibers (perpendicular to the grain).

External forces, termed *loads* (such as wind, gravity, earthquakes) produce internal resisting forces in structural members, known as *stresses. Tensile* stresses result from stretching forces, *compressive* stresses from squeezing forces. When a member is loaded so that the force is acting more or less perpendicular to its length—as in the case of a beam—the resulting stress is called *bending.* In bending, the member develops compressive stresses in the upper part and tensile stresses in the lower part of its cross section.

Strength values of wood are determined by testing small, clear specimens in bending, compression, shear, etc., and assigning *basic stresses*—that is, basic strength values—to the different species. Since small, clear pieces are seldom used in construction, lower *working stresses* (also called allowable unit stresses) must be derived to take into account the effect of knots, slope-of-grain, checks or other strength-reducing characteristics that reduce the strength of a member to a value less than for clear wood.

Thus, the *working stress* for a grade of lumber is determined by the effect of the maximum size and most unfavorable position of any strength-reducing characteristics permitted in the grade. Working stresses for bending, "Fb," and tension parallel to grain, "Ft,"; horizontal shear, "Fv;" compression both perpendicular to the grain, "Fc⊥," and parallel to the grain, "Fc" and modulus of elasticity, "E," of three selected species are shown in Figure W14a. The strength properties of wood loaded in bending, tension, shear and compression are discussed briefly below.

The allowable unit stresses shown in Figure 14a are based on *normal duration of loading.* The National Design Specification for Wood Construction defines this condition as follows: "Normal load duration contemplates fully stressing a member to the tabulated design value by the application of the full maximum normal design load for a duration of approximately 10 years (either continuously or cumulatively) and/or the application of 90% of this full maximum normal load continuously throughout the remainder of the life of the structure, without encroaching on the factor of safety." Values for normal loading generally are used for designing members to carry *dead loads* (such as floor or roof construction) and *live loads,* due to occupants, equipment and furnishings. For loads of brief duration, such as those produced by snow, earthquake or wind, higher working stresses can be used than for normal loads. Figure W14b shows the ratios of working stresses for various durations of loading in relation to normal loading, which is used as the base—1.0.

Bending Wood develops high fiber strengths in bending. When a wood beam, loaded as in Figure W15a, deflects, bending stresses are produced in the fibers of the member. If the load is great enough to produce stresses that exceed the fiber strength of the wood, the

FIG. W14b **ADJUSTMENT OF ALLOWABLE STRESSES FOR DURATION OF LOADING**

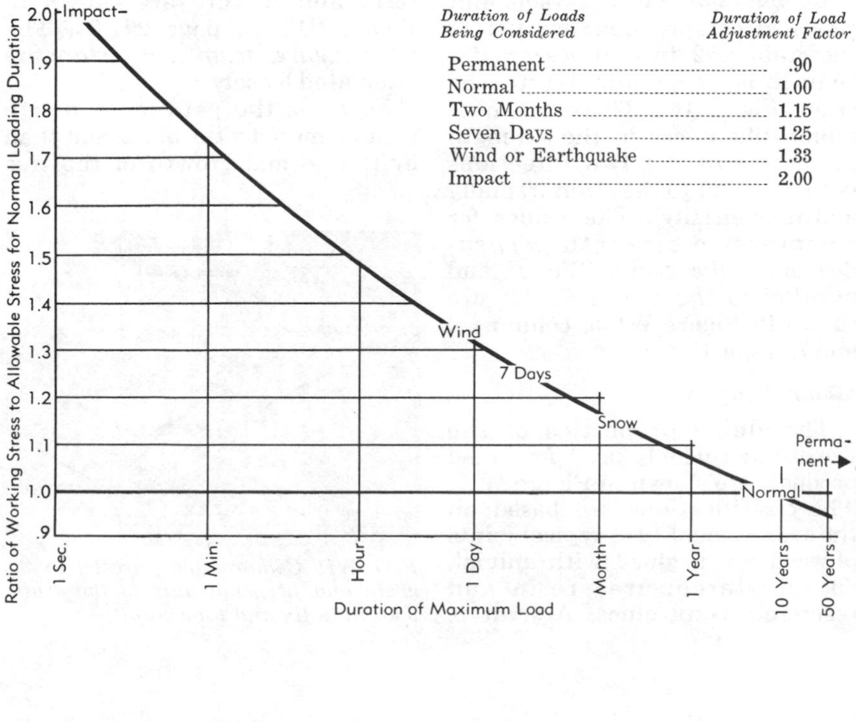

Duration of Loads Being Considered	*Duration of Load Adjustment Factor*
Permanent	.90
Normal	1.00
Two Months	1.15
Seven Days	1.25
Wind or Earthquake	1.33
Impact	2.00

beam will fail by breaking (Fig. W15b). The stresses produced in the fibers are greater in magnitude the farther they are from the central axis. If a fiber is located twice as far from the central axis as another fiber, it will have twice the stress. Thus, in a beam, bending stresses are maximum in the outermost fibers (called the *extreme fiber*) at the top and bottom of the beam. In beams subjected to bending, the fiber stresses in bending should be held below the working stress values, "Fb," for the grade of wood used (Fig. W14a, col. 3).

Horizontal Shear A beam also tends to fail by dropping down between supports (Fig. W15c) as a result of *vertical shear,* the tendency for one part of the beam to move vertically with respect to an adjacent part. In wood, beams are more likely to fail due to the tendency of the top fibers in the beam to move horizontally with respect to the bottom fibers. This produces stresses in fibers called *horizontal shear* (Fig. W15d). The horizontal shearing stresses are maximal at each end of the beam along the central (neutral) axis. In beams the fiber stress in horizontal shear should be held below the working stress values, "Fv," for the grade of wood used (Fig. W14a, col. 5).

Modulus of Elasticity Stiffness of a material is measured by its modulus of elasticity, "E" (Fig. W14, col. 8). For a beam (Fig. W15d), this measure of wood's stiffness determines the beam's resistance to deflection. Modulus of elasticity is also used in computing the load on long columns, since such columns depend on stiffness to resist buckling.

Machine grading techniques based on modulus of elasticity permit greater utilization of wood's strength properties (see page 201-27).

Tension Theoretically, tensile strength *parallel* to the grain is the strongest property of wood, but in practice, it is also greatly affected by knots, splits and checks. Since the allowable unit stress in *tension parallel to grain,* "Ft," is a value applied to the entire cross section (less holes) of a piece of wood, it must reflect an *average* internal resistance to loading which is much more affected by wood characteristics than a value applied to a narrow band of the cross-section, such as the extreme fiber in bending stress, "Fb." For most species, the allowable Ft values are approximately 2/3 of Fb values. Wood is quite weak in tension *perpendicular* to the grain, permitting the fibers to be pulled apart under heavy loading. Hence, wood generally should not be loaded so as to create this type of stress and allowable values for tension perpendicular to grain usually are not tabulated.

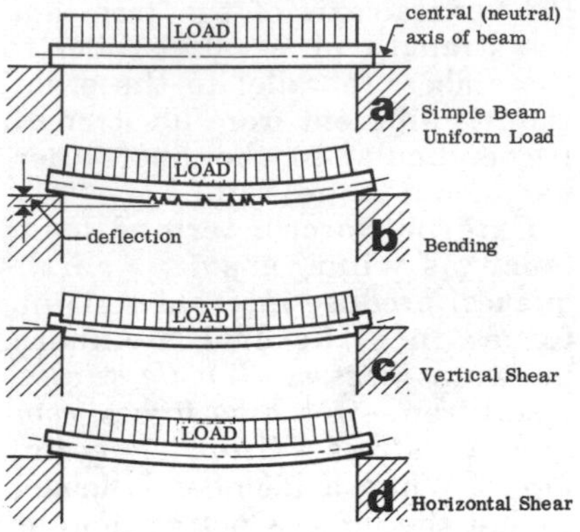

FIG. W15 Stresses produced in a beam subjected to bending

Compression The allowable unit stress in compression parallel to the grain is 2 to 5 times greater than it is perpendicular to the grain (Fig. W16). There are only minor differences in the strength properties in the two directions perpendicular to the grain (radially and tangentially). The values for compressive strength *perpendicular to the grain,* "Fc⊥," and *parallel to the grain,* "Fc," are shown in Figure W14a, columns 6 and 7, respectively.

Gluing Properties

The gluing properties of the woods most widely used for glued products are shown in Figure W17. The classifications are based on the average quality of typical joints of wood when glued with animal, casein, starch, urea resin and resorcinol resin glues. A joint is considered to be glued satisfactorily when its strength is approximately equal to the strength of the wood.

Whether it will be easy or difficult to obtain a satisfactory joint depends upon the density of the wood used, its structure, the presence of extractives or infiltrated materials in the wood and the kind of glue. In general, hardwoods are more difficult to glue than softwoods, and heartwood is more difficult than sapwood. Several species vary considerably in their gluing characteristics with different glues (Fig. W17).

Non-structural Properties

Many unique characteristics of wood result in its selection as a material for many non-structural uses. These non-structural properties include decorative characteristics, weathering qualities, natural decay resistance, and thermal expansion and conductivity.

Decorative Characteristics Color, figure patterns, texture and luster; ability to take filler, stains, bleaches and other finishes, as well as the method of cutting or sawing —all influence the decorative value of wood. The figure patterns of some of the fine hardwood veneers are shown in Figure P35 on page 201-42. Examples of softwood patterns and texture are shown in Figure P12 on page 201-33. The terms *figure, grain* and *texture* are often used loosely.

Figure is the pattern or design created more by the abnormal than by the normal growth of the tree

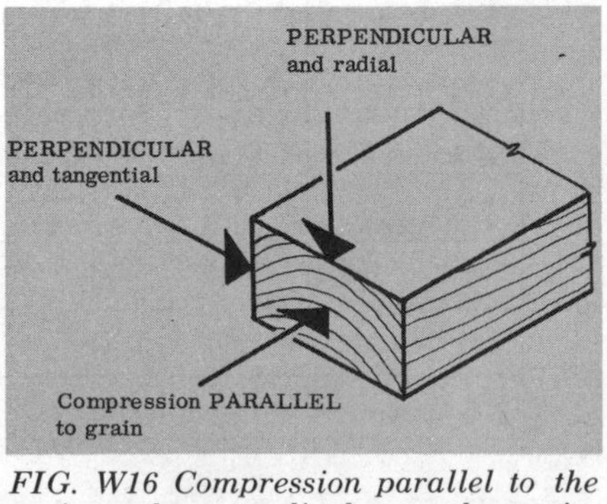

FIG. W16 Compression parallel to the grain and perpendicular to the grain, both radially and tangentially

FIG. W17 **GLUING PROPERTIES OF SELECTED WOOD SPECIES**

GLUEABILITY*	HARDWOODS	SOFTWOODS
EXCELLENT	Aspen Chestnut Cottonwood Black willow Yellow poplar	Baldcypress Western redcedar White fir Western larch Redwood Sitka spruce
GOOD	Alder Basswood Butternut Rock elm Hackberry	Eastern redcedar Douglas fir West Coast hemlock Northern white pine Southern yellow pine Ponderosa pine
SATISFACTORY	White ash Black cherry Soft maple Oak Pecan Sycamore Tupelo Black walnut	Alaska cedar
POOR	Beech Birch Hickory Hard maple	

*Ratings based on the ability of the wood to be glued under a wide range of conditions and with a variety of glues.

(wavy or swirly) and is incorrectly used synonymously with grain.

Grain is often used in referring to the annual rings (fine grain and coarse grain), but it is also used to indicate the direction of the fibers (straight grain, vertical grain); and finishers refer to the relative size of the pores (open grained, closed grain) which determines the need for a filler.

Texture, though often used synonymously with grain, usually refers more to the finer structure and the visual dimension of depth in the wood than to the annual growth rings.

In plain-sawed boards and rotary-cut veneer, the annual rings often form ellipses and parabolas making striking figures. In quarter-sawed boards and sliced veneers, the growth rings form stripes, and in species such as oak the rays add conspicuous figure. Examples of plain- and quarter-sawed boards are shown in Figure L21, page 201-22. In open-grained hardwoods (such as oak, mahogany, walnut) and sliced veneers, the appearance can be varied greatly by using fillers of different colors, and in softwoods the annual growth layers can be accentuated by applying stain.

Weathering Without a protective coating, weathering of exposed boards may involve: (1) change in color, ranging from light gray with a silvery sheen (in cedar and cypress) to dark gray with no sheen (in Douglas fir and redwood); (2) roughening and checking of the surface except in cedar, cypress and redwood, on which weather checks are inconspicuous; and (3) warping tendencies, starting with cedar, cypress and redwood, in which this tendency is slight, to a distinct tendency in most of the other softwoods and a pronounced tendency in most of the hardwoods. *To resist warping, the width of an exposed, unpainted board should not exceed 8 times its thickness.*

Thermal Expansion Like most materials, wood expands when heated and contracts when cooled. Except in special applications such as buildings kept dry but subject to wide temperature range, the thermal expansion and contraction of wood is ignored since it is much smaller than the swelling and shrinkage due to moisture content variations occurring under normal exposure conditions.

Thermal Conductivity The thermal conductivity of wood is influenced by several of its properties, the most important being specific gravity and moisture content. Lighter weight woods are better insulators, and insulation value increases as the moisture content decreases. Because of its good insulating characteristics, wood fiber is a principal constituent used in insulating fiberboards.

Natural Decay Resistance Natural toxic substances deposited in the cell structure of certain species make the heartwood resistant to decay or termite attack. Decay resistant species are: bald cypress (tidewater red), cedars, redwood, black locust and black walnut. Termite resistant species are: redwood, bald cypress (tidewater red) and eastern red cedar.

Naturally durable heartwood of the above species may be specified for all conditions of exposure except for members embedded in the ground to support permanent structures. For maximum durability, 100% heartwood of resistant species selected during grading should be used. For such species as redwood and cedar, several grading authorities—Redwood Inspection Service, West Coast Lumber Inspection Bureau and Western Wood Products Association—have *foundation grades* selected from 100% heartwood with a high toxic extractive content. Where wood is embedded in the ground for the support of permanent structures, it should be pressure-treated as explained below.

FUNGI AND INSECT HAZARD

When lumber is used under continuously dry conditions and is adequately protected from attack by insects and fungi, preservative treatment is unnecessary. Preventive measures against these hazards during the manufacturing and merchandising stages are well established and widely practiced by reputable lumber producers. Serious decay or insect problems in buildings result most often from poor construction techniques.

Decay

Microscopic plants (fungi) cause decay, molds and stains. Fungus growth can develop in wood only under the following conditions: (1) an adequate food supply of organic material, which for some forms of fungi is the wood itself; (2) mild temperatures ranging from 41° to 104°F; (3) sufficient oxygen, which is always present unless the wood is completely below the ground water line or completely submerged in water; and (4) a wood moisture content in excess of 20%.

"Dry Rot" Most decay *(rot)* occurs in wood when its moisture content is above the fiber saturation point (30% m.c.). If wood is maintained below a 20% m.c. (typical of air-dried wood), it will not decay. Wood that is continuously water soaked will not decay. Reference is often made incorrectly to "dry rot." This term is incorrect because the fungi must have access to water in order to function; and even though the wood may be dry when the "rot" is detected, it must first have been wet in order to decay. Since in the later stages of decay any wood-destroying fungus seriously reduces the strength of wood, decay is limited in structural lumber by most grading rules.

White Pocket Also known as *white speck,* this characteristic is caused by a relatively harmless fungus in living softwood trees that ceases to develop when a log containing it is cut into lumber and put into normal use. It is as harmless as other natural characteristics normally found in lumber, such as knots, shake or wane. It will not spread in any individual piece and will not spread from one piece to another. Such lumber is serviceable, and certain amounts of white speck generally are accepted by grading authorities in all but the higher grades of lumber (Fig. W18).

Molds and Stains

Molds and stains are caused by fungi which do not destroy the wood. Stained and molded stock is practically unimpaired for use except where appearance is the limiting factor.

The discoloration caused by molds is largely superficial and often is easily brushed or surfaced off the piece. Powdery surface growths vary from white or light colors to black.

The discoloration caused by stains penetrates into the sapwood and cannot be removed by surfacing. Stains appear as specks, spots, streaks or patches that vary most commonly from bluish to bluish-black and brown.

FIG. W18 White pocket (white speck) is permitted in certain lumber grades and is serviceable and acceptable for many uses.

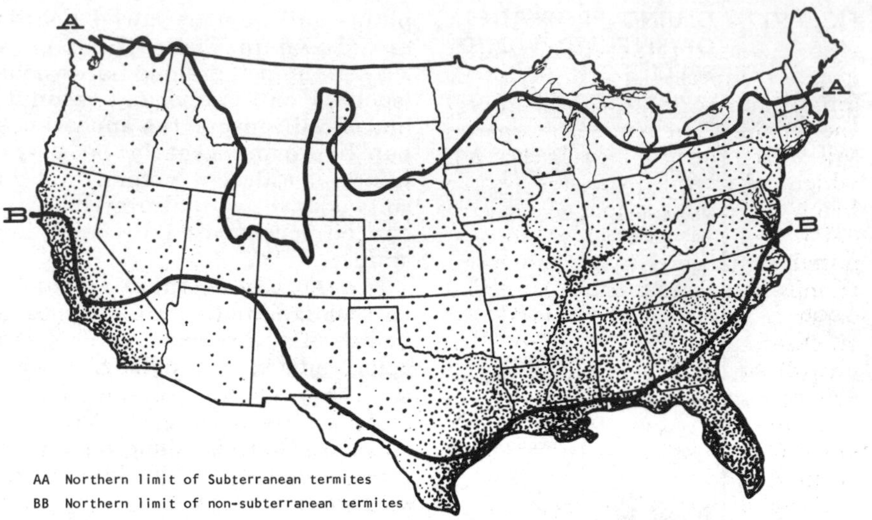

FIG. W19 General relative hazard of termite infestation. Subterranean termites have been found as far north as line A-A; the northern limit of non-subterranean termites is indicated by line B-B.

Insect Damage

Holes caused by insects, classified as pinholes and grub holes, may develop in standing trees, unseasoned or seasoned lumber. During grading such damage is taken into account. However, the strength of wood damaged by insects *after installation* cannot be estimated visually from its appearance and, if strength is an important factor, every piece found containing insect holes should be removed and replaced.

Many of the precautions taken to prevent decay are also effective against various insects that attack and destroy wood.

Subterranean Termites This type of termite accounts for approximately 95% of all termite damage in the United States. As Figure W19 illustrates, the hazard is greater where mild temperatures favor their development.

Subterranean termites (Fig. W20) live in nests and develop their colonies underground, but build tunnels through the earth to get at wood or other cellulose material such as fiberboard, fabrics and paper. Termites must have a constant source of moisture or they die. They become most numerous in moist, warm soil containing an abundant supply of food.

Subterranean termites are not carried into buildings in lumber but establish themselves by entering from ground nests after the building has been constructed, using three basic routes: (1) They may attack wood in direct contact with the ground (porch stairs, trellises and siding installed too close to grade); (2) They may enter through cracks and voids as small as 1/32″ (slabs, piers or foundation walls); and (3) They may build mud tubes over materials they cannot go through (Figs. W21, W22, W23). During certain seasons, winged forms (Fig. W20) swarm (fly) from the nest and, if successful, develop new colonies.

The control of subterranean termites (see Division 300, Methods and Systems) may consist of one of the following protective methods or more than one method used in combination: (1) metal shields, when properly installed, which isolate termites from the wood or force them to build tubes around the shield; (2) chemically treated soil, which forms a barrier through which the termites will not pass to reach the construction; (3) chemically treated, pressure-impregnated wood, which is not subject to termite attack; (4) the use of poured concrete foundations, provided no cracks greater than 1/64″ are present; (5) the use of poured reinforced concrete caps, at least 4″ thick, on unit masonry foundations, provided no cracks greater than 1/64″ occur.

Regardless of the protective method used, wherever possible construction should permit visual inspection.

Non-subterranean Termites This type of termite has been found only south of line "B-B" shown in Figure W19. Their total damage is much less than subterranean termites, but they may present a problem in the areas where they do occur. Also called *drywood termites* because of their ability to live in damp or dry wood without moisture or contact with the ground, they often are recognized by slightly compressed pellets of partially digested wood pushed through surface openings. Where this type is found, lumber should be inspected carefully to see that it is not infested before arrival at the site. When construction is underway during the swarming season, lumber should be inspected for infestation.

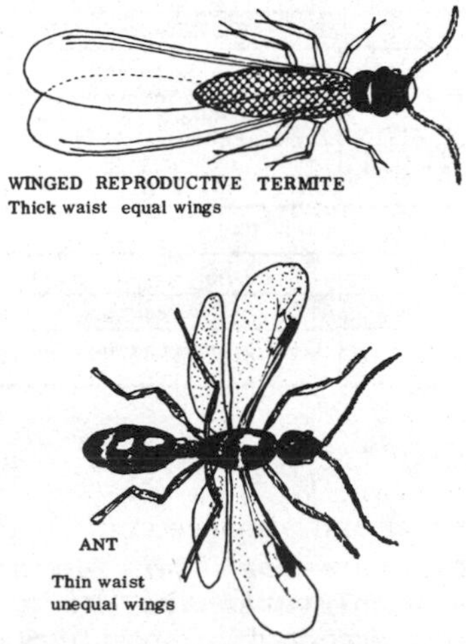

FIG. W20 Ants are often mistaken for termites. Subterranean termites may be identified by their thick waists and wings of equal length. Ants are thin-waisted and have unequal-length wings.

WOOD PRESERVATIVE TREATMENTS

Preservative treatments protect wood against decay and insect attack. Pressure treatments, in which preservative chemicals are applied under pressure to obtain maximum penetration, afford the greatest protection for lumber and plywood against decay or termite attack. Non-pressure treatments should be restricted for use only for protection of exterior millwork not in contact with the ground.

Wood preservatives may be divided into three major classes: (1) water-borne preservatives, (2) oil-borne preservatives, and (3) creosote and solutions containing creosote. (Fig. W24).

Water-borne Preservatives

Water-borne preservatives are the type most often used to pressure-treat lumber and plywood for residential construction because they leave the wood clean, odorless and easy to paint. However, two types have proven to be very durable and can be recommended for wood to be used in ground contact supporting permanent building structures. These two are ammoniacal copper arsenate and chromated copper arsenate. The latter is available in three different formulations according to AWPA Standard FDN, but each is considered to be equivalent in performance. All water-borne preservatives are relatively odorless and are paintable provided the wood has been dried to the same moisture content required for untreated wood. These preservatives, placed in contact with interior finish, will not cause discoloration except when a wet-process finish, such as stucco or plaster, is applied without a water-proof backing paper.

Oil-borne Preservatives

Lumber treated with oil-borne preservatives is used in any type of installation except those in contact with salt water.

FIG. W21 Subterranean termite infestation of foundation plates. Improper clearance between wood siding (removed for photo) and ground permitted termite entry.

FIG. W22 Termites build mud tubes to permit access from ground to wood. A series of tubes are shown here on a concrete foundation pier. Note that tubes have been built around metal shield at top of pier.

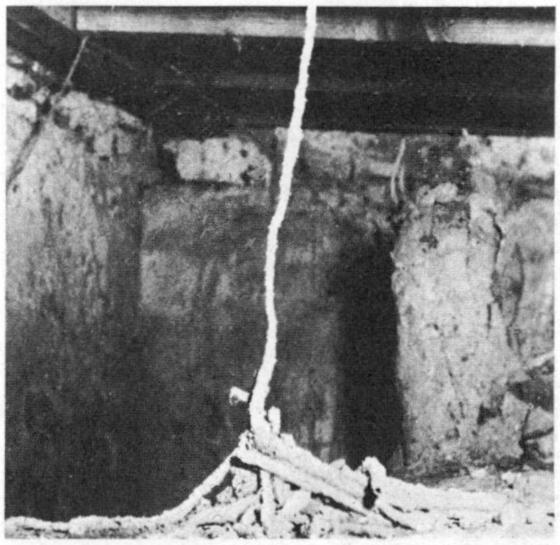

FIG. W23 Tubes shown here have been built from wood above down to the ground. Tubes can be built up foundation walls or posts many feet from the ground to wood.

Penta (Pentachlorophenol) This preservative is highly toxic to both fungi and insects, insoluble in water and thus permanent. It is the most widely used oil-borne preservative and, depending upon

FIG. W24 SUMMARY OF WOOD PRESERVATIVE TREATMENTS

PRESSURE TREATMENT

WATER-BORNE SOLUTIONS[1]

Industry Standard	Uses	Description
AWPB LP-2—Softwood Lumber, Timber and Plywood, Pressure Treated with Water-borne Preservatives for Above Ground Use	All construction uses (framing lumber, sheathing, siding and trim) where wood should be protected from moderate decay hazard & insect attack.	Treated wood is clean, bright and odorless, readily painted and glued with most adhesives. The preservative is permanently fixed in the wood and will not leach out when exposed to weather.
AWPB LP-22—Softwood Lumber, Timber and Plywood, Pressure Treated with Water-borne Preservative for Ground Contact Use	All construction uses when exposed to severe insect attack and decay hazard. Also where wood is used in contact with the ground such as: fence posts, steps, decks, etc.	
AWPB-FDN—Softwood Lumber, Timber and Plywood, Pressure Treated with Water-borne Preservatives for Ground Contact Use	All parts of wood foundation systems for residential & light construction placed directly in the ground.	
AWPA C-23 Residential and Commercial Building Poles, Pressure Treated with Water-borne Preservatives	Poles for use in ground contact as a building foundation and column support against both lateral and vertical forces.	

OIL-BORNE SOLUTIONS (PENTACHLOROPHENOL)[2]

Industry Standard	Uses	Description
AWPB LP-3—Softwood Lumber, Timber and Plywood, Pressure Treated with Light Petroleum Solvent-Penta Solution for Above Ground Use	All construction uses (framing lumber, sheathing, siding and trim) where wood should be protected from moderate decay hazard & insect attack.	Treated wood is clean, bright and relatively odorless, paintable, can be glued with special adhesives. Suitable for indoor and outdoor use. All petroleum solvent must be evaporated before painting and to eliminate odor.
AWPB LP-33—Softwood Lumber, Timber and Plywood, Pressure Treated with Light Petroleum Solvent-Penta Solution for Ground Contact Use	All construction uses where exposed to severe insect attack and decay hazard. Also where wood is used in contact with the ground such as: Fence posts, steps, decks, etc.	
AWPB LP-4—Softwood Lumber, Timber and Plywood, Pressure Treated with Volatile Petroleum Solvent (LPG)-Penta Solution for Above Ground Use	All construction uses (framing lumber, sheathing, siding and trim) where wood should be protected from moderate decay hazard & insect attack.	
AWPB LP-44—Softwood Lumber, Timber and Plywood, Pressure Treated with Volatile Petroleum Solvent (LPG)-Penta Solution for Ground Contact Use	All construction uses where exposed to severe insect attack and decay hazard. Also where wood is used in contact with the ground such as: Fence posts, steps, decks, etc.	
AWPB CP-33—Residential and Commercial Building Poles, Pressure Treated with Pentachlorophenol in Light Hydrocarbon Solvent	Poles for use in ground contact as a building foundation and column support against both lateral and vertical forces.	
AWPB CP-44—Residential and Commercial Building Poles, Pressure Treated with Pentachlorophenol in Volatile Petroleum Solvent (LPG)	Poles for use in ground contact as a building foundation and column support against both lateral and vertical forces.	
AWPA LP-7—Softwood Lumber, Timber and Plywood, Pressure Treated with Solvent-Penta for Above Ground Use	Farm buildings, fences, bridges, etc. Good for outdoor applications because of good weathering characteristics.	Treated wood varies from light to dark brown, surface slightly oily, odor of oil & penta remain for several months after treatment.[4]
AWPA C-23—Softwood Lumber, Timber and Plywood, Pressure Treated with Solvent-Penta for Ground Use	All construction uses where exposed to severe insect attack and decay hazard. Also where wood is used in contact with the ground as in fence posts, steps, decks, etc.	
AWPA C-23—Residential and Commercial Building Poles, Pressure Treated with Pentachlorophenol	Poles for use in ground contact as a building foundation and column support against both lateral and vertical forces.	
AWPA C-23—Residential and Commercial Building Poles, Pressure Treated with Creosote and Creosote Solutions	Creosote poles are widely used for pole-platform construction. They are suitable for residential pole-frame construction, but their use should be generally limited to those exterior poles that do not pass through enclosed, living space.	Creosoted poles range from dark brown to black, blending well with earth colors and natural surroundings. Because their surfaces are oily, painting is difficult at best and a characteristic smoky odor is detectable for some time following treatment.

CREOSOTE AND CREOSOTE SOLUTIONS[3]

Industry Standard	Uses	Description
AWPB LP-5—Softwood Lumber, Timber and Plywood, Pressure Treated with Creosote or Creosote Coal-Tar Sclutions for Above Ground Use	Farm buildings, fences, bridges, etc. Good for outdoor applications because of good weathering characteristics. Ok for marine use (wharfs, docks, etc.)	Treated wood varies from dark brown to black, surface oily & creosote odor remains for many months after treatment.[4]
AWPB LP-55—Softwood Lumber, Timber and Plywood, Pressure Treated with Creosote or Creosote Coal-Tar Solutions for Ground Contact Use	All construction uses when exposed to severe insect attack and decay hazard. Also where wood is used in contact with the ground as in fence posts, steps decks etc. Ok for marine use.	

NON-PRESSURE TREATMENT

OIL-BORNE, WATER REPELLENT PRESERVATIVE SOLUTION

Industry Standard	Uses	Advantages
NWMA IS-4—Water-Repellent Preservative, Non-Pressure Treatment for Millwork	Exterior millwork: precut window, door and trim components.	Clear, odorless and easily painted.[5]

1. Treated wood can be painted and glued, after redrying; not recommended for hardwoods.
2. Suitable for treatment of hardwoods as well as softwoods; not recommended for marine use.
3. Suitable for marine use; generally not paintable and retains odor for long periods; finish materials may be stained by contact with treated wood.
4. Generally not paintable; should not be used for subflooring or in contact with materials subject to staining (plaster, wallboard, etc.) Retains odor for a long period when enclosed in a structure.
5. Should not be used in contact with the ground or for conditions of severe insect attack and decay hazard.

the solvent, the treated lumber varies in color from dark brown to colorless.

Creosote

Creosote is the most widely used of all preservatives. Creosote and solutions containing it are used where protection against wood-destroying organisms is of first importance, where painting is not required and a slight odor is not objectionable. Wood may be protected in severe exposures with high retentions and in less severe exposures with lower retentions.

Creosote-Coal Tar Mixtures These mixtures are the most widely accepted preservatives for marine or salt water installations and are ideally suited for treating piling used for shore dwellings.

Creosote-Petroleum Mixtures These mixtures are used where

FIG. W25 Heavy joists, planks and fence posts are loaded on tram cars to be pressure treated.

FIG. W26 Lumber is subjected to various combinations of temperature and pressure in steel cylinders.

FIG. W27 Here the retort has been opened after pressure charge has driven the preservative into the wood.

economy is of first importance, but they should not be used for marine installations.

Treatment Methods

Methods of preservative treatment are of two types: (1) pressure processes and (2) non-pressure processes (Fig. W24).

Pressure Processes Pressure preserving processes may be classified as *full cell* or *empty cell.* Both provide the most dependable means of ensuring uniform penetration and distribution of preservative. Timber to be treated by a pressure process is loaded on tram cars (Fig. W25) and rolled into a long steel cylinder (Fig. W26) which is sealed (and in the full cell process subjected to a vacuum to remove air from the wood). Then the cylinder is filled with preservative under various combinations of temperature and pressure to force the preservative deep into the wood. After specified penetration and retention are achieved, the charge is removed (Fig. W27) and, in the case of water-borne preservatives, is air or kiln dried, leaving the protecting chemicals diffused in the wood.

Full-Cell Process is used when high retentions are required for creosote or creosote mixtures, such as in marine installations. The full-cell process also is used for treating with water-borne preservatives and fire-retarding chemicals. Preservative is not only absorbed by the cell walls, but the cell cavities also are left full. This process leaves the maximum amount of preservative in the wood but, when used with creosote or its solutions, will not leave as clean a surface as the other processes.

Empty-Cell Process subjects the wood to atmospheric or high initial air pressures prior to introducing and injecting the preservative. When pressure is released, the expanding entrapped air expels the excess preservative from the wood cell cavities, thus reducing the net retention of preservative. This method provides treatment of the cell walls but leaves the cell cavities empty. Thus it yields a drier product, gives deeper and more uniform penetrations, and permits maximum penetration for any specified weight. An empty-cell process should be used for all oil treatments if the specified retentions can be obtained by the process.

Non-pressure Processes Wood treated with non-pressure processes should not be used in contact with the ground or under severe conditions of decay or termite attack, because the preservative is not sufficiently impregnated in the wood to afford adequate protection. Non-pressure treatment can be effective for exterior millwork such as precut door, window and trim components, where the wood is generally dry but may be exposed to moisture or wetting intermittently.

Water-repellent volatile petroleum solvents containing toxic chemicals such as pentachlorophenol or zinc napthenate commonly are used.

Lumber treated with petroleum solvent solutions should be seasoned to a maximum moisture content of 19% or less as necessary for the intended use. Surfaces which are cut or milled after treatment should be retreated with brush application of preservative before final assembly or installation. Vacuum and brief dip non-pressure treatments generally provide clean, paintable surfaces.

Vacuum Processes utilize a vacuum or partial vacuum to exhaust air from the cells and pores of the wood, the preservative being forced into the wood by atmospheric pressure.

Brief Dip Processes (Three-minute Immersion) consist simply of completely submerging the wood in an open tank of the preservative solution for a period of at least three minutes.

Standards of Quality

The American Wood Preservers Association (AWPA) is a technical and research organization serving the wood preserving industry. AWPA developed and promulgated *generalized* specifications for preservative formulations, methods of treatment and minimum chemical retention, which are the recognized industry standards. A sister organization, the American Wood Preservers Institute (AWPI) performs the promotional and educational functions for the industry. AWPI has developed more detailed requirements for *specific* products and applications, based on the AWPA standards. Originally, AWPI also administered a quality control and certification program, in accordance with its own standards, then known as the AWPI Quality Standards. More recently a separate organization, the American Wood Preservers Bureau (AWPB), was established to administer the certification program and the AWPI standards became known as AWPB Quality Standards.

Certification of Quality—Pressure-Treated AWPB administers a program of quality control and certification of products treated in accordance with AWPB Quality Standards. The program consists of periodic sampling, and laboratory analysis of products from the licensee's plants; so long as products conform to applicable AWPB standards, they may be identified with the AWPB Quality Stamp (Fig. W28).

Certification of Quality—Non-pressure Treated The National Woodwork Manufacturers Association (NWMA) administers a testing and plant inspection program, resulting in certification and issuance of its seal (Fig. W29). The NWMA seal is assurance that the product conforms to NWMA IS-4 Water-Repellent Preservative Non-pressure Treatment for Millwork.

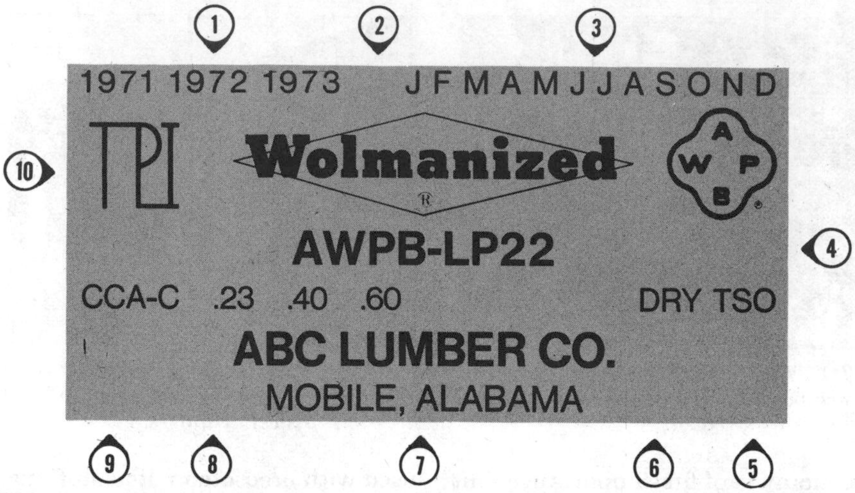

FIG. W28 The AWPB Quality Stamp provides visual assurance that a product has been pressure-treated according to the appropriate AWPB Standard. Stamp provides the following data: 1) year of treatment—earliest year shown; 2) preservative used; 3) month of treatment—first letter following blank; if no blank, then the month is Jan.; 4) governing AWPB Standard; 5) TSO means "treating service only"—treating plant is not responsible for grade or quality; 6) DRY means 19% or less m.c.; 7) name and location of treating plant; 8) chemical retention—largest number is lbs./cu. ft. required; 9) chemical description of preservative—chromated copper arsenate type C; and 10) local quality control agency which samples the products on behalf of AWPB.

FIG. W29 The National Woodwork Manufacturers Association (NWMA) seal certifies non-pressure preservative treatment of millwork in accordance with NWMA Industry Standard IS-4.

FIRE-RETARDANT TREATED WOOD

The ability of building materials to resist fire is becoming increasingly important. Protection against fire can be afforded to wood through pressure impregnation of inorganic salts which react chemically at temperatures below the ignition point of wood. The reaction causes the combustible vapors normally generated in the wood to break down into non-flammable water and carbon dioxide.

Fire-retardant treated wood has been marketed since World War II. In recent years products have been improved and better control has been established to assure uniformity. Products tested and approved by the Underwriters' Laboratories, Inc. which bear its label are recognized by model building code authorities and federal agencies as suitable for structural use in fire-resistive and noncombustible construction.

After treatment, all plywood and lumber, 2″ nominal or less, shall be dried to a moisture content of 19% or less (designers may want to specify a lower moisture content for fire retardant treated wood to be used for mill work, cabinets, office paneling, and other special uses).

Except where exterior-grade treatment complying with rain testing is specified, fire-retardant treated wood shall not be used where it will be exposed directly to the weather.

LUMBER

Lumber is the designation given to a large number of products produced in the saw and planing mill. Lumber is manufactured by sawing, resawing and passing logs lengthwise through a standard planing machine, cross-cut to length and matched. (Veneers, barrel staves and hammer handles, for example, are not lumber.) *Yard* lumber includes: (1) *boards* used for sheathing, siding, flooring, paneling, trim and patterned millwork; (2) *dimension* stock used for light framing, joists, planks, and roof and floor decking; (3) *timbers* used for posts, beams and stringers.

This subsection discusses the manufacture, classification and grading of lumber generally intended for light construction purposes.

MANUFACTURING PROCESS

The manufacture of lumber begins in the forest. After trees are felled, the logs are transported to the lumber mill for processing (Fig. L1). Following manufacture by approximately 34,000 active sawmills, some 5,000 wholesalers locate lumber and resell it to tens of thousands of wood-using plants and to over 34,000 retail lumber yards that distribute about 75% of all lumber to the ultimate user. Many retail yards and some wholesalers maintain a planing mill and other equipment to satisfy individual user requirements.

Logging Practice

Modern logging methods make it possible to transport practically the entire trunk of a tree from the forest to the mill. Portable power saws have largely supplanted hand-propelled saws for felling and cutting trees into logs (Fig. L2). Depending upon the terrain and other variable factors, logs are removed from the forests by trucks, tractors, trains or by fast, powerful cable logging systems. The logs are grouped at a landing where mobile or other types of loaders place them onto trucks for the haul to the mill (Figs. L3, L4, L5).

Sawmill Practice

Operations vary from mill to mill, depending on the size of logs, the rate of output, the type of machinery and the end-product desired; but in the main, the pattern of operations is as follows:

Once the logs reach the sawmill they may be stored in a pond until ready for processing. Logs floating in the pond are directed against the bull-chain conveyor for their trip into the mill (Fig. L6). Logs generally are sent first to a debarker. Debarking equipment peels, scrapes or blasts (using high-pressure water jets) the bark from the log, revealing its shape, knots and irregularities to the sawyer; debarkers also provide chipped waste that is bark-free (Fig. L7). The log is then placed on a carriage and fed into the head saw (Fig. L8), which may be either a circular saw or a band saw. This main saw mechanism is called the *headrig* and is the point where the log is broken down into rough cants, tim-

FIG. L1 Aerial view of a large lumber mill. Here, logs are stored in a pond until processed.

FIG. L2
Woodsman fells a giant tree by using a powered chain saw.

FIG. L3 Logs being skidded to a landing where they will be loaded for transport to the sawmill.

FIG. L4 At the landing, logs are loaded aboard trailer trucks for transport to the sawmill.

FIG. L5 At the sawmill a log unloader can remove the entire load from a truck in one bite.

FIG. L6 Workman directs logs against bull chain. High-pressure water jets wash grit away from logs.

FIG. L7 The "pencil sharpener" action of a mechanical debarker removes the bark from the log.

FIG. L8 Here, the carriage of the head-rig moves the log back and forth across a band saw.

FIG. L9 A gang saw turns out stacks of rough-cut boards in a single operation.

FIG. L10 From the head saw, boards pass through the edger, which produces square edges.

Fig. L11 The first grading and sorting operations of lumber occur on the "green chain."

FIG. L12
Stacks of lumber stand outside dry kilns waiting to be dried.

bers, planks and boards. Some mills employ a gang saw as the head saw, consisting of as many as 40 straight saws, or use one in combination with the band or circular head saw to break large cants rapidly into boards and dimension lumber (Fig. L9).

The next step is the edger (Fig. L10), a series of small saws that rip the lumber into the desired width and remove the rounded edge. The edger also rips wide pieces into narrow widths. Edged lumber moves onto the trimmer, where the saws trim off the rough ends and defects to produce the most desirable lengths.

The final step in the sawmill is called the *green chain,* where the edged and trimmed pieces are sorted visually according to grade, species and size (Fig. L11).

While automatic machinery handles much of the sawmill process, plants still depend heavily upon the judgment of skilled workmen at each step during the manufacture. Full utilization of each log depends on the knowledge and skill of the *sawyer, edgerman* and *trimmerman* working as a team.

From the green chain, the lumber may be sent to the seasoning yard or dry kiln (Fig. L12), or to remanufacturing plants for resawing, dressing, ripping, planing or other treatment.

Seasoning Practice

Lumber seasoning is the process of reducing the moisture content of the wood. Kiln drying and air drying are the methods used. In either case, the advantages to be gained over unseasoned, or *green,* lumber include: increased strength, reduction in shrinkage and checking and warping in service, less susceptibility to fungus attack in storage, and improvement in the capacity to receive pressure preservative treatment. Kiln drying can further reduce the moisture content to almost any desired value, can result in even greater reduction in weight, hastens the drying time over that required in air seasoning, and kills stain, decay fungi or insects present in the wood. Often, in the case of hardwoods, air drying in outside yards is combined with kiln drying. Following seasoning, the lumber is sent to the planing mill for sizing and surfacing, after which it is graded, tallied and prepared for shipment.

Under the American Softwood Lumber Standard (Product Standard PS 20-70), adopted in 1970, mills can fabricate dimension lumber in either the green or seasoned state provided that green lumber is surfaced to slightly larger sizes, to compensate for eventual shrinkage (Fig. L17b). The Standard also requires that lumber up to, but not including, 5″ thick be grademarked S-DRY if it was surfaced at a moisture content of 19% or less; if it was surfaced at a higher moisture content it must be marked S-GRN. Although not required by the Standard, lumber milled at 15% m.c. is identified in most species

FIG. L13 Stacks of hardwood lumber stand air-drying in yard of a Mississippi River sawmill.

FIG. L14 The planing mill cuts and surfaces lumber to "dressed sizes." Shown here is a planer.

FIG. L15 Lumber is loaded aboard freight cars for shipment. Indoor rail siding keeps it under cover.

as MC15; in grading Southern pine, the term KD (kiln-dried) is used in place of MC15.

Kiln-Dried Lumber In the dry kiln (Fig. L12), heat, humidity and air circulation are controlled to reduce the moisture content to any practical percentage. With few exceptions, the appearance grades of lumber intended for exterior and interior finish are kiln dried, and certain mills kiln dry dimension and other yard grades. Specifying *kiln-dried* lumber does not necessarily assure that it is *dry,* since drying may be terminated in the kiln at any time. The moisture content desired should always accompany the words "kiln dried" unless the moisture content established by the grading rules is suitable for the use intended. This is usually 6% to 12% in the finish grades of softwoods and hardwoods, and 15% or 19% maximum in the common yard grades of softwoods.

Air-Dried Lumber This method dries lumber by evaporation when exposed to the atmosphere, either outdoors or in unheated storage. Ordinarily, dimension and lower grades of softwood lumber used for framing are sent to a yard for air drying but, in some cases are shipped unseasoned. Structural timbers generally are not held long enough to be seasoned, though some drying may take place between sawing and time of shipment.

The moisture content of thoroughly air-dry lumber averages from 12% to 15% in the United States as a whole, varying widely from as low as 6% in summer in the arid Southwest to 24% in winter in the Pacific Northwest. Grading rules governing the various softwood species establish maximum moisture content for lumber grades specified to be seasoned, whether by air drying or kiln drying.

Hardwoods are commonly classified according to grade and size at the time of sawing, with all stock then sent to the air-drying yard (Fig. L13). After air drying, it may be kiln dried at the mill or shipped to a remanufacturing plant where it is kiln dried before being produced into flooring, interior finish, cabinet work or other finished products.

Unseasoned Lumber Lumber exceeding 2″ in thickness is rarely seasoned at the mill. Lumber 3″ and 4″ thick requires longer periods of air drying to achieve 19% m.c.; in most species, except Southern pine, such thicker lumber develops too many defects (end splitting, checking, etc.) when it is kiln-dried to the "dry" moisture content. This thicker lumber therefore is fabricated in the green state and generally is not available in the retail lumber yard with the S-DRY grademark. However, such lumber frequently achieves an acceptable moisture content of 19% or less after air drying at the lumber yard. In the absence of a grademark, a measured check can be used: if the piece reasonably approximates the thickness and width requirements of the Standard for dry lumber (Fig. L17b), it is likely to have an acceptable moisture content. An electric moisture meter, of course, provides the most reliable estimate of moisture content.

The Planing Mill

After seasoning, the lumber goes to the planing mill, where it is surfaced and worked to patterns. Equipment varies from a few machines for surfacing and resawing in some mills to complete fabricating operations in others (Fig. L14). Dried, rough-sawed products are converted into boards, dimension, flooring, ceiling, partition, siding, casing and other finished stock. After going through the planing mill the lumber may be removed to storage sheds to await shipment, but most of it usually is moved directly to cars and loaded for immediate shipment to a wholesale or retail yard or to a woodworking and industrial plant (Fig. L15).

Reforestation

The lumber industry has had to adapt production to its source of supply—trees. Raw material requirements must be geared to the volume of timber which the forests can grow. This has been accomplished by greater wood research and utilization, coupled with the application of forest conservation principles such as tree farming.

There are two basic approaches to timber harvesting and management: *clearcutting* and *selective cutting*. Clearcutting means harvesting all trees in a given area of the forest at one time. The lumber industry prefers the clearcutting practice for many species because it is more efficient and results in quicker regeneration. Conservationists, on the other hand, charge that this method is unsightly, causes erosion and destroys wildlife habitats. Selective cutting implies taking only the older, slower growing and defective trees, leaving behind enough trees for natural reseeding, wildlife shelter and erosion control. This country's forests are a national resource and it is essential that scientific harvesting methods be applied which take into consideration all environmental factors. This is particularly important in U.S. National Forests, where 25% to 30% of all softwood timber is harvested and where other natural values must be preserved for the public benefit.

Foresters now are able to grow better trees and to grow them faster than can nature unassisted by man. As soon as a stand of trees is clearcut, forestry crews start planting new trees on the same areas as soon as possible. In selectively cut areas, trees are removed in such a pattern that remaining or adjacent trees will cast their seed onto the cut-over area to start the next tree crop. Nature does not always provide a good *seed year* every year, thus requiring foresters to plant seedlings by hand or by machine or cast seed over the area to start the new crop (Fig. L16).

LUMBER MARKETING AND MEASURE

Most lumber is marketed and sold by a standard unit of measurement termed the *board foot.* Some products such as moldings, which have been machined further, are sold by the *lineal foot (running foot).* Wood shingles are packaged in bundles and sold according to the surface area they are intended to cover. Posts, poles and pilings are sold by the piece or the lineal foot. Material less than 1/4″ thick is usually classified as veneer.

FIG. L16 A dense new forest covers area "clear-cut" and replanted 20 years ago. Forest will be ready for reharvesting in 60 years.

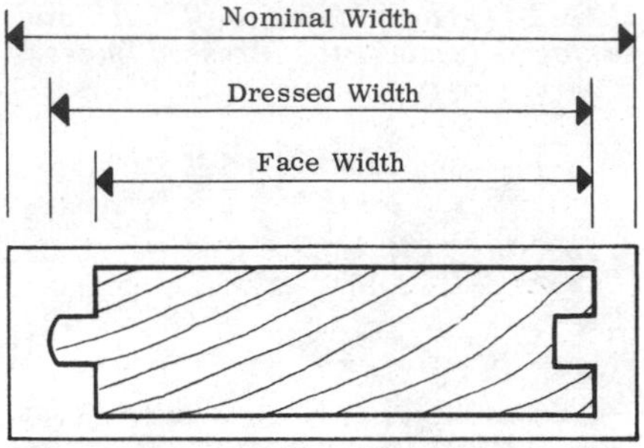

Nominal Size of lumber is its unfinished size, known commercially, as 1x6, 2x4, etc. Seasoning and planing bring it to its Dressed Size. The Face Size is that portion of piece exposed to view when in place.

FIG. L17a Nominal, dressed and face sizes of worked lumber.

FIG. L17b **AMERICAN STANDARD SIZES OF GREEN AND DRY SOFTWOOD YARD LUMBER**

Nominal Thickness or Width	Minimum Dressed Thickness or Width: Green (at over 19% m.c.)*	Minimum Dressed Thickness or Width: Dry (at 19% m.c. or less)
Boards		
1″	25/32″	3/4″
Dimension		
2″	1-9/16″	1-1/2″
3″	2-9/16″	2-1/2″
4″	3-9/16″	3-1/2″
Timbers		
5″	4-5/8″	4-1/2″
6″	5-5/8″	5-1/2″
8″	7-1/2″	7-1/4″
10″	9-1/2″	9-1/4″
12″	11-1/2″	11-1/4″
14″	13-1/2″	13-1/4″
16″	15-1/2″	15-1/4″

*Slightly different sizes apply to redwood, Western red cedar and Northern white cedar.

Sizes

Lumber is bought and sold by its nominal dimensions. The *nominal size* of lumber is its rough unfinished size, sold commercially as 1 x 6, 2 x 4, 4 x 8, etc. (Fig. L17a). The *dressed size* falls short of the nominal dimensions as a result of seasoning and surfacing. In worked lumber (Fig. L30) the *face* is that portion of the piece which will be exposed to view after it is installed (Fig. L17a).

Under the provisions of the American Softwood Lumber Standard PS 20-70, the *dressed* size of dimension lumber less than 5″ thick is related to the moisture content of the wood at the time of manufacture. Thus, lumber which is fabricated at a moisture content of over 19% (green) must be dressed to slightly larger dimensions (Fig. L17b) than lumber fabricated at a moisture content of 19% or less (dry). The objective of this requirement is to insure that both green and dry lumber achieve approximately the same *actual* dimensions (and strength) after air drying in service.

Softwood lumber usually is sold in multiples of two-foot lengths, ranging from 6′ to 24′, although many mills and lumber yards will precut members to exact dimensions to meet the demands of preassembled component construction and other special customer requirements. Hardwood lumber usually is sold by odd- and even-foot lengths or in random lengths for remanufacture into other products.

Board Measure

Foot board measure *(fbm)* is used to indicate quantities of lumber. A board foot is the amount of lumber in a piece with nominal dimensions of 1″ thickness, 12″ width and 1′0″ length. A 1 x 12 board that is 10′ long has a foot board measure of 10. (Its actual dimensions are approximately 3/4″ x 11-1/4″ x 10′.) To calculate the board feet in a piece of lumber, multiply the *thickness in*

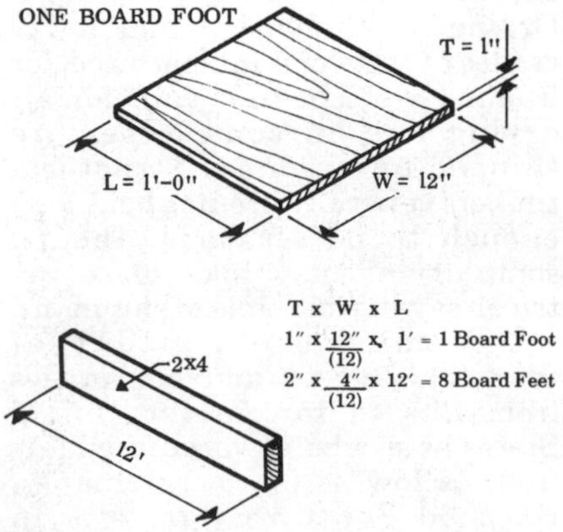

FIG. L18 Board Foot Measurement

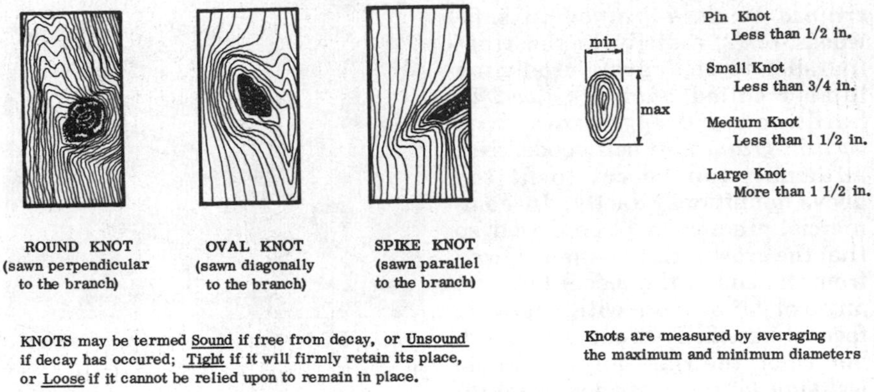

FIG. L19 Knots are limbs embedded in the tree which have been cut through during manufacture.

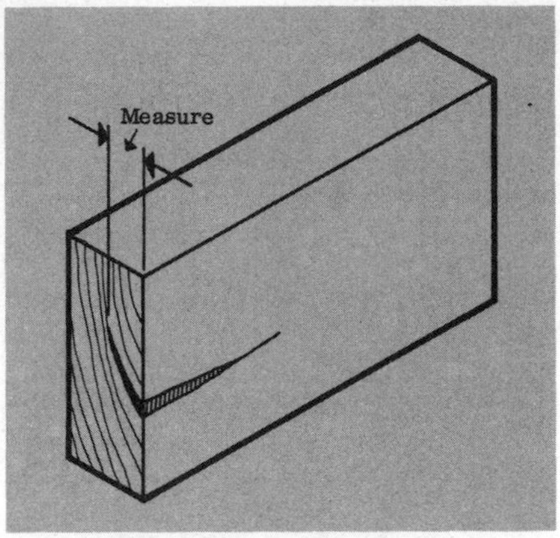

FIG. L20 Shake is a lengthwise separation between, not across, the annual rings.

inches by the *width in feet* by the *length in feet* (nominal dimensions). Since the width of a board is usually expressed in inches, it is necessary to divide by 12 to convert inches to feet (see example, Fig. L18). If the nominal thickness is less than 1″, the number of board feet equals the *width in feet* multiplied by the *length in feet.*

LUMBER QUALITY

As cut from the log, wood varies widely in quality. This results either from its natural growth characteristics or the manufacturing process. The product of the log is divided into commercial lumber grades, each having a relatively narrow range of quality.

In terms of lumber quality, or number of undesirable growth characteristics, lumber grades can be classed as *common* or *select.* Common grades are suitable for general construction purposes, whereas select grades, as the name implies, are selected to provide a surface suitable for painting or natural finishes. The bulk of yard lumber is graded as common lumber, with only boards (up to 1-1/2″ thick, 2″ and wider) available in select grades. Lumber over 1-1/2″ thick also may be graded as select but is used for fabrication into millwork and generally is not stocked as yard lumber. The word "select" in Select Structural and Select Decking does not identify a select grade—all timbers and dimension lumber grades are classed as common grades.

Quality-reducing characteristics are faults which detract from appearance, strength or general utility of the piece in the use for which it is intended. The definition of such faults is often difficult in grading because a slight or small defect in *common* lumber is often a medium or large defect in *select (finish)* lumber. In *select* lumber the defects are light, small or serious depending on the size of the piece or as they come into combination with other defects. Most of the natural and manufacturing characteristics having an effect on the quality of lumber as it is graded are briefly described below.

Natural Characteristics

Some of the characteristics that affect the appearance, grading or use of wood include knots, shakes, pitch and pitch pockets, decay, molds and stains, and insect damage. Reference should be made to the grading rules covering each species for the specific limitations of these characteristics in a given grade.

Knots *Contrary to common notion, lumber containing knots is not seriously defective for ordinary use.* Knots may be undesirable because of appearance in *select* grades, but they are characteristic of all *common* grades. (Fig. L19). Knots are restricted by size and location in grading as they do depreciate the strength value of wood used for structural purposes. The strength reduction is a result of the combined effects of the local cross-grain, produced by knots interrupting the direction of the grain, plus the checking that may develop in and around the knots during drying. On the contrary, strength reduction is not caused by the inherent inferiority of the material comprising the knots. Knots in lumber are more accurately defined as strength-reducing characteristics rather than as defects.

Shakes The effect of shakes (Fig. L20) on the end-use of the wood is largely one of appearance, since they affect only slightly the compressive and tensile strength of members. Shakes reduce the shear resistance of wood used for beams, hence limitations are imposed during grading.

Pitch Pockets The effect of pitch pockets on strength depends on their number, size and location in the piece; their effect on strength has often been overestimated. Certain restrictions are imposed in the lumber grading rules.

Decay The presence of decomposed or destroyed wood as a result of decay is restricted by grading rules in structural grades of lumber to specific proportions of the cross-section.

Molds and Stains These characteristics, even when they cannot be removed by surfacing, are quality-reducing factors only for appearance grades.

Insect Damage The effect on strength of pinholes, grub holes

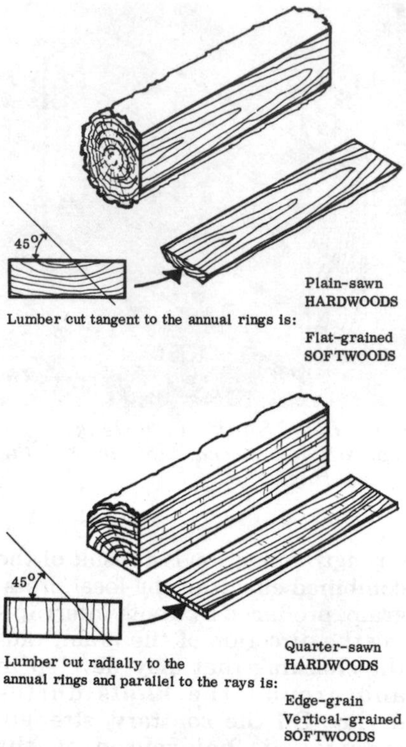

FIG. L21 Lumber sawed so that the annual rings form an angle of 45° or less with the wide faces of the piece is called "plain sawed" or "flat grain." The terms "quarter sawed," "edge-grained" or "vertical-grained" describe lumber in which the annual rings are at an angle of 45° or more.

and other insect damage is taken into account during grading. The strength characteristics of such lumber are acceptable when properly graded. Such insect damage is of primary concern in the appearance grades.

Manufacturing Characteristics

Some characteristics that affect the appearance, grading or use of lumber are the result of the manufacturing process. Manufacturing faults may be present due to sawing practice or finishing operations at either the saw or planing mill, or they may be caused during seasoning.

Effects of Sawing Theoretically, lumber is cut from a log in two distinct ways (Fig. L21): (1) tangentially to the annual growth rings, producing lumber called *plain-sawed* in hardwoods and *flat-grained* or *slash-grained* in softwoods; or (2) radially to the rings (parallel to the rays), producing lumber called *quarter-sawed* in hardwood and *edge-grained* or *vertical-grained* in softwoods. Not all lumber can be cut to fit the above definitions exactly. In commercial practice, a piece sawed so that the growth rings (when viewed from the end of the piece) form an angle of *45° or more* with the wide face is classified as *quarter-sawed,* and when the rings form an angle *less than 45°* it is classified as *plain-sawed* (Fig. L21).

For many uses, including structural applications, plain-sawed or quarter-sawed lumber is equally satisfactory, but there are certain advantages of each. These and other effects of sawing are discussed below.

Plain-sawed lumber generally is less expensive to produce because less labor and waste are involved; certain figure patterns involving growth rings and other figure sources are more conspicuously exhibited; shrinking and swelling are less in *thickness*; knots that may occur are round or oval in shape, generally have a less weakening effect and affect surface appearance less than the spike knot more common in quarter-sawed lumber; and shakes or pitch pockets, when present, extend through fewer pieces.

Quarter-sawed lumber is more costly to produce, but certain figure patterns peculiar to the radial surface are more conspicuously displayed; it shrinks and swells less in *width*; it tends to wear more evenly because the radial surface is more uniform; in some species paint adheres to the radial surface better. It also works smoother for items run to patterns or profiles, it twists and cups less and has less tendency to develop surface checks and splits in seasoning and in use.

Cross-grain results when the fibers in a piece of wood are not parallel to the edges. One type of cross-grain, called *diagonal grain,* results from sawing the log at an angle not parallel with the bark. If the center of the log is lined up with the center of the headrig carriage, the first few cuts do not cut

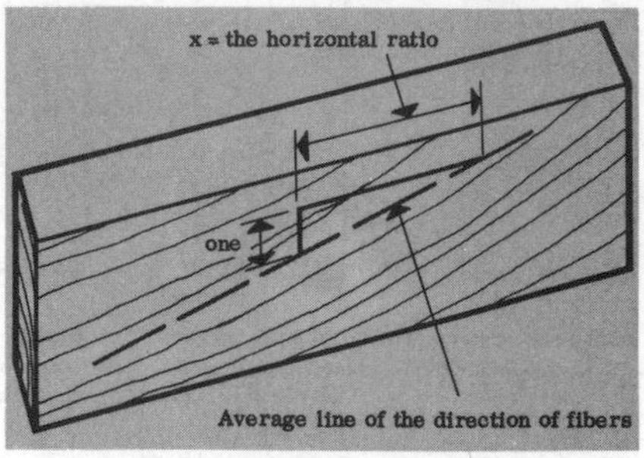

FIG. L22 Slope of grain affects strength.

the full length of the log because most logs are tapered and cross-grain may result. Cross-grain may cause warpage, end-shrinkage, rough faces, torn or lifted grain and weaker lumber.

Like knots, cross-grain has a depreciating influence on the strength of lumber due to the deviation in grain alignment. But unlike knots which are of a highly localized nature, cross-grain may involve the entire piece. The grading rules control cross-grain in structural grades by a measurement of the slope-of-grain. Figure L22 illustrates how it is measured.

Effects of Surfacing Occasionally, imperfections occur in lumber that are caused during surfacing and finishing operations. Depending on their size and location in the piece, the imperfections or blemishes resulting from manufacture defined below are limited in the grading rules so that they will

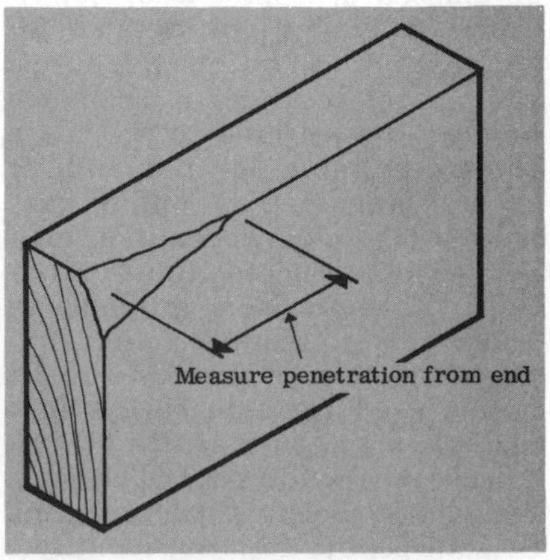

FIG. L23 Wane is the presence of bark or absence of wood.

have no more effect on the utility of the piece than other characteristics permitted in the same grade.

Torn Grain is a roughened area caused by the machine tearing out bits of the wood in dressing.

Skip is an area on a piece that fails to surface smoothly as it passes through the planer.

Mismatched Lumber results when adjoining pieces in worked or patterned lumber do not fit tightly or will not align in the same plane.

Wane is the presence of bark or the absence of wood from any cause on the edge or corner of a piece. It is restricted in grading rules because of such limitations as bearing area, nailing edge and appearance rather than direct effect on strength (Fig. L23).

Machine Burn is the darkening of the wood due to overheating by machine knives or rolls when pieces are stopped in the machine.

Effects of Seasoning The lowering in quality of lumber due to seasoning (called de-grade) is, in general, a result of unequal shrinkage which includes surface and end checking, loosening of knots and various types of warp. Seasoning de-grade can be largely eliminated by proper practice in either kiln drying or air drying, not only at the point of manufacture but also in retail yard storage and at the construction site. Lumber dried too rapidly will check and split; if dried too slowly under favorable temperature, stain or decay may develop.

The grading rules specify the amount of seasoning de-grade permitted in any of the various grades, some of which are described and illustrated below.

Warp is any deviation from a true or plane surface and includes: bow, crook, twist and cup or combinations of each (Fig. L24). In all cases it is measured from the point of greatest deviation from a straight line.

Checking is the separation of the wood normally *across the growth rings*, occurring as a surface check if present on only one surface or as a through check if it extends entirely through the piece. Checking is measured by depth and length (Fig. L25). The effect of checks is mostly one of appearance rather than strength. This is particularly true in the select grades of softwoods where large grain separations are objectionable. In structural lumber, checks exert little, if any, effect on tension and compression members. They may reduce shear resistance, and limitations in the grading rules are based on this reduction.

Splits are lengthwise grain separations extending through the piece from one surface to the other. They are measured as the penetration of the split from the end of the piece and parallel to the edge of the piece (Fig. L26).

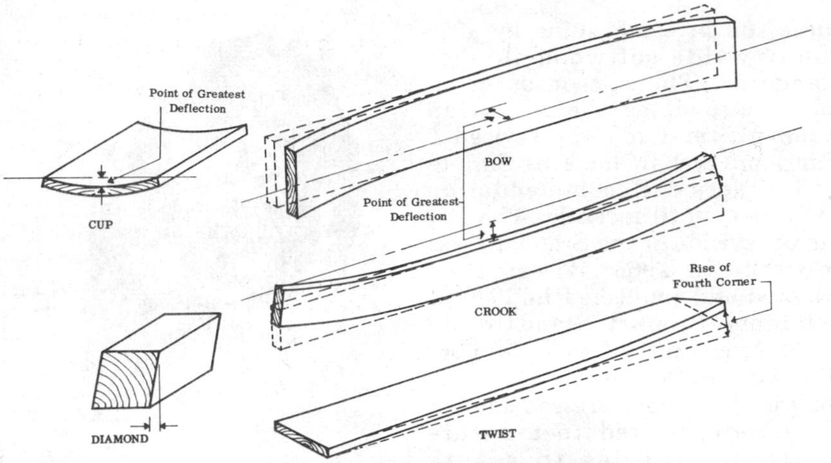

FIG. L24 The term "warp" covers several forms of distortion.

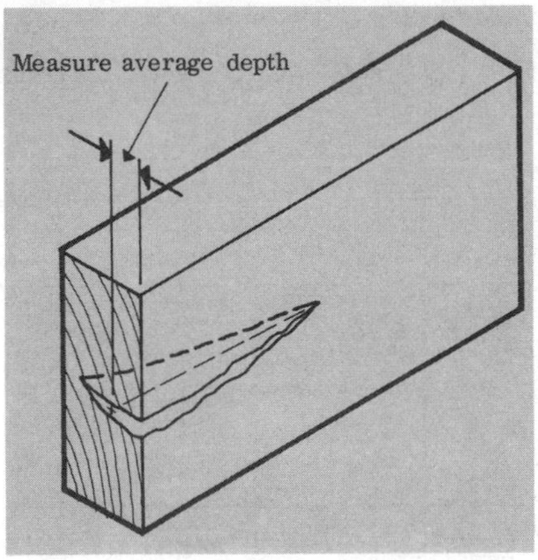

FIG. L25 Checks are separations across the grain due to seasoning.

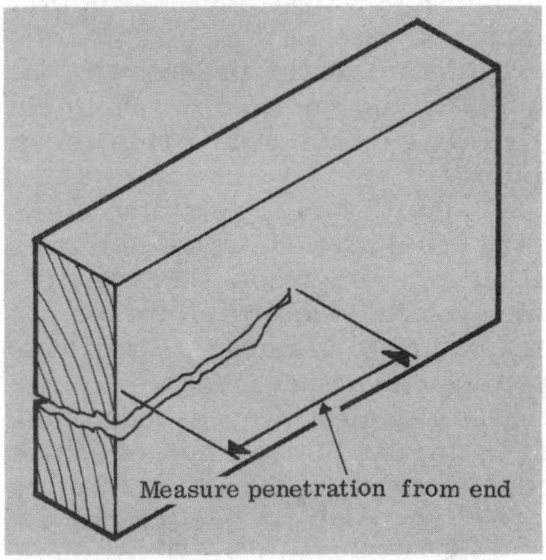

FIG. L26 Splits are lengthwise separations through the entire piece.

SOFTWOOD LUMBER

Sawmilling in the early days was entirely a local industry without need for definite standards of size and quality; manufacturers produced lumber to local customer requirements. As the lumber industry and transportation systems expanded, however, markets were no longer local. Lumber of various species originating in widely separated localities produced by different mills could not be sold and used successfully without common standards of quality.

Around 1900, a number of manufacturers' associations were formed, largely for the purpose of formulating grading rules for the various species manufactured by the producers in their region. While grading rules had much in common, they differed in detail; however, industry standardization conferences, begun in 1919, were successful in establishing by 1924 countrywide softwood lumber standards. The consensus document, known as the American Lumber Standard, was revised 5 times and was in force as Simplified Practice Recommendation SPR 16-53 until recently. The last major revision of the Standard was initiated in 1965; it was then adopted and published in 1970 as voluntary Product Standard PS 20-70, American Softwood Lumber Standard. Under the new standard, for the first time, dressed lumber sizes were related to moisture content, in order to assure reliability of actual in-service dimensions and strengths. Another major advance of PS 20-70 was the establishment of procedures for adoption of an uniform National Grading Rule for Dimension Lumber.

It should be understood, however, that the National Grading Rule for Dimension Lumber does not cover Decking Lumber, so these grade names are not the same for all associations.

"Dimension," as used in the Rule means lumber 2″ to 4″ thick and intended for use as framing members such as joists, planks, rafters, studs and small timbers. Under the new Rule, all grading authorities agree to apply the same grade names and descriptions to lumber of similar quality, regardless of species (Fig. L29). However the Rule does not restrict grading authorities from including *additional* grades of special products in their own grading rules; the Southern Pine Inspection Bureau for instance grades—and its members market—many grades not covered by the National Grading Rule. Nevertheless, the standardization adopted in 1970 considerably simplified grade recognition and the assignment of working stresses for dimension lumber.

Major Producing Regions

Three major regions produce the majority of softwood used in construction. The lumber species, approximate yearly volume of production and the association that establishes and supervises the lumber grading practices for the different species are shown for each region in Figure L27.

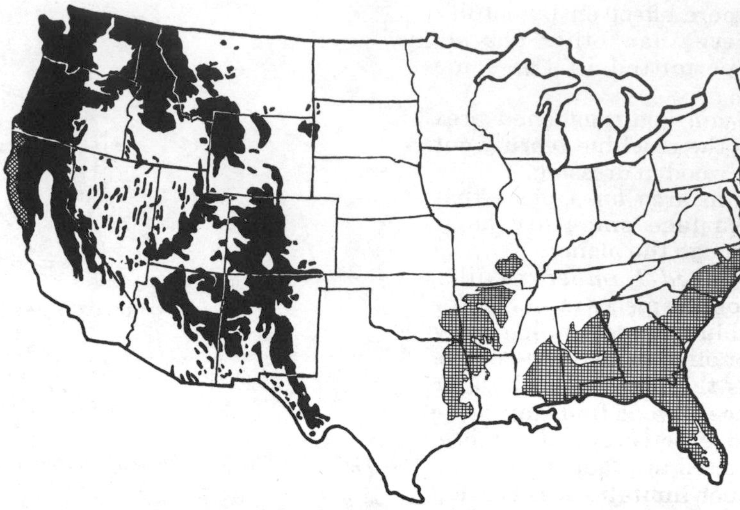

FIG. L27 **MAJOR SOFTWOOD LUMBER PRODUCING REGIONS OF THE U.S.**

KEY	REGION	SPECIES	APPROXIMATE YEARLY PRODUCTION	LUMBER ASSOCIATION	GRADING AUTHORITY
	WESTERN WOOD REGION	Douglas fir Ponderosa pine Western red cedar Incense cedar Western hemlock White fir Engelmann spruce Sitka spruce Western larch Lodgepole pine Idaho white pine Sugar pine	16.9 billion board feet	Western Wood Products Association	Western Wood Products Association West Coast Lumber Inspection Bureau
	REDWOOD	Redwood Douglas fir	2.7 billion board feet	California Redwood Association	Redwood Inspection Service
	SOUTHERN PINE REGION	Longleaf pine Shortleaf pine	6.7 billion board feet	Southern Forest Products Association	Southern Pine Inspection Bureau

Western Woods Region The western woods region is the largest of the producing areas, comprising the thirteen states of Oregon, Washington, California, Idaho, Arizona, New Mexico, Texas, Colorado, Montana, South Dakota, Nevada, Utah and Wyoming. This region produces lumber cut from ponderosa pine, Douglas fir, Engelmann spruce, Idaho white pine, incense cedar, lodgepole pine, sugar pine, Sitka spruce, western hemlock, western larch, western redcedar and white fir.

Redwood Region The redwood region is the smallest in area, encompassing but 12 northern California counties. Approximately half of the total production comprises lumber cut from redwood, the remainder being cut from other softwood species that grow in the region.

Southern Pine Region The southern pine region covers the entire South, from Texas to Virginia, with some production in Maryland and Missouri. This region produces lumber cut from shortleaf, loblolly, longleaf, slash, Virginia pine and pond pine. Commercially, these are separated and sold in two classifications called shortleaf pine (the greatest production) and longleaf pine.

STANDARDS OF QUALITY

The American Softwood Lumber Standard PS 20-70 does not

contain detailed specifications for selection or specifying of lumber; it does, however, provide the basic principles, provisions and procedures for establishing the commercial grading rules for various species.

The commercial grading rules for each region not only conform to PS 20-70 and the National Grading Rule, but also reflect the properties of individual species and recognize the specific qualities of lumber desired by users. The grading rules are detailed specifications used to facilitate the manufacture, specifying, sale and use of lumber. Their basic aim is to separate into marketable and usable classifications the wide range of lumber qualities obtained from the log. Grading is based mainly upon the quality-reducing characteristics that affect appearance, strength, durability or utility of wood.

Commercial grading rules are formulated and maintained by regional manufacturers' associations (Fig. L27), covering the species and individual products manufactured in a given producing region. These agencies publish grading rules for the species produced in their regions and maintain staffs of qualified lumber inspectors who supervise grading at the mills.

Mills which manufacture lumber according to the American Softwood Lumber Standard usually stamp each piece of dimension lumber with an appropriate grademark. Examples of official grademarks of the four chief grading authorities are shown in Figure L28. American grading authorities operate under the control of an industry group established by PS 20-70, the American Lumber Standards Committee. A comparable group, the Canadian Lumber Standards Administrative Board, authorizes grading agencies in the Canadian lumber region. Canadian lumber standards and grade names are essentially the same as those in the U.S.; hence, Canadian lumber, graded and stamped by an authorized agency is compatible with PS 20-70 and American standard engineering practice.

FIG. L28 TYPICAL GRADEMARKS FOR DIMENSION LUMBER

MILL 10 WCLB® CONST DOUG FIR S-DRY	**WEST COAST LUMBER INSPECTION BUREAU** Construction grade light framing Douglas fir maximum 19 % m.c. Mill No. 10
12 SEL DECK WWP® MC 15 INC CDR	**WESTERN WOOD PRODUCTS ASSOCIATION** Select grade decking Incense cedar maximum 15 % m.c. Mill No. 12
SPIB· No.1 KD 1850f (7)	**SOUTHERN PINE INSPECTION BUREAU** No. 1 grade light framing 1850 psi—allowable max. bending stress Mill No. 7
(50) STUD S-GRN REDWOOD RIS®	**REDWOOD INSPECTION SERVICE** Stud grade light framing Redwood over 19 % m.c. Mill No. 50

Softwood Lumber Classifications

Softwood lumber is classified by use, by size, by manufacture and by grade. A knowledge of these classifications is important in selecting and using lumber (Fig. L29).

Use The use classifications of softwood lumber include: yard lumber, and factory & shop lumber.

Yard Lumber consists of grades, sizes and patterns generally intended for ordinary construction and general building purposes. Yard lumber can be subdivided into three major categories by size: *boards, dimension lumber* and *timbers* (Fig. L29). Practically all grades in the timbers and dimension categories have working stresses assigned to them and such lumber is termed *stress grade (structural)* lumber. Working stresses (allowable unit stresses) for several important commercial species are given in Figure W14 and in the Work File. Complete tabulations for all commercially available species and grades are published by the National Forest Products Association as a Supplement to its National Design Specification (NDS) for Wood Construction. NDS constitutes the industry-recognized standard of proper engineering practice for structural wood design.

Factory and Shop Lumber is primarily for remanufacturing purposes. Since the lumber is intended for further manufacture into windows, doors, millwork and similar items, the grades are based on the amount of usable clear wood that can be cut from the piece.

Size The size classifications of softwood lumber include: boards, dimension and timbers.

Boards are pieces 1″ to 1-1/2″ thick and 2″ or more wide. They include sheathing, subflooring, roofing, finish, trim, siding and paneling. Boards less than 6″ wide may be classified as strips.

Dimension refers to pieces from 2″ thick up to, but not including, 5″ thick and 2″ or more wide. Included in this classification are joists, planks, rafters, studs and other members. These grades are intended principally for the structural elements of residences, and the chief criteria are that they adequately do the job intended for a stud, floor joist, roof rafter or other framing members.

Dimension lumber consists of *light framing, joists and planks,* and *decking.* Almost all dimension and timber lumber is *stress-graded,* with working stresses assigned to each grade. Such lumber, intended for load-bearing conditions, also is termed *structural.* Dimension lumber constitutes the bulk of common lumber.

Timbers are pieces 5″ or more in the least dimension. The timbers classification in some species is further made up of *Beams & Stringers* and *Posts & Timbers* (Fig. L29). Lumber thicker than 5″ in either face seldom is used in ordinary construction and, when stocked by a retail yard, is frequently in a green, rough (undressed) condition. When using heavy timbers, shrinkage of the assembly should be expected, and the design of the structure should be planned to minimize the effects of shrinkage. Laminated timber, made by gluing up smaller pieces that can readily be kiln dried, should be used when seasoned timbers are mandatory.

Manufacture The extent to which lumber is processed is covered in this classification and includes rough lumber, dressed lumber and worked lumber.

Rough Lumber is not dressed (surfaced) but is sawed, edged and trimmed at least to the extent of showing saw marks in the wood on four longitudinal surfaces of each piece for its overall length.

Dressed (Surfaced) Lumber is surfaced by a planing machine to attain a smooth surface and uniform size. It is designated as follows: when surfaced on one side (S1S), two sides (S2S), one edge (S1E), two edges (S2E), or a combination of sides and edges (S1S, S1S2E, S2S1E or S4S).

FIG. L29 **CLASSIFICATION OF SOFTWOOD LUMBER**

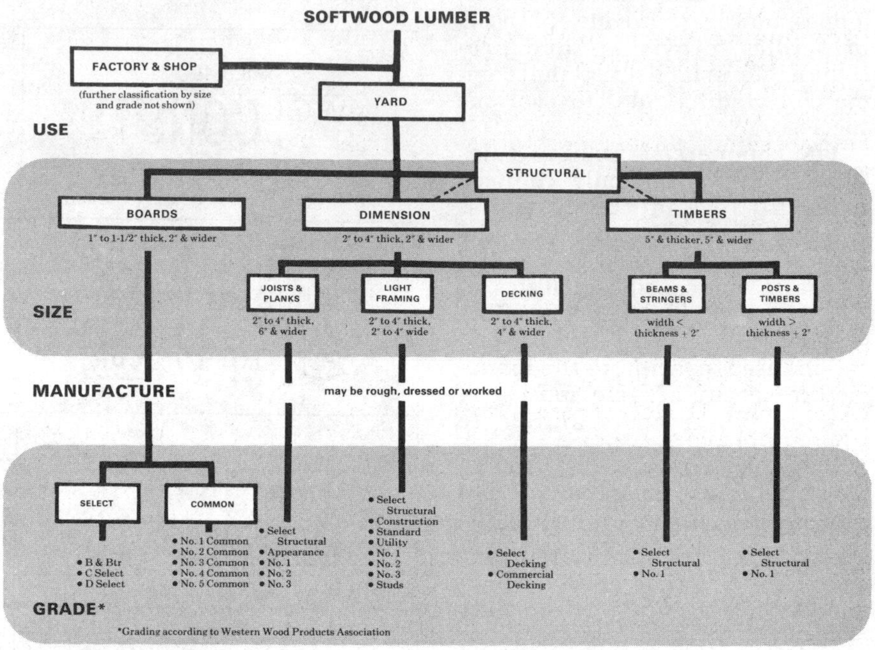

Worked Lumber, in addition to being dressed, is matched, shiplapped or patterned, or machined by running through a matching machine or molder.

Matched lumber has been worked with a tongue on one edge of each piece and a groove on the opposite edge to provide a close tongue-and-groove joint by fitting two pieces together. For end-matching, the tongue and groove are worked in the ends (Fig. L30). *Ship-lapped* lumber has been rabbeted on both edges of each piece to provide a close-lapped joint by fitting two pieces together (Fig. L30). *Patterned* lumber is shaped to a pattern or to a molded form—in addition to being dressed, matched or ship-lapped—and includes siding and other millwork.

Grades One method of classifying softwood lumber is into select

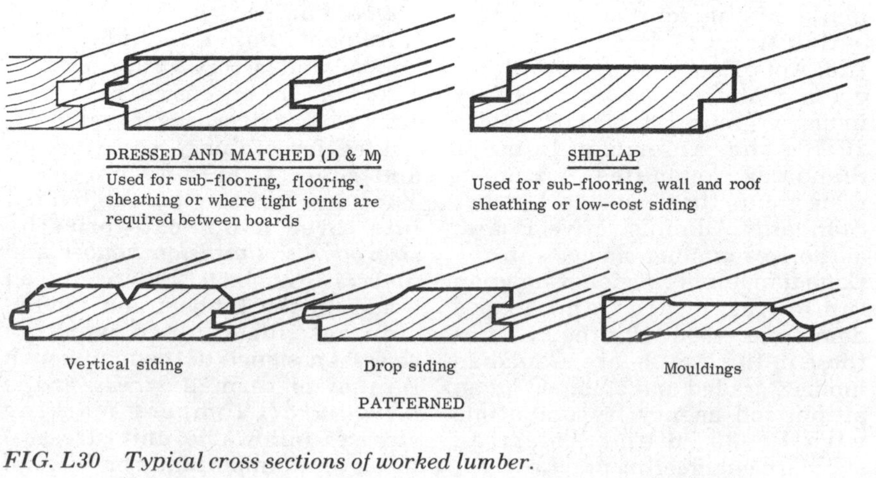

FIG. L30 Typical cross sections of worked lumber.

(finish) grades and common grades. *Select grades* have good appearance and are used where a natural or high quality paint finish is required. Generally, only boards are available in select grades as yard lumber; various worked products such as siding, paneling and flooring are manufactured usually in select grades. In most species the two highest grades, A and B, are combined and sold as "B and Better" (B&B or B&Btr) which is used for the highest quality of interior and exterior finish, trim, moldings, paneling, flooring and siding (Fig. L31). Grades C&Btr, C Select and D Select are more economical and serve many finishing purposes, where cost savings are more important than perfect appearance. *Common grades* are suitable for general construction and utility purposes.

The various grades of softwood lumber are grouped according to size and intended use into several categories, as follows:

Light framing is lumber 2″ to 4″ thick and 2″ to 4″ wide, and consists of several categories of products. Ordinary light framing grades (Construction, Standard and Utility) are intended for uses not requiring high strength, such as studs, plates and blocking. *Structural light framing* grades (Select Structural, No. 1, No. 2 and No. 3) are designed to fit engineering applications where higher bending strength ratios are needed, such as in trusses and in box beams. *Studs* (2″ to 4″ thick, 2″ to 4″ wide) are manufactured in lengths 10′ and shorter primarily for use in walls. Strength in bending is comparable to No. 3 Structural Light Framing lumber.

Structural Joists & Planks are lumber of rectangular cross section 2″ to 4″ thick and 6″ or wider, graded with respect to strength in bending when loaded either on a narrow face of a joist or on the wide face of a plank. The grades are the same as in Structural Light Framing: Select Structural, No. 1, No. 2 and No. 3.

Appearance Framing (Appearance grade) combines high strength with good appearance suitable for exposed uses. It is manufactured 2″ to 4″ thick, 2″ and wider—up to 14″ in width. This grade therefore can be used either as light framing or as joists & planks.

Decking is lumber 2″ to 4″ thick, 4″ and wider intended for exposed view as planking in ceilings and roofs. The moisture content usually is limited to 15% and the lumber is worked to a T & G pattern. It may be ordered with V or rounded edges, in striated or grooved patterns on the face. Typical grades are Select and Commercial.

Timbers which are nearly square (where the width does not exceed the thickness by more than 2″) are generally intended for use as posts and have slightly lower allowable bending strengths than rectangular beams intended for use mainly as beams. Typical grades are Select Structural and No. 1.

Most lumber is graded visually by trained inspectors. In stress-grade lumber limitations are imposed on slope-of-grain, knot placement, size and number of knots. Some structural grades in some species (for example, Douglas fir and southern pine) have additional grade requirements based on density and growth ring count.

Although most dimension lumber is graded visually, some lumber 2″ or less in thickness is now being machine stress-rated at the sawmill, allowing manufacturers to standardize and guarantee the load-bearing capacity of each piece. The more refined method of selection permits assignment of higher stresses for much of the lumber in a grade which had been subjected to visual inspection.

In machine rating, lumber is fed through a machine (Fig. L32) which subjects each piece to bending in two directions. Modulus of elasticity, "E" (measure of stiffness), is measured by instruments at 6″ intervals. The machine then electronically computes the stress grade, taking into account the interacting effect of knots, slope-of-grain, growth rate, density and moisture content. As the piece leaves the machine it is automatically stamped "MACHINE RATED," with an assigned fiber stress in bending ("f" on the grademark) and corresponding modulus of elasticity ("E") rating (Fig. L32).

The close relationship between "E" and "f" has led to the establishment of nine basic categories of stress grades for lumber 2″ or less in thickness (Fig. L33) by the Western Wood Products Association. Lumber in these nine grades can be used for most structural needs, and is especially suited for

FIG. L31 Grade stamp for boards indicating material was surfaced at 15% m.c. and is of C Select (Finish) quality.

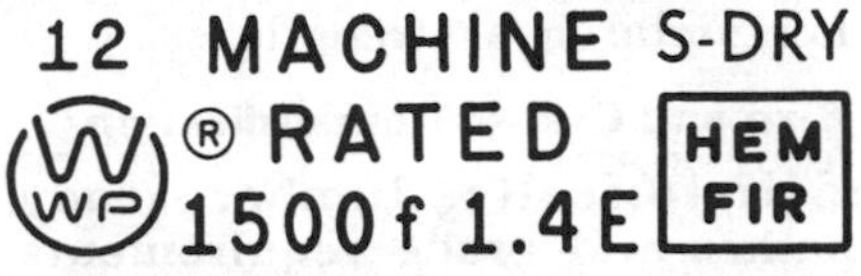

FIG. L32 Machine stress-rated lumber is automatically stamped with identifying grademark.

FIG. L33 **WWPA MACHINE STRESS-RATED (MSR) LUMBER, 2″ THICK**

Designation*	Allowable Stresses Ft (psi)	Fc (psi)
1200f - 1.2E	600	950
1500f - 1.4E	900	1200
1650f - 1.5E	1020	1320
1800f - 1.6E	1175	1450
2100f - 1.8E	1575	1700
2400f - 2.0E	1925	1925
2700f - 2.2E	2150	2150
3000f - 2.4E	2400	2400
3300f - 2.6E	2650	2650
900f - 1.0E	350	725
900f - 1.2E	350	725
1200f - 1.5E	600	950
1350f - 1.8E	750	1075
1800f - 2.1E	1175	1450

*The first number in the grade designation indicates allowable extreme fiber stress in bending (Fb) in psi; the second number is the modulus of elasticity (E) in 1,000,000 psi.

trussed rafters. Another set of five grades has somewhat lower allowable bending stresses in relation to modulus of elasticity. These special grades are intended for joists, where lower bending stresses are acceptable, since stiffness generally is the governing property. Machine stress rated lumber is also visually graded to limit certain characteristics (checks, splits, wane, warp) which might adversely affect horizontal shear strength.

Size and Grade Standardization

Historically, lumber grade names and grade requirements have been developed independently on a regional basis and, even within regions, have varied among some species. Methods of establishing stress values for stress-grade lumber also have differed among lumber producing regions, and changes in structural grading are continuing as the newer methods of machine rating are refined.

Under provisions of the American Softwood Lumber Standard PS 20-70, a National Grading Rule Committee (NGRC) was established for the purpose of developing uniform country-wide grade requirements for dimension lumber. The functions of the NGRC, according to PS 20-70, were to "establish, maintain, and make fully and fairly available grade strength ratios, nomenclature, and descriptions of grades for dimension lumber conforming to American Lumber Standards." The specifications developed by the NGRC are known as the National Grading Rule for Dimension Lumber and constitute a part of all association grading rules.

Individual association grading rules are much more extensive and detailed than the National Grading Rule, which primarily covers nomenclature and bending strength ratios for Light Framing and Joist & Plank grades. Bending strength ratios are standard relationships between the basic stresses of an ideal, *clear-wood,* piece of lumber free of strength-reducing characteristics (knots, sloping grain, splits) and the stresses developed in a particular grade of lumber in that species.

Since lumber shrinks in direct proportion to loss of moisture in drying from its green condition to the moisture content it ultimately attains in service, it is possible to establish, with reasonable tolerances, separate sizes for green and dry surfaced lumber so that both types achieve essentially the same size in service and provide equal strength and stiffness for the same spans and loading conditions.

The Supplement to the National Design Specifications, which lists working stresses for all commercial species and grades, takes into account this slight oversizing of green lumber and allows the same values for green as for dry surfaced lumber provided both are used at a moisture content of 19% or less. Higher values are permitted when lumber is surfaced at 15% m.c. and used at the same moisture content; lower values are assigned to all lumber which will be used under exposed conditions where the moisture content is likely to exceed 19% for extended periods.

HARDWOOD LUMBER

Hardwood accounts for about 25% of the total board foot production of lumber. It comes from many species and has uses ranging from small specialty items to the heaviest structural work. The predominant hardwood species are the oaks and walnut, which constitute over 50% of the total hardwood lumber production.

Areas of Growth

Most of the commercial hardwood species in the United States grow east of the Great Plains (see Fig. W1, page 201-3). The hardwoods of the West, which grow principally in California, Oregon and Washington, amount to about 2% of the total hardwood produced.

Standards of Quality

The National Hardwood Lumber Association (NHLA) publishes size and grading rules of sawed lumber. These rules, however, are used within the lumber industry to describe hardwood lumber used for remanufacture into other products and are not normally used to specify hardwood used in building construction. In the construction industry, hardwood lumber is made to special order for the specific requirements of the purchaser, and the grade descriptions and characteristics are those which are agreed upon by both buyer and seller.

Hardwood Dimension

A large proportion of the hardwood lumber produced at sawmills is used in paneling, furniture, flooring and other finished or semi-finished products. *Dimension stock,* or *hardwood dimension,* produced by hardwood sawmills is shipped to furniture, tool, piano, cabinet, and similar factories.

Hardwood *dimension* lumber is normally kiln dried and processed to a point where maximum utility is delivered to the user. It is manufactured from rough boards and flitches to the specific requirements of a particular plant or industry, in specified thicknesses, widths and lengths. It may be solid or glued-up and is classified as rough dimension, surfaced dimension, semi-fabricated dimension or completely fabricated dimension.

Rough Hardwood Dimension This consists of planks sawed and ripped to certain sizes.

Surfaced and Semi-fabricated This is rough dimension carried one or more steps further, which may include one or more of several operations such as surfacing, molding, tenoning, drum-sanding, equalizing, trimming or mitering. However, such operations do not make the product a completely fabricated one ready for assembly.

Completely Fabricated This is lumber ready for assembly into whatever type of product desired. Hardwood parts are frequently prefabricated at the sawmill site in adjoining hardwood dimension plants. Also, many hardwood flooring plants are operated in conjunction with hardwood sawmills.

The manufacture of hardwood lumber and its by-products is one of the most complex branches of the lumber industry because of the variety of uses for hardwood and the many species involved.

PLYWOOD

Plywood is an engineered panel composed of several wood veneer sheets generally placed at right angles to one another (or with veneer sheets in combination with a lumber or mineral core) and bonded together under high pressure using either a waterproof or water-resistant adhesive.

Here, the manufacture of plywood is described, as are the types, grades and sizes in which it is produced. The general characteristics of plywood and its properties are defined, and the principles governing its application are outlined along with a description of its most common uses in residential construction.

Both *decorative* and *construction* grades of plywood are manufactured. Construction plywood is used generally wherever strength is an important factor but occasionally also for paneling; decorative plywood (chiefly hardwood) is used for wall paneling, cabinet work and furniture.

CONSTRUCTION PLYWOOD

Construction plywood is manufactured primarily of softwoods, but hardwood species also are employed. However a plywood panel need not be composed entirely of the same species. Strength and stiffness requirements for inner plies are less demanding and these frequently are of a different species than the outer plies. Structural strength or stiffness is not significantly changed since the outer faces of a plywood panel provide most of its strength and stiffness lengthwise, which is the normal direction to apply the panel when used structurally.

Manufacture

The manufacturing process begins by cutting *peeler* logs, selected from the lower portion of the tree, into *blocks* about 8-1/2′ long and from 12″ to 50″ or more in diameter (Fig. P1). The blocks are placed in a giant lathe and rotated against a long knife that peels the wood into long, continuous veneer sheets, mainly 1/10″ to 3/16″ thick (Fig. P2). Veneer cut in this manner is called *rotary cut.* Veneer primarily intended for decorative plywood is sometimes sliced rather than peeled from the block (see Fig. P28), but an insignificant amount of softwood plywood is manufactured of sliced veneer.

The veneer is next conveyed to clippers, cut into desired widths, then run through long drying ovens that reduce its moisture content (to between 2% and 6%) to provide the best possible glue bond and panel stability (Fig. P3).

After the veneer is graded it goes to the glue spreaders where rotating cylinders spread glue evenly on both sides in the desired amount. The veneer is then laid-up into panels of *balanced construction* (symmetrical about the center as to thickness and generally as to species Fig. P4). Laid-up panels are inserted into a hydraulic press

FIG. P1 Selected blocks cut from peeler logs from the lower portion of a tree.

FIG. P2 At the lathe, a sharp steel blade peels the block into thin veneer.

FIG. P3 The drier-oven reduces the moisture content of the veneer.

FIG. P4 Veneers are laid-up into panels after passing through a glue spreader for uniform coverage.

FIG. P5 Laid-up panels are finally placed in a press which pressures veneers and glue into plywood panels.

FIG. P6 A careful check is made of panels for grade, appearance, accuracy of dimension and thickness.

where pressure of about 150 psi is exerted to form the veneer and glue into plywood panels (Fig. P5). After removal from the press, panels are trimmed to exact size, run through large drum sanders if they are of a sanded grade, and are finally checked for grade, appearance and accuracy of size and thickness, emerging as finished plywood ready for use (Fig. P6).

Standards of Quality

All construction plywood, except certain specialties, should conform to U. S. Product Standard PS 1-74, Construction and Industrial Plywood. This standard constitutes the recognized grading rule for construction plywood and consists of specifications to maintain a high level of quality and includes procedures for testing, grading, labeling and identification. Product Standard PS 1-74 is a consolidation of the previous Commercial Standards governing the manufacture of softwood plywood. Conformance with the requirements of the Product Standard is indicated by the grade-trademark or other suitable designation of a qualified testing and certifying agency performing the functions outlined in the Product Standard.

About 85% of all construction plywood is produced by mills belonging to the American Plywood Association. This organization maintains a quality control program to assure that plywood carrying an official grade-trademark conforms to the Product Standard. The grade-trademarks are shown in Figure WF9 of the Work File.

Timber Engineering Company (TECO) and Pittsburgh Testing Laboratories also are qualified independent testing agencies for plywood.

Plywood is graded and marked according to group of *species, type, appearance* and/or *application.*

Classification of Species

Approximately 70 species of wood used in plywood have varying strength and stiffness properties. Under Product Standard PS 1-74, these species are classed into five groups according to stiffness and other properties (Fig. P7). The group number shown in the grade stamp (Fig. WF9) refers to species used in face and back layers.

FIG. P7 **CLASSIFICATION OF SPECIES**

Group 1	Group 2	Group 3	Group 4	Group 5
Apitong	Cedar, Port Orford	Alder, Red	Aspen	Basswood
Beech, American	Cypress	Birch, Paper	Bigtooth	Fir, Balsam
Birch	Douglas Fir 2	Cedar, Alaska	Quaking	Poplar, Balsam
Sweet	Fir	Fir, Subalpine	Cativo	
Yellow	California Red	Hemlock, Eastern	Cedar	
Douglas Fir 1	Grand	Maple, Bigleaf	Incense	
Kapur	Noble	Pine	Western Red	
Keruing	Pacific Silver	Jack	Cottonwood	
Larch, Western	White	Lodgepole	Eastern	
Maple, Sugar	Hemlock, Western	Ponderosa	Black (Western Poplar)	
Pine	Lauan	Spruce		
Caribbiean	Almon	Redwood	Pine	
Ocote	Mayapis	Spruce	Eastern White	
Pine, Southern	Red Lauan	Black	Sugar	
Loblolly	White Lauan	Englemann		
Longleaf	Maple, Black	White		
Shortleaf	Mersawa			
Slash	Pine			
Tanoak	Pond			
	Red			
	Virginia			
	Western White			
	Spruce			
	Red			
	Sitka			
	Sweetgum			
	Tamarack			
	Yellow-poplar			

Basic Types

Two basic types of construction plywood are produced to meet PS 1-74—Interior and Exterior, which refer to the exposure capability of the panel. Within each type there are several grade designations based on the quality of the veneers of the panel.

Exterior Type (waterproof) These panels are made with a hot-pressed phenolic resin adhesive that becomes permanent and insoluble. Only "C" grade veneer or better is used throughout the panel. Exterior Type plywood will retain its glue bond when repeatedly wetted and dried or otherwise subjected to the weather and is intended for permanent Exterior exposure.

Interior Type (moisture-resistant) Interior Type plywood is made with "D" grade veneer or better. There are three levels of adhesive durability within the Interior Type classification: 1) Plywood bonded with Interior glue is moisture-resistant, and is intended for all interior applications where it may be temporarily exposed to the elements. 2) Plywood bonded with Intermediate glue has a lower moisture resistance than Exterior glue but greater than Interior glue. The adhesive possesses high-level bacteria, mold and moisture resistance. It is suitable for protected construction where moderate delays in providing protection may be expected or conditions of high humidity and water leakage may exist. 3) Plywood bonded with Exterior glue is intended for protected construction where maximum performance is required for protection against moisture exposure from long construction delays or other conditions of similar severity.

Appearance Grades

Each type is manufactured in several appearance grades. *Grade* designation is based on the species group and the quality of the veneer used on the face and back. Panels are built-up of veneers and are glued together at right angles to each other, using one or more of 5 veneer grades ranging from "N" (highest) through "A," "B," "C" and "D" (lowest). Veneer may be repaired to raise its quality to a higher grade. Repairs consisting of patches or plugs, usually oval or circular, along with long, narrow shims, fill voids in the veneer sheet or replace characteristics such as knots and pitch pockets.

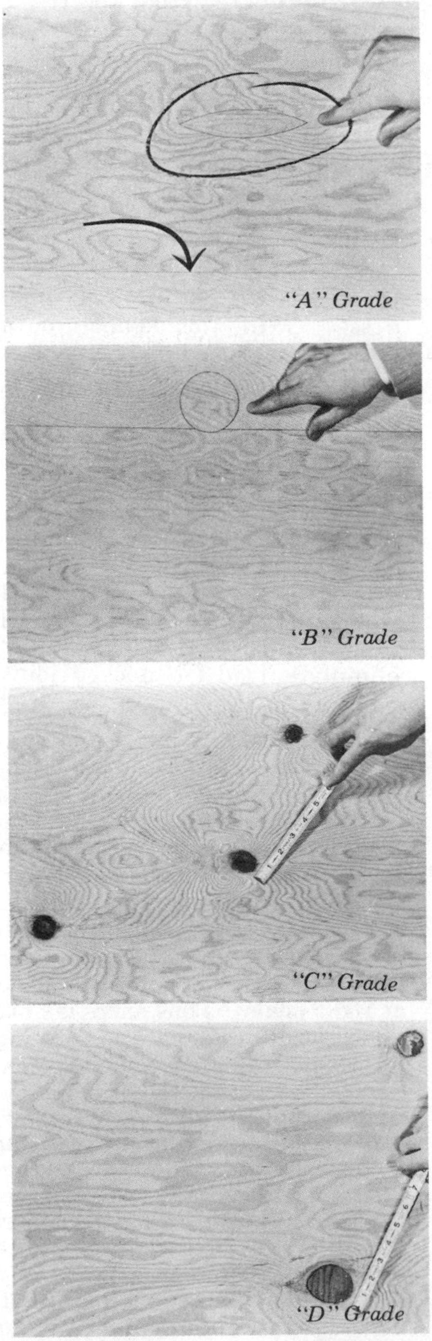

FIG. P8 Plywood veneer grades.

Veneer grades are illustrated in Figure P8 and described below.

"N" Grade. This is a specialized veneer available on special order, used for natural finish. It is select, all heartwood or all sapwood, and free of open defects with only a few small well matched repairs (not illustrated).

"A" Grade. This is a smooth *paint* grade permitting a number of neatly made repairs, plugs or patches with certain restrictions.

"B" Grade. This is a *solid surface* veneer, except for specified minor characteristics. It may have a considerable number of neatly made repairs and is suitable for concrete form work where a smooth surface is desired.

"C" Grade. This is the lowest quality of veneer permitted in any *Exterior Type* panel and is commonly used as the face of sheathing grade panels. Tight knots up to 1-1/2″, 1″ knotholes (occasional 1-1/2″ knotholes under certain conditions), splits up to 1/2″ tapering to a point and other minor defects are permitted.

"D" Grade. This is the lowest quality of veneer permitted in any *Interior Type* panel. This grade permits 2-1/2″ knotholes (occasional 3″ knotholes under certain conditions), pitch pockets 2″ up to 2-1/2″ wide and tapering splits up to 1″.

Appearance grades are listed in table form in the Wood Work File. (See the Plywood Grade-Use Guide.)

The grade of the panel is indicated by two letters, such as "A-D", describing a panel with an "A" face and a "D" back. The inner plies for any Interior Type panel are "D" or better, and "C" or better for an Exterior Type panel. Unless a decorative or other surface is specified, panels in most grades are run through large drum sanders where they are sanded smooth on both sides for appearance and balanced construction.

Engineered Grades

Wall sheathing, roof sheathing, subflooring and underlayment consume the largest volume of plywood in the construction field. Each must provide a suitable base for covering materials—whether shingles, built-up roofing, siding or flooring—and must possess stiffness and strength, nail-holding ability, durability, dimensional stability, and lateral bracing and resistance to puncture and impact.

In most buildings with conventional framing, sheathing and subflooring, no special engineering design is required if recognized standards and construction details are followed. Correct use of plywood involves selecting the appropriate Type, Grade and Identification Index Number for the spans and loads to be carried, in addition to using proper nailing.

Engineered grades are identified according to either application or the quality of the face and back veneer. Three of the engineered grades—C-D Interior, Structural (I and II), and C-C Exterior—carry an Identification Index Number (Fig. WF9). This is a set of two numbers separated by a slash. The first number indicates maximum recommended spacing of supports if the panel is used for roof sheathing. The second number indicates the maximum recommended support spacing if the panel is used for subflooring. No index number is needed for wall sheathing because the bending properties are not critical. Index numbers are based on the species group, thickness, bending strength and stiffness of face and back layers of the panel.

Engineered grades generally are left unsanded for greater stiffness and strength, as well as economy. However, certain grades, such as Underlayment, are either sanded or touch-sanded to bring adjacent panels to a uniform thickness.

Type and grade selection can be made simply and directly from the Grade-Use Guide in the Work File.

Sizes

Plywood is most readily available in panels 4′ wide by 8′ long. Mills also produce 5′ wide panels and lengths up to 12′, but they are not always stocked. Beyond these limits widths up to 10′ and any reasonable length (which has been more than 50′) can be made by *scarf-jointing* or *finger-jointing* panels together. Scarf- or finger-jointed construction consists of gluing panels either end-to-end or side-to-side continuously along a tapered or fingered cut to form one single panel (Fig. P9).

Standard nominal thicknesses for sanded panels are 1/4″, 3/8″, 1/2″, 5/8″, and 3/4″. A tolerance of ±1/32″ is permitted for unsanded and touch-sanded panels (5% for panel thicknesses greater than 13/16″), but only ±1/64″ for sanded panels.

There are always an odd number of layers for balanced construction, the minimum number being three. A layer may consist of a single thickness of veneer, or it may be a lamination of two veneers. The minimum number of layers for various thicknesses of plywood is given in Figure WF10. A diagram of how the layers are typically arranged is given in Figure P10.

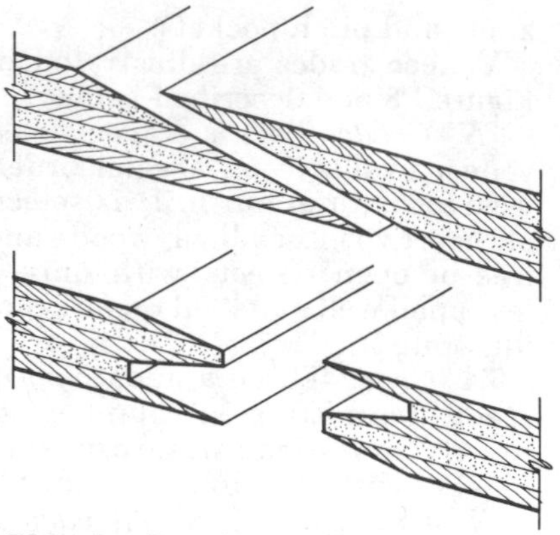

FIG. P9 Panels can be scarf-jointed (top) or finger-jointed (bottom).

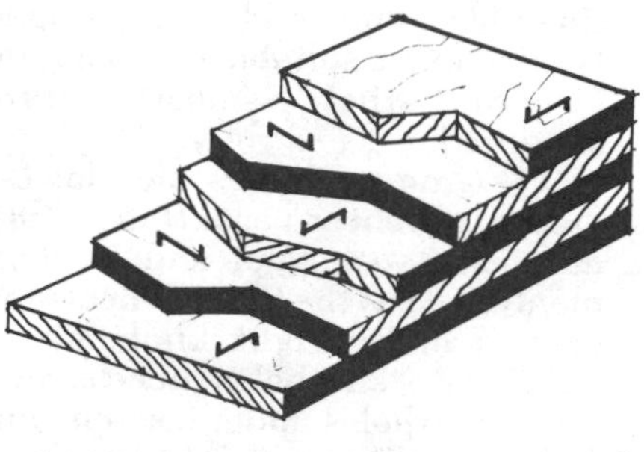

FIG. P10 Each layer is perpendicular to adjacent layers for strength and stiffness.

Specialty Plywoods

Special manufacturing processes and treatments produce a number of variations from the standard panel, each designed to meet specific use requirements.

Overlaid Plywood. This product consists of a high-grade panel of Exterior Type plywood to which resin-impregnated fiber sheets have been bonded to one or both faces. It is made in either *High Density* or *Medium Density.* The High Density face has twice as much resin (45% by weight minimum) in the overlay sheet as the Medium Density and is self-bonding to the plywood under the appropriate heat and pressures. Medium Density (17% resin by weight minimum) requires a phenolic resin adhesive to bond it to the plywood.

FIG. P11 Medium Density plywood being used for siding.

High Density has a hard, smooth, chemically-resistant, durable surface that requires no further finishing by paint or varnish. While available in several basic colors, it provides an excellent paint base and can be painted if color variation is preferred. The overlay panel is especially suitable and widely used in concrete form work for architectural finishes, highway signs and other severe exposures.

Medium Density, intended for high quality paint finishes, provides an opaque, smooth, non-glossy surface that blanks out the grain. A number of mills manufacture special Medium Density precut siding, frequently mill-primed, in widths up to 16″ (Fig. P11).

Textured Siding Panels. These plywoods are made by a number of mills under their own trade names. The panels may have a grooved surface, a rough-sawed surface, or a striated design (Fig. P12) or other textures. The panels are used for exterior siding, gable ends, and fencing, and as forms for special effects in concrete.

Paneling. Most paneling plywoods are manufactured from hardwood species (see Fig. P25). Some of the softwood species, such as California redwood, knotty pine and western red cedar, are used for interior paneling. Other manufacturers assemble softwood veneers into paneling plywood with surface textures such as embossing, striations or relief grain. Some manufacturers produce softwood panels having printed overlays simulating hardwood grains.

Tongue-and-Grooved Panels. Plywood in 1/2″, 5/8″ and 3/4″ thicknesses, with tongue-and-grooved joints, is produced by a number of mills in underlayment grades and occasionally other grades. Tongue-and-grooved panels are used to eliminate the need for blocking at joints. A specially engineered, 7-ply, 1-1/8″ thick floor panel, for use across two 48″ spans, serves as a combination subfloor and underlayment. Originally square-edged, it is now usually tongue-and-grooved (Fig. P13) to eliminate need for blocked joints.

Physical and Mechanical Properties

Aside from the characteristics of the veneer used, the most important properties of plywood derive from cross-laminated construction. Plywood utilizes the strength and dimensional stability of wood along the grain by laying alternate layers at right angles. Thus, cross-lamination provides resistance to splitting, improved dimensional stability, warp-resistance and other properties as discussed below.

Durability. In order to pass the Product Standard requirements for Exterior Type plywood, the glueline used to bond veneers is as durable as the wood itself. Exterior panels may be exposed outside without paint or other treatment and without affecting the durability of the glue bond or the life of the panels. Usually, however, plywood and other wood products

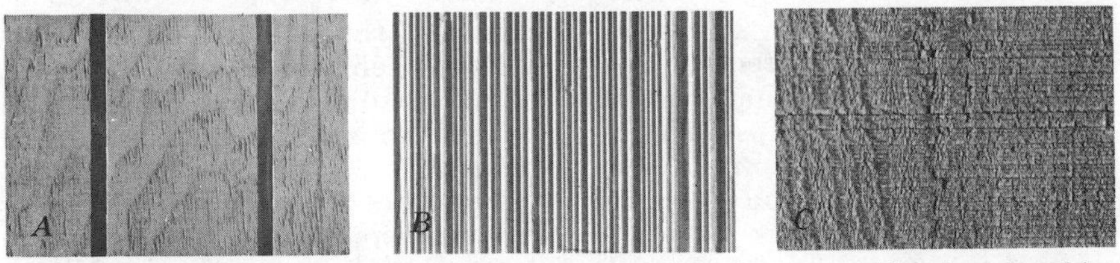

FIG. P12(a) Grooved, Exterior Type plywood ⅝″ thickness, combination sheathing and siding; grooves ¼″ deep are spaced 2″ or 4″ apart. (b) Striated, produced in Exterior and Interior Types by striating the face of the plywood with specially patterned planer. (c) Rough Sawed, saw-textured surface emphasizes character of natural wood and is especially suitable for exterior pigmented stain finishes.

are painted or stained when exposed outdoors to protect them against weathering and checking. Preservative treatments are available for plywood if it is to be used in areas of high decay or termite hazard.

Bending Strength and Stiffness. Cross-laminating of layers capitalizes on the more desirable physical and mechanical properties of wood. The strength of wood *parallel* to the grain is many times greater than that *perpendicular* to the grain. The cross-laminated construction provides longitudinal grain crosswise and improves significantly on solid wood's crosswise strength and stiffness. However, plywood panels are both stiffer and stronger along the face grain than across the face grain and should be applied with face grain perpendicular to supports for maximum resistance to bending.

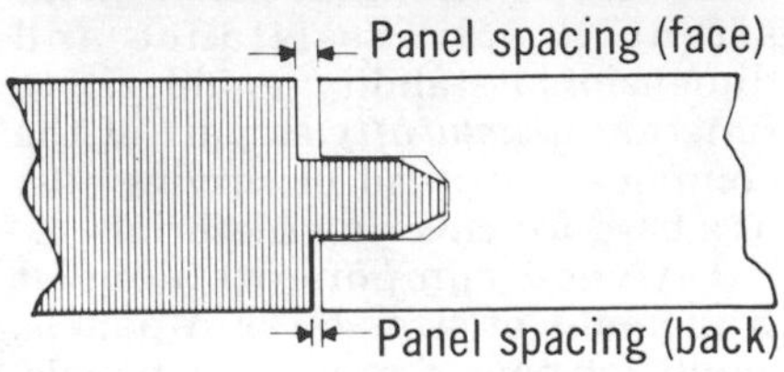

FIG. P13 Tongue-and-Grooved panels.

Racking Resistance (Rigidity). Plywood functions in its own plane as a structural bracing material when used on a wall, roof or floor. Although 5/16″ is the minimum thickness for sheathing, even 1/4″ thick plywood on a standard stud-framed wall provides twice the stiffness and more than twice the strength furnished by 1″ diagonal boards, long recognized as adequate. The racking strength of plywood and other sheathing materials is further discussed under wood framing in Division 300.

Two properties give plywood superior bracing qualities: (1) high shear strength in all directions for loads applied perpendicular to panel faces, and (2) relatively high nail-bearing strength, allowing nails to be driven as close as 1/4″ from the edge of the panel. Splitting is virtually impossible lengthwise or crosswise since no cleavage plane can exist.

Impact and Concentrated Load Resistance. Stresses resulting from impact or concentrated loading are distributed over a wide area by the cross veneers. This is an important property in sheathing, subflooring, siding and wall paneling.

Dimensional Stability. Like solid wood, plywood is subjected to dimensional changes due to variations in moisture content, temperature and other causes.

Hygroscopic Expansion and Contraction, due to water, in solid wood is generally insignificant parallel to the grain, even with large moisture content changes. *Perpendicular* to the grain, dimensional changes can be considerable (see page 201-4 Moisture Content). In plywood, any tendency for veneers to expand or contract crosswise is greatly minimized by one or more adjacent veneers running lengthwise. The total expansion both across the width and along the length of a 4′ x 8′ panel in service under normal variations in moisture content (8% to 14%) should average between 0.03″ (1/32″) and 0.05″ (1/20″).

Thermal Expansion in solid wood is about 10 times as much across the grain as it is parallel to the grain, where changes are quite small. In plywood, thermal expansion along the face is somewhat greater than in plain wood because of the influence of the crossbands. In an 8′-long Douglas fir panel (5 equal plies) the expansion caused by a 120° F rise in temperature should average approximately 0.021″ (1/50″). This is about 40% more than for plain Douglas fir wood but still insignificant for most uses. Across the grain, expansion for 120° F rise is approximately 0.029″, or only about 1/6 that for plain wood.

Warping, twisting and bowing of panels may be caused by unequal absorption of moisture along panel edges, different exposure of opposite faces to moisture, or unbalanced painting or coating of faces. Plywood that is flat at the time of manufacture will remain so unless conditions of storage or use subject it to uneven moisture content changes.

Checking may appear on any wood surface, plywood included, in the form of hairline cracks or even slightly open splits. Under severe moisture and dry conditions, such as an unpainted, unprotected plywood panel exposed to the weather, checks may in time become open cracks penetrating virtually the full thickness of the face. With moisture pick-up, the exposed veneer surface expands while the under surface glued to the crossband is held relatively fixed. The expanding outer fibers push against one another and acquire a *compression set.* As the fibers dry out and shrink they pull away from each other, creating checks of different intensity.

Remedies to control checking within acceptable limits include protection from dampness or moisture during transit and storage,

FIG. P14 Minimum plywood bending radii

PANEL THICKNESS	ACROSS GRAIN	PARALLEL TO GRAIN
1/4″	2′	5′
5/16″	2′	6′
3/8″	3′	8′
1/2″	6′	12′
5/8″	8′	16′
3/4″	12′	20′

Values shown are based on physical properties of Douglas fir.

These radii are minimums for mill-run panels of the thickness shown, bent dry. Shorter radii can be used for bending areas free of knots and short grain, and/or by wetting or steaming. (Exterior glue recommended).

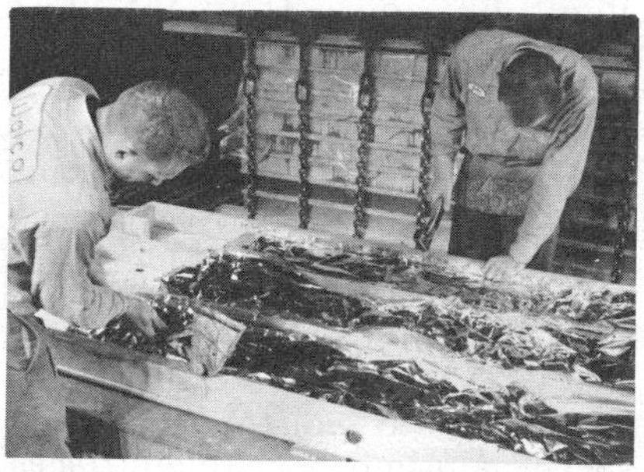

FIG. P15 Workmen insert insulation in stressed-skin panels. Note panels in press in background.

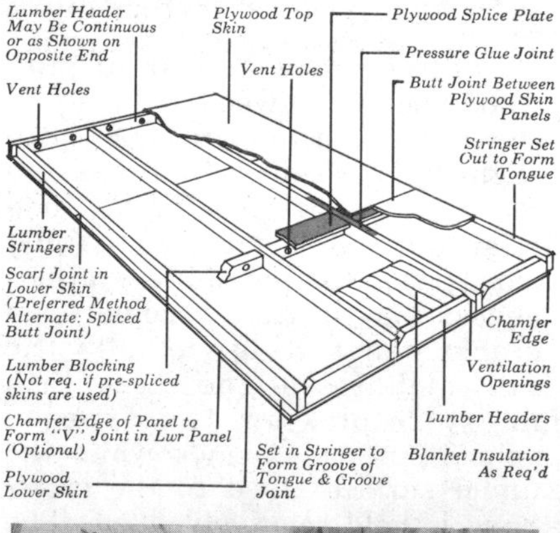

FIG. P16 Stressed-skin panels

protection during use at the job site, and the use of protective coatings such as conventional painting and water-repellent dips to minimize moisture absorption and loss.

Workability. Plywood is easily worked, cut to various sizes and shapes, and may be bent to fairly sharp radii (Fig. P14).

Fire and Decay Resistance. A high degree of resistance to either fire or decay can be imparted to plywood through pressure-impregnation treatment similar to that used for lumber (see page 201-13). Only plywood bonded with exterior glue is treated, as the adhesive used is not affected by pressure impregnation. Fire-retardant treatments have little effect on the rate of heat transfer through a panel, but virtually eliminate any fuel contribution to a fire and prevent the wood from burning by causing it to char.

FIG. P17 Sandwich panels, "honeycomb" core

FIG. P18 Pre-framed panels.

For indoors, fire-retardant paints may be applied to plywood surfaces. Certain types of paint blister or bubble, forming air pockets which insulate, retard combustion and greatly reduce flame spread along the surface.

Pre-assembled Components

The development of off-site, shop-fabricated building components has led naturally to assemblies of dimension lumber and plywood. These integral units may consist of plywood panels either nailed to precut lumber or fastened with glue, using either pressure- or nailed-glued techniques under controlled shop conditions. *Full pressure gluing,* which requires rather expensive equipment and careful control, produces excellent results (Fig. P15). *Nail gluing,* where nails furnish the pressure required for an adequate glue bond, also requires careful and precise techniques but less equipment. Nail gluing leads to satisfactory, predictable results in the laboratory but has not always been successful in the field. This presumably is due to poor gluing techniques, unfavorable temperatures, use of lumber not properly dried or surfaced, insufficient nailing for proper pressure and dried out gluelines. With proper supervision and control of these variables, field gluing can be successful.

Aside from the advantages in less erection time and reduced in-place costs that may result from the use of almost any preassembled components, plywood components derive added benfits from the plywood itself. Plywood's tensile and compressive strength, panel shear strength, split-resistance and dimensional stability enable it to function *structurally* as part of the component instead of serving just as a base for covering materials.

Plywood components consist principally of stressed-skin panels, sandwich panels, preframed panels, roof trusses, rigid frames, box beams and various curved units.

Stressed-skin Panels (more accurately, panels with stressed covers) consist of plywood sheets glued to the top and, in most cases, also the bottom faces of the longitudinal framing members so that the assembly will act integrally in resisting bending stresses (Fig. P16). Structurally, the plywood faces carry the compressive and tensile stresses induced by bending, while the framing members principally carry shear stresses. The plywood must be bonded firmly with glue to the frame in order to transfer all applied shearing stresses.

Sandwich Panels follow the general principles of stressed-skin panel design, but the faces are glued to and separated by weaker, lightweight core material such as resin-impregnated paper honeycomb (Fig. P17) or one of the plastic foam cores designed for this use. Sandwich panels possess many of the advantages of stressed-skin panels and are lighter in weight, needing no inner panel framing except, perhaps, around the perimeter. The cores provide a uniform, stabilizing support for the strong, relatively thin faces. The cores usually have little strength in flexure but have adequate compressive strength for normal floor panels. Some cores, particularly the plastic foams, also provide excellent insulation.

Preframed Panels consist of fabricated, precut lumber and plywood that are nailed or mechanically fastened together. They may be used for walls, floors or roofs (Fig. P18).

Roof Trusses are often constructed with either nailed and/or glued plywood gusset joints. Such connections permit members to be assembled without having to be overlapped at the joints. Nailed joints may be used for splicing lumber into greater lengths for developing tensile or compressive strength, moment resistance or shear strength as requirements may dictate (Fig. P19). Greater stiffness and strength can be provided when gussets are pressure-glued or nail-glued, but precision and careful techniques are essential. Typical nail-glued trusses are shown in Figure P20.

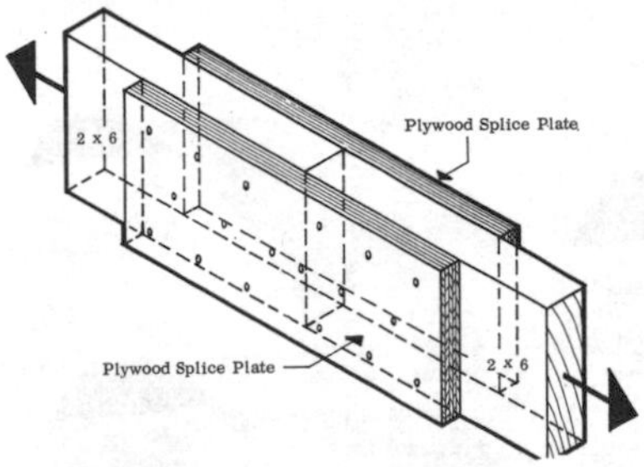

FIG. P19 Spliced joints may have plywood gusset plates on either one or both sides and are often used in truss construction.

Rigid Frames are specialized construction elements consisting of arch-shaped frames, usually comprising four pieces of straight lumber—two for posts and two for rafters. The plywood gussets, rigidly nailed or glued, connect adjacent members, transmit stresses across the joints and consolidate the frames into integral units.

Box Beams are hollow structural units consisting of two or more vertical plywood webs which are attached, usually by gluing or nailing to lumber flanges. Typical box beam cross sections are shown in Figure P21. As in a steel I-beam, the flanges carry the bending forces while the plywood webs transmit the shear. In this manner each of the materials is employed more effectively. At intervals along the beam, stiffener may be inserted between flanges and attached to the webs to distribute concentrated loads and resist web buckling.

Box beams have been used successfully for many purposes, including rather spectacular 48″ deep girders for factory roof systems and other members spanning up to 120′, as well as in many conventional applications (Fig. P21).

Although either nail gluing or nailing only may be satisfactory, pressure gluing with accurate controls is the preferred method to give maximum stiffness and strength.

FIG. P20 Nail-glued roof trusses.

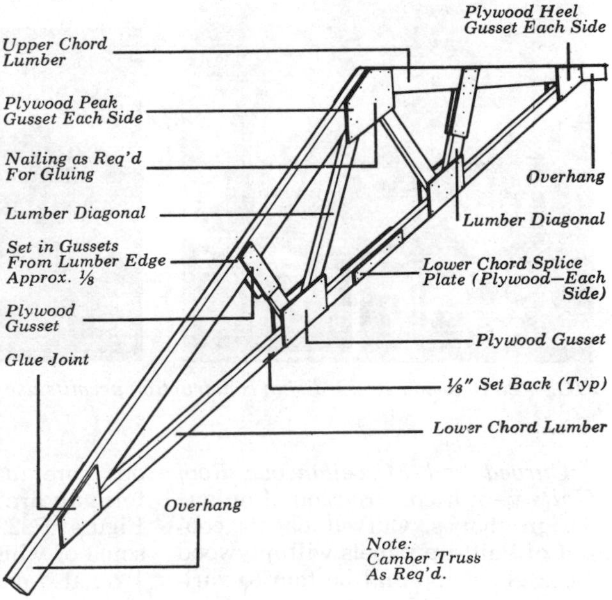

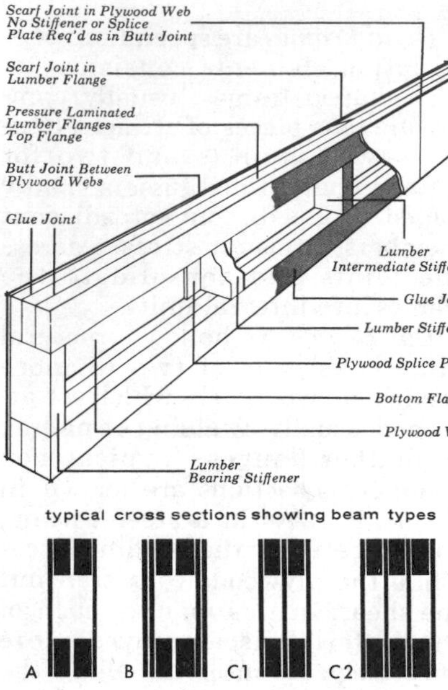

FIG. P21 Typical box beams.

FIG. P24 Box beams can be fabricated to span significant distances.

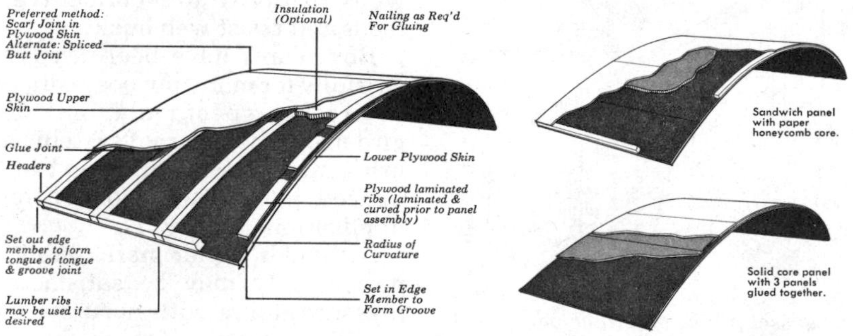

FIG. P22 Curved stressed-skin panels.

FIG. P23 Plywood stressed-skin construction permits use of many geometric shapes.

Curved and Miscellaneous Roof Units permit construction of unique design shapes. Curved panels consist of built-up panels with plywood faces glued top and bottom to various core materials. These usually form an arc of a circle as shown in Figure P22. Other design shapes, some of which are shown in Figure P23, also can be produced.

DECORATIVE PLYWOOD

The division of all plywood products into two major categories, *construction* and *decorative*, is based on the two governing Product Standards: PS 1-74 for Construction & Industrial Plywood and PS 51-71 for Hardwood & Decorative Plywood. PS 51-71 replaces Commercial Standard CS35, in keeping with the Department of Commerce program of phasing out Commercial Standards and Simplified Practice Recommendations.

Construction plywood covered by PS 1-74 is made mostly of softwood species although a few hardwood species are permitted. Decorative plywood covered by PS 51-71 consists entirely of hardwood-faced panels.

The use of softwood for interior layers of hardwood-faced panels conserves the more rare and expensive hardwood species. They are most often used as a decorative material for wall paneling, cabinet and furniture work, thin face skins in door manufacture, and face veneer for block flooring.

The species of wood used for the face veneers of panels identifies the name of hardwood plywood panels. For example, *birch plywood* has a face veneer (and, probably, back veneer) of birch.

Manufacture

Differences between the manufacture and appearance of hardwood and softwood plywood, as well as differences among hardwood plywoods, may result from: (1) the tree species; (2) the part of the tree used; (3) the cutting methods used to produce the veneer; (4) the methods used to match the veneers; and (5) the type of core construction employed.

Tree Species More than 99,000 species of hardwoods have been classified. About 240 of these are available in commercial quantities in the United States, and about 50 are widely used for hardwood plywood. Lauan, walnut, birch and gum are the most popular. Others include oak, maple, ash, cherry, elm, beech, mahogany (African and Honduras), butternut, pecan, cativo, cottonwood, sycamore, hackberry, teak and rosewood. Some of the typical hardwood veneer species, both domestic and imported, are listed in Figure P25.

Parts of the Tree The portion of the tree used greatly affects hardwood plywood figure patterns. Though most plywood comes from the trunk of the tree, some of the most interesting and valuable patterns are cut from a crotch, a burl or the stump. Veneers cut from a crotch have plume-like design. A burl produces veneers with a pattern of swirls. The stump yields rippling patterns with rich contrasts (Fig. P26).

Veneer Types The manner in which veneers are cut is important in producing various visual effects. Two logs of the same species but cut differently will have entirely different visual characteristics even though their veneer colors are similar. Five principal methods of cutting veneers are used (Fig. P27).

Rotary Cut is the same method used to produce most softwood veneers (see page 201-29). The log is mounted centrally in the lathe and turned against a razor-sharp blade, producing veneer exceptionally wide compared to that cut by other methods. Since this cut follows the log's annular growth rings, a bold, variegated ripple figure pattern is produced.

FIG. P25 **VISUAL CHARACTERISTICS OF REPRESENTATIVE HARDWOOD SPECIES**

Commercial name	Color	Type of figure
Ash, American	white to light brown	medium open grain
Basswood	white to pale brown	
Beech	white to reddish brown	
Birch	white to light reddish brown	curly grained . . . figured flat cut, plain rotary
Butternut	pale brown	leafy grain
Cativo	pale to medium brown	
Cherry, American	light to dark reddish brown	plain to rich mottle
Cottonwood	white to light grayish brown	
Elm	light brownish red	strong
Gum, Black and Tupelo	white, grayish white greenish to grayish black	
Gum, Red	pink to reddish brown	medium to highly figured
Hackberry	pale yellow, yellowish gray to light brown	
Lauan	pale grayish, yellowish brown to reddish brown	
Mahogany, African	pink to reddish brown	plain stripe to highly figured
Mahogany, Tropical American	pink to gold brown	straight to rich mottle
Mahogany, Crotch & Swirls	pink to reddish brown	moon and feather crotch . . . plain and figured swirl
Maple	white to tan	plain, curly burls
Oak, Red	pink tan to ochre	plain to flake
Oak, White	gray tan to ochre	plain to flake
Rosewood, Brazilian	pink brown and violet	wide range figure
Rosewood, East Indian	purple to straw	striped and figured
Teak	light tan, dark brown	plain, ripple, mottle, stripe
Walnut, American	soft gray brown	typical figure or stripe

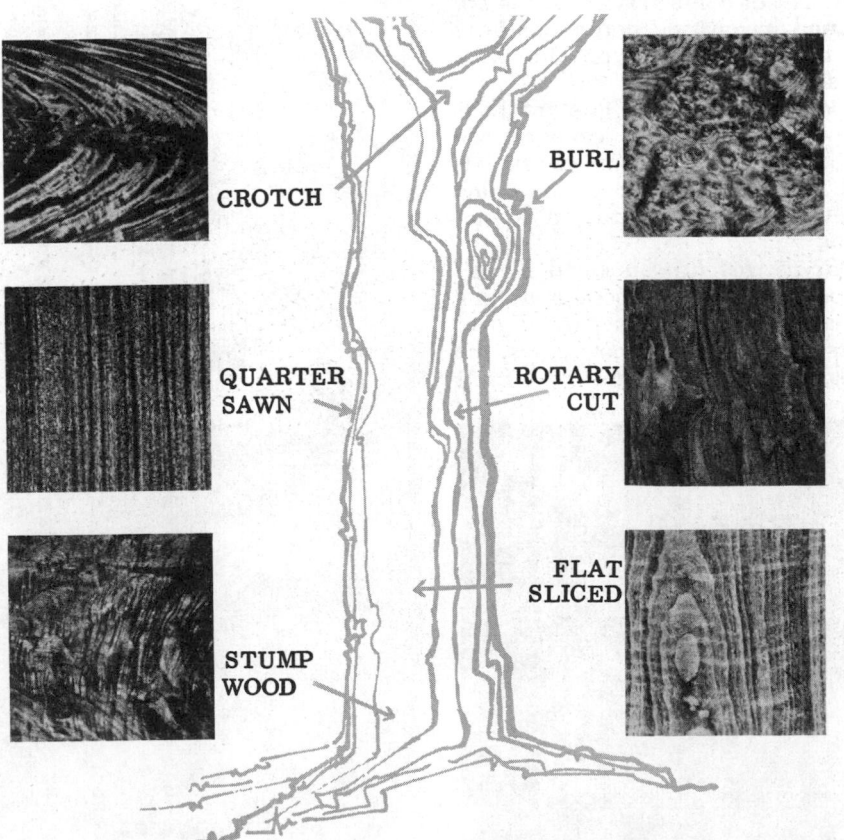

FIG. P26 Portions of tree from which various types of figure patterns are obtained. Some species produce several figure types.

Flat Slicing produces a variegated wavy figure. The half log, or flitch, is mounted with the heart side flat against the guide plate of the slicer, and the slicing is done parallel to a line through the center of the log (Fig. P28).

Quarter Slicing produces a series of stripes, straight in some woods, varied in others. The quarter log (flitch) is mounted on the guide plate so that the growth rings of the log strike the knife at approximately right angles.

Half-round Slicing shows modified characteristics of both rotary and plain sliced veneers and is a variation of rotary cutting in which segments of the log (flitches) are mounted off-center in the lathe. This results in a cut slightly across the annual growth rings.

Rift-Cut veneer is produced in various species of oak as well as other species. In these woods the ray cells which radiate from the center of the log are conspicuous. The rift or comb grain effect is obtained by cutting perpendicularly to these rays either on the lathe or slicer.

Veneer Matching The methods used to match face veneers also produce different visual characteristics. Basic and special matching methods are illustrated in Figure P29.

Book Matching is used for all types of veneers. In book matching every other sheet is turned over like the leaves of a book. Thus, the back of one veneer meets the front

FIG. P28 Flitch is held in clamps and moved against knife.

FIG. P27 Visual characteristics of veneers depend on method of manufacture.

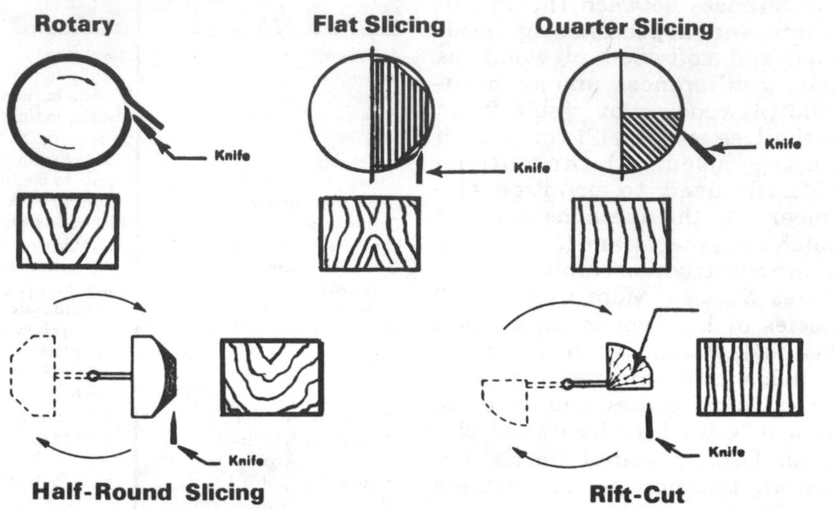

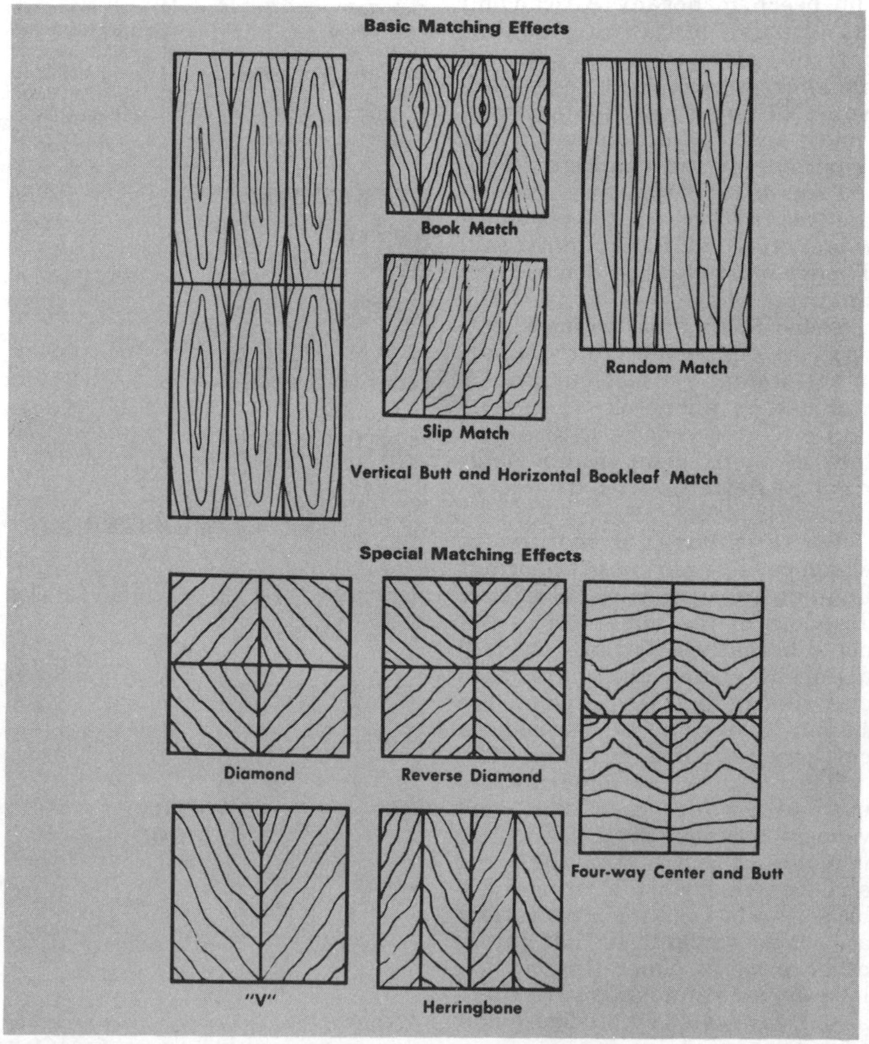

FIG. P29 Basic and special matching effects for wood veneers.

of the adjacent veneer, producing a matching joint design (also called *edge matching).*

Slip Matching results when adjacent sheets are joined side by side, without turning, to repeat the flitch figure. All types of veneer may be used, but this type of matching is common in quarter-sliced veneers.

Both book and slip-matched veneers are used in the manufacture of Premium, Good and Specialty grades of hardwood plywood and may be used for doors and matched, ungrooved wall panels.

Random Matching results when veneers are joined with the intention of creating a casual unmatched effect. Veneers from several logs may be used in the manufacture of a single panel. Random matching is generally used in prefinished V-grooved wall panels.

Vertical Butt and Horizontal Bookleaf Matching is used where the length of a flitch does not permit its fabrication into the desired height of panel. The veneer may be matched horizontally as well as vertically.

Four-way Center and Butt Matching is a special matching technique. This type of match is ordinarily applied to crotch or stump veneers since it is a most effective way of revealing the figure. Occasionally, flat-sliced veneers are matched in this manner when panel length requirements exceed the length of available veneers.

Vertical butt and horizontal bookleaf match and the other special matching types are used mostly in furniture and in custom-designed wall panel applications.

Core Construction. There are four basic types of core construction used; the innermost ply is the *core*, all other plies between the core and the face plies being crossbands. When panels are exposed on one surface only, back and face plies are of species and thickness to balance the face veneers.

Veneer Core is the method described and illustrated for softwood plywood (page 201-29). In hardwood plywood, 3, 5 or more odd number of plies are laid with grain direction of adjacent plies at right angles to each other.

Lumber Core consists of sawed lumber to which crossband plies and face veneers are glued (Fig. P30). Lumber core construction generally is used for cabinet work and furniture in which exposed edges or edge treatment such as doweled, splined or dovetailed joints are desired, or where butt hinges are to be used. The lumber core is made of narrow wood strips edge-glued together. Panels with hardwood (to match the face) banded on the edges may be specially ordered.

Particleboard Core consists of an aggregate of wood particles (also called chipboard, chipcore, particle board, etc.) fused under heat and pressure with face veneers usually glued directly to this core, although crossbanding is sometimes used. This plywood is particularly adaptable for table, desk and cabinet tops.

Mineral Core is used for fire-resistant panel construction. Veneers are bonded to a core of hard, noncombustible material.

Sizes Hardwood plywood is most commonly sold in 4′ x 8′ panels although panels are custom-made up to 40′ in length and up to 7′ in width. Stock panels may also be purchased in lengths of 10′ and 12′.

Hardwood plywood is manufactured with an uneven number of plies in thicknesses ranging from 1/8″ up to as much as 3″ in custom

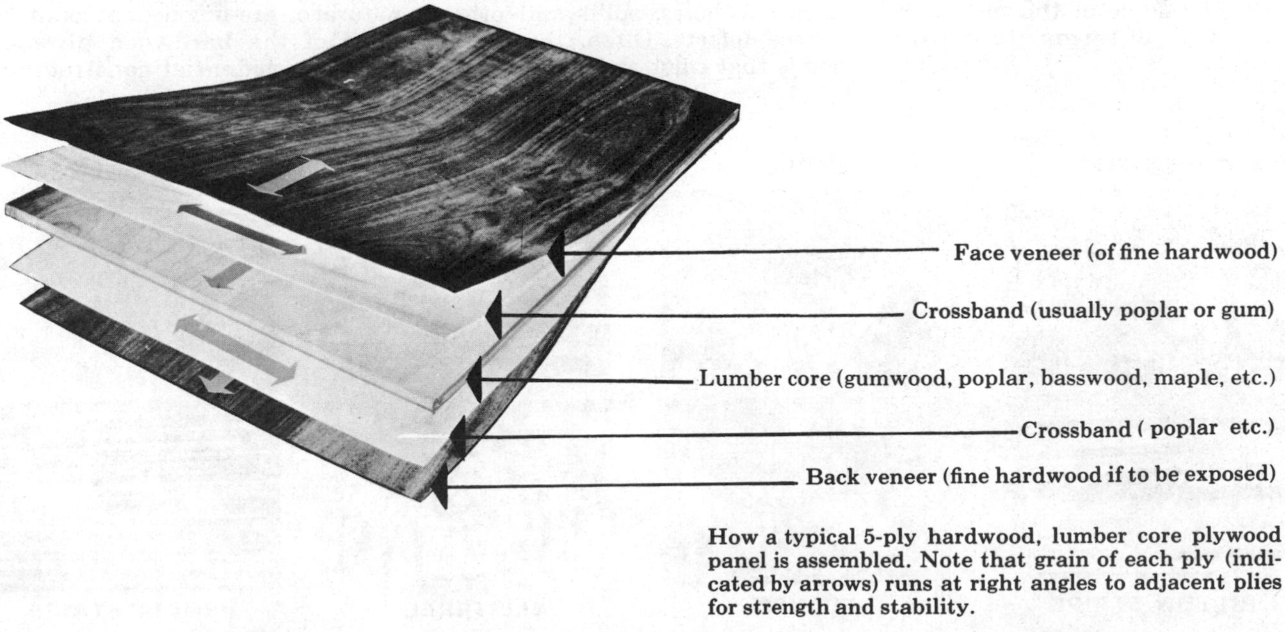

How a typical 5-ply hardwood, lumber core plywood panel is assembled. Note that grain of each ply (indicated by arrows) runs at right angles to adjacent plies for strength and stability.

FIG. P30 Typical 5-ply lumber core panel. Each ply (indicated by arrows) runs at right angles to adjacent plies.

FIG. P31 STANDARD THICKNESSES OF HARDWOOD PLYWOOD*

Plies	Thickness
3	1/8", 3/16", 1/4"
5	5/16", 3/8", 1/2"
5 and 7	5/8"
5, 7 and 9	3/4"

*Hardwood plywood is square within 1/16" measured on short dimension.

FIG. P33 CLASSIFICATION OF HARDWOOD PLYWOOD

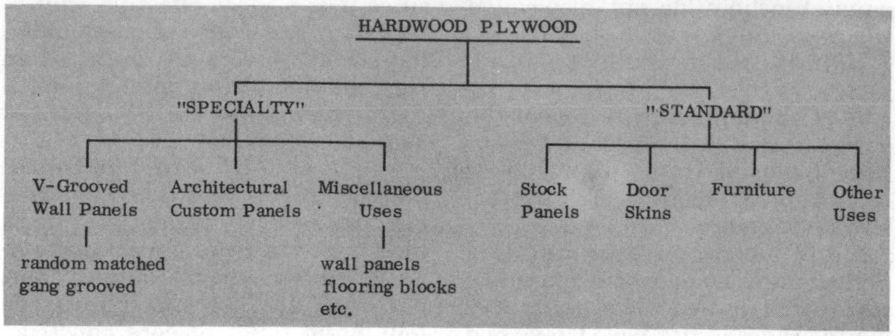

panels. The most common thicknesses produced are 1/4", 3/8", 1/2" and 3/4" in *veneer core* and 3/4" in *lumber* and *particle-board core.* The table in Figure P31 shows the most common thicknesses for varying numbers of plies.

Pattern and Figure Characteristics of Veneer

Figure refers to the highlights of "crossfire" running at right angles to grain direction. The grain character and direction are described by using the word *pattern.* Veneer men may describe figure by saying it "has a great deal of crossfire," or "has a straight or broken stripe" or "is highly figured." Figure P32 illustrates some of the most commonly used terms describing veneer.

General Classifications

Hardwood plywood falls generally into two categories: (1) "specialty" plywoods, which are not typically or necessarily manufactured to meet industry standards, and (2) "standard" plywoods produced to meet Product Standard PS 51-71 (Fig. P33).

"Specialty" Plywood There are no industry-recognized standards covering this general classification. Species, type, match, core construction and other requirements are as agreed upon by the seller and the buyer. Generally, panels containing the least amount of blemishes are sold at a higher price than those with large knots and filled knotholes, splits and other surface defects. Often, the intention is that color and grain vary.

Each manufacturer generally develops his own *grade* terminology. However, such terminology is rarely by number or letter but by an arbitrary descriptive system. For example, "aristocratic" and "homestead" are names selected to appeal to different tastes rather than to establish degrees of quality. The "aristocratic" panels might contain few blemishes and very few filled places, while the "homestead" panels would contain more knots, large filled knotholes and more "character." There may be little, if any, price differential between the so-called "aristocratic" and the "homestead" panels.

V-Grooved Wall Panels, either natural or pre-finished, account for most of the hardwood plywood used in residential construction.

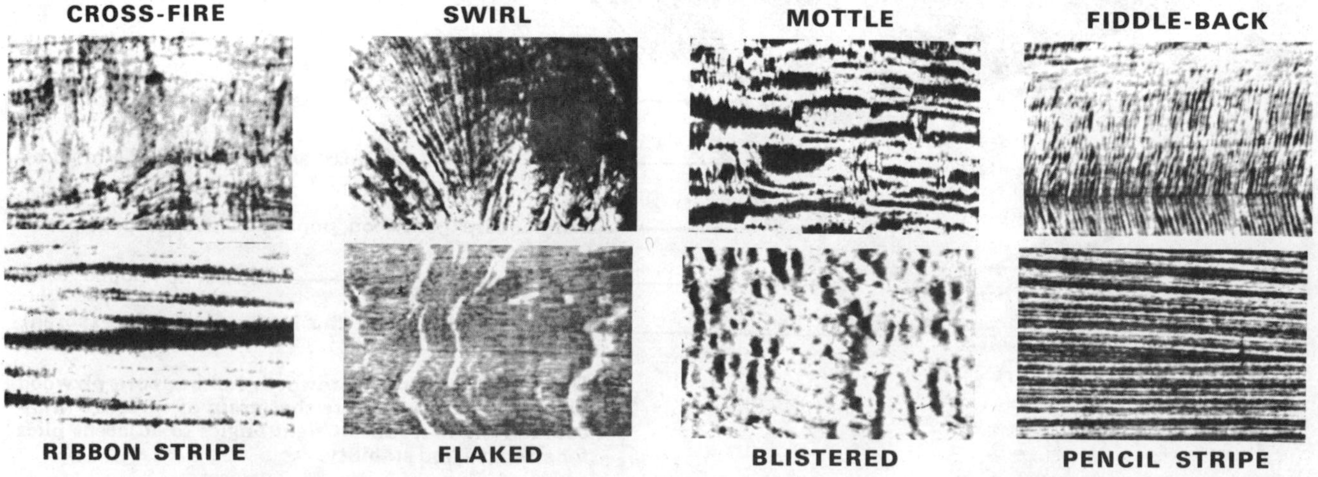

FIG. P32 Commonly used figure descriptions

There are two types of V-grooved panels: random matched, accounting for more than 75% of all wall panel production, and gang-grooved.

In random matching the individual veneer sheets, of which there may be four to nine in a 4′ x 8′ panel, are selected to simulate lumber planking. The V-grooving, at the joints of the veneer sheets, produces a mismatched panel with veneer sheets dissimilar in grain and color.

Gang-grooved panels are produced by passing one-piece or multi-piece faces under a machine with grooving knives set at certain spaced intervals. When multi-piece face veneers are used, no attempt is made to groove on the joints of the veneer. In gang-grooved panels, the spacing of grooves is identical from one panel to the next.

Architectural Custom Panels are made to special order. The desired hardwood face veneer is selected from flitch samples prior to manufacture of the plywood. The entire flitch is then fabricated by the plywood plant into panels to meet the custom specification. The only limitations of architectural custom panels are those imposed by the patterns which are available in the length and width of the selected veneer.

Miscellaneous Uses account for increasing amounts of hardwood plywood production, but are too numerous to discuss individually. Included are wall panels of varying widths (less than 4′) and flooring products such as laminated block.

"Standard" Plywood Hardwood plywood that conforms to Product Standard PS 51-71 is manufactured primarily for use as stock panels, door skins and remanufacturing in the furniture industry. The Hardwood Plywood Manufacturers Association (HPMA) maintains a quality control program which assures, if the plywood carries an official grade-trademark, that it conforms to PS 51-71 (Fig. P34).

Stock Panels, as the name implies, are fabricated and stocked by the plywood manufacturer and sold through local distribution warehouses to a number of different consumers. The character and quality of face veneers is dependent on market conditions, and buyer selection is made from the available stock at the time of purchase. Panels are typically 4′ x 8′ of varying thicknesses, such as 1/4″, 1/2″, 3/4″, and may be put to a variety of different uses.

Doorskin Plywood comprises a large percentage of hardwood veneer and is used as door *skins* to face solid and hollow core doors (see Section 202—Wood Products).

Furniture and Other Industries make extensive use of hardwood plywood. Outside the construction industry, hardwood plywood finds almost endless uses—automotive construction, ship and boat construction, aircraft and railroad cars, sporting goods, musical instruments, caskets and many other items.

"Standard" Types The word *Type* is used to indicate the panel's durability, i.e., ability of the plies to stay together under different exposure conditions. Four basic types of hardwood plywood produced to Product Standard PS 51-71 are available.

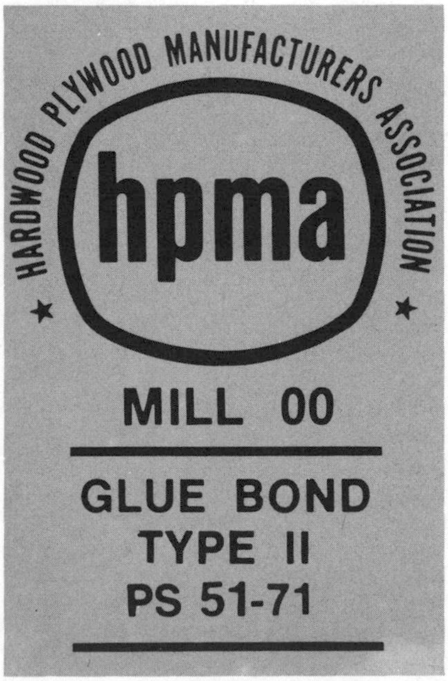

FIG. P34 The grade-trademark of the Hardwood Plywood Manufacturers Association assures conformance with the appropriate industry standard.

Technical is made with the same adhesive specifications as Type I but varies in thickness and arrangement of plies to provide approximately equal tensile and compressive strength in length and width.

Type I (Exterior) is made with waterproof adhesive and is used where it may come in contact with water and must withstand full weather exposure. Adhesives used are resins of one of the following types: phenol, resorcinol, phenol-resorcinol, melamine or melamine-urea. Veneer core construction is preferred for Type I plywood.

Type II (Interior) is made with water-resistant adhesive (urea resin) and is used where it would not be subjected to contact with water, although it should retain practically all of its strength if occasionally subjected to thorough wetting and drying. The majority of hardwood plywood is Type II.

Type III (Interior) is made with urea-resin adhesives but with higher proportions of water and/or extenders than normally used in Type II urea glue mixes. While not commonly made, it is used where it will not come in contact with any water, although it can withstand casual dampness and humidity.

"Standard" Grades The grade designates the quality of the face, back and inner plies. Hardwood veneer grades used under Product Standard PS 51-71 are designated as Premium, Good Grade, Sound Grade (#2), Utility Grade (#3), Backing Grade (#4) and Specialty Grade (SP). Figure P35 illustrates the veneer grades described below.

Premium Grade face veneers may be made from more than one piece. With most species, multi-piece faces must be book matched or slip matched. The quality of veneers is high: only a few small burls, occasional pin knots, slight color streaks and inconspicuous small patches are allowed; no other defects such as splits, shakes, worm holes or decay are permitted.

Good Grade is very similar to Premium Grade except that matching of veneer faces is not required. Sharp contrasts in color and great dissimilarity of grain and figure of two adjacent pieces of veneer in multi-piece faces are not allowed. Only a few small burls, occasional pin knots, slight color streaks and inconspicuous small patches are allowed.

Sound Grade (#2) veneers need not be matched for grain or color but must be free of open defects to provide a sound, smooth surface. This grade may contain mineral streaks, stain, discoloration, patches, sapwood, sound tight knots up to 3/4″ in average diameter, sound smooth burls up to 1″ in average diameter. Shake or decay are not permitted in Sound Grade (#2).

Utility Grade (#3) permits discolorations, stain, mineral streaks, patches, tight knots, tight burls, knotholes up to 1″ in diameter, worm holes and splits or open joints not exceeding 3/16″ wide and not extending half the length of the panel, and other minor defects. Shake and forms of decay are not permitted.

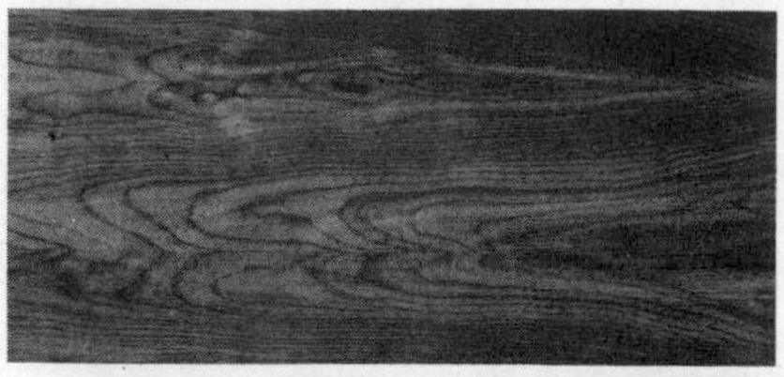

Premium Grade (for natural finish)

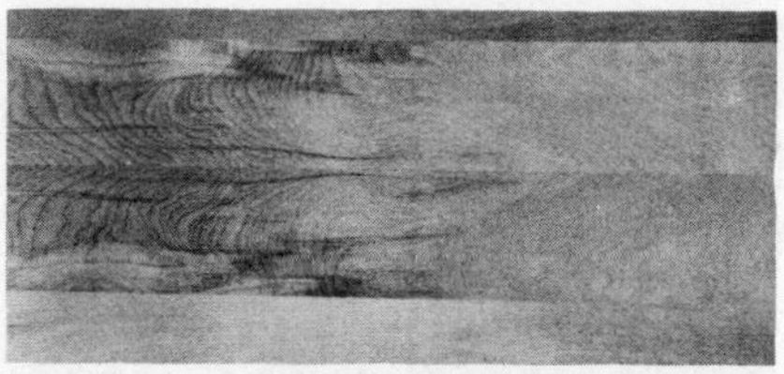

Sound Grade (#2)

Good Grade

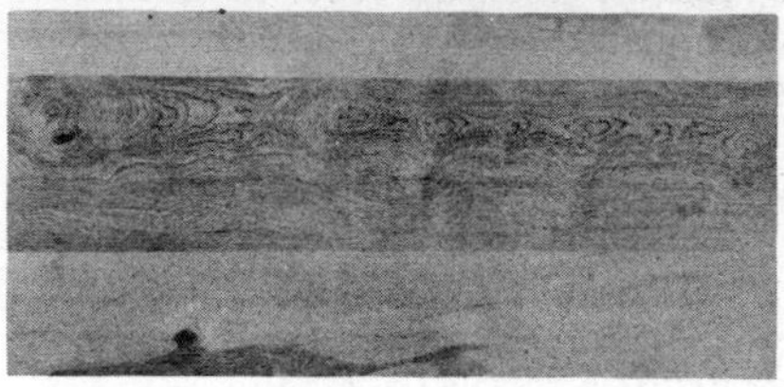

Utility Grade (#3)

FIG. P35 Appearance of "Standard" hardwood plywood grades.

Backing Grade (#4) is similar to Utility Grade except that larger-sized open defects are permitted—knotholes not greater than 3″ with a limitation on the number of knots per sq. ft., and splits up to 1″ wide, depending on the length of the split. Defects must not affect strength or serviceability of the panel.

GLOSSARY

Actual Size: See *Size.*

Air Drying: See *Seasoning.*

Allowable Unit Stress: See *Stress.*

Annual Growth Ring: The growth layer put on by a tree in a single growth year.

Back: The side opposite the face; or the poorer side of a plywood panel.

Bark: The outer, corky layer of a tree composed of dry, dead tissue.

Basic Stress: See *Stress.*

Beam: A structural member transversely supporting a load.

Beams and Stringers: See *Lumber.*

Board Foot: A measure of lumber whose nominal dimensions are 1″ thick, 12″ wide and 1′0″ long, or the equivalent.

Boards: See *Lumber.*

Bow: See *Warp.*

Cambium: The layer of tissue, one cell wide in thickness, between the bark and wood that subdivides to form the new wood and bark cells of each year's growth.

Capillaries: Thin-walled tubes or vessels found in wood.

Cellulose: The principal constituent of wood which forms the framework of the wood cells.

Check: A lengthwise separation of the wood, the greater part of which occurs across the rings of annual growth.

Core: The innermost portion of plywood, consisting of either hardwood or softwood sawed lumber, veneer or composition board.

Crook: See *Warp.*

Crossband: A layer of veneer in a plywood panel whose grain direction is at right angle to that of the face plies.

Cross Grain: See *Grain.*

Cup: See *Warp.*

Decay: The decomposition of wood substances by certain fungi.

Diagonal Stress: See *Stress.*

Dimension: See *Lumber.*

Dress: To surface lumber with a planing machine.

Dressed Size: See *Size.*

Dry Lumber: Under Product Standard PS 20-70, lumber with a moisture content of 19% or less.

Dry Rot: The term is a misnomer since all fungi require considerable moisture; however, it is loosely applied to many types of decay which, when discovered in the advanced state, permit wood to be easily crushed to a dry powder.

Edge Grain: See *Grain.*

Equilibrium Moisture Content: The moisture content at which wood neither gains nor loses moisture when surrounded by air at a given relative humidity and temperature.

Face Size: See *Size.*

Factory and Shop Lumber: See *Lumber.*

Fiber: A wood fiber is a comparatively long (1/25″ or less to 1/3″), narrow, tapering cell closed at both ends.

Fiber Saturation Point: The stage in drying (or wetting) of wood at which the cell *walls* are saturated with water, but the cell *cavities* are free of water, being approximately 30% moisture content in most species.

Figure: The pattern produced in a wood surface by annual growth rings, rays, knots and deviations from regular grain.

Flat Grain: See *Grain.*

Flitch: (a) A portion of a log sawed on two or more sides and intended for remanufacture into lumber or sliced or sawed veneer: (b) after cutting, a complete bundle of veneers laid together in sequence as they were sliced or sawed.

Gang Grooved: Plywood panels produced by passing under a machine with grooving knives set at certain intervals.

Glueline: The line of glue visible on the edge of a plywood panel. Also applies to the layer of glue itself.

Grade: The designation of the quality of a manufactured piece of wood.

Grain: The direction, size, arrangement, appearance or quality of the fibers in wood.

Cross Grain—Grain not parallel with the longitudinal axis of a piece, as a result of sawing. It may be either diagonal, spiral grain or a combination of both.

Diagonal Grain—Annual rings at right angle with the axis of a piece as a result of sawing at an angle with the bark of the tree.

Edge Grain—Lumber that has been sawed parallel with the pith of a log and approximately at right angles to the growth rings, making an angle 45° to 90° with the wide surface of the piece.

Flat Grain—Lumber that has been sawed parallel with the pith of the log and approximately tangent to the growth rings, making an angle less than 45° with the surface of the piece.

Open-Grained—Common term used for wood with large pores, such as oak, ash, chestnut and walnut.

Plain Sawed—Another term for flat grain.

Quarter Sawed—Another term for edge grain.

Vertical Grain—Another term for edge grain.

Green Lumber: Under Product Standard PS 20-70, lumber with a moisture content of over 19%. Broadly speaking, *unseasoned* lumber, not exposed to air drying or kiln drying.

Gusset Plate: Wood or metal used as a means of joining co-planar structural members. See Fig. L19, page 201-35.

Hardwood: The botanical group of trees that are broad-leaved and deciduous. The term has no reference to the actual hardness of the wood.

Heartwood: The wood extending from the pith to the sapwood, the cells of which no longer participate in the growth process of the tree. Heartwood may be impregnated with gums, resins and other materials which usually make it darker and more decay-resistant than sapwood.

High Density: See *Plywood.*

Joist: One of a series of parallel beams used to support floor and ceiling loads, supported in turn by bigger beams, girders or bearing walls.

Joists and Planks: See *Lumber.*

Kiln Drying: See *Seasoning.*

Knot: That portion of a branch or limb which has been surrounded by subsequent growth of wood.

Laminated Wood: A piece of wood built up of laminations that have been joined either with glue or mechanical fastenings.

Light Framing: See *Lumber.*

Lignin: The second most abundant constituent of wood (approximately 12% to 28%). It encrusts the cell walls and cements the cells together.

Lumber: The product of the saw and planing mill not further manufactured than by sawing, resawing and passing lengthwise through a standard planing machine, cross-cut to length and worked (see Fig. L29, page 201-26).

Beams and Stringers—Large pieces of lumber (5″ or more in thickness and 8″ or more in width), graded with respect to their strength in bending when loaded on the narrow face.

Boards—Lumber up to 2″ thick and 2″ or more in width.

Dimension—Lumber from 2″ to 4″ thick and 2″ or more in width.

Factory and Shop—Lumber intended to be cut up for use in further manufacture.

Joists and Planks—Pieces of lumber (2″ to 4″ in thickness and 4″ or more in width) graded with respect to strength in bending when loaded either on the narrow face as a joist, or wide face as a plank.

Light Framing—Lumber 2″ to 4″ in thickness and 2″ to 4″ in width.

Matched—Lumber that is edge dressed and shaped to make a close tongue and groove joint at the edges or ends when laid edge to edge or end to end.

Patterned—Lumber that is shaped to a pattern or to a molded form in addition to being dressed, matched or shiplapped, or any combination of these.

Posts and Timbers—Lumber that is approximately square, 5″ and thicker and having a width not more than 2″ greater than its thickness.

Structural—Lumber which has been machine rated or visually graded into grades with assigned working stresses (also termed *stress grade* lumber; includes most yard lumber grades, except boards).

Yard—Lumber intended for general building purposes; includes boards, dimension and timbers.

Matched: See *Lumber.*

Medium Density: See *Plywood.*

Millwork: Lumber that is shaped to a pattern or to a molded form in addition to being dressed, matched or shiplapped, or any combination of these. (Term includes most finished wood products such as doors, windows, interior trim, stairways—but not flooring or siding products.)

Moisture Content (of wood): The amount of water contained in the wood at the time it is tested, expressed as a percentage of the weight of the wood when oven-dry. (Oven-dry means wood dried in an oven to a consistent moisture content; that is, for all practical purposes, no longer holding any water.)

Nail-glued: A method of gluing in which the nails hold the wood members until the glue sets.

Nominal Size: See *Size.*

Open-grained: See *Grain.*

Overlaid: See *Plywood.*

Particleboard: A composition board consisting of distinct particles of wood bonded together with a synthetic resin or other binder.

Patterned: See *Lumber.*

Pitch Pockets: An opening extending parallel to the annual growth rings containing, or that has contained either solid or liquid pitch.

Pith: The small, soft core occurring in the structural center of a tree, branch, twig or log.

Plain Sawed: See *Grain.*

Plywood: A crossbanded assembly made of layers of veneer or with veneer in combination with a lumber core, particle board core, or other type of composition core, all joined with an adhesive.

"Standard"—Hardwood plywood produced to meet governing industry standards. Also, unsanded interior softwood plywood for sheathing and flooring.

High Density—Overlaid plywood having 40% resin by weight in the overlay sheet.

Medium Density—Overlaid plywood having 20% resin by weight in the overlay sheet.

Overlaid—Plywood in which face veneer is bonded on one or both sides with paper, resin-impregnated paper or metal.

"Specialty"—Hardwood plywood which is not typically or necessarily manufactured to meet industry standards.

Posts and Timbers: See *Lumber.*

Preframed Panels: Panels fabricated, using precut lumber and plywood.

Preservative: Any substance that will prevent, for a reasonable length of time, the action of wood-destroying fungi, insects of various kinds and similar destructive life when the wood has been properly coated or impregnated with it.

Pressure-glued: A method of gluing which places the wood members under high pressure until the glue sets.

Quarter Sawed: See *Grain.*

Rays: Strips of cellulose extending radially within the tree, storing food and transporting it horizontally in the tree (see Fig. W3, page 201-3).

Rotary Cut: See *Veneer.*

Sapwood: The living wood of pale color near the outside of the log.

Sawed Veneer: See *Veneer.*

Scarf Jointing: A joint in which the ends of plywood panels are beveled and glued together.

Seasoning: Removal of moisture from green wood in order to improve its serviceability.

Air Drying—Seasoning lumber by exposure to air usually in a yard.

Kiln Drying—Seasoning lumber in a kiln (oven) under controlled conditions of heat, humidity and air circulation.

Shake: A separation along the grain, the greater part of which occurs between the annual growth rings.

Size: *Actual Size*—The size of lumber after seasoning and dressing.

Dressed Size—Size of lumber after surfacing.

Face Size—The exposed width of a patterned piece of lumber when installed.

Nominal Size—A rough-sawed commercial size by which lumber is known and sold.

Sliced Veneer: See *Veneer.*

Softwood: The botanical group of trees that have needle- or scale-like leaves and are evergreen for the most part (cypress, larch and tamarack being exceptions). The term has no reference to the actual hardness of the wood.

"Specialty": See *Plywood.*

Springwood: The portion of the annual growth ring that is formed during the early part of the season's growth. It is usually less dense and weaker mechanically than summerwood (see Fig. W3, page 201-3).

Stain: A discoloration of wood that may be caused by such diverse agencies as micro-organisms, metals or chemicals.
Strength: A term used to describe all the properties of wood which enable it to resist different forces or loads.
Stress: Force per unit of area.
Allowable Unit Stress (Working)—Stress for use in the design of a wood member that is appropriate to the species and grade. Values for each type of stress are obtained by multiplying the basic stress (see below) for that species by the strength ratio assigned to each grade.
Basic Stress—The design stress for a clear wood specimen free from strength-reducing features, such as knots, checks and cross grain. The specimen has in it all the factors appropriate to the nature of structural timber and the conditions under which it is used except those that are accounted for in the strength ratio.
Summerwood: The portion of the annual growth ring that is formed after the springwood formation has ceased. It is usually denser and stronger mechanically than springwood (see Fig. W3, page 201-3).
Structural: See *Lumber.*
Texture: A term often used interchangeably with grain. It refers to the finer structure of the wood.
Twist: See *Warp.*
Veneer: A thin sheet of wood.
Rotary Cut—Veneer cut in a continuous strip by rotating a log against the edge of a knife in a lathe (see Fig. P27 page 201-38).
Sawed—Veneer produced by sawing.
Sliced—Veneer that is sliced by moving a log or flitch against a large knife (see Fig. P27, page 201-38).
Vertical Grain: See *Grain.*
Wane: Bark or lack of wood, from any cause, on the edge or corner of a piece of lumber.
Warp: Any variation from a true or plane surface (see Fig. L24, page 201-23).
Bow—The distortion of a board in which the face is convex or concave longitudinally.
Crook—The distortion of a board in which the edge is convex or concave.
Cup—The distortion of a board in which the face is convex or concave transversely.
Twist—A distortion caused by the turning of the edges of a board so that the four corners of any face are no longer in the same plane.
Working Stress: See *Stress.*
Yard Lumber: See *Lumber.*

We gratefully acknowledge the assistance of the following in the preparation of this text; we are indebted to them for the use of their publications as references and permission to use photographs:

American Plywood Association
American Wood Preservers Institute
American Wood Preservers Bureau
California Redwood Association
Fine Hardwoods Association
Forest Products Laboratory
Hardwood Plywood Manufacturers Association
National Hardwood Products Association
National Forest Products Association
Potlatch Forests, Inc.
Southern Forest Products Association
Southern Pine Inspection Bureau
United States Plywood Corporation
West Coast Lumber Inspection Bureau
Western Wood Products Association
Weyerhaeuser Company

WOOD 201

CONTENTS

PROPERTIES

MOISTURE CONTENT

Boards and dimension lumber 2″ or less in thickness should be specified S-DRY or MC 15 and should be identified by the appropriate grademark.

S-DRY lumber (surfaced at 19% m.c. maximum) is suitable for framing purposes in damp regions and most parts of the country except the arid regions (Fig. WF 1). In the arid regions, MC 15 lumber (surfaced at 15% moisture content maximum) is recommended, unless the lumber is allowed to air dry at the site to 15% moisture content before interior finishes are applied.

Timbers and dimension lumber over 2″ thick should be specified to have a moisture content 19% maximum, but need not necessarily be identified with the S-DRY grademark.

Most species (except for Southern pine) are not milled in the larger sizes in the dry condition, except on special order. However, heavy dimension lumber and timbers may achieve a 19% moisture content after air drying under cover at the mill or lumber yard. Verification of the moisture content with a moisture meter is recommended.

FIG. WF1

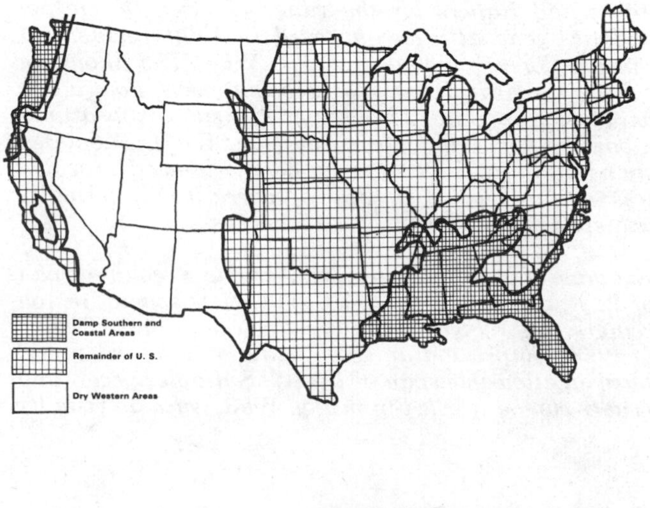

Moisture content should be measured with the type of equipment and according to procedures described in Methods of Test for Moisture Content of Wood ASTM D2016.

ALLOWABLE UNIT STRESSES

Grades of framing lumber should be suitable for the intended purpose in terms of appearance and strength. Calculated working stresses should not exceed maximum allowable working stresses tabulated in the Supplement to the National Design Specifications (NDS) for Wood Construction of the National Forest Products Association (NFPA). All engineering calculations and construction practices should conform to provisions of the NDS, latest edition.

Joists and rafters for simple frame buildings and ordinary construction can be selected by referring to Span Tables for Joists and Rafters, published by NFPA. These tables indicate the minimum modulus of elasticity, "E", and extreme fiber strength in bending, "Fb", required for various spans, lumber sizes and conditions of loading. Suitable species and grades meeting the indicated strength requirements can be selected from the tables of Allowable Unit Stresses in the Supplement to the NDS.

Abbreviated tables of Allowable Unit (Working) Stresses for representative species and for typical sizes and grades are included herein as Fig. WF3 (Visually Graded Lumber) and as Fig. WF4 (Machine Stress Rated Lumber).

MAXIMUM ALLOWABLE SPANS FOR JOISTS & RAFTERS

Abbreviated tables of Maximum Allowable Spans for Joists and Rafters for the range of strength characteristics generally encountered in light residential construction are included as Fig. WF5. The allowable span is the clear distance between supports. For rafters with a slope of 3 in 12 or greater, the span is measured along the horizontal projection (Fig. WF2). Tabulated spans are applicable to green or dry lumber, provided it is used in covered structures, where it will normally achieve a moisture content of 19% or less.

Example of floor joist selection: Assume a required span of 12'9", a live load of 40 psf and joists spaced 16" on centers. Fig. WF5 shows that a 2x8 having an "E" value of 1,600,000 psi and an "Fb" value of 1250 psi would have an allowable span of 12'-10". Suitable species and grades can be selected from Fig. WF3, (such as visually graded, dry or green Douglas Fir-Larch No. 2 grade) or Fig. WF4 (machine rated lumber, 1350f-1.8E, any WWPA species).

Example of rafter selection: Assume a horizontal projection span (Fig. WF2) of 13'0", a live load of 20 psf, dead load of 7 psf and rafters spaced 16" on centers. Fig. WF5 shows that a 2x6 having an "Fb" value of 1200 psi and an "E" value of 940,000 psi would have an allowable horizontal projection span of 13'-0". Suitable species and grades can be selected from Fig. WF3 (such as visually graded S-DRY Southern Pine No. 2) or Fig. WF4 (machine rated lumber, 1200f-1.2E, any WCLIB or WWPA species).

RAFTER SPAN, CONVERSION DIAGRAM

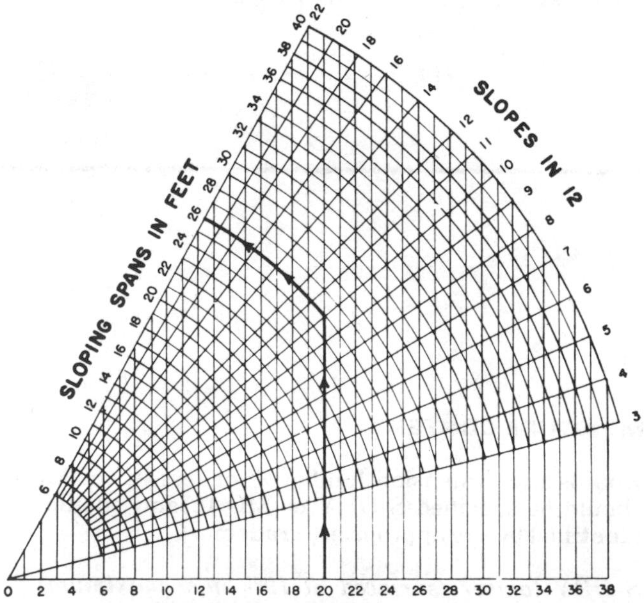

FIG. WF-2 To find the rafter span when its horizontal span and slope are known, follow the vertical line from the horizontal span to its intersection with the radial line of the slope. The diagram also may be used to determine the horizontal span when the sloping span and the slope are known, or to determine the slope when the sloping and horizontal spans are known. Example: For a horizontal span of 20 feet and a slope of 10 in 12, the sloping span of the rafter is read directly from the diagram as 26 feet. When an intermediate support is provided for rafters, the span is the actual length between top plate at eave and the intermediate support (as a knee wall). Collar ties should not be considered intermediate supports.

FIG. WF3 **ALLOWABLE UNIT STRESSES—VISUALLY GRADED LUMBER***

LIGHT FRAMING, JOISTS & PLANKS

Species and Grade	Size	Design Value in Bending "F_b" Normal Duration	Snow Loading	7-Day Loading	Modulus of Elasticity "E"	Grading Rules Agency
BALSAM FIR						
BALSAM FIR (Surfaced dry or surfaced green)						
Select Structural	2x4	1550	1780	1940	1,200,000	Northeastern Lumber Manufacturers Association
No. 1		1300	1500	1620	1,200,000	
No. 2		1100	1260	1380	1,100,000	
No. 3		600	690	750	900,000	
Appearance		1150	1320	1440	1,200,000	
Stud		600	690	750	900,000	Northern Hardwood & Pine Manufacturers Association
Construction	2x4	800	920	1000	900,000	
Standard		450	520	560	900,000	
Utility		200	230	250	900,000	
Select Structural	2x5 and wider	1350	1550	1690	1,200,000	(See notes 1 and 3)
No. 1 & Appearance		1150	1320	1440	1,200,000	
No. 2		950	1090	1190	1,100,000	
No. 3		550	630	690	900,000	
Stud		550	630	690	900,000	
CALIFORNIA REDWOOD						
CALIFORNIA REDWOOD (Surfaced dry or surfaced green)						
Clear Heart Structural	2x4	2650	3050	3310	1,400,000	
Clear Structural		2650	3050	3310	1,400,000	
Select Structural	2x4	2350	2700	2940	1,400,000	
Select Structural, Open grain		1850	2130	2310	1,100,000	
No. 1		1950	2240	2440	1,400,000	
No. 1, Open grain		1550	1780	1940	1,100,000	Redwood Inspection Service
No. 2		1600	1840	2000	1,250,000	
No. 2, Open grain		1250	1440	1560	1,000,000	(See notes 1 and 3)
No. 3		900	1040	1120	1,100,000	
No. 3, Open grain		725	830	910	900,000	
Stud		725	830	910	900,000	
Construction	2x4	950	1090	1190	900,000	
Standard		525	600	660	900,000	
Utility		250	290	310	900,000	
Clear Heart Structural	2x5 & wider	2650	3050	3310	1,400,000	
Clear Structural		2650	3050	3310	1,400,000	
Select Structural	2x5 and wider	2000	2300	2500	1,400,000	
Select Structural, Open grain		1600	1840	2000	1,100,000	
No. 1		1700	1960	2120	1,400,000	
No. 1, Open grain		1350	1550	1690	1,100,000	
No. 2		1400	1610	1750	1,250,000	
No. 2, Open grain		1100	1260	1380	1,000,000	
No. 3		800	920	1000	1,100,000	
No. 3, Open grain		650	750	810	900,000	
Stud		650	750	810	900.000	

*These "F_b" values are for use where repetitive members are spaced not more than 24 inches. For wider spacing, the "F_b" values should be reduced 13 percent. Values apply at 19 percent maximum moisture content in use.

1. When 2" lumber is manufactured at a maximum moisture content of 15 percent (grade-marked MC-15) and used in a condition where the moisture content does not exceed 15 percent the design values shown in Table W-1 for "surfaced dry or surfaced green" lumber may be increased eight percent (8%) for design value in bending "F_b", and five percent (5%) for Modulus of Elasticity "E".

2. National Lumber Grades Authority is the Canadian rules writing agency responsible for preparation, maintenance and dissemination of a uniform softwood lumber grading rule for all Canadian species.

3. Design values for Stud grade in 2x5 and wider size classifications apply to 5" and 6" widths only.

FIG. WF3 **ALLOWABLE UNIT STRESSES—VISUALLY GRADED LUMBER*** *Continued*

LIGHT FRAMING, JOISTS & PLANKS

Species	Species and Grade	Size	Design Value in Bending "F_b" Normal Duration	Snow Loading	7-Day Loading	Modulus of Elasticity "E"	Grading Rules Agency
NORTHERN WHITE CEDAR	**NORTHERN WHITE CEDAR (Surfaced dry or surfaced green)**						
	Select Structural	2x4	1350	1550	1690	800,000	Northeastern Lumber Manufacturers Association (See notes 1 and 3)
	No. 1	2x4	1150	1320	1440	800,000	
	No. 2	2x4	950	1090	1190	700,000	
	No. 3	2x4	525	600	660	600,000	
	Appearance	2x4	1000	1150	1250	800,000	
	Stud	2x4	525	600	660	600,000	
	Construction	2x4	675	780	840	600,000	
	Standard	2x4	375	430	470	600,000	
	Utility	2x4	175	200	220	600,000	
	Select Structural	2x5 and wider	1150	1320	1440	800,000	
	No. 1 & Appearance	2x5 and wider	1000	1150	1250	800,000	
	No. 2	2x5 and wider	825	950	1030	700,000	
	No. 3	2x5 and wider	475	550	590	600,000	
	Stud	2x5 and wider	475	550	590	600,000	
WESTERN CEDARS	**WESTERN CEDARS (Surfaced dry or surfaced green)**						
	Select Structural	2x4	1750	2010	2190	1,100,000	Western Wood Products Association (See notes 1 and 3) West Coast Lumber Inspection Bureau
	No. 1 & Appearance	2x4	1500	1720	1880	1,100,000	
	No. 2	2x4	1200	1380	1500	1,000,000	
	No. 3	2x4	675	780	840	900,000	
	Stud	2x4	675	780	840	900,000	
	Construction	2x4	875	1010	1090	900,000	
	Standard	2x4	500	580	620	900,000	
	Utility	2x4	225	260	280	900,000	
	Select Structural	2x5 and wider	1500	1720	1880	1,100,000	
	No. 1 & Appearance	2x5 and wider	1300	1500	1620	1,100,000	
	No. 2	2x5 and wider	1050	1210	1310	1,000,000	
	No. 3	2x5 and wider	625	720	780	900,000	
	Stud	2x5 and wider	625	720	780	900,000	
WESTERN CEDARS (NORTH)	**WESTERN CEDARS (NORTH) (Surfaced dry or surfaced green)**						
	Select Structural	2x4	1700	1960	2120	1,100,000	Nat'l. Lumber Grades Auth. (A Canadian Agency— See notes 1, 2 and 3)
	No. 1 & Appearance	2x4	1450	1670	1810	1,100,000	
	No. 2	2x4	1200	1380	1500	1,000,000	
	No. 3	2x4	650	750	810	900,000	
	Stud	2x4	650	750	810	900,000	
	Construction	2x4	850	980	1060	900,000	
	Standard	2x4	475	550	590	900,000	
	Utility	2x4	225	260	280	900,000	
	Select Structural	2x5 and wider	1450	1670	1810	1,100,000	
	No. 1 & Appearance	2x5 and wider	1250	1440	1560	1,100,000	
	No. 2	2x5 and wider	1000	1150	1250	1,000,000	
	No. 3	2x5 and wider	600	690	750	900,000	
	Stud	2x5 and wider	600	690	750	900,000	

*These "F_b" values are for use where repetitive members are spaced not more than 24 inches. For wider spacing, the "F_b" values should be reduced 13 percent. Values apply at 19 percent maximum moisture content in use.

1. When 2" lumber is manufactured at a maximum moisture content of 15 percent (grade-marked MC-15) and used in a condition where the moisture content does not exceed 15 percent the design values shown in Table W-1 for "surfaced dry or surfaced green" lumber may be increased

FIG. WF3 **ALLOWABLE UNIT STRESSES—VISUALLY GRADED LUMBER*** *Continued*

LIGHT FRAMING, JOISTS & PLANKS

DOUGLAS FIR-LARCH (NORTH)

Species and Grade	Size	Design Value in Bending "F_b" Normal Duration	Design Value in Bending "F_b" Snow Loading	Design Value in Bending "F_b" 7-Day Loading	Modulus of Elasticity "E"	Grading Rules Agency
DOUGLAS FIR–LARCH (NORTH) (Surfaced dry or surfaced green)						
Dense Select Structural	2x4	2800	3220	3500	1,900,000	Nat'l. Lumber Grades Auth. (A Canadian Agency—See notes 1, 2 and 3)
Select Structural	2x4	2400	2760	3000	1,800,000	
Dense No. 1	2x4	2400	2760	3000	1,900,000	
No. 1 & Appearance	2x4	2050	2360	2560	1,800,000	
Dense No. 2	2x4	1950	2240	2440	1,700,000	
No. 2	2x4	1650	1900	2060	1,700,000	
No. 3	2x4	925	1060	1160	1,500,000	
Stud	2x4	925	1060	1160	1,500,000	
Construction	2x4	1200	1380	1500	1,500,000	
Standard	2x4	675	780	840	1,500,000	
Utility	2x4	325	370	410	1,500,000	
Dense Select Structural	2x5 and wider	2400	2760	3000	1,900,000	
Select Structural	2x5 and wider	2050	2360	2560	1,800,000	
Dense No. 1	2x5 and wider	2050	2360	2560	1,900,000	
No. 1 & Appearance	2x5 and wider	1750	2010	2190	1,800,000	
Dense No. 2	2x5 and wider	1700	1960	2120	1,700,000	
No. 2	2x5 and wider	1450	1670	1810	1,700,000	
No. 3	2x5 and wider	850	980	1060	1,500,000	
Stud	2x5 and wider	850	980	1060	1,500,000	

DOUGLAS FIR SOUTH

Species and Grade	Size	Design Value in Bending "F_b" Normal Duration	Design Value in Bending "F_b" Snow Loading	Design Value in Bending "F_b" 7-Day Loading	Modulus of Elasticity "E"	Grading Rules Agency
DOUGLAS FIR SOUTH (Surfaced dry or surfaced green)						
Select Structural	2x4	2300	2640	2880	1,400,000	Western Wood Products Association (See notes 1 and 3)
No. 1 & Appearance	2x4	1950	2240	2440	1,400,000	
No. 2	2x4	1600	1840	2000	1,300,000	
No. 3	2x4	875	1010	1090	1,100,000	
Stud	2x4	875	1010	1090	1,100.000	
Construction	2x4	1150	1320	1440	1,100,000	
Standard	2x4	650	750	810	1,100,000	
Utility	2x4	300	340	380	1,100,000	
Select Structural	2x5 and wider	1950	2240	2440	1,400,000	
No. 1 & Appearance	2x5 and wider	1650	1900	2060	1,400,000	
No. 2	2x5 and wider	1350	1550	1690	1,300,000	
No. 3	2x5 and wider	800	920	1000	1,100,000	
Stud	2x5 and wider	800	920	1000	1,100,000	

eight percent (8%) for design value in bending "F_b", and five percent (5%) for Modulus of Elasticity "E".

2. National Lumber Grades Authority is the Canadian rules writing agency responsible for preparation, maintenance and dissemination of a uniform softwood lumber grading rule for all Canadian species.

3. Design values for Stud grade in 2x5 and wider size classifications apply to 5" and 6" widths only.

FIG. WF3 **ALLOWABLE UNIT STRESSES—VISUALLY GRADED LUMBER*** *Continued*

LIGHT FRAMING, JOISTS & PLANKS

Species	Species and Grade	Size	Design Value in Bending "F_b" Normal Duration	Snow Loading	7-Day Loading	Modulus of Elasticity "E"	Grading Rules Agency
ENGELMANN SPRUCE	**ENGELMANN SPRUCE—ALPINE FIR (ENGELMANN SPRUCE—LODGEPOLE PINE) (Surfaced dry or surfaced green)**						Western Wood Products Association (See notes 1 and 3)
	Select Structural	2x4	1550	1780	1940	1,300,000	
	No. 1 & Appearance	2x4	1350	1550	1690	1,300,000	
	No. 2	2x4	1100	1260	1380	1,100,000	
	No. 3	2x4	600	690	750	1,000,000	
	Stud	2x4	600	690	750	1,000,000	
	Construction	2x4	800	920	1000	1,000,000	
	Standard	2x4	450	520	560	1,000,000	
	Utility	2x4	200	230	250	1,000,000	
	Select Structural	2x5 and wider	1350	1550	1690	1,300,000	
	No. 1 & Appearance	2x5 and wider	1150	1320	1440	1,300,000	
	No. 2	2x5 and wider	950	1090	1190	1,100,000	
	No. 3	2x5 and wider	550	630	690	1,000,000	
	Stud	2x5 and wider	550	630	690	1,000,000	
HEM-FIR	**HEM—FIR (Surfaced dry or surfaced green)**						Western Wood Products Association (See notes 1 and 3) West Coast Lumber Inspection Bureau
	Select Structural	2x4	1900	2180	2380	1,500,000	
	No. 1 & Appearance	2x4	1600	1840	2000	1,500,000	
	No. 2	2x4	1350	1550	1690	1,400,000	
	No. 3	2x4	725	830	910	1,200,000	
	Stud	2x4	725	830	910	1,200,000	
	Construction	2x4	975	1120	1220	1,200,000	
	Standard	2x4	550	630	690	1,200,000	
	Utility	2x4	250	290	310	1,200,000	
	Select Structural	2x5 and wider	1650	1900	2060	1,500,000	
	No. 1 & Appearance	2x5 and wider	1400	1610	1750	1,500,000	
	No. 2	2x5 and wider	1150	1320	1440	1,400,000	
	No. 3	2x5 and wider	675	780	840	1,200,000	
	Stud	2x5 and wider	675	780	840	1,200,000	
IDAHO WHITE PINE	**IDAHO WHITE PINE (Surfaced dry or surfaced green)**						Western Wood Products Association (See notes 1 and 3)
	Select Structural	2x4	1550	1780	1940	1,400,000	
	No. 1 & Appearance	2x4	1300	1500	1620	1,400,000	
	No. 2	2x4	1050	1210	1310	1,300,000	
	No. 3	2x4	600	690	750	1,200,000	
	Stud	2x4	600	690	750	1,200,000	
	Construction	2x4	775	890	970	1,200,000	
	Standard	2x4	425	490	530	1,200,000	
	Utility	2x4	200	230	250	1,200,000	
	Select Structural	2x5 and wider	1300	1500	1620	1,400,000	
	No. 1 & Appearance	2x5 and wider	1100	1260	1380	1,400,000	
	No. 2	2x5 and wider	925	1060	1160	1,300,000	
	No. 3	2x5 and wider	550	630	690	1,200,000	
	Stud	2x5 and wider	550	630	690	1,200,000	

*These "F_b" values are for use where repetitive members are spaced not more than 24 inches. For wider spacing, the "F_b" values should be reduced 13 percent. Values apply at 19 percent maximum moisture content in use.

1. When 2" lumber is manufactured at a maximum moisture content of 15 percent (grade-marked MC-15) and used in a condition where the moisture content does not exceed 15 percent the design values shown in Table W-1 for "surfaced dry or surfaced green" lumber may be increased

FIG. WF3 **ALLOWABLE UNIT STRESSES—VISUALLY GRADED LUMBER*** *Continued*

LIGHT FRAMING, JOISTS & PLANKS

Species	Species and Grade	Size	Design Value in Bending "F_b" Normal Duration	Snow Loading	7-Day Loading	Modulus of Elasticity "E"	Grading Rules Agency
LODGEPOLE PINE	**LODGEPOLE PINE (Surfaced dry or surfaced green)**						
	Select Structural	2x4	1750	2010	2190	1,300,000	Western Wood Products Association (See notes 1 and 3)
	No. 1 & Appearance	2x4	1500	1720	1880	1,300,000	
	No. 2	2x4	1200	1380	1500	1,200,000	
	No. 3	2x4	675	780	840	1,000,000	
	Stud	2x4	675	780	840	1,000,000	
	Construction	2x4	875	1010	1090	1,000,000	
	Standard	2x4	500	580	620	1,000,000	
	Utility	2x4	225	260	280	1,000,000	
	Select Structural	2x5 and wider	1500	1720	1880	1,300,000	
	No. 1 & Appearance	2x5 and wider	1300	1500	1620	1,300,000	
	No. 2	2x5 and wider	1050	1210	1310	1,200,000	
	No. 3	2x5 and wider	625	720	780	1,000,000	
	Stud	2x5 and wider	625	720	780	1,000,000	
PONDEROSA PINE	**PONDEROSA PINE (Surfaced dry or surfaced green)**						
	Select Structural	2x4	1650	1900	2060	1,200,000	Nat'l. Lumber Grades Auth. (A Canadian Agency— See notes 1, 2 and 3)
	No. 1 & Appearance	2x4	1400	1610	1750	1,200,000	
	No. 2	2x4	1150	1320	1440	1,100,000	
	No. 3	2x4	625	720	780	1,000,000	
	Stud	2x4	625	720	780	1,000,000	
	Construction	2x4	825	950	1030	1,000,000	
	Standard	2x4	450	520	560	1,000,000	
	Utility	2x4	225	260	280	1,000,000	
	Select Structural	2x5 and wider	1400	1610	1750	1,200,000	
	No. 1 & Appearance	2x5 and wider	1200	1380	1500	1,200,000	
	No. 2	2x5 and wider	975	1120	1220	1,100,000	
	No. 3	2x5 and wider	575	660	720	1,000,000	
	Stud	2x5 and wider	575	660	720	1,000,000	
SPRUCE-PINE-FIR	**SPRUCE–PINE–FIR (Surfaced dry or surfaced green)**						
	Select Structural	2x4	1650	1900	2060	1,500,000	Nat'l. Lumber Grades Auth. (A Canadian Agency— See notes 1, 2 and 3)
	No. 1 & Appearance	2x4	1400	1610	1750	1,500,000	
	No. 2	2x4	1150	1320	1440	1,300,000	
	No. 3	2x4	650	750	810	1,200,000	
	Stud	2x4	650	750	810	1,200,000	
	Construction	2x4	850	980	1060	1,200,000	
	Standard	2x4	475	550	590	1,200,000	
	Utility	2x4	225	260	280	1,200,000	
	Select Structural	2x5 and wider	1450	1670	1810	1,500,000	
	No. 1 & Appearance	2x5 and wider	1200	1380	1500	1,500,000	
	No. 2	2x5 and wider	1000	1150	1250	1,300,000	
	No. 3	2x5 and wider	575	660	720	1,200,000	
	Stud	2x5 and wider	575	660	720	1,200,000	

eight percent (8%) for design value in bending "F_b", and five percent (5%) for Modulus of Elasticity "E".

2. National Lumber Grades Authority is the Canadian rules writing agency responsible for preparation, maintenance and dissemination of a uniform softwood lumber grading rule for all Canadian species.

3. Design values for Stud grade in 2x5 and wider size classifications apply to 5" and 6" widths only.

FIG. WF3 **ALLOWABLE UNIT STRESSES—VISUALLY GRADED LUMBER*** *Continued*

LIGHT FRAMING, JOISTS & PLANKS

Species and Grade	Size	Design Value in Bending "F_b" Normal Duration	Snow Loading	7-Day Loading	Modulus of Elasticity "E"	Grading Rules Agency
SOUTHERN PINE (SURFACED DRY)						
SOUTHERN PINE (Surfaced dry)						
Select Structural		2300	2640	2880	1,700,000	
Dense Select Structural		2700	3100	3380	1,800,000	
No. 1		1950	2240	2440	1,700,000	
No. 1 Dense		2300	2640	2880	1,800,000	
No. 2	2x4	1650	1900	2060	1,600,000	
No. 2 Dense		1900	2180	2380	1,600,000	
No. 3		900	1040	1120	1,400,000	
No. 3 Dense		1050	1210	1310	1,500,000	Southern
Stud		900	1040	1120	1,400,000	Pine
						Inspection
Construction		1150	1320	1440	1,400,000	Bureau
Standard	2x4	675	780	840	1,400,000	
Utility		300	340	380	1,400,000	(See note 3)
Select Structural		2000	2300	2500	1,700,000	
Dense Select Structural		2350	2700	2940	1,800,000	
No. 1		1700	1960	2120	1,700,000	
No. 1 Dense		2000	2300	2500	1,800,000	
	2x5					
No. 2	and	1400	1610	1750	1,600,000	
No. 2 Dense	wider	1650	1900	2060	1,600,000	
No. 3		800	920	1000	1,400,000	
No. 3 Dense		925	1060	1160	1,500,000	
Stud		850	980	1060	1,400,000	
SOUTHERN PINE (15% MC)						
SOUTHERN PINE (Surfaced at 15 percent moisture content-KD)						
Select Structural		2500	2880	3120	1,800,000	
Dense Select Structural		2900	3340	3620	1,900,000	
No. 1		2100	2420	2620	1,800,000	
No. 1 Dense		2450	2820	3060	1,900,000	
No. 2	2x4	1750	2010	2190	1,600,000	
No. 2 Dense		2050	2360	2560	1,700,000	Southern
No. 3		975	1120	1220	1,500,000	Pine
No. 3 Dense		1150	1320	1440	1,500,000	Inspection
Stud		975	1120	1220	1,500,000	Bureau
Construction		1250	1440	1560	1,500,000	
Standard	2x4	725	830	910	1,500,000	
Utility		300	340	380	1,500,000	(See note 3)
Select Structural		2150	2470	2690	1,800,000	
Dense Select Structural		2500	2880	3120	1,900,000	
No. 1		1850	2130	2310	1,800,000	
No. 1 Dense	2x5	2150	2470	2690	1,900,000	
	and					
No. 2	wider	1500	1720	1880	1,600,000	
No. 2 Dense		1750	2010	2190	1,700,000	
No. 3		875	1010	1090	1,500,000	
No. 3 Dense		1000	1150	1250	1,500,000	
Stud		900	1040	1120	1,500,000	

*These "F_b" values are for use where repetitive members are spaced not more than 24 inches. For wider spacing, the "F_b" values should be reduced 13 percent. Values apply at 19 percent maximum moisture content in use.

1. When 2" lumber is manufactured at a maximum moisture content of 15 percent (grade-marked MC-15) and used in a condition where the moisture content does not exceed 15 percent the design values shown in Table W-1 for "surfaced dry or surfaced green" lumber may be increased eight percent (8%) for design value in bending "F_b", and five percent (5%) for Modulus of Elasticity "E".

2. National Lumber Grades Authority is the Canadian rules writing agency responsible for preparation, maintenance and dissemination of a uniform softwood lumber grading rule for all Canadian species.

3. Design values for Stud grade in 2x5 and wider size classifications apply to 5" and 6" widths only.

FIG. WF4 **ALLOWABLE UNIT STRESSES—MACHINE RATED LUMBER***

LIGHT FRAMING, JOISTS & PLANKS

Grade Designation	Grading Rules Agency (see footnotes 1,2,3,4)	Size Classification	Design Value in Bending "F_b"			Modulus of Elasticity "E"
			Normal Duration	Snow Loading	7-Day Loading	
900f-1.0E	3	Machine rated lumber, 2x4 and wider	1050	1210	1310	1,000,000
1200f-1.2E	1,2,3,4		1400	1610	1750	1,200,000
1350f-1.3E	2,4		1550	1780	1940	1,300,000
1450f-1.3E	1,3,4		1650	1900	2060	1,300,000
1500f-1.4E	1,2,3,4		1750	2010	2190	1,400,000
1650f-1.5E	1,2,3,4		1900	2180	2380	1,500,000
1800f-1.6E	1,2,3,4		2050	2360	2560	1,600,000
1950f-1.7E	1,2,4		2250	2590	2810	1,700,000
2100f-1.8E	1,2,3,4		2400	2760	3000	1,800,000
2250f-1.9E	1,2,4		2600	2990	3250	1,900,000
2400f-2.0E	1,2,3,4		2750	3160	3440	2,000,000
2550f-2.1f	1,2,4		2950	3390	3690	2,100,000
2700f-2.2E	1,2,3,4		3100	3570	3880	2,200,000
2850f-2.3E	2,4		3300	3800	4130	2,300,000
3000f-2.4E	1,2,4		3450	3970	4310	2,400,000
3150f-2.5E	2,4		3600	4140	4500	2,500,000
3300f-2.6E	2,4		3800	4370	4750	2,600,000
900f-1.0E	1,2,3,4	See footnotes	1050	1210	1310	1,000,000
900f-1.2E	1,2,3,4		1050	1210	1310	1,200,000
1200f-1.5E	1,2,3,4		1400	1610	1750	1,500,000
1350f-1.8E	1,2,4		1550	1780	1940	1,800,000
1500f-1.8E	3		1750	2010	2190	1,800,000
1800f-2.1E	1,2,3,4		2050	2360	2560	2,100,000

*These "F_b" values are for use where repetitive members are spaced not more than 24 inches. For wider spacing, the "F_b" values should be reduced 13 percent. Values apply at 19 percent maximum moisture content in use.

1. National Lumber Grades Authority (see Footnote 2, Fig. WF3); Machine Rated Lumber, 2x4 and wider.
2. Southern Pine Inspection Bureau; Machine Rated Lumber, 2x4 and wider.
3. West Coast Lumber Inspection Bureau; Machine Rated Lumber, 2x4 and wider; Machine Rated Joists, 2x6 and wider.
4. Western Wood Products Association; Machine Rated Lumber, 2x4 and wider.

FIGURE WF-5 —**MAXIMUM ALLOWABLE SPANS FOR JOISTS & RAFTERS**

FLOOR JOISTS*

40 psf

(All rooms except those used for sleeping areas and attic floors)

DESIGN CRITERIA:
Deflection - For 40 lbs. per sq. ft. live load. Limited to span in inches divided by 360.
Strength - Live load of 40 lbs. per sq. ft. plus dead load of 10 lbs. per sq. ft. determines the required fiber stress value.

JOIST SIZE (IN)	SPACING (IN)	Modulus of Elasticity, "E", in 1,000,000 psi													
		0.8	0.9	1.0	1.1	1.2	1.3	1.4	1.5	1.6	1.7	1.8	1.9	2.0	2.2
2x6	12.0	8-6 720	8-10 780	9-2 830	9-6 890	9-9 940	10-0 990	10-3 1040	10-6 1090	10-9 1140	10-11 1190	11-2 1230	11-4 1280	11-7 1320	11-11 1410
	16.0	7-9 790	8-0 860	8-4 920	8-7 980	8-10 1040	9-1 1090	9-4 1150	9-6 1200	9-9 1250	9-11 1310	10-2 1360	10-4 1410	10-6 1460	10-10 1550
	24.0	6-9 900	7-0 980	7-3 1050	7-6 1120	7-9 1190	7-11 1250	8-2 1310	8-4 1380	8-6 1440	8-8 1500	8-10 1550	9-0 1610	9-2 1670	9-6 1780
2x8	12.0	11-3 720	11-8 780	12-1 830	12-6 890	12-10 940	13-2 990	13-6 1040	13-10 1090	14-2 1140	14-5 1190	14-8 1230	15-0 1280	15-3 1320	15-9 1410
	16.0	10-2 790	10-7 850	11-0 920	11-4 980	11-8 1040	12-0 1090	12-3 1150	12-7 1200	12-10 1250	13-1 1310	13-4 1360	13-7 1410	13-10 1460	14-3 1550
	24.0	8-11 900	9-3 980	9-7 1050	9-11 1120	10-2 1190	10-6 1250	10-9 1310	11-0 1380	11-3 1440	11-5 1500	11-8 1550	11-11 1610	12-1 1670	12-6 1780
2x10	12.0	14-4 720	14-11 780	15-5 830	15-11 890	16-5 940	16-10 990	17-3 1040	17-8 1090	18-0 1140	18-5 1190	18-9 1230	19-1 1280	19-5 1320	20-1 1410
	16.0	13-0 790	13-6 850	14-0 920	14-6 980	14-11 1040	15-3 1090	15-8 1150	16-0 1200	16-5 1250	16-9 1310	17-0 1360	17-4 1410	17-8 1460	18-3 1550
	24.0	11-4 900	11-10 980	12-3 1050	12-8 1120	13-0 1190	13-4 1250	13-8 1310	14-0 1380	14-4 1440	14-7 1500	14-11 1550	15-2 1610	15-5 1670	15-11 1780
2x12	12.0	17-5 720	18-1 780	18-9 830	19-4 890	19-11 940	20-6 990	21-0 1040	21-6 1090	21-11 1140	22-5 1190	22-10 1230	23-3 1280	23-7 1320	24-5 1410
	16.0	15-10 790	16-5 860	17-0 920	17-7 980	18-1 1040	18-7 1090	19-1 1150	19-6 1200	19-11 1250	20-4 1310	20-9 1360	21-1 1410	21-6 1460	22-2 1550
	24.0	13-10 900	14-4 980	14-11 1050	15-4 1120	15-10 1190	16-3 1250	16-8 1310	17-0 1380	17-5 1440	17-9 1500	18-1 1550	18-5 1610	18-9 1670	19-4 1780

30 psf

(All rooms used for sleeping areas and attic floors)

DESIGN CRITERIA:
Deflection - For 30 lbs. per sq. ft. live load. Limited to span in inches divided by 360.
Strength - Live load of 30 lbs. per sq. ft. plus dead load of 10 lbs. per sq. ft. determines the required fiber stress value.

JOIST SIZE (IN)	SPACING (IN)	Modulus of Elasticity, "E", in 1,000,000 psi													
		0.8	0.9	1.0	1.1	1.2	1.3	1.4	1.5	1.6	1.7	1.8	1.9	2.0	2.2
2x6	12.0	9-4 700	9-9 750	10-1 810	10-5 860	10-9 910	11-0 960	11-3 1010	11-7 1060	11-10 1100	12-0 1150	12-3 1200	12-6 1240	12-9 1280	13-1 1370
	16.0	8-6 770	8-10 830	9-2 890	9-6 950	9-9 1000	10-0 1060	10-3 1110	10-6 1160	10-9 1220	10-11 1270	11-2 1320	11-4 1360	11-7 1410	11-11 1500
	24.0	7-5 880	7-9 950	8-0 1020	8-3 1080	8-6 1150	8-9 1210	8-11 1270	9-2 1330	9-4 1390	9-7 1450	9-9 1510	9-11 1560	10-1 1620	10-5 1720
2x8	12.0	12-4 700	12-10 750	13-4 810	13-9 860	14-2 910	14-6 960	14-11 1010	15-3 1060	15-7 1100	15-10 1150	16-2 1200	16-6 1240	16-9 1280	17-4 1370
	16.0	11-3 770	11-8 830	12-1 890	12-6 950	12-10 1000	13-2 1060	13-6 1110	13-10 1160	14-2 1220	14-5 1270	14-8 1320	15-0 1360	15-3 1410	15-9 1500
	24.0	9-10 880	10-2 950	10-7 1020	10-11 1080	11-3 1150	11-6 1210	11-10 1270	12-1 1330	12-4 1390	12-7 1450	12-10 1510	13-1 1560	13-4 1620	13-9 1720
2x10	12.0	15-9 700	16-5 750	17-0 810	17-6 860	18-0 910	18-6 960	19-0 1010	19-5 1060	19-10 1100	20-3 1150	20-8 1200	21-0 1240	21-5 1280	22-1 1370
	16.0	14-4 770	14-11 830	15-5 890	15-11 950	16-5 1000	16-10 1060	17-3 1110	17-8 1160	18-0 1220	18-5 1270	18-9 1320	19-1 1360	19-5 1410	20-1 1500
	24.0	12-6 880	13-0 950	13-6 1020	13-11 1080	14-4 1150	14-8 1210	15-1 1270	15-5 1330	15-9 1390	16-1 1450	16-5 1510	16-8 1560	17-0 1620	17-6 1720
2x12	12.0	19-2 700	19-11 750	20-8 810	21-4 860	21-11 910	22-6 960	23-1 1010	23-7 1060	24-2 1100	24-8 1150	25-1 1200	25-7 1240	26-0 1280	26-10 1370
	16.0	17-5 770	18-1 830	18-9 890	19-4 950	19-11 1000	20-6 1060	21-0 1110	21-6 1160	21-11 1220	22-5 1270	22-10 1320	23-3 1360	23-7 1410	24-5 1500
	24.0	15-2 880	15-10 950	16-5 1020	16-11 1080	17-5 1150	17-11 1210	18-4 1270	18-9 1330	19-2 1390	19-7 1450	19-11 1510	20-3 1560	20-8 1620	21-4 1720

*The required extreme fiber stress in bending, "F_b", in pounds per square inch is shown below each span.

FIGURE WF-5 (Continued) **MAXIMUM ALLOWABLE SPANS FOR JOISTS & RAFTERS**

CEILING JOISTS*

20 psf

(Limited attic storage where development of future rooms is not possible. Drywall Ceiling)

DESIGN CRITERIA:
Deflection - For 20 lbs. per sq. ft. live load. Limited to span in inches divided by 240.
Strength - live load of 20 lbs. per sq. ft. plus dead load of 10 lbs. per sq. ft. determines required fiber stress value.

SIZE (IN)	JOIST SPACING (IN)	Modulus of Elasticity, "E", 1,000,000 psi													
		0.8	0.9	1.0	1.1	1.2	1.3	1.4	1.5	1.6	1.7	1.8	1.9	2.0	2.2
2x4	12.0	7-10 900	8-1 970	8-5 1040	8-8 1110	8-11 1170	9-2 1240	9-5 1300	9-8 1360	9-10 1420	10-0 1480	10-3 1540	10-5 1600	10-7 1650	10-11 1760
	16.0	7-1 990	7-5 1070	7-8 1140	7-11 1220	8-1 1290	8-4 1360	8-7 1430	8-9 1500	8-11 1570	9-1 1630	9-4 1690	9-6 1760	9-8 1820	9-11 1940
	24.0	6-2 1130	6-5 1220	6-8 1310	6-11 1400	7-1 1480	7-3 1560	7-6 1640	7-8 1720	7-10 1790	8-0 1870	8-1 1940	8-3 2010	8-5 2080	8-8 2220
2x6	12.0	12-3 900	12-9 970	13-3 1040	13-8 1110	14-1 1170	14-5 1240	14-9 1300	15-2 1360	15-6 1420	15-9 1480	16-1 1540	16-4 1600	16-8 1650	17-2 1760
	16.0	11-2 990	11-7 1070	12-0 1140	12-5 1220	12-9 1290	13-1 1360	13-5 1430	13-9 1500	14-1 1570	14-4 1630	14-7 1690	14-11 1760	15-2 1820	15-7 1940
	24.0	9-9 1130	10-2 1220	10-6 1310	10-10 1400	11-2 1480	11-5 1560	11-9 1640	12-0 1720	12-3 1790	12-6 1870	12-9 1940	13-0 2010	13-3 2080	13-8 2220
2x8	12.0	16-2 900	16-10 970	17-5 1040	18-0 1110	18-6 1170	19-0 1240	19-6 1300	19-11 1360	20-5 1420	20-10 1480	21-2 1540	21-7 1600	21-11 1650	22-8 1760
	16.0	14-8 990	15-3 1070	15-10 1140	16-4 1220	16-10 1290	17-3 1360	17-9 1430	18-2 1500	18-6 1570	18-11 1630	19-3 1690	19-7 1760	19-11 1820	20-7 1940
	24.0	12-10 1130	13-4 1220	13-10 1310	14-3 1400	14-8 1480	15-1 1560	15-6 1640	15-10 1720	16-2 1790	16-6 1870	16-10 1940	17-2 2010	17-5 2080	18-0 2220
2x10	12.0	20-8 900	21-6 970	22-3 1040	22-11 1110	23-8 1170	24-3 1240	24-10 1300	25-5 1360	26-0 1420	26-6 1480	27-1 1540	27-6 1600	28-0 1650	28-11 1760
	16.0	18-9 990	19-6 1070	20-2 1140	20-10 1220	21-6 1290	22-1 1360	22-7 1430	23-2 1500	23-8 1570	24-1 1630	24-7 1690	25-0 1760	25-5 1820	26-3 1940
	24.0	16-5 1130	17-0 1220	17-8 1310	18-3 1400	18-9 1480	19-3 1560	19-9 1640	20-2 1720	20-8 1790	21-1 1870	21-6 1940	21-10 2010	22-3 2080	22-11 2220

10 psf

(No attic storage and roof slope not steeper than 3 in 12. Drywall Ceiling)

DESIGN CRITERIA:
Deflection - For 10 lbs. per sq. ft. live load. Limited to span in inches divided by 240.
Strength - live load of 10 lbs. per sq. ft. plus dead load of 5 lbs. per sq. ft. determines required fiber stress value.

SIZE (IN)	JOIST SPACING (IN)	Modulus of Elasticity, "E", in 1,000,000 psi													
		0.8	0.9	1.0	1.1	1.2	1.3	1.4	1.5	1.6	1.7	1.8	1.9	2.0	2.2
2x4	12.0	9-10 710	10-3 770	10-7 830	10-11 880	11-3 930	11-7 980	11-10 1030	12-2 1080	12-5 1130	12-8 1180	12-11 1220	13-2 1270	13-4 1310	13-9 1400
	16.0	8-11 780	9-4 850	9-8 910	9-11 970	10-3 1030	10-6 1080	10-9 1140	11-0 1190	11-3 1240	11-6 1290	11-9 1340	11-11 1390	12-2 1440	12-6 1540
	24.0	7-10 900	8-1 970	8-5 1040	8-8 1110	8-11 1170	9-2 1240	9-5 1300	9-8 1360	9-10 1420	10-0 1480	10-3 1540	10-5 1600	10-7 1650	10-11 1760
2x6	12.0	15-6 710	16-1 770	16-8 830	17-2 880	17-8 930	18-2 980	18-8 1030	19-1 1080	19-6 1130	19-11 1180	20-3 1220	20-8 1270	21-0 1310	21-8 1400
	16.0	14-1 780	14-7 850	15-2 910	15-7 970	16-1 1030	16-6 1080	16-11 1140	17-4 1190	17-8 1240	18-1 1290	18-5 1340	18-9 1390	19-1 1440	19-8 1540
	24.0	12-3 900	12-9 970	13-3 1040	13-8 1110	14-1 1170	14-5 1240	14-9 1300	15-2 1360	15-6 1420	15-9 1480	16-1 1540	16-4 1600	16-8 1650	17-2 1760
2x8	12.0	20-5 710	21-2 770	21-11 830	22-8 880	23-4 930	24-0 980	24-7 1030	25-2 1080	25-8 1130	26-2 1180	26-9 1220	27-2 1270	27-8 1310	28-7 1400
	16.0	18-6 780	19-3 850	19-11 910	20-7 970	21-2 1030	21-9 1080	22-4 1140	22-10 1190	23-4 1240	23-10 1290	24-3 1340	24-8 1390	25-2 1440	25-11 1540
	24.0	16-2 900	16-10 970	17-5 1040	18-0 1110	18-6 1170	19-0 1240	19-6 1300	19-11 1360	20-5 1420	20-10 1480	21-2 1540	21-7 1600	21-11 1650	22-8 1760
2x10	12.0	26-0 710	27-1 770	28-0 830	28-11 880	29-9 930	30-7 980	31-4 1030	32-1 1080	32-9 1130	33-5 1180	34-1 1220	34-8 1270	35-4 1310	36-5 1400
	16.0	23-8 780	24-7 850	25-5 910	26-3 970	27-1 1030	27-9 1080	28-6 1140	29-2 1190	29-9 1240	30-5 1290	31-0 1340	31-6 1390	32-1 1440	33-1 1540
	24.0	20-8 900	21-6 970	22-3 1040	22-11 1110	23-8 1170	24-3 1240	24-10 1300	25-5 1360	26-0 1420	26-6 1480	27-1 1540	27-6 1600	28-0 1650	28-11 1760

*The required extreme fiber stress in bending, "F_b", in pounds per square inch is shown below each span.

FIGURE WF-5 (Continued) **MAXIMUM ALLOWABLE SPANS FOR JOISTS & RAFTERS**

LOW OR HIGH SLOPE RAFTERS*

20 psf (Supporting Drywall Ceiling)

DESIGN CRITERIA:
Strength - 15 lbs. per sq. ft. dead load plus 20 lbs. per sq. ft. live load determines required fiber stress.
Deflection - For 20 lbs. per sq. ft. live load. Limited to span in inches divided by 240.

RAFTER SIZE (IN)	RAFTER SPACING (IN)	Allowable Extreme Fiber Stress in Bending, "F_b" (psi). 500	600	700	800	900	1000	1100	1200	1300	1400	1500	1600	1700	1800	1900
2x6	12.0	8-6 0.26	9-4 0.35	10-0 0.44	10-9 0.54	11-5 0.64	12-0 0.75	12-7 0.86	13-2 0.98	13-8 1.11	14-2 1.24	14-8 1.37	15-2 1.51	15-8 1.66	16-1 1.81	16-7 1.96
	16.0	7-4 0.23	8-1 0.30	8-8 0.38	9-4 0.46	9-10 0.55	10-5 0.65	10-11 0.75	11-5 0.85	11-10 0.97	12-4 1.07	12-9 1.19	13-2 1.31	13-7 1.44	13-11 1.56	14-4 1.70
	24.0	6-0 0.19	6-7 0.25	7-1 0.31	7-7 0.38	8-1 0.45	8-6 0.53	8-11 0.61	9-4 0.70	9-8 0.78	10-0 0.88	10-5 0.97	10-9 1.07	11-1 1.17	11-5 1.28	11-8 1.39
2x8	12.0	11-2 0.26	12-3 0.35	13-3 0.44	14-2 0.54	15-0 0.64	15-10 0.75	16-7 0.86	17-4 0.98	18-0 1.11	18-9 1.24	19-5 1.37	20-0 1.51	20-8 1.66	21-3 1.81	21-10 1.96
	16.0	9-8 0.23	10-7 0.30	11-6 0.38	12-3 0.46	13-0 0.55	13-8 0.65	14-4 0.75	15-0 0.85	15-7 0.96	16-3 1.07	16-9 1.19	17-4 1.31	17-10 1.44	18-5 1.56	18-11 1.70
	24.0	7-11 0.19	8-8 0.25	9-4 0.31	10-0 0.38	10-7 0.45	11-2 0.53	11-9 0.61	12-3 0.70	12-9 0.78	13-3 0.88	13-8 0.97	14-2 1.07	14-7 1.17	15-0 1.28	15-5 1.39
2x10	12.0	14-3 0.26	15-8 0.35	16-11 0.44	18-1 0.54	19-2 0.64	20-2 0.75	21-2 0.86	22-1 0.98	23-0 1.11	23-11 1.24	24-9 1.37	25-6 1.51	26-4 1.66	27-1 1.81	27-10 1.96
	16.0	12-4 0.23	13-6 0.30	14-8 0.38	15-8 0.46	16-7 0.55	17-6 0.65	18-4 0.75	19-2 0.85	19-11 0.96	20-8 1.07	21-5 1.19	22-1 1.31	22-10 1.44	23-5 1.56	24-1 1.70
	24.0	10-1 0.19	11-1 0.25	11-11 0.31	12-9 0.38	13-6 0.45	14-3 0.53	15-0 0.61	15-8 0.70	16-3 0.78	16-11 0.88	17-6 0.97	18-1 1.07	18-7 1.17	19-2 1.28	19-8 1.39
2x12	12.0	17-4 0.26	19-0 0.35	20-6 0.44	21-11 0.54	23-3 0.64	24-7 0.75	25-9 0.86	26-11 0.98	28-0 1.11	29-1 1.24	30-1 1.37	31-1 1.51	32-0 1.66	32-11 1.81	33-10 1.96
	16.0	15-0 0.23	16-6 0.30	17-9 0.38	19-0 0.46	20-2 0.55	21-3 0.65	22-4 0.75	23-3 0.85	24-3 0.96	25-2 1.07	26-0 1.19	26-11 1.31	27-9 1.44	28-6 1.56	29-4 1.70
	24.0	12-3 0.19	13-5 0.25	14-6 0.31	15-6 0.38	16-6 0.45	17-4 0.53	18-2 0.61	19-0 0.70	19-10 0.78	20-6 0.88	21-3 0.97	21-11 1.07	22-8 1.17	23-3 1.28	23-11 1.39

30 psf (Supporting Drywall Ceiling)

DESIGN CRITERIA:
Strength - 15 lbs. per sq. ft. dead load plus 30 lbs. per sq. ft. live load determines required fiber stress.
Deflection - For 30 lbs. per sq. ft. live load. Limited to span in inches divided by 240.

RAFTER SIZE (IN)	RAFTER SPACING (IN)	Allowable Extreme Fiber Stress in Bending, "F_b" (psi). 500	600	700	800	900	1000	1100	1200	1300	1400	1500	1600	1700	1800	1900
2x6	12.0	7-6 0.27	8-2 0.36	8-10 0.45	9-6 0.55	10-0 0.66	10-7 0.77	11-1 0.89	11-7 1.01	12-1 1.14	12-6 1.28	13-0 1.41	13-5 1.56	13-10 1.71	14-2 1.86	14-7 2.02
	16.0	6-6 0.24	7-1 0.31	7-8 0.39	8-2 0.48	8-8 0.57	9-2 0.67	9-7 0.77	10-0 0.88	10-5 0.99	10-10 1.10	11-3 1.22	11-7 1.35	11-11 1.48	12-4 1.61	12-8 1.75
	24.0	5-4 0.19	5-10 0.25	6-3 0.32	6-8 0.39	7-1 0.46	7-6 0.54	7-10 0.63	8-2 0.72	8-6 0.81	8-10 0.90	9-2 1.00	9-6 1.10	9-9 1.21	10-0 1.31	10-4 1.43
2x8	12.0	9-10 0.27	10-10 0.36	11-8 0.45	12-6 0.55	13-3 0.66	13-11 0.77	14-8 0.89	15-3 1.01	15-11 1.14	16-6 1.28	17-1 1.41	17-8 1.56	18-2 1.71	18-9 1.86	19-3 2.02
	16.0	8-7 0.24	9-4 0.31	10-1 0.39	10-10 0.48	11-6 0.57	12-1 0.67	12-8 0.77	13-3 0.88	13-9 0.99	14-4 1.10	14-10 1.22	15-3 1.35	15-9 1.48	16-3 1.61	16-8 1.75
	24.0	7-0 0.19	7-8 0.25	8-3 0.32	8-10 0.39	9-4 0.46	9-10 0.54	10-4 0.63	10-10 0.72	11-3 0.81	11-8 0.90	12-1 1.00	12-6 1.10	12-10 1.21	13-3 1.31	13-7 1.43
2x10	12.0	12-7 0.27	13-9 0.36	14-11 0.45	15-11 0.55	16-11 0.66	17-10 0.77	18-8 0.89	19-6 1.01	20-4 1.14	21-1 1.28	21-10 1.41	22-6 1.56	23-3 1.71	23-11 1.86	24-6 2.02
	16.0	10-11 0.24	11-11 0.31	12-11 0.39	13-9 0.48	14-8 0.57	15-5 0.67	16-2 0.77	16-11 0.88	17-7 0.99	18-3 1.10	18-11 1.22	19-6 1.35	20-1 1.48	20-8 1.61	21-3 1.75
	24.0	8-11 0.19	9-9 0.25	10-6 0.32	11-3 0.39	11-11 0.46	12-7 0.54	13-2 0.63	13-9 0.72	14-4 0.81	14-11 0.90	15-5 1.00	15-11 1.10	16-5 1.21	16-11 1.31	17-4 1.43
2x12	12.0	15-4 0.27	16-9 0.36	18-1 0.45	19-4 0.55	20-6 0.66	21-8 0.77	22-8 0.89	23-9 1.01	24-8 1.14	25-7 1.28	26-6 1.41	27-5 1.56	28-3 1.71	29-1 1.86	29-10 2.02
	16.0	13-3 0.24	14-6 0.31	15-8 0.39	16-9 0.48	17-9 0.57	18-9 0.67	19-8 0.77	20-6 0.88	21-5 0.99	22-2 1.10	23-0 1.22	23-9 1.35	24-5 1.48	25-2 1.61	25-10 1.75
	24.0	10-10 0.19	11-10 0.25	12-10 0.32	13-8 0.39	14-6 0.46	15-4 0.54	16-1 0.63	16-9 0.72	17-5 0.81	18-1 0.90	18-9 1.00	19-4 1.10	20-0 1.21	20-6 1.31	21-1 1.43

*The required modulus of elasticity, "E", in 1,000,000 pounds per square inch is shown below each span. Spans are measured along the horizontal projection and loads are considered as applied on the horizontal projection.

FIGURE WF-5 (Continued) **MAXIMUM ALLOWABLE SPANS FOR JOISTS & RAFTERS**

LOW SLOPE RAFTERS*

(Roof slope 3 in 12 or less)

20 psf (No Finished Ceiling)

DESIGN CRITERIA:
Strength - 10 lbs. per sq. ft. dead load plus 20 lbs. per sq. ft. live load determines required fiber stress.
Deflection - For 20 lbs. per sq. ft. live load. Limited to span in inches divided by 240.

RAFTER SIZE (IN)	RAFTER SPACING (IN)	Allowable Extreme Fiber Stress in Bending, "F_b" (psi). 500	600	700	800	900	1000	1100	1200	1300	1400	1500	1600	1700	1800	1900
2x6	12.0	9-2 0.33	10-0 0.44	10-10 0.55	11-7 0.67	12-4 0.80	13-0 0.94	13-7 1.09	14-2 1.24	14-9 1.40	15-4 1.56	15-11 1.73	16-5 1.91	16-11 2.09	17-5 2.28	17-10 2.47
	16.0	7-11 0.29	8-8 0.38	9-5 0.48	10-0 0.58	10-8 0.70	11-3 0.82	11-9 0.94	12-4 1.07	12-10 1.21	13-3 1.35	13-9 1.50	14-2 1.65	14-8 1.81	15-1 1.97	15-6 2.14
	24.0	6-6 0.24	7-1 0.31	7-8 0.39	8-2 0.48	8-8 0.57	9-2 0.67	9-7 0.77	10-0 0.88	10-5 0.99	10-10 1.10	11-3 1.22	11-7 1.35	11-11 1.48	12-4 1.61	12-8 1.75
2x8	12.0	12-1 0.33	13-3 0.44	14-4 0.55	15-3 0.67	16-3 0.80	17-1 0.94	17-11 1.09	18-9 1.24	19-6 1.40	20-3 1.56	20-11 1.73	21-7 1.91	22-3 2.09	22-11 2.28	23-7 2.47
	16.0	10-6 0.29	11-6 0.38	12-5 0.48	13-3 0.58	14-0 0.70	14-10 0.82	15-6 0.94	16-3 1.07	16-10 1.21	17-6 1.35	18-2 1.50	18-9 1.65	19-4 1.81	19-10 1.97	20-5 2.14
	24.0	8-7 0.24	9-4 0.31	10-1 0.39	10-10 0.48	11-6 0.57	12-1 0.67	12-8 0.77	13-3 0.88	13-9 0.99	14-4 1.10	14-10 1.22	15-3 1.35	15-9 1.48	16-3 1.61	16-8 1.75
2x10	12.0	15-5 0.33	16-11 0.44	18-3 0.55	19-6 0.67	20-8 0.80	21-10 0.94	22-10 1.09	23-11 1.24	24-10 1.40	25-10 1.56	26-8 1.73	27-7 1.91	28-5 2.09	29-3 2.28	30-1 2.47
	16.0	13-4 0.29	14-8 0.38	15-10 0.48	16-11 0.58	17-11 0.70	18-11 0.82	19-10 0.94	20-8 1.07	21-6 1.21	22-4 1.35	23-2 1.50	23-11 1.65	24-7 1.81	25-4 1.97	26-0 2.14
	24.0	10-11 0.24	11-11 0.31	12-11 0.39	13-9 0.48	14-8 0.57	15-5 0.67	16-2 0.77	16-11 0.88	17-7 0.99	18-3 1.10	18-11 1.22	19-6 1.35	20-1 1.48	20-8 1.61	21-3 1.75
2x12	12.0	18-9 0.33	20-6 0.44	22-2 0.55	23-9 0.67	25-2 0.80	26-6 0.94	27-10 1.09	29-1 1.24	30-3 1.40	31-4 1.56	32-6 1.73	33-6 1.91	34-7 2.09	35-7 2.28	36-7 2.47
	16.0	16-3 0.29	17-9 0.38	19-3 0.48	20-6 0.58	21-9 0.70	23-0 0.82	24-1 0.94	25-2 1.07	26-2 1.21	27-2 1.35	28-2 1.50	29-1 1.65	29-11 1.81	30-10 1.97	31-8 2.14
	24.0	13-3 0.24	14-6 0.31	15-8 0.39	16-9 0.48	17-9 0.57	18-9 0.67	19-8 0.77	20-6 0.88	21-5 0.99	22-2 1.10	23-0 1.22	23-9 1.35	24-5 1.48	25-2 1.61	25-10 1.75

30 psf (No Finished Ceiling)

DESIGN CRITERIA:
Strength - 10 lbs. per sq. ft. dead load plus 30 lbs. per sq. ft. live load determines required fiber stress.
Deflection - For 30 lbs. per sq. ft. live load. Limited to span in inches divided by 240.

RAFTER SIZE (IN)	RAFTER SPACING (IN)	Allowable Extreme Fiber Stress In Bending, "F_b" (psi). 500	600	700	800	900	1000	1100	1200	1300	1400	1500	1600	1700	1800	1900
2x6	12.0	7-11 0.32	8-8 0.43	9-5 0.54	10-0 0.66	10-8 0.78	11-3 0.92	11-9 1.06	12-4 1.21	12-10 1.36	13-3 1.52	13-9 1.69	14-2 1.86	14-8 2.04	15-1 2.22	15-6 2.41
	16.0	6-11 0.28	7-6 0.37	8-2 0.47	8-8 0.57	9-3 0.68	9-9 0.80	10-2 0.92	10-8 1.05	11-1 1.18	11-6 1.32	11-11 1.46	12-4 1.61	12-8 1.76	13-1 1.92	13-5 2.08
	24.0	5-7 0.23	6-2 0.30	6-8 0.38	7-1 0.46	7-6 0.55	7-11 0.65	8-4 0.75	8-8 0.85	9-1 0.96	9-5 1.08	9-9 1.19	10-0 1.31	10-4 1.44	10-8 1.57	10-11 1.70
2x8	12.0	10-6 0.32	11-6 0.43	12-5 0.54	13-3 0.66	14-0 0.78	14-10 0.92	15-6 1.06	16-3 1.21	16-10 1.36	17-6 1.52	18-2 1.69	18-9 1.86	19-4 2.04	19-10 2.22	20-5 2.41
	16.0	9-1 0.28	9-11 0.37	10-9 0.47	11-6 0.57	12-2 0.68	12-10 0.80	13-5 0.92	14-0 1.05	14-7 1.18	15-2 1.32	15-8 1.46	16-3 1.61	16-9 1.76	17-2 1.92	17-8 2.08
	24.0	7-5 0.23	8-1 0.30	8-9 0.38	9-4 0.46	9-11 0.55	10-6 0.65	11-0 0.75	11-6 0.85	11-11 0.96	12-5 1.08	12-10 1.19	13-3 1.31	13-8 1.44	14-0 1.57	14-5 1.70
2x10	12.0	13-4 0.32	14-8 0.43	15-10 0.54	16-11 0.66	17-11 0.78	18-11 0.92	19-10 1.06	20-8 1.21	21-6 1.36	22-4 1.52	23-2 1.69	23-11 1.86	24-7 2.04	25-4 2.22	26-0 2.41
	16.0	11-7 0.28	12-8 0.37	13-8 0.47	14-8 0.57	15-6 0.68	16-4 0.80	17-2 0.92	17-11 1.05	18-8 1.18	19-4 1.32	20-0 1.46	20-8 1.61	21-4 1.76	21-11 1.92	22-6 2.08
	24.0	9-5 0.23	10-4 0.30	11-2 0.38	11-11 0.46	12-8 0.55	13-4 0.65	14-0 0.75	14-8 0.85	15-3 0.96	15-10 1.08	16-4 1.19	16-11 1.31	17-5 1.44	17-11 1.57	18-5 1.70
2x12	12.0	16-3 0.32	17-9 0.43	19-3 0.54	20-6 0.66	21-9 0.78	23-0 0.92	24-1 1.06	25-2 1.21	26-2 1.36	27-2 1.52	28-2 1.69	29-1 1.86	29-11 2.04	30-10 2.22	31-8 2.41
	16.0	14-1 0.28	15-5 0.37	16-8 0.47	17-9 0.57	18-10 0.68	19-11 0.80	20-10 0.92	21-9 1.05	22-8 1.18	23-6 1.32	24-4 1.46	25-2 1.61	25-11 1.76	26-8 1.92	27-5 2.08
	24.0	11-6 0.23	12-7 0.30	13-7 0.38	14-6 0.46	15-5 0.55	16-3 0.65	17-0 0.75	17-9 0.85	18-6 0.96	19-3 1.08	19-11 1.19	20-6 1.31	21-2 1.44	21-9 1.57	22-5 1.70

*The required modulus of elasticity, "E", in 1,000,000 pounds per square inch is shown below each span. Spans are measured along the horizontal projection and loads are considered as applied on the horizontal projection.

FIGURE WF-5 (Continued) **MAXIMUM ALLOWABLE SPANS FOR JOISTS & RAFTERS**

HIGH SLOPE RAFTERS*

(Roof slope over 3 in 12)

20 psf (Heavy Roof Covering)

DESIGN CRITERIA:
Strength - 15 lbs. per sq. ft. dead load plus 20 lbs. per sq. ft. live load determines required fiber stress.
Deflection - For 20 lbs. per sq. ft. live load. Limited to span in inches divided by 180.

RAFTER SIZE (IN)	RAFTER SPACING (IN)	Allowable Extreme Fiber Stress in Bending, "F_b" (psi).														
		500	600	700	800	900	1000	1100	1200	1300	1400	1500	1600	1700	1800	1900
2x4	12.0	5-5 0.20	5-11 0.26	6-5 0.33	6-10 0.40	7-3 0.48	7-8 0.56	8-0 0.65	8-4 0.74	8-8 0.83	9-0 0.93	9-4 1.03	9-8 1.14	9-11 1.24	10-3 1.36	10-6 1.47
	16.0	4-8 0.17	5-1 0.23	5-6 0.28	5-11 0.35	6-3 0.41	6-7 0.49	6-11 0.56	7-3 0.64	7-6 0.72	7-10 0.80	8-1 0.89	8-4 0.98	8-7 1.08	8-10 1.17	9-1 1.27
	24.0	3-10 0.14	4-2 0.18	4-6 0.23	4-10 0.28	5-1 0.34	5-5 0.40	5-8 0.46	5-11 0.52	6-2 0.59	6-5 0.66	6-7 0.73	6-10 0.80	7-0 0.88	7-3 0.96	7-5 1.04
2x6	12.0	8-6 0.20	9-4 0.26	10-0 0.33	10-9 0.40	11-5 0.48	12-0 0.56	12-7 0.65	13-2 0.74	13-8 0.83	14-2 0.93	14-8 1.03	15-2 1.14	15-8 1.24	16-1 1.36	16-7 1.47
	16.0	7-4 0.17	8-1 0.23	8-8 0.28	9-4 0.35	9-10 0.41	10-5 0.49	10-11 0.56	11-5 0.64	11-10 0.72	12-4 0.80	12-9 0.89	13-2 0.98	13-7 1.08	13-11 1.17	14-4 1.27
	24.0	6-0 0.14	6-7 0.18	7-1 0.23	7-7 0.28	8-1 0.34	8-6 0.40	8-11 0.46	9-4 0.52	9-8 0.59	10-0 0.66	10-5 0.73	10-9 0.80	11-1 0.88	11-5 0.96	11-8 1.04
2x8	12.0	11-2 0.20	12-3 0.26	13-3 0.33	14-2 0.40	15-0 0.48	15-10 0.56	16-7 0.65	17-4 0.74	18-0 0.83	18-9 0.93	19-5 1.03	20-0 1.14	20-8 1.24	21-3 1.36	21-10 1.47
	16.0	9-8 0.17	10-7 0.23	11-6 0.28	12-3 0.35	13-0 0.41	13-8 0.49	14-4 0.56	15-0 0.64	15-7 0.72	16-3 0.80	16-9 0.89	17-4 0.98	17-10 1.08	18-5 1.17	18-11 1.27
	24.0	7-11 0.14	8-8 0.18	9-4 0.23	10-0 0.28	10-7 0.34	11-2 0.40	11-9 0.46	12-3 0.52	12-9 0.59	13-3 0.66	13-8 0.73	14-2 0.80	14-7 0.88	15-0 0.96	15-5 1.04
2x10	12.0	14-3 0.20	15-8 0.26	16-11 0.33	18-1 0.40	19-2 0.48	20-2 0.56	21-2 0.65	22-1 0.74	23-0 0.83	23-11 0.93	24-9 1.03	25-6 1.14	26-4 1.24	27-1 1.36	27-10 1.47
	16.0	12-4 0.17	13-6 0.23	14-8 0.28	15-8 0.35	16-7 0.41	17-6 0.49	18-4 0.56	19-2 0.64	19-11 0.72	20-8 0.80	21-5 0.89	22-1 0.98	22-10 1.08	23-5 1.17	24-1 1.27
	24.0	10-1 0.14	11-1 0.18	11-11 0.23	12-9 0.28	13-6 0.34	14-3 0.40	15-0 0.46	15-8 0.52	16-3 0.59	16-11 0.66	17-6 0.73	18-1 0.80	18-7 0.88	19-2 0.96	19-8 1.04

30 psf (Heavy Roof Covering)

DESIGN CRITERIA:
Strength - 15 lbs. per sq. ft. dead load plus 30 lbs. per sq. ft. live load determines required fiber stress.
Deflection - For 30 lbs. per sq. ft. live load. Limited to span in inches divided by 180.

RAFTER SIZE (IN)	RAFTER SPACING (IN)	Allowable Extreme Fiber Stress in Bending, "F_b" (psi).														
		500	600	700	800	900	1000	1100	1200	1300	1400	1500	1600	1700	1800	1900
2x4	12.0	4-9 0.20	5-3 0.27	5-8 0.34	6-0 0.41	6-5 0.49	6-9 0.58	7-1 0.67	7-5 0.76	7-8 0.86	8-0 0.96	8-3 1.06	8-6 1.17	8-9 1.28	9-0 1.39	9-3 1.51
	16.0	4-1 0.18	4-6 0.23	4-11 0.29	5-3 0.36	5-6 0.43	5-10 0.50	6-1 0.58	6-5 0.66	6-8 0.74	6-11 0.83	7-2 0.92	7-5 1.01	7-7 1.11	7-10 1.21	8-0 1.31
	24.0	3-4 0.14	3-8 0.19	4-0 0.24	4-3 0.29	4-6 0.35	4-9 0.41	5-0 0.47	5-3 0.54	5-5 0.61	5-8 0.68	5-10 0.75	6-0 0.83	6-3 0.90	6-5 0.99	6-7 1.07
2x6	12.0	7-6 0.20	8-2 0.27	8-10 0.34	9-6 0.41	10-0 0.49	10-7 0.58	11-1 0.67	11-7 0.76	12-1 0.86	12-6 0.96	13-0 1.06	13-5 1.17	13-10 1.28	14-2 1.39	14-7 1.51
	16.0	6-6 0.18	7-1 0.23	7-8 0.29	8-2 0.36	8-8 0.43	9-2 0.50	9-7 0.58	10-0 0.66	10-5 0.74	10-10 0.83	11-3 0.92	11-7 1.01	11-11 1.11	12-4 1.21	12-8 1.31
	24.0	5-4 0.14	5-10 0.19	6-3 0.24	6-8 0.29	7-1 0.35	7-6 0.41	7-10 0.47	8-2 0.54	8-6 0.61	8-10 0.68	9-2 0.75	9-6 0.83	9-9 0.90	10-0 0.99	10-4 1.07
2x8	12.0	9-10 0.20	10-10 0.27	11-8 0.34	12-6 0.41	13-3 0.49	13-11 0.58	14-8 0.67	15-3 0.76	15-11 0.86	16-6 0.96	17-1 1.06	17-8 1.17	18-2 1.28	18-9 1.39	19-3 1.51
	16.0	8-7 0.18	9-4 0.23	10-1 0.29	10-10 0.36	11-6 0.43	12-1 0.50	12-8 0.58	13-3 0.66	13-9 0.74	14-4 0.83	14-10 0.92	15-3 1.01	15-9 1.11	16-3 1.21	16-8 1.31
	24.0	7-0 0.14	7-8 0.19	8-3 0.24	8-10 0.29	9-4 0.35	9-10 0.41	10-4 0.47	10-10 0.54	11-3 0.61	11-8 0.68	12-1 0.75	12-6 0.83	12-10 0.90	13-3 0.99	13-7 1.07
2x10	12.0	12-7 0.20	13-9 0.27	14-11 0.34	15-11 0.41	16-11 0.49	17-10 0.58	18-8 0.67	19-6 0.76	20-4 0.86	21-1 0.96	21-10 1.06	22-6 1.17	23-3 1.28	23-11 1.39	24-6 1.51
	16.0	10-11 0.18	11-11 0.23	12-11 0.29	13-9 0.36	14-8 0.43	15-5 0.50	16-2 0.58	16-11 0.66	17-7 0.74	18-3 0.83	18-11 0.92	19-6 1.01	20-1 1.11	20-8 1.21	21-3 1.31
	24.0	8-11 0.14	9-9 0.19	10-6 0.24	11-3 0.29	11-11 0.35	12-7 0.41	13-2 0.47	13-9 0.54	14-4 0.61	14-11 0.68	15-5 0.75	15-11 0.83	16-5 0.90	16-11 0.99	17-4 1.07

*The required modulus of elasticity, "E", in 1,000,000 pounds per square inch is shown below each span. Spans are measured along the horizontal projection and loads are considered as applied on the horizontal projection.

FIGURE WF-5 (Continued) **MAXIMUM ALLOWABLE SPANS FOR JOISTS & RAFTERS**

HIGH SLOPE RAFTERS*

(Roof slope over 3 in 12)

20 psf (Light Roof Covering)

DESIGN CRITERIA:
Strength - 7 lbs. per sq. ft. dead load plus 20 lbs. per sq. ft. live load determines required fiber stress.
Deflection - For 20 lbs. per sq. ft. live load. Limited to span in inches divided by 180.

RAFTER SIZE (IN)	RAFTER SPACING (IN)	Allowable Extreme Fiber Stress in Bending, "F_b" (psi). 500	600	700	800	900	1000	1100	1200	1300	1400	1500	1600	1700	1800	1900
2x4	12.0	6-2 0.29	6-9 0.38	7-3 0.49	7-9 0.59	8-3 0.71	8-8 0.83	9-1 0.96	9-6 1.09	9-11 1.23	10-3 1.37	10-8 1.52	11-0 1.68	11-4 1.84	11-8 2.00	12-0 2.17
	16.0	5-4 0.25	5-10 0.33	6-4 0.42	6-9 0.51	7-2 0.61	7-6 0.72	7-11 0.83	8-3 0.94	8-7 1.06	8-11 1.19	9-3 1.32	9-6 1.45	9-10 1.59	10-1 1.73	10-5 1.88
	24.0	4-4 0.21	4-9 0.27	5-2 0.34	5-6 0.42	5-10 0.50	6-2 0.59	6-5 0.68	6-9 0.77	7-0 0.87	7-3 0.97	7-6 1.08	7-9 1.19	8-0 1.30	8-3 1.41	8-6 1.53
2x6	12.0	9-8 0.29	10-7 0.38	11-5 0.49	12-3 0.59	13-0 0.71	13-8 0.83	14-4 0.96	15-0 1.09	15-7 1.23	16-2 1.37	16-9 1.52	17-3 1.68	17-10 1.84	18-4 2.00	18-10 2.17
	16.0	8-4 0.25	9-2 0.33	9-11 0.42	10-7 0.51	11-3 0.61	11-10 0.72	12-5 0.83	13-0 0.94	13-6 1.06	14-0 1.19	14-6 1.32	15-0 1.45	15-5 1.59	15-11 1.73	16-4 1.88
	24.0	6-10 0.21	7-6 0.27	8-1 0.34	8-8 0.42	9-2 0.50	9-8 0.59	10-2 0.68	10-7 0.77	11-0 0.87	11-5 0.97	11-10 1.08	12-3 1.19	12-7 1.30	13-0 1.41	13-4 1.53
2x8	12.0	12-9 0.29	13-11 0.38	15-1 0.49	16-1 0.59	17-1 0.71	18-0 0.83	18-11 0.96	19-9 1.09	20-6 1.23	21-4 1.37	22-1 1.52	22-9 1.68	23-6 1.84	24-2 2.00	24-10 2.17
	16.0	11-0 0.25	12-1 0.33	13-1 0.42	13-11 0.51	14-10 0.61	15-7 0.72	16-4 0.83	17-1 0.94	17-9 1.06	18-5 1.19	19-1 1.32	19-9 1.45	20-4 1.59	20-11 1.73	21-6 1.88
	24.0	9-0 0.21	9-10 0.27	10-8 0.34	11-5 0.42	12-1 0.50	12-9 0.59	13-4 0.68	13-11 0.77	14-6 0.87	15-1 0.97	15-7 1.08	16-1 1.19	16-7 1.30	17-1 1.41	17-7 1.53
2x10	12.0	16-3 0.29	17-10 0.38	19-3 0.49	20-7 0.59	21-10 0.71	23-0 0.83	24-1 0.96	25-2 1.09	26-2 1.23	27-2 1.37	28-2 1.52	29-1 1.68	30-0 1.84	30-10 2.00	31-8 2.17
	16.0	14-1 0.25	15-5 0.33	16-8 0.42	17-10 0.51	18-11 0.61	19-11 0.72	20-10 0.83	21-10 0.94	22-8 1.06	23-7 1.19	24-5 1.32	25-2 1.45	25-11 1.59	26-8 1.73	27-5 1.88
	24.0	11-6 0.21	12-7 0.27	13-7 0.34	14-6 0.42	15-5 0.50	16-3 0.59	17-1 0.68	17-10 0.77	18-6 0.87	19-3 0.97	19-11 1.08	20-7 1.19	21-2 1.30	21-10 1.41	22-5 1.53

30 psf (Light Roof Covering)

DESIGN CRITERIA:
Strength - 7 lbs. per sq. ft. dead load plus 30 lbs. per sq. ft. live load determines required fiber stress.
Deflection - For 30 lbs. per sq. ft. live load. Limited to span in inches divided by 180.

RAFTER SIZE (IN)	RAFTER SPACING (IN)	Allowable Extreme Fiber Stress in Bending, "F_b" (psi). 500	600	700	800	900	1000	1100	1200	1300	1400	1500	1600	1700	1800	1900
2x4	12.0	5-3 0.27	5-9 0.36	6-3 0.45	6-8 0.55	7-1 0.66	7-5 0.77	7-9 0.89	8-2 1.02	8-6 1.15	8-9 1.28	9-1 1.42	9-5 1.57	9-8 1.72	10-0 1.87	10-3 2.03
	16.0	4-7 0.24	5-0 0.31	5-5 0.39	5-9 0.48	6-1 0.57	6-5 0.67	6-9 0.77	7-1 0.88	7-4 0.99	7-7 1.11	7-11 1.23	8-2 1.36	8-5 1.49	8-8 1.62	8-10 1.76
	24.0	3-9 0.19	4-1 0.25	4-5 0.32	4-8 0.39	5-0 0.47	5-3 0.55	5-6 0.63	5-9 0.72	6-0 0.81	6-3 0.91	6-5 1.01	6-8 1.11	6-10 1.21	7-1 1.32	7-3 1.43
2x6	12.0	8-3 0.27	9-1 0.36	9-9 0.45	10-5 0.55	11-1 0.66	11-8 0.77	12-3 0.89	12-9 1.02	13-4 1.15	13-10 1.28	14-4 1.42	14-9 1.57	15-3 1.72	15-8 1.87	16-1 2.03
	16.0	7-2 0.24	7-10 0.31	8-5 0.39	9-1 0.48	9-7 0.57	10-1 0.67	10-7 0.77	11-1 0.88	11-6 0.99	12-0 1.11	12-5 1.23	12-9 1.36	13-2 1.49	13-7 1.62	13-11 1.76
	24.0	5-10 0.19	6-5 0.25	6-11 0.32	7-5 0.39	7-10 0.47	8-3 0.55	8-8 0.63	9-1 0.72	9-5 0.81	9-9 0.91	10-1 1.01	10-5 1.11	10-9 1.21	11-1 1.32	11-5 1.43
2x8	12.0	10-11 0.27	11-11 0.36	12-10 0.45	13-9 0.55	14-7 0.66	15-5 0.77	16-2 0.89	16-10 1.02	17-7 1.15	18-2 1.28	18-10 1.42	19-6 1.57	20-1 1.72	20-8 1.87	21-3 2.03
	16.0	9-5 0.24	10-4 0.31	11-2 0.39	11-11 0.48	12-8 0.57	13-4 0.67	14-0 0.77	14-7 0.88	15-2 0.99	15-9 1.11	16-4 1.23	16-10 1.36	17-4 1.49	17-11 1.62	18-4 1.76
	24.0	7-8 0.19	8-5 0.25	9-1 0.32	9-9 0.39	10-4 0.47	10-11 0.55	11-5 0.63	11-11 0.72	12-5 0.81	12-10 0.91	13-4 1.01	13-9 1.11	14-2 1.21	14-7 1.32	15-0 1.43
2x10	12.0	13-11 0.27	15-2 0.36	16-5 0.45	17-7 0.55	18-7 0.66	19-8 0.77	20-7 0.89	21-6 1.02	22-5 1.15	23-3 1.28	24-1 1.42	24-10 1.57	25-7 1.72	26-4 1.87	27-1 2.03
	16.0	12-0 0.26	13-2 0.34	14-3 0.43	15-2 0.53	16-2 0.63	17-0 0.74	17-10 0.85	18-7 0.97	19-5 1.09	20-1 1.22	20-10 1.35	21-6 1.49	22-2 1.63	22-10 1.78	23-5 1.93
	24.0	9-10 0.19	10-9 0.25	11-7 0.32	12-5 0.39	13-2 0.47	13-11 0.55	14-7 0.63	15-2 0.72	15-10 0.81	16-5 0.91	17-0 1.01	17-7 1.11	18-1 1.21	18-7 1.32	19-2 1.43

*The required modulus of elasticity, "E", in 1,000,000 pounds per square inch is shown below each span. Spans are measured along the horizontal projection and loads are considered as applied on the horizontal projection.

STANDARDS OF QUALITY

Softwood lumber should conform to the provisions of the American Softwood Lumber Standard PS 20-70 and the grading rules of a recognized rules writing agency approved by the American Lumber Standards Committee. Visual grading of lumber should be done according to procedures established in Methods for Establishing Structural Grades for Visually Graded Lumber ASTM D245.

SPECIES AND GRADES

The species and grades of lumber should be suitable for the intended use. Dimension lumber and timbers are graded primarily on the basis of strength properties.

Since the natural characteristics of wood which affect appearance also affect strength, the higher grades also have more severe limitations on natural characteristics and are more suitable for exposed use. A special, Appearance, grade in joist and stud sizes, combines relatively high strength with good appearance. All decking grades have one face intended for exposed use. Select Structural in timbers is suitable for exposed use, although it is graded mainly for strength. Similarly, Select Structural studs and joists have higher strength values than Appearance grades (Fig. WF 3), but have sufficiently good appearance for exposed use. Framing lumber for concealed use generally can be the lowest grade which meets the strength requirements, with some consideration given to the possibility of waste in the lowest grades.

SIZES

Dressed dimension for softwood lumber should conform to the minimum green and dry sizes of the American Softwood Lumber Standard PS 20-70 (Fig. WF7).*

WOOD PRESERVATIVE TREATMENT

Where preservative treated lumber is indicated in the paragraphs below, the lumber should be preservative treated in accordance with an appropriate AWPB or NWMA Standard (Fig. W24 Main Text, page 201-14) and each piece should be marked with the stamp of the AWPB, the NWMA or other recognized certification agency operating under procedures of ANSI Z341. Lumber with natural decay and termite resistance should carry the grademark of an inspection agency authorized by the ALSC.

In all areas of the country, lumber and wood products used in contact with the ground or embedded in the ground should be preservative *pressure* treated in accordance with an appropriate AWPB Standard.

In areas of severe termite infestation (Fig. W19 Main Text, page 201-12) and/or high decay hazard as in damp coastal areas (Fig. W8 Main Text, page 201-6) lumber and wood products should be preservative *pressure* treated under the following conditions: 1) all wood members in contact with masonry or concrete at grade, below grade or less than 8″ above grade; 2) in a crawl space, all floor joists less than 18″ above interior grade and wood girders less than 12″ above interior grade; 3) wood siding closer than 6″ to exterior finish grade.

Additional wood members and products may require preservative treatment where dictated by local experience. Soil poisoning, termite shields, monolithic concrete slab construction and other construction procedures outlined in Division 300 can minimize termite hazard. Except for structural members embedded in the ground and used as part of the foundation of permanent buildings, heartwood grades of naturally resistant species can be used in place of pressure treated lumber. Acceptable grades and species of dimension lumber are: 1) California redwood—Select Heart, Construction Heart or Foundation Grade; 2) tidewater red cypress—Heart Common; 3) Eastern red cedar—Foundation Grade.

Exterior millwork items such as doors, window units and casing trim, which are subjected to intermittent wetting and drying should be preservative treated by non-pressure methods according to NWMA IS-4, unless more effective treatment as protection against termites is indicated by local conditions.

Job-cut ends of preservative treated lumber shall be brush-coated with not less than a 3 percent solution of an approved preservative.

FIG. WF7 AMERICAN STANDARD SIZES OF GREEN AND DRY SOFTWOOD YARD LUMBER

Nominal Thickness or Width	Minimum Dressed Thickness or Width	
	Green (at over 19% m.c.)*	Dry (at 19% m.c. or less)
Boards		
1″	25/32″	3/4″
Dimension		
2″	1-9/16″	1-1/2″
3″	2-9/16″	2-1/2″
4″	3-9/16″	3-1/2″
Timbers		
5″	4-5/8″	4-1/2″
6″	5-5/8″	5-1/2″
8″	7-1/2″	7-1/4″
10″	9-1/2″	9-1/4″
12″	11-1/2″	11-1/4″
14″	13-1/2″	13-1/4″
16″	15-1/2″	15-1/4″

*Slightly different sizes apply to redwood, Western red cedar and Northern white cedar.

**Comprehensive text revisions in 1972 resulted in the elimination of Figures WF6 and WF8.*

CONSTRUCTION PLYWOOD

Construction plywood should conform to U.S. Product Standard PS 1-74 Construction and Industrial plywood.

Conformance with the Product Standard is assured by the grade-trademark of the American Plywood Association, Timber Engineering Company (TECO), Pittsburgh Testing Laboratories, or other suitable designation of a qualified testing and certifying agency. The grade-trademark used by the American Plywood Association which appears on the back or edge of the panel is shown in Figure WF9.

Plywood is graded and marked according to species group, type, appearance and construction application.

Classification of Species

Under Product Standard PS 1-74, approximately 70 wood species are classed into five groups according to stiffness (see Main Text, Fig. P7, page 201-30). The group number shown in the grade stamp refers to species used in face and back layers.

Types (Exterior, Interior)

Exterior Type plywood should be used in applications where there is permanent exposure to the weather or where unusual moisture conditions exist.

Exterior Type plywood is made with a completely waterproof glue and results in a glue bond stronger and more durable than the wood itself. No veneer used in a panel is less than grade "C".

Interior Type plywood is recommended for use only in those applications where there is a limited or controlled exposure to weather and moisture.

Interior Type plywood is made with either a highly water-resistant (Interior) glue, an Intermediate glue or completely waterproof (Exterior) glue. Structural Interior plywood is made with Exterior glue only. Veneers used in a panel are grade "D" or better.

Appearance Grades

Appearance grades are classed according to the quality of the face and back veneers of the panel. See Main Text, Fig. P8, page 201-31 for illustrations of the following veneer grades.

N Grade*—Special order "natural finish" veneer. Select all heartwood, or all sapwood, free of open defects.*
A Grade*—Best standard veneer. Smooth and paintable. May be more than one piece well jointed. Neatly made repairs are permitted.*
B Grade*—Solid surface veneer. Circular repair plugs and tight knots are permitted.*
C Grade*—Minimum veneer permitted in Exterior Type. Limited knots, knotholes, splits, and repairs permitted.*
D Grade*—Used only in Interior Type for inner plies and backs where specified.*

Engineered Grades

Engineered grades are identified according to either application or the quality of the face and back veneer. Three construction grades—C-D Interior, Structural (I and II), and C-C Exterior—carry an Identification Index Number (Fig. WF9). The first number indicates maximum recommended support spacing if the panel is used for roof sheathing; the second number indicates maximum recommended support spacing if the panel is used for subflooring. Index numbers are based on species, thickness and bending strength of the face and back layers. No index number is needed for wall sheathing because bending properties are not critical.

GRADE-USE GUIDE FOR CONSTRUCTION PLYWOOD

Plywood of the proper type, species group, grade and thickness should be used for the intended application. The grade-trademarks and general description of construction plywood panels carrying typical grade-trademarks of the American Plywood Association are shown in Figure WF10 on the following pages.

Typical Back Stamps **Typical Edge Stamp**

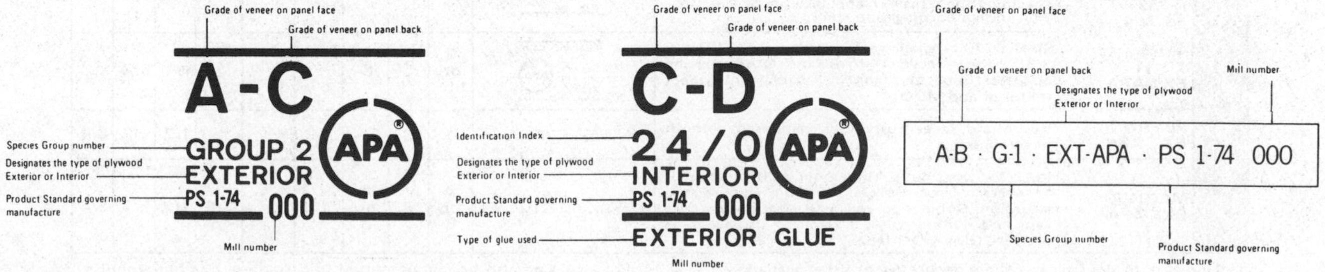

FIG. WF9 Typical grade-trademarks of the American Plywood Association

FIG. WF10 CONSTRUCTION PLYWOOD GRADE—USE GUIDE

PLYWOOD APPEARANCE GRADES

Type	Grade Designation (2)	Description and Most Common Uses	Typical (3) Grade-trademarks	Veneer Grade: Face	Veneer Grade: Inner Plies	Veneer Grade: Back	Most Common Thicknesses (inch)					
Interior Type	N-N, N-A N-B INT-APA	Cabinet quality. For natural finish furniture, cabinet doors, built-ins, etc. Special order items.	N-N G1 INT-APA PS1-74 000 / N-A G2 INT-APA PS1-74 000	N	C	N,A, or B						3/4
Interior Type	N-D-INT-APA	For natural finish paneling. Special order item.	N-D G2 INT-APA PS1-74 000	N	D	D	1/4					
Interior Type	A-A INT-APA	For applications with both sides on view, built-ins, cabinets, furniture, partitions. Smooth face; suitable for painting.	A-A G1 INT-APA PS1-74 000	A	D	A	1/4		3/8	1/2	5/8	3/4
Interior Type	A-B INT-APA	Use where appearance of one side is less important but where two solid surfaces are necessary.	A-B G1 INT-APA PS1-74 000	A	D	B	1/4		3/8	1/2	5/8	3/4
Interior Type	A-D INT-APA	Use where appearance of only one side is important. Paneling, built-ins, shelving, partitions, flow racks.	A-D GROUP 1 INTERIOR PS1-74 000 APA	A	D	D	1/4		3/8	1/2	5/8	3/4
Interior Type	B-B INT-APA	Utility panel with two solid sides. Permits circular plugs.	B-B G-2 INT-APA PS1-74 000	B	D	B	1/4		3/8	1/2	5/8	3/4
Interior Type	B-D INT-APA	Utility panel with one solid side. Good for backing, sides of built-ins, industry shelving, slip sheets, separator boards, bins.	B-D GROUP 2 INTERIOR PS1-74 000 APA	B	D	D	1/4		3/8	1/2	5/8	3/4
Interior Type	DECORATIVE PANELS–APA	Rough-sawn, brushed, grooved, or striated faces. For paneling, interior accent walls, built-ins, counter facing, displays, exhibits.	DECORATIVE B-D G1 INT-APA PS1-74	C or btr.	D	D		5/16	3/8	1/2	5/8	
Interior Type	PLYRON INT-APA	Hardboard face on both sides. For counter tops, shelving, cabinet doors, flooring. Faces tempered, untempered, smooth, or screened.	PLYRON INT-APA 000		C & D					1/2	5/8	3/4
Exterior Type	A-A EXT-APA	Use where appearance of both sides is important. Fences, built-ins, signs, boats, cabinets, commercial refrigerators, shipping containers, tote boxes, tanks, ducts. (4)	A-A G-1 EXT-APA PS1-74 000	A	C	A	1/4		3/8	1/2	5/8	3/4
Exterior Type	A-B EXT-APA	Use where the appearance of one side is less important. (4)	A-B G-1 EXT-APA PS1-74 000	A	C	B	1/4		3/8	1/2	5/8	3/4
Exterior Type	A-C EXT-APA	Use where the appearance of only one side is important. Soffits, fences, structural uses, boxcar and truck lining, farm buildings. Tanks, trays, commercial refrigerators. (4)	A-C GROUP 1 EXTERIOR PS1-74 000 APA	A	C	C	1/4		3/8	1/2	5/8	3/4
Exterior Type	B-B EXT-APA	Utility panel with solid faces. (4)	B-B G-2 EXT-APA PS1-74 000	B	C	B	1/4		3/8	1/2	5/8	3/4
Exterior Type	B-C EXT-APA	Utility panel for farm service and work buildings, boxcar and truck lining, containers, tanks, agricultural equipment. Also as base for exterior coatings for walls, roofs. (4)	B-C GROUP 2 EXTERIOR PS1-74 000 APA	B	C	C	1/4		3/8	1/2	5/8	3/4
Exterior Type	HDO EXT-APA	High Density Overlay plywood. Has a hard, semi-opaque resin-fiber overlay both faces. Abrasion resistant. For concrete forms, cabinets, counter tops, signs, tanks. (4)	HDO · A-A · G-1 · EXT-APA · PS1-74	A or B	C or C plgd	A or B			3/8	1/2	5/8	3/4
Exterior Type	MDO EXT-APA	Medium Density Overlay with smooth, opaque, resin-fiber overlay one or both panel faces. Highly recommended for siding and other outdoor applications, built-ins, signs, displays. Ideal base for paint. (4)(6)	MDO B-B G-2 EXT-APA PS1-74 000	B	C	B or C			3/8	1/2	5/8	3/4
Exterior Type	303 SIDING EXT-APA	Proprietary plywood products for exterior siding, fencing, etc. Special surface treatment such as V-groove, channel groove, striated, brushed rough-sawn and texture-embossed MDO. Stud spacing (Span Index) and face grade classification indicated on grade stamp.	303 SIDING 6-S GROUP 1 24 oc SPAN EXTERIOR PS1-74 000 APA	(5)	C	C			3/8	1/2	5/8	
Exterior Type	T 1-11 EXT-APA	Special 303 panel having grooves 1/4″ deep, 3/8″ wide, spaced 4″ or 8″ o.c. Other spacing optional. Edges shiplapped. Available unsanded, textured and MDO.	303 SIDING 6 S/W T 1-11 GROUP 2 16 oc SPAN EXTERIOR PS1-74 000 APA	C or btr.	C	C				19/32	5/8	
Exterior Type	PLYRON EXT-APA	Hardboard faces both sides, tempered, smooth or screened.	PLYRON EXT-APA 000		C					1/2	5/8	3/4
Exterior Type	MARINE EXT-APA	Ideal for boat hulls. Made only with Douglas fir or western larch. Special solid jointed core construction. Subject to special limitations on core gaps and number of face repairs. Also available with HDO or MDO faces.	MARINE A-A EXT-APA PS1-74 000	A or B	B	A or B	1/4		3/8	1/2	5/8	3/4

(1) Sanded both sides except where decorative or other surfaces specified.
(2) Can be manufactured in Group 1, 2, 3, 4 or 5.
(3) The species groups, Identification Indexes and Span Indexes shown in the typical grade-trademarks are examples only. See "Group," "Identification Index" and "Span Index" for explanations and availability.
(4) Can also be manufactured in Structural I (all plies limited to Group 1 species) and Structural II (all plies limited to Group 1, 2, or 3 species).
(5) C or better for 5 plies. C Plugged or better for 3 plies.
(6) Also available as a 303 siding.

FIG. WF10 **CONSTRUCTION PLYWOOD GRADE—USE GUIDE**

PLYWOOD ROOF DECKING [a b c d]

Panel Ident. Index	Plywood thickness (inch)	Max. span (inches)	Unsupported edge-max. length (inches)[d]	Allowable live loads (psf) — Spacing of supports center to center (inches)									
				12	16	20	24	30	32	36	42	48	60
12/0	5/16	12	12	150									
16/0	5/16, 3/8	16	16	160	75								
20/0	5/16, 3/8	20	20	190	105	65							
24/0	3/8	24	20	250	140	95	50						
	1/2		24										
32/16	1/2, 5/8	32	28	385	215	150	95	50	40				
42/20	5/8, 3/4, 7/8	42	32		330	230	145	90	75	50	35		
48/24	3/4, 7/8	48	36			300	190	120	105	65	45	35	
48/24(f)							225	125	105	75	55	40	
2•4•1	1-1/8	72	48				390	245	215	135	100	75	45
1-1/8" Grp. 1 & 2	1-1/8	72	48				305	195	170	105	75	55	35
1-1/4" Grp. 3 & 4	1-1/4	72	48				355	225	195	125	90	65	40

(a) These values apply for C-D INT-APA, C-C EXT-APA, STRUCTURAL I and STRUCTURAL II C-D INT-APA, and STRUCTURAL I and STRUCTURAL II C-C EXT-APA grades. Plywood continuous over 2 or more spans; grain of face plies across supports.

(b) Use 6d common smooth, ring-shank or spiral-thread nails for 1/2" thick or less and 8d common smooth, ring-shank, or spiral-thread for plywood 1" thick or less. Use 8d ring-shank or spiral-thread or 10d common smooth-shank nails for 2•4•1, 1-1/8" and 1-1/4" panels. Space nails 6" at panel edges and 12" at intermediate supports, except that where spans are 48" or more, nails shall be 6" at all supports. Space panel ends 1/16", and panel edges 1/8". Where wet or humid conditions prevail, double these spacings.

(c) Special conditions, such as heavy concentrated loads, may require constructions in excess of these minimums.

(d) Uniform load deflection limit: 1/180th span under live load plus dead load, 1/240th under live load only.

(e) Provide adequate blocking, tongue and grooved edges or other suitable edge support such as Plyclips when spans exceed indicated value. Use two Plyclips for 48" or greater spans and one for lesser spans.

(f) Loads apply only to C-C EXT-APA, STRUCTURAL I C-D INT-APA, and STRUCTURAL I C-C EXT-APA. Check availability before specifying.

PLYWOOD SUBFLOORING [a b c]

Panel Identification Index	Plywood Thickness (inch)	Maximum Span (d) (inches)	Nail Size & Type	Nail Spacing (inches)	
				Panel Edges	Intermediate
30/12	5/8 (h)	12 (e)	8d common	6	10
32/16	1/2, 5/8	16 (f)	8d common (g)	6	10
36/16	3/4 (h)	16 (f)	8d common	6	10
42/20	5/8, 3/4, 7/8	20 (i)	8d common	6	10
48/24	3/4, 7/8	24	8d common	6	10
1-1/8" Groups 1 & 2	1-1/8	48	10d common	6	6
1-1/4" Groups 3 & 4	1-1/4	48	10d common	6	6

(a) These values apply for C-D INT-APA, STRUCTURAL I and II C-D INT-APA, C-C EXT-APA and STRUCTURAL I and II C-C EXT-APA grades only.

(b) In some nonresidential buildings, special conditions may impose heavy concentrated loads and heavy traffic requiring subfloor construction in excess of these minimums.

(c) Edges shall be tongue and groove or supported with blocking for square-edge wood flooring, unless separate underlayment layer is installed, a minimum of 1-1/2" of lightweight concrete applied over the plywood, or finish floor is 25/32" wood strip. Minimum thickness of this underlayment layer should be 1/4" for subfloors up to 48/24, and 3/8" for thicker panels on spans longer than 24"

(d) Spans limited to values shown because of possible effect of concentrated loads. Allowable uniform loads vary, but at indicated maximum spans, floor panels carrying Identification Index numbers will support uniform live loads of more than 160 psf.

(e) May be 16" if 25/32" wood strip flooring is installed at right angles to joists.

(f) May be 24" if 25/32" wood strip flooring is installed at right angles to joists.

(g) 6d common nail permitted if plywood is 1/2".

(h) Check dealer for availability in your area.

(i) May be 24" if 25/32" wood strip flooring is installed at right angles to joists, or minimum 1-1/2" of lightweight concrete is applied over plywood.

PLYWOOD WALL SHEATHING [a b]

Panel Identification Index	Panel thickness (inch) & construction	Maximum stud spacing (inches) — Exterior covering nailed to:		Nail size (c)	Nail spacing (inches) (c)	
		Stud	Sheathing		Panel edges (when over framing)	Intermediate (each stud)
12/0, 16/0, 20/0	5/16	16	16[d]	6d	6	12
16/0, 20/0, 24/0, 32/16	3/8 & 1/2 3 ply	24	16 24[d]	6d	6	12
24/0, 32/16	1/2 4 & 5 ply	24	24	6d	6	12

(a) When plywood sheathing is used, building paper and diagonal wall bracing can be omitted.

(b) In dry conditions space panel edges ⅛", panel ends 1/16". In wet or humid conditions double spacing.

(c) Common smooth, annular, spiral-thread, or galvanized box, or T-nails of the same diameter as common nails (0.113" dia. for 6d) may be used. Staples also permitted at reduced spacing.

(d) Apply plywood with face grain across studs.

FIG. WF10 **CONSTRUCTION PLYWOOD GRADE—USE GUIDE** *Continued*

PLYWOOD ENGINEERED GRADES

Type	Grade Designation	Description and Most Common Uses	Typical Grade-trademarks (1)	Veneer Grade: Face	Veneer Grade: Inner Plies	Veneer Grade: Back	Most Common Thicknesses (inch)				
Interior Type	C-D INT-APA	For wall and roof sheathing, subflooring, industrial uses such as pallets. Most commonly available with exterior glue (CDX). Specify exterior glue where construction delays are anticipated and for treated-wood foundations. (7)	C-D 32/16 APA INTERIOR PS 1-74 000; C-D 24/0 APA INTERIOR PS 1-74 000 EXTERIOR GLUE	C	D	D	5/16	3/8	1/2	5/8	3/4
Interior Type	STRUCTURAL I C-D INT-APA and STRUCTURAL II C-D INT-APA	Unsanded structural grades where plywood strength properties are of maximum importance: structural diaphragms, box beams, gusset plates, stressed-skin panels, containers, pallet bins. Made only with exterior glue. See (6) for species group requirements. Structural I more commonly available. (7)	STRUCTURAL I C-D 24/0 APA INTERIOR PS 1-74 000 EXTERIOR GLUE	C(3)	D(3)	D(3)	5/16	3/8	1/2	5/8	3/4
Interior Type	STURD-I-FLOOR INT-APA	For combination subfloor-underlayment. Provides smooth surface for application of resilient floor covering. Possesses high concentrated- and impact-load resistance during construction and occupancy. Manufactured with exterior glue only. Touch-sanded. Available square edge or tongue-and-groove. (7)	STURD-I-FLOOR 24oc T&G 23/32 INCH APA INTERIOR 000 EXTERIOR GLUE NRB-108	C Plugged	(4)	D				19/32 5/8	23/32 3/4
Interior Type	STURD-I-FLOOR 48 O.C. (2-4-1) INT-APA	For combination subfloor-underlayment on 32- and 48-inch spans. Provides smooth surface for application of resilient floor coverings. Possesses high concentrated- and impact-load resistance during construction and occupancy. Manufactured with exterior glue only. Unsanded or touch-sanded. Available square edge or tongue-and-groove. (7)	STURD-I-FLOOR 48oc 2-4-1 T&G 1-1/8 INCH APA INTERIOR 000 EXTERIOR GLUE NRB-108	C Plugged	C(5) & D	D			1-1/8		
Interior Type	UNDERLAY-MENT INT-APA	For application over structural subfloor. Provides smooth surface for application of resilient floor coverings. Touch-sanded. Also available with exterior glue. (2)(6)	UNDERLAYMENT GROUP 1 APA INTERIOR PS 1-74 000	Plugged C	C(5) & D	D		3/8	1/2	19/32 5/8	23/32 3/4
Interior Type	C-D PLUGGED INT-APA	For built-ins, wall and ceiling tile backing, cable reels, walkways, separator boards. Not a substitute for Underlayment or Sturd-I-Floor as it lacks their indentation resistance. Touch-sanded. Also made with exterior glue. (2) (6)	C-D PLUGGED GROUP 2 APA INTERIOR PS 1-74 000	Plugged C	D	D		3/8	1/2	19/32 5/8	23/32 3/4
Exterior Type	C-C EXT-APA	Unsanded grade with waterproof bond for subflooring and roof decking, siding on service and farm buildings, crating, pallets, pallet bins, cable reels, treated-wood foundations. (7)	C-C 42/20 APA EXTERIOR PS 1-74 000	C	C	C	5/16	3/8	1/2	5/8	3/4
Exterior Type	STRUCTURAL I C-C EXT-APA and STRUCTURAL II C-C EXT-APA	For engineered applications in construction and industry where full Exterior type panels are required. Unsanded. See (6) for species group requirements. (7)	STRUCTURAL I C-C 32/16 APA EXTERIOR PS 1-74 000	C	C	C	5/16	3/8	1/2	5/8	3/4
Exterior Type	STURD-I-FLOOR EXT-APA	For combination subfloor-underlayment under resilient floor coverings where severe moisture conditions may be present, as in balcony decks. Possesses high concentrated-and impact-load resistance during construction and occupancy. Touch-sanded. Available square edge or tongue-and-groove. (7)	STURD-I-FLOOR 20oc 5/8 INCH APA EXTERIOR 000 NRB-108	C Plugged	C(5)	C				19/32 5/8	23/32 3/4
Exterior Type	UNDERLAY-MENT C-C PLUGGED EXT-APA	For application over structural subfloor. Provides smooth surface for application of resilient floor coverings where severe moisture conditions may be present. Touch-sanded. (2)(6)	UNDERLAYMENT C-C PLUGGED GROUP 2 APA EXTERIOR PS 1-74 000	Plugged C	C(5)	C		3/8	1/2	19/32 5/8	23/32 3/4
Exterior Type	C-C PLUGGED EXT-APA	For use as tile backing where severe moisture conditions exist. For refrigerated or controlled atmosphere rooms, pallet fruit bins, tanks, box car and truck floors and linings, open soffits. Touch-sanded. (2)(6)	C-C PLUGGED GROUP 2 APA EXTERIOR PS 1-74 000	Plugged C	C	C		3/8	1/2	19/32 5/8	23/32 3/4
Exterior Type	B-B PLYFORM CLASS I & CLASS II EXT-APA	Concrete form grades with high reuse factor. Sanded both sides. Mill-oiled unless otherwise specified. Special restrictions on species. Available in HDO and Structural I. Class I most commonly available. (8)	B-B PLYFORM CLASS I APA EXTERIOR PS 1-74 000	B	C	B				5/8	3/4

(1) The species groups, Identification Indexes and Span Indexes shown in the typical grade-trademarks are examples only. See "Group," "Identification Index" and "Span Index" for explanations and availability.
(2) Can be manufactured in Group 1, 2, 3, 4, or 5.
(3) Special improved grade for structural panels.
(4) Special veneer construction to resist indentation from concentrated loads, or other solid wood-base materials.
(5) Special construction to resist indentation from concentrated loads.
(6) Can also be manufactured in Structural I (all plies limited to Group 1 species) and Structural II (all plies limited to Group 1, 2, or 3 species).
(7) Specify by Identification Index for sheathing and Span Index for Sturd-I-Floor panels.
(8) Made only from certain wood species to conform to APA specifications.

FIG. WF10 **CONSTRUCTION PLYWOOD GRADE—USE GUIDE** *Continued*

PLYWOOD PANEL CONSTRUCTION

Panel Grades	Finished Panel Thickness	Number of Plys	Number of Layers
Exterior			
Marine	1/4"—3/8"	3	3
Special exterior	1/2"—3/4"	5	5
B-B concrete form	3/4"—or greater	7	7
High density overlay			
High density concrete form overlay			
Interior			
N-N, N-A, N-B, N-D, A-A, A-B, A-D			
B-B, B-D			
Structural I (C-D, C-D Plugged and			
Underlayment)	1/4"—3/8"	3	3
Structural II (C-D, C-D Plugged and	7/16"—1/2"	4	3
Underlayment)	5/8"—7/8"	5	5
	7/8"—or greater	6	5
Exterior			
A-A, A-B, A-C, B-B, B-C			
Structural I and Structural II			
Medium density and special overlays			
Interior			
(including grades with exterior glue)			
Underlayment	1/4"—1/2"	3	3
	5/8"—3/4"	4	3
	3/4"—or greater	5	5
Exterior			
C-C Plugged			
Interior			
(including grades with exterior glue)			
C-D	1/4"—5/8"	3	3
C-D Plugged	11/16"—3/4"	4	3
	3/4"—or greater	5	5
Exterior			
C-C			

HARDWOOD PLYWOOD

"Standard" hardwood plywood should meet requirements of Product Standard PS 51-71 and should carry the grade-trademark of the Hardwood Plywood Manufacturers Association certifying conformance with the Standard (Fig. WF11).

"Specialty" hardwood plywood is made in a variety of grades, textures and designs for paneling use and is not covered by any industry Standard. The manufacturers reputation is the only assurance of quality.

FIG. WF11 Grade-trademark of the Hardwood Plywood Manufacturers Association.

202 WOOD PRODUCTS

This Section 202—Wood Products now includes subsections on wood shingles and shakes, wood windows and sliding glass doors, wood doors, wood flooring, and bath and kitchen cabinets. There are many other building products made largely of wood. Discussion of those items will be included in future additions to this section.

SHINGLES AND SHAKES

The large majority of wood shingles and shakes used for roofs and sidewalls are produced from the western red cedar, a giant, slow-growing coniferous tree of the Pacific Northwest. Western red cedar has extremely fine and even grain, exceptional strength in proportion to weight, low expansion and contraction with changes in moisture content and high impermeability to water. Its outstanding characteristics are great durability and resistance to decay (Fig. SS1).

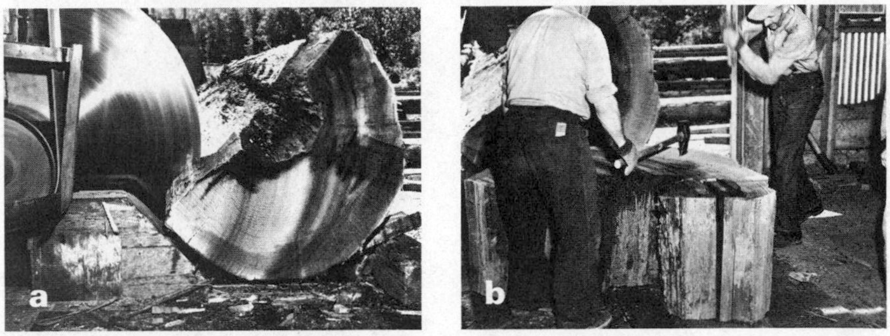

FIG. SS2 Logs are cut to lengths (a), then quartered and requartered to produce edge-grain blocks for sawing or splitting into shakes or shingles (b).

FIG. SS1 A tree grown over a fallen cedar log—no decay exists in the log after 250 years on the ground.

MANUFACTURE

Shakes are *split*, have rough split faces and either sawn or split backs, and are produced in one grade and in three types. Shingles are *sawed*, have smooth sawn faces and backs, and are produced in four grades and a variety of types.

Both shakes and shingles are made from selected cedar logs cut to proper lengths—16″, 18″ and 24″ for shingles; 18″, 24″ and 32″ for shakes. These lengths are again cut or split into the proper size blocks for the shingle cutting machine or a convenient size for splitting of shakes. In reducing the large log sections into blocks, each is quartered, split and requartered to produce blocks having a true edge grain face (Fig. SS2).

Shakes

Today, most shakes are split by machine, although some are still split by hand using a hardwood mallet and steel froe (Fig. SS3). As a result of being split, no two shakes are exactly alike. The three types of shakes produced are: *tapersplit*, *straight-split* and *handsplit-and-resawn* (Fig. SS4).

FIG. SS3 Shakes are produced by splitting cedar blocks with a mallet and froe (a), or by machine (b).

FIG. SS4 Tapersplit shakes are cut from alternate ends of the block (a); straightsplit shakes from the same end of the block (b), handsplit-and-resawn by putting handsplit material diagonally through a bandsaw (c).

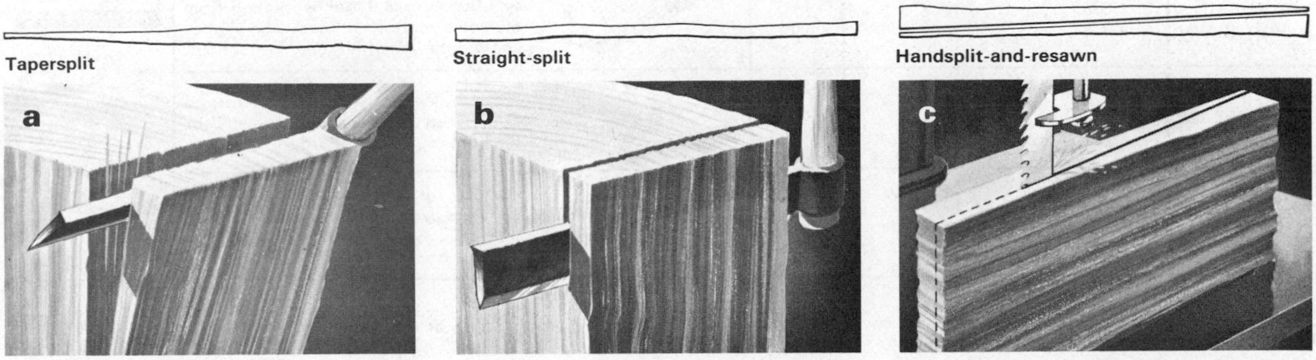

Tapersplit and straight-split shakes (Figs. SS4a & SS4b) are made by machine- or hand-riving individual shakes from the block. To obtain the butt-to-tip taper for tapersplit shakes, blocks are turned end-for-end after each split. Straight-split shakes are split from the same end of the block and hence are of uniform thickness.

Handsplit-and-resawn shakes (Fig. SS4c) are produced by splitting blocks into boards of the desired thickness. The boards are then passed diagonally through a thin bandsaw to form two shakes, each with a split face and a sawn back having thin tips and thick butts.

Packers assemble finished shakes into bundles in a standard-size frame, compressing the bundles slightly and binding them with wooden bandsticks and steel strapping (Fig. SS5).

Each type of shake may be used for roof shingles. Handsplit-and-resawn shakes have the heaviest butt lines and give the most rugged textured appearance to the roof; tapersplit and straight-split shakes give a more uniform texture. While all three types may be used for sidewalls, tapersplit and straight-split types are usually preferred because of their more uniform texture.

FIG. SS5 Grade labels beneath the bandsticks of both shakes and shingles assure that products meet or exceed quality standards.

On 8 in 12 sloped roofs or steeper, shakes are estimated to have a service life of more than 60 years. On lesser slopes, to a minimum of 4 in 12, a life of about 40 years may be expected.

Grades, Types and Sizes There is only one grade, No. 1, for cedar shakes. Shakes of this grade are 100% clear wood, graded from the split face in the case of handsplit-and-resawn shakes, and from the best face in the case of tapersplit and straight-split shakes. All are 100% heartwood, free of bark and sapwood. Tapersplit and straight-split shakes are 100% edge-grain;

FIG. SS8 **TYPES OF RED CEDAR SHAKES**

Grade	Length and Thickness	Bundles Per Square*	Weight (lbs. per square)	Description	Label
No. 1 HANDSPLIT & RESAWN	18″ x 1/2″ to 3/4″ 18″ x 3/4″ to 1-1/4″ 24″ x 3/8″ 24″ x 1/2″ to 3/4″ 24″ x 3/4″ to 1-1/4″ 32″ x 3/4″ to 1-1/4″	4 5 4 4 5 6	220 250 260 280 350 450	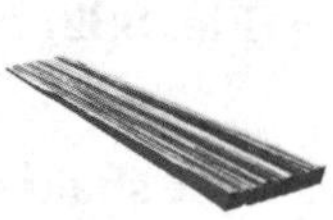 These shakes have split faces and sawn backs. Cedar blanks or boards are split from logs and then run diagonally through a bandsaw to produce two tapered shakes from each.	
No. 1 TAPERSPLIT	24″ x 1/2″ to 5/8″	4	260	 Produced largely by hand, using a sharp-bladed steel froe and a wooden mallet. The natural shingle-like taper is achieved by reversing the block, end-for-end, with each split.	
No. 1 STRAIGHT-SPLIT (BARN)	18″ x 3/8″ (True-Edge) 18″ x 3/8″ 24″ x 3/8″	4 5 5	200 200 260	 Produced in the same manner as tapersplit shakes except that by splitting from the same end of the block, the shapes acquire the same thickness throughout.	
No. 1 STARTER-FINISH	15″ x 1/2″ to 1-1/4″	—	—	 These shakes are used as the starting or underlay course at the eaves and as the final course at the ridge.	
No. 1 HIP & RIDGE	18″ x 1/2″ to 1-1/4″ 24″ x 1/2″ to 1-1/4″	—	—	 Factory cut, mitered and assembled units are produced for a 6 in 12 slope; they adjust to fit slopes between 4 in 12 and 8 in 12.	

* Generally, the number of bundles required to cover 100 sq. ft. when used for roof construction at the maximum recommended weather exposure. (See Fig. SS9).

handsplit-and-resawn shakes may have up to 10% flat grain. Nominal shake lengths are 18″, 24″ and 32″, with a maximum tolerance of 1/2″.

Shake thickness is determined by measurement of the area within 1/2″ from each edge. If corrugations or valleys exceed 1/2″ in depth, a minus tolerance of 1/8″ is permitted in the minimum specified thickness.

Shakes of random width, none narrower than 4″, are packed in straight courses in 18″ and 20″ wide frames. Curvatures in the sawn face of handsplit-and-resawn shakes should not exceed 1″ from the level plane in the length of the shake. Excessive grain sweeps on the split face are not permitted.

Starter-Finish Shakes, produced in a 15″ length, are used as a beginning or underlay course at the eave and as the final course at the ridge (Fig. SS6).

Hip and Ridge Units, factory cut, mitered and assembled, are available to fit hips and ridges, and eliminate on-the-job trimming and fitting (Fig. SS7). They are produced in 18″ and 24″ lengths, for a 6 in 12 roof slope and are adjustable to fit any slope between 4 in 12 and 8 in 12.

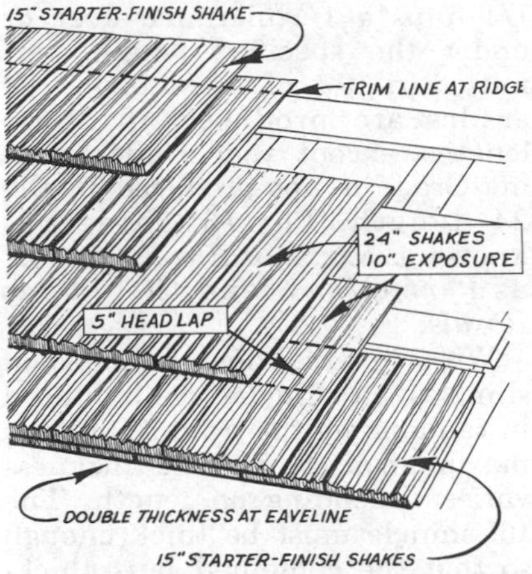

FIG. SS6 Starter-Finish Shakes are more economical for use in the initial undercourse at the eave and in the final course at the ridge.

FIG. SS7 Hip and Ridge shake units, when factory assembled, eliminate costly cutting, fitting and trimming at the job site.

A summary of types, grade and coverage of shakes is given in Figures SS8 and SS9. Methods of applying shakes to roofs and walls are described and illustrated in Division 400.

FIG. SS9 COVERAGE OF SHAKES AT VARYING WEATHER EXPOSURES

Length and Thickness	No. of Bundles	Approximate Coverage (sq. ft.) Weather Exposures											
		5-1/2″	6-1/2″	7″	7-1/2″	8″	8-1/2″	10″	11-1/2″	13″	14″	15″	16″
Handsplit-and-Resawn													
18″x1/2″ to 3/4″	4	55*	65	70	75**	80	85†	—	—	—	—	—	—
18″x3/4″ to 1-1/4″	5	55*	65	70	75**	80	85†	—	—	—	—	—	—
24″x3/8″	4	—	65	70	75	80	85	100	115†	—	—	—	—
24″x1/2″ to 3/4″	4	—	65	70	75*	80	85	100**	115†	—	—	—	—
24″x3/4″ to 1-1/4″	5	—	65	70	75*	80	85	100**	115†	—	—	—	—
32″x3/4″ to 1-1/4″	6	—	—	—	—	—	—	100*	115	130**	140	150†	—
Tapersplit													
24″x1/2″ to 5/8″	4	—	65	70	75*	80	85	100**	115†	—	—	—	—
Straight-Split													
18″x3/8″ (True-Edge)	4	—	—	—	—	—	—	—	—	—	100	106	112‡
18″x3/8″	5	65*	75	80	90	95	100†	—	—	—	—	—	—
24″x3/8″	5	—	65	70	75*	80	85	100	115†	—	—	—	—
15″ Starter-Finish Course		Use supplementary with shakes applied not over 10″ exposure											

* Recommended maximum weather exposure for 3-ply roof construction.
** Recommended maximum weather exposure for 2-ply roof construction.
† Recommended maximum weather exposure for single-coursed sidewalls.
‡ Recommended maximum weather exposure for double coursed sidewalls.

FIG. SS10 The sawyer "edges" shingles after they have been cut from cedar blocks by circular saw at his left.

Shingles

Shingles are cut from cedar blocks on an upright shingle machine (Fig. SS10). The cutting machine carries the block past a thin-gauge, razor-sharp circular saw, which slices a shingle smoothly at each stroke. Shingles cut from one block are jointed or edged and graded by one sawyer. The sawyer can readjust the block in the machine from time to time so that the highest possible grade of shingles will be produced.

The sawyer picks each shingle as it falls from the saw and places its butt firmly against a guide. Using a circular saw set at right angles to the butt, one edge of the shingle is clipped off smoothly. Flipping the shingle over, the sawyer repeats the process, making another smooth edge parallel to the one first cut. If defects are present, he trims the shingle in such a way that a narrower shingle or two shingles without defects can be obtained. Depending on the grade, the sawyer then drops the shingle into a chute which leads to the packing bins.

Shingles are graded again as they are packed into bundles in a packing frame (Fig. SS11). Bundles of shingles are held together by two bandsticks connected by metal straps. Packaged in this manner, the bundled shingles are seasoned and remain perfectly flat in the process.

Grades, Types and Sizes Shingles are marketed according to grade, thickness and type. Nominal shingle lengths are 16″, 18″ and 24″, within 1/4″ plus a 1″ tolerance over or under the specified length permitted in 10% of the shingle. All grades are produced in these lengths, except that Undercoursing Grade is not produced in the 24″ length. The 16″ shingle is known as xxxxx or *FiveX,* the 18″ as *Perfections,* and the 24″ as *Royals.*

With the exception of dimension shingles, all shingles are produced in random widths from 3″ up to a maximum of 14″. Shingle thickness varies depending on length. The 16″ shingle must be thick enough so that the combined butt thicknesses of 5 shingles will measure at least 2″. Thus, shingle thickness is designated as 5 in 2″ or 5/2″. The 18″ shingle has a thickness so that the combined butt thicknesses of 5 shingles will measure at least 2-1/4″ (5/2-1/4″); and 24″ shingles have a thickness so that 4 shingles will measure at least 2″ (4/2″). These dimensions are for unseasoned shingles measured green. If the shingles are measured after seasoning has occurred, a 3% allowance is made for shrinkage.

Shingles with extreme cross grain are not permitted in any grade.

Grade No. 1 (Blue Label) is the best or premium grade of shingle for roofs and single-coursed and the surface course of double-coursed sidewalls. Intended primarily for roofing, they are 100% edge grain, consist entirely of heartwood and contain no knots.

On 8 in 12 sloped roofs or steeper, Grade No. 1 shingles are estimated to have a service life of more than 35 years under normal climatic conditions; on roof slopes of 6 in 12, a service life of at least 25 years can be anticipated; and on 4 in 12 slopes, 20 years. On sidewalls they will have an indefinite life, usually greater than the useful life of the structure.

FIG. SS11 The packer inspects the shingles as he packs them into bundles.

Grade No. 2 (Red Label) is an excellent grade for single-coursed and the surface course of double-coursed sidewalls. It is a good grade for roofs on secondary buildings and may be used on main residence roofs when exposures are reduced approximately 20%. Shingles of this grade must be clear, or free from blemishes for 10″, 11″ and 15″ for 16″, 18″ and 24″ shingles respectively. A maximum width of only 1″ of sapwood is permissible in the first 10″. Mixed edge and flat grain are allowed.

Grade No. 3 (Black Label) is a utility grade and is used where economy is the prime consideration. It may be used as the undercourse in sidewall double-coursing, and for roofs and sidewalls of secondary buildings. These shingles may contain flat grain and sapwood, and must be 6″ clear from the butt for 16″ and 18″ shingles, and 10″ clear for 24″ shingles.

Undercoursing shingles are a lower grade than No. 3, available in 16″ and 18″ lengths. They are produced expressly for use on double-coursed sidewalls for the inner and completely concealed course (Fig. SS12a). These shingles should not be used for any purpose other than undercoursing. Some mills mix Grade No. 3 and Undercoursing shingles and these are called *Special Undercoursing* shingles.

Rebutted and Rejointed Shingles are machine trimmed for exacting parallel edges, with butts cut to exact right angles. They are also available with smooth-sanded face (Fig. SS12b). Produced in No. 1 and No. 2 Grades in 16″ and 18″ lengths, they are used on either single- or double-coursed sidewalls, where tight joints between shingles are desired.

Machine Grooved Shingles (also known as machine grooved shakes) are similar to rebutted and rejointed shingles, except the face is machine-striated to give a corrugated appearance (Fig. SS12c). They are produced in No. 1 Grade only, in 16″ and 18″ lengths, and either natural, prime-coated or finished coated. These units are used for the surface course in double-coursed sidewall applications, and are not used for roofing.

Dimension Shingles are cut to specific uniform widths where special architectural effects are desired (Fig. SS12d). They are available in 5″ or 6″ widths in 16″ lengths in No. 1 and No. 2 Grade; in 5″ or 6″ widths in 18″ lengths in No. 1 Grade; and 6″ width in 24″ lengths in No. 1 Grade. *"Fancy-butt"* shingles (Fig. SS12e) are produced from dimension shingles. The butt designs are produced by trimming the shingles on a band saw.

Hip and Ridge Units are factory-cut, mitered and assembled to fit hips and ridges of various roof slopes. These units, which reduce costs by eliminating on-the-job trimming and fitting (Fig. SS12f), are produced in either No. 1 or No. 2 Grades in 16″ and 18″ lengths for a 6 in 12 roof slope, and are adjustable to fit any slope between 4 in 12 and 8 in 12.

A summary of shingle types, grades and coverage is given in Figures SS13 and SS14. Methods for applying shingles to roofs and walls are described and illustrated in Division 400.

FIG. SS12 (a) Undercoursing shingles are used for completely concealed courses, as shown here in a double-coursed sidewall; (b) rebutted-and-rejointed shingles provide tight-fitting joints between shingles on single- or double-coursed sidewalls; (c) machine grooved shingles are used typically over low-grade undercoursing on double-coursed sidewalls and may be factory primed or finish-coated; (d) dimension shingles are uniformly sized for special design patterns on sidewalls; (e) "fancy butt" shingles, a variation of dimension shingles, have butts cut to special shapes; (f) preassembled hip and ridge units adjust to fit varying roof slopes.

FIG. SS13 COVERAGE OF WOOD SHINGLES AT VARYING EXPOSURES

Length and Thickness[2]	Approximate Coverage (sq. ft.) of Four Bundles or One Carton[1]																		
	Weather Exposures																		
	3-1/2″	4″	5″	5-1/2″	6″	7″	7-1/2″	8″	8-1/2″	9″	10″	11″	11-1/2″	12″	13″	14″	15″	15-1/2″	16″
Random-Width & Dimension 16″x5/2″	70	80	100*	110	120	140	150**	160	170	180	200	220	230	240†	—	—	—	—	—
18″x5/2-1/4″	—	72½	90½	100*	109	127	136	145½	154½**	163½	181½	200	209	218	236	254½†	—	—	—
24″x4/2″	—	—	—	—	80	93	100*	106½	113	120	133	146½	153**	160	173	186½	200	206½	213†
Rebutted-and-Rejointed 16″	—	—	—	—	50	59	63**	67	72	76	84	93	—	100†	—	—	—	—	—
18″	—	—	—	—	43	50	54	57	61**	64	72	79	—	86	93	100†	—	—	—
Machine-Grooved 16″	(normally applied at maximum exposure)													100†					
18″	(normally applied at maximum exposure)															100†			

1. Nearly all manufacturers pack 4 bundles to cover 100 sq. ft. when used at maximum exposures for roof construction; rebutted-and-rejointed and machine grooved shingles typically are packed one carton to cover 100 sq. ft. when used at maximum exposure for double-coursed sidewalls.
2. Sum of the thickness; e.g. 5/2″ means 5 butts equals 2″.

* Maximum exposure recommended for roofs.
** Maximum exposure recommended for single-coursed sidewalls.
† Maximum exposure recommended for double-coursed sidewalls.

STANDARDS OF QUALITY

The manufacture of cedar shingles and shakes should conform to the specifications of the Grading and Packing Rules of the Red Cedar Shingle & Handsplit Shake Bureau. These specifications, which meet or surpass Commercial Standards CS31 and CS199, constitute the recognized standards of the industry.

Certification of Quality

In addition to formulating specifications, a primary function of the Red Cedar Shingle and Handsplit Shake Bureau is to establish and enforce rigid quality controls covering the manufacture and grading of shingles and shakes. To insure compliance with these standards, products made by members of the Red Cedar Shingle and Handsplit Shake Bureau are inspected at frequent unannounced intervals by Bureau inspectors. Mills which conform to these standards are privileged to use the Certigrade, Certigroove, Certi-Split and Certi-Prime labels on their products.

FIG. SS14 **TYPES OF RED CEDAR SHINGLES**

Grade	Size	Bundles or Cartons* Per Square		Description		Labels
		No.	Weight			
No. 1 BLUE LABEL	24″ (Royals) 18″ (Perfections) 16″ (XXXXX)	4 bdls. 4 bdls. 4 bdls.	192 lbs. 158 lbs. 144 lbs.		The premium grade of shingles for roofs and sidewalls. These shingles are 100% heartwood, 100% clear and 100% edge-grain.	
No. 2 RED LABEL	24″ (Royals) 18″ (Perfections) 16″ (XXXXX)	4 bdls. 4 bdls. 4 bdls.	192 lbs. 158 lbs. 144 lbs.		A good grade for most applications. Not less than 10″ clear on 16″ shingles, 11″ clear on 18″ shingles and 16″ clear on 24″ shingles. Flat grain and limited sapwood are permitted.	
No. 3 BLACK LABEL	24″ (Royals) 18″ (Perfections) 16″ (XXXXX)	4 bdls. 4 bdls. 4 bdls.	192 lbs. 158 lbs. 144 lbs.		A utility grade for economy applications and secondary buildings. Guaranteed 6″ clear on 16″ and 18″ shingles, 10″ clear on 24″ shingles.	
No. 4 UNDER-COURSING	18″ (Perfections) 16″ (XXXXX)	2 bdls. 2 bdls.	60 lbs. 60 lbs.		A low grade for undercoursing on double-coursed sidewall applications.	
No. 1 or No. 2 REBUTTED-REJOINTED	18″ (Perfections) 16″ (XXXXX)	1 carton 1 carton	60 lbs. 60 lbs.		Same specifications as No. 1 and No. 2 Grades above but machine trimmed for exactly parallel edges with butts sawn at precise right angles. Used for sidewall application where tightly fitting joints between shingles are desired. Also available with smooth sanded face.	No. 1 & No. 2 labels (above)
No. 1 MACHINE GROOVED	18″ (Perfections) 16″ (XXXXX)	1 carton 1 carton	60 lbs. 60 lbs.		Same specifications as No. 1 and No. 2 Grades above; these shingles are used at maximum weather exposures and are always applied as the outer course of double-coursed sidewalls.	
No. 1 or No. 2 DIMENSION	24″ (Royals) 18″ (Perfections) 16″ (XXXXX)	4 bdls. 4 bdls. 4 bdls.	192 lbs. 158 lbs. 144 lbs.		Same specifications as No. 1 and No. 2 Grades above, except they are cut to specific uniform widths and may have butts trimmed to special shapes.	No. 1 & No. 2 labels (above)
No. 1 or No. 2 HIP AND RIDGE	18″ (Perfections) 16″ (XXXXX)	—			Same specifications as No. 1 and No. 2 Grades above; factory-cut, mitered and assembled units produced for a 6 in 12 slope and adjust to fit slopes between 4 in 12 and 8 in 12.	No. 1 & No. 2 labels (above)

* Nearly all manufacturers pack 4 bundles to cover 100 sq. ft. when used at maximum exposures for roof construction; Undercoursing, rebutted-and-rejointed and machine grooved shingles typically are packed one carton to cover 100 sq. ft. when used at maximum exposure for double-coursed sidewalls.

WINDOWS AND SLIDING GLASS DOORS

Wood was the earliest material used for the manufacture of windows, and over the years has proven its suitability for this purpose. Wood's insulating properties, ready availability, ability to take either natural or painted finishes, and ease of fabrication and repair with simple tools—all contribute to its widespread use in window manufacture. Today, wood windows account for better than half of all window production.

Recent technological innovations have further improved wood as a material for window and door construction. Kiln drying of lumber reduces shrinkage, resulting in better operation and weathertightness. Water-repellent preservative chemical treatments reduce swelling and warping, improve paint retention and increase wood's resistance to decay and insect attack under all climatic conditions. Window walls, now used in many contemporary homes, cover entire walls and consist of combinations of fixed glass for light and view ("the picture window"), and operating sash for light and ventilation (Fig. WW1). In some window walls, sliding glass doors also provide access to outdoor living spaces, in addition to light and ventilation (Fig. WW2).

FIGS. WW1 & WW2 Wood windows and wood sliding glass doors are widely used in all types of construction.

For maximum comfort at minimum heating and air conditioning costs, windows and doors must be weathertight, insulating and condensation free. Also, they must be designed and built to permit ready adaptation to varying installation conditions.

This subsection outlines the manufacture, nomenclature, standards of quality and installation of *stock* windows and sliding glass doors. Stock windows and doors are those manufactured at the mills and warehoused (stocked) in a variety of common types and sizes.

MANUFACTURE

The majority of stock windows are manufactured in accordance with specifications developed by the National Woodwork Manufacturers Association (NWMA). These specifications constitute the industry standard for windows and are known as NWMA IS-2. The following discussion of window manufacture presumes a level of quality conforming with these standards. Sliding door manufacture employs materials and methods generally similar to window manufacture. The manufacture of sliding doors is regulated by NWMA IS-3.

Materials

Better than 2/3 of all stock windows and nearly all doors are manufactured from ponderosa (western) pine. This species has excellent workability, gluability, nail-holding capacity and uniform light color suitable for natural or painted finishes. Other species commonly used are southern yellow pine, Idaho white pine, sugar pine and Douglas fir. All lumber must be sound, free from defects such as loose knots and excessive checking, and kiln dried to a moisture content of 6% to 12%. Other materials such as glass and screening are discussed below under Fabrication and Assembly.

Fabrication and Assembly

Kiln dried lumber is delivered to the millwork shop for fabrication into window and door components.

FIG. WW3 Lumber is cut by the rip saw to specified widths, excluding knotholes and other defects.

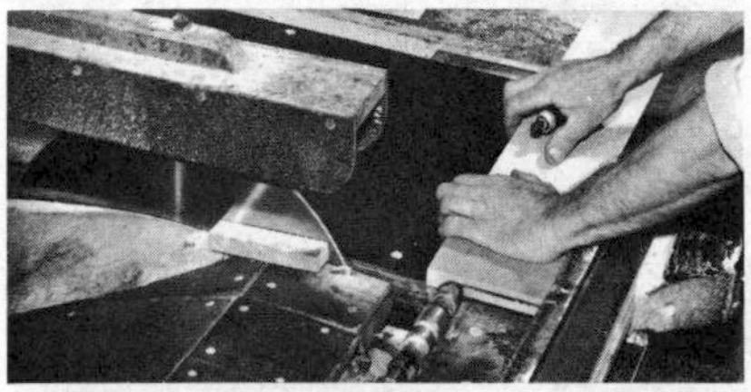

FIG. WW4 Ripped lumber is cross cut to specified lengths and each piece is appropriately marked for frame or sash component use.

FIG. WW5 The lineal milling machine accurately shapes stock to desired profile making all longitudinal grooves, channels and rabbets.

FIG. WW6 Cross milling machines cut grooves across components and shape ends for jointing.

The lumber is cut to specified lengths and widths on rip saws and crosscut saws, leaving cut stock substantially clear and without defects (Figs. WW3 & WW4). Cut lengths are then sorted and marked to indicate the window component for which they are intended. Like pieces are conveyed to molders or lineal milling machines which shape the pieces to the desired profile (Fig. WW5). The next step is cross milling, where necessary grooves and channels (dadoes, mortises and tenons, Figs. WW21 & WW22) are accurately cut across each piece (Fig. WW6) to facilitate weathertight joining of units. Milling of components is completed by routing machines which cut notches to receive recessed hardware (Fig. WW7).

When all shaping operations are completed, milled components and sometimes assembled sash are preservative treated with a solution of toxic, water-repellent agents. Non-pressure treatments (see page Wood 201-13), such as the vacuum and dip processes (Fig. WW8), are commonly used. These give the wood increased dimensional stability, water-repellent properties, and resistance to decay fungus and insect attack. Storm sash and exterior components may also be paint-primed or factory finished for additional protection.

Following preservative treatment, weatherstripping is pressure fit into grooves or surface tacked, and some hardware is installed (Figs. WW9 & WW10). Sash and frames are then assembled from components on automatic nailing machines (Fig. WW11). A continuous sander gives the assembled sash three separate sandings—coarse, medium and fine—in one trip through the machine (Fig. WW12).

The final steps consist of glazing (Fig. WW13), fitting sash to the frames and assembly with all operating hardware. Completed window units are usually provided with angle and spacer braces to prevent racking and distortion in shipment (Fig. WW14).

Weatherstripping The function of weatherstripping is to provide a seal against infiltration (leakage of

FIG. WW7 Recesses to receive hardware are cut out on routing machines.

FIG. WW8 Three-minute "penta" bath renders components water-repellent and resistant to decay and insect attack.

FIG. WW9 Weatherstripping is pressure fit into milled grooves.

FIG. WW10 Some hardware may be fastened to components before sash assembly. Sliding friction hinge is shown here.

FIG. WW11 Automatic nailing machine assembles components by driving up to 16 nails at once.

FIG. WW12 For accurate thickness and smooth finish, assembled sash are machine-sanded.

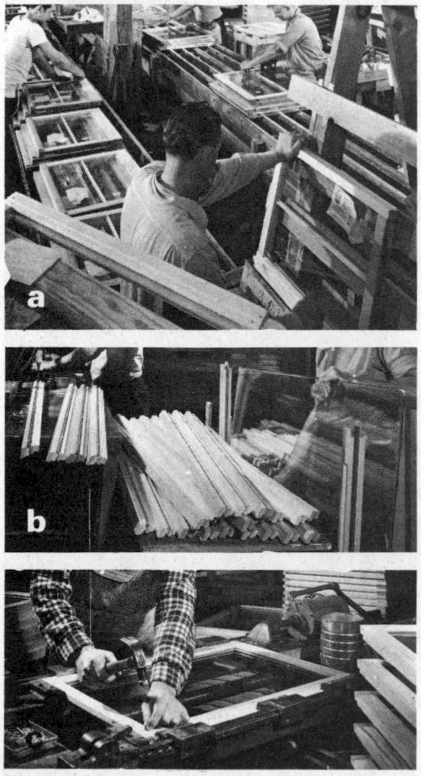

FIG. WW13 Sash may be face glazed with putty (a). In groove glazing, sash is assembled around glass (b) and squared. Pins are driven into corners completing the assembly (c).

air), and penetration of dust and windblown rain through the window. It should be incorporated on all ventilating sash to reduce infiltration substantially below the maximum established by NWMA standards. The maximum is 0.5 cu. ft. per minute per lineal foot of sash crack length, when a sample window is subjected to an air pressure equivalent to 25 mph wind. The window must have all interior trim and stops applied, be in closed position but without storm sash.

The most common weatherstripping used in wood windows is the *spring tension type* of bronze, aluminum, rigid vinyl, stainless or galvanized steel (Fig. WW15a). Extruded aluminum, cold formed galvanized steel, or rigid plastic are used as sash guides and integral weatherstripping in many double hung windows. These may be either fixed to the window frame or cushioned with plastic foam or springs to provide constant pressure against the sash stiles. In the fixed type, *woven felt* in the stiles sometimes engages a projecting fin in the sash guide for improved weathertightness (Fig. WW15b). The cushioned *compression sash guide,* in addition to weather protection often provides for easy removal of the sash (Fig. WW15c). Metal or plastic sash guides eliminate the need for finishing sash runs and the possibility of becoming clogged with paint or varnish.

Compressible bulb-type weatherstripping of vinyl or neoprene is found most often in storm sash, in screens and in sliding glass doors. It is employed by some manufacturers in main sash (Fig. WW15d). In addition to flexible vinyl, woven fabric weatherstripping also is common in sliding glass doors.

Hardware The purpose of hardware is to provide effective closure, to operate the sash, and to hold it stationary at the desired degree of openness. Hardware types range from simple hinges and locks on basement windows, to relatively complex arrangements on windows which require interior operation without removing screens. Underscreen rotary gear, push bar and lever type operators are commonly

FIG. WW14 Assembled windows are braced before packing to prevent racking and distortion in shipment.

FIG. WW15 Common types of weatherstripping.

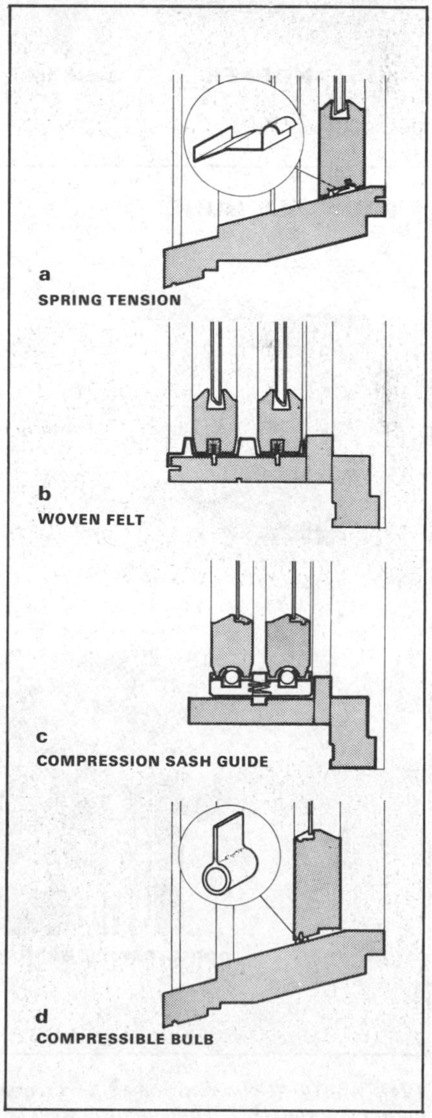

FIG. WW16 Commonly used hardware for various window types. Note in "a" & "b", sliding friction hinges provide projected action: for example, when awning sash is opened, the top rail moves downward.

used for awning windows, in combination with sliding friction hinges which provide *projected action* (Fig. WW16a). The operators generally are capable of weathertight closure, but on larger sash additional awning locks may be provided. Outswinging casement windows often utilize similar rotary gear operators, with either sliding hinges or extension hinges (Fig. WW16b); one or more casement fasteners usually are necessary to draw up relatively long sash stiles against the frame.

Double hung and horizontal sliding windows generally employ cam-action sash locks which draw the sash together at the meeting stiles (or rails) while pushing the sash outward against the frames to compress the weatherstripping. Sliding windows often are equipped with a spring-cushioned metal or plastic track at the head jamb, similar to the sash guide of the double hung window; fixed metal or plastic tracks are commonly used at the sill (Fig. WW16 c & d).

Double hung windows usually are equipped with balancing hardware which assists in raising the sash and holds it stationary in position. *Spiral spring, spring* and *coiled tape* are the most common types of balancing devices (Fig. WW17). Smaller and lighter sash may not require mechanical assistance for raising. In these, the friction fit provided by a compression sash guide may be sufficient to hold sash in place (Fig. WW16c).

Exposed hardware such as sash locks, sash lifts and various operators are available in corrosion-resistant standard (zinc, bronze, aluminum, chrome) and proprietary finishes.

Glass and Glazing Single-strength glass is limited to sizes less than 76 united inches (width plus height); and double-strength glass, to sizes less than 100 united inches. Additional information on grades and sizes of glass is given in Section 207 Glass. Most stock windows and doors are available with either insulating glass or single glazing; tinted and patterned glass usually are available on special order. The Consumer Product Safety Commission (CPSC) *Safety Standard for Architectural Glazing Materials* requires safety glazing material in sliding doors.

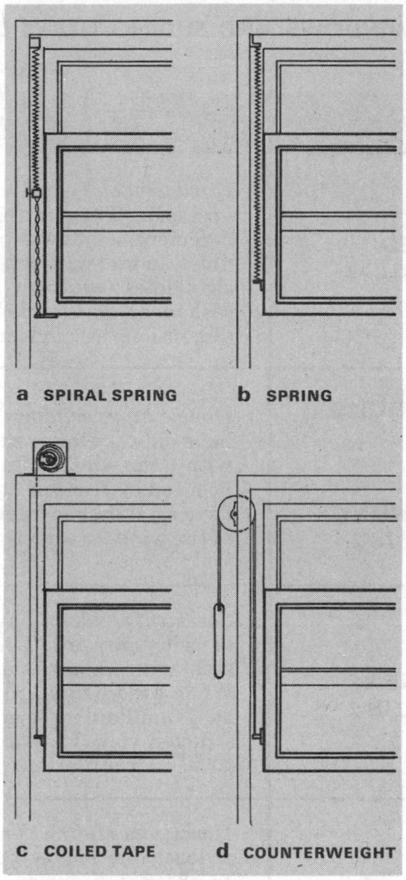

FIG. WW17 Balancing devices used for double hung windows.

Most window sash is *face glazed* in the traditional manner by bedding (back puttying) the glass in a rabbet, securing it with glazing points and sealing it with a bevel face-putty bead (Fig. WW18a). Sometimes pointing is eliminated by using adhesive bedding materials which cushion the glass, seal it and secure it in place.

Wood stop glazing replaces both pointing and face puttying (but not bedding), and provides a glass seal of greater strength, durability and better appearance (Fig. WW18b). Most sliding glass doors and large fixed windows are glazed in this manner.

Groove glazing, employed by some manufacturers, has the same advantages as wood stop glazing, but eliminates the need of handling a separate stop (Fig. WW18c). In this method, grooves in the sash are filled with glazing compound, sash members are assembled around the glass, clamped in a press and nailed or pinned together (Fig. WW13). To facilitate reglazing, one of the sash members can be removed by drawing out the pins which join it at the corners. The sash is reassembled after the new glass has been placed in the groove.

Screening Either aluminum, galvanized steel, bronze or vinyl-coated fiberglass insect screening may be used. Mesh size is limited to 18 x 14 or 18 x 16 (number of openings per inch).

Screening is secured to wood sash by rolling it into a groove and splining, by tacking or by stapling. The sash usually is "trimmed out" with a screen mold to conceal tacks, staples or splines. In metal rimmed screens, the screening is rolled into a groove of the rim and held in

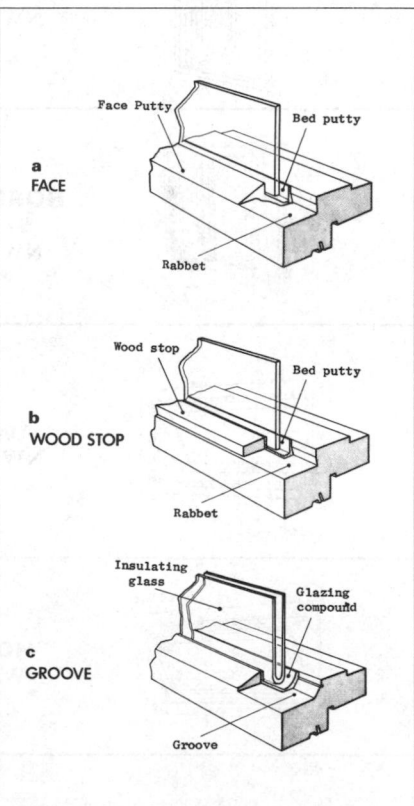

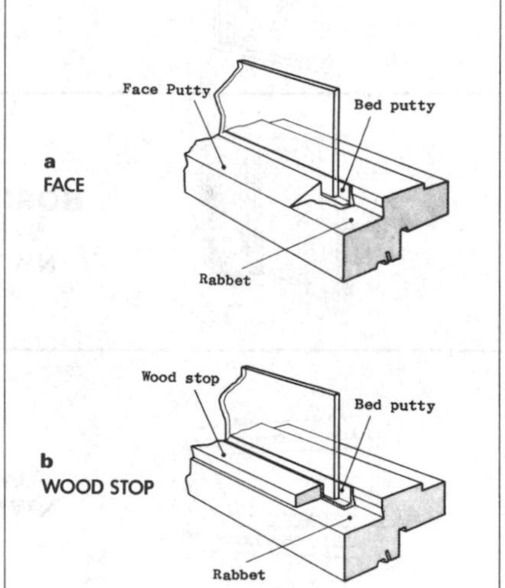

FIG. WW18 Face glazing (a), glazing compound holds glass in wood seat; wood stop glazing (b), glass is bedded in compound and secured by wood stop; groove glazing (c), glass is bedded in compound and set in continuous groove.

FIG. WW19 **SUMMARY OF WOOD WINDOWS AND SLIDING GLASS DOORS**

TYPE AND INDUSTRY STANDARD	DESCRIPTION
 **FIXED** NWMA IS-2	*Fixed windows* usually consist of a frame and glazed stationary sash. They are often flanked with double hung and casement windows, or stacked with awning and hopper units, to make up windows of custom designs. To keep the sight lines (width or height of view area) consistent, fixed sash members are made to same appropriate cross sectional dimensions as adjacent operating sash.
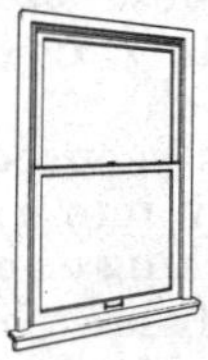 **DOUBLE HUNG** NWMA IS-2 **SINGLE HUNG** NWMA IS-2	*Double hung windows* have two operating sash; single hung have only the lower sash operative. The sash move vertically within the window frame and are maintained in the desired position by friction fit against the frame or with balancing devices. Balancing devices also assist in raising the sash. 50% of the window area is available for ventilation.
 **CASEMENT** NWMA IS-2	*Casement windows* have side-hinged sash, generally mounted to swing outward. They may contain one or two operating sash and sometimes a fixed light between the pair of sash. When fixed lights are used, a pair of casements may close on a mullion, or against themselves, providing an unobstructed view when open. Operating sash may be opened 100% for ventilation.
 **HORIZONTAL SLIDING** NWMA IS-2	*Horizontal sliding windows* have two or more sash of which at least one moves horizontally within the window frame. In three-sash design, the middle sash is usually fixed; in two-sash units, one or both sash may be ventilating. This type of window sometimes is increased in size to door proportions. Ventilating area is 50% of the window area in most designs.
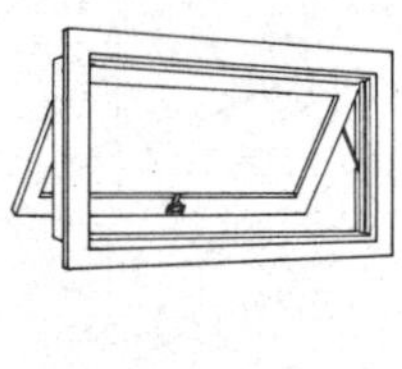 **AWNING** NWMA IS-2	*Awning windows* have one or more top-hinged, outswinging sash. Single awning sash often is combined with fixed and other types of sash into larger window units. Several sash may be stacked vertically and may close on themselves or on meeting rails which separate individual sash. When awning windows have sliding friction hinges, which move the top rail down as the sash swings out, they are said to have *projected action.* Ventilating area is considered to be 100% of operating sash area.
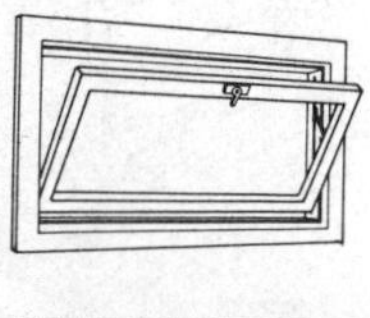 **HOPPER** NWMA IS-2	*Hopper windows* have one or more bottom-hinged, inswinging sash. Hopper sash is similar in design and operation to the awning type, and may actually be an inverted awning sash with minor hardware and weatherstripping modifications. For this reason, windows with hopper sash sometimes are referred to as awning windows. Operating sash provides 100% ventilating area.
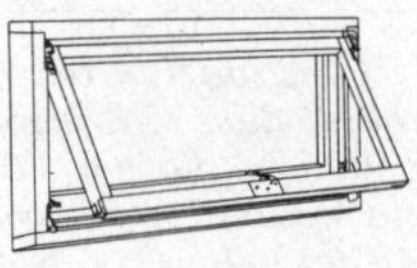 **BASEMENT** (no industry standard)	*Basement windows* generally are single sash units of simplified design intended for less demanding installations, particularly in masonry and concrete foundations. The sash may be of awning type, hopper or top-hinged inswinging (shown) type. 100% of the window is available for ventilation.

FIG. WW19 continued

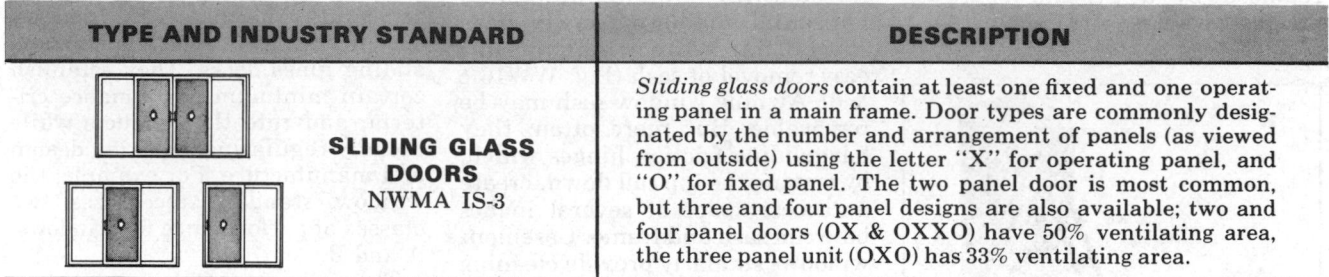

TYPE AND INDUSTRY STANDARD	DESCRIPTION
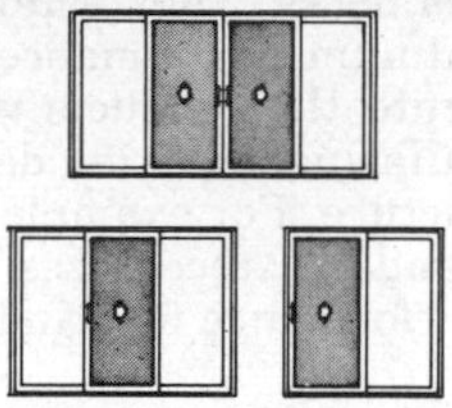 **SLIDING GLASS DOORS** NWMA IS-3	*Sliding glass doors* contain at least one fixed and one operating panel in a main frame. Door types are commonly designated by the number and arrangement of panels (as viewed from outside) using the letter "X" for operating panel, and "O" for fixed panel. The two panel door is most common, but three and four panel designs are also available; two and four panel doors (OX & OXXO) have 50% ventilating area, the three panel unit (OXO) has 33% ventilating area.

place by a vinyl or metal spline.

TYPES AND NOMENCLATURE

Wood windows may be classified according to design and manner of operation into *fixed, double (or single) hung, awning, hopper, casement, horizontal sliding* and *basement* types.

Sliding glass doors are essentially enlarged versions of the horizontal sliding window. However, their increased size, weight and use result in greater demands on frame and sash construction as well as sliding hardware. Sliding panels usually are roller supported and operate on metal tracks built into the threshold. Metal parts are "trimmed out" on the interior with wood to prevent condensation and present a uniform appearance. Door panels sometimes are steel reinforced to provide the necessary strength while retaining a slim profile.

A summary of the most common window and door types is given in Figure WW19.

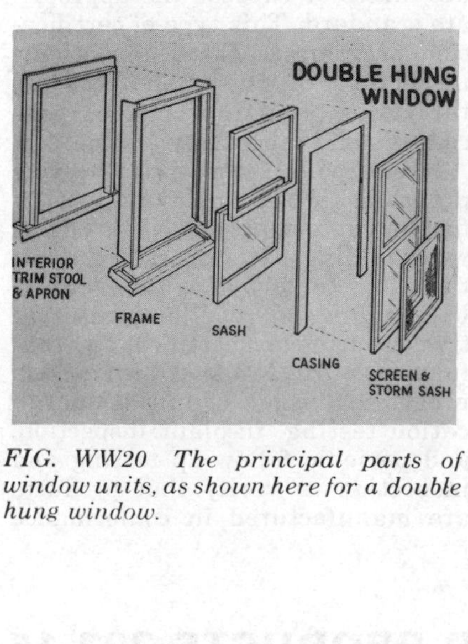

FIG. WW20 The principal parts of window units, as shown here for a double hung window.

Window and Door Components

A window unit consists of a frame, one or more sash and all necessary hardware and weatherstripping to make a complete operating unit. Storm sash and screens may be included as integrated elements of the main frame or contained in a separate combination subframe (Fig. WW20).

Sash Components The sash is made up of horizontal *rails*, vertical *stiles* and sometimes *muntins* and *bars* (Fig. WW21). A sliding glass door is composed of a main frame containing one or more

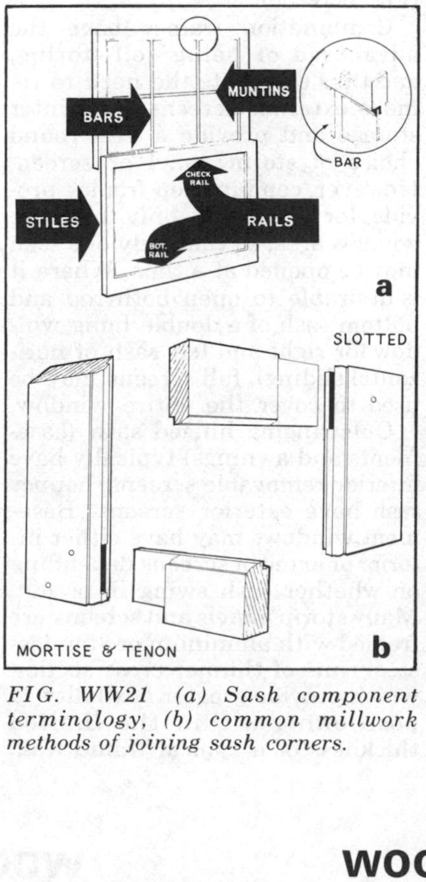

FIG. WW21 (a) Sash component terminology, (b) common millwork methods of joining sash corners.

operating and/or fixed *door panels.* Like window sash, the door panels are composed of stiles and rails and may hold either single or insulating glass. Most doors provide self-storing sliding screens in a separate track of the door frame.

Frame Components Figure WW22 illustrates construction of the traditional double-hung window. The components of other window types may vary somewhat from those illustrated, but the basic elements, consisting of a *sill, side-jambs, headjamb, stops* and *exterior casing,* are present in most window and door designs. *Blind stops* are included in some window frames between jambs and outside casings, and may be varied in thickness for proper window positioning in frame walls with 1/2" or 3/4" sheathing. *Jamb extenders* adapt windows to varying interior finish thicknesses. They are not usually a part of the window frame assembly, but may be ordered loose as optional items. *Interior trim* generally is not supplied with stock windows.

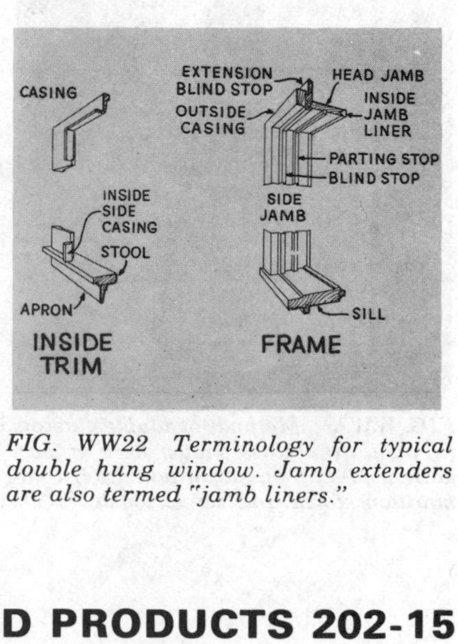

FIG. WW22 Terminology for typical double hung window. Jamb extenders are also termed "jamb liners."

Typically it is selected separately and in some installations may be replaced by a gypsum board or plaster reveal.

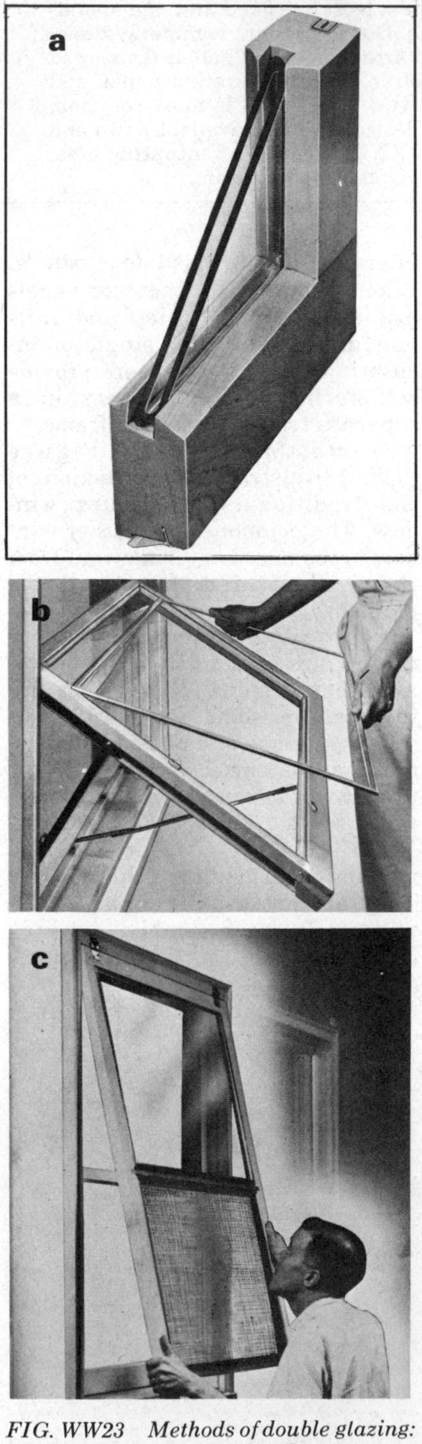

FIG. WW23 Methods of double glazing: (a) insulating glass shown here groove-glazed; (b) clip-on storm panel; (c) combination storm and screen sash.

Cleaning and Maintenance To make exterior surfaces accessible for cleaning, double hung and horizontal windows usually have special hardware which permit easy removal of sash (Fig. WW16 c & d). Awning window sash may be removable. But more often, they have sliding friction hinges which, by moving the top rail down, create an access space of several inches between sash and frame. Casement windows similarly provide cleaning accessibility by sliding or extension hinges. The in-swinging operation of hopper sash makes both surfaces automatically accessible from the inside. Most sash not ordinarily considered removable may be removed with a few simple tools.

Storm Sash and Screens For greater heat control, any window sash may be glazed with insulating glass (Fig. WW23a), or provided with *storm sash.* Clip-on metal rimmed *storm panels* commonly are used with hinged sash such as casements, awnings and hoppers (Fig. WW23b). Sliding and double hung sash may be equipped with either clip-on storm panels or an external *combination frame* which holds both storm sash and screens (Fig. WW23c).

Combination frames have the advantage of being self-storing, and thus eliminate the need to remove external screens for winter storage and provide a year-round choice of storm panel or screen. However, combination frames provide for screening only half the window area, so that only one sash may be opened at a time. Where it is desirable to open both top and bottom sash of a double hung window (or right and left sash of horizontal sliding), full screens may be used to cover the entire window.

Outswinging hinged sash (casements and awnings) typically have interior removable screens; hopper sash have exterior screens. Basement windows may have either interior or exterior screens depending on whether sash swing in or out. Many storm panels and screens are framed with aluminum or stainless steel rims of thinner cross section which provide greater flexibility in positioning within the limited thickness of a typical frame wall.

STANDARDS OF QUALITY

NWMA standards IS-2 and IS-3 constitute the accepted industry specifications for wood windows and sliding glass doors. They establish certain minimum performance criteria, and rate the products while broadly regulating material, design and manufacture. For example, the window standard recognizes two classes of performance for windows: A and B.

The following tests are run on windows: air infiltration, water infiltration and wind loading. Both classes have the same air infiltration requirement, while for water infiltration, class A windows are tested under more severe conditions than class B. Class A windows are subjected to a load twice that of class B windows. Other items covered by the standards are moisture content of the lumber, preservative treatment, screening and glass requirements. Appropriate industry standards for wood windows and sliding glass doors are shown in Figure WW19.

Certification of Quality

Industry standards themselves do not provide for inspection, testing and certification of products. These functions are performed by certification programs, sponsored by the National Woodwork Manufacturers Association (NWMA) or by independent testing laboratories. Issuance of a label (Fig. WW24) assures that the design, manufacture and tested performance of the product meets or exceeds the appropriate standard. This type of certification program is based on periodic inspection of the manufacturing facilities, quality control procedures and laboratory testing.

NWMA offers three certification programs. Wood windows are certified for conformance to NWMA IS-2; wood sliding patio doors are certified for conformance to NWMA IS-3; and water-repellent preservative treating is certified for conformance to NWMA IS-4. Each program consists of initial qualification testing, in-plant inspection, and periodic follow-up testing and inspection to verify that products are manufactured in conformance

FIG. WW24 This National Woodwork Manufacturers Association (NWMA) seal certifies wood windows in accordance with NWMA IS-2.

FIG. WW25 This National Woodwork Manufacturers Association (NWMA) seal certifies wood sliding glass doors in accordance with NWMA IS-3.

with the applicable standards. Conformance to the standards results in certification and issuance of seal (Fig. WW26). This seal is rubber-stamped on various parts of the sash and frame, or imprinted on labels affixed to the product. The seal indicates the fabricator of the product and the industry standard under which the product was produced, and is evidence of compliance with the standard.

INSTALLATION *(See also Division 400—Applications & Finishes)*

Even the best manufactured windows and sliding glass doors will not perform satisfactorily if they are not installed properly. Although window units are commonly squared and braced before shipping, they may be distorted in transit and should always be checked for squareness and equal spacing of jambs prior to installation. Braces should be renailed if necessary and left in place until the window has been installed and is ready for trimming. The frame should be plumbed and fastened securely in the rough opening. Adequate shimming should be provided so as not to distort the frame during nailing. The spaces between finished frame and rough frame should be completely filled with insulation. Drip caps and/or flashing over the cap or exterior casing are strongly recommended to minimize the possibility of leakage at the headjamb. The entire unit should be well caulked at the juncture of the exterior casing and the wall finish, preferably after both elements have been prime-coated, but prior to finish painting.

Where windows and doors are shipped with interior stops loose, it is essential that these be installed carefully for a snug, weathertight fit against sash. Weatherstripping and hardware should be adjusted for proper operation and closure, according to the manufacturer's instructions. Interior trim should be applied carefully to seal the wall cavity between the rough frame and the finish frame.

All windows and doors must be protected with a prime coat as soon after arrival at the building site as possible, and stored out of the weather in a well ventilated space. In finishing, it is important to keep paint or varnish off weatherstripping and interior finish hardware.

Metal or plastic sash guides of double hung and sliding windows should not be painted. Exterior hinges and storm sash hardware may be painted for improved corrosion resistance. Glazing compound invariably should be painted, lapping paint slightly over the glass on both face putty and back putty runs.

In addition to installation instructions, most manufacturers of stock windows provide simple homeowner's operating and maintenance recommendations. For continued trouble-free and effective operation, these should be followed.

FIG. WW26 This National Woodwork Manufacturers Association (NWMA) seal certifies preservative treatment of millwork in accordance with NWMA IS-4.

This page deliberately left blank.

Exterior doors provide for access into the home with control of weather, privacy and safety. They also may facilitate ventilation and may introduce light and view to the interior (Fig. D1).

Interior doors control passage, view and sound between interior living spaces, and may screen storage or utility areas from view. Interior doors also may be used as disappearing or movable partitions, permitting more flexible use of interior space (Fig. D2).

Because of its ready availability, easy workability and insulating properties, wood was the earliest material used in the manufacture of doors. Before the advent of plywood, doors were assembled from a number of solid wood members. The methods used to conceal joints resulted in the familiar panel treatment of surfaces. Later designs substituted thinner wood or plywood panels in a supporting framework to make interior *panel* doors, and glass to make *sash* doors.

Since World War II, mass production of plywood has permitted the introduction and widespread use of the *flush* door. Laminating plywood faces to a wood, mineral or organic core provides a dimentionally stable, economical door. Today, flush doors account for approximately 2/3 of wood doors used in all types of construction.

The following discussion outlines the variety of types and uses, methods of manufacture, standards, certification of quality and installation of wood doors.

FIG. D1 Contemporary sash and panel doors are available in a variety of styles.

FIG. D2 Accordion folding doors divide interior spaces into separate areas.

TYPES AND USES

Wood doors may be classified according to method of construction into: (1) *flush doors,* which generally consist of plywood or hardboard face panels bonded to solid or hollow cores; (2) *stile-and-rail* doors, composed of vertical and horizontal members enclosing wood, glass or louver inserts, creating panel, sash, storm, screen or louver designs; and (3) *accordion folding* doors, assembled from narrow slats which give a long, drapelike appearance.

Figure D3 summarizes the common door types, typical sizes and applicable industry standards which regulate their construction and manufacture. Standards for various door types are based on species of wood and include three major categories: (1) hardwood veneered, (2) ponderosa pine, and

FIG. D3 **SUMMARY OF WOOD DOORS**

TYPE AND INDUSTRY STANDARD		TYPICAL SIZES	DESCRIPTION
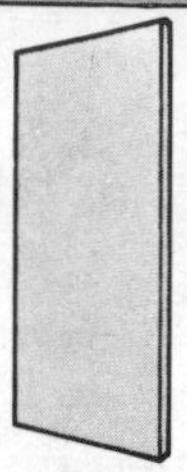	**FLUSH** Wood Flush Doors NWMA IS-1	*Hollow:* 1-3/8″ & 1-3/4″ thick 6′8″ & 7′0″ high 1′6″ thru 3′0″ wide *Solid:* 1-3/8″, 1-3/4″ & 2-1/4″ thick 6′8″ & 7′0″ high 2′4″ thru 3′6″ wide	Constructed from *hollow* or *solid*, wood or composition cores supporting face panels of wood veneers, hardboard or plastic laminates. Presents flush smooth appearance unless interrupted by glass lights or louver openings.
	**PANEL** Ponderosa Pine Doors NWMA IS-5 Douglas Fir FHDA 6-77	1-3/8″ & 1-3/4″ thick 6′8″ & 7′0″ high 1′6″ thru 3′4″ wide	Assembled from stiles and rails (vertical and horizontal components), which frame and support one or more panels. Has characteristic paneled appearance resulting from raised and recessed areas, patterned molds, etc.
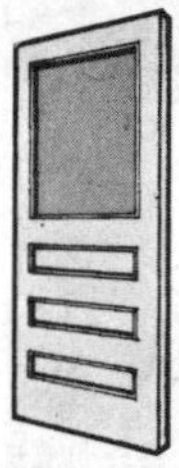	**SASH** Ponderosa Pine Doors NWMA IS-5 Douglas Fir FHDA 6-77	1-3/8″ & 1-3/4″ thick 6′8″ & 7′0″ high 2′0″ thru 3′6″ wide	Similar in construction and appearance to panel door, except that one or more panels are replaced with glass. Completely glazed doors without panels are *casement* (french) doors.
	**STORM & SCREEN** Ponderosa Pine Doors NWMA IS-5 Douglas Fir FHDA 6-77	1-1/8″ thick 6′7″, 6′9″ & 7′1″ high 2′6″ thru 3′0″ wide	Lighter, thinner stile-and-rail construction supporting screening (*screen doors*), glass panels (*storm doors*), or interchangeable screen and storm panel inserts (*combination doors*).
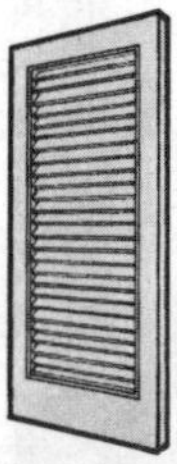	**LOUVER** Ponderosa Pine Doors NWMA IS-5 Douglas Fir FHDA 6-77	1-1/8″ & 1-3/8″ thick 6′6″, 6′8″ & 7′0″ high 1′3″ thru 3′0″ wide	Composed of stile-and-rail frame with integral louver construction, mortised into stiles or vertical dividing bars.
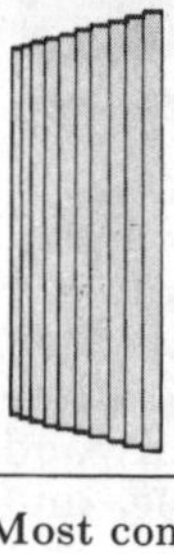	**ACCORDION FOLDING** (no industry standard)	6′8″ to 10′0″ high Practically unlimited length	Assembled from narrow wood strips or single wood slats, 3-1/2″ to 5″ wide, with fabric, plastic or metal hinges, resembling long drapelike doors.

*Most commonly used species for type of door.

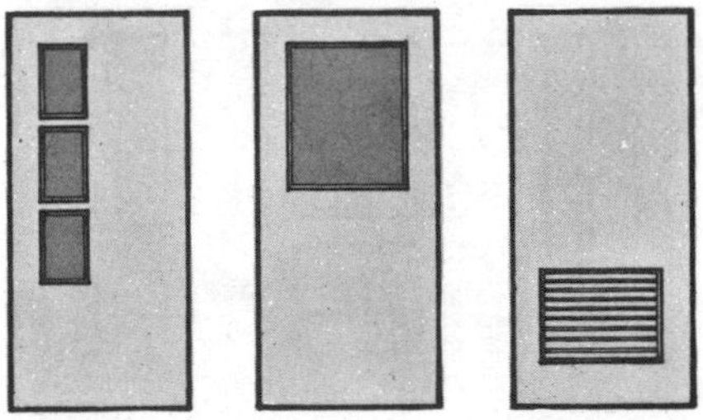

FIG. D4 Flush doors with glass and louver inserts.

(3) old growth Douglas fir, sitka spruce and western hemlock. Accordion folding doors are made of a variety of woods and are not presently covered by industry standards.

FLUSH DOORS

In addition to providing the desired surface appearance, face panels of flush doors act as *stressed skins* to impart strength, rigidity and stiffness to the door. Face panels may be hardwood veneer, plywood, high-pressure plastic laminate, or hardboard. Wood face panels are usually of two or three plies with a total thickness of at least 1/12". Door construction is described as 3-, 5-, or 7-ply by counting the total number of plies in both face panels and counting the core as one ply.

Flush doors are available with glazed lights and louver inserts. The U.S. Consumer Product Safety Commission (CPSC) "Safety Standard for Architectural Glazing Materials" requires that all glazing materials in doors be safety glazing material. In addition, industry standards limit the size of the cutout to a maximum of 40% of the door area (Fig. D4).

Solid Core

Solid core construction increases the dimensional stability of the door, provides thermal insulation, and may add sound isolation and fire resistance. Solid core doors are commonly used as exterior and entrance doors. In some applications, doors with a specific fire rating may be required, such as the door to a garage or heater room, or apartment corridor doors. These doors must carry a fire door label from an approved testing agency, such as Underwriters Laboratory

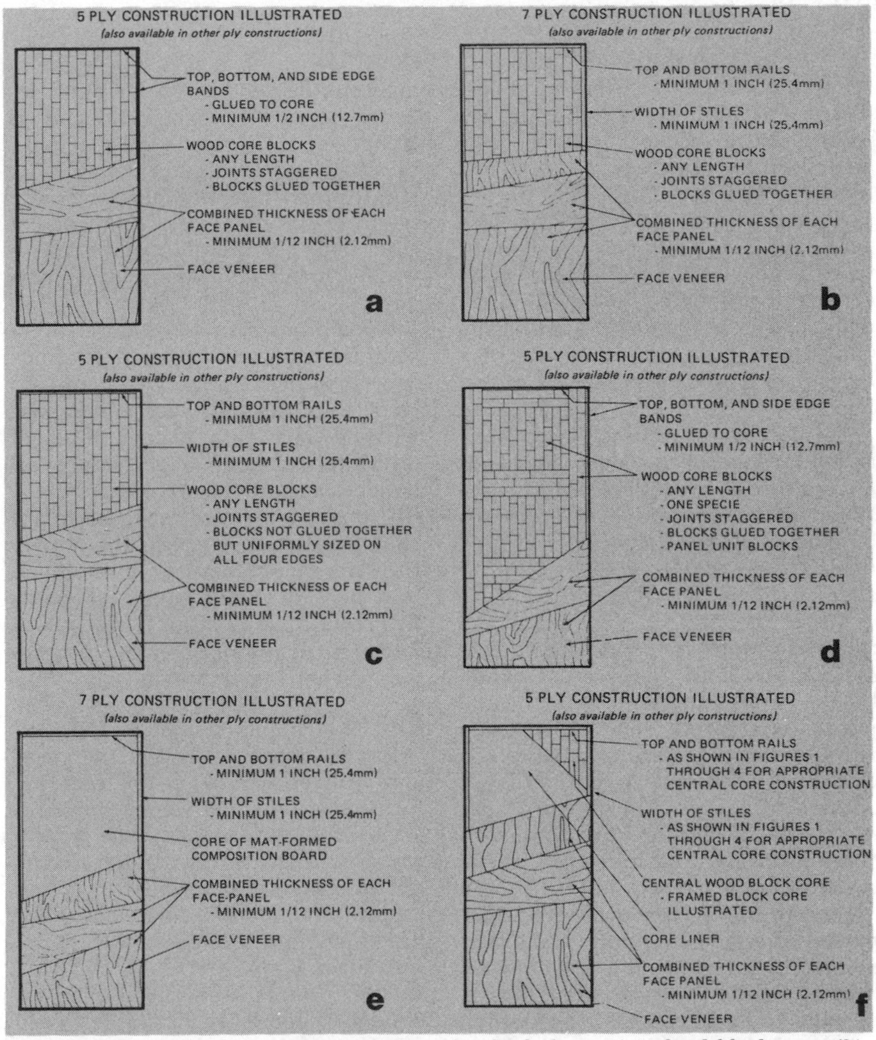

FIG. D5 Typical solid core constructions for flush doors: (a) glued block core; (b) framed block glued core; (c) framed block non-glued core; (d) stile and rail core; (e) mat-formed composition board core; (f) wood block, lined core.

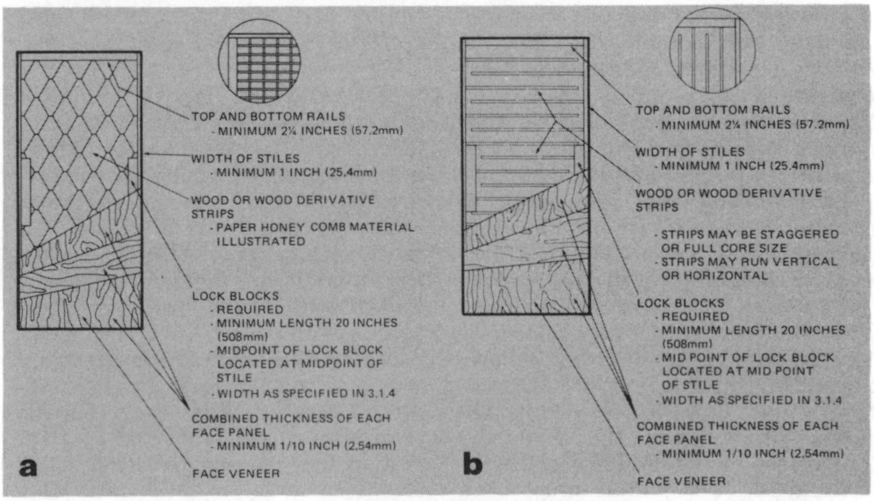

FIG. D6 Typical hollow core constructions for flush doors; 7-ply construction illustrated: (a) mesh or cellulor core, (b) ladder core.

(UL). Fire ratings for doors range from 20 minutes to 1-1/2 hours depending on materials and constructions. Many regular solid core doors are capable of a 20 minute fire rating. Local codes should be consulted for specific requirements and applications.

Solid core doors provide sufficient sound isolation for most residential uses. Where sound isolation is more critical, properly gasketed flush doors with a Sound Transmission Class (STC) rating of up to 42 may be required (see page Sound Control 106-12).

Solid cores may be of wood, particleboard, mineral composition, or a combination of these materials. Wood cores are made of strips or blocks of wood forming a *glued block core, framed block core,* or *stile and rail core* (Fig. D5).

Glued Block Core doors are the most common type of solid core construction, and consist of wood blocks bonded together with the end joints staggered. The stiles and rails of this door are bonded to the core, and the entire assembly is sanded to a smooth, uniform thickness. The face panels are bonded to this assembly, and must be a minimum of 1/12″ thick.

Framed Block Core doors are comprised of a stile and rail frame surrounding an interior core of wood blocks (which may or may not be bonded to each other), all of which are bonded to the face panels. Face panels must be a minimum of 1/12″ thick.

Stile and Rail Core doors consist of wood blocks bonded into stile and rail panel units. These units are assembled to comprise the core of the door and are bonded to the face panels. Again, face panels must be a minimum of 1/12″ thick.

Particleboard Core doors are constructed of one or more pieces of particleboard, each the full thickness of the core, which may or may not be edge or end glued together. The stiles and rails of the door may or may not be bonded to the core, depending on the type of construction. Open spaces between core pieces or between the core and frame are controlled by industry standards. Face panels must be a minimum of 1/12″ thick.

Mineral Composition Core doors are normally not available in other than fire rated doors. These doors are constructed similar to particleboard core doors, except the core is comprised of inorganic, non-combustible materials formed into a rigid slab. Special blocking and framing is normally required in this door to accommodate hardware, glazed openings or other cutouts.

Wood Block, Lined Core doors are a combination of the constructions described previously. This door consists of a central wood block core with a 3/16″ or thinner liner of particleboard or other material bonded to each face to comprise the full core thickness. Face panels are then bonded to this assembly and to the stile and rail frame. These doors may provide added sound isolation, fire resistance or other performance features depending on the properties of the liner material.

Hollow Core

Flush doors with hollow cores commonly are used for interior locations and may be used as exterior doors if bonded with waterproof adhesives. They generally do not provide as much heat or sound insulation as solid core doors, but are suitable where these factors are not critical.

Face panels like those used on solid core doors are available. Wood face panels must be at least two plies and must have a total thickness of at least 1/10″. The basic hollow core constructions are *mesh (cellular) core* and *ladder core* (Fig. D6).

Mesh (Cellular) Core doors consist of wood or wood derivative material joined, interlocked or woven to form a mesh or grid throughout the core area. Honeycomb cores of expanded paper or other material are the most common types of hollow cores. Face panels are bonded to the core and to the stiles, rails and lock blocks of the door.

Ladder Core doors are comprised of wood or wood derivative strips placed horizontally or vertically throughout the core area. Face panels are bonded to these strips and to the stiles, rails and lock blocks of the door.

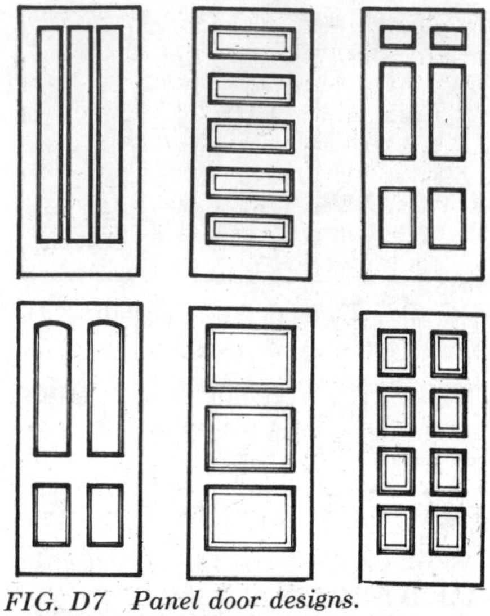

FIG. D7 Panel door designs.
FIG. D8 Sash door designs.

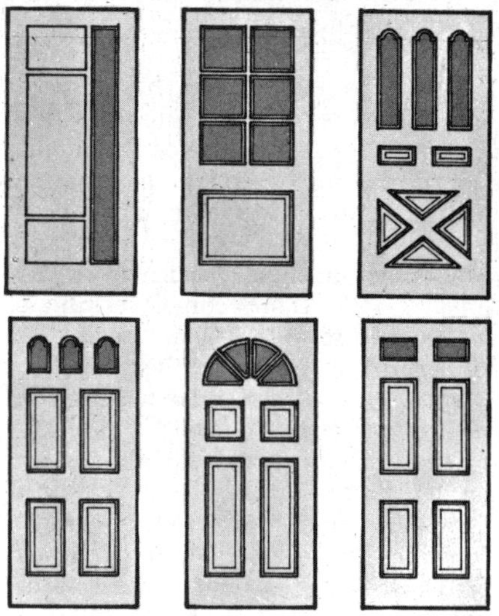

Lock blocks are provided in hollow core doors for lock mortising and backset up to 2-5/8″. Doors with special blocking may be obtained for other hardware mounting.

Stile-and-Rail Doors

Stile-and-rail doors derive their strength mainly from a supporting framework of vertical members (stiles) and horizontal members (rails) interlocked and glued together. Stiles and rails may be solid or veneered and are usually of softwoods.

The chief categories of stile-and-rail doors are: *panel, sash, louver, storm, screen* and *combination.*

Panel and Sash Panel doors consist of stile-and-rail frames enclosing flat plywood or raised wood panel fillers. They are available in a variety of designs (Fig. D7) and may be used for either interior or exterior application.

Sash doors are similar to panel doors in appearance and construction, but have one or more glass lights replacing the wood panels. They are used often for main entrance and exterior service doors (Fig. D8). Casement (french) doors are fully glazed sash doors and are referred to as *rim* doors when they have a single large glass panel and as *divided light* doors when the glass is divided into smaller lights (Fig. D9).

Stiles and rails are assembled with glued dowel or blind mortise-and-tenon construction (Fig. D10a & b). Together with mullions, muntins and bars, they frame the glass lights and/or wood panels (Fig. D10c). The contoured profiles, grooves or rabbets (*sticking*) milled on the edges of these members hold the glass or wood panels (Fig. D10d).

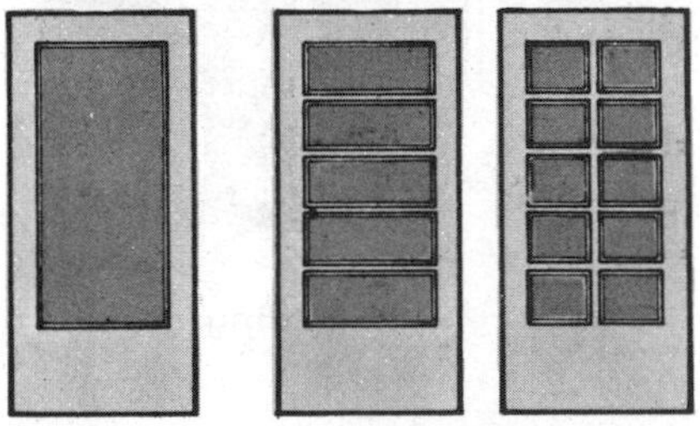

FIG. D9 Casement (french) doors.

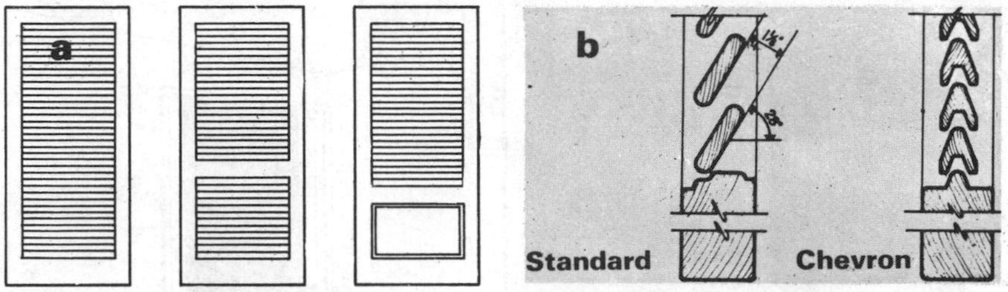

FIG. D11 Typical louver doors: (a) door designs, (b) slat styles.

Louver Louver doors are made of horizontal slats contained within stile-and-rail frames. The slats usually are blind-mortised. They are available in several common designs and two basic slat styles, *chevron* and *standard* (Fig. D11).

These doors are used in interior locations primarily as visual screens and where sound privacy is not required. Appropriate uses include closet doors, space dividers and closures around laundry areas, where ventilation and dissipation of moisture is desirable.

Storm, Screen and Combination Storm doors are lightly constructed glazed doors used with exterior doors to improve weather resistance. Screen doors are of similar construction with screening, which provides ventilation while excluding insects. Combination doors combine the functions of both storm and screen doors with interchangeable screen and glass panels (Fig. D12).

Some *self-storing* combination doors are equipped with a screen and a pair of glass panels which operate in a vertical track similar to a double hung window. Either ventilation or weather protection is possible without removing panel inserts, but the ventilating area available is limited to one half that of a full screen door. Glass jalousie inserts in combination doors have similar self-storing advantages with practically no loss in ventilating area.

FIG. D12 Storm and screen doors give added weather protection and ventilation.

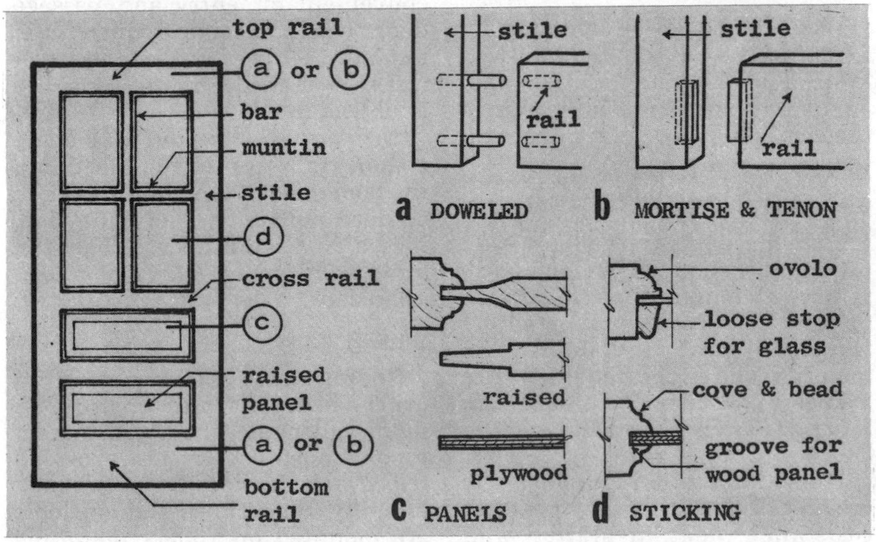

FIG. D10 Typical construction of stile-and-rail doors.

Accordion Folding Doors

Accordion folding doors are made from either wood slats or wood strips, connected with cord or fabric tapes, forming flexible drapelike doors of almost unlimited length (Fig. D13). The door generally is designed so that it will fold to a width less than the thickness of a normal interior partition.

Solid wood slats usually are connected to each other with continuous metal, vinyl or nylon fabric hinges. *Woven* doors are commonly of the basketweave type in which vertical strips of 3/8″ to 1″ wide are interwoven with nylon reinforced vinyl tape. *Corded* doors are made

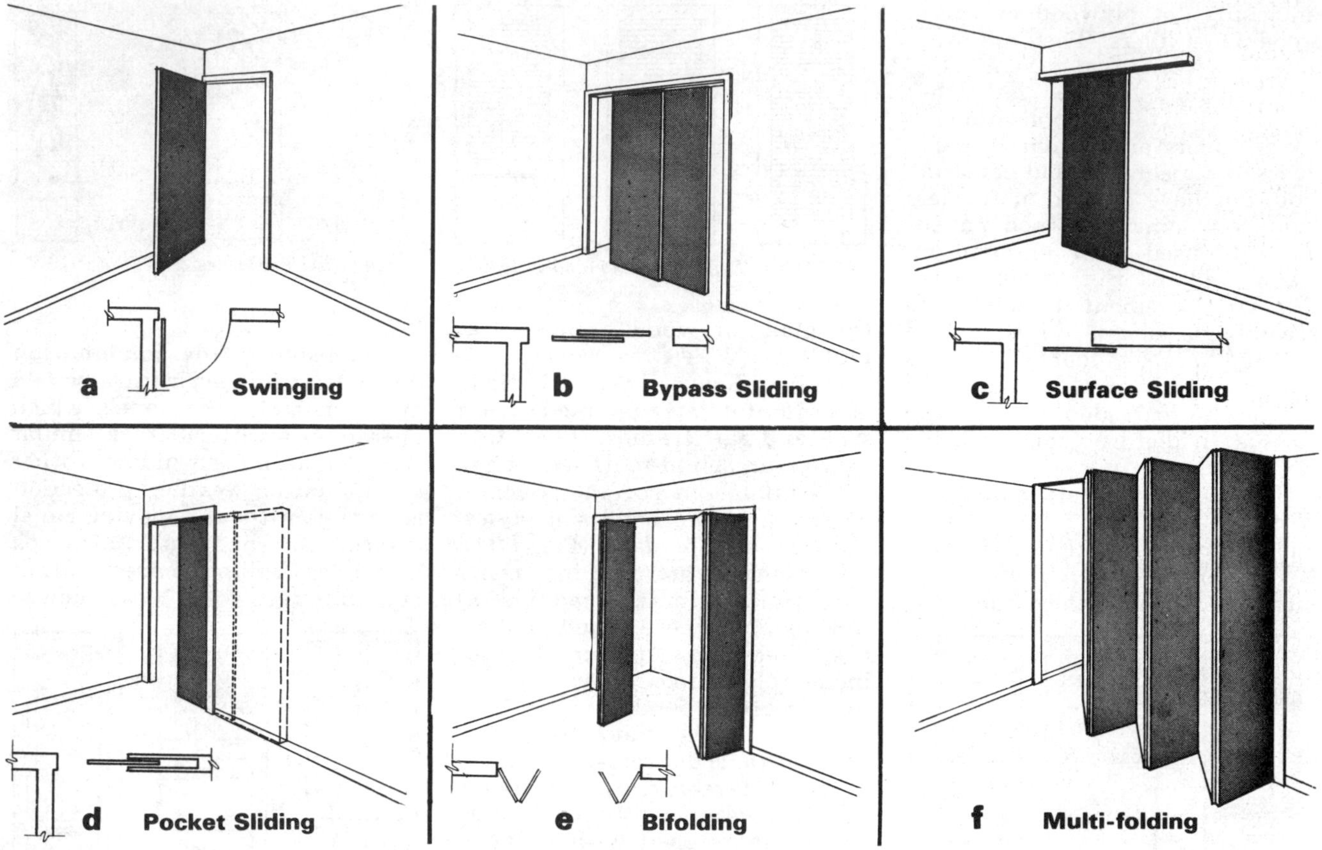

FIG. D14 Door operation: swinging (a), sliding (b, c & d), folding (e & f).

with similar wood slats connected with cotton cord. Accordion doors usually are suspended from ceiling-mounted tracks and operate on nylon rollers or glides.

These doors are primarily visual screens, for locations such as

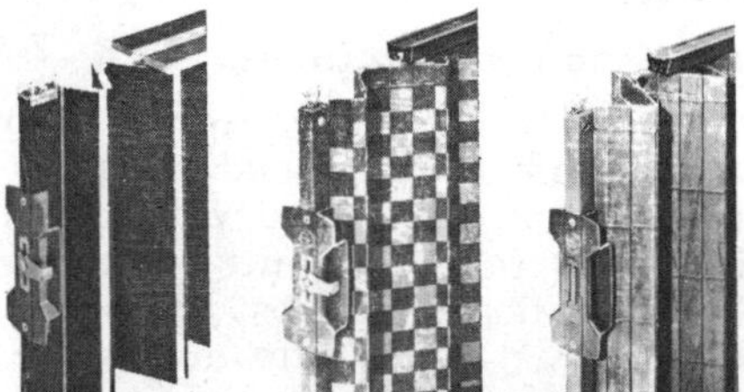

FIG. D13 Typical accordion folding doors.

closets, space dividers, laundry enclosures and where air circulation is desirable, and are not suitable where sound isolation and privacy are prime requirements. The folding operation also makes them useful where door swings are objectionable.

DOOR OPERATION

Wood doors can be operated by *swinging, sliding* or *folding* (Fig. D14). The use and location of a particular door type determines the most suitable manner of operation.

Swinging

Swinging doors typically operate on hinges secured to the side jambs, although they may also operate on pivots supported by the head jamb and floor. The majority of residential doors are *hinged* (Fig. D14a). *Pivot* hardware is used mainly for double-acting doors.

Hinged operation provides the greatest degree of security, weather resistance, heat insulation and sound isolation. Where these factors are particularly important, doors may be weatherstripped or gasketed for increased effectiveness.

Swinging door operation is most convenient for entry and passage doors which receive frequent use, but may be used in any location unless door swing into usable space is objectionable.

Door swing is referred to as being either right- or lefthand. When the door opens to you and the knob or door pull is on the right side, it is a righthand door. When the knob is on the left side, it is a lefthand door.

Sliding

Sliding doors are suspended on overhead tracks and operate on nylon rollers attached to the door. Floor guides usually are provided to prevent the door from swinging laterally. Sometimes sliding doors are mounted on rollers operating on floor tracks, but these are not common.

Sliding doors are suitable for visual screens as in closets and storage areas where *bypass sliding* doors of flush, panel or louver design are common (Fig. D14b). *Surface sliding* and *pocket sliding* doors, which operate in a similar manner, may be used for passage doors (Fig. D14c & d). Biparting surface or pocket sliding doors are used for openings 3′ or wider. For extremely large openings, multi-track installations permit stacking of several sliding doors at one or both ends.

Sliding operation ordinarily does not give a high degree of privacy, sound isolation or weather resistance, unless custom hardware or pre-engineered door units are used. The chief advantage of sliding operation is that it eliminates door swings which might interfere with the use of interior space.

Folding

Folding doors are hung on overhead tracks with nylon rollers or glides, similar to sliding doors. Some types also require a floor track. Folding operation, like sliding, is generally appropriate for locations requiring visual screening primarily.

Flush, louver and panel doors are often used for closets in single or paired *bifold* arrangements (FIG. D14e), or as space dividers and movable partitions with special hardware for *multi-folding* action and stacking (Fig. D14f). *Accordion* folding doors are used as closet doors or as space dividers and may be pocketed or end-stacked within the opening.

MANUFACTURE

The following discussion of door manufacture presumes conformance with industry standards.

Materials

The common species of hardwoods for door manufacture are oak, ash, birch, cherry, walnut, basswood and lauan. Softwood species are ponderosa (western) pine, old growth Douglas fir, sitka spruce and western hemlock.

The majority of flush doors are made with hardwood veneered

FIG. D15 Implanted blanks core for hollow door is made by feeding wood spirals into an assembled frame.

FIG. D16 Cellular core construction is made by gluing paper honeycomb into an assembled frame.

FIG. D17 Hollow flush doors are cold pressed to assure proper bonding of face panels to cores.

FIG. D18 Hot pressing sets the glueline of solid core doors bonded with thermosetting adhesives.

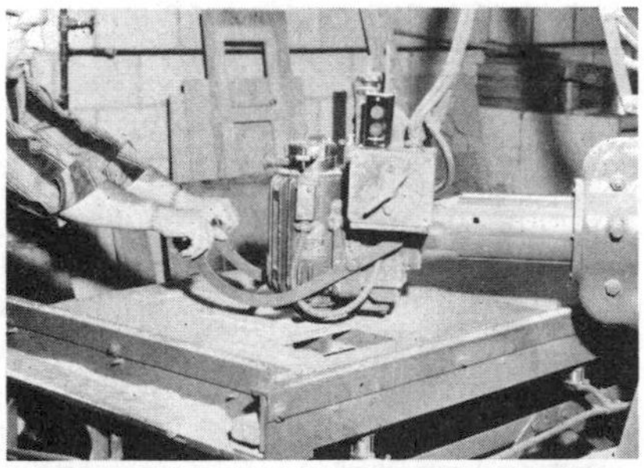

FIG. D19 Flush doors are routed for glass lights and louver openings.

FIG. D20 Muntins and bars are fitted and installed before glazing.

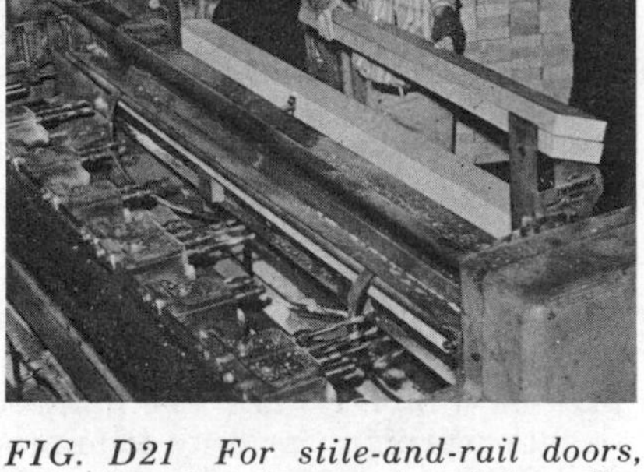

FIG. D21 For stile-and-rail doors, holes for dowels are bored in stiles automatically fed from a hopper.

FIG. D22 An endless feeder chain carries rails or stiles through a molding machine to form sticking on edges.

FIG. D23 One machine bores holes, inserts glue and drives dowels.

FIG. D24 After being cut to size, panels are shaped to make raised designs.

FIG. D25 Here, stile-and-rail doors are sanded to a smooth finish.

FIG. D26 After glazing, stops are accurately mitered and nailed in place.

FIG. D27 Doors are mortised for locksets on an automatic routing machine.

faces. Most stile-and-rail doors are manufactured from ponderosa pine, but some also are produced from other softwoods. Hardwood veneered stile-and-rail doors and softwood veneered flush doors are available, though not common. Accordion folding doors are made from both hardwoods and softwoods.

All lumber used in door manufacture is kiln-dried to a moisture content of 6% to 12% before being assembled into a door.

Solid wood cores for flush doors should contain only one species of wood which has a specific gravity of .42 or less. Composition core materials range in density up to about 37 lbs. per cu. ft. at the same moisture content.

Hardwood veneered flush doors for *exterior* use should be made with waterproof (Type I) adhesives such as thermosetting melamine urea or phenolic resins. *Interior* flush doors and most stile-and-rail doors are made with water-resistant (Type II) adhesives, such as casein (sometimes fortified with urea resins and formaldehyde), capable of resisting a limited number of wetting cycles (see Plywood, page Wood 201-41).

Fabrication and Assembly

Fabrication varies somewhat depending on the type of door.

Flush Doors Stiles, end rails, lock rails and lock blocks for hollow core flush doors are cut from kiln-dried, surfaced lumber. Stiles and rails are joined either with corrugated metal fasteners or by machined dovetail joints. The stile-and-rail frame is laid up in adhesive with one face panel. Cores, lock blocks and sometimes lock rails then are set and glued in place. Cores may be ladder, mesh or implanted blanks (Figs. D15 & D16). Assembly is completed by bonding the second skin to the frame and the core.

Hollow core flush doors usually are bonded with water-resistant adhesives which do not require heat to set. These doors are cold pressed (Fig. D17) and are allowed to cure overnight, prior to trimming and further working.

Most solid core doors are made with *thermosetting* adhesives. Continuous block cores are assembled from wood blocks, glued together and cured. After additional air drying for 24 hours, face panels are adhesive-bonded to the core and the entire assembly is *hot-pressed* to set the glueline (Fig. D18).

Further operations on both solid and hollow core doors may include routing for louver and glass openings and installing muntins and bars prior to glazing (Figs. D19 & D20).

Stile-and-rail Doors After surfacing kiln-dried lumber to exact thickness, boards are cut to proper widths and lengths. Pieces for stiles are bored for dowels on a boring machine (Fig. D21); a molding machine shapes the sticking on stile and rail edges (Fig. D22); and a double-end tenoner cuts rails to exact lengths, coping both ends to fit the sticking on the stiles. Another machine bores holes in rails, inserts glue and drives the hardwood dowels into the holes (Fig. D23).

For panel doors, wood or plywood panels are cut to size and may be shaped to form raised panel designs (Fig. D24). Stiles and rails are assembled with water-resistant adhesives around the panels and clamped in a jig. When the glue has set, doors are sanded on both sides to a smooth finish (Fig. D25).

Stile-and-rail frames, and sometimes panels, muntins and bars for sash doors, are fabricated and assembled in a similar manner. Glass is cut to size and bedded in glazing compound, and stops are accurately mitered and nailed in place (Fig. D26).

Prefabrication

Doors may be further worked and finished in the plant in varying degrees, to reduce job-site preparation and installation. They may be seal-coated to prevent moisture absorption and soiling, or may be completely surface-finished. *Prefinished* doors often are *prefit* to exact opening dimensions so that the finished edges will not be removed by job-fitting. Doors for prefabricated frames sometimes are machined for locks and hinges (Fig. D27), and bifold doors may

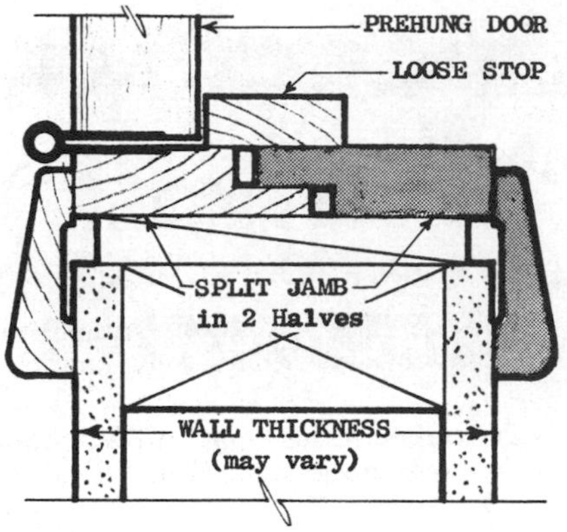

FIG. D28 Split jamb sections with trim applied may be assembled from each side of the opening and fit walls of varying thickness. Stops are then applied.

have all operating hardware installed.

Prehung door units are available assembled complete with frames, trim and hardware. The frames are of two-piece *split* design, which provides for approximately 1/2″ adjustment for varying wall thicknesses (Fig. D28). The stop may be part of the frame or left loose to be applied during installation. Frames are braced at jamb bottoms and carefully packed to prevent racking and distortion in shipment.

STANDARDS OF QUALITY

NWMA and FHDA standards constitute the recognized standards of quality for the wood door industry and are based on the type of wood and door construction as follows: Wood Flush Doors, NWMA IS-1; Ponderosa Pine Doors, NWMA IS-5; Douglas Fir, Sitka Spruce and Western Hemlock Doors, FHDA 6. Flush doors are covered by NWMA IS-1; panel, sash, louver, storm and screen doors are covered by NWMA IS-5 and FHDA 6 (Fig. D3).

In addition to establishing minimum requirements for material, design, construction and waterproof performance under test conditions, NWMA and FHDA standards also provide a basis for quality grading of wood doors.

Grades

Doors are graded according to the exposed faces and whether they are intended for natural finishes such as stain or varnish, or opaque finishes such as paint or enamel.

Wood Flush Doors (NWMA IS-1) Solid and hollow core flush doors are available in *premium, good, sound* and *specialty* hardwood grades, depending on the quality of face panels, and in *plastic* and *hardboard* grades. Flush doors for exterior use should be made with waterproof (Type I) adhesives. All other door types commonly are made with water-resistant (Type II) adhesives unless otherwise specified.

Premium Grade face panels are intended for natural finish and are made of tight and smoothly cut veneers, free from knots, wormholes, splits, shakes and torn grain. When the face consists of more than one piece, the pieces are of approximately equal width and matched at the joints for color and grain.

Good Grade face panels are made of veneers essentially similar to premium grade, but need not be matched at joints for color and grain if sharp contrasts between adjacent pieces are avoided. This grade also is intended for natural finish.

Sound Grade face veneers are intended for painting and may contain a variety of color, grain and surface defects which will not be visible after two coats of paint are applied.

Specialty Grade doors with architectural faces, matched-grain faces for special purposes, special veneer selections, or mis-matched faces, need not necessarily conform to previously mentioned grades. This grade may permit more characteristics and more defects; it may also have fewer characteristics and more defects than those set forth in other grades. The grade description and characteristics are as agreed upon by the purchaser of the door and the door manufacturer.

Plastic Grade doors have high pressure plastic laminate face panels. These doors usually are prefit to exact dimensions.

Hardboard Grade doors have hardboard face panels and are intended primarily for painting.

Ponderosa Pine Doors (NWMA IS-5). Grades are premium and standard, and are typically made with water-resistant adhesives.

Premium is made from stock practically clear of defects and is suitable for natural finishing.

Standard is made from lumber stock that may contain stains and various other minor defects and is intended for paint finishing.

Old Growth Douglas Fir, Sitka Spruce and Western Hemlock Doors (FHDA 6-77). These softwood doors usually are made with water-resistant adhesives. Waterproof ad-

FIG. D29 (a) The National Woodwork Manufacturers Association (NWMA) seal on doors assuring conformance with the appropriate industry standard; (b) facsimile of red plastic plug indicating waterproof Exterior Type I adhesive; plug is located within 3″ of corner on rail edge.

hesives are used in bonding vertical grain veneered exterior doors and may be used in flat plywood panels when specified. Grades are *select*, suitable for natural finishing, and *standard*, intended for paint finishing.

FIG. D30 The National Woodwork Manufacturers Association (NWMA) seal certifying preservative treatment of millwork in accordance with NWMA IS-4.

Certification of Quality

Doors which conform to the industry standards may be grademarked and/or labeled by the manufacturer, when requested by the buyers. Grademarks usually are stamped or branded on the edge of the door; labels are written statements with similar information attached to the door.

NWMA administers an independent certification program for hardwood veneered doors based on continuous, impartial inspection and testing of doors. The NWMA seal assures that the materials, construction and manufacture of doors conform to applicable industry standards (Fig. D29a). A red plastic plug in the top edge of the door indicates a door bonded with waterproof Exterior Type I adhesives (Fig. D29b).

Although Commercial Standards do not require it, exterior softwood doors should be water-repellent preservative treated for increased dimensional stability and resistance to decay, fungus and insect attack. The NWMA administers a certification program for preservative-treated millwork products which includes softwood doors. The NWMA seal is assurance of adequate preservative treatment in accordance with NWMA IS-4, Water Repellent Preservative Non-Pressure Treatment for Millwork (Fig. D30).

INSTALLATION *(See also Division 400, Application & Finishes)*

The satisfactory performance of doors depends on proper handling, storage, hanging and finishing.

Handling and Storage

Unfinished doors should be handled with clean gloves to prevent soiling. They should be stored flat (not on edge) on a level surface in a clean, dry, well ventilated space. They should not be stored in excessively hot, dry or humid areas, but should be conditioned for several days to the average prevailing local humidity before hanging or finishing. If doors appear warped, they should be stacked flat under uniformly distributed weights to restore flatness.

Hanging

Finished door frames should be square and plumb, and doors should be fitted in the openings with a total clearance of approximately 3/16″ in both horizontal and vertical dimensions to prevent binding under humid conditions. Doors may be trimmed for fitting, but should not be cut to smaller nominal sizes. If a substantial amount of the perimeter is removed, the construction balance of the door could be destroyed and warping could result. Glass and louver openings in flush doors should not be more than 40% of the door area, and openings should not be closer than 5″ to the door edge. The height of openings in a hollow core door should not exceed half of the door, and face skins should be supported with blocking, glued in place, to frame the opening.

Hardware When mortising for locks and other hardware, care should be taken not to impair the strength of the door by cutting away too much of the supporting construction.

All exterior doors should be hung with a minimum of three hinges per door, properly aligned so the door will not twist out of shape. Interior doors under 7′0″ in height may be hung with two hinges.

Flashing When exterior doors swing out with no overhead protection from the elements, the top edge of the door should be protected from moisture with a flashing cover. Flush doors with glass or louver openings should also be protected with flashing at the bottom of the openings.

Finishing

It is advisable to seal coat doors as soon as they are delivered to the job site to prevent undue moisture absorption and soiling. Finishing should be delayed until doors have been trimmed and fitted in the openings. All door surfaces, including top, bottom and side edges, should be sealed against moisture with two coats of paint, varnish or sealer before final hanging.

Door surfaces should be clean and dry before applying the finish. Soil marks and other surface defects should be removed by light sanding. Finishing in excessively humid weather should be avoided, and the proposed finish should be tested on a sample of the same species of wood to determine its appearance and suitability.

WOOD FLOORING

The widespread use of wood flooring in residential construction can be attributed to its distinctive natural appearance, excellent wearing qualities, moderate cost, underfoot comfort and ease of installation and maintenance. In a modern home, wood flooring can be used in almost all rooms above grade. Improved methods of protecting wood from dampness have encouraged its use in the kitchen and in rooms below grade as well.

Because of its ready availability and the relative ease of working it with hand tools, early flooring was made mostly of softwood. This flooring often was in the form of thick *planks* 3-1/2″ to 10″ wide, just as they came from the sawmill. Subsequently, improvements in manufacturing and seasoning methods brought about the development of carefully machined, narrower hardwood *strips* (3-1/4″ and narrower), which were easier to install.

Hardwood strip flooring has been and continues to be the most popular type (Fig. 1a). More recently, homeowner demand for floors simulating the early colonial random-width plank installations has encouraged the production of plank products (Fig. 1b). Various *parquet (pattern)* types, such as blocks and uniform-length strips, are being produced where more formal flooring is desired (Fig. 2).

Changing conditions in the building industry have created a demand for factory-finished flooring suitable for service immediately after installation. Such prefinished flooring is available in practically every style—strip, plank and block—and in a variety of woods.

A characteristic property of wood that must be given special attention in flooring installations is its tendency to shrink and swell as its moisture content changes. Through modern research and manufacturing methods, products with improved dimensional stability have been developed. However, individual products vary somewhat in their sensitivity to moisture and should be selected with a view to the conditions of use.

This subsection describes various types of hardwood and softwood floorings suitable for residential use and highlights their special properties. It summarizes available grades, sizes and current standards of quality, providing a basis for the selection and specifying of wood flooring.

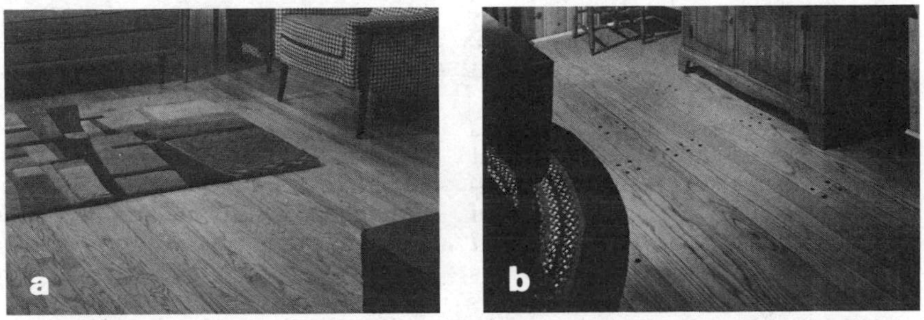

FIG. 1 Contemporary and traditional interiors are complemented by a) strip and b) plank wood floors.

MATERIALS

The terms *hardwood* and *softwood,* popularly applied to the two major groups of trees cut for lumber, actually have no bearing on the degree of hardness of the wood. In fact, many softwoods are much harder than some of the hardwoods. The terms are used primarily to distinguish the botanical characteristics of the trees. Arbitrarily, trees having broad leaves are known as hardwoods, while *coniferous* trees—those bearing needles and cones—are known as softwoods.

About 12 types of woods are regularly manufactured into flooring. Of these, the hardwoods ac-

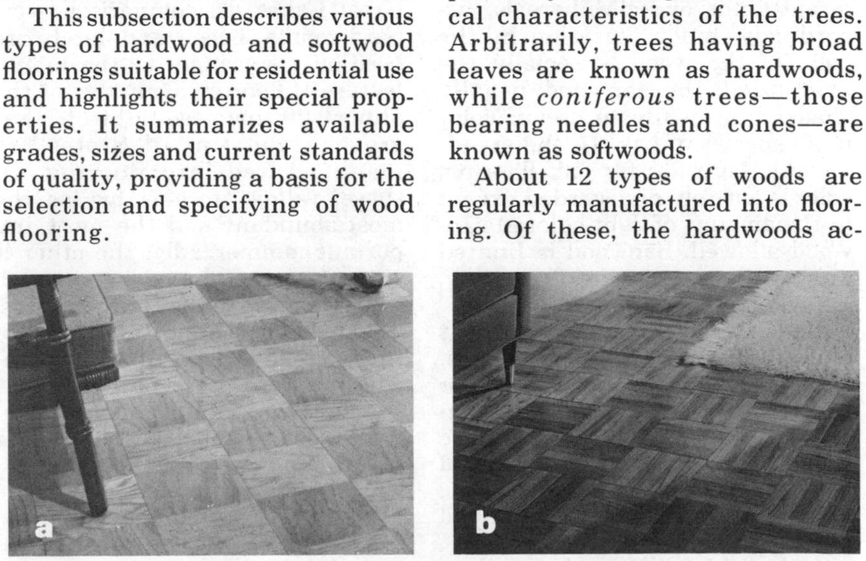

FIG. 2 Pattern (parquet) floors can be created with a) laminated plywood blocks or b) solid wood slats either preassembled into blocks or individually laid in mastic.

count for about 80% of all residential wood flooring. The greater popularity of hardwoods can be attributed to their appearance and, in the species used, substantially greater hardness and wear resistance.

Hardwoods

Of all hardwood flooring produced in a recent typical year, oak of various commercial species supplied about 92%, as compared with 6% for maple; the balance consisted largely of beech, birch, pecan and several other hardwoods in limited quantities.

Oak There are about 20 species of oak in the United States that are considered commercially important in lumber production. Of these, about half are classed as *red oak* and half as *white oak*. As growing trees, the several species within each group are readily distinguishable, but in lumber form the differences are fairly inconspicuous. Hence, precise separation of the various species within each group of oaks is impractical and unnecessary in flooring manufacture.

Oak is lumbered throughout the southern, eastern and central states, in forests of the Atlantic Plain and the Appalachian Mountains. In these regions, species of oak grow under a wide range of climatic conditions in many different kinds of soil. There is, accordingly, much variation in the color of the wood, especially the heartwood; the sapwood usually shades from white to cream color in all species of oak. In the standard grading rules for oak flooring, color is entirely disregarded except in the amount of light-colored sapwood allowed. Sapwood is limited only in the top grade of flooring, *clear* grade.

Red oak and white oak are about equal in mechanical properties. Both make a very satisfactory floor of attractive appearance when properly finished. A special feature of white oak is the prominence of large rays that make an interesting flake pattern in quarter-sawed flooring.

Although grading rules do not differentiate between red oak and white oak, practically all manufacturers supply either all-red or all-white oak flooring except in the lowest grades. Red oak flooring generally is higher in price and more uniform in color than white oak.

Maple Maple flooring is made from sugar maple, logged largely in the Northeast, the Appalachians and the lake states—Minnesota, Michigan and Wisconsin. In the lumber trade, sugar maple is known as *hard maple* or *rock maple*. This species is extremely strong, hard and abrasion-resistant, making it particularly suitable for hard use locations, such as factories and gymnasiums, as well as residences.

The so-called soft maples—silver maple, red maple and bigleaf maple—are not so hard, heavy or strong as hard maple and therefore are not used commonly for flooring.

The heartwood of both sugar maple and black maple is light reddish-brown, and the sapwood, which in mature trees is several inches thick, is creamy white, slightly tinged with brown. The contrast in color between heartwood and sapwood in maple is much less pronounced than in oak, and the standard grading rules permit natural color variation in the wood.

Beech and Birch These woods are lumbered in the Northeastern part of the country and around the Great Lakes. In comparison with hard maple, beech and birch are used only sparingly in the manufacture of flooring. Only two of the almost 20 species of birch that grow in the United States are manufactured into flooring. Of these, *yellow birch* is by far the most abundant and the most important commercially; the other is *sweet birch*. Only one species of *beech* is native to the United States.

The heartwood of all three of these species is reddish-brown, with a slight variation in color among them. Similar slight variations exist also in the color of the sapwood of the three species, which is of a lighter shade than the heartwood. As in maple flooring, the varying color of the natural wood is an accepted characteristic in grading beech and birch flooring.

Softwoods

In a recent typical year, over 50% of the softwood flooring produced was southern pine, over 40% Douglas fir and the balance in order of volume as follows: western hemlock, eastern white pine, ponderosa pine, western larch, eastern hemlock, redwood, spruce, cypress and the true firs. Western larch is similar to Douglas fir in strength properties and often is sold in mixture with Douglas fir of the northern interior region of the western states. Ponderosa pine, eastern white pine and redwood are softer than is desirable where wear is a prime factor. However, the formidable decay resistance of redwood in the all-heartwood grade has prompted its use for porch flooring.

Southern Pine Southern pine is a commercial name applied to a group of yellow pines that grow principally in the southeastern states. The group includes longleaf, shortleaf, loblolly and slash pines and several others of minor importance. Except in dimension lumber and structural timbers, no differentiation between the species is made in marketing the products of this group.

The wood of all southern pines is much alike in appearance. The sapwood and heartwood often are different in color, the former being yellowish-white and the latter a reddish-brown. However, the contrast in color between sapwood and heartwood in southern pine generally is not conspicuous in a finished floor, and the standard grading rules permit sapwood in all grades of southern pine flooring unless otherwise specified.

When flooring of uniform color is essential, the standard flooring specifications can be amended to require all-sap-face stock (for a light color) or all-heart-face material (for a reddish-brown color). However, special selection of stock for color increases the cost somewhat over the established grade.

Douglas Fir Red fir, yellow fir, coast Douglas fir and Oregon pine are other names by which Douglas fir is known in the western part of the United States and Canada, where it grows. Douglas fir occupies the same important position in the western and Pacific Coast states as southern pine does in the southeastern states.

The sapwood of Douglas fir is creamy white. The heartwood is reddish-brown, and as in southern pine, the contrast in color between the two is not so pronounced as to be objectionable in a finished floor. Pieces containing both heartwood and sapwood are permitted in all grades.

West Coast Hemlock Western hemlock grows along the Pacific Coast from northern California to Alaska and as far inland as northern Idaho and northwestern Montana. The bulk of hemlock lumber being produced comes from Oregon, Washington and California and is referred to commercially as West Coast hemlock. Both the heartwood and sapwood of western hemlock are almost white with a pinkish tinge and with very little contrast, although the sapwood may sometimes be lighter in color.

Western hemlock has light, clear color and good finishing qualities, which account for its use in moderate wear areas, such as bedrooms, where good appearance is the principal requirement. Western hemlock flooring is relatively free from warping and is easy to cut and nail but is not so hard and wear-resistant as Douglas fir and larch.

MANUFACTURE

Wood flooring is made in three basic styles: *strip, plank* and *block* (Fig. 3). The following discussion of manufacture is limited to strips and planks, as blocks may be made in many different ways. Methods of manufacture of hardwood flooring differ from those of softwood flooring.

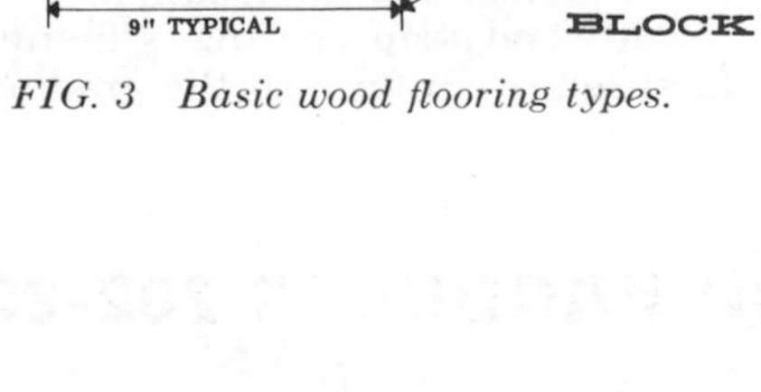

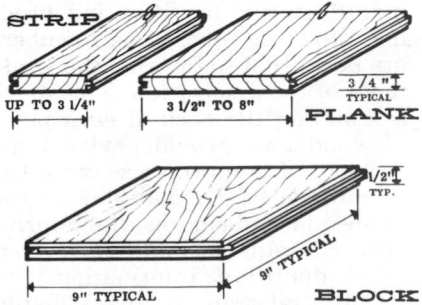

FIG. 3 Basic wood flooring types.

Hardwoods

The production of hardwood flooring begins at the sawmill, where logs are rough-sawed into boards, planks and timbers. Each log is cut in such a way as to produce the best grade of lumber in each thickness. The thicker pieces containing the highest percentage of clear lumber usually are reserved for furniture and millwork.

The common nominal thickness of lumber for flooring is 1″, from which the popular 3/4″ strips and planks are milled. After air-drying in a yard for several months, rough-sawed lumber is kiln-dried to reduce the moisture content to a range of 6% to 9%. Sawmill operations are described and illustrated in detail starting on page 17 of Section 201 Wood.

Sizing and Dressing Rough-sawed lumber often is air-dried at the sawmill, but kiln-drying is normally performed at the flooring plant. After cooling for at least 24 hours, the lumber is ready to enter the flooring mill. The first step consists of running the lumber through a ripsaw, which cuts the boards into the desired widths (Fig. 4). Lumber which is to be tongue-and-grooved (T&G) is cut wider than the face width of the flooring, so as to compensate for the tongue edge and for the material taken off in planing the edges.

In some flooring mills, the pieces then are put through a planing machine, which dresses (smooths) the top and bottom surfaces. Then, either dressed or in rough state, the pieces proceed on conveyor belts to cut-off saws (Fig. 5), where they are cut to eliminate natural defects which would be undesirable in finish flooring. The pieces emerge in random lengths as a result of this cutting operation. Next, processed strips are conveyed to a machine known as a side-matcher, which mills them to the characteristic T&G profile.

Side- and End-Matching The side-matching machine smooths the top surface accurately, cuts one or two channels in the bottom side, a tongue on one edge and a groove in the other (Fig. 6). The pieces then continue on a belt conveyor to end-matching machines, where one end of each strip is grooved and the other is tongued.

FIG. 4 Rough-sawed lumber is cut lengthwise on a ripsaw.

FIG. 5 Natural defects are eliminated on the cut-off saw.

FIG. 6 Top face is smoothed and sides are tongue-and-grooved on side-matching machine.

FIG. 7 Each piece is carefully inspected and graded prior to packaging.

Grading and Bundling Next, the pieces go to grading tables, where they are separated by grade, color and length (Fig. 7). Each piece is trade- and grade-marked, and for conventional flooring assortments the pieces of each grade are separated and bundled according to length in multiples of 1 ft. An allowance of 6″ under and 6″ over the nominal length is permitted in each bundle, so that a 4′ bundle might contain pieces 3′6″ to 4′6″. However, the most popular flooring size, 3/4″ x 2-1/4″, also is bundled in an assortment termed *nested flooring*, which includes various lengths ranging from 9″ to 102″.

Those flooring pieces which are to be prefinished undergo further treatment, including fine sanding (Fig. 8), finishing and final waxing and polishing. Bundles then are shipped to distributors in sealed boxcars or trucks which afford protection from the weather and moisture. The quantity of lumber per bundle and total required can be calculated according to Fig. 9. To determine the number of board feet necessary for a specific installation, multiply the square footage of the area to be covered by the factor shown under *required quantity* in Figure 9. For example, a 10′ x 12′ room has an area of 120 sq. ft.; 3/4″ x 2-1/4″ strip flooring has a multiple factor of 1.383; 120 sq. ft. times 1.383 equals 165.96 bd. ft. In round numbers, the 10′ x 12′ room installation would require 166 bd. ft. of 3/4″ x 2-1/4″ flooring.

FIG. 9 **QUANTITY CALCULATIONS FOR STRIP FLOORING**

STRIP DIMENSIONS		PIECES PER BUNDLE[4]	BUNDLE QUANTITY (bd. ft.)[5]		REQUIRED QUANTITY (bd. ft.)[6]	
Actual[1, 2]	Counted[3]					
3/4″ x 3¼″	1″ x 4″	12	4		1.29	
3/4″ x 2¼″	1″ x 3″	12	3	times	1.383	times
3/4″ x 2″	1″ x 2¾″	12	2.75	the	1.425	the
3/4″ x 1½″	1″ x 2¼″	12	2.25	average	1.55	area to
15/32″ x 2″	1″ x 2½″	18	3.75	bundle	1.30	be
15/32″ x 1½″	1″ x 2″	18	3	length	1.383	covered
11/32″ x 2″	1″ x 2½″	24	5	(ft.)	1.30	(ft.)
11/32″ x 1½″	1″ x 2″	24	4		1.383	

1. Dimensions are from grading rules of National Oak Flooring Manufacturers Association (NOFMA); the Maple Flooring Manufacturers Association (MFMA) recognizes standard thicknesses of 25/32″ in 1½″, 2¼″ & 3¼″ lengths and 41/32″ thickness in 2¼″ & 3¼″ lengths.
2. Nominal thickness for 15/32″ and 11/32″ flooring is ½″ and ⅜″ respectively; ¾″ is the same.
3. Equivalent dimensions used for calculating bd. ft. quantity.
4. 1″ x 4″ and 1″ x 3″ pine flooring commonly is packaged 6 pieces to the bundle.
5. Multiplier is equal to counted size times number of pieces per bundle divided by 12.
6. Multiplier makes allowance for side-matching, end-matching and normal waste.

Softwoods

The production of softwood flooring begins at the sawmill, where the *green* (unseasoned) logs are rough sawed. The resulting lumber then is seasoned either by air-dry-kiln-dried to meet moisture requirements of the grading rules. Kiln-dried flooring requires a maximum of 12% moisture content. Actual moisture content at time of installation usually averages about 9%-10%.

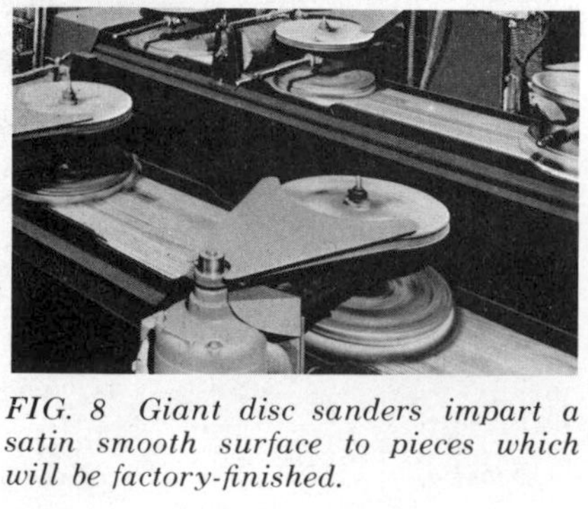

FIG. 8 Giant disc sanders impart a satin smooth surface to pieces which will be factory-finished.

After the flooring has been seasoned, it goes directly to the planing mill, where it is manufactured to the desired profile in one operation. Square-end flooring is then trimmed and packaged in equal lengths six to a bundle. End-matched flooring bundles may contain pieces from 1′ to 8′ in length laid end-to-end and stacked in one or two tiers of six courses each.

Standards of Quality

Of the three basic types of flooring produced, strip flooring and block flooring are regulated by industry-wide specifications. There are no generally accepted standards for plank flooring, but most manufacturers market products in appearance grades similar to those established for strip flooring. Three standards exist for block flooring: Federal Specification NNB-350 Wood Floor Block (Hardwood; Solid, Laminated, Slat), ANSI 010.2-1975 Laminated Block Flooring, and NBS-PS 27-70 Mosaic Parquet Hardwood Slat Flooring.

Strip flooring is manufactured generally to comply with the grading rules of the particular trade association governing the type of wood involved. The National Oak Flooring Manufacturers Association (NOFMA) promulgates and regulates the production of oak flooring through its Official Flooring Grading Rules. Since some NOFMA members also produce maple, beech, birch and pecan flooring, this association also promulgates grading rules for these woods. However, the majority of maple, beech and birch flooring is produced by members of the Maple Flooring Manufacturers Association (MFMA), which promulgates its own grading rules for these species.

Certification of Quality

All strip flooring must be kiln-dried, grade-marked and trade-marked in order to comply with NOFMA and MFMA grading rules. The two trade associations operate certification programs based on in-plant inspections of their members' manufacturing facilities and products, and only qualifying members are permitted to imprint the trade mark of the association. This mark is found on the back of each piece of wood and provides visual assurance of conformance with the grading rules. In addition to the grade and trade marks, the imprint often contains the mill number and other identifying information.

Softwood strip flooring is manufactured to conform to the grading rules of the Southern Pine Inspection Bureau (SPIB), the Western Wood Products Association (WWPA) and the West Coast Lumber Inspection Bureau (WCLIB). These associations do not require trade-marking or grade-marking under the grading rules.

STRIP FLOORING

Hardwood flooring is manufactured commonly to a *standard pattern* (Fig. 10), producing flooring which is side-matched and end-matched. The top face is generally slightly wider than the bottom, so that when the strips are driven tightly together, the upper edges make contact, but the lower edges are slightly separated. Standard T&G strip flooring usually is installed over wood subfloors or wood sleepers by *blind* (concealed) nailing at the intersection of the tongue and shoulder (Fig. 11).

Strip flooring is generally either hollow-backed (Figs. 12 a & b) or scratch-backed (Fig. 12 d). The most widely used standard pattern is 3/4″ thick and has a face width of 2-1/4″, but other widths and thicknesses are available. The strips are random length, and the proportion of short pieces depends on the grade.

Another pattern of strip flooring used to a limited degree has square edges and a flat back. (Fig. 12 c). This is generally thinner than the standard pattern and is usually installed by face nailing. The nails are driven and set so the matching

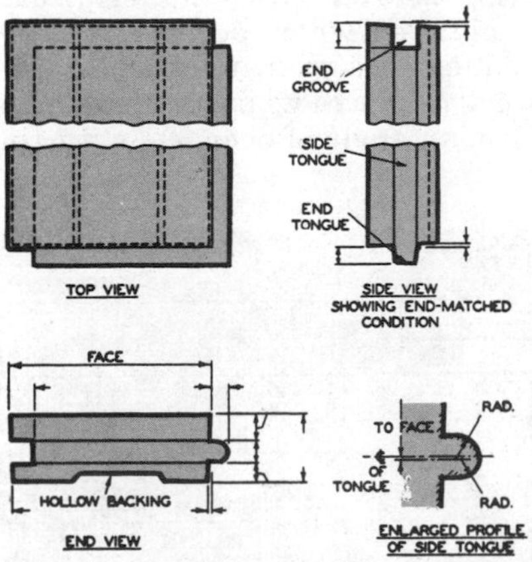

FIG. 10 Detail dimensions of typical ("standard pattern") 25/32″ T&G flooring.

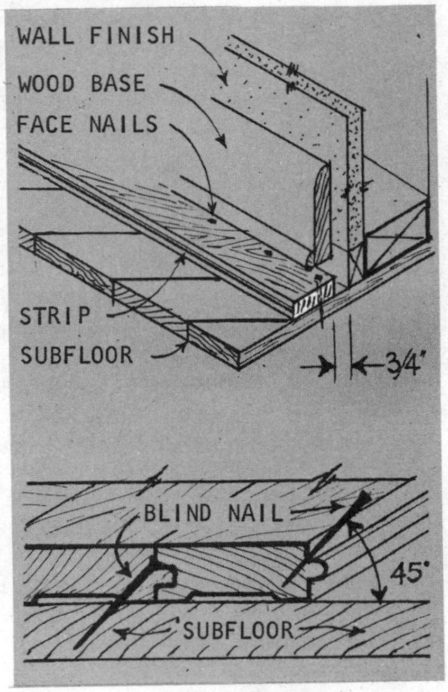

FIG. 11 Strip flooring application: a) first and last strips are face nailed; b) others are blind-nailed.

filler used in finishing will fill and conceal the holes.

While standard strip flooring generally is installed by blind nailing to wood subfloors or sleepers (Fig. 13), uniform short lengths of T&G flooring (*single slat* flooring) also can be installed with *mastic* (paste-like adhesive). Products intended for mastic installation generally have flat backs for greater contact with the adhesive and have a recess milled into the lower surface below the tongue to reduce the possibility of mastic being forced up through the joint (Fig. 12 e & f). Flooring intended for mastic installation may have T&G edges or may be grooved for insertion of a hardwood or metal spline.

Softwood flooring is made either end-matched (Fig. 12 a) or with plain ends (Fig. 12 e). It may have a wide hollow back like the 3/4″ standard pattern hardwood flooring (Fig. 12 a), a double groove like the thinner standard pattern flooring (Fig. 12 b) or a single V-shaped groove in the back (Fig. 12 d).

Standard flooring grades are based almost wholly on appearance. That is, they exclude or severely limit such defects as knots, wormholes and the like in the higher grades and permit increasing sizes and numbers of these characteristics in the lower grades. Natural variations in color generally are not limited except that, in the clear grade of oak, the amount of the lighter colored sapwood is restricted.

Traditionally, hardwood flooring has been packaged and sold in bundles containing pieces of more or less the same length, 6″ shorter or longer than the nominal length. The shortest pieces permitted per bundle and percentage of short lengths permitted in each grade are regulated by the established grading rules and are summarized in Figures 16 through 18.

NOFMA grading rules have permitted marketing 3/4″ x 2-1/4″ nested flooring, which is laid end-to-end continuously to make approximately 8′ long bundles. A nested bundle is four tiers high and three courses wide or six tiers high by two courses wide, each course being 7-1/2′ to 8-1/2′ long. Individual pieces may range from 9″ to 102″, but both the proportion of short pieces (9″ to 18″) and the minimum average length are regulated by the grading rules and depends on the grade and species.

The short pieces included in nested flooring are not detrimental to an installation. Experiments have demonstrated that even if the subflooring were omitted and a 9″ piece of strip flooring were placed so that no part of it rested on a support (Fig. 14), it could support almost

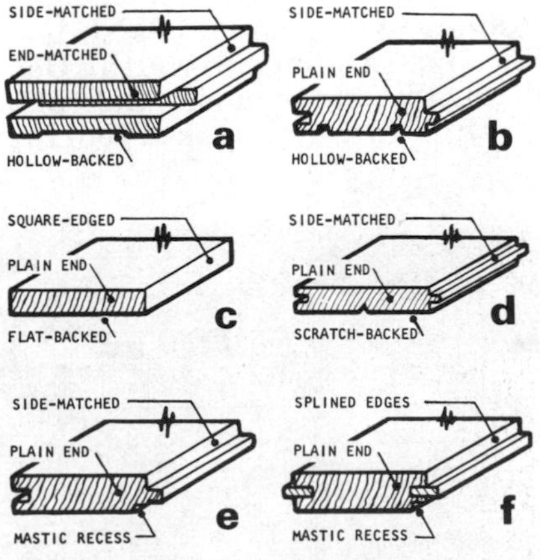

FIG. 12 Flooring patterns: (a,b,c&d) for installation with nails or screws; (e&f) for mastic installation.

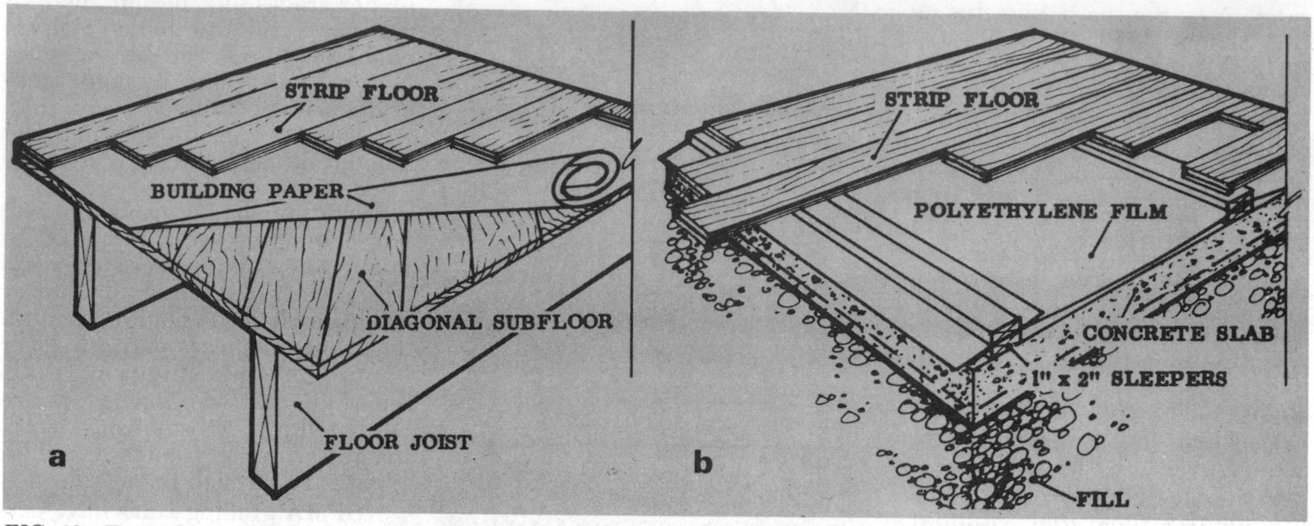

FIG. 13 Typical installation methods: a) wood subfloor and b) concrete slab. Over wood subfloor 1" x 2" sleepers 16" o.c. also can be used; over slab, 3/4" plywood over poly film can be installed as subfloor.

2-1/2 times the concentrated load under the leg of a concert grand piano. This represents an adequate safety factor for what may be assumed to be the most critical loading condition in residential use. Installations over subfloors, or those involving longer pieces, produce lower stresses and certainly are adequate.

Oak

Oak is regularly manufactured into plain-sawed and quarter-sawed flooring (Fig. 15). Most of the oak flooring manufactured is plain-sawed, the lower priced of the two. Quarter-sawed oak is characterized by a striking grain and flake pattern and by greater dimensional stability, which minimizes shrinking and swelling in width. Oak flooring is graded under the rules of the National Oak Flooring Manufacturers Association (NOFMA), as outlined in Fig. 16.

The most common oak flooring thickness for residential use is 3/4". Thinner flooring is available for use over existing floors or for light service conditions. Flooring thicker than 3/4" is intended for heavy service conditions and is seldom used in the home.

Although the great bulk of oak strip flooring is end- and side-matched, some square-edged strip flooring is made. This is intended primarily for industrial, institutional and commercial use to facilitate replacement of damaged areas, which would be more difficult with tongue-and-grooved flooring. Prefinished oak flooring is made in 4 grades: Prime, Standard, Standard & Better, and Tavern and is available in all-red or all-white oak selections. Prime grade is available on special order only.

Maple, Beech and Birch

Flooring of these species is graded both under the rules of the Maple Flooring Manufacturers Association (MFMA) and under the rules of NOFMA, as shown in Figs. 17 & 18. Although the grade names are sometimes slightly different, the requirements under both sets of rules are similar. Both generally disregard normal color variations.

These woods are very dense and hard in plain-sawed as well as quarter-sawed stock; hence, the grading rules make no special requirements in this respect. However, quarter-sawed material can be obtained when specified. MFMA defines *edge-grain* (quarter-sawed) stock as pieces whose annual rings form an angle of over 30° with the

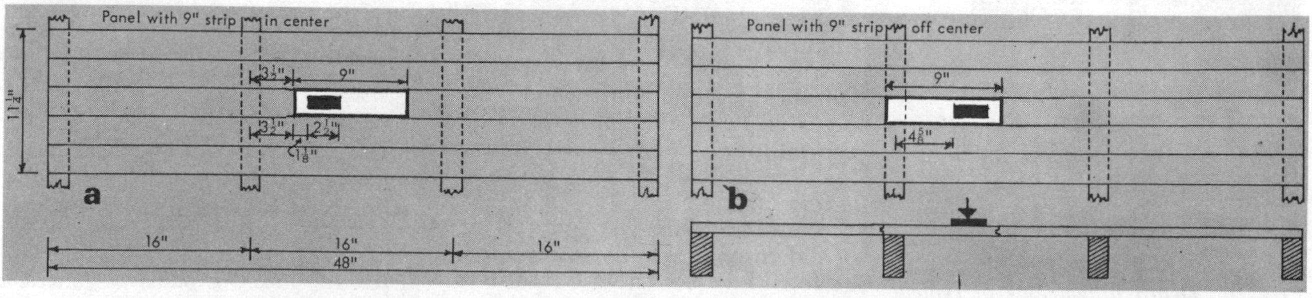

FIG. 14 Short piece of nested flooring (shaded), resting entirely on T&G's of adjacent strips, supported a concentrated load of 845 lbs or more (a); while piece partly resting on joist, more than 915 lbs (b).

face. This definition is slightly more permissive than the standard definition requiring an angle of 45° or more (Fig. 15).

Southern Pine

Southern pine is graded under the rules of the Southern Pine Inspection Bureau (SPIB), as shown in Fig. 19. It may be specified in side- and end-matched flooring and also in square-end flooring that is side-matched only. Regardless of the pattern to which it is worked, all southern pine flooring is available in both *flat-grain* and *edge-grain* stock (Fig. 15). However, edge-grain material is in limited supply available through special order and may be hard to obtain. Edge-grain material in this species must have at least six annual rings per inch across the face, in addition to the standard requirement of grain direction of over 45°.

A class of southern pine flooring intermediate between edge-grain and flat-grain is known as *near-rift* flooring. This material has annual rings that form an angle of less than 45° with the face, but the rings are so close to this position that each piece shows at least six annual rings anywhere across the face.

Douglas Fir and Other Western Softwood Species

With the exception of the B & Better Grade, which is manufactured only from *vertical-grain* material, Douglas fir flooring in the upper grades is regularly manufactured in both vertical-grain (VG) and flat-grain (FG) stock (Fig. 15). The term "vertical-grain" used in these grading rules is almost synonymous with "edge-grain" used for southern pine. However, Douglas fir Grade C requires four rings per inch and only B & Better Grade requires at least six rings per inch of face width, as does southern pine. Vertical-grain Douglas fir flooring is harder and possesses better wearing qualities than does flat-grain material. Douglas fir

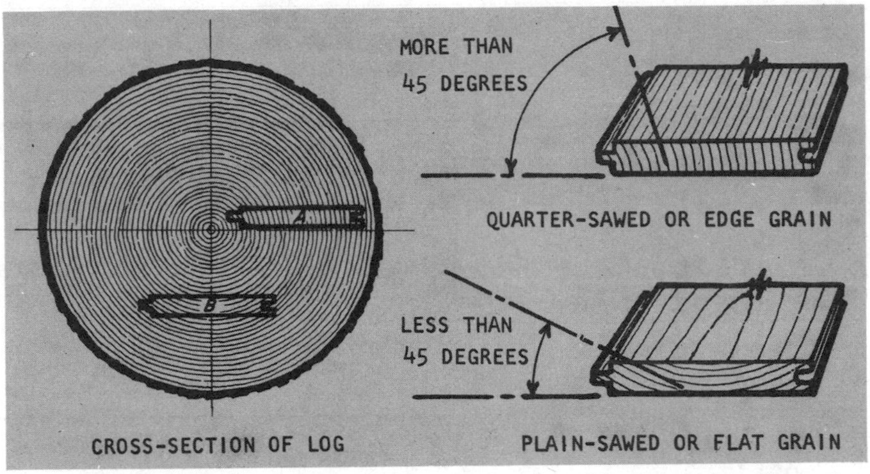

FIG. 15 The angle which the annual rings make with the surface determines whether wood is a) quarter-sawed (edge-grain) or b) plain-sawed (flat-grain).

FIG. 16 **OAK FLOORING GRADES ACCORDING TO NOFMA[1]**

TYPE OF WOOD	GRADE	DESCRIPTION	LENGTH OF STRIPS[2]	THICKNESS & WIDTH
PLAIN SAWN OR QUARTER SAWN[3]	Clear	Best appearance, best grade. Face practically clear, admitting an average of 3/8" bright sap. Color not considered.	Bundles 1¼ ft. and up; average 3¾ ft.	3/4" x 3¼" x 2¼"[2] x 2" x 1½" 1/2" x 2"[3] x 1½" 3/8" x 2"[3] x 1½"
	Select	Face may contain burls, small streaks, pinworm holes, slight imperfections in working and small tight knots that do not average more than 1 to every 3 ft.		
	Select & Better[3]	A combination of Clear and Select grades.		
PLAIN SAWN	No. 1 Common	Lays good residential floor and may contain varying wood characteristics such as flags, heavy streaks and checks, worm holes, knots and minor imperfections in working.	Bundles 1¼ ft. and up; average 2¾ ft.	
	No. 2 Common	May contain sound natural variations of the forest product and manufacturing imperfections. Red and White may be mixed. Furnishes an economical floor suitable for homes, general utility use, or where character marks and contrasting appearance are desired.		

1. Data are condensed from grading rules of the National Oak Flooring Manufacturers Association (NOFMA) and are for T&G flooring. All graded flooring is kiln-dried, grade-marked and trade-marked. For square-edge flooring see current NOFMA grading rules.
2. Also available in nested flooring.
3. Limited Supply—available on special order only.

FIG. 17 MAPLE, BEECH & BIRCH FLOORING GRADES ACCORDING TO MFMA[1]

TYPE OF WOOD	GRADE	DESCRIPTION	LENGTH OF STRIPS	THICKNESS & WIDTH
HARD MAPLE BEECH BIRCH	First Grade	Face practically free from all defects; variations in color are not considered a defect.	2 to 8 ft.; not over 40% under 4 ft., nor over 20% of 2 ft. stock	41/32″ x 3¼″ x 2¼″ x 1½″ 33/32″ x 3¼″ x 2¼″ x 1½″ 25/32″ x 3¼″ x 2¼″ x 1½″
	Second and Better Grade	Permits tight, sound knots and slight imperfections in dressing, but lays without waste.	1¼ to 8 ft.; not over 48% under 4 ft., nor over 22% of 2 ft. stock, not more than 3% of 1¼ to 1½ ft. lengths	
	Third Grade	Must be of such character as will lay and give a serviceable floor.	1¼ to 8 ft.; not over 70% under 4 ft., nor 35% of 1¼ ft. stock	

1. Data are condensed from grading rules of the Maple Flooring Manufacturers Association (MFMA). All graded flooring is kiln-dried, grade-marked, and trade-marked. Unless otherwise specified, flooring is side- and end-matched. For square-edged flooring and for special sizes, see current MFMA grading rules.

FIG. 18 MAPLE, BEECH, BIRCH & PECAN FLOORING GRADES ACCORDING TO NOFMA[1]

TYPE OF WOOD	GRADE	DESCRIPTION	LENGTH OF STRIPS	THICKNESS & WIDTH
MAPLE BEECH[2] BIRCH[2]	First Grade	Face practically free from all defects; variations in color are not considered a defect.	2 ft. and up; not over 33% of 2 to 3 ft. lengths	¾″ x 3¼″ x 2¼″ x 2″ x 1¼″ ½″[3] x 3¼″ x 2¼″ x 2″ x 1½″ ⅜″[3] x 3¼″ x 2¼″ x 2″ x 1¼″
	Second Grade	Permits tight sound knots and slight imperfections in dressing, but lays without waste.	2 ft. and up; not over 45% of 2 to 3 ft. lengths	
	Third Grade	Must be of such character as will lay and give a good serviceable floor.	1¼ ft. and up; not over 65% of 1¼ to 3 ft. lengths	
PECAN	First Grade	Face practically free from defects.	2 ft. and up; not over 25% of 2 to 3 ft. lengths	
	Second Grade	Permits tight sound knots or their equivalent, pinworm holes, streaks, light stain and slight imperfections in working, but provides a sound floor without cutting.	1¼ ft. and up; not over 40% of 1¼ to 3 ft. lengths	
	Third Grade	Must be of such character as will lay and give a good serviceable floor.	1¼ ft. and up, not over 60% of 1¼ to 3 ft. lengths	

1. Data are from grading rules of the National Oak Flooring Manufacturers Association (NOFMA). All graded flooring is kiln-dried, grade-marked and trade-marked. Unless otherwise specified, flooring is hollow-backed and side- and end-matched. For square-edged flooring and for special sizes, see current NOFMA grading rules.
2. Species in limited supply; special order only.
3. Limited supply; special order only.

flooring generally is tongue-and-grooved but not end-matched. Flooring of this wood is graded under the rules of the West Coast Lumbermen's Inspection Bureau (WCLIB) and Western Wood Products Association (WWPA), as shown in Fig. 20 & 21.

PLANK FLOORING

Flooring boards wider than 3-1/4″ are referred to as *planks*, the term *strips* being generally reserved for the narrower pieces. However, many random plank installations include both plank- and strip-sized boards, and manufacturers often refer to such assortments as plank flooring. Strip and plank floors are installed in end-to-end random pattern, with end joints staggered, using nails and/or screws.

Plank flooring is available end- and side-matched in the same

FIG. 19. SOUTHERN PINE FLOORING GRADES ACCORDING TO SPIB[1]

TYPE OF WOOD	GRADE	DESCRIPTION	PLAIN-END	END-MATCHED	THICKNESS & WIDTH
FLAT-GRAIN	B&B	Supreme quality, generally clear, and admits: small surface checks, not over 1/32" wide or 4" long; firm red heart, on not over 5% of face; one sound intergrown pin knot, 1/2" or less, in any 4 ft. length in 4 to 20 ft. or one in each piece if under 4 ft. long; one very small closed pitch pocket and one pitch streak in any 4 ft. of length in 4 to 20 ft. or one in each piece if under 4 ft. long (diffused pitch not permitted); light stain only, on not over 15% of face; light warp.	Standard practice is to furnish an assortment of standard stock lengths of 10 ft. to 16 ft., 18 ft. or 20 ft., except percentages of total board-foot quantities in lengths under 10 ft. permitted as 5% of 8 ft. and/or 9 ft. boards lengths	Standard lengths shall be 1 ft. to 8 ft. or longer; if lengths are not specifically restricted 15" shall be minimum. Flooring shall be bundled in one or two tiers of 6 courses each of 4 ft. & 8 ft. lengths or longer in multiples of 1 ft.; lengths of 3 ft. 6" or less may be in straight bundles of each nominal length or less; each course shall be within 6" of nominal length of bundle determined by averaging to the nearest foot.	1¼" x 5⅛" x 4⅛" x 3⅛" x 2⅛" 1" x 5⅛" x 4⅛" x 3⅛" x 2⅛" x 1⅛" ¾" x 5⅛" x 4⅛" x 3⅛" x 2⅛" x 1⅛" 9/16" x 5⅛" x 4⅛" x 3⅛" x 2⅛" x 1⅛" 7/16" x 5⅛" x 4⅛" x 3⅛" x 2⅛" x 1⅛" 5/16" x 5⅛" x 4⅛" x 3⅛" x 2⅛" x 1⅛"
	C	Choice quality, reasonably clear, and admits medium surface checks, not over 1/32" wide and between 4" and 10" long; firm red heart, on not over 25% of face; average of not more than 6 pin holes (1/16" or less) per surface ft.; sound or firm and tight knots not over ¾", except one between ¾" and 1½" in widths over 5"; light pitch only; pith, ¾" or less; slight shakes, skips or splits; light stain, medium on not over 15% of face; wane not more than ⅛" deep or ½" wide; medium warp.	Standard practice as in B&B, except 5% of 6 ft. and/or 7 ft., and 7% of 8 ft. and/or 9 ft.; any unfilled portion of 6 ft. and/or 7 ft. may be filled in with 8 ft. and/or 9 ft. lengths		
	C&Btr	Combination of quality consisting of B&B and C grades of finish.			
	D	A good quality grade which permits larger and more numerous characteristics than C.	Standard practice as in B&B, except 3% of 4 ft. and/or 5 ft., 6% of 6 ft. and/or 7 ft., and 8% of 8 ft. and/or 9 ft.; any unfilled portion of a shorter board length may be filled w/a longer length	Same as B&B, C, and C&Btr, except if lengths are not specifically restricted 12" shall be minimum.	
	No. 2	Low-cost serviceable flooring which admits decay holes and knots if they do not cause over 10% waste in any piece. Waste area shall be confined to one cutting area in 12 ft. and shorter, and two cutting areas in 14 ft. and longer; usable portions after cutting allowance must be at least 24" long.			
EDGE-GRAIN (vertical-grain)	B&B, C&Btr, C and D	Same as for flat-grain, except lumber has an average of at least 6 annual rings per in. across its face at each point in length, and the annual rings form an angle of 45° or more with the face. When the angle becomes less than 45° at any point, the lumber may be classed as flat-grain or near-rift, as the case may be.	Same as for flat-grain in applicable grade		
NEAR-RIFT	B&B, C&Btr, C and D	Same as for flat-grain except lumber either has fewer rings per in. than required in edge-grain but otherwise conforms to the edge-grain provision, or has an average of at least 6 annual rings per in. across the face at each point in length and has only one edge of the grain on the face but forms an angle of less than 45° with the face side.	Same as for flat-grain in applicable grade		

1. Data are condensed from the grading rules of the Southern Pine Inspection Bureau (SPIB). Graded flooring may be air-dried or kiln-dried; end-matched or plain-end; flat-grain or mixed grain and either scratch backed or hollow-backed. Grade-marking and trade-marking are optional.

FIG. 20 **DOUGLAS FIR, WESTERN HEMLOCK, WESTERN RED CEDAR, WHITE FIR, AND SITKA SPRUCE ACCORDING TO WCLIB[1]**

TYPE OF WOOD	GRADE	DESCRIPTION	LENGTH OF STRIPS	THICKNESS & WIDTH
VERTICAL-GRAIN (edge-grain) FLAT-GRAIN or MIXED GRAIN	C&Btr[2]	Fine appearance grade, with many pieces absolutely clear. May contain: medium stained sapwood, heart stain firm, two small seasoning checks; light or torn raised grain; an occasional very light skip; light warp in occasional pieces; two small, tight knots or equivalent; tongue 1/16" narrow; a 3" cutout 4 ft. or more from either end of pieces 12 ft. and longer is permissible in 50% of the shipment.	4 to 16 ft. or longer. Not less than 90% 8 to 16 ft., nor more than 3% 4 to 5 ft.	9/16" x 3-1/8" 3/4" x 2-1/8" x 3-1/8" x 5-1/8" 1" x 2-1/8" x 3-1/8" x 5-1/8"
	D	A serviceable grade which may contain: stained wood; medium splits; seasoning checks; limited holes; torn or raised grain; large pitch streaks; medium warp; four fixed knots 1" or equivalent; four medium pockets or equivalent; narrow tongue at least 1/16" in width; firm white specks 25% of face; a 3" cutout 4 ft. or more from either end of pieces 12 ft. and longer in 10% of shipment, resultant pieces shorter than 8 ft. must be free of holes.	4 to 16 ft. or longer. Not less than 80% 8 to 16 ft.	
	E	Recommended for subfloors; contains larger and more numerous characteristics than higher grades.	4 ft. or longer	

1. Data are condensed from the grading rules of the West Coast Lumbermen's Inspection Bureau (WCLIB) based on 4" wide and 12 ft. long piece. All graded flooring is kiln-dried; edges are T&G; backs are either partially surfaced, hollow-backed or scratch-backed. Grade-marking and trade-marking are optional. Special sizes are available on request.
2. Unless otherwise specified, pieces are VG, FG or MG at shipper's option. C&Btr VG must present a vertical appearance with not less than 4 annual rings per in.; D VG pieces may have an angle of grain from vertical to 60° from vertical.

thicknesses as strip flooring. Plank flooring most commonly is made of white or red oak 3/4" thick and 3-1/2" to 8" wide. It is available prefinished in dark or light shades and graded according to NOFMA rules for prefinished flooring (Fig. 22).

Some plank flooring is designed to simulate the appearance of the random-width planks commonly used in colonial homes. These rustic floors had a distinctive appearance resulting from the hardwood pegs used to secure the planks to the subfloor. Contemporary plank flooring is installed with wood screws in the face of each plank, and the screws are covered with wood plugs to simulate the pegs. Narrower, strip-sized boards which can be secured by blind nailing sometimes have decorative pegs inserted during their manufacture and are sold as plank flooring.

Because the cross-grain dimension in plank flooring is greater than in strip flooring, it is affected more by variations in atmospheric humidity. To avoid unsightly loose joints, plank edges are often eased or beveled to accentuate rather than conceal the joints.

Plank flooring is sometimes made in 3-ply laminated form, with a center cross ply. This form is more dimensionally stable in width than the solid form, making it possible to install the plank floors with tight joints. Plank flooring generally is sold by the sq. ft. rather than by the bd. ft., as strip flooring is sold, and 5% waste allowance is generally recommended in estimating required quantities.

PARQUET FLOORING

Parquet (pattern) floors consist of individual strips of wood *(single slats)* or larger units *(blocks)* installed to form a decorative geometric pattern (Fig. 23). The blocks may be made by laminating several hardwood veneers or by tying a number of solid hardwood pieces into a unit to facilitate installation. Unlike strip and plank flooring, parquet flooring usually is installed with mastic.

Oak is by far the most predominant species used in all types of pattern flooring. Other species, such as maple, walnut, cherry and East Indian teak, also are available. Sometimes a mixture of hardwoods, such as hickory, ash, elm, pecan, sycamore, beech and hackberry, are used at random in a single block.

Block Flooring

Block flooring is manufactured in several basic types. In *unit block* (also known as *solid unit block*), short lengths of strip flooring are joined together edgewise to form square units (Fig. 23a). In *laminated* (plywood) block, three or more plies of veneer are bonded with adhesive to obtain the desired thickness (Fig. 23b). *Slat block* (sometimes called *mosaic parquet hardwood slat)* flooring utilizes narrow slats or "fingers" of wood preassembled into larger units to facilitate installation (Fig. 23c).

Most unit and laminated block flooring is tongued on two adjoining or opposing edges and grooved on the other two to assure alinement between adjoining blocks. Some manufacturers produce square-edged blocks, while others include grooves on all four block edges and furnish splines for insertion between adjoining blocks. Both types are designed to be installed with mastic over a wood subfloor or concrete slab. Prefinished blocks usually have eased or beveled edges.

Although Federal Specification NN-B350b contains requirements for all three types of block flooring, it is most suitable as a standard for unit block flooring. ANSI 010.2-1975 is the preferred standard for laminated hardwood block flooring. For slat block flooring, the American Parquet Association (APA) and NBS-PS 27-70 Mosaic Parquet Hardwood Slat Flooring grading rules reflect more accurately current manufacturing practice and hence constitute the accepted industry standard. These standards and appearance grades for each type of block flooring are summarized in Fig. 24.

Unit Block Unit blocks typically consist of several 3/4″ T&G strips all laid parallel, or alternating in each quarter of the block checkerboard fashion (Fig. 25). Consequently, typical block sizes are multiples of the strip width used (Fig. 26). Common strip widths are 1-1/2″, 2″ and 2-1/4″, resulting in typical block dimensions of 6″, 6-3/4″, 7-1/2″, 8″, 9″, 10″ and 11-3/4″. The individual strips are held together with wood or metal splines embedded in the lower surface.

Laminated Block This type of

FIG. 21 **WESTERN SOFTWOODS FLOORING GRADES ACCORDING TO WWPA**[1]

TYPE OF WOOD	GRADE	DESCRIPTION	LENGTH OF STRIPS	THICKNESS & WIDTH
VERTICAL GRAIN FLAT GRAIN MIXED GRAIN	Superior Finish VG, FG or MG	Highest grade of finish lumber, with many pieces absolutely clear. Recommended for interior trim and cabinet work with natural or enamel finishes where finest appearance is important. Piece may contain: light cup and crook; medium stained wood; short splits; and small seasoning checks well scattered; slight torn or raised grain; 1 light skip on edge or back; 2 small sound tight knots or light pitch not over ½ face; wane on reverse side ½ the thickness, ⅛ the width for ¼ the length in occasional pieces; slight chip marks; 3″ cutout at end or more than 3 ft. from end permitted in pieces of otherwise high appearance. Cutouts are restricted to pieces 12 ft. and longer and 5% of the item.	Standard lengths are 4 ft. and longer; with 3% of 4 ft. and 5 ft. and 7% of 6 ft. and 7 ft. permitted	
	Prime Finish VG, FG or MG	Exhibits fine appearance; suitable where finishing requirements are less exacting. Piece may contain: medium cup & crook; stained wood; short splits; scattered seasoning checks and pin holes; medium torn grain; skips; 4 fixed knots 1″ or equivalent; 4 medium pockets; medium pitch over ⅔ of face area or less; white speck, wane; medium chip marks; 3″ cutout at the end or more than 3 ft. from end, restricted to 12 ft. and longer and 10% of the item.	Standard lengths are 4 ft. and longer; with 20% of 4 ft. and 7 ft. permitted	¾″ x 2⅛″ x 3⅛″ x 5⅛″ 1″ x 2⅛″ x 3⅛″ x 5⅛″ 1½″ x 5″ x 6¾″ x 8¾″ 2½″ x 5″ x 6¾″ x 8¾″
	C Select	Fine appearance grade, recommended for all finishing uses. Piece may contain: medium stain over ⅓ face of an occasional face; small, scattered surface seasoning checks; very light cup, crook and torn or raised grain; light skip on one edge and back; two small, tight knots, light pitch two small pockets.	Standard lengths	
	D Select	Has many of the fine appearance features of C; many pieces have a finish appearance on face and larger more numerous characteristics on the back. Piece may contain: medium stain over entire face; small, scattered seasoning checks on face, medium on back; 4 small, fixed knots; 4 small pockets; medium pitch over ⅔ of face or less; short split on one end; 4″ cutout either directly on an end or more than 2 ft. from end, restricted to pieces 12 ft. and longer and 10% of the item.	Standard lengths	

1. Data are condensed from Western Wood Products Association (WWPA) grading rules. All graded flooring is kiln-dried; edges are T&G; backs are either partially surfaced, hollow-backed or scratch-backed; grade-marking and trade-marking are optional. Special sizes are available on request.

FIG. 22 **PREFINISHED STRIP AND PLANK GRADES ACCORDING TO NOFMA**

TYPE OF WOOD	GRADE	DESCRIPTION	LENGTH
OAK[1]	Prime[2]	Excellent appearance, natural color variation; limited character marks, unlimited sap.	1¼ ft. and up; minimum average length 3½ ft.
	Standard	May contain wood characteristics which can be filled; should lay a sound floor without cutting.	1¼ ft. and up; minimum average length 2¾ ft.
	Standard & Better	Combination grade; no piece lower than Standard Grade.	1¼ ft. and up; minimum average length 3 ft.
	Tavern	Rustic appearance; all wood characteristics of species; economical and serviceable after filling and finishing.	1¼ ft. and up; minimum average length 2¼ ft.
	Tavern & Better[2]	Combination of Prime, Standard and Tavern Grades.	1¼ ft. and up; minimum average length 3 ft.
BEECH[3] PECAN[3]	Tavern & Better	Combination of Prime, Standard and Tavern Grades.	1¼ ft. and up; minimum average length 3 ft.

1. Red & white separated—graded after finishing.
2. Special order only.
3. Limited supply available through special order only.

block is made typically 15/32″ thick in 9″ squares but other sizes, such as 8″ and 8-1/2″, also are available from some manufacturers. Appearance grades established in ANSI 010.2-1975 Laminated Hardwood Block Flooring are Prime and Standard (Fig. 24).

Because of its cross-laminated construction in three to five plies, shrinking and swelling of individual blocks is minimized. This type of wood flooring has good dimensional stability and often is recommended for damp locations such as slabs-on-ground.

Adhesives used in the manufacture of laminated block flooring should be capable of resisting the temperature and humidity variations to which the flooring may be subjected. Melamine-urea resin adhesive gives good results at moderate cost and represents the typical adhesive type used for a majority of the laminated block flooring produced.

Highly water-resistant adhesives such as phenols, resorcinols and melamines may also be used if the application warrants it; however, the use of these glues adds to the cost of the product. Adhesives which create high strength dry bonds but are adversely affected by moisture are not recommended for the manufacture of laminated block flooring.

Slat Block Appearance grades are based on grading rules developed by the American Parquet Association (APA) and the National Bureau of Standards Product Standard PS 27-70 for Mosaic Parquet Hardwood Slat Flooring, which also checks compliance with the rules through periodic inspections of member plants. These products are suitable for installation in mastic over concrete surfaces, both above and on-grade. The basic components of these products are solid, 5/16″ thick, generally square-edged slats of hardwood, 3/4″ to 1-1/4″ wide and 4″ to 7″ long, assembled into basic squares. These in turn are factory-assembled checkerboard fashion, with the grain in each adjoining square reversed, into larger flooring blocks up to 30″ long and/or wide (Fig. 27).

There are several types of slat block flooring produced. Some are assembled into panels which are held together with a facepaper; these generally are 9-1/2″, 18″ or 19″ squares and are marketed unfinished. Others are made up of single basic 6″ T&G squares, which are held together by mechanical attachments and are generally factory finished. Still others are assembled into panels held together with a backing material such as a textile webbing, asphalt-saturated felt or other type of felt or non-woven interfacing. These products generally

FIG. 24 **BLOCK FLOORING TYPES AND GRADES**

FLOORING TYPE & REFERENCE STANDARD	GRADE NAME 1st	2nd	3rd
UNIT BLOCK, FS NN-B350 Unfinished	Clear	Select	No. 1 Common
Prefinished	Prime	Arabseque & Better	
LAMINATED BLOCK, ANSI 010.2-1975 Unfinished or Prefinished	Prime	Standard	
SLAT BLOCK, APA rules & NBS-PS27-70 Unfinished	Select and Better	Select	Rustic
Prefinished	Natural and Better	Natural	Cabin

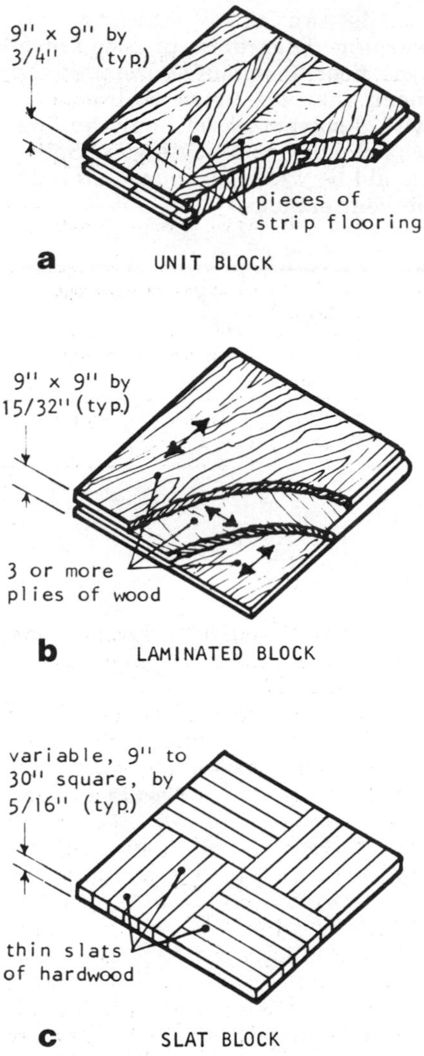

FIG. 23 Typical construction and dimensions of basic block flooring forms: a) unit block, b) laminated block and c) slat block. Slat blocks often consist of 16 or more basic squares.

FIG. 26 TYPICAL UNIT BLOCK DIMENSIONS*

STRIP WIDTH	STRIP THICKNESS 1/2"	3/4"	33/32"
1½"	7½" 9"	7½" 9"	9"
2"	8" 10"	8" 10"	
2¼"		6¾" 9" 11¼"	9" 11¼"

*All dimensions are for square blocks.

are 9-1/2", 11" or 12" squares; they may be square-edged or grooved and splined, unfinished or pre-finished.

Single Slat Parquet

Many decorative motifs can be created with *single slat* parquet flooring (Fig. 28), herringbone being the most common. This type of flooring consists of side- and end-matched strips, all carefully machined to uniform lengths ranging from 6" to 18". Typical thicknesses are 3/4" and 33/32"; common face widths are 1-1/2", 2" and 2-1/4". Since this product is essentially short lengths of strip flooring, all types of woods, grades and sizes available in strip flooring are available in single slat parquet.

MATERIALS HANDLING

To maintain proper moisture content, flooring products should not be transported or unloaded in rain, snow, or excessively humid weather. Flooring should not be delivered to the construction site until the building is enclosed, concrete and plaster work is completed and all building materials are dry. In winter, an interior temperature of 65° to 70° should be maintained for at least 5 days before flooring is delivered. It is recommended that flooring be stored for several days in the rooms where it will be installed to allow flooring to become acclimated to local conditions.

In crawl space construction, adequate cross ventilation should be

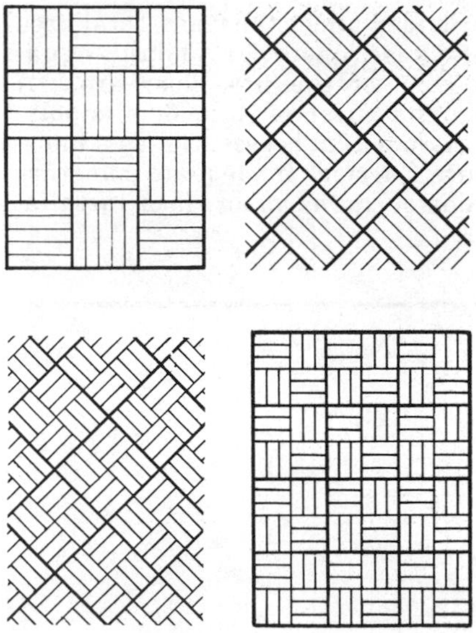

FIG. 25 Typical unit block patterns; heavy lines show individual blocks, laid in square or diagonal patterns.

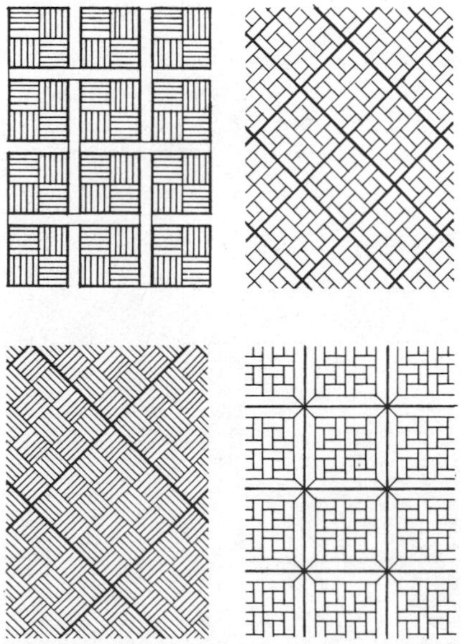

FIG. 27 Typical slat block patterns. Pickets forming borders may be part of block or separate.

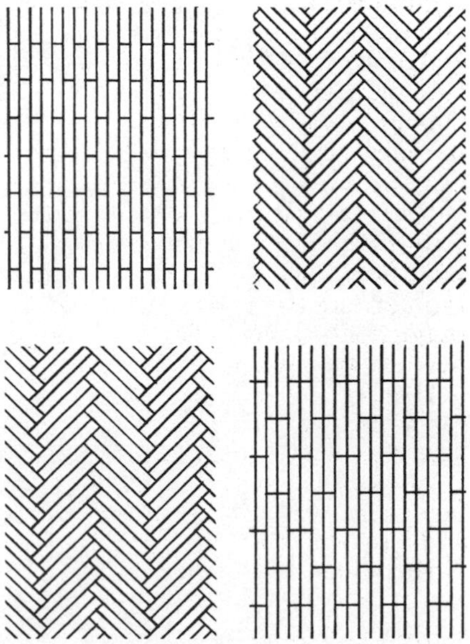

FIG. 28 Popular single slat patterns achieved with uniform-length precut wood strips.

provided under the floor. Total area of these openings should equal 1-1/2% of the first floor area. A ground cover of 4 to 6 mil. polyethylene film is essential as a moisture barrier. Inadequate moisture control can harm any floor installation.

MAINTENANCE

Properly finished wood floors are durable and easy to maintain. It is best to vacuum or dust mop floors once a week. Wood floors should never be washed or wet mopped because water seepage between boards can cause staining and warping. Regardless of floor finish type, flooring manufacturers recommend that the finish be protected with a solvent-based liquid buffing wax-cleaner or a paste wax. Spills should be wiped up immediately to prevent damage.

BATH & KITCHEN CABINETS

The origins of today's bath and kitchen cabinets can be traced back at least four centuries. In the 16th century, *ambries* (cupboards) were used to keep the food away from rats and other vermin. Ambries consisted of closed food storage areas above open shelving for pewter utensils.

More extensive built-in cabinets emerged in the 17th century. The space between the chimney breast and the side wall sometimes was filled with open and closed shelves, giving the effect of a built-in kitchen cabinet.

Wood has been and continues to be the main material used in the manufacture of bath and kitchen cabinets. However, the growing scarcity and cost of wood may result in greater use of plastic and metal components in household cabinets.

Striving to meet the demand for new cabinets caused by the post-World War II housing boom, custom cabinetry almost disappeared and mammoth new plants emerged that can make 500 to 5,000 units a day on an eight hour shift.

Vast and revolutionary innovations in manufacturing techniques accompanied this shift from custom-made cabinets to mass-produced ones. More sophisticated milling machines, presses, sanders, laminating devices, conveyors and finishing equipment have been developed to speed cabinet production.

This subsection broadly outlines the kinds of prefabricated cabinets currently manufactured, typical cabinet construction and assembly, applicable standards of quality, current certification program, proper kitchen design and installation. Except for typical sizes and design recommendations, the discussion of kitchen cabinets is applicable to bathroom cabinets as well.

CABINET TYPES

Bath and kitchen cabinets are manufactured in different styles (colonial, contemporary, traditional, provincial, Mediterranean and Spanish) and a variety of finishes (walnut, ebony, white, nutmeg, fruitwood, cherry, pecan, birch and oak). Common species of hardwood used in cabinet manufacture are ash, oak, birch, maple and walnut; softwood species include ponderosa (western) pine, Douglas fir and western hemlock (Fig. 1).

Kitchen cabinets may be classed according to function as: *base cabinets, wall cabinets* and *miscellaneous cabinets.* The latter category includes items such as broom cabinets, oven cabinets, "lazy susans", fruit and vegetable bins, tray holders and other special pieces. Optional accessories may include tray dividers, cutlery drawers, roll-out shelves, wire-mesh doors, spice shelves, cup shelves and other convenience items. *Bath cabinets* (vanities) are identical to kitchen base cabinets in construction but differ in height and depth. The variety of stock

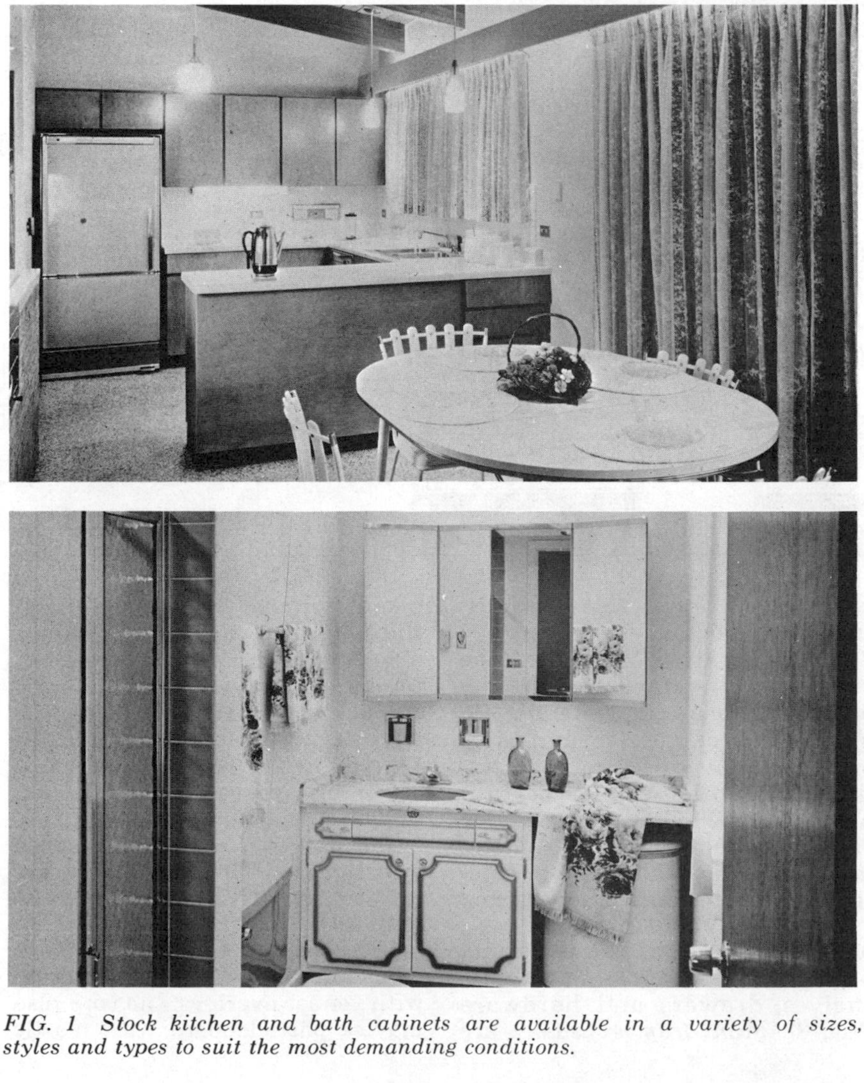

FIG. 1 Stock kitchen and bath cabinets are available in a variety of sizes, styles and types to suit the most demanding conditions.

FIG. 2 SUMMARY OF KITCHEN AND BATH CABINET TYPES AND SIZES

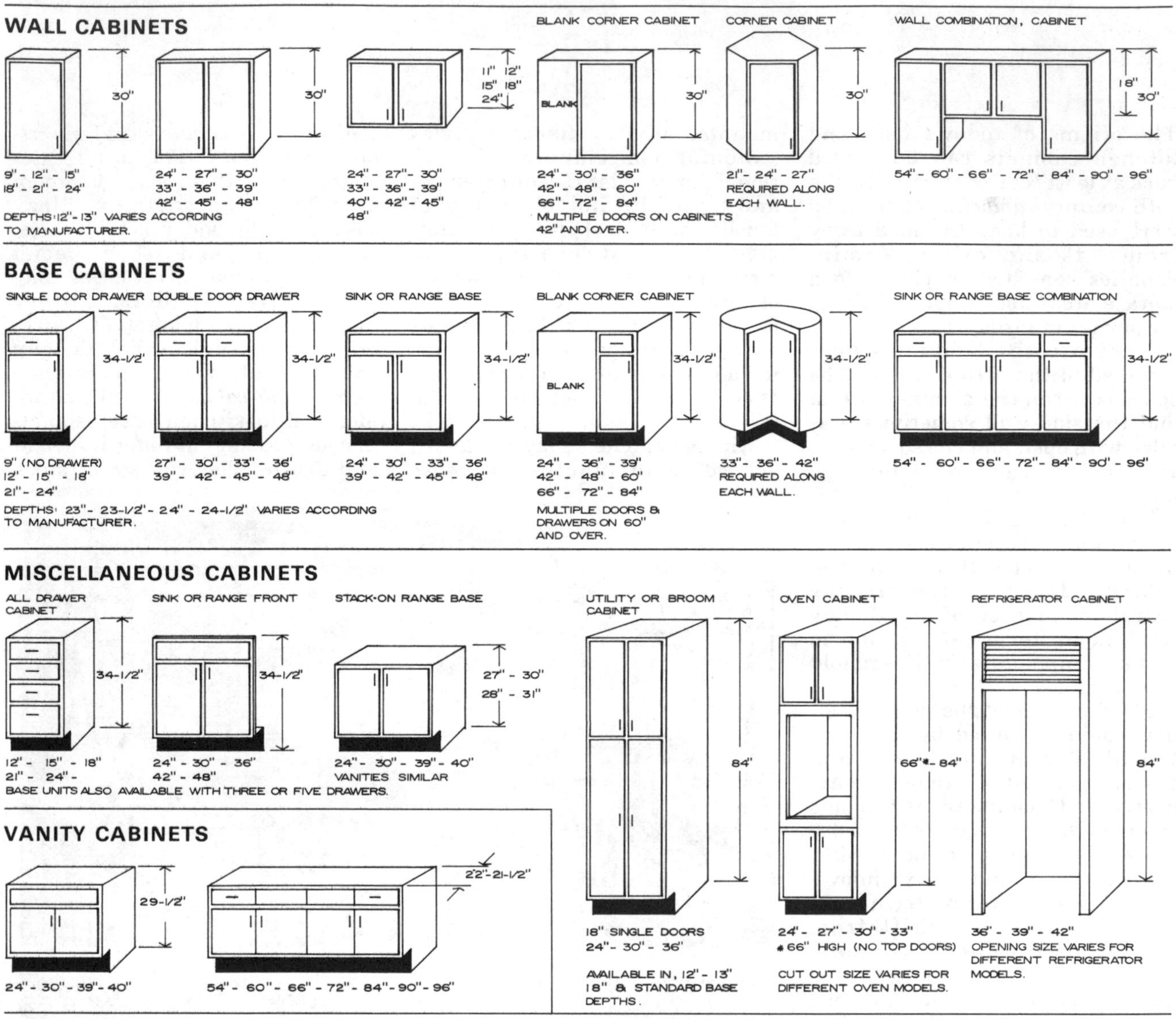

cabinet types and sizes is illustrated in Figure 2.

Industry standards regulate cabinet construction only in the broadest terms and generally do not specify particular materials or components. So long as the finished cabinets meet certain performance requirements for endurance and quality the manufacturer is free to choose the materials best suited for his operation.

The main components of cabinets typically are front frames, end panels, doors, backs, bottoms, shelves, drawers and hardware (Fig. 3). *Front frames* usually are made of hardwood, 1/2″ to 3/4″ thick. Rails, stiles and mullions are doweled (or mortise-and-tenoned), as well as glued and stapled for rigidity (Fig. 4). *End panels* typically consist of 3/16″ or 1/4″ hardwood plywood, glued to 3/4″ side frames; or 1/2″ and thicker plywood, without frames. These frequently are tongue-and-grooved into the front frames. *Doors* are 3/4″ to 7/8″ thick and may be hollow, or may have a particleboard, wood, or high density fiber core. Sometimes the core is plastic with veneer overlays; the core also may be particleboard with plastic faces or the entire door may be plastic.

Backs generally are 1/8″ hardboard or 3-ply plywood. *Bottoms and tops* are 3- or 5-ply hardwood plywood, 1/4″ to 1/2″ thick, dadoed (let) into sides and interlocked into hanging rails of wall cabinets. *Shelves* are lumber, plywood or particleboard, 1/2″ to 3/4″ thick with rounded front edges; plywood or particleboard shelves may be banded with hardwood front edges. *Drawers* frequently have hardwood lumber sides and backs, and plywood bottoms. There is as wide a variation in drawer front styling as there is in door design.

Drawer construction also varies depending on cabinet quality. In high quality cabinets, the sides are connected to the front and back with multiple *dovetail* joints, in others with *lock shouldered* or *square shouldered* joints (Fig. 5). Drawer bottoms are dadoed into the sides, front and back for more rigid construction. Some drawer units are molded plastic (except for the front), with rounded inside corners for ease of cleaning.

MANUFACTURE

There are two basic kinds of manufacturing operations: the integrated plant and the assembly plant. In the integrated plant, wood is processed from raw material (logs) to the finished products (cabinets) with minor purchases of plywood, particleboard, plastic components and hardware. Such facilities have vast milling operations for the many wood-processing steps needed to produce the final products. In the assembly plant, raw materials have already been milled either completely or partially and the production is keyed mainly to assembly.

Cabinet Assembly

In a typical assembly operation, various departments process components simultaneously and feed them to production lines.

In the front frame assembly department, previously machined stiles, rails and mullions are first

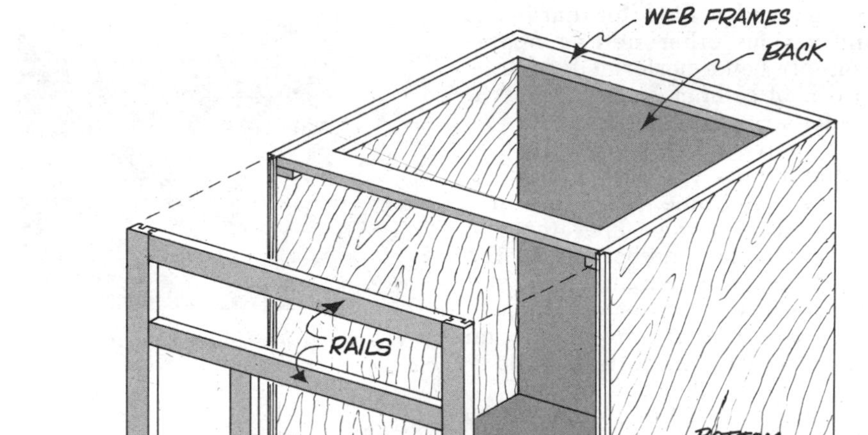

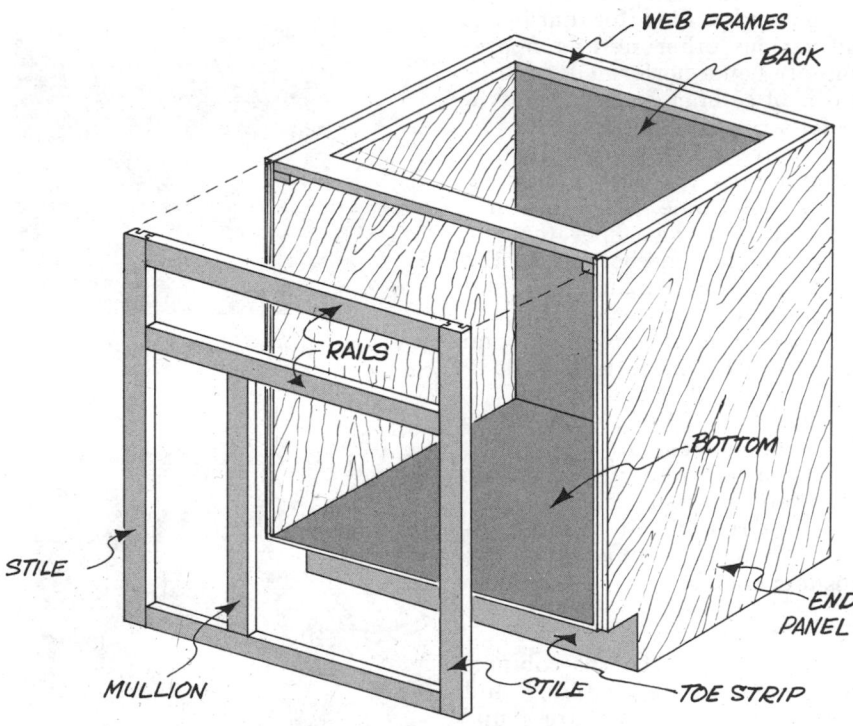

FIG. 3 Typical base cabinet construction, showing component parts.

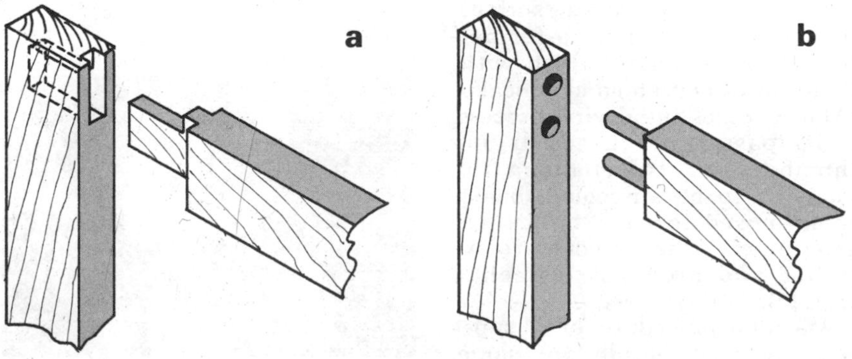

FIG. 4 Rails and stiles of frames can be joined by: a) mortise and tenon, or b) dowel joint.

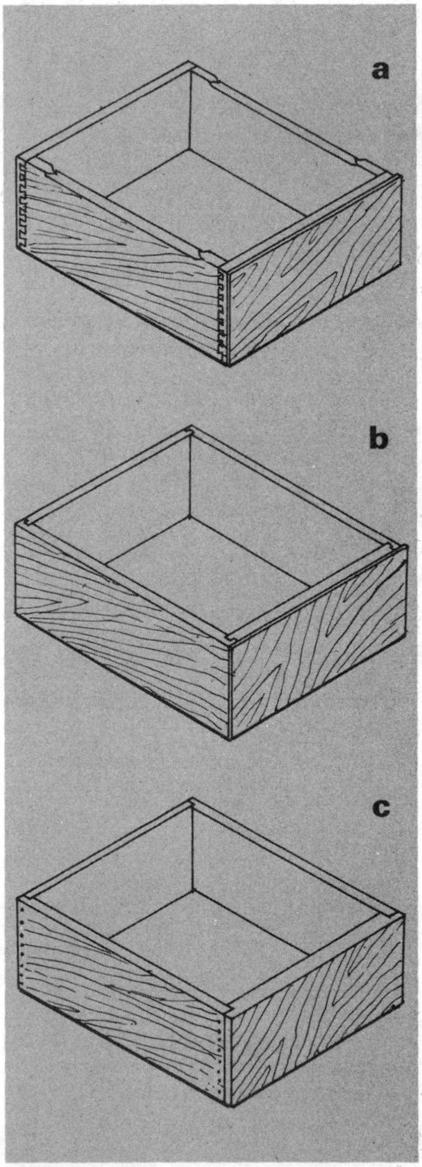

FIG. 5 Drawer construction: a) multiple dovetail; b) lock shouldered, and c) square shouldered joints.

FIG. 6 Front frames are assembled and glued in air press.

FIG. 7 Frames are passed through double sanding machine.

FIG. 8 End panels are grooved for shelves; sometimes grooves are omitted and shelves are made adjustable on pins or clips.

FIG. 9 Cabinet stiffness is increased when shelves are let into sides.

assembled, then placed into an air-press for squaring and gluing (Fig. 6). Staples are inserted to hold the frame snugly until the glue has dried. After a day's drying, the frames are sanded on both sides (Fig. 7) and moved to another section where holes are drilled for hardware.

At the same time, end panels for wall and base cabinets are being milled elsewhere in the plant. Special machines make lips on both front and back edges and run grooves for shelves (Fig. 8).

All the while, doors are being sanded and drilled for hardware and various other milling operations are being performed, such as sawing of cabinet ends and notching for strips.

In many assembly lines, the first operator puts the side panels on the cabinet frame. Then the unit is moved to the next operator who installs the top and bottom. In the third step, the hanging rails and the shelves are inserted (Fig. 9). Then the unit is conveyed to an operator who installs the back (Fig. 10).

In the next phase, glue blocks are installed and putty is applied to hide surface imperfections (Fig. 11). Then the whole cabinet is air-cleaned, inspected and hand-sanded before passing on to the finishing booths.

Finishing

In the finishing booths cabinets are stained, sealed, sanded and finished (Fig. 12). To "dress up" plain door panels, frequently they are run through a graining machine that applies a rich natural-looking grain finish to the surface. In many plants a urea-formaldehyde finish is used which withstands most household chemicals.

Then comes the drying process. Units pass through gas-drying chambers and travel along the conveyor to be air-cooled, before being moved to the "up-fitting" department, where doors are installed and hardware attached (Fig. 13).

Assembly procedure for all types of cabinets is basically the same. However, in base cabinet assembly, presses are used in the

FIG. 10 Back panels strengthen and seal cabinets; note glue blocks at front frame.

FIG. 11 Minor imperfections are repaired prior to finishing.

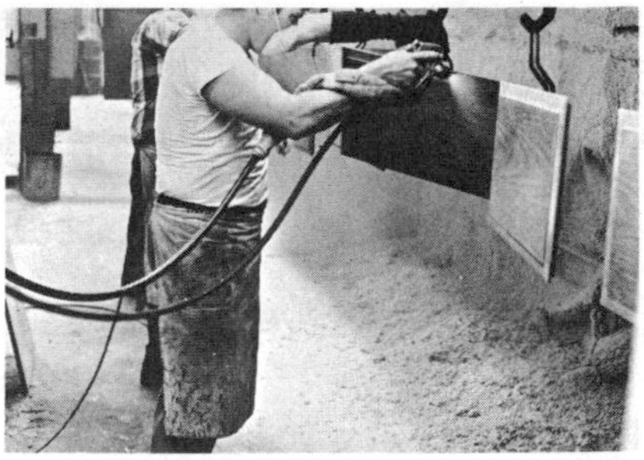

FIG. 12 Doors are sprayed with two or more coats of protective and decorative finish.

FIG. 13 Hardware is installed as cabinets move on assembly line.

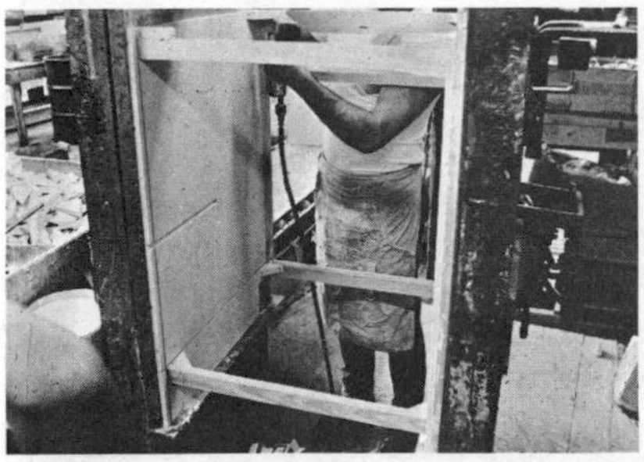

FIG. 14 To increase rigidity of base cabinets corner blocks are installed top and bottom.

FIG. 15 Glue is sprayed evenly on counter blank in preparation for post-forming plastic laminate.

FIG. 16 The NKCA Certification Seal can be found on inside face of cabinet door or drawer.

initial stages to square the units while corner and glue blocks are installed. The bottom shelf and back are installed before the unit moves down the line for further processing (Fig. 14). An additional step in base cabinet assembly is the installation of drawers and guides.

Some plants also manufacture plastic laminate countertops. In a countertop operation, the back-splash, top and front piece are assembled first into a wood blank. Then adhesive is applied to the wood surface (Fig. 15) and a plastic laminate sheet is laid over the top and pressed into intimate contact with the surface. If the countertop blank has curved surfaces such as a rounded front edge and coved backsplash, the plastic sheet is first heated, then *postformed* (molded) to the blank surface by a special press.

STANDARDS OF QUALITY

The National Kitchen Cabinet Association's (NKCA) "Minimum Construction and Performance Standards for Kitchen and Vanity Cabinets" have been the nationally recognized standards for bath and kitchen cabinets since 1968.

In 1970, the NKCA Standards were adopted by the American National Standards Institute as an American National Standard, titled ANSI A161.1—Recommended Minimum Construction and Performance Standards for Kitchen Cabinets.

The Standards apply to manufactured, factory-finished bath and kitchen cabinets made primarily of wood but including in part other non-metallic materials such as plastics, hardboards and particleboards. The Standards emphasize performance rather than construction, allowing considerable freedom in the selection of materials and methods of fabrication so long as the final products meet stringent testing procedures for endurance and quality.

Cabinet Construction

ANSI and NKCA Standards include 18 specifications covering materials and construction standards for bath and kitchen cabinets. The Standards require that all lumber and plywood parts be kiln dried to a moisture content of 12% or less at the time of fabrication. Other requirements included in the Standards are as follows: All units, except sink, oven, refrigerator and drawer cabinets should be enclosed with sides, backs and bottoms. The countertop must provide enclosure from the top on base cabinets, but wall cabinets should be provided also with individual tops.

Swinging doors should have a catch or other device to hold doors closed. When installed, adjacent cabinets and doors should be in proper alignment with each other. Doors should be of balanced (warp-resistant) construction and operate freely. Miscellaneous hardware such as drawer slides, shelf standards, brackets and rotating shelf hardware must support the design loads and operational functions described in the Standards. Hardware should comply with Cabinet Hardware Standard 201 of the Builders Hardware Manufacturers Association.

Certification of Quality

Any manufacturer of prefinished, factory-engineered kitchen cabinets may apply for participation in the NCKA Standards and Certification Program. Participants in the Standards and Certification Program need not be members of the Association.

To become a participant in the Program a cabinet manufacturer must enter into an agreement with the NKCA, which spells out the conditions under which he may affix the Certification Seal to his cabinets. The NKCA Certification Seal (Fig. 16) offers assurance to the public that cabinets bearing it have been manufactured in conformance with the NKCA Standards.

Testing and Procedures

By terms of the agreement with the manufacturer, unannounced visits can be made by an NKCA inspector to the plant twice a year. The inspector may pick one base cabinet and one wall cabinet from the assembly line or storage area and have them tested to determine compliance with the Standards. All testing under the Certification Program is done by designated members of the American Council of Independent Laboratories Inc.

Any licensee whose cabinets do not comply with the Standards is subject to exclusion from the Program unless he takes corrective action within 60 days.

To qualify for the Certification Seal, cabinets must meet performance tests which simulate

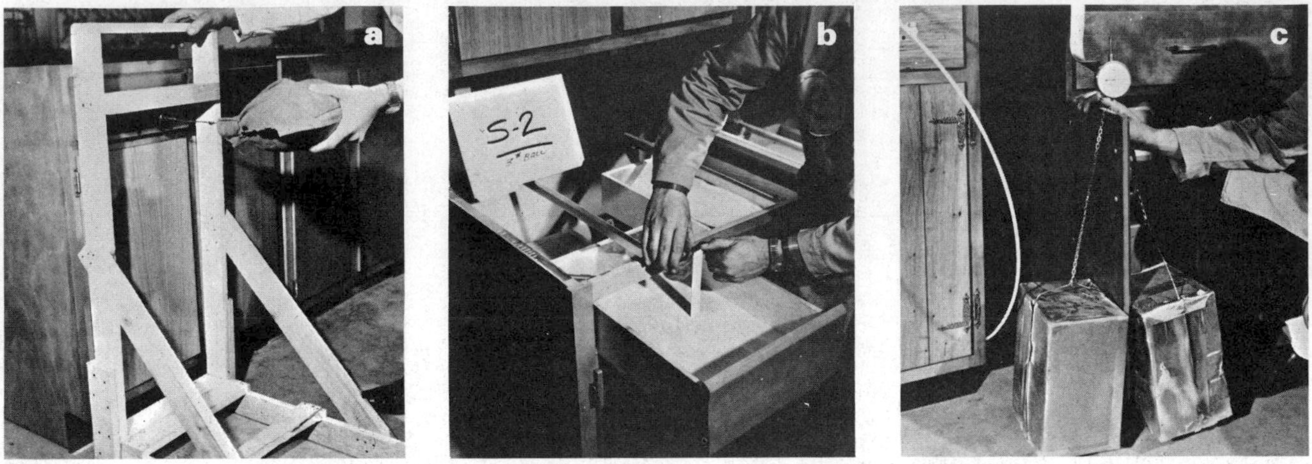

FIG. 17 NKCA cabinet performance tests include: a) dropping 10 lb. sand bag against closed door; b) dropping 3 lb. ball in open drawer, and c) racking open door with two 22.5 lb. weights. Except for dents, there must be no visible damage and units must be operable.

actual long-term conditions of use in the bathroom or kitchen. These include six tests for structural integrity (Fig. 17) and four tests for exterior cabinet finish. Door and drawer opening tests measure the results of ten years of normal use. Finish tests are designed to duplicate at least five years of normal usage.

KITCHEN DESIGN

The work triangle principle in the arrangement of appliances and kitchen cabinets is basic to good kitchen design (Fig. 18). Proper kitchen planning results in greater efficiency and comfort for the housewife. Some basic rules of thumb will be helpful in planning the kitchen layout.

No two basic appliances should be less than 4′ apart as measured from their midfronts. The recommended distances between appliances are: sink to refrigerator, 4′ to 7′; sink to range, 4′ to 6′; range to refrigerator, 4′ to 9′. Total distance between appliances along the triangle should not exceed 22′ (Fig. 18). Recommended working heights and clearance for different cabinets are shown in Figure 19.

A sufficient amount of counter space should be provided along each major appliance: 15″ next to the handle side of the refrigerator for setting out supplies; 36″ at the right of the sink for stacking dishes before washing; 20″ at the left of the sink for stacking clean dishes (may be omitted if dishwasher is included); 24″ beside the range for serving dishes; 18″ alongside a separate oven for hot pans; and at least 36″ of mixing and food preparation space, preferably near the sink. Sometimes, several of the above functions may be combined into a single counter if they are located side by side. Whenever two or more counters are combined, the multiple use counter should equal the longest recommended counter, plus 12″. For instance, the 36″ at the right of the sink, the 18″ along the refrigerator and 36″ mixing area can be combined into one counter 48″ long (36″ plus 12″).

The work triangle may be compact or elongated, according to the overall shape of the space. The most compact arrangement, with ample cabinet space and without through-traffic occurs in a *U-shaped* kitchen (Fig. 20a). But many excellent kitchens are *L-shaped* (Fig. 20b) or of the open *corridor* type (Fig. 20c). A *one-wall* arrangement also can be satisfactory (Fig. 20d) for smallish kitchens, but total distance between appliances usually becomes excessive when liberal amounts of cabinet space are provided. In a one-wall kitchen, the total length of cabinets should

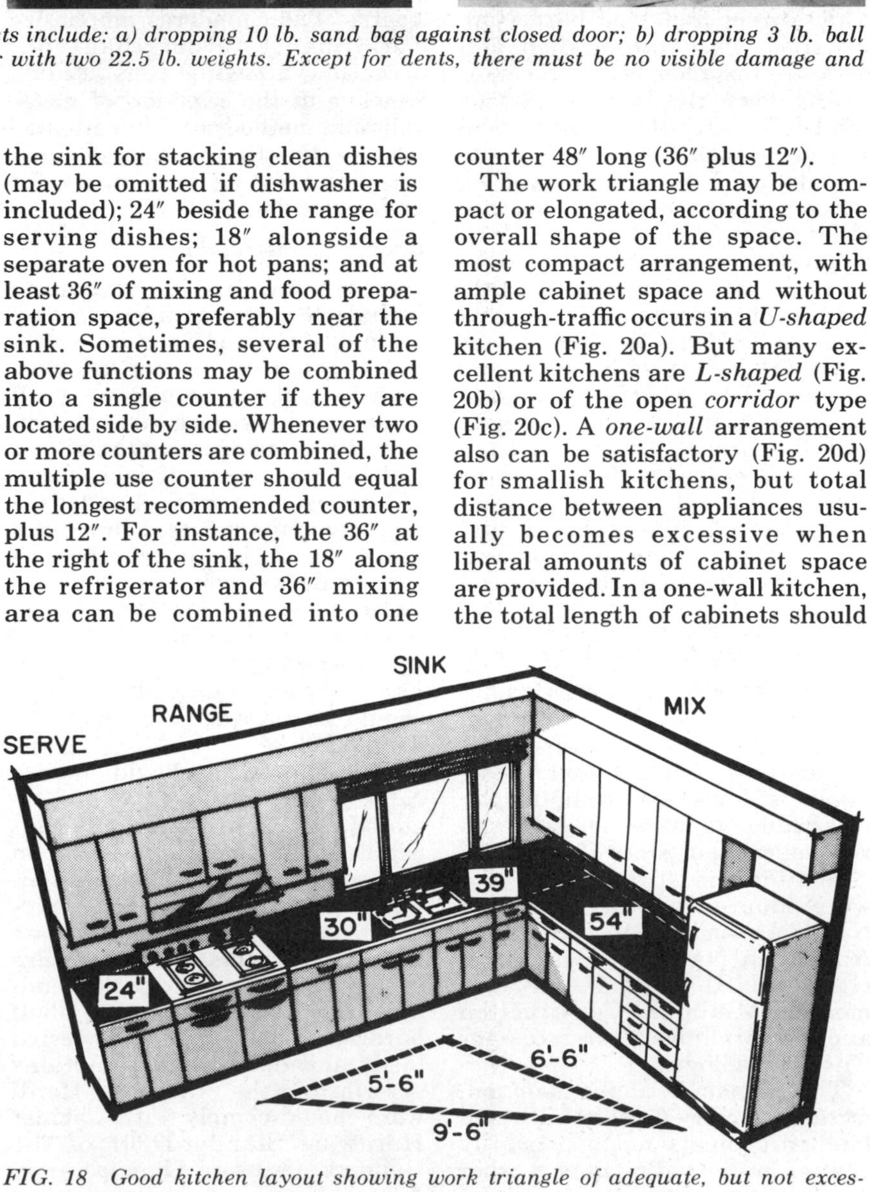

FIG. 18 Good kitchen layout showing work triangle of adequate, but not excessive dimensions.

FIG. 19 **RECOMMENDED WORK HEIGHTS AND CABINET CLEARANCES**

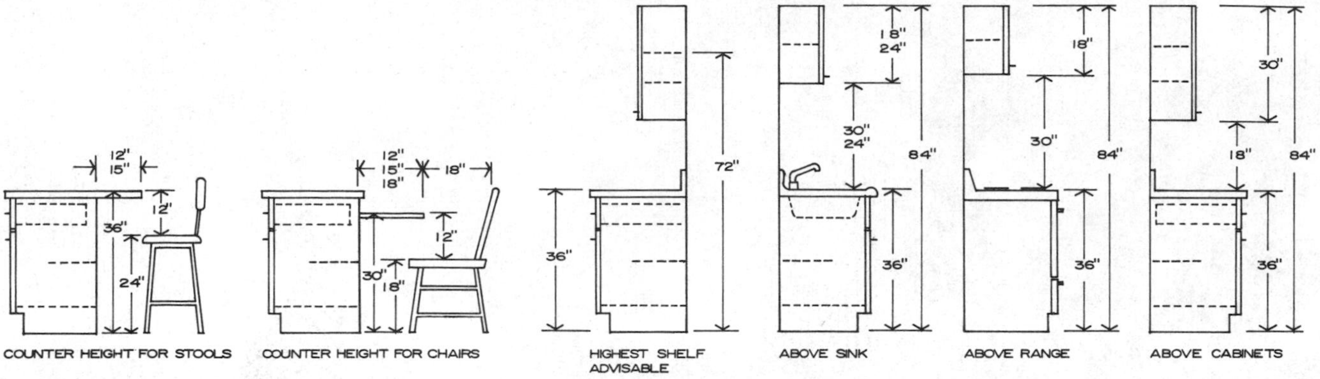

not exceed 22′ and the sink should be in the middle position, between the range and refrigerator.

In a long kitchen, it is preferred that the refrigerator should be at the end where the groceries are brought in and the range at the end nearest the dining center. Normal work flow is considered to be from refrigerator to sink to range in a counterclockwise direction.

Peninsulas or islands of cabinets can be used to improve the shape or traffic pattern and to provide the convenience of more storage space.

INSTALLATION

Appliances and cabinets should be installed level, plumb and true, in accordance with manufacturer's instructions. Cabinets should be attached with screws (not nails) to studs or other framing members. Wall cabinets should likewise be attached with screws to supporting framing rather than merely to the interior finish.

Wall or floor irregularities can interfere with proper attachment. To assure a plumb and level installation, high spots should be removed, or low spots should be shimmed. Wall base and chair rail should be removed behind cabinets to assure a flush fit.

Installation should begin in one corner. After the base corner unit is installed, other units are added using "C" clamps to connect adjacent cabinets for proper alignment. T-nuts are recommended to fasten cabinets together. Each cabinet should be checked front to back and across the front edge with a level, as it is attached to the wall. The front frame should be plumb and top should be level. Base cabinets should be attached with screws into wall studs, preferably through the back frame. After base cabinets are secured in place, the countertop is installed and the toestrip is trimmed out with resilient or wood base.

Wall cabinets should be installed next, again beginning with a corner unit. Screws should be inserted through hanging rails built into backs of cabinets at both top and bottom. Screws should not be tightened immediately; after cabinets are leveled and plumbed, and doors are perfectly aligned, the screws can be tightened.

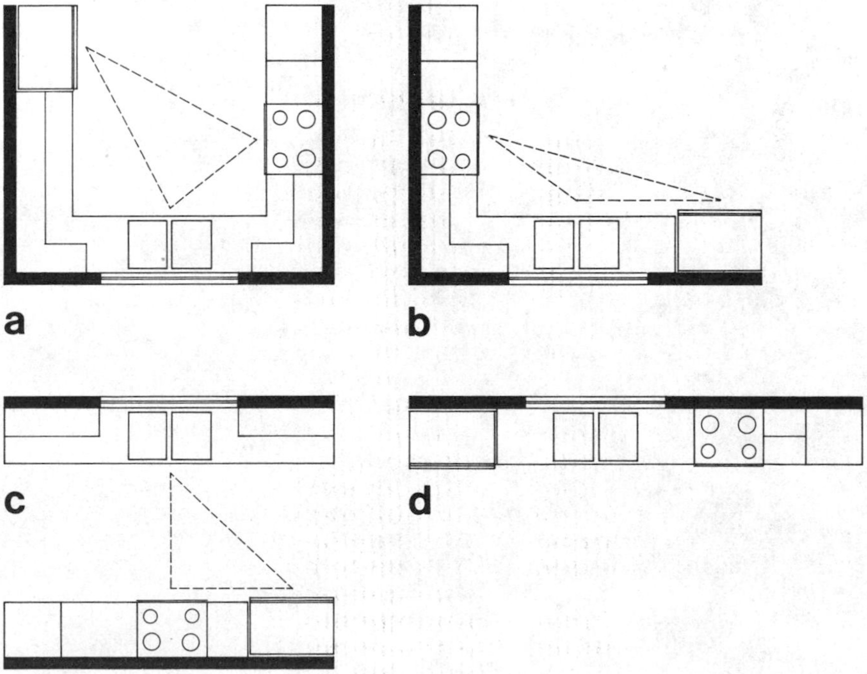

FIG. 20 Basic kitchen layouts: a) U-plan; b) L-plan; c) corridor type, and d) one-wall arrangement.

WOOD PRODUCTS 202

CONTENTS

WINDOWS AND SLIDING GLASS DOORS

STANDARDS OF QUALITY

Wood windows and sliding glass doors of suitable quality should meet or exceed appropriate industry standards shown in Figure WF-W1.

The NWMA standards constitute the accepted industry specifications for wood windows and sliding glass doors. They broadly regulate the material, design and manufacture of windows, and establish required performance criteria.

Wind Resistance

The selection of sliding glass doors for high wind zones (Fig. WF-W2) and exposed locations should be based on independent laboratory performance test reports covering infiltration, penetration of windblown rain, and wind load capability, or on a history of satisfactory performance under local conditions.

The NWMA window standard has two performance classes (A and B), based on water infiltration and wind loading. Class A windows must meet more stringent performance requirements for these criteria.

In high wind zones, Class A windows should be specified.

Preservative Treatment

Wood components of windows and sliding glass doors should be preservative-treated with toxic, water-repellent solutions in accordance with NWMA IS-4, Water-Repellent Preservative Non-Pressure Treatment for Millwork.

Preservative treatment provides resistance to decay, insect attack and differential moisture absorption, which may cause twisting or warping of wood components. This treatment minimizes intermittent swelling and shrinking due to varying atmospheric conditions or contact with water.

CERTIFICATION OF QUALITY

The National Woodwork Manufacturers Association (NWMA) and several independent laboratories administer a program of inspection and testing which results in the certification of windows which conform to standards. The NWMA or similar certification seal (Fig. WF-W3) is assurance that the design, manufacture and tested performance of a window meet or exceed the specifications of applicable industry standards.

The NWMA also administers a program of inspection and testing of windows, doors and other millwork products, which results in the certification of products which conform to NWMA IS-4. The NWMA seal (Fig. WF-W4) is assurance that the preservative solution, methods and effectiveness of treatment meet or exceed requirements of NWMA IS-4.

HANDLING AND STORAGE

Units should be handled carefully to avoid soiling prior to finishing, and should be prime coated as soon as possible after arrival at the building site. They should be stored out of the weather in a dry, well ventilated space to minimize excessive moisture loss or absorption. Installation should be avoided while wet concrete or plaster is drying.

INSTALLATION

Frames should be checked for squareness and straight-

FIG. WF-W1 INDUSTRY STANDARDS FOR WINDOWS AND DOORS

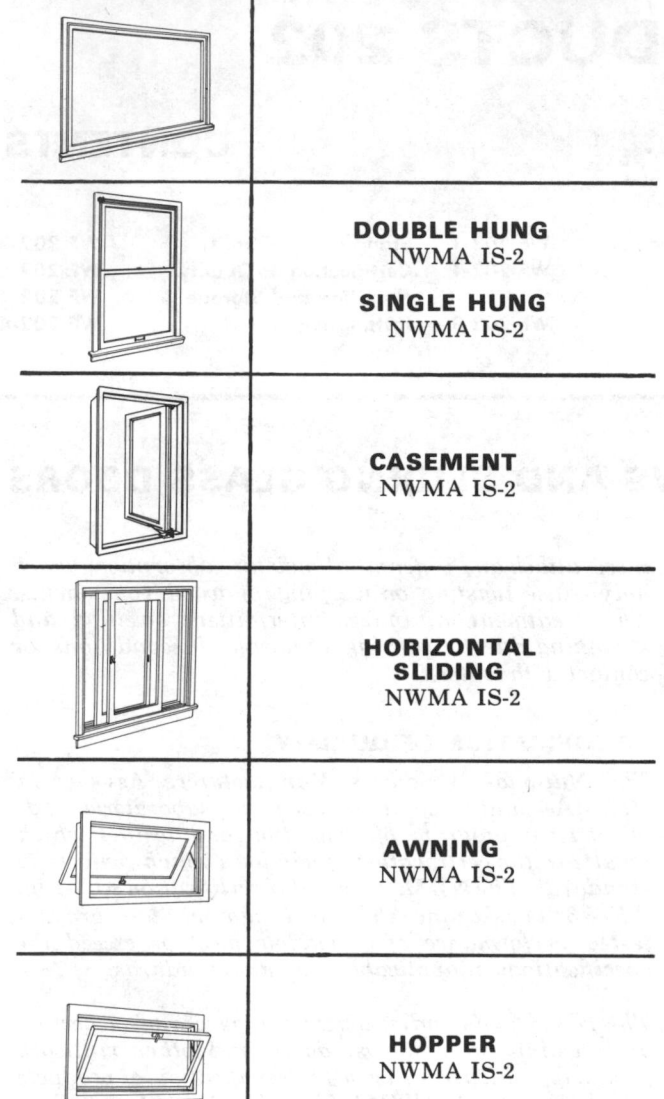

	**DOUBLE HUNG** NWMA IS-2 **SINGLE HUNG** NWMA IS-2
	**CASEMENT** NWMA IS-2
	**HORIZONTAL SLIDING** NWMA IS-2
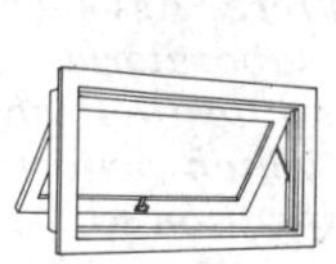	**AWNING** NWMA IS-2
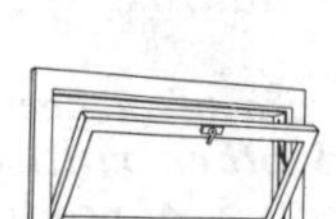	**HOPPER** NWMA IS-2
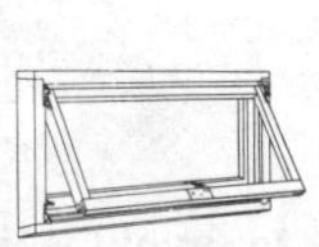	**BASEMENT** (no industry standard)
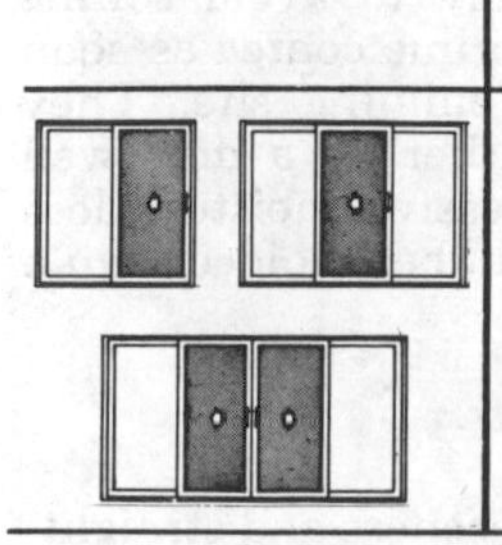	**SLIDING GLASS DOORS** NWMA IS-3

FIG. WF-W2 WIND ZONE VELOCITIES

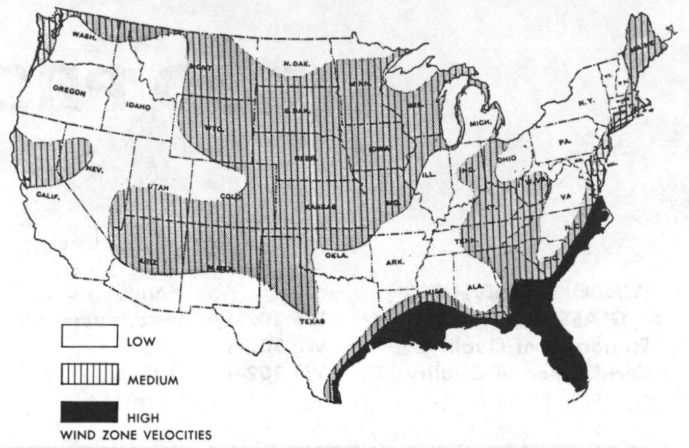

ness of jambs prior to installation. Braces installed to prevent distortion in shipment should be renailed, if necessary, and left in place until the frame is ready for trim.

Adequate shimming should be provided in the space between rough and finished frames. If the construction is insulated, this space should be filled with insulation. Interior trim should be fitted carefully to seal this space from the interior.

Flashing or drip caps are strongly recommended. The juncture of exterior finish and casing should be caulked prior to finish painting. Putty beads on both face and back of sash should be painted, lapping paint slightly over glass. Metal or plastic balancing devices, sash guides and tracks should not be painted.

See Division 400—Applications & Finishes for additional discussion of installation procedures.

QUALITY NWMA 000 CERTIFIED
WOOD WINDOW UNIT
CLASS
CONFORMS TO NWMA I.S.-2

FIG. WF-W3 Certification seal such as that of the NWMA, gives assurance that window unit conforms to applicable standards.

QUALITY NWMA 000 CERTIFIED
WATER-REPELLENT PRESERVATIVE
CONFORMS TO NWMA I.S.-4

FIG. WF-W4 The NWMA seal is stamped or die-pressed on preservative treated components and assemblies.

DOORS

STANDARDS OF QUALITY

Wood doors of suitable quality should meet or exceed appropriate industry standards shown in Figure WF-D1.

NWMA and FHDA Standards constitute the recognized standards of quality for the wood door industry and are based on the woods used in their manufacture. Most flush doors are covered by NWMA IS-1 Wood Flush Doors; stile-and-rail doors (panel, sash, louver, storm and screen) are covered by NWMA IS-5 Ponderosa Pine Doors, and FHDA 6-77, Old Growth Douglas Fir, Sitka Spruce and Western Hemlock Doors. Accordion folding doors utilize a variety of wood species and presently are not regulated by industry standards.

In addition to establishing minimum requirements for material, design, construction and performance under moisture conditions, industry standards also provide a basis for quality grading of wood doors.

Grades

The species and grades of wood doors should be suitable for their intended appearance and method of finishing (Fig. WF-D2).

Wood doors are graded according to the quality and appearance of exposed surfaces, limiting the type and extent of defects permitted. Flush door grade regulations cover face veneers only. Stile-and-rail doors are graded according to lumber stock used in stiles and rails and lumber or plywood in panels (see Main Text, page 202-26).

FIG. WF-D2 **SUMMARY OF DOOR GRADES**

Commercial Standard	Grade	Suitable Finish
Hardwood Veneered Doors, NWMA IS-1	Premium	Natural
	Good	Natural
	Sound	Painted
	Specialty	Any
	Plastic	None
	Hardboard	Painted
Ponderosa Pine Doors, NWMA IS-5	Premium	Natural
	Standard	Painted
Old Growth Douglas Fir, Sitka Spruce & Western Hemlock, FHDA 6-77	Select	Natural
	Standard	Painted

*Although suitable for natural finish, these doors often are painted.

FIG. WF-D1 **SUMMARY OF WOOD DOORS**

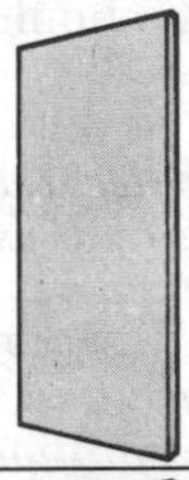

FLUSH
Wood Flush Doors NWMA IS-1

PANEL
Ponderosa Pine Doors NWMA IS-5
Douglas Fir FHDA 6-77

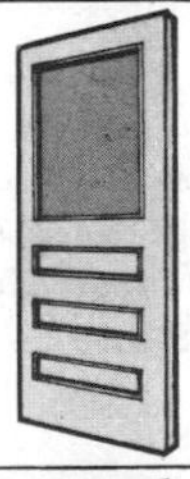

SASH
Ponderosa Pine Doors NWMA IS-5
Douglas Fir FHDA 6-77

STORM & SCREEN
Ponderosa Pine Doors NWMA IS-5
Douglas Fir FHDA 6-77

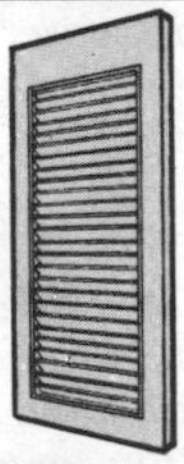

LOUVER
Ponderosa Pine Doors NWMA IS-5
Douglas Fir FHDA 6-77

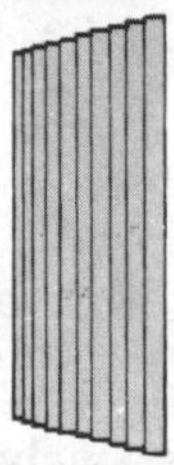

ACCORDION FOLDING
(no industry standard)

Moisture Resistance

Flush doors intended for exterior use should be bonded with waterproof (Type I) adhesives. Interior flush doors and all types of stile-and-rail doors (both for interior and exterior use) should be bonded with water-resistant (Type II) adhesives capable of resisting a limited number of wetting cycles. Where usually severe moisture conditions are expected in either interior or exterior locations, all doors should be bonded with waterproof (Type I) adhesives.

Industry standards establish performance criteria under laboratory test conditions for waterproof and water-resistant adhesives. Established performance requirements for water-resistant (Type II) adhesives are adequate to insure structural integrity of flush and stile-and-rail doors subject to average moisture conditions in interior locations. The same adhesives are suitable for normal exterior exposures of stile-and-rail doors, which typically are made with doweled stile-and-rail frames and do not depend entirely on adhesive bond for structural integrity. Flush doors in exterior locations are more dependent on durable adhesive bond, and waterproof (Type I) adhesives used must be capable of withstanding repeated severe moisture conditions, as established by performance standards.

Preservative Treatment

Wood components of stile-and-rail doors should be preservative treated with toxic, water-repellent solutions in accordance with NWMA IS-4, Water Repellent Preservative Non-Pressure Treatment for Millwork.

CERTIFICATION OF QUALITY

The National Woodwork Manufacturers Association (NWMA) administers a program of inspection and testing resulting in certification of wood doors which conform to industry standards. The NWMA seal (Fig. WF-D3) is assurance that the materials, construction and tested performance of a wood door meet or exceed the requirements of applicable industry standards. Doors bonded with waterproof (Type I) adhesives suitable for exterior use may be further identified by a small red plastic plug set in the top edge of the door (see Main Text, page 202-27).

The NWMA also administers a program of inspection and testing resulting in certification of wood doors which conform to NWMA IS-4. The NWMA seal (Fig. WF-D4) is assurance that the preservative solution, methods and effectiveness of treatment meet or exceed requirements of NWMA IS-4.

HANDLING AND STORAGE

Unfinished doors should be handled carefully to prevent soiling. They should be stored flat (not on edge) on a level surface in a clean, dry, well ventilated space. They should not be stored in excessively hot, dry or humid areas, and should be conditioned for several days to the average prevailing local humidity before hanging or finishing. If doors appear warped, they should be stacked flat under uniformily distributed weights to restore flatness.

INSTALLATION

Finished door frames should be square and plumb, and doors should be fitted in the openings with a total clearance of approximately 3/16″ in both horizontal and vertical dimensions to prevent binding under humid conditions. Doors may be trimmed for fitting, but should not be cut down to smaller nominal sizes.

When mortising for locks and other hardware, care should be taken not to impair the strength of the door by cutting away too much of the supporting construction.

All exterior doors and interior doors 7′0″ in height and over should be hung with a minimum of three hinges per door, properly aligned so the door will not twist out of shape.

Finishing

Doors should be seal-coated as soon as they are delivered to the job site to prevent undue moisture adsorption and soiling. Finishing should be delayed until doors have been trimmed and fitted in the openings. All door surfaces, including top, bottom and side edges, should be sealed against moisture with two coats of paint, varnish or sealer before final hanging.

Door surfaces should be clean and dry before applying the finish. Soil marks and surface defects should be removed by light sanding. Finishing in excessively humid weather should be avoided. The proposed finish should be suitable for the door grades used (Fig. WF-D2).

FIG. WF-D3 The National Woodwork Manufacturers Association (NWMA) seal stamped on the bottom or top of the door assures conformance with the applicable industry standard.

FIG. WF-D4 The NWMA seal is stamped or die-pressed on preservative treated components and assemblies.

203 CONCRETE

CONCRETE

INTRODUCTION

Concrete, one of the most important building materials, is used in almost every type and size of architectural and engineering structure. In residential construction, major use is in the substructure of the building for foundation walls and footings; in flat concrete work including floor and garage slabs, exterior stairs and terraces, sidewalks and driveways, curbs and streets; and for other site structures such as septic tanks and drainage swales. Recent developments in fiber reinforced concrete, using both glass and steel fibers as well as the increased use of precast concrete components, point to the growing use of concrete in the superstructure of buildings as well.

Concrete of a *suitable quality* means that which will perform satisfactory service in the use for which it is intended. To do so, it must possess four essential functional properties: *strength* to carry superimposed loads; *watertightness* to prevent water penetration; *durability* for wear and weather-resistance; and *workability* to insure proper handling, placing, finishing and curing. These properties can all be achieved with good materials and careful workmanship.

Required surface treatments may range from smooth troweled surfaces for finished floors to highly decorative geometric patterns for use on terrace, wall and other surfaces. The plastic quality of concrete offers the opportunity of almost unlimited design possibilities of form, pattern and texture.

The in-place cost of concrete as well as the long-term cost of maintenance are dependent largely on proper care in handling, placing, finishing and curing. Skillful selection and combining of ingredients in the mix are necessary to achieve the most economical concrete. But the typical in-place cost of concrete is 3 to 6 times or more the cost of materials alone, and no mix is better than the workmanship it receives in the field.

The essence of producing concrete of *suitable quality* can be summarized in one paragraph:

Sound materials must be selected, carefully proportioned and combined, using a *low* water-cement ratio, into the correct mixture with a consistency that can be worked into place to provide the required surface finish, followed by adequate curing and protection.

This section explains how to produce concrete of *suitable quality.*

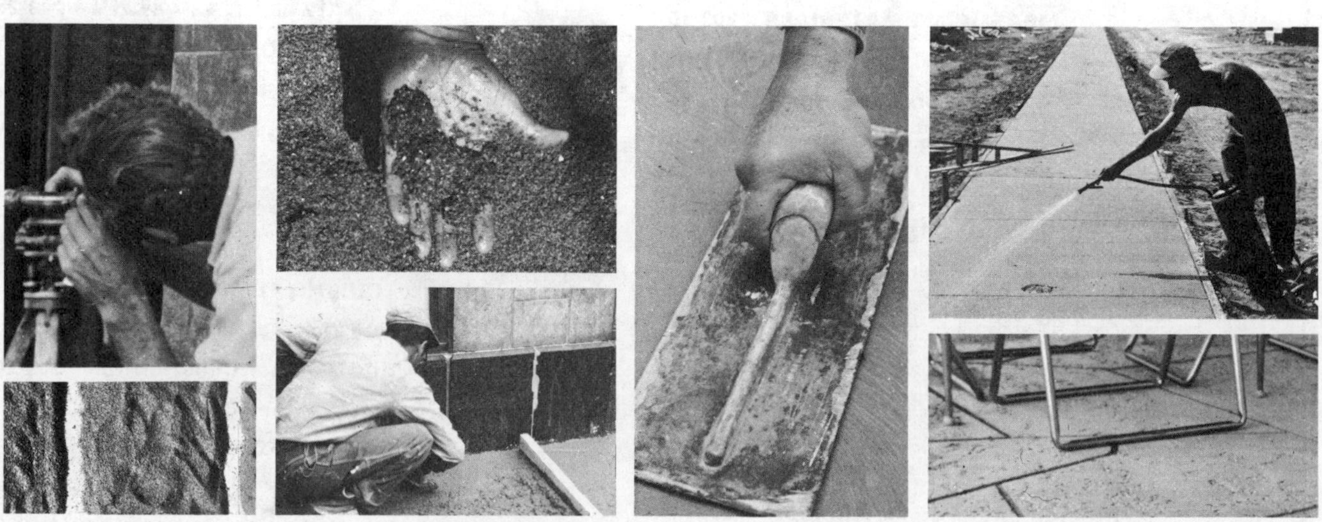

MATERIALS

Concrete is made by combining four materials into a mixture of basically two parts: aggregate and cement paste. The paste, composed of portland cement and water, binds the fine aggregate (sand) and coarse aggregate (gravel or crushed stone) into a rock-like mass when the paste hardens. The concrete hardening (setting) results from a chemical reaction between the cement and water. The portland cement, fine and coarse aggregates should conform to the requirements of the specifications of the American Society for Testing Materials (ASTM).

PORTLAND CEMENT MANUFACTURE

Portland cement is a finely pulverized material consisting principally of compounds of lime, silica, alumina and iron. It is manufactured from selected materials under closely controlled processes (Fig. 1). The process starts by mixing limestone or marl with such other ingredients as clay, shale or blast furnace slag in proper proportions and then burning this mixture in a rotary kiln at a temperature of approximately 2700° F to form a clinker. The clinker is cooled and then pulverized; a small amount of gypsum is added to regulate the setting time of the cement. The finished product, *portland cement,*

FIG. 1 Steps in the manufacture of portland cement, illustrating both dry and wet processes.

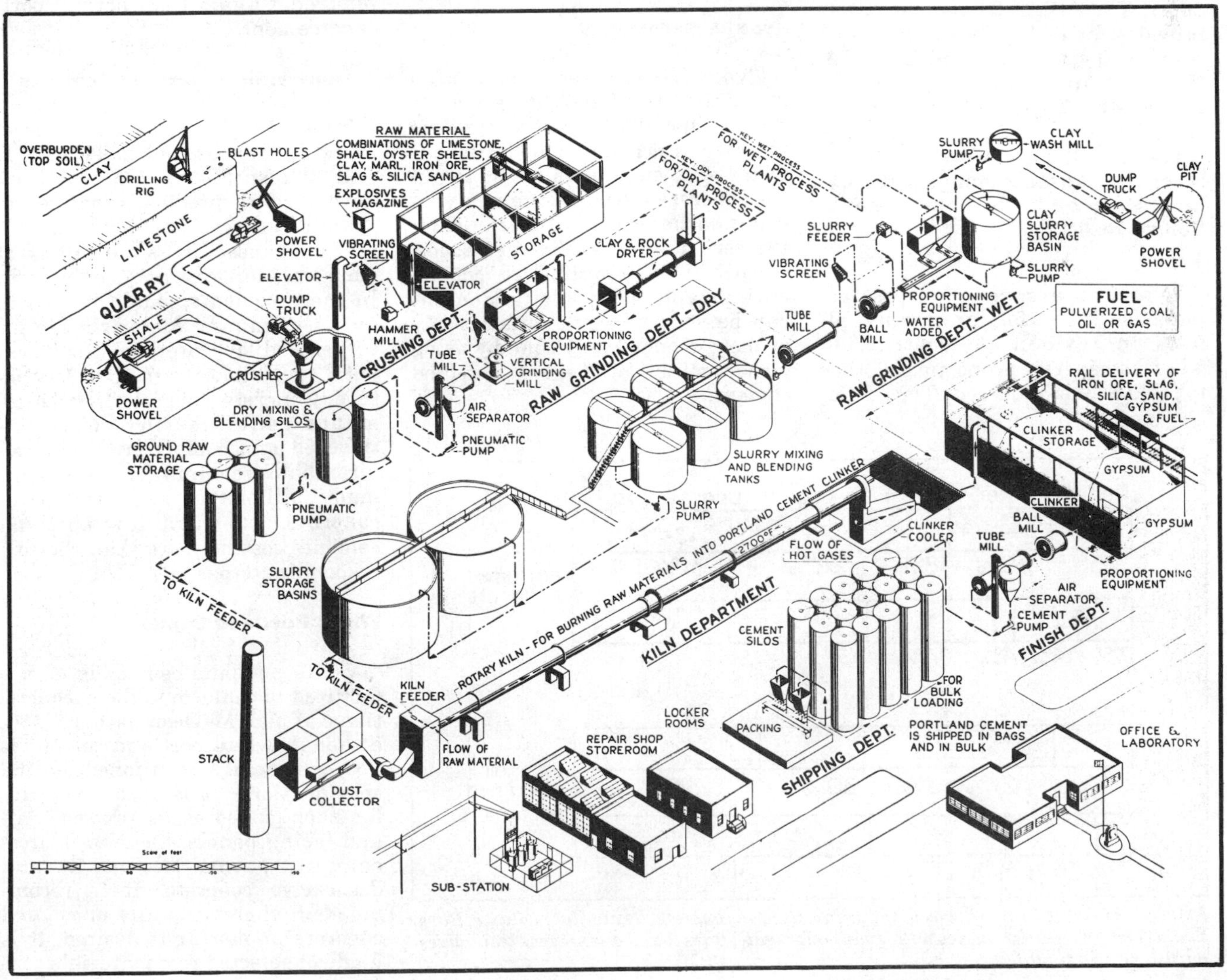

is so fine that nearly all of it will pass through a sieve with 40,000 openings to the square inch.

When portland cement is mixed with water, a paste is formed which first sets (becomes stiff) and then hardens into a solid mass. The setting and hardening is called "hydration" and is brought about by *chemical* reactions between the cement and water. The reaction is *not* simply one of evaporation or drying.

Portland cement got its name from the Isle of Portland, in the English Channel, whose grey rock the early concretes resembled.

TYPES OF PORTLAND CEMENT

Portland cements are made to meet different physical and chemical requirements for specific purposes. The ASTM Designation C150 provides for eight types of portland cement. Type I Normal; Type IA Normal, Air-entraining; Type II Moderate; Type IIA Moderate, Air-entraining; Type III High Early Strength; Type IIIA High Early Strength, Air-entraining; Type IV Low Heat of Hydration; Type V Sulfate Resisting.

Type I

Type I is a general purpose cement suitable for practically all uses in residental construction when the special properties of other types are not required. Type I Cement should not be used where the concrete will be in contact with high sulfate soils or will be subject to excessive temperatures during hydration.

Type II, Normal

Type II is used where precaution against moderate sulfate attack is important, as in drainage structures where sulfate concentrations in groundwaters are higher than normal. Type II generates less heat of hydration and at a slower rate than Type I; this moderate heat of hydration reduces temperature rise which is especially important when concrete is placed in warm weather in structures of considerable mass, such as large piers and heavy retaining walls.

Type III, Moderate

Type III is used when high strengths are desired at very early periods, usually a week or less. It is used when it is desirable to remove forms as soon as possible or to put the concrete into service quickly; in cold weather construction to reduce the period of protection against low temperatures required to control curing; and when high strengths can be secured more satisfactorily or more economically than by using richer mixes of Type I portland cement.

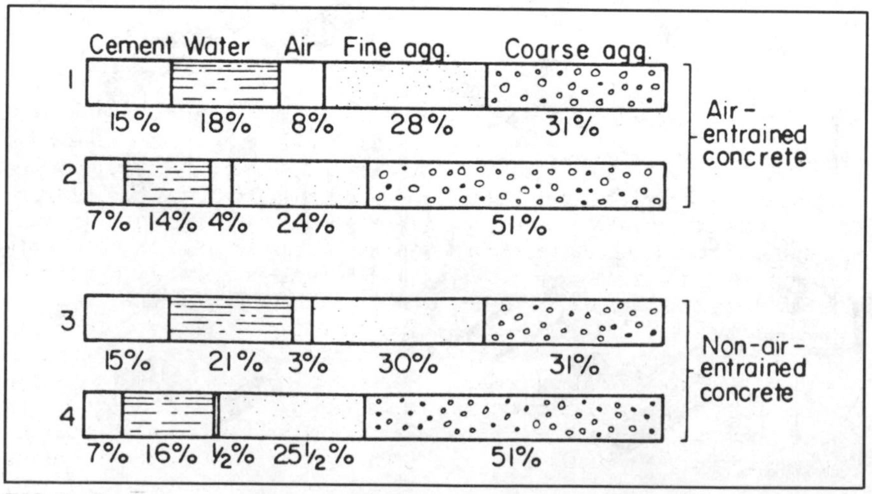

FIG. 2 Range in proportions of materials used in concrete, by absolute volume. Bars 1 and 3 represent rich mixes with small aggregate. Bars 2 and 4 represent lean mixes with large aggregate.

Type IV, Low heat

Type IV is a special cement for use where the amount and rate of heat generated during hydration must be kept to a minimum. The development of strength is also at a slower rate. It is intended for use in large masses of concrete (e.g., large gravity dams) where temperature rise resulting from the heat generated during hardening is a critical factor.

Type V, Sulfate Resisting

Type V is a special cement intended for use only in construction exposed to severe sulfate action, such as in some western states having soils and waters of high alkali content. It has a slower rate of strength gain than normal portland cement.

Air-entraining Portland Cements

Types IA, IIA, and IIIA—correspond in composition to Types I, II and III respectively and have been developed to produce concrete resistant to severe frost action and to scaling caused by applications of chemicals for snow and ice removal. In these cements, small quantities of air-entraining materials are incorporated by intergrinding them with the clinker during manufacture. Concrete made with these cements contains tiny, well-distributed and completely separated air bubbles so minute that there are many billions in a cubic foot of concrete. Air-entraining portland cements cost no more than the corresponding types in ASTM C150.

White Portland Cement

White portland cement is manufactured to conform to the specifications of ASTM Designation C150, although white portland cement is not specifically mentioned in the specifications. It is used primarily for such purposes as precast wall and facing panels, terrazzo, stucco, cement, paint, tile grout and decorative concrete. It is recommended wherever white or colored concrete or mortar is desired. It is made of selected raw materials con-

FIG. 3 **CONCRETES MADE WITH LIGHTWEIGHT AGGREGATES**

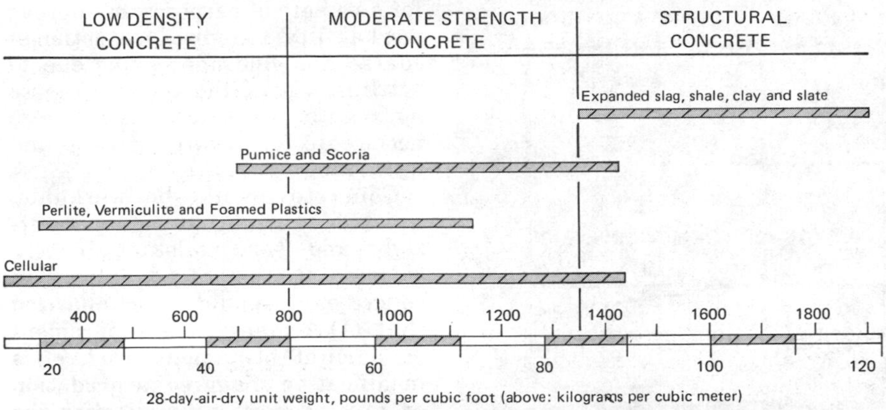

FIG. 4a Well-graded sand before and after separation into various sizes. Particles vary from dust to those approximately 1/4".

FIG. 4b Sand which lacks particles larger than 1/16" (left) and after separation (right) into four sizes. More cement is required when sand is fine. This is not a good concrete sand.

taining negligible amounts of iron and manganese oxide, the substances that give concrete its grey color.

Portland Blast-furnace Slag Cement

In these cements, granulated blast-furnace slag (a product obtained by rapidly chilling or quenching molten slag in water, steam or air) of selected quality is interground with portland cement. They should meet the latest requirements of ASTM Designation C595, Type IS and Type IS-A, air-entraining. The ASTM specifications require the slag constituent to be from 25% to 65% of the total weight of the cement.

Portland-pozzolan Cement

In these cements, pozzolan (a material consisting of siliceous, or siliceous and aluminous material) is blended with ground portland cement. They should meet the latest requirements of ASTM Designation C595, Type IP and Type IP-A, air-entraining. The ASTM specifications require the pozzolan constituent to be from 15% to 40% of the total weight of the cement.

WATER

Mixing water should be clean and free from oil, alkali or acid. In general, water that is fit to drink is suitable for mixing with cement. However, water with excessive quantities of sulfates should be avoided even though potable.

AGGREGATES

Fine and coarse aggregates constitute roughly 60% to 70% of the volume of the concrete (Fig. 2), and both the cost and the quality of the concrete are affected by the kind of aggregates selected. They should be obtained from reliable dealers whose materials are of a quality known to make suitable concrete, and they should meet the requirements of the Standard Specifications for Concrete Aggregates, ASTM C33. These specifications place limits on allowable amounts of damaging substances and also cover requirements as to grading, strength and soundness of normal weight aggregate.

The most commonly used aggregates, sand, crushed stone and air-cooled blast-furnace slag, produce normal weight concrete weighing about 135 lbs. to 160 lbs. per cu. ft. Aggregates of expanded shale, clay, slate and slag are used to produce structural lightweight concrete weighing roughly 85 lbs. to 115 lbs. per cu. ft. Other lightweight materials, such as pumice, scoria, perlite, vermiculite, and diatomite, are used to produce insulating concretes ranging in weight from 15 lbs. to 90 lbs. per cu. ft. (Fig. 3).

Fine Aggregate

Fine aggregate consists of sand or other suitable fine material. A good concrete sand will contain particles varying *uniformly* in size from very fine up to 1/4" (Fig. 4a). In well graded sand the finer particles help fill the spaces between the larger particles. Sufficient fine particles are necessary for good workability and smooth surfaces. An excess of fines requires an increase in the amount of cement-water paste to achieve a desired strength, thus increasing cost. (If the cement content is not increased, less strength will result.) A poor concrete sand is shown in Figure 4b.

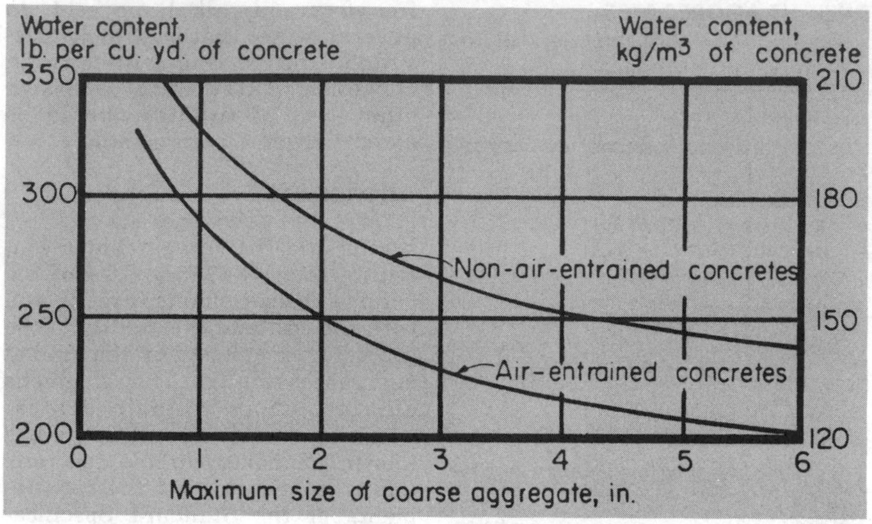

FIG. 6 The water requirement for concrete of a given consistency decreases as the maximum size of coarse aggregate increases.

Coarse Aggregate

Coarse aggregate consists of gravel, crushed stone or other suitable materials larger than 1/4″ (Fig. 5). Coarse aggregate that is sound, hard and durable is best suited for making concrete. If it is soft or flaky or wears away rapidly, it is unsatisfactory.

All aggregates, fine and coarse, should be clean and free of loam, clay or vegetable matter, since these foreign particles prevent the cement paste from properly binding the aggregate particles together. Concrete containing these objectionable materials will be porous, have low resistance to weathering, low strength and could develop surface defects such as popouts.

The maximum size of coarse aggregate, up to 1-1/2″ in size, may be used, for example, in a thick foundation wall or footing. In walls, the largest piece should never be more than 1/5 the thickness of the finished wall section. For slabs, the maximum size should be approximately 1/3 the thickness of the slab. The largest piece of aggregate should never be larger than 3/4 the width of the narrowest space through which the concrete will be required to pass during placing. This is usually the space between the reinforcing bars or between the bars and the forms.

Coarse aggregate is well graded when particles range uniformly from 1/4″ up to the largest size that may be used on the kind of work to be done. Use of the maximum allowable particle size usually results in less drying shrinkage and more economical concrete.

The maximum size of coarse aggregate used in concrete has a bearing on economy. More water usually is required for small-size aggregates than for large sizes. The water required for a slump of 3″ to 4″ is shown in Figure 6 for a wide range of coarse aggregate sizes. For a given water-cement ratio, the amount of cement required decreases as the maximum size of coarse aggregate increases. However, the increased cost of obtaining or handling aggregates larger than about 2″ may offset the savings in cement.

The most economical concrete results when the fine and coarse aggregates are uniformly graded, since this requires the least amount of cement paste to surround the aggregate and fill all the spaces between particles (Fig. 7).

ADMIXTURES

Admixtures, (materials other than portland cement, water and aggregates) added to the batch of concrete immediately before or during its mixing, are sometimes used for a variety of purposes: to improve workability; to reduce separation of coarse and fine aggregates due to settling out of the heavier coarse aggregate; to entrain air; or to accelerate or retard setting and hardening.

Concrete should be workable, finishable, strong, durable, watertight and wear resistant. In considering the use of admixtures in concrete, it should be emphasized that: (1) A change in type of cement or amount of cement used, or a modification of aggregate gradation or mix proportions may offer the surest and most economical approach to the desired objectives; (2) many admixtures affect more than one property of concrete, sometimes hurting desirable properties; (3) the effect of some admixtures is changed by such things as wetness and richness of mix, by aggregate gradation and by character and length of mixing; (4) some admixtures will not react the same with all cements, even of the same type.

FIG. 5 This is how a well-graded coarse aggregate looks before (a) and after being separated (b) into three sizes. (b) From left to right: 1/4″ to 3/8″, 3/8″ to 3/4″, 3/4″ to 1-1/2″. (a) Note how the small pieces fit among larger ones in the mixed aggregate.

Accordingly, admixtures should be used with caution since specific effects that will result from their use can seldom be predicted accurately.

The three types of admixtures most frequently used are discussed below.

Accelerators

Accelerators increase the rate of early strength development in concrete to: (1) reduce the waiting time for finishing operations to be started; (2) permit earlier removal of forms and screeds; (3) reduce the required period for curing in certain types of work; (4) advance the time a structure can be placed in service; (5) partially compensate for the slow gain in strength of the concrete even with proper protection during cold weather; (6) reduce the period of protection required for initial and final set in emergency repair and other work.

Under most conditions the common accelerators cause an increase in the drying shrinkage of concrete. In many cases, it must be decided whether to use an admixture, increase the cement content, use high-early-strength cement, provide greater protection or longer curing period, or use any combination of these.

Calcium Chloride generally is used to accelerate the time of set and to increase the rate of strength gain. It should meet the requirements of the Standard Specifications for Calcium Chloride, ASTM D98. The amount used should never exceed 2% by weight of cement. Greater amounts may cause rapid stiffening, increase drying shrinkage and corrode reinforcement. Calcium chloride always should be added in *solution* as part of the mixing water to insure uniform distribution. If it is added in dry form, all of the dry particles may not be completely dissolved during mixing. The undissolved lumps can cause popouts or dark spots in hardened concrete. Calcium chloride should never be considered as an antifreeze. To lower the freezing point of concrete appreciably would require the use of so much calcium chloride that the concrete would be ruined. Instead, protective measures should be taken to prevent the concrete from freezing by insulated covers and forms, and heating the materials and the surrounding air.

The use of calcium chloride or admixtures containing soluble chlorides is *not recommended* under the following conditions: (1) in prestressed concrete because of possible corrosion hazards; (2) in concrete containing embedded aluminum, i.e. conduit, since serious corrosion can result, especially if the aluminum is in contact with embedded steel and the concrete is in a humid environment; (3) in concrete exposed to soils or water containing sulfates; (4) in floor slabs intended to receive dryshake metallic finishes; and (5) in hot weather generally.

Retarders

Admixtures having a retarding effect on the set of cement in concrete are used to overcome the accelerating effect that temperature has on setting during hot weather, and to delay early stiffening action of concrete placed under difficult conditions. In addition, retarder solutions are sometimes applied directly to the surface of the concrete to retard the set of a surface layer of mortar so that it can be readily removed by brushing, thus exposing the aggregate and producing textures surface effects.

Because most retarders also act as water reducers, they are frequently called water-reducing retarders. Retarders may also entrain some air in concrete. Many chemicals are mentioned in current literature as having a retarding influence on the normal setting time of portland cement. Some of these have been found variable in action, retarding the set of certain cements and accelerating the set of others. Unless experience has been gained with a retarder to determine the extent of its effect on the setting time and other properties of the concrete, its use as an admixture should not be attempted without technical advice and/or advance experimentation with the cement and other concreting materials involved.

Air-entraining Agents ASTM C260

Air-entraining admixtures used in concrete will, at little or no additional cost, improve the workability and durability and, in the case of exposed flatwork, will produce a concrete resistant to severe frost action and the effects of applications of salt for snow and ice removal.

Properly proportioned air-entrained concrete contains less water per cubic yard than non-air-entrained concrete of the same slump (see page 203-13 for discussion of "Slump Test"), and has better workability. This results in a more solid weather resistant, blemish-free surface. It can be handled and placed with less segregation of materials and less tendency to bleed (the appearance of excess water rising to the surface). These properties indirectly aid in promoting durability by increasing the uniformity of the concrete.

When using air-entraining portland cement or admixtures containing air-entraining agents, care must be taken that the quantity of water in the concrete mix is adjusted to maintain the desired slump. Such mix designs and adjustments should be made only by a qualified technician.

Because of its proven record, both in the laboratory and the field, and because of its increased workability and durability, air-entrained concrete is strongly recommended for all concrete work regardless of exposure conditions. Where freezing and thawing are encountered, and where water tightness is desired, its use should be required. Whether by use of air-entraining portland cement or admixture, a proper per-

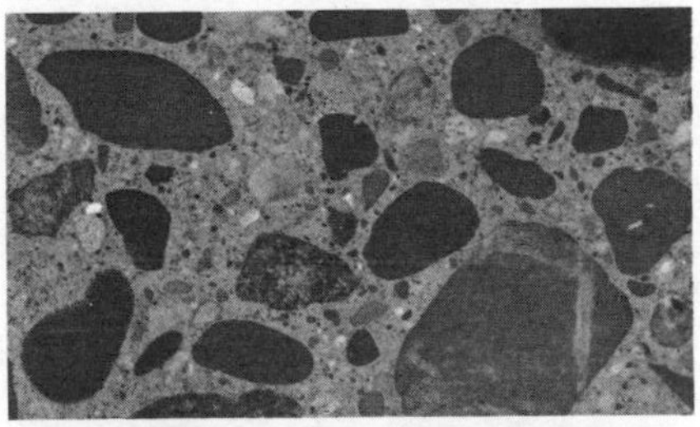

FIG. 7 Cross-section of concrete. Cement paste completely surrounds each aggregate particle and fills all spaces between particles. The quality of this paste largely establishes the quality of the concrete.

centage of entrained air in the mix should be obtained, depending upon the maximum size of aggregate as shown in the Work File.

Specifications and methods of testing air-entraining admixtures are given in ASTM C260. Air-entraining additions must meet requirements of ASTM C226.

REINFORCEMENT

Often steel reinforcing is required to increase the tensile strength of concrete or reduce volume change due to drying shrinkage. Reinforcing should conform to the applicable specifications of the ASTM listed in the Work File.

Protection of reinforcing steel is required from both corrosion and fire (see Work File for the covering thicknesses of concrete that should be provided over reinforcement).

The correct mixture of materials will determine the desirable qualities of the *plastic* concrete—consistency, workability and finishability; as well as the essential properties of the *hardened* concrete—strength, durability, watertightness, wear-resistance and economy.

In properly made concrete, each particle of aggregate, no matter how large or small, is surrounded completely by the cement-water paste, and all spaces between aggregate particles are filled completely with paste. This is illustrated in Figure 7, which shows a section cut through hardened concrete. The aggregates are considered as inert materials while the paste is the cementing medium which binds the aggregate particles into a solid mass. The quality of the concrete, therefore, is greatly dependent on the quality of the paste, and it is important that the paste itself have the strength, durability and watertightness required by the job.

WATER-CEMENT RATIO

The *water-cement ratio* is the ratio, by weight, between these two ingredients. Formerly, this propor-

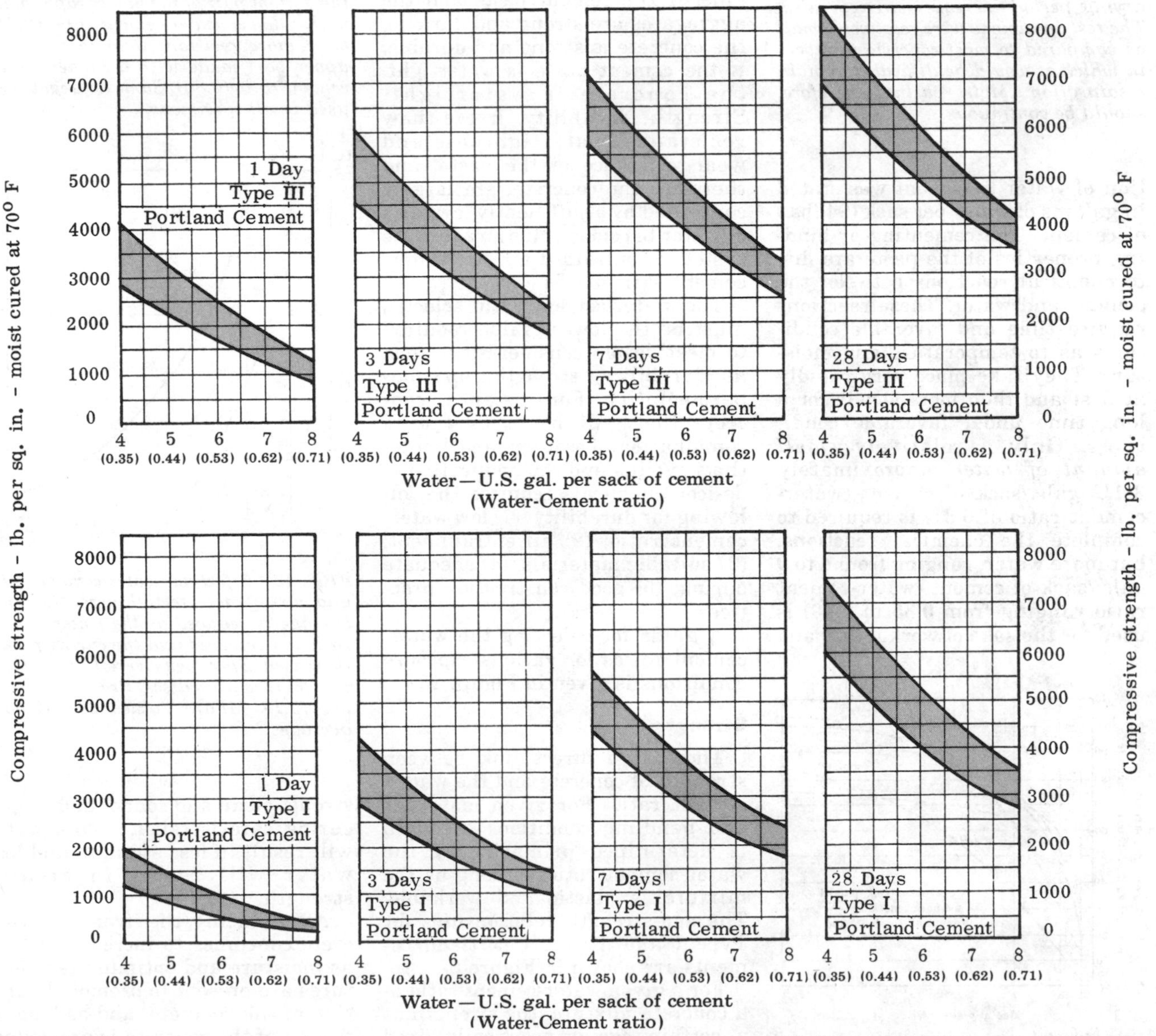

FIG. 8 Age-compressive strength relation for Types I and III portland cements. A large majority of the tests for compressive strength made by many laboratories, using a variety of materials complying with ASTM specifications, is the area within the banded curves.

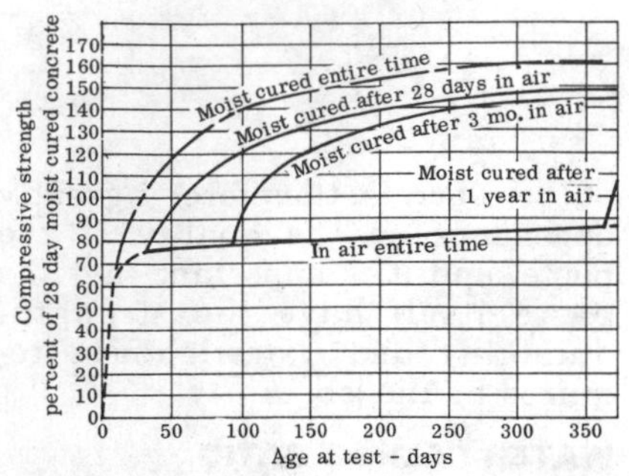

FIG. 9 Strength of concrete continues to increase as long as moisture is present to promote hydration of the cement. Note that resumption in moist curing after drying period increases strength also. The test specimens were relatively small as compared to most concrete members in which it would be difficult to obtain resaturation. Moist curing, therefore, should be continuous.

tion of water to cement was stated in gallons of water per sack (94 lbs.) of cement. The cementing or binding properties of the paste are due to chemical reactions between the cement and water. These reactions require time and favorable conditions as to temperature and moisture. They take place very rapidly at first and then more slowly for a long time under favorable conditions. Only a relatively *small amount of water,* approximately 3-1/2 gals./sack of cement (water-cement ratio of 0.31) is required to complete the chemical reactions; but more water, ranging from 4 to 9 gals./sack of cement (water-cement ratio ranging from 0.35 to 0.80) is used for the sake of workability and

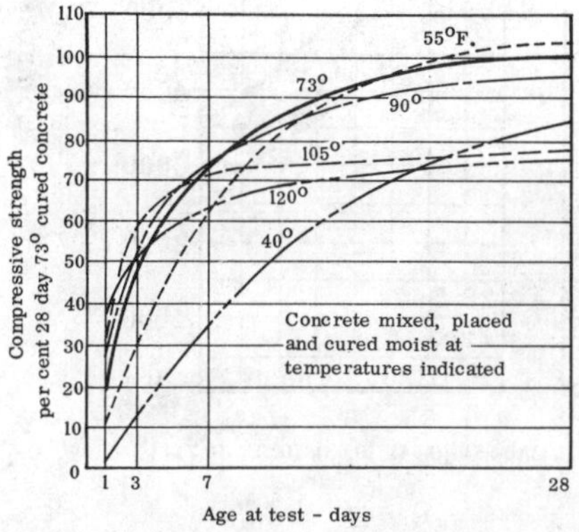

FIG. 10 The strength and other properties of concrete are affected by the temperature.

placeability, and with more water more aggregate can be used with resulting economy.

For successful results, a proper proportion of water to cement is essential. If too much water is added (a high water-cement ratio), the paste becomes thin or diluted and will be weak and porous when it hardens. Not enough water may result in a mix that cannot be properly placed and finished. Cement paste made with the correct amount of water has strong binding qualities, is watertight and durable. If the cement paste and the aggregates are strong and durable, the concrete is strong and durable. If the cement paste is watertight, the concrete is watertight. Strength, durability, freeze-thaw resistance, watertightness and wear-resistance of the paste, and therefore the concrete, are largely controlled by a sufficiently low ratio of water to cement plus an adequate system of entrained air, the water-cement ratio.

The water-cement ratio selected must be the lowest value required to meet design considerations such as durability, strength, and impermeability. For instance, concrete that will be exposed to a combination of wet-dry and freeze-thaw cycling and the application of de-icer chemicals requires the following for durability: (1) low water-cement ratio; (2) air-entrainment; (3) suitable materials; (4) adequate curing; (5) good construction practices.

A guide for selecting the water-cement ratios for various exposure conditions is given in Figure 13.

Strength

There is a direct link between strength of concrete and the water-cement ratio. For given materials and handling conditions, strength is determined primarily by the water-cement ratio, as long as the mixture is plastic and workable. The age-strength relationships for Type I and Type III portland cements are shown in Figure 8.

For a given water-cement ratio in a concrete mixture, the strength at a certain age is practically fixed (assuming that the mixture is plastic and workable, aggregates are strong, clean and sound, and proper curing care is taken). More water will result in less strength and less water will result in greater strength.

FIG. 11 The durability of concrete exposed to freezing and thawing is affected by the quality of the paste. Specimens in the bottom part of the photograph, made with a lower water-cement ratio (0.64), were more resistant than those in the upper part made with a higher water-cement ratio (0.80). Same aggregate was used in all specimens.

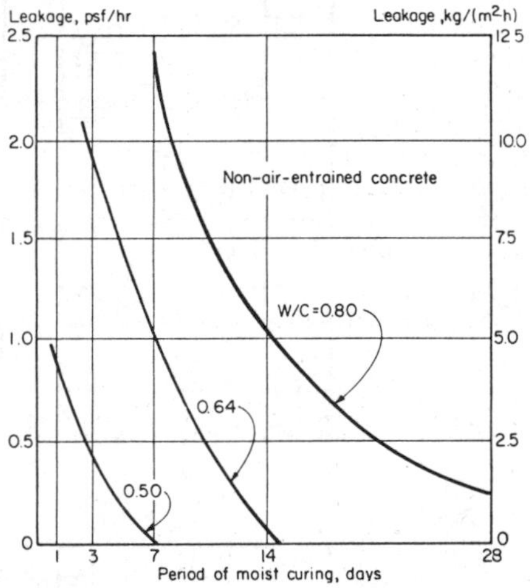

FIG. 12 Effect of water-cement ratio and curing on watertightness. Note that leakage is reduced as the water-cement ratio is decreased and the curing period increased. Specimens were 1 x 6 in. (25 x 150 mm) mortar discs. Pressure was 20 psi (0.14 MPa). Values are 48-hour averages.

After placing, the strength of concrete continues to increase as long as moisture and optimum temperatures are present to promote hydration of the concrete, and as long as drying of the concrete is prevented. When the concrete is permitted to dry, the chemical reactions between the cement and water cease and lower strengths will result than if the concrete is kept moist. Therefore, it is desirable to keep concrete constantly moist as long as possible after placing (Fig. 9). The temperature at which the concrete is made and cured also affects the rate of the chemical reactions between the cement and water, 73° F being the normal curing temperature. At temperatures above 73° F, the increasing strength during curing is higher than the normal curve the first few days, but much lower at later periods (Fig. 10). Concrete made and cured at 40° F would have lower than normal strengths early and attain full strength at later ages.

Durability

There is a direct link between the durability of the concrete and the relative quantities of water and cement in the mixture. The most destructive of the natural forces of weathering is freezing and thawing action while concrete is wet or moist. This is due to the expansion of water as it is converted into ice. If the paste in the concrete is made with a low water-cement ratio, it will be much more resistant to damage by freezing and thawing than if a higher water-cement ratio is used. This is demonstrated by the concrete cubes in Figure 11, which have been subjected to 70 cycles of freezing and thawing while water saturated. Those in the upper portion made with a water-cement ratio of 0.80 show much more disintegration than those in the lower portion made with a water-cement ratio of 0.64. Furthermore, introduction of entrained air into the concrete will significantly increase its resistance to freezing and thawing damage and the application of de-icer salts.

Watertightness

Impervious concrete requires a watertight or impermeable cement paste, and as with strength and durability, there is a direct link between the watertightness of the concrete and the relative quantities of water and cement in the mixture.

Tests show that the permeability or watertightness of the paste is dependent on the amount of mixing water used and the extent to which the chemical reactions between the cement and water have progressed. The results of subjecting discs made with cement mortar (i.e., portland cement, fine aggregate and water) to 20 psi water pressure are shown in Figure 12.

In these tests, mortar cured moist for 7 days had no leakage when made with a water-cement ratio of 0.50. There was considerable leakage with mortars made with the higher water-cement ratios. Also, in each case leakage became less as the length of curing period was increased. Mortar discs with a water-cement ratio of 0.80 leaked, even though they were moist-cured for one month.

Air entrainment improves watertightness by allowing reduction of the water-cement ratio. To be watertight, concrete must also be free from cracks and honeycombing.

As a result of the test work that has been done and the experience that has been gained in the field, definite recommendations can be made regarding the maximum amount of mixing water that should be used for various construction applications (Fig. 13).

FIG. 13 RECOMMENDED CONCRETE MIXTURES FOR VARIOUS EXPOSURE CONDITIONS

<table>
<tr><th rowspan="2">Concrete Construction Element</th><th rowspan="2">Exposure Condition[1]</th><th rowspan="2">Minimum Compressive Strength @ 28 days,[2] plant or transit (psi)</th><th colspan="2">Practical Water-cement Ratio by Weight</th><th rowspan="2">Normal Maximum Coarse Aggregate Size[5] (in.)</th><th rowspan="2">Minimum Cement[6] Content (lbs. per cu. yd.)</th><th rowspan="2">Air-entrainment by Volume (%)</th><th rowspan="2">Slump (in.)</th></tr>
<tr><th>Non-air entrained concrete</th><th>Air-entrained concrete</th></tr>
<tr><td rowspan="3">Foundations, basement walls and slabs not exposed to weather</td><td>Severe</td><td rowspan="3">3000[3]</td><td rowspan="4">See footnote 3</td><td rowspan="3">0.55</td><td rowspan="3">1</td><td>564</td><td rowspan="3">See footnote 6</td><td rowspan="12">5" maximum for hand methods of strike-off and consolidation; 3" maximum for mechanical strike-off and consolidation.</td></tr>
<tr><td>Moderate</td><td rowspan="2">520</td></tr>
<tr><td>Negligible</td></tr>
<tr><td rowspan="3">Foundations, basement walls, exterior walls and other concrete work exposed to weather</td><td>Severe</td><td>3500[3]</td><td>0.45</td><td rowspan="3">1</td><td>564</td><td>6 to 8</td></tr>
<tr><td>Moderate</td><td>3000</td><td>0.58</td><td>0.50</td><td rowspan="2">520</td><td>5 to 7</td></tr>
<tr><td>Negligible</td><td>2500</td><td>0.67</td><td>0.55</td><td>See footnote 6</td></tr>
<tr><td rowspan="3">Concrete exposed to sulfate attack</td><td>Severe</td><td rowspan="2">5000[3]</td><td rowspan="3">See footnote 4</td><td rowspan="3">See footnote 4</td><td rowspan="3">1</td><td>564</td><td>6 to 8</td></tr>
<tr><td>Moderate</td><td rowspan="2">520</td><td>5 to 7</td></tr>
<tr><td>Negligible</td><td>5000</td><td>See footnote 6</td></tr>
<tr><td rowspan="3">Driveways, garage floors, walks, porches, patios and stairs exposed to weather</td><td>Severe</td><td>4000[3]</td><td rowspan="2">See footnote 3</td><td>0.45</td><td rowspan="3">1</td><td>564</td><td>6 to 8</td></tr>
<tr><td>Moderate</td><td>3000[3]</td><td>0.50</td><td rowspan="2">520</td><td>5 to 7</td></tr>
<tr><td>Negligible</td><td>2500</td><td>0.67</td><td>0.55</td><td>See footnote 6</td></tr>
</table>

[1]See for severity of exposure, regional weathering areas.
[2]With most materials, the water-cement ratios shown will provide average strengths greater than required.
[3]Use air-entrained concrete only.
[4]Proportions should be established by the trial batch method.
[5]Maximum size of coarse aggregate should not be larger than 1/5 the narrowest dimension between forms, nor larger than 3/4 the minimum clear spacing between reinforcing bars.
[6]2% to 3% air entrainment, to improve cohesiveness and reduce bleeding, is recommended but not essential.

SELECTING THE MIX

Unless concrete mixtures are established by tests in advance of the construction work, the concrete mixtures shown in Figure 13 should be used. It is first necessary to determine the exposure condition of the concrete construction from the regional weathering areas shown in Figure 14. Areas shown as *severe exposure* are subject to many freeze-thaw cycles per year and, in the case of flat concrete work, the application of deicer chemicals; *moderate exposure* areas are subject to few freeze-thaw cycles per year and no application of deicer chemicals; *negligible exposure* areas have no freeze-thaw cycles per year and deicer chemicals applied.

When the exposure condition of the concrete construction element is known, it is possible to determine the recommended *minimum compressive strength* and the maximum permissible *water-cement ratio* to obtain the specified minimum compressive strength. The minimum cment content is included in addition to the lowest maximum water-cement content is included in addiquirements ensure satisfactory finishability, improved water resistance of slabs, and suitable appearance of vertical surfaces. The percentage of air entrainment for various maximum aggregate sizes is also given; slumps should be in the ranges below.

For example, to select the minimum compressive strength for basement walls not exposed to weather, to be built in the Chicagoland area (that is, in severe weathering region, Figs. 13 and 14) would require concrete strength of 3000 psi, with a maximum water-cement ratio of 0.46, and 6% to 8% air-entrained concrete. The minimum cement content likely to be required to achieve that strength at the recommended air-entrainment level is 565 lbs. per cu. yd.

While 2500 psi concrete can satisfactorily support typical vertical and lateral loading in a basement wall, the lower strength concrete will be less watertight than the concrete recommended herein. The higher strength concrete provides a margin of safety against the possibility of foundation creacking and leaking.

FIG. 14 **REGIONAL WEATHERING AREAS***

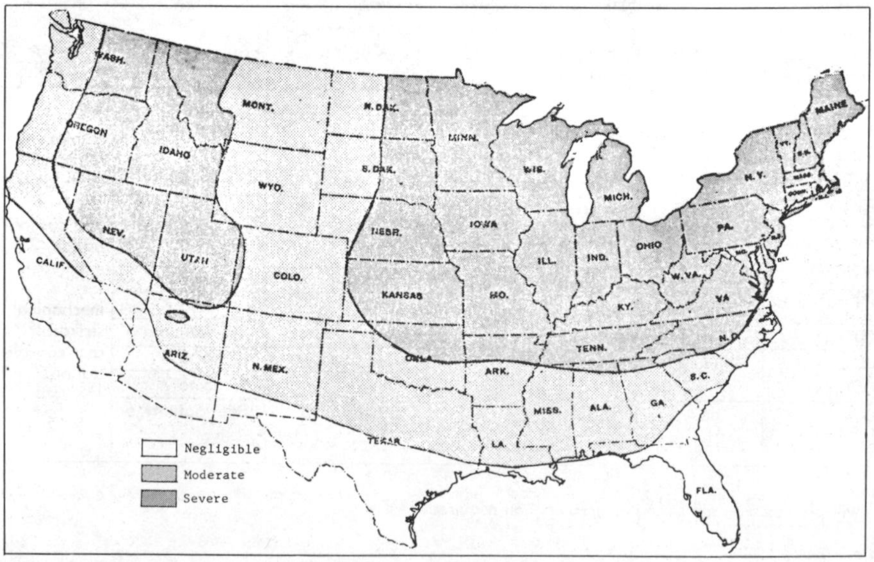

*Areas are based upon ASTM C62 weathering index map adjusted to conform with an experience survey by HUD and the Portland Cement Association. Local conditions may be more or less severe than indicated by regional classification.

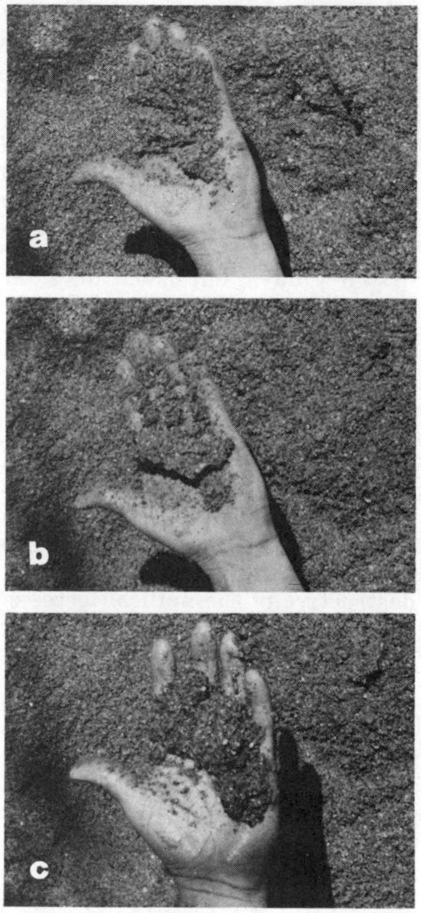

FIG. 15 (a) Damp sand; (b) wet sand; (c) very wet sand.

Allowance for Water in the Aggregate

In selecting the proper water-cement ratio, it is important to note that most fine aggregates contain some water. Therefore, allowance must be made for this in determining the amount of water to be added to the mix. A simple test for determining whether sand is "damp," "wet" or "very wet" is to press some together in your hand. If the sand falls apart after your hand is opened, it is damp (Fig. 15a); if it forms a ball which holds its shape, it is wet (Fig. 15b); if the sand sparkles and wets your hand, it is very wet (Fig. 15c).

The most practical procedure is to make trial mixes. On large jobs where many materials or unusual materials may be under consideration, or where strength is particularly important, the trials may be full-size batches checked in the field. The first trial mix may be selected on the basis of experience or established relationships such as those given in Figure 15. This table has been developed from experience and data from several sources and indicates the suggested proportions of all the various ingredients which should result in good workability.

CONSISTENCY OF THE MIX

With a given amount of cement paste more aggregate is used in stiff mixes than in more fluid mixtures; consequently, the stiff mixes are more economical in cost of the materials. Stiff mixes require more labor in placing, however, and when the mixture is too stiff the additional cost of placing may offset any savings in materials. Mixtures always should be of a consistency and workability suitable to allow the concrete to be worked into the angles and corners of forms, and around reinforcement without permitting the materials to segregate or excess free water to collect on the surface. Thin members and heavily reinforced members require more fluid mixtures than large members or flat surfaces containing little reinforcing.

A "plastic" concrete is one that is readily molded and yet will change its form only slowly if the mold is immediately removed. Mixtures of plastic consistency are required for most concrete work. Concrete of such consistency does not crumble but flows sluggishly without segregation. Thus, neither very stiff, crumbly mixes nor very fluid, watery mixes are of plastic consistency (Figs. 16 and 17).

The ease or difficulty in placing concrete in a particular location is referred to as "workability." A stiff but plastic mixture with large aggregate would be workable in a large open form but not in a thin wall with closely spaced reinforcement.

Slump Test (ASTM C143)

The slump test may be used as a rough measure of the consistency of concrete. This test is not to be considered as a measure of workability, nor should it be used to compare mixes of entirely different proportions or containing different kinds of aggregate. Any change in slump on the job indicates changes have been made in grading, or proportions of the aggregate or in water content. The mix should be corrected immediately to get the proper consistency by adjusting amounts and proportions of sand and coarse aggregate, care being taken not to change the total amount of water specified for mixing with each sack of cement. For the various

FIG. 17
Stiff, medium and wet mixtures of concrete. For foundation walls, pavements, floors and work of like character, stiff consistency is recommended. The medium mix will be found suitable for floors, slabs and beams. The wet mix may be required for very thin, heavily reinforced walls.

FIG. 16

a. *A concrete mixture in which there is not sufficient cement-sand mortar to fill all the spaces between coarse aggregate particles. Such a mixture will be difficult to handle and place, and will result in rough honeycombed surfaces and porous concrete.*

b. *A correct amount of cement-sand mortar. With light troweling, all spaces between aggregate are filled with mortar. Note appearance on edge of pile. This is a good workable mixture and will give maximum yield of concrete.*

c. *A concrete mixture in which there is an excess of cement-sand mortar. While such a mixture is plastic and workable and will produce smooth surfaces, the yield of concrete will be low and uneconomical. Such concrete also is likely to be porous.*

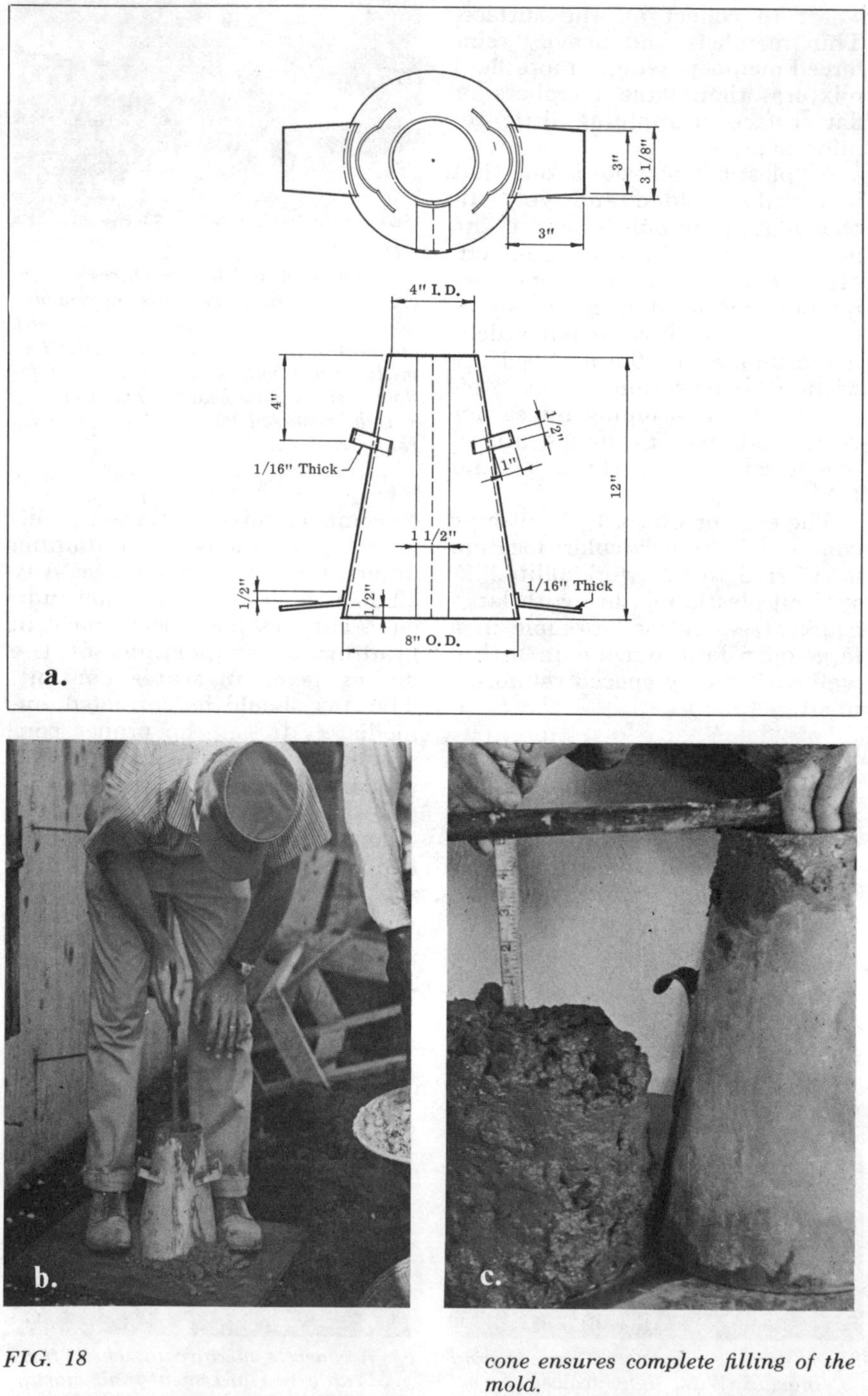

FIG. 18

a. Mold for slump test.

b. The slump test shows consistency of concrete. Here, rodding concrete in cone ensures complete filling of the mold.

c. Slump is measured from rod laid across the top of the slump cone. This amount of slump indicates a medium-wet concrete mixture.

applications in residential construction, the recommended slump varies from 1″ to 6″ (see Fig. 13, column 3).

In making the slump test, the test specimen is made in a mold (slump cone) of 16-gage galvanized metal as the form shown in Figure 18a. The base and top are open. The mold is provided with foot pieces and handles as shown.

When the slump test is made, the concrete sample is taken immediately prior to placing in the forms. The mold is placed on a flat surface, such as a smooth plank or slab of concrete, and is held firmly in place by standing on the foot pieces while filling it with concrete. The mold is filled with concrete to about 1/3 of its height. Then the concrete is puddled with 25 strokes of a 5/8″ rod about 24″ long, bullet pointed at the lower end (Fig. 18b). The filling is completed in 2 more layers, each layer being rodded 25 times and each rod stroke penetrating into the underlying layer. After the top layer has been rodded, it should be struck off with a trowel so that the mold is filled exactly. The mold is removed by gently raising it vertically immediately after being filled.

The slump of the concrete is measured, as shown in Figure 18c, immediately after the cone is removed. For example, if the top of the slump pile is 5″ below the top of the cone, the slump for this concrete is 5″.

Compression Test (ASTM C31)

This test is to determine if the concrete has the specified compressive strength. Field control specimens of various ages determine the rate of strength gain and the effectiveness of job site curing.

In making the compressive strength field test, a sample of the concrete is taken at three or more regular intervals throughout the discharge of the entire batch, except that samples are not to be taken at the beginning or end of discharge. The batch of concrete thus sampled is noted as to its location in the work, the air temperature and any unusual conditions that might be occuring at the time.

The compressive test specimen is made in a cylindrical mold that is watertight to prevent loss of water. Standard cylindrical molds are 6″ in diameter by 12″ in length if the coarse aggregate does not exceed 2″ in nominal size. The mold is filled in three layers. Each layer is puddled with 25 strokes of a 5/8″ round steel rod about 24″ long with bullet-pointed tip. Reinforcing rods or other tools should *not* be used as the puddle rod. After the top layer has been rodded, the surface of the concrete is struck off with a trowel and covered with a glass or metal plate to prevent evaporation.

Standard procedures provide for curing the specimens either in the laboratory or in the field. Laboratory curing gives a more accurate indication of the potential quality of the concrete. Field-cured specimens may give a more accurate interpretation of actual strength in the structure or slab, but they offer no explanation as to whether any lack of strength is due to error in proportioning, poor materials or unfavorable curing conditions. On some jobs both methods are used, especially when the weather is unfavorable, in order to interpret the tests properly. The laboratory test result is the one that prevails.

MEASURING MATERIALS

If uniform batches of concrete of proper proportions and consistency are to be produced, it is essential that all ingredients be measured accurately for each batch. The effects of the water-cement ratio on the qualities of concrete make it just as necessary to measure the water as the other ingredients. A troublesome factor is the effect of the varying amounts of moisture nearly always present in the aggregate, particularly in natural sand. The amount of free moisture introduced in the mixture with the aggregates must be determined and allowance made for it if accurate control is to be obtained.

Measuring Cement

If sacked cement is used, the batches of concrete should be of such size that only full sacks are used. If partial sacks of cement are used, they should be weighed for each batch. It is not satisfactory to divide sacks of cement on the basis of volume. Bulk cement (unsacked) always should be weighed for each batch.

Measuring Water

Dependable and accurate means for measuring the mixing water are essential. Portable mixers generally are equipped with water tanks and measuring devices that are fairly accurate when properly operated. The measuring device most generally used operates on the principle of the siphon. The tank is filled, and the desired amount of water is siphoned off. When the water level reaches the point at which the bottom of the siphon is set, the seal is broken and the water is shut off automatically.

Measuring Aggregates

Measurement of aggregates by volume cannot be depended upon except under most careful supervision. A small amount of moisture in fine aggregate, nearly always present, causes the aggregate to bulk or fluff up as indicated in Figure 19. The amount of bulking varies with the amount of moisture present and the grading; fine sands bulk more than coarse sands for a given amount of moisture. Note in Figure 19 that a fine sand with 5% moisture content bulks or increases in volume almost 40% over its dry volume. On a large proportion of jobs the sand contains nearly the amount of moisture that produces the maximum bulking. Only small changes in moisture can cause ap-

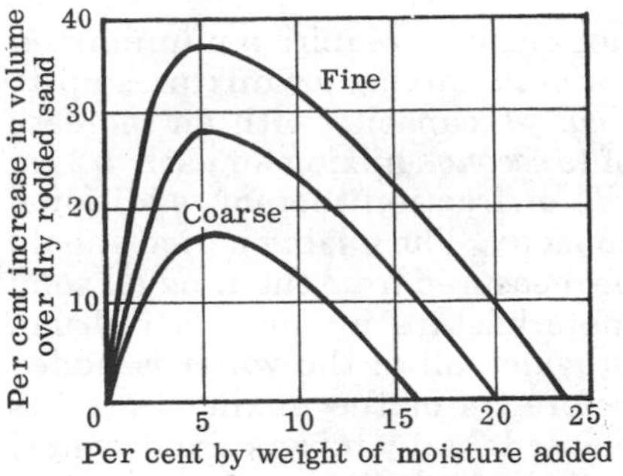

FIG. 19
The surface moisture in fine aggregate causes considerable bulking, the amount of bulking varying with the amount of moisture and the grading of the aggregate.

preciable changes in the amount of bulking. This explains why it is impossible to give accurate estimates of trial mixes by volume. The suggested trial mixes shown in Figure 15 are for a 3″ slump and will vary depending on aggregate density and gradation, as well as size of sand. Hence, the care exercised in ready-mix batching and the need for careful control, trial mixing and testing if field mixing is necessary. *Measurement of aggregates by weight is the recommended practice and should be required on all important work.*

To reduce segregation of aggregates to a minimum and to give uniformity from batch to batch, it is desirable to weigh the coarse aggregate in two or more sizes, especially if the maximum size exceeds 1″. Generally, the ratio of the maximum size particle to the minimum size for coarse aggregate separations should not exceed 2 to 1 for materials larger than 1″, and not exceed 3 to 1 for finer materials. Thus, 1-1/2″ aggregate would be separated into 1/4″ to 3/4″ and 3/4″ to 1-1/2″ sizes.

MIXING

All concrete should be mixed thoroughly until it is uniform in appearance with all ingredients uniformly distributed. The time required for thorough mixing depends on several factors. Specifications usually require a minimum of *1 minute* mixing for mixtures up to *1 cu. yd.* capacity with an increase of *15 seconds* mixing for each *1/2 cu. yd.*, or fraction thereof, additional capacity. The mixing period should be measured from the time all solid materials are in the mixer drum, provided all of the water is added before 1/4 of the mixing time has elapsed. Many mixers are provided with time devices; some of these can be set for a given mixing time and locked so that the batch cannot be discharged until the designated time has elapsed.

Mixers should not be loaded above their rated capacity and should be operated at approximately the speeds for which they are designed. If the blades of the mixer become worn or coated with hardened concrete, the mixing action will be less efficient. Badly worn blades should be replaced, and hardened concrete should be removed before each run of concrete.

Under usual operating conditions, up to about 10% of the mixing water should be placed in the drum before the aggregates and cement are added. Water should then be added uniformly with the dry materials, leaving about 10% to be added after all other materials are in the drum. When heated water is used during cold weather, this order of charging may require some modification to prevent flash setting of the cement. In this case, addition of the cement should be delayed until most of the aggregate and water have intermixed in the drum.

Ready-Mixed Concrete

In most areas ready-mixed concrete is purchased directly from a central plant. The Standard Specifications for Ready-Mixed Concrete (ASTM C94) require that the hauling be done in agitator trucks or transit mixer trucks operated at agitator speed.

In some ready-mixed operations the materials are dry-batched at the central plant and then mixed enroute to the job in truck mixers. In another procedure, the concrete is mixed in a stationary mixer at the central plant only sufficiently to intermingle the ingredients, generally about 1/2 minute. The mixing then is completed in a truck mixer enroute to the job.

Truck mixers consist essentially of a mixer with separate water tank and water measuring device mounted on a truck chassis. They usually are made with capacities of 1 to 5 cu. yds. Agitator trucks are similar but without water.

ASTM C94 requires that the concrete must be delivered and discharged from the truck mixer or agitator truck within *1-1/2 hours* after introduction of the water to the cement and aggregate.

Remixing Concrete

Appreciable initial set of concrete does not take place ordinarily until 2 or 3 hours after the cement is mixed with water. Fresh concrete that is left standing tends to dry out and stiffen somewhat before the cement sets. Such concrete may be used if upon remixing it becomes sufficiently plastic that it can be completely compacted in the forms. However, *adding water to make the mixture more workable (re-tempering) should never be allowed,* for this lowers the quality just as would a larger amount of water in the original mixing.

HANDLING AND PLACING

Each step in handling, transporting and placing the concrete should be controlled carefully to maintain *uniformity* within the batch and from batch to batch so that the complete structure has uniform quality throughout. It is essential to avoid separation of the coarse aggregate from the mortar or of water from the other ingredients. Segregation at the point of discharge from the mixer can be avoided by providing a down pipe at the end of the chute so that the concrete will drop vertically into the center of the receiving bucket, hopper or form.

Concrete is handled and transported by many methods such as chutes, push buggies, buckets handled by cranes and by pumping through a pipeline. The method of handling and transporting concrete and the equipment used should not place a restriction on the consistency of the concrete. This should be governed by the placing conditions. If placing conditions permit the use of a stiff mix, the equipment should be designed and arranged to facilitate handling and transporting such a mix.

PREPARATION

Before placing concrete, the subgrade should be properly prepared and the forms and reinforcing should be erected as required for the structure. Subgrades should be trimmed to specified elevations, uniformly compacted, and should be moist when the concrete is placed (Fig. 20). A moist subgrade is especially important to prevent too rapid extraction of water from the concrete when pavements, floors and similar work are being placed in hot weather.

FIG. 20 Subgrade should be moist when concrete is placed.

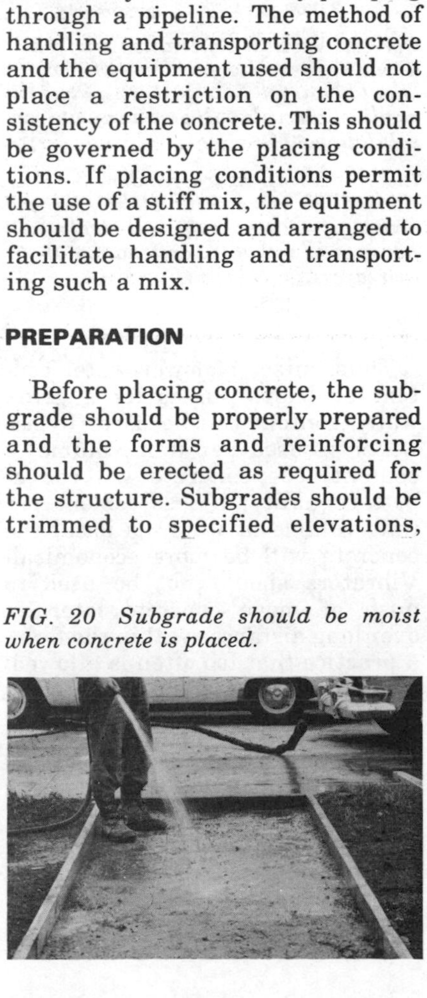

In cold weather, the subgrade must not be frozen. Snow, ice and debris must be removed from within the forms before concrete is placed.

If concrete is placed on rock, all loose material must be removed and cut faces should be nearly vertical or horizontal rather than sloping.

Forms should be clean, tight and adequately braced and constructed of material that will impart the desired texture to the finished concrete. Care should be taken to see that sawdust, nails and other debris are removed from the spaces to be concreted. To facilitate form removal, forms should be treated with a release agent such as oil or lacquer. For architectural concrete the agent should be a nonstaining lacquer or emulsified stearate. Wood forms should be moistened before placing concrete, otherwise they will absorb water from the concrete and swell.

Reinforcing steel should be clean and free of loose rust or mill scale at the time concrete is placed. Any coatings of hardened mortar should be removed from the steel.

PLACING

Concrete should be placed as nearly as practicable in its final position. It should not be placed in large quantities at a given point and allowed to run or be worked over a long distance in the form. This practice results in segregation because the mortar tends to flow out ahead of the coarser material. It also results in sloping work planes or potential weak bonding between successive layers of concrete. In general, concrete should be placed in horizontal layers of uniform thickness, each layer thoroughly compacted before the next is placed. Layers should be 6″ to 20″ thick except that thinner layers may be preferred if the forms are narrow and heavily reinforced. Each layer must be placed before the previous one stiffens.

Drop chutes will prevent spattering of mortar on reinforcement and forms. If the placement can be completed before the mortar dries, drop chutes may not be needed. Height should be limited so as to minimize

FIG. 21

PLACING IN TOP OF NARROW FORM

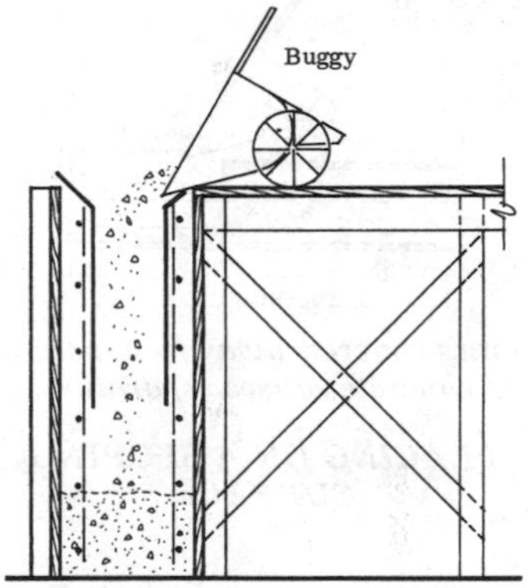

CORRECT

Separation is avoided by discharging concrete into hopper feeding into drop chute. This arrangement also keeps forms and steel clean until concrete covers them.

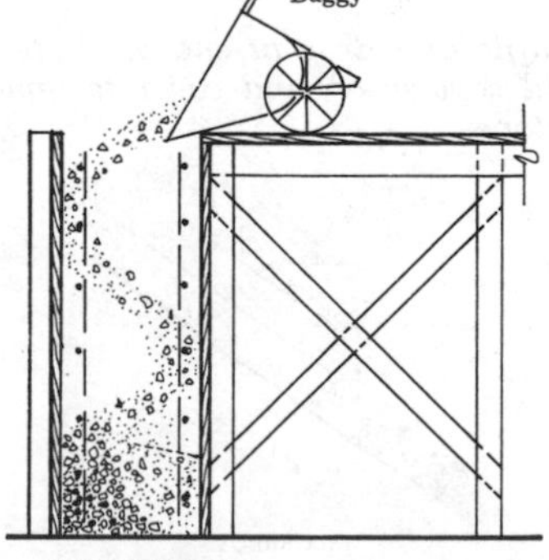

INCORRECT

Permitting concrete from chute or buggy to strike against form and richochet on bars and form faces causes separation and honeycomb at the bottom.

the likelihood of separation. (Fig. 21).

SLABS

In slab construction, placing of the concrete should be started at the far end of the work so that each batch will be dumped against previously placed concrete, not away from it (Fig. 22). The concrete should *not* be dumped in separate piles and then leveled and worked together. If stone pockets occur, some of the excess large particles should be removed and distributed to areas where there is more mortar present to surround them.

FIG. 22

PLACING SLAB CONCRETE

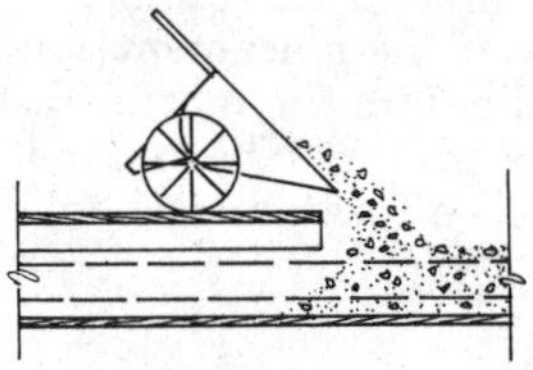

CORRECT

Concrete should be dumped into face of previously placed concrete.

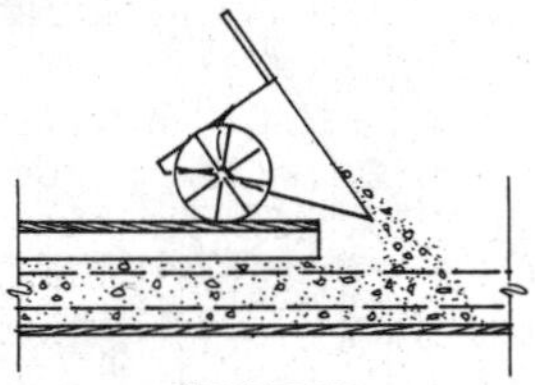

INCORRECT

Dumping concrete away from previously placed concrete causes separation.

PLACING ON A SLOPING SURFACE

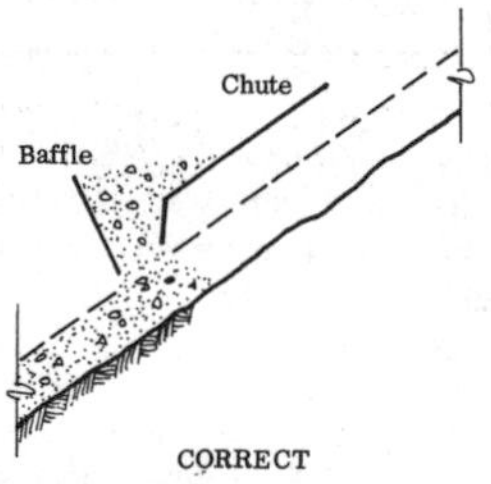

CORRECT

A baffle and drop at end of chute will avoid separation and concrete remains on slope.

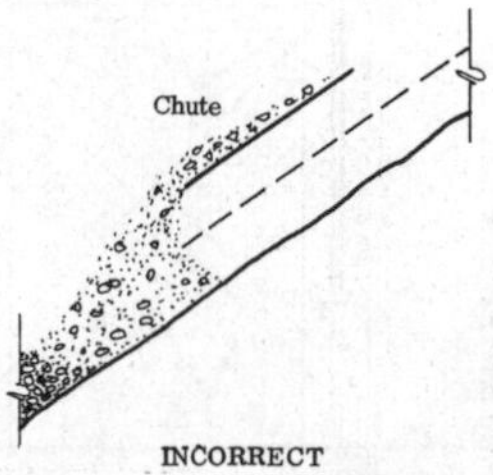

INCORRECT

Discharging concrete from free end chute onto a slope causes separation of rock which goes to bottom of slope. Velocity tends to carry concrete down the slope.

Order of Placing

In walls, the first batches should be placed at either end of the section; the placing should then progress toward the center. This same procedure should be used for each layer. In large flat open areas, the first batches should be placed around the perimeter. In all cases, the procedure should prevent water from collecting at the ends and corners of forms, and along form faces.

Compacting

Medium to high slump concrete should be compacted and worked into place by spading or puddling. Spades or long sticks, long enough to reach the bottom of the forms and thin enough to pass between the reinforcing steel and forms, should be used. Low to medium slump concrete should be compacted using mechanical vibrators either within the concrete or on the forms.

Either method should eliminate stone pockets and large bubbles of air, consolidate each layer with that previously placed, completely embed reinforcing and fixtures, and bring just enough fine material to the faces and top surfaces to produce the proper finish.

Vibrating

Mechanical vibration of itself does not make concrete stronger, more watertight or more resistant to deteriorating forces. It does permit the use of stiffer, harsher mixes.

When concrete is vibrated, the internal friction between the coarse aggregate particles is temporarily destroyed and the concrete behaves as a liquid; it settles in the forms under the action of gravity and the entrapped air bubbles rise more easily to the surface. Friction is reestablished as soon as vibration stops.

FIG. 23

WHEN CONCRETE MUST BE PLACED IN A SLOPING LIFT

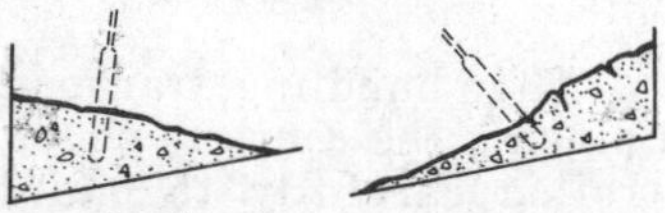

CORRECT INCORRECT

Correct: Start placing at bottom of slope so that compaction is increased by weight of newly added concrete. Vibration consolidates the concrete.

Incorrect: When placing is begun at top of slope, the upper concrete tends to pull apart, especially when vibrated below because this starts flow and removes support from concrete above.

SYSTEMATIC VIBRATION OF EACH NEW LIFT

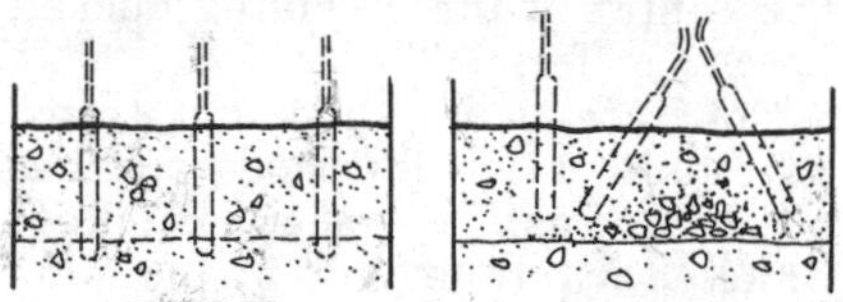

CORRECT INCORRECT

Correct: Vertical penetration of vibrator a few inches into previous lift (which should not yet be rigid) at systematic regular intervals will give adequate consolidation.

Incorrect: Haphazard random penetration of the vibrator at all angles and spacings without sufficient depth will not assure intimate combination of the two layers.

Thus, mixes of lower water content or leaner mixes for a given water content can be used. If less water is used per unit volume of cement, the concrete will be of better quality; if less cement per unit volume of water is used, the concrete will be more economical. Vibrators should not be used to push or move concrete laterally over long distances within the form, a practice that too often is allowed. The concrete should be deposited as near its final position as possible. It should be distributed in layers and then vibrated (Fig. 23). Some hand spading or puddling may be necessary along with the vibration to secure smooth surfaces and reduce pitting on formed surfaces.

Precautions should be taken not to over-vibrate to the point that segregation results. This is to be guarded against especially if the concrete is wetter than necessary for vibration. Precaution and judgment must be used to be sure that complete consolidation is secured without segregation, and that no areas are missed. Sufficient vibration usually is indicated by a line of mortar along the forms and by the submerging of the coarse aggregate particles in the mortar.

Placing on Hardened Concrete

When fresh concrete is placed on hardened concrete, it is desirable to secure good bond and a watertight joint. To accomplish these results, the hardened concrete should be fairly level, reasonably rough, clean and moist, and some aggregate particles should be exposed. Any laitance or soft layer of mortar should be removed from the top surface of the hardened concrete. (An appreciable thickness of such laitance indicates that segregation and bleeding have occurred. It should be prevented by using a stiffer mix or more fine material, particularly in the upper part of the lift.)

On floor slabs to be built in two courses, the top of the lower course may be broomed just before it sets with a stiff fiber or steel broom. The surface shoud be level, scored and free of laitance. It must then be protected and thoroughly cleaned just before the second course is placed.

When new concrete is to be placed on old concrete, the old concrete must be roughened and cleaned thoroughly. In most cases, it is necessary to remove the entire surface to expose a new surface satisfactory for bonding.

Hardened concrete should be moistened thoroughly before new concrete is placed on it. Where the concrete has dried out, it is necessary to saturate it for several days. There should be no pools of water, however, when the new concrete is placed.

Where concrete is to be placed on hardened concrete or on brick, a layer of mortar on the hard surface is necessary to provide a cushion against which the new concrete can be placed; this prevents stone pockets and secures a tight joint. The mortar should be made with the same content as the concrete and should have a slump of about 6″. It should be placed to a thickness of 1/2″ to 1″ and should be well worked into the irregularities of the hard surface. In two-course floor construction, a coat of cement and water paste of the consistency of thick paint should be brushed into the hard surface just before the second course is placed.

JOINTS

Good jointing practice is one way of ensuring crack-free concrete slabs. Joints are necessary in concrete work to allow for movement and volume changes; the handling, placing and finishing of conveniently sized areas, and the separation or isolation of independent elements.

As concrete sets or hardens, the excess mixing water is lost through evaporation and through hydration of the cement particles. This initial loss of water in the early life of the concrete section results in *shrinkage* that is of greater magnitude than any subsequent increase in the size of the hardened mass due to temperature or moisture content rise. The amount of thermal movement in concrete can be figured using its approximate coefficient of expansion (and contraction) of 0.0000055″ per inch per degree F. For example, an expansion of 0.66″ in 100′ of length for a 100° F rise in temperature can be expected when concrete is unrestrained; the same amount of contraction occurs with a 100° F drop. (Concrete placed at 50° F in the spring and rising to 150° F during summer exposure would be an extreme example of concrete subject to a 100° F temperature rise.) However, concrete shortens about 0.72″ per 100′ while drying from its saturated condition at placing to an average hardened state (moisture content equilibrium with air at 50% relative humidity). This shrinkage slightly *exceeds* the expansion caused by such an extreme increase in temperature as 100° F. Furthermore, in most outdoor applications concrete reaches its maximum moisture content (and moisture *expansion*) during the season of low temperature (and thermal *contraction*). Thus, the volume changes due to moisture and temperature variations frequently tend to offset each other. There is, therefore, no need for "expansion joints" in residential construction. The common reference to "expansion" is misleading because it infers that an increase in size after placement must be allowed for. *Concrete is at its greatest mass when it is placed.*

Control Joints

Shrinkage is inevitable as water eventually dries from freshly placed concrete and, consequently, large areas can develop jagged irregularly spaced shrinkage lines and cracks. By anticipating shrinkage and installing joints to limit areas and control where cracking occurs, the concrete work can be made more attractive, serviceable and free from unsightly and expected random cracking. Such joints are called control joints. These joints provide a break or a reduction in slab thickness and thus create a weakened section that induces movement and cracking to occur at that location. When the concrete shrinks, the cracks in these joints can open slightly, thus preventing irregular and unsightly random cracks.

The maximum spacing between control joints depends on slab thickness, shrinkage potential of the concrete, curing environment, and absence or presence of distributed reinforcement.

FIG. 24 Jointing plan for typical floor on ground. The double line around the perimeter indicates a keyed construction joint.

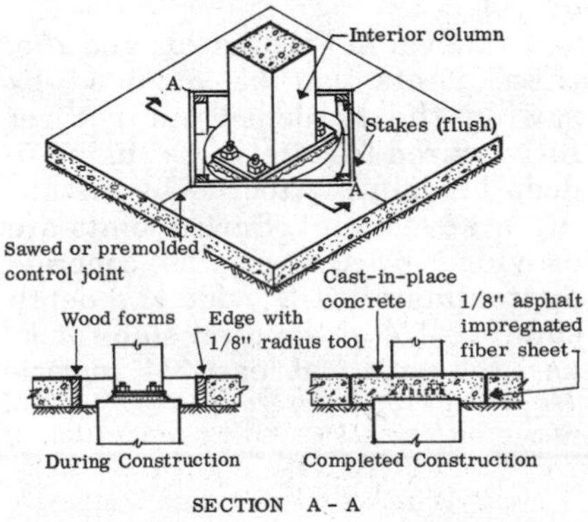

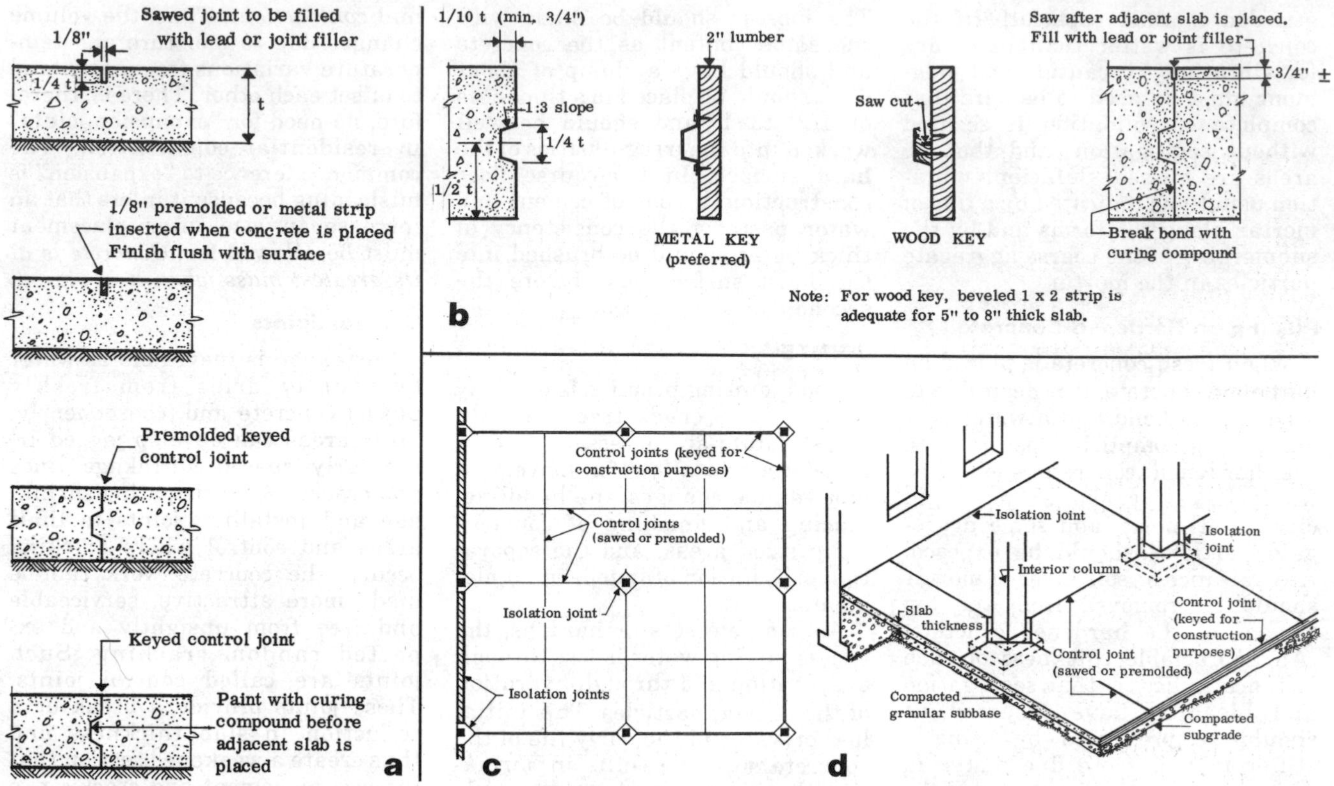

FIG. 25 (a) Several types of control joints for floors on ground. (b) Keyed joint detail for floors on ground. Using too large a key may cause a crack parallel to the joint; if the key is too small, it may shear off, resulting in loss of load transfer capacity. The horizontal saw cut permits the wood key strip to swell slightly without damaging the concrete, and it allows for easy removal of the strip. The joint should be filled with lead or joint filler, depending on traffic conditions. (c) Isolation joint for interior column. Column is boxed out so that the corners meet the control joints. This eliminates cracks radiating from column corners. Circular fiberboard forms could be used in place of the square wood forms. (d) Components of a floor on ground.

Control joints spacing should not exceed maximum intervals of from 10′ to 15′ (Fig. 24). A rule of thumb for plain concrete slabs is that joint spacing (in feet) should not exceed 2 slab thicknesses (in inches) for unreinforced concrete made with coarse 3/4″ maximum aggregate. Thus, a 4″ plain slab would require control joints at intervals not to exceed 8′.

As shown in Figures 25a and 25b, these joints can be provided by sawing the hardened but not yet fully cured concrete; by using a deep bit jointing tool or by installing a keyed joint. Sawed joints are provided by cutting the concrete approximately 1/8″ wide to a depth equal to 1/4 of the total slab thickness (generally at least 3/4″ in residential slab work), but at least equal to the maximum size of the aggregate. Tool joints should be of similar minimum depth.

Construction Joints

In large construction work, construction joints are necessary because of the inability to handle, place and finish a large area in one operation. By installing construction joints, concrete can be conveniently and practically placed in several operations without any appreciable loss in appearance or performance of the complete job. Although they may double in use as control joints or isolation joints, they must allow no vertical movement in the completed floor and are often of "keyed" design as shown in Figure 25b.

Isolation and Separation Joints

Isolation joints are often necessary to separate (i.e., prevent the bonding of) one concrete section from another, or from another material or structural part so that one can move independently from the other. They are usually formed by installing 1/8″ thick (or slightly larger) asphalt impregnated fiber sheets in floors at columns, footings and at the junctures between floors and walls. Figure 25c shows how isolation joints are formed around a column. Similar consideration around the perimeter of a slab should be given as shown in Figure 25d.

FINISHING

GENERAL PROCEDURE

Concrete may be finished in several ways depending on the esthetic effect desired. General finishing procedures are summarized first, followed by techniques to produce special surface finishes.

Bleeding

Generally, all the dry materials used in making quality concrete are heavier than water. Thus, shortly after placement, these materials will have a tendency to settle to the bottom and displace the mixing water to the surface. This occurs more easily with non-air-entrained concrete.

It is of utmost importance that the first operations of placing, screeding (striking off excess concrete) and darbying (bringing surface to true grade with enough mortar to produce the desired finish) be performed before any bleeding takes place.

The concrete should not be allowed to remain in wheelbarrows, buggies or buckets any longer than is absolutely necessary. It should be dumped and spread as soon as possible and struck off to proper grade, then immediately screeded, followed at once by darbying.

If any operation is performed on the surface while the bled water is present, serious *scaling, dusting* or *crazing* of the hardened concrete can result. This point cannot be over-emphasized and is the basic rule for successful finishing of concrete surfaces. (A further explanation of the causes and cures of scaling, crazing and dusting is found under "Surface Defects".)

Screeding

The surface is struck off or rodded by moving a straight-edge back and forth with a sawlike motion across the top of the forms and screeds. A small amount of concrete always should be kept ahead of the straight edge to fill in low spots and maintain a plane surface. (Fig. 26).

Darbying

After the concrete has been struck off or rodded, it is smoothed with a darby to level any raised spots and fill depressions (Fig. 27). Long handled floats of either wood or metal are used sometimes instead of darbies to smooth and level the concrete surface (Fig. 28).

Edging

When all bled water and water sheen have left the surface and the concrete has started to stiffen, other finishing operations such as edging may be started (Fig. 29). Edging rounds off the formed edge of the slab to prevent chipping or damage. The edger should be run back and forth until a finished edge is produced. All coarse aggregate particles should be covered, and the edger should not leave too deep an impression in the top of the slab (otherwise the indentation may be difficult to remove with subsequent finishing operations).

Jointing

Immediately following edging, the slab is jointed. Control or contraction joints are cut in the slab with the cutting edge or bit of the jointing tool to control cracking tendency in the concrete that may be due to shrinkage. In sidewalk and driveway construction, tooled joints usually are spaced at intervals equal to the width of the slab (Fig. 30), but no more than 6′ intervals for sidewalks and 15′ to 20′ intervals for driveways. Control joints, should extend to a depth of 1/4 the slab thickness. If the slab is to be grooved only for decorative purposes, jointers having shallower bits may be used.

It is good practice to use a straight board as a guide when making the groove in the concrete slab, and the tooled joints should be perpendicular to the edge of the slab. The same care must be taken in running joints as in edging, for

FIG. 26 SCREEDING

FIG. 27 DARBYING (with Darby)

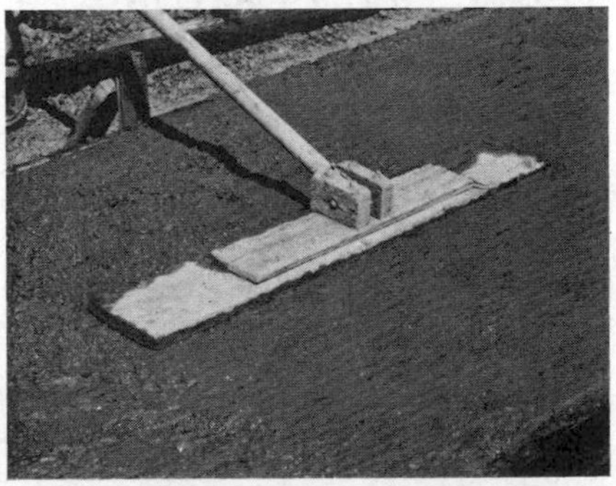

FIG. 28 DARBYING (with long handled float)

FIG. 29 EDGING

FIG. 32 FLOATING

FIG. 33 POWER FLOATING

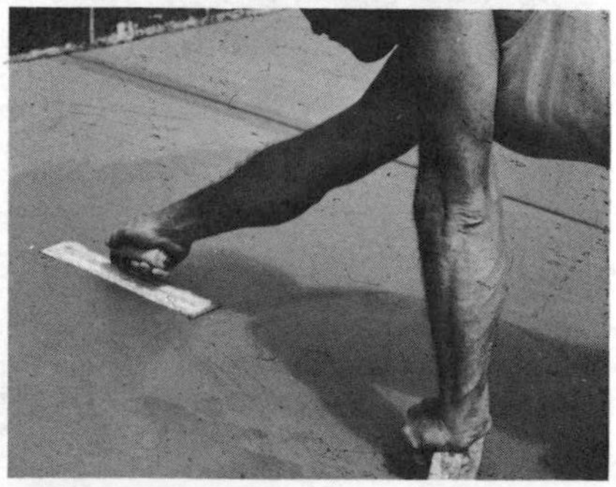

FIG. 34 TROWELING

FIG. 35 POWER TROWELING

a tooled joint can add to or detract from the appearance of the finished slab.

On large concrete flat surfaces it may be more convenient to cut joints with an electric or gasolene driven power saw fitted with an abrasive or diamond blade (Fig. 31). Cutting of joints must be done as soon as the concrete surface is firm enough not to be torn or damaged by the blade (within 4 to 12 hours), and before random shrinkage cracks can form in the concrete slab.

Floating

After edging and hand jointing operations, the slab should be floated. Many variables—concrete temperature, air temperature, relative humidity, wind and other factors—make it difficult to set a definite time to begin floating. This knowledge comes only through job experience.

The purpose of floating is threefold: (1) to embed large aggregate just beneath the surface; (2) to remove slight imperfections, humps and voids to produce a level or plane surface; (3) to consolidate mortar at the surface in preparation for other finishing operations.

FIG. 30 JOINTING

FIG. 31 JOINTING (with a saw)

Aluminum or magnesium floats should be used, especially on air-entrained concrete. This type of metal float greatly reduces the amount of work required by the finisher because the float slides more readily over the concrete surface, has a good floating action and forms a smoother surface texture than the wood float. A wood float tends to stick to and "tear" the concrete surface.

The marks left by the edger and jointer should be removed by floating (Figs. 32 and 33) unless such marks are desired for decorative purposes, in which case, the edger or jointer should be rerun after the floating operation.

Troweling

Immediately following floating, the surface should be steel-troweled (Figs. 34 and 35). (It is customary for the cement mason using hand tools to float and then steel-trowel an area before moving his knee boards.) If necessary, tooled joints and edges should be rerun before and after troweling to maintain uniformity or to remove kinks.

The purpose of troweling is to produce a smooth, hard surface. For the first troweling, whether by power or by hand, the trowel blade must be kept as flat against the surface as possible. If tilted or pitched at too great an angle, an objectionable "washboard" or "chatter" surface will result. For first troweling, a new trowel is not recommended. An older trowel which has been "broken-in" can be worked quite flat without the edge digging into the concrete. The smoothness of the surface can be improved by timely additional trowelings. There should be a lapse of time between successive trowelings to permit the concrete to increase its set. As the surface stiffens, each successive troweling should be made by a smaller-sized trowel to exert sufficient pressure for proper finishing.

Broom Finish

The steel-troweled surface leaves the concrete very smooth. However, such surfaces become quite slippery when wet. They can be slightly roughened to produce a non-slip surface by brushing or brooming the surface. The brushed surface is made by drawing a broom over the surface of the slab after steel troweling. Notice in Figure 36 that the workman is drawing the broom right over the edge joint, but the joint is not being marred. This indicates that the concrete surface has been steel-troweled properly and is hard enough. (Other types of surface finishes such as exposed aggregate are discussed under "Special Surface Finishes.")

FINISHING AIR-ENTRAINED CONCRETE

The microscopic air bubbles contained in air-entrained concrete tend to hold all the materials, including water, in suspension. This type of concrete requires less mixing water and still has good workability with the same slump. Since there is less water and it is held in suspension, little or no bleeding occurs. With no bleeding there is no waiting for the evaporation of free water from the surface. Floating and troweling should be started as soon as the slab can support the cement finisher and

FIG. 36 BROOM FINISH

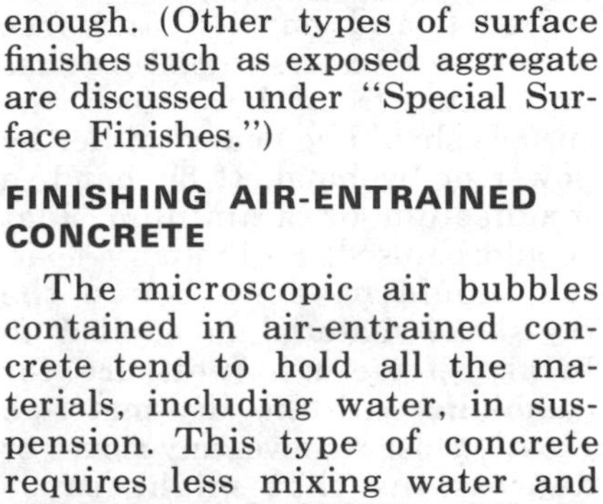

equipment.

As with regular concrete if floating is done by hand, the use of an aluminum or magnesium float is essential. A wood float drags and greatly increases the amount of work necessary to accomplish the same result. If floating is done by power, there is practically no difference between finishing procedures for air-entrained and non-air-entrained concrete, except that floating may be started sooner on the air-entrained concrete.

Major horizontal surface defects and failure are caused by finishing operations performed while bled water or excess surface moisture is present. Better results are generally accomplished, therefore, with air-entrained concrete.

SPECIAL SURFACE FINISHES

Due to the plastic quality of concrete, many surface finishes can be applied. The surface may be scored or tooled with a jointer in decorative and geometric patterns. Some of the more common finishes are discussed here.

Exposed Aggregate

An exposed aggregate surface is often chosen for any area where a special textural effect is desired (Fig. 37). If the surface is ground and polished, it is suitable especially for such places as entrances, interior terraces and recreation rooms.

Selection of aggregates is highly important, and test panels should be made before the job is started. Colorful gravel aggregate quite uniform in size—usually ranging from 1/2″ to 3/4″—is recommended. Avoid flat, sliver shaped particles or aggregate less than 1/2″ in diameter because they become dislodged during exposing operations. Exposing the aggregate used in ordinary concrete generally is unsatisfactory since this will not necessarily reveal a high percentage of coarse aggregates.

FIG. 37

EXPOSED AGGREGATE SURFACE

A 5-1/2 to 6 sack concrete with a *maximum slump of 3″* should be used. Immediately after the slab has been screeded and darbied, the selected aggregate should be scattered by hand and evenly distributed so that the entire surface is completely covered (Fig. 38a). The initial embedding of the aggregate usually is done by patting with a darby or the flat side of a 2 x 4 (Fig. 38b). After the aggregate is quite thoroughly embedded and as soon as the concrete will support the weight of a mason on kneeboards, the surface should be hand floated using a float or darby (Fig. 38c). This operation should be performed so thoroughly that all aggregate is embedded entirely just beneath the surface. The grout should completely surround and slightly cover all aggregate, leaving no holes or openings in the surface.

Shortly following this floating, a reliable retarder may be sprayed or brushed over the surface follow-

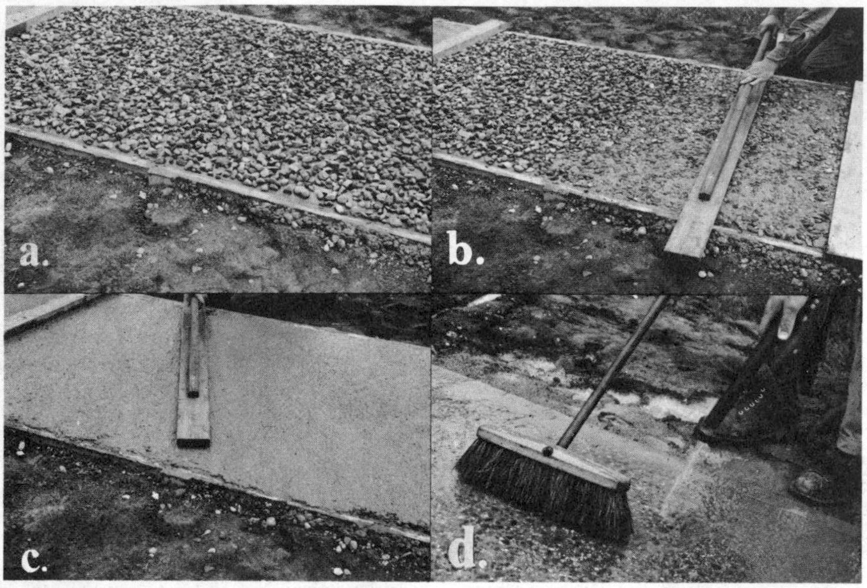

FIG. 38
a. The selected, decorative aggregate is distributed to cover the entire surface.
b. The initial embedding of the aggregate.
c. Floating the surface to completely cover the aggregate with grout.
d. Exposing the aggregate is done by simultaneously brushing and hosing with water.

ing the manufacturer's recommendations. On small jobs a retarder may not be necessary. Retarders generally are used on large jobs for better control of exposing operations. Where a retarder has been used, exposing the aggregate usually is done some hours later by brushing and hosing with water. However, the manufacturer's recommendations should be followed closely.

Whether or not a retarder has been used, the proper time for exposing aggregate is quite critical. It should be done as soon as the grout covering the aggregate can be removed by simultaneously brushing and hosing (Fig. 38d), without overexposing or dislodging the aggregate. If during exposing it is necessary for masons to move about on the surface, kneeboards should be used gently. If possible, this should be avoided because of the risk of breaking aggregate bond.

For interior areas or where a smooth surface is desired, no retarder is used and exposure of the aggregate is accomplished entirely by grinding. This may be followed by polishing which will give a surface similar to terrazzo.

Because the aggregate completely covers the surface, tooled joints in this type of work are quite impractical. Decorative or control joints are accomplished best by sawing. Control joints should be cut from 4 to 12 hours after the slab is placed and should be at least 1/5 the depth of the slab. A small-radius edger should be used before and immediately after the aggregate has been embedded to provide a more decorative edge to the slab. Another method of providing control joints is to install permanent strips of redwood before placing concrete.

In another method of placement, a top course containing the special aggregate and usually 1″ or more thick is specified.

As in all concrete work, exposed aggregate slabs should be cured thoroughly.

Dry-Shake Color Surface

This is a colored concrete surface that may be used for interior areas, terraces, decorative walks, driveways or any areas where a colored surface is desired.

It is made by applying a dry-shake material that may be purchased ready for use from various reliable manufacturers. Its basic ingredients are mineral oxide pigment (no other should be used), white portland cement and specially graded silica sand or fine aggregate. Job selecting, proportioning and mixing of a dry-shake material is not recommended.

After the concrete has been screeded and darbied and free water and excess moisture has evaporated from the surface, the surface should be floated either by power or by hand. If by hand, a magnesium or aluminum float should be used. Preliminary floating should be done before the dry-shake material is applied to bring up enough moisture for combining with this dry material. Floating also removes any ridges or depressions that might cause variations in color intensity. Immediately following this floating operation, the dry-shake material is shaken *evenly* by hand over the surface. If too much color is applied in one spot, non-uniformity in color and possibly surface peeling will result.

The first application of the colored dry-shake should use about 2/3 of the total amount needed (in lbs./sq. ft. as specified). In a few minutes, this dry material will absorb some moisture from the plastic concrete and should then be thoroughly floated into the surface, preferably by a power float. Immediately following this, the balance of the dry-shake should be distributed evenly over the surface, thoroughly floated again and made a part of the surface, taking care that a uniform color is obtained.

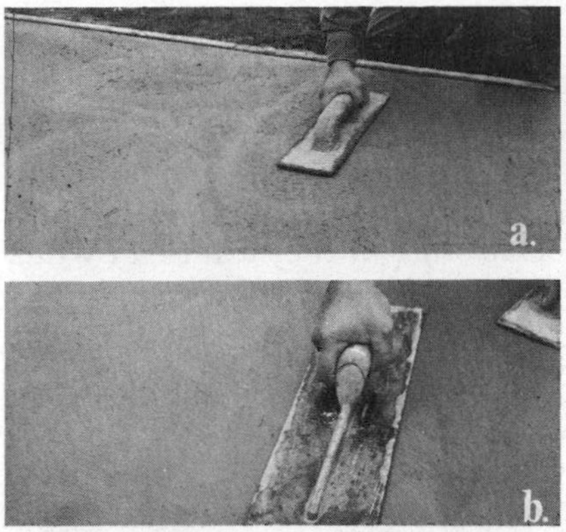

FIG. 39
a. *Swirl design on a concrete surface made with an aluminum float.*
b. *Swirl design on a concrete surface made with a steel finishing trowel.*

FIG. 40
a. *KEYSTONE OR TRAVERTINE FINISH*
b. *RANDOM-SCORED KEYSTONE FINISH*

All tooled edges and joints should be "run" before and after the applications.

Shortly after the final floating operation, the surface should be power troweled. If done by hand, then troweling should *immediately* follow the final floating. After the first troweling, whether by power or by hand, there should be a lapse of time (the length depending upon such factors as temperature and humidity) to allow the concrete to increase its set. All concrete may be troweled a second time to improve the texture and also produce a denser, harder surface.

For exterior surfaces, a second troweling usually is sufficient. Then a fine, soft-bristled pushbroom may be drawn over the surface to produce a roughened texture for better traction under foot. For interior surfaces, a third hard troweling could be specified. This final troweling should be done by hand to eliminate any washboard or trowel marks, and it will produce a smooth, dense, hard-wearing surface.

Color slabs, as with other types of freshly placed concrete, must be cured thoroughly.

After thorough curing and surface drying, interior surfaces may be given at least two coats of special concrete floor wax (also available from various reliable manufacturers) containing the same mineral oxide pigment used in the dry-shake. Care should be taken to avoid any staining, such as by dirt or foot traffic, during the curing or drying period and before waxing.

Swirl Design

This non-skid surface texture can be produced on a slab using a magnesium or aluminum float, or a steel finishing trowel. When a float is used, the finish is called a swirl float finish, and when a trowel is used, a swirl trowel finish.

After the concrete surface has been struck off, darbied, floated and steel-troweled, it is ready to be given either the swirl float or swirl trowel finish. The float should be worked flat on the surface in a semi-circular or fan-like motion. Pressure applied on the float with this motion will give a rough-textured swirl design as shown in Figure 39a.

With the same motion, using a steel trowel held flat, the cement mason can obtain a finer textured swirl design on the concrete surface (Fig. 39b). Moist curing of the slab is the final operation.

Keystone Finish

This is a special finish which has a travertine-like texture. It can be used for a terrace, garden walk, driveway, perimeter around a swimming pool or any location where an unusually decorative flat concrete surface is desired. *It should not be used in areas subjected to freezing weather.*

After the concrete slab has been struck off, darbied and edged in the usual manner, the slab is broomed with a stiff-bristled broom to insure bond when the finish (mortar coat) is applied.

The finish coat is made by mixing 1 sack of white portland cement and 2 cu. ft. of sand with about 1/2 lb. of color pigment (usually yellow is used to tint the mortar coat, but any mineral oxide color may be used). Care must be taken to keep the proportions exactly the same for all batches. Enough water is added to make a soupy mixture having the consistency of thick paint.

The mortar is placed in pails and thrown vigorously on the slab with a dash brush to make an uneven surface with ridges and depressions. The ridges should be about 1/4″ to 1/2″ high. The surface is allowed to harden enough to permit use of kneeboards.

The slab then is troweled with a steel trowel to flatten the ridges, leaving the slab surface smooth in some places and rough or coarse grained in the low spots. Depending upon the amount of troweling done on this mortar coat, many interesting textures can be produced (Figs. 40a and 40b). The slab can then be scored into random geometrical designs before curing.

SURFACE DEFECTS

There are many causes for concrete surface defects. Some of the major defects, their causes and the construction techniques which should be used to prevent them are described here. Many of them stem from improper curing as well as errors in finishing or mixing.

Scaling

Scaling is the breaking away of the hardened concrete surface of the slab to a depth of about 1/6″ to 3/16″ and usually occurs at an early age of the slab.

Early cycles of freezing and thawing of the surface of newly placed concrete can cause scaling. A favorable temperature must be maintained long enough to prevent injury (see "Curing Temperatures," page 203-27). Cycles of freezing and thawing, and applications of de-icing salts on non-air-entrained concrete also can cause scaling. Thus, air-entrained concrete is recommended for all severe exposure conditions.

Any one of the finishing operations performed while free excess water or bled water is on the surface causes a segregation of the surface fines (sand and cement) and also brings a thin layer of neat cement, clay and silt to the surface, leaving another layer of nearly clean washed sand which is not bonded to the concrete under it. To prevent scaling from this cause, water should be allowed to evaporate from the surface or be forced to evaporate by fans or blower-type heaters, or it should be removed by dragging a rubber garden hose over the surface before finishing operations begin.

Crazing

Crazing is the occurrence of numerous fine hair cracks in the surface of a newly hardened slab due to surface shrinkage. These cracks form an overall pattern similar to a crushed eggshell.

Crazing can be the result of rapid surface drying usually caused by either high air temperatures, hot sun or drying winds, or a combination of any of these. It can be prevented by using some form of curing with water since water will maintain or lower the concrete surface temperature. (See "Hot Weather Construction," page 203-28).

Premature floating and troweling when there is an excess amount of moisture at the surface, or while the concrete is still too plastic, brings an excess amount of fines and moisture to the surface. A rapid loss of this moisture will cause shrinkage at the surface which may result in crazing. Floating and troweling should be delayed until excess moisture has evaporated from the surface, and the concrete has started its initial set. Excess moisture can be avoided by reducing the slump and using air-entrained concrete.

Overuse of the vibrating screed, darby or bull float also may contribute by working an excess of mortar to the surface which tends to cause additional surface shrinkage.

Dusting

Dusting is the appearance of a powdery material at the surface of a newly hardened concrete slab.

An excess of harmful fines (clay or silt) in the concrete mix with the sand and cement at the surface can result in dusting. This emphasizes the need to use clean and well-graded coarse and fine aggregates.

Premature troweling and floating, in particular, mix excess surface water with the surface fines, thereby weakening the strength of the cement paste. Troweling should be delayed until all free water or excess moisture have disappeared and concrete has started its initial set.

When carbon dioxide (e.g., from open salamanders or gasolene engines) comes into contact with the surface of plastic concrete, a reaction takes place which impairs proper hydration. Such fumes should be vented to the outside and sufficient fresh air ventilation provided.

Condensation occasionally occurs on the surface before floating and troweling operations have been completed, usually in the spring and fall when materials have become cold due to low night temperatures. If possible, this condition should be anticipated, and the concrete should be heated or at least hot water should be used for mixing. If this is impossible, use blower-type heaters to lower the humidity directly over the slab, and fans to increase the circulation of the air. If heaters and fans are not available, open all the windows and doors to increase circulation of air. While condensation is still present, floating and troweling operations should be held to a minimum, and the surface should not be given a second troweling.

The use of neat cement as a dry-shake should not be permitted, nor should any mixture of concrete and *fine* sand be used as a dry-shake. Since condensation may occur for several hours while the concrete is beginning to harden, *emergency* measures may have to be taken in order to finish the slab: A dry mixture of one part portland cement and one part *well-graded concrete sand* well mixed and evenly and lightly distributed over the surface, followed at once by troweling. There should be no second troweling since additional condensation may take place after the first troweling.

Winter-protection heaters may lower the relative humidity around the concrete excessively and inhibit proper hydration of the cement. Water jackets should be placed on heaters to increase the relative humidity by evaporation, and moist curing methods should be employed. The location of heaters should be changed periodically so that no area will be subjected to an extreme or harmful amount of heat.

Proper curing for the specified length of time is essential. Concrete which is not cured often will be weak and the surface easily worn by foot traffic.

CURING AND PROTECTION

Curing of concrete is one of the most important construction operations but often is one of the most neglected. Concrete properly mixed, carefully placed and correctly finished will result in a poor job if proper curing operations are not followed.

PURPOSE OF CURING

The strength and watertightness of concrete improve with age as long as conditions are favorable for continued hydration of the cement. Other qualities such as resistance to freezing, thawing and weathering are affected similarly. The improvement is rapid at early ages but continues more slowly for an indefinite period as long as moisture and favorable temperatures are present.

Fresh concrete contains more than enough water for complete hydration of cement, but under most job conditions much of this water will be lost by evaporation unless certain precautions are taken. Hydration proceeds at a much slower rate when temperatures are below normal, and there is no chemical action when the temperature is near 14° F. Therefore, concrete should be protected so that moisture is not lost during the early stages of hardening and it should be kept at a temperature that will promote hydration and also protect against injury from subsequent construction activities.

CURING METHODS

Concrete can be kept moist by a number of methods such as leaving forms in place, sprinkling and ponding, using moisture-retention covers (or a seal coat applied as liquid which hardens to form a thin membrane).

Leaving Forms in Place

Forms left in place are of great assistance in retaining moisture. In hot, dry weather wood forms will dry out and should be kept moist by sprinkling. In all cases, exposed surfaces must be protected from moisture loss.

Sprinkling

Where concrete is kept moist by sprinkling (Fig. 41), care should be taken to prevent drying of the surfaces between applications of water. Alternate cycles of wetting of green concrete are conducive to crazing or cracking of the surface. A fine spray of water applied continuously provides a more constant supply of moisture and is better than copious applications of water with periods of drying between.

FIG. 41 SPRINKLING

Ponding

On flat surfaces, such as pavements, sidewalks and floors, the ponding method sometimes is used. A small dam of earth or other water-retaining material is placed around the perimeter of the surface, and the enclosed area is kept flooded with water. Ponding gives a more constant condition than does sprinkling.

Curing Covers

Moisture Retaining Covers such as burlap are sometimes used. Care should be taken to cover the entire concrete surface including exposed sides of members (and the sides of pavements and sidewalks where the forms have been removed). The covers should be kept constantly moist, enough to provide a film of moisture on the concrete surface.

Watertight Covers: Watertight paper is used on floors and other horizontal areas. It should be nonstaining and strong enough to withstand the wind and abrasive action of workmen walking over it. Seams should be overlapped several inches and covered with glued tape.

Polyethylene films often are used as watertight curing covers.

Sealing Compounds: Sealing curing compounds are available in black, colorless or white pigmented coatings. Some of them are applied in one coat, but two coats will give better results. The application should be made immediately after the concrete has been finished. If there is any delay, the concrete should be kept moist until the application is made. In extremely hot weather it is advisable to cover the slab with water for 12 hours before using curing compounds. On formed surfaces, such as beams or columns, the forms should be removed, the concrete sprayed lightly with water and then the sealing compound applied. Where the membrane must be protected against traffic, it should be covered with at least 1″ of sand or earth, or by some other means. Such a protective cover should not be placed sooner than 24 hours after the sealing compound.

CURING TEMPERATURES

The temperature affects the rate of the chemical reactions between cement and water. Consequently, temperature affects the rate at which concrete hardens, increases in strength and improves in other qualities.

Cold Weather Construction

In cold weather it is often necessary to heat the materials, cover the fresh concrete or provide a heated enclosure. The hydration of the cement causes some heat to be generated and, if this heat is retained, it raises the temperature of the concrete. Usually in cold weather the concrete should have a temperature at the time of placing of 50° F to 70° F. In no case should the material be heated to the point where the temperature of the fresh concrete is above 70° F. This results in built-in thermal stresses.

Fresh concrete should not be placed on a frozen subgrade. When such a subgrade thaws, there is likely to be uneven settlement and cracking of the concrete member it supports. The insides of forms, reinforcing steel and embedded fixtures should be free of ice at the time concrete is placed. A thin layer of warm concrete should not be placed on cold, hardened concrete for the thick upper layer will shrink as it cools, and the lower layer will expand as it warms. Bond failure will result, and this is to be guarded against especially in placing concrete floor finishes.

Mild Weather Construction

In relatively mild weather, that is when the temperature is generally above 40° to 45° F with only short periods below this range, heating only the mixing water usually will provide the desired temperatures in the concrete.

After concrete is in place, it should be kept at a favorable temperature long enough to avoid injury by exposure to the atmospheric temperature. In general, specifications require that the air surrounding the concrete after placing and when using normal portland cement (Type I) be maintained at 70° F or above for the first 3 days, or above 50° F for the first 5 days. Temperatures of 70° F for 2 days or 50° F for 3 days should be maintained when high-early-strength concrete (Type III) is being used.

In mild weather, a covering of tarpaulins may be enough protection. A layer of straw covered with tarpaulins will protect against more severe conditions, but an enclosure of tarpaulins or other watertight material with artificial heat may be necessary to provide proper protection in many cases. To prevent rapid drying of the concrete, especially near the heating elements, it may be necessary to elevate these elements and protect the concrete near them with sand kept wet constantly.

Hot Weather Construction

In extremely hot weather, extra care is required to avoid high temperatures in the fresh concrete and to prevent rapid drying of the newly placed concrete. Keeping aggregate stockpiles moistened with cool water will help in lowering the temperature. Mixing water should be chilled in very hot weather by refrigeration or by using ice as part of the mixing water. The ice should be melted by the time the concrete leaves the mixer.

Subgrades on which concrete is to be placed should be saturated some time in advance and then sprinkled just ahead of the placing operation. Wood forms should be wetted thoroughly if they have not been treated otherwise. There should be no delay in placing concrete, and it should be screeded and darbied at once. Temperature covers such as burlap kept constantly wet should be placed over fresh concrete immediately after darbying. When ready for final finishing, only a small section immediately ahead of the finishers should be uncovered, then recovered after final finish with the cover kept wet. Any delays in finishing air-entrained concrete in hot weather usually leads to the formation of a surface that is difficult to finish without leaving ridges.

The fresh concrete should be shaded as soon as possible after finishing, and moist curing should be started as quickly as possible without marring the surface. Concrete surfaces should be kept *constantly* wet during the curing period which should continue for at least 7 days.

FORM REMOVAL

The advantages of leaving forms in place as long as possible for curing and protection of concrete have been discussed above, but it is desirable sometimes to remove the forms as soon as possible. Patching and repairing of formed surfaces should be done as early as possible and this, of course, requires removal of the forms. It is often necessary also to remove forms quickly to permit their immediate re-use.

In any case, the forms should not be removed until the concrete has attained sufficient strength to insure structural stability and to carry both the dead load and any construction loads that may be imposed on it. The concrete should be hard enough so that the surfaces will not be injured in any way when reasonable care is used in removing forms.

GLOSSARY

Accelerator Admixture added to concrete to hasten its set and increase the rate of strength gain (the opposite of retarder).

Admixture Material (other than portland cement, water or aggregate) added to concrete to alter its properties (i.e., accelerators, retarders, air-entraining agents).

Aggregate Hard, inert material mixed with portland cement and water to form concrete. Fine aggregate: pieces less than and including 1/4″ in diameter; coarse aggregate: pieces larger than 1/4″ in diameter.

Air-entrained concrete Concrete containing minute bubbles of air up to approximately 7% by volume.

Alkali A soluble mineral salt present in some soils.

Alumina Aluminum oxide, constituent of ordinary clays.

Bleeding Appearance of excess water rising to the surface shortly after placing of concrete.

Calcium Chloride An accelerator added to concrete to hasten setting (not to be considered an antifreeze).

Cement A binding agent capable of uniting dissimilar materials into a composite whole.

Concrete A composite material made of portland cement, water and aggregates and, perhaps, special admixtures.

Consistency The relative ability of freshly mixed concrete to flow as measured by the slump test.

Construction joint A joint placed in concrete to permit practical placement of the work section by section.

Control joint A joint placed in concrete to form a plane of weakness to prevent random cracks from forming due to drying shrinkage and temperature changes.

Crazing Numerous fine hair cracks in the surface of a newly hardened slab.

Curing Process of keeping concrete moist for an extended period after placement to insure proper hydration and subsequent strength and quality.

Cylinder test A laboratory test for compressive stress of a field sample of concrete (6″ in diameter by 12″ in length).

Darby A tool used to level freshly placed concrete.

Darbying Smoothing the surface of freshly placed concrete with a darby to level any raised spots and fill depressions.

Dusting The appearance of a powdery material at the surface of a hardened concrete slab.

Edging The finishing operation of rounding off the edge of a slab to prevent chipping or damage.

Fill The sand, gravel or compacted earth used to bring a subgrade up to a desired level.

Fineness modulus A measure of the average size of aggregate calculated by passing aggregate through a series of screens of decreasing size.

Flash set Undesirable rapid setting of cement in concrete.

Floating A slab finishing operation which embeds aggregate, removes slight imperfections, humps and voids to produce a level surface, and consolidates mortar at the surface.

Form Temporary structure erected to contain concrete during placing and initial hardening.

Footing The base of a foundation or column wall used to distribute the load over the subgrade.

Graded aggregate Aggregate containing uniformly graduated particle size from the finest fine aggregate size to the maximum size of coarse aggregate.

Graded sand A sand containing uniformly graduated particle sizes from very fine up to 1/4″.

"Green" concrete Freshly placed concrete.

Grout A liquid mixture of cement, water and sand of pouring consistency.

High-early-strength cement Cement used to produce a concrete that develops strength more rapidly than normally.

Hydration The chemical reaction of water and cement that produces a hardened concrete.

Inert Having inactive chemical properties.

Isolation joints A joint placed to separate concrete into individual structural elements or from adjacent surfaces.

Joints See *control joint, construction joint, isolation joint.*

Laitance A soft, weak layer of mortar appearing on the top of a horizontal surface of concrete due to segregation or bleeding.

Marl Any soil or rock containing calcium carbonate (limestone).

Monolithic Concrete Concrete placed in one continuous pour without construction joints.

Mortar A mixture of portland cement, water and sand.

Neat cement A mixture of cement and water (no aggregates).

Normal portland cement See *portland cement,* Type I.

Placing Act of putting concrete in position (sometimes incorrectly referred to as pouring).

Plastic concrete Easily molded concrete that will change its form slowly only if the mold is removed.

Ponding Curing method for flat surfaces whereby a small earth dam or other water-retaining material is placed around the perimeter of the surface and the enclosed area is flooded with water.

Portland cement A hydraulic cement produced by pulverizing clinker consisting essentially of hydraulic calcium silicates and usually containing one or more of the forms of calcium sulfate as an interground addition.
- Type I and IA—Used in general construction when special properties of other types are not required.
- Type II and IIA—Used in general construction where moderate heat of hydration is required.
- Type III and IIIA—Used when high-early-strength is required.
- Type IV—Used when low heat of hydration is required.
- Type V—Used when high sulfate resistance is required.

Portland Blast Furnace cement Cement made by grinding not more than 65% of granulated blast furnace slab with at least 35% of portland cement.

Portland-pozzolan cement Cement made by blending not more than 50% pozzolan (a material consisting of siliceous or siliceous and aluminous material) with at least 50% of portland cement.

Precast concrete Concrete components which are cast and cured off-site or in a factory before being placed into their position in a structure.

Prestressed concrete Concrete subjected to compressive forces by the pre-stretching (or stressing) of high-strength steel within, which develops greater strength and stiffness.

Puddling The compacting or consolidating of concrete with a rod or other tool.

Ready-mixed concrete Concrete mixed at a plant or in trucks enroute to the job and delivered ready for placement.

Reinforcing Bars Steel placed in concrete to take tensile stresses.

Retarder An admixture, added to concrete to retard its set.

Salamander Portable stove used to heat surrounding air.

Scaling The breaking away of the hardened concrete surface of the slab (to a depth of about 1/16″ to 3/16″) usually occurring at an early age of the slab.

Screed A wood or metal templet to which a concrete surface is leveled.

Screeding Striking off excess concrete in finishing operation of concrete slab work.

Scoring Partial cutting of concrete flat work for the control of shrinkage cracking. Also used to denote the roughening of a slab to develop mechanical bond.

Segregation Separation of the heavier coarse aggregate from the mortar or of water from the other ingredients of a concrete mix during handling or placing.

Shale Laminated clay or silt compressed by earth overburden. Unlike slate, it splits along its bedding planes.

Shrinkage Decrease in initial volume due to the removal of moisture from fresh concrete. May also refer to decrease in volume due to subsequent decreases in temperature or moisture content.

Silica Silicon dioxide (SiO_2), occurring as quartz, a major constituent of sand, sandstone and quartzite.

Slab Structural A suspended, self-supporting, reinforced concrete floor or roof slab.

Slab-on grade—A non-suspended, ground-supported concrete slab, often reinforced, of one of the following types: Edge supported—Slab rests atop the perimeter foundation wall. "Floating"—Slab terminates at the inside face of the perimeter foundation wall and is said to "float" independent of the foundation wall. Monolithic—Slab and foundation wall formed into one integral mass of concrete. Also called "slab-thickened edge."

Slag A waste glass-like product generally from steel furnace.

Slag cement See *portland blast furnace cement.*

Slump A measure of the consistency of concrete mix (in inches).

Slump test Method of measuring slump by means of filling a conical mold, removing it, then measuring the "sag" or slump of the sample.

Stucco Portland cement, water, sand and possibly a small quantity of lime (portland cement plaster), along with, perhaps, other aggregates, used on exterior surfaces.

Subgrade The earth surface upon which concrete is placed.

Swale A low, flat depression to drain off storm water.

Terrazzo A floor topping made of marble chips set in cement mortar, ground smooth and polished.

Troweling A slab finishing operation which produces a smooth, hard surface.

Vibrating Mechanical method of compacting concrete.

Vibrator A mechanical tool which vibrates at a speed of 3,000 to 10,000 rpm and is inserted into wet concrete or applied to formwork to compact concrete.

Voids Air spaces between pieces of aggregate within a cement paste.

Water-cement ratio The ratio, by weight, of water and cement, or the amount of water (gallons) used per sack of cement (94 pounds) in making concrete. It is an index to strength, durability, watertightness and workability.

Workability Relative ease or difficulty with which concrete can be placed and worked into its final position within forms and around reinforcing.

We gratefully acknowledge the assistance of the Portland Cement Association for the use of its publications as references and permission to use photographs.

CONCRETE 203

CONTENTS

MATERIALS

Each of the materials listed below should conform to the current requirements of the specifications of the American Society for Testing Materials (ASTM) shown.

PORTLAND CEMENT

TYPE I, II, III, IV, or V — **ASTM C150**
TYPE IA, IIA, or IIIA (Air-Entraining) — **ASTM C175**

ASTM C150	USE	ASTM C150
TYPE I	In general construction when special properties of other types are not required.	TYPE IA*
TYPE II	In general construction where moderate heat of hydration is required.	TYPE IIA*
TYPE III	When high-early-strength is required.	TYPE IIIA*
TYPE IV	When low heat of hydration is required.	
TYPE V	When high sulfate resistance is required.	

**The letter "A" after the number designates air-entraining portland cement.*

Portland cement should be free of all lumps when used. If lumps cannot be pulverized between the thumb and finger, the portland cement should not be used.

When concrete is subjected to freezing and thawing, the use of air-entraining portland cement or an air-entraining agent to produce air-entrained concrete should be required.

Air-entraining concrete is strongly recommended for all concrete work, due to its increased durability, workability and proven record.

WATER

Mixing water should be clean and free from deleterious amounts of acids, alkalis or organic materials.

Water that is fit to drink generally is suitable for making concrete.

AGGREGATES — ASTM C33

All aggregates should be clean and free of foreign particles such as loam, clay or vegetable matter.

Aggregates should be uniformly graded from the finest particles of sand up to the largest piece of coarse aggregate.

The most economical concrete results when the fine and coarse are uniformly graded, since this requires the least amount of cement paste to surround the aggregate and fill all the spaces between the particles.

Fine Aggregate—Gradation of sand should be uniform from very fine up to 1/4″ for good workability and smooth surfaces.

Excess of fines reduces strength or requires an increase in the amount of cement to attain the desired strength.

(Continued) MATERIALS

Coarse Aggregate—Use the largest allowable particle size of coarse aggregate within the following limits:

Walls	No larger than 1/5 wall thickness
Slabs	No larger than 1/3 slab thickness

In no case, larger than 3/4 the width of the narrowest space through which concrete must pass during placing.

ADMIXTURES

Specific effects resulting from the use of admixtures seldom can be predicted accurately. Since a change in the type of cement or amount of cement, or a modification of aggregate gradation or mix proportions, or a change in curing procedure can often produce the same end result more economically, admixtures should be used with caution, if used at all.

Accelerators: ***Calcium Chloride*** **ASTM D98**

Calcium chloride, if used, should be added in solution *(never in dry form)* in an amount not to exceed 2 lbs. per sack of cement.

Calcium chloride is an accelerator and is not an antifreeze. If used in amounts to lower appreciably the freezing point of concrete, the concrete would be ruined.

Retarders

Use of a retarder should not be attempted without technical advice or advance experimentation with the cement and other materials to determine its effect on the properties of the concrete.

Air-Entraining Agents **ASTM C260**

Careful control is essential in producing air-entrained concrete, and its use should be restricted to plant- or transit-mixed batches.

Air-entrained concrete should be made with admixtures conforming to ASTM C260 (or with air-entraining portland cement—ASTM C175) to obtain the percentage of air shown here:

NOMINAL MAXIMUM SIZE COARSE AGGREGATE	TOTAL AIR CONTENT PERCENTAGE BY VOLUME OF CONCRETE
3/8″	6 to 10
1/2″	5 to 9
3/4″	4 to 8
1″	3-1/2 to 6-1/2
1-1/2″	3 to 6

REINFORCING

Deformed Steel Bars	**ASTM A615**
Billet Steel Bars	**ASTM A615**
Rail Steel Bars	**ASTM A615**
Welded Steel Wire Fabric	**ASTM A185**
Cold Drawn Steel Wire	**ASTM A82**

Any reinforcement should be clean and free of loose rust prior to the placing of concrete.

Protection of reinforcing steel should be provided using concrete of thicknesses shown in the accompanying table.

Concrete Protection Over Reinforcing Steel

LOCATION AND USE		MINIMUM COVER
Against ground without forms		3″
Exposed to weather or ground but placed in forms	Greater than 5/8″ diameter bars	2″
	Less than 5/8″ diameter bars	1-1/2″
Slabs and walls (no exposure)		3/4″
Beams, girders, columns (no exposure)		1-1/2″

MIXTURES

SELECTING THE MIX

Unless concrete mixtures are established by tests in advance of the construction work, the concrete mixtures shown in the table below should be used. It is first necessary to determine the exposure condition of the concrete construction from the regional weathering areas shown on the map. Areas shown as severe exposure *are subject to many freeze-thaw cycles per year and, in the case of flat concrete work, the application of deicer chemicals;* moderate exposure *areas are subject to few freeze-thaw cycles per year and no application of deicer chemicals;* negligible exposure *areas have no freeze-thaw cycles per year and no deicer chemicals.*

When the exposure condition of the concrete construction element is known, it is possible to determine the recommended *minimum compressive strength* and the maximum permissible *water-cement ratio.* The minimum cement content is included in addition to the lowest maximum water-cement ratio. Minimum cement requirements ensure satisfactory finishability, improved water resistance of slabs, and suitable appearance of vertical surfaces. The percentage of air-entrainment for various maximum aggregate sizes is also given; slumps should be in the ranges shown.

For example, to select the minimum compressive strength for basement walls not exposed to weather, to be built in the Chicagoland area would require concrete strength of 3000 psi, with a maximum water-cement ratio of 0.46, and 6% to 8% air-entrained concrete. The minimum cement content likely to be required to achieve that strength at the recommended air-entrainment level is 564 lbs. per cu. yd.

While 2500 psi concrete can satisfactorily support typical vertical and lateral loading in a basement wall, the lower strength concrete will be less watertight than the concrete recommended herein. The higher strength concrete provides a margin of safety against the possibility of foundation cracking and leaking.

The water-cement ratio selected must be the lowest value required to meet design considerations such as strength, durability, freeze-thaw resistance, watertightness and wear-resistance.

REGIONAL WEATHERING AREAS*

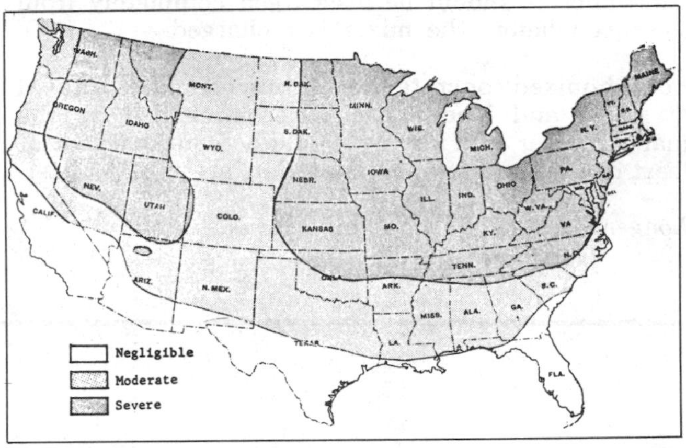

*Areas are based upon ASTM C62 weathering index map adjusted to conform with an experience survey by HUD and the Portland Cement Association. Local conditions may be more or less severe than indicated by regional classification.

RECOMMENDED CONCRETE MIXTURES FOR VARIOUS EXPOSURE CONDITIONS

Concrete Construction Element	Exposure Condition[1]	Minimum Compressive Strength @ 28 days,[2] plant or transit (psi)	Practical Water-cement Ratio by Weight: Non-air entrained concrete	Practical Water-cement Ratio by Weight: Air-entrained concrete	Normal Maximum Coarse Aggregate Size[5] (in.)	Minimum Cement[6] Content (lbs. per cu. yd.)	Air-entrainment by Volume (%)	Slump (in.)
Foundations, basement walls and slabs *not exposed to weather*	Severe	3000[3]	See footnote 3	0.55	1	564	See footnote 6	5″ maximum for hand methods of strike-off and consolidation; 3″ maximum for mechanical strike-off and consolidation.
	Moderate					520		
	Negligible							
Foundations, basement walls, exterior walls and other concrete work *exposed to weather*	Severe	3500[3]		0.45	1	564	6 to 8	
	Moderate	3000	0.58	0.50		520	5 to 7	
	Negligible	2500	0.67	0.55			See footnote 6	
Concrete *exposed to sulfate attack*	Severe	5000[3]	See footnote 4	See footnote 4	1	564	6 to 8	
	Moderate					520	5 to 7	
	Negligible	5000					See footnote 6	
Driveways, garage floors, walks, porches, patios and stairs *exposed to weather*	Severe	4000[3]	See footnote 3	0.45	1	564	6 to 8	
	Moderate	3000[3]		0.50		520	5 to 7	
	Negligible	2500	0.67	0.55			See footnote 6	

[1]See map for severity of exposure, regional weathering areas.
[2]With most materials, the water-cement ratios shown will provide average strengths greater than required.
[3]Use air-entrained concrete only.
[4]Proportions should be established by the trial batch method.
[5]Maximum size of coarse aggregate should not be larger than 1/5 the narrowest dimension between forms, nor larger than 3/4 the minimum clear spacing between reinforcing bars.
[6]2% to 3% air entrainment, to improve cohesiveness and reduce bleeding, is recommended but not essential.

(Continued) **MIXTURES**

MIXING

The concrete should be mixed thoroughly until the color is uniform and there is a uniform distribution of the materials.

The concrete should be discharged completely from the mixer before the mixer is recharged.

For job-mixed concrete the mixture should be rotated at speed and load capacity recommended by the manufacturer, and mixing should be continued for at least one minute after all materials are in the mixer.

Longer mixing gives more uniform results.

Ready-mixed Concrete ASTM C94

The concrete should be delivered and discharged within 1-1/2 hours after water is added to the cement and aggregate.

Remixing

Concrete less than 1-1/2 hours old that has stiffened or dried out slightly when left standing may be remixed and used if it can be completely compacted in the forms. Adding water to make a mixture more workable (re-tempering) should never be permitted.

Adding water lowers quality just as would a larger amount of water in the original mixing.

HANDLING AND PLACING

Handling and placing techniques should insure uniformity within the batch and from batch to batch. Equipment for conveying the concrete from mixer to place of final deposit should do so without loss or separation of the materials so as to provide practically a continuous flow of concrete at the point of delivery.

PREPARATION

The subgrade or forms should be moistened in hot weather to prevent extraction of water from the concrete mix and be free of ice or snow in cold weather. Formed surfaces should be tight and well braced.

PLACING

Concrete should be placed as nearly as practicable in its final position to avoid segregation due to rehandling or flowing. It should not be allowed to run or be worked over long distances. Concrete should not be allowed to drop freely more than 3′ or 4′. When placed in layers they should be uniform, compacted before the next layer is placed, and in thicknesses of 6″ to 12″. On flat surfaces fresh concrete should be placed into previously placed concrete.

To prevent water from collecting at the end and in corners of forms and along form faces, the order of placing should be as follows: on walls, start at ends and work toward center; on slabs, start around the perimeter.

After placing, all concrete should be compacted by rodding, spading or similar means, or mechanical vibrators.

Concrete should not be over-vibrated.

Proper compaction: (1) eliminates stone pockets and air bubbles; (2) consolidates each layer with that previously placed; (3) embeds reinforcing and other fixtures; and (4) brings enough fine materials to the faces and surfaces to produce the desired finish.

JOINTS

Control joints should be provided to prevent irregular random shrinkage cracking. Joints in sidewalk construction should be spaced at intervals generally equal to the width of the slab, but not to exceed 6′. Joints in driveway construction should be spaced generally equal to the width of the slab, but never more than 10′ to 15′ intervals.

On large slab areas the control joints should be provided so that the slab is divided into approximately square panels, with a maximum spacing of 15′ intervals, depending on thickness.

Control joints may be formed by tooling, sawing or installing keyed joints.

Tooled or sawn control joints should be made at a depth equal to either 1/5 of the slab thickness or at least equal to the largest size of coarse aggregate, whichever is larger.

(Continued) HANDLING AND PLACING

Construction Joints should be provided to subdivide the work into smaller sections when areas are too large to permit proper handling, placing or finishing operations.

Isolation Joints should be provided between floors and walls and floors and columns, and elsewhere whenever adjacent surfaces are required to move independent of each other.

Isolation joints should be formed using 1/8″ thick asphalt impregnated fiber sheets or other acceptable material less than 1/4″ thick.

FINISHING

PROCEDURE

All of the first finishing operations of placing, screeding and darbying should be performed before any excess water is forced to the surface *(bleeding).*

If any of the above operations are performed while excess or bled water is present, serious scaling, dusting or crazing of the surface will result.

The concrete should not be allowed to stand but should be placed as soon and as continuously as possible, then immediately screeded *(striking off excess concrete),* followed at once by darbying *(bringing the surface to true grade with enough mortar to produce the desired finish).*

When all bled water and water sheen has left the surface and the concrete has started to stiffen, edging may be performed and control joints may be cut.

The surface should then be floated.

Proper floating: (1) embeds large aggregate; (2) removes humps and voids to produce a level surface; and (3) consolidates the mortar at the surface.

Immediately following floating, the surface should be steel-troweled to produce a smooth, hard surface. If a smooth surface is not desired, the surface should be roughened slightly by brushing to produce a non-slip finish.

FINISHING AIR-ENTRAINED CONCRETE

When air-entrained concrete is used, floating and troweling should be started sooner (since little or no bleeding occurs).

The use of an aluminum or magnesium float is essential, since a wood float drags and greatly increases the labor necessary to produce comparable results.

CURING AND PROTECTION

To promote hydration, concrete should be protected to prevent loss of moisture and should be kept at a favorable temperature.

CURING METHODS

Concrete should be protected from early drying by leaving the forms in place; by covering exposed surfaces with wet sand, burlap, polyethylene film or other materials; by sprinkling; or by a liquid seal coat that hardens to form a thin membrane. Curing should be started as soon as possible without marring the surfaces and should be continued as long as possible—at least 5 days when normal portland cement (Type I) is used and 2 days when high-early-strength portland cement (Type III) is used.

CURING TEMPERATURES

Cold Weather Construction

Whenever the temperature of the surrounding air is below 40° F, the water and aggregate should be heated so that the temperature of the mixed concrete at the time of placing is between 50° F to 70° F. It should then be kept above 70° F for 3 days or above 50° F for 5 days when normal portland cement (Type I); or above 70° F for 2 days or 50° F for 3 days when high-early-strength portland cement (Type III) is used.

(Continued) CURING AND PROTECTION

All reinforcement, forms and subgrade with which concrete is to come in contact should be free from frost.

No dependence should be placed on salt or other chemicals for the prevention of freezing

Mild Weather Construction

When the temperature is generally above 40° F to 45° F with only short periods below this range, the mixing water should be heated or other steps taken to raise the temperature of the concrete to 70° F at the time of placing. Protection to provide the same favorable temperatures stated above for "Cold Weather Construction" is required.

Hot Weather Construction

In extremely hot weather, care should be taken to avoid high temperatures in the fresh concrete and to prevent rapid drying after it is placed. Coarse aggregate stockpiles may be cooled by sprinkling, and mixing water may be refrigerated. Ice may be used as part of the mixing water so long as it is melted by the time the concrete leaves the mixer. The subgrade and forms should be sprinkled prior to placing *(so as not to absorb water from the mix)*. Moist curing should be started immediately—as soon as surfaces are hard enough to resist marring—and should be kept constantly wet to avoid alternate wetting and drying during the curing period for at least 7 days.

FORM REMOVAL

Forms should not be removed until the concrete has attained sufficient strength to insure structural stability and to carry both the dead load and any construction loads that may be imposed on it. The concrete should be hard enough so that the surfaces will not be injured in any way when reasonable care is used in removing forms.

205 MASONRY

MASONRY

INTRODUCTION

Clay masonry, one of the oldest manufactured building materials, and *concrete* masonry are both used extensively in construction. Usually built into wall assemblies, the compressive strength, durability, fireproof character, color, texture and pattern of masonry also make it a suitable material for floor surfacing, exterior paving, decorative screens and other uses. Lightweight aggregates, produced from clay, shale and other products, have made possible lighter units as well as the fabrication of masonry into larger panels. Higher-bond-strength mortar, has enhanced prefabricated brick masonry panel construction. New developments in sealants, adhesives and other related fields assure continued use of this time-proved material.

Masonry of suitable quality must perform satisfactory service in the use for which it is intended. Exterior masonry walls must possess at least three essential functional properties: *strength* to carry applied loads, *watertightness* to prevent rain penetration and *durability* for wear and weather-resistance. Walls may also have to resist the transmission of heat, sound and/or fire.

When masonry is exposed to view, esthetic properties determine in part its suitability. Appearance is strongly affected by: *color*, resulting from both units and mortar; *texture*, resulting from either the units or method of laying; and *pattern*, resulting from the type of bond and the size of units used.

Assuming the wall has been correctly designed (see Division 300—Masonry Walls), the essence of producing masonry construction of suitable quality can be summarized in one paragraph:

Use sound materials that are handled, stored and prepared properly prior to construction. Select the type of mortar suited to the use and exposure of the wall, then proportion and mix the mortar constituents correctly. Use proper workmanship to insure full mortar joints, particularly the head joints, and follow by proper tooling of the mortar joints. Protect masonry from water penetration and freezing during construction and insure against penetration after construction by installing flashing and providing weep holes for drainage.

This section, together with Section 343—Masonry Walls, explains how to produce masonry of suitable quality.

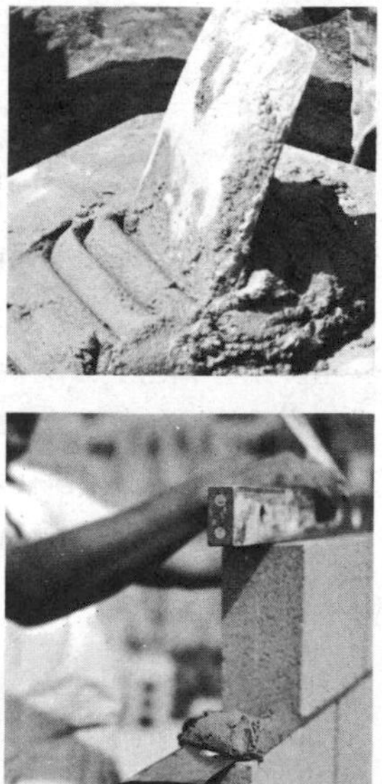

This section describes two broad classifications of masonry materials: (1) *clay* masonry units and (2) *concrete* masonry units.

CLAY MASONRY

As used here, clay masonry refers to burned clay units used in construction as an integral part of the structure and includes: brick, all types of hollow tile and terra cotta. It does not include ceramic tile, flue lining, drain tile and other similar ceramic products (see Section 442—Ceramic Tile Finishes).

Brick and Tile

Solid masonry units (brick) and hollow masonry units (tile) are composed of burned clay, shale, fire clay or mixtures of these, formed into the desired shape while plastic and then burned in a kiln. The raw materials and methods of manufacturing clay masonry have a marked influence on the properties of the units in a finished wall. Hence, a general understanding of both is essential to the selection and use of brick and tile.

Raw Materials To satisfy production requirements, clays must have plasticity which permits them to be shaped or molded when mixed with water, and then must have sufficient tensile strength to maintain their shape after forming. When subjected to rising temperatures in a kiln, the clay particles must fuse. Clay used to manufacture brick and tile has three principal forms with similar chemical compositions but having different physical characteristics:

(1) *Surface Clays* are found near the surface of the earth and are of sedimentary character; (2) *shales* are clays that have been subjected to high pressures until they have hardened almost to the form of slate; and (3) *fire clays* are mined at deeper levels than other clays, have refractory qualities (resistance to high heat as in fire brick), and normally have fewer impurities and more uniform chemical and physical properties than shales or surface clays.

Chemically, all three are compounds of silica and alumina with varying amounts of metallic oxides and other impurities. In addition to acting as fluxes (which promote fusion at lower temperatures) metallic oxides play a strong role in influencing the color of finished units.

Although manufacturers attempt to minimize variations in the properties of raw materials by mixing clays from different locations in the pit and from different pits as well as by making compensations during manufacturing processes, slight variations in the properties of finished units are quite normal.

Manufacture

In general, the manufacturing process consists of: (1) winning and storage of raw materials; (2) preparing raw materials; (3) forming units; (4) drying and glazing; (5) burning and cooling; (6) drawing and storing finished units (Fig. 1).

Winning (mining) The clays are taken from open pits or underground mines and are stored in several storage areas to permit blending. Blending produces more uniformity, helps to control color and permits some control over the suitability of raw material for manufacturing a given type of unit. If the raw clay is in large lumps it usually undergoes preliminary crushing before storage.

Preparation The clay is first crushed to break up large chunks and remove stones, after which huge grinding wheels, weighing 4 to 8 tons each, revolve in a circular pan to pulverize and mix the material. Most plants then screen the clay, passing it through an inclined vibrating screen to control particle sizes.

Forming Tempering, the first step in the forming process, produces a homogeneous, plastic mass ready for molding. It is most commonly achieved by adding water to the clay in a mixing chamber (pug mill), which contains one or two revolving shafts with blades. Three

FIG. 1 DIAGRAMMATIC REPRESENTATION OF MANUFACTURING PROCESS

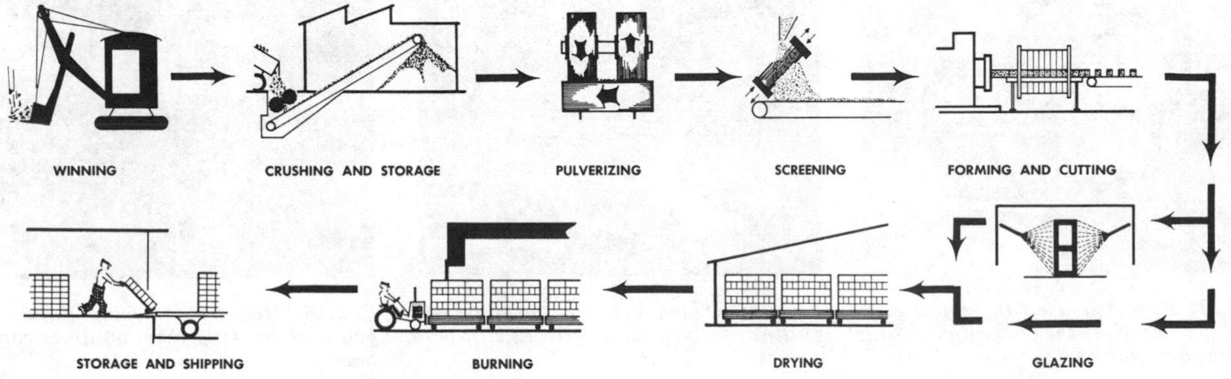

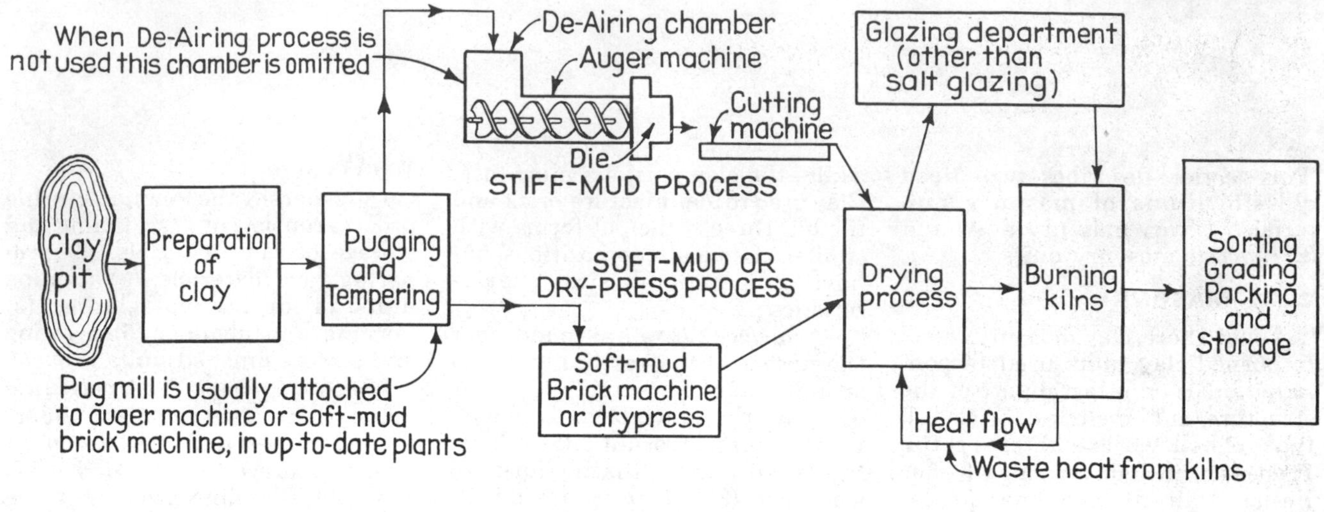

FIG. 2 MANUFACTURING FLOW CHART

methods are used to form brick and tile: the *stiff-mud, soft-mud* and *dry-press* processes (Fig. 2).

Stiff-Mud Process is the most commonly used method, all structural tile and a great many brick being made this way (Fig. 3). Clay is mixed only with sufficient water (12% to 15% moisture by weight) to produce plasticity. After thorough mixing (pugging), the tempered clay goes through a de-airing machine in which a partial vacuum is maintained. De-airing removes air holes and bubbles, giving the clay increased workability, plasticity and resultant greater strength.

Next the clay column is forced through a die (extruded) and passes through an automatic cutter (Fig. 4). Cutter-wire spacings and die sizes must be calculated carefully to compensate for normal shrinkage during drying and burning. As the clay column leaves the die, various textures may be applied.

Upon leaving the cutting table, individual units are deposited and inspected on a continuously moving belt. Suitable units are taken off and placed on dryer cars (Fig. 5) while imperfect units are returned to the pug mill.

Soft-Mud Process is used only for the production of brick and is particularly suitable for clays which contain too much natural water (20% to 30% moisture by weight) for extrusion in the stiff-mud process. Clays are mixed and then formed into units in molds. To prevent clay from sticking, the molds are lubricated with sand or water. When sand is used the brick are called sandstruck (Fig. 8i); if water is used they are called waterstruck brick (Fig. 8h).

Dry-Press Process is adaptable for clays of very low plasticity. Clays are mixed with a minimum of water (up to 10% moisture by weight), are then formed in steel molds under pressure from 500 to 1500 psi.

Drying When wet clay units come from molding or cutting machines, they contain from 7% to 30% moisture, depending on the forming method. Before the burning process begins, most of this water is evaporated in dryer kilns at temperatures ranging from 100° to 300° F. During drying time (varying from 24 to 48 hrs., depending on the type of clay), heat and humidity must be regulated carefully to avoid excessive cracking of the units.

Glazing Glazing is an additional operation used in the production of some facing tile and some brick. Ceramic glazing is a specialized, controlled procedure with two basic variations: high-fired and low-fired

FIG. 3 Pug mill used in typical stiff-mud process. Note the clay column being forced through the die.

FIG. 4 Shown here are forming and cutting of a clay column by the stiff-mud process.

FIG. 5 Hackers removing units from belt and stacking them on dryer cars in background.

FIG. 6 (a) Tunnel kiln; (b) Periodic kiln.

glazing. *High-fired* glazes are sprayed on units before or after drying and then kiln-burned at normal firing temperatures. *Low-fired* glazes are used to obtain colors which cannot be produced at high temperatures. They are applied after the unit has been burned to maturity and cooled, then are refired at lower temperatures. Spray glazes are composed of several mineral ingredients that fuse in a glass-like coating at a given temperature. Glazes are available in almost any color.

Burning Burning requires from 2 to 5 days, depending on kiln type and other variables. The chief types of kilns are *tunnel* and *periodic* kilns (Figs. 6a & 6b), fueled by natural gas, oil or coal.

Dried units are set in periodic kilns in a pattern that permits free circulation of hot kiln gases. In a tunnel kiln units are loaded on special cars which pass through various temperature zones as they travel through the tunnel.

Burning may be divided into six general stages: water-smoking (evaporating free water), dehydration, oxidation, vitrification, flashing and cooling. All except flashing and cooling are associated with rising temperatures in the kiln. Although the actual temperatures will differ with the clay or shale, water-smoking takes place at temperatures up to about 400° F; dehydration about 300° to 1800°F; oxidation from 1000° to 1800° F; and vitrification from 1600° to 2400° F.

The rate of temperature change must be controlled carefully, and kilns are equipped to provide a constant check on the firing process. Near the end of the burning process, the units may be *flashed* to produce different colors and color shadings that vary with different types of clays. Flashing is completed by creating a reducing atmosphere (insufficient oxygen for complete combustion) in the kiln near the end of the burn.

After the temperature has reached the maximum, the cooling process begins. Proper cooling requires 48 to 72 hrs. in periodic kilns, but in tunnel kilns the cooling period seldom exceeds 48 hrs. The rate of cooling has a direct effect on color and because excessive, rapid cooling will cause cracking and checking of the units, cooling must be controlled carefully.

Drawing This is the process of unloading a kiln after cooling (Fig. 7), and it is at this stage that units are sorted, graded and taken to a storage yard or loaded directly into rail cars or trucks in preparation for delivery.

Properties of Brick and Tile

The properties of finished units are dependent on characteristics of the raw materials and the effects of the manufacturing process.

Compressive Strength The clay, the method of manufacturing and the degree of burning affect compressive strength. With some exceptions, the plastic clays used in the stiff-mud process have higher compressive strengths when burned than clays used in the soft-mud or dry-press processes. For a given clay and method of manufacture, higher compressive strengths are associated with higher degrees of burning. The compressive strength of brick varies from 1500 psi to more than 20,000 psi, due mainly to the wide variation in the prop-

FIG. 7 Unloading the kiln.

FIG. 8 TYPICAL BRICK TEXTURES

a. Smooth

b. Matt—vertical markings

c. Matt—horizontal markings

d. Rugs

e. Barks

f. Stippled

g. Sandmold

h. Waterstruck

i. Sandstruck

erties of the clays used.

Absorption The absorption of water is also dependent on the clay, manufacturing method and degree of burning. Plastic clays and higher degrees of burning generally produce units having lower absorption. Generally, the stiff-mud process produces units with lower absorption than either the soft-mud or dry-press processes. Suction (the initial rate of absorption) of brick is caused by pores or small openings in burned clay that act as capillaries and tend to draw or suck water into the unit. Suction has little bearing on the transmission of water through the brick itself, but it does have an important effect on the bond between brick and mortar. Maximum bond strength and minimum water penetration are obtained when the suction of the brick *at the time of laying* does not exceed 20 grams of water (0.7 oz.) per minute. Brick having suction of more than 20 grams per minute should be wetted prior to laying in order to reduce the suction. (See page 205-29 for field test to determine suction.)

Durability The durability of clay products results from fusion during burning. *The only action of weathering that has any significant effect upon burned clay products is alternate freezing and thawing in the presence of moisture.* Where the annual average precipitation exceeds 20″ and alternate freezing and thawing is common, the ability of brick or tile to resist freezing and thawing without disintegration is very important. For brick and tile produced from the same raw material and by the same method of manufacture, either high compressive strength or low absorption are fairly accurate measures of suitable resistance to freezing and thawing.

Natural Color Burned clays produce a wide color range—from the pure tones of pearl grays or creams through buff, and to a descending scale of reds down to purple, maroon and gun metal black. The chemical composition of the clay, the burning temperatures and the method of burning control color. Of all the oxides commonly found in clays, iron probably has the greatest effect on color. Clay containing iron in practically any form will burn red when exposed to an oxidizing fire owing to the formation of ferrous oxide. When burned in a reducing atmosphere (flashing), the same clay will take on a purple cast.

For the same raw materials and methods of manufacture, lighter colors are associated with underburning, producing the *salmon* brick from clays which burn red. Underburned brick is softer, more absorptive and has decreased compressive strength. Overburning produces the *clinker* brick which is dark red to black in red clays and dark speckled brown in the buff clays. Burning at higher temperatures tends to produce a harder brick, decreases absorption and increases compressive strength.

Since, with few exceptions, the natural colors of clay products are mixtures of shades rather than pure colors, the accepted practice in specifying color is to require the shipment to match an approved sample panel of several units.

By using applied coatings, such as glazes or non-lustrous finishes, or by introducing chemicals into the kiln that vaporize and combine with the clay, color effects can be produced in almost the complete range of the spectrum.

Texture Textures are produced by the dies or molds used in forming. Smooth texture results from pressure exerted by the steel die. Many textures may be applied by attachments that cut, scratch, roll, brush or otherwise roughen the surface as the clay column leaves the die. The degree of texture ranges from fine through medium to coarse. The principal textures are *smooth, matt* with vertical or horizontal markings, *rugs, barks, stippled, sandmold, waterstruck* and *sandstruck* (Fig. 8).

Size Variation Masonry units vary in size as a result of shrinkage. Clays shrink from 4.5% to 15% during drying and burning, and allowances are made in the die size and in the length of cut to compensate for it. The principal problem is not so much the total shrinkage as the variation in shrinkage of the clay.

Fire shrinkage increases with higher temperatures, in turn producing darker shades. As a result, when a wide range of colors is desired, some variation between the size of the dark and light units is inevitable. Although manufacturers attempt to control shrinkage to obtain units of uniform size, variations in raw materials and kiln temperatures make absolute uniformity impossible; therefore specifications allow permissible variations in size.

FIG. 9 **CLASSIFICATION OF STRUCTURAL CLAY PRODUCTS**

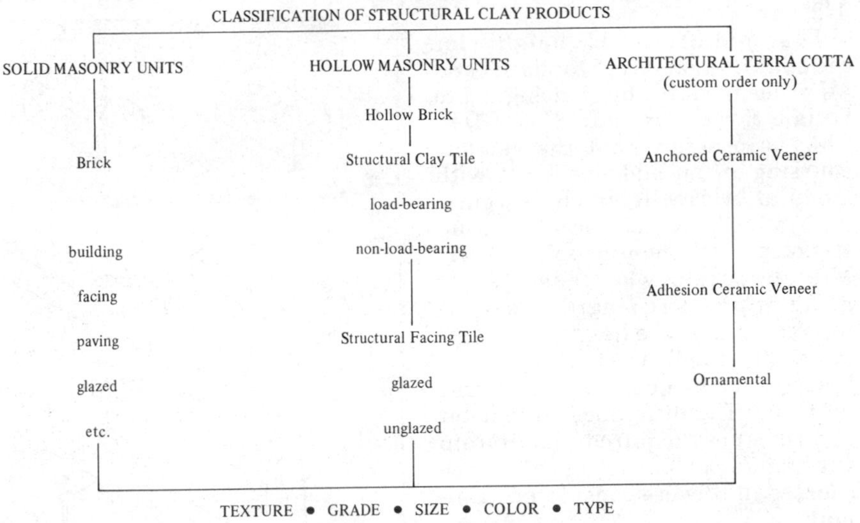

Classifications of Clay Products

Structural clay masonry units are classified (Fig. 9) as: solid masonry, hollow brick, hollow masonry and architectural terra cotta. The many individual products are qualified further as to size, grade, type, color and texture. (See Glossary, 205-36, for definition of products shown in Figure 9 not described below.)

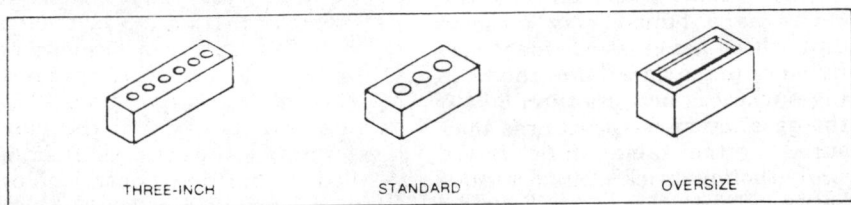

Solid Masonry (Brick) Whenever the term brick is used, it means a solid masonry unit. Cored brick, ranging from those with large cores to those with as many as 21 small "pencil" cores, are solid masonry units if the core area does not exceed 25% of the total cross sectional area of the unit. The cores are introduced to promote even drying during burning and to reduce the weight. Brick has been produced in many unusual and special shapes. Except for the non-modular, Three-inch, Standard and Oversize units shown in Figures 10 and 11, most brick is produced in *modular* sizes having nominal dimensions based on the four-inch module.

Modular coordination defines a construction system in which components or products are sized to fit together easily with minimum cutting and fitting required in the field. The use of modular coordination in the design and construction of masonry buildings not only increases productivity, but can also result in significant cost savings.

The common designations of modular brick are shown in Figure 12 and the sizes typically produced by the industry listed in Figure 13. Few manufacturers produce all sizes shown, so it is important to determine the availability of specific sizes in a given locality in the predesign stage.

The *nominal* dimensions of modular brick are equal to the *manufactured* dimensions plus the thickness of the mortar joint for which the unit is designed. In general,

FIG. 11 **SIZES OF NON-MODULAR BRICK**

Unit Designation	Manufactured Size, in. t	h	l
Three-inch[1]	3	2⅝	9⅝
	3	2¾	9¾
Standard	3¾[2]	2½	8
Oversize	3¾[2]	2¾[2]	8

[1]Sizes shown are "Kingsize." Other 3" brick are "Big John," "Jumbo," "Scotsman" and "Spartan." Originally used as a veneer unit, it is also used to construct 8" cavity and grouted walls.

[2]Since thickness varies, the designer should check with the brick manufacturer, if other than a running bond is desired.

FIG. 10 **NON-MODULAR BRICK**

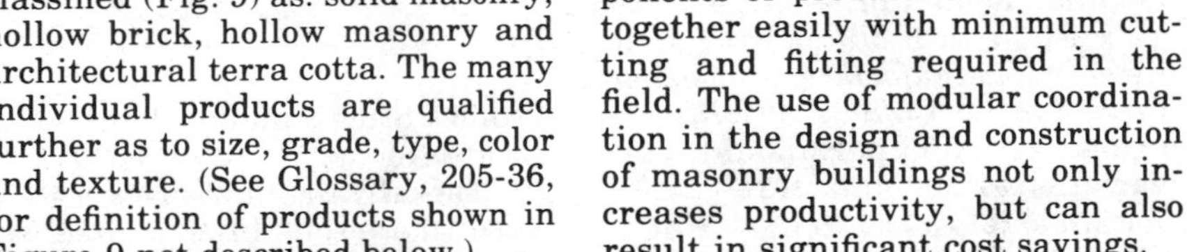

joint thicknesses are either 3/8″ or 1/2″.

The manufactured height for Standard, modular Standard and all other modular brick designed to be laid three courses in 8″ is 2-1/4″. This is to assure that the vertical coursing in an addition built with modular brick will match the coursing in an existing building constructed with non-modular brick. Manufacturers who converted to modular production agreed that the manufactured face height would remain at 2-1/4″ even though the length and thickness would become modular. The differences in mortar bed thickness required to maintain the modular coursing of three courses in 8″ were considered minimal.

Brick unit dimensions in the figures are shown in the order of thickness first, followed by the face dimensions (height and length). It is recommended that specifications for brick follow this standard order of listing brick dimensions. (Example: The *nominal* dimensions of a modular Roman brick are 4″ x 2″ x 12″; if laid with 3/8″ mortar joints, its actual *manufactured* dimensions are 3-5/8″ x 1-5/8″ x 11-5/8″ with allowance for permissible size variations in manufacture.)

Hollow Masonry Units Hollow masonry units include hollow brick and hollow clay tile. The many types of hollow clay tile are classified as either *structural clay tile* or *structural facing tile*. The cores, cells or hollow spaces exceed 25% of the gross cross sectional area of the unit. When designed to be laid with the cells in a horizontal plane, the unit is called *side construction* or *horizontal cell* tile; if designed to be laid with cells vertical, the unit is called *end construction* or *vertical cell* tile. The size, number, shape and thickness of cells will vary greatly depending on the manufacturer.

Hollow brick is defined by ASTM C-652 as a hollow clay masonry unit whose net cross-sectional area in every plane parallel to the bearing surface is not less than 60% of the gross cross-sectional area measured in the same place. Essentially, hollow brick units are cored between 25% (the upper limit for "solid" brick units) and 40%, i.e., 60% solid.

Generally, hollow brick can be used where solid brick is used; it is the same size and looks the same on the outside as solid brick. However, it is not recommended that solid brick design standards be applied to hollow brick construction. Hollow brick structures should be individually designed according to accepted engineering practice.

FIG. 12 MODULAR BRICK NOMINAL DIMENSIONS

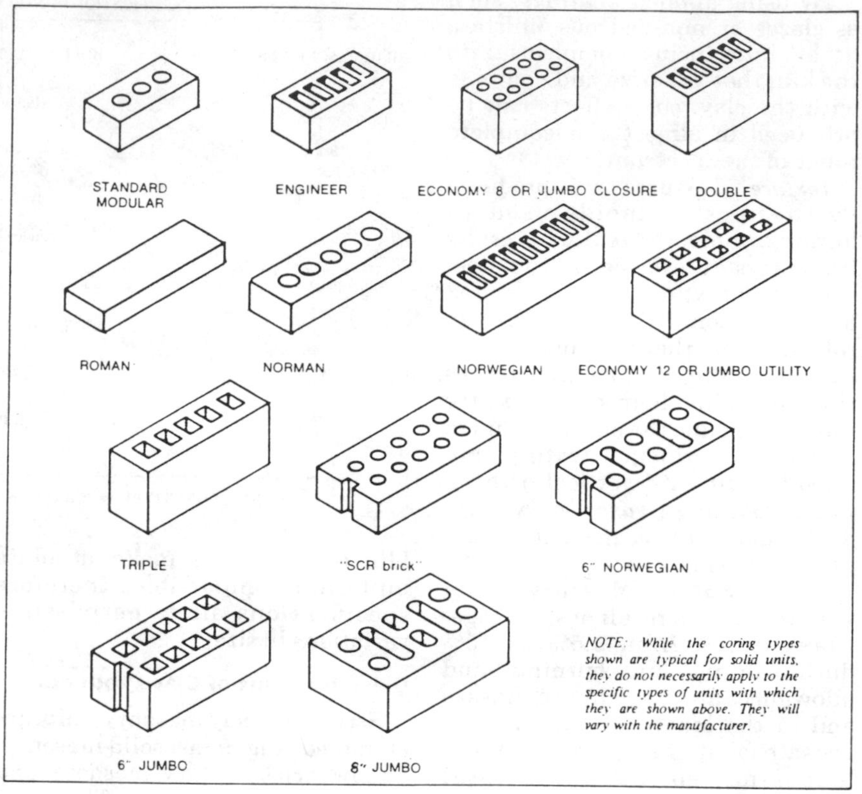

FIG. 14 HOLLOW BRICK NOMINAL DIMENSIONS

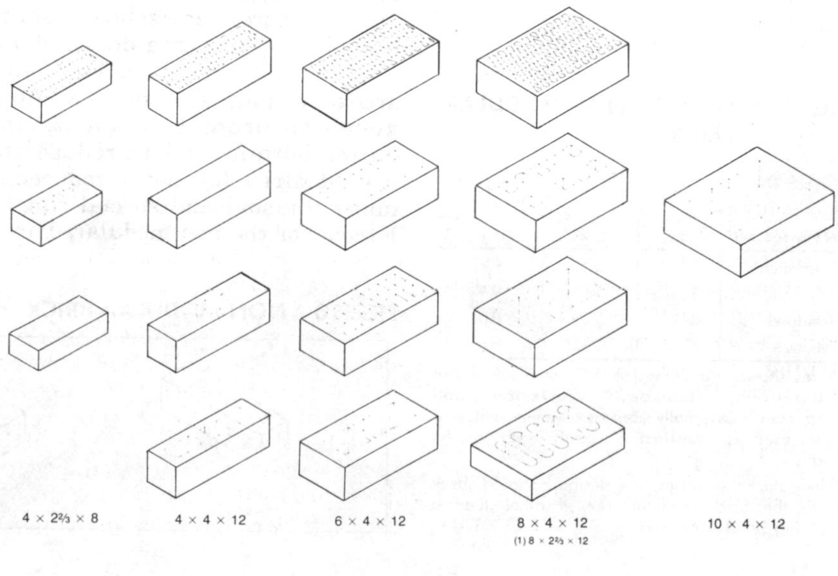

FIG. 13 **SIZES OF MODULAR BRICK[1]**

Unit Designation	Nominal Dimensions, in. t	h	l	Joint Thickness in.	Manufactured Dimensions in. t	h	l	Modular Coursing in.
Standard Modular	4	2 2/3	8	3/8	3 5/8	2 1/4	7 5/8	3C = 8
				1/2	3 1/2	2 1/4	7 1/2	
Engineer	4	3 1/5	8	3/8	3 5/8	2 13/16	7 5/8	5C = 16
				1/2	3 1/2	2 11/16	7 1/2	
Economy 8 or Jumbo Closure	4	4	8	3/8	3 5/8	3 5/8	7 5/8	1C = 4
				1/2	3 1/2	3 1/2	7 1/2	
Double	4	5 1/3	8	3/8	3 5/8	4 15/16	7 5/8	3C = 16
				1/2	3 1/2	4 13/16	7 1/2	
Roman	4	2	12	3/8	3 5/8	1 5/8	11 5/8	2C = 4
				1/2	3 1/2	1 1/2	11 1/2	
Norman	4	2 2/3	12	3/8	3 5/8	2 1/4	11 5/8	3C = 8
				1/2	3 1/2	2 1/4	11 1/2	
Norwegian	4	3 1/5	12	3/8	3 5/8	2 13/16	11 5/8	5C = 16
				1/2	3 1/2	2 11/16	11 1/2	
Economy 12 or Jumbo Utility	4	4	12	3/8	3 5/8	3 5/8	11 5/8	1C = 4
				1/2	3 1/2	3 1/2	11 1/2	
Triple	4	5 1/3	12	3/8	3 5/8	4 15/16	11 5/8	3C = 16
				1/2	3 1/2	4 13/16	11 1/2	
SCR brick [2]	6	2 2/3	12	3/8	5 5/8	2 1/4	11 5/8	3C = 8
				1/2	5 1/2	2 1/4	11 1/2	
6-in. Norwegian	6	3 1/5	12	3/8	5 5/8	2 13/16	11 5/8	5C = 16
				1/2	5 1/2	2 11/16	11 1/2	
6-in. Jumbo	6	4	12	3/8	5 5/8	3 5/8	11 5/8	1C = 4
				1/2	5 1/2	3 1/2	11 1/2	
8-in. Jumbo	8	4	12	3/8	7 5/8	3 5/8	11 5/8	1C = 4
				1/2	7 1/2	3 1/2	11 1/2	

[1]Available as solid units conforming to ASTM C216 or ASTM C62, and sometimes as hollow brick conforming to ASTM C652.

Technological improvements in masonry construction and lower production, transportation and installation costs have given hollow brick advantages over solid brick in both speed of erection and economy of construction.

Engineered brick masonry design and construction have created a trend toward larger brick units. Typical hollow brick units are shown in Figure 14. The greater core area makes bricks lighter and permits easier handling of larger sizes; larger sizes permit faster, more economical construction. For example: three 8″ x 4″ x 12″ units produce a full square foot of 8″ thick wall, compared to thirteen units of standard size brick for a square foot of similar 8″ wall.

Many home builders are using 4″ hollow brick as an economical alternative to multi-whythe solid brick on solid brick veneer construction.

Structural clay tile is produced as load-bearing and non-load-bearing. A few of the shapes and sizes are shown in Figure 16.

Structural facing tile is produced as glazed or unglazed. Glazed tile is produced from a high grade, light-burning fire clay to which a ceramic glaze has been applied. The glaze fuses into a glass-like coating on the unit during burning. Unglazed tile is produced from light- or dark-burning clays and shales and may have smooth or rough textural finishes. The nominal modular sizes of structural facing tile are shown in Figure 15 and Figure 16 illustrates a few of the common and special shapes that are produced.

Architectural Terra Cotta Architectural terra cotta is a product made to order as a veneer for both exterior and interior walls and for ornamentation.

Clay Masonry— Material Specifications

Standard specifications for the numerous types and grades of brick and tile have been developed by the American Society for Testing & Materials (ASTM) and are widely used to establish standards of quality. It is recommended that the appropriate ASTM specifications be included by reference in all specifications for brick, hollow brick, structural clay tile and/or structural facing tile.

ASTM specifications for brick, hollow brick and structural clay tile cover two or more *grades* based on the weathering index (Fig. 17). Specifications for facing brick, hollow brick and ceramic glazed structural facing tile cover requirements for two or more *types* based on color and size variations. Specifications for structural clay facing tile cover two *types* and two *classes*. When

FIG. 15 **NOMINAL MODULAR SIZES OF STRUCTURAL FACING TILE***

Thickness	Face Dimension in Wall: Height	Length
2, 4, 6 & 8″	4″	8″ & 12″
2, 4, 6 & 8″	5-1/3″	8″ & 12″
2, 4, 6 & 8″	6″	12″
2, 4, 6 & 8″	8″	12″ & 16″

*Nominal sizes include the thickness of the mortar joint for all dimensions.

FIG. 16 **HOLLOW CLAY TILE MASONRY UNITS**

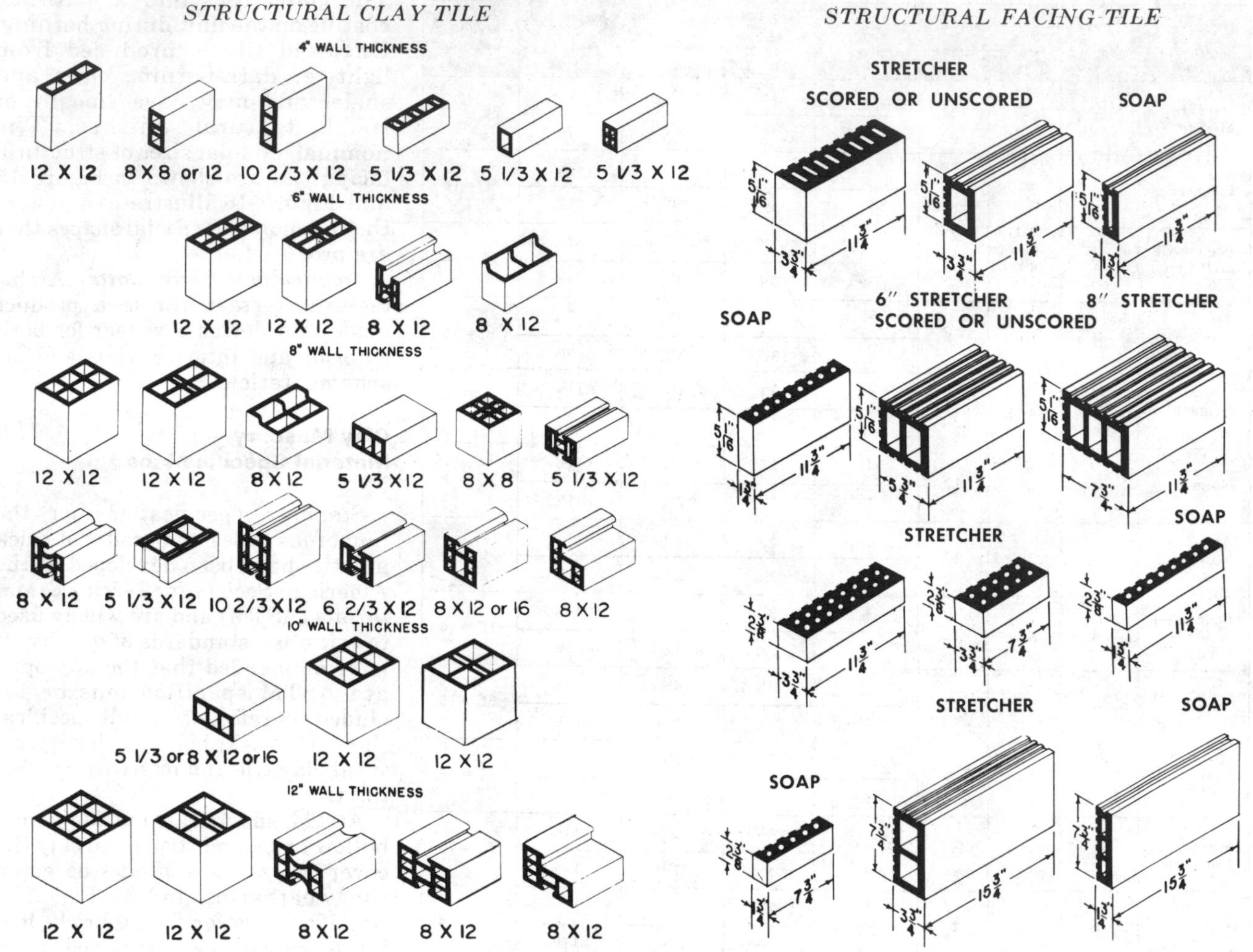

ASTM specifications are used to establish the masonry quality standards for a project, it is essential that the grade, type and class of the product be specified.

Weathering Index The effect of weathering on brick, related to the weathering index, is the product of the annual number of *freezing cycle days* and the annual *winter rainfall*.

A freezing cycle day is a day the air temperature passes above or below 32°F (0°C). The average annual number of freezing cycle days equals the difference between the mean number of days the *minimum* temperature is 32°F or below and when the *maximum* temperature is 32°F or below. Winter rainfall is the sum, in inches, of the mean monthly precipitation occurring between the first and last killing frosts in the

FIG. 17 **UNITED STATES WEATHERING INDEXES**

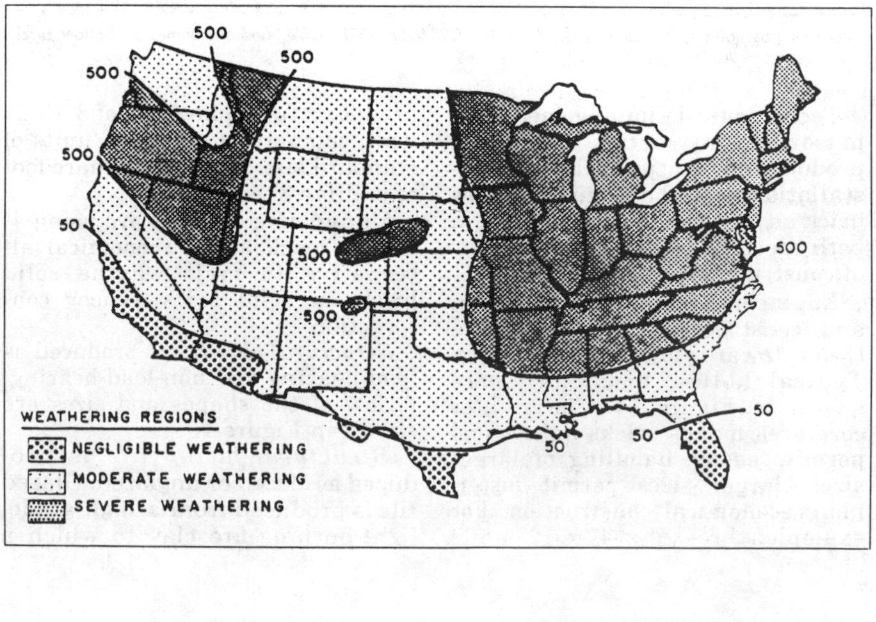

fall and spring.

A severe weathering region has an index greater than 500; a moderate region, 50 to 500; and a negligible region, less than 50. Figure 17 indicates general areas where brick is subject to severe, moderate and negligible weathering.

A summary of the physical requirements for various types of clay masonry units is given below. A guide to the selection of units based on *grade, type,* and *class,* and an explanation of each, is given in the Work File.

Building (Common) Brick ASTM C62
Covers three grades (SW, MW, NW), based on resistance to weathering. Physical requirements of each grade are given in Figure 18.

Facing Brick ASTM C216
Covers two grades (SW, MW), based on resistance to weathering, and three types (FBX, FBS, FBA), based on factors affecting appearance of the finished wall. Physical requirements are the same as for Building Brick, grades SW and MW (see Figure 18).

Hollow Brick ASTM C652
Covers two grades (SW, MW), based on resistance to weathering, and four types (HBS, HBX, HBA, HBB) based on factors affecting appearance of the finished wall. Physical requirements are the same as for Building Brick, grades SW and MW, and are given in Figure 18.

Structural Clay Load-Bearing Tile ASTM C34
Covers two grades, LBX, LB, based on strength and resistance to weathering. Physical requirements for each grade are given in Figure 19.

Structural Clay Non-Load-Bearing Tile ASTM C57
Covers one grade only (NB). The only physical requirement is a 28% limitation on maximum water absorption.

Structural Clay Facing Tile (Unglazed) ASTM C212
Covers two classes (Standard, Special Duty), based on thickness of face shells, and two types (FTX, FTS), based on factors affecting the appearance of the finished wall. Requirements for each type and class are given in Figure 20.

FIG. 18 **PHYSICAL REQUIREMENTS**

Designation	Minimum Compressive strength (brick flatwise) psi, gross area		Maximum Water Absorption by 5-hr. Boiling, %		Maximum Saturation Coefficient*	
	Avg. of 5 brick	Individual	Avg. of 5 brick	Individual	Avg. of 5 brick	Individual
Grade SW	3000	2500	17.0%	20.0%	0.78	0.80
Grade MW	2500	2200	22.0%	25.0%	0.88	0.90
Grade NW	1500	1250	no limit	no limit	no limit	no limit

*The saturation coefficient is the ratio of absorption by 24-hr. submersion in cold water to that after 5-hr. submersion in boiling water.

FIG. 19 **PHYSICAL REQUIREMENTS FOR LOAD-BEARING STRUCTURAL CLAY TILE**

GRADE	Maximum Water Absorption[1] by 1-hr. boiling, %		Minimum Compressive Strength (based on gross area)[2], psi			
			End Construction Tile		Side Construction Tile	
	Avg. of 5 Tests	Individual	Avg. of 5 Tests	Individual	Avg. of 5 Tests	Individual
LBX	16%	19%	1400	1000	700	500
LB	25%	28%	1000	700	700	500

[1] The range in percentage absorption for tile delivered to any one job should be not more than 12.
[2] Gross area of a unit should be determined by multiplying the horizontal face dimension of the unit, as placed in the wall, by its thickness.

FIG. 20 **MAXIMUM WATER ABSORPTION OF UNGLAZED STRUCTURAL CLAY FACING TILE**

TYPE	By 24-hr. Submersion in cold water, %		By 1-hr Boiling %	
	Average	Individual	Average	Individual
FTX	7%	9%	9%	11%
FTS	13%	16%	16%	19%

COMPRESSIVE STRENGTH BASED ON GROSS AREA

CLASS	End Construction Tile		Side Construction Tile	
	Minimum Avg. of 5 tests, psi	Individual Minimum psi	Minimum Avg. of 5 tests, psi	Individual Minimum psi
Standard	1400	1000	700	500
Special Duty	2500	2000	1200	1000

Facing Brick and Structural Clay Facing Tile (Ceramic Glazed) *ASTM C126*

Covers ceramic glazed facing brick and solid units as well as facing tile having a finish consisting of a ceramic color or clear glaze. Includes two grades: Grade S (select) for use with comparatively narrow mortar joints, and Grade SS (select sized or ground edge) for use where variation of face dimension must be very small; and two types: Type I (single-faced units) for use where only one finished face will be exposed, and Type II (two-faced units) for use where two opposite finished faces will be exposed. The physical requirements include minimum compressive strength, tests of finish (imperviousness, chemical resistance and crazing), and limitations on distortion and variation in dimensions.

FIG. 21 **PHYSICAL REQUIREMENTS OF PEDESTRIAN AND LIGHT TRAFFIC PAVING BRICK**[1]

Designation	Compressive Strength, flatwise, gross area, min, psi		Cold Water Absorption, max, %		Saturation Coefficient, max[2]	
Class	Avg. of 5 Brick	Individual	Avg. of 5 Brick	Individual	Avg. of 5 Brick	Individual
SX	8000	7000	8	11	0.78	0.80
MX	3000	2500	14	17	no limit	no limit
NX	3000	2500	no limit	no limit	no limit	no limit

[1]Minimum modulus of rupture values should be considered by the purchaser for uses of brick where support or loading may be severe.

[2]The saturation coefficient is the ratio of absorption by 24-h submersion in room temperature water to that after 5-h submersion in boiling water.

FIG. 22 **RAW MATERIAL SPECIFICATIONS FOR CONCRETE MASONRY UNITS**

CEMENTITIOUS MATERIALS	
Portland Cement	ASTM C150 ASTM C175 (air-entraining)
Portland-Blast Furnace Cement	ASTM 205
Natural Cement	ASTM C10
Fly Ash	ASTM C350
Hydrated Lime	ASTM C205—Type S
Siliceous & Pozzolanic	No ASTM Standards—Siliceous and pozzolanic materials should be shown by test or experience, not to be detrimental to the durability of the concrete.
AGGREGATES	
Normal Weight	ASTM C33
Lightweight	ASTM C331
ADMIXTURES	
Air-Entraining Agents	ASTM C260
Accelerators Calcium Chloride	ASTM D98
Coloring Pigments, Integral Water Repellents, etc.	No ASTM Standards—They should be previously established as suitable for use in concrete or should be shown by test or experience not to be detrimental to the durability.

Architectural Terra Cotta and Ceramic Veneer

These are custom products, and sizes and shapes are produced to meet specific job specifications.

Pedestrian and Light Traffic Paving Brick *ASTM C902*

Covers paving units to support pedestrian and light vehicular traffic in patios, walkways, floors, plazas and driveways. Three classes (SX, MX, NX), based on the severity of weather exposure, and three types (I, II, III), based on the severity of traffic exposure, are covered. Physical requirements for each class are given in Figure 21.

CONCRETE MASONRY

Concrete Masonry as used here refers to molded concrete units used primarily in construction as an integral part of the structure and includes: concrete brick, all types of hollow concrete block, slump block, split-face block and other special units. It does not include splash block, manhole block and other similar concrete products (see Division 200—Concrete Products).

Concrete Brick and Block

Solid and hollow masonry units are made from a relatively dry mixture of portland cement (and sometimes other cementitious materials), suitable aggregates, water and often admixtures. These materials are molded into the desired shape through compaction and vibration by machine, and are then cured under controlled conditions of moisture and temperature. After a period of aging, during which the required strength, moisture content and other desired properties are attained, the units are ready for use. The raw materials and methods of manufacture influence the

FIG. 23 **MANUFACTURING FLOW CHART**

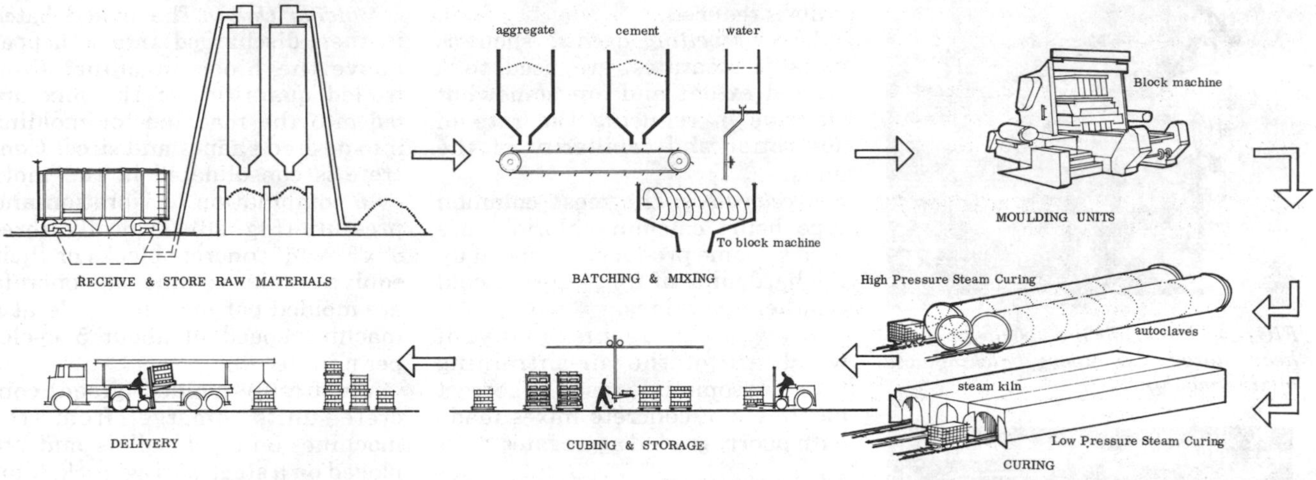

strength and other properties of the units in the finished wall, and a general understanding of both facilitates the selection and use of concrete masonry units.

Raw Materials Along with cementitious materials, water and aggregates, possible admixtures include mineral oxide coloring pigments, air-entraining materials, accelerators and water repellents.

Cementitious material in concrete masonry units consists primarily of portland cement. The cements used are mainly normal portland cement and, to a lesser extent, high-early-strength portland cement, including their air-entraining counterparts. High-early-strength cement provides greater strength during early stages of setting and curing and may reduce breakage during handling and early delivery. Portland blast-furnace slag cements, fly ash, silica flour and other pozzolanic and/or hydraulic materials also may be used as part of the cementing medium.

Pozzolanic materials, such as fly ash, possess little or no cementitious value in themselves, but have the ability to combine at ordinary temperatures with lime liberated from portland cement during hydration to form cementing compounds. The reaction is slow, contributing to strength development at ages of several months and longer. Materials such as silica flour react similarly at the elevated temperatures used in high-pressure (autoclave) curing. Fly ash and silica flour are both used as a portion of the cementing material with portland cement. Satisfactory proportions of these materials depend upon curing temperatures and type of aggregate.

Natural cement can satisfactorily replace up to 20% of the portland cement, although its use is not common.

Types of cementitious materials used in concrete masonry units and applicable ASTM specifications are shown in Figure 22.

Aggregates normally constitute about 90% of the block by weight and therefore have an important effect on the desirable properties as well as the economy of the finished units. In general, local availability primarily determines the use of any one type of aggregate.

Desirable aggregate properties include: (1) toughness, hardness and strength to resist impact, abrasion and loading; (2) durability to resist freezing and thawing and the expansion and contraction resulting from moisture and/or temperature changes; (3) uniform gradation of fine and coarse aggregate sizes to produce an economical, workable mixture and uniform appearance (the maximum size of aggregate should not exceed 1/3 of the smallest section of the block shell); and (4) cleanliness and lack of foreign particles which would impair strength and/or cause surface imperfections. The applicable ASTM specifications are shown in Figure 22.

Aggregates are classified according to their weight as: *dense or normal weight,* which include sand, gravel, crushed limestone and air-cooled slag; and *lightweight,* which include expanded shale or clay, expanded slag, coal cinders, pumice and scoria. The effect of aggregate type on weight, strength and other characteristics of concrete block is shown in Figure 33.

Admixtures are marketed for numerous uses in concrete mixes, but only a few have been found beneficial or desirable in the manufacture of concrete masonry units. The admixtures used and applicable ASTM specifications are shown in Figure 22. Admixtures should be investigated individually to determine whether they are applicable to the particular aggregate and will perform satisfactorily under individual plant conditions.

Air-entraining agents increase the plasticity and workability of a concrete mixture and distribute minute air bubbles uniformly throughout the hardened concrete, increasing its ability to withstand frost action. When used in the manufacture of concrete masonry units, air-entrainment permits greater compaction and produces a denser unit. Appearance is improved, a closer reproduction of the contours of the mold is obtained

FIG. 24 Materials are elevated to overhead storage bin above batching and mixing machine.

FIG. 25 Weigh batcher shown here positioned over the mixer.

FIG. 26 Block machine using vibration and pressure to mold units.

FIG. 27 Freshly molded concrete units are placed on steel curing racks.

and breakage of freshly molded units is reduced.

Water-repelling agents, such as metallic stearates, are used to a limited extent and are somewhat effective in reducing the rate of absorption and capillarity of the units.

Accelerators, the most common type being calcium chloride, are used by some producers to speed up the hardening of units during cold weather operations.

Workability agents, many of which are of the air-entraining type, are sometimes used to correct harshness of concrete mixes made with poorly graded aggregates.

Manufacture

While block plants vary in size, layout, equipment, type of automation and manufacturing details, the basic operations are essentially alike and include: (1) receiving and storage of raw materials, (2) batching and mixing, (3) molding units, (4) curing, (5) cubing and storage, and (6) delivery of finished units (Fig. 23).

Receiving Raw Materials Materials are delivered in bulk to the plant by truck or rail and elevated into overhead steel storage bins above the batching and mixing area (Fig. 24). At some plants, cement is delivered in special tank trucks or railroad cars and forced by compressed air through pipes directly into the overhead cement storage bin. Different aggregates are handled and stored separately to permit proper blending and proportioning.

Batching and Mixing Materials in the concrete mix are proportioned by weight with a *weigh batcher.* The weigh batcher is positioned under the storage bins and the correct amount of each material for a batch is discharged into the batcher. The weigh batcher is then positioned over the mixer and the materials are discharged into the mixer where they are mixed with a controlled quantity of water for a period of 6 to 8 minutes (Fig. 25). An electronic water meter senses the water contained in the aggregate and adjusts the amount of mixing water. Admixtures in liquid form may be added as part of the mixing water.

Molding Units The mixed batch is then discharged into a hopper above the block machine. Controlled quantities of the mix are fed into the machine for molding into desired shapes and sizes. Concrete is consolidated in the mold by a combination of vibration and pressure (Fig. 26). Typically, three 8″ x 8″ x 16″ concrete blocks or their equivalent in volume of concrete are molded per machine cycle at a machine speed of about 5 cycles per minute.

The freshly molded (green) concrete units emerge from the machines on steel pallets and are placed on a steel curing rack. This operation is termed *off bearing* and may be performed manually (Fig. 27) or with an automated rack loader.

Curing When a curing rack is filled with pallets of green units it is transported to either a steam curing kiln or autoclave for accelerated hardening. The principal curing methods used are: *low-pressure steam curing* in a kiln at temperatures of 150° to 185° F (accounting for approximately 80% of all block production), and *high-pressure steam curing* in an autoclave at temperatures of 300° to 350° F. Occasionally, where the climate is favorable, units are *moist cured* at normal temperatures of 70° to 100° F.

Low-Pressure Steam Curing, steam curing at atmospheric pressure, accelerates the hardening process using warm, moist air in the form of steam.

A typical curing cycle is shown in Figure 28 and consists of the following operations:

Holding Period—After the steam kiln has been filled with green units, the door is closed, beginning the preset or holding period. This period of delay (1 to 3 hours) prior to exposure to steam permits the units to attain initial hardening at normal temperatures (70°-100° F).

Heating Period—Following the holding period, the heating-up or steaming-up period begins. Heat is supplied by saturated steam or moist air injected into the kiln, controlled according to a predetermined time-temperature program. The temperature of the units is raised gradually at a rate not exceeding 60° F per hour, to a maximum of 150° to 165° F for units of normal weight aggregate and 170° to 180° F for units of lightweight aggregate.

Soaking Period—When the kiln air temperature reaches the predetermined maximum, the steam is shut off and the units are allowed to "soak" in the residual heat and moisture. Temperature is maintained for 12 to 18 hours or until the required strengths are developed. After soaking, an artificial drying period may be initiated by elevating the temperature in the kiln (dotted line in Fig. 28). Total curing and drying is generally accomplished within a 24-hour period.

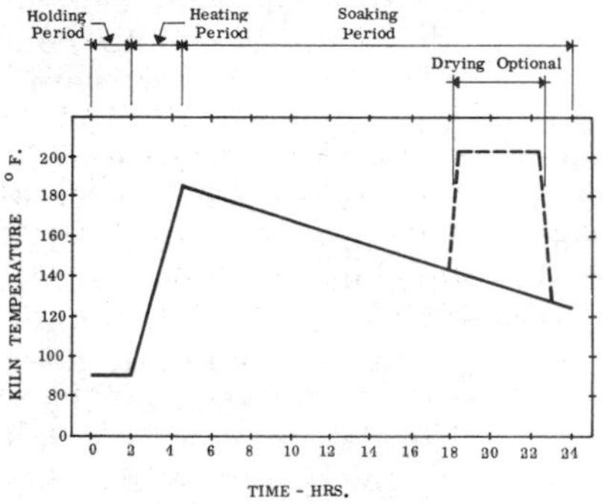

FIG. 28 Typical curing cycle.

The principal benefit of steam curing is the rapid strength gain of units, permitting their placement in inventory within hours after their molding. The compressive strength of low-pressure steam-cured block 2 to 4 days old attains 90% or more of final ultimate strength, whereas moist-cured block at normal temperatures develop strength slowly and have less than 40% of ultimate strength at this early age (Fig. 29). Low-pressure steam curing produces units of lighter color than usually obtained with moist curing.

High-Pressure Steam Curing accelerates the hardening process, using saturated steam under pressures usually of 125 to 150 psi. High-pressure curing is done in steel cylinders (autoclaves) varying in size from 6′ to 10′ in diameter and 50′ to 100′ in length (Fig. 30). Only saturated steam is used since dry, superheated steam would remove moisture from the units and prevent proper curing. The autoclave is usually designed for a gage pressure of 150 psi, which corresponds to a saturated steam temperature of 365.85° F. The temperature is the critical curing factor with the pressure necessary only to maintain proper steam quality. While many combinations of steam pressure and steam periods may produce equivalent results, a typical curing cycle consists of the following operations:

Holding Period—After molding, green units are held prior to loading in the autoclave for 2 to 5 hours to allow initial hardening.

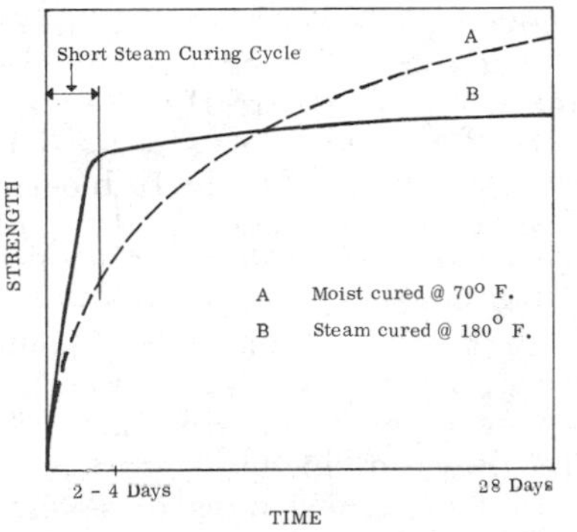

FIG. 29 Strength gain during curing.

Heating Period—Units are heated gradually in the autoclave so that full steam pressure is not attained in less than 3 hours. This is particularly desirable to improve the toughness of units made with lightweight aggregates.

Soaking Period—After a temperature of 350° F is reached, units are steamed from 5 to 10 hours (depending on the thickness of the units being cured) at constant temperature and pressure.

Blow Down—The pressure release period, following curing at constant temperature and pressure, can be made as short as feasible, 1/2 hour or less, without damage to the units. Such rapid pressure release is desirable because it facilitates rapid loss of moisture from the block without setting up shrink-

FIG. 30 Concrete masonry autoclave for high-pressure steam curing.

FIG. 31 After curing, concrete units are assembled into cubes.

FIG. 32 (a) Units being loaded on flatbed truck at plant. (b) Trucks are often mechanically equipped for unloading at the site.

FIG. 33 **PROPERTIES OF CONCRETE BLOCK MADE OF DIFFERENT AGGREGATES**

	Aggregate (Graded; 3/8" to 0.) TYPE	Density (air-dry) lbs/cu. ft.	Weight lbs/cu. ft. of Concrete	Weight 8" x 8" x 16" Unit	Compressive Strength (gross area) psi*	Water Absorption lbs/cu. ft. of Concrete	Thermal Expansion Coefficient (per°F) x 10-6
NORMAL WEIGHT	Sand and Gravel	130-145	135	40	1200-1800	7-10	5.0
	Limestone	120-140	135	40	1100-1800	8-12	5.0
	Air-cooled Slag	100-125	120	35	1100-1500	9-13	4.6
LIGHTWEIGHT	Expanded Shale	75- 90	85	25	1000-1500	12-15	4.5
	Expanded Slag	80-105	95	28	700-1200	12-16	4.0
	Cinders	80-105	95	28	700-1000	12-18	4.5
	Pumice	60- 85	75	22	700- 900	13-19	4.0
	Scoria	75-100	95	28	700-1200	12-16	4.0

*Multiply these values by 1.80 to obtain approximate corresponding values of strength of the concrete (strength of unit on net area).

age stresses (likely to develop with rapid drying in hot air at atmospheric pressure). With rapid pressure release, enough moisture is removed from normal weight aggregate blocks to bring them to a relatively stable air-dry condition soon after they have been removed from the autoclave and cooled to air temperatures. For lightweight aggregate block, some additional drying usually is needed to reach an air-dry condition.

The principal benefits of high-pressure steam curing are high early strength and greater stabilization against volume changes caused by varying moisture conditions. Units can be produced for use within 24 hours after molding which develop permanent compressive strengths at one day, equal to that of moist-cured units at 28 days. Concrete cured in an autoclave undergoes a different chemical reaction during hydration than concrete cured at atmospheric pressure, and produces more stable units with less tendency to shrink from drying or expand from wetting. The shrinkage of high-pressure steam-cured units, in drying from a saturated condition to equilibrium with air in a heated building, is about 50% less than similar moist-cured units.

Cubing and Storage When curing is complete the curing racks are removed from the steam kiln or autoclave and are conveyed to the cubing station where they are assembled into *cubes,* usually consisting of six layers of 15 to 18 blocks, 8" x 8" x 16", or an equivalent of other size units (Fig. 31). Cubes are transported to the storage area where they are placed in inventory of 3 or 4 cubes in height. Depending on the curing methods used, they remain in inventory anywhere from a few days to several weeks prior to delivery.

Delivery When ready for delivery, cubes are placed on flat-bed trucks. At the job site, cubes are picked up one or two at a time by the mechanical unloader and set at convenient site for use (Fig. 32).

Physical Properties

Like concrete, the physical properties of concrete masonry units are determined largely by the properties of the hardened cement paste and the aggregate. In general, much of the technical knowledge relating to concrete is applicable to concrete brick and block. However, differences do occur as a result of differences in mix composition and consistency, consolidation method, textural requirements, method of curing and other factors. Compared with structural concrete, concrete masonry units generally are made with a substantially lower cement factor (2.75 to 4 sacks of cement per cu. yd. versus 4 to 7 in concrete), and a much lower water-cement ratio (2 to 4 gals. per sack of cement versus 5 to 7). Aggregate is graded finer with the largest size seldom exceeding 3/8", and is often made of porous or lightweight materials. Concrete masonry units contain a relatively large volume of inter-particle void spaces (space between aggregate particles not filled with cement paste), and are cured at much higher temperatures.

Typical ranges in compressive strength, water absorption, density and other physical properties of commercial grade hollow concrete masonry units are shown in Figure 33.

Compressive Strength. The strength of concrete masonry units is difficult to predict from mix data since a simple water-cement ratio is not valid for harsh dry mixes, and because each type of aggregate exhibits different characteristics during mixing. The highest strengths are obtained from the wettest moldable mixes, but generally the maximum amount of mixing water is not used in order to cut down breakage of the freshly molded units during handling operations prior to curing. Drier mixes do not compact as well as wetter ones and therefore have lower strengths. Through experimentation, each manufacturing plant develops mixes and techniques to produce units of optimum strength compatible with other desirable properties. The compressive strength shown in Figure 33 is based on *gross-bearing area* of the concrete block, including core spaces. The compressive strength of the unit based on *net area*, excluding core space, is about 1.8 times the values shown. Solid face and veneer units (such as concrete brick and split block), when made with sand and gravel or crushed stone aggregate, will generally exceed 3000 psi in compression. The principal factors influencing compressive strength are: (1) the type and gradation of aggregate, (2) the type and amount of cementitious material, (3) the degree of compaction attained in molding the concrete units and (4) moisture content and temperature of the units at the time of testing. As shown, units made with lightweight aggregates generally exhibit lower compressive strengths than units made with heavier sand and gravel, limestone and air-cooled slags.

Tensile strength, flexural strength and modulus of elasticity of concrete masonry vary in proportion to compressive strength. Tensile strength will range generally from 7% to 10% of the compressive strength, flexural strength from 15% to 20% of compressive strength, and the modulus of elasticity from 300 to 1200 times compressive strength.

Water Absorption Absorption tests provide a measure of the density of the concrete. Absorption is a measure of the pounds of water per cu. ft. of concrete and in hollow concrete block varies over a wide range, from as little as 4 to 5 lbs. per cu. ft. for the heaviest sand and gravel units to as much as 20 lbs. per cu. ft. for the very porous, lightweight aggregate types. Solid units made with sand and gravel will normally have less than 7 lbs. per cu. ft. absorption. The absorption of the aggregate, which may vary from 1% to 5% of the dry weight of aggregate for dense aggregates to as much as 30% or more for the lightest aggregate, has strong influence on the absorption of the units.

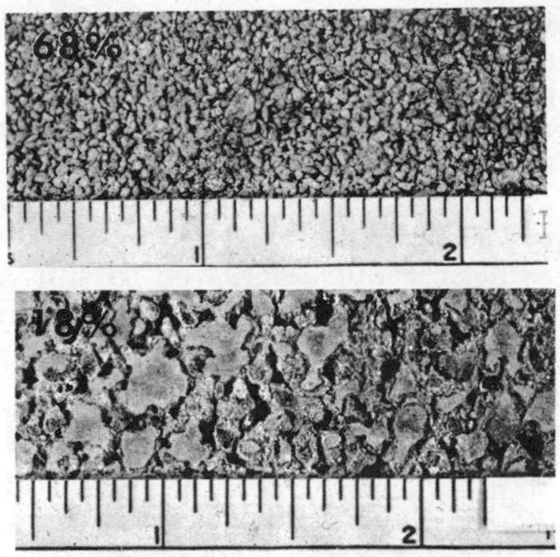

FIG. 34 Insulating values and sound absorption value vary greatly with different concrete masonry surface textures. As shown here, variation may range from fine to coarse.

The porosity and pore structure of the concrete influence other properties, such as permeability, thermal conductivity and sound absorption, but the influence is not always predictable, especially when comparing units made with different types of aggregates. High water absorption is not a desirable or purposely incorporated property, but it is accepted as a natural consequence when porosity properties such as lightness in weight, higher sound absorption or thermal insulation are desired.

A high initial rate of absorption (suction rate) results when concrete contains a large portion of relatively large, *inter-connected* pores and voids. Units with a high suction rate combined with high absorption will have high permeability to water, air and sound and may also have less resistance to frost action. On the other hand, *unconnected*, air-filled pores present in lightweight aggregate and in air-entrained cement paste impart the advantages due to porosity yet minimize permeability to water, air and sound.

An abnormally high suction rate adversely affects the structural bond of mortar to units but can be minimized by using mortars with high water retention properties.

All block intended for use in exterior walls and which will not be painted should have low absorption, and mortar joints should be carefully tooled for weathertightness.

Volume Changes Concrete masonry units undergo small dimensional variations due to changes in temperature, changes in moisture content and a chemical reaction called carbonation. Of these factors that cause dimensional changes in concrete masonry units, only moisture changes can be conveniently minimized to a significant degree. This can be done through preshrinking the units by adequate drying before they are built into the wall.

Carbonation causes irreversible shrinkage in the units as a result of a chemical reaction within the concrete when it absorbs carbon dioxide from the air. Though little test data exists, it appears that the magnitude of the change under certain conditions and over extended periods of time may approximate that due to moisture content change.

Temperature changes cause units to expand when heated and contract upon cooling. The volume changes are fully reversible, causing the unit to return to its original length after being heated and cooled through the same temperature range. The coefficient of thermal expansion for concrete block is dependent largely upon the coefficient of the aggregate, since it comprises roughly 80% or more of the concrete volume. Coefficients of thermal expansion for concrete block are shown in Figure 33.

Moisture changes cause units to expand when wetted and contract upon drying. Moisture volume changes are not quite fully reversible since during the first few cycles of wetting and drying there is a tendency for concrete to assume a permanent contraction. However, in subsequent cycles the movement is reversible.

An important factor contributing to the development of cracks in concrete masonry walls is the volume change in the units due to original drying shrinkage. If the units are laid up in the wall before they have become dimensionally stable and allowed to shrink, tensile stresses will develop wherever the wall is restrained from shrinking, and cracking can be expected. Drying shrinkage can be greatly reduced with proper curing and drying so that when placed in the wall the moisture content of the units is in equilibrium with the surrounding air; that is, dried down to the average air-dry condition to which the finished wall will be exposed in service. For a given aggregate type, a marked reduction in shrinkage with moisture change (as much as 50% reduction) can be obtained by autoclave curing compared with low-pressure curing or steam curing at normal temperatures.

The amount of residual shrinkage, or movement, a concrete unit will undergo due to moisture change after placement in a wall is the product of its shrinkage potential and its moisture loss after placement. (Residual shrinkage = shrinkage potential x moisture loss.) Shrinkage potential depends upon manufacturing technique, raw materials, age of block and other factors. Moisture loss depends upon the moisture content of the units when laid in the wall and the relative humidity and temperature of the drying environment. While measurable in the laboratory, the small magnitudes make it impractical to measure shrinkage on a routine basis as a means of demonstrating compliance with specifications. Instead, shrinkage is controlled by limiting the moisture content of the block at the time of delivery.

The correlation between moisture content and drying environment, expressed in terms of relative humidity, is shown in Figure 44. Masonry units complying with the limits for moisture content shown in Figure 45 will have a tolerable residual shrinkage; hence cracking of the wall will be minimized. The linear shrinkage classifications shown in Figure 44 depend on aggregate type and curing method. Average values of linear shrinkage are enumerated in Figure 45. *Normal weight* refers to sand-gravel, limestone and other dense aggregate block. *Lightweight* refers to all common lightweight aggregate types except pumice. Aggregate classifications are not necessarily exclusive and may overlap since many lightweight blocks contain minor amounts of normal weight aggregate.

Surface Texture Texture may be greatly varied (Fig. 34) to satisfy esthetic requirements or to suit a desired physical requirement (such as coarse texture to provide mechanical bond for direct application of plaster). With either normal weight or lightweight aggregates, units can be given practically any kind of surface texture by: (1) controlling the gradation of aggregates, (2) the amount of water used and (3) the degree of compaction at the time of molding. In addition, the surface of the unit may be ground to produce a unit to exact dimensions (Fig. 35), and the face mold can also be changed to produce numerous contours in the face shell of the unit (Fig. 36).

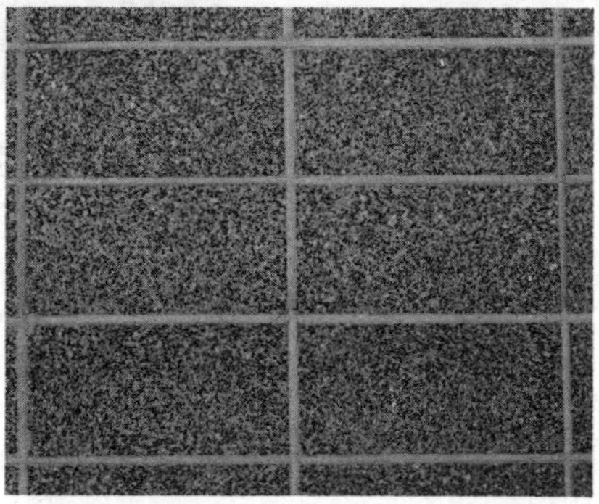

FIG. 35 Concrete masonry units may have face surfaces ground smooth to produce a desired appearance.

FIG. 36 One widely used decorative block, called Shadowal, may be laid in various patterns, thus producing a variety of effects.

Units having open surface textures offer some value in absorbing sound. Exposed concrete masonry walls built with the ordinary commercial run of block will absorb between 18% and 68% of the incident sound. Textures in Figure 34 illustrate this range. Surface finishes such as paint, which tend to close the surface pores, reduce the absorption significantly.

Color The usual aggregates and portland cements used in commercial production produce units with a range of color from off-white to gun-metal gray and from light tan to brownish shades. A wider color range, including blues, browns, buffs, greens, reds and slate grays, can be obtained using pure mineral oxide pigments singly or in combination, mixed integrally with the concrete before molding.

Classifications of Concrete Masonry Units

Concrete masonry units can be classified as *concrete brick*, con-

FIG. 37 **CLASSIFICATIONS OF CONCRETE MASONRY**

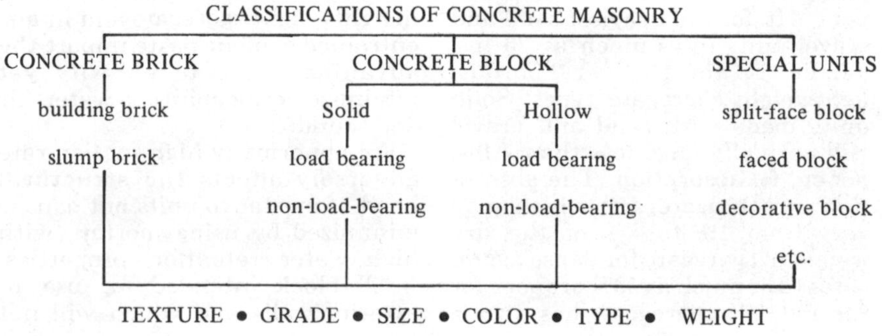

sisting of solid units; *concrete block*, consisting of both solid and hollow units; and *special units* made for a variety of purposes (Fig. 37).

Concrete Brick These units are produced completely solid or with a shallow depression called a *frog*. The purpose of the frog is to reduce weight and provide for better mechanical bond when laid in mortar. Units are sized to be laid with 3/8″ mortar joints, resulting in nominal modular dimensions of 4″ in width and 8″ in length. The thickness of vertical mortar joints is adjusted slightly so that 3 courses (3 brick and 3 bed joints) lay up 8″ in height. Some manufacturers produce the over-sized jumbo and double brick units shown in Figure 38.

Slump Brick and Block The concrete mixture used in the production of slump brick and block has a consistency that causes the units to sag or *slump* when removed from the molds, resulting in irregularly faced units varying considerably in height, surface texture and general appearance (Fig. 39). They are produced primarily to achieve special or unusual esthetic appearance.

Concrete Block Normally, concrete block is produced in three classes: (1) solid load-bearing units, (2) hollow load-bearing units and (3) non-load-bearing units.

All types may not be obtainable in every locality, but typical shapes of concrete block are shown in Figure 38. Half-length and half-height units are usually stocked for most of the units. When laid with 3/8″ mortar joints, a 7-5/8″ x 7-5/8″ x 15-5/8″ (actual dimension) unit produces nominal modular dimensions of 8″ x 8″ x 16″. Typical dimensions and component parts of a three-core corner unit are shown in Figure 40. Face shells and webs increase in thickness from the bottom of the unit to the top to permit easy removal of the core form during manufacture and to provide greater bedding area for mortar. End flanges of most units have mortar grooves 3/4″ wide by 3/16″ deep to increase transverse strength and to provide a vertical mortar key resulting in a weathertight joint.

Units shown in Figure 40 also are

FIG. 38 TYPICAL SHAPES OF CONCRETE BLOCK

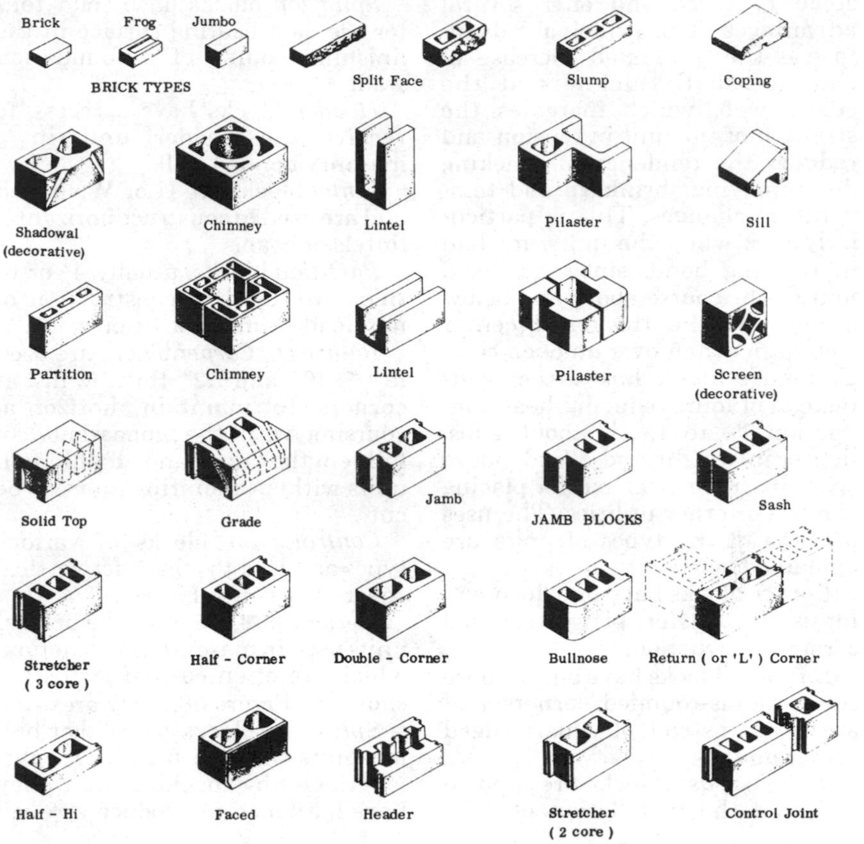

FIG. 39 Slump block produces a bold, irregular texture of varying size units.

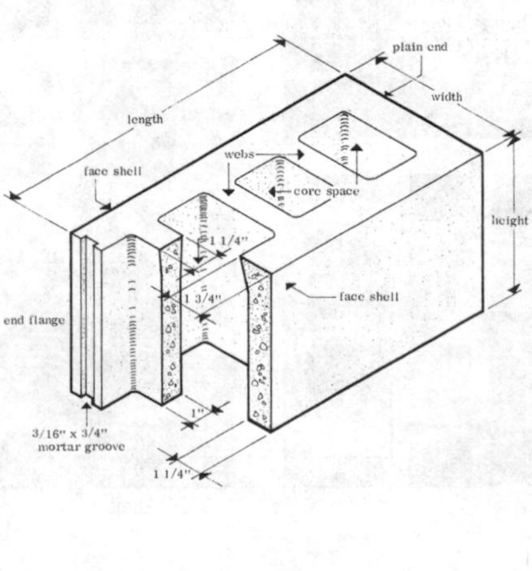

FIG. 40 Illustration of a typical 8″ x 8″ x 16″ concrete corner block.

FIG. 41 Split-face block produces a rough, natural, stone-like texture.

made with a two-core design instead of three and offer several advantages. The principal advantage is the graduated increase of the face shell thickness at the center web, which increases the strength of the unit in tension and reduces the tendency of cracking due to drying shrinkage and temperature changes. This is particularly true when the units are laid in running bond, since the head joint of the course above and below occur over the thickened center web rather than over an open core. Two-core block has three webs instead of four, reducing heat conduction 3% to 4%, is about 4 lbs. lighter in weight and affords more space in the vertical axis for placing conduit or other utilities. The uses of some of the types of units are explained below:

Corner blocks have one flush end for use in pilaster, pier or exposed corner construction.

Bullnose blocks have one or more small radius-rounded corners and are used instead of square-edged corner units.

Jamb or *sash* blocks are used to facilitate the installation of windows or other openings.

Solid top blocks have solid tops for use as a bearing surface in the finishing course of a foundation wall.

Header blocks have a recess to receive the header unit in a masonry bonded wall.

Lintel blocks are U or W shaped and are used to construct horizontal lintels or beams.

Partition blocks, usually 4″ or 6″ thick, are used in construction of non-load-bearing partition walls.

Return (L) Corner blocks are used in 6″, 10″ and 12″ thick walls at corners to maintain horizontal coursing with the appearance of full-length units and half-length units without requiring units to be cut.

Control joint blocks of various types are used to construct vertical shear-type control joints.

Special Units. Some of the special units used in masonry construction, which are often custom made, are shown in Figure 38. They are:

Split-Face blocks are solid or hollow units that are fractured (split) lengthwise by machine after they have hardened to produce a rough surface texture (Fig. 41). Color variations are obtained by introducing mineral colors and by using aggregates of different gradation and color in the concrete mixture. The units are laid in the wall with the fractured faces exposed. Units are available in random sizes (often larger than brick sizes) for broken joint patterns.

Faced blocks are units with ceramic glazed, plastic, polished or ground faces. In all cases the facing is applied in separate operations after production of the blocks. One type of glazed unit uses a glazing material which is a thermo-setting, resinous binder combined with glass silica and color pigments or colored granules cast onto each block as a semi-liquid in individual molds. Then, heat-treating sets the glazes and binds them integrally into the pores of the concrete unit.

Decorative blocks are manufactured in many different types with beveled face shell recesses that produce special appearances on wall surfaces. Different effects can be created by using the same block design in different pattern bonds.

FIG. 42. Decorative and special concrete masonry units: (a) Interior wall of double jamb block; (b) screen wall of sash block; (c & d) screen walls made from decorative block; (e) exterior wall made of standard pilaster block; (f) exterior screen made of standard solid concrete units.

Decorative and other special and custom units and their applications are shown in Figure 42.

Chimney blocks are designed in conjunction with a flue lining for use in chimney construction. *Pilaster* blocks are used in the construction of plain or reinforced pilasters and columns.

Concrete Masonry Material Specifications

Specifications of the American Society for Testing and Materials (ASTM) cover four types of concrete masonry units as summarized in Figure 43. These ASTM specifications represent the recognized industry standards for concrete masonry.

Grades Concrete masonry units intended for load-bearing construction are classified in ASTM specifications as Grade N and Grade S. Grade N units are suitable for general use, such as in exterior walls *below and above grade* that may be exposed to moisture penetration or the weather, and for interior walls and backup. Grade S units are limited to use *above grade*, in exterior walls with weather-protective coatings and in walls not exposed to the weather.

Types Either load-bearing or non-load-bearing units can be manufactured to specified limits of moisture content—designated as Type I. Type I masonry units are available in both grades N and S, and are identified as N-I or S-I. The moisture content of Type I units is regulated during manufacture to conform with limits established by governing ASTM specifications. Type II units are not restricted to a definite moisture content.

Moisture limits are established in order to minimize in-place shrinkage of units and consequent tendency to wall cracking. Limits vary depending on the average annual relative humidity, as reported by the U.S. Weather Service nearest source of manufacture (Fig. 44).

The recommended use of concrete masonry units with limited maximum moisture content is based on considerations of the change in moisture content and the shrinkage of units after installation. Where the construction is located in areas of low relative humidity and a low equilibrium moisture condition is expected, the maximum moisture content at the time of use should be lower than when the construction is located in a less arid region (Fig. 44). Also, a lower maximum moisture content is required when the shrinkage potential of the masonry unit is greater (Fig. 45).

In addition to regulating moisture content, ASTM specifications also establish requirements for strength and water absorption of masonry units (Fig. 46). The effect of moisture absorption and of other properties on masonry construction is described on page 205-16.

FIG. 43 **INDUSTRY STANDARDS FOR CONCRETE MASONRY UNITS**

Standard	Type of Unit
ASTM C55	Concrete building brick, solid veneer and split block.
ASTM C90	Hollow load-bearing concrete masonry units.
ASTM C129	Hollow non-load-bearing concrete masonry units.
ASTM C145	Solid[1] load-bearing concrete masonry units.

[1]Units with 75% or more net area.

FIG. 44 **MAXIMUM MOISTURE CONTENT OF TYPE I UNITS**

Linear Shrinkage Classification, %[1]	Max. Moisture Content, % of total absorption (avg. of 5 units)		
	Average Annual Relative Humidity, %[2]		
	Over 75	75 to 50	Under 50
Up to 0.03	45	40	35
0.03 to 0.045	40	35	30
0.045 to 0.065	35	30	25

[1] As determined by ASTM C426 Test for Drying Shrinkage of Concrete within previous 12 months.
[2] As reported by the U.S. Weather Service nearest source of manufacture.

FIG. 45 **LINEAR SHRINKAGE OF CONCRETE MASONRY UNITS**

Aggregate Type	Steam Curing Method	Average Linear Shrinkage, %[1]
Normal Weight	High Pressure	0.019
	Low Pressure	0.027
Lightweight	High Pressure	0.023
	Low Pressure	0.042
Pumice	High Pressure	0.039
	Low Pressure	0.063

[1] As determined by test, ASTM C426.

FIG. 46 **STRENGTH AND ABSORPTION REQUIREMENTS FOR CONCRETE MASONRY UNITS**

		Compressive Strength (min., psi) on Average Gross Area		Water Absorption (max., lbs/ft³) for Units of Different Densities			
				Lightweight Concrete		Medium-Weight Concrete	Normal-Weight Concrete
UNIT TYPE & STANDARD	**GRADE***	**Average of 3 units**	**Individual Unit**	**Less than 85 lbs/ft³**	**Less than 105 lbs/ft³**	**105 to 125 lbs/ft³**	**More than 125 lbs/ft³**
Hollow Load-bearing Units, ASTM C90	N-I, N-II S-I, S-II	1000 700	800 600	— 20	18 —	15 —	13 —
Solid Load-bearing Units, ASTM C145	N-I, N-II S-I, S-II	1800 1200	1500 1000	— 20	18 —	15 —	13 —
Concrete Building Brick, ASTM C55	N-I, N-II S-I, S-II	3500 2500	3000 2000	15 18	15 18	13 15	10 13
Hollow Non-loadbearing block, ASTM C129	—	350	300	—	—	—	—

*Grades S-I and S-II are limited to walls not exposed to the weather, and to exterior above grade walls protected with weather resistant coatings. Grades N-I and N-II may be used below grade as well as above; protective coatings are recommended below grade and where required above grade.

Mortar binds the individual masonry units into a wall or other building element and serves these four functions: (1) It bonds units together, sealing the spaces between them; (2) compensates for size variations in the units; (3) causes metal ties and reinforcing, if present, to act integrally with the wall; and (4) provides esthetic quality by creating shadow lines and/or color effects.

Mortar is a combination of one or more *cementitious* materials (portland cement, lime, masonry cement); a clean, well graded *sand;* and enough *water* to give a plastic, workable mixture. These materials should be proportioned to give the mortar a good balance of the following essential properties.

MORTAR PROPERTIES

Mortar is laid while plastic and then hardens. Thus it has two sets of properties: (1) those when it is in a plastic state, and (2) those after it has hardened. Plastic properties determine a mortar's construction suitability as do the properties of the hardened mortar. They both determine the suitability of the finished wall, which must be strong, durable and watertight.

For many years, the basic constituents of mortar have been portland cement, lime, sand and water. In recent years it has become commonplace to use additives in mortars which may be harmful to the hardened mortar. Since the properties of both plastic and hardened mortars depend largely on the mortar ingredients, they should conform to ASTM specifications and be properly proportioned.

Plastic Mortar Properties

Properties of mortars while plastic include *workability, water retentivity, initial flow* and *flow-after-suction.*

Workability A workable mortar is uniform, cohesive and of a consistency that makes it *usable* to the mason. A mortar is considered *workable* when particles in the mix do not segregate and when it is easy to spread, holds the weight of the units, makes alignment easy, clings to the vertical faces of masonry units and easily extrudes from the mortar joints without dropping or smearing.

Water retentivity, flow, resistance to segregation and other factors affect workability. These in turn are affected by properties of the mortar ingredients. This complex relationship makes quantitative estimates of workability difficult to obtain, and no standard laboratory test exists for measuring it.

Water Retentivity Water retention in a mortar prevents rapid loss of mixing water when the mortar contacts an absorptive masonry unit and, hence, prevents loss of plasticity. When in contact with a masonry unit of low absorption, a high degree of water retention prevents the mortar from *bleeding* moisture. Bleeding produces a thin layer of water between the mortar and the masonry unit, causing the unit to float and materially reducing bond. Water retention is measured in the laboratory by flow tests.

Initial Flow and Flow after Suction Mortar flow is determined by a simple laboratory test (Fig. 47). A truncated cone of mortar is formed on a flow table, a testing apparatus which is mechanically raised and dropped through a height of 1/2″, 25 times in 15 seconds. Under this treatment mortar will "flow," increasing its diameter. *Flow* is the ratio of the increase in diameter to the original 4″ cone base diameter, expressed as a percentage. (For example, if the diameter of the mortar mass is 8″ after the test, the *flow* is 100%.) Job conditions normally require construction mortar to have flows that range in the order of 130% up to 150%.

A similar test, flow-after-suction, indicates theoretical mortar flow after loss of water to an absorbent masonry unit. Flow of a mortar is measured and then, for 1 minute, the mortar is placed into a device which removes some water by vacuum. The mortar then is tested again for flow. These two laboratory tests give an indication of a mortar's water retention. The laboratory value of water retentivity is the ratio of flow-after-suction to initial flow, expressed in terms of percent.

FIG. 47 A laboratory test measures mortar flow by comparing the diameter of a cone of mortar before and after being rodded.

Hardened Mortar Properties

Properties of hardened mortar include *bond strength, durability, compressive strength, volume changes, appearance* and others.

Bond Strength Bond strength is perhaps the most important property of hardened mortar. Many variables affect bond, including: (1) mortar properties (type and amount of cementitious materials, water retention, air content, compressive strength); (2) type of masonry unit (surface texture, suction, moisture content); (3) workmanship (for example, pressure applied to masonry joint during laying); and (4) curing (temperature, relative humidity). The effects of some of these variables on bond are discussed below.

Air content in mortar above 12% is accompanied by a decrease in bond.

Mortar flow has a direct relationship with tensile bond strength; for all mortars, bond strength increases as flow increases.

Good workmanship is important. The time lapse between spreading mortar and placing units will affect mortar flow, particularly when assembling high suction units. For highest bond strength this time interval should be reduced to a minimum.

Because all mortar is not used immediately after mixing, some of its water may evaporate as it stands on the mortar board. The addition of water to mortar (retempering) is sometimes prohibited because it is thought to have detrimental effects. Although compressive strength will be reduced slightly if mortar is retempered, bond strength may be materially lowered if it is not. For this reason, retempering should be permitted to replace water lost by evaporation if the mortar has not begun to set. All mortar should be used within 2 hours after initial mixing.

Suction of brick, determined by test, is defined as the amount of water in grams absorbed by a surface of 30 sq. in., placed in water to a depth of 1/8" for 1 minute (Fig. 48). In practically all cases, mortar bonds best to brick whose suction is 5 to 20 grams of water at the time of laying. If brick suction exceeds 60 grams of water at the time of laying, bond may be extremely poor regardless of the type of workmanship or mortar used. Brick suction should be controlled by wetting brick prior to laying. (See page 205-29 for field test to determine if brick should be wetted prior to laying.)

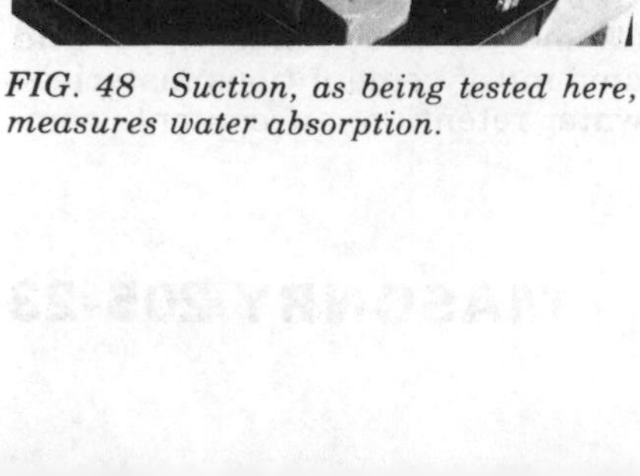

FIG. 48 Suction, as being tested here, measures water absorption.

The suction test is not applicable to concrete masonry units. Although experience indicates that concrete block at a moisture content below 15% to 20% of total absorption can draw water out of the mortar on hot, dry days, concrete block suction is usually insignificant because the units generally contain sufficient moisture to retard suction effects.

Movement will often be highly detrimental to bond. Tapping or otherwise attempting to move the unit (Fig. 49), even if done immediately after placement, can break the bond between unit and mortar, after which the mortar will not readhere well to the masonry units. Generally, units should not be realigned, tapped or moved after initial alignment; however, with low suction brick, units float and may require realignment.

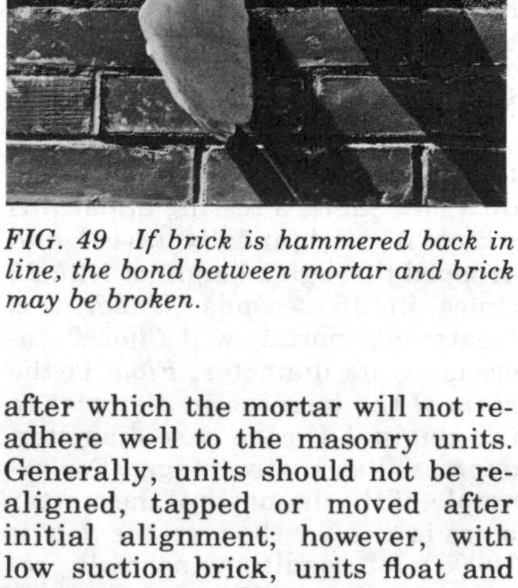

FIG. 49 If brick is hammered back in line, the bond between mortar and brick may be broken.

Durability The durability of mortar is measured principally by its ability to resist repeated cycles of freezing and thawing under natural weather conditions. Experience indicates that the durability of relatively dry masonry which resists water penetration is not a serious problem. It is generally conceded that masonry walls will stand 35 years or longer before requiring maintenance—a good indication of mortar durability.

An increase in air content may increase the durability of masonry mortar, but it will substantially decrease bond and other desirable properties. Because mortar is quite durable without air entrainment, air entraining ingredients to increase air content are *not* recommended.

Compressive Strength The compressive strength of mortar depends largely upon the amount of portland cement in the mix. Compressive strength increases with an increase in cement content of mortar and decreases with an increase in water content, and thus with increased flow.

Since there are few reports of structural distress or failures due to compressive loading, it is unimportant to use greater than moderate strength mortars for general construction. Bond strength, workability and water retentivity are considered more important than compressive strength and are normally given principal consideration in specifications.

Volume Change It is popularly believed that mortar shrinkage can be extensive and can cause leaky walls. Actually, however, maximum possible shrinkage in a tooled mortar joint containing sound mortar is so small that any resultant crack could not be seen with the naked eye. Research and field observations have shown that good materials, workmanship and design are necessary to obtain a watertight wall. Shrinkage is insignificant in mortars that have a good balance of all desirable properties and which have been tooled.

Watertightness The significant leakage of masonry walls generally is through fine cracks between the mortar and the units. Watertightness of masonry units and of mortars in common use is not a controlling factor.

Rate of Hardening The rate of hardening of mortar is the speed at which mortar develops resistance to indentation and crushing. Overly rapid hardening may interfere with the use of the mortar by the mason. Overly slow hardening may impede the progress of the work or may subject the mortar to early damage from frost action during winter. A well defined, consistent rate of hardening allows the mason to tool the joints at the same degree of hardness and thus obtain uniform joint color.

Efflorescence Mortar containing large amounts of soluble salts can

FIG. 50 Efflorescence, a white powdery mineral salt deposit appearing on masonry surfaces.

cause *efflorescence* if excessive water penetrates the masonry. Efflorescence is a white powdery substance appearing on wall surfaces (Fig. 50). It is composed of one or more water-soluble salts originally present in the mortar (or other masonry material) and is carried to the surface by movement of water which has entered the wall; it is left on the face of the masonry when the water evaporates. Methods of preventing and removing efflorescence are discussed under Workmanship, page 205-35.

Mortar Color Uniformity of joint color greatly affects the overall appearance of the masonry. Atmospheric conditions, moisture content of the masonry units and admixtures influence the shade of the mortar joints. The most important factors—uniformity of the mix and time of tooling of the mortar joint—are controllable. Careful measurement of materials and thorough mixing are important in providing uniformity from batch to batch and from day to day. If the mason tools the joint when the mortar is relatively hard, a darker color results than if he tools the joint when the mortar is relatively soft. To insure a uniform mortar color in the finished wall, the mason should tool the joints at like degrees of mortar hardness (Fig. 51).

Colored Mortars may be obtained through use of colored aggregates or suitable pigments. The use of colored aggregates is preferable. White sand, ground granite, marble or stone usually give permanent color and will not weaken the mortar. White sand, ground limestone or ground marble with white portland cement and lime should be used for white joints.

Mortar pigments must be fine enough to disperse throughout the mix, be capable of imparting the desired color when used in permissible quantities and must not react with other ingredients to the detriment of the mortar. These requirements generally are met by metallic oxides. Iron, manganese and chromium oxides, carbon black and ultramarine blue have been used successfully. Organic color should be avoided, and colors containing Prussian blue, cadmium lithopone, and zinc and lead chromates.

FIG. 51 Joints should be tooled when the mortar is of like degree of hardness to prevent variation in color.

The minimum quantity of pigment that will produce the desired results should be used, for an excess may seriously impair strength and durability. The maximum permissible quantities of most metallic oxide pigments are in the order of 10% to 15% of the cement content by weight. Caution should be exercised when using carbon black by limiting it within 2% to 3% of the cement content by weight, since greater amounts will seriously impair mortar strength.

For best results the color should be premixed with portland cement in large, controlled quantities—a service provided by some manufacturers. Premixing larger quantities will assure color more uniform than can be obtained by mixing smaller batches at the job.

MORTAR TYPES

As defined above, mortar is a mixture of one or more cementitious materials, a clean, well graded sand and as much water as possible to give a workable mix. Four mortars are described below.

Lime Mortars

Straight lime mortars (lime + sand + water) harden at a slow, variable rate, develop low compressive strength and have poor durability to the freeze-thaw cycle; but they do have high workability and high water retention. Lime hardens only upon contact with air; therefore, complete hardening occurs very slowly over a long period of time. Should cracking occur, *healing*, or the recementing of small hairline cracks, will tend to keep water penetration to a minimum.

Portland Cement Mortars

Straight portland cement mortars (portland cement + sand + water) harden quickly at a consistent rate and develop high compressive strengths with good durability to the freeze-thaw cycle, but have low workability and low water retention.

Portland Cement—Lime Mortars

To combine the advantages of both and to compensate for the disadvantages of each, portland cement-lime mortars (portland cement + lime + sand + water) are widely used. In portland cement-lime mortars, the portland cement contributes to durability, high early strength, consistent rate of hardening and high compressive strength. Lime contributes to workability, water retentivity and elasticity. Both contribute to bond strength. Sand acts as a filler, providing the most economical mix and contributing to strength. Water is the mixing vehicle which creates plastic workability and initiates the cementing action.

Masonry Cement Mortars

Masonry cement mortars (masonry cement + sand + water) are made from proprietary mortar mixes called masonry cement. The manufacturer combines into one bag portland cement with such materials as natural cement, finely ground limestone or hydrated lime, an air-entraining agent and gypsum to regulate the time of set. The principal advantage of masonry cement mortar is its convenience in mixing, and generally good workability. However, since the masonry cement standard, ASTM C-91, does not limit the additives, as does the portland cement standard, each manufacturer has the option of varying the types and extent of additives. This makes it difficult to predict the performance of masonry assemblies erected with masonry cements in general, although laboratory test results from individual manufactuers for specific cements can serve as a guide.

MORTAR MATERIALS

Standard specifications for mortar materials are contained in Specifications for Mortar for Unit Masonry, ASTM Designation C270. These specifications are based on laboratory tests, field use and are the result of over 25 years' experience.

Mortar materials should conform to the appropriate ASTM specifications given below. They are available in all localities and can be obtained at the same price as products that do not meet the specifications.

Cementitious Materials

Portland Cement (ASTM C150) Three types under ASTM C150, Type I, Type II or Type III, are used in mortar.

Air-Entraining Portland Cement (ASTM C150) is not recommended for use in mortars.

Since the amount of air entrained may be difficult to predict and since excessive air entrainment has the detrimental effect of reducing bond between mortar and masonry units or reinforcement, air-entraining cements should not be used.

Masonry Cements (ASTM C91) Two types are manufactured under C91, Type I and Type II. Only Type II masonry cement should be used in masonry mortars.

Quicklime (ASTM C5) Quicklime is essentially calcium oxide, CaO. It must be carefully mixed with water (slaked) and stored for as long as 2 weeks before use. It is always used as a mixture of lime and water having the consistency of putty.

Hydrated Lime (ASTM C207) Hydrated lime is quicklime that has been slaked before packaging, the calcium oxide having been converted into calcium hydroxide, $Ca(OH)_2$. Hydrated lime can be mixed and used without delay and therefore is much more convenient to use. Hydrated lime is available in two types, S and N. Only Type S hydrated lime is recommended for masonry mortar because ASTM specifications do not control the amount of unhydrated oxides and plasticity of Type N.

Sand (ASTM C144)

Either natural or manufactured sands may be used and should be clean, sound and well graded. For best workability, sand should include all sizes of particles ranging in size from very fine up to coarse material (not larger than 1/4″). Sand has an important effect on the workability and durability of the mortar. Sands deficient in very fine particles (less than 5% to 15% fines) produce harsh, hard-to-work mortars requiring additional cement or lime to obtain a workable mix. Sands deficient in large particles generally result in weaker mortars. Many commercially available sands do not comply with gradation limits set by ASTM C144, but gradation often can be altered easily and inexpensively by adding fine or coarse sands.

Water

If water is clean and free of deleterious acids, alkalies or organic material, it is suitable for masonry mortars.

FIG. 52 **MORTAR PROPORTIONS BY VOLUME**[1]

TYPE[2]	Portland Cement	Hydrated Lime or Lime Putty	Masonry Cement	Maximum Damp Loose Aggregate[3]	Minimum Compressive Strength 2″ Cubes @ 28 Day (psi)
M	1	1/4	—	3	2500
	or 1	—	1	6	2500
S	1	1/2	—	4-1/2	1800
	or 1/2	—	1	4-1/2	1800
N	1	1	—	6	750
	or —	—	1	3	750
O	1	2	—	9	350
K	1	4	—	15	75

[1]The weight of 1 cu. ft. of materials used is considered to be: Portland cement, 94 lbs (1 bag); hydrated lime, 40 lbs; lime putty, 80 lbs; dry sand, 80 lbs; masonry cement, weight printed on bag.
[2]Shaded figures show proportions for portland cement-lime mortar. Mortars made with masonry cement are unshaded.
[3]The damp loose aggregate should not be less than 2¼ times, nor more than 3 times the sum of the cementitious material used.

MORTAR PROPORTIONS AND USES

Selection of the type of mortar depends upon the type of masonry and its location in the building. None of the mortar types listed below will produce a mortar that rates highest in all desirable mortar properties. Adjustments in the mix to improve one property often are made at the expense of others. For this reason, the properties of each type should be evaluated and the mortar type chosen that will best satisfy the end-use requirements.

Proportions

The five types of mortars (M,S, N,O and K) recommended for the construction of unit masonry are shown in Figure 52, column 1; the proportions by volume of each type are listed in columns 2, 3, 4 and 5. The damp, loose volume of aggregate (col. 5) should be not less than 2-1/4 times, nor more than 3 times the total separate volumes of the cementitious materials used in the mortar.

Type M Mortar is a high strength mortar and has somewhat greater durability than other mortar types. It is specifically recommended for reinforced masonry below grade and in contact with earth in such use as foundations, retaining walls, walks, sewers and manholes. It is recommended specifically for reinforced masonry and where the highest compressive strength is important.

Type S Mortar is a medium-high-strength mortar recommended for use where bond and lateral strength are more important than higher compressive strength. Tensile bond strength between brick and Type S mortar approaches the maximum obtainable with cement-lime mortars. It is recommended for use in reinforced masonry and for unreinforced masonry where maximum flexural strength is required.

Type N Mortar is a medium-strength mortar recommended for general use in exposed masonry above grade where high compressive and lateral masonry strengths are not required. It is specifically recommended for chimneys, parapet walls and exterior walls subject to severe exposure.

Type O Mortar is a low-strength mortar suitable for general interior use in non-load-bearing masonry. It may be used for load-bearing walls of solid masonry where compressive stresses do not exceed 100 lbs. per sq. in., provided exposures are not severe. In general, Type O mortar should not be used where it will be subjected to freezing action.

Type K Mortar is a *very-low-strength* mortar suitable for use in interior non-load-bearing partition walls where low strength is not objectionable.

USES (General)

No single mortar type is best for all purposes. The basic rule is to *select a mortar that is only as strong (in compression) as it needs to be to satisfy the structural requirements of the wall.*

The recommended mortar types for various construction applications are given in Figure 53.

USES (Specific)

For cavity walls where wind velocities will exceed 80 mph, Type S mortars should be used. For locations where winds of lower velocity are expected, Type S or Type N may be used.

For facing tile any of the first four mortar types are acceptable. ASTM C144 allows sand particle size up to 1/4″. However, where 1/4″ joints are specified, all aggregates should pass through a No. 16 sieve (slightly less than 1/16″). For white joints, white portland cement with white aggregate should be used. As an alternate method for white or colored joints, all joints may be raked to a depth of 1/2″ and then pointed with mortar of the desired color.

For tuck pointing, only prehydrated mortars should be used. To prehydrate mortars, all ingredients should be thoroughly mixed without water, then mixed again, adding

FIG. 53 **RECOMMENDED MORTAR TYPES FOR VARIOUS CONSTRUCTION APPLICATIONS**

Construction Application	Recommended Minumum ASTM Mortar Types	Order of Relative Importance of Principal Properties		
		Plasticity[1]	Compressive Strength	Weather Resistance
Foundations, basements, walls, isolated piers[2]	M, S	3	2	1
Exterior Walls	S, N	2	3	1
Solid Masonry unit veneer over wood frame	N	2	3	1
Interior Walls—Load-bearing	S, N	1	2	3
Interior partitions—Non-load-bearing	N, O	1	—	—
Reinforced masonry (columns, pilasters, walls, beams)	M, S[3]	3	1	2

[1]Adequate workability and a minimum water retention (flow-after suction of 70%) assumed for all mortars.

[2]Also any masonry wall subject to unusual lateral loads for earthquakes, hurricanes, etc.

[3]Only portland cement-lime Type S and M mortars.

only enough water to produce a damp, unworkable mix which will retain its form when pressed into a ball. After 1 to 2 hours, sufficient water is added to make it workable.

For best results, the original mortar proportions should be duplicated. The use of prehydrated Type N mortar is recommended when the original proportions are not known.

For dirt resistance, and where medium resistance to staining is desired, aluminum tristerate, calcium stearate or ammonium stearate should be added to the construction mortar in an amount totaling 3% of the weight of the portland cement.

Where maximum dirt resistance is desired (such as facing tile walls) the construction mortar should be raked out to a depth of approximately 3/8″ and repointed with a mortar consisting of 1 part portland cement, 1/8 part lime and 2 parts graded fine (80 mesh) sand, by volume, to which has been added aluminum tristerate, calcium stearate or ammonium stearate equal to 2% of the portland cement by weight.

MORTAR MIXING

Thorough mixing is important to the development of desirable properties in any mortar. Only enough mortar as is needed for immediate use should be mixed.

All mortar should be mixed with the *maximum* amount of water that is possible to use and still produce a mortar which is workable to the satisfaction of the mason. If, after mixing, the mortar stiffens due to the loss of water by evaporation, additional water should be added and the mortar remixed. All mortar should be used within 2-1/2 hours after initial mixing and should be discarded if not.

One of the most common defects of mortar mixing is oversanding. Although sand is frequently measured by the shovel, the amount of sand that a shovel will hold varies, depending on the moisture content of the sand. Moisture in sand causes varying degrees of bulking (see Fig. 19, Section 203—Concrete).

Measurement of sand by shovel should not be permitted. The method of measuring materials should be by volume or weight so that the proportions of the mortar materials can be accurately controlled.

If the amount of sand is measured carefully and checked daily, oversanding can be avoided. Oversanded mortar is harsh and unworkable, provides a weak bond with the masonry units and tends to erode due to repeated intervals of freezing and thawing.

Except on very small jobs, machine mixing should always be used.

Machine Mixing

Machine-mixed materials should be batched in the following order: Approximately 3/4 the required water, 1/2 the sand and all of the cement are first mixed, the remainder of the sand following. The batch should be allowed to mix briefly and then water added in small quantities until workability satisfactory for the mason is attained. After all materials have been added, the mortar should be mixed a *minimum of 3 minutes and a maximum of 5 minutes.* The mixer drum should be completely empty before the next batch is recharged.

Hand Mixing

If mortar must be mixed by hand, the sand should first be spread in the mortar box, the cement spread on top of the sand, and then be mixed well with a hoe from both ends of the box. About 3/4 of the required water should then be added and mixed until all materials are uniformly damp. The water should be added in small amounts and mixing continued until workability satisfactory for the mason is attained. The batch should be allowed to stand for approximately 5 minutes and remixed thoroughly with the hoe without additional water.

Cold Weather Mixing

The temperature of the mortar when placed in the wall should be between 70° and 120° F when the outside temperatue is below 40° F. Higher temperatures may result in fast hardening, making it impossible for the mason to render good workmanship.

In cold weather, heating the mixing water is one of the easiest methods of raising the temperatures of mortar, but mixing water should not be heated above 160° F due to the danger of *flash* set when it comes in contact with the cement.

In freezing weather, moisture in the sand will turn to ice and must be thawed before the sand can be used.

Calcium Chloride The use of an admixture such as calcium chloride *to lower the freezing point* of mortar during winter construction should not be permitted. The quantity of such materials necessary to lower the freezing point of mortar to any appreciable degree would be so large that mortar strength and other desirable properties would be seriously impaired. However, *to shorten the time required for a mortar to attain sufficient strength* to resist freezing action, a calcium chloride admixture is often used. Calcium chloride should always be added to the mix in solution form and should not exceed 2% of the portland cement by weight. When masonry cements are used, the calcium chloride should not exceed 1% of the weight of the masonry cement. Calcium chloride should not be used in mortars for masonry containing joint reinforcement in order to avoid the possible effect of corrosion of the metal.

Workmanship is a primary factor in obtaining masonry of satisfactory appearance that is watertight, strong and durable. The most important part of workmanship is making the mortar joints. With solid units, all the space between units should be completely filled with mortar. With hollow units, mortar joints should also be full but the thickness of the mortar joint is normally limited to the thickness of the face shell of the unit. Laboratory tests and hundreds of observations of masonry in the field prove beyond all doubt that to obtain good masonry construction there is no substitute for completely filling mortar joints as the units are laid. Partially filled mortar joints result in leaky walls, reduce the strength of masonry from 50% to 60% and contribute to disintegration and cracking when water that can penetrate the wall expands due to cycles of freezing and thawing.

PREPARATION OF MATERIALS

Masonry units should be handled and stored off the ground and covered to prevent damage or soiling of units. Whenever possible, cementitious materials should be stored inside.

Clay Masonry

High-suction brick, if laid dry, will absorb water from the mortar before the bond has developed, thus must be thoroughly wetted before laying. In wetting brick, sprinkling is not sufficient. A hose stream should be played on the pile until water runs from all sides, after which the brick should be allowed to surface-dry before being laid. Water on the surface of the brick will cause floating on the mortar bed.

A rough but effective on-the-job method of determining whether brick should be wetted before laying is to sprinkle drops of water on their flat side. If these drops are absorbed completely in less than 1 minute, the brick should be wetted prior to laying. For a more accurate method, using a quarter and a wax pencil as guides, draw a circle on the flat side of the brick. With a medicine dropper apply 20 drops of water to the surface of the brick inside the circle. If the water is completely absorbed in less than 1-1/2 minutes, the brick should be wetted.

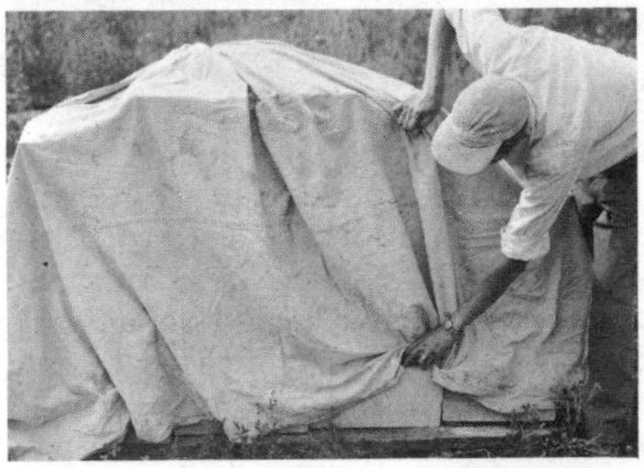

FIG. 54 Stored block should be kept dry and never wetted before laying.

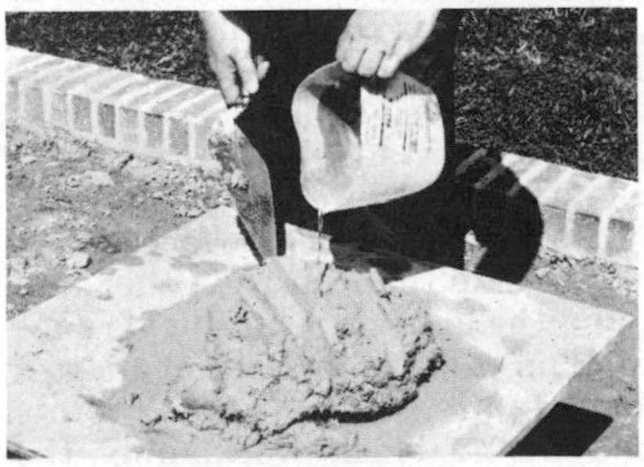

FIG. 55 Retempering is permissible to replace water lost by evaporation if mortar is less than 2 hours old.

Concrete Masonry

As specifications limit the moisture content of concrete masonry units, extra care should be taken to keep the units dry on the job. They should be stockpiled on planks or other supports, free from contact with the ground and covered for protection against wetting (Fig. 54). Concrete units should never be wetted before and during laying in the wall.

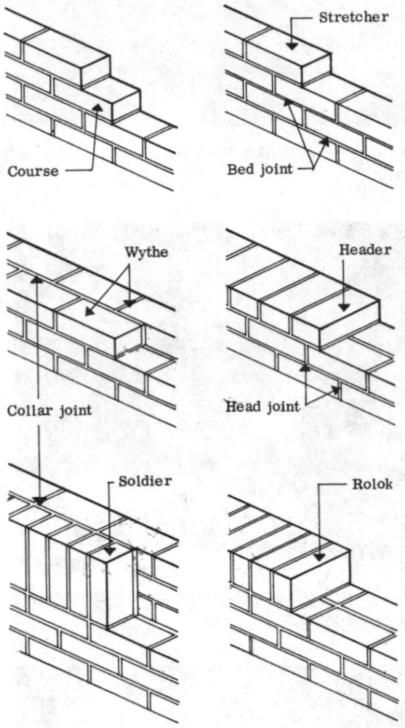

FIG. 56 Terms used to describe various positions of masonry units and mortar joints in a wall.

Mortar

Both strength and resistance to rain penetration of masonry walls are dependent to a very great degree on both the completeness of bond and the strength of the bond between the mortar and the masonry units. Water in the mortar is essential to the development of bond, and if the mortar contains insufficient water, the bond will be weak and spotty. Therefore, the mortar should be mixed with the *maximum* amount of water that is possible to use and still produce a workable mortar. Mortar that has stiffened on the mortar board because of evaporation should be retempered by the addition of water and by thorough remixing to restore workability (Fig. 55). All

FIG. 57 Spacing out units starting at corner.

FIG. 58 Buttering vertical face shells prior to laying.

FIG. 59 Positioning block. Note full vertical joint between blocks already in place.

FIG. 60 Aligning faces and leveling of blocks.

mortar should be used within 2-1/2 hours of initial mixing.

PROCEDURE

Many general workmanship techniques are applicable to the laying of solid or hollow masonry units. General procedures are shown below, as well as specific techniques applicable to either solid or hollow masonry work. Concrete block is used to illustrate the laying of hollow units. Techniques with structural clay tile and other hollow masonry are similar. Brick is used to illustrate solid masonry work. Similar techniques would be used with concrete brick and other solid units. Specific terms used to describe the positions of masonry units in a wall are shown in Figure 56.

Starting the Work

After locating the corners, the mason often spaces out the units dry (without mortar) to determine the extent to which they must be cut or joint sizes varied to accomplish accurate horizontal coursing (Fig. 57). A full, thick mortar bed is then spread for the first course and, in the case of block, furrowed with a trowel to insure plenty of mortar along the bottom edge of the face shells of the block for the first course (Fig. 58). With brick, the mortar bed should be furrowed shallow, if at all. The corner unit is laid first and carefully positioned.

Blocks have thicker face shells on one side than on the other and should be laid with the thicker face shell up to provide a larger mortar bedding area. For vertical joints in concrete block work, only the ends of the face shells are buttered. By placing several blocks on end, the mason can apply mortar to the vertical face shells of 3 or 4 blocks in one operation (Fig. 58). Each block is then brought over its final position and pushed downward into the mortar bed and against the previously laid blocks, producing full vertical mortar joints (Fig. 59). The technique of making the vertical joints (head joints) in brick is discussed below.

After several units have been laid, the mason's level is used as a straight edge to assure that they are in correct alignment, brought to proper grade and made plumb by tapping with the trowel handle (Fig. 60).

The first course of masonry should be laid carefully to ensure that it is properly aligned, leveled and plumbed. This will assist the mason in laying succeeding courses and in building a straight, true wall.

Laying Up the Corners

After the first course is laid, the corners of the wall are built first by laying up several courses higher than the center of the wall. As each course is laid at the corner, it is checked with a level for alignment (Fig. 61), for being level (Fig. 62) and for being plumb (Fig. 63). At the same time, each course is checked carefully with a level or straight edge to make certain that the faces of the units are all in the same plane to ensure true, straight walls.

In block work (except for the first course), foundations below grade and other structural elements, such as pilasters, where a full mortar bed is used on the face shells and webs, Figure 64, mortar is usually applied only to the horizontal face shells of the block. This is called face-shell mortar bedding (Fig. 65). Mortar for the vertical joints can be applied to the vertical face shells of the block to be placed or to the vertical edges of the block previously laid. Some masons butter both edges to insure well filled joints.

The use of a story- or course-pole (a board marked with the courses) provides an accurate method of finding the top of the masonry for each course (Fig. 66). Mortar joints for block should be 3/8″ thick and each course in building the corners stepped back a half block. Mortar joints for brick vary from 1/4″ to 1/2″ in thickness, depending on the size of the brick and the bonding.

Laying Between the Corners

When filling in the wall between the corners, a mason's line is stretched from corner to corner for each course and the top outside edge of each unit is laid to this line. In block work, the manner of handling or gripping the block is important. By tipping the block

FIG. 61 Aligning.

FIG. 62 Leveling.

FIG. 63 Checking for plumb.

slightly toward him, the mason can see the upper edge of the course below, thus enabling placement of the block's lower edge directly over the course below (Fig. 67). By

FIG. 64 Full mortar bedding.

FIG. 65 Face shell mortar bedding.

FIG. 66 Using a course pole.

FIG. 67 Positioning block.

rolling the block slightly to a vertical position and shoving it against the adjacent block, it can be laid to the mason's line with minimum adjustment. All adjustments to final position must be made while the mortar is soft and plastic. Any adjustments made after the mortar has stiffened will break the mortar bond.(See examples in bricklaying techniques.) By tapping lightly with the trowel handle, each block is leveled and aligned to the mason's line. The use of the mason's level between corners is necessary only to check alignment of each unit with the face of the wall.

Bricklaying Techniques

In addition to techniques of masonry workmanship described above, methods unique to bricklaying are discussed below.

Bed Joints Mortar for brick bed joints should be spread uniformly thick and furrowed only slightly (Fig. 68). As the brick is laid the excess mortar will fill the furrow and insure full bed joints. A thin bed of mortar or deep furrow will not produce sufficient mortar to fill the furrow completely and the resulting void will enable water en-

FIG. 68 Mortar furrow should be shallow.

FIG. 69 Bed joint furrow is too deep.

FIG. 70 Spread mortar over few brick only

FIG. 71 Adequate bond is developed.

tering an opening to penetrate the wall (Fig. 69).

To assure good bond, mortar should not be spread too far ahead of the actual laying of block or brick or it will stiffen and lose its plasticity (Fig. 70). This is especially true in hot weather when absorbent brick are being laid. One method of determining if bond is developing is to remove a freshly laid brick from the wall (Fig. 71). If the mortar is soft and plastic, it will stick to the brick placed on top of the bed as well as to the brick on which the bed is spread, and adequate bond will be attained.

Head Joints Head joints are more vulnerable to water penetration than bed joints since the weight of the units and wall above tend to close cracks in the bed joints. Hence, all head joints in both the face brick and back-up work should be completely filled with mortar on the end of the brick before placement (Fig. 72). When the brick is shoved into place, mortar should squeeze out at the sides and top, indicating that the head joint is completely filled (Fig. 73). An alternate method is to throw plenty of mortar on the end of the brick already in place (Fig. 74), giving the desired result shown in Figure 73 when the next brick is shoved into place. A third method is to place a full trowel of mortar on the wall (Fig. 75), shoving the next brick into this deep mortar bed and squeezing mortar out at the sides and top of the joint (Fig. 76).

Dabs of mortar spotted on both corners of the brick (Fig. 77) do not completely fill the head joints (Fig. 78). Nor can "slushing" (attempting to fill the joints from above after the brick is placed, Fig. 79) be relied upon to fill voids left in head joints (Fig. 80).

Cross Joints in header courses must also be completely filled by spreading mortar over the entire side of the header brick before it is placed (Fig. 81) and then it should be shoved into place to fill all voids (Fig. 82).

Closers Before laying the closer brick, plenty of mortar should be placed on both brick already in place (Fig. 83) and on both sides of the closer brick (Fig. 84). The

72

73

74

75

76

HEAD JOINTS
FIG. 72 Loading brick for head joint prior to laying. Mortar oozing out indicates joint is completely filled (73). Brick end may also be buttered for head joints when already in place (74). Or (75) a full trowel of mortar is thrown, then (76) brick may be shoved in place.

FIG. 77 Only spotting corners results in inadequately filled head joints (FIG. 78). Slushing from top (79) will not fill head joints (80). Note here a freshly laid brick has been removed. The slushing has not adequately filled the joint.

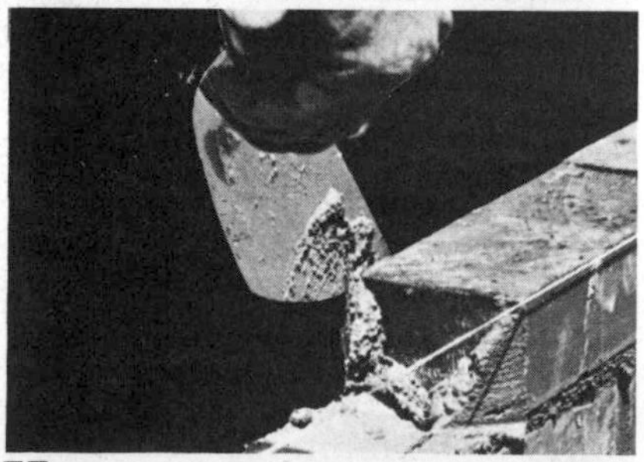
77

79

78

80

FIG. 81 & 82 CROSS JOINTS
Header brick is buttered (top), brick is shoved into place (bottom).

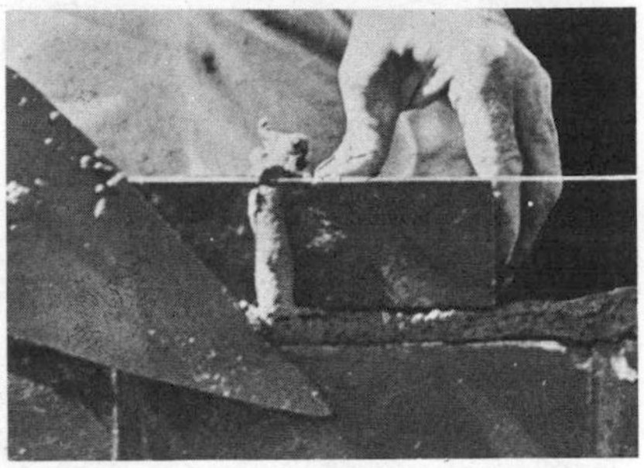

closer should then be laid without disturbing the brick already in place (Fig. 85). The method is similar in block work (Fig. 86).

FIG. 83-86 CLOSERS
Plenty of mortar is placed on brick already in place (FIG. 83), closer brick is buttered on both sides (84). Closer brick is pushed into place (85). Closer block (86).

Mortar Joints

Mortar joint finishes fall loosely into two classes, troweled and tooled joints (Fig. 87). In the troweled joint the excess mortar is simply cut off (struck) with a trowel and no further finishing is used. For the tooled joint, a special tool is used to compress and shape the mortar in the joint. Tooled joints provide the maximum protection against water penetration and should always be used in areas subjected to heavy rains, high winds and where watertight walls are mandatory.

The tooling operation compacts the mortar and forces it tightly against the masonry on each side of the joint (Fig. 88). Proper tooling is done after a section of the wall has been laid and the mortar has become "thumb-print" hard to produce joints of uniform appearance with clean, sharp lines. The jointer for tooling horizontal joints should be slightly larger than the joint and upturned on the ends to prevent gouging the mortar (Figs. 89 and 90).

The best of the troweled joints

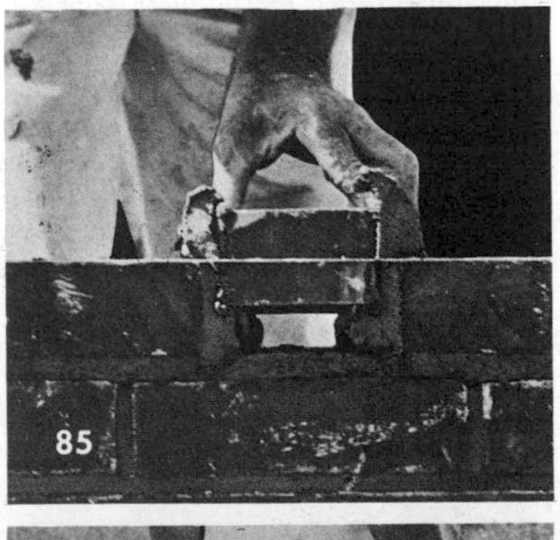

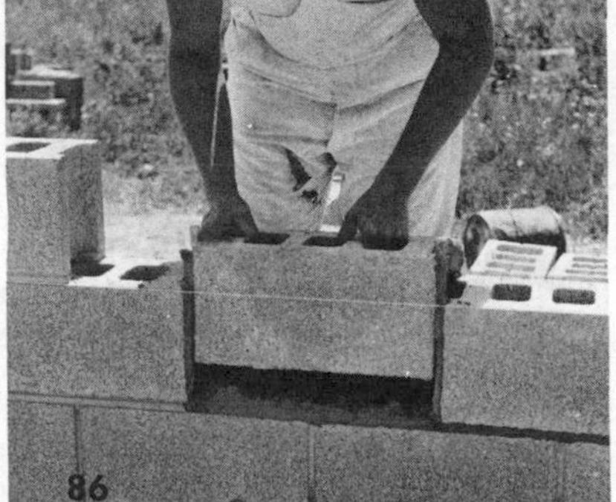

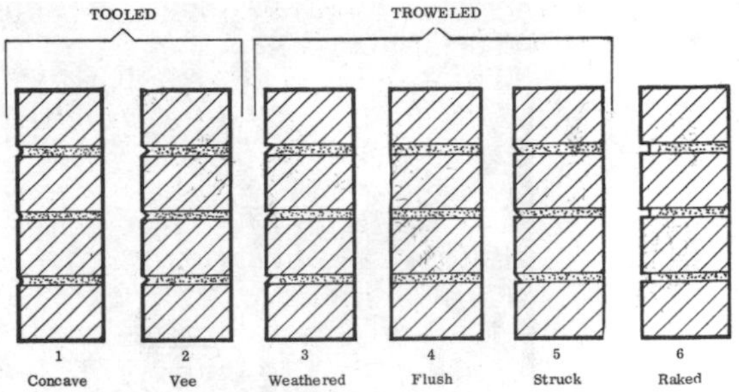

FIG. 87
TYPES OF MORTAR JOINTS

Joints before and after tooling (FIG. 88). Concave jointer for tooling block joints (89). V-jointer for tooling brick (90). Flush troweled joints (91).

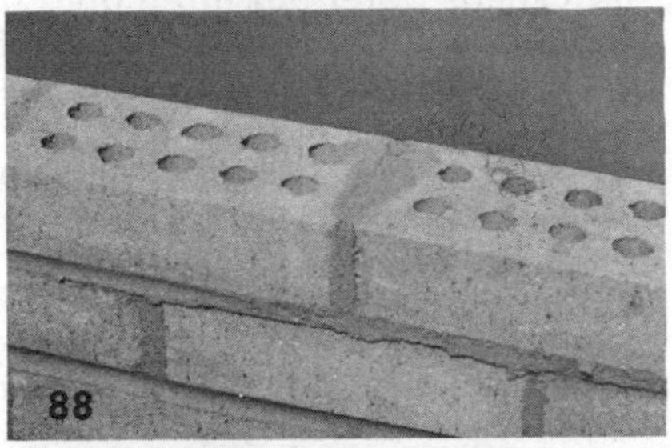

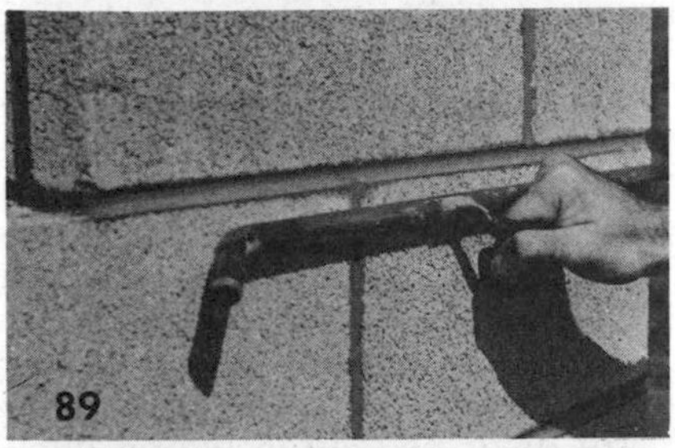

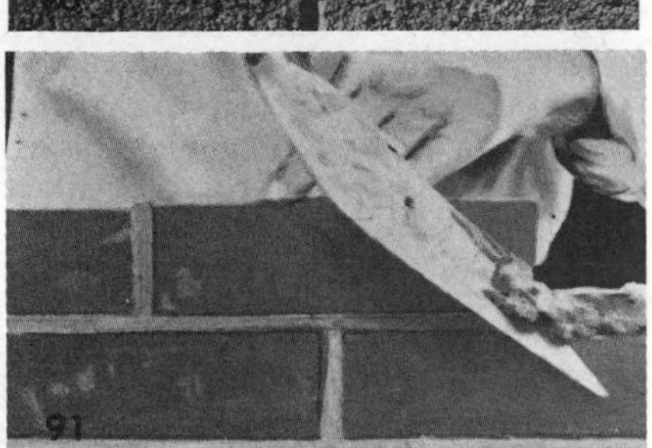

FIG. 92
Cracks due to realignment are frequently found at the corners of the wall.

FIG. 93
PARGING (back-plastering).

FIG. 94
Extruded mortar joints should be cut flush before they harden.

is a weathered joint (Fig. 87, col. 3) as it sheds water but often does not force enough mortar against the brick above the joint. The flush joint (col. 4) produces an uncompacted joint with a small hairline crack where the mortar is pulled away from the brick by the cutting action of the trowel (Fig. 91). The struck joint (col. 5) and the raked joint (which is made by removing the surface of the mortar while still soft, col. 6) produce a shelf on which water can collect.

The position of masonry units should never be shifted or realigned after they have been laid and the mortar has stiffened. Final positioning of the unit must be done while the mortar is soft and plastic, or otherwise the bond will be broken (Fig. 92). Any units which are disturbed after laying should be removed and relaid with fresh mortar.

Parging

When a wall is made up of more than one thickness of masonry, parging (back-plastering) the back of the face brick or the collar joint face of the back-up units provides additional insurance against leakage (Fig. 93). Extruded mortar joints on the back face units should be cut flush before the mortar has a chance to harden, as parging over the hardened mortar will break the bond and result in a leaky wall (Fig. 94). Slushing (filling from above) will not fill this joint completely and should not be permitted (Fig. 95). For best results, parging should be done as the walls are laid up. Careful workmanship is necessary because the pressure required to spread a parge coat of mortar on newly laid units often can break the bond between the units and their mortar joints, producing cracks that may contribute to future leakage.

FIG. 95 Here a freshly laid wythe has been removed. Note, however, slushing has not resulted in complete parging.

Basement walls in areas subject to ground water are parged on the exterior face to minimize moisture penetration (See Division 300—Masonry Walls.)

Flashing

Flashing (Fig. 96) is installed to divert to the outside any moisture that enters the masonry. Flashing should be provided under horizontal masonry surfaces such as sills and copings, at intersections of masonry walls with horizontal surfaces such as roof and chimney, over heads of openings such as doors and windows, and frequently at floor lines, depending on the type of construction (see Division 300—Masonry Walls).

FIG. 96
FLASHING

Weep Holes

Weep holes may be formed in a number of ways, but it is recommended that short lengths of sash cord be inserted by the mason in the position desired. These cords need not be removed but can be left in place to eventually disintegrate.

Protection During Construction

Unprotected, partially finished walls which may be saturated with water during a rain storm may take months to dry out, and during this drying period efflorescence may appear on the wall surfaces. Efflorescence, usually white, is a deposit on the face of the wall, caused by water entering the wall and picking up soluble salts. Acting as a vehicle, the water carries the salts to the surface of the wall, leaving a distracting appearance after evaporation.

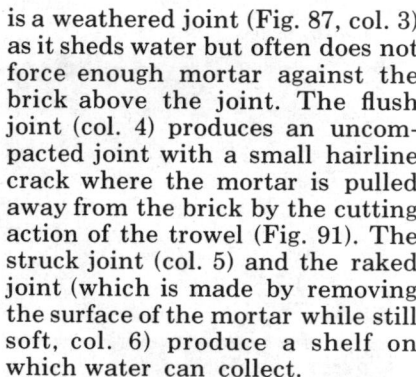

If there is a threat of precipita-

FIG. 97
Exposed top of unfinished masonry should be protected against weather.

FIG. 98
Nail or line pin holes should be filled as soon as possible.

FIG. 99
Final brushing removes loose mortar.

tion at the end of each day or shutdown period, all exposed tops of unfinished masonry should be covered with canvas or waterproof paper securely weighted down and hung over each side of the wall at least 2 feet (Fig. 97).

When building in cold weather, all materials and walls should be properly protected against freezing temperatures. Masonry should not be laid when the temperature is below 32° F on a rising thermometer, or below 40° on a falling thermometer, unless adequate precaution against freezing is provided. The masonry should not be laid on or with frozen materials, and units stored in the open or stacked near the mortar boards should be covered to prevent excessive wetting when freezing is expected. Absorptive brick should be wetted with warm water (but surface dry) just prior to laying. Masonry should be protected against freezing for at least 48 hours after being laid with the temperature on both sides of the wall maintained higher than 40° F.

Pointing and Cleaning

On completion of the work, all masonry should be carefully cleaned, removing all large particles or mortar with a putty knife or chisel. Holes in exposed joints should be pointed and filled with additional fresh mortar.

Holes left by nails or line pins should be pointed (filled) immediately while the adjacent mortar is green enough to allow adequate bond with the patch. Otherwise, water entering through nail holes may penetrate the wall (Fig. 98).

Clay Masonry. In brick work, cleaning with soap powder or other mild solutions should not be attempted until 48 hours after construction of the wall. For unglazed brick and tile (on which no green efflorescence appears) stiff brushes and water should be attempted first. If stiff brushes and water do not suffice, the surface should be thoroughly wetted with clear water and then scrubbed with a solution of acid, followed immediately by a thorough rinsing with clear water. (Green efflorescence should be removed in accordance with the manufacturer's recommendations.)

Where the removal of mortar stains requires the use of acid, muriatic (hydrochloric) acid no stronger than 1 part acid to 9 parts water may be used and should be restricted to well hardened mortar not less than 7 days old. Always test the acid solution on an inconspicuous portion to insure against adverse effects. In the case of buff or gray brick, scrubbing should be done with soap and warm water in lieu of acid. If this is not satisfactory, the brick manufacturer should be consulted for further recommendations.

Working from the top down, the wall should be wet before applying cleaning solutions, and all adjacent construction should be protected. Stiff fiber brushes should be used, cleaning only a small area at a time, followed immediately with a thorough rinsing with clean water. Efflorescence usually can be removed by dry brushing. If not, the acid wash described above may be used.

For glazed facing tile, all surfaces should be washed with soap powder and warm water, applied with a scrubbing brush, and then rinsed thoroughly with clear water. Metal cleaning tools and brushes, acid solutions or abrasive powders should not be used.

Concrete Masonry. In block work, particular care should be taken to prevent smearing mortar into the surface of the exposed concrete unit. Once hardened, embedded mortar smears can never be removed and detract from the neat appearance of the finished wall. A concrete masonry wall should not be cleaned with an acid wash, hence care must be taken to keep the wall surface clean during construction. Any mortar droppings that stick to the units should be allowed to dry before removal, as the mortar may smear into the surface of the unit if it is removed while soft. When dry and hard, most of the remaining mortar can be removed by a trowel. A final brushing removes practically all the mortar (Fig. 99).

GLOSSARY

Absorption (water) *Clay Masonry*—The weight of water a brick or tile unit absorbs when immersed in either cold or boiling water for a stated length of time, expressed as a percentage of the weight of the dry unit.

Concrete Masonry—The weight of water a concrete masonry unit absorbs when immersed in water, expressed in pounds of water per cubic foot of concrete.

Absorption Rate The weight of water absorbed when a clay brick is partially immersed for one minute, usually expressed in either grams or ounces per minute. Also called *suction* or *initial rate of absorption*.

Architectural Terra Cotta Custom-made, hard-burned, glazed or unglazed clay building units, plain or ornamental, that are machine extruded or hand molded. See *Ceramic Veneer*.

Ashlar Masonry Masonry composed of rectangular units usually larger in size than brick, with sawed, dressed or square beds, bonded with mortar. Ashlar masonry is also described according to its bond pattern, which may be coursed, random or patterned.

Atmospheric Pressure Steam Curing See *Curing (Concrete Masonry Units)*.

Backup The part of a masonry wall behind the exterior facing.

Base Course The lowest course of masonry in a wall or pier.

Batter Recessing or sloping a wall back in successive courses; the opposite of corbel.

Bed Joint Horizontal layer of mortar in which a masonry unit is laid. See *Fig. 68-71, page 205-31*.

Bond: Structural Bond Tying wythes of a masonry wall together by lapping units one over another or by connecting them with metal ties.

Tensile bond—Adhesion between mortar and masonry units or reinforcement.

Pattern Bond—Patterns formed by exposed faces of units. See *Section 343—Masonry Walls*.

Bonder See *Header*.

Brick Solid masonry unit of clay or shale, formed into a rectangular prism while plastic and burned or fired in a kiln. See *Fig. 11, page 205-7 for typical brick units*.

Acid-resistant Brick—Suitable for use where it will be in contact with chemicals and designed primarily for use in the chemical industry. Usually used with acid-resistant mortars.

Adobe Brick—Large clay brick of varying size, roughly molded and sun dried.

Building (Common) Brick—Brick for building purposes, not especially treated for texture or color. Formerly called *common* brick.

Economy Brick—Brick whose nominal dimensions are 4 by 4 by 8 in.

Engineered Brick—Brick whose nominal dimensions are 4 by 3.2 by 8 in.

Facing brick—Brick made especially for facing purposes, often with finished surface texture. They are made of selected clays or treated to produce desired color.

Fire Brick—Brick made of refractory ceramic material which will resist high temperatures.

Floor Brick—Smooth, dense brick highly resistant to abrasion, used as finished floor surface.

Hollow Brick—A masonry unit of clay or shale whose net cross-sectional area in any plane parallel to the bearing surface is not less than 60 percent of its gross cross-sectional area measured in the same plane.

Jumbo Brick—A generic term indicating a brick larger in size than the standard. Some producers use this term to describe oversize brick of specific dimensions manufactured by them.

Norman Brick—A brick whose nominal dimensions are 4 by 2-2/3 by 12 in.

Paving Brick—Vitrified brick especially suitable for use in pavements where resistance to abrasion is important.

Roman Brick—Brick whose nominal dimensions are 4 by 2 by 12 in.

"SCR Brick"—Brick whose nominal dimensions are 2-2/3" x 6" x 12". It lays up three courses to 8" and produces a nominal 6"-thick wall. Developed by the Structural Clay Products Research Foundation.

Sewer Brick—Low absorption, abrasive-resistant brick intended for use in drainage structures.

Brick, Concrete Solid concrete masonry unit, approximately a rectangular prism, usually not larger than 4" x 4" x 12".

Brick Grade Designation for durability of the unit expressed as SW for severe weathering, MW for moderate weathering, or NW for negligible weathering. See *ASTM Specifications C216, C62 and C652*.

Brick Type Designation for facing brick which controls tolerance, chippage and distortion. Expressed as FBS, FBX and FBA for solid brick, and HBS, HBX, HBA and HBB for hollow brick. See *ASTM Specifications C216 and C652*.

Buttering Placing mortar on a masonry unit with a trowel.

C/B Ratio The ratio of the weight of water absorbed by a clay masonry unit during immersion in cold water to the weight absorbed during immersion in boiling water. An identification of the probable resistance of clay brick to the action of freezing and thawing. Also called *saturation coefficient*.

Centering Temporary formwork for the support of masonry arches or lintels during construction. Also called *center*.

Ceramic Veneer Architectural terra cotta, characterized by large face dimensions and thick sections.

Adhesion Type Ceramic Veneer—Thin sections of ceramic veneer held in place without metal anchors by adhesion of mortar backing.

Anchored Type Ceramic Veneer—Thicker sections of ceramic veneer held in place by grout and wire anchors connected to backing.

Closer The last masonry unit laid in a course.

Collar Joint Interior longitudinal vertical joint between two wythes (thicknesses) of masonry. See *Fig. 56, page 205-29*.

Column A vertical member whose horizontal dimension measured at right angles to the thickness does not exceed three times its thickness.

Concrete Block See *Fig. 35, page 205-18 for typical concrete masonry units*.

Decorative Block—Various types of concrete masonry units with beveled face shell recesses which provide a special architectural appearance in wall construction (i.e.,

Shadowal block).

Faced Block—Concrete masonry units having a special ceramic, glazed, plastic, polished or ground face.

Slump Block—Concrete masonry units produced so they "slump" or sag before they harden, for use in masonry wall construction.

Split-face Block—Solid or hollow concrete masonry units that are machine fractured (split) lengthwise after hardening to produce a rough, varying surface texture.

Coping Masonry units forming a finished cap on top of an exposed pier, wall, pilaster, chimney, etc., to protect the masonry below from penetration of water from above.

Corbel Shelf or ledge formed by projecting successive courses of masonry out from the face of the wall.

Course One of the continuous horizontal layers of masonry units, bonded with mortar. One course is equal to the thickness of the masonry unit plus the thickness of one mortar joint.

Cubing The assembling of concrete masonry units into *cubes* after curing for storage and delivery. A cube normally contains 6 layers of 15 to 18 blocks (8" x 8" x 16") or an equivalent volume of other size units.

Curing (Concrete Masonry Units) Atmospheric Pressure Steam Curing—A method of curing concrete masonry units, using steam at atmospheric pressure usually at temperatures of 120° to 180°F. Also called *low-pressure steam curing.*

High-pressure Steam Curing—A method of curing concrete masonry units, using saturated steam (365°F) under presure, usually 125 psi to 150 psi. Also referred to as *autoclave* curing.

Moist Curing—A method of curing concrete masonry units, using moisture at atmospheric pressure and temperature of approximately 70°F.

Drip A projection shaped to throw off water and prevent its running down the face of the masonry surface.

Efflorescence Deposit of soluble salts, usually white in color, appearing upon the exposed surface of masonry.

Engineered Brick Masonry Masonry in which design is based on a rational structural engineering analysis.

Face Shell The side wall of a hollow masonry unit (also clay tile).

Flashing A thin, impervious sheet material, placed in mortar joints and across air spaces in masonry to collect water that may penetrate the wall and to direct it to the exterior.

Flashing The step during the burning process of clay masonry units that produces varying shades and colors in the units. See *page 205-6.*

Grout Mortar of a consistency that will flow or pour easily without segregation of the ingredients.

Hard-burned Clay products which have been fired at high temperatures. They have relatively low absorptions and high compressive strengths.

Head Joint The vertical mortar joint between ends of masonry units. See *Fig. 72-76, page 205-32.*

Header A masonry unit which overlaps two or more adjacent wythes of masonry to provide structural bond. Also called *bonder.*

High-pressure Steam Curing See *Curing.*

Joint See *Mortar Joint.*

Lintel A structural member to carry the load over an opening in a wall.

Masonry Cement Portland cement and other materials premixed and packaged, to which sand and water are added to make mortar.

Masonry Unit Natural or manufactured building units of burned clay, concrete, stone, glass, gypsum, etc.

Hollow Masonry Unit—One whose net cross-sectional area in any plane parallel to the bearing surface is less than 75% of the gross.

Modular Masonry Unit—One whose nominal dimensions are based on the 4-in. module.

Solid Masonry Unit—One whose net cross-sectional area in every plane parallel to the bearing surface is 75% or more of the gross.

Mortar A plastic mixture of one or more cementitious materials, sand and water.

Mortar Joints See *types of mortar joints, Fig. 87-91, page 205-33.*

Tooled Joint—Compressing and shaping the face of a mortar joint with a special concave or V-shaped tool.

Troweled Joint—Mortar joint which has been finished with a trowel to form a *struck* joint or a *weathered* joint. In a *raked* joint the mortar is raked out to a specified depth while the mortar is still green.

Moist Curing See *Curing (concrete masonry units).*

Moisture Content (Concrete Masonry Units) The amount of water contained in a unit, expressed as a percentage of the total absorption (i.e., concrete masonry unit at 40% moisture content contains 40% of the water it could absorb).

Parging The application of mortar to the back of the facing material or the face of the backing material. Also called *back-plastering; pargeting.*

Pier An isolated column of masonry.

Pilaster A thickened wall section or column built as an integral part of a wall.

Pointing Troweling mortar into a joint after the masonry unit is laid.

Quoin A projecting right angle masonry corner.

Reinforced Masonry Masonry containing embedded steel so that the two materials act together in resisting forces.

Retempering (of mortar) Restoring workability of mortar which has stiffened due to evaporation, by adding water and remixing.

Rowlock A brick unit laid on its face edge. Usually laid in the wall with its long dimension perpendicular to the wall face. Also spelled "rolok." See *Fig. 56, page 205-29.*

Saturation Coefficient See *C/B ratio.*

Slushing Attempting to fill vertical joints after units are laid by "throwing" mortar into the joint with a trowel from above (not recommended).

Soap A brick or tile of normal face dimensions having a nominal 2" thickness.

Soft-burned Clay products which have been fired at low temperature ranges. They have relatively high absorptions and low compressive strengths.

Soldier Masonry unit set vertically on end with face showing on the masonry surface. See *Fig. 56, page 205-29.*

Spall A small fragment removed from the face of a masonry unit by a blow or by action of the elements.

Story Pole Marked pole for measuring vertical masonry courses during construction.

Stretcher Masonry unit laid with its length horizontal and parallel with face of the masonry. See *Fig. 56, page 205-29.*

Suction The initial rate of water absorption by a clay masonry unit. See *Absorption.*

Steam Curing See *Curing (concrete masonry units).*

Structural Clay Tile Hollow masonry units composed of burned clay, shale, fire, clay or mixtures of these.

End-Construction Tile—Tile designed to be laid with axis of the cells vertical.

Facing Tile—Tile for exterior and interior masonry with exposed faces.

Side Construction Tile—Tile intended for placement with axis of cells horizontal.

Toothing Projecting brick or block in alternate courses to provide for bond with adjoining masonry that will be laid later.

Tuck Pointing Refilling defective mortar joints that have been cut out in existing masonry.

Wall Tie A header (bonder) or metal anchor that connects wythes of masonry to each other or to other materials.

Water Absorption See *Absorption.*

Water Retention (of mortar) The property of mortar that prevents the loss of water to masonry units having a high suction rate. It also prevents bleeding or water gain when mortar is in contact with units having a low suction rate. Also called *water retentivity.*

Weathering Index The product of the average annual number of freezing cycle days and the average annual winter rainfall.

Freezing Cycle Day—Any day the air temperature passes either above or below 32°F. The average number of freezing cycle days in a year equals the difference between the mean number of days the *minimum* temperature was 32°F. or below and when the *maximum* temperature was 32°F. or below.

Winter Rainfall—The sum in inches, of the mean monthly corrected precipitation occurring between the first killing frost in the fall and the last killing frost in the spring.

Wythe A continuous vertical section of masonry, one unit in thickness. Also called *withe* or *tier*. See *Fig. 56, page 205-29.*

We gratefully acknowledge the assistance of the National Concrete Masonry Association, the Portland Cement Association, the Brick Institute of America and the Louisville Cement Company. We are also indebted to them for permission to use photographs and publications as references.

MASONRY 205

CONTENTS

MATERIALS

Materials listed below should conform to the current requirements of the specifications of the American Society For Testing & Materials (ASTM).

STANDARDS OF QUALITY

The ASTM specifications for masonry do not fix the size, color or texture of wall units, but cover a range of sizes and a wide variety of colors and textures. When ASTM specifications are included in job specifications by reference to grade and type, it is recommended they be supplemented with job requirements covering size, color, texture, quality of workmanship and the specific finished appearance desired. This can be done by requesting that these variables match approved samples.

FIG. WF1 **SOLID CLAY MASONRY**

Building (Common) Brick* (Grade SW, MW, NW)	ASTM C62
GRADES (based on Weathering Index)	**USE**
SW (Severe Weathering)	For foundations or other structures below grade or in contact with earth, especially when subjected to freezing temperatures.
MW (Moderate Weathering)	For exterior walls and other exposed vertical masonry surfaces above grade.
NW (Negligible Weathering)	For all interior masonry or use as backup.

Facing Brick* (Grade SW, MW—Type FBX, FBS, FBA)	ASTM C216
GRADE (based on Weathering Index)	**USE**
SW (Severe Weathering)	For masonry in contact with earth, especially when subject to freezing temperatures.
MW (Moderate Weathering)	For exterior walls and other exposed masonry above grade.
TYPE (based on apearance of finished wall)	**USE**
FBX	For masonry where high degree of mechanical perfection, minimum variation in color range and minimum variation in size are required.
FBS	Where wider color ranges and greater size variations are permissible than are specified for FBX.
FBA	Where architectural effects are desired resulting from non-uniformity in size, color and texture of units.

*Units with core area less than 25% of unit's total cross sectional area; see Fig. 17, main text.

(Continued) **MATERIALS**

Pedestrian and Light Traffic Paving Brick* (Class SX, MX, NX—Type I, II, III) ASTM C902

CLASS (Based on Weathering)	USE
SX	Where brick may be frozen while saturated with water.
MX	Where resistance to freezing is not a factor.
NX	Interior use when an effective sealer, wax or other suitable surface coating will be applied.
TYPE (Based on traffic)	**USE**
I	Where brick will be exposed to extensive abrasion, such as in driveways and entranceways to public or commercial buildings.
II	Where brick will be exposed to intermediate traffic, such as floors, in restaurants, stores and exterior walkways.
III	Where brick will be exposed to low traffic, such as floors or patios, in single family houses.

Glazed Facing Brick* (Grade S, G—Type I, II) ASTM C126

*See Main Text for physical requirements for each class.

FIG. WF2 HOLLOW MASONRY

See Structural Clay Facing Tile (below) for description of Grades S, SS and Type I, II.

Hollow Brick* (Grade SW, MW—type HBS, HBX, HBA, HBB) ASTM C652

GRADE (Based on Weathering Index)	USE
SW (Severe Weathering)	Where a high and uniform degree of resistance to frost action and disintegration by weathering is desired and where brick may be frozen when permeated with water.
MW (Moderate Weathering)	Where a moderate and somewhat non-uniform degree of resistance to frost action is permissible and where brick is unlikely to be permeated with water when exposed to below freezing temperatures.
TYPE (Based on appearance of finished wall)	**USE**
HBS	Exposed exterior and interior walls where wider color ranges and greater variations in size are permitted than is specified for HBX.
HBX	Exposed exterior and interior walls where narrow color range and minimum permissible variation in size are required.
HBA	Where architectural effects are desired resulting from nonuniformity in size, color, and texture of units.
HBB	Where color and texture are not a consideration and greater size variation is permitted than is specified for HBX.

Structural Clay Load-Bearing Tile* (Grade LBX, LB) ASTM C34

GRADE (based on weathering)	USE
LBX	For masonry exposed to weathering and for direct application of stucco.
LB	For masonry not exposed to frost action or earth. May be used in exposed masonry if protected with a facing of 3" or more of other masonry.

Structural Clay Non-Load-Bearing Tile* ASTM C56

One grade, NB, for use in non-load-bearing walls, partitions, fireproofing and furring.

Structural Clay Facing Tile Ceramic Glazed* (Grade S, SS—Type I, II) ASTM C126

GRADE	USE
S (select)	For masonry with narrow mortar joints (1/4").
SS (select sized or ground edge)	For masonry where face dimension variation must be very small.
TYPE	**USE**
I (single-face units)	For masonry where only one finished face will be exposed.
II (two-faced units)	For masonry where two opposite finished faces will be exposed.

Structural Clay Facing Tile Unglazed* (Type FTX, FTS) ASTM C212

TYPE	USE
FTX	For exposed masonry: smooth face, low absorption and resistance to staining. Has high degree of mechanical perfection, minimum variation in color and face dimensions.
FTS	May be either smooth or rough textured, with moderate absorption and variation in dimensions, medium color range and minor surface finish defects.

*Units with core area greater than 25% of unit's total cross sectional area; see Main Text for the physical requirements for each Grade.

(Continued) **MATERIALS**

CONCRETE MASONRY UNITS[1] *(**solid** units have core area less than 25% of total cross-sectional area of the unit; **hollow** units have core area greater than 25% of the total cross-sectional area of the unit.)*

FIG. WF3 INDUSTRY STANDARDS FOR CONCRETE MASONRY UNITS

Standard	Type of Unit
ASTM C55	Concrete building brick, solid veneer and split block.
ASTM C90	Hollow load-bearing concrete masonry units.
ASTM C129	Hollow non-load-bearing concrete masonry units.
ASTM C145	Solid[1] load-bearing concrete masonry units.

[1]Units with 75% or more net area.

Concrete masonry units should conform to ASTM standards outlined in Figure WF3.

Type I concrete masonry units and building brick manufactured according to above ASTM standards should have a maximum moisture content appropriate for the local humidity conditions and the shrinkage classification of the particular masonry units (Fig. WF4).

Shrinkage classification for different units can be determined from Figure WF5.

Concrete masonry units manufactured according to ASTM standards should conform to minimum strength and maximum absorption requirements shown in Figure WF6.

FIG. WF4 MAXIMUM MOISTURE CONTENT OF TYPE I UNITS

Linear Shrinkage Classification, %[1]	Max. Moisture Content, % of total absorption (avg. of 5 units) — Average Annual Relative Humidity, %[2]: Over 75	75 to 50	Under 50
Up to 0.03	45	40	35
0.03 to 0.045	40	35	30
0.045 to 0.065	35	30	25

[1] As determined by ASTM C426 Test for Drying Shrinkage of Concrete within previous 12 months.
[2] As reported by the U.S. Weather Service nearest source of manufacture.

FIG. WF5 LINEAR SHRINKAGE OF CONCRETE MASONRY UNITS

Aggregate Type	Steam Curing Method	Average Linear Shrinkage, %[1]
Normal Weight	High Pressure	0.019
	Low Pressure	0.027
Lightweight	High Pressure	0.023
	Low Pressure	0.042
Pumice	High Pressure	0.039
	Low Pressure	0.063

[1] As determined by test, ASTM C426.

FIG. WF6 STRENGTH AND ABSORPTION REQUIREMENTS FOR CONCRETE MASONRY UNITS

		Compressive Strength (min., psi) on Average Gross Area		Water Absorption (max., lbs/ft³) for Units of Different Densities			
				Lightweight Concrete		Medium-Weight Concrete	Normal-Weight Concrete
UNIT TYPE & STANDARD	GRADE*	Average of 3 units	Individual Unit	Less than 85 lbs/ft³	Less than 105 lbs/ft³	105 to 125 lbs/ft³	More than 125 lbs/ft³
Hollow Load-bearing Units, ASTM C90	N-I, N-II S-I, S-II	1000 700	800 600	— 20	18 —	15 —	13 —
Solid Load-bearing Units, ASTM C145	N-I, N-II S-I, S-II	1800 1200	1500 1000	— 20	18 —	15 —	13 —
Concrete Building Brick, ASTM C55	N-I, N-II S-I, S-II	3500 2500	3000 2000	15 18	15 18	13 15	10 13
Hollow Non-loadbearing block, ASTM C129	—	350	300	—	—	—	—

*Grades S-I and S-II are limited to walls not exposed to the weather, and to exterior above grade walls protected with weather resistant coatings. Grades N-I and N-II may be used below grade as well as above; protective coatings are recommended below grade and where required above grade.

(Continued) **MATERIALS**

Concrete Non-Load-Bearing Block* **ASTM C129**

For use in non-load-bearing walls.

MORTAR AND MORTAR MATERIALS ASTM C270

See Mortars below for complete listing of mortar materials.

ANCHORS, TIES AND JOINT REINFORCEMENT

See Masonry Walls, Division 300.

FLASHING MATERIALS

See Masonry Walls, Division 300.

* *See Main Test for physical requirements for each grade.*

STORAGE OF MATERIALS

Cementitious mortar materials should be stored under cover in a dry place, and masonry units should be free from soil, ice and frost when laid in the wall. During freezing weather masonry units should be protected with tarpaulins or other suitable material. Structural facing tile, either glazed or unglazed, should be covered at all times.

Concrete masonry units should be protected against wetting prior to use.

Metal ties, anchors and reinforcement should be kept dry and protected from contact with soil. Before being placed, they should be free from loose rust and other coatings that will destroy or reduce bond.

MORTARS

MORTAR MATERIALS **ASTM C270**

ASTM C270 covers specifications for mortar materials which should conform to the following ASTM specifications:

Portland Cement (Type I, II, III)* **ASTM C150**

Air-Entraining Portland Cement **ASTM C150**

Air-entraining cements should not be used in mortars.

Excessive air entrainment may have effect of reducing bond between mortar and masonry units and reinforcement.

Masonry Cement (Type II) **ASTM C91**

Quicklime **ASTM C5**

Hydrated Lime (Type S) **ASTM C207**

Aggregates (Sand) **ASTM C144**

For best workability and economy, sand should be graded uniformly to include all particle sizes from very fine up to 1/4″. Sands with less than 5% to 15% very fine particles produce harsh, hard-to-work mortars requiring additional cementitious material to produce a workable mix. Sand should be measured accurately by volume or weight.

Water

Mixing water should be clean and free from deleterious amounts of acids, alkalis or organic materials.

Water that is fit to drink is generally suitable for mortar.

MORTAR COLORS

Colors should consist of inorganic compounds used in the proportions recommended by the manufacturer, but in no case exceeding 15% of the weight of the cement. Carbon black should not exceed 3% of the weight of the cement.

(Continued) **MORTARS**

ANTIFREEZE COMPOUNDS

No antifreeze liquid, salts or other substances should be used in mortar to lower the freezing point.

MORTAR PROPERTIES

The properties of different mortar types vary. The selection of the type of mortar depends upon the type of masonry and its location in the building (Fig. WF8). All mortars require adequate workability and water retention to produce mortar with suitable tensile bond strength, compressive strength, durability and appearance.

Workability

Masons consider a mortar "workable" when it will: (1) spread easily, (2) hold the weight of the units, (3) cling to the vertical face of the units, (4) make alignment easy, (5) readily extrude from the mortar joint and (6) will not drop or smear.

Water Retentivity

Water retention prevents: loss of mortar plasticity when the mortar comes in contact with units of high absorption; and "floating" of the laid unit when the mortar comes in contact with units of low absorption.

MORTAR TYPES

Portland Cement and Lime (portland cement + lime + sand + water)

Portland cement and lime produce the most predictable mortars. Portland cement contributes high compressive strength, high-early-strength and durability. Lime contributes workability, water retentivity and plasticity. Both contribute to bond strength. Sand contributes strength and, as a filler, provides the most economical mix. Water creates plastic workability and initiates the cementing action.

Masonry Cement Mortars (masonry cement + sand + water)

Masonry cement mortars are not recommended for reinforced masonry construction and may give unpredictable results for normal masonry work.

MORTAR PROPORTIONS AND USES

Proportions

Proportions of the five Types (M, S, N, O and K) of mortar recommended for use in masonry are given in Figure WF7.

MORTAR MIXING

Measurement of materials should be carefully controlled and accurately maintained.

Mortar mixing should always be done in a mechanical-

FIG. WF7 MORTAR PROPORTIONS BY VOLUME[1]

TYPE[2]	Portland Cement	Hydrated Lime or Lime Putty	Masonry Cement	Max. Damp Loose Aggregate[3]	Minimum Compressive Strength 2" Cubes @ 28 days (psi)
M	1	¼	—	3	2500
	or				
	1	—	1	6	2500
S	1	½	—	4½	1800
	or				
	½	—	1	4½	1800
N	1	1	—	6	750
	or				
	—	—	1	3	750
O	1	2	—	9	350
K	1	4	—	15	75

[1]The weight of 1 cu. ft. of materials used is considered to be: Portland cement, 94 lbs. (1 bag); hydrated lime, 40 lbs.; lime putty, 80 lbs.; dry sand, 80 lbs.; masonry cement, weight printed on bag.
[2]Shaded figures show proportions for portland cement-lime mortar type. Mortar types made with masonry cement are unshaded.
[3]The damp loose aggregate should not be less than 2¼ times, nor more than 3 times the sum of the cementitious material used.

FIG. WF8 RECOMMENDED MORTAR TYPES FOR VARIOUS CONSTRUCTION APPLICATIONS

CONSTRUCTION APPLICATION	RECOMMENDED ASTM MORTAR TYPES[1]
Foundations, basements, walls, isolated piers.	M, S
Exterior Walls.	S, N
Solid Masonry unit veneer over wood frame.	N
Interior Walls—load-bearing.	S, N
Interior Walls—non-load-bearing.	N, O
Reinforced Masonry (columns, pilasters, walls, beams).	M, S[2]

[1]Adequate workability and minimum water retention assumed for all mortars.
[2]Only portland cement-lime Type S and M mortars.

(Continued) MORTARS

batch mixer. After all materials have been added, the mortar should be mixed a minimum of 3 minutes and a maximum of 5 minutes. The consistency of mortar should be adjusted to the satisfaction of the mason, using as much water as possible for a workable mix. If the mortar begins to stiffen from evaporation or absorption of a part of the mixing water, the mortar should be retempered to restore its workability by adding water and should be re-mixed. All mortar should be used within 2-1/2 hours of initial mixing, and no mortar should be used after it has begun to set.

In cold weather the temperature of the mortar when placed should be between 70° and 120° F when the outside temperature is below 40° F. To shorten the time required for mortar to attain sufficient strength to resist freezing action, calcium chloride (ASTM D98), in an amount not to exceed 2% of the portland cement by weight, may be added to the mix in solution, never in dry form.

WORKMANSHIP

PREPARATION OF MATERIALS

Clay Masonry Units

All brick having absorption rates in excess of 20 grams of water per minute should be wetted sufficiently so that the rate of absorption when laid does not exceed this amount.

A simple test to determine if brick should be wetted is described in the Main Text, page 205-29.

Concrete Masonry Units

All concrete units should be protected against wetting prior to use. Dampening the units should not be permitted.

GENERAL RECOMMENDATIONS (ALL MASONRY)

Freezing Weather

No masonry should be laid when the temperature of the outside air is below 40° F unless provision is made to heat and maintain the temperature of the masonry materials and protect the completed work from freezing.

Protection should consist of heating and maintaining the temperature of the masonry units to at least 40° F, and maintaining an air temperature above 40° F on both sides of the masonry for a period of at least 48 hrs. if Type M or S mortar is used, and 72 hrs. if Type N or O mortar is used. These periods may be reduced to 24 hrs. and 48 hrs., respectively, if Type III (high-early-strength) portland cement is used.

Joining the Work

Where fresh masonry joins masonry that is partially set or totally set, the exposed surface of the set masonry should be cleaned, and in the case of clay masonry wetted, so as to obtain the best possible bond with the new work.

Mortar Joints and Tooling

Mortar joints that are exposed and have become "thumb-print" hard should be tooled with a round or V-shaped jointer.

The jointer should be slightly larger than the width of the mortar joint so that complete contact is made along the edges of the unit, compressing and sealing the surface of the joint. Concave or V-shaped tooled joints are recommended on exposed exterior surfaces in areas subject to hard rainstorms. On interior surfaces or where moisture penetration is not a consideration, joints may be raked, troweled or otherwise finished as desired.

Mortar joints in surfaces to be parged, plastered, stuccoed or covered with other masonry should be cut flush. Mortar joints in below-grade walls not to be parged should be tooled.

For example, slab-on-grade and crawl space foundations need not be parged when moisture penetration is not a problem.

Care should be taken to prevent smearing mortar into the finished surface of concrete units exposed to view. *Hardened, embedded mortar cannot be removed. Mortar droppings that adhere to the wall should be allowed to*

(Continued) WORKMANSHIP

dry before removal with a trowel to prevent embedment into the face of the unit.

Caulking

See Masonry Walls, Division 300.

Protection of Work

Partially completed walls should be protected from moisture penetration by covering them with a strong, waterproof membrane. The covering should overhang at least 2′ on each side of the wall and be securely anchored.

GENERAL REQUIREMENTS

Masonry should be laid plumb and true to lines. The collar joints (*vertical, longitudinal joints between two thicknesses of masonry*) in solid walls should be completely filled by parging either the face of the backing or the back of the facing.

Each unit should be adjusted to its final position in the wall while the mortar is still soft and plastic. The mason should avoid over-plumbing and pounding to realign units after they have been set into position. Where an adjustment must be made after the mortar has started to harden, the units and the mortar should be removed and replaced with fresh mortar and the disturbed units relaid.

Anchors, accessories, flashings and other items required to be built in with masonry should be built in as the work progresses.

When flashing is to be laid on or against masonry, the surface of the masonry should be smooth and free from projections which might puncture the flashing material. Weep holes spaced 24″ on center should be provided in the head joints immediately above all flashing. Weep holes should be kept free of mortar droppings.

SOLID MASONRY UNITS

Brick, including concrete brick, and other solid units should be laid with completely filled head and bed joints.

HOLLOW MASONRY UNITS

Concrete Block units should be laid with full mortar coverage on horizontal and vertical face shells, except that webs also should be bedded in all courses of piers, columns, pilasters, and in the starting course of footings and solid foundation walls, and where adjacent to cells or cavities that are to be reinforced and/or filled with mortar.

Vertical-Cell Structural Clay Tile units should be laid with divided head joints (not full) and when exposed to the weather, care should be taken to prevent continuous mortar joints through the wall.

Horizontal-Cell Structural Clay Tile units exceeding 4″ in thickness should be laid with divided bed joints. Head joints of horizontal-cell units should be placed on both sides of the tile and sufficient mortar used so that excess mortar will be squeezed out of the joints as the units are placed in position.

BONDING AND ANCHORING

See Masonry Walls, Div. 300, for discussion and recommendations covering structural bond, anchoring, wall thicknesses, lateral support, expansion and control joints.

CLEANING AND POINTING

Exposed masonry should be protected against staining by wall coverings, and excess mortar should be wiped off the surface as the work progresses. All holes in exposed masonry should be pointed, and defective joints should be cut out and repointed with mortar.

All exposed masonry should be thoroughly cleaned. *Specific recommendations for cleaning clay and concrete masonry are given in the Main Text, page 205-35.*

207 GLASS

GLASS

INTRODUCTION

Glass results from the fusion of certain common minerals at high temperatures and subsequent controlled cooling. This process causes solidification without crystallization. The chief ingredients of glass are *silica, sodium oxide* and *calcium oxide* (sand, soda and lime)—all abundant minerals throughout the United States. Glasslike mineral formations resulting from the accidental fusion of silica sand and metallic oxides by the heat of a volcano or lightning (obsidian and lightning stone) can be readily found in nature and may have served as the impetus for the invention of glass by early man.

Although man has known how to make glass for over 4500 years, the real revolution in glassmaking dates from early in the 20th century when the Fourcault and Colburn processes for mass producing large sheets of flat glass were developed. Prior to this time, glass was blown in cylinders, then cut apart and flattened to make window glass of limited size and quality.

Glass ranks with concrete, steel, aluminum and plaster as a truly 20th century material. It has exerted a tremendous influence on our technology, our modern way of life and on architectural expression. From the divided-light window of colonial days to the modern floor-to-ceiling picture window—glass technology has progressed with residential design needs and has in turn influenced them.

Just as colonial builders limited the size of windows because of high glass cost and shelter needs, today's designers continually increase glass areas as the cost of glass declines and man's ability to control his interior environment improves. Largely glass or all-glass buildings are becoming commonplace and are creating new psychological impacts on their inhabitants. The ever changing panorama of the natural outdoor environment is psychologically satisfying to most people. Possible adverse effects from over-exposure to the elements can be minimized by proper orientation and judicious selection of special glass types such as glare-reducing, heat-absorbing, reflective coated or insulating glass.

This section discusses the manufacture of glass and its generic properties, and outlines the various types of *flat glass* which find application in building construction. Additional information on uses and applications of glass is given in subsections of 202 Wood Products and 214 Aluminum Products dealing with windows and doors.

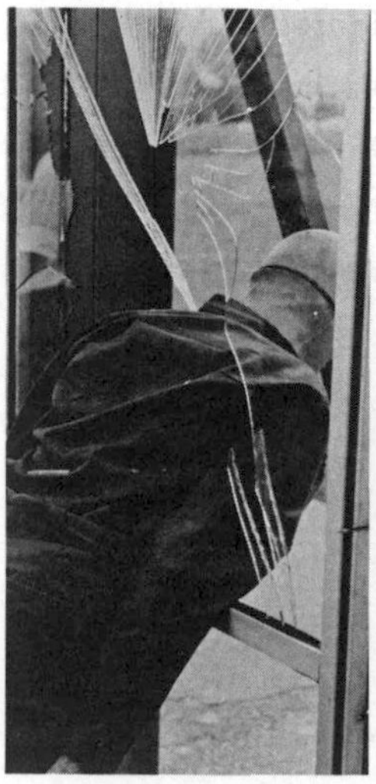

Glass is a very unusual material in terms of its internal structure. It is mechanically rigid, and thus behaves like a solid. However, most mineral solids have a crystalline structure, with the atoms arranged in a definite geometric pattern. In glass, the atoms are arranged in a random or disordered fashion, as is characteristic of a liquid. But the atoms, though arranged at random, are "frozen" in position by the rapid cooling process used to make glass. Thus glass combines some of the aspects of a solid and some of the aspects of a liquid. Because of this, it is sometimes called a *supercooled* liquid.

The physical properties of glass can be varied over an extensive range by modifying either the composition of the glass, the production techniques used, or both. In selecting an individual glass for a particular product, a combination of properties must be taken into consideration—for usually one property cannot be changed without causing a change in other properties. The following is a brief discussion of some of the properties of glass which are important in architectural applications.

MECHANICAL PROPERTIES

Mechanical properties deal with the action of forces on a material and the effects that these forces produce within the material.

One of the important mechanical properties of glass is that it is fatigue-resistant. This means that if pressure is applied to glass, causing it to bend or stretch, the glass will return exactly to its original shape when the bending or stretching force is removed (Fig. 1). Of course, if increasing force is applied, the glass will eventually break, but at any point short of breakage, it will not be permanently deformed.

The mechanical strength of glass is determined by its ability to withstand forces which cause breakage. Glass generally breaks from stretching or bending (tensile forces) and thus *tensile* strength is the chief factor used in determining its mechanical strength.

Unlike most metals, glass does not have a clearly defined tensile strength. *Theoretical* tensile strength has been estimated as high as 4,000,000 psi (lbs. per sq. in.); *actual* tensile strength of annealed glass ranges between 3,000 and 6,000 psi.

The main reason for the wide variation between theoretical and actual strength in glass is that actual strength is highly dependent upon the surface condition of the glass. No matter how carefully it is handled, glass will acquire some small nicks and scratches in the course of manufacture and later in use. These surface defects then create weak planes in the material; when a force is applied to the glass, stresses tend to concentrate at the weak planes rather than distribute themselves uniformly, as is assumed in theoretical strength calculations. Because of variations in the surface condition of different lights of glass, a group of seemingly identical glass specimens will exhibit a wide range of breaking strengths when tested. Thus, the mechanical strength of glass cannot be stated as a clearly defined figure; it must always be calculated on a probability/stress curve.

The mechanical strength of glass can be improved by both heat and chemical treatments. These processes are discussed on pages 207-9 to 207-10.

OPTICAL PROPERTIES

One of the obvious and important properties of glass is its transparency. Despite appearances, however, no glass is completely transparent. When light falls on a piece of glass, some of the light is reflected and some is absorbed. The amount of light that does pass through the glass is called *transmittance*. Commonly used clear commercial glass has a transmittance of about 85% to 90% of visible light.

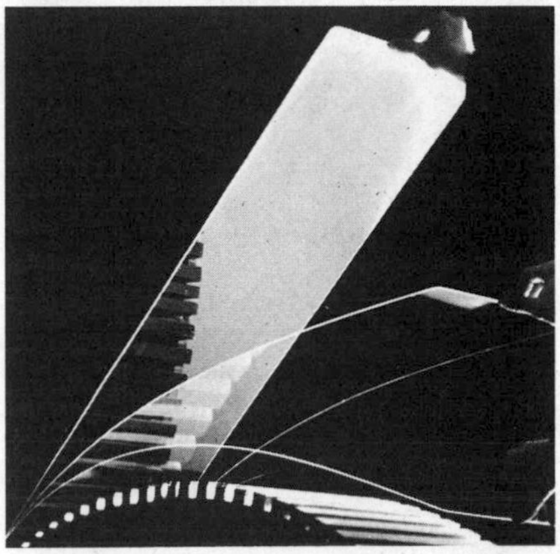

FIG. 1 The ability of glass to return to its original shape is an important quality in windows, as they are subject to repeated deflection due to wind pressure and other atmospheric conditions.

Glass may be treated in many different ways to reflect or absorb varying degrees of light. Glass which has been tinted absorbs a higher percentage of light than clear glass and, depending upon the color and thickness of the glass, may transmit anywhere between 21% and 75% of visible light. Glass which has been treated with a thin reflective coating, reflects a high proportion of light and may transmit from 8% to 50% of visible light, depending upon the nature of the coating. The metallic coating on the back of mirrors produces the maximum amount of reflectance possible, allowing no visible transmittance.

Another important optical property of glass is its fidelity (lack of distortion). When the two surfaces of a sheet of glass are perfectly parallel, the image that is seen through, or is reflected from, the glass will be free of waves or distortion (Fig. 2). However, small variations in the thickness of a

FIG. 2 The uniform thickness and flatness characteristics of plate and float glass assure distortion-free views.

sheet of glass can cause considerable distortion in the image seen. For example, in a large area of glass, a variation in thickness of only a few hundred thousandths of an inch can be quite apparent because the variation acts as a lens to magnify or diminish portions of the reflection.

The different types of flat glass (sheet, float and plate) vary in the degree of clarity they produce because of the manufacturing processes used in making them. Sheet glass is characterized by a wave-like distortion caused by the drawing process, which runs in one direction through the sheet. Float and polished plate glass have substantially flatter and more parallel surfaces than sheet glass. In commercial glazing applications plate and float glass are considered equivalent in terms of optical quality.

THERMAL PROPERTIES

The amount of heat which passes through the glass areas of a building is an important consideration in determining heating and air conditioning loads. Heat can be transferred by conduction, convection and radiation. Glass has a lower thermal conductivity than most metals, but a great deal of solar energy can be transferred directly through glass by radiation, and air currents which come into contact with glass surfaces can transfer heat by convection. Thus, overall heat loss in winter and heat gain in summer is generally greater through glass areas of a building than through ordinary masonry or wood frame walls.

However, heat gain (or loss) through windows can be controlled by several different means: insulation, reflection, absorption and shading. Special glass products such as insulating glass, reflective coated glass and tinted glass make use of these principles and can substantially reduce heat gain. In addition, shading devices such as blinds, draperies and awnings can be used with the glass to further reduce heat gain.

The thermal expansion of glass is another important consideration in the design of buildings. This factor is particularly important when using tinted or reflective-coated glasses, as these glasses absorb a higher percentage of solar energy than ordinary clear glass, and therefore will expand and contract more.

When different portions of a single pane of glass heat up at different rates, they expand at different rates, causing thermal stresses to develop within the pane. What often happens under ordinary glazing conditions is that the center portion of the pane heats up faster than the edges (which are shaded by the edge framing), causing the center portion to expand more than the cooler edge regions. This produces stress differentials within the glass, which tend to concentrate at weak points in the glass and may eventually cause breakage. Most thermal failures can be traced to minor edge damage or imperfections, as this is where weak points in a sheet of glass are most likely to occur. Hence, the best way to insure against such thermal breakage is to be sure that glass edges are strong and clean-cut.

The thermal expansion of glass also is important when two glasses or a glass and a metal are to be sealed together (as in insulating glass and types of coated glass). The rates of expansion of the two materials must be closely matched to avoid damaging differential stresses in the finished product.

CHEMICAL PROPERTIES

Glass is much more resistant to corrosion than most metals or common materials. It is non-porous and non-absorptive and thus is impervious to the common elements which affect many materials. This is made evident by the fact that glass windows, after many years of exposure to the elements, remain clear and apparently unchanged.

Nevertheless, glass will corrode and even dissolve under certain conditions. For example, hydrofluoric acid will attack soda-lime glass, causing extreme corrosion; similarly, hot concentrated alkali solutions and super-heated water will cause soda-lime glass to dissolve. These chemicals, though, usually are encountered only in laboratory situations, so that under ordinary environmental conditions, glass is considered to be an extremely inert and durable substance.

ELECTRICAL PROPERTIES

Because it offers a high resistance to the passage of electricity, glass generally is considered to be a good electrical insulator. For this reason, glass is used in making light bulbs, as well as for many other electrical products.

The progress in glass manufacture and the development of glass technology has been outstanding since the turn of the century. Today glass is manufactured in a wide variety of compositions and can be fabricated by several different techniques into a tremendous array of products.

RAW MATERIALS

Glass is manufactured in several generic types which vary according to chemical composition (Fig. 3). The most common and least expensive type is *soda-lime glass*, made from the oxides of silicon, sodium and calcium. This type of glass is used for virtually all glazing purposes—commercial and residential buildings, automobiles—and it accounts for nearly 90% of all glass production in the United States.

Soda-lime glass is composed of approximately 72% silica, 15% sodium, 9% calcium and 4% various minor ingredients. Other glass types, which have been derived from this basic glass, have slightly different ingredients. These will be mentioned only briefly here, as they are of minor importance to the construction industry.

Silica

Silica, the principal ingredient of glass, is obtained from sand and is one of the most common elements of the earth's crust. Unlike the sand found at the seashore, sand for making glass must be over 99% silica, perfectly white and not too fine. Sand of this quality usually is obtained from sandstone deposits, which are abundant in the U.S.—particularly in Pennsylvania, West Virginia, Illinois and Missouri.

Silica is the *glass former* in the glassmaking process. It can be used alone to produce glass, but it requires a very high melting temperature (3100° F) and is very difficult to form. Thus it is usually combined with other ingredients to make a more workable glass.

Sodium

Sodium for glassmaking usually is obtained from soda ash made from salt, but sodium nitrate and sodium sulfate also may be used. Sodium is added in glassmaking as a *flux*, to lower the melting point and to make a more workable mix. When sodium is added to silica, it forms a mixture that will melt at 1460° F—a considerably lower temperature than silica alone. However, silica and sodium alone form a glass which is not very durable—it dissolves slowly when in contact with water and has low resistance to chemical attack. Thus, an ingredient which will act as a *stabilizer* must be added to the glass batch.

Calcium

Calcium for glassmaking is obtained from either limestone or dolomite. It is the most commonly used stabilizer in glass manufacture, and when added to silica and sodium it produces a glass which is both durable and easily worked.

Additives

Special elements may be added to, or substituted for, the basic ingredients in glassmaking in order to produce glass with specific properties or characteristics.

Potassium Potash, a form of potassium, was used instead of soda ash in the early days of glass manufacture in the U.S. At that time wood was the principal fuel for glassmaking and potash was a by-product of wood ash. Potassium is used infrequently today because it is much more costly than sodium. Glass made with potassium is finer,

FIG. 3 **REPRESENTATIVE GLASS TYPES, THEIR COMPOSITION[1] AND USES**

Type of Glass	Sand	Soda Ash or Salt Cake	Limestone	Dolomite	Potash	Boric Acid	Litharge	Feldspar	Applications or Uses
	SiO_2	Na_2O	CaO	MgO	K_2O	B_2O_3	PbO	Al_2O_3	
100% Silica	100%[2]	—	—	—	—	—	—	—	Glass Laboratory Ware
Soda-Lime Silica[3]	68-74	11-16	8-12	1-4	—	—	—	0-4	Float, Plate, Sheet Glass
Borosilicate	65-80	2-4	3	—	0-4	13-28	—	1-3	Heat-Resistant Glasses
Lead-Alkali-Silicate	35-63	0-10	0-1	—	10-13	0-1	15-58	0-1	Radiation-Shielding Products
Optical (Flint)[4]	35	—	—	—	7	—	58	—	Ophthalmic, Lenses
Aluminosilicate	55-57	0-1	5	8-12	4	4-8	—	20-23	High Thermal Impact Glasses

[1] In addition to raw materials shown, small amounts of other elements may be added to help oxidizing, coloring, reducing, etc. Cullet (crushed glass) is also used to help melting.
[2] All figures are percentages of the ultimate oxides in glass compositions.
[3] Most important type based on volume and value; %'s given are for float, plate and sheet glass.
[4] Optical glass often contains oxides of zinc, titanium, zirconium and boron in addition to oxides shown.

FIG. 4 ADDITIVES FOR COLORING GLASS

Additive	% in glass	Color produced
Cadmium sulfide	0.03-0.1	Yellow
Cadmium sulfoselenide	0.03-0.1	Ruby & orange
Carbon & sulfur compounds		Amber
Chromium oxide	0.05-0.2	Green to yellow-green
Cobalt oxide	0.001-0.1	Blue
Copper	0.03-0.1	Ruby
Copper oxide	0.2-2.0	Blue-green
Ferric oxide	up to 4.0	Yellow-green
Ferrous oxide		Blue-green
Gold	0.01-0.03	Ruby
Iron oxide & manganese oxide	1.0-2.0 2.0-4.0	Amber
Manganese oxide	0.5-3.0	Pink-purple
Neodymium oxide	up to 2.0	Pink
Nickel oxide	0.05-0.5	Brown & purple
Selenides		Amber
Selenium		Bronze
Uranium oxide	0.1-1.0	Yellow with green fluorescence

harder and more brilliant than that made with sodium.

Lead This element is used to replace the calcium and sometimes part of the silica in soda-lime glass. It produces a soft, easy-to-melt glass, and when used in quantity, imparts a canary yellow color to the glass. Lead glass is relatively expensive, but it has excellent electrical insulating and radiation shielding properties, and is more brilliant in appearance than soda-lime glass. It is used primarily in making electrical products, objects of art and fine tableware.

Boron and aluminum These are additives used to produce types of glass which are both heat- and shock-resistant. Borosilicate glass and aluminosilicate glass, as they are called, are used primarily for laboratory ware and oven-cooking ware.

Coloring Additives There are several additives which are used specifically to produce color in glass (Fig. 4). For example, *nickel oxide* produces a tint which may range from yellow to purple depending on the glass base. *Cobalt oxide* gives an intense blue, and cobalt in combination with nickel produces a gray tone. Red glass is made with *gold, copper* or *selenium oxides,* and many other colors may be produced with other chemicals. Also, *arsenic* is sometimes added to glass in small quantities to counteract the greenish color caused by iron impurities in the glass.

Some of these colored and tinted glasses are used widely in architectural applications.

MIXING AND MELTING

Mixing and melting is the first major step in all glassmaking processes. The raw materials are granulated to nearly uniform particle size, then carefully weighed and mixed in the proper proportions. Some *cullet* (scrap glass) from a previous melt of the same kind of glass is added to the mixture, to assist in the melting process and to conserve materials.

The mixed batch is then loaded into a melting unit. The type of unit used depends upon the type and quantity of the glass needed.

Some glasses which are required in small quantities are melted in *refractory pots,* clay containers which can hold up to 3,000 pounds of glass. Up to 20 of these pots can be heated in a single furnace, thus allowing several different types of glass to be melted at the same time.

For somewhat larger quantities of glass, *day tanks* are used. Day tanks hold enough glass for a single day's operation—up to 20,000 lbs. The tank is filled with raw materials at the beginning of the operation, the glass is melted, and then all of the glass is used before the furnace is refilled.

The units most commonly used in commercial glassmaking operations are *continuous tanks,* which hold from 50 to 1,000 tons of molten glass (Fig. 5). They operate continuously, with the raw materials being fed into the loading end as

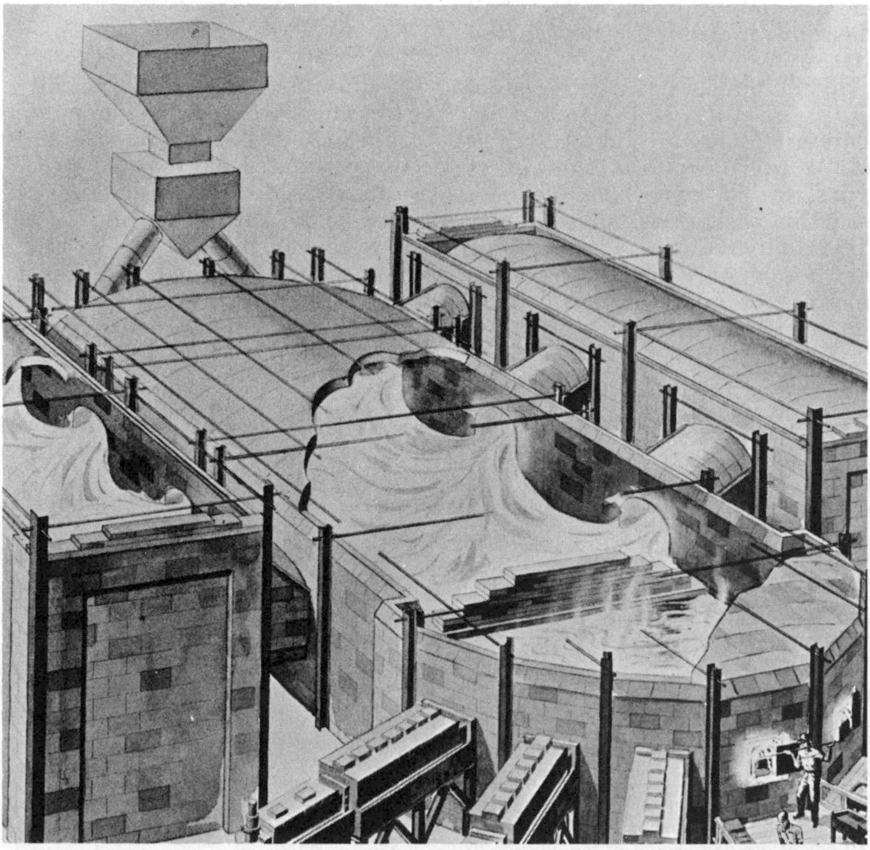

FIG. 5 A typical continuous melting tank can hold up to 1,000 tons of molten glass at a time.

rapidly as the molten glass is removed from the working end. This continuous operation can go on for a period of up to four years—depending upon how long it takes for the heat-resistant clay walls of the furnace to be worn away by the hot glass.

GLASSMAKING PROCESSES

After the molten glass leaves the melting tank, different manufacturing processes are employed to make various glass products. The following text will discuss only those processes which are used to make flat glass: (1) *drawing* for making sheet glass, (2) *rolling* for making plate and patterned glass, and (3) *floating* for making float glass.

Other processes, such as *pressing* and *foaming,* are used to make such products as glass block and cellular glass insulation, respectively. *Blowing* is of course the oldest and best known glassmaking process, but today it is primarily employed in the manufacture of containers and art objects rather than for making architectural products.

Drawing

In the early days of glass manufacture, sheet glass was made by the blowing process. The molten glass was blown into the shape of a cylinder; the cylinder was then split open, reheated and flattened to form a small sheet of glass. Unfortunately the glass made by this process could not be produced in large sheets and was of poor quality. It was characterized by uneven surfaces and longitudinal wrinkles which distorted one's vision.

The next evolution in sheet glass making was the drawing process (Fig. 6). There were three different methods used: the *Fourcault,* the *Colburn* and the *Pittsburgh* methods. All three of these methods utilized the same basic procedure.

The raw materials were first melted in a continuous melting tank. The drawing operation was then started by inserting a *bait* (steel rod) into the molten glass. Molten glass has a very high viscosity, making it sticky and highly resistant to flow. Thus, when a bait was inserted into it, the glass adhered to and was drawn up with the bait as it raised. The molten glass was drawn directly upward between jets of flame which fire-polished the surfaces and made them transparent.

Once the drawing process was started, pulling rollers kept it going. A continuous sheet was drawn from the molten glass and was pulled slowly through an *annealing lehr,* a type of continuous oven. Inside the lehr, temperatures were regulated carefully so that the glass was cooled gradually, thus avoiding excessive amounts of strain in the finished product.

After sheet glass emerged from the annealing lehr, no further finishing operations were required other than cutting to size and packing.

There were some minor differences in technique between the various methods of drawing sheet glass.

Fourcault Process This was the first process to be developed, and it was the simplest one. The glass sheet was drawn vertically from the tank to a height of about 40 ft., as described above.

Colburn Process The process was similar to the Fourcault process except that the glass sheet was drawn vertically for only about three feet and was then bent over horizontal rollers. The chief advantage of this process was that the glass sheet could be drawn through a horizontal rather than a vertical annealing lehr. The length of a vertical lehr was limited by the height to which it was practical to draw glass vertically—that is, about 30 to 40′. The horizontal lehr used in the Colburn process, however, could be up to 200′ long, thus allowing the glass to be cooled much more slowly and producing a glass that was less brittle and easier to cut. A few manufacturers still use this process today.

Pittsburgh Process In this process, the glass was drawn vertically, as in the Fourcault process. It differed from the other two processes mainly in that the drawing machine was constructed so that the sheet of glass was not contacted by center rolls until the glass was hard enough to resist being marred by the rolls.

There were several problems encountered in drawing sheet glass no matter which method was used. One problem was that the drawing process stretched the sheet causing

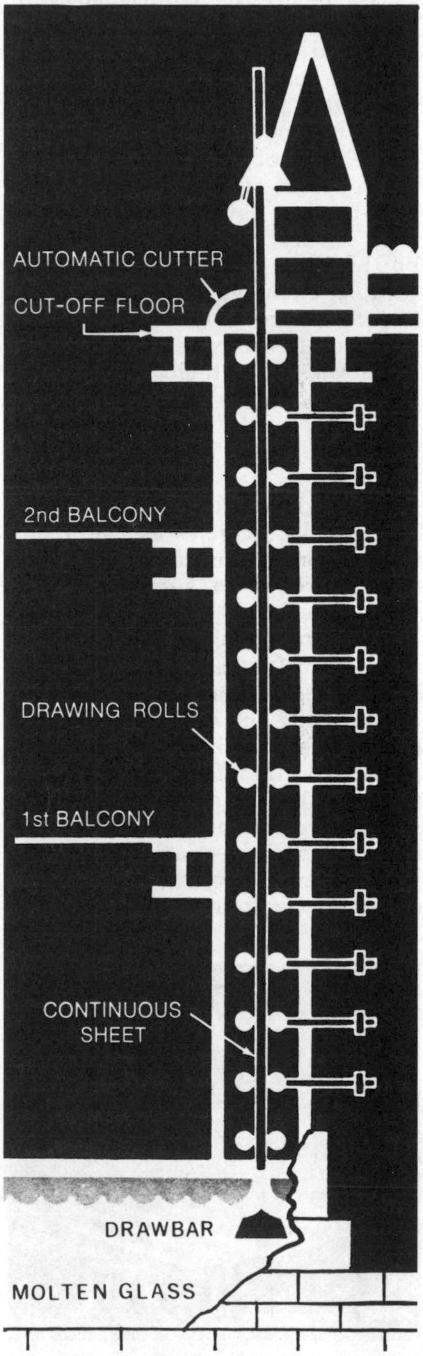

FIG. 6 The Pittsburgh Process for drawing sheet glass vertically.

"waisting-in" (variations in the width) of the glass. This was overcome by water-cooled rollers at each edge which maintained sufficient side pull to keep the width constant.

Another problem was that the drawing process caused waves and ripples to form in the rising sheet of glass, before it had cooled sufficiently to become solid. Sheet glass thus had an inherent wave or distortion, usually running in one direction. For most purposes this wave was not noticeable enough to cause a problem; however, when large pieces of glass were required or when an oblique angle of view was likely, these imperfections could be quite apparent. This largely excluded sheet glass from use for large commercial mirrors, automobile windshields and display windows, for which the superior qualities of plate or float glass were required.

Rolling

Plate glass was originally made by "casting", which was a very slow and laborious process. The glass was poured onto a casting table, then rolled flat, cooled and finally ground and polished.

Between 1922 and 1924, however, all the major glass manufacturers began using the continuous rolling process, for making both plate and patterned glass; few manufacturers still use the process today.

In the continuous rolling process, the glass first was mixed and melted in a continuous melting tank. The molten glass then flowed through an opening in the tank into the forming rolls (Fig. 7). These rolls were water-cooled, electrically driven and had precision-ground steel surfaces which could not be harmed by the extremely high temperatures of the glass. They rolled the molten glass to form a semisolid continuous ribbon of glass. The rolls could be adjusted for various thicknesses of glass, in widths up to 11′.

From the forming rolls, the ribbon of glass moved over a series of other steel rolls until it became solidified enough to enter the annealing lehr. There it was cooled gradually to remove internal strains. After leaving the lehr, the glass was cooled still further until it reached room temperature. At this point, the glass was called *rough-rolled* glass.

If a translucent effect was desired, rough-rolled glass could be used without further processing. When patterned glass was desired, the forming rolls were embossed with various decorative designs; this glass, like rough-rolled glass, did not have to be ground and polished. However, if plate glass was to be made, both surfaces still needed to be ground and polished.

Most plate glass was made by the twin-grinding method, in which the glass was ground on both sides simultaneously (Fig. 8). This method assured smooth, parallel surfaces and thus, distortion-free glass. A mixture of silica sand with water was used in the grinding process.

FIG. 7 In the rolling process, the molten glass passed between two rollers to form a continuous ribbon of glass.

FIG. 8 To produce clear plate glass, rough-rolled glass must pass under a series of large twin-grinding machines.

The sand was graded into several different sizes; the largest particles were used in the first grinding units and then the size of the particles gradually was reduced. A considerable quantity of glass was removed in the grinding process. Thus, the rough-rolled glass that was used to make plate glass was produced in a thickness greater than that desired in the finished product.

After grinding, the glass was moved on to the polishing tables, where it was passed along on a conveyor belt under a series of polishing wheels. The wheels were covered with buffs of felt, leather or a soft metal, and the polishing agent most often used was *rouge* (ferric oxide) mixed in a solution of *copperas water* (ferrous sulfate and water). After polishing, the glass was cut to size and packed for shipment.

Floating

The float process was developed by Pilkington Bros., Ltd., of England in an attempt to combine the fire-polish and low cost of sheet glass with the flatness and low distortion properties of plate glass.

The first stage in the production of float glass is the same as that used to produce sheet and plate glass—the batch is melted in a huge continuous tank. From there on, however, the process is quite different.

The molten glass leaves the tank and is floated on a bed of molten metal which is composed mostly of tin. When one liquid floats on another liquid of higher density, both surfaces of the upper liquid must become flat and parallel. Therefore, when the liquid glass moves across the float bath, it forms into a thin layer with flat parallel surfaces.

The float bath is divided into three sections (Fig. 9). In the first section, heat from the molten tin and a top heater causes the glass to float uniformly over the flat surface of the tin. In the second section, the glass receives a brilliant fire-polished surface. In the third section, the glass is cooled until it becomes sufficiently hard to be conveyed to the rollers in the lehr. Very precise control of

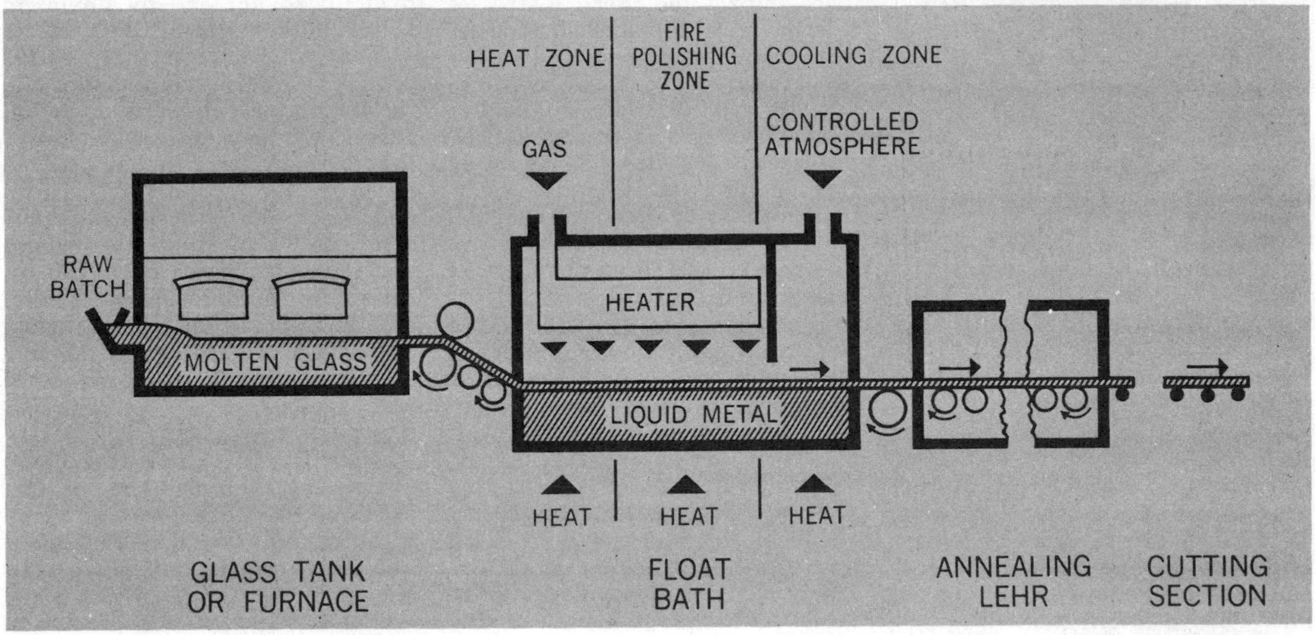

FIG. 9 The float process for making glass.

heat all along the length of the float bath is required.

After the glass leaves the float bath, it enters an annealing lehr where it is cooled gradually to insure that the glass will not contain excessive amounts of strain. It is then washed and moved to the warehouse for inspection, cutting and shipping.

The manufacture of ground and polished plate glass is rapidly being replaced by float glass manufacture because of the substantial difference in production costs. It is reasonable to expect that ground and polished plate glass will be produced in very limited quantities in the near future and will be replaced by float glass in most, if not all, commercial glazing applications.

HEAT AND CHEMICAL TREATMENTS

There are several different heat and chemical treatments which can be used to increase the strength and usefulness of flat glass products. *Annealing* relieves internal strain and facilitates fabrication. On the other hand, *tempering* and *heat strengthening* cause physical characteristics to be developed in the glass so that it cannot be fabricated afterwards. All fabrication—such as cutting, grinding, sandblasting, drilling, notching—must be performed before these heat treatments.

Annealing

All flat glass, after forming, is subjected to a controlled cooling process, *annealing,* to relieve internal strain. If a piece of glass is not annealed, parts of it will cool and contract at different rates, and when the glass approaches room temperature, it may fracture due to differential stresses throughout the sheet.

Annealing is performed in a type of continuous oven known as a *lehr.* The glass moves through the lehr on a belt and temperatures are carefully regulated throughout the length of the lehr. The glass is first raised to a high enough temperature to relieve strain developed in the forming process. Then the glass is brought down in temperature very slowly so that all parts of the sheet will cool at the same rate (Fig. 10). Thus, when the glass arrives at room temperature, it is free of internal strain and can be easily cut and processed.

Tempering

Tempering is a heat treatment used to increase the strength of various types of flat glass. The process involves heating the glass to just below the softening point and then chilling it suddenly by subjecting both surfaces to jets of cool air. This causes the surfaces to shrink and harden quickly while the interior is still fluid. The interior begins to cool and shrink next, but since the surfaces are already hardened they cannot flow and adjust to the shrinkage of the interior. This causes the surfaces and edges to be placed into a state

FIG. 10 A long ribbon of glass hardens as it cools after leaving the annealing lehr.

FIG. 11 Fully tempered glass is 3 to 5 times stronger than ordinary annealed glass of the same thickness.

of compression while the interior of the glass is in a state of tension. In properly tempered glass the opposing stresses of the surface and the interior balance each other out. The resultant glass is much stronger than ordinary flat glass because the built-in compression of the surface must be overcome before breakage can occur.

This type of glass usually is referred to as *fully tempered* glass, to distinguish it from *heat-strengthened* glass, which is only partially tempered. Fully tempered glass is 3 to 5 times more resistant to wind loads, impact and thermal stresses than ordinary annealed glass of the same thickness (Fig. 11).

Heat Strengthening

This heat treatment also is used to increase the strength of many types of flat glass. It involves heating the glass and then cooling it in a manner similar to that used in tempering. However, heat strengthening produces lower surface compression stresses in the glass than the tempering process. Thus, heat strengthened glass is only about twice as strong as annealed glass of the same thickness.

Chemical Strengthening

Glass can be strengthened by chemical treatment as well as by heat treatment. The most commonly used chemical treatment involves an exchange of ions in the surface of the glass. The glass is immersed in a molten salt bath and the large potassium ions in the salt migrate into the glass surface, replacing the smaller sodium ions there. This action crowds the surface, resulting in compression stresses similar to those produced in glass by thermal tempering.

Full temper in soda-lime glass is difficult to achieve by chemical tempering methods. A high degree of temper by chemical means generally requires careful formulation of ingredients selected to insure that glass ions interchange readily with ions in the molten salt bath. Furthermore, in chemically strengthened glass, the compressed surface layer is appreciably thinner than that in thermally tempered glass. Therefore, it is more readily weakened by surface abrasions such as cuts and scratches.

SURFACE FINISHES

The surface finish of glass depends both on the method of manufacture, and on the subsequent finishing processes to which it is subjected.

Some types of glass, such as sheet and float glass, emerge from the annealing lehr with a smooth, glossy and transparent finish, known as a *fire finish.* This type of finish is characteristic of these glassmaking processes and no further finishing operations are required. Plate glass, however, emerged from the annealing lehr with a rough, translucent surface

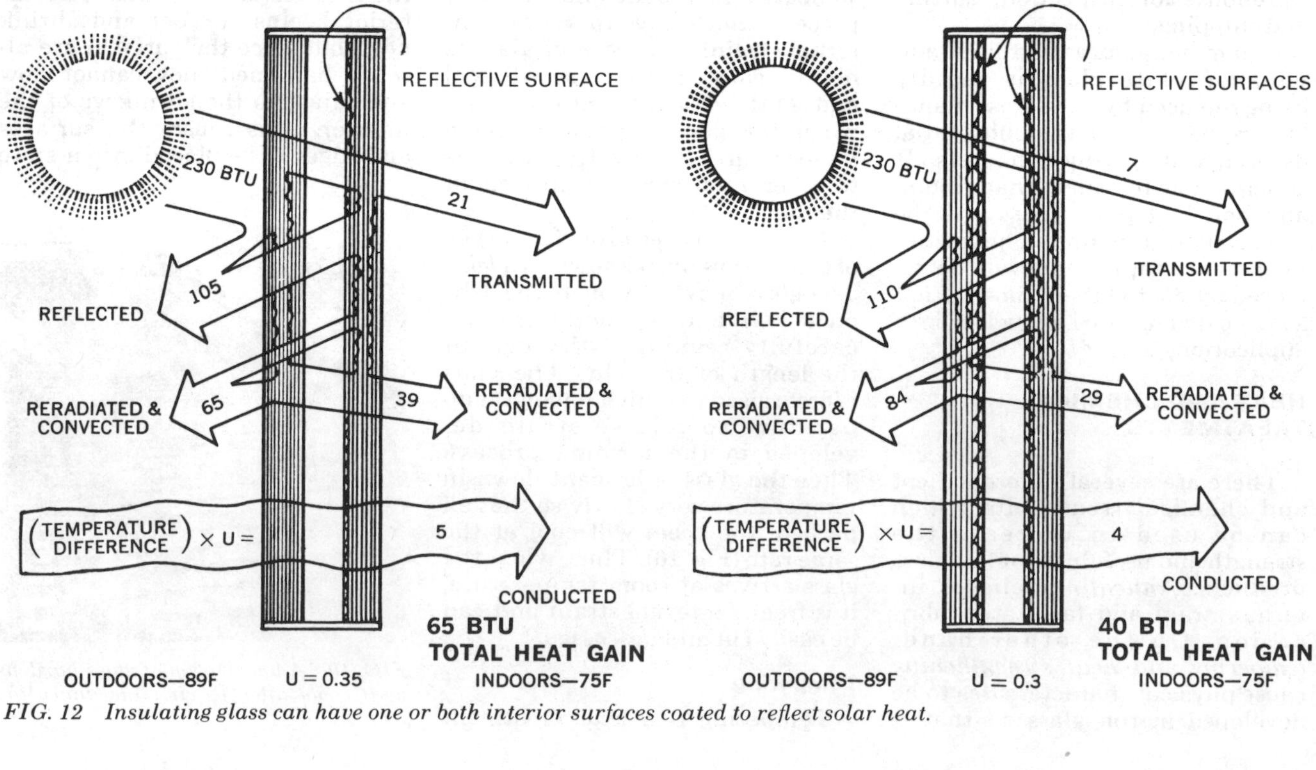

FIG. 12 Insulating glass can have one or both interior surfaces coated to reflect solar heat.

which required grinding and polishing to produce the smooth, transparent finish desired for most flat glass products.

There are many special cases, however, where a smooth, transparent glass is not essential. A translucent glass may be desired to provide privacy or to limit the amount of light transmitted into a room; a specially coated or tinted glass may be needed to reduce glare or to control the amount of solar heat transmitted into a room; in other cases, a patterned or specially treated glass may be preferred just to produce a decorative effect. Some of the surface finishes associated with these special types of glass are discussed below.

Etching

Many different degrees of transparency can be produced in glass by the etching process—from a frosted, almost opaque quality to a semi-polished, translucent appearance.

Hydrofluoric acid or one of its compounds is used in the etching process, since this is the only chemical which interacts actively with glass. The glass is dipped in or sprayed by the chemicals; the type of surface produced depends upon the glass composition, the concentration of fluorides and the time involved. If an etched design is desired, the glass is first painted with an acid-resistant chemical (the *resist*) to protect the parts of the glass outside the desired pattern. The acid eats away the unpainted surface of the glass, leaving the design.

Sandblasting

The translucent finish obtained by this method usually is rougher than that obtained by etching. Sandblasting is done by using compressed air to blow coarse, rough-grained sand against the glass surface. Often this is done through a rubber stencil to produce a decorative design.

Both etching and sandblasting reduce the strength of glass because these processes introduce surface flaws. Sandblasting, for example, reduces the strength of glass by approximately 50%.

Enameling

Translucent and solid colors can be produced on glass surfaces with vitreous enamels which are fired on at high temperatures. The firing process also imparts a partial or full temper to the glass. This type of glass is used primarily for spandrel areas and other parts of curtain wall construction.

Silvering

In the early days of mirror manufacture, silvering was done by hand, the silvering solutions being poured from a pitcher onto the flat glass surface. Some fine mirrors are still produced in this way today.

Most mirrors, however, are manufactured by an automatic process in which the silver is deposited directly on the glass as it passes on a conveyor belt under sprays of silver nitrate and tin chloride. The silvering then can be protected with a coating of shellac, varnish or paint. For almost permanent protection, an electroplated layer of copper can be applied on the silver. In order to provide true, accurate reflections, large commercial mirrors are made from plate or float glass; sheet glass is used only in the manufacture of small, inexpensive household mirrors.

Reflective Coatings

With the advent of the glass enclosed highrise, an entire new family of architectural glasses has been developed to control the transmission of solar energy into buildings. These products have a thin reflective coating applied either (1) to one surface of a single sheet of glass or (2) to an inside surface of a laminated or insulating glass unit (Fig. 12).

There are three basic processes used to apply reflective coatings to glass: (1) chemical deposition, which is accomplished by a chemical interchange at the glass surface; (2) the pyrolytic process, in which the coating material is sprayed onto heated glass and becomes fused to the glass surface; and (3) vacuum deposition, in which the glass is bombarded with metallic ions, forming an integral molecular bond with the glass.

Under appropriate lighting conditions, reflective coated glass can become a type of one-way mirror, permitting vision only to the side with the greater amount of light. During the daytime, it creates a mirror-like effect from the outside of a building (Fig. 13), while from the inside the glass is transparent. At night, the effect is reversed. Reflective coatings can be made to reflect varying degrees of light and heat, providing control over the solar energy transmitted into a building.

Tinting

Tinting is not really a type of surface finish, as it involves adding special ingredients to the basic glass batch. However, it should be

FIG. 13 Reflective glass sometimes is referred to as environmental glass because of its ability to mirror the environment.

mentioned briefly here as it is used widely to produce glass with heat-absorbing and glare-reducing qualities.

Many different ingredients can be added to glass to produce different tints. Ferrous iron, for example, imparts a bluish-green tint to glass, cobalt oxide in combination with nickel imparts a greyish tint and selenium is used to make bronze-tinted glass.

These different ingredients are added to the glass mixture at the very beginning of the melting process, and sheet, plate and float glass can all be tinted in this manner.

FLAT GLASS PRODUCTS

This section outlines the different types of flat glass and gives representative available qualities, sizes and properties of each. The multitude of available glazing products makes possible a high degree of specialization in meeting specific design requirements.

The method of glass manufacture determines the basic type of the glass—whether sheet, float or plate. By modifying the ingredients, any of these types of glass can be made either clear to provide true vision, or tinted to reduce glare and absorb solar energy. Clear or tinted glass may be coated with a reflective surface to further reduce light and heat transmission.

Where resistance to impact and thermal stresses is important, these glasses can be strengthened by heat treatments, such as tempering or heat strengthening. Glass, either clear or tinted, can be rolled with a wire mesh reinforcement to make wired glass. Finally, almost any two or more of these glasses can be combined to form a composite multilayer glass. Different types of composite glass are designed to meet specific construction requirements, such as laminated glass for safety and security, and insulating glass for thermal and sound insulation.

BASIC TYPES

The basic types of flat glass are (1) sheet glass, (2) float glass and (3) plate glass. Float glass is most commonly used; some manufacturers no longer produce the other types. In the United States, all flat glass is manufactured to meet the requirements of Federal Specification DD-G-451d.

Sheet Glass

Sheet glass was often used for ordinary glazing purposes because of its comparatively low cost. It has a brilliant fire-polished finish which resists scratching, but it lacks the parallel, distortion-free surfaces of traditionally higher-priced plate and float glass. Modern production methods have reduced float glass' cost. Some manufacturers no longer produce sheet glass. Sheet glass is divided into different types, depending upon thickness: (1) picture glass, (2) window glass, and (3) heavy sheet glass.

Picture glass is the thinnest of these types. It is used only for covering maps, photographs and other display items, and is not useful in construction. It should not be confused with window glass used for glazing picture windows.

Window glass is supplied in two thicknesses: single strength (SS) 3/32″ thick, and double strength (DS) 1/8″ thick. It comes in three different qualities as defined by Federal Specification DD-G-451d: quality AA is specially selected glass for highest grade work; quality A is select glass for superior glazing; and quality B is suitable for general glazing. Heavy sheet glass is thicker than window glass (3/16″ to 7/32″) and also comes in qualities AA, A and B (Fig. 14).

FIG. 14 **SUMMARY OF BASIC GLASS TYPES, SIZES AND USES**

Product	Quality	Nominal thickness (in.)	Nominal thickness (mm)	Maximum Area (in.)[1]	Weight (lbs./sq. ft.)	Visible light transmission (%)	Principal Uses
WINDOW GLASS	AA, A, B	SS, 3/32 DS, 1/8	2.5 3	120 u.i.[2] 140 u.i.[2]	1.22 1.63	91 91	Residential and commercial glazing
HEAVY SHEET	AA, A, B	3/16 7/32	5 5.5	84 x 120 84 x 120	2.50 2.85	90 89	Table tops, shelves, low-cost glazing
PLATE OR FLOAT	Silvering	1/4	6	50 sq. ft.	3.28	89	Mirrors and optical uses
	Mirror	1/4	6	75 sq. ft.	3.28	89	
	Glazing	1/8 1/4	3 6	74 x 120 128 x 204	1.64 3.28	90 89	Residential and commercial glazing
HEAVY PLATE OR FLOAT	Glazing	5/16 3/8 1/2 5/8 3/4 7/8	8 10 12 16 19 22	124 x 200 124 x 200 120 x 200 120 x 200 115 x 200 115 x 200	4.10 4.92 6.54 8.17 9.18 11.45	87 86 84 82 81 79	Residential and commercial glazing
ROUGH PLATE	Polished one side	17/64	7	58 x 120	3.28	88	Partitions and windows where obscurity is desired
	Rough both sides	9/32	7	124 x 240	3.70	88	

1 Maximum area varies according to producer. Larger sizes may be available from some producers.
2 U.i. = united inches, or the sum of width plus length.
3 Available from some producers, subject to factory operations.

FIG. 15 Two typical varieties of patterned glass: (a) stippled and (b) hammered.

A wave or "draw" distortion is characteristic of sheet glass, and the degree of distortion determines the grading and usefulness of the glass. Generally, the thicker the sheet, the more noticeable the distortion becomes. For this reason, heavy sheet glass is generally used for glazing in areas where a high degree of optical fidelity and clarity of vision is not required. It is used most commonly for supported table tops, shelves, and similar applications.

For best results, sheet glass should be glazed with the wave distortion running horizontally. To insure this, when specifying, the width dimension should be listed first.

Float Glass

Float glass is becoming the most widely used as production techniques are improved and it becomes available in a greater range of thicknesses and sizes. It is used today for many of the purposes which were once served only by plate glass (high-quality mirrors, display windows and automobile glazing).

Float glass may be divided into two types: *regular float* glass, in thicknesses of 1/8" to 1/4"; and *heavy float* glass, in thicknesses of 5/16" to 7/8". Regular float glass comes in three qualities: (1) silvering quality is specially selected glass which is exceptionally free from defects, intended for mirrors and optical applications; (2) mirror-glazing quality is a superior glass for mirrors and other demanding applications; and (3) glazing quality is glass used for ordinary glazing installations. Heavy float glass usually is produced in one quality only—glazing quality (Fig. 14).

Plate Glass

Like float glass, plate glass is divided into two types, depending upon thickness: *regular plate,* in thicknesses of 1/8" to 1/4"; and *heavy plate,* in thicknesses of 5/16" to 7/8". Regular plate glass comes in three qualities: silvering, mirror-glazing and glazing qualities. Heavy plate glass is produced in one quality only—glazing quality.

Rough plate glass is available on a limited basis for glazing applications where obscurity is desired. The texture pattern of rough plate may contain some surface flaws, and may vary somewhat depending upon the wearing of the forming rolls used in manufacture. It is generally available in two varieties: polished one side (17/64" thick) and rough both sides (9/32" thick), and it comes in glazing quality only (Fig. 14).

PATTERNED GLASS

Patterned glass is translucent, with some type of linear or geometric pattern embossed on one or both sides of the glass. It is produced by the same method as plate glass (rolling) but is not ground and polished.

Patterned glass is available in a wide range of textures and patterns, which vary according to manufacturer (Fig. 15). These patterns offer different degrees of light transmission and obscurity, depending upon the nature of the design and whether it is rolled into one or both surfaces of the glass. Additional obscurity may be obtained by etching or sandblasting the patterned glass, producing a

FIG. 16 **SUMMARY OF PATTERNED AND WIRED GLASS TYPES AND SIZES**

Product	Type	Nominal thickness (in.)	Nominal thickness (mm)	Maximum Area (in.)[1]	Weight (lbs./sq. ft.)	Visible light transmission (%)
PATTERNED GLASS	Floral Hammered Stippled Granular	1/8	3	60 x 132	1.60-2.10	80-90
	Ribbed Fluted[1] Striped[2]	7/32	5.5	60 x 132	2.40-3.00	80-90
WIRED GLASS	Polished (square or diamond mesh)	1/4	6	60 x 144	3.50	80-85
	Patterned (square or diamond mesh)	1/4	6	60 x 144	3.50	80-85
	Parallel wired[3]	7/32 1/4 3/8	5.5 6 10	54 x 120 60 x 144 60 x 144	2.82 3.50 4.45	80-85 80-85 80-85

[1]Maximum area varies according to producer; larger sizes may be available from some producers.
[2]These are just a few of the most common patterns available; many patterns are patented and made by one producer only.
[3]This type of wired glass does not carry Underwriter's Laboratory, Inc. fire-retardant rating; it is used mainly for decorative partitions.

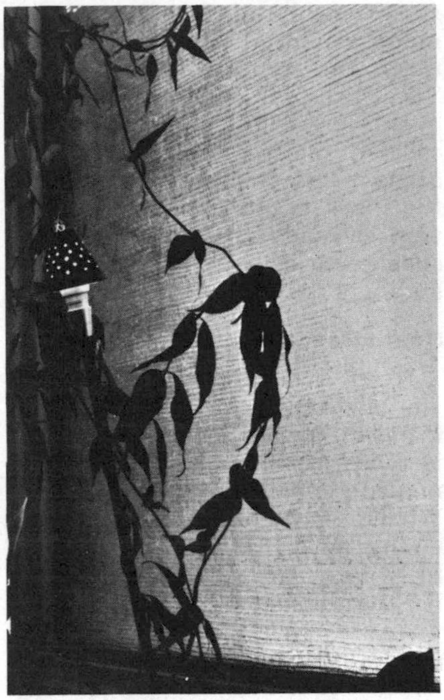

FIG. 17 Patterned glass provides visual privacy and also allows for diffused light transmission.

satin-like finish. These processes also reduce the strength of the glass, and this is a factor which should be considered when determining suitable sizes and uses.

Patterned glass is available mainly in thicknesses of 1/8″ and 7/32″, although some patterns are produced in other thicknesses on a limited basis (Fig. 16). Some patterns also are available tempered.

Patterned glass can provide visual privacy while at the same time allowing diffused light transmission (Fig. 17). It has a decorative appearance and is often used for this reason alone. Its most frequent use in the home is for tub and shower enclosures (where it should be tempered for safety); it is also used for interior partitions and exterior windows where obscured vision is desired.

WIRED GLASS

Wired glass has a wire mesh or parallel wires rolled into the center of the glass thickness. Plate glass as well as certain types of patterned glass are available wired (Fig. 18). Wired glass generally is produced in 1/4″ thickness only, although some manufacturers have other thicknesses in limited quantities.

The wire in this type of glass holds it together under low levels of impact, thereby reducing injuries from broken glass. It is considered a safety glazing material and, when properly made, meets the requirements of ANSI Standard Z97.1-1975. It is used widely for corridor and entrance doors, partitions, skylights and other potentially hazardous locations. It is widely used for fire doors and windows because of its ability to remain in place even though cracked by excessive heat. When used as a fire-resistant material, it should conform to Underwriters' Laboratories, Inc. standards for fire retardants (Fig. 16).

TINTED GLASS

Tinted glass has an admixture included in the glass batch to impart glare-reducing and heat-

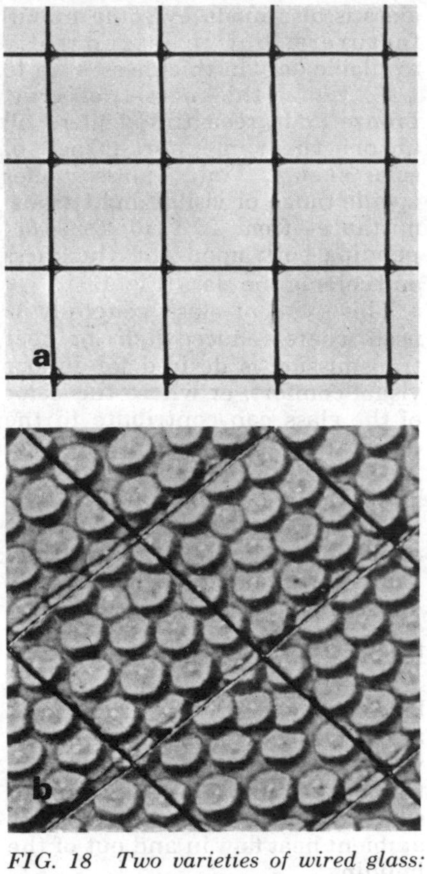

FIG. 18 Two varieties of wired glass: (a) clear plate glass with a square wire mesh and (b) patterned glass with a diamond wire mesh.

absorbing qualities to the glass. Sheet, float and plate glass all are available tinted.

Gray and bronze tinted glasses are the most commonly available types. They are generally produced in thicknesses up to 1/2″, and in glazing quality only. Green tinted

FIG. 19 **SUMMARY OF TINTED GLASS TYPES AND SIZES**

Product	Quality	Nominal thickness (in.)	(mm)	Maximum Area (in.)[1]	Weight (lbs./sq. ft.)	Visible light transmission (%)	Solar energy transmission (%)[2]	Principal Uses
BRONZE—PLATE OR FLOAT	Glazing	1/8	3	35 sq. ft.	1.64	68	65	
		3/16	5	120 x 144	2.45	58	55	
		1/4	6	128 x 204	3.27	50	46	
		3/8	10	124 x 200	4.90	37	33	
		1/2	12	120 x 200	6.54	28	24	
GRAY—PLATE OR FLOAT	Glazing	1/8	3	35 sq. ft.	1.64	62	63	Where reduced heat and light transmission is desired
		3/16	5	120 x 144	2.45	51	53	
		1/4	6	128 x 204	3.27	42	44	
		3/8	10	124 x 200	4.90	28	31	
		1/2	12	120 x 200	6.54	20	24	
GREEN—PLATE OR FLOAT	Glazing	3/16	5	96 x 128	2.45	78	55	
		1/4	6	96 x 128	3.27	74	48	

[1]Maximum area varies according to producer; larger sizes may be available.
[2]Transmission values vary slightly from one producer to another.

glass is also made by some manufacturers, but it is generally available only in thicknesses up to 1/4″. Equal thicknesses of gray, bronze and green tinted glass all absorb the same percentage of solar energy. Tinted glasses offer a wide range of visible light transmittance—from 20% to 78%—depending both upon the thickness and color of the glass (Fig. 19).

This type of glass generally is used where reduced light or heat transmission is desired for indoor visual comfort, or where the color of the glass can contribute to the design of the building. Because of its heat-absorbing characteristics, special precautions should be taken when using tinted glass to avoid damage to the edges and consequent breakage from thermal stresses.

REFLECTIVE COATED GLASS

Reflective glass is a relatively large category of glass, (sometimes referred to as *environmental* glass), which reduces the solar energy transmitted into a building. In insulating units it also minimizes ambient heat flow in and out of the building.

Reflective glasses have a transparent metal or metal oxide coating bonded to one surface of the glass. Several of the more durable coatings, such as cobalt oxide and chromium, are available on single glass as well as on insulating and laminated glasses. Other available coatings, such as copper, aluminum, nickel and gold, are less durable and can be furnished only on an inside surface of insulating or laminated glasses, where the coating is protected from weather and accidental abrasion.

A wide range of metallic colors is available with differing visual and thermal properties. Daylight reflectances range from about 10% for the more subdued coatings to about 45%, and visible light transmittance from 8% to 50%. The more common colors are neutral, golden, bronze, blue and copper. Many coatings are available also on tinted, heat-absorbing glass. When a coating is applied to such glass, it often must be heat-strengthened or fully tempered.

Coated glass products vary widely from one manufacturer to another, and certain colors and tones are offered by only one company. Consequently, individual producers should be consulted for specific product information.

HEAT-TREATED GLASS

There are two types of heat-treated glass: *fully tempered* and *heat-strengthened.* When either of these types of glass is to be used, the exact sizes desired must be specified, as the glass cannot be cut, drilled or notched after the heat treatment.

Fully Tempered Glass

Any glass 1/8″ or thicker may be tempered, except wired glass or patterned glass with deep patterns. Tempered glass may be incorporated into insulating glass or laminated glass units; heat-absorbing and heat-reflective glasses may also be tempered.

An inherent characteristic of all tempered glass is a slight deviation from flatness, particularly near the edges. The degree of deviation depends upon thickness, width, length and other factors. Usually greater thicknesses yield flatter products.

Fully tempered glass has three to five times the resistance of annealed glass to uniform loading, thermal stresses and most impact loads. When fractured, it breaks into a safe pattern of relatively small, harmless particles (Fig. 20).

FIG. 20 Fully tempered glass, when fractured, breaks into a fail-safe pattern of small, relatively harmless particles.

It is considered a safety glazing material and, when properly made, meets the requirements of ANSI Standard Z97.1-1975. It is frequently required by state or local codes in residential construction in areas where human contact with the glass is probable, such as entrances and sidelights, tub and shower enclosures, storm doors and sliding glass doors.

Heat-strengthened Glass

Heat-strengthened glass is similar to fully tempered glass, but it has only a partial heat temper. It is twice as strong as ordinary annealed glass of the same thickness. When it fails, it breaks into fragments much larger than those of fully tempered glass, and therefore it is not considered a safety glazing material. Heat-strengthened glass generally is used in applications which require a stronger product than regular annealed glass, but do not require the full strength of tempered glass.

Any type of 1/4″ or 5/16″ thick glass which can be tempered can be heat-strengthened. The most widely used form of heat-strengthened glass is *spandrel* glass. This glass has a fired-on ceramic frit on the interior surface, and it is used commonly in spandrel areas in curtainwall construction.

COMPOSITE GLASS

The two main types of fabricated glass, laminated and insulating glass, are products made by combining two or more layers of glass into a single unit.

Laminated Glass

Laminated glass consists of two or more layers of glass with interlayers of polyvinyl butyral plastic sandwiched between them, bonded together under heat and pressure to form a single unit. The thickness of the glass layers may range from 3/16″ sheet glass to 1″ plate or float glass. The plastic interlayer is usually .015″ or .030″, although it may be as thick as .090″. The glass surfaces which are to be bonded to the plastic cannot be patterned or otherwise irregular. Heat-absorbing, heat-reflecting, fully tempered, heat-strengthened and wired

FIG. 21 **SUMMARY OF LAMINATED GLASS PRODUCTS**

Product	Glass Type	Plastic interlayer (in.)	Nominal thickness (in.)	Nominal thickness (mm)	Maximum area (in.)[1]	Weight (lbs./sq. ft.)	Principal Uses
LAMINATED GLASS	SS/SS[2] Sheet	.015-.030	13/64	5	48 x 80	2.45	Partitions, doors and automotive glazing
	Lam/Lam[2] Sheet	.015-.030	15/64	6	48 x 100	2.90	
	DS/DS[2] Sheet	.015-.030	1/4	6	48 x 100	3.30	
	2 lites 1/8" plate or float	.015-.030	1/4	6	72 x 120	3.30	
	2 lites heavy plate or float	.015-.030	3/8 1/2 5/8 3/4 7/8 1	10 12 16 19 22 25	72 x 120 72 x 120 72 x 120 72 x 120 72 x 120 72 x 120	4.80 6.35 8.00 9.70 11.40 12.95	
BURGLAR-RESISTANT	2 lites 1/8" plate or float	.060-.090	5/16	8	60 x 120	3.50	Display windows
BULLET-RESISTANT	At least 4 lites of plate or float	At least 3 plies of .015	1 3/16 1 9/16 1 3/4 2	30 40 45 50	40 sq. ft. 30 sq. ft. 25 sq. ft. 22 sq. ft.	15.50 20.00 23.00 26.20	Bank teller windows and payroll booths
ACOUSTICAL STC-36[3] STC-40 STC-43	2 lites of plate or float	One or more plies of .045	9/32 1/2 3/4	7 12 19	60 x 120 60 x 120 60 x 120	3.40 6.36 9.74	Office partitions, radio and TV studios, airport glazing

[1]Maximum area varies according to producer. Larger sizes may be available from some producers.
[2]Lam indicates 7/64" sheet glass; SS, 3/32" sheet; DS, 1/8" sheet.
[3]Refers to Sound Transmission Class ratings.

glasses may all be laminated in combination with other types.

When laminated glass is broken, the fragments of glass generally adhere safely to the plastic interlayer and do not evacuate the opening. This reduces the potential hazards of flying glass and minimizes cutting injuries. All varieties of laminated glass are considered safety glazing materials and, when properly made, meet the requirements of ANSI Standard Z97.1-1975.

Most laminated glass is made of two lights of glass of varying thicknesses with a single plastic interlayer. It is used commonly for interior partitions, doors and automotive glazing. There are also several special types of laminated glass which are made for particularly demanding applications (Fig. 21).

Burglar-resistant Glass This is a laminated glass usually consisting of 2 layers of 1/8" glass with a .060" to .090" plastic interlayer. It is made primarily for use in the protection of show windows displaying valuable merchandise. It has superior resistance and serves as a major deterrent to the "smash-and-run" burglar.

Bullet-resisting Glass This is a laminated glass having four or more layers totalling 3/4" to 3" in thickness. Most manufacturers have 1-3/16", 1-9/16", 1-3/4" and 2" thicknesses listed by Underwriters' Laboratories, Inc. for resistance to medium-, high- and super-power small arms and high-power rifle arms, respectively. Thicknesses up to 7" are available on special order. This type of glass is used for bank teller windows, payroll booths and other security applications.

Acoustical Glass This is a laminated glass with one or more plastic interlayers of .045" thickness. It is particularly effective in reducing sound transmission in the frequency ranges of speech, radio and TV (250 to 4000 c.p.s. range). It is widely used for office partitions, radio and TV studios and similar applications.

Insulating Glass

Insulating glass is a factory-produced, airtight unit, consisting of two layers (lites) of glass, with a de-

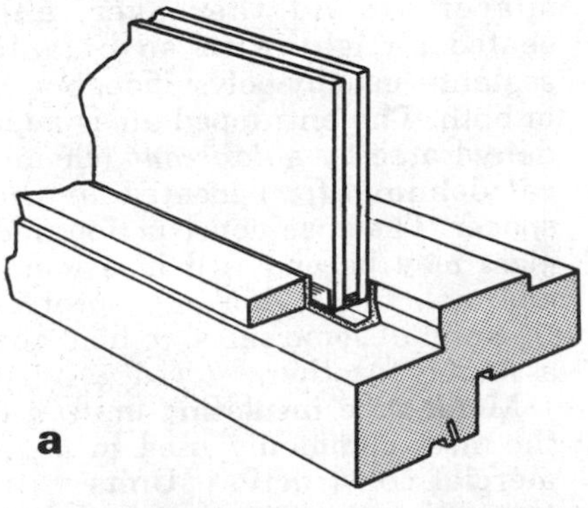

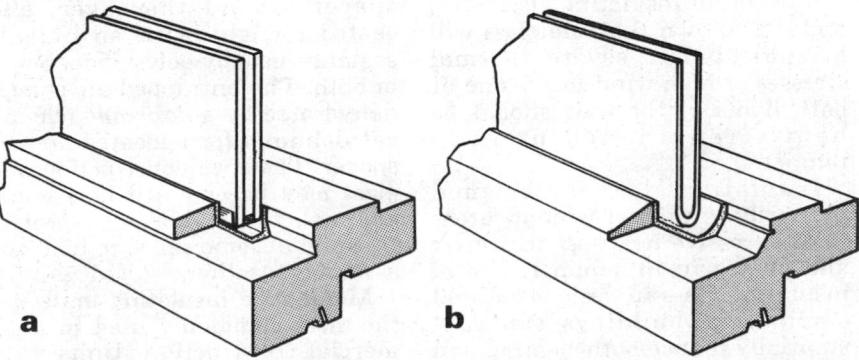

FIG. 22 Insulating glass is available with two different edge types: (a) metal edge and (b) glass edge.

FIG. 23 **MAXIMUM RELATIVE HUMIDITY CAUSING CONDENSATION ON INSULATING AND OTHER GLASS**

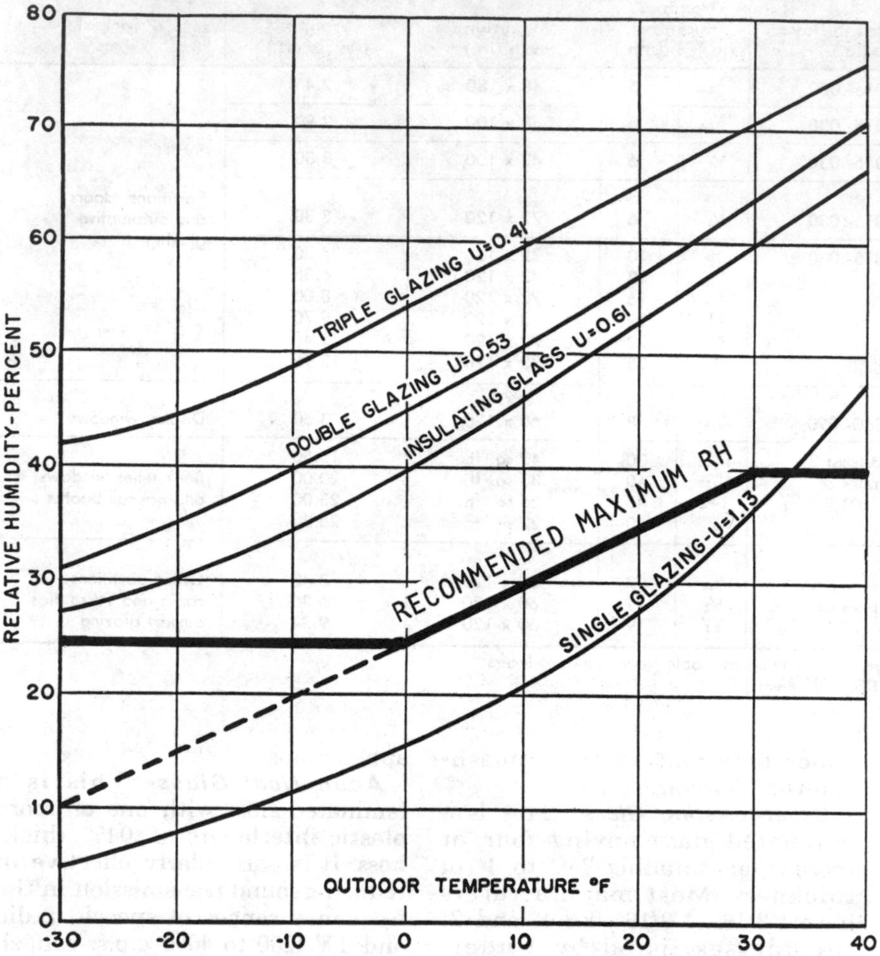

hydrated air space between. Insulating units are produced with two different edge types: metal and glass edge (Fig. 22).

Any glass type with at least one smooth surface may be used in an insulating glass unit. When heat-absorbing or reflective-coated glass is used in an insulating glass unit, and it is known that the glass will be subjected to severe thermal stresses or high wind loads, one or both lights of the unit should be heat-strengthened or fully tempered.

Insulating glass is designed primarily to protect window areas from excessive heat loss in winter and heat gain in summer. Use of insulating glass in residential and commercial buildings can substantially reduce both heating and air conditioning costs.

Insulating glass has the added advantage of allowing a much higher interior relative humidity than single glass because it minimizes condensation problems during cold weather (Fig. 23).

Metal-edge Insulating Units These consist of two lights of glass separated by a metal or rubber spacer around the edges, and sealed airtight with an organic sealant—usually polysulfide, butyl, or both. The entrapped air is kept dehydrated by a *desiccant* (chemical dehumidifier) located in the spacer. The edge construction and glass may be enclosed in a metal channel, to provide edge protection and in some cases, to hold the assembly together.

Metal-edge insulating units are the most commonly used in commercial construction. Units with 1/4″ and 1/2″ air spaces and glass thicknesses ranging from 1/8″ to 3/8″ are available on a standard basis (Fig. 24). Sealed insulating units are manufactured and stocked in a limited number of standard sizes. Special sizes can be provided by some manufacturers at slight additional cost and longer delivery time.

Glass-edge Insulating Units These consist of two lights of SS (3/32″) or DS (1/8″) sheet glass with a 3/16″ air space between. The two sheets are fused together to form a glass-to-glass sealed edge. The air space in glass edge units may be filled with dehydrated air or inert gas.

Glass-edge units are available in a more limited range of sizes and glass combinations than metal-edge units (Fig. 24). They were developed mainly to provide economical insulating glass for the smaller lights in residential and commercial buildings.

STANDARDS OF QUALITY

The most universally accepted standards within the glass industry are the Federal Specifications for glass (Fig. 25). Base glasses such as sheet, float and plate, as well as patterned, wired and tinted glasses should conform to Federal Specification DD-G-451d. This standard establishes the dimensional and optical requirements for these glazing products.

Heat-strengthened and fully tempered glass should conform to Federal Specification DD-G-1403B, which covers dimensional tolerances, strength requirements, allowable bow (warpage), and durability of ceramic colored (spandrel) glass.

Another standard which is becoming increasingly important within the glass industry is ANSI Z97.1, Performance Specifications and Methods of Test for Safety Glazing Material Used in Buildings. This standard covers the safety requirements for fully tempered, laminated and wired glass as well as rigid plastics. It is directed primarily towards assuring that safety glazings have fail-safe characteristics when broken by human impact.

The use of safety glazing in hazardous locations such as sliding

FIG. 24 SUMMARY OF INSULATING GLASS PRODUCTS

Product	Glass type and thickness (in.)	Nominal Air space (in.)	Nominal Air space (mm)	Nominal Unit thickness (in.)[1]	Nominal Unit thickness (mm)	Maximum area (sq. ft.)[2]	Weight (lbs./sq. ft.)	Principal Uses
METAL EDGE UNITS	2 lites DS sheet, plate or float, 1/8	1/4 1/2	6 12	1/2 3/4	6 19	22.2	3.27	
	2 lites plate or float, 3/16	1/4 1/2	6 12	5/8 7/8	16 22	34 41.7	4.90	Residential and commercial glazing
	2 lites plate or float, 1/4	1/4 1/2	6 12	3/4 1	19 25	50-70	6.54	
GLASS EDGE UNITS	2 lites SS sheet, 3/32	3/16	5	3/8	10	10	2.40	Smaller lights in residential and commercial glazing
	2 lites DS sheet, 1/8	3/16	5	7/16	11	24	3.20	

[1]Metal edge units with a metal channel around the edge have a unit thickness 1/16" or 1.5 mm larger than shown.
[2]Maximum area given is for units constructed with clear glass; when tinted or reflective coated glass is used, either the maximum area must be reduced, or special fabrication is required.

glass doors, storm doors and tub-shower enclosures is required by the FHA Minimum Property Standards, the major model building codes and many municipal codes. In addition, safety glazing requirements have been adopted by many states and are being considered by many others.

The main industry standard for sealed insulating units is ASTM E-6 P3 Proposed Specifications for Sealed Insulating Glass and related test methods E-6 P1 and E-6 P2, for seal longevity and seal durability, respectively. The Insulating Glass Certification Council (IGCC) founded in 1977 provides independent laboratory testing and periodic unannounced on-site plant inspections of participating manufacturers. The certification program is conducted in compliance with ANSI Z34.1—1947 (R1959), American National Standard Practice for Certification Procedures. Any sealed insulating glass manufacturer may apply for the right to use the IGCC label. Under ASTM standards, sealed insulating glass units are subjected to accelerated weathering and are tested for durability and longevity of seal, maximum initial dew point temperature and freedom of interior surfaces from fogging. Manufacturers whose products comply with ASTM E-6 P3 are assigned an IGCC number to identify the product, its performance class and date code. Approved manufacturers are permitted to use the IGCC label on the product or product information claims.

Certification of Quality—Safety Glazing

Certification programs for safety glazing were started separately by the Glass Tempering Association and the Safety Glazing Industry, Inc. several years ago. In 1972 these two programs were joined together to form the Safety Glazing Certification Council (SGCC).

The SGCC administers an independent program of testing, random sampling and product evaluation to assure that safety glazing materials comply with ANSI Standard Z97.1. Products which meet all requirements of the standard are then labelled with the SGCC certification label. There is no one universal SGCC label. Instead, each manufacturer submits his own label to SGCC for approval. This label must include reference to ANSI Z97.1, plus an SGCC identification number which indicates the plant in which the product was made. The label must be permanently affixed to the certified product at the time and place of manufacture.

The SGCC certification program is administered in accordance with American National Standard Practice for Certification Procedures, ANSI Z34.1, which provides for certification on an impartial, independent and continuing basis. All certifications guarantee compliance with the ANSI specifications for safety glazing.

FIG. 25 SUMMARY OF GLASS STANDARDS

Glass Type	Industry Standard
Sheet, plate or float (clear or tinted)	F.S. DD-G-451d
Patterned	F.S. DD-G-451d
Wired	F.S. DD-G-451d ANSI Z97.1-1975
Reflective coated	—
Heat-strengthened	F.S. DD-G-1403B
Fully tempered	F.S. DD-G-1403B ANSI Z97.1-1975
Laminated	ANSI Z97.1-1975
Insulating	ASTM E-6 P3

Manufacturers' Warranties

Specialty products such as laminated, reflective and insulating glass generally are warranted by the manufacturer against certain manufacturing defects. *Laminated* glass typically is covered for a period of five years against edge separation or defects which obstruct vision. *Reflective* glass is warranted for ten years against peeling or deterioration of the reflective coating. *Insulating* glass usually carries a warranty against any material obstruction inside the glass unit which impairs vision, such as condensation, dust or foreign particles. (The rainbow "oil slick" effect observed occasionally at certain angles is not a manufacturing defect, but the result of light refraction between accurately machined parallel glass surfaces.)

Insulating glass generally carries either a 5-year or 10-year warranty. Stock windows and sliding doors which are ordered with insulating glass are normally preglazed with 5-year quality units.

In small window units enclosed by an accurately manufactured sash, the likelihood of the seal breaking is remote, and the optical quality is not critical if the windows are used primarily for daylighting rather than viewing. For these uses insulating units with a 5-year warranty generally are acceptable.

However, larger insulating units (often used in sliding glass doors and in field-glazed fixed sash) which are intended for viewing should carry long-term assurance against annoying visual obstructions such as internal fogging. These units generally are subject to more mechanical stresses, are larger and more expensive to replace. They should be covered, therefore, by a standard 10-year warranty.

Most glass warranties provide protection against manufacturing defects, but specifically exclude damage due to faulty installation or use in other than ordinary buildings. For instance, the warranty does not apply for applications in ships, vehicles or commercial refrigeration. The manufacturer usually agrees to deliver a replacement unit without charge to the shipping point nearest the installation, but the cost of forwarding the unit and actual installation must be borne by others.

We gratefully acknowledge the assistance of the following organizations and are indebted to them for the use of their publications as references and permission to use photographs and illustrations: ASG Industries, Inc.; Architectural Aluminum Manufacturers Association; Corning Glass Works; Insulating Glass Certification Council; Libbey-Owens-Ford Co. and PPG Industries, Inc.

GLASS 207

CONTENTS

FLAT GLASS PRODUCTS

PROPERTIES

See Main Text pages 207-3 and 207-4 for discussion of glass properties.

MANUFACTURE

See Main Text pages 207-5 through 207-12 for discussion of glass manufacture.

SHEET GLASS

Sheet glass is used for most ordinary glazing purposes because it combines relatively good optical quality with low cost. It has a smooth, hard finish which resists scratching but it lacks the superior optical quality of costlier plate and float glass.

Sheet glass is available in three qualities: AA, A and B. The last two generally are specified for most glazing purposes. Of the three strength levels available, double strength (DS) is most commonly used (Fig. WF1).

Sheet glass thinner than single strength window glass should not be used for glazing purposes.

A wave (draw) distortion is characteristic of all sheet glass, and the degree of distortion determines the grading and usefulness of the glass. The thicker the glass, the more noticeable the distortion becomes.

Heavy sheet glass should not be used for glazing in areas where high optical fidelity is required.

For best results, sheet glass should be glazed with the wave distortion running horizontally. To insure this, when specifying, the width dimension should be listed first.

FLOAT AND PLATE GLASS

Float and plate glass generally are used where it is necessary to provide greater strength and better optical quality than possessed by sheet glass.

For most glazing applications, float and plate glass are considered completely interchangeable. Float and plate glass are available in silvering, mirror and glazing qualities.

Silvering quality should be used for applications requiring a high degree of optical fidelity, such as optical instruments; mirror quality should be used for high quality mirrors; glazing quality should be used for general glazing installations (Fig. WF1).

Rough plate glass is available on a limited basis for applications where light transmission as well as obscurity is desired. The textured pattern of rough plate may contain some surface flaws, and may vary somewhat depending upon the wearing of the forming rolls used in manufacture.

PATTERNED GLASS

Patterned glass is translucent, with some type of pattern embossed on one or both sides. It is available in a wide range of patterns (Fig. WF2), offering degrees of light transmission and obscurity. Additional obscurity

(Continued) FLAT GLASS PRODUCTS

FIG. WF1 SUMMARY OF BASIC GLASS TYPES, SIZES AND USES

Product	Quality	Nominal thickness (in.)	(mm)	Maximum Area (in.)[1]	Weight (lbs./sq. ft.)	Visible light transmission (%)	Principal Uses
WINDOW GLASS	AA, A, B	SS, 3/32 DS, 1/8	2.5 3	120 u.i.[2] 140 u.i.[2]	1.22 1.63	91 91	Residential and commercial glazing
HEAVY SHEET	AA, A, B	3/16 7/32	5 5.5	84 x 120 84 x 120	2.50 2.85	90 89	Table tops, shelves, low-cost glazing
PLATE OR FLOAT	Silvering	1/4	6	50 sq. ft.	3.28	89	Mirrors and optical uses
	Mirror	1/4	6	75 sq. ft.	3.28	89	
	Glazing	1/8 1/4	3 6	74 x 120 128 x 204	1.64 3.28	90 89	Residential and commercial glazing
HEAVY PLATE OR FLOAT	Glazing	5/16 3/8 1/2 5/8 3/4 7/8	8 10 12 16 19 22	124 x 200 124 x 200 120 x 200 120 x 200 115 x 200 115 x 200	4.10 4.92 6.54 8.17 9.18 11.45	87 86 84 82 81 79	Residential and commercial glazing
ROUGH PLATE	Polished one side	17/64	7	58 x 120	3.28	88	Partitions and windows where obscurity is desired
	Rough both sides	9/32	7	124 x 240	3.70	88	

1 Maximum area varies according to producer. Larger sizes may be available from some producers.
2 U.i. = united inches, or the sum of width plus length.
3 Available from some producers, subject to factory operations.

may be obtained by etching or sandblasting the glass, but these processes reduce the strength of the glass, which may limit its applications.

Patterned glass should be used where obscured vision is desired; when patterned glass is used for tub and shower enclosures, it should be tempered for safety.

WIRED GLASS

Wired glass has a wire mesh or parallel wires rolled into the center of the glass thickness. Plate glass, as well as certain types of patterned glass, are available wired (Fig. WF2).

Wired glass is used widely for corridor and entrance doors, partitions, skylights and other potentially hazardous locations. It is considered a safety glazing and fire-retardant material.

When used in fire-rated doors and windows, wired glass should conform to Underwriters' Laboratories, Inc. standards for fire retardants.

TINTED GLASS

Tinted glass has an admixture to impart glare-reducing and heat-absorbing qualities to the glass. Sheet, float and plate glass—all are available tinted (Fig. WF3).

Tinted glass should be used where reduced light or heat transmission is desired for indoor comfort.

Equal thicknesses of gray, bronze and green tinted glass absorb the same percentage of solar energy. Visible light transmittance varies with the tint and thickness of the glass.

Special precautions should be taken when glazing tinted glass to avoid damaging the edges.

REFLECTIVE COATED GLASS

Reflective glass products have a transparent metal or metal oxide coating bonded to one surface of a single light of glass, or to one or both inside surfaces of insulating or laminated glass.

Daylight reflectances range from about 10% to 45%; visible light transmittances from 8% to 50%. U-values range between 0.35 and 0.30 BTU per (hr.) (sq. ft.) (°F) for single- and double-coated insulating glass, respectively. Air conditioning and heating costs frequently can be reduced by using this type of glass.

Reflective coated glass should be used where it is desired to control solar energy gain in summer and heat loss in winter.

(Continued) FLAT GLASS PRODUCTS

HEAT-TREATED GLASS

Any glass 1/8" or thicker may be tempered or heat-strengthened, except wired glass or patterned glass with deep patterns. Heat-treated glass cannot be cut, drilled or notched after treatment.

When heat-treated glass is to be used for a project, the exact size desired should be specified.

Fully Tempered Glass

Fully tempered glass has 3 to 5 times greater resistance than annealed glass to uniform loading, thermal stresses and most impact loads. When fractured, it breaks into a safe pattern of small harmless particles, hence is considered a safety glazing material.

Fully tempered glass should be used in areas where human contact with the glass is probable, such as entrances and sidelights, tub and shower enclosures, storm doors and sliding glass doors.

Heat-strengthened Glass

Heat-strengthened glass is only twice as strong as annealed glass of the same thickness. When it fails, it breaks into fragments much larger than those of tem-

FIG. WF2 SUMMARY OF PATTERNED AND WIRED GLASS TYPES, SIZES AND USES

Product	Type	Nominal thickness (in.)	(mm)	Maximum Area (in.)[1]	Weight (lbs.sq. ft.)	Visible light transmission (%)	Principal Uses
PATTERNED GLASS	Floral Hammered Stippled Granular Ribbed Fluted Striped[2]	1/8	3	60 x 132	1.60-2.10	80-90	Partitions and windows where obscured vision is desired
		7/32	5.5	60 x 132	2.40-3.00	80-90	
WIRED GLASS	Polished (square or diamond mesh)	1/4	6	60 x 144	3.50	80-85	Fire doors and windows; partitions and skylights
	Patterned (square or diamond mesh)	1/4	6	60 x 144	3.50	80-85	
	Parallel wired[3]	7/32 1/4 3/8	5.5 6 10	54 x 120 60 x 144 60 x 144	2.82 3.50 4.45	80-85 80-85 80-85	Partitions, Doors

[1]Maximum area varies according to producer; larger sizes may be available from some producers.
[2]These are just a few of the most common patterns available; many patterns are patented and made by one producer only.
[3]This type of wired glass does not carry Underwriter's Laboratories, Inc. fire-retardant rating; it is used mainly for decorative partitions.

FIG. WF3 SUMMARY OF TINTED GLASS TYPES, SIZES AND USES

Product	Quality	Nominal thickness (in.)	(mm)	Maximum Area (in.)[1]	Weight (lbs./sq. ft.)	Visible light transmission (%)	Solar energy transmission (%)[2]	Principal Uses
BRONZE—PLATE OR FLOAT	Glazing	1/8 3/16 1/4 3/8 1/2	3 5 6 10 12	35 sq. ft. 120 x 144 128 x 204 124 x 200 120 x 200	1.64 2.45 3.27 4.90 6.54	68 58 50 37 28	65 55 46 33 24	
GRAY—PLATE OR FLOAT	Glazing	1/8 3/16 1/4 3/8 1/2	3 5 6 10 12	35 sq. ft. 120 x 144 128 x 204 124 x 200 120 x 200	1.64 2.45 3.27 4.90 6.54	62 51 42 28 20	63 53 44 31 24	Where reduced heat and light transmission is desired
GREEN—PLATE OR FLOAT	Glazing	3/16 1/4	5 6	96 x 128 96 x 128	2.45 3.27	78 74	55 48	

[1]Maximum area varies according to producer; larger sizes may be available.
[2]Transmission values vary slightly from one producer to another.

(Continued) FLAT GLASS PRODUCTS

pered glass, and it is not considered a safety glazing material.

Heat-strengthened glass should be used in applications which require moderate strength, but are not subject to accidental impact by occupants.

The most common form of heat-strengthened glass is spandrel *glass, which has a fired-on ceramic frit on the interior surface, and is used commonly in spandrel areas of curtain wall construction.*

LAMINATED GLASS

Laminated glass consists of two or more layers of glass with interlayers of polyvinyl butyral plastic sandwiched between them (Fig. WF4). Tinted, tempered, heat-strengthened and wired glasses may be combined in various combinations depending on properties desired.

When laminated glass is broken, the fragments adhere safely to the plastic interlayer and do not evacuate the opening. It may be used for interior partitioning or glazing of doors and windows for (1) impact resistance and safety, (2) security and (3) acoustical control (Fig. WF4).

INSULATING GLASS

Insulating glass consists of two lights of glass with a dehydrated air space between. Any glass type with at least one smooth surface may be used in an insulating unit (Fig. WF5).

When heat-absorbing or reflective coated glass is used in an insulating unit, and it is known that the glass will be subjected to severe thermal stress or high wind loads, one or both lights of the unit should be tempered or heat-strengthened.

Insulating glass has a U-value of 0.6 BTU per (hr.) (sq. ft.) (°F) and can substantially reduce both heating and air conditioning costs.

Insulating glass units are produced with two different edge types: (1) metal edge units, which are generally available in a wide range of sizes and glass thicknesses and (2) glass edge units, which generally are limited to the smaller sizes used in residential glazing.

FIG. WF4 SUMMARY OF LAMINATED GLASS PRODUCTS

Product	Glass Type	Plastic interlayer (in.)	Nominal thickness (in.)	Nominal thickness (mm)	Maximum area (in.)[1]	Weight (lbs./sq. ft.)	Principal Uses
LAMINATED GLASS	SS/SS[2] Sheet	.015-.030	13/64	5	48 x 80	2.45	Partitions, doors and automotive glazing
	Lam/Lam[2] Sheet	.015-.030	15/64	6	48 x 100	2.90	
	DS/DS[2] Sheet	.015-.030	1/4	6	48 x 100	3.30	
	2 lites 1/8" plate or float	.015-.030	1/4	6	72 x 120	3.30	
	2 lites heavy plate or float	.015-.030	3/8 1/2 5/8 3/4 7/8 1	10 12 16 19 22 25	72 x 120 72 x 120 72 x 120 72 x 120 72 x 120 72 x 120	4.80 6.35 8.00 9.70 11.40 12.95	
BURGLAR-RESISTANT	2 lites 1/8" plate or float	.060-.090	5/16	8	60 x 120	3.50	Display windows
BULLET-RESISTANT	At least 4 lites of plate or float	At least 3 plies of .015	1 3/16 1 9/16 1 3/4 2	30 40 45 50	40 sq. ft. 30 sq. ft. 25 sq. ft. 22 sq. ft.	15.50 20.00 23.00 26.20	Bank teller windows and payroll booths
ACOUSTICAL STC-36[3] STC-40 STC-43	2 lites of plate or float	One or more plies of .045	9/32 1/2 3/4	7 12 19	60 x 120 60 x 120 60 x 120	3.40 6.36 9.74	Office partitions, radio and TV studios, airport glazing

[1]Maximum area varies according to producer. Larger sizes may be available from some producers.
[2]Lam indicates 7/64" sheet glass; SS, 3/32" sheet; DS, 1/8" sheet.
[3]Refers to Sound Transmission Class ratings.

(Continued) **FLAT GLASS PRODUCTS**

FIG. WF5 **SUMMARY OF INSULATING GLASS PRODUCTS**

Product	Glass type and thickness (in.)	Nominal Air space (in.)	Nominal Air space (mm)	Nominal Unit thickness (in.)[1]	Nominal Unit thickness (mm)	Maximum area (sq. ft.)[2]	Weight (lbs./sq. ft.)	Principal Uses
METAL EDGE UNITS	2 lites DS sheet, plate or float, 1/8	1/4 1/2	6 12	1/2 3/4	6 19	22.2	3.27	Residential and commercial glazing
	2 lites plate or float, 3/16	1/4 1/2	6 12	5/8 7/8	16 22	34 41.7	4.90	
	2 lites plate or float, 1/4	1/4 1/2	6 12	3/4 1	19 25	50-70	6.54	
GLASS EDGE UNITS	2 lites SS sheet, 3/32	3/16	5	3/8	10	10	2.40	Smaller lights in residential and commercial glazing
	2 lites DS sheet, 1/8	3/16	5	7/16	11	24	3.20	

[1]Metal edge units with a metal channel around the edge have a unit thickness 1/16" or 1.5 mm larger than shown.
[2]Maximum area given is for units constructed with clear glass; when tinted or reflective coated glass is used, either the maximum area must be reduced, or special fabrication is required.

STANDARDS OF QUALITY

Flat glass products should conform to the appropriate industry standards shown in Figure WF6.

Ordinary sheet, float and plate glass—as well as patterned, wired and tinted glass—should conform to the requirements of Federal Specification DD-G-451d for dimensional tolerances and optical quality.

Heat-strengthened and fully tempered glass should conform to the requirements of Federal Specification DD-G-1403B, for dimensional tolerances, strength properties and allowable warpage.

Safety glazing products such as tempered, laminated and wired glass should conform to the requirements of ANSI Z97.1, for fail-safe characteristics when broken by human impact.

The main industry standard for insulating glass units is ASTM E-6 P3 Proposed Specifications for Sealed Insulating Glass and related test methods E-6 P1 and E-6 P2. The Insulating Glass Certification Council IGCC conducts a certification program in accordance with ANSI Z34.1—1947 American National Standard Practice for Certification Procedures.

IGCC provides independent laboratory testing and periodic unannounced on-site plant inspections. Participating manufacturers whose products comply with ASTM E-6 P3 are assigned an IGCC identification number which permits them to use the IGCC label on their products. The IGCC label is a manufacturer's promise to insure continuing performance through specified standards.

Where the primary function is daylighting rather than viewing and the insulating glass is relatively accessible for replacement, 5- and 10-year warranteed units can be used.

Where unobstructed vision is important and replacement costs are high, insulating units should be warranteed for a period of 10 years by a reputable, nationally advertised manufacturer.

Certification of Quality—Safety Glazing

The Safety Glazing Certification Council (SGCC) administers an independent program of testing, random sampling and product evaluation which results in

FIG. WF6 **SUMMARY OF GLASS STANDARDS**

Glass Type	Industry Standard
Sheet, plate or float (clear or tinted)	F.S. DD-G-451d
Patterned	F.S. DD-G-451d
Wired	F.S. DD-G-451d ANSI Z97.1-1975
Reflective coated	—
Heat-strengthened	F.S. DD-G-1403B
Fully tempered	F.S. DD-G-1403B ANSI Z97.1-1975
Laminated	ANZI Z97.1-1975
Insulating	ASTM E-6 P3

(Continued) **FLAT GLASS PRODUCTS**

certification of products which meet requirements of ANSI Z97.1.

All safety glazing products should be identified with a *SGCC* certification label.

The SGCC certification program is administered in accordance with American National Standard Practice for Certification Procedures, ANSI Z34.1, which provides for certification to the public on an impartial, independent, continuing basis.

209 PLASTICS

PLASTICS

INTRODUCTION

Plastics are relative newcomers to the construction field compared with traditional materials such as wood, glass and concrete. The first commercial plastic in the United States, cellulose nitrate, was developed just a little over a century ago. It was created by John Wesley Hyatt to help overcome a shortage of ivory, from which billiard balls were made.

Progress in plastics technology was slow at first. The next commercial plastic to be developed was phenol-formaldehyde, in 1909. Gradually, a few others were introduced, such as casein (1919), alkyds (1926) and polyvinyl chloride (1927). It was not until the 1930's, however, that the modern commercial plastics industry really was started; plastics were not widely used in building construction until after World War II.

Despite their newness though, plastics have become important materials in the construction field. As a group, they offer an extremely wide range of properties possessed by few other building materials: they are moisture and corrosion-resistant, lightweight, tough and easily molded into complex shapes. Furthermore, the great variety of plastic materials available—each offering slightly different advantages—makes plastics well suited for a multitude of purposes in the home.

Sheet and tile flooring is perhaps the most common use for plastics in residential construction—the familiar vinyl tile is noted for its excellent abrasion and stain resistance and good resilience. Plastic siding is becoming increasingly popular because of its good durability and freedom from maintenance; plastic-covered wood windows and doors offer similar advantages. Plastic laminates have become standard materials for counter and table tops; they are produced in many decorative designs and bright colors, and have good heat and chemical resistance. Plastics also are used in many types of buildings for break-resistant glazing, skylights and roof domes. Plastic films make excellent vapor barriers; plastic foams can be used for a wide variety of insulation purposes.

Plastics do have limitations, however, and they can create serious problems if utilized incorrectly. For satisfactory service, a suitable plastic for the intended use should be chosen. The plastic product should be correctly designed and properly integrated with other products.

This section discusses the composition and generic properties of plastics, and describes the chief applications of the most important types. For more detailed information on molecular structure and how it affects the specific properties of materials, see Section 108 Properties of Materials.

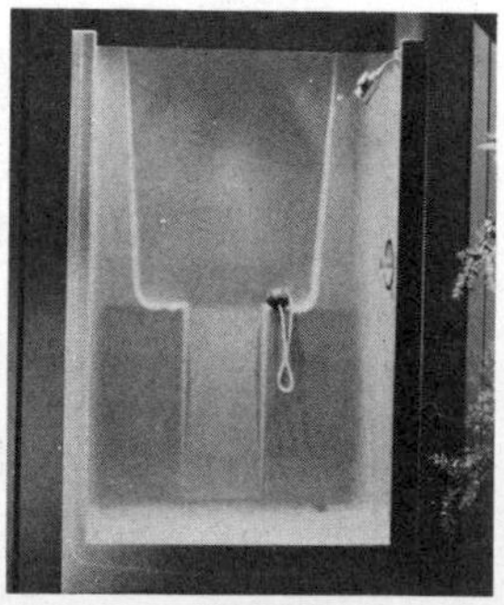

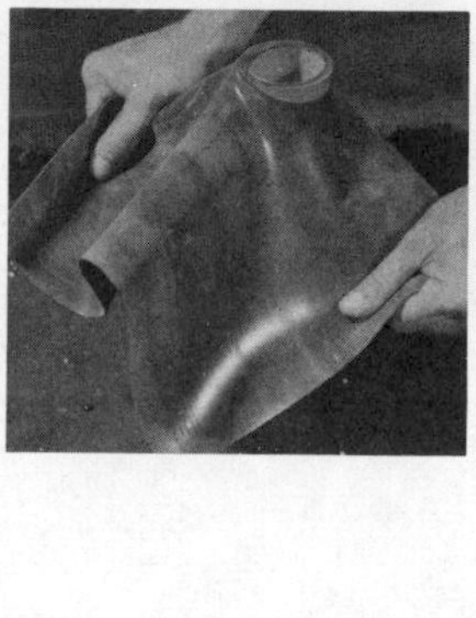

COMPOSITION

There are 20 to 30 families of materials which are classed as *plastics.* Within each of these families, a multitude of further variations and combinations are possible. The chemistry involved in the production of all of these compounds is complex, as is the terminology involved in classifying them. It is not possible to give a detailed explanation of these complexities here. However, in order to gain a basic understanding of what plastics are, how they behave and how they can be used in construction, a brief explanation of their composition, molecular structure and basic types is necessary.

Despite widely varying properties, all plastics have several characteristics in common, which warrants classifying them as a single materials group:

(1) At one stage in their formation they exhibit *plastic* behavior—that is, they will soften and can be formed into desired solid shapes, with or without the use of heat and pressure.

(2) They are made up of molecules built around a carbon atom and therefore are classed as *organic* materials. There are some exceptions to this, however: some plastics are built around a silicon atom rather than the carbon atom, yet still are considered plastics.

(3) They are *synthetic* materials; that is, they are man-made and do not occur in nature.

(4) They are *high polymers*—giant molecules made of numerous, relatively simple repeating units linked together into large aggregations.

Not all materials which possess the above characteristics are necessarily plastic. For example, synthetic rubbers have all of these characteristics, yet they are classed as rubbers, not plastics.

MOLECULAR STRUCTURE

Some plastics are soft and flexible; others are hard and brittle. Some will become soft and may even melt when exposed to intense heat; others are virtually unaffected by extremes of temperature. An examination of the molecular structure of high polymers can help to explain many of these differences in the properties and behavior of plastics.

High Polymers

Most plastics are based on carbon chemistry. This means that they are dependent on the peculiarities and properties of the carbon atom (Fig. 1). The carbon atom forms bonds with other atoms through *covalent bonding* (see Section 108, Properties of Materials). The carbon atom has a *valence* of four—that is, four points to which other atoms can attach themselves to form covalent bonds.

Atoms of other elements have different valences. For example, the hydrogen atom has a valence of one, and the oxygen atom has a valence of two. Therefore, four hydrogen atoms can attach themselves to a carbon atom (one at each valence point) to produce CH_4, the gas known as methane. Likewise, two oxygen atoms can attach themselves to a carbon atom (each oxygen atom taking up two valence points) to produce CO_2, carbon dioxide (Fig. 1).

Two adjacent carbon atoms can also attach to each other. If they attach to each other at just one valence point—creating a single bond between them—and the other valence points are taken up by other atoms, the molecule is said to be *saturated.* In other words, a saturated molecule contains only single bonds between the carbon atoms and has no free (unattached) valence points.

However, two carbon atoms may also attach to each other at two valence points, creating a double bond; or at three valence points, creating a triple bond (Fig. 2). In these cases, the molecules are said to be *unsaturated.* Under certain conditions the double or triple bonds of an unsaturated molecule can be opened up (activated) creating free valence points. Other atoms, or groups of atoms can then attach themselves to these free valence points. For example, if the double bonds in a large number of ethylene molecules are activated, they can join together to form a chain (Fig. 3), which is then known as polyethylene. Such chain formation is termed *polymerization* and is essential in the production of plastics. Only unsaturated molecules will react together to form chains, and hence, these are the only kinds of molecules which are useful in producing plastics.

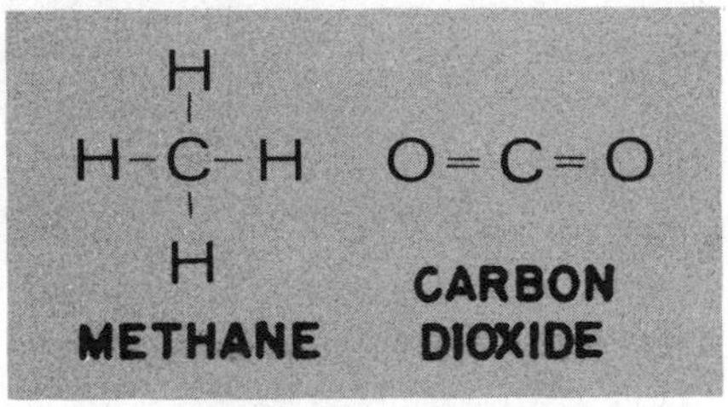

FIG. 1 Two molecules based on the carbon atom: methane and carbon dioxide.

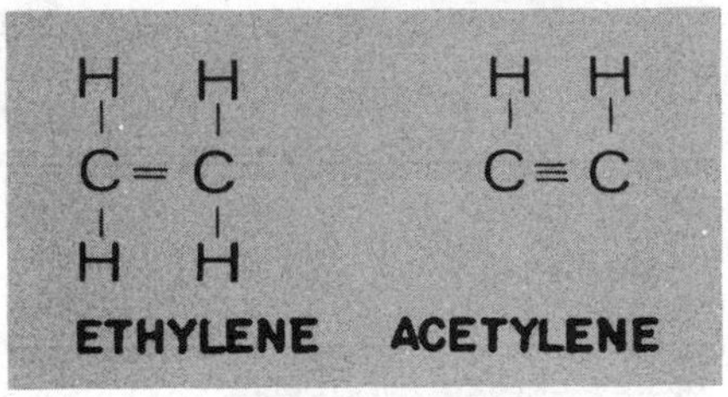

FIG. 2 Molecules can have double bonds (ethylene) or triple bonds (acetylene) between the carbon atoms; such molecules are termed unsaturated.

A single molecule such as ethylene is known as a *monomer.* Several of these molecules linked together become a *polymer*; when the number of units in the chain is very large (between several hundred and several thousand) it is called a *high polymer.* The number of monomeric units in different chains can vary widely. When chain growth does stop, free valence points are left at each end of the chain. These may be taken up by single atoms such as hydrogen, or by groups of atoms with a single net valence. In some cases, the two ends of the chain loop around and connect, thus forming a ring.

Plastics are composed of these giant chain-like molecules, and are

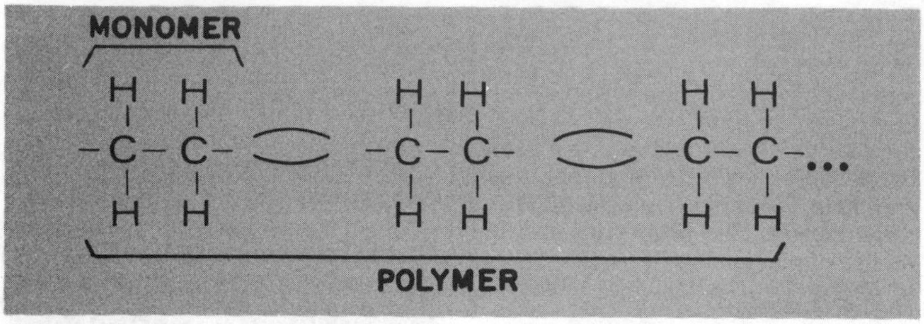

FIG. 3 Several ethylene molecules can be linked together to form polyethylene—a polymer.

therefore classed as high polymers. This is evidenced by the fact that many of the chemical names for plastics start with the prefix "poly-" which means "many" (polyethylene, polystyrene and polyvinyl chloride, for example).

Basic Categories

Plastics can be divided into two basic categories based on their molecular structure: (1) *thermoplastic* materials, which can be repeatedly softened and hardened by heating and cooling; and (2) *thermosetting* materials, which are set into permanent shape during forming, and cannot be softened again by reheating.

Thermoplastics Thermoplastic materials are made up of long chain-like molecules such as the polyethylene molecule described above. These molecules are characterized by being unattached to each other. The chemical attraction along the chain is very strong, but the attraction between adjacent chains is relatively weak. Furthermore, these chains are generally saturated, having no double or triple bonds which can be activated. This means that no free valence points can be created to cause strong attraction to adjacent chains.

At any temperature above absolute zero (−459.69°F), molecules of all materials are in constant random motion. As a result of the unconnected molecular structure of thermoplastics, the molecules can slide past each other and change shape as in a liquid. At room temperature, this random motion is slight and the plastic holds its shape. However, as the temperature rises, molecular motion increases and the attractive forces between molecules weaken, causing the material to expand and become soft and flexible.

When the plastic reaches a high enough temperature, it will flow and can then be molded into new shapes. The temperature at which this occurs is called the *softening point*, and it will vary according to the type of plastic involved. When the material is cooled again, it assumes its former stiffness and holds its new shape at normal use temperatures. With thermoplastic materials, the process of heating to soften and cooling to harden can be repeated again and again without affecting the properties of the plastic.

The configuration of the molecules in thermoplastics also affects the final properties. Some molecules are arranged in neat, continuous chains (*linear* molecules); others turn out to be highly branched. Highly branched chains do not pack together very well, whereas linear molecules can be stacked together closely. As a result, plastics composed of branched molecules tend to be softer, less strong and lower in density than those composed of linear molecules.

The molecular configuration in a given plastic can be controlled to a certain extent by the method used in its formation. For example, when ethylene is polymerized by high temperature and pressure, the molecules are often highly branched. However, ethylene can also be polymerized through the use of *catalysts,* which are substances added to the basic chemicals to speed up the chemical reaction. When polyethylene is made by this process, the molecules are generally linear. This means that variations in strength and stiffness can be produced in different batches of the same type of plastic simply by changing the method of polymerization.

There are literally hundreds of different thermoplastics. A brief description of the most important types is given on pages 209-11 to 209-14. However, a few words about the way these different types are developed may be helpful.

Starting with the ethylene molecule, which is the basic building block of many thermoplastics, substitutions can be made for one or more of the hydrogen atoms involved. Single atoms such as chlorine or fluorine, or whole groups of atoms such as the styrene monomer, can be substituted for the hydrogens (Fig. 4). For example, if one hydrogen is replaced by chlorine, a plastic known as polyvinyl chloride results; if one hydrogen is replaced by a phenyl group, polystyrene results; if one hydrogen is replaced by a methyl

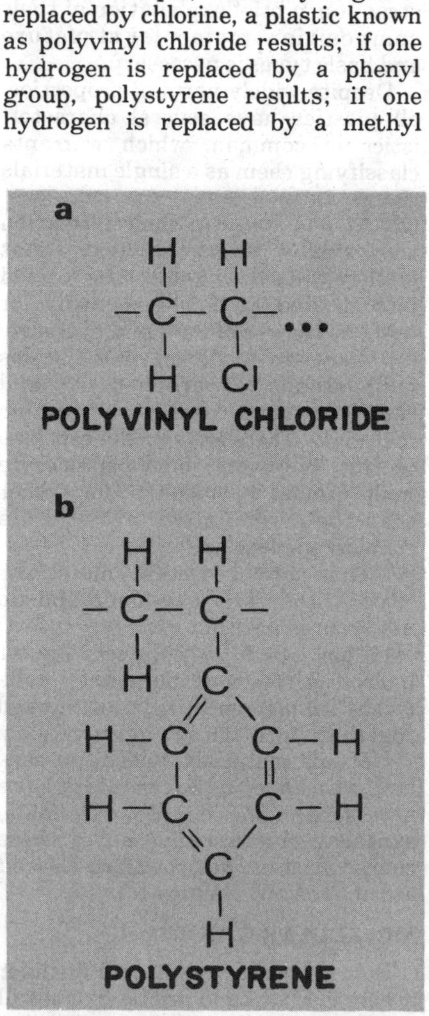

FIG. 4 Starting with the ethylene molecule, substitutions can be made for a hydrogen atom, producing (a) polyvinyl chloride or (b) polystyrene.

group (CH_3) and a second hydrogen replaced by a methoxycarbonyl group ($COOCH_3$), polymethyl methacrylate (acrylic) results.

Hundreds of such variations are possible and each variation produces a plastic with slightly different properties. A large group of plastics called *vinyls* are based on the ethylene molecule. Other types are based on different molecules. For example, *cellulosics* are based on modifications of the basic celluloid molecule.

Still other variations are possible by combining different monomeric units into the same molecular chain. These substances then are termed *copolymers*. Thus, instead of having a single repeating monomer along a polymer chain, two different monomers may alternate along the same chain. Numerous variations in the properties of plastics are made possible through such copolymerization. For example, vinyl chloride may be copolymerized with vinyl acetate, producing a substance with some of the properties of each of the original polymers (Fig. 5). Acrylonitrile-butadiene-styrene (ABS) is another copolymer which has proven to be very useful because of its unique combination of properties.

Thermosets Thermosets differ from thermoplastics in that once they are *set* (cured), they cannot be softened again. This is because during forming, they go through an irreversible chemical process which develops strong chemical bonds between adjacent molecules. Thus the molecules cannot slide past each other as they can in thermoplastic materials, but instead, form a strongly interconnected molecular structure.

One example of this process is the production of thermosetting plastics from unsaturated polyesters. Being unsaturated, the chain-like polyester molecules contain at least occasional double carbon bonds. These double bonds can be activated, creating free valence points. If other activated molecules are present in the mixture, strong crosslinks can be developed between them. For example, activated styrene molecules will react with activated polyester molecules, forming a continuous, crosslinked network. The degree of crosslinking which occurs determines the final properties of the material—that is, whether it will be resilient and rubbery, hard and glassy, or somewhere in between.

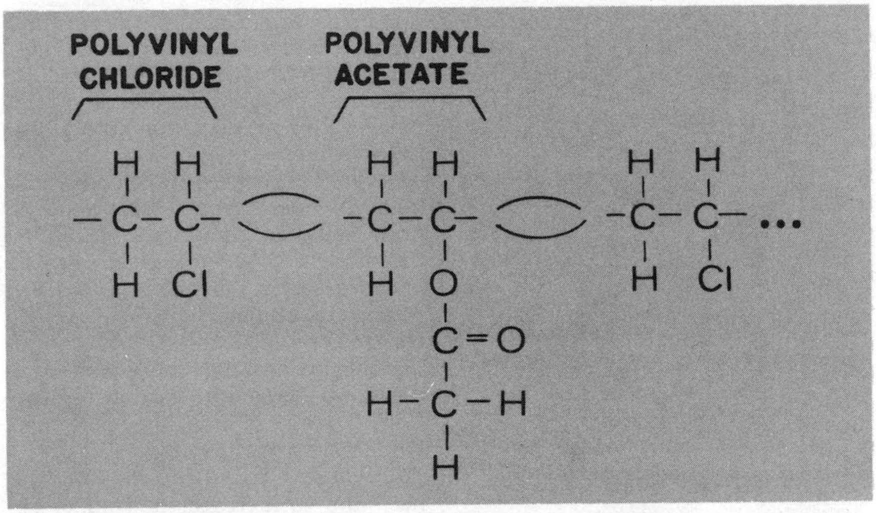

FIG. 5 Two different monomers may alternate along a polymer chain, producing a copolymer.

Another type of thermosetting reaction, termed *condensation polymerization*, produces a byproduct such as steam. An example of this is the reaction between phenol and formaldehyde, producing the thermoset known as phenol-formaldehyde. During this reaction, the phenol molecules each give up one or more hydrogen atoms and the formaldehyde molecules give up an oxygen atom. This leaves open valence points on both molecules, allowing them to join together to form a strongly interlinked structure. The leftover hydrogen and oxygen atoms join together to form water, which is given off as steam.

All thermosetting reactions are irreversible. Thus, these plastics only go through a pliable stage once and after they have set, they remain rigid and hard unless actual chemical decomposition (breaking down) occurs due to extreme heat, pressure or environmental effects.

MODIFIERS

Many plastics in the pure form are difficult to work with, expensive to produce, or lack some other property needed for a particular application. These difficulties can sometimes be overcome by the addition of several types of modifiers.

These modifiers are added to the basic plastic ingredient—the *resin*—before fabrication, in a process known as *compounding*. It is important that the different ingredients are throughly compounded into a homogeneous mass, because the uniformity of the mix contributes greatly to the final properties of the plastic.

The type and amount of modifiers used with different plastics varies greatly. Some mixes consist of 90%-95% of the basic resin, with only small amounts of modifiers. Other mixes may contain only 20%-30% of the resin, the rest being composed of various plasticizers, fillers, stabilizers or other additives.

Plasticizers

At ordinary use temperatures, many thermoplastics are inherently hard and rigid. In order to produce sheeting, tubing, film and other flexible plastic products, plasticizers must be added to these materials (Fig. 6).

Plasticizers are liquid or solid materials that are blended with the plastic resin to impart increased flexibility, impact resistance, resiliency and moldability. Basically, the plasticizer produces the same effect in a thermoplastic as high temperature does—it lessens the forces of attraction between the long chain-like molecules, allowing them to slide more freely past each other. Unfortunately, this greater flexibility also often reduces the strength

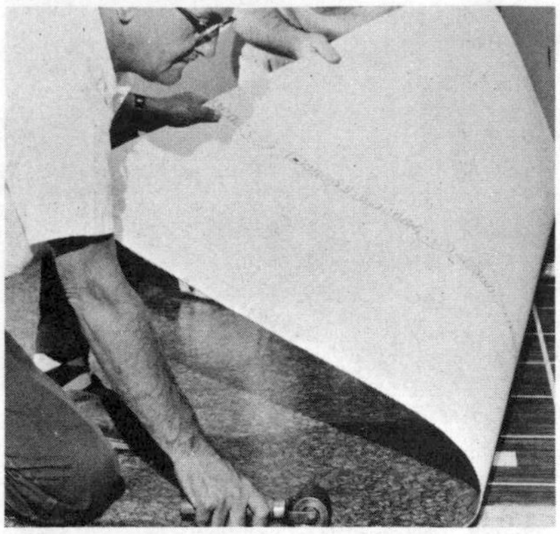

FIG. 6 The addition of plasticizers imparts flexibility to vinyl sheet flooring, making it easier to handle and install.

of the plastic and lowers its heat resistance and dimensional stability.

In order to obtain the optimum effects from the use of a plasticizer, it is important that it be compatible with the plastic resin being used. This means that the plasticizer should mix well with the resin and should not separate out during the life of the object. A poor plasticizer may migrate to the surface of the object, creating a greasy film and causing the original plastic to become hard and brittle again. Furthermore, the plasticizer should not deteriorate with age and should not cause the plastic itself to deteriorate.

Fillers

Fillers are used most commonly with thermosets, although there are some filled thermoplastics as well. Fillers generally are used to improve a particular property of the plastic, such as moldability, heat resistance or toughness. However, some fillers are used primarily to increase the bulk and therefore decrease the cost of the plastic; they are referred to as *extenders* when used for this purpose.

A wide variety of materials may be used as fillers, depending upon what properties are desired in the finished product. The amounts used are generally between 10% and 50% of the total plastic mix. Some of the more common uses for fillers are given below:

Moldability and Bulk Wood flour, made of finely ground hardwood or nut shells, is commonly added to thermosets to make a plastic which is more easily molded and less expensive to produce.

Hardness Mineral powders and metallic oxides may be added to a plastic to increase its hardness. However, if added in bulk, they also tend to increase its brittleness.

Toughness To overcome brittleness and to increase the impact resistance of a plastic, fibrous fillers commonly are added. Natural fibers such as cotton, sisal or hemp may be used, as well as synthetic fibers such as rayon, nylon or polyester.

Heat Resistance Asbestos fillers greatly increase the heat resistance of a plastic. Other inorganic fillers such as silica, clay and finely ground limestone may be used for the same purpose.

Electrical Resistance Finely ground mica or quartz may be added to increase the electrical resistance of a plastic.

Reinforcing Agents

The *reinforced plastics* are a large, very important group in the construction industry. Their strength and stiffness have been significantly increased through the addition of high-strength fillers which act as reinforcing agents. Generally, thermosetting resins such as polyester and epoxy are used in reinforced plastics. Fibrous fillers such as sisal and asbestos can be used as the reinforcing agent, but the strength of glass fiber is so superior that it has come to be used almost exclusively for this purpose.

Although a sheet of glass is quite brittle, when it is drawn into fibers, its tensile strength rises enormously. On an equal-weight basis, it is the strongest commercially available construction material, and it greatly increases the strength and stiffness of a plastic. It can be used in several different forms—as chopped strands, yarns, pressed matting or woven fabrics—depending upon the design and properties desired for the finished product.

Stabilizers

The effects of weathering on a material are an important consideration in the construction industry. Many plastics degrade when exposed to the ultraviolet rays of the sun, to oxygen and other gases in the air. As a result, small amounts of *stabilizers*, which act as antioxidants and ultraviolet light absorbers, must be added to these plastics.

Several different substances may be used as stabilizers. Carbon black is a commonly used one. When added to polyethylene, for example, it converts it from a quickly degrading material to one that stands up well to extended outdoor exposure.

Colorants

The addition of colorants to plastics is a complex subject. Some plastics, such as acrylic and polystyrene, are inherently clear and can be produced in a wide range of colors; others, such as the phenolics, are dark and opaque and therefore have a more limited color range.

Both pigments and dyes can be used in coloring plastics. Pigments are used to produce opaque colors, dyes to obtain transparent colors. The pigment or dye chosen must be compatible with all of the ingredients of the plastic mix. It should not bleed out of the plastic, causing discoloration and staining of adjacent materials. Furthermore, it should have good light and heat stability so that it will not fade or change in appearance with age.

The range of properties covered by plastics is quite broad (Fig. 7). They can be flexible or stiff, tough or brittle, transparent or opaque. However, plastics also have certain limitations, and these must be kept in mind when considering applications in the construction field. These limits can be seen most clearly when the general properties of plastics are compared with those of other structural materials such as steel, concrete, wood and glass.

STRENGTH AND STIFFNESS

Strength can be measured in terms of tension, compression and bending; tensile strength generally is the most important. Most plastics are considerably lower in tensile strength than steel, and are more on a par with wood. However, some plastic laminates and reinforced plastics approach steel in strength (Fig. 8). The highest tensile strengths are found in the *fiberglass reinforced plastics* (FRPs).

In most plastics applications stiffness is more important than strength. Stiffness is measured by the modulus of elasticity—the higher the stiffness, the greater the modulus of elasticity. Generally, plastics are less stiff than steel, glass and other structural materials (Fig. 9), and therefore deflect more than these materials under a given applied load. Thermosets are somewhat stiffer than thermoplastics, but reinforced plastics exhibit the highest modulus of elasticity—which is still only about 1/10 that of steel, or roughly in the same range as wood and concrete.

Plastics weigh considerably less than other structural materials.

FIG. 8 COMPARATIVE STRENGTH OF PLASTICS AND OTHER CONSTRUCTION MATERIALS

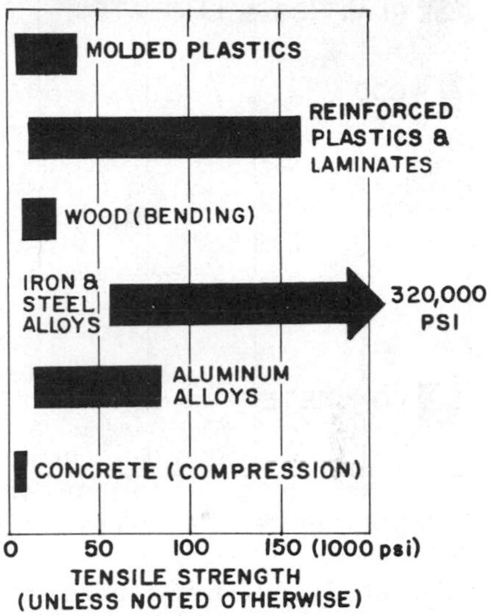

FIG. 7 IMPORTANT PROPERTIES OF SEVERAL TYPES OF PLASTICS

Plastic Type	Specific Gravity	Tensile Strength (psi)	Modulus of Elasticity (10^6 psi)	Coefficient of Thermal Expansion (10^{-6}/°F)	Thermal Conductivity (Btu/ft²/in/hr/°F)	Maximum Service Temperature (°F)
Thermoplastics:						
Acrylics	1.05-1.19	7,000-11,000	0.39-0.47	28-50	1.2-1.7	140-200
Cellulosics	1.13-1.28	2,000-9,000	—	44-111	1.1-2.3	115-220
Fluorocarbons	2.13-2.20	2,000-7,000	0-0.20	25-66	0.9-1.7	300-550
Polyethylene	0.9	1,000-5,500	0-0.35	56-195	2.3-3.6	180-275
Polystyrene	1.06	1,500-20,000	0-1.8	19-117	0.3-1.0 (foam 0.26)	140-220
Vinyls	1.18-1.67	500-9,000	0-0.4	28-195	0.9-20	120-210
Thermosets:						
Melamine	1.50	5,000-13,000	—	11-25	1.9-4.9	210-400
Phenolics	1.40-1.90	3,000-18,000	0-2.4	14-33	0.9-6.4	200-550
Reinforced Polyesters	1.50-2.10	800-50,000	0-2.0	7-56	1.2-7.2	250-450
Polyurethane	1.0-1.3	175-10,000	0-0.35	56-112	0.5-2.1 (foam 0.15)	190-250
Silicones	1.88	800-35,000	—	4-167	1.0-3.8	400-600

FIG. 9 COMPARATIVE STIFFNESS OF PLASTICS AND OTHER CONSTRUCTION MATERIALS

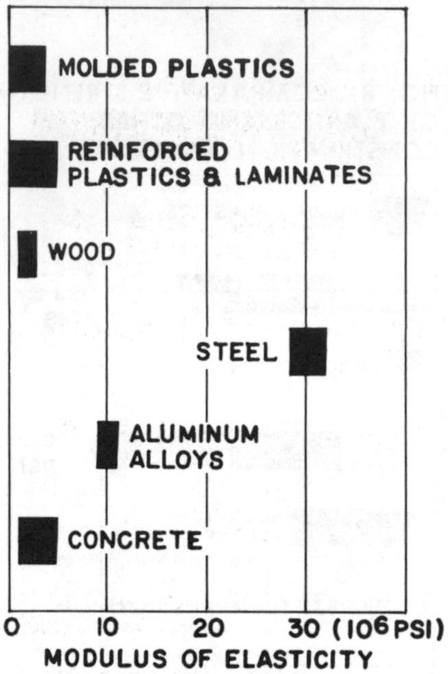

Unmodified plastic resins range from a specific gravity of about 0.9 for polypropylene to about 1.5 for polyvinyl chloride—filled or reinforced plastics are somewhat denser (Fig. 7). In contrast, steel has a specific gravity of about 7.5. Thus, when strength comparisons are made between plastics and metals on an *equal-weight* basis, plastics appear to have the advantage.

For this reason, plastics are employed in applications where strength combined with lightness is of prime importance—such as in spacecraft. In other construction, they can provide very strong, lightweight components (such as sandwich panels) which can appreciably decrease the overall weight of a structure (Fig. 10).

TOUGHNESS AND HARDNESS

One of the outstanding characteristics of many plastics is their toughness. Toughness can be defined as strength combined with resilience. This property is difficult to measure precisely, but it is generally expressed in terms of a material's *impact strength* (ability to resist impact, such as from a falling ball or a striking pendulum).

Plastics differ greatly in impact strength, and different formulations of the same plastic also show considerable variation. For example, regular polystyrene is quite brittle, whereas copolymers of styrene, such as acrylonitrile-butadiene-styrene, are noted for toughness.

Compared with other materials such as glass, concrete and wood, plastics are quite resistant to impact—a 1/8″ thick sheet of acrylic can withstand 25 times the impact of a 1/8″ sheet of double strength window glass. This is illustrated by the widespread use of plastics for break-resistant glazing and lighting fixtures.

Another property of importance is hardness (abrasion resistance). In this respect, plastics do not perform as well as glass or steel—plastic glazing scratches much more readily than glass. However, many plastics are more resistant to abrasion than wood.

CREEP

Creep (cold flow) is a permanent change in the dimensions of a material due to prolonged stress. All materials are subject to creep, but in structural materials such as steel, the rate of deformation at room temperature is so slight that it is negligible.

Because of their molecular structure, many plastics do exhibit creep at room temperature, and the amount of deformation which results can be quite large. This is particularly true of thermoplastics, whose unconnected molecules begin to slide past each other and assume new forms when an external force is applied. If the force is applied only for a short time, the plastic may recover, but if the force is long-term, deformation can be permanent and may eventually result in failure of the material.

Thermosets, laminates and reinforced plastics have greater resistance to creep, but they still cannot resist prolonged high stresses as can steel and concrete. Nevertheless, plastics can be used in self-supporting and even some load-bearing applications, if the proper formulation is chosen.

THERMAL PROPERTIES

The important thermal properties for plastics are expansion and contraction, conductivity and maximum service temperatures.

FIG. 10 Low-density polyurethane foam is injected between two strong, rigid facings in the production of this lightweight sandwich panel.

FIG. 11 COMPARATIVE THERMAL EXPANSION OF PLASTICS AND OTHER MATERIALS

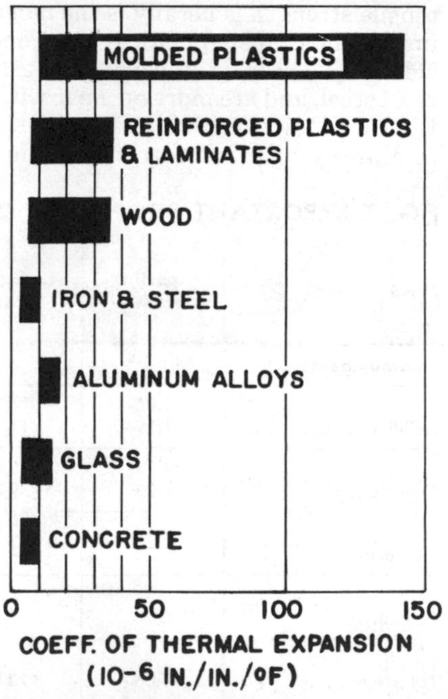

Expansion and Contraction

All building materials expand and contract with changing temperatures. As can be seen from Fig. 11, thermal expansion is much higher in plastics than in most materials. Thermosets exhibit a lower rate of expansion than thermoplastics, while reinforced plastics and laminates generally show the lowest rates of all plastics.

Changes in dimension caused by expansion and contraction can cause buckling or cracking of plastic components used in assemblies. These difficulties can be avoided through proper design of the component and through the use of expansion joints to allow for movement. Particular care must be taken when plastics are joined to metals because of the wide difference in rates of expansion of the two materials. Generally, thermosets such as phenolics or reinforced plastics work best in such applications.

Conductivity

Plastics are generally good heat insulators. They have a much lower thermal conductivity than metals. Pure, unmodified plastics also compare favorably with concrete, glass and brick—wood being the only construction material with a lower conductivity (Fig. 12).

Plastic foams, however, outperform even wood in this respect, and are among the best heat insulators available today. Polyurethane foam can have a conductivity as low as 0.15 (Fig. 7), lower even than mineral wool, which has a conductivity of 0.27. Values for other foams vary according to density, type of foaming agent used, and whether the foam is of the *open-cell* (inter-connecting) or *closed-cell* variety.

Because of these low conductivity values, plastic foams make excellent insulation materials. However, some problems have been encountered with their use in construction, due to their flammability and to the toxic fumes which they can produce when burned.

Service Temperatures

The maximum continuous service temperature of a material is the highest prolonged temperature at which it can be used without softening. Service temperatures for plastics are low compared with other building materials (Fig. 13). Many thermoplastics begin to soften at temperatures near the boiling point of water. Thermosets have better heat resistance and some can be used at temperatures up to the wood char point (380° to 400°F). Fluorocarbons have the highest continuous service temperature of unmodified plastics and can be used in nonload-bearing applications at 500°F.

Additives can substantially improve the heat resistance of plastics—thus, many filled and reinforced plastics can withstand higher temperatures than the pure, unmodified resins.

Service temperatures are important in determining the suitability of a plastic for a particular application. Most plastics can withstand the temperatures encountered in ordinary building structures; for more demanding applications, such as piping for hot water supply lines, special heat-resistant formulations may be required.

FIG. 12 COMPARATIVE THERMAL CONDUCTIVITY OF PLASTICS AND OTHER CONSTRUCTION MATERIALS

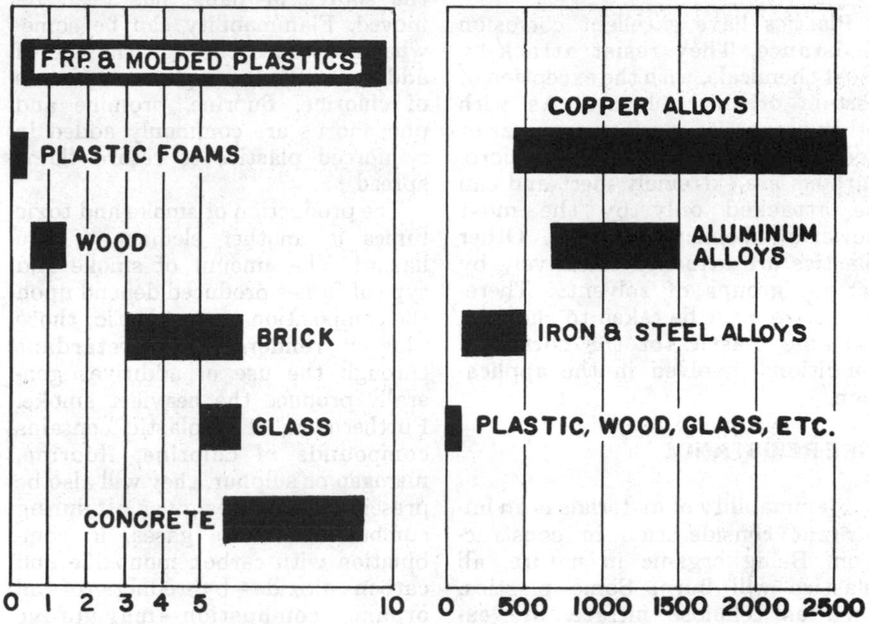

FIG. 13 COMPARATIVE SERVICE TEMPERATURES OF PLASTICS AND OTHER CONSTRUCTION MATERIALS

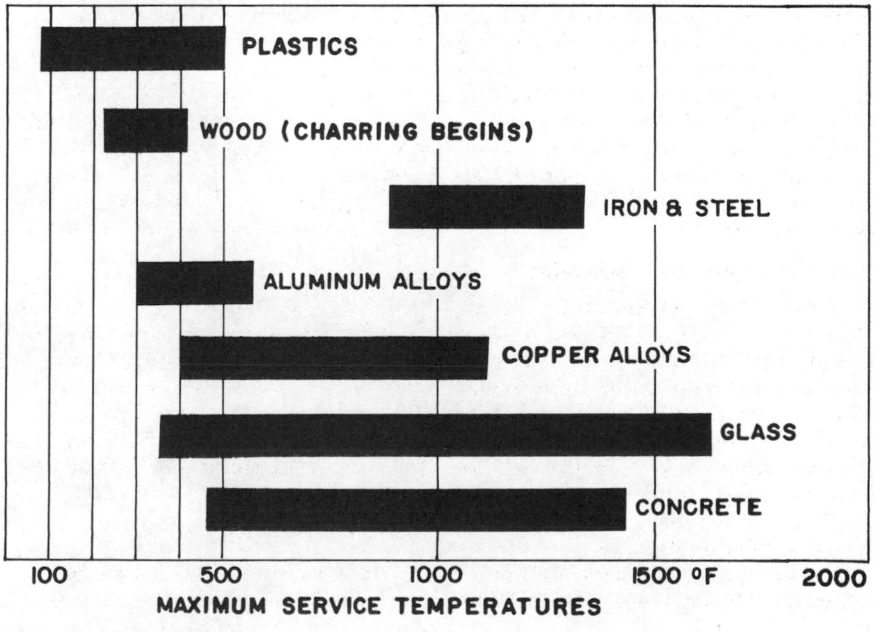

CORROSION RESISTANCE

Plastics have excellent corrosion resistance. They resist attack by most chemicals, with the exception of certain organic solvents. As with other properties, performance varies according to composition. The fluorocarbons are extremely inert and can be attacked only by the most powerful chemical reagents. Other plastics are attacked selectively by certain groups of solvents. Therefore, care must be taken to choose a suitable plastic for the corrosive conditions involved in the application.

FIRE RESISTANCE

Flammability of materials is an important consideration in construction. Being organic in nature, all plastics will burn. Some plastics, such as cellulose nitrate, are extremely flammable and are generally not permitted as construction materials by most building codes. Other plastics are slow burning, while some will stop burning after the source of flame has been removed. Flammability can be somewhat controlled through the use of additives. For example, compounds of chlorine, fluorine, bromine and phosphorus are commonly added in reinforced plastics to retard flame spread.

The production of smoke and toxic fumes is another element of fire hazard. The amount of smoke and type of fumes produced depend upon the composition of the plastic; those plastics rendered flame-retardant through the use of additives generally produce the heaviest smoke. Furthermore, if a plastic contains compounds of chlorine, fluorine, nitrogen or sulphur, they will also be present in the gases given off during combustion. These gases, in combination with carbon monoxide and carbon dioxide—byproducts of all organic combustion—may prove highly toxic.

In short, plastics are not any more flammable than wood, fabrics or other organic materials. However, because plastics are often used in place of traditional *nonflammable* materials such as glass or metals, they should be used with discretion and only after careful consideration of their potential fire hazard.

To deal with these and other problems associated with the use of plastics in construction, most of the major building codes now have chapters devoted specifically to plastics applications. Standard tests developed by the American Society for Testing and Materials (ASTM) can be used to rate plastics as to burning rate, fire endurance and flame spread. Building codes generally utilize one or more of these tests in determining which plastics are considered suitable for construction. In addition, codes often restrict the use of plastics in certain areas by special requirements such as: covering foamed plastics with noncombustible materials, spacing plastic glazing with intermittent noncombustible surfaces, or limiting the amount of plastic materials which may be used as interior finish.

MAJOR TYPES AND USES

The multitude of commercially available plastics can be broken down into several major groups. Following is a brief description of the chief characteristics and applications of the plastics most commonly used in construction. A listing of plastics and their abbreviations compiled by ASTM is given in Fig. 14.

THERMOPLASTICS

This is by far the larger group of plastics. Although thermoplastics can be softened by heating, they are dimensionally stable at normal use temperatures.

Acrylonitrile-butadiene-styrene (ABS)

These copolymers are noted for their unique combination of properties. They are an outgrowth of the high-impact polystyrenes, which are a combination of styrene and butadiene-rubber compounds. Polystyrene plastics alone are very brittle; by adding a butadiene-rubber compound, their toughness is greatly improved. The further addition of acrylonitrile to the mixture improves other properties, such as chemical resistance, rigidity and tensile strength.

FIG. 14 **ASTM ABBREVIATIONS FOR PLASTICS**

Term	Abbreviation	Thermoplastic	Thermosetting
Acrylonitrile-butadiene-styrene	ABS	x	
Acrylic:			
poly (methyl methacrylate)	PMMA	x	
Cellulosics:			
cellulose acetate	CA	x	
cellulose acetate-butyrate	CAB	x	
cellulose acetate-propionate	CAP	x	
cellulose nitrate	CN	x	
ethyl cellulose	EC	x	
Epoxy, epoxide	EP		x
Fiberglas reinforced plastics	FRP		x
Fluorocarbons:			
polytetrafluoroethylene	PTFE	x	
fluorinated ethylene propylene	FEP	x	
Melamine-formaldehyde	MF		x
Nylon:			
polyamide	PA	x	
Phenolic:			
phenol-formaldehyde	PF		x
Polycarbonate	PC	x	
Polyester	—		x
Polyethylene	PE	x	
Polypropylene	PP	x	
Polyurethane:			
urethane plastics	UP		x
Styrene:			
polystyrene	PS	x	
styrene-acrylonitrile	SAN	x	
styrene-butadiene plastics	SBP	x	
Silicone plastics	SI		x
Urea-formaldehyde	UF		x
Vinyls:			
poly (vinyl acetate)	PVAc	x	
poly (vinyl alcohol)	PVAL	x	
poly (vinyl butyral)	PVB	x	
poly (vinyl chloride)	PVC	x	
poly (vinyl fluoride)	PVF	x	

FIG. 15 Plastic piping made of ABS is light, easy to handle, and resists corrosion.

One of the largest applications for ABS plastics in the construction industry is for piping and pipe fittings (Fig. 15). They are used for water and gas supply lines, for drain, waste and vent systems. They are tough and will withstand rough usage and fairly high temperatures over extended periods of time.

ABS plastics are also used for hardware items such as door knobs, handles, latches and handrails, as well as furniture, appliance and machine parts.

Acrylics

Polymethylmethacrylate (PMMA) is the best known of the acrylic plastics. These transparent plastics are noted for their superior optical properties and can be used readily for glazing. Unlike glass, they also have the ability to "pipe" light. This means that a beam of light can be bent around a corner due to the internal reflections of the light by the outer layer of the plastic.

Acrylics also are noted for their good resistance to breakage and their outstanding weathering properties —their transparency and physical stability are not adversely affected by long outdoor exposure. However, acrylics cannot withstand high temperatures and begin to soften at 200°F. Their low abrasion resistance makes them more susceptible to scratching than glass.

Acrylics are used widely for skylights and roof domes (Fig. 16) and for glazing in public buildings where breakage hazard is high. They also are used in lighting fixtures—as translucent ceiling panels, lamp enclosures and light diffusers. Their ability to pipe light is utilized in surgical instruments and in lighted outdoor signs. Other applications include clear or translucent corrugated sheets for roofing; films or sheets bonded to wood or metal for exterior finishes; and various molded pieces of hardware.

In addition, acrylics can be dispersed as finely divided particles in a liquid or mastic medium, producing a *latex*. In this form they are used widely for making paints (see Section 216, Paints & Protective Coatings) and building sealants.

FIG. 16 Acrylic plastics can be easily molded into one-piece bubble-shaped skylights, thus minimizing leakage and eliminating bars and muntins.

FIG. 17 Air-supported structures are lightweight and portable, making them ideal for such uses as temporary construction enclosures.

Cellulosics

These plastics are based on modifications of the cellulose molecule. Cellulose acetate (CA) and cellulose acetate-butyrate (CAB) are the most common cellulosics.

Cellulosics are exceedingly tough and will withstand a great deal of rough usage. They can be produced clear or translucent, although their optical properties are not as good as those of the acrylics. CA is not suitable for extended outdoor exposure, but CAB can be formulated so as to be quite resistant to weathering.

Cellulose acetate is known primarily for its use in photographic film and recording tape. Cellulose-acetate-butyrate is used for piping and pipe fittings (primarily for gas and chemicals), as well as for outdoor lighting fixtures. Because of their toughness, both types of cellulosics are also used for a variety of hardware items, for handrails and for tool handles. Cellulosics also find use as adhesives and coating compounds.

Fluorocarbons

Polytetrafluoroethylene (PTFE) is the primary fluorocarbon used in the construction industry. It is remarkably inert to chemical attack—strong chemicals such as boiling aqua regia, chlorine or bromine do not affect it at all. Furthermore, it remains serviceable over an extensive range of temperatures (from −450° to 500°F), unmatched by most other plastics. It also has one of the lowest coefficients of friction of any known solid, which means that it has superior "anti-stick" properties. However, it is quite difficult to form and is consequently expensive to produce.

PTFE is commonly known under the trade name of *Teflon*, and it is used in particularly demanding applications such as piping for extremely corrosive chemicals at high temperatures; for low-friction slider pads to permit movement in steam lines; and as nonstick linings for pots and pans.

Nylons

Nylon is the common name for a group of plastics known as polyamides (PA). These plastics have good all-around toughness, high strength and good chemical resistance. They are also resistant to abrasion, but they are not recommended for continuous outdoor exposure.

Nylons are perhaps best known for their use in synthetic fibers and filaments. They produce high-strength fabrics which are durable and tear-resistant. These fabrics are used for sails, parachutes and many articles of clothing. In the construction industry, nylon fabrics (which may be coated or reinforced as necessary) are frequently used for air-supported structures (Fig. 17).

Because of its toughness, nylon is also used to produce molded parts such as locks, latches, rollers for sliding windows and drawers, gears and cams.

Polycarbonates

Polycarbonates (PC) are transparent plastics with exceptionally high impact strength, good heat resistance and good dimensional stability. They are fairly new plastics (developed in 1957) and are not yet widely used in construction. However, they do find use in break-resistant safety glazing, lighting fixtures, lighted signs and various pieces of molded hardware.

Polyethylene

Polyethylene (PE) is a well-known plastic made by polymerization of the ethylene molecule. It is strong, light and flexible and it retains its flexibility even at low temperatures. It is noted for its good water resistance and low moisture vapor transmission. It is produced in several different formulations (including *linear* and *branched* molecular configurations) and these may vary considerably as to strength, density and stiffness.

Polyethylene is used widely for vapor barriers in floors and walls (Fig. 18) and dampproofing of basement walls. It is also used for some types of piping (cold water, gas and chemicals), but it is not suitable for use under severe conditions involving high temperatures or extremely corrosive chemicals. Another common application of polyethylene is for wire and cable insulation.

Polypropylene

Polypropylene (PP) is similar to polyethylene in many respects (both are classed as polyolefines), but it is much stiffer and more heat-resistant than polyethylene. This makes it suitable for piping and pipe fittings which can be used under more severe conditions, such as for hot water supply lines and drainage systems. Although polypropylene lacks the flexibility of polyethylene and is also more expensive to produce, it is used for some films and sheeting, primarily in the packing industry.

Another common use for polypropylene is in the production of high-strength fibers for carpeting (Fig. 19).

FIG. 18 A polyethylene vapor barrier on the interior of a wall assembly prevents vapor condensation inside the wall.

FIG. 19 Polypropylene fiber used in indoor-outdoor carpeting has low moisture absorption and good fade resistance.

Polystyrene

Polystyrene (PS) is made from the styrene monomer. It is a transparent, water-resistant and dimensionally stable plastic. It is not affected by extended low temperatures, but it begins to soften at temperatures near that of boiling water (212°F). The main drawbacks of polystyrene are that it is quite brittle and not very weather-resistant—it becomes more brittle and yellows with outdoor exposure.

In order to improve the properties of polystyrene, it is often copolymerized with other substances. One such copolymer is high-impact polystyrene, which is produced by adding small amounts of butadiene to the styrene. This results in a much tougher plastic with greater impact resistance and greater flexibility. Another well-known copolymer of styrene is acrylonitrile-butadiene-styrene (ABS), discussed previously.

Polystyrene is used for lighting fixtures and lighted signs and also for various molded pieces of hardware. Expanded (foamed) polystyrene is used widely in the construction industry for duct and pipe insulation; insulation for freezers and walk-in refrigerators; and insulation for walls, floors and ceilings of residences and commercial buildings (Fig. 20). Polystyrene foam is also used as the core material in the manufacture of doors and sandwich panels.

Vinyls

The vinyls are a large group of plastics based on modifications of the ethylene molecule. Included in this group are such plastics as polyvinyl acetate, polyvinyl alcohol, polyvinyl butyral, polyvinyl chloride, polyvinylidene chloride and many more. These plastics exhibit a wide range of properties. They generally are characterized by good strength and toughness, fair chemical resistance and a slow rate of water absorption. They perform well at normal use temperatures as well as relatively low (freezer) temperatures, but cannot withstand prolonged exposure to high temperatures—some vinyls begin to soften at 130°F. Most types are recommended for indoor use, but some are suitable for outdoor exposure.

FIG. 20 Polystyrene foam insulation is available in blocks or boards which can be bonded directly to masonry walls.

FIG. 21 Vinyl siding, gutter and downspout systems provide an attractive, easy-to-maintain home exterior.

The most widely used vinyl in the construction industry is polyvinyl chloride (PVC). This is a very versatile plastic, with high impact resistance, good dimensional stability, abrasion resistance and good aging characteristics. It is used to produce a large variety of products such as sheet and tile flooring, gutters and downspouts, moldings, clapboards and siding, window frames, piping and drainage systems (Fig. 21). In sheet form it is used for facings in sandwich construction, and is bonded to wood or metal for exterior building finishes, doors and window frames.

PVC is also available as a foam (either rigid or flexible) and is used for the core material in sandwich construction and for cushions in seating. PVC is copolymerized with many other plastics to produce a variety of adhesives, binders for terrazzo and other floor toppings.

Some other vinyls encountered in construction are polyvinyl butyral (PVB), which is used for the interlayer in safety glass and also as a coating for upholstery fabrics; and polyvinyl acetate (PVAc), which is used in adhesives, mortars and paints.

THERMOSETS

These plastics are permanently set during forming and cannot be softened again by reheating. In general, thermosets have better heat resistance and greater stiffness than thermoplastics, but are more difficult to form and more brittle. They have a more limited range of application than thermoplastics.

Epoxies

The epoxies (EP) were introduced in the 1950's and were quickly adopted for a wide variety of uses because of their excellent adhesive strength, combined with good chemical and moisture resistance. They will adhere to almost any building material and can be used for bonding such diverse materials as metals, glass, masonry and other plastics.

Epoxies are used primarily in coating compounds and adhesives. They are often produced as two-component systems, consisting of a resin and hardener, which must be mixed together just prior to application. Because they bond well to metals, they are used as protective coatings for appliances, automobiles and piping. As adhesives, they are used in the construction of sandwich panels; for bonding facings to walls, doors and other assemblies; and for applying tile finishes. They make excellent high-strength mortars for concrete block, and for patching cracked monolithic concrete. Epoxy compounds also can be mixed with mineral aggregate or plastic chips to produce terrazzo (Fig. 22).

Although polyesters are more common, epoxies sometimes are used in the production of fiberglass reinforced plastics. Pure, unmodified epoxy cures to a hard, brittle plastic, but when reinforced with glass fibers, it results in a strong, tough material which can be used in structural and semi-structural applications.

FIG. 22 Epoxy terrazzo is a durable, low-maintenance floor finish suitable for both interior and exterior installations.

Melamines

Melamines (MF) are produced by the reaction between melamine and formaldehyde. They are classified as *amino* plastics, along with the ureas. They are hard and scratch-resistant, and can be used at temperatures approaching the boiling point of water. They also stand up well to chemical attack.

The most important application for melamine plastics is in the production of high-pressure laminates, which are used widely for counter tops and cabinet finishes (Fig. 23). They are also used as adhesives (notably for plywood) and as a protective treatment for fabrics and paper.

As molding compounds, unmodified melamines are somewhat expensive and difficult to work with; hence, they are generally mixed with mineral or wood flour fillers. In this form, they are used for molded hardware, for electrical fittings and for plastic dinnerware.

FIG. 23 Kitchen countertops employing high-pressure laminates are scratch and stain-resistant.

Phenolics

Phenolics (PF) are produced by the reaction of phenol and formaldehyde. Like many of the thermosets, pure phenolics are quite brittle and hard, and are generally mixed with fillers to improve their strength and impact resistance. Many different formulations are available, with varying properties. Phenolics have excellent heat resistance, and fiberglass reinforced types will perform well at temperatures up to 400°F. They also have excellent thermal and electrical insulating properties.

Phenolics are one of the most widely used of the thermosetting plastics. As molding compounds, they are used for electrical parts (sockets, switch boxes, circuit breakers), for a great variety of hardware items, and for appliance cabinets and parts.

Phenolic resins are important in the production of high-pressure laminates—they are used to impregnate the sheets of Kraft paper which form the base of most laminates. The resins also are used to stiffen paper honeycomb (Fig. 24), which is

FIG. 24 Paper honeycomb stiffened with phenolic resin is used in the construction of hollow core doors.

FIG. 25 Bathtub/shower units molded of fiberglas reinforced plastic have seamless one-piece construction.

then bonded to a variety of facings to produce lightweight, strong sandwich panels for walls, doors and room dividers. Adhesives and protective coatings are also made from phenolic resins.

Another major use for phenolics is as foamed insulation. In this form they are used as the core for sandwich panels, and for insulation around piping and ducts.

Polyesters

Polyester is the generic name for a large group of plastics which are used for a wide variety of purposes. Some examples of different types of polyesters are: oil modified polyesters (sometimes referred to as *alkyds*), used in paints and protective coatings; saturated polyesters, used as plasticizers in other plastic compositions; and saturated polyesters of high molecular weight, used to form fibers and film. The most important polyesters for building purposes are the unsaturated polyesters, which form the basic constituent of reinforced plastics.

The unsaturated polyesters can be crosslinked with a compatible monomer (such as styrene) to form a hard, strong thermosetting material. They have become the most widely used resin for the production of reinforced plastics because of an important combination of properties: they are easy to work with and combine readily with a variety of fillers and reinforcing agents; they cure rapidly at ordinary room temperatures without giving off any volatile by-products; and they have excellent dimensional stability, combined with high strength and toughness. Furthermore, they can be specially formulated to provide good weathering properties and good heat and flame resistance.

For maximum strength and stiffness glass fiber is used as the reinforcing agent in polyester compounds. This produces an exceptionally strong, lightweight material which can be used in many structural building applications. Some of the common uses for fiberglass reinforced polyesters are prefabricated sections for roof structures; translucent sheets for roofing and interior partitioning; window frames and sash; facings for sandwich panels; and many large, molded articles such as bathtubs, shower stalls, sinks and cabinets (Fig. 25). Products such as bathtubs and countertops may be made by molding reinforced polyester within an acrylic shell to give a combination of rigidity, strength, and attractive appearance.

Polyurethanes

The polyurethanes (UP) are a class of plastics which are known primarily for their use as low-density foams. Foams with widely varying properties can be produced from polyurethane—from soft, flexible open-cell types to tough, rigid closed-cell types. Urethane foams have excellent heat and chemical resistance and good tensile strength. They can be formulated to be fire-resistant.

Rigid urethane foams make excellent insulation material. They have the lowest thermal conductivity of any building material—lower even than mineral wool. They are used widely in construction for wall, floor and ceiling insulation (Fig. 26), for insulation around piping and ducts and as the core material in sandwich panels. Flexible urethane foams are used for cushions, upholstery and padding.

Urethane resins also are used in the production of elastomers (synthetic rubbers), adhesives, caulking and glazing compounds.

Silicones

Silicone plastics (SI) are based on the silicon atom rather than the carbon atom. They are therefore an exception to the general rule that all plastics are organic compounds.

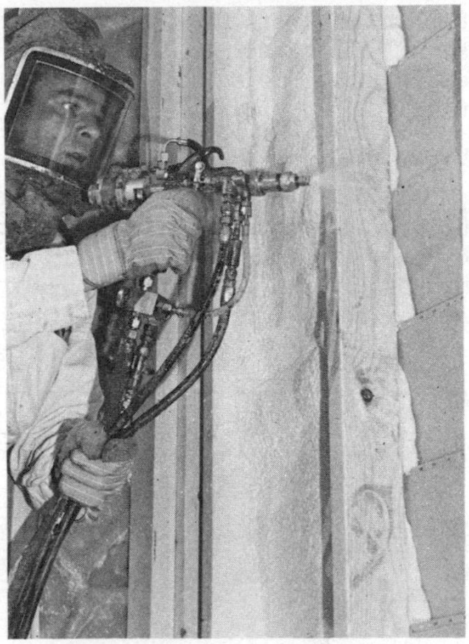

FIG. 26 Applying foamed-in-place polyurethane insulation.

FIG. 27 An application of silicone water repellent protects masonry walls from damaging weather effects.

Silicones are very stable compounds, characterized by excellent corrosion resistance and good electrical insulating properties. They can be used over a wide range of temperatures (from −80°F to 500°F), and they also stand up extremely well to most weathering conditions.

Silicones are noted for being extremely water repellent, and they are used for waterproofing masonry and brick (Fig. 27). They also make excellent sealants and are used in curtain wall construction. Small amounts of silicone are frequently added to reinforced plastics to improve the adhesion of the polyester resin to the glass fibers, and at the same time, to increase the strength and moisture resistance of the plastic.

Ureas

The most common urea plastic, urea-formaldehyde (UF), is produced by the reaction between urea and formaldehyde. Ureas are chemically related to the melamines and are classed with them as *amino* plastics.

Urea plastics have lower heat and chemical resistance than melamines, but they are more scratch-resistant and have good electrical resistance. They are used for molded hardware items (Fig. 28) and electrical fittings. Urea resins provide adhesives for use in sandwich panels, and they are also available as insulating foams.

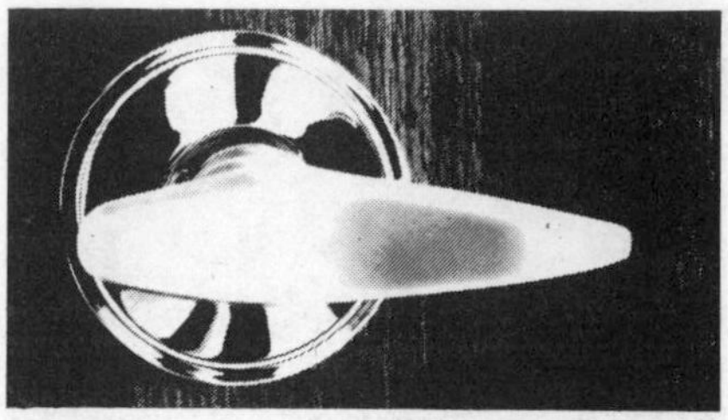

FIG. 28 Doorknobs and other hardware items can be made of urea.

GLOSSARY

Antioxidant A compound added to other substances to retard oxidation, which deteriorates plastics.
Bleeding Diffusion of a colorant out of a plastic part into adjacent materials.
Blow Molding Shaping thermoplastic materials into hollow form by air pressure and heat; usually performed on sheets or tubes.
Calendaring A process for producing film or sheeting by passing the material between revolving heated rolls.
Casting Shaping of plastic objects by pouring the material into molds and allowing it to harden without the use of pressure.
Catalyst A substance which changes the rate of a chemical reaction without itself undergoing permanent change in composition. Also, a substance used to initiate the polymerization of monomers to form polymers.
Cellular Plastic A plastic the apparent density of which is decreased substantially by the presence of numerous cells dispersed throughout its mass.
Closed Cell A cellular plastic in which there is a predominance of non-interconnecting cells.
Open Cell A cellular plastic in which there is a predominance of interconnecting cells.
Cold Flow Permanent change in dimension due to stress over a period of time, without heat.
Compounding The thorough mixing of a polymer or polymers with other ingredients such as fillers, plasticizers, catalysts, pigments, dyes or curing agents.
Compression Molding Forming plastic in a mold, by applying pressure and usually heat.
Condensation Polymerization A chemical reaction in which the molecules of two substances combine, giving off water or some other simple substance (see *Polymerization*).
Copolymer A polymer formed by the combination of two or more different monomers (see *Polymer*).
Creep See *Cold Flow*.
Crosslinking A chemical reaction in which adjacent polymer molecules unite to form a strong, three-dimensional network; usually occurs uring the curing of thermosetting plastics.
Cure To change the properties of a polymeric system into a final, more stable, usable condition by the use of heat, radiation, or reaction with chemical additives. Sometimes referred to as *set*.
Degradation A permanent change in the physical and/or chemical properties of a plastic evidenced by impairment of these properties.
Deterioration See *Degradation*.
Discoloration Any change from the initial color possessed by a plastic.
Dispersion The distribution of a finely divided solid in a liquid or solid.
Elasticity The ability to recover original size and shape after deformation.
Elastomer A material which at room temperature can be stretched repeatedly to at least twice its original length and, upon release of the stress, will return instantly and with force to its approximate original length; a rubber-like substance.
Expandable Plastic A plastic suitable for expansion into cellular form by thermal, chemical or mechanical means.
Extender A low-cost material used to dilute or extend high cost resins without appreciably lessening the properties of the original resin.
Extrusion Forcing plastic material through a shaped orifice to make rod, tubing or sheeting.
Filler A relatively inert material added to modify the strength, permanence or working properties, or to lower the cost of a resin.
Film Sheeting having a nominal thickness not greater than 0.010″.
Foamed Plastic See *Cellular Plastic*.
High Polymer A polymer of high molecular weight (see *Polymer*).
High-pressure Laminate A laminate molded and cured at pressures not lower than 1,000 psi—commonly 1,200 to 2,000 psi (see *Laminate*).
Inhibitor A substance which prevents or retards a chemical reaction; inhibitors often are added to plastic resins to prolong their storage life.
Injection Molding Forming plastic by fusing it in a chamber with heat and pressure, and then forcing part of the mass into a cooler chamber where it solidifies.
Laminate A product made by bonding together two or more layers of materials.
Latex A water suspension of fine particles of rubber or rubber-like plastics.
Linear Polymer A polymer which is arranged in a long, continuous chain with a minimum of side-chains or branches.
Modifier An ingredient added to a plastic to improve or modify its properties.
Monomer A relatively simple compound which can react to form a polymer.
Plastic A material which contains as an essential ingredient an organic substance of large molecular weight, is solid in its finished state, and, at some stage in its manufacture, can be shaped by flow. (see *Polymer*).
Plasticizer A material added to increase the workability and flexibility of a resin.
Polymer A chemical compound formed by the reaction of simple molecules into more complex molecules of higher molecular weight.
Polymerization A chemical reaction in which the molecules of a monomer are linked together to form large molecules whose molecular weight is a multiple of that of the original substance.
Postforming Forming thermosetting laminates which have already been cured into simple shapes by heat and pressure.
Reinforced Plastic A plastic material whose strength and stiffness have been upgraded by the addition of high strength fillers, such as glass fiber.
Resin The essential ingredients of a plastic mix, before final processing and fabrication of the plastic object.
Sandwich Panel A composite structural panel made of two thin, strong, hard facings bonded firmly to a core of relatively light-weight, weaker material with insulating properties.
Saturated Molecule A molecule which will not unite readily with another element or compound.
Service Temperature The maximum temperature at which a plastic can be continuously employed without noticeable reduction in strength or other properties.
Sheeting A form of plastic in which the thickness is very small in proportion to the length and width; usually refers to a product with a thickness greater than 0.010″ (see also *Film*).
Softening Range The range of temperature in which a plastic

changes from a rigid to a soft state. Sometimes referred to as *softening point.*

Stabilizer A material added to prevent or retard degradation of a plastic when exposed to sunlight or other environmental conditions.

Thermoplastic A plastic which can be repeatedly softened by heating and hardened by cooling.

Thermoset A plastic which, after curing, forms a permanently hardened product which cannot be softened again by reheating.

Transfer Molding Forming plastic by fusing it in one chamber with heat and then forcing it into another chamber where it solidifies; commonly used with thermosetting plastics.

Unsaturated Molecule A molecule which is capable of uniting with certain other elements or compounds, without creating any side products.

Vacuum Forming Shaping a heated plastic sheet by causing it to flow in the direction of reduced air pressure.

We gratefully acknowledge the assistance of the Society of the Plastics Industry, Inc., Rohm and Haas Co. and United Foam Corp. in the preparation of the text. We also acknowledge the cooperation of the following in supplying illustrations: Armstrong Cork Co., Birdair Structures, Inc., Bird and Son, Inc., Corl Corp., General Electric Co., Kawneer Co.

211 IRON AND STEEL

IRON AND STEEL

INTRODUCTION

The chief ingredient of the metals known as iron and steel is the chemical element *iron* (ferrum), from which iron-based alloys get the generic name, *ferrous metals*. Although about 5% of the earth's crust is composed of iron, it is rarely found in its metallic form in nature; it is almost always combined with oxygen (iron oxide), sulfur silicon and mixed with other minerals; it is found in rock, gravel, sand, clay and mud. In this form it is mined as iron ore, the chief source of iron. Iron and steel making involve extracting iron from the ores, combining it with carbon and other elements and then forming it into shapes.

Pure metallic iron is inherently soft, ductile and easily shaped, but too weak for most uses. The many irons and steels used in modern construction and manufacture get their strength and hardness mainly from the element carbon. Most *steels* derive the best combination of properties when carbon is present in amounts less than 1.20%.

Ferrous metals containing substantially larger amounts of carbon become so hard and brittle that they cannot be readily shaped by either hot or cold working methods. These metals are formed by being cast into molds, and are referred to as *cast iron*.

Early steel making processes could not control the impurities and carbon content sufficiently to produce metal with the properties of modern steel. Also, the rolling, drawing and heat treating processes which improve the mechanical properties of steel were unknown. Therefore, the earliest ferrous metals used in construction were essentially wrought and cast irons, shaped by hand working or casting hot metal.

The first successful North American iron works was erected in Saugus, Massachusetts in the late 1640's. Its earliest products were cooking pots and nails. Nails were so expensive, that old buildings were often burned so that the nails could be recovered.

Early steel making processes produced steel in small quantities. The first high-tonnage steel making process was discovered by an American, William Kelly, but named for Sir Henry Bessemer of England. The bessemer (pneumatic) process forced air through molten iron contained in a pear-shaped vessel. As the air rose through the iron it burned out impurities. The addition of manganese caused chemical changes which converted the iron to steel.

The Civil War slowed the development of commercial steel making in the United States until 1864. Then the bessemer process and the discovery of abundant iron ore deposits in the Lake Superior region made the United States a world leader in steel production. The modern revolution in construction, industry, commerce, transportation and communications is due in large part to the economical production of the two common metals—iron and steel.

About 20% of the total steel production is used in building construction. Nearly four tons of steel may be used in the construction and equipment of a typical home.

The name "steel" covers a variety of ferrous metals of different chemical compositions and properties. The properties of the many kinds of steel are greatly affected by alloying elements, processes of production and methods of manufacture. This section discusses iron and steel making and the resulting properties.

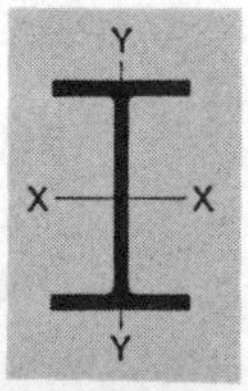

RAW MATERIAL

In this section the term "raw material" is used to designate all of the materials employed in iron and steel making up to the point of producing metal ingots used for product manufacture.

The general steps in the mining and processing of the primary raw materials, smelting into pig iron and refining into steel ingots are illustrated in Figure 1. Generally, all iron and steel making starts with the extraction of iron from its oxide form in the ores by heating in contact with coke, a form of carbon. The carbon combines separately with the oxygen and the iron of the ores, thus forming a volatile gas and a carbon-rich crude iron called *pig iron.*

The primary raw materials used to make pig iron are: (1) *iron ore,* which furnishes the iron; (2) *fuel* (coke), which furnishes the carbon for combustion and for alloying with the iron; (3) *air,* which supplies the oxygen for combustion and oxidation (burning out) of impurities (unwanted elements) in the raw materials; (4) *flux,* which provides a molten slag to separate the ash of combustion and the impurities from the molten iron; (5) *refractories,* which form the linings of furnaces, flues and vessels used to contain the molten metal.

In steel making, pig iron and metal scrap replace the ore as a primary source of iron, and electricity may replace the fuel as a source of heat. In some processes, gaseous oxygen supplements or completely replaces the combustion air. The emphasis in steel making is on reducing carbon and other impurities in pig iron originating in the raw materials.

MINING AND PROCESSING

The primary raw materials—iron ores, fuels, fluxes and refractories—are mined in areas often quite distant from iron and steel making plants. To reduce shipping costs, sometimes raw materials are processed near the mines to increase their yield and reduce bulk. They are shipped by rail and water (Fig. 2) to steel mills for direct use or

FIG. 1 Flow chart of operations in the production of steel ingot.

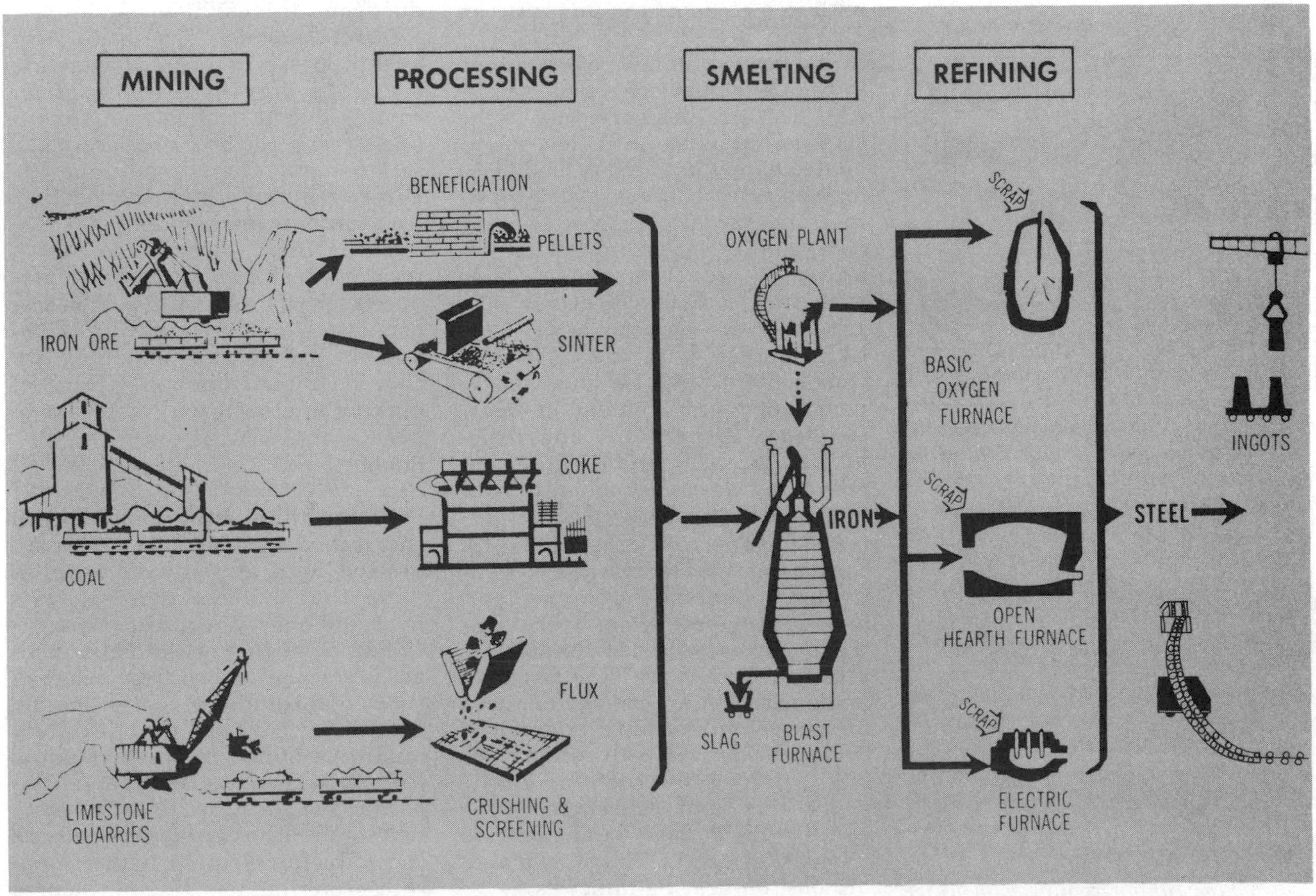

FIG. 2 Ore and limestone typically are delivered to the blast furnace (background) by boat.

FIG. 3 Typical terraced effect of open pit ore mining.

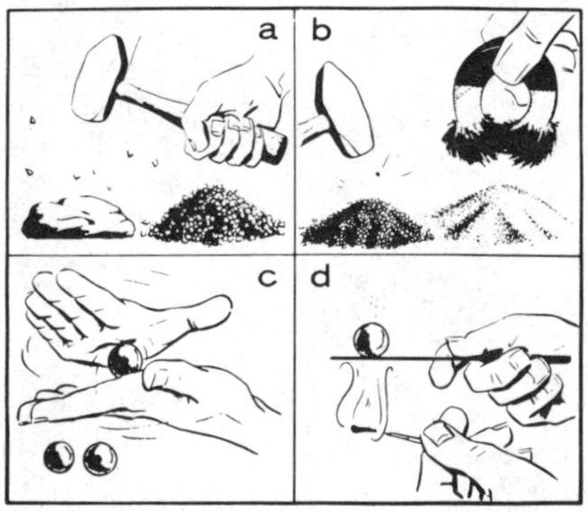

FIG. 4 Taconite is pelletized by: (a) crushing, (b) magnetic separation of ore particles; (c) rolling into pellets; (d) baking to a hard, dense consistency.

FIG. 5 Hot coke is forced from one of a battery of slot-shaped ovens into a quenching car for cooling with water sprays.

further processing in plants. Mills and plants are typically located along navigable water routes, affording low-cost transportation for bulk materials and an abundant supply of fresh water essential in steel making.

Iron Ores

Iron ore is mined by the underground or open pit methods (Fig. 3). The iron ore of the Lake Superior region dominates steel production; other deposits are mined in Alabama, New Jersey, New York, California and a few other states. Iron ore is also imported, with Canada providing more than half of all imports. The most common iron minerals are hematite (FE_2O_3), siderite ($FeCo_3$), and pyrite (FeS), with iron contents ranging up to 70%. Limonite may be considered as magnetite plus water; taconite and jasper are names referring to a wide variety of iron-bearing rocks with iron contents ranging between 22% and 30%.

The largest tonnage of ore mined in the Lake Superior region is taconite. This ore lies very close to the surface but is so hard that superheated flames are used to drill holes for explosive charges. After blasting, the taconite is processed by *beneficiation,* a treatment to improve the ore's iron content. Ores with iron contents of nearly 50% generally are shipped as is. Ores like taconite, with a 23% to 25% iron content, can be improved to nearly 66% iron content by beneficiation. Richer ores and beneficiated taconite can be further improved by direct reduction to over 90% iron content.

Beneficiation of ores generally consists of 1) grinding and concentration to improve the iron content, and 2) *agglomeration,* increasing the particles to a size suitable for blast furnace use. Two forms of agglomeration are *pelletizing,* used for jasper and taconite; and *sintering,* used to recycle iron ore dust.

Pelletizing by both magnetic separation and concentration is usually done near mine sites to avoid transporting large tonnages of non-iron-bearing minerals.

In one pelletizing process, magnetic taconite is crushed into a fine powder the consistency of flour. The powder is passed through a magnetic seperator which removes the iron particles. The iron powder then goes to a balling drum, where it is mixed with coal dust and a binder material to form pellets with 60% iron content. The pellets are baked into a hard finish in a kiln or furnace (Fig. 4).

In another pelletizing process, high grade non-magnetic ores are crushed and then concentrated in flotation cells. Flotation cells contain liquids in which the heavier iron-bearing particles sink to the bottom of the cell while the lighter particles remain suspended and are skimmed away. After filtering, the iron-bearing particles are formed into pellets and baked hard.

Sintering is a process by which iron ore dust, reclaimed from blast furnace gases and other manufacturing processes, is fused with coke and fluxes into a clinkerlike product containing 52% to 65% iron.

Direct Reduction These are various processes which concentrate ores to a very high iron content, comparable to molten iron from a blast furnace. Some ores are made into pellets, others into briquettes. Direct reduced ores with over 90% iron contents are bypassing the traditional blast furnaces in which iron for steel making has been produced for centures. The pellets and briquettes produced by direct reduction are so rich that they are charged directly into electric steel making and basic oxygen furnaces.

Fuels

Of the three main fuels used in the iron and steel industry—coal, oil and natural gas—coal supplies more than 65% of the industry's heat and energy requirements.

The chief fuel of the blast furnace is *coke,* produced from selected types of bituminous coal in coking ovens (Fig. 5). Most coal of coking quality is mined in West Virginia, Pennsylvania, Kentucky and Alabama. In addition to furnishing heat and carbon, coke also acts as the reducing element which separates iron from its oxide in the ores to make pig iron.

Fluxes

Fluxes are minerals which have an affinity for the impurities in iron ore or pig iron, combine with these impurities and separate them from the molten metal in the furnace by forming a liquid *slag*. Some fluxes are used mainly to separate the impurities, others to make the slag more fluid, thus floating it to the surface of the metal sooner. Depending on the affinity of a flux to combine chemically with certain impurities, it is classed as either *basic* or *acid*. Limestone and dolomite are examples of basic fluxes; sand, gravel and quartz rock are acid fluxes. Fluorspar is a neutral flux used mainly to make the slag more fluid.

FIG. 6 Magnesia is extracted from sea water or brines for use as a basic refractory in steel furnaces.

Refractories

Refractories are nonmetallic materials with superior heat and abrasion resistance used as linings of steel making furnaces, flues and vessels. Like the fluxes, refractories may be *acid* or *basic*, depending on their chemical reaction with impurities. Theoretically, acid and basic refractories must be used with like fluxes to minimize chemical reaction between them, but in practice this rule is not always adhered to.

A common acid refractory material is *ganister*, a form of quartzite rock obtained from quarries and mines in Pennsylvania, Wisconsin, Alabama, Utah and California. The most common basic refractory material is *magnesia*, obtained from the mineral magnesite and from sea water (Fig. 6).

Alloy Ores

The chemistry of steel requires many elements for alloying purposes, such as manganese, silicon, nickel, chromium and molybdenum. Although the United Staes is a major steel producer, it depends upon imports to furnish many of these elements. Most alloys are imported as ores and processed into useful forms called ferroalloys.

Manganese All steel uses manganese, both as a scavenger to remove oxygen from molten metal and also as an alloying element to improve hardness and wear resistance. An average of 12 pounds of manganese goes into each ton of steel.

Chromium Added in amounts of 1% to 35%, chromium provides stainless steels with resistance to heat, rust and corrosion. To other alloy steels and cast iron, it provides hardness and wear resistance. Chromium is also important as a coating metal.

Nickel Cryogenic steels, those produced at low temperatures, use nickel to improve hardness. Steels containing nickel are especially suitable for case hardening, the process of surface hardening by absorption of carbon at high temperatures.

Silicon In steel making, silicon serves as a deoxidizer and an alloying element. When used in large percentages, silicon equips cast irons to withstand highly corrosive acids.

Tungsten Tungsten provides heat resistance and is used in cutting tools of high-speed steel and tungsten carbide, an abrasion resistant material almost as hard as diamonds. It is also used as a welded deposit on parts exposed to extreme wear.

Molybdenum One of the more versatile alloying elements, molybdenum increases the hardness and corrosion resistance of steels.

Vanadium This mineral is often used in conjunction with other alloying elements: with chromium in making steel springs, with manganese in making special plates and other structural forms, and molybdenum in making certain high-temperature steels. Very small amounts (often measured in hundredths of 1% provide the desired effect in construction steels.

Boron Known widely as a cleansing agent, boron is used as a hardening agent in steel. Minute quantities are used to obtain the desired effect and these must be protected by other deoxidizers or must be added to a completely deoxidized molten metal bath.

Air and Oxygen

Oxygen and air can be considered as much a raw material as the specially prepared iron ore, coke and lime that the steel industry consumes. Air must be forced into furnaces, heated or cooled, and cleaned to control air pollution. Oxygen is a product of special plants which produce also nitrogen, and argon from ordinary air.

Water

The steel industry uses 35,000 gallons of water for each ton of raw steel produced. Often water must be cleaned both before and after use. Its primary function in steel making is as a coolant in closed systems.

IRON MAKING

Smelting is the metallurgical operation by which metal is heated and separated from the impurities with which it may be chemically combined or physically mixed in the ores. The separation of iron from oxygen with which it is chemically combined in the ore is termed *reduction*. Pig iron is made in the *blast furnace* by smelting, which extracts the iron both by reduction of its oxide and physical separation from some impurities.

Blast Furnace

The blast furnace is a cylindrical structure, approximately 120′ tall, lined with refractory brick and encased in a steel shell (Fig. 7). The charge (ore, coke and limestone) is loaded at the top and a blast of hot air is introduced near the bottom. Oxygen in the air combines with carbon in the hot coke, causing combustion and producing heat. The charge is smelted by the heat and, as the gases of combustion pass through the hot ore, another chemical reaction takes place which frees the iron from its oxide. The hot combustion gases are cleaned in "dust catchers" and recycled through brick "checkerworks" (stoves) to heat the incoming air blast (Fig. 8). The operation of the blast furnace is continuous.

The impurities in the ore and the flux combine with oxygen to form a floating slag 4′ to 5′ deep which is tapped from the furnace at approximately two-hour intervals. The molten crude iron is tapped at a lower level at about three-hour intervals; it contains approximately 93% iron and 7% extraneous elements. It is conveyed in ladle cars to *mixers*, holding furnaces, where liquid iron is mixed to equalize its chemical composition and kept hot pending further refinement.

Most of the molten iron goes to open hearth or basic oxygen steelmaking facilities. Some goes to a casting machine where it solidifies into molds, known as *pigs* (ingots).

Pig Iron Most of the iron from pig casting is made to rigid specifications which vary widely according to end use. Pig iron is also referred to as *merchant pig iron* when it is intended to be sold as an end product.

End products for pig iron vary from cast iron soil pipe to fancy dutch ovens. A merchant pig iron producer may have several hundred piles of different grades of iron and may blend a dozen or more grades of iron ore to arrive at a desired chemical composition. It is worth noting that while 30,000 cu. ft. of solids weighing just under 1,000 tons are being processed in a furnace, the chemical composition is being simultaneously controlled within fractions of 1% relative to carbon, silicon, sulfur, phosphorus and manganese.

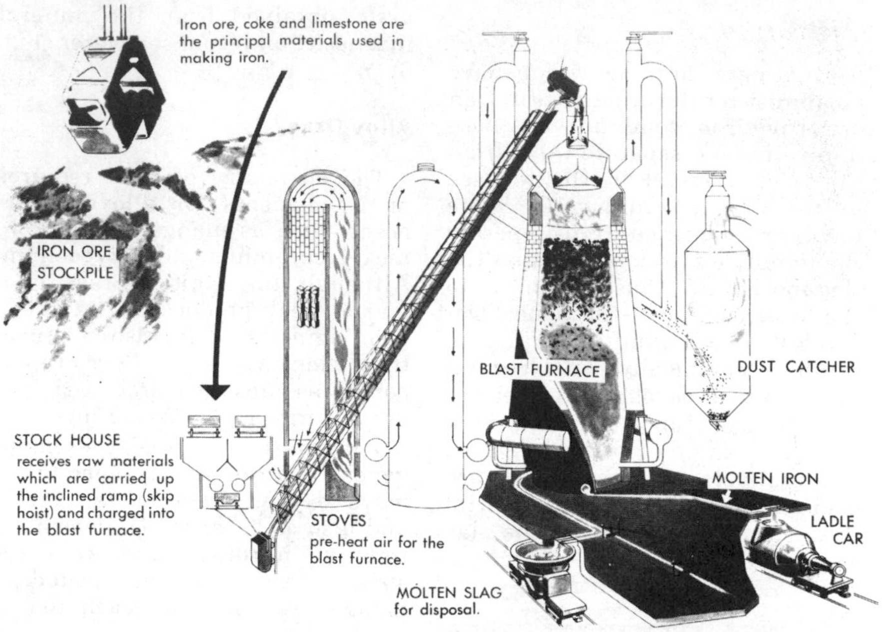
Iron ore, coke and limestone are the principal materials used in making iron.
IRON ORE STOCKPILE
STOCK HOUSE
receives raw materials which are carried up the inclined ramp (skip hoist) and charged into the blast furnace.
STOVES
pre-heat air for the blast furnace.
BLAST FURNACE
DUST CATCHER
MOLTEN IRON
LADLE CAR
MOLTEN SLAG
for disposal.

FIG. 7 Diagramatic illustration of blast furnace and auxiliaries.

FIG. 8 Overall view of a blast furnace and auxiliaries. A battery of cylindrical checkerworks appears at center left.

Foundry

A small percentage of pig iron is remelted in *foundries*, and formed into special shapes by *casting* in sand or loam molds. This process is called *iron founding*. Since no major change in chemical composition takes place when pig iron is cast, it is most often melted prior to casting with scrap metal and alloying elements to impart desirable properties. The carbon content of cast iron therefore is somewhat lower than pig iron, and ranges between 1.5% to 4.0%.

The types of cast iron produced are *gray, white, malleable, chilled, alloyed* and *nodular* cast iron. Only the first three are important in construction and will be discussed here.

Gray and White Cast Iron When molten cast iron solidifies, carbon in the iron may remain chemically combined in the form of iron carbide, or may separate out as graphite. *White cast iron* contains carbon mainly in the chemically combined form (iron carbide) which makes this metal hard and nonductile. In *gray cast iron*, the carbon is chiefly in the form of graphite flakes which impart the characteristic gray fracture and promote machinability and resistance to wear.

Malleable Cast Iron White cast iron may be made somewhat softer and more ductile, though not strictly "malleable," by *annealing* (controlled heating and cooling). The carbon in the iron carbide is converted to a more dispersed graphite form, which provides greater ductility than the flaky graphite of gray iron, or the carbide of white iron.

Direct Reduction

There are many direct reduction processes, capable of bypassing the blast furnace, presently being researched and developed. These processes are classed as (1) processes using solid reductants, (2) processes

FIG. 9 Hot "sponge ball" of iron and slag is delivered for processing into blooms.

using gaseous reductants, and (3) direct steel processes. None of these processes is ready to completely replace the efficient blast furnace.

STEEL MAKING

Steel making processes usually involve: (1) lowering the carbon content of pig iron; (2) controlling impurities in the pig iron and other raw materials; and (3) adding alloying elements to obtain desired properties.

Carbon is an important ingredient of steel, beneficial when contained within certain specified ranges which vary with the type of steel. Carbon increases steel hardness and strength up to about 1.20% content (at a lesser rate between 0.50% and 1.20%), but accompanying increases in brittleness and loss of ductility generally dictate lower contents in most steels.

Carbon, phosphorus, sulfur, manganese and silicon are classed as *impurities* when present in *uncontrolled* or *excessive* amounts. Pig iron usually contains these elements as impurities originating from the raw materials of the blast furnace. Up to certain limits or within specified ranges, the effect of these elements on steel may be beneficial (Fig. 18), and steel making processes attempt to regulate their contents within these limits or ranges. Phosphorus and sulfur impart some desirable properties, but must be limited to less than 0.05% in most steels due to their adverse effects on malleability (ability to be shaped). Silicon and manganese are strengthening elements desired in ordinary steels in the range of 0.01% to 0.35% and 0.20% to 2.00% respectively, and may be added when contained in insufficient quantities in the raw materials. All of these elements are present in *controlled amounts* in most steels. When they occur naturally, they are referred to as *residual elements;* when they are deliberately added to achieve certain properties, these and other elements are termed *alloying elements.*

Steel making processes generally start with hot pig iron delivered in ladles by overhead cranes to the steel furnace where scrap metal

FIG. 11 ANNUAL INGOT PRODUCTION*

Furnace Type	Net Tons	%
Open Hearth	20,042,934	16%
Basic Oxygen Process	77,408,157	62%
Electric & Other	27,295,262	22%
Total	124,746,353	100%

*1977 data from American Metal Market Metal Statistics 1978.

and fluxes are added. After the melting operation, the hot steel is tapped from the furnace into other ladles and moved to awaiting molds for casting into ingots or to a strand casting machine. The chief steel

FIG. 10 Hot pig iron may be converted to steel by any of the steel making furnaces illustrated below.

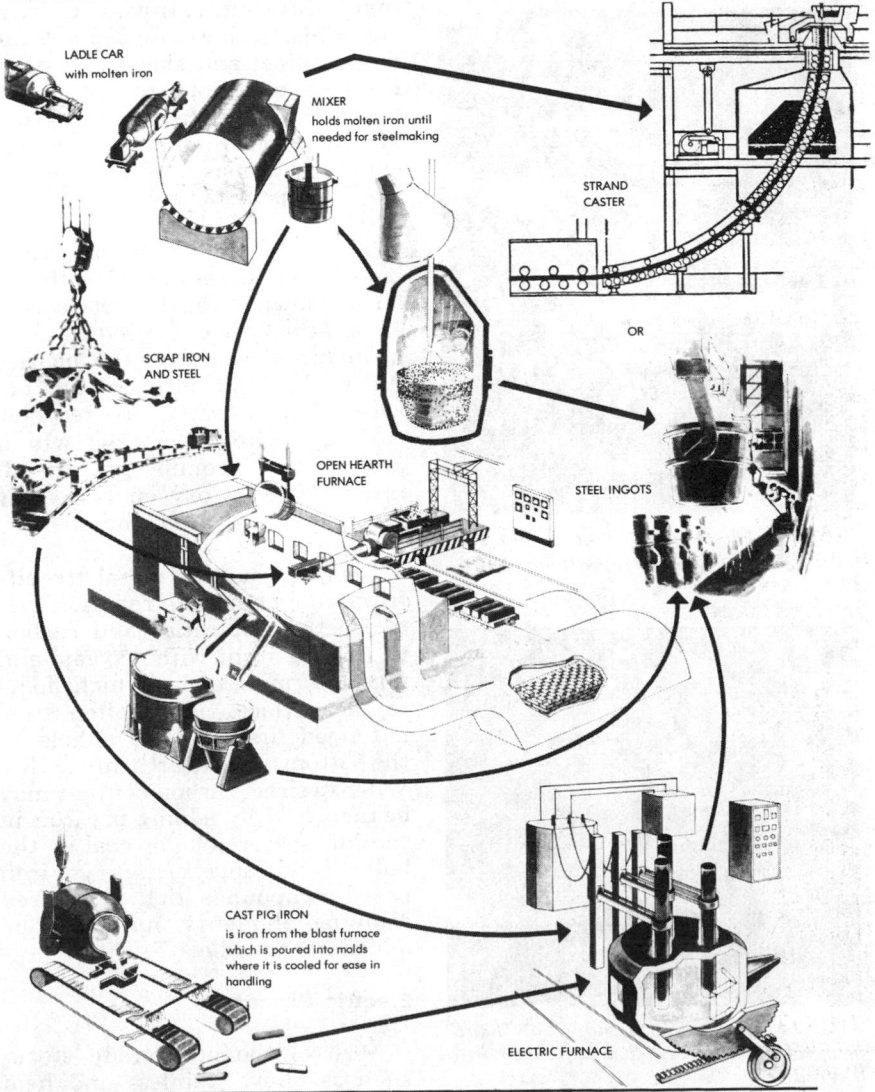

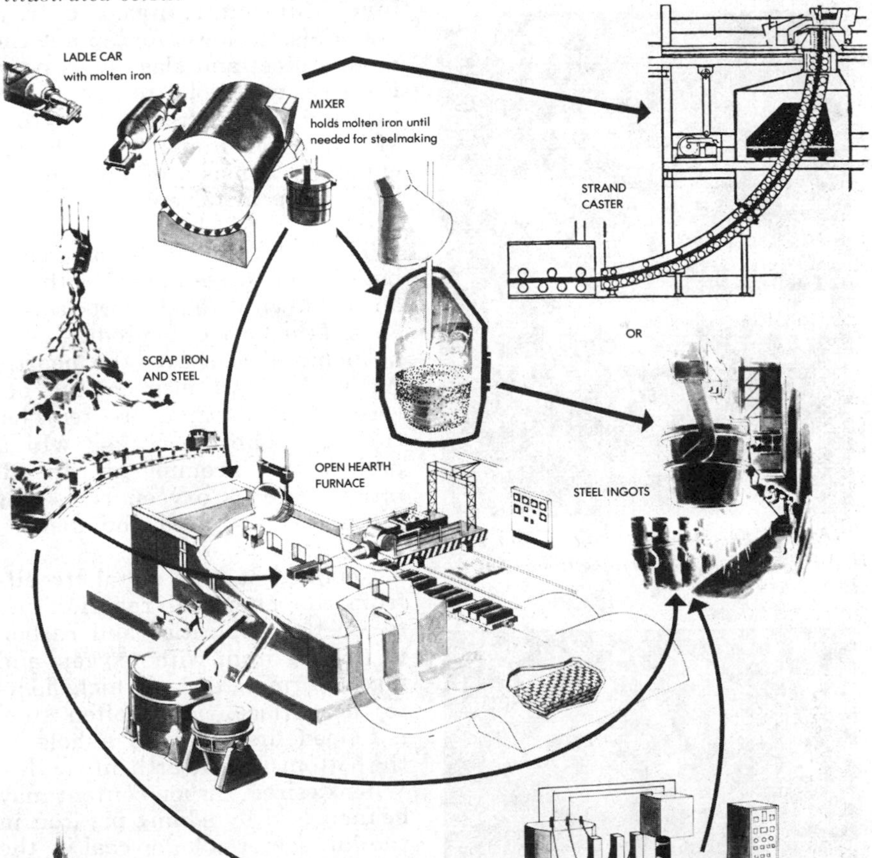

FIG. 12 Open hearth furnace: hot pig iron is poured from the ladle; charging machine (foreground) adds scrap metal and limestone flux.

FIG. 13 Electric arc furnace: hot steel is tapped from the furnace above into a ladle.

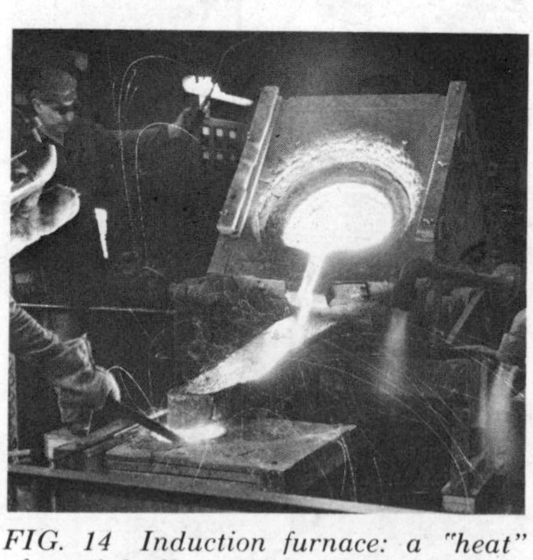

FIG. 14 Induction furnace: a "heat" of special alloy steel being poured into a mold.

making processes are *open hearth*, *electric* and *pneumatic* (Fig. 10). The volume produced by each process is illustrated in Figure 11.

Open Hearth Process

The greatest majority of open hearth furnaces are of the basic type, capable of controlling the phosphorus present in pig iron. Both ordinary carbon grades and high grade alloy steels are generally produced by this process.

The name *open hearth* is applied because the saucer-shaped hearth, or floor of the furnace, is exposed to the sweep of the flames. The furnace itself is a masonry structure lined with refractory materials arranged on two levels: the upper level for charging the raw materials, the lower for tapping the molten steel and slag by gravity. Furnaces may hold from 50 to 600 tons and produce from 25 to 50 tons of steel per hour. The charge generally consists of molten pig iron, scrap metal and fluxes (Fig. 12).

The fuel—gas, oil, tar or a combination of these—mixed with hot air, is blown through ports at the ends of the furnace to produce combustion and heat over the hearth. As in the blast furnace, the hot gases of combustion are recycled through "checkerworks" which preheat the incoming air. In addition, gaseous oxygen is usually introduced to speed the melting operation.

Refining of the metal results from the chemical reaction between the impurities and carbon in the pig iron, with oxygen and fluxes, forming a slag which floats to the surface. The molten steel is tapped first through a hole in the bottom of the hearth into ladles. When desired, carbon content may be increased by adding pig iron in the furnace or coke or coal in the ladle. Oxidizable *ferroalloys*, iron based compounds rich in desired alloying elements, may also be added in the ladle.

Electric Processes

Most of the high grade steels, such as alloy, stainless and heat

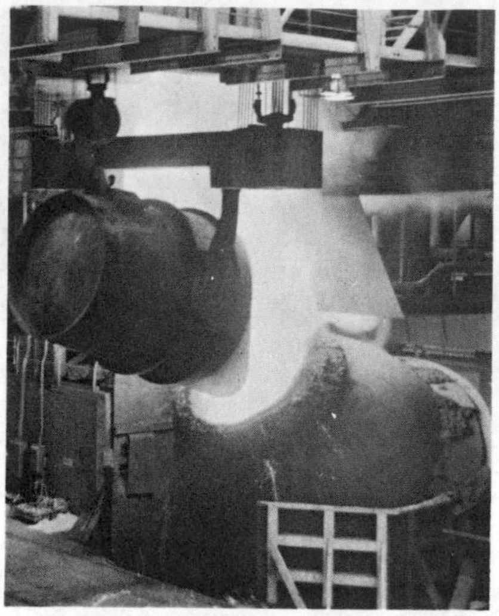

FIG. 15 Basic oxygen furnace: ladle pours hot pig iron into the mouth of a tilted furnace.

resisting steels, are produced in electric furnaces capable of developing the high temperatures and reducing conditions necessary for melting these metals. Here electrical resistance and arc radiation rather than combustion are used to produce heat, and air or oxygen are not necessary for combustion. High-purity oxygen is introduced at appropriate times to oxidize impurities, which results in better control of the oxidation process and a smaller loss of expensive alloying elements.

The charge usually consists of scrap metal, mill scale, fluxes and often alloying elements. Electric furnaces are mostly of the basic type and produce low phosphorus steel. Many grades of basic steel can be produced at a rate of 30 to 40 tons per hour in the *electric arc* furnace and may hold up to 300 tons. Special grades involving valuable alloying elements are generally produced in smaller quantities, often in the *electric induction* furnace.

Electric Arc Furnace The body of the electric arc furnace is a circular steel shell resembling a huge kettle, lined with refractory materials and mounted so that it may be tilted to pour off the molten steel. Three or more cylindrical electrodes project into the furnace from above. When high voltage current is passed through the electrodes, an electric arc is produced between them and the charge, generating heat. Many of the impurities are oxidized during the melting operation and float to the surface as slag, allowing the steel to be poured off from under it (Fig. 13).

Induction Furnace The induction furnace consists of a magnesia pot, insulated by other refractory materials and surrounded by windings of copper tubing (Fig. 14). When high voltage alternating electric current is passed through the copper tubing around the pot, an induction current is generated in the metal charge of the furnace. The resistance of the metal to the induced current creates the heat for the melting operation.

Pneumatic Process

Steel making operations which use high speed oxygen or air to oxidize impurities in the charge without fuel are *pneumatic processes.* Their chief advantages are shortened heat time and high output. With furnace capacities of 300 tons and tap-to-tap heat cycles of 45 minutes, it is possible to produce steel at a rate 6 to 10 times that of the open hearth and electric arc furnaces. The pneumatic process, called the *basic oxygen furnace,* is the predominant method of producing steel in the United States (Fig. 11).

Basic Oxygen Furnace In the oxygen process, a jet of high purity oxygen is directed onto the molten metal contained in a steel, pear-shaped vessel lined with refractory materials (Fig. 15). The charge, consisting of molten pig iron, some scrap and fluxes, is added from the top. Proportionately larger amounts of hot pig iron and lesser amounts of scrap metal are used in the oxygen furnace than in other steel furnaces.

No fuel is required since the heat of the large mass of molten pig iron is sufficient to start the chemical reaction between the oxygen, the carbon and other impurities. Oxidation in turn produces sufficient heat to continue the process without the introduction of fuels.

If the as-tapped metal has a lower carbon or manganese content than specified, *ferro-manganese* (a manganese-rich ferroalloy also containing carbon) may be added in the ladle. The carbon gives added strength and hardness, while the manganese, in addition to strength, assures proper ductility at elevated temperatures by combining with excess sulfur.

Steel Casting

No matter which steel making process is used, steel must be formed into shapes before further processing. It is either poured into *ingot molds* or put through a *strand casting machine.*

Ingots Raw steel from a furnace

FIG. 16 **CONTINUOUS CASTING**

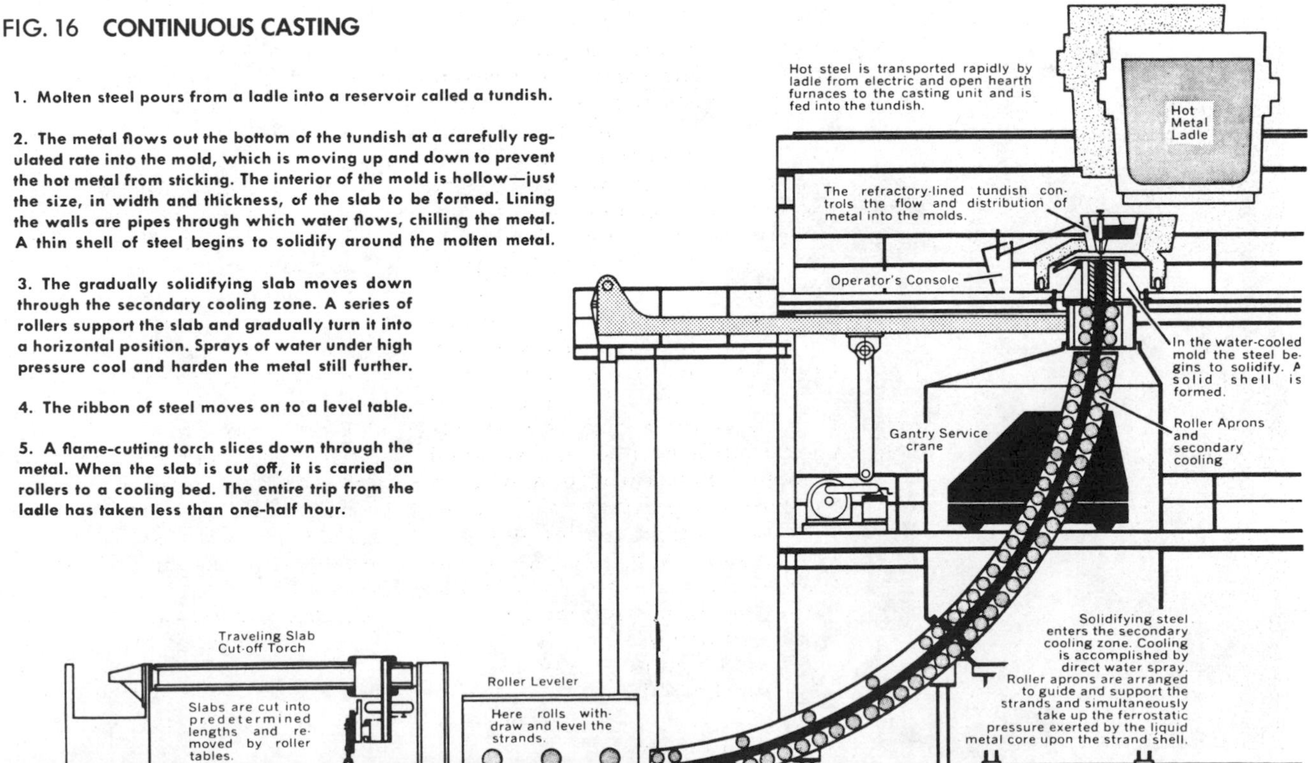
1. Molten steel pours from a ladle into a reservoir called a tundish.
2. The metal flows out the bottom of the tundish at a carefully regulated rate into the mold, which is moving up and down to prevent the hot metal from sticking. The interior of the mold is hollow—just the size, in width and thickness, of the slab to be formed. Lining the walls are pipes through which water flows, chilling the metal. A thin shell of steel begins to solidify around the molten metal.
3. The gradually solidifying slab moves down through the secondary cooling zone. A series of rollers support the slab and gradually turn it into a horizontal position. Sprays of water under high pressure cool and harden the metal still further.
4. The ribbon of steel moves on to a level table.
5. A flame-cutting torch slices down through the metal. When the slab is cut off, it is carried on rollers to a cooling bed. The entire trip from the ladle has taken less than one-half hour.
Hot steel is transported rapidly by ladle from electric and open hearth furnaces to the casting unit and is fed into the tundish.
Hot Metal Ladle
The refractory-lined tundish controls the flow and distribution of metal into the molds.
Operator's Console
In the water-cooled mold the steel begins to solidify. A solid shell is formed.
Gantry Service crane
Roller Aprons and secondary cooling
Solidifying steel enters the secondary cooling zone. Cooling is accomplished by direct water spray. Roller aprons are arranged to guide and support the strands and simultaneously take up the ferrostatic pressure exerted by the liquid metal core upon the strand shell.
Traveling Slab Cut-off Torch
Slabs are cut into predetermined lengths and removed by roller tables.
Roller Leveler
Here rolls withdraw and level the strands.
Slab Run Out Table

is typically teemed from the ladle into case iron ingot molds, which may be more than 8′ high and 3′ in diameter. The molds are coated on the inside with a releasing agent which prevents splashed metal from marring the ingot's surface quality. Often stripped ingots are moved towards furnaces called "soaking pits," while the outside is solid but the interior is still molten.

Strand Casting This process produces a continuous ribbon of steel rather than separate molds. It can result in semifinished solid products thus bypassing many of the steps necessary to produce ingots. At some point the continuous ribbon of steel is cut into desired lengths (Fig. 16).

The properties of various irons and steels are closely related to their chemical compositions and grain structure, that is, the size and arrangement of the crystalline microscopic particles making up the metal. In the initial crude refining processes, the most important factors influencing grain structure are the carbon and its chemical reaction with iron as it cools. Further refining processes, alloying elements, heat treatments and hot and cold working operations also affect grain structure. In ordinary steels, carbon primarily, as well as the other residual elements, are important. In higher grades of steel, alloying elements contribute special properties to the metal.

The essential physical property that distinguishes steel from iron is its *malleability* as it comes from the furnace and is initially cast. Other unique properties of steel are its ability to resist high stresses by deforming without breaking (toughness), and hardenability by hot and cold working. The important physical and mechanical properties of the most common iron and and steel types used in construction are shown in Figure 17. The effects of the several residual elements originating in the raw materials and of the added alloying elements are outlined in Figures 18 and 20.

CAST IRON

In casting pig iron, only slight variations in properties occur as a result of the reaction between carbon and iron, and in the resulting grain structure of the metal. Most ordinary white and gray cast irons, therefore, like pig iron, are hard and brittle, have relatively high

FIG. 17 **SELECTED PHYSICAL AND MECHANICAL PROPERTIES OF REPRESENTATIVE IRON AND STEEL TYPES**

	Metal	Specific Gravity	Coefficient of Thermal Expansion[1] (in/in/°F)	Modulus of Elasticity (psi)	Yield Point[2,3,5,6] (ksi)	Ultimate Strength[3,4,5,6] (ksi)	Ductility (Elongation in 2")[4,5]
IRONS	Gray Cast Iron	7.2	5.9×10^{-6}	15×10^6	22	30	(not ductile)
	Malleable Cast Iron	7.4	6.6×10^{-6}	25×10^6	30	54	18%
CARBON STEELS	ASTM A373	7.85	6.5×10^{-6}	29×10^6	32	58 to 75	24%
	A283	7.85	6.5×10^{-6}	29×10^6	24 to 33	45 to 60	23 to 30%
	Merchant Quality A663 Bars	7.85	6.5×10^{-6}	29×10^6	22.5 to 40	45 to 90	33 to 17%
	A675 Bars	7.85	6.5×10^{-6}	29×10^6	22.5 to 55	45 to 90	33 to 14%
	ASTM A36	7.85	6.5×10^{-6}	29×10^6	36	58 to 80	23%
HIGH STRENGTH STEELS	A572 Columbium-Vanadium	7.85	6.5×10^{-6}	29×10^6	42 to 65	60 to 80	17 to 24%
	A588	7.85	6.5×10^{-6}	29×10^6	42 to 50	63 to 70	21%
	ASTM A441	7.85	6.5×10^{-6}	29×10^6	40 to 50	60 to 70	21%
	ASTM A242	7.85	6.5×10^{-6}	29×10^6	42 to 50	63 to 70	21%
STAINLESS STEELS	AISI 410 (Martensitic)	7.75	5.5×10^{-6}	29×10^6	45	70	25%
	AISI 430 (Ferritic)	7.75	5.8×10^{-6}	29×10^6	50	75	25%
	AISI 302 (Austenitic)	8.0	9.6×10^{-6}	28×10^6	40	90	50%

[1] Within temperature range of 32° to 212°F. for stainless steels.
[2] Yield strength or yield point is measure of usable strength (i.e., maximum stress which produces deformation without increase in load).
[3] These mechanical properties of ASTM steels are specified minimums which vary according to shape and thickness of the material.
[4] Ultimate strength is measure of total reserve strength (i.e., maximum stress which produces rupture in a test specimen).
[5] Values given for stainless steels are for annealed sheet and strip only. Higher strength values are common for products and alloys hardened by cold working or heat treating.
[6] Kips per sq. in. (ksi) equals 1,000 lbs. per sq. in.

FIG. 18 **EFFECTS OF ALLOYING ELEMENTS**

Element	Common Content	Effects
CARBON	up to 0.90%	Increases hardness, tensile strength and responsiveness to heat treatment with corresponding increases in strength and hardness.
	over 0.90%	Increases hardness and brittleness; over 1.2%, causes loss of malleability.
MANGANESE	0.50% to 2.0%	Imparts strength and responsiveness to heat treatment; promotes hardness, uniformity of internal grain structure.
SILICON	up to 2.50%	Same general effects as manganese.
SULFUR	up to 0.050%	Maintained below this content to retain malleability at high temperatures, which is reduced with increased content.
	0.05% to 3.0%	Improves machinability.
PHOSPHORUS	up to 0.05%	Increases strength and corrosion resistance, but is maintained below this content to retain malleability and weldability at room temperature.
ALUMINUM	Variable	Promotes small grain size and uniformity of internal grain structure in the as-cast metal or during heat treatment.
COPPER	up to 0.25%	Increases strength and corrosion resistance.
LEAD	0.15% to 0.35%	Improves machinability without detrimental effect on mechanical properties.
CHROMIUM	0.50% to 1.50%	In alloy steels, increases responsiveness to heat treatment and hardenability.
	4.0% to 12%	In heat resisting steels, causes retention of mechanical properties at high temperatures.
	over 12%	Increases corrosion resistance and hardness.
NICKEL	1.0% to 4.0%	In alloy steels, increases strength, toughness and impact resistance.
	up to 27.0%	In stainless steels, improves performance at elevated temperatures and prevents work-hardening.
MOLYBDENUM	0.10% to 0.40%	In alloy steels, increases toughness and hardenability.
	up to 4.0%	In stainless steels, increases corrosion resistance and strength retention at high temperatures.
TUNGSTEN	17% to 20%	In tool steels, promotes hardness at high cutting temperatures; in stainless steels, smaller amounts assure strength retention at high temperatures.
VANADIUM	0.15% to 0.20%	Promotes small grain size and uniformity of internal grain structure in the as-cast metal or during heat treatment; improves resistance to thermal fatigue and shock.
TELLURIUM	Up to .05%	Improves machinability when added to leaded steels.
TITANIUM	Variable	Prevents loss of effective chromium through carbide precipitation in "18-8" stainless steels.
COBALT	17.0% to 36.0%	Increases magnetic properties of alloy steels. In smaller amounts, promotes strength at high temperatures in heat resisting steels.

compressive and low tensile strengths. However, with the proper combination of selected pig irons, scrap metals and alloying elements, cast iron can be made with increased hardness, toughness and corrosion and wear resistance. Casting also permits the formation of complex and massive shapes not readily produced by machining or rolling.

Plumbers' drainage, vent and waste pipe, ornamental railings and lamp posts are familiar cast iron products. In sanitary ware involving enameling and baking at elevated temperatures, cast iron's resistance to warping and cracking at high heat is an important advantage.

Malleable cast iron has improved breakage resistance as a result of increased toughness and ductility, and is used extensively for small articles of builders' hardware. Special *alloy* cast irons (alloying elements 1.0% to 5.0%) and *high strength* cast irons containing molybdenum, chromium and nickel are also produced for special uses. *Nodular* cast iron, produced in this country by the addition of magnesium, has greater strength, ductility, toughness, machinability and corrosion resistance.

CARBON STEELS

Carbon steels are those in which the residual elements (such as carbon, manganese, phosphorus, sulfur, silicon) are controlled, but generally no alloying elements are added to achieve special properties (Fig. 18). They contain up to 1.20% carbon, and other elements are controlled within specified limits or ranges. Over 90% of the steels manufactured into finished products are carbon steels.

Classification and Uses

In the past, steels have been classified according to carbon content as follows: (1) *soft steel*—0.20% maximum; (2) *mild steel*—0.15% to 0.25%; (3) *medium steel*—0.25% to 0.45%; (4) *hard steel*—0.45% to 0.85%; (5) *spring steel*—0.85% to 1.15%.

These broad classifications have become less useful as the need for precisely compounded steel types has increased. The following AISI and ASTM designations are more commonly used in manufacturing and construction.

AISI Designations The American Iron and Steel Institute (AISI) designation system assigns code numbers which indicate the chemical composition and the steel making process of various types of steel. For carbon steels, each four-digit code number specifies the permissible maximums or ranges for the residual elements—carbon, manganese, phosphorus, sulfur and sometimes silicon.

There are certain types of alloy steels which are designated by five numerals.

The prefix "E" designates steels made by the basic electric furnace process with special practices; the absence of a prefix indicates open hearth steels. The prefix "M" designates a series of merchant quality steels. The suffix "H" designates standard hardenability steels.

The first two digits indicate the type of steel (carbon or alloy) (Fig. 19). The last two digits of the four numeral series and the last three digits of the five-numeral series are intended to indicate the approximate mean of the carbon range. For example, in the Grade 1035, 35 represents a carbon range of 0.32% to 0.38% and in Grade E52100, 100 represents a carbon range of 0.98% to 1.10%. It is necessary, however, to deviate from this system and to interpolate numbers in the case of some carbon ranges, and for variations in manganese, sulfur or other elements with the same carbon range.

ASTM Designations Structural steels are commonly specified according to American Society for Testing and Materials (ASTM) numbers. The carbon steels typically used in building construction are ASTM A283, A663, A675, A373, and A36. These arbitrary number designations indicate certain partial chemical compositions and specified minimums for strength and ductility. ASTM specifications regulate only specific chemical elements which directly affect the fabrication and erection of steel, making it possible to produce proprietary steels of different chemical compositions conforming with the required ASTM performance standards.

Common Uses Carbon steels are used for most products in the building industry—from structural shapes, concrete reinforcing bars, sheets, plates and pipes to the smallest items of builders' hardware. Carbon steels generally possess adequate strength, hard-

FIG. 19. **AISI DESIGNATIONS**

Series Designation	ALLOY STEELS Type and Approximate Percentages of Identifying Elements
13xx	Mangenese 1.75
40xx	Molybdenum 0.20 or 0.25; or Molybdenum 0.25 and Sulfur 0.042
41xx	Chromium 0.50, 0.80 or 0.95, Molybdenum 0.12, 0.20 or 0.30
43xx	Nickel 1.83, Chromium 0.50 or 0.80, Molybdenum 0.25
46xx	Nickel 0.85 or 1.83, Molybdenum 0.20 or 0.25
47xx	Nickel 1.05, Chromium 0.45, Molybdenum 0.20 or 0.35
48xx	Nickel 3.50, Molybdenum 0.25
51xx	Chromium 0.80, 0.88, 0.93, 0.95 or 1.00
51xxx	Chromium 1.03
52xxx	Chromium 1.45
61xx	Chromium 0.60 or 0.95, Vanadium 0.13 or min. 0.15
86xx	Nickel 0.55, Chromium 0.50, Molybdenum 0.20
87xx	Nickel 0.55, Chromium 0.50, Molybdenum 0.25
88xx	Nickel 0.55, Chromium 0.50, Molybdenum 0.35
92xx	Silicon 2.00; or Silicon 1.40 and Chromium 0.70
50Bxx	Chromium 0.28 or 0.50
51Bxx	Chromium 0.80
81Bxx	Nickel 0.30, Chromium 0.45, Molybdenum 0.12
94Bxx	Nickel 0.45, Chromium 0.40, Molybdenum 0.12

B denotes Boron Steel.

Series Designation	CARBON STEELS Type and Approximate Percentages of Identifying Elements
10xx	Nonresulfurized, Manganese 1.00 per cent maximum
11xx	Resulfurized
12xx	Rephosphorized and Resulfurized
15xx	Nonresulfurized, Manganese maximum over 1.00 per cent

ness, stiffness, malleability and weldability, but poor resistance to corrosion (Fig. 18).

Occasionally, some copper and other elements may be added to improve atmospheric corrosion resistance, and by varying the chemical compositions, heat treatments and subsequent hot and cold working operations, other desirable properties can be enhanced.

Unified Numbering System Alloy and carbon steels have been assigned a designation in the Unified Numbering System (UNS) for Metals and Alloys; they were established in 1975 by the American Society for Testing and Materials (ASTM E527) and the Society of Automotive Engineers (SAE J1086). The UNS number consists of a single letter prefix followed by five digits. The letter "G" indicates standard alloy or carbon steels and "H" indicates standard hardenability steels. The first four digits usually correspond to standard AISI, ASTM or SAE steel designations and the last digit usually relates to an additional chemical requirement such as lead or boron. The last digit 6 is used to designate steels which are made by the basic electric furnace process with special practices (Fig. 20).

ALLOY STEELS

Alloy Steel is so classified when the content of alloying elements exceeds certain limits or in which a definite range of alloying elements is specified. The term "alloy steel" does not cover certain classifications such as stainless steel, tool steel, specialty steel, and others, although those steels do indeed contain specified amounts of alloying elements.

Classification and Uses

Alloy steels are generally grouped by the element or combination of elements used to obtain desired properties. *Manganese-molybdenum* steels, for example, are known for hardenability and resistance to fatigue; *silicon-manganese* steels for high relative strength and shock resistance; *nickel* steels for toughness and impact and corrosion resistance.

AISI Designations The AISI has developed a numerical designation system for identifying alloy steels. These steels are designated by a four-digit system, from 1200 to 9800, according to chemical composition (Fig. 19). Specific allowable ranges for carbon, manganese, silicon, nickel, chromium and molybdenum; limits for sulfur and phosphorus also generally are specified. The prefix "E" is used to designate *electric* furnace steel; the absence of a prefix indicates open hearth steels. The suffix "H" is sometimes used to identify steels of controlled *hardenability*, the letter "B" or "BV" between the second and third digits indicates *boron* or *boron-vanadium* alloys respectively.

The steels are grouped in series according to the most important alloying element(s), identified by the first two digits; the last two digits indicate the approximate carbon content. For example, AISI type 4023 steel indicates a molybdenum (40xx series) alloy steel with an average carbon content of 0.23% (xx23), made in the open hearth furnace.

Common Uses The alloy steels have properties essentially similar to or exceeding those of the carbon steels. In general, particular properties are improved by certain elements. For example, high silicon content gives steel excellent magnetic permeability and low core loss, making it useful in motors, generators and transformers; cobalt improves magnetic properties for such uses as permanent magnets in electrical apparatus and sealing gaskets in modern refrigerators; nickel imparts great toughness, useful in rock drilling and air hammer equipment.

Because of their higher cost, alloy steels are not generally used in construction materials. However, a particular group known as *heat-treated constructional alloy steels* have been used successfully in large structures where unusually high loads or temperatures are encountered, and are rolled in a limited variety of structural shapes. They are capable of developing very high strengths (ultimate strength up to 135,000 psi, yield point up to 100,000 psi), and show promise of increased structural use.

Alloy steels are used extensively in the manufacture of automobile and locomotive parts, excavating and heavy construction equipment, and various industrial uses. Parts of common household equipment such as vacuum cleaners, washing machines, lawn mowers and tools are sometimes made of alloy steels.

HIGH STRENGTH LOW ALLOY STEEL

High strength low alloy steels comprise a group with chemical compositions specially developed to impart better mechanical properties and greater resistance to atmospheric corrosion than are obtainable from conventional carbon structural steels.

Classification and Uses

The AISI classifies high strength low alloy steels according to chemical composition. The American Society of Testing and Materials publishes recommended practices and specifications.

ASTM Designations for Sheet Steel The most widely produced and ordered sheet steel products available are 1) *Conventional type* ASTM A607, 2) *Improved Atmospheric Corrosion Resistance,* ASTM type A606 and 3) *Improved Formability* Type ASTM A715. In general, minimum yield points vary within a range of 45 ksi to 80 ksi and minimum tensile strength from 55 ksi to 90 ksi.

Conventional Type is produced to various strength levels where atmospheric corrosion resistance and maximum formability are not a requirement.

Improved Atmospheric Corrosion Resistance Type is produced to 45 ksi or 50 ksi minimum yield points. The types of steel with enhanced corrosion resistance include type 2 with a corrosion resistance twice that of plain carbon steel and type 4 with a corrosion resistance at least four times.

Improved Formability Type is produced with special forming properties to various strength levels where maximum formability is required. This type is generally furnished for making parts too difficult for the fabricating properties of the other two types.

ASTM Designations for Steel Bars, the most frequently specified steel bars, have a minimum yield point ranging from 42 ksi to 80 ksi and a minimum tensile strength from 60 ksi to 95 ksi.

Common Uses The high strength low alloy steels are commonly used where high strength in relation to weight is important. The ASTM steels mentioned above are often used in bridge and high rise building construction. These and other high strength steels also are used extensively in various transportation vehicles, heavy construction equipment and industrial containers, and wherever low maintenance from corrosion and impact damage is important.

Certain proprietary A242 steels can be made with increased corrosion resistance, when intended for exposed architectural use. These steels form natural rust-colored, self-healing oxide coatings which inhibit further corrosion when left unpainted. They are gaining increasing use in commercial and industrial construction, but at present are not used extensively in residential work.

STAINLESS AND HEAT RESISTING STEEL

The outstanding characteristics of the two groups of steels referred to as *stainless* and *heat resisting* are implied in their name. Stainless steels possess excellent corrosion resistance at varying temperature ranges. Heat resisting steels retain their essential physical and mechanical properties at elevated temperatures. Stainless and heat resisting steels account for about 1% of total steel production.

Chromium is the alloying element mainly responsible for corrosion-resisting and heat-resisting properties, although other elements such as nickel, manganese and molybdenum also contribute to these properties. The stainless qualities of steel are derived from a self-healing chromium oxide which forms a transparent skin and prevents further oxidation. Steels with less than 4% chromium are regarded as alloy steels. Heat resisting steels normally contain 4% to 12% chromium, and stainless steels over 12%. The chief physical and mechanical properties of representative stainless steels are presented in Fig. 20.

Classification and Uses

Stainless and heat resisting steels are classified according to chromium content and internal grain structure into three major groups: *martensitic, ferritic* and *austenitic* precipitation hardening (Fig. 20).

Martensitic Steels The heat resisting and stainless steels with chromium content between 4% and 18% are known as martensitic. They may be hardened by heat treatment and are therefore sometimes referred to as the *hardenable chromium alloys.* They also have adequate cold forming characteristics and are satisfactory for hot working or forging, but must be slowly cooled or annealed after forging to prevent cracking. They are easily welded but require subsequent annealing or tempering to restore ductility.

Ferritic Steels These are stainless steels which contain from 12% to 27% chromium. They do not respond readily to heat treatment and are referred to as the *non-hardenable chromium* alloys; however, hardness may be increased somewhat by cold working. Ductility may be improved by hot or cold working followed by annealing. These steels are not welded as easily as the martensitic steels, but have greater corrosion resistance. Fair ductility after annealing allows product fabrication by a variety of processes such as cold forming, spinning and light drawing. Ferritic steels possess low strength at elevated temperatures and a low coefficient of thermal expansion.

Austenitic Steels The chromium content of the austenitic steels ranges between 16% and 26%; from 3.5% to 22% nickel is also added. These steels are characterized by a greater ability to harden as a result of cold working than the

FIG. 20 **AISI DESIGNATIONS AND CHARACTERISTICS OF STAINLESS AND HEAT RESISTING STEELS**

Grain Structure	Chief Alloying Elements	AISI Series	Representative Type	UNS Designation	Characteristics	Steel Type
MARTENSITIC (ferro-magnetic and hardenable by heat treatment)	CHROMIUM 4% to 12%	500	502	S50200	Retain their mechanical properties at high temperatures.	HEAT RESISTING STEELS
	CHROMIUM 12% to 18%	400	410	S41000	Moderate corrosion resistance, high strength and hardness.	STAINLESS STEELS
FERRITIC (ferro-magnetic and non-hardenable)	CHROMIUM 12% to 27%	400	430	S43000	Very good corrosion resistance, particularly at high temperatures.	
AUSTENITIC (non-magnetic and hardenable by cold working)	CHROMIUM 16% to 26% NICKEL 6% to 22%	300	302	S30200	Excellent corrosion resistance, high strength and ductility. Suitable for many fabrication techniques.	
	CHROMIUM 16% to 19% NICKEL 3.5% to 6% MANGANESE 5.5% to 10%	200	202	S20200	Excellent corrosion resistance at high temperatures, high strength and toughness.	
Precipitation Hardening	CHROMIUM 12% to 18% NICKEL 3% to 8.5%	—	—	S15500	Combines high strength and hardness with excellent corrosion resistance.	

ferritic types. The austenitic steels include the nickel-chromium and nickel-chromium-manganese groups of alloys. Both groups provide superior resistance to corrosion and are suited to many fabrication techniques. They have high ductility (important for deep drawing and forging operations), are easily welded, and have good oxidation resistance at elevated temperatures. A relatively high coefficient of thermal expansion is characteristic of these steels and must be considered in the design of high temperature equipment. They are also more susceptible to corrosive attack by deoxidizing sulfur gases than other stainless steels.

Precipitation Hardening Steels These are iron-chromium-nickel alloys with additional elements, which are hardenable by solution treating and aging. These steels combine high strength, hardness and corrosion resistance.

AISI Designations The heat resisting and stainless steels are grouped by the AISI into five series bearing number designations from 200 to 500 as shown in Fig. 20.

Unified Numbering System (UNS) The ASTM and the Society of Automotive Engineers developed the six digit UNS as a method of uniquely designating for each steel type. Stainless steels are identified with an "S" followed by five digits (Fig. 20).

Common Uses There are many applications for stainless steels and a wide variety of alloys from which to choose.

Martensitic Chromium Steels are suitable for applications requiring high strength, hardness and resistance to abrasion, such as steam and gas turbine parts, bearings and cutlery.

Type 410 (basic type) is one of the most widely used alloys in this category.

Ferritic Chromium .Steels are used mostly for automotive trim, applications involving nitric acid, high temperature service requiring resistance to scaling, and uses which call for low thermal expansion.

Type 405 and 430 steels are the most economical and widely used alloys in this category. Type 430 is sometimes used in building construction for gutters and downspouts, architectural trim, sills and column covers. It is also used for kitchen applications such as range hoods, range tops and appliance fronts.

Austenitic Nickel-Chromium Steels are the most numerous and popular category of stainless steels. They are sometimes referred to as the "18-8" alloys because several of the most common types (302, 303 and 304) contain approximately 18% chromium and 8% nickel. Among these, types 302 (the basic "18-8" grade) and 304 are used most often for such building applications as facias, curtain walls, store fronts, doors and windows, column covers and railings. Their relative chemical inertness and finish retention qualities also make them useful in food preparation and surgical equipment, hospital clean rooms, laboratories and for such familiar kitchen applications as sinks, countertops and appliances.

In addition to a variety of automotive and transportation equipment uses, type 301 steel is sometimes used for structural members because it work-hardens rapidly to high strength and toughness. It has good cold forming properties and is used for making roof drainage products.

Type 316 is the most resistant to attack by salt spray and corrosive industrial fumes, and is therefore used in seacoast and industrial areas where protection from such exposures is essential.

Austenitic Nickel-Chromium-Manganese Steels types 201 and 202 are similar to the nickel-chromium counterparts (types 301 and 302) and may be used interchangeably with them. In addition to automotive and food storage equipment, they are used for flatware, cooking utensils, kitchen sinks and counter tops.

Steel products are manufactured by casting in foundries, and by forging, extruding, rolling and cold drawing in mills. The products of the foundry are referred to as *cast* products, all others are *wrought* products. Wrought products account for over 90% of the total steel production, and by far the largest part of that is produced by hot rolling and cold finishing.

Steel mill products are regarded as "semifinished" when in the form of blooms, billets and slabs; and "finished" when in a form suitable for further fabrication into components or use directly in construction. Figure 21 illustrates the variety of finished mill products produced at steel mills.

CAST STEEL PRODUCTS

Castings are generally made by pouring molten steel into sand molds. Prior to casting, pig iron may be further refined and melted with scrap metal and ferroalloys at the foundry to obtain desired casting and mechanical properties.

Casting is used in preference to other forming operations when desired shapes are of a size or complexity not readily obtained by rolling or machining. Complex and massive shapes possessing great strength and impact resistance at high and low temperatures usually are cast. Iron and steel castings contribute a total of over 15 million tons to a broad range of uses in a typical year (Fig. 22). Of this, 13-1/2 million tons are in gray and malleable cast iron, 1-1/2 million tons in steel. Of the cast steel products, approximately 2/3 are carbon steels, the rest alloy and special steels.

The transportation industry relies on castings for many engine and car body parts, as does the excavating and mining equipment industry. Castings are also used extensively in many parts of steel making furnaces and rolling mills.

FIG. 22 Steel castings may range in weight from a few ounces to several hundred tons, as in this bridge saddle.

WROUGHT STEEL PRODUCTS

A small percentage of steel products, complex in shape and of irregular cross section, is manufactured by *forging*. Linear products of a constant cross section—such as bars, tubes, rods—and more complex shapes may be manufactured by *extruding*. Less complex and larger linear forms, as well as sheets, strips, rods, bars and most structural shapes used in construction are generally produced by *hot rolling* and *cold finishing*. The volume of production for various products is illustrated in Figure 23.

FIG. 21 Operations in the manufacturing sequence for steel products.

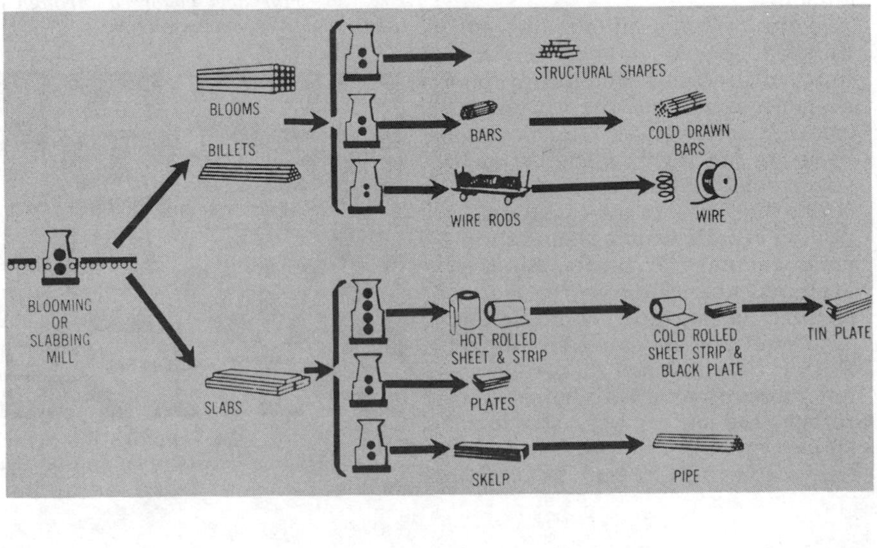

FIG. 23 ANNUAL PRODUCTION BY PRODUCT[1]

Product	Net Tons	%
Sheets & Strip[2]	43,609,360	44.5%
Bars & Tool Steel[3]	16,915,044	17.3%
Total Shapes & Plates	13,628,038	13.9%
Pipe & Tubing[3]	8,398,656	8.6%
Tin Mill Products[4]	6,100,148	6.2%
Wire & Wire Products	2,509,961	2.6%
Semifinished Shapes[5]	5,070,215	5.2%
Rails	1,703,284	1.7%
Total Steel Products	97,934,706	100.0%

[1]Figures from 1978.
[2]Includes hot rolled, cold rolled and galvanized products.
[3]Includes extruded products.
[4]Blackplate, terneplate and tinplate.
[5]Includes forgings.

Forging

Forging is a method of forming hot metal into desired shapes by pressing between heat-resistant dies. When metal is forged, toughness, strength and ductility increase significantly along the lines of flow (Fig. 24), an important advantage of this forming technique. The metal to be forged is preheated to between 2200° and 2400° F in special furnaces, and in some operations the dies are heated as well. The two major types of forging are open-die and closed-die.

Open-die In this type of forging, a large press squeezes (rather than strikes) steel between two flat surfaces. Temperature must be carefully controlled which requires frequent reheating of the steel between shaping actions. Some hydraulic open-die forging presses may accept ingots weighing several hundred tons (Fig. 25).

Closed-die In this type of forging, a hammer pounds a section of steel between "carved" dies until it reaches the desired shape. Closed-die forging most often uses *steam hammers*. *Drop hammers* develop forging force by the fall of a heavy weight or ram. In *double-action hammers* the speed of the falling weight is usually given additional impetus.

Extruding

Semifinished shapes may be converted into lengths of uniform cross section by extruding. In this process an advancing ram (Fig. 26) forces preheated, plastic metal through a tough heat-resistant die of the desired profile (Fig. 27). Presses vary in size from a small 700 ton, 2-1/2" maximum circle die press designed for fast extrusion speeds up to 25" per second, to 12,000 ton installations capable of extruding sections up to 21" in diameter at slower speeds. Stainless steel is generally extruded in shapes up to 6-1/2" in diameter.

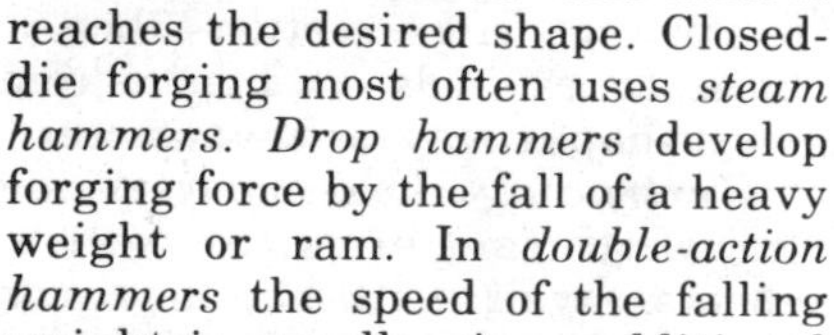

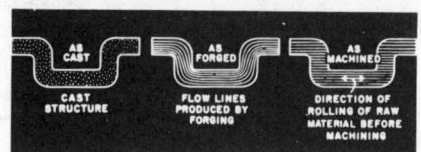

FIG. 24 Continuity of flow lines in forged metal structure results in greater strength and toughness.

FIG. 25 Hydraulic press may exert a force up to 18,000 tons and is particularly suitable for forging high strength steels.

Extruding can produce more complex sections with better surface characteristics than can be produced by rolling. It can be used to shape certain alloys without undesirable residual effects (such as excessive hardness) and is more economical than other forming methods for small quantities. However, most steel extrusions are limited to 35′ in length and 21″ in diameter.

Hot Rolling and Cold Finishing

By far the largest proportion of wrought steel products is manufactured by *hot rolling* and *cold finishing*.

Hot rolling is used to make semifinished shapes as well as some finished products. Hot rolling may be done as part of the continuous casting process on a strand caster or by using hot ingots made by traditional steel making processes.

The hot steel passes through a system of rolls which gradually imparts rectangular bloom, billet or slab shapes. These are cooled, cleaned of surface irregularities, inspected and reheated for further rolling. Semifinished shapes may be hot rolled directly into finished hot rolled products such as structural shapes, plates, sheets, strip and bars; or hot rolling may be used as an intermediate step prior to cold finishing, as in the manufacture of bar, sheet, strip and wire. The hot rolling of *finished* products is sometimes referred to as *hot finishing*.

Hot rolled products intended for cold finishing are descaled (cleaned of surface oxide scale) usually by *pickling*, which involves passing the steel through sulfuric or hydrochloric acid, followed by rinsing with hot and cold water, steam drying and usually oiling.

Cold Finishing consists of cold rolling, cold reduction and cold drawing of previously hot rolled, descaled shapes. Cold *rolling* involves passing metal at room temperature through sets of rolls to impart desired shape, finish, and mechanical properties. Cold *reduction* is a form of cold rolling which drastically reduces the thickness of a flat product (sheet, strip, blackplate, etc.) and improves the strength, surface finish and flatness. Cold reduction often hardens the metal excessively, so that it must be annealed to soften it and further *temper (skin)* rolled to obtain the proper strength and stiffness and a desired surface texture.

Cold drawing is used to make smaller or more complex bar shapes

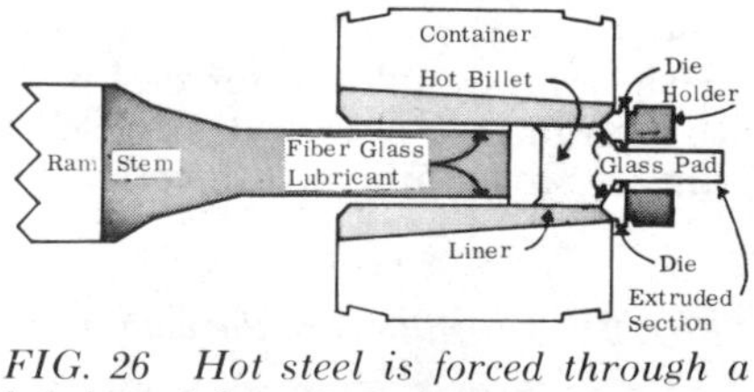

FIG. 26 Hot steel is forced through a lubricated flow extrusion press.

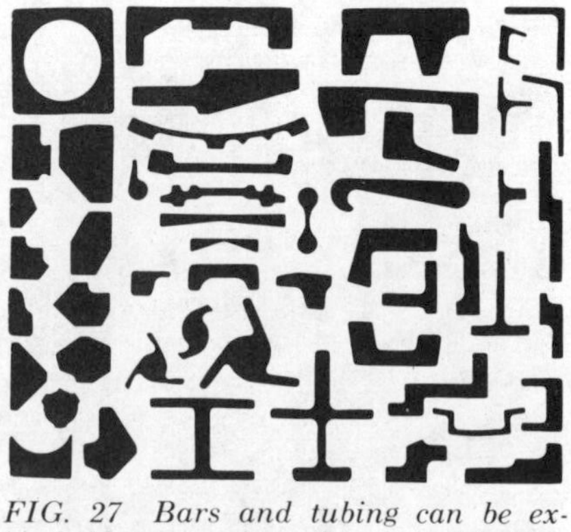

FIG. 27 Bars and tubing can be extruded into a variety of open and hollow shapes.

from hot rolled annealed bars, by pulling the bars through a hard, abrasion-resistant die. Wire is similarly cold drawn from wire rods. This operation results in improved machinability and strength, smoother finish and greater dimensional accuracy.

Slabs, Blooms and Billets These are the principal semifinished shapes from which other products are made. *Slabs* are flat rectangular shapes with a width more than twice the thickness, which is usually less than 10″. *Blooms* are rectangular shapes generally larger than 36 sq. in. *Billets* are less than 36 sq. in., more nearly square in cross section, and longer.

The type of semifinished shape rolled depends on the intended final use. Slabs are commonly used in the rolling of flat products such as plates, sheets, strip, and skelp; blooms in the rolling of rails, tube rounds and structural shapes; billets for making bars, tube rounds and wire rods (Fig. 28). *Skelp, tube rounds* and *wire rods* are also regarded as semifinished shapes. Skelp is used to make welded pipe, tube rounds for piercing into seamless pipe, and wire rods for drawing into wire.

Sheets and Strip More than a third of all steel shipped annually is the product of continuous sheet mills. The definition of these and other flat rolled products is complex and varies depending on whether they are stainless or carbon steel. Generally, the dimensions of sheets and strip lie between those of bars and plates (Fig. 29). For carbon steels, the method of rolling (whether hot or cold) is also a factor.

Hot Rolled Sheets and Strip are manufactured from semifinished slabs in sheet and strip mills. Preheated slabs are squeezed and shaped by progressive hot rolling, until the final desired thickness and width are obtained. Sheets and strip emerge from the roller tables at speeds up to 3500′ per minute and are wound into coils for shipment, or for transfer to other departments for further working.

Modern cold reducing mills may receive hot rolled steel about as thick as a half dollar and three-quarters of a mile long; two minutes later, it will be the thickness of two playing cards and more than two miles long.

Cold Reduced Sheets and Strip are made by cold reducing previously hot rolled and pickled flat products. Many sheets and strip are cold rolled to meet consumer requirements for thickness, surface finish or mechanical properties (Fig. 29). Cold reduced products may also be annealed and temper rolled or zinc or tin coated.

Blackplate is the classification for cold reduced products thinner than .0141″ and wider than 12″ (Fig. 29).

Tinplate is the classification for tin coated, flat rolled products. Like blackplate, they are generally comparable to sheet in width but thinner. Tinplate may be further reduced before coating to produce a *double reduced* product with higher strength and stiffness than the single reduced product. Tinplate is used for food & beverage containers.

Plates Flat rolled finished products thicker than sheets and strip, produced from slabs or slab ingots by hot rolling, are called plates (Fig. 29). The two principal classes of plates are *sheared* and *universal*, depending on the type of mill on which they are rolled (Fig. 30). Sheared plate mills have only horizontal rolls and produce plates with uneven edges and ends which must later be flame cut or sheared straight to desired size. Universal plate mills have vertical rolls which are capable of producing smooth, straight edges so that only the ends must be trimmed. Lighter plates may also be rolled on some of the sheet and strip mills from slabs produced in blooming or slabbing mills.

Structural Shapes Structural members, such as I-beams, H-sections, angles, channels, tees, zees and piling, are hot rolled from blooms or billets. As many as 20 to 30 passes through grooved rolls may be required to produce the desired shape (Fig. 31). The first sets of rolls drastically reduce the cross

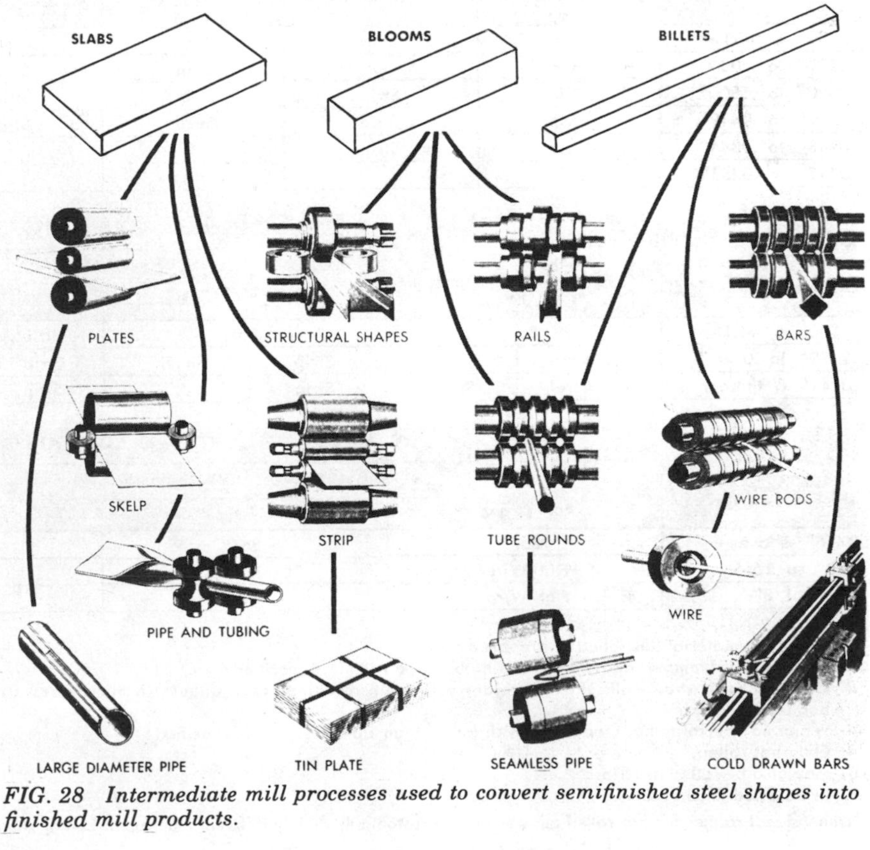

FIG. 28 Intermediate mill processes used to convert semifinished steel shapes into finished mill products.

section and impart the approximate shape; the finishing rolls gradually form the final shape and, for deep sections such as channels and angles, bend the legs into position. Wide flange shapes such as H-sections are rolled on universal mills which have vertical rolls capable of shaping the vertical surfaces.

Bars These products may be round, square, hexagonal and multifaceted long shapes, generally thicker than wire, and are produced by hot rolling and cold drawing. In hot rolling, approximately 10 to 15 roll passes taking less than 2 minutes are required to change a square hot billet to a bar. Bars over 3/4″ in diameter are usually sheared into lengths 16′ to 20′ long. Bars of smaller diameter are typically wound in coils for ease of handling and shipping. Carbon steel bars may be galvanized or treated with other corrosion-resistant coatings. Flat rectangular shapes (*flats*), generally narrower than sheets and strip, are also known as bars (Fig. 29) and may be either hot or cold rolled.

Hot rolled bars may be cold drawn to further reduce the cross section and produce complex shapes (Fig. 32) of closer dimensional tolerances and smoother finishes. Improved mechanical properties, such as greater tensile strength and hardness, also result from cold drawing. The degree of reduction is dependent on the type of metal used and the desired mechanical properties of the end product.

Wire Cold drawn products of round, square or multifaceted cross section, generally smaller than bars, are called wire. Round wire is drawn from 0.005″ up to 1″ in diameter. *Flat wire* is a cold rolled product, rectangular in shape and generally narrower than bars (Fig. 29).

Semifinished wire rods of low carbon steel may be drawn directly into wire by *single draft drawing*. For finer wire sizes or harder steels, the rods must be softened by annealing prior to continuous *multiple draft drawing* (Fig. 33) through a series of dies. Wire may also be annealed after drawing to make extremely soft *annealed wire*.

Cold drawn wire of carbon steel can be produced with extremely high tensile strengths ranging to over 500,000 psi. Corrosion resistance can be achieved by galvanizing or coating with other corrosion-resistant metals.

Tubular Products These are long, hollow metal products of round, oval, square, rectangular or

FIG. 29 GENERAL CLASSIFICATION OF FLAT-ROLLED STEEL PRODUCTS BY SIZE†

CARBON STEEL, HOT ROLLED

Thickness	Width					
	To 3-1/2″ incl.	Over 3-1/2″ to 6″ incl.	Over 6″ to 8″ incl.	Over 8″ to 12″ incl.	Over 12″ to 48″ incl.	Over 48″
.2300″ & thicker	Bar	Bar	Bar	Plate	Plate	Plate
.2299″ to .2031″	Bar	Bar	Strip	Strip	Sheet	Plate
.2030″ to .1800″	Strip	Strip	Strip	Strip	Sheet	Plate
.1799″ to .0449″	Strip	Strip	Strip	Strip	Sheet	Sheet
.0448″ to .0344″	Strip	Strip				
.0343″ to .0255″	Strip					

CARBON STEEL, COLD ROLLED

Thickness	Width						
	To 15/32″ incl.		Over 15/32″ to 12″ incl.		Over 12″ to 23-15/16″ incl.		Over 23-15/16″
.2500″ & thicker	Bar		Bar		Strip[2],	Sheet[4]	Sheet
.2499″ to .0142″	Flat Wire[1]	Strip	Strip	Sheet[3]	Strip[2],	Sheet[4]	Sheet
.0141″ & thinner	Flat Wire[1]	Strip	Strip		Strip,	Blackplate[4]	Blackplate

STAINLESS STEEL, HOT AND COLD ROLLED

Thickness	Width			
	1/4″ to 3/8″	3/8″ to 10″	10″ to 24″	24″ & Over
3/16″ & over	Bar[5]	Bar	Plate	Plate
1/8″ to 3/16″	Flat Wire[6]	Bar[6]	Strip	Sheet[7]
Up to 1/8″	Flat Wire	Strip	Strip	Sheet[7]

1. When the material has rolled or prepared edges.
2. When special temper, edge, finish or single strand rolling is specified.
3. Cut lengths or sheet coils slit from wider coils (with resulting No. 3 edges), in thicknesses .0142″ to .0821″ inclusive and widths 2″ to 12″, carbon 0.20% maximum.
4. When no special temper, edge or finish (other than Dull or Luster) is specified.
5. Only hot rolled.
6. May also be rolled as strip.
7. In polished finishes, Nos. 3, 4, 6, 7 and 8.

†Dimensional ranges for flat rolled alloy steels are customarily slightly different than for carbon or stainless steel.

FIG. 30 Hot plate is rolled back and forth in a reversing mill until desired thickness is achieved.

multifaceted cross section. Tubular products can be either pipe or tubing, the terms being applied somewhat interchangeably. Round pipe in a wide range of sizes and square or rectangular structural tubing are widely employed in building construction. Tubular products which are mechanically worked and produced on *welded* and *seamless* mills are referred to as *wrought pipe* to distinguish them from cast pipe. Wrought tubular products may be marketed in the hot or cold finished condition resulting from the mill operations as *black pipe,* or as *galvanized pipe* when hot dipped in zinc.

Welded Pipe is made chiefly by the butt weld and the electric weld processes.

In the *butt weld* process, a strip of *preheated*, semifinished skelp is continuously formed into a tubular shape between concave rolls, and the edges are joined by *mechanical pressure* during rolling (Fig. 34a). Pipe sizes 1/8″ to 4″ are commonly produced by this process.

The majority of steel pipe in this country is produced by *electric weld* processes. Preheating of the metal is not required, and localized heat necessary for joining the edges is generated during the welding process itself. Resistance welding is used to make *electric-resistance welded* (ERW) tubular products up to 20″ in diameter. In this process, a strip of *cold* metal is formed into tubular shape by rolling, and the edges are joined by heat and pressure (Fig. 34b). Welding heat is generated in the seam by the resistance of the metal to an electric current introduced by a wheel-shaped roller acting as the electrode.

For *larger diameter and thick-walled pipe,* one of several electric welding processes may be used. In most, steel strip or plates are die formed *cold* by several successive forming operations, consisting of crimping (bending the edges), U-forming, and O-forming. The last forming operations result in a nearly round shape with edges touching and ready for welding. Submerged arc and metal inert gas shielded arc (MIG) welding are often used for this purpose. The final shape and diameter is obtained by expanding the welded shell with hydraulic power against a retaining jacket.

Seamless Pipe and Tubing are made by hot piercing and hot extrusion. Piercing is accomplished by continuously feeding a hot billet between pairs of tapered rolls which spin and advance the billet over a center punch to make the pilot hole. A piercing point enlarges the hollow and a sizing mandrel shapes the final opening size and wall thickness (Fig. 35a). Most seamless products are made by hot working, although some seamless tubing is made by cold drawing a prepierced billet through a die over an internal mandrel (Fig. 35b). Stainless and alloy steel tubing up to 11″ in diameter are often made in this manner. The

FIG. 32 Irregular as well as round and rectangular bars may be easily cold drawn.

FIG. 33 Fine wire may be made by drawing rods through successively smaller die openings.

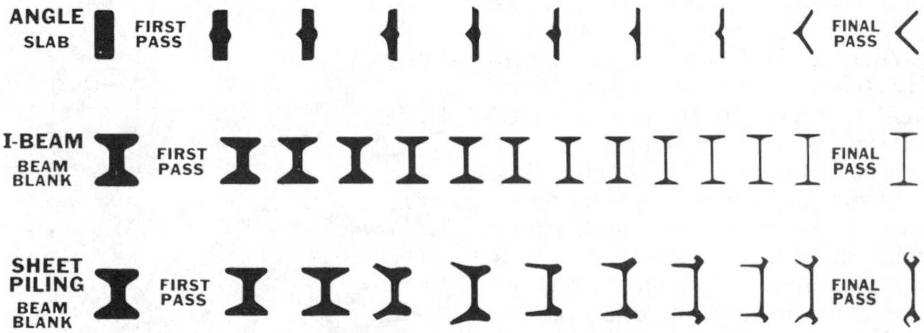

FIG. 31 Successive steps in rolling of structural shapes. Hot blanks are roughly shaped by one set of rolls, move onto other rolls which impart final shape.

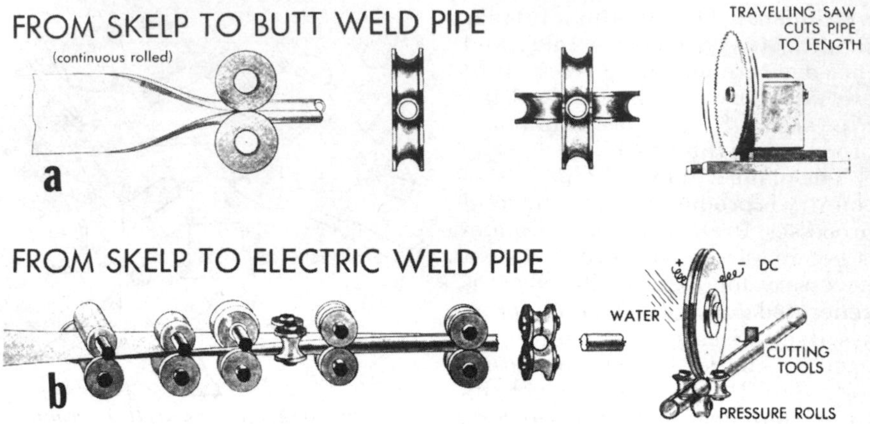

FIG. 34 Butt Weld Process (a) produces pipe up to 4" in diameter; Electric Weld Process (b) can produce pipe up to 8' in diameter.

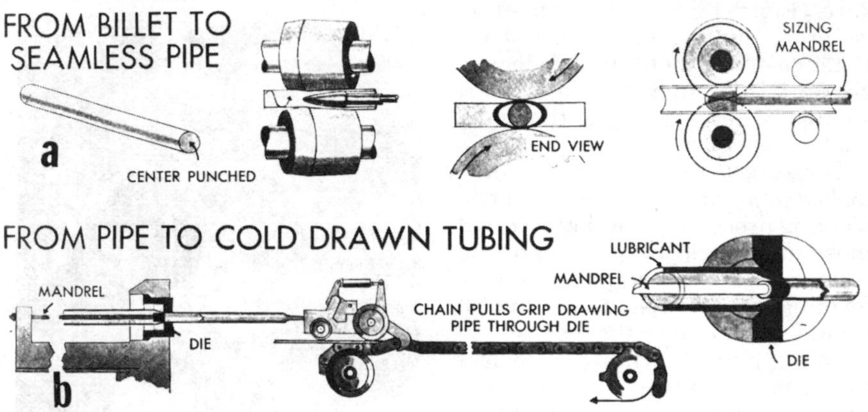

FIG. 35 (a) Most carbon steel, seamless pipe is made by hot piercing and extruding; (b) stainless and alloy steel tubing is cold drawn.

hot piercing methods can produce seamless pipe up to about 26" in diameter.

Hot extrusion produces a tubular shape by forcing hot predrilled billets through a die over an internal mandrel. This method is used primarily to make small sizes of stainless or high alloy pipe for cold reducing. It is also the most economical method of producing complex tubular shapes in small quantities.

FABRICATION

Most steels can be fabricated by hot forming, cold forming and machining. Some forming methods may be the same as those previously described; however, fabrication as used here implies reworking the products of the mills and foundries in a fabricating plant. Cold forming techniques are much more common in the manufacture of products used in construction.

Hot Forming

Hot forming operations are used on metals too thick or too hard to be formed cold. These operations include: (1) *forging*, described earlier; (2) *high temperature forming* (electro-forging), which uses heat generated by the resistance of the workpiece to an electric current passed through it by the forming roll; and (3) *high energy rate forming*, a method of hot (sometimes cold) shaping of materials by the sudden impact of a small explosive charge, a shockwave generated by an electric spark in a fluid, or sudden release of a compressed gas through a system of valves.

Cold Forming

Cold forming operations are performed on most flat-rolled and tubular products by shaping, bending or drawing the metal beyond its yield point so that a permanent set is achieved. The most common cold forming methods are *roller, stretch, shear* and *brake forming*, as well as *deep drawing*.

Roller Forming A variety of shapes (Fig. 36) are formed by passing sheets and strip through a series of roller dies which gradually impart the desired shape (Fig. 37). Many types of corrugated or ribbed decking and wall panels are made in this manner.

Stretch Forming With this process, light gauge flat products are shaped over form blocks into shallow curved shapes by pressure. The steps in stretch forming a particular shape are illustrated in Figure 38.

Shear Forming By this method, flat round blanks are formed into curved surfaces such as cylinders, domes and cones by turning the workpiece on a mandrel and shaping over a block with external pressure. *Spinning* is a type of shear forming in which the deformation and reduction in original blank thickness is not quite as severe (Fig. 39). Tank heads, television tube cones and stainless steel railing base flanges and caps are examples.

Brake Forming This is the most common method of fabricating

FIG. 36 Sheets and strip may be roller formed into a variety of shapes.

FIG. 37 Wall and roofing panels typically are roller formed from galvanized and stainless steel sheets.

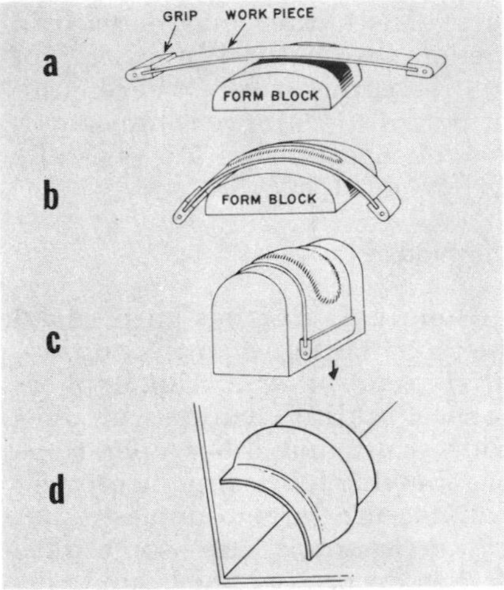

FIG. 38 In stretch forming, an oversized piece of sheetmetal is gripped (a) and bent over a form block (b) to produce desired contour (c), and then is trimmed to final shape (d).

sheetmetal components for construction. Sheets are shaped on a press brake by squeezing between a punch and die block in one or several consecutive operations (Fig. 40). Heating and air conditioning ducts and much of the architectural stainless steel storefront construction are fabricated by this method.

Deep Drawing Flat, thin-gauge blanks are often formed in a draw press between an operating punch and stationary die. Many household utensils, such as pots and pans, and commercial vessels and containers are made this way. Stainless steel can be drawn to various degrees, but the austenitic group lends itself most readily to deep drawing operations.

Machining

Most steels may be readily milled, sawed, drilled, punched sheared, tapped and reamed with appropriate equipment. Special tool steels—containing tungsten, molybdenum and chromium—that have been subjected to tempering and other heat treatments are used for this purpose. Sulfur and lead improve the machining properties of carbon steels and are added to steels requiring extensive machining. Among the stainless steels, the ferritic alloys are the easiest to machine.

PROTECTIVE COATINGS AND MECHANICAL FINISHES

Ordinary cast iron, most carbon steels and some alloy steels oxidize under ordinary atmospheric conditions. Elevated temperatures and moisture accelerate this oxidation known as atmospheric corrosion. The resulting surface scale (rust) deteriorates the metal surface and, if not prevented, progresses until the metal reverts to the oxide state found in nature. The chief function of protective coatings used on iron and carbon steel is to inhibit corrosion.

Not all ferrous metals require equal protection against progressive corrosion. Copper-bearing steels, some cast irons, and some high strength low alloy steels acquire excellent corrosion resistance from a tight surface oxide which inhibits further corrosion. Sometimes the appearance of the natural patina is objectionable and finish coatings are applied on these steels for decorative purposes. Decorative and protective coatings are used less often on stainless steels because of their superior corrosion resistance and attractive natural finishes.

Finishes for iron and steel may be classified as: (1) *mechanical*, including the as-rolled and as-drawn natural mill finish, and those imparted by further grinding, polishing or patterning; (2) *chemical*, generally consisting of cleaning or preparatory operations for further finishing; and (3) *organic* and *inorganic coatings* such as metallic, vitreous, laminated and painted finishes. Organic coatings commonly sprayed on or brush applied in the field, are treated in Section 216 Paints and Protective Coatings.

Mechanical Finishes

The *hot-rolled mill finish* (black, as-rolled finish) on carbon steel is characterized by a mill scale and rust powder which is sometimes ac-

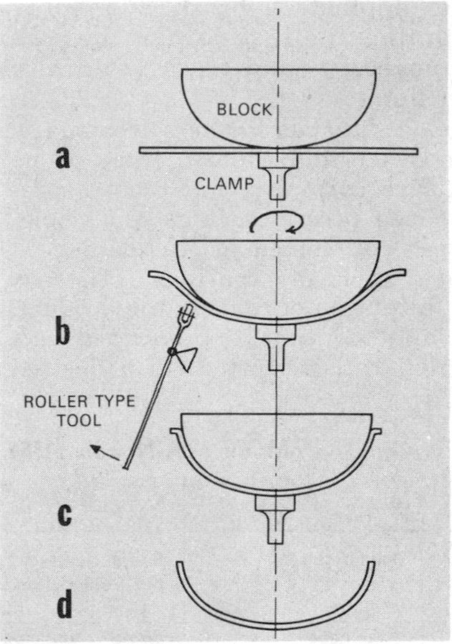

FIG. 39 In spinning, a round blank is clamped on a form block (a) and spun while pressure is applied (b), gradually imparting shape of the form block (c), then is trimmed to desired profile (d).

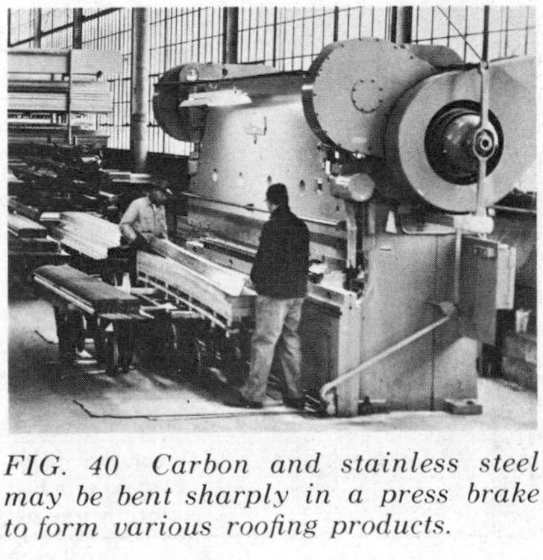

FIG. 40 Carbon and stainless steel may be bent sharply in a press brake to form various roofing products.

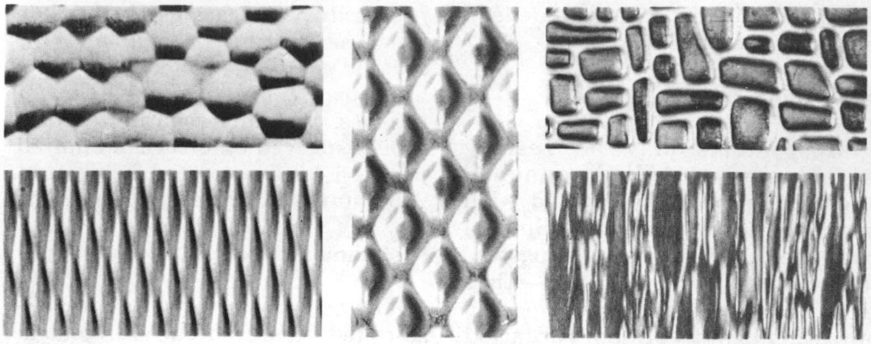

FIG. 41 Steel sheets may be stiffened and decorated with patterned finishes.

ceptable as a base for organic coatings applied by brush. For spray painting, the rust powder must be removed by sand blasting or wire brushing.

Cold finished surfaces are usually smoother and contain little or no surface corrosion, but must be degreased of oil coatings and sometimes roughened to provide a good base for organic coatings. If there is no intention of coating the product, it may be sent to a *temper mill* which rolls the steel to a desired flatness and surface quality. Cold reduced sheet and strip have a natural mill finish acceptable for many uses, and as a base for subsequent galvanizing, tinning or terne coating. *Black pipe* has a natural dark finish often left uncoated. Stainless steel sheets are produced in a variety of patterned finishes (Fig. 41) imparted by roll forming for greater stiffness, decorative effect and to reduce "oil-canning" effects (differential light reflection that emphasizes waviness of sheet surfaces in thin-gauge flat products).

Stainless Steel Sheets and Strip Because the stainless steels are inherently corrosion resistant and can be made strong and rigid in thin sheets, these products are the ones most often not protected by coatings. They are most often either *natural cold-rolled* finishes (such as Nos. 1, 2D and 2B sheet finishes), or one of the *polished* finishes (Nos. 4, 6, 7 and 8) outlined in Figure 42.

Stainless Bars, Pipe and Tubing These products are manufactured in several natural and polished finishes similar to those of sheets and strip described above. In industrial uses, they may be used in one of the natural finishes; for architectural work where uniformity of appearance is important, the polished finishes are generally preferred.

FIG. 42 **COMMON STAINLESS SHEET STEEL MECHANICAL FINISHES**

	Designation	Description	Uses
NATURAL FINISHES (unpolished)	#1 Finish	Rough dull surface resulting from hot rolling, annealing and descaling.	Generally available on heavy gauge sheets; not used in architectural applications.
	#2D Finish (#1 Strip Finish)	Smooth dull surface resulting from cold rolling, annealing and descaling.	Some architectural applications requiring low luster, as in roofing and drainage products; suitable for cold forming and further polishing.
	2B Finish (#2 Strip Finish)	Bright smooth cold-rolled surface produced either by highly polished rolls or bright annealing.	General purpose natural finish for tranportation equipment. Limited use in curtain walls and store fronts. Used for further polishing.
POLISHED FINISHES	#3 Finish	A surface produced by grinding and polishing with slightly finer abrasives than for finish #4.	General purpose natural finish for transportation equipment. Limited use in curtain walls and store fronts. Used for further polishing.
	#4 Finish	Bright polished surface produced by grinding and polishing either at the mill or fabricator plant.	Most common for architectural uses; directional grit line permits blending of fabricated or field joints with mill-polished surface.
	#6 Finish	Dull matt finish produced by brushing #4 finish.	Used in architectural applications where softer less reflective finish is desired.
	#7 Finish	A highly reflective finish produced by further polishing and buffing.	Chiefly used in architectural applications.
	#8 Finish	Mirror-like finish produced by polishing with buffing rouges until surface is free of grit lines.	Infrequent in architectural applications.

Chemical Finishes

Chemical finishes are used chiefly as cleaning and conditioning steps for other coating applications. Pickling, which removes oxide scale, and other operations for removing lubricants in product manufacture, are examples. *Conversion coatings* are sometimes used as decorative finishes on carbon and stainless steels.

Carbon Steel Conversion Coatings These are mostly chemical treatments (as with acid phosphate solution) which "convert" the chemical nature of the surface film to improve its bonding properties with paint or other applied coatings. In a recently developed decorative finish, textured steel is zinc plated and dyed using chromate conversion coatings, followed by buffing. Buffing causes the bright zinc to stand out in relief, highlighted by the color coating in the recesses.

Stainless Conversion Coatings *Surface blackening,* a common conversion coating on stainless steel, produces a decorative finish ranging from a bluish hue to dark brown or dead-black. It is produced by encouraging the growth of a thin oxide, either by controlled heat treatment or dipping in a hot chemical oxidizing bath.

Organic and Inorganic Coatings

Carbon steels are most often finished for protective and decorative purpose with organic and inorganic coatings. These may be: (1) *metallic*, applied by hot dipping, electroplating or other methods; (2) *vitreous*, fused-on glassy materials such as porcelain enamel; and (3) *laminated*, involving the adhesive application of inert plastic films.

Metallic Coatings These coatings provide protection against corrosive elements, either by: (1) acting as a "sacrificial" metal (one that is purposely permitted to corrode), as in the case of zinc coatings; or (2) by being relatively inert and corrosion-resistant (thus protecting the underlying base metal), as in coating with nickel or chromium. The metals most often used for protecting carbon steels are: zinc, tin, terne metal, aluminum, cadmium, chromium and nickel. These coating metals are most commonly applied by *hot dipping, electroplating, metalizing* and *cladding*. *Galvanizing*, coating with zinc, is usually done by hot dipping and electroplating. Aluminum coatings are generally applied by hot dipping. *Tinplating* is coating with tin, either by hot dipping or electroplating. *Bonderizing* on galvanized sheets and *galvannealing* are light metallic coatings with a suitable surface for paint coatings.

Hot Dipping is a process in which steel is immersed in a molten bath of the coating metal (Fig. 43). The most prevalent method is continuous hot-dip galvanizing which accounts for more than ten times as much as the alternative electrolytic process. The hot-dip process is used to produce heavier coatings and is the chief method of galvanizing tubular and flat rolled products. Hot dipped galvanized sheets and strip can be made with zinc coatings of designation G90 (.90 oz. per sq. ft.) which are used frequently in architectural applications. Heavier coatings of approximately 2.00 ounces per sq. ft. are made for severe corrosive applications. (Coating weights given are the combined weights on both surfaces.)

Hot dipped *aluminized* sheets are usually coated with an aluminum-

FIG_ 43 A continuous hot-dip galvanizing unit.

silicon (5% to 11%) coating known as type 1. Aluminum-silicon coatings provide excellent resistance to atmospheric and high temperature (to 1250° F) corrosion, and are extensively used in the automotive industry.

Electroplating employs an electric current and an electrolytic solution to deposit metallic coatings on steel or iron (Fig. 44). Tin, zinc and cadmium are commonly applied by this method, as well as nickel and chromium either directly or over copper. Electroplating is the chief method of making modern tinplate, and the second most important process for making galvanized products. Lighter coatings of 0.10 ounce per sq. ft. (for minimum corrosion resistance) and greater variations in coating thicknesses are possible by electroplating. The lighter coatings are normally recommended surfaces to be painted or for mildly corrosive exposures.

Metallizing consists of spraying molten metal against the surface to be coated, and is used extensively for applying zinc and aluminum coatings. The coating adheres to the base metal by a combination of mechanical interlocking and metallurgical bonding. Metallizing is the chief method of metallic coating practical for field use.

Cladding produces bimetallic "sandwich" products consisting usually of a carbon or low alloy steel core, covered with a thin sheet layer of the coating metal. Stain-

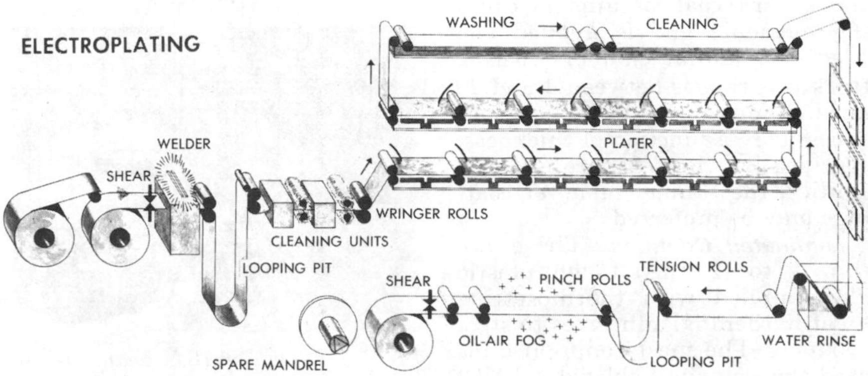

FIG. 44 Steps in the production of tinplate and galvanized products by the Electrolytic Process.

less steel and aluminum coatings are applied by hot rolling under controlled, non-oxidizing conditions. Copper may also be used for cladding steel plates by hot dipping or electroplating.

Zinc-based Paint Pretreatments are common on steel to improve the bond of subsequent paint finishes.

Bonderizing is the most common process for producing a suitable paint base on fabricated products. Zinc-coated products are dipped in hot phosphate solution, resulting in a thin (0.02 to 0.03 ounce per sq. ft.) crystalline surface film of zinc phosphate.

Vitreous Coatings Although there are several types of vitreous coatings used on metal, including glass-on-steel linings applied to hot water tanks and process piping, porcelain enamel is the type most common in architectural applications. Common uses are porcelain enamel on cast iron or steel in bath tubs, sinks, and other sanitary ware. Porcelain-enameled building panels are used in commercial and institutional buildings.

Porcelain Enamels have the same general inert abrasion- and corrosion-resistant properties as glass. They can be produced in a great variety of lightfast colors and textures. Porcelain enamel is applied in one or two coats to sheets that have been prepared by etching. In the case of the two coat system, the first enamel coat contains "adherence promoters," such as cobalt and/or nickel oxides, and is applied by spraying, dipping or flow coating, followed by firing at high temperatures. The second (and sometimes third) coat contains the coloring elements and is applied and fired in a similar manner. Coating thickness ranges between 4 and 20 mils, though durability is not necessarily a function of thickness; specifically, where flexure is a possibility, the thinner or fewer coatings may be preferred.

Laminated Coatings These are thin (2 to 12 mils) tough plastic films applied with thermosetting (heat-hardening) adhesives to steel products. The most common films used are polyvinyl-chloride (PVC) and polyvinyl-fluoride (PVF), both possessing good abrasion resistance and sufficient tenacity, toughness and flexibility to withstand severe forming operations. They have excellent resistance to weathering, good color retention and are resistant to attack by a variety of chemical agents. Although presently color selections are more limited than in porcelain enamels, the potential color variations are substantial.

JOINTS AND CONNECTIONS

Steel and iron shapes can be joined with: (1) *mechanical fastenings* such as bolts, screws and rivets; (2) *mechanically formed joints* using the workpiece itself to form the connection; (3) a variety of *welding* techniques, and (4) *adhesive bonding* methods.

The joints and connections used in the fabrication of iron and steel products are generally similar to those of other metals. The following discussion therefore will be limited to connections common in structural steel applications and installation of steel products for home construction.

Bolting, riveting and welding are often used in product manufacture but are most significant as joining methods in construction. One of the disadvantages of riveting in populated areas is the high noise level of riveting machines. For that reason, bolting and welding are finding increasing use in the erection of steel frameworks for buildings and bridges.

Mechanical Fastenings

Mechanical fastenings are available in a variety of *permanent* and *semi-permanent* types, and are used to connect metal to metal as well as to other building materials. The permanent types, such as rivets, involve permanent deformation so that destruction of the fastener or the component must take place for disassembly. The familiar and often used *semi-permanent* fastenings (screws, bolts and nails) allow disassembly without permanent damage to either the component or the fastener.

FIG. 45 A tub shell is heated in an oven prior to application of glass frit, which melts on the hot surface. Further baking produces the enamel coating.

Riveting This procedure involves the joining of components by means of a rivet with prefabricated head on one side, and field deformation of the shank to form a head on the other side. Structural rivets are generally deformed by a pneumatic or hydraulic riveter while red hot. As the rivet cools it shortens in length, drawing the components together. The resulting compressive and frictional forces between the contact surfaces are important in resisting tensile and shear stresses developed in structural joints.

Rivets are made from carbon, alloy and high-strength steels according to ASTM dimensional and chemical specifications. They are often used in shop assemblies of structural components, such as built-up columns, plate girders and beams, as well as in the field for making connections between beams, columns and girders. The number and spacing of rivets (usually in rows) are dictated by certain design criteria. Different types of heads are available (counter-sunk, flattened, etc.) to suit clearance and appearance requirements.

Bolting This is a semi-permanent mechanical method of fastening with nuts, bolts and washers. Specified tensile values must be developed in the bolts by tightening with calibrated torque wrenches or by turning the nut a specified number of times from a snug-tight position. Proper tightening is important to develop frictional resistance in the joint to counteract shearing forces. Keeping the threaded part of the bolt out of the structural parts being joined is also essential in developing the full strength of the connection.

Nuts and bolts are made in sizes ranging up to 4″ in diameter and with several types of heads (square, round, hexagonal, counter-sunk, etc.). Most bolts have square or hexagonal heads, as do the nuts, allowing the use of a wrench. Although bolts are available in both carbon and alloy steels, the use of high strength bolts for most major structural applications is becoming more prevalent (Fig. 46). Ordinary carbon steel bolts are used in securing many components of home construction, such as sill plates to foundations, beams to columns and to each other.

Stud Welding This is a method of semi-permanent fastening in which one end of the stud is arc welded by a gun-shaped unit to a steel component, and a nut is applied to the stud's threaded end (Fig. 47). It is used most often where nonstructural materials must be attached to steel, as in the application of architectural finishes to a structural framework. Stud welding is finding increasing use in composite structural design as a means of transferring stresses between structural steel and reinforced concrete to produce combined resistance to stresses. Studs are available also unthreaded or internally threaded, in a variety of shapes such as bent and rectangular eye-bolts and j-bolts.

Powder Actuated Fastenings These are permanent or semi-permanent fastenings which employ a gun powered by a small explosive charge to drive a pin into very hard and tough materials. In permanent fastenings, a specially hardened drive-pin is used to connect components in a fashion similar to nailing. Often, however, a pointed stud—threaded on one end—is driven into one of the components and assembly is completed with a nut. The availability of inexpensive guns and a variety of stud sizes and shapes makes this a useful fastening procedure in all types of construction (Fig. 48).

Threaded Fasteners and Nails The most familiar semi-permanent fastening devices consist of a variety of screws, bolts and threaded and grooved nails. They are manufactured from carbon, alloy and stainless steels in many shapes and sizes. Their uses include ordinary household applications, assemblies for product fabrication and joining of structural trim components in construction.

Expansion Bolts These are semi-permanent fasteners generally used to join materials to concrete and masonry surfaces. (They are also referred to as cinch anchors, expansion anchors and many other proprietary names.) The device consists of two or more units which develop holding power in concrete or masonry through wedge action and friction with the walls of a pilot hole by the expansion element. This action is produced by striking and/or turning the threaded machine bolt insert. Expansion bolts act in a manner similar to screws used with lead, fiber or plastic shields, but usually develop much greater withdrawal strengths (Fig. 49).

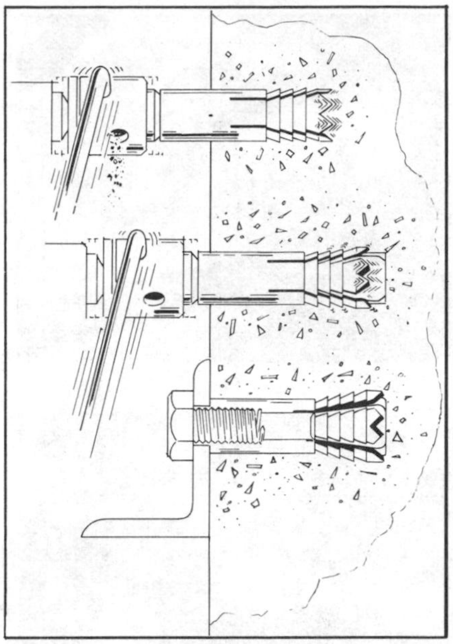

FIG. 49 Self-drilling expansion anchors speed joining of steel to concrete.

Mechanically Formed Joints Joints in thin gauge sheets are often made without the use of fasteners. Instead, a part of the material itself is used to make the connection. Two common methods are *staking* and *lockseam joining.*

Staking is a method of joining flat materials by foldover tabs inserted through matching slots in the elements to be joined. It is often used in product fabrication.

Lockseam Joining is a mechanical method of joining sheetmetals both

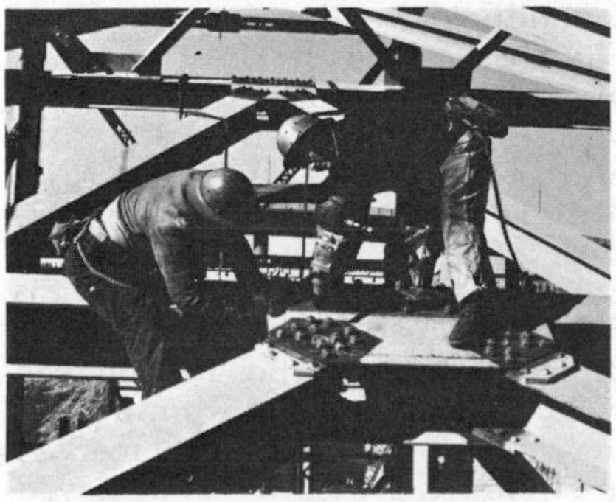

FIG. 46 Air impact wrenches speed high strength bolting of steel connections in modern buildings.

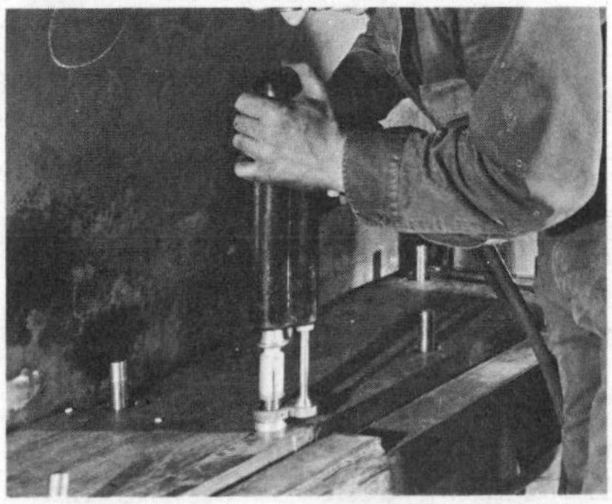

FIG. 47 Special stud welding equipment permits fast and efficient joining of many materials to steel.

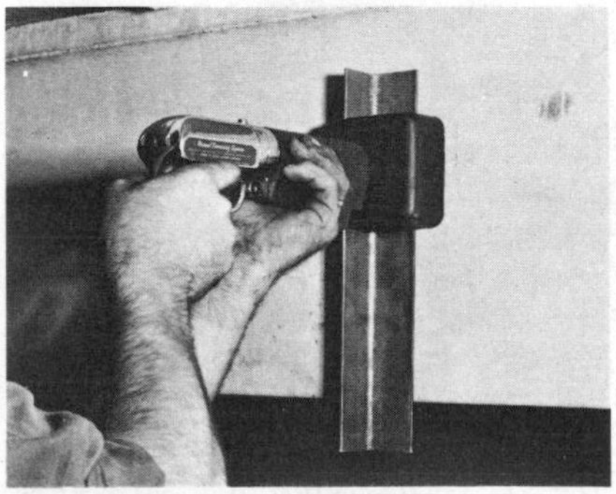

FIG. 48 Powder actuated fasteners often are used to secure steel members to concrete and masonry surfaces.

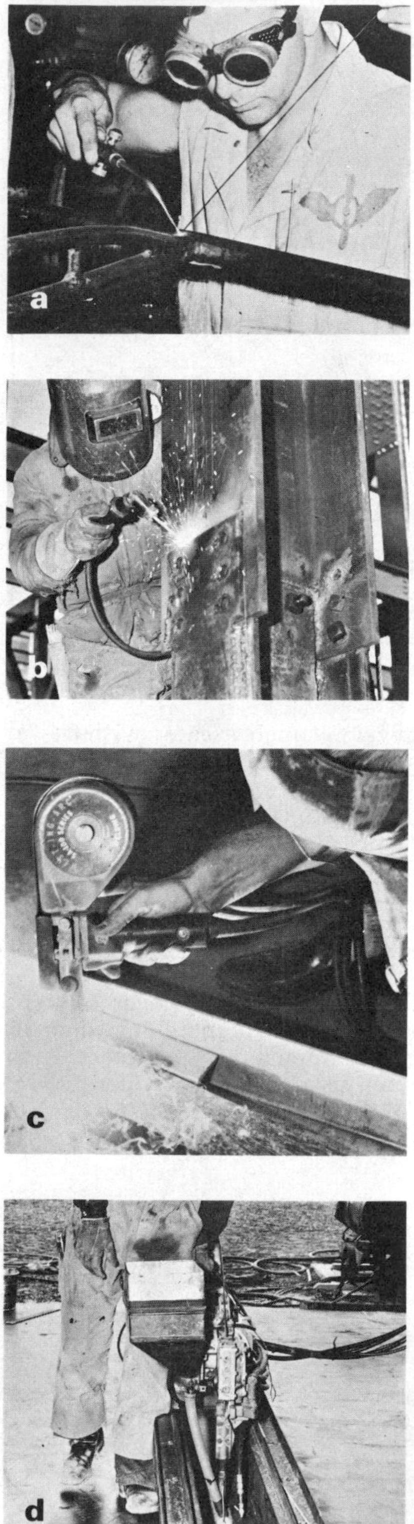

FIG. 51 *(a) Gas Welding with filler rod; (b) Shielded Metal Arc Welding; (c) Metal Inert Gas Shielded Welding (MIG); (d) Submerged Arc Welding.*

in the shop and field by bending and interlocking the elements (Fig. 50). This is a common method of fabricating air ducts for heating and air conditioning, as well as for roofing and roof drainage components.

Welding

Welding, brazing and soldering are methods of joining two or more metal components. Welded joints are made by melting the edges of the components to be joined and adding filler metal if required. Brazing and soldering generally employ non-ferrous filler metals which have melting points below those of the base metals and do not require melting of the base metal to effect a bond. When the filler metal has a melting point below 800° F the joining method is considered soldering; if above, it is brazing. Brazed connections can be as strong as welded joints, depending on service requirements. The brazing filler metal flows into a joint of any configuration by capillary attraction. Soldering and brazing are most often used in product manufacture; welding may be used either in manufacturing or structural applications.

Steel product manufacture employs welding, brazing and soldering processes similar to those of other metals. Recognizing the importance of steel as a *structural* material, the following discussion will be limited to the common welding techniques used in *structural* applications and the general effects of welding on steel.

Welding Techniques The majority of welding processes can be classified into three main groups: *gas, arc* and *resistance.* Gas welding uses a combination of oxygen and a fuel gas to produce the necessary heat to melt and bond the component edges (Fig. 51a). *Oxyacetylene* welding is a common type of *gas* welding which uses acetylene as the fuel gas.

Arc welding and resistance welding are *electric* welding processes. *Arc welding* depends on the heat of an electric arc established between an electrode and the components to be welded. In *resistance*

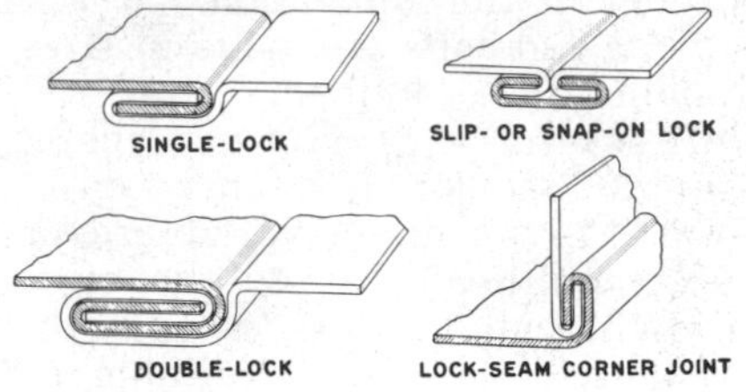

FIG. 50 *Typical methods of lockseam joining of sheetmetal components in product manufacture.*

welding, the components are butted together and heat is generated in the seam by the resistance of the metal to the passage of an electric current. An electrode in contact with the component introduces the current.

Oxygen and nitrogen in the air have a detrimental effect on the hot metal in a weld area. Most arc processes *shield* the weld area from atmospheric contamination with inert gases or granular fusible materials. *Shielded metal arc welding* (also known as *manual metal arc* or *coated stick electrode* welding) is the most common arc process (Fig. 51b). This process uses a consumable metal *stick* electrode which is gradually melted by the arc and deposits filler metal in the weld. A chemical coating on the electrode releases an inert gas which forms a shielding envelope around the weld area. This process depends on manual replacement of the stick electrode as it is consumed.

In *semi-automatic* processes using self-advancing *wire* electrodes, the wire comes off a spool at a predetermined rate and goes through a handheld welding gun. The two chief processes of this type are *metal inert-gas shielded arc* (MIG, or *gas shielded metal arc,* Fig. 51c) and *submerged arc* welding (Fig. 51d). In MIG, an inert gas or mixture of gases (argon, helium, a combination of both or carbon dioxide) from an external supply is directed through the gun on the weld area to shield it from atmospheric contaminants. In submerged arc welding, protection is achieved by burying the arc and immediate weld area under a blanket of fusible granular material. Arc heat melts the material to form a layer of protective slag over the molten metal. The slag and loose material are removed after welding is completed. In MIG, filler metal is provided by a consumable electrode; in submerged arc welding either a consumable electrode or separate filler rod may be used.

The most common welded joint types are illustrated in Figure 52. The use of welding processes with various types of steel is outlined in Figure 53.

Connections in structural work are commonly made by one of the electric arc processes and a variation of gas welding is used for cutting shapes. Gas welding is seldom used for structural connections since lower temperatures limit its use to lighter gauge, flat and tubular products. Some of the resistance processes often employed in manufacture are discussed below.

Spot Welding is a type of resistance welding where the components are joined by a series of spot welds, rather than a continuous line of weldment as in other techniques. It is produced by the resistance heating of a small area between two cylindrical electrodes under pressure (Fig. 54).

Seam Welding is actually a series of overlapping spot welds produced under pressure by a pair of disc electrodes, as in welded pipe manufacture.

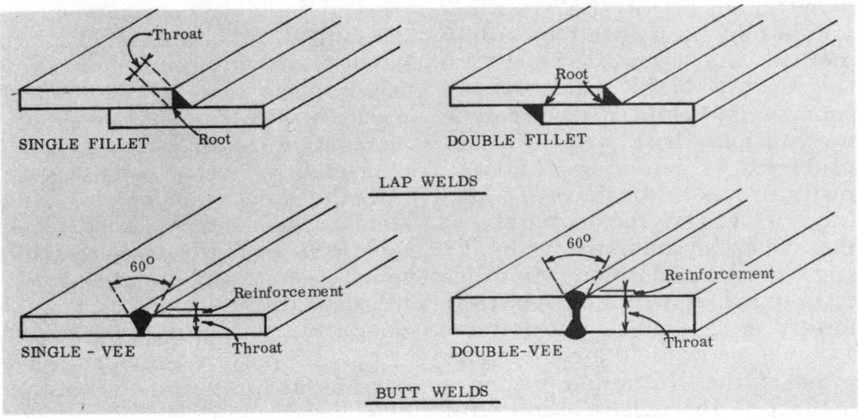

FIG. 52 Welded joints common in structural assemblies.

Projection Welding is also similar to spot welding but makes use of pre-applied projective embossments on one of the components to localize the heat and effect a weld in the desired spot.

Effects of Welding on Steel Since heat is an essential element of all welding techniques and metals respond to heat exposure in varying degrees, some steels are more weldable than others. The degree of weldability is related to the sensitivity of the metal to heat, as reflected in the soundness of the weld and the degree of undesirable side effects produced by the operation. The possible detrimental side effects are: (1) the creation of cracks in the base and weld metals; (2) changes in the physical properties of the base metal, such as decreased strength, ductility and corrosion resistance; (3) oxidation of the finished surface.

The most important element which affects the weldability of steel is the carbon content. The low carbon steels (less than 0.15%) are the most weldable; the mild steels (0.15% to 0.30% carbon) are next in weldability and can be welded by most processes up to a section thickness of 1/2″. Thicker sections require greater input of heat to effect a weld, and this may have detrimental results unless proper procedures and care are used. All ASTM structural steels have a carbon content of less than 0.35% and, with the exception of A-440, are recommended for welding.

FIG. 54 Multiple-head spot welder joins metal skin to continuous metal reinforcement of hollow metal door.

FIG. 53 WELDING PROCESSES USED TO JOIN VARIOUS TYPES OF STEEL

Type of Steel	Shielded Metal Arc	Submerged Arc	Gas Shielded Metal Arc (MIG)	Oxy-acetylene (Gas)	Spot, Seam and Projection
LOW CARBON:					
sheets	Common	Common	Common	Common	Common
plates & bars	Common	Common	Common	Occasional	Occasional
MEDIUM CARBON*:					
sheets	Common	Common	Common	Common	Common
plates & bars	Common	Common	Common	Occasional	Rare
HIGH CARBON*:					
sheets	Common	Occasional	Occasional	Common	Common
plates & bars	Common	Occasional	Occasional	Common	Not Used
LOW ALLOY*:					
sheets	Common	Common	Common	Common	Occasional
plates & bars	Common	Common	Common	Occasional	Rare
STAINLESS*:					
sheets	Common	Occasional	Common	Occasional	Common
plates & bars	Common	Common	Common	Occasional	Rare

*Preheating and/or post-heating (stress relieving) of structures may be required in the weld zones, to prevent embrittlement and excessive residual stresses on certain carbon and low alloy steels, and to stabilize in order to prevent corrosion on stainless steels.

Although all of the stainless steels can be welded by some method, the austenitic steels are the most weldable. Almost any commercial welding method can be used to join these steels without problems of soundness and porosity, or loss in toughness or ductility. However, the high rate of thermal expansion in the austenitic steels must be considered in order to control distortion. Another hazard is reducing corrosion resistance in the weld area, because some of the chromium becomes ineffective through reaction with carbon. This often occurs when stainless steels are exposed to high welding temperatures over extended periods. It can be minimized by limiting the heat input, promoting heat dissipation and selecting low-heat resistance welding methods. Annealing of the finished assembly followed by fast cooling also restores corrosion resistance.

Adhesive Bonding

Many decorative and protective laminated coatings are applied to metal base materials with special adhesives. Adhesive bonding is also used to secure thin-gauge carbon, galvanized and stainless steel sheets and foils to non-ferrous materials which provide the required rigidity, flatness, strength or insulating qualities.

Carbon steel surfacing adds abrasion and indentation resistance to softer core materials. Steel-clad honeycomb doors (Fig. 55), and steel bonded to plywood panels for column and radiator covers are common examples; galvanized or vitreous enamel coated steel sheets, bonded with rigid urethane foam to form sandwich panels, are often used in building construction.

FIG. 55 Stainless steel skins are adhesive-bonded to honeycomb cores to produce flush metal doors.

Stainless steel sheets also are bonded to rigid or insulating materials in the manufacture of many products with corrosion resistant, attractive finishes. A recent development in adhesive bonding permits the lamination of very thin stainless steel foil to kraft paper for use in packaging where additional strength and impermeability to vapor are important considerations.

The chief advantages of adhesive bonding are uniform distribution of stresses, absence of surface fasteners and economy of mass production. Ease of joining small parts not accessible by welding, and low weight in relation to strength make this a valuable structural joining method in the aircraft industry. However, high initial cost of specialized equipment, low peel and creep (flow under stress) strengths, sensitivity to certain temperatures (characteristic of low cost adhesives) at present limit their use in building construction mainly to the manufacture of panelized products.

Adhesive Types The most common adhesives used in the lamination of steel to other materials are epoxies, phenolic resins, rubber polymers modified by resins, polyvinyl acetates and chemically cured neoprenes. The most promising new adhesives for laminating steel are the polyurethanes which, with epoxies and phenolics, withstand considerable loads at a greater temperature range and come closest to the concept of "structural" adhesives. These adhesives often possess high peel strength, chemical inertness, resistance to wetting action, and shear and tensile strengths greater than the material bonded. Under certain conditions they may be used for joining structural members in building construction.

Adhesives are available in a variety of forms such as pressure-sensitive tapes, films, liquids, pastes and solids. They can be applied manually with ordinary building tools or with specialized mass production equipment which spray, roll or flow the adhesives on component surfaces. Some acquire permanent mechanical properties when heated (thermosetting adhesives); others "set" at moderate temperatures but can be softened repeatedly at high temperatures (thermoplastics). Other adhesives depend on the chemical reaction between two agents to acquire their permanent characteristics.

Although many adhesives can perform well under a given set of service conditions, there is no universal adhesive and each must be selected to meet the anticipated operating requirements. Additional discussion on adhesives also pertinent to the bonding of steel may be found in Section 221 Adhesives.

BUILDING PRODUCTS

Since residential and light construction practices evolved from experience with natural materials and local traditions, engineered steel products have entered this field gradually. Today there is a variety of iron and steel building products available for the housing industry.

Sitework Products in this category include cast iron castings for drains, manholes and guards, and corrugated pipe for drainage.

Structural These products include lightweight framing, hot-rolled products, open web steel joists and preengineered buildings.

Lightweight steel framing, consisting of cold-formed steel studs, joists and accessories, provides a non-flammable and dimensionally stable structural system. It is increasingly used for both single and multifamily housing, as well as for other types of light construction. While compatible with conventional stud or bearing wall construction, lightweight steel framing also introduces the possibility of longer spans and large-scale prefabrication (Fig. 56).

Hot-rolled sections, plate and structural pipe are commonly used in framing systems for mid- and low-rise multifamily housing. They are seldom used in single-family construction, except as the mid-span support in basements and garages. These products are also used in many buildings where noncombustibility and speed of erection are important.

FIG. 56 Prefabricated steel stud wall assembly can often be lifted in place by one or two workmen.

FIG. 57 Steel doors provide effective fire, heat and sound control through integral insulation and weatherstripping.

Open web steel joists are standard roof framing for light commercial construction, while corrugated steel floors and roof decks are often used in multi-family housing as well.

Pre-engineered buildings, typically combining rolled steel sections, open-web joists, sheet steel roofing and siding, are occasionally adapted for custom housing, but are primarily used in industrial commercial buildings.

Roofing and Siding These products consist of sidings, wall panels, rainware and roof accessories.

Residential siding, in traditional clapboard designs as well as vertical and other patterns, is an increasingly popular steel product. Standing-seam steel *roofing* has a long history of use in several housing styles. These products offer a broad choice of finishes and coatings (see pages 211-23) for durability, texture and color.

Wall panel systems, often in combination with lightweight steel framing, are used in many industrialized housing systems. However, manufactured wall panels find wider use in non-residential light construction.

Rainware describes such products as flashing, gutters, and downspouts.

Roof accessories include skylights, stacks, ventilators and roof hatches.

Doors and Windows These products include interior and exterior doors, closet doors and various windows.

Metal doors for both exterior and interior applications provide fire safety, security, durability, and dimensional stability. A metal door consists of an insulating core surrounded by steel skins and sides. Built-in magnetic weatherstripping can provide effective heat control and sound control. Designs range from smooth flush faces to variations with glazing, moldings, textured surfaces, and vinyl laminations (Fig. 57).

Bifold closet doors make use of the strength and rigidity of cold-formed sheet steel to act as a light, yet permanent, folding screen where conventional doors are not required.

Steel-framed windows offer a complete range of designs for residential and light commercial uses.

Interior Finishes These products include nonloadbearing studs and acoustical ceilings.

Nonloadbearing studs provide noncombustible, dimensionally stable framing for gypsum wallboard, plaster and other interior surfaces.

Steel acoustical ceilings are found in many types of light contruction, sometimes as an integral part of a steel deck floor-ceiling system.

Other Architectural Uses There are many steel products used in various parts of the home and in light construction:

Prefabricated stairs, such as steel pan stairs, are often found in multi-family construction. Spiral stairs are prefabricated to serve a variety of single family and low-rise multi-family uses.

Prefabricated fireplaces and flues usually offer cost, time, weight, and space advantages over conventional construction. Many designs are engineered for high performance and substantial energy savings (Fig. 58).

Kitchen cabinets in a variety of finishes and colors often coordinate with appliances to give a clean, unified appearance. Unit kitchens, for small apartments and recreation rooms, are also popular.

Mechanical Systems Iron and steel are used extensively in plumbing fixtures and piping; for hot water and steam piping; for radiators, valves and accessories; as well

FIG. 58 Prefabricated fireplaces and flues are easier to install in existing as well as new construction.

as in heating/cooling ductwork.

Plumbingware includes both cast iron and sheet steel products, such as tubs, shower and tub enclosures, toilets, bidets and sinks. They are usually finished in porcelain enamel. Stainless steel is widely used for kitchen and bar sinks.

Sheet steel ductwork, galvanized for corrosion resistance, is part of most heating and cooling systems that employ moving air. Such systems generally have fans, vents, louvers, and other sheet steel accessories.

GLOSSARY

Agglomeration Process for increasing the particle size of iron ores to make them suitable for iron and steel making.

Alloy Steel See *Steel*.

Alloying Element Element added in steel making to achieve desired properties.

Annealing See *Heat Treatment*.

Austenitic Steel See *Grain Structure*.

Beneficiation Concentrating process used to increase the iron content of ores prior to use. (See also *Agglomeration*).

Bonderizing Process to improve paint adhesion on steel by dipping lightly galvanized objects in a hot phosphate solution to form a surface film of zinc phosphate.

Carbon Steel See *Steel*.

Case Hardening Hardening of the outer skin of an iron-base alloy by promoting surface absorption of carbon, nitrogen or cyanide, generally accomplished by heating the alloy in contact with materials containing these elements, and rapid cooling.

Cast Iron High carbon iron made by melting pig iron with other iron-bearing materials, and casting in sand or loam molds; characterized by hardness, brittleness, high compressive and low tensile strengths.

Cladding Bonding thin sheets of a coating metal with desirable properties (such as corrosion resistance or chemical inertness) over a less expensive metallic core not possessing these properties. Copper cladding over steel may be applied by hot dipping; stainless steel and aluminum cladding, by hot rolling.

Coke Processed form of bituminous coal used as a fuel, a reducing agent and a source of carbon in making pig iron.

Cold Forming Forming thin sheets and strip to desired shapes at room temperature, generally with little change in mechanical properties of the metal; includes roll, stretch, shear and brake forming.

Cold Working Shaping by cold rolling, cold drawing or cold reduction at room temperature; generally accompanied by increase in strength and hardness.

Cold Drawing Shaping by pulling through a die to reduce cross-sectional area and impart desired shape; generally accompanied by increase in strength, hardness, closer dimensional tolerances and smoother finish.

Cold Reduction Cold rolling which drastically reduces sheet and strip thickness with each pass through the rolls; generally accompanied by increase in hardness, stiffness, strength and resulting in smoother finish and improved flatness.

Cold Rolling Gradual shaping between rolls to reduce cross-sectional area or impart desired shape; generally accompanied by increase in strength and hardness.

Cold Finishing Cold working which results in finished mill products.

Continuous Casting A process which receives molten metal from any form of furnace and produces semifinished products such as slabs or billets, by passing ingot teaming, stripping, soaking and rolling.

Corrosion Deleterious effect on metal surface due to weathering, galvanic action or direct chemical attack.

Galvanic Action Corrosion produced by electrolytic action between two dissimilar metals in the presence of an electrolyte.

Direct Chemical Attack Corrosion caused by a chemical dissolving of the metal.

Weathering Galvanic and/or chemical corrosion produced by atmospheric conditions.

Electric Induction Furnace See *Furnace*.

Electroplating A process which employs an electric current to coat a base metal (cathode) with another metal (anode) in an electrolytic solution.

Electrogalvanizing Electroplating with zinc, to provide greater corrosion resistance.

Ferritic Steel See *Grain Structure*.

Ferrous Alloys Composite metals whose chief ingredient is iron (*ferrum*), metallurgically combined with one or more alloying elements.

Ferroalloys Iron-based alloys used in steel making as a source of desired alloying elements.

Finished Mill Products Steel shapes which can be used directly in construction.*

Bar Hot-rolled or cold-drawn round, square, hexagonal and multifaceted long shapes, generally larger than wire in cross section. Also hot- or cold-rolled rectangular flat shapes (flats) generally narrower than sheets and strip.*

Blackplate Cold-rolled flat carbon steel products thinner than sheets and wider than strip, generally used for coating with zinc, tin or terne metal.*

Foil Cold-rolled flat product less than 24″ wide and less than 0.005″ thick.

Pipe See *Tubular Products*.

Plates Hot-rolled flat products generally thicker than sheets and wider than strip.*

Sheets Hot- or cold-rolled flat products generally thinner than plate and wider than strip.*

Strip Hot- or cold-rolled flat products generally narrower than sheets and thinner than plates.*

Structurals Hot-rolled steel shapes of special design (such as H-beams, I-beams, channels, angles and tees) used in construction.

Terneplate Blackplate which has been coated with terne metal (lead-in alloy). Also sheet metal that has been coated is referred to as terne-plate.

Tinplate Blackplate which has been coated with tin.

Tubular Products Hollow products of round, oval, square, rectangular and multifaceted cross sections. In construction, round products are generally referred to as *pipe*; square or rectangular products with thinner wall sections as *tube or tubing*.

Wire Cold finished products of round square or multifaceted cross section, generally smaller than bars; *round wire* is cold drawn, 0.005″ to less than 1″ in diameter; *flat wire* is cold rolled, generally narrower than bar.*

Flux Mineral which, due to its affinity to the impurities in iron ores, is used in iron and steel making to separate impurities in the form of molten slag.

Basic Flux Mineral (limestone, dolomite) used in basic furnaces to make *basic* (low phosphorus) steel.

Neutral Flux Mineral (fluorspar) used to make slag more fluid.

Foil See *Finished Mill Products*.

Forging See *Hot Working*.

Furnace, Blast Tall, cylindrical masonry structure lined with refractory materials, used to smelt iron ores in combination with fluxes, coke and air into pig iron.

*See General Classification of Flat-rolled Steel Products by Size, Fig. 29.

Furnace, Steel Masonry or steel structure lined with refractory materials, used to melt pig iron, scrap metal, and sometimes agglomerated ores, ferroalloys and fluxes into steel.
Basic Oxygen Suspended, tilting vessel which uses high purity oxygen to oxidize impurities in hot pig iron and other iron-bearing materials to produce low phosphorus (basic) steel.
Electric Arc Suspended, tilting kettle which melts scrap metal, ore and sometimes ferroalloys with the heat of an electric arc to produce steels of controlled chemical composition.
Electric Induction Steel encased, insulated magnesia pot in which metal scrap and ferroalloys are melted with the heat of an electric current induced by windings of electric tubing; used chiefly to produce small quantities of high grade steels such as alloy, stainless and heat-resisting steels.
Open Hearth A masonry structure with a hearth exposed to the sweep of flames in which hot pig iron, scrap metal and fluxes are melted and oxidized by a mixture of fuel and air, to produce basic or acid steel.

Galvanizing Zinc coating by electroplating or hot dipping, which produces a characteristic bright spangled finish and protects the base metal from atmospheric corrosion.

Grain Structure The microscopic, internal crystalline structure (size and distribution of particles) of a metal which affects its properties—known as austenitic, ferritic, and martensitic.
Austenitic Steels Tough, strong and non-magnetic steels. Austenitic stainless steels have a chromium content up to 25%, nickel up to 22%, and can be hardened by cold working.
Ferritic Steels Soft, ductile and strongly magnetic steels. Ferritic stainless steels usually have a chromium content between 12% and 27% and are not hardenable by heat treatment.
Martensitic Steels Can be made very hard and tough by heat treatment and rapid cooling. Martensitic stainless steels have a chromium content between 4% and 12%.

Heat Treatment Controlled heating and cooling of steels in the solid state for the purpose of obtaining certain desirable mechanical or physical properties.
Annealing Heating metal to high temperatures (1350°F to 1600°F for steel), followed by controlled cooling, to make the metal softer or change its ductility and toughness.
Quenching Rapid cooling by immersion in oil, water or other cooling medium to increase hardness.
Tempering Reheating to less than 1350°F after hardening (as by quenching) and slow cooling, to restore ductility.

Hot Dip Process Coating commonly of zinc, terne metal or aluminum by immersion in a bath of the molten coating metal.

Hot Forming Forming of hot plastic metal into desired shapes, with little change in the mechanical properties of the metal.

Hot Working Shaping hot plastic metal by hot rolling, extruding or forging, usually at temperatures above 1500°F, generally accompanied by increases in strength, hardness and toughness.
Extruding Shaping lengths of hot metal by forcing through a die of the desired profile.
Forging Shaping hot metal between dies with compression force or impact.
Hot Rolling Gradual shaping by squeezing hot metal between rolls.

Ingot Cast pig iron or steel shape made by pouring hot metal into molds.

Machinability The ability to be milled, sawed, tapped, drilled and reamed without excessive tool wear, with ease of chip metal removal and surface finishing.

Malleability The ability to be shaped without fracture either by hot or cold working.

Martensitic See *Grain Structure.*

Melting Heating metal to the molten state in a steel making furnace to control residual elements and/or add beneficial alloying elements.

Mild Steel See *Steel.*

Oxidation The chemical combination of a substance with oxygen.

Open Hearth Furnace See *Furnace.*

Pickling Removing the oxide scale formed on hot metal as it air cools, by dipping in a solution of sulfuric or hydrochloric acid.

Pig Pig iron ingot.

Pig Iron High-carbon crude iron from the blast furnace used as the main raw material for iron and steel making.
Basic Pig Iron High phosphorus iron used in basic steel making furnaces.
Bessemer (acid) Pig Iron Low phosphorus iron used in acid steel making furnaces.

Quenching See *Heat Treatment.*

Reduction Separation of iron from its oxide by smelting ores in the blast furnace.

Refining Melting of pig iron and/or other iron-bearing materials in steel furnaces to achieve desired contents of residual and alloying elements.

Refractories Non-metallic materials with superior heat and impact resistance used for lining furnaces, flues and vessels employed in iron and steel making.

Residual Elements Non-ferrous elements (such as carbon, sulfur, phosphorus, manganese and silicon) which occur naturally in raw materials and are controlled in steel making.

Scrap Metal Source of iron for iron and steel making, consisting of rolled product croppings, rejects and obsolete equipment from steel mills and foundries, and waste ferrous material from industrial and consumer products.

Slag Molten mass composed of fluxes in combination with unwanted elements, which floats to the surface of the hot metal in the furnace and thus can be removed.

Smelting Melting of iron-bearing materials in the blast furnace to separate iron from impurities with which it is chemically combined or mechanically mixed.

Steel Iron-base alloy, containing manganese, usually carbon residual and often other alloying elements, characterized by its strength and toughness; distinguished from iron by its ability to be shaped by hot and/or cold working as initially cast.
Alloy Steel Steel in which residual elements exceed limits prescribed for carbon steel, or alloying elements are added within specified ranges.
Carbon Steel Steel in which the residual elements are controlled but alloying elements are not usually added.
Heat Resisting Steel Low chromium steel with at least 4% chromium which retains its essential mechanical properties

at elevated temperatures.

High Strength Low Alloy Steel Steel with less than 1% of any alloying element, manufactured to high standards for strength, ductility, and partial chemical specifications.

Mild Steel Carbon steel with carbon content between 0.15% and 0.25%.

Stainless Steel Steel containing at least 10% chromium, with excellent corrosion resistance, strength and chemical inertness at high and low temperatures.

Strand Caster The machine which performs the continuous casting process.

Tempering See *Heat Treatment.*

Toughness Maximum ability of a material to absorb energy without breaking, as from sudden shock or impact.

Welding Creating a metallurgical bond between metals with heat and sometimes with the use of pressure and filler metal.

Arc Welding Welding methods employing an electric arc as the source of heat.

Gas Welding Welding methods employing a fuel gas (acetylene, hydrogen) and oxygen as the source of heat.

Shielded Welding Processes using gases or fusible granular materials to shield the weld area from damaging effects of oxygen and nitrogen in the air.

Shielded Metal Arc Welding (also known as manual metal arc, and stick electrode welding) Arc welding in which a flux coated metal electrode is consumed to form a pool of filler metal and a gas shield around the weld area.

Inert Gas Shielded Arc Welding Arc welding in which shielding is provided by an inert gas envelope (such as argon, helium, a combination of both or carbon dioxide) from an external supply. Filler metal is supplied by either a consumable metal electrode, as in *inert gas shielded metal arc* welding (MIG), or by a separate filler rod used with a non-consumable tungsten electrode, as in *inert gas shielded tungsten arc* welding (TIG).

Submerged Arc Welding Arc welding in which the weld area is shielded by fusible granular material which melts to protect the weld area. Filler metal is obtained either from a consumable electrode of separate filler rod.

Wrought Products Products formed by rolling, drawing, extruding and forging.

Wrought Iron Relatively pure iron, mechanically mixed with a small amount of iron-silicate slag; characterized by good corrosion resistance, weldabilty, toughness and high ductility.

We gratefully acknowledge the assistance of the following for the use of their publications as references and permission to use photographs and illustrations: American Iron and Steel Institute, National Association of Architectural Metal Manufacturers, Allegheny Ludlum Steel Corporation, Bethlehem Steel Company, A. M. Byers Company, The Ceco Corporation, Inland Steel Company, Jones & Laughlin Corporation, Lukens Steel Company, Nelson Stud Welding Division, Phillips Drill Company, Ramset Winchester-Western Division, Republic Steel Corporation, Rheem Manufacturing Company, Rigidized Metals Corporation, United States Steel Corporation, Vasco Metals Corporation, Welding Engineer Publications, Inc., Youngstown Sheet and Tube Company.

213 ALUMINUM

ALUMINUM

INTRODUCTION

Aluminum is truly a modern building material. Though first detected in 1807 as an ingredient of common clay, it was not until 1825 that the first metallic aluminum was produced; and not until 1845 was aluminum powder successfully transformed into solid particles. In 1852, due to costly production methods, pure aluminum sold for $545 per pound and was only used for the finest of jewelry and tableware.

The 1886 discovery that metallic aluminum could be produced by dissolving alumina (aluminum oxide) in molten cryolite and then passing an electric current through the solution gave birth to the modern aluminum industry. Today, through improved mining and processing techniques, the present cost of unalloyed aluminum ingot (99.5% pure) is less than $1.00 per pound.

As a building material, aluminum has been used for more than 30 years. Building and construction is the largest single market for aluminum in the U.S. (Fig. 1). In 1978 this market required over 3 billion pounds of aluminum. Products that use the largest amount include windows, doors and screens (1 billion pounds), residential siding (500 million pounds), mobile homes (300 million pounds), and awnings and canopies (150 million pounds).

A unique combination of properties makes aluminum one of the most versatile engineering and construction materials. It is light in weight, yet some of its alloys have strengths greater than that of structural steel. It has high resistance to corrosion and will not "rust red" to stain adjacent surfaces or discolor products.

It has no toxic action, has high electrical and thermal conductivities and high reflectivity to both heat and light. The metal can easily be worked to any form and readily accepts a wide variety of surface finishes. These characteristics give aluminum its extreme versatility. In the majority of applications, two or more of these characteristics come prominently into play.

One of aluminum's most important attributes is that it has met the test of time in a variety of building and construction uses for decades. For example, an aluminum cap installed on the top of the Washington Monument in 1884 was recently inspected and found to be in excellent condition. Over 11 million homes have been sided with aluminum during the past 30 years.

As with other materials, however, aluminum's properties must be understood to insure its proper use in service. This section will discuss the production of aluminum and aluminum alloys and their various properties, and methods by which aluminum products are manufactured. Individual aluminum products are discussed in Section 214.

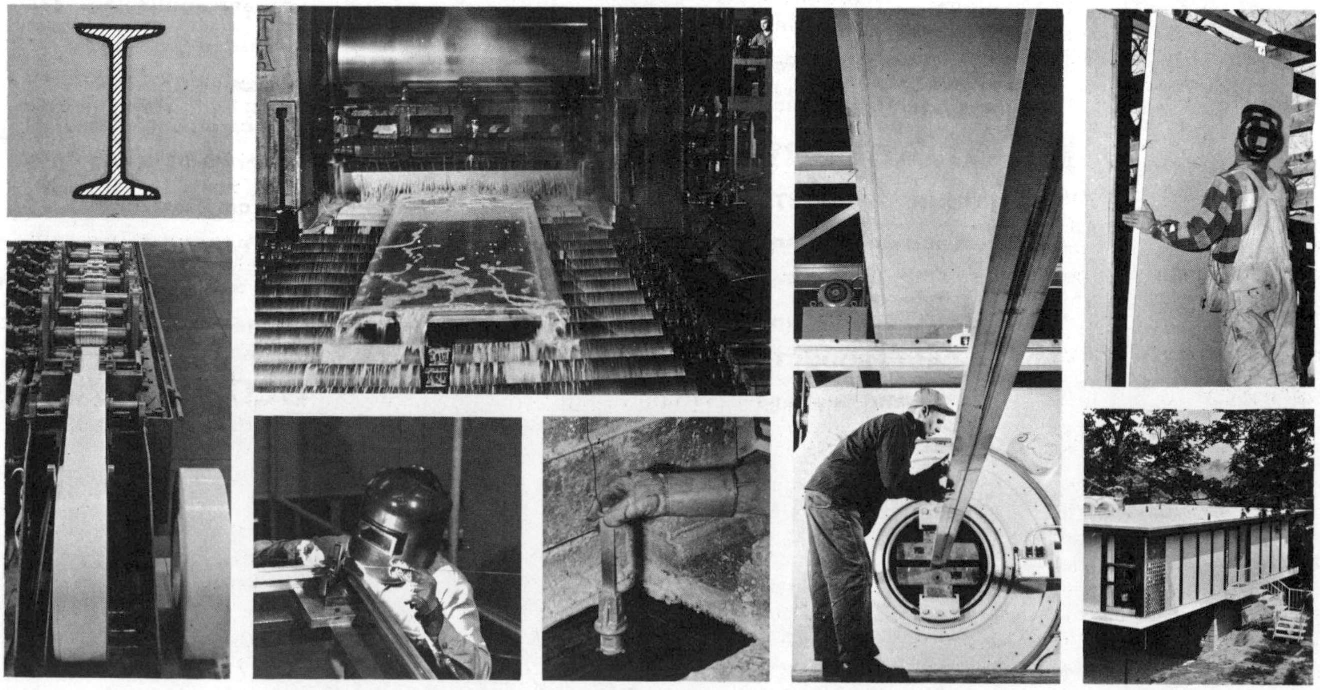

Aluminum forms 8% of the earth's crust, making it the most plentiful metal. But it is never found free in nature like gold or silver—it is always combined chemically with other elements. Though nearly all common rocks and clays contain some aluminum, it is not practical to extract aluminum unless the raw ore has an aluminum oxide content of at least 45%.

Such ores are called bauxites, after the town of Les Baux in France where one of the first bauxite ore deposits was found. Most of the world's bauxite deposits are outside of the United States, making it necessary for the United States to import most of its supply. Ninety-five percent of the bauxite mined in the United States comes from Arkansas, the remainder from small, lower grade deposits in Georgia and Alabama.

Work is now underway in the U.S. to develop methods of using domestic ores such as kaolin and alunite, as a replacement for imported bauxite. The U.S. Bureau of Mines, which is coordinating the project, believes the technology exists to extract aluminum commercially from these domestic ores.

FIG. 1 ALUMINUM CONSUMPTION BY MAJOR MARKETS

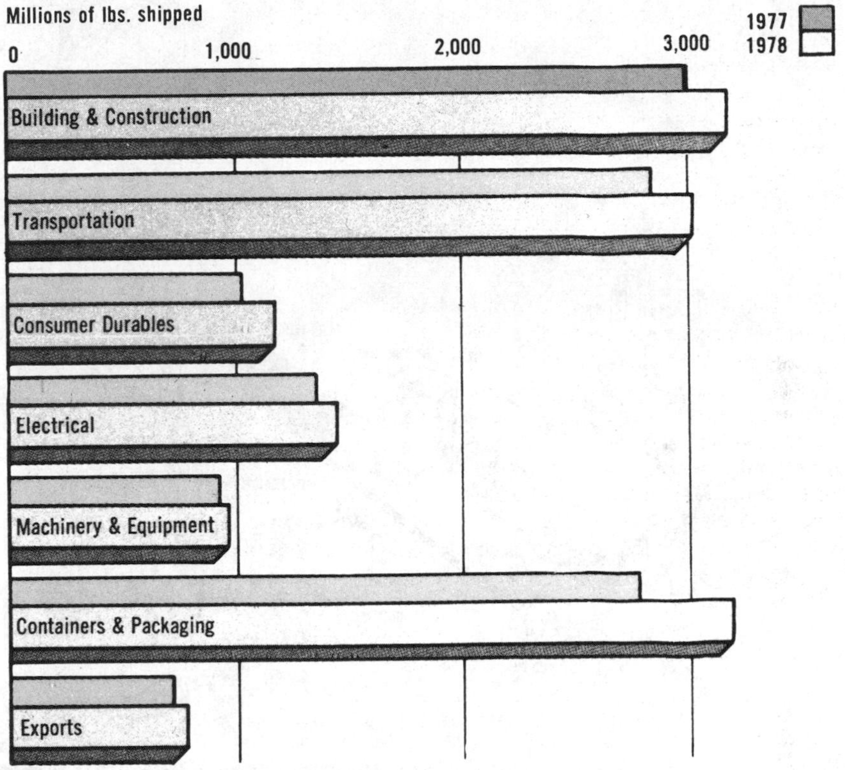

MINING

Strip or open pit mining methods are generally used to obtain the ore, which is then crushed, washed and screened. The ore is ground and then dried in kilns (large revolving steel drums) at temperatures up to 250°F). The general steps in the mining and refining of aluminum are shown in Figure 2.

REFINING

All commercial bauxite refining is done by the Bayer process which separates aluminum oxide (alumina) from the unwanted minerals.

The dried, ground bauxite is mixed with a solution of caustic soda (sodium hydroxide) which dissolves the alumina to form sodium aluminate. The silica in the bauxite reacts and precipitates out of solution. The iron oxide and other impurities are not affected chemically and, being solids, settle out.

The "green liquor" is now highly supersaturated sodium aluminate. Previously prepared hydrated alumina crystals are added to the solution. These form larger crystals, which gradually settle out of the solution. After being washed to remove any remaining traces of impurities, the concentrated aluminum hydrate crystals are roasted at temperatures of more than 2000°F. The water is thus driven off and the resulting alumina remains as a fine white powder, similar to sugar in appearance and consistency.

REDUCTION

Reduction is the electrolytic separation of the aluminum from its oxide, alumina. The general reduction procedure is described in Figure 2.

The process employs a carbon-lined vessel (pot) containing molten cryolite (sodium aluminum fluoride) in which alumina is dissolved. Metallic aluminum is separated from the alumina by passing an electric current through this solution at a temperature of about 1775°F.

Molten aluminum collects at the bottom of the pots and is tapped or siphoned off into large ladles and poured into molds to form ingots, or transferred to holding furnaces for alloying. Aluminum, as it comes from the pots, is about 99.5% pure. By additional refining, "super purity" aluminum which is about 99.99% pure can be produced.

Reduction plants built in recent years use less than 6.5 kilowatt hours per pound (kWh/lb) to convert alumina into aluminum, com-

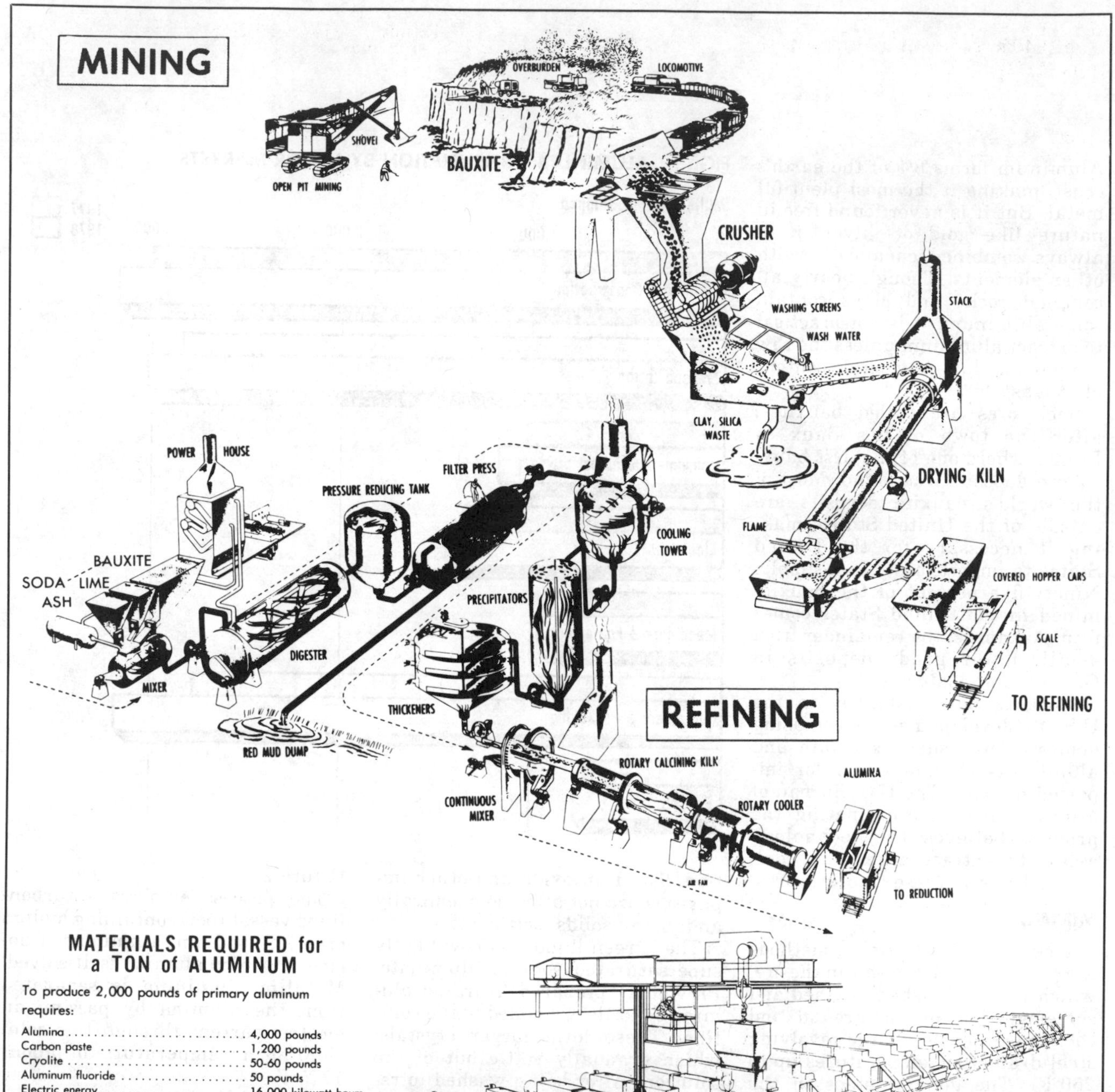

MATERIALS REQUIRED for a TON of ALUMINUM

To produce 2,000 pounds of primary aluminum requires:

Alumina	4,000 pounds
Carbon paste	1,200 pounds
Cryolite	50-60 pounds
Aluminum fluoride	50 pounds
Electric energy	16,000 kilowatt-hours

To produce 4,000 pounds of alumina needed for 2,000 pounds of primary aluminum requires:

Bauxite	8,000 pounds
Soda ash	1,000 pounds
Quicklime	160 pounds
Coal	1,000 pounds
(or 20,000-40,000 cubic feet natural gas)	

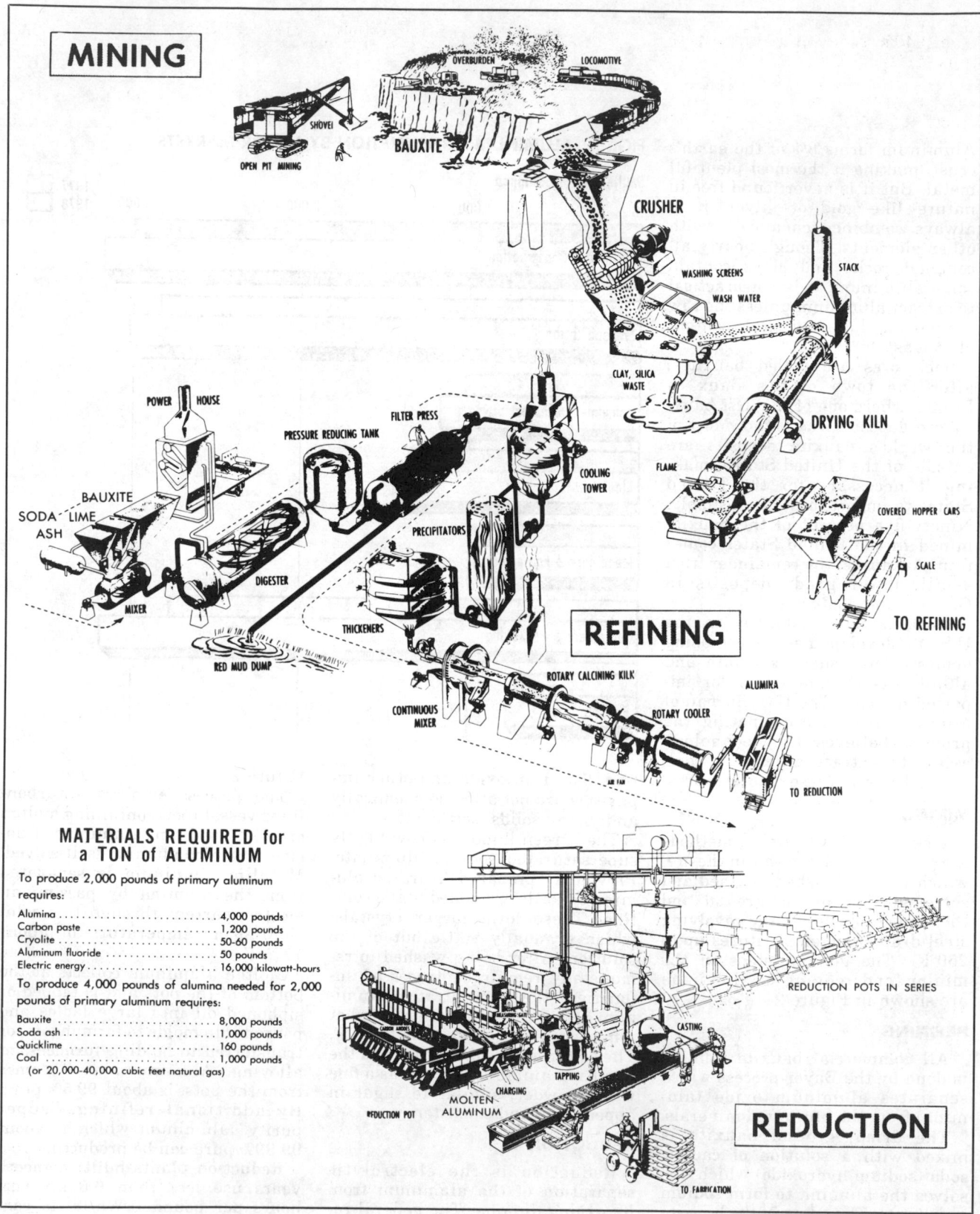

FIG. 2 The mining, refining and reduction of aluminum.

pared with about 12 kWh/lb during the 1940's. Today the average use of electricity for U.S. plants is about 8 kWh/lb. A pilot commercial plant incorporating a new process is using less than 5 kWh/lb. The industry managed to reduce its unit energy consumption by ten percent between 1972 and 1978, a goal it had earlier set for 1980.

RECYCLING

Recycled aluminum, metal which has already served a useful life in a product and is used over again, is playing an increasingly important role. More than 20% of the aluminum supply in a typical year comes from scrap.

Recycling is especially important because it saves 95% of the energy required to make aluminum from ore. Ingot that comes from the recycling process is called secondary ingot.

Old aluminum windows, doors and siding are among the many products that are being brought to scrap dealers for recycling.

ALUMINUM ALLOYS

In high purity form, aluminum has a tensile strength of about 7000 psi and is relatively soft and ductile. Most uses, however, require greater strength and other properties achieved by the addition of other elements to produce various aluminum *alloys*.

Aluminum *alloys* can be divided into two general categories: those whose strength characteristics may be improved by heat treatment (*heat-treatable* alloys), and those whose strength may be improved by mechanical methods of cold working (*non-heat-treatable* alloys).

Heat-Treatable Alloys

The initial strength of heat-treatable alloys is improved by the addition of alloying elements such as copper, magnesium, zinc, and silicon. Since these elements (singly or in various combinations) show increasingly solid solubility in aluminum with increasing temperature, it is possible to subject them to thermal treatments which will impart further strengthening.

	TEMPER DESIGNATIONS	
—F	As fabricated	
—O	Annealed	
—H	Strain hardened (wrought only)	
	—H1	Plus one or more digits* Strain hardened only
	—H2	Plus one or more digits* Strain hardened and partially annealed
	—H3	Plus one or more digits* Strain hardened, then stabilized
—W	Solution heat treated—unstable temper	
—T	Thermally treated to produce stable tempers other than —F, —O, or —H	
	—T1	Cooled from an elevated temperature shaping process and naturally aged to a substantially stable condition
	—T2	Cooled from an elevated temperature shaping process, cold worked and naturally aged to a substantially stable condition
	—T3	Solution heat treated, cold worked and naturally aged to a substantially stable condition
	—T4	Solution heat-treated and naturally aged to a substantially stable condition
	—T5	Cooled from an elevated temperature shaping process, then artificially aged
	—T6	Solution heat-treated, then artificially aged
	—T7	Solution heat-treated and stabilized
	—T8	Solution heat-treated, cold worked, then artificially aged
	—T9	Solution heat-treated, artificially aged, then cold worked
	—T10	Cooled from an elevated temperature shaping process, cold worked, then artificially aged

*Second digit indicates final degree of strain hardening, i.e., 2 is ¼ hard, 4 is ½ hard, 6 is ¾ hard, 8 is full hard

FIG. 3 In the designation of temper the letter "H" is used for non-heat-treatable alloys; "T" for heat-treatable.

The first step, called *solution heat treatment,* is an elevated-temperature process. This is followed by rapid quenching, usually in water, which momentarily "freezes" the structure and for a short time renders the alloy very workable. It is at this stage that some fabricators retain this more workable structure by storing the alloys at below-freezing temperatures until they are ready to be formed. At room or elevated temperatures the alloys are not stable after quenching, and after a period of several days at room temperature, termed *natural aging* or *room temperature precipitation,* the alloy is considerably stronger. Some alloys, particularly those containing magnesium and silicon or magnesium and zinc, continue to *age-harden* for long periods of time at room temperature.

By heating the alloy at a controlled temperature, even further strengthening is possible and properties are stabilized. This process is called *artificial aging* or *precipitation hardening.* Heat-treated alloys carry the temper designation "T" (Fig. 3).

By the proper combination of solution heat treatment, quenching, cold working and artificial aging, the highest strengths are obtained.

Non-Heat-Treatable Alloys

The non-heat-treatable (*common*) aluminum alloys have alloying elements that do not show an increase in strength with heat treatment. The initial strength of non-heat-treatable alloys depends upon the hardening effects of elements such as manganese, silicon, iron and magnesium, singly or in various combinations. These alloys can

FIG. 4 **EXAMPLES OF ALUMINUM ALLOYS OFTEN USED FOR SOME BUILDING APPLICATIONS**

GENERAL USE OF ALLOYS

APPLICATION	ALLOY NUMBER	PRODUCT
Acoustic ceiling	3003-H14	Sheet
Air duct	3003-H14, 1100-H16, Duct Stock	Sheet
Awning	3003-H14	Sheet
Builders' hardware	5050-0	Sheet
	B443.0, 356.0	Casting
	6063-T5	Extrusion
Curtain wall	3003-H14, Alclad 4043, Alclad 1235	Sheet
	B443.0	Casting
	6063-T5, 4043-F	Extrusion
Deck (corrugated)	3003-H14	Sheet
Door frame	6063-T5, 6063-T6	Extrusion
Fascia plate	6063-T42	Extrusion
	3003-H14	Sheet
Flashing	3003-0	Sheet
Flue lining	1100-H16	Sheet
Garage door	3003-H14	Sheet
Grating	6061-T6	Bar
Gravel stop	6063-T5, 6063-T42	Extrusion
Grille	B443.0	Sand Casting
	3003-H14	Sheet
	6063-T5	Extrusion
Gutter	3003-H14	Sheet
Hood	3003-H14	Sheet
Insulation	1235	Foil
Kick plate	3003-H18	Sheet
Letters	B443.0, 514.0	Sand casting
	3003-H14	Sheet
Louver	3003-H14	Sheet
	6063-T5	Extrusion
Mullion	3003-H14	Sheet
	6063-T5, 6063-T6	Extrusion
Nails	6061-T913	Wire
Railing	6063-T5, 606 1-T83	Extrusion
Relief panel (sculptural)	B443.0	Casting
Roofing	Special Roofing Alloys	Sheet
Screen	Alclad 5056-H392	Wire
Screws	2024-T4, 2017-T4	Screw Machine Stock
Shade screening	5052-H38	Sheet
Shingle	3003-H14	Sheet
Siding	3003-H14	Sheet
Structural shape	2014-T6, 6061-T6	Extrusion
	6061-T6, 2014-T6	Rolled Shape
Termite shield	3003-H14	Sheet
Terrazzo strip	3003-H14	Sheet
Threshold	6063-T5	Extrusion
Vapor barrier	1235-0	Foil
Venetian blind	5052-H18	Sheet
Weatherstrip	5052-H38	Sheet
Window	6063-T5, 6063-T6, 6061-T6	Extrusion

be strengthened by *strain-hardening, cold rolling* or other mechanical working denoted by the "H" series of temper designations (Fig. 3).

Annealing

One thermal treatment applicable to both heat-treatable and non-heat-treatable alloys and to cast and wrought products is *annealing,* an elevated temperature (600° to 800°F) treatment followed by a slow cool. The purpose of this treatment is to relieve internal stresses and to return the alloy to its softest and most ductile condition. Annealing is usually performed to condition the metal for severe forming operations. Subsequent working processes increase the mechanical properties of non-heat-treatable alloys. The mechanical properties of heat-treatable alloys may be increased by appropriate thermal treatments, after forming operations have been completed.

Cladding

To increase the corrosion resistant properties of some products or to provide for special surface appearance or preparation for additional surface treatment, an aluminum alloy or other metal coating may be metallurgically bonded to an aluminum product. Cladding of sheet and plate products is most common although some tube, rod and wire products are also clad.

Aluminum, clad with high purity aluminum or a non-heat-treatable alloy, for improved surface corrosion resistance, is designated by its four digit alloy number (explained below) preceded by the word *Alclad.* For example, Alclad 3004 is a common sheet and plate material used for building applications.

Alloy Designation Systems

Aluminum products fall into two general groups—*wrought* products and *cast* products.

Alloys and temper are designated by numerical systems for both cast and wrought products as described below. Figure 4 shows some typical uses for various alloys.

Temper Designations. A notation or temper designation is used following the alloy. The letter "H" is used for non-heat-treatable and the letter "T" for heat-treatable alloys. Some temper designations apply only to wrought products, others to cast products, but most apply to both. Additional digits after the T or H indicate variations in the basic treatment (Fig. 3).

Wrought Alloys Almost all aluminum products used in construction are made from *wrought* alloys, which are designated by a four digit index system (Fig. 5). The first digit of the number indicates an alloy group or series:

1000 Series designates aluminum of 99% or higher purity. These alloys have excellent corrosion resistance, high thermal and electrical conductivity, low mechanical properties and excellent workability. Moderate increase in strength may be obtained by strain hardening. Alloys used in electrical and chemical fields, aluminum foils for insulation and vapor barriers are in this group (1235) and alloy 1100 is frequently used for sheet metal work.

2000 Series uses copper as the principal alloying element. These alloys require heat treatment to obtain optimum properties and in the heat-treated condition their mechanical properties are similar to and may exceed those of mild steel.

Sometimes artificial aging is employed to further increase strength properties. The alloys in the 2000 Series do not have as good corrosion resistance as most other aluminum alloys and in sheet form are usually clad with high-purity alloy or a magnesium-silicon alloy of the 6000 Series. Alloy 2024 is used for fasteners. Structural shapes for construction use are sometimes made from 2014.

3000 Series alloys with manganese as the major alloying element are generally non-heat-treatable. Because only a limited percentage of manganese (up to about 1.5%) can be effectively added to aluminum, it is used as a major element in only a few alloys.

Alloy 3003 is a general-purpose alloy used for moderate-strength applications requiring good workability. In sheet form, it is used for air ducts, acoustical ceiling pans, awnings, corrugated decks, garage doors, flashings, gutters, siding, shingles and other uses.

4000 Series employs silicon as the major alloying element, which substantially lowers the melting point of the alloy without producing brittleness. Most alloys in this series are non-heat-treatable and those containing appreciable amounts of silicon become dark gray when anodic oxide finishes are applied and hence are used for architectural applications.

5000 Series uses magnesium as the major alloying element. Magnesium is one of the most effective and widely used alloying elements and when used as the major alloying element (or with manganese) the result is a moderate-to-high strength, non-heat-treatable alloy. Magnesium is more effective than manganese as a hardener and can be added in considerably greater quantities. These alloys have especially good resistance to corrosion in marine atmospheres and sea water. Shade screening, venetian blinds and weatherstripping are often made from alloy 5052.

6000 Series is composed of heat-treatable alloys containing silicon and magnesium. Though not as strong as most alloys of the 2000 or 7000 series, the 6000 series alloys possess good formability and corrosion resistance with medium strength. They may be formed in the solution heat-treated temper (T4 temper) and then reach higher strengths by artificial aging (T6 temper). This series is the most commonly used in products for residential construction—6061 for such uses as structural shapes, nails and bars; 6063 for window and door frames, hardware, louvers, tubing, and curtain walls.

7000 Series has zinc as the major alloying element which, combined

FIG. 5 The Aluminum Association has developed a four-digit system of designating wrought aluminum alloys.

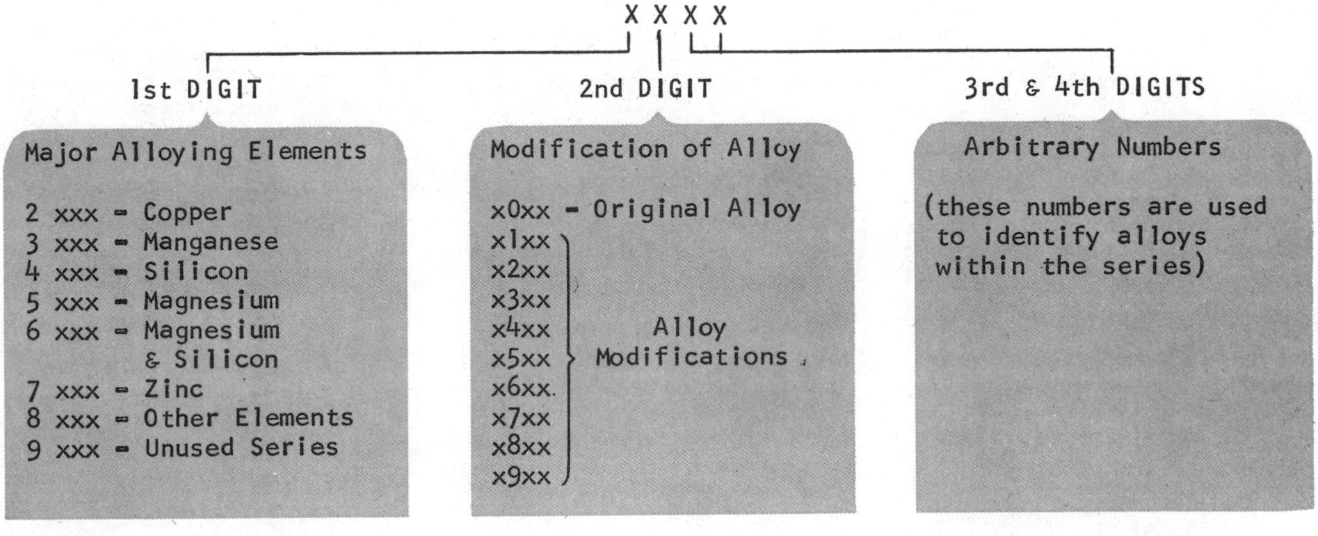

with a smaller percentage of magnesium, results in heat-treatable alloys of very high strength. Usually other elements such as copper and chromium are also added in small quantities. Among the highest strength alloys available is 7075. It is frequently used in aircraft structures and other high stressed parts.

Experimental Alloys are also designated with the groups described above, but carry the prefix "X" which is dropped when the alloy becomes standard.

Cast Alloys designations use a four-digit system similar to that for wrought alloys, shown in Figure 5.

Aluminum has a specific gravity of about 2.70, a density of approximately .170 lbs. per cu. ft., and weighs about 0.1 lb. per cu. in., as compared with 0.28 for iron and 0.32 for copper.

Pure aluminum melts at approximately 1200°F. Melting points for various alloys range from about 900°F to 1250°F.

Aluminum and aluminum alloys have relatively high coefficients of thermal expansion. For structural calculation, the coefficient of expansion is assumed to be 13 x 10^{-6}. This means, for example, that an unrestrained piece of aluminum 18′ long will expand or contract 0.259″ (approximately 1/4″) in a 100°F temperature change.

The two common metals used as electric conductors are copper and aluminum. The electrical conductivity of EC grade (electric-conductor) aluminum is about 62% that of the International Annealed Copper Standard. However, with a specific gravity of less than 1/3 that of copper, one pound of aluminum will conduct twice as much electricity as one pound of copper.

The high thermal conductivity of aluminum ("k" ranging up to 1540 for alloy 1100-0) makes it well fitted for use wherever a transfer of thermal energy from one medium to another is desired. However, high thermal conductivity is generally not required in most building material applications.

Aluminum is an excellent reflector of radiant energy through the entire range of wave lengths, from ultra-violet through the visible spectrum to infra-red (heat) waves. Aluminum's light reflectivity of over 80% accounts for its wide use in lighting fixtures, and its high heat reflectivity and low emissivity have led to its use as reflective insulation and improved performance in such applications as paint, roofing and ductwork.

Since aluminum is nontoxic, it is widely used in such products as cooking utensils, food processing equipment and aluminum foil food wrapping. Its nonmagnetic properties make it useful for electrical shielding purposes in electrical equipment and its nonsparking characteristics make it an excellent material for use around inflammable or explosive substances.

The ease with which aluminum may be fabricated is one of its most important properties. The metal can be cast, rolled to any thickness desired (down to foil thinner than paper), extruded, stamped, drawn, spun, roll-formed, hammered or forged; it can be turned, milled, bored, or machined at maximum speeds.

Almost any method of joining is applicable to aluminum and a wide variety of mechanical aluminum fasteners simplifies the assembly of products.

STRENGTH

Although pure aluminum has a tensile strength of only about 7,000 psi, by working the metal, as by cold rolling, its strength can be approximately doubled. Much larger increases in strength can be obtained by *alloying* aluminum with small percentages of one or more other metals such as manganese, silicon, copper magnesium or zinc. Like pure aluminum, the alloys are also made stronger by cold working, and heat-treatable alloys are further strengthened and hardened by *heat treatments* so that tensile strengths approaching 100,000 psi are possible (Fig. 6).

Strength decreases at elevated temperatures, although some alloys retain good strength at temperatures from 400° to 500°F. At subzero temperatures strength increases without loss of ductility so that aluminum is a particularly useful metal for low-temperature applications.

CORROSION RESISTANCE

Aluminum's high resistance to corrosion is due to the very thin inert surface film of aluminum oxide which forms rapidly and naturally in air. After reaching a thickness of about a ten-millionth of an inch, it effectively halts further atmospheric oxidation of the metal and thus protects the surface. If this natural oxide film is broken, as by a scratch, a new protective film forms immediately. The oxide film increases in thickness with temperature and remains protective to the underlying aluminum even at the melting point.

It may be necessary to protect aluminum in certain severely corrosive environments by protective coatings such as organic paints, by cladding, or by increasing the thickness and effectiveness of the oxide film by anodizing (see Finishes).

FIG. 6 MECHANICAL PROPERTIES OF REPRESENTATIVE ALUMINUM ALLOYS

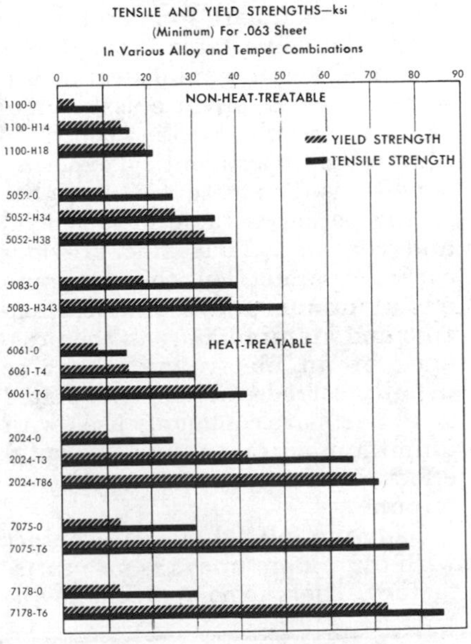

Weathering

The extent of corrosion due to weathering depends on the type and extent of contamination of the surrounding atmosphere. More corrosive elements are present in the industrial or marine areas than in rural areas.

In tests and observations (in part under the direction of the National Bureau of Standards), made over a 10-year period in different atmospheric conditions on alloys 1100, 3003 and 3004, the depth of penetration of atmospheric corrosion ranged from less than 1 mil (1/1000″) for rural areas, about 4 mils for industrial areas, and up to 6 mils in seacoast locations.

Galvanic Corrosion

When two dissimilar metals are connected by a solution which conducts electricity (an electrolyte), galvanic corrosion may occur. As in a battery, an electric current is created, and one metal corrodes away while the other is being plated. Galvanic attack can occur when moisture condenses from the air and contaminating elements act as an electrolyte. The threat of such corrosion is greatest in industrial areas because the atmospheric moisture is more contaminated, and in seacoast areas due to salt and moisture in the air.

The severity of galvanic attack varies depending on how far apart the two metals are from each other on the galvanic table (Fig. 7). For example, coupling mild steel to aluminum would produce less corrosion than would coupling copper to aluminum under identical exposure conditions. The metal that is higher in the potential table is sacrificial and corrodes. Thus zinc provides *cathodic* protection to aluminum and aluminum alloys. That is, when zinc and aluminum are in the presence of an electrolyte, the zinc usually corrodes. Thus, zinc-plated steel parts are commonly used with aluminum to prevent the harmful effect of galvanic action on the aluminum.

Cadmium-plated steel in contact with aluminum alloys is also satisfactory. There is no appreciable difference in the electrolytic potential of these two metals and hence practically no galvanic action.

Protection Against Galvanic Corrosion Aluminum is compatible with some stainless steel alloys, chromium, zinc and small areas of white bronze. Where permanent contact with other metals cannot be avoided, the risk of galvanic corrosion can be greatly reduced by painting the other metal and aluminum at the contact area with zinc chromate followed by one coat of a non-leaded paint such as aluminum paint or of a heavy-bodied bituminous paint. When the dissimilar metal cannot be painted, the aluminum may be given the insulating treatment. As an alternative, a strip of plastic or a similar insulator may be used in place of paint.

In severely corrosive atmospheres and high moisture areas the edges of the dissimilar metal joint may be sealed with compatible building mastic or caulking.

ELECTROLYTIC SOLUTION POTENTIALS*

Metal or Alloy	Potential in Volts
Magnesium	—1.73
Zinc	—1.00
520.0-T4 Alloy (Cast)	—0.92
7072 (Wrought)	—0.96
514.0 Alloy (Cast)	—0.87
Pure Aluminum and 5052-0, 1100-0, 1100-H18, 3003-H18, 6061-T6 Alloys (Wrought) and 43 Alloy (Cast)	—0.83
Cadmium	—0.82
356.0-T4 Alloy (Cast)	—0.81
2017-T4, 2024-T4 Alloys (Wrought)	—0.68
Mild Steel	—0.58
Lead	—0.55
Tin	—0.49
Brass (60-40)	—0.28
Copper	—0.20
Stainless Steel (18-8)	—0.15

*Measured in a 5.85% sodium chloride solution containing 0.3% hydrogen peroxide. The values vary somewhat depending on the particular lot of material and on the surface preparation employed.

FIG. 7 Coupling metals with dissimilar electrolytic potentials in an electrolyte results in galvanic corrosion of the metal with the highest negative potential. Degree of attack generally increases as potential difference increases.

A dangerous source of electrolytic attack may be water drainage from metals such as copper. Therefore, aluminum gutters on buildings with copper flashings must be protected by painting either the copper flashing or both the copper and aluminum.

Direct Chemical Attack

Direct chemical attack is the tendency of a chemical to dissolve the metal. Solutions of strong alkalis, sulphuric acid, hydrochloric acid, carbonates and fluorides tend to attack aluminum because they can dissolve or penetrate the protective oxide coating. Some chemical attack may occur to aluminum in contact with wet alkaline materials such as mortar, concrete and plaster, resulting in a mild etching of the aluminum surfaces. Hence, surface protection to aluminum products may be necessary during construction.

Where aluminum is *buried* in concrete, it is usually not necessary to paint the parts. However, if chlorides are present from additives or other sources in steel reinforced concrete, the aluminum parts should be coated with bituminous paint or other suitable coating.

Protection Against Corrosion When in contact with concrete, masonry or other absorbent materials under wet or intermittently wet conditions, aluminum should be protected with a coating of bituminous paint, zinc chromate primer, or a separating layer of plastic or other gasketing material. Creosote and tar coatings should not be used because of their acid contents. Where wet concrete or other alkaline material may splash against aluminum during construction, the aluminum surfaces may be protected by a coating of clear acrylic type lacquer or a suitable strippable coating if maintenance of appearance is important.

Although aluminum is inert when buried in many types of soil, some soils and the presence of stray electrical currents in the ground may cause corrosion. Therefore, buried aluminum pipe may have to be coated or otherwise protected.

Wrought mill products make up the greater part of aluminum products used in the building industry. They include plate, sheet, foil, bar, rod, wire, extruded sections, structural shapes, tube, pipe and forging stock. Mechanical properties of such materials are developed by alloying, hot working, cold working or heat treatment, or by combinations of these processes.

Cast products are produced by pouring molten metal into molds of desired shape. Cast products are manufactured in the foundry and include sand castings, permanent mold castings and die castings. The mechanical properties of some castings may be improved by heat treatment.

The general categories of wrought and cast products are outlined in Figure 8.

WROUGHT PRODUCTS

Wrought mill products are those shaped by hot and cold rolling, cold drawing, extruding and forging.

Hot and Cold Rolling

Hot rolling is performed in mills, rolling machines, through which ingot, heated to soften it, is passed until it is reduced to the desired shape (Fig. 9). Hot rolling may be continued until thicknesses of about 0.125″ are obtained. Further reduction in thickness is usually accomplished by cold rolling. Cold rolled material has a better finish and, in the case of non-heat-treatable alloys, increased strength and hardness.

Plate (minimum thickness 0.250″) Plate is made from a slab of aluminum which is hot rolled in a reversing mill (acting like a rolling pin) until reduced to a thickness of about 1″ to 3″. Further rolling operations are performed in an intermediate hot mill.

Sheet (thickness of 0.006″ to 0.249″) Sheet is rolled in a hot mill to about 0.125″. Thinner gauges are obtained by cold rolling for additional strength and better finish.

Coiled sheet is rolled to a finish gauge and width in a continuous strip up to several hundred feet in length and wound into coils. *Flat*

FIG. 8 CLASSIFICATION OF ALUMINUM MILL OPERATIONS USED TO MANUFACTURE WROUGHT PRODUCTS AND FOUNDRY OPERATIONS USED TO MANUFACTURE CAST PRODUCTS

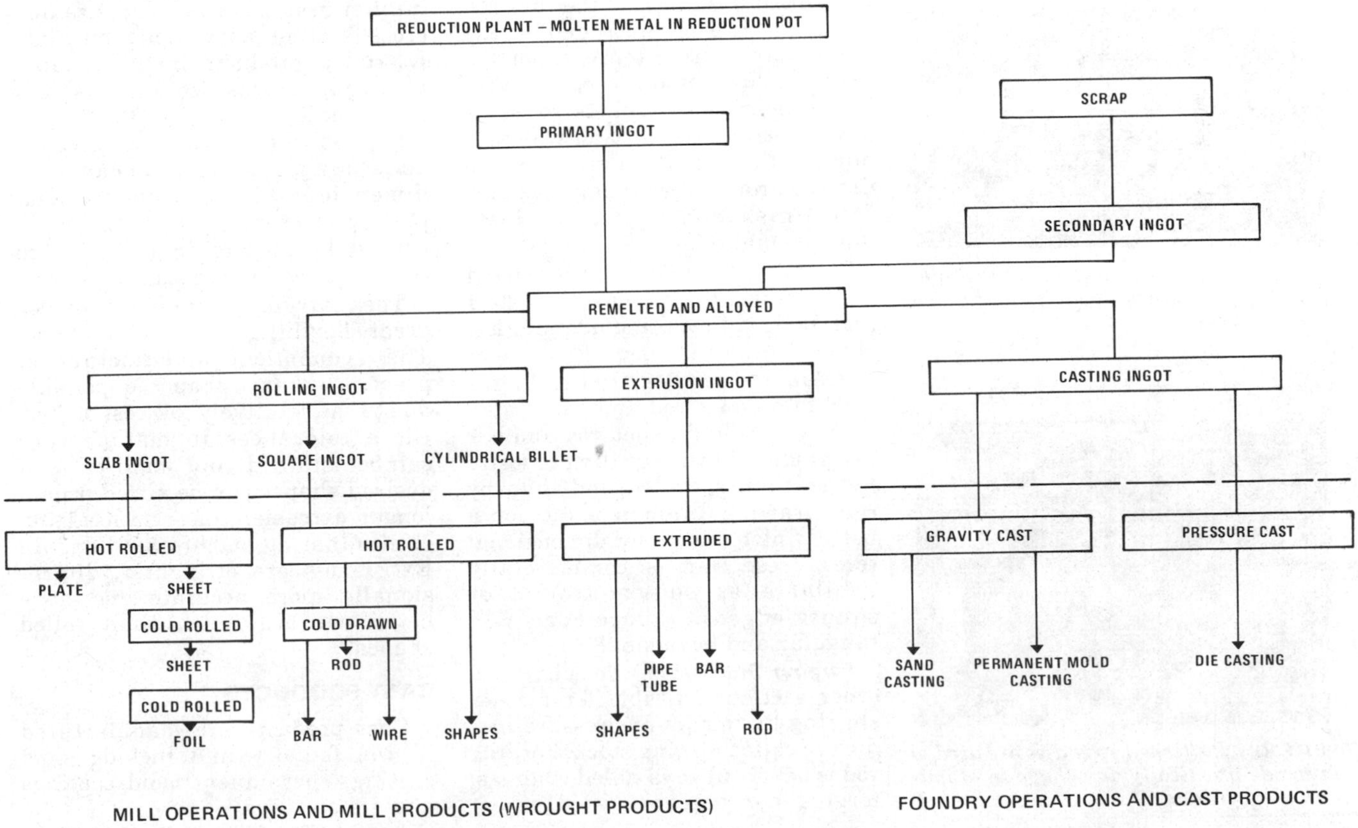

FIG. 9 Rolling machines, called mills, reduce heated ingots to desired thicknesses.

FIG. 10 Bar is hot rolled back and forth through die opening to reduce size. Note that rolls become smaller toward right end of mill.

FIG. 11 An extrusion press rams aluminum heated to a plastic state through a shaped die.

FIG. 12 Patterned steel dies are used to provide practically any shape in aluminum extrusions.

sheet is sheared to proper lengths from coiled sheet. It may be stretcher leveled for improved flatness.

Brazing Sheet is a special type of sheet which has a cladding that melts at a lower temperature than the core. Assemblies of parts made from this sheet can be brazed together by subjecting them to a temperature of about 1000°F, which does not melt the core alloy but only the cladding, joining the surfaces.

Aluminum Foil (thickness less than 0.006″), made by rolling sheet in cold rolling mills, is usually supplied in coil form. Its most common application in the building industry is for thermal insulation and vapor barriers.

Structural Shapes Though most structural shapes are extruded, some shapes may be hot rolled by a series of specially shaped rolls. The more complicated structural shapes require a greater number of roller passes to prevent undue straining of the metal in any single pass. Rolls are relatively expensive, since they require considerable metal with a high degree of machining and finishing to get the desired contour. (Hence, extruded structural shapes are often cheaper because of lower tooling costs.) Finishing operations include roller or stretch straightening and when required, the rolled shapes may be drawn through dies to improve surface finish and to obtain closer dimensional tolerances.

Standard rolled structural shapes, usually supplied in 2014 and 6061 alloys, include angles, channels and I-beams.

Rolled Rod and Bar Hot rolling is used to produce rod and bar shapes (Fig. 10). Cold finished rod and bar are produced by hot rolling to a size larger than specified and then by cold drawing through a die for a better finish and closer dimensional tolerances. Bar is commercially available as square (round or square-edged), square-edge rectangular and hexagonal.

Forging Stock Rod Rod, bar and other sections suitable for further shaping by impact or pressure in a die are called forging stock. Forging rod is nearly always rolled whereas, forging bar may be rolled or extruded, and forging shapes are usually extruded.

Redraw Rod (typically 3/8″ in diameter) This rod is suitable for further drawing to produce wire.

Cold Drawing

Drawn products are formed by pulling (drawing) metal through a die. The purpose of cold drawing is to obtain close dimensional tolerances and a better surface finish.

Wire Products Wire, which must be kept within close tolerances, is cold drawn by pulling redraw rod (3/8″ diameter) through a series of progressively smaller dies.

Tube Drawing Hollow tube sections are drawn in a process similar to drawing bar and rod, except the tube is threaded over a mandrel (shaper) positioned in the center of the die. When the tube is drawn, close tolerances both inside and out are maintained.

Extruding

Extruding is a process in which a heated aluminum ingot is pushed by hydraulic pressure through a die opening (Fig. 11). The extruded aluminum assumes the same size and cross section as the die opening. Extruded products include shapes of varying cross section, rod, bar and seamless tube (Fig. 12).

Shapes which cannot be extruded as a single section because of dimensional limitations or complexity of cross section can be obtained by fitting (mating) more than one section together.

The extruding process provides great flexibility of design shape. The economical manufacture of practically any shape is possible due to the relatively low cost of dies. Close tolerances in manufacture can be obtained, and small parts of desired shapes can be sliced from a longer extrusion, often minimizing or eliminating machining (Fig. 13). Extrusions are straighter, dimensionally more accurate and have better surface finish than rolled shapes.

CAST PRODUCTS

Cast products are manufactured in the foundry and include sand castings, permanent-mold castings and die castings.

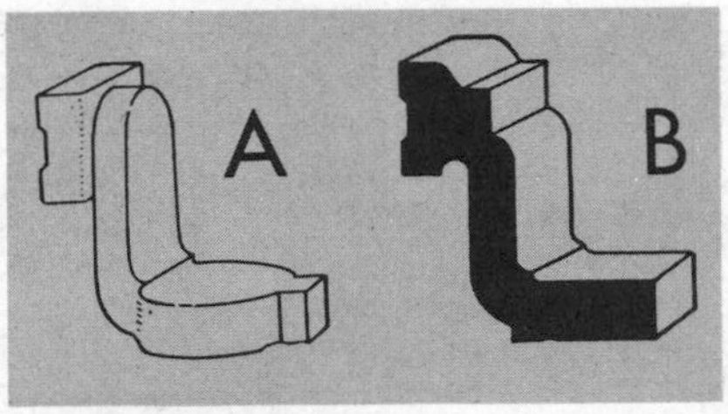

FIG. 13 Small castings, forgings or machined parts (A), may possibly be replaced with more economically produced pieces (B), sliced from an extrusion.

Sand Castings

Molds made of sand are lowest in cost and permit maximum flexibility in design changes, although labor costs are higher than for other castings. Sand castings require heavier section thicknesses than other methods, and tolerances are not as close. They are used for spandrel panels, decorative panels, lighting standard bases and general fittings.

Permanent-Mold Castings

This casting method employs metal molds with a metal core and is used where a large number of castings from the same mold is required. If a sand core is used, the casting is called a *semi-permanent* mold casting. Molten metal is poured into the mold by gravity.

Die Castings

Die castings are permanent-mold castings with the metal being forced into the mold under pressure. Die castings have smoother surfaces than permanent-mold castings, permit greater repetitive use of the molds. Die casting is generally used for the mass-production of small items such as handles, latches and other door and window fittings (Fig. 14).

FIG. 14 In die casting molten aluminum is forced into a mold. Hardware is commonly die cast (inset).

The main advantages of permanent-mold and die castings over sand castings are higher strength, better surface appearance and closer dimensional tolerances.

FABRICATION

Fabrication includes forming and machining operations performed on aluminum products and castings, after leaving the mill or foundry.

Forming

Aluminum, one of the most workable of common metals, can be fabricated into a variety of shapes by bending, stretching, rolling, drawing and forging.

Formability varies greatly with the aluminum alloy and temper.

Bending Temper and thickness greatly affect the minimum bend radius that can be used in forming. Certain aluminum alloys can be cold formed to a sharp 90° bend, others must be formed on stipulated radii.

Stretch Forming Sheets are stretched to flatten them or to form them into special shapes. The most suitable alloys for stretch forming are those with high elongation (ability to stretch without fracturing). The advantage of stretch forming is that there is hardly any tendency of the material to assume its original shape after forming stresses are removed.

Roll Forming Products are roll formed by passing strips through a series of roller dies (Fig. 15). Each successive pair of rolls shapes the metal slightly—only a small deformation is made by any single pair of rolls. No straightening is required since roll-formed products ordinarily are straight and free from distortion.

Forging High strength aluminum parts are often produced by forging (Fig. 16). Most forgings are made by hammering or pressing heated aluminum between closed steel dies. Press forging forces the metal into a die cavity of the desired shape under mechanical or hydraulic pressure. Hammer forging produces the same results by repeated blows from a drop hammer. Aluminum forgings may be heat treated after fabrication.

Since forging refines and provides desirable directional characteristics to the grain structure of the metal, improving strength and fatigue resistance, it is used to make various types of builders' hardware where strength is important.

Machining

The machining characteristics of aluminum and its alloys, especially of wrought products, are excellent. Sawing, drilling, tapping and

FIG. 15 A series of cold rolls form corrugated roof decking from embossed sheet.

FIG. 16 A large, 8000 ton hydraulic press produces forgings for use as architectural panels.

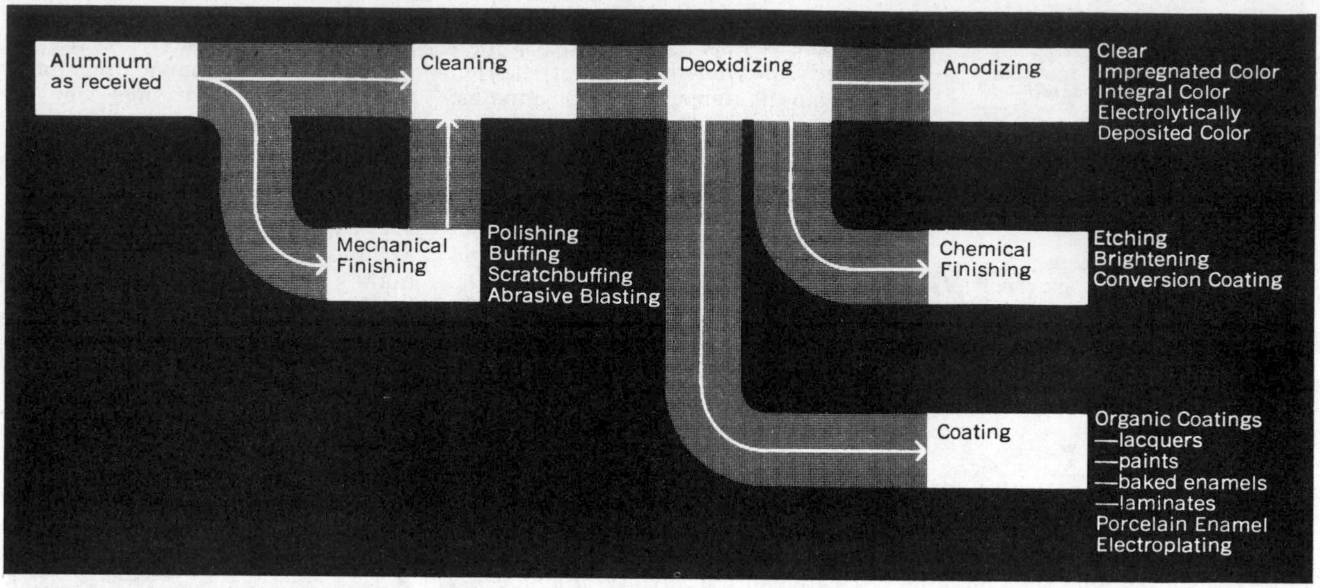

FIG. 17 Principal aluminum finishing processes.

threading methods vary little from those employed for other metals.

FINISHES

The surfaces of many aluminum products require no further finishing. The natural oxide film makes protective coatings unnecessary under ordinary conditions. However, finishes are frequently applied in order to: (1) change or improve the appearance for decorative purposes; (2) improve its abrasion or corrosion resistance; (3) provide a base for subsequent surface treatment such as painting.

Natural finishes and other common surface treatments are briefly described below (Fig. 17).

Uncontrolled Natural Finishes

Hot-rolled products show a certain amount of discoloration or darkening, while cold-finished products have a whiter, brighter surface. Extruded products have an intermediate appearance nearer that of cold-finished items and show traces of longitudinal striations (*die-lines*) caused by the extrusion process. Castings vary little in their surface appearance, except that die castings and permanent-mold castings are smoother than sand castings.

Controlled Natural Finishes

Natural finishes of cold-rolled sheet may be controlled by the degree of polishing given to the rollers of the rolling mill: *mill finish* has an uncontrolled, slightly variable finish which may not be entirely free from stains or oil; *one side bright mill finish* has a moderate degree of brightness on one side. The reverse side is uncontrolled and may have a dull, non-uniform appearance; *standard bright finish* has a relatively bright, uniform appearance on both sides, but somewhat less lustrous than the standard one side bright sheet; *standard one side bright* sheet has a uniformly bright and lustrous surface on one side, the reverse side is uncontrolled and may have a dull nonuniform appearance.

FIG. 18 Aluminum sheet and plate may be embossed providing a variety of surface textures.

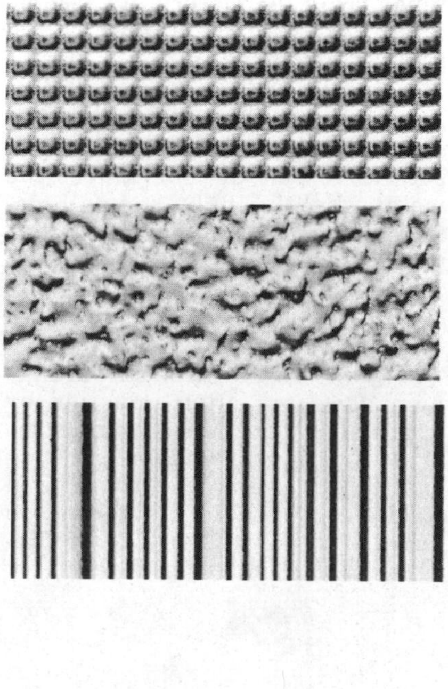

Embossed sheet is obtained by passing a mill finish sheet through a pair of rolls with a matched design or between a design roll and a plain roll. Embossed surfaces increase the stiffness and tend to "hide" wear and scratches. Patterns include wood grain, leather grain, fluted, stucco, diamond, notched, ribbed, bark and pebble designs (Fig. 18).

Mechanical Finishes

Aluminum surfaces may be mechanically treated by: *grinding* with a dry abrasive wheel, primarily for removing surface roughness from castings; *buffing* with a fine abrasive, resulting in a highly lustrous surface; *polishing* with finer abrasive than for buffing, and with a very light pressure, resulting in a mirror-like surface; *scratch-brushing* with power-driven, rotating wire brushes to give coarse or smooth lined surface textures, depending on the size of the wire bristles used; *shot-blasting* and *sand blasting* with various combinations of pressure, sand and steel shot to produce matte and course surfaces. Rough textures may be given a lacquer coat or may be anodized to minimize collection of dust.

Satin finishing is accomplished with fine wire bristles, very fine abrasives or a muslin or felt wheel to produce a soft smooth sheet. *Spin Finishing* is achieved by pressing an oily abrasive cloth against the aluminum piece while it is revolved rapidly. Subsequent applications of fine stainless steel wool and fine mesh emery give a bright, silvery finish with a pattern of concentric circles.

Highlighting provides a two-tone effect for relief surfaces such as on spandrel panels. The elevated areas are masked while the recessed areas are sand-blasted. Then the masking is reversed and the elevated areas are given a highlight effect by painting or polishing.

The Aluminum Association designations for various mechanical finishes are shown in Figure 19.

MECHANICAL FINISHES

TYPE OF FINISH	DESIGNATION*	DESCRIPTION
As Fabricated	M10	Unspecified
	M11	Specular as fabricated
	M12	Nonspecular as fabricated
	M1X	Other
Buffed	M20	Unspecified
	M21	Smooth specular
	M22	Specular
	M2X	Other
Directional Textured	M30	Unspecified
	M31	Fine satin
	M32	Medium satin
	M33	Coarse satin
	M34	Hand rubbed
	M35	Brushed
	M3X	Other
Nondirectional Textured	M40	Unspecified
	M41	Extra fine matte
	M42	Fine matte
	M43	Medium matte
	M44	Coarse matte
	M45	Fine shot blast
	M46	Medium shot blast
	M47	Coarse shot blast
	M4X	Other

* The complete designation must be preceded by AA—signifying Aluminum Association.

FIG. 19 Aluminum Association designations for various mechanical finishes.

FIG. 21 A chemical treatment spray line applies a conversion coating to extruded sections prior to painting.

Chemical Finishes

Many methods of treating aluminum surfaces employ the reaction of various chemicals on the metal. Some of the most common are discussed here. The Aluminum Association designations for various chemical finishes are shown in Figure 20.

Etch Finish A decorative, silvery white finish, also known as "frosted finish" may be produced by etching the metal with caustic soda (sodium hydroxide) or other chemical. Several methods permit partial or *design* etching to produce patterns of varying finish and colored background.

Conversion Coatings Although chemical conversion coatings are generally used on aluminum to prepare the surface for painting, some types may also be used as a final finish (Fig. 21).

Since the natural oxide film on aluminum surfaces may not provide a suitable bond for paints and other organic coatings and laminates, its chemical nature is often "converted" to improve adhesion. *Conversion films* or *conversion coatings* are generally applied by use of phosphate or chromate solutions, many of which are proprietary in nature. One of the simplest methods for paint surface preparation is to etch the surface with a phosphoric acid solution, thus providing a good mechanical bond.

Chemical Oxidizing Chemical oxide films have improved resistance to corrosion. As a base for painting, they provide additional protection in the event that the finish is scratched or broken. They are more economical and easier to apply than anodic processes (described below) which also produce oxide films, but are thinner and softer than anodic finishes. Chemical oxide coatings may be dyed, but anodic films are preferred if a superior color quality is desired.

Zincating In this process, aluminum is immersed in a zincate bath providing a very thin, firmly adherent coating of zinc. This base is often used for aluminum which is to be electroplated.

Chemical Brightening A mirror-bright surface finish can be provided on aluminum by immersing the parts in a hot chemical solution, followed by subsequent rinsing. Depending upon the alloy and its mechanical preparation, reflec-

CHEMICAL FINISHES

TYPE OF FINISH	DESIGNATION*	DESCRIPTION
Nonetched Cleaned	C10	Unspecified
	C11	Degreased
	C12	Inhibited chemical cleaned
	C1X	Other
Etched	C20	Unspecified
	C21	Fine matte
	C22	Medium matte
	C23	Coarse matte
	C2X	Other
Brightened	C30	Unspecified
	C31	Highly specular
	C32	Diffuse bright
	C3X	Other
Chemical Conversion Coatings**	C40	Unspecified
	C41	Acid chromate-fluoride
	C42	Acid chromate-fluoride-phosphate
	C43	Alkaline chromate
	C4X	Other

* The complete designation must be preceded by AA—signifying Aluminum Association.

FIG. 20 Aluminum Association designations for various chemical finishes.

FIG. 22 Assembled window frames are immersed in electrolyte tanks for anodizing treatment.

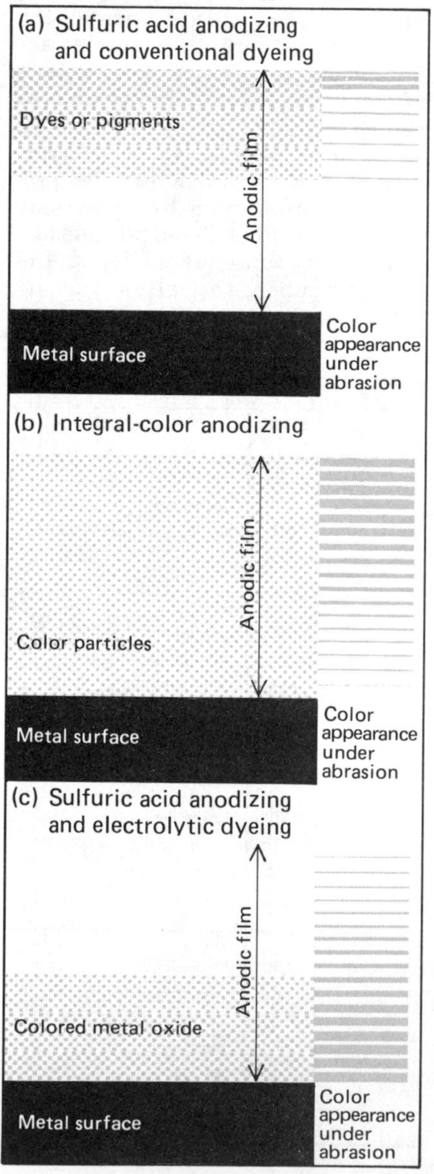

FIG. 23 Distribution of coloring elements in the anodic film of three main color anodizing processes.

tivity values up to 85% may be obtained.

Anodized Finishes

Anodized electrolytic oxide *coatings* are readily applied to aluminum alloys and are widely used. Thinner anodic "flash" *films* may be used as a pre-paint treatment.

The distinction between a coating and a film is thickness. A film is less than 0.1 mil; any thicker covering is referred to as a coating.

Ordinarily, clear anodic coatings on architectural products are preceded by either a chemical etch and/or some form of mechanical finish. Etching is the most economical and common pre-finish, imparting a frosted appearance which helps to conceal variations in the mill finish and die lines on extrusions. Mechanical finishes also impart a pleasing texture which is retained after anodizing.

If neither etching nor mechanical finishing is specified, the aluminum still must be chemically cleaned prior to anodizing to insure satisfactory results. Contaminants which are deposited on the metal surfaces during fabrication processes may affect the anodic coating and should be completely removed before anodizing.

Anodizing consists of immersing the product in an acid solution (*electrolyte*) and passing an electric current, usually of the direct type, between the metal and the solution with the aluminum serving as the anode (Fig. 22). This results in the controlled formation of a relatively thin but durable oxide or coating on the surface of the metal.

In general, such coatings are many times thicker than the natural oxide film, and may be transparent, translucent or opaque, depending on the alloys and/or electrolyte used. Anodic coatings do not change the surface texture of the metal, but they greatly increase resistance to corrosion, and provide increased resistance to abrasion.

Several anodizing processes are commonly used. The principal differences between these processes are the type of solution used for the electrolyte, the voltage and current required, and the temperature of the bath.

The *sulphuric acid process*, because of its relatively low cost, is most widely used for architectural anodizing. It produces comparatively thick, transparent and absorptive coatings which are suitable for receiving coloring dyes and pigments (see below). Thin films may be produced by this process, which are excellent pre-treatments for paint, enamel and lacquer coatings.

Color Anodizing There are now three methods of color anodizing in common use: sulfuric acid anodizing and dyeing; integral-color anodizing; and sulfuric acid anodizing and electrolytically deposited color (Fig. 23).

In sulfuric acid anodizing and conventional dyeing, the work is rinsed after it leaves the anodizing bath, and is dipped in a bath containing either an organic dye or an inorganic pigment. To set the dye, the work is dipped in a nickel acetate solution and is sealed (usually in boiling de-ionized water) to produce a clear film that provides minimum interferences with, and optimum lightfastness of, the color.

A wide range of colors is possible with this process, but not all of them are satisfactory for outdoor service. As indicated by Figure 23, the color is concentrated in the top third of the anodic film; therefore it is necessary to adjust dyeing conditions (time, temperature, dye concentration and pH) to get similar colors with different film thicknesses.

In integral-color anodizing, the coloring material is spread uniformly through the anodic film. Integral-color anodic films may be produced by one of two different methods, or by a combination of both. The first method makes use of an alloy with constituents that color the anodic film as it grows. The second method makes use of an electrolyte that yields a colored precipitate during anodizing, again coloring the anodic film as it grows.

A combination of these two methods increases the range of possible colors. The colors produced by integral-color anodizing are very light-fast but, like sulfuric acid anodizing and electrolytically deposited color, the depth of color varies with the anodic film thickness. In addition, alloy is a major variable; some colors are restricted to specific alloys.

Sulfuric acid anodizing and electrolytically deposited color concentrates the color at the base of the anodic film, Figure 23. Consequently, the color receives maximum protection from sunlight and weathering.

Sealing In most applications, the pores in the anodic film are sealed off to increase resistance to staining and corrosion and to improve color durability. Because most applications require a clean film, immersion of the work in boiling deionized water with a pH of 5.5 to 6.5 is the usual method of sealing. When maximum corrosion resistance is required, potassium dichromate is added to the water.

The thickness of anodic films and coatings range from less than 0.1 mil, in the case of primer films for receiving organic finishes, to 5.0 mils for some of the engineering hardcoats.

For architectural applications the thickness and weight of coatings are of prime importance; no coating thinner than 0.4 mil should be used even for interior purposes. Coatings for exterior parts which will be subject to contaminated atmospheres, or will not receive regular maintenance care, should have at least an 0.7 mil thickness and a minimum coating weight of 27 mg. per sq. in.

Classification of Anodic Coatings The Aluminum Association identifies the various anodic finishes under four general types: General Anodic Coatings, Protective and Decorative Coatings, Architectural Class I Coatings and Architectural Class II Coatings. The first two classifications apply to general industrial work. Therefore, only the last two, for architectural products, are discussed below. The Aluminum Association designations for anodic coatings are shown in Figure 24.

Architectural Class I Coatings are more than 0.7 mil in thickness, weigh more than 27 mg. per sq. in. and include the hardest and most durable anodic coatings available. They are suitable for exterior items which receive no regular maintenance care, and for interior architectural items subject to normal wear.

Architectural Class II Coatings range in thickness from 0.4 to 0.7 mils, with corresponding weights of from 17 to 27 mg. per sq. in. These are appropriate for interior items not subject to excessive wear or abrasion, and for exterior items such as store fronts and entrances which are regularly cleaned and maintained.

Paint Coatings

A substantial volume of aluminum is painted for both decorative and protective purposes. The most important applications are residential siding, mobile home cladding, sliding doors, awnings, canopies, and swimming pools.

The trend in painting aluminum is towards continuous processing with pretreatment, coating and curing integrated into a single operation. Roller coating enables the painting of sheet at line speeds up to 600 feet per minute.

First the bare coiled sheet is passed through a series of cleaning and surface preparation baths to provide a surface to which the paint will strongly adhere; then it is roller coated with an organic paint; and, finally, is passed through an oven where the paint is baked hard.

The result is a product that possesses an excellent combination of hardness, ductility, adhesion, and gloss; and that can undergo further fabrication without harming the painted surface. Castings and extrusions are commonly sprayed.

Electrostatic spraying, in which electrically-charged coating particles are attracted to the surface by maintaining it at the opposite charge, provides a more uniform finish than conventional spraying with less waste.

Vitreous Coatings

Porcelain enamel is a vitreous inorganic coating bonded to metal by fusion at high temperatures. A porcelain enamel coating significantly improves the existing properties of aluminum, increasing durability and corrosion resistance. Practically any texture and lightfast color is available.

Electroplating

Chromium and copper plates are the most common and easiest to apply electroplated finishes. *Chromium* is readily plated on aluminum either after zincating, or after copper, brass or nickel plating. Lustrous chrome plate is easy to obtain by buffing. Chromium plates are generally applied for decorative effect and/or resistance to abrasion.

Aluminum may be *copper plated* after the zincate treatment or by other methods including anodizing in phosphoric acid. Copper may also be electroplated after first precleaning and zincating the aluminum. Copper plates on aluminum are used where soldering operations are required, for certain electrical

ANODIC COATINGS

TYPE OF FINISH	DESIGNATION*	DESCRIPTION
General	A10	Unspecified
	A11	Preparation for other applied coatings
	A12	Chromic acid anodic coatings
	A13	Hard, wear and abrasion resistant coatings
	A1X	Other
Protective & Decorative (Coatings less than 0.4 mil thick)	A21	Clear (natural) coating
	A22	Coating with integral color
	A23	Coating with impregnated color
	A24	Coating with electrolytically deposited color
	A2X	Other
Architectural Class II (0.4 to 0.7 mil coating)	A31	Clear (natural) coating
	A32	Coating with integral color
	A33	Coating with impregnated color
	A34	Coating with electrolytically deposited color
	A3X	Other
Architectural Class I (0.7 mil and greater anodic coating)	A41	Clear (natural) coating
	A42	Coating with integral color
	A43	Coating with impregnated color
	A44	Coating with electrolytically deposited color
	A4X	Other

*The complete designation must be preceded by AA—signifying Aluminum Association.

FIG. 24 Aluminum Association designations for anodic finishes.

requirements, for a base for chromium or silver plating, and for decorative items.

Silver plating decreases electrical resistance and is sometimes used for bus-bar connections. *Tin* and *brass* plates are especially good as a base for soldering. *Zinc* and *cadmium* plates provide corrosion protection. Chemically treated *nickel* plates yield a black decorative finish.

Electropolishing

High purity aluminum may be treated by electropolishing to obtain a light reflectivity of about 85%. The metal surface is anodically smoothed to a highly polished surface suitable for use in aluminum light fixture reflectors.

Electropolished reflecting surfaces are usually given a final anodizing treatment for surface protection. Final reflectivity depends on the purity of the aluminum, the thickness of the oxide film, and the particular procedure employed.

Laminated Coatings

For special use and design applications, many other materials may be laminated to aluminum surfaces. For example, polyvinyl chloride and vinyl films are used to provide non-metallic color and long lasting finish for exterior applications such as aluminum siding. The film is adhesive bonded to the aluminum surfaces which have been suitably prepared, usually by application of a chemical conversion coating.

Temporary Protective Coatings

Waxes and lacquers applied to aluminum surfaces provide some weather protection and make cleaning easier. Aluminum so treated can usually be kept clean by wiping with a damp cloth or rinsing with clear water.

Lacquer Clear lacquer, applied by the fabricator or contractor, provides the most common type of temporary surface protection for aluminum. Methacrylate lacquers, commonly used are resistant to alkalis such as lime mortar, plaster and concrete and provide some protection against abrasion. Under normal conditions, the methacrylate coating lasts for several years and weathers by chalking rather than by unsightly peeling.

Wax Liquid waxes or wax base cleansers offer some protection against atmospheric weathering and retard grime accumulation on aluminum surfaces. However, after repeated waxings, surfaces may tend to yellow. If necessary, the wax may be removed with a solvent or mild abrasive cleaner before rewaxing.

Strippable Coatings These rather expensive coatings are usually applied by the fabricator to protect surfaces from the normal hazards encountered during construction, and are ideal for finer finishes which may be scratched or damaged more easily. Strippable coatings should be left in place until the final cleanup operation. Aluminum with such coatings should not be stored in places exposed to sunlight, since the coatings may tend to get "tacky" and difficult to remove. If they cannot be easily removed, a solvent should be used.

Tapes For protection against abrasion or staining on smaller areas, various types of tape may be used. They may be applied either by the fabricator or by the contractor.

Maintenance of Surfaces

Cleaners that may be used with aluminum vary in strength and abrasiveness. The type of cleaner used should be determined by the aluminum finish, the degree of soiling and the size of area to be cleaned.

Cleaning In general, aluminum should be washed frequently with mild cleansers rather than occasionally with abrasive, heavy-duty cleansers. Products should never be mixed. When abrasives are used, the appearance of the aluminum finish may be altered. To avoid streaking, work should be confined to the shady side of the building or done on cloudy days. Applying cleaners on sun-warmed metal or on cold days should be avoided. All cleaners should be spot tested before general use. The following cleaners are listed in increasing order of harshness.

Mild cleaners should be used on aluminum that is lightly soiled or subject to frequent cleaning. These include most soaps and detergents that can be applied with bare hands. Even mild cleaners should not be permitted to remain on the metal longer than necessary. They should be sponged on and removed with a clear water rinse.

Solvent and emulsion cleaners are effective on oil and grease stains and some types of soil. Like mild cleaners, solvents have little effect on weathered aluminum and heavy grime encrustation. Some solvents, however, can soften or dissolve certain paints and lacquers. Cleaners should be applied with one clean cloth and removed with another. Remaining residue should be removed with mild soap and water.

Abrasive cleaners may contain water, oil, wax, soap, silicones, acids or alkalines, in addition to an abrasive. They are packaged under such names as polish, cleaner, wax-polish, metal brightener or scouring powder. In general abrasive cleaners are sprinkled on a clean, damp pad and applied to the metal with a light rubbing action. A separate cloth is used to remove the cleaner. In some cleaners a wax or silicone remains which produces a shine and protects the metal somewhat.

Steel wool (size 00 or finer) or abrasive nylon pads may be used alone or with mild or abrasive cleaners. All steel wool particles must be removed since these may rust and cause stains.

Etching cleaners are strong chemicals which may change the appearance of the finish and may permanently harm the aluminum surface unless thoroughly rinsed. Cleaner concentration, application time and safety precautions suggested by the manufacturer should be followed carefully.

Cleaning work should always start from the top of the area and work down. The solution should be evenly rubbed on a small area at a time with a sponge or brush. The effectiveness of etching cleaners can be increased by using fine steel wool or nylon pads or by adding a fine abrasive. Care should be taken to avoid or to immediately wipe up any drips or spills. The cleaned area, especially cracks, grooves and holes should be thoroughly rinsed and dried as soon as the work is complete.

Special cleaners include steam cleaners and power-driven wire brushes. Steam, used alone or with a cleaning compound, is useful for removing grease or oil-based deposits and for cleaning large areas. The area to be cleaned should be spot tested.

Stainless steel or German silver wire brushes are used in cleaning large heavily weathered areas. Since power-driven wire brushes remove metal, they must be used carefully.

JOINTS AND CONNECTIONS

Practically every method of joining and connecting is used in assembling aluminum products including: mechanical fastenings, various methods of mechanically forming connections, welding (and brazing and soldering) and bonding with adhesives.

Mechanical Fastenings

A great variety of *permanent* and *semi-permanent* fasteners are used to join aluminum to itself and other materials. Permanent fastening requires destruction of the fastener or the product itself for disassembly. Semi-permanent fastening includes methods such as screwing and bolting that permit non-destructive disassemby.

Riveting This is a method of permanent mechanical fastening in which a headed pin (rivet) inserted through the pieces being joined is mechanically deformed so that a head is formed on both sides of the work, tightly gripping the pieces together. Riveting is probably the most widely used aluminum fastening method (particularly in the aircraft industry) and makes use of a great variety of rivet shapes, sizes and assembling techniques.

Aluminum rivets are typically used in joining aluminum for neater appearance and uniform finish, avoidance of galvanic corrosion, low cost and light weight.

Aluminum rivets are available in all standard designs and sizes with the usual round, brazier, flat and countersunk heads.

Rivets can be made in practically any alloy, the following alloys meet most needs: 1100, 2024, 2117, 5056, and 6053. Alloys 1100 (commercially pure aluminum) and 5056 are non-heat-treatable. The others are heat-treatable. In general, rivets to be used with parts of a given alloy should match that alloy in mechanical and physical properties.

Although standard rivets meet requirements, many special types are available including tubular and other designs for blind riveting (where the work is accessible from only one side).

Stitching Joining of various metals to other metals and other materials by wire sewing is called stitching. Aluminum's relatively low density and softness make it very suitable for stitching.

Stitching requires no backing material, pre-punching or drilling. The work need not be clamped nor the metal cleaned. Inspection is simple since it can be determined visually whether a stitch has penetrated all work and is properly clinched. Stitching is used in assembling aluminum garage doors, on sheet metal jobs such as ductwork and blower installations, and in joining aluminum to rubber, fabrics and other materials.

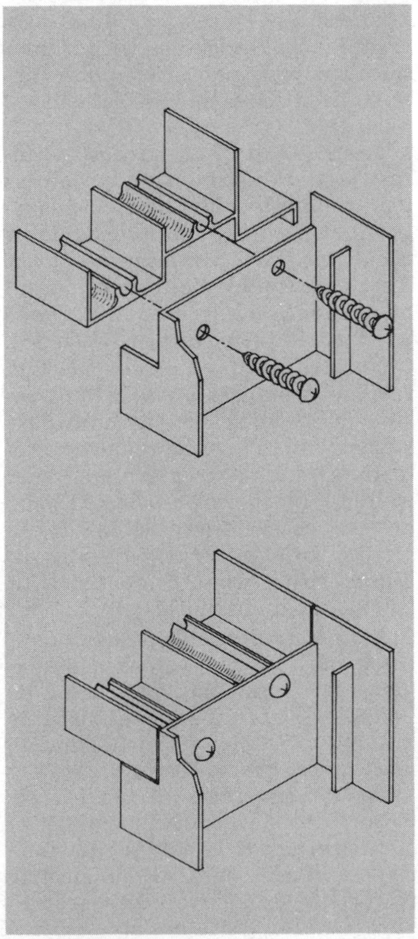

FIG. 25 Extruded window and door frame sections are often connected with screw fasteners into continuous slots.

Threaded Fasteners Semi-permanent fasteners, widely used in assembling aluminum products, include a multitude of screw and bolt types. Whatever fastener is used, however, it should be aluminum or metal compatible with aluminum.

Screws for use with aluminum are usually made from alloy 2024-T4 and are manufactured in practically any head type, size and standard thread (see Fig. 25).

Bolts and nuts and washers are usually made from alloy 2024-T4, 6061-T6 or 7075-T73, and are manufactured in practically any head style (typically hexagonal), size and standard machine thread.

Nails—Because of their freedom from corrosion and staining, aluminum nails are popular fasteners for roofing and siding made of aluminum or of non-metallic materials. The nails are manufactured from alloys 6061 and 5056 which, in this form, have tensile strengths in excess of 60,000 psi.

Special Fasteners Many other special fasteners and rivets, are available for blind fastening and other connections. Blind fasteners include pop rivets, explosive rivets and others where only one side of the work is accessible. Other special fasteners such as speed nuts and quick release fasteners are also commonly used.

Mechanically Formed Joints

Connections obtained by forming or cutting the parts to be joined in such manner that they interlock are called mechanically formed joints. Those used for aluminum work do not differ basically from such joints in other metals. Where high strength is of primary importance, other joining methods will usually be preferred. But the building industry offers many applications, such as roof work, ducts, gutters, downspouts and railings where mechanically formed joints are quite satisfactory (Fig. 26).

Staking The staked joint is commonly used in assembling extruded aluminum members. Tabs or prongs formed at the end of one member are inserted through mating holes in the member to be joined and are twisted, bent over or spun down (Fig. 27).

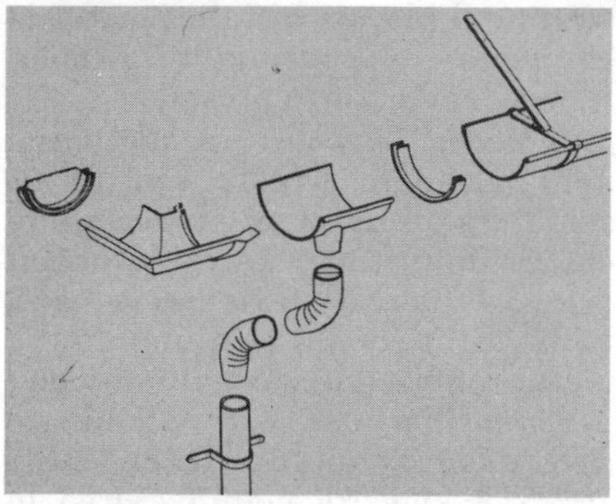

FIG. 26 Mechanically formed slip joints are used to assemble aluminum gutters and downspouts.

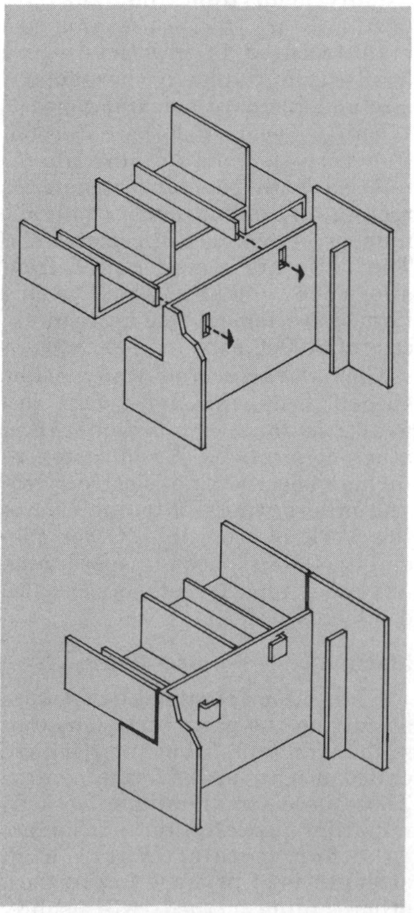

FIG. 27 Staked joints are often used in assembling frame members of windows and doors.

Welding, Brazing and Soldering

Welding (including brazing and soldering) is a method of forming a metallurgical bond between two or more members of an assembly. The selection of a welding process for a particular job depends upon many variables including conditions encountered in service, shape and thickness of parts, volume of production, and the alloys employed.

The choice of filler metal alloy for welding depends on joint design, service conditions, surface finish, welding process, alloy or alloy combinations to be joined and appearance. Therefore, filler alloys vary considerably and must be selected for each particular circumstance.

There are many welding processes suitable for use with aluminum alloys. Some of the more important processes are discussed here.

Gas Welding The earliest welding process used on aluminum is gas welding, which employs oxygen, acetylene, oxyhydrogen or other gas flame as the heat source. The gas flame melts the parent metal and a flux coated filler rod is used to complete the weld (Fig. 28).

Disadvantages include the flux residue, a potential source of corrosion or weld defects in multipass welds, which must be removed; distortion, which is greater than in arc welding; a wider heat-affected zone; and lower welding speeds. These conditions generally result in higher costs, when compared with gas shielded arc methods.

Aluminum from about 1/32″ to 1″ thick may be gas welded. Heavier material is seldom gas welded as heat dissipation is so rapid that it is difficult to apply sufficient heat to melt the metal with a torch.

Gas welding finds its greatest use today where few or occasional welds are required. It is also the obvious choice in the field, when electric power lines or portable generators for arc welding are not accessible.

Arc Welding Most forms of arc welding depend upon the production of heat from high voltage electric flow (arc) from an electrode to the parent metal (Fig. 29).

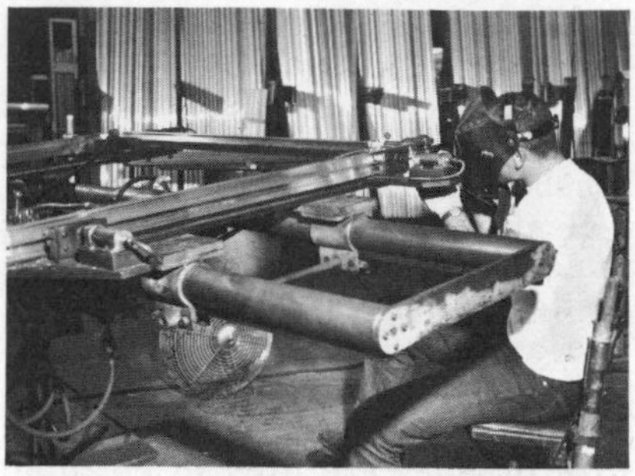

FIG. 28 Extruded window frames are connected by gas welding.

Inert Gas Shielded Arc Welding Of all the fusion welding processes the inert gas shielded arc processes are by far the most important and most widely used. In the inert gas tungsten arc (TIG) process, an arc is drawn between the work and a non-consumable tungsten electrode, the arc being shielded by a flow of inert argon or helium. Filler metal is added manually. In inert gas metal arc welding (MIG) the arc in inert gas, is drawn between an automatically fed filler wire and the work. Fluxes, used to dissolve the aluminum oxide surface film on the metal in other welding processes, are not required for these because of the action of the arc and the inert gas shield (Fig. 29).

These processes are applicable to all welding positions. Thinner gauges are commonly welded by

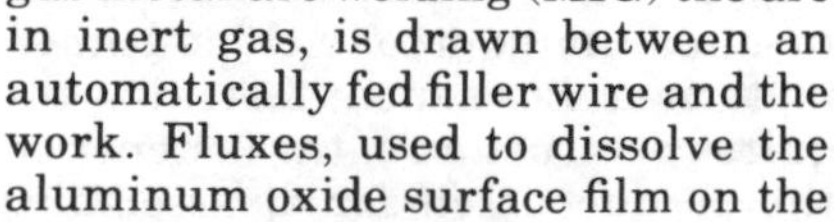

FIG. 29 Overhead welding where a pipe is joined in its final position presents no problem for inert gas shielded arc welding (MIG).

TIG. MIG is particularly appropriate for automatic welding; TIG also if filler metal is not required. Since welding speeds are high, the effect of heat input on strength is minimized.

Resistance Welding This type of welding uses an electric current conducted through the parts to be joined. Heat is generated at the juncture of the two parts (due to the resistance to the passage of heavy current) and when combined with applied pressure, fusion is produced. Resistance welding has practically no effect on the alloy temper since the weld is made so quickly that the adjoining area does not heat appreciably (Fig. 30).

Flash welding is a type of resistance welding used to butt weld together shapes such as tubes, extruded shapes, rods, bars and wire. Typical is the miter joining of aluminum window sections. In general, resistance welding is used where technically practical and where mass production is involved.

Spot and Seam Welding are resistance welding methods especially important in fabricating high-strength heat-treated sheet alloys which can be so joined with practically no loss in strength. Spot welding is widely used to replace riveting, joining sheets at intervals as required (Fig. 31). Seam welding is merely spot welding with the spots spaced so closely that they overlap, producing an air-tight joint where desired.

Brazing When the welding process uses a filler metal that is nonferrous, or an alloy with a melting point higher than 800°F, but lower than that of the materials joined, it is called brazing (Fig. 32).

In brazing, the molten metal is called *brazing filler metal*; in soldering it is called *solder*.

Brazing is employed to join millions of aluminum parts each year, which range in thickness from foil to heavy plate and castings. It is used for joints which because of their complexity of contour, limited access, or great ratio of length to part size are unsuitable.

Filler alloys for brazing aluminum include such aluminum-silicon alloys as 4047 and 4145. They have liquidus temperatures much closer to the solidus temperature of the base metal than filler alloys for brazing most other metals. Consequently, temperature control is critical, and only those aluminum alloys with relatively high melting points are brazeable—e.g., alloys 1050, 1100, 3003, 6063, and the self-aging foundry alloy Alcan 432.

Although alloy 6061 can be brazed, the melting range of this alloy is very close to that of the alloy 4047, the commonest brazing filler alloy. Alloys with high magnesium content are not considered brazeable because of poor wetting by the filler alloys. Flux is essential to remove the oxide film.

FIG. 30 In resistance welding adjoining members are fused together by the heat of an electric current.

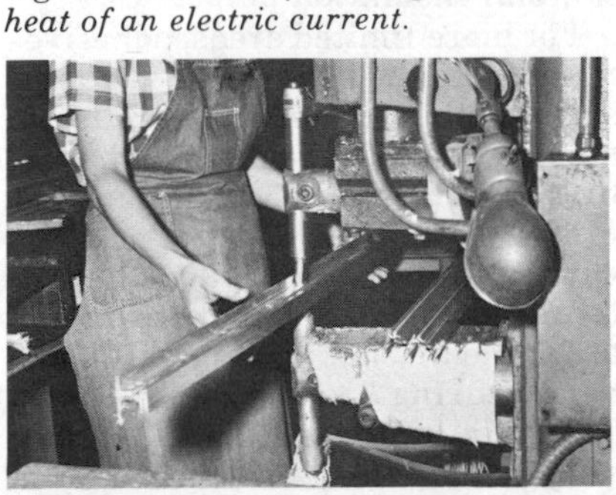

FIG. 31 Electric current flow from the electrode through the extruded members produces a spot weld.

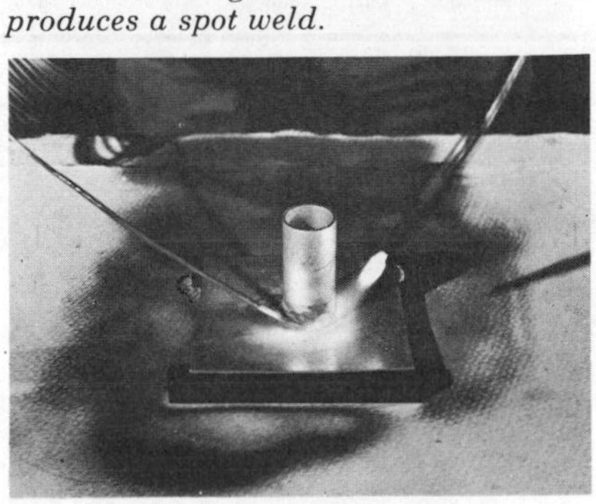

FIG. 32 In brazing the torch heats the parts being joined and the brazing alloy is provided by the filler rod.

Aluminum may be brazed by a number of methods, differing mainly in the way that heat is applied to the work. These include torch brazing, furnace brazing and salt bath brazing. *Torch brazing* is employed where the joints are accessible both for heating with the torch flame and for feeding filler metal into the joint. In *furnace brazing,* the entire work assembly is held in a furnace for one to five minutes.

Due to the capillary action of the brazing alloy, the furnace brazing process is particularly useful for inaccessible joints. In the *salt bath* process, (also known as "dip" brazing) the work is submerged in molten salt which performs as the heating medium and as the flux. Time in the salt bath varies from 30 to 180 seconds.

Whatever the process, it is essential that surfaces be thoroughly cleaned prior to brazing. For 1100 and 3003 alloys, degreasing is usually sufficient; for other alloys, an etching treatment is recommended.

Under some conditions, brazed joints develop strengths comparable to welded joints, make more economical use of filler metal, can be made faster and have a neater appearance (requiring little finishing work, if any) and brazing makes possible the joining of extremely thin sections. In general, the corrosion resistance of brazed joints is comparable to that of welded joints.

Soldering Filler metals (*solders*) with melting points under 800°F are used in soldering.

Difficulties in soldering aluminum are caused by the great affinity of aluminum for oxygen and the lack of bond between solder and the natural oxide film present on aluminum. Ordinarily a suitable flux is applied to dissolve the oxide and prevent it from reforming, or otherwise react, so that the solder may bond directly to the underlying aluminum. Another method to get below the oxide film is to mechanically abrade it down through an overlaying layer of molten solder, using the soldering iron, a wire brush or other tool.

When soldering aluminum to other metals, the aluminum surface should be tinned (pre-coated with solder) and then joined to the other metal with solder and flux. If any difficuty is encountered, the other metal surface should also be tinned first.

The essentials of soldering aluminum are removal of the surface oxide and use of a filler metal which will not encourage galvanic corrosion. Hence, appropriate fluxes are used, and higher melting solders, pure zinc and zinc-aluminum (95-5) are used where corrosion is a factor. Where corrosion resistance is less important, lower melting zinc-tin and lead-tin alloy solders are used.

In corrosive atmospheres, lacquer coatings may be used to improve the corrosion resistance of soldered aluminum.

Adhesive Bonding

Adhesive bonding utilizes adhesives for bonding aluminum to aluminum or other materials, and is gaining rapidly expanding use.

Design possibilities are limited only by extremes of stress, temperature, corrosive environment and cost. One special consideration in designing joints for the use of adhesives is that peeling stresses should be kept as low as possible. The pretreatments of aluminum used in adhesive bonding are concerned mainly with cleaning the surface. Where the adhesive bond must develop the greatest strengths, a chromic-sulphuric acid dip is often used as a chemical cleaning method.

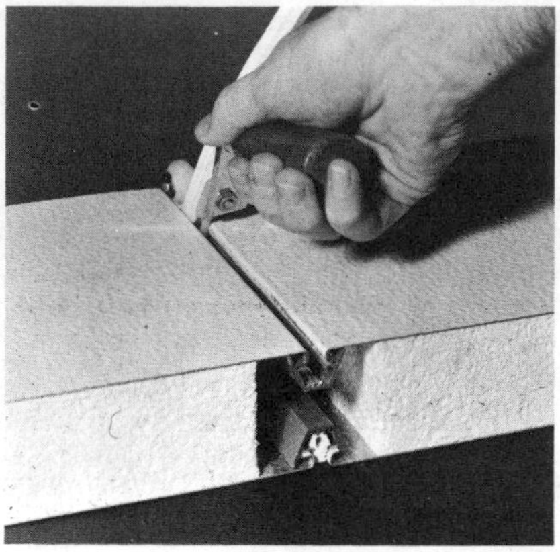

FIG. 33 Sandwich panels may be made from prefinished aluminum sheets bonded to foamed plastic cores. Panels here are joined with flexible plastic sealing strips.

Adhesives themselves may be applied by any of several techniques. Roller coating is particularly effective for high speed application of a thin coat of adhesive, as in the production of laminates. Various building materials are currently being adhesively laminated with aluminum foil to increase their thermal insulation properties.

For more limited areas, adhesives may be applied by brush, blade or spray. Certain adhesives are being adapted to mass production practice by manufacturing in film form. The film is cut into sheets, ribbons or special shapes which follow the contours of specific joint areas and assembled with the metal parts before curing.

Curing processes for adhesives vary considerably in type and form. The length of curing time ranges from three seconds to over 76 hours depending upon the type of adhesives and whether or not heat and pressure are applied to the assembled joint. Either heat, pressure, or both are used to obtain the highest bond strengths.

There is no universal adhesive. The nature of the material joined and the ultimate conditions of service determine the type of adhesive to be used. Hence, there are many proprietary adhesives developed for varying bonding processes. Natural and synthetic rubbers, polyvinyl acetates, phenolic blends and epoxies are among the commonly used adhesive bases.

Bonding of aluminum is gaining increased use in building products manufacture where plastics, cork, rubber, wood veneering and other materials are used with aluminum. The processes offer great opportunities for aluminum fabrication.

FIG. 34 A sandwich panel of aluminum bonded to a plastic core provides a complete prefinished exterior wall panel.

GLOSSARY

Aging The period of time in which a heat-treatable aluminum alloy is allowed to remain at room temperature, after heat treatment (heating and quenching), to reach a stable state of increased strength.
Age hardening: The continuing increase in strength for long periods of time of aluminum alloys after heat-treatment. Some aluminum alloys do not reach a stable condition at room temperature after a few days.
Artificial aging: Heating of the aluminum alloy for a controlled time at an elevated temperature to accelerate and increase its strength gain after heat treatment.
Alclad See Clad Alloys
Alloy Designation A numerical system used in designating the various aluminum alloys.
Alumina (AL_2O_3) Hydrated form of aluminum oxide found in bauxite.
Annealing Process in which an alloy is heated to a temperature of 600° F to 800° F and then slowly cooled to relieve internal stresses and return the alloy to its softest and most ductile condition.
Anodic Coating Surface coating applied to an aluminum alloy by anodizing.
Anodize Method of applying an electrolytic oxide coating to an aluminum alloy by building up the natural surface film using an electrical current (usually DC) through an oxygen-yielding electrolyte with the alloy serving as the anode.
Bauxite The raw ore of aluminum consisting of 45% to 60% aluminum oxide, 3% to 25% iron oxide, 2.5% to 18% silicon oxide, 2% to 5% titanium oxide, other impurities and 12% to 30% water. The ore varies greatly in the proportions of its constituents, color and consistency.
Bayer Process The process generally employed to refine alumina from bauxite.
Brazing A welding process where the filler metal is a nonferrous metal or alloy with melting point higher than 800° F but lower than that of the metals joined.
Brazing Alloy Alloy used as filler metal for brazing, generally of the 4000 Series.
Brazing Sheet Nonclad or special clad sheet for brazing purposes with the surface of the special clad sheet having a lower melting point than the core. Brazing sheet of the clad type may be clad on either one or two surfaces.
Casting Pouring molten metal into molds to form desired shapes.
Clad Alloys Alloys having one or both surfaces of metallurgically bonded coating, the composition of which may or may not be the same as that of the core, and which is applied for such purposes as corrosion protection, surface appearance, or brazing.
Alclad Sheet A clad product with an aluminum or aluminum alloy coating having high resistance to corrosion. The coating is anodic to the core alloy it covers, thus protecting it physically and electrolytically against corrosion.
Cold Working Forming at room temperature by means of rolling, drawing, forging, stamping or other mechanical methods of shaping.
Combination Process A process used to retrieve additional alumina and soda from the red mud impurities of the Bayer Process.
Common Alloy An alloy that does not increase in strength when heat-treated (non-heat-treatable). Common alloys may be strengthened by strain hardening.
Corrosion Deleterious affecting of an aluminum surface due to weathering, galvanic action and direct chemical attack.
Galvanic—corrosion produced by electrolytic action between two dissimilar metals in the presence of an electrolyte.
Direct Chemical Attack Corrosion caused by a chemical dissolving of a metal.
Weathering Corrosion (galvanic and/or chemical) produced by atmospheric conditions.
Cryolite Sodium aluminum fluoride used with alumina in the final electrolytic reduction of aluminum. Found naturally in Greenland, generally produced synthetically from alum, soda and hydrofluoric acid.
Ductile Capable of being drawn out or hammered, able to undergo cold plastic deformation without breaking.
Drawing The process of pulling material through a die to reduce the size, change the cross-section or shape, or to harden the material.
Electrolysis A chemical decomposition by the action of electric current.
Electrolyte A substance in which the conduction of electricity is accompanied by chemical decomposition.
Extrude To form lengths of shaped sections by forcing plastic material through a shaped hole in a die.
Extrusion A product formed by extruding.
Extrusion Billet A solid wrought semifinished product intended for further extrusion into rods, bars, or shapes.
Extrusion Ingot A solid or hollow cylindrical casting used for extrusion into bars, rods, shapes or tubes.
Forging The working (shaping) of metal parts by forcing between shaped dies.
Press Forging Shaping by gradually applying pressure in a press.
Hammer Forging Shaping by application of repeated blows i.e., in a forging hammer.
Heat-treatable Alloys capable of gaining strength by being heat-treated. The alloying elements show increasing solid solubility in aluminum with increasing temperature resulting in pronounced strengthening.
Hot Working Forming metal at elevated temperatures at which metals can be more easily worked (for aluminum alloys, usually in the 300° F to 400° F range).
Ingot Mass of metal cast into convenient shape for storage or transportation to be later remelted for casting or finishing by rolling, forging, etc.
Non-Heat-Treatable Alloys not capable of gaining strength by heat treatment which depend on the initial strength of the alloy or cold working for additional strength. Also

called *common* alloys.

Pots Carbon lined vessels used in the reduction process, at the bottom of which molten aluminum is collected and siphoned off.

Red Mud Solid matter impurities collected by either filtering or gravity settling of the Bayer process of aluminum refining.

Reduction The electrolytic process used to separate aluminum from aluminum oxide.

Rolling Shaping plate and sheet metal by passing metal slabs through steel rollers.

Cold Rolling Forming sheet metal by rolling, at room temperature, metal previously hot rolled to a thickness of approximately 0.125″.

Hot Rolling Shaping of plate metal by rolling heated slabs of metal. Hot rolling is usually used for work down to approximately 0.125″ thickness.

Soft Temper The state of maximum workability of aluminum obtained by annealing.

Solution Heat Treatment The first temperature raising step in the thermal treating of a heat-treatable alloy. Also called *heat treatment.*

Stretch Forming The stretching of large sheets for the purpose of flattening them or for forming them to the shape of a form block.

Strain Hardening A method of strengthening non-heat-treatable alloys by either cold rolling or other physical or mechanical working.

Temper Designation Designation (following an alloy designation number) which denotes the temper of an alloy.

Wrought Products Products formed by rolling, drawing, extruding and forging.

Bar A solid section that is long in relation to its cross-sectional dimensions, having a completely symmetrical cross-section that is square or rectangular (excluding flattened wire) with sharp or rounded corners or edges, or is a regular hexagon or octagon, and whose width or greatest distance between parallel faces is 3/8″ or greater.

Extruded Sections Rod, bar, tube or any other shape produced by the extrusion process.

Foil A solid sheet section rolled to a thickness of less than 0.006″.

Forging Stock Rod, bar or other section suitable for subsequent change in cross section by forging (working in a shaped die).

Pipe A tube having certain standardized combinations of outside diameter and wall thicknesses, commonly designated by "nominal pipe sizes" and ASA Schedule numbers.

Plate A solid section rolled to a thickness of 0.250″ or more in rectangular form and with either sheared or sawed edges.

Rod A solid round section 3/8″ or greater in diameter, whose length is greater in relation to its diameter.

Sheet A solid section rolled to a thickness ranging from 0.006″ to 0.249″, inclusive, supplied with sheared, slit or sawed edges.

Structural Shapes Solid shapes used as load bearing members. They include angles, channel sections, I beams, H beams, tees, zees and others.

Tube A hollow product, whose cross section is completely symmetrical and is round, square, rectangular, hexagonal, octagonal, elliptical, with sharp or rounded corners, and whose wall is of uniform thickness except as affected by corner radii.

Wire A solid section that is long in relation to its cross-sectional diameter having a completely symmetrical cross section that is square or rectangular (excluding flattened wire) with sharp or rounded corners or edges, or is round or a regular hexagon or octagon, and whose diameter, width or greatest distance between parallel faces is less than 3/8″.

We gratefully acknowledge the assistance of the following for the use of their publications as references and permission to use photographs and illustrations: The Aluminum Association, the Architectural Aluminum Manufacturers Association, the Aluminum Company of America, Kaiser Aluminum & Chemical Sales, Inc., the Reynolds Metals Company and Modern Metals magazine.

214 ALUMINUM PRODUCTS

This Section 214 Aluminum Products includes information on windows and sliding glass doors, storm windows and doors and siding. There are other building products made largely of aluminum which are gaining widespread use. These will be discussed in future additions to this section.

WINDOWS AND SLIDING GLASS DOORS

The use of aluminum for window construction has increased significantly since aluminum windows were introduced for industrial application in the United States about forty years ago. From this beginning, aluminum windows now account for approximately 50% of all window production. In contemporary houses, glass areas have increased until the size of windows has reached the proportions of doors, permitting aluminum windows to provide entire walls of glass, or in the case of aluminum sliding glass doors, movable walls of glass (Fig. AW1 & AW2).

FIGS. AW1, AW2—Aluminum windows and sliding glass doors are widely used in all types of construction.

MANUFACTURE

The manufacture of aluminum windows and sliding glass doors consists of fabricating wrought aluminum alloys into correct size and shape, using various techniques of die cutting, punching, notching, piercing and sawing. The aluminum members thus fabricated are assembled together with other materials—glass, glazing materials, hardware, weatherstripping, screws and other fasteners—into a finished pre-assembled component.

The workability of the aluminum metal, coupled with the ease and simplicity with which aluminum windows and sliding glass doors can be manufactured, has contributed immeasurably to their widespread availability and use. Modern machinery and production techniques produce a finished product whose parts can be fabricated to extremely close tolerances, resulting in weathertight assemblies having precision operation at relatively low cost.

Materials

Pure aluminum has very few commercial applications, but when combined with other metals—such as manganese, silicon, iron, magnesium—aluminum alloys* are formed making it useful in a variety of applications. The principal materials used in aluminum windows and sliding glass doors are *wrought* (worked into shape by mechanical means) alloys. The alloys used for windows and doors must possess strength; be resistant to corrosion; be readily worked and machined for efficient fabrication and simple assembly, and provide surfaces suitable to receive protective and decorative finishes. The wrought aluminum alloy known as 6063 (containing silicon and magnesium), which possesses good formability and corrosion resistance with medium strength, is formed by the extrusion process and artificially aged by heat treatment. After heat treatment (temper designation T5), this alloy, known commercially as 6063-T5, is commonly used for the main frame, sash, and ventilator members in both windows and sliding glass doors. Extruded shapes permit infinite possibilities of cross-sectional configuration to satisfy the requirements of glazing, weatherstripping, condensation control and assembly (Fig. AW3).

Window and door locks, hinges, handles and corners are often made from *castings* (formed by pouring or forcing molten metal into a mold) of aluminum alloy (Fig. AW4).

Other aluminum wrought alloys made by *roll forming* (passing strips of flat aluminum sheet through a series of forming rolls that bend it into a tube or other shape) also

**See Section 213 Aluminum for a detailed discussion of wrought and cast alloys, heat treating and manufacturing techniques.*

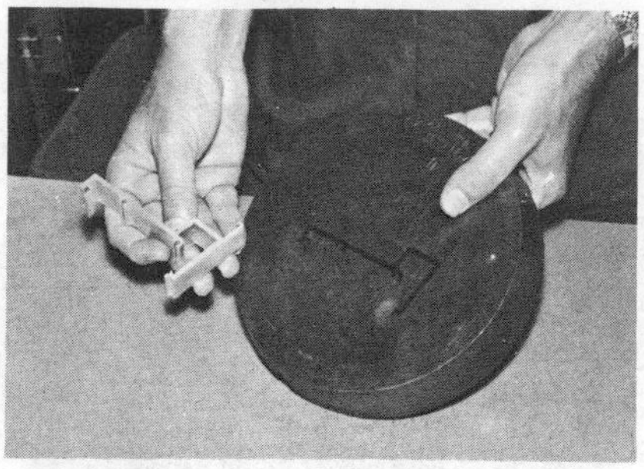

FIG. AW3—Piece of an aluminum frame member and the die through which it has been extruded.

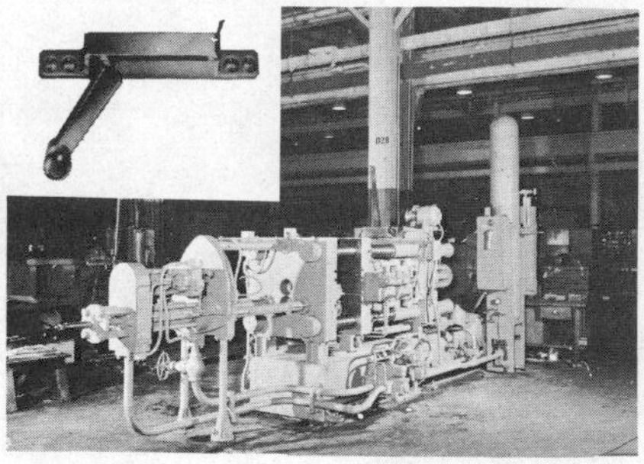

FIG. AW4—In die casting, molten aluminum is forced into a mold; hardware is commonly die cast.

FIG. AW5—Roll formers shape flat aluminum sheet into items such as glazing beads, (lower left).

have a variety of uses in aluminum window and sliding glass doors. These include such shapes as screen frames and glazing strips (Fig. AW5). Aluminum sheet is also cut and formed by stamping in a punch press where accessories such as corner locks and screen clips can be made.

Finishes The silvery sheen of the natural, *mill finish,* surface of aluminum is a common finish used on residential windows and doors. This natural oxide coating which forms on aluminum provides an adequate degree of protection against corrosion in normal atmospheres and can be readily restored by rubbing with stainless steel wool if damaged or discolored. Normal oxidation over time will convert the silvery sheet to a soft gray. Other finishes often used may be either protective or decorative.

Protective Finishes are necessary when abnormal atmosphere is present, such as found in heavy industrial and sea coast areas. Protective finishes are also necessary over certain types of decorative finishes, particularly where the surface is etched or polished.

Anodizing, an electrolytic process that provides a heavy oxide film on the surface, is a permanent protective finish against moderate abrasion and atmospheric attack. Clear anodizing, usually having a frosty appearance when an etchant is used during the cleaning cycle, is the most economical finish when corrosion protection is the primary concern.

Certain *permanent* color tones on aluminum are possible using the anodizing process. These include several shades of gray, bronze, gold, as well as black, and are produced by varying the alloy constituents, the electric current, the electrolyte and the time cycle. Practically any color is possible by impregnating the porous surface of the anodic film with dye prior to sealing. But since dye colors are not always permanent, they are not recommended for exterior use. (See Section 213 Aluminum for a description of anodizing and other finishing processes.)

Organic coatings, such as paint, that are often applied to aluminum, are generally for decorative purposes, but may also impart a certain degree of protection (See Paint Coatings, Aluminum 213).

FIG. AW6—Aluminum is extremely workable and saws easily.

FIG. AW7—Small punch press punches fastener holes.

FIG. AW8—Multiple presses punch all holes in a framing member at once.

Decorative Finishes These finishes are colored or opaque coatings (such as a backed synthetic resin enamel) similar to those used on aluminum siding. These coatings can be expected to have a life of up to 20 years when the coating is applied and baked on at the factory under controlled conditions. Organic coatings should meet AAMA 603.6 Standard, "Performance Requirements and Test Procedures for Pigmented Organic Coatings on Extruded Aluminum".

Regardless of the finish used, an aluminum window or sliding glass door is a prefinished building component and should be handled and treated with care during the construction period. The best practice is to withhold installation of these units as long as construction procedures will permit.

Fabrication

Fabrication of aluminum alloy sections into component parts of a window or sliding glass door is accomplished by sawing, drilling, punching and other traditional methods of metal working (Fig. AW6).

One of the machines used most is the small punch press (Fig. AW7). Aluminum's machinability permits the use of a variety of intricate punches and dies in the press to cut, form, shape, notch and pierce the metal. Thus, components can be joined accurately and easily to make weathertight assemblies. A multiple punch press can cut a long length of extrusion to the proper length and punch all notches, holes and slots, producing one complete framing member in a single operation (Fig. AW8).

Assembly

Materials other than aluminum—glass, plastics, steel and many others—are required to complete the assembly of windows and sliding glass doors.

Glass and Glazing Materials. Most windows and sliding glass doors for residential use have the glass installed at the factory, which eliminates glazing on the job site and allows controlled methods of glazing not feasible with onsite glazing.

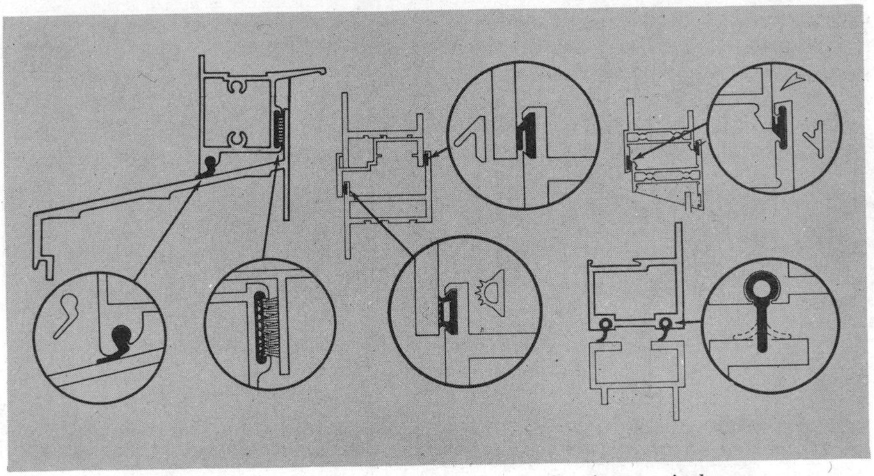

FIG. AW9—Various methods of weatherstripping aluminum windows.

Glass selection by size and grade for windows depends primarily on windload requirements and is often established by the local building code. In the absence of code requirements, recommendations of ANSI/AAMA 302.9 for windows and ANSI/AAMA 402.9 for sliding doors should be used. For sliding glass doors, in addition to wind load requirements, the glass must meet requirements for "safety type" and be in compliance with ANSI Z-97.1 or other similar standard as regulated by codes or laws.

However, other functional or other esthetic requirements often dictate the selection of different types and higher grades of glass. These include: double-glazing to minimize heat loss and heat gain; tinted glass for privacy and decorative purposes; tempered glass for additional strength; and tempered, wire or laminated glass for additional safety.

Glazing Materials for both aluminum windows and sliding glass doors include gaskets made from neoprene or polyvinyl chloride and mastic type compounds such as butyl or rubber base materials that are compatible with aluminum and do not require painting.

Weatherstripping. The function of weatherstripping is to control air infiltration, prevent rain penetration, minimize dust infiltration and withstand atmospheric conditions. It must be durable under extended usage, resist corrosion and be readily replaceable. Most residential aluminum windows and sliding glass doors are equipped with fabric pile weatherstripping (similar to that used in automobiles) or with flexible vinyl plastic, neoprene gaskets or fins (Fig. AW9). In some cases these materials are combined. Extremely thin stainless steel, which compresses to accomplish the seal, is sometimes used.

Fasteners and Hardware. A variety of fasteners and hardware are required to hold the window and door parts together and to make them operable. Although aluminum is used extensively for this purpose, other materials may also be desired or required. Other metals used with aluminum must be either compatible with aluminum or insulated from direct metal-to-metal contact with the aluminum. Otherwise, electrolytic corrosion (galvanic action) may result. Hence, when mild steel fasteners are used they must always be plated with zinc, cadmium or other suitable material. Stainless steel need not be plated and can be used in exposed locations if it is of the non-magnetic type and has a chromium content of not less than 16%. Nylon, polypropolene and other plastic materials are used for slides, bushings, bearings and similar contact surfaces.

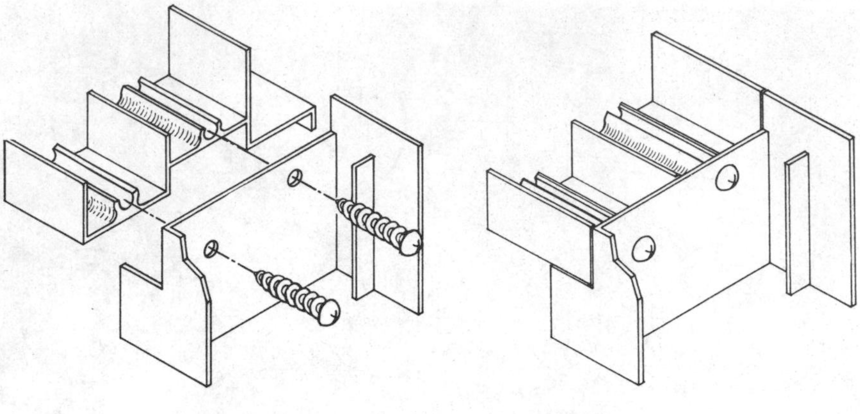

FIG. AW10—Framing members often have continuous slots to receive screws.

Frame Assembly The main frame, sash and ventilators of windows and doors are assembled by screwing or welding.

Screwed assemblies are shipped from the factory "KD" (knocked down) and later assembled by a distributor, dealer or contractor. A slot to receive the screw threads is often extruded as an integral part of the window member and holes are pre-punched to insure perfect alignment of the corner joints (Fig. AW10). Lower corners of the main frame are usually sealed with a mastic to make them watertight. Units that are completely assembled at the factory are either staked or welded together.

Welded assemblies utilize several welding methods. The flash (butt) welding method produces a continuous weld across the entire cross section of the shapes, and is made by forcing two members together at their intersection under pressure and applying a voltage which melts the parent metal (Fig. 11). Flash welding uses no filler material or flux and forms a rigid, strong and neat, watertight joint.

The inert gas welding method uses filler material and is similar to conventional gas welding used on other metals. (Fig. 12). It is seldom continuous and is generally used to join members that are of dissimilar shape. While joints made by inert gas welding can be made reasonably watertight, the use of a mastic sealing material at the lower corners may be required to prevent water leakage.

Accessories. Window and door accessories are assembled and attached during all stages of fabrication and assembly. For example, weatherstripping frequently is pulled into retaining grooves in long lengths of extruded frame members before cutting and punching takes place (Fig. 13). Operating hardware such as sash balances, pivot arms, guides and guide followers are usually assembled after the main frames are together. Accessories such as locking hardware and handles are attached by riveting, bolting, screwing, staking or interlocking with frame members. Manual rotating operators on casements, awning and jalousie windows are usually the last accessories to be assembled, sometimes being left until after installation of the unit at the job site.

Glazing One method of factory glazing utilizes a suitable glazing compound applied to the glazing leg of the window as a back-bedding material; the glass is then dropped in and held in place by aluminum or rigid plastic face stops. The stops generally snap into place to serve as trim molding and hold tension on the assembly until the glazing compound has set. Another method widely used in the factory to glaze units is to apply a channel-shaped gasket over the edge of the glass and then force the channel shape into the window or door frame around it. Flexible vinyl plastic is most often used for this application.

When windows or sliding glass doors are glazed at the job site, the manufacturer's recommendations should be followed to insure a satisfactory result.

Figure 14 shows typical steps in the assembly process.

FIG. 11 Here workman utilizes flash welding to join corners of aluminum sash frame.

FIG. 12 Here the inert gas welding method is used to join members of a window frame.

FIG. 13 Fabric pile is inserted into strip to form weather seal that is later inserted into frame (upper left).

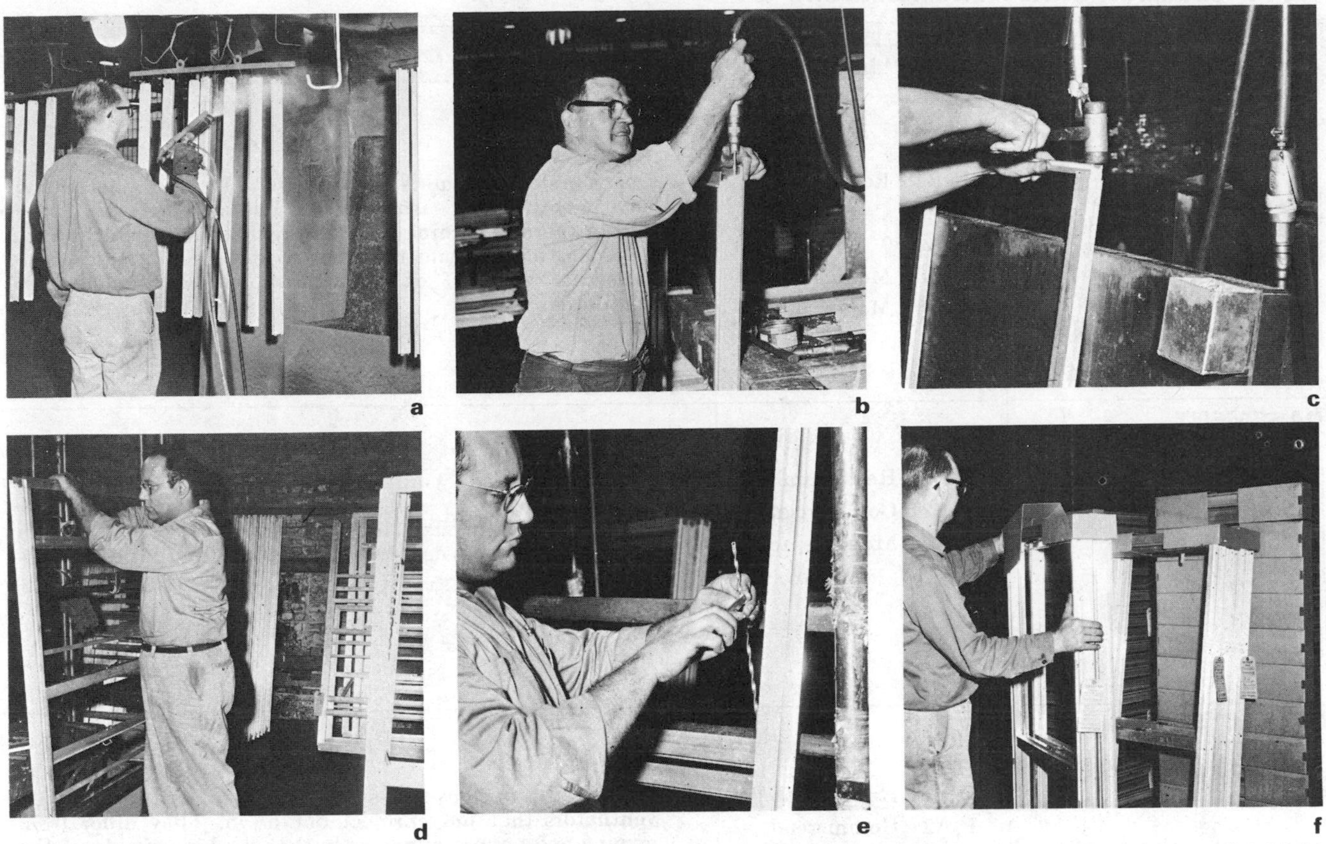

FIG. 14 Typical assembly of double-hung window: (a.) Protective coating being applied to individual members. (b.) Main frame being assembled. (c.) Sash frame is assembled. (d.) Sash being installed in main frame. (e.) Sash balance being installed. (f.) Completed assembled units being readied for shipment.

STANDARDS OF QUALITY

The design and construction of aluminum windows and sliding glass doors should conform to the specifications developed by the Architectural Aluminum Manufacturers Association (AAMA) and adopted by the American National Standards Institute (ANSI). These specifications constitute the recognized standards for the aluminum window and sliding glass door industry and include minimum provisions for frame strength and thickness, corrosion resistance, air infiltration, water resistance, windload capability and other requirements necessary to insure adequate performance. Windows are covered by Specifications for Aluminum Windows, ANSI/AAMA 302.9; sliding doors are covered by Specifications for Aluminum Sliding Glass Doors ANSI/AAMA 402.9.

The various types of aluminum windows and doors covered by ANSI/AAMA standards are illustrated and described in Fig. 15.

Window and Door Types

Aluminum windows and sliding glass doors are given designations that describe the type of window or door, the AAMA specification number under which it was produced and other information. For example, DH-B1-HP conveys the following information: The first letters describe the window or door *type* (DH means *double hung* window, C means *casement* window, SGD means *sliding glass door*, etc.); the B signifies *residential* type; the number designates the applicable *specification level;* and HP indicates a *high performance* rating.

There are six specification levels for windows: B1, B2, A2, A2.50, A3 and A4. B1 and B2 identifies windows generally suitable for single-family residences; A2 and A2.50 are commercial windows of heavier construction, with stronger hardware suitable for rental properties, apartment buildings and commercial construction; A3 and A4 specification levels designate extremely heavy windows for monumental construction.

There are four specification levels for sliding glass doors: B1 and B2 for doors meeting performance requirements for residential application; A2 for commercial applications, and A3 for monumental construction.

According to the ANSI/AAMA specifications, residential series windows and doors not rated for high performance must meet certain requirements when tested at

FIG. 15 SUMMARY OF ALUMINUM WINDOWS

TYPE	DESIGNATION & APPLICATION	DESCRIPTION
DOUBLE HUNG OR SINGLE HUNG 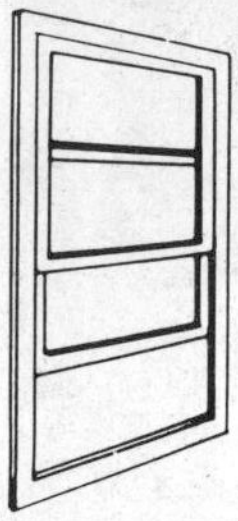	DH-B1 Residential DH-A2 Commercial DH-A2.50 Commercial DH-A3 Monumental DH-A4 Monumental	*Double-hung windows* have two operable sash; single hung have only the lower sash operable. The sash move vertically within the main frame with the assistance of mechanical balancing mechanisms that minimize the effort required to raise the sash. 50% of the window area is available for ventilation.
CASEMENT 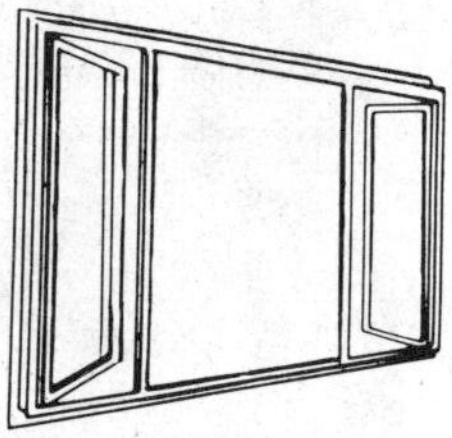	C-B1 Residential C-A2 Commercial C-A3 Monumental	*Casement windows* contain side hinged ventilators that operate individually and swing outward. They frequently have non-operating fixed lites (shown) which the ventilators close to. Casements may be open 100% for ventilation.
PROJECTED 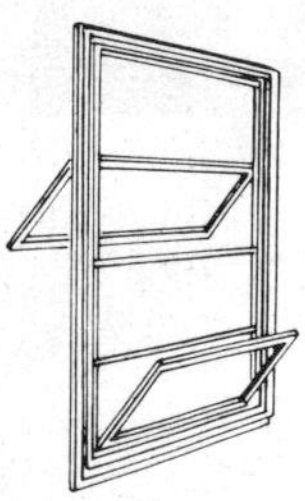	P-B1 Residential P-A2 Commercial P-A2.50 Commercial P-A3 Monumental	*Projected windows* consist of horizontally mounted ventilators that may project out or in. They differ from awning and hopper windows in that the hinged side of the ventilator moves in a track up or down when the ventilator is operated. Used commonly where large glass areas are desired and where it is desired to deflect incoming air in an upward direction.
AWNING 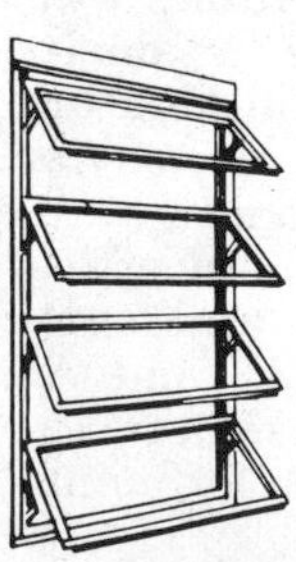	A-B1 Residential A-A2 Commercial	*Awning windows* consist of a number of top hinged ventilators operated by a single control device that swings the lower edges outward giving an "awning" effect when open. Operation is through a roto operator which also serves as the lock. Ventilating area is 100% of the window area.
HORIZONTAL SLIDING 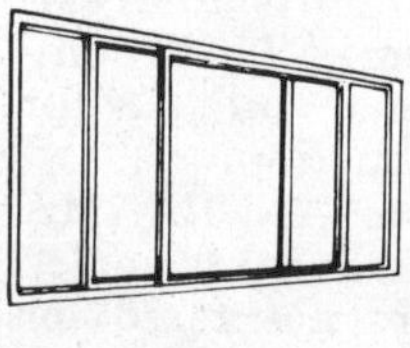	HS-B1 Residential HS-B2 Residential HS-A2 Commercial HS-A3 Monumental	*Horizontal sliding windows* have one or more operable sash arranged to move horizontally within a main frame. Ventilating area is 50% of the window area. This type of window is sometimes combined with fixed lites and constructed in large sizes reaching the proportion of a "window wall."

DESCRIPTION	DESIGNATION & APPLICATION	TYPE
Jalousie windows (louver windows) consist of a series of overlapping horizontal glass louvers which pivot together in a common frame. Ventilating area is 100% of the window area. Jalousies are not as resistant to air and dust infiltration as other types, but are used in moderate climates and as porch enclosures in all cimates. They should only be used where large heat losses or heat gains can be tolerated. Insulating effectiveness may be improved by inside storm panels, designed to be interchangeable with screens.	J-B1 Residential	**JALOUSIE** 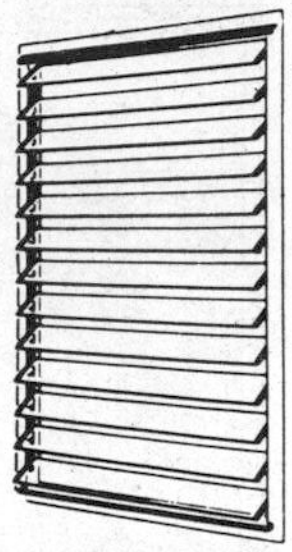
Jal-awning windows resemble an awning window in appearance but use different operating and locking mechanisms. These mechanisms are usually separate and require individual operation. Ventilating area is 100% of the window area.	JA-B1 Residential	**JAL-Awning** 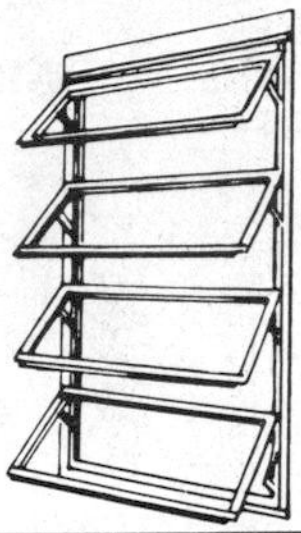
Vertical sliding windows resemble single or double hung windows in appearance but use no sash balancing devices. They are operated manually by lifting and are held in various open positions by mechanical catches that engage in the jamb or hold by friction. Ventilating area is 50% of the window area.	VS-B1 Residential	**VERTICAL SLIDING** 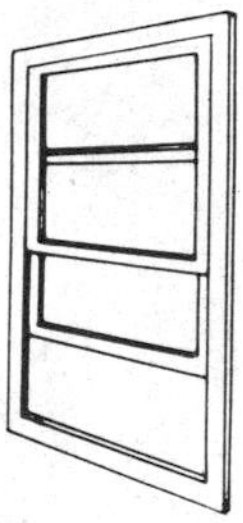
Vertically-pivoted windows consist of a ventilator sash mounted on pivots located in the center of the top and bottom main frame members. This allows the ventilator to to be reversed by rotating and permits the glass to be cleaned from the interior. This window is not used for normal ventilation and is operated solely for cleaning purposes or emergency ventilation. It is used generally in completely air conditioned high rise buildings.	VP-A2 Commercial VP-A3 Monumental	**VERTICAL PIVOT** 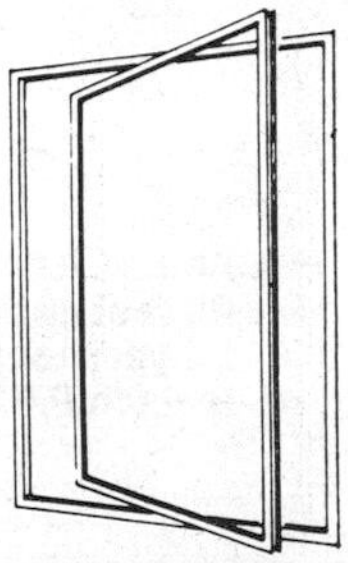
Top-hinged, in-swinging windows consist of a ventilator hinged to the main frame at the top which swings into the room. This permits the glass to be cleaned from the interior. This window is not used for normal ventilation and is operated solely for cleaning purposes and emergency ventilation. Its use is generally in completely air conditioned high-rise buildings.	TH-A2 Commercial TH-A3 Monumental	**TOP-HINGED** 

FIG. 15 continued

TYPE	DESIGNATION & APPLICATION	DESCRIPTION
SLIDING GLASS DOOR 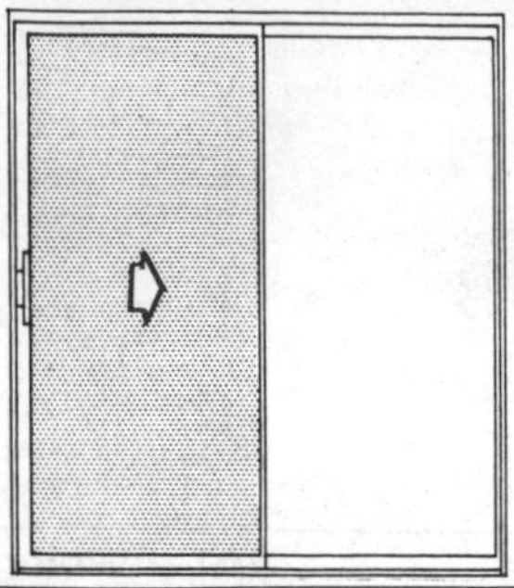	SGD-B1 Residential SGD-B2 Residential SGD-A2 Commercial SGD-A3 Monumental	*Sliding glass doors* consist of two or more framed glass panels contained in a main frame designed so that one or more of the panels move horizontally in the main frame. The most common arrangement is of 2 panels. "Single-Slide" has one operable panel, a "Double-Slide" has two. The ventilation area of a single and double slide door is 50% of the main frame area. Doors are also made in 3 and 4 panel arrangements: 3 panel arrangement can provide 1/3 or 2/3 of its area for ventilation, and 4 panel can provide 1/2 or in some cases 3/4 of its area for ventilation.

FIG. 16 WIND MAP OF THE UNITED STATES

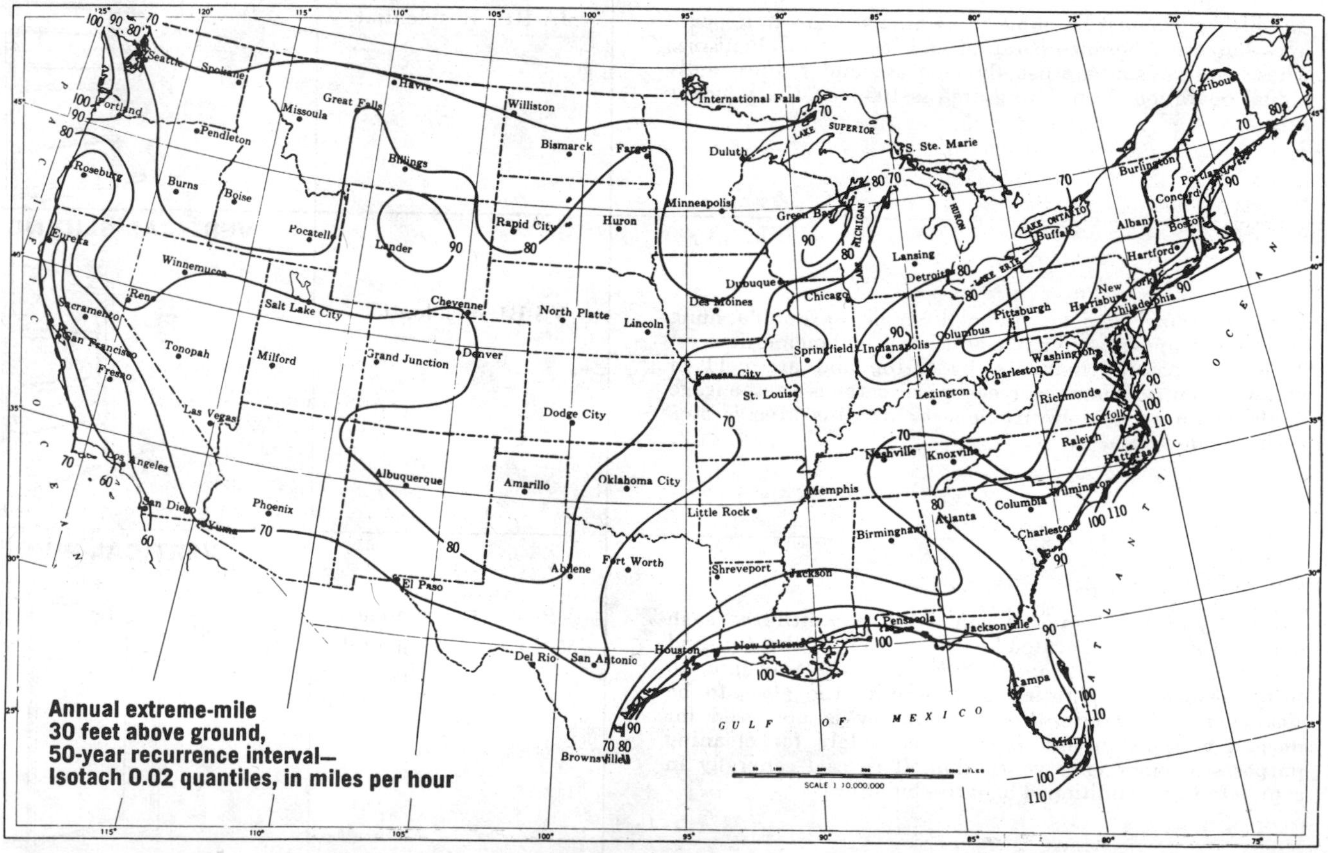

FIG. 17 **TYPES OF THERMAL-BREAK ALUMINUM WINDOWS**

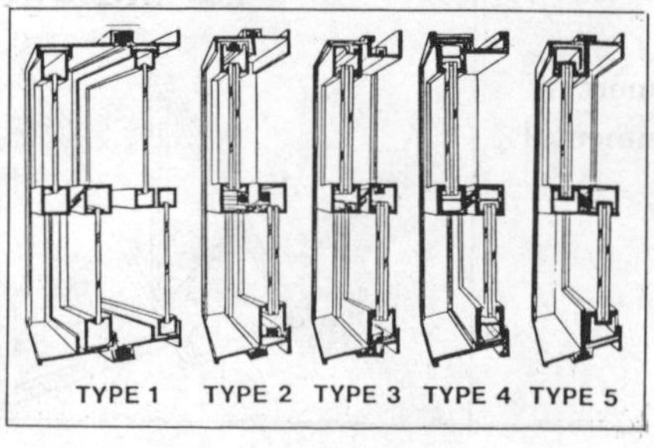

Type 1 is a "unitized" prime and storm Thermalized combination window, consisting of a single-glazed or double-glazed prime window with insulating glass, and a storm window, supplied as one unit. The frame has a built-in thermal break and the air space between the prime window and the storm window provides insulation in the usual manner provided by double-glazing.

Type 2 is a Thermalized prime window with insulating glass and a plastic thermal break in both the frame and sash.

Type 3 also uses insulating glass. It has a thermal break in the frame and uses vinyl cladding as the thermal break for the sash members.

Type 4 uses insulating glass and provides a thermal break by having vinyl cladding on both the frame and the sash members.

Type 5 uses insulating glass and a thermal break in the frame which serves a dual purpose by extending out and covering the sash members, giving them insulation at the same time.

Continued from page 7
structural pressures of 20 psf and up. The minimum test pressure is based on the effect of 70 mph winds at an elevation of 30 ft. above ground, increased by 65% to account for gusts, shape of building and a safety factor. Therefore, such windows and doors are suitable for most low-rise buildings in areas where wind velocities generally do not exceed 70 mph (Fig. 16). Since wind velocities typically are lower nearer the ground, these products can be used also in 80 mph areas at a 10 ft. elevation and in 90 mph areas at a 5 ft. elevation. (Elevations are measured to the center of the window or door, hence the 5 ft. requirement would apply to a one-story structure).

Doors and windows intended for high wind zones—in areas with winds exceeding 70 mph or at elevations over 30 ft.—should be of the HP (high performance) series.

According to ANSI/AAMA specifications, residential HP series products must meet certain requirements when tested at structural pressures of 40 psf and up. The minimum test pressure constitutes 165% of the actual wind load created by 100 mph winds at an elevation of 30 ft. above ground. In parts of the country where winds exceed 70 mph (Fig. 16), the air pressure on tall buildings can be substantially higher than the ANSI prescribed minimums. Many manufacturers produce windows and doors to meet these more demanding conditions, and AAMA certifies such products as conforming to ANSI requirements at specific wind loads.

Windows or sliding glass doors in areas of low winter design temperature (+10°F or less) should incorporate double or triple insulating glass and should have thermalized frames and sash. Thermalizing basically involves separating the aluminum parts on the warm inside portion of the window from those on the cold outside portion through the use of an insulating material such as plastic. This keeps heat from being conducted right through the frame or sash; the insulating separator is called a "thermal break" or "thermal barrier." Thermalized windows not only minimize heat conduction through the glass area, but the frame and sash as well. Figure 17 illustrates several methods of achieving thermal separation.

The Condensation Resistance Factor (CRF), as determined by AAMA 1502.6 Voluntary Standards and Tests of Thermal Performance of Residential Insulating Windows & Sliding Glass Doors, gives an indication of a window's ability to resist condensation. The higher the CRF, the less likely condensation is to occur (Fig. 18).

Certification of Quality

In addition to developing standard specifications, AAMA also sponsors a program of independent testing (Fig. 19), evaluating, inspecting and labeling aluminum

FIG. 18 **CONDENSATION RESISTANCE FACTOR**

To determine the minimum recommended CRF for your locality, find the outside winter design temperature from the map (a). Then, pick a recommended maximum indoor humidity level for that temperature from the chart (b). You can use the outside winter design temperature and the recommended humidity level with the chart (c) to find the minimum CRF. Windows with CRF ratings below 35 are not recognized as Thermalized aluminum windows.

Note that CRF ratings have nothing directly to do with heat loss or gain, or with air leakage. So, a higher CRF does not necessarily mean you will save proportionately that much more energy.

(a) WINTER OUTSIDE DESIGN TEMPERATURE

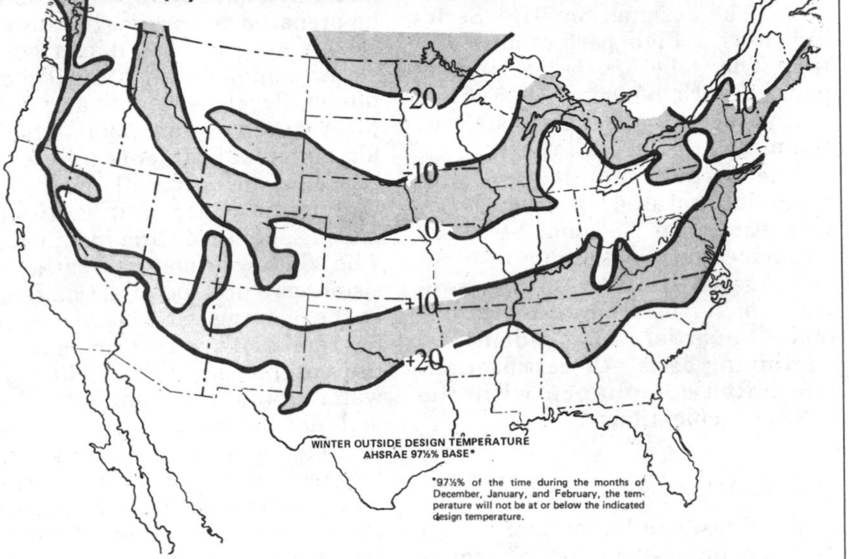
-20
-10
0
+10
+20
10
WINTER OUTSIDE DESIGN TEMPERATURE
AHSRAE 97½% BASE*
*97½% of the time during the months of December, January, and February, the temperature will not be at or below the indicated design temperature.

(b) MAXIMUM RECOMMENDED HUMIDITY LEVELS[1] [2]

OUTSIDE AIR TEMPERATURE	INSIDE RELATIVE HUMIDITY
−20°F Or Below	Not Over 15 Percent
−20°F to −10°	Not Over 20 Percent
−10°F to 0°	Not Over 25 Percent
0°F to 10°	Not Over 30 Percent
10°F to 20°	Not Over 35 Percent
20°F to 40°	Not Over 40 Percent

[1]Relative humidity levels above these are not recommended.

[2]If higher relative humidity levels are required because of special interior environmental conditions, the window manufacturer should be consulted.

(c) MINIMUM RECOMMENDED CONDENSATION RESISTANCE FACTORS (CRF)*

Outside Design Temperature	Inside Relative Humidity					
	15%	20%	25%	30%	35%	40%
−20°F	46	52	57	60	—	—
−10°F	39	46	52	56	60	—
0°F	30	39	45	52	57	61
+10°F	17	29	37	44	50	57
+20°F	0	16	25	34	40	48

*Based upon an inside air temperature of 68°F and an outside wind velocity of 15 MPH.

products to certify compliance with ANSI/AAMA Standards. The certification program is accredited by ANSI. Products which conform with the requirements of the applicable specifications are identified by the AAMA label (Fig. 20).

The manufacturer's code number, appropriate standard and maximum size tested is indicated on the label. The designation "HP" Series indicates a high performance rating. The label is 5-3/8" x 5/8", printed with blue ink on a silver background. It must be visible on the metal after installation.

The AAMA certification program is operated in accordance with American National Standard Practice for Certification Procedures ANSI Z34.1, which provides for certification to the public on an impartial, independent and continuing basis. All certifications guarantee compliance with the ANSI specifications.

INSTALLATION

The finest window or door design and the highest quality of manufacture will not compensate for poor installation workmanship. A suitable, weathertight installation, although simple, can only be assured if the aluminum window or door is installed by an experienced tradesman following a few simple guidelines and the standard instructions supplied by the manufacturer. The rough opening should be prepared to receive the unit in such a manner that it can be installed and will finish out square, plumb, level, straight and true. Most windows and doors provide for minor adjustments at the job site, but no unit will operate or weather properly if it is twisted and misaligned during installation. The window or door is weather resistant as a unit, but must be sealed in the opening with mastic, caulking compound or trim members to prevent passage of air, dust and water around the frame. Windows and doors should be supported (nailed, screwed or otherwise fastened) around the entire frame and anchored securely to the supporting construction in a firm and rigid position.

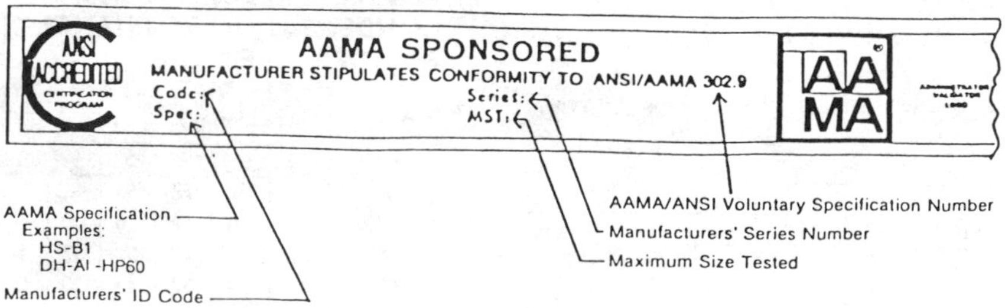

FIG. 20 The blue on silver, 5-3/8" x 5/8" AAMA label must be visible on the metal after installation. Its use on windows (example shown) and sliding glass doors (ANSI/AAMA 402.9) assures conformance with AAMA and ANSI specifications.

FIG. 19 Aluminum window and door units bearing certification seal are subjected to production controls and rigid performance tests by independent laboratories. This window unit is being tested for structural resistance to wind loading and for infiltration resistance. Other tests measure deflection of frame members under varying superimposed loads, racking, and resistance to rain penetration.

Anchoring materials and flashing should be of aluminum or material compatible with aluminum. When dissimilar materials are used they should be insulated from direct contact with the aluminum. If dissimilar metals must be used in locations where water drainage from them passes over the window or door, the dissimilar material should be painted to prevent staining of the aluminum.

An aluminum window or sliding glass door is a pre-finished building component and installation should be withheld as long as the construction schedule will permit. Once installed, normal care should be exercised to prevent damage to preserve its finish, appearance and ability to operate properly.

STORM WINDOWS & DOORS

Storm doors and windows are glazed panels or inserts which are sometimes added to prime windows and doors as an extra line of defense against the elements. Properly installed storm doors and windows provide an effective barrier to the transfer of heat (Fig. 21). They are highly recommended for use in the northern part of the country, above the 4500 winter degree-day line and are desirable down to the 2500 degree-day line. (Fig. 22). In these areas, they can significantly reduce heating costs and minimize condensation on interior windows. By preventing cold air from leaking around prime windows and doors, they also reduce drafts and improve interior comfort. In both northern and southern areas, storm windows can be retained over glass areas during the summer to help reduce heat gain through the glass and frames.

Although the term "storm doors and windows" generally refers to units intended for weather protection, it is often used collectively to include screen units as well. Accordingly, the following discussion covers both combination units, containing interchangeable glass and screen inserts, as well as single-purpose units, with fixed glass or screening. The recommendations for storm doors and windows are based largely on specifications developed by the Architectural Aluminum Manufacturers Association (AAMA), while those for screen doors and screen inserts follow specifications of the Screen Manufacturers Association (SMA).

Nearly 90% of all the storm doors and windows sold today are made of aluminum. Aluminum's popularity is based on its high resistance to corrosion and decay and its ease of

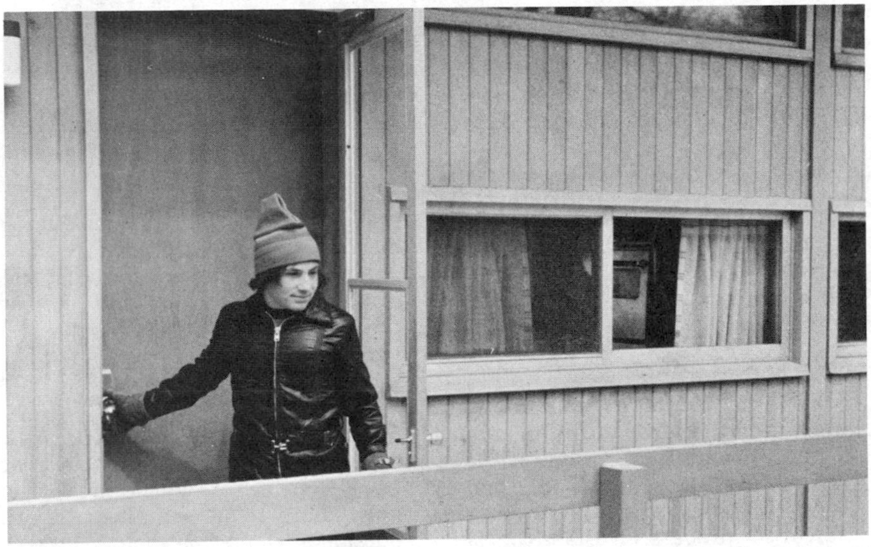

FIG. 21 Storm doors and windows conserve energy, reduce heating/cooling costs and increase comfort.

FIG. 22 **ZONES OF HEATING DEMAND (WINTER DEGREE DAYS)**

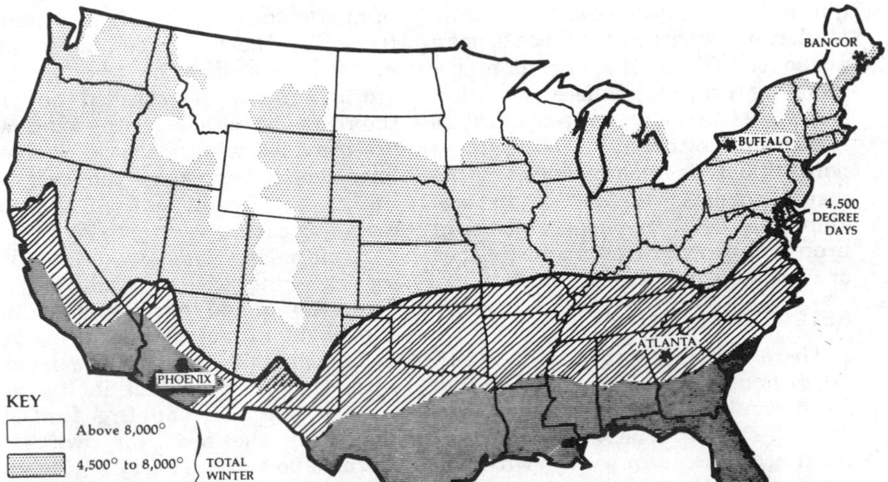

FIG. 23 SUMMARY OF STORM DOOR & WINDOW COMPONENTS

Material or Product	Standard Specification
Glass	
Annealed	FS DD-G-451c
Tempered	FS DD-G-1403A
Back Bedding Glazing Compound	
Ductile Type	AAMA 802.2
Bonding Type	AAMA 805.2
Back Bedding Glazing Tape	
Ductile Type	AAMA 804.1
Bonding Type	AAMA 806.1
Cured Rubberlike Type	AAMA 807.1
Vinyl Weatherstrip & Gasket	CS 230-60
Screening	
Vinyl Coated Glass Fibre	CS 248-G4
Aluminum Wire	CS 138-55
Fasteners	
Cadmium Plated	ASTM A-164-71
Zinc-Plated	ASTM A-165-71
Tubular Frame for Screens	SMA-1003

maintenance. Durable anodized and paint finishes minimize the need for periodic refinishing and offer a selection of colors to complement any design.

MANUFACTURE

The manufacture of aluminum storm doors and windows is similar to that of aluminum prime windows and sliding glass doors. (See pages 214-3 to 214-7 for a general description of the process.) Besides the basic aluminum sash and frame parts, many other accessory products are used in the assembly of window and door units (Fig. 23).

Storm windows and doors are made in a variety of finishes and colors. A natural *mill* finish is adequate for many moderate-cost installations, where appearance is not a prime consideration. For more demanding projects, where both corrosion resistance and appearance are important, either an *anodized* or *paint* finish is recommended. Anodized finishes can be of natural aluminum color or deeper shades of bronze or brown-black. Paint finishes are available in a variety of colors.

SELECTION

There are many types of storm doors and windows available today. Both windows and doors can be of the *self-storing* type, or may depend on a glazed *storm panel* which is applied and removed seasonally as needed.

Self-storing storm windows generally have both glass and screen inserts sliding vertically or horizontally in a frame. In self-storing *combination* (storm and screen) doors, glass and screen inserts slide vertically.

Storm Windows and Panels

The most widely used storm windows are of the triple-track and two-track type. Storm windows preferably should be installed on the outside of the prime sash. Where the prime sash swings out, as in casement and awning windows, storms may have to be on the room side.

To prevent rapid heat transfer and condensation on the interior window frame, an insulating *thermal break* should be provided between the prime window and storm window frame. A thermal break is recommended regardless of whether the storm window is installed on the room side or outside. Some windows are provided with a plastic or wood liner which serves as the thermal break. An air gap or a small wood moulding can be used instead.

Triple-track windows are intended for use mainly with vertically sliding double-hung and horizontal sliding triple-light prime windows (Fig. 24). A typical unit consists of a frame, with each insert riding in its own track. The self-storing feature eliminates the need for switching storms and screens every spring and fall. Instead, selection of glass or screen can be made easily from inside the house.

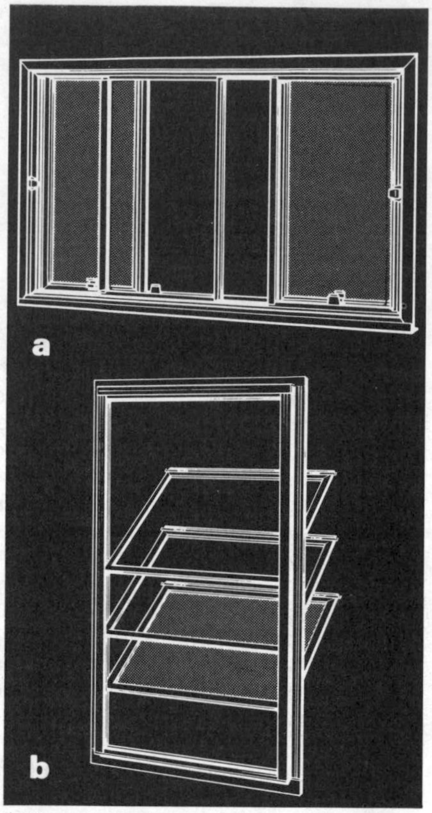

FIG. 24 Triple track windows provide complete flexibility in selection of vent or storm sash. For cleaning, horizontal sliding inserts can be lifted out (a); vertical sliding inserts pivot in (b).

FIG. 25 Storm panel covers entire jalousie windows, reducing air infiltration and heat loss.

Two-track windows are generally used with vertically sliding single-hung windows and with two-light horizontal sliders. These units also have three inserts which are stored in the frame. In vertical windows, the bottom light is usually screened; in horizontal sliders, the screen is placed right or left, depending on the prime vent sash location.

Single-track windows have two glass inserts and one screen insert, but only two inserts can be housed in the track at any one time. This makes changing from weather protection to ventilating somewhat less convenient, as one insert must always be stored elsewhere.

Clip-on Storm Panels are available for outswinging prime windows such as casements, awnings and jalousies (Fig. 25). These consist of single glass panels which are attached either to the individual sash, or across the entire prime window, including both fixed and vent sash. When the panel is clipped to the sash only, heat transfer through the glass is reduced, but heat lost through the frame and by air leakage is not affected.

The larger panels covering the entire window control heat loss more effectively, but cannot be left in place for the summer. These panels must be stored elsewhere and interchanged seasonally with screens. Storm panels should be separated from the prime window frame by an insulating gasket or weatherstripping, and should be held tightly in place by pivoting clips or other fasteners.

FIG. 26 Doors with removable inserts generally have smaller openings than self-storing doors.

Window Construction

Extrusions used for the main frame and insert frames should be of wrought aluminum alloy with a minimum tensile strength of 22,000 psi and yield strength of 16,000 psi. Reinforcing members which are not in plain view can be of cadmium, zinc-plated steel or non-corrosive materials compatible with aluminum.

Storm windows should be provided with weatherstripping to minimize air infiltration and reduce rattling. Glass inserts generally consist of a light gauge aluminum frame into which the glass is set with bedding compound, plastic channel gaskets or glazing tape. These materials should cover the glass edge continuously so that there will be no glass-to-metal contact.

Glass used for storm inserts in windows should not be less than B quality and of adequate thickness. Glass inserts in vertically sliding windows should have self-activating latches to hold the insert level in different ventilating positions. Insert frames should be designed to permit reglazing without special tools.

Screen inserts consist of insect screening stretched on an extruded or roll-formed frame. Screening can be of aluminum, vinyl-coated fiberglass or other synthetic fibers, in 14 x 18 or finer mesh.

Storm and Screen Doors

Storm doors are made either (1) with fixed glass lights, exclusively for weather protection, or (2) with both screen and glass inserts, to permit ventilation with insect control as well. In parts of the country with moderate year-round temperatures, screen doors (without glass inserts) are frequently used. Screen door construction is similar to that of storm doors. Both must be constructed rigidly, to resist sagging, bowing or rattling from repeated impact.

Many storm doors are of the combination type, providing both weather protection and insect control. Such doors are in fact stile-and-rail assemblies, framing a storm window with glass and screen inserts which can be interchanged seasonally. Doors with openings limited to the upper half of the door generally accommodate only one insert at a time, requiring that the other be stored elsewhere (Fig. 26).

Self-storing storm doors contain the equivalent of a two-track window, accommodating two inserts in one track and another in the adjacent track. The glass and screen inserts slide up and down just as they do in a vertical storm window.

Both prime doors and storm doors can be of the jalousie type. To provide ventilation and storm protection, interchangeable screen and glass panels can be clipped to the inside of the door. Clip-on glass storm panels not only provide added insulation, but also reduce air leakage between the louvers, a common problem with this type of door.

Door Construction

Storm and screen doors receive heavy usage in most homes. They are opened and closed many times a day and are subject to wind buffeting and other stresses. It is therefore important that they be durably constructed and properly installed. Rigid corner connections are particularly important to resist racking forces.

Standard door frames are 1″ or 1-1/4″ thick, with top rail and stiles 2″ to 2-1/2″ wide. The bottom rail (kickplate) varies in width depending on the size of the opening. To avoid the hazards of broken glass, storm doors should be safety-glazed with transparent plastic or with tempered, laminated or wired glass.

Doors currently are manufactured in literally thousands of types and styles, utilizing hundreds of different glass sizes. To minimize waste, it is recommended that storm windows and doors be specified so as to accommodate standard glass sizes adopted by the industry (Fig. 27). Limiting the number of glass sizes not only reduces production costs, but also facilitates replacement, particularly of tempered glass.

Construction of glass and screen inserts for doors is similar to that of windows, described above.

FIG. 27 **RECOMMENDED STANDARD STORM DOOR GLASS SIZES**

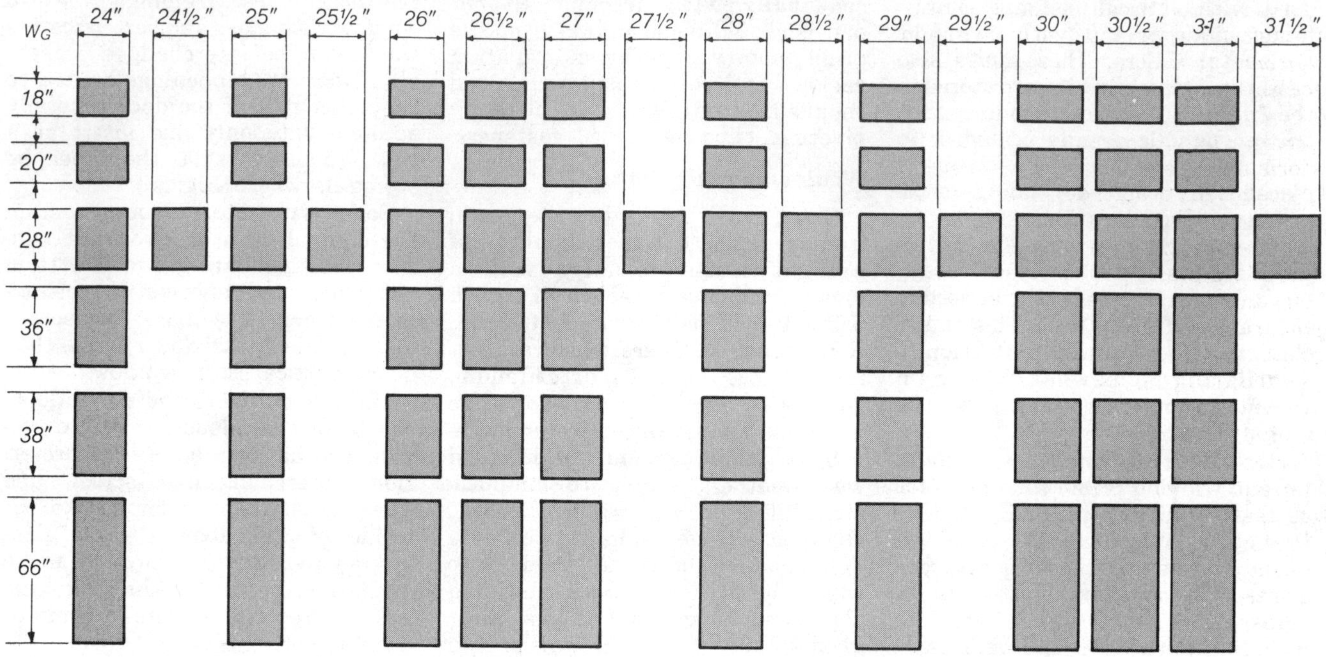

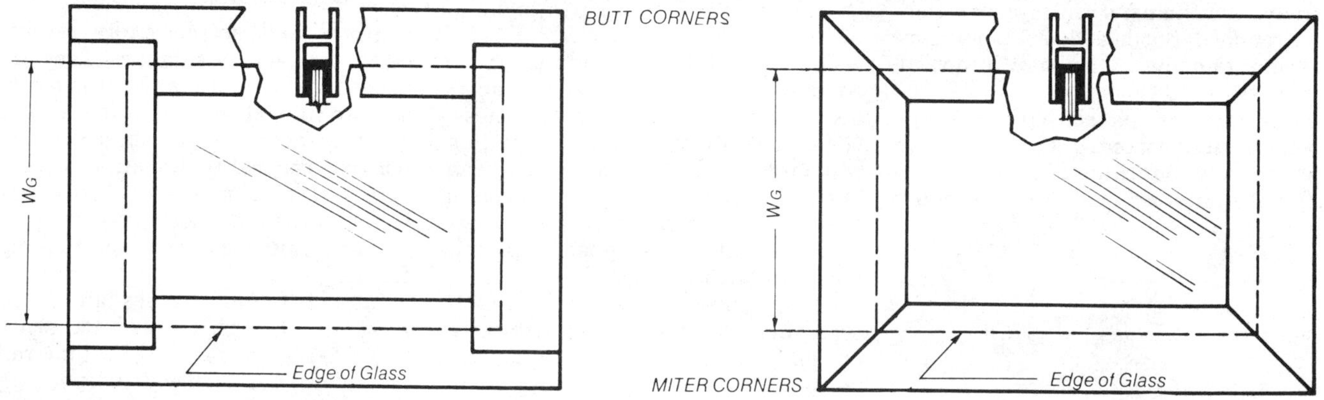

STANDARDS OF QUALITY

Aluminum storm doors and windows should conform to the specifications developed by the Architectural Aluminum Manufacturers Association (AAMA) and adopted as National Standards: (1) ANSI·1002.9 Specifications for Aluminum Vertically-Sliding and Horizontally-Operating Storm Windows for External Application and (2) ANSI 1102.7 Specifications for Aluminum Storm Doors. These constitute the recognized industry standards and establish minimum performance requirements for air infiltration, water leakage, wind load and structural strength.

Screen inserts and doors should conform to requirements developed by the Screen Manufacturers Association and likewise adopted as National Standards: (1) ANSI A201.1 Specifications for Aluminum Tubular Frame Screens for Windows and (2) ANSI A201.3 Specifications for Aluminum Swinging Screen Doors.

Certification of Quality

The AAMA sponsors an independent certification program of testing, random sampling and product evaluation to assure that aluminum storm doors and windows meet ANSI standards. Products which meet all requirements of the standards are identified with an AAMA certification label (Fig. 28).

The AAMA certification program is administered in accordance with American National Standard Practice for Certification Procedures, ANSI Z34.1, which provides for certification to the public on an impartial, independent and continuing basis.

INSTALLATION AND MAINTENANCE

Above the 4500 degree-day line (Fig. 22), storm doors or an entrance vestibule should be provided as part of the initial construction. Storm windows can be omitted in original construction, if the prime windows are equipped with insulating glass and a thermal break in the sash and frames. Existing residences in these northern areas which are not so equipped should be provided with storm doors, storm windows or panels. The economics of adding storm windows in the temperate parts of the country should be determined on a case by case basis depending on construction and energy costs.

To insure convenient, weathertight performance, storm doors and windows should be installed by experienced mechanics. Most storm windows are sold as complete units, with all accessories and hardware included with the unit. This insures smooth, convenient operation and adequate security. Installation consists simply of attaching the window frame in a previously prepared opening (Fig. 29).

Storm doors are marketed both *prehung* on an aluminum frame, with hardware attached, as well as unassembled, with frame parts and hardware supplied loose. The unassembled type provides somewhat more flexibility in adapting to special situations, but requires field assembly and attachment of suitable hardware (Fig. 30).

A spring-loaded safety chain should be provided to limit door swing and cushion the impact of a sudden opening, as by a gust of wind. A door closer is recommended to close the door when not in use. A durable latch to maintain the door in a closed position and to permit locking the door from the inside is also desirable.

Relatively little maintenance is required with aluminum storm doors and windows. Guide tracks and sash should be cleaned periodically to keep them operating smoothly. Door closers may need periodic oiling. Factory-applied paint finishes generally last for many years. However, if repainting is desired, special paints are available.

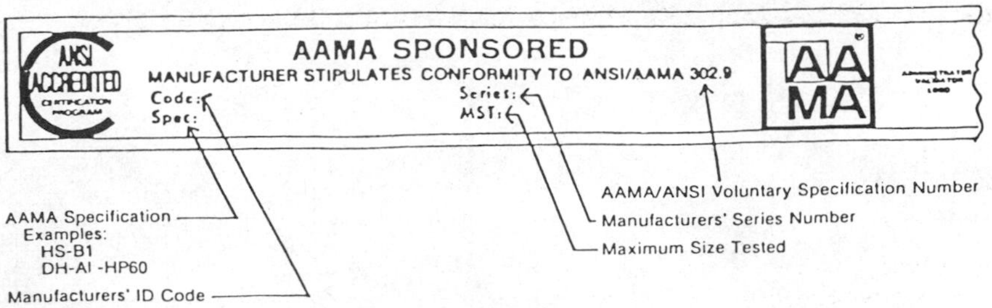

FIG. 28 The blue on silver, 5-3/8" x 5/8" AAMA label must be visible on the metal after installation. Its use on windows (examples shown), storm windows (ANSI/AAMA 1002.9) and storm doors (ANSI/AAMA 1102.7) assures conformance with AAMA and ANSI specifications.

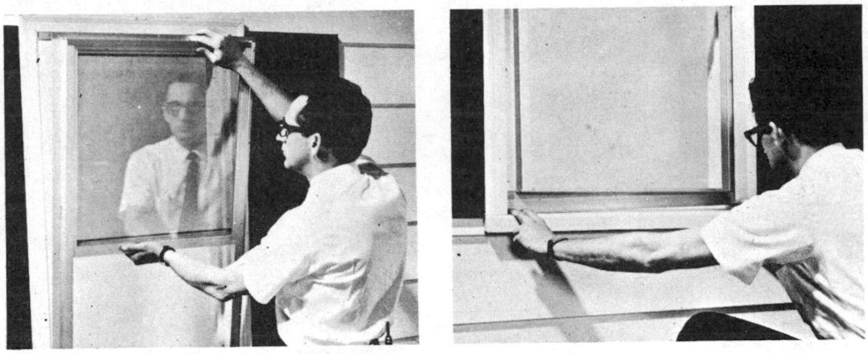

FIG. 29 Window is centered in opening (a) and secured at top; bottom expander is adjusted to sill (b) and sides screwed to jambs.

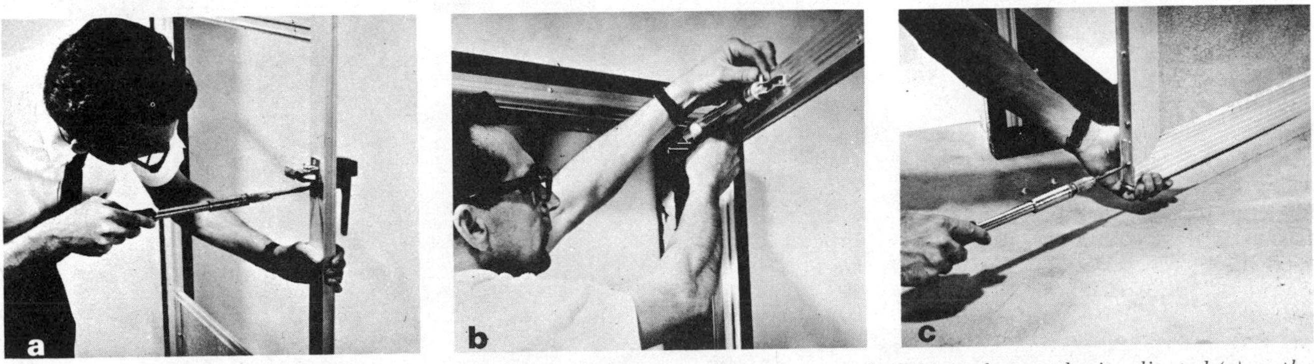

FIG. 30 After door is secured in opening, latch is installed (a); door closer is attached (b); and expander is adjusted (c) so that weather-stripping will contact sill when door is closed.

This page deliberately left blank.

Aluminum siding was first introduced to the American public following World War II. Early siding was made in the natural aluminum finish, but soon afterwards an enamel finish was applied, making it available in a choice of colors.

The prominent features of aluminum siding include minimal maintenance, good durability, light weight and high resistance to corrosion. For multiple design possibilities, aluminum siding is available in a variety of colors, textures, sizes and shapes. Although used in both new and existing constructions, aluminum now accounts for 75% of all residential re-siding projects in the country (Fig. 31).

The following recommendations are based largely on Specifications for Aluminum Siding developed by the Architectural Aluminum Manufacturers Association (AAMA) in Publication 1402.3, "Specifications for Aluminum Siding and High Performance Coatings for Aluminum Siding". The AAMA Specifications regulate the manufacture and performance of aluminum siding and constitute the recognized standard of quality for this product.

FIG. 31 Durability and ease of installation make aluminum siding a popular choice for most exterior surfaces, including walls, gable ends and soffits.

MANUFACTURE

Aluminum sheet is fabricated into horizontal and vertical panels by cold rolling. The typical alloy used in aluminum siding is known as 3003-H14, a magnesium-containing alloy distinguished by moderate strength and good workability. (For additional information on aluminum fabrication techniques, see page 213-13.)

The finishes used on aluminum siding are generally organic enamels such as vinyls, acrylics or alkyds. They are baked on at the factory after proper pretreatment of the surface.

Siding Types

Horizontal aluminum siding is available in several typical widths. The most commonly used siding type is the horizontal panel with 8″ exposure (face). Other types include the double 4″ and double 5″ exposure effects which are created by grooves bisecting a single 8″ or 10″ horizontal panel (Fig. 32a).

Vertical siding is available in both 12″ and 16″ exposures. Both exposures lend a board-and-batten effect, while the 16″ exposure is often bisected by a V-groove for additional interest (Fig. 32b).

Since aluminum, like all metals, is an effective vapor barrier, siding should be equipped with vents or weep holes to permit the escape of water vapor and prevent the accumulation of moisture in the wall.

Backerboards Aluminum siding is available with or without insulating backerboard. When the siding is backed by a suitable board, the metal gauge can be as thin as .019″. Unbacked siding should not be less than .024″ in thickness if the maximum flat area is more than 5-1/2″.

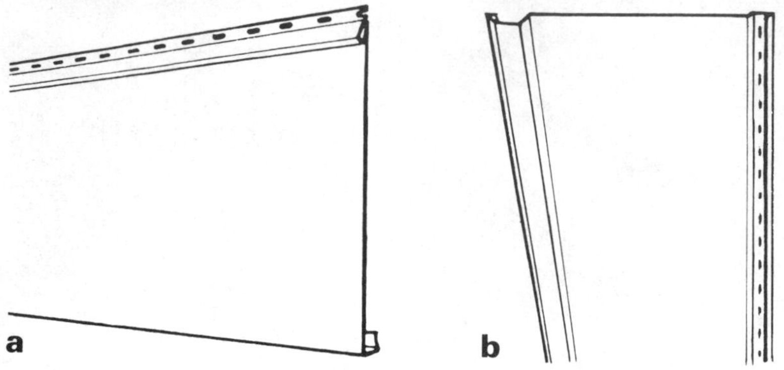

FIG. 32 Both horizontal (a) and vertical (b) siding is designed so that each successive panel will conceal fasteners and interlock with the previous panel.

FIG. 33 SUMMARY OF ALUMINUM SIDING ACCESSORIES

Material or Product	Standard Specification
Vegetable Fiber Board	FS LLL-I-535, Class G
Expanded Polystyrene	FS HH-I-524, Type I, Class 2 or Type II, Class 2
Aluminum Nails	CS263-64
Caulking Compound	FS TT-C-598

Typical backerboard materials include fiberboard or foamed plastic with or without an aluminum foil-faced kraft-paper laminate bonded to the backside. Foamed plastics should have a flame spread rating of 75 or less and should produce not more than 2500 BTU's per sq. ft. of material. The plastic backerboard should also be compatible with the aluminum substrate and its applied coatings.

Accessory Products Adhesives used to bond backerboards to siding should demonstrate sufficient moisture resistance, strength and flexibility to maintain a permanent bond under service conditions. AAMA Specifications for Aluminum Siding outline specific performance tests for the adhesives as well as for the finish coating. The AAMA Specifications also require compliance with other industry standards for accessory products employed in the manufacture or installation of siding (Fig. 33).

INSTALLATION

In refinishing projects, the quality of the finished job depends heavily on the initial step of surface preparation. Loose boards should be nailed down and irregular surfaces should be levelled. Unbacked siding cannot bridge low spots as readily as the backed type, hence furring of low spots may be required to provide adequate support. In some cases, the whole wall must be furred (Fig. 34). Good initial preparation permits siding joints to close up and avoids the wavy look of poor siding installations.

In new construction, backed siding can be applied directly to stud framing over sheathing paper. Unbacked siding should be installed only over a solid sheathing surface. When wood board sheathing is used, sheathing paper should be provided; when panel sheathing materials are applied in a way which minimizes air leakage, the sheathing paper can be omitted.

Aluminum foil laminated to kraft paper is sometimes used as sheathing paper under aluminum siding. It should be remembered that aluminum foil is an excellent vapor barrier and will trap moisture unless it is deliberately perforated to allow the escape of vapor from the wall cavity. Foil products intended for such use should be factory perforated to obtain a permeance of 5 perms or more. An efficient vapor barrier, with a permeance of 0.5 perms or less, should be used on the room side of the wall assembly. Some manufacturers also recommend the installation of two vent cups at the top and bottom of each stud space. (For more detailed information, see Moisture Control 104-15.)

Careful application of accessories such as starter strips and corner trim is essential to provide a level and square installation. Starter strips are installed first to insure proper alignment and fastening of siding (Fig. 35).

FIG. 34 Over existing walls, 3/8" × 1-1/2" or larger wood furring is applied.

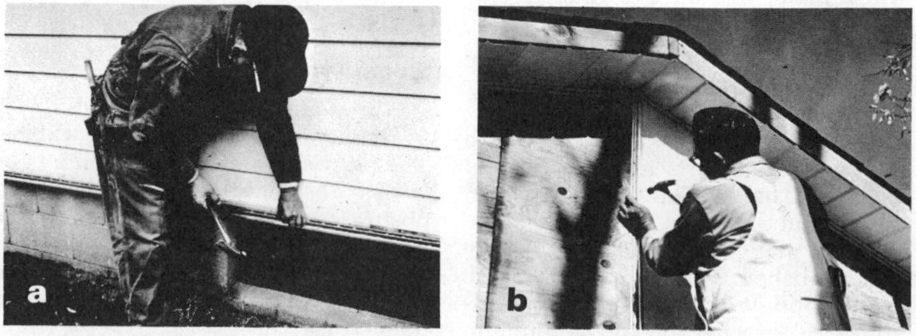

FIG. 35 Starter strip for horizontal siding (a) secures first panel; starter strip on vertical siding (b) conceals cut edge of panel on right and assures plumb installation.

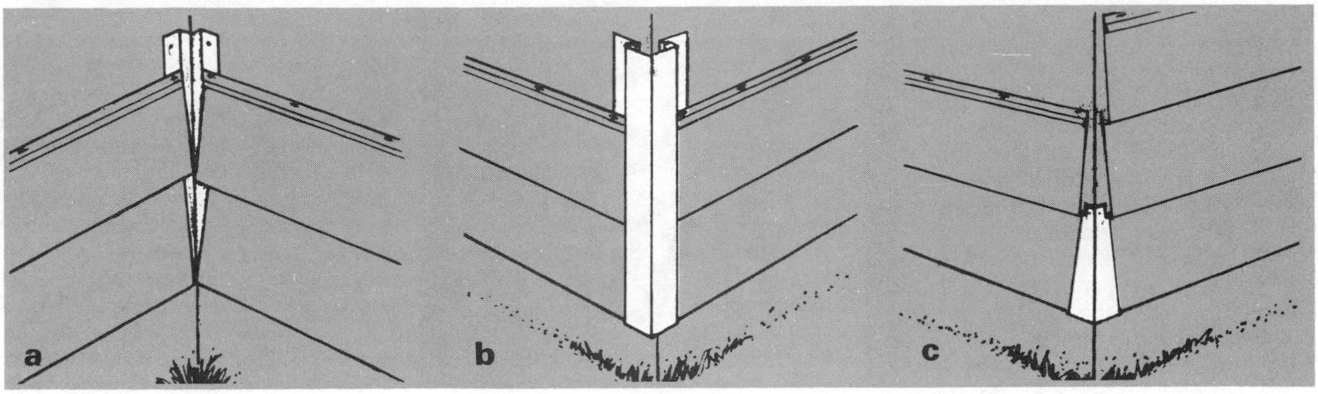

FIG. 36 Inside and outside corner trim improves appearance and, with the help of caulking, provides weather resistance as well. Corner posts (a & b) are installed ahead of siding, individual corner caps (c) after.

Continuous inside and outside corner posts are used to receive siding at corners, insets, projections and wall junctions (Fig. 36a and 36b). Since aluminum expands and contracts with temperature variations, a space should be allowed where the siding ends abut a post or any rigidly supported surface. Outside corners can also be completed with the use of outside corner caps installed after siding is in place (Fig. 36c).

FIG. 37 The first course should be pulled up and interlocked with starter strip.

FIG. 38 Panel joints should be staggered in successive courses.

In installing horizontal siding, extra care should also be used in the application of the first course, making certain that it is level and securely locked into the starter strip (Fig. 37). Each length of siding should be lapped over the preceding to create the effect of a single panel extended across the building. Factory finished ends should be exposed and cut ends lapped under the edge of adjoining panels (Fig. 38).

To fit siding around areas such as windows, eaves and doors, panels can be cut to desired size (Fig. 39). Joints should be planned so that, wherever possible, no piece is less than 20″ long.

Proper nailing is essential to successful installation. Panels should be secured with aluminum nails driven through the slotted holes in the siding at intervals of approximately 16″ along its length. Siding should be suspended on the nails, not nailed rigidly to the wall, as locked siding will not be able to expand or contract with temperature changes (Fig. 40).

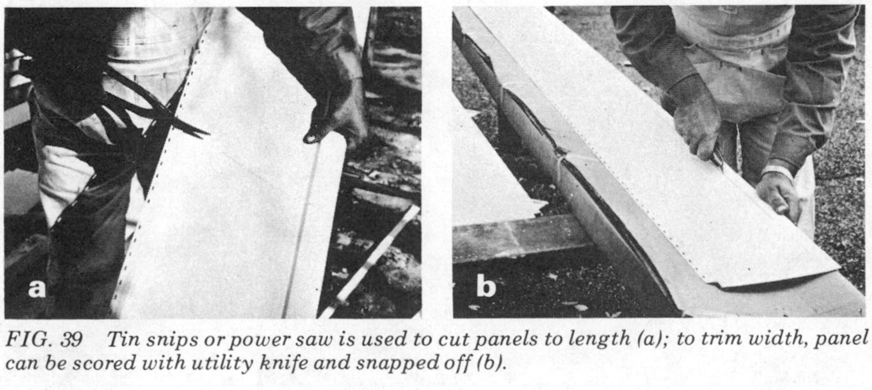

FIG. 39 Tin snips or power saw is used to cut panels to length (a); to trim width, panel can be scored with utility knife and snapped off (b).

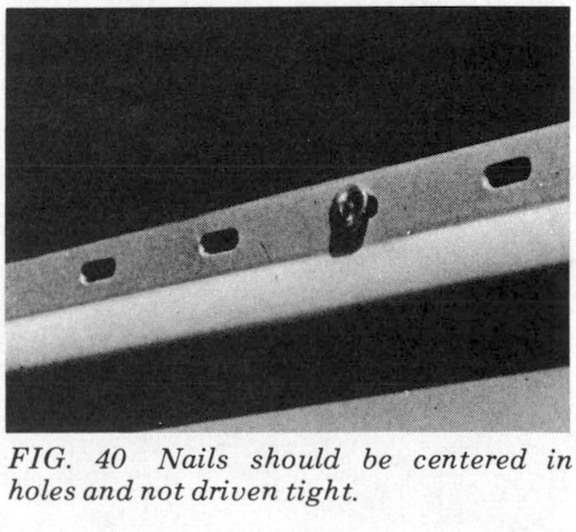

FIG. 40 Nails should be centered in holes and not driven tight.

Matching gutters, downspouts, fascia boards and soffit panels are also available (Fig. 41). Because of the natural corrosion resistance of aluminum and the durability of the baked-on enamel finish, these products are particularly well adapted to their severe exposures.

To prevent the entry of water and improve the appearance of the job, caulking should be used at all junctions where the siding abuts wood, brick, stone or other metals. Caulking compounds should be durable synthetic materials, compatible with the aluminum and paint finish.

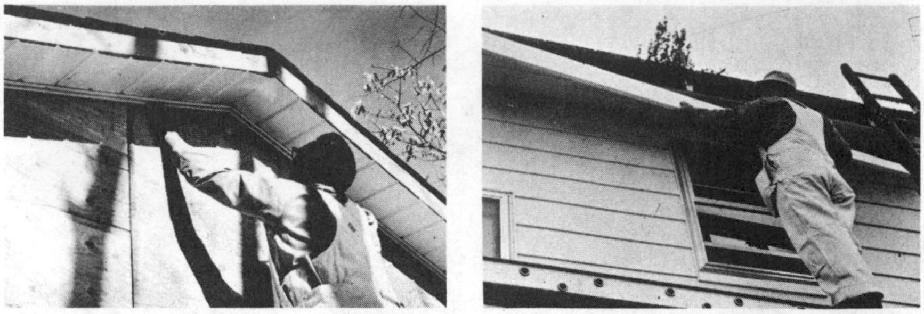

FIG. 41 Soffits panels can be of special design or standard horizontal siding cut to size (a); fascia board (b) can be scored and bent under roofing if it is wider than rafter ends.

Grounding

Although the possibility is remote, certain building codes require that aluminum siding be grounded as a precaution against electrical shock from faulty wiring. This can be done by connecting a No. 8 copper wire to a cold water pipe or a steel rod embedded in the earth (Fig. 42). Grounding allows stray electrical fault currents to be dissipated safely into the ground and eliminates shock hazard to persons touching the siding.

MAINTENANCE

Baked-on siding finishes are generally resistant to the effects of intense sunlight, snow, sleet, hail, ice or salt spray. Aside from normal rainfall, siding surfaces should be kept clean by periodic rinsings with a garden hose. Persistent stains resulting from industrial fallout, tree sap or insecticides can be removed with a non-abrasive household cleaner or special siding cleaner.

Panels damaged by storm, fire and other hazards can be replaced individually, without disturbing adjacent panels. Aluminum siding may also be repainted, if desired or when necessary due to damage. A special paint intended for aluminum surfaces is recommended, although ordinary house paint can also be used.

STANDARDS OF QUALITY

Specifications for Aluminum Siding AAMA 1402.3, developed and promoted by the Architectural Aluminum Manufacturers Association, constitute the recognized standard for this product. The Specifications establish dimensional and performance requirements for the aluminum siding, backerboards, nails, caulking and other accessory materials. Installation procedures are not covered in these specifications but should comply with manufacturer's recommendations generally provided with each package of siding.

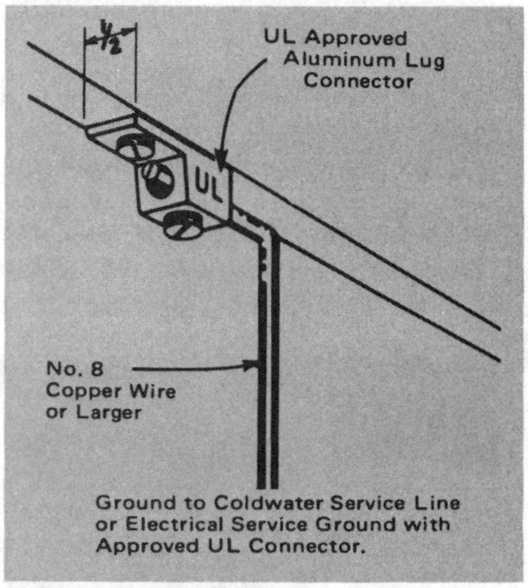

FIG. 42 Underwriters Laboratories' recommended grounding technique for aluminum siding.

ALUMINUM PRODUCTS 214

This section of the Work File, Aluminum Products 214, includes information on windows and sliding glass doors. Information on other aluminum products will be included in future Work File additions.

WINDOWS AND SLIDING GLASS DOORS

STANDARDS OF QUALITY

Aluminum windows and sliding glass doors of suitable quality for one- and two-family residential construction should meet or exceed appropriate B1 or B2 Residential specifications of the Architectural Aluminum Manufacturers Association (AAMA) shown in Figure WF1.

The AAMA specifications are also American National Standards; they are designated as ANSI A302.9 for aluminum windows and ANSI A402.9 for aluminum sliding glass doors, and constitute the recognized standards for the aluminum window and sliding glass door industry. The specifications include minimum provisions for frame member strength and thickness, corrosion resistance, air infiltration, water resistance, wind load capability and other details necessary to insure a window or door of suitable quality.

Windows and sliding glass doors of suitable quality for multiple-family dwellings, rental properties and light commercial use should meet or exceed appropriate A-2 Commercial specifications (Fig. WF1).

A-2 and A-2.50 specifications require heavier window and door construction and stronger hardware appropriate for more demanding installations. A-3 and A-4 Monumental specifications cover windows for institutional and heavy commercial use.

Windows and sliding glass doors of suitable quality for high wind load zones should meet or exceed AAMA specifications for "HP" Series windows and doors.

Both B1 and B2 specification levels are available in the wind load rated "HP" Series. "HP" Series windows and doors are designed to withstand structural pressures of 40 psf or greater which may be encountered in the high wind load zones. They also may be required in low or medium wind load zones when used in high-rise buildings, shoreline areas and other exposed locations where high wind velocities often are experienced. (See Main Text Fig. AW17 and discussion on page 214-11.)

CERTIFICATION OF QUALITY

The Architectural Aluminum Manufacturers Association (AAMA) sponsors a program of inspection and testing which results in certification of windows and sliding glass doors which conform to ANSI specifications. The AAMA label (Fig. WF2) on windows and doors is assurance that the design, manufacture and tested performance of the unit meets or exceeds appropriate ANSI specifications.

The AAMA sponsored certification program and several others are operated in accordance with the American National Standard Practice for Certification Procedures, ANSI Z34.1, which provides for certification to the public on an impartial, independent, continuing basis.

INSTALLATION

Installation of windows and doors in the rough openings should be withheld as long as construction schedules will permit. Units should be erected in place plumb, level and true, with secure anchorage to sup-

FIG. WF1 **ANSI DESIGNATIONS FOR WINDOWS AND DOORS**

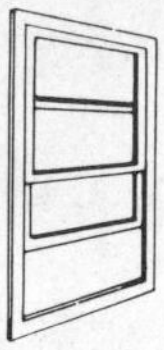	**SINGLE OR DOUBLE HUNG** DH-B1 Residential DH-A2 Commercial DH-A2.50 Commercial DH-A3 Monumental DH-A4 Monumental
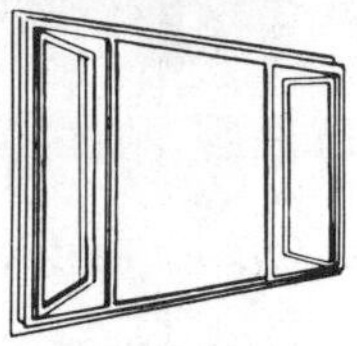	**CASEMENT** C-B1 Residential C-A2 Commercial C-A3 Monumental
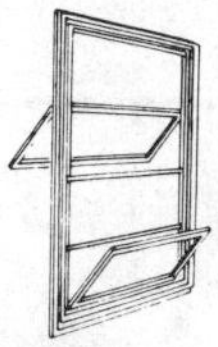	**PROJECTED** P-B1 Residential P-A2 Commercial P-A2.50 Commercial P-A3 Commercial
	**AWNING** A-B1 Residential A-A2 Commercial
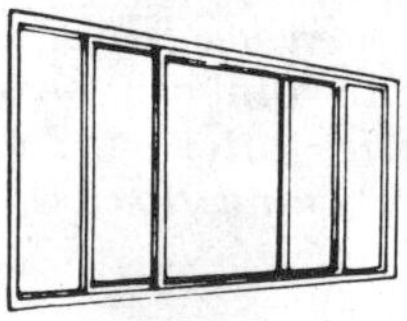	**HORIZONTAL SLIDING** HS-B1 Residential HS-B2 Residential HS-A2 Commercial HS-A3 Monumental
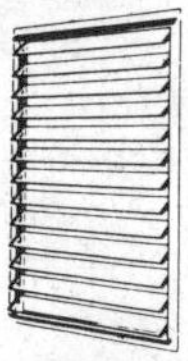	**JALOUSIE** J -B1 Residential
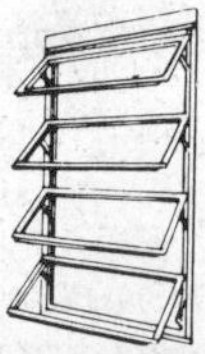	**JAL-AWNING** JA-B1 Residential
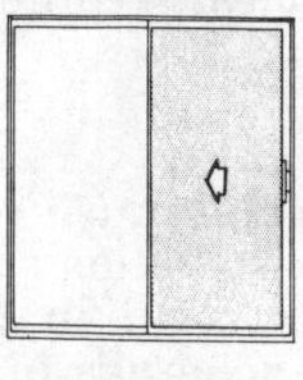	**SLIDING GLASS DOOR** SGD-B1 Residential SGD-B2 Residential SGD-A2 Commercial SGD-A3 Monumental

porting construction around the entire frame. If the construction is insulated, the space between rough opening and frame should be filled with insulation. This space should be sealed against infiltration with caulking compound and/or adequate trim.

Anchoring devices and flashing should be of aluminum or metals compatible with aluminum, such as stainless, cadmium-coated and galvanized steels. Dissimilar metals such as copper or bronze should not be used in *direct contact* with aluminum. Where they are used with aluminum, they should be insulated from the aluminum with waterproof, non-conductive materials such as neoprene, waxed papers, and coated felts. Dissimilar metals located where water may drain from them over the window or door should be painted to prevent galvanic corrosion of the aluminum. Carbon steel in similar locations should be painted to prevent rust staining.

"Dissimilar metals" are those whose electrolytic potentials are much higher or lower than that of aluminum (see page Aluminum 213-8). Galvanic corrosion of aluminum may occur in the presence of dissimilar metals and an electrolytic solution. Electrolytic solutions may form around metal parts from condensed moisture and chemicals present in the air. Waterproof non-conductors between aluminum and dissimilar metals inhibit galvanic corrosion.

Exposed aluminum surfaces of windows and sliding glass doors which may be subject to splashing with plaster, mortar or concrete masonry cleaning solutions should be covered during construction with a clear lacquer or strippable coating. Concealed aluminum surfaces in contact with intermittently wet concrete, masonry and absorbent materials (such as wood, paper and insulation) should be permanently protected. The aluminum or the adjacent material should be coated with such coatings as bituminous paint, aluminum house paint or zinc-chromate primer, to minimize the chance of chemical corrosion by acids, alkalies and salts leached out of adjacent materials. Creosote and tar coatings should not be used.

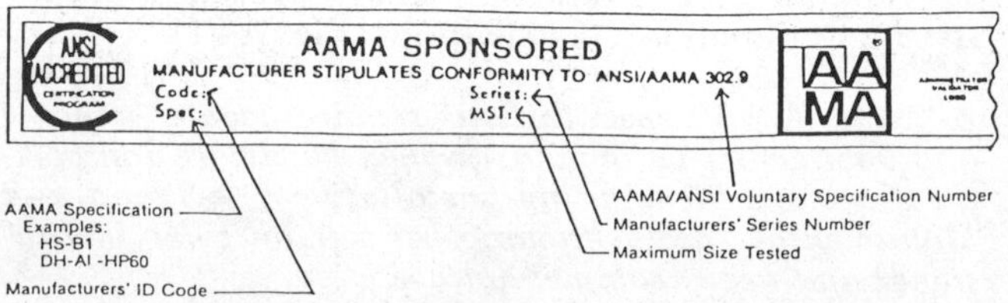

FIG. WF2 The blue on silver, 5-3/8" x 5/8" AAMA label must be visible on the metal after installation. Its use on windows (example shown) and sliding glass doors (ANSI/AAMA 402.9) assures conformance with AAMA and ANSI specifications.

216 PAINTS & PROTECTIVE COATINGS

PAINTS & PROTECTIVE COATINGS

INTRODUCTION

There is a constant demand for recommendations as to the "best" paint for a given use. However, there is rarely one "best" paint for any surface. Some of the factors which must be considered are the condition of the surface, the method of application, the curing conditions, the service expected and the relative weight of initial cost balanced against appearance and durability.

Furthermore, the entire surface-coating *system* must be considered as a unit. It is unwise and often impossible to choose a coating without regard to the surface over which it is to be applied. Though it is generally impossible to single out the best paint, it is possible to list commonly encountered surfaces and to recommend the materials most suitable for each surface.

As used in this section, the term *coating* includes all mastic or liquid-applied decorative or protective surface treatments, regardless of whether a protective film is formed as in paints and varnishes, or whether the material is absorbed into the surface, as in many stains. *Paints* are pigmented coatings forming opaque films and include enamels. *Varnishes* generally contain no pigment and form clear or translucent coatings. *Stains* may range from clear penetrating types which merely change the color of wood and leave no surface film, to heavily pigmented types which approach paints in opacity and film forming properties.

Paint is a mixture of minute solid particles known as *pigment,* suspended in a liquid medium referred to as the vehicle. The pigment generally provides hiding power and color; the vehicle combines (1) the volatile *solvent* (thinner), which assures the desired consistency for application by brush, roller or spray, and (2) the *binder,* which bonds the pigment particles into a cohesive paint film during the drying process. Some paints such as lacquers dry and harden simply by evaporation of the solvent. Most other types of paint involve chemical reactions as well as solvent evaporation.

The compositions and chief properties of various paints and coatings are described in the first subsection, *Materials.* Recommendations for use over various surfaces are given in the second subsection, *Surfaces.* Specialized terms applicable to paints and coatings are defined in the *Glossary.*

The recommendations in this section have been developed by the Scientific Committee of the National Paint and Coatings Association. The paints and coatings recommended in the following text are not all-inclusive. Individual manufacturers may have other formulations designed for specific purposes, and having some advantage over more commonly used paint and coating materials.

It also should be understood that the classification in the *Materials* subsection carries no implication of quality. Materials in any of the types described may be available in varying qualities. In the selection of paint, as with other building products, the reputation of the manufacturer is a significant factor.

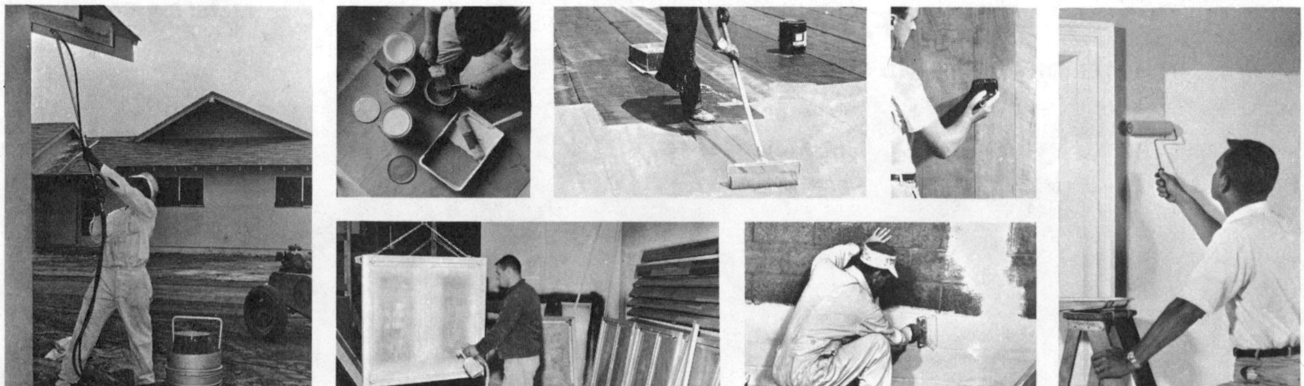

Paints and coatings listed in this section are classified by number as well as generic name. The numbers will aid in identifying and locating specific materials when mentioned elsewhere in the text. Specialized paints and coatings generally used in industrial and special-purpose applications (5000 series) are not included in this discussion.

INTERIOR PRIMERS

Primers (preparatory applications) may be classified as to their use on interior or exterior surfaces. The following materials typically are used on interior surfaces.

Wall Primers and Primer-sealers

(1110) Latex Primers and Primer-sealers These emulsion paints are recommended especially for use on gypsum board surfaces. On such surfaces, they have the property of not raising the fibers as would a solvent-thinned primer. Further advantages are their quick drying properties which allow recoating within hours, ease of equipment cleaning and excellent alkali resistance.

(1120) Alkyd Primers and Primer-sealers These usually are made of odorless alkyd, that is, the vehicle is an alkyd resin dissolved in odorless mineral spirits. They take longer to dry than the emulsion type, requiring overnight drying before recoating. They are suitable for all interior surfaces except paper-surfaced gypsum board and surfaces containing active alkali, such as partially cured or damp plaster and masonry.

(1130) Oil Primers and Primer-sealers These are primers whose vehicle is a processed drying oil, often containing some resin. Their characteristics are similar to those of alkyd primers except that drying may be slower and softer. Overnight recoating schedules still are possible. They may be thinned with odorless or ordinary mineral spirits.

(1140) Alkali-resistant Primers and Primer-sealers These primers usually are based on a butadiene-styrene copolymer or on chlorinated rubber, with one or more softer resins, such as chlorinated paraffin as plasticizers. Sometimes small amounts of drying oil may be added. Aromatic solvents often are required. Application is likely to be more difficult than for other primers. They are particularly suitable for alkaline surfaces, such as damp or partly cured plaster or masonry. They should not be expected to resist hydrostatic pressure, such as may occur below grade but may be used on above grade masonry walls.

Wood Primers

(1210) Enamel Undercoats Enamel undercoats may be of the alkyd, latex or oil-based variety and are characterized by good hiding ability and by low gloss. These undercoats produce hard, tight films which prevent the penetration of enamel coats applied over them. This property is known as *enamel hold-out.* They may be used under any pigmented interior finish but are particularly suitable under gloss and semi-gloss enamel where a smooth finish is important. Enamel undercoats are usually white but may be tinted for use under colored finish coats. Enamel undercoats contain no rust-inhibitive pigment, but they may be used as primers on metals where no rusting is expected. Alkyd undercoats dry sufficiently overnight to be painted the following day; latex undercoats may be painted in two to four hours; oil-based undercoats should dry at least 16 hours before recoating.

(1220) Clear Wood Sealers These products are developed to seal the pores in wood surfaces without impairing the natural appearance. Many sealers contain a transparent pigment to improve sealing and reduce penetration. This reduces the prominence of the grain to some extent, which usually is desirable. These products generally are used under clear finishes although they can be used under pigmented materials as well.

(1230) Paste Wood Filler This product is used to fill the pores of open-grain woods, such as oak or mahogany. A transparent and relatively coarse pigment is combined with sufficient binder to make a stiff paste, which requires thinning with a volatile solvent before application. To obtain specific effects, stain may be added. When the solvent has evaporated, the filler should be rubbed across the grain to force the filler into the pores and to remove any excess from the surface. This is followed by a cleanup wipe in the direction of the grain using a coarse cloth such as burlap.

Liquid wood fillers designed to eliminate thinning also are available. While they generally do not perform as well as paste fillers on many woods, they are adequate when extensive filling is not required. All fillers impart some color to the wood. Fillers containing stearates should not be used under urethane finishes.

Masonry Primers

(1310) Block Filler This type of

masonry primer has a relatively thick composition which permits application by brush or roller. If applied by spray, it should be brushed into the surface to insure that voids will be filled. The vehicle may be latex or a solvent-thinned material, such as an epoxy ester. The latex materials have excellent alkali resistance and may be used on damp surfaces. The solvent-thinned materials may have better chemical resistance.

(1320) Cement Grout A thin portland cement mortar often is used to give a smooth surface to rough masonry. It may be applied with brush or trowel and is available premixed with latex or in powdery form, which should be mixed with latex emulsion. Unless the mortar is made with a latex, it is very permeable and is not a good substrate for paints, especially in thin layers.

EXTERIOR PRIMERS

The following primers typically are used on exterior surfaces.

Wood Primers

(2110) Oil Primers These *house paint primers* usually contain oil to control penetration. They generally do not contain zinc oxide because the presence of zinc can cause blistering and peeling, particularly under severe moisture conditions. However, paint systems without zinc oxide are more difficult to render mold resistant.

(2120) Oil-and-resin Primers These *fortified primers* are similar to oil primers (2110), except that an alkyd or phenolic varnish is used in the vehicle. This promotes faster drying, shorter recoating time, resistance to bleeding, resistance to mold and moisture and controlled penetration.

Metal Primers

Lead pigment is toxic and coatings containing more than 0.06% lead should not be used in residential buildings or on structures with which children may come in contact.

(2210) Oil and Oleoresinous Primers

(2211) Red Lead primer (red lead in oil) is probably the oldest type of metal primer in use today because it is very reliable. The oil vehicle wets the surface well and seeps into cracks and crevices to insure protection of the metal even on poorly prepared surfaces. However, drying is comparatively slow, and the film may be too soft for some finish coats. The red lead pigment will carbonate and turn white if left exposed for too long. This primer should be covered with a finish coat within six months.

(2212) Red Lead/Mixed Pigment primers depend on the red lead for their corrosion-inhibitive properties. However, they also contain iron oxide and selected extenders which contribute to the general performance of the coating, adding hiding power, hardness and general durability. In other respects their properties are similar to those of the red lead in oil primers (2211).

(2213) Zinc Chromate (zinc yellow) primers usually are made with iron oxide as part of the pigment and with a phenolic resin varnish as the vehicle. They depend on zinc chromate for their anti-corrosive effect and generally dry harder and faster than the red lead primers.

(2220) Alkyd-based Primers

(2221) Red Lead/Mixed Pigment primers are similar to their oil-based counterpart (2212) but will dry faster and harder, making them suitable for use under harder finish coats. The surface must be cleaned more completely before application than is required for oil or oleoresinous primers (2212).

(2222) Zinc Chromate (zinc yellow) primers generally are of mixed pigment type similar to 2213, but they dry harder and faster and require more thorough cleaning of the surface. When lead-free, they often are used for priming aluminum or magnesium surfaces.

(2230) Zinc-dust/Zinc-oxide Primers These primers are available with oil, alkyd or phenolic as the vehicle. All of these primers are especially effective on galvanized steel but may be used on other metallic surfaces as well. The oil type usually is the best for general use, especially if old, rusty or weathered surfaces are to be coated. The alkyd type is for general use on clean surfaces and also is used as a heat-resistant coating.

Phenolic primers generally are recommended only when the coated surface is to be immersed in water, but they are not recommended for exterior use. When these primers are used, the surface should be given a phosphate treatment before painting.

(2240) Latex Inhibitive Primers Like most latex products, these primers are easy to apply and easy to clean up, being water-soluble before they cure. They are available in white, permitting coverage with one finish coat, usually within hours. While latex inhibitive primers will give good corrosion protection, they may show rust staining if not protected by a moisture-impervious finish coat.

(2250) Wash Primers These primers usually contain phosphoric acid and clean as well as prime the the surface. They are useful in promoting adhesion of subsequent coats of paint and frequently are used to protect freshly sand-blasted surfaces from further corrosion. Wash primers are furnished as either one- or two-part materials. One-part wash primers have the advantage of longer pot life after thinning but generally sacrifice some corrosion resistance when compared with the two-part type. It is essential that two-part formulations be used the day they are mixed.

(2260) Portland Cement Paints These are oil-based paints containing portland cement and should not be confused with cement grout (1320) or dry-powder portland cement paints (4560) designed for masonry. Portland cement paints are effective primers for galvanized steel. If white portland cement is used, a substantially white primer results, permitting tinting to shade and contributing to the ease with which the primer can be covered by finish coats.

Masonry Primers

(2310) Clear Phenolic, Alkyd and

Epoxy Coatings These are used as sealers on normally smooth masonry surfaces. If the surface is weathered, a clear sealer sometimes is recommended to stabilize the surface. If latex finish coats are to be used, the sealer should be thinned so that it will dry to a low gloss.

(2320) Silicone Water-repellent Primers These materials consist of a solution of silicone resin in a suitable solvent and are recommended when a water-repellent surface is required without change in surface appearance. Since they are transparent solutions, it is difficult to determine visually when an adequate coating has been obtained. To insure that the desired water-repellency is achieved, the spreading rate should be as recommended by the manufacturer. These materials should not be finished with other coatings until they have weathered for several years.

INTERIOR FINISH COATS

A wide range of interior paints are available with various finish surface characteristics.

Gloss Finishes

(3110) Gloss Wall Paints This term usually denotes products with oleoresinous vehicles. Their former popularity was based largely on easier application than early alkyd gloss enamels (3120). They have been largely replaced by the gloss enamels.

(3120) Gloss Enamels These often are alkyd enamels. Modern formulations offer easy application comparable to gloss wall paint (3110), with much better gloss retention and resistance to yellowing and alkaline cleaners. Latex interior enamels are available in both gloss and semi-gloss formulations; the gloss types do not produce as high a sheen as the alkyd enamels.

(3130) Floor Enamels

(3131) Alkyd and Oleoresinous enamels are formulated for relatively fast drying and for resistance to abrasion and impact. They are quite impermeable to water but are not notably resistant to alkali, such as may arise from fresh concrete. On wood or masonry, these enamels will blister and peel if moisture accumulates behind the film.

(3132) Alkali-resistant enamels usually are based on a solvent-thinned styrene-butadiene resin or chlorinated rubber. They have excellent alkali resistance but relatively poor resistance to solvents. They flow and level well but may sag on vertical surfaces. Bubbling sometimes will occur if these enamels are applied by roller on concrete floors.

(3133) Epoxy and Urethane enamels are available in one- or two-part formulations. The one-part materials are similar to alkyd and oleoresinous enamels (3131) but have somewhat improved properties, particularly abrasion resistance. The two-part materials have extremely high adhesion, abrasion resistance and resistance to water and solvents, but they require great care in surface preparation to develop the properties to the maximum extent.

(3140) Dry-fallout Spray Gloss These finishes basically are similar to gloss enamels (3120) but have a much faster acting solvent and usually a faster curing resin. They are designed for spray application only. An advantage is that much of the overspray dries quickly before reaching a surface and therefore can be removed easily.

(3150) Clear Finishes

(3151) Floor Varnishes are relatively fast-drying with good resistance to yellowing and excellent abrasion resistance. Alkyd varnishes often are paler and have better color retention than oleoresinous varnishes. However, oleoresinous varnishes may be tougher and more alkali resistant. Epoxy varnishes have good color quality and alkali resistance. Urethanes provide the best abrasion resistance but require more thorough surface preparation prior to application.

(3152) Counter Varnishes are hard-drying, print-proof films. "Bar top" varnishes should also be resistant to alcohol. Polyurethanes or polyesters often are used for this purpose.

(3153) Paneling and Trim Varnishes need not be as hard as varnish for floors or counters but should be resistant to checking and should dry tack-free. Alkyds usually exhibit the best color quality and color retention.

(3154) Shellac is a fast-drying, light-colored varnish with excellent color retention. Although abrasion resistance is poor, the finish is easily repaired. Resistance to water and alcohol is poor but resistance to petroleum solvents is good. These coatings often are used as a *wash coat* on wood before final sanding or as a clear primer or stain sealer.

Semi-gloss (Eggshell) Finishes

(3210) Semi-gloss Wall Paint This is an oleoresinous enamel with an initial intermediate gloss. It may lose its gloss after a period of time and is likely to have low resistance to grease and alkaline cleaners. It is often too soft for use on horizontal surfaces; its use is decreasing in favor of other types of semi-gloss enamels and paints.

(3220) Semi-gloss Enamel This is an alkyd-based product, usually with better gloss retention, grease and alkali resistance than semi-gloss wall paint (3210). Application may be somewhat more difficult over large areas. Gloss may be reduced rapidly the first few days after application but will then stabilize.

(3230) Semi-gloss Latex Paint And Enamel This product has the usual advantages of latex paints—ease of application and cleanup, rapid drying, little odor and nonflammability. Film leveling usually is good but hiding power may not compare favorably with alkyd products.

(3240) Dry-fallout Spray Semi-gloss This material, intended for spray application only, has the properties of its gloss counterpart (3140) but at a lower gloss level.

(3250) Clear Finishes

(3251) Flat (satin) Varnishes have an added flatting agent such as silica. They do not have the undesirable highlights associated with gloss varnishes, but they are less abrasion resistant.

(3252) Penetrating Sealers are

comparable to thin varnishes. They are applied liberally and the surplus is removed by wiping while still wet or by buffing when dry (after about 15 hours). Buffing provides better protection but involves more time and labor. These products give a satin finish with fair abrasion resistance and color retention.

Flat Finishes

(3310) Oil or Oleoresinous Flats These are seldom used today except as coating over calcimine paints. They have a predominantly limed oil vehicle which reduces penetration and permits easy application and excellent leveling. Finished appearance is excellent, but washability and resistance to yellowing is likely to be poor. Because of good hiding power in thin films, these flats sometimes are used in repainting acoustical surfaces.

(3320) Alkyd Flat Wall Paints These paints usually are superior to latex paints (3330) in hiding power and washability. Odorless flats may have very little odor during application, but an after-odor usually results as the oil portion of the resin oxidizes. Whether odorless or not, these products should be used only under conditions of good ventilation.

(3330) Latex Flat Wall Paint This is the most popular kind of interior flat wall paint. High hiding power and good washability may be built into the product, but often one is achieved at the expense of the other. Quick drying, ease of touchup and cleaning of tools and absence of flammable solvents or unpleasant odors during and after application are additional advantages. Characteristic ease of application may lead to excessive spreading rates and poor hiding ability. Even quality products may fail to form a coherent film and, hence, show poor washability if applied over a very porous surface or under conditions of low humidity or temperature extremes.

(3340) Dry-fallout Spray Flats These paints are made with fast-drying solvents and may be either of the alkyd or oleoresinous type. They are used on ceiling areas of industrial or commercial buildings where application is by spray, and where it is not practicable to protect all adjacent areas from overspray.

(3350) Fire-retardant Paints These products are made in a number of types, incorporating various properties in addition to fire retardance. They can be formulated as strictly functional or both functional and decorative coatings. Products are available with good fire retardancy after repeated washing and with all the properties of conventional interior coatings. Some of these paints are sensitive to water and high humidity and, hence, suitability for the intended purpose should be determined.

Fire retardant capability usually is expressed on the basis of *flame spread* ratings determined and listed by Underwriters' Laboratories, Inc. Flame spread, however, is not a property of the paint but depends on the number of coats, the thickness of each coat, the surface to which it is applied and other factors. The same paint may have entirely different flame spread ratings under different conditions. In particular, a paint may be noncombustible and have a low flame spread on noncombustible surfaces, but may have little fire retardancy on combustible surfaces. However, a paint with a low rating on combustible surfaces always will have a low flame spread on noncombustible surfaces. If coatings are not applied at the listed spreading rates, the flame spread rating will not apply.

(3360) Latex Floor Paints Latex floor paints have most of the application advantages of latex wall paints, as well as fairly good abrasion resistance. The absence of toxic or flammable solvents makes them useful in areas such as basements where the ventilation may be poor. They may be applied to damp (but not wet) surfaces even if alkalis are present. Continuous exposure to standing water is likely to soften the paint film.

Miscellaneous Finishes

(3411) Pigmented Stains These are stains containing insoluble pigments, generally permanent in color. If applied too heavily, they will obscure the texture, giving a painted effect to the wood. They are particularly suitable for subduing the conspicuous irregular grain effect of softwood plywood. These stains will not bleed into varnish or enamel coats applied over them.

(3412) Dye-type Stains These stains employ an organic coloring material which is soluble in the vehicle. They have better clarity and transparency than do pigmented stains (3411), and they produce brighter colors. However, they are more likely to bleed into succeeding coats, and the color is less permanent.

EXTERIOR FINISH COATS

Exterior finish coats today include many formulations with latex, alkyd and other synthetic resins.

Oil Base Finishes

(4110) Gloss House Paints Traditionally house paints were white lead in linseed oil; then various blends with zinc oxide, talc and other extenders were developed; and finally blends with various titanium dioxide pigments were introduced. These paints have been the foundation of the paint industry. Linseed oil is still the binder used in most quality oil type house paints, although modifications with other materials may provide improved properties. A number of *long-oil* alkyds (resins with 55% or more oil or fatty acid) are available which are similar to trim enamels (4310) but are softer and more durable.

Lead has been found to be harmful to health and is no longer allowed in any residential paints. Paints containing more than 0.06% lead (non-volatile content) must carry a warning label.

Zinc oxide often is included to control mold and to aid in the control of chalking; it is also a factor in regulating the hardness of the paint film. Extender pigments are used to control consistency and to develop an optimum blend of other properties.

Titanium dioxide is the most important pigment for providing hiding power in white and tinted paints and contributes to the control of chalking and fading. Some of these factors are of lesser importance with dark-colored paints which contain little or no white pigment.

A wide range of alkyd and latex based house paints are available in flat or glossy sheens. Because of excellent color retention, durability and ease of clean-up, latex exterior paints are gaining favor over the solvent thinned alkyd types.

(4120) Barn Paints These paints possess most of the characteristics of house paints (4110). They differ in that they are more likely to contain vehicle modifications, such as rosin, and they usually employ red iron oxide as the chief hiding pigment. Several other colors, such as green and black also are available. Barn paints are formulated to give uniform appearance over poorly prepared and nonuniform surfaces but lack the long-term protective qualities of house paints. They may be used also on roofs, particularly those of sheetmetal.

(4130) Shingle Stains These are pigmented products designed to provide color, only partial hiding ability, maximum penetration and some mold-resistant protection. They may contain creosote or other preservative; linseed oil and alkyd resin are the usual binders.

Oleoresinous Finishes

(4210) Trim Enamels Because of application problems and early loss of gloss, these materials largely have been superseded by alkyd enamels (4310).

(4220) Metal Paints Some paints for exterior metal are still made with oleoresinous vehicles—notably those for structural steel. They give good service and protection but usually lose their gloss earlier than do similar products made with alkyds (4320).

Alkyd Enamels

(4310) Trim Enamels Designed for use on exterior wood, these products usually are made with long oil resins. For this reason and because they usually are designed for easy brush application, they are not fast drying. Drying usually takes from a few hours to overnight. Typically, they are made in high gloss and bright colors and show good retention of color and gloss. They are not suitable where high resistance to acid, alkali or other chemicals is required. Silicone-alkyds are substantially more durable than are conventional alkyd enamels.

(4320) Equipment Enamels These are basically similar to trim enamels (4310), but because they are intended chiefly for metal, a hard-drying resin may be employed and some rust-inhibitive pigment may be added. If designed for spray application, a faster solvent also will be used and drying may be quite rapid. These products often are called *automotive enamels.*

(4330) Masonry Paints These paints are similar to the alkyd trim enamels (4310) except that they contain a higher proportion of pigment. Because very little flexibility is required of a masonry finish, these alkyd-based products perform very well except when excessive moisture and alkali are present.

Exterior Latex Paints

(4410) Latex House Paints These are intended mainly for wood and are gaining in popularity over solvent thinned alkyd paints because of excellent color retention, durability and ease of clean-up. Primers will nearly always be required over bare wood, badly weathered surfaces and bare metal. Where primers are needed, it is best to employ one specifically recommended for use with the intended finish-coat material. These products are similar to other latex paints in their general properties, and they share the advantages of ease of application, suitability for damp (not wet) surfaces and cleaning of equipment. However, they are formulated for good flexibility so that they will accommodate dimensional changes of wood.

Most latex house paints provide a matte finish and offer better color retention than oil and alkyd flats, even in light colors.

(4420) Maintenance Finish (Semi-gloss) Latex paints applied over suitably primed metal have shown good performance. However, moisture vapor permeability is fairly high, and while good resistance to pitting and rusting is obtained, some rust staining may appear unless an impermeable primer is used.

(4430) Masonry Paint Paints for masonry were the first large scale exterior use of latex systems. Because masonry surfaces show less dimensional change than do wood and metal, and also are rougher, adhesion is less of a problem. For surfaces with efflorescence, a clear or lightly pigmented solvent-thinned sealer (2310) often is recommended. Like the latex house paints (4410), these products usually give a low sheen finish.

Miscellaneous Coatings

(4510) Zone-marking Paints These paints, sometimes called *traffic paints,* are characterized by fast drying, little flow, good hiding ability, resistance to bleeding over asphalt and usually low gloss. Zone-marking paints for highway use may be reflectorized or nonreflectorized. Nonreflectorized paints are used where there is overhead lighting as, for example, on many parking lots. Reflectorized paints may be of the drop-in type where the glass beads or similar material are added after application of the paint, or of the premixed type where beads are incorporated in the paint. The drop-in type usually is more durable and has better visibility early in the life of the paint. The premixed type is easier to apply. A small amount of beads may be dropped into the premixed paint after application for early visibility.

Normal zone-marking paints are flat and therefore are not satisfactory for such interior uses as factory floors or school corridors where dirt collection is a problem. Marking paints for this purpose must have a gloss or semi-gloss finish.

(4520) Aluminum Paints

(4521) General-purpose aluminum paints may be made with aluminum powder or paste and a vehicle of

drying oil, alkyd resin or one of a number of varnishes. Quality may vary widely, depending upon the amount of aluminum, the ratio of aluminum to binder and the quality of the binder. Because of problems of stability, aluminum paints formerly were sold in a two-part container to be mixed just before use, but these problems largely have been solved and satisfactory one-part aluminum finishes now are available.

(4522) Heat-Resistant aluminum paints having the best heat-resistant characteristics generally are made with pure silicone resin. Various materials may replace part of the silicone vehicle; this results in a reduction in the temperature which the paint will withstand and a reduction in the paint's exterior durability. Another type intended for high temperature surfaces uses a small amount of vehicle (usually an oil) which burns off, leaving the aluminum melted and fused to the surface.

(4523) Aluminum Roof Coatings are usually of the bituminous type (4551).

(4530) Clear Finishes for Wood The performance of clear finishes in exterior locations often is not satisfactory because no clear finish can approach the durability of the same vehicle protected by pigment. Alkyd varnishes have good initial color and color retention but may crack and peel. Some of the newer synthetics have good durability but may darken on exposure. A long oil phenolic, *marine spar* varnish, has the best history of satisfactory performance although it also will darken and yellow with time. Penetrating coats of oil or thinned varnish should be used on the bare wood, followed by unthinned varnish. Varnishes containing ultraviolet screening agents exhibit less discoloration and generally greater durability.

(4540) Alkali-resistant Finishes These paints usually are formulated with styrene-butadiene resin or chlorinated rubber. They dry flat and ultimate deterioration of the film is by gradual chalking. These coatings often are used on masonry walls or decks where higher resistance to moisture penetration is required than can be obtained with latex paints (4430). They may be used as primers under less resistant finish coats. While their resistance to water, alkali and acid is excellent, their resistance to solvents is likely to be poor.

(4550) Roof Coatings

(4551) Bituminous roof coatings are made of special weather-resistant asphalt dissolved in a suitable solvent. They are intended primarily for use on asphalt roofing surfaces. Asbestos and other fillers may be added to prevent sagging on sloping roofs and to permit the application of relatively thick coatings. The addition of aluminum powder enables colors other than black to be obtained. For some roof coatings the asphalt is emulsified in water, permitting application to damp surfaces.

(4552) Latex (exterior) paints may be used on asbestos cement and tile roofs. Special latex paints are available for painting roll roofing. They may be used on asphalt shingles but are likely to curl them.

(4560) Portland Cement Powder Paint These paints are made from white portland cement, suitable pigments and usually small amounts of a water repellent agent. They are mixed with water just before application. Since hydration of the portland cement requires moisture, painted surfaces should be kept damp as recommended by the manufacturer. These paints are useful, low-cost finishes for rough masonry. Their principal drawback is that they are not good bases for other paints or coatings.

Fire Retardant Intumescent Coatings

These paints slow the progress of a fire by forming a protective foam. In many cases the charred foam may be scraped off and the sound substance repainted.

Heat sensitive ingredients in fire-retardant paints begin to intumesce (swell) at 300°F. The resulting foam, hundreds of times thicker than the original paint film, slows the fire, reduces smoke and delays the build-up of toxic gases.

Intumescent coatings may be used either in strategic places or throughout a building. These coatings are especially recommended for corridors or areas leading to fire exits.

They are available pigmented or clear, and are scrubbable.

This subsection outlines the broad categories of surfaces for which paints and coatings often are considered and describes the products which are suitable for each surface under varying exposure conditions. Unusual surfaces or those which usually are not painted, such as glass, underground metal or underwater concrete, are omitted in the following discussion; these surfaces usually require specialized coatings designed for the special conditions of use (5000 series) and generally can be identified by name. The products recommended in the following text are identified by generic name and by index number. (Fig. 1). The numbers may be used to locate the product description in preceding subsection.

INTERIOR WALLS AND CEILINGS

Primers typically are applied over prepared surfaces prior to subsequent finish coats. The importance of proper selection and application of primers cannot be over-emphasized. Regardless of the quality of subsequent finishing, failure of the primer will mean failure of the system.

Priming

Plaster New plaster should be primed with a latex, alkyd or oil-type primer-sealer (1110, 1120 or 1130). The latex flat finish coat (3330) may be used as a primer but if semi-gloss or gloss paints are to be used as a finish coat, the primer should be tested for *enamel holdout* (resistance to penetration by the finish coat) which results in uneven gloss. Highly alkaline or fresh, damp plaster requires an alkali-resistant primer (1140) or latex system (1110). Plaster which is wet or has a continuing flow of moisture from the back cannot be painted successfully. Textured or swirl plaster and soft, porous or powdery plaster should be given an acid wash with a solution of one pint of white household vinegar in one gallon of water. This procedure should be repeated until the surface is hard; it should then be rinsed well and allowed to dry before painting.

Gypsum Wallboard For wall-board or other paper-surfaced

FIG. 1 SUMMARY AND NUMERICAL DESIGNATIONS OF PAINTS AND COATINGS*

PRIMERS AND PRIMER-SEALERS

1000 Series	Interior Surfaces
1100	**Walls**
1110	Latex
1120	Alkyd
1130	Oil Base
1140	Alkali Resistant
1200	**Wood**
1210	Enamel Undercoats
1220	Clear Wood Sealers
1230	Paste Wood Fillers
1300	**Masonry**
1310	Block Fillers
1320	Cement Grout

2000 Series	Exterior Surfaces
2100	**Wood**
2110	Oil Base
2120	Oleoresinous
2200	**Metal**
2210	Oil & Oleoresinous
2211	Red Lead
2212	Red Lead/ Mixed Pigment
2213	Zinc Chromate
2220	Alkyd-based
2221	Red Lead/ Mixed Pigment
2222	Zinc Chromate
2230	Zinc-dust/Zinc-oxide
2240	Latex Inhibitive
2250	Wash Primers
2260	Portland Cement Paints
2300	**Masonry**
2310	Clear Phenolic, Alkyd & Epoxy
2320	Silicone Water-repellent

FINISH COATS

3000 Series	Interior Surfaces
3100	**Gloss Finishes**
3110	Wall Paints
3120	Enamels
3130	Floor Enamels
3131	Alkyd & Oleoresinous
3132	Alkali Resistant
3133	Epoxy & Urethane
3140	Dry-fallout Spray
3150	Clear Finishes
3151	Floor Varnishes
3152	Counter Varnishes
3153	Paneling & Trim Varnishes
3154	Shellac
3200	**Semi-gloss Finishes**
3210	Wall Paints
3220	Enamels
3230	Latex Paints
3240	Dry-fallout Spray
3250	Clear Finishes
3251	Flat Varnishes
3252	Penetrating Sealers
3300	**Flat Finishes**
3310	Oil & Oleoresinous
3320	Alkyd Wall Paints
3330	Latex Wall Paints
3340	Dry-fallout Spray
3350	Fire-retardant Paints
3360	Latex Floor Paints
3400	**Miscellaneous Finishes**
3411	Pigmented Stains
3412	Dye-type Stains

4000 Series	Exterior Surfaces
4100	**Oil Base Finishes**
4110	Gloss House Paints
4120	Barn Paints
4130	Shingle Stains
4200	**Oleoresinous Finishes**
4210	Trim Enamels
4220	Metal Paints
4300	**Alkyd Enamels**
4310	Trim Enamels
4320	Equipment Enamels
4330	Masonry Paints
4400	**Latex Finishes**
4410	Latex House Paints
4420	Maintenance Finishes (semi-gloss)
4430	Masonry Paints
4500	**Miscellaneous Coatings**
4510	Zone-marking Paints
4520	Aluminum Paints
4521	General Purpose
4522	Heat Resistant
4523	Roof Coatings
4530	Clear Wood Finishes
4540	Alkali Resistant
4550	Roof Coatings
4551	Bituminous
4552	Latex
4560	Portland Cement Powder Paints

5000 Series Specialized Chemical-resistant and Industrial Coatings

*Adapted from recommendations of the National Paint and Coatings Association. Lead pigment is toxic and should not be used in residential paints. Paints containing more than 0.06% lead (non-volatile content) should carry a warning label.

materials, latex primer-sealers (1110) usually are recommended. Latex emulsions are preferred, since solvent-thinned primers raise the fibers of the paper giving them an unsightly appearance. The latex finish coat (3330) generally may be used as a primer, but if semi-gloss or gloss paint is to be used as a finish coat, the primer should be tested for enamel holdout.

Brick Brick interior walls usually are somewhat porous, but if the mortar is well cured they will not be highly alkaline. They may be sealed with a latex primer-sealer (1110), or an enamel undercoater (1210) may be used with the addition of suitable oil or varnish. Latex (4430) or solvent-thinned (4540) exterior masonry paints may also be used for priming. Where efflorescence is present, it should be removed by vigorous scrubbing and the brick should be treated with a clear resin sealer (2310).

Concrete Masonry Most concrete block, particularly the lightweight types, have a very rough and porous surface. Because most paints and coatings cannot bridge the surface voids, one or two coats of block filler (1310) are recommended. By sealing the voids, block sealers reduce dirt accumulation and improve surface color uniformity. However, fillers may be omitted when above considerations are not of importance, when low-cost finishing may be required, or when the sound absorbing effect of the rough surface may be desirable.

Concrete The problems associated with painting concrete are similar in many respects to those of concrete masonry. However, the concrete surface usually is denser, less regularly formed and more likely to be contaminated with oils and other bond-breaking materials. Substances should be removed unless it can be assured that they will not affect paint adhesion. Rough or honeycombed surfaces often are treated with cement grout (1320), preferably thinned with a latex solution, particularly if the grout is to be applied in thin layers. Priming over grout should be as recommended for plaster surfaces. One or two coats of block filler (1310) may be used instead of grout. In this case, an alkali-resistant primer usually is not required.

Ferrous Metals A good rust-inhibitive primer always is desirable. It is essential for quality work, for damp locations, when water-thinned paints are used for finishing and when the finish coats do not produce an impervious film. A primer may be omitted in dry locations and if the finish coats are to be of a solvent-thinned type producing a tight film.

The slower drying oil-base primers (2210) generally will wet the surface better and give more positive results, but quick-drying primers (2220) will be more resistant to moisture penetration.

Zinc-coated Metals If applied over zinc-coated surfaces such as galvanized steel, many ordinary paints and rust-inhibitive primers will develop a layer of white, powdery material which eventually separates the paint film from the metal completely. This can be avoided by: (1) the use of a nonoxidizing coating such as some of the modern chemical-resistant coatings (5000 series); or more commonly (2) the use of a primer containing zinc dust (2230) or portland cement (2260). There are certain proprietary primers for galvanized steel (often latex) which are very effective for interior use.

Wood In many interior situations wood may be primed at the same time and with the same sealer used for plaster walls. One or two coats of paint may give a sufficiently smooth base for use under enamels, but such water-thinned paints will raise the grain and roughen the surface of many woods. Good enamel holdout is necessary under gloss or semi-gloss finish coats. Solvent-thinned wall primer-sealers (1120) provide better adhesion and flexibility. For the best enamel work, an enamel undercoater (1210) may be used either over a wall primer-sealer or directly on the wood. A clear wood sealer (1220) is recommended for natural finishes, preceded by a paste wood filler (1230), if necessary.

Acoustical Surfaces Acoustical tile or acoustical plaster rarely is painted at the time of installation but may require finishing later. Since the acoustical properties depend on a suitably porous or rough surface, a thin paint should be used so that the pores are not bridged or filled. Any of the products recommended for plaster surfaces may be used (see page 216-9) after thinning to suitable consistency. Metal acoustical surfaces should be treated as light-duty areas using one coat to avoid excessive paint buildup.

Finishing

Light-duty Areas There are some building areas which have fairly low concentrations of moisture, grease or dirt in the air and which require relatively little maintenance. These include living rooms, adult bedrooms and similar areas in the home. In such spaces a matte finish is both desirable and practical on the ceilings and most walls.

Latex flat wall paints (3330) usually are preferred because of their easy application, quick drying, easy cleanup and almost complete lack of odor. Alkyd flats (3320) can produce thicker films, have better hiding power, and give a more uniform appearance. These properties are particularly desirable on rough surfaces and when a considerable change is made in color. Both latex and alkyd paints have good resistance to soiling and marring and are relatively easy to clean.

Odorless alkyd flats are almost completely free of solvent odor during application. However, they must be used with adequate ventilation because the solvent fumes, while odor-free, may cause headaches and other physiological symptoms. Also, although these solvents may have a high flash point, there is a slight fire hazard. In occupied premises these air-drying coatings have the further disadvantage that the curing process results in odors which can persist for several days, requiring continued ventilation.

Moderate-duty Areas These are

similar to light-duty areas in most respects except that the amount of traffic or airborne dirt may dictate a more intensive cleaning schedule. Typical areas are children's rooms, entries, halls and closets. Finish coats should be chosen from the more washable flat paints available under the 3320 and 3330 groups and the semi-gloss coatings (3210, 3220 and 3230). The glossier materials generally have better wear resistance and washability but accompanying higher surface shine makes surface irregularities more conspicuous. This is particularly true when natural or artificial light strikes the surface at a sharp angle. This consideration becomes less important in smaller spaces or where the visible wall areas are broken up by interior furnishings.

Heavy-duty Areas Heavy-duty areas are those in which there may be large amounts of moisture and grease in the air, or where the amount and type of traffic requires a more durable paint film to withstand greater wear or more intensive cleaning. These areas include kitchens, recreation rooms and bathrooms in homes and apartments; and public corridors, entrance lobbies and service areas in institutional, commercial and apartment buildings.

The usual choice is a gloss (3110, 3120) or semi-gloss enamel (3210, 3220 and 3230). Where the exposure is particularly severe, specialized hard-gloss coatings can be used. The light reflectivity factors discussed above for moderate-duty areas always should be considered when selecting gloss finishes.

INTERIOR TRIM AND PANELING

The selection of primers for interior use on trim and paneling depends primarily on the type of surface material being coated and the desired finish characteristics.

Priming

Ferrous Metals Although most interior exposures are not particularly corrosive, a rust-inhibitive primer usually is desirable. Any of the metal primers, such as oil (2210), alkyd (2220) or latex (2240) may be used. The choice is dictated by the speed of drying, the amount of surface preparation and color. For very mild exposure, conventional enamel undercoats (1210) may be used, but these products have no corrosion-resistant effect and rust will spread from any break in the film.

Zinc-coated Metals Interior galvanized surfaces which are dry and do not receive heavy wear may be coated with interior latex paints (3330). However, if moisture, abrasion and repeated cleaning are expected, an inhibitive primer (2230, 2240 or 2260) generally will provide better service and is recommended.

Wood, Clear Finish Priming will depend on both the wood and type of finish desired. Trim should be primed on all surfaces before installation. Most softwoods will require a clear sealer (1220), especially those showing marked contrast between springwood and summerwood such as pine. The sealer will provide a more uniform surface over hard and soft grain and thus it is particularly recommended if the surface is to be stained.

Dense, close-grained hardwoods such as maple and birch have less need for a sealer, and the first coat may be a thinned version of the finish coat if recommended by the manufacturer. In stain systems, the products used and the order in which they are applied will depend on the effect desired. For example, applying the sealer before the stain is applied will minimize both the depth of color and the amount of grain contrast.

Open-grained hardwoods such as oak, walnut and mahogany require the use of a wood filler (1230) which also will serve as a sealer. When the grain is to be subdued, stain may be applied after the filler.

Wood, Opaque Finish Any of the several interior wall primers (1100 series) or an enamel undercoat (1210) may be used as a primer. However, water-thinned primers are apt to raise the grain of the wood. An enamel undercoat will give a smoother surface and is preferred under gloss and semi-gloss enamels, although many semi-gloss enamels are designed to be self-priming.

Finishing

The selection of finish coats is determined largely by the type of exposure expected, whether light-, moderate- or heavy-duty (see page 216-10).

Moderate- and Light-duty Areas

Clear Finishes for Wood should be selected from the 3150 group, depending on decorative and exposure requirements. For light-duty areas, satin or low-luster finishes often are used.

Opaque Finishes for Wood and Metal usually are semi-gloss enamels (3220) or latex enamels (3230). In light-duty areas the predominant interior finish used on the adjacent walls may be used.

Heavy-duty Areas

Clear Finish for Wood may be floor varnish (3151) or counter varnish (3152). Floor varnish is most resistant to abrasion, while counter varnish has good resistance to water and solvents. The selection should be made on the basis of the exposure anticipated.

Opaque Finishes for Wood and Metal usually are hard-drying enamels (3120). For special exposures, such as in laundries, a special type (3160) may be indicated.

INTERIOR FLOORS

The selection of finishes for interior floors depends primarily on the type of surface to be coated.

Wood

Clear Finishes For relatively light-duty floors, shellac (3154) may be used where color quality and color retention are very important. It has good gloss and color, but scratches easily and is not water-resistant; however, it is easily patched. Floor varnish (3151) is more resistant to wear and water but is somewhat darker in color. Penetrating sealer (3252) has a lower gloss and usually requires waxing. For heavy-duty areas such as gymnasium floors, specially designed finishes usually are required (5000 series). Open-grained woods such as oak may use a filler (1230) before coating. Some floor varnishes may be slippery when wet.

Opaque Finishes Alkyd enamels (3131) and latex floor paints (3360) generally are used for opaque floor finishes. Other fast-drying gloss enamels and paints may be used if recommended by the manufacturer. However, products not specifically recommended for floors may be too soft and slow drying, or too hard and brittle.

Concrete

Enamels suitable for wood floors (3131) generally are suitable also for concrete floors. However, alkali-resistant enamels (3132) should be used when a concrete slab is not completely cured or is placed below grade or on the ground without a vapor barrier. Even with alkali-resistant enamels, adhesion problems may be encountered if hydrostatic pressure forces water up through the floor.

Latex floor paints (3360) usually dry to a comparatively low gloss. Because they are nonflammable and have little odor, latex paints are preferred for basements and enclosed areas which are difficult to ventilate adequately. Since they are water emulsions and have excellent alkali resistance, the floor need not be completely dry before application. However, latex paints are softened by continued exposure to moisture or solvents and are not suitable for laundry and service areas where solvents are handled. For such exposures, specialized highly resistant finishes (5000 series) should be used.

Steel

A fast-drying, tough enamel (3131) applied over a corrosion-resistant primer (2222) generally is recommended for most steel floors. Floors subject to chemical agents as in industrial installations should be finished with specialized industrial coatings (5000 series).

Ceramic

Ceramic floors are seldom painted but they may be coated with an abrasion-resistant clear coating (3151) to seal the surface and facilitate cleaning.

Resilient

Resilient flooring rarely is coated when new, but coating may be required to restore appearance or change color. The coating chosen will depend on the flooring material, its condition, the adhesive used in installation and the effect desired. Specific recommendations should be obtained from the paint manufacturer for each case.

EXTERIOR PRIMING

As with the finishing of interior surfaces, the selection and application of primers are critical to the performance of the coating system.

Wood

Siding Smooth siding should be primed with a suitable wood primer of the 2100 series, compatible with the intended finish coat. The manufacturer's recommendations for the finish coat should govern the choice of a primer. Nailheads, cracks and deep surface imperfections should be puttied; caulking should be provided at perimeters of doors and windows and wherever required for weather resistance. All puttying and caulking should be done after the siding is primed.

For rough siding, shakes and shingles, any wood primer (2100 series) may be used according to finish coat recommendations; or the finish coat itself may be thinned and used as a primer if it is zinc-free and is recommended by the manufacturer. Flat house paint (4140) or shingle stains (4130) do not require special primers. Factory-primed siding does not require a field-applied primer unless the surface has been heavily damaged during shipping or application. Minor scratches, dents and raw edges should be spot-primed as recommended above.

Plywood After prolonged exterior exposure, most plywood surfaces will check and the coating will crack unless an overlaid plywood is used. Medium-density overlaid plywood with a resin-impregnated paper covering is highly recommended for an opaque paint finish. Painting recommendations for wood siding also are applicable to plywood. Textured plywood, such as grooved, rough-sawn and striated surfaces, is not intended for painting and only pigmented stains (4130) should be used.

Clear Finishes Wood surfaces intended for a natural finish usually are primed with a coat of the same clear material to be used as the finish coat. To improve penetration, some manufacturers recommend that the finish coat be thinned when used for priming. However, some products may be damaged by thinning and, hence, it should be done only when specifically recommended by the manufacturer.

Masonry

These surfaces generally do not require a special primer and the first coat can be the same material as that used for the finish coat (see Brick and Stone Masonry, page 216-14). The finish coat material may require thinning but the manufacturer's recommendations should be followed. Common brick should be sealed with a penetrating type of exterior varnish (4530) to minimize spalling and efflorescence which may cause peeling.

If the masonry surface has many large voids, a cement grout (1320) preferably thinned with latex may be applied. Sound but rough surfaces, especially in concrete block construction, may be surfaced with block filler (1310). Cast-in-place concrete should be cleaned of form oils and other bond-breaking agents unless they are guaranteed not to affect paint adhesion. Old surfaces which have become excessively chalky may be bonded or stabilized with a clear or lightly pigmented phenolic or alkyd coating (2310); but if the existing surface is soft or crumbly it should be removed to a sound layer, usually by sandblasting. Where only water repellency is required, a silicone solution may be used above grade; hot melt or cutback asphalt can be used below grade.

Metal

Ferrous Metal Proper surface preparation of iron and steel is extremely important as coating materials cannot compensate for inadequate surface preparation. Rust, mill scale, dirt, oils and old, loose paint must be removed. The standards for surface preparation recommended by the Steel Structures Painting Council should be followed.

An appropriate primer from the 2200 series should be used, depending on conditions and the finish materials intended. Oil and oleoresinous materials generally have better bonding properties and are more suitable for hand-cleaned surfaces. Alkyd and other synthetic primers dry more rapidly but require more careful surface preparation. Chemical-resistant coatings and industrial coatings (5000 series) usually are marketed as complete systems of primer and finish coat, and the manufacturer's recommendations as to the appropriate primer should be followed.

Galvanized Steel Zinc-dust/zinc-oxide (2230), latex (2240) or portland cement (2260) paints give excellent results as do certain proprietary paints designed for use on galvanized steel. Conventional exterior paints usually lose adhesion rapidly unless the surface has weathered for a considerable length of time. If the galvanized surface has weathered to the point that substantial amounts of rust are showing, the recommendations for painting ferrous metals should be followed. Certain chemical pretreatments, usually based on phosphoric acid, improve the adhesion of paints to galvanized metal. However, other acids or chemical solutions sometimes recommended for the preparation of galvanized surfaces rarely contribute to the performance of paints; actually they may be harmful and their use generally should be avoided.

Aluminum and Magnesium Zinc chromate primers (2213 or 2222) usually are recommended for these metals.

Copper, Brass, and Bronze Copper and copper-bearing metals oxidize rapidly when exposed to air and moisture. To preserve the bright metallic luster of new polished copper and brass, a clear coating should be applied promptly. Old copper and brass should be buffed or polished to a bright color, cleaned with a mild phosphoric acid cleaner and coated before the surface becomes discolored by oxidation. Specially formulated clear coatings are available which will preserve the bright luster and protect the metal from discoloration by the elements.

Copper gutters and downspouts generally are not painted or coated. The rapid forming oxide (patina) darkens the surface but protects the underlying metal from progressive deterioration. However, copper may be cleaned and coated either to match the color or to prevent the staining of adjacent surfaces. Coating is particularly recommended for copper surfaces located above aluminum products to prevent electrolytic deterioration of the aluminum by water wash from above.

EXTERIOR FINISHING

The choice of finish may be influenced by the type of exposure expected and the performance desired. High temperatures and humidity, particularly in the absence of direct sunlight, encourage the growth of mold (mildew). Therefore, it is important to choose fairly hard-drying finish coats, avoiding those types which may dry to a rough, soft surface, as these finishes will tend to trap and hold mold spores. Finish coats containing substantial amounts of zinc oxide are effective in controlling fungus growth. Various other additives are used to provide further mold resistance.

If a high degree of mold resistance is required, the primer as well as the finish coat should be designed for this purpose. However, the use of coatings containing zinc oxide is not recommended on bare wood. The degree of mold resistance is determined by the total formulation rather than by a single component, and the specific product should have adequate resistance under conditions similar to those to be encountered.

Industrial fumes may darken painted surfaces by the reaction of hydrogen sulfide with lead or occasionally mercury compounds. Paints used in areas subject to hydrogen sulfide should be free from lead and mercury.

Soot and other airborne materials may settle on and discolor the paint film. Hard-drying films will not retain dirt as tenaciously as softer films. Paints may be formulated to encourage moderate chalking of the surface which has a self-cleaning effect as the soot and dirt are carried away with the surface layer.

However, for severe weather exposures, resistance to chalking is a desirable property. Repeated exposure to rain and high winds generally will keep the surface clean but will wear away the film too quickly unless a comparatively nonchalking paint is chosen.

Wood Siding

Smooth Oil paints (4110) or exterior latex paints (4410) are preferred for whites and light tints. Oil paints have better hiding ability and will show less staining from wood resins. Latex paints are easier to apply, may be used on slightly damp surfaces and often have better color retention, particularly in light tints. Where factory-primed siding is used, the recommendations of the siding manufacturer should be followed. Clear finishes (4530) may be used where the natural appearance of the wood is desired, but their durability is much lower than that of pigmented products.

Rough Siding, Shingles and Shakes Flat house paint (4140), shingle stains (4130) or exterior latex paints (4410) usually are used for these surfaces. Latex paints generally give good service if rust-resistant nails have been used. Wood discoloration can be avoided by sealing with a solvent primer or by the use of a latex primer specially designed for resistance to staining. It should be remembered that rough surfaces have substantially greater surface area than smooth surfaces, and more paint must be applied to obtain suitable film thickness.

Wood Trim

Alkyd trim enamels (4310) should be used where high gloss and good abrasion resistance are desired, as around doors and windows. Any of the finish coats suggested above for smooth siding also may be used.

Metal

Oleoresinous paints (4220) or alkyd paints (4320) are suitable finishes over compatible primers. The alkyd paints tend to be harder and more abrasion resistant. Any of the paints recommended for wood such as alkyd enamels (4310) can be used if the surface is properly primed.

Masonry

Concrete and Concrete Masonry An alkali-resistant coating usually is required because concrete, even after some aging, contains alkali which will attack oils and oil-based materials. Latex paints (4430) or resin-based finishes (4540) also may be used. Resin-based finishes are preferred on flat areas such as ledges and window sills and areas subject to foot traffic and standing water such as patios and roofs. Other types of paint may be used on well cured concrete, on concrete that is sealed with an impervious primer or on previously painted surfaces.

Brick and Stone Masonry These materials usually do not contain alkali and almost any exterior paint may be used with satisfactory results, if the mortar is well cured. However, fresh mortar is usually strongly alkaline and paints recommended for concrete should be used on new walls. Where the natural appearance of the brick or stone is desired, a silicone-resin solution may be used to retard the penetration of water. The solution and spreading rate should be as recommended by the manufacturer. Silicone finishes should not be painted until they have weathered for at least two years.

Asbestos Cement

Asbestos cement compositions such as shingles, siding, corrugated boards and insulating sandwich panels often are supplied with a transparent water-repellent coating which should be weathered or removed before painting. For most products, trapped moisture seldom is a problem, but alkalinity may be too high for oleoresinous or alkyd types. However, the insulating core of sandwich panels may absorb moisture, releasing it later to damage the coating. It is recommended that all edges of insulating panels be sealed after cutting and before installation. Exterior latex (4430) or solvent-thinned resin (4540) paints generally are preferred.

Wood Shingles

Shingle stains (4130) are the usual choice for wood shingles. These stains are formulations which penetrate the wood and cracks rather easily. Impermeable coatings usually are not desirable because they seal in moisture and are likely to peel easily. Staining the shingles before installation usually is preferred because many areas are inaccessible after installation.

Asphalt Roofing

Asphalt shingles, roll roofing and built-up roofing rarely are coated when new, unless for decorative or heat-reflecting purposes. Most solvent-thinned paints are too brittle to be used over asphalt. The solvent will dissolve the asphalt and the paint will become stained.

Bituminous roof coatings (4551) are most commonly used, and if heat reflection is required these coatings may be obtained pigmented with aluminum or in a limited range of colors. Where white is required, certain latex paints (4552) are sufficiently flexible to give satisfactory performance. However latex paints are likely to curl shingles and are not recommended for these products. Because severe exposure conditions generally are encountered on roofs, thick coatings generally are required to give satisfactory durability.

Metal Roofing

In general, paints recommended for metal roofs follow the recommendations for other exterior metal (4220, 4320 and 4310). However, because foot traffic rarely is a consideration, softer coatings such as house paints (4110) or barn paints (4120) may be used for finish coats. As on other roofs, a relatively heavy coating is desirable for resistance to weathering. Terneplate roofs usually are painted with a special iron oxide primer and then finished as suggested above.

GLOSSARY

Abrasion Wearing away by friction.

Acid Corrosive chemical substance which attacks many common building materials, decorative finishes and coatings (see also *Alkali)*.

Adhesion The property of a paint film that enables it to stick to a surface.

Air Drying Capable of forming a solid film when exposed to air at moderate atmospheric temperatures.

Alkali Chemical substance characterized by its ability to combine with acids to form neutral salts; damaging to many paints and coatings.

Antique Finish A finish usually applied to furniture or woodwork to give the appearance of age.

Aromatic Solvents Group of organic compounds derived from coal or petroleum, such as benzene and toluene.

Baking Finish Paint or varnish which is baked at temperatures above 150°F. to dry and develop desired properties.

Binder Vehicle ingredient with adhesive qualities (e.g., linseed oil, resins) which binds the pigment and other ingredients into a cohesive film and facilitates bonding with the underlying surfaces.

Bleaching The process of restoring discolored or stained wood to its normal color or making it lighter.

Bleeding Discoloration of the finish coat by coloring matter from underlying surface or coating.

Blistering The formation of bubbles or pimples on the painted surface caused by moisture in the wood, painting before the previous coat has dried thoroughly, or excessive heat or grease under the paint.

Blushing Describes opaque lacquer that loses its gloss and becomes flat or clear lacquer that turns white or milky.

Body Used to indicate thickness or thinness of a liquid paint.

Boxing Mixing paint by pouring from one container to another several times to ensure thorough mixing.

Breathe Ability of a paint film to permit the passage of moisture vapor without causing blistering, cracking, or peeling.

Brush Tool composed of bristles set into a handle; often used to apply coatings. Bristles may be synthetic (needed for water-thinned paints) or natural, such as hog hair.

Brushability Ability or ease with which a paint can be brushed.

Brush Marks Marks of the brush that remain in the dried paint film.

Build Apparent thickness or depth of the paint after drying.

Burnishing Shiny or lustrous spots on a paint surface caused by rubbing.

Calcimine Water-base paint generally consisting of animal glue, zinc white and calcium carbonate or clay; now seldom used.

Catalyst Ingredient that speeds up a chemical reaction; sometimes used in two-component paint systems.

Caulking Compound Semidrying or slow-drying plastic material used to seal joints or fill crevices such as those around windows and chimneys.

Chalking Formation of a loose powder on the surface of a paint after exposure to the elements.

Checking Type of paint failure in which many small cracks appear in the surface of the paint.

Coating Mastic or liquid-applied surface finish—regardless of whether protective film is formed (as in paint and varnish) or whether merely decorative treatment results (as in some stains).

Colorant Concentrated color added to paints to make specific colors.

Colorfast Fade resistant.

Color Uniformity Ability of a coating to maintain a consistent color across its entire surface, particularly during weathering.

Copper Staining Usually caused by corrosion of copper screens, gutters, or downspouts washing down on painted surfaces. Can be prevented by painting or varnishing the copper.

Coverage Area over which a given amount of paint will spread and hide the previous surface; usually expressed in square feet per gallon.

Covering Power Term sometimes used as synonym for hiding power.

Cracking Type of paint failure characterized by breaks in irregular lines wide enough to expose the underlying surface.

Curing Final conversion or drying of a coating material.

Custom Color Special colors made by adding colorant to paint or by intermixing colors; permits the retailer to match a color selected by the consumer.

Cutting In Careful painting of an edge such as wall color at the ceiling line or at the edge of woodwork.

Dirt Collection The accumulation of dust, dirt, and other foreign matter on the paint surface.

Drier Ingredient included to speed the drying of paints and coatings.

Dry Dust Free Stage of drying when particles of dust that settle upon the surface do not stick to the paint film.

Drying Oil Paint vehicle ingredient such as linseed oil which, when exposed to the air in a thin layer, oxidizes and hardens to a relatively tough elastic film.

Dry Tack-Free Stage of drying when the paint no longer feels sticky when lightly touched.

Dry to Handle Stage of drying when a paint film has hardened sufficiently so the object or surface painted may be used without marring.

Dry to Recoat Stage of drying when the next coat can be applied.

Dry to Sand Stage of drying when a paint film can be sanded without the sandpaper sticking or clogging.

Durability Ability of a paint or coating to retain its desirable properties for a long time under expected service conditions.

Dye or Dyestuff Colored material used to change color with little or no hiding of the underlying surface.

Efflorescence Deposit of salts that remains on the surface of masonry, brick, or plaster when water has evaporated.

Eggshell Finish Surface sheen midway between flat and semi-gloss.

Emulsion Mixture of liquids (or a liquid and a solid) not soluble in each other, one liquid (or solid) being dispersed as minute particles in the other, base liquid, with the help of an emulsifying agent.

Enamel Paint capable of forming a very smooth, hard film, often using varnish as the vehicle; may be flat, gloss or semi-gloss.

Enamel Holdout Property of producing a tight film which prevents the penetration of subsequent enamel coats to underlying surfaces; prevents unequal absorption and uneven gloss.

Erosion Wearing away of a paint film caused by exposure.

Etch Surface preparation by chemical means to improve adhesion.

Extender Inexpensive but compatible substance which can be added to a more valuable substance to increase the volume of material without substantially diminishing its desirable properties; in paints, extender pigments improve storage and application properties.

Fading Loss of color due to exposure to light, heat, or weathering.

Feather Sanding Tapering the edge of dried paint film with sand paper.

Filler Pigmented composition for filling the pores or irregularities in a surface in preparation for finishing.

Film Thin application of paint or coating, generally not thicker than 0.010″.

Fire-retardant Paint Paint which will significantly (1) reduce the rate of flame spread, (2) resist ignition at high temperatures, and (3) insulate the underlying material so as to prolong the time required for the material to reach ignition, melting or structural weakening temperature.

Flaking Form of paint failure characterized by the detachment of small pieces of the film from the surface or previous coat of paint; usually preceded by cracking or blistering.

Flash Point Temperature at which a coating or solvent will ignite; the lower the flash point, the greater the hazard.

Flat Dull, nonreflective; opposite of gloss.

Flat Applicator Rectangular-shaped flat pad with an attached handle that is used to paint shingles, shakes, and other special surfaces.

Flat Finish Having no gloss or luster.

Flats Term applied to flat paints and coatings.

Flatting Agent Ingredient added to paints and coatings to reduce the gloss of the dried film.

Floating Separation of pigment colors on the surface of applied paint.

Flow Ability of a coating to level out and spread into a smooth film; paints that have good flow usually level out uniformly and exhibit few brush or roller marks.

Fungicide Agent that helps prevent mold or mildew growth on paint.

Glaze Used to describe several types of finishing materials.
Glazing Putty—A compound of creamy consistency applied to fill surface imperfections.
Glazing Stain—Very thin semi-transparent pigmented usually with a Vandyke brown or burnt sienna applied over a previously stained, filled or painted surface to soften or blend the original color without obscuring it.
Glaze Coat—Clear finish applied over previously coated surfaces to create a gloss finish.

Glazing Compound Dough-like material, consisting of vehicle and pigment, which retains its plasticity over a wide range of temperature and an extended period of time.

Gloss Shiny, reflective surface quality; term sometimes used broadly to include paints and coatings with these surface properties.

Grain Raising Swelling and standing up of the wood grain caused by absorbed water or solvents.

Graining: Simulating the grain of wood by means of specially prepared colors or stains and the use of graining tools or special brushing techniques.

Ground Coat Base coat in an antiquing system; applied before graining colors, glazing, or other finish coat.

Hardness Cohesion of particles on the surface, as determined by ability to resist scratching or identation.

Hiding Power Ability of a paint or coating to obscure the surface to which it is applied, generally expressed as number of sq. ft. which can be covered by one gal. of material, or number of gals. required to cover 1000 sq. ft.

Intercoat Adhesion Adhesion between two coats of paint.

Joint Tape Paper or paper-faced cotton tape used over joints between wallboard to conceal the joint and provide a smooth surface for painting.

Lacquer Coating that dries quickly by evaporation of its volatile solvent, and forms a film from its nonvolatile constituent, usually nitrocellulose; may be pigmented or clear.

Lap To lay or place one coat so its edge extends over and covers the edge of a previous coat.

Latex Paint A paint that should be thinned with water.

Leveling Ability of a film to flow out free from ripples, pockmarks, and brush marks after application.

Lifting Softening and penetration of a previous film by solvents in the paint being applied over it, resulting in raising and wrinkling.

Lightfastness See *Fading*.

Long Oil Varnish or paint vehicle with more than 55% of the resin consisting of oil or fatty acid.

Marine Varnish Varnish specially designed for immersion in water and exposure to marine atmosphere.

Masking Temporary covering of areas not to be painted.

Masking Tape A strip of paper or cloth tape, easily removable and used temporarily to cover areas that are not to be painted.

Mastic Heavy-bodied pastelike coating of high build often applied with a trowel.

Metallics Class of paints that include metal flakes.

Mil One one-thousandth of an inch.

Mildew Resistance Ability of a coating to resist the growth of molds and mildew; mildew is particularly prevalent in moist, humid, and warm climates.

Mildewcide See *Fungicide*.

Mineral Spirit Common solvent for paints and coatings derived from the distillation of petroleum.

Nap Length of fibers on a paint roller cover.

Natural Resin See *Resin*.

Nondrying Oil Oil which does not readily oxidize and harden when exposed to air.

Nonvolatile Vehicle Liquid portion of paint excepting its volatile thinner and water.

Oil Color Single pigment dispersion in linseed oil used for tinting paints.

Oil Paint See *Paint*.

Oil Stains May be penetrating and nonpenetrating. Penetrating oil stains contain dyes and resins that penetrate into the surface; nonpenetrating oil stains contain larger amounts of pigments and are usually opaque or translucent.

Oil Varnish See *Varnish*.

One-package (One-part) Formulation Paint or coating formulated to contain all the necessary ingredients in one package and generally not requiring any field additions, except pigment and thinner.

Opacity Degree of obstruction to the passage of visible light.

Opaque Coating Coating that hides the previous surface or coating.

Paint Pigmented material generally producing an opaque film—as distinguished from *varnish* and *stain* which are transparent or translucent.
Emulsion—Paint with a vehicle consisting of an oil, oleoresinous varnish or resin binder dispersed in water (see also *Emulsion*).
Latex—Emulsion paint with latex resin as the chief binder.
Oil—Paint with drying oil or oil varnish as the basic vehicle ingredient.

Paste—Paint with sufficiently concentrated pigment permitting substantial thinning before use.

Water—Paint with a vehicle which is a water emulsion, water dispersion or with ingredients that react chemically with water.

Paint Remover Compound that softens old paint or varnish and permits scraping off the loosened material.

Peeling Detachment of a dried paint film in relatively large pieces, usually caused by moisture or grease under the painted surface.

Pigment Fine, solid particles suspended in the vehicle which provide color (hiding ability) as well as other properties.

Plasticizer Substance added to impart flexibility to the hardened coating or paint film.

Polyurethane See *Resin*.

Polyvinyl Acetate See *Resin*.

Pot Life Time period after mixing of reactive components, in a two-component paint system, during which material can be satisfactorily used.

Primer First of several coats, intended to prepare the surface for the succeeding coat(s); sometimes a special product, but may be the same as the finish coat.

Primer-sealer Product formulated to possess properties of both a primer and a sealer.

Putty Dough-like material consisting of pigment and vehicle; used for sealing glass in sash or frames and for filling imperfections in wood or metal surfaces; does not retain its plasticity for extended period as does a glazing compound.

Resin Mixture of organic or synthetic compounds with no sharply defined melting point, no tendency to crystallize, soluble in certain organic solvents but not in water. It is the main ingredient of paint, binds the other ingredients together and aids adhesion to the surface. Includes acrylic, alkyd, epoxy, nitrocellulose, polyurethane, polyvinylacetate, silicone, styrene-butadiene and vinyl.

Natural Resin—May be fossil of ancient origin, such as amber, or extract of certain pine trees, such as rosin, copal and damar.

Synthetic Resin—Man-made substances exhibiting properties similar to those of natural resins; typical synthetic resins used in paints and coatings include alkyd, acrylic, latex, phenolic, urea and others.

Roller Paint applicator having a revolving cylinder covered with lambswool, fabric, foamed plastic, or other material.

Rosin Natural resin obtained from various pine trees; an ingredient of varnishes and some man-made resins.

Rust Preventive Paint or Primer First coat of paint applied directly to iron or steel structures to slow down or prevent rusting.

Sags Excessive flow, causing runs or sagging in paint film during application; usually caused by applying too heavy a coat of paint or thinning too much.

Sand Finish See *Texture*.

Sanding Surfacer Heavily pigmented finishing material used for building the surface to a smooth condition; it is sanded after drying.

Satin Finish See *Semigloss*.

Scrubbability Ability of a paint film to withstand scrubbing and cleaning with water, soap, and other household cleaning agents.

Sealer Formulation intended to prevent excessive absorption of the finish coat into a porous surface or to prevent bleeding through the finish coat.

Seeds Small, undesirable particles or granules other than dust found in a paint, varnish, or lacquer.

Self Cleaning Controlled chalking of a paint film so dirt does not adhere to the surface.

Semi-gloss Degree of surface reflectance midway between gloss and eggshell; also paints and coatings displaying these properties.

Semitransparent Degree of hiding greater than transparent but less than opaque.

Settling Paint separation in which pigments and other solids accumulate at the bottom of the container.

Shake Painter See *Flat applicator*.

Sheen Degree of luster of a dried-paint film.

Sheen Uniformity Even distribution of luster over the entire surface of an applied finish.

Shellac Coating made from special resins dissolved in alcohol.

Silicone See *Resin*.

Size Water-based formulation, with glue or starch binders, intended as a sealer over existing wall paint or plaster; now seldom used.

Skin Tough covering that forms on liquid paints and varnishes when left exposed to air.

Softness Film property displaying low resistance to scratching or indentation; opposite of hardness.

Solvent Volatile portion of the vehicle, such as turpentine and mineral spirits, which evaporates during the drying process.

Solvent-thinned Formulation in which the binder is *dissolved* in the thinner as in oil paint, rather than *emulsified* as in latex paint.

Spacking Compound Material used as a crack filler for preparing surfaces before painting.

Spar Varnish Very durable varnish designed for service on exterior surfaces.

Spatter Small particles or drips of liquid paint thrown or expelled when applying paint.

Spot Priming Method for protecting localized spots. The only areas primed are those that require additional protection due to rusting or peeling of the former coat.

Spreading Rate Measure of area which can be covered by a unit volume of material, generally expressed as sq. ft. per gal.

Stain Penetrating formulation intended primarily for wood surfaces which changes the color of the wood without obscuring the grain; depending on the amount of pigment, stains may be more or less transparent and may leave little or no surface film.

Streaking Irregular occurrence of lines or streaks of various lengths and colors in an applied film; usually caused by some form of contamination.

Styrene-butadiene See *Resin*.

Substrate Surface to be painted.

Surfacer Pigmented formulation for filling minor irregularities before a finish coat is applied; usually applied over a primer and sanded for smoothness.

System, Paint One or more coatings selected for compatibility with each other and the surface to which they are applied, as well as suitability for the expected exposure and decorative requirements.

Tack Rag Piece of loosely woven cloth that has been dipped into a varnish oil and wrung out. When it becomes tacky or sticky, it is used to wipe a surface to remove small particles of dust.

Texture Roughness or irregularity of a surface.

Texture Paint One that may be manipulated by brush, roller, trowel, or other tool to produce various effects.

Thinners Solvents used to thin coatings.

Thixotropy Property of a material that causes it to change from a thick, pasty consistency to a fluid consistency upon agitation, brushing, or rolling.

Tint Base Basic paint in a custom color system to which colorants are added to make a wide range of colors.

Touch-up Ability of a coating film to be spot repaired (usually within a few months of initial painting) without showing color or gloss differences.

Turpentine Colorless liquid, which is used as thinner for oil paints and varnishes, distilled from the products of the pine tree.

Two-package (Two-part) Formulation Paint or coating formulated in two separate packages and requiring that the two ingredients be mixed before characteristic properties can be obtained and the material can be applied.

Undercoat Prime or intermediate coat in a multicoat system.

Uniformity Not varying in gloss, sheen, color, hiding, or other property.

Varnish Coating which dries to a transparent or translucent film when exposed to air.

Oleoresinous and Oil—Varnishes containing resin and drying oil as the chief film-forming ingredients.

Polyurethane—Hard, highly abrasion- and chemical-resistant coating, often used as floor varnishes and bar top finishes.

Shellac—Fast-drying varnish consisting of lac resins, produced by the lac insect, dissolved in alcohol.

Spar—Exterior, weather-resistant varnish, based generally on long-oil phenolic resin; term originated from use on spars of ships.

Vehicle Liquid portion of paint or coating, including any ingredients dissolved in it.

Vinyl See *Resin*.

Washability Ability of a paint to be easily cleaned without wearing away during cleaning.

Water Spotting Paint appearance defect caused by water droplets.

Weathering Effect of exposure to weather on paint films.

Wrinkling Development of ridges and furrows in a paint film when the paint dries.

Yellowing Development of a yellow color or cast in white, pastels, colored, or clear finishes.

White Lead Oldest white pigment, chemically known as lead sulfate or lead carbonate.

We gratefully acknowledge the assistance of the following for use of their publications as references and permission to use photographs: Jewel Paint and Varnish Company; National Paint and Coatings Association, Inc.; Sherwin-Williams Company.

217 BITUMINOUS PRODUCTS

BITUMINOUS PRODUCTS

This Section 217 Bituminous Products now includes information on Asphalt Roofing Products. Other building products made of bituminous materials will be discussed in future additions to this section.

ASPHALT ROOFING PRODUCTS

Four basic materials—a reinforcing base which is either organic dry felt or glass fiber mat, asphalt, mineral stabilizers, and either fine or coarse surfacings—are used in the manufacture of asphalt roofing products.

In general, manufacture includes processing cellulose fibers into dry felt, saturating and coating the felt with asphalt; and then, depending on the type of finished product, surfacing the coated felt with selected mineral aggregates. When glass fiber mat is used as a reinforcing base instead of dry felt, the saturating process is bypassed. The mat goes directly from the dry looper into the coater.

RAW MATERIALS

The raw materials and some of the processing steps required to manufacture finished products are shown diagrammatically in Fig. 1.

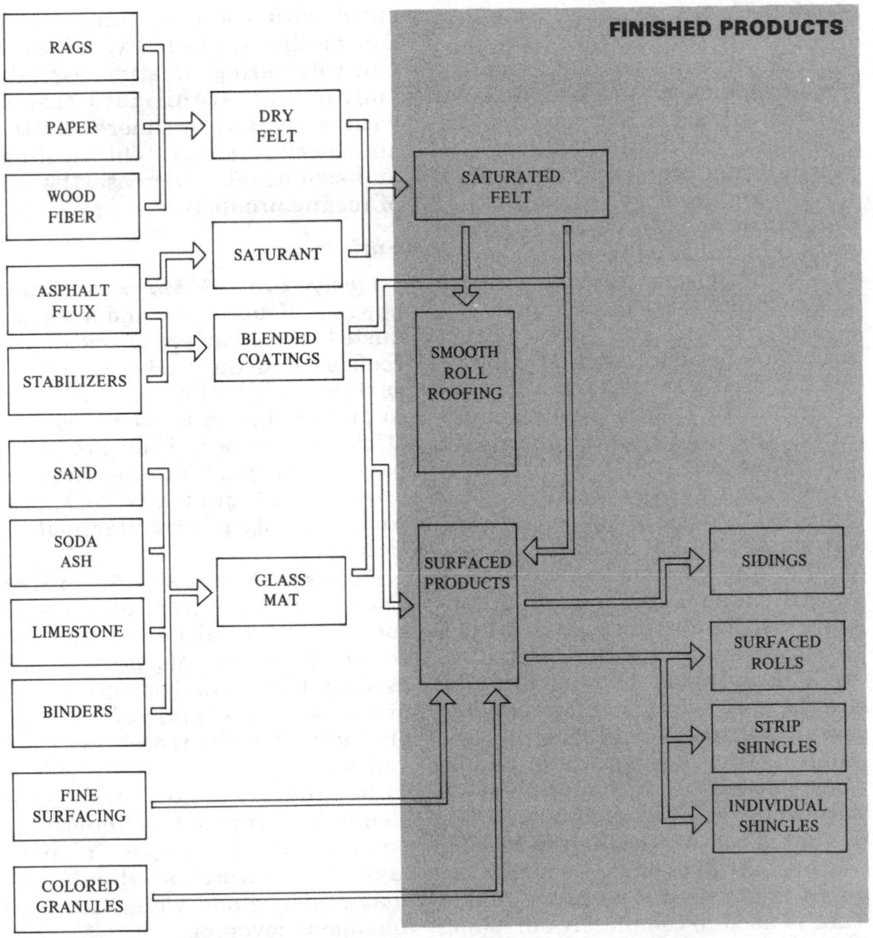

FIG. 1 Raw materials and processing steps used to make finished asphalt roofing products.

Dry Felt

Dry felt, used as a base for making underlayment, asphalt shingles, and smooth or surfaced rolls for built-up roofs, is made from combinations of cellulose fibers such as those derived from rags, paper and wood, on a machine similar to the type used for making paper (Fig. 2). The fibers are prepared by various pulping methods, then blended and proportioned to produce felt with the necessary weight, tensile strength, absorptive capacity and flexibility required to make a suitable roofing product. The felt must be able to absorb from 1-1/2 to 2 times its weight in asphalt saturants and be strong and flexible enough to withstand any strains placed on it during the manufacturing process.

As the felt comes off the end of the machine in a continuous sheet, it is cut into specified widths and wound into individual rolls from 4′ to 6′ in diameter, weighing up to a ton or more.

Glass Fiber Mat

Glass fiber mat is used as a base for asphalt shingles, coated rolls and surfaced rolls used in built up roofs. Sand, soda ash, and limestone are the raw materials which are combined to make the chopped strand glass fibers. These fibers are then mixed with a binding agent and cured to produce a mat with the proper thickness, tensile strength, tear strength, weight and flexibility required to produce a roofing product suitable for its intended use.

FIG. 2 Flow Diagram of a Typical Felt Mill.

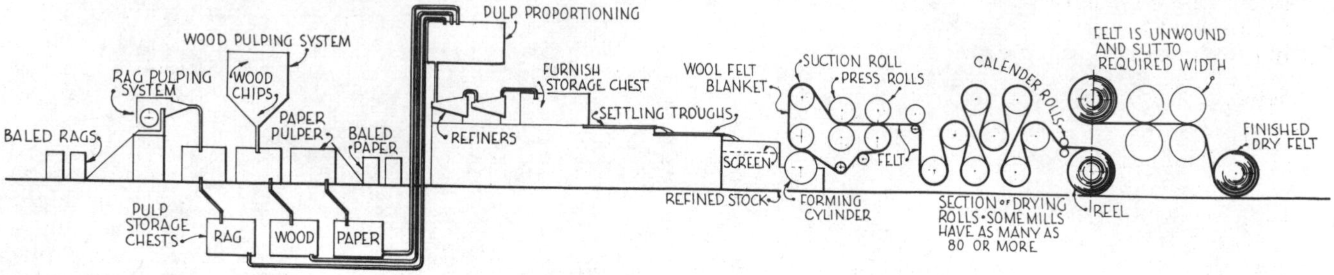

Asphalt

Asphalt used to make roofing products, known as asphalt flux, is a petroleum product obtained from the fractional distillation of crude oil. Asphalt flux is processed to produce (1) roofing grades of asphalt called *saturants* or, (2) when combined with mineral stabilizers, coating asphalts. Both forms combine with dry felt to make asphalt roofing.

The preservative and waterproofing characteristics of asphalt reside largely in certain oily constituents. In making roofing products, the body of the highly absorbent felt sheet is first impregnated (saturated) to the greatest possible extent with saturants which are oil-rich asphalts. The saturant is then sealed in with an application of a harder, more viscous coating asphalt which in turn can be further protected by a covering of opaque mineral granules. A primary difference between saturants and coating asphalts is the temperature at which they soften: the softening point of saturants varies from 100° to 160° F, that of coatings as high as 260° F. With glass fiber mat, saturant asphalt is not used. The coating mixture is used to completely surround the glass fibers and to provide a layer of coating on both sides of the mat.

Asphalt flux is also the base material used to make asphalt plastic cement, quick-setting roof adhesives, asphalt primers, other roof coatings, and adhesives.

Mineral Stabilizers

Finely ground minerals—such as silica, slate dust, talc, micaceous materials, dolomite and trap rock—called stabilizers, when combined with coating asphalts control hardness, elasticity, adhesion and weathering. Coating asphalts containing stabilizers resist weather better, are more shatter- and shock-proof in cold weather, and significantly increase the life of roofing products.

Surfacings

Finely Ground Minerals These powders, usually talc and mica, are dusted on the surface of smooth roll roofings and the backs of mineral surfaced roll roofings and shingles to prevent layers in packages from sticking together. They are not a permanent part of the finished product and gradually disappear from exposed surfaces after application.

Coarse Mineral Granules Usually natural colored slate or rock granules (either natural or colored by a ceramic process) are used on shingles and on certain roll products. These *mineral surfaced* products have increased weather and fire resistance and provide a wide range of colors and color blends. To protect the underlying asphalt from the impact of light rays, the granules should be opaque, dense and well graded for maximum coverage.

MANUFACTURE OF ROOFING

Raw materials are processed into finished roofing products in a continuous sequence (Fig. 3). The principal steps are described below.

Dry Looper

A roll of dry felt or glass fiber mat installed on the felt reel is unwound onto the dry looper. The looper provides a reservoir of felt material that can be drawn upon by the machine as circumstances demand, eliminating stoppages when a new roll must be put on the felt reel or when imperfections in the felt must be cut out.

Saturation of Felt

The felt is then subjected to a hot saturating process which eliminates any moisture and fills the felt fibers and intervening spaces as completely as possible with asphalt saturant. Glass fiber mat does not undergo this saturating process. Coating asphalt is used to fill the spaces between glass fibers.

Wet Looper

Following the saturating process, an excess of saturant usually remains on the surface of the sheet. It is held for a time on a wet looper so the natural shrinkage of the asphalt upon cooling will draw the excess into the felt, resulting in a high degree of saturation. At this point, uncoated products, such as saturated underlayment felt, move directly to the cooling looper and onto the winding mandrel for final packaging.

Coater

Material to be coated continues to the coater where coating asphalt is applied to both the top and bottom surfaces. Coating thickness is regulated by rollers, which can be brought together to reduce the amount of coating or separated to increase its thickness.

The final weight of the product is controlled by the roofing machine operator. Often machines are equipped with automatic scales that weigh the sheets during manufacture to warn the operator when the material is running over or under weight specifications. If smooth roll roofing is being made, talc or mica is applied to both sides of the sheet by spreading and pressing it through a pressure roll.

Mineral Surfacing

When mineral surfaced products such as shingles are made, granules of specified color or color combinations are applied from a hopper and spread thickly on the hot coating asphalt, and talc or mica is applied to the back (smooth) side. The sheet is then run through a series of press and cooling rollers subjecting the sheet to a controlled pressure that embeds the granules in the coating to the desired depth.

Texture

At this stage some products are textured by the pressure of an embossing roller which forms a pattern in the surface of the sheet.

Cooling Looper

The sheet then travels into the cooling or finish looper where the sheet is cooled so it can be cut to length or into shingles and packed.

Shingle Cutter

When shingles are being made the material is fed from the finish looper into the shingle cutting machine. The sheet is cut from the back (smooth) side as it passes between a cutting cylinder and an anvil roll. Shingles are then separated into stacks of the proper number for packaging. The stacks are moved to packaging equipment where the bundles are prepared for warehousing and shipment.

Roll Roofing Winder

When roll roofing is made the sheet is drawn directly from the finish looper to the winding mandrel. Here it is wound and the material is measured to length as it turns. After sufficient material accumulates, it is cut off, removed from the mandrel and wrapped for warehousing and shipment.

Care of Materials

The following simple rules should be observed in storing finished products in the warehouse and on the job to insure suitable performance: (1) roll goods should be stored on end in an upright position, and if stored in tiers, boards should be placed between the tiers to prevent damage to the ends of rolls; (2) shingle bundles should not be stored to a height of more than 4′ to eliminate the possibility of sticking or discoloration; (3) warehouse stock, particularly white and pastel shades of roofing, should be rotated by moving out of storage first the material held longest; (4) asphalt roofing stored outside should be covered for weather protection. Asphalt roofing provides a waterproof roof *after* application, but can be harmed if stored outside without protection, particularly if it is applied soaking wet.

CLASSIFICATION OF PRODUCTS

Asphalt roofing products include: (1) asphalt coatings and cements, and (2) products made on a felt base. (3) products made on a glass fiber base.

Asphalt Coatings and Cements

Asphalt coatings and cements are asphaltic materials combined with special ingredients and are either (1) mixed with suitable solvents of the cut-back type, or (2) emulsified in water. These materials are processed to various consistencies depending on their use. The materials are flammable and should never be warmed over an open fire or placed in direct contact with a hot surface. If necessary, they may be softened by placing the container, unopened, in hot water or by storing in a warm place.

Cut-back asphalt coatings and cements should be applied to a dry, clean surface and troweled or brushed vigorously to force the material into all cracks and openings and to eliminate air bubbles. Emulsions may be applied to damp or wet surfaces, but should never be applied in an exposed location if rain is anticipated within 24 hours after completion of the job.

The materials may be classified into several groups, including plastic asphalt cements, lap cements, quick setting asphalt adhesives, roof coatings, asphalt water emulsions, asphalt primers and roofing tape.

Plastic Asphalt Cements These materials are generally designated as *plastic asphalt cements* or *flashing cements.* They are processed with asbestos fibers and asphalts which are especially plastic and elastic, combined with non-oxidizing minerals, oils and solvents to produce a black plastic flashing and repair cement. They will not flow at summer temperatures nor become brittle at low temperatures. They are sufficiently elastic after setting to compensate for normal expansion and contraction of the roof deck or movement between the deck and other elements of the

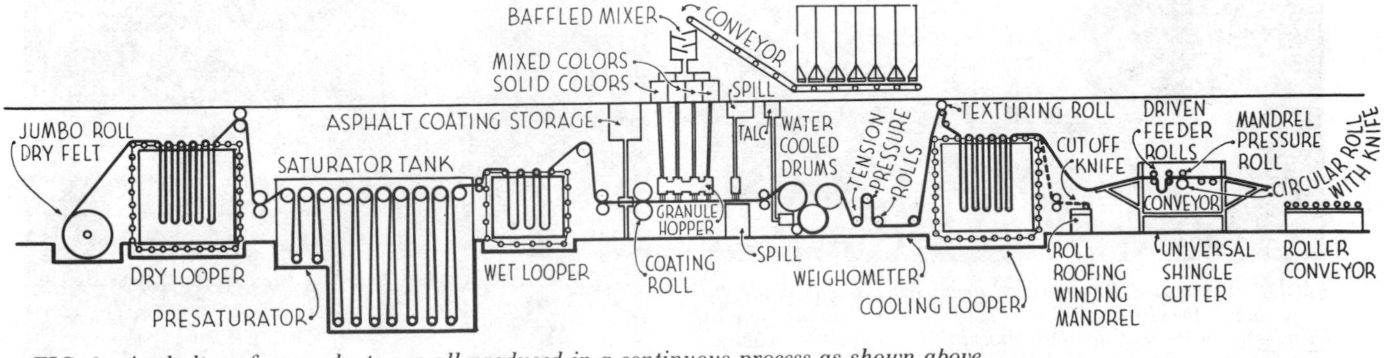

FIG. 3 Asphalt roofing products are all produced in a continuous process as shown above.

structure. These cements have excellent resistance to acids and alkalies, will not absorb water, and provide good adhesion to all surfaces. Plastic asphalt cement is used for flashing assemblies, patching, roof repairing and caulking.

Lap Cements These cements are of various consistencies but are not as thick and heavy as plastic cements. They should be used as recommended by the manufacturer. Lap cements are intended to make a watertight bond between lapping elements of roll roofing. When lap cement is applied it should be spread over the entire lapped area and any nails used to secure the roofing should pass through the cement so that the shank of the nail will be sealed where it penetrates the deck material.

Quick Setting Asphalt Adhesive Cement This material, available in brush, trowel or gun consistency, is used for cementing down the free tabs of strip shingles and for laps of roll roofing.

Quick setting asphalt adhesive cement is about the same consistency as plastic asphalt cement but is more adhesive. It contains a solvent that evaporates rapidly on exposure, causing the cement to set quickly.

Roof Coatings Roof coatings are of various types and consistencies but are usually thin enough to be applied with a spray or brush. Coatings are used to resurface existing roll roofing and built-up asphalt roofs.

Asphalt Water Emulsions This is a special type of roof coating made with asphalt, sometimes combined with other ingredients and emulsified with water. Emulsions must be stored where they will not freeze in cold weather and should be used at times when they will not be rained on for a period of at least 24 hours after being applied.

Asphalt Primer This material is very fluid and is applied with a brush or spray. It must be thin enough to penetrate rapidly into the surface pores of masonry without leaving a continuous surface film. Its purpose is to prepare masonry surfaces to receive other asphalt products and insure a good bond between the masonry and built-up roofing, plastic asphalt cement or asphalt coatings.

Roofing Tape Some manufacturers make a roofing tape of cotton fabric or other porous material saturated with asphalt. It is available in varying widths from 4″ to 36″ and is provided in rolls up to 50 yards in length. It is used with asphalt coatings, plastic cements for flashing, and in patching seams, breaks and holes in asphalt and metal roofing.

Felt or Fiberglass Base Products

Asphalt roofing and siding products made on a felt or fiberglass base may be classified broadly into three groups: (1) saturated felts, (2) roll roofing and roll siding, and (3) roofing and siding shinges.

The sizes, exposures and other information covering products typical of these three classifications are given in Figure 6. Products within each group may differ as to weight, dimension and design (for example, the weight of a 3-tab square butt strip shingles ranges from 205 lbs. to 350 lbs. per square). See Division 400—Applications and Finishes for details and recom-

FIG. 4 Saturated felts may be used: (a) for wall underlayment beneath siding; (b) for roof underlayment beneath shingles; and (c & d) for building up plies of felt to provide a waterproof membrane used on flat roofs.

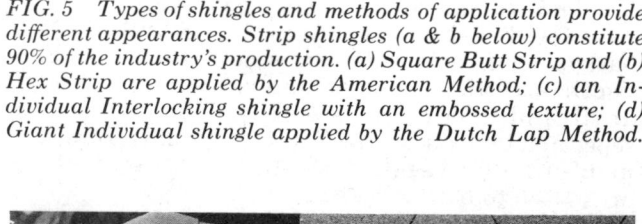

FIG. 5 Types of shingles and methods of application provide different appearances. Strip shingles (a & b below) constitute 90% of the industry's production. (a) Square Butt Strip and (b) Hex Strip are applied by the American Method; (c) an Individual Interlocking shingle with an embossed texture; (d) Giant Individual shingle applied by the Dutch Lap Method.

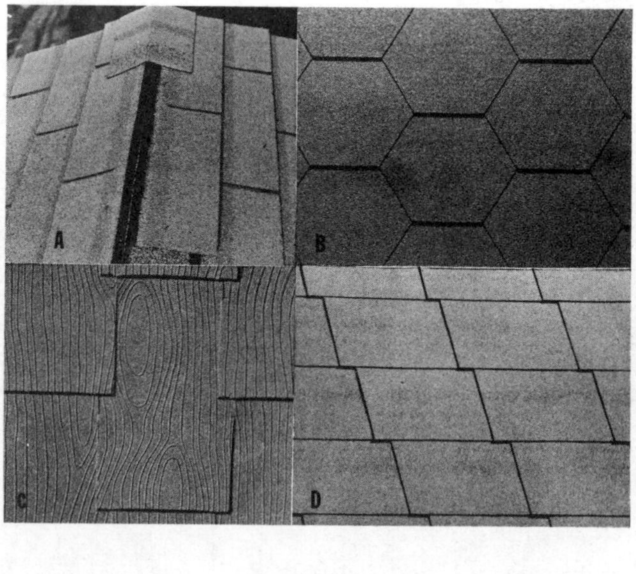

FIG. 6 **SUMMARY OF TYPICAL ASPHALT ROOFING PRODUCTS**

	PRODUCT	SHIPPING WEIGHT PER SQUARE	PACKAGES PER SQUARE	LENGTH	WIDTH	UNITS PER SQUARE	SIDE-LAP	TOP-LAP	HEAD-LAP	EXPOSURE
ROLL ROOFING	 SATURATED FELT	15 Lb 30 Lb	¼ ½	144′ 72′	36″ 36″		4″ to 6″ 4″ to 6″	2″ 2″		34″ 34″
ROLL ROOFING	 SMOOTH	65 Lb 55 Lb 45 Lb	1 1 1	36′ 36′ 36′	36″ 36″ 36″		6″ 6″ 6″	2″ 2″ 2″		34″ 34″ 34″
ROLL ROOFING	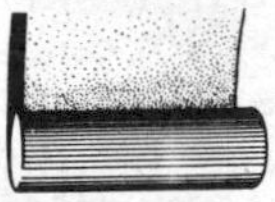 MINERAL SURFACED	90 Lb 90 Lb 90 Lb	1.0	36′	36″	1.0 1.075 1.15	6″ 6″ 6″	2″ 3″ 4″		34″ 33″ 32″
ROLL ROOFING	 PATTERN EDGE	105 Lb 105 Lb	1 1	42′ 48′	36″ 32″			2″ 2″		16″ 14″
ROLL ROOFING	 19″ SELVAGE DOUBLE COVERAGE	110 to 120	2	36′	36″			19″	2″	17″
STRIP SHINGLES	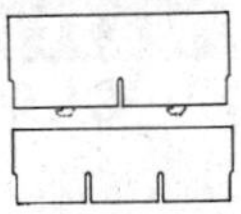 2 & 3 TAB SQUARE BUTT	205 Lb 220 Lb 225 Lb 235 Lb	3 3 3 3	36″ 36″ 36″ 36″	12″ 12″ 12″ 12″	80 80 80 80		7″ 7″ 7″ 7″	2″ 2″ 2″ 2″	5″ 5″ 5″ 5″
STRIP SHINGLES	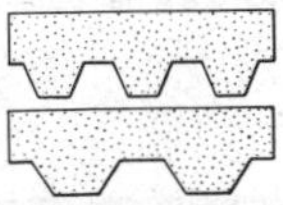 2 & 3 TAB HEXAGONAL	195 Lb	2	36″	11⅓″	86		2″	2″	5″
INDIVIDUAL	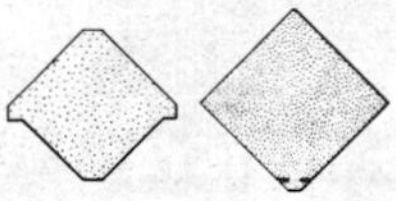 STAPLE LOCK	145 Lb	2	16″	16″	80	2½″			
GIANT INDIVIDUAL	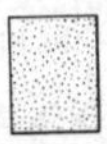 AMERICAN	330 Lb	4	16″	12″	226		11″	6″	5″
GIANT INDIVIDUAL	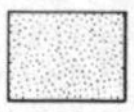 DUTCH LAP	165 Lb	2	16″	12″	113	3″	2″		10″

mended methods of installing these products.

Saturated Felts Dry felts impregated with asphalt or coal tar saturants, but otherwise untreated, are called saturated felts. They are used as underlayment for shingles, for sheathing paper and for laminations in the construction of built-up roofs (Fig. 4). Saturated felts are made in different weights; the most common are No. 15 and No. 30, weighing about 15 lbs and 30 lbs per square respectively.

Although fiberglass mats which are impregnated with asphalt but otherwise untreated may be classified under this broad category, they are not actually saturated, since glass fibers do not absorb asphalt. In reality the mat is lightly coated so that all fibers are thoroughly surrounded by asphalt. The uses for this product are mainly in built-up roofing applications.

Roll Roofing and Siding Roll roofing and siding are made by adding a coating of a more viscous, weather resistant asphalt to a felt that has first been impregnated with a saturant asphalt or to a glass fiber mat. Some roll roofing and all siding are surfaced with mineral granules providing a wide color range, while some styles are furnished in split rolls designed to give an edge pattern when applied to the roof.

Shingles All shingles are surfaced with mineral granules. There are many weights and patterns of both individual and strip shingles designed for both new construction and reroofing existing roofs (Fig. 5).

STANDARDS OF QUALITY

Asphalt shingles and sheet roofing should conform to Underwriters' Laboratories, Inc. (UL) standards. The UL specifications constitute the recognized standard for the asphalt roofing industry and include provisions for: (1) fire resistance (Fig. 7), (2) wind resistance, (3) tests of the saturation of felt to determine quantity of saturant and efficiency of saturation, (4) thickness and distribution of coating asphalts, (5) adhesion and distribution of granules, (6) weight, count, size, coloration and other characteristics of finished products before and after packaging, and (7) instructions for application.

The UL maintains a check on labeled products by making periodic factory inspections. UL inspectors are afforded the privilege of manufacturers' control laboratories and visit, inspect and analyze product samples. When necessary, they recommend to the manufacturer measures that are required to maintain the standards of the product to remain eligible to use the UL label.

Products that conform to UL standards are identified by Underwriters' Laboratories' labels similar to those illustrated in Figure 8. It is recommended that asphalt shingles conform to no less than UL55B Standard Class "C" for Asphalt Organic-Felt Sheet Roofing and Shingles. Products should carry a label as shown in Figure 8a. In areas subject to winds in excess of 75 mph*, Wind Resistant shingles are recommended. These shingles are provided with factory-applied adhesive or integral locking tabs, and the bundles carry the UL Class C label which includes the words "Wind Resistant."

* United States Weather Bureau defines hurricane winds as those of 75 mph or greater velocity.

CLASSIFICATIONS OF FIRE-RETARDANT PREPARED ROOF COVERINGS by UNDERWRITERS' LABORATORIES, INC. (UL)	
CLASS A	Includes roof coverings which are effective against severe fire exposures. Under such exposures, roof coverings of this class are not readily flammable and do not carry or communicate fire; afford a fairly high degree of fire protection to the roof-deck; do not slip from position; possess no flying brand hazard; and do not require frequent repairs in order to maintain their fire-resisting properties.
CLASS B	Includes roof coverings which are effective against moderate fire exposures. Under such exposures, roof coverings of this class are not readily flammable and do not readily carry or communicate fire; afford a moderate degree of fire protection to the roof-deck; and do not slip from position; possess no flying brand hazard; but may require infrequent repairs in order to maintain their fire-resisting properties.
CLASS C	Includes roof coverings which are effective against light fire exposure. Under such exposures, roof coverings of this class are not readily flammable and do not readily carry or communicate fire; afford at least a slight degree of fire protection to the roof-deck; do not slip from position; possess no flying brand hazard; and may require occasional repairs or renewals in order to maintain their fire-resisting properties.

FIG. 7 Underwriters' Laboratories, Inc. Standard UL 790—Test Methods for Fire Resistance of Roof Covering Materials establishes three classifications.

A

B

FIG. 8 Products conforming to UL55B Standard carry the above labels.

We gratefully acknowledge the assistance of the Asphalt Roofing Industry Bureau, Johns Manville Corporation, Koppers, and Owens-Corning Fiberglas Corporation. We are also indebted to them for permission to use photographs and publications as references.

BITUMINOUS PRODUCTS

CONTENTS

ASPHALT ROOFING PRODUCTS

MANUFACTURE

See Main Text pages 217-3 through 217-5 for description of raw materials and manufacturing processes.

COATINGS AND CEMENTS

Asphalt coatings and cements may be used in accordance with information provided in Figure WF 1.

Asphalt coatings and cements should never be warmed over an open fire or placed in direct contact with a hot surface.

These materials are flammable; if softening is necessary, the unopened container may be placed in hot water or stored in a warm place.

Asphalt coatings and cements should be applied to a dry, clean surface and troweled or brushed vigorously, forcing the material into cracks and openings to eliminate air bubbles.

Emulsions may be applied to damp or wet surface, but should never be applied in an exposed location if rain is anticipated within 24 hours after completion of the job. Emulsions must be stored where they will not freeze in cold weather.

Lap cement should be spread over the entire lapped area and any nails used to secure the roofing should pass through the cement.

Securing the nails through the cement will seal the shank of the nail where it penetrates the deck material.

FELT OR FIBERGLASS BASE PRODUCTS

Asphalt roofing and siding products made on a felt or fiberglass base may be classified broadly in three groups: (1) saturated felts, (2) roll roofing and roll siding, and (3) roofing and siding shingles. The sizes, exposures and other information covering typical products are given in Figure WF 2.

STANDARDS OF QUALITY

Asphalt shingles and sheet roofing should conform to Underwriters' Laboratories, Inc. (UL) standards.

FIG. WF1 USES AND PROPERTIES OF ASPHALT COATINGS & CEMENT PRODUCTS

PRODUCT	USES	PROPERTIES
Plastic Asphalt Cement	Flashing assemblies; patching; roof repairing; caulking	Remains elastic after setting; will not flow at summer temperatures or become brittle at low temperatures; excellent resistance to acids and alkalies; will not absorb water; good adhesion to all surfaces
Lap Cement	Watertight bond between lapping elements of roll roofing	Not as thick and heavy as plastic cements
Quick Setting Asphalt Adhesive Cement	Cementing down the free tabs of strip shingles and laps of roll roofing	Available in brush, trowel or gun consistency; same consistency as plastic asphalt cement, but more adhesive; sets quickly
Roof Coating	Resurfacing old roll roofing and built-up asphalt roofs	Thin enough to be applied with spray or brush
Asphalt Water Emulsion	Roof coating	Keep from freezing; apply when no rain is forecast for 24 hours
Asphalt Primer	Preparing masonry surfaces to receive other asphalt products; insures good bond between the masonry and built-up roofing, plastic asphalt cement or asphalt coatings	Very fluid; applied with brush or spray

(Continued) ASPHALT ROOFING PRODUCTS

The UL specifications constitute the recognized standards for the asphalt roofing industry and include provisions for: (1) fire resistance (Fig. WF 3); (2) wind resistance; (3) tests of the saturation of felt to determine quantity of saturant and efficiency of saturation; (4) thickness and distribution of coating asphalts; (5) adhesion and distribution of granules; (6) weight, count, size, coloration and other characteristics of finished products before and after packaging, and (7) instructions for application.

Asphalt shingles should conform to UL55B Standard Class C for Asphalt Organic-Felt Sheet Roofing and Shingles and should carry a UL label as shown in Figure WF 4a. In areas subject to winds in excess of 75 mph, Wind Resistant shingles should be used.

These shingles are provided with factory-applied adhesive or integral locking tabs, and the bundles carry the UL Class C Label which includes the words "Wind Resistant" (Fig. WF 4b).

CARE OF MATERIALS

Roll roofing stored outside should be protected from the weather.

Roll roofing should be stored on end in an upright position, and if stored in tiers, boards should be placed between the tiers to prevent damage to the ends of the rolls.

FIG. WF2 SUMMARY OF TYPICAL ASPHALT ROOFING PRODUCTS

	PRODUCT	SHIPPING WEIGHT PER SQUARE	PACKAGES PER SQUARE	LENGTH	WIDTH	UNITS PER SQUARE	SIDE-LAP	TOP-LAP	HEAD-LAP	EXPOSURE
ROLL ROOFING	SATURATED FELT	15 Lb 30 Lb	¼ ½	144' 72'	36" 36"		4" to 6" 4" to 6"	2" 2"		34" 34"
ROLL ROOFING	SMOOTH	65 Lb 55 Lb 45 Lb	1 1 1	36' 36' 36'	36" 36" 36"		6" 6" 6"	2" 2" 2"		34" 34" 34"
ROLL ROOFING	MINERAL SURFACED	90 Lb 90 Lb 90 Lb	1.0	36'	36"	1.0 1.075 1.15	6" 6" 6"	2" 3" 4"		34" 33" 32"
ROLL ROOFING	PATTERN EDGE	105 Lb 105 Lb	1 1	42' 48'	36" 32"			2" 2"		16" 14"
ROLL ROOFING	19" SELVAGE DOUBLE COVERAGE	110 to 120	2	36'	36"			19"	2"	17"
STRIP SHINGLES	2 & 3 TAB SQUARE BUTT	235 Lb	3	36"	12"	80		7"	2"	5"
STRIP SHINGLES	2 & 3 TAB HEXAGONAL	195 Lb	2	36"	11⅓"	86		2"	2"	5"
INDIVIDUAL	STAPLE LOCK	145 Lb	2	16"	16"	80	2½"			
GIANT INDIVIDUAL	AMERICAN	330 Lb	4	16"	12"	226		11"	6"	5"
GIANT INDIVIDUAL	DUTCH LAP	165 Lb	2	16"	12"	113	3"	2"		10"

CLASSIFICATIONS OF FIRE-RETARDANT PREPARED ROOF COVERINGS by UNDERWRITERS' LABORATORIES, INC. (UL)

CLASS A	Includes roof coverings which are effective against severe fire exposures. Under such exposures, roof coverings of this class are not readily flammable and do not carry or communicate fire; afford a fairly high degree of fire protection to the roof-deck; do not slip from position; possess no flying brand hazard; and do not require frequent repairs in order to maintain their fire-resisting properties.
CLASS B	Includes roof coverings which are effective against moderate fire exposures. Under such exposures, roof coverings of this class are not readily flammable and do not readily carry or communicate fire; afford a moderate degree of fire protection to the roof-deck; and do not slip from position; possess no flying brand hazard; but may require infrequent repairs in order to maintain their fire-resisting properties.
CLASS C	Includes roof coverings which are effective against light fire exposure. Under such exposures, roof coverings of this class are not readily flammable and do not readily carry or communicate fire; afford at least a slight degree of fire protection to the roof-deck; do not slip from position; possess no flying brand hazard; and may require occasional repairs or renewals in order to maintain their fire-resisting properties.

FIG. WF 3 Underwriters' Laboratories, Inc. Standard UL 790—Test Methods for Fire Resistance of Roof Covering Materials establishes three classifications.

A

B

FIG. WF 4 Products conforming to UL55B Standard carry the above labels.

218 GYPSUM PRODUCTS

GYPSUM PRODUCTS

INTRODUCTION

Gypsum is a common, rocklike mineral known chemically as *hydrous calcium sulfate*, usually found combined with impurities of clay, limestone and iron oxides. Gypsum has the unique property of giving up some of its chemically combined water and powdering when intensely heated (*calcined*); then it is restored chemically to the original rocklike form when water is added by forming interlocking crystals (*crystallizing*). The mixture of calcined gypsum powder and water remains plastic for a short time and can be easily shaped or molded. After it hardens by crystallizing (*sets*), it becomes an effective fire barrier which will protect supporting wood framing from combustion or steel framing from loss of strength due to high temperatures.

The properties of gypsum were known to the ancients and were utilized in construction as far back as 2000 B. C. However, early gypsum plasters such as *plaster of paris* had a very short, uncontrolled setting time and were difficult to use. Late in the 19th Century, materials capable of controlling setting time (retarders and accelerators) were added to the processed gypsum. This encouraged the use of gypsum plasters for decorative and fire-resistant purposes.

The first large-scale use of gypsum plaster was for the white palaces of Chicago's World Columbian Exposition in 1893. Shortly thereafter, the first gypsum board product was invented. This board product was intended for use as a plaster base, similar to contemporary gypsum lath, and consisted of layers of gypsum sandwiched between plies of felt paper. Further refinements in ingredients, designs and manufacturing processes resulted in the modern varieties of plaster, lath and other gypsum products used in all types of construction.

Many new, engineered gypsum products and assemblies combine the advantages of both board and plaster products to meet contemporary construction needs.

This section outlines the manufacture, properties and uses of gypsum boards, plasters and other gypsum building products. Installation of gypsum finishes is discussed in Division 400—Applications & Finishes.

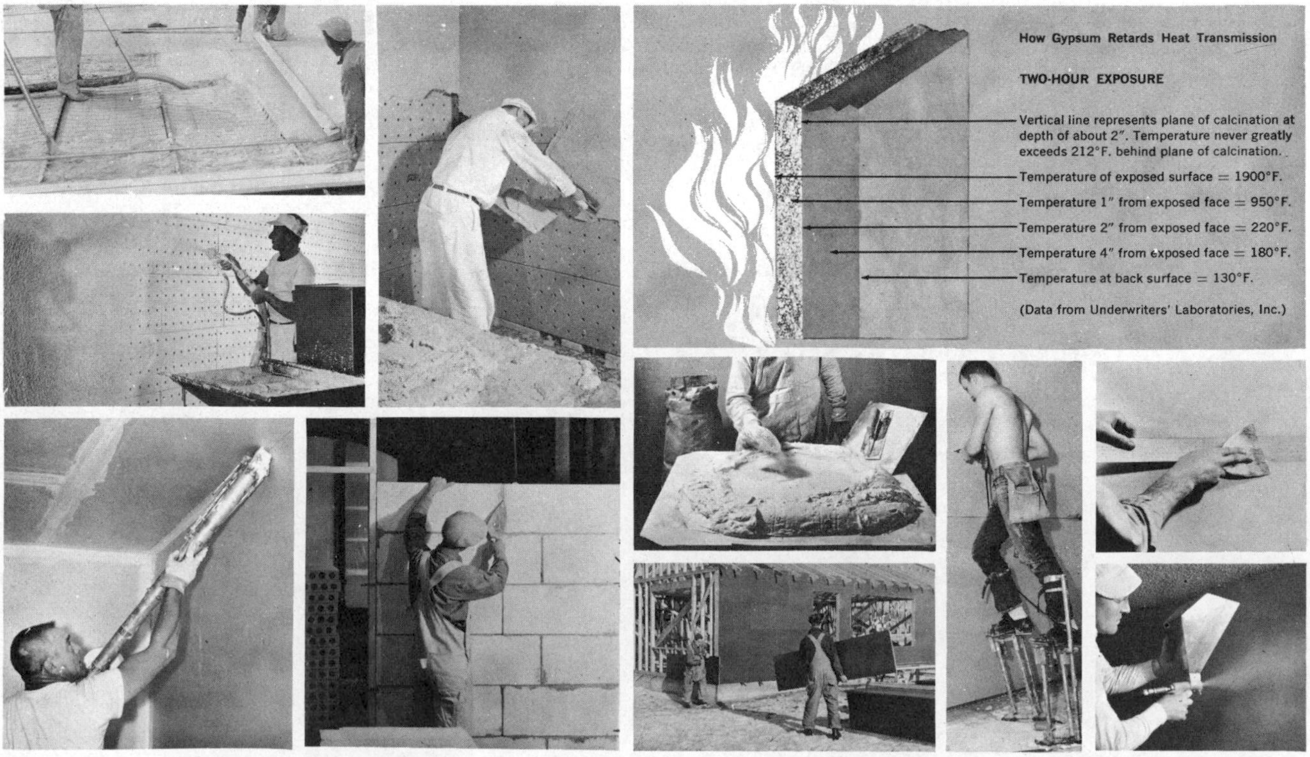

MANUFACTURE

The majority of gypsum building products can be classed as *plaster* products or *board* products. Plaster products consist principally of dry calcined gypsum and require the addition of water at the job site to develop the chemical reaction for setting. Board products, on the other hand, are prefabricated at the factory from crystallized gypsum enclosed by special paper or vinyl, and are delivered to the job site in immediately usable form. Figure 1 illustrates one manufacturing sequence for plaster and board products.

About 2/3 of all crude gypsum production (15.5 million short tons) went into building materials in a recent typical year. Gypsum products are used extensively where fire resistance, sound control and hard, dense wall surfaces are required.

The fire-resistant protection of gypsum can be described as follows: As intense heat is directed on the surface of crystalline gypsum, water in the form of steam is released when the material reaches 212° F. In addition to absorbing some of the heat and dissipating it, this conversion of water to steam prevents the gypsum from rising above 212°F. as long as steam is present. With continued exposure to heat, this action progresses inward releasing more water, and the surface becomes calcined, turning to a white chalky powder. The calcined outer area becomes a better insulator and protects the underlying uncalcined areas of gypsum and retards further calcination. Therefore, gypsum retards intense heat transfer in two ways—by releasing water and by becoming a better insulator when this water is released.

MATERIALS

The raw materials used in the manufacture of gypsum plaster products and in the cores of board products may include (1) powdered calcined *gypsum*, (2) *aggregates*, (3) *mineral* or *organic fibers*, and/or (4) *lime.*

Gypsum acts as the principal binder in gypsum products. Aggregates act as a filler and provide dimensional stability to plaster finishes or to board products. They may also provide increased hardness and fire resistance in some plasters. Wood or other organic fibers are sometimes included in plasters to control the working quality and the keying action of plasters with metal lath. Lime in the form of lime putty is added in finish coat plasters to increase bulk and improve workability. Proprietary chemical retarders and accelerators usually are introduced at the plant to control the setting time.

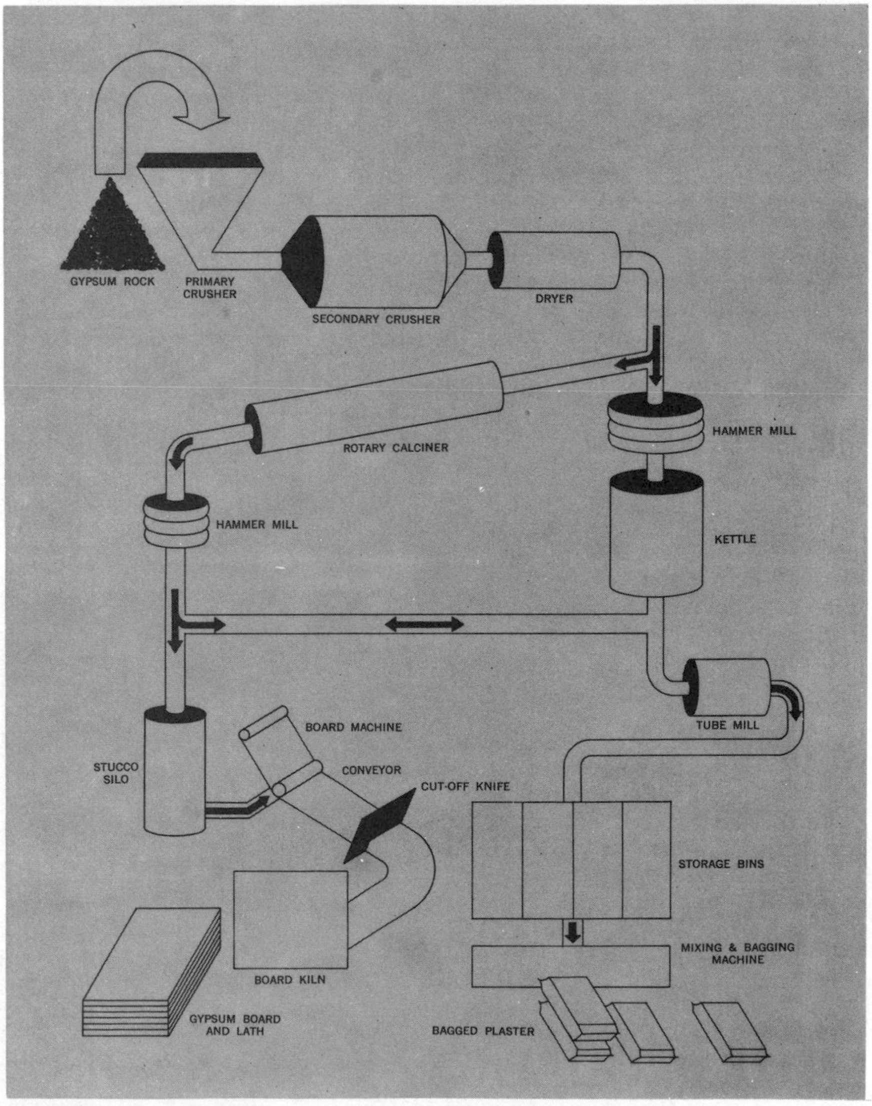

FIG. 1 Flow chart of gypsum board and plaster manufacture.

Gypsum

Gypsum is a common mineral found in many parts of the world. In its pure form it is white, though in combination with impurities it may assume a gray, brown or pinkish color. Relatively pure calcined gypsum is known as *plaster of paris*, after the huge beds which underlie the city of Paris where it was mined in abundance at the turn of the century. In the United States, it is quarried or mined throughout the country where substantial deposits are relatively accessible to transportation (Fig. 2).

After gypsum rock is mined, it is crushed to a 2″ size and is passed through a hammer mill where it is further reduced to a 1/2″ maximum particle size. In this form it is suitable for treating in a *rotary calciner*, but must be ground to a fine powder for treating in a *kettle calciner*.

The rotary calciner is an inclined steel cylinder lined with fire brick, typically 150′ long and 12′ to 15′ in diameter. As the calciner rotates, the rock gravitates to the lower end, while heat is injected to produce a temperature of about 350°F. The intense heat drives off about 3/4 of the combined water, changing the gypsum into a dehydrated form with an affinity for water.

The kettle calciner is a vertical cylinder 10′ to 12′ in diameter with internal agitators which rotate around a central shaft. As in the rotary calciner, heat is introduced near the bottom, driving off the combined water in the powdered gypsum (Fig. 1).

Aggregates

The aggregates commonly used in connection with gypsum products are *sand, vermiculite* and *perlite*.

Sand Most sands consist primarily of quartz and silica, with small amounts of mica, feldspar, clay and other minerals. Natural deposits from all over the United States are used as the main source of plastering sands. The sand is graded according to particle size and is washed if the source is known to produce impurities. Artificially produced sand (stone screenings) of the proper size and type is also suitable.

The natural characteristics of sand vary according to location and must be regulated to insure suitability as an aggregate for plaster. Standards limiting size, gradation and percentage of impurities, and testing procedures to measure these factors, have been established by the American Society for Testing and Materials and the American Standards Association. In general, rough, sharp, angular sand particles produce stronger plasters than do smooth, round ones.

Vermiculite Vermiculite consists of silica, magnesium oxide, aluminum oxides and other minerals, combined with 5% to 9% water by weight. In the United States, vermiculite ore is mined primarily in Montana and South Carolina. This micaceous mineral is composed of a series of parallel layers which expand under the pressure of water trapped between them when subjected to intense heat (Fig. 3). The name originates from the vermicular (wormlike) movement of the layers during expansion.

Processing consists of crushing the raw ore, grading according to particle size and removing impurities. Ore particles then are heated at 1600° to 2000°F. in a furnace for 4 to 8 seconds to produce expansion, and are cooled rapidly to impart pliability and toughness to the expanded particles. As the material cools it is air-lifted to remove unexpanded rock particles and to classify particle sizes. The expanded material is soft and pliable, silvery or gold in color, and contains less than 1% water by weight. Expanded vermiculite is classified into five types according to particular size. Type II, the lightest of the standard aggregates, is used in plaster products.

Perlite Raw perlite is a volcanic ore consisting primarily of silica and alumina, with 4% to 6% combined water by weight, and is quarried mostly in the western part of the United States.

Perlite ore is processed by crushing, screening and grading according to size. It then is expanded in a furnace at 1400° to 2000°F. At this temperature the glassy ore particles approach the melting point and begin to soften. A small amount of the combined water is converted to steam which "pops" the particles into frothy glass bubbles, up to 20 times the original volume (Fig. 4). After expansion, an air blast separates and grades the particles according to size.

Lime

Lime is obtained principally from limestone, a common mineral consisting of calcium carbonate (*calcite*) and/or magnesium carbonate (*dolomite*). Lime for construction purposes comes from

FIG. 2 Gypsum is quarried (left) or mined (right) throughout the country.

FIG. 3 Magnified expanded vermiculite.

FIG. 4 Perlite: crude ore (left), crushed (center), expanded (right).

dolomitic deposits in Ohio or calcitic deposits in Arizona, California, Missouri, Pennsylvania, Texas and Virginia.

Crushed limestone is calcined in a kiln at 2000°F. During the calcining process, the limestone gives off carbon dioxide and changes from the carbonate form to calcium oxide or magnesium oxide to form *quicklime.* Quicklime was used as the chief binder in lime plasters for several hundred years, before the introduction of controlled-setting gypsum plasters.

Before quicklime can be used in plastering, it must be *slaked* (combined with water) and soaked for several weeks. Slaking initiates a chemical reaction which transforms the oxide to a more stable hydroxide state. Some quicklime is sold in the natural pulverized form, but to eliminate the slaking operation, the majority is partially *hydrated* (slaked) at the mill. Hydrated lime is made in two types: *normal* (Type N), which requires 12 to 16 hours' soaking prior to use; and *special* (Type S), which can be used as soon as water is added. The addition of water to hydrated lime or quicklime results in *lime putty*, a plastic form which is combined with gypsum and water to make some finish coat plasters.

The physical and chemical properties of lime depend largely on the limestone from which it was derived. Dolomitic limestones contain a greater proportion of magnesium carbonate and produce lime plasters with better working properties than the calcitic type which contains less of this mineral. However, the magnesium oxides resulting from the calcination of magnesium carbonates are somewhat slower in hydrating. Hence, dolomitic limes often are autoclaved at the mill to assure that a minimum of unhydrated oxides remains in the manufactured lime. Lime with a maximum of 8% unhydrated oxides is suitable for finish coat plasters.

PLASTER MANUFACTURE

For basecoat plasters, calcined gypsum from the kettle is further treated in a *tube mill*, a rotary cylinder containing thousands of steel balls. The action of the steel balls on the gypsum produces a material having a finer grind, which contributes to better plasticity and workability.

Calcined gypsum at this point of manufacture sets very quickly and would be unsuitable for plastering. The type of aggregate used (perlite, vermiculite or sand), local impurities in aggregates and in water, and seasonal variations all will affect the setting time. Therefore, the amounts of retarder added vary and are formulated by manufacturers for specific locations and uses to give the desired job set, which is usually a maximum of 4 hours.

For fibered plasters, organic fibers are added to the gypsum at the mill, and ready-mixed plasters usually are combined with lightweight aggregates prior to packaging and shipping.

BOARD MANUFACTURE

As depicted by the flow chart in Figure 1, calcined gypsum without tube milling is used for the manufacture of gypsum board products. The gypsum is mixed with water to form a slurry. Special additives and glass fibers are added for highly fire-resistant gypsum boards. This slurry, which forms the core of the board, is fed onto a conveyor belt between two layers of paper. After traveling down the belt for 3 or 4 minutes, the core material sets sufficently to be cut to length (Fig. 5), and then boards are conveyed to kilns for drying. Boards not requiring additional finishes may be bundled and stacked for shipping. Others may receive special backings such as aluminum foil, decorative vinyl paper finishes, or organic texture coating prior to shipment.

FIG. 5 Streams of gypsum lath emerge from board machine on a continuous belt.

This page deliberately left blank.

BOARD PRODUCTS

Gypsum board products are panels or slabs consisting of a noncombustible gypsum core, surfaced and edged with covering material specifically designed for various uses with respect to performance, application, location and appearance.

Gypsum board products are produced for specific uses in building construction. *Wallboards* are products with prefinished surfaces or with surfaces suitable for paint finishing after appropriate fastener and joint treatment. *Sheathing* boards are used in exterior frame walls for wind bracing, as a base for finish materials, and for increased fire resistance. *Backing boards* provide a base for adhesive application of wallboards in multi-ply construction and of acoustical tiles in suspended ceilings. *Formboards* serve as both permanent forms and finished ceilings in poured gypsum concrete roof decks.

Board products are made in different sizes and thicknesses, typically with square, tongue-and-groove (T&G), tapered or beveled edges (Fig. 6). Board products are manufactured in conformance with Commercial Standards and Federal Specifications. The physical dimensions, edge types and governing standards of quality of generally available board products are summarized in Figure 7.

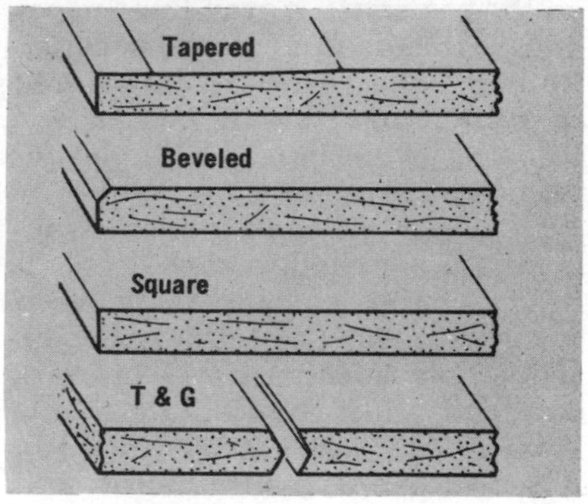

FIG. 6 Typical edge designs for gypsum board products.

WALLBOARDS

The main types of wallboards are: *regular, predecorated, foil-backed, fire-resistant (Type X)* and *water-resistant.*

Regular Wallboard

Regular wallboard is surfaced

FIG. 7 **SUMMARY OF GYPSUM BOARD PRODUCTS**

Type	Thickness (in.)	Thickness (mm)	Width	Length	Edge Detail
WALLBOARD	¼″	6.4	48″	8′, 10′, 12′	Tapered
(ASTM C36, FS SSL30d)	5/16″	7.9	48″	8′, 10′, 12′, 14′	
Regular	⅜″	9.5	48″	7′, 8′, 9′, 10′, 12′	
	½″	12.7	48″	7′, 8′, 9′, 10′, 12′, 14′	
	⅝″	15.9	48″	8′, 10′, 12′	
Foil-backed	⅜″	9.5	48″	8′, 10′, 12′	
	½″	12.7	48″	8′, 9′, 10′, 12′	
Type X	½″	12.7	48″	8′, 9′, 10′, 12′	
	⅝″	15.9	48″	8′, 9′, 10′, 12′	
LATH					Round
(ASTM C37, FS SSL30d)	⅜″	9.5	16″, 16 1/5″, 24″	4′, 8′, 12′	
Plain	½″	12.7	16″, 16 1/5″, 24″	4′	
Perforated	⅜″	9.5	16″	4′	
	½″	12.7	16″	4′	
Foil-backed	⅜″	9.5	16″	4′, 8′	
	½″	12.7	16″	4′	
Type X	⅜″	9.5	16″	4′, 8′	
BACKING BOARD	¼″	6.4	48″	8′	Square
(ASTM C442, FS SSL30d)	⅜″	9.5	48″	8′	Square
Regular	½″	12.7	24″	8′	Tongue-and-grooved
Foil-backed	⅜″	9.5	48″	8′	Square
Type X	⅝″	15.9	24″	7′	Tongue-and-grooved
	1″	25.4	24″	7′	Tongue-and-grooved
SHEATHING	⅜″	9.5	24″	8′	Tongue-and-grooved
(ASTM C79, FS SSL30d)			48″	8′, 9′	Square
Regular	4/10″	10.0	24″	8′, 9′	Tongue-and-grooved
			48″	8′, 9′	Square
	½″	12.7	24″	8′, 9′	Tongue-and-grooved
			48″	8′, 9′	Square
	⅝″	15.9	24″	8′, 9′	Tongue-and-grooved
			48″	8′, 9′	Square
Type X	⅝″	15.9	24″	8′	Tongue-and-grooved
			48″	8′, 9′	Square
FORMBOARD (ASTM C318, FS SSL30d)	½″	12.7	32″	8′, 10′	Square

on the back with a gray liner paper and on the face and along the longitudinal edges with a calendered manila paper which provides a smooth, even finish suitable for decorating.

In single-ply construction, wallboards are nailed, screwed or adhesive-bonded to supporting wood or metal studs or to furring (Fig. 8). In two-ply construction, they are generally adhesive-bonded to a layer of backing boards (Fig. 12). After appropriate fastener and joint treatment (see Division 400—Applications & Finishes). the surface may be painted or finished with wall covering or wall tile.

Most regular wallboards are produced with *tapered* edges which permit taping and finishing of joints to provide a homogeneous, smooth surface with inconspicuous joinings between individual boards (Fig. 9).

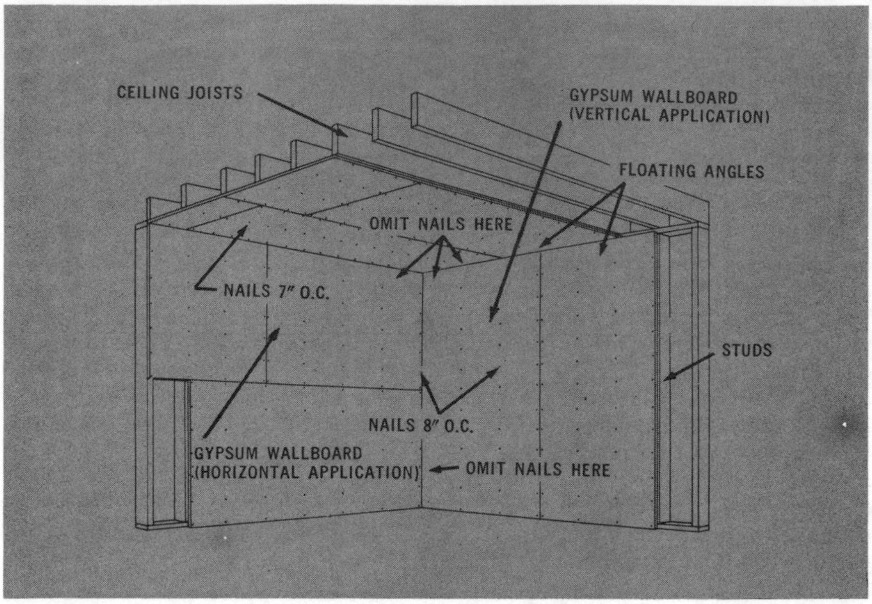

FIG. 8 Wallboard application over wood frame construction.

Predecorated Wallboard

Predecorated wallboard is essentially regular wallboard with a factory-applied decorative finish on the face of the board. The common finish materials are paper and vinyl in a variety of colors, patterns and simulated wood finishes (Fig. 10). These boards usually have accurately formed square or beveled longitudinal edges which do not require further treatment in the field. Beveled-edged and some square-edged boards simply are butted together to create a shadowline repeated at regular intervals. Square edges usually are trimmed out with matching moldings or decorative battens to conceal joint irregularities and fasteners. Some manufacturers provide a loose flap of the decorative covering material which is cemented down after installation to conceal the joint. Predecorated boards may be secured with matching prefinished nails or by adhesive bonding. A lighter weight, predecorated product, MH Board, is often used in manufactured construction, primarily mobile homes. It is available factory painted or faced with paper or vinyl.

Foil-backed Wallboard

Foil-backed wallboard is regular wallboard with a bright-finish aluminum foil bonded to the back. The foil backing serves as a vapor barrier; its use eliminates the need for a separate vapor barrier in an exterior wall assembly. When the reflective surface faces a 3/4" dead air space, it can provide increased insulating performance. Installation and edge treatment of foil-backed wallboard is the same as for regular wallboard.

Fire-Resistant (Type X) Wallboard

This board product is similar in exterior covering, appearance, edge treatment and installation to regular wallboard, but it has a core specially formulated with additives and glass fibers for greater fire-resistance. When prescribed installation methods are followed, fire-resistant assemblies of 45 and 60 minutes are possible with 1/2" and 5/8" boards, respectively, over wood frame construction. Fire-resistant wallboard also is available with an aluminum foil vapor barrier for use in exterior walls and roof/ceiling assemblies.

Water-Resistant Wallboard

Special types of wallboard are produced by some manufacturers for areas such as bathrooms, kitchens and utility rooms (Fig. 11).

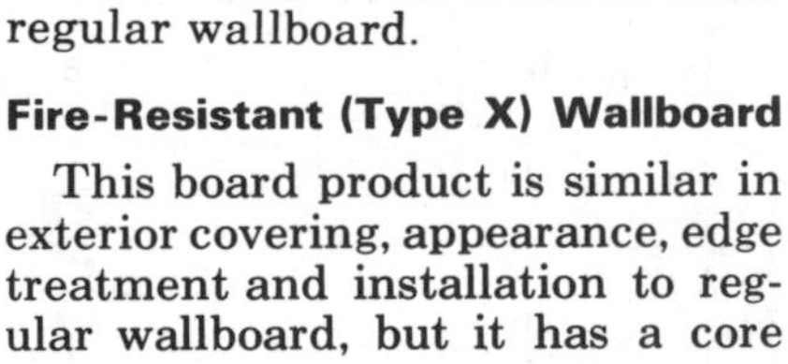

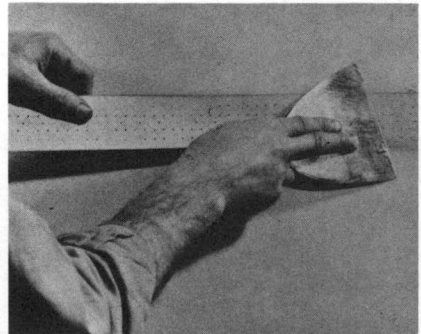

FIG. 9 Wallboard joints are finished by embedding a perforated tape in bedding compound and smoothing the tape over with several coats of topping compound.

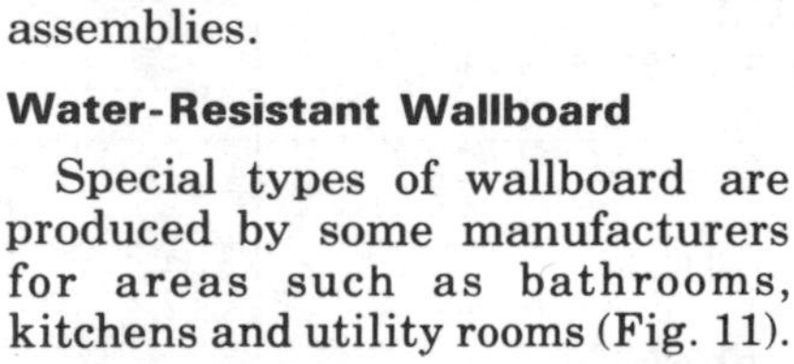

FIG. 10 Predecorated wallboard is factory-finished in a variety of textures and simulated wood patterns, and does not require further treatment in the field.

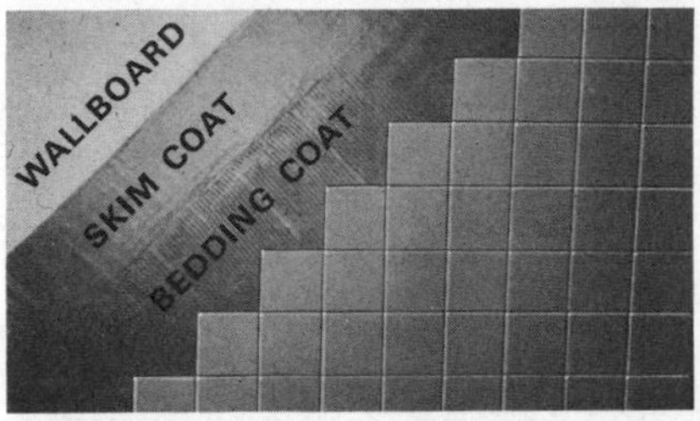

FIG. 11 In humid areas, a skim coat of waterproof adhesive should precede tile application over wallboard.

These are not recommended for high moisture areas such as saunas or steam rooms. A moderately water-resistant type is available which is suitable for paint finishing or as a base for adhesively applied ceramic, metal and plastic wall tile. Water resistance is achieved by using a multilayered covering of chemically treated paper and a gypsum core formulated with asphaltic additives. Edges usually are tapered and installation is the same as for regular wallboard.

A highly water-resistant backing board intended primarily for tiling, in areas subject to direct wetting such as in shower stalls, is described under Backing Boards.

BACKING BOARDS

These boards serve as a base to which wallboards or acoustical ceiling tiles are laminated. Figure 12 illustrates two-ply construction of wallboard bonded to backing board. Backing boards differ from regular wallboards mainly in that the paper covering on both faces and edges is a gray liner paper not suitable for decorating. Backing boards may be secured to framing with staples, screws or nails. The main types are: *regular, fire-resistant (Type X)* and *foil-backed. Coreboard,* a 3/4″ or 1″ thick, factory-laminated board, consists of two layers of regular backing board and is used in the construction of solid gypsum partitions and shaft walls.

A highly water-resistant backing board with treated paper and core is produced, and is intended primarily as a base for tile in areas subject to direct wetting, such as in shower stalls.

GYPSUM SHEATHING

Gypsum sheathing provides fire resistance, wind bracing and a base for exterior finishes in wood and metal frame construction. It usually is made of a water-resistant gypsum core completely enclosed with a firmly bonded water-repellent paper, which eliminates the need for sheathing paper. Standard 2′ x 8′ size boards, typically have V-shaped tongue-and-groove edges to shed water and are nailed horizontally across the framing members (Fig. 13). The 4′ x 8′ and 4′ x 9′ size boards have square edges and are installed vertically where improved wind bracing and infiltration resistance is desired.

FORMBOARDS

Gypsum formboards are used as permanent forms and as finished ceilings for gypsum concrete roof decking (see page 218-20). Formboards are surfaced on the exposed face and longitudinal edges with calendered manila paper specially treated to resist fungus growth. The back face is surfaced with gray liner paper as in regular wallboard. A vinyl-faced board, which provides a white, highly reflective and durable surface, is also available.

FIG. 13 Gypsum sheathing applied horizontally to wood construction.

GYPSUM LATH

Gypsum lath is a class of board products intended as a base for plastering. These products consist of a gypsum core enclosed by a multilayered, fibrous paper covering designed to assure good bond with gypsum plaster. A *suction* bond is created by tiny needlelike crystals of plaster forming in the porous paper covering. Gypsum lath may be stapled, screwed, nailed or clipped to supporting construction of wood or metal studs or to furring (Fig. 14). The main types of gypsum lath are: *plain, perforated, foil-backed* and *fire-resistant (Type X).*

Plain Lath

This is ordinary gypsum lath

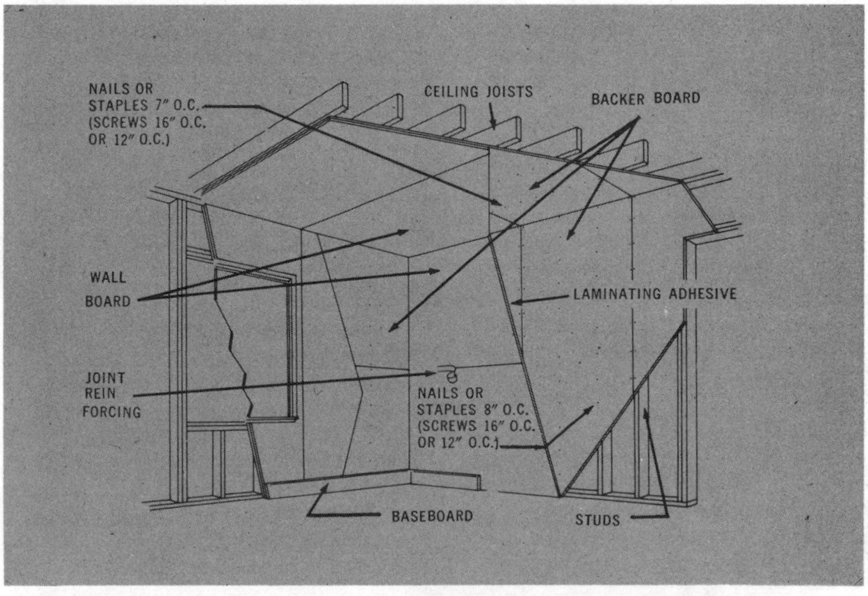

FIG. 12 Double-layer (laminated) wallboard applied to wood construction.

consisting of a gypsum core and paper covering as described above, without any of the special features incorporated in the following types.

Perforated Lath

Before the development of Type X lath, perforated lath, with 3/4″ holes drilled 4″ o.c. in each direction was used to improve the lath-to-plaster bond under fire exposure. Covering and core material are the same as for plain lath.

Foil-backed Lath

A bright-finish aluminum foil is laminated to the back of plain lath. The foil acts as a vapor barrier and can act as a reflective insulator in exterior walls and roof/ceiling assemblies if the reflective surface faces a 3/4″ dead air space.

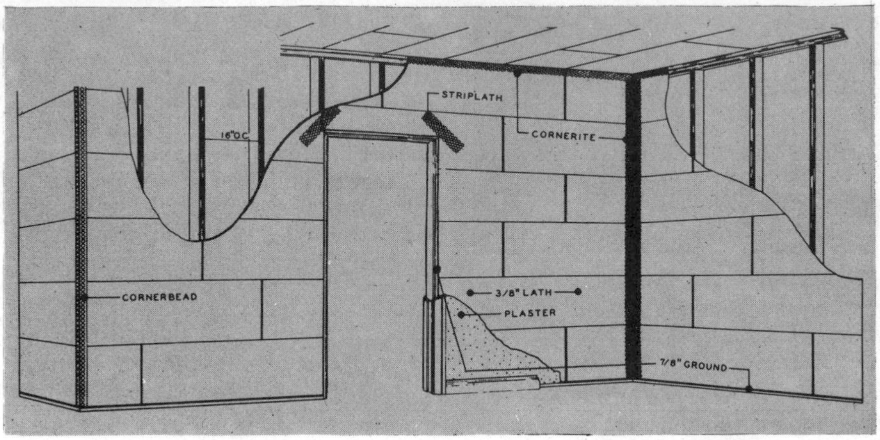

FIG. 14 Application of gypsum lath over wood frame construction.

Fire-Resistant (Type X) Lath

Special additives and glass fibers in the gypsum core of Type X lath improve its fire-resistant qualities, but covering materials are the same as for plain lath.

The discussion in this subsection is limited to a general description of the properties and uses of packaged plaster products ("bagged goods") and does not attempt to describe the many plasters which may be derived on the job for specific plastering applications. A detailed discussion of application methods and finish types of plasters is given in Division 400. For the sake of simplicity and conformance with common terminology, packaged plaster products and job-prepared plaster mixes both are referred to as "plasters."

Packaged plaster products consist primarily of powdered gypsum and additives to control setting time and working quality. These products also may contain other ingredients, such as aggregates and fibers, required for specific applications. Unfibered *neat* plasters generally require the addition of aggregates on the job. *Ready-mixed* basecoat and *prepared* gypsum finish plasters contain all of the necessary ingredients including aggregates. The addition of water to ingredients, properly proportioned either at the job site or at the mill, results in plasters suitable for application to appropriate plaster bases.

Gypsum plasters are suitable for all interior plastering in areas not subject to severe moisture conditions. They can be used in exterior locations such as for ceilings of open porches and carports or for eave soffits, where they are not directly exposed to water. They are intended for application over plaster bases such as gypsum lath, metal lath (sheet, expanded, welded or woven wire), and masonry or concrete bases. The minimum thickness and number of coats required depends on the plaster base. Three coats are required over metal lath and some gypsum lath installations. Two coats are adequate over most gypsum lath installations and masonry and concrete bases.

Three-coat work requires a *scratch coat*, a *brown coat* and a *finish coat*, each applied separately and allowed to set and partially dry. The scratch coat derives its name from the cross-raking which is performed on the wet surface to improve bond with the following brown coat (Fig. 15). In two-coat work, the scratch coat is applied as in three-coat work, but cross-raking is omitted and the brown coat is applied immediately, before the scratch coat sets. The finish coat is applied after the basecoat is set and partially dry (Fig. 16).

Plasters used for scratch and brown coats are termed *basecoat* plasters; those used for the final coat are *finish coat* plasters. The governing ASTM standards, physical and working properties of basecoat and finish coat plasters are summarized in Figure 17.

BASECOAT PLASTERS

Basecoat plasters are classed according to ingredients as *neat, wood-fibered, ready-mixed* and *bond* plasters. Basecoat plasters are formulated for hand-trowel or machine applications.

Neat Plaster

Neat gypsum basecoat plaster does not contain aggregates and generally is used where the addition of aggregates on the job is desired. It also may be obtained fibered or specially formulated for use with lightweight aggregates such as vermiculite or perlite. Fibered neat plaster is lightly fibered with cattle hair, sisal or other fine organic fibers.

Neat plaster usually is combined on the job with aggregates in proportions of 1 part plaster to 2 or 3 parts aggregate by weight. It is suitable for basecoats over metal or gypsum lath, gypsum and clay tile, concrete block and brick of moderate suction (see page Masonry 205-29). Fibered neat plaster often is used as a scratch coat over metal lath because less plaster is lost as mechanical keys are formed behind the lath. When neat plaster is used as a brown coat, sand usually is added to prevent excessive shrinkage in drying.

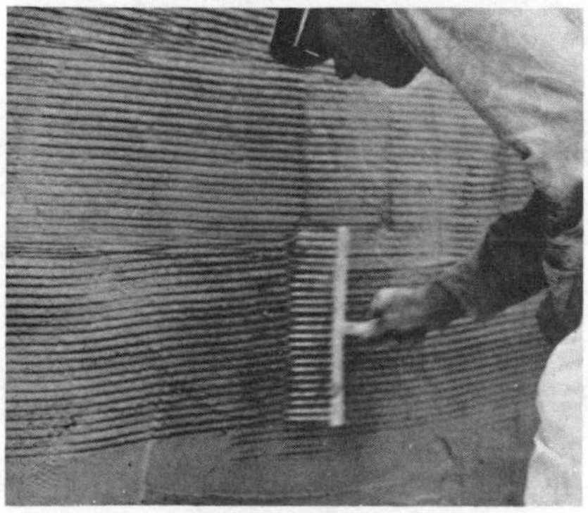

FIG. 15 Scratch coat of plaster is cross-raked for three-coat work.

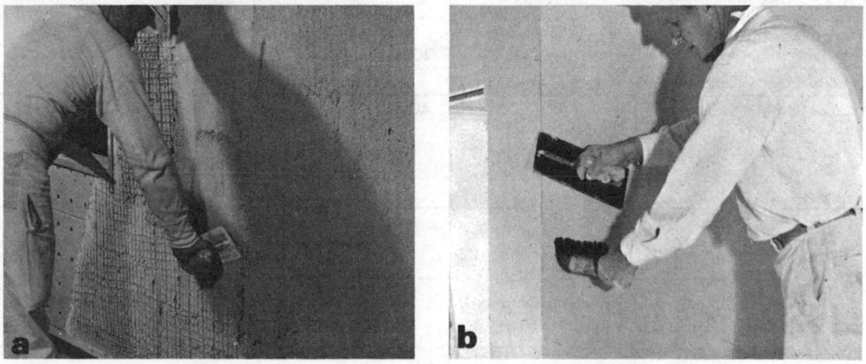

FIG. 16 Brown coat (a) is applied over scratch coat, followed by finish coat (b) in separate operations for three coat work.

Wood-Fibered Plaster

Wood-fibered plaster consists of calcined gypsum integrally mixed with selected coarse cellulose fibers which provide bulk and greater coverage. It is formulated to produce high-strength basecoats for use in highly fire-resistant ceiling assemblies. Although this plaster is often used *neat* (without aggregates), 1 cu. ft. of sand is recommended for each 100 lbs. of wood-fibered plaster to obtain a basecoat of superior hardness, strength, and added crack resistance or for use over masonry. Wood-fibered plaster may be used over the same bases as neat plaster, but it provides greater fire resistance than neat and other sanded basecoat plasters.

Ready-Mixed Plaster

As the name implies, this is a mill-prepared plaster which consists of calcined gypsum and an aggregate (usually perlite, but sometimes sand or vermiculite), only requiring the addition of water. This plaster possesses all of the features and advantages of neat plaster mixed on the job, but in addition offers mill control of the quality, gradation and proportioning of aggregates. Ready-mixed plaster is suitable for all uses where neat plaster mixes might be used.

Bond Plaster

In addition to gypsum, bond plaster contains 2% to 5% lime by weight and chemical additives which improve bond with dense nonporous surfaces such as concrete. The use of bond plaster currently is being replaced by a variety of proprietary bonding agents which permit the application of standard plasters to nonporous surfaces.

FINISH COAT PLASTERS

The finish coat is the last in the plastering operations and is intended as a base for further decorating or as a final decorative surface. Finish coat plasters usually consist of calcined gypsum, lime and sometimes an aggregate. Some finish coat plasters may require the addition of lime or sand on the job. The three basic methods of applying finish coat plasters—*trowel, float* and *spray*—largely determine surface appearance (Fig. 18).

Trowel finishes are used where a flat, smooth, nonporous surface is desired and often form the base for further decorating (Fig. 19a). When an aggregate is used, it is fine silica sand or perlite. Float finishes produce surface textures (Fig. 19b). The coarseness of the texture depends on the aggregate, mix proportions and type of float used.

FIG. 17 **GENERAL PROPERTIES OF GYPSUM PLASTERS**

BASECOAT PLASTERS						
Plaster Mix	Standard	Compressive Strength-Dry (psi) Avg.	Compressive Strength-Dry (psi) Min.	Average Tensile Strength-Dry (psi)	Setting Time	Characteristics
Neat (with 2 parts sand)	ASTM C28	900	750	175	2 to 15 hrs.[2]	Greater versatility because aggregates and/or fibers are job-mixed; less control of plaster quality.
Ready-mixed (with perlite)	ASTM C28	675[1]	400	155	1½ to 8 hrs.[2]	3½ times better thermal insulator than sanded plaster; closely controlled quality.
Wood-fibered	ASTM C28	2100 to 2500	1200	400 to 500	1½ to 9 hrs.[2]	May be used without aggregate; about 50% greater surface hardness than sanded plasters, and generally higher strength and fire resistance.
Bond	ASTM C28	2300	—	400	2 to 9 hrs.	Bonds with nonporous surfaces such as concrete.

FINISH COAT PLASTERS					
Plaster Mix	Standard	Compressive Strength—Dry (psi) Avg.	Compressive Strength—Dry (psi) Min.	Setting Time	Characteristics
Gauging (with lime putty) Regular	ASTM C28	2000 to 3000	1200	20 to 40 min.	Hard surface; good working properties.
Special High Strength	ASTM C28	variable	5000	20 to 40 min.	Hard surface; good working properties.
Gypsum Keene's Cement (with lime putty)	ASTM C61	4000 to 5000	2500	20 min. to 6 hrs.	Very hard surface, becomes harder with decreased proportions of lime; less susceptible to moisture; good working properties; can be retempered; can be polished.
Prepared	(None at present)	variable[3]	variable[3]	variable[3]	Moderately hard surface; no alkali reaction with paint; not as workable as gypsum-lime putty plasters.
Acoustical	(None at present)	variable[3]	variable[3]	variable[3]	Moderately soft; better sound absorption than conventional plasters.

[1]Not applicable to ready-mix plasters for masonry bases.
[2]Four-hour maximum for ASTM C 842 Standard Specification for Gypsum Plastering.
[3]Proprietary plaster mixes containing different ingredients which have variable effects on strength and setting time of plasters.

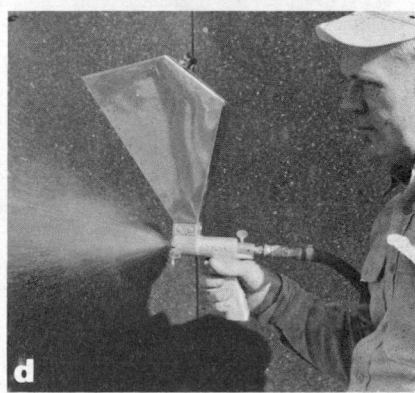

Either trowel or float finishes may be obtained with *gauging plaster, Gypsum Keene's cement, prepared gypsum finish plaster* or *veneer (thin coat) plaster. Acoustical plaster* is another mill-prepared mix which may be hand-applied but most generally today is machine-sprayed (Fig. 19c). *Molding plaster* is a type used for casting in molds and for ornamental plaster work.

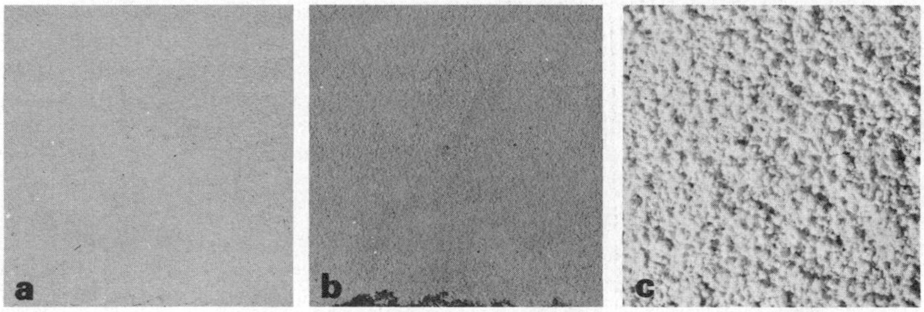

FIG. 18 Surface textures of finish coat plasters: (a) smooth trowel finish; (b) sand float finish; (c) machine-sprayed acoustical finish.

Gauging Plaster

Gauging plaster consists of coarsely ground gypsum of low consistency which soaks up mixing water readily and blends well with *lime putty* (slaked quicklime or hydrated lime). This plaster controls (gauges) the setting time of plasters when mixed with lime putty (which by itself does not set). Gauging plasters are compounded to provide either a *quick set* or a *slow set*. This eliminates the need for adding accelerators or retarders in the field, which seldom can be controlled adequately to provide the desired setting time. Gauging plaster is available as *white* plaster, which produces a brilliant, white finish when gauged with lime putty, and local plaster which may be somewhat darker in color.

Gypsum-lime putty-trowel plaster is a low-cost finish plaster resulting in a smooth surface suitable for most normal uses. The field mix consists of 100 lbs. of gauging plaster to 200 lbs. of dry hydrated lime. Although generally used without aggregates, the addition of a fine silica sand or perlite fines in an amount of at least 1/2 cu. ft. per 100 lbs. of dry gauging plaster will increase the crack and crazing resistance of the finish when used over basecoat plasters containing lightweight aggregates.

Gypsum Keene's Cement

Gypsum Keene's cement is similar to gypsum gauging plaster, but unlike ordinary gypsum plasters, Gypsum Keene's cement is burned in a kiln instead of a calciner or kettle. In this burning process, practically all of the combined water is removed, producing a "dead-burned" gypsum. The resulting material is denser than ordinary gypsums and has greater impact, abrasion and moisture resistance. Select white gypsum rock is used in the manufacture to provide the hardest white gypsum available.

Gypsum Keene's cement has an initial set of about 1-1/2 hours and a final set of approximately 4 to 6 hours. A quick-setting Gypsum Keene's cement which provides a final set of approximately 2 hours also is available.

Gypsum Keene's cement is used for gauging lime finish coat plaster, much the same as gauging plaster, but is used in preference to gauging plaster when a harder, more moisture-resistant surface is desired. Hardness is proportionate to the amount of Gypsum Keene's cement used. Both *trowel* and *float* finishes are possible with Gypsum Keene's cement.

Gypsum Keene's Cement-Lime Putty-Trowel Plaster This plaster

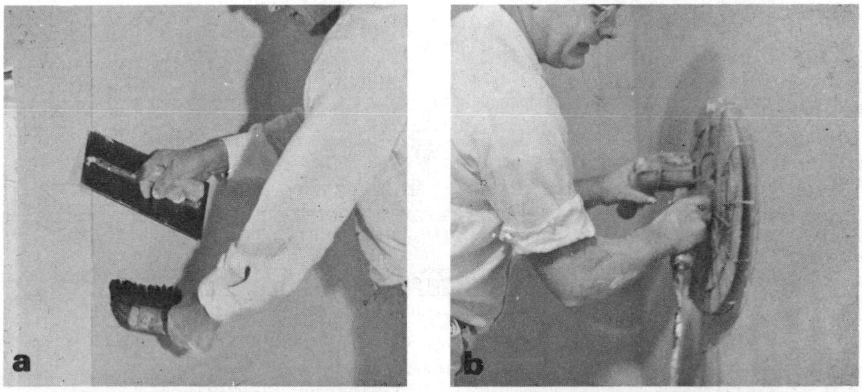

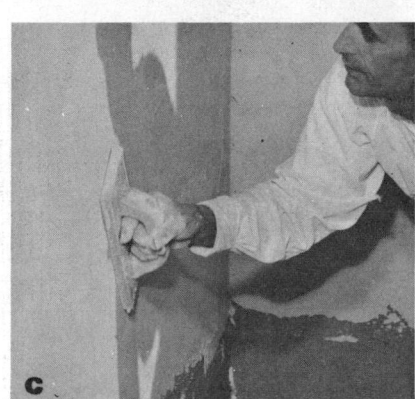

FIG. 19 Methods of applying finish coats: (a) hand trowel, (b) power trowel, (c) hand float; (d) machine sprayed.

produces a dense, hard finish and does not contain aggregates except when applied over lightweight aggregate basecoats. In this instance, at least 1/2 cu ft. of fine sand or perlite is added to each 100 lbs. of Gypsum Keene's cement and 50 lbs. of hydrated lime; an extra hard finish can be obtained by reducing the lime to 25 lbs.

Gypsum Keene's Cement-Lime-Sand Float Plaster This plaster may be used over all types of basecoat plasters where a minimum of cracking and spalling is desired, and is particularly recommended over lightweight aggregate basecoat plasters. It usually is mixed with sand and provides greater crack resistance than the trowel finish. The resulting textured surface may be painted or an integral coloring agent may be introduced in the plaster to eliminate decorating. For float finish, 150 lbs. of Gypsum Keene's cement is mixed with 100 lbs. of hydrated lime and 400 to 600 lbs. of silica sand, depending on the texture desired.

Prepared Gypsum Finish Plaster

This is a ready-mixed plaster requiring only the addition of water at the job. It consists of finely ground calcined gypsum with or without fine aggregates. It is available as white or gray *trowel* finish and *sand float, colored float* or *textured* finishes. These finishes are comparable in hardness and appearance to those obtained with job-mixed Keene's cement, but factory blending of ingredients assures somewhat better results.

Veneer (Thin Coat) Plaster

This is a high-strength plaster for application as a very thin one or two coat finish (1/16" to 1/8") over a special plaster base designed particularly for this purpose. In many parts of the country veneer plaster is the dominant type of finish because of its economy and speed of job completion.

The joints in the plaster base are treated with a glass fiber tape or reinforced with drywall paper tape and a compatible joint compound following manufacturer's recommendations. Corner beads are installed prior to plastering. Because of the composition and thinness of the plaster coat, it dries rapidly and may be decorated 24 hours after installation.

Veneer plaster can be applied as a smooth, trowel finish or as a textured, sprayed finish. Either finish is extremely hard and abrasion-resistant. Veneer plaster may be used also as a finish coat over bond plaster or as a basecoat for *gypsum-lime* and *Gypsum Keene's-lime* finishes where exceptional strength and crack resistance is desired.

Molding Plaster

A very finely ground gypsum necessary for smooth, intricate surfaces and close tolerances common in ornamental work is used in molding plaster. In most cases, the mix consists of 100 lbs. of molding plaster to 50 lbs. of dry hydrated lime.

Acoustical Plasters

These are proprietary plasters intended to increase sound absorption to reduce sound reverberation. They consist usually of ground gypsum with chemical ingredients which encourage the formation of small air bubbles to trap and absorb sound. Depending on the formulation, they may be hand- or machine-applied; however, most acoustical plasters are applied by machine. Noise reduction coefficients ranging from .40 to .60 may be obtained with these products (see page Sound Control 106-10). Surface textures may vary from fine to rough, and crack resistance is high. Indentation resistance, however, is much lower than with other plaster products and the use of acoustical plasters is limited to ceiling and upper wall areas not subject to abrasion or impact. Acoustical plasters may be left in the natural off-white or gray color or may be spray-painted.

Gypsum's inherent qualities such as incombustibility, fire resistance and relative light weight make it useful in a variety of building products. These may be classified according to common use as *partition, roof* and *ceiling* products. Like other gypsum products, they are adversely affected by moisture and should not be used under high humidity conditions or where they will be exposed directly to wetting.

This subsection discusses and illustrates engineered systems using board and plaster products, as well as other gypsum products.

PARTITION PRODUCTS

Gypsum partition products are used in *partition systems*, engineered assemblies of board and/or plaster products with special metal accessories designed for use with each system.

Partition Systems

A variety of proprietary partition systems have been developed by manufacturers of gypsum products to minimize the number of operations and speed field erection of partitions. The chief types are (1) *solid*, (2) *semi-solid*, (3) *hollow* and *demountable* (movable) systems.

Solid Partitions These may be field-erected from combinations of lath and plaster or backing boards and wallboards. Both generally utilize wood or proprietary metal ceiling and floor runners to stabilize the partitions. In lath and plaster partitions, an integral flush metal base acts as a screed. Partitions also may be trimmed out with resilient or wood base.

Lath and Plaster Partitions use self-supporting ribbed lath (Fig. 20a) or ordinary expanded mesh wired to steel bars or channels (Fig. 20b). To this base, three coats of plaster are applied in thicknesses varying from 1″ to 1-1/4″ on each side. Solid plaster partitions also may be made of 1/2″ gypsum lath, with two 3/4″ plaster faces built up of two or more coats (Fig. 20c).

Backing Board and Wallboard Partitions consist of a 1″ backing board core to which 1/2″ or 5/8″ wallboard faces are laminated with special adhesives. The core may consist of two job-laminated 1/2″ backing boards or a 1″ factory-laminated product referred to as *coreboard* (Fig. 21a). Regular wallboards usually are used requiring joint treatment and decorating, but in some instances predecorated wallboards with V-edges may be installed in a visually attractive repeat pattern. When veneer plaster finish is desired, the special plaster base designed for this type of finish is used in place of wall-

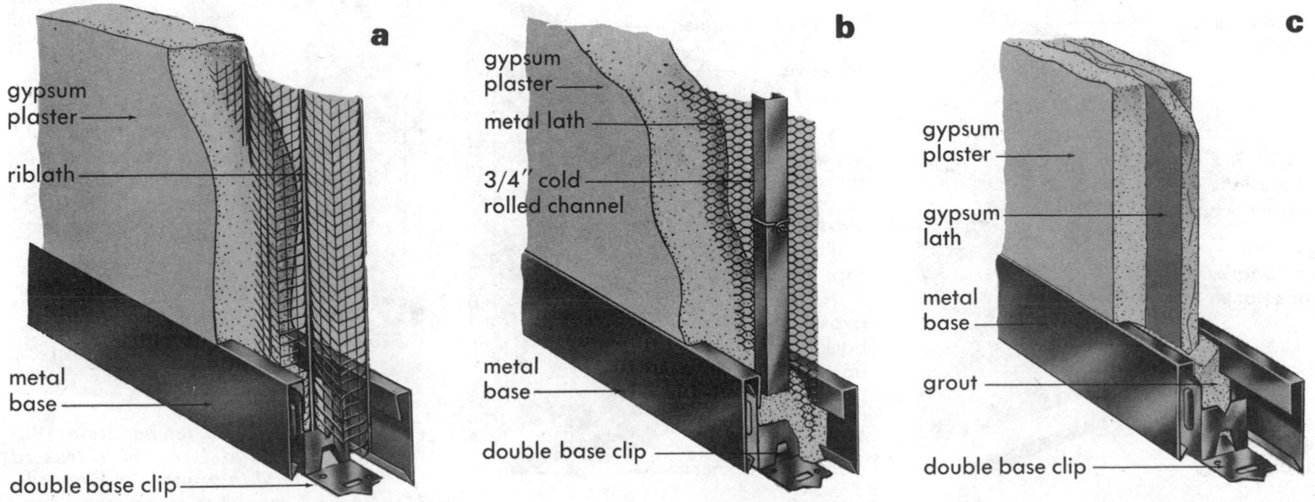

FIG. 20 Solid plaster partitions: (a) on self-supporting ribbed metal lath; (b) on stud-supported woven wire lath; (c) on gypsum lath.

board.

Double solid partitions are constructed of two separate layers of lath or backing board, separated by an air space, with plaster or wallboard faces on the exterior (Fig. 21b).

Semi-solid Partitions In semisolid partitions, the backing board core is replaced by 6″ to 8″ wide gypsum board ribs, laminated at the factory in 1″ and 1-5/8″ thicknesses (Fig. 22a). Ribs are adhesive-bonded in the field to the wallboard faces for single-ply construction or to backing boards for two-ply construction (Fig. 22b). In two-ply construction, regular or predecorated wallboard is bonded to form the finished surface. Veneer plaster finish may be used over an appropriate plaster base laminated to the ribs. Semi-solid partitions generally are not made with standard plaster finishes. Partition thicknesses vary between 2″ and 3″ depending on rib and face panel thicknesses.

Hollow Partitions These partitions are made with wood or metal studs supporting faces of plaster or wallboard. Studs usually are spaced 16″ to 24″ o.c. and are supported by wood or metal floor and ceiling runners. Wood studs normally are 3-1/2″ deep, metal channel studs are available in 1-5/8″, 2-1/2″ and 3-5/8″, and 4″ sizes (Fig. 23a). Greater wall cavities can be developed by using deeper trussed-type studs which range up to 6″ in depth (Fig. 23b), or two sets of lighter channel studs braced with wallboard (Fig. 23c).

Lath and Plaster hollow partitions have gypsum lath screwed, clipped or nailed to studs, or metal lath wired, nailed, clipped or

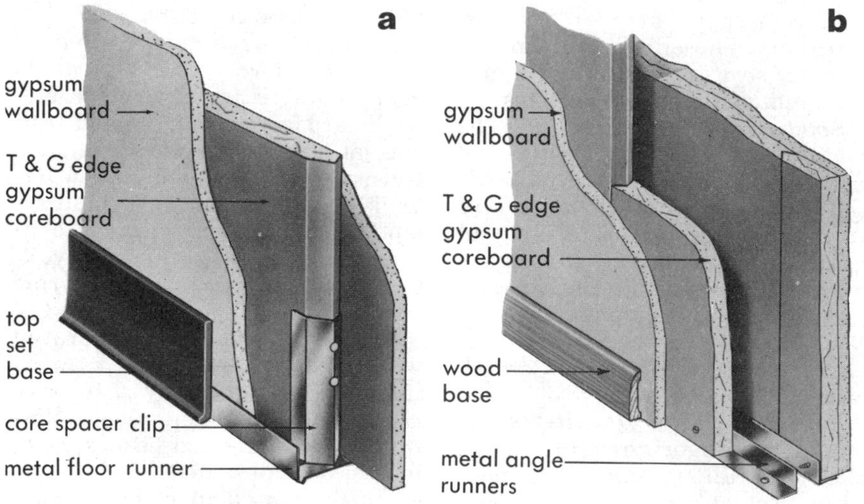

FIG. 21 Solid wallboard partitions: (a) single solid partition, and (b) double solid partition with an air space between the two faces.

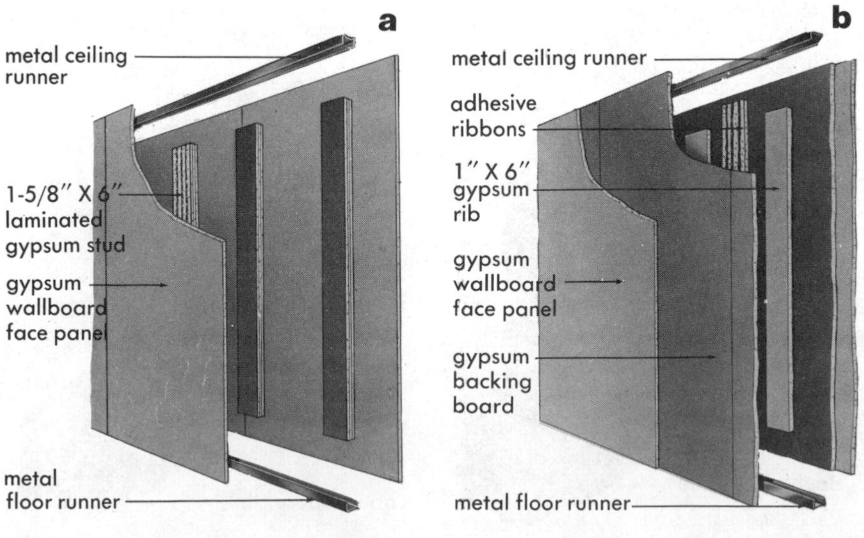

FIG. 22 Semi-solid wallboard partitions: (a) single-ply construction, and (b) two-ply construction.

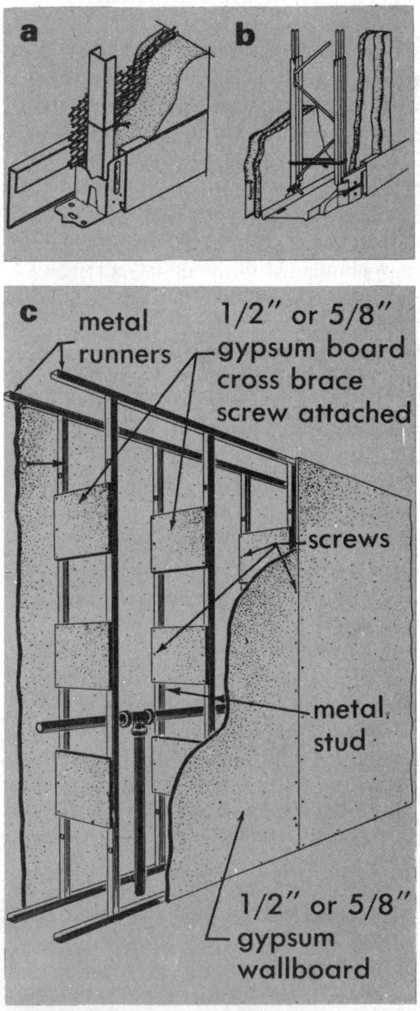

FIG. 23 Hollow plaster partitions: (a) with channel studs, (b) with trussed studs, and (c) with braced double studs. Double stud construction accommodates mechanical equipment within the cavity space.

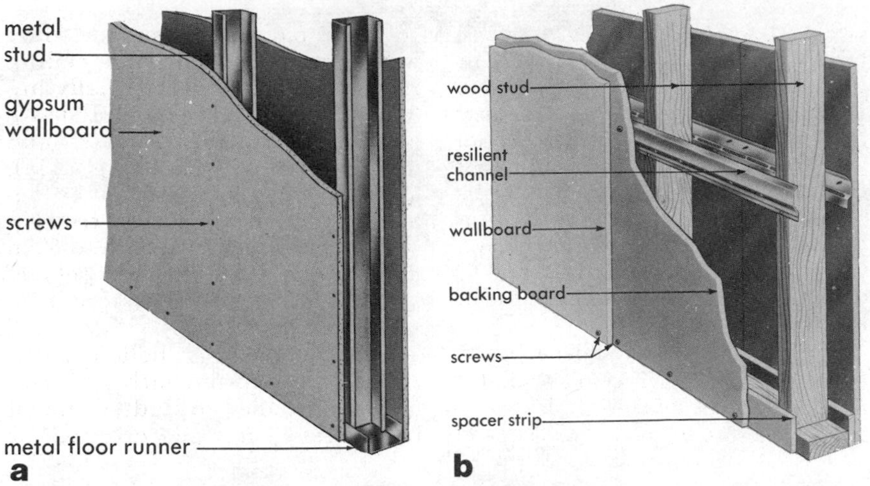

FIG. 24 Hollow wallboard partitions: (a) screwed single-ply, and (b) two-ply construction with resilient mounting for improved sound control.

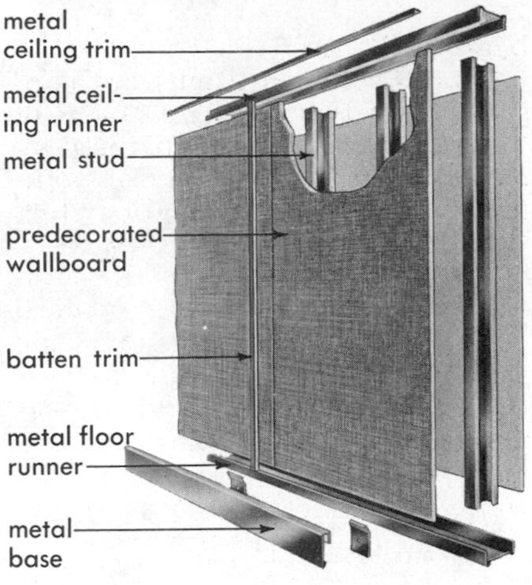

FIG. 25 Many decorative wallboards are available for demountable systems.

stapled to the studs. Two or three coats of plaster are built up over this plaster base on each face.

Wallboard hollow partitions are commonly either of single-ply or two-ply construction. For single-ply construction, wallboard is nailed or screwed directly to the studs (Fig. 24a). For two-ply construction, backing board is clipped, nailed, stapled or screwed to the studs and wallboard is adhesive-bonded to this backing with a minimum of supplemental fastening. Where soundproofing is important, resilient channels may be used in either one- or two-ply construction to mount the board (Fig. 24b). When regular wallboard is used as the exterior face, fasteners and joint treatment are required; predecorated wallboard normally is adhesive-bonded to backing board and has V-edge joints which can be left exposed.

Demountable (movable) Systems. In situations where it may be desirable to rearrange partitions, demountable panel systems frequently are used. They are designed for speedy erection and for minimum loss of components when being demounted and rearranged.

Demountable partitions systems use wallboard face panels attached to interior metal studs for stability. Various pressure fit, clip or screw systems are used to secure the wallboard panels to the studs. One system uses an H-section stud 24″ o.c., the flanges of which are pressure fit into grooves of factory-laminated, predecorated panels. In other systems, predecorated squared-edge wallboard is adhesive-bonded directly to studs 24″ o.c. and edges are secured with nails or screws. Decorative batten trim then is used to conceal fastenings at 24″ or 48″ o.c. (Fig. 25).

Shaft Walls

Shaft walls are light weight gypsum board assemblies used for elevator shafts, electrical and mechanical enclosures and stairwells.

Shaft wall assemblies with metal stud framing weigh 10 to 13 psf compared with 35 to 65 psf for masonry or masonry-and-plaster assemblies.

Shaft walls built with gypsum board are not only lighter in weight, but are easier and less expensive to install than conventional masonry construction. They can be erected floor by floor from outside the mechanical or elevator shaft; they do not delay installation of the mechanical equipment inside the shaft due to masonry scaffolding; and they do not require extensive cleanup inside the shaft upon completion.

CEILING PANELS

Two gypsum products for ceiling construction are *fire-resistant* panels and proprietary *electric-*

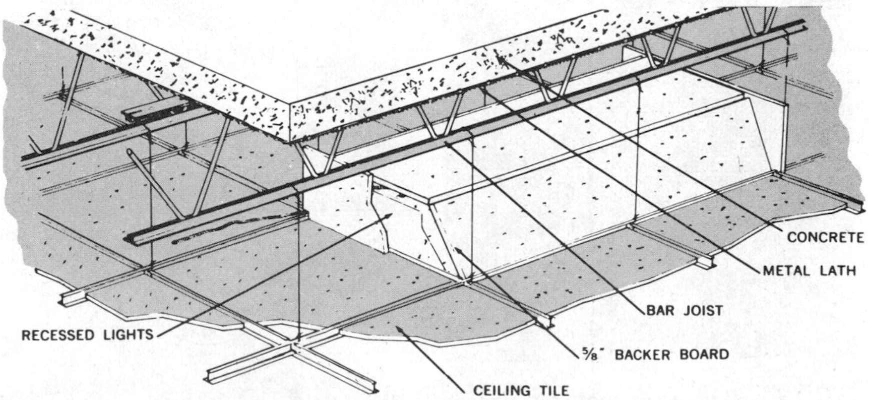

FIG. 26 Fire-resistant 5/8″ ceiling panels in T-grid suspension system provide two-hour fire protection for steel and concrete construction.

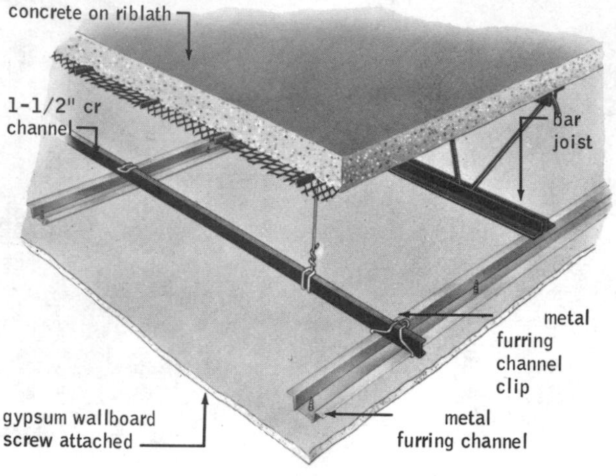

FIG. 27 Regular fire-resistant 5/8" wallboard, screwed to furring channels, provides fire protection up to three hours for steel and concrete construction.

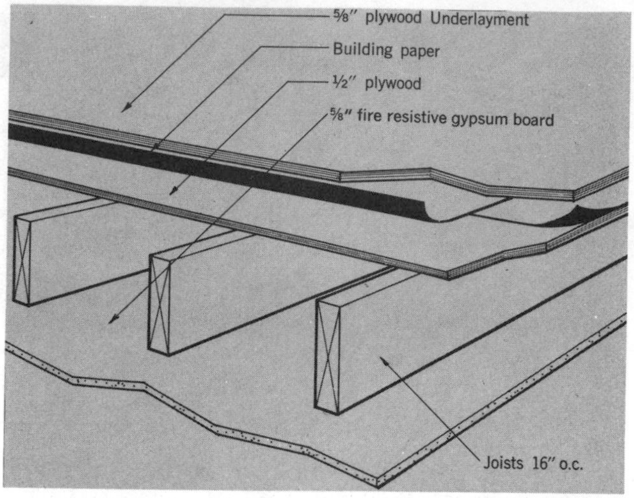

FIG. 28 Properly constructed wood joist floor/ceiling assemblies with 5/8" fire-resistant wallboard receive one-hour fire ratings.

radiant heating panels.

Fire Resistant Panels

Fire-resistant wallboards are cut into 2′ x 2′ and 2′ x 4′ panel sizes for use in some fire-rated ceilings. Common thicknesses are 3/8″, 1/2″ and 5/8″ and fire ratings up to 2 hours are possible with an approved suspension system (Fig. 26). Such ceilings also reduce sound transmission but have little sound absorptive properties. Standard 4′ x 8′ fire-resistant wallboard panels also may be used to provide fire-rated floor/ceiling assemblies for steel and wood construction (Figs. 27 and 28).

Electric-Radiant Heating Panels

Electric resistance elements may be incorporated in the wallboard itself for one-ply application, or in the backing board for two-ply construction. Heat loss calculations indicate the number of square feet of heating panels required in each space. For properly insulated residential installations, this is generally less than the total ceiling area. The areas adjacent to heating panels in each space are covered with backing board or wallboard of matching thickness. Wallboard face panels may be finished in the normal manner, but installation of acoustical tile over the system is not recommended.

Standard gypsum lath and plaster products also can be assembled into a radiant heating ceiling system by embedding heating cables

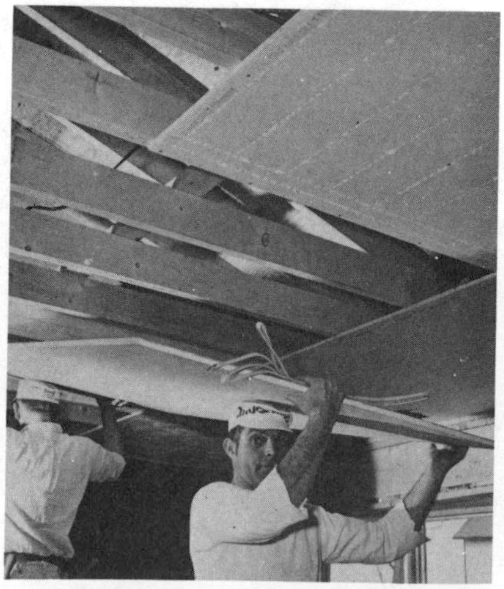

FIG. 29 Single-ply radiant panels may be nailed and finished in the same manner as regular wallboard.

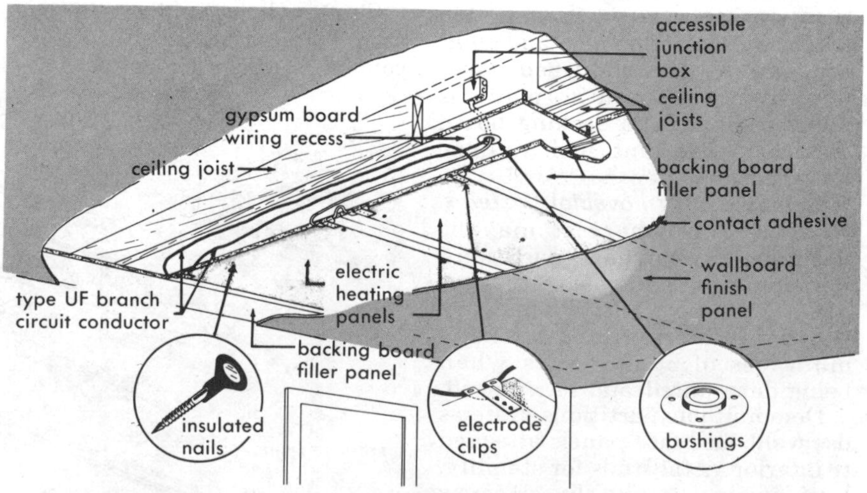

FIG. 30 Two-ply radiant system requires special accessories and job-lamination of wallboard face panels. Finishing is the same as for regular wallboard, but acoustical ceiling tile should not be used over this system.

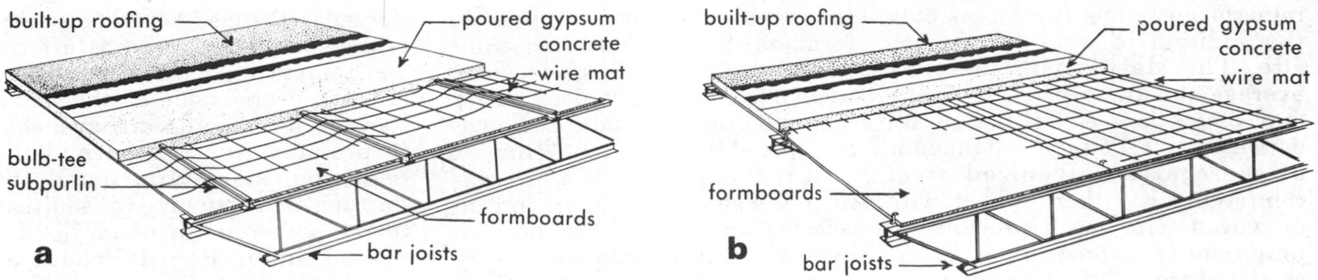

FIG. 31 Gypsum concrete roof construction: (a) with subpurlins, (b) without subpurlins.

in the plaster coats over gypsum lath.

One-ply Construction One proprietary heating panel system utilizes a single layer of 5/8″ wallboard with heating cables embedded in the core (Fig. 29). Panels come in standard 4′ x 4′, 4′ x 8′ and 4′ x 12′ sizes, but special panels, containing cables in 2/3 of the area only, allow ends and sides to be trimmed as required for particular layouts. All panels may be installed by nailing at prescribed intervals. Nailheads are concealed with compound as in other wallboards and tapered edges permit standard joint treatment.

Each panel is provided with a 12′ electric lead cable which may be connected to an electric branch circuit in a junction box. The branch circuit typically is connected to a room thermostat which regulates the panel heating cycles. Prior to electrical connection, the lead cable is taped to the back of the panel so that it will not interfere with erection.

Two-ply Construction This proprietary system uses a combination heating panel and backing board, to which a 1/4″ regular gypsum wallboard must be job-laminated (Fig. 30). The heating panel is approximately 1/2″ thick and comes in 4′ x 10′ and 4′ x 12′ sizes. It consists of a gypsum core, resistor coating and copper electrodes. The electrodes are connected to standard AC or DC house current and conduct the electricity to the resistor coating, generating heat. This coating is completely enclosed by an asbestos insulator to prevent current leaks.

The heating panels may be nailed directly to wood ceiling members or other supporting framework with special insulated nails which can penetrate the resistor coating without becoming energized. Standard panels have electrodes 1″ from each edge and may be cut to desired length, but cutting the panels to width is not possible as one of the electrodes would be removed. A special 4′ panel, however, is available with a 22″ wide resistor coating which can be cut to a minimum 2′ width.

ROOF DECKING

This category covers structural deck systems commonly used in noncombustible roof construction. The common gypsum system is *poured-in-place gypsum concrete.*

Gypsum Concrete

A gypsum concrete roof deck consists of a reinforced *gypsum concrete slab* poured on permanent formboards that may or may not be supported by subpurlins. Subpurlins span between main purlins, beams or joists. When subpurlins are not used, formboards rest on purlins, beams or joists (Fig. 31a & b).

Gypsum concrete roof decks can be installed on virtually any roof shape, size or configuration (warped, saw-tooth, curved or sloped). Primary framing may be wood, steel or concrete. The light weight of gypsum concrete, compared to portland concrete, often allows the use of a lighter structural frame. Gypsum concrete sets rapidly and the roof slab may be used as a working surface within 30 minutes after pouring.

Decks of gypsum concrete poured over gypsum or mineral fiber formboards are classified as noncombustible. When combustible wood fiber formboards are used, there generally is an insurance deficiency penalty. Gypsum concrete decks, 2″ and 2-1/2″ thick on gypsum formboards without a suspended ceiling below, qualify for 1- and 2-hour fire resistance ratings, respectively. Materials for gypsum concrete should conform to ASTM C317; materials and installation of gypsum concrete roof decks should conform to ANSI A59.1.

Gypsum Concrete Gypsum concrete is a factory-controlled mixture of gypsum and wood chips or

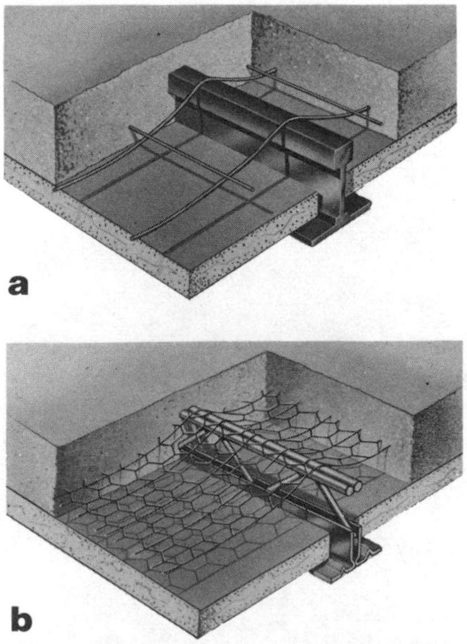

FIG. 32 Subpurlins and reinforcing: (a) bulb tee and welded wire, (b) trussed tee and woven wire.

mineral aggregate. It requires only the addition of water at the job site. The slab is poured to an average thickness of 2″ to 3-1/2″ and weighs 35 to 55 lbs. per cu. ft., depending on the type and amount of aggregate. Galvanized steel reinforcing is either welded wire or woven wire mesh providing a minimum of 0.26 sq. in. of cross-sectional area per foot of slab width (Fig. 32).

Subpurlins Steel subpurlins (rails, bulb tees or other tee sections) are welded transversely to the primary framing to support the formboards and roof slab. Subpurlins also anchor the deck against uplift forces, restrict deck movement due to temperature changes and provide lateral bracing for main roof purlins. Subpurlins vary in size, weight and shape and are selected on the basis of the span between main purlins and the required total safe load-bearing capacity of the deck.

Formboards Formboards are nonstructural components of the roof deck and usually are not considered in the load-bearing capacity of the slab. They serve as a form and remain in place to provide a functional underside of the deck. Various formboards are available to provide fire resistance, light reflection, insulation, sound absorption or economy. In addition to the gypsum formboards discussed under Board Products, wood, glass and mineral fiber boards are available. These formboard widths generally coincide with the standard 24″, 32″ and 48″ nominal subpurlin spacing.

We gratefully acknowledge the assistance of the following for the use of their publications as references and permission to use photographs and illustrations: Contracting Plasterers' and Lathers' International Association, Georgia Pacific Corporation Gypsum Association, MAC Publishers Association, Bestwall Gypsum Company, Celotex Corporation, National Gypsum Company, United States Gypsum Company.

219 ASBESTOS CEMENT PRODUCTS

ASBESTOS CEMENT PRODUCTS

Asbestos cement products, also called *mineral fiber* products, are made by combining asbestos fibers and portland cement with water and forming this mixture, under pressure, into thin, hard, rigid sheets or into pipe.

Early production of mineral fiber products in the United States, dating from 1905, was almost entirely roof shingles. From this first introduction of the material, the mineral fiber products industry has greatly improved the quality of their products which today include, in addition to roof shingles, various types of wall shingles and panel (sheet) siding, perforated and corrugated sheet material, insulating panels, various types of pipe, and many specialty products.

Mineral fiber products possess many properties that make them highly suitable as exterior finish materials and other uses. The natural finished material is completely inert, very durable and virtually maintenance free. These products are strong in compression, fire resistant, and resistant to alkalies and acids. Although weak in resistance to impact, when simple construction precautions are observed, mineral fiber products provide suitable wall and roof finishes in exposed locations under normal conditions of service.

Mineral fiber products may be divided into two classifications: *building materials* and *pipe*. This section provides a basis for understanding the general characteristics of the various products and their appropriate uses so that suitable quality in service may be achieved.

Because asebestos fibers are harmful to the body when inhaled or ingested, it is important to employ great care in installing mineral fiber products. Whenever asbestos products are drilled, sawed, shaped or sanded, the asbestos-laden dust must be prevented from contaminating the atmosphere. Manufacturers, fabricators, and installers of mineral fiber products are required to minimize the asbestos dust exposure of their workers by the Occupational Safety and Health Act (OSHA) Asbestos Standard 1910.93a.

The Occupational Safety and Health Administration (OSHA) requires that asbestos products be handled, applied, cut or otherwise worked in such a way that emission of asbestos dust into the atmosphere is controlled. A respirator, gloves, hat and other protective clothing should be worn when handling asbestos products; these cautions extend to the removal of asbestos products from their containers, as well as demolition work. Local exhaust or ventilation systems to control dust must be constructed, installed and maintained in accordance with ANSI Z9.2 for all hand and power operated tools. Surface cleaning of the work area should be done with vacuum units which have disposable dust bags. If floors must be swept with a broom, debris should be wetted to prevent

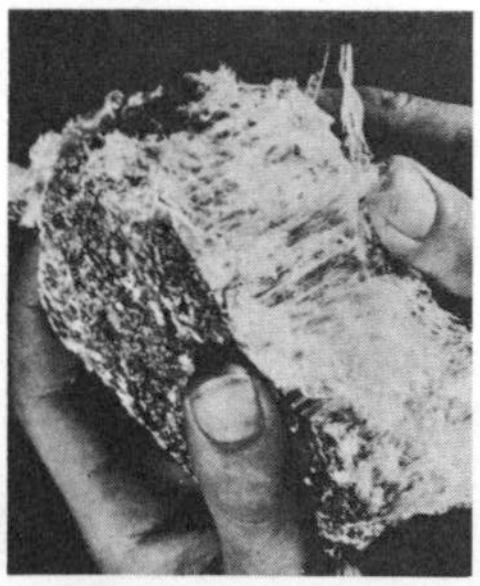

stirring up dust. Material should be wetted also when hand tools are used. Asbestos waste should be disposed of in sealed bags or closed containers.

MATERIALS

Mineral fiber building products consist essentially of asbestos fibers and portland cement, combined with water. The mixture is formed into flat, thin sheets under pressure and cured. The asbestos fibers, acting like the steel bars in reinforced concrete, impart tensile strength and flexibility to the otherwise brittle portland cement, and permit products to be produced in very thin sheet. Organic fibers are sometimes added to produce building products which are lighter in weight and have increased workability.

Asbestos

Asbestos fiber is obtained by crushing asbestos ore, principally hydrated magnesium silicate, taken from open-pit and underground mines. Fibers used in mineral fiber products vary in length from an average of 1/16″ to 3/4″, possess extremely high tensile strength (18,000 psi to over 400,000 psi), have great flexibility, and are resistant to attack by chemicals, moisture, decay and weathering. Asbestos fiber is incombustible, can be spun into yarn, woven into cloth, and formed into paper. The combination of these unique characteristics account for the commercial usefulness of asbestos.

In addition to being one of the principal constituents of mineral fiber building products and pipe, asbestos fiber is used in other building materials such as vinyl asbestos floor tile, asbestos felt, thermal insulation and as a reinforcement in plastics and other coatings. Other industrial applications include use in textiles, in parts for electrochemical cells, as a friction material in brake linings and clutch facings.

Portland Cement

Portland cement is produced by combining limestone and clay, pulverizing the mixture, and then burning it into clinkers. The clinkers are finely ground to produce the finished cement. When mixed with water, the cement hardens into a solid mass that is predictably consistent in strength and quality. The type of portland cement used in the manufacture of mineral fiber products is generally Type II conforming to ASTM C-150 (See Section 203 Concrete).

MANUFACTURING

Three methods, depending on the manufacturer and the specific product, are used to produce mineral fiber building materials: (1) Wet Mechanical Process, (2) Dry Process, and (3) Wet Formed Process. While details of production, equipment and plant size vary with manufacturers, in general,

FIG. 1 Production sequence of Wet Mechanical Process to produce mineral fiber building products.

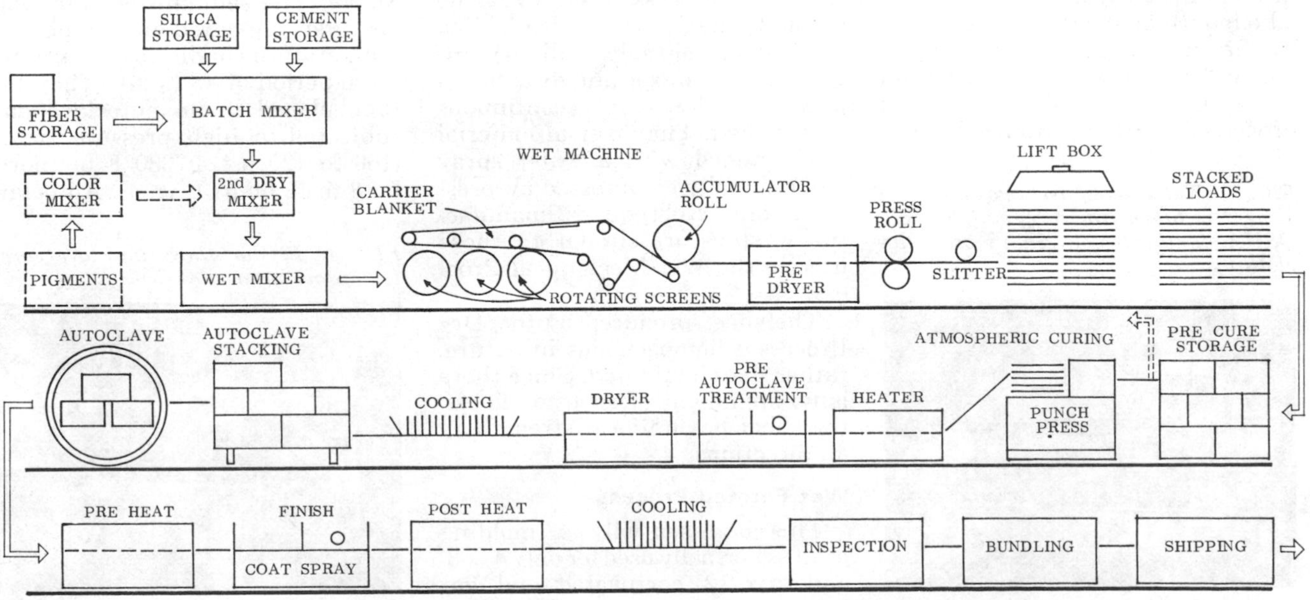

manufacture is a continuous process where asbestos fiber and portland cement are thoroughly mixed with water and formed into a plastic sheet. After the plastic sheet has hardened, it is trimmed to final size, punched with nailholes if required, and then cured. Color may be integrated with the material at the mixing stage, or it may be applied to the hardened sheet by a variety of coating processes. Textures can be embossed on the sheets while they are still in plastic form.

Wet Mechanical Process

The most commonly used method is a refinement of the original process devised in 1900, using a modified paper machine. A typical production flow diagram is shown in Figure 1. Asbestos fiber, portland cement (and silica added when products are to be cured by the autoclave method) are thoroughly dry-mixed. At the wet mixer, a controlled amount of water is added to produce a thin slurry mixed to a smooth consistency. The slurry is picked up in thin layers by rotating, cylindrical wire mesh screens, is deposited on a continuous carrier blanket in the wet machine, and then transferred to an accumulator roll. The wet solids from the slurry are permitted to build up on the accumulator roll layer by layer until the proper sheet thickness is reached. The length of sheets is determined by the circumference of the roller. The plastic sheet is removed from the accumulator roll (Fig. 2) and processed further through successive stages of manufacture, which produce the desired texture, shape, size and surface finish. Products may also be punched for nailholes or perforations. Prior to curing, discussed below, the plastic sheet may receive a texture on the accumulator roll, by a separate texture roller, or is pressed into flat or corrugated sheets.

FIG. 2 Wet sheet of asebestos fiber and cement is taken from accumulator roll. OSHA standards now require protective clothing.

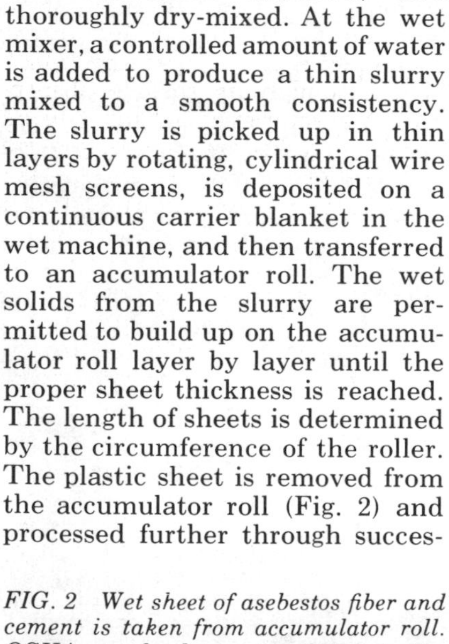

FIG. 3 Hydraulic press forms corrugated mineral fiber panels. OSHA standards now require protective clothing.

The Wet Process offers a high production rate and permits the addition of mineral pigments at the initial dry mixing stage. This process produces a built-up, laminated sheet with asbestos fibers aligned in one direction, giving the sheet significantly greater strength in the direction of fiber alignment.

Dry Process

In this process, used by some manufacturers to produce siding and roofing shingles, all dry ingredients are mixed and distributed in an even layer on a continuous moving belt. The layer of material is moistened with a water spray and gradually compressed by pressure rolls. Subsequent manufacturing steps are similar to those used in the Wet Mechanical Process.

The sheet produced by the Dry Process is homogeneous in nature, rather than laminated. Since there is no alignment of asbestos fibers, the sheet has uniform strength in all directions.

Wet Formed Process

This is essentially a molding process normally used for only 4′ x 8′ and 4′ x 12′ corrugated and flat sheets that range from 3/8″ to 2″ in thickness. A wet slurry of portland cement and asbestos fibers is poured into the mold of a hydraulic press. Hydraulic pressure forces out most of the water and compresses the mixture to the required density (Fig. 3). As in the Dry Process, the asbestos fibers are nonaligned and the sheet produced has uniform strength in all directions.

Curing

In all three manufacturing processes, curing of the semi-finished product is one of the most important steps. Any product made with portland cement becomes harder and stronger with age and proper moisture conditions. Proper curing accelerates aging, assures better dimensional stability, higher strength and other properties. Two methods are used to cure mineral fiber products: (1) *normal* curing, or (2) *autoclave curing*, using high pressure steam.

Normal Curing (Atmospheric Pressure) As the formed semifinished products leave the production line in a plastic state, they are stacked in a curing room and allowed to hydrate in air at a predetermined temperature and humidity for approximately a month. Steam at atmospheric pressure is sometimes used to provide the required warm moist air.

Autoclave Curing (High Pressure Steam) The semifinished products are first pre-cured under normal temperature conditions for a minimum period of 48 hours. They are then placed in an autoclave and subjected to high pressure steam (100 to 120 psi at 330°F or more) for 8 to 12 hours (Fig. 4). Autoclav-

FIG. 4 Products are autoclave cured to improve properties.

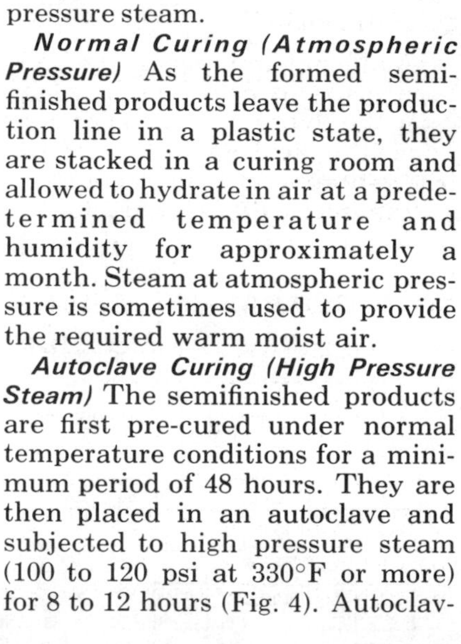

ing greatly reduces curing time, allowing products to be placed in inventory sooner.

FINISHES

Products may be unpigmented and left just as they come from the production line with a natural finish, or one of several color finishes may be applied.

Natural Finish

Natural finished products have the characteristic light gray color of portland cement products, and provide a highly suitable finish, especially for exterior use. The natural material is chemically inert, unaffected by moisture and does not require painting or other preservative treatment. The natural material provides a suitable base for paint, should color be desired.

Color Finishes

Significant improvements have been made in color stability and color durability of integral factory applied or field applied color finishes on mineral fiber products.

Integrated Pigments Mineral pigments may be introduced at the dry mixing stage during initial manufacture. Color is integrated with the base materials either throughout the entire sheet or through the topmost laminate of the sheet. A clear plastic coating may be baked over this color base.

Imbedded Granules In this factory process, pre-colored ceramic or quartz granules are embedded in the top surface of the soft plastic sheet before curing.

Paint and other Coatings *Field applied* exterior paints for use on mineral fiber are generally rubber resin base and latex masonry paints. For interior surfaces the same types of paint used on plaster surfaces are suitable.

Most *factory applied* finishes are acrylic resin coatings, in emulsion form or as solvent-thinned acrylic lacquers, and produce stable finishes. These finishes are easy to wash and resist staining.

PROPERTIES

A summary of selected properties of mineral fiber building materials is shown in Figure 5.

As described above, mineral fiber building products are produced essentially from a combination of asbestos fibers and portland cement. The strength properties of the combination far exceed those of cement or concrete materials alone. The asbestos fibers provide greatly increased tensile strength and flexibility and permit fabrication of the material into thin sheet form.

Characteristics of the finished products can be modified by: (1) increasing the ratio of asbestos fiber to portland cement, making the product more workable; (2) replacing a portion of the asbestos fiber with organic fibers, such as wood fibers, making the product lighter in weight, more flexible and easier to work; (3) increasing the length or grade of asbestos fiber, resulting in a product that is stronger and more flexible.

In general, the material is resistant to alkalies and acids, inert and durable. It may be left unfinished and is virtually maintenance free, although its surface absorption characteristics permit it to oil stain and hold dirt. However, it can be easily cleaned (see page 219-9) and, as explained above, under

FIG. 5 **SUMMARY OF SELECTED PHYSICAL PROPERTIES OF MINERAL FIBER BUILDING MATERIALS**

PROPERTY	FLAT SHEETS Rigid (Type U)	FLAT SHEETS Flexible (Type F)	SIDING	ROOFING	CORRUGATED Standard (Type A)	CORRUGATED Lightweight (Type B)
DENSITY (lbs/cu ft)	93 to 110	100 to 120	95 to 120	95 to 125	105 to 120	100 to 120
MODULUS of ELASTICITY (psi) (average of two directions)	1.35×10^6 to 2.00×10^6	1.5×10^6 to 2.4×10^6	1.3×10^6 to 3.2×10^6	1.5×10^6 to 3.2×10^6	1.2×10^6 to 3.0×10^6	1.2×10^6 to 2.3×10^6
SECTION MODULUS (in^3/Lin. ft.)	—	—	.037 to .055	.051 to .065	.90 to 1.0	.48 to .54
MOMENT OF INERTIA ($in.^4$/Lin ft.)	—	—	.0025 to .0045	.0041 to .0058	.64 to .72	.29 to .34
COMPRESSIVE STRENGTH (psi)	10,500 to 14,000	13,000 to 27,000	9,000 to 12,000	10,000 to 13,000	13,000 to 16,000* 35 to 45**	13,000 to 16,000* 10 to 13**
TENSILE STRENGTH (psi)	1,400 to 2,000	1,500 to 3,300	1,000 —	1,000 —	2,000 to 4,000	2,000 to 4,000
SHEAR STRENGTH	3,000 to 4,000	3,500 to 4,500	1,500 —	1,500 —	2,000 to 3,000	—
THERMAL COEFFICIENT OF EXPANSION (in/in/°F)	4.5×10^{-6} to 5.5×10^{-6}					4×10^{-6} to 6×10^{-6}
THERMAL CONDUCTIVITY "k" (Btu/hr/sq ft/°F/in)	—	3.4 to 4.2	3.5 to 4.5	3.5 to 4.5	3.5 to 4.5	3.5 to 4.5
COEFFICIENT OF MOISTURE EXPANSION (in/in) Dry to Wet	.0025 to .0040	.0013 to .0030	.0010 to .0018	.0004 to .0018	.0015 to .0028	.002 to .003
Normal to Wet	.0004 to .0010	.0004 to .0010	.0001 to .0006	.0004 to .0006	.0005 to .0016	.0005 to .0016

*Parallel to corrugations
**Perpendicular to corrugations

Finishes, it may be finished with a variety of paints and protective coatings.

Mineral fiber products should be carefully handled during transportation, storage and application in order to keep them free of discoloration from excess moisture (Fig. 6). They should also be temporarily wetted, enclosed or ventilated when removed from shipping containers in order to prevent the formation of asbestos dust. Care should also be exercised in installing mineral fiber products. Whenever asbestos sheet products are drilled, sawed, shaped or sanded, the dust must be prevented from contaminating the atmosphere either by dampening the product or by capturing the dust. Gloves, respirators and protective clothing should be worn when handling mineral fiber products.

FIG. 6 Mineral fiber products should be kept dry and stored flat.

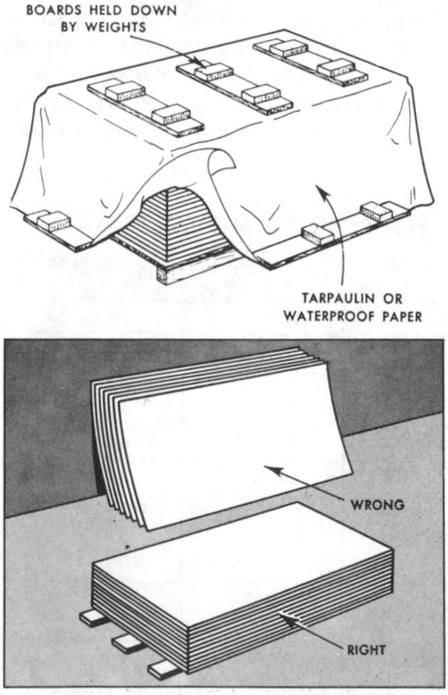

Fire Resistance

All mineral fiber building products are non-combustible. The importance of their fire resistance is indicated by the Underwriters' Laboratories, Inc. use of mineral fiber sheet as the primary standard for determining the fire rating of all other materials. Mineral fiber products are rated 0-0-0, indicating zero flame spread, zero smoke development, zero fuel contribution. By comparison, red oak is rated 100-100-100.

FIG. 7

APPROXIMATE MINIMUM BENDING RADII

THICKNESS	LONGITUDINAL	TRANSVERSE
	TYPE U	
3/16″	5′3″	7′0″
1/4″	7′0″	10′5″
3/8″	10′5″	13′0″
	TYPE F	
1/8″	2′2″	2′8″
3/16″	3′0″	4′5″

Dimensional Stability

The coefficient of *thermal* expansion (increase in length per unit of length for a temperature change of 1° F) of mineral fiber material is about that of concrete, approximately 5.0×10^{-6} inches per inch per degree F. For example, an unrestrained piece of mineral fiber 10′ long will expand or contract 0.06″ (approximately 1/16″) in a 100° F temperature change.

The material has a coefficient of *moisture* expansion of about 0.001″ to 0.003″ per inch from a dry to a saturated condition and 0.0001″ to 0.0016″ per inch from a normal condition of use to a wet condition. Maximum expansion movement in a 4′ x 8′ panel to be expected from a normal state to wet would approximate 1/16″ in width and 1/8″ in length.

Nailing

Because of the durability of mineral fiber building material and to prevent staining, nails and fasteners used in applying the material should be permanently corrosion-resistant and non-staining. Nail types and sizes for application of mineral fiber products are given in Section 400—Applications and Finishes.

Workability

Mineral fiber sheets require relatively long bending radii (Fig. 7), although curved sheets are made on special order.

FIG. 8 Roof and wall shingles are typically cut with this shingle cutter. Protective gloves should be worn.

Methods and equipment used to cut, punch and notch mineral fiber products are discussed below.

Mineral fiber products should be worked in a way which will prevent contamination of the atmosphere with asbestos dust. Power tools should have dust control attachments (Fig. 11). In using hand tools, wetting the material will prevent dust formation.

Shingle Cutters Portable shingle cutting machines (Fig. 8) may be used to save time and labor for cutting, notching and punching all sizes of siding units that are nominally 3/16″ in thickness.

Score and Snap Any sharp scoring tool can be used, but special tungsten-carbide tipped tools are more durable and do a better job more quickly. The material is scored deeply on one, preferably both sides of the sheet. The scored line is then placed along a firm

FIG. 9 Mineral fiber sheets may be snapped off after scoring. Protective gloves should be worn.

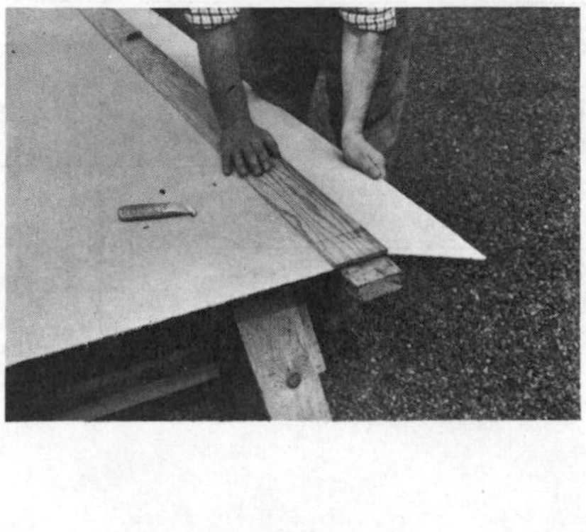

straight edge and snapped off (Fig. 9).

Sawing The best and quickest method of sawing is to use a portable power cut-off saw (Fig. 10a & b) equipped with one of the following blades:

Tungsten-Carbide Grit Discs are smooth edge, untoothed solid steel discs about 1/8″ thick with tungsten-carbide grit edge. This disc makes a quick clean cut with practically no disc breakage hazard.

Tungsten-Carbide Tipped Discs have tungsten-carbide tipped teeth and cut siding easily and quickly.

Flexible Abrasive Saw Discs are designed to eliminate the serious breakage hazard of thin abrasive discs. They are relatively flexible, thin discs made of resin saturated fabric, impregnated with abrasive grit and cut siding cleanly but slower than the discs described above. The use of thin, *rigid* grinding wheels on a power driven saw is very hazardous and is not recommended.

The most practical *hand* saw is one of the several styles of pistol-grip, coarse tooth, hack (metal) cutting saw, although a coarse tooth, cross-cut wood saw may be used in an emergency (Fig. 10c).

Drilling For easiest, most rapid drilling and long drill service-life, tungsten-carbide tipped twist drills should be used. High speed twist drills are also suitable and in an emergency any kind of twist drill can be used.

Punching An untapered metal cold punch such as a pin or drift punch, can be used, or the punch attachment on the portable shingle cutter can be used. Large holes or circular cuts can be made as illustrated in Figure 11.

Filing and Dressing To remove rough edges which may result when mineral fiber products are cut, especially when scored and snapped, a tungsten-carbide grit flat file, a coarse steel file or rasp, or a flat diagonal-grooved coarse abrasive stick should be used.

FIG. 10 Mineral fiber products may be cut by power saw (a & b) or by hand saw (c). OSHA standards now require protective clothing and dust control attachments on tools.

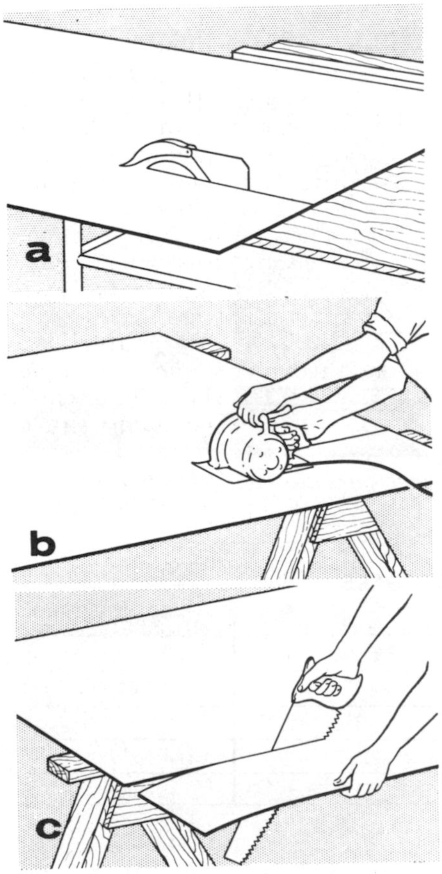

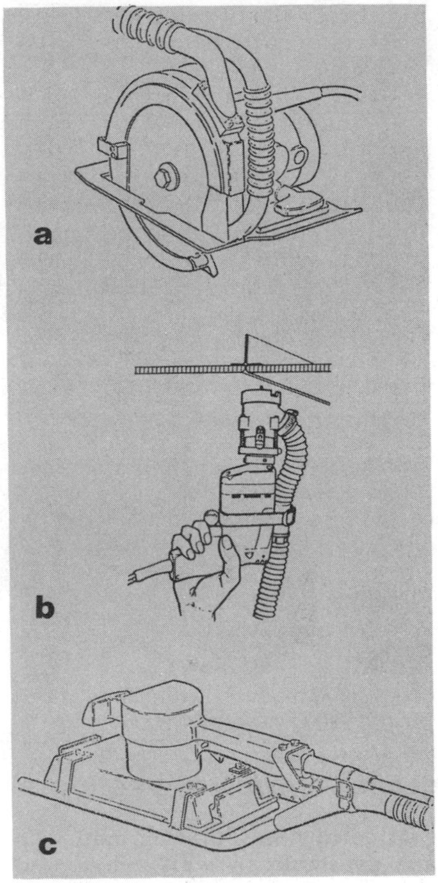

Fig. 11 Power tools can be readily equipped with vacuum attachments to capture asbestos dust: (a) saw, (b) drill (c) sander.

Paintability

Mineral fiber products generally do not require painting for protection from the weather. However, both unpainted and factory finished products can be painted satisfactorily.

Because of the alkaline content of the portland cement in mineral fiber products, paints used over these surfaces must be alkali-resistant. The paint applied must adhere well to the base and dry to a uniform appearance. Primers are used as a base coat to seal the surface and to promote adhesion of finish coats. For best appearance and maximum durability, one primer and two finish coats should be applied, although often one finish coat is sufficient. (See also Division 400 for Painting and Finishing).

FIG. 12 **MINERAL FIBER SIDING**

Sizes[1]	Exposure	Pieces per Bundle	Pieces per Square[2]
12″ x 24″	11″	18	54
9″ x 32″	8″	19	57
9-5/8″ x 32″	8-5/8″	17-17-18	52
11-3/4″ x 32″	10-3/4″	14	42
14-5/8″ x 32″	13-5/8″	11	33
14-3/4″ x 32″	13-3/4″	11	33
8-3/4″ x 48″	7-3/4″	13	39
12″ x 48″	11″	9	27

1. Three nailholes per shingle.
2. Weight per square ranges from 153 to 206 lbs.

FIG. 13 Sidings are available in many textures, styles, colors and finishes.

CLASSIFICATION OF PRODUCTS

Mineral fiber building materials are available in wall siding and shingles, roof shingles, flat sheets, insulating panels and corrugated sheets. Each is described below.

Wall Siding and Shingles

Siding products are available in a variety of styles, shapes, textures and colors (Figs. 12 & 13). Siding material textures are striated, wood grain and smooth. Finishes include: plastic coatings, plastic spray applied in the field, integral pigment with embedded mineral granules, integral pigment with clear plastic seal, vitreous ceramic and ceramic granules. Accessories, other than face nails and joint strips supplied with the siding, include matching aluminum corners, window and door molding and insulating backer board.

Mineral fiber siding and shingles should conform to ASTM C223. Recommendations for applying siding and shingles are given in Division 400.

Roofing Shingles

Similarly, roofing products come in a variety of types, shapes, textures and colors (Fig. 14 & 15). Textures are striated, wood grain and smooth. Finishes include plastic coatings and integral pigment with imbedded mineral granules. Accessories include storm anchors, starter shingles, hip and ridge shingles, and ridge roll.

Mineral Fiber roofing shingles should conform to ASTM C222. Recommendations for applying roofing shingles are given in Division 400.

Flat Sheets

Fabricated in sizes up to 4′ x 12′ in a range of thicknesses from 1/8″ to 2″, flat sheets are used both for interior and exterior applications (Figs. 16 & 17). Sheets are available in both smooth and textured surfaces and finishes of natural cement gray, preprimed or with pigmented plastic coatings. They are produced in perforated, rigid (Type U) and flexible (Type F) form. Accessories include battens (3/16″ and 1/4″ x 2″, 3″ and 4″), nails adhesives and fasteners.

FIG. 16 Flat sheets may be battened or left as large panels.

FIG. 14 A variety of mineral fiber roof shingles are available for residential and other applications.

FIG. 15 **MINERAL FIBER ROOF SHINGLES**

SIZES[1]	Exposure	Pieces per Bundle	Pieces per Square[2]
12″ x 24″	9″ to 9-1/2″	16, 20	80
14″ x 30″	6″	16	80
16″ x 8″	5″ to 7″	13	260
16″ x 16″	12″ to 13″	20, 21, 22, 23, 43,	86, 92
16-1/4″ x 16″	7″	13	130

1. 2 to 8 nailholes per shingle.
2. Weight per square ranges from 240 to 595 lbs.

FIG. 17 **MINERAL FIBER FLAT SHEETS**

TYPE	Nominal Thickness	Nominal Width	Length	Average Flexural Strength[1] (for given thickness)					
				1/8"	3/16"	1/4"	3/8"	1/2"	5/8"
TYPE F (Flexible)	1/8" to 1" (multiples of 1/8") 1-1/4" to 2" (multiples of 1/4")	48"	4', 8', 10', 12'	20	50	90	190	360	560
Type U (Utility)	3/16", 1/4", 3/8" 1/2", 5/8"	48"	4', 8', 10', 12'	—	35	65	145	260	400

1. Minimum average breaking load, lbs/6" width, span = 10".

FIG. 18 Sandwich panels are used for exterior or interior application. Workers should use dust controls on tools and wear protective clothing.

FIG. 19 **MINERAL FIBER INSULATING PANELS**

Thickness	Width	Length	Weight lb/sq. ft
11/16"	4"	6', 8', 9', 10', 12'	3.5
1-1/8"	4'	6', 8', 9', 10', 12'	3.9
1-9/16"	4'	6', 7', 8', 9', 10', 12'	4.7 to 5.4
2"	4'	6', 8', 9', 10', 12'	5.4

1. Insulation cores include wood fiberboard, foamed plastics, non-combustible inorganic expanded fibers or particles.

Mineral fiber flat sheets should conform to ASTM C220. Recommendations for applying flat sheets are given in Division 400.

Insulating Panels

Insulating panels are factory-formed by pre-assembling layers of insulating core material sandwiched between and laminated to conventional 1/8" thick flat sheets. Depending on the type of core material, the characteristic properties of the flat sheets, and the number of laminations within the panel, these *sandwich panels* may be used for exterior curtain walls, interior partitions and roof decking, and are available with square cut or rabbeted edges. They are available in natural gray, factory-primed and colored plastic finishes (Figs. 18 & 19). Mineral fiber fiberboard insulating panels should conform to ASTM C551.

Corrugated Sheets

Corrugated sheets are primarily used for industrial roof and wall applications but are used in special residential applications, such as bulkheads, to resist erosion of earth along lakesides and inland waterways. Most corrugated materials come in natural gray finish; however, some products are available in color. Accessories include ridge roll, corner roll, eave and gable trim, corrugated louver, corrugated filler strips of neoprene or rubber, fasteners and adhesives.

Standard sizes are shown in Figure 20. Asbestos-Cement corrugated sheets are produced in standard (Type A) and lightweight (Type B) thicknesses and should conform to ASTM C221.

Maintenance

In general, mineral fiber building materials require little care or maintenance. The natural-finish products do not require painting.

FIG. 20 **MINERAL FIBER CORRUGATED SHEETS**

TYPE[1]	Overall Thickness		Cross Section Minimum Thickness[2]		Weight (lbs/sq. ft.)	Maximum Span Structural[3]		Flexural Strength[4]
	Min.	Max.	(1)	(2)		Roofing	Siding	
Type A (Standard)	1-23/64"	1- 9/16"	11/32"	9/32"	3-3/4 to 4	4'6"	5'6"	500
Type B (Lightweight)	1- 3/16"	1-25/64"	11/64"	9/64"	1-3/4 to 2	2'6"	3'6"	200

1. Width: 42" (10 corrugations @ 4.2"); Length: 6" to 12'0" in increments of 6".
2. (1) crest and vale; (2) midway between crest and vale.
3. Distance from center to center of support.
4. Minimum average breaking load, lbs per ft. of width, span = 30".

They may, however, accumulate airborne dirt and are subject to normal staining.

Unfinished Surfaces

Surface dirt may be removed with soap, water and a stiff fiber brush. Tri-sodium phosphate, sodium metasilicate, or abrasive soap cleaners can also be used. Mold or mildew or fungi can be removed and/or inhibited with chlorine-containing household bleach, diluted about 10 to 1.

Copper stains may be removed with a paste consisting of one part ammonium hydroxide, 10 parts water, 16 parts talc, 4 parts ammonium chloride. Talc and ammonium chloride should be mixed thoroughly and applied with a trowel or stiff fiber brush. After 25 hours, the dried paste should be washed off.

Rust stains may be removed by washing the surface with a 5% solution of phosphoric acid or 2% solution of oxalic acid, followed by a thorough rinsing with clear water.

Wet paint should be removed immediately by a cloth soaked in turpentine or paint thinner. Dry paint spots can be scraped off with a knife blade and the area wiped with a cloth soaked in turpentine or paint thinner.

Oil and grease may be removed with a paste consisting of 2-1/2 lbs of whiting mixed with one quart of carbon tetrachloride, benzine or naphtha. The paste should be applied with a trowel and wiped off with a clean dry cloth.

Asphalt and tar stains should be wiped with a cloth soaked in turpentine or other paint thinner. If the stains do not disappear entirely, weathering will reduce their intensity.

Bloom (a surface deposit of white lime salt similar to efflorescence on brickwork) may be washed off with 5% to 10% muriatic (hydrochloric) acid solution, applied with a bristle brush or spray. Only a small area should be washed at a time, then rinsed with clear water within several minutes and allowed to dry. Full shingles should be washed to avoid spotty appearance. Surfaces which will show bloom after acid washing may be improved by dry brushing in the textured direction with a stiff fiber brush. If bloom still remains, the acid treatment may be repeated.

Finished Surfaces

Abrasive cleansers or solvents for removing dirt should not be used on mineral fiber products that have pigmented plastic or painted finishes. Products with a plastic surface coating require special treatment to remove paint stains. Commercial paint removers or high reactive solvents should not be used. Paint stains should be removed immediately. Dry paint stains may be scraped off with a knife, exercising care to prevent damage to surface coating. Touch-up paint, matching the original, is generally available from the mineral fiber product manufacturer.

We gratefully acknowledge the assistance of the Johns Manville Corporation in the Preparation of this text and in supplying photographs and illustrations.

ASBESTOS CEMENT PRODUCTS 219

CONTENTS

BUILDING MATERIALS

MATERIALS

Mineral fiber products are made by combining asbestos fibers and portland cement with water, to form into rigid sheet or pipe shapes. The Occupational Safety & Health Act (OSHA) Asbestos Standard 1910.93a governs the manufacture, working, handling and installation of asbestos products.

MANUFACTURE

See Main Text page 219-3 for description of manufacturing processes employed in the production of mineral fiber products.

FACTORY FINISHES

Products may be unpigmented and left just as they come from the production line with a natural gray finish, or one of several color finishes may be applied. Painting is not required for weather protection. The natural material is chemically inert, unaffected by moisture and does not require painting or other preservative treatment, but does provide a suitable base for paint, should color be desired.

During manufacture, color may be integrated with the base materials either throughout the entire sheet or through the topmost laminate of the sheet. A clear plastic coating may be baked over this color base.

Pre-colored ceramic or quartz granules may be embedded in the top surface of the soft plastic sheet before curing in this factory process.

Factory applied finishes are acrylic resin coatings in emulsion form or solvent-thinned acrylic lacquers; these produce stable finishes, which are easy to wash and resist staining.

PROPERTIES

Mineral fiber products are strong, hard, resistant to alkalies and acids, inert and durable (see Fig. 5 Summary of Selected Physical Properties of Mineral Fiber Building Materials, Main Text page 219-5).

Mineral fiber products should be handled carefully during transportation and application, and stored properly to keep them dry in the warehouse and on the job.

Mineral fiber products should be handled carefully during transportation and application, and stored properly to keep them dry in the warehouse and on the job. Products should be wetted for removal from shipping cartons to prevent the formation of asbestos dust. When handling and working with mineral fiber products, care should be taken to avoid contaminating the atmosphere with asbestos dust. Gloves and other protective clothing should be worn. Products should be wetted before handling and hand and power tools should be equipped with dust control devices.

They can be discolored by moisture and dampness in storage.

All mineral fiber building products are non-combustible and are rated 0-0-0, indicating zero flame spread, zero smoke development, zero fuel contribution as compared to red oak, rated 100-100-100.

The coefficient of thermal expansion (increase in length for a temperature change of 1°F) of mineral fiber material is about that of concrete, approximately 5.0 x 10^{-6} inches per inch per degree F.

The material has a coefficient of moisture expansion of about 0.0001″ to 0.0016″ per inch from a normal

(Continued) **BUILDING MATERIALS**

condition of use to a wet condition. Maximum expansion movement in a 4' x 8' panel from a normal to a wet state would approximate 1/16" in width and 1/8" in length.

Nails and fasteners used in applying mineral fiber products should be permanently corrosion-resistant and non-staining.

Mineral fiber products can be stained by rust marks from unprotected ferrous materials.

CLASSIFICATION OF PRODUCTS & STANDARDS OF QUALITY

Mineral fiber building products should conform to ASTM Specifications as shown in Figure WF 1.

Figure WF 1 also shows the variety of textures and finishes in which the products are available.

Materials also are available in a variety of types and sizes: wall siding and shingles, roof shingles, insulating panels (Fig. WF 2); flat sheets and corrugated sheets (Fig. WF 3).

FIELD PAINTING

Paint applied to unfinished mineral fiber product surfaces should adhere well to the base, should dry to a uniform appearance and should be alkali resistant.

Alkali resistant paints are required because of the high alkaline content of the portland cement in the material.

Though one finish coat often is sufficient, one primer and two finish coats should be applied for best appearance and maximum durability.

FIG. WF1 **CLASSIFICATION OF MINERAL FIBER PRODUCTS & STANDARDS OF QUALITY**

Product & ASTM Specification	Textures	Finishes
Wall Siding & Shingles ASTM C223	striated; wood grain; smooth	plastic coatings; plastic spray applied in field; integral pigment with clear plastic seal; vitreous ceramic; ceramic granules
Roofing Shingles ASTM C222	striated; wood grain; smooth	plastic coatings; integral pigment with imbedded mineral granules
Flat Sheets ASTM C220	smooth; textured	natural cement gray; pre-primed; pigmented plastic coatings
Insulating Panels ASTM C551	square cut edges; rabbeted edges	natural gray; factory-primed; colored plastic finishes
Corrugated Sheets ASTM C221	smooth	natural gray; color

FIG. WF2 **SIDING, SHINGLES AND INSULATING PANELS**

MINERAL FIBER SIDING			
Sizes[1]	Exposure	Pieces per Bundle	Pieces per Square[2]
12" x 24"	11"	18	54
9" x 32"	8"	19	57
9-5/8" x 32"	8-5/8"	17-17-18	52
11-3/4" x 32"	10-3/4"	14	42
14-5/8" x 32"	13-5/8"	11	33
14-3/4" x 32"	13-3/4"	11	33
8-3/4" x 48"	7-3/4"	13	39
12" x 48"	11"	9	27

MINERAL FIBER ROOF SHINGLES			
Sizes[3]	Exposure	Pieces per Bundle	Pieces per Square[4]
12" x 24"	9" to 9-1/2"	16, 20	80
14" x 30"	6"	16	80
16" x 8"	5" to 7"	13	260
16" x 16"	12" to 13"	20, 21, 22, 23, 43	86, 92
16-1/4" x 16"	7"	13	130

MINERAL FIBER INSULATING PANELS[5]			
Thickness	Width	Length	Weight lb/sq. ft
11/16"	4"	6', 8', 9', 10', 12'	3.5
1-1/8"	4'	6', 8', 9', 10', 12'	3.9
1-9/16"	4'	6', 7', 8', 9', 10', 12'	4.7 to 5.4
2"	4'	6', 8', 9', 10', 12'	5.4

1. Three nailholes per shingle.
2. Weight per square ranges from 153 to 206 lbs.
3. 2 to 8 nailholes per shingle.
4. Weight per square ranges from 240 to 595 lbs.
5. Insulation cores include wood fiberboard, foamed plastics, non-combustible inorganic expanded fibers or particles.

(Continued) **BUILDING MATERIALS**

FIG. WF3 FLAT AND CORRUGATED SHEET PRODUCTS

MINERAL FIBER FLAT SHEETS

Type	Nominal Thickness	Nominal Width	Length	Average Flexural Strength[1] (for given thickness) 1/8″	3/16″	1/4″	3/8″	1/2″	5/8″
Type F (Flexible)	1/8″ to 1″ (multiples of 1/8″) 1-1/4″ to 2″ (multiples of 1/4″)	48″	4′, 8′, 10′, 12′	20	50	90	190	360	560
Type U (Utility)	3/16″, 1/4″, 3/8″ 1/2″, 5/8″	48″	4′, 8′, 10′, 12′	—	35	65	145	260	400

MINERAL FIBER CORRUGATED SHEETS

Type[2]	Overall Thickness		Cross Section Minimum Thickness[3]		Weight (lbs/sq. ft.)	Maximum Span Structural[4]		Flexural Strength[5]
	Min.	Max.	(1)	(2)		Roofing	Siding	
Type A (Standard)	1-23/64″	1- 9/16″	11/32″	9/32″	3-3/4 to 4	4′6″	5′6″	500
Type B (Lightweight)	1- 3/16″	1-25/64″	11/64″	9/64″	1-3/4 to 2	2′6″	3′6″	200

1. Minimum average breaking load, lbs/6″ width, span = 10″.
2. Width: 42″ (10 corrugations @ 4.2″); Length: 6″ to 12′0″ in increments of 6″.
3. (1) crest and vale; (2) midway between crest and vale.
4. Distance from center to center of support.
5. Minimum average breaking load, lbs per ft. of width, span = 30″.

FIG. WF4 MAINTENANCE OF UNFINISHED ASBESTOS-CEMENT SURFACES

Type of Stain	Solvent	Method of Removal
Surface Dirt	soap and water; or tri-sodium phosphate; or sodium metasalicate; or abrasive soap cleaners;	use a stiff fiber brush and any of the solutions mentioned
Mold, Mildew, Fungi	chlorine-containing household bleach 10:1	
Copper Stains	ammonium hydroxide—1 part water—10 parts talc—16 parts ammonium chloride—4 parts	mix thoroughly and apply with trowel or stiff fiber brush; wash off dried paste after 24 hours
Rust Stains	5% solution phosphoric acid, or 2% solution oxalic acid	wash with solution, followed by thorough rinsing in clear water
Wet Paint	turpentine or paint thinner	soak cloth in solution and wipe immediately
Dry Paint	turpentine or paint thinner	first scrape with knife blade, then wipe with solution
Oil and Grease	paste consisting of 2½ lbs. whiting mixed with 1 quart carbon tetrachloride, benzine or naphtha	apply with trowel and wipe with clean dry cloth
Asphalt and Tar	turpentine or paint thinner	wipe with cloth soaked in solution; weathering also reduces intensity of stain
Bloom	5% to 10% muriatic (hydrochloric) acid solution	apply to small area at a time with a bristle brush or spray; then rinse in clear water within minutes; wash entire surface afterwards to avoid spots; dry brush in textured direction with stiff fiber brush.

(Continued) **BUILDING MATERIALS**

Primers are used as a base coat to seal the surface and to promote adhesion of finish coats.

Field applied exterior paints for use on mineral fiber are generally rubber resin base and latex masonry paints. For interior surfaces the same types of paint used on plaster surfaces are suitable.

MAINTENANCE

Mineral fiber building materials require little care or maintenance. Though natural finish products do not require painting, they may accumulate airborne dirt and are subject to normal staining. See Figure WF 4 for Maintenance of Unfinished Asbestos Cement Surfaces.

Mineral fiber products that have pigmented plastic or painted finishes should not be maintained with abrasive cleansers or solvents.

Products with a plastic surface coating require special treatment to remove paint stains.

Commercial paint removers or strong solvents should not be used. Paint stains should be removed immediately.

220 RESILIENT FLOORING PRODUCTS

RESILIENT FLOORING PRODUCTS

INTRODUCTION

Resilient flooring is a class of flooring products distinguished by resilience and dense, nonabsorbent surfaces. Resilience contributes to indentation resistance and quietness, while the dense surface assures durability and relative ease of maintenance. A variety of decorative colors, patterns and textures are available in sheet or tile form.

Availability of self-adhesive tile has made installation of some products easier. No-wax surfaces on both tile and sheet goods have reduced maintenance.

Linoleum, discovered a little over 100 years ago in England, was the forerunner of contemporary resilient flooring. It consisted of a blend of oxidized linseed oil binders and cork flour fillers, bonded to a burlap or felt backing. *Asphalt tile* evolved after World War I from a trowled-on mastic flooring which consisted of a mixture of asphalt, asbestos fibers and naphtha solvents.

Both linoleum and asphalt tile are no longer produced in this country, having been largely replaced by sheet vinyl and vinyl asbestos tile. The real revolution in resilient flooring is less than 30 years old and dates to the introduction of *vinyl-asbestos* and *vinyl* flooring products, whose excellent wearing properties and decorative potential substantially increased the range of resilient flooring uses. Today, a variety of resilient flooring products is available for residential and non-residential construction.

This section describes the basic types of flooring products, their manufacture and properties and selection criteria. Installation of flooring products is discussed in Division 400—Applications & Finishes.

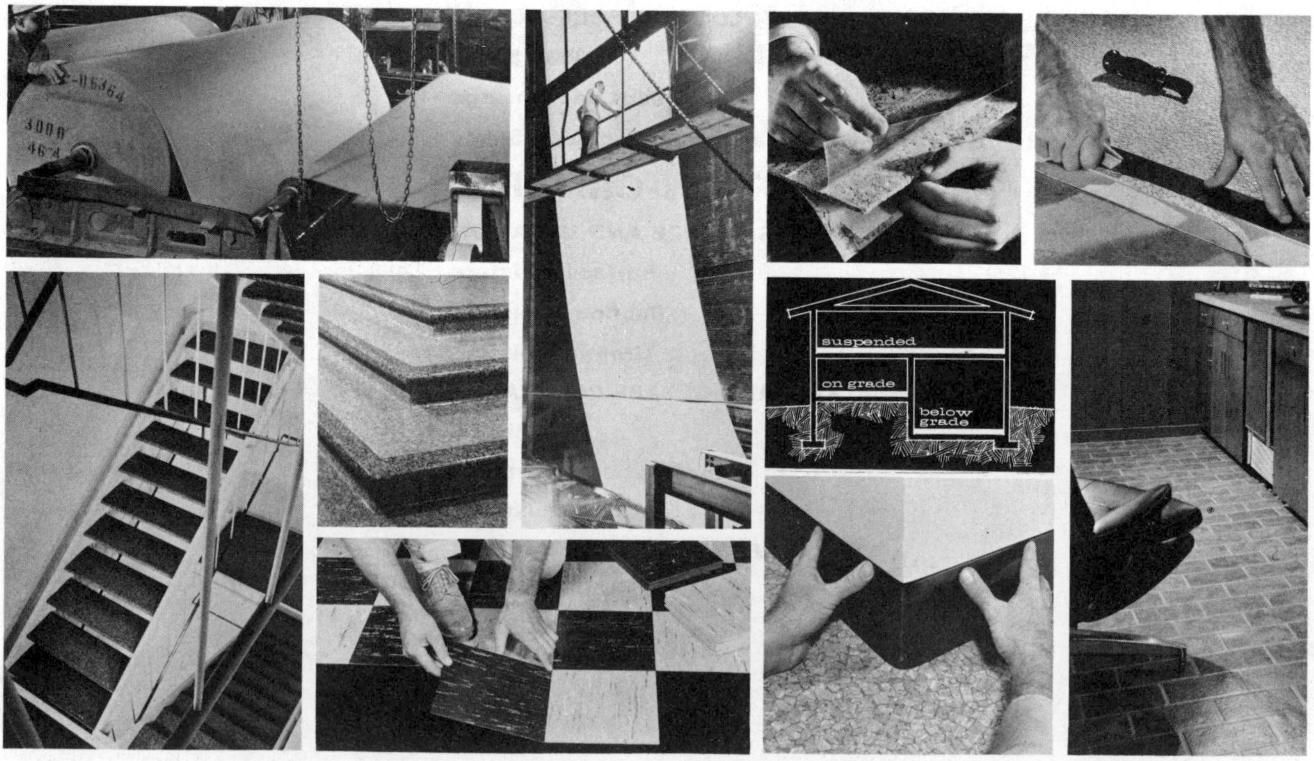

Resilient flooring is manufactured in *tile* and *sheet* forms, in different thicknesses and from a variety of ingredients. Tile sizes range from 9″ x 9″, 12″ x 12″ up to 36″ x 36″; sheet generally is made in 6′ and 12′ widths, although some products are produced in 4′6″ widths. Matching resilient accessories such as wall base, thresholds, stair treads and feature strips, are available to complete flooring installations (Fig. 1).

Resilient flooring products may be classed according to basic ingredients into (1) *vinyl,* (2) *vinyl-asbestos,* (3) *rubber* and (4) *cork.* Many resilient flooring products are manufactured to meet requirements of Federal Specifications, which constitute the industry-recognized standards of quality. In some types, the basic ingredients run the full depth of the flooring; in others, backing material adds desired physical characteristics or reduces cost. Figure 3 summarizes the ingredients and backing materials, sheet widths, tile sizes, gauges and applicable Federal Specifications.

FIG. 1 (a) Feature strips permit rectangular custom design in sheet or tile; (b) unusual curved designs are possible with sheet flooring.

VINYL-ASBESTOS TILE

Vinyl-asbestos tile is composed of asbestos fibers, ground limestone, plasticizers, pigments and *polyvinyl chloride* (PVC) resin binders. These resins permit light colors and a great variety of designs. Some tile products are available with self-adhesive backs and "no-wax" surfaces.

Ingredients are mixed under heat and pressure and rolled into blankets to which decorative chips may be added. Additional rolling and calendering produce smooth sheets of desired thickness. After waxing, sheets are cut into tiles (Fig. 2).

VINYL SHEET AND TILE

The chief ingredient of vinyl products is *polyvinyl chloride* (PVC) resin. Other ingredients include mineral fillers, pigments, plasticizers and stabilizers. Plasticizers provide flexibility; stabilizers fix the mixture to assure color stability and uniformity. Vinyl products sometimes are referred to as *flexible* vinyl to distinguish them from vinyl-asbestos products, which also are made with PVC resins and are termed *semi-flexible* vinyl.

Vinyl Sheet

Vinyl sheet products are made with a vinyl wear surface bonded to a backing. The backings may be vinyl, polymer impregnated asbestos fibers, or asphalt-saturated or resin-saturated felt. Many new vinyl products have a layer of vinyl foam bonded either to the backing or between the wear surface and the backing.

A type of sheet flooring often made with an asphalt-saturated felt backing and a thin wear surface (less than 0.014″) over an imprinted image is not suitable for permanent adhesive installation. This "floor covering" typically is made in standard rug sizes (9′ x 12′, 12′ x 12′, 12′ x 15′) and often is laid loose on the floor.

The wear surface of sheet vinyl may be either *filled* or *clear* (unfilled) vinyl (Fig. 4a & b).

Filled Vinyl In addition to PVC resins, the wear surface of filled vinyl-surfaced sheets may contain decorative vinyl chips, filler, pigments and other ingredients. Powdered vinyl resins, fillers, plasti-

FIG. 2 Vinyl-asbestos tile being inspected for quality of finish and thickness prior to packaging.

FIG. 3 **SUMMARY OF RESILIENT FLOORING PRODUCTS**

Type	Federal Specification	Backing Materials	Gauges**	Sizes	Basic Ingredients
ASPHALT TILE†					
Plain	SS-T-312B	—	1/8", 3/16"	9" x 9", 12" x 12"*	Asphaltic and/or resin binders, asbestos fibers and limestone fillers.
Greaseproof	SS-T-312B				
VINYL-ASBESTOS TILE	SS-T-312B	—	1/16", .080", 3/32", 1/8"	9" x 9", 12" x 12"*	PVC resin binders, limestone, asbestos fibers.
VINYL TILE					
Solid	SS-T-312B	—	1/16", .080" 3/32", 1/8"	9" x 9", 12" x 12" 4" x 36"*	PVC resin binders and mineral fillers.
Backed	SS-T-312B	Scrap vinyl, asbestos or rag felts	.050" to .095"	9" x 9", 12" x 12" & sizes up to 36" x 36"*	
VINYL SHEET***					
Filled Surface	L-F-475-a L-F-001641	Scrap vinyl, asbestos or rag felts	.065" to .160" (.010" to .050" wear surface)	6'0", 12'0" & 4'6"* wide	PVC resin binders and mineral fillers; clear PVC film for clear vinyl surface.
Clear Vinyl Surface			.065" to .160" (.010" to .050" wear surface)	6'0" & 12'0" wide	
RUBBER TILE	SS-B312B	—	.080", 3/32", 1/8", 3/16"	9" x 9", 12" x 12" 18" x 36"*, 36" x 36"*	Synthetic or natural rubber, mineral fillers.
CORK TILE					
Protected Surface	Interim LLL-T-00431 (Class 1)	—	1/8", 3/16", 1/4" 5/16"*, 1/2"*	9" x 9", 6" x 6"* 6" x 12"*, 12" x 12"*	Cork particles and resin binders, wax or resin finish.
Vinyl Surface	Interim LLL-T-00431 (Class 2)	—	1/8", 3/16"	9" x 9", 12" x 12"*	Cork particles and resin binders, clear vinyl finish.
LINOLEUM SHEET†			(.050" wear surface)	6'0" wide	Cork and/or wood flour with linseed oil binders.
Plain & Marbleized	LLL-F-1238A	Burlap	1/8"		
Inlaid & Molded	LLL-F-1238A	Rag felt	.090"		
	LLL-F-1238A	Burlap	1/8"		
	LLL-F-1238A	Rag felt	.090"		
Battleship	LLL-F-1238A	Burlap	1/8"		
LINOLEUM TILE†			(.050" wear surface)	9" x 9", 12" x 12"*	Cork and/or wood flour with linseed oil binders.
	LLL-F-1238A	Burlap	1/8"		
	LLL-F-1238A	Rag felt	.090"		

*Not common, available on special order from some manufacturers.
**Thinnest gauges and wear surfaces indicated are minimums suitable for adhesive installation.
***Also produced with foam backing, self-adhesive back and no-wax surface; not included in data shown.
†Asphalt and linoleum products are no longer produced in this country, but are included here for reference.

cizers, stabilizers and pigments are mixed, rolled into sheets and chopped to form vinyl chips. Vinyl chips of various colors are mixed with additional resins and spread evenly over the backing and bonded to it under high heat and pressure (Fig. 5). Minimum recommended wear surface for residential use is 0.020″ and overall thickness without foam backing is 0.065″. Foam cushioned products may range up to 3/16″ in thickness.

Clear (Unfilled) Vinyl Clear vinyl-surfaced sheets contain the decorative ingredients under a layer of clear vinyl. Because there are no fillers in the wear surface, abrasion resistance and durability are greater per unit of thickness of the wear surface than in filled vinyl-surfaced sheets. The clear vinyl wear surface should be at least 0.010″ thick. Overall gauge, without foam backing, should be at least 0.065″.

Resin-saturated felt and asbestos backings are coated with pigmented vinyl (Fig. 6), and designs are imprinted with vinyl inks. Sheet products with vinyl backings may have the design imprinted on top of the backing or on the underside of the wear surface. The clear PVC wear surface is calendered to desired thickness and laminated to the backing with heat and pressure.

Vinyl Tile

Vinyl tile may be of homogeneous *solid* composition or may be *backed* with other materials such as or-

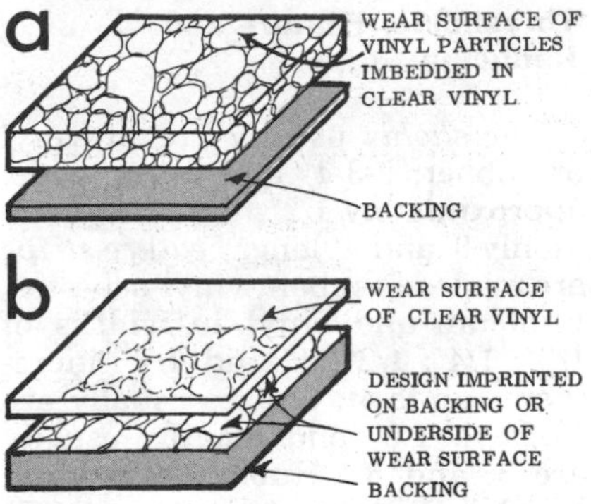

FIG. 4 Backed vinyl cross section: (a) filled vinyl, (b) clear vinyl.

FIG. 5 Vinyl ingredients are keyed in a series of presses to make vinyl sheet flooring.

FIG. 6 Felt backing is coated to form a light, smooth background for printed designs for vinyl sheet.

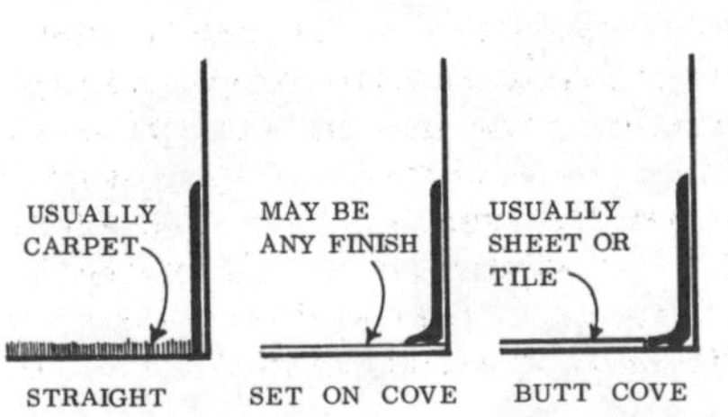

RESILIENT WALL BASE TYPES

FIG. 7 Straight, set-on and butt type wall bases are available in rubber or vinyl in 2-1/2", 4" and 6" heights.

ganic felts, asbestos fibers or scrap vinyl.

Ingredients for solid vinyl tiles are mixed at high temperature, hydraulically pressed and/or calendered into homogeneous sheets of required thickness, and sheets are cut into tile sizes. Backed products are essentially vinyl sheet flooring cut into tile sizes. Some tiles are made with a self-adhesive back.

RUBBER TILE

Natural or synthetic rubber is the basic ingredient of rubber flooring. Clay and fibrous talc or asbestos fillers provide the desired degree of reinforcement; oils and resins are added as plasticizers and stiffening agents. Color is achieved by nonfading organic pigments, and chemicals are added to accelerate the curing process.

The ingredients are mixed thoroughly and rolled into colored sheets. The sheets are calendered to uniform thickness and vulcanized in hydraulic presses under heat and pressure into compact, flexible sheets with a smooth, glossy surface. The backs then are sanded to gauge, insuring uniform thickness, and sheets are cut into tiles.

CORK TILE

Cork tile is composed chiefly of the granulated bark of the cork oak tree, native to Spain, Portugal and North Africa. Synthetic resins are added to the granulated cork, pressed into sheets or blocks and baked. Surfaces are finished with a protective coat of wax, lacquer or resin applied under heat and pressure. Sheets then are cut to tile sizes. *Vinyl* cork tile has a film of clear PVC vinyl fused to the top surface to improve durability, water resistance and ease of maintenance.

FLOORING ACCESSORIES

The most common resilient accessories used with flooring materials include: (1) *wall bases,* (2) *stair treads* and *stair nosings,* and (3) *thresholds* and *feature* and *reducing strips.* These accessories are available in more limited color

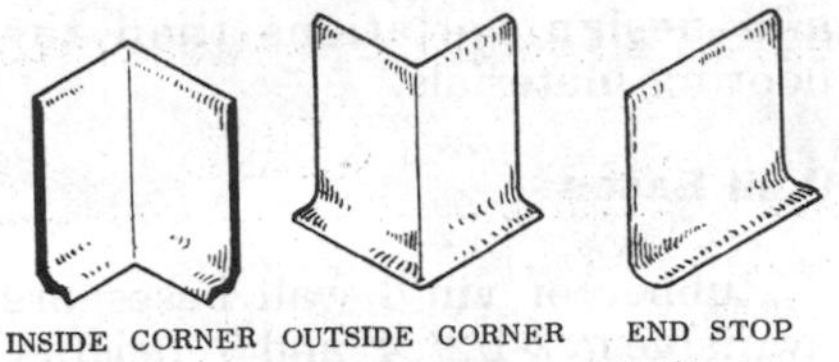

PREMOLDED BASE ACCESSORIES

FIG. 8 Set-on cove base, available in both rubber and vinyl, is the most widely used type of base.

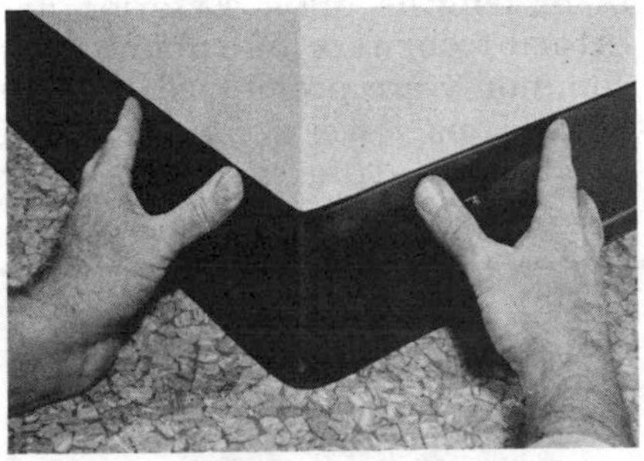

FIG. 9 Flexible vinyl base, capable of being molded around corners, is replacing premolded rubber corners.

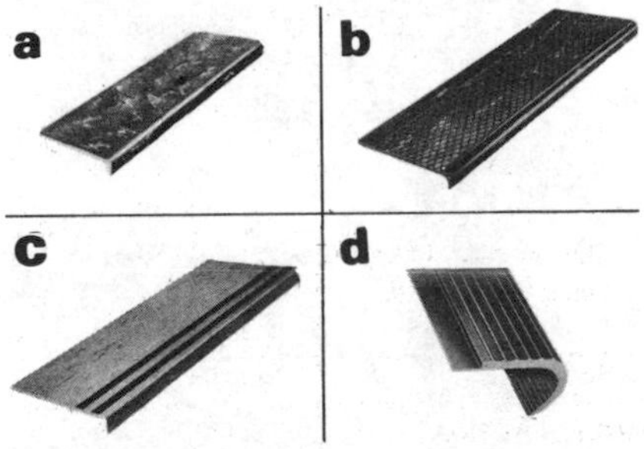

FIG. 10 Stair treads: (a) smooth, (b) textured, (c) abrasive grit. Smooth or ribbed nosings (d) may be used with sheet or tile treads.

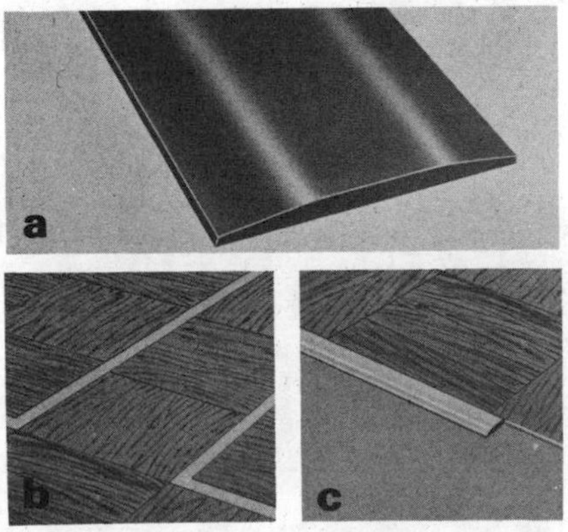

FIG. 11 Resilient accessories: (a) threshold, (b) feature strips, (c) edging strips.

and design variations than the flooring materials.

Wall Bases

Rubber or vinyl wall bases are available in 2-1/2″, 4″ and 6″ heights, in *straight* type and either *set-on* or *butt* types with coved-bottom designs (Fig. 7). Vinyl base is available in 50′ coils and in 4′ lengths. Rubber base normally is available in 4′ lengths only. Exterior and interior corners of rubber base commonly are premolded (Fig. 8). End stops with a finished edge for terminating wallbase runs are available in both rubber and vinyl. The greater flexibility of vinyl base permits field-molding around exterior corners (Fig. 9); inside corners usually are scribed and fitted on the job.

Stair Treads and Nosings

Vinyl and rubber stair treads and nosings are made with smooth or textured surfaces and may have inlaid abrasive grit strips to improve slip resistance (Fig. 10). They may be combined with standard 6′ high wall base or higher stair risers to form complete stair coverings. Lengths vary from 3′ to 12′.

Thresholds, Feature and Reducing Strips

Thresholds usually are of vinyl or rubber, 2-3/4″ and 5-1/5″ wide, approximately 1/2″ high and commonly 3′ and 4′ long. *Feature strips* are made of asphalt, vinyl-asbestos, linoleum and vinyl in widths of 1/8″, 1/4″, 1/2″, 1″ and 1/2″ increments up to 4″. Gauges usually are 1/16″ to 1/8″ and lengths usually are 3′ and 4′. Rubber and vinyl beveled *reducing strips,* 1″ to 1-1/2″ wide, 3′ to 12′ long and 1/16″ to 3/8″ thick, permit transition in finished floor heights (Fig. 11).

SELECTION CRITERIA

Suitability of a resilient flooring material for a particular location or condition depends on its physical properties. *Moisture resistance* is important in determining whether a material can be used in areas subject to surface or ground moisture. *Grease* and *alkali* resistance have a bearing on suitability for use in kitchens or over concrete, respectively. *Resilience* is related to quietness, indentation resistance and underfoot comfort.

These and other properties may be combined into ratings for *durability* and *ease of maintenance*. Moisture and alkali resistance must be considered together in determining suitable *location* with respect to grade—whether suspended, on grade or below grade (Fig. 12). Installed cost is another factor in flooring selection, but should be considered along with long-term costs which are affected by the physical properties, durability and ease of maintenance. Selection criteria for resilient flooring products are summarized in Figure 13.

MOISTURE RESISTANCE AND LOCATION

Linoleum which contains linseed oil binders needs sufficiently dry conditions so that it will not be softened from exposure to moisture. Products with organic ingredients or backings are subject to mold growth and decay. Asphalt, vinyl-asbestos, solid vinyl, and sheet vinyl with asbestos fiber backings are least affected by moisture and alkalis which may be leached by moisture from concrete floors.

The relative resistance of flooring products is reflected by recommendations for *suspended, on-grade* and *below-grade* subfloor locations (Fig. 13). Consideration is given in these recommendations to the alkali resistance of the backing materials, which in some products may vary from those of the surface. In sheet vinyl with asphalt felt backing, for instance, the vinyl surface may have excellent alkali resistance, a property not possessed by the organic felt backing.

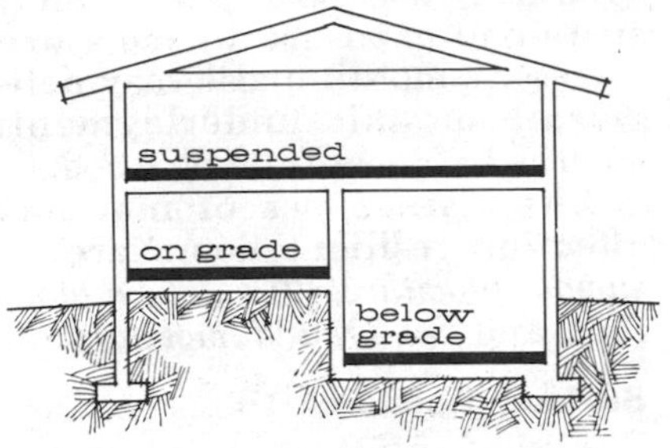

FIG. 12 Subfloor locations: on-grade level subfloors may be considered suspended when built over a ventilated crawl space.

FIG. 13 **SELECTION CRITERIA FOR RESILIENT FLOORING PRODUCTS**[1]

Type of Flooring	Location[2]	Resistance to: Grease	Alkalis	Stain[3]	Cigarette Burns	Indentation	Resilience[4]	Quietness	Ease of Maintenance	Durability
VINYL-ASBESTOS TILE	BOS	2	2	4	2	3	6	6	2	2
VINYL TILE										
Solid	BOS	1	1	1-2	1-5	1-4	2-5	2-5	3	1
Asbestos Backed[5]	BOS	1	1	1-3	3-5	2-5	2-4	2-5	3	1
Rag felt Backed[5]	S	1	2	1-3	3-5	2-5	2-4	2-5	3	1
VINYL SHEET										
Filled Surface										
Asbestos Backed[5]	BOS	1	1	1-3	3-5	2-5	2-4	2-4	2	1
Rag felt Backed[5]	S	1	2	1-3	3-5	2-5	2-4	2-4	2	1
Clear Vinyl Surface										
Asbestos Backed[5]	BOS	1	1	1-3	3-5	2-5	2-4	2-4	2	1
Rag felt Backed[5]	S	1	2	1-3	3-5	2-5	2-4	2-4	2	1
RUBBER TILE	BOS	3	3	2	1	4	2	2	3	2
CORK TILE	OS	4	4	5	4	4	1	1	5	5
VINYL CORK TILE	OS	1	3	1	5	3	3	3	1	4

[1]Numerical rating indicates rank of each product compared to other products. Highest rating is 1.

[2]B—Below grade, O—On grade, S—Suspended.

[3]Varies with staining agent.

[4]Also indicates potential underfoot comfort.

[5]Foam-cushioned products are rated highly for resilience, indentation resistance, ease of maintenance and quietness within the ranges indicated.

In addition to the effect on flooring materials directly, moisture may prevent proper bonding to the subfloor or underlayment. Water-based adhesive will not set up in the presence of water, and asphaltic emulsion adhesives eventually may be displaced under sustained exposure to moisture. Excessive moisture also may deteriorate organic underlayments such as hardboards or plywood.

The main types of moisture affecting resilient floors are (1) *surface moisture,* (2) *subfloor moisture,* and (3) *ground moisture.*

Surface Moisture

Spilled water, floor moppings and tracked-in moisture are the primary sources of surface moisture. Spilled water may be common in laundry areas, bathrooms and kitchens. Entries may be subjected to tracked-in moisture and excessive wet mopping.

Surface moisture affects flooring materials and adhesives by entering through the seams. Sheet flooring products minimize the number of seams, and when provided with moisture-resistant backings they are most suitable for use in such areas. The seams of many sheet products also may be sealed to prevent water seepage through seams.

Subfloor Moisture

Concrete subfloors and mastic underlayments mixed with water release large amounts of moisture as they cure. With lightweight-aggregate concrete (weighing less than 90 lbs. per cu. ft.), or under conditions which retard the curing process, moisture may be released over long periods of time. Field tests described in Division 400—Applications & Finishes should be used to establish that the subfloor is sufficiently dry to receive the intended product. Since it originates in the original mix of the concrete, subfloor moisture may be present in suspended as well as on-grade and below-grade subfloors.

Ground Moisture

Moisture from the ground generally is limited to on-grade and below-grade locations. Concrete floor slabs in direct contact with the ground, unless protected by a vapor barrier, may transmit moisture by capillary action. A well drained fill of sand or gravel will retard the migration of moisture.

When resilient flooring is installed on wood subfloors over crawl spaces, a vapor barrier and adequate ventilation of the subfloor space is required. A vapor barrier placed over the ground will prevent transmission of moisture from the ground. Ventilation will encourage dissipation of moisture resulting from breaks in the vapor barrier and other sources. Both crawl spaces and concrete slabs-on-ground require proper surface drainage of exterior finished grade to prevent ground water from becoming trapped under the floor (see Division 300—Methods & Systems).

ALKALI AND GREASE RESISTANCE

In residential use, the source of grease is mainly from cooking oils spilled in kitchens; alkalis may be present from various cleaner residues. If the flooring is properly protected with a floor wax or finish and spilled oils or residues are removed promptly before they seep into the seams, such temporary exposure is not a hazard, except for materials with the lowest resistance ratings (Fig. 13). Vinyl-asbestos and solid vinyl tile are not affected substantially by either grease or alkali.

RESILIENCE

Resilience is a measure of the instantaneous yielding and recovery of a surface from impact. Indentation resistance, quietness and underfoot comfort are closely related to resilience.

Indentation Resistance

In assessing indentation resistance, the momentary indentation produced from foot traffic and dropped objects are of primary importance. These impact pressures sometimes are quite high and demanding. A 105-pound girl in spike heels, for example, exerts a pressure on the floor of approximately 2000 lbs. per sq. in., while a 225-pound man with his weight spread over 3″ x 3″ heels exerts only 25 lbs. per sq. in.

Permanent indentation from heavy stationary objects, such as a piano, may be minimized by using floor protectors to distribute the load (Fig. 14). Indentation resistance of thinner flooring materials is greatly affected by the subfloor or underlayment and may be increased by selecting harder subsurface materials. Homogeneous vinyl tile and foam-cushioned vinyls have the highest indentation resistance.

Permanent indentation in some flooring types and under certain conditions cannot be entirely prevented. However, these indentations may be less conspicuous in patterned, textured and low-luster floors.

Quietness

The quietness rating indicates the effectiveness of a floor in reducing sound from foot traffic and other impact noises. While resilient floors have the capacity to soften impact sounds, they will not reduce reverberated noises originating from other sources because they have practically no sound absorption ability. Cork and foam-cushioned vinyl flooring have the

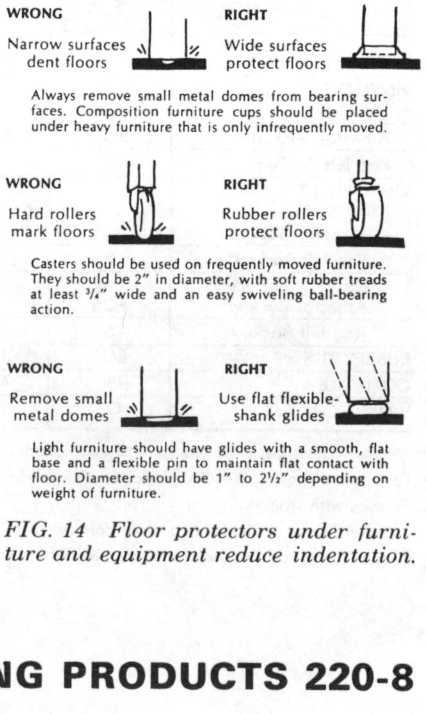
WRONG
Narrow surfaces dent floors
RIGHT
Wide surfaces protect floors
Always remove small metal domes from bearing surfaces. Composition furniture cups should be placed under heavy furniture that is only infrequently moved.
WRONG
Hard rollers mark floors
RIGHT
Rubber rollers protect floors
Casters should be used on frequently moved furniture. They should be 2″ in diameter, with soft rubber treads at least 3/4″ wide and an easy swiveling ball-bearing action.
WRONG
Remove small metal domes
RIGHT
Use flat flexible-shank glides
Light furniture should have glides with a smooth, flat base and a flexible pin to maintain flat contact with floor. Diameter should be 1″ to 2 1/2″ depending on weight of furniture.

FIG. 14 Floor protectors under furniture and equipment reduce indentation.

highest resilience and quietness ratings. Asphalt tile is at the low range of resilience and quietness.

RESISTANCE TO SUNLIGHT

The actinic rays in strong sunlight may affect some resilient floors by causing fading, shrinking or brittleness. Linoleum and vinyl products are most resistant to such deterioration. Color pigments are the critical factor in fade-resistant properties. Neutral colors show the best light resistance; pastel tones, especially yellows, blues and pinks, are least effective in retaining colors under prolonged exposure to sunlight. Cork tile has the same tendency as natural wood to fade under strong sunlight.

EASE OF MAINTENANCE

This is an appraisal of the relative expense (labor and materials) in keeping flooring at an acceptable level of cleanliness and attractiveness. Relative ratings are based on a record of experience with various flooring types as well as laboratory testing. Textured surfaces and darker colors show less scuffing and soiling and generally receive higher ease-of-maintenance ratings. No-wax vinyl products and vinyl cork tile are the easiest to maintain, followed by other vinyl products, vinyl-asbestos and linoleum.

DURABILITY

Durability reflects the ability of flooring products to retain serviceability and attractiveness over a period of time. Durability ratings are based on laboratory tests and are related to the physical properties of the material. Homogeneous vinyl tile is the most durable. Vinyl sheet flooring, rubber tile and vinyl-asbestos also are rated highly. Foam-cushioned products have good durability because the surfaces of these products absorb impact and resist abrasion.

We gratefully acknowledge the assistance of the following for the use of their publications as references and permission to use photographs: Armstrong Cork Company; Congoleum Nairn, Inc.; The Flintkote Company; Johnson Rubber Company; Mercer Plastics Company, Inc.; Resilient Floor Covering Institute.

226 CARPETING

INTRODUCTION

Since World War II, carpet production has increased almost fourfold, and carpeting has become an integral part of building construction. Approximately 70% of total carpet yardage currently is going into residential use and the balance into commercial and institutional use. Wider use of carpeting in recent years can be attributed to technological advances in carpet manufacture, increasing use of manmade fibers, and less cost.

Although carpet for the home often is selected for its comfort and decorative values, carpeting is extremely effective in reducing impact sound transmission through floor/ceiling assemblies. Therefore, its acoustical performance should be a prime consideration in flooring selection. Many common installations will provide a performance considerably higher than the minimum range of −5 to +10 on the FHA Impact Noise Reduction scale (INR), recommended for sound isolation in residential construction (see Section 106 Sound Control). Specially selected combinations of carpet and cushioning may be rated as high as +17 over wood floors, and +29 over concrete slabs.

Carpeting also is capable of absorbing sound and reducing sound reflection within a room, much like acoustical ceiling tile. The Noise Reduction Coefficient (NRC) of most carpeting ranges between .35 and .55, which compares favorably with a range of .55 to .75 for acoustical ceiling tile. Properly selected carpeting may achieve absorption coefficients equal to acoustical tile. However, carpeting is not particularly effective in controlling the transmission of airborne sound, and where this is an important consideration, special sound isolating construction should be used.

This section describes carpeting suitable for long-term *wall-to-wall* installation, as distinguished from *rugs* which usually have bound edges and are laid loose over a finished flooring. The discussion is divided into subsections outlining *Carpet Construction* according to materials and methods of manufacture, and *Selection Criteria* based on performance and standards of quality.

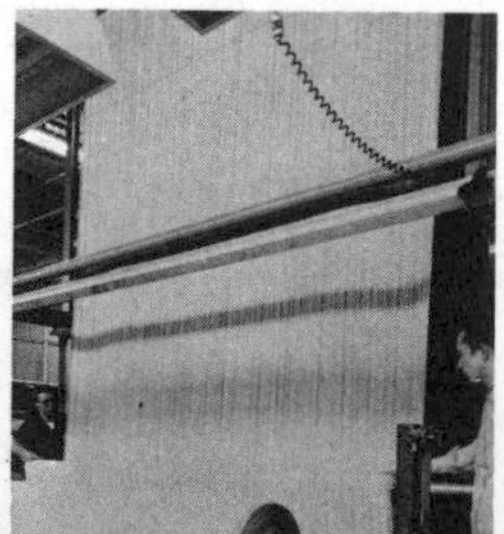

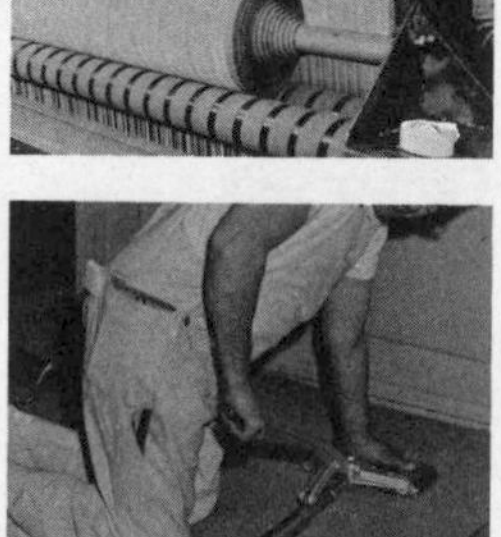

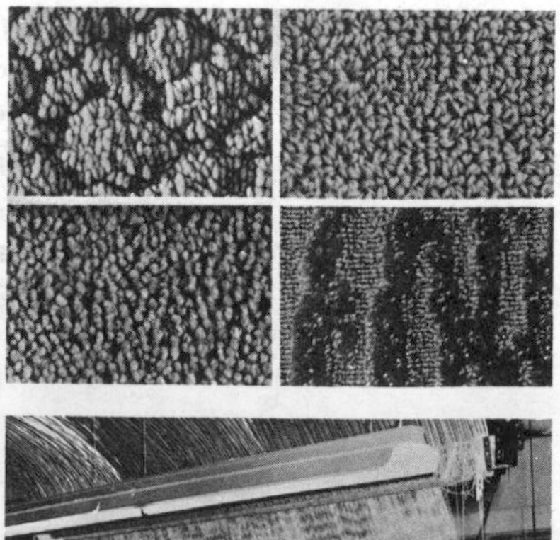

CARPET CONSTRUCTION

Most carpet consists of *pile yarns* which form the wearing surface, and *backing yarns* which interlock the pile yarns and hold them in place. Therefore, carpet can be identified according to the *carpet construction,* the method of interlocking backing and pile yarns. Comparative factors related to carpet construction, such as pile yarn weight and pile thickness, number of tufts per sq. in. and other factors may be useful in comparing carpet of similar construction.

Carpet also can be identified by the pile fibers of which the pile yarns are made, which include wool, nylon, acrylics, polyester and polypropylene. The various fibers have unique characteristics which may affect carpet performance and styling. Both carpet construction and fiber properties influence the selection of dyeing methods.

CARPET TYPES

The term *broadloom* sometimes is mistakenly thought to define a type of carpet construction, but actually it refers merely to carpet manufactured in widths over 6′. Almost all of the carpet types discussed below, with the exception of rubber-backed loomed carpet, are manufactured in widths of 9′, 12′, 15′ and sometimes 18′.

Modern carpet can be classed according to the principal methods of manufacture as *tufted, woven* and *knitted.* In general, the weaving and knitting operations interlock pile and backing yarns in a single operation, while the tufting loom stitches pile yarns onto a premade backing. Tufted carpets can be produced much faster than woven or knitted types, are less expensive to produce and now account for approximately 85% of all carpet manufactured in this country.

Punched and *flocked* carpets are two other types which recently have become available, but have not yet been fully established nor accepted for long-term wall-to-wall installation. The use of punched carpet, however, is rapidly increasing and shows considerable promise in the low-cost field. Punched carpet is made by punching loose, unspun fibers (usually polypropylene or nylon) through a woven sheet which results in a homogeneous pressed layer of fibers much like an army blanket or heavy felt. Flocked carpet is produced by spraying or electronically depositing short, loose nylon fibers on an adhesive-coated backing, similar to flocked wall coverings.

Weft Yarn

a

Stuffer Yarn

b

Chain Yarn

c

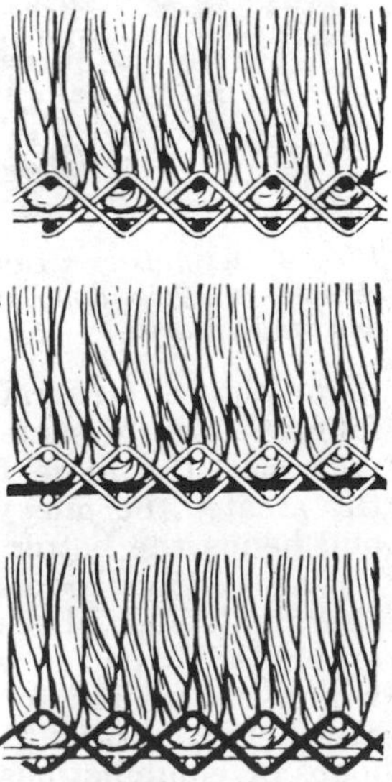

FIG. 1 Backing yarns of woven carpets identified by heavy lines: (a) also called weft shots or filling; (b) also called stuffer warp; (c) also called chain warp or binder or binder warp.

Since punched carpet is made entirely of synthetic fibers which are not subject to decay, it can readily be used on exterior as well as interior surfaces. Continuing research into flocked carpet manufacture indicates that polypropylene might be substituted for jute, presently used as backing material, thus rendering flocked carpet also suitable for exterior and high moisture locations.

Woven

Woven carpet is made on looms that interweave pile yarns and backing yarns in one operation. Backing yarns consist of *weft* (crosswise) yarns and *warp* (lengthwise) yarns. Because the tuft-forming yarns in woven carpet run lengthwise, they too are classed as warp—*face warp* or *pile warp.* Though most properly applied to woven carpet, the terms warp and weft are loosely applied to knitted carpet as well.

Weft yarns run crosswise, passing over the pile yarns (Fig. 1a). *Stuffer yarns,* a type of warp yarn, give strength and dimensional stability to the carpet (Fig. 1b). Typically, there are 2 or 3 stuffer yarns per tuft running the length of the carpet. *Chain yarns,* another type of warp, run the length of the carpet alternately passing over and under the weft yarns, forming a chain which locks the weft in place and pulls the pile yarns down tightly (Fig. 1c).

Four weaving processes are used from which woven carpets derive their names: *velvet, Wilton, Axminster* and *loomed.*

Velvet The velvet weave is the

simplest of all weaves (Fig. 2). Chain, stuffer and pile yarns are alternately raised and bound together with the weft yarns. The pile is formed by *wires* inserted between pile and warp yarns. The height of the wire determines the depth of the pile. After a row of loops has been formed over the wire and secured by the weft, the wire is withdrawn leaving a row of loops. Cut pile is created by a razor edge on the wire which cuts the loops as the wire is withdrawn.

Since the velvet loom feeds the pile yarn from a single spool, it is best suited for solid colors. However, strands of different color can be twisted together into a single yarn (moresque), producing tweeds, salt-and-pepper effects and stripes. A wide range of textural effects is possible by using uneven wires forming high and low loops.

Velvet weave can be recognized at a cut edge or open selvage where the chain warp is visible, binding all the construction yarns together in a zigzag pattern (Fig. 2).

Wilton The Wilton loom is basically a velvet loom fitted with a special *Jacquard* mechanism which utilizes punched cards, similar to computer cards, to select the various colored yarns. These cards are locked together to match the entire surface design of the carpet and are suspended in the loom. The colored pile yarns are fed from spools assembled on frames (racks) in the back of the loom, one frame for each color. As the cards pass, the Jacquard mechanism selects and lifts the appropriate pile yarns into position to be looped over the wires, while other yarns remain buried in the body of the carpet. Depending on the number of colors used, for each tuft of yarn showing on the surface, there may be as many as four strands of yarn buried in the backing, adding to the stiffness of the carpet (Fig. 3).

A carpet woven on a Wilton loom with just one colored pile yarn would have the same construction as a velvet weave and would be designated as a velvet carpet. Wilton carpets are designated as two-, three-, four- or five-frame Wiltons depending on the number of colored pile yarns used in their

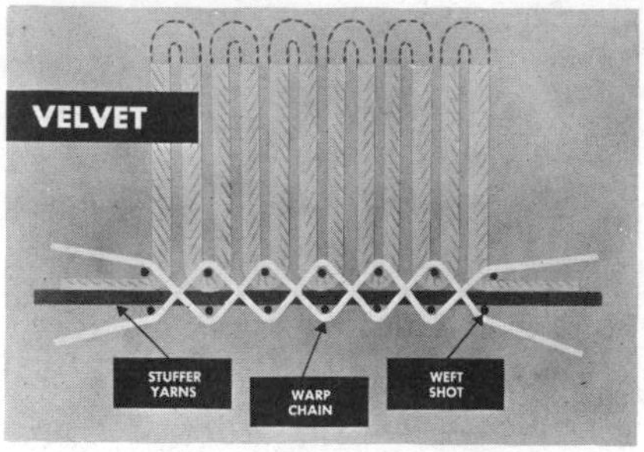

FIG. 2 Velvet weave with double chain warp and typically 2 or 3 stuffer yarns per pitch.

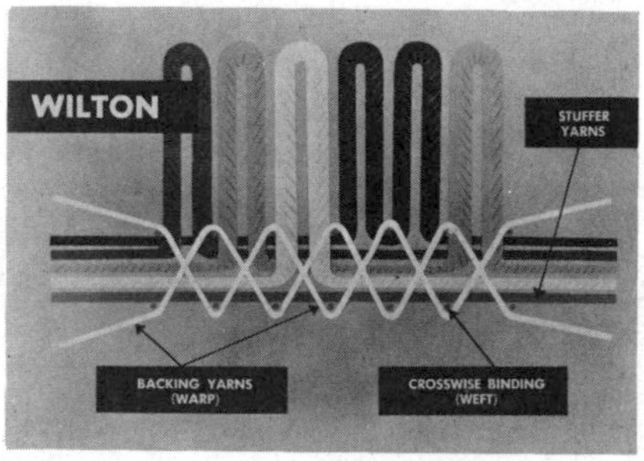

FIG. 3. Wilton weave with four frames (colors), double chain warp, two weft shots per wire and stuffer yarns

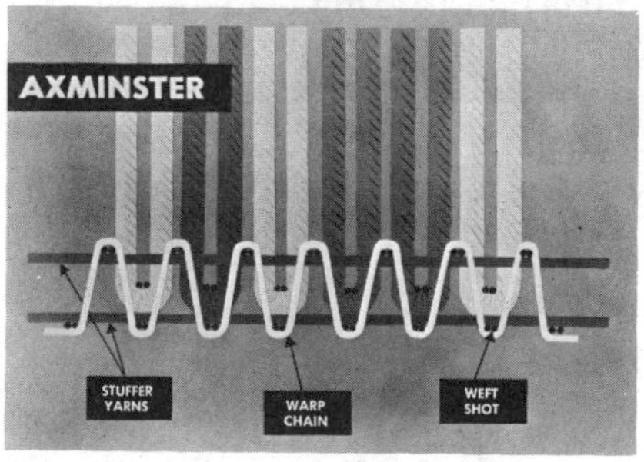

FIG. 4 Axminster weave with single chain warp, three double shots per row and stuffer yarns.

manufacture. Generally speaking, for carpets of the same density and yarn size, the more frames used, the greater the pile yarn weight, and hence the better the quality.

The Wilton loom can weave sharply delineated sculptured and embossed textures or patterns, which are created by varying the pile height, using high and low loops or combinations of cut and uncut piles. Wilton carpets with uncut loop pile are termed *round-wire Wilton.*

In multicolor Wilton weaves, the different colors of pile fibers are usually visible on the back, and the body of the carpet is thicker than that of other types. As in velvet weaves, the chain warp also is visible at a cut edge.

Axminster Axminster carpet is woven on a loom which draws pile yarns from small spools wound with yarn of various colors. The spools are locked together in a frame equal to the width of the carpet. Since each frame provides one row of tufts across the carpet and each strand of yarn provides a single pile tuft, the sequence of colors determines the finished surface pattern.

The Axminster loom simulates hand weaving because each tuft of yarn is individually inserted, and theoretically each tuft could be of a different color. This flexibility offers unlimited design possibilities and intricate patterns are produced even in lower priced carpet. Considerable delicacy and subtlety of design are possible in the higher priced carpets where the rows of tufts are closer together.

The Axminster weave is almost always a cut pile with an even pile height and elaborate patterning with many colors. The backing is so heavily ridged and sized (coated) that it usually only can be rolled easily lengthwise. The characteristic chain warp can be seen along a cut edge as in other woven carpets (Fig. 4).

Loomed Loomed carpet is a recently developed fabric intended specifically for use with a bonded rubber cushioning. It is characteristically a high-density nylon with single-level, low-loop pile about 1/8″ high, woven on a modified upholstery loom (Fig. 5). The pile back is treated with a waterproof adhesive which serves as a bond coat for the rubber backing. Sponge or foam rubber is applied to the pile fabric and the backed carpet is heated to cure the rubber. Loomed carpet generally is manufactured in 4′6″ width with a rubber backing approximately 3/16″ thick.

Tufted

Tufted carpet is manufactured by inserting tufts of pile into a premade backing by a machine which resembles a huge sewing machine with thousands of needles operating simultaneously. The *prime backing* is made from jute, kraftcord, cotton or polypropylene olefin, and is coated with a heavy layer of latex to hold the tufts permanently in place. On most tufted carpets a loosely woven, coarse jute fabric (scrim) is added as *secondary backing* to lend additional strength and dimensional stability (Fig. 6).

The variety of textures and colors in tufted carpet is increasing with innovations in tufting equipment and dyeing processes. Tufted pile can be level-loop, plush, multilevel, cut or uncut, or a combination of cut and uncut. Striated and carved effects also can be achieved. Tufted carpets are made in solids, moresques, and more recently in multicolor patterns made possible by new dyeing methods.

Tufted carpet generally can be identified on the reverse side, where the even rows of tufts punched through the prewoven backing with punched-in tufts also can be seen.

Knitted

Knitted carpet is somewhat similar to woven carpet in that the pile and backing yarns are fabricated in one operation. Unlike weaving, the knitting process loops together the backing, stitching and pile yarns with three sets of needles, in much the same way as hand knitting. A coat of latex is applied to the back as a fixative and to give additional body to the fabric (Fig. 7).

Since a single pile yarn is used in the knitting operation, as in velvet weaving, knitted carpets are usually solid colors or tweeds, although more recently patterns have become possible. Knitted carpets usually have either single- or multilevel uncut pile, but cut pile can be achieved with modification of the knitting machine.

If a knitted carpet is *grinned* (folded back) to expose the backing, it will show the continuous looping of the pile yarn from row to row, held in place by the stitching yarns. The rows of pile run in an irregularly diagonal direction with a random, homespun look, not in straight rows as in woven and tufted carpets.

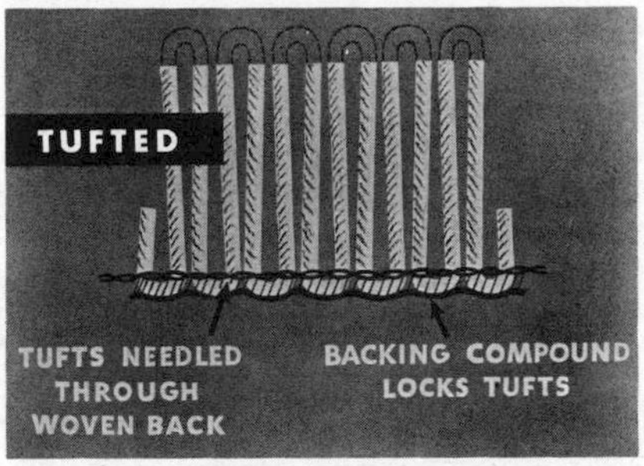

FIG. 6 Tufted carpet construction; secondary backing usually is bonded to a prime backing with latex compound.

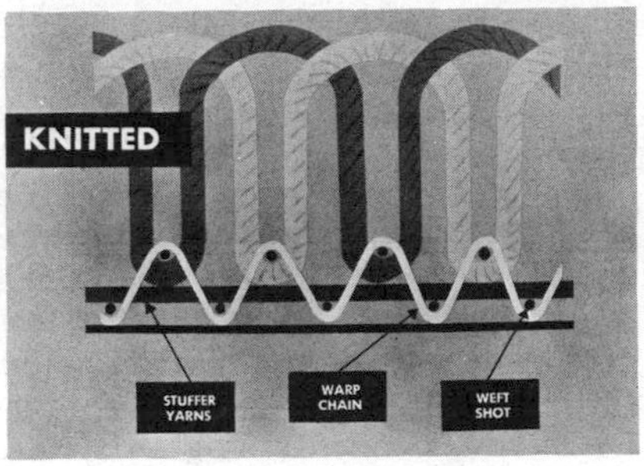

FIG. 7 Knitted carpet with typical continuously looped pile, single chain warp and stuffer yarns—all backed with latex.

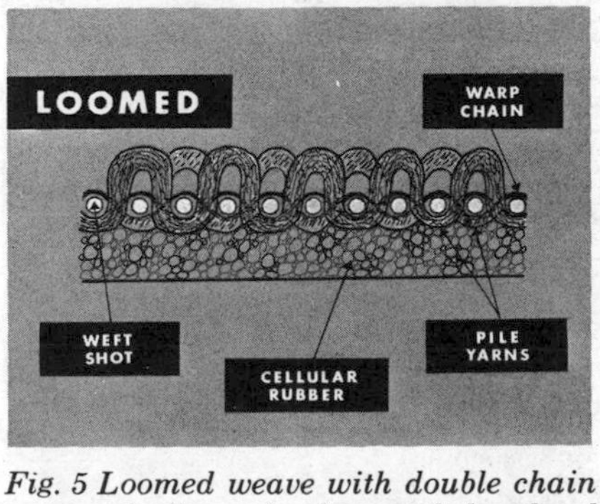

Fig. 5 Loomed weave with double chain warp, single weft shot and low-level loop pile bonded to rubber cushioning.

COMPARATIVE FACTORS

The traditional rule of thumb, "the deeper and denser (the pile), the better," often is used in making visual comparisons between carpets of similar construction and fiber type. However, such a generalization is superficial and more detailed measures of carpet construction generally are necessary to specify and to judge carpet quality properly. *Pile density* is actually a combination of *pile thickness* and *pile weight.*

Pile thickness (in inches) and pile weight (in oz. per sq. yd.) can be established accurately only by laboratory analysis, usually in accordance with standard ASTM procedures. Pile density (in oz. per cu. yd.) can be derived by a simple formula from established information about pile thickness and weight. All of these criteria are used in determining standards of quality for carpeting (see page 226-12).

Other comparative factors have been developed as a substitute for, or supplement to, the basic criteria of pile density, weight and thickness. The number of *tufts per sq. in.; pitch, gauge and needles;* and *rows, wires and stitches*—are terms which describe the spacing of individual tufts in various carpets. When considered together with *yarn size* (yarn weight) and number of *plies* (strands) twisted together to form the pile yarn, these factors provide measures of pile density and other important basic criteria relating to quality.

Pitch, Gauge and Needles

Spacing of tufts across the width of a woven carpet can be described in terms of pitch, gauge or needles. *Pitch* indicates the number of pile warp yarns (tufts) in a 27″ width of carpet and is a measure of the tightness of crosswise construction (Fig. 8). Pitch may range between 90 to 256 for velvet and Wilton carpets, and between 189 and 216 for Axminster. The higher the pitch, the tighter the weave and the greater the pile density.

In knitted and tufted carpets where there are no continuous warp lines present, *gauge* describes the actual spacing of tufts across the carpet width, expressed as a fraction such as 1/8″ 3/16″, 5/32″. For

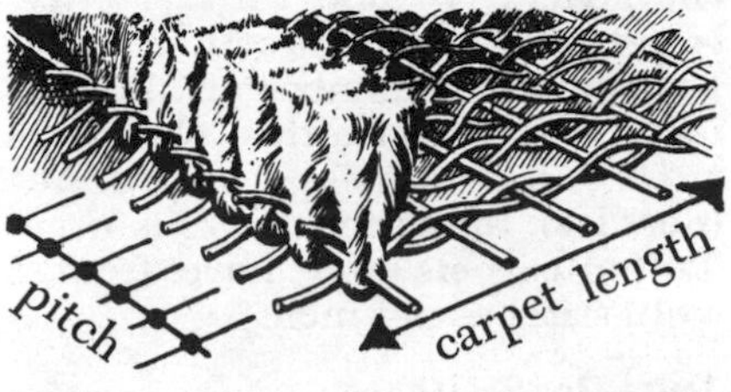

FIG. 8 Pitch, the number of tuft-forming pile yarns in a 27″ width of woven carpet.

example, 1/8″ gauge describes carpet with 8 tufts per 1″; 3/16″ gauge indicates 16 tufts in 3″ (a little more than 5 tufts per 1″).

In tufted carpets, the term *needles* often is used instead of gauge to describe the number of tufts per inch across the width. Tufted construction may vary from 4 to 13 needles (tufts) per 1″.

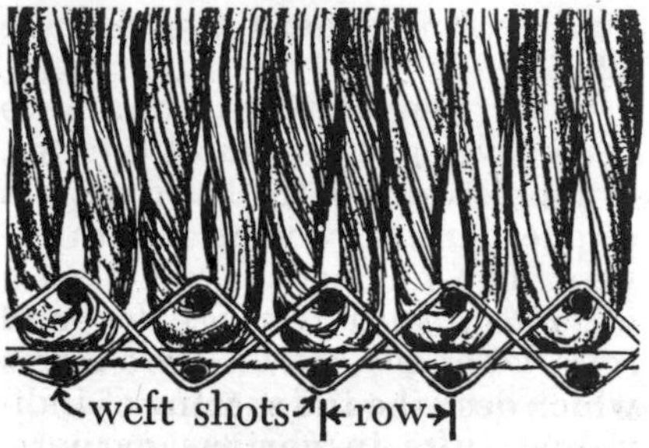

FIG. 9 Wires or rows per inch—a measure of tuft spacing and pile density lengthwise for woven carpet.

Rows, Wires and Stitches

Rows, wires and stitches are terms describing spacing of tufts lengthwise. The number of pile tufts per inch in Axminsters are described in *rows* per inch; for Wiltons and velvets in *wires* per inch (Fig. 9). Single or double weft shots are usually visible on the back of the carpet and correspond to the rows of tufts in the length of the carpet. To give an accurate indication of rows or wires per inch, it is customary to count the visible single or double weft shots in a 3″ length of carpet and express this value in rows per inch, as a fraction where necessary. Thus 17 rows in 3″ would be expressed as 5-2/3 rows or wires per inch. Woven carpets range from 4 to 11 rows or wires per inch.

In tufted carpets there are no weft yarns which can be related to the spacing of pile tufts. Hence, lengthwise construction is expressed as the number of tufts per inch. To distinguish this count from tuft count across the carpet (needles), the term *stitches* is used. Tufted carpets may have from 6 to 10 stitches per inch.

Tufts Per Sq. In.

The overall density of carpets can be computed in tufts per sq. in. from information describing both lengthwise and crosswise construction (Fig. 10), or can be determined by actually counting the tufts in a 1″ square area. This comparative method is preferred for tufted and knitted carpets, but also can be applied to woven carpet. It is helpful in comparing overall density of dissimilar carpets when different terms are used to describe lengthwise and crosswise construction.

Yarn Size and Plies

Wool fibers and synthetic staple fibers are spun into single-strand yarns. Continuous filament synthetic fibers are not spun, but emerge from the extrusion machine in bundles which are formed into a single-ply yarn. The strands of yarn may then be further twisted together into thicker yarns, each forming a single pile tuft. Depending on the number of single strands, the pile yarn is identified as *single-, two-, three- or four-ply* yarn. The size of both single-ply and multi-ply yarns can be measured by *woolen count* (yds. of yarn per oz.) or *denier* (grams per 9000 meters of yarn) and other methods depending on the spinning method employed. Synthetic fibers and yarns are most often measured according to the denier system.

Along with spacing of tufts, the yarn size and number of plies affect carpet density, resilience, compression resistance and other important performance criteria.

Backing

In woven and knitted carpets, the backing yarns and pile yarns are interlocked simultaneously during manufacture. The backing yarns typically are *jute* (sisal fiber), *kraftcord* (wood pulp fiber), *cotton, nylon* or *rayon.* Backing yarns hold the pile tufts in place and provide the necessary stiffness so the carpet will lie flat without wrinkling. In addition, they provide sufficient strength to prevent the carpet from tearing when heavy objects are moved across the surface.

Woven carpets such as velvet, Wilton and Axminster may have single or double chain yarns and several stuffer yarns per tuft running lengthwise; the number of *weft shots per wire* running across

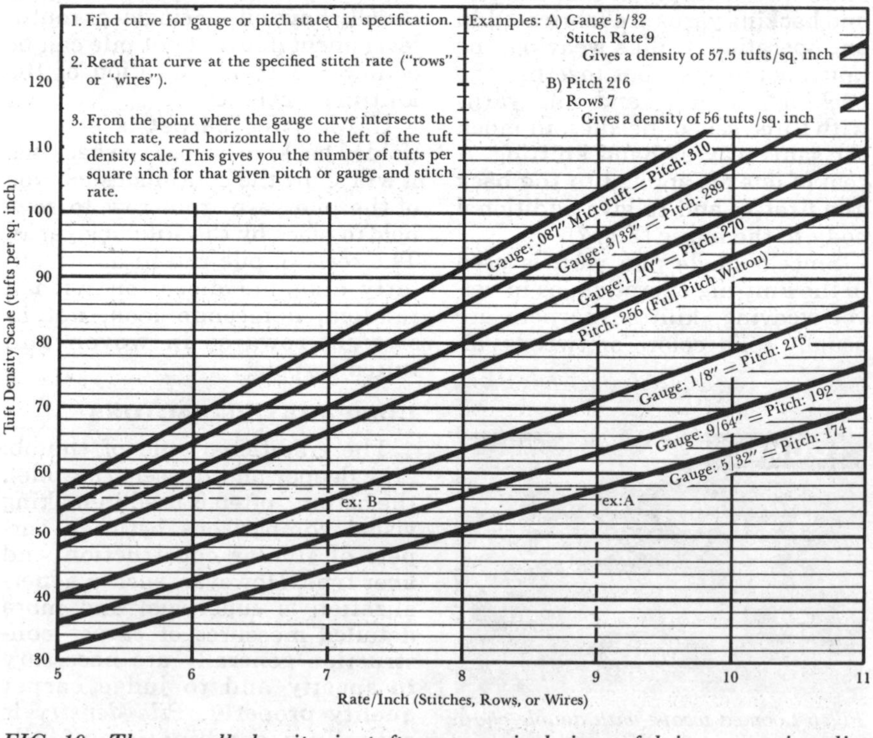

FIG. 10 The overall density in tufts per sq. inch is useful in comparing dissimilar carpets, such as tufted (ex:A) and woven (ex:B).

the carpet also may vary, resulting in such designations as 1-, 2- or 3-shot carpet. Increasing the number of backing yarns results in a stiffer, stronger backing and hence in improved carpet quality. The better Wilton and velvet carpets may be *woven through the back* by hooking the pile yarns around the bottom shot yarns, thus providing better anchorage of the pile tufts.

In tufted carpets, pile yarns are stitched into a separately prepared *prime backing*. The backing typically is a jute fabric, although woven and punched polypropylene also are used. An adhesive such as latex generally is used to secure the pile yarns in place and a *secondary backing* of jute is bonded to the prime backing for additional strength, stiffness and dimensional stability. Similar backings sometimes are used in woven and knitted carpets as well.

PILE FIBERS

Wool has been the traditional fiber used in carpet manufacture and because of its proven performance over the years remains the classic fiber to which the newer synthetic fibers often are compared. The synthetic fibers now account for approximately 75% of all carpet fiber used, and are marketed under a variety of proprietary names. The commonly used synthetic fibers can be classed generically as: *nylon, acrylic,* and *olefin. Cotton* and two other man-made fibers, *acetate* and *rayon,* have not proven suitable for long-term wall-to-wall installations and therefore are not considered here. *Polyester,* one of the newest of the synthetic fibers, shows considerable promise but has not been in use long enough to demonstrate its properties fully.

The type of fiber used does not in itself guarantee carpet quality or performance. The various fibers do have unique characteristics which may affect carpet performance and styling. Selected physical properties of pile fibers are given in Figure 11; other characteristics of fibers related to carpet performance are given in Figure 14, page 226-10.

When fibers are blended, the higher the percentage of a particular fiber, the more a carpet will reflect the characteristics of that fiber. For example, wool may be reinforced by tough nylon in a 70/30 blend, but the carpet will look and feel more like wool. Generally, at least 20% of a fiber must be used for its characteristics to be apparent.

FIG. 11 **SELECTED PROPERTIES OF PILE FIBERS**

Fiber: Type	Fiber: Length[1]	Specific Gravity	Breaking Tenacity[2]	Degrading Temperature[3]	Moisture Absorption (%)
Wool	Staple	1.32	1.17	420° - 570° F.	15 - 16
Nylon	Staple & CF	1.14	4 - 6	428° - 482° F.[4]	4 - 5
Acrylic	Staple	1.18	2 - 4	400° - 500° F.	1.3 - 2.5
Modacrylic	Staple	1.3 - 1.4	2.5 - 3.0	300° F.	.4 - 4.0
Polypropylene	CF	0.91	3 - 6	333° F.	up to .10

1. CF—Continuous Filament.
2. In grams per denier.
3. For wool, scorching to charring range; for synthetics melting temperature.
4. 428° for Type 6; 482° for Type 6,6.

Wool

The outstanding characteristic of wool is resilience which, in combination with moderate fiber strength (see tenacity, Fig. 11) and good resistance to abrasion, produces excellent appearance retention. Wool's relatively high specific gravity contributes to greater pile density.

Carpet wool is obtained from a number of countries, including Syria, Iraq, Argentina, Pakistan, New Zealand, Australia and Scotland. Wools from different regions have unique qualities, as various breeds of sheep yield different wools. Some wools are fine and lustrous, others are coarse and springy; also, fibers can vary in length from 3-1/2″ to 7″. *Woolen* yarn is made up of interlocked long and short fibers; *worsted* yarn uses only long fibers. Woolen yarns are more irregular, hence are softer and bulkier than worsted yarns. The pile of most wool carpet is made of woolen yarns spun from fibers 3-1/2″ to 5-1/2″ long. (Fig. 11).

Nylon

Today, nylon represents almost half of the total yardage used in residential carpeting and about 1/3 of the total used in the commercial field. The increased use of nylon is attributed to its lower cost, availability in many bright colors, exceptional resistance to abrasion and high fiber strength (see *tenacity,* Fig. 11).

Chemically, there are two types of nylon: *Type 6* which represents a long-chain polymer, a chemical containing 6 carbon atoms; and *Type 6,6* which results from the polymerization of two chemicals each containing 6 carbon atoms. Although the two types perform similarly as yarn, they differ slightly in their ability to be processed and dyed.

Nylon fiber is produced in two forms: *staple* nylon composed of 1-1/2″ to 6″ length fibers, which are spun into yarn much like wool or other short fibers; and *continuous filament* nylon consisting of bundles of continuous fibers which are formed into yarn without spinning.

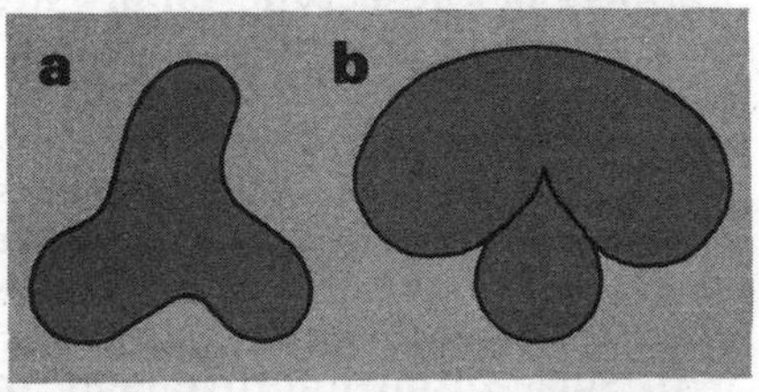

FIG. 12 Modified fiber cross sections: (a) trilobal nylon and (b) bi-component acrylic.

Both staple nylon and continuous filament nylon yarns are made in round and modified cross sections (Fig. 12). Staple fibers are mechanically crimped, while continuous

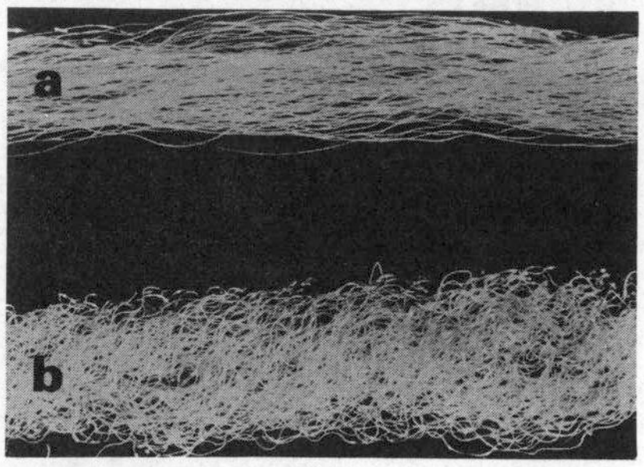

FIG. 13 Continuous filament yarn: (a) before texturizing; (b) after texturizing.

filament yarns are texturized (Fig. 13) to impart bulk and covering power to the yarns. Texturizing and crimp give the yarns or fibers an irregular alignment which further increases their bulk and covering power.

Acrylics and Modacrylics

The acrylics and modacrylics (modified acrylics) are the synthetic fibers which most closely resemble wool in abrasion resistance and texture.

Chemically, acrylic fibers are composed of 85% or more by weight of acrylonitrile. Modacrylics contain at least 35% of acrylonitrile in chemical combination with modifying materials. Acrylic fiber is produced in round, bean- and mushroom-shaped cross sections (Fig. 12b). The latter is a bi-component fiber of two filaments welded parallel to each other. Because the shrinking ratios of the two filaments are different, the fiber assumes a coiled shape in its dry equilibrium state, a property which improves texture retention.

Fibers of the acrylic family are now available only in staple form and are characterized by an appearance and high durability comparable to that of wool. Acrylics often are blended with modacrylics in commercial carpet to reduce potential flammability. However, with substantial improvement in acrylic fibers, the trend is toward 100% acrylic fiber.

Polypropylene Olefin

There are two classes of olefin, polyethylene and polypropylene; but only polypropylene has been produced in a fiber suitable for carpet construction. Polypropylene fiber consists of at least 85% propylene by weight, and is generally produced in bulked continuous filament yarn, although staple yarn also is available.

Polypropylene has the lowest moisture absorption rate of all carpet fibers, which gives it superior stain resistance as well as excellent wet cleanability (Fig. 11). Its low specific gravity results in superior covering power and high fiber strength contributes to good wear resistance.

DYEING METHODS

Color can be introduced at various stages of manufacture. For both synthetic and wool pile carpets, dyes can be added at either the fiber, yarn or carpet manufacturing stage. The dyeing method used may vary with the general properties of the fibers, the intended carpet construction and color effect desired, or may depend on special methods developed for proprietary fibers.

Solution Dyeing

Many of the synthetic fibers are colored by a dye or pigment introduced into the liquid chemical from which fibers are extruded. In this method, the dye becomes chemically a part of the fiber; therefore, solution-dyed yarns are generally colorfast and more resistant to discoloration from wet cleaning.

Stock Dyeing

Wool and synthetic staple fibers can be dyed in the raw fiber form before being spun into yarn, by being dipped into kettles of dye. This method assures uniform distribution of dye and is suitable where large quantities of a single-colored yarn are desired.

Skein and Package Dyeing

When relatively smaller quantities of colored yarn are required, dyeing often is accomplished after staple fibers are spun or continuous filament fibers are extruded into yarn. In *skein dyeing,* skeins of natural-colored yarn are immersed in vats of hot dye where they become saturated with color. For *package dyeing,* the yarn is wound on a perforated tube and the dye is forced through the perforations to soak the yarn.

Space Dyeing

Both staple and continuous filament yarns can be space dyed in alternating bands of color, generally for use in multicolored tufted carpet. In the *knit sleeve method* of space dyeing, the yarn is knitted into a fabric, print rollers apply the desired colors, then the fabric is unraveled again into yarn. In the *warp print* method, parallel yarns are fed continuously into a machine where color is applied by rollers.

Piece Dyeing and Resist Printing

Some woven and many tufted carpets are piece-dyed by immersing the entire carpet in a hot dye bath. However, the carpet need not be dyed only one color; many synthetic fibers can be chemically formulated to absorb certain compatible dyes and repulse others. Thus, it is possible to produce complex color patterns in the same dye bath. A similar technique of selective dyeing can be applied in skein dyeing. *Resist printing,* another method of selective dyeing, involves printing tufted carpet with a dye-resistant agent before piece dyeing.

Print Dyeing

The most recent development in dyeing tufted carpet is print dyeing. This method provides good dye penetration and pattern definition. With possible color variation of up to six colors, print-dyed tufted carpets compare favorably in appearance with Wiltons and Axminsters, but are cheaper to produce. The process involves silkscreening plain tufted carpet with a premetalized dye and applying an electromagnetic charge under the carpet which drives the dye deeply into the pile. The final pattern is built up by continuously processing the carpet through a series of troughs, each applying a different color.

Carpet performance cannot be attributed to any single element of carpet construction or physical property of the pile fiber. Installed performance is rather the combined effect of many variables such as surface texture and carpet construction, backing and cushioning, pile thickness, weight and density. Of these, pile weight and pile density are the most uniformly significant, regardless of fiber type or carpet construction.

Although pile fibers differ in their physical properties (Fig. 11), all fibers included in this discussion will provide adequate service when used in a carpet construction suitable for the traffic exposure (see Standards of Quality). The variables of carpet construction are so great that it is difficult to assess their individual effect on performance and selection criteria. Hence, selection criteria are outlined for fibers only (Fig. 14) and the effects of carpet construction are described where appropriate in the following discussion.

Individual selection criteria may be useful in carpet selection for special conditions of traffic or exposure to special hazards. For example, *alkali, acid* and *stain resistance* may be important in food preparation and serving areas. *Soiling resistance* and *wet cleanability* may be a factor in entry areas. *Insect and fungus resistance* may require consideration in potentially damp locations, and the possibility of *static buildup* should be considered in areas of prevailing low humidity. These and other criteria have been combined into composites such as *durability, texture retention* and *ease of maintenance* for average conditions encountered in residential use (Fig. 14).

DURABILITY

Pile weight and density, backing and cushioning firmness, the quality of installation and maintenance—all significantly affect the durability of carpeting. In addition, durability is affected by all hazards that tend to destroy the pile fibers. The resistance of fibers to these hazards is shown in ratings for resistance to abrasion, alkalis and acids, insects and fungi, and burns. The rating for durability is a composite reflecting overall resistance to destruction or loss of pile fibers, and is a measure of the service life of a carpet.

Abrasion Resistance

Abrasion resistance is the resistance of the fiber to wearing away due to foot traffic or other moving loads. It is measured in terms of the actual loss of fiber when exposed to a laboratory machine which simulates the abrasive forces due to foot traffic. Since loss of fiber occurs as a result of breaking, fiber strength is one of the physical properties which influences abrasion resistance. Nylon and polypropylene have superior fiber strength and hence excellent abrasion resistance.

Alkali and Acid Resistance

Foods and beverages are generally mildly alkaline or acid in nature; many soaps and detergents are strongly alkaline. If alkaline or acid solutions are not promptly removed when spilled, they may attack the pile fibers. Wool fibers are more vulnerable to such attacks because they have a higher moisture absorption rate and are less inert chemically than the synthetic fibers.

Insect and Fungus Resistance

The chemical composition of synthetic fibers contains no organic nutritive value, hence these fibers are immune to attack by insects (carpet beetles, moths) and fungi (decay, mold and mildew). To render wool carpet resistant to these hazards, most manufacturers treat the wool yarn in the dye bath with a chemical preservative which provides long-term protection from insect and fungus attack even after wet and dry cleaning. Since many carpets intended for interior use employ at least some organic fibers in the backing, some fungus hazard exists with all carpets, even those with synthetic fiber piles.

In general, the hazard of insect and fungus attack can be reduced be regular vacuuming and exposing the carpet to light and air. This is particularly important where large or heavy pieces of furniture make part of the carpet inaccessible to regular cleaning and ventilation. In such areas, greater insect protection can be provided for wool carpets by periodic spraying with mothproofing agents.

Burn Resistance

The synthetic fibers are flame-resistant, but they will melt and fuse when exposed to the heat of a burning cigarette or glowing ash (over 500° F). Prolonged exposure to such concentrated heat may result in complete local loss of fiber or a slightly fused spot. The fused spot is more resistant to wear and abrasion than the surrounding area, thus causing a visible discoloration in the carpet surface.

Although wool fiber will actually burn by charring, the damage from a lit cigarette usually is less severe and the charred spot will wear off more readily or can be removed with fine sandpaper. The fused spot on synthetic carpets similarly can be snipped off if it is not too severe, and, if followed by brushing, an acceptable appearance usually will result. If the burn is severe, replacement of the burnt area mav be necessary.

APPEARANCE RETENTION

Appearance retention depends on factors which change original carpet color, texture or pattern, such as fading, soiling and staining, and compression and crush resistance. Other factors common to many types of fiber and carpet construction which affect appearance without impairing serviceability are pilling, shedding, sprouting and shading (see Glossary). The ratings for appearance retention are composites of these factors.

Compression Resistance

The extent to which the pile will be compressed under heavy loads or by extensive foot traffic is termed compression resistance. For instance, of two samples tested under the same conditions, the one which compresses less is said to have better compression resistance. With the exception of polypropylene which is rated somewhat lower, all fibers are approximately equal in their ability to resist compression loads. Of greater importance in resisting such forces are the cushioning, the pile height and pile density. Looped pile construction with closely spaced tufts and tightly twisted yarns is more resistant to compression.

Compressed fibers in a small area may be restored by dampening the pile with a sponge or steam iron and brushing the pile erect. If a larger area becomes matted down by excessive traffic, it can be restored by vacuuming with a beater bar machine (Fig. 15), by using a pile lifter machine or by wet cleaning.

Crush Resistance

Crush resistance is dependent on the ability of the fibers to *recover* from short- and long-term compression loads. This property is measured by the extent to which the fiber springs back after the load has been removed, as a percentage

FIG. 14 **SELECTION CRITERIA FOR CARPET PILE FIBERS**[1]

Criteria	WOOL	NYLON	ACRYLIC	MODACRYLIC	POLYPROPYLENE
Resistance To:					
Abrasion	Good	Excellent	Good	Good	Excellent
Alkalis	Fair	Good to Excellent	Good to Excellent	Good to Excellent	Good to Excellent
Acids	Fair	Good to Excellent	Good	Good	Good to Excellent
Insects & Fungi	Excellent[2]	Excellent	Excellent	Excellent	Excellent
Burns	Good	Fair	Fair	Fair	Fair
Compression	Good	Good	Good	Good	Fair
Crushing	Excellent	Good	Good to Excellent	Good	Fair to Good
Staining	Good	Good to Excellent	Good to Excellent	Good to Excellent	Excellent
Soiling	Good	Fair to Good	Good to Excellent	Good to Excellent	Excellent
Static Buildup	Fair to Good	Fair	Good	Good	Excellent
Texture Retention	Excellent	Good to Excellent	Good to Excellent	Good	Fair to Good
Wet Cleanability	Fair to Good	Good	Good to Excellent	Good to Excellent	Excellent
Durability	Good to Excellent	Excellent	Good to Excellent	Good to Excellent	Good
Appearance Retention	Excellent	Good to Excellent	Good to Excellent	Good to Excellent	Good
Ease of Maintenance	Good to Excellent	Good	Good to Excellent	Good to Excellent	Excellent

1. In each case, criteria vary depending on the properties of the specific fiber type used, carpet construction and installation procedures.
2. When chemically treated.

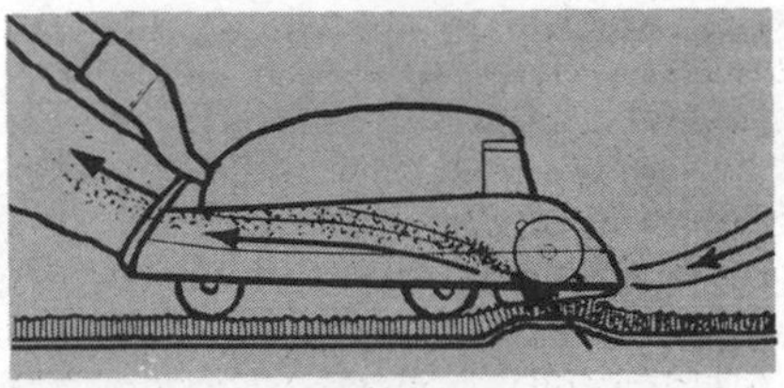

FIG. 15 The beater bar cleaner opens the pile and reaches deep-seated dirt.

of the original height. Wool and acrylic have the highest crush resistance, nylon and modacrylic moderate and polypropylene somewhat lower. Pile density, spacing of tufts and tightness of yarns affect crush resistance more than the relative rating of the fibers.

Texture Retention

The ability of carpets to retain the surface texture imparted during manufacture is related to the compression and crush resistance of the fibers, as well as the density of the pile and tightness of the yarns. Generally, all fibers have adequate texture retention, with wool rated highest and polypropylene somewhat lower than the other synthetic fibers. Within this range, special fiber types such as heat-set nylon and bi-component acrylic (mushroom cross section) have improved texture retention.

In areas where the texture has been conspicuously damaged by continuous traffic or heavy loads, the surface can be partially restored by methods suggested above under compression resistance.

EASE OF MAINTENANCE

The ease of maintenance of installed carpeting depends mainly on its resistance to soiling and staining, as well as its wet cleanability. Although dry powder cleaning products are useful for moderately soiled areas, wet cleaning (shampooing) remains the most satisfactory method for thorough in-place carpet cleaning under heavier soil conditions.

Staining Resistance

Staining is a local discoloration of the carpet and in this way differs from soiling which is more or less uniform over the carpet surface. Many foodstuffs (oil, grease, eggs, chocolate), beverages and beauty aids may cause staining. If promptly and properly treated, such stains need not be permanent. Many carpet manufacturers publish detailed recommendations for stain removal and indicate the type of solvent or cleaning agent to be used with each type of stain. Since stains are often moisture-borne, rate of moisture absorption is regarded as a factor in staining resistance. Polypropylene and acrylics have low moisture absorption and hence are rated highly for staining resistance.

The general procedure for stain removal involves: (1) promptly removing solids and blotting up liquids with absorbent materials; (2) treatment of the stain with appropriate solvent or cleaning agent, starting at the periphery and working towards the center; (3) repeated blotting of treated areas and avoidance of excessive brushing; (4) leaving a weighted cloth or paper towel over the stain area to remove most of the remaining liquid by absorption rather than air drying.

Soiling Resistance

Soiling is the result of dirt or soil particles deposited on the carpet surface by foot traffic or dust from the air. Fibers such as wool and nylon which generate higher electrostatic charges in low humidity tend to attract airborne dust and soil faster. However, the scaly surfaces of the wool fibers make soiling less apparent. Polypropylene and acrylics have higher soiling resistance because of low static buildup and low moisture absorption. Bulked fibers with modified cross sections and texturized yarns spun more loosely tend to retain dirt more tenaciously than round, tight, high-density yarns. Also, tightly woven, uncut-loop piles will keep soil near the surface where it is easier to remove; while multicolored and textured piles of medium shades will conceal soiling better than light or dark smooth plush surfaces.

Wet Cleanability

Since it is usually impractical to remove wall-to-wall carpeting for plant cleaning, the ability of carpet to respond successfully to in-place wet cleaning is an important consideration. Polypropylene which has the lowest moisture absorption rate is considered the best in this respect. Generally the synthetic fibers—with lower moisture absorption rates and inert chemical compositions—clean more readily, dry more rapidly and are less subject to mildew odors while drying than wool. However, since wool has been the traditional fiber in quality carpeting for many years, a variety of effective cleaning formulations specifically suited to wool carpeting are available.

STANDARDS OF QUALITY

There are no accepted industry-wide standards of quality for carpeting at the present time. However, the General Services Administration (GSA) and Federal Housing Administration (FHA) both have issued specifications which may become the basis for further standard development. GSA Federal Specification DDD-C-95 covers carpeting for commercial, institutional and residential use. The FHA specifications contained in Use of Materials Bulletin UM44A are related more directly to residential carpeting and are the basis for the following general recommendations.

FHA specifications do not permit the use of carpeting in baths, kitchens and service areas such as laundry, utility and furnace rooms. Remaining areas within the home or apartment such as living, dining, recreation, sleeping and dressing rooms are designated as *moderate traffic* areas. Public areas such as corridors, entrances, stairways and elevators are considered as *heavy traffic* areas.

FHA requirements for carpeting in heavy traffic areas are made by reference to Federal Specification DDD-C-95 and are summarized in Figure 16. Carpet construction requirements for moderate traffic areas are discussed below and are illustrated in Figures 17 and 18. For both types of traffic exposure, carpeting should conform to the following requirements for *pile fibers, backing* and *cushioning.*

FIG. 16 **PHYSICAL REQUIREMENTS FOR CARPETING IN HEAVY TRAFFIC AREAS**[1]

	WILTON	VELVET			TUFTED	KNITTED
	Wool or Acrylic	Wool or Acrylic	Nylon[2]	Wool or Acrylic	Wool or Acrylic	Wool or Acrylic
Pile Description	Single level loop woven thru back	Single level loop pile woven thru back	Single level loop pile woven thru back	Multilevel loop woven thru back	Single level loop	Single level loop
Tufts/sq. in.	52	60	32	34	60	26
Frames	3	—	—	—	—	—
Shots/wire	2	2	2	2	—	—
Weight oz./sq. yd.						
Pile	48	42	29	44	42	37
Total	72	60	50	64	—	58
Pile height	Min. 0.250 Max. 0.320	Min. 0.210 Max. 0.310	Min. 0.200 Max. 0.290	Min. 0.190 Max. 0.370	Min. 0.250 Max. 0.320	Min. 0.230 Max. 0.290
Chain	Cotton and/or rayon	Cotton and/or rayon	Cotton and/or rayon	Cotton and/or rayon	—	Cotton, rayon or nylon
Filling	Cotton and/or rayon, or jute	Cotton and/or rayon, or jute	Cotton and/or rayon, or jute	Cotton and/or rayon, or jute	—	Jute or kraftcord
Stuffer	Cotton, jute or kraftcord	Cotton, jute or kraftcord	Cotton, jute or kraftcord	Cotton, jute or kraftcord	—	—
Back coating oz./sq. yd.	—	8	6	8	—	—
Tuft bind (oz.)	80	80	80	80	100	14
Ply twist turns (per inch)	Min. 1.5 Max. 3.5	Min. 1.5 Max. 3.5	Min. 1.0 Max. 3.0	Min. 1.5 Max. 3.5	Min. 2.5 Max. 4.5	Min. 1.5 Max. 3.5

1. Based on Federal Specifications DD-C-95.
2. High bulk or textured continuous filament.

Carpet Construction

FHA carpet construction requirements are based on average values for pile yarn weight, pile thickness and pile density. Pile yarn weight and pile thickness are obtained from samples by tests performed according to ASTM D418 for woven carpet construction and ASTM D1486 for tufted construction.

Pile yarn weight (W) is the average weight of the pile yarn in oz. per sq. yd. when measured in accordance with specified ASTM test methods.

Pile thickness (t) is the height of the tufts above the backing in inches and similarly is established in accordance with specified ASTM test methods.

Pile density (D) is a calculated quantity in oz. per cu. yd. obtained by the formula: $D = \frac{36\ (W)}{(t)}$.

FHA establishes minimums for *pile yarn weight* (W) and *weight-density factor (WD)*, depending on the type of fiber (Fig. 17). The minimum weight-density factors (WD) given are always the minimum pile weight (W) multiplied by a pile density (D) of 2800. The FHA requirements are based on minimum weight-density factors rather than a constant pile density of 2800. If only the pile density were specified, it would establish a simple proportionate relationship between actual pile thickness and pile yarn weight which would require increasing the minimum pile weight to the same extent as the pile thickness. This requirement would be

FIG. 17 **PHYSICAL REQUIREMENTS FOR CARPETING IN MODERATE TRAFFIC AREAS**[1]

Pile Fiber	Minimum Pile Yarn Weight (W) oz./sq. yd.[2]	Minimum Weight-Density Factor (WD)[3]
Wool	25	70,000 (25 x 2800)
Acrylic		
Modacrylic		
Nylon (Staple & Filament)	20	56,000 (20 x 2800)
Polypropylene Olefin		

1. Based on FHA Use of Materials Bulletin UM-44A.
2. Computed for blended yarns by multiplying percentage of each fiber by minimum pile weight and adding the two amounts.
3. Pile Density (D)=2800, the minimum density which historically has given satisfactory service for wall-to-wall installations.

excessive for moderate traffic areas. Figure 18 illustrates calculation of the weight-density factor.

Pile Fibers

The pile yarn should be made of 100% wool, nylon, acrylic, modacrylic or polypropylene olefin fibers, or blends of these fibers in yarn (exclusive of ornamentation). Not less than 20% of any of the above fibers should be used when blended with other fibers, and such blends should not consist of more than two types of fiber. All fibers should be entirely new, not processed or reclaimed from previously knitted, woven, tufted or felted products. Wool yarn should contain at least 97% wool fibers by dry weight; synthetic fibers should be 15 denier or thicker and should contain not more than 2% chloroform soluble material.

Colorfastness Carpeting pile should be resistant to fading from wet cleaning and light exposure. Light colors should be rated as satisfactory after laboratory exposure to 20 standard fading hours and dark colors after 40 fading hours.

Backing

Backing materials should be those customarily used and accepted in the industry for each type of carpet. For tufted carpet, there should be a secondary backing reinforcement of at least 4 oz. per sq. yd., exclusive of the adhesive coating.

CUSHIONING

All carpet should be installed over cushioning to increase resilience and durability. When the cushioning is bonded to the carpet, it should meet or exceed requirements of Federal Specification DDD-C-95. Separate cushioning made of cellular rubber, jute-and-hair or hair felt should conform to GSA Specification ZZ-C-00811b and DDD-C-00123 and the minimum weight requirements for heavy and moderate traffic areas given in Figure 19.

For most carpets, cushioning is provided in the form of a separate sheet—*pad* or *padding.* In loomed carpets, cellular (foam or sponge)

FHA Weight-Density (WD) calculated for nylon or polypropylene carpet from Pile Thickness (t) and Pile Yarn Weight (W) determined by test in accordance with ASTM D-418 for woven carpet or D-1486 for tufted carpet.

Pile Thickness (t) = .25″, and

Pile Yarn Weight (W) = 20 oz./sq. yd.,

Calculated Pile Density (D) = $\frac{36W}{t}$

$$D = \frac{36 \times 20}{.25} = 2{,}880 \text{ oz./cu. yd.,}$$

Calculated Weight-Density Factor (WD)

$$WD = 20 \times 2{,}880 = 57{,}600^*$$

*Referring to Figure 17, Pile Yarn Weight (20 oz./sq. yd.) equals minimum requirement; Weight-Density Factor (57,600) exceeds minimum requirement (56,000).

FIG. 18 Weight-Density Calculation

rubber cushioning usually is bonded directly to the carpet at the mill. Currently available paddings include: felted hair, rubberized fibers and cellular rubber.

Felted Hair

The most economical conventional padding is made of felted animal hair. This padding has a waffle design to provide a skid-proof surface and improve resiliency. It is sometimes reinforced with a jute backing or with a burlap center liner. When reinforced with burlap, the hair is punched through the burlap fabric and compressed to a uniform thickness. Sizing (adhesive) sometimes is used to strengthen the bond between the fibers and the burlap core. Many manufacturers also sterilize and mothproof hair padding.

Felted padding of hair or hair-and-jute may mat down in time or may develop mildew, especially if the fibers become wet, as during cleaning. However, when properly cleaned, sterilized and treated, hair padding is suitable for reasonably dry floors at all grade levels and on conventional radiant-heated floors.

Rubberized Fibers

Some padding made of jute or hair is coated with rubber on one or both sides to hold the fibers together and provide additional resilience. Sometimes this padding has an animal hair waffle top and a jute back reinforced with a patterned, rubberized application.

Cellular Rubber

In addition to cushioning bonded directly to the carpet, foam or sponge rubber is produced in sheet form with waffle, ripple, grid or V-shaped rib designs. A scrim of burlap fabric usually is bonded to the rubber sheet to facilitate installation of the carpeting. When laid with the fabric side up, this permits a taut and even stretch of the carpet.

Rubber padding is more expensive, but it retains its resilience longer than hair padding. It is highly resistant to decay and mildew and is non-allergenic. It can be used at all grade levels, but the denser cushionings are not recommended for radiant heated floors.

FIG. 19 **PHYSICAL REQUIREMENTS FOR CARPET PADDING**[1]

Padding	Moderate Traffic		Heavy Traffic	
	Weight (oz./sq. yd.)	Thickness (in.)	Weight (oz./sq. yd.)	Thickness (in.)
Animal Hair	40	¼	50	⅜
Rubberized Jute and Hair	40	3/16	50	¼
Cellular Rubber[2]	56	¼	56	¼

1. Based on FHA Use of Materials Bulletin UM-44A
2. When bonded to carpet, should conform to Federal Specification DDD-C-95.

INSTALLATION

Carpeting may be installed over finished flooring, subflooring of concrete, plywood, particleboard, or any other suitable, smooth surface. Carpet which is not rubber-backed can be readily installed over nailable surfaces by tacking the edges or by using a *tackless strip* around the perimeter. Installations on concrete can be made by tacking if wooden plugs are provided 6″ to 9″ o.c. in the concrete slab to receive the tacks. Also, tackless strips can be installed readily on concrete with power actuated fasteners or epoxy adhesive. Regardless of the installation method, the carpet should be stretched sufficiently to eliminate wrinkles or buckles.

CERTIFICATION OF QUALITY

There is no industry-wide program for certification of carpet quality at the present time. However, most manufacturers will certify either on their invoice or by an appropriate label that their product meets or exceeds FHA requirements for moderate or heavy traffic areas.

Federal law requires that all textiles be identified with an appropriate label as to fiber content for each fiber type that constitutes 5% or more of the total. In the case of carpets consisting of backing yarns and pile yarns, only the fibers in the pile yarn need be identified. In addition, carpet samples shown to consumers must be similarly identified as to fiber content, name of manufacturer and country of origin, if imported. Several producers of synthetic fibers have agreements with carpet manufacturers which stipulate that only carpet meeting certain minimum construction requirements can display their fiber trademark.

GLOSSARY

Acrylics In carpeting, generic term including acrylic and modified acrylic (modacrylic) fibers. Acrylic is a polymer composed of at least 85% by weight of acrylonitrile; modacrylic is a polymer composed of less than 85% but at least 35% by weight of acrylonitrile.

Axminster carpet See *Woven carpets.*

Backing The carpet foundation of jute, kraftcord, cotton, rayon, or polypropylene yarn that secures the pile yarns and provides stiffness, strength and dimensional stability.

Bearding Long fiber fuzz occurring on some loop pile fabrics, caused by fibers snagging and loosening due to inadequate anchorage.

Bonded rubber cushioning Rubber or latex cushioning adhered to the carpet at the mill.

Broadloom Carpet woven on a *broad* loom in widths of 6′ or more.

Buckling Wrinkling or ridging of the carpet after installation, caused by insufficient stretching, dimensional instability or manufacturing defects.

Bulked continuous filament (BFC) Continuous strands of synthetic fiber made into yarn without spinning; often extruded in modified cross section such as multi-lobal, mushroom or bean shape, and/or texturized to increase bulk and covering power (see also *Continuous filament nylon).*

Burling Removing surface defects such as knots, loose threads and high spots to produce acceptable quality after weaving; also, filling in omissions in weaving.

Carpet General designation of fabric constructions which serve as soft floor coverings, especially those which cover the entire floor and are fastened to it, as opposed to *rugs.* (See also *Woven, Tufted, Knitted, Punched* and *Flocked* carpets.)

Chain warp See *Warp.*

Construction The method by which the carpet is made (loom or machine type) and other identifying characteristics—including pile rows per inch, pitch, wire height, number of shots, yarn count and plies, pile yarn weight and density.

Count A number identifying yarn size or weight per unit of length (or length per unit of weight) depending on the spinning system used (such as denier, woolen, worsted, cotton or jute system).

Crimping Method of texturizing staple and continuous filament yarn to produce irregular alignment of fibers and increase bulk and covering power; also facilitates interlocking of fibers, which is necessary for spinning staple fibers into yarn.

Cushioning Soft, resilient layer provided under carpeting to increase underfoot comfort, to absorb pile-crushing forces and to reduce impact sound transmission (also referred to as *underlay, lining;* see also *Padding).*

Delustered nylon Nylon on which the normally high sheen has been reduced by surface treatment.

Denier System of yarn count used for synthetic fibers: number of grams per 9,000 meters of yarn length; one denier equals 4,464,528 yards per pound or 279,033 yards per ounce.

Density See *Pile yarn density.*

Dimensional stability The ability of a fabric to retain its dimensions in service and wet cleaning.

Filling See *Weft.*

Flocked carpet Single-level velvety pile carpet composed of short fibers embedded on an adhesive-coated backing.

Fluffing See *Shedding.*

Frames Racks at the back of a Jacquard loom, each holding a different color of pile yarn. In Wilton carpets, 2 to 6 frames may be used and the number is a measure of quality as well as an indication of the number of colors in the pattern, unless some of the yarns are buried in the backing.

Frieze carpet See *Pile.*

Fuzzing Temporary condition on new carpet consisting of irregular fuzzing appearance caused by slack yarn twist, fibers snagging or breaking of yarn. Can be remedied by spot shearing.

Gauge The distance between tufts across the width of knitted and tufted carpets, expressed in fractions of an inch.

Grin Condition where the backing shows through sparsely spaced pile tufts. Carpets may be grinned (bent back) deliberately to reveal the carpet construction.

Heat-set nylon Nylon fiber which has been heat treated to retain a desired shape.

Jacquard Mechanism for a Wilton loom which uses punched cards to produce the desired color design.

Jaspe carpet See *Pile.*

Jute Strong, durable yarn spun from fibers of the jute plant, native to India and Far East, used in the backings of many carpets.

Knitted carpet Carpet made on a knitting machine by looping together backing, stitching and pile yarns with three sets of needles, as in hand knitting.

Kraftcord Tightly twisted yarn made from wood pulp fiber, used as an alternate for cotton or jute in carpet backing.

Loom Machine on which carpet is woven, as distinguished from other machines on which carpets may be tufted, flocked or punched.

Loomed carpet See *Woven carpet.*

Modacrylics See *Acrylics.*

Moresque Multicolored yarn made by twisting together two or more strands of different shades or colors.

Natural gray yarn Unbleached and undyed yarn spun from a blend of black, brown or gray wools.

Olefins Long-chain synthetic polymers composed of at least 85% by weight of ethylene, propylene or other olefin units. Currently, only polypropylene has been produced in fiber form for carpet manufacture.

Package dyeing Placing spun and wound yarn on large perforated forms and forcing the dye through the perforations.

Padding Cellular rubber, felted animal hair or jute fibers in sheet form, used as cushioning under carpet (see also *Cushioning).*

Piece dyeing Immersing an entire carpet in a dye bath to produce single- or multi-color pattern effects (see also *Resist printing).*

Pile The raised yarn tufts of woven, tufted and knitted carpets which provide the wearing surface and desired color, design or texture. In flocked carpets, the upstanding, non-woven fibers.

Cut loop pile—Pile surface in which tufts have been cut to reveal the fiber ends.

Frieze—A rough, nubby-textured carpet using tightly twisted yarns.

Jaspe—Carpet surface characterized by irregular stripes produced by varying textures or shades of the same color.

Multi-level—Texture or design created by different heights of tufts, either cut or uncut loop.

Plush—A smooth-face cut pile surface that does not show any yarn texture.

Sculptured (carved)—Surface designs created by combinations of cut and loop pile and/or variations in pile height.

Shag—Surface consisting of long twisted loops.

Stria (Striped)—A striped surface effect obtained by loosely twisting two strands of one shade of yarn with one strand of a lighter or darker shade.

Uncut loop—Pile yarns are continuous from tuft to tuft, forming visible loops.

Pile crushing Bending of pile due to foot traffic or the pressure of furniture.

Pile Height The height of pile measured from the top surface of the backing to the top surface of the pile (also referred to as *pile wire height).*

Pile setting Brushing after shampooing to restore the damp pile to its original height.

Pile yarn density The weight of pile yarn per unit of volume in carpet, usually stated in oz. per cu. yd.

Pilling Appearance defect associated with some staple fibers where balls of tangled fibers are formed on the carpet surface which are not removed readily by vacuuming or foot traffic; pills can be removed by periodic clipping.

Pitch The number of tufts or pile warp yarns in 27″ width of woven carpet.

Ply Term used to designate the number of single strands used in the finished yarn.

Polypropylene See *Olefin*

Print dyeing Screen printing a pattern on carpet by successive applications of premetalized dyes, which are driven into the pile construction by an electromagnetic charge.

Punched carpet Carpet made by punching loose, unspun fibers through a woven sheet which results in a pileless carpet similar to a heavy felt; usually consists entirelv of synthetic fibers.

Quarter width The unit of yard measure (1/4) used in referring to carpet or loom widths. Early European carpet was woven in widths of 27″ or 3/4 yds.—hence, 4/4 = 1-yard width or 3′; 12/4 width = 9′ and 16/4 width = 12′.

Repeat The distance lengthwise from one point in a figure or pattern to the same point at which it again occurs in the carpet.

Resist printing Placing a dye-resist agent on carpet prior to piece dyeing so that the pile will absorb color according to a predetermined design.

Round wire A pile wire which does not cut the pile loop; or woven carpet with an uncut loop pile.

Scrim Rough, loosely woven fabric often used as a secondary backing on tufted carpets.

Selvage The finished lengthwise edge of woven carpet that will not unravel and will not require binding or serging.

Serging A method of finishing a lengthwise cut edge of carpet to prevent unraveling; distinguished from finishing a cut end which may require binding.

Shading Bending or crushing the pile surface so that the fibers reflect light unevenly—not a defect but an inherent characteristic of some pile fabrics.

Shearing A carpet finishing operation that removes stray fibers and fuzz from loop pile, and produces a smooth level surface on cut pile.

Shedding A normal temporary condition of dislodged loose short fibers in new carpeting after initial exposure to traffic and sweeping.

Shooting See *Sprouting.*

Shot See *Weft.*

Skein dyeing Immersing batches of yarn (skeins) in vats of hot dye.

Solution dyeing Adding dye or colored pigments to synthetic material while in liquid solution before extrusion into fiber.

Space dyeing Alternating bands of color applied to yarn by rollers at predetermined intervals prior to tufting.

Sprouting Temporary condition on new carpets where strands of yarn work loose and project above the pile. Can be remedied by careful clipping or spot shearing.

Staple fibers Relatively short natural (wool) or synthetic fibers ranging from approximately 1-1/2″ to 7″ in length, which are spun into yarn.

Stock dyeing Dyeing raw fibers before they are carded (combed) or spun.

Stria (Striped) See *Pile.*

Stuffer yarn See *Warp.*

Tone-on-tone Carpet pattern made by using two or more shades of the same hue.

Tufted carpet Carpet made by inserting the pile yarns through a prewoven fabric backing on a machine with hundreds of needles (similar to a huge sewing machine).

Tufts Surface loops of pile fabric.

Underlay See *Cushioning.*

Velvet carpet See *Woven carpet.*

Vinyl foam cushioning Carpet cushioning made from a combination of foamed synthetic materials.

Warp Backing yarns running lengthwise in the carpet.
Chain Warp—Zigzag warp yarn that works over and under the shot yarns of the carpet, binding the backing yarns together.
Pile warp—Lengthwise pile yarns in Wilton carpets which form part of the backing.
Stuffer warp—Yarn which runs lengthwise in the carpet but does not intertwine with any filling (weft shot) yarns; serves to give weight, thickness and stability to the fabric.

Weaving Process of forming carpet on a loom by interlacing the warp and weft yarns.

Weft Backing yarns which run across the width of the carpet. In woven carpets, the weft shot (filling) yarns and the warp chain (binder) yarns interlock and bind the pile tufts to the backing. In tufted carpets, pile yarns which run across the carpet are also considered weft yarns.

Wilton carpet See *Woven carpet.*

Wires (Pile wire, gauge wire, standing wire) Metal strips over which the pile tufts are formed in woven carpets (see also *Round wire).*

Woolen yarn Soft, bulky yarn spun from both long and short wool fibers which are not combed straight but lie in all directions so they will interlock to produce a felt-like texture.

Worsted yarn Strong, dense yarn made from long staple fibers which are combed to align the fibers and remove extremely short fibers.

Woven carpet Carpet made by simultaneously interweaving backing and pile yarns on one of several types of looms from which the carpets derive their names.
Axminster—Carpet made on an Axminster loom, capable of intricate color designs, usually with level cut-pile surface.
Loomed—Carpet made on a modified upholstery loom with characteristic dense low-level loop pile, generally bonded to cellular rubber cushioning.
Velvet—Carpet made on a simple loom, usually of solid color or moresque, with cut or loop pile of either soft or hard twisted yarns.
Wilton—Carpet made on a loom employing a Jacquard mechanism which selects two or more colored yarns to create the pile pattern.

We gratefully acknowledge the following for the use of their publications as references and permission to use photographs and illustrations: American Carpet Institute; American Hotel & Motel Association; Commercial Carpet Corp.; E. I. duPont de Nemours & Co.; Hercules, Inc.; Mohawk Carpet Mills; Monsanto Company; Walter E. Selck & Co.; Trend Mills, Inc.; The Wool Bureau, Inc.; Wunda Weve Carpet Company, Inc.

CARPETING 226

CONTENTS

SELECTION CRITERIA

Carpet can be selected and specified broadly according to "how" it is made (construction) and "of what" it is made (pile fibers), as summarized in Selection Criteria below. Other criteria required to establish levels of suitable quality are described in the second subsection, Standards of Quality.

The term broadloom does not identify a particular type of carpet but refers merely to the typical product, which is over 6' wide. Most carpets described below, except the rubber-backed loomed type, are available in widths of 9', 12', 15' and sometimes 18'. Loomed carpet generally is made in 4'6" width, with a 3/16" attached rubber cushioning.

Carpet performance cannot be attributed to any single element of carpet construction or physical property of the pile fiber. Installed performance is rather the combined effect of many variables, such as surface texture and carpet construction, backing and cushioning, pile thickness, weight and density.

Of these, pile weight and pile density are the most uniformly significant, regardless of fiber type or carpet construction (see Main Text pages 226-5 and 226-6). However, awareness of the performance characteristics of pile fibers and the outstanding features of the several carpet construction types will aid in the selection of carpet.

CARPET CONSTRUCTION

Modern carpet can be classed according to construction as tufted, woven *and* knitted. *Four weaving processes are used from which woven carpets derive their names:* velvet, Wilton, Axminster *and* loomed *(Fig. WF1).*

Tufted carpets can be produced much faster than woven or knitted types, are less expensive to produce and now account for approximately 85% of all carpet manufactured in this country.

Comparative factors related to carpet construction, such as pile yarn weight and pile thickness, number of tufts per sq. in. and other factors, may be useful in comparing carpet of similar construction.

PILE FIBERS

Carpet also can be selected according to the pile fibers of which the pile yarns are made, which include wool, nylon, acrylics, polyester and polypropylene.

Although pile fibers differ in their physical properties, all fibers included in this discussion will provide adequate service when used in a carpet construction suitable for the traffic exposure.

A table of performance characteristics of pile fibers (Fig. WF2) may be useful in carpet selection for special conditions of traffic or exposure to special hazards.

STANDARDS OF QUALITY

There are no accepted industry-wide standards of quality for carpeting at the present time. However, the General Services Administration (GSA) and Federal Housing Administration (FHA) both have issued standards. FHA specifications contained in Use of Materials Bulletin UM44A are related more directly to residential carpeting and are the basis for the following general recommendations.

FHA requirements for carpeting in heavy traffic areas are made by reference to GSA's Federal Specification DDD-C-95 (see Fig. WF3).

(Continued) **STANDARDS OF QUALITY**

SUITABLE USES

Carpeting should not be used in baths, kitchens or service areas, such as utility, laundry or furnace rooms.

Carpeting can be used in the remaining areas within the home—living, dining, recreation, sleeping and dressing rooms—designated as moderate traffic areas, as well as in heavy duty public areas—corridors, entrances, stairways and elevators.

CARPET CONSTRUCTION

Carpeting for heavy traffic areas should conform with the detailed requirements of Fig. WF3.

Carpeting for moderate traffic areas should meet or exceed minimum *pile yarn weights* and minimum *weight-density factors* shown in Fig. WF4.

PILE FIBERS

The pile yarn should be made of 100% wool, nylon, acrylic, modacrylic or polypropylene olefin fibers or blends of these fibers (exclusive of ornamentation).

Not less than 20% of any of the above fibers should be used when blended, and such blends should not consist of more than two types of fibers.

FIG. WF1 CONSTRUCTION AND APPEARANCE FEATURES OF CARPET TYPES

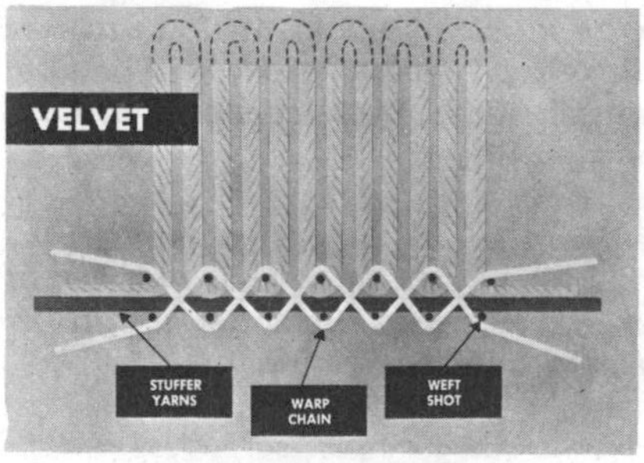

Since the velvet loom feeds the pile yarn from a single spool, it is best suited for solid colors. However, strands of different color can be twisted together into a single yarn (moresque), producing tweeds, salt-and-pepper effects and stripes. A wide range of textural effects is possible by forming high and low loops.

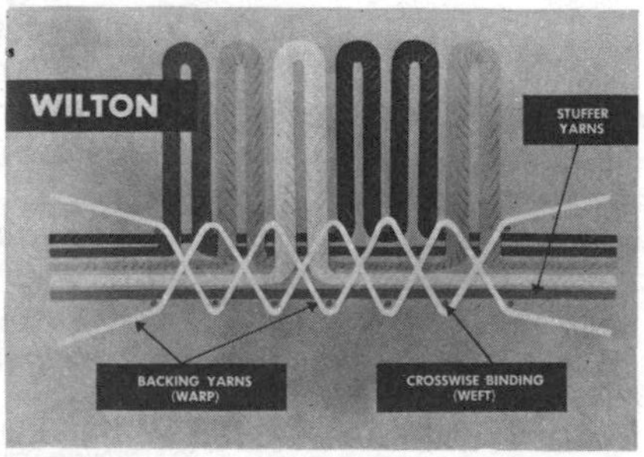

Wilton carpets are designated as two-, three-, four-, or five-frame Wiltons depending on the number of colored pile yarns used. The Wilton loom can weave sharply delineated sculptured textures or patterns, which are created by varying the pile height, using high and low loops or combinations of cut and uncut piles.

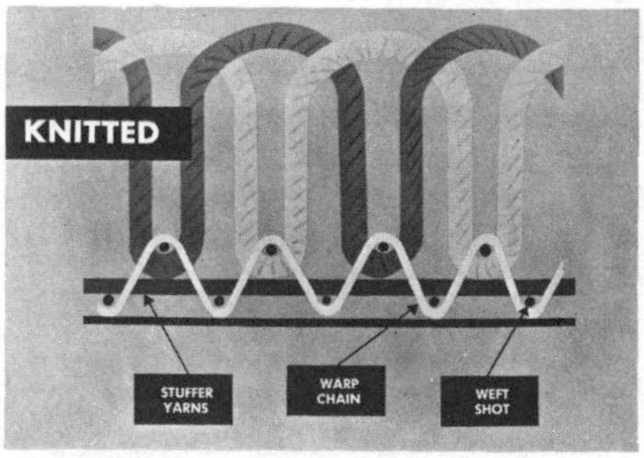

Since a single pile yarn is used in the knitting operation, as in velvet weaving, knitted carpets are usually solid colors or tweeds, although more recently patterns have become possible. Knitted carpets usually have either single- or multi-level uncut pile, but cut pile can be achieved with modification of the knitting machine.

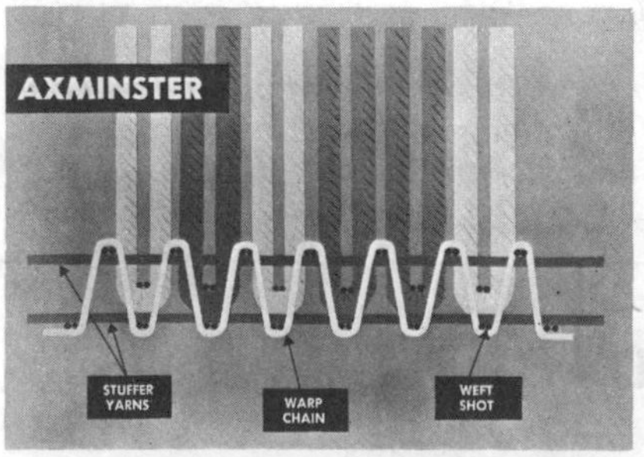

The Axminster loom simulates hand weaving because each tuft of yarn is individually inserted, and theoretically each tuft could be of a different color. This flexibility offers unlimited design possibilities and intricate patterns are produced even in lower priced carpet. The Axminster weave is almost always a cut pile with even pile height and elaborate color patterning.

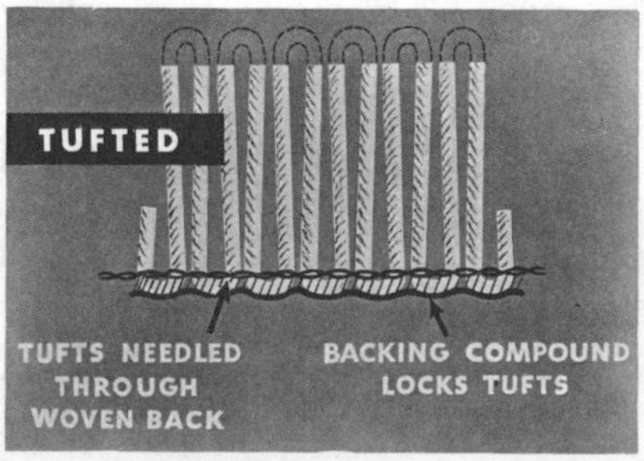

The variety of textures and colors in tufted carpet is increasing with innovations in tufting equipment and dyeing processes. Tufted pile can be level-loop, plush, multilevel, cut or uncut, or a combination of cut and uncut. Striated and carved effects also can be achieved. Tufted carpets are made in solids, moresques, and more recently in multi-color patterns.

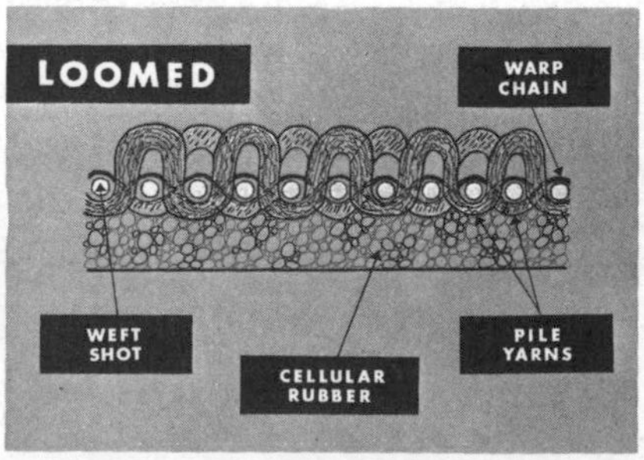

Loomed carpet is a recently developed fabric intended specifically for use with a bonded rubber cushioning. It is characteristically a high-density nylon with single-level, low-loop pile about 1/8″ high, woven on a modified upholstery loom. Loomed carpet generally is manufactured in 4′6″ width with a rubber backing approximately 3/16″ thick.

(Continued) STANDARDS OF QUALITY

Wool yarn should contain at least 97% wool fibers by dry weight; synthetic fibers should be 15 denier or thicker and should contain not more than 2% chloroform soluble material.

Pile should be resistant to fading from wet cleaning and exposure to light. Light colors should be rated as satisfactory after laboratory exposure to 20 standard fading hours and dark colors after 40 fading hours.

BACKING

Backing materials should be those customarily used and accepted in the industry for each type of carpet. For tufted carpet, there should be a secondary backing reinforcement of at least 4 oz. per cu. yd., exclusive of the adhesive coating.

CUSHIONING

All carpet should be installed over cushioning to increase resilience and durability. When the cushioning is bonded to the carpet, it should meet or exceed requirements of Federal Specification DDD-C-95.

Separate cushioning made of cellular rubber, jute-and-hair or hair felt should conform to GSA Specification ZZ-C-00811b and DDD-C-00123 and the minimum weight requirements for heavy and moderate traffic areas given in Fig. WF5.

Rubber cushioning should not be used over radiant heated floors.

CERTIFICATION OF QUALITY

There is presently no industry-wide certification program. However most manufacturers will certify either on their invoice or by an appropriate label that their product meets or exceeds FHA requirements for moderate or heavy traffic areas.

Federal law requires that all textiles be identified with an appropriate label as to fiber content for each fiber type that constitutes 5% or more of the total. In the case of carpets consisting of backing yarns and pile yarns, only the fibers in the pile yarn need be identified. In addition, carpet samples shown to consumers must be similarly identified as to fiber content, name of manufacturer and country of origin, if imported.

Several producers of synthetic fibers have agreements with carpet manufacturers which stipulate that only carpet meeting certain minimum construction requirements can display their fiber trademark.

FIG. WF2 PERFORMANCE CHARACTERISTICS OF PILE FIBERS[1]

Criteria	WOOL	NYLON	ACRYLIC	MODACRYLIC	POLYPROPYLENE
Resistance To:					
Abrasion	Good	Excellent	Good	Good	Excellent
Alkalis	Fair	Good to Excellent	Good to Excellent	Good to Excellent	Good to Excellent
Acids	Fair	Good to Excellent	Good	Good	Good to Excellent
Insects & Fungi	Excellent[2]	Excellent	Excellent	Excellent	Excellent
Burns	Good	Fair	Fair	Fair	Fair
Compression	Good	Good	Good	Good	Fair
Crushing	Excellent	Good	Good to Excellent	Good	Fair to Good
Staining	Good	Good to Excellent	Good to Excellent	Good to Excellent	Excellent
Soiling	Good	Fair to Good	Good to Excellent	Good to Excellent	Excellent
Static Buildup	Fair to Good	Fair	Good	Good	Excellent
Texture Retention	Excellent	Good to Excellent	Good to Excellent	Good	Fair to Good
Wet Cleanability	Fair to Good	Good	Good to Excellent	Good to Excellent	Excellent
Durability	Good to Excellent	Excellent	Good to Excellent	Good to Excellent	Good
Appearance Retention	Excellent	Good to Excellent	Good to Excellent	Good to Excellent	Good
Ease of Maintenance	Good to Excellent	Good	Good to Excellent	Good to Excellent	Excellent

1. In each case, criteria vary depending on the properties of the specific fiber type used, carpet construction and installation procedures.
2. When chemically treated.

(Continued) STANDARDS OF QUALITY

FIG. WF3 PHYSICAL REQUIREMENTS FOR CARPETING IN HEAVY TRAFFIC AREAS[1]

	WILTON	VELVET			TUFTED	KNITTED
	Wool or Acrylic	Wool or Acrylic	Nylon[2]	Wool or Acrylic	Wool or Acrylic	Wool or Acrylic
Pile Description	Single level loop woven thru back	Single level loop pile woven thru back	Single level loop pile woven thru back	Multilevel loop woven thru back	Single level loop	Single level loop
Tufts/sq. in.	52	60	32	34	60	26
Frames	3	—	—	—	—	—
Shots/wire	2	2	2	2	—	—
Weight oz./sq. yd.						
Pile	48	42	29	44	42	37
Total	72	60	50	64	—	58
Pile height	Min. 0.250 Max. 0.320	Min. 0.210 Max. 0.310	Min. 0.200 Max.0.290	Min. 0.190 Max. 0.370	Min. 0.250 Max. 0.320	Min. 0.230 Max. 0.290
Chain	Cotton and/or rayon	Cotton and/or rayon	Cotton and/or rayon	Cotton and /or rayon	—	Cotton, rayon or nylon
Filling	Cotton and/or rayon, or jute	Cotton and/or rayon, or jute	Cotton and/or rayon, or jute	Cotton and/or rayon, or jute	—	Jute or kraftcord
Stuffer	Cotton, jute or kraftcord	Cotton, jute or kraftcord	Cotton, jute or kraftcord	Cotton, jute or kraftcord	—	—
Back coating oz./sq. yd.	—	8	6	8	—	—
Tuft bind (oz.)	80	80	80	80	100	14
Ply twist turns (per inch)	Min. 1.5 Max. 3.5	Min. 1.5 Max. 3.5	Min. 1.0 Max. 3.0	Min. 1.5 Max. 3.5	Min. 2.5 Max. 4.5	Min. 1.5 Max. 3.5

1. Based on Federal Specifications DDD-C-95.
2. High bulk or textured continuous filament.

FIG. WF4 PHYSICAL REQUIREMENTS FOR CARPETING IN MODERATE TRAFFIC AREAS[1]

Pile Fiber	Minimum Pile Yarn Weight (W) oz./sq. yd.[2]	Minimum Weight-Density Factor (WD)
Wool		
Acrylic	25	70,000 (25 x 2800)
Modacrylic		
Nylon (Staple & Filament)	20	56,000 (20 x 2800)
Polypropylene Olefin		

1. Based on FHA Use of Materials Bulletin UM-44A.
2. Computed for blended yarns by multiplying percentage of each fiber by minimum pile weight and adding the two amounts.

FIG. WF5 PHYSICAL REQUIREMENTS FOR CARPET PADDING[1]

Padding	Moderate Traffic		Heavy Traffic	
	Weight (oz./sq. yd.)	Thickness (in.)	Weight (oz./sq. yd.)	Thickness (in.)
Animal Hair	40	¼	50	⅜
Rubberized Jute and Hair	40	3/16	50	¼
Cellular Rubber[2]	56	¼	56	¼

1. Based on FHA Use of Materials Bulletin UM-44A.
2. When bonded to carpet, should conform to Federal Specification DDD-C-95 Carpets & Rugs: Wool, Nylon, Acrylic, Modacrylic

300 METHODS & SYSTEMS

INTRODUCTION

The general objectives of the homebuilding industry are to build suitable housing of better quality, at less cost if possible, making the individual building and the total housing environment more beautiful in the process.

The selection of construction methods and systems which produce the basic house types shown in Figure 1 usually is governed by three criteria: *functional requirements, total cost* and *desired appearance.*

These basic criteria may require a consideration of: climate, site topography, initial costs, maintenance costs, building codes and zoning ordinances or other laws, availability of materials and labor, builder resourcefulness and size, buyer taste, local custom and other factors.

The selection of methods and systems is further complicated by the thousands of material and product choices available, many of which are interdependent on one another. In fact, the relationship of building elements to each other may create situations where the construction method or system is the major influence on the building design.

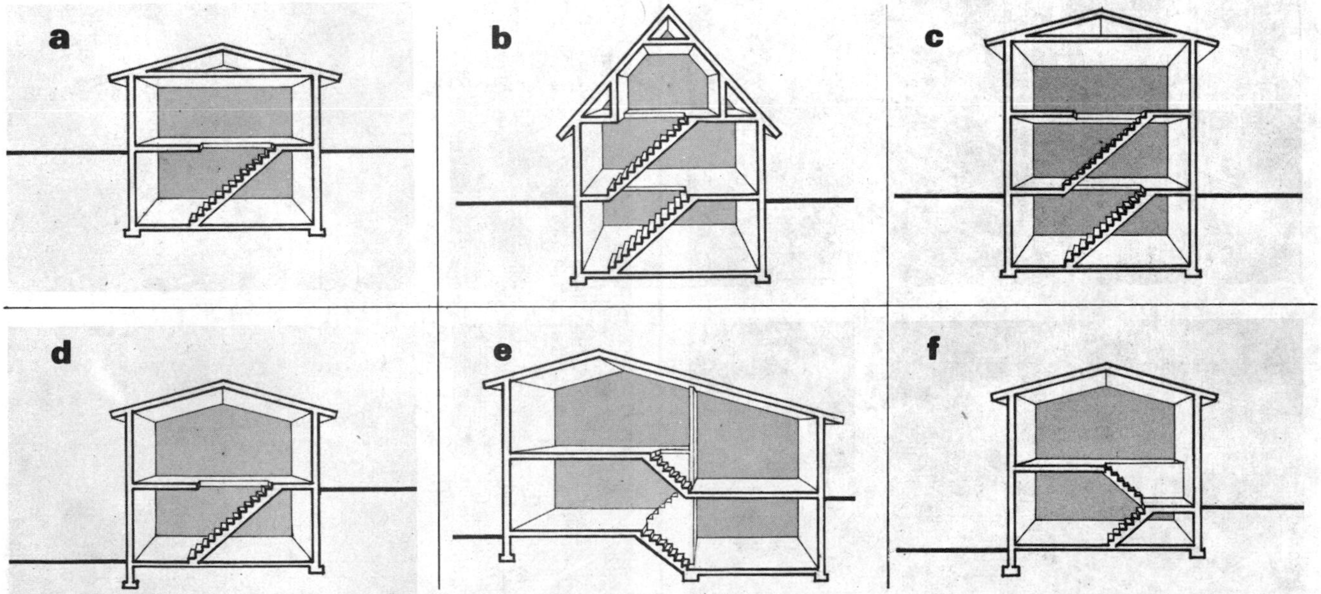

FIG. 1 **Basic House Types** *Construction methods and systems depend in part on the space enclosed. Typical residential spaces: (a)* **one-story**—*with or without basement, is more prevalent than any other basic house type; provides the most size, shape and design variations. (b)* **one-and-one-half-story**—*with or without basement, provides additional second-floor living area which varies in size with overall house dimensions and roof slopes; light, ventilation and view may be provided by using dormers. (c)* **two-story**—*with or without basement, provides maximum living areas at least cost. (d)* **bi-level**—*extremely suitable for hillside locations; provides living areas at both grade levels when connected with full flight of stairs. (e)* **split-level**—*provides distinct separation of living functions, often on three levels with an optional fourth level basement, each connected with half-flights of stairs; best suited to sloping lots; offers numerous design possibilities but can result in awkward proportions when not carefully designed. (f)* **bi-level/split-level**—*characterized by the split-level foyer between two full living levels; may provide either a sunken two-story house without basement or a raised one-story house with optional finished basement.*

While there have been outstanding experiments with a few revolutionary construction systems since World War II, most current home building practices in the United States are based on conventional light wood construction of *balloon* or *platform* framing (Figs. 2 and 3). The historic balloon frame in which each wall stud runs from foundation to roof has evolved into today's conventional platform framing. Other basic systems that are used include post and beam (Fig. 4) and masonry bearing wall construction (Fig. 5). In many areas, the use of preassembled components, such as those shown in Figure 6, is com-

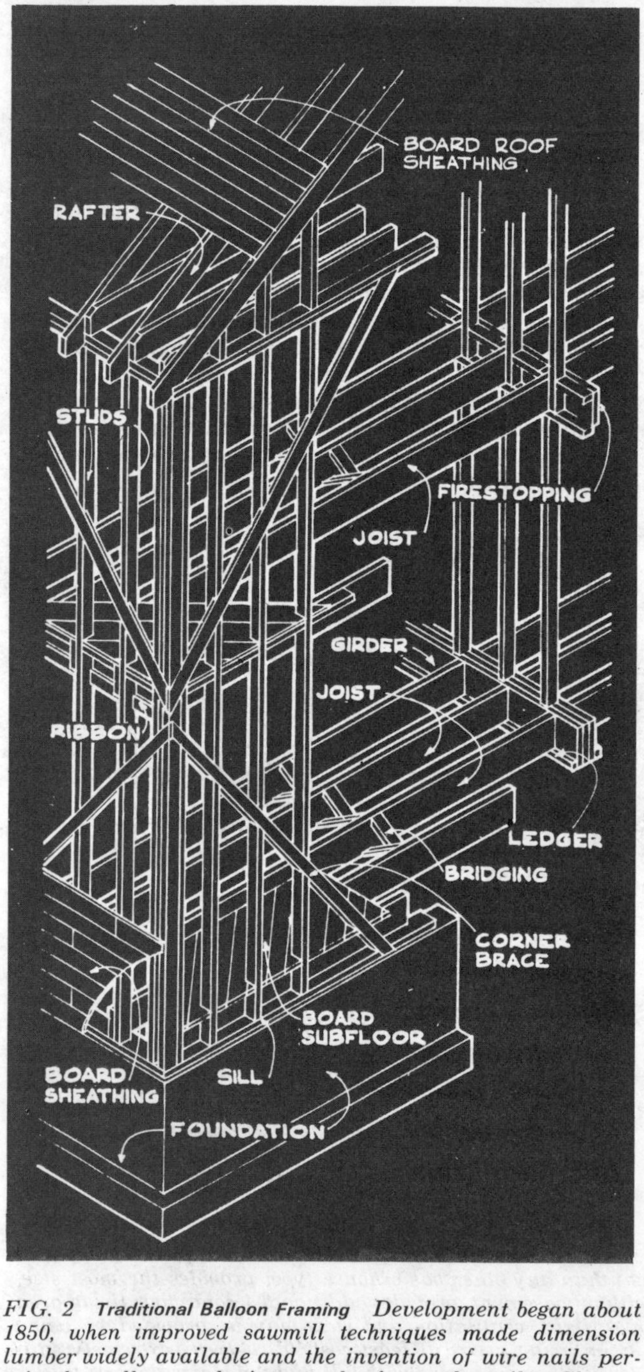

FIG. 2 **Traditional Balloon Framing** *Development began about 1850, when improved sawmill techniques made dimension lumber widely available and the invention of wire nails permitted smaller wood pieces to be fastened easily. Characterized by: individual joists, rafters and building-height studs spaced 12" to 24" o.c.; second-floor joists carried on ribbon let-into studs; wood boards for sheathing, subflooring, floor bridging and corner bracing.*

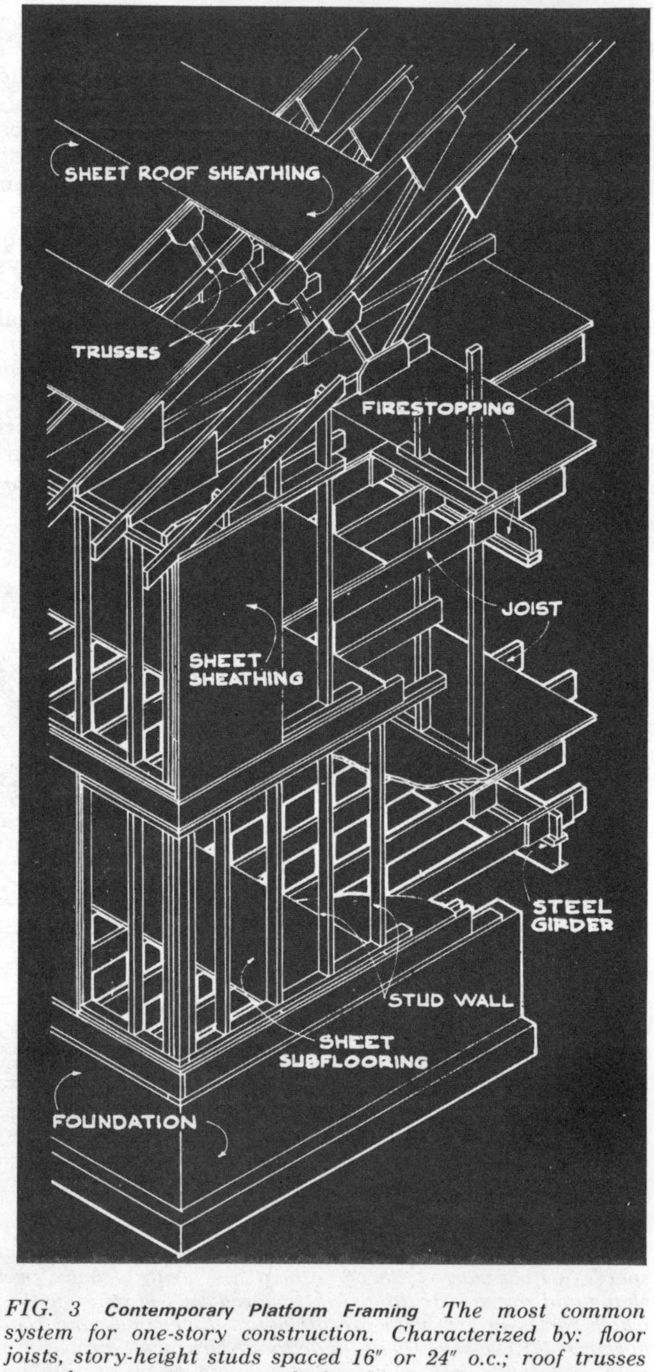

FIG. 3 **Contemporary Platform Framing** *The most common system for one-story construction. Characterized by: floor joists, story-height studs spaced 16" or 24" o.c.; roof trusses spaced up to 4' o.c.; panel sheathing such as plywood or fiberboard; unnecessary corner-bracing and floor bridging is eliminated. Floor construction provides a work platform, permitting efficient tilt-up methods; system adapts easily to panelized construction.*

mon. In the future, advanced industrialization techniques using new materials and methods may offer new construction forms far different from those typical today (Fig. 7).

Conventional wood framing still offers many advantages. As a complete construction system, it still is one of the most economical ways to build. The ease of working and fastening wood together with simple tools provides flexibility which permits job changes without extensive re-engineering. Wood framing is still the basis for most building codes and labor practices and will probably remain so for

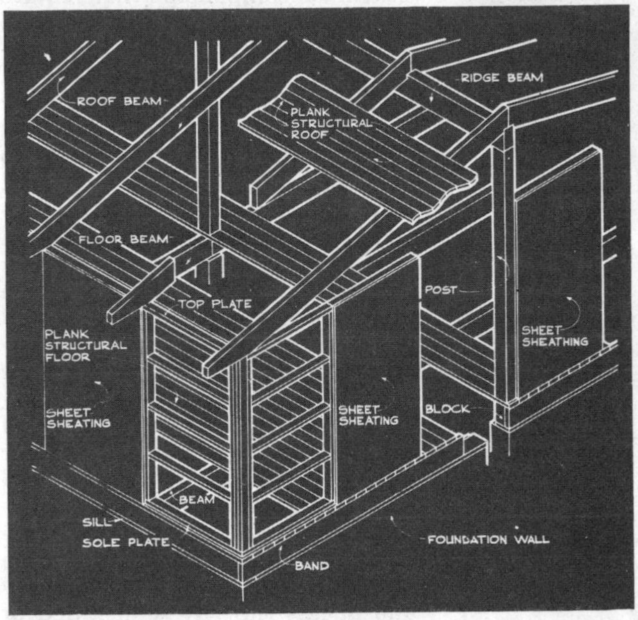

FIG. 4 **Post and Beam Framing** *Consists of posts, beams and planks. Roof and ceiling loads are distributed by planks to beams which transmit their loads directly to posts. Non-load bearing exterior walls serve only as curtain walls and to brace the frame.*

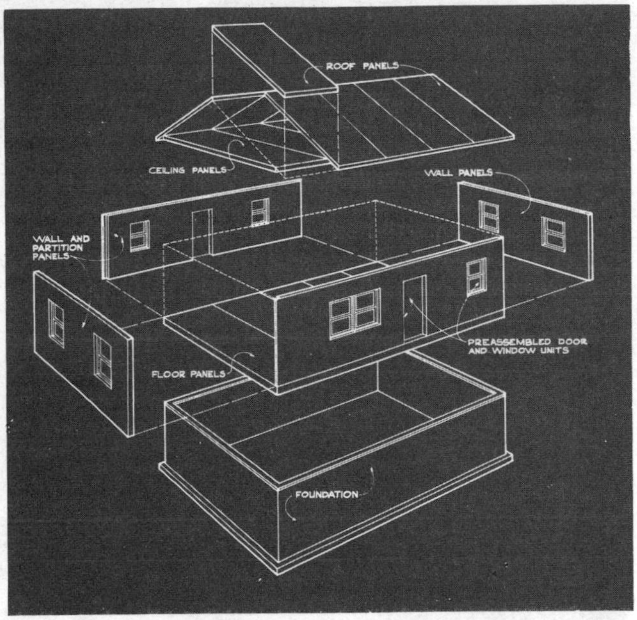

FIG. 6 **Stressed Skin Construction** *A panelized construction system for walls, floors and roofs. It consists of a "skin" (such as plywood) glued to framing members. When loaded, the skin is stressed and assists the framing members in supporting loads.*

FIG. 5 **Masonry Bearing Walls** *Many combinations of clay and concrete masonry bearing walls are possible. Shown is concrete masonry reinforced to provide resistance to lateral forces.*

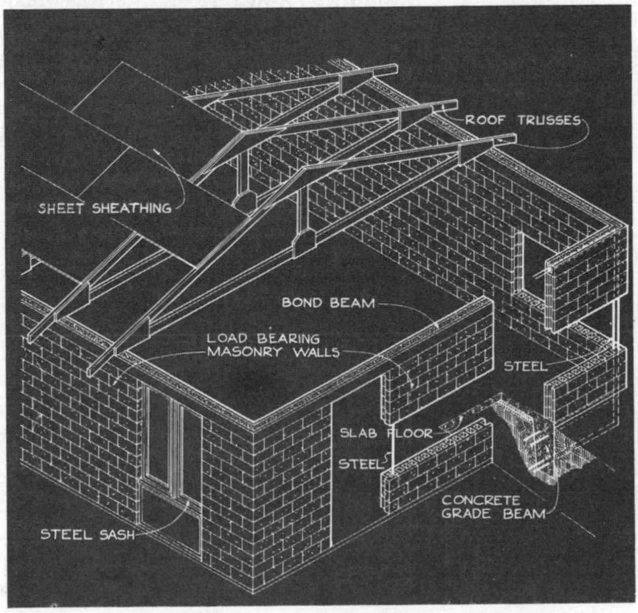

FIG. 7 **Experimental Systems** *New Materials and/or new methods such as the ones used in these experimental houses may preview construction systems of the future.*

FIG. 8 Techniques of site or factory preassembly of parts result in efficient construction methods. Preassembled components may range from small wall panels to half-house units.

Field assembly of small wall panels

Shop assembly of entire end wall

Placing field assembled floor panel

Erection of preassembled trusses

Prefabrication of kitchen-bath unit

Positioning of kitchen-bath unit

Setting large kitchen-bath-utility unit

Half-house units are fitted together

some time to come. Conventional framing is adaptable to site fabrication by the smallest builder handling each member piece by piece, as well as to off-site fabrication of individual pieces into larger preassembled components that require additional manpower or machinery for erection (Fig. 8).

Further industrialization of housing, using more and larger prefinished and prefabricated components, appears essential to help offset the rising costs of land, labor and materials. Off-site fabrication permits maximum utilization of labor and materials under factory-controlled conditions with little loss in on-site time due to bad weather. Efficiency may be increased with the use of power tools and machinery; volume purchasing of materials and stockpiling of finished parts is possible; greater convenience for workmen and better protection for finished materials is provided; and site erection of components usually can be accomplished more economically and in less time by semiskilled or even unskilled labor.

To save costs, mechanical components have been developed that combine furnace, air conditioner, water heater and electric power panel into one package. Larger mechanical components include completely furnished kitchens and bathrooms. The concept of prefinishing complete rooms has been extended to prefabricating up to half a house so that upon setting and joining two halves, an entire house is completed. Future developments may include assembling the entire unit and completely finishing it prior to site placement.

Some future building construction methods will be highly sophisticated and closely integrated systems. For instance, several integrated floor/ceiling systems available for use in commercial construction already include structure, lighting, acoustical control, heating, cooling and air distribution in one integrated system.

Components should be capable of satisfying varying design requirements, should permit simple modifications in the field in case of errors, and should be sized for ease

FIG. 9 Variety of methods and systems are used in residential construction across the country. Local climatic conditions, economic considerations and esthetic preferences may influence the selection of each.

Panelized exterior wall construction

Stressed-skin panels and trusses

Plywood vault and beam roof

Load bearing masonry walls

Wood frame and brick veneer

Plank roof on post and beam structure

Reinforced concrete masonry walls

Prefinished panel siding

Wire lath for stucco walls

Trusses 4' o.c. in research house

Prefinished exterior wall panels

Entire house fabricated off site

of shipping, storage and assembly. As the component size increases, design and construction problems increase. Design flexibility is lessened as the size of the component increases. The dimensions of large units are restricted to what can be transported physically and legally over the highways, and larger components usually require more manpower and larger erection equipment at the site.

Accordingly, the design, engineering or selection of preassembled components require judgments between size and flexibility. The most useful systems will combine the advantages of fully standardized factory-built modular units which capitalize on the inherent savings resulting from repetitive production, and those which offer the design advantages of custom fabrication in the field.

Construction practices vary widely across the country (Fig. 9). In the main, conventional methods and systems are not carefully engineered but are based on a combination of long established custom, rules of thumb and/or arbitrary building code requirements. These practices have resulted in buildings that usually have provided satisfactory performance over the years. However, when these practices are overly conservative—as they often are—they foster excessive waste and therefore higher cost. New performance standards for methods and systems are being established. These design criteria based on laboratory and field testing can materially reduce over-design, waste and cost. The homebuilding industry should continue to encourage the development of performance criteria and the design of methods and systems based on them. In addition, the industry must find the means for quicker acceptance in the marketplace of innovations as they occur.

In this division, the building is divided for convenience into major construction elements: *foundation systems, floor/ceiling systems, wall systems* and *roof/ceiling systems* (Fig. 10). Many conventional types of foundations, floors, walls and roofs are discussed and illustrated. In addition, some new or unique methods and systems are covered because they offer strong potential for future acceptance. Obviously, because of space requirements, many methods and systems cannot be described. Their omission in no way indicates or infers a lack of suitability.

An understanding of the principles and criteria on which construction methods and systems are based is necessary for the judgment in each locale and set of circumstances as to what is suitable quality. The intent of this division is to improve that understanding.

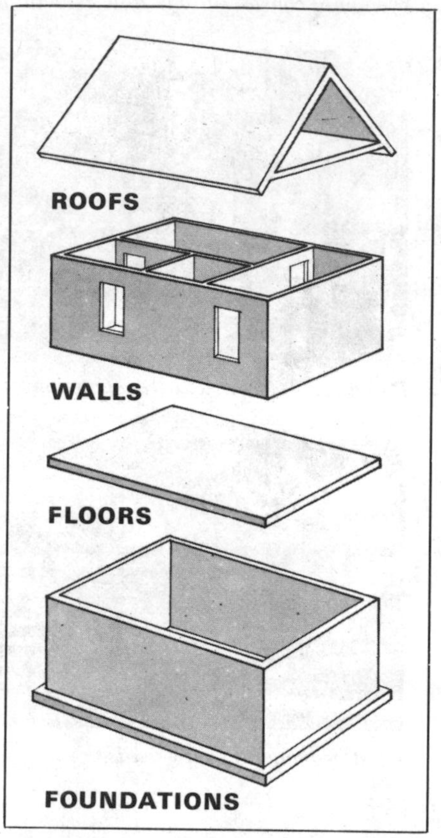

FIG. 10 For convenience in describing and illustrating the numerous combinations of residential construction methods and systems, the building may be subdivided into the four major elements shown above. These elements are discussed in appropriate sections in Division 300.

310 FOUNDATION SYSTEMS

FOUNDATION SYSTEMS

INTRODUCTION

The foundation forms the substructure of a building, supports the building's superstructure and transmits all loads to the soil. Footings, resting on the soil, receive loads from foundation walls, piers, pilasters and/or columns and must distribute the loads directly to the soil without exceeding the allowable bearing capacity of the soil.

All foundations will settle to some extent over time. The extent of settling depends on the compressibility of the underlying soil. One objective in the design and construction of foundations is to minimize this settlement, especially differential settlement, so that the structure or its use will not be impaired.

The bearing capacity and compressibility of soils can be calculated by site and laboratory tests on actual soil samples. With a knowledge of bearing capacity and the magnitude of superimposed loads, foundations can be sized using engineering analysis and settlement can be predicted. *However, this procedure is seldom the practice in residential foundation design.* Instead, residential foundations usually are sized using rules of thumb based on established practices or on the performance of foundations in a locality.

Foundations form various types of basement and basementless substructures. Basements have usable or habitable space. Basementless foundations—crawl spaces, slabs-on-ground and various types of pier and beam systems—may have either partial or no usable space.

This section describes and illustrates some of the materials and methods used to provide suitable residential foundations.

Many of the recommendations for the construction of foundations in this section are based on the standards of the American Standards Building Code requirements for Excavation and Foundations (ASA A56.1) published by the American Society of Engineers.

Specific recommendations for the construction of basements, crawl spaces, slabs-on-ground and other foundation systems are given in other individual sections of Division 300.

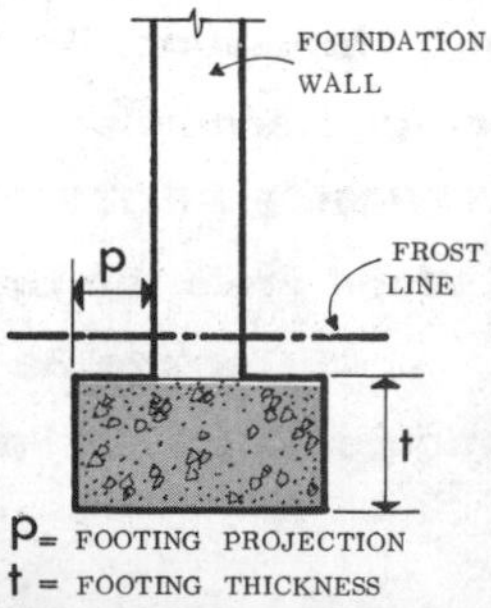

GENERAL RECOMMENDATIONS

A foundation forms the substructure of a building and is composed of several elements which may include various types of footings, walls, beams, piers, columns and pilasters (Fig. 1). The substructure must provide the support for the superstructure and distribute all building loads so that settlement will be either negligible or uniform under all parts of the building. Differential (uneven) settlement under different parts of the foundation may cause substantial problems. This uneven settlement can result in a foundation which is out of plumb or not level, and such damage as cracks in foundation walls and in finished walls and ceilings, sloping floors, binding doors and windows, and open joints in woodwork. When differential settlement is extreme, partial failure of the structural integrity of the building can result.

Many types of foundation systems are used in residential construction. Some of the more common systems are illustrated in Figure 2.

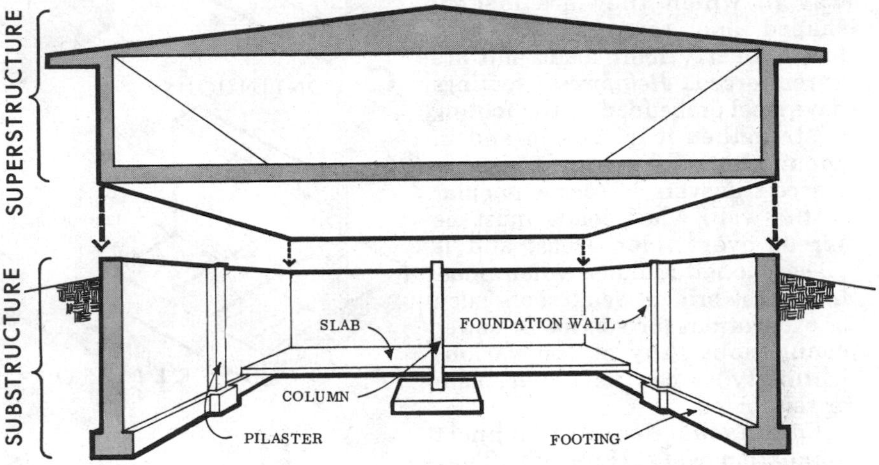

FIG. 1 The substructure of a building—its foundation—consists of load transmitting elements such as foundation walls, pilasters, piers, columns, and footings, and other elements such as grade beams and piles.

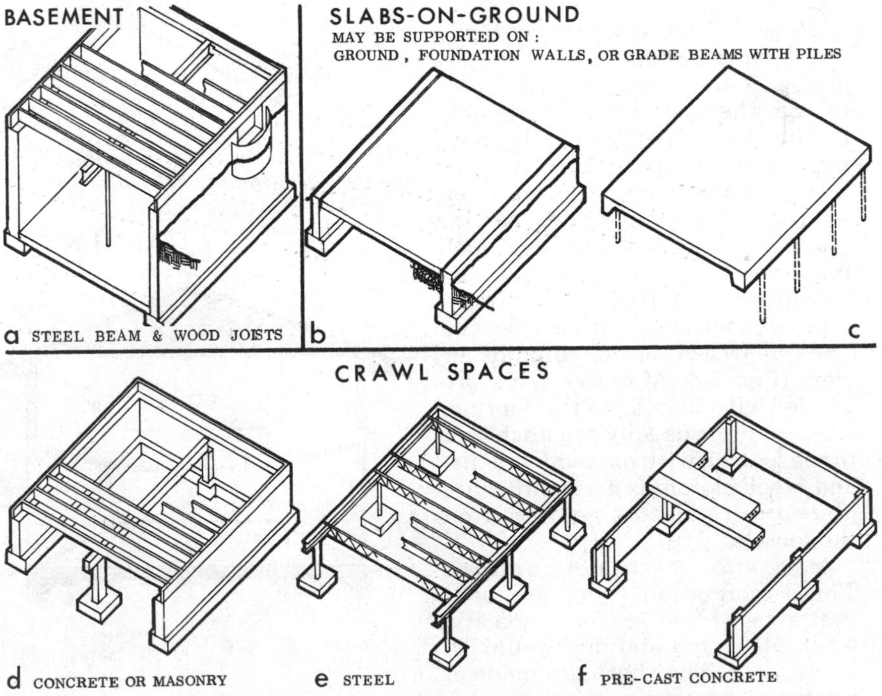

FIG. 2 Foundations may take three general forms: basements, slabs-on-ground and crawl spaces. Several of the typical methods and systems used are shown above. Slabs-on-ground may have spread footings or grade beams and piles. Basements and crawl spaces provide underfloor space for more flexible integration of mechanical, electrical and plumbing installations.

FOUNDATION TYPES

Foundations are of two general types: (1) *spread foundations*, which distribute the building loads directly over a sufficient area of soil to obtain adequate bearing capacity; and (2) *pile foundations*, which transmit the building loads through weak soils (of inadequate bearing capacity for spread foundations) to deeper layers of soil with adequate bearing value.

Spread Foundations

Spread Foundations consist of load transmitting elements—such as *walls, pilasters, columns* or *piers*—which rest on an abruptly enlarged base called a *footing*. The footings in turn spread the load directly to the supporting soil. *Grade beams* also are load transmitting elements and may rest on footings or piles. *Raft* and *mat* foundations, described below, also are classed as spread foundations.

Footings Several types of footings are illustrated in Figure 3. They usually are made of concrete and are classed according to the way in which they are loaded, shaped and constructed. *Plain* footings carry light loads and are unreinforced. *Reinforced* footings have steel embedded in the footing to strengthen it in tension and in shear (Fig. 3c). Reinforcing is placed transversely (perpendicular to the wall) when loads must be spread over wider areas, and is placed longitudinally when footings must bridge weak spots such as excavations for service or sewer connections. Any of the various footing types can be either plain or reinforced.

Continuous Footings support foundation walls (Fig. 3a). They also may be used to support several columns in a row to minimize differential settlement. When a single footing supports more than one column, it is called a *combined footing* (Fig. 3d).

Stepped Footings widen gradually at the base and distribute the load over a wider area (Fig. 3b). Although they were used commonly in the past, they generally have been replaced by reinforced footings. A footing that changes levels in stages to accommodate a sloping grade also is called a stepped footing (Fig. 3b).

Isolated Footings are independent footings that receive the loads of free-standing columns or piers (Fig. 3e). Masonry piers are bonded with mortar to the footing; steel columns usually are anchored to the footing with embedded bolts; and wood columns frequently are pinned to retain their position on the footing.

Mat and Raft Foundations These foundation types are used over soils of low bearing capacity, when other foundations would be inadequate. They both are made of concrete, heavily reinforced with steel, so that the entire foundation will act as a unit. Thus, they often are referred to as *floating foundations*.

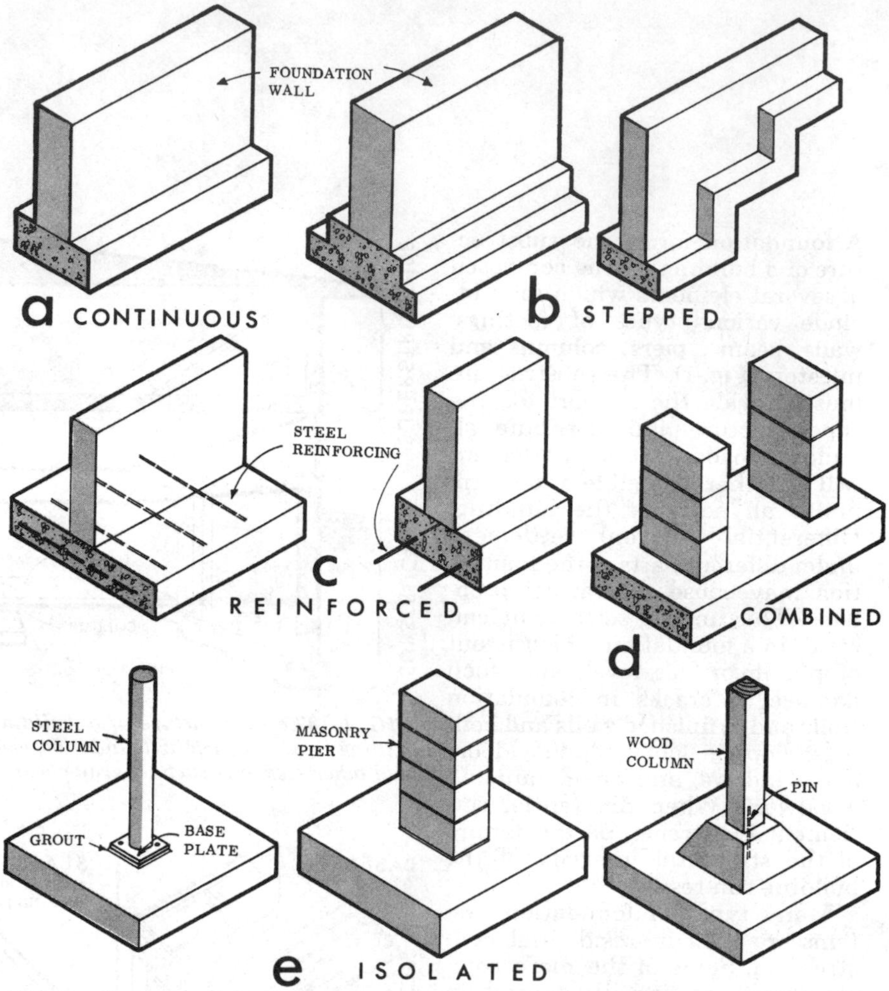

FIG. 3 Various types of footings spread building loads over soil area. Any of the types shown may be reinforced.

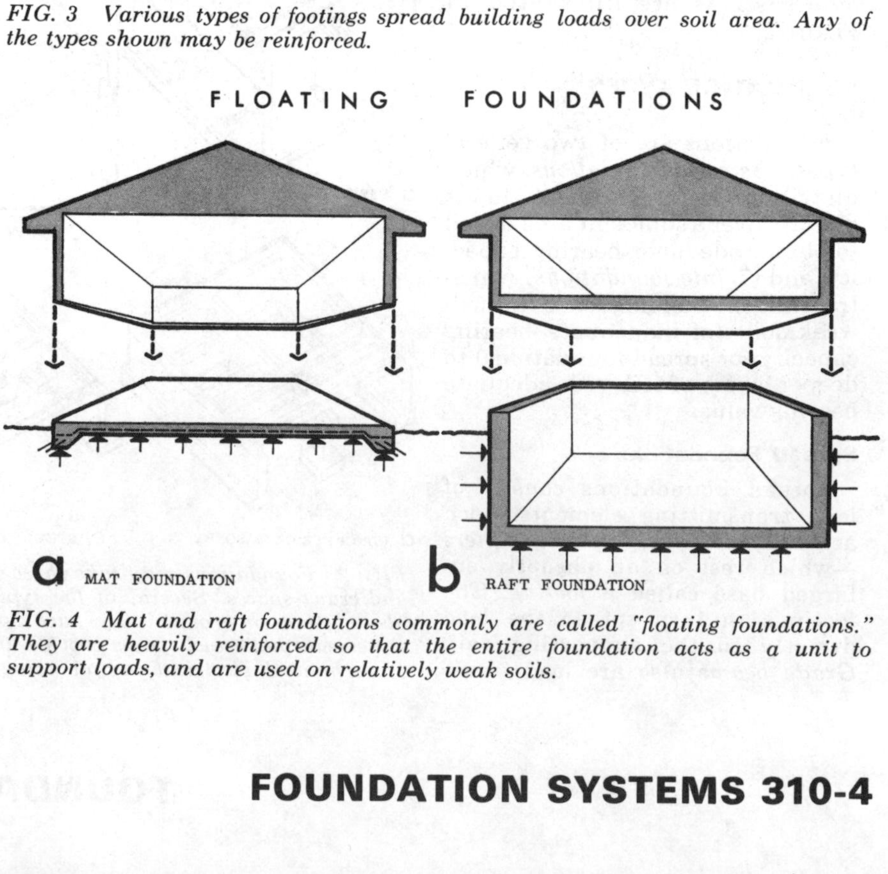

FIG. 4 Mat and raft foundations commonly are called "floating foundations." They are heavily reinforced so that the entire foundation acts as a unit to support loads, and are used on relatively weak soils.

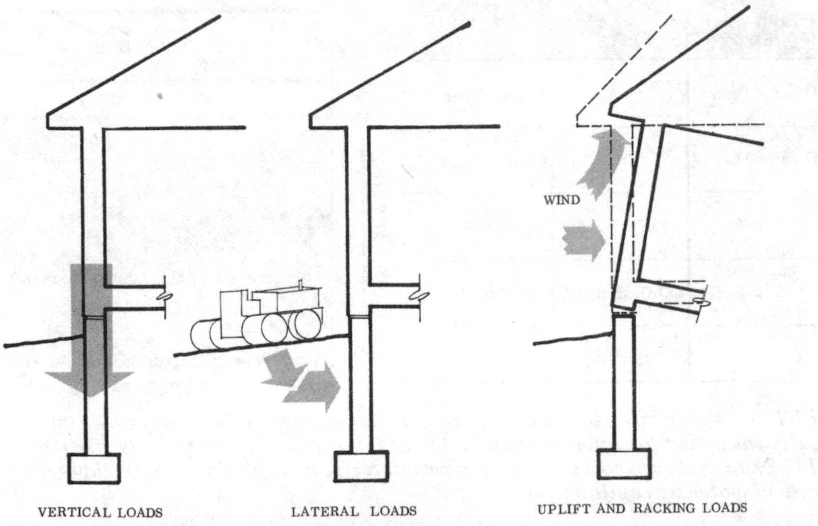

FIG. 5 Foundation walls must: (1) resist vertical loads from the weight of the building and its occupants; (2) resist lateral loads from the soil, water and other superimposed loads such as earthmoving equipment; and (3) anchor the building to the foundation.

A *mat foundation* (Fig. 4a) is a thickened slab that supports loads and transmits them as one structural unit over the entire slab and soil surface area. A *raft foundation* (Fig. 4b) consists of reinforced walls and floor (forming the raft) and is constructed by excavating until the weight of the soil removed is approximately equal to the weight of the foundation plus the weight of the superstructure. In this way, settlement is minimized on low-strength soils because the building load on the soil is no more than the weight of the excavated soil.

Foundation Walls As load transferring elements, foundation walls form an integral part of spread foundations. They must be able to support vertical loads from the superstructure; they must resist lateral (horizontal) loads from the ground and the superstructure; and they must provide anchorage for the superstructure against uplift and lateral forces caused by wind (Fig. 5).

In addition to structural functions, foundations walls must be durable, resist water and vapor penetration, provide a barrier against fire, and control air and heat flow. If the foundation encloses a habitable space, the wall must present a suitable appearance or be able to accept other finishes as well as accommodate windows, doors, and/or other openings.

Columns, Piers and Pilasters Each of these elements act in compression to transmit loads to footings. Columns and piers are free-standing; pilasters are built integrally with the wall and also may function as stiffeners to provide necessary lateral support to a wall.

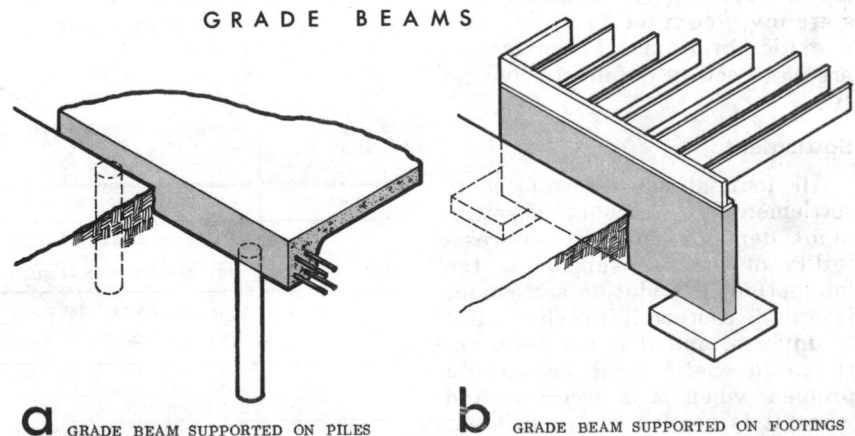

Grade Beams In contrast to a continuous footing which is supported uniformly along its length, grade beams (located approximately at grade level) receive building loads and transfer them to either spread or pile foundations. Grade beams may take many forms. The examples shown in Figure 6 include (a) a reinforced concrete beam placed integrally with the concrete slab and supported on piles, and (b) a precast reinforced wall carried on footings and supported laterally by concrete or masonry piers.

Pile Foundations

Piles are column-like units that transmit loads through poor soil to lower levels of soil having adequate bearing capacity (Fig. 7). In this sense, piles serve the same purpose as footings—they transmit loads to soil stratum capable of carrying the

GRADE BEAMS

FIG. 6 Grade beams take several forms: (a) they may be poured integrally with slabs-on-ground; or (b) they may transmit vertical loads to other foundation elements such as footings.

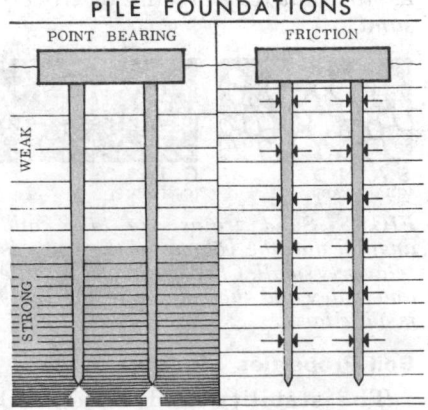

PILE FOUNDATIONS

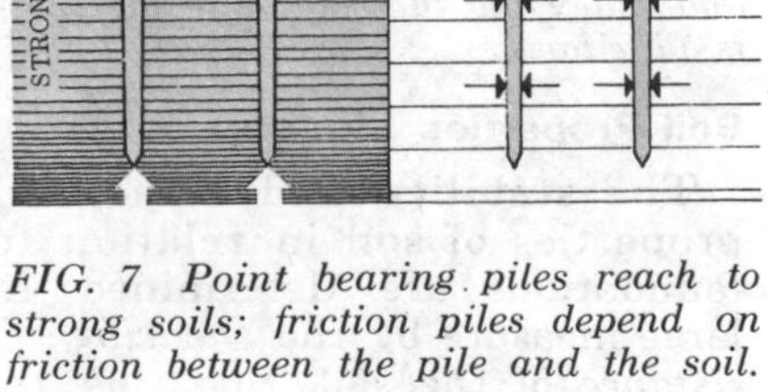

FIG. 7 Point bearing piles reach to strong soils; friction piles depend on friction between the pile and the soil.

load. Piles receive the loads from foundation walls, from isolated columns or from grade beams as shown in Figure 6a.

Pile foundations generally are (1) *point bearing* types which transmit the loads to lower, stronger soil through their points, and (2) *friction* types which develop the necessary bearing capacity through surface friction between the pile and the ground. Piles may be made of wood, concrete or steel.

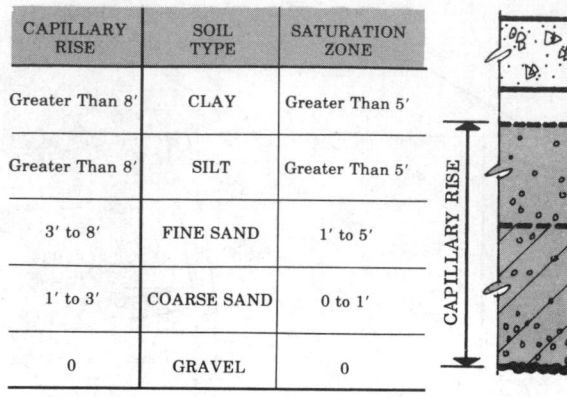

CAPILLARY RISE	SOIL TYPE	SATURATION ZONE
Greater Than 8′	CLAY	Greater Than 5′
Greater Than 8′	SILT	Greater Than 5′
3′ to 8′	FINE SAND	1′ to 5′
1′ to 3′	COARSE SAND	0 to 1′
0	GRAVEL	0

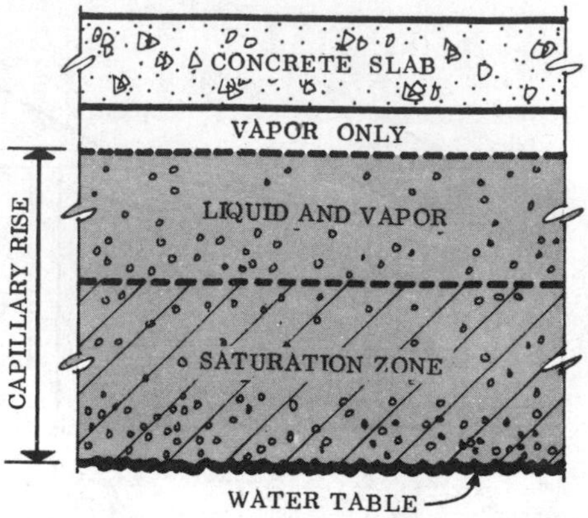

FIG. 9 Water rises in most soils by capillary action. Clays and silts may become fully saturated to almost 6′ above a water table and some water may rise more than 11′. Note that even coarse sand may permit a rise up to 3′. Coarse gravel prevents any rise of water by capillarity.

SOIL AND SETTLEMENT

Soil underlies practically every building site and usually receives the loads of low-rise building foundations. Since the structural integrity of the building is dependent on soil, a basic understanding of its nature and how it performs under load is important.

Soil Classes

Soils are classed broadly as being either granular or cohesive (Fig. 8). *Granular* soils, such as boulders, gravel and sand, consist of relatively large particles visible to the eye. In *cohesive* soils the particles are relatively small, and in some types such as fine-grained clays the particles cannot even be seen through a low-powered microscope. Silts are soils made up of particles finer than sand but coarser than clay (Fig. 10). Actual soil samples usually consist of mixtures of many soils of different particle sizes. They are described as combinations of the predominate soils such as *silty sand, clayey sand* or *gravely sand.*

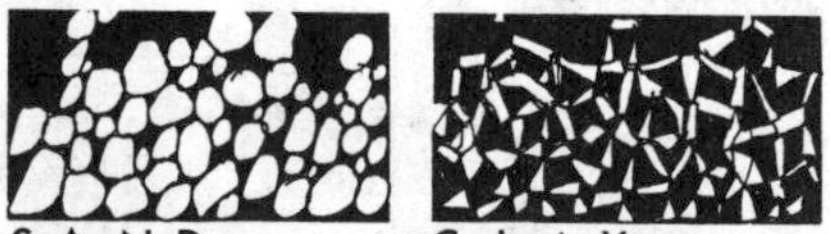

FIG. 8 Sand grains are large and angular and the volume of the voids is relatively small. Clay grains are minute and scaley and the volume of the voids is quite large.

Soil Properties

The stability and structural properties of soil in relation to foundations are determined in large measure by the *shearing resistance* of the soil. Shearing resistance in most subgrade materials is the combined effect of *internal friction* and *cohesion.*

Internal Friction This property of soil can be described as the resistance of the soil grains to slide over one another.

Cohesion The binding force that holds the soil grains together is called cohesion, and varies with the moisture content of the soil. Cohesion may be very high in clays, whereas sand and silt have little if any cohesion.

In addition to the above properties, other important characteristics of soil influencing residential foundation construction include: soil classification and composition, grain size, moisture content, density, capillarity (Fig. 9), drainage characteristics, stratification, consistency, compressibility and climatic variations. These factors are discussed in detail in Division 100.

Settlement

All foundations are subject to settlement. The amount of settlement depends on the compressibility of the soil supporting the foundation. Foundation settlement is of little concern if it is slight, if it is uniform, and if it has been anticipated. Settlement becomes a problem when it is excessive and when it is uneven under different parts of the foundation.

Differential settlement can be controlled by proportioning the size of the footings so they will transmit an equal load per unit of area to the soil. Residential loads are generally so light that footings are seldom sized in this way. In very poor soil, however, if unusual load variations exist, footing sizes should be proportioned for equal settlements.

When the soil being loaded is dense and granular, such as coarse sand or gravel, little settlement is likely and it occurs *quickly* after the load is applied (Fig. 11). When the soil is a cohesive clay, settlement is likely to be greater, but occurs *slowly*, perhaps taking several years.

The two principal causes of settlement are (1) reduction of the volume of soil voids (spaces between soil particles containing air and/or water), and (2) lateral displacement of the supporting soil.

SOIL TYPES	APPROXIMATE SIZE LIMITS OF SOIL PARTICLES
BOULDERS	LARGER THAN 3" DIAMETER
GRAVEL	SMALLER THAN 3" DIAMETER BUT LARGER THAN #4 SIEVE
SAND	SMALLER THAN #4 SIEVE[1], BUT LARGER THAN #200 SIEVE[2]
SILTS	SMALLER THAN 0.02 mm. DIA. BUT LARGER THAN 0.022 mm. DIAMETER
CLAYS	SMALLER THAN 0.002 mm. DIA.

1. APPROXIMATELY 1/4" IN DIAMETER
2. PARTICLES LESS THAN #200 SIEVE NOT VISIBLE TO THE NAKED EYE

FIG. 10 Soil types.

Reduction of Voids The reduction in the volume of the voids within soil when loaded is related to the soil's compressibility. Since sand contains large angular grains and a relatively small percentage of void spaces (Fig. 8), foundation pressures can cause little decrease in the volume of the voids. Clay, however, has a scale-like grain structure and a relatively large percentage of voids. Hence, decreases in the volume of the voids in clays subjected to foundation loads can be quite large.

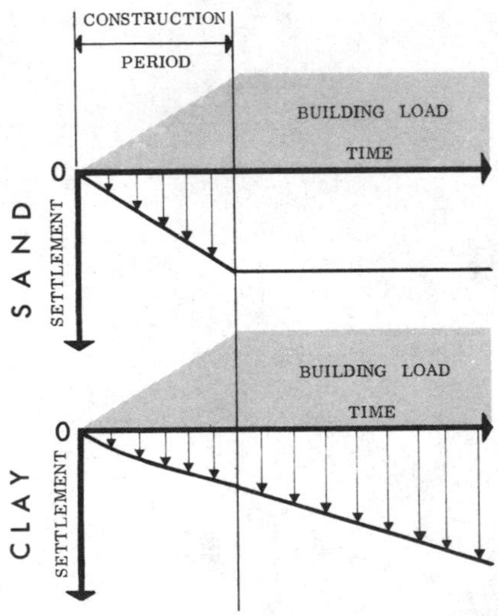

FIG. 11 Building settlement on sand tends to be quick and slight; on clay it tends to be greater and may occur over a long period of time.

If soil voids are filled with water, the water must be pressed out to allow any decrease in the volume of the voids. Sand is permeable and offers little resistance to the passage of the water, so that the total void reduction and settlement occurs almost simultaneously with the application of the load. Impermeable clays require a long time for compression (*consolidation*) to take place, since the water cannot pass out easily.

Air contained in the voids of sand offers no resistance to compression since it is expelled as soon as the load is applied. A certain amount of compression may occur instantaneously in clays if air is present in the voids, since air itself is compressible.

Lateral Displacement Lateral displacement in soil is the actual horizontal movement of soil under heavy foundation loads. Settlement due to lateral displacement is not common and generally offers no problem with normal residential foundation loads, unless the lateral support to the soil under the foundation areas is reduced or disturbed by adjoining excavation.

Soil Bearing Capacities

Since residential loads transmitted to soil usually are so small, the sizing of foundations for houses generally is based on local rules of thumb, rather than on soil tests. Building codes often list allowable bearing capacities for soils of several general classes (Fig. 12). These bearing capacities normally are conservative and rarely take into account compressibility of soils. In addition, the general descriptions of the soils are subject to different interpretations, and the use of generalized tables may lead to erroneous conclusions regarding the true load carrying capability of the soil at a specific site.

Where numerous individual house foundations are being planned, as in a subdivision, and in the case of apartments, where taller buildings produce heavier loads, soil samples and tests to determine foundation type and size should be made. Even in the case of an individual house, where foundations often are over-designed, the resulting savings in smaller foundation costs may offset the cost of soil sampling and testing.

FIG. 12 BEARING CAPACITIES OF SOILS

TYPE OF SOIL	Assumed Bearing Capacity Tons Per Sq. Ft.*
Gravels or gravel-sand mixtures, little or no fines	5
Gravel-sand-clay or gravel-sand-silt mixtures	2 to 2.5
Sands or gravelly sands, little or no fines	3 to 3.75
Sand-silt or sand-clay mixtures	2
Inorganic silts, very fine sands or inorganic clays	1

This page deliberately left blank.

This general discussion of design and construction affecting the performance of foundation systems includes: (1) a description of the loads acting on foundations and recommended minimum sizes for footings, walls, piers and other foundation elements; (2) a discussion of water and moisture problems; (3) a discussion of foundation anchorage; and (4) precautions recommended to prevent decay and termite protection of wood structural elements.

FOUNDATION SIZES

A foundation should be economical and should carry all expected loads with a reasonable factor of safety. The design of foundation elements is based on assumptions as to the kinds of loads the foundation must carry.

Loads carried by foundation walls may be divided into: (1) *dead loads*, those which act constantly, such as the weight of the superstructure and the foundation itself; and (2) *live loads*, those which are not constant, such as the weight of snow, furniture and people. These loads may be classed further as either: *vertical or lateral* (horizontal), *static* (non-moving) or *dynamic* (moving), *concentrated* at a point or *uniform* along a surface.

Dead loads usually are calculated easily by summing up the total weight of the materials used in the construction. Dead loads usually may be assumed to be uniformly distributed over the length and thickness of the foundation wall.

Live loads are variable, periodic or moving, and include lateral earth pressures on basement walls and loads due to dynamic forces such as wind, blast, earthquake and impact forces.

The type and magnitude of live loads acting on the structure vary according to locality, occupancy and other factors. Most building codes specify minimum live loads which may be used in sizing foundation walls, footings and other elements of the substructure.

Footings

When the bearing capacity of the soil is inadequate to support a foundation wall or pier alone, a footing is required to spread the loads on the supporting soil. Foundation walls built of masonry units bonded with mortar should have footings, regardless of the soil's bearing capacity.

Footing Sizes Footings should be sized for existing soil conditions and the magnitude of dead and live loads. The usual rule-of-thumb size for normal residential footings resting on soil of average bearing capacity is that the width of the footing should equal twice the thickness of the wall, and the depth of the footing should equal the wall thickness. In other words, given a wall thickness *t*, the footing width will be *2t* and its depth will be *t*. In the absence of specific building code requirements or engineering analysis, the footing sizes given in Figure 13 may be used. These sizes are based on a conventionally loaded wall footing and on soil of average bearing value (approximately 2,000 psf or greater). Footings should not be less than 6″ thick and, unless reinforced, their thickness should not be less than 1-1/2 times the footing projection.

Footing Location Footings should never be placed on frozen ground. As frozen soil thaws, excess water can transform the soil into a liquid mud, unable to support loads. Footings should be located far enough below the ground level to protect them from possible frost heaving caused by the freezing of soil moisture.

Foundations also should be extended below fill material so that footings rest on undisturbed earth. Footings can be supported on consolidated fill or unstable soil, but the foundations must be designed by engineering analysis for the existing conditions.

In areas where soils expand due to moisture changes, footings must be deep enough to be unaffected by seasonal moisture changes, or special practices may be necessary to prevent damage to the structure from heaving.

Footing Material Footings should be cast-in-place concrete, placed continuously. Concrete for unreinforced footings in foundations should have a minimum compressive strength of 2,000 psi at 28 days. Sufficient time should be allowed for the strength of the concrete to develop before subjecting a footing to loads. When concrete must be placed in cold weather, it should be placed as soon after excavation as possible and must be protected from freezing until it cures (see Section 203 Concrete).

When a large excavation has

WALL FOOTING DIMENSIONS

Based on average soil bearing value = 2000 P. S. F.

NUMBER OF FLOORS		FRAME CONSTRUCTION		MASONRY OR VENEER CONSTRUCTION	
		t″	p″	t″	p″
1	BASEMENT	6″	2″	6″	3″
1	NO BASEMENT	6″	3″	6″	4″
2	BASEMENT	6″	4″	8″	5″
2	NO BASEMENT	6″	3″	6″	4″

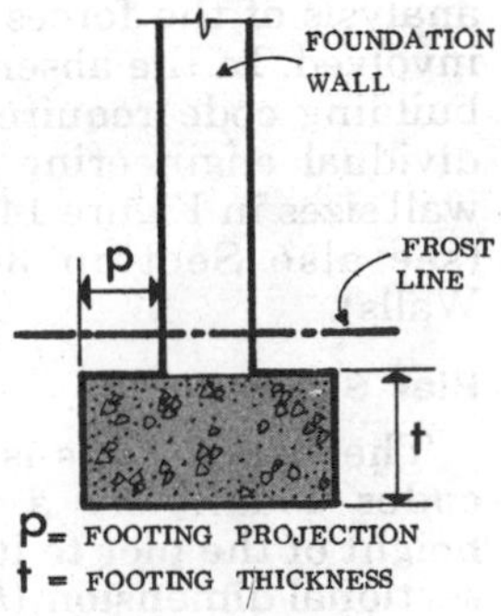

FIG. 13 Dimensions for continous wall footings.

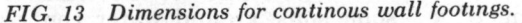

been carried deeper than desired, the depth should be refilled with concrete rather than with fill material. If it is necessary to excavate beneath a footing for a sewer or water line or other small opening, the excavation should be backfilled with compacted stone or gravel.

Reinforced Footings Concrete footings should be reinforced longitudinally with steel where soil has low bearing capacity or varies in compressibility, or where footings must span over pipe trenches or other interruptions.

When the footing projection exceeds 2/3 the footing thickness, it should be reinforced transversely with steel.

Stepped Footings Stepped footings are required on lots having sloping grades. The vertical part of the step should be constructed at least 6″ thick, have the same width as the horizontal footing, and be placed at the same time. If more than one step is necessary, the horizontal distance between steps should not be less than 2′, and the vertical step should not exceed 3/4 the horizontal distance between steps (Fig. 3).

Walls

Unreinforced concrete and concrete masonry is used extensively for foundation walls. Past experience and observation of plain concrete and masonry foundations have resulted in the development of minimum requirements governing their design. The requirements generally are empirical, particularly in designing basement walls to resist lateral earth pressures. Hence, foundation design rarely is based on theoretical analysis of the forces and stresses involved. In the absence of specific building code requirements or individual engineering analysis, the wall sizes in Figure 14 may be used (see also Section 343 Masonry Walls).

Pier Sizes

The size of piers is governed in codes usually by a ratio of the height of the pier to its least cross-sectional dimension (*h/t*), with the minimum pier section being 8″ x 12″ for solid units and 8″ x 16″ for hollow units. The unsupported heights of solid masonry piers should not exceed 10 times their least cross-sectional dimension. The same is true of piers constructed with hollow masonry units, provided that the cores are filled with mortar. Piers constructed with unfilled hollow masonry units may have unsupported heights not exceeding 4 times their least dimension.

Footings for piers, columns and masonry chimneys should be at least 8″ thick and, unless reinforced, not less than 1-1/2 times the footing projection.

FOUNDATION WALL DIMENSIONS

FOUNDATION WALL CONSTRUCTION	MAXIMUM HEIGHT OF UNBALANCED FILL H (FEET)	MINIMUM WALL THICKNESS T (INCHES)	
		FRAME CONSTRUCTION	MASONRY OR VENEER
HOLLOW MASONRY	3′	8″	8″
	5′	8″	8″
	7′	12″	10″
SOLID MASONRY	3′	6″	8″
	5′	8″	8″
	7′	10″	8″
PLAIN CONCRETE	3′	6″	8″
	5′	6″	8″
	7′	8″	8″

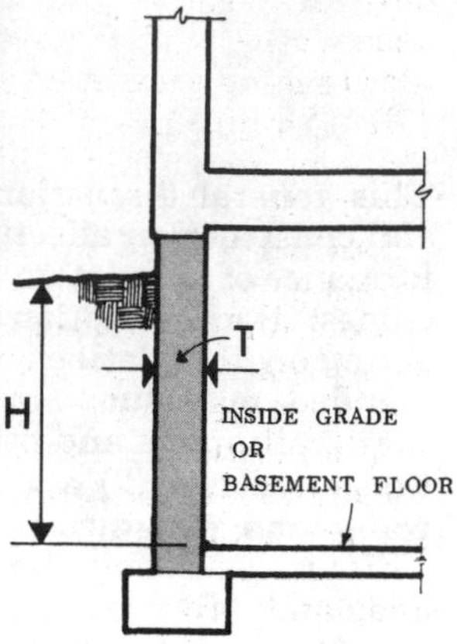

FIG. 14 Walls are sized for lateral pressures and imposed loads.

WATER AND MOISTURE PROBLEMS

Foundation design and construction must adequately solve several problems presented by water in its liquid, vapor and solid forms.

If water penetrates the walls of a basement, the livable space becomes useless. Water in a crawl space or other enclosed space may cause conditions conducive to decay, musty odors, mold growth and corrosion of some metals. Moisture below slab-on-ground can migrate through the slab, if improperly constructed, and may cause delamination of floors and may support condensation and frost accumulation at the perimeter of the walls. In some soils, ground water can rise high enough to fill heating ducts if they are embedded improperly in the concrete slab.

The two external sources of water affecting foundation design and construction are *surface water* and *ground water*. General construction methods to cope with each of these are shown in Figure 15.

Surface Water

Water that runs along the surface of the ground as a result of rain, downspout discharge, melting snow or other source should be directed away from the perimeter of the building by *protective slopes.* Protective slopes are formed by intentionally sloping the finished

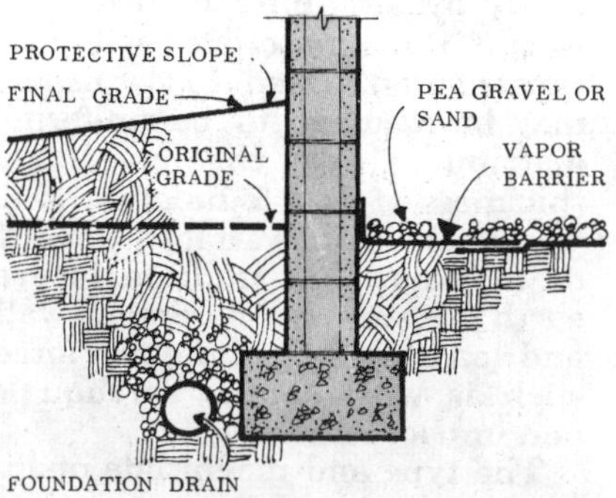

FIG. 15 General moisture control: sloped finished grade provides run-off of surface water from the foundation; vapor barrier resists rise of water vapor due to capillarity; foundation drains provide subsurface drainage.

grades around the foundation to provide positive drainage of surface water away from the building. For grassed areas, the slope should fall a minimum of 6″ in 10′, or a slope of about 5%. If the area adjacent to the foundation is paved, about a 1% slope or 1/8″ in 1′ will direct the water away.

Foundation walls enclosing basement spaces usually require additional protection by dampproofing applications of parging and/or bituminous coatings on the exterior surface below grade.

Ground Water

Most ground water problems are caused by locating foundations too close to the subsurface water table. The problem usually is caused by: (1) the temporary rise of water in the ground by seasonal variations in precipitation, and (2) the rise of water in the ground due to capillarity of the soil. When the water table is so high that the foundation would be submerged constantly, either the site should be abandoned or more extensive waterproofing methods must be used to make the foundation watertight.

Water Table Rise Seasonal fluctuations in precipitation which cause the water table to rise temporarily above the level of the foundation may be solved by subsurface drainage. The most effective means of resisting water infiltration through foundations is to collect the water and discharge it away from the site before it reaches the foundation.

Subsurface water can be collected by foundation drains placed around the foundation at or slightly above the level of the footing bottom. Drain tiles should be kept above the base of the footing to prevent possible washout of the soil beneath the footing. The drain may be constructed of tiles laid with open joints, or of perforated pipe, and should be set in a pervious gravel bed. The collected water may drain by gravity to a natural outfall at a lower elevation, or may be discharged by positive pumping from a collecting sump located within the foundation area.

Capillary Rise Vapor and water will rise in certain soils due to capillary action. In fact, moisture can move upward through the voids within some clays as much as 12′ above the water table (Fig. 9). Capillarity does not occur in coarse granular soil types such as gravel. Therefore, one method of control is to replace soil having high capillarity with granular materials.

The rise of water vapor can be controlled by covering the ground surface with a vapor barrier. The barrier should be of a material not subject to decay or insect attack. A typical vapor barrier for use under concrete slabs and in crawl spaces is 4 mil thickness polyethylene film. Vapor barrier films used as ground cover in crawl spaces generally should be covered with a ballast of sand or pea gravel to hold them in place and to prevent damage from traffic (Fig. 15).

FOUNDATION ANCHORAGE

The superstructure should be anchored to the foundation to provide resistance to lateral forces, primarily from wind. Wind pressures produce compressive, tensile (uplift), shearing and racking loads on the foundation (Fig. 16).

Overturning, Translation, Rotation

Overturning forces may produce stresses that are critical in foundations which support light superstructures such as light wood construction subjected to high wind pressures. The connections anchoring the superstructure to the foundation and the foundation itself must have adequate strength in tension to resist these forces.

When the ratio of exposed building surface to the width of the building is large, high wind forces tend to cause *translation* (lateral movement) of the superstructure. When the wind forces are not symmetrical, building *rotation* can occur. The forces tending to cause translation and rotation are transferred to the foundation by the first floor acting as a diaphragm. The connections anchoring the superstructure must develop enough shearing strength to transmit these forces to the foundation. When the first floor does not have enough rigidity to provide diaphragm action, high shear and bending stresses may be produced in the foundation walls. In this case, consideration should be given to the use of shear walls, pilasters or bond beams to stiffen the walls.

Anchorage Recommendations

Masonry Walls Sill plates of a wood joist floor system should be anchored to masonry walls with 1/2″ bolts (Fig. 17), extending at least 15″ into the filled cores of masonry units. Anchor bolts should be spaced not more than 8′ apart, with one bolt not more than 12″ from each end of the sill plate. Where earthquake design is required, maximum spacing is 6′ o.c.

Ends of floor joists (Fig. 17) should be anchored at intervals of 6′ or every fourth joist. If the floor is required to provide lateral support for the masonry wall, anchorage should be provided as shown in Figure 18.

Concrete For concrete walls, sill plates should be anchored with

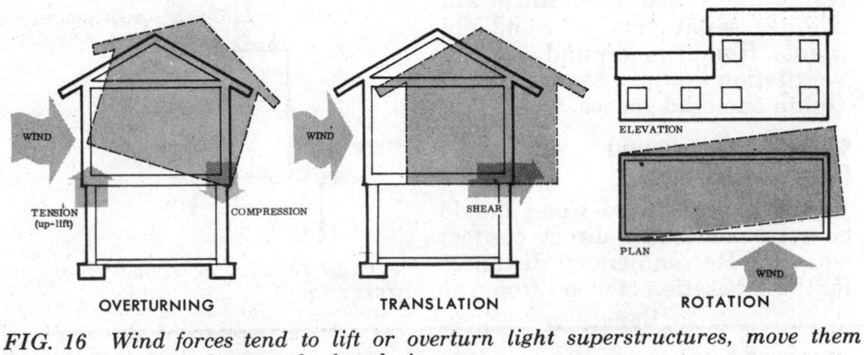

FIG. 16 Wind forces tend to lift or overturn light superstructures, move them laterally or rotate them on the foundation.

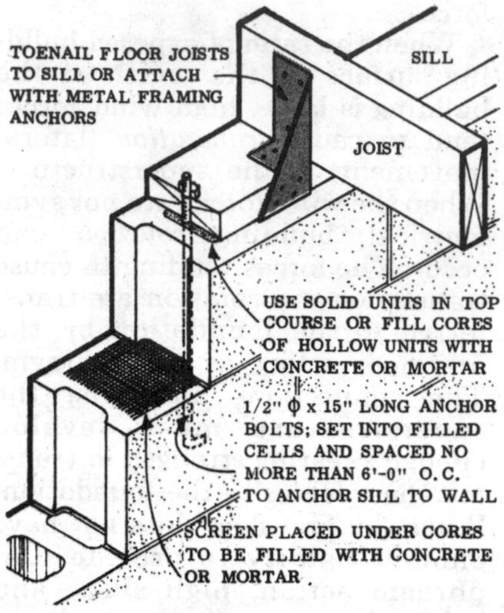

FIG. 17 Sill anchorage to masonry.

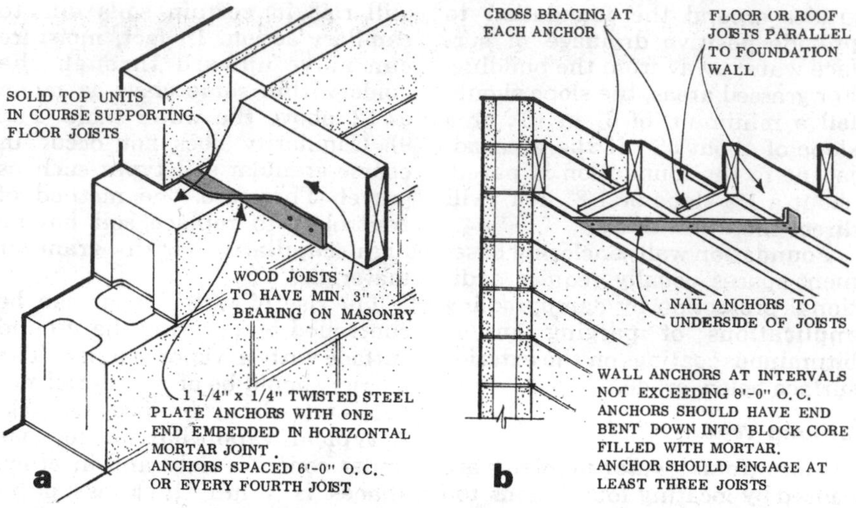

FIG. 18 Anchorage: (a) joists perpendicular to wall, (b) joists parallel.

1/2″ bolts embedded 6″, with the same maximum spacing as for masonry walls.

Hardened steel studs driven by powder actuated tools also may be used, except when earthquake design is required. Maximum spacing should not exceed 4′ o.c., with not less than 2 studs in each length of sill plate.

DECAY AND TERMITE PROTECTION

The design and construction of foundations which support buildings using wood structural floor, wall and roof systems must include precautions for the protection of the wood from the effects of decay and attack by subterranean termites (see Fungi and Insect Hazard, and Wood Preservative Treatments, Section 201 Wood).

Suitable protection of wood structural elements is provided by (1) positive site and building drainage, (2) separation of wood elements from the ground, and (3) ventilation and condensation control in enclosed spaces.

Separation of Wood From the Ground

Whenever possible, wood should be separated from direct contact with soil. Recommended distances for the separation of wood from the ground are given in Figure 19. These distances are considered necessary to: (1) maintain wood elements in permanent structures at safe moisture contents for decay protection, (2) provide a termite barrier and, (3) facilitate periodic visual inspection. When it is impossible or impractical to comply with the clearances recommended, all heartwood of a naturally durable species or pressure treated lumber should be used. However, when termites are known to exist, access space always should be provided for periodic visual inspection.

Termite Barriers

A termite barrier is any material that cannot be penetrated by termites and which drives the insects into the open where they can be detected and eliminated. Where local experience indicates that additional protection against termites is required, one or more of the following barriers is recommended, depending on the degree of hazard: (1) chemical soil treatment, (2) preservative treated lumber for all floor framing up to and including the subfloor, (3) properly installed termite shields.

Additional precautions include poured concrete foundations (provided no cracks greater than 1/64″ are present), and poured reinforced concrete caps at least 4″ thick on unit masonry foundations (provided no cracks greater than 1/64″ occur).

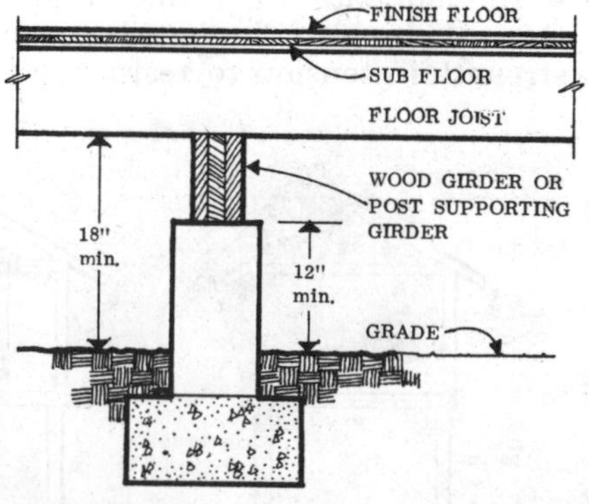

FIG. 19 Minimum separation of wood from ground.

Ventilation of Enclosed Spaces

In crawl spaces, the presence of water may produce considerable amounts of water vapor within enclosed spaces. The accumulation of water vapor may increase the possibility of decay and termite attack and result in unpleasant odors within the building. Therefore, it is important to ventilate the enclosed foundation spaces adequately to provide for the dissipation of water vapor. The amount and methods of ventilation depend on the type of space being ventilated, the degree of water vapor present, and the type of vapor barrier installed.

We gratefully acknowledge the assistance of the Building Research Advisory Board, the Federal Housing Administration, the National Concrete Masonry Association, the National Forest Products Association and the Structural Clay Products Institute for the use of their publications as references.

FOUNDATION SYSTEMS 310

CONTENTS

GENERAL RECOMMENDATIONS

FOUNDATION TYPES

Foundations are of two general types: (1) spread foundations, which distribute the building loads directly over a sufficient area to obtain adequate bearing capacity; and (2) pile foundations, which transmit the building loads through weak soils (of inadequate bearing capacity for spread foundations) to deeper layers of soil with adequate bearing value.

Most foundations in residential single-family and low-rise construction are of the spread type. (For more detailed description of foundation types see Main Text pages 310-3 through 310-5.)

SOIL AND SETTLEMENT

Spread foundations should be designed so as to support the superstructure and so as to distribute all building loads in a way which will insure that settlement will be either negligible or uniform under all parts of the building.

Uneven foundation settlement can result in defects such as cracks in finished walls and ceilings, sloping floors, and open joints in woodwork, as well as functional annoyances such as binding doors and windows. When differential settlement is extreme, partial failure of the structural integrity of the building can result.

Soils are classed broadly as either granular or cohesive. Granular soils, such as bolders, gravel and sand, consist of relatively large particles visible to the eye. In cohesive soils the particles are relatively small, and in some types such as fine-grained clays the particles cannot even be seen through a low-powered microscope. Silts are soils made up of particles finer than sand but coarser than clay (Fig. WF 1).

Important characteristics of soils influencing residential foundation construction include: grain size, moisture content, density, drainage characteristics, compressibility and climatic variations. Capillarity is another important characteristic of soils which is related to particle size and composition (Fig. WF 2).

Since foundation loads in single-family and low-rise residential construction are small, the sizing of foundations for houses generally is based on local rules of thumb, rather than on soil bearing values. However, building codes often list allowable (maximum) bearing capacities for soils which can be used as a rough guide (Fig. WF 3).

FIG. WF1 **SOIL TYPES**

SOIL TYPES	APPROXIMATE SIZE LIMITS OF SOIL PARTICLES
BOULDERS	LARGER THAN 3" DIAMETER
GRAVEL	SMALLER THAN 3" DIAMETER BUT LARGER THAN #4 SIEVE
SAND	SMALLER THAN #4 SIEVE[1], BUT LARGER THAN #200 SIEVE[2]
SILTS	SMALLER THAN 0.02 mm. DIA. BUT LARGER THAN 0.002 mm. DIAMETER
CLAY	SMALLER THAN 0.002 mm. DIA.

1. Approximately 1/4" in diameter
2. Particles less than #200 sieve not visible to the naked eye

FIG. WF2 **CAPILLARITY OF VARIOUS SOILS**

CAPILLARY RISE	SOIL TYPE	SATURATION ZONE
Greater Than 8′	CLAY	Greater Than 5′
Greater Than 8′	SILT	Greater Than 5′
3′ to 8′	FINE SAND	1′ to 5′
1′ to 3′	COARSE SAND	0 to 1′
0	GRAVEL	0

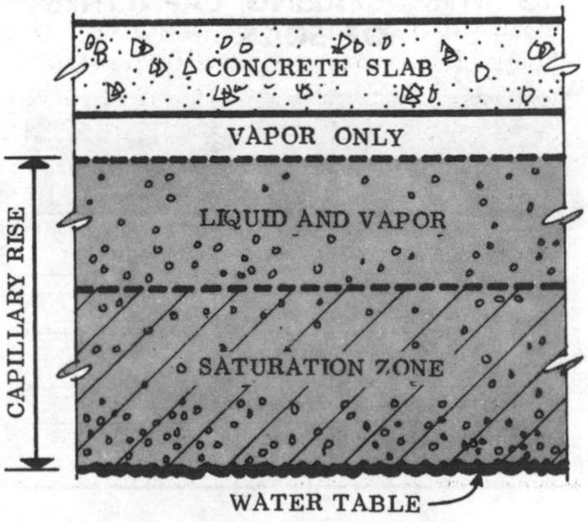

DESIGN AND CONSTRUCTION

FOUNDATION SIZES

The design of all foundation elements should be based on adequate safety factors and reasonable assumptions as to the dead and live loads which the foundation is expected to carry.

Footings

A footing should be provided whenever (1) the bearing capacity of the soil is inadequate to support a foundation wall or pier or (2) the foundation wall or pier is built of masonry units bonded with mortar.

In the absence of specific building code requirements or engineering analysis, footing sizes should be not smaller than those shown in Figure WF 4.

These footing sizes are based on a conventionally loaded wall footing and on a conservative estimate of the soil bearing value of 2,000 psf.

Footings should not be less than 6″ thick and, unless transversely reinforced, their thickness should not be less than 1-1/2 times the footing projection. When the footing projection exceeds 2/3 times the footing thickness, it should be reinforced transversely as determined by engineering analysis.

Foundations should rest on undisturbed soil, unless specifically engineered for the soil conditions. Footings should not be placed on frozen ground. Footings should be located far enough below grade level to protect them from possible ground heaving caused by the freezing of soil moisture.

In areas where soils expand due to moisture changes, footings should be deep enough to be unaffected by seasonal moisture changes, or should be specifically engineered for these conditions.

Footings should consist of monolithic cast-in-place concrete, placed continuously. Concrete for unreinforced footings in foundations should have a minimum compressive strength of 2,000 psi at 28 days. *(See Section 203 Concrete).*

When concrete must be placed in cold weather, it should be placed as soon after excavation as possible and should be protected from freezing until it cures. (See *Section 203 Concrete).*

Foundation wall and pier construction should be delayed until the footing is sufficiently cured to support the loads.

When a large excavation has been carried deeper than desired, the extra depth should be refilled with concrete rather than with fill material.

If it is necessary to excavate beneath a footing for a sewer or water line or other small opening, the excavation should be backfilled with compacted stone or gravel.

Concrete footings should be reinforced longitudinally with steel where soil has low bearing capacity or varies in compressibility, or where footings must span over pipe trenches or other bearing interruptions.

The vertical part of the step of a stepped footing should be at least 6″ thick, should have the same width as the horizontal footing, and should be placed at the same time.

FIG. WF3 **BEARING CAPACITIES OF SOILS**

TYPE OF SOIL	Assumed Bearing Capacity Tons Per Sq. Ft.*
Gravels or gravel-sand mixtures, little or no fines	5
Gravel-sand-clay or gravel-sand-silt mixtures	2 to 2.5
Sands or gravelly sands, little or no fines	3 to 3.75
Sand-silt or sand-clay mixtures	2
Inorganic silts, very fine sands or inorganic clays	1

FIG. WF4 **MINIMUM FOOTING SIZES FOR VARIOUS BUILDING TYPES***

NUMBER OF FLOORS		FRAME CONSTRUCTION		MASONRY OR VENEER CONSTRUCTION	
		t″	p″	t″	p″
1	BASEMENT	6″	2′	6″	3″
	NO BASEMENT	6″	3″	6″	4″
2	BASEMENT	6″	4″	8″	5″
	NO BASEMENT	6″	3″	6′	4″

*Based on average soil bearing value = 2000 P. S. F.

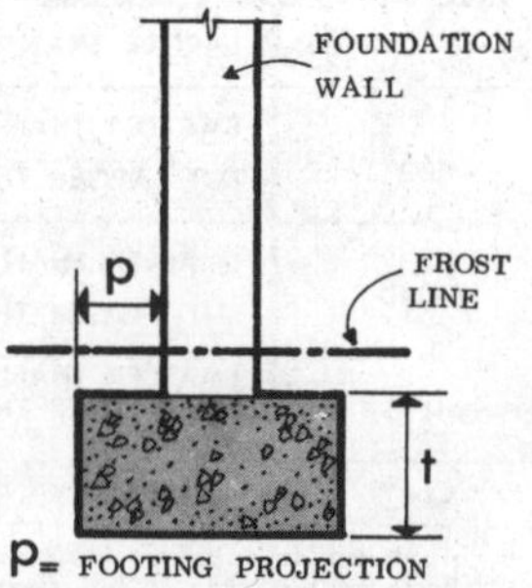

p= FOOTING PROJECTION
t = FOOTING THICKNESS

(Continued) **DESIGN AND CONSTRUCTION**

FIG. WF5 **MINIMUM FOUNDATION WALL DIMENSIONS**

FOUNDATION WALL CONSTRUCTION	MAXIMUM HEIGHT OF UNBALANCED FILL H (FEET)	MINIMUM WALL THICKNESS T (INCHES) FRAME CONSTRUCTION	MASONRY OR VENEER
HOLLOW MASONRY	3′	8″	8″
	5′	8″	8″
	7′	12″	10″
SOLID MASONRY	3′	6″	8″
	5′	8″	8″
	7′	10″	8″
PLAIN CONCRETE	3′	6″	8″
	5′	6′	8″
	7′	8″	8″

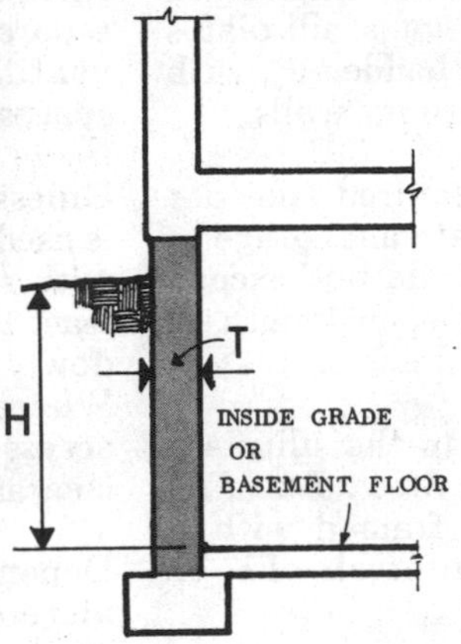

FIG. WF6 **MINIMUM CLEARANCES FOR CRAWL SPACES**

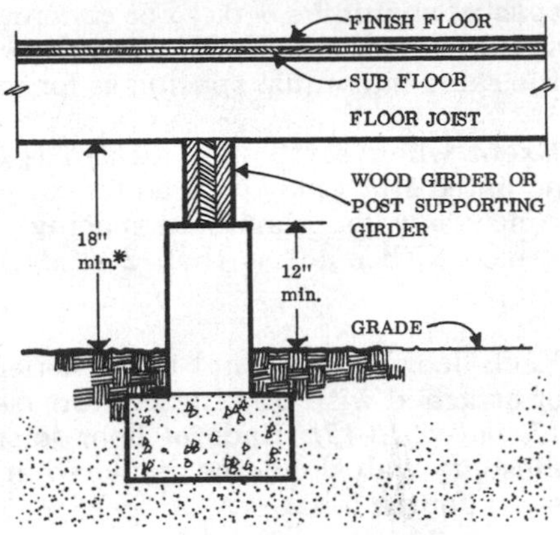

*24″ min. where mech. eqpt. is located in crawl space

If more than one step is necessary, the horizontal distance between steps should not be less than 2′, and the vertical step should not exceed 3/4 the horizontal distance between steps.

Walls and Piers

In the absence of particular building code requirements or specific engineering analysis, foundation wall sizes should not be smaller than those given in Figure WF 5 (see also Section 343 Masonry Walls).

The size of piers usually is governed in codes by a ratio of the height of the pier to its least cross-sectional dimension (h/t), with the minimum pier section being 8″ x 12″ for solid units and 8″ x 16″ for hollow units.

The unsupported heights of solid masonry piers and of piers constructed with hollow masonry units, whose cores are filled with mortar, should not exceed 10 times their least cross-sectional dimension. Piers constructed with unfilled hollow masonry units may have unsupported heights not exceeding 4 times their least dimension.

Footings for piers, columns and masonry chimneys should be at least 8″ thick and, unless reinforced, not less than 1-1/2 times the footing projection.

WATER AND MOISTURE PROBLEMS

Surface Water

Protective slopes are formed by sloping the finished grades around the foundation to provide positive drainage of surface water away from the building. For grassed areas, the slope should fall a minimum of 6″ in 10′ (approximately 5% slope). If the area adjacent to the foundation is paved, the slope should be at least 1/8″ in 1′ (approximately 1%).

Foundation walls enclosing basement spaces usually require additional protection by dampproofing applicatation, and (2) the rise of water in the ground due to capillarity of the soil (Fig. WF 2).

Ground Water

Ground water problems may be caused by: (1) a permanently high water table or the temporary rise of water table due to seasonal variations in precipitation, and (2) the rise of water in the ground due to capillarity of the soil (Fig. WF 2).

Where a high water table is expected, foundation drains should be placed all around the foundation. Drain tiles should be kept above the base of the footing to prevent possible washout of the soil beneath the footing.

The rise of water vapor can be controlled in slab-on-ground and crawl space foundations by covering the ground surface with a vapor barrier such as 4 mil polyethylene film.

FOUNDATION ANCHORAGE

Sill plates of a wood joist floor system should be anchored to masonry walls with 1/2″ bolts extending at least 15″ into the filled cores of masonry units. Anchor bolts should be spaced not more than 8′ apart, with one bolt not more than 12″ from the ends of each piece of the sill plate.

(Continued) DESIGN AND CONSTRUCTION

Where earthquake design is required, maximum spacing should be 6′ o.c. For concrete walls, sill plates should be anchored with 1/2″ bolts embedded 6″, with the same maximum spacing as for masonry walls.

Exept when earthquake design is required, powder actuated studs may be used for sill plate anchorage to concrete walls. Maximum spacing should not exceed 4′ o.c., with not less than 2 studs in each length of sill plate.

Each floor joists should be toenailed to the sill plate or attached with framing anchors *(see Main Text Fig. 17, page 310-12)*. Ends of floor joists framed with a masonry wall should be anchored at intervals of 6′ or every fourth joist.

If the floor joists are required to provide lateral support for the masonry wall, anchorage should be provided also where joists run parallel to the foundation wall. Metal anchors extending from the wall and secured to at least 3 joists should be provided at 8′ o.c. maximum *(see Main Text Fig. 18b, page 310-12)*.

DECAY AND TERMITE PROTECTION

Foundations which support wood floor, wall and roof systems should include decay and termite precautions such as: (1) positive site and building drainage, (2) separation of wood elements from the ground, and (3) ventilation and condensation control in enclosed spaces.

Unless special decay and termite resistant material is used, wood should be separated from direct contact with soil. In crawl spaces, floor joists should be at least 18″ above ground, but wood girders may project down not closer than 12″ to the ground (Fig. WF 6). Where mechanical equipment requiring occasional access is located in the crawl space, the minimum clearance from joists to ground should be 2′.

Depending on the degree of termite hazard, one or more of the following barriers should be used: (1) chemical soil treatment, (2) preservative treated lumber for all floor framing up to and including the subfloor, (3) properly installed termite shields.

Additional termite precautions include the use of cast-in-place monolithic concrete foundations and the use of poured reinforced concrete caps at least 4″ thick on hollow concrete masonry foundations. Such solid concrete walls or capped hollow walls should be free of cracks 1/64″ or greater, through which termites can penetrate to reach wood construction.

311 SLABS-ON-GROUND

SLABS-ON-GROUND

INTRODUCTION

The information and recommendations in this section apply to concrete slabs-on-ground built at or near grade level and which serve as the foundation/floor system of buildings that enclose interior spaces.

The use of concrete slabs-on-ground for residential foundation/floor systems is prevalent in all parts of the United States. The suitability of slab-on-ground construction is influenced by several factors involving the geographic location of the building, the specific building site and the construction of the superstructure. Characteristics pertaining to site preparation, topography, soil properties, climate, the type and support of the superstructure, and comfort factors such as thermal and moisture control may influence the construction of the slab. In all cases, slab construction requires concrete quality control in selecting concrete materials and mixtures, and especially in placing, finishing and curing procedures.

In this section, recommendations are given for selecting and constructing residential slabs-on-ground. If care is exercised in construction procedures to control concrete quality and if specific precautions dictated by characteristics of site, superstructure and geographic location are observed, slabs-on-ground are suitable for residential foundation/floor systems in all parts of the United States.

These recommendations are based on the following publications: Portland Cement Association's Concrete Floors on Ground; American Concrete Institute Standard 302, Recommended Practice for Concrete Floors and Slab Construction; and The Findings of the Building Research Advisory Board, National Academy of Sciences.

GENERAL RECOMMENDATIONS

Slab-on-ground foundation/floor systems are built several ways. Construction may consist (1) of a ground-supported concrete floor slab built independent of load-carrying elements, such as walls, columns or footings which transmit and distribute their loads directly to the supporting soil (Fig. 1a); or (2) of a floor slab built as an integral part of the foundation which carries all superstructure loads and transmits them to the soil (Fig. 1b). Foundation elements such as walls and footings which are common to many types of foundation systems are discussed in Section 310 Foundation Systems.

Slab-on-ground construction requirements are determined largely by site characteristics, such as drainage and soil type, and the methods used to build the slab. Slabs may be ground supported or supported structurally. They may be unreinforced (plain); lightly reinforced using welded wire fabric reinforcing; or structurally reinforced using welded wire fabric and steel reinforced bars. Four types of slabs are described below.

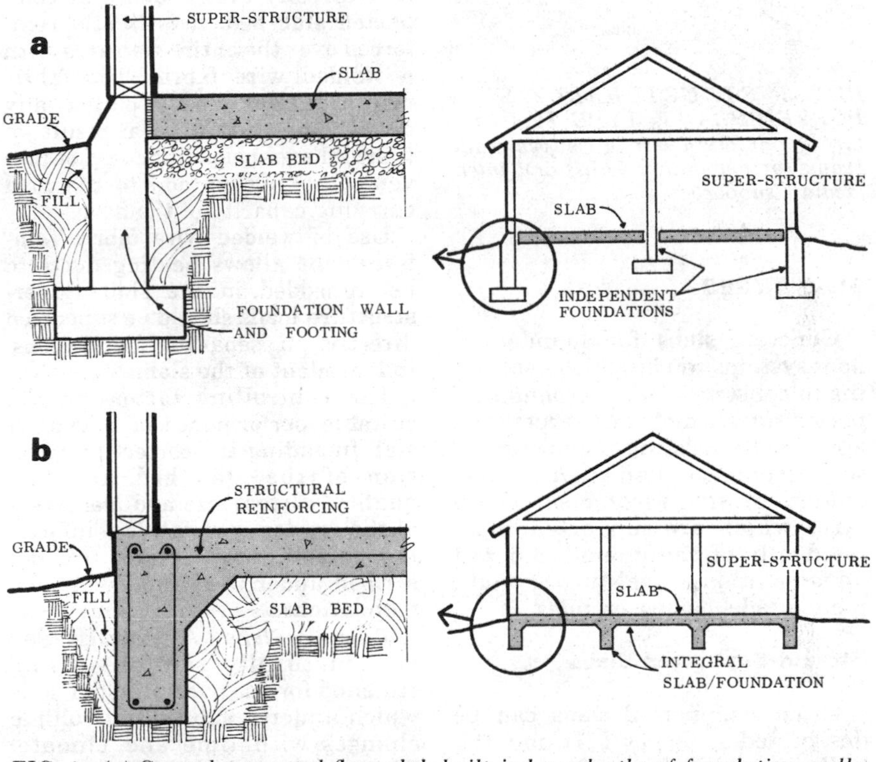

FIG. 1 (a) Ground-supported floor slab built independently of foundation walls; (b) ground-supported slab built integrally with foundation grade beam.

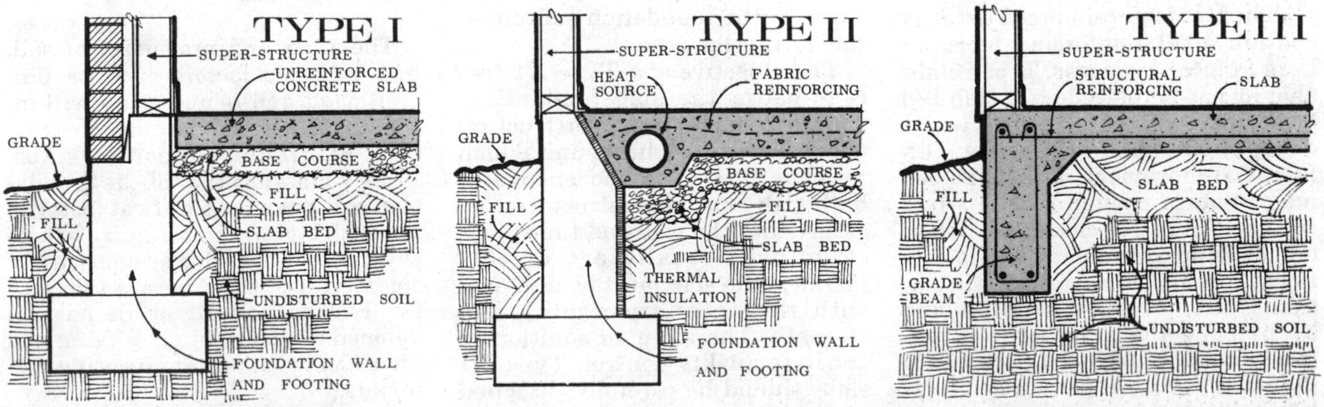

FIG. 2 GROUND-SUPPORTED SLABS: TYPE I—unreinforced; superstructure loads carried independently of slab. TYPE II—lightly reinforced to resist shrinkage cracking; superstructure loads carried independently of slab; heat source may be imbedded in slab. TYPE III—structurally reinforced and stiffened for use over problem soils; superstructure loads distributed over entire slab area.

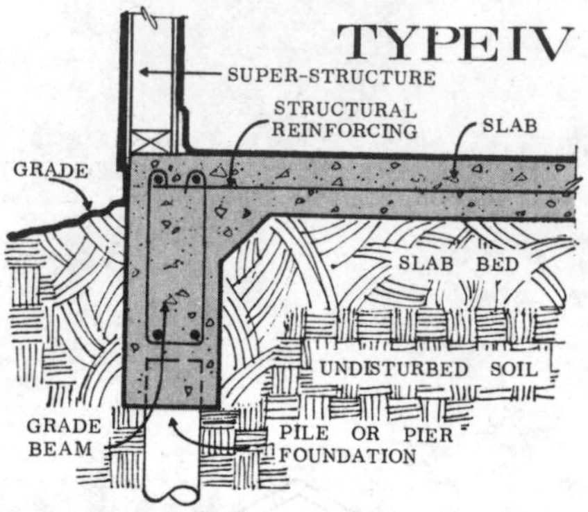

FIG. 3 STRUCTURALLY SUPPORTED SLABS, TYPE IV: structurally reinforced to span between grade beams (or foundation walls) and intermediate supports.

SLAB TYPES

Concrete slabs for foundation/floor systems can be classed according to construction as: *ground-supported slabs* which rest directly on a slab bed consisting of undisturbed soil or compacted fill and/or a base course; or *structurally supported slabs* which are supported independently of the ground and rest on foundations consisting of walls, piers, grade beams or piles.

Ground-Supported Slabs

Ground-supported slabs can be designated as Types I, II and III, as illustrated in Figure 2.

Type I Slabs A Type I slab is of at least 4" nominal thickness and cast directly on a dense or compacted slab bed. It is unreinforced but may contain steel reinforcing fabric or bars in localized areas. Type I slabs should not be placed on a slab bed that can develop significant changes in volume over time. Differential movement in the slab bed may produce enough tensile stress in the concrete to cause the slab to crack. Random cracking due to drying shrinkage is controlled by isolation joints, construction joints and by limiting the area between control joints which are located to induce cracking at preselected locations. Superstructure loads should be supported directly on foundations independent of the slab.

The controlling factors in the suitable performance of a Type I slab, including selection of the correct slab bed, concrete quality, location, control joint spacing and other details are described under Construction Recommendations.

Type II Slabs A Type II slab is of at least 4" nominal thickness and is cast directly over a dense or compacted slab bed. It is lightly reinforced over the entire slab area with a welded wire fabric. The fabric reinforcement keeps closed any cracks that form as a result of thermal expansion but does not prevent cracking, or add to the load carrying capacity of the floor.

Use of welded wire fabric reinforcement allows heating ducts to be imbedded in the slab. Superstructure loads should be supported directly on separate foundations, independent of the slab.

The controlling factors in the suitable performance of a Type II slab including the correct preparation of the slab bed, concrete quality, correct size and placement of the welded wire fabric reinforcement and other details are described under Construction Recommendations.

Type III Slabs A Type III slab is structurally reinforced and stiffened for use over problem soils which undergo substantial volume changes with time and climate. The use of foundations with spread footings is not advisable on such soils. Type III slabs receive the superstructure loads and transmit them to the foundation soil under the entire slab area.

The objective of a Type III slab is to cause the slab foundation/floor system and the superstructure to act as one monolithic unit which restricts differential movements in both slab and superstructure. To assure that the slab and the superstructure actually act in this manner, the slab must be designed with the necessary rigidity and strength. Therefore, in addition to concrete quality control, Type III slabs should be carefully designed to provide proper dimensions for stiffness and correct amount and size of steel reinforcement for strength.

Structurally Supported Slabs

Type IV slabs are structurally reinforced and can be used over very poor soils such as highly expansive clays, which are extremely sensitive to moisture, and soils of negligible bearing capacity or high in organic material content, and where Type I, II and III slabs would not be suitable (Fig. 3).

Structurally supported slabs are not dependent on the foundation soil for support. The slab bed functions only as a form to provide temporary support until the slab develops strength. Type IV slabs receive superstructure loads and transmit them to supporting grade beams, piers or intermediate foundation walls which are carried by piles or footings through the poor soil to soil of adequate bearing capacity below the level of the slab.

Type IV slabs require professional design using engineering analysis.

SITE CHARACTERISTICS

Site characteristics which may affect the design and construction of slabs-on-ground include: soil properties, moisture conditions, thermal conditions, and geographic factors which may require construction precautions to control termites or unusual conditions.

Soil Properties

The type and properties of soil, and climatic factors such as precipitation and temperature will influence the selection and construction of ground-supported slabs. Proper classification of the foundation soil becomes a critical factor in identifying *problem soils.* While several classification systems are in common use, the American Society for Testing and Materials has developed methods for classifying and identifying soils as engineering materials.*

Soil Classification The soil classification chart in Figure 4,

ASTM D 2487—Classification of Soils for Engineering purposes; ASTM D2488—Description of Soils (Visual-Manual Procedure).

FIG. 4 **SOIL CLASSIFICATION**[1]

Major Division			Group Symbols	Typical Names	Presumptive Bearing Capacity,[2] Tons Per Square Foot
COARSE-GRAINED SOILS (More than half of material is larger than the smallest particle visible to the naked eye)	Gravels (more than half of coarse fraction is larger than ¼")	Gravels with fines (appreciable amount of fines)	GW	Well-graded gravels, gravel-sand mixtures, little or no fines.	5
			GP	Poorly graded gravels or gravel-sand mixtures, little or no fines.	5
		Clean Sands (little or no fines)	GM	Silty gravels, gravel-sand-silt mixtures.	2.5
			GC	Clayey gravels, gravel-sand-clay mixtures.	2
	Sands (more than half of coarse fraction is smaller than ¼")	Clean Gravels (little or no fines)	SW	Well-graded sands, gravelly sands, little or no fines.	3.75
			SP	Poorly graded sands or gravelly sands, little or no fines.	3
		Sands with Fines (appreciable amount of fines)	SM	Silty sands, sand-silt mixtures.	2
			SC	Clayey sands, sand-clay mixtures.	2
FINE-GRAINED SOILS (More than half of material is smaller than the smallest particle visible to the naked eye)	Silts and Clays (liquid limit is less than 50)		ML	Inorganic silts, very fine sands, rock flour, silty or clayey fine sands or clayey silts with slight plasticity.	1
			CL	Inorganic clays of low to medium plasticity, gravely clays, sandy clays, silty clays, lean clays.	1
			OL	Organic silts and organic silty clays of low plasticity.	
	Silts and Clays (liquid limit is greater than 50)		MH	Inorganic silts, micaceous or diatomaceous fine sandy or silty soils, elastic silts.	1
			CH	Inorganic clays of high plasticity, fat clay.	1
			OH	Organic clays of medium to high plasticity, organic silts.	
Highly Organic Soils			PT	Peat and other highly organic soils.	

[1]Based on ASTM D2487—Classification of Soils for Engineering Purposes.
[2]National Building Code, 1976 edition, American Insurance Association.

identifies soils and divides them into coarse-grained and fine-grained soils. Each type of soil is given a descriptive name and letter symbol indicating its principal characteristics. As indicated by the typical names in Figure 4, field classification of soils includes a variety of different materials. When soil is a combination of two types, it is described as a combination of both. Thus, *clayey sand* describes a soil that is predominantly sand but which contains an appreciable amount of clay. A *sandy clay* has the properties of clay but contains an appreciable amount of sand.

Density and Consistency The most significant property of cohesionless soils is the *relative density* and is described in such terms as *very loose, loose, medium, dense* and *very dense.* Most coarse-grained soils are cohesionless. The relative density of coarse-grained soils can be estimated in the field using a standard penetration test[1] with a split-barrel sampler and power driven equipment (Fig. 5). This test is not always necessary for single sites for one- and two-family residences, but for highrise buildings and large residential developments on which extensive grading must be performed it should be required. The relative density may be estimated from the number of blows causing the sampler to penetrate 1′ and is classed from very loose to very dense (Fig. 6).

The most significant property of cohesive soils is consistency, described in such terms as *very soft, soft, medium, stiff, very stiff* and *hard.* Most fine-grained soils are cohesive. The consistency of clay soils should be determined by test from undisturbed samples using a thin-walled sampler (Fig. 5), or

1. *ASTM D1586—Penetration Test and Split-barrel Sampling of Soils.*

the assumed characteristics may not be valid.

A direct numerical value of consistency determined by laboratory test is the load per unit of area that causes failure of an unconfined soil sample in a simple compression test. The value obtained is called

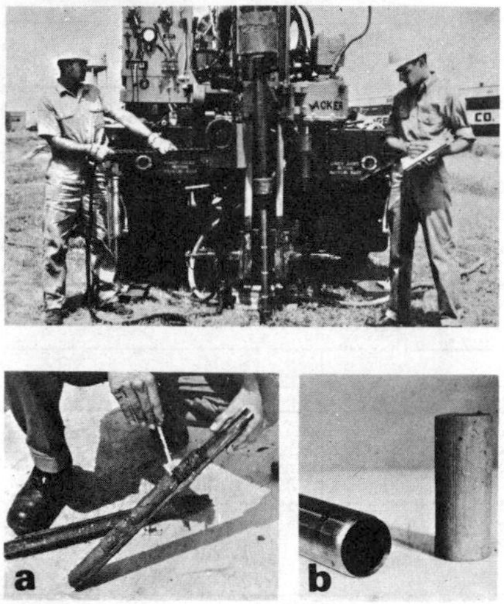

FIG. 5 Test borings to determine soil properties are made with power driven equipment and (a) a split-barrel sampler for coarse-grained soils, or (b) a thin-walled sampler for fine-grained soils.

the *unconfined compressive strength* (q_u) of the soil. Where CL, CH, OL or OH soils are encountered, this laboratory test should be performed to determine the unconfined compressive strength in order to select the type of slab used. Consistency of cohesive soils is given in Figure 7.

Purposely densifying or increasing the unit weight of a soil mass by compaction with rollers, tampers or vibrators can improve the structural properties of the soil (Fig. 26).

Plasticity If a soil within some range of water content can be rolled into thin threads, it is called plastic. All clay minerals are plastic at some range of water content—practically all very fine-grained soils contain clay minerals and therefore practically all are plastic. The degree of plasticity may be defined by the terms *fat* and *lean*. Lean clays are only slightly plastic because they usually contain a large proportion of silt or sand. The *plasticity index* (PI) is a value expressing the numerical difference between the liquid limit and the plastic limit of the soil as determined by test, such as ASTM D424, Standard Method of Test for Plastic Limit and Plasticity Index of Soils. The plasticity index is equal to the liquid limit less the plastic limit (PI = LL − PL). The Liquid Limit (LL) is the amount of moisture present when the soil changes from a plastic to a liquid state; Plastic Limit (PL) is the amount of moisture present when the soil changes from semisolid to plastic.

Any clay soil with a plasticity index greater than 15 may cause problems by expanding under changes in the moisture content of the sub grade soil, such as a rising water table.

Soil Investigation

Soil classification, descriptions defining density, consistency, plasticity and other properties of soil depend on the skill of the individual making the observation. As a result, field classification may be subject to uncertainties and inaccuracies. Soil investigation will vary with conditions. A preliminary investigation should pro-

FIG. 6 RELATIVE DENSITY OF COARSE-GRAINED SOILS

Soil Description	Rule-of-thumb Field Guide[1]	ASTM Method No. of Blows[2]	Relative Density (%)
Very Loose		0 - 4	0 - 20
Loose	Easily penetrated with ½″ reinforcing rod pushed by hand	4 - 10	20 - 40
Medium	Easily penetrated with ½″ reinforcing rod driven with 5 lb. hammer	10 - 30	40 - 60
Dense	Penetrated 1″ with ½″ reinforcing rod driven with 5 lb. hammer	30 - 50	60 - 80
Very Dense	Penetrated only a few inches with ½″ reinforcing rod driven with 5 lb. hammer	Over 50	80 - 100

1. Field guide intended as an example of one of many field procedures currently in use for indicating density and is not necessarily a preferred method.
2. Number of blows required to penetrate to a depth of 1′, as measured in accordance with ASTM D1586—Tentative Method for Penetration Test and Split-Barrel Sampling of Soils.

FIG. 7 CONSISTENCY OF UNDISTURBED COHESIVE SOILS

Soil Description	Rule-of-thumb Field Guide[1]	Unconfined Compressive Strength (q_u)[2]
Very Soft	Core (height=2 x diameter) sags under own weight	0.25
Soft	Can be pinched in two between thumb and forefinger	0.25 - 0.50
Medium	Can be imprinted easily with fingers	0.50 - 1.00
Stiff	Can be imprinted with considerable pressure from fingers	1.00 - 2.00
Very Stiff	Barely can be imprinted by pressure from fingers	2.00 - 4.00
Hard	Cannot be imprinted by fingers	4.00

1. Field guide is only an indication of soil consistency. Values of q_u are given as basic values of consistency by which field classification can be varified and should not be used for design purposes without laboratory varification.
2. q_u is unconfined compressive strength in tons/sq. ft. (not the ultimate bearing capacity of the soil for design).

vide general information about the nature and origin of existing soils without the expense of extensive soil boring and testing, and also will determine the need for and extent of further investigation.

At least one test boring at the site is recommended to determine the acceptability of the soil for ground-supported slabs. Test borings can be made with simple tools (Fig. 8) to determine soil types and the extent of each to a depth of 15′, the maximum depth of borings usually necessary to predict the performance of light residential construction.

Competent professional advice of a soil engineer experienced in a given locality may preclude the need for test borings at a single site. Also, examination should be made of structures in the immediate area for evidence of apparent settlement or expansion of the soil, provided that similar conditions of soil, topography and proposed construction prevail.

No additional soil investigation is required for single sites if the preliminary site investigation clearly indicates that the site has nonproblem soils consisting of *firm, nonexpansive* material, and that no ground or surface water problems exist, and that fill thickness greater than 3′ will not be required.

FIG. 8 Hand auger for preliminary field investigation of soils.

FIG. 9 **DESCRIPTION OF NONPROBLEM SOILS**

Soil Classification	Soil Description	Density or Consistency[1]
GW, GP	Gravels or sand-gravel mixtures, little or no fines	All densities
GM, GC	Gravels or gravel-sand mixtures, with clay or silt	Medium-dense to dense
SW, SP	Sands with little or no fines	Medium-dense to dense
SM, SC	Silty or clayey sands	Medium-dense to dense
ML	Silty or clayey fine sands	Medium-dense to dense
ML, MH	Inorganic silts	Medium-dense to dense
CL	Inorganic and silty clays of low to medium plasticity	Stiff to hard

1. See Figs. 6 & 7 for definition of terms.

Where tests reveal problem soils with an expansion or shrinkage potential, or on large developments where extensive grading and site preparation are required, an engineering study should be conducted by a qualified soils engineer to determine necessary design and construction procedures.

Nonproblem Soils Soils which are dense, coarse-grained and properly drained cause few problems for slab-on-ground construction (Fig. 9). These soils possess excellent bearing capacity because they are not subject to significant volume change. Gravels, for instance, usually occur in a dense state and are unlikely to cause any settlement problems. Usually, the only consideration is that the natural confinement of the soil not be disturbed by trenching or grading operations which would reduce the necessary lateral support.

Load-bearing capabilities of both cohesive and cohesionless soils can be improved by compaction.

When sand or silt is encountered, it should be determined whether it is dense, medium dense or loose. Loose sand or silt frequently can be compacted economically to a dense state before slab construction.

Loess Soils which are included in the ML or MH groups usually should be treated in the same manner as other soils in these groups. The loads in light residential construction generally are not sufficient to cause problems with these soils unless the slab supports masonry or other heavy concentrated loads. Heavily loaded slabs may settle unduly when loess soils become saturated and therefore proper drainage is extremely important where these soils are encountered.

Problem Soils Soils are considered problem soils if they possess less than suitable bearing characteristics for either Type I or Type II slabs. Special construction precautions are necessary with such soils as highly compressible or highly expansive clays, highly plastic soils, sands and silts which are very loose and loess soils which are heavily loaded or not well drained. If the slab is not sufficiently stiff and does not possess the necessary strength, detrimental effects to the slab and/or superstructure may result, caused by settlement or differential movement due to lack of density or change in the moisture content of the supporting soil.

Moisture Conditions

Moisture due to either surface water from precipitation or ground water, including the effects of seep-

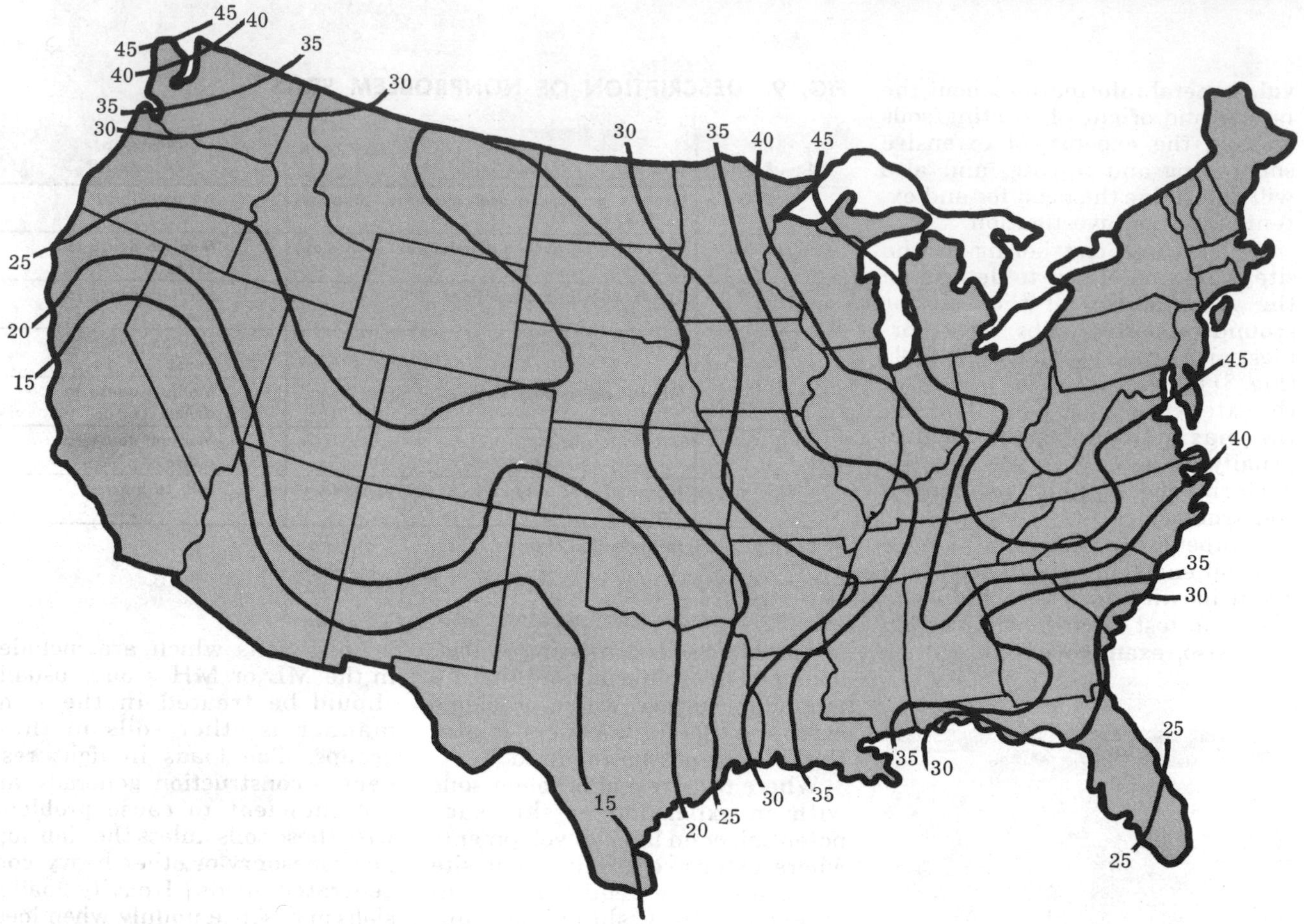

FIG. 10 Climatic ratings (C_w) used to determine acceptable soils for Types II and III slabs.

age fields: (1) may result in volume change and/or reduction of the bearing capacity of the soil; (2) may impose additional requirements for site preparation with regard to grading and drainage; and (3) may necessitate special precautions such as a base course or vapor barrier (see Moisture Control, page 311-17).

Precipitation The amount and frequency of precipitation has a primary influence on problem soils, because it may cause changes in the moisture content of soil supporting the slab. Exact values for the amount of precipitation, its seasonal variation, amount and duration of each occurence or its effect on supporting soil under the slab are difficult to measure and almost impossible to predict. However, on the basis of U.S. Weather Bureau data, a *climatic rating* (C_W) can be assigned to various areas of the United States as shown in Figure 10. These values can be used to help select the proper type of slab construction for various types of slab beds. The most important considerations are whether precipitation or the absence of it will change the moisture content of the soil intended for slab support during and after construction, and whether the change will have a detrimental effect on the slab.

When variations in moisture content occur in a type of soil that expands as it absorbs moisture and shrinks as it loses moisture, the slab it supports is subjected in turn to cycles of uplift as the soil expands and settling as the soil

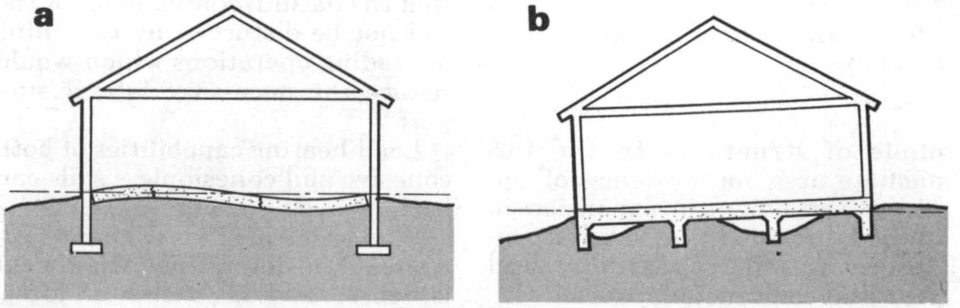

FIG. 11 Slab distortion and soil movement (highly exaggerated above): (a) Flexible slab (Type I or II) will distort with volume changes in the supporting soil. (b) Rigid slab (Type III) is designed to span uneven contours caused by volume changes in the supporting soil.

shrinks. When a period of high moisture content is followed by drought, soil moisture evaporates more rapidly around the perimeter of the slab, and soil moisture under the center of the slab—which may be due to capillary action and migration—becomes trapped and sealed from direct exposure. Moisture can be retained under the center of a slab long after extended periods of drought.

Trees and shrubbery in the immediate area of the slab perimeter also can affect the soil moisture content by providing a shield from natural precipitation and by extracting soil moisture during the growing season. Also, watering of gardens and lawns in the area of the slab perimeter can increase the soil moisture content.

When soil changes from low to high moisture content, a similar but opposite phenomenon develops. If prolonged periods of alternating drought and moisture occur, considerable differential moisture can develop in different areas under the same slab. If the soil is a type that undergoes substantial volume changes with changes in moisture, one of two conditions may result. If the slab is relatively flexible (such as Type I and Type II slabs which are unreinforced or only lightly reinforced with fabric) it will follow the uneven contour of the supporting soil that results from uneven volume change, exposing the slab to distortions and possible damage to both slab and superstructure (Fig. 11a).

On the other hand, if the slab is sufficiently rigid (such as a structurally reinforced and stiffened Type III slab) it will resist deformations even though the contour of the supporting soil is uneven. Higher soil pressures develop under the slab over the plateaus with reduced soil pressures at the valleys. The slab will be subjected to bending between these uneven contours and the soil at the plateaus may deform due to higher bearing pressures, distributing slab loads to adjacent areas (Fig. 11b). Type III slabs are designed to resist bending stresses in order to span between plateaus without damage to the slab.

Thermal Conditions

The effects of temperature, combined with moisture, may influence the depth of perimeter grade beams. The combined effects of precipitation and temperature can affect both slab selection and performance because some soils are subject to volume change called frost heave, caused by formation of ice layers when water in the soil freezes. Unless a Type III slab is used or if nonheaving soils (GW, GP, SW, SP) are present with the water level below the frost line, foundations always should be carried be-

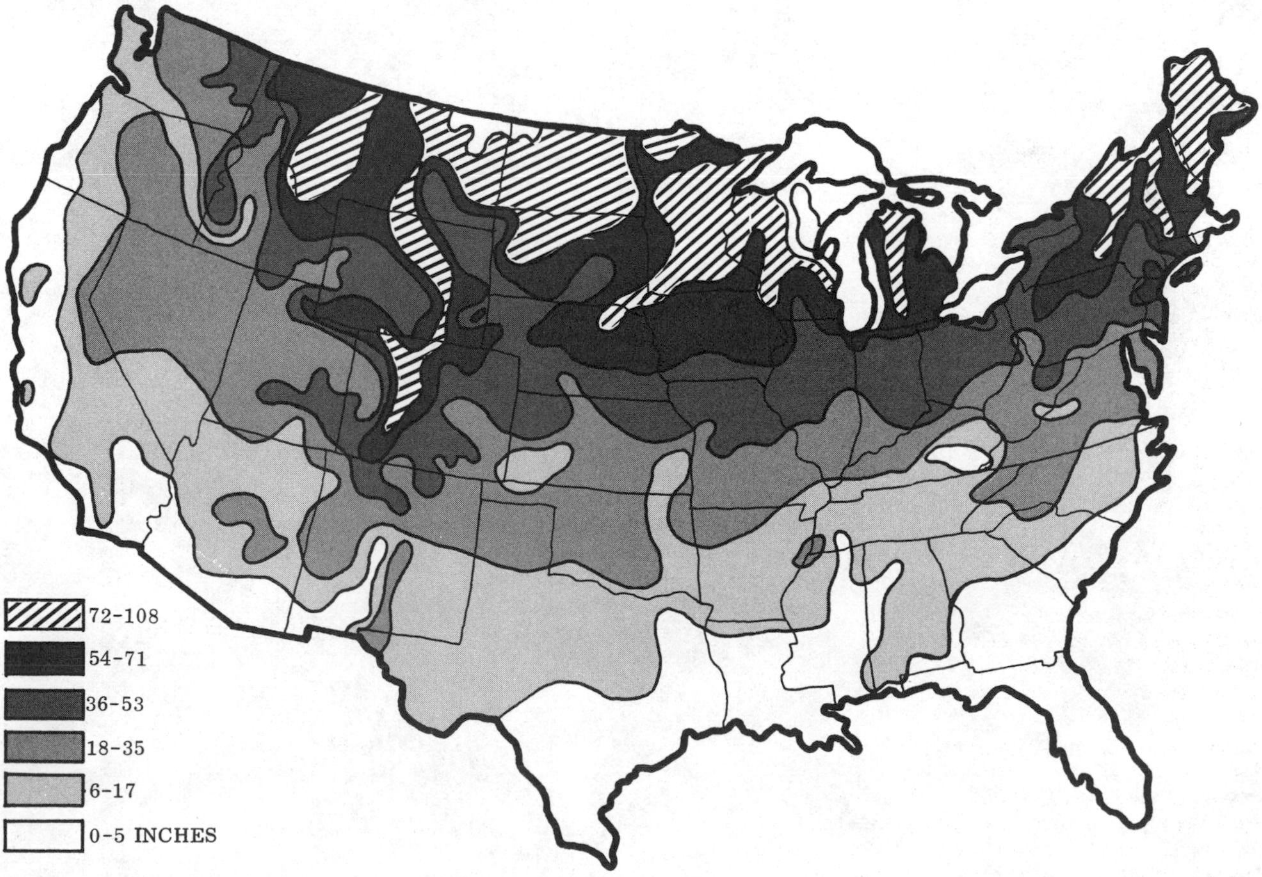

FIG. 12 Maximum depth of frost penetration in inches.

low the frost line (Fig. 12).

Temperature can affect the comfort performance of slabs, especially in cold climates where insulation is required, and often elements of the heating system are built integrally with the slab. Recommendations for insulation and heating elements in the slab are discussed under Thermal Control.

CONSTRUCTION RECOMMENDATIONS

Slab construction depends on the *slab type; site preparation,* including preparation of the slab bed, grading and backfilling; other precautions to control either *ground or surface moisture;* and *thermal control* when required. The performance of all slabs is highly dependent on the quality control of concrete materials and workmanship.

SLAB TYPES

The appropriateness of a concrete slab foundation/floor system should be based on observed soil properties, climate, functional and esthetic requirements, and economy.

Type I Slabs

Because a Type I slab is unreinforced, except for specific localized areas, the slab possesses only compressive strength and cannot tolerate appreciable amounts of tension or warping. The slab may crack from shrinkage during drying, but when used under the conditions recommended below, the cracks which occur should not open excessively nor prove detrimental to serviceability of the slab.

Slab Bed Type I slabs should be placed only on well-drained and properly graded coarse-grained soils (Fig. 13). Such soils are not affected significantly by climate and moisture content changes and should develop no appreciable volume change that would be detrimental to the slab. The soil must be capable of supporting the slab and be uniformly and adequately compacted to ensure that warping and tensile stresses which could contribute to cracking are not induced in the slab.

Type I slabs can be applied successfully to: all gravelly soils of any density (Types GW, GP); and all sandy soils with or without silts and clays (Types GM, GC, SW, SP, SM, SC) and silts (Types ML, MH) provided they are dense or medium dense. Type I slabs can be applied also to loose sandy and silty soils if the soil can be compacted to a dense or medium dense state to its entire depth before slab construction (Fig. 13).

FIG. 13 **RECOMMENDED SLAB BED FOR TYPE I SLABS**

Soil Type[1]	Density
GW, GP	All Densities
GM, GC, SW, SP, SM, SC, ML, MH	Dense or medium dense
	Loose[2]

1. See Fig. 25 for base course requirements.
2. May be loose if properly compacted to its entire depth.

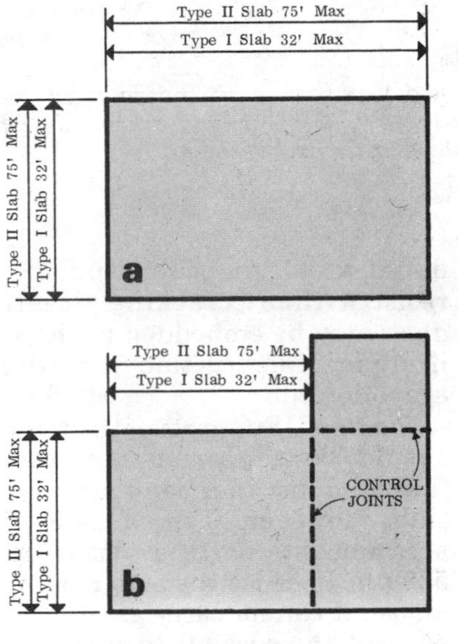

FIG. 14 Maximum dimensions for Types I and II slabs. If slab is left exposed, control joints spaced 15' to 20' apart should be provided.

Dimensions Type I slabs should be of at least 4″ uniform thickness, and rectangular or square in shape (Fig. 14a). To control shrinkage cracking during drying and subsequent thermal volume changes, no dimension of the slab perimeter should exceed 32′. However, if the slab is to be left exposed, control joints spaced 15′ to 20′ apart are recommended (Fig. 14b).

Irregularities in Shape, such as when a nonrectangular T or L shape is required, require control joints to divide the slab into squares or rectangles (Fig. 14b). If control joints are not provided, diagonal cracks which may be objectionable are likely to occur where the slab changes in shape.

Irregularities in Thickness, that would reduce the slab to less than 4″ nominal thickness, should not be permitted. Stresses caused by drying shrinkage or thermal change would tend to cause cracking at these locations. When the upper

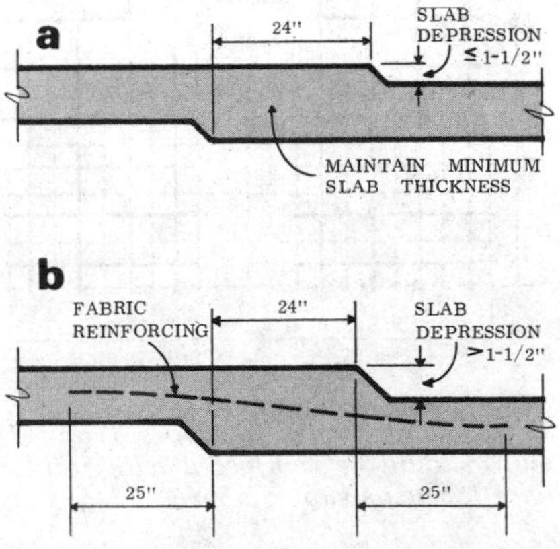

FIG. 15 (a) Minimum slab thickness should be maintained when floor depressions are required; (b) fabric reinforcement should be provided when depression exceeds 1-1/2″.

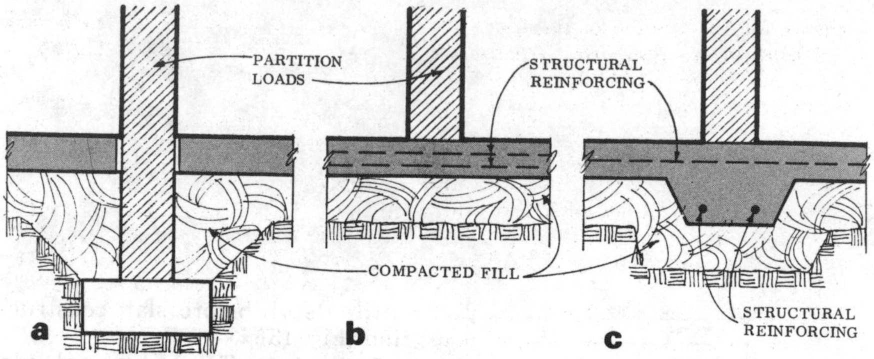

FIG. 16 For Types I and II slabs (a) heavy partition or concentrated loads should be carried on independent foundations; (b & c) alternate methods when engineering design is used to determine structural reinforcement.

surface of the slab must be lowered, such as to permit proper installation of ceramic tile or other floor finish, the uniform slab thickness should be maintained by lowering the underside of the slab beginning at least 24″ away (Fig. 15a). In addition, when the vertical displacement is greater than 1-1/2″, 6 x 6-6/6 welded wire fabric should be placed in the slab extending 25″ on either side of the point of displacement (Fig. 15b).

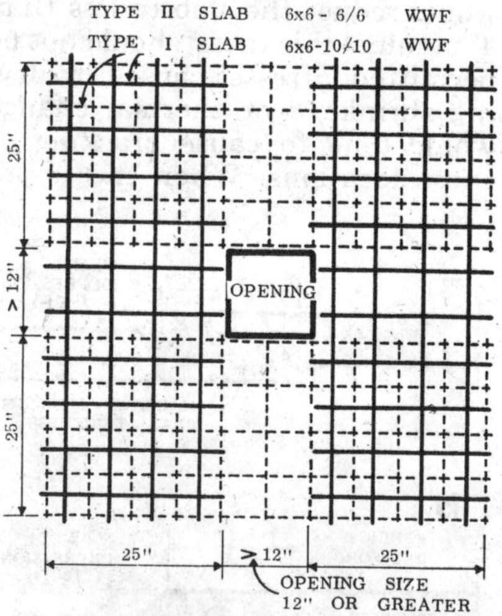

FIG. 17 Openings in Types I and II slabs should be reinforced with welded wire fabric as shown above.

Embedded Heating Elements Heating *coils or pipes* should not be embedded in Type I slabs because thermal stresses will be induced which the slab is unable to resist without cracking. Heating *ducts* may be embedded in the slab if proper construction procedures are followed.

Loading Superstructure loads should be supported directly on foundations independent of the slab. However, Type I slabs can accommodate partition loads up to 500 plf. Because the slab is unreinforced, it cannot carry greater loads without the possibility of cracking. Where this loading must be exceeded, reinforcement or special construction should be provided to minimize cracking which would result. Methods of supporting partitions or concentrated loads are illustrated in Figure 16.

Openings Openings in Type I slabs should be kept to a minimum because they can introduce stresses which are not uniform. When an opening 12″ or wider must be provided, the slab should be reinforced with 25″ wide 6 x 6-10/10 welded wire fabric around the opening (Fig. 17).

Type II Slabs

Type II slabs are lightly reinforced over their entire area to withstand slight differential movement, and to hold cracks together once they form; grading and loading are similar to Type I slabs. As with Type I slabs, the design of the slab is not influenced by the type of superstructure which is supported directly on independent foundations.

Slab Bed Some soils unsuitable for Type I slabs are acceptable for the support of Type II slabs, because the slab is lightly reinforced over its entire area. However, Type II slabs cannot be expected to resist significant differential movement of the soil. The amount of fabric reinforcement provided is based on the premise that the entire slab will remain uniformly supported by the underlying soil.

Type II slabs can be constructed on all soils suitable for Type I slabs under all climatic conditions, and these soils may be of any density, including loose (Fig. 18). In addition, Type II slabs can be applied to certain fine-grained soils (CL, OL, CH, OH) under certain climatic conditions if the *unconfined compressive strength* and *plasticity index* fall within the limits shown in Figure 18.

FIG. 18 **RECOMMENDED SLAB BED FOR TYPE II SLABS**

Soil Type[1]	Density or PI and q_u[2]	Climatic Rating[3]
GW, GP, GM, GC, SW, SP, SM, SC, ML, MH	All Densities	All
CL, OL, CH, OH	PI < 15 and $q_u/w \geq 7.5$	All
	PI > 15 and $q_u/w \geq 7.5$	$C_w \geq 45$

1. See Fig. 25 for base course requirements.
2. q_u = unconfined compressive strength; PI = Plasticity Index.
3. See Fig. 10 for Climatic Ratings (C_w).

Dimensions Type II slabs are of at least 4″ uniform thickness, and no dimension of the slab perimeter should exceed 75′ (Fig. 14a).

The same recommendations apply as for Type I slabs with respect to irregularities in thickness (Fig. 15) and shape (Fig. 14b), with the added requirement that reinforcement be continuous across control joints, or that keyed joints be provided to eliminate the possibility of vertical displacement between adjacent sections.

Reinforcement Fabric reinforcement in Type II slabs provides for control of *crack size* only. Cracks which are expected to occur should be held tightly closed by the tensile reinforcement and should not be objectionable. To control crack size,

FIG. 19 **MINIMUM FABRIC REINFORCEMENT FOR TYPE II SLABS**

Maximum Slab Dimension (feet)	Wire Spacing (inches)	Wire Gauge (number)
Up to 45	6 x 6	10/10
45 to 60	6 x 6	8/8
60 to 75	6 x 6	6/6

the slab should be provided with the minimum welded wire fabric reinforcement recommended in Figure 19. Reinforcement should be placed in the center of the slab, located midway between top and bottom of the slab and preferably set on supports such as chairs.

Embedded Heating Elements Because Type II slabs are reinforced over their entire area the welded wire fabric will hold together any cracks that form. Heating pipes or coils, as well as heating ducts, can be embedded in the slab. Construction recommendations given on page 311-18 should be followed.

Loading These slabs are expected to deflect with slight soil movement and therefore superstructure loads should be supported independently on foundations. However, like Type I slabs, Type II slabs can accommodate partition loads up to 500 plf and equivalent concentrated loads. An additional layer of reinforcement can be provided to extend at least 25" on either side of the partition or concentrated load to distribute added stresses (Fig. 16b). All heavy concentrated loads such as masonry should be supported on independent or structurally reinforced footings (Fig. 16a & 16c).

FIG. 20 **ACCEPTABLE SLAB BED FOR TYPE III SLABS**

Soil Type[1]	PI or q_u[2]	Climatic Rating[3]
CL, OL, CH, OH	PI > 15 and $q_u/w \geq 7.5$	$C_w < 45$
	$7.5 > q_u/w \geq 1.5$	All

1. See Fig. 25 for base course requirements.
2. q_u = unconfined compressive strength; PI = Plasticity Index.
3. See Fig. 10 for Climatic Rating (C_w).

Openings The same recommendations for Type I slabs apply for Type II slabs, except that the additional layer of reinforcing placed around the opening should be 6x6-6/6 welded wire fabric. The additional reinforcing around the opening will prevent the concentration of stresses which are likely

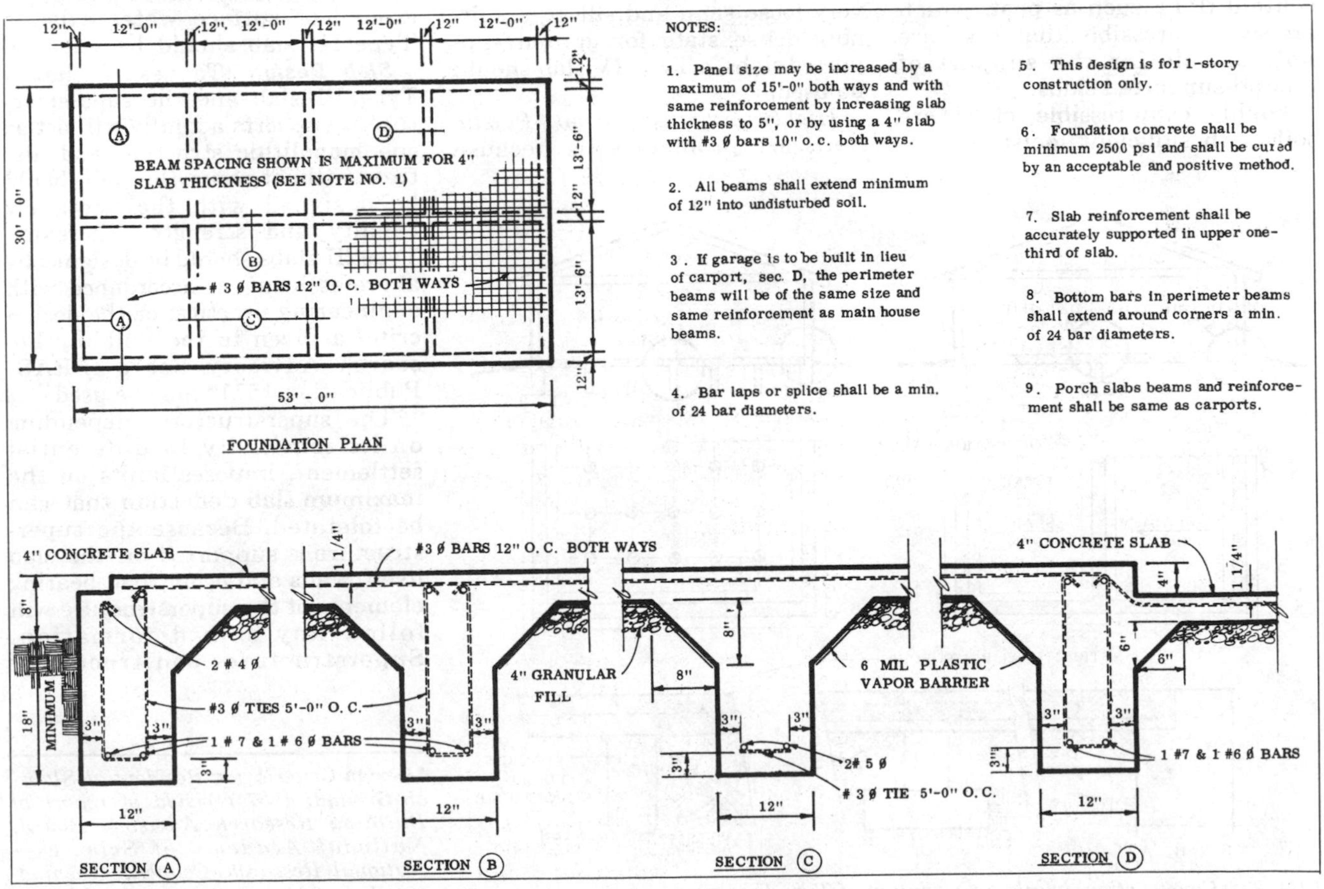
12" 12'-0" 12" 12'-0" 12" 12'-0" 12" 12'-0" 12"
BEAM SPACING SHOWN IS MAXIMUM FOR 4" SLAB THICKNESS (SEE NOTE NO. 1)
3 ø BARS 12" O. C. BOTH WAYS
30' - 0"
53' - 0"
13'-6"
FOUNDATION PLAN
NOTES:
1. Panel size may be increased by a maximum of 15'-0" both ways and with same reinforcement by increasing slab thickness to 5", or by using a 4" slab with #3 ø bars 10" o.c. both ways.
2. All beams shall extend minimum of 12" into undisturbed soil.
3. If garage is to be built in lieu of carport, sec. D, the perimeter beams will be of the same size and same reinforcement as main house beams.
4. Bar laps or splice shall be a min. of 24 bar diameters.
5. This design is for 1-story construction only.
6. Foundation concrete shall be minimum 2500 psi and shall be cured by an acceptable and positive method.
7. Slab reinforcement shall be accurately supported in upper one-third of slab.
8. Bottom bars in perimeter beams shall extend around corners a min. of 24 bar diameters.
9. Porch slabs beams and reinforcement shall be same as carports.
4" CONCRETE SLAB
#3 ø BARS 12" O.C. BOTH WAYS
2 # 4 ø
#3 ø TIES 5'-0" O. C.
1 # 7 & 1 # 6 ø BARS
18" MINIMUM
4" GRANULAR FILL
6 MIL PLASTIC VAPOR BARRIER
2# 5 ø
3 ø TIE 5'-0" O.C.
1 #7 & 1 #6 ø BARS
SECTION A
SECTION B
SECTION C
SECTION D

FIG. 21 An example of Type III slab construction.

to crack the slab at these points (Fig. 17).

Type III Slabs

Type III slabs are limited to use over problem soils. They are structurally reinforced and stiffened so that they can receive and transmit all superstructure loads to the foundation soil by distributing all loads over a larger portion of the slab. Bearing pressures on the ground are reduced and the slab and the superstructure act as one monolithic structure.

Slab Bed Because a Type III slab is structurally reinforced and stiffened, it can be used over certain problem soils such as highly compressible, highly expansive and highly plastic soils. The unconfined compressive strength, plasticity index and climatic ratings for soils on which Type III slabs should be used are given in Figure 20.

Highly Compressible Soils include those that are high in organic content (PT), such as peat, which are so compressible that they are not suitable for the support of ground-supported slabs.

Highly compressible clays are soils with sufficient moisture content to have a soft or very soft consistency or high organic content. Excessive settlement may result even under relatively light residential loading or from the weight of fill material.

FIG. 22 **ACCEPTABLE SOILS FOR TYPE IV SLABS**

Soil Type	Unconfined Compressive Strength (q_u)	Climatic Rating
CL, OL, CH, OH	$q_u/w < 1.5$	All
PT	ALL	ALL

Sands and silts which are very loose can cause serious problems beneath the slab. Fine sands and silts are particularly troublesome when very loose. Vibration of very loose sands and silts may cause a decrease in the volume of soil voids with resultant settlement, depending on the amplitude and frequency of vibration at the site. When it is not practical to compact very loose sand and silt to a suitably dense state for ground supported slabs, a Type IV slab should be used.

Highly Expansive and Plastic Soils are problem soils because they are greatly affected by climatic conditions. Any soil with a plasticity index greater than 15 may cause problems by expanding under exposure to severe climatic conditions. However, under favorable or intermediate conditions this material may be considered for Type II slabs.

Any soil with a plasticity index of 30 or higher will react vigorously to changes in moisture content and will present the most difficult design problems because of the great volume change that can be expected. Since loads associated with light residential construction are not sufficient to control such expansion by loading, extreme care must be exercised in placing light residential slabs on highly plastic soils.

Loess Soils can become problem soils when they are heavily loaded and become saturated. When slab construction is desirable over these soils and the slab must be heavily loaded but problems of ground or surface water exist, a Type III slab should be selected.

Slab Design To assure that a Type III slab and the superstructure it supports actually will act as one monolithic structure and distress will not occur, the slab should be designed with the necessary rigidity and strength. Therefore, Type III slabs should be designed by a professional in accordance with engineering practice; or the design criteria given in the Building Research Advisory Board (BRAB) Publication 1571* may be used.

The superstructure, depending on its sensitivity to differential settlement, imposes limits on the maximum slab deflection that can be tolerated. Because the superstructure is supported on the slab itself, walls and other load-bearing elements of the superstructure will follow any slab deformation. Superstructures constructed of

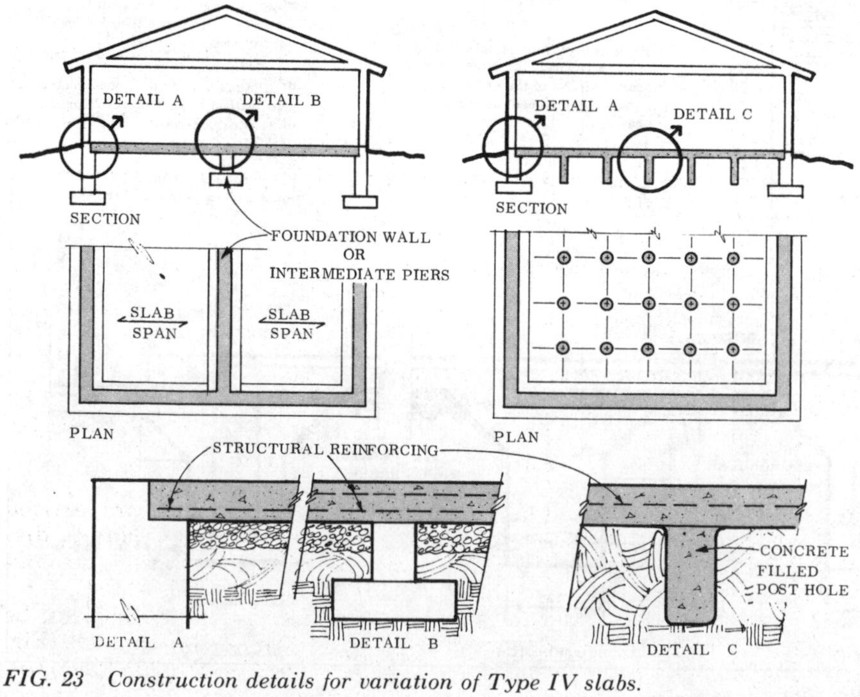

FIG. 23 Construction details for variation of Type IV slabs.

**Design Criteria for Residential Slabs-on-Ground: 1967 revision of report of Building Research Advisory Board, National Academy of Sciences—National Research Council, Washington, D.C. 20418.*

different materials tolerate deformations of differing intensities before developing undesirable effects such as cracking and distortions. For example, wood frame construction generally can sustain more deformation than masonry construction before warping and

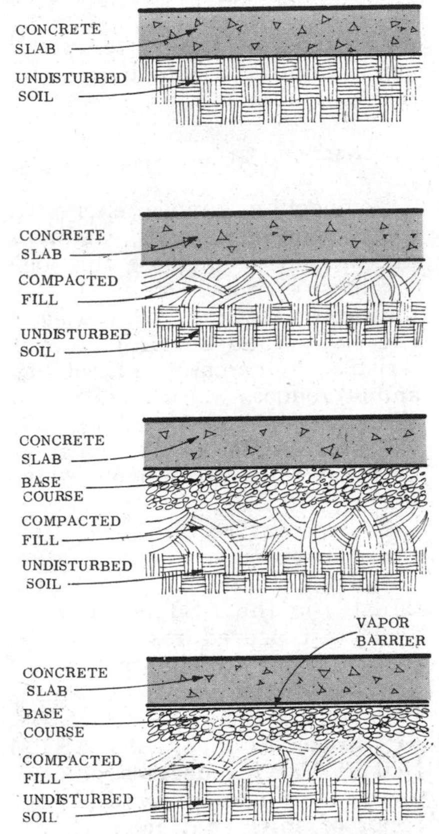

FIG. 24 Slab bed for ground-supported slabs: a base course and/or a vapor barrier may be necessary to control moisture.

cracking create mechanical and esthetic problems. Comparatively brittle cementitious materials such as concrete, stucco, plaster and mortar can tolerate only small deformations before cracking develops.

An example of a Type III slab is shown in Figure 21.

FIG. 25 **REQUIREMENTS FOR SLAB BED ACCORDING TO SOIL CLASSIFICATION[1]**

Major Division			Group Symbols	Typical Names
COARSE-GRAINED SOILS (More than half of material is larger than the smallest particle visible to the naked eye)	Gravels (more than half of coarse fraction is larger than ¼")	Clean Gravels (little or no fines)	GW	Well graded gravels, gravel-sand mixtures, little or no fines.
			GP	Poorly graded gravels or gravel-sand mixtures, little or no fines.
		Gravels with Fines (appreciable amount of fines)	GM	Silty gravels, gravel-sand-silt mixtures.
			GC	Clayey gravels, gravel-sand-clay mixtures.
	Sands (more than half of coarse fraction is smaller than ¼")	Clean Sands (little or no fines)	SW	Well graded sands, gravelly sands, little or no fines.
			SP	Poorly graded sands or gravelly sands, little or no fines.
		Sands with Fines (appreciable amount of fines)	SM	Silty sands, sand-silt mixtures.
			SC	Clayey sands, sand-clay mixtures.
FINE-GRAINED SOILS (More than half of material is smaller than the smallest particle visible to the naked eye)	Silts and Clays (liquid limit is less than 50)		ML	Inorganic and very fine sands, rock flour, silty or clayey fine sands or clayey silts with slight plasticity.
			CL	Inorganic clays of low to medium plasticity, gravely clays, sandy clays, silty clays, lean clays.
			OL	Organic silts and organic silty soils, elastic silts.
	Silts and Clays (liquid limit is greater than 50)		MH	Inorganic silts, micaceous or diatomaceous fine sandy or silty soils, elastic silts.
			CH	Inorganic clays of high plasticity fat clay.
			OH	Organic clays of medium high plasticity, organic silts.
Highly Organic Soils			PT	Peat and other highly organic soils.

Generally may be used for fill without a base course

Fill that requires a base course

Not permitted for foundation fill

1. Based on ASTM D2487—Classification of Soils for Engineering Purposes.

Type IV Slabs

Type IV slabs also receive and transmit all superstructure loads to the foundation soil, but unlike Type III slabs, they do not depend on the foundation soil for support. The slab bed functions as a form and is used only for temporary support of the slab until it develops sufficient strength to be supported on beams, piers or intermediate foundations walls. These foundation elements in turn are carried by piles or footings which distribute the loads to stable ground below the level of the slab.

Slab Bed Type IV slabs can be used over extremely poor soils (Fig. 22) which cannot be treated economically or practically to receive ground-supported slabs. These soils might include extremely expansive clays, highly organic soils, or very loose sands and silts. Foundation elements such as grade beams should be designed so that swelling of the soils beneath the slab will not lift these elements off the foundations or pull the foundations out of the ground.

Due to local practice, Type IV slabs sometimes are used over unstable soils which may not necessarily be considered *extreme* problem soils (Fig. 23). Foundation piers or footings should be designed to penetrate to a depth providing adequate support below unstable soil.

Slab Design Type IV slabs are reinforced structural slabs and should be designed by a professional in accordance with conventional engineering practice.

SITE PREPARATION

Typically, when slab-on-ground construction is intended, some preparation must be made to the subgrade, depending on the type of soil and other conditions encountered at the site. The slab bed for ground-supported slabs: (1) should provide the necessary bearing capacity for slab support; (2) should control ground moisture; and (3) should establish the proper slab elevation.

The slab bed may consist of undisturbed soil (if of the proper type, density and consistency), foundation fill and/or a base course when necessary to control capillary rise of moisture (Fig. 24). All fill material, and undisturbed soil when it is not in a suitably dense state, should be properly compacted.

Fill Materials

Compacted fills are used for residential development sites in undulating or hilly terrain to create reasonably level ground for slab construction. Soils used for grading fills and foundation fills should contain no vegetation or foreign material that would cause uneven settlement. Foundation fill and/or a base course should be provided when necessary.

Foundation Fills Areas within foundation walls also should have topsoil removed and should contain no other material that would cause the slab to settle unevenly.

Foundation fills should be used to establish the desired finished slab elevation and should be compacted to a maximum practical density. The slab bed for Type I and II ground-supported slabs should not expand or contract due to moisture. Hence, organic or active clays and silts of the OL, CH, CL, OH and PT groups are not suitable for foundation fills under these types of slabs. Soil groups GW, GP, SW and SP (Fig. 25) may be used for fill material without a base course.

Structural slabs (Type IV) which receive their support at edges and through the center by intermediate foundations require only moderate compaction because the fill material serves only as temporary support until the slab develops the necessary strength.

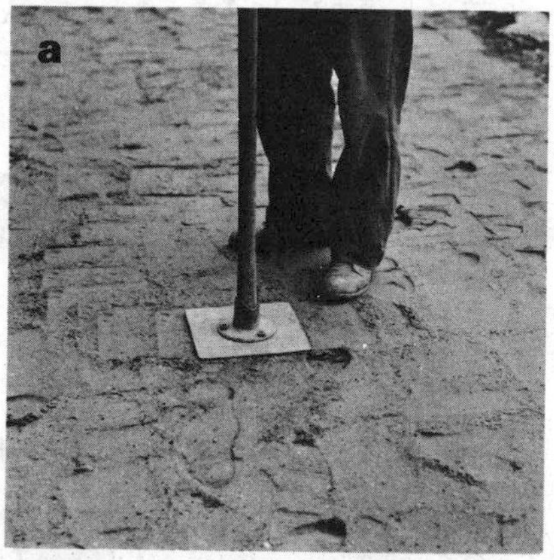

FIG. 26 Depending on the condition of the soil, the slab bed can be compacted by (a) hand tamping or (b) power driven equipment.

Compaction

Compaction applies energy to soil to consolidate it by compressing air voids to increase the soil's dry density. Proper compaction: (1) minimizes settling, (2) increases the soil's load-bearing characteristics, (3) increases soil stability, and (4) reduces water penetration. Thus, proper compaction generally will prevent slab cracking caused by differential settlement in the soil.

Compaction requirements generally are specified as a percentage of the maximum dry density obtainable in the ASTM D1557 or D698 test procedures[1]. Also the relative density can be specified for fills constructed with free-draining sand and gravel soils (GW, GP, SW, SP) in accordance with ASTM D2049[2]. Varying degrees of compaction for the many different types of soils that may be encountered cannot be stated without engineering analysis. Generally, however, clean sand fill (SW, SP) should be compacted to 95% optimum dry density and all other fine-grained, nonexpansive fills to 90% of optimum dry density in accordance with ASTM D1557. Compaction requirements for expansive soils should be determined from engineering analysis and may require somewhat less than 90% of optimum dry density.

Using suitable hand or power equipment (Fig. 26), compaction of fill materials should be performed in 4″ to 12″ layers (lifts) of soil. Re-

1. *ASTM D1557—Moisture-Density Relations of Soils Using 10-lb. Rammer and 18″ drop; ASTM D698—Moisture-Density Relations of Soils Using 5.5-lb. Rammer and 12″ Drop.*
2. *ASTM, D2049—Relative Density of Cohesionless Soils.*

gardless of the type of soil being compacted, the thinner the lift, the better the compaction.

MOISTURE CONTROL

Settlement or differential movement of slabs caused by moisture fluctuations in problem soils is discussed under Site Characteristics (page 311-4). Additional provisions for moisture control are necessary for long-term performance of concrete slab-on-ground construction. Slabs should be protected from water and water vapor to prevent damage to finish flooring materials and pipes or heating ducts embedded in the slab, and to guard against reduced effectiveness of thermal insulation. Problems may be caused either by surface or ground water.

The methods and extent of protection depend on (1) the slab elevation with respect to the elevation and slope of the finished grade; (2) location of the ground water table or other man-made water sources such as seepage fields; (3) drainage properties of the soil or fill beneath the slab; and (4) the type of finish flooring used.

Many of the moisture problems associated with slab-on-ground construction can be minimized or eliminated by proper preliminary grading, correct selection of fill or base course materials, and installation of a vapor barrier.

Grading and Drainage

The site around the structure should be graded to form protective slopes which will drain surface water away from the foundation so that surface and/or ground water will not collect under the slab (see Fig. 15, page 310-10). Surface runoff should be directed to streets or other drainage structures. Backfill adjacent to foundation walls should be uniformly compacted with hand or machine tamping equipment to at least 85% of optimum dry density.

Finish grade should slope downward in all directions away from the slab to a minimum of 6″ for a distance of 10′ or approximately a 4% to 5% slope. Paved areas adjacent to the structure having less than a 5% slope should be provided with a positive means of drainage. On hillside locations, the site should be graded to divert surface water around and away from the structure.

Slab Elevation

The elevation of the slab should be established so that the top of the slab or any element of wood construction is at least 8″ above finished grade level. If heating ducts are used in or under the slab, elevations should be established so that the bottom of the ducts are at least 2″ above finished grade, unless the ducts are of concrete, asbestos-cement or ceramic tile. Heating ducts should be installed as recommended.

Base Course

A base coarse is placed over undisturbed soil or fill when it is necessary to provide a capillary stop for water rising through the slab bed (Fig. 24). It also will provide uniform structural support for the slab, and some degree of thermal insulation against heat loss to the ground. At least a 4″ base course of limited capillarity should be provided when the foundation fill is not of limited capillarity, unless unusually favorable soil and drainage conditions exist. Material which passes a 2″ sieve and is retained on a 1/4″ sieve may be assumed to be of limited capillarity.

Soil Groups GM, GC, SM, SC, ML and MH generally require a base course; soil Groups GW, GP, SW, SP can be used without a base course (Fig. 25).

Vapor Barrier

Where no drainage or soil problems exist and in arid regions where irrigation and heavy sprinkling is not done, the vapor barrier may not be necessary. However, because it is impossible to correct moisture problems after construction and because a vapor barrier usually can be installed at nominal cost during construction, a vapor barrier generally is recommended for all slab-on-ground construction.

Where floor coverings, household goods or equipment must be protected from damage by moisture conditions, a vapor barrier should be installed under the slab (Fig. 24). The vapor barrier membrane should have a permeance of less than 0.30 perms and should resist deterioration as well as puncture from heavy traffic during construction. The vapor barrier should be installed with 6″ laps and be fitted carefully around utility and other service openings.

Vapor barrier materials suitable for slab-on-ground construction include single-layer membranes such as polyethylene sheet or multiple-layer membranes such as glass fiber reinforced waterproof paper with polyethylene film extrusion-coated onto both sides. Polyethylene sheet used over sand or firmly compacted soil under unreinforced slabs should be at least 4 mil nominal thickness; when used over gravel or crushed stone or under any structurally reinforced slab, it should be at least 6 mil nominal thickness.

THERMAL CONTROL

The following recommendations are for: (1) slab insulation, (2) slabs heated by pipes embedded in the slab, (3) slabs containing warm air ducts, and (4) slabs in dwellings for which summer cooling is to be provided.

Slab Insulation

Heat loss of floor slabs built on ground is more nearly proportional to the perimeter than the area of the slab. Heat loss to the ground below is minor. However, it is important that in areas where heating systems are required, insulation should be installed around the slab perimeter, to prohibit excessive heat loss directly to areas of low outside air or ground temperature. Uninsulated perimeter surfaces of slabs may be uncomfortably cold, may produce condensation and may result in uneconomical heating efficiency.

Insulation should be (1) virtually nonabsortive of moisture (noncapillary), (2) not permanently

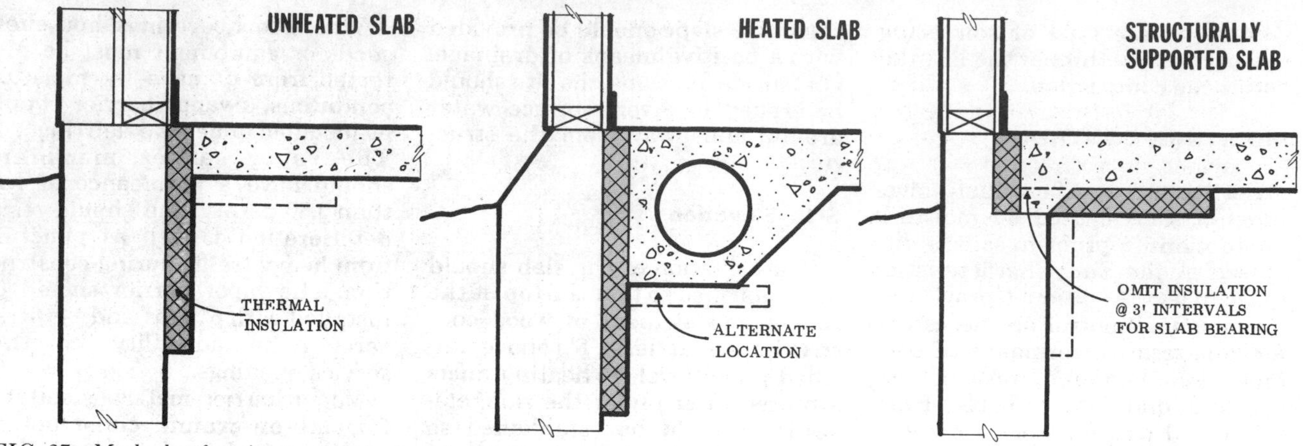

FIG. 27 Methods of perimeter insulation for slab construction; insulation can be installed either vertically or horizontally.

harmed by wetting or contact with wet concrete, (3) sufficiently rigid to resist damage and compression during placing of concrete, and (4) resistant to decay and insect damage.

Insulation of at least 1″ thickness should be provided around the perimeter of the slab; however, 2″ thickness is recommended. The insulation should be installed around the entire perimeter of the slab, extending vertically 24″ below grade level or horizontally for a distance of 24″ under the slab (Fig. 27).

Embedded Heating and Cooling Elements

Whenever pipes or ducts must pass through slabs, care should be taken to prevent pipes or ducts from breaking by isolating them from the slab. Pipes should pass through the slab vertically and be provided with space for expansion and with flexible connections if necessary to accommodate differential movement between the slab and soil.

When galvanized metal ducts or copper pipe is embedded, calcium chloride should not be used in the concrete mixture. Aluminum should not be embedded in any concrete.

Heating Ducts Heating ducts embedded in the slab may be of metal, asbestos-cement, wax-impregnated paper, ceramic tile or concrete pipe. Asbestos-cement ducts with watertight joints may be set on a sand leveling bed and be backfilled with sand to the underside of the slab. Precautions should be taken to see that the position of ducts is not disturbed during concreting.

If metal or paper ducts are used, they should be completely enclosed in at least 2″ of concrete to prevent moisture from collecting in them; joints may be taped or cement-grouted. These ducts often must be weighted so that they will not rise in the freshly poured concrete.

When the bottom of the ductwork adjacent to the slab perimeter is below exterior finish grade, the duct system should be of ceramic tile, asbestos-cement or concrete pipe. Joints which are completely or partially below exterior grade should be watertight. In areas where termite control is indicated, these joints should be resistant to termite penetration or soil treatment chemicals.

If the slab is not reinforced (Type I slab), 6x6-6/6 wire mesh reinforcement should be placed to extend 19″ on either side of the centerline of the duct or to the slab edge, whichever is closer (Fig. 28a).

Embedded Pipes and Coils Type II slabs may be heated by circulation of heated liquids through embedded pipes. Ferrous or copper pipe generally is used with approximately 2-1/2″ to 3″ of cover over the pipe (Fig. 28b). Where pipes pass through control joints or construction joints, provision should be made for possible movement across the joint. The piping should be pressure-tested before the concrete is placed and

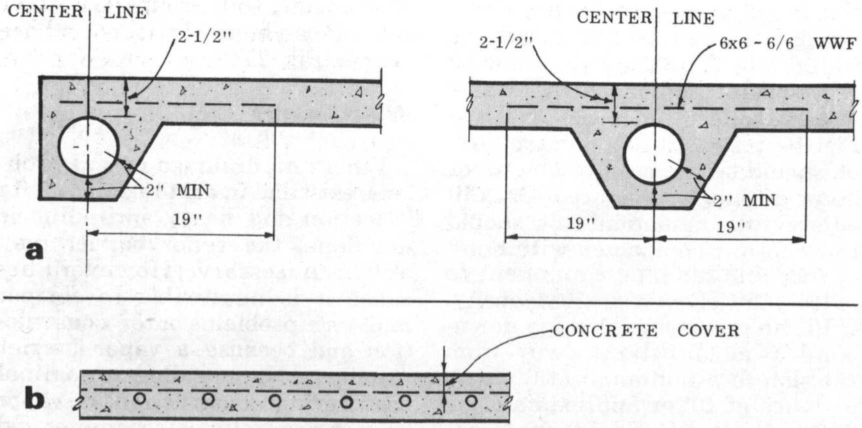

FIG. 28 Details for embedding heating sources in the slab.

air pressure should be maintained in the pipe during concreting operations; water pressure should not be used for testing. After the concrete is placed, the slab should not be heated until curing of the concrete has been completed. Slabs should be warmed gradually using luke warm liquid in the pipe to prevent the cold concrete from cracking.

Cooling Pipes The use of embedded pipes for summer cooling is not recommended with slab-on-ground construction. Under conditions of high humidity, condensation may form on the cooled slab which may damage flooring and support mold or fungus growth.

TERMITE CONTROL

In addition to the general recommendations given in Sections 201 and 310, slab-on-ground construction requires special considerations in areas where subterranean termite hazard is a significant problem (see Figure W19, page Wood 201-11). Slabs are expected to crack in varying degrees and termites can penetrate through cracks as small as 1/32″—through joints between the slab and the foundation wall, through control joints and through openings made for plumbing and conduit.

Some degree of protection of wood structural elements is provided by positive site and foundation drainage and separation of wood elements from the ground. Foundations and piers of hollow masonry units should have solid masonry caps in which all joints or voids are filled with mortar. Unless the foundation fill is of limited capillarity, at least a 4″ base course of limited capillarity and a suitable vapor barrier should be provided under the slab.

Where local experience indicates that additional protection against termites is necessary, termite barriers should be provided. Additional protection can be afforded by using pressure-treated wood which will resist decay as well.

Soil poisoning is the most effective termite barrier. The area under and around the slab should be treated, as well as all possible points of entry. Application of chemicals and precautions involved in their use should be in accordance with the manufacturer's recommendations.

The effective performance of *termite shields* depends on proper construction and workmanship and long-term integrity of the slab without cracking—which is unpredictable. Therefore, the effectiveness of termite shields is questionable.

FIG. 29 **CONCRETE MATERIALS**

Material	Standard Specification
PORTLAND CEMENT	
Types I, II, III, IV, V	ASTM C150
Types IA, IIA, IIIA (Air Entraining)	ASTM C150
Pozzolan and Blast Furnace	ASTM C595
READY-MIXED CONCRETE	ASTM C94
AGGREGATES	
Normal Weight	ASTM C33
Lightweight	ASTM C330
ADMIXTURES	
Air-entraining Agents	ASTM C260
Accelerators: Calcium Chloride	ASTM D98
REINFORCING	
Deformed Steel Bars	ASTM A305
Billet Steel Bars	ASTM A15
Rail Steel Bars	ASTM A16
Welded Steel Wire Fabric	ASTM A185
Cold Drawn Steel Wire	ASTM A82

CONCRETE QUALITY CONTROL

In addition to the effects of such factors as soil properties, slab type, control of moisture and resistance to heat transfer, careful control of concrete quality is essential to slab-on-ground construction. Unless concrete quality control measures are observed, undesirable properties in the hardened concrete may occur—particularly at the wearing surface—which may lead to a soft or dusting surface, permeable concrete, cracking or poor durability.

General recommendations for producing concrete of suitable quality are discussed in Section 203 Concrete. Reference should be made to the entire 203 Concrete section for general recommendations covering (1) materials, (2) mixtures, (3) handling and placing, (4) finishing and (5) curing. Unless specific exceptions are made in the following text, these general recommendations for quality control apply to slabs-on-ground.

Materials

In general materials used in mixing concrete should conform to the ASTM specifications given in Figure 29.

Portland Cement Concrete made with air-entrained portland cement provides increased durability, workability and improved finishing characteristics at little if any extra cost. A small amount of entrained air is recommended for all residential concrete slabs. At least 3% to 5% entrained air by volume is recommended for normal-weight aggregate concrete; concrete made with lightweight coarse and fine aggregates require an air content of 4% to 7%. For optimum quality control, air-entrained concrete should be plant- or transit-mixed.

Aggregates Figure 30 gives recommended grading for fine aggregates for residential slab-on-ground construction. The maximum size of coarse aggregates should not be more than 1/3 the thickness of unreinforced or lightly reinforced

FIG. 30 **GRADINGS OF FINE AGGREGATE FOR CONCRETE SLABS**

Sieve Size	Percent of aggregate Passing Sieve: Normal Weight	Lightweight
⅜″ (10 mm)	100	100
No. 4	95 - 100	85 - 100
No. 8	80 - 90	—
No. 16	50 - 75	40 - 80
No. 30	30 - 50	30 - 65
No. 50	10 - 20	10 - 35
No. 100	2 - 5	0 - 5

FIG. 31 **RECOMMENDED MIXTURES FOR RESIDENTIAL SLABS-ON-GROUND[1]**

Maximum Size of Coarse Aggregate[2]	Minimum Cement Content (lbs./cu. yd.)	Maximum Water/Cement By Weight		Slump	Probable Compressive Strength[3]
		Nonair-Entrained Concrete	Air-Entrained Concrete		
1½"	470	0.51	0.40	2" to 4"	3500 psi
1"	520				
¾"	540				
½"	590				
⅜"	610				

[1]For both air-entrained and nonair-entrained concretes.

[2]Normal weight aggregate; different mixes may be required for lightweight aggregate concretes.

[3]Strength refers to approximate compressive strength of cylinders made and tested according to applicable ASTM Standards at 28 days that have been continuously moist cured.

slabs, or 3/4 the minimum clear spacing between reinforcing bars and forms in structurally reinforced slabs. Generally, normal weight aggregate larger than 1-1/2" or lightweight aggregate larger than 1" are not used in residential slabs.

Mixtures The best proportions for a concrete mixture are the most economical combination of portland cement, available aggregate and other materials which will produce concrete of the required durability, abrasion resistance and strength, and which can be placed and finished properly. For slab construction, the properties of the plastic concrete (which affect placing and finishing) are equally as important as properties of the hardened concrete such as abrasion resistance, durability and strength. The plastic properties will greatly affect the quality of the top 1/16" to 1/8" of the slab surface. However, these properties are not measured easily, and there is a tendency to emphasize the importance of more readily determined properties such as compressive strength.

Recommendations for slump, strength, maximum size of coarse aggregate and minimum cement content required for each size aggregate are given in Figure 31. The cement content in the mix should be properly proportioned to produce a slab that can be easily placed and finished and in addition will possess adequate strength and durability. Mixtures with low cement content result in harsh concretes and produce slabs which tend to bleed excess water to the surface, are difficult to finish and result in poor surface characteristics. Mixtures with excessive cement content can result in increased drying shrinkage.

Excessive slump and consequent bleeding and segregation are primary causes of poor performance in concrete floor slabs. The slump range given in Figure 31 is recommended to produce slabs that can be easily placed and properly consolidated, yet will have a consistency not conducive to bleeding or segregation during placing and finishing operations.

When the appearance of the finished slab is important, different batches of concrete placed in the same floor slab should have approximately the same slump, which should not exceed 3". When lightweight concrete is placed at slumps in excess of 3", the coarser lightweight aggregate may rise to the surface and cause excessive bleeding.

Handling and Placing

Care should be taken in all handling operations to minimize the possibility of segregation. Whether the concrete is handled and placed by chute, buggy or bucket, it should be discharged as near as possible to its final position. Advance planning of suitable access will avoid the necessity of using higher slump concrete and excessive delays.

The slab bed should be completely compacted at the time of concreting. When a vapor barrier is not used and a base course is provided, a separator of building paper or other material which will withstand handling and construction traffic is recommended. The separator should be installed between the base course and the slab to prevent the fines or paste of plastic concrete from seeping down into the base course or fill.

When a vapor barrier, waterproof membrane or separator is not used, the slab bed should be moistened to minimize extraction of water from the concrete. However, at the time of placing, there should be no muddy or soft spots or free water standing on the slab bed.

Steel reinforcing, ductwork or heating coils embedded in the slab should be set and supported at the proper elevation before the concrete is placed. Chlorides should not be used in concrete in which dissimilar metals are embedded, such as reinforcing steel and heating ducts.

The concrete should not be placed faster than it can be spread, screeded and darbied, because these operations should be performed before excess water has had an opportunity to collect on the surface of the freshly placed concrete.

Finishing

Any of the following finishing operations performed with excess (bled) water on the surface will cause dusting, crazing or scaling.

With the exception of heavily reinforced structural slabs which usually are placed with mechanical vibration equipment, initial compaction of the concrete should be accomplished in the first operations of spreading, vibrating, screeding and darbying.

Screeding Of all the placing and finishing operations, screeding (striking off the surface to a predetermined grade) has the greatest effect on surface tolerances. Concrete should be screeded *immediately* after it is placed, and screed stakes used to establish surface elevation should be removed as the work progresses to avoid walking back into the screeded areas.

Darbying The purpose of darbying is to eliminate the ridges and voids left by screeding. In addition, it should slightly embed the coarse aggregate, thus preparing the surface for subsequent finishing operations of edging, jointing, floating and troweling. Darbying should *immediately* follow screeding.

A slight stiffening of the concrete is necessary after darbying before proceeding with further finishing operations. No subsequent operations should be performed until the concrete will sustain foot pressure with only about 1/4″ indentation.

Edging Usually, edging is not required for most interior slabs-on-ground, and is more commonly performed on sidewalks, driveways and steps. When the slab is to be finished with resilient flooring requiring a level subfloor, an edger should not be used. Edges at construction joints in the slab may be lightly stoned to remove irregularities after forms are stripped and before the adjacent slab is placed.

Jointing When the slab is to receive a finished floor such as resilient flooring or carpeting, control joints are not necessary as the flooring finish will cover any unsightly cracking which results from shrinkage.

Control Joints are used in exposed finished slabs to provide controlled straight cracks by reducing the effective cross section of the slab. Joints can be made by grooving the moist concrete with a jointing tool or by sawing a groove in the slab after initial set. The depth of control joints should be approximately 1/4 the nominal slab thickness or equal to at least the largest size of coarse aggregate, whichever is greater. Tensile stresses generated will be relieved at the control joint, thus reducing the likelihood of cracks occurring where they might be objectionable.

Control joints in Type I slabs which are to be left exposed should be placed to provide slab areas approximately 15′ to 20′ square in order to minimize the width of the resulting cracks. Such spacing usually permits the joints to be placed under partitions. In structurally reinforced slabs (Types III and IV) control joints usually are unnecessary.

Floating When the water sheen disappears and the concrete will support the weight of the mason, it may be floated. Floating embeds coarse aggregate, removes slight surface imperfections which may be left by edging or jointing, and consolidates mortar at the surface for troweling.

Troweling Troweling should not be performed unless the surface has been power- or hand-floated and should be performed *immediately* after floating. When hand-troweling, the mason customarily floats a small area and then trowels it before removing his kneeboards. If necessary, tooled joints and edges should be rerun before and after troweling to maintain true lines and depths. For first troweling, an old trowel is recommended which can be worked flat without digging into the concrete surface. The trowel blade should be kept as flat as possible against the surface. If the trowel blade is tilted or pitched at too great an angle, an objectionable chatter effect will result.

Surface smoothness can be improved by additional trowelings, with time allowed between successive trowelings to permit the concrete to become harder. As the surface stiffens, each troweling operation should be made with successively smaller trowels tipped at progressively higher angles so that sufficient pressure can be applied for proper finishing.

For exposed slabs, additional troweling increases the compaction of fines at the surface, giving greater density and better wear resistance. A second troweling is recommended even if the slab is to be finished with resilient flooring because it results in closer surface tolerances and better surface for the application of the flooring.

Finishing Lightweight Structural Concrete Finishing operations for slabs made with lightweight structural concrete (containing coarse aggregates of expanded clay, shale or slag) should vary somewhat from slabs made with normal weight concrete. When the surface of the concrete is worked, there is a tendency for the coarse lightweight aggregate rather than the mortar to rise to the surface. Lightweight concrete can be easily finished if the following precautions are observed: (1) The mix should be properly proportioned and should not be over- or under-sanded in an attempt to meet unit weight requirements. (2) Finishing should not be started too early and the concrete should not be over-worked or over-vibrated. A well proportioned mix can generally be placed, screeded and floated with much less effort than for normal weight concrete. Excessive darbying or floating is a principal cause of finishing problems because the heavier mortar is driven down and the coarse aggregate is brought to the surface.

Curing and Protection

Curing methods using water, such as sprinkling, ponding and the use of wet coverings, are the most effective and should be used whenever practical. Although not usually as effective as curing with wet coverings, moisture retention membranes are widely used because of their convenience. In any case, the most important consideration is that the surface of the concrete be kept uniformly wet or moist. Partial drying of any part of the surface of the slab can result in crazing and cracking. Moist curing compounds will affect the bonding properties of adhesives and should not be used on slabs intended for resilient flooring unless they are known to be compatible with the adhesive. Also, curing compounds are not recommended for slabs intended for further surface treatment.

Length of Curing Sufficient curing time always should be allowed for the slab to develop adequate strength prior to subjecting it to loading. Curing should be started as soon as it is possible to apply the curing medium without damaging the surface. Curing time and surface temperatures are summarized in Figure 32. The concrete temperature should not be allowed to fall below 50°F. In any curing procedure, rapid cooling may induce thermal cracking. Rapid cooling during the first 12 to 24 hours will be particularly critical.

FIG. 32 **LENGTH OF CURING**

Type of Portland Cement	Curing Time	Concrete Surface Temperature
Type I	5 days	70° F. or higher
	7 days	50° F. to 70° F.
Type II	3 days	50° or higher
Other than Type I or Type II	7 days	70° or higher
	14 days	50° F. to 70° F.

We gratefully acknowledge the following for the use of their publications as references and permission to use photographs and illustrations: Building Research Advisory Board, National Academy of Sciences—National Research Council; Federal Housing Administration; Portland Cement Association; Soiltest, Inc.; Soil Testing Services, Inc.

SLABS-ON-GROUND 311

CONTENTS

GENERAL RECOMMENDATIONS

The following recommendations apply to concrete slabs-on-ground built at or near grade level, which serve as the foundation/floor system to enclose interior spaces.

SLAB TYPES

Slabs for foundation/floor systems are classed either as: (1) ground-supported slabs (Types I, II and III) which rest directly on a slab bed consisting of undisturbed soil or compacted fill and/or a base course (see Main Text, page 311-3); or (2) structurally supported slabs (Type IV) which are supported independent of the ground and rest on foundations consisting of walls, piers, grade beams or piles (see Main Text, page 311-4).

SITE CHARACTERISTICS

The principal site characteristics affecting slab-on-ground design and construction are soil properties and moisture conditions.

Soil Investigation

A preliminary investigation should be made to identify and classify the foundation soil to determine the need for and extent of further investigation (Figs. WF1, WF2 & WF11).

Examination should be made of structures in the immediate area for evidence of apparent settlement or expansion of the soil, provided similar conditions of soil, topography and proposed construction prevail.

FIG. WF1 **DESCRIPTION OF NONPROBLEM SOILS**

Soil Classification	Soil Description	Density or Consistency[1]
GW, GP	Gravels or sand-gravel mixtures, little or no fines	All densities
GM, GC	Gravels or gravel-sand mixtures, with clay or silt	Medium-dense to dense
SW, SP	Sands with little or no fines	Medium-dense to dense
SM, SC	Silty or clayey sands	Medium-dense to dense
ML	Silty or clayey fine sands	Medium-dense to dense
ML, MH	Inorganic silts	Medium-dense to dense
CL	Inorganic and silty clays of low to medium plasticity	Stiff to hard

1. See Main Text Figs. 6 & 7, page 311-6, for definition of terms.

(Continued) GENERAL RECOMMENDATIONS

FIG. WF2 **RECOMMENDED SLAB BED FOR SLABS-ON-GROUND**

Slab Type	Soil Type[1]	Density[2]	Plasticity[3]	Consistency (q_u/w)[4]	Climatic Rating[5]
TYPE I	GW, GP	All	—	—	All
	GM, GC, SW, SP, SM, SC, ML, MH	Dense or Medium Dense[6]			
TYPE II	GW, GP, GM, GC, SW, SP, SM, SC, ML, MH	All	—	—	All
	CL, CH,	—	PI $<$ 15	$q_u/w \geq 7.5$	All
	OL, OH	—	PI $>$ 15	$q_u/w \geq 7.5$	$C_w \geq 45$
TYPE III	CL, CH,	—	PI $>$ 15	$q_u/w \geq 7.5$	$C_w < 45$
	OL, OH	—	—	$7.5 > q_u/w > 1.5$	All
TYPE IV	CL, CH, OL, OH	—	—	$q_u/w < 1.5$	All
	PT	—	—	All	

1. See Figure WF11 for base course requirements.
2. See Main Text, page 311-5 and Figure 6 for discussion of density of coarse-grained soils.
3. PI=Plasticity Index, determined in accordance with ASTM D424; see Main Text page 311-5 for discussion of plasticity.
4. q_u/w=Unconfined Compressive Strength/tons/sq. ft. of undisturbed sample determined in accordance with ASTM D2166; see Main Text, page 311-5 and Figure 7 for discussion of consistency of cohesive soils.
5. See Main Text, page 311-7 for Climatic Rating (C_w).
6. May be loose if properly compacted to its entire depth.

At least one test boring at the site is recommended to determine the type, extent and acceptability of the soil for ground-supported slabs. A test boring can be made with simple tools to a depth of 15′, the depth of borings usually necessary to predict the performance of slabs-on-ground for light residential construction.

No additional soil investigation should be required for single sites when the preliminary site investigation clearly indicates: (1) that the site has nonproblem soils consisting of firm, nonexpansive materials (Fig. WF1); (2) that no ground or surface water problems exist; and (3) that fill thickness greater than 3′ will not be required.

When investigation reveals problem soils with expansion or shrinkage potential, or on large developments where extensive grading and site preparation are required, an engineering study should be conducted by a qualified soil engineer to determine the necessary design and construction procedures (see Main Text, page 311-6).

Special construction procedures, such as the use of a Type III slab, should be employed over problem soils such as highly compressible or highly expansive clays, highly plastic soils, very loose sands and silts, and loess soils which are heavily loaded or not well drained.

Slabs used over problem soils should be sufficiently rigid and possess the necessary strength, or detrimental effects to the slab and/or superstructure may result.

See Main Text, page 311-4 through 311-8, for a discussion of soil properties, moisture conditions and thermal conditions which may affect problem soils, as well as other factors affecting slab-on-ground construction.

CONSTRUCTION RECOMMENDATIONS

Slab construction depends on the slab type; site preparation, including preparation of the slab bed, grading and backfilling; other precautions to control ground and surface moisture; and thermal control when required. The performance of all slabs is highly dependent on the quality of concrete materials and workmanship.

SLAB TYPES

The type of slab foundation/floor system should be based on observed soil properties, climate, functional requirements and economy.

Type I Slabs

A Type I slab is unreinforced except for specific localized areas of the slab (Fig. WF3). It possesses only compressive strength and cannot tolerate appreciable amounts of tension or warping. The slab may crack from shrinkage during drying, but when placed over the proper slab bed and when used under the conditions recommended below, cracks which occur should not open excessively nor prove detrimental to the serviceability of the slab.

Slab Bed Type I slabs should be placed only on well drained and properly graded coarse-grained soils (Fig. WF2).

These soils are not affected significantly by climate and moisture content changes and will develop no appreciable volume change that would be detrimental to the slab.

Dimensions A Type I slab should be of at least 4″ uniform thickness and be rectangular or square in shape. To minimize shrinkage during drying and subsequent thermal volume changes, no dimension of the slab perimeter should exceed 32′, unless control joints are provided (Fig. WF4a). However, if the slab is to be left exposed, control joints should be provided at closer spacing of 15′ to 20′ o.c.

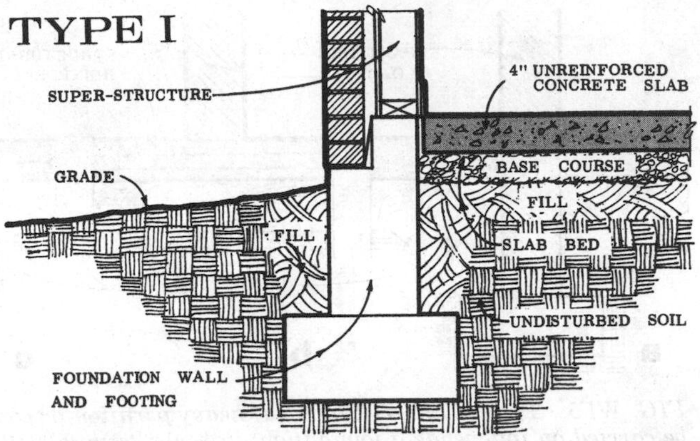

FIG. WF3 Type I Slab: unreinforced, ground-supported; superstructure loads carried independent of the slab on separate foundation elements.

When a nonrectangular T or L shape slab is required, control joints should be provided to divide the slab into squares or rectangles (Fig. WF4b).

Irregularities in thickness, that would reduce the slab to less than 4″ nominal thickness, should not be permitted. When the upper surface of the slab must be lowered to permit proper installation of flooring finish, the uniform slab thickness should be maintained by lowering the underside of the slab beginning at least 24″ away (Fig. WF4c). When the vertical displacement is greater than 1-1/2″, 6x6-6/6 welded wire fabric should be placed in the slab extending 25″ on either side of the point of displacement (Fig. WF4d).

Embedded Heating Elements Heating coils or pipes should not be embedded in Type I slabs because the

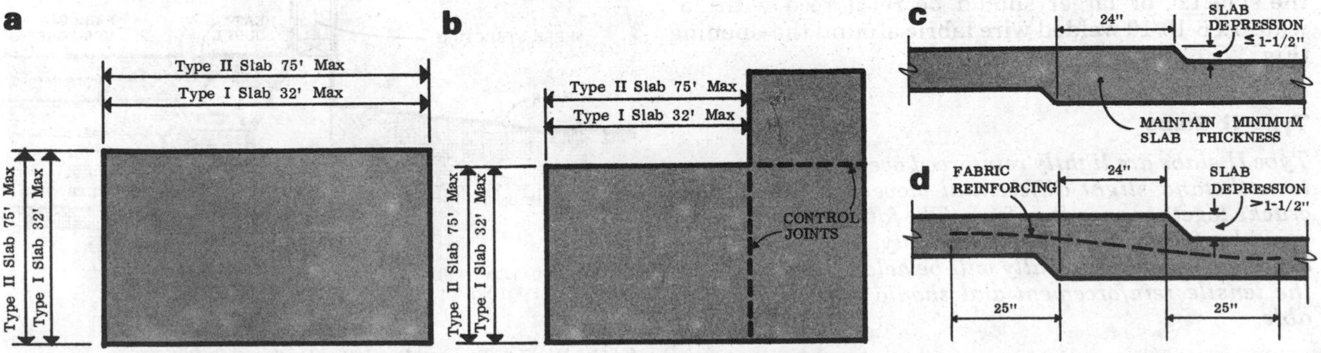

FIG. WF4 (a & b) Maximum dimensions for Types I and II Slabs. (c) Minimum 4″ slab thickness should be maintained when floor depressions are required. (d) Fabric reinforcing is required when depression exceeds 1-1/2″.

(Continued) CONSTRUCTION RECOMMENDATIONS

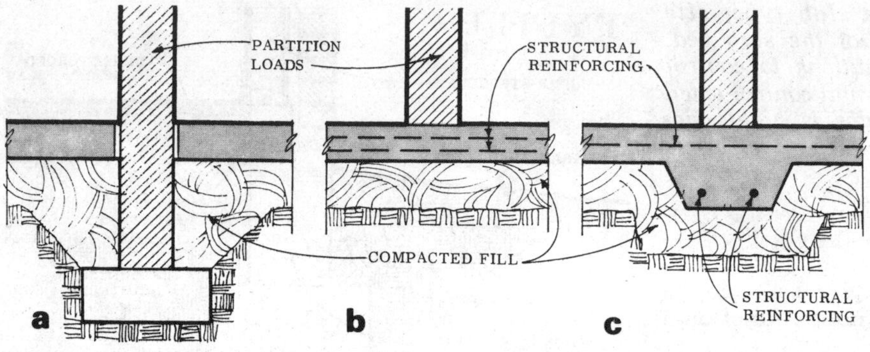

FIG. WF5 Types I and II Slabs: (a) heavy partition or concentrated load should be carried on independent foundation; (b & c) alternate methods when engineering design is used to determine structural reinforcement.

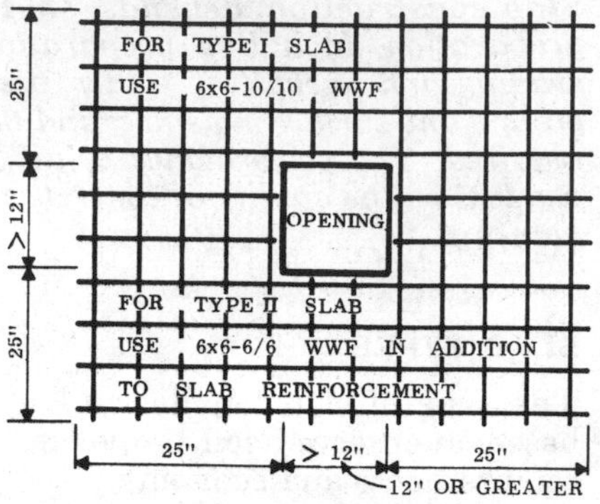

FIG. WF6 Openings larger than 12″ in Types I and II slabs should be reinforced with welded wire fabric.

slab is unable to resist induced thermal stresses without cracking.

Heating *ducts* may be embedded in the slab and should be installed using the construction procedures recommended on page WF311-9.

Loading Superstructure loads should be supported directly on foundations independent of the slab (Fig. WF3), because an unreinforced Type I slab cannot accommodate excessive loading without the possibility of cracking.

Partition loads in excess of 500 plf, or equivalent concentrated loads, should not be permitted on the slab unless special construction is provided. Where this loading must be exceeded, reinforcement or special construction methods such as illustrated in Figure WF5 should be provided.

Openings Openings which introduce nonuniform stresses should be kept to a minimum. Openings in the slab 12″ or larger should be reinforced with 25″ wide 6 x 6-10/10 welded wire fabric around the opening (Fig. WF6).

Type II Slabs

Type II slabs are lightly reinforced over their entire area to withstand slight differential movement and to hold cracks together once they form. The fabric reinforcement provides for control of crack size only. Cracks which are expected to occur normally will be held tightly closed by the tensile reinforcement and should not be objectionable.

Slab Bed Type II slabs should be constructed on the soil types recommended in Figure WF2.

Certain soils unsuitable for Type I slabs are acceptable for the support of Type II slabs because they are lightly reinforced over their entire area (Fig. WF7). However, Type II slabs cannot be expected to resist significant differential movement of the soil. The amount of fabric reinforcement provided is based on the premise that the entire slab will remain uniformly supported by the underlying soil.

Dimensions Type II slabs should be at least 4″ uniform thickness and be rectangular or square in shape. No dimension of the slab perimeter should exceed 75′ (Fig. WF4a).

When a nonrectangular T or L shape is required, control joints should be provided to divide the slab into squares or rectangles (Fig. WF4b). Reinforcement

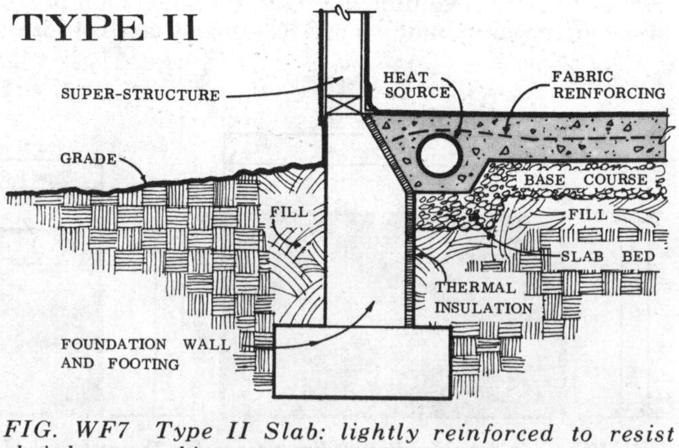

FIG. WF7 Type II Slab: lightly reinforced to resist shrinkage cracking; superstructure loads carried independent of slab; heat sources may be imbedded in slab.

(Continued) CONSTRUCTION RECOMMENDATIONS

should be continuous across control joints, or keyed joints should be provided to eliminate the possibility of vertical displacement between adjacent sections.

Irregularities in thickness that would reduce the slab to less than 4″ nominal thickness should not be permitted. When the upper surface of the slab must be lowered to permit proper installation of flooring finish, the uniform slab thickness should be maintained by lowering the underside of the slab beginning at least 24″ away (Fig. WF4c). When the vertical displacement is greater than 1-1/2″, 6x6-6/6 welded wire fabric should be placed in the slab extending 25″ on either side of the point of displacement (Fig. WF4d).

Reinforcement The amount of welded wire fabric reinforcement in Type II slabs should be as recommended in Figure WF8. Reinforcement should be placed in the center of the slab, located midway between top and bottom of the slab and preferably set on supports such as chairs.

FIG. WF8 **MINIMUM FABRIC REINFORCEMENT FOR TYPE II SLABS**

Maximum Slab Dimension (feet)	Wire Spacing (inches)	Wire Gauge (number)
Up to 45	6 x 6	10/10
45 to 60	6 x 6	8/8
60 to 75	6 x 6	6/6

Embedded Heating Elements Heating ducts, pipes and coils conducting heated liquids should be installed as recommended on page WF311-09.

Because Type II slabs are lightly reinforced over their entire area with welded wire fabric which will hold together any cracks that form, heating pipes or coils, as well as heating ducts, can be embedded in the slab.

Loading Superstructure loads should be supported on foundations independent of the slab. Partition loads in excess of 500 plf, or equivalent concentrated loads, should not be permitted on Type II slabs, unless special construction is provided such as is illustrated in Figure WF5.

Openings Openings which introduce nonuniform stresses should be kept to a minimum in Type II slabs. Openings in the slab 12″ or larger should be reinforced with an additional layer of 6 x 6-6/6 welded wire fabric around the openings (Fig. WF6).

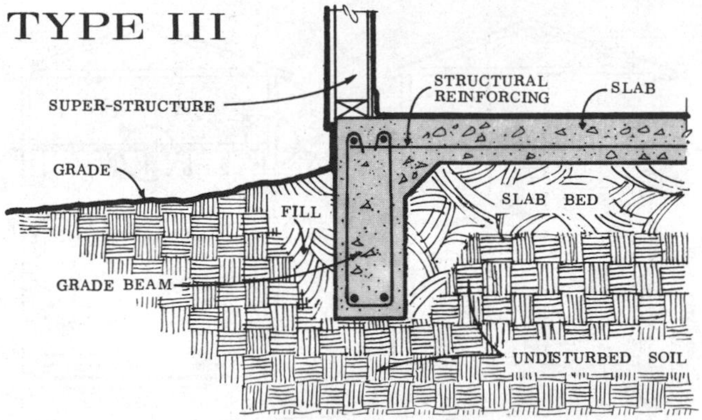

FIG. WF9 An example of Type III Slab construction.

The additional reinforcing around the openings will prevent the concentration of stresses which are likely to crack the slab at these points.

Type III Slabs

Type III Slabs are limited to use over problem soils.

Slab Bed Type III slabs should be used over certain problem soils such as highly compressible, highly expansive and highly plastic soils. The unconfined compressive strength, plasticity index and climatic ratings for soils on which Type III slabs should be used are given in Figure WF2.

Type III slabs are structurally reinforced and stiffened so they can receive and transmit all superstructure loads to the foundation soil by distributing all loads over a larger portion of the slab. Bearing pressures on the ground are reduced and the slab and superstructure act as one monolithic structure. An example of a Type III slab is illustrated in Figure WF9.

Slab Design The slab should be designed with the necessary rigidity and strength so that the slab and the superstructure will act as one monolithic unit. Type III slabs should be designed by a professional in accordance with engineering practice; or the design criteria given in the Building Research Advisory Board (BRAB) Publication No. 1571 may be used*.

**Criteria for Selection and Design of Residential Slabs-on-Ground, report No. 33 to the Federal Housing Administration by Building Research Advisory Board—National Academy of Sciences, Washington, D.C. 20418.*

(Continued) CONSTRUCTION RECOMMENDATIONS

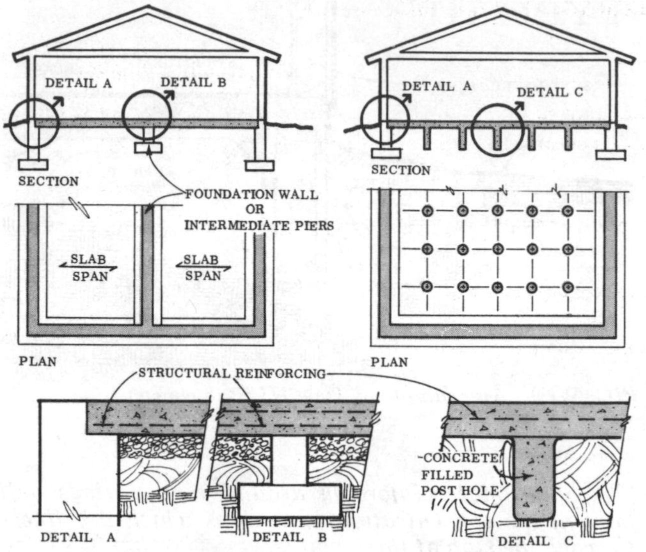

FIG. WF10 Construction details for variation of Type IV Slab.

Type IV Slabs

Type IV slabs receive and transmit all superstructure loads to the foundation soil through footings, piers or piles and therefore do not depend on the soil beneath the slab for support.

Slab Bed Type IV slabs should be used over extremely poor soils which cannot be treated economically or practically to receive ground-supported slabs (Fig. WF2).

Foundation elements such as grade beams should be designed so that swelling of the soils beneath the slab will not lift these elements off the foundation or pull the foundation out of the ground.

Foundation piers or footings should penetrate to a depth providing adequate support below unstable soil.

Due to local practice, Type IV slabs sometimes are used over unstable soils which may not necessarily be considered extreme problem soils (Fig. WF10).

Slab Design Type IV slabs should be designed by professional analysis in accordance with conventional engineering practice for reinforced structural slabs.

SITE PREPARATION

The slab bed for ground-supported slabs should: (1) provide the necessary bearing capacity for slab support; (2) control ground moisture; and (3) establish the proper slab elevation.

The slab bed should consist of either undisturbed soil of the proper type, density and consistency, or foundation fill and/or a base course.

The soil should be uniformly and adequately compacted to support the slab capably and to ensure that warping and tensile stresses which could contribute to cracking are not induced in the slab.

Fill Material

Fill material should be used to create a reasonably level ground for slab construction on residential sites where undulating or hilly terrain exist. Soils used for grading and foundation fills should contain no vegetation or foreign material that would cause uneven settlement. Areas within foundation walls also should have topsoil removed.

Foundation fills should be used to establish the desired finished slab elevation and should be compacted to a maximum practical density.

Structural slabs (Type IV) are supported at edges and through the center by intermediate foundations. They can be placed over fill material which requires only moderate compaction because the fill serves only as temporary support until the slab develops the necessary strength.

The slab bed for Types I and II ground-supported slabs should not expand or contract due to moisture. Hence, organic or active clays and silts of the OL, CH, CL, OH and PT groups should not be used for foundation fills under these slabs (Fig. WF11).

Soil Groups GW, GP, SW and SP (Fig. WF11) may be used for fill material without a base course.

Compaction

Compaction requirements should be specified in terms of a percentage of the maximum dry density obtainable in the ASTM D1557 or D698 test procedures,[1] or the relative density can be specified for fills constructed of free draining sand and gravel soils (GW, GP, SW, SP) in accordance with ASTM D2049.[2]

---

1. ASTM D1557—Moisture-Density Relations of Soils Using 10-lb. Rammer and 18″ drop; ASTM D698—Moisture-Density Relations of Soils Using 5.5-lb. Rammer and 12″ Drop.
2. ASTM D2049—Relative Density of Cohesionless Soils.

(Continued) CONSTRUCTION RECOMMENDATIONS

Fig. WF11 **RECOMMENDATIONS FOR SLAB BED ACCORDING TO SOIL CLASSIFICATION**[1]

Major Division			Group Symbols	Typical Names	Presumptive Bearing Capacity,[2] Tons Per Square Foot
COARSE-GRAINED SOILS (More than half of material is larger than the smallest particle visible to the naked eye)	Gravels (more than half of coarse fraction is larger than ¼")	Gravels with fines (appreciable amount of fines)	GW	Well-graded gravels, gravel-sand mixtures, little or no fines.	5
			GP	Poorly graded gravels or gravel-sand mixtures, little or no fines.	5
		Clean Sands (little or no fines)	GM	Silty gravels, gravel-sand-silt mixtures.	2.5
			GC	Clayey gravels, gravel-sand-clay mixtures.	2
	Sands (more than half of coarse fraction is smaller than ¼")	Clean Gravels (little or no fines)	SW	Well-graded sands, gravelly sands, little or no fines.	3.75
			SP	Poorly graded sands or gravelly sands, little or no fines.	3
		Sands with Fines (appreciable amount of fines)	SM	Silty sands, sand-silt mixtures.	2
			SC	Clayey sands, sand-clay mixtures.	2
FINE-GRAINED SOILS (More than half of material is smaller than the smallest particle visible to the naked eye)	Silts and Clays (liquid limit is less than 50)		ML	Inorganic silts, very fine sands, rock flour, silty or clayey fine sands or clayey silts with slight plasticity.	1
			CL	Inorganic clays of low to medium plasticity, gravely clays, sandy clays, silty clays, lean clays.	1
			OL	Organic silts and organic silty clays of low plasticity.	
	Silts and Clays (liquid limit is greater than 50)		MH	Inorganic silts, micaceous or diatomaceous fine sandy or silty soils, elastic silts.	1
			CH	Inorganic clays of high plasticity, fat clay.	1
			OH	Organic clays of medium to high plasticity, organic silts.	
Highly Organic Soils			PT	Peat and other highly organic soils.	

[1]Based on ASTM D2487—Classification of Soils for Enginering Purposes.
[2]National Building Code, 1976 edition, American Insurance Association.

☐ Generally may be used for fill without a base course

▒ Fill that requires a base course

■ Not permitted for foundation fill

Generally, clean sand fill (SW, SP) should be compacted to 95% of optimum dry density, and all other fine-grained, nonexpansive fills to 90% of optimum dry density in accordance with ASTM D1557.

Compaction requirements for expansive soils should be determined from engineering analysis and may require somewhat less than 90% of optimum dry density.

Backfill adjacent to foundation walls should be uniformly compacted to at least 85% of optimum dry density.

Varying degrees of compaction for the many different types of soils that may be encountered cannot be stated without engineering analysis.

Compaction of soil should be performed with suitable hand or power equipment in 4″ to 12″ layers (lifts) of soil.

Regardless of the type of soil being compacted, the thinner the lift, the better the compaction. Proper compaction: (1) minimizes settling, (2) increases the soil's load-bearing characteristics, (3) increases soil stability, and (4) reduces water penetration. Thus, proper com-

(Continued) **CONSTRUCTION RECOMMENDATIONS**

paction generally will prevent slab cracking caused by differential settlement of the soil.

MOISTURE CONTROL

See Main Text, page 311-7, for a discussion of differential movement of slabs caused by moisture fluctuations in problem soils.

Slabs should be protected from surface or ground water or water vapor to prevent damage to finish flooring materials and pipes or heating ducts embedded in the slab, and to guard against reduced effectiveness of thermal insulation.

The methods and extent of protection depend on (1) slab elevation with respect to elevation and slope of finish grade, (2) the location of the ground water table or any man-made water source such as seepage fields; (3) drainage properties of the soil beneath the slab; and (4) the type of finish flooring used.

Grading and Drainage

The site around the structure should be graded to form protective slopes which will drain surface water away from the foundation toward streets or other drainage structures. On hillside locations, the site should be graded to divert surface water around and away from the structure.

Finished grade should slope downward in all directions away from the slab to a minimum of 6″ for a distance of 10′, or approximately a 4% to 5% slope. Paved areas adjacent to the structure having less than a 5% slope should be provided with a positive means of drainage.

Slab Elevation

The slab elevation should be established so that the top of the slab or any element of wood construction is at least 8″ above finished grade level.

If heating ducts are used in or under the slab, elevations should be established so that the bottom of the ducts are at least 2″ above finished grade, unless the ducts are of concrete, asbestos-cement or ceramic tile with watertight joints. Heating ducts should be installed as recommended on page WF311-9.

Base Course

When the slab bed (foundation fill or undisturbed soil) is not of limited capillarity, at least a 4″ base course of limited capillarity should be provided to prevent the capillary rise of water through the soil. The base course also should provide uniform structural support for the slab.

Material which passes a 2″ sieve and is retained on a 1/4″ sieve is assumed to be of limited capillarity.

Generally, a base course should be provided over GM, GC, SM, SC, ML and MH soils (Fig. WF11).

Soils of the GW, GP, SW and SP groups generally do not require a base course.

Vapor Barrier

Where floor coverings, household goods or equipment must be protected from damage by moisture conditions, a vapor barrier should be installed under the slab.

The vapor barrier membrane should have a permeance of less than 0.30 perms and should resist deterioration as well as puncture from heavy traffic during construction. The vapor barrier should be installed with 6″ laps and be fitted carefully around utility and other service openings.

Polyethylene sheet used over sand or firmly compacted soil under unreinforced slabs should be at least 4 mil nominal thickness; when used over gravel or crushed stone, or under structurally reinforced slabs, it should be at least 6 mil nominal thickness.

Where no drainage or soil problems exist (such as in arid regions where heavy sprinkling and irrigation is used) the vapor barrier may not be necessary.

However, a vapor barrier generally is recommended under all slab-on-ground construction. A vapor barrier usually can be installed at nominal cost during construction, and it is often very difficult to correct moisture problems after construction.

THERMAL CONTROL

Special precautions should be taken for slabs in which heated pipes are embedded, for slabs containing warm air ducts, and in areas where heating systems are required in the dwelling.

Slab Insulation

In areas where heating systems are required, insulation should be installed around the perimeter of the slab to prohibit excessive heat loss directly to areas of low outside air or ground temperatures.

(Continued) CONSTRUCTION RECOMMENDATIONS

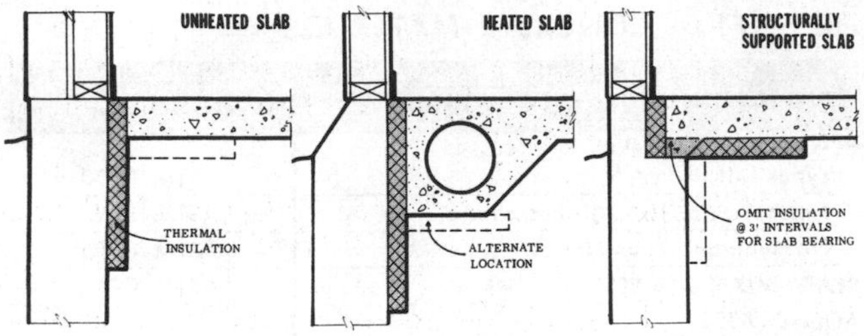

FIG. WF12 Methods of perimeter insulation for slab construction; insulation can be installed either vertically or horizontally.

Insulation should be: (1) virtually nonabsorptive of moisture (noncapillary), (2) not harmed permanently by wetting or contact with wet concrete, (3) sufficiently rigid to resist damage and compression during placing of concrete, and (4) resistant to decay and insect damage.

Insulation at least 1″ thick (preferably 2″ thick) should be provided around the perimeter of the slab. The insulation should be installed around the entire slab perimeter, extending either vertically 24″ below grade level or horizontally for a distance of 24″ under the slab.

Alternate methods of perimeter slab insulation are shown in Figure WF12.

Embedded Heating and Cooling Elements

When pipes or ducts pass through the slab, they should be isolated from the slab to prevent them from breaking. Pipes should pass through the slab vertically and be provided with space for expansion and with flexible connections if necessary to accommodate differential movement between the slab and the soil.

When galvanized metal ducts or copper pipe are embedded, calcium chloride should not be used in the concrete mixture. Aluminum should not be used in any concrete.

Heating Ducts Heating ducts embedded in the slab should be either metal, asbestos-cement, wax-impregnated paper, ceramic tile or concrete pipe.

When metal or paper ducts are used, they should be completely enclosed in at least 2″ of concrete to prevent moisture from collecting in them.

Asbestos-cement ducts with watertight joints may be set on a sand leveling bed and be backfilled with sand to the underside of the slab.

The position of ducts should not be disturbed during concrete placement.

Metal or paper ducts often must be weighted so they will not rise in the freshly placed concrete; joints may be taped or cement-grouted.

When the bottom of ductwork adjacent to the slab perimeter is below exterior finish grade, the duct system should be of ceramic tile, asbestos-cement or concrete pipe. Joints which are completely or partially below exterior grade should be watertight. In areas where termite control is required, these joints should be resistant to termite penetration or soil treatment chemicals.

If the slab is not reinforced (Type I), 6x6-6/6 welded wire fabric should be placed to extend 19″ on either side of the centerline of the duct or to the slab edge, whichever is closer (Fig. WF13).

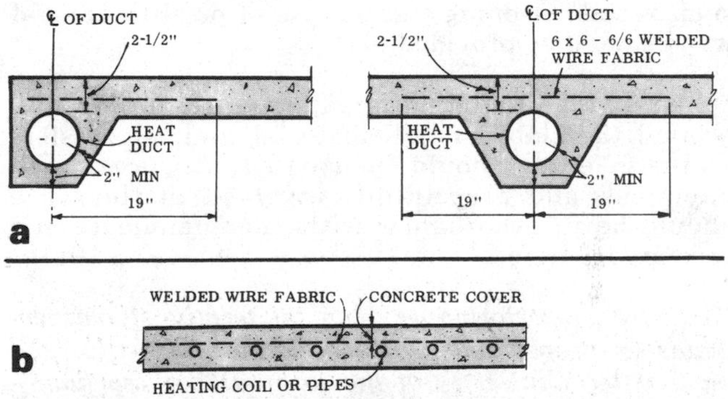

FIG. WF13 Details for (a) embedded heating ducts in Types I and II Slabs, and (b) embedded heating pipes in Type II Slabs.

Embedded Pipes and Coils Generally, heating pipes or coils embedded in Type II slabs should be ferrous or copper pipe, and should be placed with approximately 2-1/2″ to 3″ of concrete cover over the pipes.

Where piping passes through control joints or construction joints, provision should be made for possible movement across the joint.

The piping should be pressure-tested before the concrete is placed and air pressure should be maintained during concrete placement; water pressure should not be used for testing. After the concrete is placed, the slab should not be heated until concrete curing has been completed. Slabs should be warmed gradually, using lukewarm liquid in the pipe to prevent the cold concrete from cracking.

(Continued) CONSTRUCTION RECOMMENDATIONS

Cooling Elements Cooling pipes should not be embedded in concrete slabs-on-ground.

Under conditions of high humidity, condensation may form on the cooled slab which could damage flooring and support mold or fungus growth.

TERMITE CONTROL

Protection of wood structural elements should be provided by positive site and foundation drainage and separation of wood elements from the ground. Foundations and piers of hollow masonry units should have solid masonry caps in which all joints or voids are filled with mortar. Unless the foundation fill is of limited capillarity, at least a 4″ base course of limited capillarity and a suitable vapor barrier should be provided under the slab.

Where local experience indicates that additional protection against termites is necessary, termite barriers such as soil poisoning and the use of pressure-treated wood should be provided.

When soil poisoning is used, the area under and around the slab, backfill material and all possible points of entry should be treated. Application of chemicals and precautions involved in their use should be in accordance with the manufacturer's recommendations.

The effective performance of metal termite shields depends on proper construction and workmanship, and on long-term integrity of the slab without cracking—which is unpredictable. Therefore, the effectiveness of metal termite shields is questionable.

See Section 310 Foundation Systems for a discussion of termite control.

CONCRETE QUALITY CONTROL

Unless concrete quality control measures are observed, undesirable properties in the hardened concrete may occur—particularly at the wearing surface—which may lead to a soft or dusting surface, permeable concrete, cracking or poor durability.

Reference should be made to the entire Work File Section 203 Concrete for general recommendations for producing concrete of suitable quality.

Materials

Materials used in mixing concrete for slabs-on-ground should conform to the ASTM specifications given in Figure WF14.

FIG. WF14 **CONCRETE MATERIALS**

Material	Standard Specification
PORTLAND CEMENT	
Types I, II, III, IV, V	ASTM C150
Types IA, IIA, IIIA (Air Entraining)	ASTM C150
Pozzolan and Blast Furnace	ASTM C595
READY-MIX CONCRETE	ASTM C94
AGGREGATES	
Normal Weight	ASTM C33
Lightweight	ASTM C330
ADMIXTURES	
Air-entraining Agents	ASTM C260
Accelerators: Calcium Chloride	ASTM D98
REINFORCING	
Deformed Steel Bars	ASTM A305
Billet Steel Bars	ASTM A15
Rail Steel Bars	ASTM A16
Welded Steel Wire Fabric	ASTM A185
Cold Drawn Steel Wire	ASTM A82

Portland Cement A small amount of entrained air is recommended for all residential concrete slabs—at least 3% to 5% entrained air by volume for normal weight aggregate concrete, and 4% to 7% entrained air for concrete made with lightweight coarse and fine aggregates. Air-entrained concrete should be plant- or transit-mixed for optimum quality.

Concrete made with air-entrained portland cement provides increased durability, workability and improved finishing characteristics at little if any extra cost.

Aggregates The maximum size of coarse aggregate should not be more than 1/3 the thickness of unreinforced or lightly reinforced slabs, or 3/4 the minimum clear spacing between reinforcing bars and forms in structurally reinforced slabs.

Recommended grading for fine aggregate for residential slab-on-ground construction is given in Figure WF15.

FIG. WF15 **GRADINGS OF FINE AGGREGATE FOR CONCRETE SLABS**

Sieve Size	Percent of aggregate Passing Sieve	
	Normal Weight	Lightweight
⅜″ (10 mm)	100	100
No. 4	95 - 100	85 - 100
No. 8	80 - 90	—
No. 16	50 - 75	40 - 80
No. 30	30 - 50	30 - 65
No. 50	10 - 20	10 - 35
No. 100	2 - 5	0 - 5

(Continued) CONSTRUCTION RECOMMENDATIONS

Mixtures *For slab construction, the properties of the plastic concrete (which affect placing and finishing) are equally as important as properties of the hardened concrete such as abrasion resistance, durability and strength. The plastic properties will generally affect the quality of the top 1/16" to 1/8" of the slab surface. However, these properties are not measured easily and there is a tendency to emphasize the importance of more readily determined properties such as compressive strength.*

Recommendations for slump, strength, water/cement ratio, maximum size of coarse aggregate and minimum cement content required for each size aggregate are given in Figure WF16.

FIG. WF16 **RECOMMENDED MIXTURES FOR RESIDENTIAL SLABS-ON-GROUND[1]**

Maximum Size of Coarse Aggregate[2]	Minimum Cement Content (lbs./cu. yd.)	Maximum Water/Cement By Weight		Slump	Probable Compressive Strength[3]
		Nonair-Entrained Concrete	Air-Entrained Concrete		
1½"	470				
1"	520				
¾"	540	0.51	0.40	2" to 4"	3500 psi
½"	590				
⅜"	610				

[1]For both air-entrained and nonair-entrained concretes.

[2]Normal weight aggregate; different mixes may be required for lightweight aggregate concretes.

[3]Strength refers to approximate compressive strength of cylinders made and tested according to applicable ASTM Standards at 28 days that have been continuously moist cured.

The cement content in the mix should be properly proportioned to produce a slab that can be easily placed and finished and in addition will possess adequate strength and durability.

When appearance of the finished slab is important, different batches of concrete in the same floor slab should have approximately the same slump which should not exceed 3". Lightweight concrete also should not be placed at slumps in excess of 3".

Handling and Placing

All handling operations should minimize the possibility of segregation. Whether handled and placed by chute, buggy or bucket, concrete should be discharged as near as possible to its final position.

The slab bed should be completely compacted prior to the time the concrete is placed. When a vapor barrier is not used and a base course is provided, a separator of building paper or other material that will withstand handling and construction traffic should be used under the slab to prevent the fines or paste in the plastic concrete from seeping down into the slab bed.

When a vapor barrier, waterproof membrane or separator is not used, the slab bed should be moistened to minimize extraction of water from the concrete. At the time of placing, there should be no muddy or soft spots or free water standing on the slab bed.

Steel reinforcing, ductwork or heating pipes embedded in the slab should be set and supported at the proper elevation before the concrete is placed. Chlorides should not be used in concrete in which dissimilar metals are embedded, such as reinforcing steel and heating ducts.

The concrete should not be placed faster than it can be spread, straightened and darbied, because these operations should be performed before excess water has had an opportunity to collect on the surface of the freshly placed concrete.

Finishing

Any of the following finishing operations performed with excess (bled) water on the surface will cause dusting, crazing and scaling.

Screeding Screeding should be performed *immediately* after the concrete is placed. Screed stakes used to establish surface elevation should be removed as the work progresses to avoid walking back into the screeded area.

Darbying In addition to slightly embedding the coarse aggregate, darbying should eliminate ridges and voids left by screeding.

The concrete should be allowed to stiffen slightly after darbying. No subsequent operations should be performed until the concrete will sustain foot pressure with only about 1/4" indentation.

Edging When the slab is to be finished with resilient flooring requiring a smooth subfloor, edging should not be performed.

Jointing The depth of control joints should be approximately 1/4 the normal slab thickness or equal to at least the largest size of coarse aggregate.

(Continued) CONSTRUCTION RECOMMENDATIONS

For Type I slabs which are to be left exposed, control joints should be provided 15′ to 20′ apart. Control joints also should be provided at the juncture of slab intersections as recommended on page WF311-3.

When the slab is to receive a floor finish such as resilient flooring or carpeting, control joints are not necessary, as the flooring will cover any unsightly cracking resulting from shrinkage. Also, control joints usually are unnecessary in structurally reinforced slabs (Types III and IV).

Floating When the water sheen disappears and the concrete will support the weight of the mason, it should be floated to embed coarse aggregate, remove slight surface imperfections left by edging or jointing, and consolidate mortar at the surface for troweling.

Troweling The surface should be hand- or power-floated before any troweling operation is performed. If necessary, tooled joints and edges should be rerun before and after troweling to maintain true lines and depths.

Time should be allowed between successive trowelings to permit the concrete to become harder. As the surface stiffens, each troweling operation should be made with successively smaller trowels tipped at progressively higher angles so that sufficient pressure can be applied for proper finishing.

Surface smoothness can be improved by additional trowelings. For exposed slabs, additional troweling increases the compaction of fines at the surface, giving greater density and better wear resistance. A second troweling is recommended even if the slab is to be finished with resilient flooring, because it results in closer surface tolerances and a better surface for the application of flooring.

Curing

The surface of the concrete should be kept uniformly wet or moist throughout the curing period. Partial drying of any part of the surface (which can result in crazing and cracking) should not be allowed.

Curing methods using water (such as sprinkling, ponding and the use of wet coverings) are the most effective and should be used whenever practical.

Although not usually as effective as curing with water and wet coverings, moisture retention membranes are widely used because of their convenience.

Moist curing compounds should not be used for slabs intended for further surface treatment, or for slabs intended for resilient flooring unless they are known to be compatible with the adhesive, since bonding properties of adhesive or other materials may be affected.

Length of Curing Sufficient curing time always should be allowed for the slab to develop adequate strength prior to subjecting it to loading. Curing should be started as soon as the curing medium can be applied without damaging the surface. Recommended curing time and surface temperatures are summarized in Figure WF17. The concrete temperature should not be allowed to fall below 50° F., and rapid cooling during the first 12 to 24 hours should not be allowed, as thermal cracking may be induced.

FIG. WF17 **LENGTH OF CURING**

Type of Portland Cement	Curing Time	Concrete Surface Temperature
Type I	5 days	70° F. or higher
	7 days	50° F. to 70° F.
Type II	3 days	50° or higher
Other than Type I or Type II	7 days	70° or higher
	14 days	50° F. to 70° F.

314 SPECIALIZED FOUNDATION SYSTEMS

SPECIALIZED FOUNDATION SYSTEMS

This section now includes information on the all-wood preservative-treated foundation. Future additions to this section will discuss prefabricated steel and precast concrete foundation systems.

WOOD FOUNDATION SYSTEMS

The all-wood foundation system described in the following text was developed cooperatively in 1969 by the National Forest Products Association, the American Wood Preservers Institute, and the Economics and Marketing Division of the U.S. Forest Service. The foundation system has so far been utilized in more than 20,000 housing units located in 40 states.

Specific design and construction recommendations for the system are based on experience with houses utilizing preservative-treated wood foundations built in the United States and Canada. Studies conducted by the NAHB Research Foundation, Inc., on four houses of similar floor plan, three with wood foundations and one with a concrete block foundation, have demonstrated the practicality and advantages of the system.

Advantages include reduction in construction delays due to inclement weather, savings in cost and in erection time, and creation of more comfortable below-grade living area. The system is approved as a permanent foundation in homes insured by the Federal Housing Administration, the Veterans Administration and the Farmers Home Administration. It is accepted under the Basic Building Code and the Standard Building Code.

GENERAL DESCRIPTION

The all-wood foundation system is fabricated of preservative pressure-treated lumber and pressure-treated plywood. All structural elements of the substructure are wood. Foundation wall sections of nominal 2″ lumber and plywood cladding may be factory fabricated or constructed at the job site.

Footing plates for the foundation are nominal 2″ pressure-treated wood planks resting on a 4″ or thicker bed of fine gravel or crushed stone.

The system can be used for both basement and crawl space construction. The exterior of the foundation

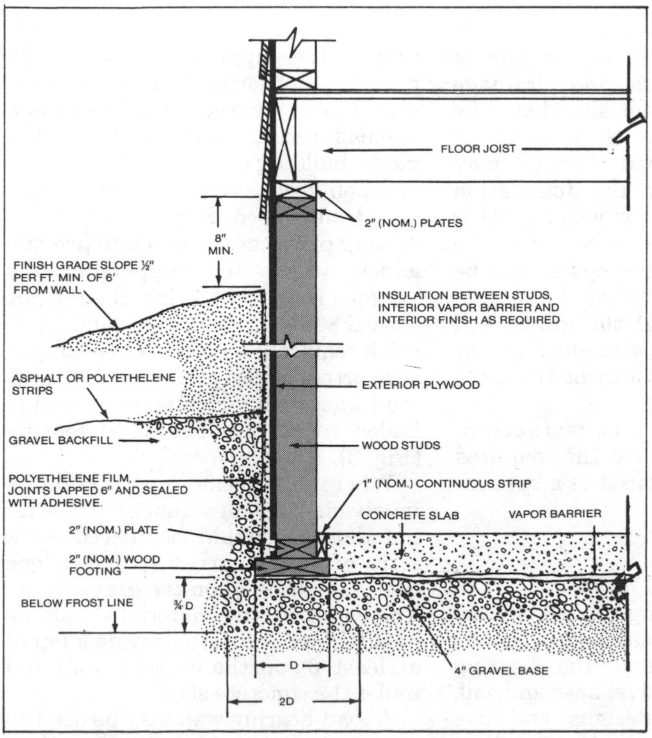

FIG. 1 Details of basement wall construction showing pressure treated framing (shaded). Insulation between studs, vapor barrier and interior finish usually are provided for habitable spaces.

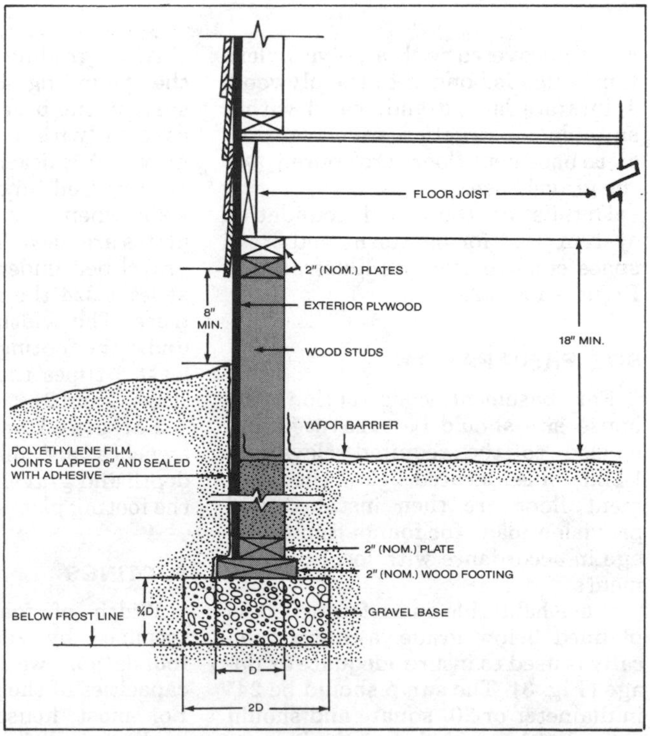

FIG. 2 Details of crawl space wall construction showing pressure-treated framing (shaded). Greater clearance under floor joists may be indicated when access to equipment in crawl space is desired.

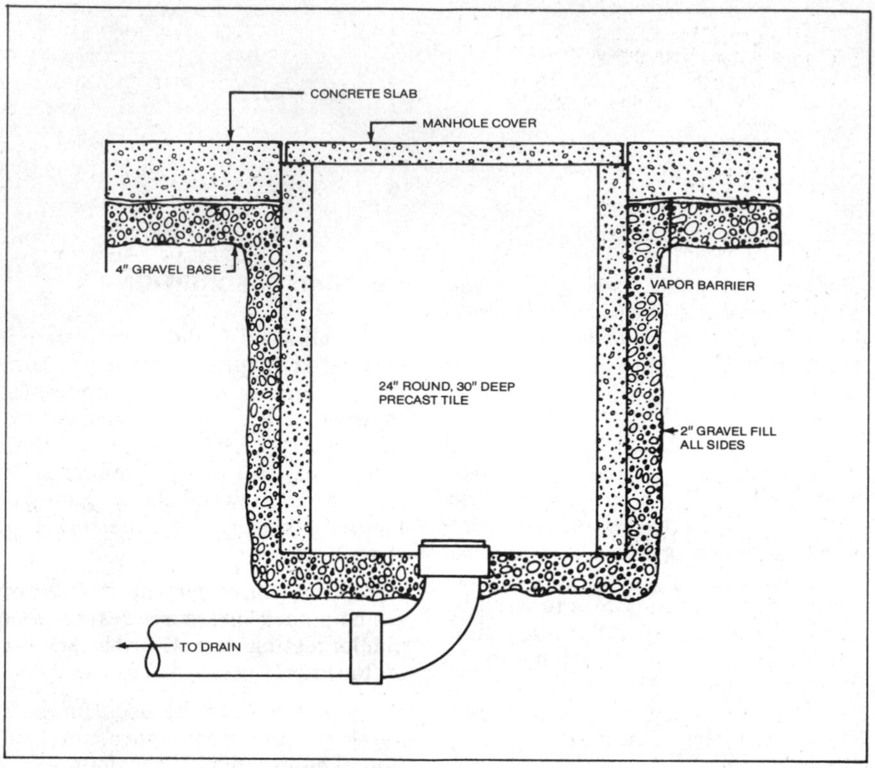

FIG. 3 Sump well and gravel bed provide effective sub-surface drainage and reduce possibility of water leakage. Separate knee-wall below grade supports brick veneer above.

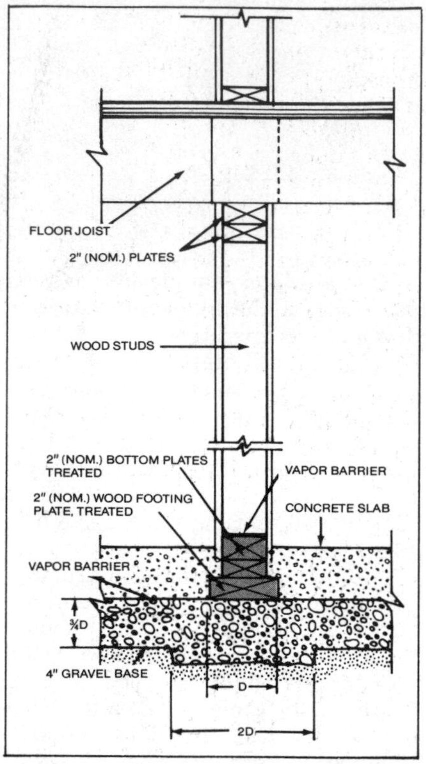

FIG. 4 Interior load bearing walls are constructed similar to exterior foundation walls.

walls is covered with a polyethylene film which is bonded to the plywood. Joints are lapped and sealed with a suitable construction adhesive. Concrete basement floors are poured over the gravel base.

Details of the wood foundation system used for basement and crawl space construction are illustrated in Figures 1 and 2.

SITE PREPARATION

For basement construction, the house site should be excavated and leveled to the required elevation. Utility lines located below the basement floor are then installed and provision made for foundation drainage in accordance with local requirements.

Since habitable space frequently is planned below grade, a sump typically is used to insure adequate drainage (Fig. 3). The sump should be 24″ in diameter or 20″ square and should extend at least 30″ below the bottom of the basement slab. The sump can be drained by gravity or by pump to a storm sewer or drainage swale.

After grading and installation of the plumbing lines and drainage system, the basement site should be covered with a 4″ thick layer of gravel. A thicker bed of gravel may be required under the foundation walls when 2″ x 8″ or wider footing plates are used. The thickness of the gravel bed under footings should be at least 3/4 the width of the footing plate. The width of the gravel bed under the footing plates should be at least 2 times the width of the plate (Figs. 1 and 2).

For crawl space construction, trenches are dug to the required depth and gravel placed as a base for the footing plates.

FOOTINGS

Width of footing plates is determined by vertical loads on the foundation wall and the bearing capacities of the gravel base and soil. For most house designs and live loading conditions, nominal 2″ x 6″ or 2″ x 8″ footings will be adequate when the bearing capacity of the gravel base is 3,000 psf (lbs. per sq. ft.) and that of the supporting soil is 2,000 psf. These bearing values are conservative compared to typical design values for stable soils allowed by the model building codes.

When brick veneer exterior construction is used, 2″ x 10″ or 2″ x 12″ footing plates are required to provide added width for supporting the veneer. Height of brick should not exceed 8′-0″. To reduce the amount of brick required, the veneer typically is supported below grade on a 2″ x 4″ stud knee-wall which rests on and is nailed to the treated wood footing (Fig. 3).

Where basement construction is involved, footings under exterior foundations should be placed on a strip of vapor barrier (polyethylene film) laid directly on the gravel base. The vapor barrier material should be of sufficient width to provide a lap of at least 6″ on the exterior wall and under the concrete slab.

A load-bearing wall may be used as center support under the first floor (Fig. 4). The wood footing and gravel base under the interior wall are designed in the same manner as the

FIG. 5 RECOMMENDATIONS FOR STRUCTURAL FRAMING OF EXTERIOR FOUNDATION WALLS

Construction	House width, ft.	Height of fill, in.	Uniform Load Conditions: Roof - 40 psf live; 10 psf dead, Ceiling - 10 psf, 1st floor - 50 psf live & dead, 2nd floor - 50 psf live & dead			Uniform Load Conditions: Roof - 30 psf live; 10 psf dead, Ceiling - 10 psf, 1st floor - 50 psf live & dead, 2nd floor - 50 psf live & dead		
			Lumber species & grade*	(Nominal) stud & plate size in.**	Stud spacing in.**	Lumber species & grade*	(Nominal) stud & plate size in.**	Stud spacing in.**
Basement - 2 stories								
	32 or less	24	D	2 x 6	16	D	2 x 6	16
		48	D	2 x 6	16	D	2 x 6	16
		72	A	2 x 6	16	A	2 x 6	16
			B	2 x 6	12	B	2 x 6	12
			C	2 x 8	16	D	2 x 8	16
			D	2 x 8	12			
		86	A†	2 x 6	12	A†	2 x 6	12
			B	2 x 8	16	B	2 x 8	16
			C	2 x 8	12	C	2 x 8	12
	24 or less	72	C	2 x 6	12			
			D	2 x 8	16	C	2 x 6	12
		86				D	2 x 8	12
Basement - 1 story								
	32 or less	24	B	2 x 4	16	C	2 x 4	16
			D	2 x 4	12	D	2 x 4	12
			D	2 x 6	16	D	2 x 6	16
		48	D	2 x 6	16	D	2 x 6	16
		72	A	2 x 6	16	A	2 x 6	16
			B	2 x 6	12	C	2 x 6	12
			D	2 x 8	16	D	2 x 8	16
		86	A†	2 x 6	12	A†	2 x 6	12
			B	2 x 8	16	B	2 x 8	16
			C	2 x 8	12	D	2 x 8	12
	28 or less	24				D	2 x 4	16
		48				B	2 x 4	12
		72	C	2 x 6	12			
		86	D	2 x 8	12			
	24 or less	24	D	2 x 4	16			
		48	B	2 x 4	12			
Crawl Space - 2 stories								
	32 or less					B	2 x 6	16
			B	2 x 6	16	D	2 x 6	12
	28 or less		D	2 x 6	12			
	24 or less					C	2 x 6	16
Crawl Space - 1 story								
	32 or less		B	2 x 4	12	A	2 x 4	16
			B	2 x 6	16	B	2 x 4	12
			D	2 x 6	12	D	2 x 6	16
	28 or less		A	2 x 4	16			
			D	2 x 6	16			
	24 or less					C	2 x 4	12

Note: Assume wall height of 8′, soil pressure of 30 pcf, and 2,000 psf soil bearing pressure

** Studs and plates in interior bearing walls supporting floor loads only must be of lumber species and grade "D" or higher. Studs shall be 2 inches by 4 inches at 16 inches on center where supporting one floor and 2 inches by 6 inches at 16 inches on center where supporting 2 floors. Footing plate shall be 2 inches wider than studs.

* Species, species groups and grades having the following minimum (surfaced dry or surfaced green):

Stress		Group A	Group B	Group C	Group D
F_b (repetitive member) psi:	2 x 6	1750	1450	1150	975
	2 x 4	2000	1650	1300	1150
F_c psi	2 x 6	1250	1050	850	700
	2 x 4	1250	1000	800	675
$F_{c\perp}$ psi:		385	385	245	235
F_v psi:		90	90	75	70
E psi:		1,800,000	1,600,000	1,400,000	1,100,000

Where indicated (†), length of end splits or checks at lower end of studs not to exceed width of piece.

exterior foundation wall footings. If a girder support is used in place of the interior wall, footings under the foundation ending walls and under the posts should be designed for concentrated loads.

FOUNDATION WALLS

Structural requirements for the foundation wall of typical one and two-story wood frame houses are shown in Figures 5 and 6.

Design Method

Lumber framing members in foundation walls enclosing a basement are designed to resist the lateral pressure of the fill, as well as the vertical forces resulting from live and dead loads on the structure. The plywood sheathing is designed to resist maximum inward soil pressures occurring at the bottom of the wall.

Soil pressures on the retaining walls are based upon "fluid pressure" of the retained earth. For most soils, the assumption of 30 pcf (lbs. per cu. ft.) of equivalent fluid pressure is generally satisfactory. With soils high in clay or fine silt and of low permeability, or with poorly drained soils, a higher equivalent fluid pressure should be used.

Foundation walls for crawl space construction are designed to resist only vertical loads when the difference between the outside grade and ground level in the crawl space is 12″ or less. Framing is covered with 1/2″ plywood to keep out vermin and provide additional stability.

Construction

Foundation walls are erected over a previously laid wood footing sized in accordance with Figure 5. It is common practice to prefabricate the foundation wall into modular sections (Fig. 7). Treated foundation wall studs are end-nailed to the treated top and bottom plates using two 16d corrosion resistant nails. Treated plywood panels, 1/2″ or thicker, are attached to the studs using corrosion resistant 8d nails spaced 6″ o.c. along panel edges and 12″ o.c. at intermediate supports. Panels generally are applied with face grain vertical and parallel to studs. Vertical joints between panels within wall sections should be located over studs (Fig. 8).

Foundation wall sections are attached to the wood footing and to adjacent sections using corrosion resistant 10d nails spaced 12″ o.c. and face-nailed through the bottom plate and through mating studs (Fig. 9). Wall panel joints should not occur over joints in the wood footing. An upper top plate should be face-nailed to the lower top plate of the foundation wall with corrosion resistant 16d nails spaced 8″ o.c. to tie the wall sections together.

An alternate method of attaching the upper top plate is to fabricate the foundation wall sections with the edge of the plywood wall panels extending 3/4″ above the top of the treated top plate. The field-applied top plate is then face-nailed to the top plate of the wall section using corrosion resistant 10d nails spaced 16″ o.c. and is secured to the overlapping edge of the plywood using 8d corrosion resistant nails spaced 6″ o.c.

Joints in the field-applied top plate should be staggered with respect to joints in the treated top plate of the wall panels. At corners and wall intersections, the field-applied upper plate of one wall is extended across the treated plate of the intersecting wall to tie the building together. The first floor is then installed following standard platform frame construction practices (Fig. 10).

When the height of backfill on a wall parallel to the joists exceeds 4′, solid bridging at intervals of approximately 4′ should be installed between the first five rows of floor joists nearest the wall. The bridging transfers the lateral load from the

FIG. 6 PLYWOOD CLADDING RECOMMENDATIONS FOR EXTERIOR FOUNDATION WALLS

Height of fill, in.	Stud spacing, in.	Plywood face grain across studs[2]			Plywood face grain parallel to studs		
		Grade[3]	Minimum thickness[1]	Identification index	Grade[3]	Minimum thickness[1]	Identification index
24	12	B	1/2	32/16	B	1/2	32/16
	16	B	1/2	32/16	B	1/2 (4, 5 ply)	32/16
48	12	B	1/2	32/16	B A	1/2[4] (5 ply) 1/2	32/16 32/16
	16	B	1/2	32/16	A B	5/8 3/4	42/20 48/24
72	12	B	1/2	32/16	A	5/8[4] (5 ply)	42/20
	16	A	1/2[4]	32/16	B	3/4[4]	48/24
86	12	B	1/2[4]	32/16	A B	5/8[4] 3/4[4]	42/20 48/24
	16	A B	5/8 3/4	42/20 48/24	A	5/8[5]	42/20

Note: Assume 30 pcf soil pressure

[1] Minimum thickness ½″ except crawl space sheathing may be ⅜″ for face grain across studs and maximum 3′ depth of unequal fill.

[2] Minimum 2″ blocking between studs required at all horizontal panel joints more than 4′ below adjacent ground level.

[3] Grade designations: A means Structural I C-D; B means C-D Exterior Glue.

[4] For this fill height, thickness and grade combination, panels which are continuous over less than three spans (across less than three stud spacings) require blocking 2′ above bottom plate. Offset adjacent blocks and fasten through studs with two 16d corrosion resistant nails at each end.

[5] For this fill height, thickness and grade combination, panels requrie 16″ above bottom plate. Offset adjacent blocks and fasten through studs with two 16d corrosion resistant nails at each end.

FIG. 7 Prefabricated foundation sections are delivered to previously prepared site—utilities laid under and footings laid over gravel bed.

FIG. 8 Section of foundation wall being erected over wood footing; sheathing extends to overlap next section at corner.

FIG. 9 Wall section will be secured to footing and adjacent section; a second top plate will tie sections together.

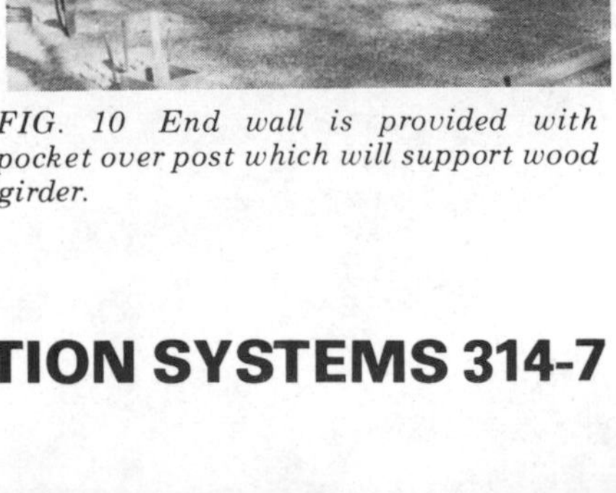

FIG. 10 End wall is provided with pocket over post which will support wood girder.

top of the foundation studs into the floor system.

Plywood panel joints in the foundation wall should provide a minimum space of 1/8″ for caulking with a suitable sealant. In both basement and crawl-space construction, the exterior side of foundation walls should be covered below grade with a 6-mil polyethylene vapor barrier. The film should be bonded to the plywood, with joints lapped 6″ and sealed (Figs. 1 and 2).

The top edge of the vapor barrier should be effectively adhered to the plywood wall with adhesive. Film areas near grade level can be protected from mechanical damage and exposure using treated plywood, brick, stucco or other covering appropriate to the architectural design. Where plywood is used for this purpose, the top edge of a 12″ wide panel of treated material is attached to the wall several inches above the finish grade level. The top inside edge of the panel should be caulked full length before the panel is fastened to the wall.

For habitable basement space, suitable insulation should be installed between studs; an interior vapor barrier and wall finish typically are applied to the stud framing.

FLOORS

For basement construction, a 6 mil polyethylene film is applied over the gravel bed, and a concrete slab at least 3″ thick is placed over the film. Placement of the film and the slab can be delayed until after the house is under roof. Backfill should not be placed against the foundation walls until after the concrete slab is in place and set and the top of the wall is adequately braced. Where height of backfill exceeds 4′, gravel should be used for the lower portion.

For crawl space construction, a 6 mil polyethylene film ground cover should be used. Film edges should be lapped and sealed. Joist floor construction should be suspended at least 18″ above the ground.

CENTER SUPPORT

Framing members in load-bearing walls acting as center support for the first floor joist system are constructed similar to exterior foundation walls (Fig. 4), except that vertical loads only need to be considered in design.

Load-bearing Wall

In basement construction, load-bearing interior walls should be made with pressure-treated bottom plates and, in areas of low termite hazard, with untreated studs and top plates. A double bottom plate is used to provide clearance between the top of the concrete slab floor and the untreated vertical framing in the wall (Fig. 4). In areas of high termite hazard, pressure-treated studs should be used. The interior wall framing can be covered both sides with gypsum wallboard or other interior finish as desired.

Interior load-bearing walls in

crawl-space construction should be made of treated framing and covered on one side with 1/2″ treated plywood.

Interior bearing walls abutting exterior foundation walls should be tied to the exterior walls using corrosion resistant 10d nails spaced 12″ o.c. by face nailing the end studs of the interior wall to the edges of the center studs in the outside walls, or to the edges or faces of blocks nailed between exterior wall studs. The upper top plate of the interior wall should be extended across the treated top plate of the exterior wall.

Interior wall sections should be attached to adjacent interior sections by face nailing through mating studs using corrosion resistant 10d nails spaced 12″ o.c.

FIG. 11 The superstructure, like the foundation, can be conventionally built or prefabricated off-site without affecting traditional appearance.

Post-and-girder Support

Where a post-and-girder center support system is used, posts should be supported on blocks of sufficient size to assure that the vertical concentrated loads do not exceed the bearing capacity of the gravel base and soil. Likewise, foundation walls supporting the center girder should be designed to carry and distribute the concentrated load from the girder to the wood footing and gravel base (Fig. 10), so that allowable gravel and soil bearing capacities are not exceeded. Design of a suitable post-and-girder support system requires engineering calculations and should not be attempted without benefit of professional advice. The main floor and superstructure can be conventionally designed and built or prefabricated off-site (Fig. 11).

Lumber and Plywood Preservative Treatment

All lumber and plywood used in the exterior foundation walls below the upper top plate and all wood footings or bottom plates placed in contact with concrete or the ground should be pressure-treated. Treatment should be in accordance with Standard AWPB-FDN of the American Wood Preservers Bureau, entitled "Standard for Softwood Lumber, Timber and Plywood Pressure Treated with Water-Borne Preservatives For Ground Contact Use In Residential And Light Commercial Foundations."

All treated lumber and plywood should bear an approved AWPB-FDN Quality Mark or that of an independent inspection agency that maintains continuing control, testing, and inspection over the quality of the product as described in the AWPF-FDN Standard and modified herein. Each piece of treated lumber should be dried to not over 19% moisture content after treatment. Southern pine lumber less than 4″ thick should contain no boxed heart and the total volume of heartwood in any piece should be limited to 20%.

To the extent practical, all fabricating, cutting, boring, and trimming of lumber and plywood should be done prior to treatment. However, where lumber and plywood is cut after treatment, it should be brush-coated with the same preservative used in the original treatment, but having a 3% solution of the initial treating solution. Preservatives can be obtained from the supplier.

SPECIALIZED FOUNDATION SYSTEMS 314

CONTENTS

GENERAL DESCRIPTION

The all-wood foundation system is fabricated of preservative pressure-treated plywood and lumber. Foundation wall sections may be factory fabricated or constructed at the job site.

SITE PREPARATION

The house site should be excavated and leveled to the required elevation, removing all topsoil and plant or organic materials. Portions of the site which will support footings should be excavated down to undisturbed soil.

A bed of porous gravel (or crushed stone) should be provided under all footings. The *depth* of the gravel bed should be at least 3/4 the width of the footing plate; the *width* of the gravel bed under the footing plates should be at least 2 times the width of the footing plate (Fig. WF1 and WF2).

Basement Construction

Where habitable space is planned below grade, a sump should be used to insure adequate foundation drainage. The sump should be 24″ in diameter or 20″ square and should extend at least 30″ below the bottom of the basement slab.

The sump can be drained by gravity or by pump to a storm sewer or drainage swale.

The basement site should be covered with a 4″ deep bed of gravel extending from footing to footing.

Crawl Space Construction

For crawl space construction, trenches should be dug to the required depth and gravel should be placed as a base for the footing plates (Fig. WF2).

FOOTINGS

Width of footing plates should be determined according to house type, structural design and soil conditions, as shown in Figure WF3. Footings should be located below the front line.

For most house designs and live loading conditions, nominal 2″ x 8″ footings will be adequate when the bearing capacity of the gravel base is 3,000 psf and that of the supporting soil is 2,000 psf. These conservative bearing values were used in calculating recommendations in Figure WF3.

Exterior brick veneer construction should be supported on footing plates 4″ wider than those shown in Figure WF3. Height of brick veneer supported on wood kneewall should not exceed 8′-0″

Where basement construction is involved, footings under exterior foundations should be placed on a strip of vapor barrier laid directly on the gravel base. The vapor barrier should provide a lap of at least 6″ on the exterior wall and under the concrete slab.

(Continued) WOOD FOUNDATION SYSTEM

Six mil polyethylene film typically is used as a vapor barrier in crawl spaces and basements of all-wood and other foundation systems.

When a load-bearing wall is used as center support under the first floor, the footing plate and gravel base under the interior wall should be designed in the same manner as the exterior foundation wall footings.

If a girder support is used in place of the interior load-bearing wall, footings under the foundation endwalls and under the posts should be designed for concentrated loads.

FOUNDATION WALLS

Foundation walls should be founded on treated wood footings bearing on gravel over undisturbed soil.

Design Method

Framing members in foundation walls should be sized and spaced as shown in Figure WF3, in accordance with house design, soil conditions and lumber group selected.

When the exterior finish grade is only 12″ higher than the grade in the crawl space, foundation walls need only be designed to resist vertical loads and soil pressure can be ignored.

When the height of fill (difference between exterior and interior grades) exceeds 12″, the plywood sheathing should be designed to resist inward soil pressures occurring at the bottom of the wall (Fig. WF4).

For most soils, the assumption of 30 pcf equivalent fluid pressure on foundation walls is generally satisfactory. For soils high in clay or fine silt and of low permeability, or for poorly drained soils, a higher pressure should be assumed.

Construction

Foundation wall studs should be end-nailed to the treated top and bottom plates using two 16d corrosion resistant nails. Treated plywood sheathing, 1/2″ or thicker, should be attached to the studs using corrosion resistant 8d nails; nails should be spaced 6″ o.c. along panel edges and 12″ o.c. at intermediate supports.

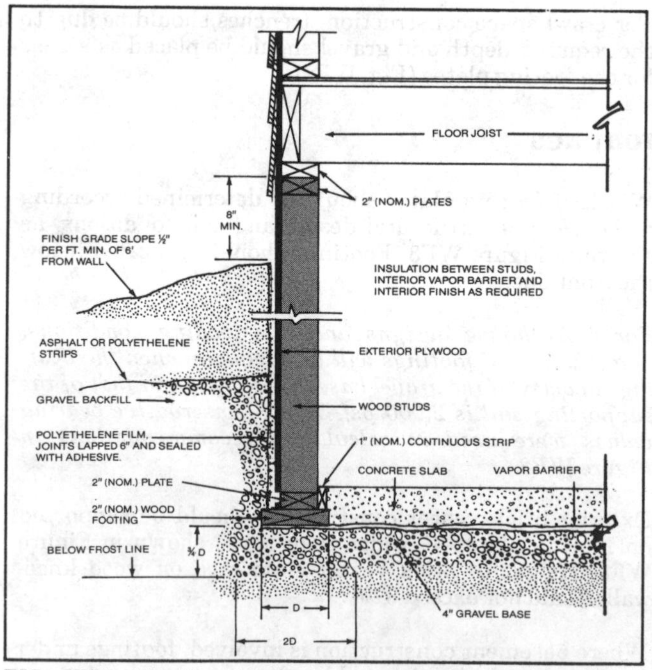

FIG. WF1 Details of basement wall construction showing pressure-treated framing (shaded). Insulation between studs, vapor barrier and interior finish usually are provided for habitable spaces.

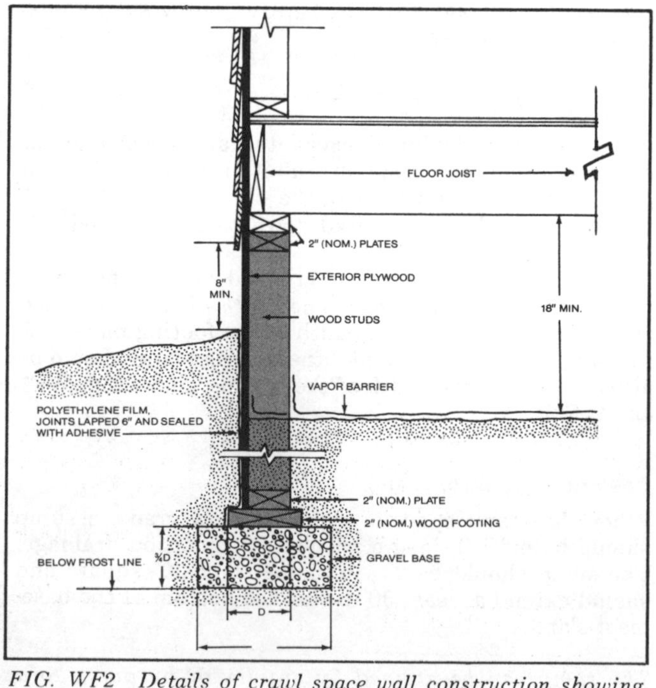

FIG. WF2 Details of crawl space wall construction showing pressure-treated framing (shaded). Greater clearance under floor joists may be indicated when access to equipment in crawl space is desired.

FIG. WF3 RECOMMENDATIONS FOR STRUCTURAL FRAMING OF EXTERIOR FOUNDATION WALLS

Construction	House width, ft.	Height of fill, in.	Uniform Load Conditions — Roof - 40 psf live; 10 psf dead; Ceiling - 10 psf; 1st floor - 50 psf live & dead; 2nd floor - 50 psf live & dead			Roof - 30 psf live; 10 psf dead; Ceiling - 10 psf; 1st floor - 50 psf live & dead; 2nd floor - 50 psf live & dead		
			Lumber species & grade*	(Nominal) stud & plate size in.**	Stud spacing in.**	Lumber species & grade*	(Nominal) stud & plate size in.**	Stud spacing in.**
Basement - 2 stories								
	32 or less	24	D	2 x 6	16	D	2 x 6	16
		48	D	2 x 6	16	D	2 x 6	16
		72	A	2 x 6	16	A	2 x 6	16
			B	2 x 6	12	B	2 x 6	12
			C	2 x 8	16	D	2 x 8	16
			D	2 x 8	12			
		86	A†	2 x 6	12	A†	2 x 6	12
			B	2 x 8	16	B	2 x 8	16
			C	2 x 8	12	C	2 x 8	12
	24 or less	72	C	2 x 6	12	C	2 x 6	12
			D	2 x 8	16			
		86				D	2 x 8	12
Basement - 1 story								
	32 or less	24	B	2 x 4	16	C	2 x 4	16
			D	2 x 4	12	D	2 x 4	12
			D	2 x 6	16	D	2 x 6	16
		48	D	2 x 6	16	D	2 x 6	16
		72	A	2 x 6	16	A	2 x 6	16
			B	2 x 6	12	C	2 x 6	12
			D	2 x 8	16	D	2 x 8	16
		86	A†	2 x 6	12	A†	2 x 6	12
			B	2 x 8	16	B	2 x 8	16
			C	2 x 8	12	D	2 x 8	12
	28 or less	24				D	2 x 4	16
		48				B	2 x 4	12
		72	C	2 x 6	12			
		86	D	2 x 8	12			
	24 or less	24	D	2 x 4	16			
		48	B	2 x 4	12			
Crawl Space - 2 stories								
	32 or less					B	2 x 6	16
			B	2 x 6	16	D	2 x 6	12
	28 or less		D	2 x 6	12			
	24 or less					C	2 x 6	16
Crawl Space - 1 story								
	32 or less		B	2 x 4	12	A	2 x 4	16
			B	2 x 6	16	B	2 x 4	12
			D	2 x 6	12	D	2 x 6	16
	28 or less		A	2 x 4	16			
			D	2 x 6	16			
	24 or less					C	2 x 4	12

Note: Assume wall height of 8′, soil pressure of 30 pcf, and 2,000 psf soil bearing pressure

** Studs and plates in interior bearing walls supporting floor loads only must be of lumber species and grade "D" or higher. Studs shall be 2 inches by 4 inches at 16 inches on center where supporting one floor and 2 inches by 6 inches at 16 inches on center where supporting 2 floors. Footing plate shall be 2 inches wider than studs.

* Species, species groups and grades having the following minimum (surfaced dry or surfaced green):

Stress		Group A	Group B	Group C	Group D
F_b (repetitive member) psi:	2 x 6	1750	1450	1150	975
	2 x 4	2000	1650	1300	1150
F_c psi	2 x 6	1250	1050	850	700
	2 x 4	1250	1000	800	675
$F_{c\perp}$ psi:		385	385	245	235
F_v psi:		90	90	75	70
E psi:		1,800,000	1,600,000	1,400,000	1,100,000

Where indicated (†), length of end splits or checks at lower end of studs not to exceed width of piece.

(Continued) WOOD FOUNDATION SYSTEM

Vertical joints between sheathing panels always should be located over studs. Panel joints should not occur over joints in the wood footing plate.

Plywood panels can be applied either with face grain perpendicular or parallel to studs.

Foundation wall sections should be face-nailed to the wood footing and to adjacent sections using corrosion resistant 10d nails spaced 12″ o.c.

An upper top plate should be face-nailed to the lower top plate of the foundation wall with corrosion resistant 16d nails spaced 8″ o.c.

An alternate method of attaching the upper top plate is to fabricate the foundation wall sections with the edge of the plywood wall panels extending 3/4″ above the top of the treated top plate. The field-applied top plate is then face-nailed to the top plate of the wall sections using corrosion resistant 10d nails spaced 16″ o.c. and is secured to the overlapping edge of the plywood using 8d corrosion resistant nails spaced 6″ o.c.

Joints in the field-applied top plate should be staggered with respect to joints in the treated top plate of the wall panels. At corners and wall intersections, the field-applied upper plate of one wall should be extended across the treated plate of the intersecting wall to tie the building together.

The first floor framing is installed according to standard platform frame construction practices.

When the height of backfill on a wall parallel to the joists exceeds 4′, solid bridging at intervals of approximately 4′ should be installed between the first five rows of floor joists nearest the wall.

Plywood panels should be spaced to provide an 1/8″ to 3/16″ joint for caulking with a suitable sealant. In both basement and crawl space construction, the exterior side of foundation walls should be covered below grade with a 6 mil polyethylene film.

The film should be bonded to the plywood, with joints lapped 6″ and sealed. The top edge of the film should be effectively adhered to the plywood wall with adhesive.

Film areas near grade level can be protected from damage by using treated plywood, brick, stucco or other exterior finish. For habitable basement space, suitable wall in-

FIG. WF4 **PLYWOOD CLADDING RECOMMENDATIONS FOR EXTERIOR FOUNDATION WALLS**

Height of fill, in.	Stud spacing, in.	Plywood face grain across studs [2]			Plywood face grain parallel to studs		
		Grade [3]	Minimum thickness [1]	Identification index	Grade [3]	Minimum thickness [1]	Identification index
24	12	B	1/2	32/16	B	1/2	32/16
	16	B	1/2	32/16	B	1/2 (4, 5 ply)	32/16
48	12	B	1/2	32/16	B A	1/2[4] (5 ply) 1/2	32/16 32/16
	16	B	1/2	32/16	A B	5/8 3/4	42/20 48/24
72	12	B	1/2	32/16	A	5/8[4] (5 ply)	42/20
	16	A	1/2[4]	32/16	B	3/4[4]	48/24
86	12	B	1/2[4]	32/16	A B	5/8[4] 3/4[4]	42/20 48/24
	16	A B	5/8 3/4	42/20 48/24	A	5/8[5]	42/20

Note: Assume 30 pcf soil pressure

[1] Minimum thickness ½″ except crawl space sheathing may be ⅜″ for face grain across studs and maximum 3′ depth of unequal fill.

[2] Minimum 2″ blocking between studs required at all horizontal panel joints more than 4′ below adjacent ground level.

[3] Grade designations: A means Structural I C-D; B means C-D Exterior Glue.

[4] For this fill height, thickness and grade combination, panels which are continuous over less than three spans (across less than three stud spacings) require blocking 2′ above bottom plate. Offset adjacent blocks and fasten through studs with two 16d corrosion resistant nails at each end.

[5] For this fill height, thickness and grade combination, panels requrie 16″ above bottom plate. Offset adjacent blocks and fasten through studs with two 16d corrosion resistant nails at each end.

(Continued) WOOD FOUNDATION SYSTEM

sulation, vapor barrier and wall finish typically are required.

FLOORS

For basement construction, a vapor barrier should be applied over the gravel bed, and a concrete slab at least 3″ thick should be placed over the barrier.

Backfill should not be placed against the foundation walls until after the concrete slab is set and the top of the wall is adequately braced. Where height of backfill exceeds 4′, gravel should be used for the lower portions of the backfill.

For crawl space construction, a 6 mil polyethylene film ground cover should be used. Film edges should be lapped and sealed. Joist floor construction should be suspended at least 18″ above the ground in the crawl space.

CENTER SUPPORT

Houses 20′ or more in width generally require a center support for economical floor framing.

Load-bearing Wall

In basement construction, load-bearing interior walls always should be made with pressure-treated double bottom plates. In areas of high termite hazard, studs also should be pressure-treated.

Interior load-bearing walls in crawl space construction should be made of pressure-treated studs; studs should be covered on one side with 1/2″ treated plywood.

Interior bearing walls abutting exterior foundation walls should be tied to the exterior walls using corrosion resistant 10d nails spaced 12″ o.c.; end studs of the interior wall should be face-nailed to studs or blocking in the exterior wall. The upper top plate of the interior wall should be extended across the treated top plate of the exterior wall.

Interior wall sections should be attached to adjacent interior sections by face-nailing through mating studs using corrosion resistant 10d nails spaced 12″ o.c.

Post-and-girder Support

Where a post-and-girder center support system is used, posts should be supported on blocks of sufficient size to assure that the vertical concentrated loads do not exceed the bearing capacity of the gravel base and soil.

Likewise, foundation walls supporting the center girder should be designed to carry and distribute the concentrated load from the girder, so that allowable gravel and soil bearing capacities are not exceeded.

Design of a suitable post-and-girder support system requires engineering calculations and preferably should be performed by a qualified professional.

LUMBER AND PLYWOOD PRESERVATIVE TREATMENT

All lumber and plywood used in the exterior foundation walls below the upper top plate and all wood footings or bottom plates placed in contact with concrete or the ground should be pressure-treated.

Treatment should be in accordance with AWPB-FDN, the American Wood Preservers Bureau *Standard for Softwood Lumber, Timber and Plywood Pressure Treated with Water-Borne Preservatives for Ground Contact. Use In Residential And Light Commercial Foundations.*

All treated lumber and plywood should bear an approved American Wood Preservers Board *Quality Mark* or the mark of an independent inspection agency which maintains continuing control, testing and inspection over the quality of the product.

Each piece of treated lumber should be dried to not over 19% moisture content after treatment. Southern pine lumber less than 4″ thick should contain no boxed heart and the total volume of heartwood in any piece should be limited to 20%.

To the extent feasible, fabricating and cutting of lumber and plywood should be done prior to preservative treatment. However, where lumber or plywood is cut after treatment, it should be brush-coated with the same preservative used in the original treatment.

315 POLE CONSTRUCTION

INTRODUCTION

The pole construction system described in the following text has evolved from millenia of human experience with tree trunks and timbers as construction materials. From prehistoric lake dwellers to 20th century farmers, men have been embedding poles in the ground and building simple elevated structures to protect themselves from unfriendly humans, ravaging animals, fierce winds and rampaging floods.

Observing some structures to be short-lived because of decay, man has learned to recognize naturally decay-resistant species such as redwood, cypress and cedar. However, it was not until post-World War II years, when pressure-treated poles became widely available, that pole construction of farm buildings and beach cottages became commonplace.

The use of chemically treated poles gained impetus from the favorable experience with a variety of houses built on the West Coast in the 1950s. These houses were erected on steep slopes, on periodically flooded lowlands, in areas subject to earthquakes—yet at reasonable cost, with a high degree of permanence.

The following technical information was abstracted from a study by the American Wood-Preservers' Institute (AWPI), done for the Department of Housing and Urban Development (HUD). Some recommendations and illustrations also were obtained from publications of the Forest Products Laboratory of the U.S. Department of Agriculture.

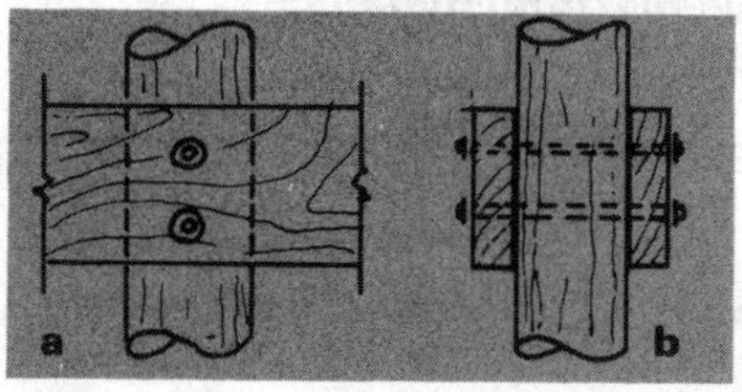

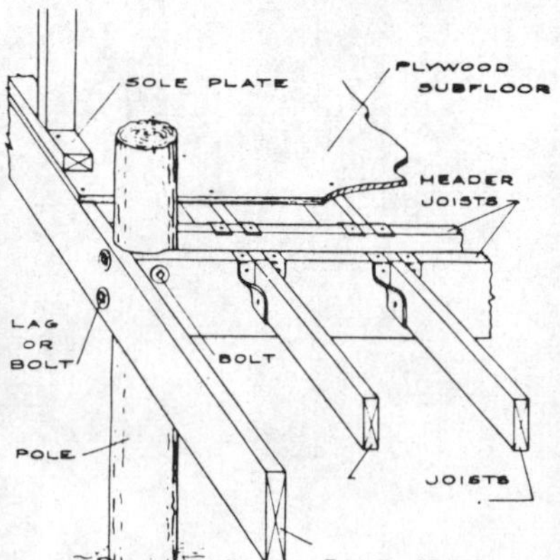

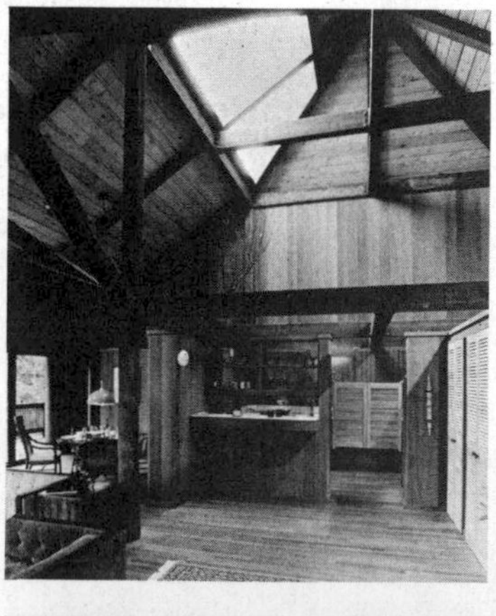

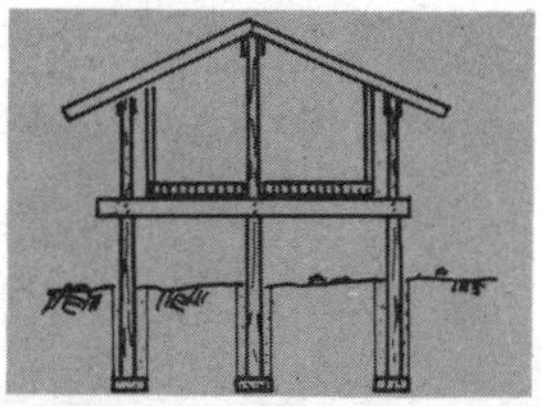

GENERAL DESCRIPTION

A pole house is a structure whose contact with the ground is through a series of decay- and termite-resistant round tree trunks (*poles*) or squared wood members (*timbers*). Poles are cheaper and longer-lasting, are more frequently used and hence give their name to the system (Fig. 1).

Although cedar, redwood and cypress have inherent decay resistance, their availability in long, thick poles is limited and their decay resistance varies with local growing conditions. Chemical treatment of Douglas fir, Southern pine, ponderosa pine and red pine, according to American Wood Preservers' Association (AWPA) specifications assures long-term resistance to infestation and in-ground durability.

FIG. 1 Treated poles are frequently combined with timbers to form the structural frame and enhance the design.

In typical pole construction, poles ranging in diameter from 6″ to 12″ are spaced 6′ to 12′ apart to support floor and roof areas up to 150 sq. ft. (Fig. 2). The poles typically are dropped into predug holes, made either with a power auger or by hand digging. Occasionally, poles are "jetted" with a stream of water or driven with a pile driver. Proper embedment of the pole in the ground and backfilling of the hole are critical to the structural soundness of the building. A concrete pad *under* the pole, or a concrete necklace *around* the pole, may be required to increase the contact area with the soil. The pad and necklace help to distribute the load over a larger area.

The poles may extend substantially above ground to form the load-bearing frame of the superstructure or may terminate at the first floor level, thus permitting conventional stud and joist construction above the pole-framed platform (Fig. 3).

Embedded poles can support the superstructure either by spreading the loads with a concrete footing or by bearing on rock or very stiff soils. *The following recommendations are made only for situations where a concrete footing pad, necklace, or backfill will be used to spread the building loads.*

Where the poles are used as point bearing or friction piles (see page 310-5 Foundation Systems), the services of an engineer should be retained. In general, Section 2907 of the Uniform Building Code should be used as the basic design guide.

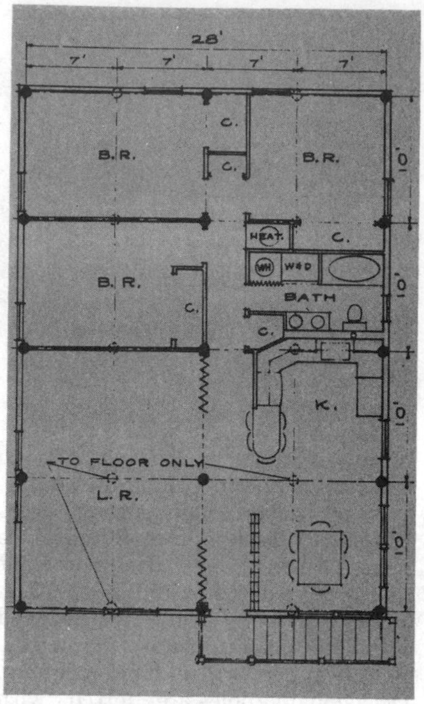

FIG. 2 Plan for small house with all poles inside house; short poles could be eliminated and poles placed outside walls if floor girders are used.

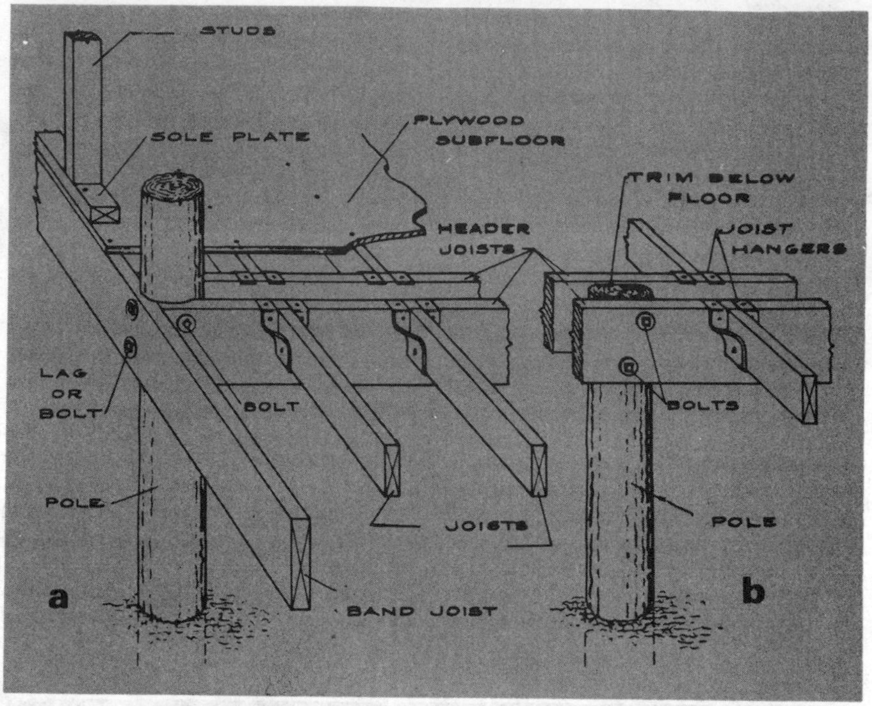

FIG. 3 In pole-frame building, poles continue up past main floor (a); in pole-platform building, poles stop at main floor (b). Lag bolts and joists are used for relatively light loads; timbers and through-bolted shear connections usually are required for efficient pole loading.

ADVANTAGES & DISADVANTAGES

The advantages of pole construction are most pronounced in locations where conventional methods are either uneconomical or do not meet the desired objectives of minimum disturbance to the terrain, freedom from flooding, or resistance to strong winds and earthquakes.

Thus, on steep slopes, pole buildings are generally more economical because they use relatively light wood members; because pole construction does not require heavy equipment for excavation, it retains existing drainage patterns and preserves ground cover (Fig. 4).

On level ground subject to periodic flooding (as in coastal areas from high winds or in river plains from spring floods), pole buildings permit raising the main level safely above flood levels. When there is adequate embedment and proper connections, rampaging waters can flow under the building with minimum damage to the structure (Fig. 5). However, all pole buildings in flood-prone areas should be carefully engineered for specific local conditions rather than the generalized guidelines of this section.

In hurricane plagued areas, whether on flat or sloping ground, pole construction permits secure anchorage of the structure from roof to footings. Proper connections and adequate embedment are particularly important in high wind areas, which are often subject to flooding.

The following precautions should be observed with pole buildings: 1) in most parts of the country, it is necessary to insulate the first floor and to provide a finish to protect the insulation; 2) in heavily wooded areas, vegetation and combustible materials should be cleared from under where the house is raised above grade only sufficiently to create a crawl space, a skirt may be desirable to keep out animals and prevent accumulation of debris under the house; and 4) the underfloor cavity should remain accessible to permit periodic inspections of the poles and the underside of the floor.

FIG. 4 Pole-house "floats" above terrain.

SITE PREPARATION

One of the greatest opportunities afforded by pole construction is the preservation of the natural features of the site. Buildings should be fitted to the terrain, varying setbacks and sideyards as dictated by drainage, views, orientation to the sun and prevailing winds. Clustering buildings is one way to reduce land and construction costs, and to shorten road and utility runs.

Drainage can be crucial on hillside sites. Pavements and roofs, road cuts and utility trenches may concentrate surface waters into destructive, erosive streams. Heavy machinery should be used as little as possible and necessary roads should be terraced along contour lines. It is generally preferable to hand-dig the holes and to remove as little of the natural vegetation as possible. Hand digging theoretically is possible for holes up to 6′ deep, but may be uneconomical except for shallow poles, as described in Special Construction on Rocky Slopes below.

On relatively flat sites, machine digging is more practical, but truck mounted augers generally are limited to 18″ hole diameters. While these hole sizes are adequate for close pole spacing in good soils, larger diameters are generally required for most soil conditions.

A minimum number of poles should be used, consistent with other design and structural requirements. The most efficient use of poles usually occurs when the vertical load per pole is about 8,000 lbs. Loads up to twice this amount require special engineering; one-half the indicated loading is generally inefficient.

Dug holes should be at least 8″ larger than the diameter of the pole, to permit accurate aligning and plumbing (Fig. 6). In most cases the hole is wider than this minimum in order to develop a bearing area of sufficient size for the loads. Specific recommendations for hole size and depth are given below.

POLE SELECTION

The main considerations in pole selection are: 1) type of treatment and 2) size and length.

Preservative treatment

Treated poles should conform to AWPA C-23, Round Poles and Posts Used in Building Construction—Preservative Treatment by Pressure Process. The standard specifies how much preservative should be retained in each pole and how deep it should penetrate into the wood. Treated poles are frequently branded to provide users with certain basic information. Information included is the manufacturer's name, location and year of treatment, pole size, length, wood species and type of preservative.

There are four main types of preservative treatment: 1) waterborne salts; 2) pentachlorophenol in light solvent; 3) pentachlorophenol in heavy oil; and 4) creosote. Each treatment results in varying degrees of these desirable properties: cleanliness, paintability, attractive color and freedom from odors.

Waterborne salts preservatives produce treated poles that are comparatively dry to the touch, paintable, and free from objectionable odor. The color is a shade of brown or green, depending upon the preservative and wood species used.

Pentachlorophenol in light solvent produces a comparatively dry surface that usually can be stained or painted. However, if painting is definitely intended, the specification should so indicate, so that especially oil-free poles are selected. There might be a slight odor until the oil has vaporized. This treatment has little effect on the natural color of the wood, but is less durable than the other treatments.

Pentachlorophenol in heavy oil generally produces an oily surface making it difficult to apply paint successfully. Colors vary from light to dark brown, and an odor is usually present for some time. Poles treated with penta in heavy oil are readily

FIG. 5 Pole substructure elevates home above natural terrain permitting unobstructed runoff of storm or flood waters.

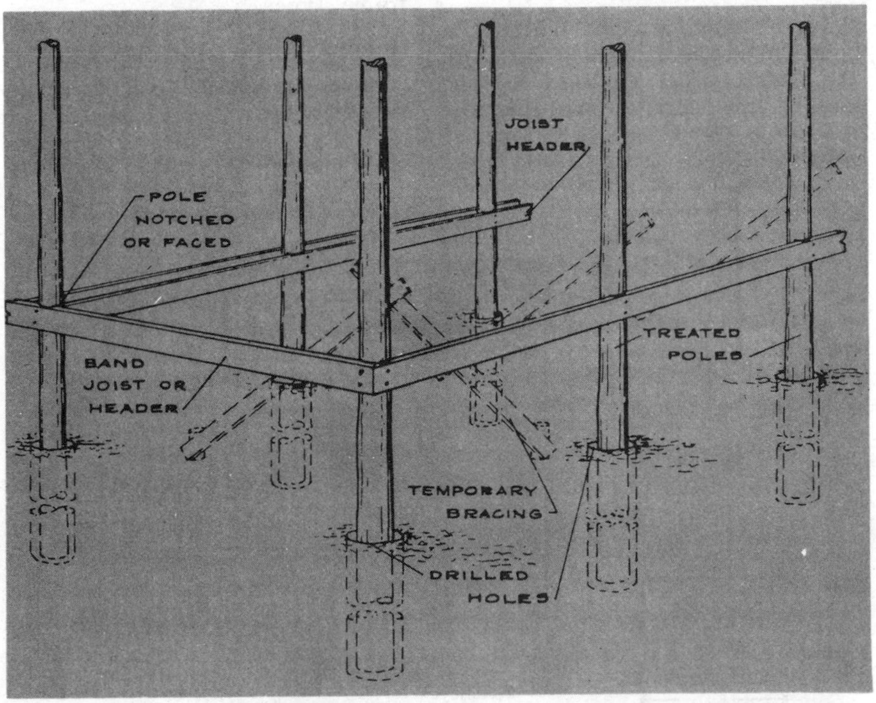

FIG. 6 Poles are aligned and plumbed, then braced temporarily, before holes are backfilled. Heavier horizontal members are likely to be used in typical home construction.

available because they are frequently used by utility companies for electric and telephone poles.

Creosote treated poles are widely available, because they too are favored by utility companies. They are generally used for pole-platform construction, but if used in pole-frame construction, they should not pass through enclosed living spaces. Creosoted poles range from dark brown to black and exude a smoky odor for a long time. Because their surface is oily, painting is not recommended.

Where penta or creosote treated poles are used, there should be no nails driven through a finish material and into the treated wood. The capillary space between the nail and the wood may draw preservative along this path and stain the finish material.

Field cuts, notches or drilled holes through the preservative zone should receive several liberal applications of a preservative to protect the wood. Field application of preservatives should be in accordance with AWPA M4, Standard for the Care of Preservative-Treated Wood Products.

CONSTRUCTION METHODS

There are two principal methods of pole construction: 1) *pole-frame*, and 2) *pole-platform*. The size of the footing and depth of embedment is partly dependent on the method of construction.

FIG. 8 SOIL CLASSIFICATION FOR POLE CONSTRUCTION

Rating	General Description	Soil Groups	Maximum Bearing Value	Lateral pressure per ft. of embedment
Good	Compact, well graded sand & gravel; hard clay or graded fine & course sand	GW, GP, SW and SP	6,000 psf	400 psf
Average	Loose gravel, medium clay or more compact compositions	GM, GC, SM, SC and SL	3,000 psf	200 psf
Below Average	Sand or clay containing large amounts of silt	CL, OL, MH, CH and OH	150 psf	100 psf

In pole-frame buildings, the main vertical members extend from below the ground surface to the roof. The poles not only support the vertical weight of the building, but give it considerable lateral resistance to the forces of wind, flood, and earthquake.

In pole-platform buildings, treated poles form the substructure, permitting conventional construction methods to be used above. Pole-platform structures generally require greater pole embedment than pole-frame types.

General Recommendations

A few general rules will avoid problems in matching conventional framing methods with pole construction. Poles should not be placed within stud walls, because it is costly to apply finishes around poles which vary in diameter and shape. The beams, girders and posts should be spaced so that standard, readily available sizes of lumber and panel materials can be used with a minimum of waste.

On level ground and on steep slopes with good soils (6,000 psf and better), non-cementitious, granular backfill can be used, provided a concrete pad or necklace is used (Fig. 7). Common backfill materials include sand, gravel, or crushed rock. Clean sand often is the least expensive and can achieve 100% compaction when properly tamped. Excavated soil

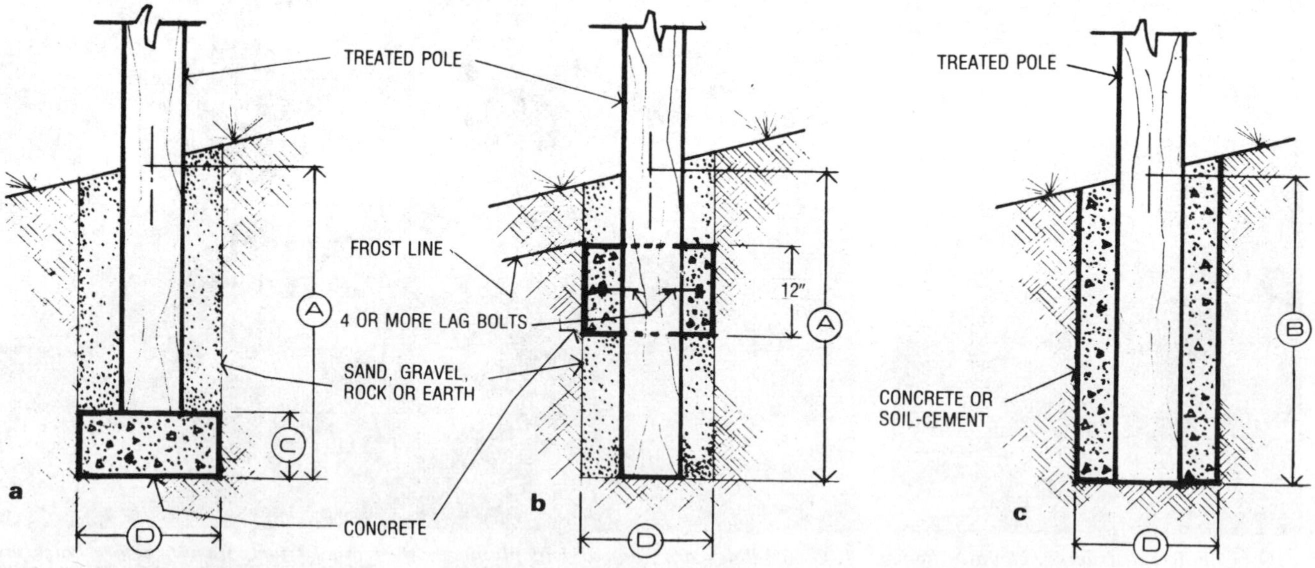

FIG. 7 Recommended pole footings: a) concrete pad; b) concrete necklace; c) concrete or soil-cement backfill. Dimensions A, B and D can be obtained from Figures 10 through 14. Pad thickness C should be determined by engineering calculations.

from the site also can be used, provided it is free of organic matter and is compacted by tamping.

Under these conditions, backfilling with concrete, or with soil-cement (excavated soil mixed with portland cement in 5 to 1 ratios), will generally reduce required hole depth by 1.5′ or 2.0′ (Fig. 7).

On steep slopes with average and below average soils (3,000 psf and lower), only concrete and soil-cement backfill is suitable. When concrete or soil-cement is used, a separate pad or necklace is not necessary.

Design Assumptions

Pole embedment depends on a variety of factors. To reduce the variables to a smaller number which could be tabulated readily, certain design assumptions have been made:

Soil bearing values fall in three general categories as shown in Figure 8.

Building height of pole-frame structure is 9′ from floor to top plate; 4′ from top plate to ridge (Fig. 9).

Embedment condition A: Each pole either rests on concrete pad or has concrete necklace and is backfilled with granular material (Fig. 9).

Embedment condition B: Each pole is completely backfilled to grade either with concrete or soil-cement (Fig. 9).

Pole-frame Buildings

Embedment depths vary according to the slope of the site, as well as the soil, pole spacing and unsupported pole height.

Flat Slope On sites with a slope of *less than 1 in 1*, all poles should be embedded and supported as recommended in Figure 10.

Steep Slope On sites with a slope of *1 in 10 and up to 1 in 1*, poles should be embedded and supported as recommended in Figures 11 and 12.

It should be noted that embedment recommendations differ for uphill and downhill poles. The uphill poles, having a short unsupported height, provide the necessary rigidity for the structure. In order to transfer this rigidity to the longer downhill poles, the floor should be designed as a diaphragm. Properly nailed conventional joists and 1/2″ or thicker plywood floors provide such diaphragm action.

Pole-platform Buildings

Embedment depths of poles vary according to the slope of the site, as well as the other factors tabulated for pole-frame buildings.

Flat Slopes On sites with a slope of *less than 1 in 10*, all poles should be embedded and supported as recommended in Figure 13.

Steep Slopes On sites with a slope of *1 in 10 and up to 1 in 1*, uphill poles should be embedded and supported as recommended in Fig. 14. Recommendations for downhill poles

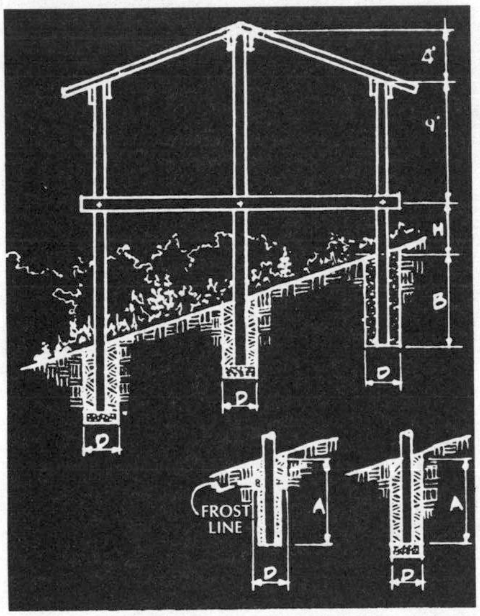

FIG. 9 Building height design assumptions and embedment conditions: a) concrete pad or necklace and granular backfill; b) concrete or soil-cement backfill. See Figure 11 for recommended values of A, B, D & H.

FIG 11 **EMBEDMENT DEPTHS FOR DOWNHILL POLES OF POLE-FRAME BUILDINGS ON SLOPES UP TO 1:1.**

Soil Strength	Embedment Depth, ft. Slope of Grade Up to 1:3	Up to 1:2	Up to 1:1
Below Average	4.5	6.0	*
Average	4.0	5.0	7.0
Good	4.0	4.0	6.0

*Embedment depth is greater than 8′, considered excessively expensive .

FIG. 10 **EMBEDMENT DEPTHS FOR POLE-FRAME BUILDINGS ON SLOPES LESS THAN 1:10.**

H, ft.	Pole Spacing, ft.	Good Soil: Embedment Depth, ft.† A	B	D, in.	Tip Size, in.	Average Soil: Embedment Depth, ft.† A	B	D, in.	Tip Size, in.	Below Average Soil: Embedment Depth, ft.† A	B	D, in.	Tip Size, in.
1½ to 3	8	5.0	4.0	18	6	6.5	5.0	24	6	*	6.0	36	6
	10	5.5	4.0	21	7	7.0	5.0	30	7	*	6.5	42	7
	12	6.0	4.5	24	7	7.5	5.5	36	7	*	7.0	48	7
3 to 8	8	6.0	4.0	18	7	7.5	5.5	24	7	*	7.0	36	7
	10	6.0	4.5	21	8	8.0	6.0	30	8	*	7.5	42	8
	12	6.5	5.0	24	8	*	6.0	36	8	*	8.0	48	8

*Embedment depth is greater than 8′, considered excessively expensive.
†Where a concrete floor slab is used at grade, embedment depths may be reduced to 70% of those shown.

FIG. 12 **EMBEDMENT DEPTHS FOR SHORTER, UPHILL POLES IN POLE-FRAME BUILDINGS ON SLOPES UP TO 1:1.**

H, ft.	Pole Spacing, ft.	Good Soil Embedment Depth, ft. A	Good Soil Embedment Depth, ft. B	Good Soil D, in.	Good Soil Tip Size, in.	Average Soil Embedment Depth, ft. A	Average Soil Embedment Depth, ft. B	Average Soil D, in.	Average Soil Tip Size, in.	Below Average Soil Embedment Depth, ft. A	Below Average Soil Embedment Depth, ft. B	Below Average Soil D, in.	Below Average Soil Tip Size, in.
1½ to 3	6	7.0	5.0	18	8	*	6.5	18	8	*	*	*	*
	8	7.5	5.5	18	9	*	7.0	24	9	*	*	*	*
	10	*	6.0	21	9	*	8.0	30	9	*	*	*	*
	12	*	6.5	24	10†	*	*	*	*	*	*	*	*
3 to 8	6	7.5	5.5	18	8	*	7.0	18	8	*	*	*	*
	8	8.0	6.0	18	9	*	8.0	24	9	*	*	*	*
	10	*	7.0	21	10†	*	*	*	*	*	*	*	*
	12	*	7.0	24	11†	*	*	*	*	*	*	*	*

*Embedment depth is greater than 8′, considered excessively expensive.
†These tip diameters may be decreased 1″, provided embedment is increased by 0.5 ft.

FIG. 13 **EMBEDMENT DEPTHS FOR POLES IN PLATFORM BUILDINGS ON SLOPES LESS THAN 1:10.**

H, ft.	Pole Spacing, ft.	Good Soil Embedment Depth, ft.† A	Good Soil Embedment Depth, ft.† B	Good Soil D, in.	Good Soil Tip Size, in.	Average Soil Embedment Depth, ft.† A	Average Soil Embedment Depth, ft.† B	Average Soil D, in.	Average Soil Tip Size, in.	Below Average Soil Embedment Depth, ft.† A	Below Average Soil Embedment Depth, ft.† B	Below Average Soil D, in.	Below Average Soil Tip Size, in.
1½ to 3	8	4.0	4.0	18	5	5.5	4.0	24	5	7.0	5.0	36	5
	10	4.5	4.0	21	5	6.0	4.0	30	5	8.0	5.5	42	5
	12	5.0	4.0	24	5	6.5	4.5	36	5	*	5.5	48	5
3 to 8	8	5.0	4.0	18	6	6.5	4.5	24	6	*	6.0	36	6
	10	5.5	4.0	21	7	7.0	5.0	30	7	*	6.5	42	7
	12	6.0	4.5	24	7	7.5	5.5	36	7	*	7.0	48	7

*Embedment depth is greater than 8′, considered excessively expensive.
†Where concrete slab is used at grade, embedment depths may be reduced to 70% of those shown.

FIG. 14 **EMBEDMENT DEPTHS FOR UPHILL POLES IN PLATFORM BUILDINGS ON SLOPES UP TO 1:1.**

H, ft.	Pole Spacing, ft.	Good Soil Embedment Dept, ft. A	Good Soil Embedment Dept, ft. B	Good Soil D, in.	Good Soil Tip Size, in.	Average Soil Embedment Dept, ft. A	Average Soil Embedment Dept, ft. B	Average Soil D, in.	Average Soil Tip Size, in.	Below Average Soil Embedment Dept, ft. A	Below Average Soil Embedment Dept, ft. B	Below Average Soil D, in.	Below Average Soil Tip Size, in.
1½ to 3	6	5.5	4.0	18	6	7.5	5.0	18	6	*	7.0	24	6
	8	6.5	4.5	18	7	*	6.0	24	7	*	8.0	36	7
	10	7.0	5.0	21	7	*	6.5	30	7	*	*	*	*
	12	7.5	5.5	24	8	*	7.0	36	8	*	*	*	*
3 to 8	6	7.0	5.0	18	8	*	6.5	18	8	*	8.0	24	8
	8	7.5	5.5	18	9	*	7.0	24	9	*	*	*	*
	10	*	6.0	21	10†	*	7.5	30	10†	*	*	*	*
	12	*	6.5	24	10†	*	*	*	*	*	*	*	*

*Embedment depth is greater than 8′, considered excessively expensive.
†These tip diameters may be decreased 1″, provided embedment is increased by 0.5 ft.

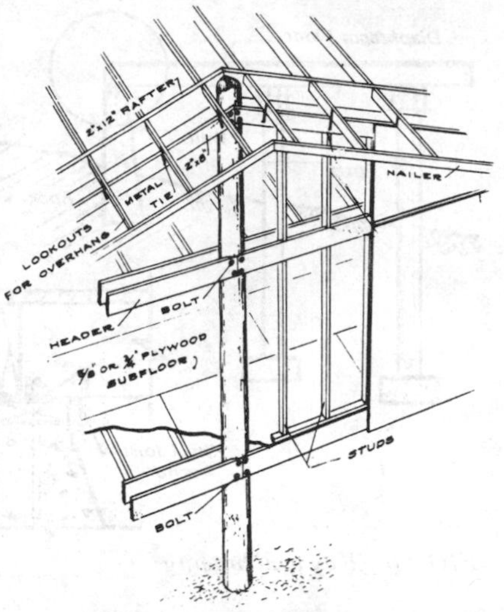

FIG. 15 Double headers at exterior end walls frame ridge-high pole and are bolted to it.

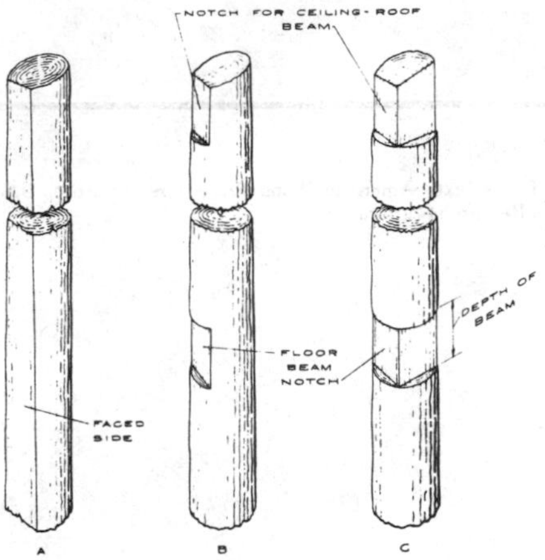

FIG. 16 Poles are faced or notched to provide better contact with beams and make suitable connections with shear connectors such as spike grid shown below.

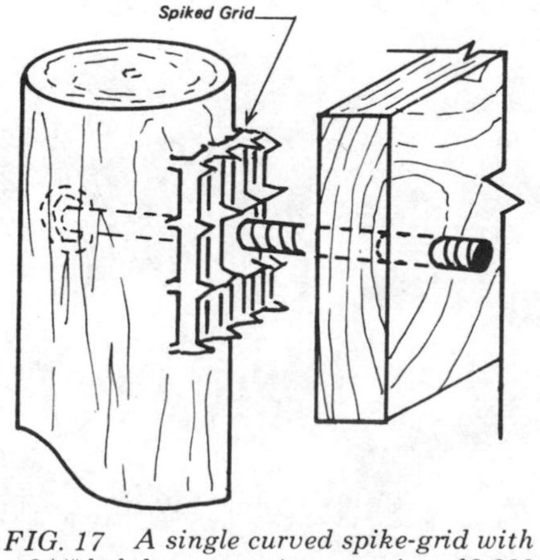

FIG. 17 A single curved spike-grid with a 3/4" bolt has a carrying capacity of 3,800 lbs; with 1" bolt, 4,100 lbs.

are given in Fig. 11. The floor should be designed as a diaphragm to transfer the rigidity of the uphill poles to the rest of the structure.

Connections

It is generally better to frame poles with pairs of beams by lag-bolting, through-bolting, or nailing to the supporting pole (Fig. 15). To make a stronger connection, poles can be faced or notched (Fig. 16). A spike-grid is preferred to squaring or notching because it minimizes the damage to the preservative-treated outside layer and increases the load-bearing value of the connection (Fig. 17).

Accepted design practice is outlined in *Pole Building Design* by American Wood-Preservers' Institute and *Timber Construction Manual* by the American Institute of Timber Construction. Strength values used in designing bolted and nailed connections should conform to the *National Design Specifications for Stress-Grade Lumber and Its Fastenings* by the National Forest Products Association. Section 2907 of the Uniform Building Code can also be used as a design guide.

SPECIAL CONSTRUCTION ON ROCKY SLOPES

The following discussion of pole bracing as an alternative to embedment is informational only, and should not be attempted without competent engineering advice.

On rocky slopes, where deep embedment as shown in Figures 11 to 15 cannot be developed without expensive machine digging or rock drilling, it is possible to use shallow hand-dug holes, provided the pole structure is adequately braced and anchored.

Alternates to deep embedment include: 1) *pole seating* and *cross-bracing;* 2) *shear walls* and *floor diaphragms.*

Pole Seating and Crossbracing

If the understructure is adequately braced, the poles need to be embedded only far enough to prevent slipping downhill. The embedment should be at least 4′ and the poles should be placed in specially dug seats in the rock. Since the rock surface is likely to be irregular, tamped earth or wetted sand should be used to make a flat bearing surface under each seated pole. If a pole seat cannot be dug into the rock, a concrete pad may be used over the rock, and the pole anchored to the pad with lag bolts or standard post anchors. The pad should be 8″ thick and of the same diameter as the hole which could be 8″ larger than the pole.

In one crossbracing method, 5/8″ or 3/4″ steel rods are fitted through drilled holes, flooded with preservative, then fastened with nuts, using beveled washers to avoid crushing the wood. Steel rod crossbracing usually is not needed between every pole; several diagonal rods often are adquate (Fig. 18a).

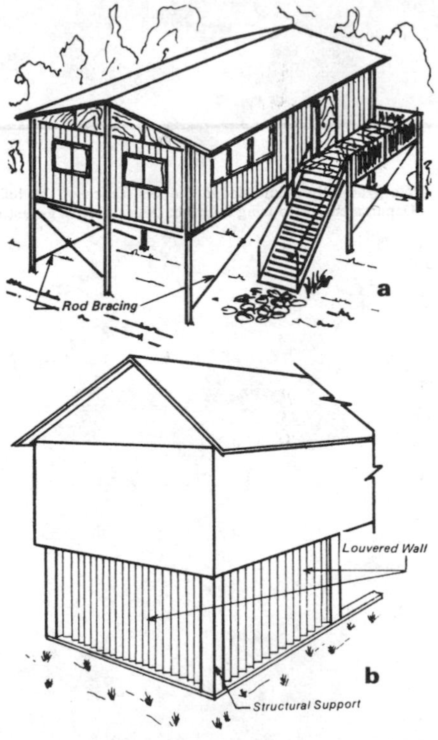

FIG. 18 Rod braced house (a) and louvered skirting around bracing (b).

FIG. 19 Shear wall. Rod bracing may still be required on columns.

The pole structure also can be knee-braced with 2″ x 6″ wood members, nailed or bolted to form a diagonal connection from the poles to the floor beams. The bracing can be left exposed or covered with "skirt" boards around the perimeter of the house. (Fig. 18b).

Shear Walls and Floor Diaphragms

A masonry or concrete fireplace block, which is supported on concrete footings at least 4′ below grade, can act as a shear wall to resist horizontal forces (Fig. 19). Similarly, a masonry or concrete utility room below the first floor can act to brace the structure and anchor it.

Keywalls also can be used to resist the horizontal forces pushing the structure down the slope (Fig. 20). For both shear walls and keywalls, floors should be designed as diaphragms to transfer the lateral rigidity of the walls to the poles. Such floors can be built of conventional wood joists and 1/2″ or thicker plywood, with nailing as recommended by the American Plywood Association.

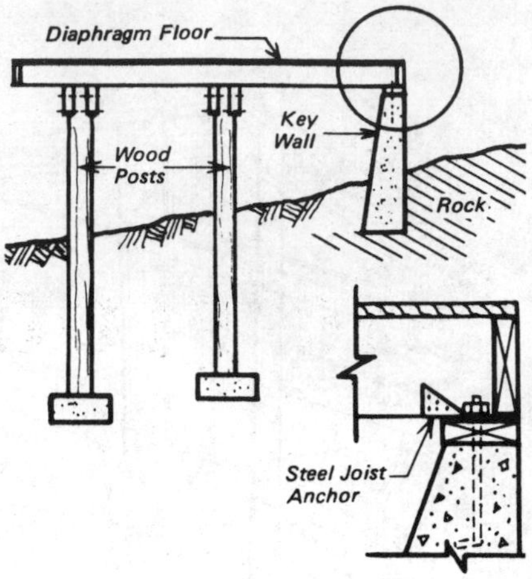

FIG. 20 Key wall bracing.

We gratefully acknowledge the assistance of the following companies, agencies and associations in preparation of this text: American Wood Preservers Institute, U.S. Department of Housing & Urban Development, Forest Products Laboratory of U.S. Department of Agriculture and AIA Research Corporation.

320 FLOOR/CEILING SYSTEMS

INTRODUCTION

In multilevel buildings, a ceiling can be an integral part of the floor construction. Hence, systems may be called floor/ceiling systems and considered one of the basic elements of the superstructure. The design, selection and construction of a floor/ceiling system considers (1) the required level of functional performance, (2) costs, and (3) the desired finished appearance. These considerations often are influenced by the systems and finishes used in other basic elements of the construction.

Functional performance requirements of a floor/ceiling system usually include: (1) adequate strength to support dead loads (the actual weight of the materials used in the floor/ceiling system and other elements in the building which it supports) and live loads (such as from occupants and furnishings); (2) provision for lateral support of walls; (3) satisfactory resistance to the transmission of airborne and structure-borne sound; (4) suitable fire resistance; (5) suitability for the application of finish materials; (6) adaptability to economical methods of assembly and erection; (7) space to accommodate heating, air conditioning, electrical and plumbing equipment; and (8) control of heat loss and the flow of water vapor.

Floor/ceiling systems must support superimposed loads safely and without excessive deflection or vibration. Design dead and live loads ordinarily are established by local building codes which require structural members to be sized so that an allowable fiber stress in bending is not exceeded when the floor/ceiling system is fully loaded. Codes usually do not limit deflection, since it does not have a bearing on structural failure and safety. Deflection limitations are established in floor/ceiling design to reduce deflection to visually acceptable limits, to assure the integrity of applied finish materials, and to limit vibration to acceptable levels of human perception. Deflection, rather than bending strength, generally is the controlling factor in the structural design of wood floor/ceiling systems.

Depending on the design of the foundation system, a first floor system may provide lateral support to foundation walls; and floor/ceiling systems also may provide lateral support for other elements of the superstructure against loads caused by wind and/or earthquake.

The ultimate costs of a floor/ceiling system include in-place costs, maintenance costs and operating costs. In-place costs are the sum of the costs of materials, labor and overhead required to assemble and/or install the floor/ceiling system. The selection of the system may affect other costs such as the cost of applying floor and ceiling finishes, constructing interior partitions and providing space to accommodate heating, air conditioning, electrical and plumbing equipment.

Maintenance costs and operating costs attributable to the floor/ceiling system (such as heat loss or gain through the construction) have an important influence on the ultimate cost of the floor and should not be overlooked if long-term economy is to be achieved.

The system is in part determined by the desired finish applications. Selection of floor/ceiling finish materials: (1) may affect the spacing and span of structural members; (2) may determine the need for ceiling furring and/or floor underlayment; and (3) even may dictate methods of construction.

Prefinished floor or ceiling surfaces may not require the application of an additional finish after the system has been placed. However, the floor should provide a working platform for finishing operations during the construction

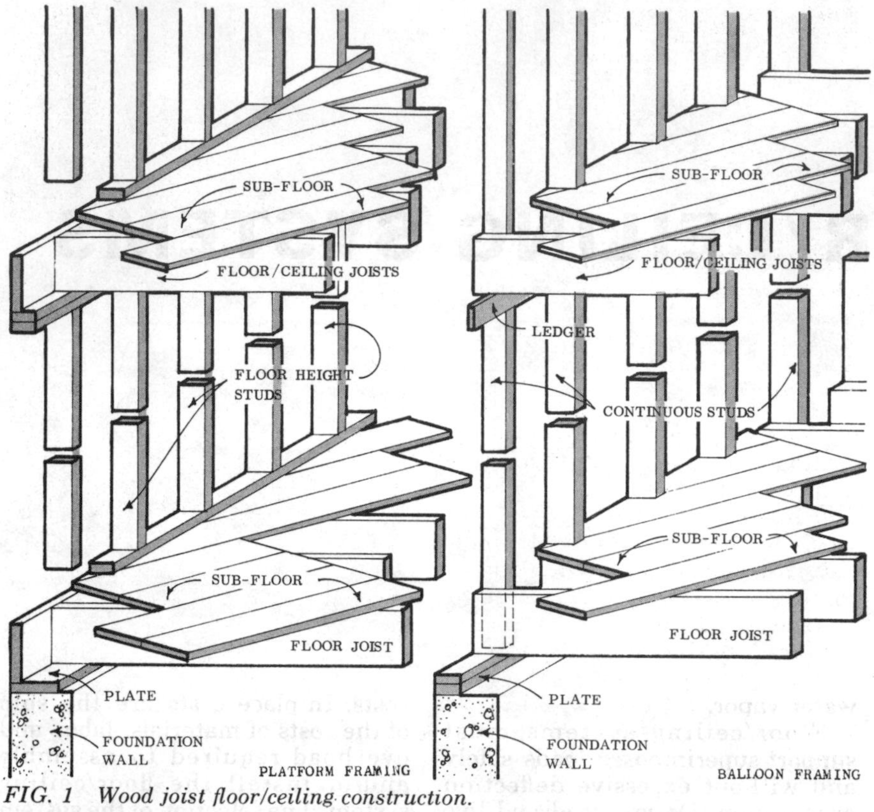

FIG. 1 Wood joist floor/ceiling construction.

FIG. 2 Wood Joists: (a) over basement foundation, center girder is steel beam; (b) joists over crawl space, center girder is built-up wood member; (c) 1" x 6" tongue-and-grooved board subfloor applied diagonally; (d) plywood subfloor over joists.

period. Therefore, a floor/ceiling system often is selected which has a subfloor or sheathing application that is later provided with finished surfaces.

Typical residential *suspended* floor/ceiling systems include: *joist* construction, *plank-and-beam* construction, *panelized* construction; and *steel* and *concrete* constructions. Each system is described briefly in this section. A more detailed discussion of each system will be found in individual sections immediately following. Concrete slab-on-ground floor constructions are discussed in Section 311.

JOIST FLOORS

Balloon-frame and platform-frame constructions typically use joist floor systems consisting of nominal 2" lumber joists spaced 12", 16" or 24" o.c. Subflooring of wood boards or plywood is applied to form a structural floor and working platform (Fig. 1). Generally, joists span roughly one-half the width of the building from the exterior walls to the center load-bearing wall or beam. Interior load-bearing walls may restrict the flexibility of room planning. However, for relatively narrow buildings (approximately 16' or less) joists may span from exterior wall to exterior wall without a center support.

Joist floor systems provide space for heating, air conditioning, electrical and plumbing equipment, and easily may be insulated to minimize heat loss and/or improve sound transmission characteristics.

Some types of floor finishes (such as hardwood flooring) may be installed directly to the subflooring; other finishes (such as resilient flooring) may require an underlayment. A variety of ceiling finishes may be applied, usually directly to the underside of the joists.

Both balloon- and platform-frame systems (see page 300-2) are readily adaptable to multistory construction. Because of its adaptability to the use of tilt-up and component construction methods, platform framing provides the greatest flexibility for the wide variety of possible design requirements in basic house types (Fig. 2).

PLANK-AND-BEAM FLOORS

Plank-and-beam systems are characterized by the use of nominal 2″ x 6″ or 2″ x 8″ tongue-and-grooved (T&G) or splined planks spanning between beams generally spaced 4′ to 8′ o.c. The planks serve as the subfloor and working platform and transmit the floor loads to fewer but larger members than in wood joist floor systems (Fig. 3).

Floor finishes such as strip wood flooring may be applied directly over the planks, but sheet and tile flooring require an underlayment. When used in multistory construction, beams and the underside of planks can be left exposed to provide the finished ceiling. However, without the application of an additional ceiling finish, a plank-and-beam floor/ceiling system may offer less resistance to sound transmission than other systems, and does not provide space for the installation of heating, air conditioning, electrical and/or plumbing equipment. Hence, plank-and-beam systems generally are used as floor systems and are suited particularly to one-story construction (Fig. 4). They permit a lower first-floor-to-grade elevation and result in a lower building height. Interior wall placement should consider the location of beams.

Plank-and-beam floor/ceiling systems often are integrated with plank-and-beam roof/ceiling systems and post-and-beam wall systems. In such design, the expression of the basic structural system largely determines the interior appearance and character.

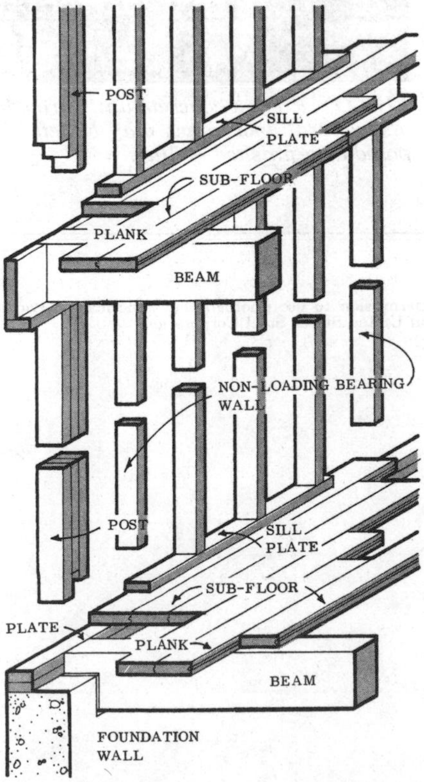

FIG. 3 Plank-and-beam construction.

FIG. 4 Plank floor systems result in a minimum finished grade-to-floor separation. In this system, laminated wood planks are a single-thickness structural and finish floor.

PANELIZED FLOORS

Panelized systems take many forms. One common system uses 1-1/8″ thick plywood panels spanning between beams spaced 4′ o.c. Panels serve as a combination subfloor and underlayment (Fig. 5). Other systems consist of simple preframed panels in which the members are precut and sheathed to form a preassembled component (Figs. 6 and 7). Frame members are nominal 2″ dimension lumber and function as floor joists. Panels may be fabricated on or off site in practically any size, with insulation and vapor barrier installed and interior finish applied.

Panelized systems also include *stressed-skin* and *sandwich* panel construction. These panels, which often require specialized equipment and quality control, are factory-built.

Stressed-skin floor panels are made with framing members to which plywood skins are bonded either by glue-nailing techniques or by adhesive applied under heat and pressure (Fig. 8). When the

FIG. 5 4′ x 8′ plywood panels 1-1/8″ thick with tongue-and-grooved edges form a combination subfloor-underlayment over supports spaced 4′ o.c.

FIG. 6 This preframed floor panel consists of plywood sheathing attached to 2″ x 4″ stiffeners which span 4′ from beam to beam.

FIG. 7 Preframed floor panel with top and bottom plywood surfaces applied to 2″ x 8″ frame members.

FIG. 8 Stressed-skin panel with plywood surfaces glued to a 2″ lumber frame: after glue is placed and plywood positioned, the entire panel is treated with heat and pressure in press at rear.

panel is loaded, both the framing members and the skin surfaces are stressed and act as an integral unit to carry loads. The stressing of the skin surfaces permits a reduction in the size of the framing members.

Sandwich panels are similar in principle to stressed-skin panels but the faces are glued to and separated by weaker lightweight core materials (Fig. 9), instead of actual framing members.

Panelized systems can be used in a variety of ways: panels may span between transverse or longitudinal beams or be centrally supported by bearing walls or beams as with other floor/ceiling systems.

FIG. 9 A sandwich panel with a honeycomb core used in floor construction.

Practically any flooring finish may be applied to panelized systems. Ceiling finishes can be field-applied or may be an integral, prefinished part of the panels.

Though panelized construction provides many of the advantages of component fabrication, the incorporation of heating, air conditioning, electrical and plumbing equipment requires careful design of the system.

STEEL AND CONCRETE FLOORS

Steel and concrete systems are not commonly used in residential construction, but the potential advantages they offer may indicate increased use in the future. These systems usually have higher in-place costs than conventional wood framing systems, but offer such advantages as longer clear spans, increased fire resistance and greater dimensional stability. Most systems generally require a minimum of on-site fabrication and permit the building to be enclosed quickly during inclement weather (Fig. 10 and 11).

Most steel and concrete floor systems can receive the same wide variety of floor and ceiling finishes applicable to wood floor/ceiling systems, and some concrete floor systems may be left with the underside exposed as the finished ceiling.

FIG. 10 Steel bar joists provide first-floor support in this all-steel foundation floor system.

FIG. 11 Precast concrete units provide a structural floor which may be left exposed as a finished ceiling.

We gratefully acknowledge the assistance of the following for the use of their publications as references and permission to use photographs: American Plywood Association, House & Home Magazine, National Forest Products Association, Practical Builder Magazine, and United States Steel Corporation.

321 WOOD FRAMED FLOORS

WOOD FRAMED FLOORS

INTRODUCTION

Modern light wood framing of the type used in typical residential construction is an outgrowth of the *balloon frame* which originated around the middle of the 19th century in Chicago. It was a revolutionary development which drastically changed the direction of the housing industry and contributed substantially to the colonization of the West. Some historians go so far as to claim that "if it had not been for the knowledge of the balloon frame, Chicago and San Francisco could never have arisen, as they did from little villages to great cities in a single year.... Contemporaries knew quite well that houses would never have sprung up with such incredible rapidity, either on the prairies or within the big cities, had it not been for this construction."

Prior to the evolution of the balloon frame, houses were built of heavy timbers, mortised-and-tennoned together and pinned with hardwood dowels. The principle of balloon framing involves the repetitive use of slender structural members—*studs* for walls, *joists* for roofs and floors—tied together with simple nailed connections. It changed house construction from a highly skilled craft to a relatively simple mechanical procedure which could be learned quickly by most handymen familiar with hand tools. For that reason, perhaps, it was regarded with suspicion and acquired its derisive name on the assumption that such a house would blow away like a balloon the first time a good wind hit it.

The development of the balloon frame is closely associated with the level of industrialization reached in America in the middle of the 19th century. Aside from the evolution of sawmill technology, the ability to mass produce cut nails cheaply generally is credited with the impetus to this revolutionary idea. When the manufacture of cut nails was started at the turn of the century, wrought nails cost 25¢ a pound. With the introduction of more sophisticated machinery, the price was reduced to 8¢ in 1820, to 5¢ in 1833 and finally to a mere 3¢ a pound by 1842.

Over the years, balloon framing was modified to permit the use of shorter (and cheaper) wall studs spanning from floor to floor rather than from foundation to roof as in the original system. This new approach made it possible first to erect a floor platform, then to nail together a full-length wall on the level, where it could be done more conveniently and faster than in an upright position. This modified framing system evolved as wood framing moved west with the early settlers and came to be known as *Western*, or *platform framing*. Although balloon framing still is being used in house construction, the majority of new homes today are platform framed.

The search for economies in labor and materials is a continuing process. As the cost of labor became a more significant factor in the total housing cost, *plank-and-beam* construction, and other floor/roof systems utilizing greater spacing of fewer and larger structural members, came into use. Such systems were more compatible with *post-and-beam* wall framing and, like the skeleton structure of tall buildings, enabled designers to treat the space between posts as a *curtain wall*, intended mainly to keep out the weather rather than resist loads.

However, despite all the innovations in light wood framing, the essential construction principles inherent in the balloon frame persist. The nail still remains an essential connective element in all light wood framing systems and the exterior covering between framing members—the *sheathing*—is relied upon to provide structural continuity to the assembly of light members.

The discussion of wood framing systems in this volume is divided into three sections: this section, 321 Wood Framed Floors; section 341 Wood Framed Walls and section 361 Wood Framed Roof/Ceilings. All are divided into two subsections, one dealing with proper framing methods; the second, with suitable application of sheathing.

Most of the material in this and related sections on wood framing has been abstracted from Agriculture Handbook No. 73 *Wood-Frame House Construction* by L. O. Anderson, of the Forest Products Laboratory, USDA, 1970.

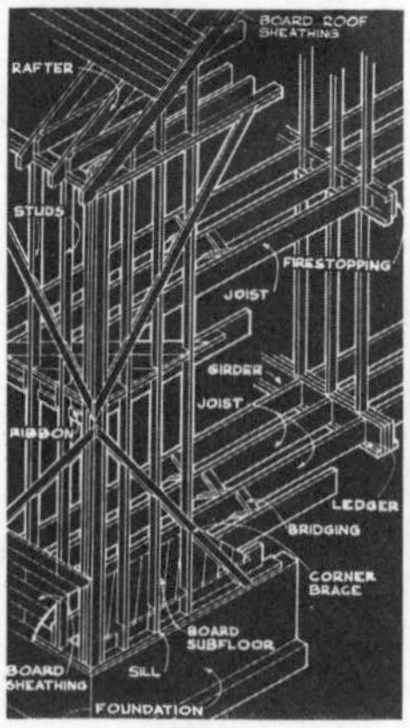

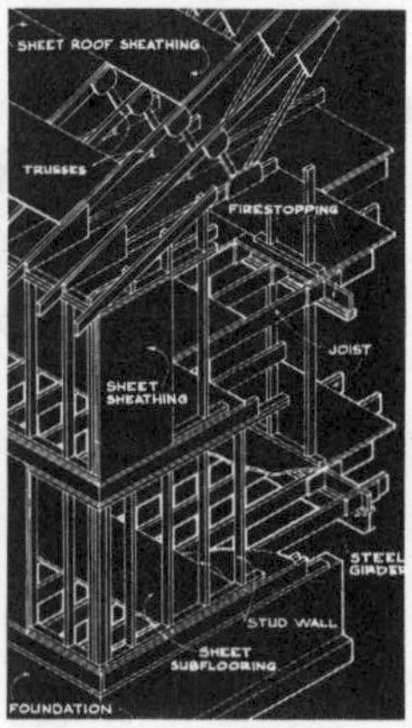

A suspended wood joist floor is part of the superstructure of a building, which includes also the walls and roof/ceiling assemblies. The construction of a joist floor is essentially the same whether the superstructure is wood frame, brick veneer or solid masonry.

The floor framing consists specifically of the *posts, beams* (or *girders*), *sill plates* and *joists.* The floor *sheathing* (subfloor) ties the various framing elements together and stabilizes the joists to prevent twisting and buckling. When these elements are assembled properly on a foundation, they form a level anchored platform for the erection of the partitions, additional floors and roof of the house. The posts and center beams, which support the inside ends of the joists, sometimes are replaced with a wood frame or masonry wall when the basement area is divided into rooms. Wood frame houses also may be constructed over a crawl space with floor framing similar to that used for a full basement. In a slab-on-ground house, wood floor framing would be limited to the upper floors.

LUMBER MATERIALS

The moisture content of beams and joists used in floor framing should not exceed 19%; however, a moisture content of about 15% is much more desirable. Dimension lumber generally can be obtained at a moisture content of 19% or less when S-DRY is specified or at 15% or less when MC 15 is specified. In arid parts of the country, lumber of higher moisture contents can be specified if the lumber is allowed to approach its equilibrium moisture content before applying interior finishes and trim such as baseboard, shoe mold and door casings.

The National Grading Rule for Dimension Lumber, established in accordance with Product Standard PS 20-70 provides a series of uniform grade designations applicable to all species. *Joists and planks* (2″ to 4″ thick and 5″ or wider) are graded as Select Structural, No. 1, No. 2 and No. 3 grades. All grades are suitable for floor joists if they have adequate strength characteristics for the intended spacing and span conditions. (See Floor Joist Span Tables, 201 Wood Work File.) *Light framing* (2″ to 4″ thick and 2″ to 4″ wide) is sorted into up to seven grades. These grades in order of decreasing strength are: Select Structural, No. 1, No. 2, Construction, No. 3, Standard and Utility. An eighth, Stud grade is similar to No. 3 grade, but is cut to 10′ or shorter lengths, is straighter and has a better nailing surface. The lower light framing grades preferably should be used for sills and plates rather than for studs and posts.

One of the important considerations in the design of a wood floor system is to equalize shrinkage of the wood framing at the exterior walls and at the center support. By using approximately the same total depth of cross-grain wood at the center girder as at the exterior wall, differential shrinkage can be minimized. This reduces plaster cracks, prevents sticking doors and avoids other inconveniences caused by uneven shrinkage. For instance, if there is a total of 12″ of cross grain wood at the foundation wall (including joists and sill plate), this should be balanced with an equal amount of cross-grain wood at the center girder. To avoid uneven shrinkage, a steel beam center support generally is preferred. If a wood girder is used, MC 15 lumber should be specified, or the joists should be supported on ledger strips rather than over the top of the girder.

NAILING PRACTICES

Of primary consideration in the construction of a house is the method used to fasten the various wood members together. These connections are most commonly made with nails, but in certain cases metal straps, lag screws, bolts, staples and adhesives may be used.

Proper fastening of frame members and covering materials provides the rigidity and strength to resist windstorms, earthquakes and other destructive hazards. Adequate nailing also is important from the standpoint of suitable performance of wood parts. In general, nails give stronger joints when driven into the side grain (perpendicular to wood fibers) than into the end grain of wood. For instance, the allowable lateral (shear) loads for nails driven into the side grain are 50% higher than those allowed when driven into the end grain. Whenever possible, connections should be designed in such a way that nails are loaded in *shear*, exerting a sideways pull on the nail, rather than in *direct withdrawal*, pulling straight out. Connections never should be designed in a way which would load an end nailed joint in direct withdrawal (Fig. 1).

Figure 2 outlines recommended nailing practices for the floor framing of a well-constructed wood frame house. Penny designations and actual sizes of common wire nails are shown in Figure 3.

When houses are located in high wind areas, they should be provided with special connections or supplemental fasteners as recommended below under the discussion for sill-plate construction.

POSTS, BEAMS AND GIRDERS

Wood or steel posts generally are used in the basement to support wood girders or steel beams. Masonry piers also can be used in basements but most commonly are employed in crawl-space houses.

Round steel posts known as pipe columns or *lally* columns can be used to support both wood girders and steel beams. Columns should be provided with a steel bearing plate at each end. Secure anchoring of the girder or beam to the post or column is essential (Fig. 4).

Wood posts for free-standing use in a basement should be of solid or built-up lumber not less than 6″ x 6″ in size. When combined with a frame wall, posts may be 4″ x 6″ to conform to the depth of the studs. Wood posts should be squared at both ends and securely fastened to the girder (Fig. 5a). The bottom of the post should rest on and be pinned to a concrete pedestal 2″ to 3″ above the finish floor (Fig. 5b). It is good practice to treat the bottom of the post with a decay preservative, or use a vaporproof covering such as heavy roll roofing over the concrete pedestal. A sheet metal covering is preferred in areas subject to termite infestation.

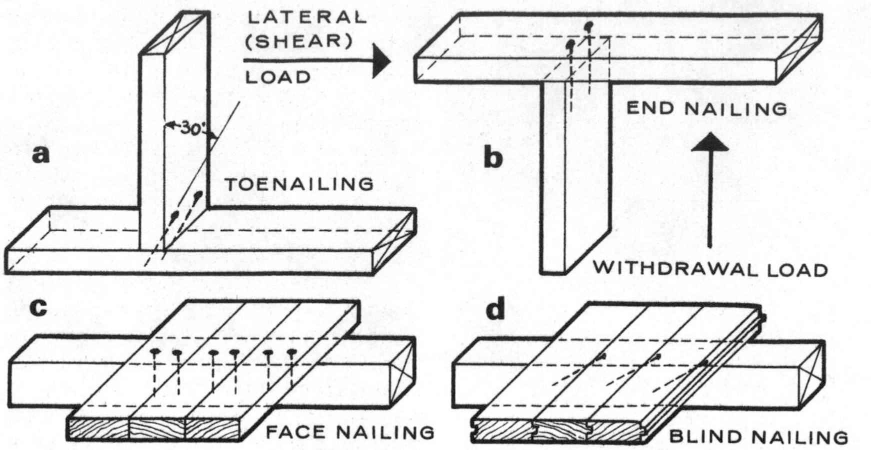

FIG. 1 Common nailing methods: (a) toenailing—slant nailing at a 30° angle to the face or edge of the attached piece; (b) end nailing—nailing into end grain of the supporting piece; (c) face nailing—nailing into side grain through the wide dimension of the attached piece; (d) blind nailing—toenailing through the edge of the plank so as to conceal the nail head.

FIG. 2 NAILING RECOMMENDATIONS FOR WOOD FRAMED FLOORS

Connection	Nailing Method, Number of Nails and Types
Members of built-up girder	Face nail 2-10d at each end; 2 rows of 10d, 32″ o.c. top and bottom, staggered
Header to floor joist	End nail 2-16d
Floor joist to sill plate or girder	Toenail 2-10d or 3-8d at each joist
Floor joists to each other over girder	Face nail 3-16d at lap over girder
Header and stringer joists to sill plate	Toenail 8d, 16″ o.c.[1]
Bridging to joists	Toenail 2-8d each end of bridging
2″ ledger strip to girder	Face nail 3-16d at each joist
Header to tail joists, less than 6′[2]	End nail 3-16d and toenail 2-10d
Trimmer to header, less than 6′[3]	End nail 3-16d and toenail 2-10d

[1] Nailing not required when sheathing laps over header or stringer and is nailed to sill plate with 8d nails 8″ o.c.
[2] Longer tail joists should be supported on framing anchors or ledger strip.
[3] Longer header (up to 10′) should be supported on joist hangers, posts or special connectors.

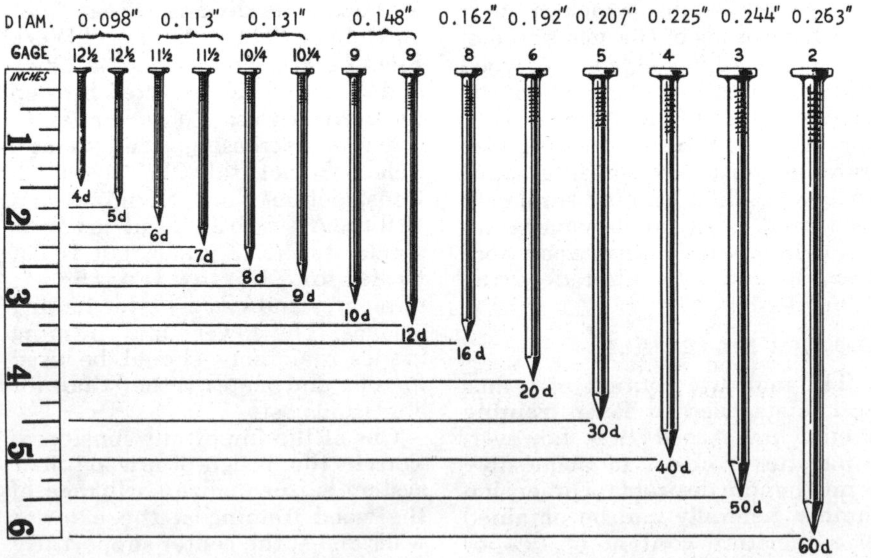

FIG. 3 Standard sizes of common wire nails. Thicker nails in sizes 10d to 60d, are known as spikes.

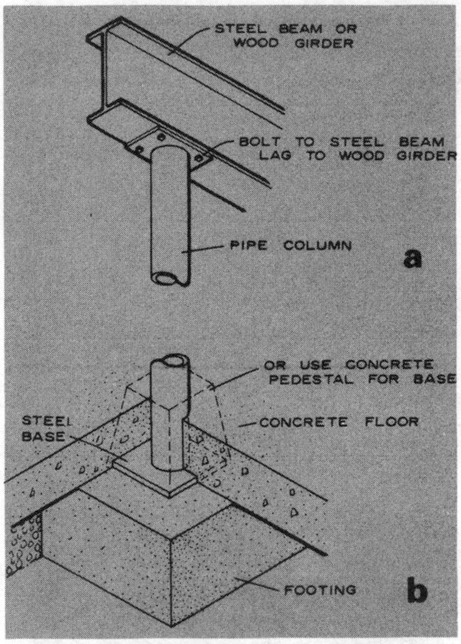

FIG. 4 (a) Steel top plate provides a means of secure anchorage to beam; (b) base plate distributes the load and anchors the steel post.

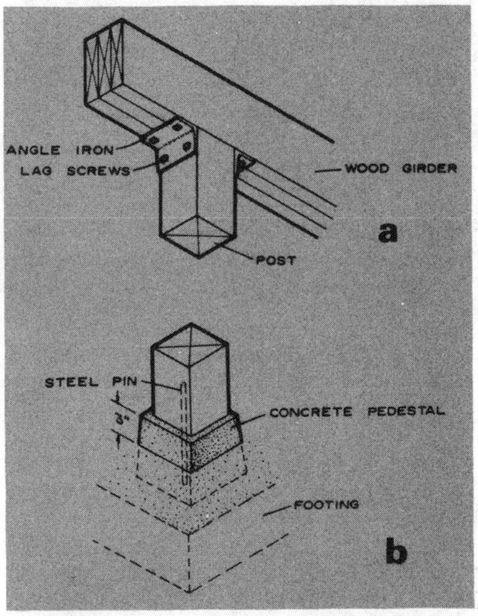

FIG. 5 Wood post should be connected to girder with metal angles or framing anchors (a); base of post should be pinned and protected against decay and termites (b).

Both wood girders and steel beams are used in present-day house construction. The standard *I-beam* and *wide flange beam* are the most commonly used steel beam shapes. Wood girders are of two types—*solid* and *built-up* (Fig. 6). The built-up beam is preferred because it can be made up from drier dimension lumber and is more dimensionally stable. Solid lumber or commercially available glue-laminated girders may be desirable where exposed in finished rooms below.

(The use of the term "beam" in connection with steel members and "girder" with wood members is based on common usage, rather than technical difference in meaning. In engineering terminology a "beam" is any member with loads applied perpendicular to its direction and a "girder" is a principal beam, frequently supporting other beams. In this broader sense joists, girders, headers and trimmers are types of beams. Although wood joists are not generally referred to as beams, both built-up and solid wood members carrying joists frequently are referred to as "beams" or "girders".)

The built-up girder (Fig. 6) usually is made up of two or more pieces of 2″ dimension lumber nailed together, the ends of the pieces joining over a supporting post. A two-piece girder may be face nailed from one side with 10d nails, two nails at the end of each piece and others driven near top and bottom in two rows 16″ o.c., staggering the nails. A three-piece girder is face nailed from each side with 20d nails, two nails near each end of each piece and others in two rows driven 32″ o.c. staggered.

Ends of wood girders should bear at least 3-1/2″ on foundation walls or pilasters. When the lumber is not preservative treated a 1/2″ air space should be provided at each end and at each side of wood girders framing into the wall (Fig. 6). In termite-infested areas, these pockets should be lined with sheet metal, extended on the inside about 3-1/2″ to form a termite shield. The girder should be installed true and level. Wood shims should not be used to level girders or beams. Where adjustment is required, slate, metal or other durable noncompressible material should be used to support the girder ends. The shim space should be grouted solid under the beam or girder.

GIRDER-JOIST CONNECTION

The simplest method of floor framing is one where the joists bear directly on the wood girder or steel beam, in which case the top of the beam coincides with the top of the anchored sill plate. This method is used when adequate headroom is provided below the girder and the dropped girder is not objectionable. It is simple to frame and has the advantage of providing ample joist cavities for ductwork, electrical and mechanical equipment.

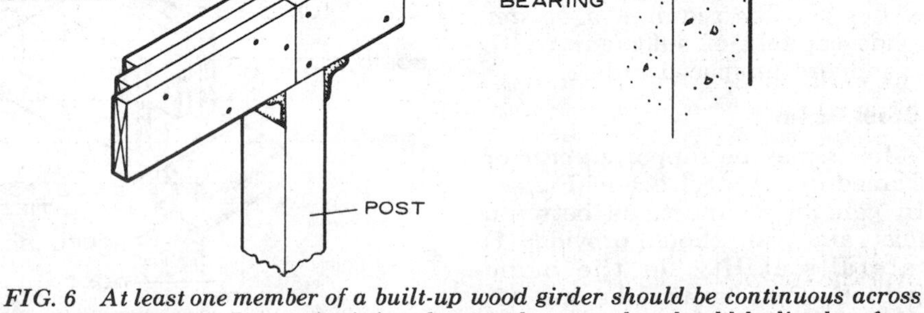

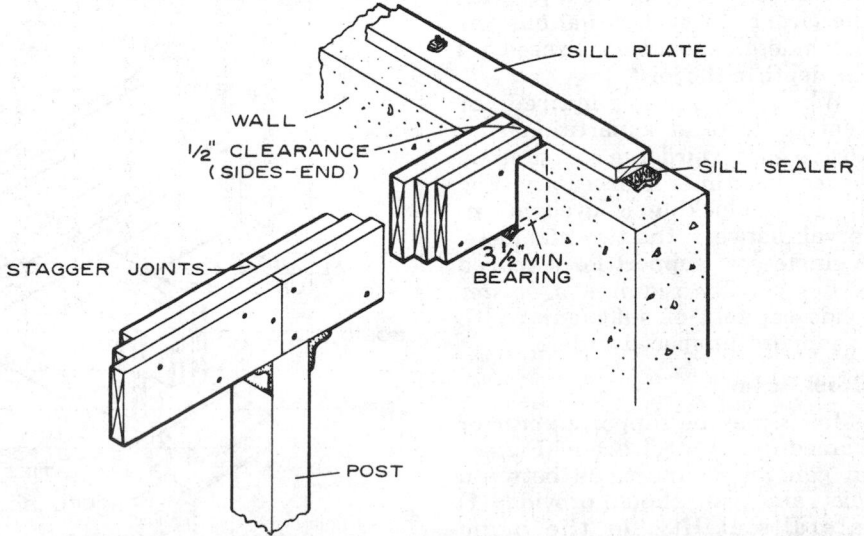

FIG. 6 At least one member of a built-up wood girder should be continuous across the supporting post. In termite infested areas, beam pocket should be lined to form a termite shield.

Wood Girder or Beam

For more uniform shrinkage and to provide greater headroom, joists can be framed into the girder with joist hangers or *ledger strips* (Fig. 7). The ledger strip should be at least 2" x 2" in size and should be face nailed to the girder with three 16d nails at each point where a joist frames into the girder.

Depending on sizes of joists and girders, joists may be supported on the ledger strip in several ways. When individual boards rather than plywood is used for subflooring, continuity across the supporting beam or girder must be provided by elements other than the subfloor. A continuous horizontal tie between exterior walls is obtained by nailing notched joists together (Fig. 7a). Instead of notching the joists, a connecting "scab" at each pair of joists can provide this continuity and also serves as a nailing area for the subfloor (Fig. 7b). A steel strap is used to tie the joists together when the tops of the beam and the joists are level (Fig. 7c). When notched joists or scabs ride over the top of the girder, it is important that a small space be allowed above the beam to provide for shrinkage of the joists. Sometimes the ledger strip is positioned higher on the girder and the joists are notched so that the joist shoulders bear on the ledger strip and the joist bottoms are level with the girder. In such situations the notch depth should not exceed 1/4 the depth of the joist.

When a space is required for heating ducts in a partition supported on the girder, a *spaced wood girder* sometimes is necessary (Fig. 8). Solid blocking is used at intervals between the two members. A single post support for a spaced girder usually requires a *bolster* (wide capital), of sufficient width to support the spaced girders.

Steel Beam

Joists may be supported *over* or framed *into* a steel beam (Fig. 9). In general, connections between joists and beam should provide: (1) lateral stability for the beam (which often is an I-beam), (2) resistance against twisting and

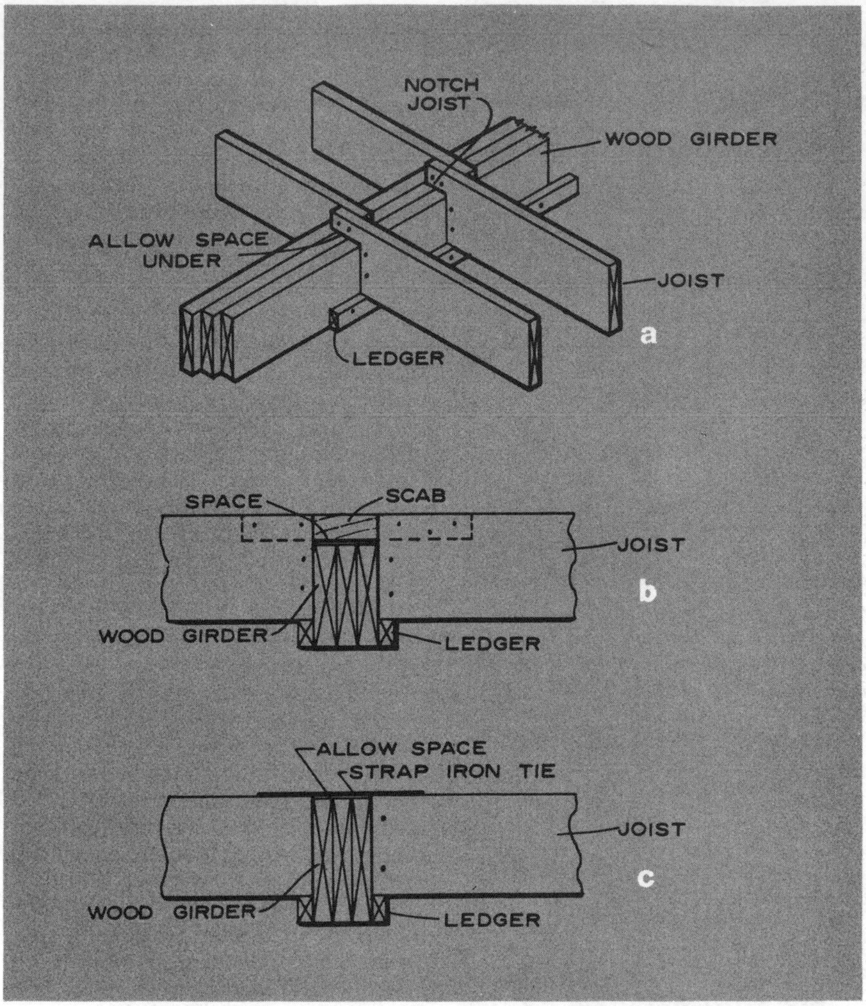

FIG. 7 Typical joist-girder connection showing how joists can be tied together, when wood board subfloor is used.

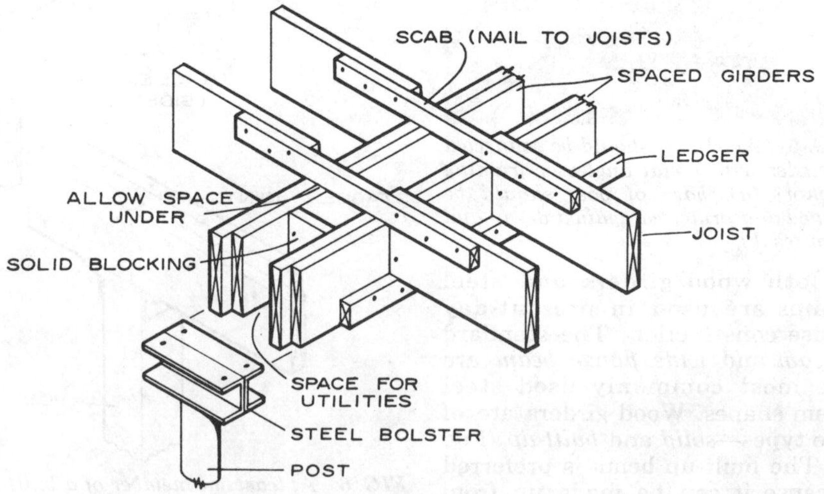

FIG. 8 Spaced girder detail showing connections to steel post and floor joists.

overturning for the joists, (3) structural continuity and diaphragm action to resist lateral pressures against foundation walls, and (4) anchorage for the entire floor assembly. This is accomplished generally by bolting a 2" x 4" or 2" x 6" plate to the upper or lower flange of the beam, on whichever the joists are supported (Fig. 9a). Each joist should be toenailed to the wood plate with two 10d or three 8d nails and the plate should be bolted 24" o.c., either through the beam web or the flange.

When it is not feasible to use a nailing plate, special clips can be nailed to the sides of the joists to develop the kind of connection described above. In Figure 9b, where a nailing plate was omitted, a metal strap tie provides continuity at each joist and anchors the floor assembly whereas the solid 2" x 4" blocking fitted between the joists keeps the joists from twisting and overturning. Since the strap ties only prevent the joists from pulling apart, the joists should be installed so the ends are in contact with the beam web to provide compressive resistance as well.

SILL PLATE CONSTRUCTION

The sill assembly over the foundation wall conforms either to platform or balloon framing methods. In all cases, the sill plate should be securely anchored to the wall and should be separated from the finish grade by at least 8". In the Gulf States and Florida, the sill plate, and all wood members within 18" of the grade, should be preservative treated against decay and termites. In areas of high humidity but where termite infestation is not a hazard, inherently decay-resistant grades of redwood, cypress or red cedar can be used. Additional information on sill plate anchorage and decay/termite protection can be found in Sections 310, 311, 312 and 313.

Platform Framing

The *box sill* assembly commonly is used in platform construction. It consists generally of a 2" *band* joist and a 2" x 4" or 2" x 6" sill plate as shown in Figure 10. The plate

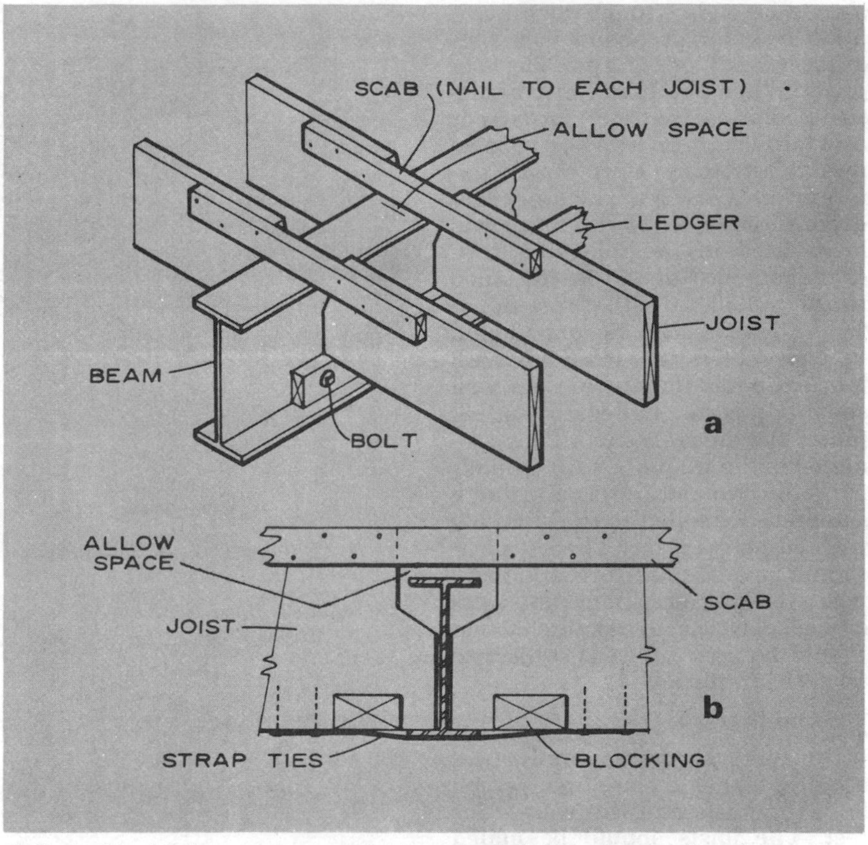

FIG. 9 Wood joists can be supported on wood ledger (a), or directly on lower flange of steel beam (b).

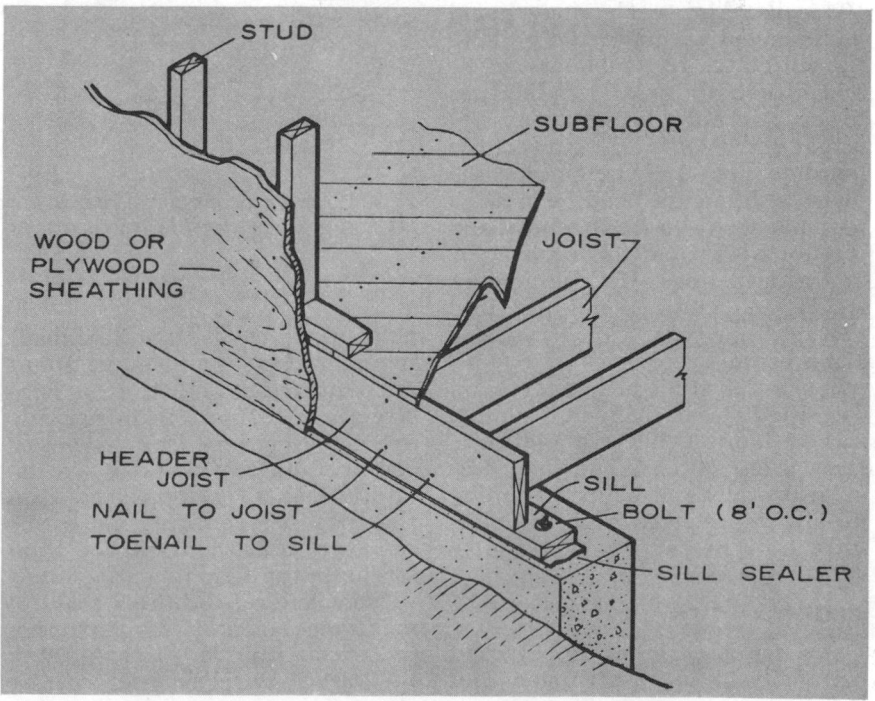

FIG. 10 Platform-framed box sill assembly; studs rest on sole plate.

should be wide enough to provide a full base for the band joist plus at least 1-1/2″ bearing for the floor joists. When the band joist is perpendicular to the floor joists it is also termed a *header*; when parallel it is designated a *stringer*. To differentiate between the two conditions, the latter terminology will be utilized in the following text. Each floor joist should be toenailed to the sill plate with three 8d or two 10d nails; the header should be end nailed to each joist with two 16d nails. If suitable rigid wall sheathing is not nailed to the sill plate, the header should be toenailed to the sill plate with 8d nails 16″ o.c. The stringer similarly should be toenailed to the sill plate with 8d nails 16″ o.c. The sill plate should be anchored to the foundation wall over a resilient sill sealer or bed of grout. A termite shield should be used in areas subject to termite infestation.

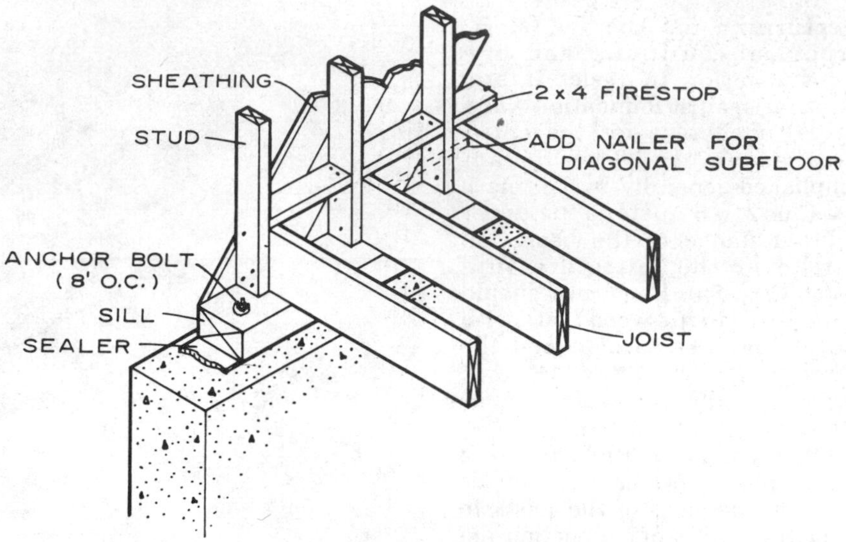

FIG. 11 Balloon-framed firestopped sill assembly; studs rest on sill.

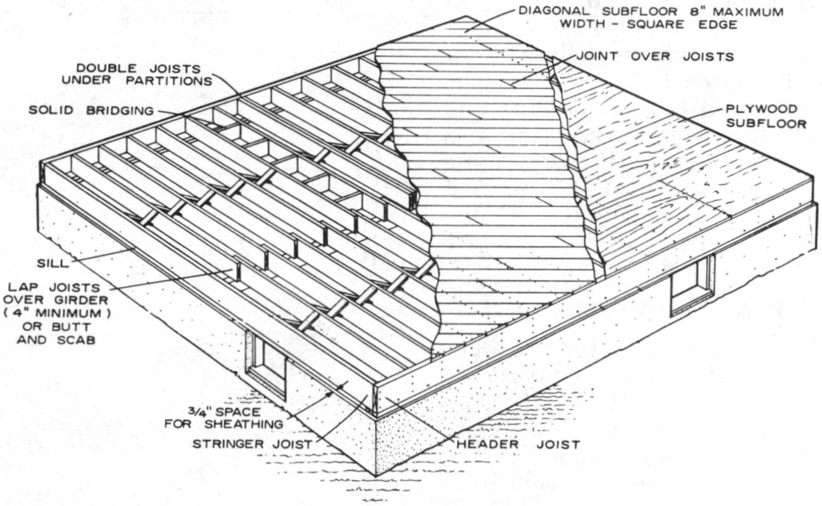

FIG. 12 Typical platform-framed floor construction, showing alternate subfloor types: plywood or wood boards. Cross-bridging is no longer necessary in residential floors.

Balloon Framing

Balloon framing generally uses a 2″ or thicker sill plate upon which both the joists and the wall studs rest. The joists should be nailed to the sill plate as in platform construction but in addition should be face nailed to each stud with two 10d nails. Also, each stud should be toenailed with two 10d nails to the sill plate. The subfloor can be laid diagonally or at right angles to the joists and a *firestop* generally is required between the studs at the floorline (Fig. 11). When a diagonal board subfloor is used, a nailing member such as a 2″ x 4″ normally is required between joists and studs at the wall lines. If this member is wide enough to extend fully into the stud cavity, it can serve also as a firestop.

Since wall studs bear directly on the sill plate in balloon framing, rather than on the joist platform, there is less potential shrinkage in exterior walls and balloon framing usually is preferred over the platform type in two-story masonry veneer houses.

FLOOR JOISTS

Floor joist lumber is selected primarily to meet strength and stiffness requirements, which depend upon the dead and live loads to be carried. For average spans encountered in residential construction, stiffness requirements are more critical than strength requirements. In other words, typical joist sizes are selected on the basis of their stiffness (resistance to deflection) rather than strength (ability to carry loads without failure). Stiffness requirements place an arbitrary maximum on deflection under loads: 1/360 of the joist span, in inches. For instance, the maximum allowable deflection for a 15′ joist span would be 180″/360, or 1/2″.

Other desirable qualities for joist lumber are good nail holding ability and freedom from warp. Wood floor joists are generally 2″ thick and 6″ to 12″ deep. The size depends upon the loading, length of span, joist spacing, and the species and grade of lumber used. Suitable floor joists can be selected by referring to tables for Allowable Unit Stress and Maximum Allowable Spans for Floor Joists in 201 Wood Work File. Joist grades should be selected for the specific design conditions. Where several combinations of joist size and spacing satisfy the given requirements, a deeper joist at greater spacing generally is preferred, provided of course the subfloor and/or ceiling finish is selected to span the greater spacing. The deeper joists usually require less lumber, less labor and result in a stiffer floor assembly.

Joist Installation

After the sill plates have been anchored to the foundation walls, the joists are laid out over the sill plates and center supports. The floor joists are spaced uniformly at 12″, 16″ or 24″ o.c. as determined by design requirements. Extra joists are added where required to support bearing partitions (Fig. 12) or to frame floor openings. Joists are braced temporarily with wood strips nailed across the top, until the subfloor is applied.

Joists with a slight crook should be placed so that the crown is on top; after the subfloor and normal floor loads are applied, the joist will tend to straighten out. The largest edge knots also should be placed on top, since the upper side of the joist is in compression and the knots will have a lesser effect on strength and deflection.

In platform construction the header joist should be fastened by nailing into the end of each joist with two 16d nails. In addition, the header joists and the stringer joists (parallel to the exterior walls) should be toenailed to the sill with 8d nails spaced 16″ o.c. (Fig. 12), unless rigid wall sheathing covers the sill plate and is nailed to it with 8d nails 8″ o.c.

Each floor joist should be toenailed to the sill and center beam or girder with two 10d or three 8d nails; joists should also be nailed to each other with at least three 16d nails when they lap over the center beam. The lap should be not less than 4″, and the overhang of either joist beyond the edge of the girder or beam should not exceed 12″. The limit on overhangs is imposed in order to reduce the possibility of lifting the subfloor resting on the overhang when a heavy concentrated load is placed on the

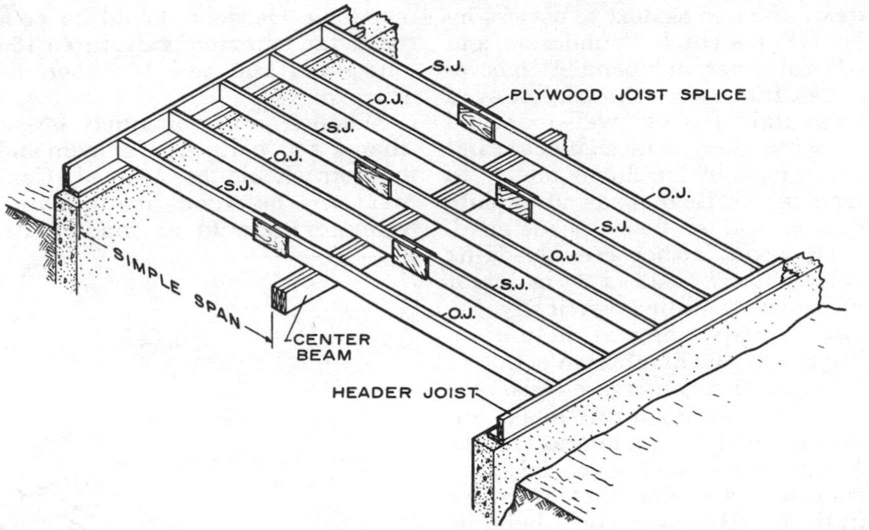

FIG. 13 In-line joist system framed with pairs of overhanging joists (o.j.) and supported joists (s.j.).

joist near its midspan. If a nominal 2″ scab is used across butt-ended joists, it should be nailed to each joist with at least three 16d nails at each side of the joint. These and other recommended nailing practices are outlined in Figure 2.

For greater lumber economy the *in-line* joist system sometimes is used in floor and roof framing when center supports are present. This system normally allows the use of joists one size smaller than in conventional floor framing. Briefly, the system consists of uneven-length joists with the long, *overhanging* joist cantilevered over the center support and connected to the *supported* joist with a metal connector or plywood splice plate (Fig. 13). In typical single-family construction, the overhang varies between about 1′10″ and 2′10″. For Douglas fir and southern pine No. 1 lumber, *total* spans (between foundation walls) of up to 22′ are possible with 2″ x 6″s; spans up to 30′ with 2″ x 8″s; and spans up to 34′ with 2″ x 10″ joists. Detailed information on this type of joist design can be obtained from a publication of the American Plywood Association titled "Concept No. 108—Design Systems for In-Line Joist Splice." Other innovative floor systems which depend on special attachment of a plywood subfloor are discussed on page 321-14.

Floor joists under all parallel bearing partitions should be doubled; if access into the partition is required for air ducts, the double joists should be spaced and solid blocking should be used between them as in spaced wood girder construction (Figs. 8 and 14).

BRIDGING

Cross-bridging consists of metal struts or wood members 1″ x 3″ or larger installed at midspan in intervals not over 8′, when the depth of joists is equal to or greater than 6 times the joist thickness (Fig. 12). Cross-bridging between joists traditionally has been used in house construction for many years, but recently has been

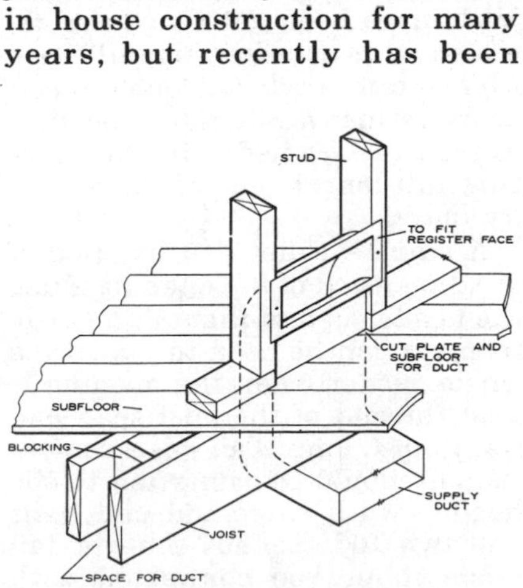

FIG. 14 Spaced joist construction under parallel bearing partition.

demonstrated as next to useless by NAHB Research Foundation and several other independent laboratories. It has been shown that even with tight-fitting, well-installed bridging, there is no significant improvement in the floor's ability to transfer vertical loads after subfloor and finish floor are installed.

However, either cross-bridging or *solid blocking* (2″ x 4″ or larger members installed vertically between joists) may be helpful in improving the diaphragm action of a floor, and may increase resistance to lateral (horizontal) loads imparted to it by foundation walls. It may be required in special situations where engineering design so indicates. However, some building codes still require cross-bridging, or solid blocking in all cases.

Solid blocking between joists should be used to provide a more rigid base for nonbearing partitions which happen to fall between joists (Fig. 12). Solid blocking should be well fitted and securely nailed to the joists. Load-bearing partitions running parallel to the joist direction always should be supported by doubled joists, either by placing the partition over a normally spaced joist which has been doubled, or by providing an additional pair of joists under the partition.

FLOOR OPENINGS

It is frequently necessary to provide openings in floors for large ducts, chimneys and stairs. Where joists have been cut away, the remaining *tail joists* usually are framed into single or double cross-joists, termed *headers*. The headers in turn are framed into the abutting full-length joists, known as *trimmers*.

A single header can be used if the dimension of the opening along the header is less than 4′; a single trimmer can be used to support a single header when the opening is near the end of the joist span and less than 4′ long. Tail joists shorter than 6′ should be connected to the headers with three 16d end nails and two 10d toenails. Longer tail joists should be connected with framing anchors or supported on a ledger strip as in joist-girder connections. Headers should be connected to trimmers with three 16d end nails and two 10d toenails (Fig. 15).

When framing openings larger than 4′ x 4′, both the trimmers and headers should be doubled. Connections between headers and trimmers should be made with framing anchors, joist hangers or ledger strips. Tail joists over 6′ long should be similarly attached to headers; shorter tail joists can be attached as recommended above for smaller openings. Nailing sequence for double headers is illustrated in Figure 16; the opening first is framed out using single

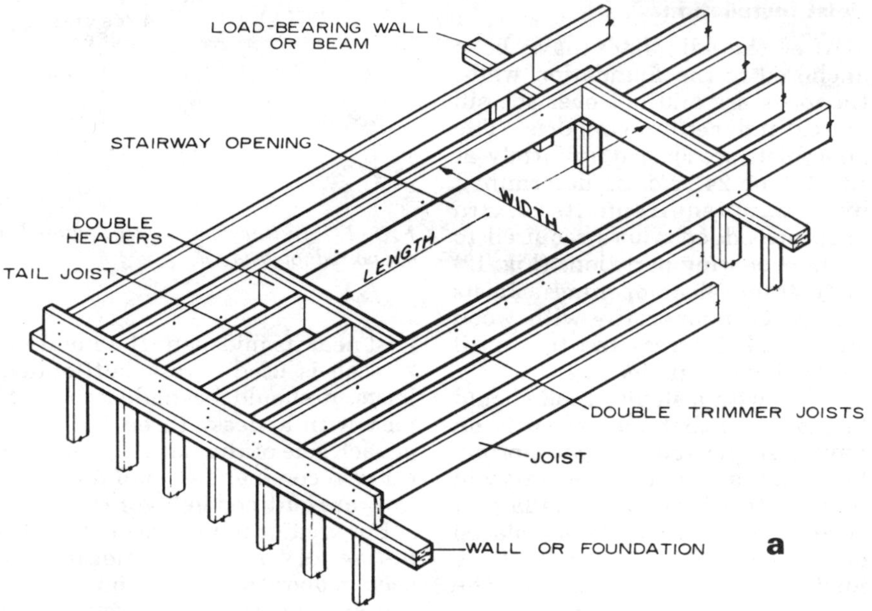

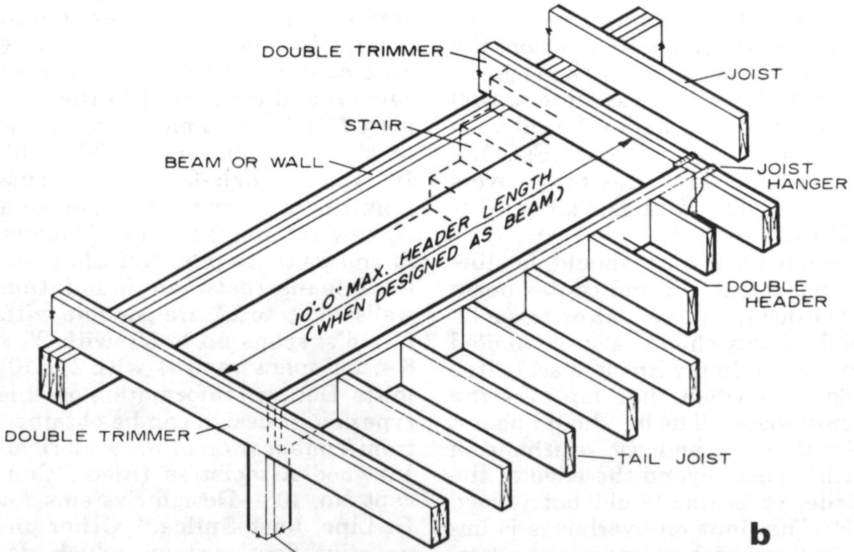

FIG. 15 Framing for stairs: (a) length of opening parallel to joists; (b) length of opening perpendicular to joists.

headers and trimmers; then the second joists are applied and face nailed to the first.

The largest floor openings are usually for stairs; for straight runs and typical floor heights, they are usually about 10′ long and 3′ to 4′ wide. The long dimension of stairway openings may be either parallel or at right angles to the joists. However, it is much easier to frame a stairway opening when its length is parallel to the joists (Fig. 15a).

When the length of the stair opening is perpendicular to the length of the joists, a long doubled header is required (Fig. 15b). A header under these conditions without a supporting wall beneath usually is limited to a 10′ length and should be designed as a beam. A load-bearing wall under all or part of this opening substantially simplifies the framing, as the joists then bear on the top plate of the wall rather than on the header.

FLOOR PROJECTIONS

The framing for floor projections beyond the lower wall such as for a bay window or balcony generally consists of extended floor joists, if they run perpendicular to the supporting wall (Fig. 17a). This extension normally should not exceed 24″ unless specifically engineered for greater projections. The joists forming each side of the bay should be doubled; nailing procedure, in general, should conform to that for stair openings. The subflooring should be extended flush with the outer framing member.

In framing the roof, rafters often are supported on a header over the bay area, in the plane of the supporting wall below. The header thus supports the roof load and transmits it to the foundation wall below, so that the wall of the bay has less load to support.

Projections parallel to the direction of the floor joists should generally be limited to small areas and extensions of not more than 24″ (Fig. 17b). In this construction, the stringers should frame into doubled joists. Joist hangers or a ledger strip can be used to frame joists into the double header, as in a joist-girder connection.

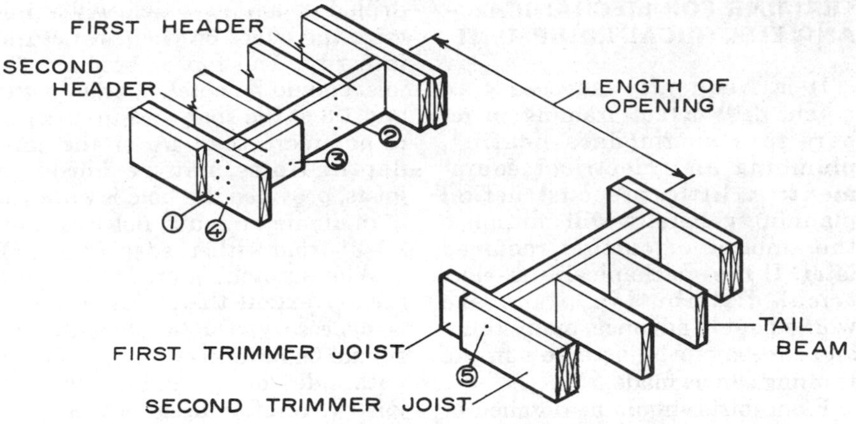

FIG. 16 Nailing sequence for floor openings: (1) nailing trimmer to first header; (2) nailing header to tail beams; (3) nailing header together; (4) nailing trimmer to second header; (5) nailing trimmers together.

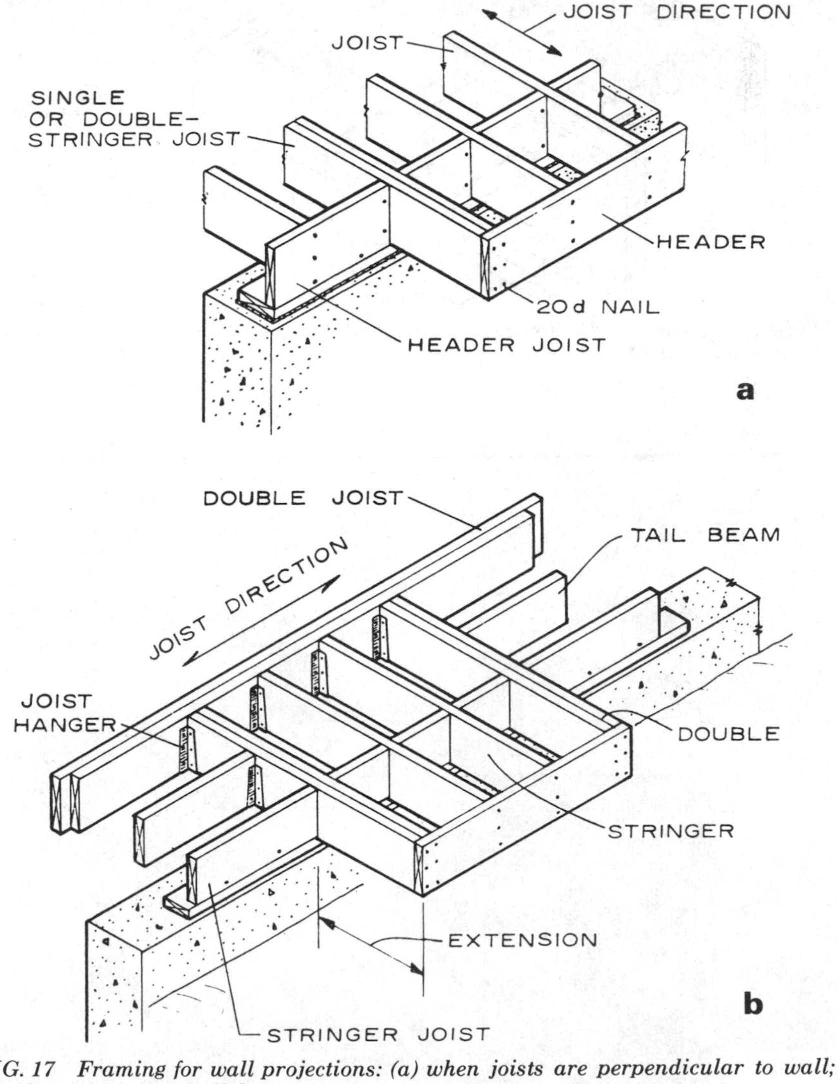

FIG. 17 Framing for wall projections: (a) when joists are perpendicular to wall; (b) when joists are parallel.

FRAMING FOR MECHANICAL AND ELECTRICAL EQUIPMENT

It is frequently necessary to notch, drill or cut framing members to accommodate heating, plumbing and electrical equipment. A little preconstruction planning generally will minimize the amount of cutting required later. If the mechanical and electrical distribution plans are worked out in advance, many time- and cost-saving adjustments in the framing can be made.

Floor joists should be notched or drilled in a way which will minimize the effect on their structural integrity. The top or bottom of a joist should be notched only in the end 1/3 of the span and to a depth of not more than 1/6 of the joist depth. Holes may be bored in joists, provided the hole is not over 2″ in diameter and is not less than 2-1/2″ from either edge (Fig. 18).

Where notching or drilling to a greater extent than allowed above is necessary, the weakened joist should be reinforced with scabs on both sides or by adding another joist. If reinforcing is not possible because of conflict with the mechanical equipment, the joist should be cut away and the area treated as a floor opening, with tail beams supported on headers.

When the plumbing wall adjacent to a bathroom also serves as a load-bearing partition, the supporting double-joists should be spaced and blocked (Fig. 19) as recommended for a spaced girder. The joist nearest the front apron of the tub similarly should be doubled to prevent excessive deflection under the weight of a tubfull of water.

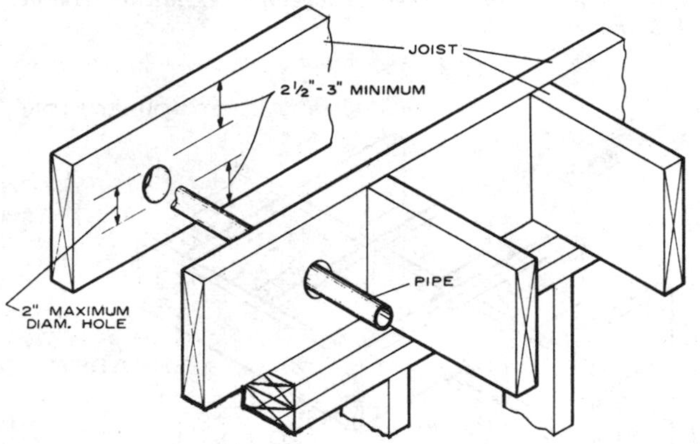

FIG. 18 Recommended limits for boring of holes in load-bearing floor joists.

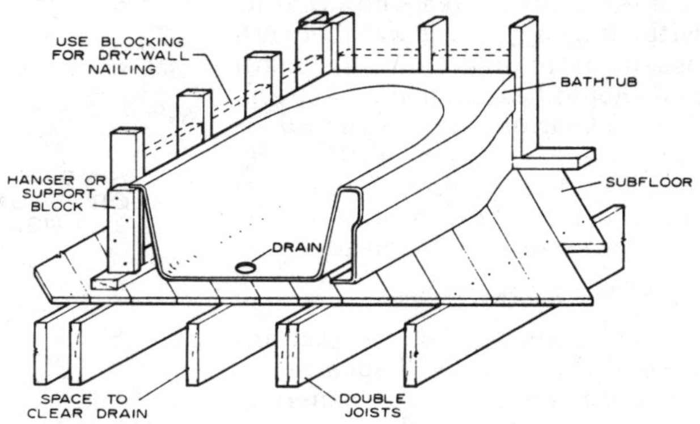

FIG. 19 Recommended framing to minimize floor deflection under tub.

FLOOR SHEATHING AND PLANKING

The structural material which spans between framing members in walls, floors or roofs is termed *sheathing.* Floor sheathing acts as the working platform and serves as a base for floor finishes, hence it is also referred as the *subfloor.* A conventional subfloor for joist framing up to 24″ o.c. is composed of nominal 1″ wood boards or 1/2″ to 3/4″ plywood. When heavier structural members such as beams are used for the framing, the support spacing can be substantially increased and heavier, 2″ thick, boards generally are utilized. When 2″ thick lumber is laid flat and is stressed in bending across its 2″ dimension it is referred to as a *plank*, and hence this type of subfloor also is called *planking.* The planking generally is tongue-and-grooved to facilitate the transfer of concentrated loads from one plank to another. In lumber grading, such T & G material is termed *decking.*

(It should be noted that the meaning of the terms "beam", "plank" and "decking" as used in lumber grading does not completely agree with that in general construction terminology. In light residential construction, 4″ thick framing members generally are referred to as "beams"; yet in lumber grading these are classed as joists and planks and the term "beams and stringers" is restricted to rectangular members 5″ and thicker. In lumber grading, "decking" means lumber 2″ to 4″ thick, 4″ and wider, with a maximum moisture content of 15% or 19% and with T & G edges; yet in common usage both installations with this special product and ordinary surfaced four-sides lumber (joists and planks) frequently are referred to as "planking," or as "decking" when used in a roof assembly, which is termed "roof deck." For instance the traditional "plank-and-beam" framing system utilizes planking which is really "decking" and beams which are really "joists", according to lumber grading terminology.)

The following discussion describes—in addition to conventional floor sheathing and planking—several *floor assemblies* which depend on a specialized type of sheathing product (2-4-1 plywood) or unconventional method of sheathing installation (gluing).

CONVENTIONAL SUBFLOOR

A conventional subfloor usually consists of either (a) nominal 1″ thick wood boards 4″ to 8″ wide or (b) plywood 1/2″ to 3/4″ thick, depending on species, type of finish floor and spacing of joists (Fig. 12). Recommended subfloor construction for specific floor finishes are given in Sections 442 Ceramic Floors and Walls, 452 Resilient Flooring and 453 Terrazzo. Subfloor construction for carpeting—with or without underlayment—is the same as for resilient flooring.

Board Subfloor

Boards may be applied either *diagonally* or at *right angles* to the joists. When the boards are placed at right angles to the joists, wood strip or plank flooring must be laid at right angles to the boards. Diagonal board installation is preferred because it permits a finish wood flooring to be laid either parallel or at right angles to the joists. Boards may have square, shiplap or T & G edges. Board ends should always rest over the joists (Fig. 12) unless the board ends are tongue-and-grooved. Subfloor boards should be nailed to each joist with two 8d nails for widths under 8″ and three 8d nails for 8″ widths. Where desired, 7d annularly threaded nails can be substituted for the 8d smooth shank nails (Fig. 20).

Generally, the joist spacing should not exceed 16″ o.c. when 25/32″ or thinner strip and plank flooring is laid parallel to the joists, or where parquet flooring is used; nor exceed 24″ o.c. when 25/32″ finish flooring is installed at right angles to the joists. For other types of finish flooring, a suitable underlayment should be installed over the subfloor, as recommended in the appropriate sections of Division 400 Applications & Finishes.

In balloon framing, blocking should be installed between ends of joists at the wall for nailing the ends of diagonal subfloor boards (Fig. 11).

FIG. 20 **RECOMMENDED NAILING OF WOOD SUBFLOORS**

Subfloor	Nailing Method, Number of Nails and Types	Nail Spacing
½″ plywood	Face nail 2-6d	6″ o.c. at panel edges and 10″ o.c. at intermediate supports
⅝″ plywood and up to 1¼″ thick	Face nail 2-8d	
1″ x 6″ boards and smaller	Face nail 2-8d	Per board at each joist
1″ x 8″ boards	Face nail 3-8d	
2″ x 6″ planks and smaller	Face nail 1-16d and toenail 1-10d	Per plank at each beam
2″ x 8″ planks	Face nail 2-16d and toenail 1-10d	
2″ x 10″ and 2″ x 12″ planks	Face nail 3-16d and toenail 1-10d	

Plywood

Plywood can be obtained in a number of grades designed to meet a broad range of end-use requirements. Under normal conditions, Standard Interior, Structural I or Structural II Interior grades are satisfactory. All Interior grades also are available with fully waterproof adhesives identical with those used in Exterior plywood. This type is useful where a hazard of prolonged exposure to moisture exists, such as in bathrooms, in floors adjacent to plumbing fixtures or where the subfloor may be exposed to the weather for long periods during construction. CC Exterior grade also is made with waterproof glue, and can also be used when unusual moisture conditions are anticipated.

All plywood suitable for subfloors has a panel identification index marking on each sheet. These markings indicate the allowable spacing of rafters and floor joists for the various thicknesses when the plywood is used as roof sheathing or subfloor. For example, an index mark of 32/16 indicates that the plywood panel is suitable for a maximum spacing of 32″ for rafters and 16″ for floor joists. Thus, the suitability of each piece of plywood for a particular job can be determined visually regardless of species.

Generally, when underlayment is intended over the subfloor, the minimum subfloor thickness for species such as Douglas fir and southern pine is 1/2″ when joists are spaced 16″ o.c., and 5/8″ thick for species such as western white pine and ponderosa pine. The same plywood thicknesses can be used also for 24″ joist spacing, when 25/32″ strip or plank flooring is installed at right angles to the joists. However, the somewhat limp feeling underfoot during construction frequently prompts selection of the next plywood thickness when joist spacing is greater than 16″ o.c.

Plywood also can serve as combined subfloor-underlayment, eliminating the need for separate underlayment under resilient floorings, carpeting and other nonstructural floor finishes. Plywood used in this manner must be tongue-and-grooved, or blocked with 2″ lumber along the unsupported edges between joists. It must also be protected during construction or completely resanded prior to flooring installation.

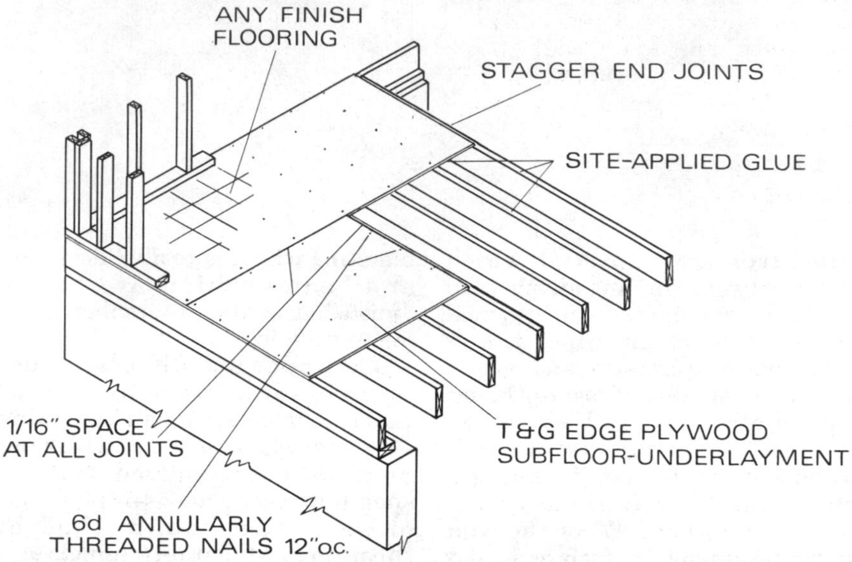

FIG. 21 The glued subfloor-underlayment system derives its strength from composite T-beam action of joists and plywood. Hence glueline between these elements must be strong and durable.

Plywood panels must be applied with face grain or long dimension perpendicular to joists if index numbers are to apply (Fig. 20); panels should be staggered so that end joints in adjacent panels break over different joists. Panels should be nailed as shown in Figure 20, spacing all panels approximately 1/32″ at end and edge joints. At intersections with vertical surfaces, at least 1/8″ space should be provided.

GLUED SUBFLOOR-UNDERLAYMENT

The glued floor system is based on newly developed elastomeric adhesives which firmly bond the plywood to the wood joists. Floor and joists are fused into an integral T-beam unit and floor stiffness is increased up to 70% when compared with conventional floor construction. Field-gluing virtually eliminates the squeaks which can result from shrinking lumber (Fig. 21).

Combining the functions of subflooring and underlayment into a single-layer floor results in material and labor economies. By gluing the floor to the joists, joists can span longer distances or, for some spans, smaller joists can be used. Field-glued floors have better creep resistance than nailed-only floors and can be installed quickly using materials and techniques readily available at the construction site.

Plywood should be Underlayment grade, 1/2″, 5/8″ or 3/4″ thick, depending on joist spacing. Although T & G plywood is used most often, square edge panels may be used if solid blocking is placed under the panel edges between joists and the plywood is also glued to the blocking. The basic joist framing is the same as for a conventional subfloor.

Plywood panels should be placed with the face grain or long dimension perpendicular to the joists, with end joints staggered and spaced 1/16″ at all joints. Before each panel is set in place, a bead of glue is applied to the joists with a caulking gun; for extra stiffness, the edge groove also may be glued. The plywood panel then is secured with 6d annularly threaded nails, spaced 12″ o.c. at all supports.

More detailed information on the glued floor system, including clear span tables, application sequence, and list of recommended adhesives, is contained in the booklet, APA Glued Floor System, published by the American Plywood Association.

2-4-1 SUBFLOOR-UNDERLAYMENT

This floor system depends on a special plywood product generally 1-1/8″ thick (for group 1 and 2 species; 1/8″ thicker for groups 3 and 4) capable of spanning up to 48″ between supports. The framing generally consists of solid, built-up, or laminated wood girders spaced 48″ o.c. and supported either on a central beam, a wall or a series of piers 48″ o.c. (Fig. 22).

The typical box sill at the foundation wall can be used when the greater separation between finish grade and finish floor is not objectionable (Fig. 22a). A more intimate relationship between finish floor and exterior grade can be achieved when the girders are supported on posts (Fig. 22b) or pocketed in the wall (Fig. 22c).

The 2-4-1 plywood panels have a core construction similar to Underlayment grade, meaning that the face ply is solid (plugged) and the ply just below the face is C grade or better, to prevent puncturing of the face under concentrated loads. The face ply can be specified sanded or touch-sanded; the latter is specified when field sanding is intended at the end of construction, just prior to installing the floor finish.

The panels are installed generally as described for other plywood subfloors: face plies perpendicular to supports, end joints staggered, all joints spaced 1/16″. Panels should be nailed with 8d annularly threaded (or 9d smooth shank) nails 6″ o.c. at all panel edges. Blocking under panel edges is not required because 2-4-1 plywood edges are tongue-and-grooved to transfer bending stresses. The system has inherent construction economies and is suitable for any type of flooring finish.

PLANK SUBFLOOR

While conventional joist framing involves support spacing at a maximum of 24″ o.c. and a 2-4-1 floor system extends the spacing to 48″, plank-and-beam framing permits spacing of supports at 6′ o.c. or more. Besides the economy in materials and labor, T & G or splined plank subfloors provide an acceptable finish surface on the underside of the floor, which can be left exposed. The disadvantage of the system, however, is that it does not provide the usual cavities for mechanical and electrical distribution equipment, and in the absence of a separate ceiling finish, relatively poor sound isolation, particularly from impact noise. For these reasons, plank floors are most suitable over unfinished spaces or where impact sound isolation from above is not an important factor. Plank-and-beam construction generally is more adapted to roof assemblies in typical residential construction.

The plank subfloor generally requires underlayment for resilient flooring and adhesively applied ceramic tile or thinset terrazzo. However it can be used as a base for carpeting with suitable cushioning (padding), if the subfloor surface is reasonably level and

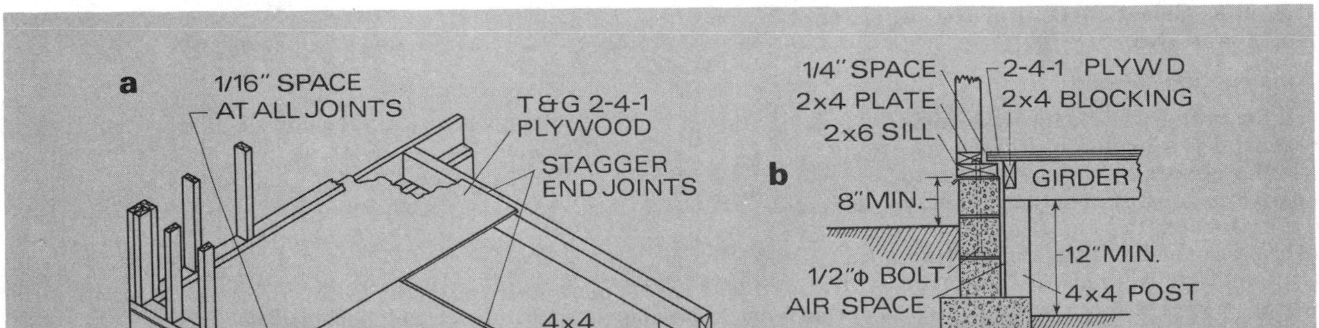

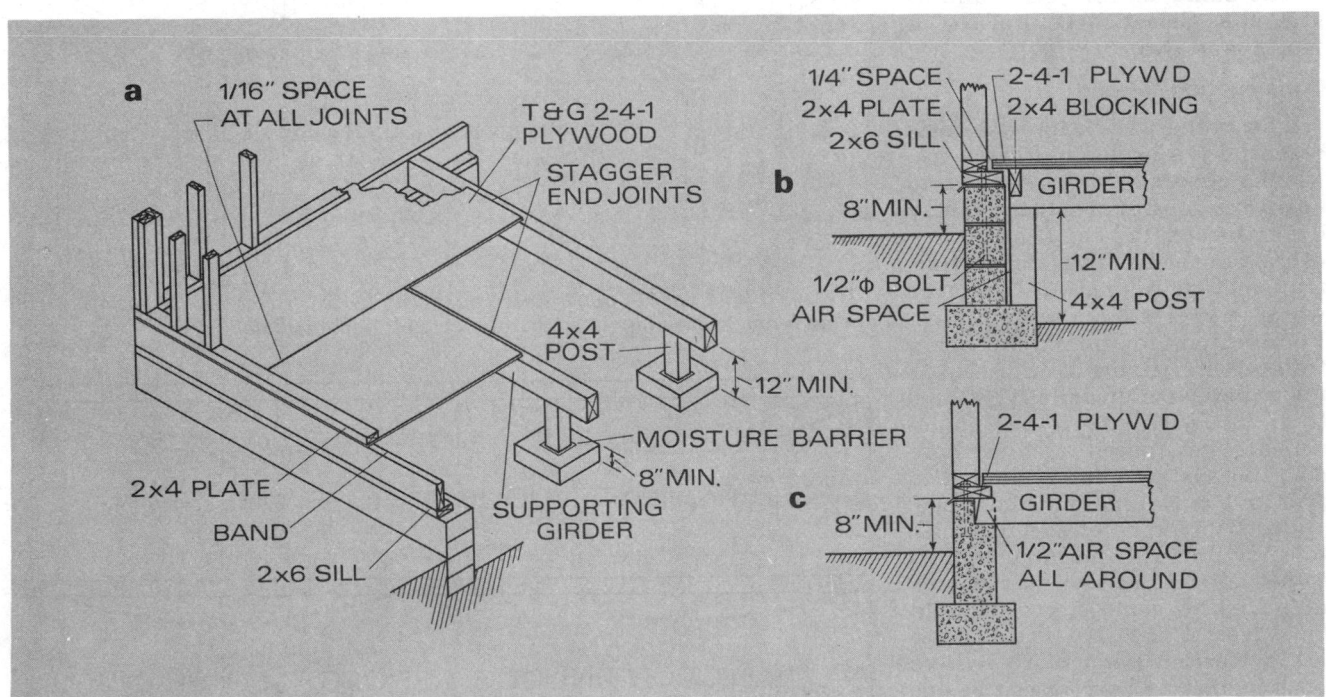

FIG. 22 Alternate framing methods for 2-4-1 subfloor-underlayment: (a) box sill on foundation wall; (b) wood post on footing; (c) pocket in foundation wall.

smooth. Heavy cushioning and deep pile carpeting, with or without underlayment, also considerably improves impact sound isolation and is recommended where sound transmission is a factor.

Plank subfloors can be built with tongue and grooved (T & G) or surfaced-four-sides (S4S) lumber. T & G lumber is graded in most species as Select Decking and Commercial Decking, both of which are suitable for subfloors. Decking is manufactured 2″ to 4″ thick and in widths ranging from 4″ to 12″. The maximum width for subfloors should not exceed 8″ at 19% moisture content; the wider boards should be used only at 15% moisture content. The thinner 2″ planks generally are made with a single tongue and groove; the 3″ and 4″ thick planks with two. Planks with grooves on both edges, for use with splines, also are available. One face usually is of better quality and is intended for exposed view, such as in a paneled ceiling. The face also may be worked to a special pattern with V-joints, rabbets or rounded edges.

S4S lumber suitable for planking is the same as for joists and is graded as Select Structural No. 1, No. 2 and No. 3.

Construction Details

The plank-and-beam system is essentially a skeleton framework. Planks generally are intended to support moderate, uniformly distributed loads. These are carried to the beams which in turn transmit their loads to posts or walls supported on foundations. Foundations for plank-and-beam framing are similar to those used for other wood framing systems (Fig. 23). With beams (and sometimes posts) spaced at 6′ or more o.c., this system is well adapted to pier and crawl space foundations.

Posts should be of adequate size to carry the load and large enough to provide full bearing for the ends of beams. In general, posts should be at least 4″ x 4″, but 4″ x 6″ posts are preferred, particularly where beams join or where spaced beams are used. Where the ends of beams abut over a post, the 6″ dimension

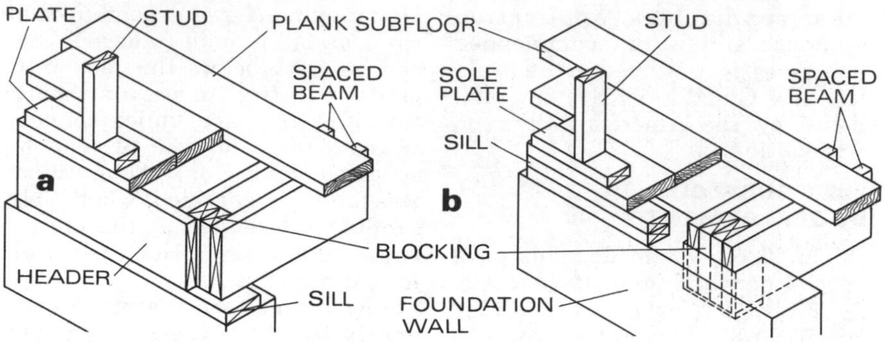

FIG. 23 Spaced beam (or girder) for plank floors can be (a) supported on foundation sill or (b) framed into wall pocket.

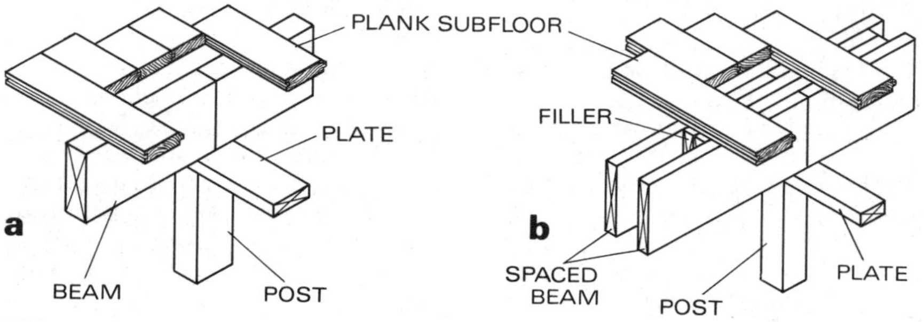

FIG. 24 Intermediate support for plank floors can be provided with (a) solid beam or (b) spaced beam over wood post.

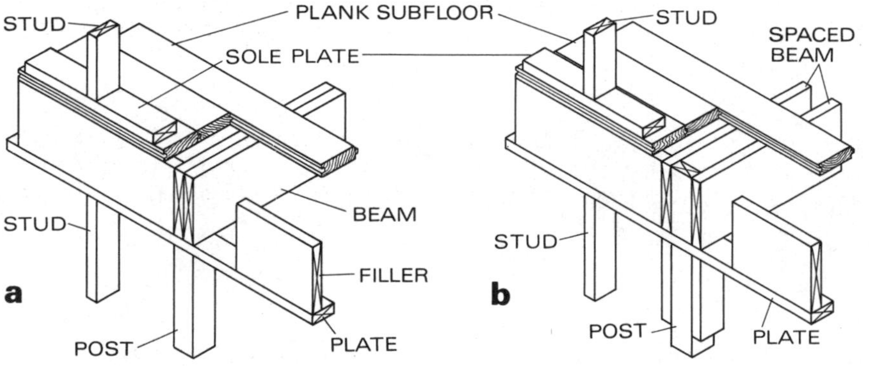

FIG. 25 Plank floor at upper floor can be supported by (a) built-up beam or (b) spaced beam, framed into exterior wall over post and top plate.

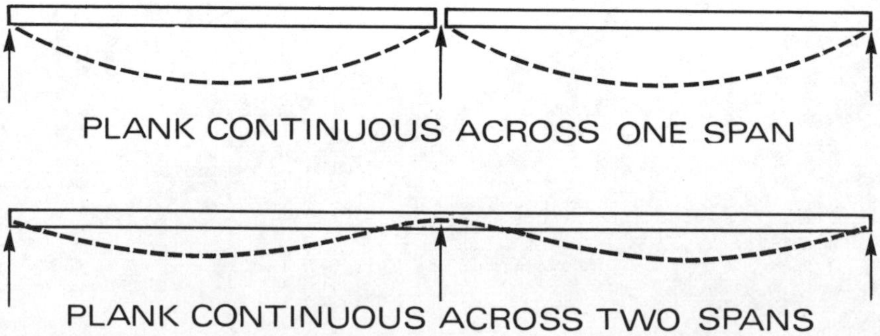

FIG. 26 Resultant deflection under similar loads is less when individual planks are long enough to reach across two spans, than in single span floors.

should be parallel to the beams (Fig. 24). Suitable beam sizes vary with grades, species, span, loading and spacing. Beams may be solid lumber, glue-laminated, or may be built up of several joists securely nailed to each other or to spacer blocks between them. The minimum required fiber stress in bending and modulus of elasticity are given in the Work File. Beams should be securely fastened to posts with framing anchors or clip angles. Details of upper floor beams, framing into an exterior frame wall, are illustrated in Figure 25.

In laying out the plank subfloor, greater advantage can be taken of the strength and stiffness of the material by making the planks continuous over more than one span. For example, using the same span and uniform load in each case, a plank which is continuous over two spans is nearly 2-1/2 times as stiff as a plank which extends over a single span (Fig. 26). Required values for fiber stress in bending and modulus of elasticity for different loading conditions of 2″ plank are given in the Work File. All planks, regardless of width, should be blind nailed to each support with one 10d nail; in addition, planks 4″ or 6″ wide should be face nailed also with one 16d nail at each support; 8″ planks should be face nailed with two 16d nails; 10″ and 12″ planks with three 16d nails.

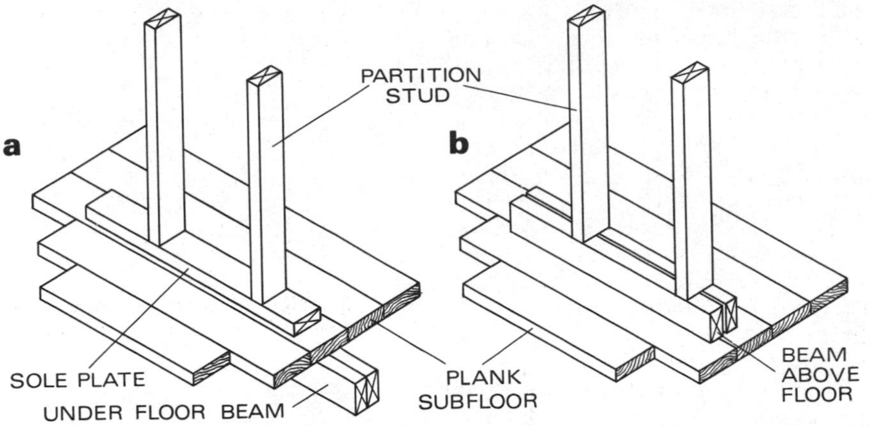

FIG. 27 Interior non-bearing partitions running parallel to the plank floor can be supported on (a) suitable underfloor beam or (b) double vertical sole plate.

A wood strip floor should be laid at right angles to the plank subfloor, using the same procedure followed in conventional construction. Where the underside of the plank is to serve as a ceiling, care is needed to make sure that flooring nails do not penetrate through the plank.

Interior partitions in the plank-and-beam system usually are non-bearing. Where bearing partitions occur, they should be placed over beams and the beams enlarged to carry the added load. If this is not possible, supplementary beams must be placed in the floor framing. Non-bearing partitions, which are parallel to the planks, should be supported on a sole plate consisting of double 2″ x 4″s set on edge and spiked together (Fig. 27b). Where door openings occur in the partition, the double 2″ x 4″s may be placed under the plank floor and framed into the beams with framing anchors (Fig. 27a). Non-bearing partitions located at right angles to the planks, do not require supplementary framing, since the partition load is distributed across a number of planks.

We gratefully acknowledge the assistance of the Forest Products Laboratory, USDA and the National Forest Products Association in the preparation of the above text and illustrations.

PARTITION
STUD

WOOD FRAMED FLOORS 321

CONTENTS

FLOOR FRAMING

LUMBER MATERIALS

Floor framing lumber generally should have a moisture content of 19% or less at the time interior finishes are installed; in arid parts of the country and where a gypsum wallboard ceiling will be secured directly to the floor joists, the framing lumber should have a moisture content of 15% or less.

Framing lumber 2″ or less in thickness generally is available in S-DRY (19% m.c. maximum) and MC-15 (15% m.c. maximum) grades. Thicker pieces may not be as readily available in the mill-stamped dry grades, but may achieve the desired moisture content by air drying under cover at the local yard or at the job site.

NAILING PRACTICE

Framing members should be connected to each other with the type of nails, number of nails and nailing method recommended in Figure WF1.

Proper nailing is essential to achieving stable, durable frame construction. Connection between framing members preferably should be made in a way which will place nails in shear rather than in direct withdrawal. Allowable shear loads for nails driven into side grain are 50% higher than for end grain.

Connections never should be made in a way which would load nails in direct withdrawal from end grain.

FIG. WF1 **NAILING OF WOOD FRAMED FLOORS**

Connection	Nailing Method, Number of Nails and Types
Members of built-up girder	Face nail 2-10d at each end; 2 rows of 10d, 32″ o.c. top and bottom, staggered
Header to floor joist	End nail 2-16d
Floor joist to sill plate or girder	Toenail 2-10d or 3-8d at each joist
Floor joists to each other over girder	Face nail 3-16d at lap over girder
Header and stringer joists to sill plate	Toenail 8d, 16″ o.c.[1]
Bridging to joists	Toenail 2-8d each end of bridging
2″ ledger strip to girder	Face nail 3-16d at each joist
Header to tail joists, less than 6′[2]	End nail 3-16d and toenail 2-10d
Trimmer to header, less than 6′[3]	End nail 3-16d and toenail 2-10d

[1] Nailing not required when sheathing laps over header or stringer and is nailed to sill plate with 8d nails 8″ o.c.
[2] Longer tail joists should be supported on framing anchors or ledger strip.
[3] Longer header (up to 10′) should be supported on joist hangers, posts or special connectors.

(Continued) **FLOOR FRAMING**

POSTS, BEAMS & GIRDERS

Wood posts should be raised on a concrete pedestal at least 3″ above basement floor and should be securely pinned to the pedestal. Posts should be installed plumb and true and securely connected to wood beam or girder. Built-up wood girders supporting floor loads should not exceed the maximum allowable spans shown in Figure WF2.

The tables indicate required lumber sizes for various conditions of loading and framing, assuming a minimum fiber stress in bending (Fb) of 1,000 psi and 1,500 psi. Species and grades meeting this requirement can be selected from Figure WF3 of 201 Wood Work File.

Solid wood beams of equivalent cross section may be substituted for built-up girders consisting of several 2″ pieces.

In termite infested areas, all wood members framing into masonry should be preservative pressure treated. When the lumber is not preservative treated, a 1/2″ air space always should be provided at ends and sides of members framing into masonry walls and the pocket should be lined with sheetmetal.

In termite infested areas, lining of the girder pocket with sheetmetal provides nominal assurance against termite attack. Use of pressure treated lumber is a more positive method of protection. Separation of the wood girder from masonry with 1/2″ air space is desirable in all areas.

FIG. WF2 **MAXIMUM ALLOWABLE SPANS FOR BUILT-UP GIRDERS SUPPORTING ONE OR TWO FLOORS**

MAXIMUM ALLOWABLE SPANS ("S") FOR GIRDERS SUPPORTING ONE FLOOR[1]

LUMBER		HOUSE WIDTHS ("L")[2]								
Fb, psi	Size	20′	22′	24′	26′	28′	30′	32′	34′	36′
1000	(2)2″ x 6″	4′-5″	4′-0″	—	—	—	—	—	—	—
	(3)2″ x 6″	5′-6″	5′-3″	5′-0″	4′-10″	4′-8″	4′-5″	4′-2″	—	—
	(2)2″ x 8″	5′-10″	5′-3″	4′-10″	4′-5″	4′-2″	—	—	—	—
	(3)2″ x 8″	7′-3″	6′-11″	6′-7″	6′-4″	6′-2″	5′-10″	5′-5″	5′-1″	4′-10″
	(2)2″ x 10″	7′-5″	6′-9″	6′-2″	5′-8″	5′-3″	4′-11″	4′-8″	4′-4″	4′-1″
	(3)2″ x 10″	9′-3″	8′-10″	8′-5″	8′-1″	7′-10″	7′-5″	6′-11″	6′-6″	6′-2″
	(2)2″ x 12″	9′-0″	8′-2″	7′-6″	6′-11″	6′-5″	6′-0″	5′-8″	5′-4″	5′-0″
	(3)2″ x 12″	11′-3″	10′-9″	10′-3″	9′-10″	9′-6″	9′-0″	8′-5″	7′-11″	7′-6″
1500	(2)2″ x 6″	5′-3″	4′-10″	4′-5″	4′-1″	—	—	—	—	—
	(3)2″ x 6″	6′-9″	6′-5″	6′-2″	5′-11″	5′-8″	5′-3″	4′-11″	4′-8″	4′-5″
	(2)2″ x 8″	7′-0″	6′-4″	5′-10″	5′-4″	5′-0″	4′-8″	4′-4″	4′-1″	—
	(3)2″ x 8″	8′-11″	8′-6″	8′-1″	7′-9″	7′-5″	7′-0″	6′-6″	6′-2″	5′-10″
	(2)2″ x 10″	8′-11″	8′-1″	7′-5″	6′-10″	6′-4″	5′-11″	5′-7″	5′-3″	4′-11″
	(3)2″ x 10″	11′-4″	10′-10″	10′-4″	9′-11″	9′-6″	8′-11″	8′-4″	7′-10″	7′-5″
	(2)2″ x 12″	10′-10″	9′-10″	9′-0″	8′-4″	7′-9″	7′-2″	6′-9″	6′-4″	6′-0″
	(3)2″ x 12″	13′-9″	13′-2″	12′-7″	12′-1″	11′-7″	10′-10″	10′-2″	9′-6″	9′-0″

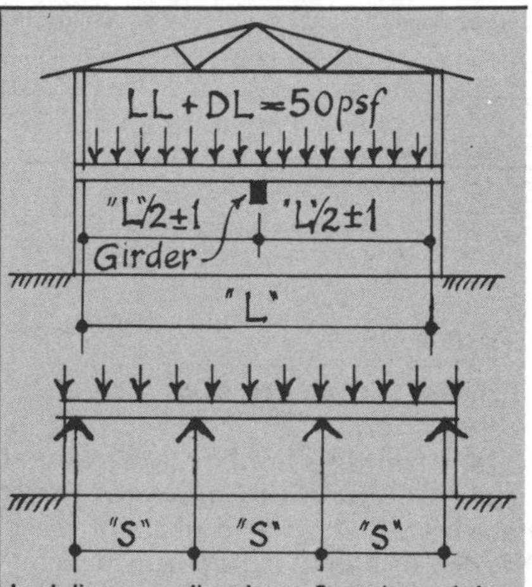

Load diagram applies also to 2 - st. house below

MAXIMUM ALLOWABLE SPANS ("S") FOR GIRDERS SUPPORTING TWO FLOORS[1]

LUMBER		HOUSE WIDTHS ("L")[2]								
Fb, psi	Size	20′	22′	24′	26′	28′	30′	32′	34′	36′
1000	(3)2″ x 8″	4′-7″	4′-2″	—	—	—	—	—	—	—
	(2)2″ x 12″	4′-9″	4′-4″	4′-0″	—	—	—	—	—	—
	(3)2″ x 10″	5′-10″	5′-4″	4′-11″	4′-7″	4′-3″	4′-0″	—	—	—
	(3)2″ x 12″	7′-1″	6′-6″	6′-0″	5′-6″	5′-2″	4′-10″	4′-6″	4′-3″	4′-0″
1500	(3)2″ x 6″	4′-2″	—	—	—	—	—	—	—	—
	(2)2″ x 10″	4′-8″	4′-3″	—	—	—	—	—	—	—
	(3)2″ x 8″	5′-6″	5′-0″	4′-7″	4′-3″	—	—	—	—	—
	(2)2″ x 12″	5′-8″	5′-2″	4′-9″	4′-5″	4′-1″	—	—	—	—
	(3)2″ x 10″	7′-0″	6′-5″	7′-6″	5′-6″	5′-1″	4′-9″	4′-6″	4′-3″	4′-0″
	(3)2″ x 12″	8′-6″	7′-9″	7′-2″	6′-8″	6′-2″	5′-9″	5′-5″	5′-1″	4′-10″

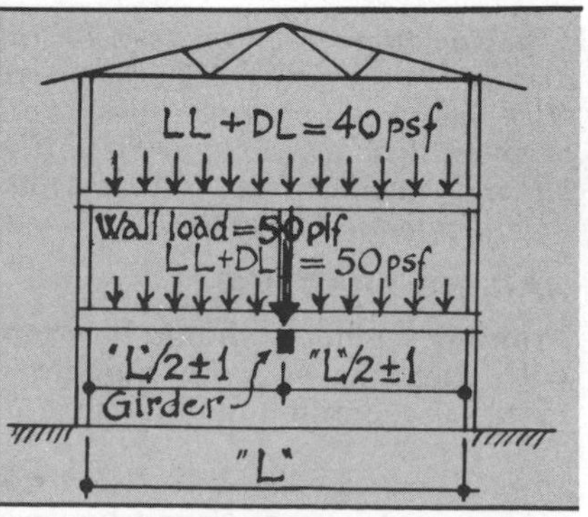

[1]Maximum spans are based on 30 psf live load for second floor and 40 psf live load for first floor; 10 psf dead load for either floor; 50 plf dead load for partition above girder.

[2]Spans given may be interpolated on a straight line basis for house widths intermediate between those shown.

(Continued) FLOOR FRAMING

Ends of wood girders should bear at least 3-1/2″ on foundation walls or pilasters. The girder should be installed level and true, using shims of non-compressible durable materials such as slate or steel. The shim space should be grouted solid.

JOIST-TO-BEAM CONNECTION

Joist-to-beam (or girder) connection should be made in a way which will provide: (1) lateral stability for the beam; (2) resistance against twisting for the joists; (3) structural continuity and diaphragm action through the floor assembly; (4) anchorage for the entire floor assembly.

Refer to Main Text for typical joist-to-girder connection details.

Wood joists should bear at least 3″ over rough surfaces such as concrete or masonry. Joists should bear at least 1-1/2″ over smooth surfaces such as metal or wood (steel beam, wood girder, foundation, sill plate).

Ledger strip at wood girder should be at least 2″x2″ nominal, securely nailed as recommended in Figure WF1. Joists supported on the ledger strip should not be notched more than 1/4 their depth at the ends. Joists framing over the top of a beam or girder should be set in-line or lapped a minimum of 4″. Overlap should be limited so that neither joist will extend beyond the edge of the beam or girder more than 12″.

SILL PLATE CONSTRUCTION

The foundation sill plate should be anchored to the foundation wall with 1/2″ anchor bolts 8′-0″ o.c. or with powder-activated fasteners 4′-0″ o.c.

For more detailed information on sill anchorage and lateral support refer to Section 310 Foundation Systems. It is important to transmit the uplift resistance developed through anchorage of the sill plate to other components of the superstructure.

The sheathing board should extend to cover the sill plate and should be nailed to it with 8d nails 8″ o.c. Where the sheathing cannot be nailed to the sill, the header and stringer joists should be toe-nailed to the sill with 8d nails 16″ o.c. and the sheathing should be nailed to these joists.

In areas subject to winds of 90 mph or greater, 18 ga. metal straps should be provided approximately 32″ o.c. One end should be turned under the sill plate and secured to it with one 8d nail, the other end extending to the studs and secured to it with three 8d nails.

FLOOR JOISTS

Floor joists should be installed with the crown up. Joists under load-bearing partitions and under heavy loads, such as heavy refrigerators and bathtub aprons, should be doubled.

BRIDGING

Bridging should be provided when the depth of joists is equal to or greater than 6 times the joist thickness (nominal dimensions). When required, bridging should be installed at 8′ intervals.

Bridging generally is not required in light residential construction, except when 2″ x 12″ or larger joists are used, or engineering calculations indicate its desirability.

FLOOR OPENINGS

Floor openings extending 4′ or more across the joists should be provided with double headers. Double trimmers should be provided to support double headers, or any opening 4′ x 4′ or larger. A double header built-up of the same size members as adjacent joists should not exceed a span of 10′.

Headers with spans over 10′ or headers supporting a load-bearing partition in addition to the adjacent floor construction should be designed as a beam, according to accepted engineering practice.

FLOOR PROJECTIONS

Floor joists cantilevered beyond the foundation wall and supporting only floor loads and a non-load bearing exterior wall should be limited to 2′.

Larger cantilevers and those intended to support load-bearing walls should be based on engineered designs.

NOTCHING & DRILLING FOR EQUIPMENT

Floor joists should be notched or drilled in a way which will minimize the effect on their structural integrity. The top and bottom of a joist should be notched only in the end 1/3 of the span and to a depth of not more than 1/6 of the joist depth. Bored holes may be located anywhere in the span, but should be not over 2″ in diameter and should be not closer than 2-1/2″ from either top or bottom edge.

Preconstruction planning of piping, wiring and ductwork will minimize the amount of notching and drilling required later.

Where notching or drilling to a greater extent than recommended above is required, the weakened joists should be reinforced with short pieces on both sides or by adding another joist. If reinforcing is not possible, because of conflict with required equipment, the joist should be cut away and the area treated as a floor opening with tail joists supported on headers.

FLOOR SHEATHING & PLANKING

CONVENTIONAL SUBFLOORS

Subfloors composed of lumber boards, plywood, or planking should be nailed to the floor framing as recommended in Figure WF3a.

Lumber boards for subfloor should have a nominal thickness of at least 1″ and a maximum nominal width of 8″. Board ends should be supported on the floor framing unless end-matched boards are used.

Boards may have square, shiplap, or T&G edges. Boards can be laid perpendicular to the joists or diagonally. Diagonal installation involves more materials and labor and is infrequently used.

Plywood should be Standard, Structural I or Structural II grade—Interior type, where no special moisture problems are anticipated. Where moisture presents a potential hazard, either Structural grade (or Standard) Interior type *with waterproof glue* or Exterior type should be used. Panels should be installed with grain direction perpendicular to joists and end-joists staggered. Adjacent panels should be spaced 1/32″; 1/8″ space should be allowed at intersection with vertical surfaces.

Plywood carrying the DFPA grademark shows the maximum permissible span for subfloor or roof deck use. Detailed information on plywood grades and maximum spans is provided in Wood 201 Work File.

Lumber for plank subfloors should have a nominal thickness of 2″; maximum nominal width should be 8″ when the moisture content is 19% or less; lumber wider than 8″ should have a moisture content of 15% or less.

PLANK SUBFLOORS

Plank floors can be built with T&G or splined lumber graded as Decking, or S4S lumber graded as Joists & Planks. Decking lumber generally is required by grading rules to be at 15% m.c. or less; Joists and Planks may be obtained S-DRY (19% m.c. maximum) or MC15, as desired.

Plank subfloors and supporting beams for plank-and-beam construction should be capable of supporting the expected design loads without exceeding the maximum allowable working stresses. Subfloors should be nailed as shown in Figure WF1.

Maximum allowable working stresses for most commercial grades and species are tabulated in the Supplement to the National Design Specifications published by the National Forest Products Association (NFPA). An abbreviated table for more common grades and species is included in Wood 201 Work File.

Wood beams supporting plank floors or roofs should not exceed the maximum allowable spans tabulated in Figure WF4. T&G 2″ plank decking should not exceed the maximum allowable spans shown in Figure WF5.

FIG. WF3a **NAILING OF WOOD SUBFLOORS**

Subfloor	Nailing Method, Number of Nails and Types	Nail Spacing
½″ plywood	Face nail 2-6d	6″ o.c. at panel edges and 10″ o.c. at intermediate supports
⅝″ plywood and up to 1¼″ thick	Face nail 2-8d	
1″ x 6″ boards and smaller	Face nail 2-8d	Per board at each joist
1″ x 8″ boards	Face nail 3-8d	
2″ x 6″ planks and smaller	Face nail 1-16d and toenail 1-10d	Per plank at each beam
2″ x 8″ planks	Face nail 2-16d and toenail 1-10d	
2″ x 10″ and 2″ x 12″ planks	Face nail 3-16d and toenail 1-10d	

FIG. WF3b **TABLE OF NAIL SIZES**

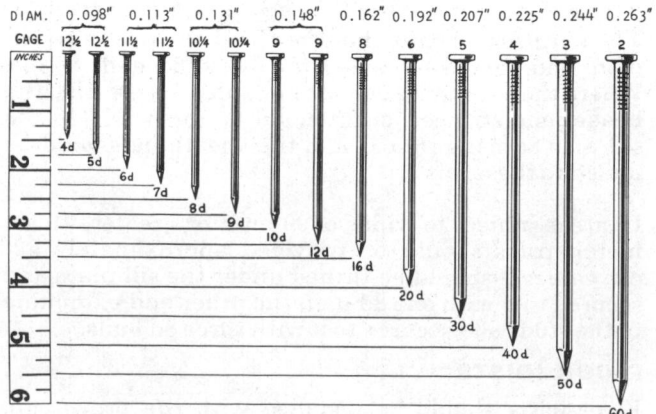

(Continued) FLOOR SHEATHING & PLANKING

FIG. WF4 **MAXIMUM ALLOWABLE SPANS FOR WOOD BEAMS SUPPORTING PLANK ROOFS & FLOORS***

LOADING CRITERIA: 20 psf live load—10 psf dead load **20 psf**

Beam Span	Beam Size	MINIMUM E (psi) AND MINIMUM Fb (psi)					
		Deflection l/240			Deflection l/360		
		Beams 6′ o.c.	Beams 7′ o.c.	Beams 8′ o.c.	Beams 6′ o.c.	Beams 7′ o.c.	Beams 8′ o.c.
10′	(2)2″ x 8″	570,000 1,030	665,000 1,200	760,000 1,370	855,000 1,030	997,000 1,200	1,140,000 1,370
	(1)4″ x 8″	485,000 880	566,000 1,030	646,000 1,175	727,000 880	847,000 1,030	969,000 1,175
	(3)2″ x 8″	380,000 685	443,000 800	506,000 915	570,000 685	667,000 800	759,000 915
11′	(2)2″ x 8″	754,000 1,245	880,000 1,450	1,005,000 1,660	1,131,000 1,245	1,320,000 1,450	1,507,000 1,660
	(1)4″ x 8″	647,000 1,065	755,000 1,245	862,000 1,420	970,000 1,065	1,132,000 1,245	1,293,000 1,420
	(3)2″ x 8″	503,000 830	587,000 970	670,000 1,105	754,000 830	880,000 970	1,005,000 1,105
12′	(2)2″ x 8″	980,000 1,480	1,144,000 1,725	1,306,000 1,970	1,470,000 1,480	1,716,000 1,725	1,959,000 1,970
	(1)4″ x 8″	840,000 1,270	980,000 1,480	1,120,000 1,690	1,260,000 1,270	1,470,000 1,480	1,680,000 1,690
	(3)2″ x 8″	653,000 985	762,000 1,150	870,000 1,315	979,000 985	1,143,000 1,150	1,305,000 1,315
	(2)2″ x 10″	472,000 910	551,000 1,060	629,000 1,210	708,000 910	826,000 1,060	943,000 1,210
13′	(2)2″ x 8″	1,245,000 1,740	1,453,000 2,025	1,660,000 2,315	1,867,000 1,740	— —	— —
	(1)4″ x 8″	1,067,000 1,490	1,245,000 1,735	1,422,000 1,985	1,600,000 1,490	1,867,000 1,735	— —
	(3)2″ x 8″	830,000 1,160	969,000 1,350	1,106,000 1,545	1,245,000 1,160	1,453,000 1,350	1,659,000 1,545
	(2)2″ x 10″	600,000 1,070	700,000 1,245	800,000 1,420	900,000 1,070	1,050,000 1,245	1,200,000 1,420
14′	(2)2″ x 8″	1,555,000 2,015	1,815,000 2,350	— —	— —	— —	— —
	(3)2″ x 8″	1,037,000 1,340	1,210,000 1,570	1,382,000 1,790	1,555,000 1,340	1,815,000 1,570	— —
	(2)2″ x 10″	749,000	874,000	998,000	1,123,000	1,311,000	1,497,000

*FLOOR beams should be selected from values given for 30 psf or 40 psf live loads, deflection limited to l/360; 30 psf live loads are suitable for sleeping rooms only. ROOF beams may be selected from any of the values given; where dead loads exceed 10 psf, select beams from table with next higher live load.

(Continued) **FLOOR SHEATHING & PLANKING**

FIG. WF4 **MAXIMUM ALLOWABLE SPANS FOR WOOD BEAMS SUPPORTING PLANK ROOFS & FLOORS*** *(Continued)*

LOADING CRITERIA: 30 psf live load—10 psf dead load **30 psf**

Beam Span	Beam Size	MINIMUM E (psi) AND MINIMUM Fb (psi)					
		Deflection Limit l/240			Deflection Limit l/360		
		Beams 6′ o.c.	Beams 7′ o.c.	Beams 8′ o.c.	Beams 6′ o.c.	Beams 7′ o.c.	Beams 8′ o.c.
10′	(1)3″ x 8″	1,020,000 1,645	1,190,000 1,920	1,360,000 2,195	1,754,000 1,430	2,047,000 1,670	— —
	(1)4″ x 8″	727,000 1,175	848,000 1,370	969,000 1,565	1,091,000 1,175	1,273,000 1,370	1,454,000 1,565
	(3)2″ x 8″	570,000 915	665,000 1,070	760,000 1,220	854,000 915	997,000 1,070	1,138,000 1,220
	(2)2″ x 10″	409,000 840	477,000 980	545,000 1,120	613,000 840	715,000 980	817,000 1,120
11′	(1)3″ x 8″	1,357,000 1,990	1,584,000 2,320	1,809,000 2,655	2,035,000 1,990	— —	— —
	(1)4″ x 8″	970,000 1,420	1,132,000 1,660	1,293,000 1,895	1,454,000 1,420	1,697,000 1,660	1,938,000 1,895
	(3)2″ x 8″	754,000 1,105	880,000 1,290	1,005,000 1,475	1,130,000 1,105	1,319,000 1,290	1,506,000 1,475
	(2)2″ x 10″	544,000 1,020	635,000 1,190	725,000 1,360	816,000 1,020	952,000 1,190	1,088,000 1,360
12′	(1)4″ x 8″	1,260,000 1,690	1,470,000 1,970	1,679,000 2,255	1,890,000 1,690	2,206,000 1,970	— —
	(3)2″ x 8″	979,000 1,315	1,142,000 1,535	1,305,000 1,755	1,469,000 1,315	1,714,000 1,535	1,958,000 1,755
	(2)2″ x 10″	708,000 1,210	826,000 1,410	944,000 1,615	1,062,000 1,210	1,239,000 1,410	1,416,000 1,615
13′	(1)4″ x 8″	1,600,000 1,985	1,867,000 2,315	— —	— —	— —	— —
	(2)2″ x 8″	1,245,000 1,545	1,453,000 1,805	1,659,000 2,060	1,867,000 1,545	2,179,000 1,805	— —
	(2)2″ x 10″	900,000 1,425	1,050,000 1,665	1,200,000 1,900	1,350,000 1,425	1,575,000 1,665	1,799,000 1,900
14′	(3)2″ x 8″	1,555,000 1,790	1,815,000 2,090	2,073,000 2,385	— —	— —	— —
	(2)2″ x 10″	1,123,000 1,650	1,310,000 1,925	1,497,000 2,200	1,684,000 1,650	1,965,000 1,975	2,245,000 2,200

*FLOOR beams should be selected from values given for 30 psf or 40 psf live loads, deflection limited to l/360; 30 psf live loads are suitable for sleeping rooms only. ROOF beams may be selected from any of the values given; where dead loads exceed 10 psf, select beams from table with next higher live load.

(Continued) FLOOR SHEATHING & PLANKING

FIG. WF4 **MAXIMUM ALLOWABLE SPANS FOR WOOD BEAMS SUPPORTING PLANK ROOFS & FLOORS*** *(Continued)*

LOADING CRITERIA: 40 psf live load—10 psf dead load **40 psf**

Beam Span	Beam Size	MINIMUM E (psi) AND MINIMUM Fb (psi)					
		Deflection Limit l/240			Deflection Limit l/360		
		Beams 6′ o.c.	Beams 7′ o.c.	Beams 8′ o.c.	Beams 6′ o.c.	Beams 7′ o.c.	Beams 8′ o.c.
10′	(1)3″ x 8″	1,360,000 2,055	1,587,000 2,400	— —	2,040,000 2,055	2,381,000 2,400	— —
	(2)2″ x 8″	1,134,000 1,710	1,323,000 1,995	1,512,000 2,280	1,701,000 1,710	1,985,000 1,995	2,267,000 2,280
	(1)4″ x 8″	969,000 1,470	1,131,000 1,715	1,291,000 1,960	1,453,000 1,470	1,696,000 1,715	1,937,000 1,960
	(2)2″ x 10″	545,000 1,050	636,000 1,225	726,000 1,400	817,000 1,050	953,000 1,225	1,089,000 1,400
11′	(2)2″ x 8″	1,509,000 2,070	1,761,000 2,415	— —	— —	— —	— —
	(1)4″ x 8″	1,293,000 1,775	1,509,000 2,070	1,723,000 2,365	1,939,000 1,775	2,263,000 2,070	— —
	(2)2″ x 10″	725,000 1,275	846,000 1,490	966,000 1,700	1,087,000 1,275	1,268,000 1,490	1,449,000 1,700
12′	(1)4″ x 8″	1,680,000 2,110	1,960,000 2,460	— —	— —	— —	— —
	(3)2″ x 8″	1,305,000 1,645	1,523,000 1,920	1,739,000 2,190	1,957,000 1,645	2,284,000 1,920	— —
	(2)2″ x 10″	944,000 1,510	1,101,000 1,760	1,258,000 2,010	1,416,000 1,510	1,652,000 1,760	1,887,000 2,010

*FLOOR beams should be selected from values given for 30 psf or 40 psf live loads, deflection limited to l/360; 30 psf live loads are suitable for sleeping rooms only. ROOF beams may be selected from any of the values given; where dead loads exceed 10 psf, select beams from table with next higher live load.

FIG. WF5 MAXIMUM ALLOWABLE SPAN FOR 2" PLANK DECKING

NOMINAL 2" RANDOM LENGTH WOOD DECKING—END-MATCHED AND CENTER-MATCHED

Modulus of Elasticity, E (1,000 psi)	MAXIMUM SPAN AND MINIMUM Fb (psi) FOR VARIOUS LOADING CONDITIONS					
	Deflection limit l/240			Deflection limit l/360		
	20 psf live load	30 psf live load	40 psf live load	20 psf live load	30 psf live load	40 psf live load
1,000	7′-10″ 620	6′-11″ 630	6′-3″ 650	6′-11″ 470	6′-0″ 480	5′-6″ 500
1,100	8′-2″ 660	7′-1″ 670	6′-6″ 700	7′-1″ 500	6′-2″ 510	5′-8″ 530
1,200	8′-4″ 700	7′-4″ 710	6′-8″ 730	7′-4″ 530	6′-5″ 540	5′-10″ 560
1,300	8′-7″ 730	7′-6″ 740	6′-10″ 780	7′-6″ 560	6′-7″ 570	6′-0″ 590
1,400	8′-10″ 780	7′-8″ 790	7′-0″ 810	7′-8″ 590	6′-9″ 600	6′-1″ 620
1,500	9′-3″ 810	8′-1″ 830	7′-4″ 850	8′-1″ 620	7′-0″ 630	6′-5″ 650
1,600	9′-3″ 850	8′-1″ 860	7′-4″ 890	8′-1″ 650	7′-0″ 650	6′-5″ 680
1,700	9′-5″ 880	8′-3″ 900	7′-6″ 930	8′-3″ 670	7′-2″ 680	6′-6″ 700
1,800	9′-7″ 920	8′-4″ 930	7′-7″ 960	8′-4″ 700	7′-4″ 710	6′-8″ 730
1,900	9′-9″ 960	8′-6″ 970	7′-9″ 1,000	8′-6″ 730	7′-5″ 730	6′-9″ 760

NOMINAL 2" WOOD DECKING—BUTT JOINTS OVER FRAMING

Plank Span	Live Load (psf)	Deflection Limit	REQUIRED MINIMUM Fb (psi) AND E (psi) FOR VARIOUS LOADING CONDITIONS							
			Type A		Type B		Type C		Type D	
			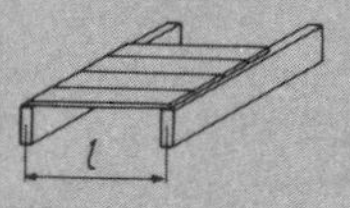		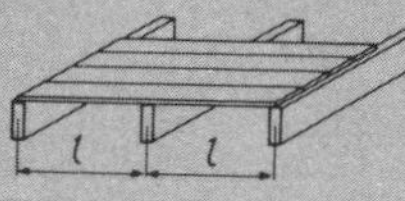		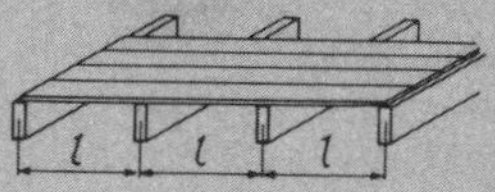		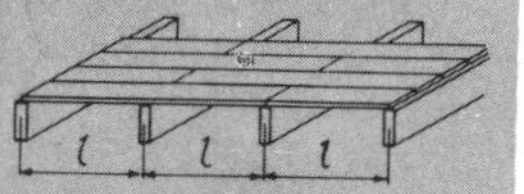	
			E, psi	Fb, psi	E, psi	Fb, psi	E, psi	Fb, psi	E, psi	Fb, psi
6′	20	l/240	576,000	360	239,000	360	305,000	288	408,000	360
		l/360	864,000	360	359,000	360	457,000	288	611,000	360
	30	l/240	864,000	480	359,000	480	457,000	384	611,000	480
		l/360	1,296,000	480	538,000	480	685,000	384	917,000	480
	40	l/240	1,152,000	600	478,000	606	609,000	480	815,000	600
		l/360	1,728,000	600	717,000	600	914,000	480	1,223,000	600
7′	20	l/240	915,000	490	380,000	490	484,000	392	647,000	490
		l/360	1,372,000	490	570,000	490	726,000	392	971,000	490
	30	l/240	1,372,000	653	570,000	653	726,000	522	971,000	653
		l/360	2,058,000	653	854,000	653	1,088,000	522	1,456,000	653
	40	l/240	1,829,000	817	759,000	817	968,000	653	1,294,000	817
		l/360	2,744,000	817	1,139,000	817	1,451,000	653	1,941,000	817
8′	20	l/240	1,365,000	640	567,000	640	722,000	512	966,000	640
		l/360	2,048,000	640	850,000	640	1,083,000	512	1,449,000	640
	30	l/240	2,048,000	853	850,000	853	1,083,000	682	1,449,000	853
		l/360	3,072,000	853	1,275,000	853	1,625,000	682	2,174,000	853
	40	l/240	2,731,000	1067	1,134,000	1067	1,444,000	853	1,932,000	1067
		l/360	4,096,000	1067	1,700,000	1067	2,166,000	853	2,898,000	1067

340 WALL SYSTEMS

INTRODUCTION

Wall systems are one of the basic elements of the superstructure. The design, selection and construction of a wall system considers (1) the required level of functional performance, (2) costs, and (3) the desired finished appearance. These considerations often are influenced by the systems and finishes used in other basic elements of the construction.

Functional performance requirements of a wall system usually include: (1) adequate structural strength to support dead loads (the actual weight of the wall and of other building elements which it supports) and live loads (such as from wind and earthquake and from live loads from other building elements supported by the wall; (2) weathertightness to control the flow of heat, moisture, air and water vapor; (3) satisfactory levels of sound and visual privacy; (4) suitable fire resistance; (5) the accommodation of heating, air conditioning, electrical and plumbing equipment; (6) suitability for the application of various finish materials; (7) adaptability to economical methods of assembly and erection; and (8) provision for the installation of doors, windows and other openings. Where there are special requirements other than the above, the conventional constructions described in this chapter may not be adequate.

Wall systems must safely support all superimposed vertical loads and resist horizontal and racking loads due to lateral forces, such as wind and earthquake, and transmit all loads to the foundation. Rarely are walls specifically engineered. Ordinarily, building codes or rules of thumb based on established practices determine materials and methods of construction.

The ultimate costs of a wall system include in-place costs, maintenance costs and operating costs. In-place costs are the sum of the costs of materials, labor and overhead required to assemble and/or install the system. The selection of a wall system may affect other costs such as installing insulation and a vapor barrier, applying siding or other finishes, and providing for window and door openings. The necessary time and costs to enclose the building and the costs of installing heating, air conditioning, electrical and plumbing equipment also may be influenced by the type of system. Maintenance costs and operating costs attributable to the wall system (such as heat loss or gain through the construction) have an important influence on the ultimate cost of the wall and should not be overlooked if long-term economy is to be achieved.

Occasionally some of the structural elements in post-and-beam systems are exposed, and in masonry construction the finished appearance may be determined by the wall construction. But generally, wall systems are covered with applied materials which provide the desired finished appearance and performance. The selection of finish materials; (1) may affect the spacing and span of the primary structural members; (2) may determine the need for wall sheathing and/or other finish bases; (3) may dictate the kind, type and physical properties of the subsurface materials, and (4) may dictate methods of construction.

The esthetic appearance of a building is influenced greatly by the location, size and type of doors, windows and other openings that form a part of the wall system. Also, the wall system should be compatible with the roof and floor systems used in the construction. Different structural systems should not be mixed indiscriminately; colors, textures, forms and details must be related if a unified appearance is to result.

Typical wall systems include conventional *stud wall* construction, *panelized* construction, *masonry* construction and *post-and-beam* construction. Each system is described briefly in this section. Section 425 discusses Stucco wall systems.

STUD WALLS

The following discussion of conventional stud wall construction refers to "stick-built" erection methods, where carpenters on the job handle individual studs rather than prefabricated panels.

Balloon-frame and platform-frame constructions (see page 300-2) typically use wood stud wall systems consisting of nominal 2″ x 4″ lumber studs, spaced 16″ or 24″ o.c. (Fig. 1). Walls may be sheathed with wood boards, plywood, fiber or gypsum boards which provide rigidity, form a weather barrier and may be necessary as a nailing base to receive exterior finishes. Some types of exterior and interior finishes can be applied directly to the studs, other finishes must be applied over sheathing or other bases.

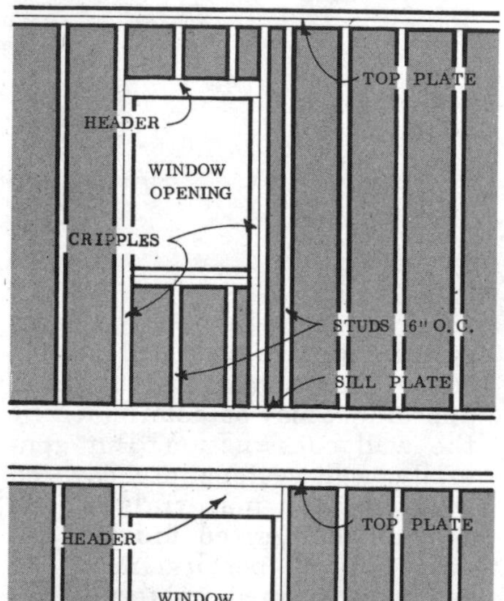

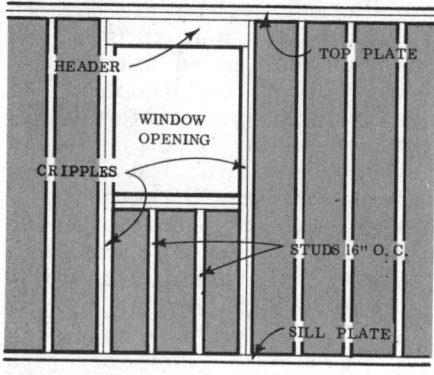

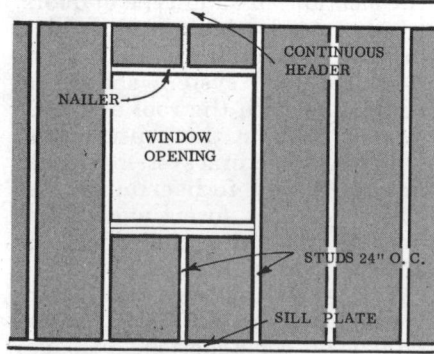

FIG. 1 Typical wood stud construction, showing framing members. Structural wall sheathing or diagonal bracing is required to resist wind and other stresses, but is not shown here.

Because of the existing body of building codes and other regulations and practices, stud spacing rarely is specifically engineered. The conventional 16″ and 24″ spacing of studs has evolved from years of established practice and is based more on accommodating the finish materials applied to the framework than on actual engineering for imposed loads. For example, in one-story residential construction, 2″ x 4″ studs could be placed at 6′ intervals and adequately support the generally imposed roof loads. But few exterior or interior finish materials can span such intervals economically. The 16″ and 24″ spacing permits the use of standard widths and lengths of panel sheathing and siding, and hence remains the convention in use.

In stud systems, roof and other vertically imposed loads are carried by the studs. Sheathing and applied finishes resist lateral forces but are not assumed to share in carrying vertical loads.

Wood stud systems lend themselves to on-site or off-site fabrication. Preframed panel wall systems are, in fact, wood stud systems fabricated in sections away from the job site. Stud wall systems provide flexibility in design and erection and require only the use of simple hand tools and fastenings. Changes can be made easily during construction, particularly with on-site fabrication. Space between studs provides for the installation of insulation and heating, air conditioning, electrical and plumbing equipment (Fig. 2). Studs may be erected by themselves or framed on the subfloor with sheathing and siding applied and windows and doors installed. The assembly then is tilted into place (Fig. 3).

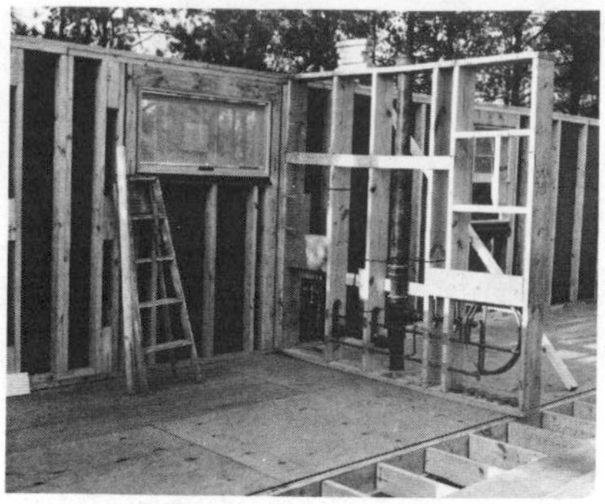

FIG. 2 Space within stud walls can be used for plumbing lines and other mechanical equipment.

FIG. 3 Walls (a) can be sheathed after studs are erected, (b) can be framed and sheathed on the subfloor; (c) insulation, siding, windows and doors may be installed before walls are erected.

Wood stud walls may be framed, erected and sheathed quickly and the building can be enclosed quickly. Interior finishing operations and mechanical installations then may take place under cover with maximum flexibility of job scheduling.

FIG. 4 Panels may be built off site in sizes conveniently handled by two men.

FIG. 5 4' wide preframed panels being erected. Studs are spaced 24" o.c. Space at the top of panels is for continuous header.

FIG. 6 Large wall panel with sheathing applied and windows installed can be fabricated off site.

Wood stud wall systems are economical and are most widely used. They provide excellent structural performance, make use of relatively inexpensive basic materials, permit the application of many finishes, and are adaptable to practically any methods of on-or off-site fabrication.

PANELIZED WALLS

Wall panel and component systems may consist of simple preframed panels using precut wood studs to which sheathing has been applied (Figs. 4 and 5). Fabrication may include the installation of insulation, vapor barriers, wiring, interior and exterior finishes, windows and doors (Figs. 6 and 7); and panels may be fabricated off site in any size that is conveniently handled.

Panelized systems also include *stressed-skin* and *sandwich* construction. These panels, which often require specialized equipment and quality control, are factory-built.

Stressed-skin wall panels usually are made of framing members to which plywood skins are bonded either by glue-nailing techniques or adhesive application under heat and pressure (Fig. 8). When the wall panel is loaded, both the framing members and the skin surfaces are stressed and act as an integral unit to carry loads. Since the plywood performs a structural function, the framing members ordinarily can be reduced in size, resulting in thinner wall sections. For example, wall and partition studs, which normally are 2" x 4", can be reduced to 1" x 3" or 2" x 3" with stressed-skin design.

Sandwich panels have exterior and interior surfaces glued to and separated by lightweight core materials which generally are insulating (Fig. 9). The panels are similar in principle to stressed-skin panels in that the surfaces become stressed and assist the core in resisting loads. A variety of materials are used in the construction of sandwich panels, such as polystyrene or urethane foam cores with aluminum, plywood or gypsum faces.

Stressed-skin and sandwich panel systems have exhibited

FIG. 7 Off-site fabrication may also include: (a) panels with exterior finish and windows installed; (b) panels with siding, windows and insulation.

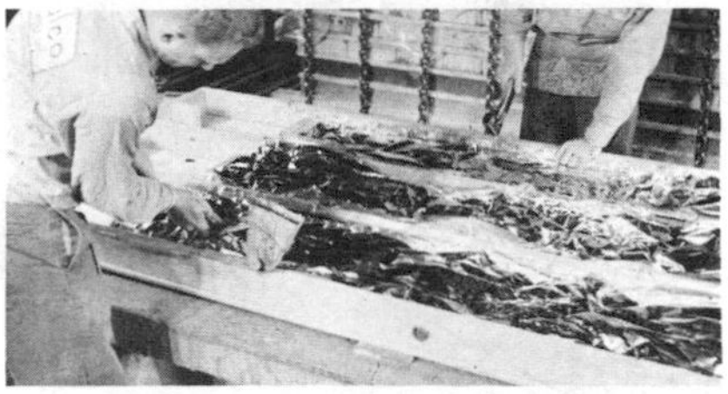

FIG. 8 Stressed-skin wall panel being fabricated. Workman at left is spreading glue along 2" x 4" frame member; man at right is inserting foil type insulation.

FIG. 9 Sandwich wall panel with prefinished exterior aluminum surface and gypsum board interior surface over a 3" foamed plastic core.

excellent structural performance and provide many of the advantages of component fabrication. However, they may require careful

FIG. 10 **MASONRY WALL TYPES**

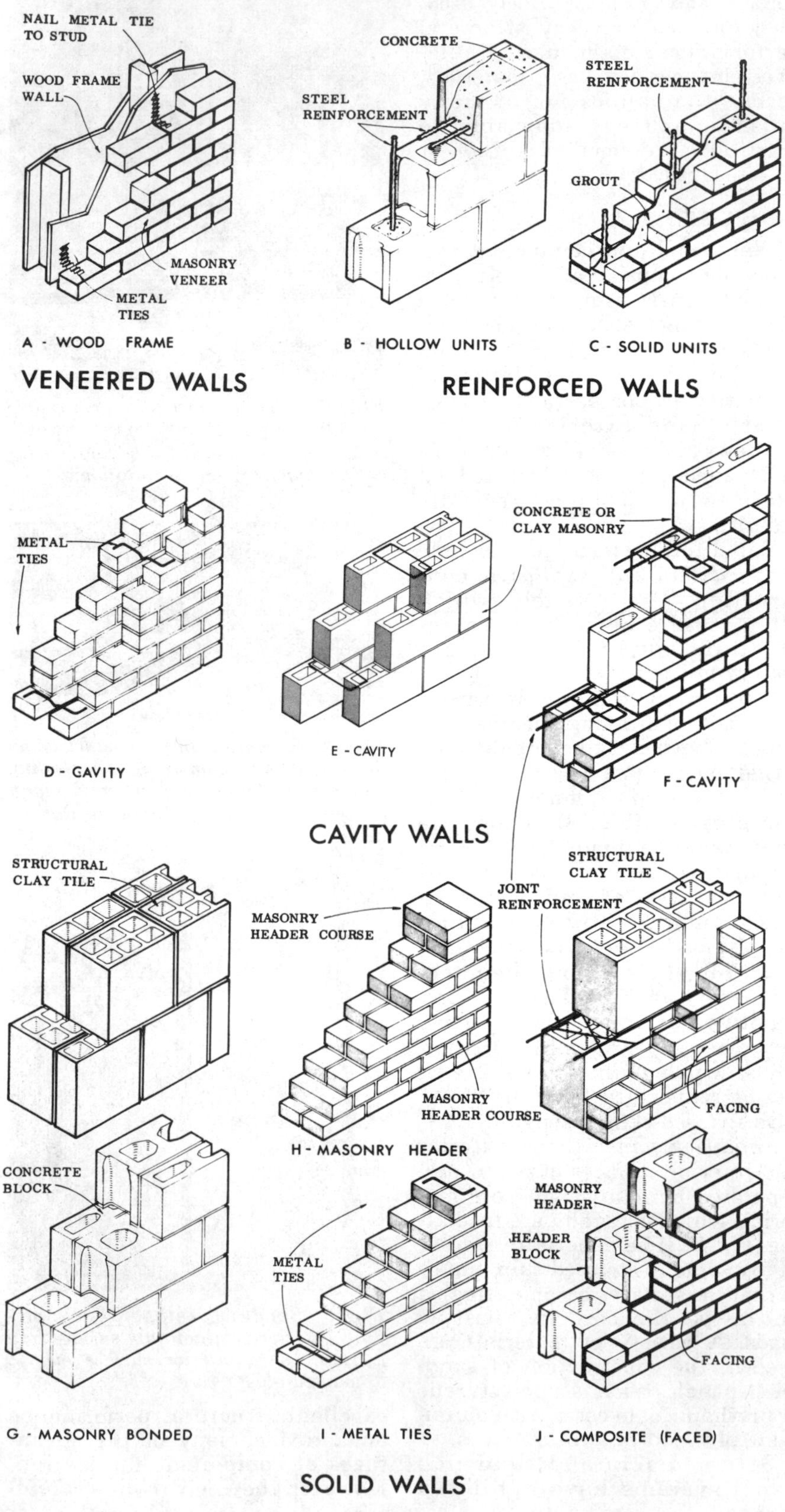

design to incorporate mechanical equipment, may require special handling and erection procedures in the field, and may not be acceptable to local codes. Therefore, they have had limited success to date.

Many local building codes specifically require 2″ x 4″ studs spaced 16″ or 24″ o.c. Some codes permit the acceptance of any system which performs as well as the system specified. Properly engineered and constructed panelized systems which do perform adequately should be accepted whenever possible.

MASONRY WALLS

Load-bearing and non-load-bearing masonry walls may be constructed of clay and/or concrete masonry units. They are classed as solid walls, cavity walls, veneered walls and reinforced walls, depending on methods of construction (Fig. 10).

In masonry veneer construction, wood framing assumes all vertical loads and the masonry performs essentially as an exterior finish surface with the basic function of providing a weather barrier to the system. Space between studs typically is used for the installation of insulation and heating, air conditioning, electrical and plumbing lines.

In solid, cavity and reinforced masonry systems, the installation of insulation and mechanical equipment may be incorporated between thicknesses of masonry or within furred spaces on the interior side of the wall.

Usually masonry walls are left exposed, but they may be finished on the exterior with stucco or other coatings. Interior surfaces can be finished with applications such as plaster or plaster board, or can be left unfinished as the desired appearance.

A masonry wall assembly should be suitable for the conditions of loading, the locality, the exposure and the type of occupancy (Fig. 11). In many areas, single-thickness masonry walls are adequate for the control of heat and moisture and may be more economical than wood stud walls. Generally, however, the selection of masonry

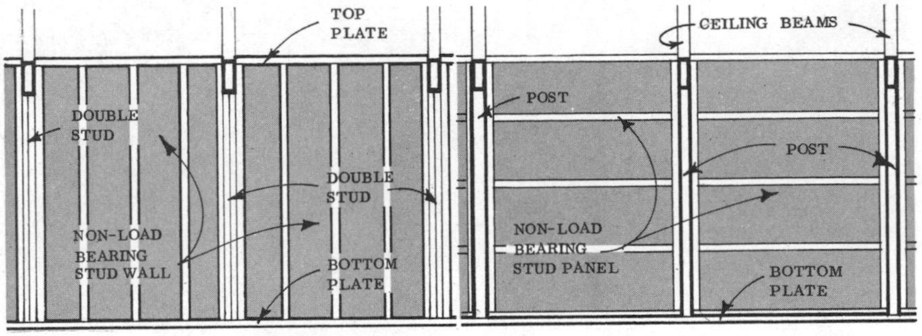

FIG. 12 Post-and-beam wall construction.

construction is based on functional performance and desired appearance.

Masonry wall systems are widely used because they provide excellent structural performance and offer long-term, low-maintenance finishes.

POST-AND-BEAM WALLS

Post-and-beam wall systems consist of posts which carry the imposed loads of floors, roofs and/or ceilings, with non-load-bearing wall sections placed between the posts (Fig. 12). Posts typically are 4″ x 4″ lumber members or are built up of several 2″ x 4″s and are generally spaced 4′ to 8′ o.c. to support roof beams. Wall areas between posts are framed, sheathed and finished similar to other wood wall systems. They serve primarily as weather barriers, and provide resistance to racking and give lateral support to the posts. Resistance to lateral forces generally is achieved with sheathing materials or diagonal bracing. Since the wall areas between posts do not carry any vertical loads, framing members may be placed horizontally to receive vertical exterior siding; and areas between posts can accommodate large windows or doors easily (Fig. 13). Glass panels often are actually set (stopped) into the posts. No headers are required, but the placement of windows and doors is determined by the spacing and location of the posts (Fig. 14).

Post-and-beam wall systems often are integrated with plank-and-beam roof/ceiling systems. Planks and beams generally are left exposed to form the finished ceiling, and posts may be partially exposed on the interior and/or exterior surfaces as part of the finished wall. In such designs, the expression of the basic structural system largely determines the interior and exterior appearance and character.

FIG. 11 (a) In some load-bearing masonry construction, floors are supported directly by masonry. (b) Single-thickness masonry walls are suitable in some localities. Here, mortar joints are left untooled. (c) Cavity walls may be used (a) to provide space for insulation, and (b) to receive plumbing and wiring runs.

FIG. 13 Here, roof loads are carried on posts spaced 6′ o.c. Wall areas between posts are combinations of window, ventilation and solid areas.

FIG. 14 Exterior walls between posts carry no roof loads. Therefore, large window and door areas without headers can be constructed.

We gratefully acknowledge the assistance of the following for the use of their publications as references and permission to use photographs: Aluminum Company of America: Dow Chemical Company; House & Home Magazine; Lumber Dealers Research Council; Practical Builder Magazine.

343 MASONRY WALLS

MASONRY WALLS

INTRODUCTION

A masonry wall assembly is a good example of the adage: "The whole is equal to the sum of its parts." The masonry units, the mortar that bonds them together, and the workmanship of the mason are all interdependent. The use of quality materials will not compensate for poor workmanship, nor will masonry constructed with the finest workmanship give satisfactory performance if poor materials are used. Above all, the wall must be properly designed for the necessary *strength, watertightness* and *durability.* Depending on its location and use, other wall properties—*resistance to heat transfer, fire resistance, sound resistance, radiation protection*—could be equally important. All of these properties can be achieved through proper design.

Past experience and observation of the performance of masonry walls has resulted in the development of certain standards and minimum requirements governing their design. These criteria are based on observation with the result that the design of masonry walls is most often established in local building codes by minimum wall thicknesses and lateral support requirements, rather than by theoretical analysis of the forces and stresses actually involved.

In this section, most of the recommendations for the design of masonry walls are based on the provisions of the American National Standards Institute (ANSI) Building Cost Requirements for Masonry ANSI A41.1—1953 (R 1970), and the American Concrete Institute (ACI) Building Code Requirements for Concrete Masonry Structures, ACI Standard 531-79. These publications have been widely accepted as representative of good practice in the design and construction of all types of masonry under average conditions of exposure and loading.

Section 205 Masonry, explains how to select sound masonry units, choose the correct mortar type for the use and exposure of a wall, and illustrates the techniques of proper masonry workmanship. This section describes the numerous types of masonry wall assemblies, their properties when built, and illustrates numerous construction details essential to satisfactory design. Taken together, these two sections explain how to produce masonry of suitable quality.

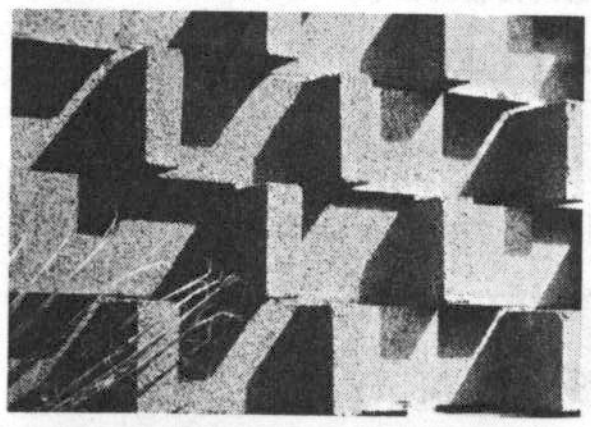

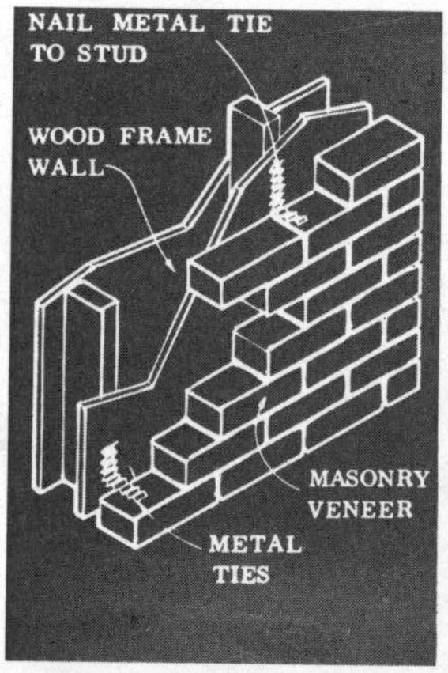

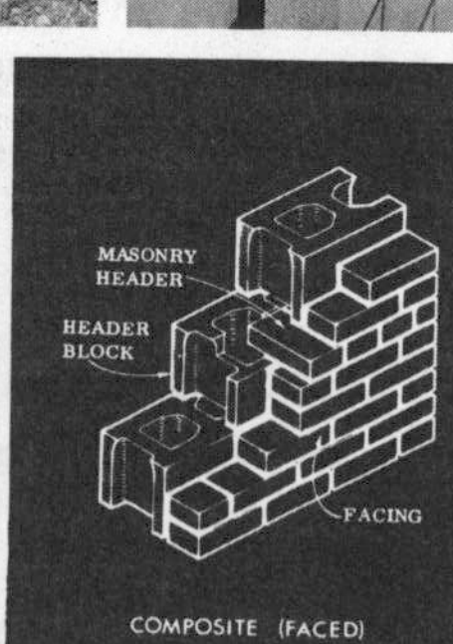

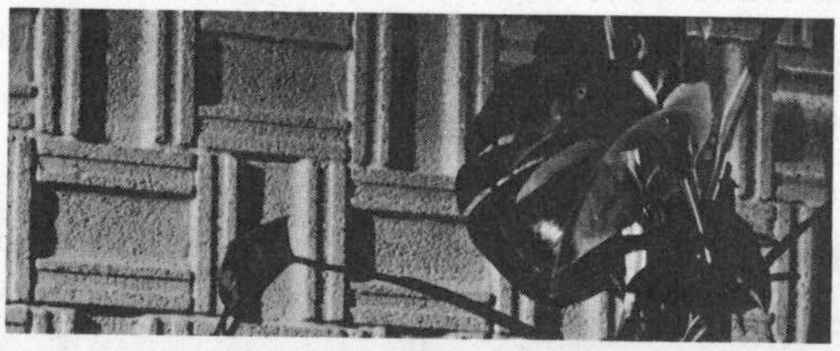

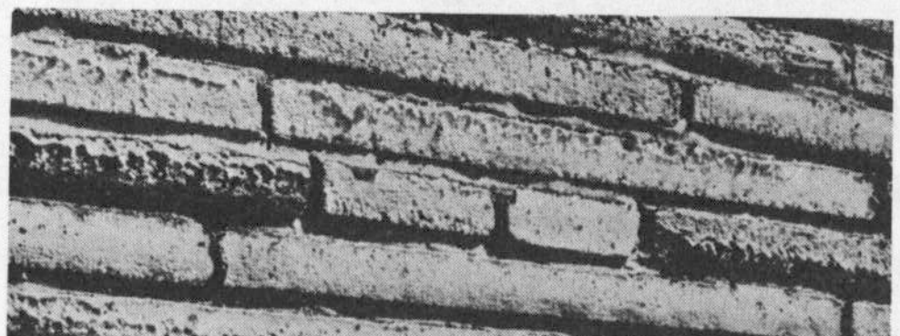

MATERIALS

Masonry *units* and *mortars* are discussed in Section 205 Masonry. In this section, *other materials* necessary for the suitable completion of masonry walls are described and illustrated.

MASONRY UNITS

The properties of solid and hollow *clay* and *concrete* masonry units are given in Masonry, pages 205-3 through 205-22.

A grade-use guide to the selection of solid and hollow *clay* and *concrete* masonry units is given in the Work File, pages WF205-1 through WF205-4.

MASONRY MORTARS

The properties of mortar types, including a discussion of how mortar materials should be proportioned and mixed are given in Masonry, pages 205-23 through 205-28.

A guide to the selection of mortar materials and for selecting different mortar types for various construction applications is given in the Work File, pages WF205-4 through WF205-6.

FLASHING MATERIALS

Flashing is installed at vulnerable locations in masonry to: (1) prevent entry of water or to collect water that might enter through cracks in the masonry, and (2) if collected, to then divert it through weep holes to the exterior.

Flashings are formed from sheet metal, bituminous fabric membranes, plastic sheet materials or a combination of these materials. (Fig. 1, 2 & 3).

Because replacement costs can exceed the initial installed costs, it is prudent to select a permanent flashing material for the original installation. The recommendations for installing flashing materials are discussed later in this section under Moisture Control, page 343-9. See Work File, page WF343-1 for specifications and ASTM designations for the flashing materials described.

Copper

Though more costly than most other flashing materials, copper is durable, available in special preformed shapes, and is an excellent moisture barrier. It is not materially affected by the caustic alkalis present in masonry mortars, hence copper can safely be imbedded in fresh mortar and will not deteriorate in continuously saturated, hardened mortar unless excessive chlorides are present. When using copper flashing, the use of chloride-base additives in mortar (such as calcium chloride to accelerate the set of mortar in cold weather) should be prohibited. One objection to copper flashing is that when exposed to weather, the copper wash may stain or discolor light-colored masonry surfaces. Where copper staining would be objectionable, *lead-coated* copper can be used.

Galvanized Steel and Lead

These materials are subject to corrosion in fresh mortar and when placed in contact with mortar, it is advisable to protect them with a bituminous coating. Galvanized steel is widely used in residential building and generally performs adequately. However, exposed galvanized steel flashing requires painting maintenance.

Aluminum

Flashings made of aluminum are subject to attack by the caustic alkalis present in fresh, unhardened mortar. Although dry, seasoned mortar will not affect aluminum, corrosion can occur if the adjacent mortar becomes wet. Therefore, aluminum will be unsatisfactory as à flashing unless coated. Aluminum flashing will not stain or discolor light-colored masonry surfaces.

Stainless Steel

Stainless steel flashings are avail-

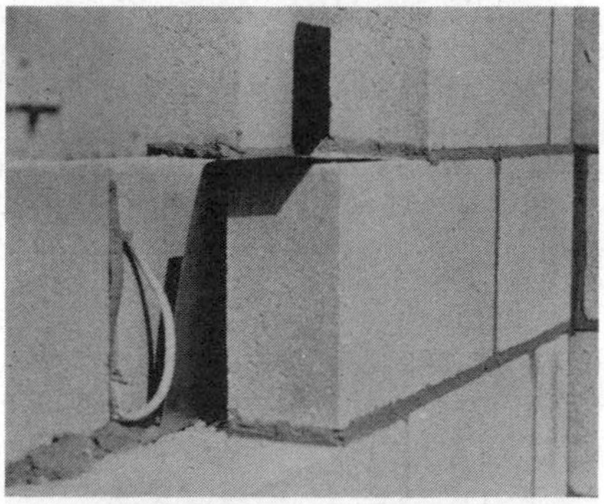

FIG. 1 Continuous thru-wall metal flashing in a cavity wall; rope wick forms a weep hole.

FIG. 2 Plastic flashing installs easily around a vent.

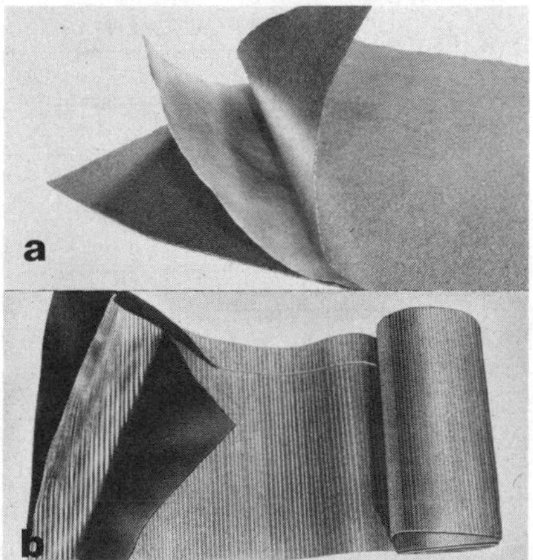

FIG. 3 Combination flashing materials: (a) copper (or aluminum) bonded to asphalt-saturated kraft paper; (b) 5-layers assembly composed of copper (or aluminum) sandwiched between asphalt-saturated cotton fabric with outer layers of ductile mastic.

able in several gages and finishes. They are durable, highly resistant to corrosion, excellent moisture barriers and are surprisingly workable. There are a number of types available, the varieties depending basically on the proportions of chromium, nickel and steel. These varieties have differing degrees of resistance to corrosion and are priced accordingly. Stainless steel should be specified by number and not by the generic term alone. (See Work File.) In addition to other favorable properties, stainless steel will not stain adjacent areas, resists rough handling and can be soldered, welded or brazed.

Bituminous Fabrics

Flashing materials composed of fabrics saturated with bitumen are used as damp checks and as low cost substitutes for metal flashing at the base of walls, heads of openings and at window sills. If they are permanently insoluble in water and are installed with unbroken skins (which usually requires extra care in the field), they can be effective although they are not as permanent as metal flashing.

Plastics

Plastic flashings are available that are tough, resilient and highly resistant to corrosion. Some plastic flashings will not withstand the corrosive effects of masonry mortars, and since the chemical compositions of plastics vary so widely, it is important not to consider *all* plastics as permanent flashing. Since recognized standards have not yet been established for plastic flashing, it is necessary to rely on performance and test records of the material or the reputation of its manufacturer in order to insure suitable quality and performance.

Combination Flashings

Combination flashings consist of materials combined to utilize the best properties of each—such as sheet metal coated with plastic or bituminous materials. The combination of materials can provide a lower cost flashing by reducing the thickness of metal required, or it may permit the use of corrodible metals which might otherwise prove unsatisfactory unless coated.

ANCHORS AND TIES

Numerous types of masonry anchors, ties and joint reinforcement are used in masonry construction (Fig. 4). Recommendations for installing these materials are discussed and illustrated later in this section under *Bonding and Anchorage,* page 343-25. See the Work File, page WF343-2 for specifications and ASTM designations for anchors, ties and joint reinforcement.

FIG. 4 MASONRY ANCHORS, TIES, AND JOINT REINFORCEMENT

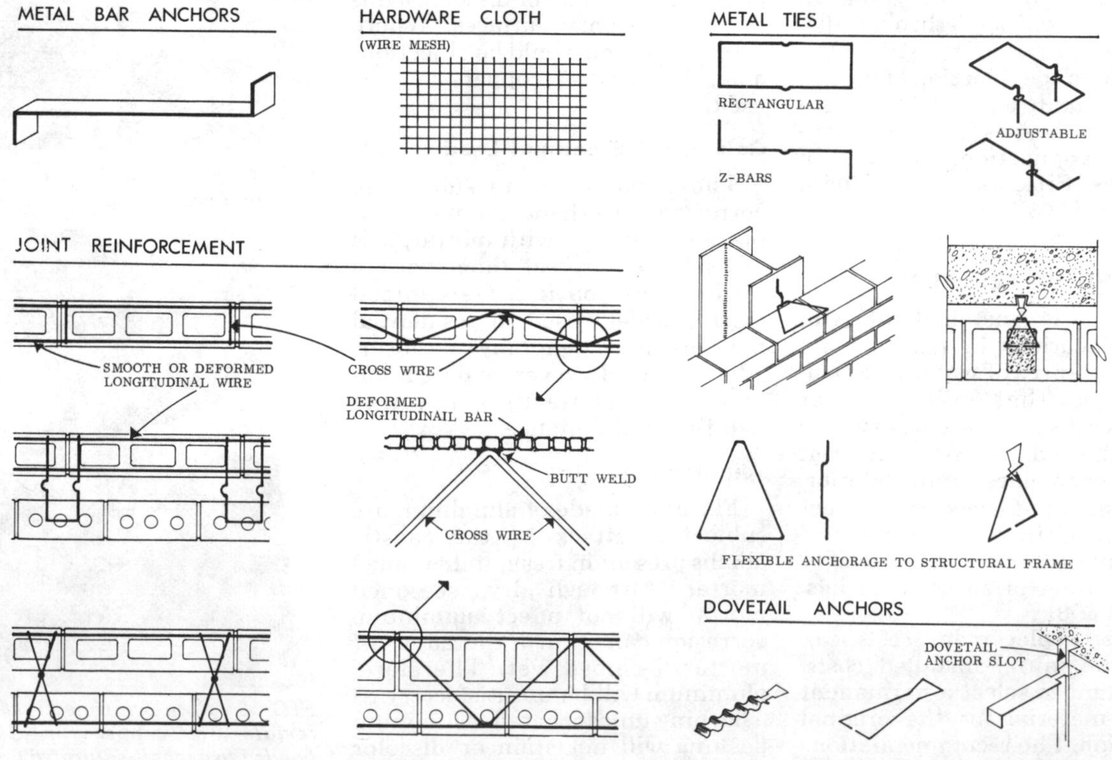

DESIGN AND CONSTRUCTION

WALL TYPES

Masonry walls may be classified as *solid* walls, *cavity* walls, *veneered* walls and *reinforced* walls (Fig. 5). No one wall assembly is suitable as a standard design for all exposures, conditions of loading or appropriate to all localities and types of occupancy. As with practically all materials and methods of construction, the final selection of a wall type is a compromise between function, economy and appearance.

Solid Walls

A *solid* wall acts as a unit to support loads and is built up of units laid close together with all joints between units filled solidly with mortar. Solid walls may be built of either solid or hollow masonry units (or both in combination) in any required thickness, and are used for either load-bearing or non-load-bearing construction. Structural bond of solid walls is provided by metal ties, masonry headers or joint reinforcement as shown in Figure 5. Walls which utilize masonry headers or have units overlapping in alternate courses (Fig. 5G). are called *masonry bonded.*

Solid walls may be further classified depending on the type of units used in their construction. They may consist of: (1) solid units of brick or concrete brick and block, (2) hollow units of concrete block, hollow brick, or structural clay tile, and (3) composite (faced) walls consisting of facing and backup units of different materials bonded so that both facing and backup are load bearing.

Typical details of 8″ solid brick walls are shown in Figure 5H&I. Solid walls bonded with masonry headers are less resistant to rain penetration than metal tied walls, and hence should be used only in areas of slight exposure (See Fig. 14). The addition of furring to the inner wall reduces through-the-wall moisture penetration and permits use in areas of severe exposure. In areas subjected to hurricanes or wind driven rains, the exterior surface of concrete block walls is usually stuccoed to resist rain penetration.

FIG. 5 TYPICAL MASONRY WALL ASSEMBLIES

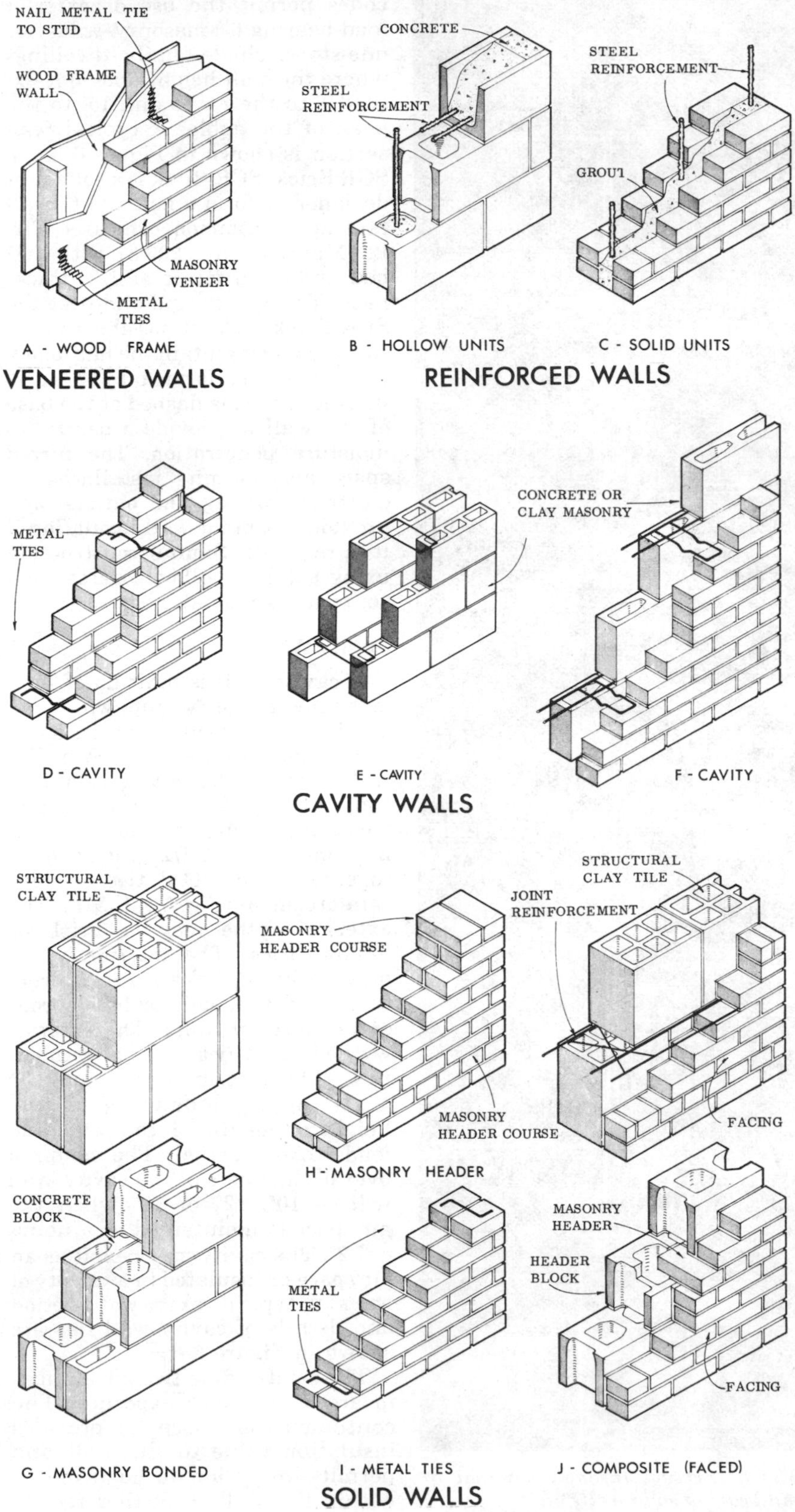

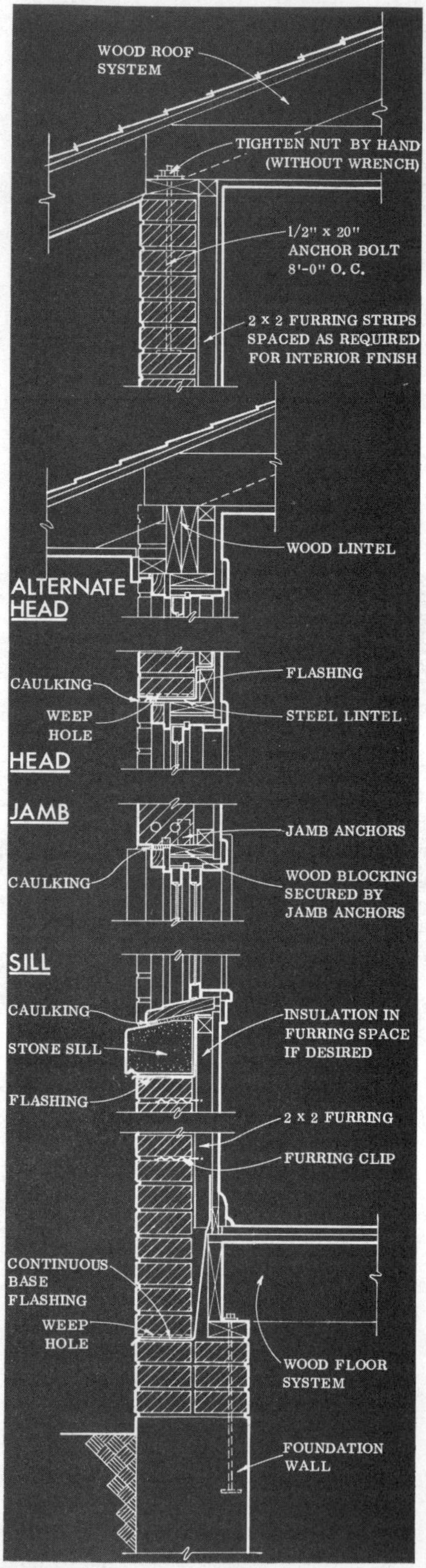

FIG. 6 Section through nominal 6" load-bearing solid wall.

All nationally recognized building codes permit the use of exterior load-bearing 6" masonry walls for one-story, single-family dwellings where the wall height does not exceed 9′ to the eaves and 15′ to the peak of the gable. A typical wall section is shown in Figure 6 using SCR Brick. SCR Brick is a solid unit designed to form a nominal 6″ wall laid up in common bond (see Fig. 12, Masonry, pg. 205-8) with full bed and head joints of 1/2″ thickness. The length and width of the SCR Brick make it possible to turn corners and maintain the half-bond. An air space, formed by 2″ x 2″ furring strips, is flashed at the base of the wall to provide a barrier to moisture penetration. The furred space also permits installation of electrical wiring and outlets, and provides room for wall insulation if it is required. Similar construction using solid or hollow concrete masonry units may also be used.

Cavity Walls

A cavity wall is built up of solid or hollow masonry units which are deliberately separated into an inner and an outer wall. A cavity wall consists of two wythes of masonry at least 4″ thick separated by a continuous air space not less than 2″ or more than 4-1/2″ wide, bonded together with metal ties or joint reinforcement (Fig. 5D, E & F). The exterior wythe may be brick or hollow brick. The interior wythe may be brick, hollow brick, structural clay tile, solid or hollow concrete masonry units. The exterior wythe is usually a nominal 4″ thickness. The interior wythe may be 4″, 6″ or 8″, depending on the height or length of the wall and loads to be carried. The nominal overall thickness of the cavity wall will be 10″, 12″ or 14″ when the air space is maintained at a nominal 2″. The cavity may be left as an air space or insulated in a variety of ways. A typical cavity wall section and details of cavity wall ties are shown in Figure 5.

The cavity offers two advantages in areas of severe exposure. The continuous air space (1) provides insulation value to the wall and permits insulation to be installed within the wall to further reduce

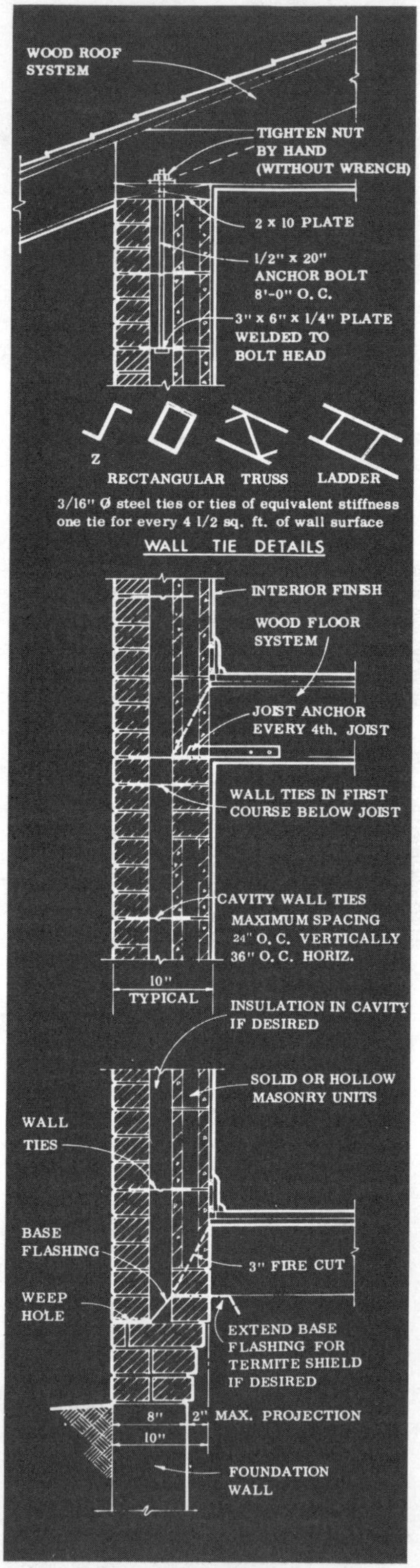

FIG. 7 Section through nominal 10″ load-bearing cavity wall.

heat transfer (See Fig. 31); and (2) acts as a barrier to moisture, eliminating any need for furring since rain penetration to the interior is practically impossible if proper flashing and weep holes are installed.

To be effective, certain precautions must be observed during construction. The cavity must be kept free of mortar droppings that could form a bridge allowing moisture to penetrate to the interior face. Two methods are illustrated in Figure 8. Weep holes must be provided in conjunction with flashing to properly drain the cavity of any water that may enter the exterior wythe and collect on the flashing. Metal cavity wall ties, designed to resist both tension and compression, are used to bond the exterior and interior wythes and should be solidly embedded in mortar so that the two wythes can act together in resisting wind or other lateral forces.

A vapor barrier is not required in a cavity wall where: (1) the cavity is insulated with fill materials such as water repellent vermiculite or silicon treated perlite, which will not retain excessive moisture; or (2) rigid board materials, such as glass fiber, foamed glass or foamed plastics, that are at least 1″ less in thickness than the cavity and are installed next to the inner wythe.

Masonry Bonded Walls Masonry bonded walls are used in areas of slight weather exposure, for economical foundation and exterior load-bearing walls.

Veneered Walls

The use of masonry units only as a facing material (veneer), without utilizing its load-bearing properties, is common in residential construction (Fig. 9). The facing of a veneered wall is attached but not bonded to act structurally with the load-bearing backup material.

Key factors in constructing brick veneer walls include correct selection of masonry materials, adequate anchorage to the back-up wall, and proper installation of flashing and weep holes.

Since the veneer is isolated from the back-up wall by an air space

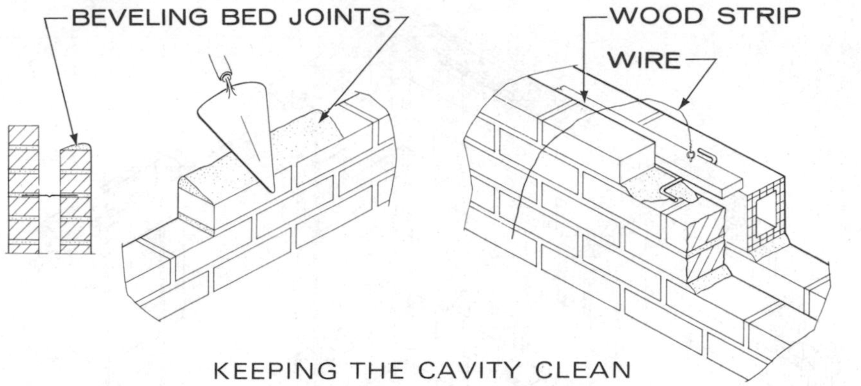

FIG. 8 (a) Left, mortar bed is beveled so little, if any, mortar will be squeezed out of bed joints into cavity; (b) a wood strip is placed on each tier of metal ties to catch mortar droppings, and raised as the work progresses.

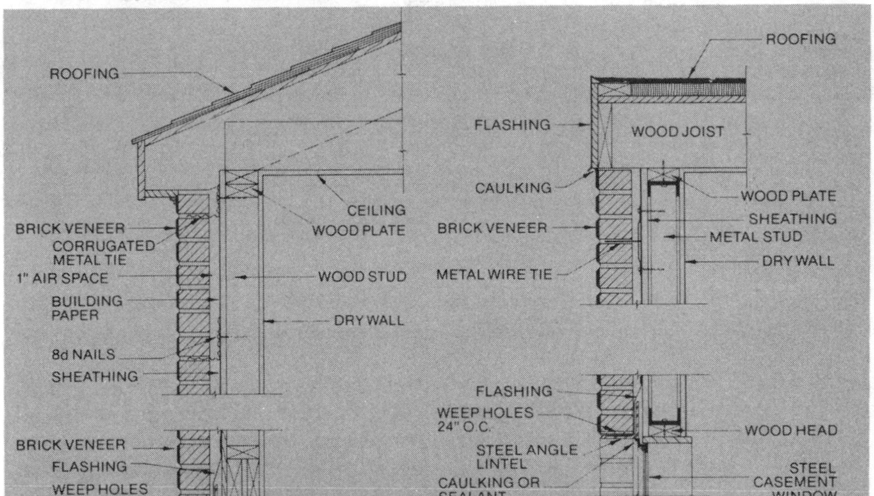

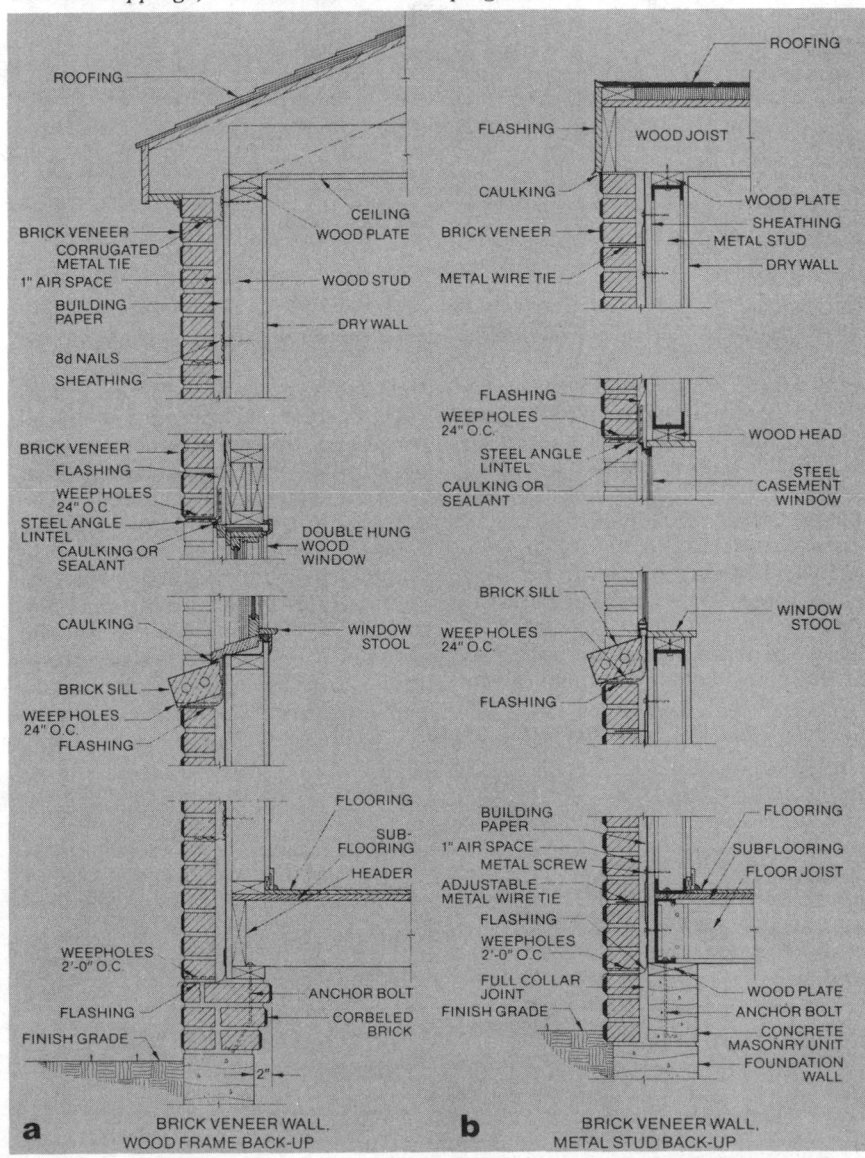

FIG. 9 Brick veneer wall construction section: (a) wood stud wall; (b) metal stud wall.

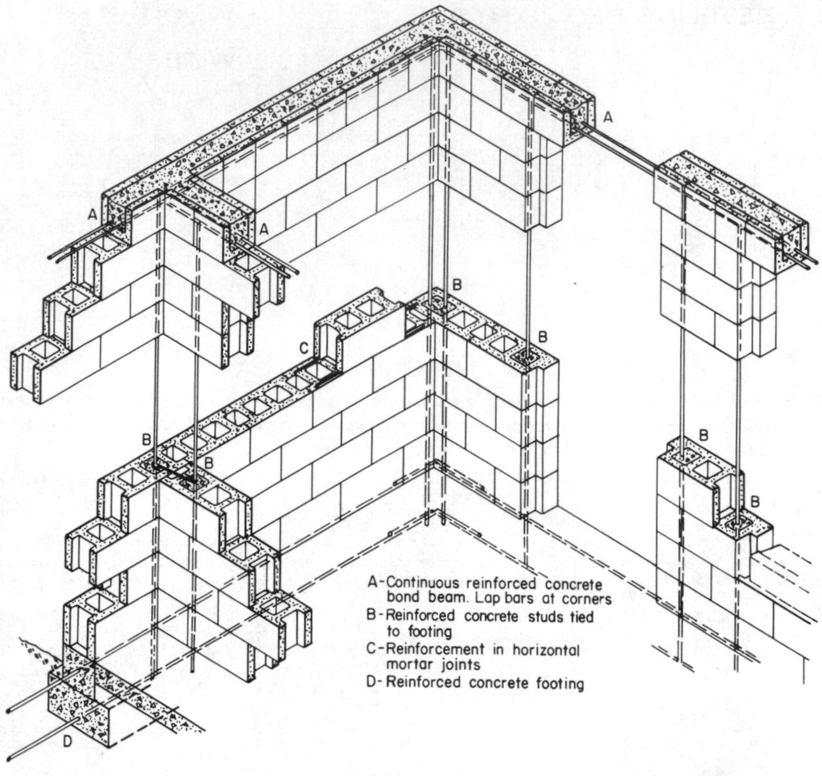

FIG. 10 Typical methods of reinforcing concrete masonry walls.

and exposed to exterior temperature extremes, Severe Weathering (SW) Grade facing brick should be used. Brick veneer walls carry only their own weight, therefore Type N mortar is recommended, which is consistent with the principle that the lowest strength mortar be used compatible with structural requirements.

In wood frame construction (Fig. 9a) the brick veneer is anchored to the back-up wall by nailing corrosion-resistant corrugated metal ties to wall framing members. In metal stud wall construction (Fig. 9b), the veneer is anchored using adjustable wire ties, screwed to the metal studs. The ties, spaced not to exceed 16″ horizontally and 24″ vertically, permit vertical and horizontal movement parallel to the plane of the wall, but resist tension and compression perpendicular to the plane of the wall.

Flashing and weep holes to provide resistance to rain penetration should be located at the bottom of the wall and at all openings.

Reinforced Walls

Using the principles employed in the design of reinforced concrete, steel reinforcement is placed and embedded with the masonry units so that the masonry has greatly increased resistance to forces that produce tensile, shear and compressive stresses. The chief use of steel reinforcement is in vertical members, such as walls and columns, and for lintels and bond beams (Fig. 5 B&C and 10).

Workmanship techniques required of the mason in building reinforced walls are the same as for non-reinforced masonry, except it is of even greater importance that all joints be completely filled with mortar and grout, and that a strong bond be developed between the mortar, steel reinforcing and masonry units.

PATTERN BONDS

The pattern formed by the masonry units and the mortar joints on the face of the wall is called *pattern bond.* Pattern bond may result either from the type of *structural* bond used or may be purely decorative. (Structural bond is the method by which individual masonry units are interlocked together, using masonry headers or metal ties, to cause the entire wall

FIG. 11 **TRADITIONAL PATTERN BONDS**

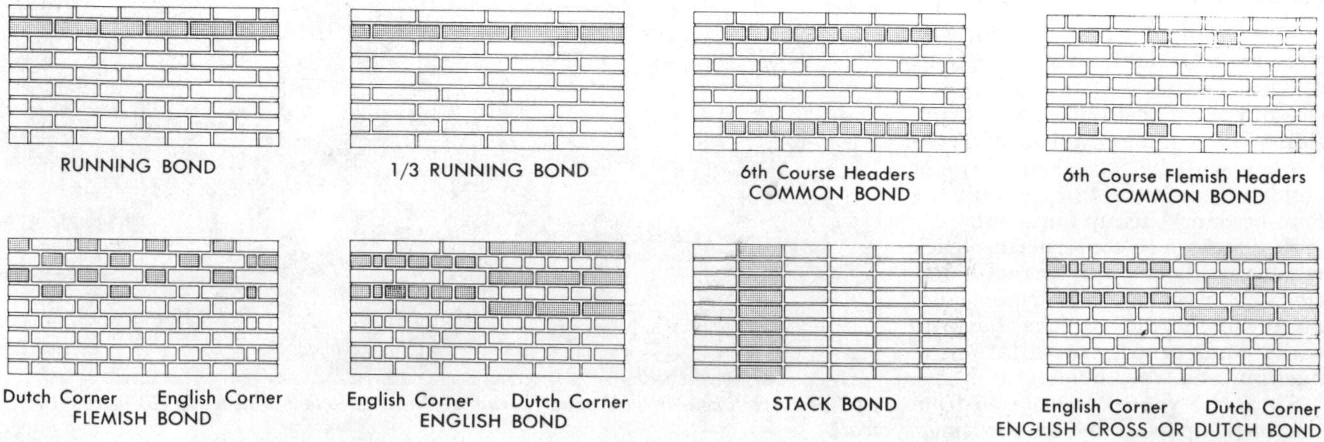

assembly to act as a single structural unit.)

Brick Pattern Bonds

Brick bond patterns commonly used are described briefly below and are illustrated in Figure 11.

Running Bond This is the simplest of the basic pattern bonds. Horizontal courses consist of all stretchers and since there are no masonry headers, metal ties are usually used for structural bond. Running bond is used largely in cavity and veneered walls of solid units and often in concrete block and clay tile walls where masonry bonding may be accomplished by wider stretcher units (Fig. 5G).

Common (American) Bond This is a variation of running bond with a course of full length headers at every 5th, 6th or 7th course, depending on the structural bonding requirements. It is important in laying out any bond pattern that the corners be started correctly. For common bond, a three-quarter brick starts each header course at the corner.

Flemish Bond This pattern is made up of alternate stretchers and headers in each course. Where the headers are not used for the structural bonding, half-brick (blind headers) may be used. Two methods are used in starting the corners. The "Dutch" corner uses a three-quarter brick to start each course and in the "English" corner 2″ or quarter brick closures must be used.

English Bond In this pattern alternate courses are composed of headers and stretchers. The joints between stretchers in all courses line up vertically. Blind headers are used except in structural bonding courses.

English Cross or Dutch Bond This is a variation of English bond. Vertical joints between the stretchers in alternate courses do not line up vertically, but center on the stretchers themselves in the courses above and below.

Stack Bond This is a pattern bond with all vertical joints aligned. Since units do not overlap, this pattern is usually bonded to the backing with rigid steel ties. In large wall areas and in load-bearing construction, it is advisable to reinforce the wall with joint reinforcement or steel rods placed in the horizontal mortar joints. The alignment of vertical mortar joints requires dimensionally accurate units, or prematched units carefully selected for size so that each vertical joint will align.

Contemporary Bonds Many of the traditional bonds have been modified by projecting and recessing units or by leaving out units to form perforated walls and screens. Contemporary brick bonds are shown in Figure 12. The effects of specific patterns can be greatly altered by varying the pattern of individual unit color and texture, joint types, joint color and by recessing or projecting individual units, courses or wythes.

Block Pattern Bonds

Many bond patterns currently used in concrete masonry are illustrated in Figure 13. Although decorative, all of the patterns drawn in Figure 13 have been shown by test to be suitable for load-bearing construction.

WORKMANSHIP

A discussion, including illustrations, of correct workmanship to obtain suitable quality masonry construction is given in Masonry, pages 205-29 through 205-35.

MOISTURE CONTROL

The principal sources of moisture in masonry are rain penetration, capillary action from contact with the ground and condensation of vapor within the masonry.

Methods used to control rain penetration in masonry walls include adequate flashing, tooled mortar joints, parging, painting, tightly caulked door and window openings, sufficient slope (wash) to readily drain all horizontal surfaces (as at sills and copings, including an overhang and drip) and adequate gutters and downspouts.

Condensation of water vapor may occur within uninsulated masonry and can be controlled by installing a vapor barrier on the warm side of the wall. A water-emulsion asphalt paint applied to the wall surface as it is constructed is one method used to provide such a barrier. Masonry insulated with proper types of insulation may not require a vapor barrier.

Flashing

Wall flashing is installed to collect moisture that may penetrate the wall and then divert it through weep holes out of the wall. Exposure of the masonry to rain, which varies greatly throughout the United States, is a basic consideration in design against mois-

FIG. 12 **CONTEMPORARY BRICK BONDS**

FIG. 13 CONCRETE MASONRY PATTERN BONDS

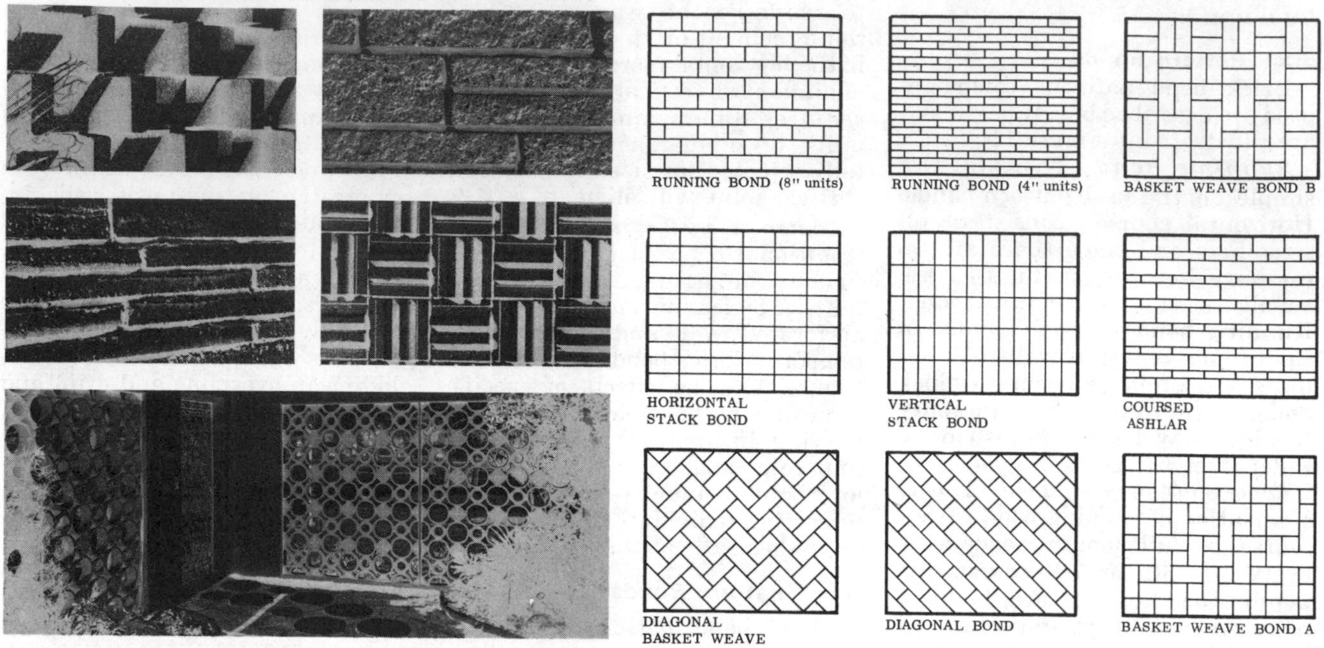

ture penetration. Exposure may be defined roughly in terms of the resultant wind pressures and precipitation maps shown in Figure 14 as: (1) *Severe*—Annual precipitation 30" or over, wind pressure 30 lbs per sq. ft. or over; (2) *Moderate*—annual precipitation 30" or over, wind pressure 20 to 25 lbs per sq. ft.; (3) *Slight*—annual precipitation less than 30", wind pressure 20 to 25 lbs per sq. ft. or annual precipitation less than 20"

The following recommendations apply to areas of *severe* exposure such as the Gulf Coast, Atlantic Seaboard and the Great Lakes; and to *moderate* exposures, typical of the Mid West. Where exposure is *slight*, internal flashing such as lintel or sill flashing, may be eliminated, and in some areas external flashing may be reduced to a minimum.

Flashing Placement The vulnerable locations in masonry where flashing is necessary or recommended are described below (See Fig. 15).

Base Flashing and Damp Checks are used at the base of the wall. Base flashing diverts to the exterior any moisture having entered above this point, and a damp check stops the upward capillary travel of ground moisture. If installed properly, metal shields, used for termite protection of wood joists, may double for these purposes. (See Division 300 for details of termite protection.)

Sill Flashing should be placed under and behind all sills. The ends of sill flashing should extend beyond both sides of the opening and turn up at least 1" into the wall. Sills should slope and project from the wall to drain water away from the building and to prevent staining. When the undersides of sills do not

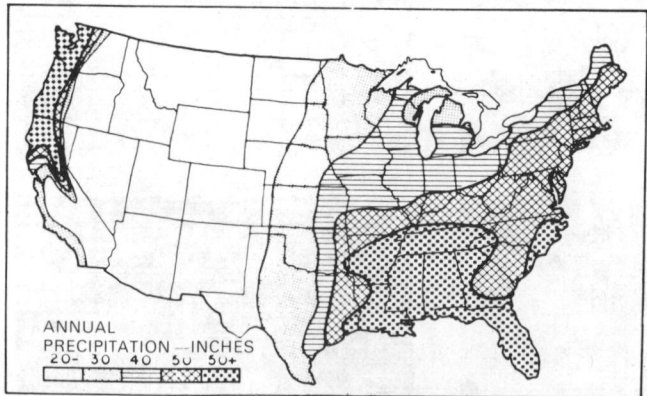

PRECIPITATION

WIND PRESSURE

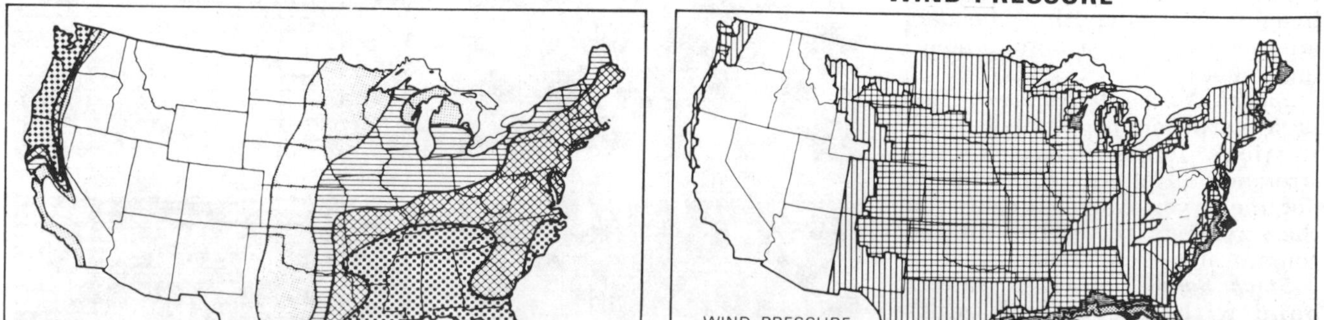

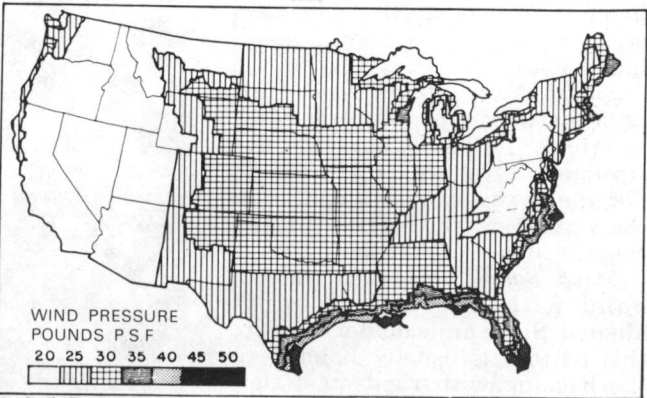

FIG. 14 Weather exposure based on precipitation and wind pressure.

slope, a drip notch should be provided or the flashing may be extended and bent down to form a drip.

Head Flashing should be placed over all openings except those completely protected by overhanging projections. At steel lintels the flashing should be placed under and behind the facing material with its outer edge bent down over the lintel to form a drip.

Roof Flashing at chimney walls consists of *base* flashing placed under the roofing and turned up against the masonry with *counter* flashing overlapping the base flashing and extended into and imbedded securely in the mortar joints.

Weep Holes Flashing should be drained to the outside. Concealed flashings in tooled mortar joints are not self-draining without weep holes, but serve rather as a trap to collect moisture and by concentrating moisture in one spot may do more harm than good. Except at window sill flashing where little water collects, weep holes should be provided every 24″ horizontally in the head joint immediately above any flashing. Weep holes may be formed by omitting the vertical mortar (head) joint at intervals of 2′, or by inserting oiled rods or pins in the head joints which are then removed when the mortar is ready for tooling. Such open weep holes provide access for insects and when placed over lintel flashing may contribute to wall staining. It is recommended that "wicks" made of 1/4″ glass fiber rope (Fig. 1) or similar inorganic material be placed in the weep holes, or 1/2″ glass fiber insulation can be placed in open head joints. When weep holes are filled with inorganic materials they should be spaced every 16″. In all cases, special care in workmanship should be taken to keep weep holes free of mortar droppings.

Flashing Installation Masonry must be relatively smooth and free of projections that might puncture flashing and destroy its effectiveness. Through-wall flashing should be placed on a thin bed of mortar with another thin mortar bed placed on top of the flashing to receive and bond the next masonry

FIG. 15 **DETAILS OF FLASHING INSTALLATIONS IN DIFFERENT MASONRY WALL CONSTRUCTIONS**

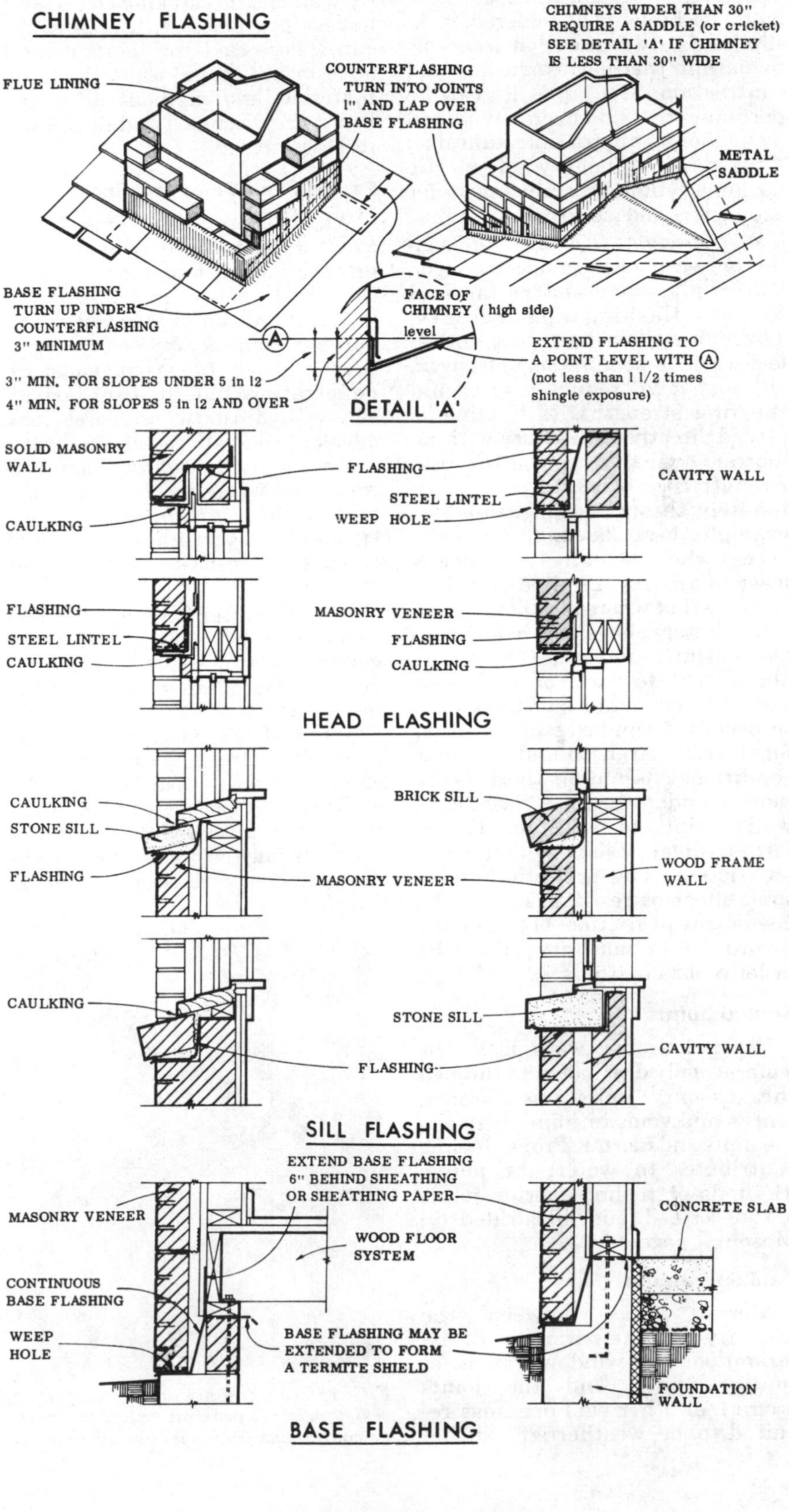

course. Flashing seams must be thoroughly bonded to prevent water penetration. Although most sheet metal flashing can be soldered, lock-slip joints are required at intervals to permit thermal expansion and contraction. Many plastics can be permanently and effectively joined by heat or an appropriate adhesive. The elastic pliability of plastic flashings eliminates the need for expansion and contraction seams.

Wall Strength The effect of flashing on wall strength depends upon mortar placement and mortar bond to both flashing and masonry. Through-wall flashing does not affect a wall's compressive strength, although it will reduce bending and shearing strengths. If flashing is placed directly on masonry with no mortar beneath it, the flexural strength is zero at that point. Fortunately, through-wall flashing (for example, base flashing) generally occurs where bending resistance is least significant. Limited test data indicate that where mortar is placed immediately above and below copper flashing, flexural strength is about 30% to 70% of unflashed walls. There is not sufficient data to permit a similar generalization for shearing strength under similar conditions. Assuming there is no mortar under the base flashing, a wall's ability to withstand lateral forces (shear resistance) depends on friction. The wall still retains some shearing resistance since the coefficient of friction between the flashing and masonry is in the order of 0.25 to 0.50.

Tooled Joints

When masonry walls leak, the water usually does not pass through the masonry units but through cracks and voids in joints between the units and mortar. Proper tooling contributes to watertight joints. Methods of tooling mortar joints are described and illustrated in Masonry, page 205-33.

Caulking Joints

Mortar joints between the masonry and the perimeter of exterior door and window frames, at control joints, and the joints around all other wall openings required to be weathertight should be cleaned out to a uniform depth of at least 3/4″ and then filled solidly with an elastic caulking compound forced into place with a pressure gun. These caulking joints should be from 1/4″ to 3/8″ wide. Caulking materials and methods of installation are discussed and illustrated in Division 400.

Parging and Dampproofing

Where subsoil conditions do not cause water to build up hydrostatic pressure against masonry foundation walls below grade, moisture penetration can be controlled by parging and *dampproofing* (coatings applied to resist moisture penetration due to capillarity). Where hydrostatic pressures may develop, concrete foundation walls are normally used rather than masonry and water penetration is controlled by *waterproofing* (continuous watertight membrane capable of resisting water under pressure).

Type M mortar parging is applied to the exterior face of the wall below grade and trowelled to a smooth, dense surface extending from the footing to 6″ above finished grade. The top of the parging should be beveled to form a wash and thickened to form a cove between the wall and the footing (Fig. 16). Coal tar or asphalt-based dampproofing products may be applied over parging to increase resistance to moisture penetration. They are available in solvent form for hot application or in emulsions for application at normal temperatures. They may be brushed, sprayed or troweled on the wall surface.

In well drained soils, one of three methods may be used: (1) parging alone may be applied in two 1/4″ thick coats of mortar; (2) one 1/4″ coat of parging followed by two brush coats of bituminous dampproofing; or (3) one heavy trowled-on coat of cold, fiber reinforced asphaltic mastic applied directly to the masonry surface.

In wet and impermeable soils, moisture control can be achieved either by (1) two 1/4″ thick coats of mortar plus two brush coats of bituminous dampproofing; or (2) one 1/4″ thick mortar application followed by one heavy troweled-on coat of cold, fiber-reinforced asphaltic mastic.

If parging is applied in two coats, the first should be roughened when partially set, hardened for 24 hours, then moistened before the second coat is applied. Parging should be moist cured for at least 48 hours and dry before dampproofing is applied. All pin holes in the dampproofing cover should be eliminated since dampproofing is only effective if it forms a continuous coating from the point above finish grade to a point below the cove at the top of the footing.

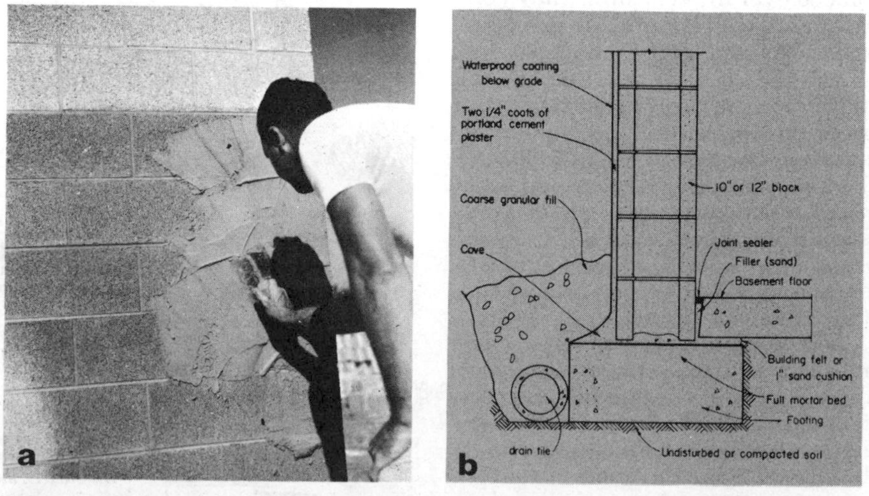

FIG. 16 (a) Masonry units in contact with the ground are parged or dampproofed to resist moisture penetration due to capillarity. (b) Parging and dampproofing should be continuous from above grade to below the cove formed at the footing.

CRACK CONTROL

Differential movements within masonry walls can produce stresses that may cause serious cracking unless precautions are taken to minimize them. Wall movements great enough to cause cracking of masonry are generally caused by one or more of the following conditions: (1) expansion or contraction of the masonry due to temperature changes; (2) expansion or contraction due to changes in moisture content of units—particularly concrete masonry units; (3) contraction in concrete masonry units due to chemical reactions called *carbonation* (see *Volume Changes,* Masonry page 205-17); (4) structural movements caused by unequal settlement of the foundation or at the point of concentration of applied loads; (5) concentration of stress that develops at wall sections where door, window and other openings occur; and (6) members built into the wall that locally restrain or prevent the wall from moving, as at the junction of floors, roofs, columns and intersecting walls.

The problem is further complicated by the fact that: (1) materials, particularly dissimilar materials as in a composite wall, may expand or contract at different rates; (2) exterior walls having cold exterior surfaces and warm interior surfaces may have tendencies to warp; (3) the degree of restraint to movement imposed by roofs, floors and intersecting walls or the effect of openings is difficult to estimate; and (4) metal lintels, windows and door frames have substantially greater coefficients of expansion than masonry and can cause relatively large differential movements.

To minimize the effects of such movements, expansion joints and control joints must be planned and built into the wall. Because so many of these variables may act singly or in combination, the location of expansion and control joints cannot be established with mathematical accuracy. The recommendations given below have evolved from experience and observation.

Expansion Joints For Clay Masonry

A clay masonry wall 100′ long

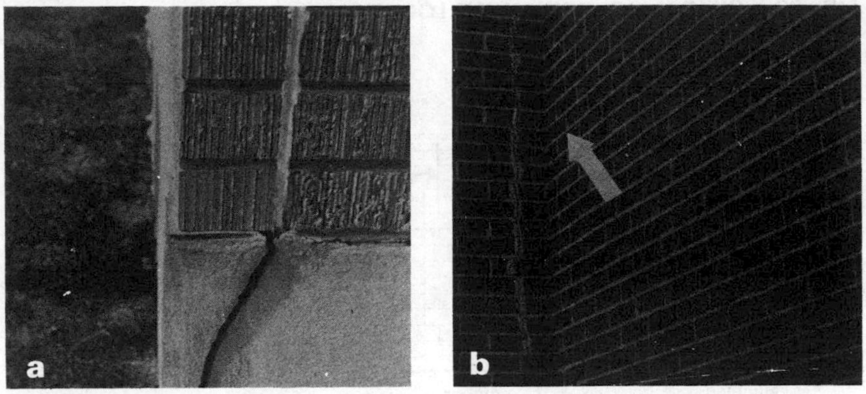

FIG. 17 Cracks in masonry due to wall movement: a) foundation, b) wall.

FIG. 18 **EXPANSION JOINT LOCATIONS IN MASONRY WALLS**

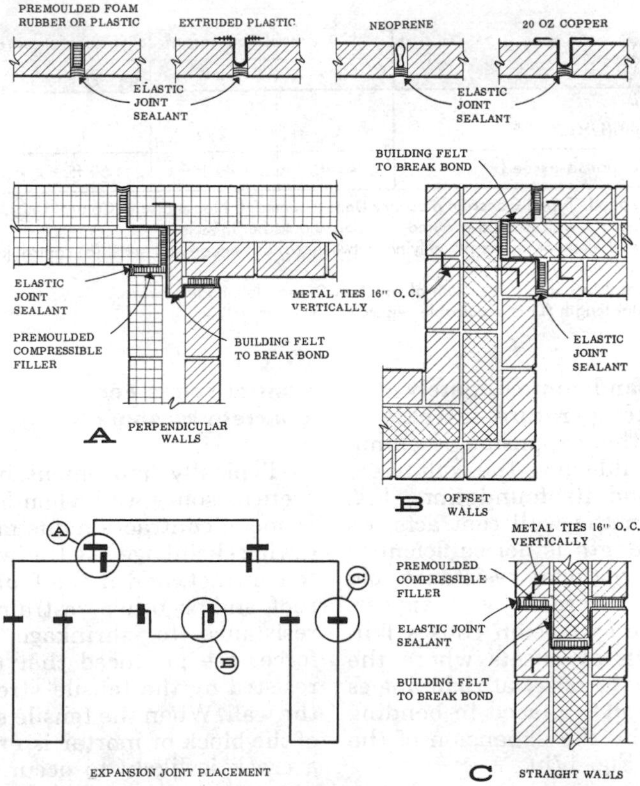

FIG. 19 **RECOMMENDED SPACING OF EXPANSION JOINTS**

Outside Temperature Ranges[1]	Maximum Length of Wall[3]			
	Unheated or Insulated		Heated not Insulated	
	Solid, ft.	Openings, ft.[2]	Solid, ft.	Openings, ft.[2]
100 and over	200	100	250	125
Less than 100	250	125	300	150

[1]The range from the lowest average temperature to the highest in degrees F.
[2]Openings 20% or more of wall area.
[3]For insulated cavity or veneer walls, distances are approximately 1/2 lengths shown.

FIG. 20 **TYPICAL CONTROL JOINT LOCATIONS**

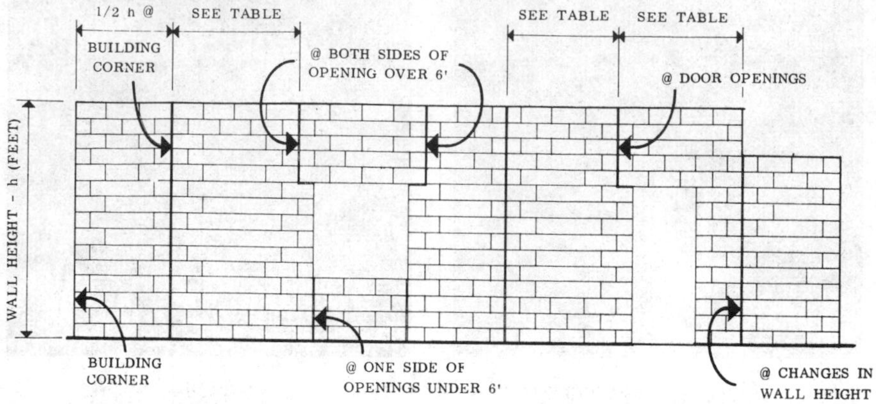

FIG. 21 **MAXIMUM SPACING OF CONTROL JOINTS IN UNREINFORCED MASONRY[1]**

Panel Dimensions[2]	Vertical Spacing of Joint Reinforcements[3]			
	None	24″	16″	8″
Ratio of panel[2] Length to height (L/H)	2	2½	3	4
Panel length (L) not to exceed	40 ft.	45 ft.	50 ft.	60 ft.

[1]Moisture controlled, Type I Concrete Masonry Units assumed. If Non-moisture Control Units, reduce joint spacing by 50%; if wall is solid-grouted, spacing should be 33% less.
[2]A panel is a wall element in one plane lying between: (a) corners or wall ends, (b) control joints or (c) a control joint and end wall.
[3]Continuous metal ties with a minimum of two No. 9 gage longitudinal wires.
[4]Maximum panel length (L) is applicable regardless of height (H).

might expand approximately 3/8″ due to a temperature increase of 100° F; if this expansion is cumulative it could cause the wall to extend beyond its foundation (Fig. 17a). When the wall contracts, its tensile strength is not sufficient to overcome frictional resistance and one or more cracks will appear. Cumulative expansion causes failure in walls at offsets where the masonry in the offset at right angles to a long wall is placed in bending and shear by the expansion of the long wall (Fig. 17b).

Expansion joints are installed to provide a complete separation through the structure. They are expensive, may be difficult to maintain and should be avoided if possible. In general, expansion joints should be located at offsets when the wall expanding into the offset is 50′ or more in length, and at junctions of walls in L, T or U shaped buildings (Fig. 18). The table in Figure 19 gives the recommended spacing.

Control Joints For Concrete Masonry

Typically, movement in a concrete masonry wall when first built is one of contraction due mainly to drying shrinkage. As the wall tries to contract, and if the foundation, roof and/or other restraint offers resistance to shrinkage, tensile forces are produced that must be resisted by the tensile strength of the wall. When the tensile strength of the block or mortar is exceeded, a crack is likely to occur.

Cracking due to shrinkage can be controlled by one or more methods: (1) moisture controlled, Type I concrete masonry units manufactured according to ASTM specifications to reduce shrinkage potential (see Fig. 44, page 205-21; (2) horizontal steel reinforcement to increase tensile resistance to cracking and to minimize the width of cracks; and (3) control joints located to accommodate wall movements.

Control joints are continuous vertical joints built into concrete masonry walls as they are laid up to minimize shrinkage cracks and to control cracks at locations where points of weakness or concentration of stresses are expected. When constructed properly, control joints should (1) permit slight wall movements to occur without causing the masonry units to crack, (2) seal the joint against the weather, and (3) stabilize the wall laterally across the joint.

Location and Spacing The layout of control joints will depend upon the following variables: the length and height of the wall, the type of wall, moisture content of units at time of laying, the spacing of joint reinforcement, the use of bond beams, architectural details, and *on the experience records as to the need for control joints in the locality where the building is built.* In the absence of specific local data or practice, the locations and spacings of control joints shown in Figures 20 and 21 are recommended.

Control joints should be placed at changes in wall height, or wall thickness junctions of load-bearing and non-load-bearing walls, at junctions of walls with columns or pilasters, where openings occur, at wall intersections and at return angles in L, T and U shaped buildings. In composite walls, where concrete masonry is used as back-up for other materials: (1) extend control joints through facing if masonry bonded; (2) control joint need not extend through facing if bonded with metal ties.

When stucco or plaster are applied directly to masonry units, control joints should extend through the stucco or plaster.

Because the temperature- and moisture-volume change of concrete masonry below grade in foundation walls is usually quite low, and because it is difficult to construct a watertight joint in masonry below grade, *control joints are not recommended for basement or foundation walls.* Where the above grade masonry is designed with control joints, movements at these joints may cause cracking to develop in the foundation walls below. To minimize this possibility, it is recommended that a continuous

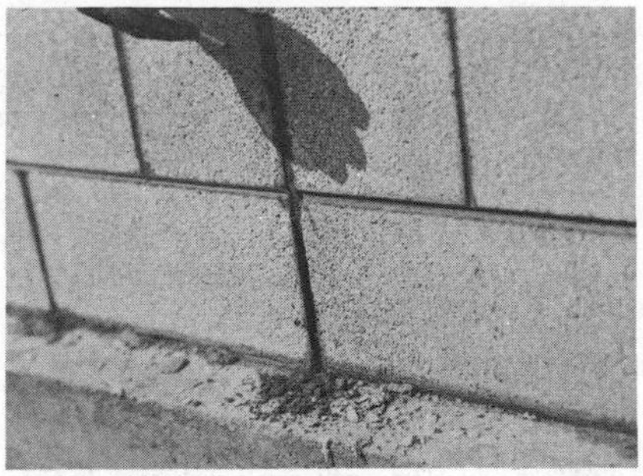

FIG. 22 Mortar in control joint is raked out to a depth of 3/4".

FIG. 23 Control joint is sealed with knife grade caulking compound.

FIG. 24 Control joint may be constructed with building paper or roofing felt inserted in end core of block.

FIG. 25 Paper or felt prevents mortar from bonding on one side of the joint and the filled core provides stability.

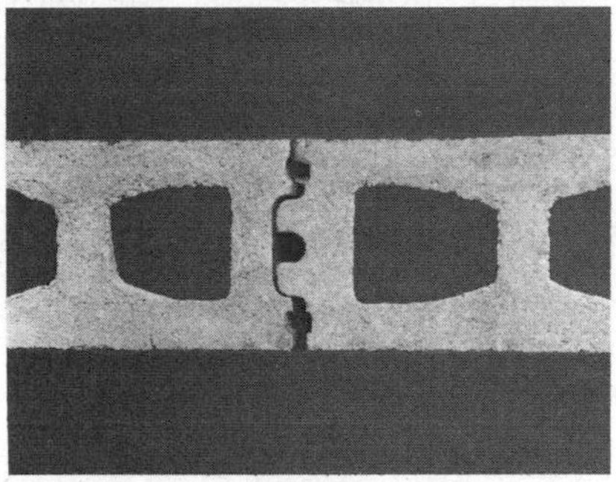

FIG. 26 Special control joint block provides lateral stability by means of tongue-and groove-shaped ends.

FIG. 27 Control joint blocks are made in full- and half-length units.

bond beam be provided at the top of the foundation walls and that the control joints in the above grade wall be stopped at the bond beam.

Constructing a Control Joint Control joints should be laid up in mortar just as any other vertical joint. If the control joint is to be exposed to the weather or to view, a recess should be provided for the caulking material by raking out the mortar to a depth of about 3/4" after the mortar has become quite stiff (Fig. 22). A pressure gun and a thin flat caulking trowel are used to force knife-grade caulking compound into the joint (Fig. 23). To keep them as unnoticeable as possible, they should be built plumb and of the same thickness as the other mortar joints.

One type of control joint can be built with regular stretcher block. Building paper or roofing felt is inserted in the end core of full (and half-length) block and extends the full height of the control joint. The core is then filled with mortar for lateral support (Fig. 24). The paper, or felt, cut to convenient lengths and wide enough to extend across the joint, prevents the mortar from bonding on one side of the joint, thus permitting the control joint to function (Fig. 25). A control joint block with tongue-and-groove shaped ends, available in many areas, provides lateral support to wall sections on each side of the control joint (Fig. 26). These control joint blocks are made in full and in half-length units (Fig. 27).

Control joints are often unsightly and require maintenance, hence their use should be limited in so far as possible. Methods used to increase the tensile strength of the wall and to improve the dimensional stability of the units will reduce the number of control joints required.

Other methods are also available to increase the crack-resistance of walls. Two-core block with thickened fillets or three-core block with thickened center core increase the tensile strength of the concrete and crack resistance of the wall approximately 50% and 25% respectively over the typical three-core block shown. Note in Figure 21, when joint reinforcement is used, control joint spacing may be increased.

COLD WEATHER CONSTRUCTION

During below-normal temperatures (40°F and below) masonry strength and mason productivity may be affected. As the air temperature falls below 40°F, more of the masonry materials must be preconditioned to achieve satisfactory masonry strength. The chemical reaction between portland cement and water in mortar is slowed or stopped at temperatures below 40°F. Hydration and strength development proceed only above 32°F and only when sufficient water is available.

Cold weather masonry construction may proceed at temperatures below freezing, provided the mortar ingredients are heated and, as the temperature drops, the newly laid masonry is maintained above freezing during the early hours after construction. Heated mortar ma-

FIG. 28 RECOMMENDED PROTECTION FOR COLD WEATHER MASONRY CONSTRUCTION[1]

Work Day Temperature	Construction Requirement[2]	Recommended Protection[2]
All Temperatures	Normal masonry procedures.	Masonry materials should be received, stored and protected in ways that prevent water damage. Cover tops of newly laid masonry with plastic or canvas to prevent water damage.
40°F to 32°F	Heat mixing water to produce mortar temperatures between 40°F-120°F. Never place masonry on frozen, snowy or ice-covered surfaces.	Cover walls and materials with plastic or canvas to prevent wetting and freezing.
32°F to 25°F	Heat mixing water and sand to produce mortar temperatures between 40°F-120°F.	Provide windbreaks for wind velocities over 15 mph. Cover walls to prevent freezing. Maintain masonry above 32°F for 16 hours using auxiliary heat or insulating blankets.
25°F to 20°F	Mortar on boards should be maintained above 40°F. Masonry units below 20°F must be heated above 20°F.	
20°F to 0°F And below	Heat mixing water and sand to produce mortar temperatures between 40°F-120°F.	Provide enclosures and supply heat to maintain masonry enclosure above 32°F for 24 hours.

[1]For unenclosed construction sites. These recommendations do not apply where the complete work area, including materials storage space, is enclosed and heated.
[2]Each requirement is in addition to the one above it.

FIG. 29 ESTIMATED *U* VALUES FOR 8″ HOLLOW CONCRETE MASONRY WALLS

Wall Details	*U* Value Based on Density of Concrete Used in Block				
	60 pcf	80 pcf	100 pcf	120 pcf	140 pcf
No insulation	0.32	0.34	0.38	0.43	0.55
No insulation, ½″ gypsum board on furring strips	0.21	0.23	0.25	0.27	0.31
No insulation, ½″ foil-backed gypsum board on furring strips	0.15	0.15	0.16	0.17	0.19
Loose fill insulation in cores	0.12	0.14	0.18	0.21	0.35
Loose fill in cores, ½″ gypsum board on furring	0.10	0.12	0.14	0.17	0.24
Loose fill in cores, ½″ foil-backed gypsum board, furring	0.08	0.10	0.11	0.12	0.16
1″ rigid glass fiber, ½″ gypsum board	0.14	0.14	0.15	0.15	0.17
1″ polystyrene, ½″ gypsum board	0.12	0.12	0.12	0.13	0.14
1″ polyurethane, ½″ gypsum board	0.10	0.10	0.11	0.11	0.12
Loose fill in cores plus 1″ rigid glass, ½″ gypsum board	0.08	0.09	0.10	0.11	0.14
Loose fill in cores plus 1″ polystyrene, ½″ gypsum board	0.07	0.08	0.09	0.10	0.12
Loose fill in cores plus 1″ polyurethane, ½″ gypsum board	0.07	0.07	0.08	0.09	0.11
R-7 blanket insulation, ½″ gypsum board, furring	0.09	0.10	0.10	0.10	0.11

terials produce mortars with performance characteristics identical to those in the normal temperature range.

Recommendations for cold weather masonry construction are discussed in Masonry, pages 205-28, -34, -35, and are summarized in Figure 28.

HEAT CONTROL

The amount of thermal insulation installed in a masonry wall depends on considerations of human comfort and the effect of total heat loss and heat gain and, hence, on the cost of heating and cooling the building. Estimated savings in both heating and air conditioning costs may be used as a guide to determine the amount of insulation that can be installed economically. In many cases, insulation may be necessary only to raise or lower the inside surface temperature of the wall to increase human comfort (see Section 105 Heat Control). Cold wall surfaces absorb radiant heat from the body and may produce a feeling of chill even though the inside air temperature is 70° F or above. This comfort factor may dictate the use of insulation which could not be justified by economical considerations alone.

Hourly heat loss or gain through an exterior wall is measured by its *U-value*. *U-value* is the amount of heat in *British Thermal Units* (BTU) transmitted in one hour per sq. ft. of wall area, for a difference in temperature of 1°F between the inside and outside air. Figure 29 gives U-values of 8″ walls with various combinations of finish and insulation. *R-value* is the reciprocal of the U-value (R=I/U). R-values are useful for estimating the effect of the individual components of a building section on the total heat flow because R-values can be directly added. See Section 105 Heat Control for more detailed discussion of U- and R-values. Figure 30 lists R-values for single-wythe concrete masonry walls. Figure 31 gives R- and U-values for some typical concrete masonry walls.

Insulating Masonry Walls

In general, the methods used to add insulation to masonry walls are illustrated in Figures 32 through 38, and include: (1) fill insulation poured in the cavity or core space of the masonry; (2) rigid insulation placed in or applied to the face of the wall; and (3) batt or blanket insulation placed in a furring space on the interior face of the wall.

Loosefill insulation used in cavity walls and the cores of hollow masonry should possess these properties: (1) insulation should not transmit moisture from exterior to interior wythe; (2) insulating efficiency should not be impaired by retained quantities of moisture; (3) loose fills should support their own weight without settlement; (4) insulation should be inorganic or have comparable resistance to decay, fire, and vermin; and (5) granular fills must be pourable in lifts of at least 4′.

Types of loose fill meeting the above criteria are: (1) water-repellent, expanded vermiculite; and (2) silicone-treated, expanded perlite. Rigid boards having the required properties include: (1) expanded or molded polystyrene, (2) expanded polyurethane, (3) rigid urethane, (4) cellular glass, (5) preformed fiber glass, and (6) perlite board.

FIG. 30 ***R* VALUES OF SINGLE-WYTHE CONCRETE MASONRY WALLS**[1]

Nominal Wall Thickness, In.	Insulation In Cells[2]	*R*-value Based on Concrete Unit Weight 60 pcf	80 pcf	100 pcf	120 pcf	140 pcf
4	Filled	3.36	2.79	2.33	1.92	1.14
	Empty	2.07	1.68	1.40	1.17	0.77
6	Filled	5.59	4.59	3.72	2.95	1.59
	Empty	2.25	1.83	1.53	1.29	0.86
8	Filled	7.46	6.06	4.85	3.79	1.98
	Empty	2.30	2.12	1.75	1.46	0.98
10	Filled	9.35	7.45	5.92	4.59	2.35
	Empty	3.00	2.40	1.97	1.63	1.08
12	Filled	10.98	8.70	6.80	5.18	2.59
	Empty	3.29	2.62	2.14	1.81	1.16

[1]*R* values do *not* include the sums of the effect of air film or surface conductance on the inside of the walls ($1/f_i = 0.68$) and on the outside ($1/f_o = 0.17$).

[2]Loose fill insulation such as perlite, vermiculite, or others of similar density.

Capacity Insulation

In warm climates and the summer months in cooler climates, the heat gain characteristics of walls are important. Heat gain in buildings with masonry walls depends on the capacity insulation—heat storage capacity—of the walls. The greater the capacity insulation, the lower the heat gain and consequently the cooler the inside temperatures. In general, heavier walls have a greater capacity insulation. They absorb and retain the heat during the day and dissipate it to the outside during the cooler hours of the night.

CLEANING MASONRY

See Masonry, page 205-35, for a discussion of methods of removing mortar, mortar stains or efflorescence from masonry wall surfaces

FIG. 31 **R AND *U* VALUES OF SOME TYPICAL CONCRETE MASONRY WALLS**

Components	Heat Resistance Values, *R* A	B	C	D1	D2	E	F1	F2
Surface film (outside)	0.17	0.17	0.17	0.17	0.17	0.17	0.17	0.17
8″ hollow concrete masonry at 100-pcf density (cores open)	1.75			1.75	1.75			
Cores filled with bulk insulation			4.85				4.85	4.85
Concrete brick at 140-pcf density		0.44				0.44		
2″ air cavity		0.97						
4″ hollow concrete masonry at 100-pcf density		1.40				1.40		
2″ bulk insulation in cavity						4.00		
Batt insulation between furring strips: 1″				3.70			3.70	
2 or 2¼″					7.00			7.00
½″ gypsum board interior finish				0.45	0.45		0.45	0.45
⅝″ plaster, lightweight aggregate		0.39				0.39		
Surface film (inside)	0.68	0.68	0.68	0.68	0.68	0.68	0.68	0.68
TOTAL resistance, *R*	2.60	4.05	5.70	6.75	10.05	7.08	9.85	13.15
U value, 1/*R*	0.384	0.247	0.175	0.148	0.100	0.141	0.102	0.076

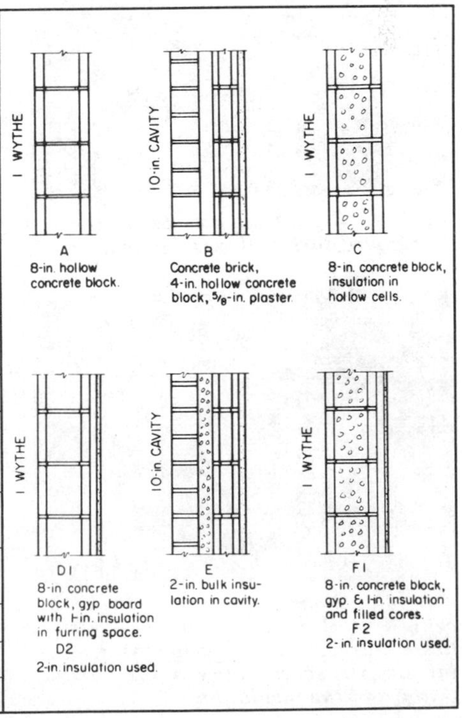

FIG. 32 Loose fill insulation being blown into an existing wall.

FIG. 33 Blanket or batt insulation installed between furring strips attached to masonry.

FIG. 34 Rigid foam polystyrene insulation board installed in a cavity wall.

FIG. 35 Glass fiber insulation board being installed in a cavity wall. Brick ties hold insulation against wythe leaving an air space between the outside wythe and the insulation.

of new construction.

PAINTING MASONRY

The cost of painting a *clay* masonry wall in most cases must be charged to appearance alone, not to increased performance or resistance to weather. The only possible exception would be cement-water paint applied to correct a leaky wall condition. In the case of *concrete* masonry, paint not only provides an attractive finish but may also be necessary to make the masonry surface watertight.

Masonry paints are of four general classifications: (1) *cement-based* paints which come in powdered form and are mixed with water before applying; (2) *water-thinned* emulsion paints which include butadiene-styrene, vinyl acrylic, alkyd, and multicolored lacquers; (3) *fill coats* which are similar to cement-based paints but contain an emulsion paint in place of some water, giving improved adhesion and tougher film than unmodified cement paints; and (4) *solvent-thinned paints* which, except for special-purpose paints and special applications, should only be used on interior masonry walls which are not susceptible to moisture penetration. Solvent-thinned paints include oil-base, alkyd, synthetic rubber, chlorinated rubber and epoxy. (See also Section 216 Paints & Protective Coatings).

FIREPLACES

The design and construction of an efficient, functional fireplace requires adherence to some basic rules concerning dimensions and the placement of the various component parts. The objectives of a correctly designed fireplace are to: (1) assure proper combustion of the fuel; (2) deliver smoke and other products of combustion up the chimney; (3) radiate the maximum amount of heat into the room; and (4) afford simplicity and fire safety in construction.

The first two of these objectives are closely related and depend mainly upon the shape and dimensions of the combustion chamber, the proper locating of the fireplace throat and the smoke shelf, and the ratio of the flue area to the area

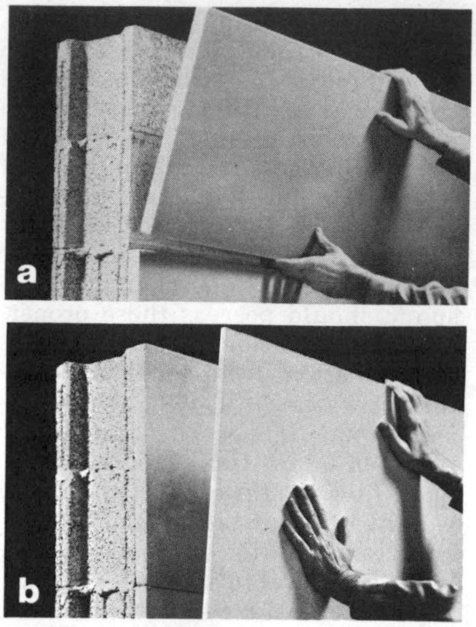

FIG. 36 (a) Insulation board is adhesive to bonded interior face of masonry; (b) gypsum board is adhesive bonded to glass fiber insulation board.

FIG. 37 Granular insulation fills a brick cavity wall; it can also be used to fill concrete block cores.

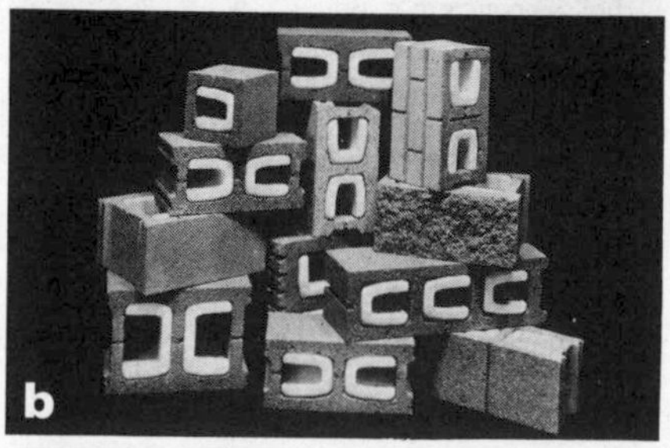

FIG. 38 (a) Insulation inserted in the face-shell of specially formed blocks; (b) specially formed insulation inserted in cores of concrete block.

of the fireplace opening. The third objective depends upon the dimensions of the combustion chamber, while the fourth depends upon the size and shape of the masonry units and their ability to withstand high temperatures without warping, cracking or deterioration. Typical details of fireplace construction are shown in Figure 39.

Size And Shape

The size of a fireplace depends not only on esthetic considerations, but also on the size of the room to be heated. For a room with 300 sq. ft. of floor area, a fireplace with an opening 30″ to 36″ wide is sufficient. Approximate dimensions for sizing the fireplace are given in Figure 40.

Combustion Chamber The shape of the combustion chamber influences both the draft and the amount of heat that will be radiated into the room. The slope at the back throws the flame forward and leads the gases with increasing velocity through the throat. The slope at the back and sides, also radiates the maximum amount of heat into the room. The dimensions for the correct wall angles and slopes given in Figures 39 & 40 may be varied slightly to correspond with brick coursing, but no major changes in dimensions should be attempted.

The combustion chamber, unless it is of the metal preformed type, should be lined with firebrick and should be laid in fireclay mortar. The back and end walls should be at least 8″ thick to support the weight of the chimney above.

Throat Because of its effect on the draft, the throat of the fireplace must be carefully designed. It should be not less than 6″ and preferably 8″ above the highest point of the fireplace opening. The sloping back should extend to the same height and form the support for the back of the damper. A metal damper should be placed in the throat and extend the full width of the fireplace opening, preferably of a design in which the valve plate will open upward and toward the back. This plate when open will form a barrier for downdrafts and will deflect them upward with the

FIG. 39 **TYPICAL FIREPLACE CONSTRUCTION DETAILS**

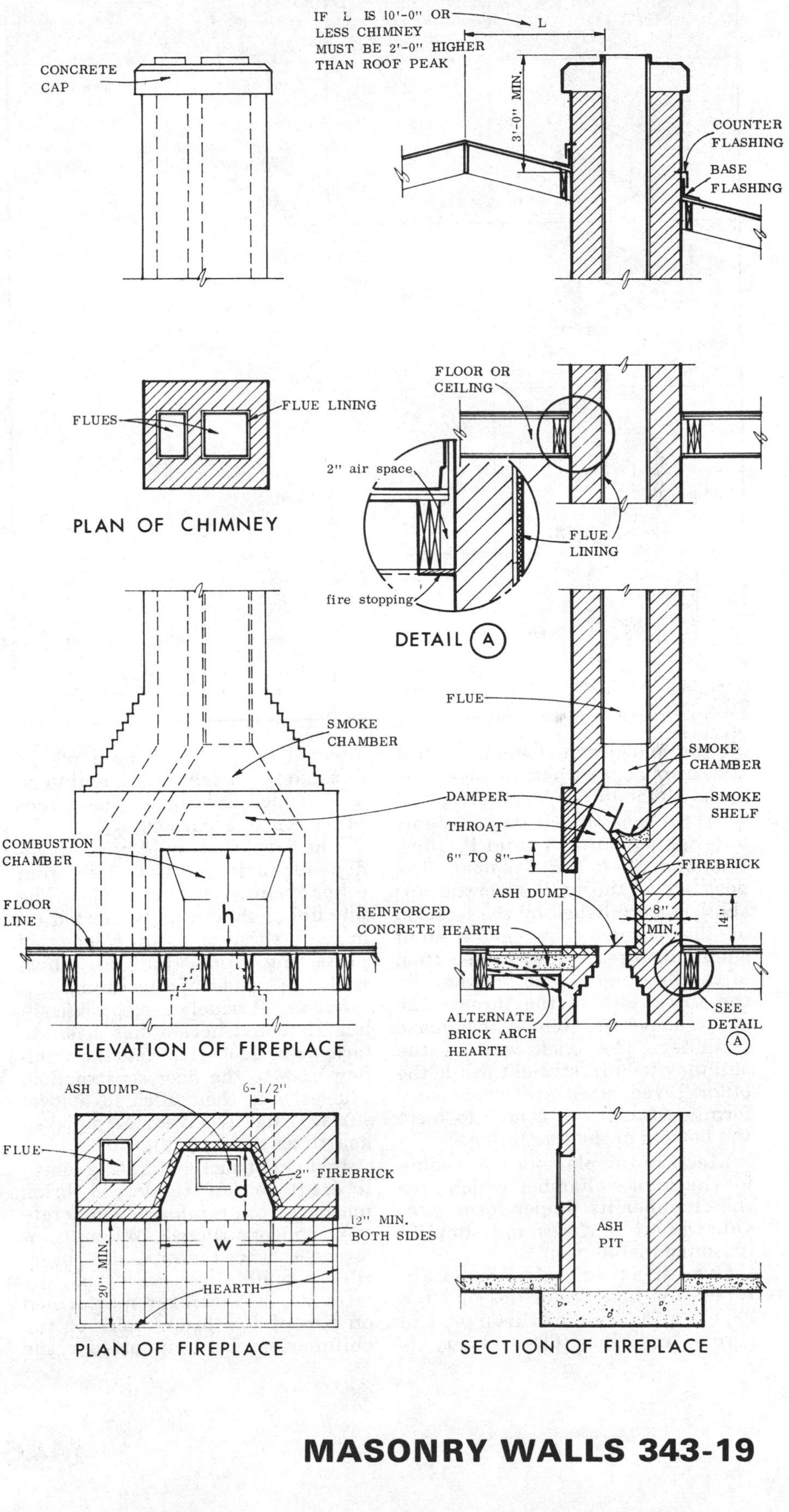

FIG. 40 **FIREPLACE TYPES AND DIMENSIONS**

TYPE	OPENING HEIGHT h"	HEARTH SIZE w" x d"	FLUE SIZE (standard modular)
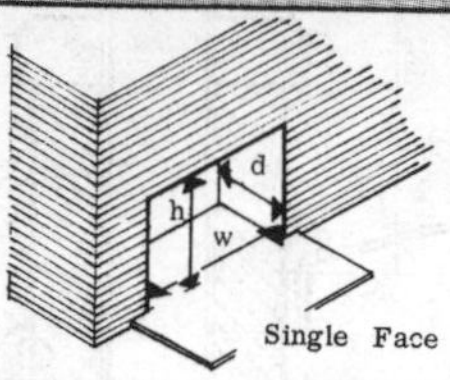 Single Face	29"	30" x 16'	12" x 12"
	29"	36" x 16"	12" x 12"
	29"	40" x 16"	12" x 16"
	32"	48" x 18"	16" x 16"
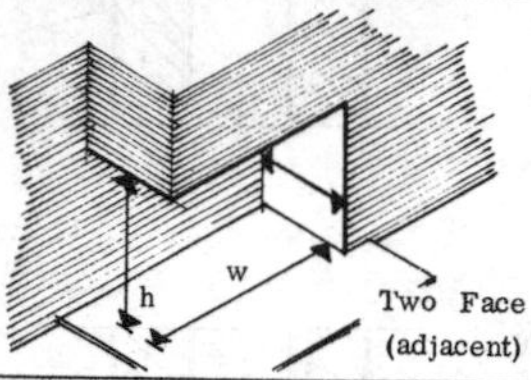 Two Face (adjacent)	26"	32" x 16"	12" x 16"
	29"	40" x 16"	16" x 16"
	29"	48" x 20"	16" x 16"
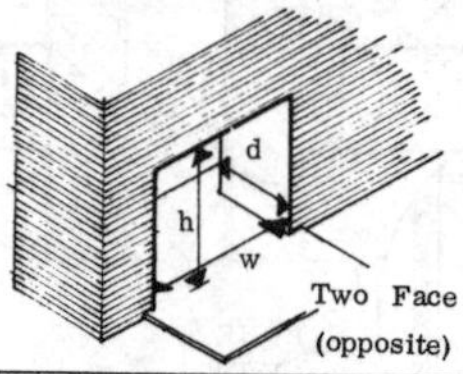 Two Face (opposite)	29"	32" x 28"	16" x 16"
	29"	36" x 28"	16" x 20"
	29"	40" x 28"	16" x 20"
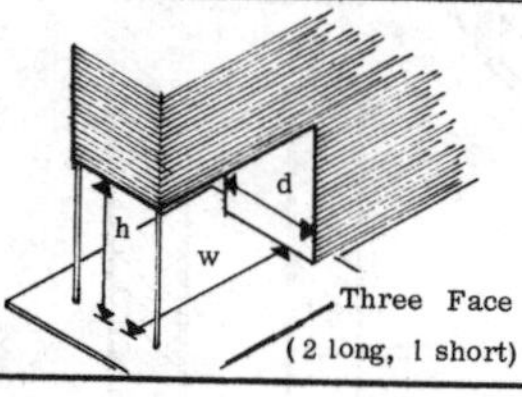 Three Face (2 long, 1 short)	27"	36" x 32"	20" x 20"
	27"	36" x 36"	20" x 20"
	27"	44" x 40"	20" x 20"

ascending column of smoke. When the fireplace is not in use, the damper should be kept closed to prevent heat loss from the room and to keep out dirt from the flue.

Smoke Shelf and Chamber The position of the smoke (downdraft) shelf is established by the location of the throat. The smoke shelf should be directly under the bottom of the flue and extend horizontally the full width of the throat. The space above the shelf is the smoke chamber. The back wall of the chimney is built straight, while the other three sides are sloped uniformly toward the center to meet the bottom of the flue lining.

Metal lining plates are available for the smoke chamber which give the chamber its proper form, provide smooth surfaces and simplify masonry installation.

Flue It is desirable to obtain relatively high velocities of flue gases and smoke through the throat and flue. The velocity is affected by both the area of the flue and the height of the chimney. The proper sizes of fireplace flues are given in Figure 40.

The fireplace should have an independent flue entirely free from other openings or connections. The flue lining should be supported on at least three sides by a ledge of protecting brick, finishing flush with the inside of the lining.

Hearth A widely used practice in hearth construction has been to form an arch of brick from the chimney base to the floor construction. The arch is then filled to a level surface and receives the finished hearth as shown in Figure 39.

Instead of such an arch, a cantilevered slab of reinforced brick masonry or reinforced concrete may be constructed. When reinforced brick masonry is used, ribbed metal lath serves as the form and is laid in the mortar bed on top of the brick walls of the chimney base and supported at the projecting end by a temporary form. On the metal lath is placed a bed of Type M mortar of sufficient depth to cover the ribs. This is followed at once by the placement of the brick on edge 1/2″ apart without mortar. The joints should then be filled with cement mortar grout and the reinforcing bars set in place, pressing them down into their proper position approximately 1-1/2″ below the top surface.

Fireplace Types

There are several types of fireplaces being used today, and the basic principles involved in their design and construction are the same. The single-faced fireplace is the most common and the oldest variety, and therefore most of the standard design information is based on this type.

The multi-faced fireplace, if used properly, may be effective and attractive. It presents certain problems as to draft and opening size which usually must be solved on an individual basis. Certain standard sizes that have been found to work satisfactorily under most conditions are given in Figure 40.

Most fireplaces are not very energy efficient. They can be made more energy efficient by the following practices: (1) Provide a source of outside air for fireplace combustion, so that heated air is not drawn out of the house. (2) Adjust the damper to the type of fire—a slow-burning log may only need a 1″ or 2″ damper opening to provide adequate combustion without smoking. (3) Provide a glass screen with an adjustable damper, so that the damper in the throat of the fireplace can be left open while the combustion air is controlled by the damper in the glass screen. When the fireplace is not in use the glass screen can be closed, preventing heated inside air from being wasted.

CHIMNEYS

Due to the large mass and weight of a chimney, it is quite possible that its foundation will have to support a greater unit load than the surrounding foundations. Hence, special attention must be given to the chimney footing and the rest

FIG. 41 SMOKE PIPE CONNECTION TO CHIMNEY

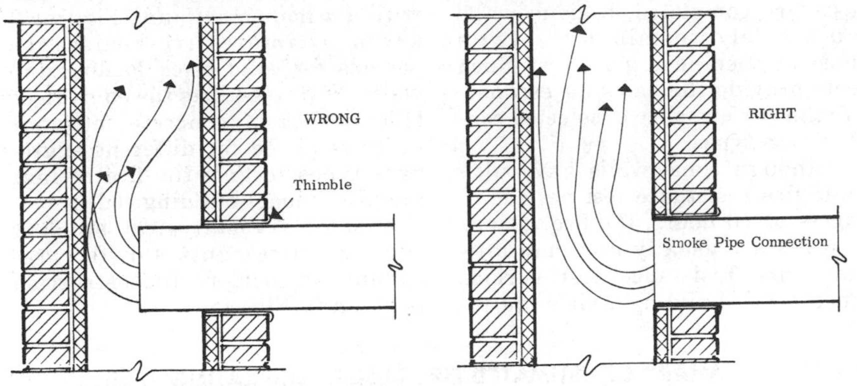

of the foundation's design so the resulting unit load on the soil under all portions of the building foundation will be nearly the same. This is done to prevent *uneven* settling of the foundation which could result in cracking of masonry or other damage to the building.

The thickness of the masonry in the base of a chimney that contains only flues may be 4″ thick, but if a fireplace is included it must be at least 8″. The ash pit in the base should be fitted with a tight cast iron door and a frame of the same material anchored securely approximately 16″ above the floor. If there is no basement, the door should be located one or two courses above the finish grade line. Flue lining is not required 8″ below the point where the lowest smoke pipe thimble or flue ring enters the flue.

Each flue must be built as a separate unit entirely free from other flues or openings. Flue walls should have all joints completely filled with mortar. In masonry chimneys with walls less than 8″ thick, liners should be separated from the chimney wall and the space between the liner and masonry should not be filled; only enough mortar should be used to make a good joint and hold the liners in position. The exposed joints inside the flue should be struck smooth. The interior surface should not be plastered.

Although unlined *fireplace chimneys* with masonry walls at least 8″ thick may perform adequately, the use of fireclay flue lining is recommended. The use of flue lining provides a smooth interior surface and hence a more efficient operation of the flue.

FIG. 43 ULTIMATE FIRE RESISTANCE PERIODS FOR LOADBEARING CLAY AND SHALE BRICK WALLS

Nominal Wall Thickness in.	Wall type	Ultimate Fire Resistance Period in Hours: Incombustible Members Framed Into Wall or No Framed-in Members — No Plaster	Plaster[1] on One Side	Plaster[1] on Two Sides	Combustible Members Framed Into Wall — Plaster No Plaster	Plaster[1] on Exposed Side
4	Solid	1¼	1¾	2½	—	—
8	Solid	5	6	7	2	2½
12	Solid[2]	10	10	12	8	9
12	Solid[3]	12	13	15	—	—
10	Cavity	5	6	7	2	2½

[1]To achieve these ratings, each plastered wall face must have at least ½-in., 1:3 gypsum-sand plaster.
[2]Based on load failure.
[3]Based on temperature rise (for non-loadbearing walls).

FIG. 42 STANDARD TIME-TEMPERATURE CURVE (ASTM E 119 FIRE TEST)

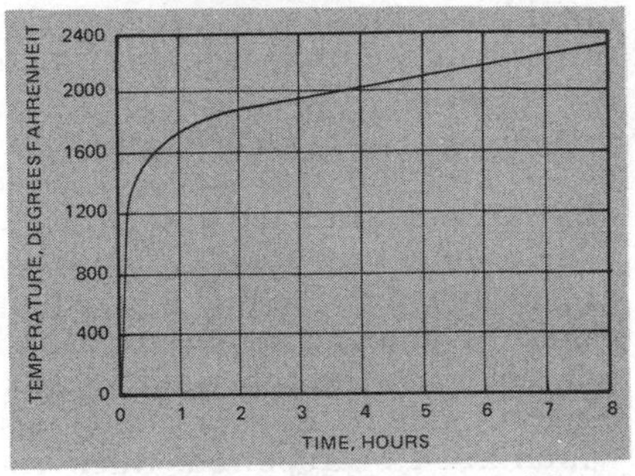

The tops of chimneys without flue linings may have to be rebuilt every few years due to the disintegrating effect of smoke and gases on the mortar. The chimney must be taken down to a point where the mortar joints are solid and a new top rebuilt.

Flue linings should be set into place before the brickwork has reached the top of the flue lining section below, and then the brick built up around them. Where offsets or bends are necessary in the chimney, they should be formed by equally mitering both ends of abutting sections of lining. This prevents reduction of the flue area, since it is important that the same effective area be maintained the whole height of the flue.

Smoke pipe connections should enter the side of the flue at a thimble or flue ring, which should be built as the work progresses. The metal smoke pipe should not extend beyond the inside face of the flue (Fig. 41). The top of the smoke pipe should not be less than 18″ below the ceiling and no wood or combustible materials should be placed within 6″ of the thimble.

At the tops of the chimneys, the flue lining should project at least 4″ above the top course or capping and should be surrounded with portland cement mortar at least 2″ thick, finished with a straight or concave slope to direct the air currents upward at the top of the flue. This wash also serves to drain water from the top of the chimney and should be constructed with a drip to keep the walls dry and clean.

Wood framing and furring members should never be placed closer than 2″ to the walls of the chimney. The space between chimney and floor members should be filled with an incombustible material providing a firestop to prevent the transfer of fire from below through the space.

After the chimney has been completed and the masonry cured, a test may be conducted on each flue. This is done by building a smudge fire at the top of the flue and, when smoke is pouring freely from the top, covering the top tightly. Any escaping smoke indicates openings which must be repaired. Such repair is usually difficult and expensive, hence it is far more satisfactory and economical to see that construction is properly executed as the work progresses.

Smoke passages and chimney flues should be kept clean. The most efficient method of cleaning a chimney flue is by means of a weighted brush or bundle of rags lowered and raised from the top. An accumulation of soot may cause a chimney fire with consequent danger of sparks igniting the roof or damage the chimney, which could permit the passage of fire.

FIRE RESISTANCE

The fire resistance of walls, floors and partitions is usually based on the Standard Methods of Fire Tests of Buildings' Construction and Materials, ASTM Designation E119. The standard fire test does not measure the fire hazard in terms of actual performance in a real fire. Rather, the standard fire test gives a comparison of the measure of performance between assemblies tested under similar conditions.

The test consists of exposing one side of a wall to a fire of controlled intensity measured by the standard time-temperature curve (Fig. 42). Immediately after firing, the hot face of the wall is subjected to a fire hose stream. The wall must withstand the fire and hose stream without passage of flame or gases, and heat transmission through the wall must be limited to less than a 250°F rise in temperature. A loadbearing wall must also be able to resist its assigned design stresses without failing. Within 72 hours after the tests are completed, bearing walls must safely sustain loads twice those applied during the test. Such tests provide the basis for estimating the performance of selected construction types.

Although some walls have ultimate fire resistance test periods in excess of 10 hours, the fire ratings of walls are usually less than ultimate fire resistance test periods since most building code requirements are in multiples of one hour with a 4 hour maximum. Figure 43 gives ultimate fire resistance periods for solid brick loadbearing walls. Figure 44 lists the equivalent thicknesses of concrete masonry units required for different aggregate types to give the fire ratings found in model building codes.

A summary of typical building code requirements for fire resistance in residential construction is given in Fig. 45.

FIG. 44 **SUMMARY OF ESTIMATED FIRE RESISTANCE RATINGS FOR CONCRETE WALLS OF EQUIVALENT THICKNESS**[1]

Concrete Masonry Units	Members Framed Into Wall or Partition						
	None or Noncombustible (in.)[2]						
	4-Hr.	3-Hr.	2-Hr.	1½-Hr.	1-Hr.	¾-Hr.	½-Hr.
Expanded Slag or Pumice Aggregates	4.7	4.0	3.2	2.7	2.1	1.9	1.5
Expanded Clay or Shale Aggregates	5.7	4.8	3.8	3.3	2.6	2.2	1.8
Limestone, Cinders, or Unexpanded Slag Aggregates	5.9	5.0	4.0	3.4	2.7	2.3	1.9
Calcareous Gravel Aggregates	6.2	5.3	4.2	3.6	2.8	2.4	2.0
Siliceous Gravel Aggregates	6.7	5.7	4.5	3.8	3.0	2.6	2.1

[1]Equivalent thickness is the solid thickness that would be obtained if the same amount of concrete contained in a hollow unit were recast without core holes.
[2]Where combustible members are framed into the wall, the wall must be of such thickness or be so constructed that the thickness of solid material between the end of each member and the opposite face of the wall, or between members set in from opposite sides, will not be less than 93% of the thickness shown in table.

FIG. 45 **TYPICAL CODE REQUIREMENTS FOR FIRE SAFE RESIDENTIAL CONSTRUCTION**

Building Component	30-Story	8-10 Story	4-5 Story	3-Story Garden
	Type I Fireproof	Type II Noncombustible	Type III Ordinary—Protected	Type IV Ordinary—Unprotected
	Required Fire Resistance Ratings (Hrs.)			
Fire Walls	4	4	4	4
Exterior Bearing Walls	4	3	3	3
Exterior Nonbearing Walls	2	2	2	2
Interior Bearing Walls	4	2	1	1
Stairs	2	2	2	1
Elevator Shaft	2	2	2	1
Pipe and Duct Shafts	2	2	2	1
Corridor Walls	2	2	2	1
Heat Plant & Boiler Room Walls	2	2	1	1

STRENGTH AND STABILITY

Under average conditions of exposure and loading, the minimum requirements of the American National Standards Institute (ANSI), Building Code Requirements for Masonry, (ANSI A41.1-1953) (R 1970) and the American Concrete Institute (ACI), Building Code Requirements for Concrete Masonry Structures, (ACI Standard 531-79) are recommended as the basis for wall design. The requirements for wall thicknesses and the provisions for unsupported wall heights and wall lengths have developed from performance records over a long time and are incorporated in most up-to-date building codes.

Many of the strength properties and stability measures for masonry walls—compressive and transverse strengths, lateral support and wall thickness, bonding and anchorage, the support of walls over openings—are discussed below.

COMPRESSIVE STRENGTH

The *ultimate* compressive strength (strength at failure) of masonry walls is closely related to the compressive strength both of the masonry units and the mortar. The quality of workmanship, thickness of mortar joints, regularity of the units' bearing surfaces, and workability of mortar are also important.

High compressive strength is developed by using both high strength units and mortars and by using proper workmanship to completely fill all joints with mortar. The *allowable* compressive stresses (strength used for designing purposes) recommended for various

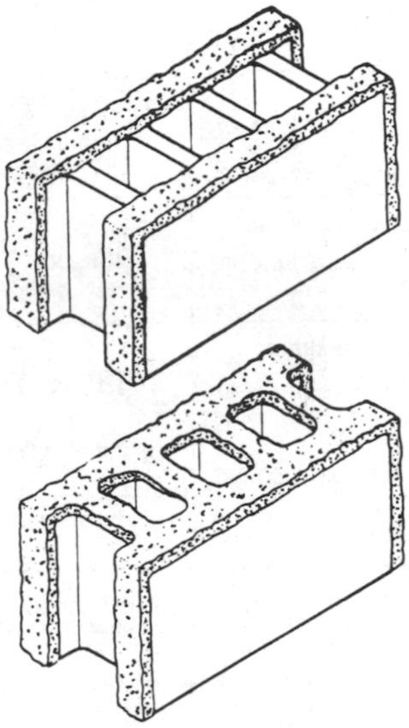

FIG. 47 Examples of full mortar bedding, bottom, and face-shell mortar bedding, top.

FIG. 46 ALLOWABLE COMPRESSIVE STRESSES IN UNIT MASONRY

TYPE OF WALL CONSTRUCTION	COMPRESSIVE STRENGTH OF UNITS in PSI	ALLOWABLE COMPRESSIVE STRESSES OVER GROSS CROSS-SECTIONAL AREA in PSI[1]				
		Type M Mortar	Type S Mortar	Type N Mortar	Type O Mortar	Type K Mortar
Solid Masonry of clay, shale or concrete brick	8000 and over	400	350	300	200	100
	4500 to 8000	250	225	200	150	100
	2500 to 4500	175	160	140	110	75
	1500 to 2500	125	115	100	75	50
Piers of Hollow Units		85	75	70	—	—
Solid Masonry of Solid Concrete Masonry Units	Type I	175	160	140	100	—
	Type II	125	115	100	75	—
Piers of Hollow Units[2]		105	95	90	—	—
Hollow Walls (cavity or masonry bonded)[3]	Solid Units:	Solid Units:				
	2500 and over	140	130	110	—	—
	1500 to 2500	100	90	80	—	—
	Hollow Units	70	60	55	—	—

[1]Mortar types, mortar proportions and the recommended mortar types for various construction applications are given on page WF205-5.
[2]Cellular spaces of hollow units filled solidly with concrete or Type M or Type S mortar.
[3]On gross cross-sectional area of wall minus area of capacity between wythes. The allowable compressive stresses for cavity walls are based upon the assumption that the floor loads bear upon but one of the two wythes. When hollow walls are loaded concentrically, the allowable stresses may be increased by 25%.

FIG. 48 **LATERAL SUPPORT REQUIREMENTS FOR MASONRY WALLS**

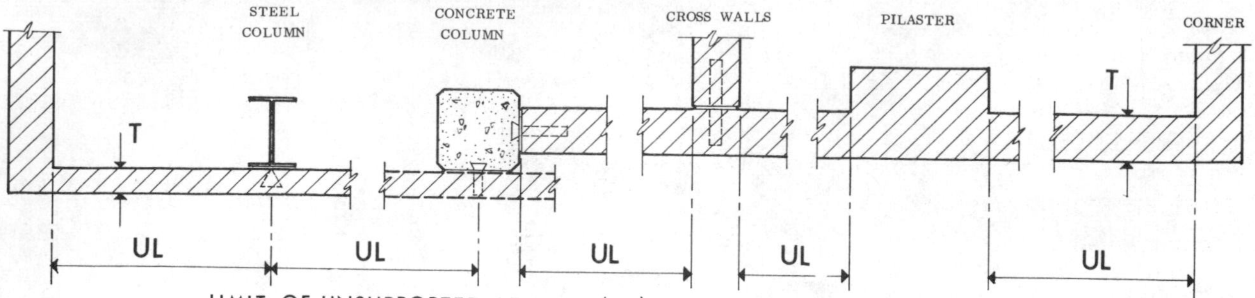

LIMIT OF UNSUPPORTED LENGTH (UL) PROVIDED BY VERTICAL SUPPORTS

WALL TYPE (Interior or Exterior)	MAXIMUM SPACING OF VERTICAL OR HORIZONTAL SUPPORT UL OR UH
LOAD BEARING Solid walls built of solid units	20 x T[1]
LOAD BEARING Hollow walls or walls built of hollow units	18 x T[2]
NON-LOAD BEARING (thickness of wall may include plaster)	36 x T

1 When laid with Type M, S, N or O mortar. If Type K mortar is used reduce spacing to 12xT.

2 Thickness of cavity walls equals sum of nominal thickness of inner and outer wythes.

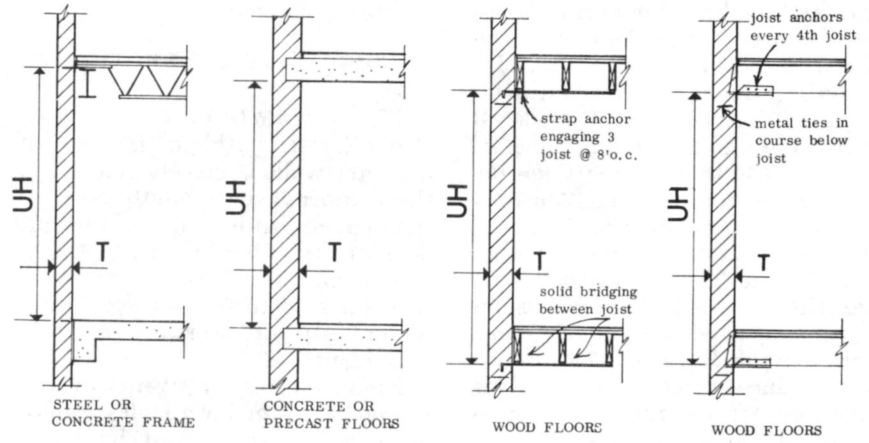

LIMIT OF UNSUPPORTED HEIGHT (UH) PROVIDED BY HORIZONTAL SUPPORTS

wall types, unit strengths and mortar types are given in Figure 46.

The principal loads producing compressive stresses in walls are the *dead* loads—the weights of the walls, floors and roofs—and the *live* loads—the weights of occupants, furniture and equipment. These loads can be determined with reasonable accuracy and once they are determined, the resultant compressive stresses developed in the walls carrying them can be computed. Such stresses should not exceed the allowable stresses given in Figure 46 for the type of wall under construction.

Actually, typical construction today does not require walls to carry large vertical loads. For example, the superimposed load on the foundation wall of a typical one-story residence rarely exceeds 15 psi. About 75% of the brick produced today have compressive strengths of over 4500 psi; concrete block, over 1000 psi. When used with typical mortars having compressive strengths in the range of 750 to 2500 psi, it is evident that most masonry walls for one-story buildings have far more compressive strength than is needed.

Tests have established the relationship between the compressive strength of walls and masonry units from which they are laid. The compressive strength of walls constructed of hollow concrete units, laid with face-shell mortar bedding, is approximately 42% of the compressive strength of the units, and 53% when laid with full mortar bedding (Fig. 47). Thus a wall built with full-mortar bedding and concrete block units with compressive strengths of 1000psi would have a compressive wall strength of 0.53 x 1000, or 530 psi. Using an allowable compressive stress of 70 psi (from Fig. 46), a wall having a strength of 530 psi will have a factor of safety of 530/70, or about 7-1/2. A factor of safety of 4 is generally considered ample for masonry wall assemblies. is generally considered ample for masonry wall assemblies.

TRANSVERSE STRENGTH

The transverse strength of a wall is a measure of its lateral stability; that is, its resistance to such lateral forces as wind pressure, earth pressure (in foundation or retaining walls), and earthquake forces. Transverse loads induce tensile stresses in the masonry that must be resisted. The lateral stability of masonry walls depends on the tensile strength of bond between mortar and units. Factors that will increase tensile bond strength will also increase transverse strength. These include: (1) the use of Type S mortar mixed with the maximum amount of water consistent with workability; (2) maintaining the suction of clay masonry units below 20 grams per minute, per 30 sq. in. (A simple field test to determine the suction of brick is described in Masonry, page 205-29); and (3) using good workmanship in which mortar joints are full.

Lateral stability of masonry walls is also increased by vertical loads which produce compressive stresses that offset tensile stresses caused by transverse loads. The addition of reinforcing steel also greatly increases lateral strength of masonry walls.

LATERAL SUPPORT OF WALLS

Masonry walls must be laterally supported or braced at certain intervals either by vertical or horizontal supports. When the limiting distance is length, vertical supports may be columns, pilasters or cross walls. When the limiting distance is height, horizontal supports may be floors, beams or roofs. The spacing of these supports is primarily dependent on the wall thickness. The spacing of supports and other lateral support requirements are summarized in Figure 48. In walls of different classes of units or mortars, the ratio of wall height or length to wall thickness should not exceed that allowed for the weakest of the combinations of units and mortars in the member.

If vertical supports are provided, there is no limit on the height of walls between floors or between floor and roof. When horizontal supports are provided, there is no limit on the length of the wall. Thus lateral supports in the form of columns, pilasters or cross walls are not required.

Lateral Support During Construction

Masonry walls are usually not designed to be freestanding, that is without support from columns, piers or crosswalls. Wind pressures can create four times as much bending stress in a new freestanding wall as in a wall which is connected to other structural elements of the building. This stress often occurs at the bottom of the wall where flashing and lack of bond in fresh mortar decreases the wall strength to resist tensile wind forces.

Bracing should be installed to resist wind pressure if the height of the wall exceeds that given in Fig. 49 for various peak wind velocities. Where bracing is used, the heights shown are the safe heights above the bracing. For example, to withstand wind gusts of 50 MPH, the freestanding height (without bracing) of a 10″ thick wall weighing 67 psi or less should not exceed 7-1/2′. The curves in the figure assume the mortar has no tensile strength and the walls are freestanding, unreinforced, ungrouted concrete masonry. In cavity walls, the thickness is assumed to be 2/3 the sum of the thicknesses of the two wythes.

WALL THICKNESSES, HEIGHTS AND LENGTHS

See Work File, page WF343-4.

BONDING AND ANCHORAGE

Two methods are generally used in the structural bonding of wall assemblies. One method, using masonry bonders is accomplished by overlapping and interlocking the masonry units. The other, and the preferred method, is the use of metal ties imbedded in connecting mortar joints (see Fig. 5).

Masonry Bonded Walls

The masonry bonded wall is based on variations of two traditional methods of bonding: (1) English bond, consisting of alternating courses of headers and stretchers; and (2) Flemish bond, consisting of alternating headers and stretchers in every course.

Overlapping stretchers, laid parallel with the length of the wall, develop longitudinal bonding strength; the headers laid across the width of the wall bond the wall transversely. Types of masonry bonded walls are shown in Figure 5, using brick (5H), overlapping stretchers (5G), brick and block or tile backup (5J).

Masonry bonded brick walls should be bonded so that not less than 4% of the wall surface is composed of headers, not exceeding 24″ spacing vertically or horizontally. In multiple unit tile or block walls, structural bonding is achieved by overlapping stretcher

FIG. 49 **MAXIMUM UNSUPPORTED HEIGHT OF NONREINFORCED, UNGROUTED CONCRETE MASONRY WALLS DURING CONSTRUCTION**

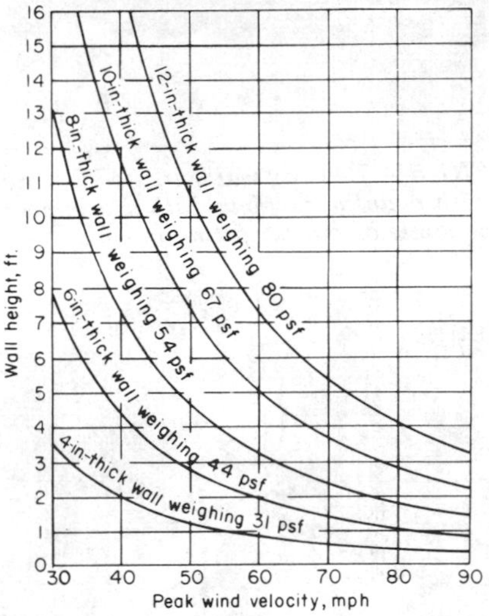

FIG. 50 (a) Effect of heavy backup unit on masonry header (top); (b) in an 8″ wall, headers are laid every 7th course to bond facing with backup.

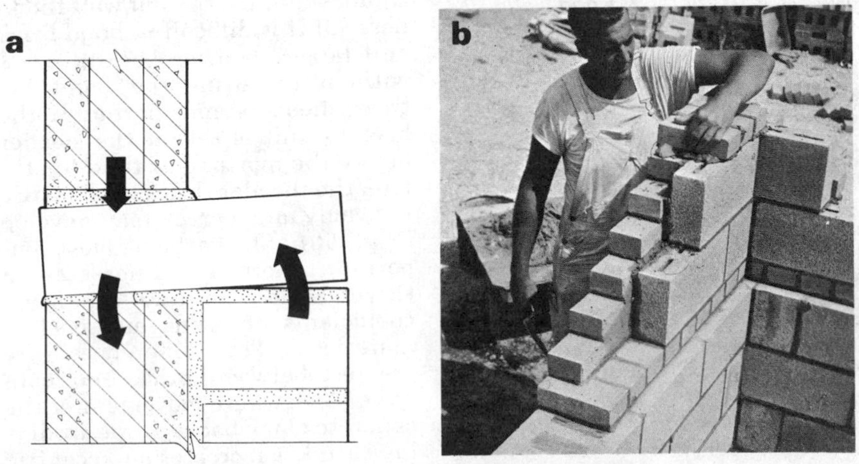

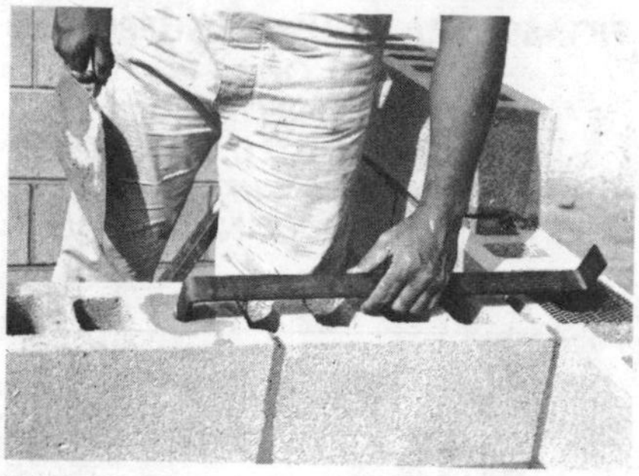

FIG. 51 Bearing walls are tied together with metal bars. Metal lath is to support concrete or mortar filling in cores.

FIG. 52 Cores are filled with concrete or mortar to embed metal tie bar ends.

FIG. 53 Metal strips of galvanized hardware cloth are used to tie non-load-bearing walls to intersecting walls.

FIG. 54 Here a complete wall with metal strips built in is tied to intersecting wall by building the metal strips into mortar joints of intersecting wall.

units rather than using header units. The stretcher units should be bonded at vertical intervals not exceeding 34″ by lapping at least 4″ over the unit below, or by lapping with units at least 50% greater in thickness at vertical intervals of 17″ (see Fig. WF1, Work File, page WF343-5).

Metal Tied Walls

Structural bonding with metal ties is used in both solid wall and cavity wall construction (Fig. 5). The usual requirements for such bonding are 3/16″ diameter steel ties (or metal ties of equivalent stiffness and strength) spaced so that at least one tie occurs for every 4-1/2 sq. ft. of wall area. The maximum vertical distance between ties should not exceed 18″ and the maximum horizontal distance should not exceed 36″. Ties in alternate courses should be staggered (see Fig. WF2, Work File page WF343-6).

Bonding with metal ties is recommended for exterior walls over the masonry bonded method because walls have greater resistance to moisture penetration and the ties allow slight differential movement between the facing and backing which may relieve stresses and prevent cracking.

Rain penetration in an 8″ masonry bonded wall may be the result of several factors: (1) It is sometimes difficult for the mason to obtain full mortar joints between brick headers, and moisture may be conveyed to the inside face of the wall since the header is continuous through the full wall thickness. (2) It is difficult to bond brick and larger, heavier backup units without impairing the bond between headers and mortar. If the backup unit is set on the header before the mortar has time to stiffen, the header brick will settle unevenly and a crack may develop (Fig. 50). (3) Perhaps most important, there are differences in thermal and moisture expansion coefficients of facing and backup materials. The differences are greatest between brick and concrete block where movement of the concrete block backup, due to drying shrinkage, creates an eccentric load on the brick header which tends to rupture the bond between headers and mortar at the external face. Thus, the desirability of providing some flexibility in the joints which connect materials of different characteristics becomes apparent.

For these reasons, cavity walls or solid metal tied walls are recommended, particularly where resistance to rain penetration is important or where wide variations in the physical characteristics of facing and backup materials exist.

Bonding and anchorage requirements for all wall types are given in the Work File, page WF343-5. Recommendations for some types of walls are summarized below.

Concrete Block Load-Bearing Walls When such walls intersect they should not be tied together in a masonry bond, except at the corners. One wall should terminate at the face of the other wall with a control joint located at this point. For lateral support, bearing walls are tied together with a metal tie bar 1/4″ thick, 1-1/4″ wide and 28″ long, with 2″ right angle bends on each end (Fig. 51). Tie bars are spaced not over 4′ apart vertically. The bends at the ends of the tie bars are embedded in cores filled with mortar (Fig. 52).

Concrete Block Non-Load Bearing Walls. Such walls are tied to other intersecting walls using strips of metal lath or 1/4″ mesh galvanized hardware cloth placed across the

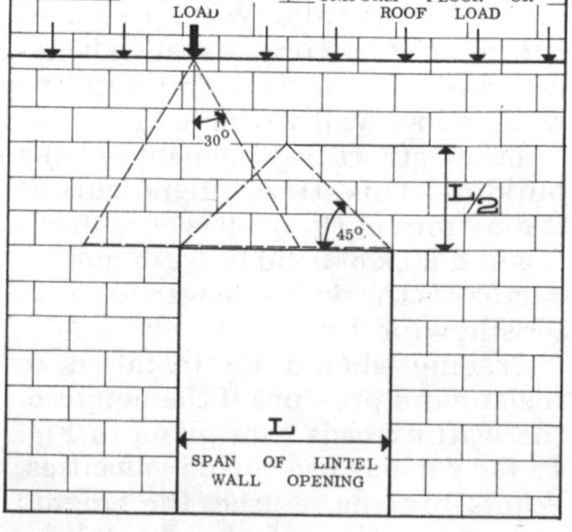

FIG. 55 Determining the loads imposed on masonry lintels.

joint between the two walls (Fig. 53). The metal strips are placed in alternate courses in the wall. When one wall is constructed first, the metal strips are built into the wall and later tied into the mortar joint of the second wall (Fig. 54). Where the two walls meet, a control joint should be provided if the joint is exposed to view.

Reinforced Masonry Walls These walls are structurally bonded by grout which is poured into the cavity (collar joint) between the wythes of masonry. The grout core also serves to contain and bond the reinforcing steel.

FIG. 56 **TYPICAL LINTELS FOR CLAY AND CONCRETE MASONRY**

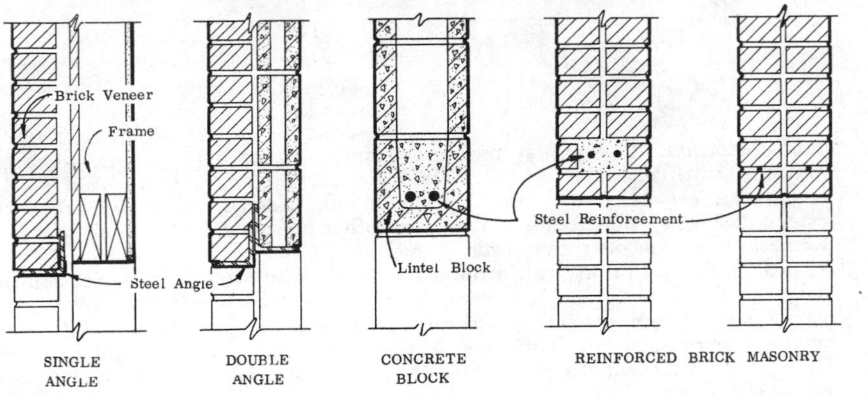

LINTELS AND ARCHES

The design of the structural member to support masonry over openings should not be slighted. Although rule-of-thumb methods generally result in over-design, arbitrary selection of steel shapes or reinforcing steel may result in inadequate sizing, resulting in structural cracks that cause subsequent leaking and an unsightly appearance.

The dead weight of a masonry wall which must be supported over openings can safely be assumed as the weight of a triangular section whose height is 1/2 the clear span of the opening (Fig. 55). Arching action of the masonry above the top of the opening may be counted upon to support the additional weight. Any concentrated load over an opening may be distributed along a wall length equal to the base of the triangle whose sides make an angle of 30° with the vertical from the point of application of the load. The horizontal thrust resulting from any arching action must be resisted by sufficient mass in the adjoining wall.

The most common method of supporting masonry over an opening is by steel angles or structural shapes (Fig. 56). Reinforced masonry lintels where the steel is completely protected from the elements have certain advantages: lower maintenance costs result because the steel is completely protected from the elements, eliminating periodic painting; lower initial costs result because savings may be effected since less steel is required.

FIG. 57 **TYPICAL BRICK ARCHES**

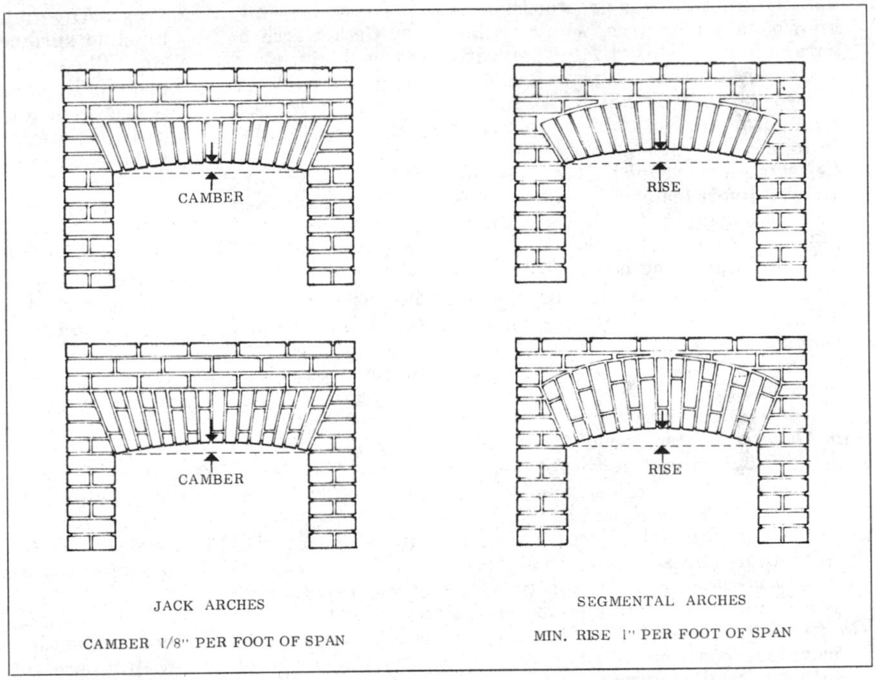

Various forms of rounded (segmental) and flat (jack) arches are also used to span openings. Figure 57 illustrates typical brick arch construction.

PIERS AND PILASTERS

Piers are free-standing columns; pilasters are columns that are bonded to and are built as integral parts of the wall. Both carry concentrated vertical loads. In addition, pilasters often provide necessary lateral support to walls.

The unsupported heights of solid masonry piers should not exceed 10 times their least cross-sectional dimension. The same is true of piers constructed with hollow masonry units provided that the cells are filled with mortar. Piers constructed with unfilled hollow masonry units may have unsupported heights not exceeding four times their least dimension.

Actual dimension The actual measured dimension of a masonry unit. (See *nominal dimension*).

Anchor A piece or assemblage, usually metal, used to attach parts (e.g., plates, joists, trusses or other masonry) to masonry or masonry materials.

Area wall The retaining wall around basement windows below grade.

Arch A curved compressive structural member, spanning openings or recesses; also built flat.

Back Arch A concealed arch carrying the backing of a wall where the exterior facing is carried by a lintel.

Jack Arch One having horizontal or nearly horizontal upper and lower surfaces. Also called *flat* or *straight arch*.

Major Arch Arch with spans greater than 6 ft and equivalent to uniform loads greater than 1000 lb per ft. Typically known as Tudor arch, semicircular arch, Gothic arch or parabolic arch. Has rise to span ratio greater than 0.15.

Minor Arch Arch with maximum span of 6 ft and loads not exceeding 1000 lb per ft. Typically known as jack arch, segmental arch or multicentered arch. Has rise to span ratio less than or equal to 0.15.

Relieving Arch One built over a lintel, flat arch, or smaller arch to divert loads, thus relieving the lower member from excessive loading. Also known as *discharging* or *safety arch*.

Trimmer Arch An arch, usually a low rise arch of brick, used for supporting a fireplace hearth.

ASTM American Society for Testing Materials.

Bearing wall A wall that supports a vertical load in addition to it own weight.

Blocking A method of bonding two adjoining or intersecting walls, not built at the same time, by means of offsets whose vertical dimensions are not less than 8 inches.

Bond beam Course or courses of a masonry wall grouted and usually reinforced in the horizontal direction. Serves as horizontal tie of wall, bearing course for structural members or as a flexural member itself.

Bond course The course consisting of units which overlap more than one wythe of masonry.

Capacity insulation The ability of masonry to store heat as a result of its mass, density and specific heat.

Cavity wall A wall built of masonry units so arranged as to provide a continuous air space within the wall (with or without insulating material), and in which the inner and outer wythes of the wall are tied together with metal ties.

Centering Temporary formwork for the support of masonry arches or lintels during construction. Also called *center(s)*.

Chase A groove or continuous recess built in a masonry wall to accommodate pipes, ducts or conduits.

Composite wall A multiple-wythe wall in which at least one of the wythes is dissimilar to the other wythe or wythes with respect to type or grade of masonry unit or mortar.

Compressive strength The maximum compressive load (in lbs. per sq. in.) that a masonry unit will support divided by the *gross* cross-sectional area of the unit in sq. in. (Where compressive strength is based on the *net* cross-sectional area it should be specifically stated.)

Corbel A shelf or ledge formed by projecting successive courses of masonry out from the face of the wall.

Curtain wall An exterior non-load-bearing wall in skeleton frame construction, not wholly supported by girders or beams at each story. Such walls may be anchored to columns, spandrel beams or floors.

Damp course A course or layer of impervious material which prevents capillary entrance of moisture from the ground or a lower course. Often called *damp check*.

Dead load The weight of all permanent and stationary construction or equipment included in a building. (See *live load.)*

Engineered brick masonry Masonry design based on accepted structural engineering analysis.

Exterior wall Any outside wall or vertical enclosure of a building other than a party wall.

Faced wall A wall in which facing and backing are of different materials and are bonded together to exert common action under load.

Facing Any material, forming a part of the wall, used as a finished surface.

Field The expanse of wall between openings, corners, etc., principally composed of stretchers.

Fire division wall Any wall which subdivides a building so as to resist the spread of fire. It is not necessarily continuous through all stories to and above the roof.

Fireproofing Any material or combination of materials built to protect structural members so as to increase their fire resistance.

Fire resistive materials See *non-combustible material.*

Fire wall Any wall which subdivides a building to resist the spread of fire and which extends continuously from the foundation through the roof.

Flashing Sheet metal or other suitable material built into the wall for the purpose of (1) collecting any water that may penetrate the wall, and (2) to divert such moisture to the exterior (see Fig. MW1, page 343-3).

Foundation wall A load-bearing wall below the floor nearest to exterior grade serving as a support for a wall, pier, column, floor or other structural part of a building.

Furring A method of finishing the interior face of a masonry wall to provide space for insulation, prevent moisture transmittance, or to provide a level surface for finishing.

Gross cross-sectional area The total area of a section perpendicular to the direction of the load, including areas within cores or cellular spaces.

Grounds Nailing strips placed in masonry walls as a means of attaching trim or furring.

Header course A continuous bonding course of header brick.

Hollow wall A wall built of masonry units arranged to provide an air space within the wall between the facing and backing wythes.

Joint reinforcement Steel wire, bar or fabricated reinforcement which is placed in horizontal mortar joints.

Lateral support (of walls) Means whereby walls are braced *either* horizontally by columns, pilasters or crosswalls, or vertically by floor and roof constructions (see Fig. MW36, page 343-24).

Lead The section of a wall built up and racked back on successive courses. A line is attached to leads as a guide for constructing a wall between them.

Live load The total of all moving and variable loads that may be placed upon or in a building. (See *dead load*).

Masonry bonded hollow wall A hollow wall in which the

facing and backing are bonded together with solid masonry units.

Net cross-sectional area The gross cross-sectional area of a section minus the area of cores or cellular spaces.

Nominal dimension The dimension greater than the *actual* masonry dimension by the thickness of a mortar joint, but not more than 1/2″.

Noncombustible Any material which will neither ignite nor actively support combustion in air at a temperature of 1200° F when exposed to fire.

Non-load-bearing wall A wall which supports no vertical load other than its own weight.

Panel wall An exterior non-load-bearing wall in skeleton frame construction, wholly supported at each story.

Parapet wall That part of any wall entirely above the roof line.

Parging The process of applying a coat of cement mortar to masonry.

Partition An interior wall, one story or less in height.

Party wall A wall used for joint service by adjoining buildings.

Perforated wall A wall which contains a considerable number of relatively small openings. Often called *pierced* wall or *screen* wall.

Plain masonry Masonry without reinforcement, or reinforced only for shrinkage or temperature changes.

Prefabricated brick masonry Masonry construction fabricated in a location other than its final inservice location in the structure. Also known as preassembled, panelized and sectionalized brick masonry.

Retaining wall Any wall subjected to lateral pressure other than wind pressure, or a wall built to support a bank of earth.

Shear wall A wall which resists horizontal forces applied in the plane of the wall.

Solar screen A perforated masonry wall used as a sunshade.

Solid masonry wall A wall built of solid or hollow masonry units laid contiguously, with joints between units *completely* filled with mortar.

Spandrel wall That part of a panel curtain wall above the head of a window in one story and below the sill of the window in the story above.

Stack Any structure or part thereof which contains a flue or flues for the discharge of gases.

Story That portion of a building included between the upper surface of any floor and the upper surface of the floor next above, except that the topmost story should be that portion of a building included between the upper surface of the topmost floor and the ceiling or roof above. Where a finished floor level directly above a basement or cellar is more than 6 feet above grade such basement or cellar is considered a story.

Veneered wall A wall having a face of masonry units or other weather-resisting noncombustible materials securely attached to the backing, but not so bonded as to intentionally exert common action under load.

Veneer A single facing wythe of masonry units or similar materials securely attached to a wall for the purpose of providing ornamentation, protection or insulation, but not bonded or attached to intentionally exert common action under load.

Wall tie A bonder or metal piece which connects wythes of masonry to each other or to other materials.

Cavity wall tie A rigid, corrosion-resistant metal tie which bonds two wythes of a cavity wall. It is usually steel, 3/16 in. in diameter and formed in a "Z" shape or a rectangle.

Veneer wall tie A strip or piece of metal used to tie a facing veneer to the backing.

Waterproofing Prevention of moisture flow through masonry due to water pressure.

Weep holes Opening placed in mortar joints of facing materials at the level of flashing, to divert to the exterior any moisture collected on the flashing.

Wythe A continuous vertical section of masonry one masonry unit in thickness. Also called *withe* or *tier*.

We gratefully acknowledge the assistance of the Federal Housing Administration, the National Concrete Masonry Association, the Portland Cement Association, the Brick Institute of America, Vermiculite Institute, A. A. Wire Products Company, American Cynamid Company, Dow Chemical Company and Owens-Corning Fiberglas Corporation for the use of their publications as references and permission to use photographs and illustrations.

MASONRY WALLS 343

CONTENTS

MATERIALS

MASONRY UNITS

See Work File, Masonry.

MORTAR AND MORTAR MATERIALS

See Work File, Masonry.

FLASHING MATERIALS

See Fig. MW1, Main Text.

Flashings are built into the masonry to collect water that may enter the wall and then divert it through weep holes to the exterior. Suitable flashing materials must possess at least four important properties: (1) imperviousness to moisture penetration; (2) resistance to corrosion caused either by exposure to the atmosphere or the caustic alkalis present in mortars; (3) toughness to resist puncture, abrasion or other damage during installation; and (4) formability so that the flashing can be worked to the correct shape and retain this shape after installation.

METAL FLASHING

Copper Sheet (Type ETP or FRTP) **ASTM B152**

Exposed to weather	16 oz. per sq. ft.
Completely concealed	10 oz. per sq. ft.

Light colored masonry surfaces which may be subject to stain and discoloration from the copper wash can be flashed with lead-coated copper (ASTM B-101).

Aluminum Sheet

Exposed to weather	0.019″ thick-16,000 psi tensile strength
Completely concealed	0.015″ thick-14,000 psi tensile strength

Aluminum is attacked by caustic alkalies present in fresh unhardened mortar and should be protected from corrosion with a bituminous coating.

Galvanized (Zinc Coated) Steel **ASTM A93**

Exposed to weather	26 gage
Completely concealed	28 gage

Galvanized steel, when exposed, requires periodic maintenance and should be accessible for inspection and painting.

Stainless Steel (No. 2B Finish) **AISI Type 304**

Exposed to weather	28 gage
Completely concealed	30 gage

COMBINATION FLASHING

These flashings are made by coating sheet metal with various materials. The coating materials commonly used include bitumen, kraft paper, bituminous satu-

(Continued) **MATERIALS**

rated cotton fabrics, glass fiber fabrics and/or combinations of these materials. The sheet metals commonly used for combination flashing are: Copper in weights per sq. ft. of 3 oz., 5 oz. or 7 oz.; and aluminum of 0.004″ or 0.005″ thickness.

While no recognized standards exist for combination flashing, the weights of the copper and the thicknesses of the aluminum sheetmetal given above when coated both sides with any of the coatings listed above and when installed with unbroken skins provide suitable concealed flashing.

BITUMINOUS AND PLASTIC FABRIC FLASHING

Fabric saturated with bitumen or plastic fabrics are suitable as flashing if they are permanently insoluble in water and are installed with unbroken skins.

Careful workmanship is required to install bituminous and plastic fabric flashings. They are often inadvertently damaged during installation by sharp trowels or when heavy masonry units are dragged across them. When used for concealed flashing, thin fabrics may sag and not retain the desired shape as a result of changes in temperature or poor workmanship.

No industry standards exist for fabric flashing. Use of fabric flashing for either exposed or concealed work should be based on performance records and test data of the material and on the reputation of its manufacturer.

ANCHORS, TIES AND REINFORCEMENT

See Fig. 4, Main Text.

Anchors, ties and reinforcement should be copper or zinc coated steel or be of non-corrodible metal having total strength equivalent to the types listed below and conform to the following specifications: Copper Coated Wire (Grade 30HS) ASTM B227, Zinc Coated Wire (Class 2) ASTM A116, Zinc Coated Anchors & Ties (Class B) ASTM A153.

Wire Mesh Ties

See Figs. 53 & 54, Main Text.

Wire mesh ties or hardware cloth for anchoring non-load-bearing partitions to intersecting masonry should be at least 15 gage coated steel wire, 1/2″ mesh, 12″ minimum in length, and 1″ less in width than the wall thickness.

Corrugated Metal Ties

Corrugated metal ties for anchoring masonry veneer or load-bearing partitions to intersecting masonry should be 22 gage coated sheet steel, not less than 7/8″ wide and 6″ in length.

Dovetail Anchors

Dovetail anchors for use with embedded slots or inserts in concrete should be 16 gage coated sheet steel, 7/8″ wide and long enough to permit the ends of the tie to be embedded in the outer face-shell mortar beds of hollow units or the center of mortar beds in solid units.

Rigid Steel Anchors

See Figs. 51 & 52, Main Text

Rigid steel anchors for anchorage of load-bearing walls and fire walls should be 1-1/4″ by 1/4″ steel with ends turned up not less than 2″, not less than 16″ long for 8″ walls or not less than 24″ long for 12″ walls.

Joint Reinforcement

Joint reinforcement for use in horizontal courses of masonry should be fabricated from zinc-coated cold-drawn steel wire (ASTM A82) consisting of at least two smooth or deformed longitudinal wires 9 gage or larger, weld connected with 12 gage or larger cross wires. Distance between welded contacts of cross wires should not exceed 6″ for smooth longitudinal wire or 16″ for deformed longitudinal wire. The width of joint reinforcement should permit it to be embedded in the mortar joint not closer than 5/8″ to masonry face surface.

Cavity Wall Ties

See Fig. 7, Main Text

Ties for use in cavity walls should be 3/16″ diameter, coated steel wire, formed either in a rectangular shape not less than 4″ wide (for use with hollow units having cells vertical), or in a "Z" shape with 2″ legs (for use with solid units or hollow units having cells horizontal). The length of ties should be long enough to permit the ends of the tie to be embedded in the outer face-shell mortar beds of hollow units or the center of mortar beds of solid units.

Joint reinforcement comprising 9 gage or larger cross wires may be used instead of metal ties provided the cross wires are spaced to render strength and stiffness per square foot of wall area equivalent to that of 3/16″ diameter "Z" ties used at the maximum spacing.

DESIGN AND CONSTRUCTION

Faced Wall Ties

Ties for use in faced walls should meet the same requirements given above for cavity wall ties.

WALL TYPES

Masonry walls may be classified as solid walls, cavity walls, veneered walls and reinforced walls (see Fig. 5, Main Text). Each type may be constructed of either solid or hollow masonry units. Solid and cavity walls are built to be either load-bearing or non-load-bearing; veneered walls are non-load-bearing; reinforced walls are load-bearing. Descriptions of each type are given in the Main Text, pages 343-5 through 343-8, including illustrations of traditional and contemporary pattern bonds.

Since no one wall type is suitable as a standard design for all exposures, conditions of loading or appropriate to all localities or types of occupancy, the final selection of a wall can only be made after weighing all the requirements of wall function, cost and appearance.

WORKMANSHIP

The workmanship techniques and skills required of the mason to produce masonry of suitable quality are described and illustrated in the Main Text, Masonry, pages 205-29 through 205-35.

MOISTURE CONTROL

Moisture in masonry may result from : (1) Rain penetration, (2) capillary action, and (3) condensation. Rain penetration is controlled by flashing, parging, dampproofing, caulking, painting, tooling of mortar joints and other construction details such as sloping sills or projections that keep direct rainfall off the face of the masonry. Moisture due to capillary action resulting from direct contact of the masonry with the ground is controlled by dampproofing. Moisture from condensation of vapor within the wall is controlled by installing a vapor barrier or by insulating the wall properly (see Main Text.

FLASHING

In areas of *severe* or *moderate* exposure (see Fig. 14, Main Text, flashing should be installed at the following ing vulnerable locations: (1) At the base of the wall; (2) under the sill of wall openings; (3) over the heads of wall openings, and (4) where masonry penetrates the roof surface (*see Fig. 15, Main Text, page 343-11).*

Weep holes should be provided in the head joint immediately above the flashing, except below window sills. *(Weep holes are normally omitted below window sills, since little water normally collects on the flashing.)*

PARGING

Exterior masonry basement walls subject to moisture penetration should be parged and dampproofed *(see Main Text, page 343-12).*

The vertical longitudinal joints (collar joints) in solid walls should be completely filled by parging with the type of mortar used in laying the units. *(See Fig. 93, Main Text, Masonry).*

CAULKING

The perimeter of wall openings, exposed control joints, and the intersection of masonry with other materials should be caulked.

PAINTING

Cement-water paint, applied in two coats, may be required to make exterior concrete masonry surfaces watertight or to correct a leaky clay masonry wall condition. *(See Division 400—Painting and Finishing).*

TOOLED MORTAR JOINTS

Where exposed masonry is required to be watertight, mortar joints should be tooled. *Methods of tooling are described and illustrated in the Main Text, Masonry, page 205-33.*

CRACK CONTROL

To minimize the possibility of cracking in masonry walls: (1) Expansion Joints in clay masonry should be built into the wall at the locations and spacing shown in Figs. 18 & 19, Main Text. (2) Control Joints in concrete masonry should be built into the wall at the locations and spacing shown in Fig. 20, Main Text.

When constructed properly, control joints provide a continuous vertical joint that will: (1) permit slight wall movement without cracking the masonry units; (2) seal the joint against weather, and (3) stabilize the wall laterally across the joint by means of a shear key.

COLD WEATHER CONSTRUCTION

Masonry under construction in cold weather should be protected. *(See Main Text, & Work File, Masonry; Fig. 28, and Main Text, Masonry Walls).*

(Continued) DESIGN AND CONSTRUCTION

FIREPLACES AND CHIMNEYS

For maximum performance, fireplaces and chimneys should be built to conform to the dimensions given in Figures 39 & 40 Main Text.

Each fireplace should be vented by a separate flue.

Masonry chimneys should extend at least 2′ above any part of the roof or roof ridge when within 10′ of the chimney.

When the combustion chamber is lined with fire brick at least 2″ in thickness, the total thickness of the combustion chamber wall, including the lining, should not be less than 8″ in thickness.

Fireplace walls should be separated from combustible construction as follows: (1) Framing members, 2″ air space. The air space should be fire stopped at floor level with extension of ceiling finish, strips of asbestos board or other noncombustible material; (2) subfloor and flooring, 3/4″ air space. Trim in contact with back of fireplace should be of noncombustible material or should be fire-retardant wood; (3) Wall sheathing, 2″ air space; (4) Furring strips, 1″, except that furring may be applied directly to masonry at sides or corners of fireplace.

When the floor construction and floor finish is of combustible material, the hearth should extend at least 16″ in front of the fireplace opening and at least 8″ on each side of the opening. Combustible materials should not be placed within 8″ of the top or side edges of the fireplace opening.

STRENGTH AND STABILITY

All masonry should be supported on other masonry, concrete or steel. Wood girders or wood framing should not be used for support.

ALLOWABLE COMPRESSIVE STRESSES

Masonry construction should be designed and constructed so that the allowable compressive stresses given in Fig. 46, Main Text, are not exceeded for the type of wall construction, kinds of units and types of mortar used.

When composite walls or other structural masonry elements are composed of different kinds or types of units or mortars, the maximum compressive stress should not exceed the allowable stress for the weakest of the combinations of either the masonry unit(s) or the mortar type(s) of which the member is composed.

WALL THICKNESSES

Load-Bearing Walls

Load-bearing walls in residence buildings three stories or less in height may be 8″ in thickness when not over 35′ in height and the roof is designed to impart no horizontal thrust.

Except where earthquake design is required, exterior walls in one-story residence buildings may be 6″ in thickness when (1) not over 9′ in height; (2) the height to the peak of the gable does not exceed 15′, and (3) girders and concentrated loads are supported on integral piers or pilasters not less than 8″ x 12″.

Interior bearing partitions in dwellings not more than 1-1/2 stories or 20′ in height may be 6″ in thickness.

Load-bearing walls exceeding 35′ in height should be at least 12″ in thickness for the uppermost 35′ of their height, and should be increased 4″ in thickness for each successive 35′ increase in height.

Cavity walls built of solid masonry units not more than 25′ in height above the support of the wall may be 10″ in thickness. The facing and backing of cavity

(Continued) STRENGTH AND STABILITY

walls should each have a thickness of at least 4″ and the cavity should not be less than 2″ nor more than 3″ in width.

Non-Load-Bearing Walls

Non-load-bearing walls on the exterior may be 4″ less in thickness than required for bearing walls, but not less than 8″ in thickness except where 6″ walls are permitted in one-story residences.

LATERAL SUPPORT

Masonry walls must be laterally supported (braced) by either vertical OR horizontal supports.

Walls should not vary in thickness between their lateral supports *(exception is made in residence buildings not over two stories in height, where vertical recesses may be built into 8″ walls if the recess is not more than 4″ deep nor occupies more than 4 sq. ft. of wall area.)*

Ratio of Wall Height or Wall Length to Thickness

The ratio of the unsupported wall height to wall thickness, OR the ratio of unsupported wall length to wall thickness should not exceed the values given in Figures 48 & 49, Main Text, for various wall types.

Methods of Support

Lateral support may be obtained by cross walls, piers or buttresses when the limiting distance is measured horizontally OR by floor and roof when the limiting distance is measured vertically. Sufficient bonding and/or anchorage should be provided to transfer the load from the wall to its supports.

Piers

The unsupported height of piers should not exceed 10 times their least dimension. The unsupported height of piers built with hollow masonry units should not exceed 4 times their least dimension unless the cellular spaces are filled solidly with Type M or Type S mortar.

Support During Erection

Where walls may be exposed to high winds during erection, they should not be built higher than the heights shown in Figure 49, unless adequately braced or until provision is made for prompt installation of permanent bracing at the floor or roof level.

BONDING AND ANCHORAGE

All masonry walls should be structurally bonded with masonry headers, metal ties or joint reinforcement.

Masonry Headers

In walls built with solid units, there should be not less than one header course in each 7 courses; OR masonry headers should compose not less than 4% of the wall surface, should extend not more than 4″ into the backing, and should be spaced not more than 24″ apart vertically or horizontally.

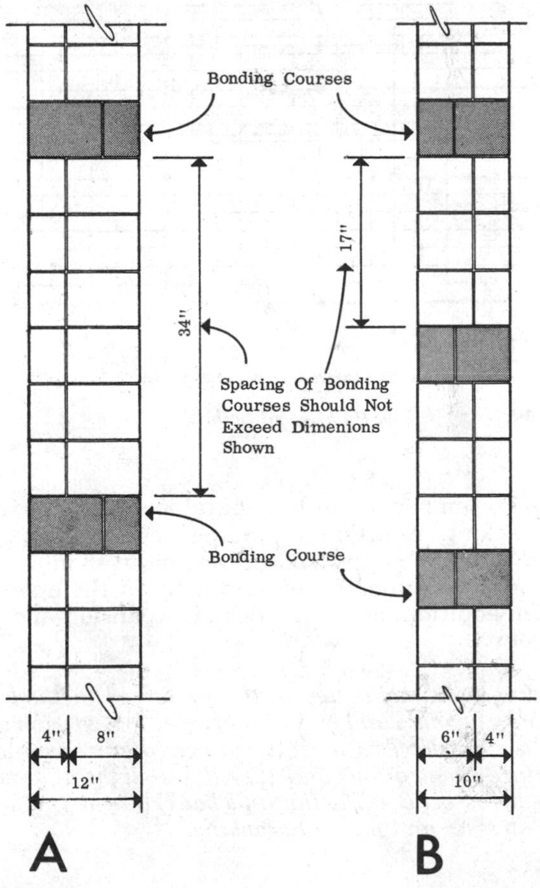

FIG. WF-1

Where two or more hollow units make up the wall thickness they should be bonded at maximum vertical intervals of 34″ by lapping a stretcher course at least 3-3/4″ over the course below, or by lapping at vertical intervals not exceeding 17″ with units at least 50% greater in thickness than the units below (see Fig. WF1).

(Continued) STRENGTH AND STABILITY

Metal Ties

Metal ties in walls solid built with solid or hollow units, should be used for each 4-1/2 sq. ft. of wall surface. Ties in alternate courses should be staggered. The distance between adjacent ties should not exceed 18″ vertically nor 36″ horizontally (see Fig. WF2). In cavity walls the spacing of the ties should not exceed 24″ vertically nor 36″ horizontally. Ties

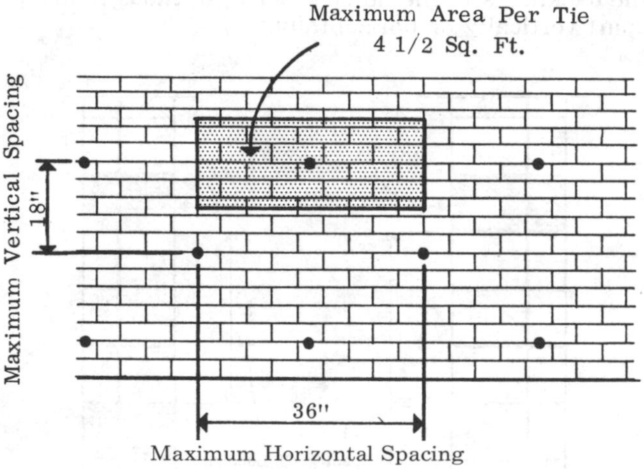

FIG. WF2 Solid walls of solid or hollow units.

should be embedded in horizontal joints of the facing and backing. Additional bonding ties should be provided at all openings, spaced not more than 3′ apart around the perimeter and within 12″ of the openings, and, in addition, on both sides of expansion and control joints.

Bonding with metal ties is the preferred method over masonry headers and is recommended for exterior walls because: (1) it offers a wall with greater resistance to moisture penetration, and (2) allows slight differential movements between facing and backing which may relieve stresses and prevent cracking.

Joint Reinforcement

To control cracks in concrete masonry walls, continuous joint reinforcement, spaced not more than 16″ apart vertically, should be installed in walls exceeding 20′ in length. *(See Fig. 20, Main Text, for effect of joint reinforcement on control joint spacing.)*

When joint reinforcement is embedded in horizontal mortar joints, it should be not closer than 5/8″ to masonry face surface and should be continuous except through control joints and expansion joints. Lengths of joint reinforcement should be lapped at least 6″ and the lap should contain at least one cross wire of each length. To compensate for loss of reinforcement at wall openings, the joint reinforcement should be installed in the first and second bed joints immediately above and below openings and should extend at least 24″ beyond the end of sills and lintels.

Exterior Walls

Masonry walls and partitions should be securely bonded or anchored where they abut or intersect walls, floors and columns if these are depended upon for lateral support.

Exterior walls facing against or abutting concrete members should be anchored to the concrete by the use of dovetailed anchors inserted in slots built into the masonry *(see Fig. 4, Main Text)*. Anchors should be spaced not more than 18″ vertically and 24″ horizontally.

Load-Bearing Walls

When two bearing walls meet or intersect and the courses are built up together, the intersections should be bonded by laying at least 50% of the units at the intersection in a toothed masonry bond.

When intersecting bearing walls are carried up separately, the walls should be tied together with rigid steel anchors embedded in the horizontal mortar joint at vertical spacings not exceeding 32″ with the hooked ends projecting into and embedded in cores of joints filled with mortar *(see Figs. 51 & 52, Main Text)*; or the vertical joint can be regularly toothed at 8″ maximum offsets and rigid steel anchors provided in horizontal joints spaced not more than 4′ apart vertically.

Except in areas subject to earthquakes, it is recommended that concrete block bearing walls not be tied together in a toothed masonry bond except at corners. Instead, intersecting walls should be carried up separately, anchored with metal anchors and when the intersection is exposed to view, a control joint should be installed.

The bearing course of masonry under structural members such as floor joists, concrete slabs and under lintels should be of solid masonry units at least 4″ in thickness or hollow masonry with cells filled solidly with mortar.

(Continued) STRENGTH AND STABILITY

Non-Load-Bearing Walls

Non-bearing partitions abutting or intersecting other walls or partitions should be anchored with corrugated metal tie anchors at vertical intervals of not more than 3′ or by masonry bonding in alternate courses.

Intersecting non-load-bearing concrete block walls should be bonded together as specified above for concrete block load-bearing walls, except that non-load-bearing partitions 6″ or less in thickness may be anchored with wire mesh ties, hardware cloth or joint reinforcement, spaced not more than 16″ vertically *(See Figs. 53 & 54, Main Text).*

Faced Walls

Facing and backing should be bonded with either metal ties or masonry headers as described above under "Masonry Headers" and "Metal Ties".

Cavity Walls

Cavity walls should be bonded with metal ties as described above under "Metal Ties".

Veneered Walls

Masonry veneer should be attached to the backing by corrugated metal ties. One tie should be used for each 2 square feet of wall area, and the distance between adjacent metal ties should not exceed 24″ either vertically or horizontally. Ties should be embedded at least 2″ in a horizontal joint of the facing. Additional bonding ties, spaced not more than 3′ apart around the perimeter and within 12″ of the opening, should be provided at all openings and on both sides of expansion or control joints. *(See Fig. 9, Main Text).*

LINTELS AND ARCHES

The masonry above openings should be supported by lintels of metal or reinforced masonry or by masonry arches. Lintels should bear on the wall not less than 4″ at each end and should be of sufficient stiffness to carry the superimposed load without deflection of more than 1/360 of the clear span. Masonry lintels and arches should be designed in accordance with engineering practice.

PROPERTIES OF SELECTED MASONRY WALLS

The following contains data which permits comparison of several masonry wall properties.

For a given wall thickness, the properties shown include: (1) *Fire Rating,* measured in hours; (2) *Sound Transmission Class* (STC) measured in decibels; (3) *Heat Transmission,* measured by the wall's U-value in BTU's transmitted through one sq. ft. of wall in one hour for each degree Fahrenheit of temperature difference between air on the warm side and the air on the cool side of the wall; (4) *Wall Weight,* measured in lbs. per sq. ft. and (5) *Estimated Wall Costs,* measured in dollars per sq. ft.

The estimates shown permit only approximate relative cost comparison as determined by a 1978 survey of masonry contractors in the 16 county Chicago metropolitan area. Estimates are based on the labor and materials to construct an 8 ft. x 32 ft. running bond wall section without offsets or openings. Other factors affecting cost include: (1) construction of any openings, offsets, corners, piers; structural bonding or different patterns; (2) design considerations; (3) site conditions; (4) seasonal variations; and (5) geographic location.

For additional information regarding specific properties of masonry walls, see Section 105, Heat Control and Section 106, Sound Control.

PROPERTIES OF SELECTED MASONRY WALLS*

*"Estimated Cost Per Square Foot" values are intended as an index of relative cost for various masonry wall assemblies and are not to be used for cost estimating purposes. These and other masonry wall properties may be useful as selection criteria in determining walls of suitable quality for the desired functions.

4" WALLS

(Tooled 1 Side, unless noted)

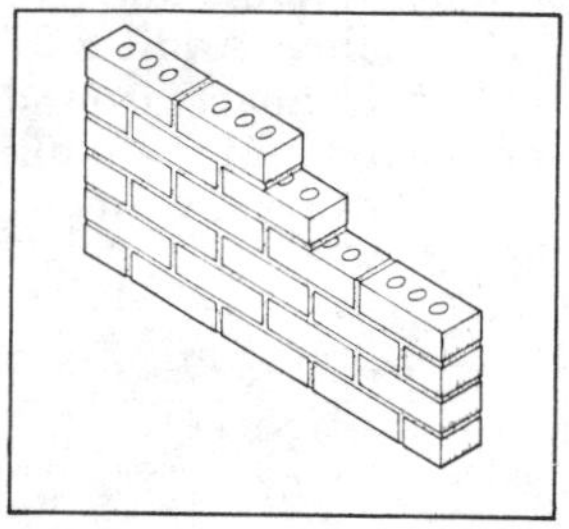

FACE BRICK (2-1/4" x 8")

Fire Rating (hr.): 1
Sound Transmission Class (DB): 45
U Value (BTU/Sq. Ft. Hr. – Deg. Fahr.): .... .76
Wall Weight (lb./Sq. Ft.): 38
ESTIMATED COST PER SQUARE FOOT: 4.89

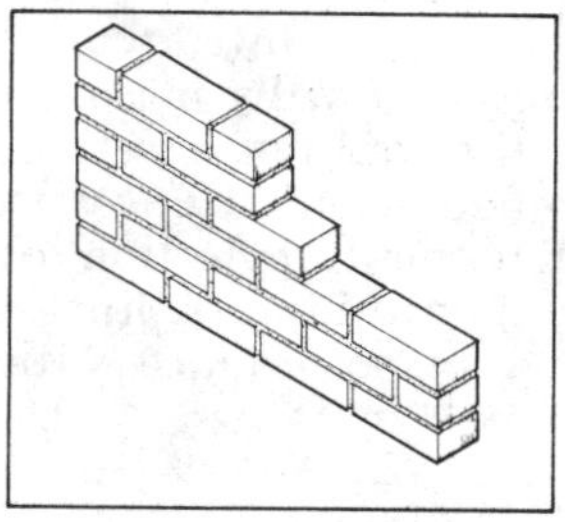

COMMON BRICK (2-1/8" x 8")

Fire Rating (hr.): 1
Sound Transmission Class (DB): 34
U Value (BTU/Sq. Ft. Hr. – Deg. Fahr.): .... .76
Wall Weight (lb./Sq. Ft.): 38
ESTIMATED COST PER SQUARE FOOT: 3.70

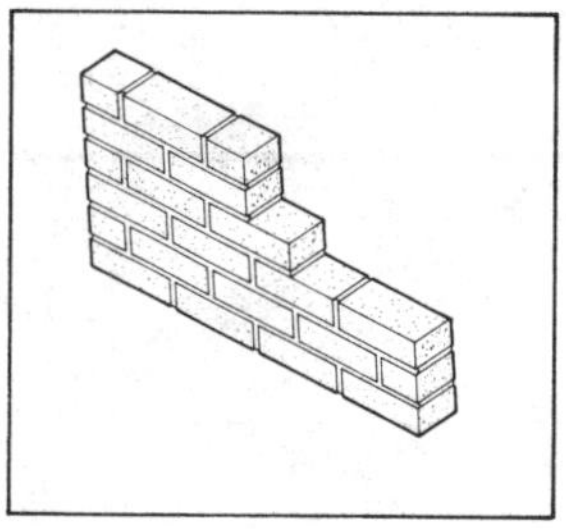

CONCRETE BRICK (2-1/4" x 7-5/8")

Fire Rating (hr.): 1
Sound Transmission Class (DB): 37
U Value (BTU/Sq. Ft. Hr. – Deg. Fahr.): .... .67
Wall Weight (lb./Sq. Ft.): 35
ESTIMATED COST PER SQUARE FOOT: 3.78

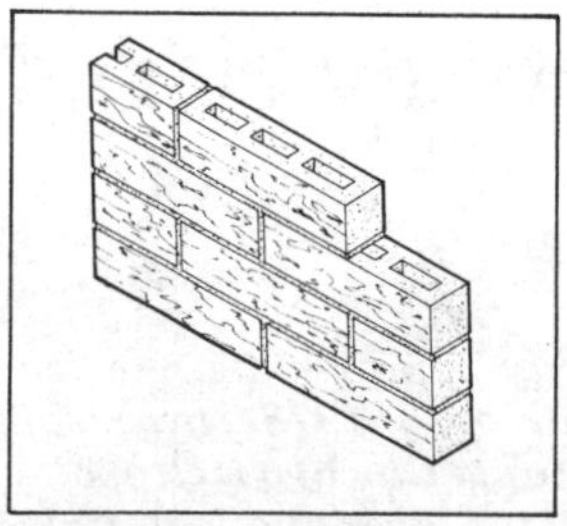

HEAVYWEIGHT CONCRETE "SLUMP" BLOCK (4" x 16" Face Size)

Fire Rating (hr.): .... -1
Sound Transmission Class (DB): 36
U Value (BTU/Sq. Ft. Hr.– Deg. Fahr.): .... .46
Wall Weight (lb./Sq. Ft.): 33
ESTIMATED COST PER SQUARE FOOT: 3.94

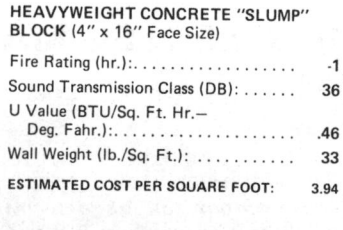

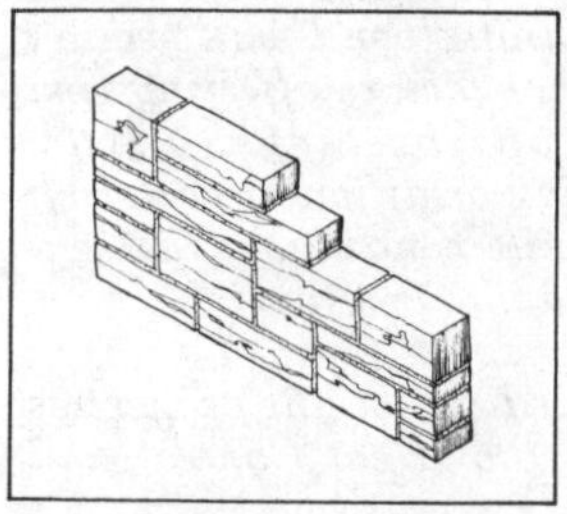

LANNON STONE

Fire Rating (hr.): 1
Sound Transmission Class (DB): 46
U Value (BTU/Sq. Ft. Hr.– Deg. Fahr.): .... .68
Wall Weight (lb./Sq. Ft.): 50
ESTIMATED COST PER SQUARE FOOT: 8.65

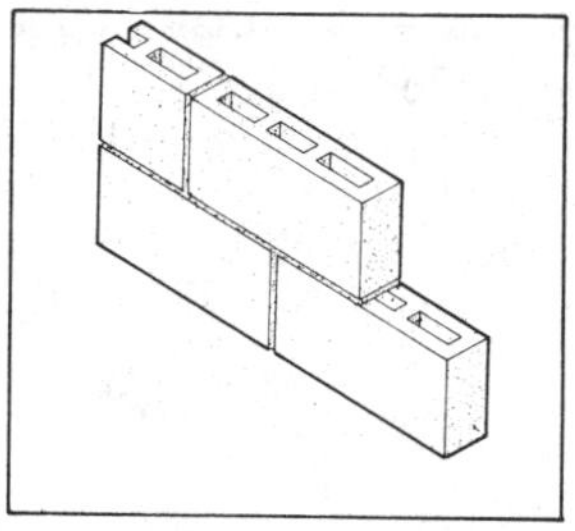

LIGHTWEIGHT CONCRETE BLOCK

Fire Rating (hr.): 1
Sound Transmission Class (DB): 40
U Value (BTU/Sq. Ft. Hr.– Deg. Fahr.): .... .43
Wall Weight (lb./Sq. Ft.): 24
ESTIMATED COST PER SQUARE FOOT: 2.37

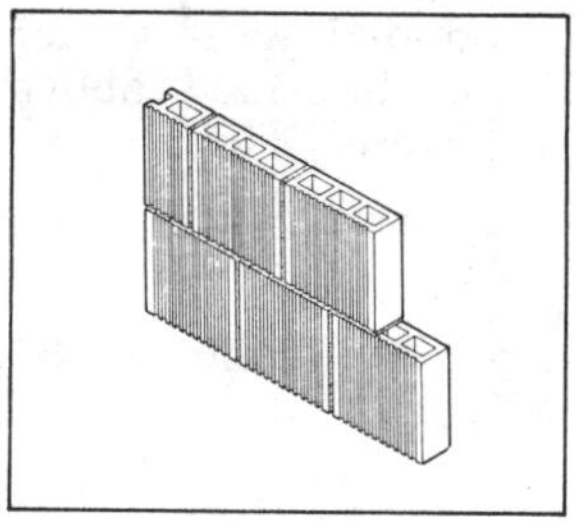

CLAY PARTITION TILE (12" x 12" Face Size) (Trowel Cut Joint)

Fire Rating (hr.): (Plaster 2 Sides)..... 1
Sound Transmission Class (DB): 28
U Value (BTU/Sq. Ft. Hr.– Deg. Fahr.): .... .55
Wall Weight (lb./Sq. Ft.): 21
ESTIMATED COST PER SQUARE FOOT: 1.96
(No Plaster)

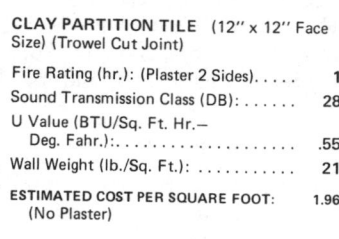

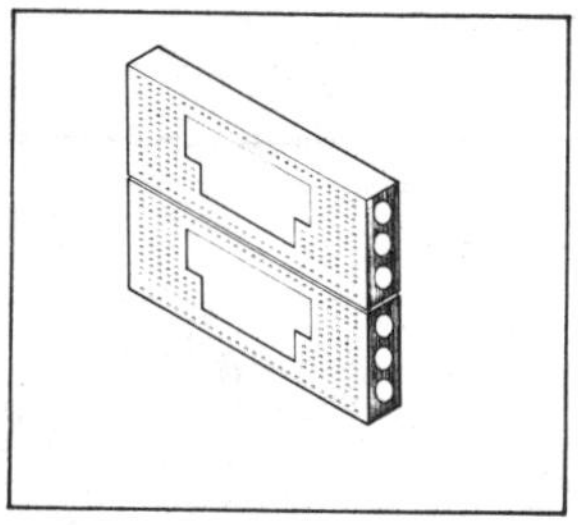

GYPSUM PARTITION TILE
(12" x 30" Face Size) (Trowel Cut Joint)

Fire Rating (hr.): (Plaster 2 Sides)..... 4
Sound Transmission Class (DB): 42
U Value (BTU/Sq. Ft. Hr.– Deg. Fahr.): (Plaster 2 Sides)....... N.A.
Wall Weight (lb./Sq. Ft.): (Plaster 2 Sides)........ 26
ESTIMATED COST PER SQUARE FOOT: 1.89
(No Plaster)

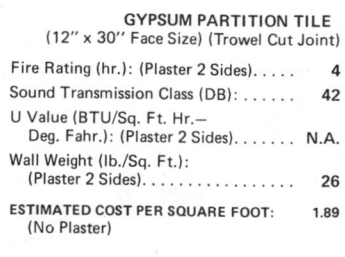

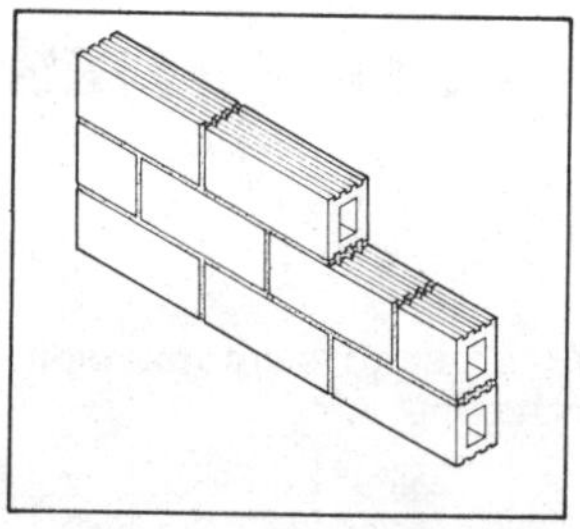

GLAZED TILE (6 TC Series and 6 TCD Series – Tooled 2 Sides)

Fire Rating (hr.): .... -1
Sound Transmission Class (DB): 32
U Value (BTU/Sq. Ft. Hr.– Deg. Fahr.): .... .40
Wall Weight (lb./Sq. Ft.): 27
ESTIMATED COST PER SQUARE FOOT
1 Side Glazed: 7.60
2 Sides Glazed: 8.90

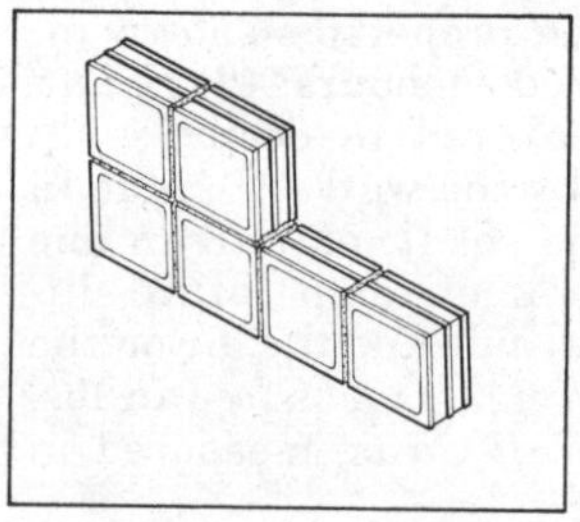

GLASS BLOCK* (8" x 8") (Tooled 2 Sides)

Fire Rating (hr.): .... –
Sound Transmission Class (DB): 27
U Value (BTU/Sq. Ft. Hr.– Deg. Fahr.): .... .56
Wall Weight (lb./Sq. Ft.): 20
ESTIMATED COST PER SQUARE FOOT: 11.77

*6" x 6" and 12" x 12" Glass Block are also available.

(continued) PROPERTIES OF SELECTED MASONRY WALLS

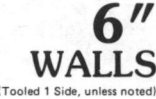

6" WALLS
(Tooled 1 Side, unless noted)

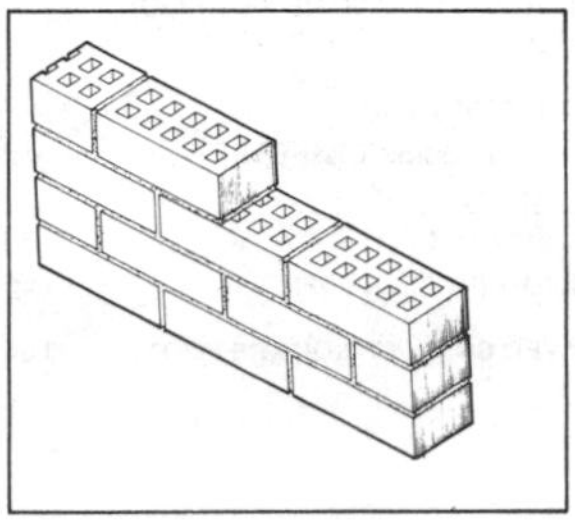

BRICK, THRU WALL UNITS
(3-1/2" x 11-1/2" Face Size)

Fire Rating (hr.): 2+
Sound Transmission Class (DB): 41
U Value (BTU/Sq. Ft. Hr.–Deg. Fahr.): .66
Wall Weight (lb./Sq. Ft.): 42
ESTIMATED COST PER SQUARE FOOT: 6.50

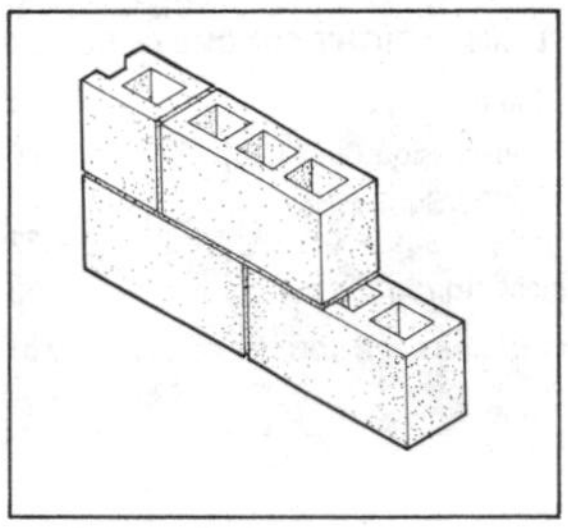

LIGHTWEIGHT CONCRETE BLOCK

Fire Rating (hr.): 1
Sound Transmission Class (DB): 45
U Value (BTU/Sq. Ft. Hr.–Deg. Fahr.): .38
Wall Weight (lb./Sq. Ft.): 26
ESTIMATED COST PER SQUARE FOOT: 2.60

8" WALLS
(Tooled 1 Side, unless noted)

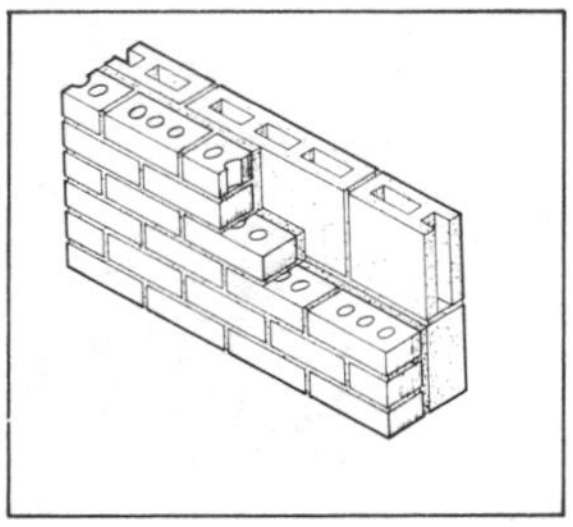

4" FACE BRICK & 4" LIGHTWEIGHT CONCRETE BLOCK

Fire Rating (hr.): 3+
Sound Transmission Class (DB): 48
U Value (BTU/Sq. Ft. Hr.–Deg. Fahr.): .33
Wall Weight (lb./Sq. Ft.): 53
ESTIMATED COST PER SQUARE FOOT: 6.70

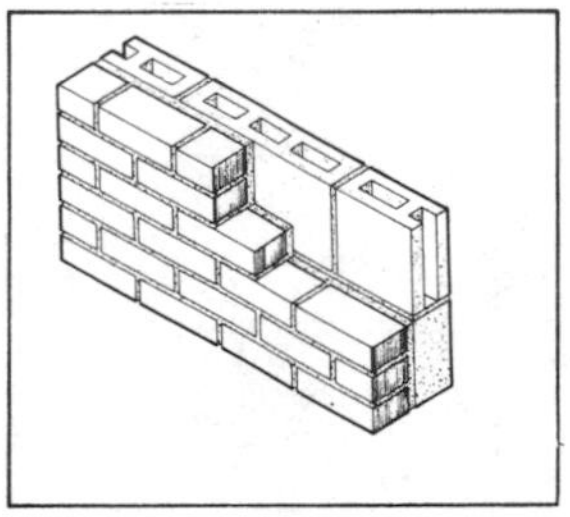

4" COMMON BRICK & 4" LIGHTWEIGHT CONCRETE BLOCK

Fire Rating (hr.): 3+
Sound Transmission Class (DB): 48
U Value (BTU/Sq. Ft. Hr.–Deg. Fahr.): .33
Wall Weight (lb./Sq. Ft.): 53
ESTIMATED COST PER SQUARE FOOT: 5.56

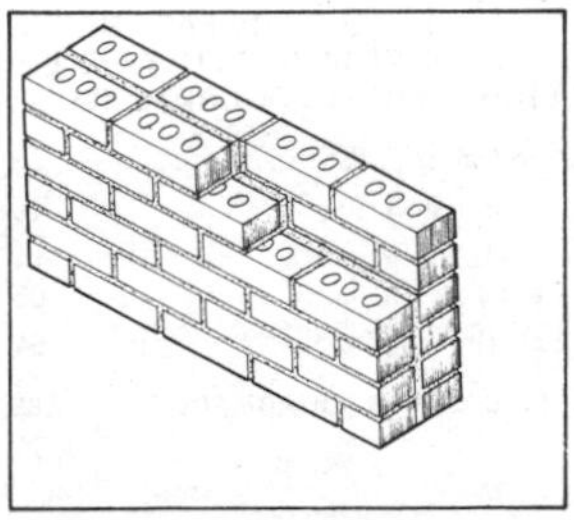

4" FACE BRICK & 4" FACE BRICK-TOOLED 2 SIDES

Fire Rating (hr.): 4
Sound Transmission Class (DB): 49
U Value (BTU/Sq. Ft. Hr.–Deg. Fahr.): .54
Wall Weight (lb./Sq. Ft.): 62
ESTIMATED COST PER SQUARE FOOT: 9.33

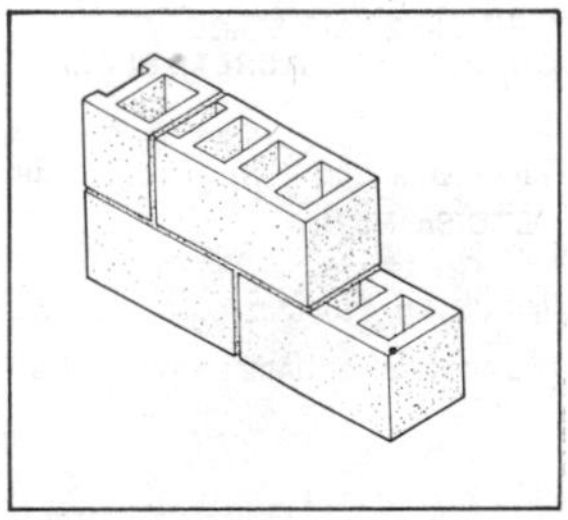

LIGHTWEIGHT CONCRETE BLOCK
(Tooled 2 Sides)

Fire Rating (hr.): 3
Sound Transmission Class (DB): 46
U Value (BTU/Sq. Ft. Hr.–Deg. Fahr.): .35
Wall Weight (lb./Sq. Ft.): 32
ESTIMATED COST PER SQUARE FOOT: 2.92

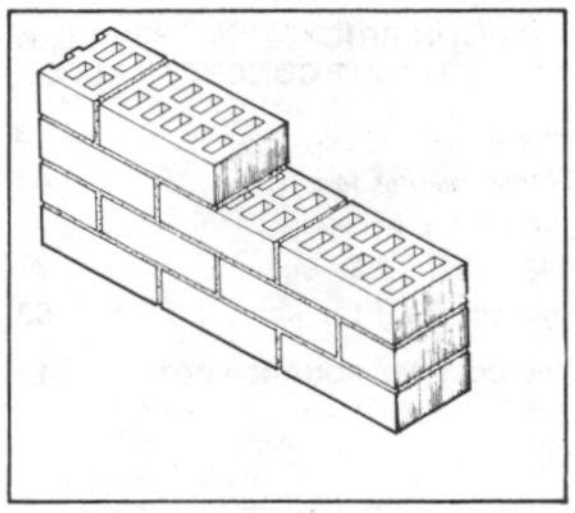

BRICK, THRU WALL UNITS
(3-1/2" x 11-1/2" Face Size)

Fire Rating (hr.): 3
Sound Transmission Class (DB): 50
U Value (BTU/Sq. Ft. Hr.–Deg. Fahr.): .47
Wall Weight (lb./Sq. Ft.): 61
ESTIMATED COST PER SQUARE FOOT: 7.65

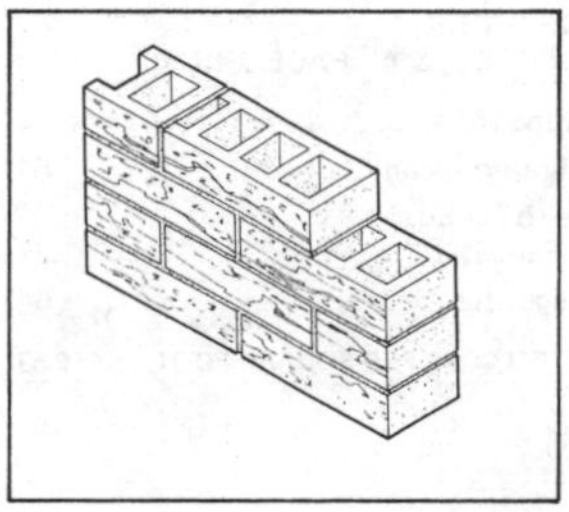

HEAVYWEIGHT CONCRETE "SLUMP" BLOCK
(4" x 16" Face Size)

Fire Rating (hr.): 3
Sound Transmission Class (DB): 44
U Value (BTU/Sq. Ft. Hr.–Deg. Fahr.): .38
Wall Weight (lb./Sq. Ft.): 45
ESTIMATED COST PER SQUARE FOOT: 4.53

(continued) PROPERTIES OF SELECTED MASONRY WALLS

10″ WALLS
(Tooled 1 Side, unless noted)

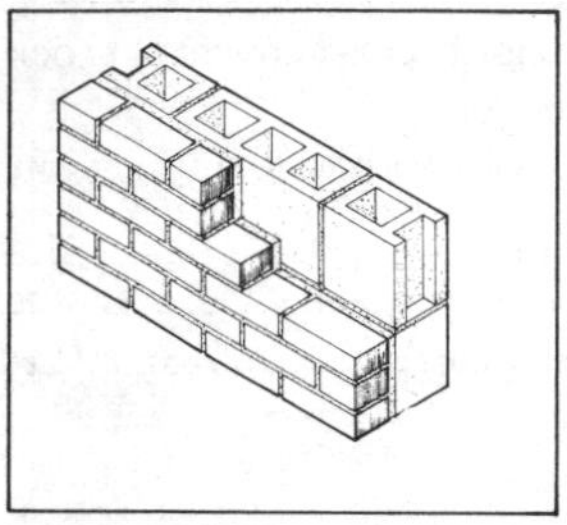

4″ FACE BRICK & 6″ LIGHTWEIGHT CONCRETE BLOCK

Fire Rating (hr.): 4
Sound Transmission Class (DB): 50
U Value (BTU/Sq. Ft. Hr.– Deg. Fahr.): .32
Wall Weight (lb./Sq. Ft.): 62
ESTIMATED COST PER SQUARE FOOT: 7.01

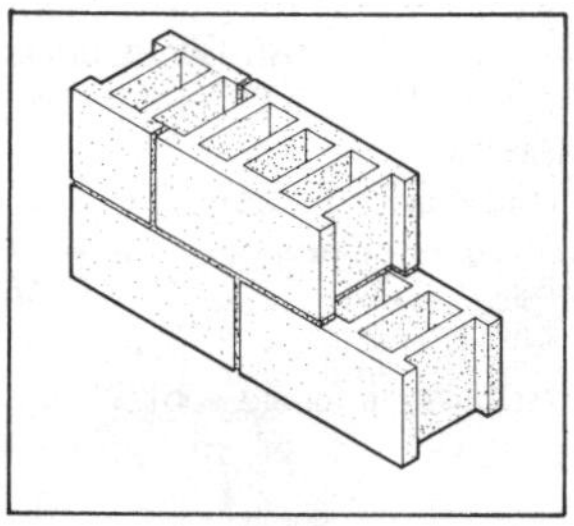

LIGHTWEIGHT CONCRETE BLOCK
(Tooled 2 Sides)

Fire Rating (hr.): 3
Sound Transmission Class (DB): 48
U Value (BTU/Sq. Ft. Hr.– Deg. Fahr.): .33
Wall Weight (lb./Sq. Ft.): 39
ESTIMATED COST PER SQUARE FOOT: 3.34

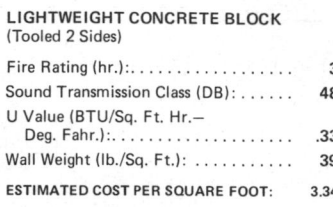

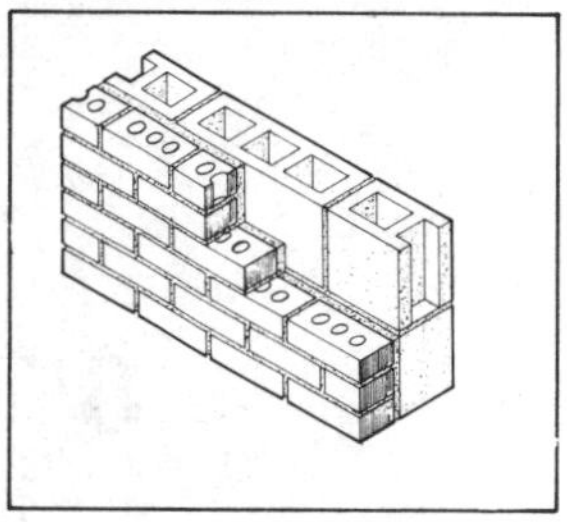

4″ COMMON BRICK & 6″ LIGHTWEIGHT CONCRETE BLOCK

Fire Rating (hr.): 4
Sound Transmission Class (DB): 50
U Value (BTU/Sq. Ft. Hr.– Deg. Fahr.): .32
Wall Weight (lb./Sq. Ft.): 62
ESTIMATED COST PER SQUARE FOOT: 5.90

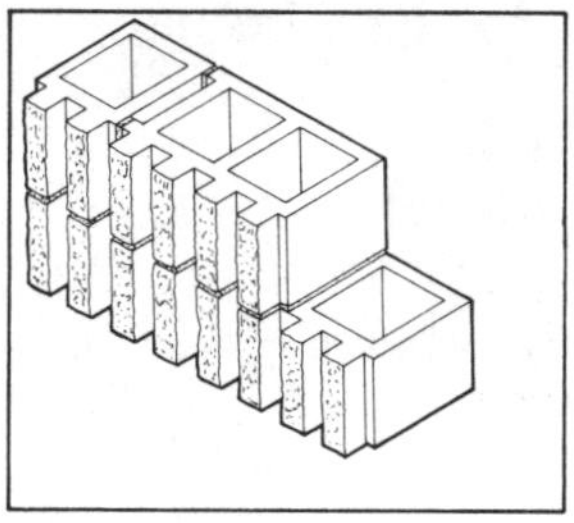

FLUTED (4 Flutes)
HEAVYWEIGHT CONCRETE BLOCK
(Tooled 2 Sides)

Fire Rating (hr.): 2
Sound Transmission Class (DB): 50
U Value (BTU/Sq. Ft. Hr.– Deg. Fahr.): .49
Wall Weight (lb./Sq. Ft.): 61
ESTIMATED COST PER SQUARE FOOT: 4.77

10″ CAVITY WALLS
(With Flashing, Tooled 2 Sides)

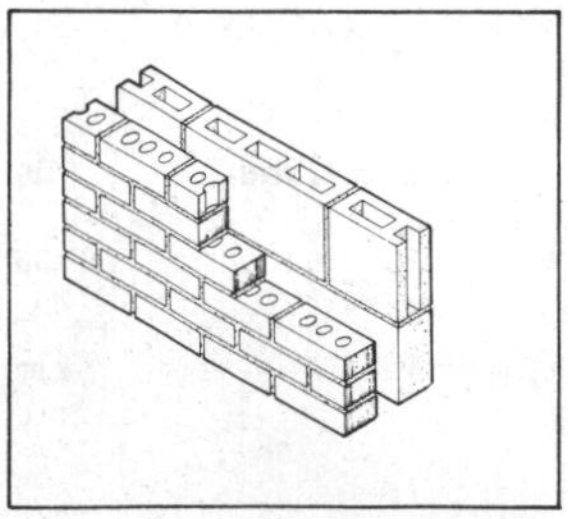

4″ FACE BRICK, 2″ AIR SPACE, & 4″ LIGHTWEIGHT CONCRETE BLOCK

Fire Rating (hr.): 3
Sound Transmission Class (DB): 48
U Value (BTU/Sq. Ft. Hr.– Deg. Fahr.): .27
Wall Weight (lb./Sq. Ft.): 53
ESTIMATED COST PER SQUARE FOOT: 6.95

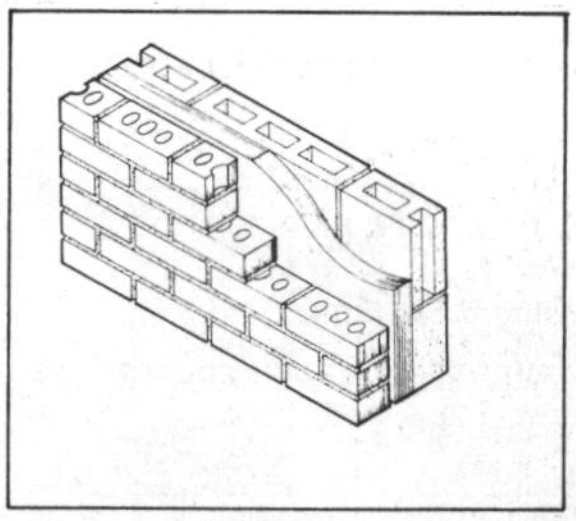

4″ FACE BRICK, 2″ URETHANE INSULATION, & 4″ LIGHTWEIGHT CONCRETE BLOCK

Fire Rating (hr.): 3
Sound Transmission Class (DB): 49
U Value (BTU/Sq. Ft. Hr.– Deg. Fahr.): .05
Wall Weight (lb./Sq. Ft.): 54
ESTIMATED COST PER SQUARE FOOT: 7.85

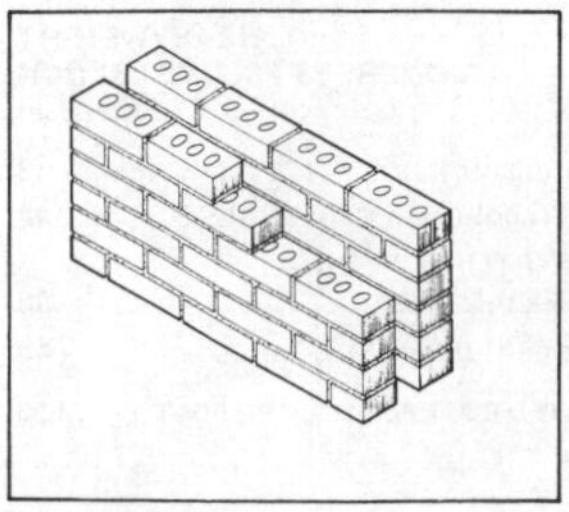

4″ FACE BRICK, 2″ AIR SPACE, & 4″ FACE BRICK

Fire Rating (hr.): 4
Sound Transmission Class (DB): 51
U Value (BTU/Sq. Ft. Hr.– Deg. Fahr.): .37
Wall Weight (lb./Sq. Ft.): 64
ESTIMATED COST PER SQUARE FOOT: 9.57

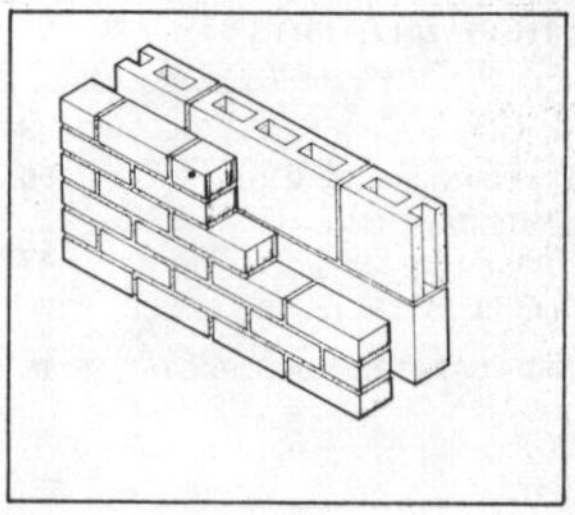

4″ COMMON BRICK, 2″ AIR SPACE, & 4″ LIGHTWEIGHT CONCRETE BLOCK

Fire Rating (hr.): 3
Sound Transmission Class (DB): 48
U Value (BTU/Sq. Ft. Hr.– Deg. Fahr.): .27
Wall Weight (lb./Sq. Ft.): 53
ESTIMATED COST PER SQUARE FOOT: 5.81

(continued) PROPERTIES OF SELECTED MASONRY WALLS

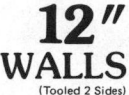

12" WALLS
(Tooled 2 Sides)

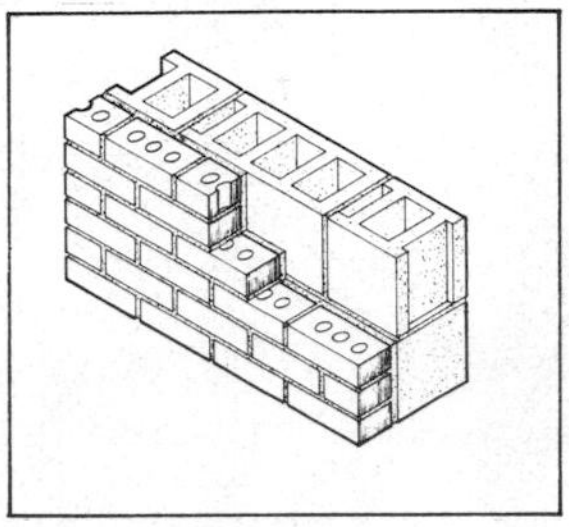

4" FACE BRICK & 8" LIGHTWEIGHT CONCRETE BLOCK

Fire Rating (hr.): 4
Sound Transmission Class (DB): 50+
U Value (BTU/Sq. Ft. Hr.–Deg. Fahr.): .29
Wall Weight (lb./Sq. Ft.): 72
ESTIMATED COST PER SQUARE FOOT: 7.25

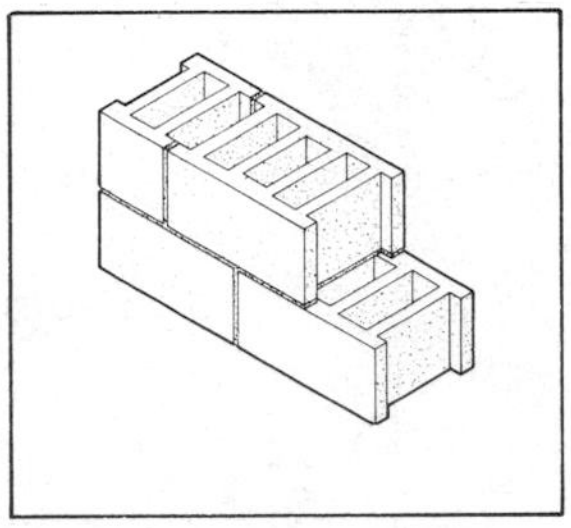

LIGHTWEIGHT CONCRETE BLOCK

Fire Rating (hr.): 3-4
Sound Transmission Class (DB): 49
U Value (BTU/Sq. Ft. Hr.–Deg. Fahr.): .31
Wall Weight (lb./Sq. Ft.): 47
ESTIMATED COST PER SQUARE FOOT: 3.76

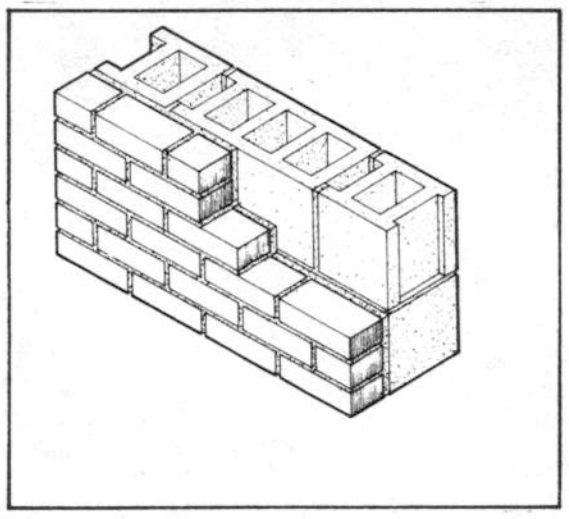

4" COMMON BRICK & 8" LIGHTWEIGHT CONCRETE BLOCK

Fire Rating (hr.): 4
Sound Transmission Class (DB): 50+
U Value (BTU/Sq. Ft. Hr.–Deg. Fahr.): .29
Wall Weight (lb./Sq. Ft.): 72
ESTIMATED COST PER SQUARE FOOT: 6.12

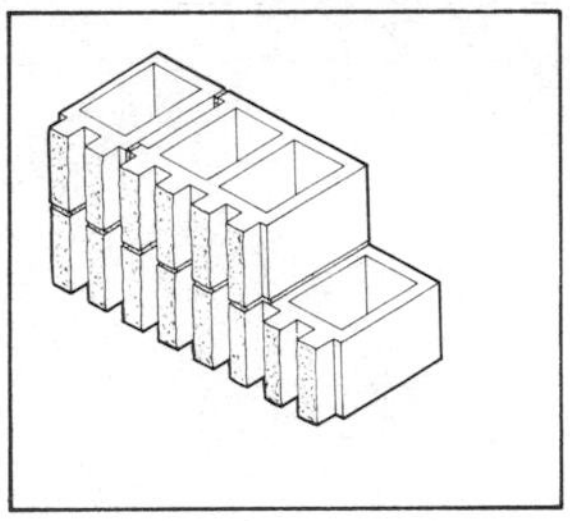

FLUTED (4 Flutes) **HEAVYWEIGHT CONCRETE BLOCK**

Fire Rating (hr.): 3-4
Sound Transmission Class (DB): 54
U Value (BTU/Sq. Ft. Hr.–Deg. Fahr.): .47
Wall Weight (lb./Sq. Ft.): 73
ESTIMATED COST PER SQUARE FOOT: 5.27

BRICK PAVING
(Basket Weave Pattern, 4.5 per Sq. Ft.)

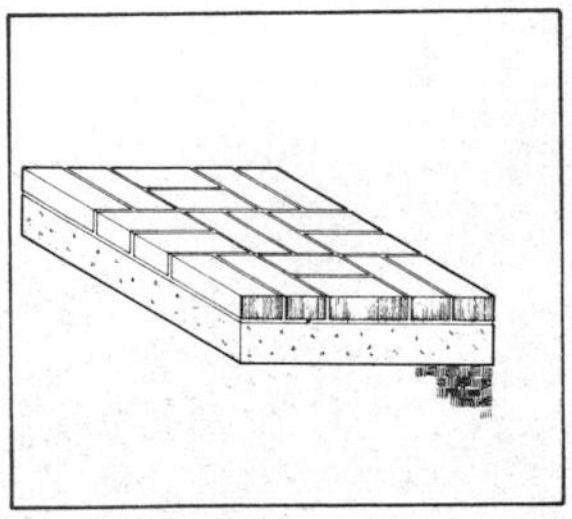

MORTAR SET—BED & ALL JOINTS
(Type "M" Mortar)

Weight (lb/Sq. Ft.): 29
ESTIMATED COST PER SQUARE FOOT: 6.20

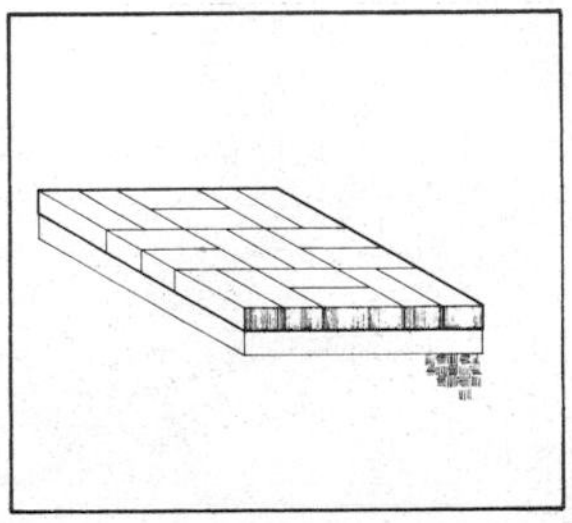

DRY SET & BUTTED
(Sand Swept Into Joints)

Weight (lb./Sq. Ft.): 27
ESTIMATED COST PER SQUARE FOOT: 5.00

360 ROOF/CEILING SYSTEMS

INTRODUCTION

In many cases, ceiling construction is an integral part of the roof construction. The entire system more correctly can be called the roof/ceiling system and considered one of the basic elements of the superstructure.

The design, selection and construction of a roof/ceiling system considers (1) the required level of functional performance, (2) costs, and (3) the desired finished appearance. These considerations often are influenced by the systems and finishes used in other basic elements of the construction.

Functional performance requirements of a roof/ceiling system usually include: (1) adequate structural strength to support dead loads (actual weight of the roof/ceiling system) and live loads (such as from snow, ice and wind); (2) the control of the flow of heat, water, air and water vapor; (3) satisfactory levels of sound privacy; (4) suitable fire resistance; (5) suitability for the application of finish materials; (6) adaptability to economical methods of assembly and erection; and (7) accommodation of heating, air conditioning, electrical and plumbing equipment.

The ultimate costs of a roof/ceiling system include in-place costs, maintenance costs and operating costs. In-place costs are the sum of the costs of materials, labor and overhead required to assemble and/or install the roof/ceiling system. The selection of a roof system may affect other costs such as the cost of installing insulation, applying roofing finishes, providing interior wall support, and the necessary time to enclose the building. Maintenance costs and operating costs attributable to the roof/ceiling system (such as heat loss or gain through the construction) have an important influence on the ultimate cost of the system and should not be overlooked if long-term economy is to be achieved.

Visually and structurally, the form of a roof is determined by the roof/ceiling system. But generally, the system is covered with applied materials which provide the desired finished appearance and performance. Selection of the finish materials: (1) may affect the spacing, span and slope of structural members; (2) may determine the need for roof sheathing and/or underlayment; and (3) even may dictate methods of construction.

The roof/ceiling system should be compatible with the wall and the floor/ceiling systems used in the construction. Different structural systems should not be mixed indiscriminately; colors, textures, forms and details must be related if a unified appearance is to result.

Typical residential roof/ceiling systems include: *truss* construction, *joist-and-rafter* construction, *joist* construction, *plank-and-beam* construction, and *panelized* construction. A brief discussion of these roof/ceiling systems is given in this section. A more detailed discussion of each system will be found in individual sections immediately following.

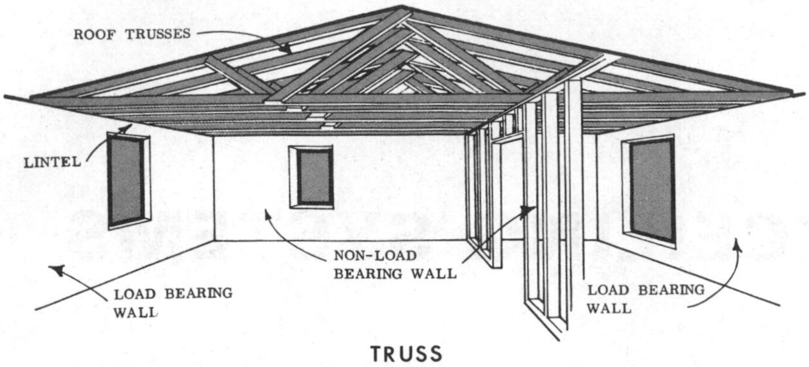

FIG. 1 Wood truss roof/ceiling construction.

TRUSS ROOFS

Trusses for residential construction are composed of a number of wood framing members (chords) arranged in a framework of triangles to form a supporting structural element for both a roof and a ceiling (Fig. 1). Individual members, generally 2″ x 4″ and 2″ x 6″, are connected by wood or metal gusset plates, metal gang-nail plates or metal ring connectors (Fig. 2). The truss acts as a unit to support roof and ceiling loads and to span the entire width of the building between exterior walls. This clear span allows interior walls to be non-load bearing and permits almost complete flexibility in placing walls (Fig. 3).

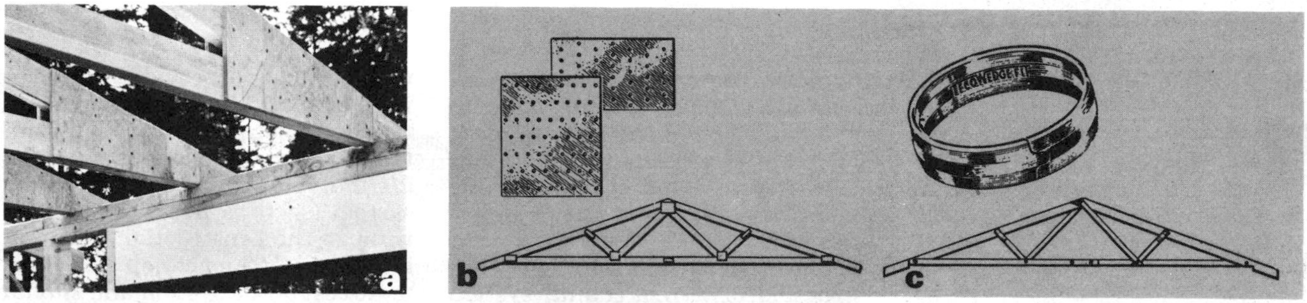

FIG. 2 "Heel joints" often are made with plywood gussets connecting top and bottom chords. Plywood gussets can be nailed (a) and may be glued. Metal plates (b) often are used and ring connectors (c) also can be used.

FIG. 3 Trusses provided clear spans and eliminate the need for interior load-bearing walls. Here, the entire truss system has been furred with 1″ x 4″s to receive acousticaltile,and wiring has been started prior to partition framing.

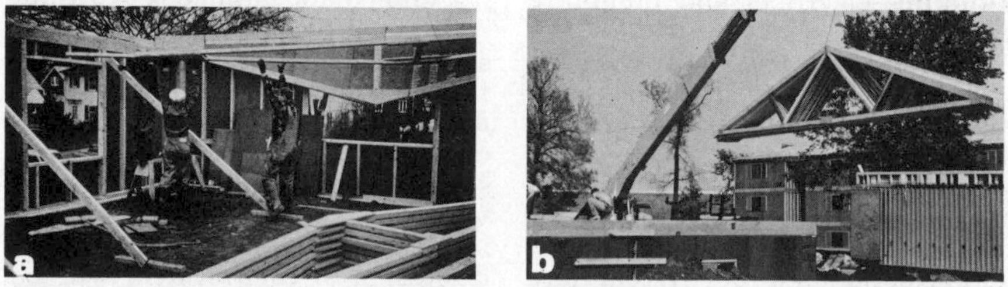

FIG. 4 (a) Individual trusses are positioned easily by hand. (b) Crane equipment can lift a stack of trusses directly from truck to roof.

Because of the structural action within a truss, individual members primarily resist compression and tension forces. Bending stresses are relatively small. Hence, truss members may be considerably smaller than their counterparts in joist-and-rafter systems which resist bending forces primarily. Trusses require stress-rated lumber and must be built from engineered designs. They commonly are spaced from 24″ to 48″ o.c.; and like other preassembled components, they can be erected easily, permitting the building to be enclosed quickly (Fig. 4).

Truss roof systems may be designed for flat (Fig. 5) or sloped roofs to receive membrane or shingle roofing finishes. Plywood or board sheathing typically is fastened to the truss to receive the roofing.

These systems provide overhead space for the installation of insulation and heating, air conditioning, electrical and plumbing equipment.

Almost any interior finish may be applied to the lower chord member at the ceiling line of the truss. Plaster bases and plaster board finishes commonly are applied directly to the trusses; with truss spacing greater than 24″ o.c. and with other materials, it may be necessary to apply nailing strips or furring strips across the trusses to receive the interior finish.

Trusses are economical and are widely used (Fig. 6). They provide excellent structural performance, make efficient use of materials, are easy to assemble and to erect, and provide maximum interior flexibility.

FIG. 5 Flat roof/ceiling construction is possible with truss systems.

FIG. 6 (a) Trusses can be assembled in the field on simple "jig" patterns using only hand tools, but they often are preassembled in shops with press equipment. (b) These trusses have been designed to provide second-floor space (NAHB Research House).

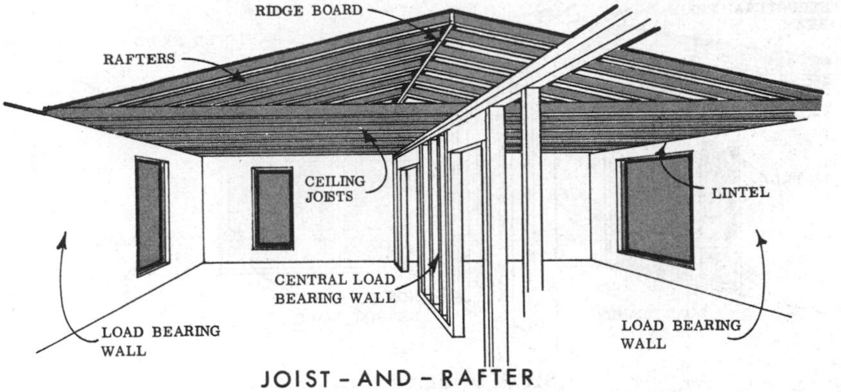

FIG. 7 Wood joist-and-rafter roof/ceiling construction.

JOIST-AND-RAFTER ROOFS

Joist-and-rafter systems consist of nominal 2″ lumber ceiling joists and roof rafters spaced 12″, 16″ or 24″ o.c. Typical construction for joist-and-rafter roofs is shown in Figure 7. Joists span roughly one-half the width of the building from the exterior walls to a center load-bearing wall or beam. The rafters are supported on exterior walls and bear against each other and/or a ridge board at the high point of the roof (Fig. 8). When loaded, the ridge tends to lower and the rafters exert a thrusting force which tends to spread the outside walls. This thrust is resisted by the ceiling joists acting as ties from wall to wall.

Joist-and-rafter construction generally requires longer erection time than truss construction and may expose materials longer to possible weather hazards.

Interior load-bearing walls somewhat restrict the flexibility of room planning. However, for relatively short spans (approximately 16′ or less) load-bearing interior walls or beams generally are not required. If the house plan form includes a number of wings of different spans, joist-and-rafter framing may be more economical than truss construction.

These systems provide overhead space for the installation of insulation and mechanical equipment; and in addition, with adequate rafter slopes, economical attic storage or expansion space can result (Fig. 9).

Interior and exterior finishes may be applied as for truss roofs.

FIG. 8 Joists and rafters are framed entirely on-site. Different gable spans are used here. Gables meet at a "valley." Shorter rafters from valley to ridge are called "jack rafters." High gable to the right encloses attic or story-and-a-half expansion space.

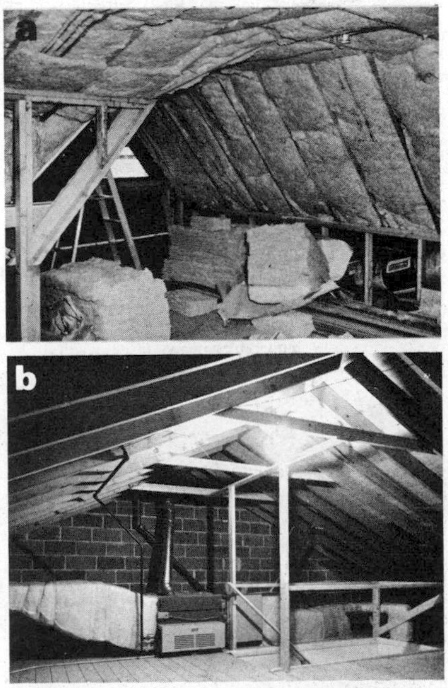

FIG. 9 Joist-and-rafter systems can provide space for attic storage, future expansion and mechanical equipment.

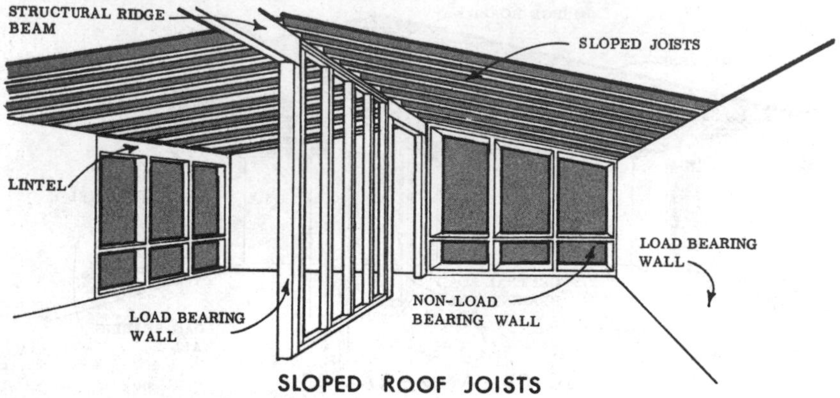

FIG. 10 Wood joist roof/ceiling construction.

FIG. 11 Roof joists here span from exterior walls to center ridge beam.

FIG. 12 Joist systems can create interiors with high, sloped ceilings. Since end walls are non-load bearing, large gable-end glass areas may be used.

JOIST ROOFS

In this roof/ceiling system, the functions of ceiling joists and roof rafters are provided for by roof joists alone (Fig. 10). These joist members generally are of nominal 2″ dimension lumber spaced 12″, 16″ or 24″ o.c. Joists may have a clear span between exterior walls or they may be centrally supported on a load-bearing partition or structural beam (Fig. 11). When the roof is sloped rather than flat, the construction also may be called a ridge-supported rafter system which often is used to achieve an interior sloped ceiling effect (Fig. 12). Roof sheathing materials may be attached directly to the top of the joists, and ceiling finishes applied directly to the underside.

Ductwork can be accommodated in the joist space if it is run parallel to the joists. When it is necessary to run ductwork perpendicular to the joists, it is common to use dropped-ceiling areas over hallways. Wiring can be concealed in the joist space and the system is relatively easy to insulate. Interior and exterior finishes are applied in the same manner as for truss and joist-and-rafter roofs.

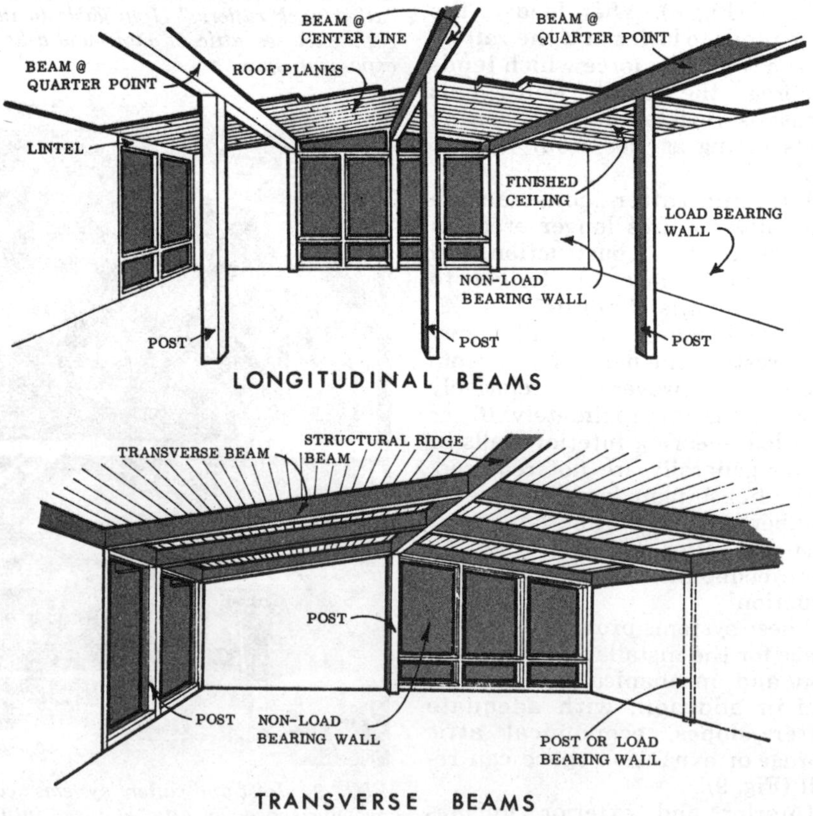

FIG. 13 Wood plank-and-beam roof/ceiling construction.

PLANK-AND-BEAM ROOFS

Plank-and-beam roofs typically are used on one-story buildings and for low-slope and flat roofs. These systems are characterized by the use of 2″ x 6″ and 2″ x 8″ tongue-and-grooved (T&G) or splined planks spanning between either longitudinal or transverse beams (Fig. 13). Beam spacing and thickness of planking are determined by imposed loads, spans and strength of the materials. Nominal 2″ plank generally may be used on spans up to approximately 8′.

Longitudinal beams span parallel to the long dimension of the building and are supported on end walls or columns. In buildings up to approximately 36′ wide, the beams typically are placed at quarter points. This divides the width of the building into two equal spans for planking on each side of the center ridge beam. Planks may be supported along the exterior sides by walls or beams. Interior planning flexibility is slightly more restricted than when transverse beams are used because of the greater number of beam supports required.

Transverse beams, generally spaced 6′ to 8′ o.c., span perpendicular to the long dimension of the building. They may be supported by load-bearing walls and/or beams along the ridge or center of the building.

The system may be used with wood stud exterior walls and is often used with post-and-beam exterior walls (Fig. 14).

FIG. 14 Longitudinal beams here are carried by posts, walls and another beam (beneath the ridge).

FIG. 15 The transfer of roof loads from planks to beams permits the use of large glass areas without headers.

FIG. 16 Planks, beams and posts often are part of interior finish. Note ventilation panels used above glass areas.

The plank ceiling and roof deck does not provide overhead space for the convenient concealment of ductwork and wiring. If the system requires insulation, it is necessary to use rigid insulation applied to the top of the deck—which in turn may limit the selection of roofing finishes.

Plank-and-beam systems, in contrast to truss and joist-and-rafter roof systems, require different design and construction methods and care in erection, since the finished appearance generally is an integral part of the system (Fig. 15). The planks, beams and columns generally are left exposed as the finished interior surface. (Fig. 16).

PANELIZED ROOFS

Panelized systems may consist of simple preframed, precut and sheathed panels (Fig. 17). Nominal 2″ dimension frame members of the panels function as rafters or ceiling joists. Panels may be fabricated on or off site in any size that is conveniently handled.

Panelized systems also include *stressed-skin* and *sandwich panel* construction. These panels, which often require specialized equipment and quality control, are factory-built.

Stressed-skin roof panels are made with framing members to which plywood *skins* are bonded either by glue-nailing techniques or by adhesive applied under heat and pressure (Fig. 18). When the roof panel is loaded, the skin and the framing members act as an integral unit to carry loads. The stressing of the skin surfaces permits a reduction in the size of the framing members.

Sandwich panels are similar in principle to stressed-skin panels but the faces are glued to and separated by weaker lightweight core materials (Fig. 19), instead of actual framing members.

Panelized systems may be used in a variety of ways to form sloped or flat roofs and ceilings (Fig. 20). Panels may span between transverse or longitudinal beams or can be centrally supported by bearing walls or beams as in joist roofs.

FIG. 17 This preframed roof panel has been assembled on the ground, partially sheathed, and lifted into position.

FIG. 18 Stressed-skin roof panels with plywood faces being installed over glue-laminated beams 8′ o.c.

FIG. 19 Sandwich panel with foamed plastic core, plywood upper surface and aluminum-faced gypsum ceiling finish.

FIG. 20 Vaulted stressed-skin panels can create unusual roof/ceiling forms.

Membrane or shingle roofing finishes may be applied to panelized systems. Interior finishes may be field-applied or may be an integral, prefinished part of the panel.

Though panelized construction provides many of the advantages of component fabrication, incorporation of overhead mechanical equipment requires careful design of the system.

We gratefully acknowledge the assistance of the following for the use of their publications as references and permission to use photographs: American Plywood Association, House & Home Magazine, National Forest Products Association, Practical Builder Magazine, Western Wood Products Association.

411 SHINGLE ROOFING

SHINGLE ROOFING

INTRODUCTION

The primary function of shingles on a sloped roof is to provide a surface which *sheds water.* In contrast, flat or very low sloped roofs are covered with built-up roofing which provides a *waterproof membrane.* Due to the large amount of exposed surface, shingle roofs also contribute greatly to the architectural appearance of the building, not only in terms of shape or form, but because shingles also add color, texture and/or pattern to the roof surface.

In addition to cost, other criteria to be considered in selecting the kind of shingles and the method of application to be used include: the slope of the roof, expected service life of the roofing, wind resistance, fire resistance, and local climate. Depending on the color and kind of shingle selected, total heat loss or gain of the roof may also be affected.

This section describes the materials and methods recommended to provide a suitable roof finish of shingles. Beginning with General Recommendations typical of all shingle roofs, the text is divided into sub-sections giving more specific recommendations for roof finishes of asbestos-cement shingles, asphalt shingles, wood shingles and handsplit wood shakes. Other roof finishes are described in other sections of Division 400.

GENERAL RECOMMENDATIONS

All shingles are applied to roof surfaces in some overlapping fashion to shed water. Accordingly, shingles are suitable for use on any roof that has sufficient slope to insure good drainage.

In addition to shingle units, many accessory materials are required to prepare the roof deck and to apply the shingles. Accessories usually required include *sheathing, underlayment, flashing, roofing cements, eaves flashing, drip edge,* and *roofing nails (or fasteners).* Depending on the kind of shingle, other accessories, such as starter shingles, hip and ridge units may also be required. While most materials and typical methods of applying shingle roofs are discussed in the following text, the instructions and recommendations of the shingle manufacturer should always be consulted to insure the most suitable performance.

In this section, the roof construction is assumed to be correctly and adequately ventilated; a practice that is essential to prevent condensation of water vapor. Condensation, the actual cause of many unjust complaints of roof leakage, can be eliminated as a hazard by providing sufficient ventilation in attic areas or enclosed air spaces beneath the roof (see Division 100—Moisture Control).

NOMENCLATURE

Definitions of both roof and roofing terminology are shown below to assist in understanding the text and figure illustrations.

Roof Terminology

Several common types of sloped roofs are illustrated in Figure 1. These simple shapes are often complicated by intersecting walls or dormers, other intersecting roofs and by projections through the roof such as chimneys, plumbing vents and roof ventilators.

The composite drawing in Figure 2 names the elements of a sloped roof.

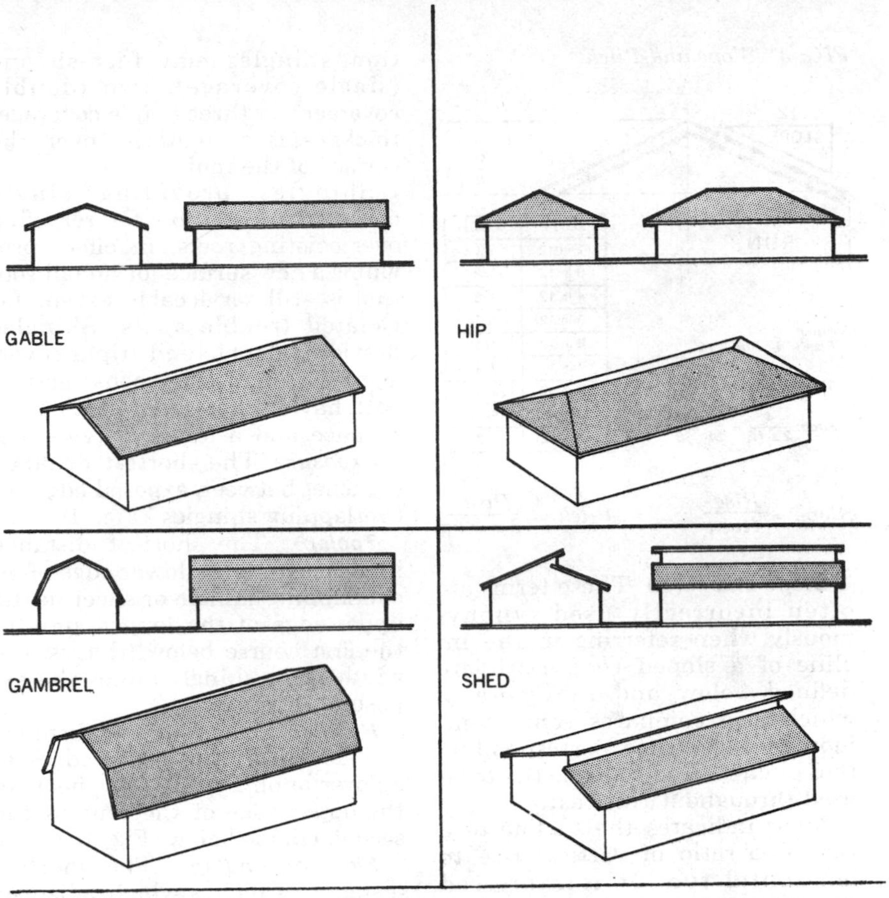

FIG. 1 Shingles applied to sloped roofs shed water. In contrast, flat roofs require a watertight membrane of built-up roofing.

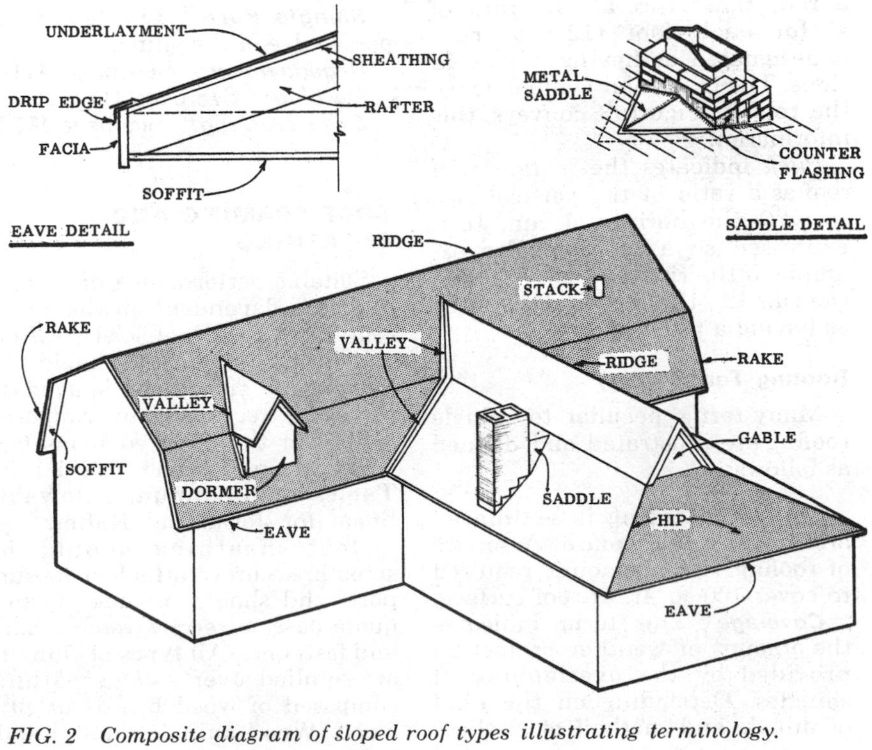

FIG. 2 Composite diagram of sloped roof types illustrating terminology.

FIG. 3 Slope and Pitch.

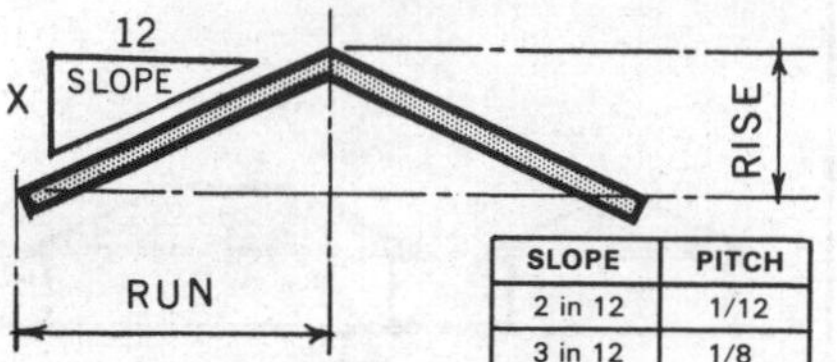

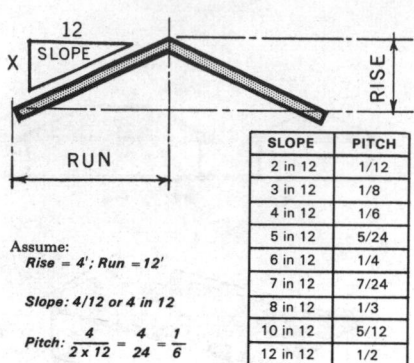

SLOPE	PITCH
2 in 12	1/12
3 in 12	1/8
4 in 12	1/6
5 in 12	5/24
6 in 12	1/4
7 in 12	7/24
8 in 12	1/3
10 in 12	5/12
12 in 12	1/2

Assume:
Rise = 4'; Run = 12'

Slope: 4/12 or 4 in 12

Pitch: $\frac{4}{2 \times 12} = \frac{4}{24} = \frac{1}{6}$

$$Slope = \frac{Rise}{Run;} \qquad Pitch = \frac{Rise}{2 \times Run}$$

Slope and Pitch These terms are often incorrectly used synonymously when referring to the incline of a sloped roof. Both are defined below and in Figure 3, which also compares some common roof slopes to corresponding roof pitches. Roof *slope* is the term used throughout this text.

Slope indicates the incline of a roof as a ratio of vertical rise to horizontal run. It is expressed sometimes as a fraction but typically as "X" in 12. For example, a roof that rises at the rate of 4″ for each foot (12″) of run, is designated as having a *4 in 12* slope. The triangular symbol above the roof in Figure 3 conveys this information.

Pitch indicates the incline of a roof as a ratio of the vertical rise to *twice* the horizontal run. It is expressed as a fraction. For example, if the rise of a roof is 4′ and the run 12′, the roof is designated as having a pitch of 1/6.

Roofing Terminology

Many terms peculiar to shingle roofing are illustrated and defined as follows:

Square Roofing is estimated and sold by the square. A square of roofing is the amount required to cover 100 sq. ft. of roof surface.

Coverage This term indicates the amount of weather protection provided by the overlapping of shingles. Depending on the kind of shingle and method of application, shingles may furnish one (single coverage), two (double coverage), or three (triple coverage) thicknesses of material over the surface of the roof.

Shingles providing single coverage are suitable for reroofing over existing roofs, in effect providing a new surface for an old roof that is still serviceable except for isolated trouble spots. Shingles providing double and triple coverage are used for new construction, both having increased weather resistance and a longer service life.

Exposure: The shortest distance in inches between exposed edges of overlapping shingles (Fig. 4).

Toplap: The shortest distance in inches from the lower edge of an overlapping shingle or sheet to the upper edge of the lapped unit in the first course below (that is, the width of the shingle minus the exposure) (Fig. 4).

Headlap The shortest distance in inches from the lower edges of an overlapping shingle or sheet, to the upper edge of the unit in the second course below (Fig. 4).

Side- or Endlap: The shortest distance in inches which adjacent shingles or sheets horizontally overlap each other (Fig. 4).

Shingle Butt: The lower exposed edge of the shingle.

Underlayment: See page 411-5.
Flashing: See page 411-6.
Eaves Flashing: See page 411-7.

FIG. 4 Roofing Terminology

E = Exposure
TL = Toplap
HL = Headlap
SL = Sidelap
W = Width for Strip Shingles or Length for Individual Shingles

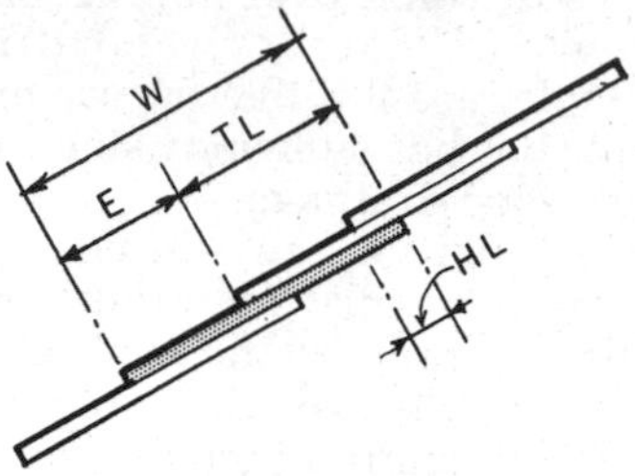

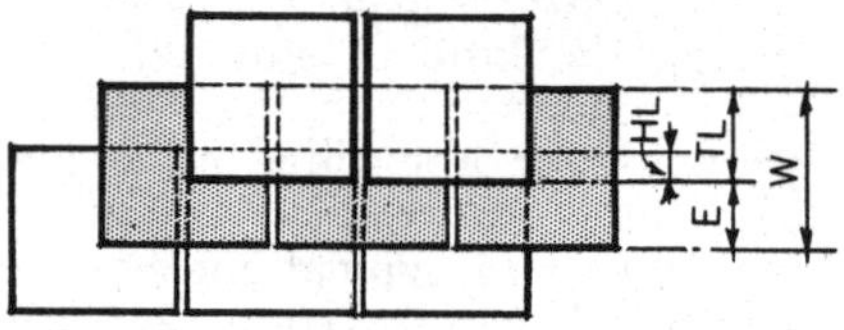

AMERICAN METHOD

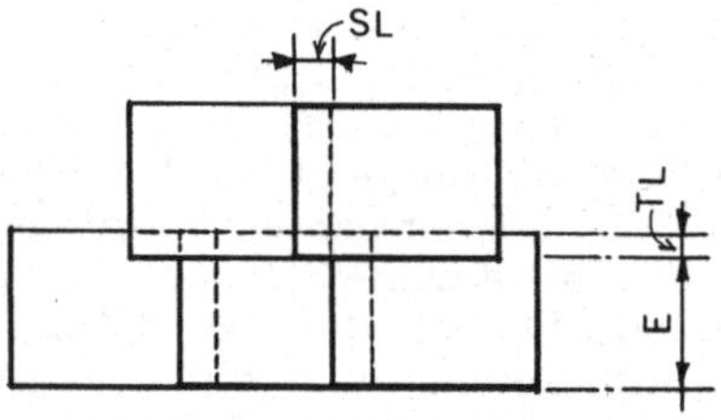

DUTCH LAP METHOD

ROOF FRAMING AND SHEATHING

Suitable performance of shingle roofing is dependent on the entire roof structure. Roof joists, rafters and other supports should be adequately sized and spaced to carry the necessary superimposed loads over the required spans (see Work File, page Wood WF201-1, for Tables of Maximum Allowable Spans for Joists and Rafters).

Roof sheathing should be smooth, securely attached to supports and should provide an adequate base to receive roofing nails and fasteners. All types of shingles are applied over *solid* sheathing composed of wood boards or plywood. Wood shingles and hand-split wood shakes are also applied over *spaced* sheathing.

Lumber Sheathing

Solid sheathing of boards should be seasoned to moisture content shown in Figure W8, page Wood 201-6. Boards should preferably be tongue-and-grooved (T&G) sheathing, nominal 1″ thickness and usually not over 6″ nominal width to minimize shrinkage. The boards should be tightly matched and securely face- and edge-nailed with two 2-1/8″ x 0.109″ (7d) annularly threaded or 2-1/2″ x 0.131″ (8d) common wire nails at each framing member. Maximum spacing of supports for nominal 1″ lumber boards is 24″ on center (o.c.).

Spaced sheathing of nominal 1″ x 3″, 1″ x 4″ or 1″ x 6″ boards may be used in blizzard-free areas (generally where the outside design temperature is warmer than 0° F) to receive the application of wood shingles and wood shakes. Seasoning and face-nailing of spaced boards should be the same as described for solid sheathing.

Plywood Sheathing

Plywood sheathing should be of the proper Interior or Exterior Construction Grade (see Work File, page Wood WF201-16) and should conform to the spans and thicknesses given in Figure 5.

Plywood 1/2″ thick or less should be applied with 1-3/4″ x 0.120 (5d) annularly threaded nails or 2″ x 0.113″ (6d) common wire nails; thicker plywood should be applied with 2-1/8″ x 0.109″ (7d) annularly threaded nails or 2-1/2″ x 0.131″ (8d) common wire nails. (1-1/2″ x 0.135″ helically threaded nails also may be used.) Nails should be spaced 6″ o.c. around panel edges and 12″ o.c. at other points of bearing on framing members.

Non-Wood Sheathing

When gypsum, concrete, fiberboard or sheathing materials other than wood or plywood are recommended by their manufacturers for use as roof sheathing under shingles, the sheathing manufacturer should specify the type of fastening required to attach sheathing to supports.

ROOF SLOPE LIMITATIONS

Free and effective drainage of water from sloped roof surfaces is essential for long service from shingles. As the roof slope decreases, the runoff of water is slower and the susceptibility to leakage from wind-blown rain or snow being driven up underneath the butt-end of the shingle increases. Accordingly, the roof slope governs, or may impose limitations on, the following: (1) shingle selection, (2) shingle exposure, (3) underlayment requirements, (4) eaves flashing requirements, and (5) methods used in shingle application.

In this text, when a reduction in the slope of the roof requires precautions other than normal practice, the roof is referred to as being *Low Slope;* otherwise, the term *Normal Slope* is used.

The minimum slope recommended for each kind of shingle is shown in Figure 6. Slope limitations for underlayment and eaves flashing requirements vary with each kind of shingle and are discussed below. Normal and Low Slope application methods are discussed in the following sub-sections for each kind of shingle.

UNDERLAYMENT

Roof underlayment performs several functions: (1) it protects the sheathing from moisture absorption until shingles can be applied; (2) it provides important additional weather protection by preventing the entrance of wind driven rain below the shingles onto the sheathing or into the structure; (3) in the case of asphalt shingles, it prevents direct contact between shingles and resinous areas in wood sheath-

FIG. 5 **PLYWOOD ROOF SHEATHING**[1,2]

Identification Index	Plywood Thickness	Maximum Rafter or Joist Spacing (inches o.c.)			
		Asphalt Shingles and Wood Shingles & Shakes		Asbestos-Cement Shingles	
		Edges Blocked[3]	Edges Unblocked	Edges Blocked[3]	Edges Unblocked
16/0	5/16″, 3/8″	16″	—	—	—
20/0	5/16″, 3/8″	20″	—	—	—
24/0	3/8″, 1/2″	24″	24″	16″	—
30/12	5/8″	30″	24″	24″	16″
32/16	1/2″, 5/8″	32″	24″	24″	16″
36/16	3/4″	36″	30″	32″	24″
42/20	5/8″, 3/4″, 7/8	42″	32″	32″	24″
48/24	3/4″, 7/8	48″	32″	42″	28″

1. Plywood continuous over two or more spans, grain of face plys across supports; Structural I and II, Standard and C-C Exterior Construction Grades only.
2. For 1/2″ plywood or less, use 6d common or 5d threaded nails; for 1″ plywood or less use 8d common or 7d threaded nails.
3. Unsupported edges of sheathing should be blocked using wood blocking, tongue-and-grooved edges (5 ply, 1/2″ or thicker) or special corrosion-resistant metal H clips. Use 2 clips for spans 48″ or greater, 1 clip for lesser spans.

FIG. 6 **SUMMARY OF MINIMUM ROOF SLOPE RECOMMENDATIONS**

Types of Roofing	Normal Slope	Low Slope[1]
Asbestos Cement Shingles	5 in 12	3 in 12[2]
Asphalt Shingles	4 in 12	2 in 12[3]
Wood Shakes	4 in 12	3 in 12[4]
Wood Shingles	5 in 12	3 in 12[5]

1. Where eaves flashing is required, it should be cemented.
2. Requires double underlayment.
3. Strip shingles only; requires double underlayment and Wind Resistant shingles or cemented tabs.
4. Requires solid sheathing, underlayment in addition to interlayment and reduced weather exposure.
5. Requires reduced weather exposure.

ing which, because of chemical incompatability, may be damaging to the shingles.

Underlayment should be a material such as asphalt saturated felt which has low vapor resistance. Materials such as coated felts or laminated waterproof papers which act as vapor barriers should not be used. Such materials may permit moisture or frost to accumulate between the underlayment and the surface of the roof sheathing.

Underlayment requirements for different kinds of shingles for various roof slopes are summarized in Figure 7.

Underlayment should be applied over the entire roof as soon as the roof sheathing has been completed with at least a 2″ toplap at all horizontal joints and a 4″ sidelap at end joints in the underlayment (Fig. 8). The underlayment should be lapped 6″ from both sides over all hips and ridges. Only sufficient fasteners need be used to hold the underlayment securely in place until shingles are applied. Shingles should not be applied over wet underlayment.

FIG. 7 SUMMARY OF UNDERLAYMENT RECOMMENDATIONS FOR SHINGLE ROOFS

Type of Roofing	Sheathing	Type of Underlayment	Normal Slope		Low Slope	
Asbestos-Cement Shingles	Solid	No. 15 asphalt saturated asbestos (inorganic) felt, OR No. 30 asphalt saturated felt	5/12 and up	Single layer over entire roof	3/12 to 5/12	Double layer over entire roof[1]
Asphalt Shingles	Solid	No. 15 asphalt saturated felt	4/12 and up	Single layer over entire roof	2/12 to 4/12	Double layer over entire roof[2]
Wood Shakes	Spaced	No. 30 asphalt saturated felt (interlayment)	4/12 and up	Underlayment starter course; interlayment over entire roof	Shakes not recommended on slopes less than 4/12 with spaced sheathing	
	Solid[3,5]	No. 30 asphalt saturated felt (interlayment)	4/12 and up	Underlayment starter course; interlayment over entire roof	3/12 to 4/12[4]	Single layer underlayment over entire roof; interlayment over entire roof
Wood Shingles	Spaced	None required.	5/12 and up	None required	3/12 to 5/12[4]	None required
	Solid[5]	No. 15 asphalt saturated felt	5/12 and up	None required[6]	3/12 to 5/12[4]	None required[6]

1. May be single layer on 4 in 12 slope in areas where outside design temperature is warmer than 0° F.
2. Square-Butt Strip shingles only; requires Wind Resistant shingles or cemented tabs.
3. Recommended in areas subject to wind driven snow.
4. Requires reduced weather exposure.
5. May be desirable for added insulation and to minimize air infiltration.
6. May be desirable for protection of sheathing.

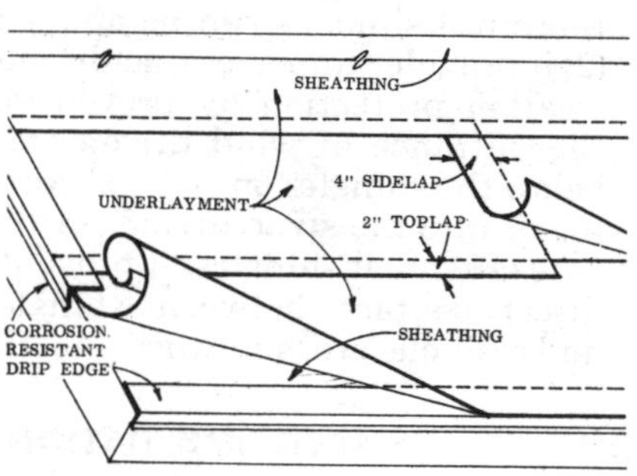

FIG. 8 Application of Underlayment

FLASHING

Roofs are often intersected by other roofs, adjoining walls and projections through the roof, creating opportunities for leakage. Special construction, called flashing, must be installed at these locations to weatherseal the roof. Care must be used to make all flashing watertight and water-shedding.

Valley Flashing

The joint formed by two sloping roofs meeting at an angle is called a valley. Drainage concentrates in the valley, causing runoff toward and along the joint, and is therefore especially vulnerable to leakage. An unobstructed drainage way must be provided with enough capacity to carry away the water rapidly.

Chimney Flashing

The chimney is usually built on a foundation separate from the foundation that supports the structure and both are normally subject to some differential settling. To permit differential movement between chimney and roof structure, without damage to the waterseal, it is necessary to secure *base flashings* to the roof deck and secure *counter flashing,* which is bent down over the base flashing, to the masonry.

When chimneys project through the roof below the ridge, a saddle (See Fig. 2) should be constructed between the back face of the chimney and the roof. The purpose of the saddle is to prevent the accumulation of snow and ice behind the chimney and to deflect water down around the chimney.

Flashing Materials

Corrosion resistant metal used for flashing valleys, chimneys and other roof intersections should be at least 26 gauge galvanized steel, 0.019″ thickness aluminum or 16 oz. copper. (See pages Masonry Walls 343-3 & 343-4 for a discussion of these and other flashing materials).

Additional requirements and typical methods of flashing are described in the following subsections for each kind of shingle.

ROOFING CEMENTS

These include *plastic asphalt cements, lap cements, quick setting asphalt adhesives, roof coatings* and *primers.* They are used for installing eaves flashing, for flashing assemblies, for cementing tabs of asphalt shingles and laps in sheet materials, and for roof repairing. Each is discussed in Section 217 Bituminous Products.

The type and quality of materials and methods of application on shingle roofs should be as recommended by the manufacturer of the shingle roofing.

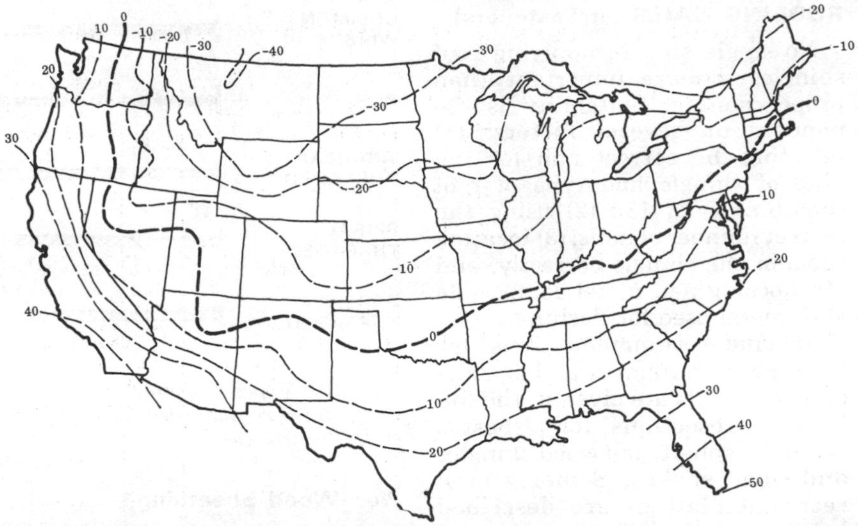

FIG. 9 OUTSIDE DESIGN TEMPERATURES

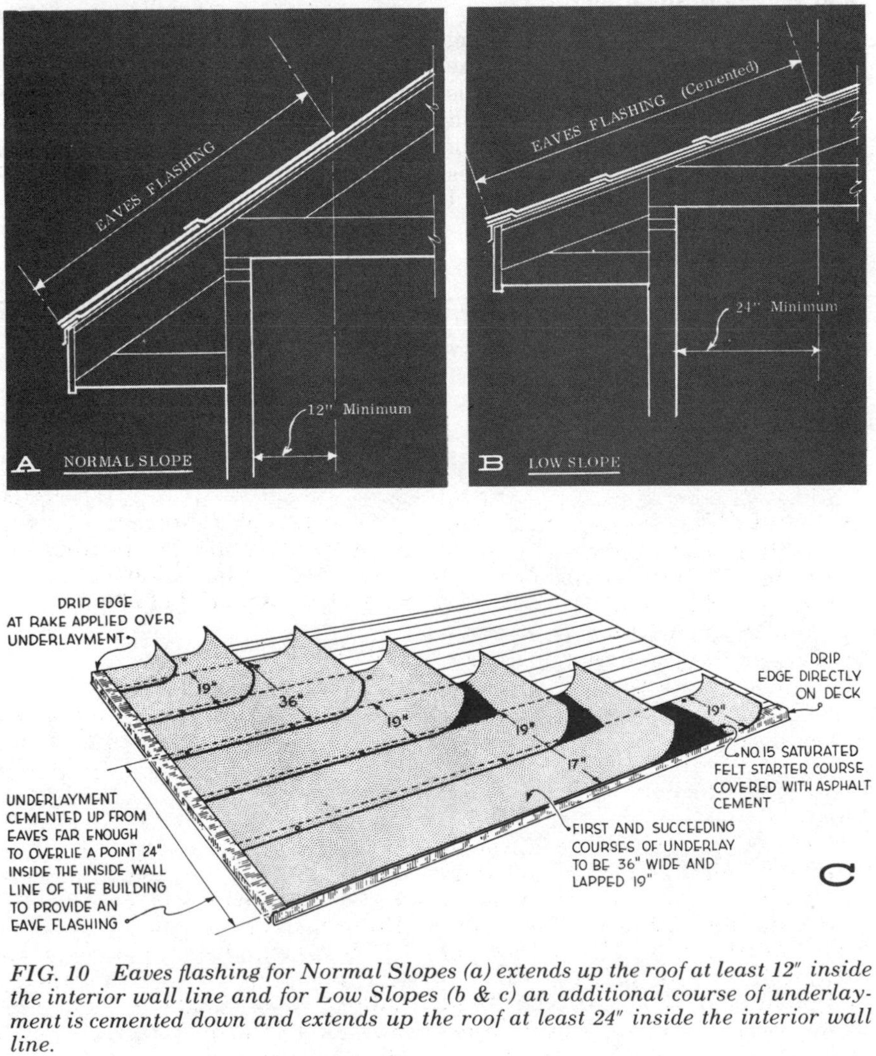

FIG. 10 Eaves flashing for Normal Slopes (a) extends up the roof at least 12" inside the interior wall line and for Low Slopes (b & c) an additional course of underlayment is cemented down and extends up the roof at least 24" inside the interior wall line.

EAVES FLASHING

Eaves flashing is recommended in areas where the outside design temperature is 0° F or colder (Fig. 9), or wherever there is a possibility of ice forming along the eaves. This protection is important to avoid leaks and moisture penetration caused by water backing up under shingles behind dams of ice and snow that collects at the eaves.

Different methods of flashing are used to prevent leakage from this cause, depending on the slope of the roof, the severity of icing conditions and the kind of shingle.

For Normal Slopes, eaves flashing is typically formed by a course of smooth or mineral surfaced coated roll roofing applied over the underlayment. It extends up the roof to cover a point at least 12" inside the interior wall line of the building (Fig. 10a).

For Low Slopes or in areas subject to severe icing, a double layer of underlayment should be applied over the entire roof. The two layers should be cemented together with asphalt cement up to the roof to a point that is at least 24" inside the interior wall line of the building (Fig. 10b).

Specific recommendations for eaves flashing are described in the following sub-sections for each kind of shingle.

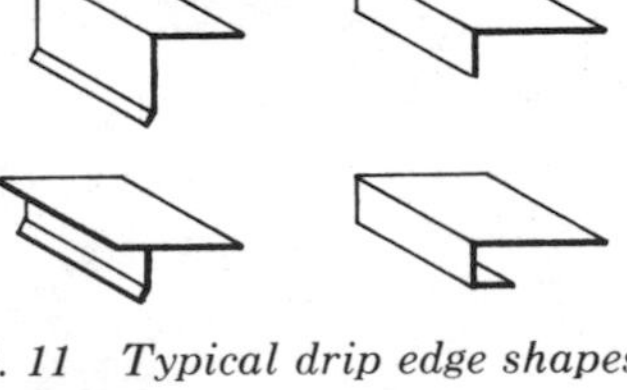

FIG. 11 Typical drip edge shapes.

DRIP EDGE

A continuous corrosion-resistant, non-staining material such as 26 gauge galvanized steel, formed to provide a drip, and nailed along the eaves and rakes is recommended for most shingle roofs. Drip edges are designed and installed to protect the edges of the deck and prevent leaks at this point by allowing water to drip free of underlying eave and cornice construction. Typical shapes of preformed drip edges are shown in Figure 11.

Drip edge is applied to the sheathing and *under* the underlayment at the eaves, but *over* the underlayment up the rake. (Fig. 12).

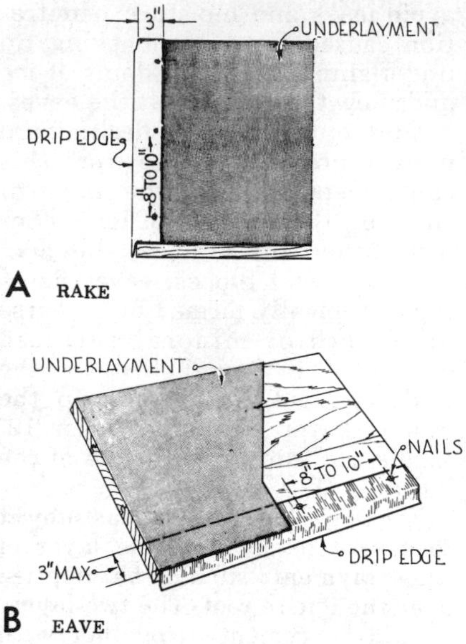

FIG. 12 Application of drip edge at rake (a) and eave (b).

ROOFING NAILS (or Fasteners)

No single step in applying roof shingles is more important than proper nailing. Suitability is dependent on several factors: (1) selecting the correct nail for the kind of shingle and type of roof sheathing (Fig. 13); (2) using the correct number of nails; (3) locating them in the shingle correctly, and (4) choosing nail metal compatible with metal used for flashings.

Specific recommendations for the type, size, number and spacing of roofing nails are given in the following subsections for asbestos cement, asphalt, and wood shingles and wood shakes. Some general recommendations are described below.

Lumber Boards, Plank Decking and Plywood Sheathing

Roofing nails should be long enough to penetrate through the shingle and through lumber boards or plywood sheathing. They should penetrate at least 1″ into plank decking. Nails for applying shingles over plywood sheathing should have threaded shanks.

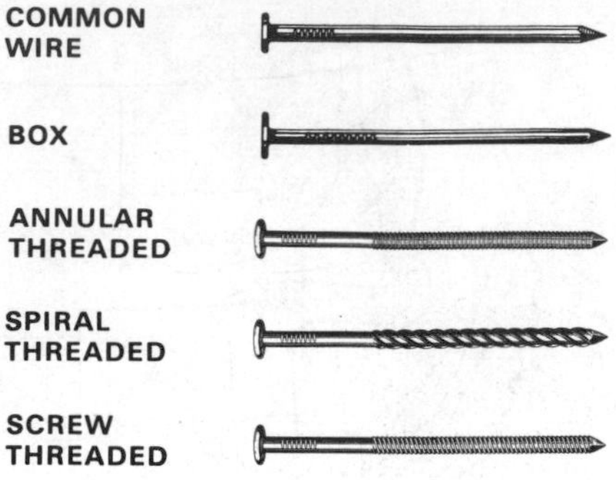

FIG. 13 Types of smooth and threaded shank nails recommeded for application of shingle roofing.

Non-Wood Sheathing

When gypsum, concrete, fiberboard or sheathing materials other than wood are used for the roof sheathing, special fasteners and/or special details for fastening are often necessary to provide adequate anchorage of the shingle roofing. Recommendations of the manufacturer of the non-wood deck material should be followed to insure his responsibility for the suitable performance of the roof.

In 1964, the Mineral Fiber Products Bureau (formerly called the Asbestos-Cement Products Association), whose members manufacture asbestos cement building materials, changed the terminology of their products. Asbestos cement roof shingles are now designated as *mineral fiber* roof shingles. "Mineral fiber" is the term used throughout the following text.

The following text describes materials and methods recommended to provide a suitable roof covering of mineral fiber shingles over both new and existing construction.

Reference should be made to the entire preceding subsection, General Recommendations, pages 411-1 through 411-8, for: (1) definition of terminology; (2) limitations imposed by the slope of the roof; (3) roof sheathing thicknesses, spans and nailing schedules; and (4) typical accessory materials—underlayment, flashing, roofing cements, eaves flashing, drip edge and roofing nails—recommended for preparation of the roof for the application of shingles. Unless specific exceptions are made in the following text, the General Recommendations apply for each kind of shingle and method of application.

MATERIALS

Mineral fiber roof shingles are manufactured in four basic types: (1) Individual units and (2) Multiple units, for application by the American Method; (3) Dutch Lap units and (4) Ranch Style units, for application by the Dutch (or Scotch) Method. Each type of shingle is applied in a distinct way giving a different appearance to the roof. Starter units, hip and ridge shingles and ridge roll are also manufactured of mineral fiber.

Shingles are made with prepunched holes correctly located to receive concealed roofing nails. Additional holes are provided in Dutch Lap and Ranch Style units to receive storm anchors.

MINERAL FIBER ROOF SHINGLES

Type	Width	Height	Thickness	Exposure	Weight per square
Individual Unit	8″ 9″	16″ 18″	5/32″ 1/4″ 1/4″	7″ 7″ 8″	350 lbs. 585 lbs. 570 lbs.
Multiple Unit	30″	14″	5/32″	6″	440 lbs.
Dutch Lap	16″	16″	5/32″	12″ x 13″	265 lbs.
Ranch Style	24″	12″	5/32″	20″ x 9″	255 lbs.

FIG. MF1 Products of individual manufacturers may vary slightly from the dimensions shown.

Mineral fiber shingles can be discolored by moisture and dampness while still in packages and until applied. Prior to application on the job, they should be kept clean and dry by complete protection from weather.

Mineral fiber roof shingles should conform to the requirements of ASTM C222. Data covering the four types is summarized in Figure MF1. See Section 219 Asbestos Cement Products for additional information covering manufacture and properties.

Nails and Fasteners

Roofing nails for applying mineral fiber shingles should be corrosion-resistant, such as hot dipped galvanized steel, aluminum or stainless steel, needle or diamond pointed, with large flat heads approximately 3/8″ diameter. Shanks should not exceed 0.140″ in overall outside diameter, so as not to exceed the diameter of prepunched holes in the shingles. Shanks may be smooth or threaded; however, threaded nails are preferred to increase holding power. Aluminum nails should have screw threads of approximately 12-1/2° thread angle; galvanized steel nails, if threaded, should have annular threads.

Solid Lumber Sheathing Nails for application of shingles over lumber sheathing should be approximately 1-1/4″ x 0.120″, preferably with threaded shanks.

Plywood Sheathing Nails for application of shingles over plywood sheathing *should have threaded shanks* and should be approximately 1-1/4″ x 0.120″, or long enough to penetrate through the sheathing.

Plank Decking and Non-wood Sheathing See Roofing Nails (or Fasteners), page 411-8.

Storm Anchors In addition to nails, storm anchors are used to apply Ranch Style and Dutch Lap units. Storm anchors should be copper, aluminum or lead, approximately 3/4″ long with a flat base, having sufficient ductility to permit repeated clinching and bending upward without cracking the shank (Fig. MF2).

The number, spacing and pattern of nails and fasteners are described below under Application of Shingles.

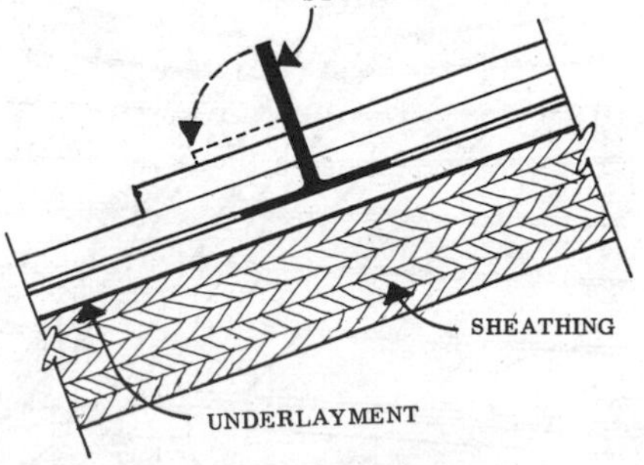

FIG. MF2 Storm anchors secure Dutch Lap and Ranch Style shingles applied by the Dutch Method.

PREPARATION FOR SHINGLES

Slope Limitations

Limitations imposed by the roof slopes are shown in Figure 6, page 411-5. Mineral fiber shingles are applied by Normal Slope Methods on roofs with slopes of 5 in 12 or greater. Using Low Slope Methods, they may be used on slopes as low as 3 in 12.

Mineral fiber shingles are not recommended on roof slopes less than 3 in 12.

Roof Sheathing

Sheathing types, thicknesses, spans and nailing requirements are discussed on page 411-4.

Underlayment

Underlayment material should be unperforated No. 15 asphalt saturated asbestos (inorganic) roofing felt, or No. 30 asphalt saturated organic felt. Coal tar saturated felt, which may stain the shingles, should not be used. Underlayment should be applied over the entire roof as described on page 411-5 and as shown below under Application of Shingles.

Cant Strips

When the American Method is used with Individual and Multiple unit shingles, a 1/4″ x 2″ wood strip is applied along the eaves to pitch (cant) the starter strip at the proper angle (Figs. MF4 & MF6).

Drip Edge

A corrosion-resistant non-staining drip edge should be installed along the eaves and rakes (see Fig. 12, page 411-8). Drip edge is applied over the cant strips when they are used.

Eaves Flashing

In areas where the outside design temperature is 0° F or colder (Fig. 9, page 411-7), or where the possibility of ice forming along the eaves could cause a backup of water, eaves flashing should be installed as described below.

Normal Slope (5 in 12 or greater) The eaves flashing is formed by cementing an additional strip of underlayment over the first underlayment course. A continuous layer of plastic asphalt cement is applied, at the rate of 2 gals. per square, from the eave to cover a point at least 12″ inside the interior wall line of the building (Fig. 10a, page 411-7) and is covered with the additional ply of underlyament.

Low Slopes (3 in 12 to 5 in 12) A cemented eaves flashing for low slopes is formed just as described above for normal slopes. However, the flashing should extend from the eave to cover a point at least 24″ inside the interior wall line (Fig. 10b, page 411-7).

Flashing

See General Recommendations, page 411-6.

Valley Flashing The *open* or *closed* method may be used to construct valley flashing as illustrated in Figure MF3.

Chimney and Side Wall Flashing When shingles butt against vertical surfaces such as sidewalls, dormers or chimneys, individual flashing units of corrosion-resistant metal should be inserted between each course of shingles. The metal flashing units should be folded from a width of 19″ material and have a standing leg 6″ high and a flat leg 4″ wide on the shingle. The length of the flashing should be 3″ greater than the exposure of the shingles. Sidewall exterior finish materials should cover the standing leg. In the case of masonry chimneys, a stepped counter flashing of corrosion-resistant metal should cover the standing leg of the flashing units and be trimmed off just above the surface of the shingles (see Fig. MW13, Masonry Walls, page 343-11).

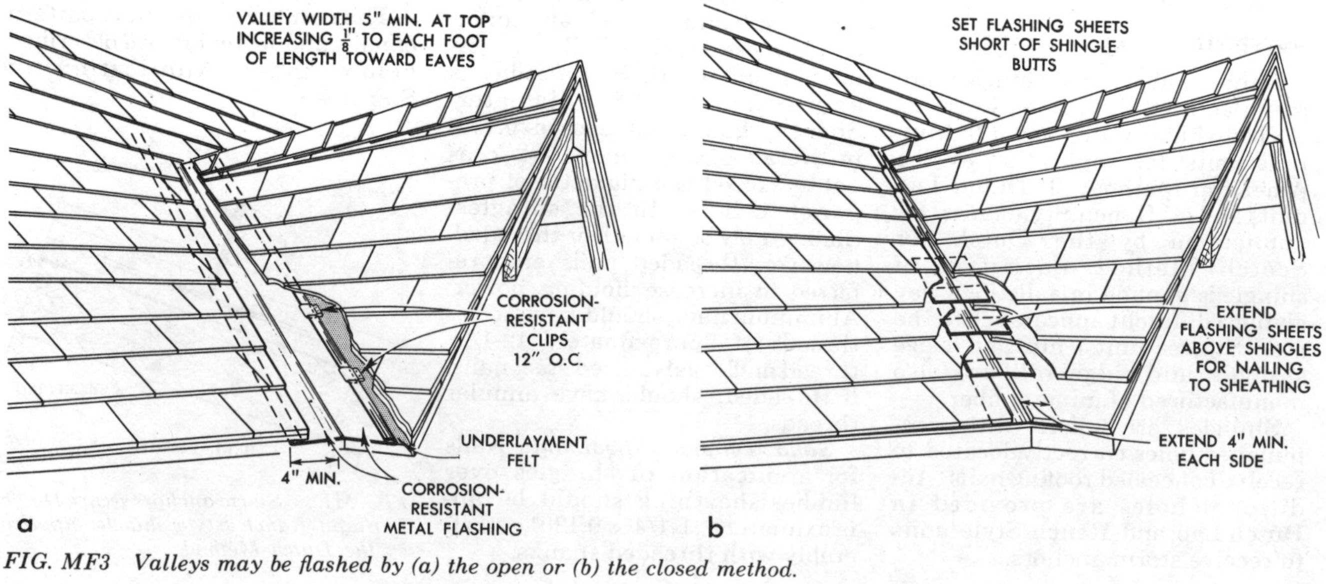

FIG. MF3 Valleys may be flashed by (a) the open or (b) the closed method.

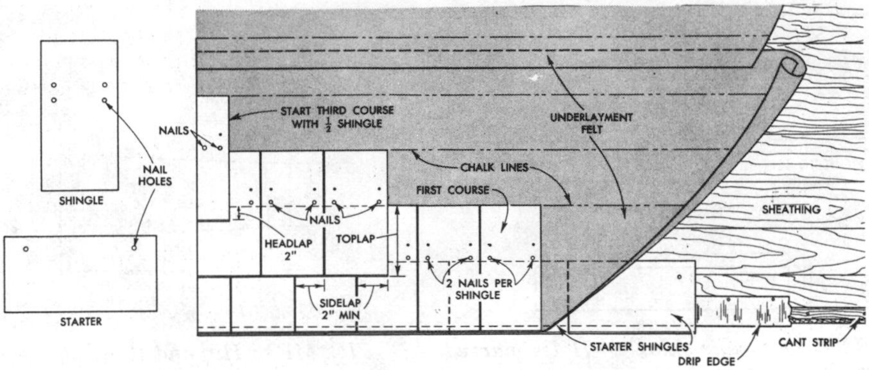

FIG. MF4 Application of Individual shingles with straight butt line.

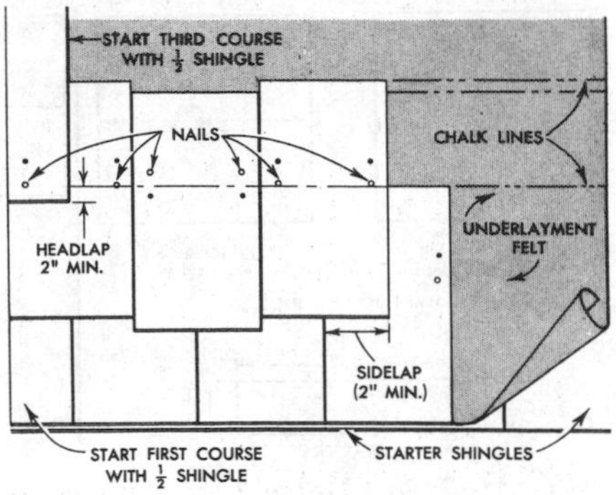

FIG. MF5 Staggered butt line can be achieved with Individual shingles by varying the exposure.

APPLICATION OF SHINGLES (New Construction)

The *American Method* is used to apply Individual and Multiple shingles without a sidelap. The *Dutch Method* with side- and top-lap is used to apply Dutch Lap and Ranch Style shingles. In both methods the underlayment felt should be rolled back and starter shingles applied directly on top of the metal drip edges previously nailed. Starter shingles should overhang the rakes and the eaves approximately 3/4″. When starters match the color of the field shingles and are used without a gutter, they are usually applied with the colored side down for better appearance from below.

For illustrations and discussion of tools and methods used to cut, punch or notch mineral fiber roof shingles, see Section 219 Asbestos Cement Products.

American Method (Normal Slope)

Individual Shingles (Straight Butt Line) After replacing the underlayment over the starter shingles, the first shingle course is applied starting with a half shingle overhanging the rake approximately 3/4″ and in line at the butt with the starter shingles. Each shingle should be applied with nails through two of the prepunched holes. The second course is started with a full shingle and the third course, like the first course, starts with a half shingle. Chalklines should be marked on the underlayment for each shingle course for accurate alignment (Fig. MF4).

Individual Shingles (Staggered Butt Line) Individual shingles have two sets of prepunched holes to permit a staggered application. The first course is applied as described above for straight butt line. In starting the second course, the exposure of every second shingle is reduced by dropping it down and nailing through the upper set of nailholes. Staggering can be varied depending on the appearance desired, but is limited to the distances between the upper and lower holes. The third course starts with a half shingle and alternating shingles should be staggered as in the previous course.

Individual shingles with either straight or staggered butt line should have a minimum headlap and sidelap of 2″ (Figs. MF4 & MF5).

Multiple Unit Shingles After replacing the underlayment over the starter shingles, the first shingle course is applied starting with a half shingle overhanging the rake approximately 3/4″ and in line at the butt with the starter shingles. The second course is started using a full shingle and the third course, like the first course, starts with a half shingle. Chalklines are not necessary if the visual alignment design of the shingles is properly utilized (Fig. MF6). Overdriving of nails causes the butts of the shingles to cock up and should be avoided to prevent damage to the shingles.

Because the design of Multiple unit shingles may vary among manufacturers, application instructions of the manufacturer for nailing, positioning and alignment should be followed.

FIG. MF6 Application of Multiple Unit shingles.

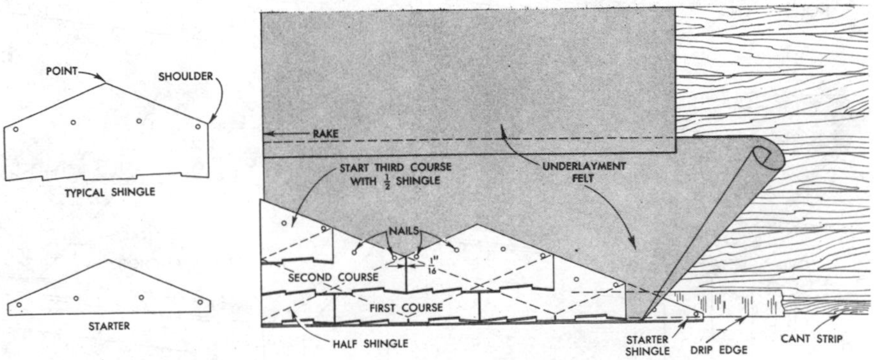

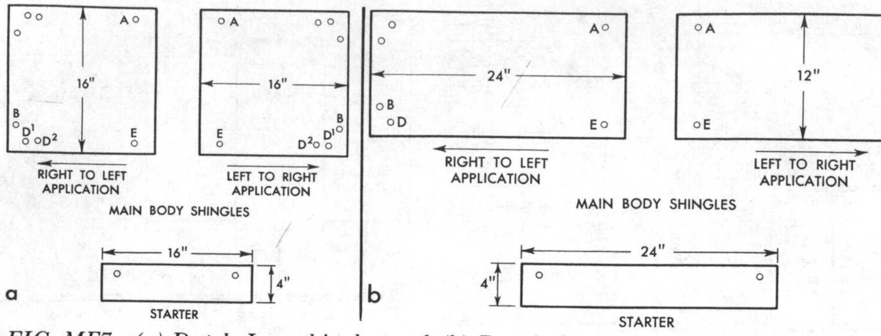

FIG. MF7 (a) Dutch Lap shingles and (b) Ranch Style shingles. Holes marked A and B are for nails; holes D and E are for storm anchors.

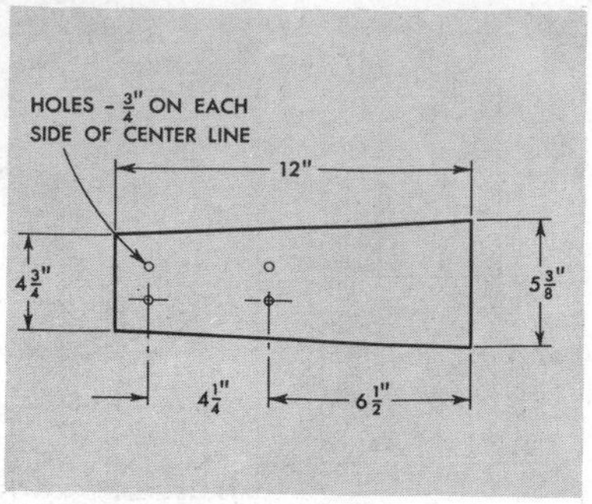

FIG. MF9 Hip and Ridge shingles.

Dutch Method (Normal Slope)

The Dutch Method is used to apply Dutch Lap shingles (Fig. MF7a) and Ranch Style shingles (Fig. MF7b). The following application methods refer to shingles having a 4″ sidelap.

The starter shingles are applied as shown in Figure MF8a & 8b. After the felt is replaced over the starter shingles, a piece 4″ wide cut from a full shingle is applied to form a rake starter. A storm anchor, "A", is inserted from below and the rake starter is secured in place using two nails (Fig. MF8c). Over the rake starter, a full-size shingle is applied by first inserting storm anchor "B" and positioning the shingles so that the storm anchor "A" appears through the free corner (Fig. MF8d). Storm anchor "A" is clinched and the shingle secured with two nails (Fig. MF8e). Each succeeding shingle is applied in the same manner, inserting first a storm anchor in one corner, and then engaging the other corner over the previous storm anchor. The second shingle course is started with a shingle 4″ less than full width, punched with a new hole as required to hold the shingle along the rake, and is completed using full shingles. The third course is started with a shingle 8″ less than full width; the fourth course with a shingle 12″ less than full width, and so on up the rake. Automatic alignment of the shingles is usually provided by the location of the prepunched holes.

Hips and Ridges

Before application of hip and ridge shingles (Fig. MF9), wood nailing strips equal in thickness to the roof shingle assembly should be applied at the hips and ridges to provide a flat surface. Hips or ridges should then be covered with two thicknesses of underlayment

FIG. MF8 Application of Dutch Lap and Ranch Style shingles.

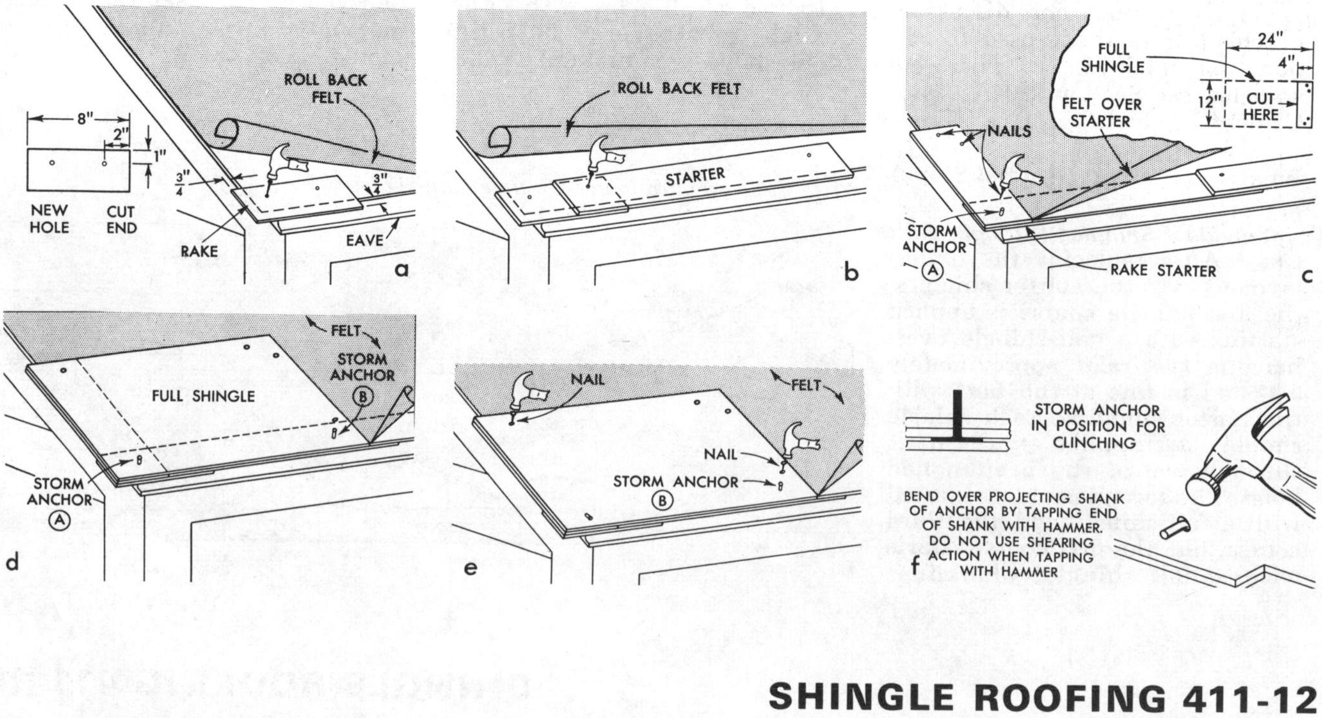

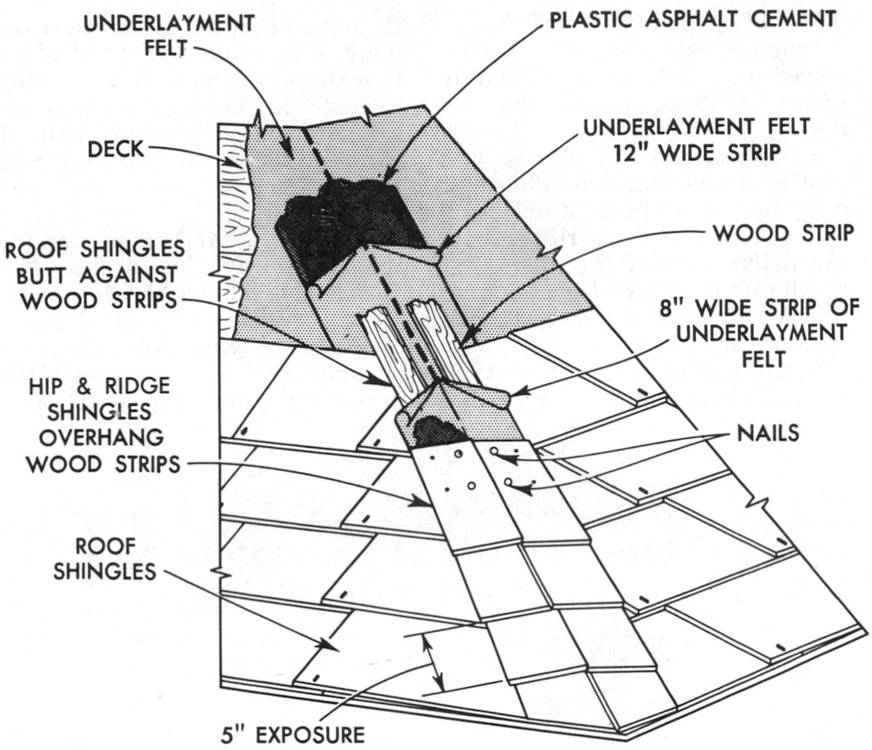

FIG. MF10 Application method for finishing hips and ridges.

as shown in Figure MF10, or a double thickness of 8″ wide felt may be adhered with plastic asphalt cement over the nailing strips. The application of hip and ridge shingles is started with a half-length shingle and covered with a full-length shingle. Thereafter, full-length shingles are applied with a maximum of 8″ exposure (depending on the design of the shingle used) by nailing each shingle through two nailholes. Alternating shingles are lapped on either side of hip or ridge. Finally, the ridge joint should be pointed with plastic asphalt roofing cement.

Ridge Roll

A tapered half-round ridge roll is used in some localities (Fig. MF11). Prior to applying ridge roll, a wood nailing strip of adequate height is installed along the apex of all hips and ridges. The nailing strip is then covered with a double thickness of underlayment. Whenever possible, the application of ridge roll starts at the end of the ridge and is laid in a direction opposite to that of the prevailing winds; at hips, ridge roll is started at eaves. The first section of ridge roll is placed with the large end at the starting point overhanging roof shingles 1/4″. Each end is fastened to the nailing strip by nailing at the head end through a corrosion-resistant clip designed for this purpose. The next section of ridge is placed to overlay the starter at least 2″. The clip is bent back to fasten the exposed end of the next ridge roll section. This method is continued until the entire hip or ridge is covered. All ridge pieces should be embedded in asphalt plastic cement and all exposed nails should be pointed with plastic cement. The opening of end units may be closed with colored cement mortar or with a closure consisting of a half-round piece cut from a shingle that is nailed or screwed into the end of the nailing strip.

Low Slope Method
(3 in 12 to 5 in 12)

Methods for applying shingles on roof slopes less than 5 in 12 but not less than 3 in 12 are the same as described above for Normal Slopes, 5 in 12 and over, with certain exceptions with regard to underlayment.

For slopes 3 in 12 up to 4 in 12, two layers of underlayment should be applied over the entire roof. For slopes 4 in 12 up to 5 in 12, one layer of underlayment is sufficient, except two layers are required in areas where the outside design temperature is 0° F or less.

When eaves flashing is required on Low Slope roofs, the additional layer should be cemented at the eaves as described on page 411-7.

FIG. MF11 Ridge roll may be used on ridge as alternate to ridge shingles.

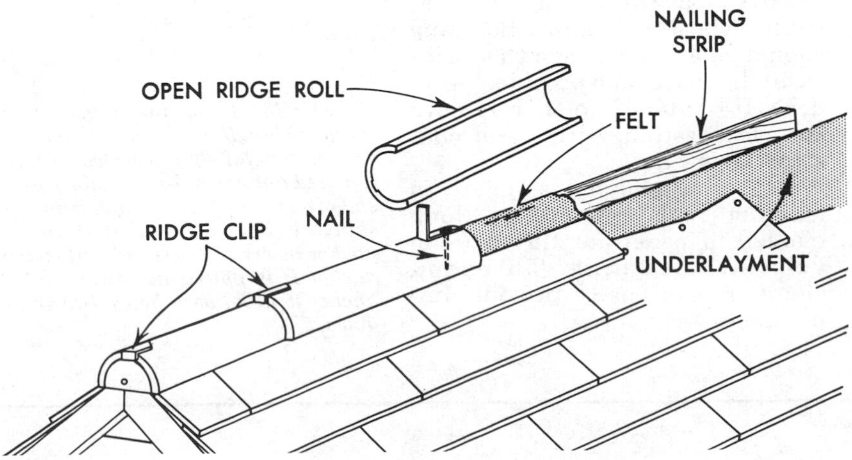

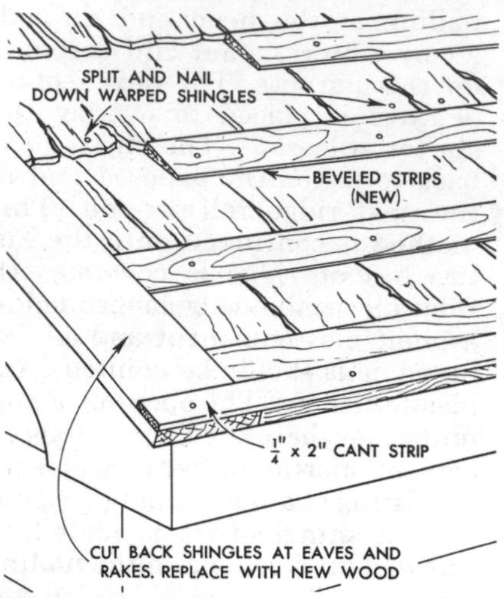

FIG. MF12 Preparation of existing roof prior to application of new shingles.

REROOFING EXISTING CONSTRUCTION

When mineral fiber shingles are applied to an existing roof, the surface must be made as smooth as possible. Shingles may be applied directly over old asphalt shingles provided the sheathing and roof framing members are sound. Loose or curled wood shingles should be split and nailed down, and missing shingles replaced. Beveled wood strips should be installed below the butts of each course of shingles (Fig. MF12).

If the existing wood shingles fail to retain nails with at least the force of 40 lbs per fastener, the shingles should be removed and the existing sheathing prepared as in the case of new construction. If old wood shingles were applied over spaced sheathing, new sheathing should be fitted between the strips to fill the spaces and provide a solid deck. It is often simpler to remove the old sheathing strips and construct a new deck.

Nails for applying shingles to existing roofs should be long enough to penetrate through the existing sheathing. Threaded shank roofing nails provide improved holding power.

Replacing Damaged Shingles

Damaged shingles may be removed and replaced as shown in Figure MF13 and as described below.

The damaged shingle is first shattered and removed. The nails may then be withdrawn or cut off by hooking a slater's ripper over the nails and striking the offset handle of the ripper with a hammer (Fig. MF13a). A new shingle is then positioned.

In the case of American Method of application, a hole is drilled through the joint between the overlying shingles and the head of the new shingle. One nail is installed through this hole and a corrosion-resistant metal flashing strip 3" wide is placed over the nailhead and under the overlying shingle (Fig. MF13b).

For Dutch Method of application, the new shingles should be anchored in place with two spots of roofing cement placed on the head of the underlying shingle and by engaging the storm anchors. Nailing is not required (Fig. MF13c & d).

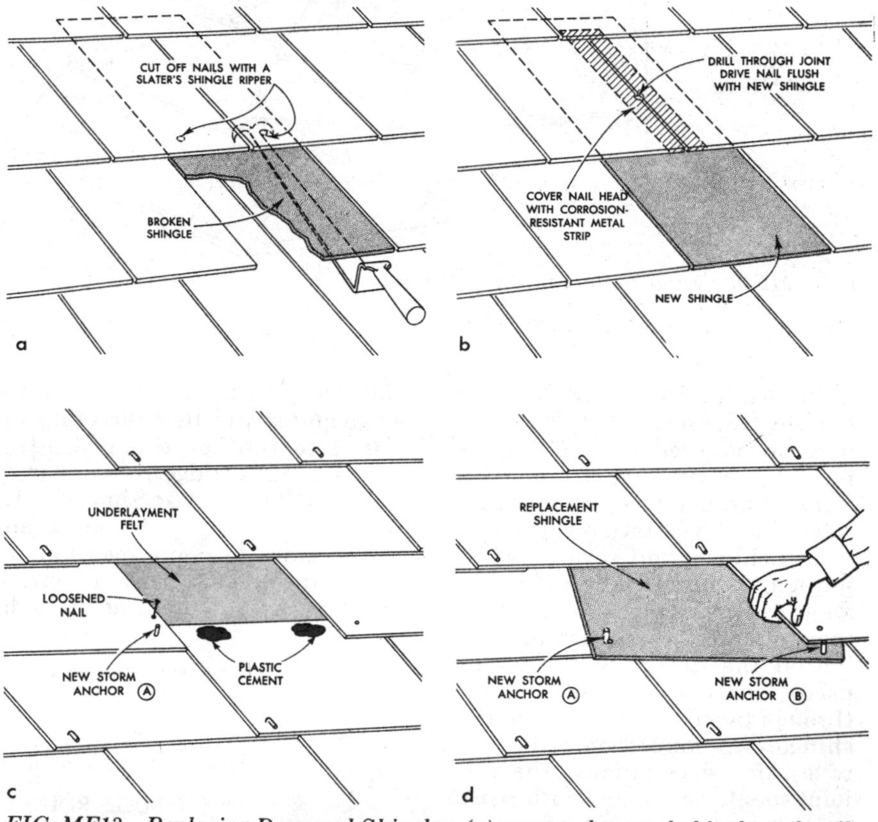

FIG. MF13 Replacing Damaged Shingles: (a) remove damaged shingle and nails; (b) new shingle is nailed through side joint and the nail covered with corrosion resistant metal strip inserted in lap; (c) in Dutch Method, the damaged shingle is removed and the nails and storm anchor sheared off by engaging them with the hook end of a slater's ripper and striking offset on handle with hammer. Nail in exposed shingle is loosened so that sheared storm anchor can be removed and new storm anchor inserted. Loosened nail is driven home and plastic cement applied; (d) storm anchor B is placed in new shingle. Overlapping shingle is raised to allow storm anchor to align with holes. Shingle is embedded in cement and storm anchor bent down.

ASPHALT SHINGLES

The following text describes materials and methods recommended to provide a suitable roof covering of asphalt shingles over both new and existing construction.

Reference should be made to the entire first sub-section, General Recommendations, pages 411-1 through 411-8, for: (1) definition of terminology; (2) limitations imposed by the slope of the roof; (3) roof sheathing thicknesses, spans and nailing schedules; and (4) typical accessory materials—underlayment, flashing, roofing cements, eaves flashings, drip edge, and roofing nails—recommended for preparation of the roof for the application of shingles. Unless specific exceptions are made in the following text, the General Recommendations apply for each kind of shingle and method of application.

MATERIALS

Asphalt roof shingles are manufactured in three basic kinds of units: (1) *Strip shingles* of the Square Butt or Hexagonal type, (2) *Individual* shingles of the interlocking type or staple-down type; and (3) *Giant Individual* shingles for application by either the American or Dutch Lap Methods.

For use in high wind areas, *wind resistant* strip shingles have tabs with factory-applied adhesive or integral locking tabs. Lightweight individual shingles applied by the Dutch Lap Method are intended primarily for reroofing existing roofs.

Each kind of shingle is applied in a distinct way, giving a different appearance to the roof. Starter units, hip and ridge shingles and valley roll are also manufactured to match asphalt shingles.

Shingles should conform to no less than the Underwriters' Laboratories, Inc. (UL) Standard UL55B Class C and each bundle of shingles should be identified with the UL Class C Label. In areas subject to winds in excess of 75 mph*, Wind Resistant shingles are recommended. These shingles are provided with factory-applied adhesive or integral locking tabs, and the bundles carry the UL Class C label and include the words "Wind Resistant."

Data covering types of asphalt shingles is summarized in Figure AR1. See Section 217 Bituminous Products for additional information covering manufacture and physical properties.

*United States Weather Bureau defines hurricane winds as those of 75 mph or greater velocity.

FIG. AR-1 **ASPHALT ROOF SHINGLES**

	SHINGLE TYPE*	SHIPPING WEIGHT PER SQUARE	PACKAGES PER SQUARE	LENGTH	WIDTH	UNITS PER SQUARE	SIDE-LAP	TOP-LAP	HEAD-LAP	EXPOSURE
STRIP SHINGLES	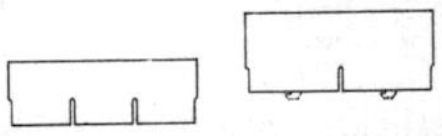 2 & 3 TAB SQUARE BUTT	235 Lb	3	36″	12″	80		7″	2″	5″
		205 Lb	3	36″	12″	80		7″	2″	5″
		220 Lb	3	36″	12″	80		7″	2″	5″
		225 Lb	3	36″	12″	80		7″	2″	5″
	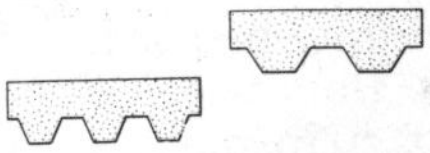 2 & 3 TAB HEXAGONAL	195 Lb	3	36″	11⅓″	86		2″	2″	5″
INDIVIDUAL	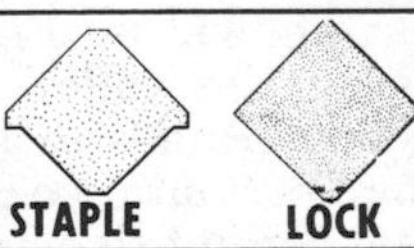 STAPLE LOCK	145 Lb	2	16″	16″	80	2½″			
GIANT INDIVIDUAL	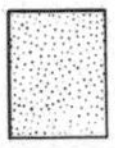 AMERICAN	330 Lb	4	16″	12″	226		11″	6″	5″
	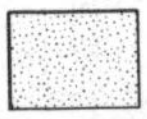 DUTCH LAP	165 Lb	2	16″	12″	113	3″	2″		10″

*see page Bituminous Products 217-8 for standards of quality and UL labels.

Nails and Fasteners

Nails (Fig. AR2) for applying asphalt roofing should be corrosion-resistant such as hot-dipped galvanized steel or aluminum with sharp points and flat heads, 3/8″ to 7/16″ in diameter. Shanks should not exceed 0.135″ or be less than 0.105″ in overall outside diameter (10 to 12 gauge wire). Shanks may be smooth or threaded; however, threaded nails are preferred for increased holding power. Aluminum nails should have screw threads of approximately 12-1/2° thread angle. Galvanized steel nails, if threaded, should have annular threads. Nail lengths typically required are given in Figure AR3.

The number, spacing and pattern of nails are described below under Application of Shingles.

FIG. AR2 Asphalt Shingle Nails: (a) smooth, (b) annular threaded, and (c) screw threaded.

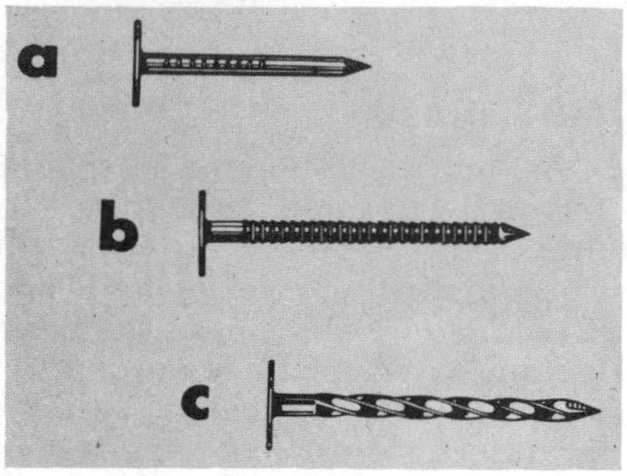

FIG. AR3 **ASPHALT SHINGLE NAIL LENGTHS**

Application	1″ Sheathing	3/8″ Plywood
Strip or Individual Shingle (new construction)	1-1/4″	7/8″
Over Asphalt Roofing (reroofing)	1-1/2″	1″
Over Wood Shingles (reroofing)	1-3/4″	—

PREPARATION FOR SHINGLES

Limitations imposed by the slope of the roof are shown in Figure 3, page 411-4. All asphalt shingles may safely be applied using normal application methods on roofs with slopes of 4 in 12 or greater. Square-butt strip shingles may be used on slopes as low as 2 in 12 when applied using the Low Slope Methods described below under Application of Shingles.

Roof Sheathing

Sheathing types, thicknesses, spans and nailing requirements are discussed on page 411-4. If lumber boards are used, they should not be over 6″ in nominal width. Wider boards may swell or shrink in width enough to buckle or distort the shingles. Badly warped boards and those containing excessively resinous areas or loose knots should be rejected. If any of these defects appear after the sheathing has been applied, they should be covered with 26 gauge galvanized steel patches before the underlayment is placed.

Underlayment

Underlayment should be No. 15 asphalt saturated felt, applied over the roof as described on page 411-5 and as shown below under Application of Shingles. No. 30 asphalt felt is too stiff to be used for underlayment and may cause wrinkling of shingles applied over it.

Eaves Flashing

In areas where the outside design temperature is 0° F or colder (see Fig. 9, page 411-7), or where there is a possibility of ice forming along the eaves and causing a backup of water, eaves flashing is recommended.

Normal Slope (4 in 12 or greater) A course of 90 lb. mineral surfaced roll roofing or not less than 55 lb. smooth roll roofing is installed at the eaves over the underlayment. It should overhang the underlayment and drip edge from 1/4″ to 3/8″, and should extend up the roof to cover a point at least 12″ inside the interior wall line of the building (Fig. 10a, page 411-7). When the eave overhang requires the flashing to be wider than 36″, the necessary horizontal lap joint is cemented and located *outside* the exterior wall line of the building.

Low Slope (as low as 2 in 12) *Square Butt Strip shingles* may be applied on slopes as low as 2 in 12. For slopes less than 4 in 12 down to 2 in 12, or in areas subject to severe icing, a double layer of underlayment should be used over the entire roof. Eaves flashing may be formed by cementing the two layers of underlayment together with asphalt cement up the roof to a point that is at least 24″ inside the interior wall line of the building (Fig. 10b, page 411-7).

The eaves flashing is formed by applying a continuous layer of plastic asphalt cement, at the rate of 2 gals. per 100 sq. ft., to the surface of the underlayment starter course before the first full course of underlayment is applied. Cement is also applied to the 19″ underlying portion of each succeeding course which lies within the eaves flashing area, before placing the next course. It is important to apply the cement uniformly with a comb trowel, so that at no point does underlayment touch underlayment when the application is completed. The overlying sheet is pressed firmly into the entire cemented area (Fig. 10c, page 411-7).

Drip Edges

A corrosion resistant, non-staining drip edge should be installed along the eaves and rakes (see Fig. 12, page 411-8).

Flashing

See General Recommendations, page 411-6.

Valley Flashing The *open* or *closed* method may be used to construct valley flashing. A valley underlayment strip of No. 15 asphalt saturated felt, 36″ wide, is applied first as shown in Figure AR4, centered in the valley and secured with enough nails to hold it in place. The horizontal courses of underlayment are cut to overlap this valley strip a minimum of 6″. Where eaves flashing is required, it is applied over the valley underlayment strip.

Open Valleys may be flashed with the metals described under Flashing Materials, page 411-6, or with 90 lb. mineral surfaced asphalt roll roofing in a color to match or to contrast with the roof shingles. The

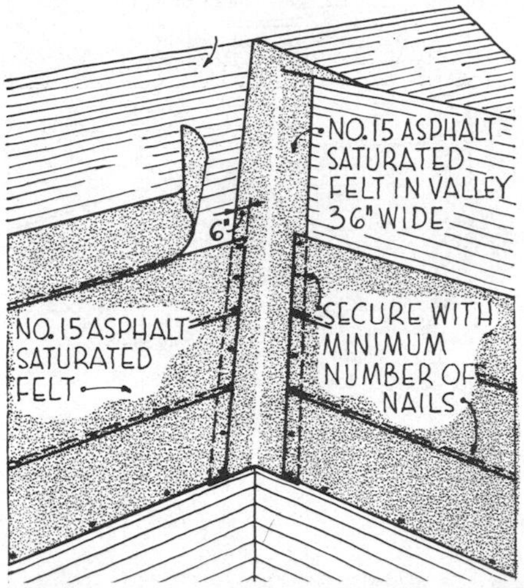

FIG. AR4 Underlayment is centered in the valley prior to applying flashing.

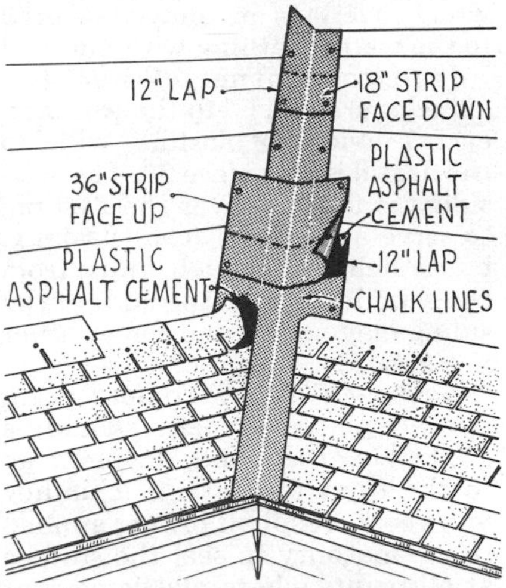

FIG. AR5 Open valley flashing using mineral surfaced roll roofing.

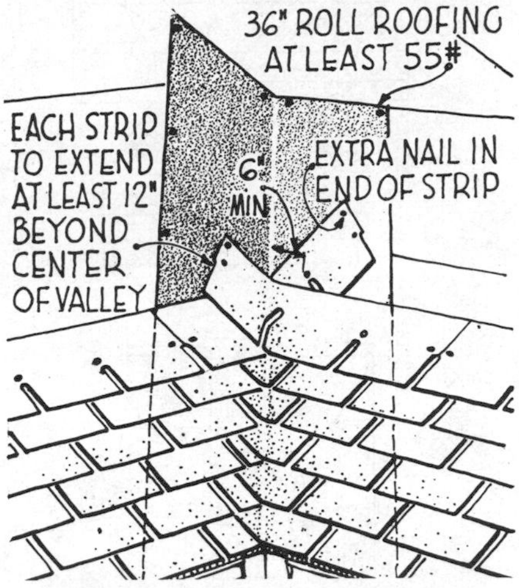

FIG. AR6 Closed valley flashing using woven strip shingles.

method is illustrated in Figure AR5. An 18″ wide strip of mineral surfaced roll roofing is placed over the valley underlayment, centered in the valley with the surfaced side down, having the lower edge cut to conform to and be flush with the eaves flashing. When necessary to splice the material, the ends of the upper segments overlap the lower segments 12″ and are secured with asphalt plastic cement. Only enough nails are used in rows 1″ in from each edge to hold the strip smoothly in place.

Another strip, 36″ wide, is placed over the first strip, centered in the valley with the surfaced side up, being secured with nails and lapped if necessary just as the underlying 18″ strip.

Before shingles are applied, a chalkline is snapped on each side of the valley for its full length. The lines should start 6″ apart at the ridge and spread wider apart (at the rate of 1/8″ per foot) at the eave. The chalklines serve as a guide in trimming the last shingle units to fit the valley, insuring a clean, sharp edge. The upper corner of each end shingle is clipped as shown in Figure AR5, to direct water into the valley and prevent water penetration between courses. Each shingle is cemented to the valley lining with asphalt cement to insure a tight seal. No exposed nails should appear along the valley flashing.

Closed (Woven) Valleys can be used only with strip shingles. Individual shingles cannot be used because nails might be required at or near the center of the valley lining. This method has the advantage of doubling the coverage of the shingles throughout the length of the valley, increasing weather resistance at this vulnerable point.

A valley lining made from a 36″ wide strip of 55 lb. (or heavier) roll roofing is placed over the valley underlayment and centered in the valley (Fig. AR6).

Valley shingles are laid over the lining either by (1) applying them on both roof surfaces at the same time, weaving each course in turn over the valley, or (2) covering each surface first to the point approximately 36″ from the center of the valley and weaving the valley shingles in place later.

In either case, the first course at the valley is laid along the eaves of one surface over the valley lining, extending it along the adjoining roof surface for a distance of at least 12″. The first course of the adjoining roof surface is then carried over the valley on top of the previously applied shingle. Succeeding courses are then laid alternately, weaving the valley shingles over each other.

The shingles are pressed tightly into the valley and nailed normally (see Nailing . . ., page 411-21), except no nail should be located closer than 6″ to the valley center line and two nails are used at the end of each terminal strip.

Vertical Wall Flashing When the rake of a roof abutts a vertical wall, the most satisfactory method of protecting the joint is to use stepped metal base flashing applied over the end of each course of shingles (Fig. AR7).

The metal flashing is rectangular in shape, approximately 6″ long and 2″ wider than the exposed face of the shingles. With strip shingles laid 5″ to the weather, it would be 6″ x 7″. The flashing is bent with one end extending 2″ onto the roof underlayment, the remainder extending up the wall surface. Each flashing piece is placed just up-roof from the exposed edge of the shingle

FIG. AR7 Vertical wall flashing using individual metal flashing units.

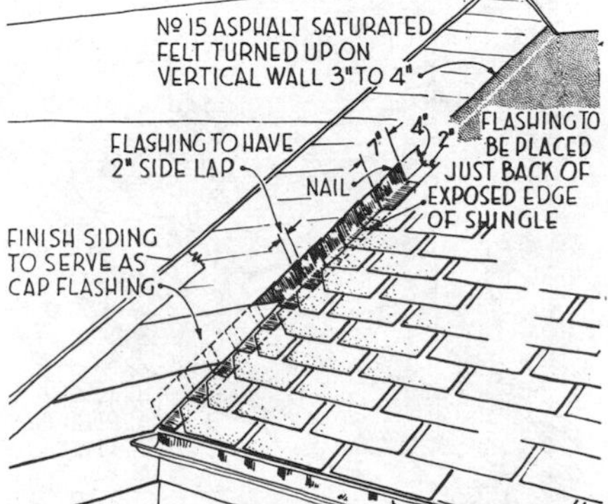

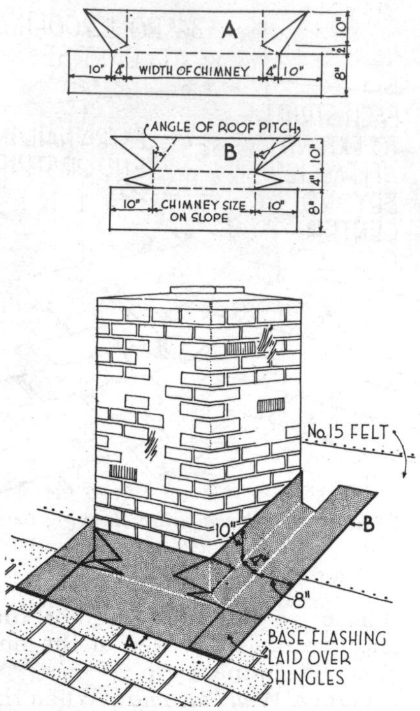

IG. AR8 Patterns and application chimney base flashing.

G. AR9 Patterns and application of idle (cricket) flashing.

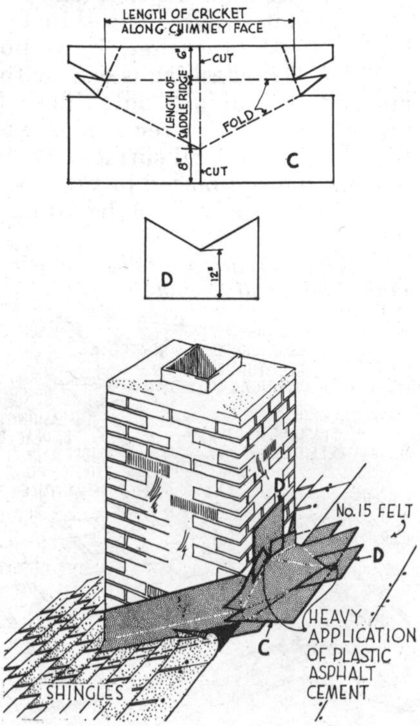

which overlaps it, and is secured to the wall sheathing with one nail in the top corner. When the shingles are laid 5″ to the weather, each 7″ width of flashing will lap the next flashing piece 2″. The wall siding is finished over the flashing to serve as counter (cap) flashing, but is held far enough away from the shingles so that the ends of the siding may be painted for weather protection.

Chimney Flashing Before flashings are placed, shingles should be applied over the underlayment up to the front face of the chimney and a coat of asphalt primer applied to the masonry to seal the surface at all points where plastic cement will be applied.

Base Flashing may be 90 lb. mineral surfaced roll roofing. The chimney front, cut to pattern "A" in Figure AR8, is applied first by securing the lower section over the shingles in a bed of asphalt plastic cement. The vertical upper section is secured against the masonry with plastic cement and nails driven into the mortar joints; the triangular ends are bent around the corners of the chimney and cemented into place.

The base flashing for chimney sides, cut to pattern "B", are applied as shown. Each is secured to the deck with plastic cement and to the brickwork with plastic cement and nails; the triangular ends of the upper section being turned around the chimney corners and cemented in place over the front base flashing. The base flashing covering the saddle is cut to pattern "C" (Fig. AR9) and secured in plastic cement over the underlayment. The standing portion extends up the masonry 6″ to 12″ and is capped with a piece cut to pattern "D", bedded in plastic cement. A second piece cut to pattern "D" is set tightly in plastic cement and centered over that portion of the saddle flashing, extending up the roof. This provides added protection at the point where the ridge of the saddle joins the roof.

Stepped flashing using sheet-metal (as described under Vertical Wall Flashing) or roll roofing can be used as alternate methods of base flashing. Base flashing made

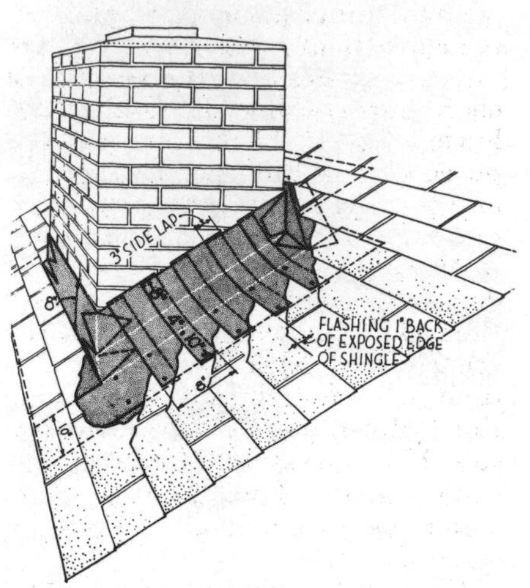

FIG. AR10 Alternate method of base flashing using roll roofing.

FIG. AR11 Metal counter (cap) flashing must turn down over base flashing to complete the flashing assembly.

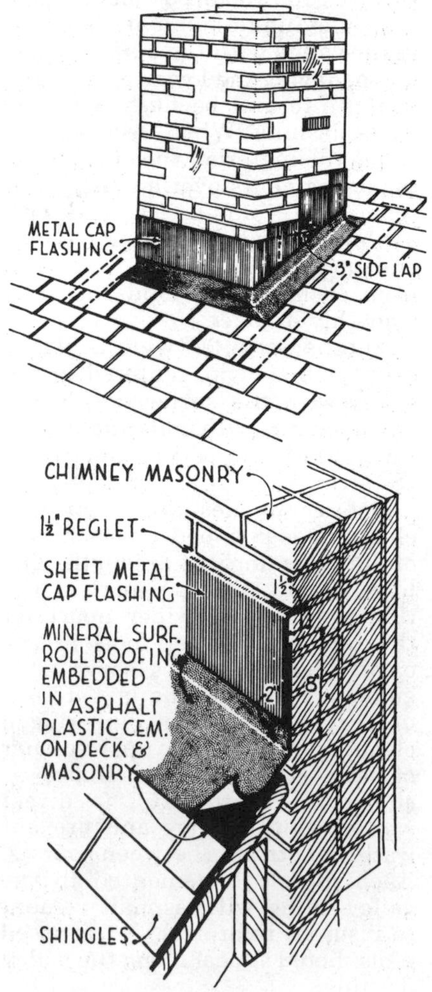

of 90 lb. mineral surfaced roll roofing consists of nailing 8″ x 22″ rectangular pieces over the end tab of each course of shingles, holding the lower edge 1″ back from the exposed edge of the covering shingle, and then bending it up against the masonry and securing it with plastic cement (Fig. AR10). The nails driven through the lower edge of the flashing into the roof sheathing are covered with plastic cement, used to secure the end shingle to the flashing, and by the shingle itself. This operation is repeated for each course. The flashing units should be wide enough to lap each other at least 3″.

Counter Flashing of metal must be applied over base flashing to complete the flashing installation (Fig. AR11). One method of securing counter flashing to masonry is to rake out the mortar joint to a depth of 1-1/2″ and the bent back edge of the flashing is inserted in the raked joint between bricks. The joint is then refilled with portland cement mortar or plastic cement, and the flashing bent down over the base flashing.

The counter flashing on the chimney front is one continuous piece. On the sides and rear, the sections are cut to conform to the locations of brick joints and pitch of roof, lapping each other at least 3″.

Stack Flashing Pipes projecting through the roof require special flashing methods. Corrosion-resistant metal sleeves which slip over stacks have adjustable flanges to fit any roof slope (Fig. AR12), or asphalt products may be used successfully.

The roofing is applied up to the point where the stack projects and is fitted around it (Fig. AR14). When metal flashing is used, the sleeve is fitted down over the stack, adjusted to the slope of the roof, and the top of the sleeve turned into the stack opening (Fig. AR13). When asphalt products are used, an oval opening is cut in a rectangular flange of 55 lb. or heavier roll roofing. The flange should be large enough to extend 4″ below, 8″ above and 6″ on each side of the stack when installed. The flange is slipped over the pipe and

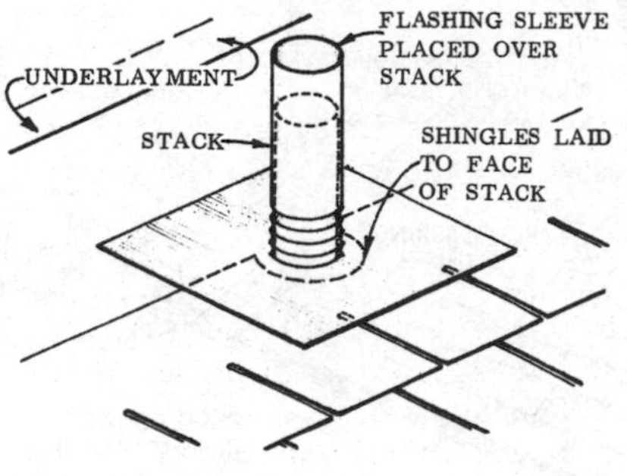

FIG. AR12 Adjustable metal flashing sleeves fit over stacks.

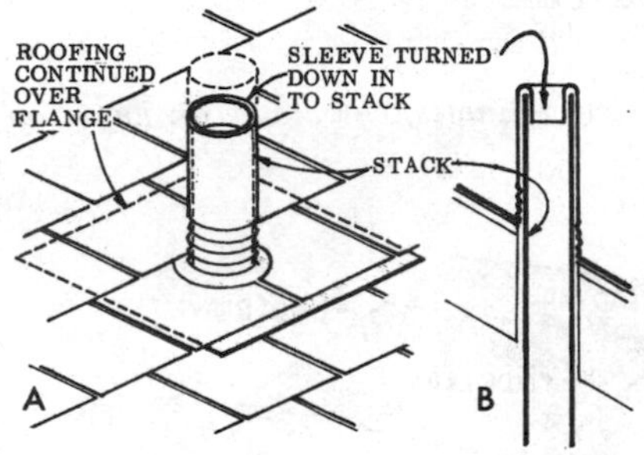

FIG. AR13 Shingles are laid and metal sleeve is turned into top of stack.

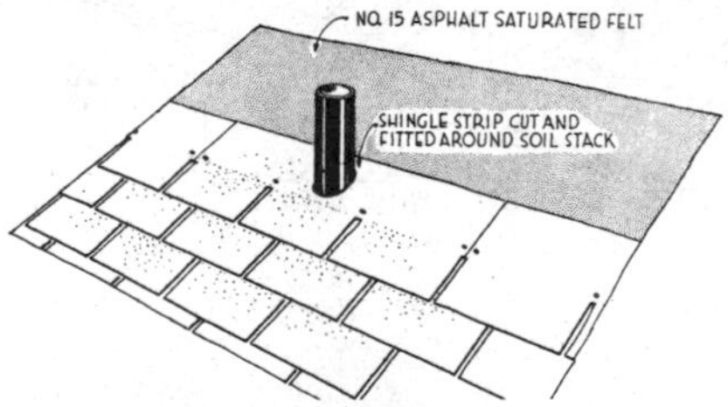

FIG. AR14 Shingles are laid to fit around stack.

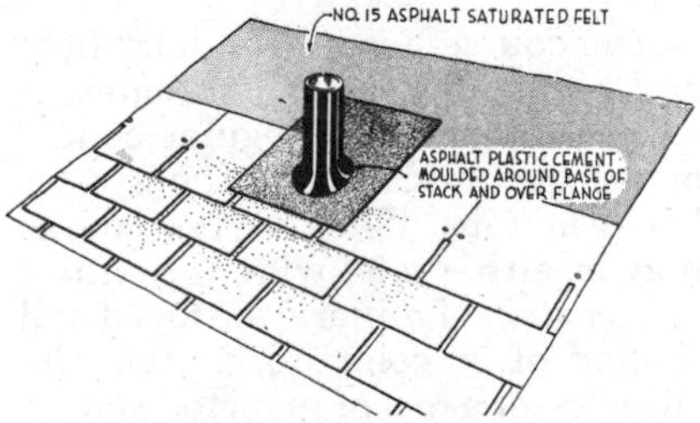

FIG. AR15 Roll roofing flange and plastic asphalt cement collar form flashing.

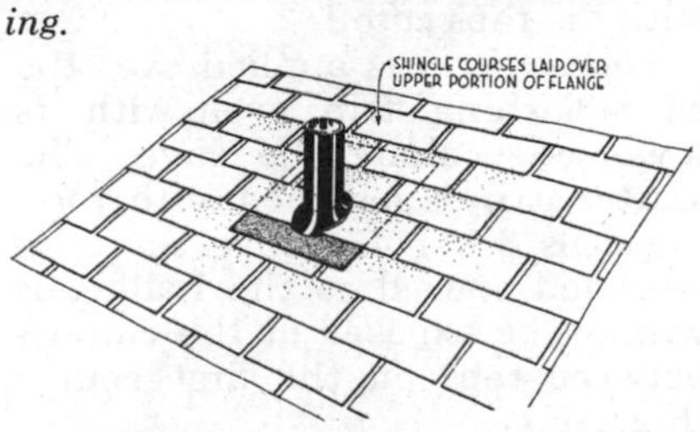

FIG. AR16 Shingles applied over flange complete flashing assembly.

applied flat on the roof. A collar of plastic cement is formed around the pipe, extending up the pipe and out over the flange about 2″ (Fig. AR15). The cement should be vigorously and thoroughly troweled in place to give proper adherence and eliminate all air pockets. The shingles are continued up the roof, fitted around the stack, and bedded in plastic cement where they overlay the flange at the side and back of the stack (Fig. AR16).

APPLICATION OF SHINGLES

Shingles are laid so they overlap and cover each other to shed water. The extent of overlapping depends on their shape and method of application and furnishes single, double and sometimes triple coverage over different areas of the roof. *Square Butt Strip* shingles, some *interlocking* shingles and *giant individual* shingles applied by the American Method, which provide either double or triple coverage, are suitable for new construction.

Strip and Individual hexagonal shingles and *giant individual* shingles, applied by the Dutch Lap Method, provide only single coverage and are used to reroof existing construction.

Regardless of the method of application, on new construction shingles should only be applied to a tight roof deck providing a suitable nailing base, with underlayment, drip edge and flashings in place. Chimneys should be completed and counter flashing installed. Stacks and other equipment requiring openings through the roof should be in place with provision for counter flashing where necessary; preferably, gutters should be hung.

Strip Shingles (Normal Slope)

On small roofs, strip shingles may be laid from either rake. On roofs 30′ and longer they should be started at the center and worked both ways from a vertical line to assure more accurate vertical alignment and to provide for meeting and matching above a projection such as a dormer or chimney.

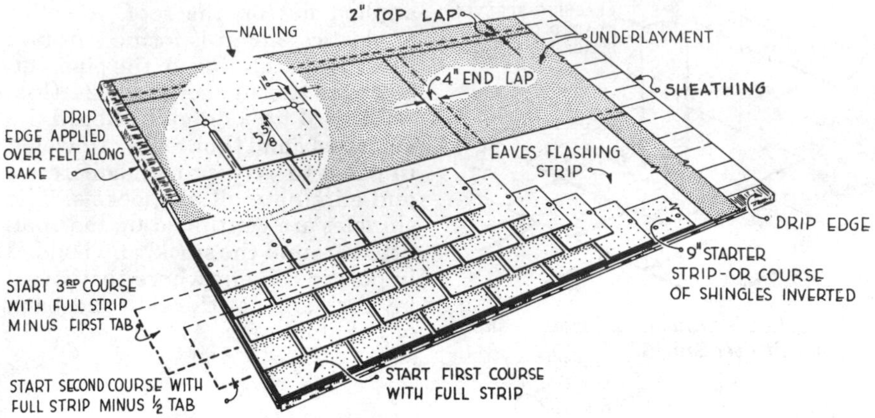

FIG. AR17 Application of Square Butt Strip Shingles, cutouts break joints on halves.

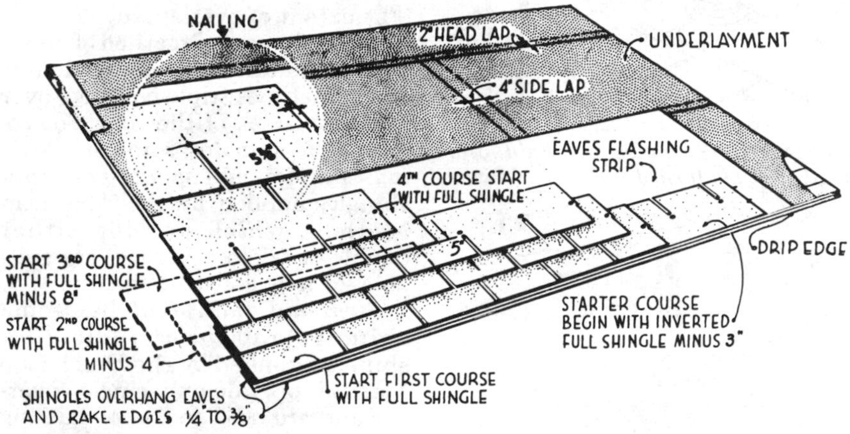

FIG. AR18 Application of Square Butt Strip Shingles, cutouts break joints on thirds.

Where a roof surface is broken, shingles should be started from a rake and laid toward a dormer or into a valley. Where unbroken, they should be started at the rake that is most visible. If both rakes are equally visible, shingles are started in the center and worked both ways.

Minor variations in the dimensions of asphalt shingles are unavoidable. Variations seldom exceed ± 1/4″ in either length or width in a 12″ x 36″ strip shingle, and usually will be no more than ± 1/8″. However, to control the placement of shingles so that cutouts will be accurately aligned horizontally, vertically and diagonally, it is recommended that horizontal and vertical chalklines be used to control alignment.

Starter Course The purpose of a starter course, for both square-butt and hexagonal pattern shingles, is to back up the first regular course of shingles and fill in the spaces between tabs. The starter course may be either a 9″ wide (or wider) *starter strip* of mineral surfaced roll roofing of a color to match the shingles, or a row of inverted *starter course* shingles or strips made from the top lap portion of the shingles with the tabs cutoff.

Starter Strip is applied over the eaves flashing strip, even with its lower edge along the eave. The starter strip is fastened with roofing nails 3″ or 4″ above the eave edge and spaced so the nailheads will not be exposed at the cutouts between tabs on the first course (Fig. AR17).

Starter Course Shingles are applied with tabs facing up the roof over the eaves flashing strip. Nails are spaced so they will not be exposed at the cutouts of the first course (Fig. AR18). This method is preferred for hexagonal type strip shingles to insure a satisfactory color match at the spaces between tabs along the eaves (Figs. AR20 & AR21).

First and Succeeding Courses (Square-Butt Shingles) The first course is started with a full shingle, while succeeding courses are started with full or cut strips depending upon the pattern desired. The major variations for *square-butt strip* shingles are as follows:

Cutouts Break Joints on Halves when the second course is started with a full shingle cut 6″ short (1/2 of a tab), the third with a full shingle with the entire first tab removed, the fourth with 1-1/2 tabs cut from the shingle and so on, causing the cutouts to be centered on the tabs of the course below (Fig. AR17). This pattern can also be made with less cutting by starting the third course with a full shingle.

Cutouts Break Joints on Thirds when the second course is started with a full shingle cut 4″ short (1/3 of a tab), the third course with a full shingle cut 8″ short (2/3 of a tab), and the fourth with a full shingle (Fig. AR18). The shingles are placed so that the lower edges of the butts are aligned with the top of the cutouts of the underlying course.

Random Spacing is achieved by removing different amounts from the starting tab of succeeding courses in accordance with the following general principles: (1) the width of any starting tab should be at least 3″; (2) cutout centerlines of any course should be located at least 3″ laterally from the cutout centerlines in both the course above and the course below; (3) starting tab widths should be varied sufficiently so the eye will not follow a cutout alignment.

Starting the first course with a full length strip, Figure AR19, illustrates the length of the starting tab at the rake of each succeeding course in order to produce suitable random spacing.

First and Succeeding Courses (Hex Shingles) Hexagonal strip shingles (Figs. AR20 & AR21) permit no spacing variations as do square-butt strips. Application begins with a roll roofing starter course or inverted shingles. The first course starts with a full strip, and second and succeeding courses with a full strip less 1/2 tab and full strips, alternating as shown. Each course is applied so that the lower edge of the tabs are aligned with the top of the cutouts of the preceding course.

Vertical Application Research findings* show that strip shingles may be applied faster by laying one 36″ strip of shingles in vertical courses from the eave to ridge, rather than laying courses horizontally from rake to rake. This method results in an equally weathertight roof at labor cost reductions as high as 25%. This method is recommended for solid colored shingles, since color blends laid in this manner may develop a "striped" effect on the roof. Most manufacturers, however, recommend against application of any shingles in a vertical manner. Such instructions are actually printed on some manufacturers' shingle packages.

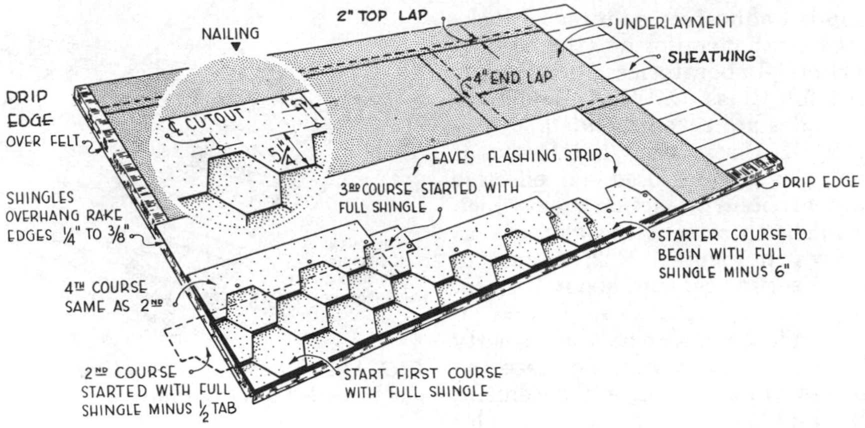

FIG. AR20 Application of Two-Tab Hexagonal Strip Shingles.

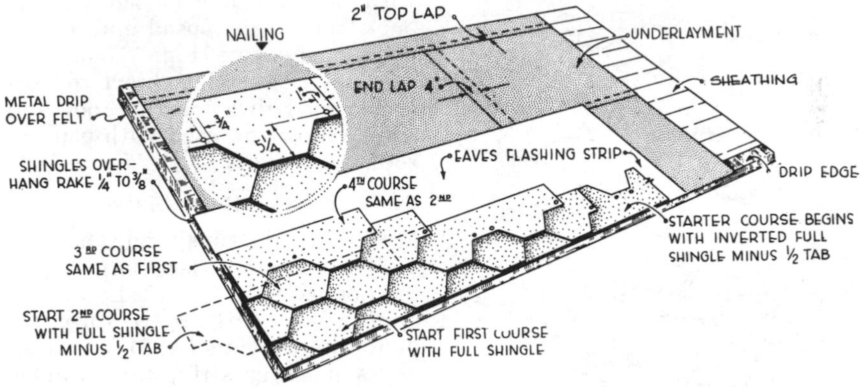

FIG. AR21 Application of Three-Tab Hexagonal Strip Shingles.

Nailing and Fastening Strip Shingles The number of nails per strip and the placing of the nails are vital for suitable roof application. To prevent buckling, each shingle should be in perfect alignment before driving any nail. Nailing should be started at the end nearest the shingle last applied, proceeding to the opposite end, with nails driven straight to avoid cutting the fabric of the shingle with the edge of the nailhead. The nailhead should be driven flush, not sunk below the surface of the shingle. If a nail fails to penetrate the sheathing, an additional nail should be used in a new location.

*NAHB Research Foundation.

FIG. AR19 Random spacing of Square Butt Strip Shingles.

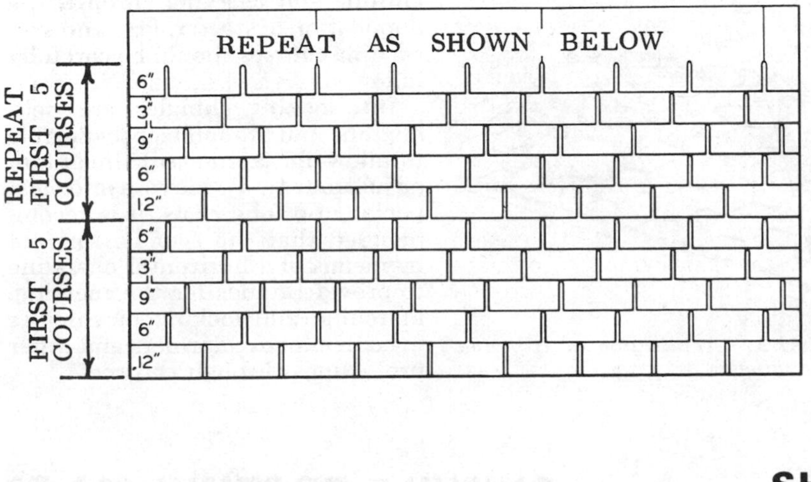

Three-Tab Square-Butt Shingles require 4 nails for each three-tab square butt strip. When applied with a 5″ exposure, the nails are placed 5/8″ above the top of the cutouts and located horizontally with one nail 1″ back from each end and one nail on the centerline of each cutout (Figs. AR17 & AR18).

Two and Three Tab Hexagonal Shingles require 4 nails per strip, located in a line 5-1/4″ above the exposed butt edge and horizontally as follows: (1) for two-tab shingles, one nail 1″ back from each end of the strip and one nail 3/4″ back from each angle of the cutouts (Fig. AR19); (2) for three-tab shingles, one nail 1″ back from each end and one nail centered above each cutout (Fig. AR20).

Windy Locations require additional protection. In areas where wind velocities are 75 mph or greater*, shingles manufactured with factory

*United States Weather Bureau defines hurricane winds as those of 75 mph or greater velocity.

applied adhesives or integral locking tabs that conform to the Underwriters' Laboratories, Inc. Standard for Class C Wind Resistant Shingles are recommended.

If UL Class "C" Wind Resistant Shingles are not used, an effective way to obtain protection against high winds is to cement the *free tabs* as shown in Figure AR22. A spot of quick setting cement, about 1 sq. in. in area, for each tab is pressed on the underlying shingle with a putty knife or caulking gun. The free tab is then pressed against the cement. In applying the cement, shingles should not be bent back further than necessary.

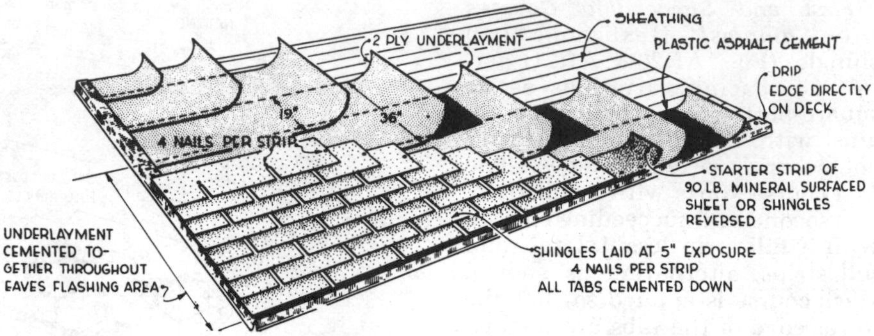

FIG. AR24 Low Slope Application of Strip Shingles over double underlayment and cemented eaves flashing.

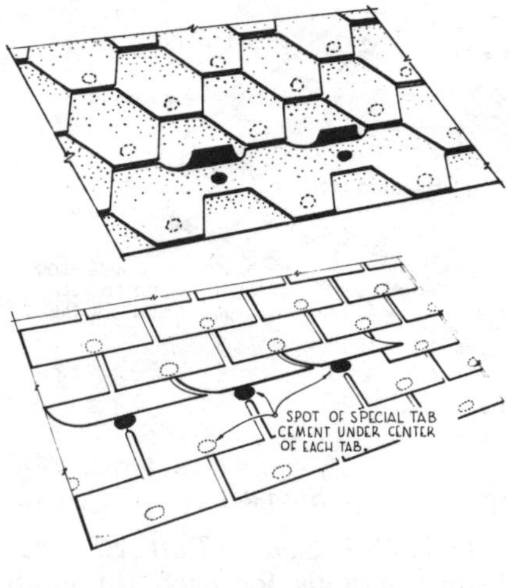

FIG. AR22 Free tabs may be cemented down for wind protection.

Three tab square butt strip shingles should receive six nails instead of four in high wind areas. One nail should be placed at each end, as described previously, with two nails 5/8″ above and positioned just to either side of the centerline of each cutout.

Hips and Ridges Hips and ridges may be finished by using hip and ridge shingles furnished by the manufacturer or by cutting pieces at least 9″x 12″ either from 12″x 36″ square butt shingle strips, or from mineral surfaced roll roofing of a color to match the shingles. They are applied by bending each shingle lengthwise down the center with equal exposure on each side of the hip or ridge. In cold weather the shingles should be warmed before bending. They are applied beginning at the bottom of a hip or at one end of a ridge with a 5″ exposure. Each shingle is secured with one nail at each side, 5-1/2″ back from the exposed end and 1″ up from the edge (Fig. AR23).

Metal ridge roll subject to corrosion may discolor the roof and should never be used with asphalt roofing products.

Strip Shingles (Low Slope)

Square tab strip shingles may be used on roof slopes less than 4 in 12 but not less than 2 in 12. Low slope application methods require: (1) double underlayment, (2) cemented eaves flashing strip, (3) shingles provided with factory applied adhesives conforming to the Underwriters' Laboratories, Inc. Standard for Class C Wind Resistant Shingles, or each free tab of square butt strip shingles should be cemented as shown in Figure AR22.

Application of strip shingles over double underlayment and cemented eaves flashing is shown in Figure AR24. Any shingle arrangement described under normal slope application methods may be used.

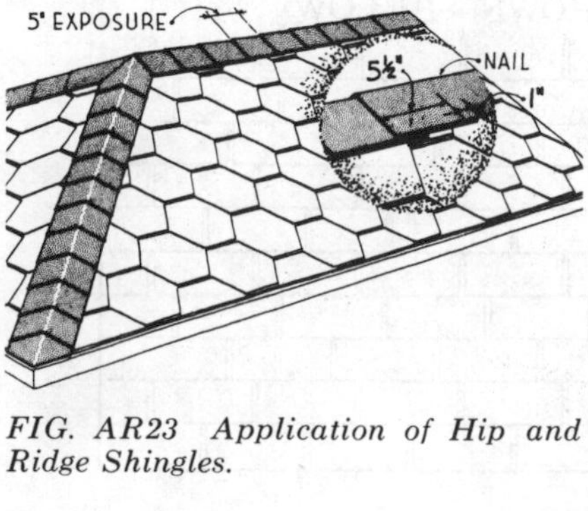

FIG. AR23 Application of Hip and Ridge Shingles.

Interlocking Shingles

Interlocking (lock-down) shingles are designed to provide resistance to strong winds. They have integral locking devices that vary in detail but which can be classified into five general groups as shown in Figure AR25. Types 1, 2, 3 and 4 are individual shingles, while type 5 is a strip shingle generally having two tabs per strip. They do not require use of either field or factory-applied adhesives, although use of cement may occasionally be necessary along rakes and eaves where the locking devices may have to be removed.

Interlocking shingles are used for both new construction and reroofing. The roof slope should not be less than the minimum specified by the shingle manufacturer. The individual shingles (Types 1 through 4) are intended for roof slopes of 4 in 12 and greater.

Starter, First and Succeeding Courses Due to the diversity of designs available, the manufacturers' instructions should be studied and the directions for beginning and carrying through the application of starter, first and succeeding courses should be carefully followed.

Interlocking shingles are self-aligning, but are sufficiently flexible to allow for a limited amount of adjustment. To save time, especially on long roofs, it is recommended that the roof be squared by means of a horizontal chalkline to provide guides for the meeting, matching and locking of shingles in courses above dormers and other projections through the roof.

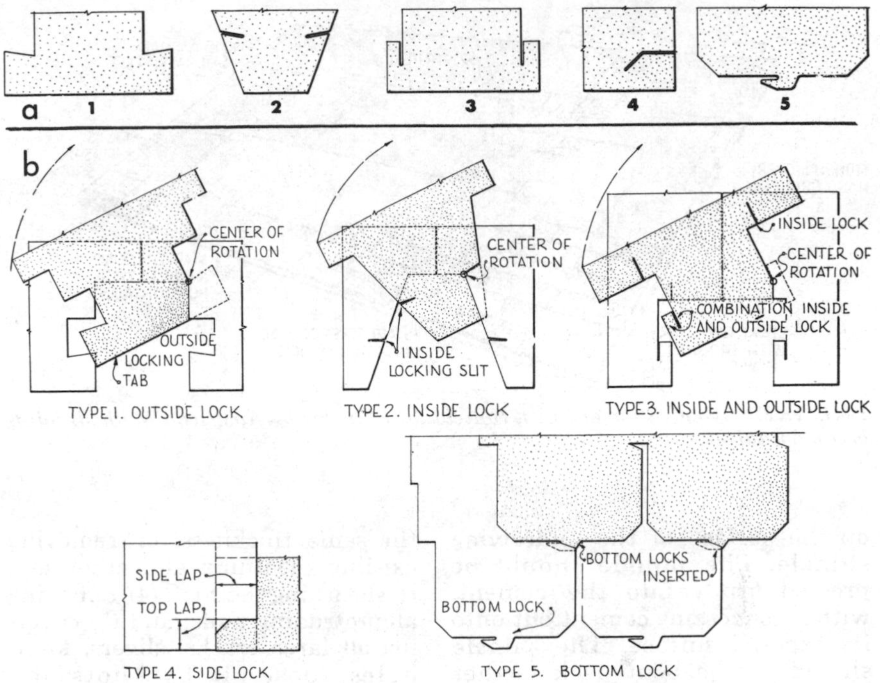

FIG. AR25 (a) Interlocking Shingle tabs, (b) methods of locking shingles.

Interlocking The interlocking devices are accurately manufactured to insure a definite space relationship between each shingle. The locking devices should be engaged correctly. Methods of locking the different shingle types are shown in Figure AR25.

Nailing Interlocking Shingles The location of the nails is essential to the performance of the locking device. The shingle manufacturer's instructions, supplied with each square of roofing, specify precisely where the nails should be located to ensure the most suitable results.

Cementing Along eaves and rakes, some styles of interlocking shingles require removal of part or all of the locking device. When shingles are interlocked, wind pressures are transmitted from each shingle to the adjacent shingle. To prevent excessive wind damage, any shingle at the rake or eaves having part or all the locking device removed should be cemented down or face-nailed.

Hips and Ridges The procedure for finishing hips and ridges is the same as for strip shingles (Fig. AR23).

Giant Individual Shingles (American Method)

The American Method of application may be used for new construction or for reroofing on roof slopes 4 in 12 or greater. Horizontal and vertical chalklines should always be used to insure good alignment. This is especially important when a dormer or other obstruction intervenes between two ends of a roof, as it insures that courses above the obstruction will meet and match.

Starter Course The starter course is laid even with the lower edge of the eaves flashing strip with the starter shingles placed horizontally in contact with each other. Each shingle is secured with 3 nails (Figs. AR26 & AR27).

First Course All shingles, except those in the starter course, are laid with the long dimension parallel to the rake and exposed 5″ to the weather. The first course consists of full shingles, with the first shingle nailed flush with the eaves and rake edges of the starter course. Succeeding shingles are nailed flush with the eave edge of the starter course and are placed not more than 3/4″ apart.

Second and Succeeding Courses *Joints Break On Thirds* when the second course begins with an 8″ wide shingle and continues with full shingles spaced not more than 3/4″ apart. The third course begins with a 4″ wide shingle and the fourth with a full shingle, this sequence being repeated to the ridge (Fig. AR26).

Breaking Joints on Halves is an alternate arrangement which starts the second course at the rake with a 6″ wide shingle, and the third course again with a full shingle, repeating the sequence to the ridge (Fig. AR27).

Nailing and Fastening Giant Individual Shingles Each shingle requires 2 nails, located 5-1/2″ to 6″ up from the lower exposed edge of the shingle and 1-1/2″ in from the outer edge.

Windy Locations Where wind velocities are 75 mph or greater, shingles require additional fastening. A spot of quick setting cement, not less than 1 sq. in. in area, is placed under the center of the exposed portion of each shingle and

FIG. AR26 Application of Giant Individual Shingles by American Method, joints break on thirds.

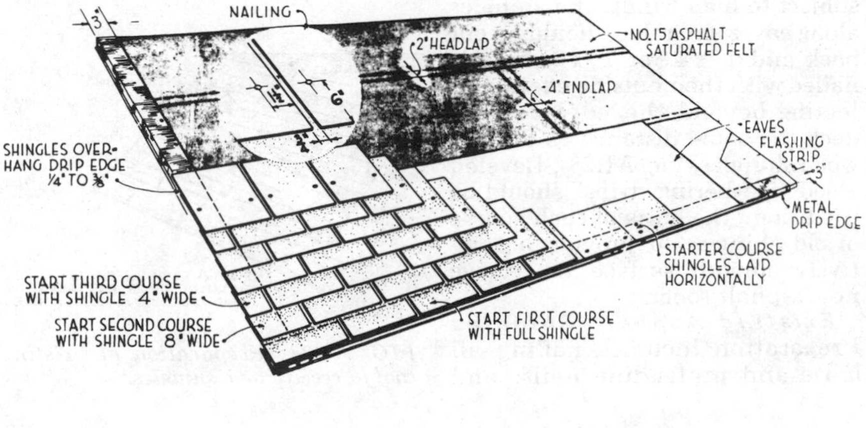

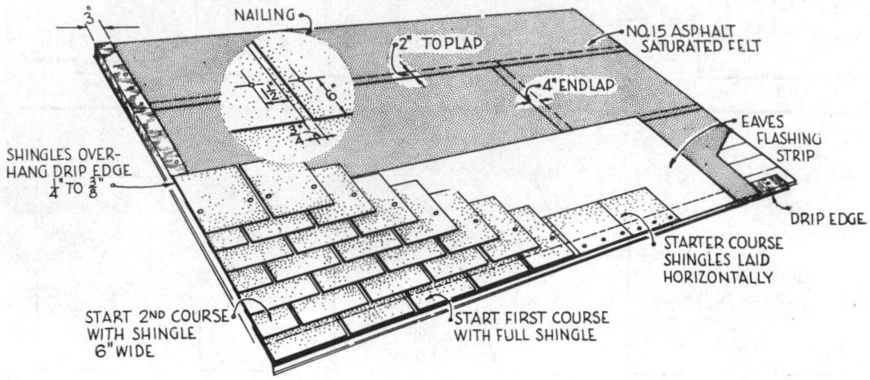

FIG. AR27 Application of Giant Individual Shingles by American Method, joints break on halves.

on the surface of the underlying shingle. The shingle should be pressed firmly into the cement, without squeezing cement out onto its exposed surface. The shingle should not be bent back farther than necessary. Along the rakes, shingles are cemented to the drip edge as well as to each other.

Hips and Ridges are finished in the same manner as are strip shingles (Fig. AR23).

REROOFING EXISTING CONSTRUCTION

When a roof is to be reroofed, existing wood shingles, asphalt shingles or roll roofing need not be removed before applying a new asphalt roof, provided: (1) the strength of the existing roof framing is adequate to support the new dead and live loads; (2) the existing roof sheathing is sound and will provide adequate anchorage for nails used to apply new roofing.

If either the framing or sheathing is inadequate, the old roofing, regardless of type, must be removed to correct the defects.

Existing Roofing to be Removed

When the old roofing must be removed, the roof should be prepared for new roofing by: (1) repairing and/or reinforcing existing roof framing where required to provide adequate strength; (2) removing existing sheathing that is damaged, and replacing with new sheathing; (3) filling all spaces between boards with securely nailed wood strips of the same thickness, or removing existing sheathing altogether and re-sheathing the roof; (4) removing all protruding nails; and (5) covering all large cracks, slivers, knotholes, loose pitchy knots and excessively resinous areas with sheetmetal securely nailed to the sheathing.

Existing Roofing to Remain

If inspection indicates that the existing roofing need not be removed, the surface of the existing roofing should be prepared to receive new roofing, as described below for various types of roofs.

Existing Wood Shingles Preparation includes: removing all loose or protruding nails, and renailing shingles in a new location; nailing down all loose shingles; splitting all badly curled or warped shingles and nailing down the segments; replacing missing shingles with new ones. When shingles and trim at eaves and rakes are badly weathered and if the location is subject to high winds, the shingles along eaves and rakes should be cut back and 1″x4″ or 1″x6″ boards nailed with their outside edges projecting beyond the edges of the deck the same distance as did the wood shingles (Fig. AR28). Beveled wood "feathering strips" should be used along the butts of each course of old shingles to provide a relatively smooth surface to receive new asphalt roofing.

Existing Asphalt Shingles Preparation includes nailing all loose and protruding nails, and removing and replacing all badly weathered edging strips.

Existing Roll Roofing Preparation includes splitting all buckles and nailing segments down smoothly, and removing all loose and protruding nails.

Existing Built-Up Roofs If the slope of the roof is not less than 2-in-12 and if the roof sheathing is sound and can be expected to provide good nail holding power, new shingles can be applied. Old slag, gravel or other coarse surfacing materials should first be removed, leaving the surface of the underlying felts smooth and clean. New asphalt shingles are applied directly over the felts as described under Strip Shingles (Low Slope), page 411-22.

Underlayment (Reroofing)

When existing roofing must be removed before applying new material, underlayment is required over the entire roof, as prescribed for new construction.

When new shingles can be applied directly over old roofing, no underlayment is required as the old roofing adequately serves the same purpose.

Drip Edge and Eaves Flashing Strip (Reroofing)

When renewing an existing wood shingled roof, drip edges and eaves flashing strip should be installed as for new construction (Fig. AR29 and AR30).

Flashing (Reroofing)

Existing metal flashing, unless perforated or badly deteriorated, may be left in place and re-used.

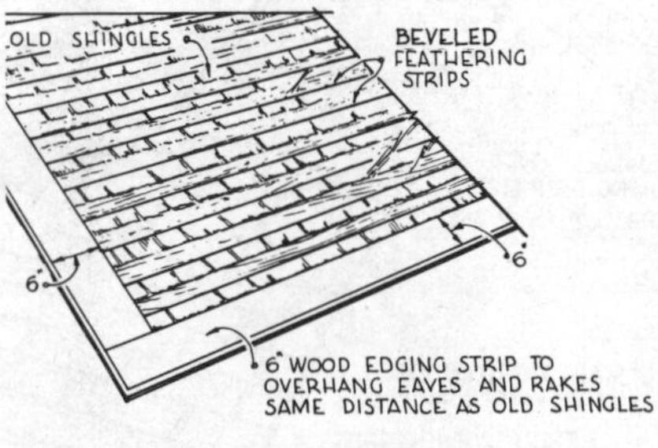

FIG. AR28 Preparation of existing roof to receive new shingles.

Flashing Against a Vertical Wall When new asphalt shingles are being applied over an old roof, the joint between a vertical wall and the rake of an abutting roof may be treated as shown in Figure AR30.

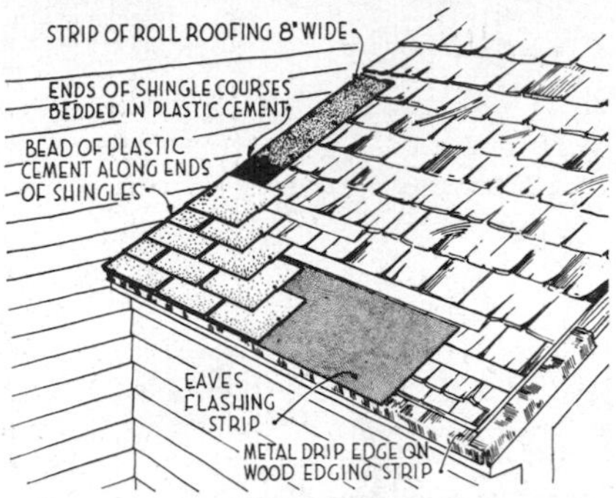

FIG. AR30 Wall Flashing (reroofing).

A strip of smooth roll roofing 6″ or 8″ wide is applied over the shingles abutting the wall surface, using a row of nails approximately 4″ o.c. along each edge. As the work proceeds, the strip is covered with asphalt plastic cement and the end of each course of new shingles is firmly secured by bedding it in the cement. Improved appearance and a tight joint can be achieved by using a caulking gun to apply a final bead of cement between the ends of the shingles and the siding.

FIG. AR29 Application of drip edge. Installation at rake (a) and installation at eave (b).

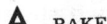

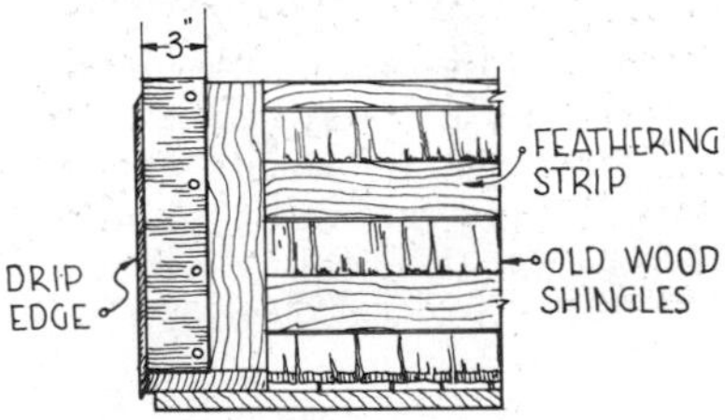

A RAKE

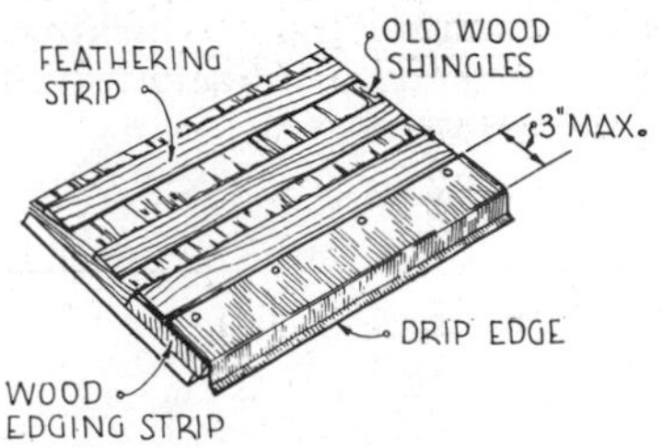

B EAVE

Chimney Flashing Figure AR31 illustrates a method of chimney flashing using existing flashing.

Application of Shingles (Reroofing)

After the existing roof has been prepared, the method of application of the various types of shingles are similar to those described above for new construction.

Since giant individual shingles, applied by the Dutch Lap Method, and individual hexagonal shingles are intended primarily for reroofing, these application methods are described below.

Giant Individual Shingles (Dutch Lap Method) This application method results in a single coverage roof intended primarily for reroofing and should not be applied on roof slopes less than 4-in-12.

Nails and Special Fasteners are used for the Dutch Lap Method. Each shingle, after the first course, is secured with 2 nails and 1 fastener. Nail types are the same as for new construction, but longer in length (see Fig. AR3). When laying the shingles from left to right, the nails should be located in the upper lefthand and lower righthand corners, 1″ in from the side, top and bottom edges (Fig. AR32). These positions are reversed when laying from right to left.

Fasteners may be non-corrodible wire staples applied with a stapler, a special metal clip furnished by the manufacturer of the shingle, or by a spot of special asphalt cement. The fasteners are used to secure the exposed lower corner of the shingle to the overlapped portion of the adjacent shingle in the same course. The exposed corner should not be nailed to the deck, except along the eave and rake edges of the deck.

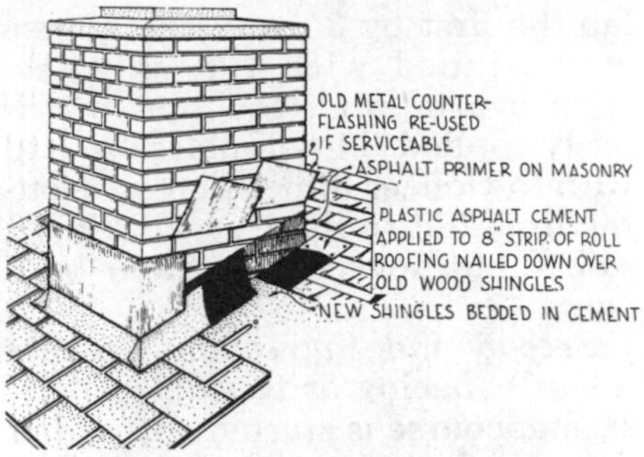

FIG. AR31 Chimney Flashing (reroofing). Old metal counter flashing may be reused (if still serviceable). It should extend down over new base flashing made by applying plastic asphalt cement over end of shingles and against masonry and covering with a strip of mineral surfaced roofing.

First Course is started at the lower corner of the roof with a piece of shingle 3″ wide and 12″ high, laid flush with one nail in each corner, 1″ in from each edge. The first shingle, cut to a length of 15″, is placed at the corner of the eave and rake to cover and be flush with the outer edges of the 3″ x 12″ unit, and is secured with exposed nails (Fig. AR32). The second shingle, full size, is placed to over-

FIG. AR32 Application of Giant Individual Shingles by the Dutch Lap Method.

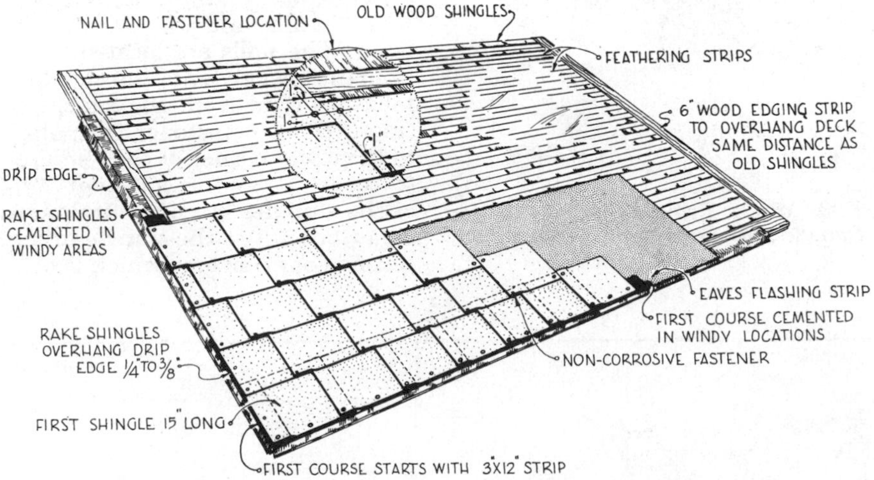

lap the first by 3″, and the course is continued with full shingles, each overlapping the one previously applied. The shingles are laid with the longer dimension horizontal, from left to right or from right to left, but never from center both ways.

Second and Succeeding Courses are self-spacing as to sidelap. The second course is started with a full shingle aligned with the vertical exposed edge of the second shingle in the first course, and placed with a toplap of 2″. The left end is trimmed even with the rake edge of the eaves flashing strip. The course is then continued with full shingles maintaining a 2″ toplap and a 3″ sidelap. All rake and eave shingles are secured with exposed nails.

Windy Locations (areas with wind velocities of 75 mph or greater) require a strip of asphalt cement applied along the rakes and eaves, extending back from the edges of the deck approximately 6″. The eave and rake shingles are embedded in the cement in addition to being secured with exposed nails.

Individual Hex Shingles Individual hexagonal shingles are of two types. Both are relatively lightweight shingles intended primarily for reroofing over old roofing, and should not be applied on slopes less than 4-in-12. They are applied by nailing with the lower corner secured either by a staple (Fig. AR33a) or with an integral locking device (AR33b).

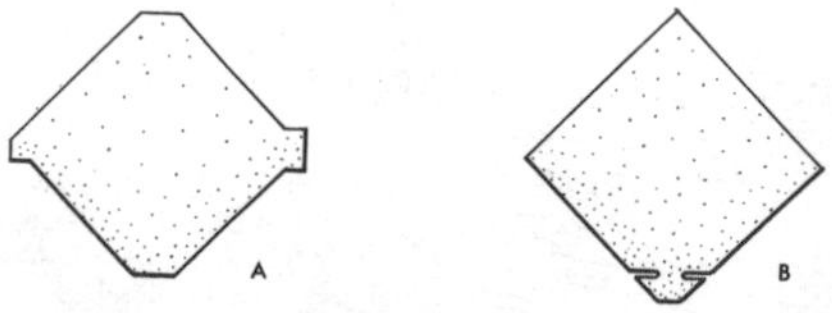

FIG. AR33 Typical Individual Hex Shingles; (a) staple type, (b) lock type.

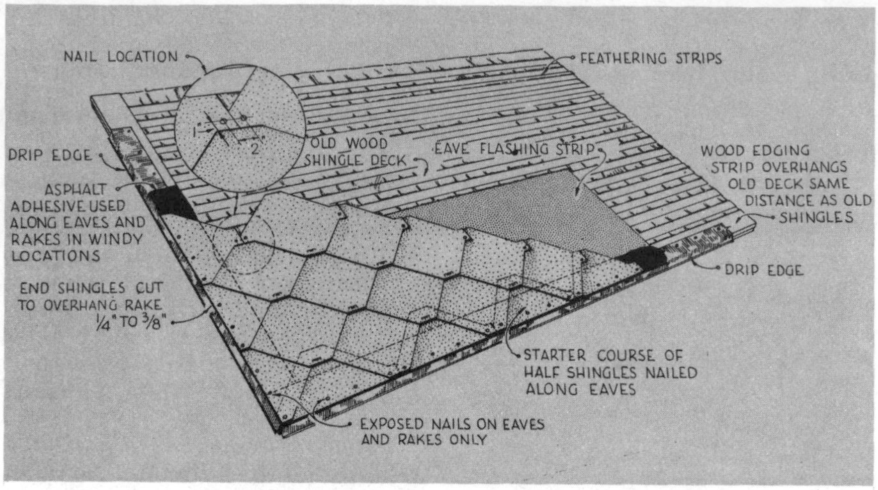

FIG. AR34 Application of Individual Staple Type Hex Shingles.

FIG. AR35 Application of Individual Lock Type Hex Shingles.

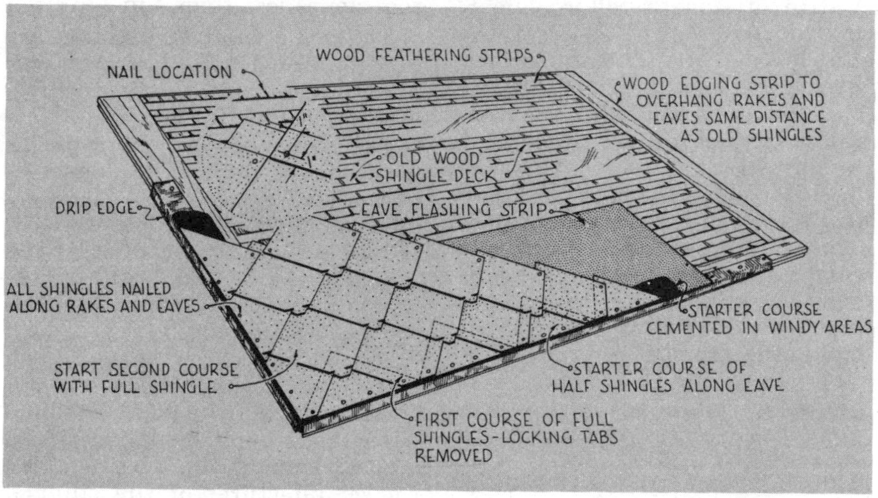

Nails are the same type used for new construction, but longer in length (see Fig. AR3). Each staple-type shingle should be secured with 2 nails and one staple fastener. The nails are located 1″ up from the lower edge of the shoulder tabs and 1″ in from the side edge (Fig. AR34). A fastener is applied to the lower corner of each shingle to secure it to the adjacent tabs of the shingles in the course next below but not to the shingle in the second course below, which lies on the deck. The tab should never be nailed down. A stapling device that does not penetrate the complete shingle assembly should be used so that the roofing will not be vulnerable to leaks from driving rains.

Each lock-type shingle is secured with 2 nails, one at each shoulder (Fig. AR35). The lower corner is locked by inserting the locking tab under the sides of adjacent shingles in the course next below.

WOOD SHAKES

The following text describes materials and methods recommended to provide a suitable roof covering of handsplit wood shakes over both new and existing construction.

Reference should be made to the entire first sub-section, General Recommendations, pages 411-1 through 411-8, for: (1) definition of terminology; (2) limitations imposed by the slope of the roof; (3) roof sheathing thicknesses, spans and nailing schedules; and (4) typical accessory materials—underlayment, flashing, roofing cements, eaves flashing, drip edge, roofing nails—recommended for preparation of the roof for the application of shingles and shakes. Unless specific exceptions are made in the following text, the General Recommendations apply.

MATERIALS

Wood shakes are manufactured in three types of units: (1) handsplit-and-resawn, (2) tapersplit, and (3) straightsplit. Shakes are produced in three lengths—18″, 24″ and 32″. Each is 100% clear wood and 100% heartwood. Both tapersplit and straightsplit shakes, produced largely by hand, are of 100% edgegrain. Handsplit-and-resawn shakes have split faces and sawn backs and may include up to 10% flat grain. Factory assembled hip and ridge units are available.

Each bundle of shakes should carry the official grade-marked label of the Red Cedar Shingle & Handsplit Shake Bureau. Data covering types of shakes are summarized in Figure SK1. See Section 202 Wood Products for additional information covering manufacture and properties.

Nails

Nails for applying shakes should be corrosion-resistant such as hot-dipped galvanized steel, or aluminum. Common wire nails, or even electro-galvanized nails, may shorten the life of a shake roof and should not be used. Nail shanks may be smooth or threaded. Threaded nails provide increased holding power, although smooth shank nails are used in the majority of applications over wood sheathing and give suitable service. Nails for applying shakes to plywood sheathing should be threaded. If threaded, galvanized steel nails should have annular threads; aluminum nails should have screw threads of approximately 12-1/2° thread angle. Box nails, having smaller diameters and longer lengths than common nails, are also suitable. For new roofs, nail lengths typically required are 6d (2″) for shakes with 3/4″ to 1-1/4″ butt thickness, 5d (1-3/4″) for shakes with 1/2″ to 3/4″ butt thickness. Nails two-penny (2d) sizes larger should be used to apply hip and ridge units. For re-roofing, nails should be long enough to penetrate through the existing sheathing.

The number, spacing and pattern of nails are described below under Application of Shakes.

PREPARATION FOR SHAKES

Slope Limitations and Exposure

Roof slope limitations are discussed on page 411-5.

The maximum recommended exposure for double coverage roofs is 13″ for 32″ shakes, 10″ for 24″ shakes and 7-1/2″ for 18″ shakes. A triple-coverage roof is achieved by reducing these exposures to 10″ for 32″ shakes, 7-1/2″ for 24″ shakes and 5-1/2″ for 18″ shakes. Approximate roof coverage for shakes using different exposures is shown in Figure SK2.

Shakes are recommended on slopes 4 in 12 or steeper. Suitable installations have been achieved on slopes as low as 3 in 12 by taking the following extra precautions: (1) reducing exposures to provide triple coverage and (2) using solid sheathing with an underlayment of No. 30 asphalt saturated felt applied over the entire roof with No. 30 asphalt saturated felt interlayment between each course. (See Application of Shakes below).

FIG. SK1 **TYPES OF RED CEDAR SHAKES**

Grade	Length and Thickness	Bundles Per Square	Weight (lbs. per square)		Description	Label
No. 1 HANDSPLIT & RESAWN	18″ x 1/2″ to 3/4″ 18″ x 3/4″ to 1-1/4″ 24″ x 1/2″ to 3/4″ 24″ x 3/4″ to 1-1/4″ 32″ x 3/4″ to 1-1/4″	4 5 4 5 6	220 250 280 350 450		These shakes have split faces and sawn backs. Cedar blanks or boards are split from logs and then run diagonally through a bandsaw to produce two tapered shakes from each.	
No. 1 TAPERSPLIT	24″ x 1/2″ to 5/8″	4	260	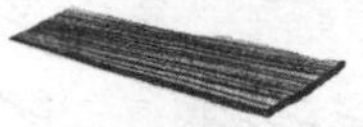	Produced largely by hand, using a sharp-bladed steel froe and a wooden mallet. The natural shingle-like taper is achieved by reversing the block, end-for-end, with each split.	
No. 1 STRAIGHT-SPLIT (BARN)	18″ x 3/8″ 24″ x 3/8″	5 5	200 260		Produced in the same manner as tapersplit shakes except that by splitting from the same end of the block, the shapes acquire the same thickness throughout.	

FIG. SK2 **ROOF COVERAGE OF SHAKES AT VARYING WEATHER EXPOSURES**

Length and Thickness	Type of Shake	No. of Bundles	Approximate Coverage (sq. ft.) Weather Exposures								
			5-1/2″	6-1/2″	7″	7-1/2″	8″	8-1/2″	10″	11-1/2″	13″
18″ x 1/2″ to 3/4″	Handsplit-and-Resawn	4	55*	65	70	75**	—	—	—	—	—
18″ x 3/4″ to 1-1/4″	Handsplit-and-Resawn	5	55*	65	70	75**	—	—	—	—	—
24″ x 1/2″ to 3/4″	Handsplit-and-Resawn	4	—	65	70	75*	80	85	100**	—	—
24″ x 3/4″ to 1-1/4″	Handsplit-and-Resawn	5	—	65	70	75*	80	85	100**	—	—
32″ x 3/4″ to 1-1/4″	Handsplit-and-Resawn	6	—	—	—	—	—	—	100*	115	130**
24″ x 1/2″ to 5/8″	Tapersplit	4	—	65	70	75*	80	85	100**	—	—
18″ x 3/8″	Straight-Split	5	65*	—	—	—	—	—	—	—	—
24″ x 3/8″	Straight-Split	5	—	65	70	75*	—	—	—	—	—
15″ Starter-Finish Course			Use supplementary with shakes applied not over 10″ exposure								

* Recommended maximum weather exposure for 3-ply roof construction.
** Recommended maximum weather exposure for 2-ply roof construction.

Roof Sheathing

Sheathing types, thicknesses, spans and nailing recommendations are discussed on page 411-4.

Shakes may be applied over either spaced or solid sheathing, depending on the climatic conditions of the region.

In areas not subject to wind-driven snow, the use of spaced sheathing, such as nominal 1″ x 4″ or 1″ x 6″ boards, is suitable (Fig. SK3). The center-to-center spacing of boards should be equal to the exposure of the shake, but not more than 10″.

In areas subject to wind-driven snow (generally where the outside design temperature is 0° F or colder, see Fig. 9, page 411-7), solid sheathing of nominal 1″ boards or plywood is recommended.

Interlayment (and Underlayment)

Interlayment strips and the underlayment course used at eaves should be No. 30 asphalt saturated felt. The installation of interlayment between each shake course, essential to provide a baffle against infiltration of wind-driven rain or snow, is described below under Application of Shakes.

Eaves Flashing

In areas where the outside design temperature is 0° F or colder (see Fig. 9, page 411-7), or where there is a possibility of ice forming along the eaves and causing a backup of water, eaves flashing is recommended. In these areas shakes should be applied over solid sheathing.

On slopes 4 in 12 or steeper, the eaves flashing is formed by applying an additional course of No. 30 asphalt saturated felt over the underlayment starting course at the eaves. The eaves flashing should extend up the roof to cover a point at least 24″ inside the interior wall line of the building. When the eave overhang requires the flashing to be wider than 36″, the necessary horizontal joint is cemented and located outside the exterior wall line of the building. (see Fig. 10a, page 411-7).

For slopes 3 in 12 up to 4 in 12, or in areas subject to severe icing, eaves flashing may be formed as described above for 4 in 12 slopes, except that the double layer of No. 30 asphalt saturated underlayment is cemented. The eaves flashing is formed by applying a continuous layer of plastic asphalt cement, at the rate of 2 gals. per 100 sq. ft., to the surface of the underlayment starter course before the second layer of underlayment is applied. Cement is also applied to the 19″ underlying portion of each succeeding course which lies within the eaves flashing area, before placing the next course. It is important to apply the cement uniformly with a comb trowel, so

FIG. SK3 Spaced Sheathing: (a) center to center spacing equals exposure; (b) maximum recommended exposure provides 2-ply roof. A 3-ply roof is achieved by reducing exposures.

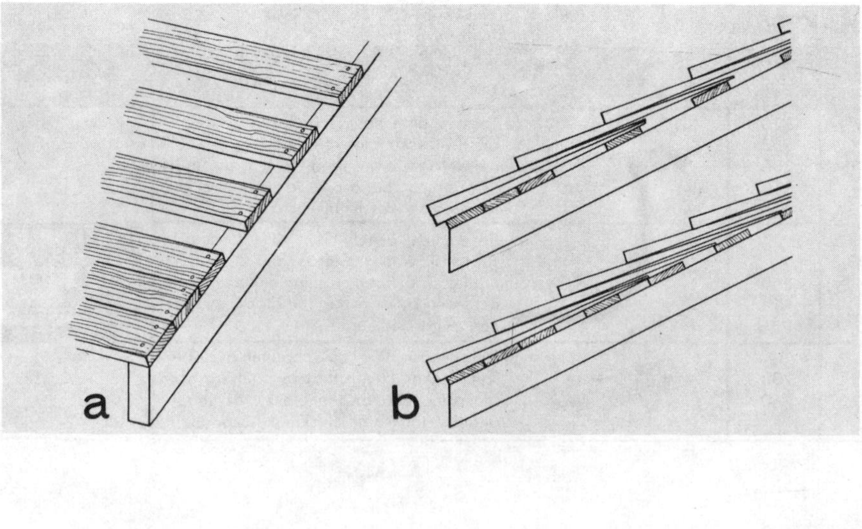

that at no point does underlayment touch underlayment when the application is completed. The overlying sheet is pressed firmly into the entire cemented area.

Drip Edge

Wood shakes should extend out over the eave and rake a distance of 1″ to 1-1/2″ to form a drip. Accordingly, corrosion-resistant drip edge is unnecessary at these points as the starter course of shakes and the rake shakes accomplish the same purpose.

Flashing

General recommendations for flashing are discussed on page 411-6.

Unless special precautions are taken (Fig. SK4), copper flashing materials are not recommended for use with red cedar shakes. Premature deterioration of the copper may occur when the metal and wood are in intimate contact in the presence of moisture.

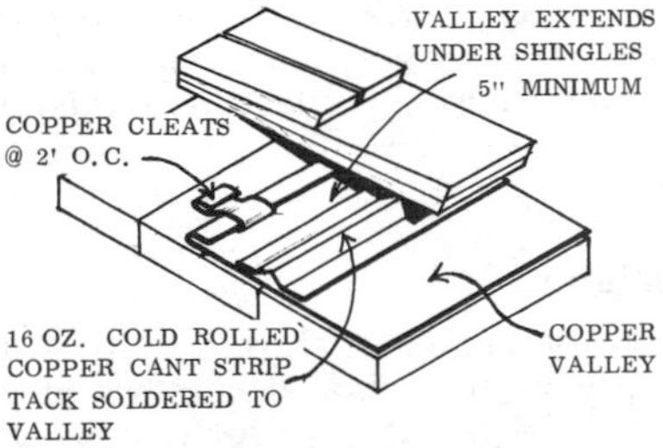

FIG. SK4 Surface contact of copper with red cedar must be minimized.

Valley Flashing The *open* method, using roofing felt and sheetmetal, or the *closed* method, using hand-fitted shakes, may be used to construct valley flashing. For longest service life, the open method is highly recommended.

Open Valleys are first covered with a valley underlayment strip of No. 30 asphalt saturated felt, at least 20″ wide, centered in the valley and secured with enough nails to hold it in place (Fig. SK5). Metal flashing strips, 20″ wide, are then nailed over the valley underlayment. If galvanized steel flashing is used, it should be preferably 18 gauge (but not less than 26 gauge), preferably center-crimped, and painted on both surfaces. It is desirable to edge-crimp valley sheets by turning the edges up and back approximately 1/2″ toward the valley centerline. This provides an additional water-stop, directing water down the valley. Shakes laid to finish at the valley are trimmed parallel with the valley to form a 6″ wide gutter.

Closed Valleys are first covered with a 1″ x 6″ wood strip, nailed flat into the saddle and covered with roofing felt as specified above for *open* valleys. Shakes in each course are edge-trimmed to fit into the valley, then laid across the valley with an undercourse of metal flashing having a 2″ headlap, and extending 10″ under the shakes on each side of the saddle.

FIG. SK5 Method for constructing open valley flashing.

Chimney and Sidewall Flashing When shakes butt against vertical surfaces such as sidewalls, dormers or chimneys, individual flashing units of corrosion-resistant metal should be inserted between each course of shakes. One leg of the flashing should extend at least 6″ under the shake with the up-turned leg covered by the vertical wall finish or by metal counter flashing, as shown in Figure SK6.

On new construction, chimney flashing should be built in by the mason (Fig. SK7). If masonry work

FIG. SK6 Base flashing units are laid with each shake course and counter flashed.

FIG. SK7 Base flashing and counter flashing detail at chimney.

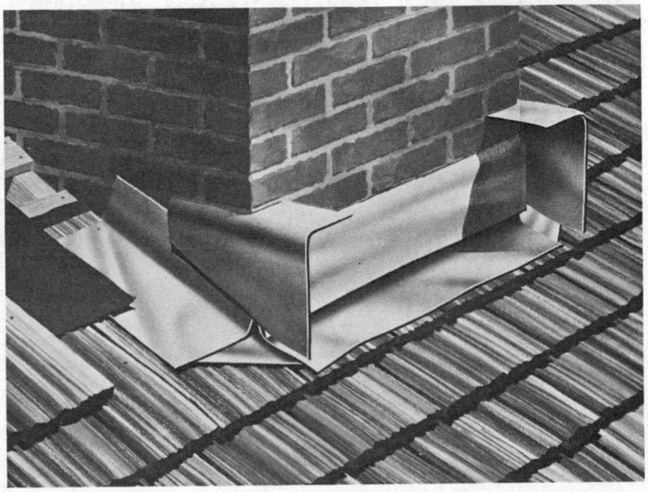

is completed, flashing should be inserted between bricks to a depth of 3/4″ by removing mortar, and filling over the flashing with bituminous mastic.

If flashing is galvanized steel, it should be painted on both surfaces before it is placed. Flashing strips which must be bent to sharp angles should be painted after bending; and if crimped to form a waterstop in the centers of valleys, the crimped surface should be given a double coat of paint.

APPLICATION OF SHAKES (New Construction)

Along the eave line, a 36″ wide starter strip of No. 30 asphalt saturated felt underlayment should be laid over the sheathing. The starter course of shakes at the eave line should be doubled, using an undercourse of 24″, 18″ or 15″ shakes, the latter being made expressly for this purpose. Striking visual effects can be achieved by using a double undercourse at the eave line. After each course of shakes is applied, an 18″ wide strip of No. 30 asphalt saturated felt interlayment should be applied over the top portion of the shakes extending onto the sheathing (Fig. SK8). The bottom edge of the interlayment should be positioned at a distance above the butt equal to twice the exposure. For example, if 24″ shakes are being laid at 10″ exposure, the bottom edge of the felt should be positioned at a distance 20″ above the shake butts; the strip will then cover the top 4″ of the shakes and extend 14″ onto the sheathing.

Individual shakes should be spaced approximately 1/4″ to 3/8″ apart to allow for possible expansion due to moisture absorption. The joints between shakes should be offset at least 1-1/2″ in adjacent courses. The joints in alternate courses also should be kept out of direct alignment when a 3-ply roof is being built.

When straight-split shakes, which are of equal thickness throughout, are applied the "froe-end" of the shake (the smoother end from which it has been split) should be laid undermost.

The application of shakes can be speeded by the use of a special shingler's or lather's hatchet (Fig. SK9). The hatchet is used for nailing, carries an exposure gauge and has a sharpened heel which expedites trimming.

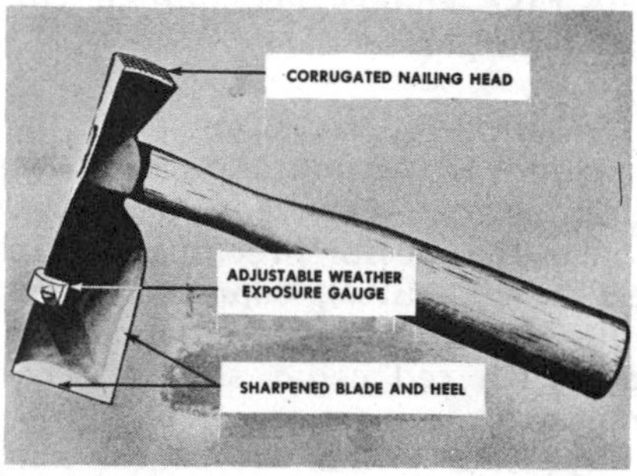

FIG. SK9 Special shingler's hatchet speeds application of both wood shakes and shingles. Nailing, measuring the correct exposure, and cutting and trimming may all be done with this one tool.

FIG. SK8 Application of shakes over spaced sheathing. Solid sheathing may also be used.

Nailing

Only two nails should be used to apply each shake, regardless of its width. They should be placed approximately 1″ in from each edge, and from 1″ to 2″ above the buttline of the succeeding course. Nails should be driven until the heads meet the shake surface but no further (Fig. SK10).

Hips and Ridges

The final shake course at the ridge line, as well as shakes that terminate at hips, should be secured with additional nails and should be composed of smoother textured shakes. A strip of No. 30 asphalt saturated felt, at least 12″ wide, should be applied over the crown of all hips and ridges, with an equal exposure of 6″ on each side. Prefabricated hip and ridge units can be used, or the hips and ridges may be cut and applied on the site. In site-construction of hips, shakes approximately 6″ wide are sorted out. Two wooden

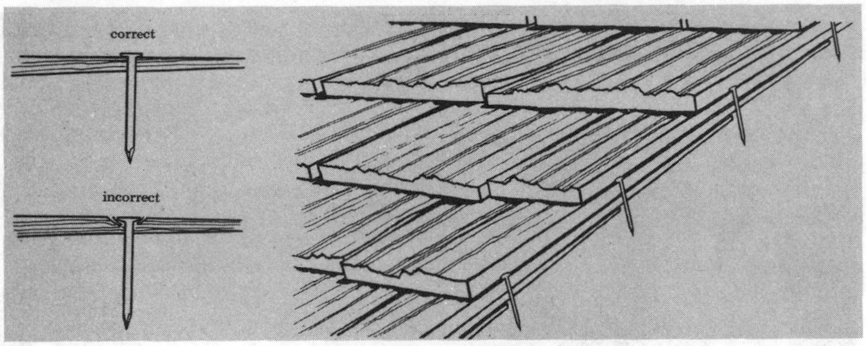

FIG. SK10 Nails should be correctly placed and nailheads should not break the shake surface.

straightedges are tacked on the roof, 6″ from the centerline of the hip, one on each side. The starting course of shakes should be doubled. The first shake on the hip is nailed in place with one edge resting against the guide strip. The edge of the shake projecting over the center of the hip is cut back on a bevel. The shake on the opposite side then is applied and the projecting edge cut back to fit. Shakes in the following courses are applied alternately in reverse order (Fig. SK11). Ridges are constructed in a similar manner. Exposure of hip and ridge shakes normally is the same as the shakes on the roof.

Ridge shakes are laid along an unbroken ridge that terminates in a gable at each end, should be started at each gable end and terminate in the middle of the ridge. At that point, a small saddle is face-nailed to splice the two lines. The first course of shakes should always be doubled at each end of the ridge.

FIG. SK11 Hips and Ridges. Starter course should be doubled as shown.

Application Variations

Graduated Exposures A variation in roof appearance may be achieved with handsplit shakes by reducing the exposure of each course from eaves to ridge. This requires shakes of several lengths.

Such a graduated-exposure roof is shown in Figure SK12, and is built by starting at the eaves with 32″ shakes, laid 13″ to the weather. One-third of the roof area is covered with shakes of this length, reducing the exposure gradually to about 10″. The next third of the area is covered with 24″ shakes, laid with a 10″ exposure and diminishing to about 8-1/2″. The roof is completed with 18″ shakes, starting with 8″ exposure and diminishing to 5″.

Staggered Lines Irregular and random roof patterns can be achieved by laying shakes with butts placed slightly below or above the horizontal lines governing each course (Fig. SK13). If an extremely irregular pattern is desired, longer shakes may be interspersed over the roof with their butts several inches lower than the course lines.

Mixtures of fairly smooth and very rough shakes will produce a more rugged appearance than a roof composed entirely of rough shakes.

Tilted Strip at Gables Dripping water may be eliminated by inserting a single strip of bevel siding the full length of each gable rake with the thick edge flush with the sheathing edge (Fig. SK14). The resulting inward pitch of the roof surface diverts water away from the gable edge.

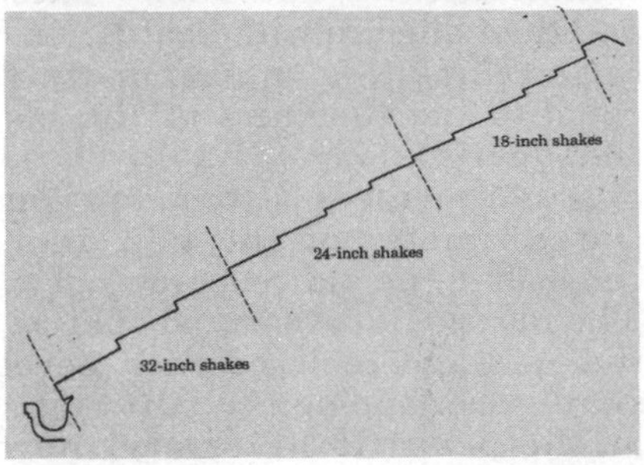

FIG. SK12 A roof with graduated exposure is created by using all three shake sizes.

FIG. SK13 Staggered butt lines produce varied appearance.

FIG. SK14 Beveled strip also provides decorative termination at gable.

REROOFING EXISTING CONSTRUCTION

Structures which require new roof coverings, either for appearance or serviceability, may be covered with any of the three types of handsplit cedar shakes. Except over slate, tile or asbestos-cement shingles, shakes may be applied directly over the existing roof.

When the existing roof is to be over-roofed with shakes, a 6″ wide strip of the old roofing is removed along the eaves and gable rakes, and then filled in with boards having a thickness approximately equal to the thickness of the old roof section (Fig. SK15). These boards provide a strong base at the perimeter of the new roof, concealing the old roof from view. The old ridge covering should be removed and replaced with bevel siding, overlapping the butt edges at the peak. If necessary, new flashing should be installed over existing flashing. Nails for applying a new roof should be long enough to penetrate the underlying existing sheathing.

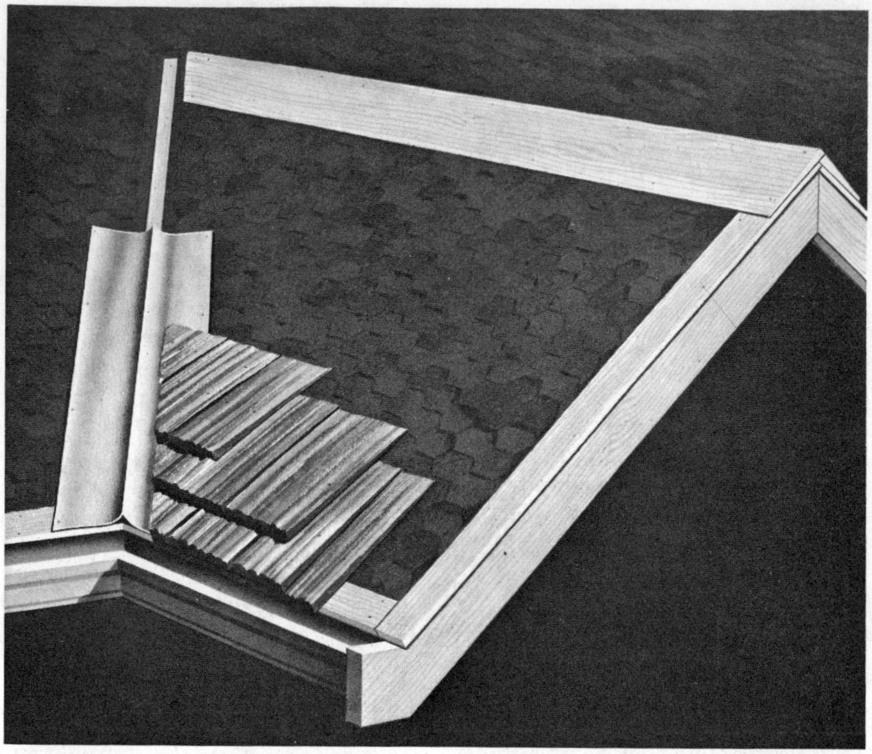

FIG. SK15 Preparation of existing roof to receive application of new wood shakes or shingles.

WOOD SHINGLES

The following text describes materials and methods recommended to provide a suitable roof covering of wood shingles over both new and existing construction.

Reference should be made to the entire first sub-section, General Recommendations, pages 411-1 through 411-8 for: (1) definition of terminology; (2) limitations imposed by the slope of the roof; (3) roof sheathing thicknesses, spans and nailing schedules; and (4) typical accessory materials—underlayment, flashing, roofing cements, eaves flashing, drip edge, and roofing nails—recommended for preparation of the roof for the application of shingles. Unless specific exceptions are made in the following text, the General Recommendations apply.

MATERIALS

Wood shingles are manufactured in 24″, 18″ and 16″ lengths conforming to three grades: No. 1 (Blue Label), No. 2 (Red Label) and No. 3 (Black Label). Preformed, factory-built hip and ridge units are available.

Each bundle of wood shingles should carry the official grade-marked label of the Red Cedar Shingle & Handsplit Shake Bureau. Data covering wood shingles are summarized in Figure WS1. See Section 202 Wood Products for additional information covering manufacture and properties.

Nails

Nails for applying shingles should be corrosion-resistant, such as hot-dipped galvanized steel, or aluminum. Common wire nails, or even electro-galvanized nails, may shorten the life of the shingle roof and should not be used. Nail shanks may be smooth or threaded. Threaded nails provide increased holding power, although smooth shank nails are used in the majority of applications over wood sheathing and give suitable service. Nails for applying wood shingles to plywood sheathing should be threaded. If threaded, galvanized steel nails should have annular threads; aluminum nails should have screw threads of approximately 12-1/2° thread angle. Box nails, having smaller diameters and longer lengths than common nails are also suitable.

Nail dimensions typically required are given in Figure WS2. Nails two-penny (2d) sizes larger should be used to apply hip and ridge units. For over-roofing, nails should be long enough to penetrate through the existing sheathing; the 5d size (1-3/4″) is normally adequate.

The number, spacing and pattern of nails are described below under Application of Shingles.

FIG. WS2
NAIL SIZES RECOMMENDED FOR APPLICATION OF WOOD SHINGLES

SIZE	LENGTH	GAUGE	HEAD	SHINGLES
3d*	1-1/4″	14-1/2	7/32″	16″ & 18″
4d*	1-1/2″	14	7/32″	24″
5d**	1-3/4″	14	7/32″	16″ & 18″
6d**	2″	13	7/32″	24″

*3d and 4d nails are used for new construction.
**5d and 6d nails are used for over roofing.

PREPARATION FOR SHINGLES

Slope Limitation and Exposure

Roof slope limitations are discussed on page 411-5. The exposure of wood shingles is dependent on the slope of the roof.

Standard exposures of 5″, 5-1/2″ and 7-1/2″ for shingle lengths of 16″, 18″ and 24″ respectively are used on slopes 5 in 12 or greater. Four bundles of shingles will cover 100 sq. ft. (1 square) of roof area when the shingles are applied at standard exposures. Using reduced exposures, wood shingles may be

FIG. WS1 **TYPES OF RED CEDAR SHINGLES**

Grade	Size	Bundles Per Square	Weight Per Square	Description		Labels
No. 1 BLUE LABEL	24″ (Royals) 18″ (Perfections) 16″ (XXXXX)	4 bdls. 4 bdls. 4 bdls.	192 lbs. 158 lbs. 144 lbs.		The premium grade of shingles for roofs and sidewalls. These shingles are 100% heartwood, 100% clear and 100% edge-grain.	
No. 2 RED LABEL	24″ (Royals) 18″ (Perfections) 16″ (XXXXX)	4 bdls. 4 bdls. 4 bdls.	192 lbs. 158 lbs. 144 lbs.		A good grade for all applications. Not less than 10″ clear on 16″ shingles, 11″ clear on 18″ shingles and 16″ clear on 24″ shingles. Flat grain and limited sapwood are permitted.	
No. 3 BLACK LABEL	24″ (Royals) 18″ (Perfections) 16″ (XXXXX)	4 bdls. 4 bdls. 4 bdls.	192 lbs. 158 lbs. 144 lbs.		A utility grade for economy applications and secondary buildings. Guaranteed 6″ clear on 16″ and 18″ shingles, 10″ clear on 24″ shingles.	

used on slopes as low as 3 in 12. They are not recommended on slopes less than 3 in 12.

On 4 in 12 slopes, exposures should be reduced to 4-1/2", 5" and 6-3/4", for 16", 18" and 24" shingles respectively. On 3 in 12 slopes they should be reduced further to 3-3/4", 4-1/4" and 5-3/4" for 16", 18" and 24" shingles respectively, which will assure four layers of shingles over the roof area.

Shingle exposure should not be increased beyond a point equivalent to the length of the shingle minus 1", divided by 3. This will provide at least triple coverage (three layers of wood at every point) to insure complete freedom from leakage in heavy wind driven rain or snow storms. Approximate coverage of 4 bundles of shingles at varying exposures is shown in Figure WS3.

Roof Sheathing

Sheathing types, thicknesses, spans and nailing recommendations are discussed on page 411-4.

Either spaced or solid sheathing may be used with wood shingles. If solid sheathing is preferred, either nominal 1" boards or plywood may be used. When spaced sheathing, such as nominal 1" x 3" or 1" x 4" boards, is used, it is applied so that the space between boards is not greater than the width of the boards themselves.

Underlayment

Underlayment typically is not required between shingles and spaced or solid sheathing, but may be desirable for the protection of sheathing and to insure against air infiltration.

When underlayment is desired, No. 15 asphalt saturated felt may be used.

Drip Edge

Wood shingles should extend out over the eave and rake a distance of 1" to 1-1/2" to form a drip. Accordingly, corrosion-resistant drip edge is unnecessary at these points as the shingle starter course and rake shingles accomplish the same purpose.

FIG. WS3
ROOF COVERAGE OF WOOD SHINGLES AT VARYING EXPOSURES

Length and Thickness*	Approximate Coverage (sq. ft.) of Four Bundles — Weather Exposures 3-1/2"	4"	4-1/2"	5"	5-1/2"	6"	6-1/2"	7"	7-1/2"
16" x 5/2"	70	80	90	100**	—	—	—	—	—
18" x 5/2-1/4"	—	72-1/2	81-1/2	90-1/2	100**	—	—	—	—
24" x 4/2"	—	—	—	—	—	80	86-1/2	93	100**

* Sum of the thickness e.g. 5/2" means 5 butts = 2"
** Maximum exposure recommended for roofing

Eaves Flashing

In areas where the outside design temperature is 0° F or colder (Fig. 9, see page 411-7), or where there is a possibility of ice forming along the eaves and causing a backup of water, eaves flashing is recommended. Sheathing should be applied *solidly* above the eave line to cover a point at least 24" inside the interior wall line of the building.

For 4 in 12 slopes, the eaves flashing is formed by applying a double layer of No. 15 asphalt saturated felt to cover this section of solid sheathing. When the eave overhang requires the flashing to be wider than 36", the necessary horizontal joint between the felt strips is cemented and located outside the exterior wall line.

For slopes 3 in 12 up to 4 in 12, or in areas subject to severe icing, eaves flashing may be formed as described above for 4 in 12 slopes, except that the double layer of No. 15 asphalt saturated underlayment is cemented. The eaves flashing is formed by applying a continuous layer of plastic asphalt cement, at the rate of 2 gals. per 100 sq. ft., to the surface of the underlayment starter course before the second layer of underlayment is applied. Cement is also applied to the 19" underlying portion of each succeeding course which lies within the eaves flashing area, before placing the next course. It is important to apply the cement uniformly with a comb trowel, so that at no point does underlayment touch underlayment when the application is completed. The overlying sheet is pressed firmly into the entire cemented area. (Fig. 10, pg. 411-7).

Flashing

General recommendations for flashing are discussed on page 411-6.

Unless special precautions are taken (see Fig. SK4, page 411-29), copper flashing materials are not recommended for use with red cedar shingles. Premature deterioration of the copper may occur when the metal and wood are in intimate contact in the presence of moisture.

Valley Flashing Only the *open* method should be used to construct valley flashing; butting of shingles together in the valley at the centerline to form a closed valley is not recommended.

On slopes up to 12 in 12, metal valley sheets should be wide enough to extend at least 10" on each side of the valley centerline (Fig. WS4). On roofs of steeper slope, narrower sheets may be used extending on each side of the centerline of the valley for a distance of at least 7".

To avoid restricting the delivery of a valley and to prevent splashing of water flowing from a steep slope on the shingles of a lower slope, the edges of the shingles on the lower slope should be lined up at least 1" farther back from the centerline of the valley than those on the steeper slope. (As an alternative, a vertical ridge, or water-stop, can be made by crimping the metal up in the center of the valley.) It is desirable to edge-crimp valley sheets by turning the edges up and back approximately 1/2" toward the valley centerline. This provides an additional water-stop, directing water down the valley. The open portion of the valley should be at least 4" wide, but valleys may taper from a width of 2″ where they start and increase at the rate of 1/2″ per 8′ of length, to a wider width as they descend.

In areas where the outside design temperature is 0° F or colder, underlayment should be installed under metal valley sheets.

Chimney and Sidewall Flashing See page 411-29.

APPLICATION OF SHINGLES

The first course of shingles at the eaves should be doubled or tripled and should project 1″ to 1-1/2″ beyond the eaves to provide a drip.

The second layer of shingles in the first course should be nailed over the first layer to provide a minimum sidelap of at least 1-1/2″ between joints (Fig. WS5). If possible, joints should be "broken" by a greater margin. A triple layer of shingles in the first course provides additional insurance against leaks at the cornice. No. 3 grade shingles frequently are used for the starter course.

Shingles should be spaced at least 1/4″ apart to provide for expansion. Joints between shingles in any course should be separated not less than 1-1/2″ from joints in the adjacent course above or below, and joints in alternate courses should not be in direct alignment.

When shingles are laid with the recommended exposure, triple coverage of the roof results (Fig. WS5).

When the roof terminates in a valley, the shingles for the valley should be carefully cut to the proper miter at the exposed butts and should be nailed in place first so that the direction of shingle application is *away* from the valley. This permits valley shingles to be carefully selected and insures shingle joints will not break over the valley flashing.

The application of wood shingles can be speeded by the use of a special shingler's or lather's hatchet (Fig. SK9, page 411-30). The hatchet is used for nailing, carries an exposure gauge and has a sharpened heel which expedites trimming.

Nailing

To insure that shingles will lie flat and for maximum service, only two nails should be used to secure each shingle. Nails should be placed not more than 3/4″ from the side edge of shingles, at a distance of not more than 1″ above the exposure line (Fig. WS5). Nails should be driven flush but not so that the nailhead crushes the wood.

FIG. WS6 Application of Hips and Ridges.

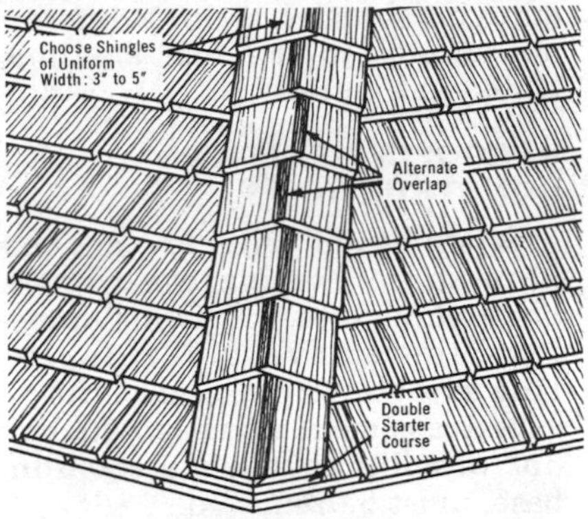

Hips and Ridges

Hips and ridges should be of the modified "Boston" type with protected nailing. Either site-applied or preformed factory-constructed hip and ridge units may be used. Nails should be increased at least two sizes larger than the nails used to apply the shingles.

Construction of the hips and ridges is shown in Figure WS6. Hips and ridges should begin with a double starter course.

Gable Rakes

Shingles should project 1″ to 1-1/2″ over the rake. End shingles can be canted to eliminate drips as shown in Figure SK14, page 411-31.

FIG. WS4 Methods for constructing open valley flashing. Closed valley flashing is not recommended.

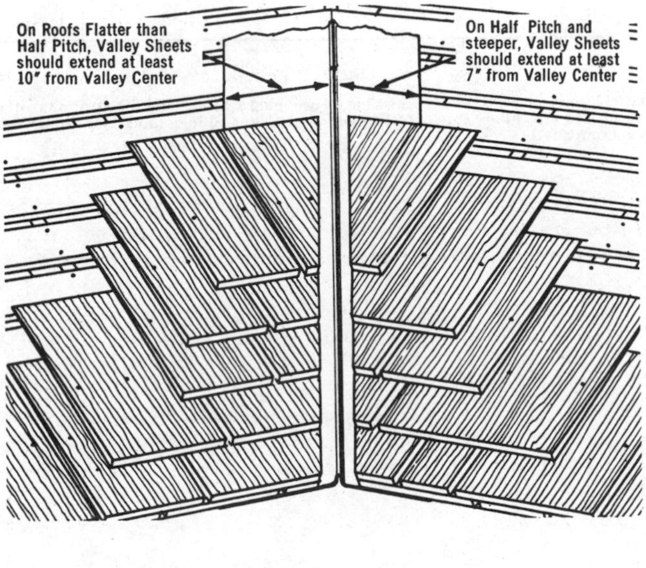

FIG. WS5 Application of wood shingles over spaced (or solid) sheathing.

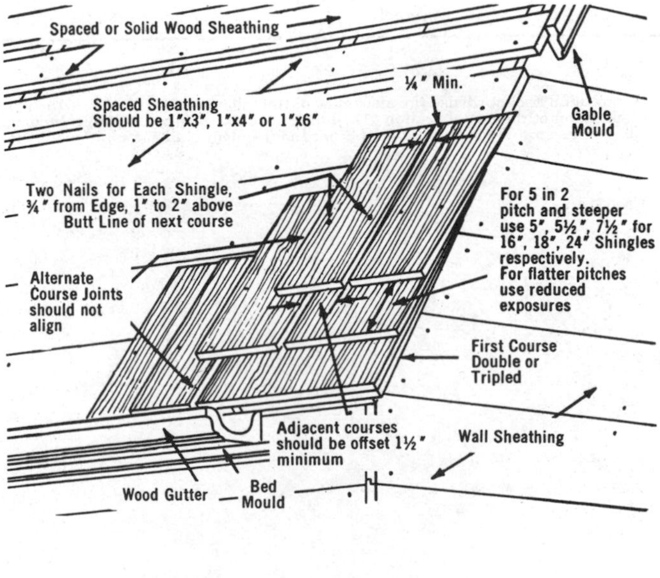

Alternate Application Methods

Several alternate methods of applying shingles, giving a different appearance to the roof are illustrated in Figure WS7.

REROOFING EXISTING CONSTRUCTION

When wood shingles are applied to an existing wood or asphalt shingle roof, it is not necessary to remove old shingles. However, existing slate, tile or asbestos-cement shingles, which do not provide an adequate nailing base, must be removed.

The first course of shingles at the eaves is removed just below the butts of the shingles in the second course. Old shingles should also be cut back and removed along the gable rakes (Fig. SK15, page 411-32). The removed shingles along the eaves and gable rake edges are replaced with boards having a thickness approximately equal to the thickness of the old roof section. These boards provide a strong base at the perimeter of the new roof, concealing the old roof from view. Ridge shingles should be replaced with strips of bevel siding, laid with the thin edges down the roof.

If necessary, new flashing should be installed over existing flashing.

New shingles are applied by identical methods used in new construction. If the existing roof sheathing is spaced, it is not necessary for all new nails to strike the sheathing.

FIG. WS7 ALTERNATE SHINGLE APPLICATIONS

(7a)
Thatch. Shingles are positioned above and below a hypothetical course line, with deviation from the line not to exceed 1".

(7b)
Serrated. Courses are doubled every third, fourth, fifth or sixth course. Doubled courses can be laid butt edge flush or with slight overhang.

(7c)
Dutch Weave. Shingles are doubled or superimposed at random throughout the roof area.

(7d)
Pyramid. Two extra shingles, narrow shingle over a wide one, are superimposed at random.

We gratefully acknowledge the assistance of the following for the use of their publications as references and permission to use photographs and illustrations in the preparation of the entire Section 411 Shingle Roofing: Asphalt Roofing Industry Bureau, Mineral Fiber Products Bureau, Red Cedar Shingle and Handsplit Shake Bureau, Federal Housing Administration, Independent Nail Corporation, W. H. Maze Company.

SHINGLE ROOFING 411

CONTENTS

GENERAL RECOMMENDATIONS

Materials typically required or which may affect the application of shingles include: sheathing, underlayment, flashing, roofing cements, eaves flashing, drip edge and roofing nails. Specific recommendations for these materials are given below for each kind of shingle roofing.

See Main Text, page 411-3, for definitions of terminology.

The instructions and recommendations of the shingle manufacturer should always be consulted to insure suitable performance.

ROOF FRAMING

See Work File (Table of Contents) page Wood WF201-1 for Maximum Allowable Spans for Joists and Rafters.

SHEATHING

Sheathing should be securely attached to supports, form a smooth surface to receive the roofing materials and provide an adequate nailing base.

Lumber

Boards may be tongue and grooved, shiplapped or square edge. Spacing of supports should not exceed 24″ on center. Joints in roof sheathing should occur over supports unless end-matched boards are used. End-matched boards may break joints between supports provided each piece used bears on at least two rafters or joists, and end-joints in adjacent boards are staggered. Boards should be nailed to supports with at least two 7d annularly threaded or 8d common wire nails at each bearing.

Moisture content of boards should be as shown in Figure W8, Main Text page Wood 201-6.

Solid sheathing of wood boards should have a minimum actual thickness of 5/8″ and should usually not exceed 6″ nominal width, to minimize shrinkage.

Spaced sheathing of nominal 1″ x 3″, 1″ x 4″ or 1″ x 6″ boards should have a minimum actual thickness of 3/4″.

Non-Wood Sheathing

When materials other than lumber boards or plywood are recommended by their manufacturers for use as roof sheathing under shingles, the sheathing manufacturer should specify the type of fastening required to attach sheathing to supports.

(Continued) GENERAL RECOMMENDATIONS

FIG. WF1 PLYWOOD ROOF SHEATHING[1,2]

Identification Index	Plywood Thickness	Maximum Rafter or Joist Spacing (inches o.c.)			
		Asphalt Shingles and Wood Shingles & Shakes		Asbestos-Cement Shingles	
		Edges Blocked	Edges Unblocked	Edges Blocked	Edges Unblocked
16/0	5/16", 3/8"	16"	—	—	—
20/0	5/16", 3/8"	20"	—	—	—
24/0	3/8", 1/2"	24"	24"	16"	—
30/12	5/8"	30"	24"	24"	16"
32/16	1/2", 5/8"	32"	24"	24"	16"
36/16	3/4"	36"	30"	32"	24"
42/20	5/8", 3/4", 7/8"	42"	32"	32"	24"
48/24	3/4", 7/8"	48"	32"	42"	28"

1. Plywood continuous over two or more spans, grain of face plys across supports; Structural I and II, Standard and C-C Exterior Construction Grades only.
2. For 1/2" plywood or less, use 6d common or 5d threaded nails; for 1" plywood or less use 8d common or 7d threaded nails.
3. Unsupported edges of sheathing should be blocked using wood blocking, tongue-and-grooved edges (5 ply, 1/2" or thicker) or special corrosion-resistant metal H clips. Use 2 clips for spans 48" or greater, 1 clip for lesser spans.

Plywood

Plywood sheathing should be of the proper Interior or Exterior Construction Grade (see page Wood WF201-16), and should conform to the spans and thicknesses in Figure WF1. Exterior Type should be used when sheathing will be exposed to weather such as eave overhangs without soffits.

ROOF SLOPE LIMITATIONS

The minimum roof slopes are given below for each kind of shingle roofing.

Roof slope governs, or may impose limitations on: (1) shingle selection, (2) shingle exposure, (3) underlayment requirements, (4) eaves flashing requirements, and (5) methods used in shingle application. (See Fig. 6 Main Text, page 411-5).

UNDERLAYMENT

Underlayment recommendations are given below for each kind of shingle roofing.

The purpose of underlayment is to: (1) protect sheathing from moisture absorption until shingles are applied; (2) provide additional weather protection by preventing the entrance of wind driven rain below the shingles onto the sheathing or into the structure; and (3) in the case of asphalt shingles, to prevent direct contact between shingles and resinous areas in wood sheathing, which, because of chemical incompatability, may be damaging to the shingles.

FLASHING

Typical methods for flashing are described and illustrated in the Main Text for each kind of shingle roofing.

Exposed flashing materials should be sheetmetal such as 26 gauge galvanized steel, 0.019" thickness aluminum or 16 oz. copper.
Additional requirements are given below for each kind of shingle roofing.

EAVES FLASHING

Eaves flashing is recommended in areas where the outside design temperature is 0° F or colder (Fig. 9, Main Text, page 411-7), or where there is a possibility of ice forming along the eaves.

Eaves flashing is important to avoid leaks and moisture penetration caused by water backing up under shingles at the eaves behind dams of ice and snow.

Eaves flashing recommendations are given below for each kind of shingle roofing.

DRIP EDGE

Drip edges are designed and installed to protect the edges of the deck at the eaves and rakes, and help prevent leaks at these points by allowing water to drip free of underlying eave and cornice construction.

Drip edge should be continuous of a material such as 26 gauge galvanized steel.

Drip edge recommendations are given below for each kind of shingle roofing.

ROOFING NAILS

Nail suitability is dependent on (1) selecting the correct nail for the kind of shingle and type of roof sheathing; (2) using the correct number of nails; (3) locating them in the shingle correctly, and (4) choosing nail metal compatible with metal used for flashings.

Lumber Boards, Plank Decking and Plywood Sheathing

Roofing nails should be long enough to penetrate through lumber boards or plywood sheathing. They should penetrate at least 1" into plank decking. Nails for applying shingles over plywood sheathing should have threaded shanks.

(Continued) GENERAL RECOMMENDATIONS

Non-Wood Sheathing

When gypsum, concrete, fiberboards or sheathing materials other than lumber boards or plywood are used for the roof sheathing, special fasteners and/or special details for fastening are often necessary to provide adequate anchorage of the shingle roofing. Recommendations of the manufacturer of the non-wood sheathing material should be followed to insure his responsibility for suitable performance of the roof.

ASBESTOS CEMENT SHINGLES

Unless specific exceptions are made in the following text, the General Recommendations apply.

MATERIALS

Shingles should conform to the requirements of ASTM C222. Size, thickness and maximum exposure should comply with Fig. WF2.

FIG. WF2 **MINERAL FIBER ROOF SHINGLES**

Type	Width	Height	Thickness	Exposure	Weight per square
Individual Unit	8″	16″	5/32″	7″	350 lbs.
			1/4″	7″	585 lbs.
	9″	18″	1/4″	8″	570 lbs.
Multiple Unit	30″	14″	5/32″	6″	440 lbs.
Dutch Lap	16″	16″	5/32″	12″ x 13″	265 lbs.
Ranch Style	24″	12″	5/32″	20″ x 9″	255 lbs.

Nails and Fasteners

Nails should be corrosion-resistant, have needle or diamond points, large flat heads at least 3/8″ in diameter, with smooth or threaded shanks. Nails should have a minimum length of 1-1/4″ and the shank diameter should not exceed 0.140″, so as not to exceed the diameter of prepunched holes in shingles. Threaded nails should be used for plywood sheathing, and are preferable for lumber board sheathing.

Storm anchors used with Dutch Lap or Ranch design should be copper, aluminum or lead, approximately 3/4″ long with flat base. For spacing and pattern of nails and storm anchors, see Main Text, Application of Shingles, page 411-11.

PREPARATION FOR SHINGLES

Slope Limitations

Asbestos cement shingles are applied using normal application methods on roofs with slopes 5 in 12 or steeper. They may be applied on slopes as low as 3 in 12, if: (1) underlayment is doubled, and (2) eaves flashing is cemented.

Sheathing

Sheathing should be applied to form a solid surface.

Drip Edge

A continuous drip edge is recommended at the eaves and rakes.

Underlayment

Underlayment should be No. 15 asphalt saturated asbestos (inorganic) felt *OR* No. 30 asphalt saturated felt:

Normal Slope (5 in 12 or steeper)	Low Slope (3 in 12 to 5 in 12)
Single layer over entire roof.	Double layer over entire roof. May be single layer on 4 in 12 slope in areas where outside design temperature is warmer than 0° F.

(Continued) ASBESTOS CEMENT SHINGLES

Eaves Flashing*

Normal Slope (5 in 12 and steeper)	Low Slope [1] (3 in 12 to 5 in 12)
Additional course of underlayment at eaves cemented over first underlayment layer. Eaves flashing extends up the roof to cover a point at least 12″ inside interior wall line[2].	Additional course of underlayment at eaves cemented over first underlayment layer. Eaves flashing extends up the roof to cover a point at least 24″ inside interior wall line[2].

1. Or areas subject to severe icing.

2. See Fig. 10a, b & c, Main Text, page 411-7.

Flashing

See Main Text, page 411-10.

APPLICATION OF SHINGLES

Recommended methods for applying asbestos cement shingles for new and existing construction are described and illustrated in Main Text, page 411-11.

** Eaves flashing is recommended in areas having outdoor design temperature of 0°F or colder (See Fig. 9, Main Text, page 411-7) OR where there is possibility of ice forming at the eaves.*

ASPHALT SHINGLES

Unless specific exceptions are made in the following text, the General Recommendations apply.

MATERIALS

Shingles should conform to no less than Underwriter's Laboratories, Inc. (UL) Standard UL 55B Class C. Each bundle of shingles should be identified with a UL Label. (Fig. WF3a).

In areas subject to winds in excess of 75 mph, the UL Label should carry the words "Wind Resistant" (Fig. WF3b), or each free shingle tab should be cemented using quick setting asphalt cement.

Shingles used for new construction should provide double coverage and weigh at least 235 pounds per square.

Sizes, maximum exposure and minimum lap should comply with Fig. WF4 & Fig. AR1, Main Text, pg. 411-15.

Nails and Fasteners

Nails should be corrosion-resistant, have sharp points, large flat heads 3/8″ to 7/16″ in diameter, with smooth or threaded shanks. Nails should have a minimum length of 1-1/4″ for new construction, longer for reroofing. Shanks should be 0.105″ to 0.135″ in diameter.

FIG. WF3 **UL LABELS FOR SHINGLES**

Threaded nails should be used for plywood sheathing, and are preferable for lumber board sheathing.

Square Butt Strip shingles should be nailed with at least 4 nails per strip; Individual shingles should be nailed with at least 2 nails per shingle. For spacing and pattern of nails, see Main Text Application of Shingles, page 411-19.

PREPARATION FOR SHINGLES

Slope Limitations

All types of asphalt shingles may be applied using normal application methods on roofs with slopes 4 in 12 or steeper. Square Butt Strip shingles may be applied on slopes as low as 2 in 12, if: (1) underlayment is doubled, (2) shingles are "Wind Resistant" (or each free shingle tab is cemented), and (3) eaves flashing is cemented.

(Continued) ASPHALT SHINGLES

Sheathing

Sheathing should be applied to form a solid surface. If lumber boards are used for sheathing they should not exceed 6″ in nominal width. Where large (over 1″ diameter) loose knots, knot holes or excessively resinous areas occur in applied sheathing, they should be covered with sheet metal patches prior to roofing.

Drip Edge

A continuous drip edge is recommended at the eaves and rakes.

Underlayment

Underlayment should be No. 15 asphalt saturated felt:

Normal Slope (4 in 12 or steeper)	Low Slope (2 in 12 to 4 in 12)
Single layer over entire roof.	Double layer over entire roof.

Flashing

See Main Text, page 411-16.

Eaves Flashing*

Normal Slope (4 in 12 or steeper)	Low Slope [1,2] (2 in 12 to 4 in 12)
Single course of 90 lb. mineral surfaced roll roofing, OR 55 lb. smooth roll roofing at eave applied directly over underlayment. Eaves flashing extends up the roof to cover a point at least 12″ inside the interior wall line[3].	Double layer of No. 15 underlayment cemented at eaves. Eaves flashing extends up the roof to cover a point at least 24″ inside the interior wall line[3].

1. Or areas subject to severe icing.
2. Low Slope is applicable for Square Butt Strip shingles of the "Wind Resistant" type or (when each free shingle tab is cemented).

Other types of asphalt shingles may be recommended by the manufacturer for use on slopes less than 4/12.
3. See Fig. 10a, b & C, Main Text, page 411-7.

APPLICATION OF SHINGLES

Recommended methods for applying asphalt shingles for new and existing construction are described and illustrated in the Main Text, page 411-19.

** Eaves flashing is recommended in areas having outdoor design temperature of 0°F or colder (See Fig. 9, Main Text, page 411-7) OR where there is possibility of ice forming at the eaves.*

FIG. WF4 **ASPHALT ROOF SHINGLES***

	SHINGLE TYPE*	SHIPPING WEIGHT PER SQUARE	PACKAGES PER SQUARE	LENGTH	WIDTH	UNITS PER SQUARE	SIDE-LAP	TOP-LAP	HEAD-LAP	EXPOSURE
STRIP SHINGLES	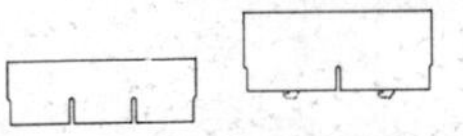 2 & 3 TAB SQUARE BUTT	235 Lb	3	36″	12″	80		7″	2″	5″
		205 Lb	3	36″	12″	80		7″	2″	5″
		220 Lb	3	36″	12″	80		7″	2″	5″
		225 Lb	3	36″	12″	80		7″	2″	5″
	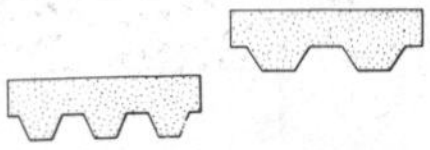 2 & 3 TAB HEXAGONAL	195 Lb	3	36″	11⅓″	86		2″	2″	5″
INDIVIDUAL	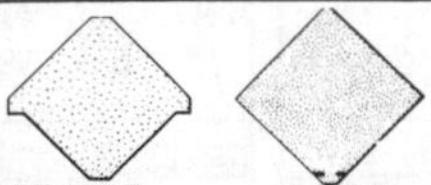 STAPLE LOCK	145 Lb	2	16″	16″	80	2½″			

*For Giant Individual type shingles, see Figure AR1, Main Text, page 411-15.

WOOD SHAKES

Unless specific exceptions are made in the following text, the General Recommendations apply.

MATERIALS

Each bundle of shakes should carry the official grade-marked label of the Red Cedar Shingle & Handsplit Shake Bureau.

Sizes, maximum exposure and grades should comply with Figs. WF5 & WF6.

Nails and Fasteners

Nails should be corrosion-resistant. Common wire nails or electro-galvanized nails should not be used. Nail lengths are given in Main Text, page 411-27. Only two nails should be used to secure each shake regardless of its width. For spacing and pattern of nails, see Main Text, Application of Shakes, page 411-30.

PREPARATION FOR SHAKES

Slope Limitations

Shakes are recommended on slopes 4 in 12 or steeper. They may be applied on slopes as low as 3 in 12 if: (1) solid sheathing is used, (2) underlayment is applied over entire roof, and (3) exposure is reduced to provide triple coverage. See Fig. WF6 for maximum exposure to provide double and triple coverage.

Sheathing

Shakes may be applied over either spaced or solid sheathing.

FIG. WF6
MAXIMUM EXPOSURE FOR WOOD SHAKES

TYPE	LENGTH	DOUBLE COVERAGE	TRIPLE COVERAGE
Hand-Split and Resawn Shakes	18″	7½″	5½″
	24″	10″	7½″
	32″	13″	10″
Tapersplit Shakes	24″	10″	7½″
Straight-Split Shakes	18″		5½″
	24″		7½″

In areas not subject to wind-driven snow, sheathing such as nominal 1x4 or 1x6 boards may be spaced equal to the exposure of the shake, but not more than 10″.

In areas subject to wind-driven snow (generally where the outside design temperature is 0°F or colder), solid sheathing of lumber boards or plywood should be used.

Drip Edge

Shakes should extend 1″ to 1-1/2″ beyond eaves and rakes. Drip edge is unnecessary.

FIG. WF5 **RED CEDAR SHAKES**

Grade	Length and Thickness	Bundles Per Square	Weight (lbs. per square)		Description	Label
No. 1 HANDSPLIT & RESAWN	18″ x 1/2″ to 3/4″ 18″ x 3/4″ to 1-1/4″ 24″ x 1/2″ to 3/4″ 24″ x 3/4″ to 1-1/4″ 32″ x 3/4″ to 1-1/4″	4 5 4 5 6	220 250 280 350 450		These shakes have split faces and sawn backs. Cedar blanks or boards are split from logs and then run diagonally through a bandsaw to produce two tapered shakes from each.	
No. 1 TAPERSPLIT	24″ x 1/2″ to 5/8″	4	260		Produced largely by hand, using a sharp-bladed steel froe and a wooden mallet. The natural shingle-like taper is achieved by reversing the block, end-for-end, with each split.	
No. 1 STRAIGHT-SPLIT (BARN)	18″ x 3/8″ 24″ x 3/8″	5 5	200 260		Produced in the same manner as tapersplit shakes except that by splitting from the same end of the block, the shapes acquire the same thickness throughout.	

(Continued) WOOD SHAKES

Interlayment (and Underlayment)

Interlayment and underlayment each should be No. 30 asphalt saturated felt.

Normal Slope (4 in 12 or steeper)	Low Slope (3 in 12 to 4 in 12)
Spaced or solid sheathing: Underlayment starter course at eaves and interlayment between each shake course over entire roof.	**Solid Sheathing:** Single layer of underlayment over entire roof and interlayment between each shake course over entire roof. Shakes are not recommended on slopes less than 4 in 12 with spaced sheathing.

Flashing

See Main Text, page 411-29. Copper flashing is not recommended for use with red cedar shakes unless special precautions are taken as shown in Fig. SK4, Main Text, page 411-29.

Eaves Flashing*

Normal Slope (4 in 12 or steeper)	Low Slope [1] (3 in 12 to 4 in 12)
Additional course of underlayment at eaves applied over underlayment starter course. Eaves flashing extends up the roof to cover a point 24″ inside the interior wall line[2].	Additional course of underlayment at eaves cemented over first underlayment. Eaves flashing extends up the roof to cover a point 24″ inside the interior wall line[2].

1. Or areas subject to severe icing.

2. See Fig. 10a, b & C, Main Text, page 411-7.

APPLICATION OF SHAKES

Recommended methods for applying wood shakes for new and existing construction are described and illustrated in the Main Text on page 411-30.

**Eaves flashing is recommended in areas having outdoor design temperature of 0°F or colder (See Fig. 9, Main Text, page 411-7) OR where there is possibility of ice forming at the eaves.*

WOOD SHINGLES

Unless specific exceptions are made in the following text, the General Recommendations apply.

MATERIALS

Each bundle of shingles should carry the official grade-marked label of Red Cedar Shingle & Handsplit Shake Bureau.

Sizes, maximum exposures and grades should comply with Figs. WF7 and WF8.

FIG. WF7 **RED CEDAR SHINGLES**

Grade	Size	Bundles Per Square	Weight Per Square		Description	Labels
No. 1 BLUE LABEL	24″ (Royals) 18″ (Perfections) 16″ (XXXXX)	4 bdls. 4 bdls. 4 bdls.	192 lbs. 158 lbs. 144 lbs.		The premium grade of shingles for roofs and sidewalls. These shingles are 100% heartwood, 100% clear and 100% edge-grain.	
No. 2 RED LABEL	24″ (Royals) 18″ (Perfections) 16″ (XXXXX)	4 bdls. 4 bdls. 4 bdls.	192 lbs. 158 lbs. 144 lbs.		A good grade for all applications. Not less than 10″ clear on 16″ shingles, 11″ clear on 18″ shingles and 16″ clear on 24″ shingles. Flat grain and limited sapwood are permitted.	
No. 3 BLACK LABEL	24″ (Royals) 18″ (Perfections) 16″ (XXXXX)	4 bdls. 4 bdls. 4 bdls.	192 lbs. 158 lbs. 144 lbs.		A utility grade for economy applications and secondary buildings. Guaranteed 6″ clear on 16″ and 18″ shingles, 10″ clear on 24″ shingles.	

(Continued) **WOOD SHINGLES**

FIG. WF8
MAXIMUM EXPOSURE FOR WOOD SHINGLES

Roof Slope	Shingle Lengths 16″	18″	24″
5 in 12 or steeper	5″	5-1/2″	7-1/2″
4 in 12	4-1/2″	5″	6-3/4″
3 in 12	3-3/4″	4-1/4″	5-3/4″

Nails and Fasteners

Nails should be corrosion-resistant. Common wire nails or electro-galvanized nails should not be used. Nail lengths are given in Fig. WS2, Main Text, page 411-33. Only two nails should be used to secure each shingle regardless of its width. For spacing and pattern of nails see Main Text Application of Shingles, page 411-35.

PREPARATION FOR SHINGLES

Slope Limitations

Shingles are recommended on slopes 5 in 12 or steeper using standard exposures of 5″, 5-1/2″ and 7-1/2″ for shingle lengths of 16″, 18″, and 24″ respectively. Shingles may be applied on slopes as low as 3 in 12 if exposure is reduced as shown in Fig. WF8.

Sheathing

Wood shingles may be applied over either spaced or solid sheathing. The space between spaced boards should not exceed the width of the boards.

Drip Edge

Shingles should extend 1″ to 1-1/2″ beyond eaves and rakes. Drip edge is unnecessary.

Underlayment

Underlayment is not required, but may be desirable for the protection of the sheathing.

Eaves Flashing*

Sheathing should be applied solidly above the eave line to cover a point at least 24″ inside the interior wall line of the building:

Normal Slope (4 in 12 or steeper)	Low Slope [1] (3 in 12 to 4 in 12)
Double layer of No. 15 asphalt saturated felt at eaves, over the section of solid sheathing, applied to cover a point at least 24″ inside the interior wall line[2].	Double layer of No. 15 asphalt saturated felt cemented over the section of solid sheathing, applied to cover a point at least 24″ inside the interior wall line[2].

1. Or areas subject to severe icing.

2. See Fig. 10a, b & c, Main Text, page 411.7.

Flashing

See Main Text, page 411-29—Copper flashing is not recommended for use with red cedar shingles unless special precautions are taken as shown in Fig. SK4, Main Text, page 411-29.

APPLICATION OF SHINGLES

Recommended methods for applying wood shingles for new and existing construction are described and illustrated in the Main Text on page 411-35,

** Eaves flashing is recommended in areas having outdoor design temperature of 0°F or colder (See Fig. 9, Main Text, page 411-7) OR where there is possibility of ice forming at the eaves.*

421 SHINGLE SIDING

SHINGLE SIDING

INTRODUCTION

Individual shingle units are applied to exterior walls in an overlapping fashion to shed water and resist weather penetration. On interior walls they are used for esthetic purposes. In either case, shingles may contribute color and/or texture to a wall surface and when properly applied and maintained, they provide a suitable exterior wall finish offering long service.

Depending on the type of shingle and method of application selected, esthetic effects range from wall surfaces having relatively smooth, uniform coursing to those with random coursing which give a heavily textured appearance. Color can be achieved by a variety of factory or field applied coatings, or by leaving the material in its natural color.

This section describes and illustrates the materials and methods recommended to provide a suitable wall siding finish using shingles. The text is divided into General Recommendations typical of any shingle siding application, followed by subsections giving specific recommendations for asbestos cement (mineral fiber) shingles, wood shakes and wood shingles. Other wall sidings are described in other sections of Division 400.

Wall shingles shed water, resist weather penetration, offer a variety of textural and color finishes, and if maintained give long service. Accordingly, shingles are a suitable siding material for exterior wall surfaces.

In addition to shingle units, several accessory materials usually are required to prepare the wall surface for the application of shingles. Depending on the type of shingle and method of application, accessory materials may include: *wall sheathing, sheathing paper, backer board, undercoursing, channel* and *corner moldings, flashing, caulking* and *siding nails.* While most materials and typical methods of applying shingle siding are discussed in the following text, the instructions and recommendations of the shingle manufacturer always should be consulted to assure the most suitable performance.

In this section, the exterior wall is assumed to be insulated correctly and adequately where climatic conditions require it. It also is assumed that precautions have been taken to prevent condensation of water vapor within the wall. The hazard of condensation within walls normally can be eliminated by providing a suitable vapor barrier on the warm side of the insulation, as close to the interior wall surface as possible (see Division 100, Moisture Control).

NOMENCLATURE

Shingle siding terminology is described and illustrated below to assist in understanding the text and illustrations.

Shingles: Relatively small individual siding units which overlap each other to provide weather protection. They typically are applied to a nailing base, such as sheathing or horizontal nailing strips, which supports the shingles between structural framing members.

Single Coursing: A method of applying shingles without the use of an undercourse (Fig. 1). In the case of wood shingles, this method results in concealed nails and a smaller exposure than is found in double-coursed walls.

Double Coursing: A method of applying shingles over an undercourse of lower grade shingles or other suitable material (Fig. 2). Nails are exposed as a result of face (butt) nailing and greater shingle exposures are possible than with single coursing. Double coursing usually results in greater wall coverage at lower costs.

Backer Board: An undercoursing material, typically fiberboard, used beneath siding materials such as mineral fiber shingles (Fig. 2b). It adds insulation value, increases resistance to impact, and provides a heavier shadowline at the butt than shingle siding applied directly without the use of a backing material.

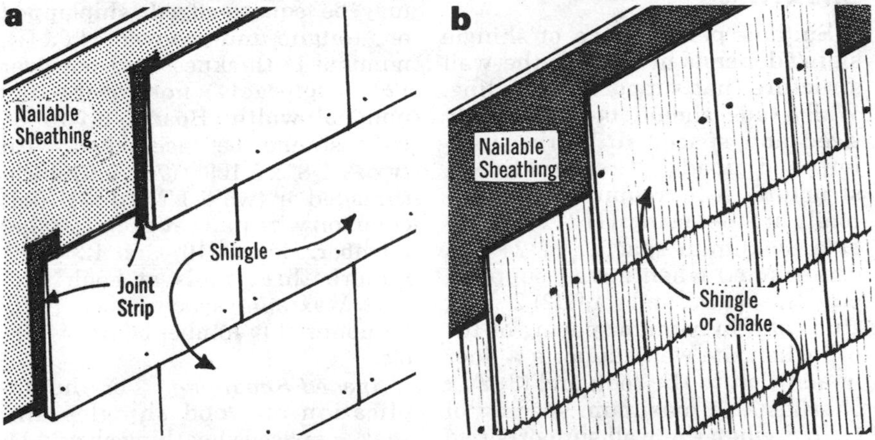

FIG. 1 Single-coursed walls have shingles applied without the use of an undercourse: (a) mineral fiber siding, (b) wood shakes or shingles.

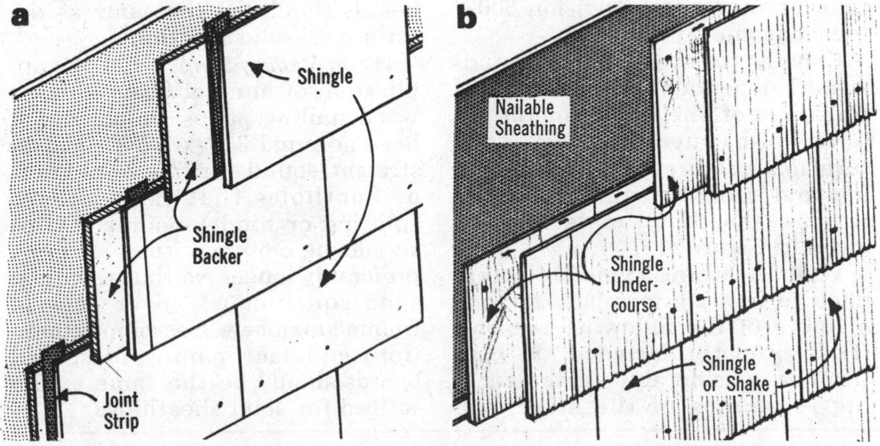

FIG. 2 Double-coursed walls have shingles applied over an undercourse: (a) mineral fiber siding, (b) wood shakes or shingles.

Exposure: The shortest distance in inches between exposed edges of overlapping shingles (Fig. 3).

Toplap: The shortest distance in inches from the lower edge of an overlapping shingle to the upper edge of the lapped unit in the first course below; that is, the width of the shingle minus the exposure (Fig. 3).

Headlap: The shortest distance in inches from the lower edges of overlapping shingles, to the upper edge of the unit in the second course below (Fig. 3).

Shingle Butt: The lower, exposed edge of the shingle.

Sheathing: See below.

Sheathing Paper: See page 421-5.

Flashing: See page 421-5.

WALL FRAMING AND SHEATHING

Suitable performance of shingle siding depends in part on the wall structure that supports the siding. Wall studs should be adequately sized and spaced to carry superimposed loads. Conservatively, structural load requirements normally are met using: (1) 2″ x 4″ studs at a maximum spacing of 24″ on center (o.c.) when a wall supports only the ceiling and roof; (2) 2″ x 4″ studs at a maximum spacing of 16″ o.c. when the wall supports a floor in addition to the roof; and (3) 2″ x 6″ studs at a maximum spacing of 16″ o.c. when the wall supports two floors in addition to the roof. Smaller members and/or wider spacings may be possible under certain conditions (see Division 300—Methods and Systems).

Usually, the spacing of wall studs is more dependent on the type and thickness of sheathing and/or interior finish materials than on structural load requirements. Stud spacing and nailing requirements for various sheathing materials are given in Figure 4.

Wall sheathing typically is applied to framing members for one or more of the following reasons: (1) if properly fastened, it contributes considerably to the racking resistance of the structural frame*; (2) it contributes insulation which reduces heat transmission and resists infiltration of wind and moisture; and (3) it functions in many cases as a nailing base for exterior finish materials.

When wall sheathing is used as a nailing base, it should be smooth and securely attached to its supports. Shingles may be applied over nailable solid sheathing consisting of wood boards, plywood or nailable fiberboard. They may be applied also to wood nailing strips over almost any type of sheathing. Wood shakes and shingles can be applied also over spaced sheathing.

Lumber Sheathing

Solid Sheathing Boards for solid sheathing should be seasoned to the moisture content shown in Figure W8, page Wood 201-6. Boards may be square-edged, shiplapped or tongue-and-grooved (T&G), nominal 1″ thickness and not over 12″ (preferably not over 10″) nominal width. Boards up to 8″ wide should be face-nailed with two 2-1/8″ x 0.109″ (7d) annularly threaded or two 2-1/2″ x 0.131″ (8d) common wire nails at each framing member. Boards 10″ and 12″ wide require three nails at each support. Maximum spacing of supports for nominal 1″ lumber boards is 24″ o.c.

Spaced Sheathing For the application of wood shingles and shakes, spaced sheathing should be at least nominal 1″ x 3″, 1″ x 4″ or 1″ x 6″ boards. Seasoning conditions and face-nailing of spaced boards should be the same as described for solid sheathing.

Wood Nailing Strips For the application of mineral fiber shingles, wood nailing strips should be at least nominal 3/8″ x 3-5/8″ boards, straight, sound and free from knots or knotholes that might cause splitting or hinder nailing. Strips should be not less than 4′ long, preferably longer, so that they can span continuously over several framing members. Seasoning conditions and face nailing of spaced boards should be the same as described for solid sheathing.

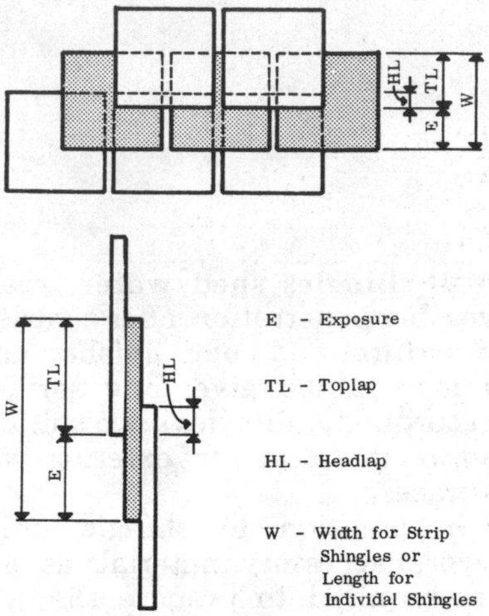

FIG. 3 Shingle Siding Terminology.

Plywood Sheathing

Plywood sheathing should be of the proper Interior or Exterior Construction Grade (see Work File, page Wood WF201-16), and should conform to the thicknesses and stud spacing given in Figure 4.

Plywood 1/2″ thick or less should be applied with 1-3/4″ x 0.120 (5d) annularly threaded nails or 2″ x 0.113″ (6d) common wire nails. (1-1/2″ x 0.135″ helically threaded nails also may be used.) Nails should be spaced 6″ o.c. around panel edges and 12″ o.c. at intermediate points of bearing. Staples can be used if of 16-gauge galvanized wire with 3/8″ crown width, and long enough to provide 1″ penetration into the stud. They should be spaced 4″ o.c. around panel edges and 8″ o.c. at intermediate points of bearing.

Nail-Base Fiberboard Sheathing

Nail-base fiberboard sheathing should be homogeneous high-density structural insulating board conforming to Insulation Board Institute (IBI) Specification No. 2 and ASTM D2277. Sheathing should be applied with its long dimension parallel to the framing members (vertically) with sufficient bearing for satisfactory nailing along all edges. Approximately 1/8″ should be left between adjoining boards and at the ends of boards.

**Tests by the Division of Building Research-National Research Council of Canada indicate that interior wall finishes such as gypsum board or plaster provide about 4 times as much resistance to racking as does lumber board sheathing applied horizontally with corner bracing. Hence, the National Building Code of Canada does not require sheathing except when exterior finish materials require it for a nailing base.*

FIG. 4 **WALL SHEATHING REQUIREMENTS**

Sheathing	Thickness	Maximum Stud Spacing
Wood Boards[1]	3/4″	24″
	11/16″	16″
Fiberboard	1/2″	24″[2]
	25/32″	16″ or 24″[2]
Nail-base Fiberboard[3]	1/2″	16″ or 24″
Gypsum Board	1/2″	24″[2]
Plywood[4]	5/16″	16″
	3/8″	16″ or 24″
	1/2″	24″

1. Wall bracing not required if boards are applied diagonally.
2. Wall bracing required.
3. Wall bracing not required when 4″ x 8″ sheets are installed vertically and nailed 3″ o.c. around perimeter and 6″ o.c. at intermediate supports.
4. Wall bracing not required.

The sheathing should be applied with large-headed roofing nails, staples or other approved fasteners. The nails should be spaced 6″ o.c. applied to intermediate framing members first. Fasteners then are applied 3″ o.c. around all four edges and held to a minimum of 3/8″ from the edge. Fasteners should be large-headed, 1-1/2″ x 0.120″ (4d), plain or annularly threaded roofing nails. The fastener heads should be driven flush with the surface of the sheathing. When applied as described above, supplemental diagonal corner bracing is not necessary for 1/2″ sheathing fastened to studs spaced 16″ or 24″ o.c.

Non-Wood Sheathing

When gypsum board, fiberboard or sheathing materials other than wood, plywood or nail-base fiberboard sheathing are recommended by the manufacturer for use as wall sheathing under shingles, the sheathing manufacturer should specify the type of fastening required to attach sheathing to supports. In the case of 1/2″ gypsum board, 1-1/2″ x 0.120″ plain or annularly threaded roofing nails should be spaced 8″ o.c. at intermediate supports and 4″ o.c. along panel edges.

Non-lumber sheathing material such as rigid fiber- and gypsum boards, rated by the manufacturer as "sheathing grade," are not adequate nailing bases. When this type of sheathing is used, siding should be applied by nailing *through* the sheathing and into framing members or furring strips. Unless the sheathing is nail-base fiberboard, which requires the use of special shingle nails as shown in Figure MF3, page 421-8, application methods such as the Wood Nailing Strip Method should be used as described under Application of Siding, page 421-11.

Wall Sheathing Paper

Sheathing paper is applied directly to studs to resist the infiltration of air and moisture when sheathing is not used and to provide moisture protection for lumber boards used for sheathing. Sheathing paper is not necessary over plywood, fiberboard or treated gypsum sheathing.

Sheathing paper should be a material such as No. 15 or No. 30 asphalt-saturated felt which has low vapor resistance. Materials such as coated felts or laminated waterproof papers which have high vapor resistance and act as vapor barriers should not be used. Such materials may permit moisture or frost to accumulate as a result of moisture condensation between the sheathing paper and the surface of the wall sheathing.

When sheathing paper is used, it should be applied over the entire wall as soon as the lumber board sheathing has been completed, with at least a 4″ toplap at all horizontal joints and a 4″ sidelap at end joints. It should be lapped 6″ from both sides around all corners. Only sufficient fasteners need be used to hold the sheathing paper securely in place until shingles are applied. Shingles should not be applied over wet sheathing paper.

FLASHING

Walls may contain window or door openings, intersect with sloped roof surfaces, or terminate in exposed locations against other materials, creating opportunities for leakage. Flashing must be installed at these locations and be made watertight and water-shedding to weatherseal the wall. Exposed flashing should be corrosion-resistant metal.

Flashing Materials

Metal flashing should be at least 26 gauge galvanized steel, 0.019″ aluminum, or 16 oz. copper. (See pages Masonry Walls 343-3 & 343-4 for a discussion of these and other flashing materials.)

Additional requirements and typical methods of flashing are described in the following subsections.

Flashing At Openings

Heads, sills and jambs of openings should be flashed. However, in areas not subject to wind driven rain, head flashing may be omitted when the vertical height between the top of finish trim of openings and the bottom of eaves soffits is equal to or less than 1/4 of the eave overhang.

Heads and Jambs of Openings, Wood Frame Walls Metal flashing should be installed extending from at least 2″ above trim and turned down over the outside edge of drip caps forming a drip. Where sheathing paper is omitted, flashing should extend up behind the sheathing.

For unsheathed walls, jambs should be flashed with a 6″ wide strip of metal, 3 oz. copper-coated paper, or 6 mil polyethylene film.

Flashing At Wall and Roof Intersections

Sheetmetal flashing should be installed at all horizontal intersections and at vertical intersections when the exterior finish material does not provide a self-flashing joint. (See also pages 411-10, 411-17, 411-29 for additional information on sidewall flashing.)

CAULKING

A non-shrinking caulking compound, either in white or matching color, should be used to weather-seal all joints where siding abuts wooden trim, masonry or other projection.

SIDING NAILS (and Fasteners)

No single step in applying shingle siding is more important than proper nailing. Suitability is dependent on several factors: (1) selecting the correct nail for the kind of shingle and type of wall sheathing; (2) using the correct number of nails; (3) locating them in the shingle correctly; and (4) choosing nail metal compatable with metal used for flashing.

Specific recommendations for type, size, number and spacing of nails are given in the following subsections for mineral fiber shingles and wood shingles and wood shakes. Some general recommendations are given below.

Lumber Boards and Plywood Sheathing

The nails should be long enough to penetrate through the shingle and through the lumber boards or plywood sheathing. Nails for applying shingles over plywood sheathing should have threaded shanks.

Non-Wood Sheathing

When gypsum board, fiberboard or sheathing materials other than lumber or plywood are used for wall sheathing, special fasteners and/or special details for fastening are necessary to provide adequate anchorage of the shingle siding. Recommendations of the manufacturer of the non-wood material should be followed to assure his responsibility for satisfactory performance of the wall.

In 1964, the Mineral Fiber Products Bureau (formerly called the Asbestos Cement Products Association), whose members manufacture asbestos cement building materials, changed the terminology of their products. Asbestos cement wall shingles and siding are now designated *mineral fiber* wall shingles and siding. "Mineral fiber" is the term used throughout the following text.

This subsection describes materials and methods recommended to provide a suitable wall finish of mineral fiber shingles over both new and existing construction.

Reference should be made to the entire preceding subsection, General Recommendations, pages 421-3 through 421-6 for: (1) definition of terminology, (2) wall sheathing recommendations, including sheathing thickness, spans and nailing schedules; and (3) typical accessory materials—sheathing paper, flashing, caulking and siding nails—recommended for preparation of the wall and for the application of shingles. Unless specific exceptions are made in the following text, the General Recommendations apply for each kind of shingle and method of application.

FIG. MF1 MINERAL FIBER SIDING

Sizes[1]	Exposure	Pieces per Bundle	Pieces per Square[2]
12" x 24"	11"	18	54
9" x 32"	8"	19	57
9-5/8" x 32"	8-5/8"	17-17-18	52
11-3/4" x 32"	10-3/4"	14	42
14-5/8" x 32"	13-5/8"	11	33
14-3/4" x 32"	13-3/4"	11	33
8-3/4" x 48"	7-3/4"	13	39
12" x 48"	11"	9	27

1. Three nailholes per shingle.
2. Weight per square ranges from 153 to 206 lbs.

MATERIALS

Mineral fiber siding is usually supplied in rectangular units having straight or wavy butt edges. Units range in size from 8" to 16" in width and from 24" to 48" in length and are designed to be applied with a minimum nominal toplap of 1". Depending on their size, the units range up to approximately 3/16" in thickness (Fig. MF1).

Siding units are manufactured with prepunched nailholes correctly sized and located to receive exposed face nails. The holes are aligned the proper distance above the lower edge of the units to establish the recommended amount of toplap (Fig. MF2). For standard application, face nails driven through these holes and resting on the top edge of the next lower course serve as guides for the correct amount of toplap, exposure and unit alignment.

Mineral fiber shingles can be discolored by moisture and dampness while still in packages and until applied. Prior to application on the job, they should be kept clean and dry by complete protection from weather. Siding should not be applied on exterior walls over wet sheathing or wet sheathing paper.

Mineral fiber siding materials should conform to the requirements of ASTM C223. Data covering types of siding are summarized in Figure MF1. See Section 219 Asbestos Cement Products for additional information covering manufacture and properties.

Nails

Face nails should be long enough to penetrate into and hold securely in the nailing base. Face nails normally are furnished with the siding by the manufacturer to insure the use of the proper kind and type. All nails and fasteners should be permanently corrosion-resistant and stain-resistant. Only those furnished with the siding or recommended by the siding manufacturer should be used.

A summary of face nail types recommended for use in applying mineral fiber siding is given in Figure MF3. The number, spacing and pattern of nails are described under Application of Siding.

PREPARATION FOR SIDING

Wall Sheathing

Siding should be applied to surfaces that are smooth and dry, and which will provide adequate support and an adequate nailing base. Mineral fiber siding can be applied over lumber, plywood, nail-base fiberboard, or other suitable types of rigid sheathing. Sheathing material thicknesses, spans and nailing requirements are discussed on page 421-4.

If wood boards are used for sheathing, they preferably should be not over 8" in width, free from loose knots or large knotholes, and have tongue-and-grooved or shiplapped edges. Warped boards that would prevent the siding from lying flat should not be used.

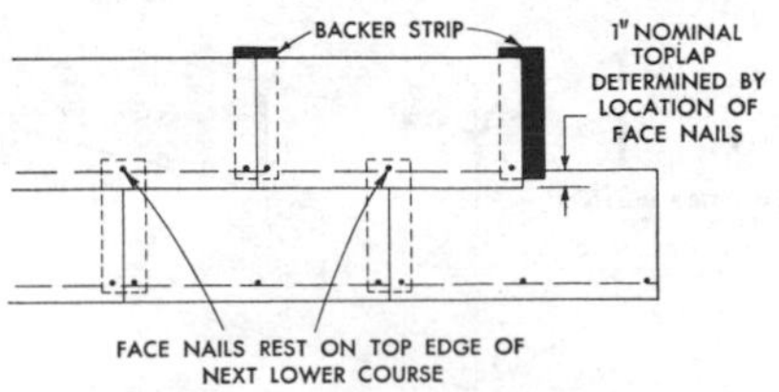

FIG. MF2 Prepunched nailholes in mineral fiber siding help insure proper toplap, alignment and minimize damage to siding during application.

NAIL TYPES FOR USE WITH MINERAL FIBER SIDING

Sheathing Type	NEW CONSTRUCTION Nailed Directly to Sheathing or Wood Nailing Strips	Nailed Through Fiberboard Shingle Undercoursing
Board Lumber	1-1/8″ Screw Thread 1-7/16″ Screw Thread 1-1/2″ Annular Thread	2″ Annular Thread
Plywood	1-1/8″ Screw Thread	1-3/4″ Screw Thread
Wood Nailing Strip	1-1/8″ Screw Thread	1-3/4″ Screw Thread
Nailbase Fiberboard	1-3/8″ Special Annular Thread	2″ Special Annular Thread
RE-SIDING (over existing wood siding or shingles)		
Nailing Base	**Nail Types**	
Sound Board Lumber	1-3/4″ or longer, annular thread, screw thread or helical thread	
Wood Nailing Strip	1-1/8″ Screw Thread	

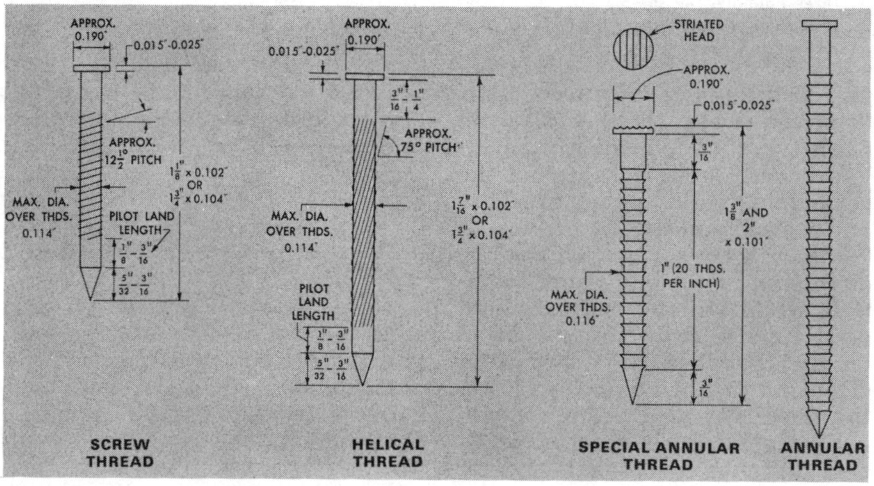

FIG. MF3 Face nails for mineral fiber siding normally are furnished by the siding manufacturer to insure the use of the proper nails.

Sheathing Paper

Sheathing paper for use over lumber sheathing should be water-repellent yet vapor porous (permeable to water vapor), such as No. 15 asphalt-saturated felt. Coal tar saturated felts which may stain the siding should not be used. Vapor barrier membranes should not be used for sheathing paper between siding and sheathing. Sheathing paper is discussed on page 421-5.

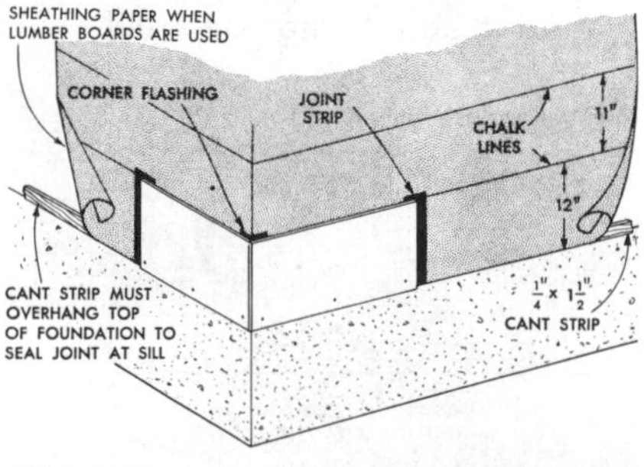

FIG. MF4 Flashing strips are applied behind each vertical joint between mineral fiber siding units. The Direct Application Method is illustrated.

Joint Flashing Strips

Asphalt (not coal tar) saturated and coated, water-repellent joint flashing strips are supplied by the manufacturer with all types of mineral fiber siding, and are applied behind each vertical joint between siding units (Fig. MF4).

Cant Strips

A 1/4″ x 1-1/2″ wood strip should be nailed level along the bottom edge of the sheathing so that it overhangs the top of the foundation wall a sufficient amount to seal the joint between the top of the foundation, the wood sill and the bottom of the sheathing (Fig. MF4). The bottom edge of the siding should align with the bottom edge of the cant strip. This strip gives the necessary cant (pitch) to the first course of shingles, and provides solid bearing to prevent breakage.

Flashing

See General Recommendations, page 421-5.

APPLICATION OF SIDING (New Construction)

Depending on the type of sheathing, mineral fiber shingles may be applied by several methods: (1) *Direct* Application to nailable sheathing; (2) *Shingle Backer* Method or (3) *Channel* Method of application over nailable sheathing or directly over framing members without sheathing; and (4) *Wood Nailing Strip* Method for use over any type of sheathing or directly over framing members without sheathing.

The methods described are based on applying 12″ x 24″ siding units from left to right on new construction. For applying other sizes of siding, the manufacturers' instructions should be followed.

Careful layout should establish the location of the first horizontal course and assure that all courses will meet and match at corners and are level *all around the building.* The total number of horizontal courses to complete one wall should be laid out, and any adjustments should be made so that the top courses at the eaves will not be too narrow. This may require uniform adjustments in reducing the exposure of each shingle course. A level chalkline should be marked all around the building to fix the location of the top edge of the first course of siding. Additional horizontal chalklines should mark the top of each succeeding course. The lines should be spaced to provide the desired exposure (Fig. MF4).

For illustrations and discussion of tools and methods used to cut, punch or notch mineral fiber wall shingles, see Section 219 Asbestos Cement Product.

Direct Application Method This method is used for sheathing types which provide a nailing base, such as wood boards, plywood or nail-base fiberboard sheathing.

First and Odd Numbered Courses A full size siding unit is started at the lefthand corner of the wall. This unit should be carefully applied plumb, level and in alignment with the chalkline because it guides the laying of all other units.

The correct face nails for the type of sheathing and application method being used are supplied by the manufacturer with the siding as described in Figure MF3. Nails should be driven snug but not too tight.

Before the last nail is driven at the righthand end of the unit, a felt joint flashing strip is inserted in place and is secured with the last nail. Joint flashing strips always should be used, centered behind the butted joint between siding units, and with the lower end overlapping the cant strip or head of the next lower course.

Application of full size siding units is continued in the first course with the top edges aligned to the chalkline. The last unit in a course or at an opening should not be less than 6″ wide, and if smaller spaces will remain to be filled, a few inches should be cut from the units applied earlier in the course. Necessary holes for face nails in less than full size pieces of siding should be punched. Ends of units should be butted tightly end to end with no space in between. Wedge-shaped spaces between ends of units indicate incorrect application.

Second and Succeeding Even Numbered Courses In succeeding courses of 12″ x 24″ siding, vertical joints should break on "halves"; thus, the second course is started with a half unit. When using 32″ long units, vertical joints also should break on "halves." When 48″ long units are used, they should break on "thirds" and the second course is started with a 2/3 unit. Some bundles of siding contain the necessary shorter pieces for starting courses.

A partial unit is started at the lefthand corner with its head edge aligned to the chalkline and its lower (butt) edge overlapping the head of the next lower course the correct distance to provide the necessary toplap between courses. A nail is inserted in a face nail hole and the siding positioned so that the shank of the nail is resting on the top edge of the next lower course, thus establishing the proper amount of toplap. The unit then is nailed in place with a felt joint flashing strip installed as described above. The course is continued with full size units.

The face nails described in Figure MF3 should always be used. The nails should penetrate completely through the sheathing. When the siding is applied directly to nominal 1″ wood board sheathing or 5/16″ and thicker plywood without wood nailing strips or shingle backer, the standard threaded face nail is used.

Shingle Backer Method This method uses rigid insulating fiberboard as a backer for the shingles and may be used over wood boards, plywood or high-density fiberboard sheathing, since they provide a nailing base for both the shingle backer and the siding (Figs. MF5, MF6 & MF7). Special methods may be used to apply siding directly over framing members.

The rigid, water-resistant, insulating fiberboard shingle backer should be at least 5/16″ thick and should underlay the full width and length of each course of siding completely. The shingle backer should be approximately 1/4″ narrower than the siding. The siding is applied with its head edge flush with the head edge of the shingle backer and its butt edge extended down and beyond the butt edge of the underlying shingle backer to

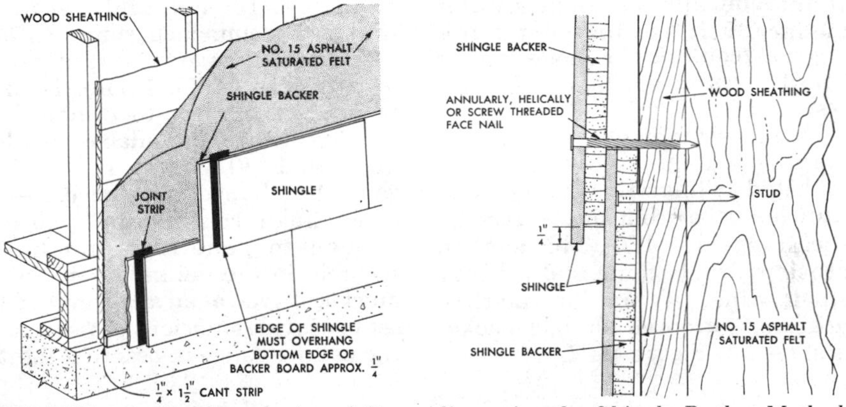

FIG. MF5 Application of mineral fiber siding using the Shingle Backer Method.

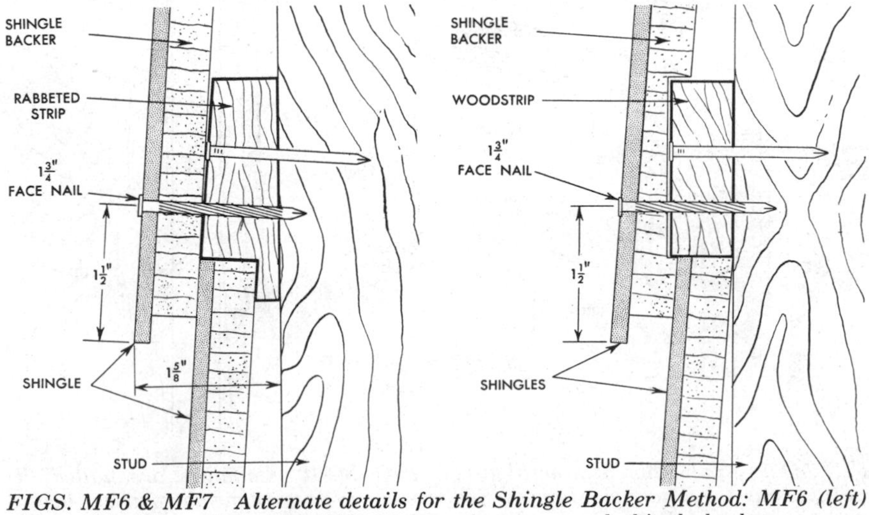

FIGS. MF6 & MF7 Alternate details for the Shingle Backer Method: MF6 (left) using rabbetted nailing strip; MF7 (right) using a grooved shingle backer.

provide a drip edge. Shingle backer should be at least 4′ long and applied with staggered joints in succeeding courses. The vertical joints of shingle units should not occur over the ends of the shingle backer. Shingle backers are applied with their butt edges overlapping the head of the lower course of siding approximately 1/4″ less than the headlap used between courses of siding.

The shingle backer and the siding should be secured to the wall with face nails driven through the face nail holes in the siding and into the underlying sheathing. Shingle backer is held in position before applying siding by nailing each panel about 3″ from the top edge with two or three 1-1/4″ galvanized roofing nails. The type and length of the face nails should be as described in Figure MF3.

Succeeding courses of siding should be applied in a similar manner. Felt flashing joint strips are not required if shingle backer board is water-repellent and stain-resistant.

Direct attachment to framing members is possible using a rabbeted nailing strip with a special shingle backer that also serves as sheathing (Fig. MF6). This method provides firm backing and a heavy shadowline. A similar method utilizing a grooved shingle backer is shown in Figure MF7.

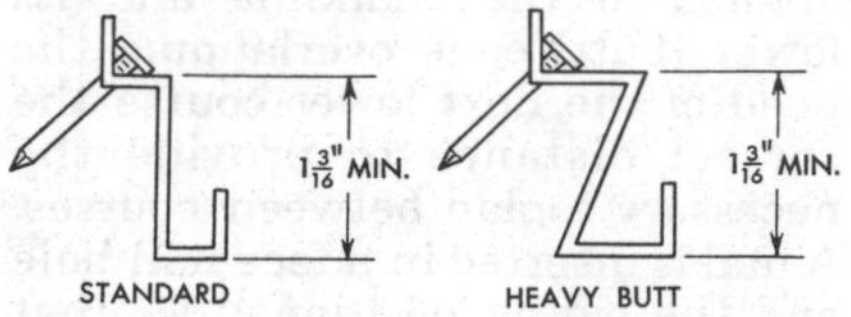

FIG. MF8 Typical profiles of metal channel moldings for installing mineral fiber siding by the Channel Method.

Channel Method This method uses one of several styles of metal channel molding designed for securing mineral fiber siding to nailable sheathing or directly to framing members. Metal channel should be corrosion-resistant and stain-resistant metal of sufficient thickness to provide adequate support and rigidity. Moldings usually are prepainted to match siding color and also provide either a standard or heavy butt shadow (Fig. MF8).

The 1-3/16″ channel permits standard prepunched siding to be used when units are reversed head to butt (Fig. MF9). Channels are secured at 16″ o.c. to either the framing member or nailable sheathing with 10 (0.135″) to 12 (0.106″) gauge hot-dipped galvanized steel, large head roofing nails, long enough to penetrate and hold securely in the nailing base. Nails must be driven at an angle and well set to provide sufficient pressure on the nailing flange to assure a tight, vibration-free assembly (Fig. MF10).

Channel molding should have a vertical nailing flange at least 3/8″ high that can bear against the face of the sheathing, and should be strong and rigid enough to secure the siding to the wall sheathing.

Direct attachment to framing members requires siding either 10″, 24″, 32″ or 48″ long, wide enough to span from center to center of studs. The face nail holes must be punched within 3/8″ to 1/2″ from the vertical edge of the siding to assure nailing into the stud when two siding units are butted together. Intermediate face nail holes must be provided for nailing each stud, and face nails should be long enough to penetrate at least 1″ into the stud.

When this method of nailing is used, studs must be located accurately on 16″ centers wherever possible. Warped studs should not be used.

Where it is not possible to nail directly into a stud, either a twist nail or a self-clinching nail (Fig. MF11) should be used to attach the corner of the siding to the non-lumber sheathing. The *twist nail* is a slender copper or aluminum nail with flat head and medium needle point, designed for twist clinching; that is, manual twisting of the exposed shank. The *self-clinching nail* is designed so that

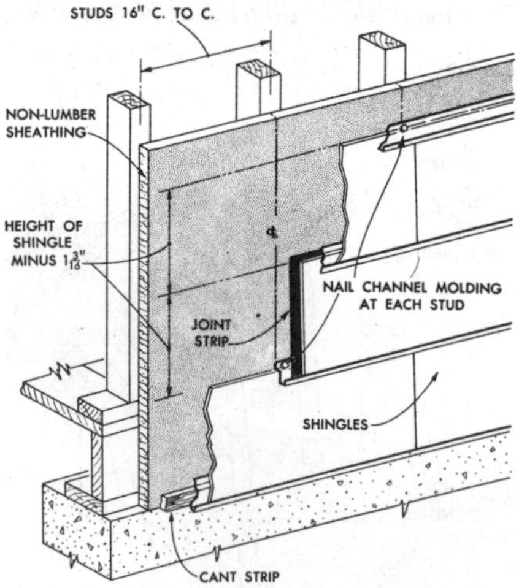

FIG. MF9 Application of mineral fiber siding using the Channel Method. See detail in Figure MF10.

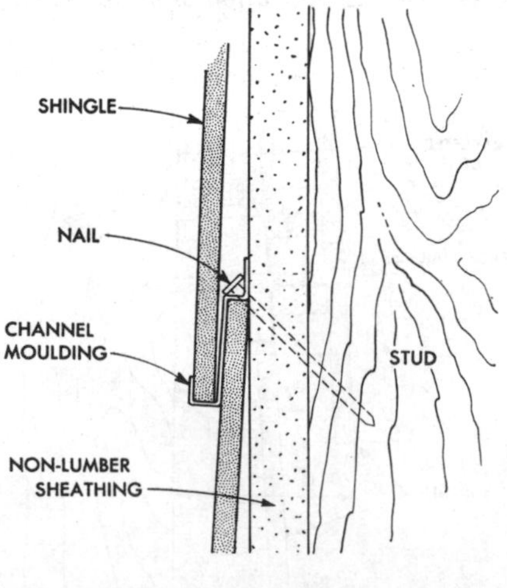

FIG. MF10 Channels are nailed directly to framing members when sheathing does not provide a nailing base.

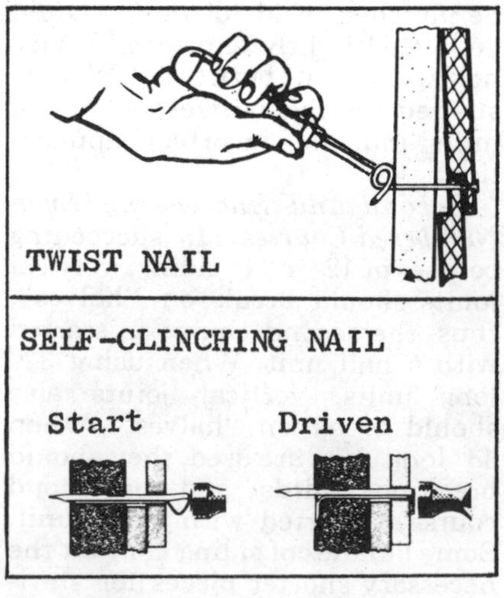

FIG. MF11 Special nail types are available for attaching siding to non-nailable sheathing.

its shank clinches automatically while the nail is fully driven.

The inside of exterior walls should be left exposed for final inspection to assure that all siding nails are properly driven into the studs and that twist nails or self-clinching nails are properly clinched.

Wood Nailing Strip Method The Wood Nailing Strip Method may be used over lumber boards, plywood or any type or thickness of non-lumber sheathing that does not have adequate nail holding power, because the strips rather than the non-lumber sheathing serve as the nail base. No special nails, fasteners or channels are required, and the standard face nails shipped with the siding are used.

In addition to providing maximum security for the siding and economy in application, this method produces a heavy shadowline along the butt edge of each course of siding.

Wood nailing strips 3/8″ x 3-5/8″ are placed to overlay the head of the lower course of siding in an amount equal to the desired toplap less 1/4″ (Fig. MF12). This firmly clamps the head of the course of siding to the wall. At each framing member, the wood strip is nailed through the sheathing and to each underlying stud with two 2″ x 0.113″ (6d) or longer annularly threaded nails, penetrating the stud at least 1″. The wood strip should end over framing members; joints in the strips should be staggered in succeeding courses. Vertical joints between shingle units should not occur over the end joints of wood strips.

The next course of siding then is applied with its butt edge placed over the wood strip with the lower edge of the siding extending down over the lower edge of the strip not more than 1/4″, to provide a drip. The siding then is nailed through the face nail holes to the wood nailing strips with the threaded face nails supplied with the siding.

Corner Finishing

Where wall surfaces join at *inside corners*, the courses of siding are generally butted, although metal molding or wood corner strips may be used. *Outside corners* can be finished by any one of the following three methods:

Metal Corner Finish Several styles of individual stain-resistant metal corners and finish moldings are available (Fig. MF13). They should be applied as recommended by the manufacturer.

Wood Corner Board Finish Corners finished with wood strips can be made of previously primed or painted nominal 1″ lumber applied along one corner. A 1″ x 4″ is applied along one corner so that its side edge projects a distance beyond the corner equal to the thickness of the board applied on the adjoining wall. A 1″ x 3″ then is applied with its edge butting against the back of the 1″ x 4″ first applied. Thus, the wood corner strips will appear the same width on each wall (Fig. MF14). When wood nailing strips or shingle backer, or both, are used, wood corner boards thicker than nominal 1″ lumber or blocking may be required.

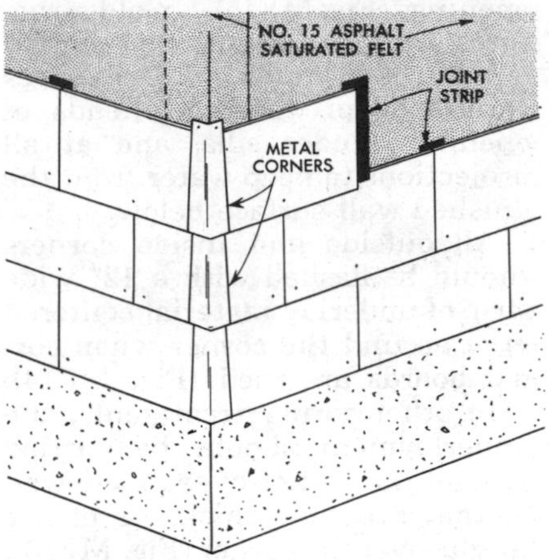

FIG. MF13 Mineral fiber siding may be weathersealed at corners using metal corners especially designed for this purpose.

Lapped or Woven Corner Finish *(for white or gray shingles only)* For lapped adjoining courses at outside corners, a whole shingle is placed temporarily against each adjoining wall, so that shingles meet and join at the corner. The adjoining vertical edges of the siding units are scribed and cut at the proper angle to form a lapped or woven corner joint, and edges are dressed with a file. Corner joints should be lapped to the left and right in alternate courses. Courses on adjoining walls should meet and match at corners.

Flashing at Openings and Corners

Metal flashing should be applied over all exposed door and window

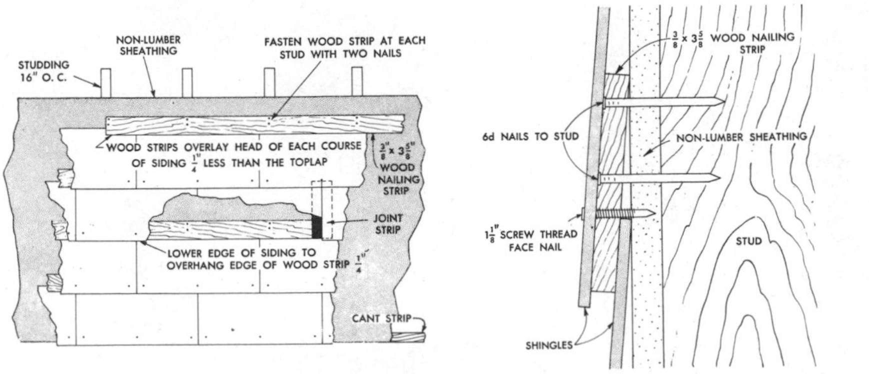

FIG. MF12 Application details of mineral fiber siding using the Wood Nailing Strip Method.

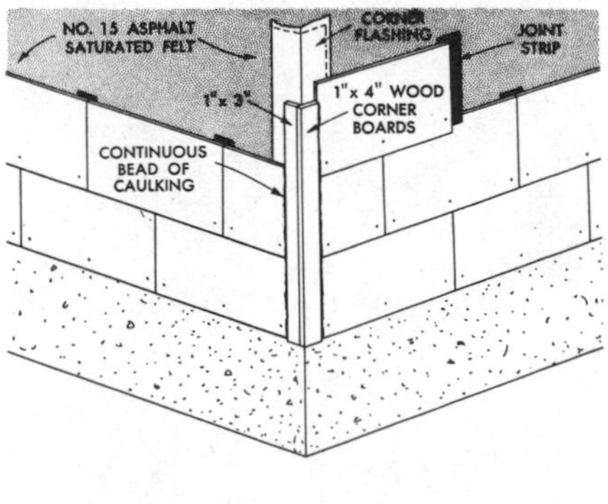

FIG. MF14 Alternate corner detail using wood corner boards.

openings (Fig. MF15). On old structures, the flashing must be renewed if not in good condition. Drip edges should be provided at heads of openings, under sills, and at all projections to keep water from the finished wall surface below.

All outside and inside corners should be flashed with a 12″ wide strip of underlay material centered in or around the corner when corner boards are used (Fig. MF14).

In other corner treatments, the underlayment should be carried around the corner of each sidewall so that a double thickness of felt results over the corner (Fig. MF13).

RE-SIDING EXISTING CONSTRUCTION

The old wall to be covered should be repaired to provide a smooth, substantially true surface having adequate nailholding capacity and the ability to support the new siding. When the existing exterior finish consists of relatively thin bevel siding applied directly to widely spaced framing members

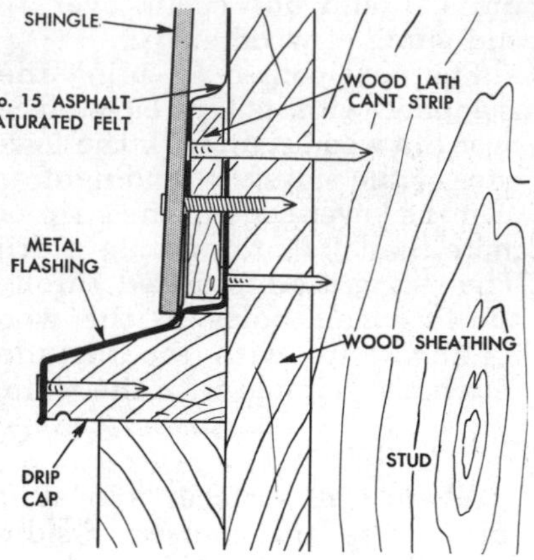

FIG. MF15 Detail for head flashing over openings.

and does not provide an adequate nailing base for the new siding, the walls should be stripped at nailing locations with beveled wood strips. The face nails described in Figure MF3 should be used for application of new siding.

Re-siding Over Non-Lumber Sheathing When mineral fiber siding is to be applied to a structure that originally was sheathed with one of the various kinds of non-nailable sheathing, and the existing siding to be covered will not provide an adequate nailing base, the only recommended attachment method is the Wood Nailing Strip Method (see page 421-11). Otherwise, all existing siding and sheathing should be removed down to the existing framing members and replaced with new sheathing as well as new siding.

Re-Siding Over Stucco When new siding is to be applied over old stucco, the best practice is to remove the old stucco, nails and lath, and any protruding element, and apply the new siding directly on the existing wood lumber sheathing. However, if the stucco is in reasonably good condition, or is removed to bare studs, siding can be applied by the Wood Nail Strip or Metal Channel Methods described above under Application of Siding, modifying the steps as necessary to meet any special conditions.

WOOD SHAKES AND SHINGLES

The following text describes materials and methods recommended to provide a suitable finish with wood shakes or wood shingles over both new and existing construction.

Reference should be made to the entire first subsection, General Recommendations, pages 421-3 through 421-6, for: (1) definition of terminology; (2) wall sheathing recommendations, including sheathing thickness, spans and nailing schedules; and (3) typical accessory materials—sheathing paper, flashing, caulking and siding nails—recommended for preparation of the wall and for the application of shingles. Unless specific exceptions are made in the following text, the General Recommendations apply for each kind of shake and shingle and method of application.

MATERIALS

Shakes

Wood shakes are manufactured in one grade and three types: (1) *handsplit-and-resawn*, (2) *taper-split* and (3) *straight-split*. Shakes are produced in three lengths—18″, 24″ and 32″. Each should be 100% heartwood and both taper-split and straight-split shakes have 100% edge grain. Handsplit-and-resawn shakes have split faces and sawn backs and may include up to 10% flat grain.

Data covering types of shakes are summarized in Figure SS1. The label shown in Figure SS1 is assurance that the shakes meet the quality standards of the Red Cedar Shingle & Handsplit Shake Bureau. See Section 202 Wood Products for additional information covering manufacture and properties.

Shingles

Wood shingles are manufactured in 16″, 18″ and 24″ lengths conforming to four grades: No. 1 (Blue Label), No. 2 (Red Label), No. 3 (Black Label), and Undercoursing Grade. In addition, shingles are available in Rebutted-and-Rejointed, Dimension and Machine-Grooved types.

Data covering wood shingles are summarized in Figure SS2. The labels shown in Figure SS2 are assurance that the shingles meet the quality standards of the Red Cedar Shingle & Handsplit Shake Bureau. See Section 202 Wood Products for additional information covering manufacture and properties.

Nails

Nails for applying shakes and shingles should be corrosion-resistant, such as hot-dipped galvanized steel, or bright or colored aluminum-alloy. Common wire nails or even electro-galvanized nails may shorten the life of the shake or shingle wall and should not be used. Nails may have smooth or threaded shanks and medium or blunt diamond points. Threaded nails provide increased holding power, although smooth shank nails are used in the majority of applications over wood sheathing and give suitable service. Nails for applying wood shingles to plywood sheathing should be threaded. If threaded, galvanized steel nails should have annular threads; aluminum nails should have screw threads with approximately 12-1/2° thread angle. The length of nails should be such that they penetrate completely through the sheathing.

FIG. SS1 **TYPES OF RED CEDAR SHAKES**

Grade	Length and Thickness	Bundles Per Square*	Weight (lbs. per square)		Description	Label
No. 1 HANDSPLIT & RESAWN	18″ x 1/2″ to 3/4″ 18″ x 3/4″ to 1-1/4″ 24″ x 3/8″ 24″ x 1/2″ to 3/4″ 24″ x 3/4″ to 1-1/4″ 32″ x 3/4″ to 1-1/4″	4 5 4 4 5 6	220 250 260 280 350 450		These shakes have split faces and sawn backs. Cedar blanks or boards are split from logs and then run diagonally through a bandsaw to produce two tapered shakes from each.	CERTI-SPLIT Handsplit Red Cedar Shakes NUMBER 1 GRADE RED CEDAR SHINGLE & HANDSPLIT SHAKE BUREAU
No. 1 TAPERSPLIT	24″ x 1/2″ to 5/8″	4	260	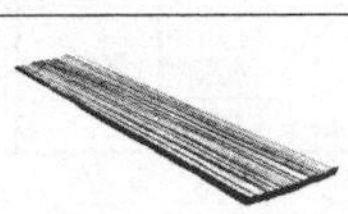	Produced largely by hand, using a sharp-bladed steel froe and a wooden mallet. The natural shingle-like taper is achieved by reversing the block, end-for-end, with each split.	
No. 1 STRAIGHT SPLIT (BARN)	18″ x 3/8″ (True-Edge) 18″ x 3/8″ 24″ x 3/8″	4 5 5	200 200 260	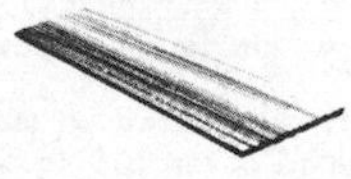	Produced in the same manner as taper-split shakes except that by splitting from the same end of the block, the shapes acquire the same thickness throughout.	

* Generally the number of bundles required to cover 100 sq. ft. when used for roof construction at the maximum recommended weather exposure.

FIG. SS2 TYPES OF RED CEDAR SHINGLES

Grade	Size	Bundles or Cartons* Per Square: No.	Bundles or Cartons* Per Square: Weight	Description		Labels
No. 1 BLUE LABEL	24″ (Royals) 18″ (Perfections) 16″ (XXXXX)	4 bdls. 4 bdls. 4 bdls.	192 lbs. 158 lbs. 144 lbs.		The premium grade of shingles for roofs and sidewalls. These shingles are 100% heartwood, 100% clear and 100% edge-grain.	CERTIGRADE Red Cedar SHINGLES BLUE 1 LABEL RED CEDAR SHINGLE & HANDSPLIT SHAKE BUREAU
No. 2 RED LABEL	24″ (Royals) 18″ (Perfections) 16″ (XXXXX)	4 bdls. 4 bdls. 4 bdls.	192 lbs. 158 lbs. 144 lbs.		A good grade for most applications. Not less than 10″ clear on 16″ shingles, 11″ clear on 18″ shingles and 16″ clear on 24″ shingles. Flat grain and limited sapwood are permitted.	CERTIGRADE Red Cedar SHINGLES RED LABEL RED CEDAR SHINGLE & HANDSPLIT SHAKE BUREAU
No. 3 BLACK LABEL	24″ (Royals) 18″ (Perfections) 16″ (XXXXX)	4 bdls. 4 bdls. 4 bdls.	192 lbs. 158 lbs. 144 lbs.		A utility grade for economy applications and secondary buildings. Guaranteed 6″ clear on 16″ and 18″ shingles, 10″ clear on 24″ shingles.	CERTIGRADE Red Cedar SHINGLES BLACK 3 LABEL RED CEDAR SHINGLE & HANDSPLIT SHAKE BUREAU
No. 4 UNDER- COURSING	18″ (Perfections) 16″ (XXXXX)	2 bdls. 2 bdls.	60 lbs. 60 lbs.		A low grade for undercoursing on double-coursed sidewall applications.	RED CEDAR SHINGLE BUREAU UNDERCOURSING GRADE THIS GRADE of Red Cedar Shingles is RECOMMENDED FOR UNDERCOURSING OF DOUBLE-COURSED SIDEWALLS ONLY Red Cedar SHINGLES
No. 1 or No. 2 REBUTTED- REJOINTED	18″ (Perfections) 16″ (XXXXX)	1 carton 1 carton	60 lbs. 60 lbs.		Same specifications as No. 1 and No. 2 Grades above but machine trimmed for exactly parallel edges with butts sawn at precise right angles. Used for sidewall application where tightly fitting joints between shingles are desired. Also available with smooth sanded face.	No. 1 & No. 2 labels (above)
No. 1 MACHINE GROOVED	18″ (Perfections) 16″ (XXXXX)	1 carton 1 carton	60 lbs. 60 lbs.		Same specifications as No. 1 and No. 2 Grades above; these shingles are used at maximum weather exposures and are always applied as the outer course of double-coursed sidewalls.	CERTIGROOVE CEDAR SHAKES NUMBER 1 GRADE RED CEDAR SHINGLE BUREAU
No. 1 or No. 2 DIMENSION	24″ (Royals) 18″ (Perfections) 16″ (XXXXX)	4 bdls. 4 bdls. 4 bdls.	192 lbs. 158 lbs. 144 lbs.		Same specifications as No. 1 and No. 2 Grades above, except they are cut to specific uniform widths and may have butts trimmed to special shapes.	No. 1 & No. 2 labels (above)

* Nearly all manufacturers pack 4 bundles to cover 100 sq. ft. when used at maximum exposures for roof construction; Rebutted-and-Rejointed and Machine Grooved singles typically are packed one carton to cover 100 sq. ft. when used at maximum exposure for double-coursed sidewalls.

FIG. SS3 COVERAGE OF WOOD SHINGLES AT VARYING EXPOSURES

Length and Thickness[2]	Approximate Coverage (sq. ft.) of Four Bundles or One Carton[1] — Weather Exposures																		
	3-1/2″	4″	5″	5-1/2″	6″	7″	7-1/2″	8″	8-1/2″	9″	10″	11″	11-1/2″	12″	13″	14″	15″	15-1/2″	16″
Random-Width & Dimension																			
16″ x 5/2″	70	80	100	110	120	140	150*	160	170	180	200	220	230	240**	—	—	—	—	—
18″ x 5/2-1/4″	—	72½	90½	100	109	127	136	145½	154½*	163½	181½	200	209	218	236	254½**	—	—	—
24″ x 4/2″	—	—	—	—	80	93	100	106½	113	120	133	146½	153*	160	173	186½	200	206½	213**
Rebutted-and-Rejointed																			
16″	—	—	—	—	50	59	63*	67	72	76	84	93	—	100**	—	—	—	—	—
18″	—	—	—	—	43	50	54	57	61*	64	72	79	—	86	93	100**	—	—	—
Machine-Grooved																			
16″	(normally applied at maximum exposure)													100**					
18″	(normally applied at maximum exposure)															100**			

1. Nearly all manufacturers pack 4 bundles to cover 100 sq. ft. when used at maximum exposures for roof construction; Rebutted-and-Rejointed and Machine Grooved shingles typically are packed one carton to cover 100 sq. ft. when used at maximum exposure for double-coursed sidewalls.
2. Sum of the thickness; e.g. 5/2″ means 5 butts equal 2″.

* Maximum exposure recommended for single-coursed sidewalls. ** Maximum exposure recommended for double-coursed sidewalls.

Nail lengths typically required are given in Figure SS4 and are illustrated in Figure 13, page 411-8. The number, spacing and pattern of nails for shakes and shingles are described below under Application of Siding.

Fiberboard Nail-base Sheathing For applying shingles and shakes to nail-base sheathing, nails should be aluminum-alloy, annularly threaded, 1-3/8″ (4d) and 2″ (6d) x 0.099″, with slightly countersunk 5/32″ head and long needle point.

PREPARATION FOR SIDING

Exposure Recommendations

The exposure of shakes and shingles is dependent on their length and whether the walls are to be single- or double-coursed. For single coursing, the weather exposure of the shakes and shingles should never be greater than half the length of the shingle minus 1/2″, so that two layers of wood occur over the entire surface of the wall.

Shingles Maximum recommended weather exposures for single-coursed wall construction are 7-1/2″ for 16″ shingles, 8-1/2″ for 18″ shingles, and 11-1/2″ for 24″ shingles. When double-coursed, 16″ shingles can be laid at weather exposures up to 12″, 18″ shingles up to 14″, and 24″ shingles up to 16″. Approximate wall coverage for shingles using different exposures is shown in Figure SS3.

Shakes Maximum recommended weather exposures for single-coursed wall construction are 8-1/2″ for 18″ shakes, 11-1/2″ for 24″ shakes, and 15″ for 32″ shakes. When double-coursed, 18″ shakes can be laid at weather exposures up to 14″, and 24″ shakes up to 20″. When straight-split shakes are double-coursed, maximum weather exposure is 16″ for 18″ shakes and 22″ for 24″ shakes. Approximate wall coverage for shakes using different exposures is shown in Figure SS5.

Wall Sheathing

Sheathing types, thicknesses, spans and nailing recommendations are discussed on page 421-4.

Sheathing Paper

To provide moisture protection when lumber boards are used for sheathing, No. 15 or No. 30 asphalt-saturated felt should be applied over the entire wall as described on page 421-5. When spaced (open) sheathing is used, sheathing paper is applied first to the framing members (Fig. SS11).

FIG. SS4 Recommended minimum nail lengths for wood shakes and shingles[1].

Type	Nail Lengths[2] Double-Coursed Sidewalls	Single-Coursed Sidewalls
SHINGLES	1-3/4″ (5d)	1-1/4″ (3d)
SHAKES[3]	2-1/2″ (8d)	1-1/2″ (4d)

1. For application over plywood sheathing, nails should be threaded.
2. Nails should be long enough to penetrate through sheathing.
3. Because thickness and exposure of shakes vary, larger nails may be necessary to obtain suitable penetration in the sheathing.

APPLICATION OF SIDING

The two basic methods used to apply shakes and shingles are: (1) *single coursing* and (2) *double cours-*

FIG. SS5 COVERAGE OF SHAKES AT VARYING WEATHER EXPOSURES

Length and Thickness	No. of Bundles	Approximate Coverage (sq. ft.) Weather Exposures 5-1/2″	6-1/2″	7″	7-1/2″	8″	8-1/2″	10″	11-1/2″	13″	14″	15″	16″
Handsplit-and-Resawn													
18″ x 1/2″ to 3/4″	4	55	65	70	75	80	85*	—	—	—	—	—	—
18″ x 3/4″ to 1-1/4″	5	55	65	70	75	80	85*	—	—	—	—	—	—
24″ x 3/8″	4	—	65	70	75	80	85	100	115*	—	—	—	—
24″ x 1/2″ to 3/4″	4	—	65	70	75	80	85	100	115*	—	—	—	—
24″ x 3/4″ to 1-1/4″	5	—	65	70	75	80	85	100	115*	—	—	—	—
32″ x 3/4″ to 1-1/4″	6	—	—	—	—	—	—	100	115	130	140	150*	—
Tapersplit													
24″ x 1/2″ to 5/8″	4	—	65	70	75	80	85	100	115*	—	—	—	—
Straight-Split													
18″ x 3/8″ (True-Edge)	4	—	—	—	—	—	—	—	—	—	100	106	112**
18″ x 3/8″	5	65	75	80	90	95	100*	—	—	—	—	—	—
24″ x 3/8″	5		65	70	75	80	85	100	115*	—	—	—	—

* Recommended maximum weather exposure for single-coursed sidewalls.
** Recommended maximum weather exposure for double coursed sidewalls.

ing. When single-coursed (Fig. SS6), units are applied in the same manner as are wood roof shakes and shingles (see Section 411 Shingle Roofing) except that greater weather exposures are employed. When double-coursed (Fig. SS7), units are applied at much greater weather exposures over a course of either No. 3 Grade or Undercoursing Grade shingles.

Using either method, at least two layers of shingles (double coverage) occur at every point on the wall surface. Except where specifically noted, the application methods described and illustrated below are applicable to both shakes and shingles.

Laying Out

The exposure and number of of courses are determined by measuring the height of the dominant wall from a point 1″ below the top of the foundation to the top of the windows, underside of soffits, or other vertical reference (Fig. SS8).

The height is divided into equal parts—as near as possible to the maximum recommended weather exposure of the shingle or shake. This measurement and the number of courses are transferred to a story pole (Fig. SS9), or may be established with chalklines all around the building. Exposure of courses can vary slightly to line up with the bottom and top of openings and at the eave. Story-pole marks or chalklines should be made after sheathing paper has been applied (Fig. SS10).

Single Coursing

The first course at the foundation should be double-coursed. Each shake or shingle should be secured with the correct nails (see Figure SS4), driven about 3/4″ from each edge and not to exceed 1″ above the buttline of the succeeding course. A third nail is driven in the center of individual shakes or shingles wider than 8″. Joints between shingles in any one course should be separated at least 1-1/2″ from the joints in adjacent courses, and joints in alternate courses should not be in direct alignment.

If fiberboard or gypsum sheathing is used, shingles and shakes should be applied over 1″ x 3″ nailing strips (Fig. SS11) spaced according to shingle exposure, or special fasteners designed for this purpose should be used.

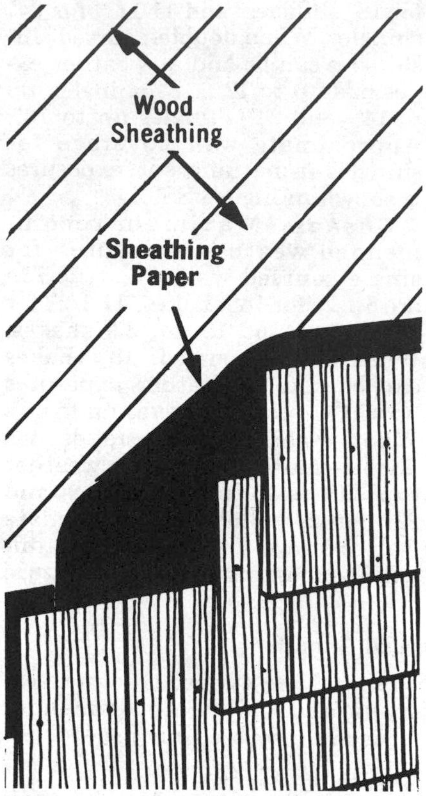

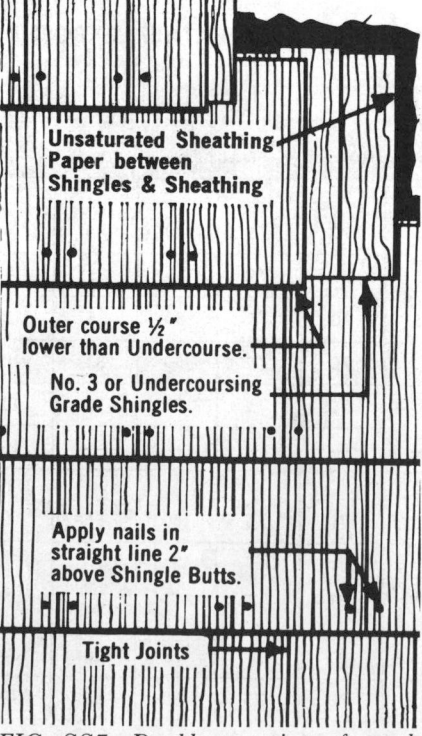

FIG. SS6 Single coursing of wood shake or shingle sidewalls.

FIG. SS7 Double coursing of wood shake or shingle sidewalls.

Double Coursing

The bottom course is laid with two under coursings and one outer course. This triple coursing eliminates the need for a costly drip cap, and makes the bottom course as thick as the succeeding courses (Fig. SS12).

A straight-edge, such as shiplap or 1″ x 4″ lumber, should be tacked tightly along a chalkline snapped between the bottom story-pole marks. Undercourse shingles, spaced approximately 1/8″ apart, are laid on the upper lip of the straightedge and fastened with a single staple near the top (Fig. SS13).

The outer course then is applied, laid tight without spacing. As each double course is laid, a chalked string is stretched from opposite corner marks to indicate the level of the next course and relocate the position of the straight-edge.

Each shingle or shake should be nailed with a minimum of two nails (Fig. SS4), applied 2″ above the buttline and 3/4″ from the edge of each shingle or shake. On shakes wider than 8″, additional nails should be applied not more than 3″ apart.

Corners

Outside corners may be lapped alternately. One outer course should protrude slightly past the corner, butting the shake on the other side of it. The protruding corner then is trimed with a knife and plane and touched-up with stain (Fig. SS14). Metal corner molding designed for this purpose also may be used.

Inside corners should be fitted with a lumber strip approximately 1-1/2″ square to receive the undercourse and outer shakes or shingles with a minimum of fitting. The strip should be stained after application.

Re-siding Exterior Walls

Shingles or shakes can be applied either directly to existing

FIG. SS8

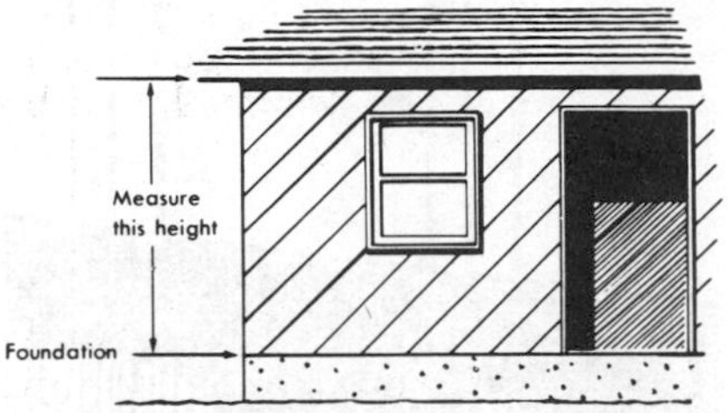

FIG. SS9

FIG. SS10

FIG. SS11

FIG. SS12

FIG. SS13

FIGS. SS8 through SS13, Double-coursed Sidewalls. Exposure of courses is determined from the wall height (SS8); courses are laid out around the building using a storypole or chalkline (SS9 & SS10). For spaced sheathing, nailing strips are applied over sheathing paper (SS11). Two layers of undercourse shingles are used for the first course (SS12). Straight-edge forms guide for nailing undercoursing and exposed courses (SS13).

siding or to new sheathing; building paper should be applied over old siding and other steps are the same as described above (Fig. SS15). Around windows and doors, before application of courses, molding strips to which shakes and shingles should be joined should be nailed on the face of old casings, flush with outer edges.

For re-siding over stucco, nailing strips first should be nailed over the wall to provide a good base for nailing double coursing (Fig. SS16). The horizontal strips should be spaced with centers 2″ above butt-lines of each course to receive the nails.

FIG. SS14 Corner treatments for shakes: (a) jointed inside corner, (b) woven outside corner, (c) mitered inside corner, (d) mitered outside corner. Corner treatments for shingles: (e) overlapping outside corner, (f) mitered inside corner using metal flashing in the joint.

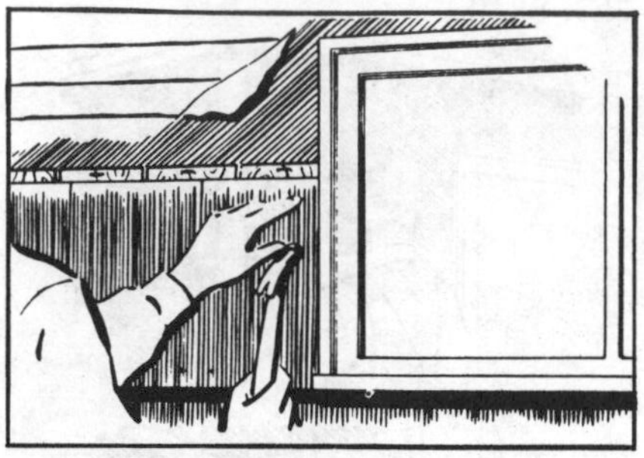

FIG. SS15 Re-siding with shingles is similar to new wall construction.

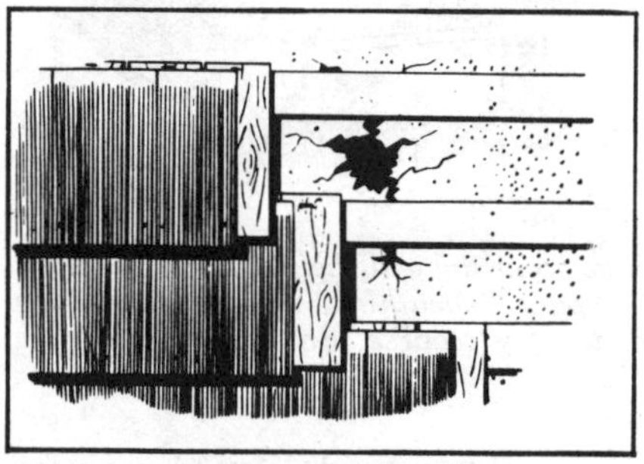

FIG. SS16 Horizontal nailing strips must be used over non-nailable sheathing.

The lath strip application illustrated in Figure SS17 should be used to apply shingles and shakes over insulating sheathing. Lath nailed at the studs provides a nailing base for double-coursing nails. As an alternate for attaching to non-wood sheathing, 3/8″ x 1-5/8″ treated or decay-resistant lath may be used. Lath should be spaced according to shingle exposure with butts of undercoursing shingles resting on the lath. The lath should be nailed to the studs through the sheathing with 2-1/8″ x 0.109″ (7d) annularly threaded or 2-1/2″ x 0.131″ (8d) common wire nails. The outer course of shakes or shingles is nailed to the lath with small-headed corrosion-resistant annularly threaded nails penetrating the lath. The butts of the outer course should project approximately 1/2″ below the lath (Fig. SS17).

Joints should not be closer than 1-1/2″ to joints of undercoursing in the same course. Joints in the following courses should be staggered at least 1-1/2″.

FIG. SS17 Lath strips are used with non-nailable sheathing for double-coursed sidewalls.

We gratefully acknowledge the assistance of the following for the use of their publications as references and permission to use photographs and illustrations in the preparation of the entire Section 421 Shingle Siding: Mineral Fiber Products Bureau, Red Cedar Shingle & Handsplit Shake Bureau, Federal Housing Administration, ES/Products, Inc., Independent Nail Corporation, W. H. Maze Company.

SHINGLE SIDING 421

CONTENTS

GENERAL RECOMMENDATIONS

Depending on the type of shingle and method of application, accessory materials typically required or which may affect the application of shingle siding include: wall sheathing, sheathing paper, backer board or undercoursing, channel and corner moldings, flashing, caulking and siding nails. Specific recommendations for these materials are given below for each kind of shingle siding.

The instructions and recommendations of the shingle manufacturer always should be consulted to insure suitable performance.

WALL FRAMING AND SHEATHING

Wall studs should be adequately sized and spaced to carry superimposed loads.

Usually the spacing of wall studs is more dependent on the type and thickness of sheathing and/or interior finish materials than on structural load requirements.

Sheathing, properly applied, can provide (1) resistance to racking, (2) additional weather protection, and (3) a nailing base for exterior siding.

Lumber Sheathing

Boards may be tongue-and-grooved, shiplapped or square edge. Spacing of framing members should not exceed 24″ on center (o.c.). Joints in lumber sheathing should occur over studs unless end-matched boards are used. Joints in end-matched boards may fall between studs provided each board bears on at least two studs and end-joints in adjacent boards are staggered. Boards 4″, 6″ and 8″ wide should be nailed to supports with at least two 2-1/8″ x 0.109″ (7d) annularly threaded or 2-1/2″ x 0.131″ (8d) common wire nails at each bearing; three nails should be used in 10″ and 12″ wide boards.

Moisture content of boards should be as shown in Figure W8, Main Text, page Wood 201-6.

Sheathing paper should be used over lumber board sheathing.

Plywood Sheathing

Plywood sheathing should be of the proper Interior or Exterior Construction Grade (see page Wood WF201-16), and should conform to the spans, thicknesses and nailing recommendations in Figure WF1.

Nail-Base Fiberboard Sheathing

Nail-base fiberboard sheathing should be homogeneous high-density structural insulating board conforming to Insulation Board Institute (IBI) Specification No. 2 and ASTM D2277.

Nail-base fiberboard sheathing should be applied vertically with large-headed, 1-1/2″ x 0.120″ (4d) plain or annularly threaded roofing nails, spaced 3″ o.c. around panel edges and 6″ o.c. at intermediate supports.

Non-Wood Sheathing

When gypsum board, fiberboard or sheathing materials

FIG. WF1 **PLYWOOD WALL SHEATHING**[1]

Identification Index	Plywood Thickness	Maximum Stud Spacing Shingles Attached to:		Fasteners	
		Stud	Sheathing	Nails[2]	Staples[3]
12/0, 16/0, 20/0	5/16″	16″	16″[4]		1¼″
16/0, 20/0, 24/0	3/8″	24″	16″ 24″[4]	6d Common[5]	1⅜″
24/0, 32/16	1/2″	24″	24″		1½″

1. Plywood continuous over 2 or more spans. Building paper and diagonal wall bracing can be omitted.
2. Spaced 6″ o.c. around panel edges and 12″ o.c. at other points of bearing.
3. Spaced 4″ o.c. around panel edges and 8″ o.c. at other points of bearing.
4. Apply plywood with face grain across studs.
5. Or 5d annularly threaded nails.

other than wood, plywood or nail-base fiberboard sheathing are recommended by the manufacturer for use as wall sheathing under shingle siding, the sheathing manufacturer should specify the type of fastening required to attach the sheathing to supports.

Non-lumber sheathing material such as rigid fiber- and gypsum boards, rated by the manufacturer as "sheathing grade," are not adequate nailing bases for shingle siding. When this type of sheathing is used, siding should be applied by nailing through the sheathing and into framing members or nailing strips.

SHEATHING PAPER

Sheathing paper should be a material such as No. 15 asphalt-saturated felt which has low vapor resistance.

Sheathing paper is used over wood board sheathing or is applied directly to studs when sheathing is not used, to resist infiltration of air and moisture.

FLASHING

Exposed flashing materials should be sheetmetal such as 26 gauge galvanized steel, 0.119″ thickness aluminum or 16 oz. copper. Additional requirements may be given below for each kind of shingle siding.

Methods for flashing are described and illustrated in the Main Text for each kind of shingle siding.

CAULKING

A non-shrinking caulking compound, either in white or matching color, should be used to weatherseal all joints where siding abuts wooden trim, masonry or other projection.

SIDING NAILS

Nail suitability is dependent on (1) selecting the correct nail for the kind of shingle and type of wall sheathing; (2) using the correct number of nails; (3) locating the nails in the shingle correctly; and (4) choosing nail metal compatible with metal used for flashing.

Lumber Boards and Plywood Sheathing

Nails should be long enough to penetrate the siding and extend through lumber boards or plywood sheathing. Nails for applying shingles over plywood sheathing should have threaded shanks.

Non-Wood Sheathing

When gypsum board, fiberboard or sheathing materials other than lumber, plywood or nail-base fiberboard are used for wall sheathing, special fasteners and/or special details for fastening are necessary to provide adequate anchorage of the shingle siding. Recommendations of the manufacturer of such non-nailable sheathing material should be followed to insure his responsibility for suitable performance of the finished wall.

ASBESTOS CEMENT SHINGLES

Unless specific exceptions are made in the following text, the General Recommendations apply.

MATERIALS

Shingles should conform to the requirements of ASTM C223. Size and maximum recommended weather exposure are given in Figure WF2.

Nails

All nails and fasteners should be permanently corrosion-resistant and stain-resistant. Only those furnished with the siding, or recommended by the siding manufacturer, should be used.

A summary of face-nail types recommended for use in applying mineral fiber siding is given in Figure WF3. For number, spacing and pattern of nails, see Main Text, Application of Siding, page 421-8.

PREPARATION FOR SIDING

When sheathing is used with mineral fiber siding, it should be smooth sheathing of lumber, plywood, nail-base fiberboard or other type of rigid sheathing which provides an adequate nailing base.

If wood boards are used for sheathing, they preferably should be not over 8″ in width, free from loose knots or large knotholes, and have tongue-and-grooved or shiplapped edges.

Sheathing Paper

Sheathing paper should be used over lumber board sheathing and should be at least No. 15 asphalt-saturated felt. Coal tar saturated felts which may stain the siding should not be used.

FIG. WF2 **MINERAL FIBER SIDING**

Sizes[1]	Exposure	Pieces per Bundle	Pieces per Square[2]
12″ x 24″	11″	18	54
9″ x 32″	8″	19	57
9-5/8″ x 32″	8-5/8″	17-17-18	52
11-3/4″ x 32″	10-3/4″	14	42
14-5/8″ x 32″	13-5/8″	11	33
14-3/4″ x 32″	13-3/4″	11	33
8-3/4″ x 48″	7-3/4″	13	39
12″ x 48″	11″	9	27

1. Three nailholes per shingle.
2. Weight per square ranges from 153 to 206 lbs.

Joint Flashing Strips
Asphalt (not coal tar) saturated and coated, water-repellent joint flashing strips supplied by the manufacturer with all types of mineral fiber siding should be applied behind each vertical joint between siding units.

Cant Strips
A 1/4″ x 1-1/2″ wood cant strip is used to pitch the first course of shingles at the proper angle, provides solid bearing to prevent breakage, and can be used to seal the joint between the top of the foundation, the wood sill and the bottom of the sheathing.

APPLICATION OF SIDING
Recommended methods for applying mineral fiber shingles for new and existing construction are described and illustrated in the Main Text, pages 421-8 through 421-12.

FIG. WF3 NAIL TYPES FOR USE WITH MINERAL FIBER SIDING

NEW CONSTRUCTION

Sheathing Type	Nailed Directly to Sheathing or Wood Nailing Strips	Nailed Through Fiberboard Shingle Undercoursing
Board Lumber	1⅛″ Screw Thread 1⁷⁄₁₆″ Screw Thread 1½″ Annular Thread	2″ Annular Thread
Plywood	1⅛″ Screw Thread	1¾″ Screw Thread
Wood Nailing Strip	1⅛″ Screw Thread	1¾″ Screw Thread
Nailbase Fiberboard	1⅜″ Special Annular Thread	2″ Special Annular Thread

RE-SIDING (over existing wood siding or shingles)

Nailing Base	Nail Types
Sound Board Lumber	1¾″ or longer, annular thread, screw thread or helical thread
Wood Nailing Strip	1⅛″ Screw Thread

WOOD SHAKES AND SHINGLES

Unless specific exceptions are made in the following text, the General Recommendations apply.

MATERIALS
Each bundle of shakes or shingles should carry the official grade-marked label of the Red Cedar Shingle & Handsplit Shake Bureau. Sizes and grades should comply with Figures WF4 and WF5 (see next page).

Nails
Nails should be corrosion-resistant. Common wire nails or electro-galvanized nails should not be used. Nailing recommendations are given in the Main Text, Application of Siding, page 421-15.

PREPARATION FOR SIDING

Exposure Recommendations
The exposure of shakes and shingles is dependent on their length and whether the walls are to be single-course or double-coursed.

For single coursing, the exposure should not be greater than half the length of the shake or shingle minus 1/2″, so that two layers of wood occur over the entire surface of the wall.

Maximum recommended weather exposures for shakes and shingles are given in Figure WF6.

FIG. WF6 MAXIMUM EXPOSURE FOR WOOD SHAKES & SHINGLES

Type	Length	Single Coursing	Double Coursing
Handsplit & Resawn Shakes	18″	8-1/2″	14″
	24″	11-1/2″	20″
	32″	15″	—
Tapersplit Shakes	24″	11-1/2″	20″
Straight-Split Shakes	18″	8-1/2″	16″
	24″	11-1/2″	22″
Random-Width & Dimension Shingles	16″	7-1/2″	12″
	18″	8-1/2″	14″
	24″	11-1/2″	16″
Rebutted & Re-jointed Shingles	16″	7-1″2″	12″
	18″	8-1/2″	14″
Machine-Grooved Shingles	16″	—	12″
	18″	—	14″

Wall Sheathing
When solid sheathing is used, shingles and shakes should be applied over smooth sheathing of lumber, plywood, nail-base fiberboard, or other type of rigid sheathing which provide an adequate nailing base. For spaced sheathing, boards should be at least nominal 1″ x 3″, with spacing of boards on center equal to shingle exposure.

APPLICATION OF SIDING
Recommended methods for applying wood shakes and shingles for new and existing construction are described and illustrated in the Main Text, see page 421-15.

FIG. WF4 TYPES OF RED CEDAR SHINGLES

Grade	Length and Thickness	Bundles Per Square*	Weight (lbs. per square)	Description	Label
No. 1 HANDSPLIT & RESAWN	18" x 1/2" to 3/4" 18" x 3/4" to 1-1/4" 24" x 3/8" (sidewall use only) 24" x 1/2" to 3/4" 24" x 3/4" to 1-1/4" 32" x 3/4" to 1-1/4"	4 5 4 4 5 6	220 250 260 280 350 450	 These shakes have split faces and sawn backs. Cedar blanks or boards are split from logs and then run diagonally through a bandsaw to produce two tapered shakes from each.	
No. 1 TAPERSPLIT	24" x 1/2" to 5/8"	4	260	 Produced largely by hand, using a sharp-bladed steel froe and a wooden mallet. The natural shingle-like taper is achieved by reversing the block, end-for-end, with each split.	CERTI-SPLIT Handsplit Red Cedar Shakes NUMBER 1 GRADE RED CEDAR SHINGLE & HANDSPLIT SHAKE BUREAU
No. 1 STRAIGHT-SPLIT (BARN)	18" x 3/8" (True-Edge) 18" x 3/8" 24" x 3/8"	4 5 5	200 200 260	 Produced in the same manner as taper-split shakes except that by splitting from the same end of the block, the shapes acquire the same thickness throughout.	

FIG. WF5 TYPES OF RED CEDAR SHAKES

Grade	Size	Bundles or Cartons* Per Square		Description	Labels
		No.	Weight		
No. 1 BLUE LABEL	24" (Royals) 18" (Perfections) 16" (XXXXX)	4 bdls. 4 bdls. 4 bdls.	192 lbs. 158 lbs. 144 lbs.	 The premium grade of shingles for roofs and sidewalls. These shingles are 100% heartwood, 100% clear and 100% edge-grain.	CERTIGRADE Red Cedar SHINGLES BLUE 1 LABEL
No. 2 RED LABEL	24" (Royals) 18" (Perfections) 16" (XXXXX)	4 bdls. 4 bdls. 4 bdls.	192 lbs. 158 lbs. 144 lbs.	 A good grade for most applications. Not less than 10" clear on 16" shingles, 11" clear on 18" shingles and 16" clear on 24" shingles. Flat grain and limited sapwood are permitted.	CERTIGRADE Red Cedar SHINGLES RED LABEL
No. 3 BLACK LABEL	24" (Royals) 18" (Perfections) 16" (XXXXX)	4 bdls. 4 bdls. 4 bdls.	192 lbs. 158 lbs. 144 lbs.	 A utility grade for economy applications and secondary buildings. Guaranteed 6" clear on 16" and 18" shingles, 10" clear on 24" shingles.	CERTIGRADE Red Cedar SHINGLES BLACK 3 LABEL
No. 4 UNDER-COURSING	18" (Perfections) 16" (XXXXX)	2 bdls. 2 bdls.	60 lbs. 60 lbs.	 A low grade for undercoursing on double-coursed sidewall applications.	RED CEDAR SHINGLE BUREAU UNDERCOURSING GRADE This Grade of Red Cedar Shingles is recommended for undercoursing of double-coursed sidewalls only Red Cedar SHINGLES
No. 1 or No. 2 REBUTTED-REJOINTED	18" (Perfections) 16" (XXXXX)	1 carton 1 carton	60 lbs. 60 lbs.	 Same specifications as No. 1 and No. 2 Grades above but machine trimmed for exactly parallel edges with butts sawn at precise right angles. Used for sidewall application where tightly fitting joints between shingles are desired. Also available with smooth sanded face.	No. 1 & No. 2 labels (above)
No. 1 MACHINE GROOVED	18" (Perfections) 16" (XXXXX)	1 carton 1 carton	60 lbs. 60 lbs.	 Same specifications as No. 1 and No. 2 Grades above; these shingles are used at maximum weather exposures and are always applied as the outer course of double-coursed sidewalls.	CERTIGROOVE CEDAR SHAKES NUMBER 1 GRADE RED CEDAR SHINGLE BUREAU
No. 1 or No. 2 DIMENSION	24" (Royals) 18" (Perfections) 16" (XXXXX)	4 bdls. 4 bdls. 4 bdls.	192 lbs. 158 lbs. 144 lbs.	 Same specifications as No. 1 and No. 2 Grades above, except they are cut to specific uniform widths and may have butts trimmed to special shapes.	No. 1 & No. 2 labels (above)

*Nearly all manufacturers pack 4 bundles to cover 100 sq. ft. when used at maximum exposures for roof construction; Rebutted-and-Rejointed and Machine Grooved shingles typically are packed one carton to cover 100 sq. ft. when used at maximum exposure for double-coursed sidewalls.

425 STUCCO

INTRODUCTION

In most parts of the country, *stucco* is the term applied to exterior plastering; the corresponding interior application is simply referred to as *plaster* or *plastering.* However, on the West Coast, only the final finish coat is referred to as stucco, and this textured or colored coat may be used in interior locations as well.

Exterior plastering (stucco) involves considerations of hazards such as water penetration, corrosion of reinforcement and accessories, and stresses due to wide variations of temperature and humidity which are not normally present in interior applications. Exterior applications therefore require practices and precautions which are somewhat different from interior applications.

Because of the exposure to direct wetting and high moisture conditions, stucco requires cementitious materials such as *portland cement, masonry cement* and *plastic cement* which are not adversely affected by such conditions. *Lime* historically was used as the chief binder in stucco finishes. It is now added to portland cement in the field or is included during manufacture in some of the other cements to improve plasticity.

Both masonry and plastic cements consist chiefly of portland cement and include sufficient amounts of plasticizers which makes the field addition of lime unnecessary. Finishes made with plastic cement or masonry cement or with portland cement mixed with less than 50% by volume of lime are called *portland cement stucco.* When lime is added to portland cement in amounts greater than 50% by volume, the finish is referred to as *lime-cement* (or *portland cement-lime) stucco.*

Stucco finishes have many of the desirable properties of concrete, and when properly applied are hard, strong and decorative. They can be formed to any shape, can be finished with a wide variety of surface textures and colors, and are resistant to fire, weather and fungus attack. Stucco finishes are used predominantly in the southern, southwestern and western parts of the country, but have proven durable and economical in both warm and cold climates.

The stucco membrane is in effect a thin concrete slab, built up with tools and methods associated with plastering. Therefore, the discussion of the properties and the behavior of concrete found in Section 203 Concrete and the description of plastering methods in Section 432 Plaster Finishes will be helpful in understanding recommendations given in the following subsections, *Materials* and *Applications.*

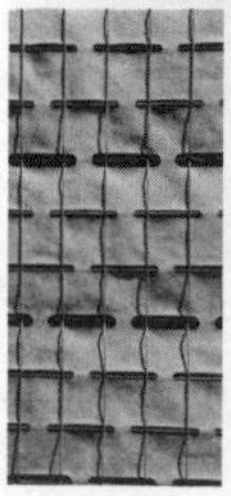

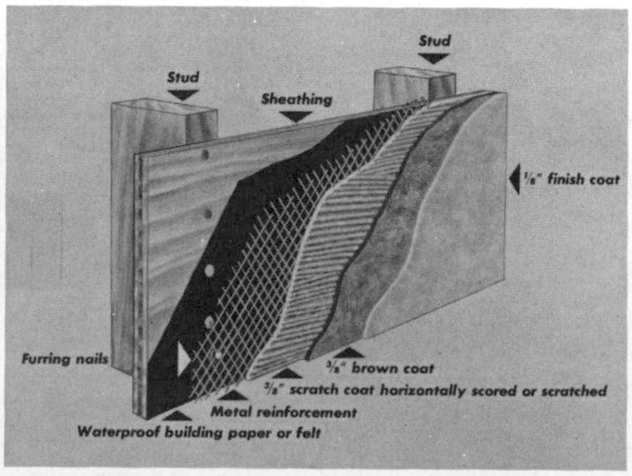

MATERIALS

The basic ingredients of stucco are essentially the same as those of concrete, mortars and plasters, and are classed broadly as *aggregates, cementitious materials, water* and sometimes *admixtures.* Aggregates act as inert fillers and provide strength; cementitious materials provide the binder; water, together with cementitious materials, forms the paste and initiates the reaction that binds the aggregates together into a solid mass; and admixtures sometimes are added to improve plasticity or cohesiveness of the mix, or to retard or to accelerate the setting time.

Accessory materials commonly employed in the application of stucco finishes are *backing paper* (or felt), *metal reinforcement, flashings* and various *metal accessories* for edge protection and crack control.

Although portland cement and lime-cement mixes are used primarily for exterior applications, they can be used as well in high-moisture interior locations, such as laundries, saunas and tub-shower enclosures. The discussion in this section deals primarily with exterior finishes, but many of the principles and recommendations given also apply to interior portland cement and lime-cement plasters.

TERMINOLOGY

The following definitions relating to stucco finishes are given to assist in understanding the text.

Base Coat The first stucco layer in two-coat work; or either the scratch or brown coat in three-coat work.

Brown Coat The second stucco layer in three-coat work which provides additional strength and a suitably true and plane surface for the application of the finish coat.

Control Joint A prefabricated metal accessory intended to relieve shrinkage, temperature or structural stresses in the stucco finish, thus minimizing cracking.

External Corner A projecting angle formed by abutting walls or vertical surface and soffit (not to be confused with *exterior,* meaning exposed to the weather).

Finish Coat The final decorative stucco layer in either two-coat or three-coat work.

Internal Corner An enclosed angle of less than 180° formed by abutting walls or the juncture of walls and the ceiling (not to be confused with *interior,* meaning protected from the weather).

Scratch Coat The first stucco layer in three-coat work which embeds the reinforcement and provides a suitably rigid and roughened (scratched) surface for the following coat.

Stucco Finish As generally used in this text, all elements of the stucco membrane, such as the several coats of stucco, metal reinforcement, accessories and backing paper, when required. *Stucco finish* is a term also applied to factory-prepared mixes for application as finish coats.

Stucco Mix Mortar consisting of properly proportioned quantities of cementitious materials, aggregate, water and sometimes pigments, plasticizers and other admixtures.

Three-coat Work Application of stucco in three separate layers (scratch, brown and finish coats) totaling at least 7/8" in thickness.

Two-coat Work Application of stucco in two layers (base coat and finish coat) totaling at least 5/8" in thickness.

FIG. 1 **SUMMARY OF STUCCO MATERIALS**

Material	Standard Specification
Cementitious Material	
Portland Cement	
Types I, II, III	ASTM C150
Types IA, IIA, IIIA	ASTM C175
Masonry Cement	ASTM C91
Plastic Cement	ASTM C150
Lime, Hydrated*	
Special Finishing Normal Finishing	ASTM C206
Mason's, Types S & N	ASTM C207
Quicklime	ASTM C5
Aggregates	
Sand	ASTM C144
Lightweight	ASTM C332
Metal Reinforcement	FS-QQ-L-101a
Metal Accessories	ANSI A42.3
Backing Paper	
Roofing Felt	FS-HH-R-595
Building Paper	FS-UU-B-790

*Both Special and Normal limes are available as air-entraining types (SA and NA).

CEMENTITIOUS MATERIALS

The cementitious materials used in stucco finishes are portland cement, masonry cement, plastic cement and lime. These products should conform to the requirements of the standard specifications given in Figure 1.

Portland Cement

The types of portland cement suitable for stucco (Fig. 1) are used also in concrete and are discussed in detail on page Concrete 203-4.

Type I portland cement generally is used unless special properties possessed by Types II and III are required. Types IA, IIA and IIIA (air entraining) portland cement provide improved workability, resistance to bleeding (concentration of water and fines at the surface) and to damage from freezing.

Type I (Normal) Type I generally is used for most stucco work. It is available in natural gray or white. The white cement is particularly suitable for finish coats with integral coloring or colored aggregate.

Type II (Moderate Heat) Type II portland cement has greater resistance to decomposition from the action of sulphates and lower heat of hydration. Resistance to sulphate attack may be a useful property when stucco is applied to brick or block masonry containing soluble sulphate materials.

Type III (High Early Strength) This portland cement develops strength faster in the first few days after application. Therefore, Type III may be desirable in cold weather to minimize the hazard of freezing damage and to speed application of successive coats of stucco.

White Portland Cement Truer colors can be obtained for finish coats by using a white portland cement manufactured from specially selected ingredients. Although portland cement is inherently moisture resistant, white portland cement is available also with moisture resistant additives and is marketed as *waterproof* white portland cement. These products must contain lesser amounts of iron oxides, but generally should conform to the requirements of ASTM C150 for Type I portland cement.

Masonry Cement

Masonry cement consists primarily of portland cement with varying amounts of such materials as natural cements, finely ground limestone, air entraining agents and sometimes lime. It is used extensively for stucco finishes in some parts of the country, particularly in the southwestern states.

Plastic Cement

Plastic cement is produced primarily in the southwestern United States. It is similar in properties and formulations to Types I and II portland cement, but may contain larger amounts of air entraining agents, insoluble residues, and plasticizing agents not in excess of 12% of the volume. With the exception of the abovementioned variations, plastic cement should conform to the requirements of ASTM C150, for Type I and Type II portland cement.

Lime

Building limes can be classed as: (1) *quicklime,* which requires slaking and soaking for approximately 1 to 2 weeks to hydrate oxides and to make a plastic putty usable in mortars or plasters; and (2) *hydrated lime,* which can be used either dry or as lime putty after brief soaking, depending on the type. Hydrated limes include mason's hydrated lime, intended primarily for masonry mortars, and finishing hydrated lime intended for stucco and plastering uses; both are available in *Type S (special), Type SA (special air-entraining), Type N (normal)* and *Type NA (normal air-entraining).*

Building limes may be manufactured from calcitic limestone, naturally high in calcium carbonate, or from dolomitic limestone which normally contains substantial amounts of magnesium carbonate. When magnesium carbonates are calcined during manufacture, magnesium oxides result. These oxides produce a highly workable, plastic putty but take longer to hydrate than the calcium oxides in calcitic limes. If the oxides are not completely hydrated during soaking or during the manufacture of hydrated lime, the hazard exists that they will be present in the stucco finish. Under conditions of high humidity they could hydrate in the stucco finish, causing expansion and consequent cracking.

Therefore, dolomitic limes often are autoclaved at the mill to assure that a minimum of unhydrated oxides remains in the manufactured hydrated lime. *Type S* hydrated lime conforming to ASTM C206 must contain not more than 8% unhydrated oxides, regardless of the raw materials used.

Type N hydrated lime also produced to the requirements of ASTM C206, has no limitation on unhydrated oxides. However, when calcitic limestone is used in its manufacture, the resulting product (identified as calcitic or high-calcium finishing hydrated lime) consists of at least 95% calcium oxide, which hydrates readily during manufacture. Some oxides may remain unhydrated, but these can be expected to hydrate substantially during the required soaking period for Type N limes of 12 to 16 hours. Often, high calcium finishing lime will meet ASTM limitations for unhydrated oxides for Type S lime, but may not develop the degree of plasticity specified for this lime within the time limit established by ASTM.

Quicklime, which is not hydrated at the mill, must be slaked and soaked on the job, during which time hydration of oxides takes place. The type of raw material affects the degree and speed of hydration possible during the normal soaking period of 1 to 2 weeks. Quicklime identified as calcitic or high calcium can be expected to result in greater hydration of oxides.

Type S (special) finishing hydrated lime conforming to ASTM C206 requires no preliminary soaking and can be mixed with other ingredients before water is added. Because it is easy to use and because unhydrated oxides are limited during manufacture, Type S lime most often is used and specifically is recommended for stucco finishes. Type N hydrated finishing lime also under ASTM C206, and quicklime are acceptable if they are identified as high-calcium or calcitic limes, and provided they have been properly slaked and/or soaked as recommended by the manufacturer.

AGGREGATES

The aggregate constitutes most of the volume of the stucco mix. One cu. ft. of a typical stucco mix contains approximately 0.97 cu. ft. of aggregate and 0.25 to 0.33 cu. ft. of cementitious material. Therefore, it is important that the aggregate be clean, sound and well graded with particles ranging from fine up to a maximum size of 1/8″.

The best and most practical mix results when the aggregate size is proportioned so that a minimum of water-cement paste is required to coat the aggregate and fill the voids between particles. Either an excess of fine or coarse particles can result in a mix containing too much cement or too much water, either of which can be detrimental to the stucco finish (see page 425-13).

FIG. 2 **RECOMMENDED GRADING FOR SAND AGGREGATE**

Sieve Size	Percent of Aggregate by Weight Retained on Sieve	
	Minimum	Maximum
No. 4	—	0
No. 8	—	10
No. 16	10	40
No. 30	30	65
No. 50	70	90
No. 100	95	100
No. 200	97	100

Sand

Sand aggregate for stucco should consist of natural or manufactured sand which is clean, granular and free from deleterious amounts of loam, clay, soluble salts or vegetable matter. It should be well graded within the limits given in Figure 2 and, with the exception of grading, should conform to the requirements of ASTM C144.

Lightweight Aggregate

Expanded shale, clay, slate and slag sometimes are used where sand is not available locally. Vermiculite and perlite have been used experimentally for stucco where lightweight, insulating properties and/or fire resistance are prime factors. Proportioning of these materials should be determined by laboratory test or by making test panels. Thermal insulating properties, gradation and unit weight for lightweight aggregate should conform to the requirements of ASTM C332.

PREPARED FINISH COAT

Several factory-prepared mixes *(stucco finishes)* are manufactured for application over basecoats. These prepared mixes contain all the necessary finish-coat ingredients, with the exception of water. They are available in white or a variety of colors and, in general, provide more satisfactory results than those obtained by the addition of pigments in the field.

WATER

Water for mixing and curing should be clean and free from deleterious amounts of oils, acids, alkalies, salts or organic substances that could adversely affect proper hydration or cause corrosion of metal reinforcement and accessories.

The presence of water promotes *hydration* of portland cement and *carbonation* of lime—the chemical processes by which these materials harden. However, more water generally is added to the mix than is necessary to provide for hydration and/or carbonation. The excess is required to bring the mix to a workable consistency. The amount of water added for workability depends on the characteristics of the cementitious material, gradation of aggregate, suction of the supporting construction, and drying conditions.

ADMIXTURES

Admixtures such as asbestos, animal, vegetable or glass fibers sometimes are added to the mix to improve plasticity and are termed *plasticizers*. These plasticizers can be added to the first (scratch) coat where greater cohesiveness of the mix is required, particularly for ceilings and soffits, and to reduce the amount of mix lost through openings in metal reinforcement when backing paper is not used. Asbestos fibers also may be used for machine-applied basecoats to prevent segregation of the aggregate and to provide lubrication in the hose.

Asbestos fibers should be clean and free from harmful amounts of any substance such as unhydrated oxides that may be injurious to the stucco ingredients. Animal hair or vegetable fibers should be 1/2″ to 2″ in length, free from grease, oil, dirt or other impurities.

Other admixtures intended to accelerate or to retard setting time generally are not recommended, but may be used where local practice indicates satisfactory results. When used, admixtures should not reduce the compressive strength of the mix by more than 15% below the strength of a comparable mix without admixtures.

BACKING PAPER

Installations with metal reinforcement often require a backing paper: (1) to isolate stucco from the supporting construction, (2) to prevent loss of the mix through lath openings, (3) to facilitate complete embedment of the lath, and (4) to resist water penetration through breaks in the stucco membrane. When the stucco is hand applied, backing paper is not required generally on horizontal surfaces protected from direct wetting.

The paper should have a vapor permeance of at least 5 perms to permit stucco drying and to prevent moisture from becoming trapped within the wall. Waterproof and vapor permeable products used for sheathing paper, such as 15 lb. asphalt-saturated roofing felt and building paper conforming to FS-UU-B-790-Class D, are suitable for this purpose. Backing paper may be factory-attached to metal reinforcement or can be installed separately. *Absorptive* paper attached to some types of metal reinforcement is not water resistant and is not suitable for exterior applications where weather protection is required.

METAL REINFORCEMENT

The most common commercially available types of metal reinforcement are *expanded metal lath* and

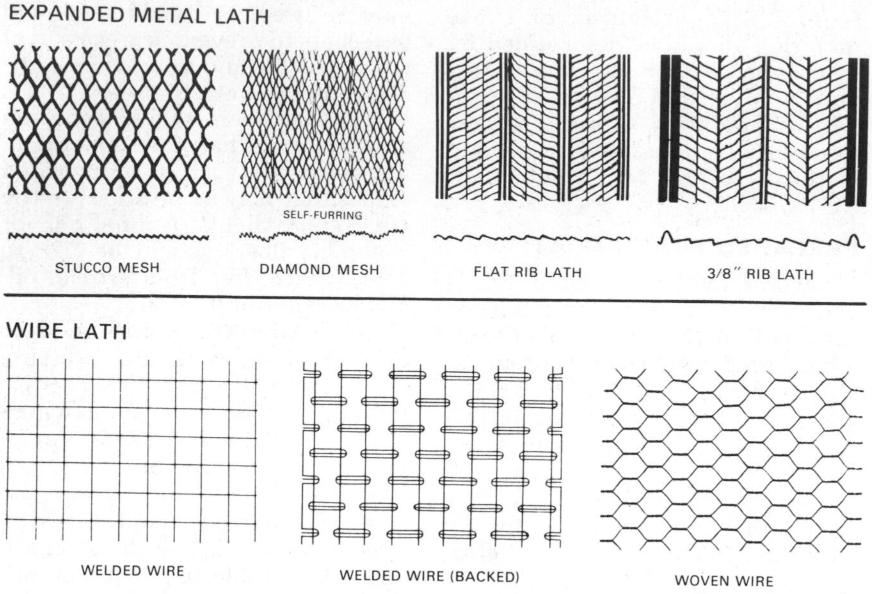

FIG. 3 Metal reinforcement suitable for stucco finishes; all but stucco mesh and 3/8″ rib lath are available in plain, self-furring and/or paper-backed types.

wire lath (Figs. 3 & 4). Openings in the lath should be large enough to permit the scratch coat to be forced through the openings and to embed the reinforcement, thereby preventing corrosion. Metal reinforcement should conform to the standard specifications given in Figure 1 and the weight requirements given in Figure 4. Self-furring reinforcement should be provided with devices to keep the back of the lath at least 1/4″ away from the supporting construction.

Expanded Metal Lath

This type of reinforcement is formed from light gauge steel sheets that have been cut in a regular pattern and expanded (stretched) to form diamond-shaped openings (Fig. 3). Both expanded zinc-coated steel sheet and cold-rolled steel sheet, coated with corrosion-resistant paint after fabrication, are used.

Diamond Mesh Plain, self-furring and paper-backed types of diamond mesh are produced in sheet form. The plain type generally should be installed over backing paper with furring fasteners. Diamond mesh is produced in weights of 2.5 and 3.4 lbs. per sq. yd.; only the heavier 3.4 weight is suitable for stucco use.

The self-furring type is dimpled to hold it away from the surface to which it is attached. It can be nailed directly to smooth, reinforced concrete, masonry or other solid surface to receive stucco.

Paper-backed diamond mesh, with paper strips attached to the back surface, is produced primarily for machine application of stucco. Spaces are provided between paper strips to permit fastening to studs or other supporting construction.

Stucco Mesh While similar in pattern to diamond mesh, stucco mesh has larger openings and is designed primarily as a base for stucco over solid surfaces. It should be installed with backing paper and furring fasteners.

Rib Lath This type of expanded metal lath consists of integrally formed V-shaped ribs which provide greater stiffness and permit wider spacing of supports (Fig. 9). Laths with 1/8″, 3/8″ and 3/4″ ribs are produced (Fig. 4). The 1/8″ rib lath, referred to as *flat rib lath,* generally requires furring fasteners and can be used for most stucco applications. It is available in weights of 2.75 and 3.4 lbs. per sq. yd., but only the heavier is acceptable as a stucco base.

The *3/8″ rib lath* does not require furring but is not recommended for severe weather exposure because it is practically impossible to encase the lath with stucco mix. However, in milder climates and in locations protected from direct wetting, such as soffits, the 3/8″ rib lath can be used. Its chief advantage is the ability to provide a stucco base for support spacing up to 24″ o.c. (Fig. 9).

Wire Lath

Wire lath is produced as *woven* and *welded* wire fabric. Wire lath should be fabricated from copper-bearing galvanized cold drawn steel wire of the appropriate gauge (Fig. 4). Wire lath is used exten-

FIG. 4 METAL REINFORCEMENT FOR STUCCO FINISHES

Type	Weight lbs/sq. yd.	Opening Size	Dimensions
Expanded Metal Lath			
Diamond Mesh[1]	3.4	5/16″ x 3/8″	27″ x 96″
Stucco Mesh	1.8 & 3.6	1 3/8″ x 3 1/8″	48″ x 99″
Flat Rib Lath	3.4	5/16″ x 3/8″	24″ x 96″ & 27″ x 96″
3/8″ Rib Lath	3.4 & 4.0	5/16″ x 3/8″	27″ x 96″ & 27″ x 99″
Wire Lath			
Woven Wire[1]	1.7 (18 ga.)	1″ Hex.	3′ x 150′ (rolls)[2]
	1.4 (17 ga.)	1 1/2″ Hex.	
Welded Wire[1]	1.4 (16 ga.)	2″ x 2″	3′ x 150′ (rolls)[2]
	1.4 (18 ga.)	1″ x 1″	
	1.9 (16 ga.)	1 1/2″ x 2″	

1. Available in plain, self-furring and paper-backed types.
2. Paper-backed and self-furring types also available in sheet form.

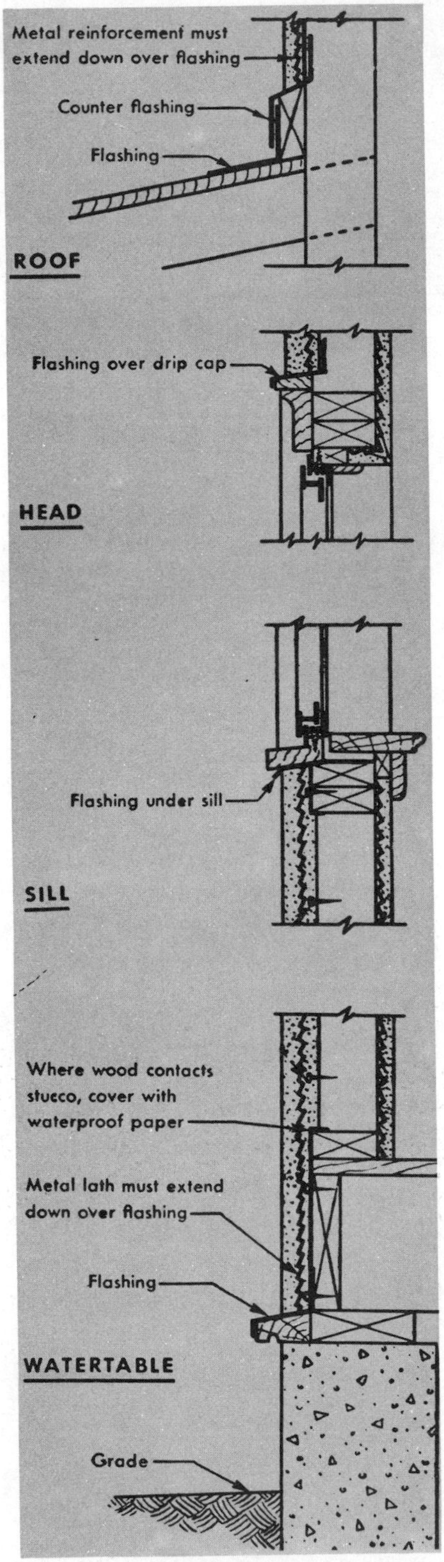

FIG. 5 Flashing details for critical locations; metal drip screeds generally are substituted for watertable and flashing at foundation.

sively as a base for stucco finishes.

Woven Wire Fabric Hexagonal woven wire fabric *(stucco netting)* is made by interweaving wire into hexagonal openings 1″ to 2-1/4″ wide (Fig. 3). The plain type is nailed to supporting construction with furring nails over backing paper. Woven wire fabric also is available in self-furring type, with or without backing paper, which eliminates the need for furring fasteners or separate installation of backing paper, depending on the type of lath used. Minimum wire size for 1″ openings should be 18 gauge, and 17 gauge for 1-1/2″ openings (Fig. 4).

Welded Wire Fabric This type of wire lath consists of a grid formed by welding steel wire at intersections into 1″ x 1″ and 2″ x 2″ squares. It is available with backing paper and stiffener wires and in self-furring type. The fabric should be galvanized, coated with corrosion-resistant paint after welding, or fabricated from galvanized wire. Wire should be at least 16 gauge for fabric with 2″ x 2″ openings, and at least 18 gauge for fabric with 1″ x 1″ openings.

FLASHING

Properly installed flashing to prevent water from entering the construction is essential to durable and satisfactory service of stucco surfaces. Flashing should be provided in all locations where the hazard of water penetration exists (Fig. 5). Water entering behind the stucco finish could cause corrosion of metal reinforcement and accessories and damage to interior finishes. Only corrosion resistant materials, such as galvanized steel, copper, stainless steel and suitable plastics, should be used for flashing (see page Masonry Walls 343-3). Metal reinforcement should be installed over flashing because these materials do not provide adequate bond for stucco.

METAL ACCESSORIES

In residential construction, wood trim normally covers the edges of the stucco finish. When wood trim is not used, exposed edges requiring protection from damage should be trimmed with metal *casing beads* (Fig. 6a). *Control joints* (Fig. 6b) are incorporated to break

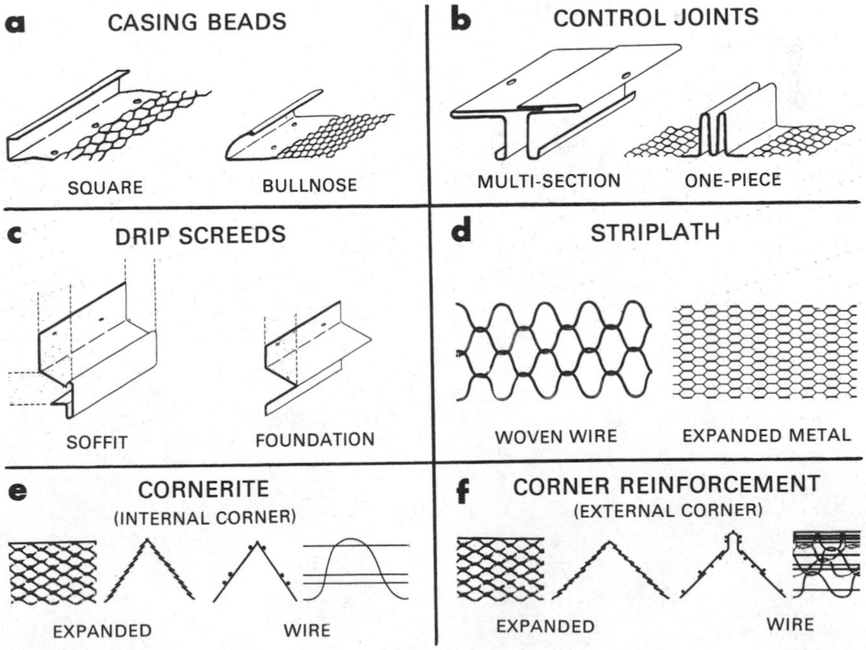

FIG. 6 Typical accessories for stucco finishes. Casing beads (a) are available with or without expanded flanges; external corner reinforcement (f) also is available with solid metal nosing (cornerbead) in white alloy zinc. Other accessories are produced to meet special conditions.

large areas into self-contained smaller panels, and *drip screeds* are used as terminating lines at the bottom of walls (Fig. 6c). Other accessories such as *striplath* and *cornerite* (Fig. 6d & e) are used for reinforcing flat surfaces and internal corners, respectively. External corners often are provided with *corner reinforcement* (Fig. 6f) when the metal lath is not flexible enough to be bent around corners.

Corner reinforcement without a solid nosing is designed to permit complete embedment of the metal within the stucco finish. However, *cornerbead* made with a solid nosing which remains exposed can be used, provided it is made of white alloy zinc (99% pure zinc). Casing beads should be formed from metal not less than 24 gauge; other accessories should be at least 26 gauge. All exposed metal accessories (not completely embedded in the stucco) should be not less corrosion resistant than white alloy zinc.

The primary considerations for the application of stucco finishes are: (1) preparation of supporting construction and stucco base, (2) selection and proportioning of the stucco mix, (3) method of application, (4) curing and (5) precautions to control cracking.

SUPPORTING CONSTRUCTION

Stucco can be applied directly to suitable concrete and masonry surfaces, and to wood and metal framing if properly prepared with backing paper and metal reinforcement. All supporting construction should be capable of supporting design loads without excessive deflection or without introducing stresses that might cause cracking of the finish. The calculated deflection of ceiling members should not exceed 1/360 of the span under full design load. Headers or lintels of adequate size should be provided over openings and all construction should be true and plumb or level.

Supporting construction should be firm and well braced, but also should be capable of accommodating structural movement where it is expected to occur. Concrete and masonry construction should be provided with control joints and expansion joints as dictated by good design practice. Control joints in the stucco should be placed directly over these joints in the supporting construction and wherever required to relieve stresses and minimize cracking (see page 425-17).

STUCCO BASES

Stucco bases are constructions or surfaces which develop adequate bond for the direct application of stucco. Bond can be provided by *mechanical key* or by *suction*. Mechanical keying results from the interlocking of the stucco mix with openings or projections in the base surface; suction develops bond by drawing part of the stucco paste into minute pores in the base surface by capillary action. Over metal reinforcement, bond is developed by mechanical keying; over concrete and masonry, bond is developed by either mechanical keying or suction, or both.

Stucco should not be applied directly to gypsum products such as gypsum plaster, gypsum masonry or gypsum lath, because these products generally do not provide adequate bond for portland cement mixes. Backing paper and metal reinforcement should be installed over such materials before stucco is applied. Wood lath, fiber insulating lath or sheetmetal lath should not be used as a base for stucco.

Masonry

When stucco is to be applied directly to masonry, mortar joints in the masonry should be struck flush or slightly raked. The masonry surface should be clean, sound and firm and should not contain efflorescence, grease, waterproofing compounds, paint or other substances detrimental to good bond. Efflorescence, oil and grease can be removed by washing with a 10% solution of muriatic acid and water. Other coatings or substances that will adversely affect bond should be removed with appropriate solvents and rinsed clean.

Most masonry, such as medium-hard common and face brick, standard weight concrete block and many types of stone, generally provide adequate suction and often mechanical bond. The surface can be tested for suction by spraying it with clean water to see how quickly the moisture is absorbed. If the water is readily absorbed, good suction is likely to result; if water droplets form, the surface does not provide adequate suction and bond will depend on mechanical key. It can be determined by observation whether the surface is sufficiently rough to assure mechanical keying with the stucco.

On surfaces which absorb moisture too quickly, such as soft common brick and lightweight concrete block, the stucco will quickly stiffen and become difficult to apply properly. Excessive suction can be controlled by fine spraying (not soaking) the masonry with several applications of water immediately prior to application of the first stucco coat. The surface should appear slightly damp, but there should be no free water present when the stucco is applied. When stucco is machine-applied, it may not be necessary to dampen the surface because the mix usually contains more water than for hand application.

Hard-burnt brick, glazed concrete block and some types of stone may provide poor or nonuniform bond. If suction is not uniform over the entire surface, parts of the wall may draw more moisture than others, and the final finish will be spotty in color. Also, nonuniform bonding is likely to cause cracking because the stucco surface will move with the supporting base in the areas of good bond and independent of the base in areas of poor bond.

If adequate or uniform bond cannot be obtained, the surface should be prepared with metal reinforcement, a dash coat or a bonding agent (see below). If metal reinforcement is used, it is recommended that backing paper also be used to isolate the stucco membrane from the base and prevent dissimilar suction. When stucco is to be applied as a continuous finish over dissimilar rigid bases, a control joint or a strip of metal reinforcement should be placed over the juncture of the dissimilar materials.

On chimneys, stucco should not be applied directly to the masonry, but should be applied over metal reinforcement. However, backing paper should not be used, because heat might cause deterioration. Flashings should be installed under the chimney cap and at junctures with the roof to prevent water penetration (see page Masonry Walls 343-11).

Concrete

When applying stucco directly to concrete, the surface should be inspected carefully for laitance, form oil or other substances which might impair bond. Sandblasting can be used to remove form oil, curing compounds and other foreign materials. Smooth or slick spots on fresh concrete due to form oil also can be removed by scrubbing with strong soap and water and allowing the concrete to dry to restore suction. However, it is recommended that form oil not be used for concrete forms when direct application of stucco is intended.

Rough lumber can be used for forms, or the freshly placed concrete can be lightly sandblasted to roughen the surface immediately after forms are stripped. Hardened, smooth concrete may require more extensive preparation such as chipping, acid etching, or roughening with a bush hammer. A bonding agent or a dash coat also may be used where roughening of the surface is impractical.

FIG. 7 Dash coat is used to provide bond for concrete or masonry surfaces.

Dash Coat If it is doubtful that good bond can be obtained on concrete or masonry supporting construction, it is sometimes advisable to use a dash bond coat. The mix for a dash coat should consist of 1 part by volume of portland cement and from 1 to 2 parts sand. A small straw broom or long-fibered stiff brush is dipped in the mix and the material is splattered on the wall with a quick throwing motion (Fig. 7). The dash coat should not be troweled and should be allowed to set before the stucco base coat is applied. If there is a delay in applying the base coat, the dash coat should be kept moist until the next coat is applied.

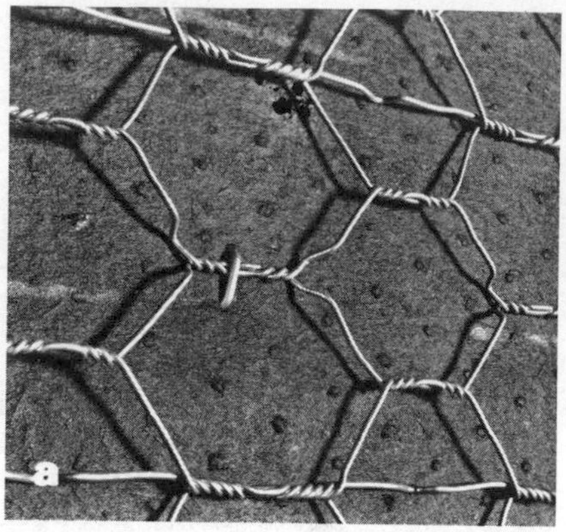

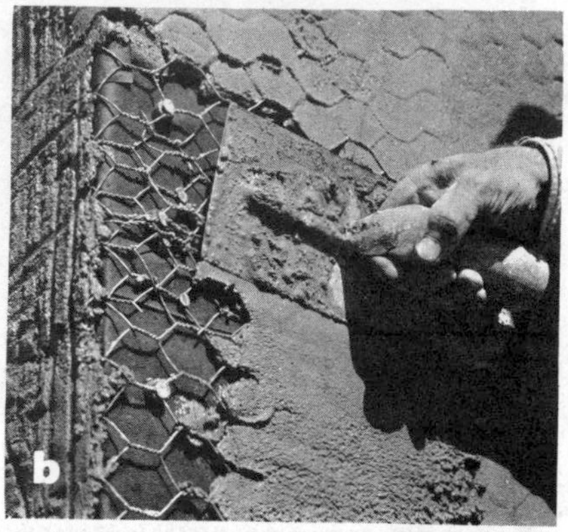

FIG. 8 (a) Self-furring wire lath is crimped every 3"; (b) Fiber spacers on furring nails keep plain wire lath 1/4" away from the backing paper.

Bonding Agent Another method of obtaining good bond on various concrete and masonry surfaces is to apply a waterproof bonding agent recommended by the manufacturer for exterior use. The supporting surface should be structurally sound and rigid, and free from laitance, dust, grease and all materials soluble in water such as water-based paints or glue.

Acid Etching A solution of 1 part muriatic acid and 6 parts water should be used for acid etching. Before applying the acid solution, the surface first should be wetted so that the acid will only act on the surface; several applications may be necessary. After treatment, the concrete should be thoroughly washed with water to remove all traces of the acid and be allowed to dry to restore suction.

Metal Reinforcement

Metal reinforcement should be used when stucco is applied over: (1) wood frame construction, either sheathed or unsheathed; (2) steel frame construction; (3) concrete and masonry providing unsatisfactory bond; and (4) all flashings and chimneys.

Stucco applied over metal reinforcement acts as a thin reinforced slab which, though supported by the underlying construction, is not in sufficiently rigid contact to permit direct transfer of stresses. The reinforcement should be

FIG. 9 RECOMMENDED SUPPORT SPACING FOR METAL REINFORCEMENT[1]

Reinforcement	Weight (lbs/sq. yd.)	Support Spacing
Diamond Mesh	3.4	16" o.c.
Stucco Mesh[2]	1.8 & 3.6	16" o.c.
Flat Rib Lath	3.4	19" o.c.[3]
⅜" Rib Lath	3.4 & 4.0	24" o.c.
Woven Wire	1.7 (1" hex.)	16" o.c.
	1.4 (1½" hex.)	16" o.c.
Welded Wire	1.4 (1" x 1")	16" o.c.
	1.4 (2" x 2")	16" o.c.
	1.9 (1½" x 2")	16" o.c.[4]

1. Support spacing applicable to both wood and metal framing for both vertical and horizontal surfaces, unless otherwise noted.
2. Not recommended for vertical surfaces over metal framing or for horizontal surfaces.
3. Spacing of metal framing for horizontal surfaces should not exceed 13½" o.c.
4. Spacing may be increased to 24" o.c. when stiffener wires are provided 6" o.c.

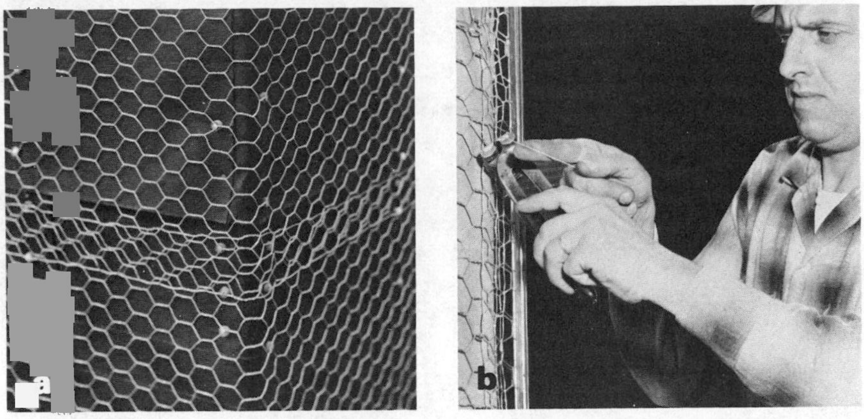

FIG. 10 (a) External corners can be reinforced by bending lath around the corner. (b) Zinc cornerbead makes a strong, neat corner.

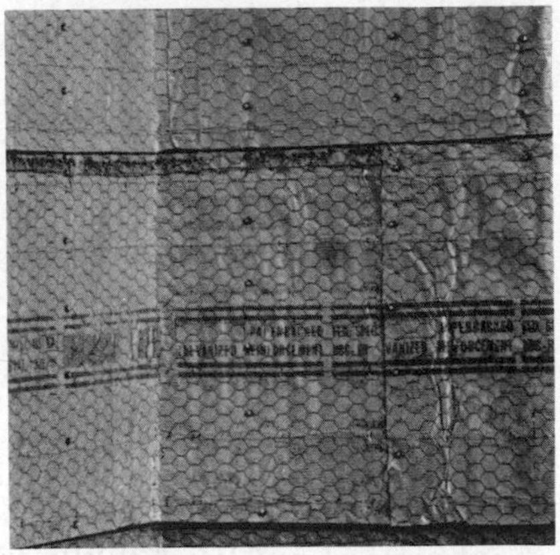

FIG. 11 Woven wire lath is readily bent to conform to internal corner.

attached securely to supporting construction and stretched tightly between supports to eliminate slack. Loosely attached reinforcement generally results in an uneven thickness of stucco, with the thinner and weaker sections more likely to crack. At least a 1/4″ space should be maintained between the reinforcement and the supporting construction by use of either furring fasteners or self-furring reinforcement (Fig. 8). The metal reinforcement should be of suitable type and weight for the support spacing (Fig. 9) and should be attached with fasteners recommended for the particular supporting construction (Fig. 12).

Generally, the reinforcement should be installed with the long dimension perpendicular to supports and end laps staggered, forming a continuous network of metal over the entire surface. Expanded metal lath should be lapped 1/2″ at sides and 1″ at ends, and edge ribs can be nested. Wire lath should be lapped at ends and sides at least one full mesh but not less than 1″. Generally, lower sheets should be lapped over upper sheets at horizontal laps; however, paper-backed reinforcement should be applied shingle fashion with upper sheets sandwich-lapped over the lower sheets with metal on metal.

External corners can be formed with cornerbead, with prefabricated corner reinforcement, or by bending the metal reinforcement around the corner (Fig. 10). Internal corners can be reinforced with cornerite wired to the reinforcement or the metal reinforcement itself can be bent to conform to the angle (Fig. 11).

FIG. 12 FASTENING RECOMMENDATIONS FOR REINFORCEMENT OVER WOOD SUPPORTS[1]

Reinforcement	Nails		Staples	
	Type[2]	Maximum Spacing	Type	Maximum Spacing
Diamond Mesh	1½″, 11 ga. barbed, 7/16″ head	6″ o.c.	⅞″ leg, 16 ga. ¾″ crown	6″ o.c.
Stucco Mesh[3]		6″ o.c.		6″ o.c.
Flat Rib Lath		6″ o.c.		6″ o.c.
⅜″ Rib Lath		6″ o.c.[4]	1¼″ leg, 16 ga., ¾″ crown	4½″ o.c.
Woven & Welded Wire		6″ o.c.	⅞″ leg, 16 ga. 7/16″ crown	6″ o.c.

1. Applicable to both vertical and horizontal surfaces, unless otherwise noted.
2. 1″ roofing nails suitable on vertical surfaces for all but ⅜″ ribbed lath. 1½″ 12 ga., ⅜″ head furring nails suitable to attach welded and woven wire on vertical surfaces.
3. Not recommended for horizontal surfaces.
4. Maximum spacing 4½″ o.c. for horizontal surfaces using 2″ nails.

Wood Framing

In wood frame construction, stucco should be applied over backing paper and metal reinforcement. The reinforcement should be attached with the long dimension across the studs with corrosion-resistant nails or staples, sized and spaced as shown in Figure 12. Nails generally should provide at least 7/8″ penetration into supporting construction on walls and 1-3/8″ penetration on ceilings. Aluminum nails react chemically with fresh portland cement and should not be used. For most types of reinforcement, 1″ or 1-1/2″ nails are generally suitable. However, 3/8″ rib lath usually is nailed at the rib and 1-1/2″ or 2″ nails are recommended to assure the minimum penetration for walls and ceilings.

Wood frame construction falls into two categories: *unsheathed (open-frame)* and *sheathed* construction. Generally, studs should be of at least 2″ x 4″ dimension, spaced not more than 24″ o.c. when supporting the ceiling and roof only, and not more than 16″ o.c. when supporting one or two floors. Studs in exterior walls supporting more than two floors should be at least 2″ x 6″ dimension, spaced not more than 16″ o.c.

Unsheathed Construction The structural frame should be properly braced, by let-in 1″ x 4″ or larger diagonal bracing or by other means, to resist racking forces. Paper-backed reinforcement can be applied directly to unsheathed construction. If backing paper is applied separately, a soft annealed steel *line wire*, 18 gauge or heavier, should be stretched across the faces of the studs in horizontal strands about 6″ apart. The wire should be stretched taut by first nailing or stapling it at every 5th stud, then securing it to the center of intermediate studs and stretching it tightly by raising and lowering the attachments (Fig. 13a). Backing paper then is nailed over the line wire with edges lapped at least 3″. The wire will prevent the paper from sagging, and a more uniform thickness of stucco less likely to crack results.

Application of the flexible reinforcement should start at least one full stud away from any corner and should be bent around the corner to avoid a joint at the corner. If the reinforcement is too rigid to be bent easily, prefabricated corner reinforcement should be used. Horizontal laps should be wired at least once between each stud space but at spacings not more than 9″ o.c., and vertical laps should be made over supports. The bottom of a stucco panel should be terminated with a metal drip screed. The screed will help to support the stucco membrane and, if properly located, will permit drainage of any water that might penetrate the stucco (Fig. 14).

As the stucco is applied by trowel or machine, it should be forced through the openings in the reinforcement against the paper backing to assure at least 90% coverage of the reinforcement. An alternate method sometimes used to assure complete embedment of the reinforcement is to eliminate the backing paper and the reinforcement is back plastered between supports. The back-plaster coat should be at least 3/8″ thick and should be applied only after the scratch coat has hardened sufficiently so that pressure will not break the stucco keys. The surface should be uniformly dampened before application and moist-cured for at least 24 hours after application.

Sheathed Construction Lumber board, plywood or other suitable panel sheathing should be applied horizontally and be fastened securely to each stud (Fig. 13b). Lumber boards used for sheathing should not be more than 8″ in width; they should be fastened with at least 2 nails at each support to minimize the possibility of damage from warping. The sheathing should be covered with backing paper, lapped at least 3″, followed by metal reinforcement fastened through the sheathing into the wood supports. Paper-backed reinforcement can be attached directly over the sheathing. The reinforcement should be returned at least 6″ around corners. Vertical laps over supports and horizontal laps should be nailed or stapled 6″ o.c. into the sheathing. Unless the reinforcement is self-furring, nails should be of the furring type; staples can be used for self furring reinforcement.

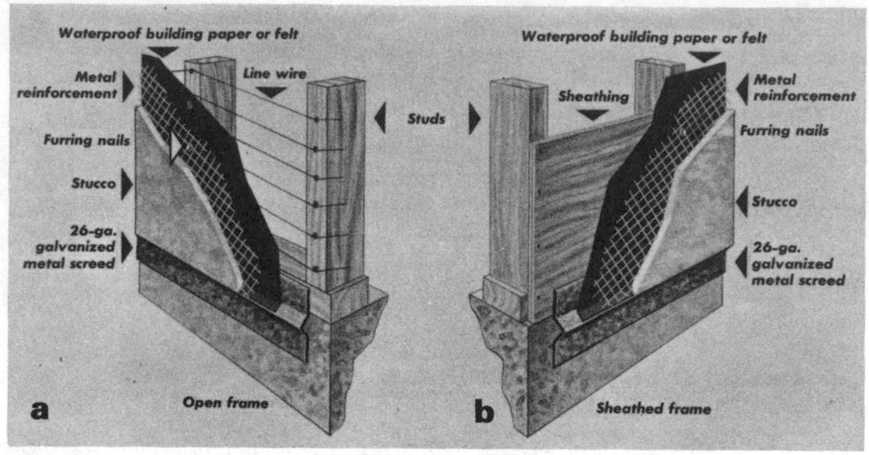

FIG. 13 (a) Unsheathed (open frame) construction; (b) sheathed construction. Line wire is not required with paper-backed reinforcement.

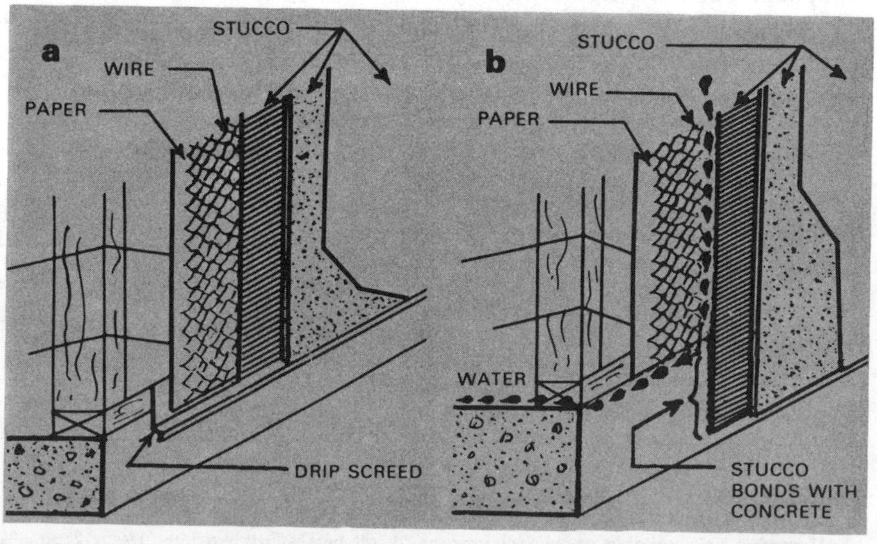

FIG. 14 (a) Properly installed drip screed allows drainage and reduces the possibility of water penetration into the wall. (b) No drip screed and bond between concrete and stucco may result in moisture backup.

Metal Framing

The use of stucco finishes over metal framing is not common in low-rise residential construction. However, the thermal insulating and fire-resistant properties of lightweight aggregates such as perlite and vermiculite have encouraged the use of stucco in multistory buildings. Lightweight aggregates are not recommended for exterior applications, and are used primarily for a back-plaster coat. Stucco applied over metal framing provides noncombustible construction and the back plaster thickness can be varied to produce fire ratings up to 4 hours.

Assemblies consisting of steel studs, furring channels and/or reinforcing rods should be properly spaced and securely fastened according to the specialty manufacturer's recommendations. The recommended spacing of metal supports for various types of reinforcement closely parallels that of wood (Fig. 9).

Concrete and Masonry

Concrete and masonry surfaces which do not provide adequate bond and which cannot be economically prepared for direct application of stucco (see page 425-9) require the use of metal reinforcement. The surface should be covered with backing paper and reinforcement should be attached with suitable fasteners spaced 6″ o.c. vertically and 16″ o.c. horizontally. Horizontal and vertical laps should be secured with fasteners or wire ties 6″ o.c. Self-furring reinforcement can be attached with powder-actuated fasteners or concrete stub nails; plain reinforcement should be attached with concrete furring nails. All nails should be long enough to provide at least 3/8″ penetration into the concrete or masonry.

MIXES

Generally, stucco is applied either in two or three coats, depending on the supporting construction. Three-coat application, used over metal reinforcement, consists of a *scratch coat*, a *brown coat* and a *finish coat*—each applied separately and allowed to set partially. In *two-coat* application, generally used over concrete and masonry bases, a single *base coat* takes the place of the scratch and brown coats, and is followed by a *finish coat.* The scratch coat and brown coat in three-coat work sometimes also are referred to as base coats.

Several recommended mixes, classified according to plaster groups (C, L, F and P) are outlined in Figure 15. The letter designations broadly indicate the type of mix: C = (portland) cement, L = lime, F = finish coat and P = plastic cement. The selection of a suitable stucco mix for a specific coat depends in part on the bonding properties of the stucco base (Fig. 16).

Generally, recommended proportions are 1 part cement (including lime when used) to at least 3 and not more than 5 parts by volume of damp, loose sand. Since wide variations in sand quality and gradation exist throughout the country, the correct proportioning of sand to cement should be determined from trial mixes or previous experience with local aggregate sources.

Proper gradation of aggregate from fine to coarse particles (Fig. 2) is highly important to producing a durable stucco finish. If the mix contains an excess of fine particles, more water-cement paste is required to coat the larger surface area of the particles. When an excess of coarse particles is present, more paste is required to fill the spaces between particles.

A lack of adequate water-cement paste is demonstrated by poor workability. The addition of either too much water or too much cement to improve workability results in a stucco finish subject to high drying shrinkage and cracking. Also, sand fines often contain a high percentage of clay particles which shrink excessively upon drying; while an excess of coarse particles produces a harsh

FIG. 15 RECOMMENDED STUCCO MIXES*

Group	Portland Cement	Lime	Masonry Cement Type II	Plastic Cement	Sand
C	1	0 to ¼	—	—	3 to 4
C	1	—	1	—	6 to 7½
C	1	¼ to ½	—	—	4 to 6
L	1	½ to 1¼	—	—	4½ to 9
L	—	—	1	—	3 to 4
F	1	1¼ to 2	—	—	5 to 10
P	1	—	—	1	6 to 10
P	—	—	—	1	3 to 4

*Proportions are given in parts by volume.

FIG. 16 STUCCO MIX SELECTION GUIDE

Stucco Base	Recommended Groups for Stucco Coats: Scratch	Brown[1]	Finish[2]
Low absorption (placed concrete, dense clay brick)	C, P	C, L, P	L, F, P
High Absorption (concrete unit masonry, clay brick & structural tile)	L, P	L, P	L, F, P
Metal Reinforcement[3]	C, P	C, L, P	L, F, P

1. Use as base coat in two-coat work over concrete and masonry bases.
2. Finish coat may be factory-prepared "stucco finish."
3. Over any type of supporting construction.

mix which is difficult to apply. Therefore, aggregate should be selected carefully and mix proportions should not be relied on to overcome deficiencies in aggregate quality or gradation.

A good mix can be recognized by its workability: ease of troweling, good adhesion to the base and sufficient cohesiveness so that the mix will not sag on vertical surfaces. Once the mix has been determined, uniformity in proportioning of materials from batch to batch will help to assure uniform suction and color.

The materials for all coats should be thoroughly mixed. Dry materials should be mixed to a uniform consistency before water is added. A power mixer is recommended for uniform blending of materials (Fig. 17), and it should be operated for at least 5 minutes after all ingredients are in the mixer. Materials should be mixed only in quantities which can be used within 2-1/2 to 3-1/2 hours; remixing to restore plasticity is permissible, but retempering with additional water should be avoided.

When the stucco is applied by machine, the amount of water should be adjusted to produce a mix of proper consistency for the length of hose, temperature and humidity conditions, and type of base. When measured with a standard 2″ x 4″ x 6″ slump cone, the slump should be from 2-1/2″ to 4″ at the mixer and from 2″ to 3-1/4″ at the nozzle. The hose length from the machine to the working surface should be as straight and as short as possible, and generally not longer than 200′.

Masonry and Plastic Cement Mixes

When masonry cement and plastic cement are used, the addition of lime or plasticizers generally is not necessary as plasticizing materials are included during manufacture. The use of these cements simplifies field mixing because only sand and water need to be added. Either masonry cement or plastic cement can be combined with portland cement in equal volumes when greater strength is required (Figs. 15 and 16). Plastic and masonry cements alone or in combination with portland cement can be used for any of the three stucco coats.

FIG. 17 A power mixer combined with pump facilitates accurate mixing and machine spraying of stucco.

FIG. 18 Protective enclosures in mild climates prevent rapid loss of moisture and facilitate curing.

Portland Cement Mixes

Portland cement should be supplemented with lime, plastic cement or masonry cement to impart suitable workability and plasticity to the mix. For base coats, portland cement can be mixed with up to 1-1/4 parts lime, or in equal proportions with masonry or plastic cements. For the finish coat, 1 part portland cement can be supplemented with an equal part of plastic cement or 1/2 to 2 parts lime (Figs. 15 & 16). Truer colors will be obtained if white portland cement and light-colored sand are used for field-mixed finish coats.

Plasticizers such as asbestos, animal, vegetable or glass fibers can be added to mixes for the first coat, and only in quantities sufficient to obtain required plasticity or cohesiveness. The kind and exact amount of plasticizing agent will vary with the type of aggregate, stucco base and manner of application. The amount of asbestos fibers used should not exceed 3% of the cementitious material by weight. The best combination of ingredients should be determined by trial mix prior to application.

STUCCO APPLICATION

Stucco should not be applied in freezing weather or to surfaces containing frost. The temperature of the stucco should be maintained at 50° F or higher during application and for at least 48 hours after the finish coat is applied. In the case of falling temperatures, the stucco should be protected against freezing by heated enclosures (Fig. 18, see also page 425-16).

Hand application of stucco involves traditional tools and practices proven successful over centuries of use. However, machine application incorporates additional advantages. The wet mix can be delivered under pressure to scaffolding in multistory buildings (Fig. 18); application is speeded, and obvious lap and joint marks, sometimes difficult to avoid with hand application, are eliminated. Machine application results in a more uniform texture and deeper and darker colors can be produced than are possible with hand application. Also, some textures can be obtained best by spraying. For these reasons, the finish coat sometimes is applied by machine over a hand-applied base coat.

Three-Coat Work

Three coats totalling at least 7/8″ are required over all types of metal reinforcement. The first coat, the *scratch coat,* provides the mechanical keying with the reinforcement and a rigid surface for the following coat. The *brown coat* provides additional thickness and strength and assures a level surface for the application of the final decorative *finish coat.* Generally, each coat should be moist-cured before the following coat is applied (see

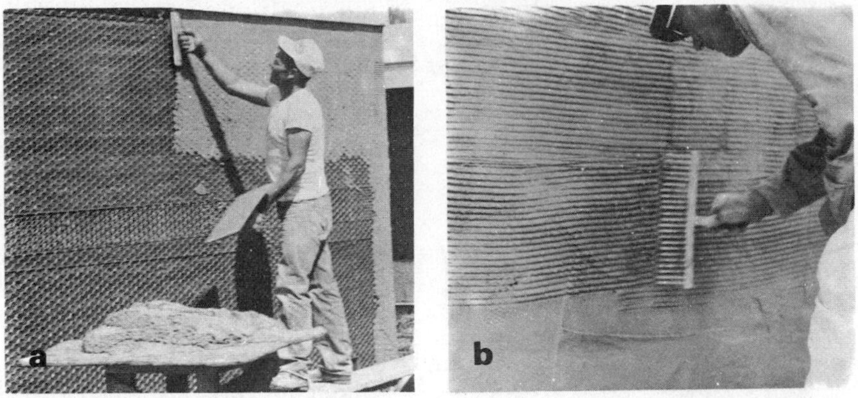

FIG. 19 Scratch coat: (a) mix is spread on reinforcement with a trowel; (b) the surface is raked with a scratcher.

Curing, page 425-16).

Scratch Coat The first coat should be at least 3/8″ thick when measured from the rigid base or the backing paper. It should be applied evenly, with sufficient pressure to squeeze the mix through the reinforcement to form mechanical keys (Fig. 19a). The space between the metal and the backing should be filled as thoroughly as possible so that the metal will be sufficiently embedded. The coverage on the surface should be such that no more than 10% of the surface of the reinforcement remains exposed. As soon as the scratch coat has become firm, its entire surface should be cross-raked *(scratched)* to provide good mechanical bond for the brown coat (Fig. 19b). On walls, horizontal cross raking generally is recommended. Regardless of whether application is by hand or by machine, cross raking is done manually.

A dash bond coat over a rigid base can be used as the first coat of three-coat application. The dash-coat surface is sufficiently rough so that cross raking is not required. The dash coat should be moist cured for 12 to 24 hours before the brown coat is applied.

Brown Coat After the scratch coat has been moist-cured and is sufficiently firm to carry the weight of the next coat, the brown coat should be applied. If the scratch coat becomes dry, it should be dampened again to control suction before the brown coat is applied. The brown coat should be approximately 3/8″ thick, bringing the thickness of both coats to at least 3/4″ (Fig. 20a).

When applying the brown coat, if possible, an entire wall panel should be covered without stopping. Interruptions in the work should be made at screeds, control joints, corners, pilasters, belt courses, doorways, openings or other lines where changes in appearance, which might show through the finish coat, will not be obvious.

The brown coat should be *rodded* with a straightedge and then *darbied* to impart a level, uniformly rough texture. With machine application, these operations are performed by hand (Fig. 20b). The same type of darby or float should be used for the entire surface to assure uniformity of texture. Deep scratches or surface defects should be smoothed out, as such irregularities may cause unequal suction or may show through to the finish coat. External corners should be formed with a straight edge when corner reinforcement is not used (Fig. 20c).

After moist curing, the surface should be allowed to dry uniformly before the finish coat is applied (see Curing, page 425-16).

Finish Coat The brown coat should be uniformly dampened if the finish coat is to be hand-applied. Dampening is not required for machine-applied finishes. The

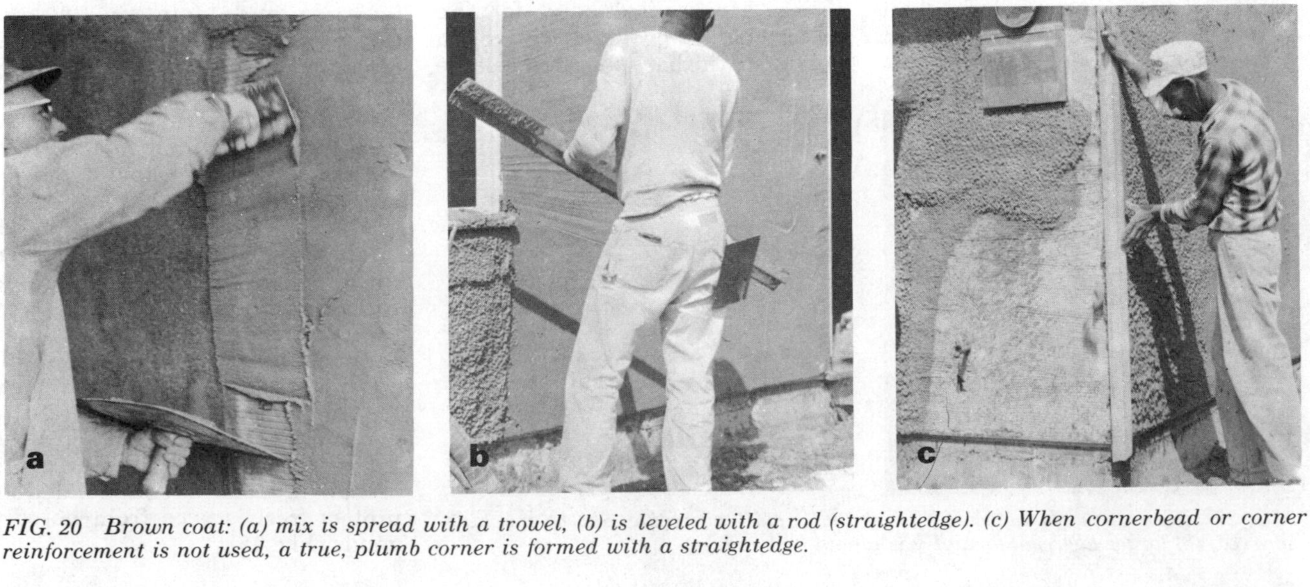

FIG. 20 Brown coat: (a) mix is spread with a trowel, (b) is leveled with a rod (straightedge). (c) When cornerbead or corner reinforcement is not used, a true, plumb corner is formed with a straightedge.

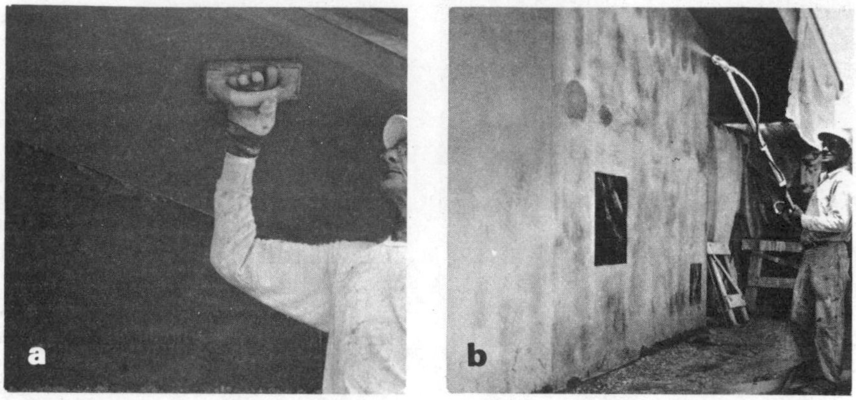

FIG. 21 *The finish coat can be hand-floated (a), or sprayed by machine (b).*

finish coat should be at least 1/8″ thick, bringing the total stucco finish to at least 7/8″ thickness (Fig. 21a). When machine applied, the sprayed finish coat often is made in two or more consecutive applications (Fig. 21b). The mix for the first application is generally of thinner consistency and is applied primarily to insure complete coverage of the brown coat. The successive applications are somewhat thicker and are intended to develop the finished texture.

Factory-prepared stucco finishes assure greater uniformity of color in the finish coat. A variety of decorative surface finishes are possible with both machine and hand application (Fig. 22, also see page 425-19). The finish coat should be moist-cured if a plain stucco is used and is intended for painting. Prepared stucco finishes often contain curing agents and moist curing is not necessary; in fact, it is specifically discouraged by some manufacturers because it may cause streaking of the color surface. The finish coat should be maintained at the proper temperature for at least 48 hours after application (see page 425-16).

Two-Coat Work

Over rigid concrete or masonry bases, two-coat application is acceptable, but three coats should be used where increased thickness and/or strength is required. The *base coat* and *finish coat* should total at least 5/8″ in thickness.

Base Coat The surface of the concrete or masonry should be dampened uniformly to assure uniform suction. When a chemical bonding agent is used, dampening of the surface is not required. The base coat is applied approximately 3/8″ thick and should not be scratched as in three-coat application. Mechanical bond for the finish coat is provided by rough floating of the base coat in the same manner as the brown coat in three-coat work (Fig. 23).

When using machine application, the base coat is built up with several applications, with the final application being rodded and darbied manually.

Finish Coat The finish coat should be approximately 1/4″ thick, and can be applied in a variety of textures by hand or machine, as with three-coat application.

CURING

Like concrete, stucco requires sufficient water and favorable temperatures to facilitate the process of hydration by which the cement gains strength and hardens to a solid mass. The large exposed area of stucco relative to its volume encourages rapid loss of moisture by evaporation. *Moist curing* of each coat attempts to replace this moisture and maintain an adequate supply for the chemical reaction to continue. An ordinary hose nozzle can be adjusted to provide a fine spray for curing; a steady stream which could erode the stucco surface should not be used.

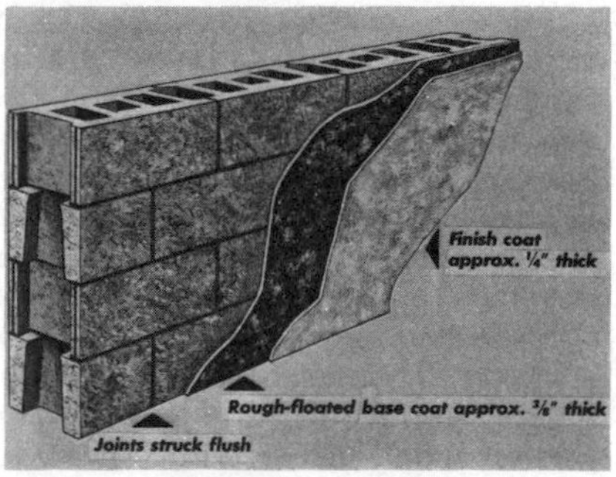

FIG. 23 *Two coats of stucco are adequate over a suitably rough concrete or masonry surface.*

The act of *moist curing* does not infer spraying at set intervals, but rather periodic observation and spraying when the surface appears to be drying out. Since evaporation is increased with increased air circulation and higher temperatures, more frequent spraying generally is required in breezy, warm weather. Also, wind breaks or hangings are recommended to protect fresh stucco from hot dry winds (Fig. 18).

FIG. 22 *Rock dash (marblecrete) finish: (a) Small aggregate can be machine-sprayed; (b) larger aggregate usually is placed by hand.*

FIG. 24 **RECOMMENDED CURING SCHEDULE FOR STUCCO FINISHES**

Application[1]	Moist Curing[2]	Total Setting Time[3]
Scratch Coat	12 to 24 hours	At least 48 hours (between coats)
Brown Coat	12 to 24 hours	At least 7 days (between coats)
Finish Coat	12 to 24 hours[4]	At least 48 hours (after application)

1. In two-coat work, requirements for base and finish coats are the same as for brown and finish coats, respectively.
2. Moist curing should be delayed until scratch and brown coats are sufficiently set to prevent erosion, and for 12 to 24 hours for finish coat.
3. At least 50° F. temperature should be maintained during and after application.
4. Not recommended for prepared colored "stucco finishes."

In addition to adequate moisture, proper hydration requires that the stucco mix and each stucco coat be maintained at temperatures of at least 50° F. The finish coat should be maintained at this temperature for at least 48 hours after application. In cold weather, therefore, it may be necessary to heat the mixing water and provide a heated enclosure around the work area.

Curing Procedures

While it is generally accepted in the industry that adequate water to assure hydration is essential in producing quality stucco finishes, there is no general agreement as to exact curing procedures. The curing recommendations given in Figure 24 are based on the premise that *deliberate moist curing* assures a greater measure of success than does dependance on favorable climatic conditions or on water contained in the mix.

In some parts of the country, no deliberate moist curing is performed, and each coat is applied as soon as the previous coat is sufficiently hard to support the following coat. This method, sometimes called *doubling back*, relies on the protective nature of each covering coat to retain the moisture and facilitate curing of the underlying coat.

Another recommended practice involves application of coats at not less than 24 hour intervals, with continuous moist curing between coats and after the finish coat is applied.

The curing method chosen should be based on accepted local practice which has a demonstrated history of satisfactory performance. All the various curing procedures attempt either to maintain the previous coat uniformly damp, or to permit it to become uniformly dry. This is particularly true of the base (or brown) coat, because uniformity of color in the finish coat is dependent on equal suction in the base coat.

The long waiting period between base and finish coats recommended in Figure 24 assures that the brown coat will be uniformly dry. When shorter intervals are allowed, it is often necessary to keep the base coat damp by continuous moist curing, until the finish coat is applied. Unless a rigid schedule for rapid successive applications can be maintained, the need for continuous moist curing may be more demanding than the recommended short curing period and subsequent drying delay.

CRACK CONTROL

The proper design of supporting construction, selection of materials and application of stucco finish—all can substantially minimize or eliminate undesirable cracking. Cracks can develop in the stucco finish from: (1) shrinkage stresses, (2) structural movement, (3) restraints due to penetration of mechanical or electrical equipment or intersecting walls or ceilings, or (4) weak sections in the finish due to thin spots, openings, or changes in materials in the supporting construction.

Shrinkage Stresses

As the wet stucco mix hydrates and dries, it also tends to shrink, setting up stresses in the stucco membrane. If shrinkage stresses exceed the tensile strength of the stucco membrane, cracking results. It is desirable, therefore, to achieve the full strength of the stucco mix and to minimize the degree of shrinkage. Proper moist curing and protection from temperature extremes allows the stucco to dry slowly and uniformly, thus increasing strength and reducing shrinkage. Shrinkage stresses also are minimized by proper proportioning of mixes and by proper placement of control joints.

Proportioning Mixes The chief factors in proportioning mixes which affect the balance between tensile strength and shrinkage stresses in the finished stucco are the cement content, the water content, and the aggregate quality and gradation.

Rich mixes, high in cement content, generally result in stucco finishes with higher initial shrinkage, balanced to some extent by increased strength. Excessive water in the mix increases the volume of minute voids in the stucco as water evaporates or is used up in hydration, and results in both higher shrinkage and lower strength.

Poorly graded sand requires the addition of more water or more cement, or both, to obtain suitable workability, either of which can be detrimental and cause shrinkage cracking (see page 425-13).

Control Joints When stucco is applied directly to masonry or concrete and no reinforcement is used, control joints in the stucco should coincide with joints in the base. Additional control joints to resist shrinkage cracking in the stucco finish are not required, because the underlying base will resist stresses due to shrinkage.

When stucco is applied to metal reinforcement (over any type of supporting construction) control joints should be installed in the stucco finish not more than 18′ apart, forming panels not larger than 150 sq. ft.

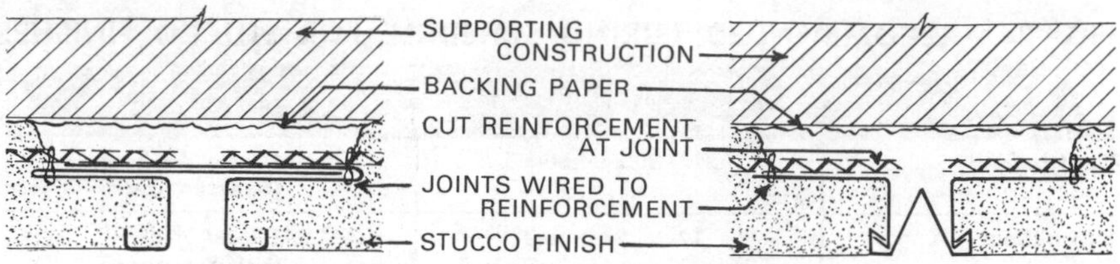

FIG. 25 Multi-section control joints effectively relieve stresses due to drying shrinkage, structural movement and temperature variations.

For vertical joints in walls and for all joints in protected soffits, either one-piece or multi-section control joints can be used (Fig. 5b). The multi-section control joint is particularly effective in relieving stresses caused by shrinkage and differential structural movement in the supporting construction (Fig. 25). For horizontal joints, a watertight control joint designed for the purpose should be used (Fig. 26).

Structural Movement

Structural movement may occur from such causes as foundation settlement, excessive deflection of structural members due to overloading, racking due to wind forces and expansion and contraction of materials caused by temperature and humidity variations. Good structural design attempts to minimize these movements where possible; and specially designed joints are used to localize movements which cannot be reduced to negligible proportions.

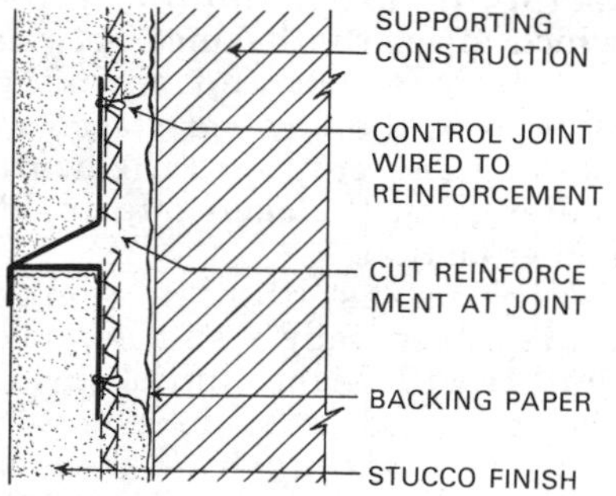

FIG. 26 Specially designed horizontal control joints provide weathertightness as well as crack control.

Relief joints and control joints are used to localize movement and minimize stresses within the stucco finish. Relief joints at ceilings and walls should be used where the stucco is intersected (Fig. 27) or is penetrated by structural elements such as beams, columns or load bearing walls (Fig. 28). Location of control joints is dependent on the type of supporting base.

Rigid Bases Where stucco is applied directly to a concrete or masonry base, firm and continuous support of the stucco membrane will tend to resist shrinkage, temperature and impact stresses in the stucco finish. However, the finish will be subjected to structural movements in the base and is likely to crack wherever excessive stress concentrations develop. It is particularly important, therefore, in this type of installation to provide proper control joints in the base (see page Masonry Walls 343-14, Fig. MW18). Metal control joints should be placed in the stucco finish directly over each underlying joint in the base. Control joints also should be placed in the stucco finish over any construction, isolation or expansion joint which occurs in the base and at the juncture of dissimilar materials.

Metal reinforcement Stucco applied over backing paper and metal reinforcement acts as a thin concrete slab. Because the stucco finish is somewhat isolated from the supporting construction, it is not subjected to structural stresses to the same degree as stucco applied directly to rigid bases. However, such a thin reinforced slab is subject more to shrinkage, impact and temperature stresses because it is not supported and restrained by an underlying rigid base. Of these, shrinkage stresses are the most critical and involve the greatest movement. Therefore, control joints located as recommended for shrinkage purposes also will relieve stresses occasioned by other forces acting on the stucco membrane.

Weakened Sections

Normal stresses caused by shrinkage, humidity or temperature variations are evenly distributed in stucco of uniform thickness, but tend to be higher at the thin spots. If the stucco finish is applied over an irregular, rigid base, or over sagging metal reinforcement, the resulting finish will be of unequal thickness, creating weakened sections. Such concentration of stresses often results in cracking at these weakened points. To minimize this hazard, the stucco should be applied over a level base in uniform thickness.

Weakened sections also are created at corners of cutouts and openings in the stucco finish. Metal reinforcement will resist the concentration of stresses at the corners and generally prevents random

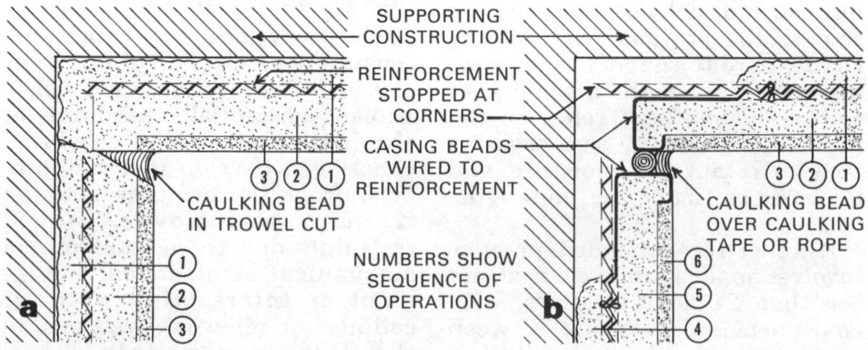

FIG. 27 Perimeter relief at internal corner between wall and soffit: (a) using trowel cut, (b) formed with casing beads.

diagonal cracking. When reinforcement is not used and stucco is applied directly to a rigid base, strip reinforcement (striplath) can be installed diagonally at each corner and will act similarly to prevent cracking.

As in concrete, weakened sections can be introduced deliberately at planes where cracking would not be objectionable. A trowel cut adjacent to a door or window frame will result in a straight, inconspicuous crack rather than objectionable random cracking. Often, the trowel cut can be concealed by trim; where trim is not intended, the cut can be left exposed if neatly formed and caulked.

Restraints

Free movement of the stucco membrane generally is restrained at corners and wherever the membrane is penetrated by fixtures or structural elements. Cracking due to restraint is not common at external corners, particularly if control joints are provided in each plane as recommended for shrinkage control.

Cracking is more frequent at internal corners formed by the intersection of load-bearing elements (columns, walls, pilasters) and non-load-bearing walls, and at the juncture of walls and suspended or furred ceilings. At such corners, *perimeter relief* should be provided by omitting cornerite, stopping the metal reinforcement in each plane and including relief joints which will permit independent movement in each plane.

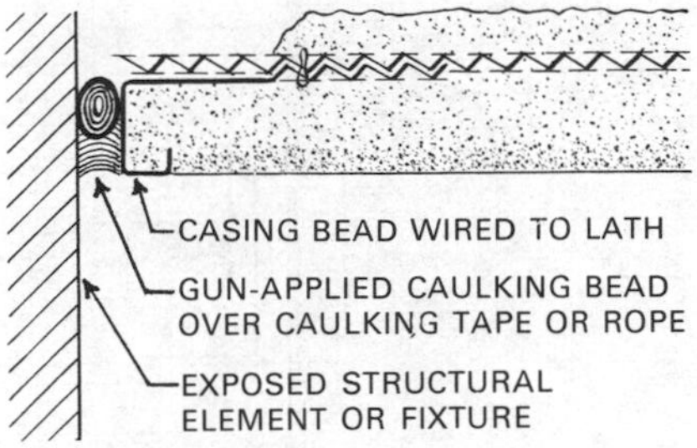

FIG. 28 Perimeter relief joint for use where structural elements or fixtures penetrate the surface.

Relief joints can be formed by a pair of casing beads (Fig. 27b) or a trowel cut (Fig. 27a). The casing beads should be wired to the reinforcement, not rigidly nailed to the supporting construction. The corner should be caulked with a nonhardening resilient compound, preferably after initial shrinkage has taken place.

Where a stucco surface abuts or is penetrated by a structural element or a fixture, a relief joint should be provided. The stucco should terminate with a casing bead; the space between the bead and the structural element or fixture should be caulked (Fig. 28).

SURFACE FINISHES

In addition to providing the desired decorative surface, the finish selected also affects the weather resistance and maintenance requirements of the stucco installation. Smooth troweled stucco finishes generally are not recommended because they tend to make surface irregularities more conspicuous. The rougher textures generally will conceal slight color variations, lap joints, uneven dirt accumulation and streaking.

The range of possible stucco textures is practically limitless. Brick, stone and travertine marble can be simulated by skilled applicators; many other finishes can be produced by hand or machine, depending on the desired texture. The finish may be natural or integrally colored so that painting is

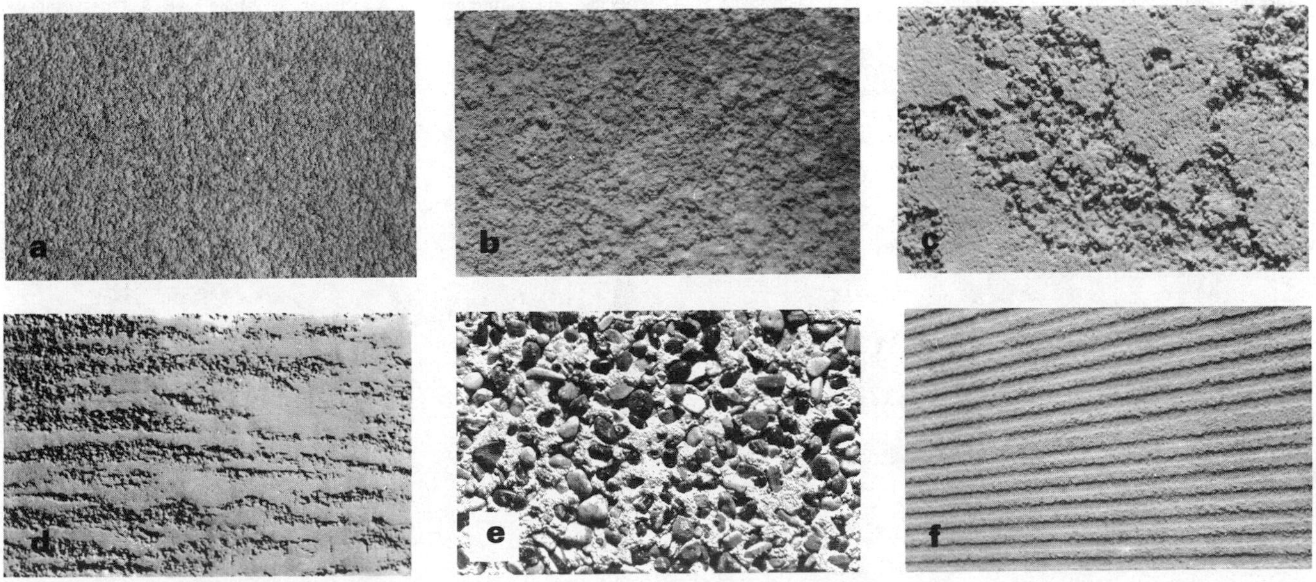

FIG. 29 Typical stucco textures: (a) **float finish,** *made with carpet or rubber-faced float; (b)* **wet-dash,** *coarse aggregate mix is dashed on with a brush; (c)* **dash-troweled,** *high spots of dashed surface are troweled smooth; (d)* **stippled-troweled,** *surface is stippled with broom, then high spots are troweled smooth; (e)* **rock-dash** *or* **pebble-dash,** *decorative pebbles are thrown by hand or machine against wet surface; (f)* **combed,** *surface is formed with a notched template.*

not required. Colored finishes can be effected either by using a prepared *stucco finish*, by adding pigment to the mix, or by embedding colored aggregate in the stucco surface. Manufacturers of colored stucco finishes issue application recommendations which should be followed carefully.

Several of the common textures are illustrated in Figure 29. The character of these textures can vary substantially with each applicator. Therefore, a standard sample should be produced at the beginning of the work, and as the work progresses, it should be checked with the sample for conformity.

RESURFACING

Existing concrete, masonry and stucco surfaces which are firm and sound and which can provide adequate bond by suction or mechanical key are suitable for direct application of stucco. Painted or coated surfaces can be sandblasted or otherwise treated to remove the coating. Smooth concrete surfaces can be etched or treated with a bonding agent.

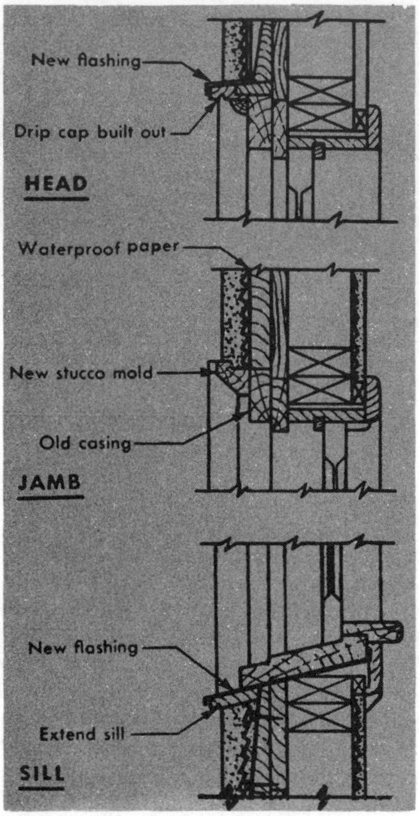

FIG. 30 Resurfacing details for stucco over wood frame construction.

When none of these methods is practical or effective, backing paper and metal reinforcement should be applied over the surface as for new construction.

Either one-, two- or three-coat work may be necessary, depending on the type of base. One finish coat may be sufficient over substantially sound, level stucco with good bonding properties. If the existing stucco is irregular or requires removal of loose material, a base and finish coat are required. Two coats similarly are required over concrete and masonry surfaces suitable for direct application.

When metal reinforcement is used, three coats should be applied. The procedures for application of the several coats are the same as for new construction.

When new stucco is applied over an existing stucco surface, consideration should be given to all points where water might enter behind the new finish (Fig. 30). Often drip caps and flashings over openings must be extended and new window sills and stucco molds installed to project beyond the new stucco finish.

We gratefully acknowledge the assistance of the following for the use of their publications as references and permission to use photographs and illustrations: California Lathing and Plastering Contractors Association, Contracting Plasterers' and Lathers' International Association, Keystone Steel & Wire Company, MAC Publishers Association, Metal Lath Association, National Lime Association, Portland Cement Association.

STUCCO 425

CONTENTS

MATERIALS

See Main Text, page 425-3, for definitions of terminology.

Materials used for stucco finishes should conform to the standard specifications given in Figure WF1.

CEMENTITIOUS MATERIALS

The cementitious materials used in stucco finishes include portland cement, masonry cement, plastic cement and lime. See page WF425-6 for recommended mixes.

Portland Cement

Portland cement should be used in combination with other more plastic materials such as masonry cement, plastic cement or lime.

Type I portland cement generally is used for stucco mixes. Type II, because of its greater resistance to sulphates and greater heat of hydration, can be used over brick or concrete block containing soluble sulphates and during cold weather. Type III develops high early strength and can be used when there is a hazard of freezing. Air entraining portland cement Types IA, IIA and IIIA are recommended for resistance to deterioration from freeze-thaw cycles.

Masonry and Plastic Cements

Masonry and plastic cements consist mostly of portland cement, but contain sufficient quantities of plasticizers, added during manufacture, so that either can be used with only the addition of mixing water and aggregate.

Lime

Lime should be used only in combination with portland cement. Type S (special) finishing hydrated lime is specifically recommended for stucco mixes.

Quicklime and Type N (normal) finishing or Mason's hydrated limes can be used provided they are designated as calcitic or high calcium types by the manufacturer, and provided they are slaked and/or soaked for the required length of time.

Quicklime should be slaked and soaked for a period of 1 to 2 weeks prior to use. Type N lime should be soaked for a period of 12 to 16 hours prior to use.

FIG. WF1 SUMMARY OF STUCCO MATERIALS

Material	Standard Specification
Cementitious Material	
Portland Cement	
Types I, II, III	ASTM C150
Types IA, IIA, IIIA	ASTM C175
Masonry Cement	ASTM C91
Plastic Cement	ASTM C150
Lime, Hydrated*	
Special Finishing Normal Finishing	ASTM C206
Mason's, Types S & N	ASTM C207
Quicklime	ASTM C5
Aggregates	
Sand	ASTM C144
Lightweight	ASTM C332
Metal Reinforcement	FS-QQ-L-101a
Metal Accessories	ANSI A42.3
Backing Paper	
Roofing Felt	FS-HH-R-595
Building Paper	FS-UU-B-790

*Both Special and Normal limes are available as air-entraining types (SA and NA).

(Continued) MATERIALS

Restrictions on the use of building limes are intended to assure either (1) that a minimum of unhydrated oxides are present in the manufactured hydrated product, or (2) that the ingredients are known to hydrate readily during normal slaking and soaking periods. The presence of unhydrated oxides in the stucco finish under conditions of high humidity may initiate a cycle of hydration, causing expansion and consequent cracking of the stucco surface.

FIG. WF2 RECOMMENDED GRADING FOR SAND AGGREGATE

Sieve Size	Percent of Aggregate Retained on Sieve	
	Minimum	Maximum
No. 4	—	0
No. 8	—	0
No. 16	10	40
No. 30	30	65
No. 50	70	90
No. 100	95	100

AGGREGATE

Sand should be natural or manufactured aggregate which is clean, granular and free from deleterious amounts of loam, clay, soluble salts or vegetable matter. Sand should be well graded within the limits given in Figure WF2.

Proper grading of aggregate is essential to producing suitable stucco finishes. An excess of either fine or coarse aggregate particles can result in a mix containing too much cement or too much water, both of which are detrimental to the quality of the stucco finish.

PREPARED FINISH COAT

The manufacturer's recommendations for stucco finishes should be followed carefully.

Factory-prepared mixes (stucco finishes) are available in a variety of colors and textures for decorative finish coats. All ingredients, except water, are included in the mix, and better results generally are obtained for decorative finishes than with field-mixed formulations.

WATER

Water for mixing and curing should be clean and free from deleterious amounts of oils, acids, alkalis, salts and organic materials.

ADMIXTURES

Admixtures such as plasticizers, retarders or accelerators should not be used in quantities which will reduce the compressive strength of the mix by more than 15%, as compared with a similar mix without admixtures.

Plasticizers such as asbestos, animal, vegetable or glass fibers should be free from impurities which may be injurious to the stucco. Asbestos fibers should not exceed 3% by weight of the cementitious materials. Animal hair and vegetable fibers should be 1/2″ to 2″ long.

METAL REINFORCEMENT

Metal reinforcement should conform to the weight requirements given in Figure WF3.

Self-furring reinforcement should be designed to keep the back of the reinforcement at least 1/4″ from the backing paper or stucco base. Openings in the reinforcement should be large enough to facilitate embedment of the metal in the stucco mix.

FIG. WF3 METAL REINFORCEMENT FOR STUCCO FINISHES

Type	Weight lbs/sq. yd.	Opening Size	Dimensions
Expanded Metal Lath			
Diamond Mesh[1]	3.4	5/16″ x 3/8″	27″ x 96″
Stucco Mesh	1.8 & 3.6	1 3/8″ x 3 1/8″	48″ x 99″
Flat Rib Lath	3.4	5/16″ x 3/8″	24″ x 96″ & 27″ x 96″
3/8″ Rib Lath	3.4 & 4.0	5/16″ x 3/8″	27 x 96 & 27″ x 99″
Wire Lath			
Woven Wire[1]	1.7 (18 ga.)	1″ Hex.	3′ x 150′ (rolls)[2]
	1.4 (17 ga.)	1 1/2″ Hex.	
Welded Wire[1]	1.4 (16 ga.)	2″ x 2″	3′ x 150′ (rolls)[2]
	1 4 (18 ga.)	1″ x 1″	
	1.9 (16 ga.)	1 1/2″ x 2″	

1. Available in plain, self-furring and paper-backed types.
2. Paper-backed and self-furring types also available in sheet form.

(Continued) MATERIALS

Expanded metal lath and wire lath should be fabricated from low carbon steel protected by galvanizing, or from cold-rolled steel coated with corrosion-resistant paint.

BACKING PAPER

Backing paper should be a waterproof paper or felt having a vapor permeance of at least 5 perms.

Roofing felt (conforming to FS HH-R-595) and building paper (conforming to FS UU-B-790, Class D) are waterproof but vapor permeable, and generally are used.

FLASHING

Flashing should be corrosion-resistant material, such as copper, galvanized or stainless steel, or suitable plastics (see page Masonry Walls WF343-1, for specific recommendations).

Exposed aluminum flashing is attacked by alkali in the stucco finish. Aluminum flashing may be used in concealed locations when protected with a bituminous coating or when laminated between bituminous coated kraft paper.

METAL ACCESSORIES

Metal accessories which will be completely embedded in the stucco (such as cornerite, strip-lath and prefabricated corner reinforcement) should be manufactured from galvanized low-carbon steel or painted cold-rolled steel.

Exterior metal accessories designated to be left partially or completely exposed (such as control joints, drip screeds, casing beads and cornerbeads) should be manufactured from metal not less corrosion resistant than white alloy zinc.

APPLICATION

SUPPORTING CONSTRUCTION

All supporting construction should be capable of supporting design loads without excessive deflection or without introducing stresses that might cause cracking of the finish. The calculated deflection of ceiling members should not exceed 1/360 of the span under full design load. Headers or lintels of adequate size should be provided over openings and all construction should be true and plumb or level.

Supporting construction should be firm and well braced, but also should be capable of accommodating structural movement where it is expected to occur.

Concrete and masonry construction should be provided with control joints and expansion joints as dictated by good design practice. Control joints in the stucco should be placed directly over these joints in the supporting construction and wherever required to relieve stresses and minimize cracking (see page WF425-8).

STUCCO BASES

Stucco should be applied only to stucco bases which develop adequate bond for the direct application of stucco.

Bond can be provided by mechanical key or by suction. Over metal reinforcement, bond is developed by mechanical keying; over concrete and masonry, bond is developed by either mechanical keying or suction, or both.

Stucco should not be applied directly to gypsum products such as gypsum plaster, gypsum masonry or gypsum lath. Backing paper and metal reinforcement should be installed over such materials before stucco is applied.

Wood lath, fiber insulating lath or sheet metal lath should not be used as a base for stucco.

Masonry

When stucco is to be applied directly to masonry, mortar joints in the masonry should be struck flush or slightly raked. The masonry surface should be clean, sound and firm and should not contain efflorescence, grease, waterproofing compounds, paint or other substances detrimental to good bond.

Most masonry, such as medium-hard common and face brick, standard weight concrete block and many types of stone, generally provide adequate suction and often mechanical bond. The surface can be tested to determine

(Continued) **APPLICATION**

suction by spraying it with clean water. If the water is readily absorbed, good suction is likely to result; if water droplets form, the surface does not provide adequate suction and bond will depend on mechanical key. If the surface is sufficiently rough, it will provide mechanical key.

On surfaces which absorb moisture too quickly, such as soft common brick and lightweight concrete block, excessive suction should be controlled by fine spraying (not soaking) the masonry with several applications of water immediately prior to application of the first stucco coat.

If adequate or uniform bond cannot be obtained, the surface should be prepared with metal reinforcement, a dash coat or a bonding agent (see below).

If metal reinforcement is used, it is recommended that backing paper also be used to isolate the stucco membrane from the base and prevent dissimilar suction.

When stucco is to be applied as a continuous finish over dissimilar rigid bases, a control joint or a strip of metal reinforcement should be placed over the juncture of the dissimilar materials.

Concrete

When applying stucco to concrete, substances such as laitance, form oil or other substances should be removed from the surface.

Rough lumber can be used for forms, or the freshly placed concrete can be lightly sandblasted to roughen the surface immediately after forms are stripped. Hardened, smooth concrete may require more extensive preparation, such as chipping, acid etching, or roughening with a bush hammer. A bonding agent or a dash coat also may be used where roughening of the surface is impractical, see Main Text, page 425-10.

Metal Reinforcement

Metal reinforcement should be used when stucco is applied over: (1) wood frame construction, either sheathed or unsheathed; (2) steel frame construction; (3) concrete and masonry providing unsatisfactory bond; (4) all flashings; and (5) chimneys.

At least a 1/4″ space should be maintained between the reinforcement and the supporting construction by use of either furring fasteners or self-furring reinforcement. The metal reinforcement should be of suitable type and weight for the support spacing (Fig. WF4) and should be attached with fasteners recommended for the particular supporting construction (Fig. WF5).

Generally, the reinforcement should be installed with the long dimension perpendicular to supports and end laps staggered, forming a continuous network of metal over the entire surface. Expanded metal lath should be lapped 1/2″ at sides and 1″ at ends, or ends and sides can be nested. Wire lath should be lapped at ends and sides at least one full mesh but not less than 1″. Generally, lower sheets should be lapped over upper sheets at horizontal laps; however, paper-backed reinforcement should be applied shingle fashion with upper sheets lapped over the lower sheets.

External corners should be reinforced with cornerbead, with prefabricated corner reinforcement or by bending the reinforcement around the corner.

Internal corners should be reinforced with cornerite or by bending the reinforcement to conform to the angle, unless a suitable relief joint is provided.

Wood Framing

In wood frame construction, stucco generally should be applied over backing paper and metal reinforcement. The reinforcement should be attached with the

FIG. WF4 RECOMMENDED SUPPORT SPACING FOR METAL REINFORCEMENT[1]

Reinforcement	Weight (lbs/sq. yd.)	Support Spacing
Diamond Mesh	3.4	16″ o.c.[3]
Stucco Mesh[2]	1.8 & 3.6	16″ o.c.
Flat Rib Lath	3.4	19″ o.c.[3]
⅜″ Rib Lath	3.4 & 4.0	24″ o.c.
Woven Wire	1.7 (1″ hex.)	16″ o.c.
	1.4 (1½″ hex.)	16″ o.c.
Welded Wire	1.4 (1″ x 1″)	16″ o.c.
	1.4 (2″ x 2″)	16″ o.c.
	1.9 (1½″ x 2″)	16″ o.c.[4]

1. Support spacing applicable to both wood and metal framing for both vertical and horizontal surfaces, unless otherwise noted.
2. Not recommended for vertical surfaces over metal framing or for horizontal surfaces.
3. Spacing of metal framing for horizontal surfaces should not exceed 13½″ o.c.
4. Spacing may be increased to 24″ o.c. when stiffener wires are provided 6″ o.c.

(Continued) **APPLICATION**

long dimension across the studs with corrosion-resistant nails or staples, sized and spaced as shown in Figure WF5. Nails generally should provide at least 7/8″ penetration into supporting construction on walls and 1-3/8″ penetration on ceilings.

For most types of reinforcement, 1″ and 1-1/2″ long nails are generally suitable. However, 3/8″ riblath usually is nailed at the rib and 1-1/2 or 2″ nails are recommended to assure the minimum penetration for walls or soffits.

Studs should be of at least 2″ x 4″ dimension, spaced not more than 24″ o.c. when supporting the ceiling and roof only, and not more than 16″ o.c. when supporting one or two floors. Studs in exterior walls supporting more than two floors should be at least of 2″ x 6″ dimension, spaced not more than 16″ o.c.

Unsheathed Construction The structural frame should be properly braced by let-in 1″ x 4″ or larger diagonal bracing or by other means, to resist racking forces.

Paper-backed reinforcement can be applied directly to unsheathed construction. If backing paper is applied separately, a soft annealed steel line wire, 18 gauge or heavier, should be stretched across the faces of the studs in horizontal strands about 6″ apart. Backing paper then should be nailed over the line wire with edges lapped at least 3″.

Application of flexible reinforcement should start at least one full stud away from any corner and the reinforcement should be bent around the corner to avoid a joint at the corner. If the reinforcement is too rigid to be bent easily, cornerite or cornerbead should be used. Horizontal laps should be wired at least once between each stud space but at spacings not more than 9″ o.c., and vertical laps should be made over supports. The bottom of a stucco panel should be terminated with a metal drip screed.

As the stucco is applied by trowel or machine, it should be forced through the openings in the reinforcement against the paper backing to assure at least 90% coverage of the reinforcement.

When the construction is back-plastered, (see Main Text, page 425-12) the back-plaster coat should be at least 3/8″ thick and should be applied only after the scratch coat has hardened sufficiently so that pressure will not break the stucco keys.

Sheathed Construction Lumber board, plywood or other suitable panel sheathing should be applied horizontally and should be fastened securely to each stud. Lumber boards used for sheathing should not be more than 8″ in width and should be fastened with at least 2 nails at each support. The sheathing should be covered with backing paper lapped at least 3″, followed by metal reinforcement fastened through the sheathing into the wood supports.

Paper-backed reinforcement should be attached directly over the sheathing. The reinforcement should be returned at least 6″ around corners. Vertical laps over supports and horizontal laps should be nailed or stapled 6″ o.c. into the sheathing. Unless the reinforcement is self-furring, nails should be of the furring type; roofing nails and staples should be used only for self-furring reinforcement.

FIG. WF5 FASTENING RECOMMENDATIONS FOR REINFORCEMENT OVER WOOD SUPPORTS[1]

Reinforcement	Nails		Staples	
	Type[2]	Maximum Spacing	Type	Maximum Spacing
Diamond Mesh	1½″, 11 ga. barbed, 7/16″ head	6″ o.c.	7/8″ leg, 16 ga. ¾″ crown	6″ o.c.
Stucco Mesh[3]		6″ o.c.		6″ o.c.
Flat Rib Lath		6″ o.c.		6″ o.c.
3/8″ Rib Lath		6″ o.c.[4]	1¼″ leg, 16 ga. ¾″ crown	4½″ o.c.
Woven & Welded Wire		6″ o.c.	7/8″ leg, 16 ga. 7/16″ crown	6″ o.c.

1. Applicable to both vertical and horizontal surfaces, unless otherwise noted.
2. 1″ roofing nails suitable on vertical surfaces for all but 3/8″ ribbed lath. 1½″ 12 ga., 3/8″ head furring nails suitable to attach welded and woven wire on vertical surfaces.
3. Not recommended for horizontal surfaces.
4. Maximum spacing 4½″ o.c. for horizontal surfaces using 2″ nails.

(Continued) APPLICATION

Metal Framing

Assemblies consisting of steel studs, furring channels and/or reinforcing rods should be properly spaced and securely fastened according to the specialty manufacturer's recommendations.

Concrete and Masonry

Concrete and masonry surfaces which do not provide adequate bond and which cannot be economically prepared for direct application of stucco should be covered with backing paper and reinforcement. Reinforcement should be attached with suitable fasteners spaced 6″ o.c. vertically and 16″ o.c. horizontally. Horizontal and vertical laps should be secured with fasteners or wire ties 6″ o.c. All nails should be long enough to provide at least 3/8″ penetration into the concrete or masonry.

MIXES

Generally, stucco is applied in either two or three coats, depending on the supporting construction. Three-coat application, used over metal reinforcement, consists of a scratch coat, a brown coat and a finish coat—each applied separately and allowed to set partially. In two-coat application, generally used over concrete and masonry bases, a single base coat takes the place of the scratch and brown coats, and is followed by a finish coat. The scratch coat and brown coat in three-coat work sometimes also are referred to as base coats.

Stucco mixes should be proportioned as recommended in Figure WF6. Mix selection should be determined by the type of base and coat for which it is intended (Fig. WF7).

FIG. WF6 RECOMMENDED STUCCO MIXES*

Group	Portland Cement	Lime	Masonry Cement	Plastic Cement	Sand
C	1	0 to ¼	—	—	3 to 4
C	1	—	1	—	6 to 7½
C	1	¼ to ½	—	—	4 to 6
L	1	½ to 1¼	—	—	4½ to 9
L	—	—	1	—	3 to 4
F	1	1¼ to 2	—	—	5 to 10
P	1	—	—	1	6 to 10
P	—	—	—	1	3 to 4

*Proportions are given in parts by volume.

The materials for all coats should be thoroughly mixed. Dry materials should be mixed to a uniform consistency before water is added. Only enough water should be added to assure workability. A power mixer is recommended for uniform blending of materials, and it should be operated for at least 5 minutes after all ingredients are in the mixer. Materials should be mixed only in quantities which can be used within 2-1/2 to 3 hours; remixing to restore plasticity is permissible, but retempering with additional water should be avoided.

When the stucco is applied by machine, the amount of water should be adjusted to produce a mix of proper consistency for the length of hose, temperature and humidity conditions, and type of base. When measured with a standard 2″ x 4″ x 6″ slump cone, the slump should be from 2-1/2″ to 4″ at the mixer and from 2″ to 3-1/4″ at the nozzle. The hose length from the machine to the working surface should be as short as possible, and generally not longer than 200′.

STUCCO APPLICATION

Stucco should not be applied in freezing weather or to surfaces containing frost. The temperature of the stucco should be maintained at 50°F. or higher during application and for at least 48 hours after the finish coat is applied. In the case of falling temperatures, the stucco should be protected against freezing by heated enclosures (see Curing, page WF425-7).

Three-Coat Work

Over all types of metal reinforcement, three coats of stucco totaling at least 7/8″ should be applied.

Scratch Coat The first coat should be at least 3/8″ thick when measured from the rigid base or the back-

FIG. WF7 STUCCO MIX SELECTION GUIDE

Stucco Base	Recommended Groups for Stucco Coats		
	Scratch	Brown[1]	Finish[2]
Low absorption (placed concrete, dense clay brick)	C, P	C, L, P	L, F, P
High Absorption (concrete unit masonry, clay brick & structural tile)	L, P	L, P	L, F, P
Metal Reinforcement[3]	C, P	C, L, P	L, F, P

1. Use as base coat in two-coat work over concrete and masonry bases.
2. Finish coat may be factory-prepared "stucco finish."
3. Over any type of supporting construction.

(Continued) **APPLICATION**

ing paper. It should be applied evenly, with sufficient pressure to squeeze the mix through the reinforcement to form mechanical keys. The space between the metal and the backing should be filled as thoroughly as possible so that the metal will be sufficiently embedded. The coverage on the surface should be such that no more than 10% of the surface of the reinforcement remains exposed. As soon as the scratch coat has become firm, its entire surface should be cross-raked *(scratched)* to provide good mechanical bond for the brown coat. Regardless of whether application is by hand or by machine, cross raking is done manually.

Brown Coat The brown coat should be applied after the scratch coat has been moist cured and is sufficiently firm to carry the weight of the next coat. If the scratch coat becomes dry, it should be dampened again to control suction before the brown coat is applied. The brown coat should be approximately 3/8″ thick, bringing the thickness of both coats to at least 3/4″.

When applying the brown coat, if possible, an entire panel should be covered without stopping. Interruptions in the work should be made at screeds, control joints, corners, pilasters, belt courses, doorways, openings or other lines where changes in appearance, which might show through the finish coat, will not be obvious.

The brown coat should be rodded with a straightedge and then darbied to impart a level, uniformly rough texture. The same type of darby or float should be used for the entire surface to assure uniformity of texture. Deep scratches or surface defects should be smoothed out.

FIG. WF8 RECOMMENDED CURING SCHEDULE FOR STUCCO FINISHES

Application[1]	Moist Curing[2]	Total Setting Time[3]
Scratch Coat	12 to 24 hours	At least 48 hours (between coats)
Brown Coat	12 to 24 hours	At least 7 days (between coats)
Finish Coat	12 to 24 hours[4]	At least 48 hours (after application)

1. In two-coat work, requirements for base and finish coats are the same as for brown and finish coats, respectively.
2. Moist curing should be delayed until scratch and brown coats are sufficiently set to prevent erosion, and for 12 to 24 hours for finish coat.
3. At least 50° F. temperature should be maintained during and after application.
4. Not recommended for prepared colored "stucco finishes."

External corners should be formed with a straightedge when corner reinforcement is not used. After moist curing the surface should be allowed to dry uniformly before the finish coat is applied (see Curing, page WF425-6).

Finish Coat The brown coat should be uniformly dampened if the finish coat is to be hand-applied. The finish coat should be at least 1/8″ thick, bringing the total stucco finish to at least 7/8″ thickness.

The finish coat should be moist-cured unless a prepared stucco finish is used.

Prepared stucco finishes often contain curing agents and moist curing is not necessary; in fact, moist curing is specifically discouraged by some manufacturers because it may cause streaking of the color surface.

The finish coat should be maintained at the proper temperature for at least 48 hours after application (see page WF425-6).

The finish coat can be applied in a variety of textures by hand or machine, as with three-coat application (see Main Text, page 425-19).

Two-Coat Work

Over rigid concrete or masonry bases, at least two coats of stucco, totaling a minimum of 5/8″ thickness, should be applied.

Three coats should be used where increased thickness and/or strength is required.

Base Coat The surface of the concrete or masonry should be dampened uniformly to assure uniform suction, except when a chemical bonding agent is used. The base coat should be applied approximately 3/8″ thick and should be rough floated.

Finish Coat The base coat should be uniformly dampened if the finish coat is to be hand-applied. The finish coat should be approximately 1/4″ thick.

CURING

The several coats of stucco should be moist cured as recommended in Figure WF8.

The curing recommendations given in Figure WF8 are based on the premise that deliberate moist curing assures a greater measure of success than does dependence on favorable climatic conditions or on water contained in the mix. (See Main Text, page 425-16).

(Continued) **APPLICATION**

Other accepted local curing practices which have a demonstrated history of satisfactory performance may be used in place of the recommendations given in Figure 8.

All of the various curing procedures attempt either to maintain the previous coat uniformly damp, or to permit it to become uniformly dry. This is particularly true of the base coat, because uniformity of color in the finish coat is dependent on equal suction in the base coat.

CRACK CONTROL

The proper design of supporting construction, selection of materials and application of stucco finish—all can substantially minimize or eliminate undesirable cracking. Cracks can develop in the stucco finish from (1) shrinkage stresses, (2) structural movement, (3) restraints due to penetrating equipment and structural elements, or intersecting walls and ceilings, or (4) weak sections due to thin spots and openings in the supporting construction.

Shrinkage Stresses

Rich mixes containing substantially more portland cement than recommended in Figure WF6 should not be used. Water should be added only in sufficient quantities to make a workable mix.

Both excessive water and rich mixes contribute to high drying shrinkage and cracking hazard.

When stucco is applied over metal reinforcement, control joints in the stucco finish should be provided not more than 18′ o.c. forming panels not larger than 150 sq. ft.

Control joints permit shrinkage of the stucco finish and relieve stresses without cracking.

Structural Movement

Control joints should be provided in the stucco finish directly over all control joints, expansion joints and isolation joints which occur in the supporting construction. Control joints also should be provided at the juncture of dissimilar materials in the supporting construction.

The stucco finish cannot be expected to resist stress concentrations caused by differential movement in the supporting construction. Control joints relieve these stresses and minimize cracking by allowing the stucco finish to move with the supporting construction.

Relief joints should be provided at all locations where the stucco finish is intersected, penetrated or abutted by structural elements such as load-bearing walls, columns and beams.

Relief joints minimize cracking by allowing structural elements to move independent of the adjacent stucco finish.

Weakened Sections

Stucco should be applied over a level base in uniform thickness.

If stucco is applied over an irregular rigid base or over sagging reinforcement, the resulting membrane will be of uneven thickness. The thinner spots constitute weakened sections where higher stress concentrations are likely to cause cracking.

When stucco is applied directly to a rigid base, corners of window and door openings should be reinforced with striplath secured to the base diagonally at all corners.

Restraints

Relief joints should be provided at internal corners formed by (1) vertical structural elements (columns, walls, pilasters) and nonload-bearing stucco walls, and (2) suspended or furred ceilings and either load-bearing or nonload-bearing walls.

At such corners, cornerite should be omitted and reinforcement should not be continued around the corners, but rather stopped to permit independent movement in each plane (see Main Text, Figs. 27 & 28).

Relief joints should be provided where mechanical fixtures or other equipment penetrate through the stucco finish. Relief joints formed by trowel cuts or casing beads should be tightly caulked with an elastic nonhardening compound. Casing beads should be wired to metal reinforcement, not rigidly nailed to the supporting construction.

SURFACE FINISHES

An approved sample of the desired color, texture and/or design should be produced at the beginning of the work to serve as a standard of comparison as the work progresses.

A great variety of surface finishes can be produced in stucco. See Main Text, page 425-19 and Figure 29 for a discussion of typical stucco textures.

431 GYPSUM BOARD FINISHES

GYPSUM BOARD FINISHES

INTRODUCTION

Gypsum wallboard construction originated about 1915, when small, 3/8″ thick sheets were developed primarily for remodeling. In the following decades, erection and joint treatment techniques were gradually evolved which improved surface appearance and extended the use of wallboard to all types of construction. However, it was not until after World War II, with greater emphasis on total system or assembly performance that gypsum wallboard gained widespread use.

Gypsum wallboard finishes commonly are referred to as *drywall* finishes because they have initial low moisture content and require little or no water during application. These factors contribute to their dimensional stability and minimize undesirable effects on other construction components. Although plywood, hardboard and fiberboard sometimes are classed as "drywall", the term most often is applied to gypsum wallboard.

This section discusses the materials and methods recommended to provide suitable gypsum wallboard finishes for various assemblies. *General Recommendations* for materials and methods applicable to both single-ply and multi-ply constructions are discussed in the first subsection. The following subsections provide more specific information for *Single-ply Construction* and *Multi-ply Construction.*

Only general recommendations for sound isolation, fire resistance, thermal insulation and moisture resistance are included. More detailed information of assembly performance with respect to these criteria is given in appropriate Work File sections of Division 100—Criteria & Principles. The manufacture, properties and uses of gypsum board and other gypsum products are discussed in Section 218 Gypsum Products.

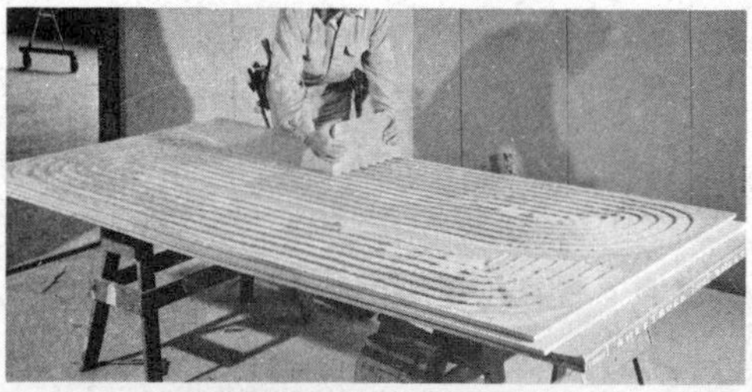

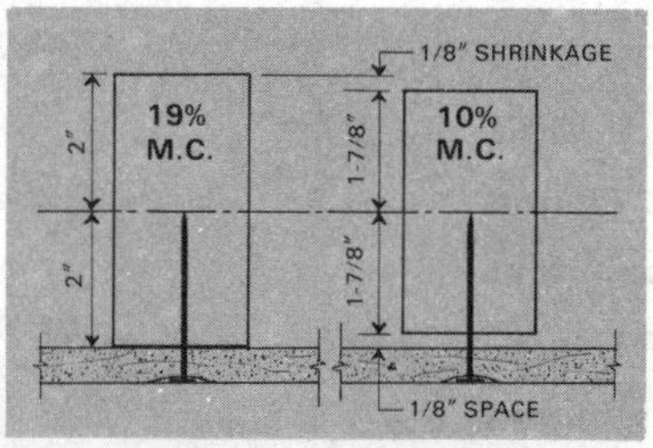

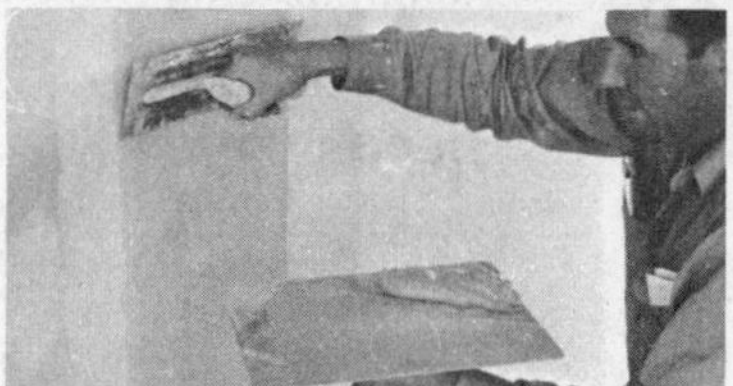

GENERAL RECOMMENDATIONS

Gypsum board finishes consist of one or more plies of *gypsum board* having incombustible gypsum cores with surfaces of paper or other sheet material. The face plies of these finishes consist of either *plain* wallboard intended for decorating with paint, wall covering or wall tile, or *pre-decorated* wallboard. Both plain and predecorated wallboard provide economical finishes capable of efficient erection with different structural systems.

The majority of gypsum board finishes in residential construction utilize *plain* wallboard requiring fastener and joint treatment and decorating. *Predecorated* wallboard constitutes a small portion of the total use and involves somewhat different installation methods. It is usually applied directly to studs with color-matched nails or is laminated to a base layer of gypsum backing board. Joints either are left exposed, treated with special moldings or battens, or in some proprietary systems concealed with a loose flap of the decorative material.

The following discussion is concerned primarily with plain wallboard finishes unless predecorated wallboard is mentioned specifically.

In addition to other functional requirements, walls and ceilings often demand specific performance with respect to sound isolation, fire resistance, thermal insulation and moisture control. A variety of gypsum boards and engineered systems are manufactured today to meet specific needs. However, performance is closely related to the properties, arrangement and installation of construction elements comprising the wall or ceiling assembly.

To provide finishes of suitable quality, consideration should be given to: (1) preparation of supporting construction; (2) selection of gypsum board and accessory materials, (3) job conditions; (4) application of wallboard, including joint and fastener treatment; and (5) special constructions where sound isolation, fire resistance, thermal insulation or moisture resistance are important.

The Gypsum Association publishes the Recommended Specifications for the Application and Finishing of Gypsum Board GA 216. It is the currently recognized industry standard and the basis for many of the following recommendations.

TERMINOLOGY

An understanding of the terminology applicable to gypsum wallboard finishes is basic to the following discussion.

Adhesive, Drywall Adhesives specifically intended for the application of gypsum wallboard may be classed according to use as: *laminating* adhesive for bonding layers of gypsum board; *stud* adhesive for attaching wallboard to wood supports; *contact* adhesive for bonding layers of gypsum board or bonding gypsum board to metal studs.

Back Blocking A single-ply installation procedure for reinforcing butt-end and/or edge joints to minimize surface imperfections such as cracking and ridging.

Butt-end Joint Joint in which mill- or job-cut (exposed core) board ends or edges are butted together.

Corner Cracking Cracks occurring in the apex of inside corners, such as between adjacent walls or at walls and ceilings.

Corner Floating Installation procedure which eliminates some mechanical fasteners at interior corners and permits sufficient movement of boards to eliminate corner cracking.

Cross Furring Furring members installed perpendicular to framing members.

Crown, Joint The maximum height to which joint compound is applied over a wallboard joint.

Edges Board extremities which are paperbound and run the long dimension of the board as manufactured (Fig. 1).

Ends Board extremities which are mill- or job-cut exposing the gypsum core, and run the short di-

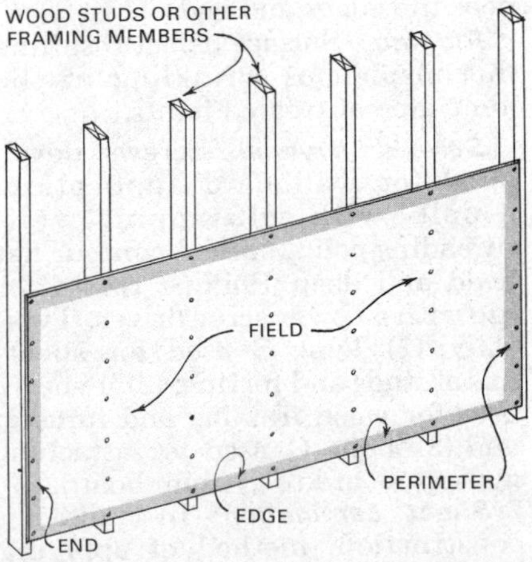

FIG. 1 Plain wallboard edges usually are tapered; backing board edges are either square or tongue-and-grooved.

mension of the board as manufactured (Fig. 1).

Fastener Treatment Method of concealing fasteners by successive applications of compound until a smooth wallboard surface is achieved.

Featheredging (Feathering) Tapering joint compound to a very thin edge to assure inconspicuous blending with the adjacent wallboard surface.

Field The surface of the board exclusive of the perimeter (Fig. 1).

Horizontal Application Board applied with *edges perpendicular* to supporting members such as studs, joists, channels or furring strips (Fig. 2a).

Joint Beading See ***Ridging***

Joint Treatment Method of reinforcing and concealing wallboard joints with tape and successive layers of compound.

Nailpopping Surface defect resulting in a conspicuous protrusion of compound directly over the nailhead (see page 431-14).

Nails, Drywall Nails suitable for wallboard application. Typically: bright, coated or chemically treated, low-carbon steel nails with flat, thin, slightly filleted and countersunk head approximately 1/4″ in diameter, and medium or long diamond point. Annularly threaded nails (GWB-54) and smooth or deformed shank nails should conform to ASTM C-514.

Nail Spotting See ***Fastener Treatment***

Perimeter Surface of the board near the edges and ends (Fig. 1).

Ridging Surface defect resulting in conspicuous wrinkling of the joint tape at treated joints.

Screws, Drywall Screws developed for wallboard application, usually with self-tapping, self-threading point, special-contour flat head and deep Phillips recess for use with a power screwdriver. Typically: (1) Type S used for sheet-metal studs and furring; (2) Type W used for wood framing and furring; and (3) Type G used for attaching gypsum board to gypsum board.

Sheet Lamination In multi-ply construction, method of applying adhesive to the entire surface to be bonded.

Strip Lamination In multi-ply construction, method of applying adhesive in parallel strips spaced 16″ to 24″ apart.

Strip Reinforcing In single-ply construction, installation procedure in which strips of gypsum board are applied to framing members to reinforce wallboard joints and provide a base for adhesive application.

Taping Applying joint tape over embedding compound in the process of joint treatment.

Vertical Application Boards applied with *edges parallel* to supporting members (Fig. 2b).

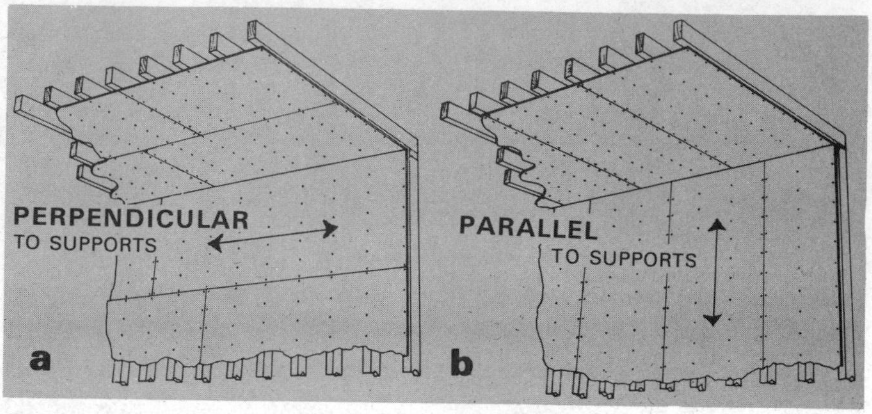

FIG. 2 (a) Horizontal and (b) vertical application of wallboard to wood frame construction by the single nailing method.

SUPPORTING CONSTRUCTION

Gypsum board can be applied directly to wood, metal, masonry or concrete supporting construction which is capable of supporting design loads and which provides a firm, level, plumb and true base. Generally, the deflection of ceiling members supporting gypsum board finishes should not exceed 1/240 of the span under full design load. Headers or lintels should be provided over openings to support structural loads and special construction should be provided where required to support wall-hung equipment and fixtures.

Wood and metal supporting construction usually consists of: (1) self-supporting *framing* members such as wall studs, ceiling joists and roof trusses; and/or (2) *furring* members such as wood strips and metal channels which are supported by underlying construction of framing, concrete or masonry.

Wallboard can be applied directly to above-grade interior masonry or concrete. However, exterior or below-grade walls of solid masonry or concrete generally require furring to separate the wallboard from possible moisture in the wall.

All framing and furring should be accurately aligned in the same plane and should be spaced not to exceed the maximum recommended for the wallboard thickness. Window and door frames should be of appropriate depth for the intended wall thickness. When installing mechanical and electrical equipment, consideration should be given to the thickness of the gypsum board finish and design of trim components such as cover plates, registers and grilles.

Wood Framing

Framing lumber should be of the proper grade for the intended use, graded by a recognized grading authority. The moisture content of framing lumber is important in the performance of gypsum board finishes. It should be as near as possible to the equilibrium moisture content for the particular area and should not exceed 15% at the time the gypsum board is installed (see Work File, page Wood WF201-1).

In light residential construction, wood studs are typical framing members for walls and partitions; wood joists and trusses are the most common ceiling and roof framing members. These framing members generally are spaced 16″ or 24″ o.c., which permits direct application of most gypsum board. However,

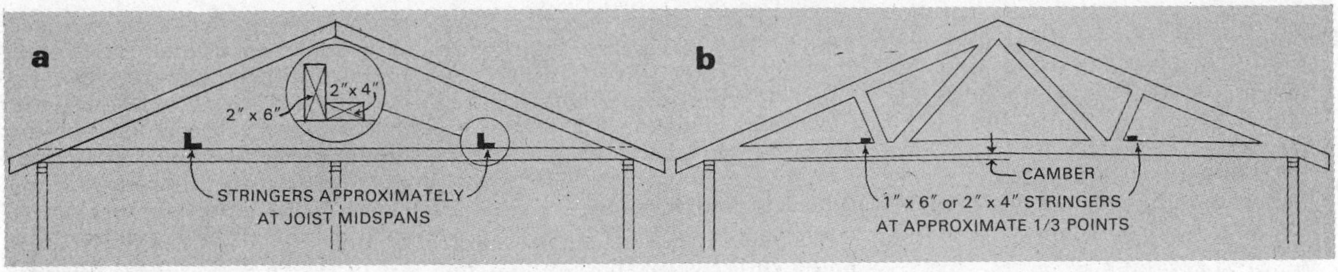

FIG. 3 (a) Joist spacing can be regulated with 2" x 4" stringers at midpoints; joists can be leveled with 2" x 6" stringers set on edge; (b) When cross furring is not used, 1" x 6" or 2" x 4" stringers at 1/3 points are recommended to regulate truss spacing; additional 2" x 6" stringers also may be used for leveling as shown in 3a.

furring may be required when member spacing exceeds the maximum recommended for the particular board thickness or when special performance such as increased sound isolation is desired.

Studs Wood studs in load-bearing partitions should be of nominal 2" x 4" dimension or larger. In nonload-bearing or staggered-stud partitions, nominal 2" x 3" studs may be used, spaced 16" o.c. when openings are provided and 24" o.c. when no openings are included. Backup members should be provided at all interior corners for support and/or as a nailing base for the gypsum board. Warped or crooked studs should not be used.

Joists Ceilings intended for the direct application of gypsum board should have all joists evenly spaced with bottom faces aligned in a level plane. Excessively bowed or crooked joists should not be used, but joists with slight crown may be used if they are installed with crown up. Slightly crooked or bowed joists can be straightened and leveled by nailing stringers approximately at midspan (Fig. 3a).

Trusses Irregularities in spacing and leveling often are magnified in wood truss construction because relatively small members are used for bottom chords and larger spans are involved than in joist construction. Cross furring generally is recommended to remedy these problems. However, when gypsum board is nailed directly to the bottom chords, the chords can be aligned and leveled by using stringers approximately at third points continuously across the entire ceiling. Built-in camber to compensate for future deflection of bottom chords also is recommended (Fig. 3b).

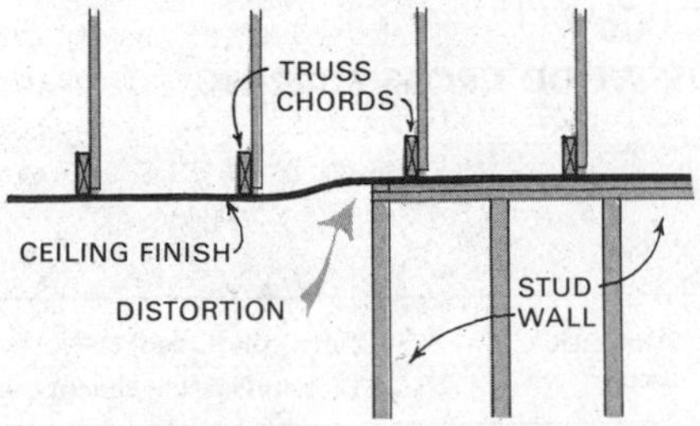

FIG. 4 In truss construction, ceiling distortion may result if substantial loads are introduced after interior partitions are erected.

With trussed roof construction, exterior walls and the ceiling sometimes are finished before interior partitions are erected and finished. When this method is used, the roofing and other construction elements which would increase roof loads should be installed before interior partitions are erected. If substantial roof loads are introduced after partitions are installed, ceiling distortion may result near interior partitions as trusses deflect (Fig. 4).

Metal Framing

Metal framing is common in high-rise residential construction where noncombustible construction is required. Nonload-bearing partitions typically are framed with metal studs; ceilings usually are framed with open-web joists or light beams.

Metal Studs Typically, metal studs are 25 gauge cold-formed steel, electrogalvanized to resist corrosion and conforming to ASTM C-645. This gauge is compatible with most drywall sheetmetal screws used for fastening gypsum board to metal. Proprietary nailable studs also are available but are not used as frequently (Fig. 5).

A variety of systems have been developed by gypsum board and metal specialty manufacturers to meet requirements for sound isolation and fire resistance. Many systems are designed for ease of erection and demountability with a minimum of labor and waste of materials. Systems should be selected on the basis of desired performance criteria and should be

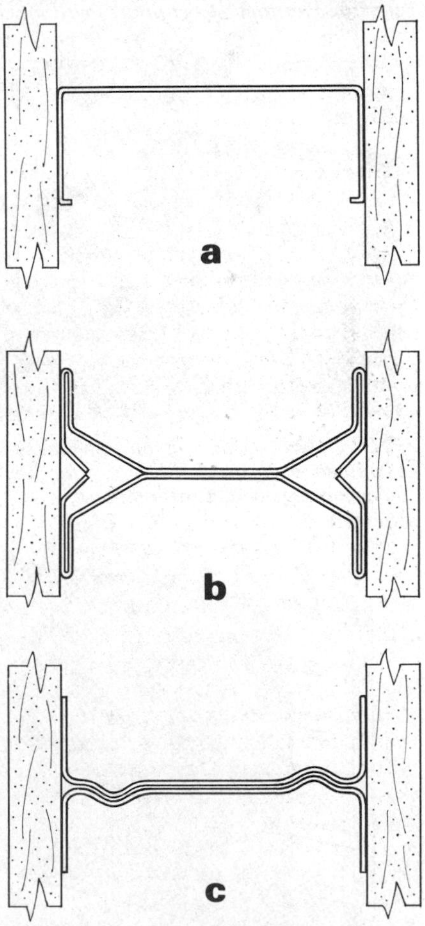

FIG. 5 Metal studs: (a) screw stud for drywall sheetmetal screws, (b) nailable studs for special ring-shank nails, or (c) nailable stud for smooth shank nails.

installed in accordance with the manufacturer's recommendations. Stud spacing should not exceed the maximum spans recommended for single-ply and multi-ply construction.

Open-web Joists These members generally are not designed to receive wallboard directly and often are spaced more than 24″ o.c. Hence, it is necessary to provide suitable furring or suspension systems at appropriate spacing for attaching wallboard (Fig. 6, see also Metal Furring, below).

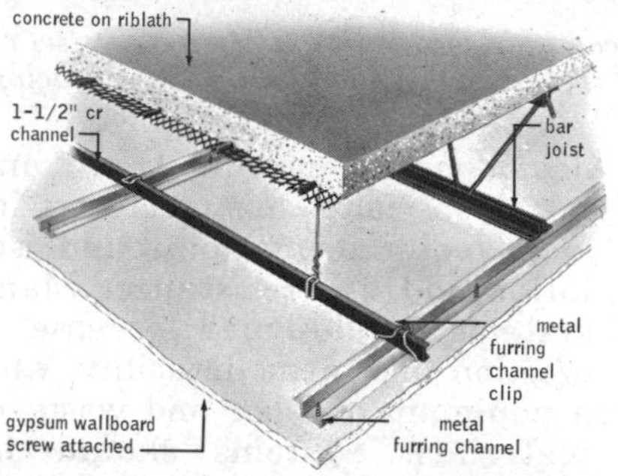

FIG. 6 Open-web joists usually require a grillage of cold-rolled channels and furring channels to support wallboard.

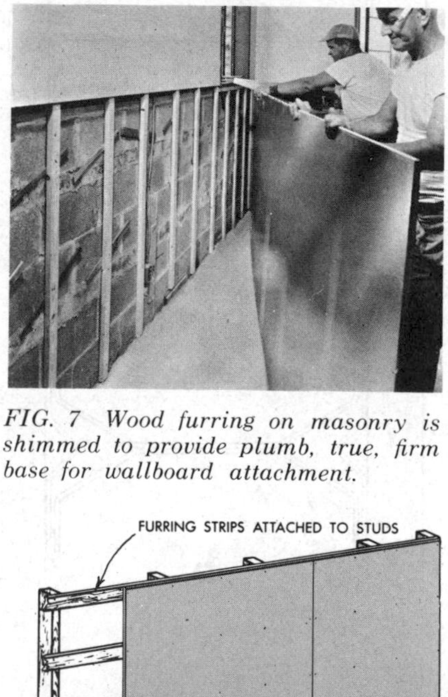

FIG. 7 Wood furring on masonry is shimmed to provide plumb, true, firm base for wallboard attachment.

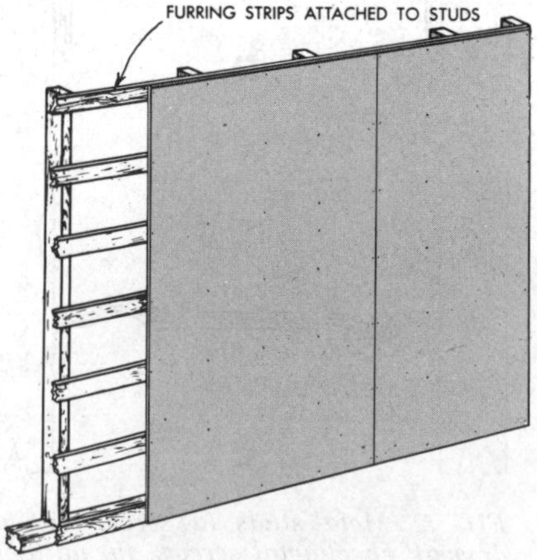

FIG. 8 Gypsum board attached to wood cross furring.

Concrete and Masonry

Wallboard finishes can be laminated directly to above-grade, interior masonry and concrete if the surface is sufficiently dry, smooth, plumb and true. Above-grade, exterior cavity walls also may be suitable for direct lamination if the cavity is properly insulated when required. However, predecorated wallboard with a decorative film which is highly impermeable to vapor generally should not be laminated directly to concrete or masonry, because moisture may become trapped within the gypsum core of the board.

Rough or protruding edges should be removed and depressions greater than 4″ in diameter and 1/8″ deep should be filled with grout or similar filler. Mortar joints should be cut flush with the face of the wall. Surfaces should be free from form oils, curing compounds, loose particles, dust, grease or efflorescence to assure adequate bond. It is essential that concrete be allowed to cure for at least 28 days before wallboard is applied.

Exterior and below-grade walls and horizontal surfaces subject to moisture or excessive heat loss, or any surface which cannot be prepared readily for adhesive lamination should be furred to provide a suitable base for attaching the wallboard (Fig. 7).

Furring

Wood or metal furring may be necessary: (1) to provide a suitably plumb, true and/or properly spaced supporting construction; (2) to eliminate capillary transfer of moisture in exterior or below-grade masonry and concrete walls, and to minimize the likelihood of moisture condensation on interior wall surfaces; (3) to improve thermal insulating properties of exterior walls; (4) to increase assembly thickness to accommodate mechanical equipment; or (5) to improve sound isolation by resilient mounting.

Wood Furring Wood furring strips most commonly are used with wood frame and masonry or concrete construction where noncombustibility or decay resistance is not a consideration. When original framing is of irregular spacing or the spacing is too great for the intended wallboard thickness, *cross furring* should be applied perpendicular to framing members (Fig. 8). Recommendations for wood cross furring are given in Figure 9.

Cross furring smaller than 2″ x 2″ is not sufficiently stiff to permit the wallboard to be nailed without excessive hammer rebound, and may result in loose board attachment. Thinner cross furring is acceptable for screw-attached wallboard because little or no impact is involved when screws are driven. Thinner furring also may be acceptable if the wallboard is applied with stud adhesive and nailing is limited to locations where the furring is fully supported by underlying framing.

When the primary purpose of furring is to increase the wall or ceiling assembly thickness, 1″ x 2″ or larger *parallel furring* can be applied directly over the primary framing, nailed 16″ o.c. with nails providing penetration recommended for cross furring (Fig. 9).

FIG. 9 MINIMUM REQUIREMENTS FOR WOOD CROSS FURRING[1]

Location	Support Spacing (in.)	Support Spacing (mm)	Furring	Nails: Spacing	Nails: Penetration
CEILINGS	16″ o.c.	406 o.c.	2″ x 2″	2 at each support	⅞″ (22 mm) (threaded)
	24″ o.c.	610 o.c.			1¼″ (32 mm) (smooth shank)
WALLS	16″ o.c.	406 o.c.	1″ x 2″	1 at each support	¾″ (19 mm) (threaded or smooth shank)
	24″ o.c.	610 o.c.	1″ x 3″		

[1]Wallboard screw attached; if nailed, 2″ x 2″ furring should be used in all cases.

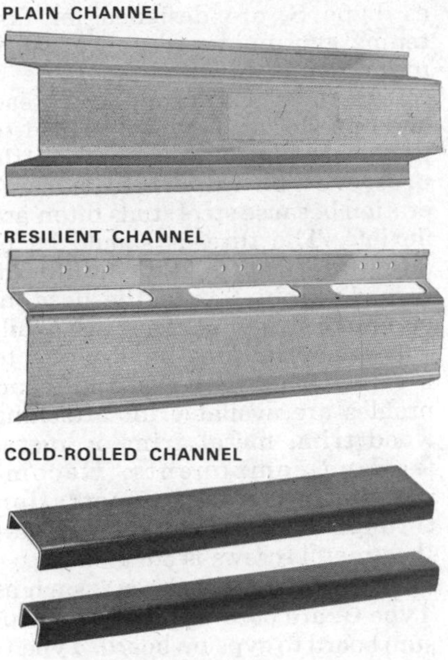

FIG. 10 Metal furring channels.

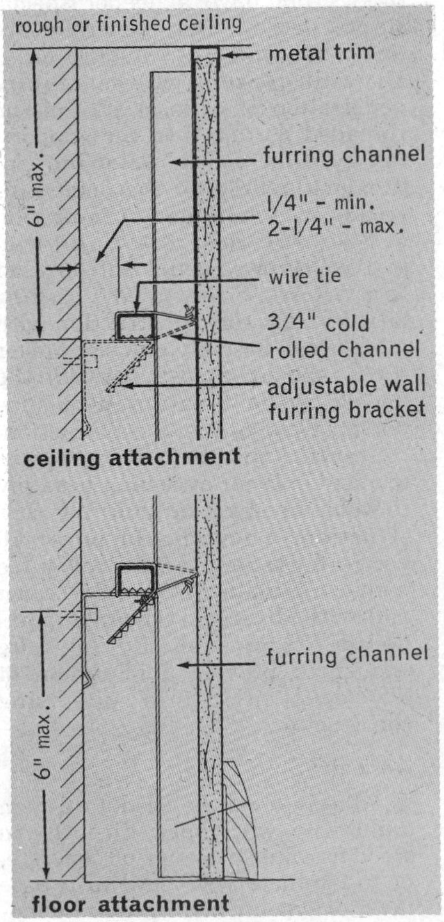

FIG. 11 Adjustable brackets have notched surfaces to permit plumb and true installation of channel grid.

Furring strips over masonry or concrete can be secured with cut nails, helically threaded concrete nails or powder actuated fasteners of appropriate size. Maximum fastener spacing should be 16″ o.c. for 1″ x 2″ strips and 24″ o.c. for 2″ x 2″ strips or larger.

Metal Furring Metal furring of various types can be used with all types of supporting construction. It is used in preference to wood furring when noncombustible or sound isolating assemblies are required. Metal furring members may be either *cold-rolled channels* or *drywall channels* (Fig. 10).

Cold-rolled Channels are multipurpose members used for both plaster and gypsum board furred partitions and in most types of suspended ceiling assemblies. These channels usually are of 16 gauge steel with either a galvanized or a black asphaltum finish and are available in sizes ranging from 3/4″ to 2″ wide, in 16′ to 20′ lengths. Because the gauge is not designed to receive drywall sheetmetal screws, the channels generally are wire-tied to supporting construction. Cold-rolled channels are used primarily as a supporting grid for lighter drywall channels to which wallboard can be screw-attached (Fig. 6).

Drywall Channels usually are roller-formed of 25 gauge electrogalvanized steel. Most drywall channels are designed for wallboard attachment with self-drilling sheetmetal screws, but nailable channels similar to nailable studs also are available. Channels usually are installed perpendicular to framing with a 1-1/4″ screw or a 5d nail at each support. Over masonry or concrete surfaces, they may be installed horizontally or vertically, whichever is more convenient, with fasteners as described for wood furring.

A variety of proprietary clips, runners and adjustable brackets are available with engineered metal furring systems to facilitate rapid erection over irregular masonry walls (Fig. 11). The manufacturer's specific recommendations for fasteners, spacing, accessories and installation should be consulted.

Drywall channels can be either plain or resilient. *Plain channels* are

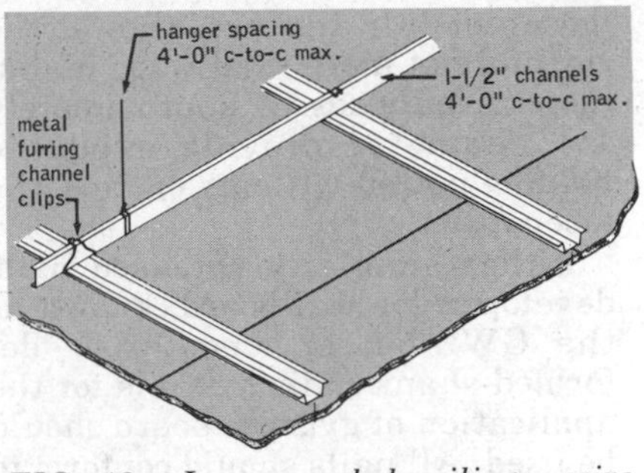

FIG. 12. In suspended ceilings, wire hangers are used to support and level the channel grillage.

used primarily with unit masonry and concrete wall assemblies which possess inherent sound isolation because of their mass and are not significantly improved by resilient mounting if they are sealed and airtight. In suspended ceiling assemblies, a grillage composed of cold-rolled channels and plain drywall channels is used to support the wallboard finish (Fig. 12). If sound or vibration isolation is desired, resilient hangers can be wired into the suspension system to isolate the ceiling from framing.

Resilient channels are used over both metal and wood framing (see Fig. 27, page 431-17). In addition to improving sound isolation, they help to isolate the gypsum board from structural movement, thus minimizing the possibility of joint cracking.

ACCESSORY MATERIALS

Accessory materials for the application of gypsum board include mechanical fasteners and adhesive for board attachment, compounds and tape for joint and fastener treatment, and metal edge and corner trim to protect exposed edges and exterior corners.

Mechanical Fasteners

Nails and screws are used to apply both single- and multi-ply finishes. Clips and staples are limited to attaching the base ply in multi-ply construction.

Nails A variety of nails are used for applying gypsum board. Nails can be bright, coated or chemically treated. Shanks may be either smooth or annularly threaded, generally with medium or long diamond point. The nailhead should be flat or slightly concave, thin at the rim and not more than 5/16″ in diameter. Nailheads of approximately 1/4″ diameter provide adequate holding power without cutting the facepaper.

Either annularly threaded nails developed for wallboard, known as the GWB-54, or smooth- or deformed-shank nails suitable for the application of gypsum board should be used. All nails should conform to ASTM C-514. Casing nails and common nails have heads that are too small in relation to the shank or are too thick and should not be used.

Nails should be of the proper length for the wallboard thickness (Fig. 13). Generally recommended penetration into supporting construction for smooth shank nails is 7/8″. Annularly threaded nails provide more withdrawal resistance, require less penetration and generally minimize nailpopping. However, for fire-rated construction, 1″ or more penetration usually is required and the longer smooth shank nails generally are used.

Screws Gypsum board can be fastened to both wood and metal supporting construction with drywall screws. The usual finish for drywall screws is a zinc phosphate coating with baked-on linseed oil. These screws typically are self-drilling, have self-tapping threads and flat Phillips recessed heads for use with a power screwdriver. A special contour head design makes a uniform depression free of ragged edges and fuzz.

Screws pull the board tightly to the framing without damaging the board, thus minimizing fastener surface defects due to loose board attachment. The three basic types of drywall screws for wood, sheetmetal and gypsum construction and recommended penetration for various uses are shown in Figure 14. Other specially designed screws are available for attaching wood or metal trim and for wallboard attachment to heavier gauge loadbearing studs. The use of screws generally is acceptable in fire-rated constructions.

Drywall Wood Screws, such as Type W, are designed for fastening to wood framing or furring. Type W screws are diamond-pointed to provide efficient drilling action through both gypsum and wood and have a specially designed thread for quick penetration and increased holding power. Recommended minimum penetration into supporting construction is 5/8″, but in two-ply application when the face layer is being screw-attached, the additional holding power developed in the base ply permits reducing the penetration into supports to 1/2″. Type W screws are available in 1-1/4″ length. (Drywall sheetmetal screws, more readily available in longer sizes, may be substituted in two-ply construction.)

Drywall Sheetmetal Screws, such as Type S, are designed for fastening gypsum board to 25 gauge metal studs or furring. Type S screws have a self-tapping thread and mill-slot drill point designed to penetrate sheet metal with little pressure—an important consideration because steel studs often are flexible. The threads should be of adequate depth and should be turned within 1/4″ of the head to eliminate stripping. They are available in several lengths, from 1″ to 2-1/4″. Other lengths and head profiles are available for attaching wood trim, metal trim or metal framing components. Recommended minimum penetration through sheetmetal for drywall sheetmetal screws is 3/8″.

Drywall Gypsum Screws, such as Type G, are used for fastening gypsum board to gypsum board. Type G screws are similar to Type W screws, but have a deeper special thread design. They generally are available in 1-1/2″ length only. Drywall gypsum screws require penetration of at least 1/2″ of the threaded portion into the supporting gypsum board. Allowing approximately 1/4″ for the point, this results in a minimum penetration of 3/4″. For this reason, drywall gypsum screws should not be used to attach wallboard to 3/8″ backing board. In two-ply construction with a 3/8″ thick base ply, nails or longer screws should be used to provide the necessary penetration into supporting wood or metal construction.

Staples Staples are recommended only for attaching base ply to wood members in multi-ply construction. They should be of 16 gauge flattened galvanized wire with a minimum 7/16″ wide crown and with divergent sheared bevel points. Staples should be long enough to provide a minimum of 5/8″ penetration into supporting construction.

FIG. 13 **NAILS FOR GYPSUM BOARD APPLICATION**[1]

Drywall Nails	Recommended Penetration (in.)	Recommended Penetration (mm)	Board Thickness (in.)	Board Thickness (mm)	Nail Length (in.)	Nail Length (mm)
Annularly Threaded[2] ASTM C-514 	¾″	19	⅜″	9.5	1⅛″	29
			½″	12.7	1¼″	32
			⅝″	15.9	1⅜″	35
Smooth Shank[3] ASTM C-514 	⅞″	22	⅜″	9.5	1¼″	32
			½″	12.7	1⅜″	35
			⅝″	15.9	1½″	38

[1]For fire-rated constructions, building code and/or manufacturer's specifications should be followed.
[2]Typically, GWB-54: ¼″ diameter head, 0.098″ diameter shank.
[3]Typically, ¼″ diameter head, 0.099″ diameter shank, with dished head.

Adhesives

Adhesives can be used to attach single-ply wallboard directly to wood framing, masonry or concrete, or to laminate wallboard to a base layer of gypsum board, sound deadening board or rigid foam insulation. They generally are used in combination with nails or screws

FIG. 14 TYPICAL SCREWS FOR GYPSUM BOARD APPLICATION[1]

Drywall Screws[2]	Recommended Penetration	Screw Length (in.)	Screw Length (mm)	Application
For fastening to wood ASTM C-894 	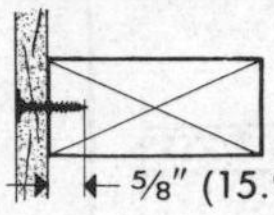 5/8" (15.9 mm) min.	1¼"	32	Single-ply or base ply application
	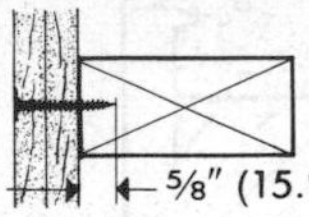 5/8" (15.9 mm) min.	1⅝"	41	Face ply application in two ply construction
For fastening to sheetmetal ASTM C-646 	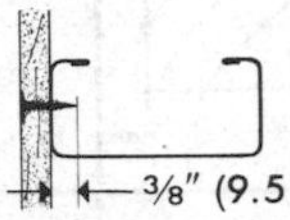 3/8" (9.5 mm) min.	1"	25.4	Single-ply or base-ply application
	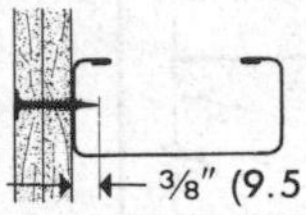 3/8" (9.5 mm) min.	1⅝"	41	Face ply application in two ply construction
For fastening to gypsum board ASTM C-893 	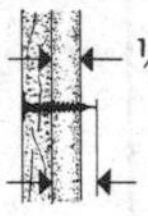 ½" (12.7 mm) min. ¾" (19 mm) min.	1½"	38	Gypsum board to gypsum board in multi-ply construction
	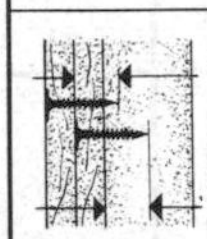 ¾" (19 mm) min. ¾" (19 mm) min.			

[1]Generally acceptable for fire-rated construction.
[2]Screws shown are Types W, S & G, respectively; other drywall screws also are available.

which provide temporary or permanent supplemental support.

With the increased use of adhesives for applying gypsum board finishes, new adhesives are being developed to meet specific installation conditions. Hence, adhesive formulations vary and manufacturers' recommendations should be consulted in each case. Adhesives for applying wallboard finishes may be classed as: *stud adhesives, laminating adhesives* and *contact adhesives.*

Stud Adhesives Adhesives formulated for attaching single-ply wallboard to wood supporting construction are classed as *stud adhesives.* They generally are rubber or asphaltic base, are of caulking consistency that will bridge framing irregularities, and are applied with a gun in a continuous bead. Solvent-base stud adhesives should not be used near an open flame or in poorly ventilated areas. *Structural* stud adhesives should conform to ASTM C-557 which establishes rigid standards for strength, bridging ability, aging and other qualities necessary for long-term structural performance. Fewer permanent supplemental fasteners are required for structural stud adhesives than are required for adhesive-nail attachment with *nonstructural* stud adhesives.

Laminating Adhesives These generally are casein-base adhesives in powder form to which water is added. Laminating adhesives can be used to laminate gypsum boards to each other and to suitable masonry or concrete surfaces, but are not intended for adhesive-bonding to wood framing or furring. In addition to the several adhesives specifically formulated for laminating purposes, some powder embedding compounds (see Joint Compounds, below) can be used when mixed at the recommended consistency. Although some stud adhesives and contact adhesives also can be used for laminating purposes, they should not be used unless specifically recommended by the manufacturer.

Only as much adhesive should be mixed as can be used within the working time specified by the manufacturer. Water used for mixing should be at room temperature and clean enough to drink. Adhesive usually is applied over the entire board area with a suitable spreader and requires temporary support or supplemental fasteners until the adhesive develops sufficient bond strength.

Contact Adhesives These adhesives consist of a synthetic rubber base, sometimes with resins added, in either solvent or emulsion formulations. Contact adhesives can be used to laminate gypsum boards to each other and to bond wallboard to metal studs in demountable partitions. The adhesive is applied by roller, spray gun or brush in a thin, uniform coating to both surfaces to be bonded. For most contact adhesives some open time usually is required before surfaces can be joined and bond can be developed. To assure maximum possible adhesion between mating surfaces, the face board should be impacted over its entire surface with a suitable tool such as a rubber mallot.

The advantages of contact adhesive are immediate bond, excellent long-term strength, resistance to fatigue, and no temporary or permanent fasteners required in the field of the board. One disadvantage is the inability of the adhesive to bridge irregularities in the mating surfaces, thus creating discontinuity in the bond film. Another disadvantage is that most of these adhesives do not permit repositioning of the boards once contact between surfaces has been made. A slip sheet of polyethylene film or tough building paper can be used to facilitate gradual bonding of surfaces as the slip sheet is withdrawn.

Modified Contact Adhesive has been developed which retains the advantages of contact adhesive while removing some disadvantages. This adhesive combines good long-term strength, bridging ability and sufficient immediate bond to permit erection with a minimum of temporary fasteners. In addition, the adhesive has an open assembly time of about 1/2 hour before bond develops, during which time boards can be repositioned. Modified contact adhesive is intended for attaching wallboard to all types of supporting construction, including framing, furring, gypsum board and rigid insulating board.

Joint Tape and Joint Compounds

Joint tape and joint compound are used to produce a wallboard finish that is smooth, monolithic in appearance, with inconspicuous joinings between boards and no visible fasteners. Joint compound and tape should conform to ASTM C475 Treatment Materials for Gypsum Wallboard.

Joint Tape Tape used for joint reinforcement typically is a strong fibered tape with chamfered edges. The special paper resists tensile stresses across the joint as well as longitudinally.

Joint Compounds These may be classed according to function and properties as (1) *embedding compound,* used for embedding and bonding the joint tape; (2) *topping compound,* used for final smoothing and leveling over joints and fasteners; and (3) *all-purpose compound,* which combines features of both embedding and topping compounds.

These compounds either are packaged in powder form to be mixed with water on the job, or are premixed by the manufacturer. With the exception of quicksetting embedding compounds, these compounds harden by evaporation of the water and can be kept in a wet mix state for several days if covered with a moist cloth or a thin layer of water. Both premixed and job-mixed compounds should be protected from freezing.

Embedding Compound typically is a casein-based adhesive, similar to laminating adhesive. It is available either premixed or as a powder which is mixed with water. In addition to embedding tape, it is used as a first coat for concealing fasteners and edge and corner trim. Some embedding compounds also can be used for laminating gypsum board in multi-ply construction when

FIG. 15 **STANDARD DRYWALL METAL TRIM SHAPES**

USASI Designation	Trim Profile	Application
CORNERBEAD[1] CB—100 x 100 CB—118 x 118 CB—114 x 114 CB—100 x 114 CB—PF (paper flange steel corner combination bead)	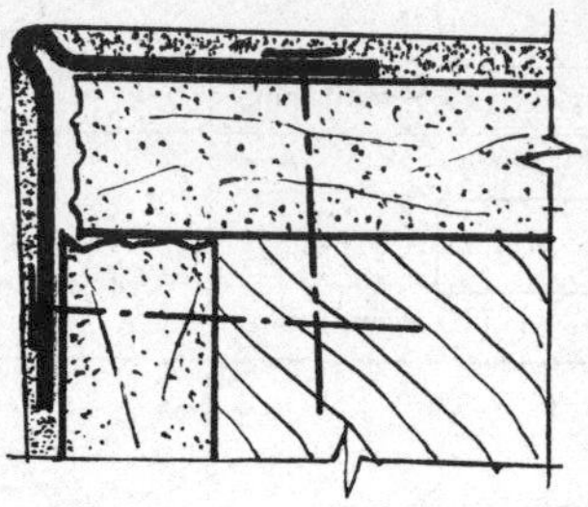	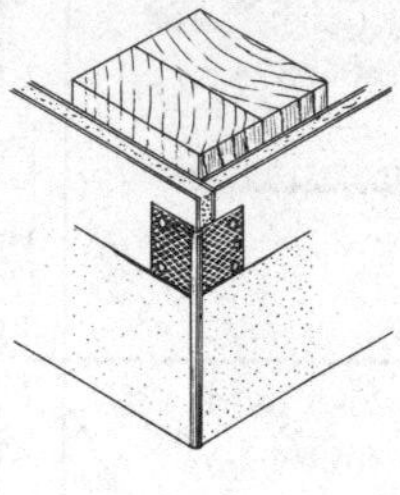
U BEAD[2] U-38 U-12 U-58 U-34	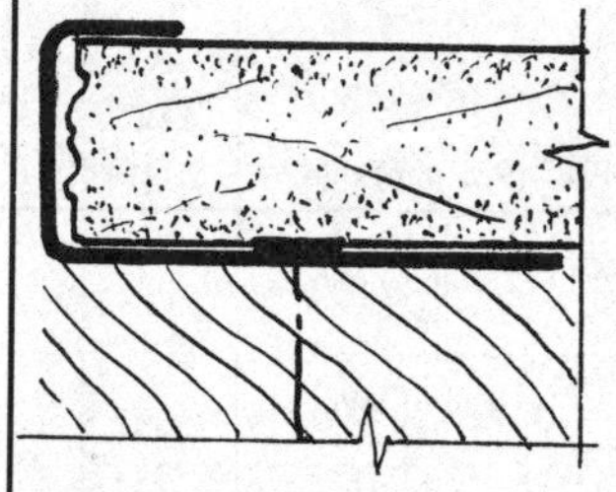	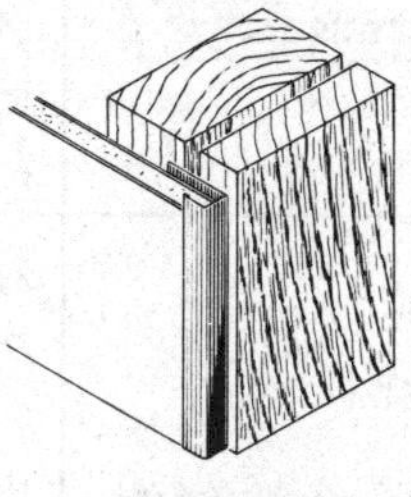
L BEAD[2] LB-38 LB-12 LB-58 LB-34	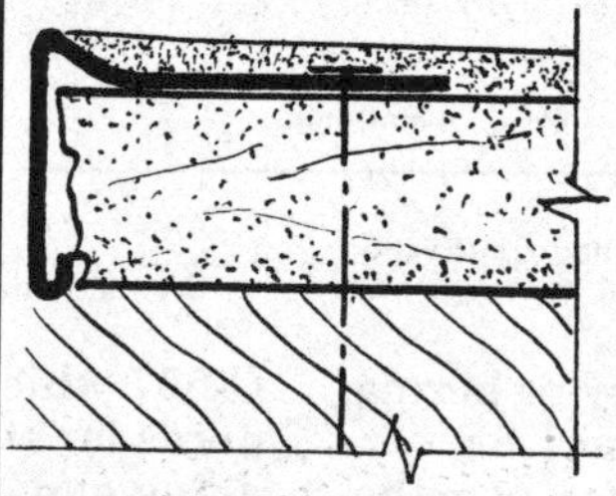	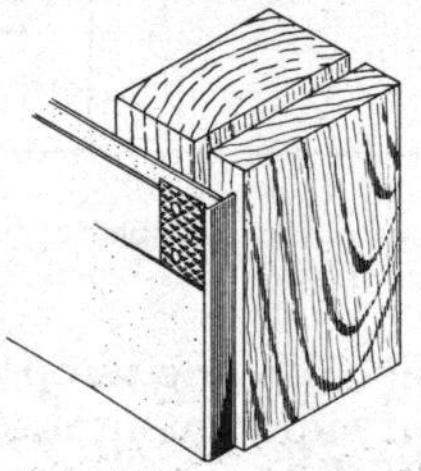
LK BEAD[3 4] LK-S LK-B	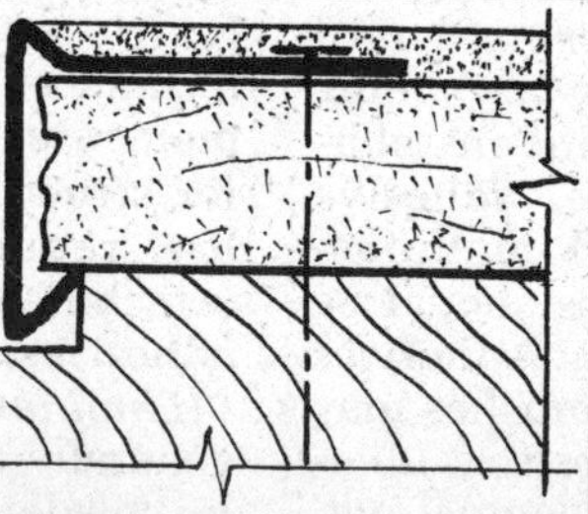	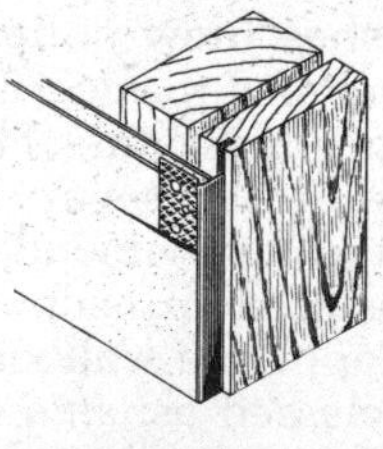
LC BEAD[2] LC-38 LC-12 LC-58 LC-34	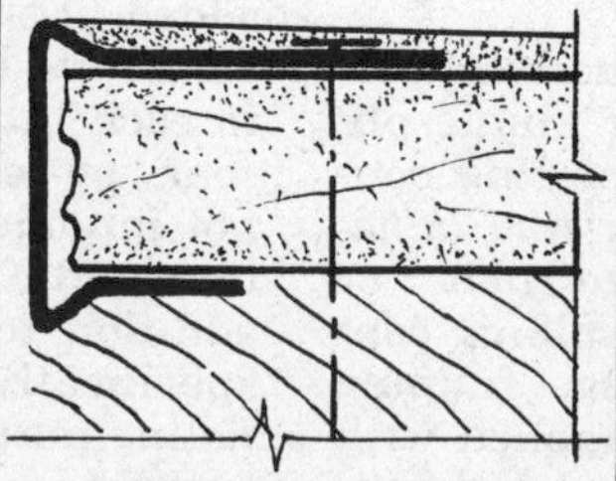	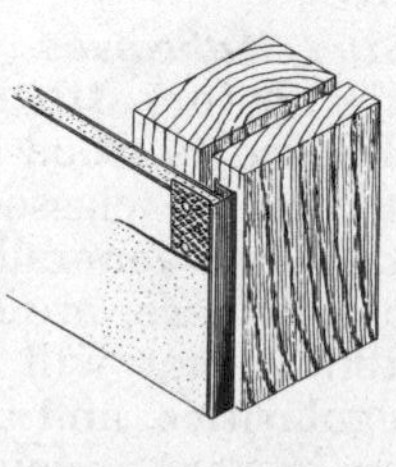

[1]Numbers indicate width of flanges: e.g.—114 x 114 has two 1¼″ flanges.
[2]Numbers indicate thickness of board intended: e.g. U-12 is U Bead intended for ½″ wallboard.
[3]Letter (S) indicates square nose; letter (B), bull or round nose.
[4]For use with kerfed jamb.

mixed at the consistency recommended by the manufacturer for that purpose. Usually, embedding compound requires at least overnight drying before following coats can be applied. However, quick-setting compounds which dry in 2 to 4 hours also are available which permits three-coat application of embedding and topping compounds in one day.

Topping Compound is primarily a surface filler used to conceal and smooth over embedded tape, fasteners and trim. It is a casein or casein-vinyl formulation available either in premixed or powder form. It bonds well with joint tape and compound, gypsum board and fasteners; it sands easily and provides a surface with sufficient "tooth" and suction for painting.

All-purpose Compound, available from some manufacturers, combines the characteristics of both adhesive and filler. It is a convenient formulation available in powder or premixed form, in either a machine or hand tool consistency. All-purpose compound can be used for embedding tape, topping over tape, finishing over metal trim and concealing fasteners. However, it should not be used for laminating gypsum boards unless specifically recommended by the manufacturer.

Edge and Corner Trim

Exposed wallboard edges and ends generally are protected by wood or metal trim. Wood casing trim typically is used at window and door frames; wood or resilient wall base is used to protect wallboard edges at the floor. Metal trim, such as corner beads and casing beads, may be used in similar locations when inconspicuous edge protection is desired without featuring the trim as a design element.

Trim shapes are available for different conditions and varying wallboard thicknesses. The standard shapes and designations are illustrated in Figure 15. On outside corners, metal cornerbeads are used to protect the wallboard from damage and provide neat, slightly rounded corners, difficult to obtain with tape and joint compound alone. Casing beads and cornerbeads with metal flanges generally are nailed or screwed to supporting construction approximately 6″ o.c. (Fig. 15). Beads with paper flanges are held in place by embedding the flanges in joint compound.

JOB CONDITIONS

Temperature and humidity affect the performance of joint treatment materials, the scheduling of joint finishing operations, the quality of the joint, and sometimes the bonding properties of adhesives.

During the winter season, when the outside temperature is less than 55° F., interior finishes should not be installed unless the building is completely enclosed and has a controlled heat of 55° to 70° F. This temperature should be maintained for 24 hours before installation, during installation and until a permanent heating system is in operation.

During summer months, the building need not be completely glazed, but all materials and installation should be protected from the weather.

Ventilation should be provided to eliminate excessive humidity. If the building is glazed, windows can be kept partially open to provide air circulation. In enclosed areas without natural ventilation, temporary air circulators should be used. Under slow drying conditions, additional drying time between coats of joint compound should be allowed. Drafts during hot, dry weather should be avoided to prevent the joint compound from drying too rapidly.

Delivery of gypsum board should coincide as closely as possible with the installation schedule. Boards should be stored on the job flat, inside under cover. Stacking long lengths on short lengths should be avoided to prevent the longer boards from breaking. For short periods of time, boards may be placed vertically against the framing with the long edges of the boards horizontal. All materials should remain in their wrappings or containers until ready for use.

FIG. 16 **SUMMARY OF GYPSUM BOARD PRODUCTS[1]**

Type	Thickness (in.)	Thickness (mm)	Width	Length	Edge Detail
WALLBOARD	¼″	6.4	48″	8′, 10′, 12′	
(ASTM C36, FS SSL30d)	5/16″	7.9	48″	8′, 10′, 12′, 14′	
	⅜″	9.5	48″	7′, 8′, 9′, 10′, 12′	
Regular	½″	12.7	48″	7′, 8′, 9′, 10′, 12′, 14′	
	⅝″	15.9	48″	8′, 10′, 12′	Tapered
Foil-backed	⅜″	9.5	48″	8′, 10′, 12′	
	½″	12.7	48″	8′, 9′, 10′, 12′	
Type X	½″	12.7	48″	8′, 9′, 10′, 12′	
	⅝″	15.9	48″	8′, 9′, 10′, 12′	
BACKING BOARD	¼″	6.4	48″	8′	Square
(ASTM C442, FS SSL30d)	⅜″	9.5	48″	8′	Square
Regular	½″	12.7	24″	8′	Tongue-and-grooved
Foil-backed	⅜″	9.5	48″	8′	Square
Type X	⅝″	15.9	24″	8′	Tongue-and-grooved
	1″	25.4	24″	7′	Tongue-and-grooved

[1]Recommended stock items per Simplified Practice Recommendation R266-63; other sizes and edge details are stocked by some manufacturers and distributors or are available on special order.

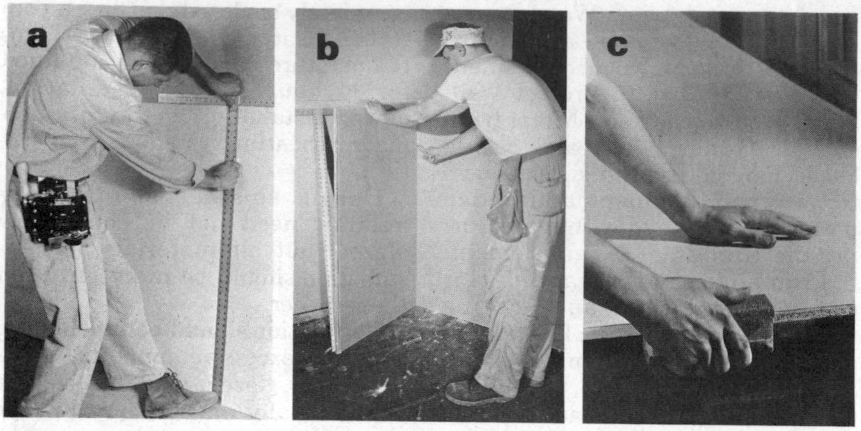

FIG. 17 (a) Facepaper is scored; (b) gypsum core is snapped along score line and back paper is cut; (c) rough edges can be smoothed with coarse sandpaper or a rasp.

APPLICATION OF GYPSUM BOARD

Gypsum board finishes may be either single-ply using wallboard alone, or multi-ply using wallboard over a base ply of backing board or sound deadening board. Figure 16 summarizes typical dimensions, edge details and recognized standards of quality for gypsum board products. More detailed discussion of these and other gypsum board products is given in Section 218 Gypsum Products. Maximum recommended support spacing for various wallboard thicknesses is given in the following subsections for single-ply and multi-ply constructions.

Wallboard installed perpendicular to supporting members results in a stiffer finish surface and hence is preferred. Whenever possible, boards should be applied to span ceilings and walls to avoid butt-end joints which are difficult to finish inconspicuously. When butt-end joints do occur, they should be staggered and located as far from the center of walls and ceilings as possible so they will be less conspicuous. With the exception of face layers in two-ply construction, board ends and edges which are parallel to supporting members should fall on these members.

Successful application of gypsum board finishes depends on proper measuring, cutting and fitting, board attachment and joint and fastener treatment.

Measuring, Cutting and Fitting

All measurements should be taken accurately at the intended ceiling or wall location for each edge or end of the board. Accurate measuring usually will indicate irregularities in framing and trimming so that allowances can be made in cutting.

Straight-line cuts across the full width or length of the board are made by scoring through the facepaper into the gypsum core (Fig. 17a). The core of the board then is snapped and the back paper is cut (Fig. 17b). Cut edges should be smoothed with a rasp or trimmed with a sharp knife (Fig. 17c).

Ceiling panels should be installed first, and all panels should be cut so that they fit easily into place. Joints should be loosely butted without forcing the boards into position. Tapered edges should be placed next to tapered edges and square job- or mill-cut ends should be butted to square ends. Square ends should not be placed next to tapered edges because excessive build-up of compound over the tapered edge results in a conspicuous ridge due to differential shrinkage.

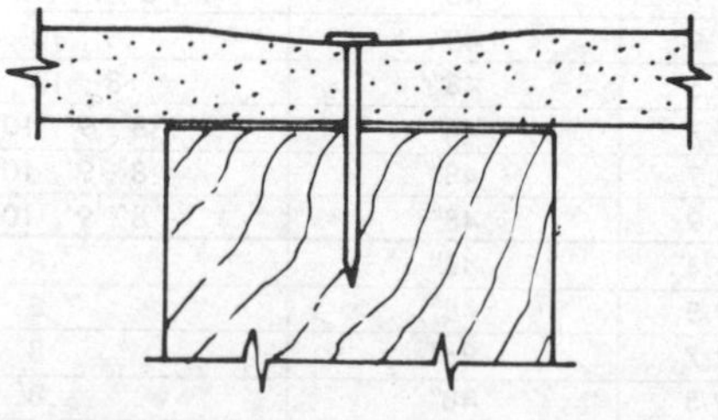

FIG. 18 Nails should be driven perpendicular with a crown headed hammer which forms a uniform depression not more than 1/32" deep.

Board Attachment

Wallboard and backing board can be attached to supporting construction with mechanical fasteners, with adhesive or by a combination of adhesive and mechanical fasteners. Specific attachment methods are discussed in the following subsections for single-ply and multi-ply construction.

Mechanical Fasteners Nails and screws commonly are used to attach gypsum board in both single- and multi-ply installations; however, in two-ply construction, staples and clips can be used to attach the base ply. Adhesive, when used, generally is supplemented with mechanical fasteners.

All fasteners should be of a type recommended for the intended method of application. Ordinary wood screws, sheetmetal screws or common nails are not designed to hold the board tightly and countersink neatly and should not be used. Fasteners should be placed at least 3/8" from board edges and ends, and application should start in the middle of the board and proceed outward toward the perimeter. Fasteners should be driven as nearly perpendicular as possible while the board is held in firm contact with supporting construction. Nails should be driven with a crown-headed hammer forming a uniform depression around the nailhead. Fastener heads should not be seated deeper than about 1/32" below the board surface and particular care should be taken not to break the facepaper (Fig. 18).

Adhesive-Bonding Adhesive can be used to secure single-ply wallboard to wood framing and furring, masonry and concrete or to laminate face ply to a base layer of gypsum board, sound deadening board or rigid foam insulation. The surface to be bonded should be free from dirt, grease, oil or other foreign material.

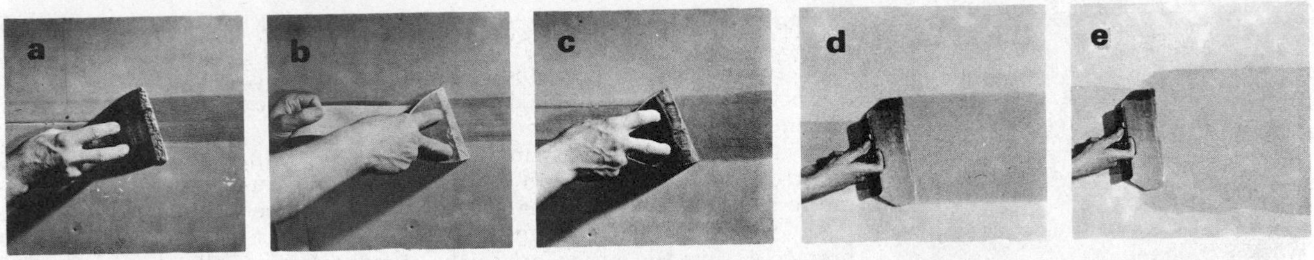

FIG. 19 Finishing knives of varying widths are used to spread embedding compound, embed joint tape, and apply successively wider coats of topping compound.

JOINT AND FASTENER TREATMENT

After wallboard has been erected, specific areas require treatment to achieve a smooth homogeneous appearance. Panel joints in the same plane and interior corners require taping. Exterior corners and exposed edges require metal corner and edge trim. Both taped and trimmed areas, as well as fasteners require finishing with compound.

Joint Taping and Finishing

A minimum of three coats of joint compound is recommended for all taped joints—an embedding coat to bond the tape and two finishing coats over the tape. Each coat should be allowed to dry thoroughly so that the surface can be readily sanded and so that the compound will obtain maximum shrinkage before the next coat is applied.

Depending on atmospheric conditions, 12 to 24 hours may be required for each successive coat to dry, unless quicksetting compound is used. The second and third coats should be lightly sanded when dry to remove surface irregularities and blend evenly with the adjacent wallboard. Care should be taken not to raise the fibers on the facepaper, as this may result in uneven surface texture and spotty appearance.

Flush Joints Embedding compound is spread into the depression formed by the tapered edges of adjacent boards (Fig. 19a) and over all butt-end joints. The tape is centered over the joint and smoothed to avoid wrinkling. The tape then is pressed into the compound by drawing the knife along the joint with enough pressure to squeeze out excess compound (Fig. 19b). This excess compound may be removed or redistributed as a skim coat over the tape (Fig. 19c). When the embedding coat is dry, a second coat of compound is applied with the edges feathered 2″ to 4″ beyond the tape edges (Fig. 19d). Over the dry second coat, a third coat of compound is spread with edges feathered 2″ to 4″ beyond the second coat to blend smoothly with the wallboard surface (Fig. 19e). Joints can be treated in a similar manner by machine application of tape and joint compound (Fig. 20).

Interior Corner Joints Compound is applied to both sides of interior corners, the tape is folded along the center crease and embedded snugly in the corner to form a right angle (Fig. 21). Excess compound is removed and the surfaces are subsequently finished with additional coats of compound as

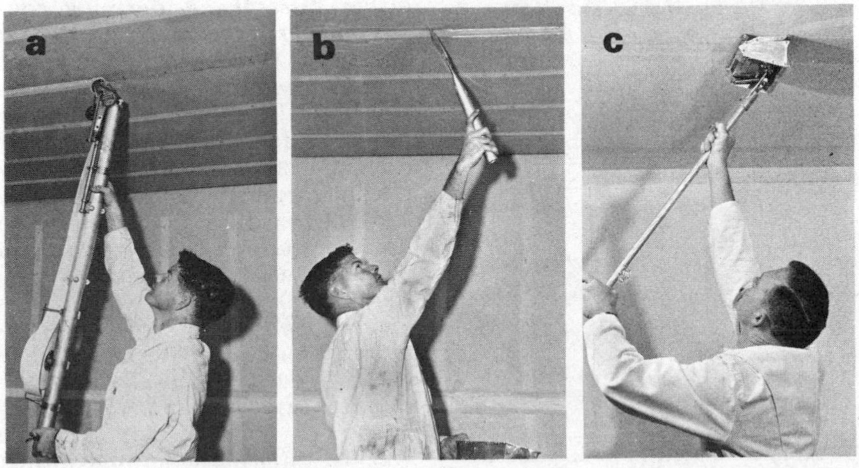

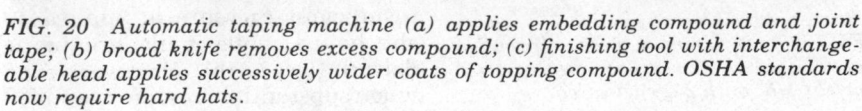

FIG. 20 Automatic taping machine (a) applies embedding compound and joint tape; (b) broad knife removes excess compound; (c) finishing tool with interchangeable head applies successively wider coats of topping compound. OSHA standards now require hard hats.

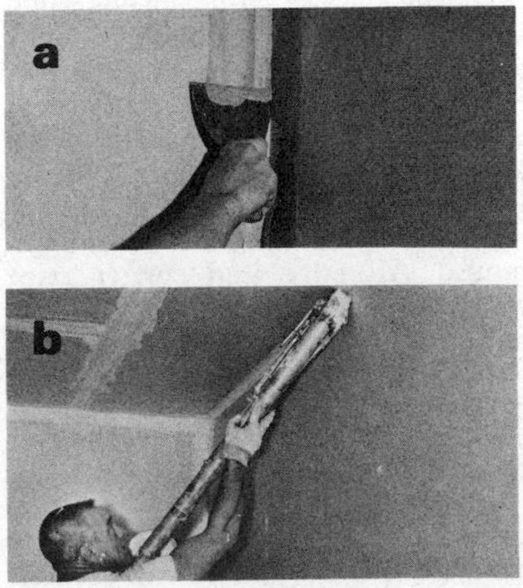

FIG. 21 (a) each side of interior corner can be treated separately with a finishing knife, or (b) both sides can be treated at once with a corner tool.

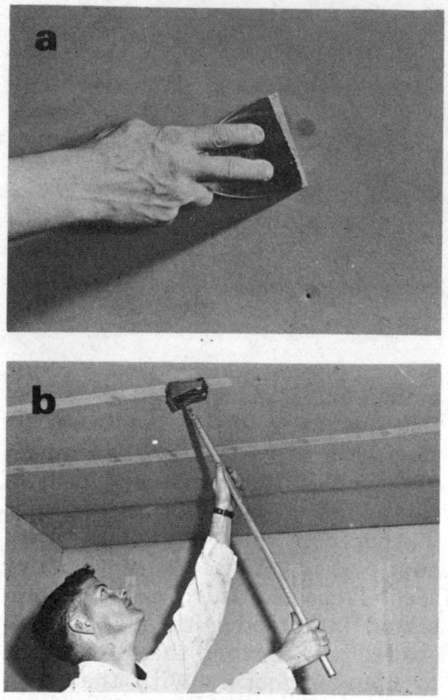

FIG. 22 Slightly recessed screw or nail heads are treated with (a) a finishing knife, or (b) an automatic spotter. A hard hat should be worn.

described for flush joints. Hand tools also are available for corner finishing.

Edge and Corner Trimming

All trim shapes except metal U bead are installed after the wallboard is erected (Fig. 15). Metal cornerbeads and casing beads such as L, LC and LK, are installed with the flange outward and are nailed, crimped or screwed 5″ o.c. through the gypsum board into supporting construction. Metal cornerbeads with paper flanges are secured by embedding the flanges in joint compound. All edge and corner trim except U bead require finishing with joint compound to obtain a smooth surface and/or conceal fasteners. Methods and sequence of three-coat application are the same as for fastener treatment.

Unlike other trims, U bead is fastened through the back flange first, then the wallboard is erected with edges inserted in the trim. Finishing with compound is not required because fasteners are concealed.

Fastener Treatment

A first coat of embedding compound usually is applied over fastener heads at the same time joints are taped. Fasteners may be treated (spotted) using an automatic spotter or a finishing knife (Fig. 22). The knife should be pressed against the surface to finish level and remove excess compound. Second and third coats of compound generally are applied over the fasteners when similar coats are applied to taped joints, edges and corners. Each successive coat should be feathered slightly beyond the one previously applied, sanding between coats as required. Each coat should be thoroughly dry before the next coat is applied, and care should be taken not to raise the fibers in the facepaper when sanding.

FINISH DEFECTS

Improper installation or adverse job conditions may result in surface imperfections such as *nailpopping, shadowing, cracking* at corners and *ridging* at joints. Such finish defects can be minimized or completely eliminated if their causes are understood and appropriate precautions are taken.

Nailpopping

Nailpopping is a surface defect in which compound is pushed out over the nailhead beyond the surface of the board. It is due primarily to loose boards caused by wood shrinkage and/or excessively long nails, and improper nailing.

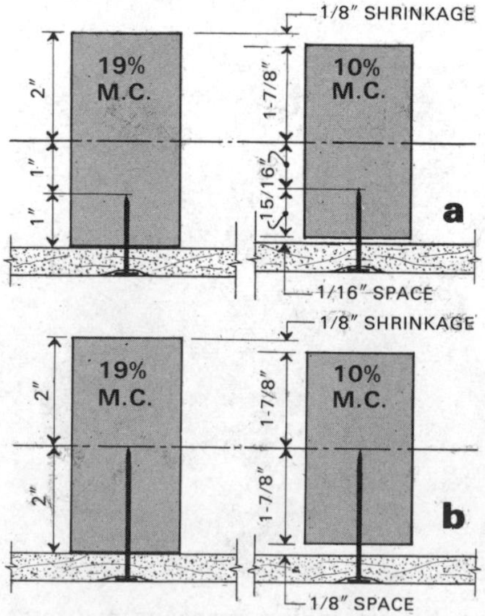

FIG. 23 As stud shrinks, separation may be (a) 1/16″ when nail penetrates 1″ or (b) 1/8″ with 2″ penetration.

Wood Shrinkage Wood shrinks across the grain as its moisture content is reduced (see page Wood 201-4). As a piece of lumber shrinks it tends to expose the shank of a nail driven into the edge (Fig. 23). If shrinkage is substantial and nails are too long, separation between board and framing member can result. A separation as small as 1/32″ can result in an objectionable nailpop. It is essential that framing and furring lumber have as low an initial moisture content as possible, that lumber be kept dry during storage and installation, and that nail lengths do not substantially exceed those recommended for specific applications. Framing and furring lumber should be at the equilibrium moisture content for the area and should not exceed 15% moisture content. Nail penetration should be as near as possible to those recommended in Figure 13. Annularly threaded nails require less penetration than smooth shank nails and tend to minimize nailpopping.

Improper Nailing When gypsum board is improperly nailed, loose boards may result due to loose nails, loose attachment or punctured facepaper.

Loose Nails can result when nails either miss the underlying construction or enter it at an angle near the edge (Fig. 24a). Such nails eventually will be loosened by vibration until they protrude beyond the finish surface. To guard against such defects, the location of supporting members should be marked carefully on the wallboard prior to nailing, nails should be driven perpendicular to the surface into the member, and nails which have missed the member should be removed promptly. After nails have been removed, nailholes should be dimpled and treated with compound.

Loose Attachment results when insufficient pressure is applied to the board while nails are being driven (Fig. 24b). However, even when substantial pressure is applied

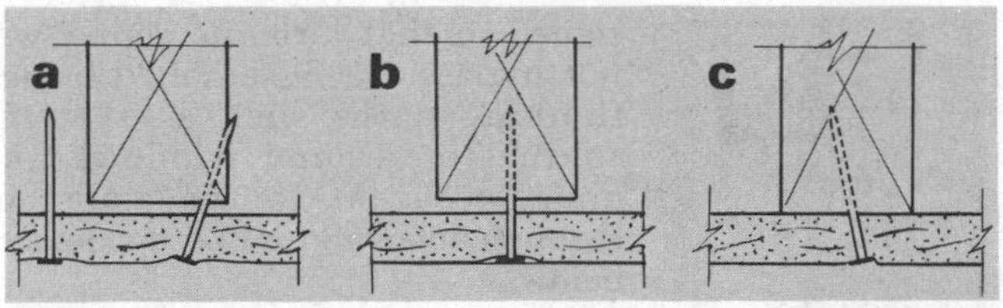

FIG. 24 Nailpopping can result from (a) loose nails which miss framing members, (b) loose boards, or (c) obliquely driven nails which puncture facepaper.

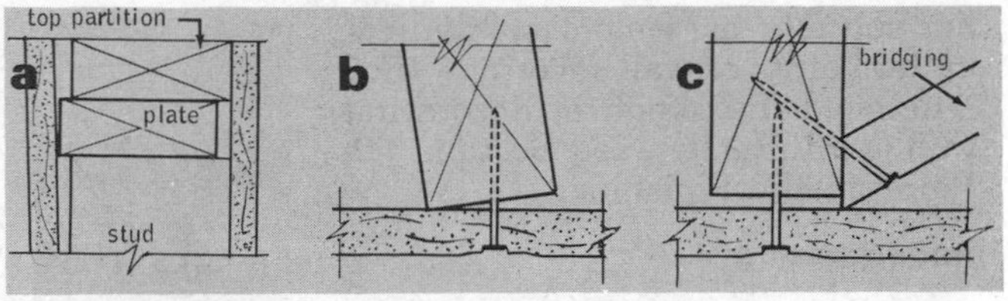

FIG. 25 Punctured facepaper and loose boards can be caused by (a) misaligned members, (b) twisted members, (c) projecting blocking.

properly it may not be possible to produce snug contact with supporting members if the board is cut too large and must be forced into place, or if nailing improperly starts at the perimeter and proceeds inward. To prevent loose board attachment, panels should be cut to fit easily into place, pressure should be applied while nailing, and proper nailing sequence from the middle of the panel outward should be observed. Use of screws and double nailing also tends to minimize loose boards.

Punctured Facepaper can be caused by improper nailhead design, obliquely driven nails (Fig. 24c), excessive and repeated impact during nailing, or extremely dry facepaper. When the paper is punctured, the gypsum core is easily crushed and offers less resistance to loosening of the board from normal impact and vibration. In hot, dry periods, the facepaper may become more brittle, increasing the hazard of puncturing. This hazard can be minimized by increasing the air humidity during storage and erection of the wallboard.

Other common causes of punctured facepaper are misaligned or twisted members and projections caused by improperly installed blocking or bracing (Fig. 25). Intimate contact between wallboard and supporting members is impossible in such cases, and hammer impact causes the board to rebound which often ruptures the paper. Such defects should be corrected before the wallboard is erected. Use of screws, adhesive, or two-ply construction will minimize problems resulting from such defects.

Shadowing

Shadowing is simply a surface discoloration, but may appear similar to nailpopping when viewed at a distance. Shadowing is the result of an accumulation of dust or dirt caused by increased thermal conductivity and moisture condensation over the nailhead. It occurs commonly on roof/ceiling or exterior wall assemblies which are inadequately insulated and in regions with great indoor-outdoor temperature variations. Shadowing can be remedied by periodic washing or decorating; it can be avoided by proper insulation design and use of laminated two-ply construction. Where a permanent ventilating system is employed, adequate air filtration and humidity control will be helpful.

Cracking and Ridging

Corner cracking may result from differential structural stresses at interior corners where adjacent walls or walls and ceilings meet. Joint ridging often results when joints are treated during cold or highly humid weather, and occurs more frequently when joint compound is applied too thickly and is feathered over a wide area. These finish defects are not commonly encountered in multi-ply construction.

In single-ply construction, cracking at interior angles can be minimized by corner floating (see page 431-23); ridging can be substantially reduced by backblocking or strip reinforcing (see page 431-24). A bullnose-edge wallboard recently has been developed specifically to overcome the ridging problem in single-ply construction (see page 431-23).

DECORATING

The wallboard installation should be reinspected prior to decorating, particularly at joints and fasteners, and imperfections should be sanded or repaired. It is essential that joint compound be thoroughly dry before any form of decoration is applied. Depending on temperature and humidity conditions, 24 hours or more of good drying time generally is required.

Wallboard surfaces can be decorated with emulsion or oil-base paints or with paper, fabric and vinyl wall coverings. It is necessary to prepare the wallboard surface with a sealer and/or primer. A good *sealer* will seal the pores and lay down the fibers in the paper surface so that suction and texture differences between paper and joint compound will be equalized; it also will seal in surface impurities sometimes present in the wallboard surface. A good primer will provide uniform texture and will conceal color or surface variations. Some sealers incorporate the properties of primers as well.

Oil-base paints, lacquers and enamels tend to raise the fibers in the facepaper and hence require a sealer and/or primer. Sealers are used under wall coverings primarily to facilitate later removal without marring the wallboard surface and to provide an adequate base for redecorating.

Most emulsion paints such as latex, resin and casein do not raise the fibers in the facepaper. Some of the better latex paints have inherently good sealing and priming properties and require no special preparation of the wallboard. However, applying a pigmented emulsion sealer as a first coat or using at least two coats of emulsion paint generally will provide a durable, attractive finish if the wallboard has been properly installed and joints, fasteners and trim have been properly treated.

Glue size, shellac and varnish are

not suitable as sealers or primers under paints or wall coverings. (For a detailed discussion of decorating wallboard finishes, see Section 441 Painting and Finishing.)

Redecorating

If nailpopping and other surface defects occur after decorating, it is advisable to wait until one full heating cycle has been completed before repairing and redecorating.

SPECIAL CONSTRUCTIONS

Most special constructions intended to provide improved sound, fire, moisture and thermal control are predicated on gypsum board finishes with complete joint treatment and panel caulking. Material used for caulking should be a nonhardening, nonshrinking, resilient sealant. In each case, the manufacturer's specific installation recommendations should be followed for individual products and/or systems.

Sound Isolating Constructions

The properties of sound and principles of room acoustics and sound isolation are discussed in Section 106 Sound Control. For detailed information on specific assemblies, see the Work File, 106 Sound Control.

Transmission of airborne and structure-borne sound can be reduced in wall and ceiling assemblies by: (1) controlling airborne sound through flanking paths, (2) separating the supporting structural construction of each finish surface; (3) resilient mounting of finishes with channels, clips or sound deadening board, (4) including sound absorbing materials in the assembly cavity, and (5) increasing the weight of the surface construction.

Controlling airborne sound through flanking paths is essential to airborne sound isolation in any assembly. Failure to observe normal construction and design precautions can seriously impair the effectiveness of other sound control methods and the performance of the entire assembly. Resilient-mounting systems and separated structural framing systems are the most effective, most practical sound control methods individually capable of bringing lightweight framed as-

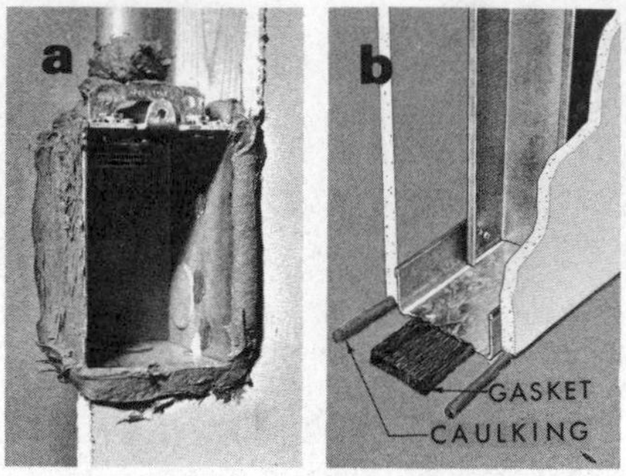

FIG. 26 Sound transmission can be reduced by caulking with acoustical sealant at critical points such as electrical boxes and partition runners.

semblies to generally acceptable STC and IIC ratings. Sound absorbing materials and increased weight generally are used in combination for their cumulative effect.

Airborne Sound Through Flanking Paths Airborne sound transmission can be reduced by constructing an airtight assembly and by avoiding back-to-back location of recessed fixtures such as medicine cabinets, and electrical, telephone, television and intercom outlets which perforate the gypsum board surface. In addition, any opening for such fixtures and related piping outlets should be caulked (Fig. 26). The entire perimeter of a sound isolating partition should be caulked at wallboard edges, and under runners to make it airtight.

Assemblies intended for sound isolation should not contain heating or air conditioning ducts or ventilating outlets and should be designed so that airborne sound will not bypass the assembly through flanking paths such as through ceilings or around windows and doors adjacent to the walls, or through floors and into crawl spaces below.

Separated Construction Separated construction systems can be developed with wood framing by using separate sets of studs, each supporting a surface of one or more plies of gypsum board. This is accomplished by staggering the studs on the same top and bottom plates, or by placing the studs on separate plates (Fig. 27). By using separated wall framing systems, kitchen cabinets, ceramic tile, lavatories and other wall-hung fixtures can be installed in the conventional manner, which is not possible on walls with resilient mounting. Separated walls also can be developed with all-gypsum double-solid and semi-solid partitions (see page 431-28).

Resilient Mounting Gypsum board finishes can be mounted on walls using resilient channels or resilient clips or over sound deadening board. Kitchen cabinets, lavatories, ceramic tile, medicine cabinets and other fixtures should not be supported on resiliently mounted walls, as the added weight and fastenings may "short out" the construction acoustically. On ceilings, resilient channels attached to wood joists or resilient hangers wired to steel joists in suspension systems can be used.

Resilient Channels are used in

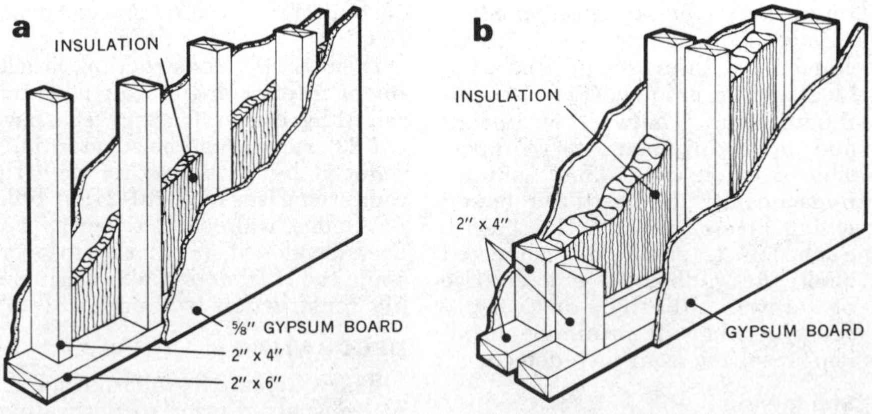

FIG. 27 Separated construction: (a) staggered studs on single plate; (b) double row of studs using separate plates.

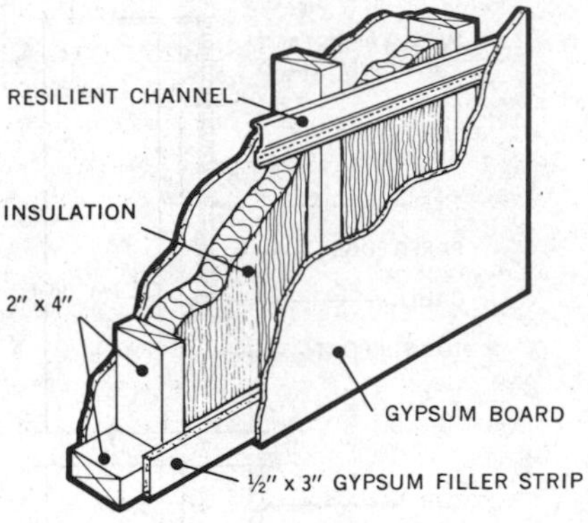

FIG. 28 Resilient channel mounting substantially improves sound isolation in wood frame construction.

both wall and ceiling assemblies to isolate the wallboard surface from the structural framing. Channels generally are screw attached perpendicular to framing members (Fig. 28). In ceiling assemblies, nails used to attach channels to wood joists may work loose as the framing lumber dries and cause floor squeaking; therefore, screw attachment is preferred.

Resilient channels can support limited loads without losing their acoustical effectiveness, and therefore are not intended for two-ply construction. Care should be taken to provide the proper size cutouts for all mechanical and electrical outlets and caulking with a nonhardening, resilient sealant is required around all outlets to keep the wall airtight and flexible at the same time.

Resilient Clips can be used to attach the base ply of two-ply finishes

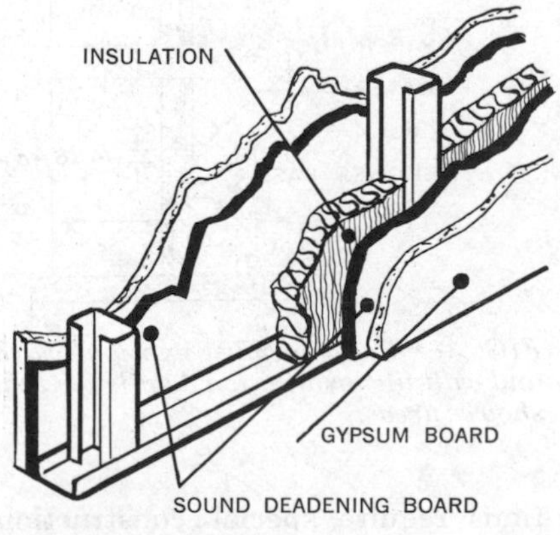

FIG. 29 Resilient mounting with sound deadening board improves sound isolation in metal frame construction.

to wood studs, but are not recommended for ceiling application. However, proprietary nail-on clips are available which are designed for use with auxiliary 1" thick wood furring for single-ply construction.

Sound Deadening Board used in residential construction includes: (1) wood- (or cane) based fiberboard, (2) mineral (including glass) fiberboard, (3) plastic foamboard, and (4) regular gypsum board. Sound deadening board is manufactured in sizes 1/4" to 5/8" thick, 4' wide and 8' to 9' long. It is used as a base layer to which wallboard is attached and generally is installed parallel to framing members with end joints staggered on alternate studs (Fig. 29). Gypsum wallboard can be applied either parallel or perpendicular to supporting construction and is generally laminated to the sound deadening board. However, nailing or screwing through the sound deadening board into the framing construction is required to satisfy fire ratings, but will reduce sound isolating effectiveness.

Wood fiberboard and plastic foamboard are used over wood construction or where noncombustible materials are not required. Mineral fiberboard is used with metal framing where noncombustible materials are required.

Sound Absorbing Materials Mineral fiber insulating batts and blankets may be used in the cavity space of assemblies to improve assembly performance in resisting sound transmission. The full height of the cavity should be filled with batts or blankets carefully fitted behind electrical outlets, blocking and fixtures, and around cutouts necessary for plumbing lines.

Insulating batts and blankets may be paper- or vapor barrier-faced with flanges and are installed by stapling within the stud space. Stapling flanges across the stud face is not recommended because wrinkled flanges may prevent intimate board-to-stud contact. In metal stud and laminated all-gypsum board partitions, the blankets can be attached to the back of the gypsum board. Batts and blankets without facings required in noncombustible construction, are installed by friction fit within the stud space.

Fire-Resistant Construction

Gypsum board is a noncombustible building material which has excellent fire-resistant properties (see page Gypsum Products 218-3). Gypsum board made of specific composition containing fibrous glass reinforcement (Type X) has greater fire resistance than regular gypsum board.

Fire resistance ratings of gypsum board assemblies can be improved by: (1) use of special fire-resistant Type X board, (2) increasing the gypsum board thickness and in some cases (3) filling the stud cavity with mineral wool. Unless all of the materials used in the tested assembly classify as *noncombustible,* the term *combustible* often is used after the assigned hourly classification.

Ordinary 2" x 4" wood stud partitions, when finished with 5/8" Type X wallboard on both sides and installed as specified, receive a one-hour combustible rating. Wood stud partitions have been rated two-hours when two layers of 5/8" Type X board are used on each side. Metal stud partitions may receive similar ratings with two-ply faces when special construction elements mentioned above are included. Wood joist floor/ceiling assemblies with single-layer wallboard generally are limited to one-hour ratings. However, steel-and-concrete assemblies protected by a suspended gypsum board ceiling generally result in higher ratings.

Another criterion for fire hazard of exposed surface materials is the rate of surface flame spread. The flame spread rating of gypsum board is under 20, based on a flame spread of 0 for asbestos-cement board and 100 for red oak. Gypsum board's low flame spread rating more than satisfies most interior finish requirements. (For more detailed discussion of the principles of fire resistance, See Section 107 Fire Resistance.)

Thermal Insulating Construction

In certain regions of the country, exterior wall and ceiling assemblies may require thermal insulating construction to minimize heat gain or loss through the assembly. In such areas a vapor barrier commonly is required to retard the migration of moisture and condensation within the assembly. (For a discussion of heat and moisture control, see Division 100—Criteria & Principles.)

Typical materials used to increase thermal resistance of assemblies with gypsum board finishes include: (1) flexible fiber insulation (batts and blankets); (2) loose fill insulation; (3) rigid plastic insulation; (4) reflective foil faced air space; and (5) air spaces provided by furring over solid materials such as masonry and concrete.

Vapor control can be provided as an integral part of the gypsum board or as part of the insulation. Foil-backed gypsum board may provide some insulating value when combined with a dead air space as well as an effective vapor barrier and can be used with or without other insulating materials. Rigid plastic foam and reflective foil insulation are materials with inherently low vapor permeability and generally do not require a separate vapor barrier. Flexible insulations such as batts and blankets are available with or without vapor barriers. Loose fill insulation requires either a separate vapor barrier, such as foil or polyethylene film, or an integral barrier provided by insulating gypsum board.

Two important considerations should not be overlooked in insulating design: (1) if reflective foil is to provide effective insulating value, a dead air space of at least 3/4″ should be provided, and (2) when a vapor barrier is used, it should be located on the warm side of the insulation. The installation of insulating and vapor control materials which are independent of the gypsum board finish essentially is the same as for other finishes. Only wood and metal furring and rigid plastic insulation which provide support for the gypsum board finish require special consideration in this section (see page 431-6 for a discussion of wood and metal furring).

Rigid plastic insulation is intended primarily for exterior and below-grade concrete or masonry walls. Rigid insulating foamboards are made from expanded polystyrene or polyurethane, in panels 1/2″ to 3″ thick up to 4′ wide and 12′ long. The foam boards may be adhesive-bonded to the masonry or concrete surface using methods and adhesives specifically recommended by the insulation manufacturer. They should be fitted together so that there will be no gaps between the boards which could create *cold spots* or allow moisture migration. Gypsum board is then bonded to the rigid foam-insulation with an approved adhesive or screw attached to special channels provided with the foam-board. When adhesive is used, the wallboard should be held in place by temporary bracing or double-headed nails which later can be removed after sufficient bond strength develops. Supplemental fasteners are required.

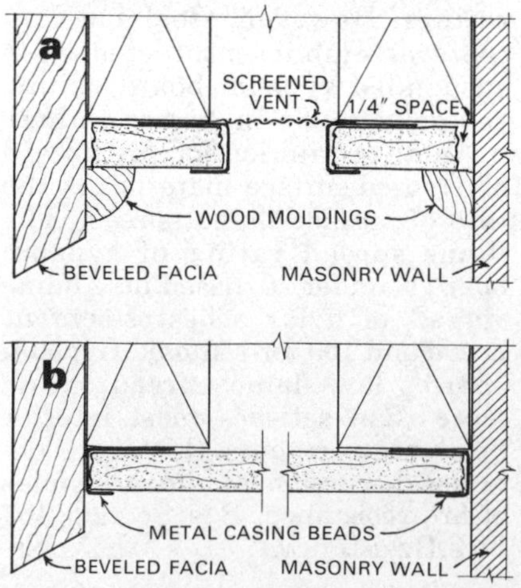

FIG. 30 Wallboard can be protected from damp exterior surfaces by (a) air spaces and wood moldings, or (b) by metal trim.

Moisture-Resistant Construction

Most gypsum board is sensitive to moisture and should not be used in locations subject to direct wetting or continuous high humidity. However, it can be used in limited exterior locations such as eave soffits, porch and carport ceilings when properly finished and painted, and in interior locations such as tub-showers and laundry areas when protected with wall tile. Such locations require special construction details and precautions.

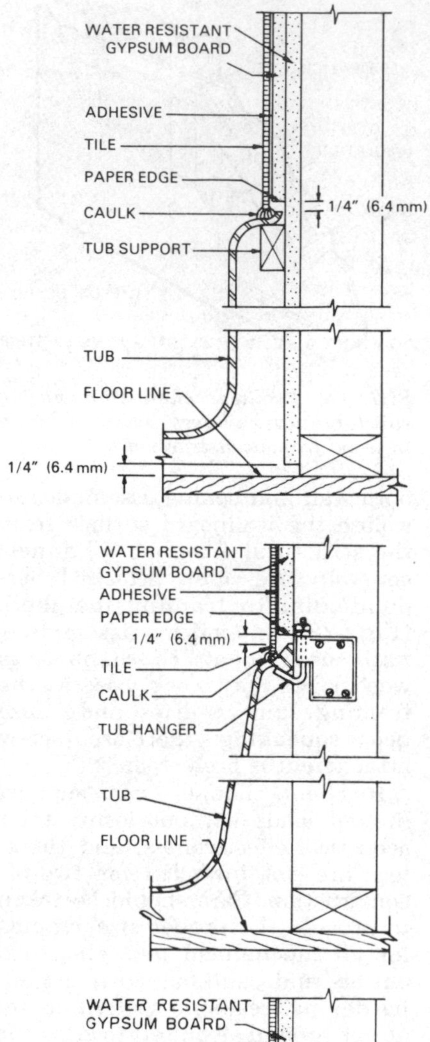

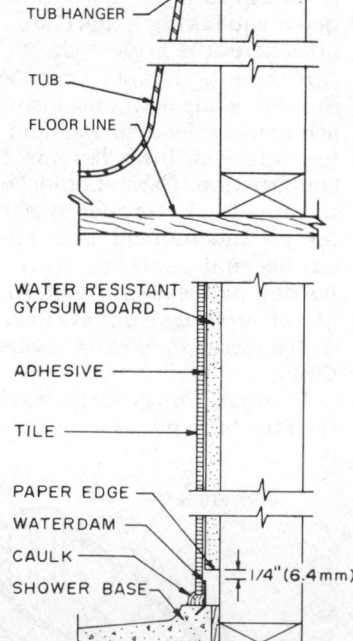

FIG. 31 Recommended gypsum board and wall tile installation details for tub-shower areas.

Exterior Locations Regular or water-resistant wallboard used on horizontal surfaces in exterior locations should be at least 1/2″ thick. Facia boards at wallboard soffits should be designed to form a drip edge to prevent water runoff across the wallboard surface and to protect the wallboard edge (Fig. 30). In addition, edges and ends at the perimeter of the board should be protected by metal edge trim. Unless protected by metal trim or other water stop, wallboard edges should be spaced at least 1/4″ from abutting masonry. Exposed wallboard surfaces and metal trim should be painted with two coats of exterior paint. Adequate ventilation of eaves, ceilings and attic spaces should be provided (see Section 360 Roof/Ceiling Systems).

Interior Location Special water-resistant wallboard or backing board is recommended for all interior walls subject to high humidity, except saunas and steam rooms. Moisture resistant gypsum board should not be used on ceilings. This type of board is particularly recommended as a base for wall tile in tub-shower areas, although regular wallboard will perform satisfactorily if properly installed. Foil-backed wallboard should not be used in high humidity interior areas because the foil backing may prevent moisture from being dissipated, thus trapping it within the board core. But where a high humidity area is located on an exterior wall, adequate insulation and a vapor barrier on the warm side of the wall are necessary to prevent condensation in the wall cavity.

Tub-Shower Areas Wallboard in tub-shower areas should be at least 1/2″ thick. Stud spacing should not exceed 24″ o.c.; if stud spacing exceeds 16″ o.c., blocking should be installed not more than 4′ o.c. between all studs, with the first row of blocking approximately 1″ above the top of the tub or shower receptor. The gypsum board should be applied perpendicular to framing members with paperbound edges parallel to the top of the tub or shower receptor and spaced at least 1/4″ above it. (Fig. 31). Generally, gypsum board should be fastened with nails spaced not more than 8″ o.c. or screws spaced not more than 12″ o.c. However, when ceramic tile over 5/16″ thick is intended, nails should be spaced 4″ o.c. and screws 8″ o.c.

Special water-resistant backing board or wallboard should be installed as recommended by the manufacturer or the Gypsum Association. When regular wallboard is used, joints between adjacent boards, including those at corners, should be taped and treated as for painting. Before installing tile, a water-resistant sealer should be applied to the entire wallboard surface, including treated joints and corners. When tile adhesive is used as a sealant, it should be spread separately from the bonding coat in a continuous, uniform coating approximately 1/16″ thick, covering all cut ends and all cutouts for fixtures. A Type I, highly water resistant organic adhesive should be used to apply ceramic wall tiles.

The tile finish should cover all surfaces as illustrated in Figure 32. The space between tile and tub rim, shower receptor or subpan top should be caulked with a continuous bead of sealant (Fig. 31). To prevent this caulked joint from opening due to settlement, it is recommended that the tub be supported at the walls on metal hangers or vertical blocking nailed to the studs. Waterproof shower receptor pans or subpans should have an upstanding leg or flange at least 1″ higher than the water dam or threshold in the entry to the shower.

Electric Panel Heating

Electric panel heating can be developed utilizing backing board, electric heating cables, a filler material and wallboard. The backing board base layer should be attached perpendicular to furring or framing members with nails, screws or staples spaced 7″ o.c. Electric heating cables should be securely attached to the backing board, maintaining at least 1-1/4″ clearance either side of the joist centerline and 4″ to 6″ around the perimeter of the ceiling at the walls (Fig. 33).

All inspection and testing of the electrical installation should be performed before the face layer is applied. The maximum cable temperature should not exceed 125° F. at any time.

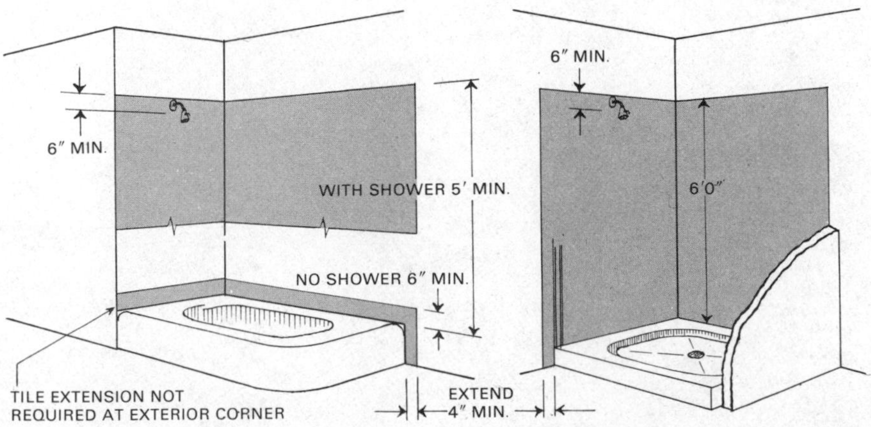

FIG. 32 Recommended tile coverage of gypsum board in tub-shower areas is indicated by shaded areas.

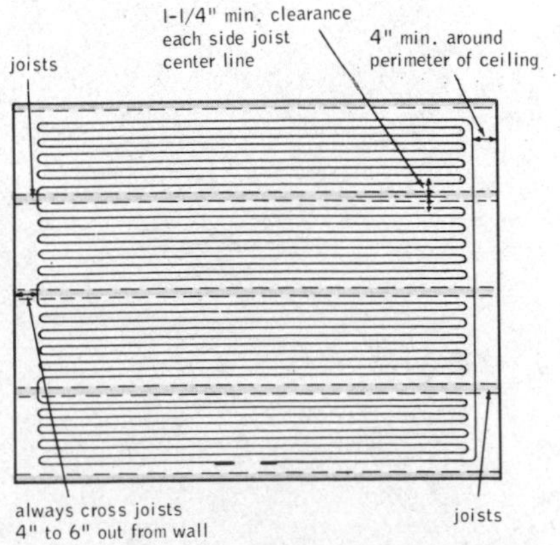

FIG. 33 Typical cable layout for electric panel heating system.

A filler material consisting of sanded gypsum plaster or embedding compound and sand mixed according to the manufacturer's recommendations is used to completely embed the heating cables. The filler material should be leveled to provide complete contact with the back of the face layer. Kiln-dried wood furring may be installed directly under and parallel to the framing members to facilitate leveling of the filler material and provide a firm base for attaching the face layer.

Filler material consisting of embedding compound and sand possesses inherent adhesive properties and the face layer requires only limited nailing as in adhesive-nail attachment. With such fillers, the wallboard face layer is applied immediately after the filler is spread and is nailed 16″ o.c. along supports. When the filler does not serve as an adhesive, it should be allowed to dry completely; then a thin coat of embedding compound is applied to provide continuous filler-to-wallboard contact, and the wallboard is nailed 7″ o.c. along supports. At the ceiling perimeter, nails should be kept 8″ to 10″ from the walls to prevent striking the heating cable where it crosses the joists.

All filler material and adhesive should be completely dry before wallboard joints and fasteners are treated. Filler material, when specifically formulated for the purpose, can be troweled on as the finish material instead of a wallboard face layer.

The system should not be operated until filler material and joint compound is completely dry, which may require as much as a week of good drying conditions or two weeks in cold, damp weather. Proper insulation design is important to the efficient operation of any electric panel heating. Thermal insulating batts used in the ceiling above the panel heating system should be unfaced friction-fit mineral or glass wool.

For a discussion of other electric panel heating systems, see page Gypsum Products 218-17.

SINGLE-PLY CONSTRUCTION

Single-ply gypsum board construction commonly is used in light residential construction and consists of a single layer of wallboard permanently attached to supporting construction. This subsection discusses the assemblies, methods and precautions recommended to obtain suitable wallboard finishes with single-ply construction. However, where a specific degree of fire resistance or sound isolation is required, the general recommendations for special constructions (page 431-15) and the manufacturer's detailed installation instructions should be followed.

Reference should be made to the entire preceding subsection, General Recommendations, pages 431-3 through 431-20, for a discussion of: (1) terminology; (2) supporting construction; (3) accessory materials—mechanical fasteners, adhesives, edge and corner trim, joint compound and joint tape; (4) job conditions, (5) application of gypsum board—fastening, adhesive-bonding, joint and fastener treatment; (6) finish defects; and (7) special constructions.

Sound ratings ranging from 35 to 50 STC can be achieved with single-ply construction, the higher values requiring special construction systems such as resilient mounting, separated construction and/or sound absorbing materials. Fire ratings for single-ply framed partitions generally are limited to one hour.

Because of the large number of fasteners in the surface of the board and the greater dependence on accurately aligned supporting construction, single-ply finishes are more susceptible to surface problems such as joint ridging corner cracking and nailpopping. However, with the selection of proper supporting construction and methods of attachment, attention to ridging, cracking and nailpopping precautions and care in joint treatment, it is possible to produce visually attractive, durable finishes.

FIG. 34 **RECOMMENDED SUPPORT SPACING FOR SINGLE-PLY CONSTRUCTION**

Wallboard Thickness				Maximum Support Spacing	
(in.)	(mm)	Location	Application[1]	(in.)	(mm)
3/8″	9.5	Ceilings[2,3]	Perpendicular	16″ o.c.	406
		Walls	Perpendicular or parallel	16″ o.c.	406
1/2″ & 5/8″	12.7 & 15.9	Ceilings[3]	Perpendicular	24″ o.c.	610
			Parallel	16″ o.c.	406
		Walls	Perpendicular or parallel	24″ o.c.	610

[1]Direction of wallboard edges relative to supports.
[2]3/8″ ceilings should not support insulation.
[3]For ceilings to receive a water base spray texture finish, use 5/8″ board applied perpendicular to 24″ supports or 1/2″ perpendicular to 16″ supports.

CEILING AND WALL ASSEMBLIES

Framing or furring members should be spaced as recommended in Figure 34, regardless of the method of attachment. With the exception of 3/8″ wallboard on ceilings, boards can be applied either perpendicular or parallel to supporting members. Perpendicular application generally is preferred because it provides greater strength and shorter joint lengths.

Wallboard edges and/or ends generally should occur over supporting members; however, ends in end-floated construction need not fall over members. Butt-end joints should be staggered with respect to each other on the same side of a partition and with respect to joints on the opposite side of the partition.

Wood Framing and Furring

Wallboard typically is attached to wood furring and/or framing with nails or screws or by a combination of adhesive and nails. Resilient metal furring can be used with wood framing where increased sound isolation is desired.

Screw Attachment Fewer fasteners are required when screws are used to attach wallboard, and the number of fasteners requiring treatment and possible surface defects are reduced. Drywall wood screws providing at least 5/8″ penetration into supports should be used, spaced as shown in Figure 35.

Nail Attachment It is particularly important that the wallboard be drawn tightly against the framing or furring members so that the board will not move on the nail shank. Wallboard can be attached by either the single- or double-nailing method in accordance with GA 216. In general, nail spacing is based on requirements for fire resistance and on intimate-board-to-stud contact to reduce possible surface defects, rather than on structural support.

Single Nailing can provide satisfactory results if the general recommendations and above precautions are carefully followed. Nails should be spaced 7″ o.c. on ceilings and

FIG. 35 **SINGLE-PLY SCREW ATTACHMENT TO WOOD SUPPORTS**

Support Spacing		Screw Spacing[1]			
		Walls		Ceilings	
(in.)	(mm)	(in.)	(mm)	(in.)	(mm)
16″ o.c.	406 o.c.	16″ o.c.	406 o.c.	12″ o.c.	305 o.c.
24″	610 o.c.	12″ o.c.	305 o.c.	12″ o.c.	305 o.c.

[1]Spacing applicable to both field and perimeter of board.

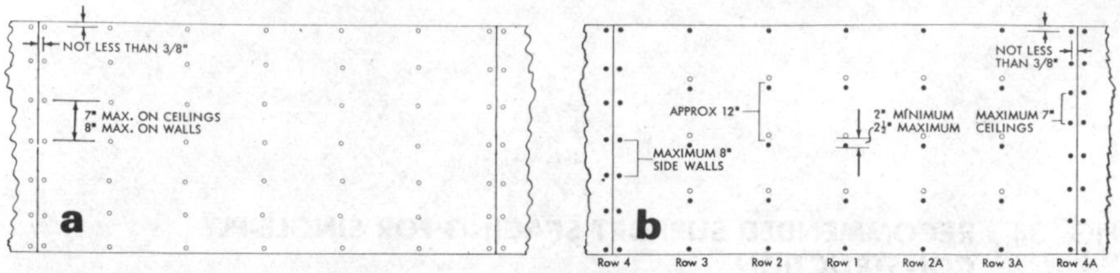

FIG. 36 (a) Single-nailing method; (b) double-nailing method: first nails shown by black dots are installed starting with Row 1, then Row 2 and 2A, etc., working outward. Second nails shown by circles are installed in same row sequence.

8″ o.c. on walls along supports (Fig. 36a).

Double Nailing assures consistently better board-to-stud contact and fewer defects than single nailing. One set of nails 12″ o.c. is driven first in the field of the panel, followed by a second set of nails 2″ from each of the nails first applied. Edges and ends falling over supports should be nailed 7″ o.c. on ceilings and 8″ o.c. on walls (Fig. 36b). While the total number of nails per panel is about 10% higher when double nailing, the number of nail spots that must be concealed is 20% less than for single nailing.

Adhesive-Nail Attachment Compared to conventional nailing, adhesive-nail attachment has the advantage of reducing the number of nails in the middle of the panel by at least 50%—consequently reducing the labor required for fastener treatment. The intimate bond between wallboard and supporting members results in stronger, stiffer assemblies and fewer fastener defects. Although screws can be used in place of nails, their use is not common because good board-to-stud contact is assured by the adhesive and the number of face fasteners is reduced to a minimum.

Stud adhesive should be applied with a caulking gun in accordance with the manufacturer's recommendations. A straight bead of adhesive is applied to the face of members supporting the field of the panel (Fig. 37a). Where two adjacent panels join over a supporting member two beads should be applied (Fig. 37b). When a predecorated wallboard joint is to be left untreated, two parallel beads of adhesive should be applied, one near each edge of the member (Fig. 37c). Generally, the volume of the bead should be such that when the wallboard is applied, the adhesive will spread to an average width of 1″ at least 1/16″ thick.

Care should be taken in applying the adhesive to prevent it from being squeezed out at the joints. Adhesive should not be applied to members which are not required for wallboard support, such as diagonal bracing, blocking and plates.

The number and spacing of supplemental nails (or screws) varies depending on the adhesive properties.

Nonstructural Stud Adhesives require fasteners in the field as well as at the perimeter of panels for wall and ceiling installation. Fastener spacing varies with the type of fastener, support spacing and load-bearing conditions (Fig. 38).

Structural Stud Adhesives conforming to ASTM C557 require only perimeter fasteners for wall application. The fasteners should be spaced 16″ o.c. for edges or ends that fall parallel to supports, and at each support for edges or ends that fall perpendicular to supports. For ceiling application, the perimeter is nailed as for walls, but additional fasteners are required in the field spaced 24″ o.c. (Fig. 39).

When installing predecorated wallboard with beveled or finished edges not requiring joint treatment, it often is desirable to avoid fasteners at the vertical joints. For these and plain wallboard applied vertically on walls, fasteners may be omitted along panel edges if appropriate means such as prebowing or temporary bracing are used to insure contact with the adhesive until it develops full bond strength. For plain wallboard, temporary bracing or nailing should be allowed to remain in place and joint or nail treatment should not be started for at least 24 hours after installation.

Metal Framing and Furring

When proprietary nailable metal studs and channels are used, the manufacturer's recommendations for fastening should be followed. Resilient channels generally are screwed perpendicular to wood framing members through predrilled holes. Wallboard should be attached to metal framing and furring with appropriate drywall sheet-metal screws, spaced not more than 12″ o.c. along supports for both walls and ceilings. For fire-rated construction, wallboard generally is applied parallel to supports with screws spaced 8″ o.c. along edges

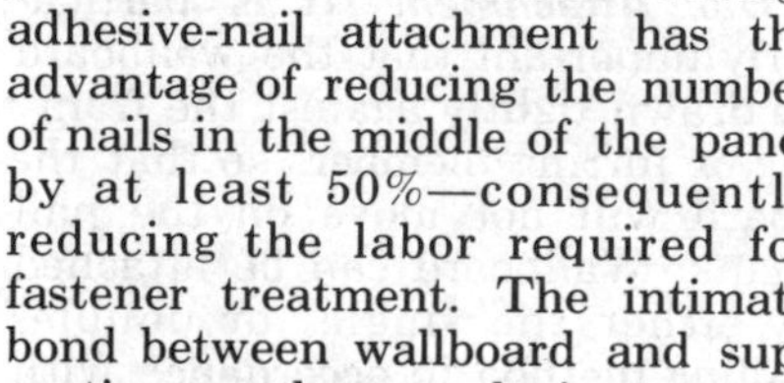

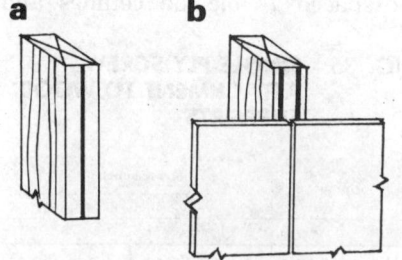

FIG. 37 Beads of stud adhesive are applied (a) straight under field of board, or (b) parallel under joints.

FIG. 38 RECOMMENDED FASTENER SPACING FOR NONSTRUCTURAL ADHESIVES

Location	Support Spacing		Fastener Spacing[1]			
			Nails		Screws	
	(in.)	(mm)	(in.)	(mm)	(in.)	(mm)
Ceilings	16″ o.c.	406 o.c.	16″ o.c.	406 o.c.	16″ o.c.	406 o.c.
	24″ o.c.	610 o.c.	12″ o.c.	305 o.c.		
Load-bearing Walls	16″ o.c.	406 o.c.	16″ o.c.	406 o.c.	24″ o.c.	610 o.c.
	24″ o.c.	610 o.c.	12″ o.c.	305 o.c.	16″ o.c.	406 o.c.
Nonload-bearing Walls	16″ o.c.	406 o.c.	24″ o.c.	610 o.c.	24″ o.c.	610 o.c.
	24″ o.c.	610 o.c.	16″ o.c.	406 o.c.		

[1]Applies to both field and perimeter of the board.

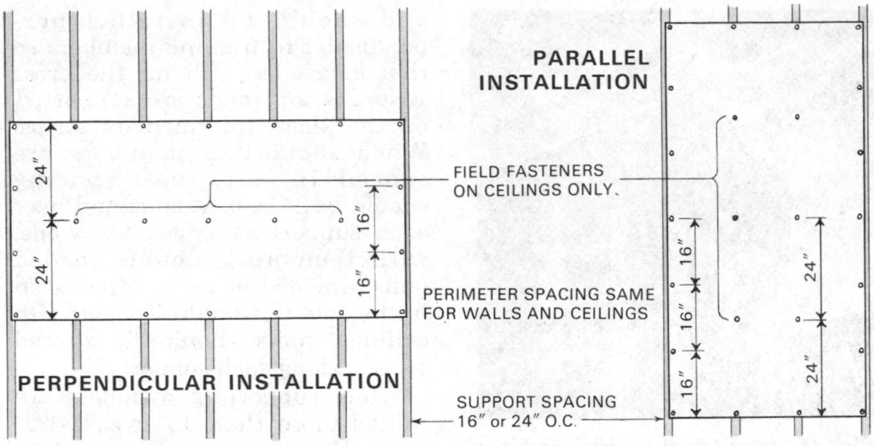

FIG. 39 Recommended supplemental fastener spacing for structural adhesives.

and 12" o.c. in the field.

Concrete and Masonry

Generally, concrete and masonry construction are furred before applying gypsum board. However, wallboard can be laminated directly to interior and some above-grade exterior masonry (see page 431-6), if the surfaces are sufficiently dry, smooth, plumb and true (Fig. 40). Some laminating adhesives and stud adhesives are suitable for this purpose, but individual manufacturers' recommendations vary and should be consulted to assure proper selection and application. While adhesive is setting, temporary support should be provided generally with threaded stub nails or cut nails.

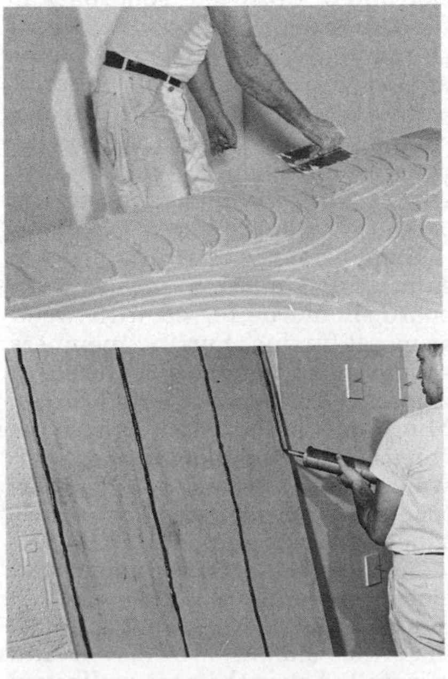

FIG. 40 Laminating adhesive is applied with a notched spreader, stud adhesive is applied in beads with a caulking gun.

CRACKING AND RIDGING PRECAUTIONS

Improper installation or adverse job conditions may cause surface imperfections such as cracking at corners and ridging at joints. Corner cracking may result from differential structural stresses at inside corners where walls and ceilings meet. Joint ridging often occurs when joints are treated during cold or humid weather; it occurs more frequently when joint compound is not properly applied.

Corner floating is intended to minimize cracking at inside corners. *Back blocking* and *strip reinforcing* are installation procedures which substantially reduce ridging with wood frame construction. However, two-ply finishes also are effective for this purpose and often may be found more practical. A proprietary bullnose-edge wallboard recently has been developed specifically to overcome the problem of ridging in single-ply construction. It utilizes a special prefill joint compound with nonshrinking and fast setting properties which develops high initial bond strength (Fig. 41).

Corner Floating

Corner floating can be used with either screw or nail attachment and consists of omitting some fasteners at interior corners (Fig. 42). When corners are floated, wallboard should be applied to the ceiling first. It should be fitted snugly at both wall-to-wall and ceiling-to-wall intersections. Framing members or blocking should be used at corners to provide solid backup even though they may not be required to receive fasteners.

At ceiling-to-wall intersections, the last row of fasteners should be omitted on the wall so that nails start 8" from the intersection. On ceilings, where ceiling joists are parallel to the intersection, nailing should start at the intersection; where the joists are perpendicular, nails should start 7" from the intersection (Fig. 42). For interior corners at the meeting of walls, the last row of fasteners is omitted from the board first applied only, and the overlapped board is nailed in the conventional manner. When double nailing, the same unrestrained clearances at corners should be maintained as for single nailing described above.

Back Blocking

Back blocking is a system of reinforcing joints against stresses which cause surface defects, par-

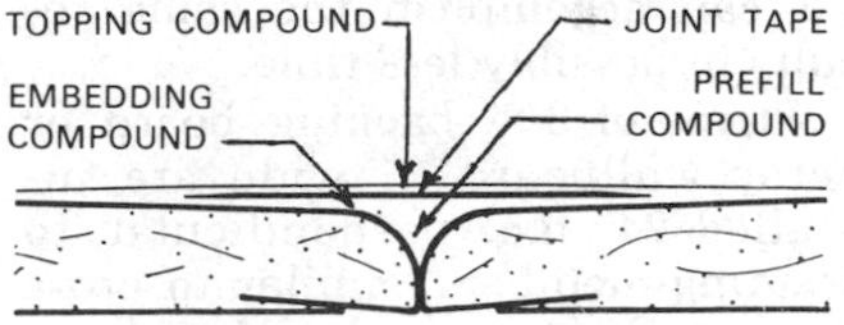

FIG. 41 Bullnose-edge wallboard requires prefilling joint channels before standard joint treatment is applied.

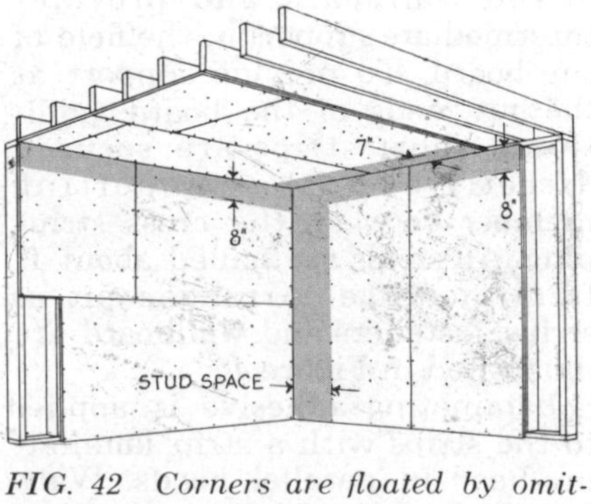

FIG. 42 Corners are floated by omitting fasteners at the intersections of walls and/or ceilings, as indicated by shaded areas above.

ticularly ridging. Simple back blocking consists of laminating wallboard blocks between framing members. To reinforce the wallboard ends by back blocking, it is necessary to *float* the ends by locating end joints midway between framing members (Fig. 43). However, because ends are not tapered, the need for relatively high crowns and wide feathering at such joints still may cause ridging when joint treatment is performed under adverse weather conditions. *End-joint tapering* can be used in conjunction with floating to reduce the need for high crowns and wide feathering at joints (Fig. 44).

Back blocking and associated constructions most often are performed on ceilings because imperfections are more conspicuous on such large, unobstructed, obliquely lighted surfaces.

Strip Reinforcing

This method of reinforcing joints against ridging offers many of the advantages of two-ply construction at reduced cost. By combining adhesive lamination with mechanical fastening, the number of nails in the face ply is reduced. Although it results in thicker assemblies than does back blocking, strip reinforcing can accomplish the same results in possibly less time.

Strips of 3/8″ backing board or scrap wallboard 8″ wide are installed 24″ o.c. perpendicular to framing members, similar to cross furring. Strips may be 4″ wide at the base of the partition and at the partition-ceiling intersection. This cross stripping supports the edges of the wallboard and provides intermediate support in the field of the board. To provide support at the butt ends of the board, additional short strips are secured directly over the supporting member between the cross stripping. All strips are nailed about 1″ from the edge. Arrangement of strips, fasteners and wallboard are illustrated in Figure 45.

Laminating adhesive is applied to the strips with a strip laminating head in parallel beads. Wallboard should be 1/2″ or thicker and should be installed perpendicular to framing members so that long edges fall on the cross stripping and ends are supported on the short intermediate strips. When supporting members are spaced 16″ o.c., nail spacing should be 24″ o.c. at ends and over each support at edges for walls, with temporary double-headed nails provided at every other stud in the field until adhesive sets. On ceilings, nails should be spaced 24″ o.c. along each support.

When supporting members are spaced more than 16″ o.c., 1-1/2″ drywall gypsum screws are required in the gypsum board strips midway between framing members for both walls and ceilings as shown in Figure 45. Appropriate drywall nails providing generally recommended penetration should be used to attach both stripping and wallboard.

FIG. 43 In end floating, backing blocks are installed with adhesive beads perpendicular to end joints.

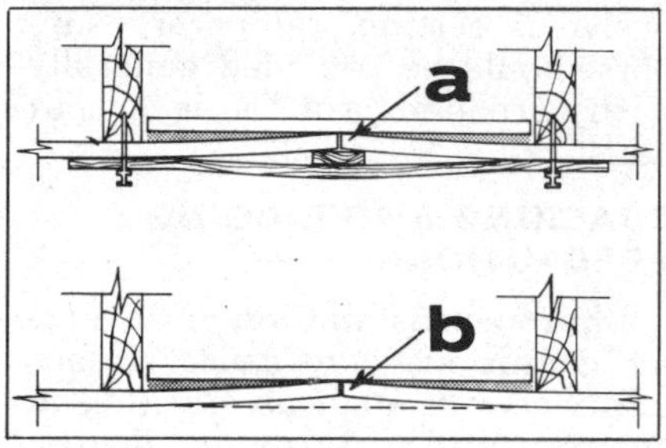

FIG. 44 End tapering butt-end joints: (a) nail-attached wood strips force ends upward; (b) depression is formed after strips are removed.

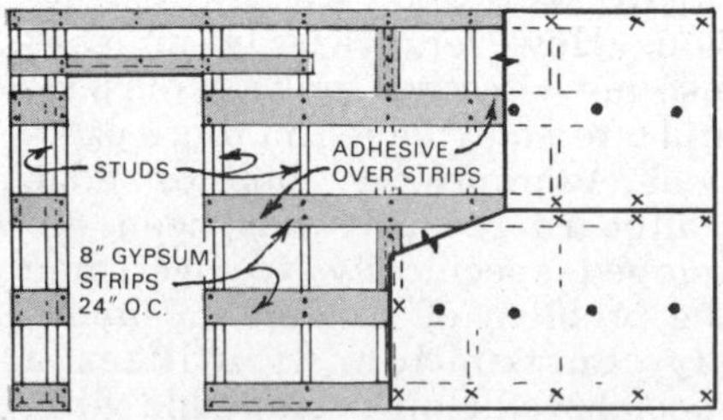

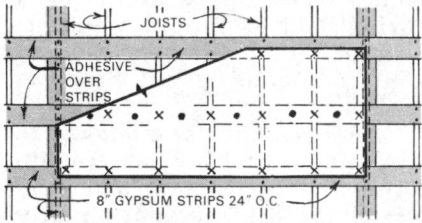

CEILINGS: PERIMETER, FIELD NAILS (X) AT EACH SUPPORT

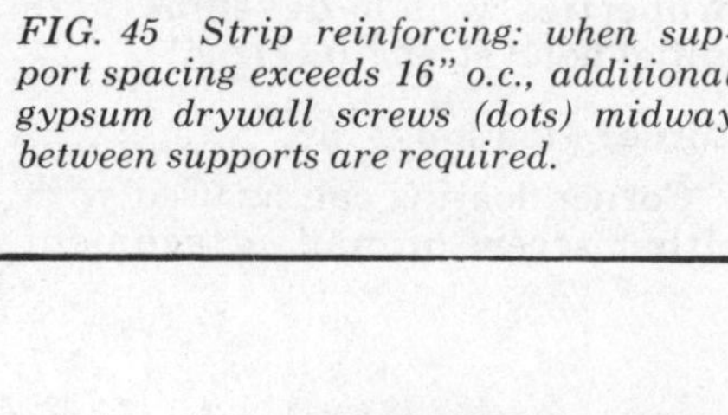

WALLS: PERIMETER NAILS (X) ONLY AT EACH SUPPORT

FIG. 45 Strip reinforcing: when support spacing exceeds 16″ o.c., additional gypsum drywall screws (dots) midway between supports are required.

RESURFACING EXISTING CONSTRUCTION

Wallboard may be used to provide a new finish on existing walls and ceilings of wood, plaster, masonry or wallboard. If the existing surface is sufficiently smooth and true after appropriate preparation and if it provides solid backup for the new wallboard without shimming, 1/4″ wallboard can be applied by adhesive, nail or screw application. Nails and screws should be long enough to provide recommended penetration into supporting members under the existing finish (see Figs. 13 & 14).

Existing surfaces which are not sufficiently smooth to receive wallboard should be furred with wood or metal furring. Furring members should be shimmed as required to provide a suitably level base. Minimum wallboard thickness for various support spacings and installation methods should be as recommended for new construction over furring.

Surface trim for mechanical and electrical equipment such as switch plates, outlet covers and ventilating grilles should be removed and reinstalled over the new wallboard finish.

Multi-ply finishes are particularly suitable where increased sound isolation and/or fire resistance are desired, and generally provide better surface quality than single-ply construction.

Reference should be made to the entire first subsection, General Recommendations, pages 431-3 through 431-20, for a discussion of: (1) terminology, (2) supporting construction; (3) accessory materials—mechanical fasteners, adhesives, edge and corner trim, joint compounds and joint tape; (4) job conditions; (5) application of gypsum board—fastening, adhesive-bonding, joint and fastener treatment; (6) finish defects; and (7) special constructions.

Multi-ply finishes consist of two or more layers of gypsum board and include (1) *supported* ceiling and wall assemblies, and (2) *self-supporting* all-gypsum partitions. Face layers usually are laminated to a base layer of wallboard, backing board or sound deadening board, with only a limited number of fasteners to assure adhesive bond. Multi-ply construction, therefore minimizes fastener treatment and surface defects such as shadowing and nailpopping. It further provides inherent joint-reinforced construction and reduces the incidence of joint ridging.

Sound ratings up to 60 STC and two-hour fire ratings are possible with some two-ply framed assemblies and multi-ply all-gypsum partition assemblies. Framed load-bearing wall assemblies with two-ply finishes receive somewhat lower fire ratings than nonload-bearing assemblies.

On framed ceilings, the use of two-ply gypsum board finishes alone generally does not provide substantially higher fire and impact sound ratings than does comparable single-ply construction. Selection of two-ply finishes for ceilings often is based on improved surface quality rather than on fire or sound ratings.

SUPPORTED CEILING AND WALL ASSEMBLIES

The gypsum board finishes discussed under this heading are of two-ply construction and may be supported by wood or metal framing and furring, concrete or masonry. The maximum spacing for framing and furring depends mainly on the base-ply thickness and whether the base ply is attached perpendicular or parallel to supporting construction (Fig. 46). Generally, support spacing of 24″ is acceptable when the base ply is 1/2″ or thicker; support spacing 16″ o.c. is sometimes required when 3/8″ base ply is used. The same maximum support spacing is applicable to both wood and metal framing and furring.

Base-Ply Attachment

The base ply of supported wall and ceiling finishes generally consists of wallboard or backing board. Both are available in regular, insulating and fire-resistant boards, but backing board more frequently is used for the base ply because it is more economical. Sound deadening board may be used where increased sound isolation is required. Fastener size and spacing for applying sound deadening board varies for different fire and sound-rated constructions and the manufacturer's recommendations should be followed.

The base ply can be attached to supporting framing or furring with nails, staples or screws. A method utilizing resilient clips is limited to walls. Base-ply attachment for all-gypsum partitions is discussed under Self-supporting Gypsum Partitions below.

Wood Framing and Furring Fastener spacing for nails, screws and staples varies primarily with the method of face-ply attachment (Fig. 47).

Nail Attachment of base ply should be with suitable drywall nails, long enough to provide the recommended penetration into supporting construction. When the face ply is to be laminated, the base ply should be nailed as recommended for single-ply construction. Double nailing generally is not used because nailheads will be covered by the face ply and fastener treatment is not a consideration.

FIG. 46 **SUPPORT SPACING FOR TWO-PLY CONSTRUCTION**[1]

Thickness of Base Ply (in.)	Thickness of Base Ply (mm)	Location	Application of Base Ply[2]	Maximum Support Spacing (in.)	Maximum Support Spacing (mm)
3/8″	9.5	Ceilings[4, 5]	Perpendicular	16″ o.c.	406 o.c.
		Walls	Perpendicular or Parallel	24″ o.c.[3]	610 o.c.
1/2″ & 5/8″	12.7 & 15.9	Ceilings[5]	Perpendicular	24″ o.c.	610 o.c.
		Walls	Perpendicular or Parallel	24″ o.c.	610 o.c.

[1]Unless noted, installation with mechanical fasteners with or without adhesive.
[2]Direction of board edges relative to supports.
[3]16″ o.c. if no adhesive used between plies.
[4]Two-ply 3/8″ ceilings should not support insulation.
[5]For two-ply ceilings to receive a water base spray texture finish, use 5/8″ board applied perpendicular to 24″ supports or 1/2″ perpendicular to 16″ supports.

When the face ply is to be attached with mechanical fasteners, fewer nails are required in the base ply because the nails in the face ply will support the base ply as well.

The base ply on walls generally is installed parallel to framing members, except when predecorated wallboard is intended as the face ply. On ceilings, application of the base ply perpendicular to framing members is preferred. At inside corners it is recommended that only the overlapping base ply be nailed and fasteners be omitted from the face ply so as to create a floating corner more capable of resisting structural stresses (Fig. 48). The base ply generally need not be nailed at exterior corners as support will be provided when the face ply and/or corner trim is nailed.

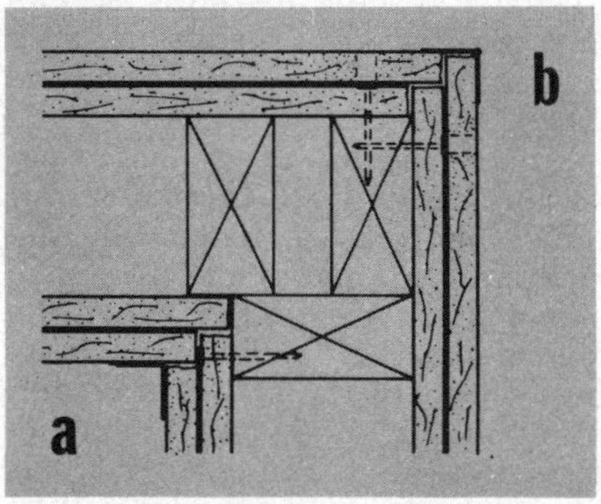

FIG. 48 Floating corners: (a) base ply is nailed only one side of interior corner; (b) temporary nails in face ply at exterior corner may be countersunk.

Screw Attachment of the base ply to wood framing or furring should be with appropriate drywall wood screws providing at least 5/8″ penetration into supporting construction. Spacing of screws depends on the method of face-ply attachment and supporting construction (Fig. 47). Inside corners should be floated as described for nailing.

Staple Attachment generally is applicable to wood supporting construction only. Staples should be 1″ long for 3/8″ thick base ply and increase proportionately for thicker boards to provide at least 5/8″ penetration into supporting members. Recommended spacing for staples is given in Figure 47. Staples generally should be driven with the crown perpendicular to the edges of the board, except where edges fall over supports they should be driven parallel to the edges (Fig. 49). When driven, the crowns should bear tightly against the board without breaking the facepaper.

FIG. 47 **BASE-PLY ATTACHMENT TO WOOD SUPPORTING CONSTRUCTION[1]**

	Nail Spacing		Screw Spacing		Staple Spacing	
Location	Laminated Face Ply	Nailed Face Ply	Laminated Face Ply[2]	Screwed Face Ply	Laminated Face Ply	Nailed or Screwed Face Ply
Walls	8″ (203 mm) o.c.	16″ (406 mm) o.c.	16″ (406 mm) o.c.	24″ (610 mm) o.c.	7″ (178 mm) o.c.	16″ (406 mm) o.c.
Ceilings	7″ (178 mm) o.c.					

[1]Supports spaced 16″ or 24″ o.c. unless noted; fastener spacing applies to perimeter and field fastening along supports.
[2]12″ o.c. for both ceilings and walls when supports are spaced 24″ o.c.

Metal Framing and Furring The base ply generally is attached to metal framing or furring with drywall sheetmetal screws. When proprietary nailable studs or channels are used, the manufacturer's fastening recommendations should be followed.

For walls, screw spacing should be 12″ o.c. in the field and 8″ o.c. at edges when the face ply is laminated, and 16″ o.c. throughout the board when the face ply is attached with screws. Screws should be long enough to penetrate metal studs or furring by at least 3/8″.

Two-ply ceiling finishes on metal framing or furring are not commonly used because they do not substantially improve fire resistance or impact sound isolation. They do provide increased airborne sound isolation and may be used where this is an important consideration.

Concrete and Masonry These assemblies have inherently good fire resistance and sound isolation which two-ply finishes do not substantially improve. However, when single-ply wallboard cannot be laminated directly to masonry or concrete and the walls must be furred, two-ply finishes may be desired for superior surface finish. The base ply is applied to such walls as described above for wood or metal furring.

Face Ply Attachment

Joints in the face ply should be offset from joints in the base ply by at least 10″. Wallboard can be applied either horizontally or vertically, whichever results in least waste of materials and shorter joint lengths. On walls, wallboard applied horizontally generally

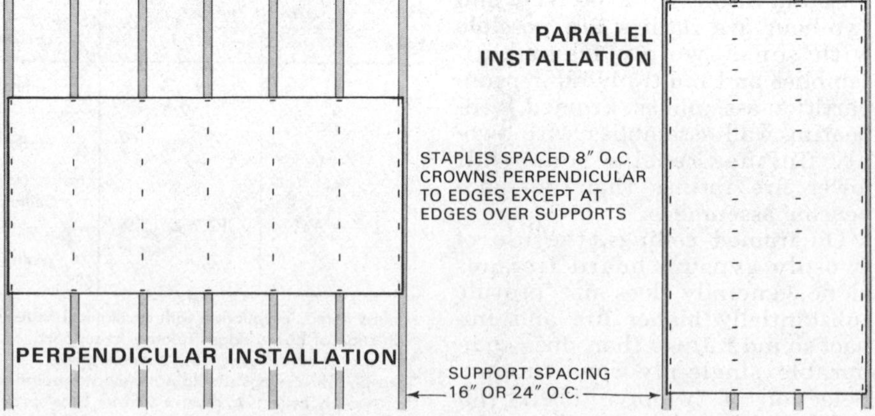

FIG. 49 Staple attachment of base ply to wood supporting construction.

results in shorter and less conspicuous joints and therefore is preferred. However, predecorated wallboard which is to be left with joints untreated and plain wallboard which is to be trimmed with battens usually are installed vertically.

Nail and Screw Attachment When the face ply is attached with mechanical fasteners and no adhesive between plies, the maximum spacing and minimum penetration recommended for nails and screws in single-ply construction should be observed in accordance with GA 216. However, the specifications for some proprietary sound- and fire-rated partition assemblies may differ.

Adhesive Attachment The face ply typically is attached to gypsum board and sound deadening board by *sheet lamination* or *strip lamination.* Generally, strip lamination is specified for systems with sound deadening board.

Sheet lamination involves covering the entire back of the face ply with laminating adhesive using a notched spreader or other suitable tool (Fig. 50). Size and spacing of notches varies depending on the adhesive. The wallboard should be erected using moderate pressure, and adhesive squeezed out at the joints should be promptly removed. In strip lamination the face ply is applied vertically and adhesive is spread in vertical ribbons spaced 16″ to 24″ o.c., depending on adhesive and assembly (Fig. 51).

Supplemental mechanical fasteners or temporary wood bracing are required to assure intimate adhesive bond between base and face plies. When structural ceiling members are used to anchor temporary wood bracing, ceiling installation should be delayed until after the wall finish is installed.

Temporary Fasteners may be used when the face ply is laminated to gypsum board in wood construction and a fire rating is not required. These may be double-headed nails, stucco furring nails or drywall nails, generally spaced to provide face support approximately 24″ o.c. in each direction. All nails should be long enough to assure proper penetration into supporting construction. Nails should be driven through gypsum board washers or other rigid material to facilitate removal without damaging the face ply. After nails are removed, nailholes should be dimpled and finished with compound as for permanent nails.

FIG. 50 Sheet lamination: notched spreader or laminating box (shown) can be used to apply adhesive.

FIG. 51 Strip lamination: laminating head applies adhesive beads in parallel strips to the base ply.

Permanent Fasteners generally are required in the field as well as in the perimeter of the face ply in fire-rated assemblies. Nail spacing varies depending on the tested assembly construction and is not related to whether adhesive is used between plies.

In sound-rated partitions where fire resistance is not a consideration, the face ply generally is applied vertically over sound deadening board, with permanent mechanical fasteners at top and bottom of boards. Intermediate fasteners may be omitted along vertical joints and in the field if temporary bracing is provided until adhesive has developed sufficient bond strength.

Some permanent mechanical fasteners in each laminated wallboard panel are recommended, particularly on ceilings. Usually, these can be spaced at the perimeter of the board where they will be concealed by joint treatment and/or trim.

SELF-SUPPORTING GYPSUM PARTITIONS

These are nonload-bearing, noncombustible multi-ply partitions, fabricated on the job almost entirely from gypsum boards. Inner plies consist of backing board in rib or sheet form, to which wallboard face plies are laminated. Unlike the ceiling and wall assemblies described above, which depend on wood, metal, concrete or masonry construction to support the surface finish, gypsum partitions are entirely self-supporting and require only floor and ceiling runners to stabilize the partition. Runners should be attached securely to ceiling and floor construction with fasteners spaced 24″ o.c. and should be carefully caulked at partition perimeters and under runners to close air leaks. Maximum height of partitions is limited by the design and dimensions of components; hence the manufacturer's recommendations should be consulted.

Gypsum partitions are classed according to assembly construction as: (1) *semi-solid,* utilizing backing board ribs; or (2) *solid,* with coreboard (thicker backing board) inner plies. All gypsum board components such as ribs, coreboard, backing board and wallboard typically are installed in vertical position.

Semi-solid Partitions

These partitions are constructed with ribs 1″ to 1-5/8″ thick and 6″ to 8″ wide, with single- or two-ply faces (Fig. 52). In single-faced partitions, the face ply is attached directly to the ribs. Partitions with two-ply faces consist of ribs, a backing board base ply and wallboard face ply.

Ribs are cut 6″ shorter than wallboard and can be laminated to the face ply on one side of the partition either before or after it is erected. Perimeters of door and window openings should be reinforced with additional ribs. Adhesive can be spread on the erected face ply or on the ribs, keeping it 1″ from the rib edges. The rib-reinforced panels are secured to wood or metal runners at top and bottom with appropriate drywall screws spaced 12″ o.c.

The ribs on the erected panels then are spread with adhesive, facing panels are installed and screw-attached to runners with screws spaced 12″ o.c. To assure adequate bond at ribs, 1-1/2″ drywall gypsum screws should be provided at each rib 12″ from the top and bottom and spaced not more than 36″ o.c.

For partitions with two-ply faces, the ribs and backing board base plies are assembled and installed as for single-ply faces. Face plies then are sheet laminated as for two-ply supported partitions using supplemental drywall gypsum screws.

Separated rib construction to improve sound isolating effectiveness can be provided in either single- or two-ply ribbed partitions. Installation procedure is essentially the same as for ordinary ribbed partitions, but twice as many ribs are required, spaced 12″ o.c. and staggered on opposite sides of the partition (Fig. 52b).

Solid Partitions

Solid gypsum partitions are an extension of the concept of semi-solid partitions and consist of a solid coreboard to which wallboard faces are laminated on each side (Fig. 53). The vertical joints in the face ply should be offset at least 3″ from joints in the coreboard. The face ply typically is attached by sheet lamination with supplemental drywall gypsum screws. Double- or triple-solid partitions separated by air spaces provide increased fire resistance and sound isolation (Fig. 53b).

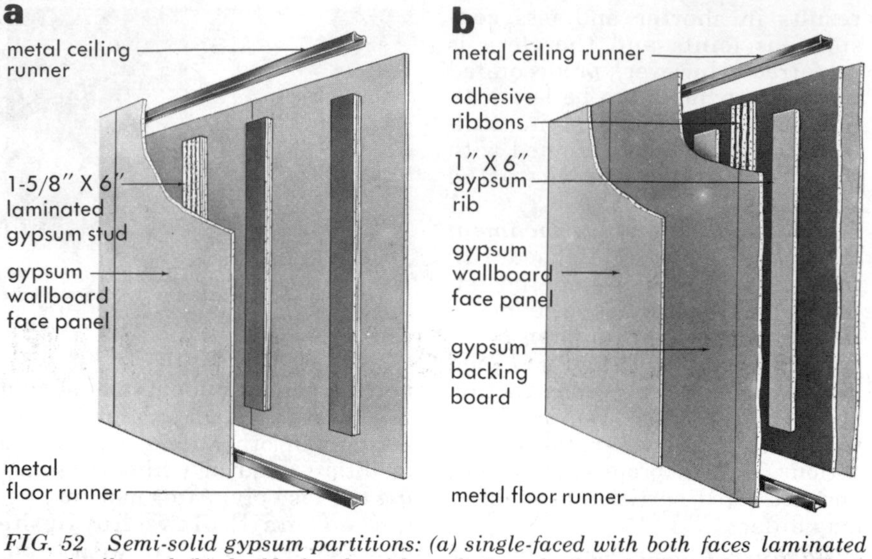

FIG. 52 Semi-solid gypsum partitions: (a) single-faced with both faces laminated to each rib; and (b) double-faced, with each two-ply face laminated to alternate ribs forming separated rib construction.

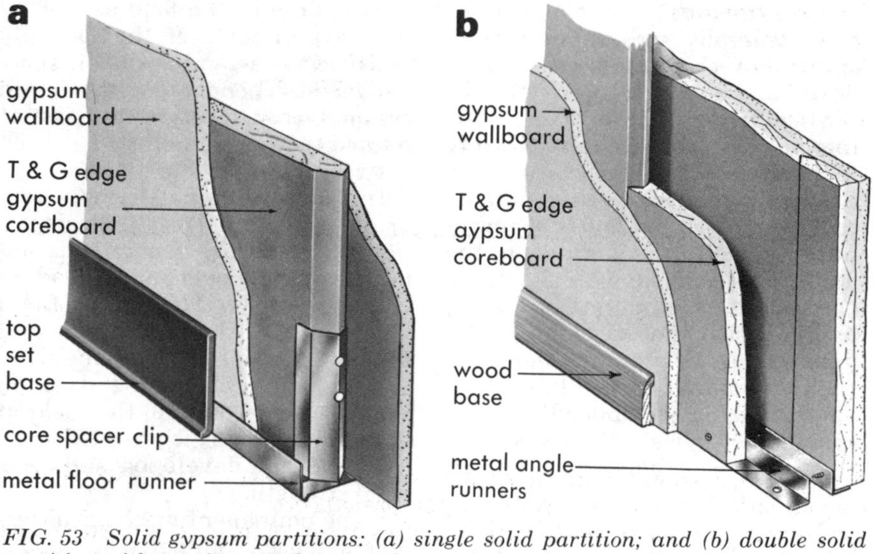

FIG. 53 Solid gypsum partitions: (a) single solid partition; and (b) double solid partition with an air space between the two faces.

We gratefully acknowledge the assistance of the following for the use of their publications as references and permission to use photographs and illustrations: Gypsum Association, Gypsum Drywall Contractors International; Celotex Corporation, Flintkote Company, Georgia-Pacific Corporation, Johns-Manville Corporation, Kaiser Gypsum Company, National Gypsum Company, United States Gypsum Company.

GYPSUM BOARD FINISHES 431

CONTENTS

GENERAL RECOMMENDATIONS

To provide gypsum board finishes of suitable quality, consideration should be given to: (1) preparation of supporting construction; (2) selection of gypsum board and accessory materials; (3) job conditions; (4) application of gypsum board, including joint and fastener treatment; and (5) special constructions where sound isolation, fire resistance, thermal insulation and moisture control are important.

Recommended Specifications for the Application and Finishing of Gypsum Board (GA 216), published by the Gypsum Association, constitutes the recognized industry standard and forms the basis for many of the following recommendations. These recommendations primarily apply to plain wallboard finishes unless predecorated wallboard is specifically mentioned.

See Main Text page 431-3 for a definition of terminology.

See Main Text, page 431-24. for a discussion of resurfacing existing construction.

SUPPORTING CONSTRUCTION

Wood, metal, concrete or masonry supporting construction intended for the application of gypsum board finishes should be capable of supporting design loads and should provide a firm, level and true base. Deflection of ceiling framing members should not exceed 1/240 of the span under full design loads.

Adequate headers or lintels should be provided over openings to support structural loads and special construction should be provided where required for wall-hung equipment and fixtures.

Spacing of framing or furring members should not exceed that recommended for the intended wallboard thickness and installation method.

Exterior and below-grade concrete and masonry and framing members which do not conform to the above requirements should be furred.

(continued) GENERAL RECOMMENDATIONS

Wood Framing

Framing lumber should be of the proper grade for the intended use. The moisture content of lumber should not exceed 15% at the time of installation and should be as near as possible to the equilibrium moisture content for the particular area (see page Wood WF201-1).

Wood studs in load-bearing partitions should be of 2″ x 4″ dimension or larger. In nonload-bearing or staggered-stud partitions, 2″ x 3″ studs may be used but the spacing should not exceed 16″ o.c. when openings are included or 24″ o.c. when no openings are included in the partition.

Ceiling joists should be regularly spaced with bottom faces aligned in a level plane. Slightly crooked joists should be installed crown up.

Stringers at midspans can be used to correct minor irregularities in spacing and to level wood joist ceilings (see Main Text, Figure 3a, page 431-5).

Because relatively small members are used for bottom chords in wood truss construction and larger spans are involved, spacing often is irregular and bottom chords are not properly aligned.

Generally, wood trusses should be cross furred with 2″ x 2″ or larger furring strips to provide a properly spaced and level base for attaching wallboard; or stringers should be nailed over the bottom chords at approximate 1/3 points of the span (see Main Text, Figure 3b, page 431-5).

Metal Framing

Metal framing systems should be selected on the basis of desired performance criteria and should be installed in accordance with the manufacturer's recommendations.

Open-web joists generally are not designed to receive wallboard directly; hence, suitable furring or suspension systems should be provided for wallboard attachment.

Concrete and Masonry

Generally, wallboard can be laminated directly to interior, above-grade masonry and concrete, and exterior, above-grade cavity walls if the cavity is properly insulated when required.

Concrete and masonry surfaces intended for direct lamination of wallboard should be dry, smooth, plumb, true and free from loose particles, efflorescense or surface films.

Concrete surfaces should be allowed to cure for at least 28 days before wallboard is laminated. Masonry walls should have flush or tooled joints. Rough or protruding edges should be removed and depressions greater than 4″ in diameter and 1/8″ deep should be filled with grout or mortar.

Exterior and below-grade walls and other surfaces subject to moisture penetration or excessive heat loss, or surfaces which cannot readily be prepared for adhesive lamination, should be furred to provide a suitable base for wallboard attachment.

Furring

Wood or metal furring may be necessary: (1) to provide a suitably plumb, true and/or properly spaced supporting construction; (2) to eliminate capillary transfer of moisture in exterior or below-grade walls, and to minimize the likelihood of moisture condensation on interior wall surfaces; (3) to improve thermal insulating properties for exterior walls; (4) to increase assembly thickness to accommodate mechanical equipment; or (5) to improve sound isolation by resilient mounting.

Wood Furring Furring strips over masonry and concrete surfaces should be 1″ x 2″ or larger. They should be properly shimmed and secured with appropriate fasteners spaced 16″ o.c. for 1″ thick strips and 24″ o.c. for 2″ thick strips.

When framing is of irregular spacing or the spacing is too great for the intended wallboard thickness, cross furring perpendicular to framing members should be applied in accordance with the recommendations given in Figure WF1.

Cross furring smaller than 2″ x 2″ is not sufficiently stiff to permit the wallboard to be nailed without excessive hammer rebound, and may result in loose board attachment. Thinner cross furring is acceptable for screw-attached wallboard because little or no impact is involved when screws are driven. Thinner furring also may be acceptable if the wallboard is applied with stud adhesive and nailing is limited to locations where the furring is fully supported by underlying framing.

Parallel furring applied directly over primary wood framing should be 1″ x 2″ or larger and should be nailed 16″ o.c. with nails providing penetration as

(Continued) GENERAL RECOMMENDATIONS

recommended for cross furring (Fig. WF1).

Metal Furring Cold-rolled and drywall furring channels should be selected on the basis of required performance criteria and should be installed in accordance with the manufacturer's recommendations.

Cold-rolled channels generally are of 16 gauge steel, are wire-tied to supporting construction and are used primarily as a supporting grid for lighter drywall channels.

Drywall channels are of 25 gauge steel to which wallboard can be screw-attached, and can be used over all types of supporting construction. Special resilient drywall channels are available for sound isolating, framed assemblies.

ACCESSORY MATERIALS

Accessory materials for the application of gypsum board include: mechanical fasteners, adhesives, joint tape and compounds, and edge and corner trim.

Mechanical Fasteners

Nails Annularly-threaded or smooth shank nails used for the application of gypsum board should be as recommended in Figure WF2. Nails should conform to industry standards given in this table.

Unless specifically required for fire-rated assemblies, longer nails which result in substantially greater penetration into supporting construction should not be used.

Annularly threaded GWB-54 nails conforming to ASTM C-514, provide increased holding power with less penetration, minimize nailpopping, and are widely recommended.

Casing nails and common nails should not be used because the diameter of the head is too small in relation to the shank.

Screws Drywall screws providing the minimum penetration recommended in Figure WF3 should be used to attach gypsum board. Screws should conform to industry standards shown in this table.

Drywall screws have self-drilling points, self-tapping threads and flat heads designed to drive slightly below the board surface without creating ragged edges. Type W, Type S and Type G screws are typical drywall screws intended for wood, sheetmetal and gypsum supporting construction, respectively. Other proprietary

FIG. WF1 MINIMUM REQUIREMENTS FOR WOOD CROSS FURRING[1]

Location	Support Spacing (in.)	Support Spacing (mm)	Furring	Nails: Spacing	Nails: Penetration
Ceilings	16" o.c.	406 o.c.	2" x 2"	2 at each support	7/8" (22 mm) (threaded)
	24" o.c.	610 o.c.			1¼" (32 mm) (smooth shank)
Walls	16" o.c.	406 o.c.	1" x 2"	1 at each support	¾" (19 mm) (threaded or smooth shank)
	24" o.c.	610 o.c.	1" x 3"		

[1]Wallboard screw attached; if nailed, 2" x 2" furring should be used in all cases.

FIG. WF2 NAILS FOR GYPSUM BOARD APPLICATION[1]

Drywall Nails	Recommended Penetration (in.)	Recommended Penetration (mm)	Board Thickness (in.)	Board Thickness (mm)	Nail Length (in.)	Nail Length (mm)
Annularly Threaded[2] ASTM C-514	¾"	19	⅜"	9.5	1⅛"	29
			½"	12.7	1¼"	32
			⅝"	15.9	1⅜"	35
Smooth Shank[3] ASTM C-514	⅞"	22	⅜"	9.5	1¼"	32
			½"	12.7	1⅜"	35
			⅝"	15.9	1½"	38

[1]For fire-rated contructions, building code and/or manufacturer's specifications should be followed.
[2]Typically, GWB-54: ¼" diameter head, 0.098" diameter shank.
[3]Typically, ¼" diameter head, 0.099" diameter shank, with dished head.

FIG. WF3 SCREWS FOR GYPSUM BOARD APPLICATION[1]

Drywall Screws[2]	Recommended Penetration (in.)	Recommended Penetration (mm)	Screw Length (in.)	Screw Length (mm)	Application
For fastening to Wood ASTM C-894	⅝"	15.9	1¼"	32	Single-ply or base ply application
			1⅝"	41	Face ply application in two ply construction
For fastening to sheetmetal ASTM C-646	⅜"	9.5	1"	25.4	Single-ply or base ply application
			1⅝"	41	Face ply application in two ply construction
For fastening to gypsum board ASTM C-893	¾"[3]	19[3]	1½"	38	Gypsum board to gypsum board in multi-ply construction

[1]Generally acceptable for fire-rated construction.
[2]Screws shown are Types W, S & G, respectively; other drywall screws also are available.
[3]Allows for ¼" point plus ½" threaded shank; hence acceptable only for ½" or thicker board.

(continued) GENERAL RECOMMENDATIONS

screws also are suitable when used according to manufacturers' recommendations.

Ordinary wood or sheetmetal screws should not be used for the application of gypsum board.

Staples Staples used in multi-ply construction to attach the base ply should be 16 gauge flattened galvanized wire, with a minimum 7/16″ crown and divergent sheared bevel points. Staples should provide at least 5/8″ penetration into supporting construction.

Adhesives

Application of gypsum board with stud adhesives, laminating adhesives and contact adhesives for single-ply and multi-ply construction should be in accordance with the specific recommendations of the adhesive manufacturer.

See Main Text, page 431-8, for a discussion of adhesives for the application of gypsum board finishes.

Joint Tape and Joint Compound

Joint tape and joint compound should conform to ASTM C-475, Treatment Materials for Gypsum Wallboard.

See Main Text, page 431-10, for a discussion of joint tape and joint compounds.

Edge and Corner Trim

Exposed wallboard edges and ends and exterior corners should be protected with wood or metal trim.

Typical metal trim shapes are illustrated in Figure WF4.

JOB CONDITIONS

Temperature and humidity affect the performance of joint treatment materials, scheduling of joint finishing operations, the quality of the joint, and sometimes the bonding properties of adhesives.

During the winter season when the outside temperature is less than 55° F., interior finishes should not be installed unless the building is completely enclosed and has a controlled heat of 55° to 70° F. This temperature should be maintained for 24 hours before installation, during installation and until a permanent heating system is in operation.

Adequate ventilation should be provided at all times,

FIG. WF4 STANDARD DRYWALL METAL TRIM SHAPES

USASI Designation	Trim Profile	Application
CORNERBEAD[1] CB—100 x 100 CB—118 x 118 CB—114 x 114 CB—1 x 114 CB-PF (paper flange steel corner combi nation bead)	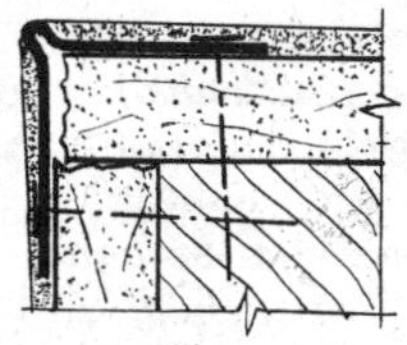	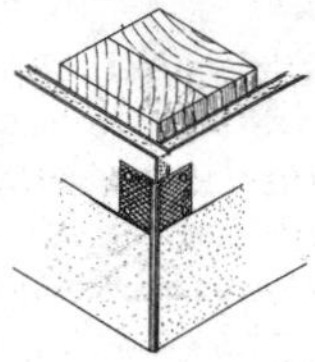
U BEAD[2,3] U-38 U-12 U-58 U-34	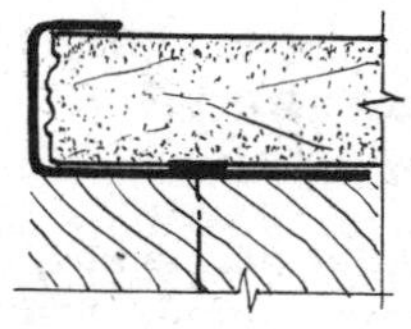	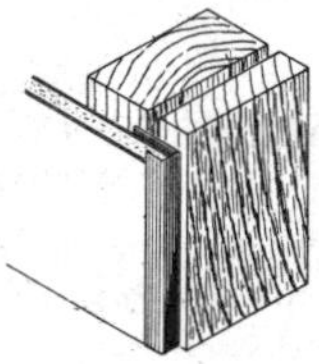
L BEAD[2,3] L-38 L-12 L-58 L-34	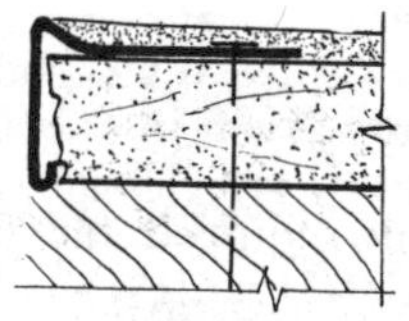	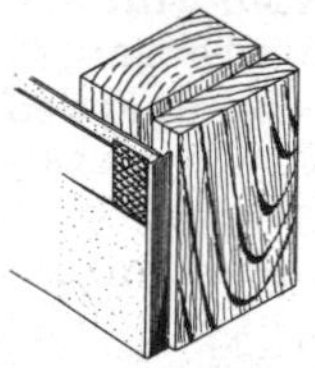
LK BEAD[3,4] LK-S LK-B	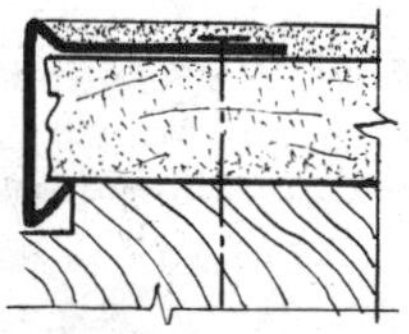	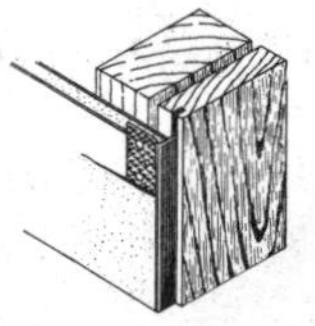
LC BEAD[2] LC-38 LC-12 LC-58 LC-34	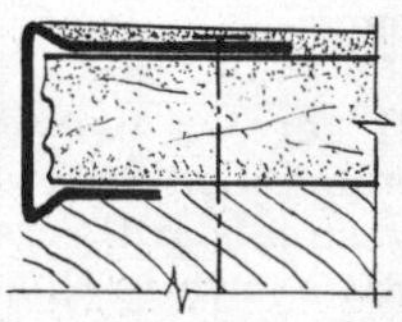	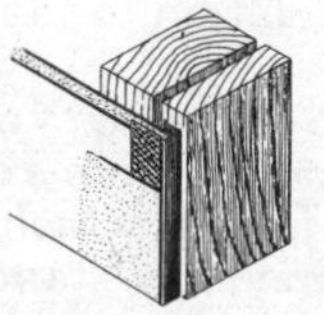

[1] Numbers indicate widths of flanges: e.g. CB—114 x 114 has two 1¼″ flanges.
[2] Numbers indicate thickness of board intended: e.g. U-12 is U Bead intended for ½″ wallboard.
[3] Letters S & B indicate: square-nose (S), or bull or round nose (B).
[4] For use with kerfed jamb.

(continued) GENERAL RECOMMENDATIONS

but particularly during joint and fastener treatment. Drafts should be avoided over newly treated surfaces during hot, dry weather to prevent joint compound from drying too rapidly.

Delivery of gypsum board should coincide as closely as possible with the installation schedule, and board should be stored flat, inside, under cover. Boards and other material should remain in their wrappings or containers until ready to be used.

APPLICATION OF GYPSUM BOARD

Gypsum board should conform to the ASTM and/or Federal Specifications given in Figure WF5.

Support spacing should not exceed that recommended for the gypsum board thickness in single-ply or multi-ply construction (see Figs. WF8 and WF13).

Wallboard should be applied in lengths to fit walls and ceilings with a minimum of butt-end joints. Where butt-end joints do occur, they should be staggered and located as far as possible from the center of ceilings and walls.

With the exception of the face layer in two-ply construction, board ends and edges which are parallel to supporting members should fall over these members.

Measuring, Cutting and Fitting

All measurements should be taken accurately. Boards should be cut to fit easily without being forced into position. Tapered edges should be placed next to tapered edges, and square-cut ends should be butted to square-cut ends.

Board Attachment

Specific recommendations for attaching gypsum board are given in the following subsections.

Mechanical Fastening Fasteners should be of the type recommended for the intended method of application.

Ordinary wood screws, sheetmetal screws or common nails are not suitable because they are not designed to hold the board tightly and countersink neatly.

Application of fasteners should start in the middle of the board and proceed outward. Fasteners should be spaced at least 3/8″ from board ends and edges. They should be driven as nearly perpendicular to the surface as possible while the board is held in firm contact with supporting construction. Nails should be driven with a crown-headed hammer so that a uniform depression will be formed around the nailhead without breaking the face paper. Screw heads should be set below the board surface not more than 1/32″.

Adhesive Bonding Because adhesive formulations and properties vary, the manufacturer's specific recommendations should be consulted in each case.

FIG. WF5 SUMMARY OF GYPSUM BOARD PRODUCTS[1]

Type	Thickness (in.)	Thickness (mm)	Width	Length	Edge Detail
WALLBOARD	1/4″	6.4	48″	8′, 10′, 12′	Tapered
(ASTM C36, FS SSL30d)	5/16″	7.9	48″	8′, 10′, 12′, 14′	
	3/8″	9.5	48″	7′, 8′, 9′, 10′, 12′	
Regular	1/2″	12.7	48″	7′, 8′, 9′, 10′, 12′, 14′	
	5/8″	15.9	48″	8′, 10′, 12′	
Foil-backed	3/8″	9.5	48″	8′, 10′, 12′	
	1/2″	12.7	48″	8′, 9′, 10′, 12′	
Type X	1/2″	12.7	48″	8′, 9′, 10′, 12′	
	5/8″	15.9	48″	8′, 9′, 10′, 12′	
BACKING BOARD	1/4″	6.4	48″	8′	Square
(ASTM C442, FS SSL30d)	3/8″	9.5	48″	8′	Square
Regular	1/2″	12.7	24″	8′	Tongue-and-grooved
Foil-backed	3/8″	9.5	48″	8′	Square
Type X	5/8″	15.9	24″	8′	Tongue-and-grooved
	1″	25.4	24″	7′	Tongue-and-grooved

[1]Recommended stock items per Simplified Practice Recommendation R266-63; other sizes and edge details are stocked by some manufacturers and distributors or are available on special order.

(continued) **GENERAL RECOMMENDATIONS**

Surfaces to receive adhesive should be free from dirt, grease, oil or other foreign material. Solvent-based adhesives should be kept away from open flames and should be used in well-ventilated areas.

JOINT AND FASTENER TREATMENT

All taped joints, fastener heads and metal trim should be treated with at least three coats of joint compound. Each coat should be allowed to dry thoroughly before the next coat is applied. Compound should be applied in thin, successively wider coats, featheredging to blend smoothly with the surrounding board surface. The second and third coats should be lightly sanded to remove surface irregularities, taking care not to roughen the fibers in the wallboard facepaper.

Depending on atmospheric conditions, 12 to 24 hours may be required for each successive coat to dry, unless quicksetting compound is used.

Joint Taping and Finishing

All visible joints in the surface and at interior corners of the wallboard should be taped and finished. Joint tape should be adhered to the wallboard with an embedding joint compound or a compound specifically recommended for the purpose by the manufacturer.

Topping compound, all-purpose compound or embedding compound may be used as cover coats over the tape.

Edge and Corner Trimming

All exposed exterior corners and ends and edges of the wallboard should be protected with appropriate wood or metal trim (Fig. WF4), and properly finished with compound when required.

U bead is installed before the wallboard is erected and does not require finishing with compound. Other types of metal casing beads and cornerbeads are installed after the wallboard is erected and require finishing with three coats of compound as recommended above.

Fastener Treatment

All exposed fastener heads and holes left by temporary nails should be treated with three coats of joint compound as recommended above.

FINISH DEFECTS

Many installation procedures and precautions are designed to eliminate surface defects such as nailpopping, shadowing, ridging and cracking. For a detailed discussion of the causes of such defects and remedies, see Main Text, pages 431-14 and 431-23.

DECORATING

Joint compound should be thoroughly dry before decorating is started. A suitable sealer or sealer-primer should be applied to the wallboard when a flexible wall covering or an oil base paint, a lacquer or an enamel is intended.

The need for a sealer-primer under emulsion paints depends on the paint formulation and the manufacturer's specific recommendations should be consulted.

SPECIAL CONSTRUCTIONS

The manufacturer's specific installation recommendations should be followed for individual products and/or systems.

Material used for caulking should be a nonhardening, nonshrinking, resilient sealant.

Most special constructions intended to provide improved sound, fire, moisture and thermal control are predicated on gypsum board finishes with complete joint treatment and panel caulking.

Sound Isolating Construction

Sound isolating assemblies should be made airtight to reduce airborne sound transmission. Back-to-back location of recessed fixtures such as medicine cabinets and electrical, telephone, television and intercom outlets should be avoided. Such fixtures should be staggered on opposite sides of the assembly and openings for such fixtures or related piping should be caulked. The entire perimeter of sound isolating partitions should be caulked.

Assemblies intended for sound isolation should not contain heating and air conditioning ducts or ventilating outlets and should be designed so that airborne sound will not bypass the assembly through flanking paths such as through ceilings or around windows and doors adjacent to the walls, or through floors and into crawl spaces below.

Resilient Mounting Heavy wall finishes such as ceramic tile and wall hung fixtures, such as lavatories, kitchen cabinets and bookcases, should not be supported on resiliently mounted walls.

If such fixtures or heavy wall finishes are supported on resiliently mounted walls, the added weight and fastenings may adversely affect the acoustical performance of the construction.

(continued) GENERAL RECOMMENDATIONS

On ceiling assemblies, it is recommended that the channels be screw-attached. If nails are used, they may work loose as the framing lumber dries and may cause floor squeaking.

Sound absorbing batts and blankets with flanges should be attached to the sides of wood studs and joists or to the back of the gypsum board in metal framed and solid gypsum partitions. Flanges should not be attached across the faces of wood studs and joists.

Fire-Resistant Constructions

Where a specific degree of fire resistance is required for gypsum board assemblies, applicable building code requirements should be followed. The manufacturer's or testing laboratory's specifications for materials and installation for fire-rated constructions should be followed.

For a discussion of fire-rated constructions, see Main Text, page 431-17.

Thermal Insulating Construction

When foil-backed wallboard is used, a dead air space of at least 3/4″ should be provided adjacent to the reflective side of the wallboard within the assembly cavity for thermal value.

Foil-backed wallboard should not be used: (1) as the backing material for ceramic tile; (2) as the face ply in laminated two-ply construction; (3) for laminating directly to masonry or concrete; or (4) in connection with electric heating cables in electric panel heating.

Thermal insulating batts and blankets should be attached as recommended for sound absorbing batts and blankets, above.

When a vapor barrier is used, it should be located on the warm side of the insulation.

Rigid plastic insulation should be installed in accordance with the manufacturer's recommendations and should not be laminated to exterior masonry or concrete with a surface temperature less than 45° F.

Generally, at least 24 hours' drying time should be allowed before wallboard is laminated to the rigid plastic insulation.

Moisture-Resistant Construction

Exterior Locations Gypsum board applied to exterior locations should be at least 1/2″ thick and should not be exposed to direct wetting. Wallboard ends and edges should be protected by metal trim or should be isolated from contact with potentially damp surfaces such as concrete and masonry by an air space of at least 1/4″ (see Main Text, page 431-18, Fig. 30).

Facia boards at wallboard soffits should be designed to form a drip edge to prevent water runoff across the wallboard surface and protect the wallboard edge. Adequate ventilation should be provided for attic spaces and eaves above wallboard soffits (see Section 360 Roof/Ceiling Systems).

Exposed wallboard surfaces and metal trim should be painted with two coats of exterior paint.

Interior Locations Water-resistant wallboard is recommended for all interior areas subject to high humidity, such as bathrooms and laundry and utility areas, but is not recommended for high moisture areas, such as saunas or steam rooms.

Foil-backed wallboard should not be used in high humidity areas, such as bath and shower enclosures as the foil backing may prevent moisture from being dissipated, thus trapping it within the board core.

Where a high humidity area is located on an exterior wall, adequate insulation and a vapor barrier on the warm side of the insulation should be provided to prevent condensation in the wall cavity.

Tub-shower Areas *Water-resistant gypsum wallboard and proprietary highly water resistant backing board are particularly recommended as a base for wall tile in tub-shower areas, although regular wallboard will perform satisfactorily if properly installed and prepared to receive tile.*

Wallboard in tub-shower areas should be at least 1/2″ thick, supported on framing members spaced not more than 24″ o.c. If spacing exceeds 16″ o.c., wood blocking should be installed between studs approximately 4′ o.c., with the first row of blocking located approximately 1″ above the top of the tub or shower receptor. Wallboard should be applied perpendicular to the framing members with paperbound edges parallel to the top of the tub or shower receptor and spaced approximately 1/4″ above it. This space should be filled with a continuous bead of waterproof sealant.

For ceramic tile 5/16″ thick or less, wallboard should be secured with nails spaced not more than 8″ o.c. or screws spaced not more than 12″ o.c. For ceramic tile

(continued) GENERAL RECOMMENDATIONS

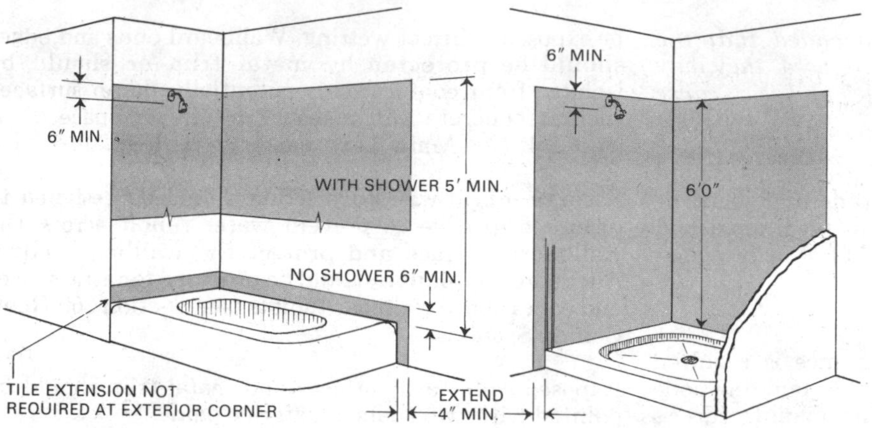

FIG. WF6 Recommended tile coverage of gypsum board in tub-shower areas.

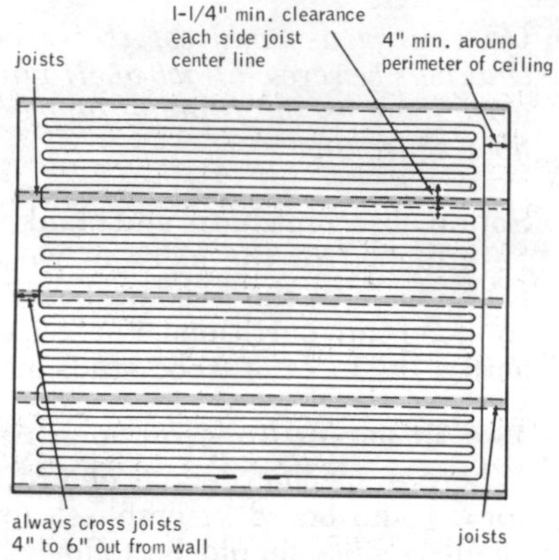

FIG. WF7 Typical cable layout for electric panel heating system.

thicker than 5/16″, nail spacing should not exceed 4″ o.c. and screw spacing should not exceed 8″ o.c.

Water-resistant gypsum wallboard and backing board should be prepared for wall tile in accordance with the manufacturer's recommendations.

For regular wallboard, joints between adjacent boards and at corners should be taped but should not receive finishing coats. Before tile is installed, a water-resistant sealer should be applied to the entire wallboard surface. When tile adhesive is used as the sealer, it should be spread separately from the bonding coat in a continuous uniform coating, approximately 1/16″ thick, covering all cut ends and cutouts for fixtures. A Type I, highly water resistant organic adhesive should be used to apply ceramic wall tiles.

See Main Text, page 431-18, Figure 31, for recommended gypsum board and wall tile installation details for tub-shower areas.

Wall tile should cover all gypsum board surfaces as illustrated in Figure WF6. The space between tile and top of the tub or shower receptor should be caulked with a continuous bead of waterproof sealant.

Electric Panel Heating

The manufacturer's detailed recommendations for specific electric panel heating systems should be consulted.

Electric panel heating can be developed utilizing backing board, electric heating cables, a filler material and wallboard in accordance with the following recommendations.

Backing board should be attached perpendicular to supporting members with mechanical fasteners providing the generally recommended penetration into supporting members and spaced 7″ o.c.

Electric heating cables should be securely attached to the backing board in accordance with the cable manufacturer's recommendations, maintaining at least 1-1/4″ clearance on either side of each joist centerline and 4″ to 6″ clearance around the perimeter of the ceiling at the walls (Fig. WF7).

All inspection and testing of the electrical installation should be performed before the face layer is applied. Maximum cable temperature should not exceed 125° F.

A suitable filler material should be used to cover heating cables completely. Filler material should be leveled to provide continuous contact with the face layer. If the filler is to serve as an adhesive, the wallboard should be applied immediately after the filler is spread and should be attached with nails spaced 16″ o.c. along supports. When the filler does not serve as an adhesive, it should be allowed to dry completely. Then, laminating adhesive should be applied and the wallboard should be nailed 7″ o.c. along supports. Nails should be kept 8″ to 10″ from the perimeter of the ceiling at walls to prevent striking the cables where they cross the joists.

Insulating (foil-backed) wallboard should not be used as the face layer.

For a discussion of electric panel heating systems see Main Text page 431-19 and page Gypsum Products 218-17.

SINGLE-PLY CONSTRUCTION

Unless specific exceptions are made in the following text, the General Recommendations apply. The following recommendations are for normal conditions. Where a specific degree of sound isolation, fire resistance, thermal insulation or moisture resistance is required, the general recommendations for Special Constructions and the manufacturer's detailed installation specifications should be followed.

FIG. WF8 RECOMMENDED SUPPORT SPACING FOR SINGLE-PLY CONSTRUCTION

Wallboard Thickness (in.)	Wallboard Thickness (mm)	Location	Application[1]	Maximum Support Spacing (in.)	Maximum Support Spacing (mm)
3/8"	9.5	Ceilings[2,3]	Perpendicular	16" o.c.	406 o.c.
		Walls	Perpendicular or parallel	16" o.c.	406 o.c.
1/2" & 5/8"	12.7 & 15.9	Ceilings[3]	Perpendicular	24" o.c.	610 o.c.
			Parallel	16" o.c.	406 o.c.
		Walls	Perpendicular or parallel	24" o.c.	610 o.c.

[1]Direction of wallboard edges relative to supports.
[2]3/8" ceilings should not support insulation.
[3]For ceilings to receive a water base spray texture finish, use 5/8" board applied perpendicular to 24" supports, or 1/2" board perpendicular to 16" supports.

CEILINGS AND WALL ASSEMBLIES

Wood and metal supporting members (framing or furring) should be properly spaced for the intended wallboard thickness and direction of application as recommended in Figure WF8.

With the exception of 3/8" wallboard on ceilings, which is applied perpendicular to supports, wallboard may be applied either perpendicular or parallel to supports.

FIG. WF10 SINGLE-PLY SCREW ATTACHMENT TO WOOD SUPPORTS

Support Spacing (in.)	Support Spacing (mm)	Fastener Spacing[1] Walls (in.)	Walls (mm)	Ceilings (in.)	Ceilings (mm)
16" o.c.	406 o.c.	16" o.c.	406 o.c.	12" o.c.	305 o.c.
24" o.c.	610 o.c.	12" o.c.	305 o.c.	12" o.c.	305 o.c.

[1]Spacing applicable to both field and perimeter of board.

However, perpendicular application generally is preferred because it provides greater strength and shorter joint lengths.

Except in end-floated construction, wallboard ends and or edges should occur over supporting members. Butt-end joint should be staggered with respect to each other on the same side of the partition and with respect to joints on the opposite side of the partition.

Wood Framing and Furring

Nail Attachment Wallboard should be nailed to wood supporting construction by either the single- or double-nailing method in accordance with GA 216 (Fig. WF9), with appropriate nails providing the recommended penetration into supporting construction (Fig. WF2).

Double nailing results in better board-to-stud contact and fewer nail spots requiring treatment.

Screw Attachment Wallboard should be attached to wood framing or furring with appropriate drywall wood screws (such as Type W), long enough to provide at least 5/8" penetration into supporting members and spaced as shown in Figure WF10.

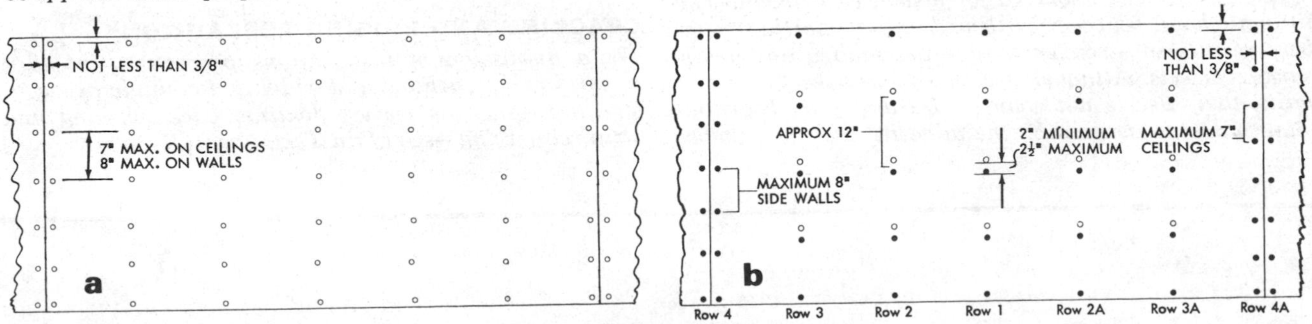

FIG. WF9 (a) Single nailing method; (b) double-nailing method: first nails shown by black dots are installed starting with Row 1, then Row 2 and 2A, etc., working outward. Second nails shown by circles are installed in same row sequence.

(continued) SINGLE-PLY CONSTRUCTION

Fewer fasteners are required when screws are used to attach wallboard, and the number of fasteners requiring treatment and possible surface defects are reduced.

Adhesive-Nail Attachment Stud adhesive should be applied to the faces of supporting members in sufficient volume to provide a cross section of adhesive approximately 1/16" x 1" when wallboard is applied. Care should be taken in spreading the adhesive to prevent it from being squeezed out at the joints when the wallboard is applied. Adhesive should not be applied to members not required for wallboard support such as diagonal bracing, blocking and plates.

For structural stud adhesives conforming to ASTM C-557, fastener spacing should be as recommended in Figure WF11. For nonstructural stud adhesives (which do not conform to ASTM C-557) fastener spacing should be as recommended in Figure WF12.

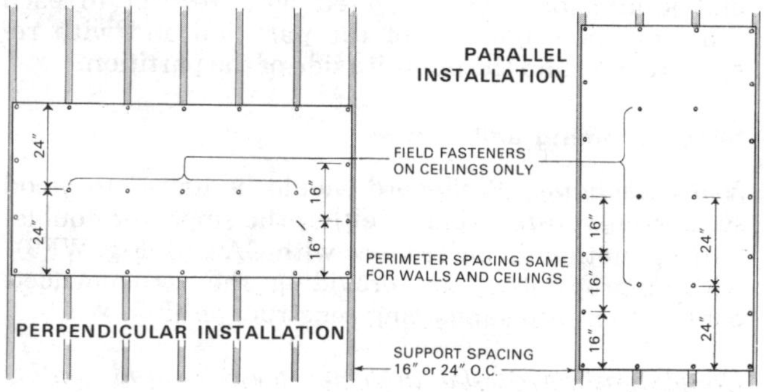

FIG. WF11 Recommended supplemental fastener spacing for structural adhesives.

Compared to conventional nailing, adhesive-nail attachment has the advantage of reducing the number of nails in the middle of the panel by at least 50%—consequently reducing the labor required for fastener treatment. The intimate bond between wallboard and supporting members results in stronger, stiffer assemblies and fewer surface defects. Although screws can be used instead of nails, their use is not common because good board-to-stud contact is assured by the adhesive and the number of face fasteners is reduced to a minimum.

FIG. WF12 **RECOMMENDED FASTENER SPACING FOR NONSTRUCTURAL ADHESIVES**

Location	Support Spacing (in.)	Support Spacing (mm)	Fastener Spacing[1] Nails (in.)	Nails (mm)	Screws (in.)	Screws (mm)
Ceilings	16" o.c.	406 o.c.	16" o.c.	406 o.c.	16" o.c.	406 o.c.
	24" o.c.	610 o.c.	12" o.c.	305 o.c.		
Load-bearing Walls	16" o.c.	406 o.c.	16" o.c.	406 o.c.	24" o.c.	610 o.c.
	24" o.c.	610 o.c.	12" o.c.	305 o.c.	16" o.c.	406 o.c.
Nonload-bearing Walls	16" o.c.	406 o.c.	24" o.c.	610 o.c.	24" o.c.	610 o.c.
	24" o.c.	610 o.c.	16" o.c.	406 o.c.		

[1]Applies to both field and perimeter of the board.

Metal Framing and Furring

Suspended furred ceilings, metal furring over concrete, masonry and wood construction and primary metal framing should be installed as recommended by the manufacturer.

Drywall furring channels are applied perpendicular to wood framing members, secured with one 1-1/4" screw or one 5d nail at each support.

Wallboard should be attached to suitable metal framing or furring channels with drywall sheetmetal screws spaced not more than 12" o.c. on both walls and ceilings.

Concrete and Masonry

Interior and above-grade concrete and masonry which are intended for direct lamination of wallboard should be dry, smooth, plumb and true. Wallboard should be bonded with appropriate adhesives according to the manufacturer's recommendations. Temporary bracing or fasteners should be provided until the adhesive has developed sufficient bond strength.

CRACKING AND RIDGING PRECAUTIONS

For a discussion of installations procedures intended to minimize cracking and ridging in single-ply construction—such as corner floating, back blocking and strip reinforcing—see Main Text, page 431-23.

MULTI-PLY CONSTRUCTION

Unless specific exceptions are made in the following text, the General Recommendations apply. The following recommendations are intended for normal conditions; where a specific degree of sound isolation, fire resistance, thermal insulation or moisture resistance are required, the general recommendations for special constructions and the manufacturer's detailed installation instructions should be followed.

Multi-ply finishes generally provide better surface quality than single-ply construction, and are particularly suitable where increased fire-resistance and/or sound isolation are desired.

SUPPORTED CEILING AND WALL ASSEMBLIES

The following recommendations generally apply to two-ply gypsum board finishes supported by wood or metal framing or furring.

Wood or metal framing and furring members should be properly spaced for the intended base-ply thickness and direction of application as recommended in Figure WF13.

FIG. WF13 **SUPPORT SPACING FOR TWO-PLY CONSTRUCTION**[1]

Thickness of Base Ply (in.)	Thickness of Base Ply (mm)	Location	Application of Base Ply[2]	Maximum Support Spacing (in.)	Maximum Support Spacing (mm)
3/8"	9.5	Ceilings[4,5]	Perpendicular	16" o.c.	406 o.c.
		Walls	Perpendicular or Parallel	24" o.c.[3]	610 o.c.
1/2" & 5/8"	12.7 & 15.9	Ceilings[5]	Perpendicular	24" o.c.	610 o.c.
		Walls	Perpendicular or Parallel	24" o.c.	610 o.c.

[1]Unless noted, installation with mechanical fasteners with or without adhesive.
[2]Direction of board edges relative to supports.
[3]16" o.c. if no adhesive used between plies.
[4]Two-ply 3/8" ceilings should not support insulation.
[5]For two-ply ceilings to receive a water base spray texture finish, use 5/8" board applied perpendicular to 24" supports or 1/2" board perpendicular to 16" supports.

Base-Ply Attachment

Wood Framing and Furring Wallboard or backing board base ply should be attached to wood supporting construction with appropriate nails, screws or staples, spaced as recommended in Figure WF14.

When the face ply is to be attached with mechanical fasteners and adhesive is not used, fewer fasteners are required in the base ply because the fasteners in the face ply will support the base ply as well.

When staples are used, they should be driven with crown perpendicular to the paperbound edges; except where edges fall over supports, they should be driven parallel to the edges.

Metal Framing and Furring For walls, the base ply should be attached with drywall sheetmetal screws spaced 12" o.c. in the field and 8" o.c. at edges when the face ply is to be laminated, and 16" o.c. throughout the panel when the face ply is to be screw-attached.

Two-ply ceiling finishes on metal supporting construction do not provide substantially better surface quality, impact sound isolation or fire resistance than single-ply finishes screw-attached to metal furring. When two-ply ceiling finishes on metal supports are desired for other properties, the manufacturer's specific recommendations for proprietary systems should be consulted.

Concrete and Masonry *These assemblies have inherently good fire resistance and sound isolation which two-ply finishes do not substantially improve. When two-ply construction is used primarily for improved surface finish over furred walls, the base ply is attached as recommended for wood or metal furring, above.*

Face-Ply Attachment

The face ply should be installed in a direction which will result in the shortest and least conspicuous joints. All joints in the face ply should be staggered and offset at least 10" with respect to joints in the base ply.

FIG. WF14 **BASE-PLY ATTACHMENT TO WOOD SUPPORTING CONSTRUCTION**[1]

Location	Nail Spacing: Laminated Face Ply	Nail Spacing: Nailed Face Ply	Screw Spacing: Laminated Face Ply[2]	Screw Spacing: Screwed Face Ply	Staple Spacing: Laminated Face Ply	Staple Spacing: Nailed or Screwed Face Ply
Walls	8" (203 mm) o.c.	16" (406 mm) o.c.	16" (406 mm) o.c.	24" (610 mm) o.c.	7" (178 mm) o.c.	16" (406 mm) o.c.
Ceilings	7" (178 mm) o.c.					

[1]Supports spaced 16" or 24" o.c. unless noted; fastener spacing applies to perimeter and field fastening along supports.
[2]12" o.c. for both ceilings and walls when supports are spaced 24" o.c.

(continued) MULTI-PLY CONSTRUCTION

When the face ply is attached with mechanical fasteners to supporting construction and no adhesive is used, mechanical fasteners should be spaced as for single-ply construction (Figs. WF9 and WF10).

When the face ply is laminated to the base ply, supplemental temporary nails or bracing should be provided until the adhesive develops bond strength. Temporary nails or bracing should be spaced to provide face support not more than 24″ o.c. in each direction.

Some permanent fasteners in each laminated wallboard panel are recommended, particularly on ceilings. Usually, these can be spaced at the perimeter of the board where they will be concealed by joint treatment and/or trim. Permanent fasteners in the field as well usually are required for fire-rated assemblies.

If the base ply is sound deadening board or 3/8″ gypsum board, fasteners providing suitable penetration through the base ply into supporting construction should be used for temporary or permanent support.

SELF-SUPPORTING GYPSUM PARTITIONS

The manufacturer's detailed recommendations for materials and installation of semi-solid and solid all-gypsum partitions should be followed. Partition design and type should be selected on the basis of height limitations, sound isolation and fire resistance, as well as other pertinent criteria.

For a discussion of self-supporting gypsum partitions, see Main Text, page 431-27.

441 PAINTING AND FINISHING

PAINTING AND FINISHING

INTRODUCTION

The main purpose of paints and finishes is to beautify and protect the underlying surface. The paint film serves as a protective shield between the construction materials and the elements which attack and deteriorate them. Typical causes of paint failure are sunlight, temperature variations, salt water, water vapor, decay, mildew, chemicals and abrasion. When regularly renewed, painting offers long-term protection that extends the useful life of the materials and the structure.

Paints may serve other functions as well. Painting is an essential part of sanitary programs for hospitals, as well as for commercial, institutional and home kitchens and washrooms. Gloss and semi-gloss paint coatings provide smooth, non-absorptive surfaces that can be washed easily and kept free of dirt and germs.

White and light-tinted paints reflect light, brighten rooms and increase visibility. Flat paints soften and evenly distribute interior illumination. The functional use of color creates comfortable living and working conditions in the home, shop and office.

Color-coded painting frequently is used to enhance public comfort, safety and efficiency. Certain colors are stimulating, others are relaxing, and yet others universally are associated with potentially dangerous situations.

This section discusses recommended methods of applying opaque film-forming coatings *(painting)*, as well as clear and translucent materials such as varnishes and stains *(finishing)*. The information is abstracted from the Paints and Protective Coatings Manual of the Army, Navy, and Air Force compiled by the National Bureau of Standards. Additional discussion of materials for painting and finishing and the surfaces for which they are suitable can be found in Section 216 Paints & Protective Coatings.

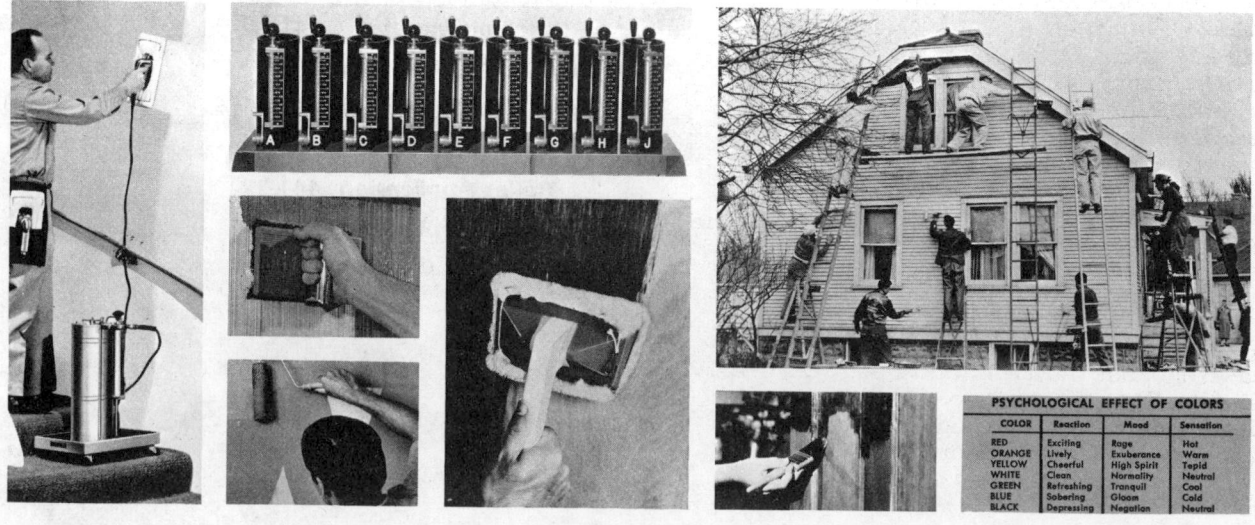

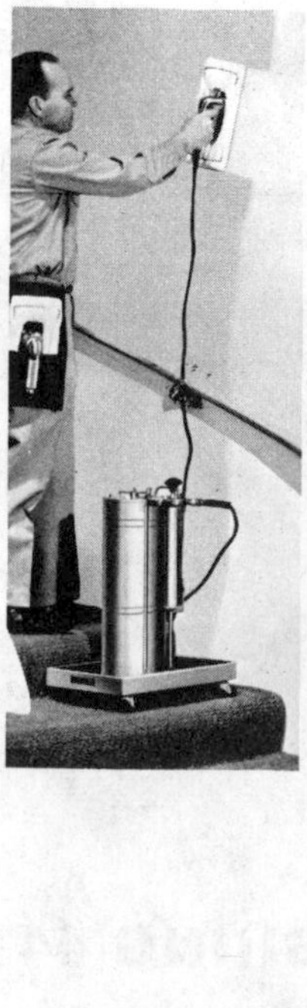

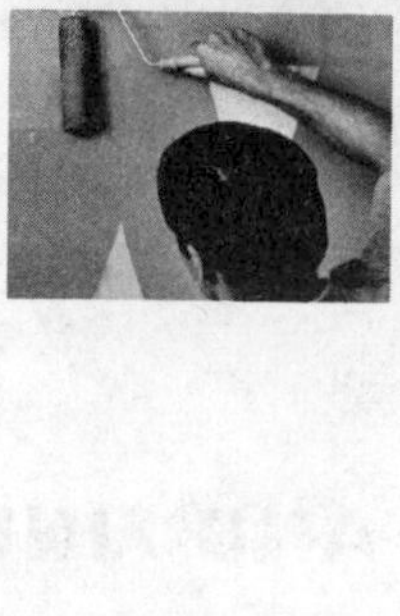

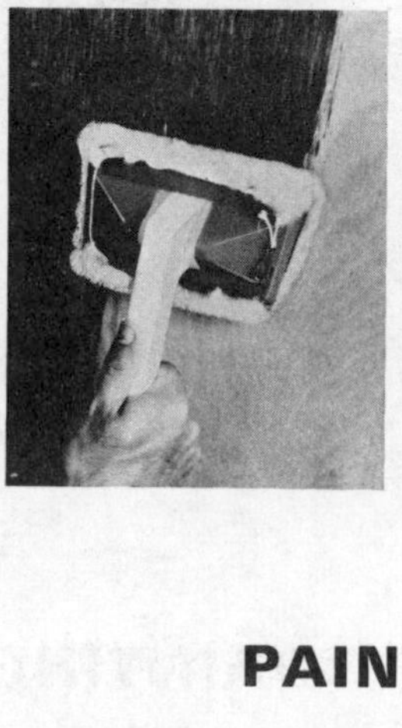

PSYCHOLOGICAL EFFECT OF COLORS

COLOR	Reaction	Mood	Sensation
RED	Exciting	Rage	Hot
ORANGE	Lively	Exuberance	Warm
YELLOW	Cheerful	High Spirit	Tepid
WHITE	Clean	Normality	Neutral
GREEN	Refreshing	Tranquil	Cool
BLUE	Sobering	Gloom	Cold
BLACK	Depressing	Negation	Neutral

GENERAL RECOMMENDATIONS

Satisfactory painting and finishing depends on: 1) proper selection of paint or coating materials; 2) adequate preparation of materials and surfaces, and 3) suitable application methods and procedures.

MATERIALS

Proper storage conditions and stocking procedures are essential to avoid material damage, explosion hazards and application problems.

Storage

Paint materials should be stored in dry and well-ventilated areas maintained at a temperature range of 50°F to 90°F. Lower temperatures generally increase material viscosity and make it necessary to temperature-condition materials before use. Freezing temperatures may permanently damage water-based paints.

If the paint material is sensitive to heat, temperatures over 100°F may bring about a chemical reaction within the container. This generally results in increased viscosity to the point of gelation, and the paint becomes unusable. Also, overheating may cause covers to blow off, creating a serious fire hazard.

Other hazards to be considered are high humidity, which may cause containers to corrode and labels to deteriorate, and poor ventilation, which may allow excessive concentration of solvent vapors that are both toxic and combustible.

Stocking

Paint materials should be rotated so that oldest stocks are used first. It is desirable to make paints ready for use by conditioning at the proper temperature and mixing thoroughly. Using leftover paint should be avoided but, if necessary to do so, such material should be strained and temperature-conditioned again as recommended below for unopened containers.

Paint material should be stored in full, tightly sealed containers. Required quantities should be carefully estimated so as to have little or no material left at the end of a job. It is safer generally to discard small quantities than to use paint that has skinned. When saved, leftover paint should be placed in smaller containers, filled full, and tightly sealed. All information from the original container should be transferred to the new container.

CONDITIONING AND MIXING

Unless stored at the recommended temperature, paint materials should be temperature-conditioned at 65°F to 85°F for at least 24 hours before use. After this time, paints should be mixed, thinned, or tinted as specified, and strained if necessary.

Paints consist of two principal components—the solid pigment and the liquid vehicle. Suspension agents are included in most paints to minimize settling of solids. However, when stored for long periods, separation of the pigment and vehicle may take place. The purpose of mixing is to reblend settled pigment with the vehicle, to eliminate lumps, skins or other detriments to proper application.

Mixing

Paint materials always should be mixed prior to issuance or delivery to the job. Mixing can be done either by manual or mechanical methods, but the latter is preferred to insure maximum uniformity. Some hand stirring should be done just prior to use and periodically during application. Regardless of the procedure used, caution is recommended to avoid the incorporation of an excess of air through overmixing. Figure 1 outlines the recommended mixing procedures for different coating types.

FIG. 1 RECOMMENDED MIXING PROCEDURES FOR DIFFERENT COATING TYPES

COATING TYPE	MIXING EQUIPMENT	REMARKS
Enamel, semigloss or flat oil paints	Manual, propeller or shaker	Mix until homogeneous
Water-based latex paints	Manual, propeller or shaker	Use extreme care to avoid air entrapment
Clear finishes, such as varnishes and lacquers	Manual	Generally require little or no mixing
Extremely viscous finishes, such as coal tar paints	Propeller	Use extreme care to avoid air entrapment
Two-package metallic paints, such as aluminum paint*	Propeller	Add small amount of liquid to paste and mix well; slowly add remainder of vehicle, while stirring, until coating is homogeneous
Two-Component Systems	Manual, propeller or shaker	Mix until homogeneous; check label for special instructions

*With metallic powder, first make into a paste with solvent, then proceed as recommended above.

It is important to mix ready-mixed paints *prior* to, as well as after introducing thinners or other additives. It is best to use the same conditioning procedures for all components of multi-component paint materials before mixing. Manufacturers' label directions regarding proper mixing should be followed as closely as possible.

Mechanical Mixers The two most common mechanical mixers either vibrate the full sealed container, or utilize propellers that are inserted into the paint. Vibrating shakers are used for full containers up to 5 gallons. Propeller mixers are simple hand drill attachments and generally are used on the job for containers 1 quart or larger (Fig. 2).

Manual Mixing This method is more time-consuming than mechanical mixing, but can be used when no mechanical equipment is available, or for materials that had been recently well-mixed. The paint should be stirred from the bottom up, in order to dislodge any settled matter on the bottom or lower sides of the container. Properly mixed paint should have a completely blended appearance with no evidence of varicolored swirls or lumps at the top, indicating unmixed pigment solids or foreign matter.

When two or more cans of the same color are to be used in the same area, the paint may be "boxed" by pouring back and forth from one container to the other until uniform. This procedure minimizes the visual difference when switching from one container to the next.

Tinting

On-the-job tinting should be avoided as a general practice, in order to minimize errors and to eliminate color-matching problems at a later date. One exception is the practice of tinting intermediate coats for identification and to assure complete coverage under the finish coat. Tinting should be done with care, using only colors which are known to be compatible, and preferably adding not more than 4 oz. per gallon of paint. Use of more than 8 oz. per gallon should be avoided unless specifically approved by the paint manufacturer. Tinting of chalking-type exterior paints is not recommended except for identification of intermediate coats.

Many paint manufacturers employ tinting systems for the addition of colorants to their base paints. The colorants are formulated to precise standards of tinting strength and are packaged in tubes or in bulk for use with a colorant dispensing machine at the paint shop (Fig. 3). Each color in the system has an identifying number and mixing formula indicating the proportions of base paint and colorants to be combined. Thus, any color can be reproduced accurately at a later time without guesswork.

Tinting Colors There are two types in general use: colors-in-oil and universal tinting colors.

Colors-in-oil are limited to use with standard oil-base paints, alkyd resins, chlorinated rubbers and other solvent systems. They should not be used with the other synthetics or with water-thinned paints.

Universal tinting colors are used in the same manner as colors-in-oil. They are more compatible with a wide variety of paint materials. Many can be used with both solvent-thinned and water-thinned paints.

Tinting Procedure Tinting should be done by experienced personnel, in the paint shop or at the paint store. The paint should be at application consistency before tinting. Ingredients should be compatible and fluid, so as to blend readily. Mechanical agitation is important to insure uniform color distribution but overmixing should be avoided. When trying to achieve a paint color not included in a manufacturers' tinting system, the paint sample should be allowed to dry before it is compared with the color chip used for reference. When the desired color has been achieved, it is important to maintain a written record of the tinting formula; the container should be identified by name or number, and a spot of paint applied to the cover for visual reference.

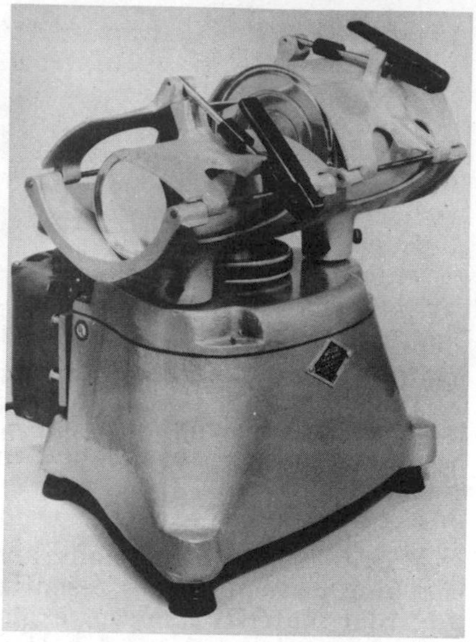

Fig. 2 Mechanical mixer can handle cans from 1/4 pint to one gallon in size.

Fig. 3 Machine dispenses carefully measured quantities of colorant into container placed under the turntable.

Straining

Paint from newly-opened containers normally does not require straining. However, if there is evidence of skins, lumps, color flecks or foreign materials, the paint should be strained through a fine sieve or commercial paint strainer. To avoid clogging of the nozzle, straining is recommended whenever the paint is to be applied by spray gun.

Thinning

Paints usually are formulated and packaged at the proper viscosity for application by brush or roller without thinning. However, thinning frequently is required and recommended by the manufacturer for spray application. The arbitrary addition of thinners should be avoided; unnecessary or excessive thinning may result in an inadequate film thickness, and may reduce the durability of the applied coating.

Thinning should be performed by competent personnel using only compatible thinning agents recommended in label instructions, or contract specifications. Thinning should not be used as a means of improving brushing or rolling properties of cold paint materials; instead, materials should be conditioned to suitable temperature (65°F to 85°F) before use.

PREPARATION OF SURFACES

Proper surface preparation prior to painting is essential to achieve maximum coating life. The best paint will not perform effectively if applied on a poorly prepared surface. The initial cost of adequate surface preparation is rewarded by longer coating life and improved appearance. The selection of surface preparation method is dependent upon: 1) nature and condition of surface to be painted; 2) type of exposure; 3) surface texture and/or appearance desired, and 4) material and labor costs.

Material surfaces often contain contaminants such as dirt, grease, rust and moisture which reduce adhesion and cause blistering, peeling, flaking or cracking of the paint film. Many surface defects which also adversely affect adhesion are: metal burrs and irregular weld areas, wood knots and torn grain, powdery concrete and masonry, and old paints in various stages of deterioration. In general, sharp edges, irregular areas, cracks and holes create conditions conducive to early paint failure.

Because iron and steel are important building materials which corrode readily unless properly prepared and painted, special attention is given in the following discussion to suitable preparation methods for these and other metals. In all cases involving the ferrous metals, painting should be started as soon as possible after the surface has been prepared. The suitability of treatments for various material surfaces is given in Figure 4.

FIG. 4 RECOMMENDED PREPARATION OF VARIOUS SURFACES

PREPARATION METHOD	WOOD	METALS		CONCRETE & MASONRY
		STEEL	OTHER	
MECHANICAL				
Hand Tool Cleaning[1]	OK	OK	OK	OK
Power Tool Cleaning	OK[2]	OK	—	OK
Flame Cleaning	—	OK	—	—
Brush-Off Blast Cleaning	—	OK	OK	OK
Other Blast Cleaning	—	OK	—	—
CHEMICAL AND SOLVENT				
Solvent Cleaning	OK	OK	OK	—
Alkali Cleaning	—	OK	—	OK
Steam Cleaning	—	OK	—	OK
Acid Cleaning	—	OK	—	OK
Pickling	—	OK	—	—
PRETREATMENTS				
Hot Phosphate	—	OK	—	—
Cold Phosphate	—	OK	—	—
Wash Primers	—	OK	OK	—
PRIMERS	—	—	—	OK
SEALERS	OK	—	—	—
FILLERS	OK	—	—	OK

1. Also suitable for gypsum wallboard and plaster
2. Sanding only recommended

Mechanical Treatments

Metals can be prepared for painting by cleaning with hand tools, power tools, flame, and abrasive blast. The use of goggles and gloves is essential for blast and flame cleaning and is recommended for the other methods as well. Dust respirators may be required also.

Hand Tool Cleaning This method will remove only superficial contaminants, such as rust and mill scale and loosely adhering paint. It is primarily recommended for spot-cleaning in exposures where corrosion is not a serious factor. Since manual cleaning removes only surface contamination it is essential to use primers which will thoroughly "wet" the surface and penetrate to the base metal.

Before cleaning is started, the surface should be treated with solvent cleaners to remove oil, grease, dirt and chemicals. Rust scale and heavy buildup of old coatings should then be removed with impact tools such as chipping hammers, chisels and scalers. Loose mill scale and non-adhering paint can be removed with wire brushes and scrapers. Caution should be exercised to avoid deep marking or scratching the surface by the tools used.

Power Tool Cleaning This method provides faster and more adequate surface preparation than the hand tool method but is less economical than blasting for effective large area cleaning. Power tools can be used for removing small amounts of tightly adhering contaminants which are difficult to dislodge with hand tools.

Power tools are driven either electrically or pneumatically and include a variety of attachments for the basic units. Chipping hammers are used for removing tight rust, mill scale and heavy paint coats. Rotary and needle scalers are handy for removing rust, mill scale and old paint from large metallic and masonry areas. Wire brushes can be used for removing loose mill scale, old paint, weld flux, slag and dirt deposits. Grinders and sanders are useful in removing old paint, rust or mill scale on small surfaces and for smoothing rough surfaces (Fig. 5).

As with hand tools, care is recommended to avoid cutting too deeply into the surface, since this may result in burrs that are difficult to protect satisfactorily. When using wire brushes, care should be taken to avoid polishing the metal surface, which may prevent adequate coating adhesion. Power tool cleaning should be preceded by solvent or chemical treatment; painting should be started and completed as soon after cleaning as possible.

Flame Cleaning This method is satisfactory for both new and maintenance work. Oil and grease should be removed prior to flame cleaning both for safety and adequacy of preparation. Wire brushing normally follows flame cleaning to remove loose matter. A high velocity oxyacetylene flame is used in this treatment. Extreme caution and adequate ventilation are necessary to prevent fires.

The paint generally is applied while the surface is still warm, thereby speeding up drying time and making it possible to paint when air temperatures are somewhat below 50°F. However, the painting operation should be sufficiently removed from the cleaning operation while solvent thinned paints are used to avoid a fire hazard.

Blast Cleaning This method involves the high velocity impact of abrasive particles such as sand on the surface. The abrasive is discharged, either wet or dry, under pressure. In the wet system the water—and sometimes a rust inhibitor—either is mixed with the abrasive in the pressure tank or is introduced into the blast stream near the nozzle (Figs. 6 and 7).

Metal surfaces which have been blast cleaned should be prime coated the same day, since rust forms rapidly on newly cleaned surfaces. Metal or synthetic shot, grit or similar abrasives are used where recovery of the abrasive is possible, otherwise sand generally is used. The abrasive should be of a size sufficient to remove surface contamination without pitting the surface, or creating extreme peaks and valleys. Pitted surfaces require thicker paint films for adequate protection and high peaks represent areas of possible paint failure.

As in other mechanical cleaning, surface grease or oil should be removed by solvent cleaning prior to blasting. Dry blasting should be avoided if temperature of the steel is less than 5°F above the dew point, to prevent condensation and subsequent rusting.

Blast cleaning is the most effective of mechanical treatments and often is used on metal structures but also may be used, with caution, on masonry surfaces. The depth to which the surface is abraded can be varied depending on the intended paint system and severity of exposure. There are four methods of blast cleaning; the most effective methods, reaching down to, or near, the untarnished "white metal" require more labor, hence are costlier (Fig. 8).

White metal blast provides for the complete removal of all rust, mill scale and other contaminants from the surface and is the preferred technique in blast cleaning. It is used for coatings which must withstand exposure to very corrosive atmospheres and where a high initial cost of surface preparation is warranted.

Near-white metal blast surfaces are nearly free of all contaminants but will show streaks and discolorations distributed across the "white metal" surface. This technique is 10% to 35% cheaper than white metal blasting and has proven adequate for many protective coatings in moderately severe exposures.

Commercial blast describes the removal of superficial contaminants such as loose scale and rust. This method results in a surface quality generally adequate for the majority of paint systems under normal exposure conditions.

Brush-off blasting is a relatively low-cost method, used to remove old finishes, loose rust and mill scale. Brush-off blasting is not suitable where severe corrosion is prevalent, but is intended as a more efficient substitute for manual and power tool cleaning. Brush-off blasting also is used for the removal of loose or deteriorated paint from masonry.

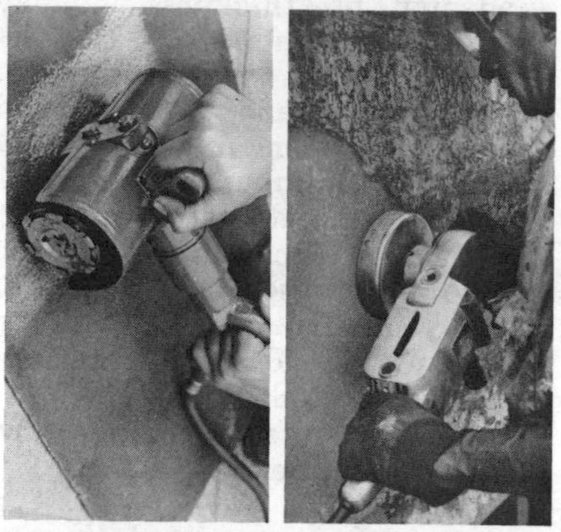

Fig. 5 Power cleaning tools generally include toothed discs for chipping and wire brushes for grinding.

Fig. 6 In wet blast cleaning, a high-speed jet of water dislodges and carries off contaminants.

Fig. 7 Dry blast cleaning is used where the resulting dust is not objectionable.

FIG. 8 RATES OF BLAST CLEANING*

METHOD	sq. ft. per hour
White-Metal	100
Near-White	175
Commercial	370
Brush-Off	870

*Approximate cleaning rates using 100 PSI with 5/16 in. nozzle.

Chemical and Solvent Treatment

Chemical cleaning methods usually are more suited to paint-shop application, while mechanical methods are more practical for use in the field. With the exception of blast cleaning, chemical cleaning is both more efficient and more effective than mechanical methods.

The use of goggles and rubber gloves generally is recommended and additional protective clothing is required where acid or alkaline solutions are used. Adequate ventilation is essential during chemical and solvent cleaning. Respirators should be used when operating in confined areas; proper fire precautions should be taken when flammable cleaners are used.

Solvent Wiping and Degreasing Solvent cleaning is a procedure used for dissolving and removing surface contaminants such as oil, grease, dirt and paint remover residues, prior to painting or mechanical treatment. It is best to remove soil, cement spatter and other dry materials first with a wire brush. The surface then should be scrubbed with brushes or rags saturated with solvent. Other effective methods include immersing the work in the solvent or spraying solvent over the surface.

The solvent quickly becomes contaminated, so it is essential that several clean solvent rinses be applied to the surface. Clean rags should be used for rinsing and wiping dry. Mineral spirits is an effective solvent for cleaning under normal conditions. Toxic solvents and solvents with low flash points constitute health and safety hazards. Use of solvents with flash points below 100°F, or solvents with a maximum safe concentration of less than 100 parts per million is not recommended. Rags should be placed in fireproof containers after use.

Alkali Cleaning This method is more efficient, less costly and less hazardous than solvent cleaning, but is more difficult to carry out. It is suitable for most metals, except stainless steel and aluminum. Alkali cleaner solutions are applied at relatively high temperatures (150°F to 200°F) since cleaning efficiency increases with temperature. Alkalis attack oil and grease, converting them into soapy residues that wash away with water.

Commercial cleaners contain other active ingredients which aid in removing surface dirt and other contaminants such as mildew. These cleaners also are effective in removing old paint by breaking up the dried vehicle. The most commonly used alkaline cleaners are *trisodium phosphate, caustic soda* and *silicated alkalis.* For steel, these cleaners should contain 0.1% chromic acid or potassium dichromate to prevent corrosion. Cleaners can be applied by brushing, scrubbing, spraying or the work can be immersed in soak tanks.

After cleaning, thorough rinsing is necessary to remove the soapy residue as well as all traces of alkali. Rinsing also avoids the possibility of chemical reaction with the applied paint. The rinse water should be hot (175°F) and preferably should be applied under pressure. Universal pH test paper should be used to insure that the surface is neutral and no free alkali is present on the surface after rinsing.

Steam Cleaning This method involves the use of steam or hot water under pressure, sometimes with a detergent included for added effectiveness. The steam or hot water removes oil and grease by emulsifying or diluting them, then flushing them away with water. When steam is used on old paint, the vehicle swells so that it loses its adhesiveness and is easily removed. Under ordinary conditions steam or hot water alone may be used to remove dirt deposits, soot and grime. In severe cases, wire brushing and/or brush-off blast cleaning may be necessary to augment the steam cleaning.

Acid Cleaning This method is used to remove oil, grease, dirt and other surface contaminants, and is suitable for cleaning iron, steel, concrete and masonry. It should not be used on stainless steel or aluminum surfaces.

Iron and steel surfaces are treated with solutions of phosphoric acid containing small amounts of solvent, detergent and wetting agent. Unlike alkali cleaners, acid cleaners also remove light rust and etch the surface to assure better adhesion of applied coatings. There are many types of metal acid cleaners and rust removers, each formulated to perform a specific cleaning job. The most common formulations are based on *phosphoric acid* solutions. The procedures usually involve applying the cleaner by brush or rag, allowing time for it to act, thorough rinsing, wiping and/or air drying. Smaller objects, particularly in the shop, can be immersed in hot cleaner and followed by several rinses.

Concrete and masonry surfaces can be washed with a solution of muriatic (hydrochloric) acid to remove efflorescence and laitance, to clean the surface, and to improve its bonding properties. Efflorescence is a white, powdery deposit often found on masonry surfaces. Laitance is a fine gray powder which sometimes develops on the surface of improperly finished concrete. Paints applied over efflorescence or laitance may loosen prematuraly and result in early coating failure. It is best to remove as much as possible of these deposits using a stiff fiber or wire brush; putty knives or scrapers also may be used.

Oil and grease should be removed prior to acid cleaning, either by solvent wiping or by steam or alkali cleaning. To start acid cleaning, the surface first should be thoroughly wetted with clean water, then scrubbed with a 5% solution (by weight) of muriatic acid, using a stiff fiber brush. In extreme cases, up to a 10% muriatic acid solution may be used, and it may be allowed to remain on the surface up to 5 minutes before scrubbing. Work should be done on small areas, not greater than 4 sq. ft. in size. Immediately after the surface is scrubbed, the acid solution should be rinsed completely from the surface by thoroughly sponging or flushing with clean water. The surface should be rinsed immediately to avoid formation of salts on the surface which are difficult to remove.

This procedure also may be used to etch hard and dense concrete surfaces with poor surface bonding properties. The etching procedure breaks up the "glaze" so that the paint will adhere to the surface. To determine whether etching is required, a few drops of water are poured on the surface. If the water is quickly absorbed, the surface can be readily painted; if the drops remain on the surface, etching is recommended.

Surface glaze may be removed also by rubbing with an abrasive stone, light sandblasting, or allowing the surface to weather for 6 to 12 months. The surface also may be etched with a solution of 3% zinc chloride and 2% phosphoric acid; this solution is not washed off but allowed to dry to produce a paintable surface.

Concrete and masonry surfaces generally are alkaline in nature and it may be necessary to use acid cleaning methods to neutralize these surfaces before applying alkali-sensitive coatings. The presence of free alkali may be detected by testing with pH paper.

Pickling This method is used only in the mill or shop to remove mill scale, rust and rust scale from iron and steel products. Sulfuric, hydrochloric, nitric, hydrofluoric and phosphoric acid are used individually or in combination. Pickling usually is accomplished by immersing work in tanks, followed by several intermediate rinses to remove acids and salts. The final rinse usually is performed with a weak alkali solution to retard rusting and lingering acid solution.

Paint Removers Paint and varnish removers generally are used on small areas. Solvent-type removers or solvent mixtures are selected according to the type and condition of the old finish as well as the nature of the surface. Removers are available as flammable or non-flammable types, and in liquid or semi-paste consistency. While most paint removers require scraping to remove the softened paint, some removers loosen the paint sufficiently to permit flushing off the debris with steam or hot water.

To retard evaporation some removers contain paraffin wax, which often remains on the surface after the paint has been removed. It is essential that this residue be removed from the surface prior to painting, to insure proper bonding of the paint coat. The wax can be removed with mineral spirits or a solvent specifically recommended by the manufacturer. Most removers are toxic to a degree and some are flammable. Consideration should be given to adequate ventilation and fire control whenever they are used.

Metal Pretreatments

Pretreatments of metal surfaces improve coating adhesion, reduce underfilm corrosion and extend coating life. Of the several pretreatment methods, hot phosphate treatments are limited to the shop or plant; cold phosphate and wash primers can be applied in the field as well.

Hot Phosphate Treatments These treatments utilize zinc or iron phosphate solutions to form crystalline deposits on the surface of the metal. They greatly increase the bond and adhesion of applied paints while reducing underfilm corrosion. Zinc phosphate generally produces the best results and is most widely used.

It usually is necessary to apply thicker paint coats over the zinc phosphate coatings if a gloss finish is desired, because the heavier phosphate coatings absorb considerably more paint. If a higher gloss finish is desired, iron phosphate is preferred to the zinc phosphate pretreatment, because it produces a finer crystalline structure and hence a thinner film. Hot phosphate treatments are very effective in insuring good coating adhesion and tight bonds.

Cold Phosphate Treatments These treatments are produced with a mixture of phosphoric acid, wetting agent, water-miscible solvent and water. An acid concentration of about 5% to 7% (by weight) may be adequate when the area is not exposed during application to high temperatures, direct sunlight, or high wind velocities. Such adverse environmental conditions cause rapid evaporation and consequent high acid concentration.

When a dry, grayish-white powdery surface develops within a few minutes after application, the acid has reacted properly and has the proper dilution. If a dark color develops and the surface is somewhat sticky, the acid probably is too concentrated. In such cases, if the area is small, wiping with damp rags may bring about the desired appearance; otherwise, the surface should be rinsed with water and re-treated with a more dilute solution.

Although cold phosphate treatments may produce a crystalline deposit on the metal surface, the density of the deposit is not as great as the hot phosphate treatment; therefore paint adhesion may be less effective. The procedures used for cold phosphating are adaptable to field use on large or small structures.

Wash Primers These formulations result in a form of cold phosphatizing. They are said to perform more efficiently than the standard cold phosphating treatments and are generally replacing them in field use. Wash primers derive their name from the method of applying them in very thin, "wash", coats. They usually contain polyvinyl butyral resin, phosphoric acid, and a rust inhibitive pigment such as basic zinc chromate or lead chromate. Wash primers develop extremely good adhesion to blast-cleaned or pickled steel, and provide a sound base for topcoating. They also may be used to improve adhesion of coatings to hard-to-paint surfaces, such as galvanized or stainless steel and aluminum.

Primers, Sealers, and Fillers

Primers often are applied on masonry to seal chalky surfaces and to improve adhesion of water-based topcoats. Sealers are used on wood to prevent resin exudation or *bleeding*. Fillers help to produce a smooth finish on open grain wood and rough masonry surfaces.

Primers Ordinary latex paints do not adhere well to chalky masonry surfaces. To overcome this problem, "modified" latex paints can be used or the surface should be treated with an oil-based primer before the latex paint is applied. The entire surface should be vigorously wire brushed by hand or power tools, then dusted to remove all loose particles and chalk residue. The primer then should be brushed on freely to assure effective penetration and should be allowed to dry. The primer should not be used as a finish coat, for it is not as weather-resistant.

Knot Sealers Sealers are used on bare wood to prevent resin bleeding through newly applied paint coatings. Freshly exuded resin, while still soft, may be scraped off with a putty knife and the affected area may be cleaned with alcohol. Hardened resin may be removed by scraping or sanding. Since the sealer is not intended for use as a priming coat, it should be used only when necessary, and applied only over the affected area. When a previously painted softwood lumber surface becomes discolored over knots, the knots should be covered with sealer before new paint is applied.

Fillers These are used on porous wood, concrete, and masonry to fill the pores and to provide a smoother finish coat.

Wood fillers are used on open-grained hardwoods. In general, those hardwoods with pores larger than in birch *(open grain)* should be filled (Fig. 9). When filling is necessary, it should be done 24 hours after any staining operations. If staining is not warranted, natural (uncolored) filler should be applied directly to the bare wood. When desired, the filler may be colored with some of the stain in order to accentuate the grain pattern of the wood.

To apply, the filler should be thinned first with mineral spirits to a creamy consistency, then liberally brushed *across* the grain, followed by light brushing *along* the grain. It should be allowed to stand 5 to 10 minutes until most of the thinner has evaporated, and the finish has lost its glossy appearance. Before it has a chance to set and harden, the filler should be wiped off *across* the grain, using burlap or other coarse cloth, rubbing the filler into the pores of the

FIG. 9 **FINISHING CHARACTERISTICS OF MAJOR WOOD SPECIES**

WOOD SPECIES	GRAIN TYPE	REMARKS
HARDWOODS		
Ash	Open	Requires filler
Aspen	Closed	Paints well
Basswood	Closed	Paints well
Beech	Closed	Paints poorly; varnishes well
Birch	Closed	Paints and varnishes well
Cherry	Closed	Varnishes well
Chestnut	Open	Requires filler; paints poorly
Cottonwood	Closed	Paints well
Cypress	Closed	Paints and varnishes well
Elm	Open	Requires filler; paints poorly
Gum	Closed	Varnishes well
Hickory	Open	Requires filler
Mahogany	Open	Requires filler
Maple	Closed	Varnishes well
Oak	Open	Requires filler
Teak	Open	Requires filler
Walnut	Open	Requires filler
SOFTWOODS		
Alder	Closed	Stains well
Cedar	Closed	Paints and varnishes well
Fir	Closed	Paints poorly
Hemlock	Closed	Paints fairly well
Pine	Closed	Variable depending on grain
Redwood	Closed	Paints well

wood while removing the excess. The operation is finished by stroking along the grain with clean rags, making sure that all excess filler is removed.

Knowing when to start wiping is important; wiping too soon may pull the filler out of the pores, while allowing the filler to set too long may make it difficult to wipe off. A simple test for dryness consists of rubbing the finger across the surface: if the filler forms into a ball, it is time to wipe; if the filler slips under the pressure of the finger, it is still too wet for wiping. The filler should be allowed to dry for 24 hours before applying any finish coats.

Masonry fillers can be applied by brush or spray to bare concrete, stucco, concrete block, or other masonry surfaces, both new and old. The surface should be rough and sound, free of loose and powdery materials. If the surface voids are relatively large, it is preferable to apply two coats of filler, rather than one heavy coat. In large areas, the filler can be sprayed on, then troweled smooth. One to 2 hours drying time should be allowed between filler coats, and painting should be delayed 24 hours after the final filler coat.

REPAIR OF SURFACES

It is essential that surfaces intended for painting be clean, sound and uniform in texture. Consideration should be given to: repairing or replacing damaged materials or surfaces; replacing broken windows and loose putty or glazing compound; filling cracks, crevices and joints with caulking compound or sealants; and patching cracks or holes in wood, masonry and plaster.

Caulking and Sealing Materials

Caulking compounds generally have an oil or oleoresinous base and are used in relatively fixed joints subject to very limited movement. Sealants, on the other hand, are resin-based, elastomeric (rubber-like) compounds, intended for use in expansion joints or other movable joints. (See Section 222 Caulking and Sealants for more detailed discussion of these products.)

Caulking Compounds These compounds should be used to fill joints and crevices around doors and windows in wood, brick, concrete and other masonry surfaces. They are supplied in a *gun grade* suitable for use with caulking gun and a *knife grade* which can be spread with a putty knife. Caulking compounds tend to dry on the surface but remain soft and tacky inside. The exposed surface of the caulking bead should be painted each time the surrounding area is painted to help extend its life.

Sealants In critical locations where a weathertight joint is essential at all times, sealants should be used. Sealants are preferred to caulking compounds, because of their better adhesion and greater elasticity, even at low temperatures. Although they are more expensive than caulking compounds, their greater reliability and durability often is worth the difference in cost. Sealants are available as one-component and two-component types as well as in preformed tapes.

Putty This is usually an oil-based material used to fill nail holes, cracks and imperfections in wood surfaces. Putty generally is supplied in bulk form and is applied with a putty knife. It should be applied after priming to prevent the wood from drawing the oils out. It dries to a harder surface than caulking compound and should not be used for joints or crevices (Fig. 10). Special glazing putty is used to install small lights in wood and metal sash.

Glazing Compounds These preparations are used both for bedding and face glazing of windows. Glazing compounds set firmly but retain a limited flexibility, generally greater than that of putty. Glazing compounds have a life expectancy of approximately 5 to 20 years, when protected by periodic painting over the exposed surface.

Patching Materials

Cracks, holes and crevices in masonry, plaster, wallboard and wood may be filled with various patching materials. These are supplied either ready for use, or as a dry powder to which water is added.

When using patching materials on masonry, plaster or wallboard, the crack should first be opened with a putty knife or wall scraper so that weak material is removed and the patching compound can be forced in completely. The compound should be applied with a putty knife or trowel, depending on the size of the hole. Then the surface should be smoothed off slightly convex, to allow for shrinkage.

Portland Cement Grout A soupy mixture of portland cement, sand and water *(grout)* can be used to repair cracks in concrete and masonry. Synthetic resins such as latex or epoxy may be added to improve bonding qualities and to permit application in thin layers. Hydrated lime sometimes is added to lengthen its working time. Portland cement grout should be moist-cured for several days or longer.

Plastic Wood This filler is suitable for filling gouges and nail holes. It is applied in a manner similar to putty. After the plastic wood has completely dried, it should be sanded until smooth.

Patching Plaster This preparation is used for repairing large areas in plaster. It is similar to ordinary plaster except that it hardens more quickly. It is supplied as a powder to be mixed on the job with water.

Spackle This compound is used to fill cracks and small holes in plaster and wallboard. It is easy to apply and to sand smooth after it hardens. It is supplied both as a paste and as a powder.

Joint Compound Although intended primarily for sealing joints between wallboards, it can also be used to repair large cracks. It is supplied as a powder and is used in conjunction with joint tape which gives it added strength.

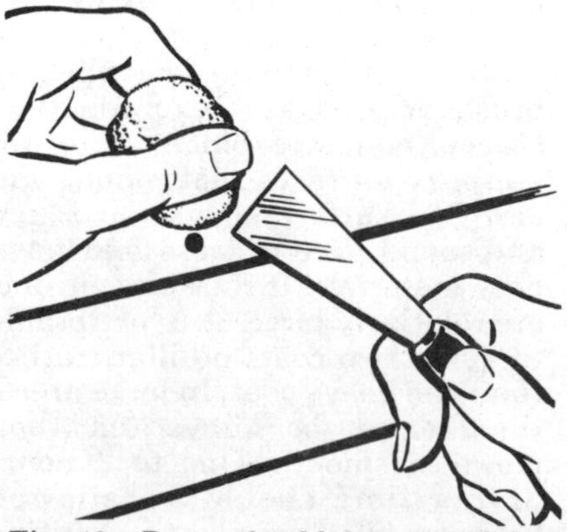

Fig. 10 Putty should be pressed into nailholes forming convex surface to allow for shrinkage.

PAINT APPLICATION

The most common methods of applying paint are by *brush, roller* and *spray*. Dip and flow coat methods generally are limited to shop work. Of the three methods designed for field use, brushing is the slowest, and spraying is the fastest. One man can cover up to 1000 sq. ft. per day by brushing, 2000 to 4000 sq. ft. by rolling, and up to 12,000 sq. ft. by spraying. Coatings applied by brush may leave brush marks in the dried film; rolling leaves a stippled effect, while spraying yields the smoothest finish, if done properly. The choice of method is based on many additional factors such as: 1) environmental conditions; 2) type of surface and coating, and 3) appearance of finish desired.

General surroundings may prohibit the use of spray application because of possible fire hazards or potential damage from overspray. Adjacent areas not intended to be coated should be masked when spraying is performed. If the time spent in masking preparations is extensive, it may offset the advantage of the speed of spraying operations.

Roller coating is efficient on large flat surfaces. Corners, edges and odd shapes, however, generally require brushing. Spraying is especially suitable for large surfaces, but it can be used also for round or irregular shapes. Brushing is ideal for small surfaces or for "cutting-in" corners and edges. Dip and flow coat methods are appropriate for volume production painting of small items in the shop.

Rapid-drying lacquer type products such as vinyls, generally are sprayed. Application of such products by brush or roller is extremely difficult, especially in warm weather or outdoors on breezy days.

Job Conditions

To obtain optimum performance from a coating there are certain basic application procedures which should be followed, regardless of the type of equipment used. Cleaned, pretreated surfaces should be prime-coated within the specific time limits recommended.

It is important that surface and air temperatures are between 50°F and 90°F for water-thinned coatings, and 45°F to 95°F for other coatings, unless the manufacturer specifies otherwise. The paint material should be maintained at a temperature of 65°F to 85°F at all times. Paint should not be applied when the temperature is expected to drop to freezing before the paint has dried. Wind velocity should be below 15 miles per hour and relative humidity below 80%.

Masonry surfaces that are damp (not wet) may be painted with latex or cementitious paints. Otherwise, the surface should be completely dry before painting. Paints should be applied at recommended spreading rates. When successive coats of the same paint are used, each coat should be tinted a different color to assure complete coverage with each coat. Sufficient time should be allowed for each coat to dry thoroughly before the finish coat is applied. The finish coat should be allowed to dry for as long as practicable before the space is occupied or put into service.

Brush Application

The quality of brush applied finish depends largely on the choice of suitable equipment and application methods.

Equipment Brushes, like any other tools, should be of good quality and maintained in good working condition at all times. Brushes can be identified by the type of bristle used: *natural*, *synthetic* or *mixed*.

Natural brushes are suitable for application of most paints except latex. They are produced primarily from horsehair and hog bristle. Hog bristle is unique in that each bristle end forks out like a tree branch, a characteristic termed *flagging*. This permits more paint to be carried on the brush and leaves finer brush marks. Fine brush marks tend to flow together more readily resulting in a smoother finish (Fig. 11).

Horsehair bristles are used in cheaper brushes and are not recommended for fine quality work

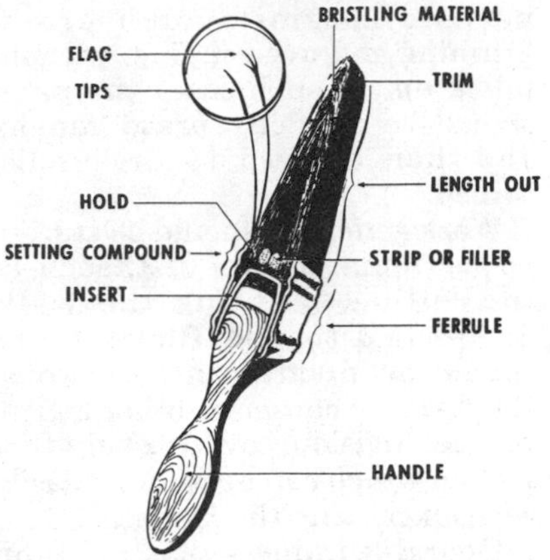

Fig. 11 Typical paint brush construction.

because the bristles quickly become limp. The ends do not flag, they hold far less paint and do not spread it as well. Brush marks left in the coating tend to be coarse and do not level out as smoothly.

Some brushes contain a mixture of hog bristle and horsehair, and their quality depends upon the percentage of each type used. Animal hair is utilized in very fine brushes for special purposes. Oxhair blends, for example, produce good varnish brushes; squirrel and sable are ideal for stripping, lining, lettering and free-hand art brushes.

New natural bristle brushes can be improved by soaking for 48 hours in linseed oil, followed by thorough cleaning in mineral spirits. This procedure makes the bristles more flexible and swells the bristles in the ferrule of the brush. Swelling results in a better grip so that fewer bristles are apt to work loose when the brush is used.

Synthetic brushes most often are made of nylon fibers. By artificially kinking the fibers and "exploding" the ends, manufacturers have increased the paint load nylon can carry, and have reduced the coarseness of brush marks. Because of importing difficulties, nylon is steadily replacing hog bristle.

Nylon generally is superior to horsehair, but being a soluble synthetic, it is unsuitable for applying lacquer, shellac and other products containing strong solvents which might soften or dissolve the bristles. Because water does not cause any appreciable swelling of nylon bristles, nylon is especially recommended for use with latex paints. Nylon brushes are ready to use when received, without any special conditioning.

It is important to use the right size brush for the job, particularly one that is not too large. A large area does not necessarily get painted faster with an oversize brush. If the brush is too large for the job the user is likely to tire faster and apply the paint unevenly.

Brushes are further identified by their shape, size and the specific painting jobs for which they are intended.

Wall brushes are flat and square-edged, ranging in width from 3″ to 6″; they can be used for painting large continuous surfaces, either interior or exterior.

Sash and trim brushes are available in four shapes: flat square-edged; flat angle-edged; round, and oval (Fig. 12a). These brushes range in width from 1-1/2″ to 3″ or varying diameters from 1/2″ to 2″ and should be used for painting window frames, sash, narrow boards, and either interior or exterior trim surfaces. For fine-line painting, the edge of the brush is often chisel-shaped to make precise edging easier.

Enameling and varnish brushes are flat, square-edged or chisel-edged, and available in widths from 2″ to 3″. The select, fine bristles are somewhat shorter than in other brushes in order to spread the relatively high-viscosity gloss finish in a smooth, even film.

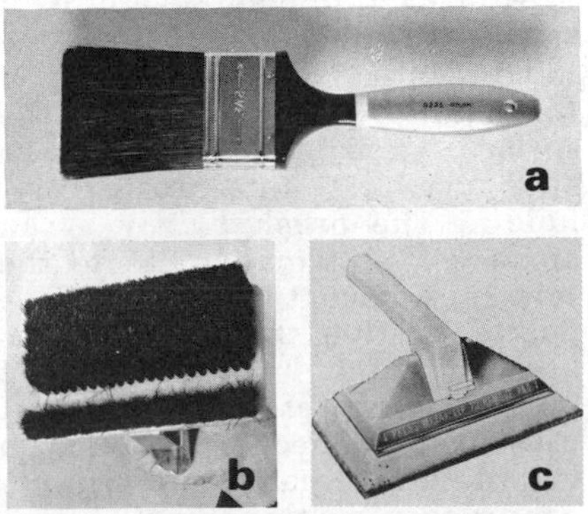

Fig. 12 Brushes are made in different shapes, sizes and types to suit materials and conditions: a) sash and trim brush, b) flat brush and c) pad applicator.

Stucco and masonry brushes have the general appearance of flat wall brushes and are available in widths ranging from 5″ to 6″. Bristles can be hog or nylon; the latter are preferred for rough surfaces because of their higher abrasion resistance.

Flat Brushes (Fig. 12b) typically are made of 1-1/4 in. nylon bristles arranged in two parallel groups, 3 x 7 in. and 1/2 x 7 in., separated by a 1 inch space. The use of short bristles in this format aids in the rapid application of paint on rough surfaces, such as wood shakes and concrete blocks. The narrow bristle area improves paint application in grooves, joints and under overhangs of siding, shakes and window sills.

Pad Applicators (Fig. 12c) are made of a foam pad, covered with napped fabric about 3/16 in. deep. The applicator has a spring loaded handle so that it remains parallel to the surface being painted. The handle is threaded to accept an extension pole. The pad is recommended for applying exterior latex paints to siding, shakes and wood fencing. The pad applicator can be used also to apply stains to shakes, shingles and paneling by substituting a lambswool pad. Pile depths of 1/2, 3/4 and 1-1/4 in. are available for surfaces of increasing roughness.

Application The brush should be dipped into the paint up to one-half of the bristle length, then withdrawn and tapped against the inside of the bucket to remove excess paint (Fig. 13).

After each dipping, several light strokes should be made in the area to be painted, to transfer most of the paint from the brush to the surface. Then the paint should be spread evenly and uniformly, holding the brush lightly at an angle of approximately 45° to the surface. Adjacent areas should be painted so that the brush strokes are completed by sweeping the brush into the wet edge of the paint previously applied. This helps to eliminate lap marks and provide a more even coating. Finally, light cross-brushing will smooth the surface and eliminate brush or sag marks. Very fast-drying finishes may not permit much cross-brushing and overlapping, without piling up. In such cases the paint should be applied, spread rapidly and then allowed to dry undisturbed.

Work generally should be started on the topmost area first, such as the ceiling, painting the walls downward to the floor. Work should be divided into sections, starting at a corner or other logical vertical division, covering only that section which can be reached easily without moving the ladder.

Doors, windows, casings and other trim should be painted after walls and ceilings or other major surfaces are completed. (A possible exception might be a situation where scaffolding is moved along from area to area.) Facia boards, moldings and other surfaces with narrow leading or indented edges should be painted first, then followed with the surrounding continuous surfaces.

It is important that corners and edges be painted so that the stroke is completed by sweeping off the corner or edge. It is best to use the edge of the brush and twist it slightly if necessary to cover rough surfaces, instead of poking the brush into corners or crevices.

Roller Application

The quality of a roller applied finish depends on the choice of suitable equipment and application methods.

Equipment A paint roller consists of a replaceable fiber cover on a cylindrical cage revolving on a shaft the extension of which forms a handle. The cover may be 1-1/2″ to 2-1/4″ inside diameter, and usually 3″, 4″, 7″ or 9″ long. Special rollers are available in lengths from 1-1/2″ to 18″. Roller fabric and nap length should be suited to the type of paint and texture of the surface to be painted. Special rollers are used for painting pipes, fences and other hard-to-reach areas (Figs. 14 & 15).

Rollers are available with lambs wool, mohair, polyester and modacrylic covers. Cover material selection depends on the type of surface and paint intended, as shown in Figure 16.

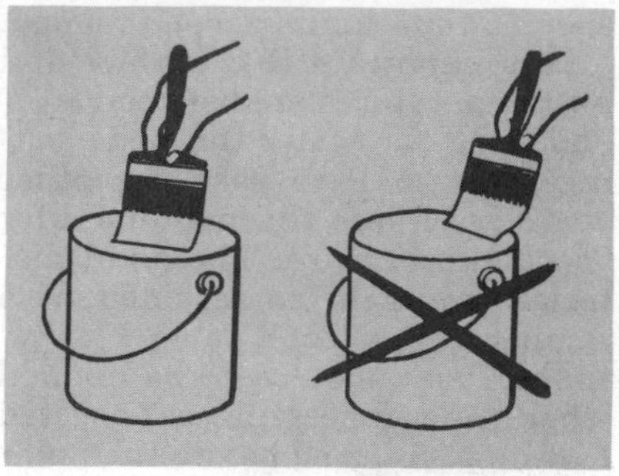

Fig. 13 Excess paint should be removed by tapping, not by wiping the brush against the side of the can.

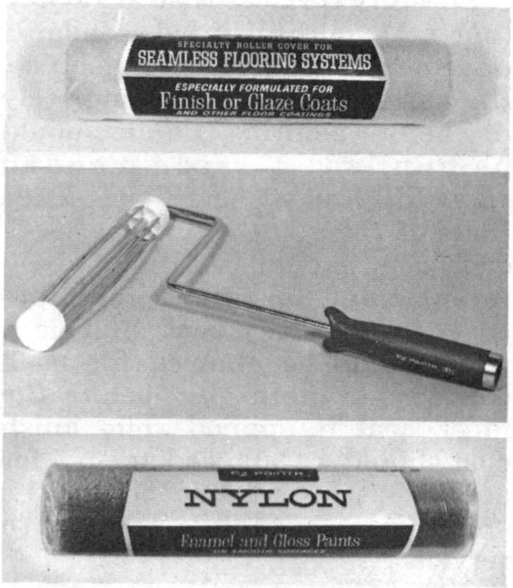

Fig. 14 Parts of a typical roller: handle and interchangeable covers.

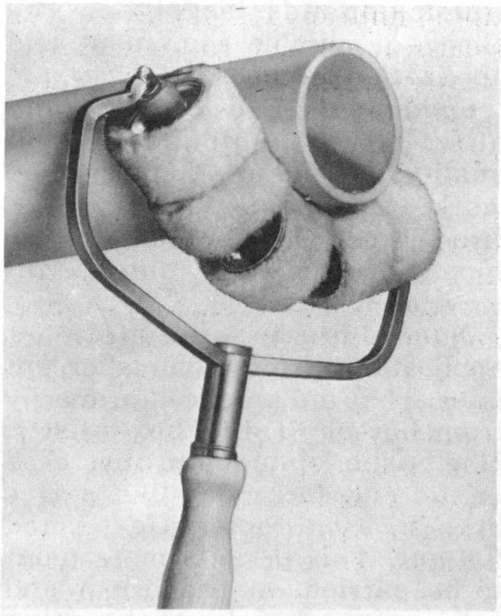

Fig. 15 Special rollers for painting pipes flex to adjust to pipe diameter.

Lambs wool is the most solvent-resistant cover material used and is available in nap lengths up to

FIG. 16 **ROLLER COVER SELECTION GUIDE**

TYPE OF PAINT	COVER MATERIAL AND NAP LENGTH		
	SMOOTH SURFACES[1]	SEMI-SMOOTH SURFACES[2]	ROUGH SURFACES[3]
Aluminum	polyester, 1/4″ to 3/8″	modacrylic, 3/8″ to 3/4″	modacrylic, 1″ to 1 1/4″
Alkyd enamel or semigloss	modacrylic[5], 1/4″ to 3/8″	modacrylic, 3/8″ to 3/4″	
Enamel undercoat	modacrylic[5], 1/4″ to 3/8″	modacrylic, 3/8″ to 3/4″	
Epoxy coatings	lambs wool pelt[5], 1/4″ to 3/8″	lambs wool pelt, 1/2″ to 3/4″	lambs wool pelt, 1″ to 1 1/4″
Exterior house paint:			
Latex over wood	polyester, 1/4″ to 3/8″	modacrylic, 3/8″ to 3/4″	
Latex over masonry	modacrylic, 1/4″ to 3/8″	modacrylic, 3/8″ to 3/4″	modacrylic, 1″ to 1 1/4″
Oil or alkyd over wood	polyester, 1/4″ to 3/8″	modacrylic, 3/8″ to 3/4″	
Oil or alkyd over masonry	modacrylic, 1/4″ to 3/8″	modacrylic, 3/8″ to 3/4″	modacrylic, 1″ to 1 1/4″
Floor enamels	modacrylic[5], 1/4″ to 3/8″	modacrylic, 3/8″ to 3/4″	modacrylic, 1″ to 1 1/4″
Interior wall paint:			
Alkyd or oil	modacrylic, 1/4″ to 3/8″	modacrylic[4], 3/8″ to 3/4″	modacrylic, 1″ to 1 1/4″
Latex	modacrylic, 1/4″ to 3/8″	modacrylic[4], 3/8″ to 3/4″	modacrylic, 1″ to 1 1/4″
Masonry sealer	mohair, 3/16″ to 1/4″	modacrylic[4], 3/8″ to 3/4″	modacrylic[4], 1″ to 1 1/4″
Metal primers	modacrylic, 1/4″ to 3/8″	modacrylic[4], 3/8″ to 3/4″	
Varnish—all types	modacrylic[5], 1/4″ to 3/8″		

(1) Smooth Surfaces: hardboard, smooth metal, smooth plaster, drywall.
(2) Semi-smooth Surfaces: sand-finish plaster and drywall, light stucco, blasted metal, semi-smooth masonry.
(3) Rough Surfaces: concrete block, brick, heavy stucco, wire fence.
(4) Also lambs wool pelt: 1/2″ to 3/4″ on semi-smooth surfaces, 1″ to 1 1/4″ on rough surfaces.
(5) Also mohair, 3/16″ to 1/4″.

1-1/4″. It is recommended for synthetic finishes, for application on semi-smooth and rough surfaces. However, it mats badly in water, and is not recommended for water-based paints such as latex.

Mohair is made primarily of Angora hair. It is quite solvent-resistant and is supplied in 3/16″ and 1/4″ nap length. It is recommended for use on smooth surfaces and for synthetic enamels, but can be used also with water-based paints.

Modacrylic fiber has low absorption and excellent resistance to water. It is best for most conventional solvent and water-based paints, except those which contain strong solvents such as ketones. It is available in all nap lengths from 1/4″ to 1-1/4″.

Rayon fabric generally is not recommended because it mats in water and frequently produces unsatisfactory results.

Application The paint should be poured into the tray to about one-half of its depth. Then the roller should be moved back and forth along the ramp to fill the cover completely and uniformly with paint (Fig. 17).

As an alternative to using the tray, a specially designed galvanized wire screen can be used inside a 5-gallon paint container. This screen attaches to the wall and remains at the correct angle for loading and spreading paint on the roller (Fig. 18).

The first load of paint on a roller should be worked out on a newspaper to remove entrapped air from the roller cover. As the roller is passed over a surface, thousands of tiny fibers continually compress and expand, metering out the coating and wetting the surface. Unlike brushing and spraying, roller application is less susceptible to variation in painter ability.

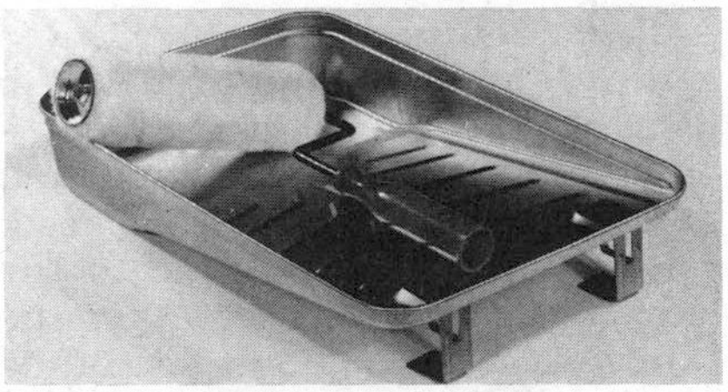

Fig. 17 Typical tray used in roller-painting average size jobs, such as home interiors.

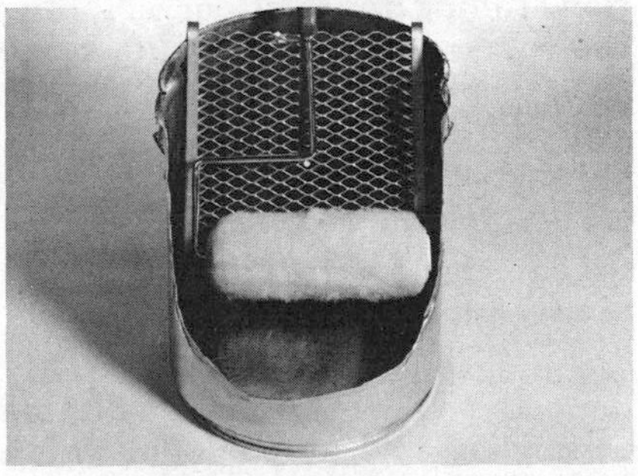

Fig. 18 For larger jobs, 5-gallon container with inside screen may be used in place of tray.

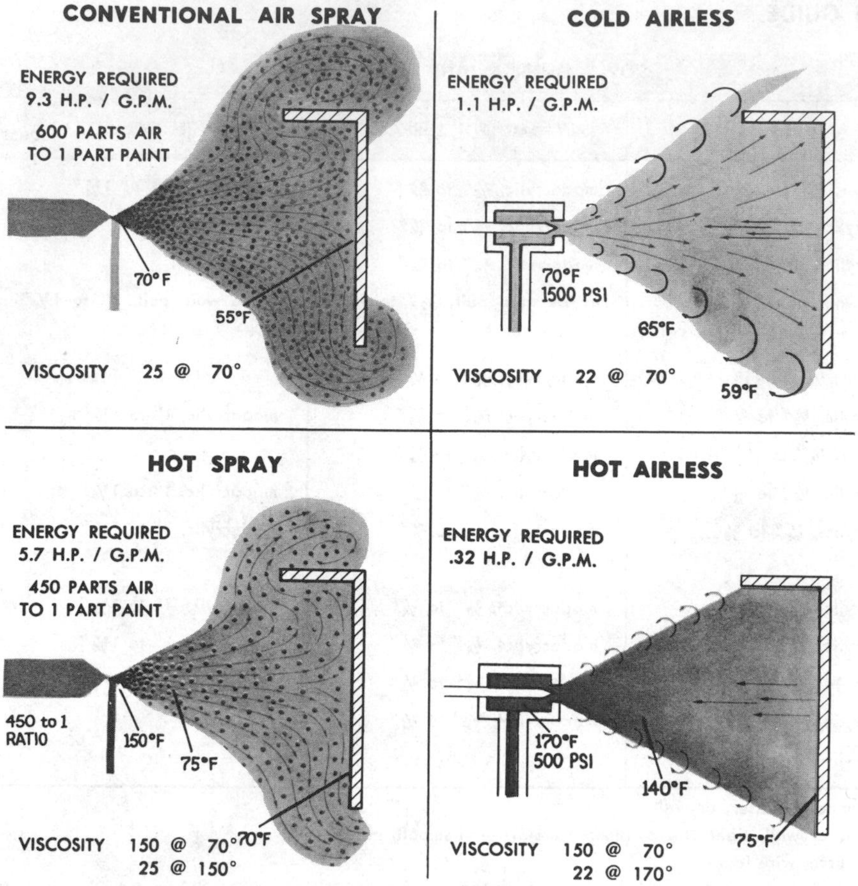

Fig. 19 Basic paint spraying methods: airless methods have the advantage of less overspray and better coverage.

The junctures between walls and ceilings, molding and wood trim usually are painted first, using a 3″ wall brush. Then the major wall or ceiling surfaces are covered, rolling as close as possible to the trim and corners to maintain the same texture. The paint should be applied by working from the dry area into the just-painted area. Rolling completely in one direction, rolling too fast, or spinning the roller at the end of the stroke should be avoided. Final strokes should be rolled with minimal pressure, to pick up any excess paint on the surface.

Spray Application

Successful spray application depends on selection of suitable equipment and proper application techniques.

Equipment Spray equipment is available in four general types from which the methods derive their names (Fig. 19).

Cold air (conventional) spraying involves forcing the paint materials through an adjustable gun nozzle with compressed air, at normal air temperatures. The compressed air feeds the paint under pressure from the "pot" and atomizes it at the nozzle. Controls on the gun make it possible to vary the air pressure, spray pattern and amount of paint delivered (Fig. 20).

Cold air spraying is the least expensive and most common spray technique used today. Wide use of this method has led to the development of equipment suitable for practically any type of coating material and painting problem. One disadvantage of this method is that it tends to create excessive overspray because of the high ratio of air to paint used.

Cold airless spraying uses hydraulic rather than air pressure to force the paint through the nozzle. The equipment is similar to that used for conventional spraying except that the compressor operates a hydraulic pump and the air hose is omitted. The paint is atomized by forcing it through a specially shaped orifice at a pressure of 1500 to 3000 psi. This pressure produces a high speed stream of the coating material at the orifice. The high speed, coupled with the sudden release of pressure, causes atomization without the use of compressed air; thus there is no air turbulence to deflect the paint or to cause overspray as in the conventional method.

The absence of compressed air also reduces rebounding of the paint in crevices and corners, thus providing more uniform coverage. Airless spraying usually permits the use of products with a higher viscosity; less thinner is required and desired film thickness can be achieved faster. The need for a single hose leading into the gun makes it easier to handle. The lack of overspray offers still two other advantages: cleanup is easier and masking is minimized. However considerable caution must be exercised because of the high operating pressures.

The hot spray technique can be adapted to either conventional or airless spray painting, but most often is used with air spray. The paint tank and hose are heated to raise the paint temperature to 130°F to 180°F. Introducing heat lowers paint viscosity thereby reducing the quantity of solvent needed. The resultant coating has a higher solids content and produces greater film thickness per coat. The use of heat also allows application at lower pressure with less overspray and rebounding, as well as more uniform application at low temperatures. Only materials specially formulated for hot spray application should be used.

Application Most materials that can be brushed or roller applied, also can be sprayed. Exceptions include very thick or stringy materials, some textured materials and some rubber-base coatings. Control of the spray operation requires control over a

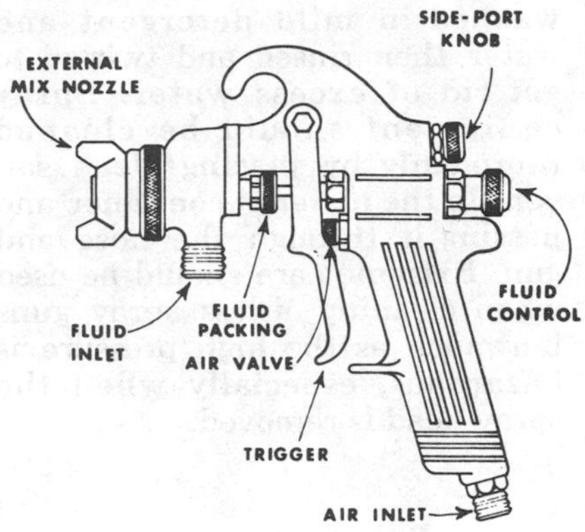

Fig. 20 Parts of a typical spray gun.

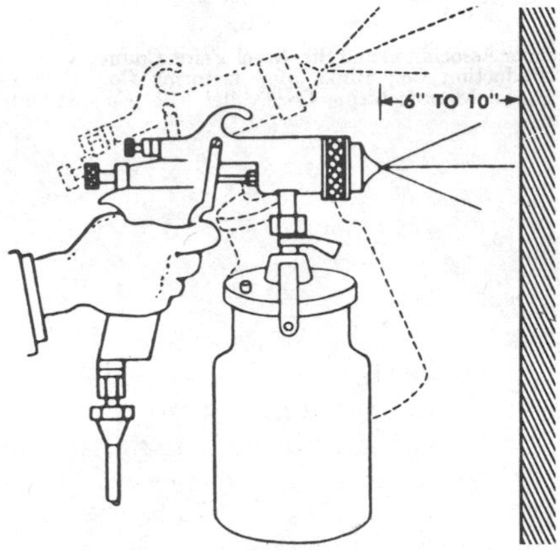

Fig. 21 For uniform finish, spray gun always should be held perpendicular to surface.

number of variables.

Paint viscosity should be low enough to permit proper atomizing but high enough to apply without running or sagging. Viscosity should be adjusted only when necessary and then according to manufacturers' instructions. Excessive thinning can result in excessive overspray, runs and sags, and inadequate surface protection.

Air and material pressures should be the lowest practicable, which will result in a quality finish with good flow-out and minimum overspray. When the material is heavy or viscous, when the fluid hose is too long, or when a more rapid rate of application is required—it may be necessary to increase the material hose pressure. In this case, it may also be necessary to increase the air pressure since it is important to maintain a proper ratio of material pressure to atomizing air pressure.

The optimum spray gun pressure varies with the nozzle, type of paint and surface; it generally must be determined by experimentation. However at a temperature of 70°F with a material hose length of 25′, the following air pressures are suggested for initial settings: 1) lacquers 40-45 lbs.; 2) enamels 35 lbs., and 3) alkyd flats 25-30 lbs.

The spray pattern can be varied from round to fan-shaped. As the width of the pattern is increased, the paint flow should be correspondingly increased to maintain the same coverage (Fig. 20).

The spray gun should be held 6″ to 10″ from the surface. Holding the gun too far away causes *dusting*, a phenomenon in which the paint solvent evaporates in mid-air and the coating hits the surface in a nearly dry state. Tilting the gun causes the paint to be more heavily applied in one part of the spray pattern than another (Fig. 21).

It is suggested that a swinging free arm motion be used, feathering out at the end of the stroke by pulling the gun trigger after beginning the stroke and releasing it before the stroke is completed (Fig. 22). For external corners, the spraying should stop several inches from the corner, then the gun should be swept up and down along the edge so as to cover both sides at the time (Fig. 23a). Internal corners should be painted as illustrated in Figure 23b.

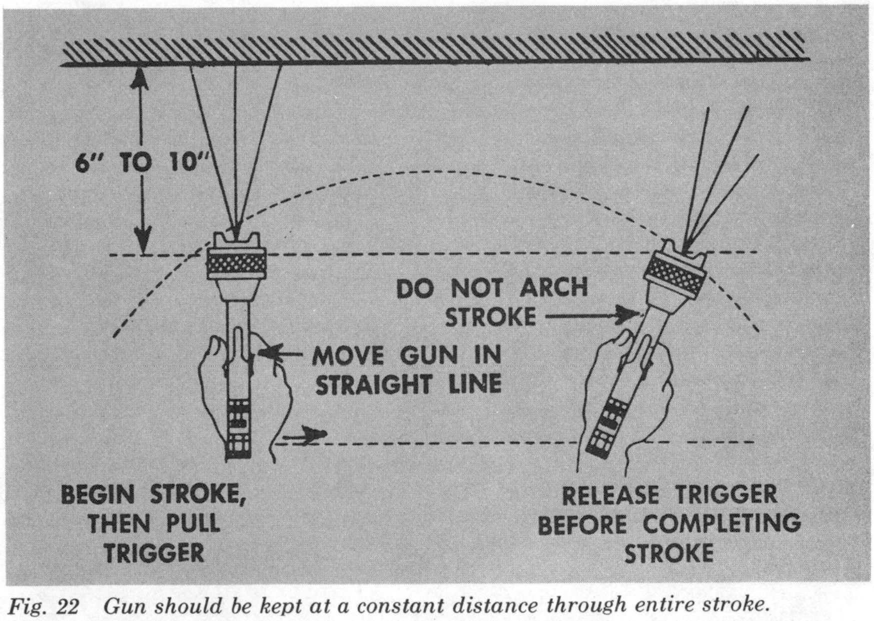

Fig. 22 Gun should be kept at a constant distance through entire stroke.

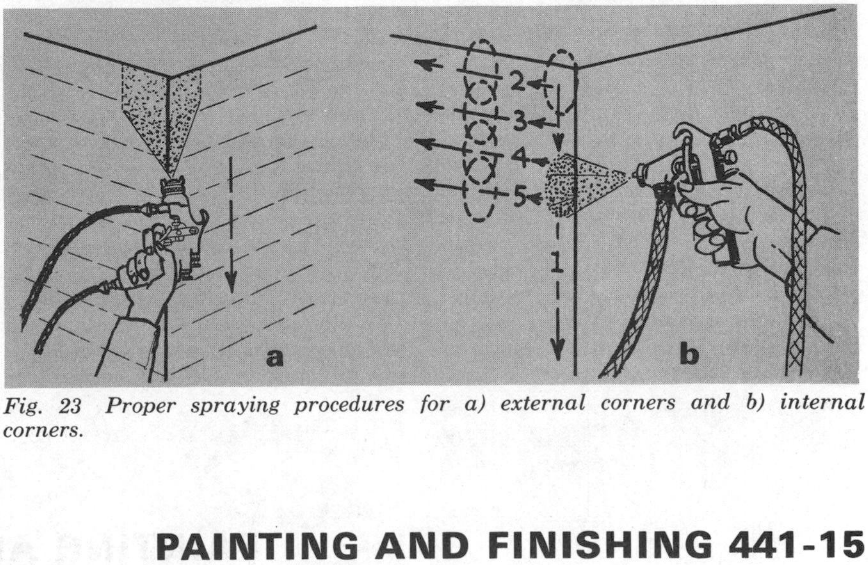

Fig. 23 Proper spraying procedures for a) external corners and b) internal corners.

Cleanup

It is essential that all tools and equipment be cleaned immediately after use, before the paint has a chance to harden. It is suggested that as much paint as possible be removed by brushing out on waste paper or other disposable surface; then tools should be cleaned thoroughly with a compatible solvent. Rollers and brushes should be cleaned two or three times in fresh solvent until no paint is noticeable, then wiped clean and dry. After cleaning, brushes should be washed with mild detergent and warm water, rinsed in clear water, and twirled to get rid of excess water. They then should be combed and placed in brush-keepers or wrapped in paper to dry flat.

Rollers should be similarly washed in mild detergent and water, then rinsed and twirled to get rid of excess water. Spray equipment should be cleaned thoroughly by placing clean solvent in the material container and passing it through the hose and gun. Extreme care should be used when cleaning airless spray guns inasmuch as the high pressure is hazardous, especially when the spray head is removed.

We gratefully acknowledge the assistance of the National Bureau of Standards, National Paint Varnish & Lacquer Association and the Jewel Paint Company in the preparation of this section. We are indebted to the following for supplying photographs and illustrations: Air Reduction Co.; Binks Manufacturing Co.; Clemco-Clementina Ltd.; Robert C. Collins Co.; The DeVilbis Co.; E-Z Paintr Corporation; H & G Industries, Inc.; Harbil Manufacturing Co.; Miller Mfg. Co.; Nordson Corporation; Partek Corporation of Houston; Tennant Co.; Von Schrader Manufacturing Co.

PAINTING & FINISHING 441

CONTENTS

GENERAL RECOMMENDATIONS

MATERIALS

For a description of typical paints and coatings and the surfaces for which they are suitable, see Main Text pages 216-1 through 216-16.

All paints or other coatings should be standard commercial brands with a history of satisfactory use under conditions equal to or similar to the conditions present in the area concerned.

The paint or coating selected should be designed and recommended by the manufacturer for the specific use proposed. The printed application instructions should clearly identify the suitability of the material for the type of exposure (exterior or interior), the type of surface to be covered (wood, metal, masonry, concrete or plaster) and the type of service to which the paint will be subjected (exposure to moisture, frequent washing or heavy traffic).

Storage

Paint materials should be stored in dry and well-ventilated areas maintained at a temperature range of 50°F to 90°F.

Freezing temperatures may permanently damage water-based paints. If the paint material is sensitive to heat, temperatures over 100°F may bring about a chemical reaction within the container, resulting in increased viscosity to the point of gelation, making the paint unusable. Also, overheating may cause covers to blow off, creating a serious fire hazard.

Other hazards to be considered are high humidity, which may cause containers to corrode and labels to deteriorate, and poor ventilation, which may allow excessive concentration of solvent vapors that are both toxic and combustible.

Stocking

Paint materials should be rotated so that the oldest stocks are used first. Required quantities of paint should be carefully estimated so as to have little or no material left at the end of a job. Using leftover paint should be avoided.

When saved, leftover paint should be placed in small containers, filled full and tightly sealed. All information from the original container should be transferred to the new container. It is safer generally to discard small quantities than to use paint that has skinned.

CONDITIONING AND MIXING

Unless stored at the recommended temperature, paint materials should be temperature-conditioned at 65°F to 85°F for at least 24 hours before use. After this time, paints should be mixed, thinned or tinted as specified, and strained if necessary, all according to manufacturers' recommendations.

Paints consist of two principal components—the solid pigment and the liquid vehicle. Suspension agents are included in most paints to minimize settling of solids. However, when stored for long periods, separation of the pigment and vehicle may take place. The purpose of mixing is to reblend settled pigment with the vehicle, to eliminate lumps, skins or other detriments to proper application.

Mixing

Paint materials always should be mixed prior to issuance or delivery to the job. Caution is recommended to avoid the incorporation of excess air through over-

(Continued) **GENERAL RECOMMENDATIONS**

FIG. WF 1 **RECOMMENDED MIXING PROCEDURES FOR DIFFERENT COATING TYPES**

COATING TYPE	MIXING EQUIPMENT	REMARKS
Enamel, semigloss or flat oil paints	Manual, propeller or shaker	Mix until homogeneous
Water-based latex paints	Manual, propeller or shaker	Use extreme care to avoid air entrapment
Clear finishes, such as varnishes and lacquers	Manual	Generally require little or no mixing
Extremely viscous finishes, such as coal tar paints	Propeller	Use extreme care to avoid air entrapment
Two-package metallic paints, such as aluminum paint*	Propeller	Add small amount of liquid to paste and mix well; slowly add remainder of vehicle, while stirring, until coating is homogeneous
Two-Component Systems	Manual, propeller or shaker	Mix until homogeneous; check label for special instructions

*First make into a paste with solvent, then proceed as recommended above.

mixing. Some hand stirring should be done just prior to use and periodically during application.

Mixing can be done either by manual or mechanical methods but the latter is preferred to insure maximum uniformity (Fig. WF1). See Main Text page 441-4 for discussion of various mixing procedures.

Tinting

Tinting should be done by experienced personnel, in the paint shop or at the paint store. On-the-job tinting should be avoided as a general practice, except in tinting intermediate coats for identification and to assure complete coverage under the finish coat.

Avoiding on-the-job tinting minimizes errors and eliminates color-matching problems at a later date.

Tinting should be done with care, using only colors which are known to be compatible, and preferably adding not more than 4 oz. per gallon of paint, unless specifically approved by the manufacturer. Tinting of chalking-type exterior paints is not recommended except for identification of intermediate coats.

Colors-in-oil should be used only with standard oil-base paints, alkyd resins, chlorinated rubbers or other solvent systems. They should not be used with other synthetics or with water-thinned paints.

Universal tinting colors are used in the same manner as colors-in-oil. They are more compatible with a wide variety of paint materials. Many can be used with both solvent-thinned and water-thinned paints.

Thinning and Straining

The arbitrary addition of thinners should be avoided. Thinning should be performed only with known compatible thinning agents or those specifically recommended by the manufacturer. Thinning should not be used as a means of improving brushing or rolling properties of cold paint materials. Instead, materials should be conditioned to suitable temperature (65°F to 85°F) before use.

Unnecessary or excessive thinning may result in an inadequate film thickness and may reduce the durability of the applied coating.

PREPARATION OF SURFACES

All surfaces should be suitably prepared prior to painting or finishing.

The best paint will not perform effectively if applied on a poorly prepared surface. Proper preparation is essential to achieve maximum coating life. See Figure WF2 and Main Text page 441-5 for recommended methods of surface preparation.

FIG. WF 2 **RECOMMENDED PREPARATION OF VARIOUS SURFACES**

PREPARATION METHOD	WOOD	METALS		CONCRETE & MASONRY
		STEEL	OTHER	
MECHANICAL				
Hand Tool Cleaning[1]	OK	OK	OK	OK
Power Tool Cleaning	OK[2]	OK	—	OK
Flame Cleaning	—	OK	—	—
Brush-Off Blast Cleaning	—	OK	OK	OK
Other Blast Cleaning	—	OK	—	—
CHEMICAL AND SOLVENT				
Solvent Cleaning	OK	OK	OK	—
Alkali Cleaning	—	OK	—	OK
Steam Cleaning	—	OK	—	OK
Acid Cleaning	—	OK	—	OK
Pickling	—	OK	—	—
PRETREATMENTS				
Hot Phosphate	—	OK	—	—
Cold Phosphate	—	OK	—	—
Wash Primers	—	OK	OK	—
PRIMERS	—	—	—	OK
SEALERS	OK	—	—	—
FILLERS	OK	—	—	OK

1. Also suitable for gypsum wallboard and plaster.
2. Sanding only recommended.

(Continued) **GENERAL RECOMMENDATIONS**

Mechanical Treatments In Field

Mechanical treatments generally are intended to remove corrosion, mill scale or old paint. Solvent cleaning to remove surface oil or grease is recommended prior to other treatments. The use of goggles and gloves is essential for blast and flame cleaning and is recommended for the other methods as well. Good ventilation generally is desirable; dust respirators also may be required.

Hand tool cleaning should be performed to remove loose mill scale and non-adhering paint but should avoid deep scarring of the surface.

Power tool cleaning should be utilized to remove tightly adhering contaminants which cannot be removed with hand tools, but should avoid scarring the surface with chipping hammers or polishing the surface with wire brushes.

Flame cleaning should be performed with extreme caution and should be followed by wire brushing Paint should be applied while the surface is still warm.

Blast cleaning should be performed with an abrasive suitable to remove surface contamination without pitting the surface. Surface should be prime coated the same day.

Chemical and Solvent Treatment In Shop

The use of goggles and rubber gloves generally is recommended and additional protective clothing is required where acid or alkaline solutions are used. Adequate ventilation is essential during chemical and solvent cleaning. Respirators should be used when operating in confined areas; proper fire precautions should be taken when flammable cleaners are used. See Main Text page 441-7 for further discussion.

Metal Pretreatments

Hot phosphate treatments are limited to the shop or plant. Cold phosphate and wash primers can be applied in the field as well. See Main Text page 441-8 for discussion of various methods and procedures.

Primers, Sealers and Fillers

Application of the finish coat always should be preceded by a suitable primer, filler, sealer or a base coat of the same paint as the finish coat which performs the above functions.

Primers often are applied on masonry to seal chalky surfaces and to improve adhesion of water-based topcoats. Sealers are used on wood to prevent resin exudation or bleeding. Fillers help to produce a smooth surface on open grain wood and rough masonry surfaces (Fig. WF3).

FIG. WF 3 **FINISHING CHARACTERISTICS OF MAJOR WOOD SPECIES**

WOOD SPECIES	GRAIN TYPE	REMARKS
HARDWOODS		
Ash	Open	Requires filler
Aspen	Closed	Paints well
Basswood	Closed	Paints well
Beech	Closed	Paints poorly; varnishes well
Birch	Closed	Paints and varnishes well
Cherry	Closed	Varnishes well
Chestnut	Open	Requires filler; paints poorly
Cottonwood	Closed	Paints well
Cypress	Closed	Paints and varnishes well
Elm	Open	Requires filler; paints poorly
Gum	Closed	Varnishes well
Hickory	Open	Requires filler
Mahogany	Open	Requires filler
Maple	Closed	Varnishes well
Oak	Open	Requires filler
Teak	Open	Requires filler
Walnut	Open	Requires filler
SOFTWOODS		
Alder	Closed	Stains well
Cedar	Closed	Paints and varnishes well
Fir	Closed	Paints poorly
Hemlock	Closed	Paints fairly well
Pine	Closed	Variable depending on grain
Redwood	Closed	Paints well

REPAIR OF SURFACES

It is essential that surfaces intended for painting or finishing should be clean, sound, smooth and uniform in texture.

Consideration should be given to: repairing or replacing damaged materials or surfaces; replacing broken windows and loose putty or glazing compound; filling cracks, crevices and joints with caulking compound or sealants; and patching cracks or holes in wood, masonry and plaster. Products and materials suitable for this purpose are described in Figure WF4.

PAINT APPLICATION

The most common methods of applying paint are by brush, roller and spray. The choice of method is based on factors such as: (1) environmental conditions; (2) type of surface and coating; and (3) appearance of finish desired. See Main Text page 441-10 for further details.

Job Conditions

Surface and air temperatures during application should be between 50°F and 90°F for water-thinned

(Continued) GENERAL RECOMMENDATIONS

FIG. WF 4 **MATERIALS FOR REPAIR OF SURFACES**

MATERIAL	WHERE USED	CHARACTERISTICS
Caulking Compound	Filling joints and crevices around doors and windows in concrete, wood, brick, block and other masonry surfaces	Tends to dry on the surface but remains soft and tacky inside; Exposed surface of caulking bead should be painted each time surrounding area is painted to help extend its life (5-15 years)
Sealants	Filling joints in locations where movement occurs and a weathertight joint is essential at all times	Better adhesion and greater elasticity than caulking compound, even at low temperatures. Life expectancy 15-30 years
Putty	Filling nail holes, cracks, and imperfections in wood surfaces; installing small lights in wood and metal sash; not used for joints or crevices	Dries to a harder surface than caulking compound
Glazing Compound	Bedding and face glazing of windows	Sets firmly but retains limited flexibility, generally greater than putty. Life expectancy of 5-15 years when protected by periodic painting over exposed surface

coatings, and 45°F to 95°F for other coatings, unless the manufacturer specifies otherwise. Paint should not be applied when the temperature is expected to drop to freezing before the paint has dried. Wind velocity should be below 15 mph and relative humidity below 80%.

Surfaces generally should be completely dry before painting except masonry surfaces which may be damp (not wet) if painted with latex or cementitious paints.

Workmanship

When successive coats of the same paint are used, each coat should be tinted a different color to assure complete coverage with each coat. Sufficient time should be allowed for each coat to dry thoroughly before the finish coat is applied. The finish coat should be allowed to dry for as long as practicable before the space is occupied or put into service.

The line of demarcation between different colors or finishes should be straight and neat. Previously finished surfaces should be protected by suitable masking or covering. Stain or paint splatter from adjacent surfaces shall be removed.

Unless otherwise recommended by manufacturer, a minimum of two coats should be used. Painted exterior wood surfaces should have a total thickness of 4.0 mils or more (Fig. WF5).

FIG. WF 5 **DRY FILM THICKNESS OF APPLIED PAINT COATINGS**

COVERAGE RATE (SQ. FT. PER GAL.)	APPROX. DRY FILM THICKNESS—FINISH COATS (MILS)
450	2.25
500	2.00
550	1.75
600	1.50
650	1.30
700	1.10
750	1.00

442 CERAMIC TILE FINISHES

INTRODUCTION

Ceramic tile is made from nonmetallic minerals fired at high temperatures (from the Greek, *keramos* —potter's clay) and manufactured in modular unit sizes (tiles) which facilitate installation. The term "ceramic tile" includes several products of varying dimensions, properties and appearance: 1) *glazed wall tile,* typically 4¼″ x 4¼″ units used mainly on walls; 2) *ceramic mosaic tile,* usually 1″ x 1″ units intended mostly for floors but used also on walls; 3) *paver tile,* floor units similar to mosaic tile but generally 4″ x 4″ or larger; 4) *quarry tile,* typically 6″ x 6″ natural clay units for floors. Some glazed wall tile are sufficiently abrasion-resistant to make them suitable for floors as well, while all floor tile can be used on walls. All of these products possess good or superior abrasion-resistance, ability to take on a variety of patterns and colors and a high degree of resistance to moisture.

These properties make ceramic tile uniquely suited for areas exposed to severe foot traffic, intermittent or continuous contact with water, or corrosive chemicals. The development of simplified installation procedures and increased variety of decorative tiles has greatly extended ceramic tile use. Today, ceramic tile is being specified for just about every interior area of residential buildings as well as in commercial, institutional and industrial buildings. Many types of tile also are suited for exterior use.

The origins of ceramic tile date back 5,000 years to the beginning of civilization, as evidenced by burnt natural clay tiles found by archeologists in early Egyptian and Babylonean excavations. Decorative and functional clay tiles were also an essential part of later Greek and Roman architecture. The Italian town of Faenza is credited with developing, in the 14th century, what is now known as *faience* tile—a highly decorative tile with a textured irregular hand-made appearance. A century later, Holland started production of *delft* tile, characterized by a figure or landscape in blue or violet.

In the U.S., production of ceramic tile was spurred by the introduction of the *dust-press* method at the end of the 19th century. Samuel Keys is credited with first producing floor and wall tile in Pittsburgh in 1867; by 1937 there were 52 manufacturers of ceramic tile in the U.S. It is generally agreed that more improvements in manufacturing and installation techniques have been made by the U.S. industry during the past 25 years than in the preceding 70-odd centuries.

Much of the progress is due to the research and educational activities of the Tile Council of America (TCA). In the early 50's, TCA sponsored research resulting in the development of several adhesives and mortars which eliminated the need for thick setting beds, for laborious pre-soaking of tile and for extended curing. To assure satisfactory tile installations, TCA: 1) licenses manufacturers of mortars and grouts according to Council formulations; 2) promulgates standards for the manufacture and installation of ceramic tile—resulting in a series of ANSI Standards; and 3) annually publishes its *Handbook for Ceramic Tile Installation.*

Most of the discussion and installation details in this section have been abstracted from TCA's Handbook. This section is divided into three subsections: *General Recommendations, Installation on Floors* and *Installation on Walls.*

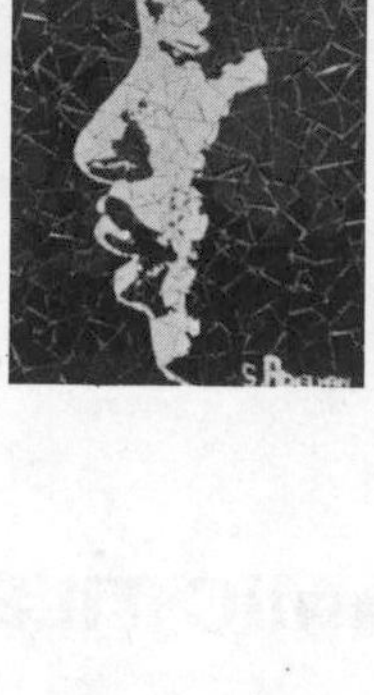

GENERAL RECOMMENDATIONS

Like most other thin non-structural finishes, ceramic tile depends on a sound, rigid and dimensionally stable backing surface. Installation of ceramic tile on floors resembles terrazzo (Fig. 1), involving either an unbonded, isolated, setting bed (thickbed method), or direct bonding to the backing (thinset method). As in terrazzo, direct bonding is made possible by newly-developed synthetic resins possessing greater bonding power than conventional portland cement. However, the terminology used by the ceramic tile industry differs somewhat from that in the terrazzo industry. The following text will generally employ the terms familiar to the tile industry.

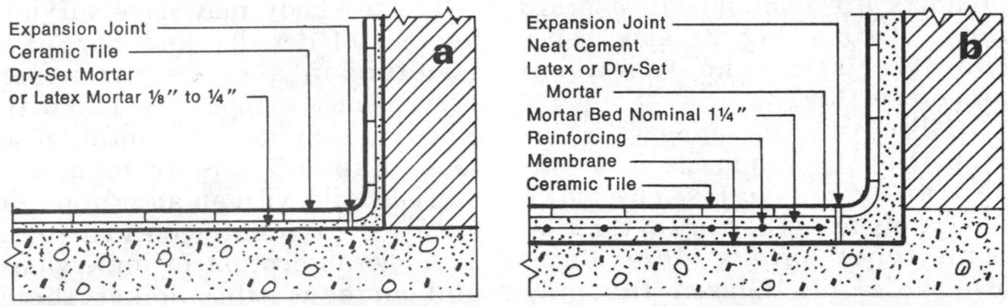

FIG. 1 Basic ceramic tile installation methods: (a) thinset method—tile bonded to the backing; (b) conventional (thickbed) method—tile and mortar setting bed isolated from the backing by membrane.

TERMINOLOGY

The following terms are basic to an understanding of ceramic tile installation:

Adhesive Prepared organic materials, ready for use with no further addition of liquid or powder, which cure or set by evaporation; distinguished from mortars by the absence of silicious fillers (sand) which are included in mortars either at the plant or in the field.

Backing Suitable surface for the application of tile, such as a structural subfloor or wall surface.

Ceramic Tile A thin surfacing unit, made from clay or a mixture of clay and other ceramic materials, having either a glazed or unglazed face, and fired to a temperature sufficiently high to produce specific physical properties and characteristics.

Ceramic Tile Finish (also "tilework") All of the elements normally installed by the tile contractor, from the backing to the face of the tile. This may include just a tilesetting product, tile and grout (Fig. 1a) or a mortar bed, reinforcement and cleavage membrane, as well as the above items (Fig. 1b).

Mortar Bed Installation A thickbed method of installing tile over a 3/4″ to 1-1/4″ portland cement-and-sand mortar bed, using neat portland cement (water-cement paste) as the bond coat (Fig. 1b).

Grout Any formulation used to fill the joints between tiles; may be cementitious, resinous or a combination of both.

Cleavage (Isolation) Membrane A membrane such as saturated roofing felt, building paper, or 4 mil polyethylene film, installed between the backing and the mortar bed to permit independent movement of the tile finish.

Mortar, Leveling or Setting Type A mixture of portland cement, sand, and sometimes lime, used as a bed into which the tile is installed (setting bed), or as a coat to produce a plumb and level surface (leveling coat), so that subsequent coats can be applied in a uniform thickness.

Mortar, Bonding Type Any of a variety of formulations used to bond the tile to the backing or the mortar bed. Formulations may be mainly cementitious, such as commercial cement mortar; resinous, such as epoxy mortar; or a combination of both, as in latex-portland cement mortar. Also see *adhesives.*

Neat Portland Cement Unsanded mixture of portland cement (p.c.) and water used as a bond coat in conventional thickbed installation.

Thickbed Installation See *Installation.*

Thinset Installation A method of bonding tile with a thin layer (1/16″ to 1/4″) of special mortar or adhesive to a suitable backing or to a properly cured mortar bed (Fig. 1a).

MATERIALS

Materials other than tile involved in ceramic tile installation include: 1) *mortars* and *adhesives* for setting tile; 2) *grouts* for filling the joints; and 3) *related materials* to complete the installation, such

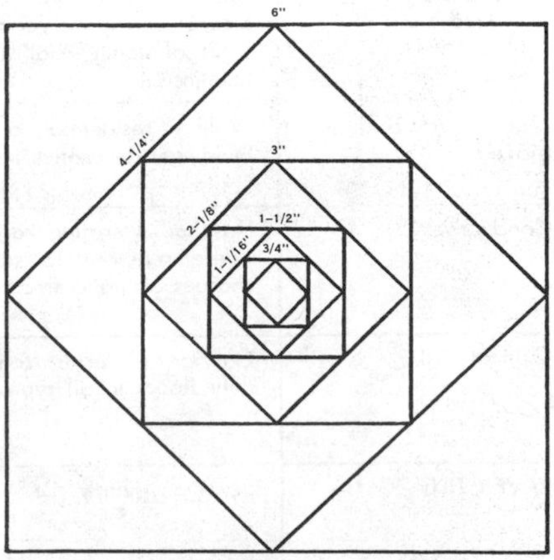

FIG. 2 Derivation of nominal tile sizes of less than 36 sq. in.

as portland cement, reinforcing metal, lath or mesh, cleavage membrane and several others.

Ceramic Tile

Ceramic tile is made from clay or from clay mixed with other ceramic materials; oxides may be included for glaze coloring. Ceramic tile products are available in a broad range of sizes (Fig. 2), appearance characteristics and functional properties (Fig. 3).

In discussing the appearance or properties of ceramic tile it is convenient to discuss the *face* separately from the *body* of the tile. Some types have a glazed face, others do not. Glazed tile may have an opaque decorative glaze over a light-colored body, or a transparent glaze which reveals the natural color of the body. Unglazed tile usually is a hard, dense, unit of homogenous composition throughout, including the face (example—some mosaic tile and all quarry tile).

The tile body may have varying degrees of density and porosity, resulting in a greater or lesser perviousness to moisture. Tile with a body which has an absorption of less than 0.5% is termed *impervious;* tile with an absorption of 0.5% to 3% is identified as *vitreous.* The low absorption of impervious and vitreous tiles makes them suitable for exterior installation subject to freezing and to interior installations subject to continuous immersion in water as in swimming pools. *Semivitreous* tile has an absorption of 3% to 7%; *nonvitreous,* 7% to 18%. Semivitreous and nonvitreous tiles are not suitable for exposure to freezing or immersion in water.

There are only two grades of tile recognized by the Tile Council of America (TCA) and the American National Standards Institute (ANSI): *standard grade* and *seconds.* Standard grade tile is as perfect as it is commercially practicable to manufacture. Individual tiles are harmonious in

FIG. 3 SUMMARY OF CERAMIC TILE PRODUCTS

Tile & Description	Suitable Uses	Limitations	Nominal Sizes
GLAZED WALL TILE			
Non-vitreous body, matte or bright glaze	All interior walls, in residential, commercial and institutional buildings.	Not suitable for floors or areas subject to freezing.	5/16″ thick: 4¼″ x 4¼″, 6″ x 4¼″, 6″ x 6″ Scored, Octagon, Hexagon and other special shapes.
Non-vitreous body crystalline glaze	All interior walls, as well as light duty residential floors and countertops.	Not suitable for areas subject to freezing.	
Vitreous body, matte or bright glaze	Interior walls, countertops, refrigerator and chemical-tank linings and exterior areas subject to freezing.	Not suitable for floors.	
CERAMIC MOSAIC TILE			
Porcelain body, unglazed	All uses, interior or exterior, floors and walls; especially suitable for bathrooms, kitchens and swimming pools.	None except limitations imposed by installation method and grout.	¼″ thick: 1″ x 1″, 2″ x 1″, 2″ x 2″
Porcelain body, glazed	Interior or exterior walls in residential, commercial and institutional buildings; decorative inserts with above tile.	Not suitable for floors.	
Natural clay body, unglazed	Swimming pool runways; floors of porches, entrances and game rooms in the home; floors and walls of commercial, institutional and industrial buildings.	Exterior use requires special frostproof body.	
Natural clay body, glazed	Walls in residential, commercial and institutional buildings; decorative inserts with above tile.	Exterior use requires special frostproof body.	
Conductive	Hospital operating rooms and adjoining areas, where required to safely dissipate dangerous charges of static electricity.	Requires special installation procedures according to NFPA No. 56A.	
QUARRY TILE	Exterior or interior; from moderate to extra heavy duty floors in all types of construction.	Recommended for interior use and exteriors not subject to freezing.	½″ thick: 3″ x 3″, 4″ x 4″, 6″ x 3″, 6″ x 6″, 8″ x 4″ and special shapes.
PAVER TILE	Same as quarry tile.	None	⅜″ thick: 4″ x 4″ ½″ thick: 4″ x 4″, 6″ x 6″, 8″ x 4″

*Sizes shown are those considered typical by the governing industry Standard, ANSI A137.1; many more sizes and shapes are produced by tile manufacturers. Special purpose tiles which combine properties of several of the above types also are produced.

MANUFACTURER'S NAME
AND ADDRESS

GRADE SEAL

STANDARD GRADE

THIS TILE COMPLIES WITH
RECOMMENDED STANDARD SPECIFICATIONS
FOR CERAMIC TILE TCA 137.1-1976

MANUFACTURER'S NAME
AND ADDRESS

GRADE SEAL

SECONDS

THIS TILE COMPLIES WITH
RECOMMENDED STANDARD SPECIFICATIONS
FOR CERAMIC TILE TCA 137.1-1976

FIG. 4 Seal format prescribed by TCA 137.1-1976 for ceramic tile products.

color, although they may vary in shade, and they are free from spots and defects visible from a distance of more than 3′. Tile graded as "seconds" may have minor blemishes and defects not permissible in standard grades, but is free from structural defects or cracks.

TCA 137.1-1976 constitutes the recognized industry standard for the manufacture, testing and labeling of ceramic tile. According to this standard, tile should be shipped in sealed cartons with the grade of contents indicated by grade seals. (Quarry tile is shipped in cartons with the grade printed on the carton in lieu of grade seals.) The seal consists of a strip of paper, not less than 2″ by 4″, applied over the opening of the container in such a way that the container cannot be opened without breaking the seal. Seals have a distinctive coloring: *blue* for standard grade, *yellow* for seconds (Fig. 4).

All ceramic tile used where appearance is an important factor should be standard grade. Additional evidence of compliance with the governing standard, TCA 137.1 is recommended, in the form of a manufacturer's written certification, termed a Master Grade Certificate.

The broad field of ceramic tile products is grouped by TCA 137.1 into four major types: 1) glazed wall tile; 2) mosaic tile; 3) quarry tile; and 4) paving tile. Many other *special purpose* tile products combining the properties of the basic type also are manufactured. Figure 3 summarizes the generally available types, sizes and uses of ceramic tile. A variety of trim shapes are produced, such as caps, corners and bases, in matching colors, both for thinset and conventional installation (Fig. 5).

Glazed Wall Tile This tile generally is 1/4″ to 5/16″ thick, 4-1/4″ x 4-1/4″ or larger; it is available in various sizes, in plain and scored surfaces, in bright, matte and crystalline finishes, in hand-decorated and sculptured designs (Fig. 3). The governing standard defines glazed wall tile as one "not required or expected to withstand impact or be subject to freezing and thawing conditions". Many types of glazed tile are suitable also for flooring, in areas subject to moderate foot traffic, as in most residential and some commercial areas. However, TCA does not cover this type of tile and manufacturer's specifications should be relied on to determine suitability for the intended exposure. Glazed tile suitable for flooring usually has a heavier glaze, with a matte or crystalline finish.

Most wall tile is made with an impervious glaze over a *non-vitreous* (absorbent) body and is

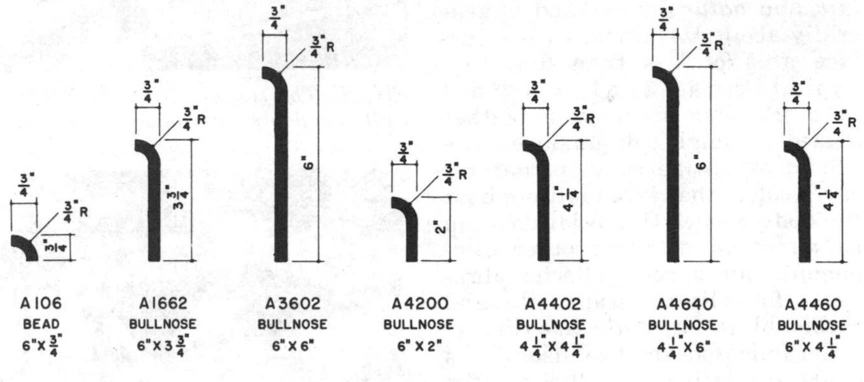

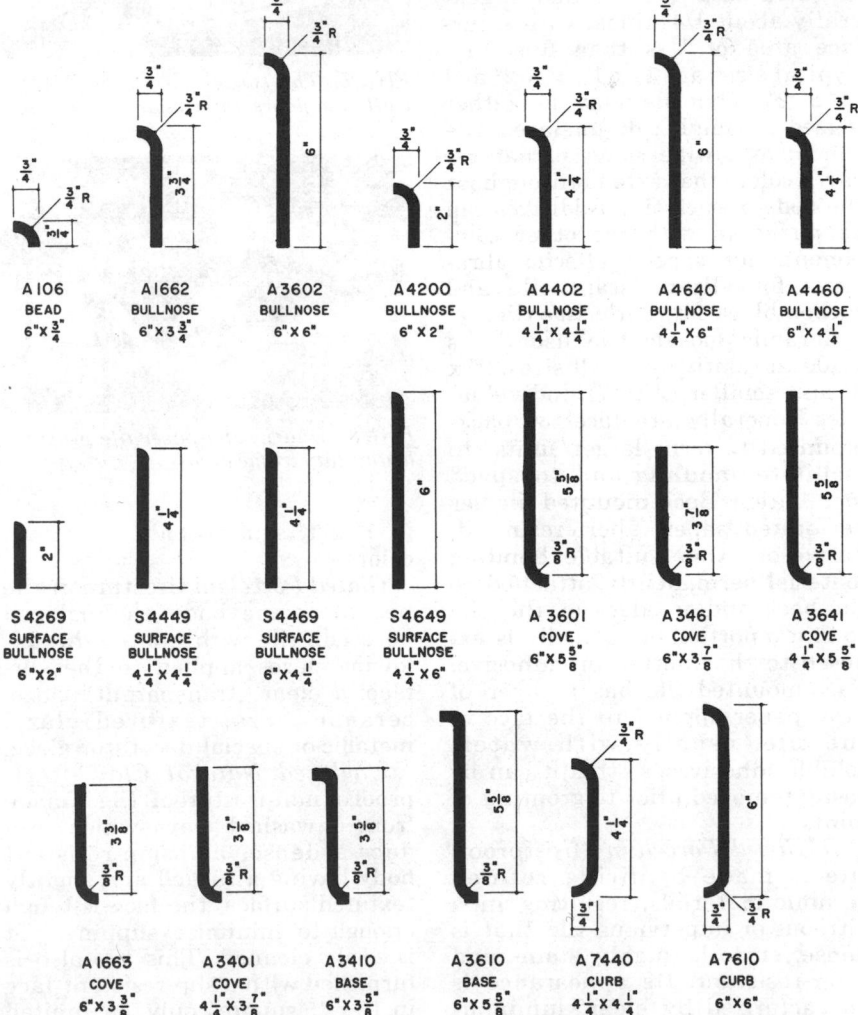

FIG. 5 Standard wall tile trim shapes. Similar bullnose, cove and base shapes are available for other types of tile (see ANSI A137.1 or manufacturer's literature).

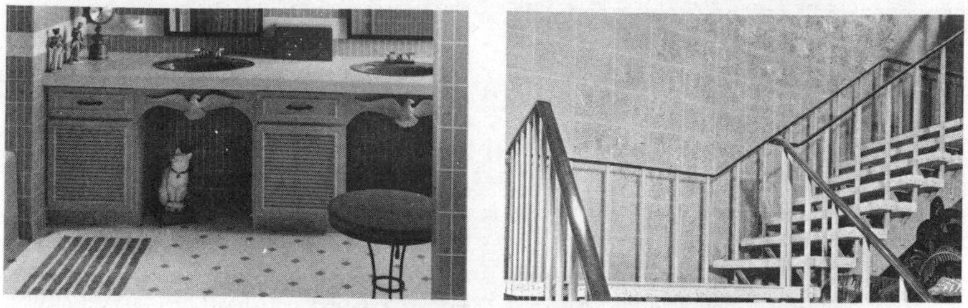

FIG. 6 Glazed tile can be used for floor and countertops as well as walls.

FIG. 7 The range of ceramic mosaic tile uses is virtually limitless: decorative walls and floors, indoors and outdoors.

FIG. 8 Quarry and paver tile provide attractive low-maintenance floor finishes in heavy-duty traffic areas.

intended for interior use. A small amount of this type of tile also is made with an *impervious* or *vitreous* body suitable for more demanding exposure to moisture and/or freezing. All wall tile is glazed, the majority with a *bright* (shiny) or *matte* (dull) finish. Some tile is made with other special glazes which are more appropriate for light residential traffic than bright or matte glazes.

Most wall tile is made with a cushioned-edge profile, but square-edge tile also is available. Soap dishes, paper holders, towel bar posts and other accessories for bathroom uses are available in matching colors.

Ceramic Mosaic Tile This tile is available in two body types—*porcelain* and *natural clay*—and is generally about 1/4″ thick, with a surface area of less than 6 sq. in. Typical sizes are 1″ x 1″, 1″ x 2″ and 2″ x 2″. The tile may be either glazed or unglazed; unglazed tile can have a wide range of natural earth colors that extend throughout the body of each tile. Additives can be mixed in with the other components for special effects: abrasives for slip-resistant tile and carbon black for *conductive* tile.

Ceramic mosaic tile usually is made in relatively small sizes, 2″ x 2″ and smaller (Fig. 7). Individual tiles generally are face- or back-mounted to form larger units, to facilitate handling and to speed installation. Back-mounted tile has perforated paper, fiber, resin adhesive or other suitable bonding material permanently attached to the back and/or edges of the tile so that a portion of each tile is exposed to the mortar or adhesive. Face-mounted tile has a layer of kraft paper applied to the face of the tile, usually with water-soluble adhesives, so that it can be easily removed prior to grouting of joints.

Unglazed Porcelain (frostproof) tile is made by firing refined ceramic materials, resulting in a vitreous or impervious tile that is dense, smooth, highly stain- and wear-resistant. Its appearance is characterized by clear, luminous colors or granular blends. It is also available with an abrasive surface in 1″ x 1″ squares only, in certain colors.

Glazed Porcelain (frostproof) tile has the same body as unglazed porcelain tile with one of the following glazes applied to the surface: a clear (transparent) glaze, ceramic glaze, textured glaze, metallic or special decorator glaze.

Unglazed Natural Clay (frostproof or non-frostproof) tile is made from unwashed clays which produce a dense, abrasion-resistant body having a rugged and slightly textured surface; the face is tough enough to minimize slipping, yet is easily cleaned. This tile also is furnished with a slip-resistant face in 1″ x 1″ squares only, in limited colors.

Glazed Natural Clay (frostproof or non-frostproof) has the same body as unglazed natural clay tile with one of the following glazes applied to the surface: a clear, ceramic, metallic or decorator glaze.

Conductive Tile is an impervious ceramic tile with additives which minimize electrostatic build-up and explosion hazard in rooms where explosive gas mixtures are used—hospital rooms and laboratories. Both the tile and the installation should comply with the National Fire Protection Association Bulletin No. 56A.

Quarry Tile This is an unglazed tile, with a surface area of 9 sq. in. or more, 1/2″ thick, made by the extrusion process from natural clay or shale (Fig. 8). Typical tile sizes are 3" x 6", 6" x 6", 8" x 4", 9" x 6", 9" x 9", in hexagonal and other special shapes. Tile thicker than 3/4", made of the same materials by the same process, is termed *packing house tile*.

Paver Tile As the name implies, pavers are used primarily for floors and are similar in appearance to quarry tile. They are similar to mosaic tile in composition and physical properties, but are relatively thicker (1/2" typical) and larger (6 sq. in. or more). Typical tile sizes are 4" x 6", 6" x 6" and 8" x 4" (Fig. 8).

Tilesetting Products

Commercially available tilesetting products possess a variety of special properties, each intended to make them suitable for setting tile over certain backings or under given conditions. Figure 9 summarizes the several common tilesetting products and the governing industry standards.

Neat portland cement (cement-water paste) can be used as a bond coat for tile only over a portland cement mortar setting bed, while the cement is still plastic. Most thinset tilesetting materials, such as dry-set and latex-portland cement mortar, also can be used to bond ceramic tile to a mortar bed after it has cured for a minimum of 24 hours. While the mortar bed is still plastic, dry-set mortar can be used as 1/16" thick bond coat. When organic adhesives are used, the mortar setting bed must be fully cured and dry. To insure a satisfactory installation in the proper plane, the mortar bed—whether left plastic or allowed to harden—preferrably should be installed by the tilesetting contractor.

Resinous tilesetting materials, such as epoxies and furans offer greater bonding strength and chemical resistance than cementitious mortars. However, special tilesetting skills are required, and both labor and materials are likely to be appreciably costlier than with cementitious mortars.

Portland Cement Mortar This is a mixture of portland cement, sand, water and sometimes hydrated lime. The proportions vary depending on whether the mortar is intended for a leveling bed or a setting bed, whether it is to be used on a vertical or horizontal surface. The neat cement coat with which tile is applied is not considered a mortar, but rather a *bond coat.*

Portland cement mortar is the traditional material for installing ceramic tile. It is suitable for most surfaces and ordinary types of installation. The relatively thick bed, 3/4" to 1" on walls and 3/4" to 1-1/4" on floors, facilitates accurate slopes and true planes in the tile finish.

FIG. 9 SUMMARY OF TILESETTING PRODUCTS

Type	Description	Industry Standards: Product	Industry Standards: Installation	Limitations
PORTLAND CEMENT MORTAR	Field mix of p.c., sand and water for setting bed; mix p.c. and water for bond coat. Interior or exterior.	ASTM C-150 (P.C.)	ANSI A108.1	Requires 1¼" setting bed, resulting in additional thickness and weight. Requires pre-soaking of most glazed wall tile.
DRY-SET MORTAR[1]	Prepared mix[2] of p.c., sand and certain additives which impart water retentivity and eliminate moist-curing. High resistance to moisture, impact and freezing. Interior and exterior.	ANSI A118.1	ANSI A108.5	Three days of moist curing before grouting recommended for heavy duty floors.
LATEX-PORTLAND CEMENT MORTAR	Factory mix of p.c. and sand to which liquid latex resins are added in the field to impart greater bonding strength and resilience. Interior or exterior.	ANSI A118.4	ANSI A108.5	Not suitable for wood, metal, hardboard or particleboard backings.
EPOXY MORTAR	Two-part system of 1) epoxy resin and finely graded sand and 2) hardener. High bond strength and chemical resistance. Interior or exterior.	ANSI A118.3	ANSI A108.6	Requires special installation skills. Material and labor costs relatively high. Not suitable for walls.
EPOXY ADHESIVE	Two-part system of 1) epoxy resin and 2) hardener. Good bond strength and chemical resistance. Interior or exterior. A three-part system containing portland cement powder is also available.	NONE	NONE	Requires special skills. Installation cost high, but lower than epoxy mortar.
FURAN MORTAR	Two-part system of 1) furan resin and 2) hardener. Highest chemical resistance and installation cost. Interior or exterior.	NONE	NONE	Same as epoxy mortar.
ORGANIC (MASTIC) ADHESIVE	Prepared organic material, ready-to-use without addition of liquid or powder. Adequate bond strength and resilience for light duty on floors and most walls. Interior use only.	ANSI A136.1	ANSI A108.4	Not suitable for exterior use or interior use in continuous contact with water. Applied 1/16" thick; requires smooth backing.

1. Conductive dry-set mortar—product specifications, ANSI A118.2; installation specifications, ANSI A108.7.
2. Requires addition of water in the field.

A portland cement mortar bed should be reinforced with metal lath or mesh, whenever it is backed with a cleavage membrane. Suitable backings for mortar bed installations are: brick or block masonry, monolithic concrete, concrete/glass-fiber reinforced backerboard, wood boards, plywood, foam insulation board, gypsum wallboard and gypsum plaster; on walls, open stud framing or furring can be used, always in conjunction with metal lath.

Dry-set Mortar This is a mixture of portland cement, sand and resinous additives which impart water retentivity. Improved retentivity prevents rapid loss of water to the backing surface and improves bond.

Dry-set mortar is suitable for use over a variety of surfaces. It has excellent impact resistance, is nonflammable, is suitable for exterior work and does not require presoaking of absorptive tile, as required with most cementitious mortars. For these reasons it is the preferred mortar for absorptive glazed tile.

Dry-set mortar is available either as an unsanded mortar or as a factory-sanded mortar. Presanded mortars, to which only water needs to be added, are strongly recommended. Dry-set mortar is not affected by prolonged contact with water and does not form a water barrier. It is limited in truing or leveling the work of other trades to about 1/4″ thickness, but can be used in a layer as thin as 1/16″, when the mortar bed is still plastic.

Suitable backings include brick and concrete masonry, monolithic concrete, concrete/glass-fiber reinforced backerboard, cut-cell expanded polystyrene or rigid closed-cell urethane insulation board, gypsum wallboard, cured p.c. mortar, ceramic tile and marble.

Latex-Portland Cement Mortar This is a mixture of portland cement, sand and a special latex additive, which imparts greater flexibility to the mortar. The uses of latex-p.c. mortar are similar to those of dry-set mortar; it is applied over a cured mortar bed in a coat 1/8″ to 1/4″ thick. Since latex formulations vary, manufacturer's directions should be followed.

Epoxy Mortar Epoxy mortar is available in a two part mortar consisting of an epoxy resin and a hardener, or a less expensive three part system consisting of an epoxy resin, a hardener and portland cement powder. After the parts are mixed, a chemical reaction sets in and the mortar must be used within a limited time (*working time*). Epoxy mortar is suitable for use where chemical resistance or high bond strength is an important consideration. Acceptable backings include concrete, plywood, steel plate and ceramic tile. The mortar is applied in a layer 1/16″ to 1/8″ thick. Working time, bond strength, water-cleanability before cure and chemical resistance vary with manufacturer. Epoxy mortar frequently is used in combination with epoxy grout, resulting in a highly durable water- and chemically-resistant installation.

Epoxy Adhesive Like epoxy mortar, this too is a two-part system consisting of an epoxy resin and hardener. However, a different formulation makes this product somewhat less costly than epoxy mortar, as well as less chemically-resistant. It is intended for thinset application of tile on walls, floors and counters, where high bond strength and ease of applications are the overriding considerations. Although chemical resistance is lower than for epoxy mortars, it is substantially better than that of organic adhesives.

Furan Mortar This two-part mortar system, consisting of a furan resin and hardener, is suitable where chemical resistance is the most critical consideration. It is used primarily on floors, in laboratories and industrial plants. Acceptable subfloors include concrete, steel plate and ceramic tile. Furan mortar typically is used in combination with furan grout.

Organic Adhesive (Mastic) This is a prepared organic material, ready-to-use with no further addition of liquid or powder, which hardens by evaporation. Organic adhesives are used in a layer 1/16″ thick and are suitable for floors, walls and counters, where surfaces are sufficiently true and properly prepared in accordance with adhesive manufacturer's directions. Suitable backings include monolithic concrete, gypsum wallboard, a p.c. mortar bed, gypsum plaster, cement-asbestos board, brick, ceramic tile, marble, and plywood.

Organic adhesives eliminate presoaking of tile and are suitable for most residential and commercial uses. They can be used for the walls of residential tub and shower areas, but they are not suitable for high moisture areas such as swimming pools or for exterior installations. Their chief advantage is low cost and a flexible bond. Bond strength varies greatly among numerous brands available. Solvents in some adhesives may be irritating to the skin and sometimes are flammable.

Grouts

Grouting materials for ceramic tile are available in many forms to meet the requirements of the different kinds of tile and types of exposures. Cementitious grouts consist of portland cement modified to provide specific qualities such as whiteness, uniformity, hardness, flexibility and water retentivity. Resinous grouts (epoxies and furans) possess special properties such as high bond strength and chemical resistance, but are harder and costlier to apply. Figure 10 summarizes commonly available tile grouts and governing standards, where they exist.

Commercial Cement Grout This is a mixture of portland cement and other ingredients, resulting in a water-resistant, uniformly colored material. Commercial cement grout is the most commonly specified grout for tile *walls* subject to ordinary use. Soaking of non-vitreous wall tile is required to prevent loss of moisture to the absorptive tile body. The moisture in the grout is essential to promote good bond and develop full grout strength. To assure an adequate supply of moisture for the hardening process (hydration), damp curing of the grout is necessary. Commercial cement grout usually is white, but colored grouts also are available.

Sand-Portland Cement Grout This is an on-the-job mixture of normal portland cement, fine graded sand and sometimes hydrated lime. It is used for ceramic mosaic tile, with quarry tile and paver tile. *Mixes should be 1 part portland cement to 1 part sand for joints up to 1/8" wide; 1 to 2 for joints up to 1/2" wide; and 1 to 3 for joints over 1/2" wide. Up to* 1/5 part lime may be added for improved workability. Pre-soaking of non-vitreous wall tile is necessary. Damp-curing of the grouted surface is required for the same reasons as in commercial cement grout installations.

Dry-set Grout This is a prepared mixture of portland cement and sand, with additives to improve water retentivity. Dry-set grout has the same properties as dry-set mortar and is suitable for grouting walls subject to ordinary use. This grout eliminates soaking of tile, although dampening sometimes is desirable under very dry conditions.

Latex Grout This is a mixture of one of the three preceding products, consisting mainly of portland cement with a latex additive to increase stain resistance and resilience of the joints. Latex grout is suitable for all installations subject to ordinary use. It is less rigid and less water-permeable than sand-p.c. grout, allowing for more movement and greater exposure to moisture. This grout therefore is particularly suitable in tub and shower areas.

Silicone Rubber Grout This is an engineered elastomeric grout system for interior use employing a single component non-slumping silicone rubber which, upon curing, is resistant to staining, moisture, mildew, cracking, crazing, and shrinking. It adheres tenaciously to ceramic tile, cures rapidly, withstands exposure to moisture as well as sub-freezing temperatures and hot, humid conditions. It should not be used on kitchen counter tops or other food preparation surfaces, since status under FDA Regulations has not been determined.

Mastic Grout This is a prepared one-part formulation which can be used directly from the container. Mastic grout is designed for use with glazed non-vitreous wall tile and hardens by a chemical process which does not require damp-curing. It is more flexible and stain-resistant than regular sand-cement grout and is available in a variety of colors to complement the tile color.

Epoxy Grout Like epoxy mortar, this is a two-part system consisting of an epoxy resin and a hardener. It is made in several formulations, each intended for a

FIG. 10 **SUMMARY OF TILE GROUTING PRODUCTS**[1]

Type	Description	Limitations
COMMERCIAL PORTLAND CEMENT GROUT	Prepared mix[2] of p.c. and other ingredients resulting in a water-resistant, uniformly colored material; usually white, but available in colors. Commonly used for tile walls and floors subject to ordinary use. Suitable for interior or exterior.	Requires pre-soaking of most glazed wall tile and damp curing of grouted surface.
SAND-PORTLAND CEMENT GROUT	Field mixture of p.c., fine graded sand (sometimes lime), and water. Most commonly used grout for ceramic mosaic tile. Suitable for floors and walls, interior or exterior.	Same as for commercial cement grout.
DRY-SET GROUT	Prepared mix[2] of p.c., sand and additives which provide water retentivity & eliminate need for pre-soaking of glazed wall tile. Damp curing may develop greater strength in portland cement grouts. Generally used on walls, interior or exterior.	Dampening of tile surface prior to grouting may be required under extremely dry conditions.
LATEX-P.C. GROUT	Prepared mix[2] of any of the above formulations, plus latex additives. It is less rigid and less permeable than regular cement grout. The addition of latex to portland cement grouts eliminates the necessity of damp curing for 72 hours. Can be used on walls or floors, interior or exterior.	Same as for dry-set grout.
SILICONE RUBBER GROUT	Ready-to-use one-part caulking-type material which cures from moisture vapor in the air. Used mainly for glazed tile walls. Flexible, mildew-resistant, stain-resistant.	Bonds only to clean, dry surfaces. Must be used with proper adhesive to prevent staining.
MASTIC GROUT	Ready-to-use one-part formulation which does not depend on availabiity of water for bond or curing. Suitable for most interior walls and light duty floors. More flexible & stain-resistant than cementitious grouts; contains agent to eliminate mildew growth.	Not suitable for exterior use or for surfaces in continuous contact with water.
EPOXY GROUT	Ready-to-use two-part system consisting of epoxy resin and hardener. Formulation for wall tile omits coarse fillers which might scratch glaze; coarse silica fillers generally are included in grout intended for quarry tile and pavers. Highly resistant to chemical staining, hence used for industrial and kitchen floors, countertops.	Requires special skills for proper application. Material cost and labor higher than with other grout systems.
FURAN RESIN GROUT	Ready-to-use two-part system consisting of furan resin and hardener. Highest chemical resistance; used primarily in industrial installations with quarry tile and pavers.	Same as for epoxy grout.

1. No national industry standards have been established for most tile grouting products—except ANSI A118.3 for epoxy grout.
2. Requires addition of water in the field.

specific ceramic tile. AAR-II is the designation of an acid- and alkali-resistant, two-part formulation developed by the Tile Council of America and manufactured under license from that organization. It is highly stain-resistant and impervious, and is used typically in conjunction with epoxy mortar or epoxy adhesive. Its use requires special skills and usually is more expensive.

For glazed wall tile and mosaic tile, epoxy grout is formulated without coarse fillers which might scratch the glaze. It can be used on walls, floors and counters subject to food staining.

For quarry tile and pavers, epoxy grout often contains a coarse silica filler. This grout is especially formulated for industrial and commercial installations where chemical resistance is of paramount importance. This grout has excellent bonding characteristics.

Furan Resin Grout This is a two-part grout consisting of a furan resin and hardener. It is intended for quarry tile and pavers, mainly in industrial areas requiring a maximum of chemical-resistance. It is generally used in conjunction with furan mortar.

Related Materials and Products

Installation of ceramic tile involves a variety of related materials and products, described below and summarized in Figure 11.

Portland Cement Normal (Type I) portland cement typically is used for mortar setting beds, for leveling beds, for the neat cement bond coat and for field-mixed cement grout. When a white grout is intended, it is recommended that *white* portland cement be used in the mortar bed as well.

Sand Sand should be clean, free from impurities and well-graded. When *fine* sand is specified, it should pass through a 16-mesh screen.

Hydrated Lime This material is used in the portland cement-sand mortar mix to make it more plastic for application on walls, and sometimes on floors. Type S (special) lime should be specified, to assure a dimensionally stable mortar bed.

Sealants Sealants for use in expansion joints and control joints should be suitable for the exposure expected. Conditions which deteriorate sealants are extreme temperature variations, direct sunlight, repeated stretching and compression, frequent wetting and drying or immersion in water. The sealant can be a single-component or a two-component type. Color of sealant should be selected to match or blend with the tile or adjacent materials. When used on floors, it must be hard enough to withstand occasional or repeated foot traffic. It should have a *Shore A* hardness index of not less than 25 for joints subject to occasional foot traffic; in high traffic areas the hardness index should be not less than 35.

Metal Reinforcement Reinforcing for a mortar bed typically consists of metal lath on walls and welded wire fabric on floors. Welded wire fabric should be one of the following: 2″ x 2″ mesh—16/16 wire; 3″ x 3″ mesh—13/13 wire; or 1-1/2″ x 2″ mesh—16/13 wire. Metal lath should be expanded lath weighing not less than 2.5 lb. per sq. yd., or sheet lath weighing not less than 4.5 lb. per sq. yd.

Cleavage Membrane The membrane should be one of the following: 1) 15 lb. asphalt (or 13 lb. coal-tar) saturated roofing felt; 2) reinforced waterproof duplex asphalt paper; 3) polyethylene film at least 4 mils thick; or 4) troweled-on membrane waterproofing layer.

FIG. 11 SUMMARY OF RELATED MATERIALS & PRODUCTS

Material or Product	Industry Standard
PORTLAND CEMENT	
Grey or White, Type 1	ASTM C-150
SAND	
for Concrete	ASTM C-33
for Mortar	ASTM C-144
HYDRATED LIME	
Type S	ASTM C-206 ASTM C-207
SEALANTS	
Two-part	TT-S-00227E
One-part	TT-S-00230C
Silicone Rubber	TT-S-001543A
METAL REINFORCEMENT	
Welded Wire Fabric	ANSI A-42.3
Expanded Metal Lath	ANSI A-42.4
CLEAVAGE MEMBRANE	
Polyethylene Film	ASTM C-171
Asphalt Roofing Felt	ASTM D-226
Coal Tar Roofing Felt	ASTM D-227
Reinforced Asphalt Paper	FS UU-P264

JOB CONDITIONS

Materials should be delivered in the manufacturer's original sealed containers and should be stored under cover to prevent damage or contamination. Tile generally should be set and grouted when the temperature is at least 50°F (10°C) and rising. The mortar manufacturer's recommendations should be checked to determine if a higher temperature is required or when a lower temperature is permissible. Backing surfaces and supporting construction should be inspected to determine their suitability prior to starting any work.

SUPPORTING CONSTRUCTION

A tile installation is only as good as the supporting construction—that is, the assembly of structural members and backing (subfloor or wall surface.) If the supporting construction deflects, expands or contracts, these movements will tend to be passed on to the tile finish.

This is particularly true in bonded installations. Except in small areas, a bonded tile finish cannot absorb these movements and the internal stresses are likely to result in cracks. It is essential, therefore, to provide supporting construction with a minimum amount of movement; where movement is anticipated, expansion and/or control joints should be provided in the supporting construction and extended into the tile.

The problem of movement is less critical with mortar bed installations isolated from the backing by a cleavage membrane. Although the membrane permits some independent movement in the supporting construction, the movement which can be tolerated still is limited, and dimensionally stable, rigid construction is desirable.

The backing is that part of the supporting construction to which the tile is bonded: it can be a subfloor, wall finish or mortar leveling coat. The backing must likewise be

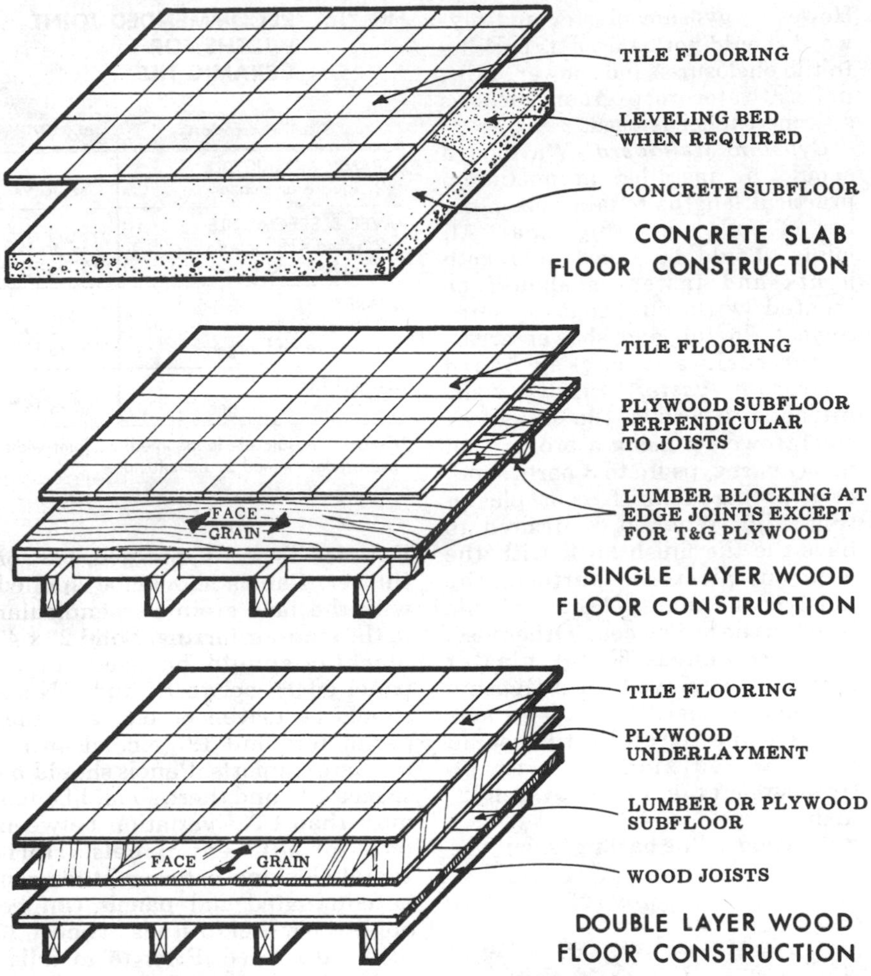

FIG. 12 Typical supporting construction for ceramic tile floors.

rigid and dimensionally stable. In thinset installations, it is important that the backing be sound, true to line, plumb, and sloped or level, as designed. Also, the backing should be free of any foreign substances or loose materials which might weaken the bond with the tile finish. When a mortar bed is used, trueness of the backing is less critical, but where slopes are intended, they should not be developed by varying the mortar bed thickness; slopes should be built into the backing.

Floors

Floors should be engineered so that the maximum deflection under full load does not exceed 1/360 of the span. {For instance, on a span of 15′ (180″), the deflection should be not more than 180/360, or 1/2″}. The most common floor assemblies encountered in residential buildings involve a concrete or wood subfloor. Additional information on suitable preparation for various floor surfaces and for specific tilesetting methods is given in the following subsections.

Concrete Subfloor Tile can be installed on a concrete subfloor either by the thinset or the portland cement mortar bed method (Fig. 12a). For thinset bonding directly to the floor slab, the surface should be steel troweled and fine broomed, with no variations in the plane or slope exceeding 1/8″ in 10′. For portland cement installations a variation of 1/4″ in 10′ is acceptable. Concrete slabs should be thoroughly moist-cured before tile application; liquid curing compounds, or other coatings which may reduce bond should not be used. Slabs should not be saturated with water at the time of installation. To prevent cracking, the slab should be properly reinforced, concrete quality should be controlled, and suitable joints should be provided in the slab.

Wood Subfloor Regardless whether a portland cement mortar bed or thinset method is intended, supporting joists should be spaced not over 16″ o.c. For thickbed installations, a single-layer subfloor or 1″ nominal boards or 5/8″ plywood is acceptable (Fig. 12b). For thinset installations with most tilesetting products (except epoxy mortar) a double floor is required, consisting of a subfloor as described above and an underlayment (Fig. 12c). The underlayment should be Exterior type CC-plugged plywood, 1/2″ thick for residential use; 5/8″ thick for more demanding installations, as in commercial buildings. When epoxy mortar is used in resi-

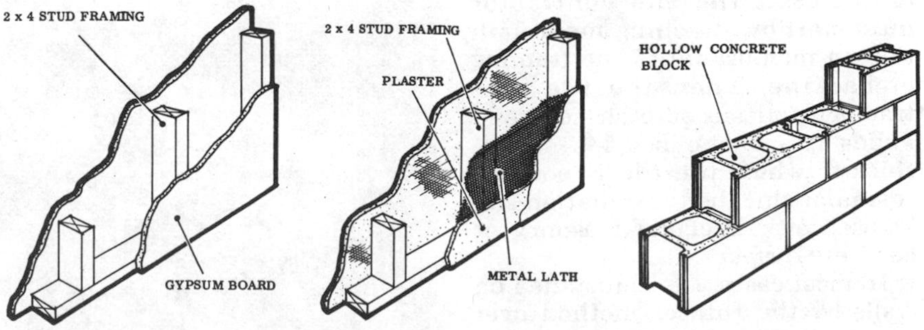

FIG. 13 Typical supporting construction for ceramic tile wall finishes. In areas subject to wetting, a p.c. mortar bed should be used in place of gypsum plaster. Concrete/glass-fiber reinforced backerboard can also be used where gypsum board is indicated here.

dential construction, the underlayment can be omitted if a 5/8″ plywood subfloor is used and longitudinal joints between panels are supported with solid wood blocking 2″ x 4″ or larger.

A 1/4″ space should be left between the underlayment and all vertical surfaces such as walls; or projections such as pipes or posts. For organic adhesives, 1/16″ should be left between underlayment ends and edges; for epoxy mortar, 1/4″ should be left at ends and edges of panels and these joints should be filled with epoxy. Adjacent underlayment edges should not be more than 1/32″ above or below each other.

Concrete/glass fiber-reinforced backer board may be used in lieu of a portland cement mortar bed over a 1/2″ thick plywood subfloor. This backer board is designed for use with ceramic tile floors and is available in various sizes and approximately 1/2″ (7/16″) thick. It can be nailed or screwed to wood joists through 1/2″ plywood subflooring. Ceramic tile can be bonded to it with either dry-set or latex-portland cement mortars. Adhesives are not recommended for use with this product.

Walls

Wall surfaces should be sound, plumb and true, with square corners. For thinset application the solid backing usually is installed by others. The backing surface should not vary from a true plane by more than 1/8″ in 8′. For portland cement mortar bed installations, either a solid backing or open framing is suitable (Fig. 13). In either case, the tile contractor must start by attaching metal lath (over a membrane) to the framing or backing. Then the tile contractor applies a scratch coat and builds up a mortar bed 1/4″ to 3/4″ thick to which the tile is bonded. A similar thickbed installation can be used over irregular masonry or concrete surfaces.

In most cases, tile is installed on walls by the thinset method over gypsum wallboard, gypsum plaster, plywood, concrete/glass fiber-reinforced backing board or other dimensionally stable solid backing. However, gypsum plaster and plywood should not be used as backing in tub enclosures and shower stalls, unless waterproof grout such as silicone rubber is used.

Gypsum Wallboard Wallboard should be installed in maximum practical lengths to minimize number of end joints (Fig. 13a). All joints should be taped, and both joints and fasteners should be treated with one coat of joint cement. In tub and shower areas, water-resistant backer board

Gypsum Plaster Plaster walls intended to receive tile should be steel troweled, using a brown coat mix (1 part gypsum to 3 parts sand) or a prepared finish coat plaster (Fig. 13b). When it is desired to have the tile finish flush with the adjacent plastered surface, the finish coat is omitted and the tile is set on the brown coat. Otherwise, a full thickness 3-coat plaster surface is employed. (See Gypsum Products, page 218-11). The surface which is to receive tile should be troweled smooth with no trowel marks or ridges over 1/32″ high.

Plywood The backing should be Exterior Type CC-plugged, 3/8″ or thicker. Panels should be applied with the face grain perpendicular to the studs or furring. Solid 2″ x 4″ blocking should be used under panel edges between studs. Nails should be driven 6″ o.c. at panel perimeters and 12″ o.c. at intermediate supports. Panels should be spaced 1/8″ and there should be not more than 1/32″ variation between faces of adjacent panels. Horizontal blocking between studs can be eliminated and panels can be applied parallel to studs, when 1/2″ plywood is used. Prior to installation, all plywood edges should be sealed with an exterior primer or aluminum paint.

Concrete/Glass Fiber-reinforced Backing Board. A backer board designed for use with ceramic tile in wet areas around tubs and showers and on floors. Available in various sizes and approximately 1/2″ thick, this material can be nailed or screwed in place over wood or metal studs, or over a 1/2″ thick plywood subfloor. Ceramic tile can be bonded to it with either dry-set or latex-portland cement mortars. It can be used in place of metal lath, portland cement scratch coat and mortar bed.

Masonry & Cement Suitable backings include monolithic concrete, concrete/glass fiber-reinforced backing board, brick masonry and concrete block masonry (Fig. 13b). For thinset installations, the wall surface should be clean, sound, free of paint, efflorescence,

FIG. 14 Tile should be laid so that it is symmetrical with respect to accessories and end tiles are larger than half size.

FIG. 15 **RECOMMENDED JOINT WIDTHS FOR CERAMIC TILE**

Tile Description	Joint Width
CERAMIC MOSAIC TILE* 2″ square or smaller	1/32″ to 1/8″
PAVER & SPECIAL TILE 2 3/16″ to 4 1/4″ square 6″ square and larger	 1/16″ to 1/4″ 1/4″ to 1/2″
QUARRY TILE 3″ to 6″ square 6″ square and larger	 1/8″ to 3/8″ 1/4″ to 1/2″
GLAZED TILE 3″ square and larger	1/8″ to 1/2″

*Ceramic mosaic tile is mounted, so joint widths are predetermined by manufacturer.

cracked or chipped areas. When latex-p.c. or dry-set mortars are to be used, smooth concrete or brick surfaces should be roughened by sandblasting or bush-hammering to assure good bond.

For mortar bed installations, the surface qualities of the backing are not as critical. If a leveling bed is to be used, the wall surface can be more irregular than if the tile is bonded directly to the wall, but it should be free of surface coatings and loose material which might affect the bond. When a cleavage membrane is used, surface contamination or irregularity is not important, but the wall should be structurally sound and dimensionally stable.

TILE LAYOUT AND SETTING

The area to be covered with tile should be checked for squareness and midlines should be established in both directions. The tile pattern should be laid out so that a tile is either butted to the midline or bisected by it. The most desirable method is the one which produces larger than half-tiles around the perimeter of the area.

Tiles should be cut with a suitable tool, rather than split. Wall tile usually is cut with a tile cutter; mosaics are cut with "nippers"; quarry tile may be cut with a water-cooled masonry saw or heavy-duty cutter. Tiles should be closely fitted to projections, smoothing all ragged edges with a carborundum stone. All joint lines should be straight, plumb, level and of even width. Accessories in the tilework should be evenly spaced, properly aligned with tile joints and set in the proper plane (Fig. 14). When the tile is not of the self-spacing type, joint widths should be approximately as shown in Figure 15. Special setting frames or grids are becoming popular for setting this type of tile more quickly and accurately.

The tile usually is set into a previously spread layer of mortar, adhesive or neat cement (Fig. 16). Sometimes it is more convenient to brush a coat of the tilesetting material on the back of the tile. Alignment of joints should be checked frequently for trueness and squareness (Fig. 17). After the tile is set, the mesh or paper of face-mounted tile should be soaked off and removed as recommended by the manufacturer. Mortar or adhesive also should be removed from the face of the tile with a recommended solvent. If strings or ropes were used to space tile, they should be removed as soon as the mortar or adhesive has set sufficiently to hold the tile firmly in place.

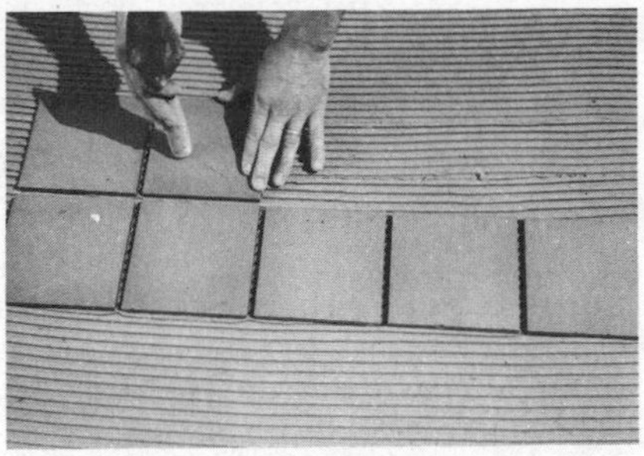

FIG. 16 Quarry tile is laid in a bond coat of dry-set mortar and tapped in place with the trowel handle.

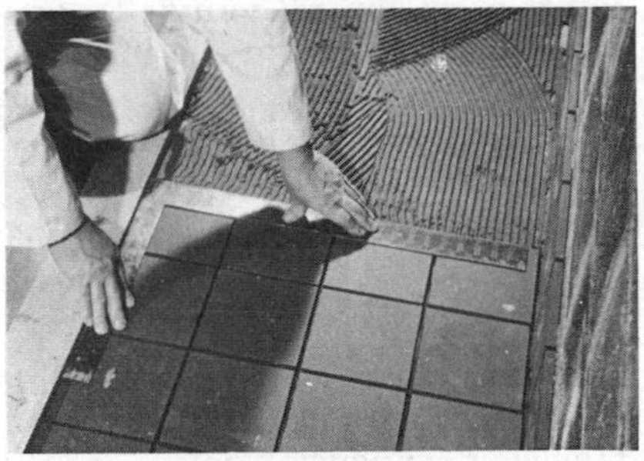

FIG. 17 Quarry tile is spaced to provide 1/4" to 1/2" wide joints. A straightedge and metal square are used periodically to check joint alignment and squareness.

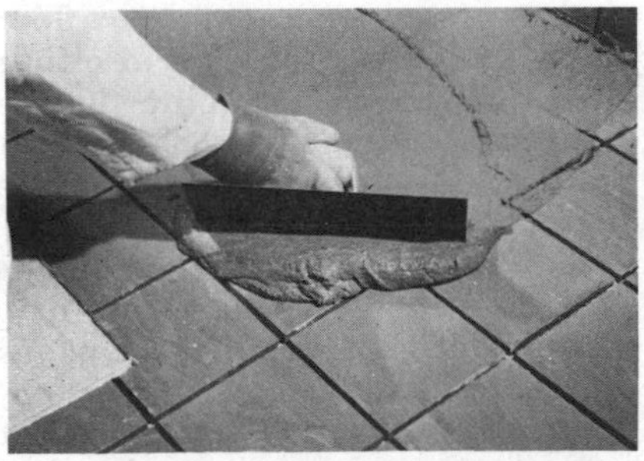

FIG. 18 Joints are filled with p.c. grout by troweling diagonally across the joints. Kneeling boards prevent damage to newly-laid tile.

FIG. 19 Excess grout is wiped from the tile surface with burlap or rough-textured cloth.

GROUTING AND CURING

Grout should be applied to the face of the tile by troweling diagonally across the joints (Fig. 18). Enough grout should be forced into each joint, so that the joint is filled down to the mortar. Joints of square-edged tile should be filled flush with the tile surface; joints of cushion-edged tile should be tooled to the depth of the cushion. The finished grout surface should be uniform in color, smooth and free of pinholes or low spots. The tile surface should be washed and sponged thoroughly, then polished with a dry clean cloth or burlap (Fig. 19).

Glazed wall tile usually requires soaking prior to setting with p.c. mortar. If glazed tile has been set with a material other than p.c. mortar and was not presoaked prior to installation, the tile should be soaked after setting, if a p.c. grout is to be used. Soaking is not required for other types of grout.

Commercial cement and p.c. grout always should be damp-cured for 72 hours. Drywall grout need not be damp-cured, except when used: 1) on floors; 2) on either floors or walls exposed to periodic wetting; or 3) in exterior locations. Damp-curing can be accomplished by periodically applying a spray mist to the tile surface, or by covering with polyethylene film after initial spraying.

CONTROL AND EXPANSION JOINTS

To insure an attractive crack-free installation, suitable joints should be provided in the backing and/or in the tile finish.

Joints in Backing

Control joints are provided in cementitious materials such as concrete and stucco in order to localize shrinkage cracks due to loss of moisture by hydration and evaporation. When control joints are provided, expansion joints usually are not necessary because shrinkage exceeds the likely thermal expansion (see 203 Concrete and 425 Stucco).

In brick masonry shrinkage is minimal, hence thermal movement is critical and expansion joints must be provided (see 343 Masonry Walls). Where differential settlement or movement of adjacent elements is expected, isolation joints should be used. Expansion joints frequently act also as isolation joints or as control joints. To conform to usage in the tile industry, all joints *in the tile finish* will be referred to as "expansion joints" in the following discussion.

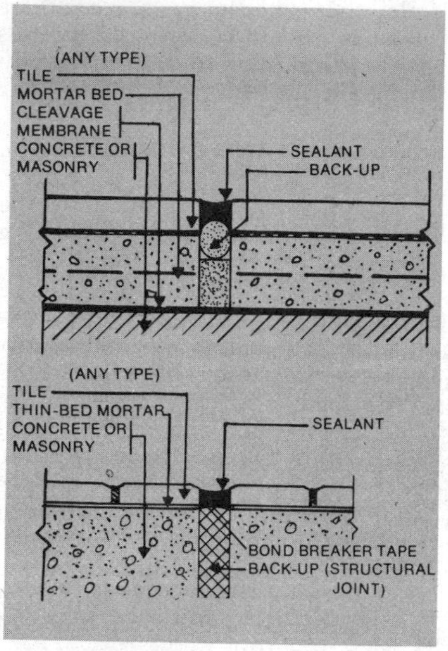

FIG. 20 Expansion joints in the tile finish should always be provided over structural joints in the backing. Additional joints may be required in isolated thickbed installations.

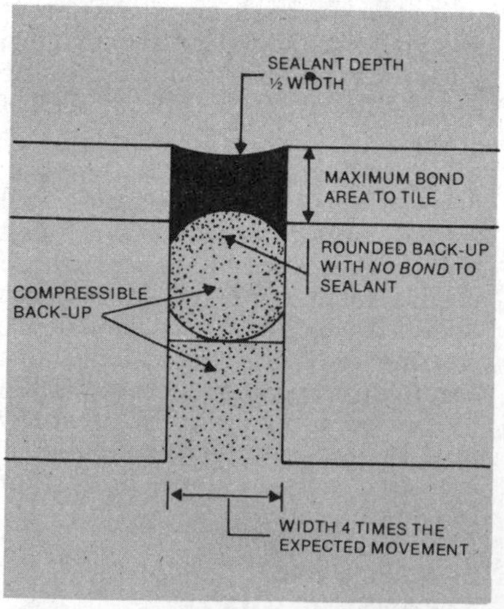

FIG. 21 Proper sealant depth, good bond at the sides and bond-free contact with backup strip are essential to long-term effective performance.

Joints in Tile Finish

Expansion joints are not required in areas less than 12′ wide and over old, well cured concrete. However, expansion joints in the tile finish should always be provided: 1) directly over expansion, isolation, construction or control joints in the backing; 2) where the tile finish abuts projecting surfaces; and 3) at proper intervals in floor areas measuring over 24′ in either direction. Expansion joints in the tile finish should never be narrower than the underlying joint in the backing (Fig. 20).

Tile edges to which the sealant will bond should be clean and dry. Sanding or grinding of these edges is recommended to obtain maximum sealant bond. Some sealant manufacturers recommend priming of edges. When a primer is used, care should be taken to keep it off the tile face. Sealant depth should be approximately 1/2 the width. Compressible back-up strips should be used in the joint to prevent bonding of the sealant to the joint filler and to insure a proper sealant profile (Fig. 21).

Exterior Locations Exterior installations should be provided with expansion joints sized as follows: at least 3/8″ wide for joints 12′ o.c.; at least 1/2″ wide joints 16′ o.c. These minimum joint widths should be increased 1/16″ for each 15°F of temperature differential (range between summer high and winter low) over 100°F. Decks exposed to the sky in the northern USA should have 3/4″ wide joints 12′ o.c.

Interior Locations Thickbed tile installations in interior areas should have expansion joints in the tilework spaced 24′ to 36′ each way. Joints for interior quarry tile and paver tile should be the same width as the grout joints, but not less than 1/4″. Joints for ceramic mosaic tile and glazed wall tile preferrably should be 1/4″ wide but never less than 1/8″. Large thinset installations may benefit from ap-

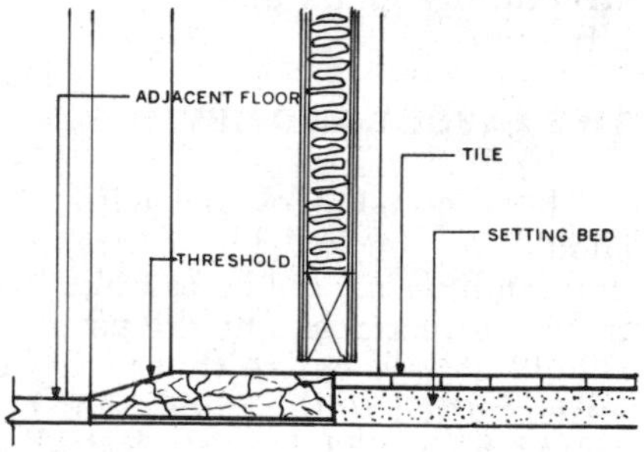

FIG. 22 In doorways, high end of thresholds should be located under door.

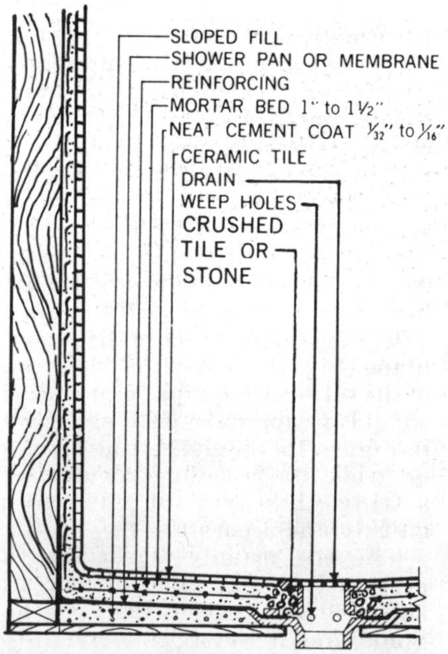

FIG. 23 In shower stall, inclusion of a membrane, crushed stone and weepholes assure drainage of water which might seep under tile finish.

propriately located expansion joints, but no generalizations can be drawn. Joints always are desirable at offsets and changes in backing construction.

THRESHOLDS

When it is not practical to depress the subfloor for a conventional mortar bed installation and adjacent finishes are subsequently thinner, a transition is required. A marble or slate threshold located in a doorway, or in a cased opening, usually serves this purpose. These thresholds are set with the same bonding agents as ceramic tile (Fig. 22).

The threshold may be of other materials when installed by different flooring contractors. Resilient flooring contractors typically use rubber or vinyl thresholds; carpenters are likely to use a hardwood threshold in connection with wood strip flooring; carpeting might be butted to tile with a metal or vinyl edge strip.

To protect the edge of tile from damage a hard material such as marble is preferred. This is not as critical in residential as in commercial installations.

SHOWER RECEPTORS

Residential shower receptors can be either precast terrazzo, plastic or field-installed ceramic tile. Tile can be installed over wood or concrete subfloors; a mortar bed must be used over wood subfloors. In either case, a waterproof membrane or pan must underlie the tile finish and it must be sloped 1/4″ per foot to the drain. The sides of the membrane or pan should be turned up the wall at 5″ (Fig. 23).

Although only the mortar bed method is recommended for shower receptors, the slope should be developed in the backing rather than in the mortar bed. This assures positive runoff of any water which might penetrate the tile finish. Crushed stone or tile around the shower drain and weepholes in the drain permit this occasional water to seep into the drainage system (Fig. 23).

When a concrete subfloor is used,

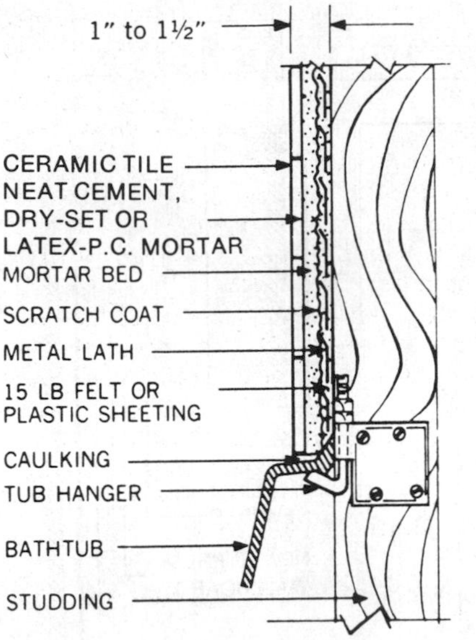

FIG. 24 Caulking the space between the tile finish and tub apron with a suitable sealant assures watertight joint.

the subfloor itself should be sloped to the drain. With a wood subfloor, a separate p.c. mortar fill should be used to create the slope. The fill can vary from 1/2″ to 1-1/4″ in thickness and preferrably should include a latex additive. The latex improves the bond strength of the p.c. mortar and makes it possible to apply the mortar in thin layers.

The waterproof membrane or pan should be installed over the sloped fill, followed by the tile finish set in a mortar bed. Any of the commercial grouts, except mastic, are suitable for grouting the tile joints. Figure 24 illustrates proper detailing of tiled tub-shower areas; Figure 25 shows recommended tile coverage of gypsum board in bathing areas.

TILE OVER OTHER FINISHES

Ceramic tile can be installed readily over existing tile or over existing floor and wall finishes, such as resilient flooring, hardwood flooring, painted concrete, gypsum plaster and wallboard. Ideally, existing finishes or coatings should be completely removed so that the tile can be placed on the original backing. If this is impractical, a new backing surface can be installed for thinset application, or the tile can be installed by the conventional mortar bed method. If the existing finish is considered suitable for thinset application of tile, it is imperative that the mortar or adhesive manufacturer's recommendations be consulted. Figure 26 illustrates several tile-over-tile installation possibilities.

Thickbed installations present no special problems over existing finishes. For thinset installations, it is important to insure that the backing surface is free of contaminants, such as dirt, dust, efflorescence, oil or grease. The maximum variation in backing surface should not exceed the dimensions indicated in Supporting Construction, page 442-10. Painted surfaces which are glossy or scaly

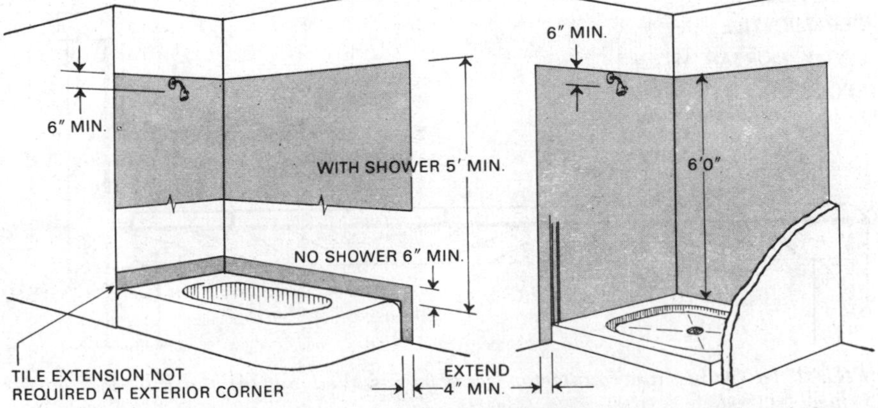

FIG. 25 Minimum recommended tile coverage of gypsum board in tub and shower areas. Gypsum boards should be of the water-resistant type.

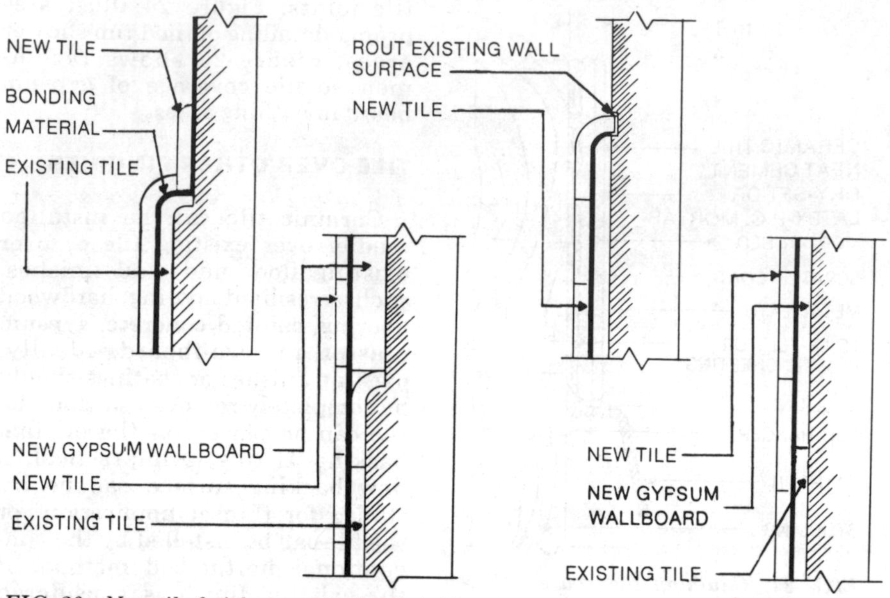

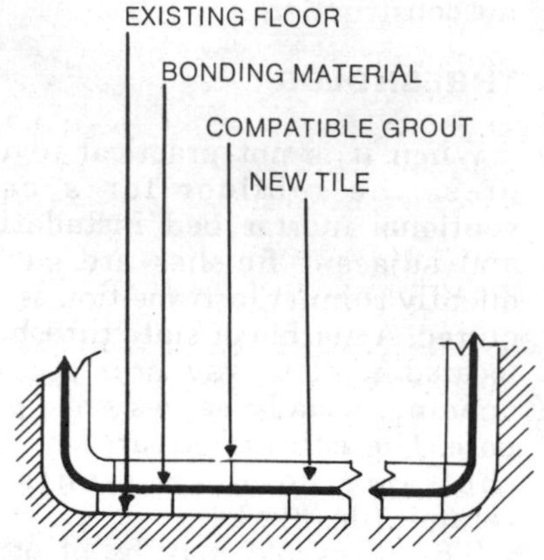

FIG. 26 New tile finishes can be bonded directly to the existing tile surface or over a suitable new backing such as gypsum board.

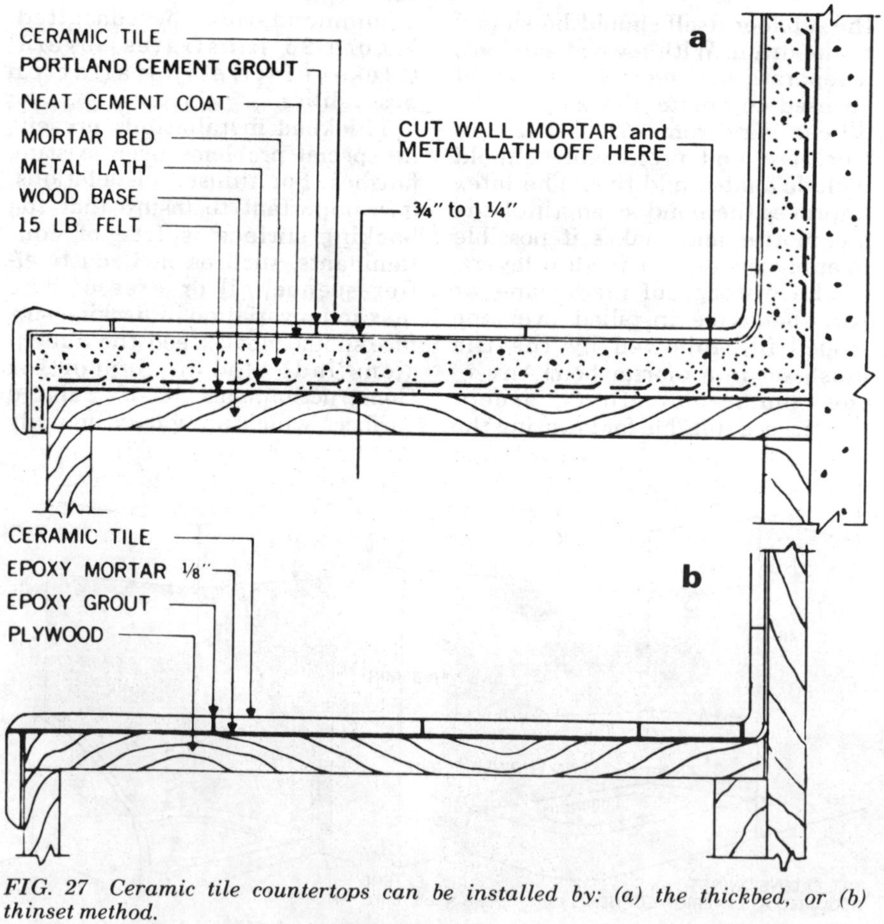

FIG. 27 Ceramic tile countertops can be installed by: (a) the thickbed, or (b) thinset method.

should be treated by sandblasting or chemical paint remover. The adhesive or mortar manufacturer should be consulted to determine whether a sealer or primer is needed. Where the structural floor assembly is capable of carrying the weight and the old floor surface is questionable, the more reliable isolated thickbed installation should be used.

COUNTERTOPS

Either mosaic, quarry or glazed wall tile can be used as a durable finish for countertops. The tile is installed typically over a wood backing either by the thickbed or thinset method (Fig. 27).

An epoxy grout or a special stain-resistant grout recommended for that purpose by the manufacturer should be used. The backing for a mortar bed installation should be 1″ x 6″ boards or 3/4″ exterior plywood. Boards should be spaced 1/4″ apart (Fig. 27a). Plywood should be provided with saw cuts 6″ to 8″ o.c. to prevent warping. For thinset installation, only 3/4″ exterior plywood should be used. A 1/4″ gap should be provided between plywood panels and the gap should be closed with a batten strip from below (Fig. 27b).

INSTALLATION ON FLOORS

Selection of the most suitable installation method depends on the *performance level* that is desired, the type of supporting construction and whether the installation is interior or exterior.

INSTALLATION PERFORMANCE LEVELS

The type of traffic (wheeled or foot traffic, impact loads) and the contaminants (water, grease, alkalis, chemicals) to which the floor will be subjected are a prime consideration in the selection of a suitable performance level. Once the performance level has been determined, an appropriate installation method can be selected. A description of the several performance levels and specific installation details for various performance levels are given in the Work File.

In addition to the installation method, consideration should be given also to: (1) the tile selected with respect to the wear properties of the surface; (2) fire-resistance properties of the tile finish including the entire wall or floor assembly; (3) acoustical properties (especially Impact Noise Rating) of the floor assembly; and (4) slip-resistance of the floor tile.

Unglazed tile (mosaic, quarry and paver tile) generally is used for floor surfacing and will give satisfactory wear for all the performance levels listed. If, however, a decorative glazed tile or an especially soft-body unglazed tile is used, it should be approved for use on floors by the tile manufacturer. In determining suitability of glazed tile for the intended use, color pattern, surface texture and glaze hardness should be considered. All glazed tile will show traffic wear in time.

INTERIOR CONCRETE

An interior concrete slab, not being subjected to the drastic temperature and humidity fluctuations encountered outdoors, offers a complete range of tile installation methods. The tile flooring can be set over a *bonded* mortar bed, an *isolated* mortar bed, or directly over a properly finished slab with a *thinset* mortar or adhesive. The specific procedure for installing tile depends mostly on the tilesetting material selected. Furan resins generally are limited to industrial installations, hence are not discussed in the following text.

Portland Cement Mortar

Neat portland cement can be used as the bond coat only when the concrete surface is still plastic and suitably rough to bond with the neat cement. Since this condition is difficult to achieve on a structural slab surface, a separate mortar *setting bed* with a suitable surface must be installed by the tile contractor.

When a setting bed is used, it should be isolated from the structural slab with a cleavage membrane, so that stresses induced in the slab are not transmitted to the tile finish. If a concrete slab-on-grade is made up of a properly engineered concrete mix, with suitable reinforcement, adequate finishing and curing, a well drained granular base, and proper jointing—the likelihood of excessive movement is minimized and there is no need to isolate the tile finish. Under these conditions, tile can be applied also by the thinset method directly to the slab or over the bonded setting bed.

Although the precautions required to assure a stable concrete backing appear extensive, they are inherent in all good concrete work. It is generally cheaper to comply with good practice than to cope with excessive slab movement, regardless of installation method.

Glazed semivitreous tile should be dampened by placing it on a wet cloth or in a shallow pan prior to setting. No free moisture should remain on the backs of the tile when set. Vitreous and impervious tile, such as mosaic tile and quarry tile, do not require soaking.

A thin layer of neat cement paste, 1/32″ to 1/16″ thick, should be troweled over the setting bed or on the back of the tile. Another way is to dust a thin layer of dry portland cement, 1/32″ to 1/16″ thick over the setting bed; this bond coat should be worked lightly with a trowel or brush until damp immediately prior to setting the tile.

The tile should be set in position and tamped firmly into the mortar, to obtain maximum contact and a strong bond (Fig. 28a). All tile surfaces should be brought to a true level at proper elevation. Tamping

FIG. 28 Mounted ceramic mosaic tile is: (a) laid in neat cement bond coat; (b) after tamping in place, paper is sponged off; (c) grout is troweled into joints and excess is removed from surface with clean cloth.

and leveling should be completed within one hour after placing the tiles.

When installing face-mounted ceramic mosaic tile, the paper should be wetted within one hour, using no more water than necessary for removing the paper and glue (Fig. 28b). Any tiles that are out of line or not level should be adjusted before the mortar assumes initial set.

Sand-portland cement grout typically is used with quarry tile, pavers and ceramic mosaic tile. For quarry and pavers, the grout generally consists of 1 part portland cement to 2 parts sand and 1/5 part hydrated lime by volume. For ceramic mosaic tile, a mix of 1 part p.c. to 1 part finely screened sand is used (Fig. 28c). First, filling and pointing is done with neat cement paste. Joints in paver, quarry and cushion-edged glazed tile should be tooled (compressed) with a pointing tool, so as to firmly bond a layer of pointing mortar to the grouting mortar and the tile.

All excess grout should be removed from the surface with a clean cloth or burlap. Before the grout sets, all voids and gaps should be filled. When portland cement-based grouts are used, the entire installation should be damp-cured for 72 hours. For all grouting materials, manufacturer's instructions should be followed.

The tile surface should be thoroughly cleaned while the grout is still plastic. Muriatic acid may be used to clean glazed tile; sulfamic acid can be used to clean unglazed tile 10 days after setting. Joints should be soaked before acid is applied and flushed thoroughly with water after cleaning. All metal and enameled surfaces should be protected with grease or other strippable coating.

Isolated Setting Bed The cleavage membrane should be spread over the slab and should be folded at edges and ends to form a locked joint. Metal reinforcement should be lapped at least one full mesh, and should be supported approximately in the middle of the setting bed. The mesh should be cut to fit accurately and kept free of vertical surfaces. Figure 29 illustrates typical isolated bed construction.

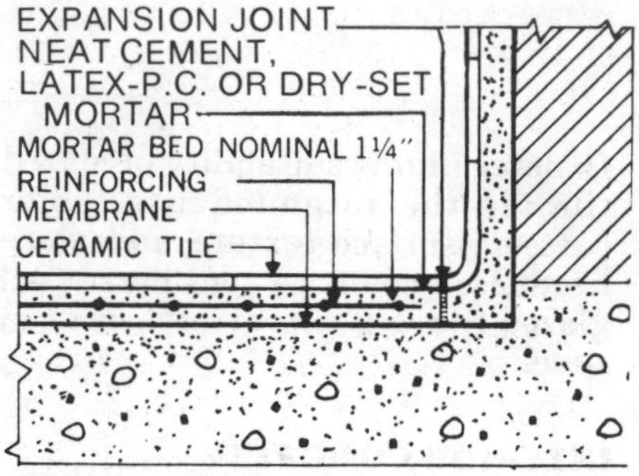

FIG. 29 Metal reinforcement always should be used in isolated thickbed installations.

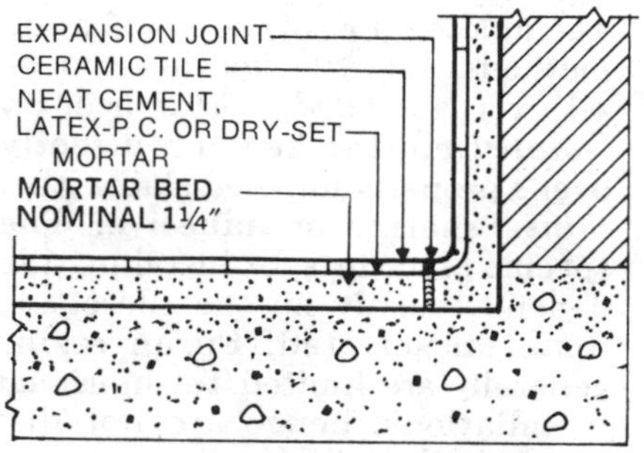

FIG. 30 When the mortar bed is bonded to the backing, reinforcement can be omitted.

The setting bed mix should consist of 1 part portland cement, 5 parts dry sand (or 6 parts damp sand) and 1/10 part hydrated lime by volume. Only enough water should be added to the mix to obtain the desired consistency. The setting bed should be applied 3/4″ to 1-1/4″ thick, tamped firmly and screeded to a level surface.

Bonded Setting Bed The bonded bed method is used infrequently because it does not possess the chief advantage of unbonded installations—separation from stresses in the concrete subfloor. Its chief advantage over thinset methods is the ability to produce a smooth finish over a rougher subfloor and reliance on familiar, readily available materials such as portland cement and sand. Bonded setting bed construction is similar to isolated construction except that the cleavage membrane and reinforcement can be omitted (Fig. 30).

The mortar bed can be bonded with a synthetic bonding compound or neat portland cement. For the latter, a fairly rough, wood floated, surface is required. The slab should be thoroughly saturated with water before the neat cement and mortar is spread. The mortar mix, grout mix and general installation procedures for a bonded setting bed are identical with those described above for an isolated setting bed.

Latex-P.C. and Dry-Set Mortars

These mortars generally are used to bond the tile directly to a concrete slab (Fig. 31), or to a cured mortar leveling bed—hence the slab should be designed to assure a minimum of movement. The slab should be well cured, free of cracks, waxy or oily films and curing compounds. The slab surface should be steel troweled and fine broomed. Dry-set or latex-p.c. mortar should be mixed as recommended by the manufacturer. Mortar should be troweled in a layer approximately 1/4″ thick for mosaic tile and 3/8″ to 1/4″ thick for quarry and paver tile. The tile should be tamped into the mortar to insure a level surface and intimate contact.

Paper should be removed from face-mounted tile within one hour of setting, using a minimum of water. Grouting should be delayed for at least 24 hours, or until the tile is firmly set. Suitable grouts include: dry-set grout, latex grout, sand-p.c. grout, commercial cement grout and mastic grout. Dry-set and p.c. grouts should be damp-cured for 72 hours.

Organic Adhesives

Since these products are recommended for residential light traffic only, they are used infrequently over a concrete slab, except in small areas and over relatively smooth plane surfaces. For organic adhesives, which are applied only 1/16″ thick, the slab surface should be *steel-troweled* and fine broomed and the surface should not vary by more than 1/16″ in 3′.

Tile installation with organic adhesives requires the same general precautions as with other thinsetting products (Fig. 31). The tile should be tamped into place to assure maximum contact with adhesive. Any of the available grouts can be used, but latex-p.c. grout, epoxy and mastic grout are most commonly used with organic adhesives.

Epoxy Mortar and Adhesive

The performance and working properties of epoxy adhesives are similar to those of epoxy mortars. Most of the following discussion concerning epoxy mortars applies to adhesives as well.

Epoxy mortar is likely to be used in conjunction with an epoxy grout to provide chemical resistance. These properties are most desirable in laboratories, industrial plants and in food preparation areas. However, the high bond strength of the mortar, combined with the general stain-resistance of the grout, make this a preferred system for many hard service areas (such as entries, public restrooms, lobbies, restaurants), particularly when combined with quarry tile. Since the mortar is installed in a layer 1/4″ to 3/8″ thick, it can absorb minor variation in the subfloor surface—not over 1/4″ total in 10′. A wood-floated concrete surface is suitable (Fig. 32).

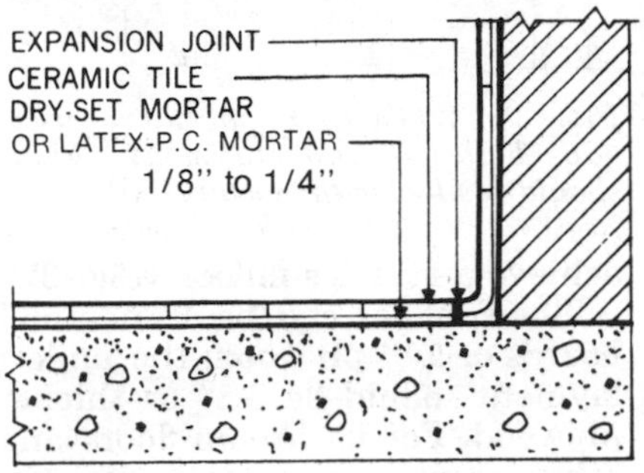

FIG. 31 Dry-set and latex p.c. mortar can be used for thin-setting of tile directly to concrete slab.

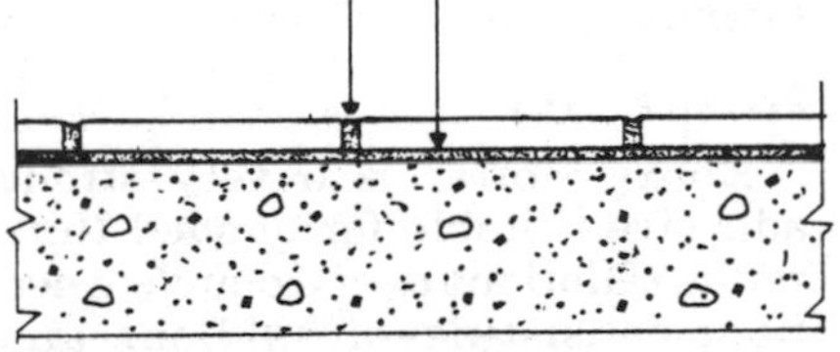

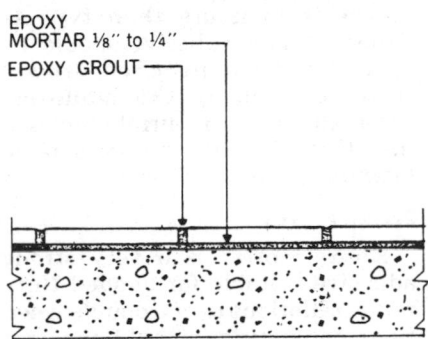

FIG. 32 Epoxy mortar and grout possess high bond strength and chemical resistance. Epoxy adhesive can be substituted in less demanding situations.

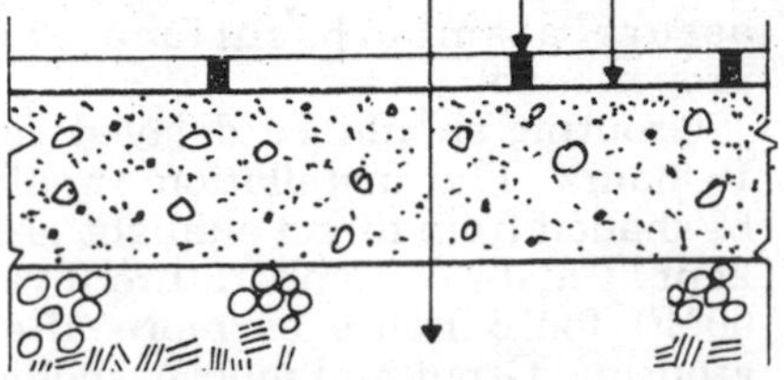

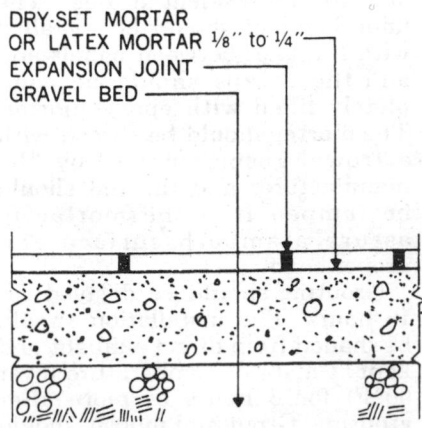

FIG. 33 Dry-set mortar can be used for thin-setting of tile in both interior and exterior locations.

The epoxy mortar should be of the same color as the grout. The tile should be tamped into the setting bed, taking care not to smudge the face of the tile. Mortar should be immediately removed from the face, as suggested by the manufacturer.

Grouting should be delayed at least 16 hours. The grout should be forced into the joints with a squeegee or rubber trowel so as to completely fill them. The installation should be shaded from direct sun and kept at a relatively even temperature (over 60°F) for at least 8 hours after grouting, to facilitate curing. All traffic should be kept off the floor for at least 40 hours. Light traffic can be permitted after this period; heavy traffic, after 7 days.

Epoxy grout can be used also with other tilesetting mortars. Where chemical resistance at the tile surface is desirable but high bond strength with the backing is not essential, epoxy grout can be used with latex or dry-set mortars. These cementitious mortars generally produce adequate bond and are less expensive than epoxy mortar.

EXTERIOR CONCRETE

Except for installations over built-up roofing or waterproofing membranes, exterior tile installations are likely to be bonded directly to the concrete slab (Fig. 33). Over a suitably finished concrete surface, dry-set mortar can be used in a coat 1/4″ to 3/8″ thick to bond the tile directly to the slab.

If the structural slab is not suitable for direct application of tile, a mortar setting bed can be used; this method assures a suitable surface for the setting of tile (Fig. 34). Since the mortar bed is installed by the tile contractor, he can follow-up with a neat p.c. bond coat while the bed is still plastic. For exterior tilework, it is important to consult with latest industry standards and tile manufacturer's recommendations, particularly in freezing areas.

Expansion joints extending through the tile finish and the mortar bed should be provided at intervals of 12′ to 16′ each way (Fig. 34). Expansion joints should be provided also where the tile finish abuts restraining vertical surfaces and where expansion, isolation or control joints are provided in the slab. Both the structural slab and the tile surface should be sloped to the drain. See page 442-13 for additional discussion of expansion joints.

WOOD SUBFLOORS

Wood subfloors usually consist of wood joists and plywood or board sheathing. Unless stiffness is carefully considered in the design, the

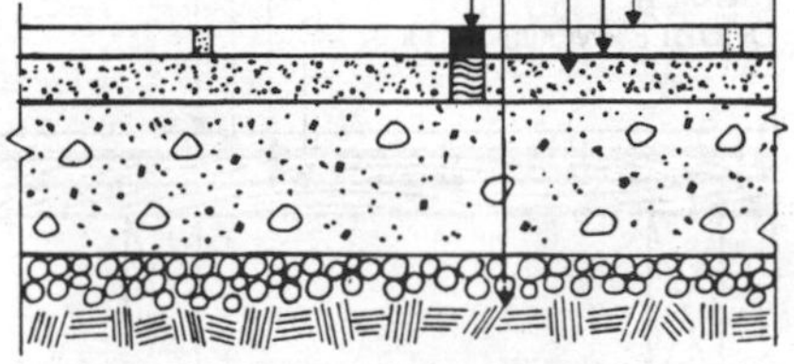

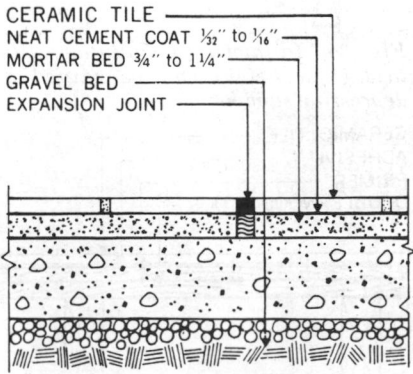

FIG. 34 When a mortar bed is used, expansion joints should be provided 12′ to 16′ o.c. as well as over joints in the backing.

joists are likely to deflect and vibrate more than massive structural systems such as concrete and/or steel; the boards and plywood, being hygroscopic, also are likely to expand and contract with humidity variations.

It is important, therefore, for the tile installation to be either completely isolated from the subfloor so as to permit differential movement, or to be bonded to the subfloor with a tilesetting material which can resist the movement without breaking the bond. When the installation depends on a very strong bond and no setting bed is

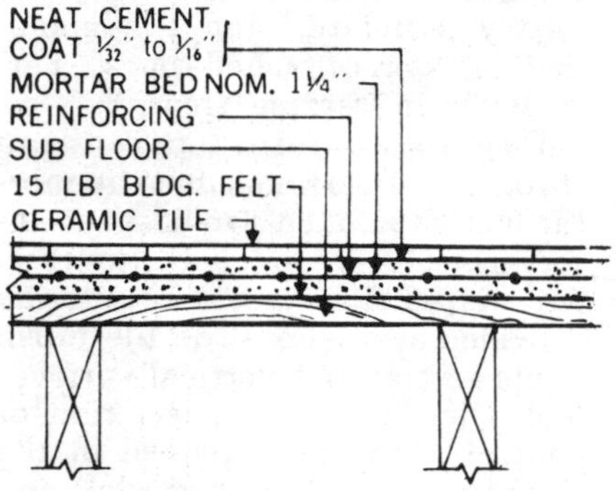

FIG. 35 A single-layer wood subfloor can be used under a reinforced isolated mortar bed.

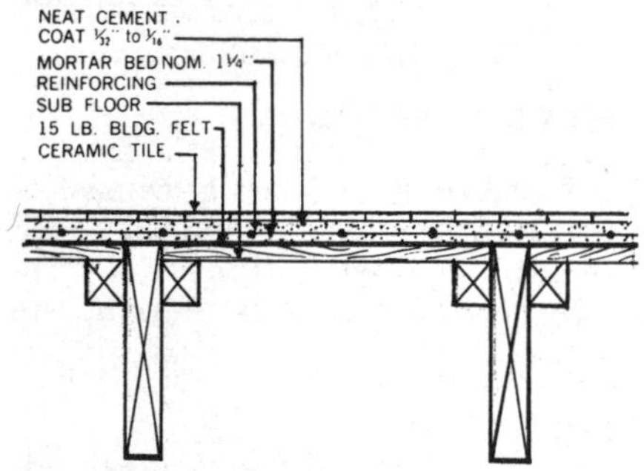

FIG. 36 In thickbed installations, 5/8" to 3/4" in height can be gained by depressing subfloor.

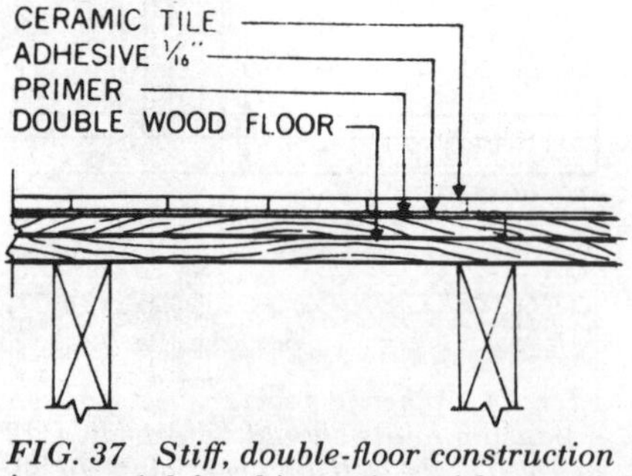

FIG. 37 Stiff, double-floor construction is essential for thin-setting with organic adhesive.

used, the grout also must be strong enough to resist the internal stresses transmitted to the tile joints.

A typical *isolated* conventional installation consists of a reinforced mortar setting bed and a latex, dry-set or neat cement bond coat—all over a cleavage membrane. A typical *bonded* thinset installation utilizes epoxy mortar or epoxy adhesive; in relatively small areas, such as residential bathrooms, tile also can be bonded to a plywood subfloor with organic adhesives. Although this material is not as strong as epoxy, it is adequate to resist the smaller stresses developed in a small, bathroom-sized area.

Portland Cement Mortar

The actual procedure of installing a mortar setting bed is the same as for concrete subfloors. The grout can be any of the several commercially available formulations, as well as sand-p.c. grout. The complete tile installation, including the mortar bed should be approximately 1" to 1-1/2" thick (Fig. 35).

Since many interior floor finishes are much thinner than 1-1/2", a transition problem develops where a tile finish abuts a thinner finish. To minimize this problem, it is possible to depress the subfloor between the joists, thus gaining 5/8" to 3/4" in height (Fig. 36). The subfloor, usually plywood, is supported on cleats level with the tops of the joists. (Depressing the subfloor below the tops of the joists and leveling the joists is not recommended). This method of tile installation is somewhat more durable in high moisture areas, such as bathrooms, than the thinset method. It is also more resistant to impact due to the mass of the mortar bed underlaying the tile finish.

Organic Adhesive

This thinset installation is suitable for all areas in the home, except those subject to excessive moisture, such as in a tiled sunken tub or in a swimming pool. It will not withstand heavy traffic or impact loads. Double floor construction generally is recommended to

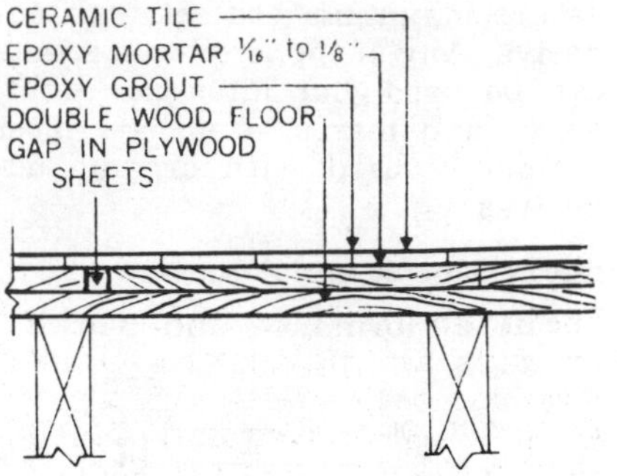

FIG. 38 When epoxy mortar is used, gaps between underlayment panels should be filled with mortar.

achieve a stiffer subfloor (Fig. 37).

The subfloor can be 1" nominal boards or 5/8" plywood; the underlayment should be 3/8" or thicker plywood. For bathroom floors and other areas subject to occasional moisture, type I adhesive should be used. This type is capable of withstanding more prolonged exposure to moisture than type II adhesive. The adhesive should be spread with a notched trowel recommended by the manufacturer and the tile firmly pressed into the adhesive. Grout can be latex-p.c., mastic or epoxy.

Epoxy Mortar

Epoxy mortar is a very strong adhesive suitable for bonded thinset installations over a double wood floor consisting of subfloor and underlayment (Fig. 38). The subfloor should be 5/8" plywood or 1" nominal boards; the underlayment should be 1/2" or thicker plywood (5/8" in non-residential uses). The underlayment should be installed with 1/4" gaps between panels and and these gaps should be completely filled with epoxy mortar. The mortar should be spread with a trowel recommended by the manufacturer and the tile should be tamped into the mortar to assure a smooth surface and proper bond.

Grouting should be delayed for 16 hours. The installation should be shaded from direct sunlight and kept at a stable temperature (over 60°F) for 8 hours or more after grouting. Grout and mortar should be cleaned off the face of the tile as soon as possible and before it hardens.

INSTALLATION ON WALLS

The ceramic products typically used on walls are glazed wall tile and ceramic mosaic tile. Quarry tile and paver tile are infrequently used as wall finishes. Where the wall surface is sufficiently sound, true, clean and free of deleterious coatings, tile can be bonded to the wall by the thinset method. The surface should not vary more than 1/8″ in 8 ft. if thinset materials are to be used; not more than 1/4″ in 8 ft. for portland cement mortar beds.

Wall surfaces which are irregular, chipped, cracked or coated with materials which would impede bond, require the use of a mortar bed. The mortar setting bed must be installed generally over a base coat of mortar (*scratch coat*) applied to metal lath over a cleavage membrane.

The scratch coat consists of 1 part portland cement, 5 parts damp sand and 1/5 part hydrated lime. It is normally applied by a plastering contractor, but when so specified, can be installed by the tile contractor. The float coat generally is applied by the tile contractor to assure a plastic surface for setting the tile with neat cement.

A solid backing is not essential when the tile is to be installed by the mortar bed method. Over wood or metal studs, the backing can be omitted and the membrane and metal lath can be attached directly to the studs. The lath-reinforced mortar bed generally is rigid enough by itself to support the tile without excessive deflection between the studs; to assure adequate vertical rigidity the studs should be at least 2-1/2″ deep at 16″ o.c., or 3-1/2″ deep when spaced 24″ o.c.

The materials, installation methods and suitable backings for tile finishes vary depending on whether the installation is interior or exterior. Exterior installations are much more critical because of exposure to moisture and temperature variations.

INTERIOR SURFACES

Suitable backings for tile finishes include masonry and concrete walls, as well as plywood, gypsum board and plaster over furring or stud framing. In tub and shower areas, the thinset method is acceptable, but it is essential that the backing be inherently water-resistant or properly prepared to render it so. Dry-set mortar can be used as a bond coat for thinset application in sunken tubs and pools over a p.c. mortar bed.

Portland Cement Mortar

Traditional installations on walls typically require a preliminary *scratch coat* followed by a mortar bed (two-coat method). The scratch coat can be omitted when the wall provides a solid backing for the application of metal lath, mortar and tile (one-coat method). The tile actually is set in a separate *bond coat*, but it usually is not counted in describing the installation.

The one-coat method can be used over existing wall finishes which are not suitable for thinset application because of surface conditions (Fig. 39). It is also a preferred method of tiling tub and shower areas in new work, especially in the western states. The backing can be masonry, concrete, plaster, wallboard plywood or any other solid wall surface. The single bed coat is 3/8″ to 3/4″ thick and consists of 1 part p.c., 1/2 part hydrated lime and 5 parts damp sand.

The two-coat method, involving both a scratch coat and a bed coat, is used most frequently over open framing of furring or studs (Fig. 40). On metal studs, the overall thickness of scratch coat and mortar bed should not exceed 1″; on wood studs overall mortar thickness should not be over 1-1/2″. The mortar mix is the same as for one-coat installations. The scratch coat should be allowed to cure for at least 24 hours and should be dampened thoroughly before the

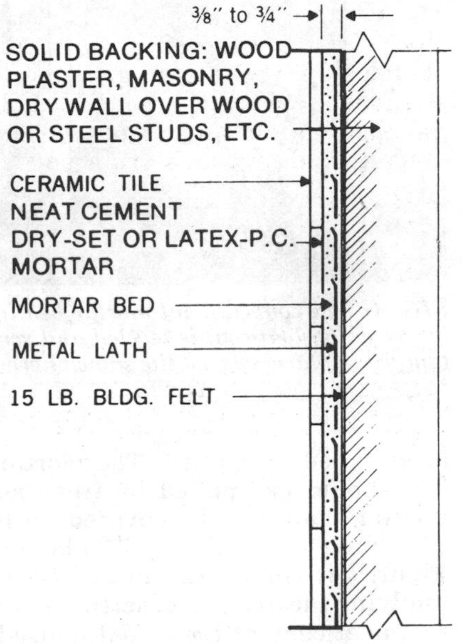

FIG. 39 One-coat method can be used over a solid backing which is too uneven for thin-setting of tile.

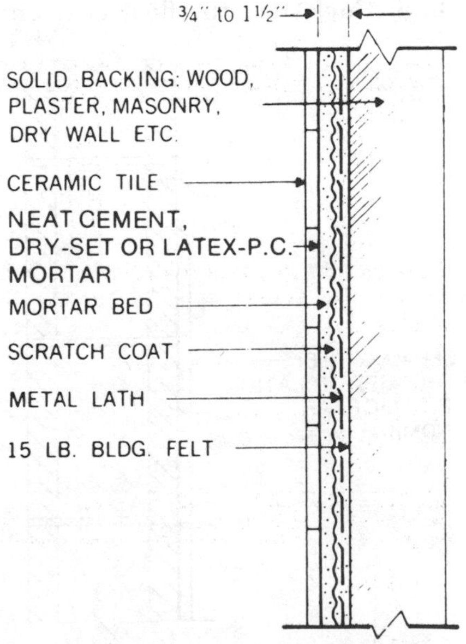

FIG. 40 Two-coat method can be used over open framing, as well as over solid backing.

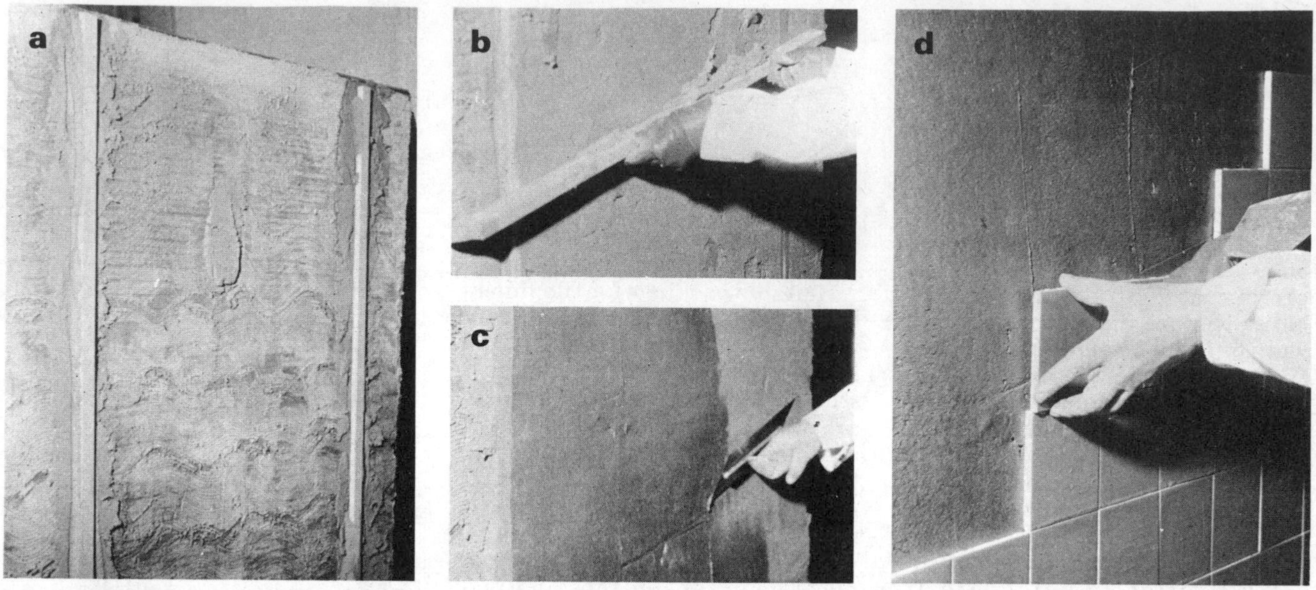

FIG. 41 In conventional mortar bed installations: (a) scratch coat is applied to metal lath and temporary screeds are inserted; (b) float coat is leveled and screeds are removed; (c) neat cement coat is troweled on; and (d) tile is applied and tamped in place. Glazed tile shown here must be soaked before setting.

mortar bed is applied. The mortar bed should be applied in areas no greater than may be covered with tile while the bed is still plastic. Figure 41 illustrates the steps in applying tile over a p.c. setting bed.

A neat cement bond coat is used for setting tile with both the one- and two-coat methods. The bond coat should be troweled 1/32″ to 1/16″ thick over the float coat and the tile should be tamped into it. The portland cement for the bond coat should be ordinary grey or white, depending on whether the grout is to be grey or white. The grout can be neat portland cement, p.c. mixed with equal part sand, latex or epoxy. Where commercial portland cement or p.c. grout is used, the finish should be cured for at least 72 hours.

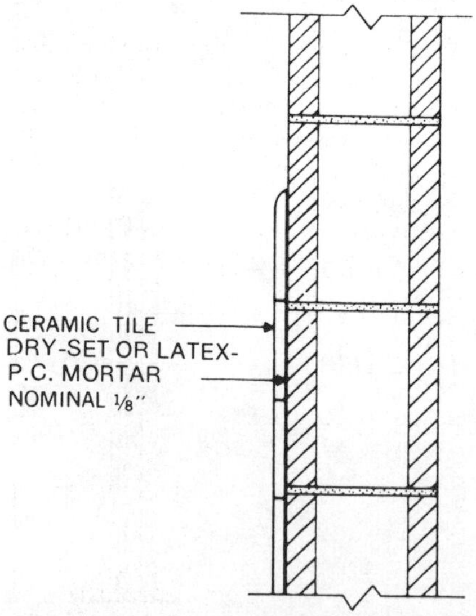

FIG. 42 For thin-setting of tile directly to masonry, surface should be smooth with flush-troweled joints.

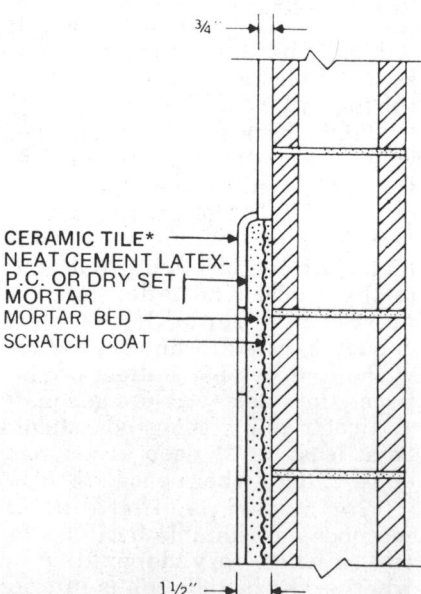

Dry-set and Latex-p.c. Mortars

These thinset mortars are suitable for any sound, clean and true surface such as gypsum plaster, wallboard, monolithic concrete or concrete and brick masonry (Fig. 42). They also can be used over a bonded mortar bed when the bed is used mainly for leveling an irregular solid surface (Fig. 43), or when a full-thickness mortar backing on open framing has been installed by others (Fig. 44). When such a mortar backing is used, the backing should be allowed to cure for several days until it is firm.

The backing surface should be dampened just before spreading mortar. Mortar should be troweled on in appropriate thickness (3/32″ to 1/4″ maximum) in an area which can be readily covered with tile before the mortar starts skinning over (Fig. 45a). The tile generally should be tamped into place to obtain at least 80% coverage; in tubs, showers and other high-moisture areas 100% coverage should be obtained (Fig. 45b). Any

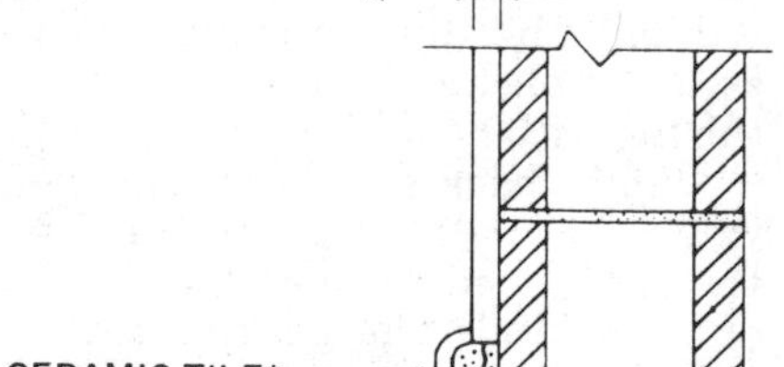

FIG. 43 Over an uneven masonry surface, a leveling bed should be applied first. Tile can be set with neat cement, or thinset mortar after curing.

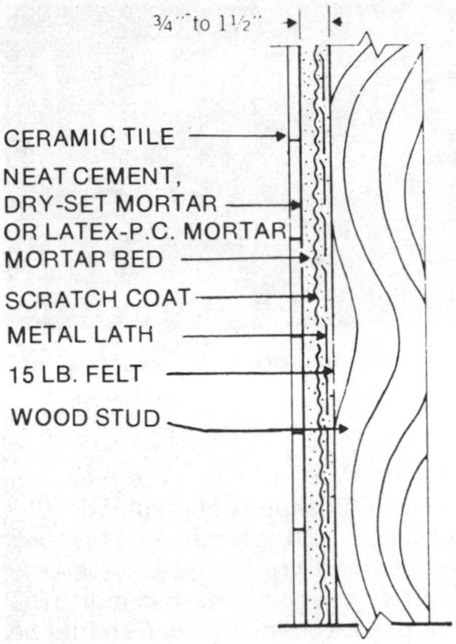

FIG. 44 Open-framing requires two-coat mortar bed over metal lath. Except with dry-set mortar, curing of the bed is required for thinset applications.

grout can be used, but in selecting a suitable type, consideration should be given to the type of exposure (Fig. 45c). In residential installations, sand-p.c., commercial portland cement, drywall grout and mastic grout are most likely.

Organic Adhesive

The use of organic (mastic) adhesive is limited to light-duty interior areas and over smooth surfaces such as plaster, wallboard, cement-asbestos board or plywood (Fig. 46). It can be used also over concrete or masonry if the surface is relatively smooth (maximum variation 1/8″ in 8′). Type I adhesive should be used for installations requiring *prolonged* water-resistance, such as in tub and shower areas, and on bathroom floors. Type II adhesive can be used in areas subject to occasional or *intermittent* wetting. However, neither type is suitable for installations requiring continuous contact with water such as in sunken tubs or pools. Manufacturer's recommendations should be followed to determine which type is best for the specific use intended, or whether a primer is required.

The adhesive should be spread with a notched trowel recommended by the manufacturer, leaving no bare spots (Fig. 47a). The tile should be tamped into the adhesive to assure maximum contact (Fig. 47b). The *average* contact area of tile removed for inspection should be not less than 75%, and no *individual* tile or mounted assembly should have less than 40% coverage of adhesive. Adjustments should be made while the adhesive is still plastic. Grouting should be delayed for at least 24 hours to permit solvent evaporation from the adhesive. Joints should be grouted with a product suitable for the conditions. Mastic, drywall or latex grouts commonly are used with organic adhesives (Fig. 47c).

EXTERIOR SURFACES

Tile installations on exterior walls require the same precautions regarding movement which have been discussed under exterior floors. The supporting construction, including the backing, must be designed to accommodate movement due to temperature and/or moisture variations. This means primarily the inclusion of suitable control, isolation and expansion joints; also, inclusion of proper metal reinforcement in concrete and masonry; careful control of mortar and concrete mixes; sound engineering design and good workmanship (see 203 Concrete and 311 Slabs-on-Ground).

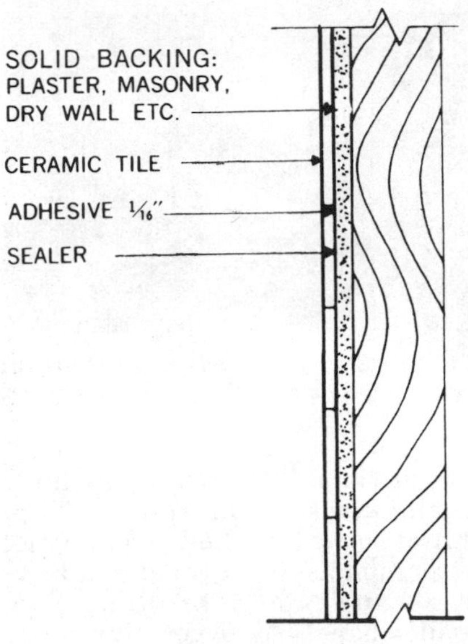

FIG. 46 Organic adhesives can tolerate limited moisture exposure (such as in shower stalls) provided a water-resistant adhesive and backing are used and board edges are sealed.

Exterior tile can be applied either by the conventional mortar bed or the thinset method (Figs. 40 & 41). Chief factors in determining the most suitable method are: 1) condition of the backing, whether suitable for direct bonding; 2) dimensional stability of the backing; and 3) provisions for crack control.

On large areas exposed repeatedly to direct wetting, the

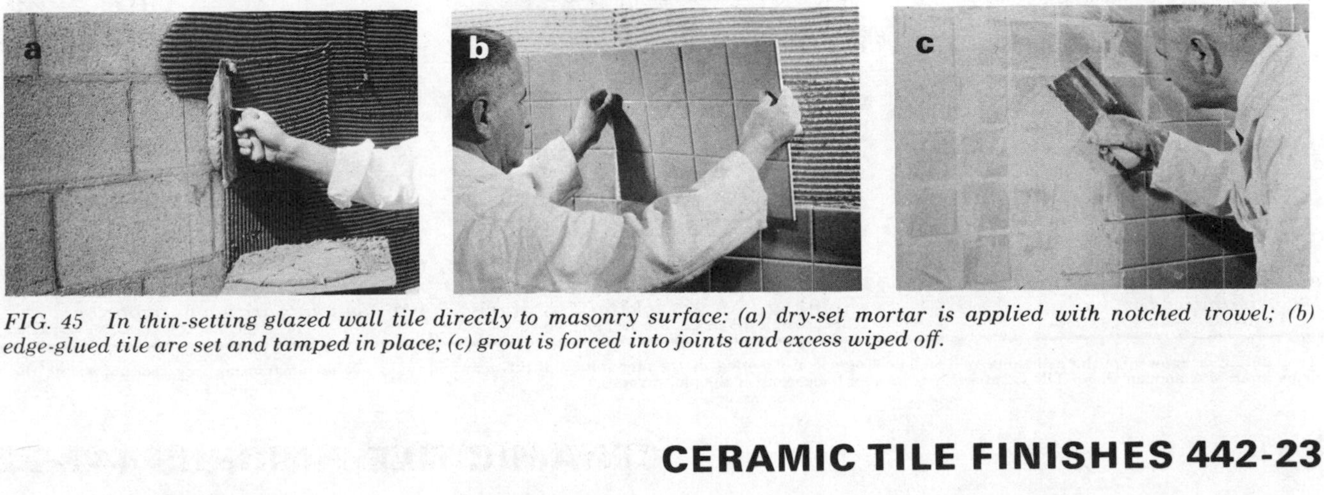

FIG. 45 In thin-setting glazed wall tile directly to masonry surface: (a) dry-set mortar is applied with notched trowel; (b) edge-glued tile are set and tamped in place; (c) grout is forced into joints and excess wiped off.

FIG. 47 Over previously applied organic adhesive: (a) edge-mounted tile is applied; (b) tamped in place with "rubber float" trowel; (c) same trowel is used to force grout into joints and remove excess grout.

conventional *isolated* method usually gives better results. The thinset *bonded* method generally is more reliable in covered outdoor areas and when broken into smaller panels by decorative trim, battens or expansion joints. However, tile can be thinset over an isolated mortar bed with dry-set mortar.

Regardless of installation methods, expansion joints should be provided in the tile finish over all underlying structural joints, at the tile perimeter, and against any projecting surface. In isolated installations, expansion joints in the tile finish should be provided at 12′ to 16′ o.c. Where these expansion joints do not coincide with joints in the backing, they should be provided in addition to the underlying joints. In a thinset installation over

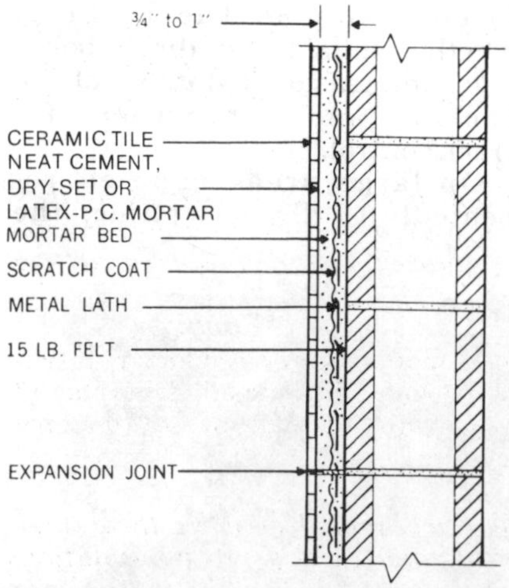

FIG. 48 Expansion joints 12′ to 16′ o.c. are essential in exterior isolated thick-bed installations on walls.

a suitable stable backing, the additional joints in the tile finish are not as essential because the tile finish is intended to move with the backing.

In all exterior tile installations, it is essential to provide suitable flashings, overhangs or copings to prevent entry of water into, or behind, the tilework. Exterior walls of habitable spaces should include a suitable vapor barrier on the warm side of the wall to prevent condensation in the wall cavity. Additional information on the design and construction of expansion joints is given in General Recommendations, page 442-03.

Portland Cement Mortar Bed

There is no substantial difference in installing tile over a mortar bed outdoors, from the procedure described for interior locations (see page 442-14). The entire installation should be backed by cleavage membrane. Metal lath should be firmly attached to the backing and cut at all expansion joints (Fig. 48).

A scratch coat consisting of 1 part p.c., 5 parts damp sand and 1/5 part lime usually is installed by a plastering contractor. The mortar bed—consisting of 1 part p.c., 5 parts damp sand and 1/2 part hydrated lime—normally is installed by the tile contractor. The overall thickness of both coats should be 3/4″ to 1″ thick. The mortar bed mix can be varied up to 1 part p.c., 1 part lime and 7 parts damp sand. However, extremely rich mixes (too much portland cement) should be avoided, because they result in greater drying, shrinkage and cracking. Similarly, a total mortar bed thickness in excess of 1″ should be avoided.

Over a still plastic mortar bed neat cement can be used for setting tile. If dry-set or latex-p.c. mortar is to be used, the mortar bed should be cured for at least 24 hours, but longer curing periods up to 10 days are preferred. Suitable grouts include commercial cement, dry-cure

Thinset Installation

Waterproof backings suitable for bonding of tile with thinset mortars include: monolithic concrete, concrete and brick masonry, structural clay tile and a cured p.c. mortar bed (Fig. 49). Latex-p.c. and dry-set mortar typically are used for setting tile. Either latex or dry-set grout can be used for filling tile joints. Damp curing for at least 72 hours is essential.

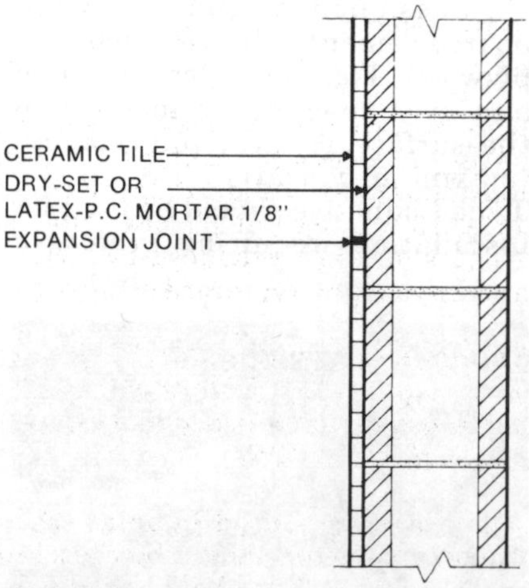

FIG. 49 Exterior thinset applications directly over masonry are expected to move with the masonry, hence expansion joints are not as important.

We gratefully acknowledge the assistance of the Tile Council of America in the preparation of this text and many of the illustrations. We also appreciate the cooperation of American Olean Tile Company, Inc. in supplying many of the photographs.

CERAMIC TILE FINISHES 442

CONTENTS

GENERAL RECOMMENDATIONS

MATERIALS

All tile and accessory products should be delivered in the manufacturer's original containers and should be stored under cover to prevent water damage or contamination.

Tile

Glazed wall tile, ceramic mosaic tile, quarry tile and paver tile should conform to requirements of the Recommended Standard Specifications for Ceramic Tile TCA 137.1. Special purpose tile not covered by TCA 137.1 should be produced by reputable manufacturers and have a proven record of success in the intended use.

Suitability of a tile product for a particular use depends on a variety of factors such as appearance, size, perviousness to moisture, abrasion resistance and other properties (Fig. WF1).

All ceramic tile used in areas where appearance is an important factor should be Standard grade, as defined in TCA 137.1. Tile should be shipped in sealed cartons with the grade of contents imprinted on the carton or indicated by a grade seal.

For quarry tile, the grade generally is imprinted on the carton; most other tile products have a grade seal consisting of a strip of paper applied in such a way that it must be broken to open the carton. The seal for Standard grade is blue; for Seconds, yellow.

Tilesetting Products

P.c. mortar, dry-set mortar, latex-p.c. mortar, epoxy mortar and organic adhesive should conform to appropriate industry standards summarized in Figure WF2.

Tilesetting products not covered by industry standards can be used if they are produced by reputable manufacturers and have a proven record of success in the intended use.

Installations with p.c. mortar, dry-set mortar, latex-p.c. mortar, epoxy mortar and organic adhesive should conform to recommendations of appropriate industry standards summarized in Figure WF2.

Installations with products not covered by these standards should follow manufacturer's recommendations and good industry practice.

Grouts

Epoxy grout should conform to requirements of ANSI

(Continued) GENERAL RECOMMENDATIONS

FIG. WF1 SUMMARY OF CERAMIC TILE PRODUCTS

Tile & Description	Suitable Uses	Limitations	Nominal Sizes
GLAZED WALL TILE Non-vitreous body, matte or bright glaze	All interior walls, in residential, commercial and institutional buildings.	Not suitable for floors or areas subject to freezing.	
Non-vitreous body crystalline glaze	All interior walls, as well as light duty residential floors and countertops.	Not suitable for areas subject to freezing.	5/16″ thick: 4¼″ x 4¼″, 6″ x 4¼″, 6″ x 6″ Scored, Octagon, Hexagon and other special shapes.
Vitreous body, matte or bright glaze	Interior walls, countertops, refrigerator and chemical-tank linings and exterior areas subject to freezing.	Not suitable for floors.	
CERAMIC MOSAIC TILE Porcelain body, unglazed	All uses, interior or exterior, floors and walls; especially suitable for bathrooms, kitchens and swimming pools.	None except limitations imposed by installation method and grout.	
Porcelain body, glazed	Interior or exterior walls in residential, commercial and institutional buildings; decorative inserts with above tile.	Not suitable for floors.	¼″ thick: 1″ x 1″, 2″ x 1″, 2″ x 2″
Natural clay body, unglazed	Swimming pool runways; floors of porches, entrances and game rooms in the home; floors and walls of commercial, institutional and industrial buildings.	Exterior use requires special frostproof body.	
Natural clay body, glazed	Walls in residential, commercial and institutional buildings; decorative inserts with above tile.	Exterior use requires special frostproof body.	
Conductive	Hospital operating rooms and adjoining areas, where required to safely dissipate dangerous charges of static electricity.	Requires special installation procedures according to NFPA No. 56A.	
QUARRY TILE	Exterior or interior; from moderate to extra heavy duty floors in all types of construction.	Recommended for interior use and exteriors not subject to freezing.	½″ thick: 3″ x 3″, 4″ x 4″, 6″ x 3″, 6″ x 6″, 8″ x 4″ and special shapes.
PAVER TILE	Same as quarry tile.	None	⅜″ thick: 4″ x 4″ ½″ thick: 4″ x 4″, 6″ x 6″, 8″ x 4″

*Sizes shown are those considered typical by the governing industry Standard, ANSI A137.1; many more sizes and shapes are produced by tile manufacturers. Special purpose tiles which combine properties of several of the above types also are produced.

A118.3. Other grouting products should be used in accordance with manufacturer's specific instructions.

Figure WF3 summarizes general properties and limitations of the main types of grout.

Related Materials and Products

Other products used in installing ceramic tile finishes should conform to appropriate industry standards summarized in Figure WF4.

Installation of ceramic tile involves a variety of related materials and products, described below and summarized in Fig. WF4.

White portland cement should be used in the mortar bed when a white grout is intended for filling tile joints.

Normal (Type I) portland cement typically is used for mortar setting beds, for leveling beds, for the neat cement bond coat and for field-mixed cement grout.

Sand should be clean, free from impurities and well-

(Continued) GENERAL RECOMMENDATIONS

graded. When fine sand is specified, it should pass through a 16-mesh screen.

Freedom from impurities is essential to a strong, durable tile finish.

Type S (special) hydrated lime should be used to assure a dimensionally stable mortar bed.

Lime is used in the portland cement-sand mortar mix to make it more plastic for application on walls, and sometimes on floors.

Sealants for use in expansion joints and control joints should be suitable for the exposure expected.

Conditions which deteriorate sealants are extreme temperature variations, direct sunlight, repeated stretching and compression, frequent wetting and drying or immersion in water.

When used on floors, the sealant should have a *Shore A* hardness index of not less than 25 for joints subject to occasional foot traffic; in high traffic areas the hardness index should be not less than 35.

The sealant can be a single-component or a two-component type. Color of sealant should match or blend with the tile or adjacent materials.

Isolated (unbonded) mortar setting beds should be reinforced with suitable metal lath, wire fabric or mesh.

Welded wire fabric may be one of the following: 2" x 2" mesh—16/16 wire; 3" x 3" mesh—13/13 wire; or 1-1/2" x 2" mesh—16/13 wire. Welded wire fabric generally is used for mortar bed reinforcement on floors.

Metal lath may be expanded lath weighing not less than 3.4 lb. per sq. yd., or sheet lath weighing not less

FIG. WF2 SUMMARY OF TILESETTING PRODUCTS

Type	Description	Industry Standards		Limitations
		Product	Installation	
PORTLAND CEMENT MORTAR	Field mix of p.c., sand and water for setting bed; mix p.c. and water for bond coat. Interior or exterior.	ASTM C-150 (P.C.)	ANSI A108.1	Requires 1¼" setting bed, resulting in additional thickness and weight. Requires pre-soaking of most glazed wall tile.
DRY-SET MORTAR[1]	Prepared mix[2] of p.c., sand and certain additives which impart water retentivity and eliminate moist-curing. High resistance to moisture, impact and freezing. Interior and exterior.	ANSI A118.1	ANSI A108.5	Three days of moist curing before grouting recommended for heavy duty floors.
LATEX-PORTLAND CEMENT MORTAR	Factory mix of p.c. and sand to which liquid latex resins are added in the field to impart greater bonding strength and resilience. Interior or exterior.	ANSI A118.4	ANSI A108.5	Not suitable for wood, metal, hardboard or particleboard backings.
EPOXY MORTAR	Two-part system of 1) epoxy resin and finely graded sand and 2) hardener. High bond strength and chemical resistance. Interior or exterior.	ANSI A118.3	ANSI A108.6	Requires special installation skills. Material and labor costs relatively high. Not suitable for walls.
EPOXY ADHESIVE	Two-part system of 1) epoxy resin and 2) hardener. Good bond strength and chemical resistance. Interior or exterior. A three-part system containing portland cement powder is also available.	NONE	NONE	Requires special skills. Installation cost high, but lower than epoxy mortar.
FURAN MORTAR	Two-part system of 1) furan resin and 2) hardener. Highest chemical resistance and installation cost. Interior or exterior.	NONE	NONE	Same as epoxy mortar.
ORGANIC (MASTIC) ADHESIVE	Prepared organic material, ready-to-use without addition of liquid or powder. Adequate bond strength and resilience for light duty on floors and most walls. Interior use only.	ANSI A136.1	ANSI A108.4	Not suitable for exterior use or interior use in continuous contact with water. Applied 1/16" thick; requires smooth backing.

1. Conductive dry-set mortar—product specifications, ANSI A118.2; installation specifications, ANSI A108.7.
2. Requires addition of water in the field.

(Continued) GENERAL RECOMMENDATIONS

FIG. WF3 SUMMARY OF GROUTING PRODUCTS[1]

Type	Description	Limitations
COMMERCIAL PORTLAND CEMENT GROUT	Prepared mix[2] of p.c. and other ingredients, resulting in a water-resistant, uniformly colored material; usually white, but available in colors. Commonly used for tile walls and floors subject to ordinary use. Suitable for interior or exterior.	Requires pre-soaking of most glazed wall tile and damp curing of grouted surface.
SAND-PORTLAND CEMENT	Field mixture of p.c., fine graded sand (sometimes lime), and water. Most commonly used grout for ceramic mosaic tile. Suitable for floors and walls, interior or exterior.	Same as for commercial cement grout.
DRY-SET GROUT	Prepared mix[2] of p.c., sand and additives which provide water retentivity & eliminate need for pre-soaking of glazed wall tile. Damp curing may develop greater strength in portland cement grouts. Generally used on walls, interior or exterior.	Damping of tile surface prior to grouting may be required under extremely dry conditions.
LATEX-P.C. GROUT	Prepared mix[2] of any of the above formulations, plus latex additives. It is less rigid and less permeable than regular cement grout. The addition of latex to portland cement grouts eliminates the necessity of damp curing for 72 hours. Can be used on walls or floors, interior or exterior.	Same as for dry-set grout.
SILICONE RUBBER GROUT	Ready-to-use one-part caulking-type material which cures from moisture vapor in the air. Used mainly for glazed tile walls. Flexible, mildew-resistant, stain-resistant.	Bonds only to clean, dry surfaces. Must be used with proper adhesive to prevent staining.
MASTIC GROUT	Ready-to-use one-part formulation, which does not depend on availability of water for bond or curing. Suitable for most interior walls and light duty floors. More flexible & stain-resistant than cementitious grouts; contains agent to eliminate mildew growth.	Not suitable for exterior use or for surfaces in continuous contact with water.
EPOXY GROUT	Ready-to-use two-part system consisting of epoxy resin and hardener. Formulation for wall tile omits coarse fillers which might scratch glaze; coarse silica fillers generally are included in grout intended for quarry tile and pavers. Highly resistant to chemical staining, hence used for industrial and kitchen floors, countertops.	Requires special skills for proper application. Material cost and labor higher than with other grout systems.

[1]No national industry standards have been established for most tile grouting products—except ANSI A118.3 for epoxy grout.
[2]Requires addition of water in the field.

than 4.5 lb. per sq. yd. Metal lath generally is used for mortar bed installations on walls.

A cleavage membrane should be used under all isolated (unbonded) tile finishes.

The membrane may be one of the following: 1) 15 lb. asphalt (or 13 lb. coal-tar) saturated roofing felt; 2) reinforced waterproof duplex asphalt paper; 3) polyethylene film at least 4 mils thick; or 4) troweled-on membrane waterproofing layer.

JOB CONDITIONS

Tile should be set and grouted when the temperature is at least 50° F (10° C) and rising.

The mortar manufacturer's recommendations should be checked to determine if a higher temperature is required, or when a lower temperature is permissible.

Backing surfaces and supporting construction should be inspected to determine their suitability prior to starting any work.

SUPPORTING CONSTRUCTION

Supporting construction should be rigid, sound and dimensionally stable. Where movement is anticipated, expansion and/or control joints should be provided in the supporting construction and extended through the tile finish.

(Continued) GENERAL RECOMMENDATIONS

The problem of movement is less critical with mortar bed installations, which generally are isolated from the backing by a cleavage membrane. Although the membrane permits some independent movement in the supporting construction, the extent of movement which can be tolerated still is limited, and dimensionally stable, rigid construction is desirable.

Floors

Floors should be engineered so that the maximum deflection under full load does not exceed 1/360 of the span.

Concrete subfloor For thinset bonding directly to the floor slab, the surface should be steel troweled and fine broomed, with no variations in the plane or slope exceeding 1/8″ in 10′. For portland cement mortar bed installations, the variation should not exceed 1/4″ in 10′.

Concrete slabs should be properly reinforced, concrete quality should be controlled, and suitable joints should be provided in the slab. Slabs should be thoroughly moist-cured before tile is applied; liquid curing compounds, or other coatings which may reduce bond should not be used. Slabs should not be saturated with water at the time of tile application.

Wood Subfloor The floor assembly should consist of joists spaced not over 16″ o.c. and a subfloor of 5/8″ or thicker plywood or 1″ nominal boards.

For mortar bed installations, a single-layer subfloor of 1″ nominal boards is acceptable. For thinset installations with most tilesetting products (except epoxy mortar) double floor construction is required; a double floor consists of either a board or plywood subfloor, and an underlayment.

The underlayment should be Exterior type CC-plugged plywood, generally 1/2″ thick for residential use, and 5/8″ thick for more demanding installations, as in commercial buildings. Adjacent underlayment edges should not be more than 1/32″ above or below each other.

When epoxy mortar is used in residential construction, the underlayment can be omitted if a 5/8″ plywood subfloor is used and longitudinal joints between panels are supported with solid wood blocking 2″ x 4″ or larger.

A 1/4″ space should be left between the underlayment and all vertical surfaces such as walls or pipes. For organic adhesives, 1/16″ should be left between underlayment ends and edges; for epoxy mortar, 1/4″ should be left at ends and edges of panels and these joints should be filled with the mortar.

Concrete/glass fiber-reinforced backing board may be used in lieu of a portland cement mortar bed over 1/2″ thick plywood subfloor. This backer board is designed for use with ceramic tile floors and is available in various sizes approximately 1/2″ thick. It can be nailed or screwed to wood joists through 1/2″ plywood subflooring. Ceramic tile can be bonded to it with either dry-set or latex-portland cement mortars. Adhesives are not recommended for use with this product.

Walls

Wall surfaces should be sound, plumb and true, and should not vary from a true plane by more than 1/8″ in 8′.

Gypsum Wallboard Wallboard should be installed in maximum practical lengths to minimize number of end joints. All joints should be taped, and both joints and fasteners should be treated with at least two coats of joint compound. In tub and shower areas, water-resistant gypsum backing board should be used for thinset installations. Joints should be taped and filled with one coat (maximum 4″ wide) of joint compound.

Gypsum Plaster Plaster walls intended to receive tile should be steel troweled, using a brown coat mix (1

FIG. WF4 SUMMARY OF RELATED MATERIALS & PRODUCTS

Material or Product	Industry Standard
PORTLAND CEMENT	
Grey or White, Type 1	ASTM C-150
SAND	
for Concrete	ASTM C-33
for Mortar	ASTM C-144
HYDRATED LIME	ASTM C-206
Type S	ASTM C-207
SEALANTS	
Two-part	TT-S-00227E
One-part	TT-S-00230C
Silicone Rubber	TT-S-001543A
METAL REINFORCEMENT	
Welded Wire Fabric	ANSI A-42.3
Expanded Metal Lath	ANSI A-42.4
CLEAVAGE MEMBRANE	
Polyethylene Film	ASTM C-171
Asphalt Roofing Felt	ASTM D-226
Coal Tar Roofing Felt	ASTM D-227
Reinforced Asphalt Paper	FS UU-P264

(Continued) **GENERAL RECOMMENDATIONS**

part gypsum to 3 parts sand) or a prepared finish coat plaster.

When it is desired to have the tile finish flush with the adjacent plastered surface, the finish coat is omitted and the tile is set on the brown coat. Otherwise, a full thickness 3-coat plaster surface is employed.

The surface which is to receive tile should be troweled smooth with no trowel marks or ridges over 1/32″ high.

Plywood The plywood should be Exterior type CC-plugged, 3/8″ or thicker. Panels should be applied with the face grain perpendicular to the studs or furring. Solid 2″ x 4″ blocking should be used under panel edges between studs. Nails should be driven 6″ o.c. at panel perimeters and 12″ o.c. at intermediate supports.

Panels should be spaced 1/8″ and there should be not more than 1/32″ variation between faces of adjacent panels. If horizontal blocking between studs is to be eliminated and panels are installed vertically, then plywood should be at least 1/2″ thick. All plywood edges should be sealed with an exterior primer or aluminum paint prior to installation.

Concrete/Glass Fiber-reinforced Backing Board *This is backing board designed for use with ceramic tile in wet areas. It is available in various sizes approximately 1/2″ thick, and can be nailed or screwed in place over wood or metal studs. Ceramic tile can be bonded to it with either dry-set or latex-portland cement mortars. It can be used in place of metal lath, portland cement scratch coat and mortar bed.* Adhesives should not be used over concrete/glass fiber-reinforced backing board.

Brick, Block and Concrete For thinset bonded installations, the wall surface should be clean, sound, free of paint, efflorescence, cracked or chipped areas. When latex-p.c. or dry-set mortars are to be used, smooth concrete or brick surfaces should be roughened by sandblasting or bush-hammering to assure good bond. For mortar bed application, metal lath should be secured to a sound surface.

TILE LAYOUT AND SETTING

Tile should be laid out symmetrically and so as to produce larger than half-tiles around the perimeter of the area to be covered. Individual tiles should be cut with a suitable tool and should be closely fitted to projections.

All joint lines should be straight, plumb, level and of even width. Accessories in the tilework should be evenly spaced, properly aligned with tile joints and set in the proper plane. When the tile is not of the self-spacing type, joint widths should be as shown in Figure WF5.

FIG. WF5 **RECOMMENDED JOINT WIDTHS FOR CERAMIC TILE**

Tile Description	Joint Width
CERAMIC MOSAIC TILE*	
2″ square or smaller	1/32″ to 1/8″
PAVER & SPECIAL TILE	
2 3/16″ to 4 1/4″ square	1/16″ to 1/4″
6″ square and larger	1/4″ to 1/2″
QUARRY TILE	
3″ to 6″ square	1/8″ to 3/8″
6″ square and larger	1/4″ to 1/2″
GLAZED TILE	
3″ square and larger	1/8″ to 1/2″

*Ceramic mosaic tile usually is mounted, so joint widths are predetermined by manufacturer.

GROUTING AND CURING

Enough grout should be forced into each joint, so that the joint is filled completely, down to the mortar. Joints of square-edged tile should be filled flush with the tile surface; joints of cushion-edged tile should be tooled to the depth of the cushion.

The finished grout surface should be uniform in color, smooth and free of pinholes or low spots. The tile surface should be washed thoroughly after grouting, then polished with a dry clean cloth.

Glazed wall tile should be soaked prior to setting with p.c. mortar. When glazed tile has been set with a material other than p.c. mortar and was not presoaked prior to installation, the tile should be soaked after setting, if a p.c. grout is to be used.

Commercial cement and p.c. grout always should be damp-cured for at least 72 hours. Drywall grout should be damp-cured, when used: (1) on floors; (2) on floors or walls exposed to periodic wetting; or (3) in exterior locations.

Damp-curing can be accomplished by periodically applying a spray mist to the tile surface, or by covering for 3 days with polyethylene film after initial and daily spraying.

(Continued) **GENERAL RECOMMENDATIONS**

CONTROL AND EXPANSION JOINTS

Joints in Backing

The backing should be provided with suitable construction, control, isolation or expansion joints, as required to relieve internal stresses and to minimize random cracking.

Joints in Tile Finish

Expansion joints in the tile finish always should be provided: (1) directly over expansion, isolation, construction or control joints in the backing; (2) where the tile finish abuts projecting surfaces; and (3) at proper intervals in floor areas measuring over 24′ in either direction. Expansion joints in the tile finish should not be narrower than the underlying joint in the backing.

Expansion joints are not required in areas less than 12′ wide and over old, well cured concrete.

Sealant depth in a joint should be approximately 1/2 the joint width. Compressible back-up strips should be used in the joint to prevent bonding of the sealant to the joint filler and to insure a proper sealant profile.

Tile edges to which the sealant will bond should be clean and dry. Sanding or grinding of these edges is recommended to obtain maximum sealant bond. Some sealant manufacturers recommend priming of edges. When a primer is used, care should be taken to keep it off the tile face.

Exterior Locations Exterior installations should be provided with expansion joints sized as follows: at least 3/8″ wide for joints 12′ o.c.; at least 1/2″ wide for joints 16′ o.c.

These minimum joint widths should be increased 1/16″ for each 15° F of temperature differential (range between summer high and winter low) over 100° F. Decks exposed to the sky, located above 40° N latitude, should have 3/4″ wide joints 12′ o.c.

Interior Locations Mortar bed installations in interior areas should have expansion joints in the tilework spaced 24′ to 36′ each way. Joints for interior quarry tile and paver tile should be the same width as the grout joints, but not less than 1/4″. Joints for ceramic mosaic tile and glazed wall tile preferably should be 1/4″ wide but never less than 1/8″.

Large thinset installations may benefit from appropriately located expansion joints, but no generalizations can be drawn. Joints always are desirable at offsets and changes in backing construction.

THRESHOLDS

A suitable threshold should be provided in doorways and in cased openings if the tile finish on one side is substantially higher or lower.

A marble or slate threshold typically is used; these materials can be set with the same mortars or adhesives as ceramic tile.

SHOWER RECEPTORS

Over a wood subfloor, a tiled shower receptor should be installed only by the mortar bed method. A waterproof membrane should underlie the tile finish and it should be sloped 1/4″ per foot to the drain. The sides of the membrane should be turned up the wall 5″.

When the subfloor is concrete, the subfloor itself should be sloped to the drain. Over a wood subfloor, a separate p.c. mortar fill should be used to create the slope under the membrane.

TILE OVER OTHER FINISHES

For thinset installations, the backing surface should be free of contaminants such as dirt, dust, efflorescence, oil or grease. The maximum variation in backing surface should not exceed the dimensions indicated in Supporting Construction (page WF442-4).

Mortar bed installations involving metal reinforcement are less demanding and present no special problems over existing finishes.

Painted surfaces which are glossy or scaly should be treated by sandblasting or chemical paint remover. The adhesive or mortar manufacturer should be consulted to determine if a sealer or primer is needed.

COUNTERTOPS

The backing for a mortar bed installation should be 1″ x 6″ boards or 3/4″ exterior plywood. Boards should be spaced 1/4″ apart. Plywood should be scored with saw cuts 6″ to 8″ o.c. to prevent warping.

For thinset (epoxy) installation, only 3/4″ exterior plywood should be used. A 1/4″ gap should be provided between plywood panels and the gap should be closed with a batten strip from below.

INSTALLATION ON FLOORS

The two basic methods of ceramic tile installation— thinset *and* mortar bed*—are illustrated in Figure WF6.*

INSTALLATION PERFORMANCE LEVELS

The type of traffic (wheeled or foot traffic, impact loads) and the contaminants (water, grease, alkalis, chemicals) to which the floor will be subjected are prime factors in establishing the performance level *(Fig. WF7). Once the performance level has been determined, an appropriate* installation method *can be selected (Fig. WF8).*

INTERIOR CONCRETE

Because interior concrete slabs are not subject to drastic temperature and humidity fluctuations, the tile finish can be bonded *directly to the slab by the thinset method or by the mortar bed method. However, when a mortar bed is used, it is generally reinforced with wire mesh and a cleavage membrane is used to* isolate *the tile finish from stresses in the slab.*

If a concrete slab-on-grade is made up of a properly engineered concrete mix, with suitable reinforcement, adequate finishing and curing, a well drained granular base, and proper jointing—the likelihood of excessive movement is minimized and there is no need to isolate the tile finish. Under these conditions, tile can be applied safely by the thinset method directly to the slab or over a bonded mortar bed.

Portland Cement Mortar Bed

Mortar bed thickness should not exceed 1-1/4″. The tile should be tamped firmly into the mortar to obtain maximum contact. All tile surfaces should be brought to a true level at proper elevation. Tamping and leveling should be completed within one hour after placing the tiles.

When installing face-mounted ceramic mosaic tile, the paper should be wetted within one hour, using no more water than necessary for removing the paper and glue. Any tiles that are out of line or level should be adjusted before the mortar assumes initial set.

Excess grout should be removed from the surface. Before the grout sets, all voids and gaps should be filled. When portland cement-based grouts are used, the entire installation should be damp-cured for 72 hours. When other grouting materials are used, manufacturer's instructions should be followed.

The tile surface should be thoroughly cleaned while the grout is still plastic. Muriatic acid may be used to clean glazed tile; sulfamic acid can be used to clean unglazed tile 10 days after setting. Joints should be soaked before acid is applied and should be flushed thoroughly with water after cleaning. All metal and enameled surfaces should be protected from the acid with a strippable coating.

Latex-p.c. and Dry-set Mortars

The slab should be well cured, free of cracks, waxy or oily films and curing compounds. The slab surface should not vary by more than 1/4″ in 10′ and should be steel troweled and fine broomed finished.

Mortar should be mixed as recommended by the manufacturer. Mortar should be troweled in a layer approximately 1/4″ thick for ceramic mosaic tile and 1/4″ to 3/8″ thick for quarry and paper tile. The tile should be tamped into the mortar to insure a level surface and intimate contact.

Paper should be removed from face-mounted tile within one hour of setting, using a minimum of water. Grouting should be delayed for at least 24 hours, or until the tile is firmly set.

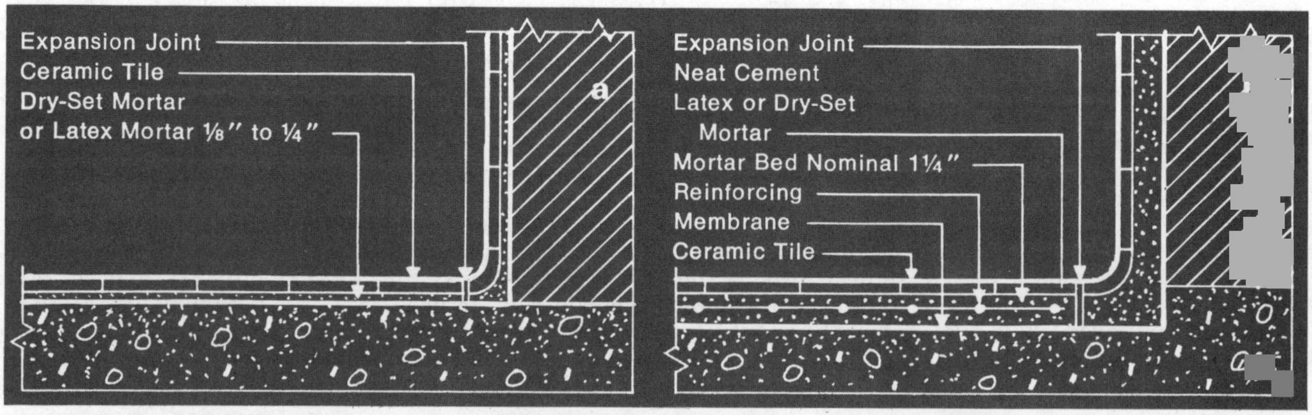

FIG. WF6 Basic ceramic tile installation methods: (a) thinset method—tile bonded to the backing; (b) conventional (thick-bed) method—tile and mortar setting bed isolated from the backing by membrane.

(Continued) INSTALLATION ON FLOORS

Organic Adhesives

For organic adhesives, the slab surface should be *steel-troweled* and fine broomed and the surface should not vary by more than 1/16″ in 3′.

Surface requirements are more demanding for organic adhesives because they are applied in a bond coat only 1/16″ thick. Most other tilesetting materials are applied in thicker coats, hence rougher, wood-floated, surfaces are adequate.

FIG. WF7 INSTALLATION PERFORMANCE LEVEL RATINGS*

Area Description	Type of Use or Traffic	Performance Level
Office space, commercial reception areas, public spaces in restaurants & stores, corridors, shopping malls:	a) Light high-heel traffic b) Heavy high-heel traffic	Moderate Heavy
Kitchens:	a) Residential b) Commercial c) Institutional	Light or Moderate Heavy Extra Heavy
Toilets, bathrooms:	a) Residential b) Commercial c) Institutional	Light or Moderate Moderate or Heavy Heavy
Hospitals:	a) General b) Kitchens c) Operating rooms	Heavy Extra Heavy Heavy
Exterior decks:	a) Roof decks, large on-grade patios b) Walkways, small patios	Extra Heavy Heavy or Extra Heavy

*PERFORMANCE-LEVEL RATINGS DEFINED:
LIGHT: Normal residential foot traffic; occasional 300 pound loads on soft rubber wheels.
MODERATE: Better residential, normal commercial and light institutional use; 300 pound loads on rubber wheels and occasional 100 pound loads on steel wheels.
HEAVY: Heavy high-heel traffic; 200 pound loads on steel wheels, 300 pound loads on rubber wheels.
EXTRA HEAVY: High impact service, as in institutional kitchens and industrial work areas; 300 pound loads on steel wheels.

FIG. WF8 SUITABLE METHODS FOR VARIOUS PERFORMANCE LEVELS

Performance Level	Installation				Remarks
	Subfloor	Tile	Mortar or Adhesive	Grout	
LIGHT	Wood	Ceramic Mosaic or Quarry	Organic	Latex-p.c.	Low cost; O.K. for home bath, entry
	Wood	Ceramic Mosaic	Epoxy	Sand-p.c.[1]	Costlier; OK. for kitchen, dining rm.
	Existing tile finish	Ceramic Mosaic or Quarry	Epoxy	Mastic or Latex-p.c.	Residential renovation
MODERATE	Wood	Ceramic Mosaic	P.c. mortar bed	Sand-p.c.[1]	O.K. for depressed wood subfloor
	Wood	Ceramic Mosaic or Quarry	Epoxy	Epoxy	Very costly; very durable even over wood subfloor
	Concrete	Ceramic Mosaic	Dry-set	Sand-p.c.[1]	Moderate cost; smooth surface
	Concrete	Ceramic Mosaic	Dry-set	Latex-p.c.	Moderate cost; eliminates damp curing of grout
HEAVY[3]	Concrete	Ceramic Mosaic or Quarry	Dry-set[2]	Epoxy	Costlier; good chemical resistance
	Concrete	Ceramic Mosaic or Quarry	P.c. mortar bed	Sand-p.c.[1]	Moderate cost; bed assures smooth surface
	Concrete	Ceramic Mosaic or Quarry	Dry-set	Sand-p.c.[1]	Low cost; good all purpose floor
	Concrete	Ceramic Mosaic or Quarry	Epoxy	Epoxy	Very costly; most durable and easy to maintain

[1]Damp cured for 3 days after grouting.
[2]Damp cured for 3 days before grouting.
[3]All installations suitable for EXTRA HEAVY performance when packing house tile is substituted for mosaic tile.

(Continued) **INSTALLATION ON FLOORS**

Epoxy Mortar (or Adhesive)

The subfloor surface should be steel troweled and fine broomed and should not vary over 1/4″ in 10′.

The epoxy mortar (or adhesive) should be of the same color as the grout. The tile should be tamped into the mortar bed, taking care not to smudge the face of the tile.

Grouting should be delayed at least 16 hours. The grout should be forced into the joints so as to completely fill them.

The installation should be shaded from direct sunlight and kept at a relatively even temperature (over 60° F) for at least 8 hours after grouting. All traffic should be kept off the floor for at least 40 hours. Light traffic can be permitted after this period; heavy traffic, after 7 days.

Where chemical resistance at the tile surface is desirable but high bond strength with the backing is not essential, epoxy grout can be used with latex or dry-set mortars. These cementitious mortars generally produce adequate bond and are less expensive than epoxy mortar.

EXTERIOR CONCRETE

Except for installations over built-up roofing or waterproofing membranes, exterior tile finishes are likely to be bonded directly to the slab. Therefore, it is important that the normal slab and tile movement due to shrinkage and temperature changes be accommodated with suitable control or expansion joints.

Expansion joints extending through the tile finish and the mortar bed should be provided at intervals of 12′ to 16′ each way. Expansion joints should be provided also where the tile finish abuts restraining vertical surfaces and where expansion, isolation or control joints are provided in the slab. Both the structural slab and the tile surface should be sloped to drain.

Exterior finishes can be installed over a p.c. mortar bed, or with thinset mortars such as dry-set and latex-p.c. Installation procedures are the same as described for interior concrete.

WOOD SUBFLOORS

Wood subfloors expand and contract with moisture changes. It is important therefore for the tile installation to be either completely isolated from the subfloor or bonded to it with a mortar which can resist the differential movement.

Portland Cement Mortar

Mortar bed installations on wood subfloors should be isolated from the subfloor with a cleavage membrane.

A mortar setting bed is installed over a wood subfloor the same way as over a concrete subfloor. The grout can be any of the several commercially available formulations, as well as sand-p.c. grout. The complete tile installation, including the mortar bed is approximately 1″ to 1-1/2″ thick.

The subfloor should not be depressed below the tops of the joists.

To gain 5/8″ to 3/4″ in finish height, a plywood subfloor can be installed between rather than over the floor joists. The subfloor should be supported on cleats nailed to the sides of joists so that the subfloor and joists form a plane, level, surface.

Organic Adhesive

Organic adhesive should be used only over a double floor, including a subfloor and a plywood underlayment at least 3/8″ thick. This adhesive should not be used in areas subject to excessive moisture (sunken tubs or swimming pools) and heavy traffic (commercial and public areas). For bathroom floors and other areas subject to occasional moisture, type I adhesive should be used.

Type I adhesive is capable of withstanding more prolonged exposure to moisture than type II adhesive. The adhesive should be spread with a notched trowel recommended by the manufacturer and the tile firmly pressed into the adhesive.

Epoxy Mortar (or Adhesive)

Epoxy mortar (or adhesive) generally should be used over a double floor, including a subfloor and a plywood underlayment at least 1/2″ thick. The underlayment should be installed with 1/4″ gaps between panels and these gaps should be completely filled with epoxy mortar.

In residential use the underlayment can be omitted if the subfloor is 5/8″ or thicker plywood, installed perpendicular to joists, with edges blocked.

Grouting should be delayed for 16 hours. The installation should be shaded from direct sunlight and kept at over 60° F for 8 hours or more after grouting. Grout and mortar should be cleaned off the face of the tile as soon as possible and before it hardens.

INSTALLATION ON WALLS

INTERIOR SURFACES

Suitable backings for tile finishes include masonry and concrete walls, as well as plywood, gypsum board and plaster, over furring or stud framing. In tub and shower areas, the thinset method is acceptable, but it is essential that the backing be inherently water-resistant or properly prepared to render it so.

Portland Cement Mortar Bed

The one-coat mortar method should be used only over a solid backing. The single (bed) coat should be 3/8″ to 3/4″ thick and should consist of 1 part p.c., 1/2 part hydrated lime and 5 parts damp sand.

The one-coat method can be used over existing wall finishes which are not suitable for thinset application because of surface conditions. It is also a preferred method of tiling tub and shower areas in new work, especially in the western states. The two-coat method, involving both a scratch coat and a bed coat, is used most frequently over open framing or furring.

On metal studs, the overall thickness of scratch coat and bed coat should not exceed 1″; on wood studs overall mortar thickness should not be over 1-1/2″. The scratch coat should be allowed to cure for at least 24 hours and should be dampened thoroughly before the float coat is applied. The float coat should be applied in areas no greater than may be covered with tile while the bed is still plastic.

Dry-set and Latex-p.c. Mortars

The backing surface should be dampened just before spreading mortar. Mortar should be troweled on in appropriate thickness (3/32″ to 1/4″ maximum) in an area which can be readily covered with tile before the mortar starts skinning over. The tile should be tamped into place to obtain at least 80% coverage; in tubs, showers and other high-moisture areas 100% coverage should be obtained.

Any grout can be used, but in selecting a suitable type, consideration should be given to the type of exposure. In residential installations, sand-p.c., commercial cement, drywall grout and mastic grout are most likely.

Organic Adhesive

Type I adhesive should be used for installations requiring more *prolonged* water-resistance, such as on walls of tub and shower areas. Type II adhesive should be used in areas subject to occasional or *intermittent* wetting.

However, neither type is suitable for installations requiring continuous contact with water such as in sunken tubs or pools.

Tile should be tamped into the adhesive to assure maximum contact. The *average* contact area of tile removed for inspection should be not less than 75%, and no *individual* tile or mounted assembly should have less than 40% coverage of adhesive. Grouting should be delayed for at least 24 hours to permit solvent evaporation from the adhesive.

EXTERIOR SURFACES

Suitable backings for exterior tile finishes include brick and block masonry, monolithic concrete and other moisture resistant, stable construction. On large areas exposed to direct wetting, the mortar bed isolated *method usually gives better results. The thinset* bonded *method generally is more reliable in covered outdoor areas and when broken into smaller panels by decorative trim, battens or expansion joints.*

Regardless of installation methods, expansion joints should be provided in the tile finish over all underlying structural joints, at the tile perimeter, and against any projecting surface. In isolated installations, expansion joints in the tile finish should be provided in addition to those over underlying joints.

In a thinset installation over a suitable stable backing, the additional joints in the tile finish are not as essential because the tile finish is intended to move with the backing.

Suitable flashings, overhangs or copings to prevent entry of water into the tilework should be provided in all exterior installations. Exterior walls of habitable spaces should include a suitable vapor barrier on the warm side of the wall to prevent condensation in the wall cavity.

Portland Cement Mortar Bed

The lath and mortar bed should be backed by a cleavage membrane. Metal lath should be firmly attached to the backing and cut at all expansion joints. Scratch coat should consist of 1 part p.c., 5 parts damp sand and 1/5 part lime; mortar bed should consist of 1 part p.c., 5 parts damp sand and 1/2 part hydrated lime. The overall thickness of both coats should be 3/4″ to 1″ thick.

(Continued) **INSTALLATION ON WALLS**

The mortar bed mix can be varied up to 1 part p.c., 1 part lime and 7 parts damp sand. However, use of too much portland cement should be avoided because this results in greater drying shrinkage and cracking. Similarly, a total mortar bed thickness in excess of 1" should be avoided.

Neat cement should be used for setting tile only over a still plastic mortar bed. If dry-set or latex-p.c. mortar is to be used the mortar bed should be cured for 24 hours or longer.

Curing periods up to 10 days are preferred. Suitable grouts include commercial cement, drywall, latex and conventional sand-p.c. All of these grouts require damp-curing at least 72 hours for exterior use.

Thinset Installation

Backings for direct bonding with thinset mortars should be smooth, sound and dimensionally stable. Suitable control and expansion joints should be provided to minimize internal stresses. Tile finish should be cured for 72 hours or longer.

If the masonry or concrete backing is too irregular for direct bonding, a p.c. leveling coat can be applied first and the tile applied to the new surface. Latex p.c. and dry-set mortar typically are used for exterior thin-setting of tile. Either latex of dry-set grout can be used for filling tile joints.

452 RESILIENT FLOORING

RESILIENT FLOORING

INTRODUCTION

Resilient flooring typically is used where economical, dense, nonabsorbent wear surfaces are desired. Depending on the product, resilient floor finishes offer a wide range of decorative effects and may provide underfoot comfort, good durability and relative ease of maintenance. Specific products with enhanced properties are available to suit most functional requirements in residential construction.

Suitability of a flooring product is related to the product's physical properties and intended use. In addition to initial installed cost, ease of maintenance and durability are important considerations in determining suitability. These factors in turn depend on subfloor conditions, proper installation and maintenance.

This section describes the materials and methods recommended for the installation of permanently bonded resilient floors. Temporary floor coverings of thinner gauge and less durable wear surfaces are not discussed. These often are laid on the floor loose or are only partially cemented and are not considered permanent installations. The text is divided into subsections beginning with *General Recommendations* typical of most resilient floors, followed by subsections giving more specific recommendations for the application of *Sheet Flooring* and *Tile Flooring.*

The manufacture, physical properties and selection criteria of sheet and tile products are discussed in Section 220 Resilient Flooring Products.

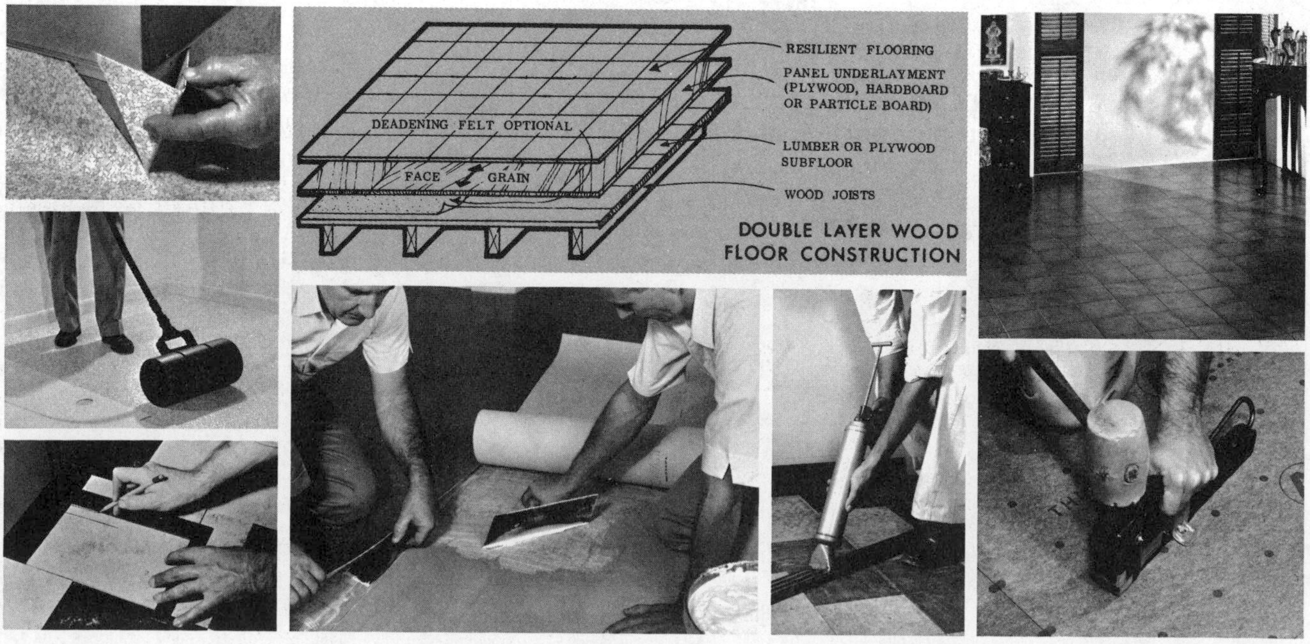

GENERAL RECOMMENDATIONS

The installation of all resilient flooring, whether sheet or tile, depends on (1) proper job conditions, (2) adequate subfloor construction, and (3) proper selection of underlayment, adhesives and flooring material. The following subsection discusses general recommendations for the installation of flooring in new and existing construction.

Because of the variety of proprietary flooring materials, underlayments and adhesives, the individual manufacturers' recommendations always should be consulted to insure proper use in each case.

TERMINOLOGY

An understanding of terminology is basic to the discussion of flooring installation.

Subfloor The structural material or surface which supports floor loads and the finish floor, and serves as a working platform during construction. If the subfloor is sufficiently dense, smooth, stiff, dimensionally stable and possesses adequate bonding properties, resilient flooring may be applied directly without the use of underlayment.

Subfloor Location The selection of resilient flooring products and adhesives is determined largely by subfloor location—whether below grade, on grade or suspended (Fig. 1). These designations are based mainly on the relative likely exposure to moisture and alkali, which adversely affect many flooring materials and adhesives.

Underlayment A mastic or panelboard material installed over the subfloor to provide a suitable base for resilient flooring when the subfloor does not possess the necessary properties for direct application of the flooring.

Hardboard is a rather dense panelboard manufactured of wood fibers with the natural lignin in the wood reactivated to serve as a binder for the wood fibers.

Particleboard is a panelboard manufactured of wood chips and/or flakes or other ligno-cellulosic material together with a synthetic resin adhesive.

Lining Felt An asbestos or asphalt felt specifically manufactured for use under resilient flooring.

Primer A brushable solvent-base asphaltic preparation recommended as a first coat over porous or dusty concrete floors and panel underlayments; intended to seal the pores and improve the bond with asphaltic adhesives used for the installation of asphalt and vinyl-asbestos tile.

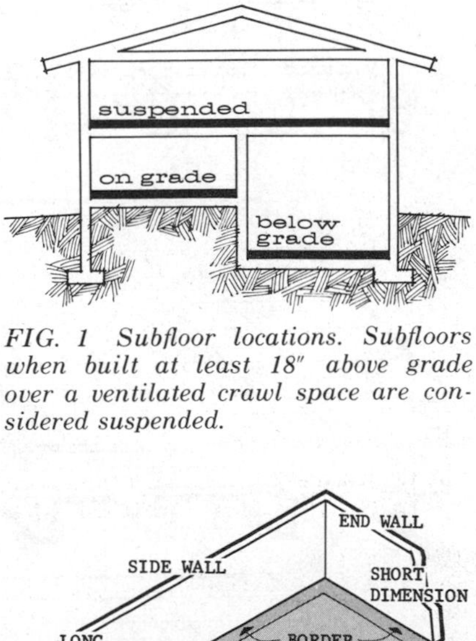

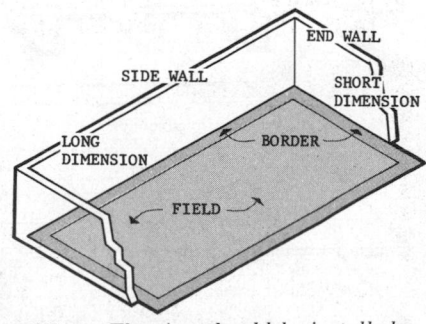

FIG. 1 Subfloor locations. Subfloors when built at least 18" above grade over a ventilated crawl space are considered suspended.

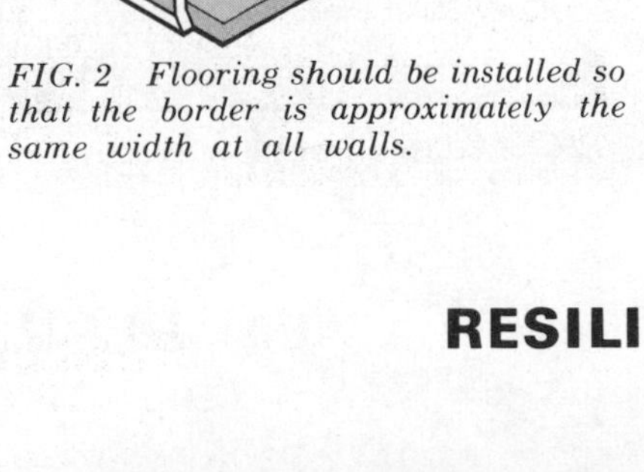

FIG. 2 Flooring should be installed so that the border is approximately the same width at all walls.

Sealer A solution of equal parts of wax-free shellac and denatured alcohol; recommended as a first coat over existing stripwood floors from which finish has been removed; intended to seal the wood pores, prevent excessive moisture absorption and provide a dimensionally stable base for direct application of lining felt or resilient flooring.

Solvent-base Adhesives and primers consisting of cementitious binders and fillers *dissolved* in a volatile hydrocarbon carrier such as alcohol or cutback.

Emulsion Adhesives and mastic underlayments consisting of cementitious binders and fillers *suspended* in a liquid carrier such as water.

Installation Terms

The following terms may be used to describe the area to be covered (Fig. 2) and some of the operations involved in the application of resilient flooring.

Endwall The wall along the short dimension of the room.

Sidewall The wall along the long dimension of the room.

Field The middle portion of the flooring installation, exclusive of the border.

Border The flooring at the perimeter of the room adjacent to the walls, which is installed separately from the field.

Flashing Bending up sheet material against a wall or a projection, either temporarily for the purpose of fitting, or permanently so as to form a one-piece resilient base.

Scribing A method of transferring the profile of an obstruction, projection or flooring edge to a piece of flooring so that it can be accurately cut and fitted.

JOB CONDITIONS

Installation of finish flooring should be delayed until all other interior work, with the possible exception of door trim and touch-up painting, has been completed. If it is necessary to install flooring sooner, the finished floor should be protected with plain, undyed unsaturated building paper. Door trim may be installed prior to flooring if it is shimmed up with scraps of flooring to permit later installation of the flooring under the trim.

Care should be taken to prevent freezing and damage to materials during delivery, handling and storage. Materials should be stored on a smooth, level surface in a warm, dry place. Sheet flooring should not be stored for long periods of time and the rolls should be stored vertically, on end. Tile flooring should be stacked not more than five cartons high.

A temperature of at least 70° F. and not over 90° F. should be maintained for 48 hours before installation, during installation and for 48 hours after installation. A minimum temperature of 55° F. is recommended thereafter, and adequate ventilation should be provided at all times. Ventilation is particularly important when using solvent-based adhesives and primers which are highly volatile and flammable.

FLOOR CONSTRUCTION

The two main types of subfloor typically used as a base for flooring in residential construction are wood and concrete. All subfloors, whether covered with underlayment or not, should be firmly fastened and structurally adequate to carry the intended loads without excessive deflection.

Figure 3 illustrates *wood joist* and *concrete slab* floor construction suitable for the application of resilient flooring. Plank-and-beam construction is similar to the wood joist construction illustrated, but the subfloor is 2″ or thicker and structural members are larger and spaced further apart. Underlayment is recommended over plank subfloors.

Concrete slab floors may be supported on ground or suspended above grade. Basement floors are concrete slabs supported on ground below grade.

Good site drainage is required for all foundation types, regardless of grade, and therefore the exterior finish grade should be sloped away from the foundation to divert surface water (see Fig. 15, page Foundation Systems 310-10).

Wood Subfloors

Suspended wood floors above grade usually are sufficiently removed from excessive external moisture sources, but floors over crawl spaces at or near grade may require the crawl space to be ventilated and the ground inside the crawl space to be covered with a suitable vapor barrier (see Work File, 313 Crawl Spaces).

Wood floors may be of single-layer construction consisting of a combination subfloor/underlayment, or of double-layer construction composed of separate subfloor and underlayment. Minimum recommended subfloor thickness for various joist spacing and nailing requirements are outlined in Figure 4. The use of thicker subfloors will result in stiffer floors.

Combination Subfloor/Underlayment Wood subfloors suitable for resilient flooring without underlayment should consist of rigid structural panel material such as plywood. Single lumber board construction is not acceptable for the direct application of resilient flooring.

For normal moisture conditions, sanded Interior Underlayment grade is suitable; however, for unusual moisture conditions, such as encountered in bathrooms and

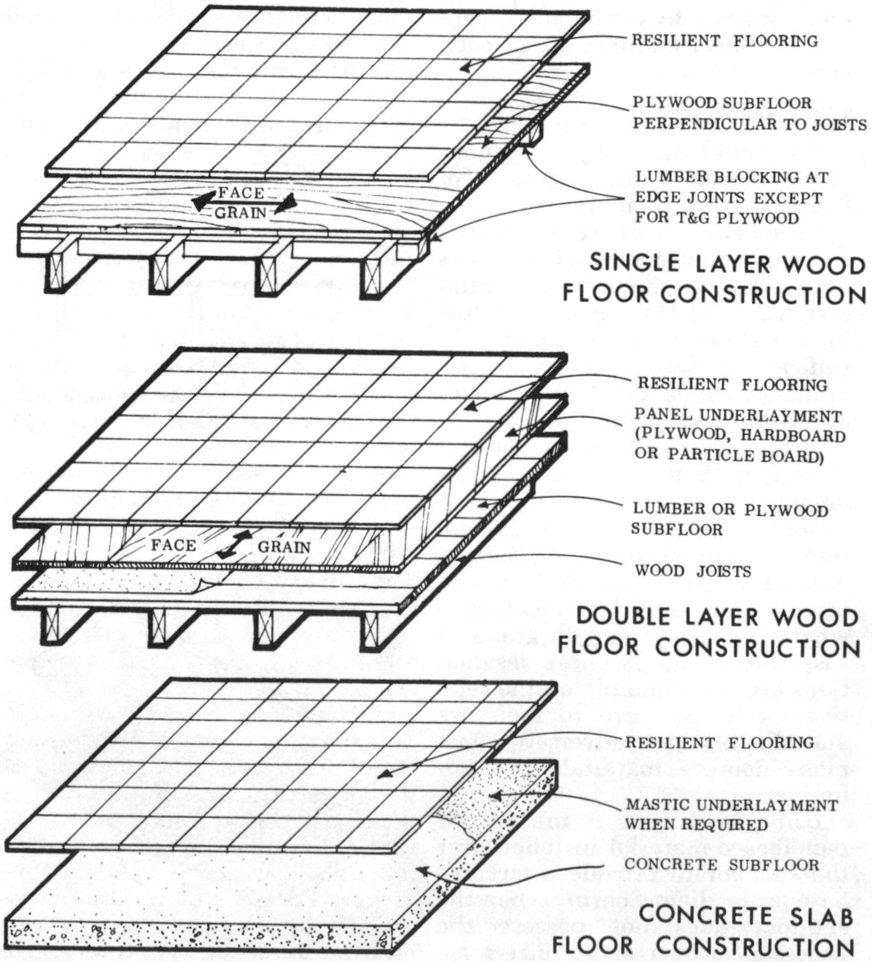

FIG. 3 Typical subfloor construction for the application of resilient flooring.

FIG. 4 WOOD SUBFLOOR CONSTRUCTION[1]

Combination Subfloor/Underlayment[2]		Subfloor[4]	Maximum Support Spacing	Subfloor with Underlayment[5]	
Nails[3]	**Nail Spacing**			**Nails[3]**	**Nail Spacing**
2" (6d) annularly threaded	6" o.c. around panel edges; 10" o.c. at intermediate supports	1/2" Plywood	16" o.c.	2" (6d) common	6" o.c. around panel edges; 10" o.c. at intermediate supports
		5/8" Plywood	20" o.c.	2 1/2" (8d) common	
		3/4" Plywood	24" o.c.		
2 1/2" (8d) annularly threaded	6" o.c. at panel edges and intermediate supports	1 1/8" Plywood (2-4-1)	48" o.c.	—	—
—	—	3/4" lumber boards less than 6" wide	16" o.c.	2 1/4" (7d) annularly threaded	2 per board at each support
—	—	3/4" lumber boards wider than 6" and less than 8" wide	16" o.c.		3 per board at each support

1. Based on 40 psf live load.
2. Blocking at longitudinal edges is required unless 5/8" or thicker plywood with T&G edges is used.
3. Common nails may be substituted for annularly threaded nails by increasing the size 1/4" (1d).
4. Plywood thicknesses given are for Group 1 Softwoods in accordance with U. S. Product Standard PS1. For Groups 2 and 3, increase thickness 1/8"; for Group 4, increase thickness 1/4" (see Work File, page Wood WF201-16).
5. Blocking not required if underlayment joints are staggered with respect to subfloor joints.

laundry areas, C-C (Plugged) Exterior, Underlayment grade bonded with Exterior glue, or a sanded grade of Exterior type plywood should be used (see Work File, page Wood WF 201-16).

The plywood panels should be installed with the surface grain perpendicular to the joists and end joints of the panels centered over the joists (Fig. 3). The longitudinal edges should be supported on 2" x 3" or heavier blocking. However, plywood with tongue-and-grooved edges 5/8" and 3/4" thick, and the 1-1/8" material known as "2-4-1" under the American Plywood Association grading rules (Fig. 5), may be used without blocking. Tongue-and-grooved 1/2" plywood is not recommended for single-layer construction without blocking.

Subfloor panels should be spaced approximately 1/32" at joints and approximately 1/8" at all intersection with vertical surfaces, to allow for expansion from humidity changes. Nailing should start in the middle of the panel and proceed outward to the edges, with nailheads set flush or slightly countersunk. Perimeter nailing should be located not more than 3/8" from the edges. Panel joints should be sanded smooth and flush prior to the flooring installation. It is important to protect single layer floors from damage during construction. However, the difficulty of furnishing adequate protection during construction may warrant the use of unsanded or touch-sanded plywood and then power sanding the entire floor area just before the flooring is installed.

Subfloor with Underlayment This type of construction is often desirable because the subfloor provides a convenient working platform during most of the construction period and installation of the underlayment can be scheduled more closely to the installation of the flooring.

Subfloors to be covered with underlayment may be lumber board or an unsanded plywood of adequate thickness, nailed as recommended in Figure 4. An Interior Standard grade of plywood is suitable for normal moisture conditions. For unusual moisture conditions, an Exterior C-C grade or an Interior Standard grade with Exterior glue should be used. Blocking is not required if joints in the underlayment are staggered. The subfloor is installed as recommended for single-layer construction.

The moisture content of lumber boards used for subflooring should be as recommended in the Work File, Figure WF1, page Wood WF 201-1.

Suitable panel underlayments and their installation are discussed on page 452-6.

Concrete Subfloors

Concrete subfloors should have a density of at least 90 lbs. per cu. ft. If lightweight concrete of lower density has been used, it should be refinished with a 1" topping of a regular concrete mix, bonded to the existing concrete with a bonding agent such as polyvinyl-acetate compound.

Subfloors on or below grade should be separated from the ground with a suitable vapor barrier such as polyethylene film (see Work File. 311 Slabs-on-Ground).

Concrete subfloors generally provide a suitable surface for resilient flooring, but require precaution against excessive moisture. The moisture content of concrete floors should be checked, particularly when lightweight concrete is used which is slow in curing. In some instances, dampness can be observed under stationary objects on a floor on or below grade, but most often moisture cannot be detected readily and tests should be performed to determine if the floor is sufficiently dry to receive the flooring.

Concrete slabs which have been

FIG. 5 Single-layer 1-1/4" T&G plywood can be installed on joists spaced 48" o.c. without blocking, but should be protected during construction.

treated with curing, hardening or separating compounds also should be tested for adequate bond.

Suitability Tests The following tests may be performed to determine acceptable moisture levels and bonding properties of concrete subfloors.

Relative Humidity Test is intended for testing new suspended concrete floors for the installation of moisture-sensitive products such as linoleum, cork, and rag felt-backed vinyl. A relative humidity meter is placed on the concrete next to an interior wall or column. Depending on atmospheric conditions, the meter should read 40% to 65%. The meter is then covered with an 18″ square of polyethylene film and the edges sealed to the floor with adhesive or tape. When substantial temperature fluctuations are expected, the meter should be protected with several layers of burlap or insulating batts. If there are no leaks in the polyethylene film, the reading should stabilize after 24 hours on a normal slab and after 72 hours on a thick, massive slab. A sufficiently dry slab will register a reading of 80% or less. If the reading is over 80%, the slab should be allowed to cure further and the test should be repeated.

Mat Moisture and Bonding Test is designed for checking moisture or surface bonding properties of concrete floors to establish suitability for installation of: (1) rubber tile, solid vinyl tile and vinyl sheet with asbestos fiber backing *on* or *below grade;* and (2) any resilient flooring on slabs from which paint, oil, concrete hardeners or curing compounds have been removed.

Linoleum or vinyl sheet mats approximately 36″ square are applied with two bands of adhesives in several locations on the subfloor. One band should be a water-soluble adhesive, the other a water-resistant latex adhesive. The mats are then sealed to the floor with adhesive or industrial tape (Fig. 6). They should remain in place for 72 hours, then be removed and the adhesive examined. If there is too much moisture present, the water-soluble adhesive will be partly or completely dissolved and the water-resistant adhesive will be stringy and will provide little bond. If excessive moisture is still indicated after further curing of the concrete and repeated testing, the above-mentioned flooring types should not be used and consideration should be given to the use of asphalt or vinyl-asbestos tile which have greater moisture resistance. However, for these products the primer test also should be performed.

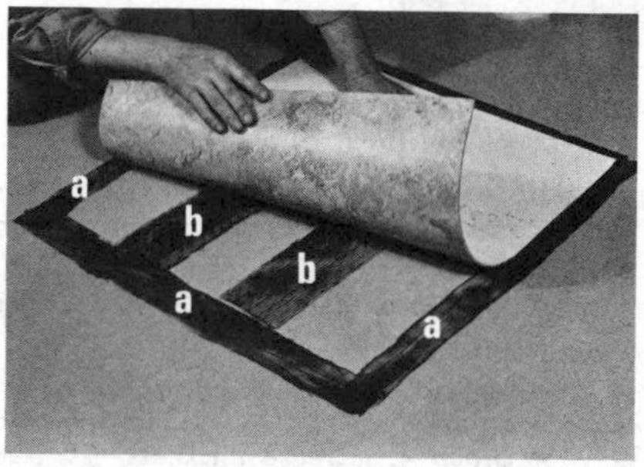

FIG. 6 The mat moisture and bonding test for suspended concrete subfloors: (a) water-resistant latex adhesive, (b) water-soluble adhesive.

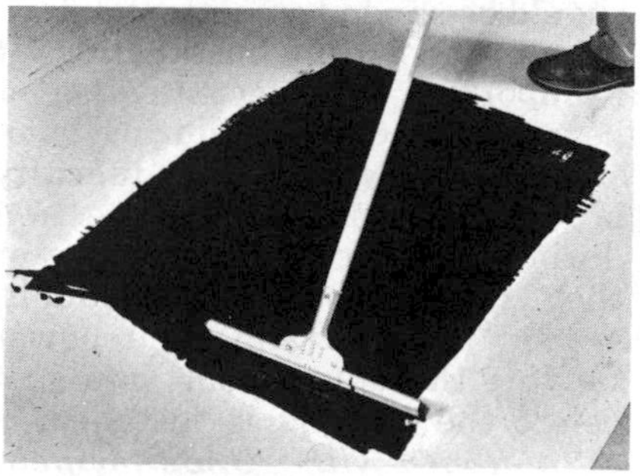

FIG. 7 The primer test determines moisture content of concrete slabs before installing asphalt or vinyl-asbestos tile.

Primer Test determines suitability for the installation of asphalt and vinyl-asbestos tile over suspended, on-grade and below-grade concrete floors. Patches of asphaltic primer are applied in several places on the floor and are allowed to dry for 24 hours (Fig. 7). If the primer does not readily peel and sticks to the floor when scraped with a putty knife, the subfloor is dry enough to receive asphalt or vinyl-asbestos tile.

Surface Preparation Expansion joints, scored joints, cracks and depressions should be filled with latex floor patching compound or mastic underlayment to obtain a smooth surface. Poorly finished or spalled concrete with an irregular surface may require grinding or smoothing with mastic underlayment. Curing or hardening compounds may leave surface films which often adversely affect adhesive bonding. If the mat bonding test reveals insufficient bonding properties, the film should be removed by power wire brushing or grinding with a terrazzo grinder. An asphaltic primer is required on dusty or porous concrete floors for the installation of asphalt and vinyl-asbestos tile.

UNDERLAYMENTS

Underlayments are used over subfloors to provide a smooth, level, hard and clean surface to receive flooring products. In general, panel underlayments are used over wood subfloors; mastic underlayments are applied over concrete subfloors (Fig. 3).

Panel

Panel underlayments may be used over any structurally adequate suspended subfloor such as plywood or lumber boards. The three common types of panel underlayment are *plywood, hardboard* and *particleboard.*

Stiffer floors result when joints in the underlayment are staggered with respect to joints in a plywood subfloor and blocking can be omitted. Over a board subfloor, underlayment joints should not fall over joints in the floor boards (Fig. 8). To allow for expansion from humidity changes, underlayment panels should be spaced approximately 1/32″ at joints and approximately 1/8″ at all intersections with vertical surfaces. End joints in the underlayment panels should be staggered ashlar fashion (Fig. 8).

Nailing should start in the middle of the panel and proceed outward to the edges, with nailheads set flush or slightly countersunk. Prior to the installation of flooring, panel joints should be sanded smooth and flush.

Minimum underlayment thicknesses are given below for each material; however, greater thicknesses may be warranted where increased stiffness is desired.

Plywood A sanded Interior Underlayment grade, at least 1/4″ thick, generally is suitable for plywood underlayment. In areas subject to excessive moisture, such as in bathrooms and laundry areas, an Exterior C-C (Plugged) or Underlayment grade with Exterior adhesive should be used. Over a plywood subfloor, underlayment may be installed with the face grain either parallel or perpendicular to the floor joists. However, if the underlayment is installed with the face grain perpendicular to the floor joists a stiffer floor assembly results. Over a board or plank subfloor, plywood should be installed with the face grain perpendicular to the direction of the floor boards (Fig. 8), and perferably with long edges over joists. Plywood underlayment should be installed with the fasteners and spacing recommended in Figure 9.

Hardboard Hardboard underlayment is suitable to bridge small irregularities and to provide a smooth surface over an otherwise structurally sound wood floor (Fig. 4). Underlayment grade hardboard is an untempered service type, surfaced one side. It is manufactured in a nominal 1/4″ (0.215″ actual) thickness, in 3′ x 4′ and 4′ x 4′ panels, and should conform to CS251 and the specifications of the American Hardboard Association. Installation of panels is the same as for other panel underlayment using fasteners and spacing recommended in Figure 9.

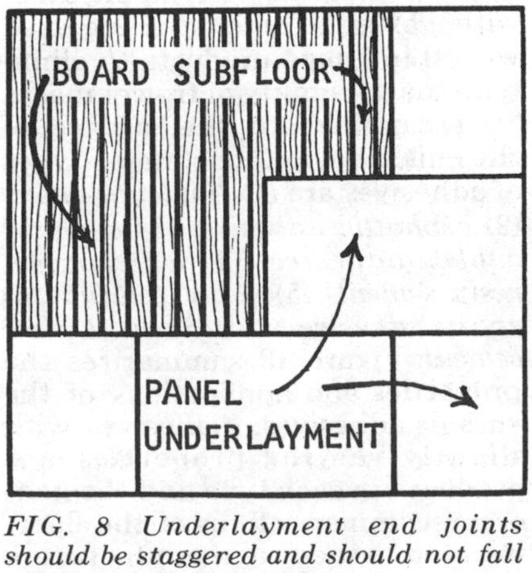

FIG. 8 Underlayment end joints should be staggered and should not fall over joints in the subfloor.

FIG. 9 FASTENING RECOMMENDATIONS FOR PANEL UNDERLAYMENT

Underlayment	Fasteners[1] Nails[2]	Staples[3]	Fastener Spacing
PLYWOOD			
¼″	1″ (2d)	⅞″	*Nails:* 8″ o.c. each direction in middle of panel, 6″ o.c. around panel perimeter. *Staples:* 6″ o.c. each direction in middle of panel, 3″ o.c. around panel perimeter.
⅜″	1¼″ (3d)	1⅛″	
½″	1¼″ (3d)	1¼″	
⅝″	1½″ (4d)	1½″	
¾″	1½″ (4d)	1⅝″	
HARDBOARD[4]			
¼″	1″ (2d)	⅞″	*Nails and Staples:* 6″ o.c. each direction in middle of panel, 3″ o.c. around panel perimeter.
PARTICLEBOARD[5]			
⅜″	1¼″ (3d)	1⅛″	*Nails:* 10″ o.c. each direction in middle of panel, 6″ o.c. around panel perimeter. *Staples:* 6″ o.c. each direction in middle of panel, 3″ o.c. around panel perimeter.
⅝″	1½″ (4d)	1½″	
¾″	1½″ (4d)	1⅝″	

1. Fastener lengths shown are minimum to provide ¾″ penetration for nails and ⅝″ for staples. If longer fasteners are used, penetration should not exceed 1″.
2. Annularly threaded; common nails may be used by increasing the sizes ¼″ (1d).
3. Galvanized divergent chisel, 16 gauge, ⅜″ crown width.
4. Nominal thickness, actual thickness 0.215″.
5. Moisture content as shipped from mill should not exceed 7% average.

Particleboard This type of underlayment should not be used in areas subject to excessive moisture, and its moisture content as shipped from the mill should not exceed an average of 7%. Particleboard underlayment should be at least 3/8″ thick and should conform to CS236 and the National Particleboard Association "Specifications for Mat-formed Wood Particleboard for Floor Underlayment." Particleboard underlayment generally is manufactured in 4′ x 8′ panels and in thicknesses ranging from 3/8″ to 3/4″. Nailing recommendations for various panel thicknesses are given in Figure 9.

Some resilient flooring manufacturers do not recommend the use of particleboard as an underlayment material. However, many particleboard manufacturers issue a guarantee covering the entire resilient flooring installation if the flooring and underlayment are installed to their specifications. If particleboard is used, it is recommended that it be a product backed by a manufacturer's guarantee and that the installation specifications of the manufacturer be followed.

Mastic

Suitable mastic underlayments contain a chemical binder such as latex, asphalt or polyvinyl-acetate resins and portland, gypsum or aluminous cement. Mixtures consisting of powdered cement and sand to which only water has been added function only as crack fillers and when applied in thin coats break down under traffic.

Latex underlayments are most suitable for applications requiring a thin layer (3/8″ or less), for skim coating and patching where the fill must be featheredged (Fig. 10).

FIG. 10 Latex underlayment can be feathered to a thin edge when patching subfloor depressions.

Skim coating does not attempt to raise the elevation of the floor, but merely smooths over surface irregularities.

Asphaltic and polyvinyl-acetate underlayments are used for installations requiring thicker underlayments (more than 3/8″). All mastic underlayments should be troweled smooth and true, with not more than 1/8″ variation from a straight line in 10′. Subfloors should be free of wax, oil or protective surface coatings before applying mastic underlayment.

Latex Latex underlayment consists of a mixture of a synthetic latex emulsion, a cementitious material such as portland cement, fine sand and water. The underlayment should be mixed in accordance with the manufacturer's recommendations and should be troweled on in layers not over 1/8″ thick and may be feathered to a thin edge (Fig. 10). The final thickness should be built up in several applications and should not exceed 3/8″. Any rough surfaces should be leveled with a sanding machine after the mastic has dried completely. Latex underlayment is a quick-setting compound and all splashed material should be cleaned off promptly.

Asphaltic solvent-base primers and adhesives used over latex underlayment for the installation of asphalt or vinyl-asbestos tile should be allowed adequate setting time before installing the tile.

Asphaltic This type of underlayment is a mixture of an asphaltic emulsion, a cementitious material such as portland cement, fine sand and water. Proper proportioning varies, depending on ingredients and thickness of the underlayment, and the manufacturer's instructions should be followed. Asphaltic underlayment generally is not recommended for installations less than 3/8″ in thickness.

The concrete subfloor should be primed with a thin solution of the same asphaltic emulsion being used in the underlayment mixture. The mastic is poured uniformly over the entire surface using wood screeds to establish the desired thickness. A straightedge is used to level the mastic over the screeds, after which the surface is troweled to a smooth finish. The setting time of the fill should not be accelerated with excessive heat, as this may cause cracking or crazing. Where it is necessary to trowel the mastic to a featheredge, the floor should be given a brush coat of undiluted asphaltic emulsion to prevent brittleness after the mastic has dried.

Solvent-base asphaltic primers and adhesives should not be applied over asphaltic underlayments; asphaltic emulsion or latex adhesives should be used.

Polyvinyl Acetate This underlayment is formed by mixing portland cement and sand with a specially compounded polyvinyl-acetate (PVAc) resin emulsion. The PVAc liquid also should be applied to the concrete subfloor to improve bond with the underlayment. When a fill of 1/2″ or thicker is required, a portland cement concrete "topping" may be used without PVAc resins, but the PVAc liquid still should be used as a bonding agent on the concrete subfloor.

This type of underlayment may be used with any adhesive and flooring type, and installation is similar to that described for asphaltic underlayment.

LINING FELT

Some manufacturers recommend the use of lining felt over panel underlayment to minimize the possibility of joints opening and tile splitting due to underlayment movement. However, this is not a significant hazard if the underlayment and/or subfloor is sufficiently thick and is properly installed. Of greater significance is the fact that indentation resistance of flooring materials is substantially reduced by lining felt, which imposes limitations on its use.

Asbestos lining felt is used in conjunction with some vinyl sheet products over nonporous subfloors, such as old resilient floors, terrazzo, ceramic tile, and concrete that has been treated with curing agents. The liner's purpose is to provide a means for the water in the adhesive to be absorbed, which allows the adhesive to set up and dry.

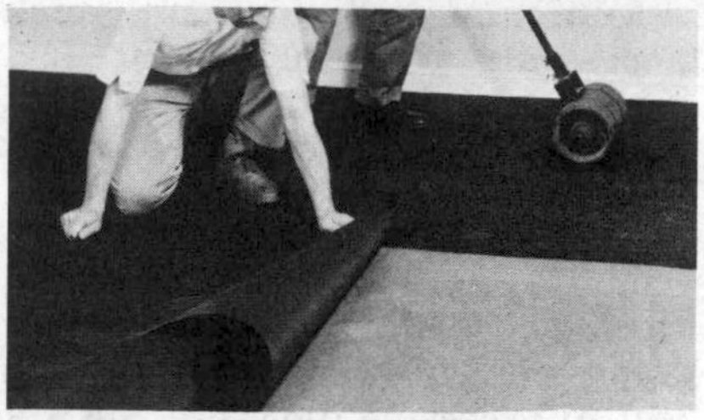

FIG. 11 Felt strips are cemented and rolled separately.

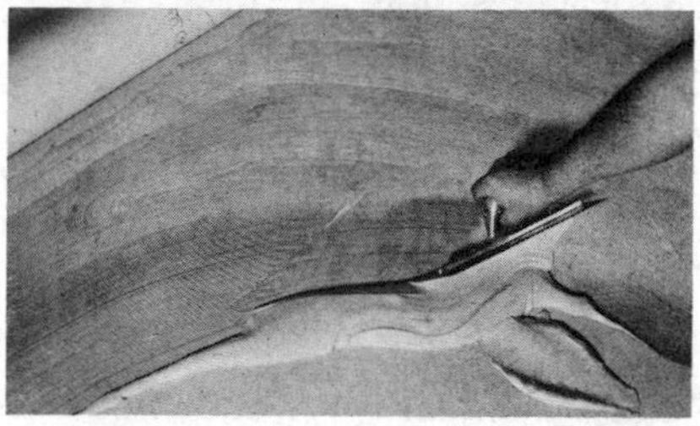

FIG. 12 Latex adhesive is gradually spread to a uniform thickness by a swirling motion with a notched trowel.

Old resilient flooring, backing, or lining felt should not be sanded. These products may contain asbestos fibers which are considered hazardous when inhaled.

ADHESIVES

Most adhesives used for the installation of resilient flooring are of a troweling grade for application with a notched trowel (Fig. 12). Some types also are made of a more fluid consistency for application with a brush. These often are used with thin gauge products to eliminate the possibility of trowel marks "telegraphing" to the surface of the finished floor. The major types of adhesives are (1) *linoleum paste,* (2) *asphaltic adhesive,* (3) *asphalt-rubber adhesive,* (4) *waterproof-resin cement,* (5) *latex cement,* (6) *epoxy adhesive* and (7) *wall base cement.* Figure 13 summarizes the properties and applications of the various adhesives. Adhesives with slightly varying properties are produced in each type and the general recommendations of the flooring and adhesive manufacturers should be followed.

FIG. 13 **SUMMARY OF ADHESIVES FOR RESILIENT FLOORING**[1]

Adhesive	Type	Flooring	Installation	Precautions
LINOLEUM PASTE	Clay: Water-base Troweling grade	Lining felt All sheet flooring Vinyl tile with backing Rubber Tile Cork & Vinyl Cork Tile	Over suspended wood or concrete subfloors, panelboard and latex underlayments; install flooring immediately.	Do not use on or below grade, or over suspended subfloors subject to moisture, or for solid vinyl, asphalt, and vinyl-asbestos tile. Remove with damp cloth.
ASPHALT[2] Cutback Asphalt	Asphalt: Solvent-base Troweling grade	Vinyl-asbestos tile	Over all concrete subfloors and panelboard underlayment primed with asphaltic primer. Allow to set 30 minutes; install tile within 4 to 18 hours as recommended.	Do not use under or over lining felt or asphaltic underlayments. Combustible. Remove with fine steel wool and soapy water.
Asphalt Emulsion	Clay and asphalt: Water-base Troweling grade	Vinyl-asbestos tile 1/16″ and thicker	Over all subfloors and over lining felt. Allow to set 30 to 60 minutes; install tile within 8 to 24 hours as recommended.	Do not use to install lining felt. Keep from freezing. Remove with fine steel wool and soapy water.
ASPHALT-RUBBER	Asphalt and rubber: Water-base Brush & troweling grades	Vinyl-asbestos tile	Overall subfloors and over lining felt. Allow to set 30 minutes before installing tile, or as recommended.	Use brushing grade with 1/16″ vinyl-asbestos tile; keep from freezing. Remove with fine steel wool and proprietary cleaner.
WATERPROOF RESIN	Resin: Solvent-base Troweling grade	Vinyl with rag felt, vinyl or rubber backing Rubber Tile Cork Tile Solid vinyl	Over all suspended subfloors; install flooring within 15 minutes.	Do not use on or below grade, for asphalt or vinyl-asbestos tile or to install lining felt. Remove with fine steel wool and proprietary cleaner. Combustible.
LATEX[3]	Rubber: Water-base Brush and troweling grades	Vinyl with asbestos, or rubber backing Rubber Tile Solid Vinyl Tile Lining Felt All sheet flooring	Over all subfloors; install flooring immediately.	Keep from freezing. Remove with fine steel wool and soapy water.
EPOXY	Resin and catalyst: Troweling grade	Solid vinyl and rubber tile on or below grade	Over all types of subfloors. Mix only as much as can be spread within 30 minutes and covered within 2½ hours, or as recommended.	Guard against tile slipping while adhesive sets.
WALL BASE Synthetic Rubber Cement	Synthetic rubber resin: Solvent-base Brush grade	Vinyl cove base Metal nosings and edges All resilient tile except asphalt and vinyl-asbestos to metal surfaces	Apply to both wall and material; install material within 20 minutes.	Combustible. Remove with proprietary remover.
Standard Cove Base Cement	Solvent-base Troweling grade	Vinyl and rubber cove base	Over dry walls above grade; install base within 15 minutes.	Combustible. Remove with fine steel wool and soapy water.

[1]Formulations and working properties of adhesives vary and manufacturer's specific recommendations should be followed.
[2]May be used over latex underlayment if sufficient setting time is allowed.
[3]May be used to install asphalt and vinyl-asbestos tile, however asphaltic adhesives more commonly are used because of less cost.

Linoleum Paste

Linoleum paste is a low-cost adhesive typically used for installing linoleum, vinyl sheet flooring and lining felt. It is a water-based adhesive which remains water-soluble even after hardening and should not be used on concrete slabs on or below grade or in other areas subject to moisture.

Flooring should be installed immediately after spreading the adhesive, then rolled with a 100 lb. roller to assure proper bond. The paste dries to a firm film within 24 hours and becomes hard within a few days. The rigidity of the hardened adhesive restrains sheet flooring dimensionally and prevents curling.

Asphaltic Adhesive

These are low-cost, water-resistant adhesives used exclusively for installing vinyl-asbestos. They may be used on all types of below-grade, on-grade and suspended subfloors. Asphaltic adhesives can be used with most underlayments, except that cutback asphalt adhesive and solvent-base primer should not be used over emulsion-based underlayment. Asphaltic adhesives retain their bonding qualities while the tile gradually conforms to minor surface irregularities. They eventually harden to provide a durable water- and alkali-resistant bond. Asphaltic adhesives are either *solvent-base* or *water-emulsion.*

Solvent-base These adhesives consist of a solution of asphalt in hydrocarbon solvents and are available in brushable primers or thicker adhesive formulations. *Cutback asphalt* adhesive, the most common of the solvent-base asphaltic adhesives, is of a troweling grade consistency. Because the volatile solvents are flammable and mildly toxic, these adhesives should not be used near an open flame and the area should be well ventilated. The adhesive should be allowed to set approximately 30 minutes before applying tile. The working time ranges from 4 to 18 hours.

Water-emulsion These adhesives consist essentially of tiny asphalt particles suspended in water, and become water-resistant upon drying. They are of troweling grade consistency and are spread with a notched trowel. The adhesive should be allowed to set for 30 to 60 minutes before installing tile, but working time ranges from 8 to 24 hours. This permits application to relatively large areas, overnight setting of adhesive and installation of tile the following day.

Asphalt-Rubber Adhesive

This is an asphalt and rubber water-emulsion used primarily for the installation of vinyl-asbestos tile. The working characteristics of this adhesive are similar to those of asphaltic adhesive, and it can be used over any type of subfloor or underlayment.

Waterproof-Resin Adhesive

Most adhesives of this type are resins in an alcohol solvent. They have approximately the same application and working characteristics as linoleum paste, but are not soluble in water. However, they are adversely affected by alkaline moisture and are unsuitable for use over concrete in contact with the ground on or below grade. With the exception of asphalt and vinyl-asbestos tile, waterproof-resin cements may be used for applying most resilient flooring in suspended locations requiring waterproof bond. The volatile alcoholic content makes these adhesives flammable, and they should not be used near an open flame or in poorly ventilated areas.

Latex Adhesive

Latex cements consist of rubber resins and water, and are extremely moisture- and alkali-resistant after they set. On drying, the adhesive turns to a rubbery film which eventually hardens and becomes insoluble in water. Latex cements generally provide good adhesion and may be used for installing all types of resilient flooring on any type of subfloor or underlayment. They are relatively higher in cost and normally are not used for installing vinyl-asbestos and asphalt tile where cheaper asphaltic adhesives also are suitable. Latex cements more commonly are used for installing rubber tile, vinyl tile and asbestos-backed sheet products on or below grade. Flooring may be installed immediately or as recommended by the manufacturer.

Epoxy Adhesive

Epoxy adhesives are extremely moisture-resistant formulations with high bonding strength and excellent durability. However, the relative high cost, difficulty in spreading and excessive slipping of tiles immediately after installation presently moderate the use of these adhesives.

Epoxy adhesives are mixed on the job and consist of a two-part mixture—epoxy resins and a catalyst which activates the mixture. A chemical reaction begins as soon as the components are mixed and the adhesive must be used within 2 to 3 hours. However, formulations vary and the manufacturer's recommendations should be followed closely.

Wall Base Cement

The two basic adhesives used to secure resilient wall base are *standard cove base cement* and *synthetic rubber cement.* Both are solvent-base formulations which set as the volatile solvents escape from the mix; both are flammable and should not be used near an open flame or in poorly ventilated areas.

Standard Cove Base Cement This low-cost cement typically is used for installing resilient wall base and premolded corners. It develops bonding power approximately 10 minutes after application to either the wall surface or the wall base (Fig. 14).

Synthetic Rubber Cement This is a *contact* adhesive which must be applied to both the wall and the resilient base. Bond is developed as soon as the two surfaces touch. Synthetic rubber cement also is used typically for installing metal or resilient nosings, edges and binding strips. It is particularly useful in bonding resilient accessories in the field where movement after fitting would be undesirable.

RESILIENT ACCESSORIES

The most common resilient accessories used with flooring materials include: *wall base, stair tread, stair nosings, thresholds, feature strips* and *reducing strips* (Fig. 15). These accessories are available in more limited color and design variations than the flooring materials. Installation of these accessories is described in the following subsections.

REFLOORING EXISTING CONSTRUCTION

Both sheet and tile products may be installed over existing finished floors which have been suitably prepared to provide a firm, smooth and stable base for the new flooring. The main types of floor finishes encountered in residential reflooring are: (1) stripwood and wood block or plank; (2) concrete; (3) ceramic tile, terrazzo or marble. New flooring may be installed over old resilient flooring provided the old flooring is not cushioned or deeply embossed. For specific recommendations, the manufacturer of the new flooring should be consulted.

All existing floors intended for reflooring should be inspected to assure a firm base free of loose particles. Existing wax, oil, grease films and most paints should be removed. Surfaces should be smooth and reasonably level, or leveled with an appropriate underlayment when necessary. Recommendations for flooring and adhesive selections with respect to grade locations should be adhered to as in new construction.

Wood Floors

A suitable panel underlayment is recommended over all existing wood floors. However, sheet products may be applied directly to double-layer floor construction consisting of a lumber board or plywood subfloor and a finish stripwood floor of tongue and grooved boards at least 3/4" thick and not wider than 3", if the existing floor is sufficiently smooth, level, firm and well seasoned.

When panel underlayment is not used, protective film on wood floors may be removed with a power sander or a blowtorch. Mild chemical solutions in water should not be used because the large quantity of water required for scrubbing and rinsing may cause the boards to warp. After a wood floor has been sanded, one brushed coat of floor sizing consisting of equal parts of a wax-free shellac and denatured alcohol should be applied to seal the wood. This prevents excessive moisture absorption from adhesive and other sources, and minimizes the possibility of boards warping in the future.

Other double wood floors of block and plank and single-layer plank floors should be covered with underlayment consisting of at least 1/4" plywood, 1/4" hardboard or 3/8" particleboard.

Concrete Floors

Most paint films can be removed from concrete floors by power grinding, burning with a blowtorch or scrubbing with a solution of trisodium phosphate and water. After the paint has been removed, the floor should be scrubbed with a strong soapy solution, rinsed with clear water and allowed to dry thoroughly. Household paint removers should not be used because they react adversely with many adhesives.

Paints with a chlorinated rubber or resin base often are tightly bonded to the concrete surface and, if the mat bonding test reveals good adhesion, may be left in place.

Ceramic Tile, Terrazzo, Marble

Ceramic tile, terrazzo and marble floors should be cleaned with a scrubbing machine using soapy water and clean sharp sand. A floor grinder used dry with 4-1/2 grit opencoated paper may be preferred to avoid delay due to drying. Badly fitted joints and cracks should be leveled with latex underlayment or a patching compound.

MAINTENANCE

A resilient floor can be damp-mopped immediately, and should be polished within 48 hours. However, it should not be given a thorough cleaning until 4 to 5 days after installation. The floor then should be washed with a mild solution of a

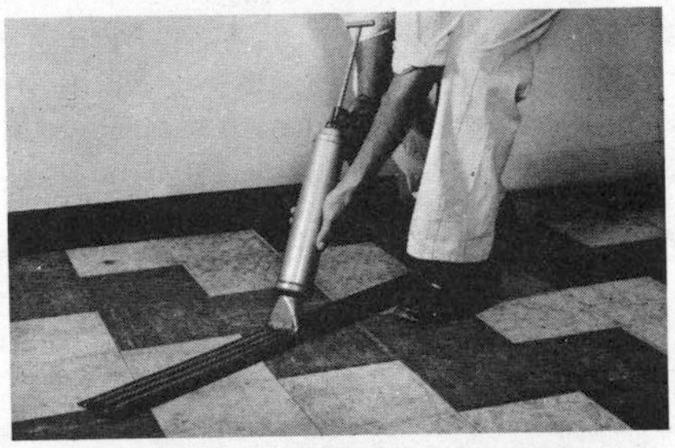

FIG. 14 Standard cove base cement may be applied with a notched spreader or by cartridge type gun shown here.

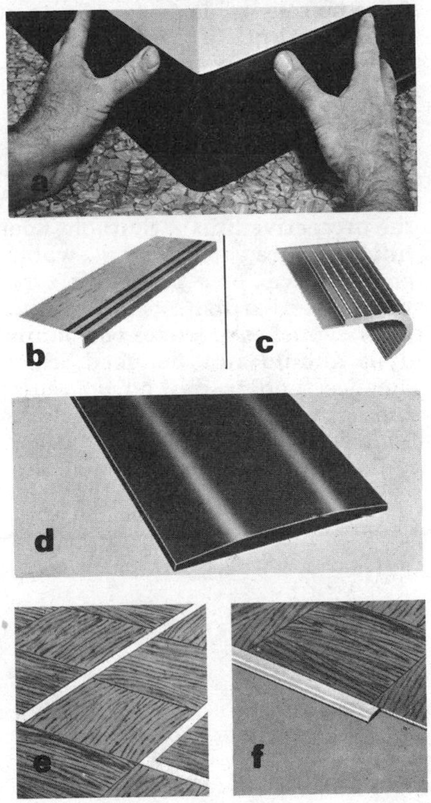

FIG. 15 Resilient accessories: (a) wall base, (b) stair tread, (c) stair nosing, (d) threshold, (e) feature strip, and (f) edging strip.

nonabrasive, nonalkaline cleaner or soap, and thoroughly rinsed with cold water. It should be allowed to dry, then be waxed with a floor finish recommended by the flooring manufacturer for the type of floor. A soft cotton mop or wax applicator should be used, applying the floor finish as thinly as possible.

No-wax tile and sheet floors should be cleaned the same way as other resilient floors; however, waxing is not necessary. Because they have no renewable surface finish, it is important to protect no-wax floors from severe abrasion, such as from gritty dirt and furniture without floor protectors.

Resilient floors should not be mopped too often and should be waxed often enough to maintain the protective finish. Suitable floor finishes are usually a water-emulsion wax or a polymeric type. Lacquer, varnish, shellac and plastic finishes or waxes containing dyes should not be used. Such finishes tend to darken or yellow the pattern and often are hard to remove.

Old wax should be stripped often enough to prevent build-up in light traffic areas. Most manufacturers have proprietary removers for each type of finish, and to assure easy removal it is desirable to use the same type of floor finish between strippings. Eventually foot traffic may reduce the gloss on no-wax floors; it may then be treated with an appropriate product to restore the gloss.

The daily care of a floor that has received one or more applications of floor finish consists of brushing with a soft broom or dry mop to remove surface dust. Spilled oils, grease and other substances should be wiped up immediately to prevent damage to the floor. An occasional damp mopping may be required to remove dried-on dirt. Sweeping compounds containing oil, dyes or abrasives should not be used. Floors in entrances should be protected by throw rugs.

Resilient floors should be protected from heavy concentrated loads with proper floor rests and protectors (Fig. 16). Stationary household furniture should rest on flat glass or plastic cups. Casters on rolling fixtures should be flat, as wide as possible and of a nonabrasive material. Chair legs should be equipped with flat floor rests which are large enough to distribute the load without permanently indenting the floor.

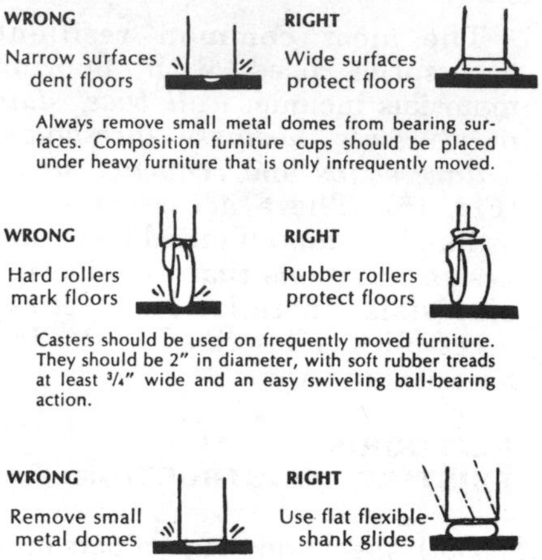
WRONG
Narrow surfaces dent floors
RIGHT
Wide surfaces protect floors
WRONG
Hard rollers mark floors
RIGHT
Rubber rollers protect floors
WRONG
Remove small metal domes
RIGHT
Use flat flexible-shank glides

Always remove small metal domes from bearing surfaces. Composition furniture cups should be placed under heavy furniture that is only infrequently moved.

Casters should be used on frequently moved furniture. They should be 2" in diameter, with soft rubber treads at least 3/4" wide and an easy swiveling ball-bearing action.

Light furniture should have glides with a smooth, flat base and a flexible pin to maintain flat contact with floor. Diameter should be 1" to 2 1/2" depending on weight of furniture.

FIG. 16 Floor rests and floor protectors are essential to good maintenance.

The following text describes the materials and methods recommended for a suitable installation of sheet flooring primarily over new floor construction. Reflooring existing construction is discussed on page 452-11.

Reference should be made to the entire preceding subsection, General Recommendations, pages 452-03 through 452-12, for a discussion of (1) terminology, (2) proper job conditions, (3) adequate subfloor construction, and (4) accessory materials—underlayments, adhesives, lining felt and resilient accessories—required for flooring installation. Unless specific exceptions are made in this subsection, the General Recommendations apply and complement the following discussion.

MATERIALS

The most common type of resilient flooring available in sheet form is *vinyl*. The manufacture, physical properties and selection criteria for vinyl products are discussed in Section 220 Resilient Flooring Products. Ingredients, gauges and standards of quality are summarized in Figure 17; selection criteria are outlined in Figure 18.

FIG. 17 SUMMARY OF SHEET FLOORING PRODUCTS

Type	Federal Specification	Backing Materials	Gauges[1]	Sizes	Basic Ingredients
LINOLEUM SHEET[4]			(.050" wear surface)		
	LLL-F-1238A	Burlap	1/8"		
Plain & Marbleized	LLL-F-1238A	Rag felt	.090"		Cork and/or wood flour
Inlaid & Molded	LLL-F-1238A	Burlap	1/8"	6'0" wide	with linseed oil
	LLL-F-1238A	Rag felt	.090"		binders
Battleship	LLL-F-1238A	Burlap	1/8"		
VINYL SHEET[2]					
Filled Surface	L-F-475-a L-F-001641	Scrap vinyl, asbestos or rag felts	.065" to .160" (.010" to .050" wear surface)	6'0", 12'0" & 4'6"[3] wide	PVC resin binders and mineral fillers; clear PVC film for clear vinyl surface.
Clear Vinyl Surface			.065" to .160" (.010" to .050" wear surface)	6'0" & 12'0" wide	

[1]Thinnest gauges and wear surfaces indicated are minimums suitable for adhesive installation.
[2]Also produced with foam backing, not included in data shown.
[3]Not common, available on special order from some manufacturers.
[4]Linoleum sheet is no longer produced in this country, but is included here for reference only.

FIG. 18 SELECTION CRITERIA FOR SHEET FLOORING PRODUCTS[1]

Type of Flooring	Location[2]	Resistance to Grease	Alkalis	Stain[3]	Cigarette Burns	Inden-tation	Resil-ience[4]	Quiet-ness	Ease of Mainte-nance	Durability
VINYL SHEET[5]										
Filled Surface										
Asbestos, vinyl & rubber backing	BOS	Excellent	Excellent	Excellent to Good	Good to Poor	Good to Fair	Good to Fair	Good to Fair	Good	Excellent
Rag felt backing	S	Excellent	Good	Excellent to Good	Good to Poor	Good to Fair	Good to Fair	Good to Fair	Good	Excellent
Clear Vinyl Surface										
Asbestos, vinyl & rubber backing	BOS	Excellent	Excellent	Excellent to Good	Good to Poor	Good to Fair	Good to Fair	Good to Fair	Good	Excellent
Rag felt backing	S	Excellent	Good	Excellent to Good	Good to Poor	Good to Fair	Good to Fair	Good to Fair	Good	Excellent

[1]Criteria vary depending on the properties and thicknesses of wear surface and backing material; performance may be further affected by properties of adhesives, underlayments or subfloors. See page Resilient Flooring Products 220-7 for a discussion of selection criteria.
[2]B-Below grade, O-On grade, S-Suspended.
[3]Varies with staining agent.
[4]Also indicates potential underfoot comfort.
[5]Foam-cushioned and no-wax products are rated highly for resilience, quietness and resistance to indentation.

PREPARATION FOR FLOORING

The effectiveness of a flooring installation greatly depends on the proper selection and preparation of the elements which make up the installation. *Subfloors* which are to receive flooring directly must be firm, smooth and dense and possess good bonding properties. When these properties are lacking, it may be necessary to prepare the subfloor by either grinding, sanding or with the use of underlayment. *Adhesives* should be selected which are compatible with all elements, including the flooring itself. All elements of the installation must be suitable for the intended location—whether suspended, on grade or below grade.

Subfloor

Vinyl sheet products with asbestos fiber backing may be installed on suspended, on-grade and below-grade subfloors. Rag felt-back products are subject to decay and alkali attack in the presence of moisture and are limited to suspended locations. Linoleum should be installed on suspended floors only.

Concrete subfloors in all locations should be tested for moisture and bonding properties to insure suitability for the proposed flooring and adhesive (see Suitability Tests, page 452-6).

Underlayments

The need for underlayment is determined by the condition of the

FIG. 19 SUMMARY OF ADHESIVES FOR SHEET FLOORING

FLOORING	SUBFLOOR AND LOCATION: WOOD (Suspended)	CONCRETE (Suspended)	CONCRETE (on- and below grade)
VINYL* (Rag felt backing)	Waterproof-Resin	Waterproof-Resin	Not Recommended
VINYL* (Vinyl or rubber backing)	Latex Waterproof-Resin	Latex Waterproof-Resin	Latex
VINYL* (asbestos backing)	Latex	Latex	Latex

*Foam-cushioned products are installed by specially formulated adhesives recommended by flooring manufacturer.

subfloor. For a discussion of underlayments and recommended uses, see General Recommendations, page 452-6).

Adhesives

The adhesives generally recommended for use with sheet products in various locations are summarized in Figure 19. See General Recommendations, page 452-8 and Figure 13 for a discussion of adhesives and their properties.

Lining Felt

Asbestos lining felt should be used in the installation of certain sheet flooring products on nonporous floors as described in General Recommendations 452.

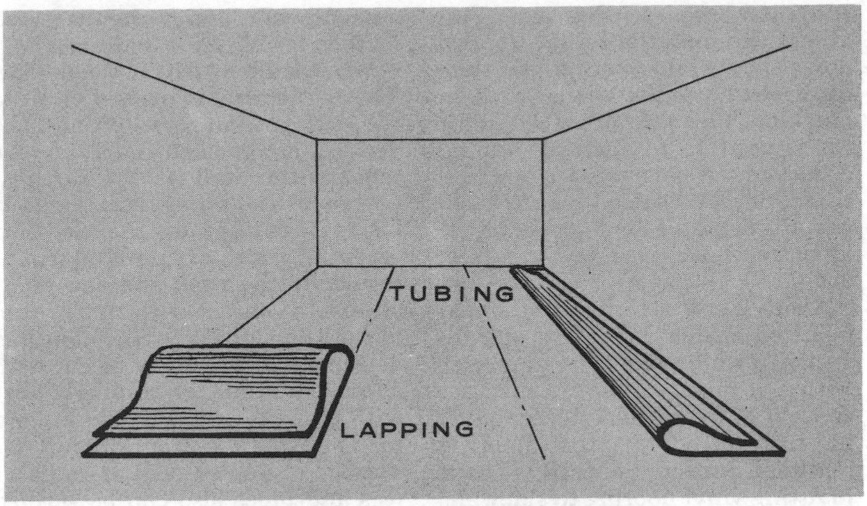

FIG. 21 Methods of handling and adhesive bonding sheet flooring.

APPLICATION OF FLOORING

The installation of sheet materials should be planned to minimize the number and total length of seams. Necessary seams should be placed in inconspicuous locations, out of the path of heavy foot traffic. In a rectangular room, running the flooring strips parallel to the sidewalls generally results in an economical installation with a minimum of seams. However, sheet flooring installed directly over stripwood floors should run perpendicular to the floor joints and may result in seams running the short dimension, parallel to endwalls (Fig. 20).

Installation of sheet flooring consists generally of *fitting and cutting, adhesive bonding* and *seam treating.*

Fitting and Cutting

Many sheet flooring products must be kept rolled face out until time of installation. Some manufacturers warn against rolling sheet flooring face in because it may set up internal stresses that can result in residual growth of the sheet flooring long after it is installed. When fitting adjacent strips which have a definite geometric repeat pattern, the pattern should be aligned and matched. Sheet flooring with random overall patterns should be reversed (turned end for end) to minimize differences in graining and shading.

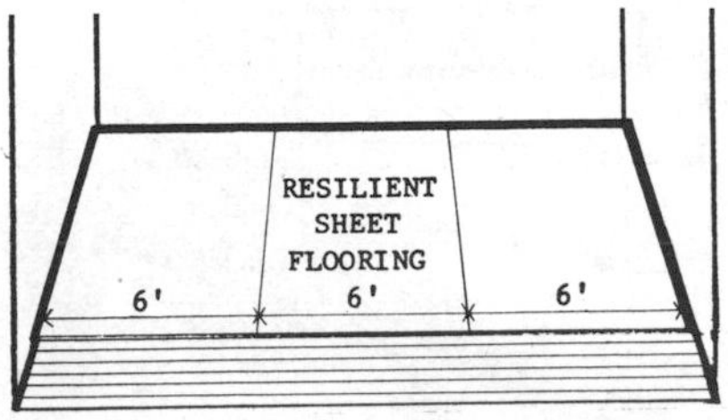

FIG. 20 Sheet flooring installed parallel to sidewalls will result in a minimum of seams; sheet flooring installed over stripwood floors, however, should always run perpendicular to the floor joists.

Methods of cutting and fitting sometimes are related to the intended method of bonding the sheet goods. The adhesive application is facilitated by folding the strip back on itself, thus working only half of the strip at a time. The strip is folded back along its length in the *tubing method*, and back along its width in the *lapping method* (Fig. 21).

Sheet flooring generally is cut and fitted to walls and to each other by *knifing, scribing* and *seam cutting.*

After squaring the end of the roll (Fig. 22), strips are cut 3″ longer than the distance between endwalls to provide excess material for fitting to wall recesses or irregularities. The first strip is placed in position with the side edges butted against the sidewall and the ends of the material are flashed (bent up) against the endwalls. The material is scribed along the length of the sidewalls with dividers or a scribing tool (see *Scribing* below) and cut to fit snugly to the wall surface. Ends which have been flashed against the endwalls also may be scribed or may be trimmed by cutting away small pieces at a time until the desired fit is achieved.

Knifing This method of end-fitting involves some waste and is used generally with the less expensive felt-backed products. After a strip of flooring is fitted in place against a sidewall or adjacent strip, excess material is cut away gradually to fit the endwalls or other vertical surfaces.

Safety cuts are made at various points to allow the material to be fitted into difficult locations without cracking. One or more safety cuts should be made first wherever the material tends to buckle as it is flashed against the wall. When the flooring has to be fitted into a corner before it can be knifed at the walls, the *inside corner safety*

FIG. 22 The end of the roll is squared with a straightedge and square.

cut is the simplest (Fig. 23a). This allows the material to be dropped down snugly into a corner and then be flashed up the walls without cracking. The *outside corner safety cut* is used to fit flooring around a curved or cornered projection (Fig. 23b). A *crossover safety cut* permits bending the flooring under an obstacle such as a radiator (Fig. 23c).

Scribing When flooring must be fitted against a vertical surface such as a wall, column or stair riser with irregular corners, the exact contour of the vertical surface must be transferred to the flooring material. Scribing generally is used in fitting vinyl flooring to sidewalls and endwalls, and is the preferred method for all sheet flooring.

When scribing with *dividers* (Fig. 24), the flooring is pushed close to the wall or the projection and the dividers are opened so that one leg touches the wall at the furthest point and the other rests about 1″ inside the edge of the flooring. Keeping the dividers perpendicular to the wall, and one point always in contact with it, the profile is transferred to the flooring. The material then may be cut with a linoleum knife, or for vinyl sheet with a notched blade knife, and further fitted in place. Cutouts for isolated obstacles such as radiator legs and pipes also can be scribed with a *scribing tool* in a similar manner (Fig. 25).

Pattern Scribing is used when the size of the room or wall projections make it difficult to scribe the flooring itself. Complex profiles are first scribed to a piece of felt or heavy paper, checked for fit and trimmed; then they are transferred to the flooring (Fig. 26). This assures a better fit, particularly desirable when wall base is not used or when the base is a straight type which will not cover minor irregularities in scribing.

End Scribing, when the flooring is tubed, is generally similar to that described for sidewalls. However, when the lapping method is used, each strip must be partly cemented before scribing. The unfitted end of the sheet is lapped back and, working towards the endwall from the center of the room, adhesive is applied to within 3′ of the endwall. The sheet is laid into the adhesive and rolled. The loose, uncemented end of the material then is scribed to the endwall by drawing a mark perpendicular to the edge of the material 9″ to 12″ from the wall and extending it from the material onto the subfloor. The material then is buckled and butted against the wall line (Fig. 27). The perpendicular distance between the line on the subfloor and the line on the material is measured with the scriber. With the scriber set at this distance, the material is scribed to the wall line and cut.

Seam Cutting To make a neat,

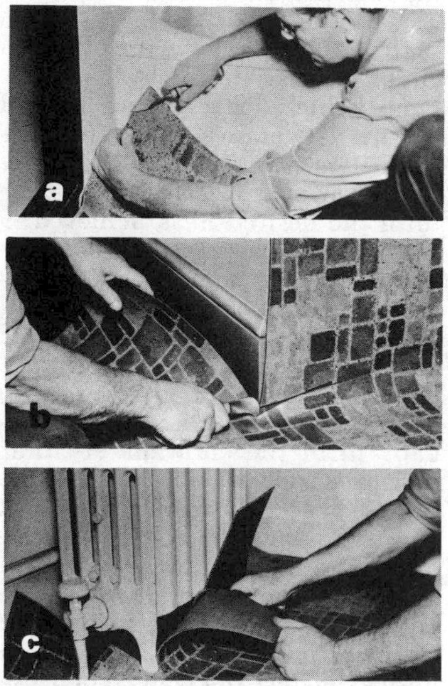

FIG. 23 Safety cuts facilitate fitting: (a) inside corner, (b) outside corner, (c) crossover safety cut.

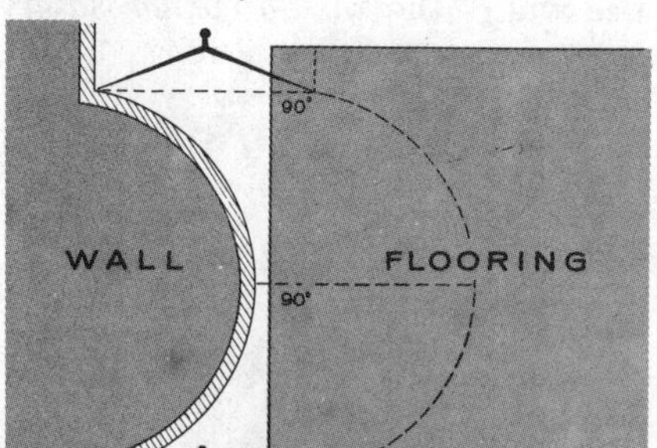

FIG. 24 When scribing with dividers, the dividers should be kept perpendicular to the wall surface.

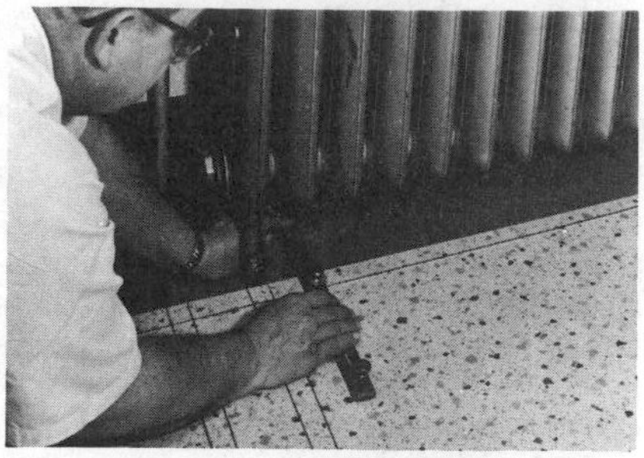

FIG. 25 An 18″ scriber permits scribing under obstacles which would be inaccessible with dividers.

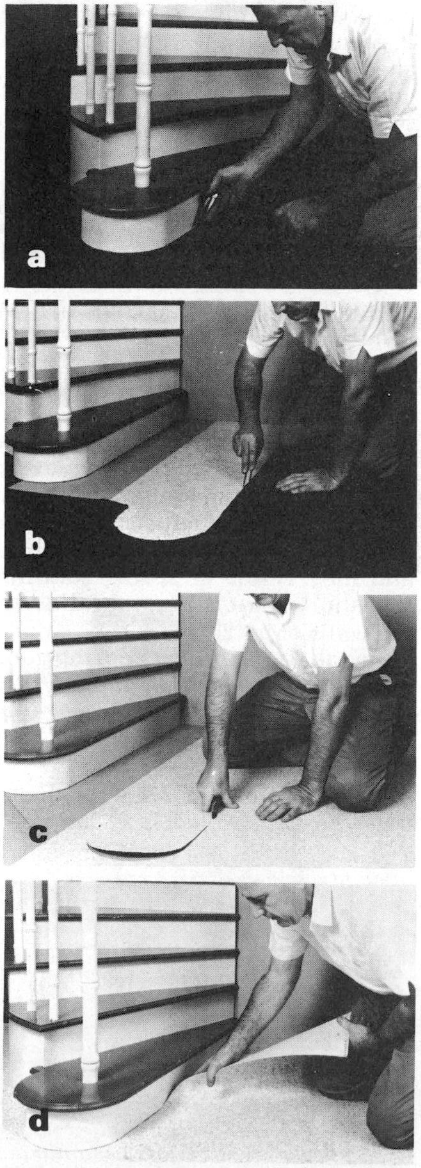

FIG. 26 Pattern scribing: (a) felt pattern is scribed, (b) is transferred to the flooring, (c) flooring is cut to the pattern, and (d) is checked for fit.

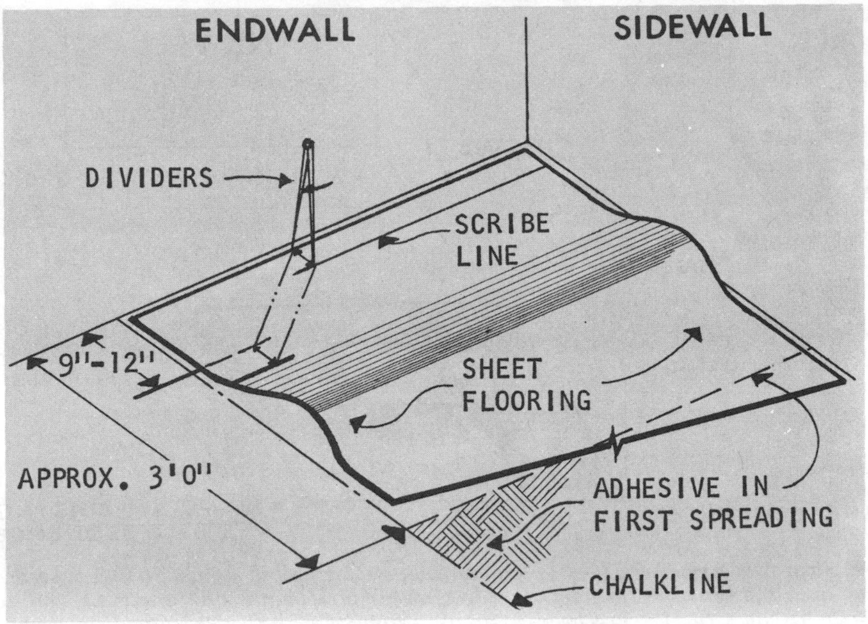

FIG. 27 When lapping, each strip is partly cemented before end scribing.

inconspicuous seam between adjacent strips of flooring, it is necessary to cut away the selvage (factory-cut edge). Sheet widths are generally 1/2″ greater than the nominal width to permit this excess material to be cut away. On goods with a geometric repeat pattern, the amount of selvage cut away cannot be varied if the pattern at the seam is to match and be aligned properly. In goods with random overall pattern, the selvage to be trimmed is less critical and may vary (Fig. 28).

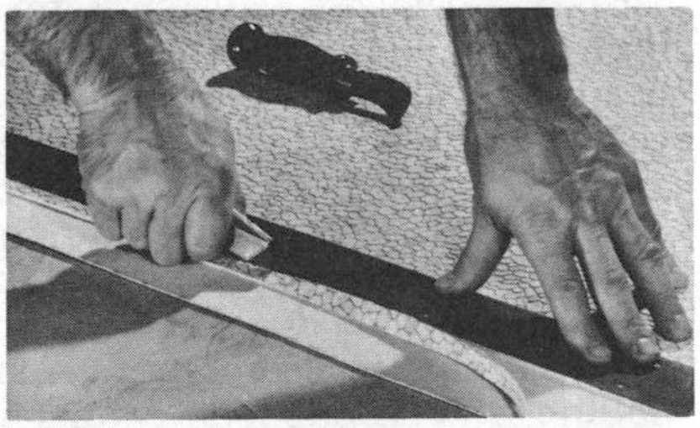

FIG. 28 The selvage is trimmed with a straightedge and either a linoleum or a straight blade utility knife.

Seams between adjacent strips are made chiefly by *underscribing, single cutting* or *double cutting*. To prevent misalignment due to shifting, strips should be partially or completely bonded to the floor before seams are cut. When using the single or double cutting methods, adhesive generally is kept 4″ or more from the seam until both edges are cut. This prevents adhesive from being pulled up when the lower edge (selvage) is removed. Both in straightedging and in seam cutting, a piece of scrap flooring should be inserted under the cut line to prevent the backing from fraying. The knife should be held as low as possible with the blade perpendicular to the floor surface.

Underscribing involves the use of a recess scriber to transfer the exact edge profile from the previously straightedged strip to the adjacent strip. As it is guided along the finished edge, the tool leaves a score mark on the overlapped adjacent strip directly above. The scribe mark then is used as a guide for cutting (Fig. 29).

Underscribing is the preferred way of making seams for flooring with filled vinyl wear surfaces. The trimmed edge can be cemented down completely before underscribing, which often minimizes the number of operations involved in fitting and cementing the flooring.

Single Cutting (Fig. 30) is commonly used for making seams in patterned flooring where the pattern must be carefully aligned. The straightedged strip is laid over the untrimmed strip, overlapping exactly as required to fit the pattern. The lower sheet then is scored with the linoleum knife guided along the upper trimmed edge. The lower strip then is cut along the score line.

Double Cutting (Fig. 31) is similar to single cutting except that neither edge is pretrimmed and the cut is made through both strips at the same time. Using a straightedge as a guide, the upper sheet is cut in the middle of the overlapped edges, scoring the sheet below. The lower sheet is then cut along the score line. Double cutting is an older method and is not used as commonly as the underscribing and single cutting methods. However, double cutting

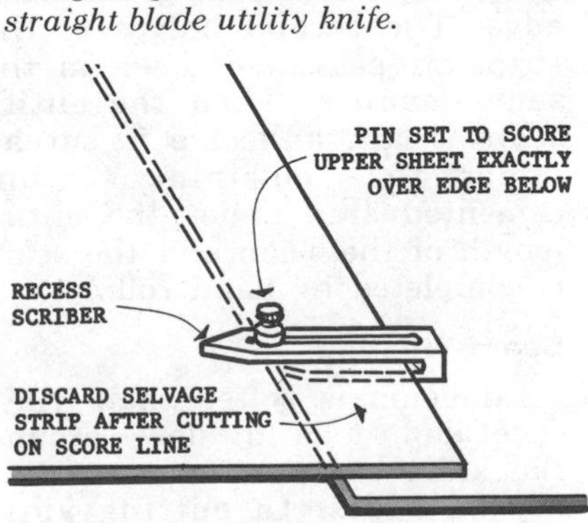

FIG. 29 Underscribed seam.

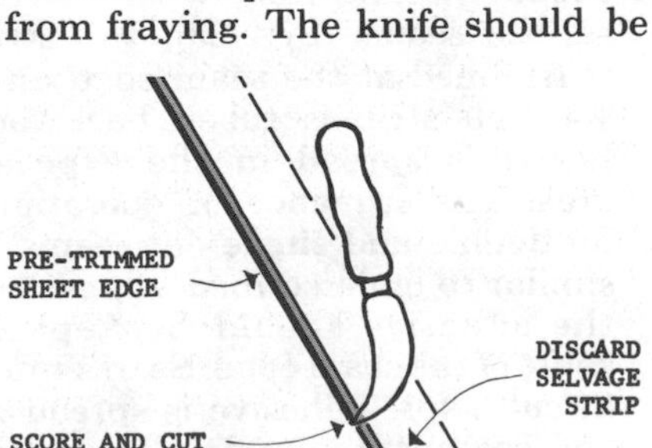

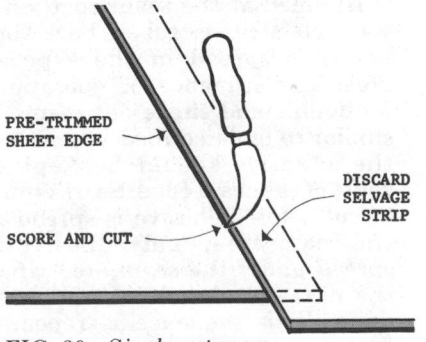

FIG. 30 Single-cut seam.

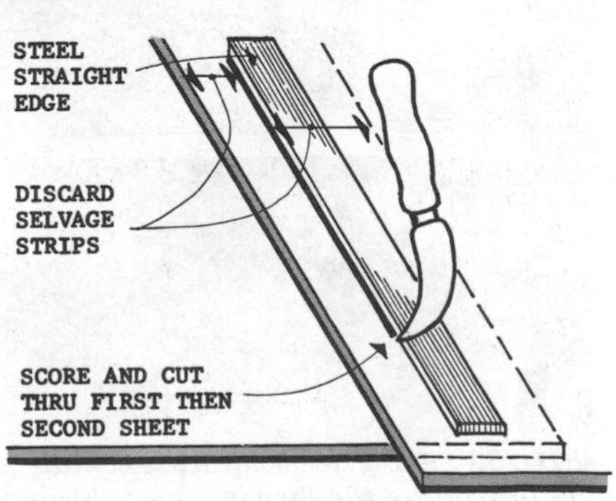

FIG. 31 Double-cut seam.

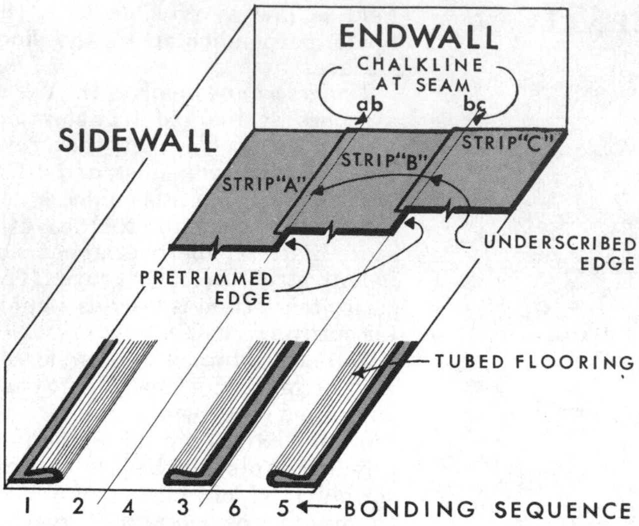

FIG. 33 Tubing with underscribed seams: Strip A is scribed to side- and endwalls, edge "ab" is trimmed, area 1 is bonded, then area 2. Strip B is lapped over Strip A, end-scribed, edge "bc" is pretrimmed and area 3 is bonded; then edge "ab" is underscribed, cut and area 4 is bonded. Strip C is installed as Strip B after being scribed to side- and endwalls.

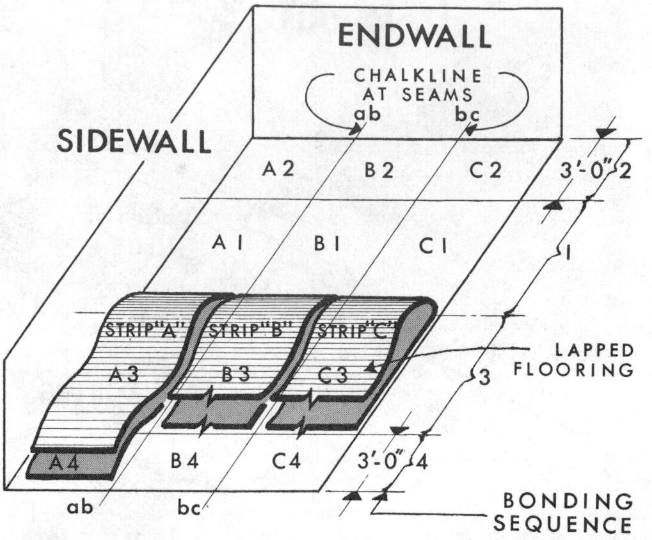

FIG. 34 Lapping with underscribed seams: Strip A is scribed to sidewall, edge "ab" is pretrimmed and areas A1 and A3 are bonded; areas A2 and A4 are bonded after end-scribing to walls. Strip B is lapped over A and entire strip is installed as Strip A; edge "ab" of Strip B is then underscribed and cut. Strip C is installed as Strip B after scribing to sidewall.

may be used with some clear (unfilled) vinyl-surfaced sheet which are not as easily scored with an underscriber.

Adhesive Bonding

To facilitate spreading adhesive prior to bonding, the flooring usually is folded back by either tubing or lapping (Fig. 21). The sequence of cutting and fitting, adhesive spreading and cementing operations depends on whether the flooring is installed by tubing or lapping. Lapping generally is the preferred method, but tubing is used where wall projections or the shape of the room make lapping difficult or impossible. Tubing also is recommended for heavy gauge vinyl sheet and for burlap-backed linoleum.

Adhesive is spread with a notched trowel, spreader or brush, depending on the type of adhesive. As each strip is pasted down, the seam should be smoothed with a hand roller and the entire strip should be rolled with a 100 lb. or heavier roller (Fig. 32).

Tubing Generally, the first strip is scribed to the sidewall and endwalls prior to cementing down. For tubed installations with underscribed seams (Fig. 33), the strip is trimmed at the seam edge also. As each strip is tubed back, adhesive is spread in the exposed area. The sequence of operations for double and single cut seams is similar to underscribed seams, but the adhesive should be kept 4″ short of the seam edge. Seams must be cut before adhesive is spread in the seam area, and adhesive is spread under the seam area when the next tubed half is cemented down. This sequence is repeated for subsequent strips.

Lapping Depending on the size of the room and working time of the adhesive, spreading may be combined across the width of the room and more than one strip may be worked at a time. For lapped installations with underscribed seams (Fig. 34), the first strip is scribed to the sidewall, and starting at the center of the room adhesive is spread under strips to within 3′ of the endwall. The strips are cemented down and rolled. Each strip is end-scribed and the remaining area is spread with adhesive, cemented down and rolled.

The sequence of operations for double and single cut seams is similar, but adhesive in each spreading is kept 4″ from the seam edge. The second halves of the strips are cemented down in the same manner. Then the entire seam is cut, adhesive is spread under the remaining 8″ uncemented area under the entire length of the seam and the seam is completed by hand rolling.

Seam Treating

Linoleum is softer than vinyl sheet and tends to "flow" around the knife blade without creating ragged edges. In cutting vinyl sheet, however, burrs often are

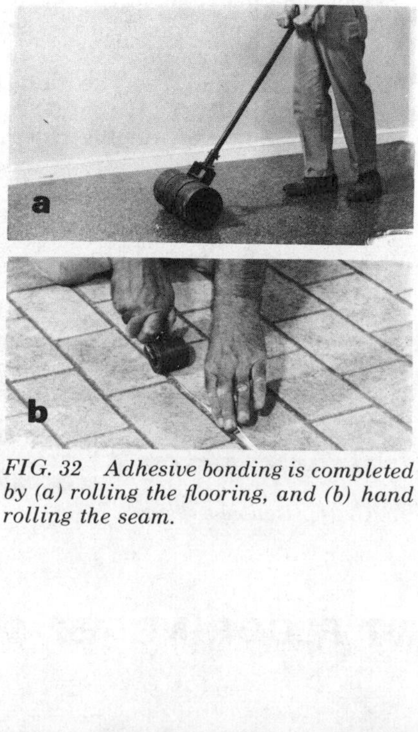

FIG. 32 Adhesive bonding is completed by (a) rolling the flooring, and (b) hand rolling the seam.

FIG. 35 Flashed cove base with (a) outside metal corner, (b) inside metal corner, (c) metal binding strip, (d) wax cove fillet.

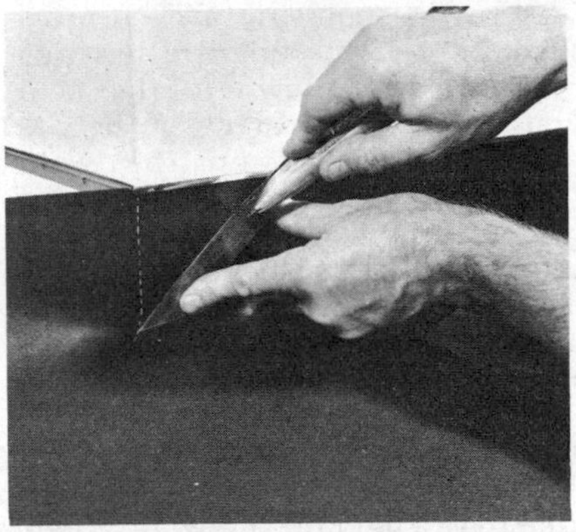

FIG. 36 Felt pattern is scribed to the binding strip and 45° V cutout is formed for outside corner of flashed base.

created on the cut edges and must be removed to make the seams inconspicuous. After hand rolling the seams, burred edges may be removed with the back of a linoleum knife moved in a "chisel" motion along the seam; or a hammerhead or scrap of flooring doubled in half (face out) may be rubbed along the burrs while the seam is kept damp to avoid scuffing.

An *electric iron* also may be used to treat seams: a 1″ wide strip of heavy-duty aluminum foil is laid over the seam (dull side down) and rubbed to make the burrs show through. The foil then is pressed with an electric iron held at an angle and moved in repeated strokes toward the mechanic. When the foil has cooled it is removed, and the seam is cleaned with a damp cloth and a light abrasive cleaner.

Tension Floors

Tension floor installation is a unique method of perimeter bonding recommended for certain sheet flooring products. Flooring may be installed by this method over well-bonded existing resilient floors over concrete and wood subfloors on all grade levels. Special adhesives allow the floor to be bonded only at the room perimeter, seam lines and at columns or fixtures.

The two types of adhesive used are a two-part epoxy and a special solvent-based mastic. Installation method varies slightly with type of adhesive used, flooring type and manufacturer's instructions. As in all resilient floor installations, the subfloor surface must be free of moisture, dust, paint, wax, oil or grease; the subfloor should be structurally sound and smooth.

A tension floor installation should be allowed to dry 4 to 5 days before cleaning. Floor maintenance generally is the same as for other resilient floors. However, when moving heavy objects such as appliances, a piece of plywood or particleboard should be placed under the object to prevent tearing or buckling of the flooring material.

Cove Base Flashing

Sheet flooring may be trimmed out with standard set-on cove base (see Resilient Accessories, page 452-11), or the material may be extended partly up the wall to make an integral flashed cove base (Fig. 35). To make the cove profile at the meeting of the wall and floor, a 7/8″ wax fillet strip is cemented or a wood cove molding is nailed in place. A binding strip then is nailed or cemented to the wall at the desired height to protect the top of the base. The flooring may be softened with a torch to facilitate molding it against the fillet strip. In heavy traffic areas, inside and outside metal corners also may be desirable (Fig. 35). Similar metal endstops are available for closing base ends at door openings without casing trim.

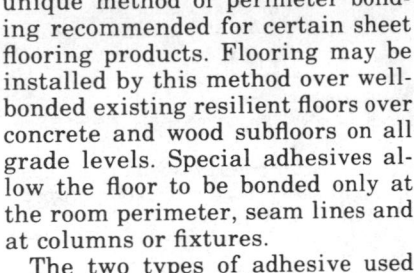

FIG. 37 Outside one-piece corner: (a) V cutout is transferred from the pattern to the flooring before installation; (b) corner piece is cut out separately with a square; (c) before fitting matching corner piece, a narrow strip of backing is removed at the fold.

In rooms requiring a decorative sheet border, the full width strips may be pasted down in the field first; then the narrower border strips are flashed up the wall, scribed against binding strips and cemented in place. After the border strip has been cut to proper width, inside and outside corners without metal corner pieces may be scribed and installed.

Where no decorative borders are used, it is necessary to flash the ends and edges of the flooring strips used in the field up the wall. When working with such larger pieces, it is difficult to cut flooring accurately by fitting and cutting in place. Therefore, it is necessary to make a felt pattern of the entire floor and base, carefully scribing and cutting around projections to the binding strips (Fig. 36). The pattern is transferred to the sheet flooring which then can be readily cut, fitted and cemented in place by either the tubing or lapping method. Inside corners can be made by scribing as shown in Figure 54 for resilient base. Figure 37 illustrates the preferred method of making an outside corner. A less durable corner can be made simply by mitering the flashed material at the corner.

A 100 lb. roller may be used to roll the floor; while a small hand roller should be used for the flashed cove base. Care should be taken not to allow the edge of the heavy roller to ride over the fillet as it may damage the installation.

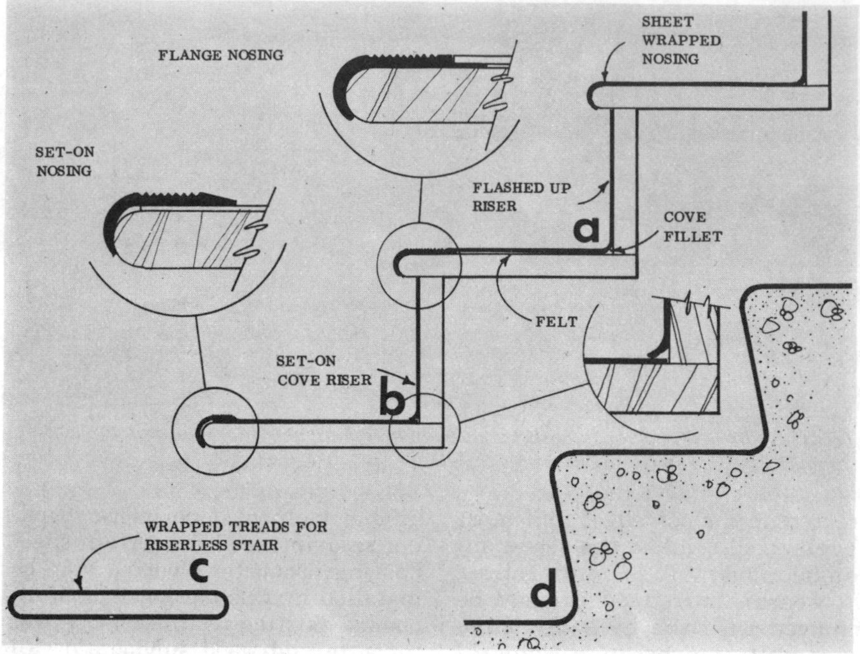

FIG. 38 Stair coverings with sheet flooring: (a) one-piece tread and riser; (b) sheet tread and set-on cove riser; (c) wrapped riserless wood tread; (d) continuous risers and treads on concrete.

STAIR COVERING

Resilient sheet materials can be used to provide a suitable covering over wood and concrete stairs. The sheet material can be flashed from the tread up the riser (Fig. 38a) or used with a set-on coved rubber or vinyl riser (Fig. 38b). The nosing of a wood tread may be wrapped with the sheet flooring or protected by a separate metal or rubber nosing. Treads on riserless wood stairs may be completely wrapped with sheet materials (Fig. 38c). Formed concrete stairs can be covered with continuous runs of sheet flooring on both treads and risers (Fig. 38d).

Methods of installing sheet stair covering are similar to those described for sheet flooring. Accurate layout, scribing, fitting and proper selection of adhesive are essential.

TILE FLOORING

The following text describes the materials and methods recommended to provide a suitable tile flooring installation primarily over new floor construction.

Reference should be made to the entire first subsection, General Recommendations, pages 452-3 through 452-12, for a discussion of: (1) terminology, (2) proper job conditions, (3) adequate subfloor construction, and (4) accessory materials—underlayments, adhesives, lining felt and resilient accessories—required for flooring installations. Unless specific exceptions are made in this subsection, the General Recommendations apply and complement the following discussion.

MATERIALS

Resilient floor tile is available in *viny, vinyl-asbestos, rubber* and *cork.* Linoleum and asphalt tile are no longer produced in this country. Tile products range in size from 9" x 9" up to 36" x 36", in various gauges and in varying colors and patterns. Some tiles are made in no-wax finishes and with self-adhesive backs. Ingredients, gauges and standards of quality are summarized in Figure 39; selection criteria are given in Figure 40.

PREPARATION FOR FLOORING

The effectiveness of a flooring installation greatly depends on the proper selection and preparation of the elements which make up the installation. *Subfloors* which are to receive the flooring directly must be firm, smooth and dense, and possess good bonding properties. When the subfloor does not possess these qualities, it may be necessary

FIG. 39 **SUMMARY OF TILE FLOORING PRODUCTS**

Type	Federal Specification	Backing Materials	Gauges**	Sizes	Basic Ingredients
ASPHALT TILE†	SS-T-312B	—	1/8", 3/16"	9" x 9", 12" x 12"*	Asphaltic and/or resin binders, asbestos fibers and limestone fillers.
Plain					
Greaseproof	SS-T-312B				
VINYL-ASBESTOS TILE	SS-T-312B	—	1/16", .080", 3/32", 1/8"	9" x 9", 12" x 12"*	PVC resin binders, limestone, asbestos fibers.
VINYL TILE					PVC resin binders and mineral fillers.
Solid	SS-T-312B	—	1/16", .080" 3/32", 1/8"	9" x 9", 12" x 12" 4" x 36"*	
Backed	SS-T-312B	Scrap vinyl, asbestos or rag felts	.050" to .095"	9" x 9", 12" x 12" & sizes up to 36" x 36"*	
RUBBER TILE	SS-T-312B	—	.080", 3/32", 1/8", 3/16"	9" x 9", 12" x 12" 18" x 36"*, 36" x 36"*	Synthetic or natural rubber, mineral fillers.
CORK TILE					
Protected Surface	Interim LLL-T-00431 (Class 1)	—	1/8", 3/16", 1/4" 5/16"*, 1/2"*	9" x 9", 6" x 6"* 6" x 12"*, 12" x 12"*	Cork particles and resin binders, wax or resin finish.
Vinyl Surface	Interim LLL-T-00431 (Class 2)	—	1/8", 3/16"	9" x 9", 12" x 12"*	Cork particles and resin binders, clear vinyl finish.
LINOLEUM TILE†	LLL-F-1238A	Burlap	(.050" wear surface) 1/8"	9" x 9", 12" x 12"*	Cork and/or wood flour with linseed oil binders
	LLL-F-1238A	Rag felt	.090"		

*Not common, available on special order from some manufacturers.

**Thinnest gauges and wear surfaces indicated are minimums suitable for adhesive installation.

†Asphalt and linoleum products are no longer manufactured in this country, but are included here for reference.

FIG. 40 SELECTION CRITERIA FOR TILE FLOORING PRODUCTS[1]

Type of Flooring	Location[2]	Resistance to Grease	Resistance to Alkalis	Resistance to Stain[3]	Resistance to Cigarette Burns	Indentation	Resilience[4]	Quietness	Ease of Maintenance	Durability
VINYL-ASBESTOS	BOS	Good	Good	Fair	Good	Good	Poor	Good	Good	Good
VINYL										
Solid	BOS	Excellent	Excellent	Excellent to Good	Good to Poor	Excellent to Fair	Good to Poor	Good to Poor	Good	Excellent
Asbestos, vinyl & rubber backing	BOS	Excellent	Excellent	Excellent to Good	Good to Poor	Good	Good to Fair	Good to Poor	Good	Excellent
Rag felt backing	S	Excellent	Good	Excellent to Good	Good to Poor	Good to Poor	Good to Fair	Good to Poor	Good	Excellent
RUBBER	BOS	Good	Good	Good	Excellent	Excellent	Good	Good	Poor	Good
CORK	S	Fair	Fair	Poor	Fair	Fair	Excellent	Excellent	Good	Poor
VINYL CORK	S	Excellent	Good	Good	Poor	Good	Good	Good	Good	Fair

1. Criteria vary depending on properties and thickness of wear surface and backing material material; performance may be further affected by properties of adhesives, underlayments or subfloors. See page Resilient Flooring Products 220-7 for a discussion of selection criteria.
2. B-Below grade, O-On grade, S-Suspended.
3. Varies with staining agent.
4. Also indicates potential underfoot comfort.

to prepare the surface by grinding or sanding or with the use of *underlayment*. *Adhesives* should be selected that are compatible with all elements, including the flooring itself. All elements also should be suitable for the intended locations—whether suspended, on grade or below grade.

Subfloor

Solid vinyl, asbestos-backed vinyl, vinyl-asbestos and rubber tile may be installed on subfloors in suspended, on-grade and below-grade locations. Cork, vinyl-cork, and rag felt-backed vinyl tile should be installed on suspended subfloors only.

Concrete subfloors in all locations should be tested for moisture and bonding properties to insure suitability for the proposed flooring and adhesive (see Suitability Tests, page 452-6).

Underlayment

The condition of the subfloor will determine the need for underlayment. For a discussion of underlayments and recommended uses, see General Recommendations, page 452-6. Mastic underlayments should be compatible with intended adhesives and primers as discussed below.

Adhesives

The adhesives recommended for tile products in various locations are summarized in Figure 41. For a discussion of adhesive types and their properties, see General Recommendations and Figure 13, page 452-8.

Solvent-base asphaltic adhesives and primers used for the installation of asphalt and vinyl-asbestos tile should not be used over asphaltic underlayments. For such installations, asphaltic emulsions should be used. When cutback asphalt, which is a solvent-base adhesive is used over latex underlayment, sufficient time must be allowed before installing tile. Porous or dusty concrete floors and panel underlayments intended for the installation of asphalt or vinyl asbestos tile should be primed with an asphaltic primer before applying adhesive. Some tiles are made with self-adhesive backs.

INSTALLATION OF FLOORING

The first step in the installation of tile flooring is to make a trial design layout of the tiles. The tiles then are cemented down according to the best layout and walls are trimmed out with resilient or other base.

Design Layout

In determining design layout, it is important that tile joints do not fall over joints in the subfloor or in the underlayment below. Because few walls are seldom exactly parallel, it is necessary to center and square off the room (Fig. 42). Tiles may be laid in either *square* or *diagonal* designs. Over existing stripwood floors, a diagonal installation is preferred.

Square Designs Tile joints for square designs run parallel to the walls. The trial layout is started by snapping chalklines to divide the room into four equal quadrants

FIG. 41 **SUMMARY OF ADHESIVES FOR TILE FLOORING**

FLOORING	SUBFLOOR AND LOCATION		
	WOOD (Suspended)	CONCRETE (Suspended)	CONCRETE (on and below grade)
VINYL (solid)	Waterproof Resin Latex	Waterproof Resin Latex	Latex Epoxy
VINYL (rag felt backing)	Linoleum Paste Waterproof Resin	Linoleum Paste Waterproof Resin	Not Recommended
VINYL (asbestos backing)	Latex Linoleum Paste	Latex Linoleum Paste	Latex
RUBBER	Linoleum Paste Waterproof Resin	Waterproof Resin Linoleum Paste	Latex Epoxy
VINYL-ASBESTOS	Asphalt Emulsion Asphalt Cutback Asphalt Rubber	Asphalt Emulsion Asphalt Cutback Asphalt Rubber	Asphalt Emulsion Asphalt Cutback Asphalt Rubber
CORK	Linoleum Paste Waterproof Resin	Linoleum Paste Waterproof Resin	Not Recommended

(Fig. 43). If the floor area is extremely large or job conditions prevent working the entire floor area, each quadrant may be subdivided further. Full tiles should be laid starting from the center working toward the walls. The first trial may result in unequal border tiles at sidewalls and endwalls (Fig. 43a).

To equalize border tiles at walls, the horizontal row of tiles may be shifted so that it is bisected by the chalklines as shown in Figure 43b. In general, layout lines should be shifted so that approximately equal border tiles of at least one half a tile or wider will result.

Checkerboard patterns may be made by alternating tiles of two different colors, or alternating the grain direction in directional tiles. Either a tile which matches the field or a contrasting one-color sheet or tile can be used for the border.

For a *hollow square* design, another common two-color pattern, it is important to lay out the trial design so that a full *repeat* of the pattern will occur at all walls (Fig. 44). Such patterns commonly are used with one-color borders to emphasize the design in the field.

Diagonal Designs Tile joints are run at 45° angles to the walls for diagonal designs. A trial row of tiles is laid out along each centerline and diagonally to the corner to determine the resulting condition at the border (Fig. 45a).

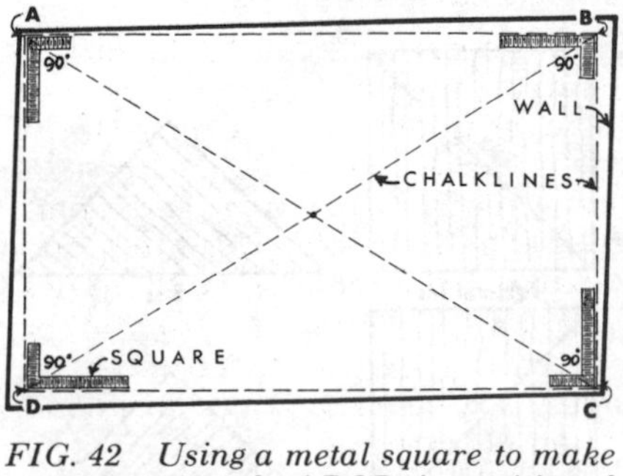

FIG. 42 Using a metal square to make corners, rectangle ABCD is positioned so that wall irregularities are evenly distributed at borders. Diagonals should be equal if rectangle is absolutely square.

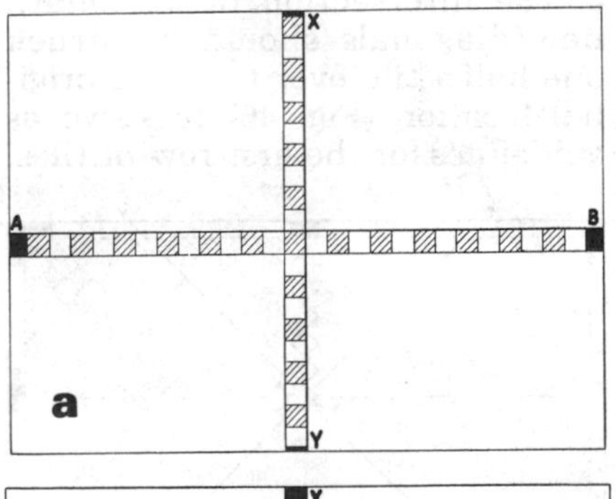

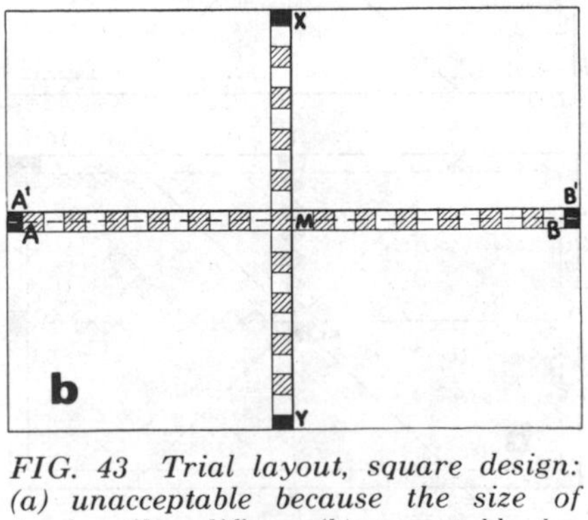

FIG. 43 Trial layout, square design: (a) unacceptable because the size of border tiles differs; (b) acceptable because border tiles are equal and wider than one-half tile.

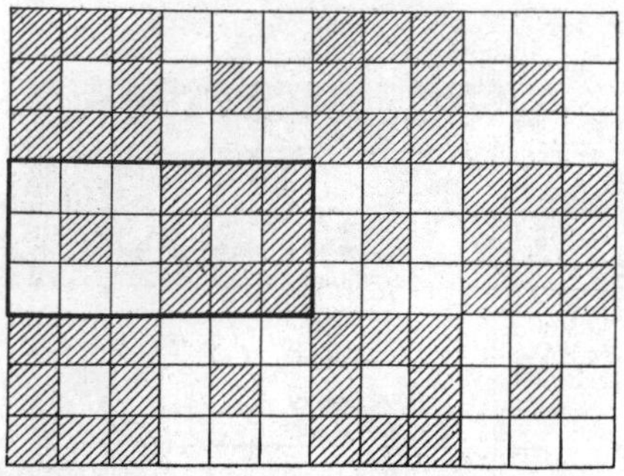

FIG. 44 "Hollow square" design; repeat unit indicated by dark outline.

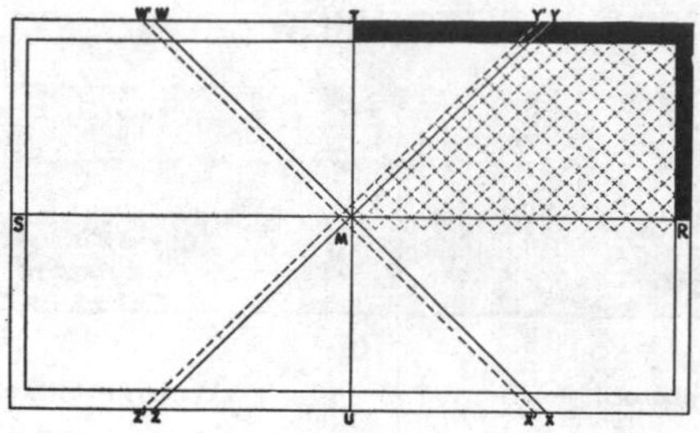

FIG. 46 New chalkline yz may be required as a guide for first row of tiles. As tile is installed clockwise, alignment may be checked against diagonals.

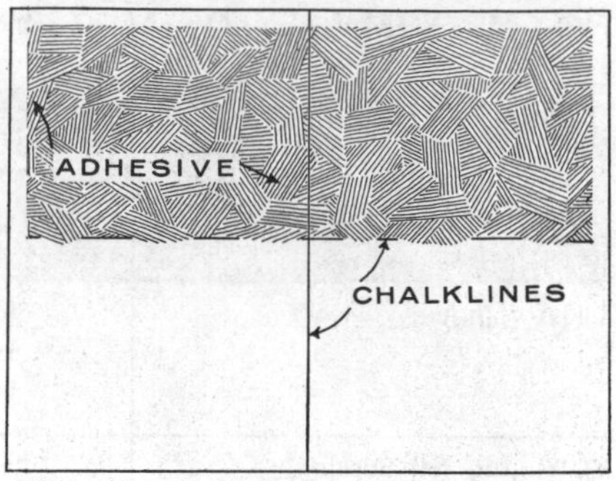

FIG. 48 Guidelines should not be obscured while adhesive is spread.

Two-color designs, requiring a border laid on the square, also must be checked to assure that the half tiles along each of the walls will be of the same color. To achieve this, it may be necessary to shift tiles from the trial position shown in Figure 45a, so that tiles will abut rather than be bisected by diagonal lines (Fig. 45b). The proper starting position of tiles at the center of the room is determined by the dimensions of the room and must be established in each case by trial layout.

If it is determined that the design should start with single tiles at the intersection of diagonals, new diagonals should be struck one half a tile over from the original location (Fig. 46) to serve as guidelines for the first row of tiles.

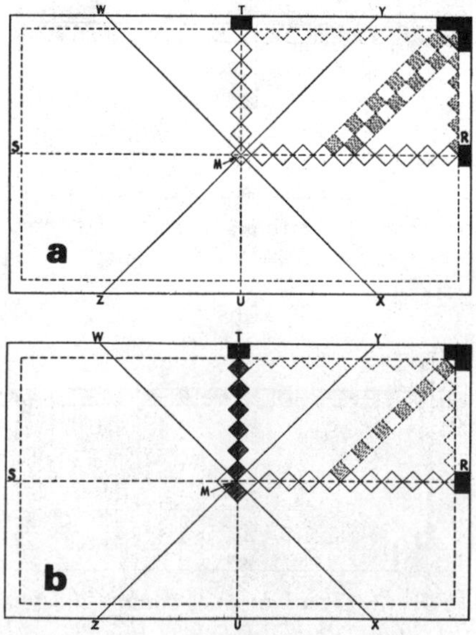

FIG. 45 Diagonal design, trial layout: (a) unacceptable, (b) acceptable, all half tiles at the border are of the same color.

One-color designs without a border present no special problem and the starting point need only be adjusted to assure that the layout will not result in excessively narrow tiles at the walls. In one-color installations with tiles having a directional grain, it is important that the half tiles near the border fit properly into a basketweave pattern. If the tiles are cut properly with respect to grain, tiles at both right and left sides will result which fit the pattern (Fig. 47).

Adhesive Bonding

To allow sufficient working space to handle materials, only half or part of the floor area, depending on the working time of the adhesive, should be covered with adhesive at one time. Care should be taken in applying the adhesive so that chalklines will not be completely concealed. This may be done by leaving ends and intermediate points on the chalkline uncovered (Fig. 48). Lines should be transferred to the flooring as they become covered.

Adhesive such as cutback asphalt develops holding power by setting before tiles are applied. However, with some adhesives such as waterproof-resin and epoxy, the possibility of tiles slipping may be a problem. Temporary lattice strips nailed at the chalklines will prevent the tiles from sliding out of place during installation (Fig. 49).

Tiles should be installed with tight, straight and carefully aligned joints in both directions. Each tile should be butted to the preceding tiles and lowered in place without sliding (Fig. 50).

Installation on the square with a two-color checkerboard pattern

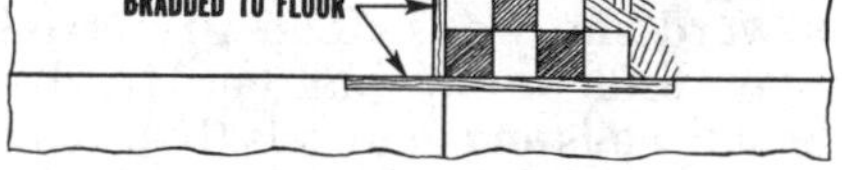

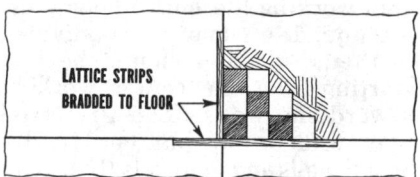

FIG. 49 Starting tiles can be restrained from slipping by nailing lattice strips as shown above.

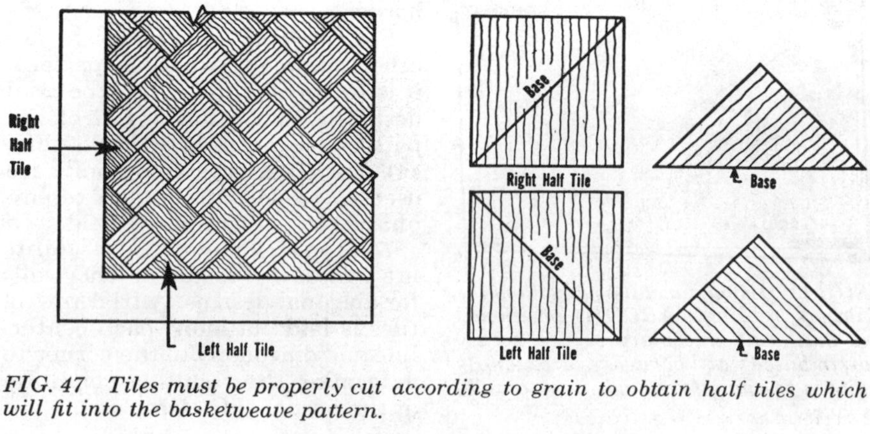

FIG. 47 Tiles must be properly cut according to grain to obtain half tiles which will fit into the basketweave pattern.

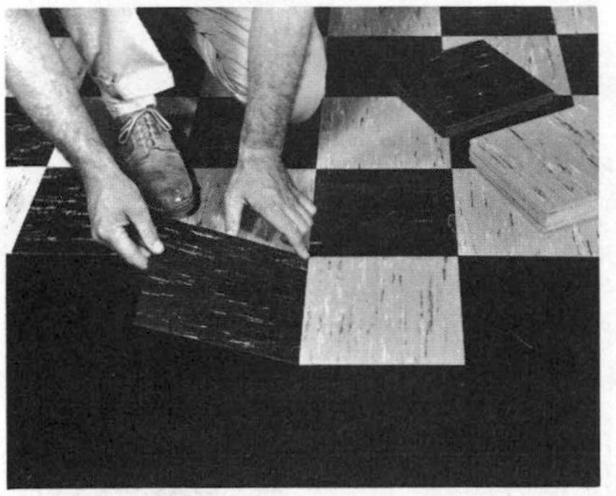

FIG. 50 Tiles should be installed by butting each tile against the others without sliding it into place so as not to force adhesive up into the joint.

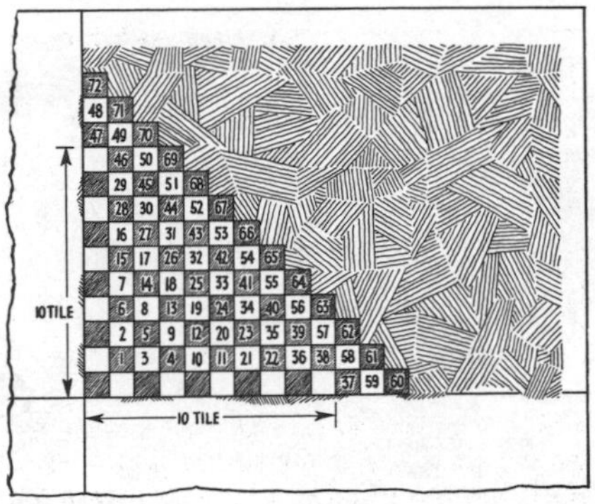

FIG. 51 Checkerboard square designs can be installed more quickly by laying tiles of one color in diagonal rows as numbered.

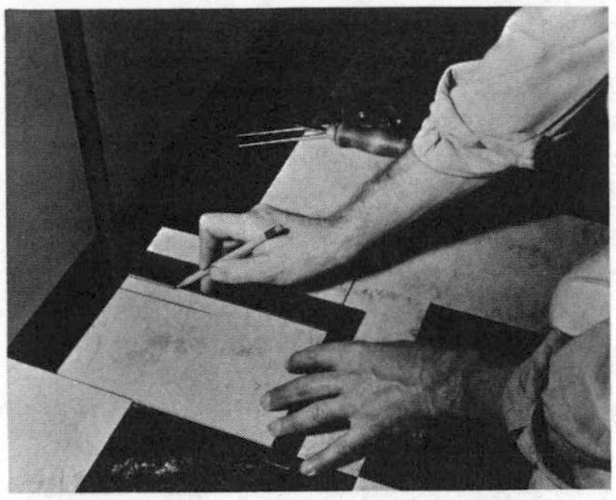

FIG. 52 Square is used to position cutout for pipe, circle cutout is scribed with dividers, the tile then is cut and slid into place.

can be speeded by starting with a row of ten tiles along each centerline, and laying tiles of one color at a time in diagonal rows (Fig. 51).

For diagonal designs, tiles may be cemented down in one quadrant at a time, starting at the center, laying the tiles along the diagonal and proceding clockwise around the room (Fig. 46).

Fitting and Cutting

Border tiles may require scribing and fitting against walls and obstructions (Fig. 52). They are scribed essentially as described for sheet flooring using dividers or a scribing tool to transfer the contour of the vertical surface (Fig. 24).

Radiators resting directly on tile floors may cause tiles to curl or shift from the heat. Metal discs (washers or nuts) should be inserted in cutouts in the tiles to support the radiator legs (Fig. 53).

Most resilient tiles can be cut at room temperature with a linoleum or straight blade knife. However, it is easier to cut vinyl-asbestos and asphalt tile by heating the tile and the knife with a blowtorch or a heat gun.

Wall Base

The two main types of wall base used with resilient tile are *set-on* and *butt-cove* base (Fig. 14). Most tile installations are made with set-on base because it is easier and more economical to install. This type of base is set on top of the finished floor and requires less accurate fitting of both flooring and base. Both set-on and butt-cove base are available in rubber and vinyl materials.

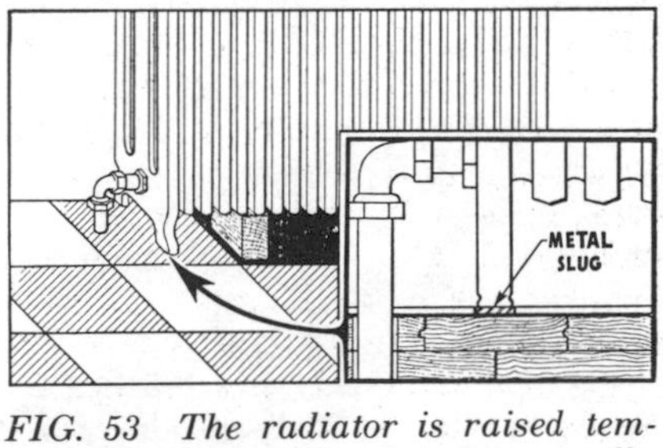

FIG. 53 The radiator is raised temporarily, tile is slipped under to scribe legs, metal discs are placed in cutouts in tiles to support the legs.

Rubber Base This type of base is manufactured in straight runs and in premolded outside and inside corners. These facilitate turning corners with a minimum of field labor, the only fitting required being at butted joints.

Vinyl Base The greater pliability of vinyl base permits molding it around corners and therefore it is manufactured mainly in straight runs. Inside corners are made by notching a *V* in the toe of the base at the fold and heating it to increase pliability. The base may then be formed and pressed into the corner. Inside corners also can be made by butting the first piece squarely in the corner and scribing the second piece around it (Fig. 54).

Outside corners are formed by first cutting away about half the thickness from the back of the material at the point of bend (Fig. 55a). At the back of the cove heel, additional material should be cut away approximately 1″ to each side

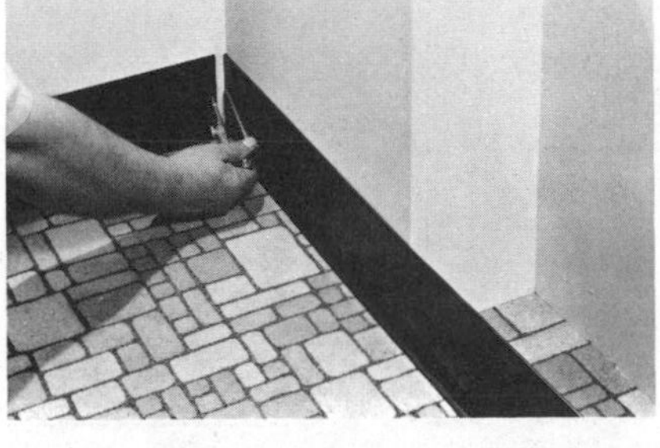

FIG. 54 Inside base corner: right side is butted to corner and bonded; left side is scribed to right side and bonded.

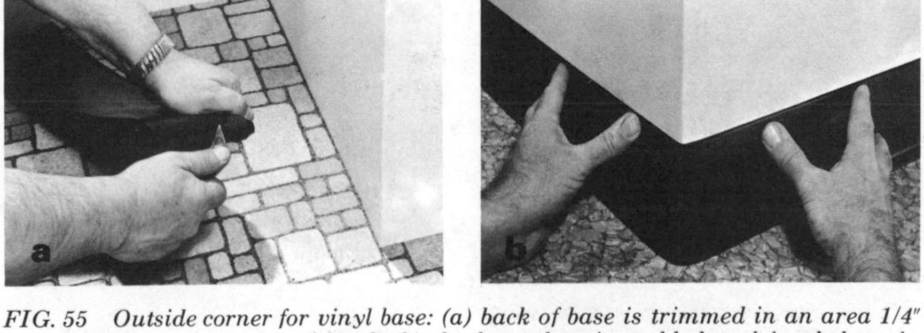

FIG. 55 Outside corner for vinyl base: (a) back of base is trimmed in an area 1/4″ to 1″2″ wide at the point of bend; (b) the base then is molded and bonded to the outside corner.

of the bend. The base then can be molded readily around the corner and set in place with standard base cement or synthetic rubber cement (Fig. 55b). Synthetic rubber cement is preferred because it provides immediate holding power and prevents movement from contraction as bond is developed.

STAIR COVERING

Resilient tiles can provide an attractive wearing surface on wood and concrete stairs. Tile treads usually are installed with rubber, vinyl or metal nosings (Fig. 56). Risers for tile treads may be of matching tile or a one-piece set-on coved riser similar to wall base. Matching one-piece treads with premolded nosings also are available from some manufacturers (Fig. 56). Methods of installation are similar to those described for tile floors. Accurate layout, scribing, fitting and proper selection of adhesives are essential.

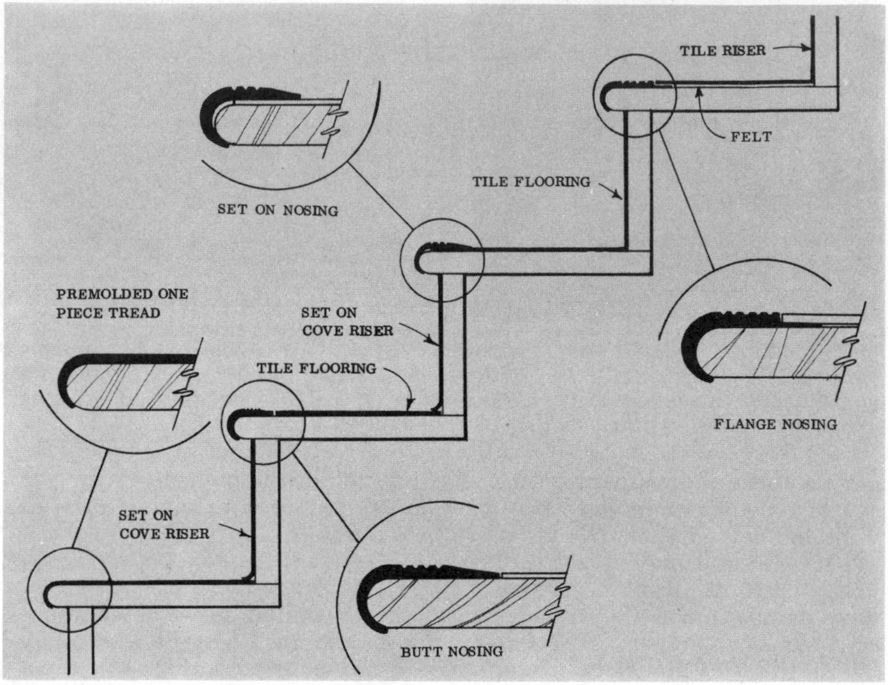

FIG. 56 Stair coverings with tile flooring and resilient accessories.

We gratefully acknowledge the assistance of the following for the use of their publications as references and permission to use photographs: Armstrong Cork Company; Congoleum Nairn, Inc.; The Flintkote Company; Johnson Rubber Company; Johns-Manville Corp.; Kentile, Inc.; Mercer Plastics Company, Inc. The Ruberoid Company.

RESILIENT FLOORING 452

CONTENTS

GENERAL RECOMMENDATIONS

See page 452-3 for definitions of terminology.

Because of the variety of proprietary flooring materials, underlayments and adhesives, the individual manufacturer's recommendations should be consulted to insure proper use in each case.

JOB CONDITIONS

A temperature of at least 65° F. and not over 90° F. should be maintained for 48 hours before installation, during installation and 48 hours after installation. A minimum temperature of 55° is recommended thereafter. Adequate ventilation should be provided at all times.

FLOOR CONSTRUCTION

Resilient flooring and adhesives should be suitable for the intended subfloor location—whether suspended, on grade or below grade (Fig. WF 1).

The subfloor and supporting structural members should be adequate to carry the intended dead and live loads without excessive deflection.

The exterior finish grade should be sloped away from the foundation to divert surface water (see Fig. 15, page Foundation Systems 310-10).

Concrete and wood subfloors may be suitable for direct application of resilient flooring or may require the use of underlayment.

Wood Subfloors

Suspended wood floors over crawl spaces should be protected against dampness. A suitable vapor barrier and adequate ventilation should be provided (see Section 313 Crawl Spaces).

FIG. WF1 Subfloor locations. Subfloors when built at least 18" above grade over a ventilated crawl space are considered suspended.

(Continued) **GENERAL RECOMMENDATIONS**

Suspended wood floors above grade usually are sufficiently removed from excessive external moisture.

Plywood and lumber board subfloors should conform to the minimum thickness for various joist spacing and nailing recommendations given in Figure WF 2.

The moisture content of lumber boards and framing members should be as recommended in Figure WF 1, page Wood WF 201-1. Single lumber board construction should not be used for the direct application of resilient flooring.

Plywood panels should be installed with the surface grain perpendicular to the joists, and end joints of panels centered over joists. Longitudinal edges should be supported on 2″ x 3″ or heavier blocking unless: (1) tongue-and-grooved plywood 5/8″ or thicker is used, or (2) a suitable underlayment is installed with joints staggered relative to joints in the subfloor.

Plywood panels should be spaced approximately 1/32″ at joints and approximately 1/8″ at all intersections with vertical surfaces, to allow for expansion from humidity changes. Nailing should start at the middle of the panel and proceed outward to the edges, with nailheads set flush or slightly countersunk. Perimeter nailing should be located not more than 3/8″ from panel edges.

Combination Subfloor/Underlayment Plywood subfloors, without separate underlayment, under normal moisture conditions, should be sanded Interior Underlayment grade. For unusual moisture conditions, such as in bathrooms and laundry areas, C-C (Plugged) Exterior, Underlayment grade bonded with Exterior adhesive, or a sanded grade of Exterior type plywood should be used (see page Wood WF 201-16).

The subfloor should be adequately protected from damage during construction and panel joints sanded smooth and flush prior to the flooring installation; or, the entire surface should be completely sanded prior to installation.

The difficulty of adequately protecting the subfloor during construction may warrant the use of an unsanded or a touch-sanded plywood and power sanding the entire floor area just before the flooring is installed.

Subfloor with Underlayment Plywood subfloors to be covered with underlayment should be an unsanded

FIG. WF2 **WOOD SUBFLOOR CONSTRUCTION**[1]

Combination Subfloor/Underlayment[2] Nails[3]	Combination Subfloor/Underlayment[2] Nail Spacing	Subfloor[4]	Maximum Support Spacing	Subfloor with Underlayment[5] Nails[3]	Subfloor with Underlayment[5] Nail Spacing
2″ (6d) annularly threaded	6″ o.c. around panel edges; 10″ o.c. at intermediate supports	½″ Plywood	16″ o.c.	2″ (6d) common	6″ o.c. around panel edges; 10″ o.c. at intermediate supports
		⅝″ Plywood	20″ o.c.	2½″ (8d) common	
		¾″ Plywood	24″ o.c.		
2½″ (8d) annularly threaded	6″ o.c. at panel edges and intermediate supports	1⅛″ Plywood (2-4-1)	48″ o.c.	—	—
—	—	¾″ lumber boards less than 6″ wide	16″ o.c.	2¼″ (7d) annularly threaded	2 per board at each support
—	—	¾″ lumber boards wider than 6″ and less than 8″ wide	16″ o.c.		3 per board at each support

1. Based on 40 psf live load.
2. Blocking at longitudinal edges is required unless ⅝″ or thicker plywood with T&G edges is used.
3. Common nails may be substituted for annularly threaded nails by increasing the size ¼″ (1d).
4. Plywood thicknesses given are for Group 1 Softwoods in accordance with U.S. Product Standard PS1. For Groups 2 and 3, increase thickness ⅛″; for Group 4, increase thickness ¼″ (see Work File, page Wood WF201-16).
5. Blocking not required if underlayment joints are staggered with respect to subfloor joints.

(Continued) GENERAL RECOMMENDATIONS

Interior Standard grade. For unusual moisture conditions, a C-C Exterior grade or an Interior Standard grade with Exterior adhesive should be used.

Subfloor with underlayment often is the preferred construction because the subfloor provides a convenient working platform during most of the construction period and installation of the underlayment can be scheduled more closely to the installation of the flooring.

Concrete Subfloors

The density of concrete subfloors intended for direct application of flooring should not be less than 90 lbs. per cu. ft. Lightweight concrete slabs of lower density should be refinished with a 1″ topping of a regular concrete mix, chemically or mechanically bonded to the slab.

Subfloors on or below grade should be separated from the ground with a suitable vapor barrier (see Section 311 Slabs-on-Ground).

The moisture content and surface bonding properties of concrete subfloors should be tested to determine suitability for the intended flooring and adhesive.

Lightweight concrete is slow in curing and retains moisture longer. Slabs which have been treated with curing, hardening or separating compounds may have poor bonding properties. See Main Text, page 452-6 for a discussion of relative humidity and mat moisture and bonding tests.

Poorly finished, spalled or irregular concrete should be patched or refinished with mastic underlayment. Curing, hardening or separating compounds should be removed if test reveals inadequate adhesion.

UNDERLAYMENTS

When the subfloor is not sufficiently smooth, level and hard, or does not possess adequate bonding properties, it should be covered with underlayment.

Panel underlayments generally are used over wood subfloors; mastic underlayments generally are used over concrete subfloors. Some manufacturers do not accept particleboard underlayment.

Panel Underlayment

Panel underlayment should be at least 1/4″ plywood, 1/4″ hardboard or 3/8″ particleboard.

Panels should be spaced approximately 1/32″ at joints and approximately 1/8″ at all intersections with vertical surfaces. End joints in the underlayment should be staggered ashlar fashion.

Over board subfloors, underlayment joints should not fall over joints in the board floor.

Over plywood subfloors, stiffer floors result when joints in the underlayment are staggered with respect to joints in the subfloor and blocking can be omitted.

Panel underlayments should be installed with the fasteners and spacing recommended in Figure WF 3.

Nailing should start in the middle of the panel and proceed outward to the edges, with nailheads set flush or slightly countersunk. Prior to installation of flooring, panel joints should be sanded smooth and flush.

Plywood For normal moisture conditions, sanded Interior Underlayment grade should be used. For areas subject to excessive moisture, such as in bathrooms and laundry areas, a C-C (Plugged) Exterior grade or Underlayment grade bonded with Exterior adhesive should be used (see page Wood WF 201-16).

FIG. WF3 FASTENING RECOMMENDATIONS FOR PANEL UNDERLAYMENT

Underlayment	Fasteners[1] Nails[2]	Staples[3]	Fastener Spacing
PLYWOOD			*Nails:* 8″ o.c. each direction in middle of panel, 6″ o.c. around panel perimeter. *Staples:* 6″ o.c. each direction in middle of panel, 3″ o.c. around panel perimeter.
1/4″	1″ (2d)	7/8″	
3/8″	1 1/4″ (3d)	1 1/8″	
1/2″	1 1/4″ (3d)	1 1/4″	
5/8″	1 1/2″ (4d)	1 1/2″	
3/4″	1 1/2″ (4d)	1 5/8″	
HARDBOARD[4]			*Nails and Staples:* 4″ o.c. each direction in middle of panel, 3″ o.c. around panel perimeter.
1/4″	1″ (2d)	7/8″	
PARTICLEBOARD[5]			*Nails:* 10″ o.c. each direction in middle of panel, 6″ o.c. around panel perimeter. *Staples:* 6″ o.c. each direction in middle of panel, 3″ o.c. around panel perimeter.
3/8″	1 1/4″ (3d)	1 1/8″	
5/8″	1 1/2″ (4d)	1 1/2″	
3/4″	1 1/2″ (4d)	1 5/8″	

[1]Fastener lengths shown are minimum to provide 3/4″ penetration for nails and 5/8″ for staples. If longer fasteners are used, penetration should not exceed 1″.

[2]Annularly threaded; common nails may be used by increasing the sizes 1/4″ (1d).

[3]Galvanized divergent chisel, 16 gauge, 3/8″ crown width.

[4]Nominal thickness, actual thickness 0.215″.

[5]Moisture content as shipped from mill should not exceed 7% average.

(Continued) GENERAL RECOMMENDATIONS

Over board subfloors, panels should be installed with the face grain perpendicular to the direction of the floor boards, preferably with long edges centered over joists.

Over plywood subfloors, panels may be installed with face grain either parallel or perpendicular to the floor joists. However, a stiffer floor assembly results if panels are installed with face grain perpendicular to floor joists.

Hardboard Hardboard underlayment should conform to CS251 and the specifications of the American Hardboard Association. It should be installed with the smooth side up, in panels not exceeding 4′ x 4′.

Particleboard Particleboard underlayment should conform to CS236 and the National Particleboard Association's "Specifications for Mat-formed Wood Particleboard for Floor Underlayment." Particleboard should not be used in areas subject to excessive moisture, and its moisture content as shipped from the mill should not exceed an average of 7%.

Some resilient flooring manufacturers do not recommend the use of particleboard as an underlayment material. However, many particleboard manufacturers issue a guarantee covering the entire resilient flooring installation if the flooring and underlayment are installed to their specifications. If particleboard is used, it is recommended that it be a product backed by a manufacturer's guarantee and the installation specifications of the manufacturer be followed.

Mastic Underlayment

Mastic underlayment should be mixed and applied in accordance with the manufacturer's recommendations. The underlayment should be troweled level and smooth, with not more than 1/8″ variation from a straight line in 10′ in any direction.

Latex underlayment generally should be built up in layers not more than 1/8″ thick and is used for skim coating and fills not more than 3/8″ thick. For thicker fills, asphaltic and polyvinyl-acetate underlayments are used.

Solvent-base asphaltic primers and adhesives should not be applied over asphaltic underlayments. Over latex underlayment, sufficient open time should be allowed for solvent-base adhesives to set before flooring is installed.

Polyvinyl-acetate underlayment is suitable with all primers and adhesives.

ADHESIVES

Adhesives should be compatible with the intended flooring, underlayment and location. Adequate ventilation should be provided for volatile, flammable adhesives.

See Main Text, page 452-8 and Figure 13 for a discussion of adhesives and their properties.

RESILIENT ACCESSORIES

For a description of resilient accessories such as wall base, stair treads and nosings, thresholds, feature strips and reducing strips, see Main Text page 452-11 and page Resilient Flooring Products 220-6.

REFLOORING

Existing wood, concrete, ceramic tile, terrazzo or marble floors may be suitable for reflooring. Most manufacturers permit installation of new resilient flooring over well-bonded existing sheet or tile, provided the old floor is not deeply embossed or foam-backed.

Existing resilient flooring or lining felt should not be sanded. If existing surface is irregular, underlayment should be installed.

MAINTENANCE AND PROTECTION

Thorough cleaning of a new resilient floor should be delayed for at least 24 hours after installation. The floor then should be washed with a mild solution of nonabrasive, nonalkaline cleaner or soap and rinsed clean. It should be waxed with a floor finish recommended by the manufacturer often enough to maintain a protective coating.

Legs of heavy furniture and other concentrated loads should be supported on floor protectors. Throw rugs or mats should be positioned at entrances to prevent outside dirt and moisture from damaging the floor.

SHEET FLOORING

Unless specific recommendations are made in the following text, the General Recommendations apply.

MATERIALS

Sheet flooring products should conform to applicable Federal Specifications and minimum gauges given in Figure WF 4. Products should be selected according to the type of subfloor and location—whether suspended, on grade or below grade.

Linoleum and rag felt-backed products should be installed on suspended subfloors only.

Vinyl sheet products with asbestos fiber backing may be installed on suspended, on-grade and below-grade subfloors.

Selection criteria are summarized in Main Text, page 452-13, Figure 17.

ADHESIVES

Adhesives should be compatible with the intended flooring, underlayment, lining felt and subfloor location. The recommendations of the flooring manufacturer should be followed.

Figure WF 5 outlines generally recommended adhesives for various types of sheet flooring. See Main Text, page 452-8 and Figure 13 for a discussion of adhesives and their properties.

FIG. WF4 SUMMARY OF SHEET FLOORING PRODUCTS

Type	Federal Specification	Backing Materials	Gauges[1]	Sizes	Basic Ingredients
LINOLEUM SHEET[4]			(.050″ wear surface)	6′0″ wide	Cork and/or wood flour with linseed oil binders
Plain & Marbleized	LLL-F-1238A	Burlap	1/8″		
	LLL-F-1238A	Rag felt	.090″		
Inlaid & Molded	LLL-F-1238A	Burlap	1/8″		
	LLL-F-1238A	Rag felt	.090″		
Battleship	LLL-F-1238A	Burlap	1/8″		
VINYL SHEET[2]					PVC resin binders and mineral fillers; clear PVC film for clear vinyl surface.
Filled Surface	L-F-475-a L-F-001641	Scrap vinyl, asbestos or rag felts	.065″ to .160″ (.010″ to .050″ wear surface)	6′0″, 12′0″ & 4′6″[3] wide	
Clear Vinyl Surface			.065″ to .160″ (.010″ to .050″ wear surface)	6′0″ & 12′0″ wide	

[1]Thinnest gauges and wear surfaces indicated are minimums suitable for adhesive installation.
[2]Also produced with foam backing, not included in data shown.
[3]Not common, available on special order from some manufacturers.
[4]Linoleum sheet is no longer produced in this country, but is included here for reference only.

FIG. WF5 SUMMARY OF ADHESIVES FOR SHEET FLOORING*

Subfloor and Location	Flooring		
	VINYL (rag felt-backed)	VINYL (vinyl or rubber-backed)	VINYL (asbestos-backed)
WOOD or CONCRETE (Suspended)	Waterproof-Resin	Latex Waterproof-Resin	Latex
CONCRETE (on and below grade)	Not Recommended	Latex	Latex

*Foam-cushioned products are installed by specially formulated adhesives recommended by the flooring manufacturer.

(Continued) SHEET FLOORING

LINING FELT

Lining felt should be used under burlap-backed linoleum.

APPLICATION OF FLOORING

The installation of sheet flooring should be planned to minimize the number and total length of seams. If possible, seams should be placed in inconspicuous locations, out of the path of heavy foot traffic.

Cutting and Fitting

Factory-cut edges (selvages) should be trimmed, and the flooring accurately fitted to the walls and other vertical surfaces. Seams between adjacent strips should be made straight, snug and inconspicuous. Seams should not be cut until adjacent strips are partly or completely secured in position with adhesive. (Tension floor installation requires that seaming be done first.)

Adhesive Bonding

Adhesive should be spread under each strip in the area revealed as the strip is lapped or tubed back on itself. Only enough area should be covered in each spreading as can be covered with flooring within the bonding time recommended by the adhesive manufacturer.

As each part of the flooring is cemented down, it should be rolled with a 100 lb. or heavier roller. Seams should be tightly bonded, rolled with a hand roller and properly treated to remove all burred edges.

For a discussion of recommended methods of adhesive bonding and seam treating, see Main Text page 452-18.

TILE FLOORING

Unless specific exceptions are made in the following text, the General Recommendations apply.

MATERIALS

Tile products should conform to applicable Federal Specifications and minimum gauges given in Figure WF 6. Products should be selected according to type of subfloor and location—whether suspended, on grade or below grade.

Cork, vinyl-cork, rag felt-backed vinyl and linoleum tile should be installed on suspended floors only.

Solid vinyl, asbestos-backed vinyl, asphalt, vinyl-asbestos and rubber tile may be installed on suspended, on-grade and below-grade subfloors.

Selection criteria are summarized in Main Text, page 452-22, Fig. 40.

ADHESIVES

Adhesives should be compatible with the intended flooring, underlayment and subfloor location. The recommendations of the flooring manufacturer should be followed.

Figure WF 7 outlines generally recommended adhesives for various types of tile flooring. See Main Text, page 452-8 and Figure 13 for a discussion of adhesives and their properties.

Concrete subfloors intended for the application of vinyl-asbestos and asphalt tile should be given the primer test to determine suitability for asphaltic adhesives.

Dusty or porous concrete subfloors and panel underlayments intended for the installation of asphalt or vinyl-asbestos tile should be primed with an asphaltic primer before applying asphaltic adhesive.

(Continued) **TILE FLOORING**

FIG. WF6 SUMMARY OF TILE FLOORING PRODUCTS

Type	Federal Specification	Backing Materials	Grade Location[1]	Gauges[2]	Sizes	Basic Ingredients
ASPHALT TILE[4] Plain	SS-T-312B	—	BOS	1/8″, 3/16″	9″ x 9″, 12″ x 12″[3]	Asphaltic and/or resin binders, asbestos fibers and limestone fillers.
Greaseproof	SS-T-312B					
VINYL-ASBESTOS TILE	SS-T-312B	—	BOS	1/16″, .080″, 3/32″, 1/8″	9″ x 9″, 12″ x 12″[3]	PVC resin binders, limestone, asbestos fibers.
VINYL TILE Solid	SS-T-312B	—	BOS	1/16″, .080″ 3/32″, 1/8″	9″ x 9″, 12″ x 12″ 4″ x 36″[3]	PVC resin binders and mineral fillers.
Backed	SS-T-312B	Scrap vinyl, Asbestos felt Rag felt	BOS S	.050″ to .095″	9″ x 9″, 12″ x 12″ & sizes up to 36″ x 36″[3]	
RUBBER TILE	SS-T-312B	—	BOS	.080″, 3/32″, 1/8″, 3/16″	9″ x 9″, 12″ x 12″ 18″ x 36″[3], 36″ x 36″[3]	Synthetic or natural rubber, mineral fillers.
CORK TILE Protected Surface	Interim LLL-T-00431 (Class 1)	—	S	1/8″, 3/16″, 1/4″ 5/16″[3], 1/2″[3]	9″ x 9″, 6″ x 6″[3] 6″ x 12″[3], 12″ x 12″[3]	Cork particles and resin binders, wax or resin finish.
Vinyl Surface	Interim LLL-T-00431 (Class 2)	—	S	1/8″, 3/16″	9″ x 9″, 12″ x 12″[3]	Cork particles and resin binders, clear vinyl finish.
LINOLEUM TILE[4]	LLL-F-1238A	Rag felt	S	(.050″ wear surface) 1/8″	9″ x 9″, 12″ x 12″[3]	Cork and/or wood flour with linseed oil binders.
	LLL-F-1238A	Burlap	S	.090″		

[1]B-Below grade, O-On grade, S-Suspended.
[2]Thinnest gauges and wear surfaces indicated are minimums suitable for adhesive installation.
[3]Not common, available on special order from some manufacturers.
[4]Not made in this country any longer, but included here for reference.

FIG. WF7 SUMMARY OF ADHESIVES FOR TILE FLOORING

Subfloor and Location	Flooring		
	VINYL-ASBESTOS	VINYL (solid and asbestos-backed) and RUBBER	VINYL (rag felt-backed), CORK
WOOD or CONCRETE (Suspended)	Asphalt Emulsion Asphalt Cutback Asphalt Rubber	Waterproof-Resin Latex	Waterproof-Resin
CONCRETE (on and below grade)	Asphalt Emulsion Asphalt Cutback Asphalt Rubber	Latex Epoxy	Not Recommended

(Continued) TILE FLOORING

Solvent-base asphaltic adhesives and primers used for the installation of asphalt and vinyl-asbestos tile should not be used over asphaltic underlayments. When solvent-base asphaltic adhesive is used over latex underlayment, sufficient setting time for the adhesive should be allowed before installing tile.

LINING FELT

Lining felt should not be used for the installation of 1/16″ vinyl-asbestos tile.

APPLICATION OF FLOORING

The tile pattern should be laid out so that joints do not fall over joints in the subfloor or underlayment. Tiles should be accurately scribed and fitted against the walls and other vertical surfaces. Radiator legs should be supported on metal discs or washers inserted in cutouts in the tile.

Border tiles adjacent to walls should be at least one half a tile wide and of approximately equal size around the perimeter of the room.

Adhesive should be spread only in as large an area as can be covered with tile within the bonding time recommended by the adhesive manufacturer.

Tiles should be installed with tight, straight and carefully aligned joints in both directions. Each tile should be butted to the preceding tiles and lowered into place without sliding to prevent the adhesive from being forced up into the joint.

453 TERRAZZO

TERRAZZO

INTRODUCTION

The term terrazzo is derived from the Italian "terrassa" (terrace on the roof) because many early installations were made in upper-level outdoor living areas. The origins of terrazzo date back to the beginnings of the Christian era, having evolved from *mosaics,* the art of placing colorful pieces of stone, vitreous enamel or marble (tesserae) in decorative designs on walls and floors. Earliest terrazzo installations in Roman times were made of a mixture of crushed brick in lime mortar, which was ground and polished after it hardened. Later, the decorative chips salvaged from mosaic work were introduced in the outdoor terraces to create highly colorful and durable floor finishes.

Terrazzo has been used for centuries as an economical and durable outdoor-indoor finish and traditionally has been defined as a form of mosaic flooring made by embedding small pieces of marble in mortar and polishing. In modern times, the mortar (matrix) for terrazzo finishes generally has been portland cement and the decorative chips were limited to marble. More recently, new synthetic resinous binders have been developed and other hard rocks capable of being polished are used.

Because of the expense involved in hand assembly of decorative chips in complex designs for large areas, mosaic work survives today primarily as custom inserts in terrazzo installations. The tesserae usually are assembled in the shop and mounted on paper; then the assembly is installed in a mortar setting bed in the desired location. After the paper is removed, joints are grouted and the entire installation is ground and polished.

Grinding and polishing are distinctive operations which differentiate terrazzo from seamless floorings which also utilize resinous binders. A notable exception to this general rule is *rustic* portland cement terrazzo which is not ground but is washed to reveal the decorative chips. In rustic terrazzo, selected colorful gravels sometimes are used instead of marble chips.

Terrazzo is a durable, low-maintenance floor finish, generally suitable for exterior and interior locations. It can be cast in place or precast in a variety of shapes and colors. In precast form, its chief use in residential construction has been for shower receptors, window sills and stair treads. More decorative products intended for kitchen counters, lavatory tops and tables are marketed under the name of 'art marble' or 'cultured marble'. These precast terrazzo products involve installation methods different from those associated with cast-in-place terrazzo floor finishes, and are not included in this discussion.

This section is divided into subsections outlining the recommended *materials* and *applications* for terrazzo floors. Since a large proportion of terrazzo floors utilize portland cement mortars and are installed over concrete surfaces, it is recommended that the general properties of concrete (Section 203) and recommended practices for slabs-on-ground (Section 311) be reviewed.

NTMA COLOR PLATE DESIGNATIONS ACCORDING TO CHIP SIZE

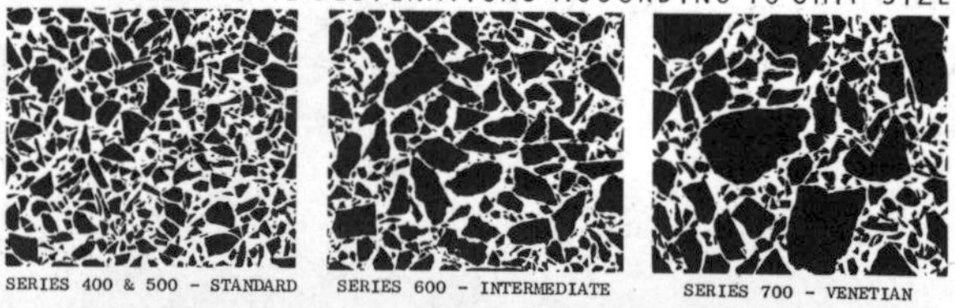

SERIES 400 & 500 - STANDARD | SERIES 600 - INTERMEDIATE | SERIES 700 - VENETIAN

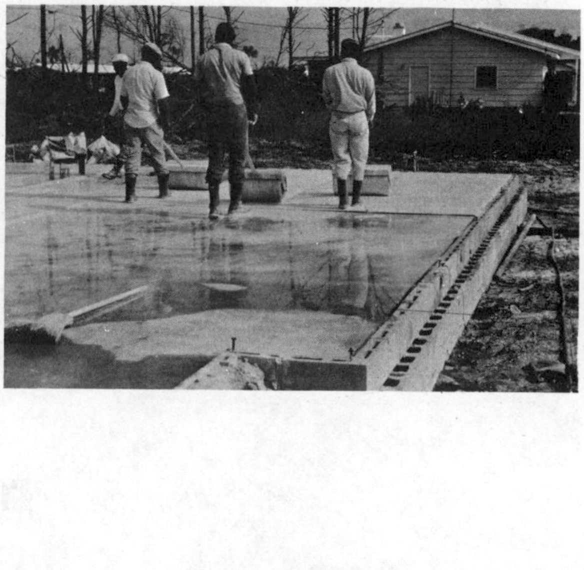

MATERIALS

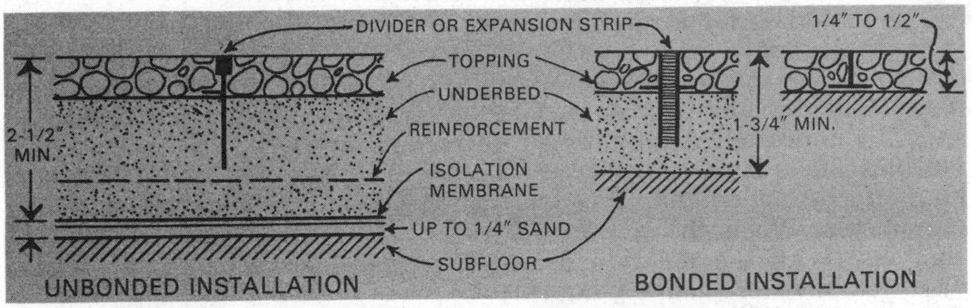

FIG. 1 Basic construction elements for bonded and unbonded installations.

The decorative wear layer of terrazzo finishes is termed the *topping*. In some installations, the topping is placed directly on a wood, metal or concrete subfloor; in others, an intervening portland cement underbed is required. When an underbed is used, it may be rigidly *bonded* to the subfloor, or it may be *unbonded* by separating it from the subfloor with an isolation membrane (Fig. 1). When no underbed is used, the topping always is bonded to the subfloor. Other terms relating to terrazzo installations are illustrated in Figure 1 and are described under Terminology.

The basic ingredients of terrazzo toppings are *binder materials* and *decorative chips*. Other materials which may be required are: (1) *pigments;* (2) accessories such as *divider strips* and *expansion strips* for decorative purposes and crack control; (3) *metal reinforcement* for tensile strength and shrinkage control; (4) *isolation membranes* to prevent the transfer of stresses from subfloor to underbed and topping; and (5) *curing compounds* and *sealing materials* to maintain the quality of the terrazzo finish.

Materials for terrazzo finishes should conform to the applicable industry standards given in Figure 2.

TERMINOLOGY

The following terms are basic to terrazzo finishes and will aid in understanding this discussion.

Binder Cementitious or resinous material which gives the matrix adhesive and other important physical properties.

Divider Strips All-metal or plastic-top metal strips provided in the terrazzo finish to control cracking due to drying shrinkage, temperature variations and minor structural movements; also are used for decorative purposes and convenience in placing the topping.

Expansion Strips Double divider strips separated by resilient material and provided generally for the same purpose as divider strips, but where a greater degree of structural movement is expected.

Isolation Membrane Membrane such as asphalt-saturated roofing felt, building paper or polyethylene film installed between subfloor and underbed to prevent bond and permit independent movement of each.

Matrix Topping mortar consisting of binders, and sometimes pigments and inert fillers, which fills the spaces between chips and binds them into a homogeneous mass.

Subfloor Structural material or surface intended to serve as a working platform during construction and to support the finish floor and design loads.

Terrazzo Finish All elements installed by the terrazzo contractor from subfloor to finished surface, such as sand bed, isolation membrane, underbed and topping.

Thinset Method of installing relatively thin toppings (1/4" to 1/2") directly over any suitable subfloor, generally possible with resinous binders only.

Topping Decorative wear layer consisting of marble chips embedded in a suitable matrix and requiring grinding, polishing or washing to form the finished terrazzo surface.

Underbed Layer of nonstructural portland cement mortar sometimes used over the subfloor to provide a suitable base for portland cement terrazzo and to minimize cracking.

BINDERS

The binders used in terrazzo toppings can be classed as *cementitious* and *resinous*. The cementitious binders are the older, traditional terrazzo materials, whereas the resinous binders are the result of relatively recent developments in synthetic organic chemistry since World War II.

Cementitious

These binders include portland cement and magnesium oxychloride cement (magnesite). Portland cement still is the most common material for terrazzo finishes; magnesite is seldom used today as a finish, but rather as a mastic underlayment under other floor finishes.

Portland Cement Either white or natural (gray) Type I portland cement may be used for terrazzo finishes. Natural portland cement is less expensive and generally acceptable if a vivid or light colored matrix is not essential to the color scheme. White portland cement is more expensive but offers a lighter background for displaying the decorative chips and produces clearer, truer colors with mineral pigments. Type IA air-entraining portland cement is recommended for exterior applications subject to freezing and thawing.

Magnesium Oxychloride Cement Terrazzo toppings of 1/2" thick magnesium oxychloride cement can be installed directly over most sound subfloors. Because portland cement terrazzo requires an underbed over subfloors other than concrete, magnesite terrazzo was useful in remodeling work before the advent of the resinous binders. However, magnesite has lower resistance to moisture, abrasion and chemicals than do most resinous binders and therefore largely has been replaced with these more durable finishes. Magnesite without decorative aggregate still is used as a mastic underlayment for thinset resinous terrazzo and other finishes over existing wood floors.

The materials for magnesite terrazzo are supplied in two parts—a dry mix and magnesium chloride flakes. The dry mix is composed of specially processed magnesium oxide plus fillers such as sand, wood flour, crushed stone and other inert ingredients. When the dry mix is combined with the chloride flakes dissolved in water (gauging liquid), the magnesium oxide reacts with the magnesium chloride to produce magnesium oxychloride cement. This plastic cement sets within a few hours and binds the entire mass into a hard finish.

Over sound wood floors, the magnesite is bonded with metal or wire lath; over concrete, a chemical bonding agent is used. Detailed installation recommendations are contained in USA Standard A88.6, Specifications for Terrazzo Oxychloride Composition Flooring and Its Installation.

Resinous

Resinous binders possess most of the desirable properties of magnesite and portland cement binders and have greater strength, chemical resistance and abrasion resistance. In addition, the ability of resinous binders to bond with most subfloor materials and to be applied in very thin toppings (1/4" to 1/2") makes them suitable for many installations where portland cement cannot be used. However, substantially higher material costs increase overall installed cost and limit the economic use of resinous terrazzo to installations where their special properties are specifically required.

Synthetic binders currently available for terrazzo finishes include *catalyst-cured* resins and *latex (emulsion)* resins. Because of the relatively high cost of resins and the desire to impart special properties, these resins are combined with various additives and fillers which together form the *matrix*.

Catalyst-cured Resins These are two-component liquid systems which consist of the basic resin and a catalyst curing agent. After the two liquids are combined, a dry mix consisting of mineral fillers, silica, pigment and marble chips is added to form the terrazzo mix.

FIG. 2 **SUMMARY OF TERRAZZO MATERIALS**

Material	Standard
Cementitious Binders	
Portland Cement, Type I	ASTM C-150
Magnesium Oxide	ASTM C-275
Magnesium Chloride	ASTM C-276
Resinous Matrixes	
Epoxy	NTMA*
Polyester	MIL-F-52505 (MO) & NTMA*
Latex	MIL-D-3134F & NTMA*
Decorative Chips	NTMA*
Aggregate	
Sand	ASTM C-33
Magnesite Fine Aggregate	USAS A88.6
Pigments	NTMA*
Metal Reinforcement	
Welded Wire Fabric	ASTM A-185
Isolation Membranes	
Roofing Felt	FS-HH-R-595
Building Paper	FS-UU-B-790
Polyethylene Film	CS 238
Curing Compounds	ASTM C-309

*Specifications of the National Terrazzo and Mosaic Association

Epoxy matrixes exhibit superior overall chemical resistance and adhesion, good abrasion resistance and impact strength and negligible curing shrinkage. Among their disadvantages are relatively high cost and toxicity to the skin during application and before curing.

Polyester matrixes are characterized by good chemical resistance, particularly to acids of high concentration. Polyester resins are resistant to water and weather, are strong, durable and possess good abrasion resistance.

Some of the disadvantages of polyester resins are somewhat lower resistance to alkali than epoxy and stronger odor during installation and curing period.

Latex Resins These are synthetic rubber-like (latex) resins suspended in water. The liquid is blended with a dry mix of dehydrating powders, fillers, pigments and marble chips to produce the topping mix. The latex resins commonly used for terrazzo matrixes are polyvinyl chloride, neoprene and acrilic.

Polyvinyl Chloride and Neoprene matrixes have excellent working qualities and are odor-free during and after installation. They adhere well to typical subfloor materials and remain flexible after curing—an important property for finishes over less rigid subfloors such as wood. Both matrixes exhibit excellent chemical resistance and nonslip properties.

These matrixes are somewhat slower curing than polyester and epoxy and hence take longer to install. Another disadvantage is that they are not generally available in white.

Acrylic matrixes have most of the desirable properties of the other latex formulations but are not as flexible. However, they cure quickly and are available in lighter colors at slightly higher cost.

DECORATIVE CHIPS

Decorative chips in terrazzo toppings consist principally of marble. Chips of various types and colors are quarried and are carefully selected to avoid off-color or contaminated material. Originally, marble was defined as a metamorphic rock formed by the recrystallization of limestone. However, both geologically and commercially, marble has been redefined to include all *calcareous* (limestone) rocks such as onyx and travertine, as well as attractive serpentine rocks, capable of being ground and polished. Granite, which is not a marble, is extremely hard and weather resistant and often is used for heavy-duty industrial and rustic terrazzo.

Domestic and imported chips are available in a wide range of colors for terrazzo use. Chips are graded by number according to size as shown in Figure 4. For example, a #1 chip retained on a 1/8" screen and passing a 1/4" screen will be between 1/8" and 1/4" in size. Producers normally supply #0, #1 and #2 as separate sizes; some produce #3 and #4 as separate sizes. Larger sizes are frequently grouped as #3-4 mixed, #5-6 mixed, #7-8 mixed and #4-7 mixed.

The appearance and minimum thickness of toppings are determined largely by the size and type of chips used. Toppings are identified on this basis as: *standard, intermediate, Venetian* and *Palladiana* (Fig. 4). standard, intermediate and Venetian use crushed chips which are mixed with the resinous or portland cement matrix and spread uniformly in the topping.

Palladiana toppings consist of cut slabs of marble about 1/2″ thick and are set by hand in the desired pattern. Spaces between slabs are filled with a standard topping mix. Palladiana toppings are custom finishes similar to mosaics.

The National Terrazzo and Mosaics Association and some producers of chips publish standard color plates as an aid in the selection and specifying of decorative chips and color pigments. To assure uniformity, it is recommended that marble chips and color combinations be designated by plate number (Fig. 5). However, the finished terrazzo should not be expected to duplicate the color plate exactly, because slight variations in color and veining are likely to occur between different shipments of chips. It is recommended that physical samples be prepared as a standard of comparison at the beginning of each project.

FINE AGGREGATE

In terrazzo finishes the term aggregate refers to material such as sand and fillers, and does not include decorative chips. Fine silica sand and fillers generally are included in the proprietary formulations of resinous matrixes. Sand aggregate is one of the basic ingredients of the mortar underbed, but it is not used in the portland cement topping. Sand should be well graded, natural or manufactured aggregate which is clean, granular and free of deleterious amounts of clay, salts or organic matter.

Fine aggregate used in magnesium oxychloride cement normally is included in the proprietary dry mix and should conform to USA Standard A88.6. Abrasive aggregate no longer is recommended for nonslip interior surfaces. Abrasive strips have proven more effective and are discussed on page 453-14.

FIG. 4 **GRADING OF DECORATIVE CHIPS FOR TERRAZZO TOPPINGS**

<table>
<tr><th colspan="4">Type of Topping & Minimum Thickness</th><th>Chip Number</th><th>Passes Screen (in.)</th><th>Retained on Screen (in.)</th></tr>
<tr><td rowspan="9">3/4″ Venetian</td><td rowspan="5">5/8″ Intermediate</td><td rowspan="3">5/8″ Standard* 1/2″ Thinset</td><td rowspan="2">1/4″ Thinset</td><td>0</td><td>1/8</td><td>1/16</td></tr>
<tr><td>1</td><td>1/4</td><td>1/8</td></tr>
<tr><td></td><td>2</td><td>3/8</td><td>1/4</td></tr>
<tr><td colspan="2"></td><td>3</td><td>1/2</td><td>3/8</td></tr>
<tr><td colspan="2"></td><td>4</td><td>5/8</td><td>1/2</td></tr>
<tr><td colspan="3"></td><td>5</td><td>3/4</td><td>5/8</td></tr>
<tr><td colspan="3"></td><td>6</td><td>7/8</td><td>3/4</td></tr>
<tr><td colspan="3"></td><td>7</td><td>1</td><td>7/8</td></tr>
<tr><td colspan="3"></td><td>8</td><td>1 1/8</td><td>1</td></tr>
</table>

*Palladiana toppings consist of broken marble slabs set in a standard chip mix and should not be less than 1/2″ thick.

PIGMENTS

Terrazzo toppings may be the natural color of the binder or can be treated with pigments to produce a wide variety of colors. Pigments should be stable, nonfading mineral or synthetic formulations, compatible with the binder.

ACCESSORIES

Accessories for terrazzo installations include *divider strips, expansion strips* and *base beads. Metal* accessories commonly are made from half-hard brass or white alloy zinc (approximately 99% zinc). Plastic-top strips and beads in various colors with brass, zinc and galvanized steel supports also are available. Type 300 stainless steel strips are produced for use in exterior terrazzo; aluminum strips are used primarily for resinous terrazzo. The standard length of strips and beads is 6′0″ but longer lengths are available on special order.

Only plastic-top strips and beads should be used in conductive floors to prevent the accumulation of dangerous electrostatic charges (see page 453-14). In exterior locations, white alloy zinc accessories are not suitable because a chemical reaction often causes blooming (a white chalky deposit on the strips). In such exterior locations, all-brass, stainless steel or zinc-supported plastic-top accessories are suitable. Aluminum strips should not be used with portland cement terrazzo because the aluminum is subject to attack by alkalis present in the portland cement topping mix.

Divider Strips

Divider strips are installed: (1) to localize shrinkage cracking in portland cement terrazzo, (2) to assure uniformity in topping thick-

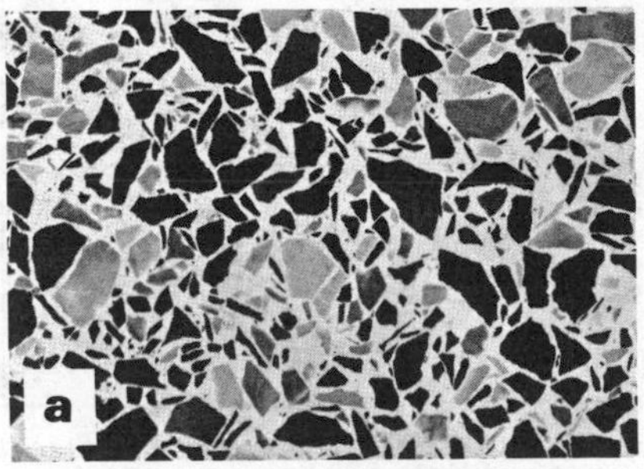

PLATE 470 70% Dark Red 30% Blue-Gray

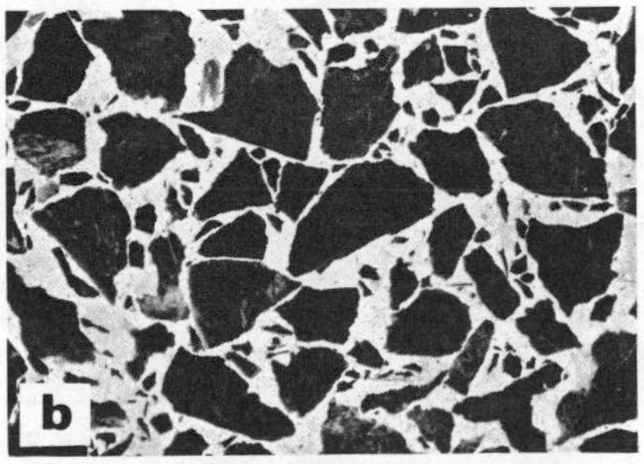

PLATE 712 100% Dark Green #1, #2, #3/4, #5/6, #7/8

FIG. 5 NTMA color plates: (a) Standard topping is identified by proportions of colored chips; (b) Venetian plate shows chip size as well as color.

ness, (3) to act as decorative elements, (4) to permit accurate placement of the different colors forming the floor pattern, and (5) to act as construction joints, subdividing the floor into convenient working areas.

A variety of divider strips are available to suit particular installations. The thickness of the top of the strip (exposed face) can range from 20 gauge up to heavy 1/2″ thick bars; 18 gauge is the recommended minimum. Divider strips with a 1/8″ or wider face usually have a heavy plastic or metal top section and a thin metal bottom section (Fig. 6). For a topping thickness of 5/8″, a 1-1/4″ deep divider is recommended; 1-1/2″ deep strips are recommended for venetian toppings 3/4″ and thicker. For thinset installations, special angle or T-shaped strips 1/4″ to 1/2″ deep are used (Fig. 6). Other special strips are available for wall base and for edgings at the juncture with other flooring materials (Fig. 6).

Expansion Strips

Provision for crack control is necessary at lines in the floor where substantial structural movement can be expected, particularly over beams and bearing walls and over expansion joints or isolation joints in the underlying subfloor. Expansion strips manufactured for this purpose consist of a pair of divider strips separated by a resilient material such as neoprene (Fig. 6). Expansion strips generally are manufactured in types and sizes similar to divided strips.

Base Beads

All-metal or plastic-top base beads are used to terminate terrazzo wall base on the wall surface. They are made in several profiles and to accommodate vertical terrazzo thicknesses varying from 1/2″ to 1-1/2″.

METAL REINFORCEMENT

Metal reinforcement generally is required for the underbed in most installations, except where the underbed is bonded to a suitable concrete subfloor. Underbed reinforcement should be corrosion-resistant welded wire fabric of at least 16 gauge thickness with wires spaced not more than 2″ o.c. Thin underlayments for resinous toppings may use welded or woven wire weighing at least 1.4 lbs per sq.yd. or expanded metal lath weighing at least 2.5 lbs per sq.yd.

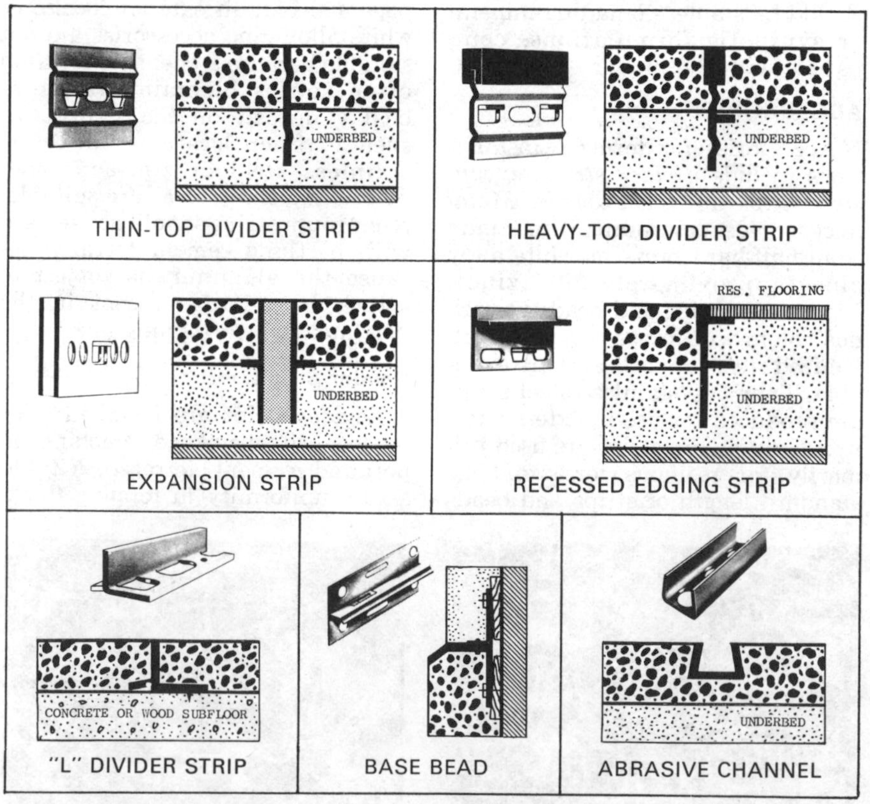

FIG. 6 Accessories: Shallow T- and L-shaped (angle) strips for installation without underbed are available in thin-top, heavy-top and edging profiles.

ISOLATION MEMBRANE

An isolation membrane is required in all unbonded installations to separate the underbed from the subfloor. The membrane may be 15 lb. asphalt-saturated roofing felt, water resistant building paper or 4 mil polyethylene film.

CURING MATERIALS

Curing is necessary for portland cement toppings to develop optimum wear properties. The materials suitable for this purpose include polyethylene film, nonstaining (nonasphaltic) water-resistant building paper and clean water. Spray-on curing compounds are recommended for curing terrazzo surfaces. However, spray-on compounds should not be used to cure concrete slabs which are to receive a thinset, chemically bonded or monolithic topping, as they may prevent bond.

SEALERS AND PROTECTIVE FINISHES

Cementitious matrixes are more porous than resinous matrixes and require greater protection from staining and chemical deterioration. The marble chips which make up more than 70% of the surface possess varying degrees of resistance to staining and deterioration. A penetrating sealer capable of internally sealing the pores in the matrix and in the marble chips should be used. These are generally proprietary formulations. Only products specified by the manufacturer for the particular purpose should be used, and they should be applied in accordance with the manufacturer's recommendations. See page 453-15 for further discussion of sealers and protective finishes.

Suitable terrazzo finishes require attention to: (1) general recommendations common to most installation methods; (2) specific application procedures for particular methods, and (3) various finishing procedures for special surface finishes.

Depending on whether the topping and/or underbed are adhered to the subfloor, terrazzo systems can be classed as *bonded* or *unbonded*. Bonded systems include thinset, monolithic, chemically bonded and bonded-underbed methods. *Thinset* installations consist of a thin (1/4″ to 1/2″) resinous topping usually placed directly over any sound wood, metal or concrete subfloor (Fig. 7a). In *monolithic* installations, a somewhat thicker (5/8″ or thicker) portland cement topping is applied directly over a rough-finished concrete subfloor (Fig. 7b). If the concrete subfloor is an existing, smooth-troweled concrete slab, a portland cement topping can be *chemically bonded* with a bonding agent (Fig. 7c). Monolithic and chemically bonded installations depend on the subfloor to resist shrinkage stresses and conspicuous cracking in portland cement toppings. However, *bonded-underbed* installations provide more positive crack control by creating weakened planes at each divider strip (Fig. 7d).

Unbonded installations provide the greatest degree of crack control and are preferred wherever structural movement is expected. Separation between subfloor and underbed can be accomplished by an isolation membrane alone, or with a membrane and sand cushion (Fig. 7e). When a sand cushion is used, it is called a floating installation.

GENERAL RECOMMENDATIONS

The installation of terrazzo involves (1) preparation of the subfloor; and depending on the method of installation, (2) placement of the topping, strips and underbed when used.

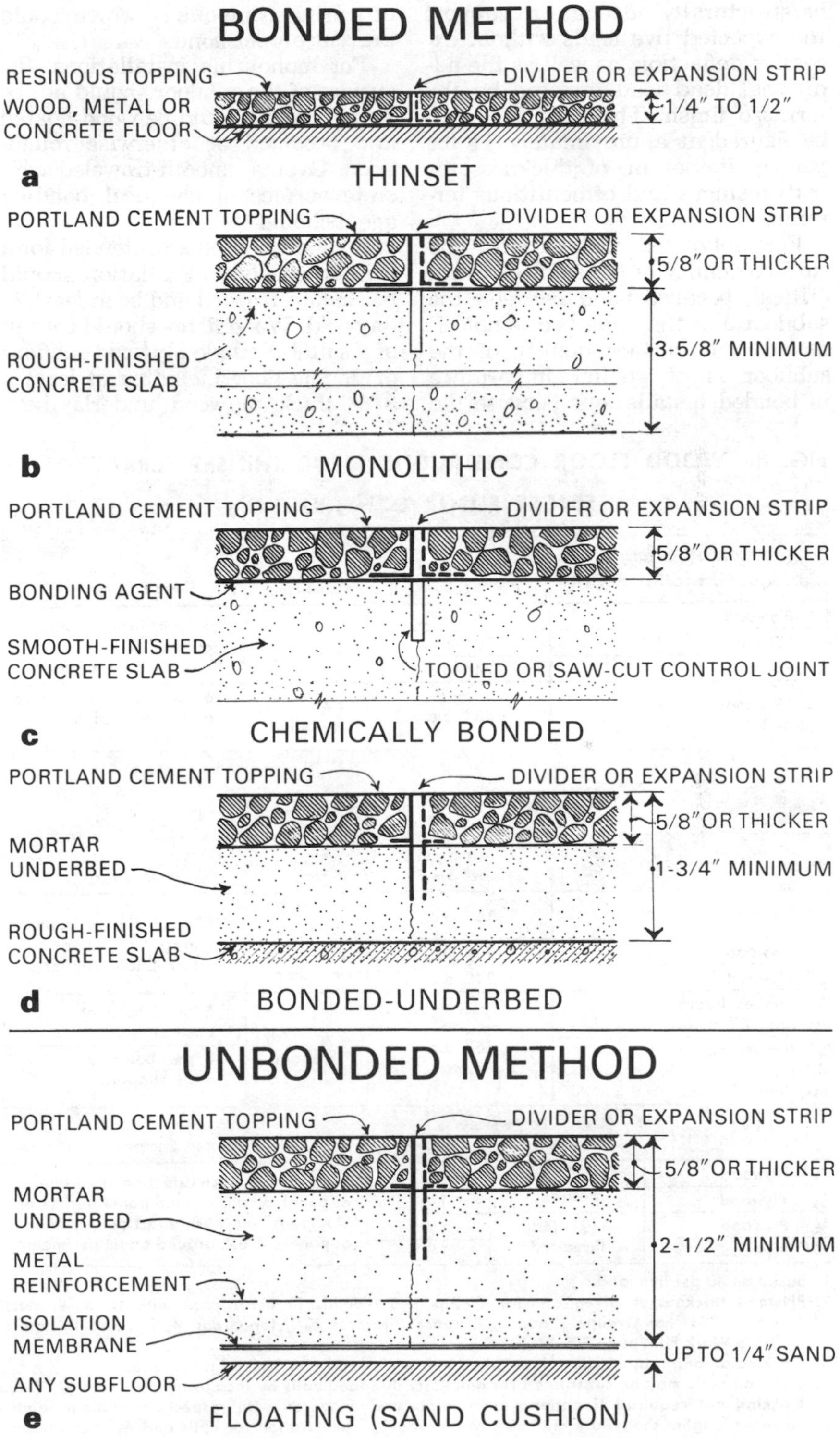

FIG. 7 Recommended details for basic terrazzo installation methods.

Subfloor Preparation

Regardless of the installation method, the subfloor should be sound and free of loose material. It should be structurally adequate to support the expected live loads without excessive deflection, as well as the additional dead load imposed by the terrazzo finish. The dead load can be figured at approximately 11 lbs per sq. ft. per in. of thickness for both resinous and cementitious terrazzo.

For unbonded installations, the surface quality of the subfloor is not critical, because bond between the subfloor and the underbed is not effected. The surface quality of the subfloor is of greater importance in bonded installations because the quality of the installation depends on adequate bond with the subfloor. Hence, for all bonded installations, the subfloor should be free of coatings such as paint, wax and sealing or curing compounds which could prevent proper bond.

For monolithic installations, the surface of the subfloor should not be troweled but should be wood-floated and broomed or otherwise roughened. Over a smooth-troweled concrete surface, a chemical bonding agent should be used.

A concrete subfloor intended for a thinset resinous installation should be smooth-trowled and be at least 21 days old. Wood floors should consist of a suitable single plywood subfloor or double floor including at least a 3/8″ thick plywood underlayment (Fig. 8). Plywood should be Exterior Type CC-Plugged grade or better. The recommendations of the manufacturer of the resinous matrix should be followed for subfloor preparation. Over most subfloor materials, the topping can be adhesively bonded and priming of the subfloor is required to achieve bond. Over plywood some manufacturers recommend direct adhesive bonding after plywood edges have been bound with a special glass fiber tape. Many latex manufacturers prefer that mechanical bond be achieved by wire or expanded metal lath over asphalt-saturated felt, secured to the subfloor 6″ o.c. along supports.

The subfloor for monolithic, chemically bonded and thinset installations should conform to the profile of the finished terrazzo surface, whether level or pitched, to insure that the topping will be of uniform thickness. The subfloor surface should be in a straight plane; there should be no more than 1/8″ variation when tested with a 10′ straightedge. The topping mix should not be used to fill voids or to develop pitch, because excessive variations in topping thickness are conducive to cracking. Where the existing subfloor is not sufficiently smooth or does not have the desired profile, it should be leveled or pitched with a suitable mastic underlayment such as latex or magnesite. Over stripwood floors, these underlayments should be installed at least 3/8″ thick using a 15 lb. felt and expanded metal or wire lath secured to the subfloor 6″ o.c. along supports.

Monolithic and chemically bonded portland cement toppings require control joints in the subfloor as discussed under Strip Placement.

FIG. 8 WOOD FLOOR CONSTRUCTION FOR THINSET TERRAZZO[1]

SINGLE FLOOR CONSTRUCTION

Combination Subfloor/Underlayment[2,3]	Maximum Support Spacing	Nails[4]	Nail Spacing
⅝″ Plywood	16″ o.c.	2″ (6d) threaded	6″ o.c. around panel edges; 10″ o.c. at intermediate supports.
¾″ Plywood	20″ o.c.		
⅞″ Plywood	24″ o.c.		
1⅛″ Plywood (2-4-1)	48″ o.c.	2½″ (8d) threaded	6″ o.c. at panel edges and intermediate supports.

DOUBLE FLOOR CONSTRUCTION

Subfloor[2]	Maximum Support Spacing	Nails	Nail Spacing
½″ Plywood	16″ o.c.	2″ (6d) common	6″ o.c. around panel edges; 10″ o.c. at intermediate supports.
⅝″ Plywood	20″ o.c.	2½″ (8d) common	
¾″ Plywood	24″ o.c.		
¾″ lumber boards less than 6″ wide	16″ o.c.	2¼″ (7d) threaded	2 per board at each support.
¾″ lumber boards wider than 6″ and less than 8″ wide			3 per board at each support.

Underlayment[2,5]	Fasteners[6] Nails[4]	Fasteners[6] Staples[7]	Fastener Spacing
⅜″ Plywood	1¼″ (3d) threaded	1⅛″	Nails: 8″ o.c. each direction in middle of panel, 6″ o.c. around panel perimeter. Staples: 6″ o.c. each direction in middle of panel, 3″ o.c. around panel perimeter.
½″ Plywood		1¼″	
⅝″ Plywood	1½″ (4d) threaded	1½″	
¾″ Plywood		1⅝″	

1. Based on 40 psf live load.
2. Plywood thicknesses given are for Group 1 Softwoods in accordance with U. S. Product Standard PS1. For Groups 2 and 3, increase thickness ⅛″; for Group 4, increase thickness ¼″ (see Work File, page 201-6).
3. Blocking at longitudinal edges is required unless plywood with T&G edges is used.
4. Common nails may be substituted for annularly threaded nails by increasing their size ¼″ (1d).
5. Blocking not required if underlayment joints are staggered with respect to subfloor joints.
6. Fastener lengths shown are minimum to provide ¾″ penetration for nails and ⅝″ for staples. If longer fasteners are used, penetration should not exceed 1″.
7. Galvanized divergent chisel, 16 gauge, ⅜″ crown width.

Underbed

An underbed is necessary primarily in unbonded installations but may be required in some bonded installations. The underbed can be relatively thin because it is not a structural slab but functions as a surface preparation for the topping. The thickness of the underbed varies depending on the installation method (Fig. 7d & e).

The underbed should be reinforced with welded wire fabric in all unbonded installations and in

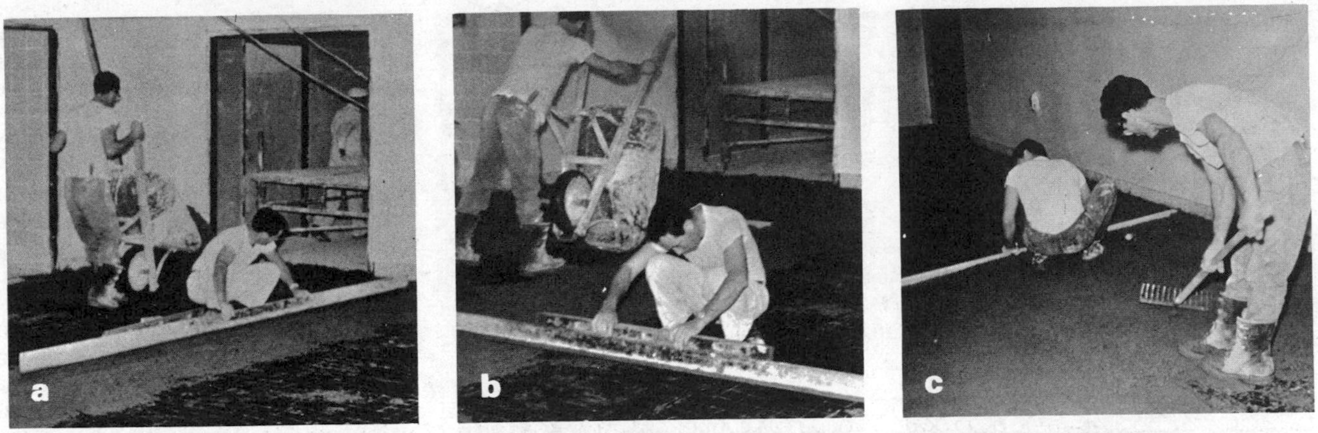

FIG. 9 *Underbed installation: (a) mortar is placed in windrows and leveled to form screeds; (b) windrows are leveled in transverse direction; (c) mortar is placed between windrows and leveled.*

FIG. 11 RECOMMENDED STRIP PLACEMENT

Matrix	Installation Method	Divider Strip Spacing[1]	Expansion Strip Spacing[1]
Portland Cement	Unbonded	6′ o.c. maximum	As needed[2]
	Bonded-underbed		
	Chemically Bonded	15′ to 20′ o.c.—Panels not over 300 sq. ft.[3]	30′ o.c. maximum[3]
	Monolithic		
Resinous	Thinset	As desired for decorative or convenience purposes	30′ o.c. maximum[4]

1. Locate divider strips or expansion strips on beams and girder lines wherever possible.
2. Provide expansion strips over all expansion joints in the subfloor or at other joints where movement is expected.
3. Divider strips and expansion strips can be alternated or expansion strips can be used at 15′ o.c.
4. Recommended for large areas or long corridors; otherwsie not required.

bonded installations over wood and metal subfloors. Unless the underbed is bonded, an isolation membrane should be spread over the subfloor lapping 4″ at all ends. Metal reinforcement should be placed over the membrane, lapping adjacent sheets at least two mesh spaces, except where expansion strips occur. No conduit or pipe should be located within the minimum recommended underbed thickness.

The mix for the underbed mortar should consist of one part portland cement and 4 to 5 parts sand, mixed with sufficient water to make a workable mortar. The surface over which the underbed is placed should be sufficiently wet to prevent water from being drawn out of the mix. Isolation membranes need only be uniformly dampened; concrete slabs should be thoroughly saturated by overnight soaking.

The underbed mortar should be placed and screeded (Fig. 9) to a height which will permit installation of the selected topping thickness at the desired finish floor elevation.

Strip Placement

When an underbed is used, conventional divider strips (Fig. 6) should be installed while the underbed is still semi-plastic. Standard 1-1/4″ or 1-1/2″ strips should be set in the desired pattern with the top of the strip at the finished floor elevation. The strips then should be troweled firmly along the edge to assure adequate anchorage in the underbed (Fig. 10).

In many bonded installations an underbed is not used and the terrazzo is applied directly to the subfloor. In monolithic and chemically bonded installations, it usually is not possible to install strips while the concrete subfloor is still plastic

FIG. 10 *Strip placement: (a) long strips are placed first and aligned with a straightedge; (b) cross strips are installed next; (c) each strip is leveled and troweled firmly along its edge to seat firmly in the underbed.*

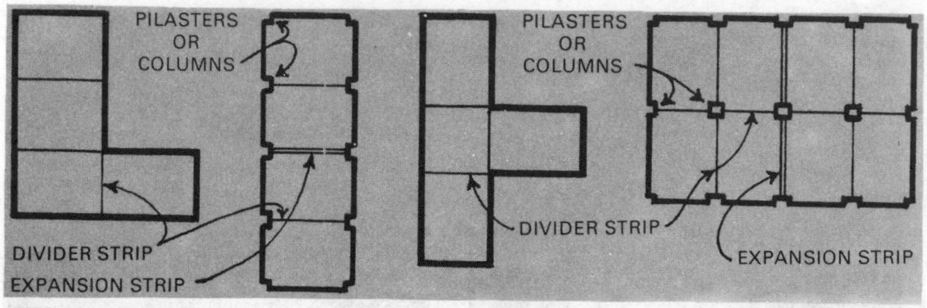

FIG. 12 For monolithic and chemically bonded installations, strips should be spaced 15′ to 20′ o.c. In all installations, locating strips at column lines is recommended.

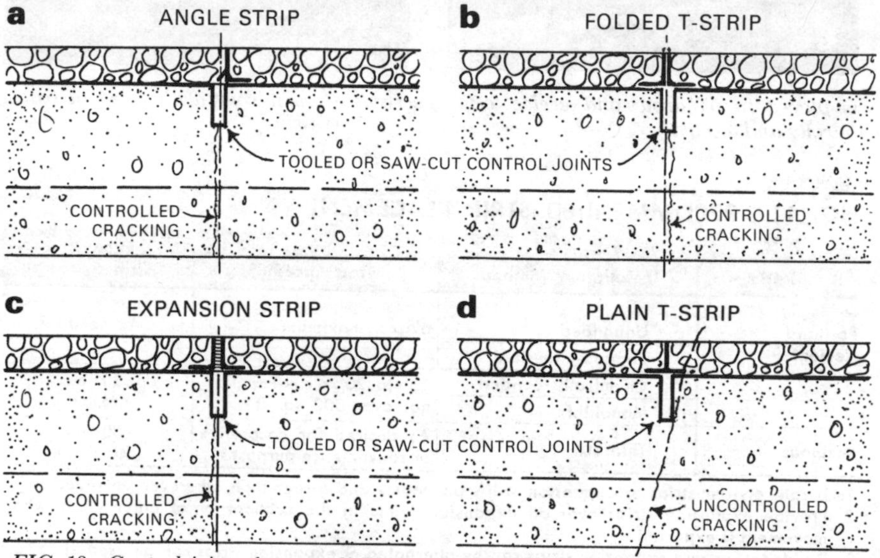

FIG. 13 Control joints for monolithic and chemically bonded installations: (a, b & c) inconspicuous controlled cracking is encouraged adjacent to strip; (d) uncontrolled cracking results from improper installation.

because the subfloor often is installed by a different contractor. For such installations, angle or T-shaped strips (Fig. 6) are secured to the concrete with adhesive or mechanical fasteners.

Generally, divider strips should be provided wherever a discontinuity in the subfloor exists, such as at all construction, isolation and control joints and changes in materials. Expansion strips can be used at these locations if desired, but they are specifically recommended at structural expansion joints where a greater degree of movement is anticipated. Divider or expansion strips should be used over beams and girders supporting the subfloor, because tensile stresses are likely to develop due to negative bending moments. Spacing of divider and expansion strips for various methods of installation is given in Figure 11.

Divider Strips In resinous terrazzo installations, divider strips are not required for control of shrinkage cracking but often are desirable for decorative purposes and convenience in placing and finishing of the topping. In thinset installations the strips assure uniform topping thickness and conserve material. They also minimize the appearance of waviness in large areas

In portland cement terrazzo installations using an underbed, strips should be spaced not more than 6′ o.c. When a portland cement topping is placed directly over a concrete slab, the slab should be provided with cut or tooled control joints 15′ to 20′ o.c. in each direction, forming panels not more than 300 sq. ft.; length of panels should not exceed the width by more than 50%. Divider strips should be placed at all such control joints, forming similar panels in the topping (Figs. 12 & 13). In residential slab-on-ground construction where a minimum of stripping is desired, control joints and divider strips can be placed in doorways and adjacent to partitions where they will be concealed by base trim (Fig. 14).

In all cases, divider strips should be continuous from wall to wall and preferably should form right angle intersections rather than acute angles. If a terrazzo wall base is used, the strips should be carried up to the base screed. When the terrazzo is carried beyond the base to form a wainscot, strips should be continued from the base in a straight line up the wall. The recommended strip spacing for portland cement terrazzo wainscoting is 3′ o.c. maximum; strip spacing for resinous wainscoting is not critical.

Expansion Strips Generally, greater consideration should be given to the need and location of divider and expansion strips in the bonded systems because the terrazzo finish will tend to move with the subfloor. In monolithic, chemically bonded and thinset installations, ex-

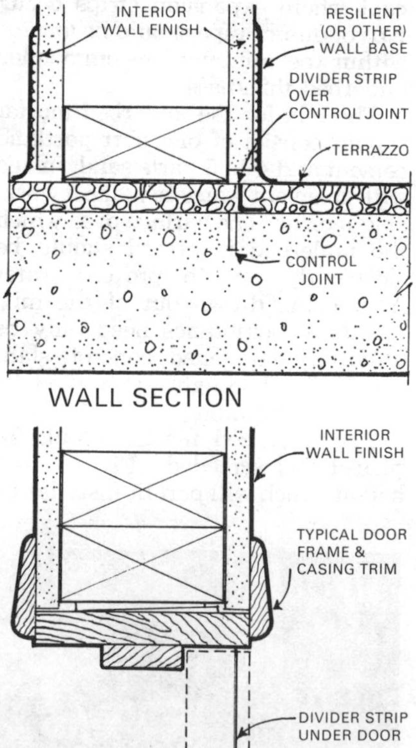

FIG. 14 Recommended strip placement for monolithic and chemically bonded terrazzo over slab-on-ground where strips are not a design feature.

pansion strips should be provided approximately 30′ o.c. where unobstructed large areas or long corridors are involved. In monolithic and chemically bonded installations, expansion and divider strips can be alternated 15′ o.c. (Fig. 12).

Unbonded and bonded-underbed installations will generally have a sufficient number of divider strips distributed over the floor so that expansion strips are not required except over expansion joints in the subfloor.

Topping

The following recommendations are primarily for portland cement terrazzo with regular finish. For recommendations concerning resinous terrazzo, see Thinset Method, page 453-13; for palladiana and rustic installations see Special Surface Finishes, 453-14.

The steps involved in installing terrazzo toppings can be grouped according to sequence of operations as: (1) mixing, (2) placing the topping, (3) preliminary finishing and (4) final finishing and protection.

Mixing A power mixer should be used for portland cement topping, using carefully measured and consistently proportioned ingredients. A conventional portland cement mix should consist of 200 lbs. of marble chips per 94 lb. bag of cement. When pigment is used, it should be measured by weight (not volume) and should be added to the dry mix. After thorough mixing of dry ingredients, water should be added at the ratio of 5 to 5-1/2 gallons per bag of cement. Mixing is best accomplished in a power mortar mixer.

FIG. 15 Slush coat of neat cement is broomed into the surface just before the topping mix is placed.

FIG. 16 Topping is spread and troweled even with the divider strips.

Resinous binder formulations vary among manufacturers and hence proprietary mixing and installation recommendations should be carefully followed.

Placing Over an underbed, installation of the topping should be delayed for 12 to 24 hours to permit the underbed to harden. Immediately before the topping is placed, a slush coat of neat cement, pigmented to match the topping mix, should be broomed into the underbed (Fig. 15). The various colored mixes with decorative chips then are placed in the appropriate panels according to the desired pattern. The topping then is troweled even with the tops of divider strips which act as screeds for leveling the surface (Fig. 16).

To assure even distribution and adequate chip content, the surface usually is seeded with additional chips of the type contained in the mix (Fig. 17). Chips should be wetted, sprinkled over the surface and then embedded by additional troweling. Then the entire area is rolled with a heavy roller until all excess water has been brought to the surface (Fig. 18). The surface again should be troweled smooth and uniform, revealing the tops of divider strips.

Before finishing operations are started, portland cement toppings should be wet-cured for at least 3 to 7 days either by ponding, by covering with watertight covers, or by spraying with a curing compound (see page Concrete 203-24).

Preliminary Finishing After proper curing, the preliminary finishing operations of *grinding, grouting* and *curing* may be performed.

The surface should be ground using a terrazzo grinding machine, first with No. 24 or finer abrasive stones, followed by a second grinding with No. 80 or finer stones (Fig. 19). The grinding operations should be performed while the surface is

FIG. 17 Additional chips are seeded and troweled into the surface where necessary for uniform appearance.

FIG. 18 Rolling compacts the topping mix, brings up excess water and creates a harder, more durable surface.

FIG. 19 After proper curing, the surface is flooded and is wet-ground to reveal the decorative chips.

covered with water.

After grinding, the surface should be thoroughly rinsed with sufficient water to remove dust and loose fine particles. Then a skim coat of neat cement grout of the type used in the topping mix (including pigment to match the topping) should be troweled over the entire surface filling all voids.

The installation then should be wet-cured again for at least 3 days using a watertight cover or curing compound. The compound or cover should be allowed to remain in place as protection against physical damage or staining until final finishing is performed.

Final Finishing These operations include *polishing, cleaning* and *sealing*. A floor machine with No. 80 or finer stones should be used for polishing. The polishing operation should continue until the grout skim coat is removed (except in voids and depressions) to reveal the decorative chips in approximately 70% to 75% of the area.

A solution of neutral liquid cleaner should be used according to the manufacturer's recommendations to clean the polished floor. It should then be rinsed with clean water and be allowed to dry. The cleaned, polished terrazzo surface finally should be treated with a suitable sealing compound. These compounds also are proprietary formulations and should be applied in accordance with manufacturer's recommendations.

Unless all work on the premises is completed, the finished terrazzo should be protected with a covering such as heavy building paper until ready for use.

INSTALLATION METHODS

The general recommendations for subfloor preparation, installation of underbed, strips and topping should be supplemented with recommendations provided for each individual installation method.

Unbonded

This method provides the greatest degree of isolation between subfloor and underbed, permits independent movement by each and hence minimizes the possibility of surface cracking. An unbonded installation is suitable over any tight, structurally sound subfloor in interior or exterior locations. In exterior locations the unbonded system generally is limited to concrete supporting construction because wood or metal would require decay- or surface corrosion-resistant protection afforded by a bonded finish.

Unbonded installations require the use of a reinforced underbed to support the topping. The combined thickness of underbed and topping should be at least 2-1/2″ (Fig. 7e). Separation between subfloor and underbed can be provided by an isolation membrane alone, or by a membrane spread over a layer of sand 1/16″ to 1/4″ deep. Over a smooth subfloor a double layer of polyethylene film similarly will provide isolation between surfaces. Membranes should be lapped at least 4″ at all joints.

The metal reinforcement should be placed over the membrane, lapped at joints at least two mesh spaces and wired together. After some mortar has been poured in place, the reinforcement should be lifted to approximately 1″ above the membrane. Underbed, strips and topping should be installed as indicated under general recommendations.

Bonded

The chief advantages of bonded installations are smaller dead load and lesser finish thickness. Most bonded methods allow the topping to be placed directly over the subfloor without an intervening underbed. However, these methods do not generally develop the degree of crack control obtained when an underbed is used with conventional divider strips. Therefore, the bonded-underbed method (Fig. 7d) often is used if the greater dead load and thickness can be accommodated. Since the difference in thickness between bonded-underbed and unbonded methods is only 3/4″ (9 lbs per sq. ft), the unbonded method always should be considered because it is most effective in reducing cracking hazard.

In the principal bonded methods —monolithic, chemically bonded and thinset—the relatively thin topping by itself cannot resist the various stresses acting on it. Adequate bond between topping and the subfloor is necessary to resist stresses across the full depth of the floor construction and, therefore, special care should be taken to remove foreign substances which might impair bond.

Also, the topping should be of uniform thickness to prevent concentration of stresses and consequent cracking at thin spots. It is important that the subfloor be leveled or pitched to conform with the desired finished floor profile.

Bonded-underbed This method can be used over a metal or concrete subfloor and provides a measure of crack control for portland cement terrazzo not available in other bonded installations. The total recommended thickness of underbed and topping is 1-3/4″ for bonded-underbed as compared to 2-1/2″ for unbonded installations.

The crack control advantage of this method is derived from the relatively frequent spacing of divider strips. The maximum spacing of 6′ o.c. for conventional divider strips induces unnoticeable stress relieving cracks along the strips. The minimum overall terrazzo thickness results from the normal strip depth (1-1/4″ or 1-1/2″) plus a 1/4″ tolerance for subfloor irregularity.

The underbed should be bonded properly to the subfloor. Natural bond can be obtained between the portland cement underbed and a suitably roughened concrete slab; over a smooth-troweled concrete slab, a chemical bonding agent can be used to bond the underbed. Over a metal deck, adequate bond can be expected without special treatment if the surface is free of corrosion, pickling acids, oil or other foreign substances. Additional bond is gained by the keying action of the underbed with depressions in a cellular metal deck. Corrugated and ribbed metal decks generally depend on composite action of the deck and a reinforced structural concrete slab. In such floor assemblies, the top of the concrete slab constitutes the subfloor and topping should be installed by the monolithic or chemically bonded method.

Underbed and topping generally are installed as described under

General Recommendations.

Monolithic This term refers to a portland cement topping installed over a concrete subfloor where natural bond is developed between the portland cement surfaces. Monolithic installations can be used in interior as well as exterior locations.

The concrete subfloor should be reinforced properly and be at least 3-5/8″ thick. The surface should be brushed with a wire broom or otherwise roughened to remove laitance and assure good mechanical bond with the topping. As soon as the concrete is sufficiently hard to resist tearing with a jointing tool or power saw, control joints should be installed in the slab approximately 15′ to 20′ apart (Fig. 13). The surface should be protected from damage and dirt with suitable coverings until the topping is placed.

If the subfloor has not been adequately protected from dirt, paint, plaster or other construction residue, the surface should be thoroughly scrubbed with detergents and rinsed clean. Then the subfloor should be saturated with water by ponding overnight. Excess water should be removed and a slush coat of neat portland cement, pigmented to match the topping, should be broomed into the surface immediately before the topping mix is applied. The topping should be placed, finished and cured as described under General Recommendations.

Chemically Bonded This method of installing portland cement topping is intended for smooth-troweled concrete subfloors. For this reason it often is suitable for older concrete floors not originally prepared for a terrazzo topping. Where a terrazzo topping is planned in the initial design, a roughened subfloor surface is preferred.

The subfloor surface should be inspected, and where random cracks exist or control joints are not provided 15′ to 20′ o.c., such joints should be installed with a power saw. Divider strips or expansion strips should be secured at all control joints.

Chemically bonded installations depend on a bonding agent such as epoxy, epoxy-polysulfide or polyvinyl-chloride resin. The subfloor surface should be free of all coatings such as paint, curing compounds or wax and the bonding agent should be applied in accordance with the manufacturers recommendations.

The topping should be placed, finished and cured as described under General Recommendations.

Thinset Using a resinous binder, it is possible to install very thin toppings over practically any sound, firm subfloor such as wood, metal or concrete. Epoxy and polyester toppings should be at least 1/4″ thick. When installed over concrete, the subfloor surface should be smooth-troweled. Latex toppings normally require an installed thickness of at least 3/8″ and hence either a troweled or broomed finish is acceptable.

Although some manufacturers allow resinous toppings 1/8″ thinner than those recommended above, industry concensus favors the slightly heavier installations. This is particularly true of epoxy and polyester, where 1/8″ thickness would require decorative aggregate the size of sand, and an acceptable minimum thickness over subfloor irregularities would be difficult to achieve.

Thinset terrazzo finishes should not be installed on slab-on-ground construction subject to excessive moisture or high alkaline content. To test for moisture, a piece of polyethylene film approximately two feet square should be taped down in several locations and be allowed to remain for 8 hours. Any visible moisture accumulation under the mat is an indication of excessive moisture in or under the slab. Under such conditions, the alkalinity of the moisture also should be tested and the manufacturer should be consulted.

Divider strips are not required with resinous binders for shrinkage control, but should be provided where there is a discontinuity in the subfloor, such as at all isolation joints, control joints and changes in materials. Divider strips may be provided for decorative purposes or as construction joints at the termination of a pour. Since wall base often is installed as a separate operation, a strip should be provided at the toe of the base. In large areas or long corridors expansion strips 30′ o.c. should be provided. Closer spacing of expansion or divider strips may be preferred for decorative purposes, to minimize the illusion of waviness or to insure uniform topping thickness.

Manufacturer's recommendations should be followed with respect to the use of primers and application of the topping mix. Generally, the dry components such as fillers and chips are mixed first, then added to the one-part or two-part liquid resin. The mix is poured onto the subfloor, then spread and leveled with a trowel. The setting time before grinding depends on the specific resin formulations. Epoxy and polyester toppings can be ground the day after application. Latex toppings usually require 3 to 5 days setting time before grinding.

For residential use, sealing of the surface often is not required, because the resinous matrixes are quite dense and resistant to most staining agents. However, the marble chips vary in their stain resistance and depending on anticipated traffic conditions, a sealer may be desirable. Manufacturer's recommendations should be consulted in each case.

Vibrated

Traditional terrazzo installations can develop surface imperfections due to portland cement shrinkage, and the appearance of the finished floor sometimes suffers from uneven chip distribution.

To overcome these problems, the Terrazzo Industry Research Institute, a branch of NTMA, has developed a special vibrating machine which improves the performance and appearance of most terrazzo floor installations including rustic and polyacrylate thinset.

The machine has a series of rollers, some of which are attached to a vibrator. On the first pass, the rollers align the chips with flat surfaces up. Then, the vibrating rollers push the chips below the surface and evenly distribute them, making it possible to use more chips and less water. Although the vibrating machine can use stiffer mixes, chemical additives often improve the workability lost by reducing the water/cement ratio.

The reduction of water, use of more chips and the thorough vibration generally result in a denser, harder terrazzo surface with little shrinkage cracking and few pinholes.

REFINISHING EXISTING FLOORS

New terrazzo finishes can be installed over practically any type of existing floor finish. The unbonded method can be used if the subfloor and structural system are capable of supporting the additional dead load (28 lbs per sq. ft./for 2-1/2″ thick finish) and the new finished floor height is not objectionable. Since no bond with the existing floor is intended, relatively little preparation of the surface is necessary and the type of existing finish is not critical. An irregular or slightly pitched surface can be leveled with a sand cushion under the isolation membrane.

If the existing structural and architectural conditions limit the finish thickness and the dead load, one of the bonded methods should be considered. The monolithic method which requires a roughened subfloor for bond generally infrequently is used. The chemically bonded and thinset methods most often are suitable for refinishing existing floors. The chemically bonded topping is thicker, heavier and requires a concrete subfloor, but it is not limited as to chip size. The lighter weight resinous thinset toppings can be used over most subfloors, including wood.

As in new construction, the condition of the subfloor or of the old finish surface is of critical importance for thinset resinous toppings. Surface coatings such as paint, wax and sealing compounds should be removed with appropriate solvents, sanding or grinding.

The topping can be installed directly over properly cleaned marble tile, quarry tile, slate or other hard tile product. Resilient tile and sheet flooring, as well as adhesive, should be removed. The surface should be ground smooth to expose the subfloor prior to installing the topping. It is recommended that the chemical bonding agent or resinous binder be tested first in a small area to insure proper bond with the prepared surface.

Existing stripwood and lumber board floors should be covered with suitable mastic or plywood underlayment at least 3/8″ thick. Manufacturers' recommendations should be consulted to determine the proper type of underlayment and installation procedure for each type of resin matrix.

FIG. 20 SURFACE FINISHES FOR VARIOUS INSTALLATION METHODS

Installation Method	Binder	Chip Size	Surface Finish
Thinset	Polyester & Epoxy	#0 & #1	Regular Conductive
	Latex	Standard	
Monolithic Chemically Bonded	Portland Cement	Standard Intermediate Venetian	Regular Rustic
Bonded-Underbed Unbonded	Portland Cement	Standard Intermediate Venetian	Regular Rustic Conductive

SPECIAL SURFACE FINISHES

Terrazzo floors can be described according to (1) binder (portland cement or one of the resins), (2) installation method (unbonded or one of the bonded methods), (3) chip size and type (standard, intermediate, Venetian or Palladiana) and (4) surface finish (regular, rustic or conductive) (Fig. 20).

Terrazzo installations in homes, apartments, offices, stores, building lobbies and other locations of normal foot traffic require no special additives or surface treatment. The usual binder formulations previously described, after normal grinding and polishing, provide adequate resistance to abrasion, staining, discoloration and slipping hazard. These *regular* terrazzo finishes are used in the great majority of installations.

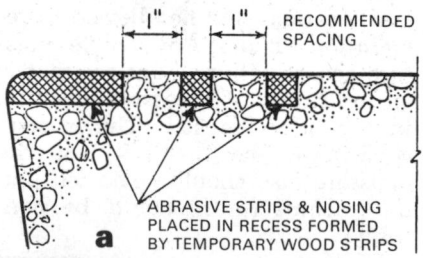

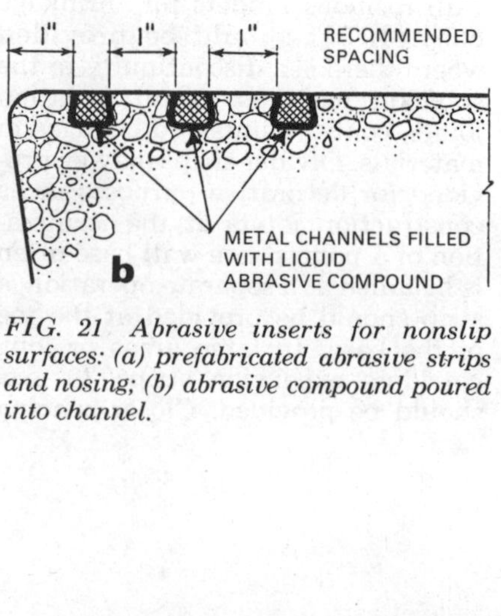

FIG. 21 Abrasive inserts for nonslip surfaces: (a) prefabricated abrasive strips and nosing; (b) abrasive compound poured into channel.

Special surface finishes have been developed for situations where unusual hazards exist. In exterior locations, a greater degree of protection from slipping hazard is afforded by *rustic* terrazzo. In hospitals and laboratories where oxygen is used and potentially explosive atmospheres exist, *conductive* terrazzo will prevent electrostatic sparks. For locations subject to damage from acids, such as hospitals, laboratories and industrial buildings, many resinous matrixes are suitable and are replacing the specialized *acid-resistant* portland cement formulations once recommended. Where greater slip resistance than that provided by regular terrazzo is required for interior locations, as on ramps and stairs, regularly spaced abrasive strips are recommended (Fig. 21). Some resinous matrixes can be formulated to provide inherently good slip-resistant properties. *Abrasive* terrazzo, containing abrasive aggregates in the mix or on the surface is no longer recommended because its slip resistance has proven to be temporary. The regular and special surface finishes can be used with most installation methods (Fig. 20).

Rustic

Rustic (washed) terrazzo is particularly suitable for nonslip and decorative exterior surfaces. Either a weather-resistant marble, quartz or granite chip or decorative gravel generally is used. Although resinous binders can be used, they generally are limited to decorative wall installations. Portland cement is the common material for rustic terrazzo floors, and in areas subject to freezing, portland cement with an air-entraining agent is recommended. Rustic terrazzo should not be used where the surface will be subjected to de-icing salts and chemicals.

After the topping is placed, rolled and leveled as described under General Recommendations, the surface is washed with a stream of water from a hose. The water pressure should be sufficient to remove the paste to a depth of about 1/16″ between chips or gravel. The surface then should be scrubbed with a solution of 1 part muriatic acid to 10 parts water to remove particles of paste clinging to the aggregate. A final rinsing with clean water removes the acid solution and neutralizes the surface. A light grinding may be desirable for exposed chips in pool areas, and the floor may be sealed in accordance with manufacturer's directions, but usually no further operations are necessary.

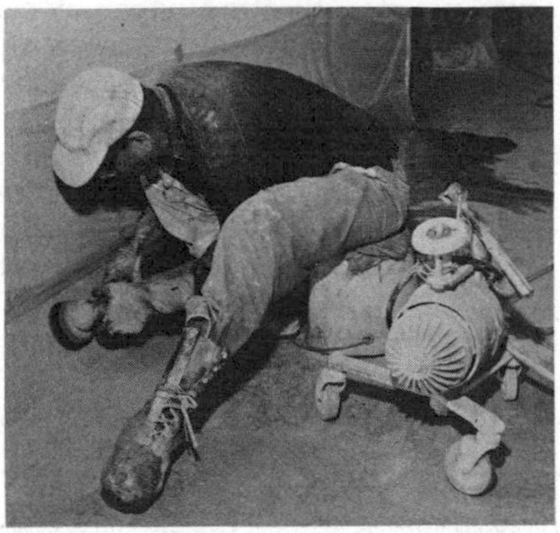

FIG. 24 Base grinding: small hand-held wheels are used to grind and polish terrazzo base and wainscoting.

Conductive

Conductivity in the terrazzo finish is achieved by adding a chemical powder, acetylene black, dissolved in isopropyl alcohol. For portland cement terrazzo the acetylene black is added at the job to the topping and underbed mixes. Resinous matrixes usually have the conductive ingredients included during manufacture. Wire mesh reinforcement and plastic-topped divider strips are required to insure uniform conductivity within the terrazzo finish. Proportions of acetylene black, alcohol, portland cement and chips, as well as many installation details and testing procedures are regulated by the National Fire Protection Association Code No. 56.

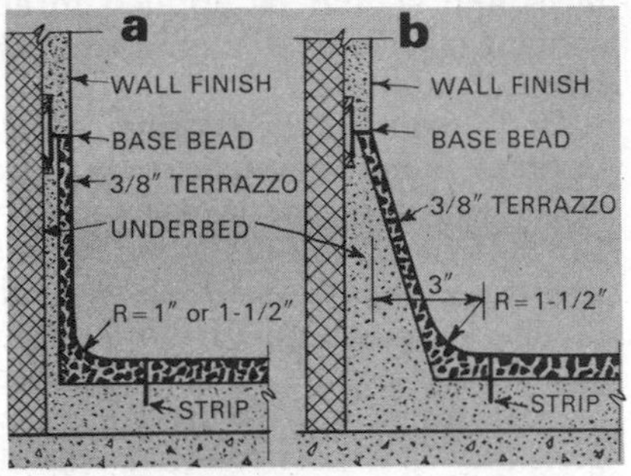

FIG. 22 Coved wall base: (a) straight type and (b) splayed type. Underbed thickness may be varied creating recessed, flush (shown) or projecting base.

SPECIAL ELEMENTS

Terrazzo finishes are not limited to flooring or horizontal surfaces. Cast-in-place terrazzo can be carried up the wall 4″ to 12″ to form a coved

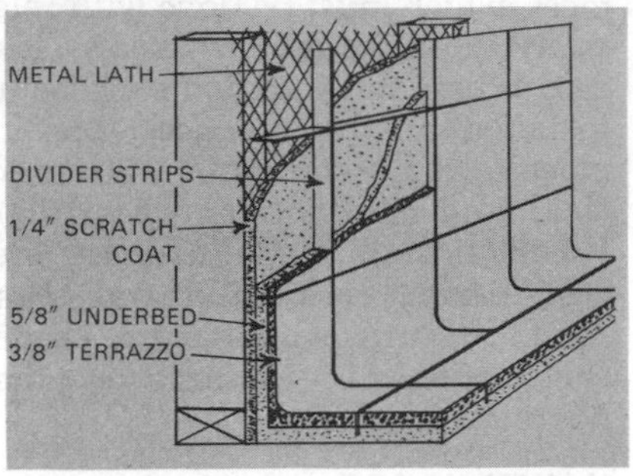

FIG. 23 Wainscot can be installed over studs and metal lath (shown) or over any masonry base which will provide adequate bond for portland cement mortar.

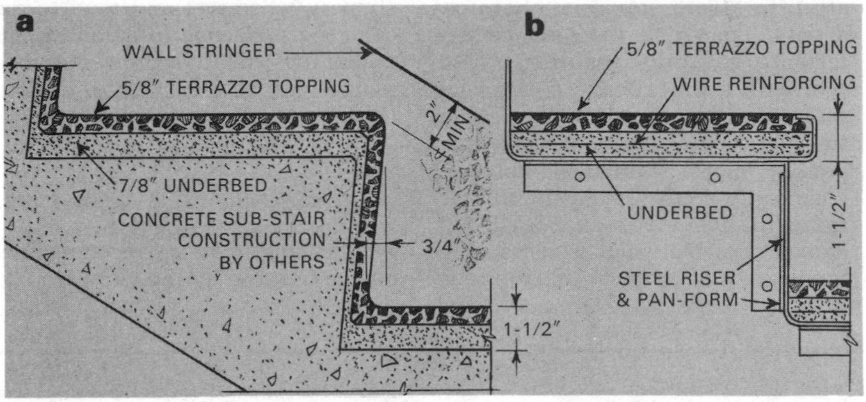

FIG. 25 Cast-in-place stairs: (a) continuous treads and risers over concrete substair; (b) treads cast in steel pan-forms.

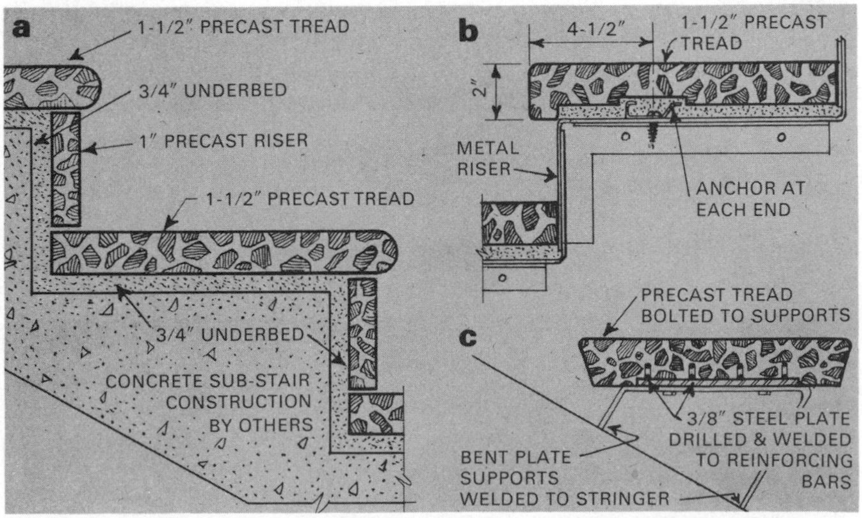

FIG. 26 Precast stair treads and risers can be used with (a) concrete or (b & c) metal stair construction.

wall base (Fig. 22) or can be continued to form a protective wainscot (Fig. 23). Full-height terrazzo finishes are possible, though generally are not practical because of the labor involved in grinding vertical surfaces, which must be done with relatively small hand-held wheels (Fig. 24). Where terrazzo floors are being installed on the premises, cast-in-place terrazzo stairs often are practical and economical (Fig. 25). Where time is of the essence, precast treads, risers and wall base (Fig. 26) are available for use in conjunction with terrazzo or other flooring materials.

MAINTENANCE

Portland cement matrixes generally are fairly porous and require protection from deterioration and staining from various chemical agents. Therefore, portland cement terrazzo generally requires protection with a proprietary penetrating sealer capable of internally sealing the pores. Resinous matrixes are denser, more stain resistant and chemically inert. However, all terrazzo finishes consist of at least 70% stone or marble which possess varying degrees of porosity and stain resistance. Resinous terrazzo may require similar protection depending on the chip type and conditions of use. When used, sealers should be of a type specifically recommended by the manufacturer for terrazzo surfaces and should be applied in accordance with the manufacturer's recommendations.

Both resinous and cementitious terrazzo floors are sufficiently hard and abrasion resistant over their entire surface. Therefore, surface coatings such as wax which provide abrasion protection for softer floor finishes are not required. The effect of foot traffic on unwaxed terrazzo generally is a positive one, tending to polish and to impart further luster to the surface. Surface coatings also generally are not recommended because they tend to make the floor slippery when wet and tend to discolor the surface in low-traffic areas, unless periodically removed and replaced with new coatings of uniform thickness.

Cleaning

Terrazzo should be cleaned periodically by damp mopping; only a neutral liquid cleaner recommended for the purpose by the manufacturer should be used. Soaps and scrubbing powders containing detergents, water-soluble inorganic salts or crystallizing salts should not be used. Sweeping compounds containing oil may stain and permanently discolor the terrazzo surface. Dry buffing after each mopping (and as required between moppings) will remove surface dirt and restore luster.

Terrazzo, like any floor surface, is susceptible to staining. Stains should be treated as soon as possible because they are more difficult to remove after they have set. The type of stain should be determined so that the proper remover can be selected. Stain removers fall into three general categories: (1) *solvents,* such as carbon tetrachloride, which dissolve grease, chewing gum and lipstick; (2) *absorbents,* such as chalk, talcum powder, blotting paper or cotton, which absorb fresh grease and moist stains; and (3) *bleaches,* such as household ammonia, hydrogen peroxide, acetic acid or lemon juice, which discolor stains. Manufacturers of terrazzo sealers and cleaners also produce stain removal kits which can be used successfully if recommendations are followed carefully.

We gratefully acknowledge the assistance of the following for use of their publications as references and for permission to use photographs: American Terrazzo Strip Company; Boyardee Tile Company; John Caretti & Company; Crossfield Products Company; H. B. Fuller Company; Grazzini Brothers and Company; Mid-States Terrazzo and Tile Co.; National Terrazzo & Mosaics Association; Portland Cement Association; Selby, Battersby & Co.; Terrazzo Marble and Supply Company.

TERRAZZO 453

CONTENTS

MATERIALS

Terrazzo finishes consist of a topping applied either directly to the subfloor or over a portland cement underbed. The basic ingredients of terrazzo toppings are binder materials and decorative chips. Other materials which may be required are: mineral or synthetic pigments; accessory materials such as divider strips and expansion strips; metal reinforcement; isolation membranes; and curing and sealing materials.

See Main Text, page 453-3, for definitions of terminology.

Materials for terrazzo finishes should conform to the applicable industry standards given in Figure WF1.

BINDERS

Either cementitious or resinous binders can be used in terrazzo toppings. Cementitious binders are the older, traditional materials, whereas resinous binders are the result of more recent developments in synthetic organic chemistry.

Cementitious

These binders include portland cement and magnesium oxychloride cement (magnesite).

Type I portland cement still is the most common material for terrazzo finishes. Natural (gray) portland cement is less expensive and generally is acceptable if a vivid or light colored matrix is not essential to the color scheme. White portland cement offers a lighter background for displaying the decorative chips and produces clearer, truer colors with mineral pigments.

Type IA air-entraining portland cement is recommended for exterior applications subject to freezing and thawing.

FIG. WF1 **SUMMARY OF TERRAZZO MATERIALS**

Material	Standard
Cementitious Binders Portland Cement, Type 1	ASTM C-150
Magnesium Oxide	ASTM C-275
Magnesium Chloride	ASTM C-276
Resinous Matrixes Epoxy	NTMA*
Polyester	MIL-F-52505 (MO) & NTMA*
Latex	MIL-D-3134F & NTMA*
Decorative Chips	NTMA*
Aggregate Sand	ASTM C-33
Magnesite Fine Aggregate	USAS A88.6
Pigments	NTMA*
Metal Reinforcement Welded Wire Fabric	ASTM A-185
Isolation Membranes Roofing Felt	FS-HH-R-595
Building Paper	FS-UU-B-790
Polyethylene Film	CS 238
Curing Compounds	ASTM C-309

*Specifications of the National Terrazzo and Mosaic Association

(Continued) **MATERIALS**

Magnesite now is not often used because it has lower resistance to moisture, abrasion and chemicals than do most resinous binders. Magnesite without decorative aggregate still is used as a mastic underlayment for thin-set terrazzo and other finishes over existing wood floors.

Materials and application for magnesite toppings should conform to the requirements of USA Standard A88.6, Specifications for Terrazzo Oxychloride Composition Flooring and Its Installation.

Resinous

Epoxy and polyester matrixes are two-part formulations (a basic resin and a catalyst) to which a dry mix (fillers, pigment, additives and decorative chips) is added. These resins are distinguished by good chemical resistance and adhesion, good abrasion resistance and impact strength.

Materials and application for epoxy, polyester and latex terrazzo should conform to the requirements of the National Terrazzo & Mosaics Association (NTMA) Specifications, as well as the manufacturer's recommendations.

Latex matrixes (polyvinyl chloride, neoprene and acrylic binders) are one-part formulations consisting of the liquid resin emulsion added to a dry mix (fillers, pigments, additives and decorative chips). The latex matrixes exhibit good chemical resistance, nonslip properties and are relatively easy to install.

DECORATIVE CHIPS

Marble and other decorative chips for various topping types should not exceed the sizes shown in Figure WF2.

Producers normally supply #0, #1 and #2 chips as separate sizes; some supply #3 and #4 as separate sizes. Larger sizes are frequently grouped as #3-4 mixed, #5-6 mixed and #4-7 mixed.

To assure uniformity of color and texture, it is recommended that decorative chips and pigment combinations be designated by standard color plate number (see Main Text, Fig. 5, page 453-5) and that physical samples be prepared as a standard of comparison at the beginning of each project.

FINE AGGREGATE

Sand used for underbed mortar should be a well graded natural or manufactured aggregate which is clean, granular, free from clay, salts or other deleterious materials.

FIG. WF2 GRADING OF DECORATIVE CHIPS FOR TERRAZZO TOPPINGS

Type of Topping & Minimum Thickness	Chip Number	Passes Screen (in.)	Retained on Screen (in.)
3/4" Venetian; 5/8" Intermediate; 5/8" Standard*, 1/2" Thinset; 1/4" Thinset	0	1/8	1/16
	1	1/4	1/8
3/4" Venetian; 5/8" Intermediate; 5/8" Standard*, 1/2" Thinset	2	3/8	1/4
3/4" Venetian; 5/8" Intermediate	3	1/2	3/8
	4	5/8	1/2
3/4" Venetian	5	3/4	5/8
	6	7/8	3/4
	7	1	7/8
	8	1 1/8	1

*Palladiana toppings consist of broken marble slabs set in a standard chip mix and should not be less than 1/2" thick.

Fine aggregate for magnesite and resinous terrazzo generally is included in the proprietary prepared mix.

PIGMENTS

Pigments used for coloring terrazzo matrixes should be stable, nonfading mineral or synthetic formulations, compatible with the binder.

ACCESSORIES

For a discussion of accessories for terrazzo installations (divider strips, expansion strips and base beads), see Main Text, page 453-5.

Metal accessories commonly are made of half-hard brass or white alloy zinc; aluminum and Type 300 stainless steel accessories also are available. Plastic-top strips and beads are available in various colors with brass, zinc and galvanized steel supports. The standard length of strips and beads is 6'0", but longer lengths may be ordered specially.

The thickness of metal accessories should be at least 18 gauge.

In exterior locations, all-brass, stainless steel or zinc-supported plastic-top accessories should be used. White alloy zinc accessories should be used in interior locations only.

Aluminum accessories should not be used with portland cement terrazzo.

(Continued) MATERIALS

Only plastic-top accessories should be used for conductive terrazzo floors.

Divider Strips

Over an underbed, for topping thickness of 5/8", a 1-1/4" deep divider strip should be used; 1-1/2" deep strips are recommended for toppings 3/4" and thicker. For thinset, monolithic and chemically bonded installations, special angle or T-shaped strips equal to the desired topping thickness should be used.

METAL REINFORCEMENT

For a portland cement underbed, metal reinforcement should be corrosion-resistant welded wire fabric of at least 16 gauge thickness, weighing at least 1.4 lbs. per sq. yd.

Metal reinforcement for bonding resinous and magnesite toppings or mastic underlayments to wood subfloors should be woven or welded wire fabric weighing at least 1.4 lbs. per sq. yd., or expanded metal lath weighing 2.5 lbs. per sq. yd.

ISOLATION MEMBRANES

Isolation membranes should be 15 lb. asphalt-saturated roofing felt, water-resistant building paper or 4 mil polyethylene film.

CURING MATERIALS

Suitable curing materials (such as polyethylene film, nonstaining (nonasphaltic) waterproof paper, liquid curing compounds and clean water) should be used for curing portland cement toppings.

Liquid curing compounds should not be used to cure concrete slabs to which terrazzo toppings are to be applied directly.

SEALERS AND PROTECTIVE FINISHES

Sealers and protective finishes should be products specifically recommended by the manufacturer for use over the specific terrazzo type and should be applied in accordance with the manufacturer's recommendations.

APPLICATION

Suitable terrazzo finishes require attention to: (1) general recommendations common to most installation methods, (2) specific application procedures for particular methods, and (3) finishing procedures for special surface finishes.

Terrazzo installations should conform to the minimum thicknesses given in Figure WF3.

GENERAL RECOMMENDATIONS

The installation of terrazzo involves (1) preparation of the subfloor, and depending on the method of installation, (2) placement of the topping, strips and underbed when used.

Subfloor Preparation

Regardless of the installation method, the subfloor should be firm, sound and free of loose material. For bonded installations, the subfloor also should be free of coatings such as paint, wax and sealing or curing compounds which could prevent proper bond.

The subfloor should be structurally adequate to support the expected live loads without excessive deflection, as well as the additional dead load imposed by the terrazzo finish.

The dead load can be figured at approximately 11 lbs. per sq. ft. per in. of thickness for both resinous and cementitious terrazzo.

A concrete subfloor intended for a thinset resinous installation (Fig. WF3a) should be smooth-troweled and be at least 21 days old. Wood floors should consist of a suitable single plywood subfloor or a double floor including at least a 3/8" thick plywood underlayment (Fig. WF4). Plywood should be Exterior Type CC-Plugged grade or better. The recommendations of the manufacturer of the resinous matrix should be followed for subfloor preparation.

For monolithic installations (Fig. WF3b), the surface of the subfloor should not be troweled but should be wood-floated and broomed or otherwise roughened. Over a smooth-troweled concrete surface (Fig. WF3c), a chemical bonding agent should be used.

The subfloor for monolithic, chemically bonded and thinset installations should conform to the profile of the

(Continued) **APPLICATION**

finished terrazzo surface, whether level or pitched, to insure that the topping will be of uniform thickness. The subfloor surface should be in a straight plane, and there should be no more than 1/8″ variation when tested with a 10′ straightedge. The topping mix should not be used to fill voids or to develop pitch.

Where the existing subfloor is not sufficiently smooth or does not have the desired profile, it should be leveled or pitched with a suitable mastic underlayment compatible with the topping material.

Underbed

The underbed should be reinforced with welded wire fabric in all unbonded installations and in bonded installations over wood and metal subfloors. In unbonded installations, an isolation membrane should be spread over the subfloor lapping 4″ at all edges and ends. Metal reinforcement should be placed over the membrane, lapping adjacent sheets at least two mesh spaces, except where expansion strips occur.

The underbed mix should consist of one part portland cement and 4 to 5 parts sand, mixed with sufficient water to make a workable mortar.

The underbed should be placed and screeded to a height which will permit installation of the selected topping thickness at the desired finish floor elevation.

Electrical conduit or other piping should not be located within the minimum recommended underbed thickness (Fig. WF3d & e).

The surface over which the underbed is placed should be sufficiently wet to prevent water from being drawn from the mix. Isolation membranes need only be uniformly dampened; concrete slabs should be thoroughly saturated by overnight soaking.

Strip Placement

Divider strips are installed: (1) to localize shrinkage cracking in portland cement terrazzo, (2) to assure uniformity in topping thickness, (3) to act as decorative elements, (4) to permit accurate placement of the different colors forming the pattern and (5) to act as construction joints subdividing the floor into convenient working areas.

Expansion or divider strips should be provided wherever a discontinuity in the subfloor exists, such as at all construction joints, isolation joints and control joints and at changes in materials. Generally, divider and expansion strips should be provided for various installation methods as recommended in Figure WF5.

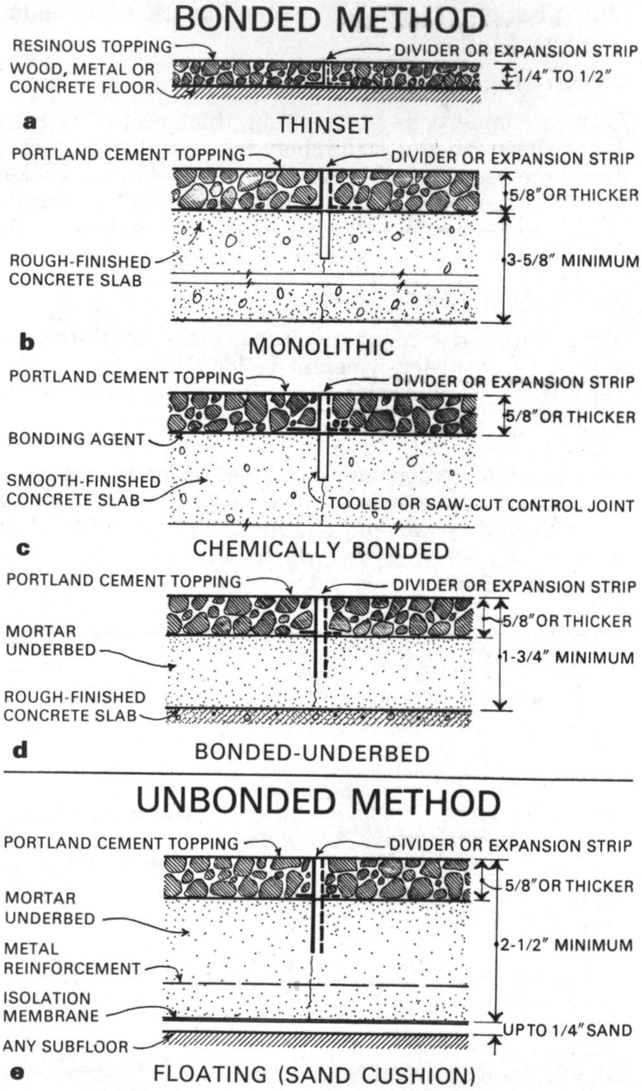

FIG. WF3 Recommended details for basic terrazzo installation methods.

Whenever possible, divider or expansion strips should be located over beams and girders supporting the subfloor, where high tensile stresses are likely to develop due to negative bending moments.

Divider Strips In portland cement terrazzo installations with an underbed, divider strips should be spaced not more than 6′ o.c. When a portland cement topping is placed directly over a concrete slab (Fig. WF3b & c), the slab should be provided with cut or tooled control joints 15′ to 20′ o.c. in each direction. Joints should form panels no larger than 300 sq. ft. and the length of panels

(Continued) **APPLICATION**

FIG. WF4 **WOOD FLOOR CONSTRUCTION FOR THINSET TERRAZZO**[1]

SINGLE FLOOR CONSTRUCTION

Combination Subfloor/Underlayment[2,3]	Maximum Support Spacing	Nails[4]	Nail Spacing
⅝″ Plywood	16″ o.c.	2″ (6d) threaded	6″ o.c. around panel edges; 10″ o.c. at intermediate supports.
¾″ Plywood	20″ o.c.		
⅞″ Plywood	24″ o.c.		
1⅛″ Plywood (2-4-1)	48″ o.c.	2½″ (8d) threaded	6″ o.c. at panel edges and intermediate supports.

DOUBLE FLOOR CONSTRUCTION

Subfloor[2]	Maximum Support Spacing	Nails	
½″ Plywood	16″ o.c.	2″ (6d) common	6″ o.c. around panel edges; 10″ o.c. at intermediate supports.
⅝″ Plywood	20″ o.c.	2½″ (8d) common	
¾″ Plywood	24″ o.c.		
¾″ lumber boards less than 6″ wide	16″ o.c.	2¼″ (7d) threaded	2 per board at each support.
¾″ lumber boards wider than 6″ and less than 8″ wide			3 per board at each support.

Underlayment[2,5]	Fasteners[6] Nails[4]	Fasteners[6] Staples[7]	Fastener Spacing
⅜″ Plywood	1¼″ (3d) threaded	1⅛″	Nails: 8″ o.c. each direction in middle of panel, 6″ o.c. around panel perimeter. Staples: 6″ o.c. each direction in middle of panel, 3″ o.c. around panel perimeter.
½″ Plywood		1¼″	
⅝″ Plywood	1½″ (4d) threaded	1½″	
¾″ Plywood		1⅝″	

1. Based on 40 psf live load.
2. Plywood thicknesses given are for Group 1 Softwoods in accordance with U. S. Product Standard PS1. For Groups 2 and 3, increase thickness ⅛″; for Group 4, increase thickness ¼″ (see Work File, page 201-6).
3. Blocking at longitudinal edges is required unless plywood with T&G edges is used.
4. Common nails may be substituted for annularly threaded nails by increasing their size ¼″ (1d).
5. Blocking not required if underlayment joints are staggered with respect to subfloor joints.
6. Fastener lengths shown are minimum to provide ¾″ penetration for nails and ⅝″ for staples. If longer fasteners are used, penetration should not exceed 1″.
7. Galvanized divergent chisel, 16 gauge, ⅜″ crown width.

should not exceed the width by more than 50%. Divider strips should be placed at all such control joints forming similar panels in the topping (see Main Text, Figs. 12 & 13, page 353-10).

In all cases, divider strips should be continuous from wall to wall and preferably should form right angle intersections rather than acute angles. If a terrazzo wall base is used, the strips should be carried up to the base screed. When the terrazzo is carried beyond the base to form a wainscot, strips should be continued from the base in a straight line up the wall.

The recommended divider strip spacing for portland cement terrazzo wainscoting is 3′ o.c. maximum; strip spacing for resinous wainscoting is not critical.

Expansion Strips Where unobstructed large areas or long corridors are involved in monolithic, chemically bonded and thinset installations, expansion strips should be provided approximately 30′ o.c.

In monolithic and chemically bonded installations, expansion strips and divider strips can be alternated 15′ o.c. Expansion strips generally are not required in these installations if the subfloor is continuous, homogeneous and 1000 sq. ft. or less in area.

Topping

The following recommendations are primarily for portland cement terrazzo with regular finish. For recommendations concerning resinous terrazzo, see Thinset

(Continued) APPLICATION

Method, page WF453-7; for Palladiana and rustic installations, see page WF453-8.

Mixing A conventional portland cement mix should consist of 200 lbs. of marble chips per 94 lb. bag of cement. When pigment is used, it should be measured by weight (not volume) and should be added to the dry mix. After thorough mixing of dry ingredients, water should be added at the rate of 5 to 5-1/2 gals. per bag of cement.

Placing Over an underbed, the topping should not be placed for 12 to 24 hours to permit the underbed to harden. Immediately before the topping is placed, a slush coat of neat cement, pigmented to match the topping mix, should be broomed into the underbed.

The surface should be seeded with additional chips where necessary to assure adequate chip content and even distribution. Chips should be wetted, sprinkled over the surface and then embedded by additional troweling. The entire area then should be rolled with a heavy roller until all excess water has been brought to the surface. The surface again should be troweled smooth and uniform revealing the tops of the divider strips.

Before finishing operations are started, portland cement toppings should be wet-cured for 3 to 7 days.

Toppings may be cured either by ponding, by covering with watertight covers or by spraying with a curing compound.

Preliminary Finishing After curing, the surface should be ground using a terrazzo grinding machine, first with No. 24 or finer abrasive stones, followed by a second grinding with No. 80 or finer stones. Grinding operations should be performed while the surface is covered with water.

After grinding, the surface should be thoroughly rinsed with sufficient water to remove dust and loose fine particles. Then a skim coat of neat cement grout of the type used in the topping mix (including pigment to match the topping) should be troweled over the entire surface, filling all voids.

The installation then should be wet-cured again for at least 3 days. The curing compound or watertight cover should be allowed to remain in place as protection against physical damage or staining until final finishing is performed.

Final Finishing A floor machine with No. 80 or finer stones should be used for polishing. The polishing operation should continue until the skim coat is removed (except in voids and depressions) to reveal the decorative chips in approximately 70% to 75% of the area.

The polished floor should be cleaned with a neutral liquid cleaner, in accordance with the manufacturer's recommendations. It then should be rinsed with clean water and be allowed to dry. The cleaned, polished terrazzo surface finally should be treated with a suitable sealing compound.

Unless all work on the premises is completed, the floor should be protected with a covering such as reinforced building paper until ready for use.

FIG. WF5 RECOMMENDED STRIP PLACEMENT

Matrix	Installation Method	Divider Strip Spacing[1]	Expansion Strip Spacing[1]
Portland Cement	Unbonded	6′ o.c. maximum	As needed[2]
	Bonded-underbed		
	Chemically Bonded	15′ to 20′ o.c.—Panels not over 300 sq. ft.[3]	30′ o.c. maximum[3]
	Monolithic		
Resinous	Thinset	As desired for decorative or convenience purposes	30′ o.c. maximum[4]

1. Locate divider strips or expansion strips on beams and girder lines wherever possible.
2. Provide expansion strips over all expansion joints in the subfloor or at other joints where movement is expected.
3. Divider strips and expansion strips can be alternated or expansion strips can be used at 15′ o.c.
4. Recommended for large areas or long corridors; otherwise not required.

INSTALLATION METHODS

The general recommendations for subfloor preparation, installation of underbed, strips and topping should be supplemented with recommendations provided for each individual installation method.

Unbonded

This method provides isolation between subfloor and underbed, permitting independent movement by each, and hence minimizes the possibility of surface cracking. An unbonded installation is suitable over any structurally sound subfloor in interior or exterior locations.

Unbonded installations should include a reinforced underbed to support the topping. The combined thickness of the underbed and topping should be at least 2-1/2″ (Fig. WF3e). Separation between subfloor and underbed should be provided by an isolation membrane alone or by a membrane spread over a layer of sand

(Continued) **APPLICATION**

1/16″ to 1/4″ deep. Membranes should be lapped at least 4″ at all ends and edges.

The metal reinforcement should be placed over the membrane, lapped at joints at least 2 mesh spaces and wired together. The reinforcement should be positioned within the underbed approximately 1″ above the membrane.

Bonded

The chief advantages of bonded installations are smaller dead load and lesser finish thickness. Most bonded methods (except bonded-underbed) allow the topping to be placed directly over the subfloor without an intervening underbed. However, these methods do not generally develop the degree of crack control provided by an underbed with conventional divider strips.

Bonded-underbed *The bonded-underbed method is suitable over metal or concrete subfloors and provides a measure of crack control for portland cement terrazzo surpassed only by the unbonded installation.*

Unless natural bond can be obtained between the portland cement underbed and a suitably roughened concrete slab, a chemical bonding agent should be used. For proper bond, a metal deck should be free of corrosion, pickling acids, oil or other foreign substances.

Monolithic *This term refers to a portland cement topping installed over a concrete subfloor where natural bond is developed between the subfloor and topping. Such installations can be used in interior as well as exterior locations.*

The concrete subfloor for monolithic installations should be properly reinforced and be at least 3-5/8″ thick. The surface should be brushed with a wire broom or otherwise roughened to remove laitance and assure good mechanical bond with the topping.

As soon as the concrete is sufficiently hard to resist tearing with a joining tool or power saw, control joints should be installed in the slab approximately 15′ to 20′ apart. Divider or expansion strips should be installed at all control joints in the subfloor. The surface of the slab should be protected from damage and dirt with suitable coverings until the topping is placed.

Chemically Bonded *The chemically bonded method of installing portland cement topping is intended for smooth-troweled concrete subfloors. For this reason it often is suitable for older concrete floors not originally prepared for terrazzo topping. Chemically bonded installations depend on a bonding agent such as epoxy or polyvinyl chloride.*

The subfloor surface should be free of all coatings such as paint, curing compounds or wax, and the bonding agent should be applied in accordance with the manufacturer's recommendations.

The subfloor should be inspected, and where random cracks exist or control joints are not provided 15′ to 20′ o.c., such joints should be installed with a power saw. Divider strips or expansion strips should be secured at all joints in the subfloor.

Thinset *High-strength, resinous binders are used for thinset installations, making it possible to install very thin toppings over practically any sound, firm subfloor such as wood, metal or concrete.*

Epoxy and polyester toppings should be at least 1/4″ thick. When installed over concrete, the subfloor should be smooth-troweled. Latex toppings should be at least 3/8″ thick; either a troweled or broomed subfloor finish is acceptable.

Thinset finishes should not be installed on slab-on-ground construction subject to excessive moisture or high alkaline content.

To test for moisture, pieces of polyethylene film approximately 2′ square should be taped down in several locations and be allowed to remain for 8 hours. Visible moisture accumulation under the mats is an indication of excessive moisture in or under the slab. Under such conditions, the alkalinity of the moisture should also be tested and the manufacturer should be consulted.

Divider strips should be provided where there is a discontinuity in the subfloor, such as at all isolation joints, control joints and at changes in materials. In large areas or long corridors, expansion strips 30′ o.c. are recommended.

Divider strips are not required for resinous binders to control shrinkage, but they may be provided for decorative purposes and for convenience in placing the topping. When wall base is installed as a separate operation, a strip should be provided at the toe of the base.

The manufacturer's recommendations should be followed with respect to the use of primers, mixing of ingredients, application of the topping, grinding and sealing of the surface. Over wood subfloors, the manufacturer's specific recommendations should be consulted

(Continued) **APPLICATION**

for details of taping plywood joints and for use of mastic underlayment, wire or metal reinforcement.

Sealing of the surface often is not required for residential use, because the resinous matrixes are quite dense and resistant to most staining agents. However, decorative chips vary in their stain resistance and, depending on anticipated traffic, a sealer may be desirable.

Refinishing Existing Floors

See Main Text, page 453-13, for a discussion of refinishing existing floors.

SPECIAL SURFACE FINISHES

The binder formulations previously described, after normal grinding and polishing, usually provide adequate resistance to abrasion, staining, discoloration and slipping hazard. These regular terrazzo finishes are used in the great majority of installations. Special surface finishes have been developed for situations where unusual hazards exist. The regular and special surface finishes can be used with most installation methods (Fig. WF6).

Rustic

Rustic (washed) terrazzo is particularly suitable for nonslip and decorative exterior surfaces. Either weather-resistant decorative chips or decorative gravel can be used. Portland cement is the common binder material and in areas subject to freezing, portland cement with an air-entraining agent is recommended. Rustic terrazzo should not be used where the surface will be subjected to de-icing salts and chemicals.

After the topping is placed, rolled and leveled as described under General Recommendations, the surface should be washed with a stream of water to remove the mortar to a depth of about 1/16″ between chips or gravel. The surface then should be scrubbed with a

FIG. WF6 SURFACE FINISHES FOR VARIOUS INSTALLATION METHODS

Installation Method	Binder	Chip Size	Surface Finish
Thinset	Polyester & Epoxy	#0 & #1	Regular Conductive
	Latex	Standard	
Monolithic Chemically Bonded	Portland Cement	Standard Intermediate Venetian	Regular Rustic
Bonded-Underbed Unbonded	Portland Cement	Standard Intermediate Venetian	Regular Rustic Conductive

solution of muriatic acid, rinsed with water and sealed.

Conductive

Conductive floors are used to prevent electrostatic sparks in potentially explosive atmospheres. For portland cement terrazzo, conductivity is achieved by adding acetylene black dissolved in isopropyl alcohol to the topping and underbed mixes. Resinous matrixes usually have the conductive ingredients included during manufacture. Wire mesh reinforcement and plastic-top divider strips are required to insure uniform conductivity within the terrazzo finish.

Installation and testing procedures for conductive terrazzo floors should conform to the National Fire Protection Association Code No. 56.

SPECIAL ELEMENTS

See Main Text, page 453-15, for a discussion of special elements such as precast and cast-in-place coved wall base, wainscot, and stair treads and risers.

MAINTENANCE

Terrazzo finishes consist of at least 70% decorative chips which possess varying degrees of porosity and stain resistance. Portland cement matrixes generally are fairly porous and require a penetrating sealer to resist deterioration and staining by various chemical agents.

When used, sealers should be of a type specifically recommended by the manufacturer for terrazzo surfaces and should be applied in accordance with the manufacturer's recommendations.

Both resinous and cementitious terrazzo are sufficiently hard and abrasion resistant over their entire surface, so that surface coatings such as wax are not required.

Terrazzo should be cleaned periodically by damp mopping; only a neutral liquid cleaner recommended for the purpose by the manufacturer should be used. Soaps and scrubbing powders containing detergents, water-soluble inorganic salts or crystallizing salts should not be used. To remove dirt and restore luster, the surface should be dry-buffed between moppings. Stains should be treated as soon as possible with an appropriate stain remover.

Manufacturers of terrazzo sealers and cleaners also produce stain removal kits which can be used successfully if recommendations are followed carefully.

512 INDIVIDUAL SEWAGE DISPOSAL

INDIVIDUAL SEWAGE DISPOSAL

INTRODUCTION

Population movement within the United States continues to be from rural to metropolitan areas. Because of the difficulty of immediately providing municipal sewerage systems for this new growth, individual sewage disposal systems continue to be an important method of sewage disposal where they are acceptable. It is estimated that almost 50 million persons are served by more than 15 million individual sewage disposal systems in the United States. Of more importance even, roughly one-fourth of the new homes are being constructed with such systems.

Disposal of human and other domestic wastes in rural areas has been accomplished traditionally with a *cesspool,* a lined pit which permitted liquids to seep into the ground and which retained solids to be pumped out periodically. In urban areas such a disposal system is considered a major health hazard and is not generally permitted. Instead, the *septic sewage disposal system* has evolved which is relatively safe and nuisance free. Such a system consists generally of a *septic tank,* where the sewage is retained long enough to separate and decompose the solids, coupled with some type of *soil absorption system* (absorption field, seepage pit or absorption bed) where the liquid effluent from the tank gradually seeps into the soil.

The sewage is treated both in the tank and in the soil, as "anaerobic" (oxygen-free) bacteria decompose the organic matter into simpler, harmless compounds. However, the anaerobic decomposition of organic matter is accompanied by foul odor and the presence of pathogenic (disease-producing) organisms, hence the septic effluent cannot be discharged on the ground surface, lake or stream, without creating a nuisance or health hazard.

Properly designed and maintained soil absorption systems practically eliminate the health and nuisance hazard, but still pose problems in certain areas. In tight soils, an absorption system may not be possible, or the land area required may be too large. Recently developed individual home "package" plants use an electric motor to pump air through the sewage in the tank enabling "aerobic" (oxygen-dependent) bacteria to break down the organic matter more completely before it leaves the tank. Aerobic treatment results in a purer, relatively odorless effluent, which manufacturers claim can be safely discharged into any running stream or creek, provided it has been first chlorinated to kill pathogenic organisms. Most health authorities however cite the uncertainty of maintenance, and hence of effluent quality, with individual home "package" systems and recommend disposal of the effluent by soil absorption.

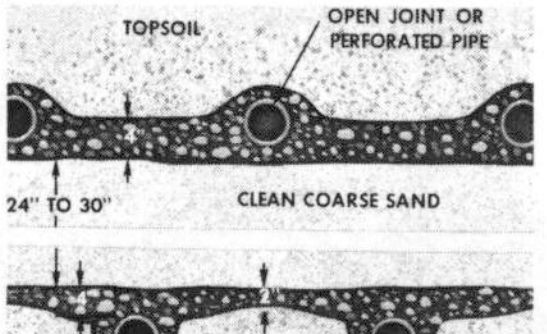

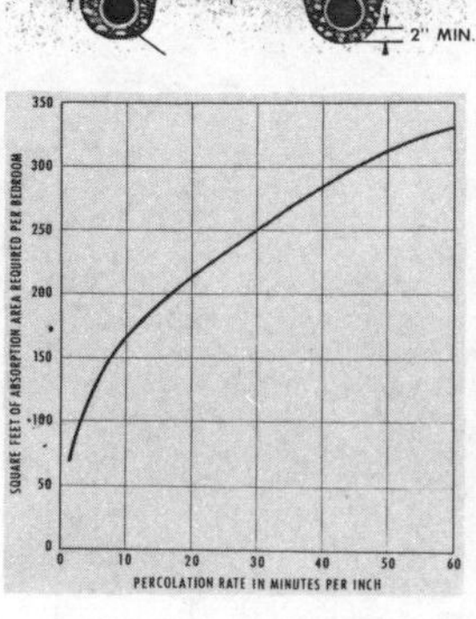

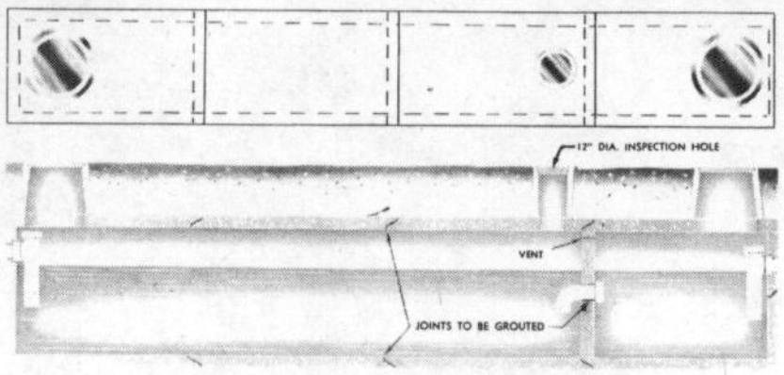

SEPTIC TANK SYSTEMS

A sewage treatment plant—whether individual or large community type—works on the same principle as decay and oxidation does in nature, except that the action is more concentrated. In a treatment plant, favorable conditions are created for decay bacteria to break down organic solids into chemically simpler water-soluble compounds: phosphates and nitrates. In the natural cycle, limited amounts of phosphates and nitrates are produced and these become the nutrients for other growing things. Plants absorb such nutrients through the roots and, with the help of sunlight, recombine them with carbon dioxide from the air to form new living tissue. In sewage treatment, large amounts of phosphates and nitrates are produced and flushed away by the effluent to be absorbed by the soil or discharged into a body of water.

Phosphates and nitrates are relatively harmless when released into the soil or water in limited quantities. However when introduced into a lake or stream in excessive amounts, they spur the growth of small aquatic plants (algae) which use up valuable dissolved oxygen as they proliferate and die off in large quantities. The over-enrichment of a body of water with excessive nutrients, proliferation of algae and reduction of oxygen content are characteristic of the *eutrophication* process which gradually destroys aquatic life and renders the water useless for recreation and water supply. Lake Erie is an example of a valuable body of fresh water which is dying of over-enrichment, largely from domestic sewage treatment plants.

The removal of solids and the partial biological treatment which takes place in a septic tank is known as *primary treatment.* Both individual home "package" plants, and most municipal and community sewage plants, provide so-called *secondary treatment,* producing relatively odorless but nutrient-rich effluents. Hence, the discharge of effluents from individual package plants into bodies of water should be discouraged and sewage plants should be equipped to remove or neutralize the nutrients (termed *tertiary treatment)* before discharging into lakes or streams.

Proper disposal of human excreta (body wastes) is a major factor in influencing the health of individuals, where public sewers are not available. Many diseases, such as dysentery, infectious hepatitis, typhoid and paratyphoid, and various types of diarrhea are transmitted from one person to another through the fecal contamination of food and water, largely due to the improper disposal of human wastes.

Safe disposal of domestic wastes is necessary to protect the health of the community and to prevent the occurrence of nuisances. To accomplish satisfactory results, such wastes must be disposed so that they: (1) will not contaminate any drinking water supply and will not pollute the waters of any bathing beach, shellfish breeding ground, or stream used for water supply or for recreation; (2) will not be discharged on or near the surface where they would be accessible to insects, rodents or other small animal carriers which may come into contact with food or drinking water; (3) will not be accessible to children and give rise to a nuisance due to foul odor or unsightly appearance; (4) will not overburden any body of water with excess nutrients, or degrade its quality below officially adopted water quality standards.

Despite the widespread use of individual systems, it should be emphasized, however, that connection to an adequate public sewerage system is the most satisfactory method of sewage disposal. Where connection to a public sewer is not feasible, and when a considerable number of residences are to be served, consideration should be given next to the construction of a community sewerage system and treatment plant. Specific information on this matter can be obtained generally from the local health authority.

The discussion and recommendations presented in the following subsection on septic tank systems have been abstracted from a publication of the Bureau of Community Environmental Management, U.S. Public Health Service (HEW), entitled "Manual of Septic Tank Practice", 1967 edition.

The discussion of septic systems is intended as an aid in understanding and implementing the recommendations contained herein or the requirements of local authorities. These recommendations are made with the full understanding that they may differ in some instances from other authoritative sources such as FHA Minimum Property Standards, National Plumbing Code or BOCA Basic Plumbing Code. Where local authorities use these documents as reference standards or have their own requirements, such local regulations should supersede the following recommendations.

TERMINOLOGY

Key terms relating to individual sewage disposal are defined below and illustrated in Figure 1.

Absorption Bed A wide trench exceeding 36″ in width containing a minimum of 12″ of clean, coarse aggregate and a system of 2 or more distribution pipes through which treated sewage may seep into the surrounding soil. (Also termed *seepage bed.)*

Absorption Field An arrangement of absorption trenches through which treated sewage is absorbed into the soil. (Also termed *disposal field.)*

Absorption Trench A trench not over 36″ in width, containing a minimum of 12″ of clean, coarse aggregate and a distribution pipe, through which treated sewage is allowed to seep into the soil.

Building Drain The lowest part of a house drainage system which receives the discharge from soil, waste and other drainage pipes *inside the building* and conveys it to the building sewer beginning 3′ outside the building wall.

Building Sewer That part of a drainage system *outside the building* which connects the building drain to a public sewer, private sewer, individual sewage disposal system or other point of disposal. (Also termed *house sewer.)*

Cesspool A covered and lined underground pit used as a holding tank for domestic sewage and designed to retain the organic matter and solids, but to permit the liquids to seep through the bottom and sides. Cesspools are almost universally prohibited in this country and are not acceptable as a means of sewage disposal.

Disposal Field See Absorption Field.

Distribution Line Open joint or perforated pipe intended to permit soil absorption of effluent.

Dry Well A covered and lined underground pit, similar to a seepage pit, but intended to receive water free of organic matter, such as from roof drains, floor drains or laundry tubs. Used as an auxiliary to a septic system to avoid overloading the septic tank absorption system. (Also called *leaching well or pit.)*

Effluent Partially treated liquid sewage flowing from any part of the disposal system—septic tank or absorption system.

Impervious Soil A tight cohesive soil such as clay which does not allow the ready passage of water.

Individual Sewage Disposal System A combination of a sewage treatment plant (package plant or septic tank) and method of effluent disposal (soil absorption system or stream discharge) serving a single dwelling.

Lateral A branch of the absorption field, consisting of either (1) the length of distribution line between overflow pipes (Fig. 9) or (2) the length of distribution line between the tee or cross fitting and the farthest point in a closed loop field (Fig. 7).

Pervious Soil Usually a granular soil, such as sand or gravel, which allows water to pass readily through it.

Scum A mass of sewage matter which floats on the surface of the sewage in a septic tank.

Scum Clear Space In a septic tank, distance between the bottom of the scum mat and the bottom of the outlet device (tee, ell or baffle).

Seepage Bed See Absorption Bed.

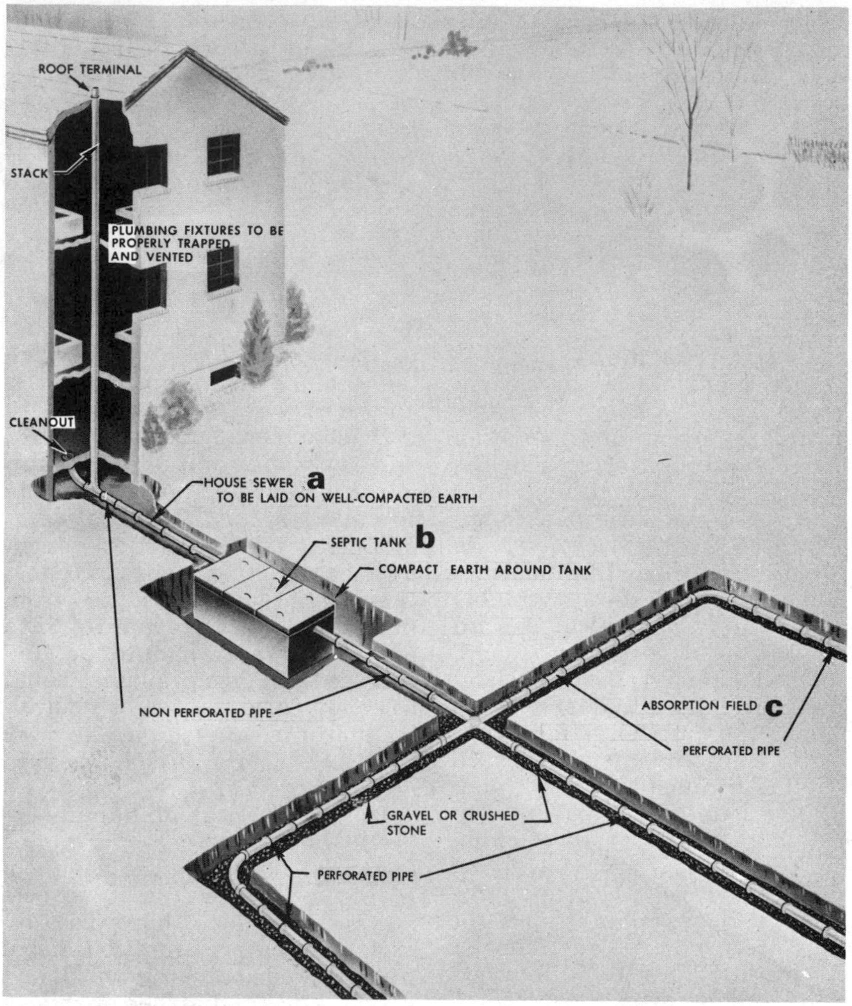

FIG. 1 Typical septic sewage disposal system consists of (a) house sewer, (b) septic tank, and (c) absorption field. Solids are separated and retained in airtight tank, liquid is passed to absorption field and gases are vented through house sewer and vent stack.

Seepage Pit A covered underground pit with concrete or masonry lining designed to permit partially treated sewage to seep into the surrounding soil. (Also called *dry well* in some parts of the country.)

Septic (Sewage) Disposal System A system for the treatment and disposal of domestic sewage by means of a septic tank and a soil absorption system.

Septic Tank A watertight, covered receptacle designed to receive the discharge of sewage from a building sewer; to separate solids from the liquid, digest organic matter and store digested solids through a period of detention; and to allow the clarified liquids to discharge for final disposal.

Serial Distribution A combination of several absorption trenches, seepage pits, or absorption beds arranged in sequence so that each is forced to utilize the total effective absorption area before liquid flows into the succeeding component (Fig. 9).

Sewage Liquid waste containing organic (animal or vegetable) matter in suspension or solution, as well as liquids containing chemicals in solution.

Sludge The accumulated solids which settle out of the sewage, forming a semi-liquid mass on the bottom of the septic tank.

Sludge Clear Space In a septic tank, the distance between the top of the sludge and the bottom of the outlet device (Fig. 22).

Soil Absorption Field See Absorption Field.

Soil Absorption System Any system that utilizes the soil for subsurface absorption of treated sewage, such as an absorption trench, absorption bed or seepage pit.

SUITABILITY OF SOIL

The first step in the design of a septic sewage disposal system is to determine whether the soil is suitable for the absorption of septic tank effluent and, if it is, how much absorption area is required. In general, three conditions must be met: (1) The percolation time should be within the range specified in Figure 2; that is, not over 60 minutes for absorption fields and beds, and not over 30 minutes for seepage pits; (2) The bottom of the trench or seepage pit should be at least 4′ above the maximum seasonal elevation of the ground water table; (3) Rock formations or other impervious strata should be at a depth greater than 4′ below the bottom of the trench, bed or pit.

Unless these basic conditions can be satisfied, the site is unsuitable for a soil absorption system.

Visual Examination

The water absorption of soils can be determined best by percolation tests, but a number of clues help in estimating the likely characteristics of a particular soil. These include the soil maps and descriptions of the U.S. Department of Agriculture; on site visual inspection of soil texture and color, and the record of experience in adjacent areas with similar soils.

Soil particle size governs pore size which, in turn, influences rate of water movement. Sandy and gravely soils generally have large particles and large pores, while clay and silt soils have small particles and small pores (although total void area may be larger). In some soils, especially clay and silt types, there may be clumping of basic particles into larger particles giving the soil a well-developed structure. A lump of structured soil will break apart with little force along well-defined cleavage planes. Such structured clay soil is likely to have better water absorption characteristics than one with little or no structure.

Another clue to its absorptiveness is the color of the soil. Bright, uniform, reddish-brown to yellow colors throughout a soil profile indicate favorable oxidation conditions, good air and water movement, and desirable absorption properties.

In some cases, examination of road cuts, stream embankments or building excavations will give useful information. Well drillers' logs can also be used to obtain data on ground water and subsurface conditions. In some areas, subsoil strata vary widely in short distances, and borings must be made at the site of the system. If the subsoil appears suitable, as judged by the visual characteristics described above, percolation tests should be made at points and elevations selected as typical of the area in which the soil absorption system will be located.

FIG. 2 **REQUIRED ABSORPTION AREA IN VARIOUS SOILS**

PERCOLATION RATE (TIME IN MIN. FOR WATER TO FALL 1")	MIN. AREA PER BEDROOM (SQ. FT.)
2 or less	85
3	100
4	115
5	125
10	165
15	190
30	250
45	300
60	330

Percolation Tests

Subsurface exploration generally is necessary to determine soil conditions: the depth of pervious material, the presence of rock formations and level of ground water. Augers with extension handles often are used for making such investigations (Fig. 3).

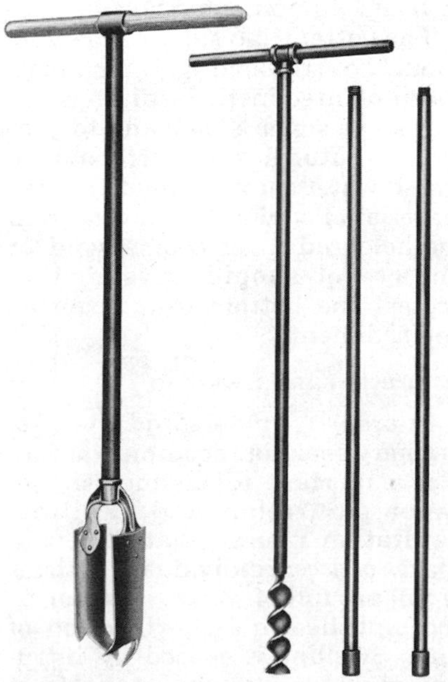

FIG. 3 Augers with extension handles can be used to bore deep test holes.

The percolation tests help to determine the acceptability of the site and establish the type and size of the soil absorption system. Figure 2 shows the square feet of absorption or seepage area recommended for soils with various percolation rates, as well as the slowest percolation rate acceptable for each type of soil absorption system.

Test Holes

The length of time required for percolation tests varies depending on the type of soil. In most cases, the test holes should be kept filled with water for at least 4 hours, (preferably overnight) before measuring percolation rate. The purpose of the preliminary soaking of the test holes is to obtain percolations when the soil is soaked and saturated.

At least six separate test holes should be made; these should be spaced uniformly over the proposed soil absorption area and should reach down to the depth of the proposed absorption system. The holes can be 4″ to 12″ in diameter; to save time, labor and volume of water required per test, shallow holes can be bored with a 4″ auger (Fig. 4a). Deeper holes frequently required for seepage pits may have to be dug by shovel or heavy equipment (Fig. 4b).

The bottom and sides of the hole should be scratched with a knife or sharp-pointed instrument, in order to remove smeared soil and to provide a natural soil surface into which water may percolate. Loose material should be removed from the hole and 2″ of coarse sand or fine gravel should be added to protect the bottom from scouring and sediment.

Saturation and Swelling

In order to understand the preliminary soaking recommendation, it is important to distinguish between saturation and swelling. Saturation means that the void spaces *between* individual particles of soil are full of water; this can be accomplished in a short period of time. Swelling is caused by intrusion of water *into* the particles of soil; this is a slow process, and is the chief reason for recommending the overnight soaking period, especially in clay soils.

The suggested test procedures are intended to insure that the soil is given ample opportunity to swell and to approach its condition during the wettest season of the year. Thus, the test should give comparable results in the same soil, whether made in a dry or in a wet season. In sandy soils containing little or no clay, the swelling procedure is not essential, and the test may be made after the water from one filling of the hole has completely seeped away.

Percolation-Rate Measurement

In conducting the test, the hole should be filled with clear water to a minimum depth of 12″ over the gravel. In most soils, it is necessary to refill the hole by supplying a surplus reservoir of water, possibly by means of an automatic syphon, to keep water in the hole for at least 4 hours and preferably overnight (Figs. 4 and 5).

If water remains in the test hole after the overnight swelling period, the depth should be adjusted to approximately 6″ over the gravel. The drop in water level should be measured from a fixed reference point at 30-minute intervals over a 4-hour period. The drop which occurs in the final 30-minute period should be used to calculate the percolation rate. For instance, a drop of 2-1/2″ in 30 minutes produces a percolation rate of 30/2.5 or 12 minutes per inch.

If no water remains in the hole after the overnight swelling period, more water should be added to bring the depth of water in the hole to approximately 6″ over the

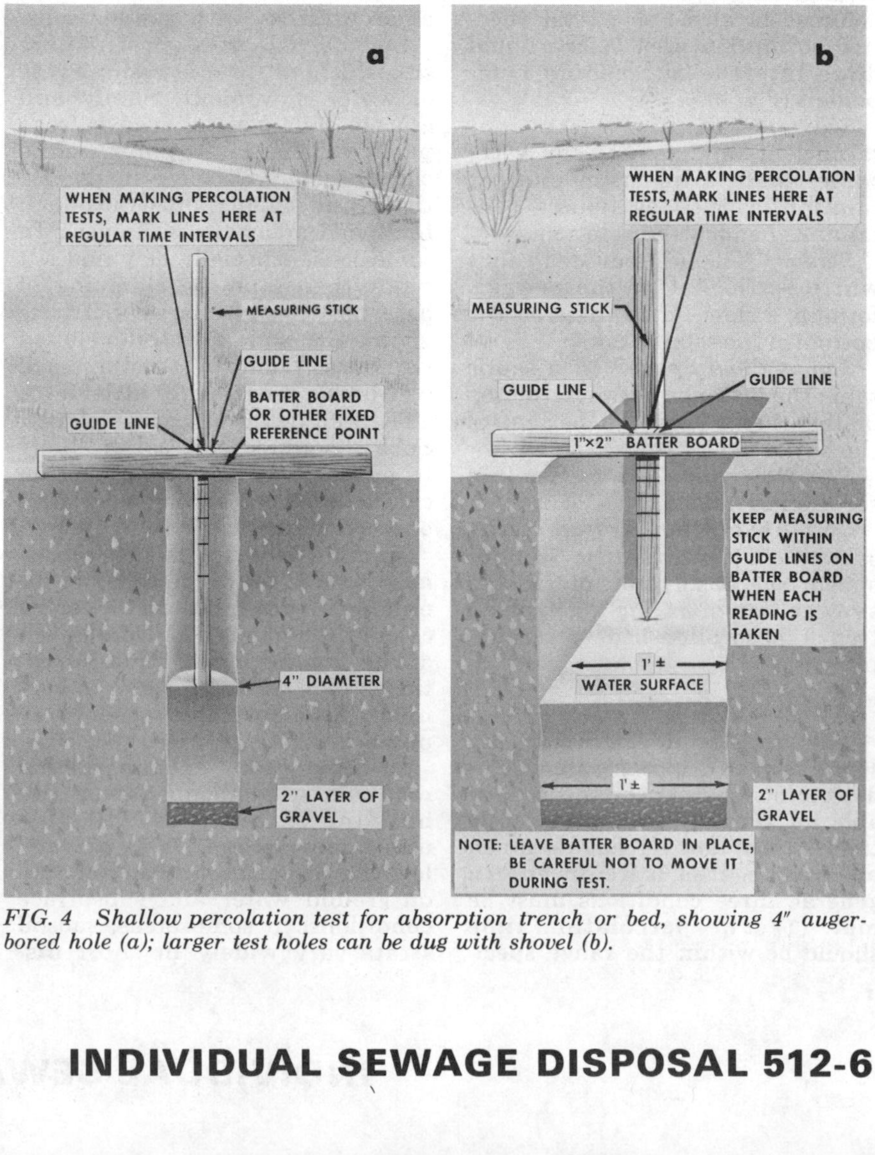

FIG. 4 Shallow percolation test for absorption trench or bed, showing 4″ auger-bored hole (a); larger test holes can be dug with shovel (b).

gravel. From a fixed reference point, the drop in water level should be measured at 30-minute intervals for 4 hours, refilling up to 6″ over the gravel as necessary. The drop that occurs during the final 30-minutes should be used to calculate the percolation rate.

In sandy soils (or other soils in which the first 6″ of water seeps away in less than 30 minutes, after the overnight swelling period), the measurements should be taken at 10-minute intervals and the test run for one hour. The drop that occurs during the final 10 minutes is used to calculate the percolation rate.

SOIL ABSORPTION SYSTEMS

From the results of the percolation tests it is possible to determine whether a septic system is feasible and which type of soil absorption system can be used for the area in question. As shown in Figure 2, only soils with a percolation rate of 30 minutes per inch or less are suitable for seepage pits; if the rate is greater than 60 minutes, the soil is not suitable for any type of absorption system.

When a soil absorption system is usable, three types of design may be considered: *absorption field, absorption bed* or *seepage pit.*

The selection of the absorption system is dependent to some extent on the location of the system in the area under consideration. A safe distance must be maintained between the site and any source of water supply. Since the distance that pollution will travel underground depends upon numerous factors, including the characteristics of the subsoil formations and the quantity of sewage discharged, no specified distance would be absolutely safe in all localities. In general, the horizontal separation between elements of the soil absorption system and other site improvements should not be less than that outlined in Figure 6.

Seepage pits should not be used in areas where water supplies are obtained from shallow wells, or where there are limestone formations and sinkholes through which the effluent may easily contaminate the ground water.

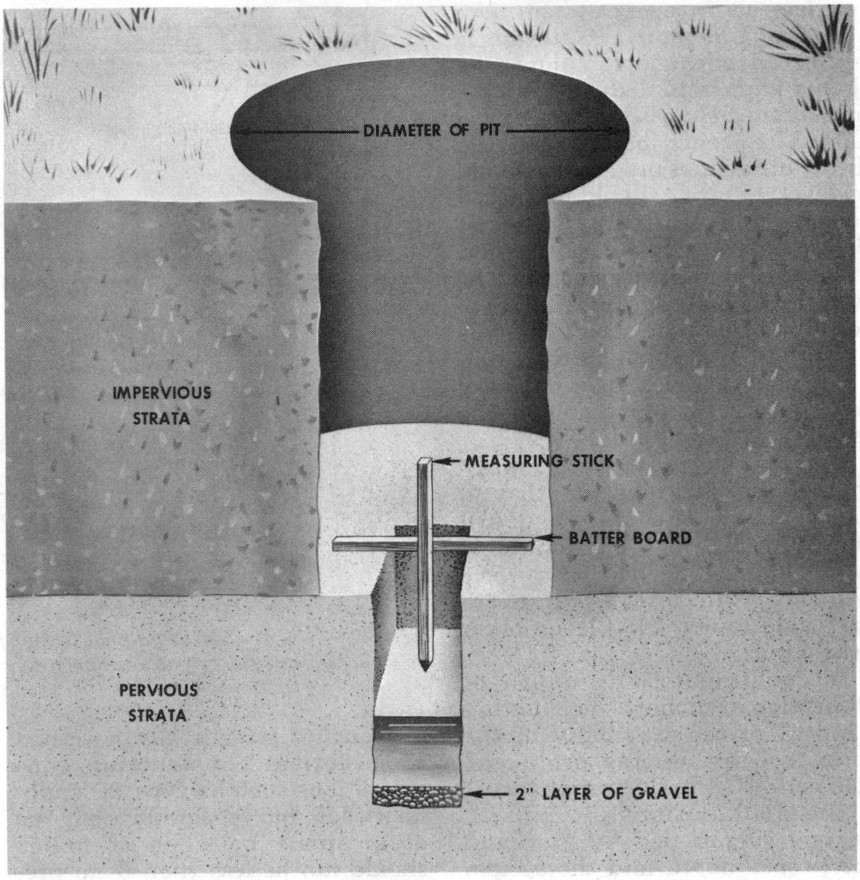

FIG. 5 Percolation tests for seepage pits must be made in deep test holes dug down to the level of the proposed pit bottom.

Absorption Field

An absorption field consists of a series of absorption trenches inside of which are laid effluent distribution lines. The distribution line may consist of 1′ lengths of 4″ agricultural drain tile, 2′ to 3′ lengths of vitrified clay sewer pipe (both laid with open joints), or perforated plastic pipe. In areas having unusual soil or water characteristics, local experience should be reviewed before selecting piping materials.

The individual laterals (branches) should not be over 100′ long and there should be at least

FIG. 6 MINIMUM CLEARANCES FOR COMPONENTS OF A SEPTIC SEWAGE DISPOSAL SYSTEM[1]

FROM,	TO WELL[2]	TO WATER LINE[3]	TO FOUNDATION WALL	TO PROPERTY LINE	TO STREAM
Building Sewer	50′	10′	—	—	50′
Septic Tank	50′	10′	5′	10′	50′
Absorption Field	100′	25′	20′	5′	50′
Absorption Bed	100′	25′	20′	5′	50′
Seepage Pit	100′	50′	20′	10′	50′

[1] Based on recommendations of U.S. Public Health Service, HEW.
[2] And to suction line between well and pressure tank.
[3] Pressure type.

two laterals in every system. Use of more and shorter laterals is preferred because if something should happen to disturb one line, most of the field will still be serviceable. The trench bottom and tile distribution lines preferably should be level. Laterals should be spaced so as to provide at least 6′ of undisturbed earth between trenches.

Many different designs may be used in laying out subsurface absorption fields. The choice depends on the size and shape of the available disposal area, the capacity required and the topography of the area (Figs. 7 and 9).

Design To provide the minimum required gravel depth and earth cover, the trench depth should be at least 24″. Additional depth may be needed for surface contour adjustment, for extra aggregate under the tile or other design purposes.

In considering the depth of absorption trenches, the question is raised of the possibility of the lines freezing during prolonged cold periods. Freezing rarely occurs if the distribution line is adequately surrounded by gravel at the proper depth and the system is kept in continuous operation. However pipes under driveways or other surfaces which are usually cleared of snow should be close-jointed and insulated or should be installed at a greater depth.

The required absorption area per bedroom is determined by the results of the soil percolation tests, and may be obtained from Figure 2. (Ideally, the house site should be large enough to allow room for an additional system if the first one fails or if bedrooms are added later.) Thus for a 3-bedroom house on a lot where the minimum percolation rate was 15 minutes per 1″, the necessary absorption area would be 3 x 190 sq. ft. per bedroom, or 570 sq. ft. For trenches 2′ wide, the required total length of trench would be 570 ÷ 2, or 285 linear feet. If this were divided into 3 laterals as if for sloping ground (Fig. 9), the length of each lateral would be 285 ÷ 3, or 95′. In a closed loop system on level ground (Fig. 7), each of the half loops could be 100′, and the middle lateral 85′. The spacing of trenches is generally governed by practical construction considerations such as type of equipment, soil and topography, but the clear space between trenches should not be less than 6′ to prevent short circuiting (flow between trenches).

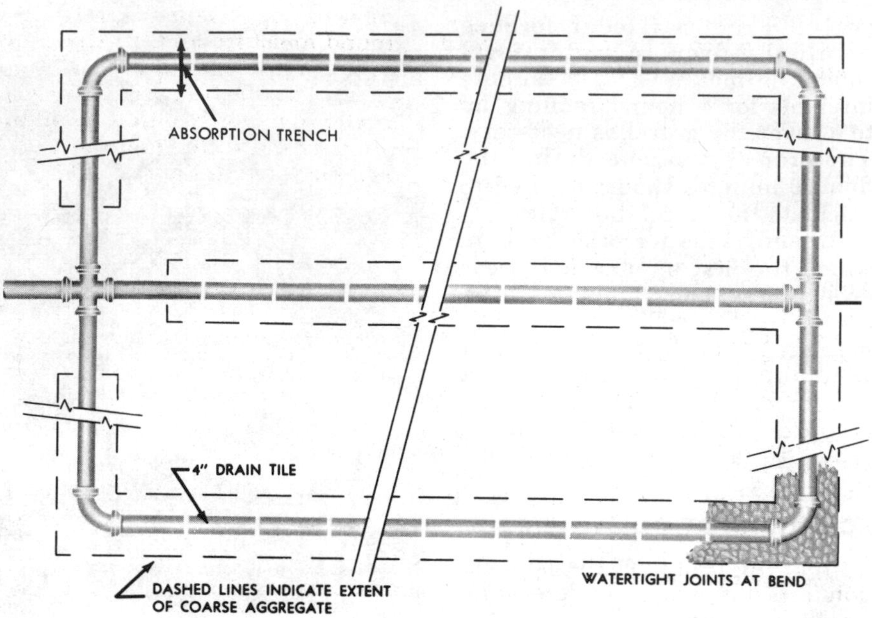

FIG. 7 Typical absorption field layout for level ground. Length of lateral should be figured from start of absorption trench (dashed lines) to furthest point.

In the example cited, trenches are 2′ wide, there are 3 trenches, and 2 spaces between trenches at 6′ each. The total width of trenches and spaces is 6′ plus 12′, equals 18′. Required area for the absorption field is 18′ times 95′, equals 1710 sq. ft., plus additional land required to provide the clearances recommended in Figure 6.

Construction Careful construction is important in obtaining a satisfactory soil absorption system. Attention should be given to the protection of the natural absorption properties of the soil. Care must be taken to prevent sealing of the surface on the bottom and sides of the trench. Trenches should not be excavated when the soil is wet

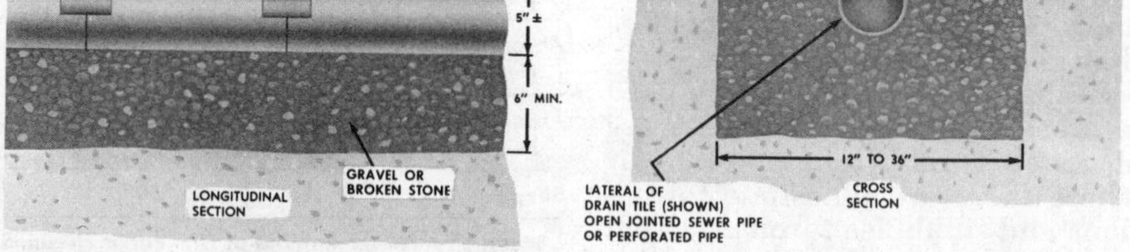

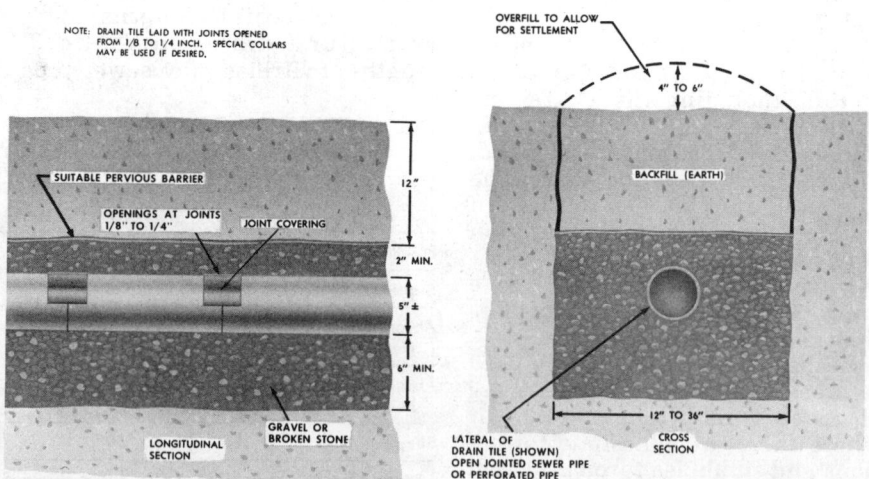

FIG. 8 Details of typical absorption trench.

enough to smear or compact easily. Soil moisture is right for safe working only when a handful will mold with considerable pressure. Open trenches should be protected from surface runoff to prevent the entrance of silt and debris. If it is necessary to walk in the trench, a temporary board laid on the bottom will reduce compaction. If smearing and compaction does occur, surfaces should be raked to a depth of 1″, and loose material removed, before the gravel is placed in the trench.

The distribution pipe should be laid in a level trench of sufficient width and depth (Fig. 8), and should be surrounded by clean, graded gravel or rock, broken hard-burned clay brick or similar aggregate, ranging in size from 1/2″ to 2-1/2″. Cinders, broken shell and similar material are not recommended, because they are usually too fine and may lead to premature clogging. The coarse aggregate should extend from at least 2″ above the top of the pipe to at least 6″ below the bottom of the pipe. If open joint drain tile is used, the upper half of each joint opening should be covered with a strip of 15 lb. roofing paper, as shown in Figure 8.

The top of the stone should be covered with untreated building paper, a 2″ layer of hay or straw, or similar pervious material to prevent the stone from becoming clogged by the earth backfill. An impervious covering should not be used, as this interferes with *evapotranspiration* (absorption and release of moisture by plants) at the surface. Although generally not figured in the calculations, evapotranspiration often is an important factor in the operation of absorption fields and beds.

Open joint tile generally is laid with the opening covered by strips of roofing paper, but drain tile connectors, collars, clips or similar accessories are of value in obtaining uniform spacing, and protection of tile joints. Such accessories commonly are made of galvanized iron, copper or plastics.

Tree and shrub roots sometimes penetrate into the tile and block the flow of effluent. Root problems can be prevented by using a liberal amount of gravel or stone over and under the tile. Roots generally seek ground moisture and, in the small percentage of cases where they become troublesome, there is usually some explanation involving changing moisture conditions. For example, at a residence which is used only during the summer, roots are most likely to penetrate when the house is uninhabited, or when moisture immediately below or around the gravel becomes less plentiful than during the period when the system is in use. In general, trenches constructed within

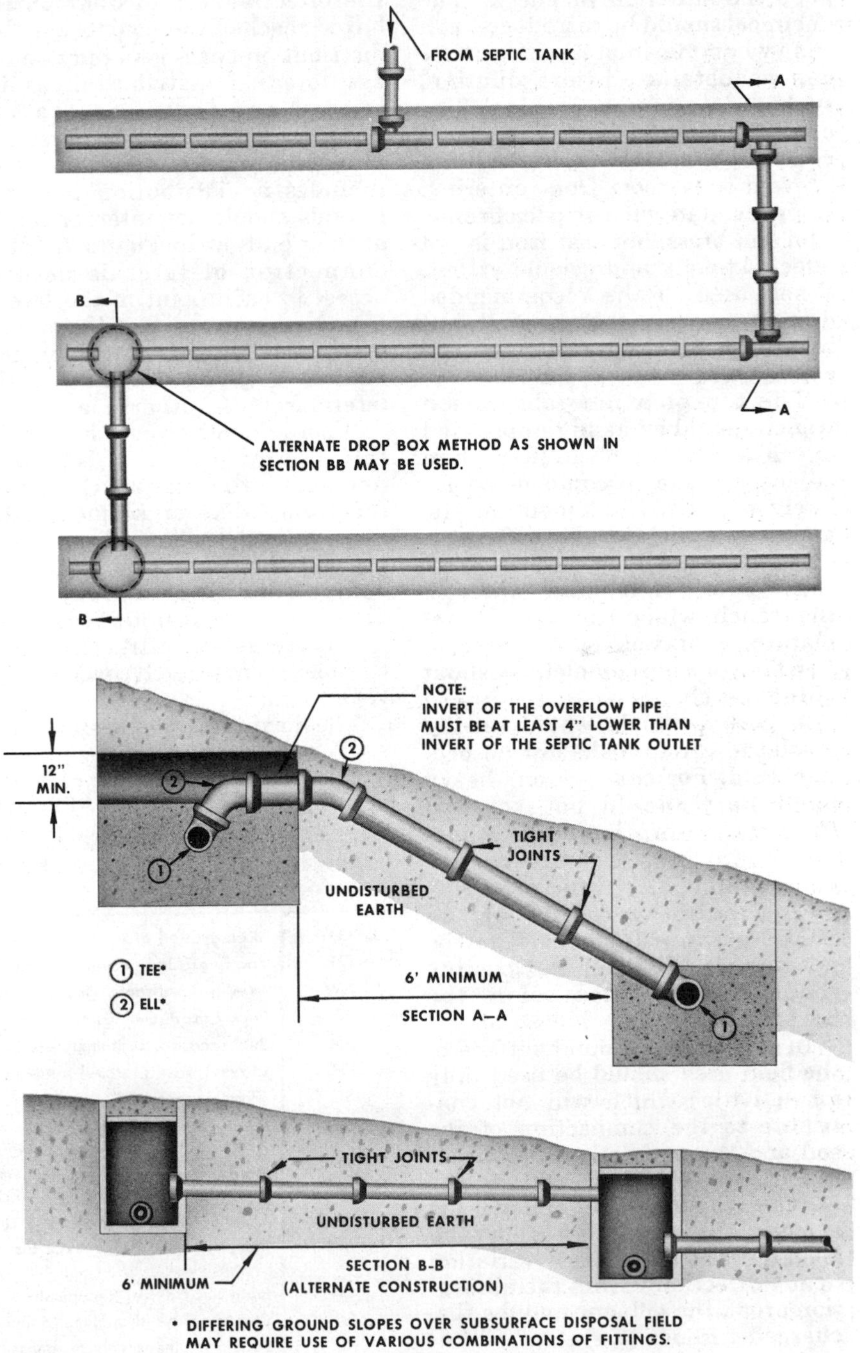

FIG. 9 Typical absorption field layout for sloping ground.

10′ of large trees or dense shrubbery should have at least 12″ of gravel or crushed stone beneath the tile.

Where clogging due to root penetration is suspected or anticipated, about 2 lbs. or 3 lbs. of copper sulfate crystals flushed down the toilet bowl once a year usually will keep the situation in check. The chemical should be introduced at a time when maximum contact time can be obtained before dilution, such as late in the evening. The chemical destroys any roots it comes in contact with, but will not prevent new roots from entering. Copper sulfate will corrode chrome, iron and brass, but cast iron is not affected to any appreciable extent. When used in the recommended dosage, copper sulfate will not interfere with operation of the septic tank.

The top of a new absorption trench should be hand tamped and overfilled 4″ to 6″ higher than adjacent grade to compensate for likely settling. Settlement of the fill in the trench to a lower level is very undesirable because it may cause collection of storm water in the trench, which can lead to premature saturation of the absorption field or even complete washout of the trench.

A heavy vehicle would readily crush the tile in a shallow absorption field. For this reason, heavy machinery should be excluded from the absorption field unless special provision is made to support the weight. Machine tamping or hydraulic backfilling of the absorption trenches should not be permitted. All machine grading should be completed before the field is laid.

In general the ground surface of the field area should be used only for activities which will not contribute to the compaction of the soil and the consequent reduction in porosity.

Effluent Distribution Where the slope of the ground is essentially level (does not exceed 6″ variation in any direction within the absorption area), the effluent may be discharged through a system of laterals connected at their extremes to form a continuous closed loop (Fig. 7). Such a system permits the effluent to flow into all the tile lines whenever there is a discharge from the septic tank.

Laterals traditionally were connected to the house sewer through a *distribution box* and the ends were not interconnected. A Public Health Service study for FHA on the effectiveness of distribution boxes, reached the conclusion that distribution boxes generally do not assure equal distribution as intended and therefore are not needed for single-family systems. However where local practice includes a distribution box, the laterals should be interconnected at their ends as in Figure 7. Interconnection of laterals assures access to each point in the lateral from two sides, so that if one part of the lateral became damaged, effluent would be able to reach the lateral from the other side.

Where sloping ground is used for the absorption field, it is usually necessary to construct a small temporary dike or surface water diversion ditch above the field, to prevent the absorption field from being eroded by surface runoff. The dike should be maintained or the ditch kept free of obstructions until the field becomes well covered with vegetation.

The maximum ground slope suitable for soil absorption systems should be governed by local factors affecting the erosion of the ground used for the absorption field. Excessive slopes which are not protected from surface water runoff or do not have adequate vegetation cover to prevent erosion should be avoided. Generally, ground having a slope greater than one vertical to two horizontal should be investigated carefully to determine if satisfactory from the erosion standpoint. Near banks, the horizontal distance from side of the trench to the ground surface should be 2′ or such larger dimension which would prevent lateral flow of effluent and breakout on surface.

In sloping ground each lateral should be connected to the next by a closed overflow line laid on an undisturbed section of ground, as shown in Figure 9. The arrangement is such that all effluent is discharged to the first trench until it is filled; excess liquid then is carried by means of the closed overflow line to the next succeeding or lower trench. In that manner, each portion of the subsurface system is used in succession.

Design and construction procedures for sloping ground fields generally are the same as for level ground. The bottom of each trench and its distribution line should be

FIG. 10 **REQUIRED WALL AREA FOR SEEPAGE PITS**[1]

SYMBOL	CHARACTER OF SOIL	MINIMUM WALL AREA PER BEDROOM
GW	Well-graded gravels, gravel-sand mixtures, little or no fines.....	50
GP	Poorly graded gravels or gravel-sand mixtures, little or no fines..	50
SW	Well-graded sands, gravelly sand, little or no fines............	75
SP	Poorly graded sands or gravelly sands, little or no fines........	75
SM	Silty sand, sand-silt mixtures..............................	125
GC	Clayey gravels, gravel-sand-clay mixtures..................	150
SC	Clayey sands, sand-clay mixtures...........................	150
ML	Inorganic silts and very fine sands, rock flour, silty or clayey fine sands or clayey silts with slight plasticity............	300
CL	Inorganic clays of low to medium plasticity, gravelly clays, sandy clays, silty clays, lean clays........................	350
OL	Organic silts and organic silty clays of low plasticity...........	UNSUITABLE FOR SEEPAGE PITS
MH	Inorganic silts, micaceous or diatomaceous fine sandy or silty soils, elastic silts..........................	
CH	Inorganic clays of high plasticity, fat clays..........	
OH	Organic clays of medium to high plasticity, organic silts.........	
Pt	Peat and other highly organic silts........................	

[1] According to FHA's Uniform Soils Classification System.

level with the laterals generally following surface contours. Adjacent trenches may be connected either with the overflow line or a drop box arrangement. The invert (bottom) of the first overflow line should be at least 4″ lower than the invert of the septic tank outlet (Fig. 9).

Overflow lines should be 4″, tight-joint pipe with direct connections to the distribution lines in adjacent trenches or to a drop box arrangement. Care must be exercised in constructing overflow lines to insure at least 6′ of undisturbed earth between trenches. The overflow line should rest on undisturbed earth and backfill should be carefully tamped.

Absorption Beds

An absorption bed is a variation of the typical trench used in absorption fields. The basic design is essentially the same as for an absorption trench, but it is generally much wider than 36″ and contains two or more distribution lines, 4′ to 6′ apart. Absorption beds have the advantages of making more efficient use of the land and of modern earth moving equipment employed at most housing projects.

The bottom absorption area for absorption beds is the same as for trenches in absorption fields (Fig. 1). The bed should have a minimum depth of 24″ below natural ground level to provide a minimum of 12″ gravel and 12″ earth backfill. The gravel should extend at least 2″ above and 6″ below the distribution line. The bottom of the bed and bottom of distribution line should be level. Distribution lines should be spaced not more than 6′ apart and not more than 3′ from the bed sidewall. At least 2 distribution lines should be used in a bed. When more than one bed is used there should be a minimum of 6′ of undisturbed earth between adjacent beds, and the beds should be connected "in series" as illustrated in Figure 9 for absorption fields in sloping ground.

Seepage Pits

Seepage pits, being considerably deeper than absorption trenches or beds, present a greater hazard of contaminating ground waters than other absorption methods. Since pits handle a larger amount of effluent in a limited land area, the possibility of clogging and saturation is greater—increasing the hazard of effluent breaking through to the surface.

Pits generally should not be used where absorption trenches or beds are feasible. Where the top 3′ or 4′ of soil is not porous enough for an absorption field, but is underlaid with suitable granular material, seepage pits constitute an acceptable substitute. Seepage pits may also be used where the land area is too small, too irregular or heavily wooded to accommodate trench or bed absorption systems.

In deep pits, it is important that the required pit capacity be computed on the basis of percolation tests made in each vertical stratum penetrated (Fig. 5). Soil strata in which the percolation rates are in excess of 30 minutes per inch should not be included in computing the absorption area. The weighted average of the results should be computed to obtain an average percolation rate. In shallow pits where the soil is relatively uniform throughout the depth of the pit, the required absorption area can be established by soil experts according to visual analysis of the soil type (Fig. 10). In other cases, the absorption area should be established by percolation tests as for other absorption methods.

Design Figure 2 gives the absorption area requirements per bedroom for soils with different percolation rates. The absorption area of the seepage pit is the vertical wall area of the actual excavated pit (not the pit lining). Only the area of pervious strata below the inlet should be figured; no allowance should be made for impervious strata or for bottom area. Having established the required area, Figure 11 may be used to select the desired dimensions of circular or cylindrical seepage pits.

To demonstrate a typical pit capacity calculation, let us again assume that the absorption system is to be designed for a 3-bedroom home on a lot where the percolation rate is 1″ in 15 minutes. Assume also that the water table does not rise above 17′ below the ground surface, that seepage pits with effective depth of 10′ can be provided, and that the house is in a locality where it is common practice to install seepage pits of 5′ diameter (i.e., 4′ to the outside walls, which are surrounded by about 6″ of gravel).

According to Figure 2, 570 (3 x 190) sq. ft. of absorption area would be needed. Referring to Figure 11, it can be determined that a 5′ round pit 10′ deep would provide 157 sq. ft. of wall area. Dividing 570 by 157, we get 3.6, that is, 3 pits 10′ deep and one 6′. If it is desired to make all pits the same size, four 5′ round pits 9′ deep can be selected from Figure 11, providing 4 x 141 or 564 sq. ft. of wall area—close enough to the total required. Small diameter, relatively deep pits, require the re-

FIG. 11 **EFFECTIVE WALL ABSORPTION AREAS (SQ. FT.) OF ROUND SEEPAGE PITS**

PIT DIAMETER	DEPTH BELOW INLET									
	1 FT.	2 FT.	3 FT.	4 FT.	5 FT.	6 FT.	7 FT.	8 FT.	9 FT.	10 FT.
3′	9.4	19	28	38	47	57	66	75	85	94
4′	12.6	25	38	50	63	75	88	101	113	126
5′	15.7	31	47	63	79	94	110	126	141	157
6′	18.8	38	57	75	94	113	132	151	170	188
7′	22.0	44	66	88	110	132	154	176	198	220
8′	25.1	50	75	101	126	151	176	201	226	251
9′	28.3	57	85	113	141	170	198	226	254	283
10′	31.4	63	94	126	157	188	220	251	283	314
11′	34.6	69	104	138	173	207	242	276	311	346
12′	37.7	75	113	151	188	226	264	302	339	377

moval of less soil than larger pits with equivalent wall area, but may be harder to excavate. When this is the case, larger pits can be selected—for instance, two 10′ round pits 9′ deep, or two 13′ round pits 7′ deep (Fig. 11).

Where more than one pit is provided, they can be connected in series or in parallel. However, they should be separated by a distance equal to 3 times the diameter of the largest pit; for pits over 20′ in depth, the minimum distance between pits should be 20′ (Fig. 12). The area of the lot on which the house is to be built should be large enough to maintain this distance between the pits while still allowing room for additional pits if the first ones should fail. If this cannot be done, other sewerage facilities should be explored.

Construction The permeability of the soil may be damaged during excavation. Digging in wet soils should be avoided as much as possible; cutting teeth on mechanical equipment should be kept sharp. Bucket auger pits should be reamed to a larger diameter than the bucket; all loose material should be removed from the excavation.

Pits should be backfilled with clean gravel to a depth of 1′ above the pit bottom or 1′ above the reamed ledge to provide a sound foundation for the lining. Lining materials for seepage pits can be clay or concrete masonry units in brick, block or ring shapes. Rings should have weep holes or openings to provide for seepage. Concrete brick and block should be laid dry (unmortared) with staggered, slightly open joints; hollow concrete block frequently is laid on the side with cores horizontally to facilitate soil absorption. Standard clay brick should be laid flat to form a 4″ wall. The outside diameter of the lining should be at least 1′ smaller than the excavation diameter, to provide at least 6″ (preferably 1′) of annular space to be filled with coarse gravel to the top of the lining (Fig. 13).

Either a corbelled brick dome or a flat concrete cover can be used over the pit. Corbelled bricks should be laid either in cement mortar or have a 2″ covering of concrete. If a flat cover is used, a prefabricated type is preferred, and it should be reinforced to be equivalent in strength to an approved septic tank cover. The cover should rest on undisturbed ground 12″ beyond the lining. A 9″ capped opening in the pit cover should be provided for pit inspection.

Connecting lines should be of a sound durable material, the same as used for the house-to-septic tank connection. All connecting lines should be laid on a firm bed of undisturbed soil throughout their length. The grade of a connecting line should be at least 2% (1/4″ in 1′). The pit inlet pipe should extend horizontally at least 1′ into the pit with a tee or ell to divert flow downward to prevent washing and eroding of the sidewalls.

Abandoned seepage pits should be filled with earth or gravel to prevent cave-ins.

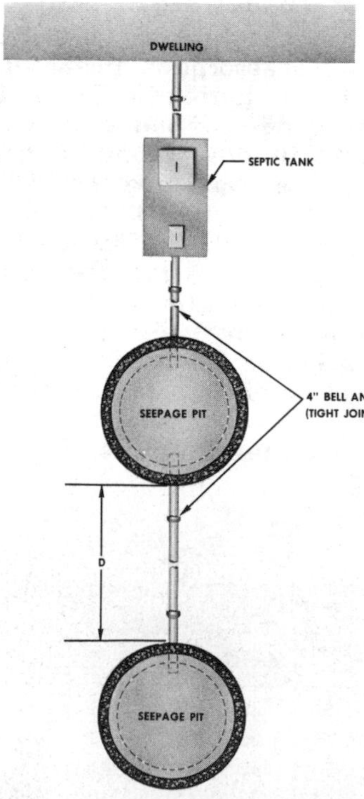

FIG. 12 Typical septic tank system with 2 seepage pits connected in series. Distance "D" between pits should be 3 times pit diameter and at least 20′ for pits deeper than 20′.

SEPTIC TANKS

The septic tank is the first component in a private sewage disposal system, which receives the sewage from the building sewer (Fig. 1). Despite its physical position in the system, it is discussed here—after absorption systems—because the determination must first be made whether the soil is suitable for a subsurface absorption system, and whether adequate area is available for the type of system intended. Except for the choice of tank capacity and component materials to be made by the system designer, the options are relatively few. Interior design of the tank is generally a matter of local practice and if approved by the local health board can be specified with confidence.

Functions

A septic tank performs three basic functions: (1) removal of solids; (2) sludge and scum storage, and (3) biological treatment. The chief function is separation of solids, but the other functions are essential also, if the system is to be more than a holding tank, requiring frequent pumping. Biological treatment of the organic matter contained in the sludge and scum assures that only a small portion of solids—undigestable matter—remains in the tank permanently between pumpings.

Removal of Solids Clogging of soil with tank effluent varies directly with the amount of suspended solids in the liquid. As sewage from a building sewer enters a septic tank, its rate of flow is reduced so that larger solids sink to the bottom or rise to the surface. These solids are retained in the tank, where most of them are gradually broken down by bacteria into water soluble components, which do not clog the soil.

Sludge and Scum Storage The heavier sewage solids settle to the bottom of the tank, forming a blanket of sludge. The lighter solids, including fats and greases, rise to the surface and form a layer of scum. A considerable portion of the sludge and scum is liquefied through decomposition or digestion. During this process, gas is

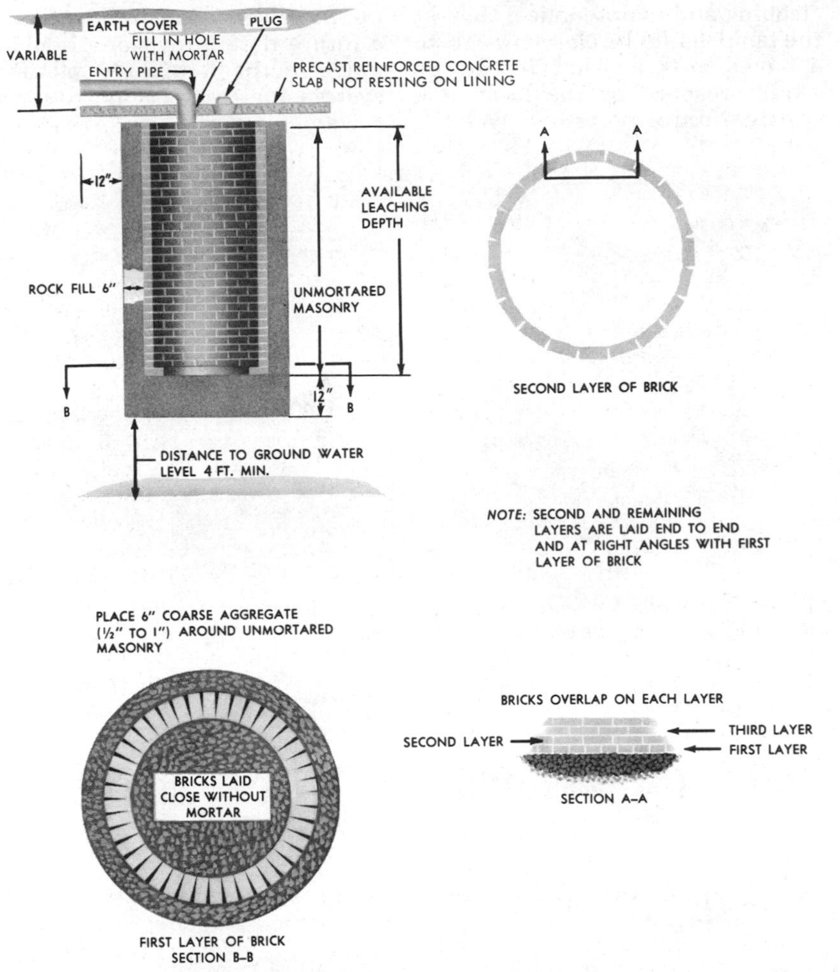

FIG. 13 Round seepage pits may be constructed of unmortared concrete block or clay brick masonry. Brick is laid normally, block is laid sideways, with cores facing out.

liberated from the sludge, carrying a portion of the solids to the surface, where they accumulate with the scum. Ordinarily, they undergo further digestion in the scum layer, and a portion settles again to the sludge blanket on the bottom. The settling action is retarded by grease in the scum layer and by gasification in the sludge blanket.

Biological Treatment Solids and liquid in the tank are subjected to decomposition by bacterial and natural processes. The bacteria are of a variety called anaerobic (oxygen-free) which thrive in the absence of oxygen. This decomposition or treatment of sewage under oxygen-free conditions is termed "septic," hence the name of the tank.

Contrary to popular belief, septic tanks do not accomplish a high degree of bacterial removal. Although the sewage undergoes treatment while passing through the tank, pathogenic (disease-producing) organisms are not removed. Hence, septic tank effluents cannot be safely discharged on the surface, into a ditch, lake or stream. This, however, does not detract from the value of the tank. As previously explained, a tank's primary purpose is to condition the sewage to prevent rapid clogging of the subsurface absorption system.

Further treatment of the effluent, including the removal of pathogenic organisms, is effected by percolation through the soil. These organisms slowly die out as their nutrients are used up and as they enter the unfavorable environment afforded by the soil.

Tank Location

Although septic tanks are supposed to be watertight when installed, some leakage may develop, from openings in the inlet and outlet pipes or cracks in the tank, as the installation ages. Hence, septic tanks should be located where they are not likely to contaminate any well, spring or other source of water supply. Underground contamination may travel for considerable distances, usually in the direction of the water table gradient—that is, from a higher to a lower water table area. In general, water tables follow the ground surface contours; hence, septic tanks preferably should be located downhill from private wells or natural springs.

However, locating the tank at the lower surface elevation than the well is no guarantee against possible contamination. Sewage from soil absorption systems occasionally contaminates wells having higher surface elevations. This happens because elevations of subsurface absorption systems are almost always higher than the level of water in wells which may be located nearby, and contamination may travel downward to the water bearing stratum (Fig. 14). It is

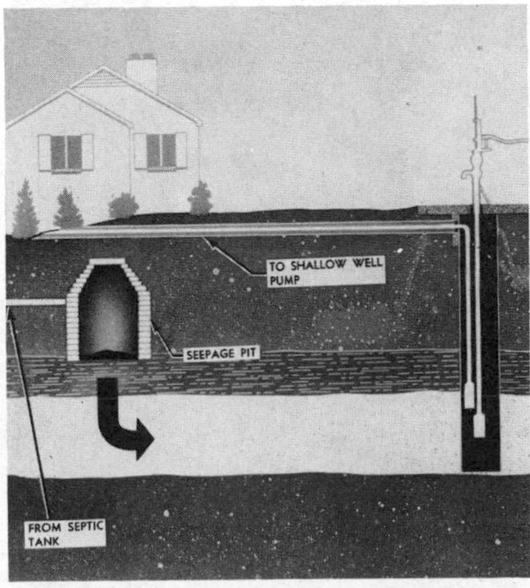

FIG. 14 A seepage pit can contaminate private water supply, if it is inadequately separated from the well or the water bearing stratum.

necessary, therefore, to rely upon horizontal as well as vertical distances for protection. Tanks should never be closer than 50′ from any source of water supply; greater distances are preferred where possible (Fig. 6).

The septic tank should not be located within 5′ of any building foundation, as structural damage may result during construction or seepage may enter the basement. The tank should not be located in swampy areas, nor in areas subject to flooding. Ideally, the tank should be located where the largest possible area will be available for the absorption system. Consideration should also be given to the location from the standpoint of cleaning and maintenance; that is, the tank should be close enough to a paved area so that it can be readily reached by the hose of a septic pumping truck. Where public sewers may be installed at a future date, provision should be made in the household plumbing system for connection to such a sewer.

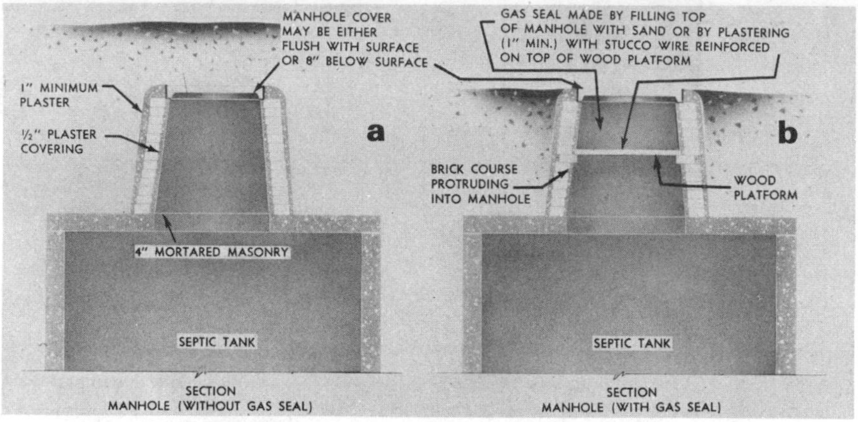

FIG. 17 Manholes for septic tank access should be provided either with (a) 8″ of fill over cover, or (b) gas seal under the manhole cover.

FIG. 15 RECOMMENDED SEPTIC TANK CAPACITIES[1]

NUMBER OF BEDROOMS	MINIMUM TANK CAPACITY (GAL.)
3 or less	900
4	1,000
Additional Bedrooms	Add 250 gal./bedroom

[1] Provides for use of garbage grinder, dishwasher, automatic clothes washer, and other household appliances.

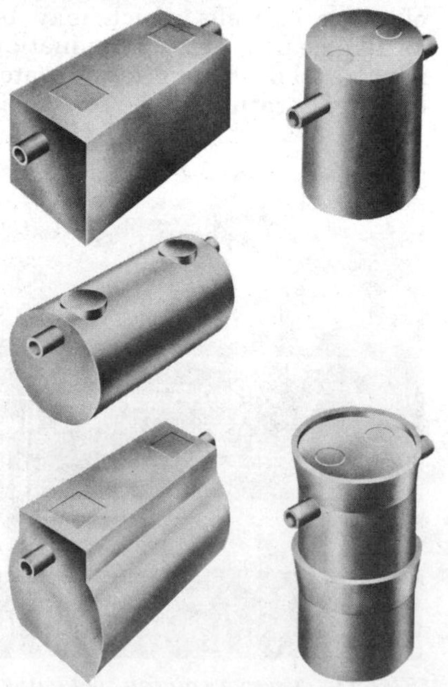

FIG. 16 Septic tanks typically are precast concrete or welded sheet steel in several basic shapes.

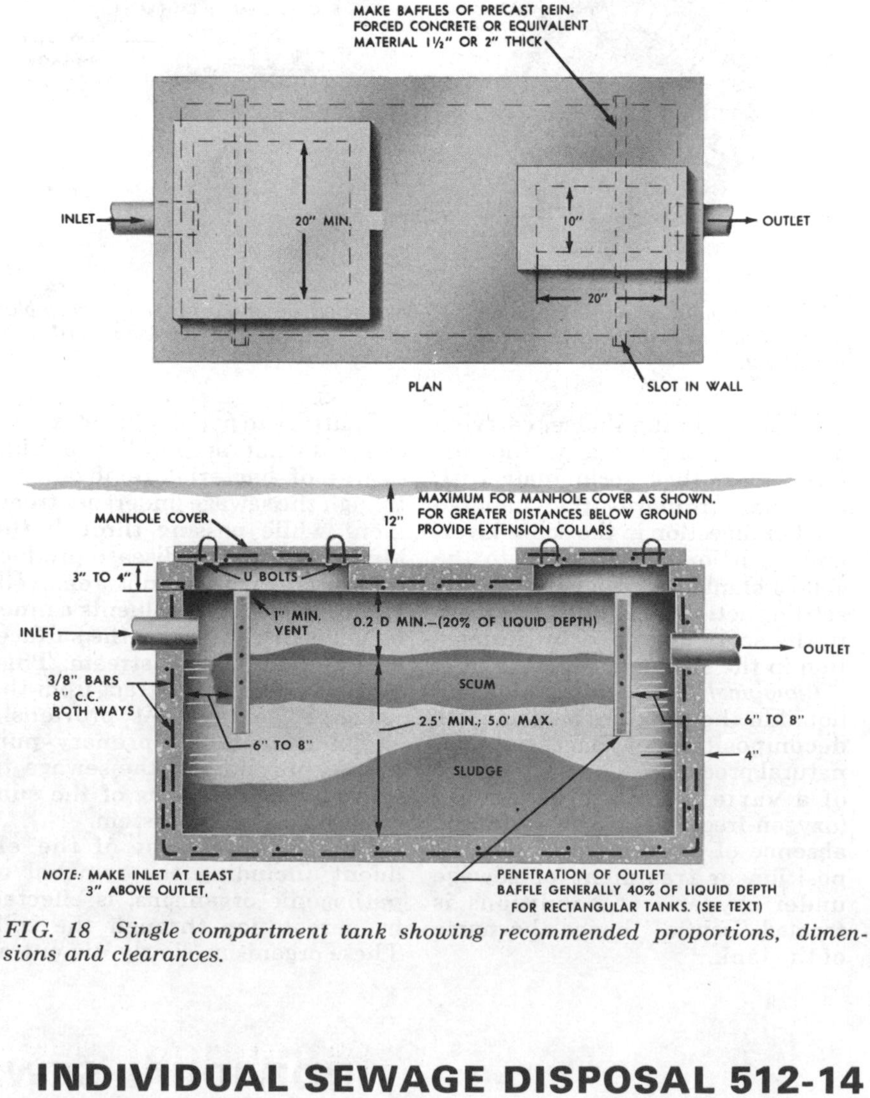

FIG. 18 Single compartment tank showing recommended proportions, dimensions and clearances.

Tank Design

Grease interceptors (grease traps) usually are not required for household sewage disposal systems. However, when provided, the discharge from a waste disposer (garbage grinder) should never be passed through them. The septic tank capacities recommended in this discussion are sufficient to receive the grease and solids normally discharged from a typical home.

It is generally advisable to have all sanitary wastes from a household discharge to a single septic disposal system. For household installations, it is also usually more economical to provide a single disposal system than two or more with the same total capacity. Normal household waste, including that from the laundry, bath and kitchen, should pass into a single system.

Capacity Adequate capacity is one of the most important considerations in septic tank design. Larger tank capacities increase the time that sewage is retained in the tank, assuring better removal of solids and biological treatment. With improperly sized tanks, the hazard exists of overloading the tank to the point that inadequately treated sewage will run into the absorption system and clog it. The liquid capacities recommended in Figure 15 allow for the use of all household appliances, including automatic washing machines, dishwashers and waste disposers. They also insure enough excess capacity to retain the sewage a sufficient time to settle out and digest solids.

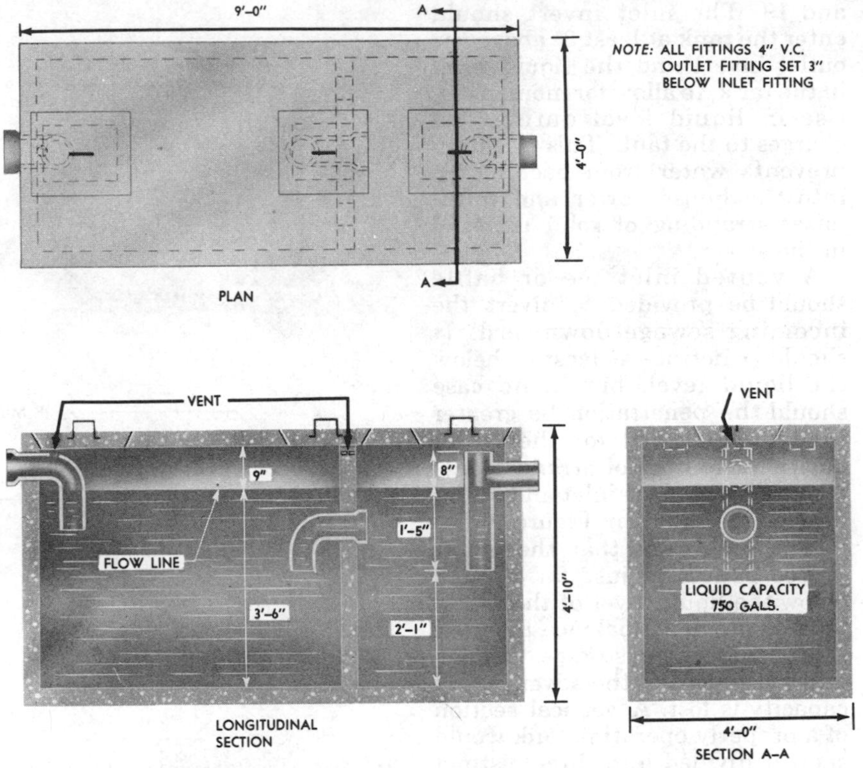

FIG. 19 Typical design of double compartment precast concrete tank. Dimensions shown are for 750 gallon tank no longer considered adequate for the average home.

Materials and Construction

Septic tanks should be watertight and constructed of corrosion- and decay-resistant materials, such as concrete, coated metal, vitrified clay, heavyweight concrete masonry units or hard burned bricks. Precast and cast-in-place reinforced concrete tanks are common in all parts of the country. Prefabricated steel tanks meeting Commercial Standard CS177 are acceptable in some states.

Special attention should be given to job-built tanks to insure watertightness and prevent contamination. Heavyweight concrete block should be laid on a solid foundation and all mortar joints should be well filled. The interior of the tank should be surfaced with two 1/4" thick coats of portland cement-sand plaster. Some typical septic tanks are illustrated in Figure 16.

Precast concrete tanks should have a minimum wall thickness of 3", and should be adequately reinforced to facilitate handling. When precast slabs are used as covers, they should be watertight, have a thickness of at least 3", and be adequately reinforced, so as to be capable of supporting a dead load of 300 lbs. per sq. ft.

Backfill and Access Backfill around septic tanks should be made in thin layers thoroughly tamped in a manner that will not produce undue strain on the tank. Backfill may be consolidated with the use of water, provided the backfill is placed and wetted in layers and the tank is first filled with water to prevent floating.

Adequate access should be provided to each tank compartment for inspection and cleaning. Both the inlet and outlet devices should be accessible. Access to each compartment can be provided by either a removable cover or a manhole at least 20" wide. Where the top of the tank is located more than 18" below the finished grade, manholes and inspection holes should extend to approximately 8" below the finished grade (Fig. 17a), or can be extended to finished grade if a *gas seal* is provided to keep odors from escaping. A gas seal is made by filling the top of the manhole with sand or by sealing with a minimum of 1" of portland cement plaster over an intermediate wood platform (Fig. 17b). In most instances, the extension can be made using clay or concrete pipe, but proper attention must be given to the accident hazard involved when manholes are extended close to the ground surface.

Tank Construction Typical single and double compartment tanks are illustrated in Figures 18 and 19. The inlet invert should enter the tank at least 3″ above the outlet invert and the liquid level in the tank, to allow for momentary rise in liquid level during discharges to the tank. This free drop prevents water from backing up into the house sewer and minimizes stranding of solid material in the sewer.

A vented inlet tee or baffle should be provided to divert the incoming sewage downward. It should penetrate at least 6″ below the liquid level, but in no case should the penetration be greater than that allowed for the outlet device. A number of arrangements commonly used for inlet and outlet devices are shown in Figure 20.

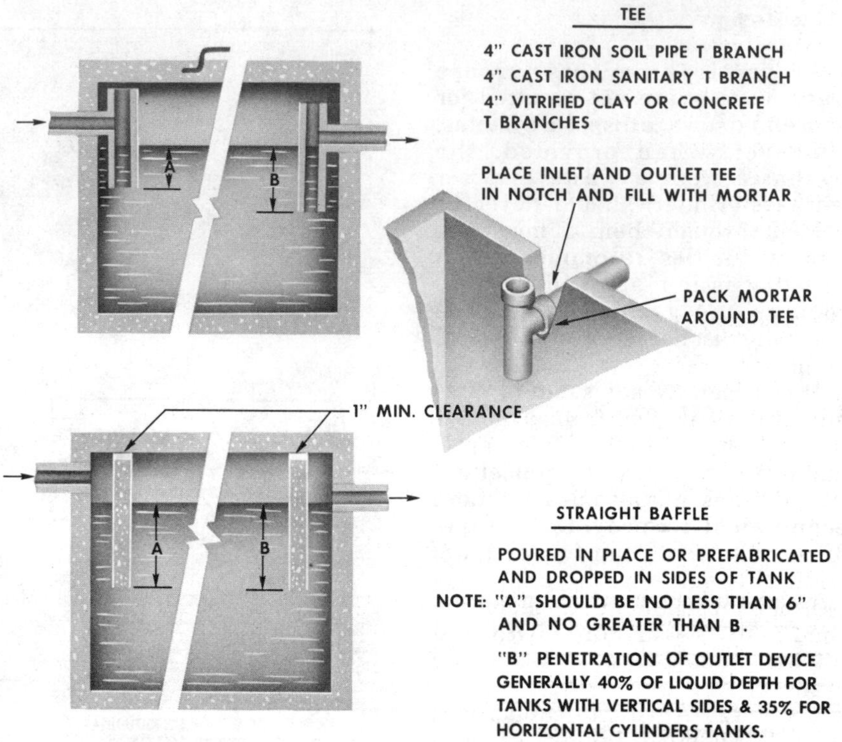

FIG. 20 Acceptable inlet and outlet devices for septic tanks.

It is important that the outlet device penetrate just far enough below the liquid level of the septic tank to provide a balance between sludge and scum storage volume; otherwise, part of the advantage of capacity is lost. A vertical section of a properly operating tank would show it divided into three distinct layers; scum at the top, a middle zone free of solids, and a bottom layer of sludge. The outlet device retains scum in the tank, but at the same time, it limits the amount of sludge that can be accommodated without scouring, which results in sludge discharging in the effluent from the tank. Field studies of sludge accumulation in septic tanks indicate that the outlet device should generally extend to a distance below the surface equal to 40% of the liquid depth. For horizontal, cylindrical tanks, this should be reduced to 35%. For example, in a horizontal cylindrical tank having a liquid depth of 42″, the outlet device should penetrate 14.7″ (42 x .35) below the liquid level.

The outlet device should extend above the liquid line to approximately 1″ from the top of the tank. The space between the top of the tank and the baffle allows gas to pass off from the tank into the house vent (Fig. 1).

The shape of a septic tank does not affect its operation. Available data indicate that, for tanks of a given capacity, shallow tanks function as well as deep ones. However, it is recommended that the smallest plan dimension be at least 24″. Liquid depth may range between 30″ and 60″. If the tank is deeper than 72″, the additional depth should not be considered in calculating tank capacity.

Adequate capacity above the liquid line is required to provide for that portion of the scum which floats above the liquid. Approximately 30% of the total scum accumulates above the liquid line. In addition to the provision for scum storage, 1″ clearance between top of inlet and outlet tees (or baffles) and top of tank usually is provided to permit free passage of gas back to the inlet and house vent stack.

For tanks having straight vertical sides, the distance between the top of the tank and the liquid line should be equal to approximately 20% of the liquid depth. In horizontal cylindrical tanks, an area equal to approximately 15% of the total circle should be provided above the liquid level. This condition is met if the liquid depth (distance from outlet invert to bottom of tank) is equal to 79% of the diameter of the tank.

Although a number of arrangements are possible, the term "compartments" as used here, refers to a number of units in series. These can be either separate units linked together, or sections enclosed in one continuous shell (Fig. 19) with watertight partitions separating individual compartments.

A single compartment tank will give acceptable performance, but a two compartment tank, with the first compartment equal to one-half to two-thirds of the total volume, generally provides better suspended solids removal. Compartments should be at least 24″ in the smallest dimension and have a liquid depth ranging from 30″ to 72″.

Installation Inspection A septic disposal system should be tested and inspected before it is used. Prompt inspection before backfilling should be required by local regulations, even where plans for the disposal system have been checked before issuance of a building permit. The septic tank should be filled with water and allowed to stand overnight to check for leaks. The soil absorption system should be carefully inspected before it is covered to be sure that the system is installed according to plans and good trade practice. Backfill material should be free of large stones and other deleterious material and should be overfilled a few inches to allow for settling.

Tank Operation and Maintenance

Septic tanks should be cleaned before too much sludge or scum is allowed to accumulate. If either the sludge or scum approaches too closely to the bottom of the outlet device, particles will be scoured into the absorption system and will clog the system. When this happens, liquid may back up in the plumbing fixtures or may break through to the ground surface. When an absorption system is clogged in this manner, it is not only necessary to clean the tank, but it also may be necessary to construct a new absorption system.

The tank capacities recommended in Figure 15 assure a reasonable period of good operation between cleanings. There are wide differences in the rate that sludge and scum accumulate; for example, in one case out of 20, the tank will reach the danger point, and should be cleaned, in less than 3 years. *Tanks should be inspected at least once a year and cleaned when necessary.*

Tank Inspection Although it is difficult for most home owners, actual inspection of sludge and scum accumulations is the only way to determine definitely when a given tank needs to be pumped. When a tank is inspected, the depth of sludge and scum should be measured in the vicinity of the outlet baffle. The tank should be cleaned if either: (1) the bottom of the scum mat is within approximately 3″ of the bottom of the outlet device, or (2) sludge comes within the limits specified in Figure 21.

Scum can be measured with a stick to which a weighted flap has been hinged, or with any device that can be used to feel out the bottom of the scum mat (Fig. 22).

FIG. 21 **MAXIMUM ALLOWABLE SLUDGE ACCUMULATION**

TANK CAPACITY (GAL.)[1]	LIQUID DEPTH			
	2.5 FT.	3 FT.	4 FT.	5 FT.
	MINIMUM SLUDGE CLEAR SPACE[2]			
750	5″	6″	10″	13″
900	4″	4″	7″	10″
1,000	4″	4″	6″	8″

[1] Tanks smaller than listed require more frequent cleaning.
[2] See Figure 22.

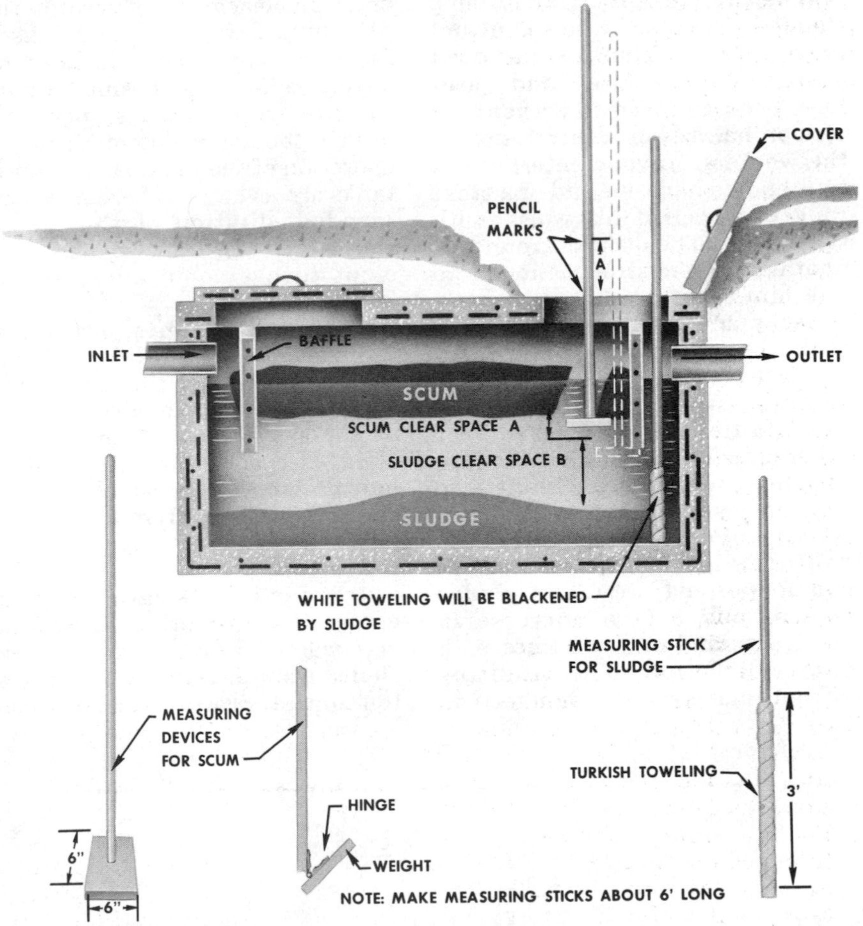

FIG. 22 Procedure for measuring sludge and scum accumulation; tank should be cleaned when space A is 3″ or less or when space B is less than specified in Figure 21.

The stick is forced through the mat, the hinged flap falls into a horizontal position, and the stick is raised until resistance from the bottom of the scum is felt. With the same tool, the distance to the bottom of the outlet device can be found.

A long stick wrapped with rough, white toweling and lowered to the bottom of the tank will show the depth of sludge and the liquid depth of the tank (Fig. 22). The stick should be lowered behind the outlet to avoid scum particles. After several minutes, if the stick is carefully removed, the sludge line can be distinguished by sludge particles clinging to the toweling.

Tank Cleaning In most communities where septic tanks are used, there are firms which conduct a business of cleaning septic tanks. The local health department can make suggestions on how to obtain this service. *Tanks should not be washed or disinfected after pumping.* A small residual of sludge should be left in the tank to assure the retention of anaerobic bacteria which carry on the biological treatment process.

When a large septic tank is being cleaned, care should be taken not to enter the tank until it has been thoroughly ventilated and gases have been removed to prevent explosion hazards or asphyxiation of the workers. Anyone entering the tank should have one end of a stout rope tied around his waist, with the other end held above ground by another person strong enough to pull him out if he should be overcome by any gas remaining in the tank.

Effect of Chemicals Operation of septic tanks is not improved by the addition of disinfectants or other chemicals. Some proprietary products which are claimed to "clean" septic tanks contain sodium hydroxide or potassium hydroxide as the active agent. Such compounds may result in sludge bulking and a large increase in alkalinity, and may interfere with bacterial action. The resulting effluent may severely damage soil structure and cause accelerated clogging, even though some temporary relief may be experienced immediately after application of the product. Properly designed and maintained septic systems do not require the aid of any chemicals or other additives, and their use should be avoided.

The harmful effects of ordinary household chemicals generally are overemphasized. Small amounts of chlorine bleach, introduced at the plumbing fixtures, may be used for odor control and will have no adverse effects. The small quantities of lye or caustics, normally used in the home, do not harm the operation of the tank. If the septic tanks are as large as herein recommended, dilution of the lye or caustics in the tank will be sufficient to overcome any possible harmful effects.

Soaps, detergents, bleaches, drain cleaners or other material normally used in the home will have no appreciable adverse effect on the system if used in moderation. The advice of responsible officials should be sought before chemicals arising from a hobby or home industry are discharged in the systems.

Precautions Drainage from garage floors or other sources of oily waste always should be excluded from the tank. Roof drains, foundation drains and drainage from other sources producing large volumes of water should not be piped into the septic tank or absorption area. Even intermittent large flows will stir up the contents of the tank and carry some of the solids into the outlet line; the absorption system following the tank will likewise become flooded or clogged, and may fail.

Toilet paper substitutes should not be flushed into a septic tank. Paper towels, newspaper, wrapping paper, rags and sticks may not decompose in the tank, and are likely to cause clogging of the absorption system. Fibrous materials such as hair are a frequent cause of clogging both of the plumbing and disposal system and should not be allowed to enter the drains. Strainer cups on bathroom lavoratories and tubs will help avoid this problem.

Waste brines from household water softener units have no adverse effect on the action of the septic tank, but may cause a slight shortening of the life of an absorption system installed in a structured clay soil.

A chart showing the location of the septic tank and absorption system should be placed at a suitable location in dwellings served by such a system. The charts also should acquaint home owners of the necessary inspection and maintenance, thus forestalling failures.

INDIVIDUAL SEWAGE DISPOSAL 512

CONTENTS

SEPTIC TANK SYSTEMS

To protect the public health and avoid nuisances individual disposal systems should be designed so that they: (1) will not contaminate any drinking water supply and will not pollute the waters of any bathing beach, shellfish breeding ground, or stream used for water supply or for recreation; (2) will not be discharged on or near the surface where they would be accessible to insects, rodents or other small animal carriers which may come into contact with food or drinking water; (3) will not be accessible to children and give rise to a nuisance due to foul odor or unsightly appearance; (4) will not overburden any body of water with excess nutrients, or degrade its quality below officially adopted water quality standards.

The following recommendations may differ in some instances from authoritative sources such as FHA Minimum Property Standards, National Plumbing Code or BOCA Basic Plumbing Code. Where local authorities use these documents as reference standards or have their own requirements, such local regulations should supersede these recommendations.

SUITABILITY OF SOIL

The first step in the design of a septic sewage disposal system is to determine whether the soil is suitable for the absorption of septic tank effluent and, if it is, how much absorption area is required.

The area intended for a soil absorption system should meet the following conditions: (1) The percolation time should be not over 60 minutes for absorption fields and beds, and not over 30 minutes for seepage pits (Fig. WF1); (2) The bottom of the trench, bed or seepage pit should be at least 4′ above the maximum seasonal elevation of the ground water table; (3) Rock formations or other impervious strata should

FIG. WF1 **REQUIRED ABSORPTION AREA IN VARIOUS SOILS**

PERCOLATION RATE (TIME IN MIN. FOR WATER TO FALL 1″)	MIN. AREA PER BEDROOM (SQ. FT.)
2 or less	85
5	125
10	165
15	190
30 [1]	250
45	300
60 [2]	330

[1] Soil not suitable for seepage pit if percolation rate is over 30 min.
[2] Soil not suitable for any absorption system if percolation rate is over 60 min.

FIG. WF2 **MINIMUM CLEARANCES FOR COMPONENTS OF A SEPTIC SEWAGE DISPOSAL SYSTEM**[1]

FROM	TO WELL[2]	TO WATER LINE[3]	TO FOUNDATION WALL	TO PROPERTY LINE	TO STREAM
Building Sewer	50′	10′	—	—	50′
Septic Tank	50′	10′	5′	10′	50′
Absorption Field	100′	25′	20′	5′	50′
Absorption Bed	100′	25′	20′	5′	50′
Seepage Pit	100′	50′	20′	10′	50′

[1] Based on recommendations of U.S. Public Health Service, HEW.
[2] And to suction line between well and pressure tank.
[3] Pressure type.

(Continued) **SEPTIC TANK SYSTEMS**

be at a depth greater than 4′ below the bottom of the trench, bed or pit; (4) The area should be large enough to provide the required absorption area in trench and bed bottoms or pit sidewalls as shown in Figure WF1.

Test Holes

At least six separate test holes should be made; these should be spaced uniformly over the proposed soil absorption area and should reach down to the depth of the proposed absorption system.

The holes can be 4″ to 12″ in diameter; to save time, labor and volume of water required per test, shallow holes can be bored with a 4″ auger. Deeper holes frequently required for seepage pits may have to be dug by shovel or heavy equipment.

The bottom and sides of the hole should be scratched with a knife or sharp-pointed instrument, in order to remove smeared soil and to provide a natural soil surface into which water may percolate. Loose material should be removed from the hole and 2″ of coarse sand or fine gravel should be added to protect the bottom from scouring and sediment.

Percolation-Rate Measurement

The test hole should be filled with clean water to a minimum depth of 12″ over the gravel. Except in sandy soils, water should be kept in the hole for at least 4 hours and preferably overnight.

After the overnight swelling period, water should be added to bring the level to approximately 6″ over the gravel. The drop in water level should be measured from a fixed reference point at 30-minute intervals over a 4-hour period. The drop which occurs in the final 30-minute period should be used to calculate the percolation rate.

In sandy soils (or other permeable soils in which the first 6″ of water seeps away in less than 30 minutes, after the overnight swelling period), the measurements should be taken at 10-minute intervals and the test run for one hour. The drop that occurs during the final 10 minutes is used to calculate the percolation rate.

The suggested test procedures are intended to insure that the soil is given ample opportunity to swell and to approach its condition during the wettest season of the year. Thus, the test should give comparable results in the same soil, whether made in a dry or in a wet season.

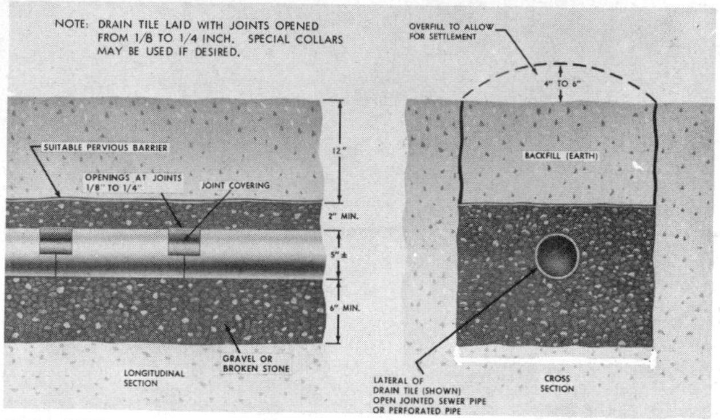

FIG. WF3 Construction of typical absorption trench and bed.

In sandy soils containing little or no clay, the swelling procedure is not essential, and the test may be made after the water from one filling of the hole has completely seeped away.

SOIL ABSORPTION SYSTEMS

When a soil absorption system is feasible, three types of design may be considered: absorption field, absorption bed or seepage pit.

The horizontal separation between elements of the soil absorption system and other site improvements should not be less than that shown in Figure WF2.

Seepage pits should not be used in areas where water supplies are obtained from shallow wells, or where there are limestone formations and sinkholes through which the effluent may contaminate the ground water.

Absorption Field

The individual *laterals* (branches) should not be over 100′ long and there should be at least two laterals in every system. The minimum absorption area of trench bottoms should be as shown in Figure WF1 for various percolation rates. The trench bottom and tile distribution lines should be level. Laterals should be spaced so as to provide at least 6′ of undisturbed earth between trenches. The trench depth should be at least 24″ (Fig. WF3).

Additional depth may be needed for surface contour adjustment, for extra aggregate under the tile or other design purposes.

The distribution pipe should be at least 4″ and should be laid in a level trench of sufficient width and depth

(Continued) **SEPTIC TANK SYSTEMS**

(Fig. WF3). The pipe should be surrounded by clean, graded gravel or rock, broken hard-burned clay brick or similar aggregate, ranging in size from 1/2″ to 2-1/2″. The coarse aggregate should extend from at least 2″ above the top of the pipe to at least 6″ below the bottom of the pipe. If open joint rather than perforated drain tile is used, the upper half of each joint opening should be covered with a strip of 15 lb. roofing paper, as shown in Figure WF3.

Open joint tile also can be used with drain tile connectors, collars, clips or similar accessories which assure uniform spacing, and protection of open joints. Such accessories commonly are made of galvanized iron, copper or plastics.

The top of the stone should be covered with a vapor-pervious barrier such as untreated building paper, a 2″ layer of hay or straw, to permit vapor movement and yet prevent the stone from becoming clogged by the earth backfill. The soil over a new absorption trench should be hand tamped and overfilled 4″ to 6″ higher than adjacent grade to compensate for likely settling.

Settlement of the fill in the trench to a lower level is undesirable because it may cause pooling of storm water in the trench, which can lead to premature saturation of the absorption field.

In general the ground surface of the field area should be used only for activities which will not contribute to the compaction of the soil and the consequent reduction in porosity.

Where the slope of the ground is essentially level (does not exceed 6″ variation in any direction within the absorption area), the absorption field should consist of several laterals connected at their extremes to form a continuous closed loop.

Such a system permits the effluent to flow into all the tile lines whenever there is a discharge from the septic tank. Interconnection of laterals assures effluent access to each point in the lateral from two sides, so that if one part of the lateral became damaged, effluent would be able to reach the lateral from the other side.

In sloping ground, laterals should be arranged "in series" and each should be connected to the next by a closed overflow line laid on an undisturbed section of ground.

The series arrangement is such that all effluent is discharged to the first trench until it is filled; excess liquid then is carried by means of the closed overflow line to the next succeeding or lower trench.

Overflow lines should be 4″, tight-joint pipe with direct connections to the distribution lines in adjacent trenches or to a drop box arrangement. Overflow lines should be laid on undisturbed ground, taking care to minimize the disturbance to the soil between trenches.

Generally, ground having a slope greater than 50% should be investigated carefully to determine if satisfactory from the erosion standpoint. The maximum ground slope suitable for soil absorption systems is governed by local factors affecting the erosion of the ground used for the absorption field. Excessive slopes which are not protected from surface water runoff or do not have adequate vegetation cover to prevent erosion should be avoided.

Near banks, the horizontal distance from side of trench to ground surface should be 2′, or such larger dimension which would prevent lateral flow of effluent and breakout on surface.

Absorption Beds

An absorption bed is a variation of the typical trench used in absorption fields. The basic design is essentially the same as for an absorption trench, but the absorption bed is generally much wider than 36″ and contains two or more 4″ distribution lines, 4′ to 6′ apart.

The minimum bottom absorption area for absorption beds should be as shown in Figure WF1. The bed should have a minimum depth of 24″ below natural ground level and should provide a minimum of 12″ gravel and 12″ earth backfill. The gravel should extend at least 2″ above and 6″ below the distribution line. The bottom of the bed and bottom of distribution line should be level.

Distribution lines should be spaced not more than 6′ apart and not more than 3′ from the bed sidewall. At least two distribution lines should be used in each bed. When more than one bed is used there should be a minimum of 6′ of undisturbed earth between adjacent beds, and the beds should be connected in series.

Seepage Pits

Pits generally should not be used where absorption trenches or beds are feasible. Where the top 3′ or 4′ of soil is not porous enough for an absorption field, but is underlaid with suitable granular material,

(Continued) SEPTIC TANK SYSTEMS

FIG. WF4 **WALL AREAS (SQ. FT.) OF SEEPAGE PITS**

PIT DIAMETER	DEPTH BELOW INLET									
	1 FT.	2 FT.	3 FT.	4 FT.	5 FT.	6 FT.	7 FT.	8 FT.	9 FT.	10 FT.
5′	15.7	31	47	63	79	94	110	126	141	157
6′	18.8	38	57	75	94	113	132	151	170	188
7′	22.0	44	66	88	110	132	154	176	198	220
8′	25.1	50	75	101	126	151	176	201	226	251
9′	28.3	57	85	113	141	170	198	226	254	283
10′	31.4	63	94	126	157	188	220	251	283	314

FIG. WF5 **RECOMMENDED TANK CAPACITIES**[1]

NUMBER OF BEDROOMS	MINIMUM TANK CAPACITY (GAL.)
3 or less	900
4	1,000
Additional Bedrooms	Add 250 gal./bedroom

[1] Provides for use of garbage grinder, dishwasher, automatic clothes washer, and other household appliances.

seepage pits constitute an acceptable substitute. Seepage pits may also be used where the land area is too small, too irregular or heavily wooded to accommodate trench or bed absorption systems.

Soil strata in which the percolation rates are in excess of 30 minutes per inch should not be included in computing the absorption area. Where the soil generally has a percolation rate in excess of 30 minutes per inch, seepage pits should not be used. The minimum absorption area of pit sidewalls should be as shown in Figure WF1.

Having established the required area, Figure WF4 may be used to select the desired dimensions of circular or cylindrical seepage pits. Where more than one pit is provided, they can be connected in series or in parallel.

Multiple pits should be separated by a distance equal to 3 times the diameter of the largest pit; for pits over 20′ in depth, the minimum distance between pits should be 20′. The area of the lot on which the house is to be built should be large enough to maintain this distance between the pits while still allowing room for additional pits if the first one(s) should fail.

Seepage pits should be backfilled with clean gravel to a depth of 1′ above the bottom to provide a sound foundation for the lining.

Lining materials for seepage pits can be clay or concrete masonry units in brick, block or ring shapes. Rings should have openings to provide for seepage. Concrete brick and block should be laid dry (unmortared) with staggered, slightly open joints; hollow concrete block frequently is laid on the side with cores horizontally to facilitate absorption. Standard clay brick should be laid with open joints to form a 4″ wall.

The outside diameter of the lining should be at least 1′ smaller than the excavation diameter, to provide at least 6″ (preferably 1′) of annular space to be filled with coarse gravel to the top of the lining.

Either a corbelled brick dome or a flat concrete cover can be used over the pit. Corbelled bricks should be laid either in cement mortar or have a 2″ covering of concrete. If a flat cover is used, a prefabricated type is preferred, and it should be reinforced to be equivalent in strength to an approved septic tank cover.

The cover should rest on undisturbed ground 12″ beyond the lining. A 9″ capped opening in the pit cover should be provided for pit inspection.

Connecting lines should be of a sound durable material, the same as used for the house drain. All connecting lines should be laid on a firm bed of undisturbed soil throughout their length. The connecting line should be sloped at least 2% (1/4″ in 1′). The pit inlet pipe should extend horizontally at least 1′ into the pit with a tee or ell to divert flow downward to prevent washing and eroding of the sidewalls.

SEPTIC TANKS

Tank Location and Size

Septic tanks should be located where they are not likely to contaminate any well, spring or other source of water supply. They preferably should be located downhill from private wells or natural springs. Tanks should never be closer than 50′ from any source of water supply, such as a well or stream; separation from other building elements should be as shown in Figure WF2. The tank should not be located in swampy areas, nor in areas subject to flooding. Septic tank capacities should be as shown in Figure WF5.

The liquid capacities recommended in Figure WF5 allow for the use of all household appliances, including automatic washing machines, dishwashers and waste disposers. They also insure enough excess capacity to

(Continued) **SEPTIC TANK SYSTEMS**

retain the sewage a sufficient time to settle out and digest solids.

Materials and Construction

Septic tanks should be watertight and constructed of corrosion- and decay-resistant materials, such as concrete, coated metal, vitrified clay, heavyweight concrete masonry units or hard-burned bricks.

Precast and cast-in-place reinforced concrete tanks are common in all parts of the country. Prefabricated steel tanks meeting Commercial Standard CS177 are acceptable in some states.

Special attention should be given to job-built tanks to insure watertightness and prevent contamination. Heavyweight concrete block should be laid on a solid foundation and all mortar joints should be well filled. The interior of the tank should be surfaced with two 1/4″ thick coats of portland cement stucco.

Precast concrete tanks should have a minimum wall thickness of 3″, and should be adequately reinforced to facilitate handling. When precast slabs are used as covers, they should be watertight, have a thickness of at least 3″, and be adequately reinforced, so as to be capable of supporting a dead load of 300 lbs. per sq. ft.

Backfill around septic tanks should be made in thin layers thoroughly tamped in a manner that will not produce undue strain on the tank. Backfill may be consolidated with the use of water, provided the backfill is placed and wetted in layers and the tank is first filled with water to prevent floating.

Adequate access (at least 20″ wide) should be provided to each tank compartment for inspection and cleaning. Both the inlet and outlet devices should be accessible. Where the top of the tank is located more than 18″ below the finished grade, manholes and inspection holes should extend to approximately 8″ below the finished grade or to finished grade if a gas seal is provided to keep odors from escaping.

A gas seal is made by filling the top of the manhole with sand or by sealing with a minimum of 1″ of portland cement over an intermediate wood platform.

The inlet invert should enter the tank at least 3″ above the outlet invert and the liquid level in the tank. A vented inlet tee or baffle should be provided to divert the incoming sewage downward. The tee or baffle should penetrate at least 6″ below the liquid level, but in no case should the penetration be greater than that allowed for the outlet device. Tee or baffle arrangement, clearances, minimum dimensions and proportions should be as shown in Figure WF6.

The outlet device generally should extend to a distance below the surface equal to 40% of the liquid depth. For horizontal, cylindrical tanks, this should be reduced to 35%. The outlet device should extend above the liquid line to approximately 1″ from the top of the tank.

It is recommended that the tank width be at least 24″; liquid depth may range between 30″ and 60″. If the tank is deeper than 72″, the additional depth should not be considered in calculating tank capacity. In a compartmented tank, each compartment should be at least 24″ wide and have a liquid depth ranging from 30″ to 72″.

For tanks having straight vertical sides, the distance between the top of the tank and the liquid line should be equal to approximately 20% of the liquid depth. In horizontal cylindrical tanks, an area equal to approximately 15% of the total circle should be provided above the liquid level.

The septic disposal system should be tested and inspected before it is backfilled and used. The septic tank should be filled with water and allowed to stand overnight to check for leaks. Thereafter, tanks should be inspected at least once a year and cleaned when necessary.

The tank capacities recommended assure a reasonable period of satisfactory operation between cleanings. There are wide differences in the rate that sludge and scum accumulate; for example, in one case out of 20, the tank will reach the danger point, and should be

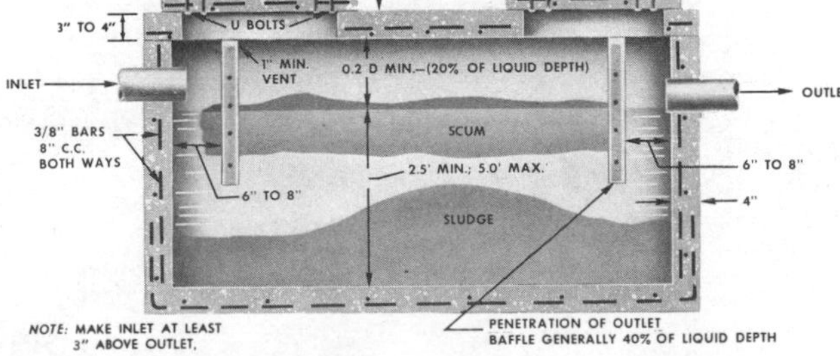

FIG. WF6 Single compartment tank showing recommended proportions, dimensions and clearances.

(Continued) **SEPTIC TANK SYSTEMS**

cleaned, in less than 3 years.

A septic tank should be cleaned if either: (1) the bottom of the scum mat is within approximately 3″ of the bottom of the outlet device, or (2) the sludge clear space is reduced to the minimum limits specified in Figure WF7.

Refer to Main Text page 512-17 for methods of measuring sludge accumulation and general maintenance recommendations.

FIG. WF7 **MAXIMUM ALLOWABLE SLUDGE ACCUMULATION**

TANK CAPACITY (GAL.)[1]	LIQUID DEPTH			
	2.5 FT.	3 FT.	4 FT.	5 FT.
	MINIMUM SLUDGE CLEAR SPACE			
750	5″	6″	10″	13″
900	4″	4″	7″	10″
1,000	4″	4″	6″	8″

[1] Tanks smaller than listed require more frequent cleaning.

521 WIRING

INTRODUCTION

A good home wiring system is a necessity. Even though a full supply of electricity is made available by the power company, the extent to which it can be effectively used in the home is determined by the interior wiring. Starting from the *service entrance,* where electricity enters the house, it is the responsibility of the home's wiring system to deliver a full supply of electricity to every outlet. In a properly planned system each *circuit* and *outlet* is designed to serve a specific purpose, and *switch* controls are located with both convenience and safety in mind.

Electrical appliances are popular aids to homemaking. They are constantly being improved to provide higher standards of performance and convenience. The resulting increases in wattage are reflected in greater demands upon the home wiring system. A few years ago a 550-watt, six-pound hand iron was considered satisfactory; today, 1,000-watt irons are commonplace. Electric ranges have increased from 6,500 watts until now 12,000-watt ranges are typical, and double-oven ranges with even higher ratings are not uncommon.

The number of appliances in homes is increasing. Work-saving and time-saving appliances, like the automatic washer, electric clothes dryer, dishwasher and food waste disposer, are no longer considered luxuries. Many families today enjoy the convenience of home freezers and the comfort of air conditioning. Electric heating is rapidly being accepted for residential use in many parts of the country.

All of these innovations are placing increased demands on the wiring system. When loads outstrip the capacities of the wiring system, operating efficiency suffers. When voltage drop becomes excessive in the circuit, it results in poor lighting and inefficient appliance operation. A 5% voltage loss, for example, produces a 10% loss of heat in any appliance or a 17% loss of light from an incandescent lamp.

Average home consumption of electrical power has increased tenfold in the past forty years. Despite increased electrical generation by nuclear plants, hydroelectric stations and other non-fossil means, four-fifths of our electrical production today is still based on coal, oil and gas. Spiraling electrical production puts a severe strain on these non-renewable fossil resources and frequently results in adverse environmental conditions. The current national commitment to energy conservation and pollution abatement places an additional responsibility on the designer—to make the wiring system not only convenient and efficient, but to discourage waste as well.

The first subsection, Design & Planning, presents recommendations for *planning* an adequate electrical wiring system. Criteria of good practice for *installing* residential wiring systems are given in the second subsection, Methods & Systems.

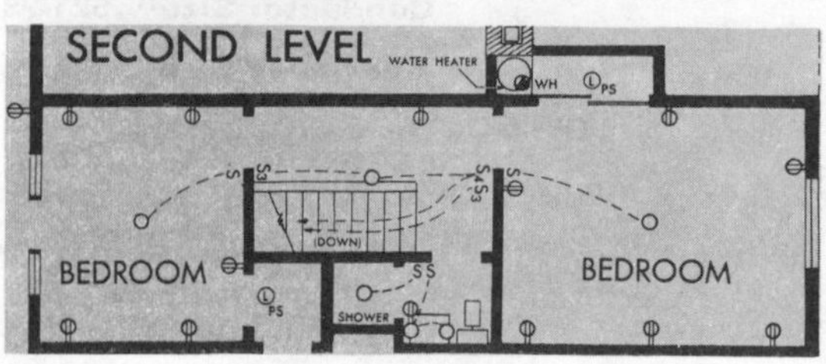

During the period 1937—1978, the average annual consumption of electrical energy in residences has risen from 805 to between 8,800 to 8,850 kilowatt-hours. In some areas of the country the average consumption is over 10,000 kilowatt-hours and is still rising. The electrical industry had studied this trend back in 1954 and evolved a set of farsighted recommendations which were adopted in 1958 as USA Standard C91.1-1958 Requirements for Residential Wiring.

The following recommendations are based chiefly on this Standard, with only minor revisions to conform with the 1978 National Electrical Code and HUD Minimum Property Standards. The purpose of these recommendations is to provide design criteria which will assure not only an efficient and convenient system, but one which will minimize physical obsolescence. They thus supplement the requirements of the National Electrical Code, whose chief emphasis is safety.

GENERAL RECOMMENDATIONS

A fundamental prerequisite of any adequate wiring installation is conformity to the applicable safety regulations. The approved industry standard for electrical safety is the National Electrical Code, developed and published by the National Fire Protection Association (See p. 602-6). Usually, local municipal ordinances and state laws are based upon the National Electrical Code or make it a part of their requirements by reference.

Inspection service to determine conformance with safety regulations is available in most communities from the local building official. Where available, a certificate of inspection should be obtained. In the absence of inspection

FIG. 1 **GRAPHIC ELECTRICAL WIRING SYMBOLS***

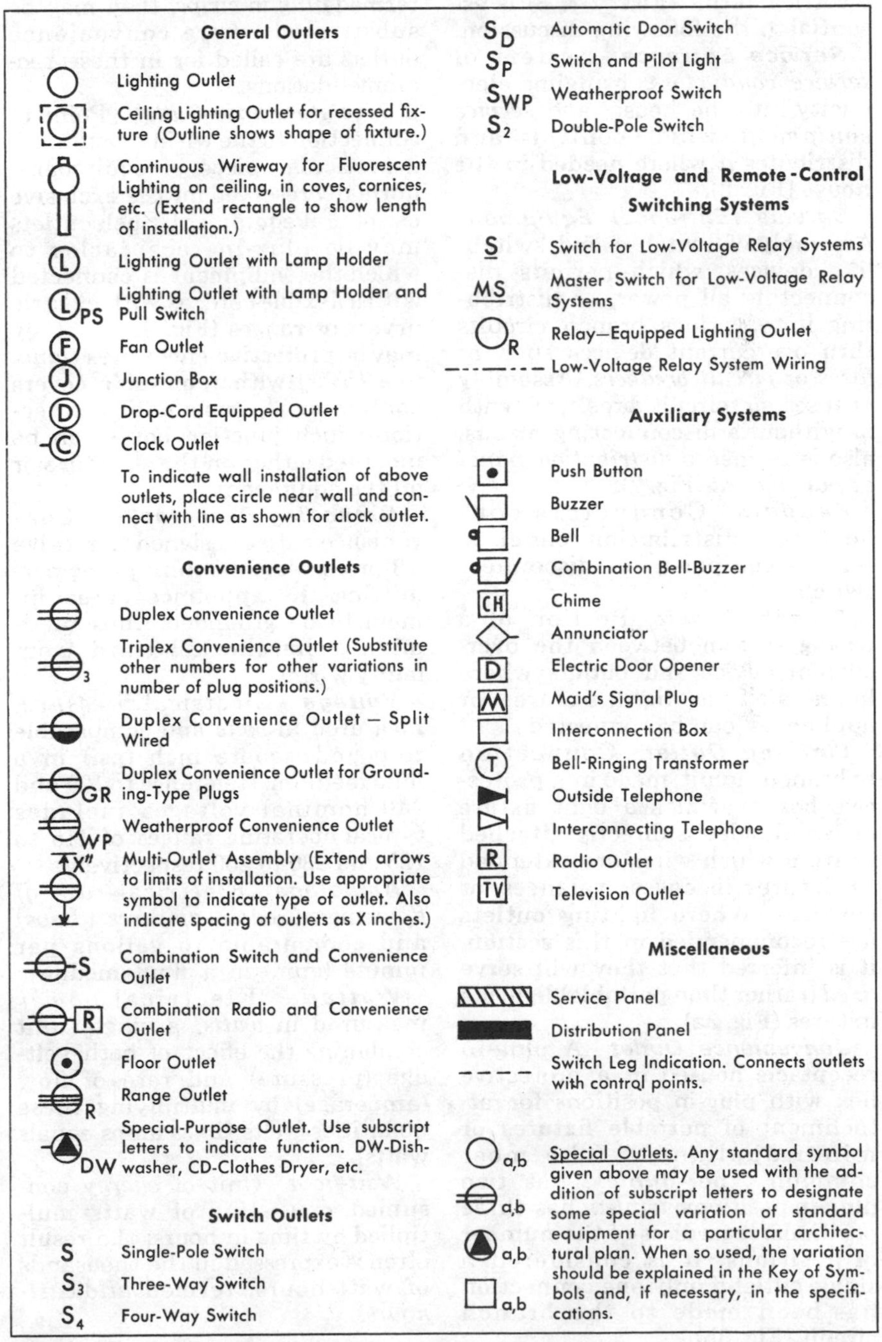
General Outlets
Lighting Outlet
Ceiling Lighting Outlet for recessed fixture (Outline shows shape of fixture.)
Continuous Wireway for Fluorescent Lighting on ceiling, in coves, cornices, etc. (Extend rectangle to show length of installation.)
L
Lighting Outlet with Lamp Holder
L PS
Lighting Outlet with Lamp Holder and Pull Switch
F
Fan Outlet
J
Junction Box
D
Drop-Cord Equipped Outlet
C
Clock Outlet
To indicate wall installation of above outlets, place circle near wall and connect with line as shown for clock outlet.
Convenience Outlets
Duplex Convenience Outlet
3
Triplex Convenience Outlet (Substitute other numbers for other variations in number of plug positions.)
Duplex Convenience Outlet — Split Wired
GR
Duplex Convenience Outlet for Grounding-Type Plugs
WP
Weatherproof Convenience Outlet
X"
Multi-Outlet Assembly (Extend arrows to limits of installation. Use appropriate symbol to indicate type of outlet. Also indicate spacing of outlets as X inches.)
S
Combination Switch and Convenience Outlet
R
Combination Radio and Convenience Outlet
Floor Outlet
R
Range Outlet
DW
Special-Purpose Outlet. Use subscript letters to indicate function. DW-Dishwasher, CD-Clothes Dryer, etc.
Switch Outlets
S
Single-Pole Switch
S3
Three-Way Switch
S4
Four-Way Switch
SD
Automatic Door Switch
SP
Switch and Pilot Light
SWP
Weatherproof Switch
S2
Double-Pole Switch
Low-Voltage and Remote-Control Switching Systems
S
Switch for Low-Voltage Relay Systems
MS
Master Switch for Low-Voltage Relay Systems
R
Relay—Equipped Lighting Outlet
Low-Voltage Relay System Wiring
Auxiliary Systems
Push Button
Buzzer
Bell
Combination Bell-Buzzer
CH
Chime
Annunciator
D
Electric Door Opener
M
Maid's Signal Plug
Interconnection Box
T
Bell-Ringing Transformer
Outside Telephone
Interconnecting Telephone
R
Radio Outlet
TV
Television Outlet
Miscellaneous
Service Panel
Distribution Panel
Switch Leg Indication. Connects outlets with control points.
a,b
Special Outlets. Any standard symbol given above may be used with the addition of subscript letters to designate some special variation of standard equipment for a particular architectural plan. When so used, the variation should be explained in the Key of Symbols and, if necessary, in the specifications.

*Adapted from USA Standard Y32.9-1962 Graphic Electrical Wiring Symbols for Architectural and Electrical Layout Drawings.

service, an affidavit should be obtained from the electrical contractor attesting to conformity with the applicable safety regulations.

Terminology

A clear understanding of standard electrical symbols (Fig. 1) and electrical terms (Figs. 2 & 3) is essential to the following discussion.

Service Entrance System of *service conductors,* bringing electricity into the house, and *service equipment,* which controls and distributes it where needed in the house (Fig. 2).

Service (Entrance) Equipment Assembly of switches and switch-like devices, which permits disconnecting all power, or distributing it to various branch circuits thru overcurrent devices such as *fuses* or *circuit breakers.* Assembly of fuses or circuit breakers, with or without a disconnecting means, also is termed a *distribution panel* or *panelboard* (Fig. 2).

Feeders Conductors connecting a distribution panel to service equipment or disconnect switch.

Branch Circuit Portion of a wiring system between the overcurrent device and outlets where loads, such as light fixtures or appliances, can be connected.

Lighting Outlet Connection to branch circuit, made in a protective box, to which a light fixture or lampholder is directly attached or from which wires are extended to fixtures in coves, valances or cornices. Where lighting outlets are recommended in this section, it is inferred that they will serve *fixed* (rather than portable) lighting fixtures (Fig. 3a).

Convenience Outlet A plug-in receptacle housed in a protective box with plug-in positions for attachment of portable fixtures or minor appliances. The most common type, *duplex,* has two plug-in positions; *triplex,* has three positions; regardless of the number of positions, it is considered a single outlet if only one connection has been made to the branch circuit (Fig. 3b).

Split-Receptacle Outlet Duplex receptacle which has been wired so that each plug-in position is on a separate 120 volt circuit or two separate line wires of a single 3-wire 240 volt circuit (Fig. 8 & 11).

Multi-Outlet Assembly A series of plug-in receptacles, usually spaced at regular intervals and contained in a protective raceway (channel). Such assemblies also are termed *plug-in strips;* they may be substituted where convenience outlets are called for in these recommendations.

Special-Purpose Outlet Point of connection to the wiring system for a particular piece of equipment, normally reserved for the exclusive use of that equipment. Such outlets may be plug-in receptacles to which the equipment is connected with a flexible cord, as with electric dryers or ranges (Fig. 3c); or they may be protective enclosures *(junction boxes)* with removable covers containing "permanent" connections; such junction boxes can be mounted either on the structure or on the equipment.

Grounding Receptacle Convenience outlet designed to receive a 3-prong plug, the third prong permitting the appliance or equipment to be grounded, thus minimizing the shock hazard from faulty wiring.

Voltage Electrical *pressure,* measured in *volts* and comparable to pounds/square inch (psi) in a fluid medium. Reference to 120 and 240 nominal voltages includes typical operating ranges of 115 to 125 and 230 to 250 respectively.

Amperage Electrical *rate of flow,* measured in *amperes (amps)* and comparable to gallons per minute (gpm) in a fluid medium.

Wattage Electrical *power* measured in *watts,* a single unit combining the effect of both voltage (pressure) and rate of flow (amperage) by multiplying these quantities (volts times amps equals watts).

Watt-hour Unit of *energy* consumed, consisting of watts multiplied by time in hours; the result often is expressed in the thousands of watt-hours, termed *kilowatt-hours.*

Outlets and Switches

Lighting Outlets The amount and type of illumination required should be fitted to the various seeing tasks carried on in the home. In living areas there should also be suitable lighting for recreation and entertainment. In many instances, best results are achieved by a blend of lighting from both fixed and portable luminaires. Good home lighting, therefore, requires thoughtful planning and careful selection of permanent fixtures, portable lamps and other lighting equipment.

Convenience Outlets All outlets on 15 and 20 amp general purpose circuits should be of the grounding type, minimizing the hazard of shock from short circuits.

Multiple-Switch Controls All

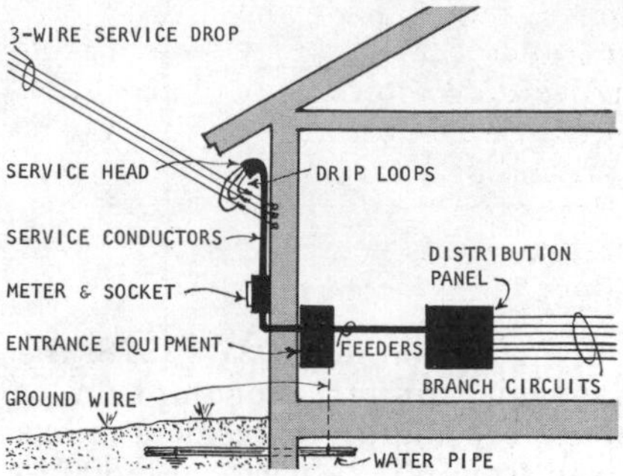

FIG. 2 Service equipment must include main disconnect (switch), but distribution panel may or may not be included within the same box.

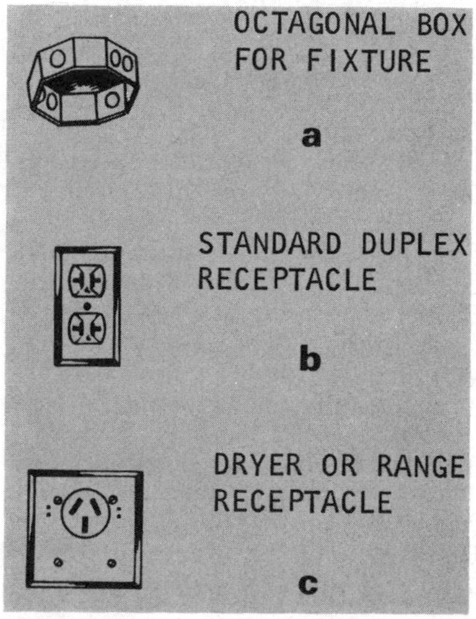

FIG. 3 Types of outlets: (a) lighting, (b) convenience and (c) special purpose.

spaces which have more than one entrance for which wall-switched lighting is recommended should be equipped with multiple-switch controls at each *principal entrance.* (Principal entrances are those commonly used for entry to a room when going from a lighted to an unlighted condition. For instance, a door from a living room to a porch is a principal entrance to the porch. However, this door would not necessarily be considered a principal entrance to the living room unless the front entrance to the house is through the porch.)

If this recommendation would result in the placing of switches controlling the same light within ten feet of each other, one of the switch locations may be eliminated. Also, where rooms may be lighted from more than one source (as when both general and supplementary illumination are provided from fixed sources), multiple switching is required for one set of controls only, usually for the general illumination.

Number of Outlets Where recommendations are based on the number of linear or square feet, the number of outlets should be determined by dividing the total linear or square footage by the recommended amount per outlet. If the resulting number contains a fraction larger than .5, it should be increased to the next larger number. The floor area is to be computed from the outside dimensions of the house or the centerline dimensions of individual rooms.

Dual-Purpose Rooms Where a room is intended to serve more than one function, such as a combination living-dining room or a kitchen-laundry, the following outlet recommendations are separately applicable to specific areas. In a living-dining room for instance, convenience outlets in the dining area should be of the split-receptacle type while ordinary outlets may be used in the rest of the room. Lighting outlet provisions may be combined in any manner that will assure both general overall illumination as well as local illumination of work surfaces. In considering locations for *wall switches,* the area is considered as a single room.

Dual Functions of Outlets When two different recommendations require an outlet at the same location, only one need be installed. In such instances, particular attention should be paid to additional wall-switch controls which may be necessary.

For example, a lighting outlet in an upstairs hall may be located at the head of the stairway, thus satisfying both a hall lighting outlet and a stairway lighting outlet recommendation with a single outlet. Stairway provisions will necessitate multiple-switch control of this lighting outlet both at the head and foot of the stairway. In addition, the multiple-switch control rule given above, when applied to the upstairs hall, may require a third point of control elsewhere in the hall because of its length.

Special Conditions

The following recommendations are necessarily general in nature and are intended to apply to most situations encountered in the typical home or apartment. Where unusual design or uncommon construction methods make it impossible to conform with these recommendations, the following rules of thumb will serve as a guideline for modified provisions.

Structural Compatibility The wiring installation should be fitted to the structure. If compliance with a particular recommendation would require alteration of doors, windows or structural members, alternate provisions should be made. For example, the extensive use of window walls may make it necessary to alter the recommended outlet or switch heights, or to resort to surface wiring.

Advance Planning If certain facilities or functions are indicated as future additions—such as a basement recreation room — initial wiring should be so arranged that none of it need be replaced or moved when the ultimate plan is realized. Final wiring for the future addition may be left to a later date if desired. If air conditioning is not included in the initial construction, provisions should be made for a future central system or window units.

Multi-Family Dwellings These recommendations may be applied also to dwelling units of multi-family dwellings. However, they must be supplemented with good engineering judgment as to wiring adequacy in common-use spaces. Particular attention should be paid to limiting voltage drop in feeders to individual dwelling units in order to assure proper operation of appliances.

OUTLET LOCATIONS

Where lighting outlets are mentioned in this section, their types and locations should conform with the lighting fixtures or equipment to be used. Unless a specified location is stated, lighting outlets may be located anywhere within the area under consideration to produce the desired lighting effects.

Convenience outlets preferably should be located near the ends of wall space, rather than near the center, thus reducing the likelihood of being concealed behind large pieces of furniture. Unless otherwise specified, outlets should be located approximately 12″ above the floor line.

Wall switches normally should be located at the latch side of doors or at the traffic side of arches and within the room or area where the outlets are located. Some exceptions to this practice are: (1) the control of exterior lights from indoors; (2) the control of stairway lights from adjoining areas when stairs are closed off by doors at head or foot; (3) the control of lights from the access space adjoining infrequently used areas, such as storage areas. Wall switches are normally mounted at a height of approximately 48″ above the floor line but may be mounted lower where it is desired to align them with door knobs and to put them within reach of younger children.

Exterior Entrances

Lighting One or more wall-switched lighting outlets should be located at the front and service entrances. Where a single wall outlet is desired, location on the latch

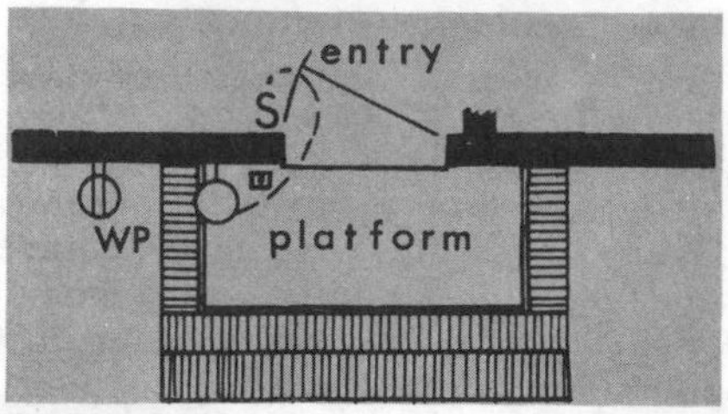

FIG. 4 Locating the light at the latch side of the door assures illumination of the lock and the traffic path.

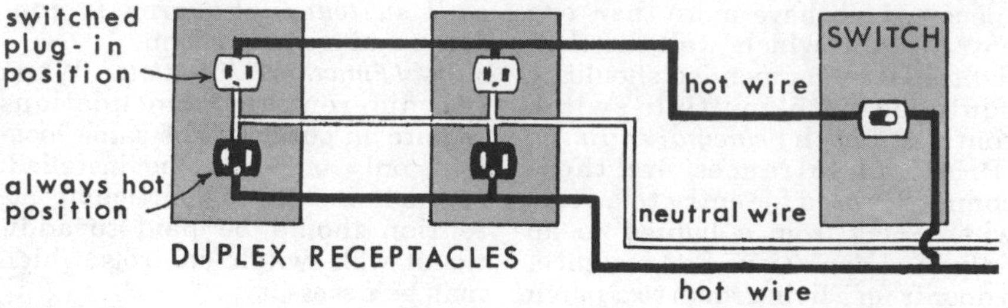

FIG. 8 The use of two "hot" wires in a duplex split-receptacle permits switching one side, while the other remains always hot.

side of the door is preferable (Fig. 4).

It is recommended that wall-switched lighting outlets be installed at other entrances. The principal lighting requirements at entrances are the illumination of steps leading to the entrance and of faces of people at the door. Switched outlets in addition to those at the door are often desirable for post lights to illuminate terraced or broken flights of steps or long approach walks (Fig. 5).

Convenience A weatherproof outlet, preferably near the front entrance, located at least 18″ above the entrance platform should be provided (Fig. 4). It is recommended that this outlet be controlled by a wall switch inside the entrance for convenient operation of outdoor decorative lighting. Additional waterproof outlets along the exterior of the house are recommended to serve decorative garden treatments (Fig. 5) and for the use of appliances or electric garden tools, such as lawn mowers and hedge trimmers.

FIG. 5 Exterior lighting can enhance the architecture and light the way to the main entrance.

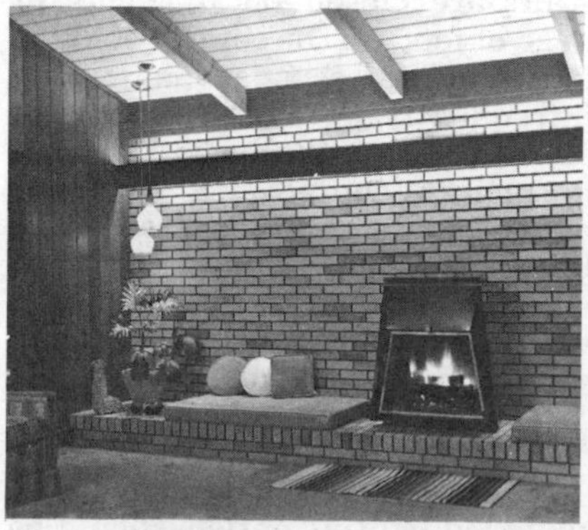

FIG. 6 Valences direct light both up and down for dramatic effects; coves direct light mostly up, cornices mostly down.

Living and Recreation Rooms

Lighting Some means of general illumination in living and recreation rooms is essential. This lighting may be provided by ceiling or wall fixtures, by lighting in coves, valances or cornices, or by portable lamps (Fig. 6). Wall-switch controlled outlets in locations appropriate to the intended lighting method should be provided (Fig. 7).

These recommendations also apply to sun rooms, enclosed porches, television rooms, libraries, dens and similar areas. Outlets for decorative accent lighting, such as picture illumination and bookcase lighting, also should be considered.

Convenience Convenience outlets should be placed so that no point along the floor line in any usable wall space is more than 6′ from an outlet in that space. Where floor to ceiling windows prevent meeting this requirement with ordinary convenience outlets, other suitable means such as floor outlets or multi-outlet assemblies can be used (Fig. 7).

If general illumination is to be provided from portable lamps, in lieu of fixed lighting, then at least two wall-switched separate plug-in positions should be planned. These can be provided with two switched regular duplex outlets or one switched plug-in position in each of two *split-receptacle* outlets (Fig. 8).

If regular convenience outlets are wall-switch controlled and intended for portable lighting, then

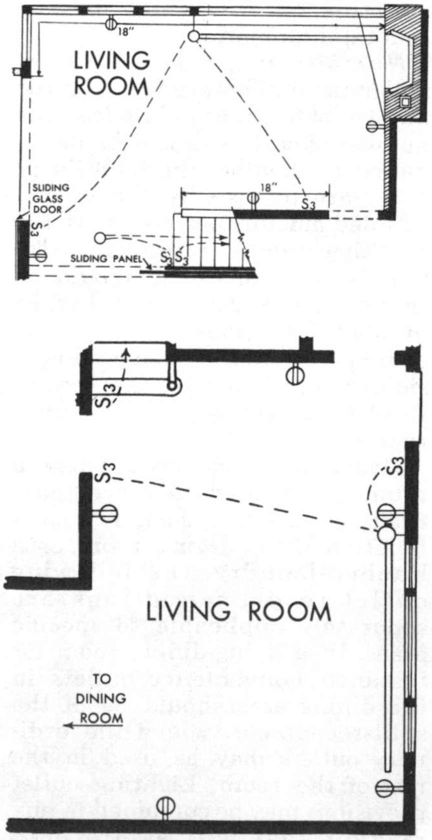

FIG. 7 Valence and cornice lighting often are used over windows, and outlets have to be placed carefully so they will be concealed by the facia board.

it is recommended that additional unswitched convenience outlets be provided, in order not to limit the use of portable equipment such as radios and clocks in that location. Also, a single convenience outlet should be installed in combination with the wall switch at one or more of the switch locations for the use of a vacuum cleaner or other portable appliances.

One or more outlets for entertainment equipment such as television, radio and record player should be provided at bookcases, shelves and other suitable locations. In rooms with conventional fireplaces, an outlet may be desirable in or near the mantel shelf.

Dining Areas

Lighting Each dining room, or dining area combined with another room, or breakfast nook, should have at least one wall-switch controlled lighting outlet. Such an outlet normally is located over the probable location of the dining or breakfast table to provide direct illumination of the area (Figs. 9 & 10).

Convenience Convenience outlets should be placed so that no point along the floor line in any usable wall space is more than six feet from an outlet in that space. When it is intended to place the dining or breakfast table against a wall, one of the recommended outlets should be located in that wall, just above table height (Fig. 10).

Where open counter space is to be built-in, an outlet should be provided above counter height for the use of portable appliances. All convenience outlets in the dining area should be of the split-receptacle type, providing increased capacity at each outlet (Fig. 11).

FIG. 9 A pull-down ceiling lamp and cornice lighting provide effective low-key illumination in the dining room.

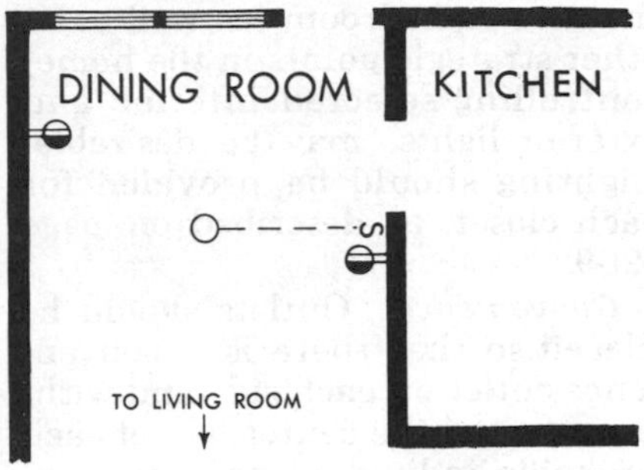

FIG. 10 A hanging ceiling lamp often is located in the center of the dining table, not necessarily the center of the room.

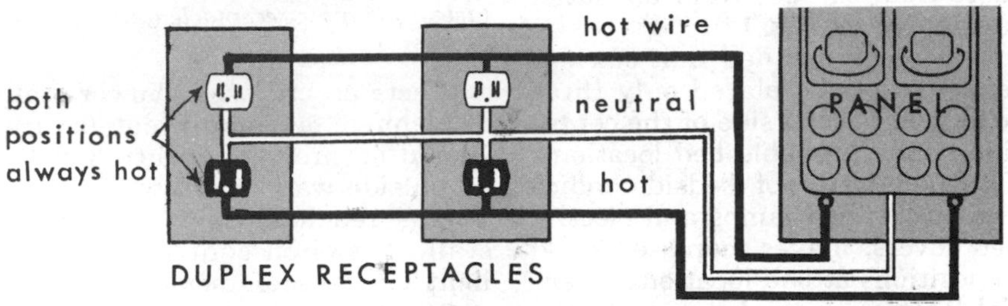

FIG. 11 Split-receptacles minimize overloads and double the capacity of duplex receptacles by connecting each plug-in position to a separate hot wire. Where local codes do not permit this type of installation, both plugs on each receptacle can be wired to the same circuit, and alternate receptacles wired to different circuits.

Bedrooms

Lighting Good general illumination is essential in the bedroom. This shall be provided from a fixed ceiling fixture or from lighting in valances, coves or cornices (Fig. 6). Wall-switch controlled outlets should be placed in locations appropriate to the lighting method selected (Fig. 12).

Light sources located over full-length mirrors, or in the bedroom ceiling, directly in front of the clothes closet, may serve as general illumination. A master-switch in

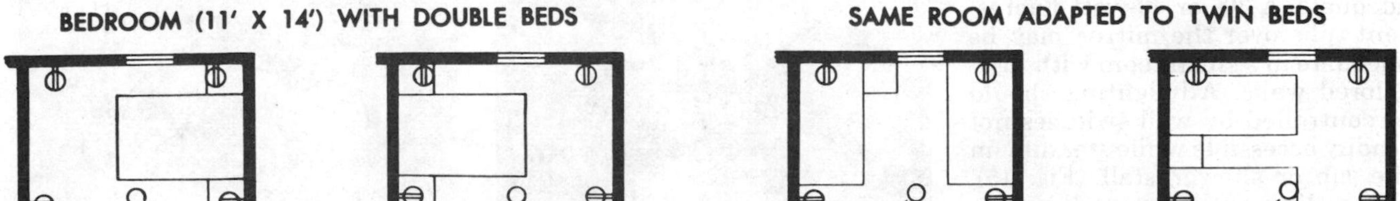

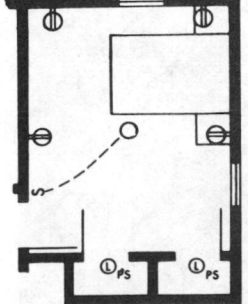

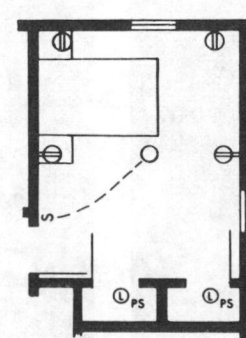

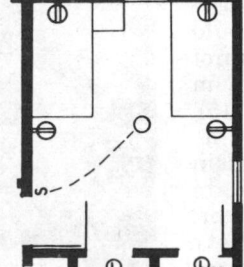

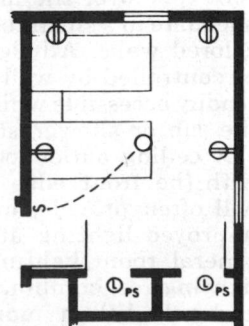

FIG. 12 Good electrical planning anticipates various furniture arrangements, as shown in these alternate bedroom layouts. A ceiling light should be placed where it will look and work best, not necessarily in the center of the room.

the master bedroom (as well as at other strategic points in the home) controlling selected interior and exterior lights, may be desirable. Lighting should be provided for each closet, as described on page 521-9.

Convenience Outlets should be placed so that there is a convenience outlet on each side and within six feet of the center line of each probable individual bed location. Additional outlets should be placed so that no point along the floor line in any other usable wall space is more than six feet from an outlet in that space (Fig. 12).

It is recommended that convenience outlets be placed only three to four feet either side of the center line of the probable bed locations. The popularity of bedside radios and clocks, bed lamps and electric bed covers, makes increased plug-in positions at bed locations essential. Triplex or quadruplex convenience outlets are recommended at these locations (Fig. 13a).

It is also recommended that a receptacle be provided at one of the switch locations for the use of a vacuum cleaner, floor polisher or other portable appliances (Fig. 13b).

Bathrooms and Lavatories

Lighting Illumination of both sides of the face when at the mirror is essential. This can be accomplished with diffused lighting from a luminous ceiling, or with more concentrated light from several sources (Fig. 14). A *single concentrated* light source, either on the ceiling or the wall, generally is not adequate. A 30- or 40-watt fluorescent tube over the mirror may be adequate in a small room with light colored walls. All lighting should be controlled by wall switches not readily accessible while standing in the tub or shower stall (Fig. 15).

A ceiling outlet located in line with the front edge of the basin will often provide, in addition to improved lighting at the mirror, general room lighting and safety lighting for combination shower and tub. When more than one mirror location is planned, equal consideration should be given to the lighting in each case.

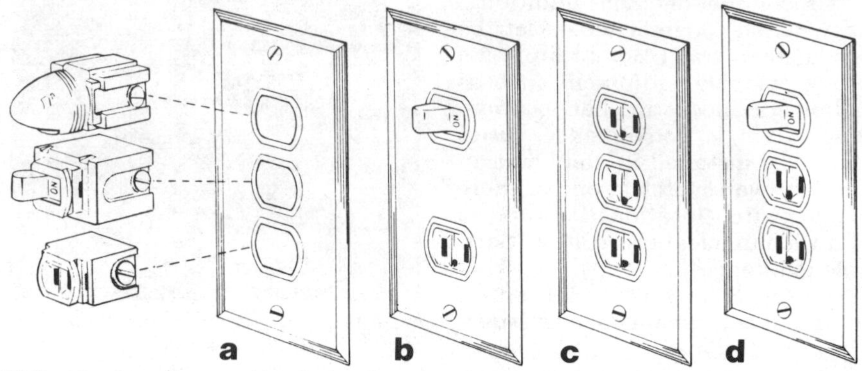

FIG. 13 Interchangeable devices can be included in the same outlet box in various combinations: (a) pilot light, switch and receptacle, (b) switch and receptacle, (c) triplex receptacle and (d) switch and duplex receptacle.

Where an enclosed shower stall is planned, a vapor-proof fixture should be provided, controlled by an outside wall switch which cannot be reached from within the stall. A switch-controlled night light also is recommended.

Convenience One outlet near the mirror, three to five feet above the floor, should be provided. It is recommended that an outlet be installed at each separate mirror or vanity space, and also at any space that might accommodate an electric towel dryer, electric razor or other small electrical equipment. A receptacle which is a part of a bathroom lighting fixture is not suitable for this purpose unless it is rated at 15 amps and wired with at least 15-amp rated wires (Fig. 15).

Special Purpose A wall-switched, built-in ventilating fan

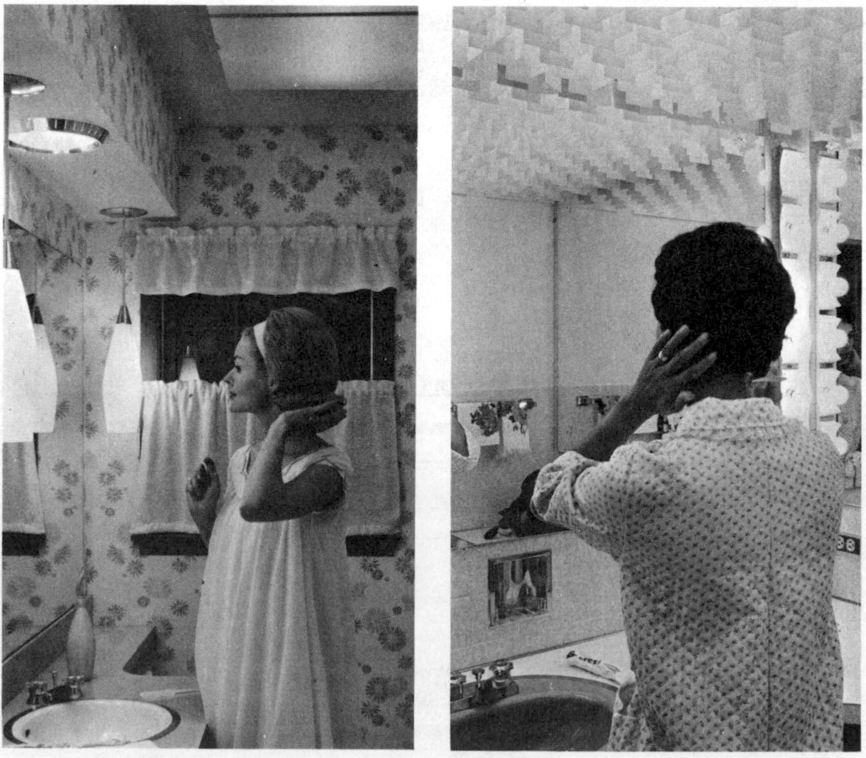

FIG. 14 Adequate bathroom lighting may consist of (a) multiple light sources on both sides of the mirror or (b) diffused illumination from a luminous ceiling.

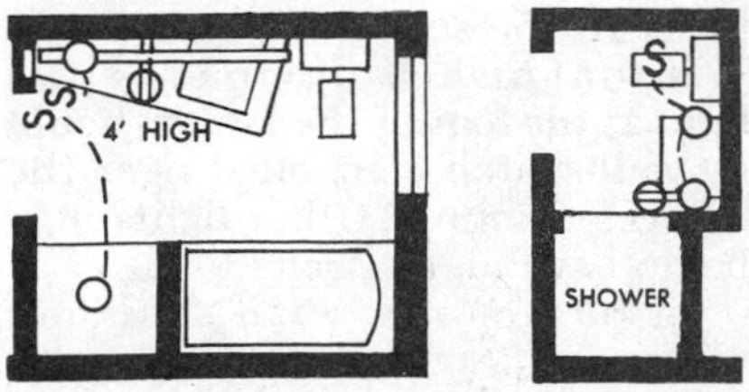

FIG. 15 In a small bathroom, a shower light may not be necessary; placing the switch out of reach from the shower stall may be difficult.

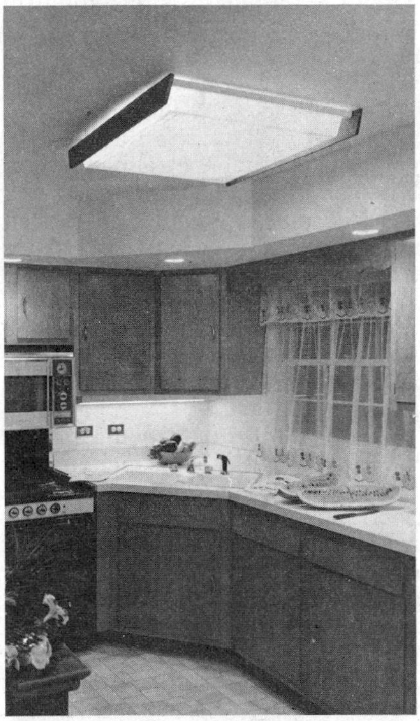

FIG. 16 Ceiling luminaire provides overall illumination, while soffit downlights and undercabinet fluorescent strips highlight work areas.

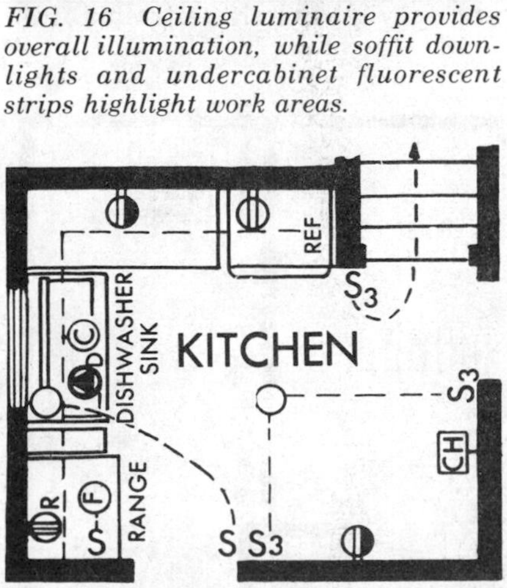

FIG. 17 Outlets in the kitchen must be carefully located and sized for the intended equipment. Convenience outlets should be of the split-receptacle type.

capable of providing a minimum of eight air changes per hour is recommended. Such a fan is generally required where natural ventilation is not provided thru windows or skylights. Also, in many parts of the country an electric space heater is desirable for increased comfort during bathing.

Kitchen

Lighting Wall-switched outlets for general illumination and for lighting at the sink are essential. Lighting design should provide for illumination of the work areas, sink, range, counters and tables (Fig. 16).

Undercabinet lighting fixtures within easy reach may have pull chain, push button or other local-switch control. Consideration should also be given to outlets to provide inside lighting of deep cabinets.

Convenience The planning should include: one outlet for the refrigerator; one outlet for each 4 linear feet of work-surface frontage, with at least one outlet to serve each separate work surface or planning desk. Work-surface outlets should be located approximately 44″ above finish floor level (Fig. 17). If work or dining table is planned next to a wall, this wall space should have one outlet located just above table level.

An outlet is recommended at any wall space that may be used for ironing or for an electric roaster. Convenience outlets in the kitchen, other than that for the refrigerator, should be of the split-receptacle type (Fig. 11).

Special Purpose Often required: one outlet each for a range, ventilating fan, dishwasher and waste disposer.

A special recessed receptacle with hanger is recommended for an electric clock. The clock should be located so as to be easily visible from all parts of the kitchen. It is also desirable to provide for a food freezer either in the kitchen or in some other convenient location.

Laundry Areas

Lighting Outlets for fixed lights should be installed to provide illumination of work areas, such as laundry tubs, sorting tables, washing, ironing and drying centers. At least one outlet in the room should be wall-switch controlled. For laundry trays in unfinished basements at least one ceiling outlet, centered over the trays, is recommended (Fig. 18).

Convenience At least one convenience outlet should be provided, preferably near the sorting table. If properly located, one of the special-purpose outlets (the washer outlet, for example) may also serve as a convenience receptacle for such items as laundry hot plate or sewing machine. Convenience outlets in the laundry area should be of the split-receptacle type (Fig. 11).

Special Purpose One outlet for each of the following pieces of equipment should be provided: automatic washer; hand iron or ironer; and clothes dryer. Outlets for a ventilating fan and clock are desirable. Requirements for electric water heater should be obtained from the local power company.

Closets

Lighting Generally, one outlet for each closet should be provided. Where shelving or other conditions make the installation of lights within a closet ineffective, outlets in the adjoining space should be so located as to provide light within the closet. Wall-switches or automatic door switches are preferred, but pull-switches are acceptable (Fig. 12).

Halls

Lighting These recommenda-

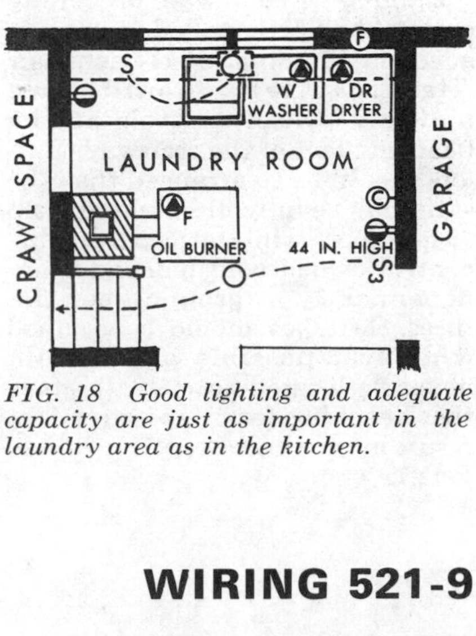

FIG. 18 Good lighting and adequate capacity are just as important in the laundry area as in the kitchen.

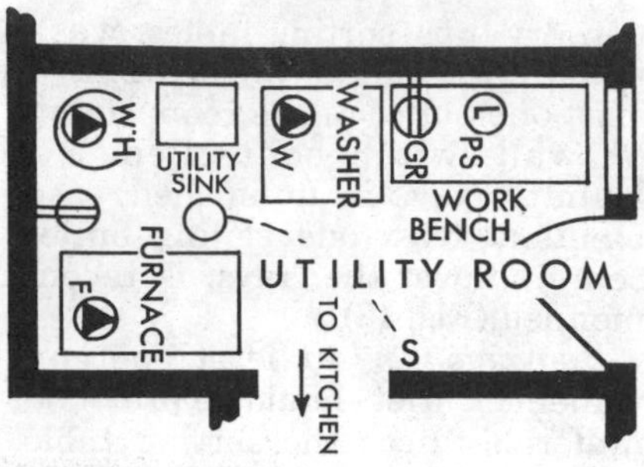

FIG. 19 Proximity to service equipment and use of exposed wiring make the cost of extra outlets in the utility room relatively low.

tions apply to passage halls, reception halls, vestibules, entries, foyers and similar areas. Wall-switch controlled outlets should be installed for proper illumination of the entire area. Particular attention should be paid to irregularly shaped spaces. It is recommended that a switch-controlled night-light be installed in any hall serving bedrooms.

Convenience One outlet for each 15 linear feet of hall, measured along the center line, should be provided. Each hall over 25 sq. ft. in floor area should have at least one outlet. In reception halls and foyers, convenience outlets should be placed so that no point along the floor line in any usable wall space is more than 10 feet from an outlet in that space.

A receptacle at one of the switch outlets is recommended for convenient connection of maintenance equipment such as vacuum cleaner, floor polisher or electric broom.

Stairways

Lighting Fixed wall or ceiling outlets shall be installed to provide adequate illumination of each stair flight. Outlets should have multiple-switch controls at the head and foot of the stairway. Outlets should be so arranged that the stair may be fully illuminated from either floor, while also permitting control of bedroom hall lights independently of ground-floor fixtures. Switches should be grouped whenever possible and should never be located so close to steps that reaching for a switch might result in missing a step.

These recommendations apply to any stairway serving finished rooms at both ends. For stairways to unfinished basements or attics, see appropriate headings below.

Convenience Outlets generally are not required, except at intermediate landings of larger than normal size, where a decorative lamp, night light or portable equipment might be used.

Utility Room

Lighting Lighting outlets should be placed to illuminate furnace area, and work area if planned. At least one lighting outlet should be wall-switch controlled (Fig. 19).

Convenience At least one convenience outlet should be provided, preferably near the furnace or near the planned work-bench location (Fig. 19).

Special Purpose Provide outlets for each piece of mechanical equipment requiring electrical connections such as furnace, boiler, water heater, water pump or compressor.

Basement

Lighting Lighting outlets should be placed to illuminate designated work areas, the foot of the stairway, enclosed spaces and specific pieces of equipment, such as furnace, pump or work-bench. Additional outlets for open (unpartitioned) basement space should be provided at the rate of one outlet for each 150 sq. ft. (Fig. 20).

In unfinished basements the light at the foot of the stairs should be wall-switch controlled near the head of the stairs. Other lights may be pull-switch controlled.

In basements with finished rooms, with garage space, or with other direct access to outdoors, the stairway lighting provisions apply.

It is suggested that a pilot light be installed in conjunction with the switch at the head of the stairs, when the basement is infrequently visited or when the stair layout makes it difficult to determine visually whether the light is on.

Convenience At least two convenience outlets should be provided in the open basement area. If a work bench is planned, one of these outlets should be placed at the specific location selected for this purpose (Fig. 20).

Additional convenience outlets may be desired for basement laundry, dark room, hobby area and for appliances such as dehumidifier or portable space heater.

Special Purpose Provide an outlet for each piece of mechanical equipment requiring electrical connections such as furnace, boiler, water heater, water pump or compressor. An outlet for a food freezer also should be considered.

Accessible Attic

Lighting Planning should include one outlet for general illumination, wall-switch controlled from

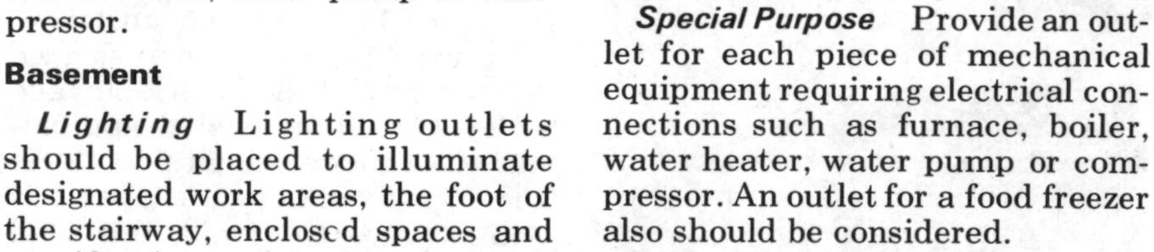

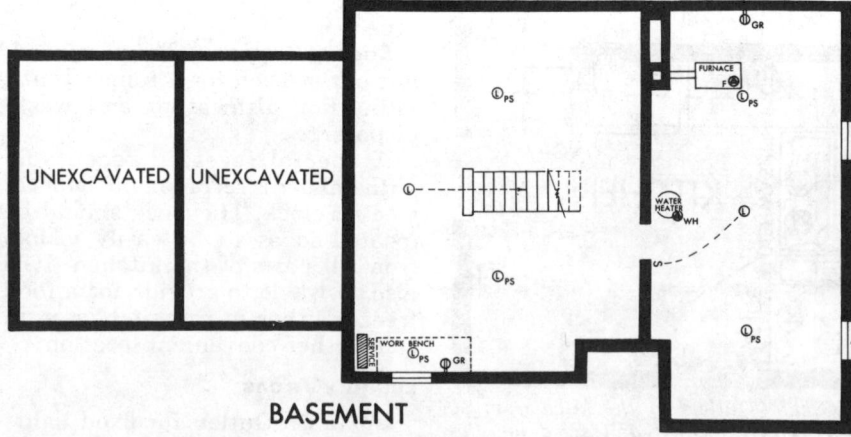

FIG. 20 Open basement wiring requirements are minimal, but future remodeling possibilities should be considered.

the foot of the stairs. The installation of a pilot light in conjunction with the switch controlling the attic light is recommended. When no permanent stairs are installed, this light may be pull-switch controlled, if located over the access door. Where it is contemplated that an unfinished attic will be developed into habitable rooms, the attic-lighting outlet should be multiple switch controlled at the top and the bottom of the stairs. There should be one outlet for each enclosed space.

Convenience A convenience outlet is desirable for portable equipment and for "trouble lights" capable of reaching into distant dark corners. If an open stairway leads to future attic rooms, provide a junction box with direct connection to the distribution panel, for future extension to convenience outlets and lights when rooms are finished.

Special Purpose A summer cooling fan, multiple switch controlled from convenient points throughout the house often is desirable.

Porches

Lighting Each porch, breezeway or other similar roofed area of more than 75 sq. ft. in floor area should have at least one wall-switched lighting outlet. Large or irregularly shaped areas may require two or more outlets. Multiple-switch control should be installed at entrances when the porch is used as a passage between the house and garage (Fig. 21).

Convenience At least one convenience outlet should be provided for each 15 linear feet of wall bordering porch or breezeway. Outlets should be of the weatherproof type if exposed to the elements. It is recommended that all outlets be controlled by a wall switch inside the house. Since this area often is used as an outdoor dining area, it is also recommended that at least one outlet be of the split-receptacle type.

Terraces and Patios

Lighting One or more fixed outlets on the building wall, or on a post centrally located in the area, is recommended for the purpose of providing general illumination. Such outlets should be wall-switch controlled just inside the house door opening onto the area.

Convenience At least one weatherproof outlet for each 15 linear feet of house bordering terrace or patio should be provided. It is recommended that these outlets be wall-switch controlled from inside the house and be located at least 18" above patio level.

Garage or Carport

Lighting At least one wall-switched ceiling outlet is recommended for one- and two-car garages. If the garage has no covered access from the house, provide one exterior outlet, multiple-switch controlled from garage and residence.

If the garage is to be used also as work area or laundry, provisions appropriate to these uses should be made. Additional interior outlets often are desirable even if no specific additional use is planned for the garage. For long driveways, additional illumination, such as by post lighting, is recommended. These lights should be wall-switch controlled from the house (Fig. 21).

Convenience At least one outlet should be provided for one- and two-car garages and carports (Fig. 21).

FIG. 22 Exterior lighting assures safety and creates dramatic views from inside and outside the house.

Special Purpose If a food freezer, work bench, or automatic door opener is planned, outlets appropriate to these uses should be provided.

Exterior Grounds

Lighting Floodlights often are desirable for illumination of surrounding grounds. Outlets may be located on the exterior of the house or garage, or on appropriately placed posts (Figs. 21 & 22). All outlets should be switch controlled from within the house. Multiple- and master-switch control from strategic points also is desirable.

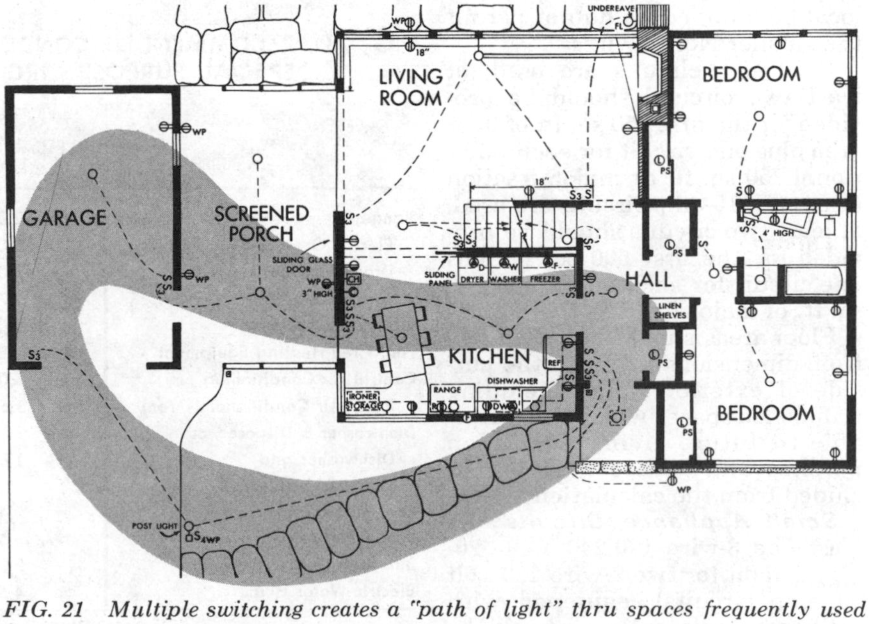

FIG. 21 Multiple switching creates a "path of light" thru spaces frequently used for passage.

CIRCUIT RECOMMENDATIONS

In addition to the *branch circuits* providing power for lights, appliances, fixed and portable equipment, a system may include *feeder circuits* serving secondary distribution panels near heavy-use locations.

Branch Circuits

A residential wiring system usually includes three types of branch circuits—general purpose, small appliance and special purpose. Each circuit involves consideration as to voltage, amperage and number and size of conductor wires.

General-Purpose General-purpose circuits supply all lighting outlets throughout the house and all convenience outlets except the convenience outlets in spaces designated for dining, cooking and laundering. All general-purpose circuits should be wired with suitable conductors: not smaller than No. 14 AWG (American Wire Gauge) if protected by 15 amp overcurrent device and No. 12 AWG for 20 amp device. Since the No. 14 conductors are more limited in circuit length (30 ft.) and current-carrying capacity (1,725 watts), No. 12 conductors with 20 amp devices are highly recommended for all general purpose circuits. Some local building codes do not permit the smaller No. 14 wires.

If 20 amp circuits are used, at least two circuits should be provided for the first 750 sq. ft. of floor area plus one circuit for each additional 750 sq. ft. or major fraction thereof. If 15-amp circuits are used, at least two circuits should be provided for the first 500 sq. ft. plus one circuit for each additional 500 sq. ft. or major fraction thereof.

Floor areas should be calculated from dimensions taken to the outside of exterior walls, including unfinished spaces which are adaptable to future living use; open porches and garage may be excluded from the calculation.

Small Appliance Circuits At least one 3-wire 120-240 Volt, 20-amp circuit (or two 2-wire 120 volt 20 amp circuits), equipped with split-receptacle outlets, should be provided for all areas regularly intended for cooking, dining or laundering.

The use of 3-wire circuits for supplying convenience outlets in the locations mentioned is an economical means of increasing the capacity and offers practical operating advantages. Such circuits provide greater capacity at individual outlet locations and lessen voltage drop in the circuit. For maximum effectiveness, the upper half of all receptacles should be connected to the same side of the circuit (Fig. 11). A separate 20-amp circuit should be provided for each appliance or piece of equipment rated over 1400 watts.

Special Purpose Circuits Separate circuits should be provided and conductors should be sized appropriately for the specific equipment tabulated in Fig. 23.

Spare circuit positions should be provided in the distribution panel for at least two future 20-amp 2-wire 115-volt circuits in addition to those initially installed. If the branch circuit or distribution panel is installed in a finished wall, raceways should be extended from the panel to the nearest accessible unfinished space for future use.

The majority of appliances for residential use are made for 110 to 125 volts. However, there is a growing tendency among manufacturers to make fixed appliances for use on 230-250 volt circuits, because of improved operating efficiency. It is recommended that the higher voltage be used in those cases where a choice exists.

Electrical heating and cooling systems are being increasingly used in some areas. These systems generally are custom designed to suit the particular house and local climatic conditions. Adequate wiring and service capacity criteria for such purposes are not included here and should be determined by consultation with local power company.

Feeder Circuits

Consideration should be given to the use of separate subpanels served by appropriate-size feeders, near heavy-use locations throughout the house, such as at electric range or clothes dryer.

SERVICE REQUIREMENTS

Service should be 3-wire 120-240 volt 100-amp minimum. Service entrance conductors should be insulation-covered and sized as required by the National Electric Code.

Such 100-amp service generally is sufficient to provide for lighting

FIG. 23 RECOMMENDED CONDUCTORS FOR SPECIAL PURPOSE CIRCUITS

ITEM	TYPICAL WATTAGE	RECOMMENDED CONDUCTOR CAPACITY		
		Amps	Wires	Volts
Range, or	12,000*	50	3	120/240
Electric Oven and	5,000	30	3	120/240
Drop-In Cooktop	7,000	40	3	120/240
Automatic Washer	700	20	2	120
Electric Clothes Dryer	5,000*	30	3	120/240
Fuel-Fired Heating Equipment	800	20	2	120
Central Air Conditioning, or	5,000*	40	3	120/240
Room Air Conditioner (1 Ton)	1,500	20	2	120
Dishwasher & Disposer, or	2,700	20	3	120/240
Dishwasher and	1,800	20	2	120
Waste Disposer	900	20	2	120
Food Freezer	400	20	2	120 or 240
Water Pump	500	20	2	120 or 240
Bathroom Space Heater	2,000	25	2	240
Electric Water Heater	4,500*	30	3	240

*Typical wattages are shown; verify with manufacturer and/or local power company.

needs in a typical home for lighting, appliances and fixed equipment totaling 18,000 watts (Fig. 23).

For larger homes or where the total load is larger than 18,000 watts, the local power company should be consulted. Generally, 150-amp service is adequate for the conditions described above, plus three to five tons of air conditioning. A larger, 200-amp, service usually may be required when electric space heating is employed.

Inadequate service conductors and service equipment have been the most serious obstacles to the use of new equipment and appliances in the home. To provide for probable increases in electrical use, even when electric heating or cooling are not contemplated, it is suggested that service conductors be oversized to No. 1/0 AWG; accordingly, service equipment should be of sufficient rating to match the carrying capacity of these wires.

Because of the many major uses of electricity in the kitchen requiring individual-equipment circuits, it is recommended that the main service equipment or a subpanel be located near or on a kitchen wall. Such a location usually is close to the laundry, thus minimizing long circuit runs and wiring costs (Fig. 24).

SPECIAL CONSIDERATIONS

Additional facilities commonly considered a part of the wiring system and installed by the electrical trade include: entrance signals, radio and TV antenna outlets, telephone outlets, fire alarm and internal communication (intercom) systems.

Entrance Signals

Signal push buttons should be provided at each commonly used entrance. These should be connected to the door chime and/or buzzer giving distinctive signals for the main and service entrances. Electrical supply for entrance signals usually is obtained from a low-voltage transformer installed near the service equipment.

In smaller homes the door chime generally is installed in the kitchen. In larger homes the chime should be installed in a central location, where it may be heard throughout the house. In a multi-level house extension signals may be necessary to insure being heard in quarters located away from the main entry level. Entrance-signal conductors should be no smaller than the equivalent of a No. 18 AWG copper wire, and preferably should be installed in conduit.

Communication

For the larger house or one with accommodations for a resident servant, additional signal and communication devices deserve consideration. Dining-room-to-kitchen call signals include an under-the-table floor treadle or push-button connected to an annunciator in the kitchen. More extensive systems may consist of push-button stations in each bedroom, living room and other habitable rooms; or an intercommunication system (with or without radio or music system) operated from a master control panel in the master bedroom, recreation room or kitchen and connected to remote stations throughout the house.

Home Fire Alarm

An automatic fire-alarm protection often is desirable, with detectors at least in the storage space and boiler or furnace area. The alarm bell and test button should be installed in the master bedroom or other suitable location in the sleeping area. Because of the increased possibility of failure due to causes originating in another system, operating power supply should be from a separate and independent transformer, rather than from the entrance-signal or other multiple-use transformer.

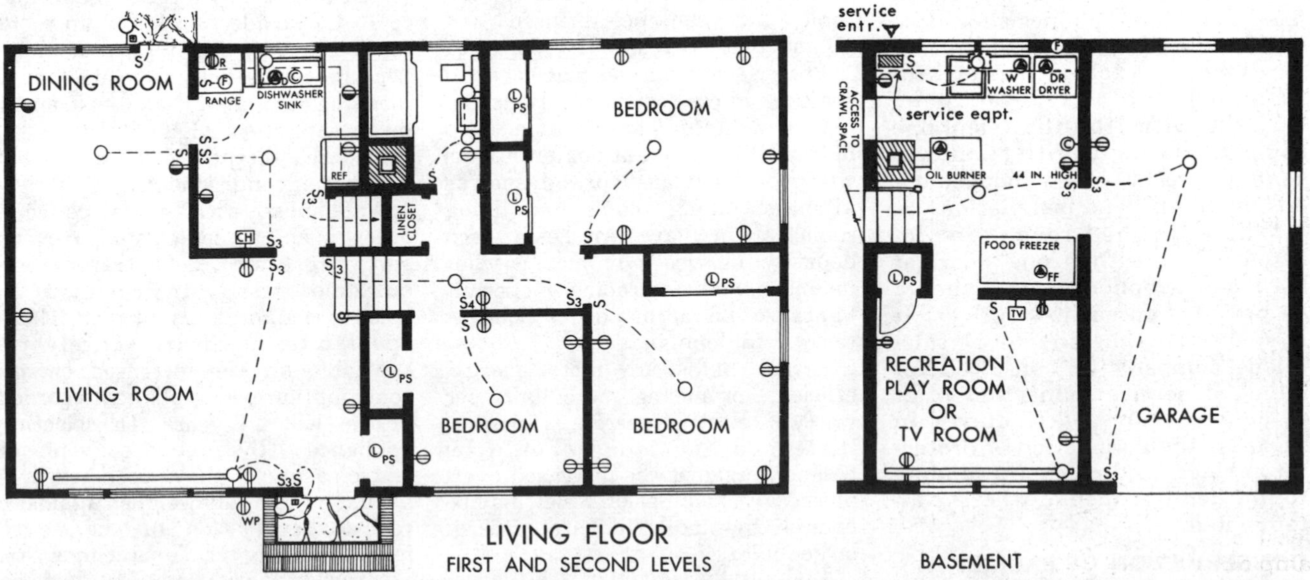

FIG. 24 In this house, a second distribution panel is unnecessary, because the service equipment is centrally located with respect to cooking, laundry and heating equipment.

Television & Radio

Provisions for a television antenna consist mainly of a nonmetallic outlet box at each television set location, connected to the attic by a suitable transmission line such as 300-ohm lead-in wire. The transmission line should be unshielded in a UHF service area and sufficiently slack to reach any reasonable location of the antenna in the attic. If the transmission line is shielded, the shield should be grounded. Where an unshielded transmission line is used, grounding should be in accordance with the National Electrical Code and the line should be kept away from pipes and other wires. A convenience outlet should be provided adjacent to each television antenna outlet. Antenna systems with more than two outlets generally require a booster amplifier, except in very strong signal areas.

In most areas, AM radio receivers used purely for the reception of broadcast communication do not require antenna and ground connections. For FM reception, provisions similar to those for television should be included.

Telephone

At least one convenient fixed telephone connection should be provided, but outlets for additional fixed or portable phones should be considered. Suggested additional locations are: kitchen, master bedroom, den and recreation room. Advance wiring by the telephone company during construction will generally result in an inconspicuous, more efficient installation.

Where finished rooms are located on more than one floor, at least one telephone outlet should be provided on each such floor. It is suggested that the local telephone company be consulted for details of service connection prior to construction, particularly in regard to the installation of protector cabinets and raceways leading to finished basements, where contemplated.

CONSERVATION OF ENERGY

The proliferating use of electrical equipment in the home causes increased demands not only on home wiring systems, but on national energy resources as well. Residential use of energy in 1977 accounted for nearly 22% of the total consumption of energy in this country. Energy usage per appliance has increased considerably for each type of appliance since 1959. Appliances that were once luxuries are now commonplace; the number of homes with color televisions has risen from 5% in 1964 to 81% in 1979.

Conservation of energy in the home is both a worthwhile national goal and a money-saving consumer objective. By careful design of the wiring system and judicious selection of equipment, it is possible to minimize energy waste, without sacrificing comfort and convenience.

Lighting

A primary purpose of a home wiring system is lighting, which is responsible for nearly 8% of home energy consumption. The use of lighting in the home is extensive, ranging from necessary illumination to enhancing interior design. However, misplaced or excessive illumination wastes electricity and money.

An entire room does not require the same high level of illumination, if close-up activity is ordinarily restricted to specific areas. For instance, lighting should be concentrated on areas such as the study desk, kitchen counter and dining table, but can be diffused and dimmer over the rest of the room.

Proper selection and placement of switches in convenient locations can reduce wasteful consumption from lights left on unintentionally. Rooms with more than one doorway, such as kitchens, dining rooms and living rooms, should have switches at each doorway. The use of dimmer switches in dining and living rooms will permit lights to be adjusted for various tasks or for lounging.

Larger bulbs are more energy-efficient, producing more light per watt of electric energy. However, increased efficiency should be balanced against the increased overall energy consumption which may be caused by indiscriminate use of larger bulbs.

Light bulbs should not be larger than required for the specific task or design effect. General service incandescent bulbs have a rated life of approximately 750 to 1,000 hours. For hard-to-reach areas such as roof overhangs and post lights, "long life" bulbs with a life expectancy of 2,500 hours can be used. While ordinary incandescent bulbs have a life span of 750 to 1,000 hours, fluorescent tubes are rated for 10,000 hours or longer. What's more, fluorescent tubes produce about 80 lumens per watt, or 4 times as much light as a standard incandescent bulb.

Appliances

Selection of energy-efficient appliances can reduce operating costs and minimize demands on the wiring system. Before the purchase of any appliance, the *life-cycle cost* (the long-range, lifetime, operating and maintenance cost) of the product should be determined. Buying an energy-efficient appliance may be initially more costly, but a significant saving can be made on subsequent electrical bills and energy demands.

The Energy Efficiency Ratio (EER) established by the National Bureau of Standards can be helpful in determining the relative efficiency of certain household appliances. The EER of air conditioners, for example, is determined by dividing the number of BTUs per hour by the number of watts needed for operation. The EER for air conditioners will range from 5 to 10 for smaller units and up to 12 for larger units. The higher the number, the more efficient the machine—meaning less energy and cost spent on operation. An EER of 7 or more is considered acceptable for household window air conditioners.

Appliances with extra conveniences, such as "self-cleaning" ovens and "frost-free" refrigerators, require more energy and cost more to operate than ordinary models. These special features are largely responsible for the increased energy consumption per appliance recorded in the last 20 years. In selecting appliances, the extra convenience and pleasure of the special features should be carefully weighed against a realistic evaluation of the added initial and long-term operating costs. The local power company can be helpful in making an economic life-cycle cost analysis and in providing other energy conserving tips.

The generally accepted mental picture of electricity is based on the theoretical structure of the atom. The atom can be visualized as consisting of a *nucleus,* and one or more *electrons,* moving in orbits around it (Fig. 25). Groups of electrons moving in neighboring orbits can be thought of as belonging to the same orbital shell. Depending on the number of electrons, there can be as many as seven such shells.

The nucleus consists of one or more positively charged *protons,* and one or more electrically neutral particles called *neutrons.* Each orbital electron has a negative charge, equal in magnitude and opposite to the positive charge of a nuclear proton.

In a neutral atom, the negative and positive charges balance out. But electrons in the outer shell can be dislodged and become "free" electrons, separated from the parent atom, which is then left with a net positive charge. A free electron may then attach itself to another neutral atom, lodging on its outer shell and creating a negative charge on the atom (Fig. 26). Single atoms or clusters of atoms with extra or fewer electrons are termed *ions.*

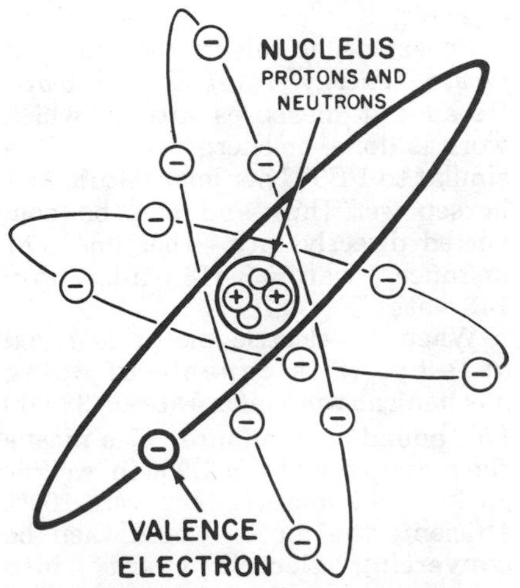

FIG. 25 Valence electrons, moving in furthest orbit from the nucleus, have the least attraction to it and can be dislodged to create electron flow—electricity.

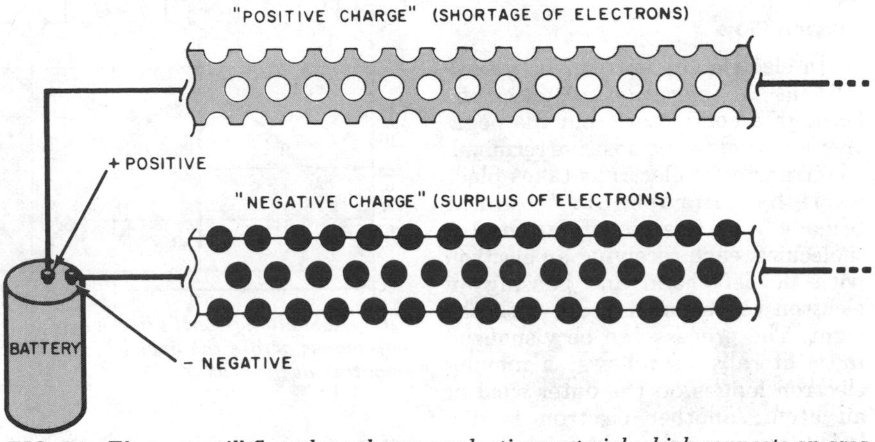

FIG. 26 Electrons will flow through any conductive material which connects an area with a surplus of electrons to an area with a shortage of electrons, such as the two terminals of a battery.

FUNDAMENTALS OF ELECTRICITY

A *battery* is the simplest source of electricity and is made up of one or more *cells.* A cell consists usually of a carbon rod (positive electrode), a piece of copper or zinc (negative electrode) and a chemical compound (electrolyte). The electrolyte breaks down readily into positively and negatively charged particles (ions) which attach themselves to the electrodes, imparting a negative or positive charge. The current-generating action can be illustrated by the *voltaic wet cell,* which uses a solution of sulfuric acid as the electrolyte (Fig. 27). In a *dry cell,* such as a flashlight battery, the zinc casing is used as the negative electrode, while a special paste acts as the electrolyte between the casing and the carbon rod (positive electrode).

The battery usually produces small amounts of electricity at low voltage. For larger amounts of power at higher voltage, an electric generator is used. Unlike the battery, which converts *chemical* energy into electric power, the generator utilizes *kinetic* (motion) energy. At a generating plant, the kinetic energy usually comes from a steam turbine. The steam is created by burning coal, gas or oil, or from the heat of a nuclear reactor. The steam turbine

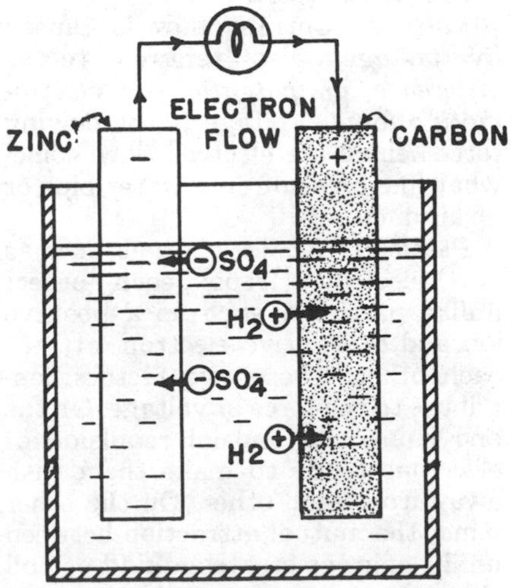

FIG. 27 The electrolytic solution (H_2SO_4) in a wet cell breaks down into positive (H_2) and negative (SO_4) ions. While the SO_4 ions carry electrons to the zinc (Zn) electrode, they also react with it, changing it into $ZnSO_4$.

turns the generator armature which has loops of wires wound around it. As the armature turns between the poles of a magnet, the loops cut across the invisible lines of magnetic force between the poles. This movement causes current to flow through the wire loops of the armature (Fig. 28).

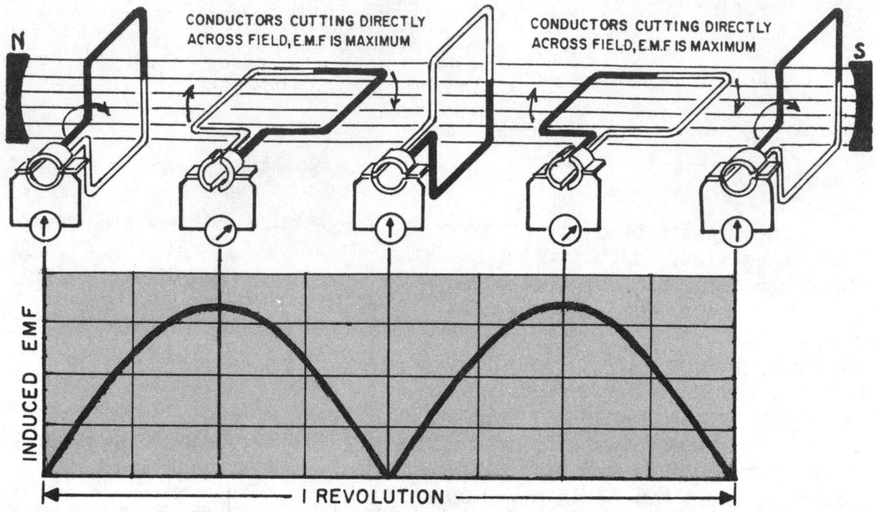

FIG. 28 Voltage (EMF) in the loop of a d-c generator armature is highest when that loop moves across the lines of force between the poles of a magnet. Using more coils smooths out the peaks.

Current Flow

An electric current can be visualized as a migration of electrons through a conductor from the negative terminal to the positive terminal. The transfer of electrons takes place as if by microscopic "bucket brigade"—a line-up of atoms or molecules, each receiving an electron with the left hand and passing an electron to the next atom with the right. The process can be visualized more literally as follows: a moving electron lodges on the outer shell of an atom; another electron is dislodged as a result, migrating to an adjacent atom. The action is repeated over and over again like a chain reaction (Fig. 26).

In order to keep the electrons flowing, there must be a continuous uninterrupted *circuit* (path) from the negative to the positive terminal. If the path is interrupted, electron flow stops, resulting in an *open circuit.* If a short cut is inadvertently created, permitting a sudden surge of electrons through an easier path, then we have a *short circuit.*

Voltage Current flow is caused by *voltage,* also referred to as *difference of potential* or *electromotive force.* Voltage is the driving force behind the electron flow, somewhat like *pressure* in a water pipe or an air duct.

Particles with like charges, such as two electrons, repel each other; unlike particles, such as a positive ion and a negative electron, attract each other. These two factors contribute to the force of voltage. On the one hand, the mutual repulsion of electrons tends to make them rush away from each other. On the other hand, the mutual attraction between unlike charges creates a kind of pull that draws electrons toward the positively charged atoms.

As stated earlier, electrons actually flow from a negatively charged to a positively charged body. However, the older notion that electricity flows from positive to negative terminal still is used in many textbooks and electrical diagrams as a "convention", a kind of accepted fiction or customary code. If electricity is assumed to flow from positive to negative terminal, then current must be composed of positive charged particles. This concept is contrary to established electron flow theory, but is useful in certain calculations.

Amperage The rate at which electrons, or current, flows through a conductor is measured in *amperes.* The speed with which electricity is transmitted is about the same as the speed of light, 186,000 miles per second. That is, if a current is started at one end of a line 186,000 miles long, it will be felt at the other end one second later. Rate of flow is different from speed because it takes into account "current density"—that is, the number of electrons moving along from atom to atom. Thus, *amperage* measures the *quantity* of electricity passing through the circuit *per unit of time.*

The quantity of electricity without reference to time is measured in *coulombs.* A coulomb can be defined roughly as the number of electrons conducted past a point. The number is unimaginably large—6,250,000,000,000,000,000 electrons. Defined more specifically, one ampere equals one coulomb of electricity passing a point in the circuit every second.

Resistance Low-temperature research has shown that some metals become super-conductive at temperatures near absolute zero (−459.6°F). But at normal use temperatures, the internal structure even in the best conductors opposes the flow of electric current and converts some current into heat. This internal friction-like effect is termed *resistance* and is measured in *ohms.* An ohm is that resistance which allows one ampere (one coulomb per second) to flow, when pushed by a pressure of one volt.

Power The electrical unit of *power* (energy rate) is the *watt.* Because it measures rate at which work is done, or energy is used, it is similar to 1 BTU per hour (Btuh) or 1 horsepower (hp) and can be converted directly into either one. For instance, 1 watt = 3.413 Btuh, 1 hp = 746 watts (Fig. 29).

When an electric motor is rated at 1 hp, it is capable of doing mechanical work at a rate of 33,000 foot-pounds per minute. This means the motor could lift a 1,000 lb. weight 33 ft. every minute. If it were 100% efficient, the motor would then be converting electric energy into mechanical work at a rate of 746 *watts.* In electric heating, electricity is converted into heat rather than mechanical work (Fig. 29).

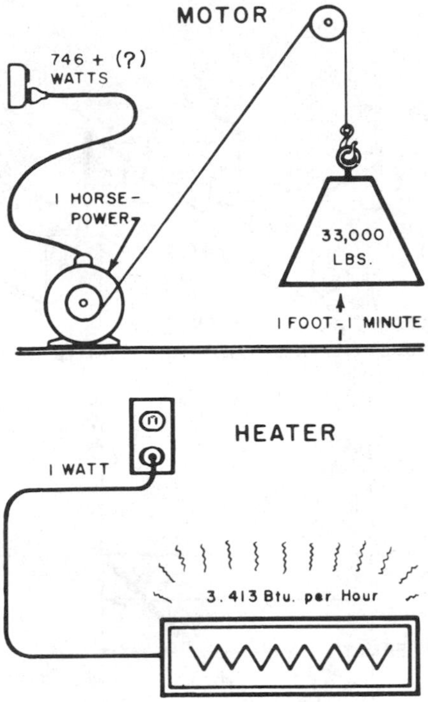

FIG. 29 Electric energy can be converted into kinetic (motion) or thermal (heat) energy.

Stating the relationship in electrical terms, one gets the basic *equation for power, W = EI,* where wattage (W) equals voltage (E) times amperage (I). The equation says, in effect: the heavier the flow rate (amperes) at a given supply pressure (volts)—the higher the rate (watts) at which energy is being supplied and used.

An electric utility, for example, might use this equation to find out the load demand on the distribution lines in watts or *kilowatts* (1,000 watts). Or, a designer may wish to determine the wattage of an electric heater whose rating is 22 amperes. Multiplying 22 amperes times 240 volts gives the answer, 5,280 watts (about 5.3 kilowatts).

Ohm's Law

Voltage, amperage and resistance in an active circuit are interrelated. The greater the resistance, the greater the voltage required to produce a given amperage. Current and resistance together, then, imply the required voltage pressure, and that is stated in *Ohm's Law:* $E = IR$, where voltage (E) equals amperage (I) times resistance (R). Ohm's Law makes it possible to determine one of these values, if the other two are known (Fig. 30).

Combining the power equation with Ohm's equation, it is possible to derive many equations for the relationship between voltage, wattage, amperage and resistance (Fig. 31). The power equation as stated above is $W = EI$; substituting IR for E (Ohm's equation), we get $W = (IR)I = I^2R$. In other words, wattage (W) equals amperage (I) squared times resistance (R). This equation helps determine the power used in an electric heater or the power lost in transmitting electricity. In a resistance heater, the power is transformed into useful heat, but in distribution wires it is an undesirable waste of energy.

For instance, a No. 14 copper wire, 100 feet long, has a resistance of about 0.25 ohm. Maintaining a current of 10 amperes at 120 volts, what would be the line loss in the conductors? The line loss (Fig. 31) is obtained from the power equation; $W = I^2R$, or $(10)^2$ amps $\times$ 0.25 ohms = 25 watts. The total power would be $EI = W$, or 120 volts times 10 amps equals 1,200 watts. The 25 watt line loss would constitute a little over 2% of the total power, which is well within the recommended maximum line loss of 3%.

Alternating and Direct Current

So far it has been convenient to discuss electrical principles in terms of *direct* rather than *alternating* current, its more common commercial form. Direct current (d-c) always flows in one direction in a circuit: namely from the negative to the positive terminals of the source. Alternating current (a-c) also flows from negative to positive terminals but the terminals change polarity quickly, causing a reciprocating (back-and-forth) surge of electricity along the wire.

Alternating current is produced by an a-c generator (Fig. 32) designed so that the voltage and current reverse direction periodically. In the commonly used 60-cycle power supply, one cycle of a complete rise and fall of voltage in each direction takes place in 1/60 second. The voltage of a 120-volt direct current is actually 120 volts at the generator and should not drop below 115 volts at the load. However, because alternating current consists of rapidly changing values from zero to 170 volts, the average or useful voltage appears as 120 volts to the power consuming device (Fig. 33). Thus, the effect of both types of current is equalized and

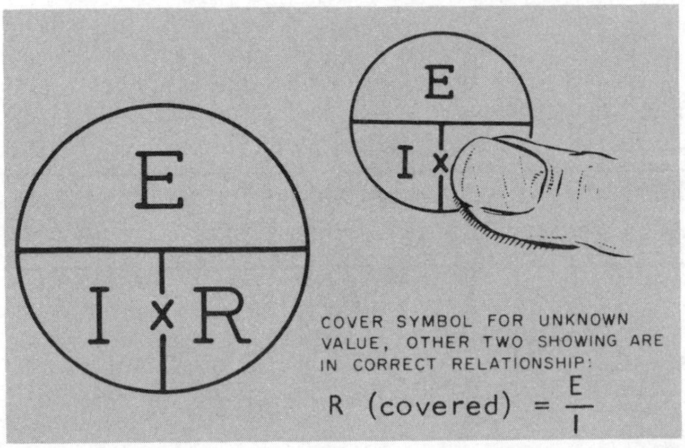

FIG. 30 Ohm's law gives relationship between voltage (E), amperage (I) and resistance (R).

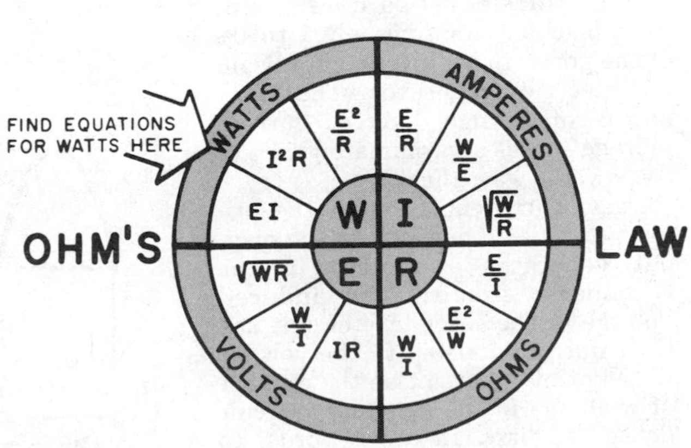

FIG. 31 Any one of the values in the center of the wheel can be found, if any other two are known.

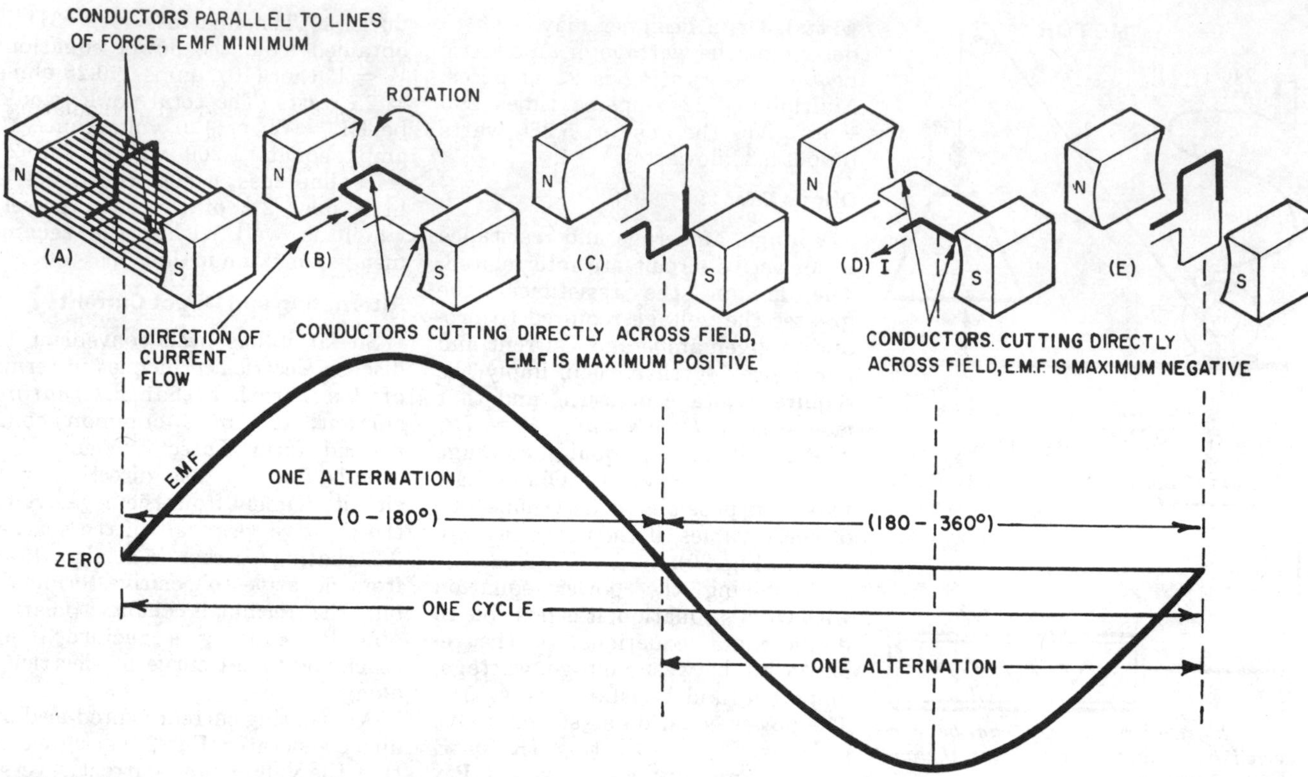

FIG. 32 In an a-c generator, voltage (EMF) peaks twice in one cycle, but alternate peaks are of opposite polarity, as current flow reverses in each alternation.

the above equations can be used for both a-c and d-c calculations.

Alternating current is now used almost exclusively in residential and commercial wiring because it provides greater flexibility in voltage selection and simplicity of equipment design. When alternating current is used, the voltage can be stepped up (increased) or stepped down (decreased) through the use of a *transformer* (Fig. 34). An alternating current transformer is a very efficient machine, wasting only 1 to 4% of the power in heat. By comparison, an electric generator—the only means of changing direct current voltage—loses approximately 25% of its power in waste heat.

Passing current through the primary circuit on one side of the magnet generates a current in the secondary circuit. The induced voltage in the secondary circuit has the same relationship to the voltage in the primary circuit as the number of windings in the secondary circuit do to the first. In other words, to double the voltage, it is merely necessary to double the windings in the secondary circuit; to triple the voltage, triple the windings; to reduce the voltage by 50%, use half the number of windings.

TRANSMISSION AND SERVICE

Electricity from a generating station typically is transmitted at *high voltage* to substations within the city, then to transformers near the users. There it is converted to *line voltage* (120 and 240 volts) for use in small commercial buildings, apartments and homes.

Primary Distribution

Most power is generated in power plants at 12,000 volts. For transmission over long distances, the voltage is stepped up by transformers to 115,000 volts or more, depending on the distance. Very high voltage is used because more power can be transmitted with a lower

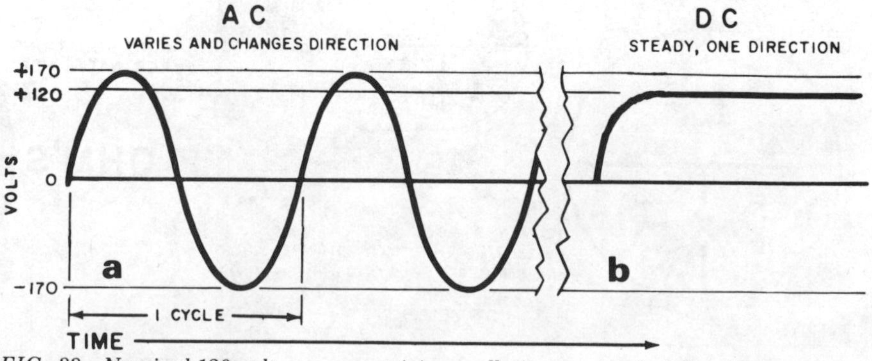

FIG. 33 Nominal 120-volt a-c current (a) actually is generated with voltage peaks of 170 volts to approximate the effect of 120-volt d-c current (b).

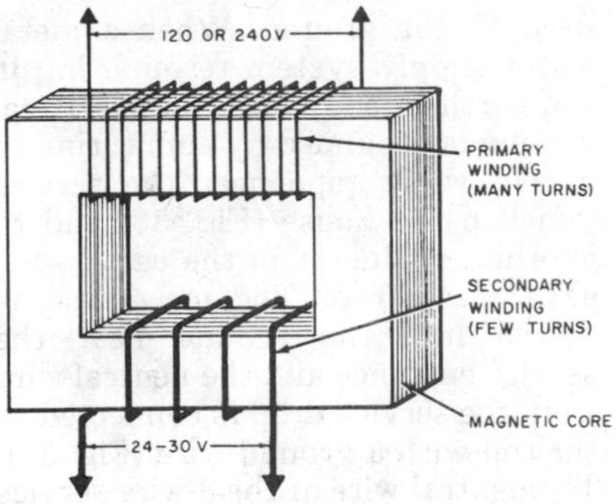

FIG. 34 Transformer is used to step up or step down voltage. To reduce voltage from 120 to 30 volts would require four times as many turns in the primary as in the secondary winding.

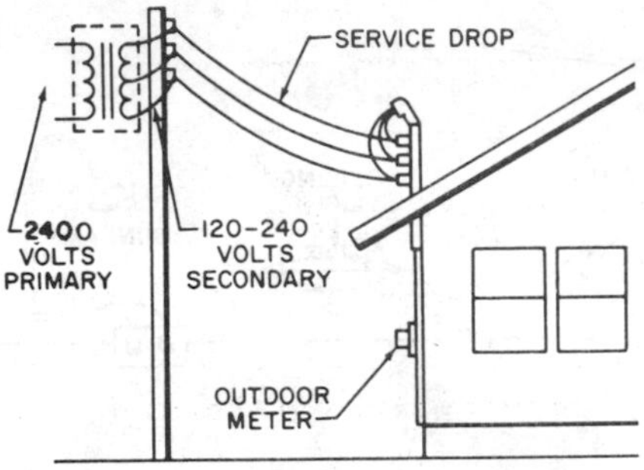

FIG. 35 Pole mounted transformer steps down high voltage to line voltage. Service drop should be at least 10' above ground.

power loss in the lines. Utility companies generally use the highest voltage consistent with safety and economy in their transmission lines.

In order to avoid hazards from high voltage transmission lines in built-up areas, high voltage is reduced at substations to either 12,000/7,200 volts or 4,160/2,400 volts. Smaller transformers, generally mounted on poles (Fig. 35), set on the ground or in underground vaults, further transform current to line voltage (120/240 volts).

Single-phase Service

Electric heating equipment and most large, fixed appliances such as water heaters, electric ranges and clothes driers, require 240 volts because they operate more efficiently at the higher voltage. However, general lighting and smaller, movable appliances such as dishwashers, refrigerators and toasters operate adequately on 120 volts. To satisfy both needs, 120 and 240 volts must be available. The dual voltage is provided in each block or group of blocks by a 240 volt, center-tapped secondary transformer (Fig. 36). This transformer makes both 120-volt and 240-volt power available through a 3-wire electric service entrance.

The three-wire *service drop* runs from the utility pole to the house. A 3-wire circuit consists of two *hot wires*, each carrying 120 volts and a *neutral wire*, having no voltage. Items requiring 240 volts are connected between the two hot wires; loads requiring 120 volts can be connected between either hot wire and the neutral (Fig. 36).

The neutral wire is connected to the ground, either before it enters the house (Fig. 37) or to a water pipe inside the house (Fig. 38). The two hot wires are generally color-coded with black and/or red insulating jackets. The neutral wire (also termed the ground*ed* wire) is usually white. The neutral wire is continuous throughout the system, is never interrupted by fuses or switches and generally is not connected to a hot wire.

For a residential or commercial wiring system, the earth is generally

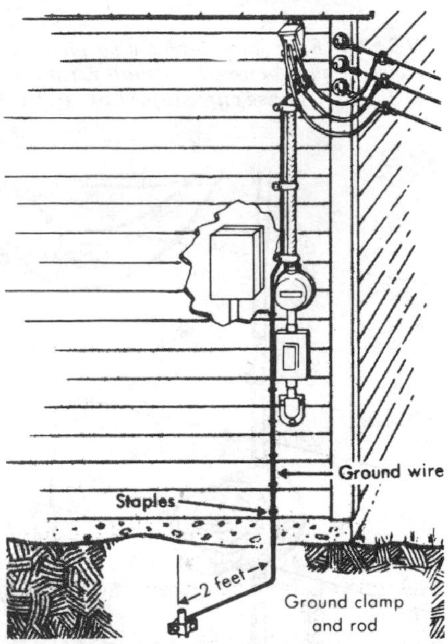

FIG. 37 The ground rod should be solid ½" copper or ¾" galvanized pipe, 8' long.

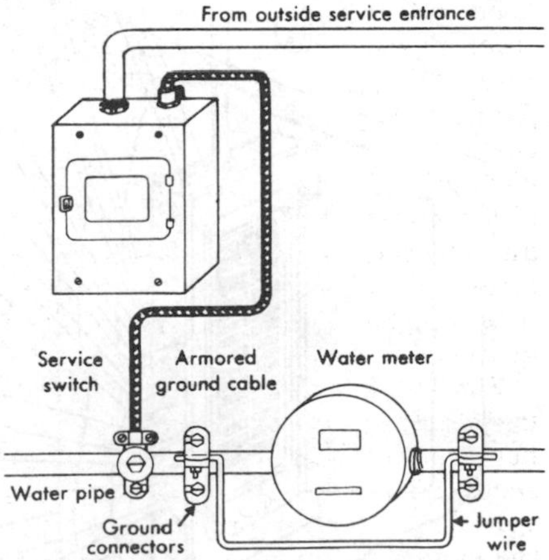

FIG. 38 Electrical system is grounded by connecting neutral bar in service panel to water pipe on street-side of meter. When connecting on house-side of meter, jumper wire should be used around meter, valves and unions.

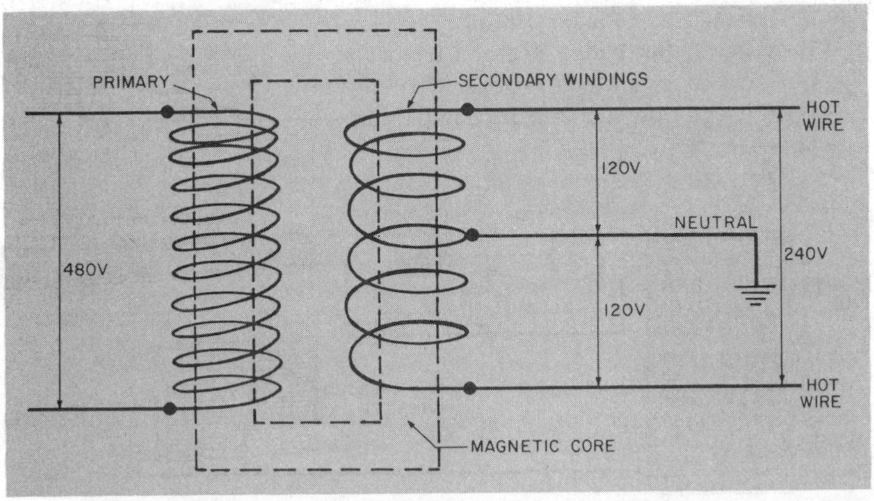

FIG. 36 Center-tapped transformer can convert two-wire 480-volt (or higher) current to three-wire 120/240 volt service.

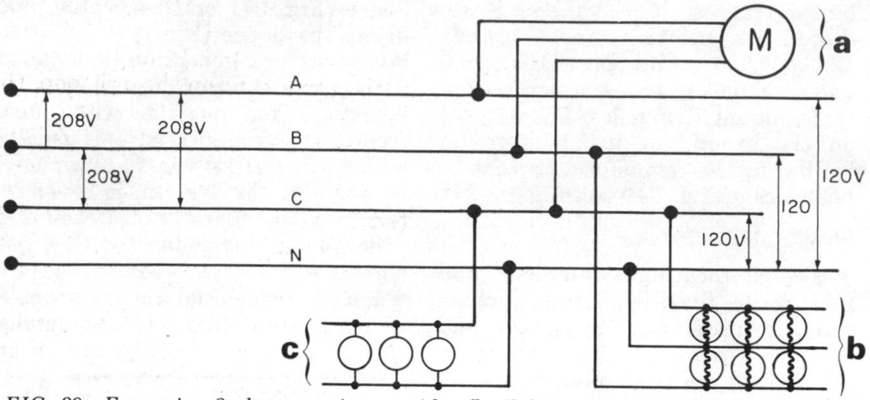

FIG. 39 Four wire, 3-phase service provides flexibility in serving different loads: (a) 3-wire, 3-phase circuit for motor, (b) 3-wire, 1-phase circuit for kitchen appliances and (c) 2-wire, 1-phase current for lights and receptacles.

used as the ground. When a metal water supply system terminating in the earth is available, the electrical system is grounded by connecting to a cold water pipe near the service panel in the house (Fig. 38) and to another conductor in the earth such as a metal rod. The metal rod is driven into the ground near the service entrance and the neutral wire from the service drop is connected to the rod with a ground wire (Fig. 37). The neutral wire of the 3-wire service is similarly grounded at the utility pole.

Three-phase Service

Single-phase 240-volt service is

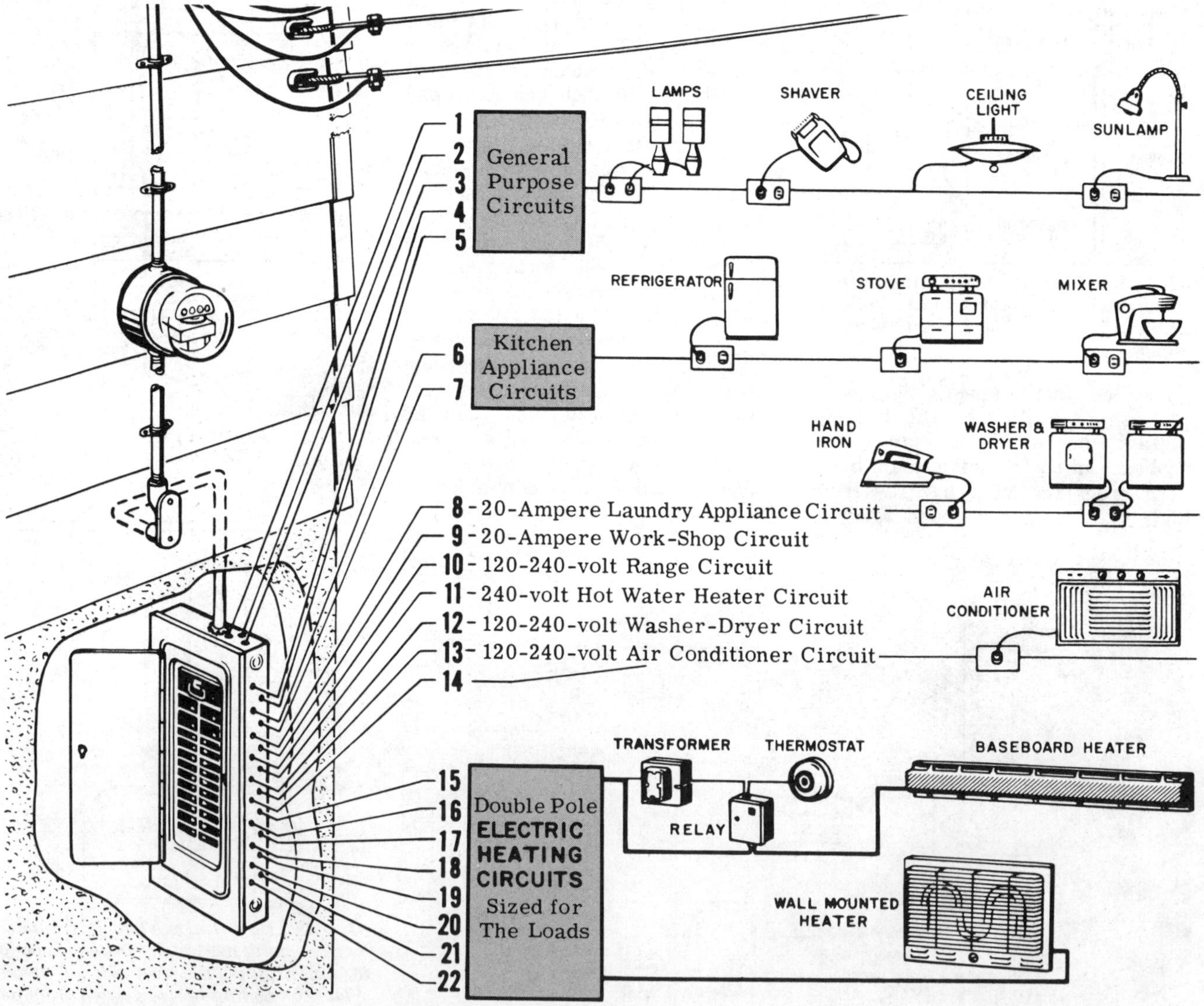

FIG. 40 The service equipment (panel) is the heart of the home wiring system; from here current is distributed all over the house and current flow is regulated to prevent overloads.

used almost universally for residences in this country. However, four-wire *three-phase* service is available for industrial and commercial applications and occasionally is used also in residences. This system is particularly suitable for electrical heating/cooling installations, where large motors are involved.

Three-phase service consists of three hot wires and a neutral wire. The voltage in each hot wire is out of phase (out-of-step) with the others by 1/3 of a cycle, as if produced by three different generators. Connections made between any of the two hot wires produces nominal 208-volt current; connections made between either of the hot wires and the neutral results in nominal 120-volt current. Although the service has four wires, branch circuits may have two, three, or four wires, depending on the type of equipment served (Fig. 39).

A three-phase system should be planned so that the total load is about equally divided between the three phases. It should be remembered also that *three* wires are hot (rather than two, as in 3-wire one-phase service) and all three require overload protection.

OVERLOAD PROTECTION

The utility service drop usually is connected to the customer's service entrance conductors at the building, forming a *drip loop,* to prevent water from entering the conductor cable. From the drip loop, the conductors go first to a *watt-hour meter,* which keeps a record of how much power is consumed, and finally to the *service equipment.* The service equipment is the origin of all the branch circuits which supply the house electrical needs (Fig. 40).

The service equipment provides both 120-volt and 240-volt current for the general purpose, kitchen appliance and special equipment circuits. In addition, line-voltage circuits can be used to power transformers which supply low-voltage circuits.

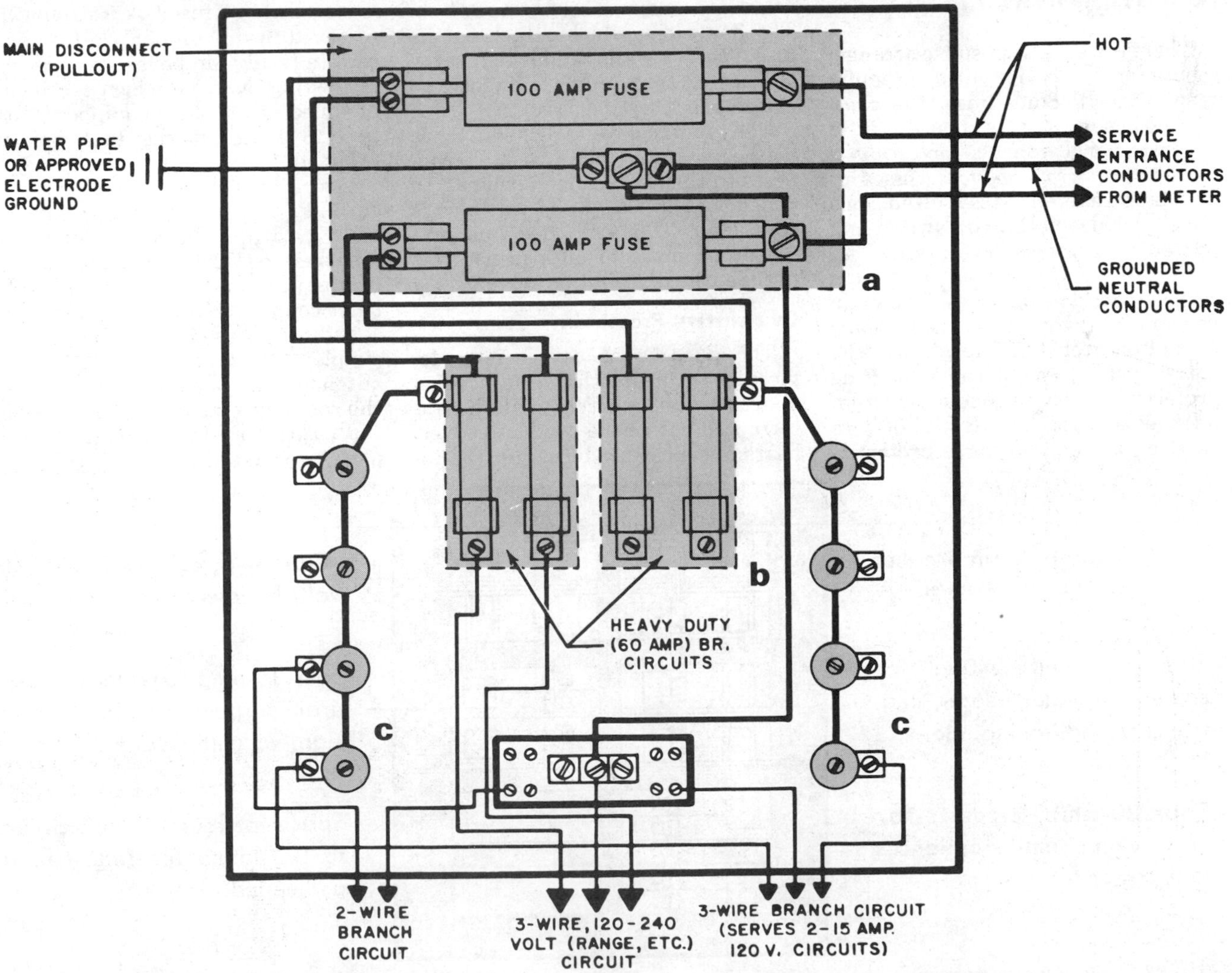

FIG. 41 This service panel consists of a fused 100-amp pullout switch (a) serving as main disconnecting means; two 60-amp pullouts (b) protecting heavy-duty circuits for electric heating and air conditioning; and 8 plug fuses (c) for appliances, equipment and lighting circuits.

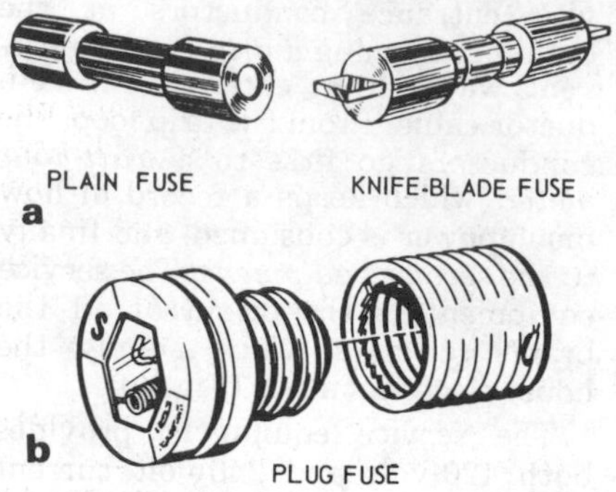

FIG. 42 Fuses are of the (a) cartridge type and (b) plug type. Type S adaptor makes it impossible to insert a 30-amp fuse in a 20-amp socket, thus avoiding dangerous overloads.

Service Equipment

There are several possible arrangements for the service entrance equipment (panel). Sometimes, the *main disconnect switch* is housed in a separate panel from the circuit *overcurrent protection devices* (fuses or circuit breakers). Most often, one panel includes both main switch and overcurrent devices.

Figure 41 illustrates a typical service panel, using fuses. The main disconnect switch provides a means of deliberately de-energizing the entire system, and at the same time protects the service conductors from unexpected overloads. Each hot wire entering the equipment is protected by a 100-amp cartridge fuse in the main disconnect switch, while the neutral wire is grounded. The distribution system includes two 60-amp 120/240 volt circuits protected by cartridge fuses, as well as 20-amp 120-volt circuits with plug fuses.

A 100-amp service such as in Figure 41 is adequate for general lighting and several household appliances typical of most residences. However, there are too few circuits and the mains are too small to supply an electric space heating system as well. For whole-house electric space heating, a 200-amp service generally is required.

At upper right in Figure 41, the service conductors enter the panel and connect to the terminals which serve the main disconnect switch. The hot wires, which attach to the disconnect switch terminals, carry the current through the main fuses to the branch fuses and circuits.

The grounded neutral conductor is neither switched nor fused. It connects to the neutral bus terminal, to which the service *ground* wire also is connected. This grounded bus provides terminals for all load neutrals in the entire distribution system.

Overcurrent Protection

The purpose of overcurrent devices is to insure that in case of a short circuit, current will be interrupted before the conductor overheats and possibly burns up the insulation. The older, more traditional device is the *fuse;* the newer, more typical device is the *circuit breaker.*

Fuses These protective devices have a fusible (meltable) wire, designed to melt and break when current flow exceeds the rated capacity of the circuit. Thus the circuit is opened and current flow is stopped before it becomes a hazard. Circuits of 30 amps or less can use *plug fuses;* for heavier currents, *cartridge fuses* are indicated.

Plug fuses are obtainable in sizes from 0 to 30 amperes (Fig. 42). The National Electrical Code requires that *type S* fuses be used for all plug-fused installations. The *type S* fuse uses a slightly different adaptor for each ampere rating; once the adaptor is installed in a fuse box it cannot be removed readily and only a fuse of a certain rating can be inserted. Existing wiring systems which are being modified should be equipped with type S fuses, if the fuse box is retained.

Cartridge fuses are made in all current ratings and voltage ratings from zero up. Fuses from 0 to 60 amperes use the *ferrule* type connection; above 61 amperes the *knife-blade* connection is used (Fig. 42).

Circuit breakers These are simple switch-like devices which automatically open the circuit when the rated current is exceeded—as in the case of a short circuit. Unlike plug fuses and most cartridge fuses,

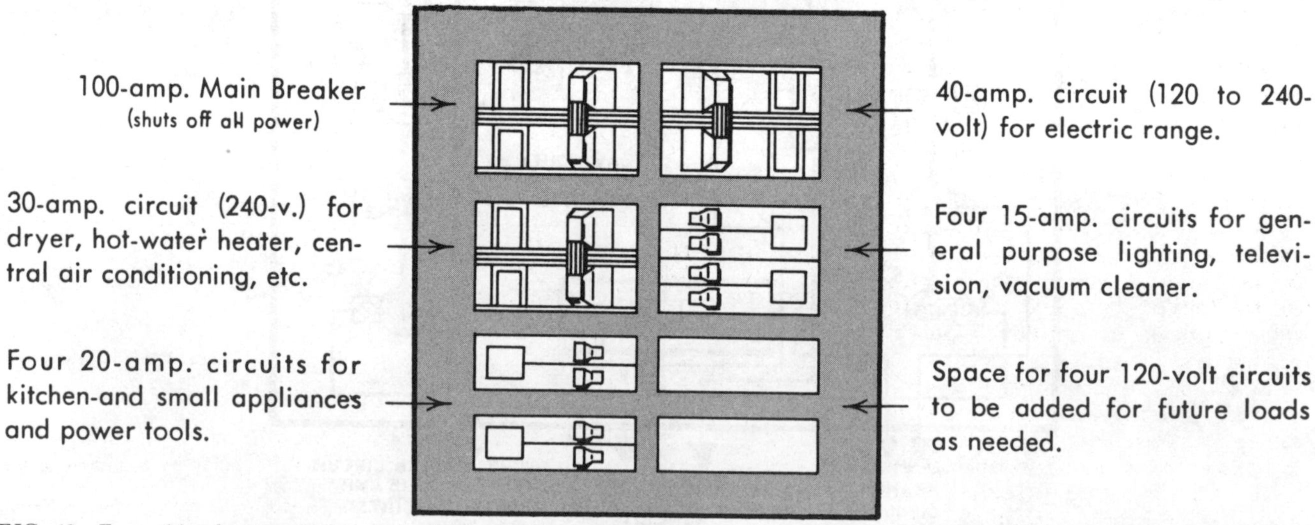

FIG. 43 Face of fuseless service panel, including 100-amp main disconnect breaker and smaller load circuit breakers. Breakers for 120-volt 3-wire or 240-volt circuits are connected in pairs, so each hot wire is protected. (The 1978 National Electrical Code permits up to 6 parallel main disconnects, so the panel could also be wired to provide separate protection for a clothes dryer, water heater, range, etc., without an additional system main disconnect.)

circuit breakers do not damage themselves when they *trip* (open). To reset (close) the breaker, it is simply necessary to push the toggle into the "on" position. Most small circuit breakers are of the thermal-magnetic type, meaning that when the breaker is tripped, the thermal element is heated. It may be necessary, therefore, to wait about a minute for the element to cool, before the breaker can be reset.

Circuit breakers are available in a wide range of sizes for many applications. Small breakers for 15, 20, 30 and 40-amp branch circuits are widely used today in service entrance equipment for residential and commercial buildings. All of these can be combined in a 100-amp service panel, including a larger 100-amp breaker for the main switch (Fig. 43).

Grounding

There are two types of grounding involved in a home wiring installation: (a) system grounding and (b) equipment grounding. *System grounding* refers to connecting the current-carrying neutral wire to the grounding terminal in the main switch, which in turn is connected to a water pipe. The neutral wire also is termed the ground*ed* wire. *Equipment grounding* refers to the use of a ground*ing* wire (or the metal raceway) to connect the metal frames of equipment and appliances back to the service cabinet and ground. The wire connecting the service cabinet to the water pipe is termed the *ground* wire.

Unlike the grounded (neutral) wire, the grounding wire and the raceway carry no current, except when a short circuit in the equipment or circuit wiring causes a "fault" current to flow. Under these conditions, equipment grounding minimizes shock hazard to occupants by providing a low-resistance path for the fault current and by opening the circuit overcurrent device in case of excessive flow.

In an electrical system involving conduit, tubing or metallic cable, metal boxes are used and equipment or fixtures permanently connected at the box become automatically grounded. Portable equipment with cords can be grounded through grounding receptacles and three-prong grounding plugs. In either case a grounding wire, color-coded green, is included in the cord or cable connecting the box or receptacle to the frame of the equipment.

When a non-metallic sheathed cable system is used, a bare grounding wire often is included to tie the system to the ground. Permanently connected equipment has its green grounding wire attached to the bare copper wire in a junction box. Grounding receptacles housed in non-metallic boxes have a special terminal for the grounding copper wire. Portable equipment cords include a green grounding wire which contacts the grounding system through the round or U-shaped prong of the plug.

Equipment grounding provides adequate safety from shocks, except in high hazard areas such as near pools, in bathrooms and outdoor locations. Receptacles closer than 10′ from the edge of a pool should not be installed at all. Poolside receptacles at distances from 10′ to 15′, as well as all outdoor receptacles, should be equipped with special devices capable of opening the circuit when even a small amount of current is flowing through the grounding system. These devices are called *ground fault circuit interrupters* (GFCI's) and are set to open the circuit when the ground fault current exceeds 5/1,000 of an ampere. The use of GFCI's is required by the NEC, because the severity of electrical shock is greatly increased when standing on wet ground.

FIG. 44 **PROPERTIES OF COMMONLY USED COPPER CONDUCTORS**

Conductor			Maximum Current (Amp)[4]	
Size (AWG, MCM)[1]	One-half Natural Size	Diameter (In.)[3]	Conductor type: RUW, T, TW	Conductor type: RH, RUH, THW, RHW, THWN
14	●	.0641	15	15
12	●	.0808	20	20
10	●	.1019	30	30
8	●	.1285	40	45
6	●	.184	55	65
4	●	.232	70	85
2	●	.292	95	115
1	●	.332	110	130
0	●	.372	125	150
00	●	.418	145	175
000	●	.470	165	200
0000	●	.528	195	230
350 MCM[2]	●	.681	260	310

[1] Sizes 14 to 8 are solid wires. Sizes 6 to 2 are 7-strand cables. Sizes 1 to 0000 are 19-strand cables. Size 350 MCM is a 37-strand cable.
[2] MCM is the designation of wire size in thousands of circular mils (350 MCM = 350,000 circular mils).
[3] Exclusive of insulation
[4] Based on ambient temperature of 86°F, (30°C); not more than 3 conductors in raceway, cable or direct burial. Where number of conductors exceeds 3, maximum current given should be derated as follows: 4-6 cond.—80%; 7-24 cond.—70%; 25-42 cond.—60%; 43 or more—50%.

WIRING METHODS

Electric current is carried and distributed by *conductors.* Technically speaking, conductors are the metallic cores which actually conduct the electricity, but in practice the entire wire, including the insulating cover, is termed the conductor. Conductors generally are protected from mechanical damage by an outer sheath. When the conductors and the protective sheath are fabricated together, the assembly is referred to as *cable.* Conductors can also be inserted on the job into *raceways*—specially built channels into which wires are placed or pulled.

Conductors

Conductors consist of an aluminum or copper core, covered by an outer coating of insulating material. The conductors for building wiring are available in American Wire Gauge (AWG) sizes, ranging from No. 14 to 4/0, and in larger sizes from 250 MCM (thousand circular mils) to 2,000 MCM (Fig. 44). Copper and aluminum are the typical conductor materials; aluminum generally is limited to the larger wire sizes. Due to its lower conductivity, aluminum wire smaller than No. 12 AWG (rated 15 amps) is not made. The same types of insulation are used for both copper and aluminum conductors.

Insulations used on electrical conductors are designated by basic-type letters identifying the material, such as rubber (R), thermoplastic (T), asbestos (A). Since the insulation generally determines the suitability of the conductor to certain exposures—such as moisture, heat or mechanical damage—conductors are identified by the material and properties of the insulating cover (Fig. 45).

Insulation thickness varies with wire size and may also vary with type of material, application or other code limitation. In addition to the basic-type letter derived from the material, other letters may be added to indicate modified or additional characteristics. For example, when an "H" is added to the basic "R" letter, the resulting type RH wire may be used to 167°F. maximum temperature instead of 140°F., which is the limit for RF wire (previously designated simply "R"). A double "H" indicates resistance up to 190°F., while a "W" indicates moisture resistance.

Cables

Cables are combinations of two or more conductors, factory-encased in a protective covering to safeguard them from mechanical damage. They can have metallic coverings as in *armored cable;* plastic or cambric (varnished cotton) jackets as in *non-metallic sheathed cable* (Fig. 46).

Armored Cable This cable (sometimes referred to as "BX") is a factory assembly of insulated conductors inside a flexible metallic covering. It may be run in the voids of masonry block or tile walls where the walls are not exposed to excessive moisture. It should not be run below grade; a transition should be made from armored cable to metal conduit or tubing when supplying a basement appliance.

Armored cable should be securely fastened by approved sleeves, staples or straps, spaced not more

Trade Name	Type Letter	Maximum Operating Temp.	Insulation	Applications and/or Limitations	Outer Covering
Heat-Resistant Rubber	RH	167 F	Heat-Resistant Rubber	Dry Locations	Moisture-Resistant, Flame-Retardant, Non-Metallic
	RHH	194 F			
Moisture- and Heat-Resistant Rubber	RHW	167 F	Heat- and Moisture-Resistant Rubber	Dry and Wet Locations	Moisture-Resistant, Flame-Retardant, Non-Metallic
Latex Rubber	RUH	167 F	90% Unmilled Grainless Rubber	Dry Locations	Moisture-Resistant, Flame-Retardant, Non-Metallic
	RUW	140 F		Dry and Wet Locations	
Thermoplastic	T	140 F	Flame-Retardant, Thermoplastic Compound	Dry Locations	None
Moisture-Resistant Thermoplastic	TW	140 F	Flame-Retardant, Thermoplastic Compound	Dry and Wet Locations	None
Moisture- and Heat-Resistant Thermoplastic	THW	167 F	Flame-Retardant, Moisture- and Heat-Resistant Thermoplastic	Dry and Wet Locations	None
Heat-Resistant Thermoplastic	THHN	194 F	Flame-Retardant, Heat-Resistant, Thermoplastic	Dry Locations	Nylon jacket
Moisture- and Heat-Resistant Thermoplastic	THWN	167 F	Flame-Retardant, Moisture- and Heat-Resistant, Thermoplastic	Dry and Wet Locations	Nylon jacket

than 4′-6″ apart. The cable should be supported within 12″ of an outlet box or other termination. When the box is mounted on equipment which needs to be moved for servicing occasionally, the unsupported cable length can be 24″.

Armored cable should be cut away to allow at least 6″ of the conductors into the box. The remaining armored portion should be clamped to the box. When the armor is cut back to provide free conductors, care should be exercised to prevent damaging the conductor insulation. Also, a fiber insulating bushing should be inserted in the end, between the armor and the conductor insulation. This bushing should remain visible within the box.

Armored cable should be handled and bent with care. If it is bent too sharply, the convolutions of the armor can damage the conductor insulation, resulting in possible short circuits or leakage to ground.

Non-metallic Sheathed Cable This cable (sometimes referred to as "Romex") consists of two or more insulated conductors having an outer sheath of moisture resistant, non-metallic material (Fig. 46). The conductor insulation is rubber, neoprene, thermoplastics, or a moisture resistant, flame retardant fibrous material.

Non-metallic sheathed cable should be securely fastened by approved staples, straps or other supports, in such a manner that the cable will not be damaged. The supports must not be more than 4′-6″ apart. Generally the cable must be fastened within 12″ of each termination. In concealed spaces of existing or prefabricated buildings where such support is impractical, the cable may be fished between points of access. When run at an angle to joists, the cable should either be supported on a running board secured to the joists, or run through holes bored through the joists.

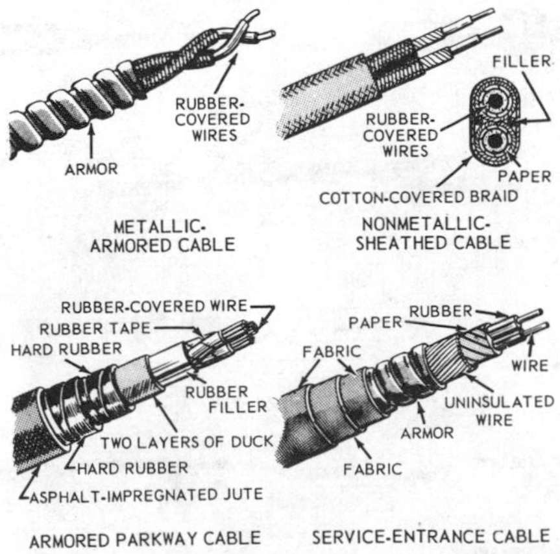

FIG. 46 Common types of electrical cable.

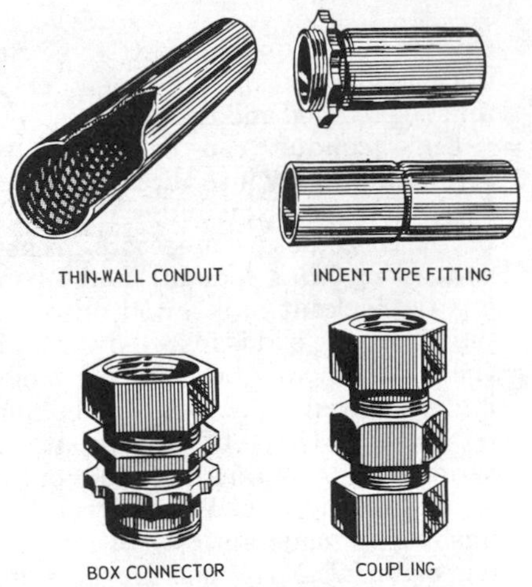

FIG. 47 Electrical metallic tubing (EMT, or "thinwall" conduit) and fittings.

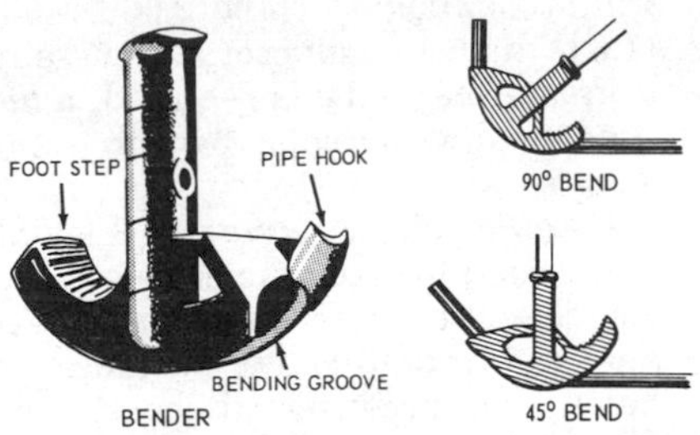

FIG. 48 EMT bender assures smooth curves, without pinching the interior of the pipe.

Care should be taken not to damage the outer sheath of the cable during installation. No bend should have a radius less than 5 times the diameter of the cable.

There are two types of non-metallic sheathed cable: type NM and type NMC.

Type NM cable has conductors which satisfy the requirements for type RH, RW, T and TW wire, and a sheath which is flame-retardant and moisture-resistant. Type NM is limited to use in dry locations such as air voids in masonry or tile walls not exposed to excessive moisture.

Type NMC conductors must satisfy the same requirements for type NM, but the sheath must be not only flame-retardant and moisture-resistant, but also fungus-resistant and corrosion-resistant. Type NMC may be used in damp or corrosive locations, as well as dry areas.

Raceways

Raceways are protective enclosures designed for placement of conductors in the field. Included in the general term are *electrical metallic tubing, rigid* and *flexible conduit* and *wireways*.

Electrical Metallic Tubing (EMT) This electrical pipe (also called "thin-wall conduit") may be used for both concealed and exposed work (Fig. 47). It is the most common type of raceway used in single family, low-rise residential and commercial construction. With proper joints and connections it may be installed in wet, as well as dry, locations.

Tubing normally is made in 1/2″ to 2″ nominal trade size. The actual internal diameter is slightly greater than the nominal size. Connectors and couplings for EMT are secured to the tubing by a setscrew or dimples made on the job in the collar of the fitting and tube after assembly. Watertight compression fittings also are available.

Tubing is frequently selected for residential wiring because it is light, easy to handle, and requires no special threading tools. However, it requires more skill to bend than rigid conduit, and should be performed with a special type of bender to prevent flattening (Fig. 48). The radius of the inner edge of a field bend should not be less than shown in Figure 49.

FIG. 49 **MINIMUM RADIUS OF FIELD-BENT CONDUIT & TUBING**

Size of Conduit (in.)	Radius to Center of Conduit (in.)
1/2	4
3/4	4-1/2
1	5-3/4
1-1/4	7-1/4
1-1/2	8-1/4
2	9-1/2
2-1/2	10-1/2
3	13
3-1/2	15
4	16
4-1/2	20
5	24
6	30

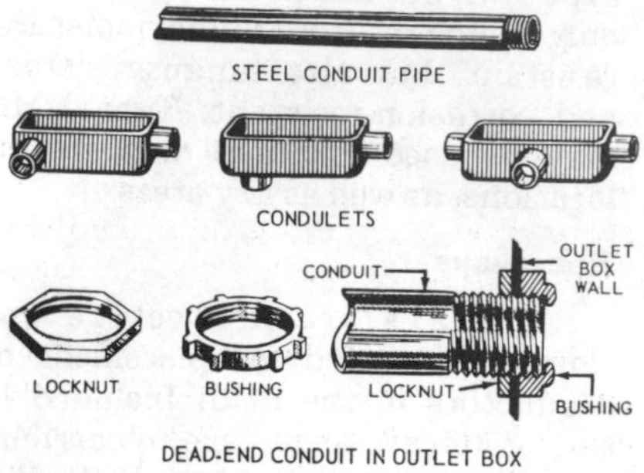

FIG. 50 Rigid ("heavywall") conduit, condulets and fittings.

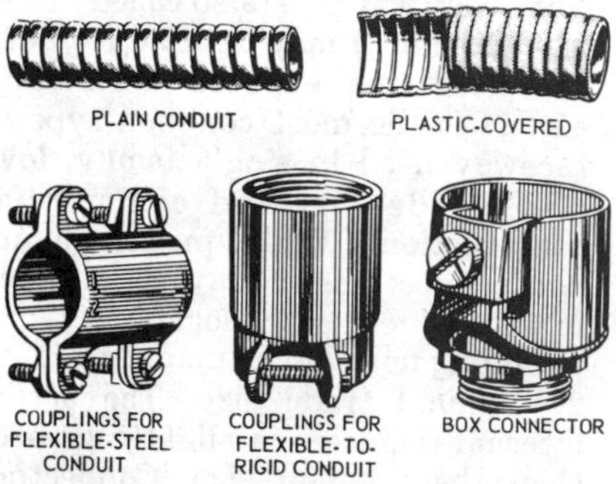

FIG. 51 Flexible metal conduit and fittings.

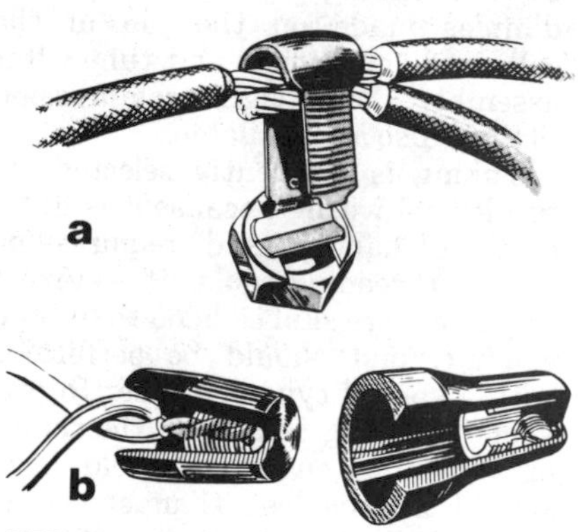

FIG. 52 Solderless connectors: (a) split bolt connector used for overhead wire connections and (b) wire nuts, for connections in approved accessible boxes.

The cut ends of conduit or electrical metallic tubing should be *reamed* (smoothed) to remove rough edges. A combination coupling should be used to make a transition from flexible conduit to electrical metallic tubing or rigid conduit.

Rigid Metal Conduit This conduit (also termed "heaviwall conduit") resembles plumbing pipe and is shipped in standard lengths of 10′ (Fig. 50). A coupling is furnished for one end of each length of rigid conduit. Both ends are reamed and threaded when shipped. If it is cut during installation, the new ends must be reamed and threaded.

Rigid conduit can be joined by screwing directly into threaded hubs, by locknuts and bushings (Fig. 50), or by threadless connectors. When connecting with locknut and bushing the locknut is screwed onto the pipe, then the end is inserted through the knockout in a box and the bushing is screwed on. When the bushing is seated, the outer locknut is screwed tightly against the outer surface of the box. When a conduit bushing is made entirely of insulating material, a locknut must be installed on both sides of the enclosure.

Rigid metal conduit is made in trade sizes of 1/2″ to 6″. The 6″ size will accommodate three 2,000 MCM (1.63″ diam.) conductors. Generally, if greater capacity is required, a *bus* type conductor (solid bar) in a bus duct is used.

Flexible Metal Conduit Flexible metal conduit (sometimes referred to as "Greenfield") is similar in appearance to armored cable, but does not have the pre-inserted conductors (Fig. 51). It cannot be used where it will be subject to corrosive gases or excessive moisture. For the most part it must conform to the same regulations as rigid conduit with respect to rewiring existing conduit, reaming cut edges, and other code requirements.

Flexible conduit connectors are fastened to the conduit by setscrews or clamps, and to the box by an expanding mechanism, a locknut, or a threaded hub. Regulations for bending and fastening flexible conduit are the same as for armored cable. Flexible conduit should be installed taut and then fastened securely to facilitate pulling conductors with a fish tape.

Non-metallic Conduit This conduit is made of non-metallic material that is resistant to moisture and chemical atmospheres. It is flame retardant, resistant to crushing, resistant to distortion from heat, low temperatures and sunlight. It can be used underground in floors, walls and ceilings.

Wireways Wireways are metal troughs with hinged or removeable covers in which conductors are placed after the wireway has been installed. Because they must be accessible throughout their entire length, wireways can be used only for exposed work. Splices can be made

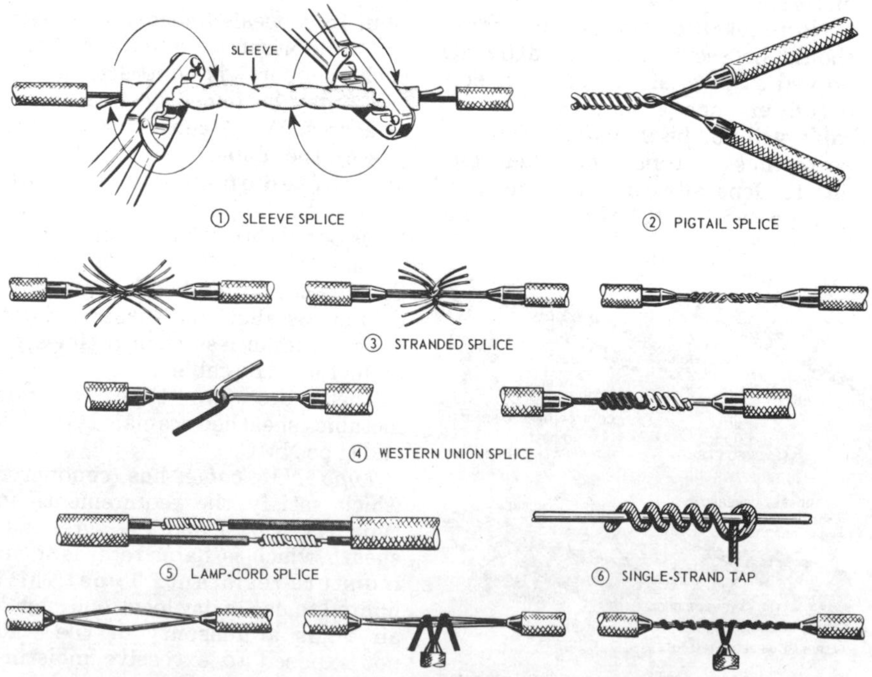

FIG. 53 Typical wire joints before insulation is applied. Various proprietary connectors for splicing and tapping conductors eliminate need for reinsulating joint.

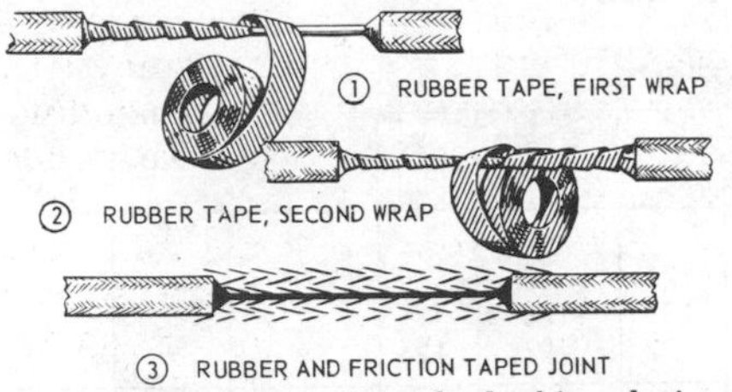

FIG. 54 Traditional method of insulating joint, using rubber tape and friction tape. Plastic electrical insulating tape can substitute for both.

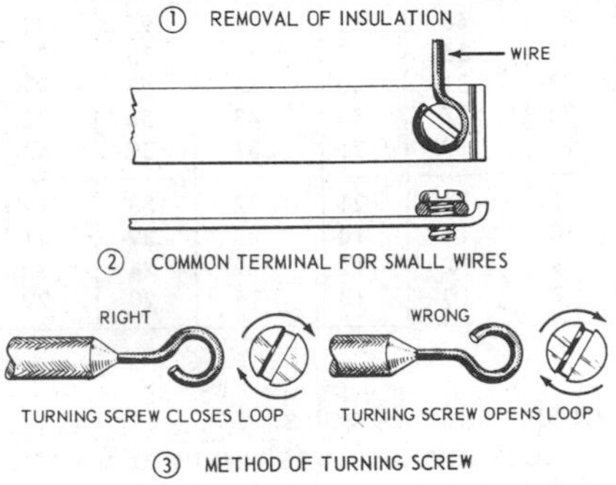

FIG. 55 Terminal connection: conductor loop should be turned clockwise, in same direction as set screw is tightened.

and laid into the wireway if the total conductor and splice cross-sectional area does not exceed 75% of the area of the wireway and is accessible. Otherwise percentage of fill is limited to 20% of the cross-sectional area of the wireway.

GENERAL RECOMMENDATIONS

The wiring equipment, materials and methods should conform to the latest edition of the National Electrical Code and be acceptable to the local utility and building authority.

Connections

Conductors should be spliced with approved solderless connectors (Fig. 52) or by brazing, welding or soldering. Splices should be made mechanically secure before soldering (Fig. 53). Bare ends of conductors should be covered after splicing with an insulation equal to that on the conductors (Fig. 54). All connections should be made in an accessible device such as a junction box, switch box or fixture base.

Splices should never be made outside a raceway and then pulled into it. They should be made in a wireway which has a removable cover or in an accessible box, which can be opened for inspection or repairs. Each conductor run should be complete from one opening to the next before conductors are pulled. If conductors are inserted before the raceway is completed, they can be damaged during the process of making up the raceway.

Terminal connections should be made securely to prevent accidental loosening (Fig. 55). Connections should be made without damaging the conductors. Terminals for securing more than one conductor should be approved for that purpose.

Outlet, junction and switch boxes used in wiring installations should provide adequate space for the conductors. Conductors should be sized adequately to minimize line losses and voltage drop. Voltage drop between the distribution panel and the load should be limited to less than 3%.

Both raceway joints and conductor connections should be made strong and secure. A mechanically secure raceway connection is one in which the joint is as strong as the raceway itself. An electrically secure connection will carry the full short-circuit, that is, the entire return current, if one of the hot wires should accidentally be grounded. Such accidental ground will then open the circuit by blowing the fuse or tripping the circuit breaker. A poorly connected system can result in excessive heating at conductor joints and possible fire hazard.

An excessive number of bends in conduit can result in damage to the conductors when they are pulled through a conduit. A run of rigid conduit or electrical metallic tubing should not contain more than the equivalent of four quarter bends (360° total), including those bends located at the fitting outlet.

Because of the different characteristics of copper and aluminum, pressure terminals or pressure connectors should be marked as suitable for mixing copper and aluminum conductors, when this combination is used.

FIG. 56 MAXIMUM LENGTH OF TWO-CONDUCTOR RUNS*

	MAXIMUM LENGTH OF RUN (FT.)																		
CONDUCTOR SIZE, AWG	5 amp	10 amp	15 amp	20 amp	25 amp	30 amp	35 amp	40 amp	45 amp	50 amp	55 amp	70 amp	80 amp	95 amp	110 amp	125 amp	145 amp	165 amp	195 amp
14	274	137	91																
12		218	145	109															
10			230	173	138	115													
8					220	182	156	138											
6								219	193	175	159								
4									309	278	253	199							
3										350	319	250	219						
2											402	316	276	232					
1												399	349	294	254				
0												502	439	370	319	280			
00													560	470	403	355	305		
000														588	508	447	385	339	
0000															641	564	486	421	360

*To limit voltage drop to 3% at 240 volt. For 120 volt, multiply length by 0.50. (Example: Find maximum run for #10 conductor carrying 30 amp at 120 volt . . . 115 × 0.5 = 57.5 ft.).

FIG. 57 **MAXIMUM NUMBER OF CONDUCTORS IN ONE CONDUIT OR TUBING**

CONDUCTOR		CONDUIT TRADE SIZE												
Type Letters	Size (AWG)	1/2"	3/4"	1"	1-1/4"	1-1/2"	2"	2-1/2"	3"	3-1/2"	4"	4-1/2"	5"	6"
TW, T, RUH, RUW	14	9	15	25	44	60	99	142						
	12	7	12	19	35	47	78	111	171					
	10	5	9	15	26	36	60	85	131	176				
	8	3	5	8	14	20	33	47	72	97	124			
RHW and RHH (without outer covering), THW	14	6	10	16	29	40	65	93	143	192				
	12	4	8	13	24	32	53	76	117	157				
	10	4	6	11	19	26	43	61	95	127	163			
	8	1	4	6	11	15	25	36	56	75	96	121	152	
TW, T, THW, RUH, RUW	6	1	2	4	7	10	16	23	36	48	62	78	97	141
	4	1	1	3	5	7	12	17	27	36	47	58	73	106
	3	1	1	2	4	6	10	15	23	31	40	50	63	91
	2	1	1	2	4	5	9	13	20	27	34	43	54	78
	1		1	1	3	4	6	9	14	19	25	31	39	57
RHW and RHH (without outer covering)	0		1	1	2	3	5	8	12	16	21	27	33	49
	00		1	1	1	3	5	7	10	14	18	23	29	41
	000		1	1	1	2	4	6	9	12	15	19	24	35
	0000			1	1	1	3	5	7	10	13	16	20	29

Conductor Size

Adequately sized wires for specific loads can be selected from Figure 56. The table gives a direct reading of maximum distance between distribution and heating unit for different loads. For example, consider a 3,000-watt 240-volt heater which is 75′ from the distribution center. First, watts should be converted to amps: I = W/E = 3,000/240 = 12.5 amps; Checking in Fig. 56 under the 15-amp column heading (the next higher load), one can find the maximum allowable distance from the source—91 feet. This is the maximum distance in which the current will not exceed the permissible voltage drop of 3%. Reading across from 91 to the left margin shows that the smallest permissible wire size is No. 14 AWG. Actually, the voltage drop in this case will be about 2%. This lower figure is preferable because the lower the voltage drop, the closer the heater will come to full output.

Raceway Size

The number and size of conductors which can be placed in a raceway is limited not only by the size of the interior opening, but also by certain code restrictions.

The major factor limiting the number of conductors that may be put into raceways is *fill*—the percentage of total cross-sectional area of the raceway filled with conductors. Maximum fill is 40% when three or more non-lead covered conductors are installed in conduit or tubing.

The number of conductors of one size which can be placed inside a metal conduit or tubing should not exceed the recommendations of Figure 57. When using several conductors of unequal size in the same raceway, it is necessary to determine the total cross-sectional area of all conductors (including insulation), then select the raceway size on the basis of percentage of fill. In such cases, allowable percentage of the cross-sectional area of the raceway will be found in Figure 58.

LOW-VOLTAGE CIRCUITS

In house wiring systems, *low voltage* refers to a-c power supplied by step-down transformers from the 120/240 volt *line voltage* service (Fig. 59). Although the National Electrical Code treats anything less than 50 volts as low voltage, in practice the term usually refers to circuits of less than 30 volts.

Less insulation is required for this type of system and the NEC, as well as many local building codes, permit low-voltage wiring without protective raceways. In residential construction, low-voltage wiring is used for bell or buzzer systems, intercoms, control wiring for heating/cooling systems and other uses requiring relatively small amounts of power.

Because low-voltage wiring is generally used for low-power systems, the conductors are quite small, No. 18 AWG or smaller. In situations requiring multiple and remote switching, low-voltage systems can be used effectively to turn light fixtures or other equipment on and off. The smallness of the wire and elimination of the raceway requirement, makes this wiring method considerably less costly than 120-volt installations in tubing or conduit. The equipment is fed from the line voltage circuit, as in standard installations, but the switch opening or closing the circuit is activated by a small, low-voltage power relay.

FIG. 58 **MAXIMUM FILL OF CONDUIT OR TUBING**

	NUMBER OF CONDUCTORS				
TYPE OF CONDUCTORS	1	2	3	4	over 4
Plain (not lead-covered)	53%	31%	40%	40%	40%
Lead-covered	55%	30%	40%	38%	35%

We gratefully acknowledge the assistance of the Commonwealth Edison Company, Chicago, Illinois, in revising the original American National Standard C91.1-1958 text. We also acknowledge the cooperation of the following in supplying specific information and illustrations: Honeywell, Inc., American National Standards Institute, Chicago Lighting Institute and Sears, Roebuck and Company.

CONTENTS

DESIGN & PLANNING

The following recommendations are based chiefly on USA Standard C-91.1-1958 with minor revisions to conform with the 1978 National Electrical Code and HUD Minimum Property Standards. The recommendations are intended to provide design criteria which will assure not only an efficient and convenient electrical system, but one which will minimize physical obsolescence. They thus supplement the requirements of the National Electrical Code, whose chief emphasis is safety. (See Main Text page 521-4 for definition of common electrical terms used in this section.)

GENERAL RECOMMENDATIONS

The entire system should be installed in accordance with requirements of the National Electrical Code, published by the National Fire Protection Association, and local ordinances and laws.

Outlets and Switches

The number and type of lighting outlets should be fitted to the various seeing tasks carried on in the home.

All outlets on 15- and 20-amp general purpose circuits should be of the grounding type, minimizing the hazard of shock from short circuits.

All spaces which have more than one entrance and for which wall-switched lighting is recommended should be equipped with multiple-switch controls at each principal entrance.

If this recommendation would result in the placing of switches controlling the same light within 10′ of each other, one of the switch locations may be eliminated. Principal entrances are those commonly used for entry to a room when going from a normally lighted to an unlighted condition.

Wiring should be so arranged as to minimize the future replacement or reworking necessary to implement any intended future additions.

OUTLET LOCATIONS

Lighting outlets should be located to conform with the lighting effects and fixtures to be used.

Wall switches should be located at the latch side of doors or at the traffic side of arches, and within the room or area where the outlets are located. Unless otherwise specified, switches should be mounted approximately 48″ above the floor.

Switches may be mounted lower when it is desired to align them with other architectural elements, such as door knobs, or to put them within reach of younger children.

Unless otherwise specified, convenience outlets should be mounted approximately 12″ above the floor. Convenience outlets in living rooms, bedrooms, and dining areas should be placed so that no point along the floor line in any usable wall space is more than 6′ from an outlet in that space.

Convenience outlets preferably should be located near the ends of wall space, rather than near the center, to reduce the likelihood of being concealed behind large pieces of furniture.

Exterior Entrances

One or more wall-switched lighting outlets should be located at the front and service entrances.

A weatherproof convenience outlet, located at least 18″ above the entrance platform, should be provided.

Living and Recreation Rooms

Outlets for general illumination, as well as for specific task lighting, should be provided. General illumination outlets should be wall-switch controlled.

Illumination may be provided by ceiling or wall fixtures, by lighting in coves, valances or cornices, or by portable lamps.

(Continued) **DESIGN & PLANNING**

Convenience outlets should be placed so that no point along the floor line in any usable wall space is more than 6′ from an outlet in that space.

If, in lieu of fixed lighting, general illumination is to be provided from portable lamps, then at least two separate wall-switched plug-in positions should be provided.

These can be provided with two switched regular duplex outlets or one switched plug-in position in each of two split-receptacle outlets.

One or more outlets for entertainment equipment such as television, radio and record player should be provided at bookcases, shelves or other suitable locations.

Dining Areas

Each dining room, or dining area combined with another room, or breakfast nook, should have at least one wall-switch controlled lighting outlet.

All convenience outlets in the dining area should be of the split-receptacle type, providing for connection to two appliance circuits.

Bedrooms

General illumination should be provided from either ceiling or wall outlets, controlled by one or more wall switches.

Outlets should be placed so that there is a convenience outlet on each side and within 6′ of the center line of each probable individual bed location.

It is often desirable to place a convenience receptacle at one of the switch locations, for the use of a vacuum cleaner, floor polisher or other portable appliance.

Bathrooms and Lavoratories

Lighting sources at the mirror should be capable of illuminating both sides of the face. Wall switches should be located so as not to be readily accessible while standing in the tub or shower stall.

One convenience outlet near the mirror, 3′ or more above the floor, should be provided.

A receptacle which is a part of a bathroom lighting fixture is not suitable for this purpose unless it is rated at 15 amps and wired with at least 15-amp rated wires.

A wall-switched or timer-operated built-in ventilating fan capable of providing a minimum of 8 air changes per hour should be provided where no natural ventilation is included.

A fan capable of moving 50 to 55 cubic feet of air per minute will assure 8 air changes per hour in a 5′ x 10′ bathroom. In many parts of the country an electric space heater is desirable for increased comfort during bathing.

Kitchen

Lighting design should provide for general illumination of the work areas, sink, range, counters and tables.

The following convenience outlets should be provided: one outlet for the refrigerator; one outlet for each 4 linear feet of work-surface frontage; and at least one outlet to serve each separate work surface or planning desk.

Convenience outlets should be located approximately 8″ above the work surface. Convenience outlets in the kitchen, other than that for the refrigerator, should be of a split-receptacle type.

One outlet each for a range, ventilating fan, dishwasher and waste disposer is often required. A special recessed receptacle with hanger is recommended if an electric clock is planned.

Laundry and Laundry Areas

Outlets for fixed lights should be installed to provide illumination of work areas, such as laundry tubs, sorting tables, washing, ironing and drying centers. At least one outlet in the room should be wall-switch controlled.

Convenience outlets in the laundry area should be of the split-receptacle type.

One outlet for each of the following pieces of equipment may be required: automatic washer; hand iron or ironer; and clothes dryer. Outlets for a ventilating fan and clock are desirable.

Requirements for electric water heater should be obtained from the local power company.

Closets

Generally, one lighting outlet for each closet should be provided. Where shelving or other conditions make the installation of lights within a closet ineffective, outlets in the adjoining space should be so located as to provide light within the closet.

(Continued) **DESIGN & PLANNING**

Wall switches or automatic door switches are preferred, but pull-switches are acceptable.

Halls

Wall-switched fixed fixtures should be installed for proper illumination of the entire area with particular attention paid to irregularly shaped spaces.

A switch-controlled night light may be desirable in bedroom halls, reception halls, vestibules, entries and foyers.

One convenience outlet for each 15 linear feet of hall, measured along the center line, should be provided. Each hall over 25 sq. ft. in floor space should have at least one outlet.

In reception halls and foyers, convenience outlets should be placed so that no point along the floor line in any usable wall space is more than 10′ from an outlet in that space.

A receptacle at one of the switch outlets is recommended for convenient connection of maintenance equipment such as vacuum cleaner, floor polisher or electric broom.

Stairways

Fixed wall or ceiling lighting outlets should be installed to provide adequate illumination of each stair flight. Outlets should have multiple-switch controls at the head and foot of the stairway.

Outlets should be so arranged that the stair may be fully illuminated from either floor, while also permitting control of bedroom hall lights independently of ground-floor fixtures.

Switches should be grouped whenever possible and should be located far away from the head or foot of a stair to avoid the hazard of tripping while reaching for a switch.

These recommendations apply to any stairway serving finished rooms at both ends. For stairways to unfinished basements or attics, see appropriate headings below.

Utility Room

Lighting outlets should be placed to illuminate the furnace area, and work area if planned. At least one lighting outlet should be wall-switch controlled.

At least one convenience outlet should be provided, preferably near the furnace or near the planned workbench location.

Outlets should be provided for each piece of mechanical equipment requiring electrical connections such as furnace, boiler, water heater, water pump or compressor.

Basement

Lighting outlets should be placed to illuminate designated work areas, the foot of the stairway, enclosed spaces and specific pieces of equipment, such as furnace, water pump or workbench. Additional outlets for open (unpartitioned) basement space should be provided at the rate of one outlet for each 150 sq. ft.

In unfinished basements the light at the foot of the stairs should be wall-switch controlled near the head of the stairs.

Other lights may be pull-switch controlled. In basements with finished rooms, garage space, or other direct access to outdoors, the stairway lighting provisions apply.

At least two convenience outlets should be provided in the open basement area. Outlets should be provided for each piece of mechanical equipment requiring electrical connections such as furnace, boiler, water heater or water pump.

Additional convenience outlets may be desired in a basement equipped with workbench, laundry, dark room or hobby area and for appliances such as a dehumidifier or portable space heater.

Accessible Attic

One lighting outlet should be included for general illumination and it should be wall-switch controlled from the foot of the stairs. Where it is contemplated that an unfinished attic will be developed into habitable rooms, the lighting outlet should be multiple-switch controlled at the top and the bottom of the stairs. There should be one outlet for each enclosed space.

In an unfinished attic a convenience outlet is desirable for portable equipment and for "trouble lights" capable of reaching into distant corners. If an open stairway leads to future finished rooms, a junction box to the distribution panel should be provided for extension to future convenience and lighting outlets.

Porches

Each porch, breezeway or other similar roofed area of more than 75 sq. ft. in floor space should have at least one wall-switched lighting outlet. Multiple-

(Continued) **DESIGN & PLANNING**

switch controls should be installed at entrances when the porch is used as a principal passage between the house and garage.

At least one convenience outlet should be provided for each 15 linear feet of wall bordering a porch or breezeway. Outlets should be of the weatherproof type if exposed to the elements.

It is recommended that all exterior outlets be controlled by a wall switch inside the house. Since this area often is used as an outdoor dining area, it is also recommended that at least one outlet be of the split-receptacle type.

Terraces and Patios

General illumination should be provided by one or more fixed lighting outlets place in a central location such as on the building wall, or on a light post. Such outlets should be wall-switch controlled from inside the house, adjacent to the door which opens onto the area.

At least one weatherproof convenience outlet for each 15 linear feet of house bordering the terrace or patio should be provided. These outlets should be wall-switch controlled from inside the house and should be located at least 18″ above the patio level.

Garage or Carport

At least one wall-switched ceiling lighting outlet should be provided in any garage space. If the garage has no covered access from the house, an exterior outlet, multiple-switch controlled from garage and from house, should be provided.

CIRCUIT RECOMMENDATIONS

Branch Circuits

General purpose circuits should be wired with suitable conductors: not smaller than No. 14 AWG (American Wire Gauge) copper wire if protected by 15-amp overcurrent device and No. 12 AWG wire for 20-amp device.

Since the No. 14 wires are more limited in circuit length (30 ft.) and current-carrying capacity (1,725 watts) No. 12 wires with 20-amp devices are highly recommended for all general purpose circuits. Some local building codes do not permit use of the smaller No. 14 wires.

If 20-amp circuits are used, at least two circuits should be provided for the first 750 sq. ft. of floor area plus one circuit for each additional 750 sq. ft. or major fraction thereof. If 15-amp circuits are used, at least two circuits should be provided for the first 500 sq. ft. plus one circuit for each additional 500 sq. ft. or major fraction thereof.

Floor areas should be calculated from dimensions taken to the outside of exterior walls, including unfinished spaces which are adaptable to future living use; open porches and the garage may be excluded from the calculation.

Two 2-wire 120 volt 20-amp receptacle circuits (or one 3-wire split receptacle circuit where permitted by local code requirements) should be provided for all areas regularly intended for cooking, dining or laundering. A separate 20-amp circuit should be provided for each appliance or piece of equipment rated over 1400 watts.

The use of 3-wire circuits for supplying convenience outlets is an economical means of increasing the capacity at specific locations and reducing the voltage drop in the circuit. However, two separate receptacle circuits may have to be used where local codes do not permit split receptacles.

Separate circuits should be provided and conductors should be sized appropriately for the specific equipment tabulated in Fig. WF 1.

FIG. WF1 RECOMMENDED CONDUCTORS FOR SPECIAL PURPOSE CIRCUITS

ITEM	TYPICAL WATTAGE	RECOMMENDED CONDUCTOR CAPACITY		
		Amps	Wires	Volts
Range, or	12,000*	50	3	120/240
Electric Oven and	5,000	30	3	120/240
Drop-In Cooktop	7,000	40	3	120/240
Automatic Washer	700	20	2	120
Electric Clothes Dryer	5,000*	30	3	120/240
Fuel-Fired Heating Equipment	800	20	2	120
Central Air Conditioning, or	5,000*	40	3	120/240
Room Air Conditioner (1 Ton)	1,500	20	2	120
Dishwasher & Disposer, or	2,700	20	3	120/240
Dishwasher and	1,800	20	2	120
Waste Disposer	900	20	2	120
Food Freezer	400	20	2	120 or 240
Water Pump	500	20	2	120 or 240
Bathroom Space Heater	2,000	25	2	240
Electric Water Heater	4,500*	30	3	240

*Verify wattages with manufacturer and/or local power company.

(Continued) DESIGN & PLANNING

Spare circuit positions should be provided in the distribution panel for at least two future 20-amp, 2-wire, 115-volt circuits in addition to those initially installed. If the branch circuit or distribution panel is installed in a finished wall, raceways should be extended from the panel to the nearest accessible unfinished space for future use.

The majority of appliances for residential use are made for 110 to 125 volts. However, because of improved operating efficiency, there is a growing tendency among manufacturers to make fixed appliances for use on 230 to 250 volt circuits. It is recommended that the higher voltage be used in those cases where a choice exists.

The use of electrical heating and cooling systems is increasing in some areas. These systems generally are custom designed to suit the particular house and local climatic conditions. Adequate wiring and service capacity criteria for such purposes are not included here and should be determined by consultation with local power company.

SERVICE RECOMMENDATIONS

Service for each home or dwelling unit should be 3-wire 120/240 volt 100-amp minimum. Service entrance conductors should be covered with suitable insulating materials and sized as required by the National Electrical Code.

Such 100-amp service generally is sufficient to provide for the power needs in a typical home for lighting, appliances and fixed equipment totaling 18,000 watts. For larger homes or where the total load is larger than 18,000 watts, the local power company should be consulted.

Generally, 150-amp service is adequate for the conditions described above, plus three to five tons of air conditioning. A larger 200-amp service usually may be required when electric space heating is employed as well.

Because of the many major uses of electricity in the kitchen requiring individual equipment circuits, it is recommended that the main service equipment or a sub-panel be located near or on a kitchen wall. Such a location usually is close also to the laundry, thus minimizing long circuit runs and wiring costs.

SPECIAL CONSIDERATIONS

For recommendations on entrance signals, communication systems, fire alarms, television and radio antennas, see Main Text pages 521-12 through 521-14.

METHODS & SYSTEMS

FUNDAMENTALS OF ELECTRICITY

For a detailed explanation of electrical principles and current properties, see pp. 521-15 through 521-18 of Main Text. Relationship between voltage, wattage, amperage and resistance are given in Figure WF2.

TRANSMISSION AND SERVICE

Electricity from a generating station typically is transmitted at high voltage *to substations within the city, then to transformers near the users. Here it is converted to* line voltage *(120/240 volts) for use in small commercial buildings, apartments and homes.*

Single-phase Service

Single-phase 120/240 volt service is used almost universally for residences in this country. Electric heating equipment and most large, fixed appliances such as water heaters, electric ranges and clothes driers, require 240 volts. General lighting and most portable appliances such as washing machines, refrigerators and toasters operate adequately on 120 volts.

The entire wiring system should be grounded by connecting the neutral wire to a water pipe or directly to the earth. Ground connection to a water pipe should be made on the street side of a water meter; if connected on the house side, a jumper cable should be used around the meter.

If a metal water pipe cannot be used, a metal rod is driven into the ground 2′ from the house near the service entrance; the neutral wire from the service drop is connected to the rod with a ground wire.

Three-phase Service

Four-wire three-phase *service is most suitable for industrial and commercial applications and occasionally is used in residences. This system is particularly suited for electrical heating/cooling installations, where large motors are involved.*

A three-phase system should be planned so that the total load is about equally divided between the three phases. All *three* wires should be provided with overload protection.

DISTRIBUTION AND OVERLOAD PROTECTION

Service Equipment

The service equipment should include a main disconnecting means and a distribution system, protected by an overcurrent device for each circuit.

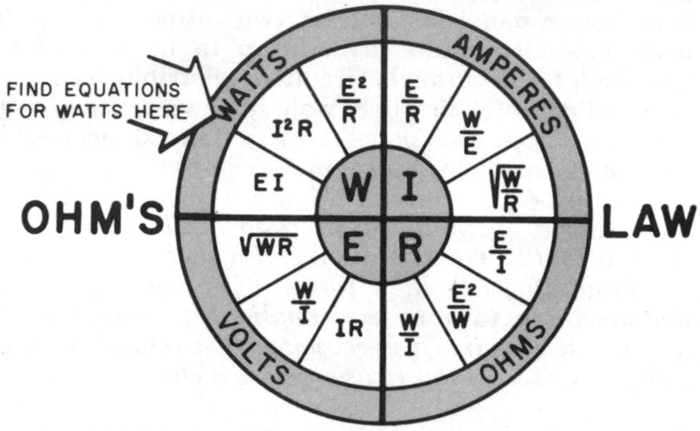

FIG. WF2 Ohm's Law and derivative formulas make it possible to find any of the values in the middle of the wheel, if any other two are known.

Overcurrent Protection

Overcurrent devices should be selected not to exceed the rated current-carrying capacity of the circuit conductors.

Overcurrent devices insure that, in case of a short circuit, current will be interrupted before the conductor overheats and possibly burns up the insulation. Overcurrent devices can be of the fuse or circuit breaker type.

Type S plug fuses should be used for fused installations up to 30 amps.

For fused installations larger than 30 amps cartridge fuses can be used.

Circuit breakers should be clearly marked with rated amperage and be capable of being switched on or off by hand.

Most circuit breakers are of the thermal-magnetic type, meaning that when the breaker is tripped, the thermal element is heated. It may be necessary, therefore, to wait for the element to cool, before the breaker can be reset manually.

Grounding

There are two types of grounding involved in a home wiring installation: (a) system grounding, *connecting the current-carrying neutral wire to the grounding terminal in the main switch and (b)* equipment grounding, *connecting the metal frames of equipment and appliances back to the main switch.*

Metal-enclosed appliances and equipment used in damp areas or within 5′ of water pipes should be grounded by

(Continued) METHODS & SYSTEMS

appropriate connection through a grounding wire and/or the raceway system back to the service ground.

Unlike the grounded (neutral) wire, the grounding wire or raceway normally carries no current, except when a short circuit in the equipment or circuit wiring causes a fault current to flow. Under these conditions, equipment grounding minimizes shock hazard to occupants by providing a low-resistance path for the fault current and by opening the circuit overcurrent device.

Receptacles should not be installed closer than 10′ to the edge of a pool. Poolside receptacles at distances from 10′ to 15′, as well as all outdoor receptacles, should be equipped with *ground fault circuit interrupters* (GFCI's).

GFCI's open the circuit when the ground fault current exceeds 5/1,000 of an ampere. The use of GFCI's is required by the NEC, because the severity of electrical shock is greatly increased when standing on wet ground.

WIRING METHODS

Conductors

Conductor sizes are designated by AWG (American Wire Gauge) number and MCM (thousand circular mills) sizes as shown in Figure WF3.

The wiring system should be designed so that the maximum current carried by conductors will not exceed amperages shown in Figure WF3.

Conductors also are designated by the type of insulation and/or outer covering, as indicated in Figure WF4.

Conductors should have an insulating jacket and outer covering suitable for the anticipated temperature and moisture conditions.

Cables

Cables are combinations of two or more conductors, factory-encased in a protective covering to safeguard them from mechanical damage. Residential wiring may involve armored cable ("BX") or non-metallic cable ("Romex").

All cable should be securely fastened by approved sleeves, staples or straps, spaced not more than 4-1/2′ apart. The cable should be supported within 12″ of an outlet box or other termination. When the box is mounted on equipment which may require moving for servicing, the unsupported cable length should be not greater than 24″.

FIG. WF3 PHYSICAL AND ELECTRICAL PROPERTIES OF COMMONLY USED COPPER CONDUCTORS

Conductor			Maximum Current (Amp)[4]	
Size (AWG, MCM)[1]	One-half Natural Size	Diameter (In.)[3]	Conductor type: RUW, T, TW	Conductor type: RH, RUH, THW, RHW, THWN
14	●	.0641	15	15
12	●	.0808	20	20
10	●	.1019	30	30
8	●	.1285	40	45
6	●	.184	55	65
4	●	.232	70	85
2	●	.292	95	115
1	●	.332	110	130
0	●	.372	125	150
00	●	.418	145	175
000	●	.470	165	200
0000	●	.528	195	230
350 MCM[2]	●	.681	260	310

[1] Sizes 14 to 8 are solid wires. Sizes 6 to 2 are 7-strand cables. Sizes 1 to 0000 are 19-strand cables. Size 350 MCM is a 37-strand cable.
[2] MCM is the designation of wire size in thousands of circular mils.
[3] Exclusive of insulation
[4] Based on ambient temperature of 86°F, (30°C); not more than 3 conductors in raceway, cable or direct burial. Where number of conductors exceeds 3, maximum current given should be derated as follows: 4-6 cond.—80%; 7-24 cond.—70%; 25-42 cond.—60%; 43 or more—50%.

Armored cable may be run in the air voids of masonry block or tile walls where the walls are not exposed to excessive moisture. Armored cable should not be buried in concrete, should not be run below grade, or in areas subject to excessive moisture.

Armored cable should be handled and bent with care, to avoid pinching the conductors.

Non-metallic sheathed cable type NM should be used only in dry locations.

(Continued) **METHODS & SYSTEMS**

FIG. WF4 CONDUCTOR DESIGNATIONS AND PROPERTIES

Trade Name	Type Letter	Maximum Operating Temp.	Insulation	Applications and/or Limitations	Outer Covering
Heat-Resistant Rubber	RH	167 F	Heat-Resistant Rubber	Dry Locations	Moisture-Resistant, Flame-Retardant, Non-Metallic
	RHH	194 F			
Moisture- and Heat-Resistant Rubber	RHW	167 F	Heat- and Moisture-Resistant Rubber	Dry and Wet Locations	Moisture-Resistant, Flame-Retardant, Non-Metallic
Latex Rubber	RUH	167 F	90% Unmilled Grainless Rubber	Dry Locations	Moisture-Resistant, Flame-Retardant, Non-Metallic
	RUW	140 F		Dry and Wet Locations	
Thermoplastic	T	140 F	Flame-Retardant, Thermoplastic Compound	Dry Locations	None
Moisture-Resistant Thermoplastic	TW	140 F	Flame-Retardant, Thermoplastic Compound	Dry and Wet Locations	None
Moisture- and Heat-Resistant Thermoplastic	THW	167 F	Flame-Retardant, Moisture- and Heat-Resistant Thermoplastic	Dry and Wet Locations	None
Heat-Resistant Thermoplastic	THHN	194 F	Flame-Retardant, Heat-Resistant, Thermoplastic	Dry Locations	Nylon jacket
Moisture- and Heat-Resistant Thermoplastic	THWN	167 F	Flame-Retardant, Moisture- and Heat-Resistant, Thermoplastic	Dry and Wet Locations	Nylon jacket

Type NMC cable may be used in damp or corrosive locations, as well as dry areas.

The outer sheath of non-metallic cable should not be damaged during installation. No bend should have a radius less than 5 times the diameter of the cable.

Raceways

Raceways are protective enclosures designed for insertion of conductors in the field. The major raceway types are: (1) electrical metallic tubing (EMT), (2) rigid metal conduit, (3) flexible metal conduit and (4) wireways.

The cut ends of raceways should be smoothed to remove rough edges. Raceways should be securely connected and rigidly supported.

Electrical metallic tubing (EMT), also called "thinwall conduit", is made in sizes from 1/2" to 4" and may be used for both concealed and exposed work. It is the most common type of raceway used in single family, low-rise residential and commercial construction. With proper joints and connections it may be installed in damp, as well as dry, locations.

In bending metallic tubing, care should be taken not to flatten the pipe. The radius of the inner edge of a field bend should not be less than shown in Figure WF5.

Rigid conduit is made in sizes 1/2" to 6" and is shipped in standard lengths of 10'. A coupling is furnished for one end of each length of rigid conduit. Both ends are reamed and threaded when shipped.

Flexible metal conduit (sometimes called "Greenfield") is similar in appearance to armored cable, but does not contain pre-inserted conductors. It is available in sizes ranging from 3/8" to 4" and is sold in coils up to 250'.

Flexible conduit should not be used in areas subject to

(Continued) **METHODS & SYSTEMS**

FIG. WF5 MINIMUM RADIUS OF FIELD-BENT CONDUIT AND TUBING

Size of Conduit (in.)	Radius to Center of Conduit (in.)
1/2	4
3/4	4-1/2
1	5-3/4
1-1/4	7-1/4
1-1/2	8-1/4
2	9-1/2
2-1/2	10-1/2
3	13
3-1/2	15
4	16
4-1/2	20
5	24
6	30

excessive moisture or corrosive gases, unless plastic-covered.

Flexible conduit should be supported at intervals not greater than 4-1/2′ and within 12″ of a box. Radius of bends should not be less than 5 times the conduit diameter.

Wireways are metal troughs with hinged or removeable covers in which conductors are placed after the wireway has been installed. Because they must be accessible throughout their entire length, wireways can be used only for exposed work.

Splices made in a completely accessible wireway should have a cross-sectional area not in excess of 75% of the area of the wireway. If the wireway is not fully accessible at the splice location, then the percentage of fill should not exceed 20% of the cross-sectional area of the wireway.

GENERAL RECOMMENDATIONS

The wiring equipment, materials and methods should conform to the latest edition of the National Electrical Code and be acceptable to the local utility and building code authorities.

Connections

Conductors should be spliced with approved solderless connectors, by brazing, welding or soldering. Splices should be made mechanically secure before soldering. Bare ends of conductors should be covered after splicing with an insulation equal to or better than that on the conductors. All connections should be made in an accessible device such as a junction box, switch box or fixture base.

Splices should not be made outside a raceway and then pulled into it. Splices should be made in a wireway which has a removable cover, or in an accessible box, which can be opened for inspection or repairs. Each raceway run

FIG. WF6 MAXIMUM LENGTH OF TWO-CONDUCTOR RUNS*

	MAXIMUM LENGTH OF RUN (FT.)																		
CONDUCTOR SIZE, AWG	5 amp	10 amp	15 amp	20 amp	25 amp	30 amp	35 amp	40 amp	45 amp	50 amp	55 amp	70 amp	80 amp	95 amp	110 amp	125 amp	145 amp	165 amp	195 amp
14	274	137	91																
12		218	145	109															
10			230	173	138	115													
8					220	182	156	138											
6								219	193	175	159								
4									309	278	253	199							
3										350	319	250	219						
2											402	316	276	232					
1												399	349	294	254				
0												502	439	370	319	280			
00													560	470	403	355	305		
000														588	508	447	385	339	
0000															641	564	486	421	360

*To limit voltage drop to 3% at 240 volt. For 120 volt, multiply length by 0.50. (Example: Find maximum run for #10 conductor carrying 30 amp at 120 volt . . . 115 × 0.5 = 57.5 ft.).

(Continued) METHODS & SYSTEMS

FIG. WF7 MAXIMUM NUMBER OF CONDUCTORS IN ONE CONDUIT OR TUBING

CONDUCTOR		CONDUIT TRADE SIZE												
Type Letters	Size (AWG)	1/2"	3/4"	1"	1-1/4"	1-1/2"	2"	2-1/2"	3"	3-1/2"	4"	4-1/2"	5"	6"
TW, T, RUH, RUW	14	9	15	25	44	60	99	142						
	12	7	12	19	35	47	78	111	171					
	10	5	9	15	26	36	60	85	131	176				
	8	3	5	8	14	20	33	47	72	97	124			
RHW and RHH (without outer covering), THW	14	6	10	16	29	40	65	93	143	192				
	12	4	8	13	24	32	53	76	117	157				
	10	4	6	11	19	26	43	61	95	127	163			
	8	1	4	6	11	15	25	36	56	75	96	121	152	
TW, T, THW, RUH, RUW	6	1	2	4	7	10	16	23	36	48	62	78	97	141
	4	1	1	3	5	7	12	17	27	36	47	58	73	106
	3	1	1	2	4	6	10	15	23	31	40	50	63	91
	2	1	1	2	4	5	9	13	20	27	34	43	54	78
	1		1	1	3	4	6	9	14	19	25	31	39	57
RHW and RHH (without outer covering)	0		1	1	2	3	5	8	12	16	21	27	33	49
	00		1	1	1	3	5	7	10	14	18	23	29	41
	000		1	1	1	2	4	6	9	12	15	19	24	35
	0000			1	1	1	3	5	7	10	13	16	20	29

should be complete from one opening to the next before conductors are pulled.

If conductors are inserted before the raceway is completed, they can be damaged during the process of making up the raceway.

Terminal connections should be made securely to prevent accidental loosening. Connections should be made without damaging the conductors. Terminals for securing more than one conductor should be approved for that purpose.

Outlet, junction and switch boxes used in wiring installations should provide adequate space for the conductors.

Conductors should be sized to minimize line losses and voltage drop. Voltage drop between the distribution panel and the load should not exceed 3%.

A run of rigid conduit or electrical metallic tubing should not contain more than the equivalent of four quarter bends (360° total), including those bends located at the fitting outlet.

An excessive number of bends in conduit can result in damage to the conductors when they are pulled through a conduit.

Conductor Size

Conductors should be adequately sized for the design loads and length of run according to Figure WF6.

FIG. WF8 MAXIMUM FILL OF CONDUIT OR TUBING FOR CONDUCTORS

TYPE OF CONDUCTOR	NUMBER OF CONDUCTORS				
	1	2	3	4	over 4
Plain (not lead-covered)	53%	31%	40%	40%	40%
Lead-covered	55%	30%	40%	38%	35%

Figure WF6 gives a direct reading of maximum distance between distribution panel and load for different conductor sizes. Sizes are calculated on the basis of a maximum 3% voltage drop between panel and load. In selecting a wire size for a known load, the permitted length of run should be greater than the anticipated distance from panel to load.

Raceway Size

The maximum fill of raceways should not exceed 40% when 3 or more non-lead covered conductors are installed. The number of conductors of one size which can be placed inside a metal conduit or tubing should not exceed the recommendations of Figure WF7.

When using several conductors of unequal size in the same raceway, the cross-sectional area of all conductors (including insulation), should not exceed the maximum percentage of fill given in Figure WF8.

539 PASSIVE HEATING & COOLING

PASSIVE HEATING & COOLING

INTRODUCTION

For thousands of years primitive peoples intuitively designed their dwellings to take advantage of the local climate. In each region a style of building evolved that kept buildings warm in winter and cool in summer with a minimum of energy. This was accomplished *passively,* that is, through the use of the sun, wind and other natural phenomena.

In the early Sinaqua and Anasazi Indian cliff dwellings of Arizona, known today as "Montezuma's Castle", an overhanging rock brow protected southern exposures against the high summer sun and kept interiors cool. In winter, the rock brow did not shut out the low afternoon sun, allowing the rock face to warm up. Since rock is a good conductor and storer of heat, it retained the heat and kept the dwellings warm during the night.

The Pueblo Indians of New Mexico also knew how to build energy efficient dwellings more than 1,000 years ago. By facing their structures south and using high, small, deeply cut windows, they were able to delay the entry of heat into the interior of their structures until needed—late in the day when temperatures started dropping. Exterior surfaces were painted a light color to reflect the sun's rays. They carefully crowded their dwellings side by side and on top of one another to reduce temperature fluctuations in both summer and winter and cooked in rooms separated from living spaces to avoid unwanted heat gain in summer.

Descendents of the Pilgrims also evolved a suitable style of building for their Northeastern region. In the cold winters of New England, keeping warm meant having efficient heat sources and tight buildings to retain the heat. The typical New England "saltbox" included warming stoves and fireplaces at the center of the house and a minimum of exterior wall surface. A north-sloping roof provided protection from icy winds and allowed snow to build up for added insulation against the cold. Most windows faced south and the exterior was painted a dark color to absorb as much solar radiation as possible.

Studies of these and other intuitive responses to local climate have laid the groundwork for a more widespread development of similar naturally-adaptive heating and cooling techniques. An increasing number of architects and engineers continue to refine and adapt these concepts into innovative and energy efficient buildings. Their work has resulted in buildings which more effectively collect, store and distribute solar energy for heating, and which better utilize natural climatic factors for cooling. Such naturally-adaptive heating and cooling systems require relatively little external energy for their operation and are termed *passive.* Solar heating or cooling installations with sophisticated heat transport, control and storage methods, known as *active* solar systems, are beyond the scope of this section.

The following subsections will discuss the elements of passive heating and cooling design and will provide examples of how those elements have been integrated into systems.

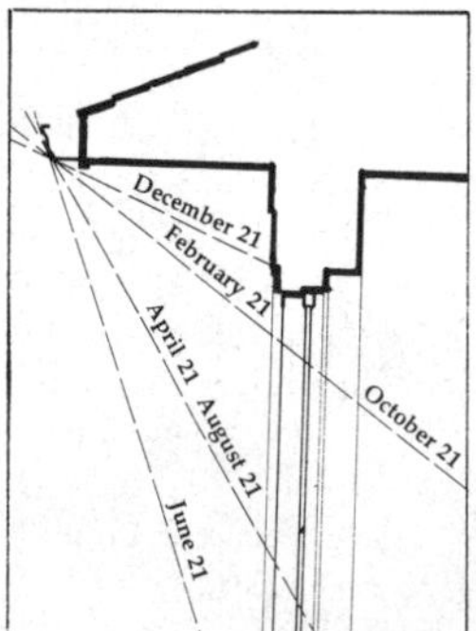

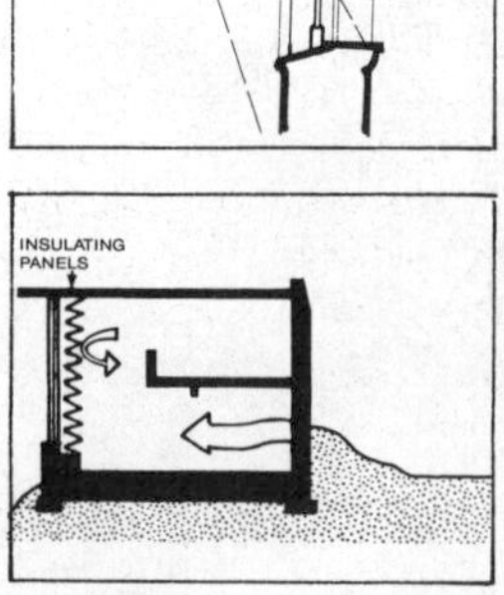

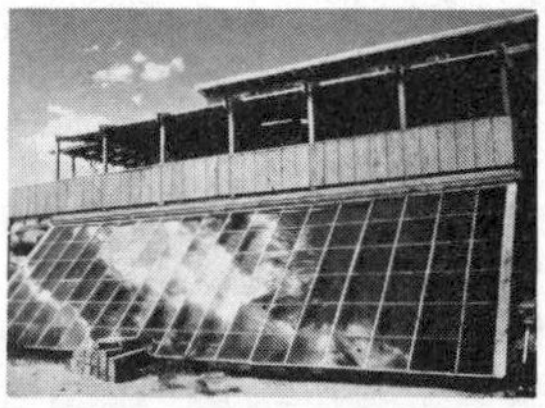

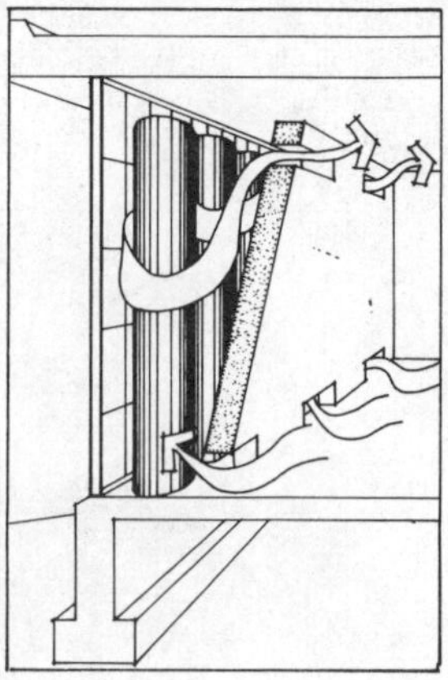

PASSIVE SOLAR PROPERTIES

All buildings are exposed to the warming rays of the sun and the cooling influences of natural breezes. When properly designed and oriented, they can take advantage of the local climate for heating and cooling, with reduced reliance on mechanical systems (Fig. 1). With such an approach, windows, skylights, walls, roofs, floors and the natural features of the site regulate heat flow into or out of the building. Conduction, convection, radiation and gravity provide the means to move the heat through the space.

Passive solar systems are distinguished from active systems in that they use few or no power-consuming fans and pumps, and because the collection, storage and distribution facilities are integrated with the building design. In active systems, the collection and storage elements are manufactured products which can be placed in, on or near the building and tied to it with pipes or ducts.

In general, all effective passive buildings share the following basic qualities: (1) they should have an energy efficient envelope which will not leak heat in winter or gain excessive heat in summer; (2) they should have an adequate amount of solar collection area, in the form of windows, skylights, glass wall, water bags or other solar heat traps; and (3) they should be capable of storing heat for later use, either in the structural elements or in specially provided facilities.

FIG. 1 Early Pueblo Indian dwellings provide examples of passive heating and cooling.

FIG. 2 AVERAGE DAILY SOLAR RADIATION IN LANGLEYS*

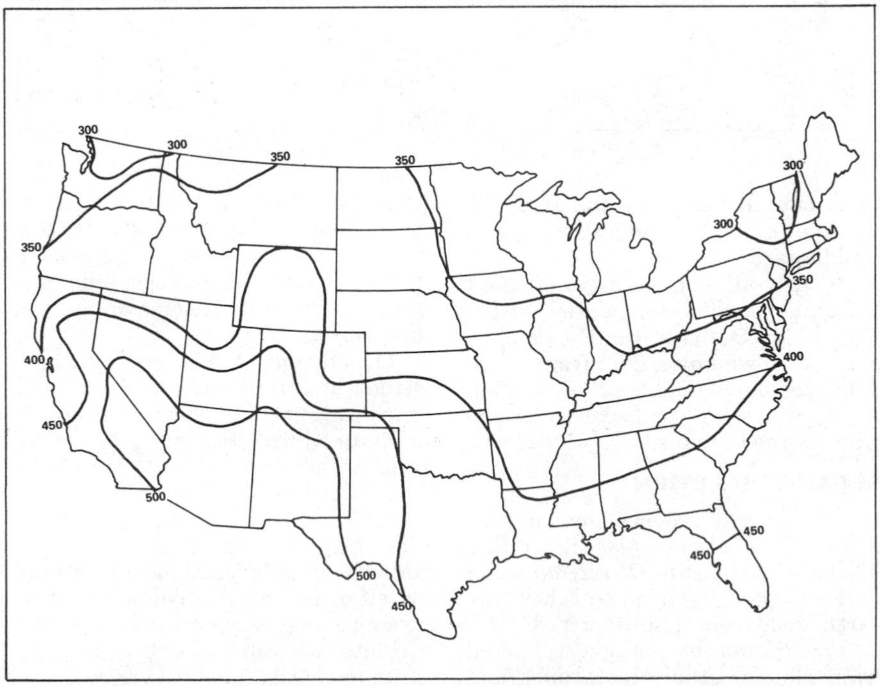

*Multiply langleys shown by 3.69 to obtain BTUs per sq. ft.

BUILDING ENVELOPE

A passively heated structure can only collect and store a limited amount of heat. With the exception of a few sunny regions, more heat is required for modern comfort than the passive system can deliver. The extra heat is generally supplied by some supplementary source, such as wood stoves, fireplaces or a conventional heating system. The recommendations of Chapter 105 Heat Control are intended to mini-

FIG. 3 CLEAR DAY INSOLATION FOR 40°N LATITUDE*

21st of:	Surface Perpendicular to Sun	Horiz. Surface	South Facing Surface at Tilt Angle of:**				
			30°	40°	50°	60°	90°
January	2182	948	1660	1810	1906	1944	1726
February	2640	1414	2060	2162	2202	2176	1730
March	2916	1852	2308	2330	2284	2174	1484
April	3092	2274	2412	2320	2168	1956	1022
May	3160	2552	2442	2264	2040	1760	724
June	3180	2648	2434	2224	1974	1670	610
July	3062	2534	2409	2230	2006	1728	702
August	2916	2244	2354	2258	2104	1894	978
September	2708	1788	2210	2228	2182	2074	1416
October	2454	1348	1962	2060	2098	2074	1654
November	2128	942	1636	1778	1870	1908	1686
December	1978	782	1480	1634	1740	1796	1646

*All values in BTUs per sq. ft.
**Angle measured from horizontal

FIG. 4 AVERAGE ANNUAL PERCENTAGE OF POSSIBLE SUNSHINE

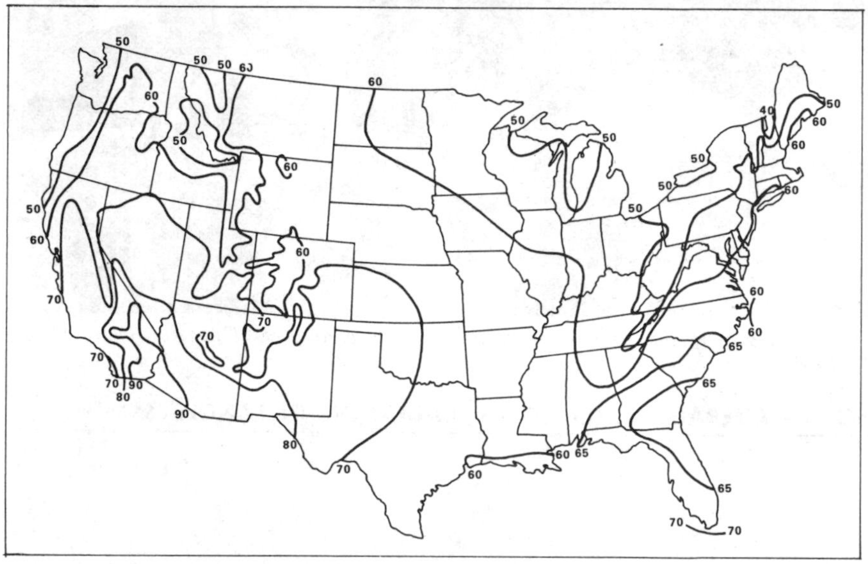

mize overall heat loss and to make passive principles applicable to such buildings.

In general, an energy efficient building should be insulated to optimum levels, have tightly caulked and sealed windows, door frames and sillplates, have double or triple glazing and as few nonsolar-collecting windows as possible.

SOLAR COLLECTION

Solar energy reaches the earth as *direct* (parallel) or *diffuse* (non parallel) rays of the sun. *Direct radiation* is that energy which reaches the earth relatively unobstructed. *Diffuse radiation* is sunlight reflected from clouds, atmospheric dust and obstructions such as adjacent buildings. The relative proportion of direct and diffuse radiation, as well as total radiation, varies from region to region, making a particular solar system more feasible in some areas than in others.

The amount of solar energy which strikes a surface is termed *insolation* (incident solar radiation) and is measured in *langleys* (ly). The amount of energy contained in 1 ly is equal to 1 calorie per sq. cm. or 3.69 BTUs per sq. ft. of surface. The map in Figure 2 shows the amount of daily sunshine which falls in all parts of the country, when averaged over a typical year.

Figure 3 tabulates the *average clear day insolation* on various surfaces, taking into account both diffuse and direct radiation. To account for cloudy days, the value for average clear day isolation should be multiplied by the appropriate percentage of possible sunshine from the map in Figure 4. The map shows the average percentage of available sunshine on a yearly basis. Maps for individual months are also available from the American Society of Heating, Refrigerating, and Air Conditioning Engineers (ASHRAE).

In designing a passive building, building shape, orientation and solar collection area should be given careful consideration.

Shape & Orientation

The building should be shaped and sited so that it is exposed to the maximum amount of sunshine in winter and a minimum amount in summer. In northern latitudes, that means buildings should expose as much surface to the south as possible. A building in Pennsylvania, for example, located at latitude 40° N, receives nearly three times as much solar radiation on its south face in January as it does in June; due south is therefore the optimum orientation for a collector surface (Fig. 5).

East and west sides, on the other hand, receive two and one half times more insolation in summer than in winter making them inappropriate for heat collection in winter. At latitudes below 35° N, houses are exposed to even more sun on the south side in winter while east and west walls receive two to three times more sun in summer than the south walls.

A square building is never the optimum form for solar collection at any latitude, unless local site conditions dictate such a shape. Buildings which are elongated on the north-south axis are even less efficient in winter and summer. The optimum shape is usually a form elongated along the east-west direction.

Solar Collection

Most solar energy reaches the earth in two forms—*visible light* (short-wave radiation) and *infrared light* (long-wave radiation). It is only the short-wave radiation, however, which can pass through glass; long-wave radiation is absorbed and conducted or radiated away. Short-wave radiation which passes through the

FIG. 5 **DAILY SOLAR HEAT GAIN FACTORS FOR 40°N LATITUDE***

Orientation	Jan	Feb	Mar	Apr	May	Jun	Jul	Aug	Sep	Oct	Nov	Dec
N	118	162	224	306	406	484	422	322	232	166	122	98
NNE	123	200	300	400	550	700	550	400	300	200	123	100
NE	127	225	422	654	813	894	821	656	416	226	132	103
ENE	265	439	691	911	1043	1108	1041	903	666	431	260	205
E	508	715	961	1115	1173	1200	1163	1090	920	694	504	430
ESE	828	1011	1182	1218	1191	1179	1175	1188	1131	971	815	748
SE	1174	1285	1318	1199	1068	1007	1047	1163	1266	1234	1151	1104
SSE	1490	1509	1376	1081	848	761	831	1049	1326	1454	1462	1430
S	1630	1626	1384	978	712	622	694	942	1344	1566	1596	1482
SSW	1490	1509	1370	1081	848	761	831	1049	1326	1454	1462	1430
SW	1174	1285	1318	1199	1068	1007	1047	1163	1266	1234	1151	1104
WSW	828	1011	1182	1218	1191	1179	1175	1188	1131	971	815	748
W	508	715	961	1115	1173	1200	1163	1090	920	694	504	430
WNW	265	439	691	911	1043	1108	1041	903	666	431	260	205
NW	127	225	422	658	813	894	821	656	416	226	132	103
NNW	123	200	300	400	550	700	550	400	300	200	123	100
HOR	706	1092	1528	1924	2166	2242	2148	1890	1476	1070	706	564

*BTUs per sq ft per day: 21st of each month on a vertical surface except as noted

glass eventually strikes an interior surface and is transformed into an equivalent amount of heat energy. Part of this heat energy is conducted and convected away, the remainder is radiated away in the form of infrared (long-wave) radiation, sometimes called *thermal radiation.* This thermal radiation will not penetrate the glass and is therefore trapped inside the space.

This phenomenon is termed the *greenhouse effect.* It is the reason why a car parked in the sun on a fall day can become hot and stuffy inside and why ordinary windows can act as solar collectors.

The type of glass used for solar collection can have a significant effect on the efficiency of the system. All types of glass—*clear, heat-absorbing* or *reflecting*—lose about the same amount of heat by conduction. However, the three differ widely in the amount of heat transmitted and trapped under the glass. Figure 6 lists the percentage of heat gain for various combinations of single and double glass, as it would typically be used—vertical windows with common design temperatures of 90° F outside to 75° F inside in summer and 25° F outside to 70° F inside in winter.

Clear single glass is the most effective heat gainer but loses the most by conduction. Clear double glazing has comparable percentage of gain, and loses less by conduction.

In climates where preventing summer heat gain is important, heat absorbing or reflecting glass on south exposures should be considered, but only if external shading or screening are not practical. Because reflective

FIG. 6 **HEAT GAIN THROUGH GLASS***

Glass Type	% Heat Gain	
	Summer	Winter
SINGLE GLAZING		
Clear	97	68
Heat-absorbing	86	41
Reflective	58	19
DOUBLE GLAZING		
Clear outside Clear inside	83	68
Clear outside Heat-absorbing inside	74	52
Clear outside Reflective inside	50	42
Heat-absorbing outside Clear inside	42	28
Reflective outside Heat-absorbing inside	31	17

*Figures taken at 40°N latitude. Based on average summer and winter sun angle and vertical glass surface.

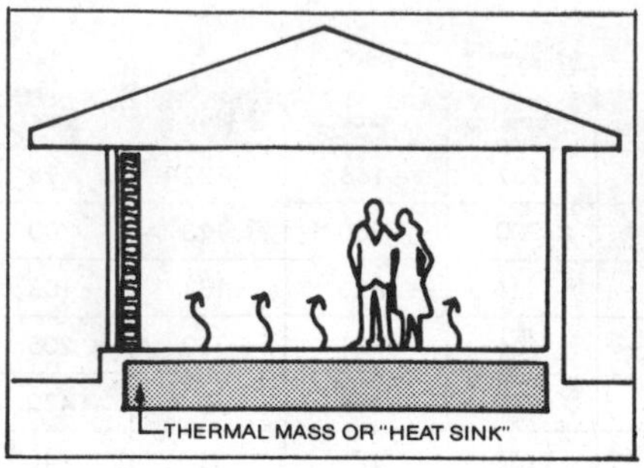

FIG. 7 A thick concrete slab can store and radiate heat for occupant's comfort.

glass reduces heat gain year-around, it is desirable only where the cooling energy costs are substantially higher than heating costs.

Although skylights and windows are the primary means of passive heat collection, they also are responsible for the greatest winter heat loss and summer heat gain. Passive systems usually employ some insulating or shading technique to prevent unwanted heat loss or gain through the glass collector. Some more traditional approaches to reducing excessive window heat loss and heat gain are described in Chapter 105 Heat Control.

STORAGE

The more glass exposed to the sun, the more heat will be trapped behind it. Windows used as passive solar collectors have the disadvantage that even on very cold days, they can quickly overheat the space they serve. In order to soak up the excess collected heat which can cause overheating, and to retain heat for nighttime or cloudy day use, floor and wall surfaces can be utilized as "heat sinks" (Fig. 7).

Materials such as water, concrete, brick, tile or stone are ideal because they can absorb and store large quantities of heat within a relatively small volume. The collected heat is released by radiation to colder surfaces and the air to maintain a relatively constant temperature. The greater the temperature differential between the mass and its surroundings, the faster the heat is released. Also, the greater the heat storage capacity (*thermal mass*) of the materials in the building, the longer it will take for the room air to overheat and the longer a room can remain comfortable without heat from an auxiliary source.

The thermal mass preferably should be located in or near the direct sun. In that case, a thermal storage capacity of 30 lbs. of water, or 150 lbs. of masonry should be used for every 1 sq. ft. of south glass. Thermal mass located elsewhere should be increased four times.

Summer Cooling

On summer nights, when the outside air is at its lowest temperature, introduction of that air into the house will carry excess heat away from interior floor and wall surfaces. In the process, those surfaces will be cooled and will therefore be able to absorb and store excess heat trapped in the building the following day. This reduces summer heat buildup within the building and aids cooling.

Heat Storage Capacity

All materials vary in their capacity to store heat. The two most common measures of that ability are *specific heat* and *heat capacity*.

Specific heat This term represents the number of BTUs required to raise 1 lb. of material 1° F. For example, water has a specific heat of 1.0 which means that 1 BTU is required to raise the temperature of 1 lb. of water 1° F. Since 1 gallon of water weighs 8.4 lbs., 8.4 BTUs are required for a 1° F rise in temperature.

Various materials absorb different amounts of heat while undergoing the same temperature rise. For example, it takes 100 BTUs to heat 100 lbs. of water 1° F, but only 22.5 BTUs are required to heat 100 lbs. of aluminum 1° F. Thus the specific heat of aluminum is .225. Water therefore has a specific heat approximately four times greater than aluminum.

The specific heats of various building materials are listed in Figure 8.

Heat Capacity This term represents the amount of heat needed to raise 1 cu. ft. of a material 1° F. Dense solid materials are effective as heat storage devices because their density permits considerable weight to be contained in each cubic foot of volume. Such materials are therefore often used as storage devices even though water has a greater specific heat. Concrete, for example, has a specific heat of approximately a quarter of water, but because of its density, it has a heat capacity of more than half that of water (Fig. 8).

FIG. 8 HEAT STORAGE CAPABILITIES OF BUILDING MATERIALS

Material	Specific Heat BTUs per lb. per 1°F	Density lb per cu. ft.	Heat Capacity BTUs per cu. ft. per 1°F
Water (40°F)	1.00	62.5	62.5
Steel	0.12	489	58.7
Cast Iron	0.12	450	54.0
Copper	0.092	556	51.2
Aluminum	0.214	171	36.6
Basalt	0.20	180	36.0
Marble	0.21	162	34.0
Concrete	0.22	144	31.7
Asphalt	0.22	132	29.0
Ice (32°F)	0.487	57.5	28.0
Glass	0.18	154	27.7
White Oak	0.57	47	26.8
Brick	0.20	123	24.6
Limestone	0.217	103	22.4
Gypsum	0.26	78	20.3
Sand	0.191	94.6	18.1
White Pine	0.67	27	18.1
White Fir	0.65	27	17.6
Clay	0.22	63	13.9
Asbestos Wool	0.20	36	7.2
Glass Wool	0.157	3.25	0.51
Air (75°F)	0.24	0.075	0.018

PASSIVE SYSTEMS

Passive solar heating or passive cooling systems are the result of adaptive design, not of mechanics. They cannot be grouped according to engineering specifications or performance standards but must be categorized according to the design approach they utilize. This, in turn, makes generic discussion of passive design more difficult because to date most designs have been the products of their inventor's ingenuity, not of established standards. What has emerged however, are three distinct design approaches within which most passive systems can be grouped: *direct gain, indirect gain* and *isolated gain.*

DIRECT GAIN

The direct gain concept is the most common passive solar building solution and has many historic precedents, including the cliff dwellings of the Sinaqua and Anasanzi Indians of Arizona. In simple terms the process can be described as: sun to living space, to thermal mass (Fig. 9). Solar radiation is collected in the living space and then stored in a thermal storage mass, such as massive floors or walls. Thus, the living space is heated by the solar radiation penetrating the windows and it serves as a "live-in" collector, storage, and distribution facility.

The basic requirements for the direct gain building include: (1) a large south-facing glazed collector area with the living space exposed directly behind; (2) a large floor and wall thermal mass exposed to solar radiation; and (3) a method for protecting the collector/windows and storage mass from heat loss to the exterior. The examples which follow illustrate how some designers have applied the concept of direct gain solar heating to building design.

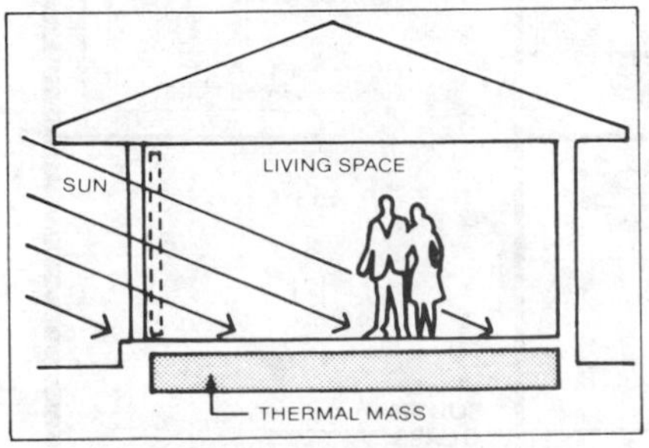

FIG. 9 Direct gain passive structures are "live-in" solar collection/storage systems.

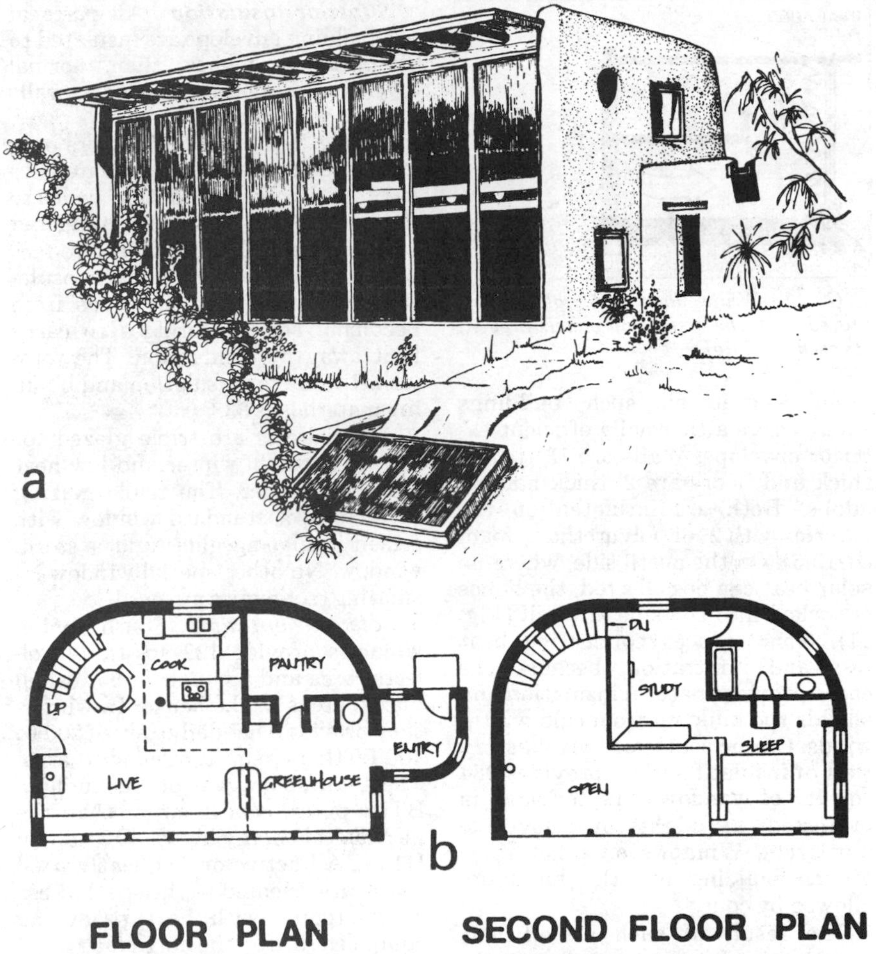

FIG. 10 Wright house: elevation (a) and floor plan (b).

Wright House

One of the best known direct gain passive houses is the Wright house in Santa Fe, New Mexico, where heating degree days (DDH) number 6063 with an average winter temperature of 25° F and an average summer temperature of 70° F. This updated version of the Pueblo adobe dwelling has been monitored and shown to provide 85% of the owner's heating requirements through direct passive gain (Fig. 10). The rest is provided by several wood-burning stoves.

The percentage of solar contribution is due in part to the fact that interior air temperatures are allowed to fluctuate between 60° F and 80° F with considerable heat stratification between upper and lower portions of the interior. Such conditions are typical of direct gain structures because large glazed areas are necessary to admit enough solar heat to maintain comfort into the night. However, use of large glazed areas may result in a rapid rise of temperature during consecutive hours of sunshine because the thermal mass may not be large enough to soak up all the incoming solar energy.

Insulating panels which can be lowered from the ceiling to block off the windows permit a degree of control over excess radiation in summer and heat loss in winter. The wide range of air temperatures is made more acceptable by the nearly constant wall temperatures. These moderate the effect of the air, making occupants comfortable at higher than normal temperatures in summer, and lower than normal temperatures in winter.

The house follows the most basic

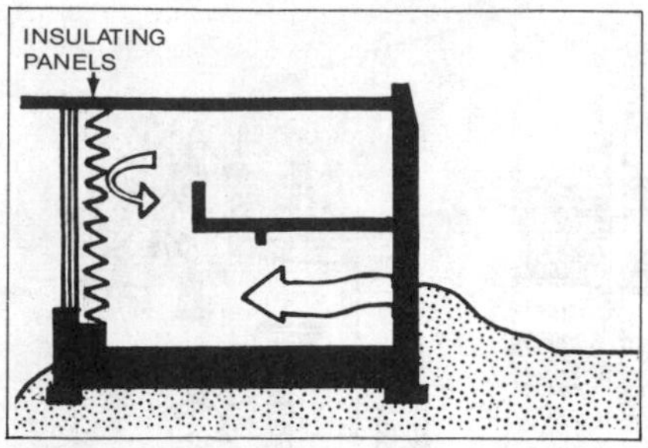

FIG. 11 While floor and wall radiate stored heat, the insulating panels prevent its escape at night.

premise—that all such buildings should have a thermally efficient exterior envelope. Walls are 13″ to 17″ thick and floors are 2′ thick natural adobe. Both are insulated on the exterior with 2″ of polyurethane foam (*R*-12.5). On the north side, where no solar heat can be collected, the house is tucked into the side of a hill (Fig. 11). This reduces conductive heat loss and infiltration, because the earth provides natural insulation and shields the building from cold winter winds. On the south face, a two-story wall of insulating glass provides 384 sq. ft. of window/collector and is protected at night by moveable insulation. Windows on other faces of the building are the minimum allowed by code.

The result of such an efficient envelope is that the house loses only about 13,000 BTUs per hr. at the local winter design temperature of 7° F. A similarly sized house insulated to the latest HUD Minimum Property Standards would lose approximately 23,500 BTUs per hr.

Illinois Low-Cal House

Architects at the University of Illinois Small Homes Research Council have developed plans for a more conventional approach to the concept of direct gain passive solar heating. Their 1,500 sq. ft. *Low-Cal House* has been computer-analyzed and shown to require only 1/3 of the energy needed for a house of equal size built to 1974 HUD Minimum Property Standards and located in an area with a heating season of between 4,500 and 8,000 DDH (Fig. 12).

It utilizes no new technology or design techniques but depends on superior insulation and solar orientation.

Superior Insulation All parts of the building envelope are insulated to significantly higher than normal levels including ceilings, floors, walls and window glazing.

Ceiling *R*-value is 38.5, floors over unheated spaces are insulated to *R*-19, perimeter foundation walls to *R*-10. Exterior walls are staggered studs spaced 8-1/2″ from face to face to permit the addition of extra insulation and to prevent the studs from becoming thermal bridges to carry heat away by conduction. The total wall *R*-value for insulation and building materials is 33.

All windows are triple glazed for low heat loss in winter and low heat gain in summer. The triple glazing consists of a standard window with sealed insulating glass and a storm window. No other special window insulating techniques are used.

Solar Orientation South facing windows provide 122 sq. ft. of collector area and account for 85% of all glazed areas in the house. In winter, they provide a net daily gain of 200 to 400 BTUs per sq. ft. of window area. The total solar gain of 14.3 million BTUs per season is larger than the mechanical energy input of 13 million BTUs. All active or frequently used rooms are oriented to the south. Less active rooms with heat producing equipment are on the north face.

No special massive interior finishes are employed as "heat sinks" to store the collected heat. The high levels of insulation are relied on to retain the heat absorbed by the air, furnishings and interior surfaces. Overhangs and deciduous trees are used to shade the windows from excess summer heat gain but are carefully calculated to work with window placement for at least 90% exposure to the sun from October 21 to February 21 (Fig. 13).

Pitkin County Air Terminal

The Pitkin County Air Terminal in Aspen, Colorado is one of the largest passively solar heated structures in the U.S. It contains 16,000 sq. ft. of terminal space organized into three pods that are staggered and linked

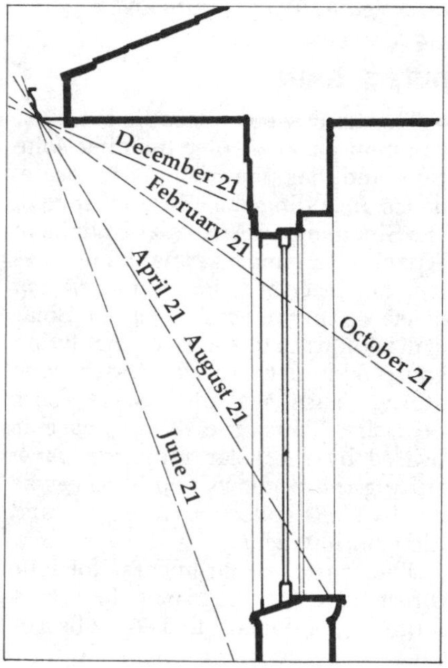

FIG. 13 Overhang keeps out high summer sun, but lets in low winter sun.

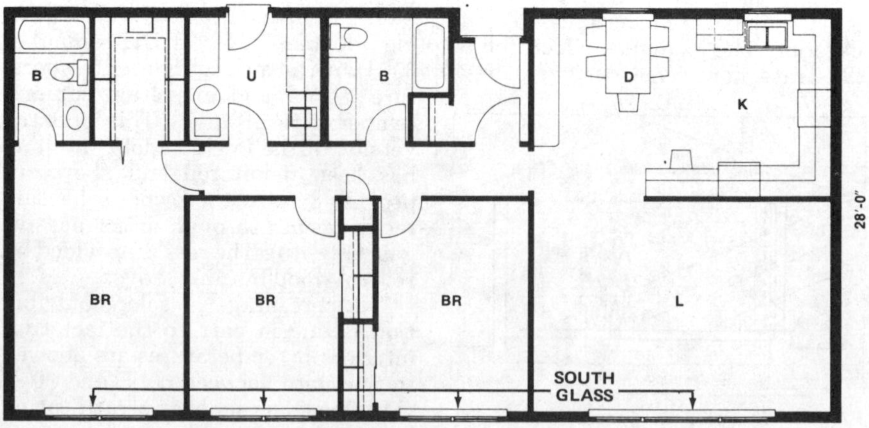

FIG. 12 Basic Low Cal plan for use on a north facing lot.

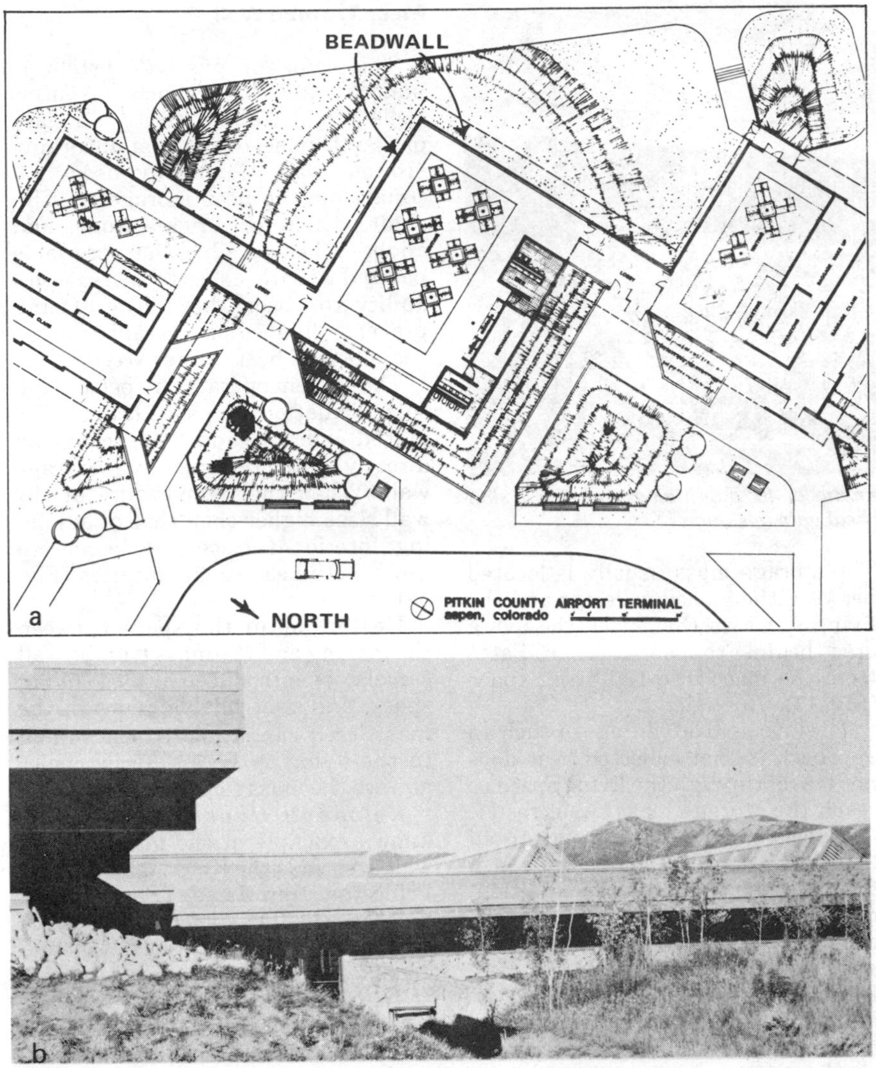

FIG. 14 Pitkin County Airport: floor plan (a) and elevation (b).

together to achieve maximum solar orientation (Fig. 14). The Aspen area has approximately 8,948 DDH per season and no cooling requirement. The basic elements of the passive solar system are abundant southern glazed windows and 1,750 sq. ft. of interior thermal mass, and a well-insulated earth sheltered building envelope.

Southern glazing Solar radiation is permitted to enter the building through 750 sq. ft. of vertical double glazed windows and 1,750 sq. ft. of horizontal skylights orientated to the south. The glass is protected from excessive night and cloudy-day heat loss by two ingenious inventions, "Skylid" and "Beadwall." Both are commercially available and are adaptable to buildings of residential scale. Large overhangs also limit unwanted summer heat gain.

Beadwall consists of a pair of translucent plastic panels separated to create a 3″ wide cavity (Fig. 15). On sunny days the clear panels admit solar radiation; at night, polystyrene beads are blown into the cavity to create an insulated wall and reduce heat loss. When sunlight is again available for collection, the beads are removed by vacuum and deposited in a concealed holding chamber until they are needed.

In summer the system can be reversed so that the cavity is filled with the insulating beads during daylight hours to prevent unwanted heat gain. At night, the beads are removed to permit the warm interior to radiate its heat to the cooler night air.

Skylids are insulated louvers that are placed inside the building behind skylights. The louvers automatically pivot open during sunny weather and close by themselves at night and on cloudy days. In the summer, the louvers are kept closed by a manual device. When closed, the Skylid is a fairly effective thermal barrier, with an approximate R-value of 4.

In winter the louvers open and close automatically by gravity, as freon moves between two cannisters mounted on the inner and outer faces. When the sun heats up the air around the outer cannister, the freon inside evaporates and flows to the inner one where it condenses to counterbalance

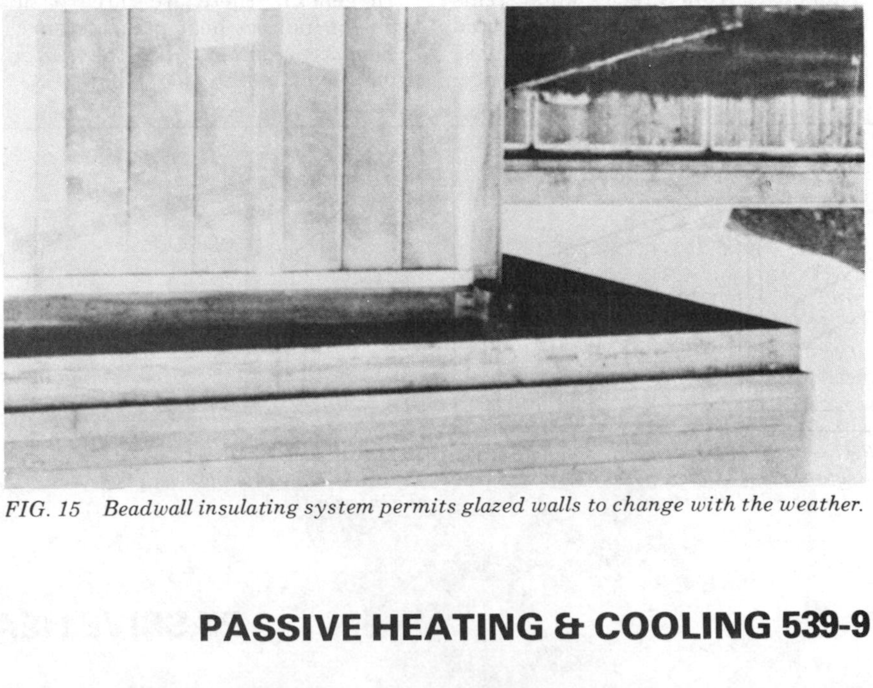

FIG. 15 Beadwall insulating system permits glazed walls to change with the weather.

FIG. 16 *Because dry earth 2 to 3 ft. down remains at 40° F or higher year round, sloping earth berms reduce heat loss in winter and heat gain in summer.*

the louvers and causes them to fall open. When the sun disappears and the air around the outer cannister cools off, a reverse flow automatically closes the louvers.

Thermal Mass The necessary thermal storage capacity is provided by grout filled concrete block interior partitions and a thickened concrete slab floor near the Beadwalls.

Building Envelope In addition to optimum levels of rigid form board and flexible insulation, the building is further protected by earth berms (mounds) wherever possible (Fig. 16). The result is increased thermal resistance and decreased infiltration through the wall.

INDIRECT GAIN

Indirect gain passive solar systems differ from direct gain systems in that they first deposit collected heat into a thermal storage mass before it is introduced into the space. The storage mass usually is located between the glazed collector and the living space so that when the living area begins to cool, heat radiates from the mass into the living space (Fig. 17).

The principal advantage of such an approach is that collected heat does not travel through the living space to reach the storage mass. As a result, more heat can be collected and stored before the space becomes too warm for occupant comfort. In addition, radiant heat flow into living spaces can more readily be controlled in such systems than in direct gain systems. Placing moveable insulation on the interior face is one typical regulatory technique. When the space begins to overheat it can be moved across the wall to reduce heat flow.

In general, there are three basic types of indirect heat gain systems: the *mass trombe wall*, the *water trombe wall*, and the *roof pond*.

Mass Trombe Wall

This type of passive system is named after its French inventor, Felix Trombe, and consists of a glass or plastic panel over a heat absorbing storage wall. Commonly used materials for the heat storage wall include concrete, brick, stone and adobe. When possible, the material is painted a dark color to increase its ability to absorb heat. The glazing prevents immediate reradiation of the collected heat to the exterior.

The system operates by permitting solar radiation to pass through the glazed collector and be transmitted directly into the massive storage wall. When the temperature of the wall rises higher than that of adjoining interior surfaces, it begins to radiate its heat to the interior (Fig. 18).

Heated air in the space between the glazing and the mass trombe wall can also be introduced into the living space, and controllable vents in the mass itself can regulate heat supply to the living space and draw cooler air into the mass for heating.

Kelbaugh House One widely known example of the mass trombe wall system is the Kelbaugh House in Princeton, New Jersey (Fig. 19). The area normally experiences 5,100 heating degree days (DDH) and usually receives approximately 55% of the possible sunshine. The house contains 2,100 sq. ft. of living space, with most rooms abutting the thermal storage wall for radiant heat transfer.

Figure 20 illustrates the solar heating system of the house. The storage element of the system is a trombe wall constructed of 15″

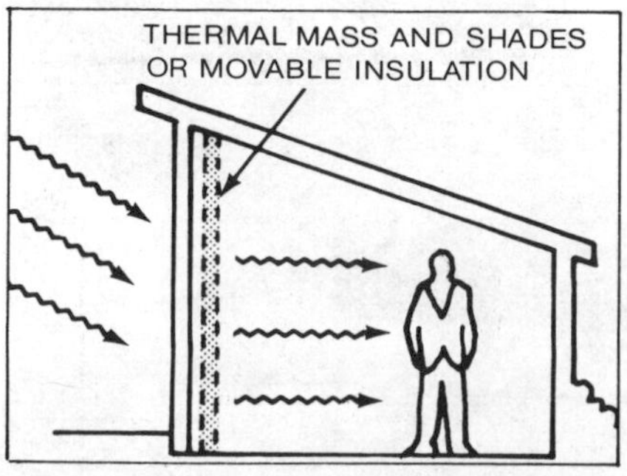

FIG. 17 *Thermal mass and movable insulation or shades permit greater control over solar radiation.*

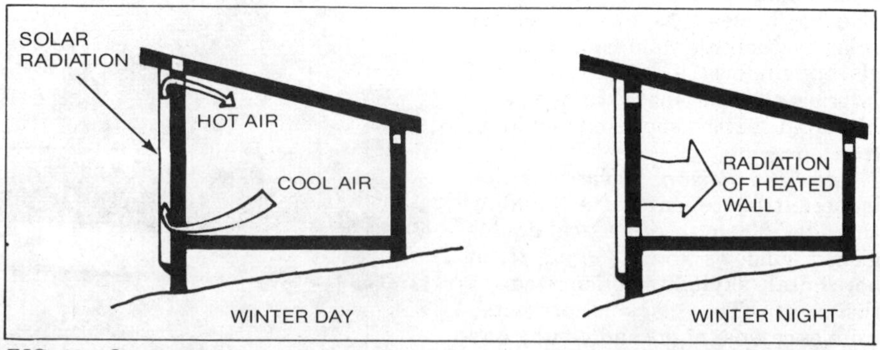

FIG. 18 *On winter days, air is heated by the sun (a); at night, the wall radiates heat to the interior (b).*

FIG. 19 South elevation and solar collector of the Kelbaugh house.

concrete and painted with a special black coating. In front of the wall are two sheets of double-strength window glass. The total collection area is approximately 600 sq. ft. Vents in the wall at the top and bottom of each floor level are part of a natural convection loop which permits circulation of the air.

A lean-to greenhouse which acts sympathetically with the solar wall was added to grow ornamental and edible plants. Its thick concrete floor is painted black to collect and store heat to heat the cellar as well as the greenhouse.

The design also provides substantial summer cooling. Four fans at the eaves exhaust heat which builds up at the southern wall and ventilates the entire house by pulling air across the rooms from windows on the north wall. By ventilating the concrete wall at night, the wall is cooled to absorb heat from the rooms on the following day. The greenhouse is shaded in the summer by slat blinds drawn over the glass and also by two large deciduous trees.

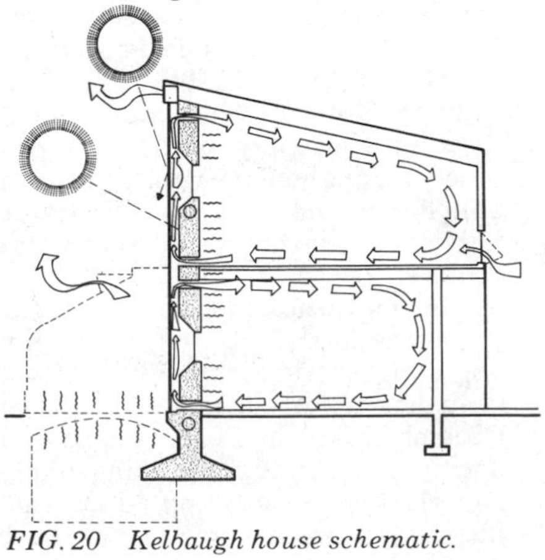

FIG. 20 Kelbaugh house schematic.

Temperatures normally fluctuate 5° to 10° F during a 24 hr. cycle with some stratification between upper and lower levels. Daytime settings were between 60° F and 64° F while the usual nighttime setting was 58° F. The seasonal low indoor temperatures were 58° F for the lower level and 62° F for the upper level, and the corresponding highs were 70° F and 75° F. The estimated lower level average was about 63° F, while that for the upper level was 67° F.

For the first winter of operation, the design heat loss was calculated to be 27,500 BTUs per DDH, or 13 BTUs per sq. ft. per DDH. The passive solar heating system provided 76% of the necessary heat.

Modifications made between the first and second winters improved the system so that the calculated heat loss for the season was reduced to 24,000 BTUs per DDH and the passive system supplied 84% of the heat.

Water Trombe Wall

The trombe wall concept of solar collection and storage can be adapted to utilize water as the storage mass (Fig. 21). Water is an effective storage medium because it has a higher specific heat than most solid massive materials and can therefore store more heat per cubic foot.

The principal disadvantage of water filled trombe walls is that containing a liquid is more difficult than containing a solid such as concrete. Steel drums are the most common containers; however, some designers have experimented with concrete block walls in which the voids are filled with vinyl bags containing water. One house in Los Alamos, New Mexico uses 2,000 water filled wine bottles as the storage wall and one manufacturer has developed a translucent fiberglass reinforced cylindrical column container (Fig. 22).

The containers can be left exposed to the interior or embedded in some massive material, such as adobe or concrete. In the David Wright house and several other houses, the containers have been covered with adobe to form window seats.

Convection currents within the water rapidly transfer the heat from the collection surface to the entire volume of water. Heat is therefore delivered to the interior spaces at a faster rate than with mass trombe walls. In addition, the collector surface operates at a lower temperature so that less heat is lost back through the glazed collector before it is absorbed.

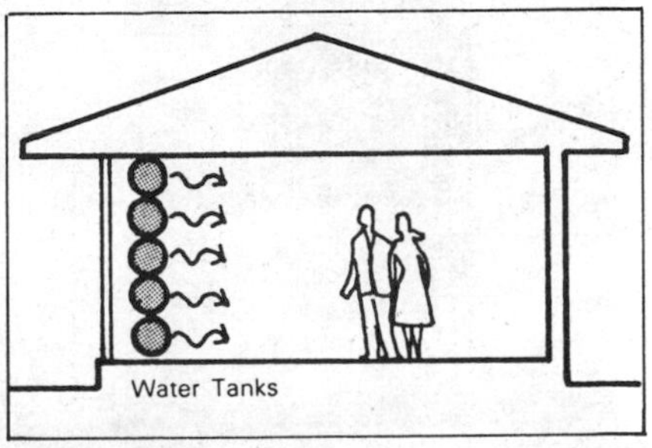

FIG. 21 Water filled drums radiate heat to interior.

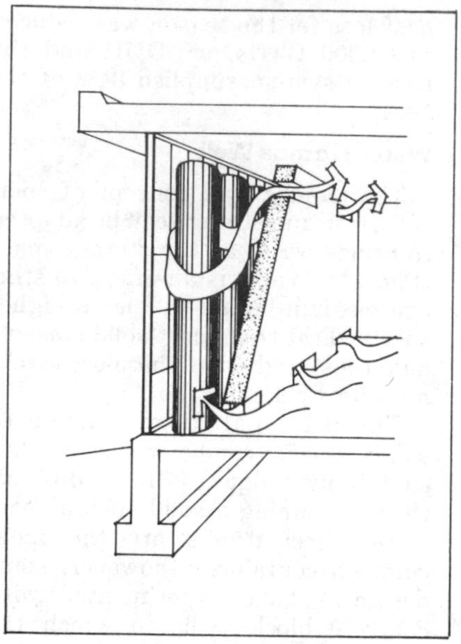

FIG. 22 Water filled cylinders provide adequate storage at a relatively low cost.

Baer House One building which uses the water trombe wall is the Baer house in Corrales, New Mexico. The system, called "Drumwall" by its inventor, Steve Baer, utilizes water filled oil drums stacked in vertical racks, placed between the interior space and a glass covered wall. It functions in a climate of 4,348 DDH and 1,150 cooling hours over 80° F (Fig. 23).

The drums are oriented to face south and are painted black on the exterior face to maximize heat absorption. In all, there are four south facing walls with approximately 100 sq. ft. of collector per wall. The total area of blackened drum face equals 260 sq. ft.

The collector surfaces are covered at night or on cloudy days by insulating shutters which are hinged at the base of each wall. Each shutter lies flat on the ground during sunny days with its aluminum inner surface reflecting additional sunlight onto the drums. At night the shutter is hoisted back into the vertical position by a hand crank and nylon rope.

The shutters are made of cardboard and paper honeycomb, filled with urethane foam and covered with an aluminized surface. The shutters have an *R*-value of 10. Some effectiveness is lost, however, because of air leaks around the shutter edges.

On sunny mid-winter days, these drums absorb approximately 1,400 BTUs per sq. ft. of glass. Convection currents within the drums rapidly distribute this heat to the remaining volume of water, thereby keeping the solar collection surfaces relatively cool. Water temperatures normally remain below 100° F. The drums distribute the heat to the rooms by radiation and natural convection. Room temperatures can be regulated by a sliding thermal curtain located behind the Drumwall.

In spaces too distant from the water trombe wall, solar heat enters the building through skylights protected by Skylids and is stored in interior wall-mounted drums.

Although the house has only 400 sq. ft. of collection surface for 2,000 sq. ft. of floor area, air temperature usually remains between 63° and 70° F during the winter. The 5,000 gallons of water in the drums, together with the concrete slab floor and adobe partitions, moderate extreme temperature swings.

If daily temperature swings of 10° F are permitted, the collected solar heat can provide 75% of the total heating requirements. Auxiliary heat is supplied by two wood stoves, but less than 1 cord of wood is used each winter.

In summer the shutters are closed during daylight hours and opened at night so that the drums will release stored heat to the cooler night-time air. With the shutters closed during the day, the cool water in the drums acts as a "sink" for excess interior heat. The reflective aluminum skin of the entire structure also helps to shed the hot desert sun.

FIG. 23 Efficiency of the "drumwall" is increased by reflecting surfaces.

Roof Pond

A variation of the trombe wall utilizes horizontal water containers on the roof of a building, to collect and store the sun's energy. Heat is collected by exposing the ponds to the sun by day, and covering them with insulation at night to prevent the collected heat from escaping into the cool night air. The solar heat collected in roof ponds is radiated directly to the rooms through a metal ceiling. Thus a continuous, evenly distributed supply of heat is provided throughout the building (Fig. 24). The system is capable of delivering heat when the water temperature drops as low as 70° F.

A prime disadvantage of roof pond collectors is that heat from the low angle winter sun is difficult to collect on a horizontal surface. On a clear December day in Pennsylvania, for example, the sunlight striking a horizontal roof is less than half that hitting an equal area of a south

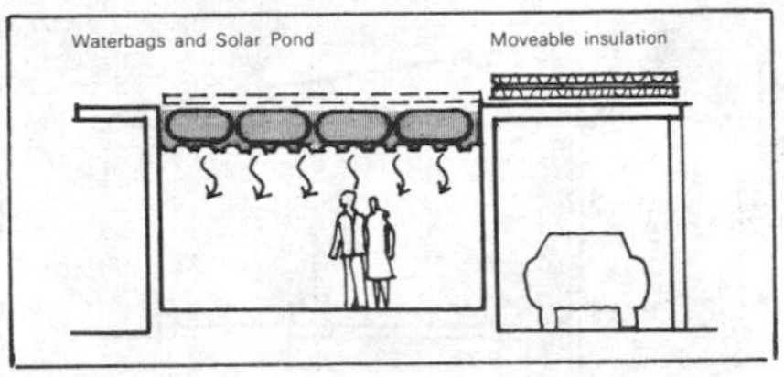

FIG. 24 Roof ponds store up solar heat in daytime to heat interior at night.

facing wall.

To compensate for this deficiency, more pond area must be provided. Stratification of the heat within the roof pond itself is another problem, as the warmer, lighter water remains near the top of the pond at night, losing heat to the outside air. The cooler, denser water falls to the bottom, just above the rooms, where heat is needed the most. An additional problem is that in many climates, the roof must be protected against snow accumulation which could block the sun from the water bags.

In general, roof pond collectors are better suited to the warmer south and southwestern states, where the sun climbs higher in the sky on a winter day, and snow build-up is not a threat. Roof ponds are also well suited for summer cooling, which is of primary importance in such climates.

Insulated panels are used to cover the roof pond on summer days—to keep the sun from warming the ponds, while the cool water absorbs excess heat from the rooms below. At night the panels are automatically removed and the roof pond radiates this heat to the night air. Heat stratification within the roof pond *aids* this cooling process because the coldest water lies just above the rooms.

Atascadero House The most successful application of the roof pond approach to passive heating and cooling has been with the "Skytherm" system invented by Harold Hay. His house located in Atascadero, California, has utilized this system since 1973 with no supplementary heating or cooling (Fig. 25). Normal interior temperatures have been maintained between 68° F and 82° F while outside temperatures ranged from 32° F to 115° F. Winter severity is represented by 2,970 DDH.

The single story house has 1,100 sq. ft. of heated floor area and an adjoining carport and patio. The roof pond covers all 1,100 sq. ft. of the house. Nine movable insulated panels of *R*-15 rigid polyurethane foam slide automatically over the roof ponds on winter nights or are stacked three-deep above the patio and carport on sunny winter days.

Water is contained in 8′ x 38′ transparent plastic bags that are arranged in four troughs running north and south on the roof. With a total volume of 7,000 gallons, the 8″ depth of water has the heat capacity of approximately 16″ of concrete but the weight of only 4″. The waterbed-like bags lay on an earthquake-proof steel deck supported by interior and exterior concrete block walls. A plastic liner between the water bags and the roof deck protects against leaks and rainwater seepage. Above the water bags, an inflatable transparent plastic cover protects them from the sun's ultraviolet radiation and helps trap the solar heat.

FIG. 25 Water filled bags are located behind the fascia.

ISOLATED GAIN

Isolated gain passive systems utilize collectors and storage masses which are thermally isolated from the living space. The concept therefore allows collector and storage to function independently while the building can draw heat as required.

Isolated gain systems can generally be grouped into two categories: those which use a greenhouse area, or similar structure (often called a *sunspace*) to collect and store heat and those which utilize the thermosiphon principle to move collected heat.

Sunspace

Using a sunspace allows maximum solar collection in this space (and consequent overheating) without affecting the comfort within the primary living space. The collected heat can be admitted as needed through doors, windows or dampers.

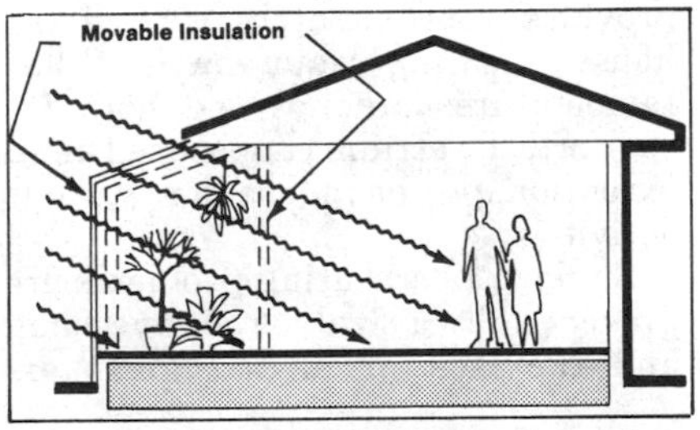

FIG. 26 Insulation is lowered to reduce heat loss and raised to allow solar collection through this greenhouse.

An atrium, sun-porch, greenhouse and sunroom, all represent potential examples of a sunspace (Fig. 26). In general, this technique requires a glazed collector space which must be both attached yet distinct from the living space, it must have the proper southern exposure, and the collector space must contain or be thermally linked to a solar storage mass for heat retention and later distribution.

Erwin House A typical example of the isolated gain concept is the Erwin house in Nacogdoches, Texas

FIG. 27 South elevation, Erwin house.

where 1,800 DDH and approximately 2,050 cooling hours over 80° F are typical. The 2,275 sq. ft. of living space are wrapped around a south facing atrium which can be thermally connected or separated as desired by 400 sq. ft. of windows and glass patio doors (Fig. 27). Direct radiation enters the 725 sq. ft. of glass exposure, is absorbed by the atrium's dark brown tile floor and stored in the supporting 8″ thick concrete slab.

As heat is required, doors are opened to allow heated atrium air to enter the living space. The system provides approximately 90% of the house's heating requirement. When temperatures are not too high for comfort, the atrium can be used as an extension of the house for various activities.

Summer overheating problems are prevented by a sizable roof overhang and the lack of angled glass exposures or skylights. The southern atrium is also designed to act as a summer wind scoop to draw air through and out small north windows to naturally ventilate and cool. The 2″ x 6″ stud construction accommodates *R*-19 insulation; the windows are double glazed, and the garage and utility areas have been located on the west perimeter to prevent summer heat gain and to give added thermal protection in winter. The northern face has a low profile with berming and minimal window openings to reduce heat loss.

Thermosiphon

The second type of isolated gain system permits heat to build up within a roof, wall or room-size collection area before drawing it off to storage or living areas. Thermosiphon (convection), the natural rise of heated air or fluid, is the usual means of transport (Fig. 28).

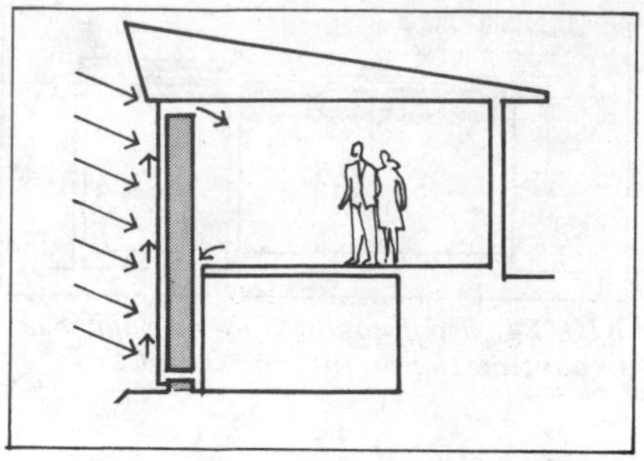

FIG. 28 Thermosiphoning wall delivers heat by natural convection and radiation.

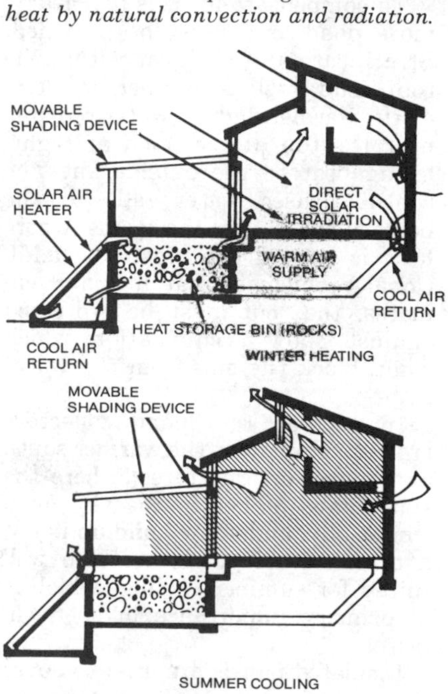

FIG. 30 Thermosiphon principles can be utilized for heating or cooling.

In the typical installation, a glazed collector is located between the direct sun and the living space, and is distinct from the building structure. In this space, a convection heat flow occurs when air heated by the sun rises naturally into an appropriately placed living space or storage mass, causing cooler air or liquid to fall again. A continuous heat gathering circulation is thereby established.

Davis House A typical example of the thermosiphon system is the Davis house in Corrales, N.M. (Fig. 29). The basic elements of this passive solar house include 420 sq. ft. of thermosiphoning collectors, single glazed and located in front of the house below a large rock bed storage bin that supports the south-facing patio. The Corrales area typically

FIG. 29 Davis house lean-to collector.

experiences 4,348 DDH and 1,500 cooling hours over 80° F.

Air ducts lead to and from the house so that with the aid of manually operated dampers, solar heated air can flow up to heat the rock bed storage, while cooler air returns to the bottom of the collector for continuous natural circulation (Fig. 30).

For heat distribution to be effective, the area of contact between storage and house is as critical as the spatial arrangement of the house. This 1,000 sq. ft. house is arranged with an open interior plan and a loft. Heat is distributed through the house by natural convection once the house dampers to storage are opened, with cooler air returning back to the collectors during the day or back to the solar heated storage at night. In addition, direct gain heating through roof monitors contribute to the system and adobe walls provide additional thermal storage to reduce interior temperature swings.

Approximately 75% of the heat and 100% of the cooling are provided by the system.

601 LAND SURVEY & DESCRIPTION

LAND SURVEY & DESCRIPTION

INTRODUCTION

Every legal description of property is based upon a land survey. In order to prepare a legal description of a parcel or tract of land, someone at some time must have measured and marked the land—made a survey—upon which the description is based.

The subject of land descriptions and surveys is centuries old. The earliest records of man refer to skillful measurements and calculations with respect to land, but it is impossible to assign the birth of the science of surveying to any particular year or even country. The Chinese, at an early date, and the ancient Egyptians practiced the art of surveying. In Egypt, it appears that it was necessary every spring to re-establish corners and boundary lines obliterated by floods of the Nile River.

Today, the transfer of land ownership and the mortgaging of property require legal land descriptions by which the location and boundaries of land parcels can be determined. Descriptions must be referenced to established systems of field marks and measures. Hence, a general knowledge of survey systems is necessary to understand the description of land in writing.

This section describes the several general systems of survey and types of legal land descriptions used in the United States today.

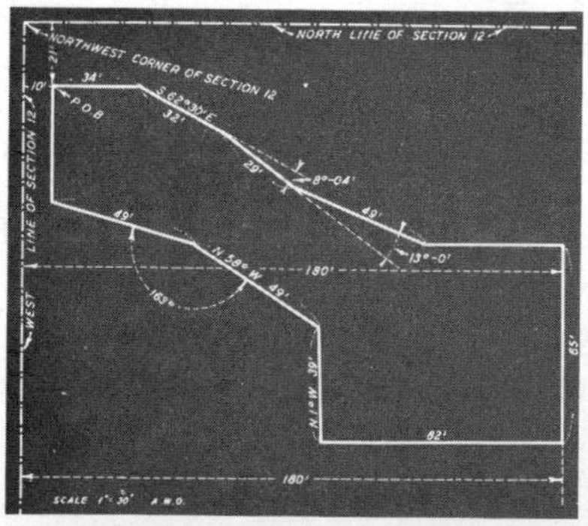

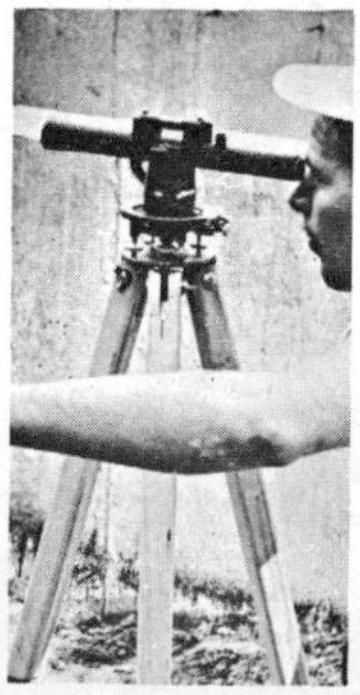

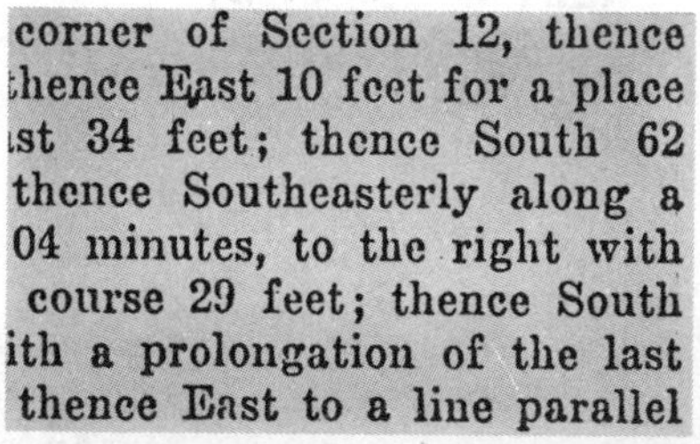
corner of Section 12, thence
thence East 10 feet for a place
st 34 feet; thence South 62
thence Southeasterly along a
04 minutes, to the right with
course 29 feet; thence South
ith a prolongation of the last
thence East to a line parallel

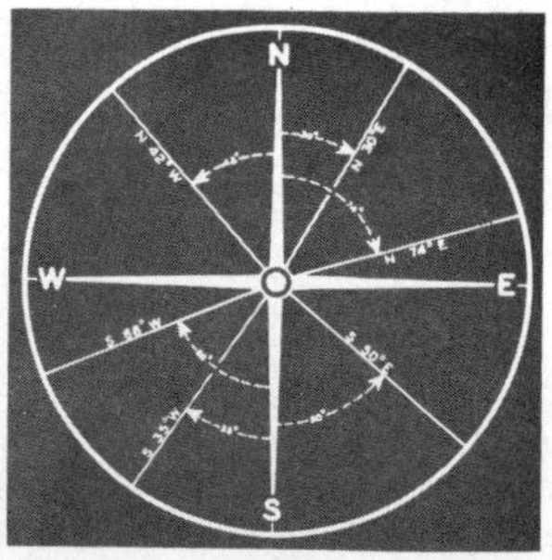

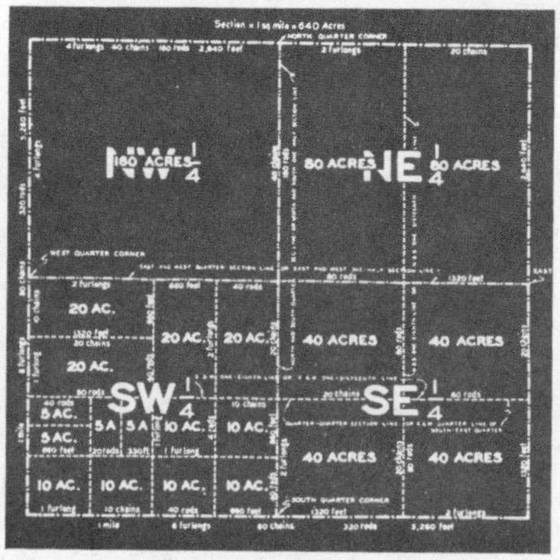

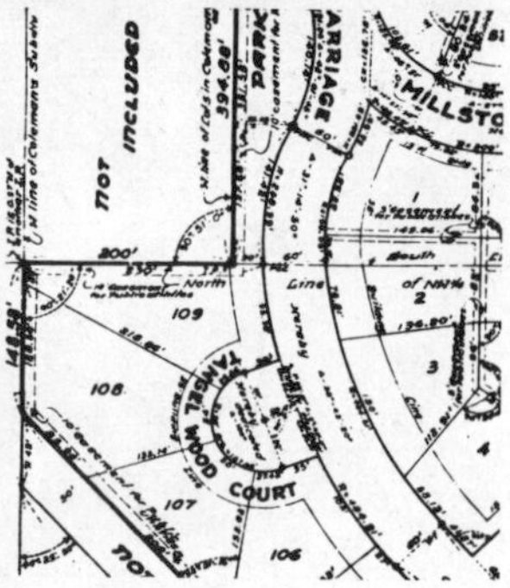

A survey is the measure and marking of land, accompanied by maps and field notes which describe the measures and marks actually made in the field. The lengths and directions of boundary lines are established between reference points on the ground. The reference points may be natural marks, such as rivers, or they may be stone, concrete or other artificial markers located and set in the field by a surveyor. Natural points, noted as permanent, and permanent markers set in the field by surveyors are called *monuments.*

Today, survey measure of lengths commonly is made in feet and decimal parts of feet. However, terminology based on measures no longer used is encountered often in surveys and legal descriptions. For example, early surveys were based on lengths such as *chains* and *links.* Surveyors actually made field measurements with a metal chain made of 100 links. The chain was 66′ long, and 80 chains equaled one statute mile (5,280′). Some of the various measures used and their equivalents are shown in Figure 1.

Two general systems of survey are used in the United States: the *metes and bounds system,* evolved from Colonial days, which is based on tracing the boundary lines surrounding an area; and the *Rectangular System* of the Federal Government Survey used in most of the states, which is based on a modified grid of north-south and east-west lines.

LAND MEASURES & EQUIVALENTS		
1 link	=	7.92 inches
1 chain	=	100 links
1 chain	=	66 feet
1 chain	=	4 rods
1 rod	=	16-1/2 feet
1 rod	=	25 links
1 furlong	=	10 chains
1 furlong	=	40 rods
1 mile	=	80 chains
1 mile	=	320 rods
1 mile	=	8 furlongs
1 section	=	1 square mile
1 section	=	640 acres
1 acre	=	160 rods
1 acre	=	43,560 square feet

FIG. 1 Early surveys were made using measures such as "chains" and "rods." Various equivalents in measures are shown here.

Though the following discussion of survey systems may appear to be a rather exact basis upon which land can be described, it is not. Inaccuracies in surveying, which may or may not be avoidable, may present difficulty and disagreement in the legal description of land. Some of the uncertainties in land description based on problems inherent in surveying are discussed on page 601-11.

METES AND BOUNDS SYSTEM

Survey by metes and bounds (Figs. 2 & 7) consists of beginning at a known point and "running out" the boundaries of the area by *courses* (directions) and *distances* (lengths), and fixing natural or artificial monuments at the corners. The place of beginning (P.O.B.) must be a known point that can be readily identified. The point must be established and witnessed so that it can be relocated with cer-

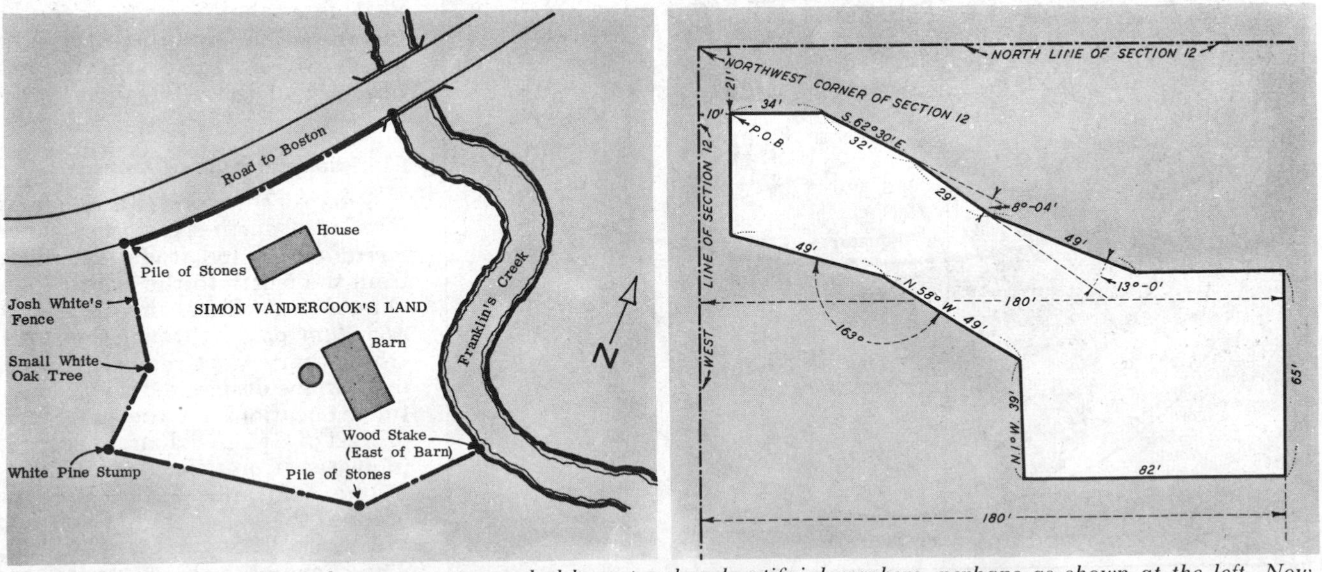

FIG. 2 Early metes and bounds surveys were marked by natural and artificial markers, perhaps as shown at the left. Now surveys are established more carefully with accurate distances and angles. In the example shown at the right, the starting point of the survey (P.O.B.) has been referenced to the rectangular system.

tainty if the *monument* which identifies it is destroyed or removed. The survey description "must close"; that is, if the courses and distances of the description are traced in order, one must return to the starting point. An example of a modern legal description based on a metes and bounds survey is shown in Figure 7.

The metes and bounds system is one of the oldest known manners of surveying and describing land. The system evolved from the early practice of "settlement first, survey afterwards." Land being bought, claimed or inherited would be identified by known natural points such as trees and rivers, and artificial points such as roads, structures or other manmade marks.

For example, in 1784, when Simon VanderCook sold his Massachusetts land, the survey upon which a legal description was based started at a "white pine stump," proceeded north to a "small white oak tree" and continued with a series of measures by "chains and links" between markers such as trees, stakes and piles of stones, finally returning to the "middle of the stump where it first begun" (Fig. 2).

Early metes and bounds surveys and descriptions often were based on monuments that lacked permanency. Surveying of large and irregular tracts of land without regard to any system or uniformity, and the failure of the surveyors to make their survey notes a matter of public record, created situations that gave rise to frequent boundary line disputes and litigation.

Today, metes and bounds surveys do not have the uncertainty as to place of beginning that existed in Colonial days. Present day surveys and descriptions may refer to government survey section lines and corners (see Rectangular System) as monuments, in addition to permanent artificial marks placed by surveyors.

Metes and bounds surveys and descriptions still are used in 20 states, which include the states of the New England area, the Atlantic Coast states (except Florida), and Hawaii, Kentucky, Tennessee, Texas, Virginia and West Virginia. Also, the system may be used anywhere to survey areas which are irregular in size and shape. A discussion of legal description by the metes and bounds method is found on page 601-9.

RECTANGULAR SYSTEM

After the Revolutionary War, the federal government found itself with vast tracts of undeveloped and uninhabited land, with few natural characteristics suitable for use as monuments in metes and bounds descriptions. It was necessary to devise a new standard system of describing land so that parcels could be located readily and permanently for land office sales. A committee headed by Thomas Jefferson evolved a plan, which the Continental Congress adopted in 1785, for dividing the land into a series of rectangles. This plan, designated the *Rectangular System* (also called the *Government System*) of survey, is in use today in the other 30 of the 50 United States (Fig. 4).

Meridians and Base Lines

A map of the world (Fig. 3) shows a series of north-south lines called *meridians* of longitude extending from the North to the South Pole. One of these lines, the *Greenwich Meridian*, passes through the Royal Observatory at Greenwich, England and is designated as 0° longitude. Locations on the globe east or west of Greenwich are measured in degrees from that meridian.

The reference lines running around the globe due East and West are called *parallels* of latitude. The *Equator*, which circles the center of the globe, is designated as 0° latitude. Locations on the

LINES OF LONGITUDE AND LATITUDE (Meridians & Parallels)

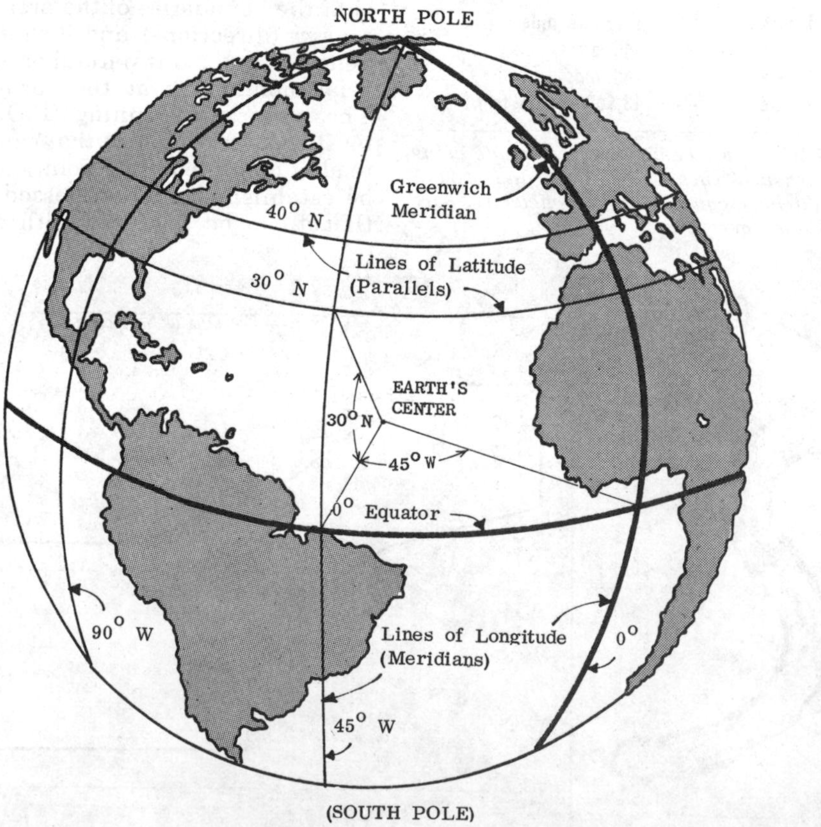

FIG. 3 The rectangular system of survey is based on a modified grid of meridians and parallels measured from the Greenwich Meridian and the Equator.

FIG. 4 **MERIDIANS AND BASE LINES OF THE UNITED STATES RECTANGULAR SURVEYS**

Rectangular System is used in states shown unshaded, Hawaii and Alaska. Each principal meridian and its base line is shown on the map with location by longitude and latitude in table below; shaded areas use metes and bounds surveys.

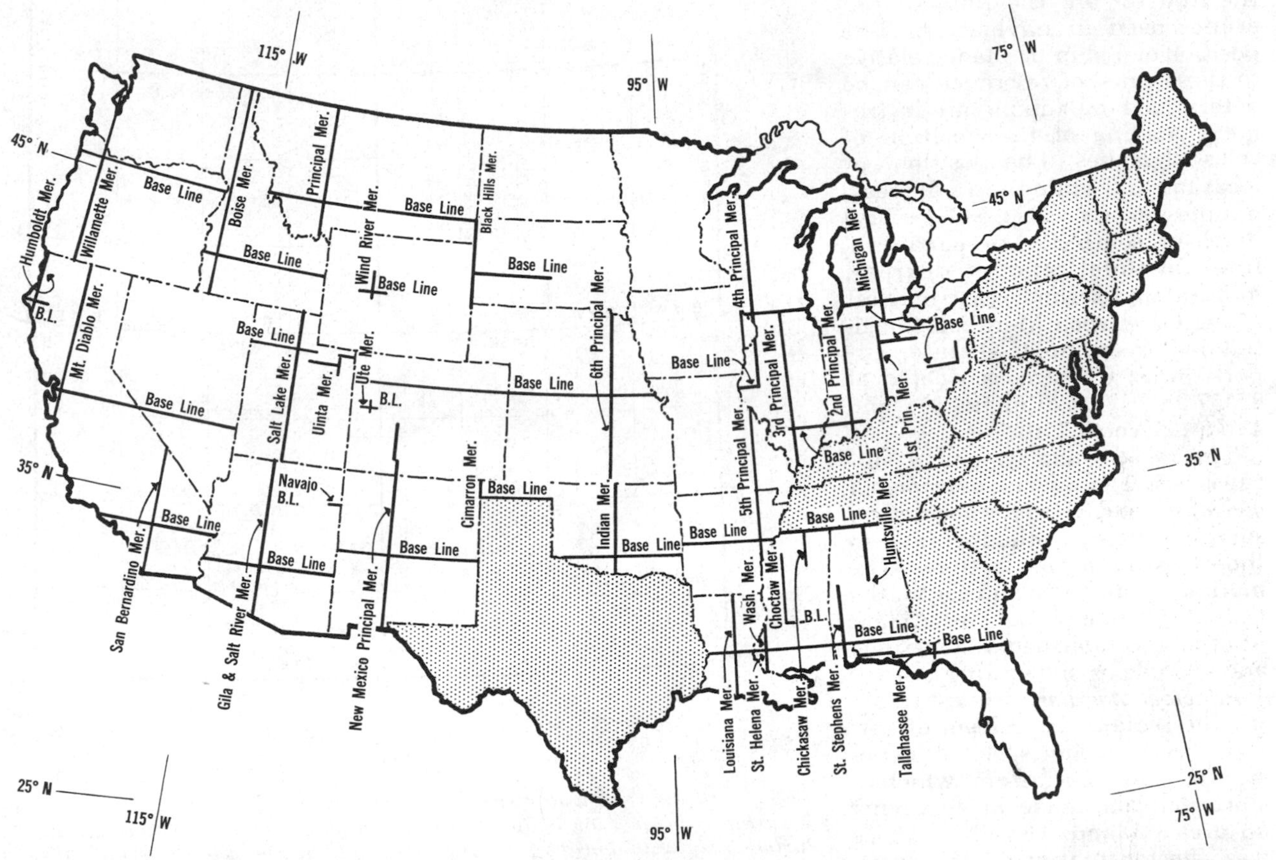

Meridians.	Governing surveys (wholly or in part) in States of—	Longitude of principal meridians west from Greenwich. ° ′ ″	Latitude of base lines north from Equator. ° ′ ″	Meridians.	Governing surveys (wholly or in part) in States of—	Longitude of principal meridians west from Greenwich. ° ′ ″	Latitude of base lines north from Equator. ° ′ ″
Black Hills	South Dakota	104 03 00	44 00 00	Navajo	Ariz. & N. Mex.	108 32 45	35 45 00
Boise	Idaho	116 24 15	43 22 31	New Mex. Principal	New Mexico	106 52 41	34 15 25
Chickasaw	Mississippi	89 15 00	34 59 00	New Mex. Principal	Colorado	106 53 36	
Choctaw	Mississippi	90 14 45	31 54 40	Principal	Montana	111 38 50	45 46 48
Cimarron	Oklahoma	103 00 00	36 30 00	Salt Lake	Utah	111 54 00	40 46 04
Copper River	Alaska	145 18 42	61 49 11	San Bernardino	California	116 56 15	34 07 10
Fairbanks	Alaska	147 38 33	64 51 49	Second Principal	Ill. & Ind.	86 28 00	38 28 20
Fifth Principal	Ark., Iowa, Minn., Mo., N. Dak., & S. Dak.	91 03 42	34 44 00	Seward	Alaska	149 21 53	60 07 26
First Principal	Ohio & Indiana	84 48 50	41 00 00	Sixth Principal	Colo., Kans., Nebr., S. Dak. & Wyo.	97 23 00	40 00 00
Fourth Principal	Illinois[1]	90 28 45	40 00 30	St. Helena	Louisiana	91 09 15	31 00 00
Fourth Principal	Minn. & Wisc.	90 28 45	42 30 00	St. Stephens	Ala. & Miss.	88 02 00	31 00 00
Gila and Salt River	Arizona	112 18 24	33 22 33	Tallahassee	Florida	84 16 42	30 28 00
Humboldt	California	124 07 11	40 25 04	Third Principal	Illinois	89 10 15	38 28 20
Huntsville	Ala. & Miss.	86 34 45	35 00 00	Uinta	Utah	109 57 30	40 26 20
Indian	Oklahoma	97 14 30	34 30 00	Ute	Colorado	108 33 20	39 06 40
Louisiana	Louisiana[2]	92 24 15	31 00 00	Washington	Mississippi	91 09 15	31 00 00
Michigan	Mich. & Ohio	84 22 24	42 26 30	Willamette	Ore. & Wash.	122 44 20	45 31 00
Mount Diablo	Calif. & Nev.	121 54 48	37 51 30	Wind River	Wyo.	108 48 40	43 01 20

1. Numbers are carried to fractional township 29 north in Ill., and are repeated in Wisc., beginning with south boundary of the State; range numbers are in regular order. 2. Latitude doubtful; is to be verified.

Note: East boundary of Ohio, known as "Ellicott's Line," longitude 80°32′20″, was the first reference meridian, with township numbers counting from Ohio River, and range numbers in regular order. Township and range numbers within U. S. military land in Ohio are counted from south and east boundaries of tract.

globe north and south of the Equator are measured in degrees from that parallel of latitude.

The Greenwich Meridian and the Equator are the lines of reference used in navigation. The position of a ship or plane relative to these lines of reference can be determined by taking an instrument reading of the position of celestial bodies. The position, or location, is stated in degrees, minutes and seconds (see Glossary) north or south of the equatorial line (latitude); and in degrees, minutes and seconds east or west of the Greenwich Meridian (longitude). For example, O'Hare Airport tower, Chicago, is located at 87°57′28″ West Longitude and 41°59′10″ North Latitude.

The rectangular system of survey is based on a series of *principal meridians* and *base lines* located by surveyors using methods similar to those used in navigation. The meridians run north and south, the base lines run east and west. Their position is established by longitude and latitude as shown in Figure 4.

Principal Meridians In establishing the rectangular system of survey, surveyors first selected a substantial landmark from which a start could be made in surveying an area of land. Usually a place was selected that could be identified readily, such as the mouth of a river, and wherever possible a monument was placed. From this point the surveyors ran a line due North and South and designated it as the *principal meridian* for that particular state or area. The location of the principal meridian (its longitude) was fixed by a reading measured in degrees, minutes and seconds west of the Greenwich Meridian.

As territories were opened and surveyed by the government, principal meridians were established for each area opened. Some of the principal meridians were referred to by number, such as the *First, Second, Third* Principal Meridian. The Third Principal Meridian, for example, is the line located 89°10′ 15″ west of Greenwich and extends from the mouth of the Ohio River to the northern boundary of Illinois.

DIVISION OF LAND BY RECTANGULAR SURVEY SYSTEM

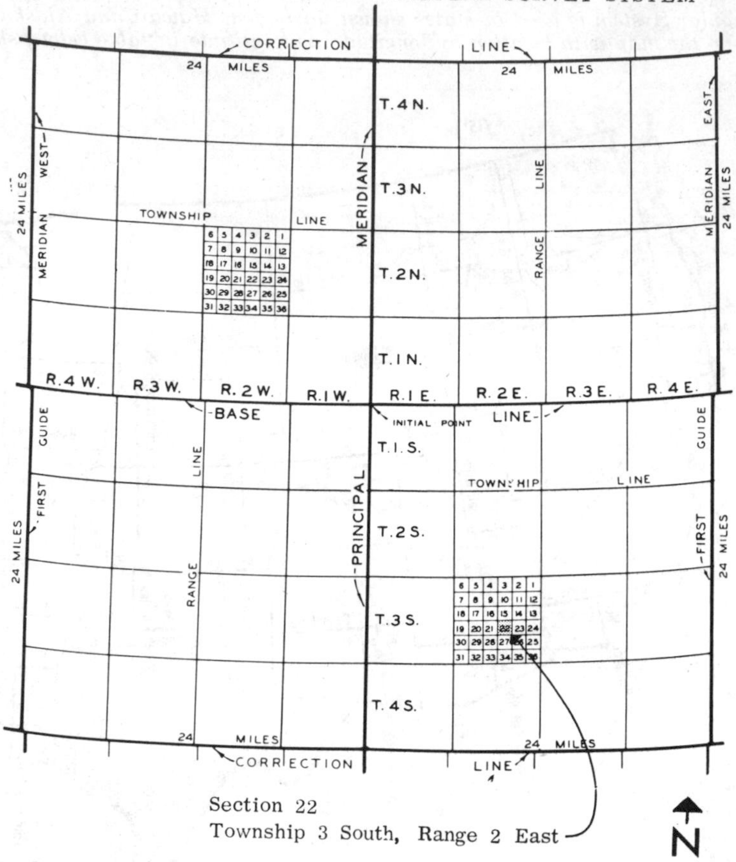

FIG. 5 In the rectangular survey system, 24 mile squares are divided into 16 townships and each township is divided into 36 sections. (Sections are subdivided further, see Fig. 6). Locations of land areas are referenced to the principal meridian and its base line.

Other principal meridians were named for the states, such as the Michigan Meridian which covers the survey of that state. The 35 principal meridians and the 32 base lines established in the United States are shown in Figure 4.

Base Lines After the principal meridian for a particular area was fixed, a point on the meridian was selected from which a line was run at right angles, due East and West. This line was designated as the *base line* for the area, and its location (its latitude) was established in degrees, minutes and seconds north of the equatorial line (Fig. 4).

In certain instances the base line for a new territory was established by extending the base line of an adjoining territory, previously surveyed. For example, the base line for both the Second and Third Principal Meridians is a parallel of latitude 38°28′20″ north of the equatorial line (Fig. 4).

Correction Lines All meridians meet at the North and South Poles. They are not parallel lines. This fact is not observable from a point on the earth's surface without the aid of surveying instruments, but an accurate measure of a midwestern area 6 miles square would show its north line to be about 50′ shorter than its south line. To compensate for the convergence of the meridians, it was necessary to establish additional reference lines to measure equal east-west distances. East and west lines, parallel to the base line, were located at intervals of 24 miles north and south of the base line. These lines were designated as *correction lines.*

Guide Meridians After the east-west correction lines were set, *guide meridians* running due North and South at 24 mile intervals on each side of the principal meridians were established. Guide meridians extend from the base line to the first correction line, and then from correction line to correction line. The guide meridians and the correction lines form squares approximately 24 miles on each side as shown in Figure 5.

Township and Range Lines

The 24 mile squares were then divided into 16 smaller tracts by township lines and range lines. *Township lines* were run east and west (parallel to the base line) at 6 mile intervals. *Range lines* were run north and south at 6 mile intervals. This cross-hatching resulted in a grid of squares, called *townships*, approximately 6 miles on each side (Fig. 5).

Two reference numbers were assigned to each square—a *township* number and a *range* number. Rows of squares (tiers) were num-

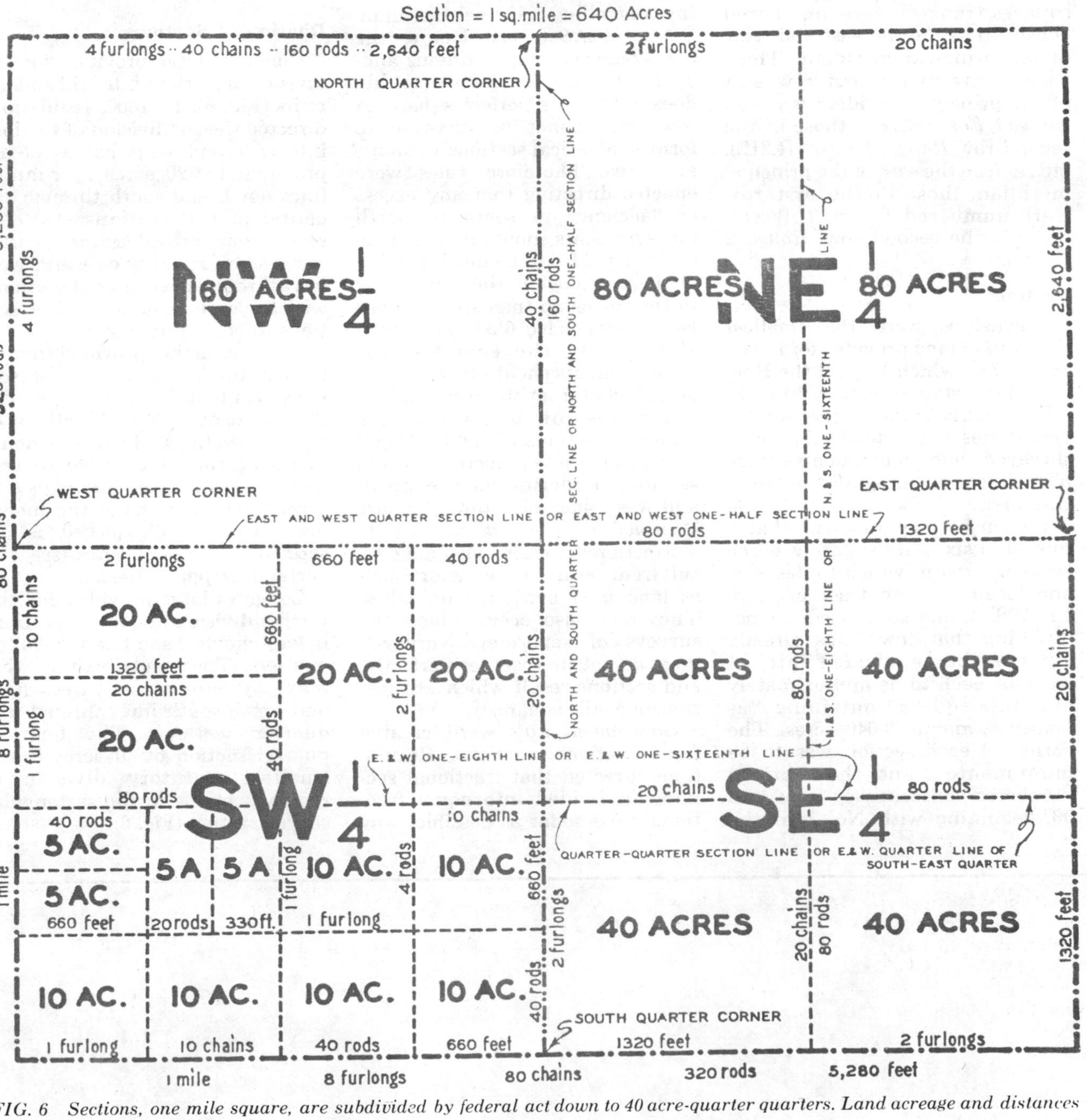

FIG. 6 Sections, one mile square, are subdivided by federal act down to 40 acre-quarter quarters. Land acreage and distances within the section are shown here.

bered consecutively to the north or south of the base line. Thus, each square in the first row north of the base line is called *Township 1 North* (T.1N.); each in the second row, *Township 2 North* (T.2N.), etc. Similarly, each square in the first row south of the base line is called *Township 1 South*; each in the second row, *Township 2 South*, etc. In the same manner, rows of squares (ranges) were numbered consecutively to the east and west of the principal meridian. Thus, each square in the first row east of the principal meridian is called *Range 1 East* (R.1E.), those in the second row *Range 2 East* (R.2E.), etc., and on the west of the principal meridian, those in the first row were numbered *Range 1 West*, those in the second row, *Range 2 West*, etc.

Sections

Townships were the smallest divisions of land provided for in the act of 1785, which created the Rectangular system of survey. In making the early surveys, the outside boundaries of the townships were surveyed and monuments were placed at every mile on the township lines.

It soon became apparent that a township six miles square was too large an area in which to describe and locate a given tract of land. In 1796, Congress passed an act directing that townships already surveyed be subdivided into 36 *sections*, each to be approximately one mile square containing "as nearly as may be" 640 acres. The corner of each section was to be monumented, and the sections numbered consecutively from 1 to 36, beginning with No. 1 in the northeast corner of the township, ending in the southeast corner with No. 36, as shown in Figure 5. This manner of numbering sections in a township has continued to this date. A given tract of land can be located within a particular square mile by giving the section number, the township number north or south of the base line, and the range number east or west of the controlling principal meridian.

Fractional Sections Due to the convergence of the meridians and/or to other causes, each township does not form a perfect square. A township cannot be surveyed to form 36 identical sections in shape and area. Therefore, rules were enacted directing that any excess or deficiency in south to north measurements should be added to or deducted from the north portion of the sections in the north row in the township (namely, Sections No. 1 through No. 6). Any excess or shortage in the east to west measurements should be added to or deducted from the west portion of the west row in the township (namely, Sections 6, 7, 18, 19, 30 and 31). Accordingly, such adjusted sections bordering on the north and west sides of a township are classified as *fractional sections*.

Fractional sections also may result from geographical factors such as land area bordering on lakes. They may also occur where the surveys of separately surveyed areas do not tie together perfectly, and sections result which are less than one mile square.

Government Lots were created from fractional sections. Regulations directed that fractional sections be divided into equal fractional parts so far as possible. Any remaining portion was to be divided into parts called *Government Lots* and numbered consecutively. The intent of the regulations was to divide fractional sections into as many regular parts as possible according to their relative position in the section, with any remaining portions divided into lots of approximately the same size.

Division of Sections

Congress later provided for the division of sections into still smaller units (Fig. 6). In 1800, regulations directed the subdivision of sections into east and west halves of approximately 320 acres by running lines north and south through the center of the sections. In 1805 regulations provided for quarter sections by running east and west lines through the center of sections, with the corners of all quarter sections to be monumented.

This act further provided that all corners marked in the public surveys would be accepted as the proper corners of the sections or quarter sections. That is, monumented corners set by the government surveyors would *stand as true corners* whether or not the monument was actually placed as described in the field notes (see Uncertainties, page 601-11).

Congress later provided for the further division of quarter sections in *half-quarters* and finally *quarter-quarters*. These subdivisions were made by running east-west lines and north-south lines through the quarter sections. The quarter-quarter section of 40 acres is the smallest statutory division of regular sections in the rectangular survey system (Fig. 6).

The objective of a legal description is to identify, locate and define a specific piece of land and distinguish it from all other tracts of land. Legal descriptions in the United States may be based on: (1) the metes and bounds survey system, (2) the rectangular (government) survey system, and (3) reference to recorded maps and plats. It is not uncommon to find a combination of these methods used in one description. For example, a lot in a subdivision may be described by reference to a *recorded plat* of a land parcel, which in turn is described by a *metes and bounds* survey within the framework of the *rectangular system.*

METES AND BOUNDS DESCRIPTION

A metes and bounds description consists of a series of statements describing each portion of the boundary around the parcel of land being described. Starting from a *place of beginning* (P.O.B.), the direction and distances of the property lines are *called* (stated in the description), until the perimeter has been traced around the entire property, returning to the starting point. Each leg of the perimeter is described first by direction, then by length (Fig. 7).

Direction

When directions (courses) agree with the major compass directions, they are so named. Courses other than due North, South, East or West (cardinal points) are called *angular courses.* Angular course direction is stated in degrees, minutes and seconds as an angular deviation east or west of due North or South. (Fig. 8). (There is one exception to this convention explained below under Prolongation.)

The direction always is written with the general statement North or South, followed by the angle of

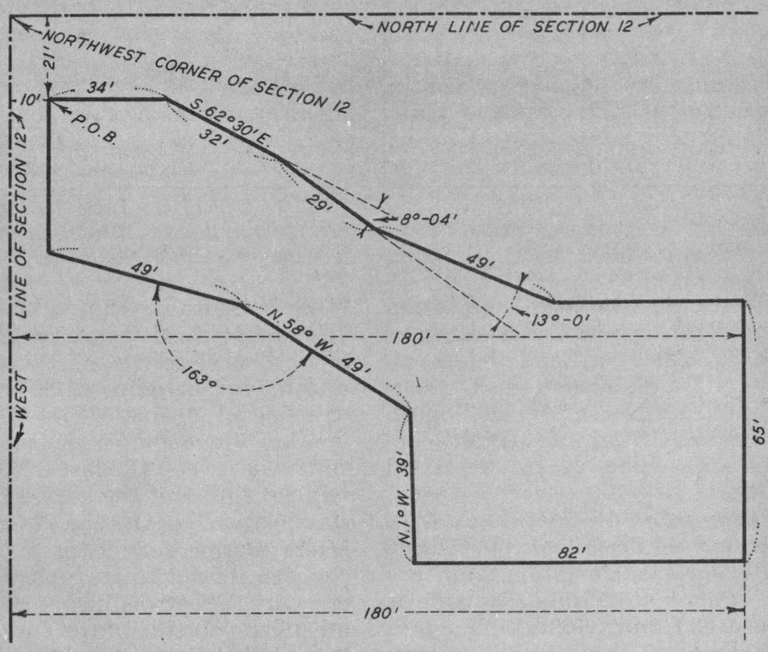

Commencing at the North West corner of Section 12, thence South along the section line 21 feet; thence East 10 feet for a place of beginning; thence continuing East 34 feet; thence South 62 degrees, 30 minutes, East 32 feet; thence Southeasterly along a line forming an angle of 8 degrees, 04 minutes, to the right with a prolongation of the last described course 29 feet; thence South 13 degrees, 0 minutes, to the left with a prolongation of the last described line a distance of 49 feet; thence East to a line parallel with the West line of said section and 180 feet distant therefrom; thence South on the last described line a distance of 65 feet; thence due West a distance of 82 feet to a point; thence North 1 degree West 39 feet; thence North 58 degrees West a distance of 49 feet; thence Northwesterly along a line forming an angle of 163 degrees as measured from right to left with the last described line a distance of 49 feet; thence North to the place of beginning.

FIG. 7 A metes and bounds description traces the perimeter of a land parcel. Starting from a reference point, in this case tied to a rectangular survey, the direction and distances of segments of the boundary are described in order.

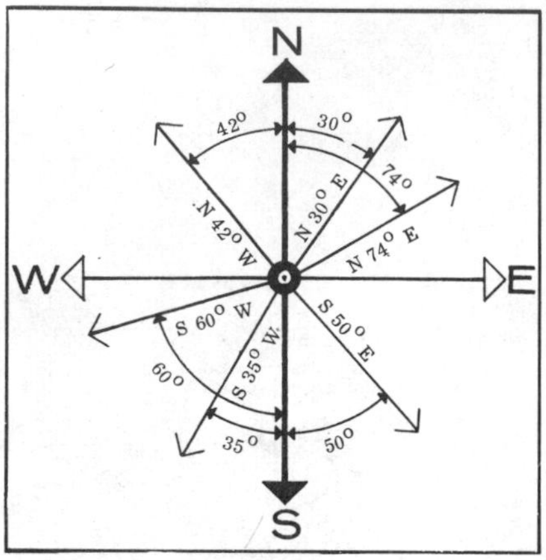

FIG. 8 Boundary line directions are designated by their eastward or westward deviation from North or South.

eastward or westward deviation from North or South. For example: "North 30 degrees East" defines an angular course direction 30° eastward of due North. A call "South 42 degrees 15 minutes West" defines a direction 42°15′ westward of South.

Prolongation Sometimes course direction is described by stating the amount of deviation from the preceding course line. An angle is measured from an extension, a *prolongation*, of the preceding line, and might be stated as: "thence Northerly 30 degrees to the left of an extension of the last described line." The third and fourth legs of the description shown in Figure 7 are examples of prolongation.

Curved Lines

A curved line sometimes is referred to in a description as:

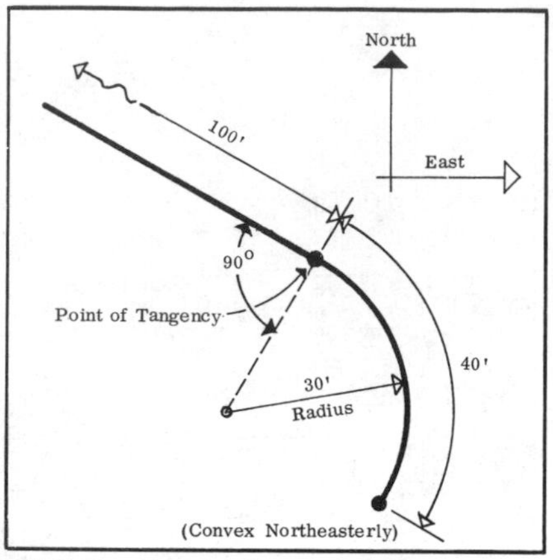

FIG. 9 Curved lines are described by giving the radius and length of the arc, and the direction of the curve.

"thence South 80 degrees East 100 feet to a point of tangency; thence along a curve convex northeasterly and having a radius of 30 feet; a distance of 40 feet" (Fig. 9). The point of tangency, or the point of contact between the last described straight line and the curve, is the starting place of the curve. Convex refers to the outside of a curve; concave, if used, refers to the inside of a curve. "Convex northeasterly" means the outside of the curve lies to the northeast.

Distance

Distance generally is stated in feet and decimal parts of feet. Sometimes it may be stated in feet and inches or fractions of feet, and rarely in chains, rods and links (see Fig. 1). Measure is made from point to point along straight lines and along the curve of a curved line.

RECTANGULAR SYSTEM DESCRIPTIONS

As a result of land being surveyed by the rectangular system, large unmarked areas of land are reduced to series of small squares; basic lines of survey are established by astronomical measurements; permanent monuments have been located on the ground by the surveyors to establish quarter-quarter section corners; and careful survey notes of the description and location of all monuments have been made and filed with the government.

Any tract of land in the government survey is described legally by reference to its appropriate principal meridian and base line, township and range numbers, section number, and, if necessary, its location within the section.

Accordingly, such a description for a parcel of land in Cook County, Illinois might read: "the South West quarter of the North East quarter of Section 6, Township 39 North, Range 13 East of the Third Principal Meridian" (Fig. 10).

Government lots (see page 601-8) are described by identifying the lot number and the quarter section in which the lot is located: for example, "Government Lot 1 in the North West quarter of the fractional Section 19, Township 38 North, Range 11 East of the Third Principal Meridian."

The rectangular system often is used to describe the boundaries of a subdivision in which individual

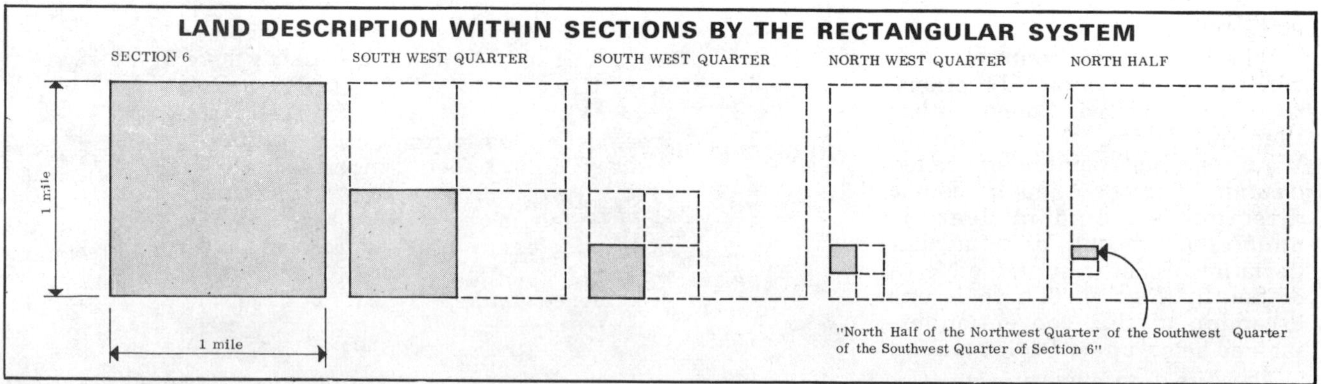

FIG. 10 In understanding a description such as "the North half of the North West quarter of the South West quarter of the South West quarter of Section 6," it is somewhat easier to locate the tract of land within the section by reading the description in reverse; that is, "Section 6, South West quarter, South West quarter, North West quarter, North half."

lots may then be described by reference to a recorded plat or map. The plat is prepared by a surveyor and recorded in the appropriate county records.

REFERENCE TO RECORDED PLAT

Curved street patterns may produce irregular shaped blocks and lots. The entire subdivision may be described within the rectangular system or by a metes and bounds description, but it may be impractical to use these systems to describe individual blocks or lots. Therefore, many states provide for the legal description of property by the reference to plats properly recorded, generally in the office of the appropriate county.

A surveyor surveys the tract and prepares a plat. The tract or subdivision may be divided into blocks and then into lots. Lots and blocks may be given consecutive numbers or letters enabling identification of a particular area of land within the subdivision by a lot and/or block number. For further identification it is common practice to give the subdivision a name. An example of a recorded subdivision plat is shown in Figure 11.

The surveyor's plat is captioned with the legal description of the land. It shows all boundary lines, all necessary monuments and dividing lines for blocks, lots and streets, and the numbering and dimensions of each lot.

Easements and restrictions also may be indicated on the plat. The plat then is certified by the surveyor and signed by the owners. Upon approval by the proper authorities (such as municipal or county zoning boards, planning and building commissions), it is recorded generally in the appropriate county recorder's office.

When recorded, the lots in a subdivision easily may be identified legally by reference to lot and block numbers, the subdivision name, and the section, township and range in which the entire subdivision boundaries are described within a rectangular survey or metes and bounds description.

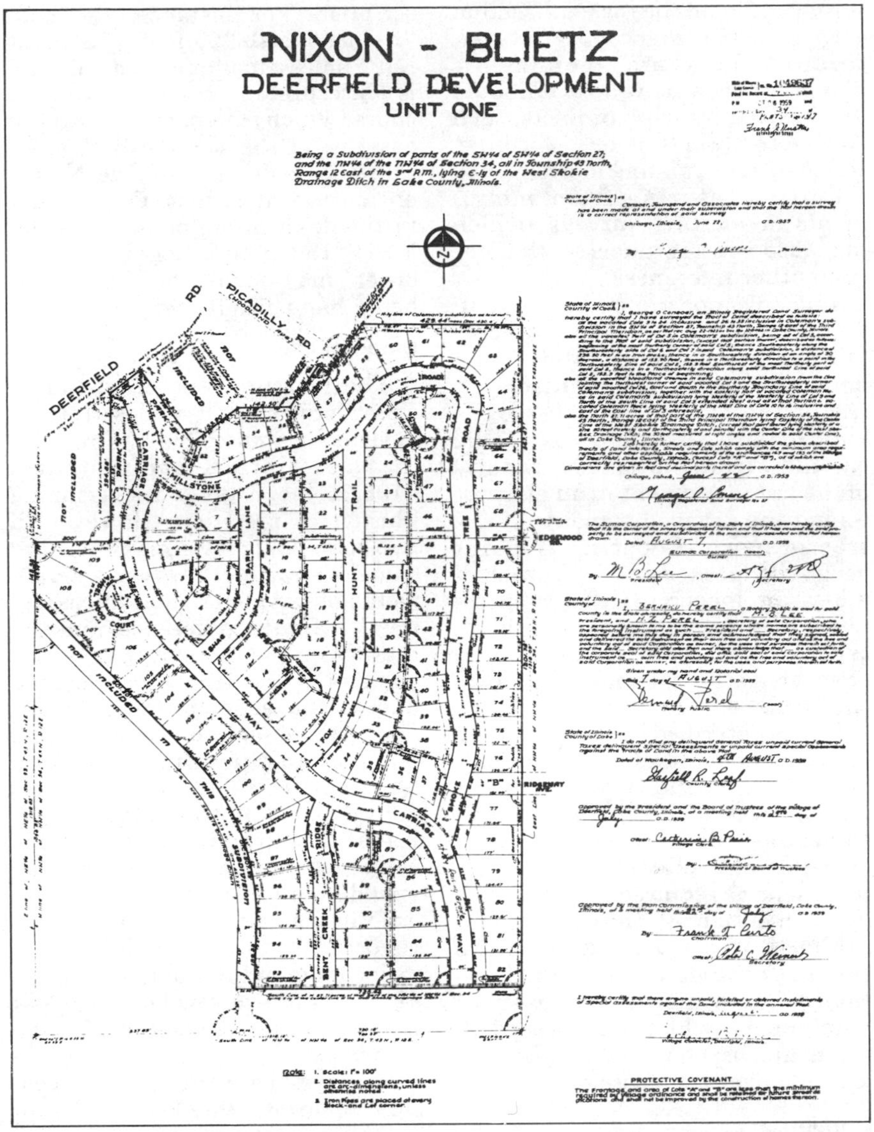

FIG. 11 Land often is described by reference to a recorded plat which shows an entire subdivision, lot dimensions, easements and other information. This example includes surveyor's certification and plan commission and county approvals. The note at top center ties the subdivision to the rectangular system, and as noted (upper right), this plat has been filed with the Lake County Recorder, plat book 34, page 147. Any lot may be described by number within the "Nixon-Blietz Deerfield Development, Unit #1" as recorded.

UNCERTAINTIES

Making a survey is not the exercise of pure science by a technician, in the manner a mathematician would solve a problem, with an exact answer to be arrived at inevitably by any surveyor. Mistakes, errors and judgments may result in conflicts and uncertainties in the legal description of land. Whether a recorded subdivision plat, a rectangular survey, or a metes and bounds description is involved, there may be differing opinions as to the precise location of the boundary lines of a tract of land. The exact location of a point on the land can be a matter of opinion and subject to interpretation.

Location of Markers

While making the original government surveys, upon which re-surveys should be based, section corners were marked. However, through the years, monuments may have been moved or uprooted during earth-moving projects, such as road construction, or in farming. Relocating a missing corner may require measuring from monuments in existing surveys (which may also contain uncertainties) or from other references.

The disappearance of monuments constantly causes uncertainties in surveying. Metes and bounds surveys monumented to a tree or a post which no longer exists may require the surveyor to dig carefully in the probable vicinity of the lost monument in an attempt to find the decayed remains of the tree or post. Locating a point originally marked by the corner of a barn no longer existing may require inquiry among older inhabitants of an area for information concerning the original location of the barn.

Roads described in an old survey may have been abandoned and offer little or no trace of their former location. Even though title companies may have unrecorded evidence of such roads, surveying judgment is required to fix a line in substitution of the lost road.

Many other uncertainties are caused by conjecture as to the location of lakes and rivers used as monuments, which have receded, dried up, or changed course and location over the years.

Convergence

Sixteen townships, each roughly six miles square, were laid out in original government surveys in areas approximately 24 miles square (see page 601-7). This was a practical but not perfect effort to adjust for the convergence of meridians of longitude as they travel north. Sections were marked by lines spaced at one mile intervals. These lines were run parallel to the range line closest to the principal meridian. This causes a divergence in direction which becomes progressively more pronounced in each successive tier of sections. Therefore, an ambiguity may result when a cardinal point is used as a course in a legal description. For instance, the call, "thence North 150 feet to a point ...," unless monumented, usually is interpreted as indicating a course which runs *parallel* with the east line of the section. If the section line deviates from true North, an uncertainty exists. Other clues in the description, or perhaps following the description in reverse order, may result in the course being more firmly defined.

Inaccuracies

The measurement of distances called for in some legal descriptions is virtually impossible. As written in a description, the distance, "126.01 feet" gives an assuring illusion of precision. However, if two surveying crews measured the same distance, they very rarely would observe the same measurement. Minute differences in the tautness of the measuring tape and changes in the length of the tape due to the effect of temperature fluctuations are unavoidable. A point in mathematical definition has neither length, breadth, nor depth. The ".01" of a foot is equivalent to a little less than 1/8″. It would be amazing if two surveyors would locate this invisible point at exactly the same place. Yet a surveyor must express an opinion as to its location and the opinions of individual surveyors will differ unless the point is monumented in the first place.

In surveying a tract, it is necessary to locate the boundaries of abutting tracts as shown in any earlier surveys and to reconcile the perimeter lines of the tract being surveyed with them. The surveyor must check back through the records to find the legal descriptions of these abutting tracts.

Section corners and other marks in the rectangular system sometimes were placed incorrectly. Early government surveyors sometimes used crude tools and methods. Their measuring instrument was a 66′ chain made of 100 metal links. Weather changes resulted in expansion or contraction of these chains, and wear and use caused them to stretch. In some instances, surveyors would count the revolutions of the wheel of the vehicle in which they rode to measure long, straight stretches. Distances also were "stepped off" by the measured stride. Even though these measurements were far from accurate, markers were placed accordingly. Such monuments are considered legal points of reference, regardless of their accuracy in being placed.

In addition to errors in distance, the perimeter lines of a given section sometimes deviate from the true East-West or North-South courses. In locating the south boundary of a tract described as the "North 10 acres of the North West quarter" of a section in which the north line deviates from a true East and West course, some judgment is required. The south line might be considered to be parallel with the north line of the section, or it might run due East and West. Again, the surveyor must search for physical evidence, such as a stake set by a prior surveyor or a fence indicating a line of occupation, in order to decide which line was intended.

Courses may show an appreciable error in direction. For example, a metes and bounds course shown as "South 7 degrees West" in a prior survey might be observed to be actually "South 6 degrees West," indicating that an error could have been made by an earlier surveyor, or that faulty equipment was used.

Mistakes, such as careless observations and inaccurate measurements, are random and unpredictable. And though the best surveying instruments are subject to a certain degree of error, these errors are predictable and can be allowed for in observing measurements. Again, however, this calls for some human judgment.

Recorded Plats

When an individual lot is described by reference to a recorded subdivision plat, the plat together with all information and data shown thereon is part of the lot description. It is possible that the lot size may not be *exactly* as shown on the plat. The incorporation by reference is of the *entire plat* and *all data* appearing thereon. This includes not only dimensions and expressions of quantity and area, but also the notations of the courses and distances of the lines of the subdivision and of the identity and locations of the monuments marking the boundaries.

When this information is incorporated into a legal description, it is, in effect, a metes and bounds description and is subject to rules of interpretation, should any conflict or discrepancy be contained in the data of the plat. For example, monuments prevail over courses and distances, courses prevail over distances, and expressions of quantity or area are generally the least reliable.

It is possible that the overall recorded measurements of a block may not equal the recorded widths of the lots in the block. If such an error is found, the widths of some or all of the lots must be adjusted accordingly. If the entire row of lots is vacant and unmonumented, the overage or deficiency could be distributed among all of the lots. Lots marked by the subdivider's stakes and sold with reference to them probably would escape adjustment. If improvements have been erected close to the boundaries of a lot as platted, without a surveyor checking the location, disputes may arise.

Errors and mistakes, fairly common when land values were low, have led to boundary disputes when land values greatly increased.

As discussed, plats are subject to errors and mistakes which may result in uncertainties and conflicts. The plat, drawn by the surveyor from field notes, and the field notes are *secondary* evidence of the surveyor's acts. The *primary* evidence consists of the monuments placed in the ground to mark lines and corners.

The systems of survey upon which legal descriptions are based are as precise as is possible. There is always the possibility of uncertainties in separate surveys, and mistakes may occur in the writing and rewriting of legal descriptions. Descriptions may *identify* land parcels adequately, but surveys are necessary to actually *locate* boundary lines.

The laws governing questions of legal description and survey are complex and include federal, state and local regulations. Court decisions and opinions pertaining to land disputes may be based on any of these regulations deemed appropriate, or on principles dating back to the real estate laws of England. This section is to acquaint the reader with the basic systems of survey and forms of legal descriptions. The complexities of resolving uncertainties in either land survey or legal land description are a matter of real estate law and interpretation, and hence are outside the scope of the *Construction Lending Guide.*

GLOSSARY

Angular Measure The deviation between two lines which meet at a point, expressed in degrees, minutes and seconds.
Degree Unit of angular measure equal to the angle contained within two radii of a circle which describe an arc equal to 1/360 part of the circumference of the circle; also used to define an arc equal to 1/360 part of the circumference of a circle.
Minute Unit of angular measure equal to 1/60 of a degree.
Second Unit of angular measure equal to 1/60 of a minute.

Base Line Parallel of specified latitude, used in the rectangular survey system serving as the main east-west reference line with a principal meridian for a particular state or area.

Call In surveyor's language, the statement or mention of a course and/or distance.

Cardinal Points The four major compass headings of North, East, South and West.

Correction Lines East-west reference lines used in the rectangular survey system, located at 24-mile intervals to the north and south of a base line.

Course Compass direction from one reference point to the next for each leg of a metes and bounds survey.
Angular Course Compass direction in degrees, minutes and seconds, stated as a deviation eastward or westward from due North or South; used in metes and bounds surveys and descriptions.

Degree See *Angular Measure*

Equator See *Parallel*

Fractional Section See *Section*

Guide Meridian See *Meridian*

Great Circle Line described on a sphere by a plane bisecting the sphere into equal parts. The Equator is a great circle, as are pairs of opposing meridians.

Latitude Position of a point on the earth's surface north or south of the Equator, stated as an angular measure (degrees, minutes and seconds) of the meridian arc contained between that point and the Equator.

Legal Description A written identification of the location and boundaries of a parcel of land. A legal description may be based on a metes and bounds survey, the rectangular system of survey, or it may make reference to a recorded plat of survey.

Longitude Position of a point on the earth's surface east or west of the Greenwich Meridian, stated as an angular measure (degrees, minutes and seconds) of the arc on the Equator contained between a meridian passing through that point and the Greenwich Meridian.

Metes and Bounds System of land survey and description based on starting from a known reference point and tracing the boundary lines around an area.

Meridian Imaginary north-south line on the earth's surface described by a great circle arc from the North Pole to the South Pole. All points on a meridian are of the same longitude.
Greenwich (Prime) Meridian The meridian passing through the Royal Observatory at Greenwich, England, and designated as the starting line (0°) for measuring east and west longitude.
Principal Meridian Meridian of specified longitude, used in the rectangular survey system, serving as the main north-south reference line for a particular state or area.
Guide Meridians North-south reference lines located at 24-mile intervals east and west of a principal meridian.

Minute See *Angular Measure.*

Monument Permanent reference point for land surveying whose location is recorded; either a manmade marker or a natural landmark.

Parallel Imaginary east-west line on the earth's surface, consisting of a circle on which all points are equidistant from one of the poles. All points on a parallel are the same latitude.
Equator The parallel circling the middle of the earth, all points of which are equidistant from both North and South Poles; designated as the starting line (0°) for measuring north or south latitude.

Plat A map of surveyed land showing the location and the boundaries and dimensions of the parcel.
Recorded Plat A plat which is recorded at an appropriate governmental office, usually the county recorder's office. The recorded plat, in addition to location notes and boundary line layout may contain information such as restrictions, easements, approvals by zoning boards and planning commissions, and lot and block numbers for a subdivision.

P.O.B. (Place of Beginning) Starting point of a metes and bounds survey or description.

Range Lines North-south reference lines used in the rectangular survey system, located at 6-mile intervals between guide meridians.

Rectangular (Government) Survey System Land survey system based on geographical coordinates of longitude and latitude; originally established by acts of Congress to survey the lands of public domain and now used in 30 states.

Second See *Angular Measure*

Section An area of land used in the rectangular survey system, approximately 1 mile square, bounded by section lines. The rectangular system provides for the further subdivision of sections into halves, quarters and quarter-quarters.
Fractional Section Any "adjusted" section of land generally containing less (sometimes more) than 1 square mile. The deficiency (or excess) may be the result of the convergence of meridians, the presence of bodies of water, or uncertainties in surveying.

Section Lines North-south reference lines used in the rectangular survey system, parallel to the nearest range line to the east, and east-west lines parallel to the nearest township line to the south; these lines divide townships into 36 approximately equal squares called sections.

Survey The measure and marking of land, accompanied by maps and field notes which describe the measures and marks made in the field.

Township An area of land, used in the rectangular survey system, approximately 6 miles square, bounded by range lines and township lines.

Township Lines East-west reference lines used in the rectangular survey system located at 6-mile intervals between correction lines.

We gratefully acknowledge the assistance of the Chicago Title and Trust Company for the use of their publications as references and permission to use illustrations.

602 CODES

CODES

Introduction

The planning, construction, location and use of buildings are regulated by a variety of laws enacted by local, state and federal governments. These statutes and ordinances—which include building, plumbing, electrical and mechanical codes—are intended to protect the health, safety and general welfare of the public.

Such laws are not new. The ancient code of the Babylonian emperor, Hammurabi, dating back to about 1800 B.C., is often cited as the first recorded building code. It provided severe penalties for construction practices that violated the health or safety of citizens—if a building collapsed killing the occupants, the architects and/or builders were put to death! This ancient code, based on the idea that the strong should not injure the weak, sets the principle also for today's construction regulations: that is, the public has a right to be protected from the harmful acts of others.

Modern building regulation evolved over time, starting in the early 19th century with the adoption of fire laws in a number of large cities. These laws prohibited the construction of wooden buildings within certain congested parts of the city. About the same time, cities also began to adopt health regulations to improve the living conditions of the poor—the forerunners of today's minimum housing codes. Some courts, however, resisted the enactment of such laws as infringements on personal property rights.

As the validity of fire and health laws slowly became established, courts began to accept governmental control of *all* aspects of building construction involving the health and safety of the public. Today, practically all cities of any size regulate the planning, construction and installation of building systems through a variety of laws and ordinances.

Despite the fact that the right to regulate building construction constitutionally rested in the state, before 1960 few state governments exercised that right. Usually the states delegated to local governments "the power to regulate buildings to protect the public health, safety and welfare."

Lately, many states are taking a very active role in building regulation. Over half the states have now adopted some form of statewide building regulation concerned with the construction of "industrialized" buildings, mobile homes and/or conventional construction.

Recent federal legislation in the areas of occupational health and safety, environmental protection, pollution controls, and consumer protection, coupled with increased state legislative activity, offer evidence that the building regulatory process is to become ever more complex.

Regulatory codes incorporate many recognized industrial specifications and standards. Section 101 Construction Standards discusses some of the various types and purposes of industry standards and the organizations active in establishing them.

This section describes some problems in the code regulatory system and how model codes can help solve these problems. It also describes efforts of the major building officials' organizations and the numerous federal, state and local groups who are working to improve the code regulatory system.

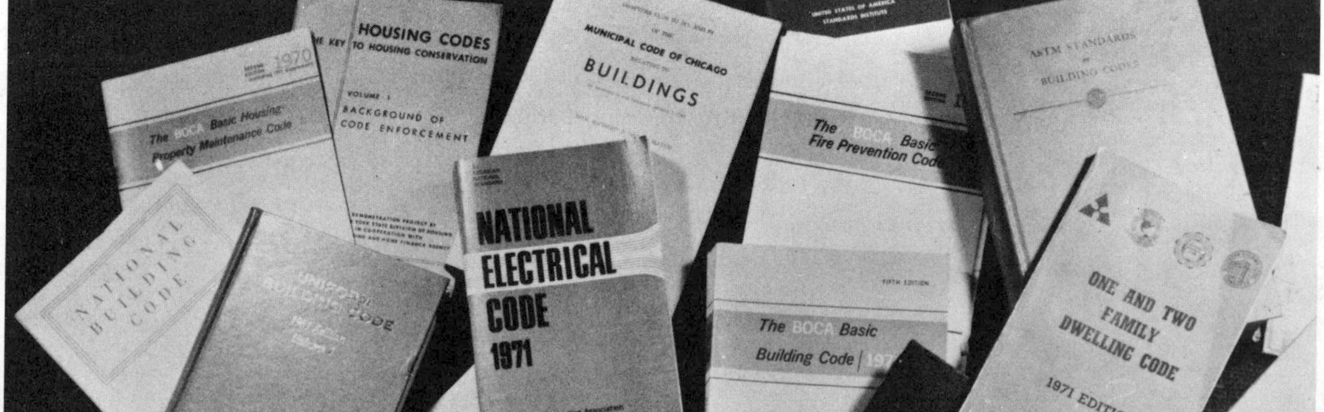

Building codes establish minimum requirements to protect the public health, safety and welfare. They *do not* necessarily contain criteria which assure efficient, convenient or adequately equipped buildings.

A building code establishes requirements for the construction and occupancy of buildings. It contains standards of performance and specifications for materials, methods and systems. Codes also cover structural strength, fire resistance, adequate light and ventilation, and other considerations determined by the design, construction, alteration and demolition of buildings. *A code becomes a law when it is adopted by a municipality as a public ordinance.* Local communities may write their own codes or may legally adopt other codes, such as state building codes or one of the *model codes* described in this subsection.

ADOPTION

The homebuilding industry is enmeshed in an extraordinary network of established building codes which attempt to assure that building construction will be safe. They generally accomplish that purpose; but codes and code administration are criticized widely as being restrictive to home building progress by retarding the acceptance of new and improved uses of materials and methods, and thereby unnecessarily increasing the costs of construction. In some instances, the adoption of improved codes has stimulated better building practices. But the existing complex and chaotic building code situation is recognized as one of the problems facing the homebuilding industry today because of the use of "specification" type codes, the lack of code uniformity, the multiplicity of codes, the slow response of codes to change and inadequate performance standards.

SPECIFICATION AND PERFORMANCE CODES

A *specification* type building code establishes building construction requirements by reference to particular materials and methods. Many unrevised or out-of-date codes are based on techniques of balloon framing. (See Methods & Systems, page 300-2.) Such a code may establish outside wall construction requirements by specifying, for example, 2″ x 4″ wall studs spaced 16″ on center, with 1″ board sheathing applied diagonally. A builder seeking to use wall construction with 24″ stud spacing (which is structurally *sound* and *safe*), or another type of sheathing that is perhaps even more adequate than the diagonal sheathing, would find that his methods do not meet the specifically stated code requirements.

A *performance* type code does not limit the selection of methods and systems to a single type, but establishes the requirements of performance for building elements. Such a code establishes design and engineering criteria without reference to specific methods of construction. For example, an outside wall would be required to support certain loads and forces, perhaps meeting stated insulating and permeability requirements. Any system performing as the code requires would be acceptable, regardless of materials and methods used.

The use of true performance codes in all of the communities of this country would be ideal but is impractical. Permitting the house and elements of the house to be built with any method as long as certain performance criteria are met is idealistically fine and proper but practicably unworkable. Local code administration would have to be in the hands of extraordinarily competent people equipped to interpret performance criteria and evaluate any proposed methods, uses or systems.

A workable solution lies somewhere between the pure *performance* and the *specification* codes. Codes may be considered performance codes if they adequately provide for the acceptance of alternate methods and systems.

Local Administration

Generally, the local administration and enforcement of codes is by a building inspector or engineer who has the authority to approve materials and methods which may not be directly referenced in the code. Qualified people are necessary to administer a building code properly. No matter how good a code may be, it must be enforced by someone experienced, informed and objective. Most builders do not complain about the careful and competent inspector who is consistent and tough, but complain about the part-time inspector who does not understand construction and may be arbitrary and inconsistent. The competent code enforcer who knows his job knows when the letter of the code should prevail and when subjective interpretation should be made.

However, the local code administrator is faced continually with the difficulties of judgment and may well argue that his job is to check compliance. An analogy which is often drawn, and which is reasonably correct, is that the local policeman cannot be asked to make judgments in exercising his function, but is charged only with determining compliance to the law. This suggests that the determination of performance criteria and judgment as to whether methods and systems do perform suitably must be made by technically qualified people, not at the local level. The model code groups, discussed later, are better qualified to offer this type of service at this time.

The model code groups, supported by building officials themselves, have performed a great service to the homebuilding industry. But model codes have not fully solved the problems of code uniformity. Only about one-fourth of the cities using model codes have adopted the model without some modification. The National Association of Home Builders points out: "The model codes are developed by qualified people and, in general, are up to date, but over 30% of the changes made 'to suit local conditions' come from local codes prepared 20 or more years ago."

The modification of model codes by the local community often is influenced by self-interest groups. If local communities would not change the model codes when they are being enacted into local statutes, a major step forward would be achieved.

Lack of Uniformity

Since establishing and enforcing building codes are local functions, the home builder who works in more than one community often is faced with a frustrating variety of requirements. This lack of uniformity in building codes complicates the job of both the builder and the manufacturer.

In developing products, the manufacturer must consider the problems of winning acceptance by literally thousands of local building code administrators. He cannot concern himself only with performance and public acceptance. Much effort has been made to unify the codes used by communities, and considerable improvement has been gained through the local adoption of model codes.

The National Association of Home Builders, the U.S. League of Savings Associations and other progressive trade associations have long encouraged the creation of a single model code for one and two family dwellings. Four organizations (American Insurance Association; Building Officials and Code Administrators International, Inc.; International Conference of Building Officials; and Southern Building Code Congress) cooperated in developing such a code.

In 1971 the four code groups adopted a consensus code titled the One and Two Family Dwelling Code. This model code contains requirements for building, planning and construction, including specifications for heating, cooling and plumbing. The One and Two Family Electrical Code published by the National Fire Protection Association is also included in the One and Two Family Dwelling Code by reference.

Publication of this single national model code constituted a significant step toward uniform minimum regulations but it remains for the code to be recognized and adopted by state and local governments.

Multiplicity of Codes

If the builder were dealing with a single code, his problem would be greatly simplified. However, in addition to local building codes which tell him how the structure must be assembled, he also may have to satisfy a number of additional codes covering a variety of subjects such as plumbing, electrical wiring, traffic, utilities, health and sanitation, land planning, building occupancy and zoning.

The lack of code uniformity and the multiplicity of additional regulations result in a state of code complication which generally forces the builder to design and construct his homes to the most conservative and most expensive criteria. He often is unable to improve his techniques because of the restrictive nature of any one of the codes.

Slow Response to Change

Codes often are criticized for failing to recognize new materials and methods. Judging new products on the basis of performance criteria must be in the hands of technically qualified and equipped organizations.

This function is one in which the model code groups can exercise great leadership and provide a major stimulus to progress. The groups have responded reasonably to progressive change. However, by their very size, they are subject to the many pressures that deter progress, and sometimes have been slow in acting on proposed changes. For example, it took almost four years for all of the major code groups to accept the NAHB's proposal to eliminate floor bridging, a proposition well documented through extensive research and testing.

Inadequate Performance Standards

The homebuilding industry is made up of so many diverse industries and serves such a wide proliferation of interest groups, no single group of performance criteria, testing and standards exists. In selecting materials or determining the suitability of materials and methods, the specifier and the builder make reference to ASTM Specifications, Federal Specifications, American Standards, Commercial Standards, FHA Minimum Property Standards, a variety of trade association certification programs, grademarks and trademarks, Underwriters' Laboratories labels and many other construction standards. (See Section 101 Construction Standards.)

Though manufacturers themselves have substantial commitments in research and testing facilities, no single agency charged with the development of performance criteria and testing presently is recognized by all of the groups interested in homebuilding.

As long as codes continue to complicate builders' operations and deter manufacturers from developing new materials and products, unnecessary costs are added to homebuilding.

MODEL CODES

To help fill the need for workable codes, four major organizations of industry and professional groups and several states have developed building codes which may be adopted into law for use in local communities. The codes commonly are referred to as the *model codes.*

Model codes have been widely accepted by local communities and now are used by more than 80% of the communities with a population over 10,000. The organizations which prepare model codes also provide for the continuous updating necessary to include recommendations based on industry research and the development of new materials and methods.

The National Plumbing Code, The Uniform Plumbing Code, Uniform Heating and Comfort Cooling Code, and the National Electric Code also have been developed to cover the special concerns of public health and

FIG. 1 MODEL CONSTRUCTION CODES AND SPONSORING ORGANIZATIONS

BUILDING CODES	
Basic Building Code	Building Officials and Code Administrators International, Inc.
National Building Code	American Insurance Association
Southern Standard Building Code	Southern Building Code Congress
Uniform Building Code	International Conference of Building Officials
One and Two Family Dwelling Code	All of the above
ELECTRICAL CODES	
National Electrical Code	National Fire Protection Association
One and Two Family Electrical Code	National Fire Protection Association
ELEVATOR CODES	
Safety Code for Elevators, Dumbwaiters and Escalators	American National Standards Institute
Safety Code for Manlifts	American National Standards Institute
FIRE PREVENTION CODES	
Basic Fire Prevention Code	Building Officials and Code Administrators International, Inc.
Fire Prevention Code	American Insurance Association
HOUSING CODES	
Basic Housing—Property Maintenance Code	Building Officials and Code Administrators International, Inc.
Housing Code	American Public Health Association
Southern Standard Housing Code	Southern Building Code Congress
Uniform Housing Code	International Conference of Building Officials
PLUMBING CODES	
Basic Plumbing Code	Building Officials and Code Administrators International, Inc.
National Plumbing Code	American Society of Mechanical Engineers
Southern Standard Plumbing Code	Southern Building Code Congress
Uniform Plumbing Code	International Association of Plumbing and Mechanical Officials
MISCELLANEOUS CODES	
Boiler and Unfired Pressure Vessel Code	American Society of Mechanical Engineers
Flammable Liquids Code	National Fire Protection Association
Safety Code for Mechanical Refrigeration	American Society of Heating, Refrigerating, & Air-Conditioning Engineers, Inc.
Life Safety Code (formerly Building Exits Code)	National Fire Protection Association

*See Construction Standards Appendix, page 101-10 for addresses of model code sponsoring organizations.

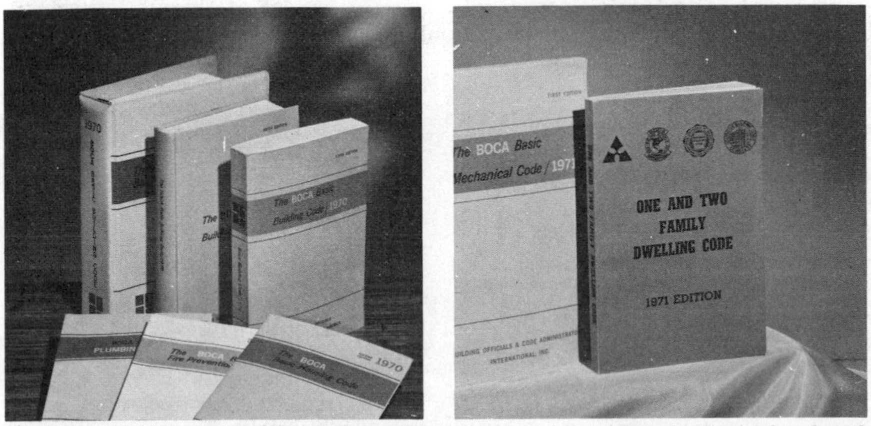

FIG. 2 BOCA publishes several model codes, which are periodically updated and revised (a). The One and Two Family Dwelling Code is a national consensus code distributed by BOCA and other code groups (b).

safety as affected by plumbing, heating, cooling and wiring of buildings. Many local communities also have adopted these codes, since some of the model codes do not cover these subjects.

The nationally recognized model codes and sponsoring organizations are shown in Figure 1.

Basic Building Code

The Building Officials and Code Administrators International, Inc. (BOCA) developed the Basic Building Code back in 1950. Substantially revised, the Basic Building Code is still widely used. BOCA also publishes the Basic Fire Prevention Code, Basic Housing-Property Maintenance Code, Basic Plumbing Code, Basic Mechanical Code and the One and Two Family Dwelling Code (Fig. 2).

National Building Code

The first model building code was introduced in 1905 by the National Board of Fire Underwriters, now known as the American Insurance Association. The organization was concerned primarily with public protection against fire hazards, and this original code was designed to guide communities in setting up fire safety standards. With subsequent additions, however, it laid the groundwork for the development of building codes throughout the country. Now known as the National Building Code, an abbreviated edition also is available for communities needing a less extensive code. The National Building Code does not cover the subjects of plumbing and wiring.

Southern Standard Building Code

The Southern Building Code Conference drafted the Southern Standard Building Code in 1945. Now used extensively in the southern states, it was designed to meet special problems in that region (Fig. 3). The Southern Building Code Congress (SBCC) also publishes the Southern Standard Plumbing Code, Southern Standard Gas Code, Southern Standard Mechanical Code, Southern Standard Housing Code and the One and Two Family Dwelling Code. The latter is the same consensus code adopted by the other code groups.

FIG. 3 The Southern Standard Building Code, shown above, covers building planning and construction. It is used mainly in the southern states.

Uniform Building Code

In 1927, the International Conference of Building Officials (initially the Pacific Coast Building Officials Conference) published its Uniform Building Code. Since that time, the code has gained wide national acceptance, particularly in the West (Fig. 4). The International Conference of Building Officials (ICBO) also publishes the Uniform Building Code Standards, Uniform Mechanical Code, Uniform Housing Code, Uniform Code for the Abatement of Dangerous Buildings, Uniform Sign Code, Uniform Fire Code, Uniform Plumbing Code, and One and Two Family Dwelling Code.

National Plumbing Code

The National Plumbing Code has been developed as an American National Standard, ANSI A40.8-1955, with the cooperation of several engineering, trade and industry organizations (Fig. 5).

The purpose of the National Plumbing Code is to protect the public health. The code, therefore,

FIG. 5 The National Plumbing Code often is incorporated by reference into local building codes.

FIG. 4 The Uniform Building Code of the International Conference of Building Officials is used mostly in the western states.

covers the proper design, installation and maintenance of plumbing systems based on principles of sanitation and safety, and not necessarily on efficiency, convenience or adequacy for good service or future expansion of system use. Standards for materials and fixtures are based largely on other industry specifications such as Commercial Standards, American Standards and ASTM Specifications.

Uniform Plumbing, Uniform Heating & Comfort Cooling Codes

In 1929, the Western Plumbing Officials Association established the first Uniform Plumbing Code, and since that time has added the Uniform Heating and Comfort Cooling Code. These codes have been developed in the interest of public health and safety and cover the design, installation and maintenance of plumbing, heating and air conditioning systems. Materials and equipment standards are based largely on other industry specifications and standards. These codes have gained widespread use in the western states, as well as in other communities across the country.

FIG. 6 Local building codes may incorporate the National Electric Code by reference.

National Electric Code

The National Electric Code was produced by the National Fire Protection Association and has been adopted as an American National Standard, ANSI C1-1971 (Fig. 6).

The purpose of the National Electric Code is to assure the safeguarding of persons and of buildings and their contents from hazards arising from the use of electricity for light, heat, power and other purposes. The code contains basic minimum provisions for safety which, with proper maintenance, will result in installations free from hazard, but not necessarily efficient, convenient or adequate for good service or future expansion of electrical use.

The Code makes reference to many other industry specifications such as Underwriters Laboratories' labels and requirements. It is revised periodically and is adopted by reference into many building codes.

ENERGY CODES

Energy consumed in buildings in the United States amounts to about one-third of all energy consumed. Building energy consumption in 1978 has been increasing at a rate of 2.23% per year. Slowing the rate of increase through increased efficiency in building systems design can significantly reduce overall energy demand.

The National Energy Plan calls for substantial decreases in energy consumption through conservation. One element of the conservation program is enforcement of thermal efficiency standards and insulation requirements in new and renovated buildings by state and local governments.

The National Conference of States on Building Codes and Standards (NCSBCS) in 1973 recognized the need for a model national, uniform energy conservation code. NCSBCS requested the National Bureau of Standards (NBS) to develop standards that could be used by all states for energy-efficient design in new buildings. NBS completed these standards in 1974 but they were considered too complex for most state and local code enforcement officials to administer. As a result NCSBCS requested the American Society of Heating, Refrigerating and Airconditioning Engineers, Inc. (ASHRAE) to translate the NBS standards into enforceable language.

In 1975, ASHRAE published its Standard 90-75, Energy Conservation in New Building Design. ASHRAE Standard 90-75 was the first nationally recognized standard for energy efficient design applicable to all building types. It was, however, written for use by design engineers. State and local practitioners still needed a document that could be administered through the traditional code enforcement system. In 1976, NCSBCS translated ASHRAE Standard 90-75 into the *Model Code for Energy Conservation in New Building Construction.* Under a grant from the Energy Research and Development Administration (now the Department of Energy), NCSBCS worked with building officials from all parts of the country who administer model codes so that the *Model Code for Energy Conservation* would be applicable to all geographic areas in the country and could be incorporated easily into existing codes. The NCSBCS *Model Code for Energy Conservation in New Building Construction* was published in January 1977.

The Energy Policy and Conservation Act (EPCA), became law in 1975. The law provided for substantial grants to those states that developed and implemented statewide energy conservation plans aimed at reducing state wide energy consumption 5% by 1980. State energy conservation plans had to include at least five elements: 1) mandatory lighting efficiency standards for public buildings; 2) programs to promote the availability and use of car pools and van pools; 3) mandatory standards and policies related to procurement practices of state and local governments; 4) a traffic law which permits a right turn on red after stopping; and 5) mandatory thermal energy standards and insulation requirements for new and renovated buildings.

In 1976, Congress passed the Energy Conservation and Production Act (ECPA). Title III of ECPA contained the following provisions: 1) the Department of Housing and Urban Development (HUD) was directed to develop performance-oriented thermal energy efficiency standards; 2) HUD is to monitor the progress of state and local governments in implementing the standards; 3) if a substantial amount of new housing is not constructed in conformance with the standards, HUD can recommend to the Congress that the standards be made mandatory. If Congress concurs:

(1) All construction assisted by federal money will have to be consistent with the standards. This will include federal grants, loans, and loan guarantees and will apply to all Savings and Loan Associations that are federally insured.

(2) Local governments will not be required to adopt the standards into their building code. HUD and the states will be responsible for insuring that all proposed construction is in conformance with the standards. However, local governments that elect to adopt the standards will be reimbursed for the cost of the certification.

(3) If a local government incorporates the federal standards into its codes, then no further approvals will be required. Money will be available to communities to assist them in incorporating and enforcing the federal standards.

The U.S. Department of Energy and HUD are currently working on the development of Building Energy Performance Standards (BEPS) mandated by Congress under ECPA in 1976. Until those standards are developed, the NCSBCS Model Code for Energy Conservation seems to be the most popular document influencing energy-conscious building design. The NCSBCS Model Code defines criteria and regulations to control the heat lost and gained by the exterior building envelope, increases the efficiency of the heating, ventilating and airconditioning systems, and encourages energy-efficient lighting design.

CODE ADVANCEMENT

The growing complexity of building design and construction and the subsequent increase of regulatory controls in the last decade has induced several organizations to improve the effectiveness of the regulatory process with better code documents, administration and enforcement.

National Institute of Building Sciences

The Housing and Community Development Act of 1974 authorized the creation of the National Institute of Building Sciences (NIBS).

The Institute gives the United States a new national center for (1) assembly, storage and dissemination of technical data and related information on construction; (2) development and promulgation of nationally recognized performance criteria, standards, test methods and other evaluative techniques, and (3) evaluation and pre-qualification of existing and new building techniques.

The Institute was initiated by the federal government with the assistance of the National Academy of Science, National Academy of Engineering Research Council.

Council of American Building Officials

In 1972 the governing bodies of three national building officials organizations—BOCA, ICBO, SBCC—established the Council of American Building Officials (CABO).

CABO's policies are established by a board of trustees composed of representatives of the governing bodies of the member organizations. Between annual meetings of the board of trustees, the Council is managed by a nine-member executive committee; the executive directors of the three organizations serve as ex-officio members.

Examples of CABO's work include: (1) advancing the One and Two Family Dwelling Code as a recognized national standard through the procedures of the American National Standards Institute. (2) sponsoring the development of model performance standards that will compliment the requirements of model codes; (3) sponsoring a national research activity that provides a single approval agency for manufacturers of building components, systems or materials; (4) developing and adopting a model mechanical, plumbing, and fire prevention code as the standard for all three organizations.

Model Code Standardization Council

The Model Code Standardization Council (MCSC) is composed of representatives of the four organizations that promulgate and publish model codes: AIA, BOCA, ICBO and SBCC. The purpose of the organization, as the name suggests, is to foster more uniformity in the language and format of the model code documents published by each organization.

Work of the MCSC presently underway includes standardization of the code definitions and chapter format; uniform classifications of types of construction; and uniform classifications of use and occupancy.

An example of the work of MCSC is illustrated by the committee to develop a List of Reference Standards for Model Codes. When approved by the voting members of MCSC, the List of Reference Standards can be adopted for inclusion in any one of the model codes.

The objectives of MCSC Standards Committee are to: (1) develop and maintain a list of current standards for model codes; (2) investigate conflicts that exist between the lists of standards referenced in each of the model codes; and (3) encourage the adoption, without amendment, of the recommended list of current standards at the federal, state and local levels.

An industry standard, to be recommended by MCSC, must meet the following criteria: (1) be necessary for enforcement of existing provisions of one of the model codes; (2) be written in mandatory language following the format of one of the national standards writing groups such as ASTM or ANSI; (3) be recognized as the authoritative standard for the subject covered; (4) serve as a primary standard used in enforcement and not a secondary standard referenced by a primary standard; (5) not be in conflict with the intent of code requirements for which they will serve as reference; and (6) be readily available.

Associated Major City Building Officials

Building code officials from the nation's 30 largest cities created in 1974 an organization called Associated Major City Building Officials (AMCBO). Its purpose is to exchange ideas, experience and information on the growing complexities of code administration and enforcement in densely populated urban areas. Public Technology Inc. (PTI), a Washington-based organization, serves AMCBO in a secretariat capacity.

PTI was established in 1971 as a non-profit, tax-exempt corporation to help apply new technology for the benefit of state and local governments. It was organized and underwritten by six public interest groups: Council of State Governments, International City Management Association, National Association of Counties, the National Governor's Conference, the National League of Cities and the U.S. Conference of Mayors.

A representative of AMCBO is appointed to serve, along with the 50 state delegates, on the National Conference of States on Building Codes & Standards.

National Conference of States on Building Codes & Standards

The National Conference of States on Building Codes and Standards (NCSBCS) was formed in 1967. The organization is composed of state delegates, designated by their respective governors, to represent the 50 states in discussions and programs pertaining to building regulation. Serving NCSBCS in a secretariat capacity is the Office of Building Standards and Codes Services (Center for Building Technology, National Bureau of Standards) located in Washington, D.C.

NCSBCS was established to: (1) provide a forum for states to discuss and develop solutions to code regulatory problems; (2) promote the adoption and administration of uniform building codes and standards that regulate construction *in* and *between* states; (3) create an effective voice for state input on the committees of nationally recognized standards-writing organizations; (4) support comprehensive training and educational programs for code enforcement personnel; and (5) foster cooperation and encourage innovation between code administration officials and the design, manufacturing, and consumer interests affected by the code regulatory system.

Examples of NCSBCS' work are the development of model state legislation for (1) a Mobile Home Standard, (2) the Education and Certification of Code Enforcement Officers, and (3) a Statewide Construction Standard.

National Academy of Code Administration

The National Academy of Code Administration (NACA) was incorporated in 1970. Its prime objective is to develop code administration as a recognized profession in the United States. A board of trustees, selected from groups affected by regulatory codes, guides development of the Administration.

NACA plans to coordinate the numerous code administration educational programs that exist throughout the country. Additional objectives include the following: (1) sponsoring continuing education programs for code administrators and code enforcement officers; (2) developing a college curriculum for code administrators; (3) establishing professional standards and certifying the professional standing of those who meet the standards; (4) conducting research, creating new knowledge and developing the capability to support the professional development of those who have chosen code administration as a career.

The Center for Building Technology —National Bureau of Standards

The Center for Building Technology of the National Bureau of Standards (CBT-NBS) provides services to other federal agencies, state and local government and the building community through laboratory and technical field support. CBT disseminates research findings through publications and conferences.

CBT's Office of Building Standards and Codes Services provides technical assistance to private and public organizations which promulgate building standards and codes. CBT also promotes the performance concept in building construction and procurement. A newly created office within CBT deals with energy conservation.

National Fire Protection Association

The National Fire Protection Association (NFPA) was organized in 1896 to promote the science and advance the methods of fire protection. NFPA is a non-profit educational organization which publishes and distributes various publications on fire safety—including model codes, materials standards, and recommended practices.

These technical materials, aimed at minimizing losses of life and property by fire, are prepared by NFPA Technical Committees and are adopted at the annual meeting of the Association. All are published in the *National Fire Codes,* a 10 volume compilation of NFPA's official technical material. The National Electrical Code is a part of Volume 5.

603 METRICATION

METRICATION

INTRODUCTION

Metrication is the process of change from the customary (inch-pound) measurement system to the modern metric system. The customary system is the collection of English and other non-SI metric units currently used in the United States. The metre and six other standardized base units are the foundation for the metric system used throughout the industrialized world.

The General Conference on Weights and Measures (CGPM), an international treaty organization, developed the International System of Units (SI) in 1960. SI is the basis for worldwide standardization of metric measurement units. It is a universal, rational, coherent, international and preferred measurement system which is derived from earlier decimal metric systems but supersedes them.

Use of the metric system in the United States was legalized by Congress in 1866, but was not made mandatory. The Metric Conversion Act of 1975 committed the U.S. to a voluntary conversion to SI, as interpreted or modified by the Secretary of Commerce.

A timetable for the changeover to SI units has been established by the construction industry. The target date for metric conversion—M Day—is January 1, 1985. The application of SI units and new preferred numeric values will simplify measurements in construction and reduce errors in calculations.

SI has fewer units than the customary system. One unit length, the metre (m), and its decimally related multiples and submultiples, such as kilometre (km) and millimetre (mm), will replace a variety of customary units. The replaced units include the mile, furlong, chain, link, rod, fathom, yard, foot, hand, inch and mil. In addition, the change to the metre will correct the difference between the standard foot and the survey foot (which is two parts per million longer), by replacing both.

Many SI numbers are already in use: the second, ampere, candela, watt, volt, ohm, farad, coulomb and lumen.

It is thought that metrication will change the entire measurement base of the construction industry including guidelines, codes, standards, drawings, specifications and associated documents.

The building industry is most likely to use SI units for linear measurement, area volume and mass.

A world renowned expert on metrication and one of its most ardent advocates, Hans J. Milton, has also been a prolific writer on the subject. The material in this section has been largely abstracted from the National Bureau of Standards (NBS) publication 938, *Recommended Practice for the Use of Metric (SI) Units in Building Design and Construction* authored by Milton while on loan to the NBS by the Australian government.

This section describes the history of the metric system; the properties of the SI system, its applications and proper usage. Factors by which customary units may be converted to SI are found in 603 Metrication Work File.

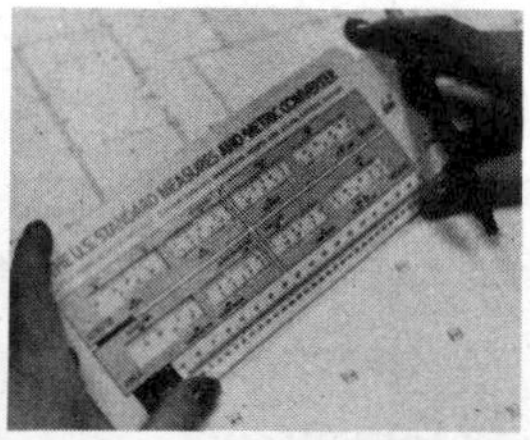

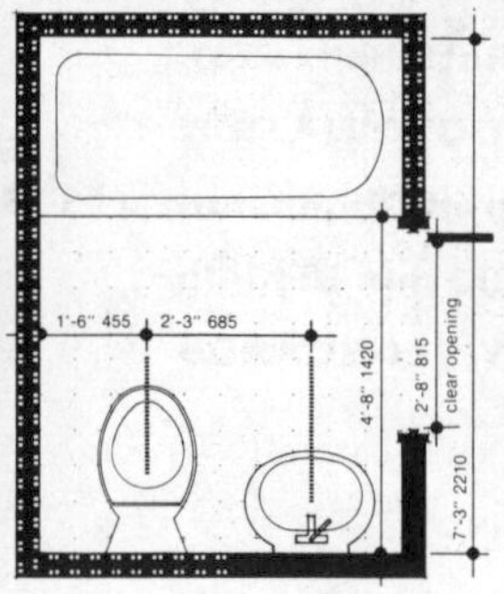

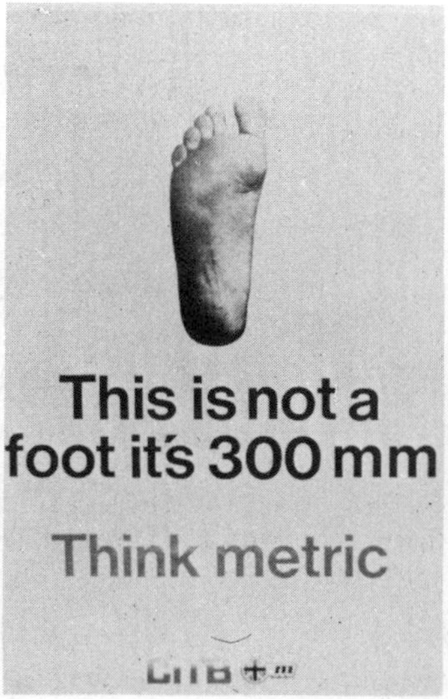

Fractional Inch	Dec. Equiv.	Millimetres
1/64	0.0156	0.397
1/32	0.0313	0.794
3/64	0.0469	1.191
1/16	0.0625	1.588
5/64	0.0781	1.984
1/12	0.0833	2.117
3/32	0.0938	2.381

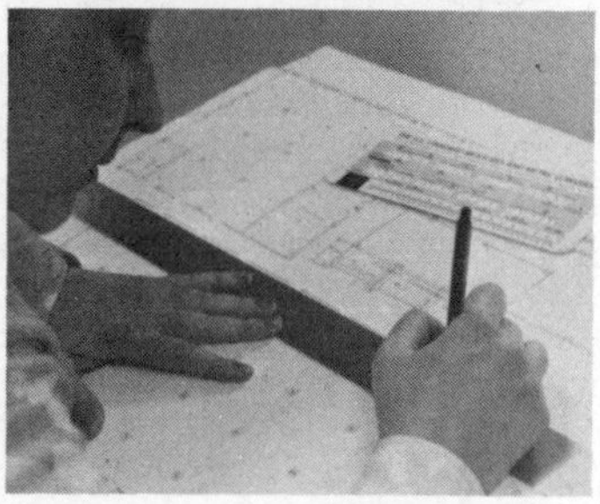

Metrication will not only force a complete revision in the accepted units of measurement and calculation, but also offer a logical opportunity to revise standard product sizes, performance criteria and methods of testing.

There are concerns expressed about the change to metric: manufacturers worry about the cost of retooling and introducing new product sizes; people worry about learning many new names and numbers.

Experience in other countries has shown that metrication takes less time and costs less than estimated. The cost of metrication occurs once; the benefits last forever. Approximately one-third of the contracts awarded to the top United States construction firms have been for work outside the U.S. American know-how is one of the exports not usually figured into foreign trade discussions. The change to SI will help the U.S. maintain a role of technical leadership and keep it from lagging behind the metric world.

FIG. 1 **SOME METRIC RECOGNITION POINTS**

Quantity	Unit	Description
Linear Measurement	1 mm	approximate diameter of a paper clip wire
	25 mm	vertical dimension of an ordinary U.S. postage stamp; nearest equivalent to 1 inch
	100 mm	international cigarette length; basic metric building module; nearest metric preferred dimension to 4 inches (1.6% less)
	600 mm	height of three courses of concrete blocks including mortar joints
	2 m	approximate height of a standard door opening (6′8″)
Area Measurement	500 mm²	face area of an ordinary U.S. postage stamp [20 mm x 25 mm]
	1 m²	1 square metre: approximate area of a shower base
Volume Measurement	1 L	1 liter: 5.7% more than a U.S. quart; new soft drink bottle size
Mass [Weight]	1 g	approximate weight of a paper clip or a dollar bill; artificial sweetener package size
	5 g	mass [weight] of a nickel [5¢ coin]
	1 kg	1 kilogram: the base unit of mass in SI; mass of water in a cube with 100 mm sides (approximately 2.2 pounds)
Temperature	0°C	freezing point of water (32°F)
	20°C	comfortable thermostat setting in winter (68°F)
	37°C	normal body temperature (98.6°F)
	100°C	boiling point of water (212°F)

TERMINOLOGY

A lifetime using the customary system has given people a feel for the length of an inch or a foot. New associations or recognition points need to be developed in order to be comfortable using SI (Fig. 1).

The following terms are basic to an understanding of metrication and SI:

Metric System A measurement system developed in France during the 1790s, based on the metre and a series of other fundamental units which have been standardized. The system also contains a set of decimal prefixes (milli—1/1000; kilo—1000 times) which can be attached to reference units to alter their magnitude.

Coherent Unit System System in which relations between units contain as numerical factor only the number "one" or "unity", because all derived units have a unity relationship to the constituent base and supplementary units. For example, one litre is equal to one thousandth of one cubic metre (0.001 m³) and (when filled with water) has a mass of one kilogram.

SI—The International System of Units The modern metric system is a coherent unit system developed in 1960 by the General Conference on Weights and Measures (CGPM), an international treaty organization. SI has been adopted by all countries which have changed to metric measurement since 1960. Countries which were using metric systems before 1960 are changing those units that have been superseded by the SI system.

Metrication A term coined in Britain to describe metric conversion, any process of change from customary measurement to SI, including the planning and coordination necessary for the change.

Exact Conversion The change from a customary value to its *precise* metric equivalent, generally expressed to a number of decimal places. For example, 24″ is exactly 609.6 mm (Fig. 23).

Soft Conversion A change in description only, but without physical change; generally a soft conversion is the rounding of an exact conversion to a more workable numerical value. For example, 24″ is approximately 610 mm (Fig. 23).

Hard Conversion A physical change from an existing numerical value to a new and "preferred" metric value, requiring a definite change in physical characteristics. For example, 24″ ideally should be changed to an even 600 mm, 1.6% smaller (Fig. 23).

Rationalization The selection of the most rational, preferred and economical alternatives, after re-

FIG. 2 **CUSTOMARY AND SI BASE UNITS**

Quantity	Customary	SI
Length	foot	metre (m)
Mass (Weight)	pound	kilogram (kg)
Time	second	second (s)
Temperature	degree Fahrenheit	kelvin (K)
Electric Current	ampere	ampere (A)
Luminous Intensity	candela	candela (cd)
Amount of a Substance	—	mole (mol)

FIG. 4 **CULTURAL VARIABILITY OF THE FOOT AS A UNIT OF LINEAR MEASURE**

Country	Local Name	Metric Equivalent, mm
Mexico	Pie	279
Netherlands	Voet	283
Poland	Stopa	288
Sweden	Fot	296
Switzerland	Fuss	300
Great Britain	Foot	305
Germany	Rheinisher Fuss	313
Denmark	Fod	314
Austria	Wiener Fuss	316
Norway	Fod	318
France	Pied	325
Italy	Piede Librando	513

search into the technical implications and an appraisal of the economic impact. For example, after rationalization, Canada reduced the number of reinforcing bar sizes from 11 to 8.

Dimensional Coordination The systematic application of preferred and related dimensions in the design of buildings and the manufacture and positioning of building components, assemblies, and elements.

Modular Coordination Dimensional coordination based on the international building module, which has a dimension of 100 mm, slightly less than 4″.

Quantity The measureable attribute of a physical phenomenon. There are *base units* for seven quantities and *supplementary units* for two quantities.

All other quantities are derived from these base and supplementary units. For example, the metre is a base unit; the radian is a supplementary unit; the pascal is a derived unit.

Unit Reference value of a given quantity as defined by CGPM resolution. There is only one unit for each quantity in SI. For example, the metre is the base unit for length.

Preferred (Metric) Sizes Sizes for building components or assemblies preferred over others—normally, selected multiples of the basic 100 mm module.

M-Day A date selected by an industry after which all activities should be carried out in SI metric units. In construction, M-Day is January 1, 1985, after which all *new* projects should be constructed in SI units.

Metric Board The 17 member United States Metric Board, established in the Metric Conversion Act of 1975 to coordinate the voluntary conversion to the metric system.

The American National Metric Council (ANMC) A private and self-supporting national organization, established in 1973 to provide assistance in the conversion to metric measurement through coordination, planning and information services to its membership and all segments of society in the U.S.

HISTORY OF MEASUREMENT

Measurement is a comparison between an unknown quantity and a known quantity. It is also a ratio of how many times a known thing can be divided into something unknown. Measurement tells how large, how many, how fast, how much, of almost anything imaginable.

A unified system of measurement is a means of communication. In the history of mankind, various number systems and representational symbols have been used. Though remnants exist of systems based on the numbers 5, 12 and 20, only systems based on the numbers 2, 10 and 60 are still in wide use.

SI uses the base-10 (decimal) system for seven designated base elements: length, mass (which is closely related to weight), time, heat, electricity, light and the amount of a substance. Figure 2 compares the names for the base units in SI and the customary system.

It took many centuries to develop accurate units of measurement. The first units of measurement were based on parts of the body. People used the human hand, or the human foot (Fig. 3); they also used the length of a furrow made by a plow, or the distance a man could walk in a day. Since it is harder and it takes longer to walk uphill than downhill, the Chinese even made an uphill mile shorter.

All of these forms of measurement were inexact. The size of a human hand or foot varied among individuals. Different people plowed different furrows and walked different distances in a day. There was a need for a unified

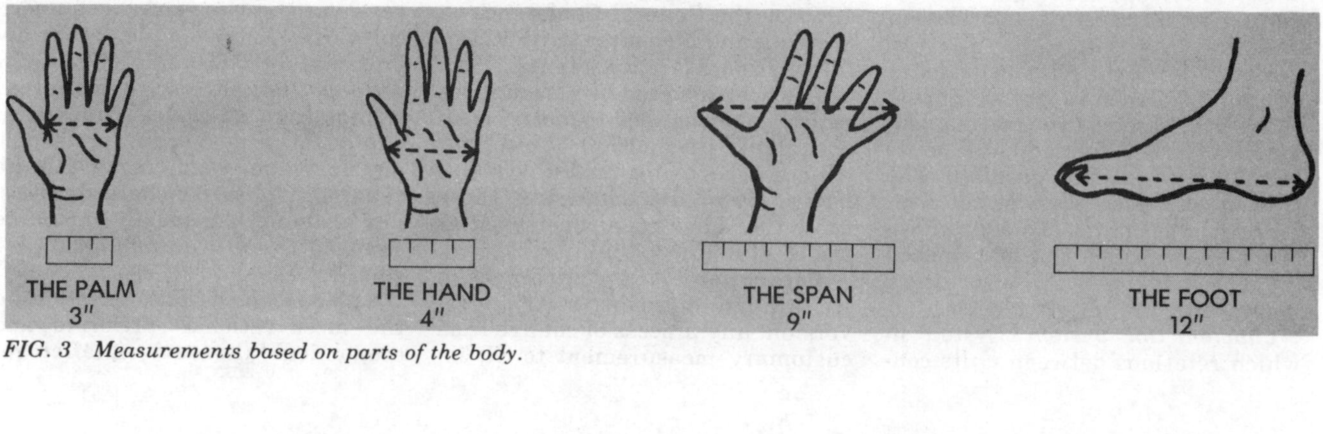

FIG. 3 Measurements based on parts of the body.

standard system of weights and measures—one that everybody could agree on.

The Romans established a unified system of weights and measures. They made a bronze rod of specific length and called it a *pes*, Latin for foot; they made a bronze weight called a *libra*, Latin for pound. Both the pes and libra were kept under guard in a temple and used as standards of measurement throughout the Roman Empire. The U.S. customary abbreviation for pound, *lb.*, is derived from the Latin libra.

After the fall of the Roman Empire, units of measurement once again became localized and confusion existed when different regions tried to communicate numerically.

Noses, Thumbs and Barleycorns

Local measurements often were based on the king's body measurements and differed in various countries or times (Fig. 4).

The U.S. customary system uses the yard as a standard unit of measurement. Legend says that King Henry I of England established by royal decree the length of the yard as 36″, the distance from Henry's nose to his thumb (Fig. 5).

Another arbitrary and unrelated unit of the customary system is the system of *thirteens*, based on *barleycorns*, used to measure shoes. A barleycorn is approximately 1/3″. Shoe sizes vary about 1/3″ from size to size; for example, a size 7 shoe is 1″ smaller than a size 10. After size 13, shoes start over again at 1.

Number Systems in Current Use

Sexagesimal System Based on the number 60, it was first used by the Babylonians and survives in the subdivision of time and angular measurement in the customary system. For instance, the hour has 60 minutes; the circle has 360 degrees. Both minutes and degrees can be further subdivided in 60 seconds.

Binary System This system is based only on numerals 0 and 1; it is used in electronic computers, because it can represent yes/no or true/false decision logic.

Decimal System This base-10 system, upon which SI is founded, was probably derived from man's use of his fingers to count. The Greeks and Romans first used a decimally based word structure for numbers, then adopted letter numerals for their representation. *Roman numerals*, used exclusively in Europe until 1100 A.D., were gradually replaced by *Arab numerals* of Hindu-Arabic origin, which were much simpler to use.

The *Roman Numeral System* consists of quinary (base-five) symbols: I equals one; V equals five; X equals 10; L equals 50; C equals 100; D equals 500; M equals 1000. In-between numbers are formed by addition or subtraction to the seven basic symbols. When a smaller number precedes a larger, it is subtracted; when the smaller follows the larger number, it is added. Thus, 1978 is written MCMLXXVIII: C before M means nine hundred; all other numbers retain their absolute values and are added.

The Hindu-Arabic System which we use today expresses all numbers by combinations of 10 symbols, each having an *absolute* value and a position value (Fig. 9).

Decimal position notation relates each number's position value as a power of ten. For example, 10^1 equals 10, 10^2 equals 100, 10^3 equals 1000. Numbers to the right of the decimal are expressed in negative powers of ten. For example, 10^{-1} equals 0.1, 10^{-2} equals 0.01 and 10^{-3} equals 0.001. Each multi-digit number is written as a linear combination of powers of ten. For example:

$456 = 4 \times 10^2 + 5 \times 10^1 + 6;$

$37.5 = 3 \times 10^1 + 7 + 5\ (\times 10^{-1}).$

THE METRIC SYSTEM & SI

The scientific evolution of the 18th and 19th centuries, the industrial revolution of the 19th century and the desire for more international trade accelerated the movement towards a universal system of weights and measures.

The French National Assembly adopted the metric system in 1795, under pressure from leading scientists of the era. The system was based on a unit of length called the *metre* (or *meter*) derived from the Greek word *metron*, meaning measure. The metre is expressed in *geodetic* terms, that is, measures related to the size and shape of the earth.

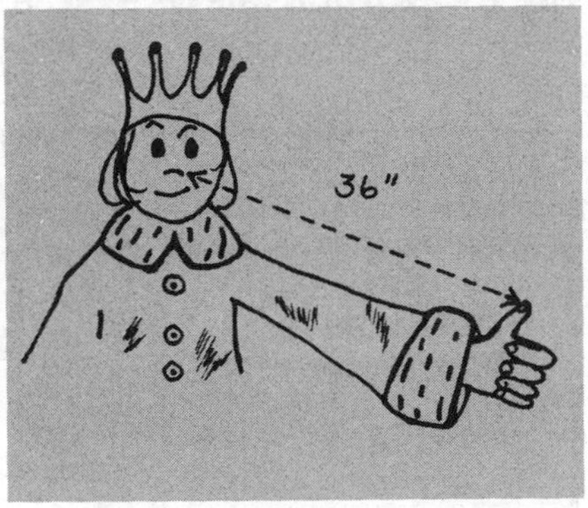

FIG. 5 The yard is believed to have been established as a unit of measure by royal decree of King Henry I.

The French chose as their standard unit of length the ten-millionth part of an imaginary line running through Paris from the Equator to the North Pole. Eventually, that distance was marked on a platinum/iridium bar which became the international standard (prototype). It was kept at the International Bureau of Weights and Measures in Sèvres, France.

Additional units in the system—length, area, volume, capacity and mass—were related decimally to the metre through the properties of water. For example, one *litre* represents one thousandth of one *cubic metre*. When filled with water at its maximum density, a litre has a mass (weight) of one thousand grams or one *kilogram*.

Science and technology demanded a means of precisely defining the metre in a way that could be accurately reproduced without having to refer to the prototype. In 1960, the General Conference on Weights and Measures (CGPM) introduced a new definition based on the electromagnetic emission of a certain krypton atom (Fig. 8).

Adoption of SI

The decimal metric system has been accepted by 90% of the world's nations. Paradoxically, the United States, the first nation to introduce a system of decimal currency in 1785, is one of the last nations to adopt SI. It shares this dubious distinction with Brunei, Burma and

FIG. 6 RELATIONSHIP OF BASE, DERIVED AND SUPPLEMENTARY SI UNITS

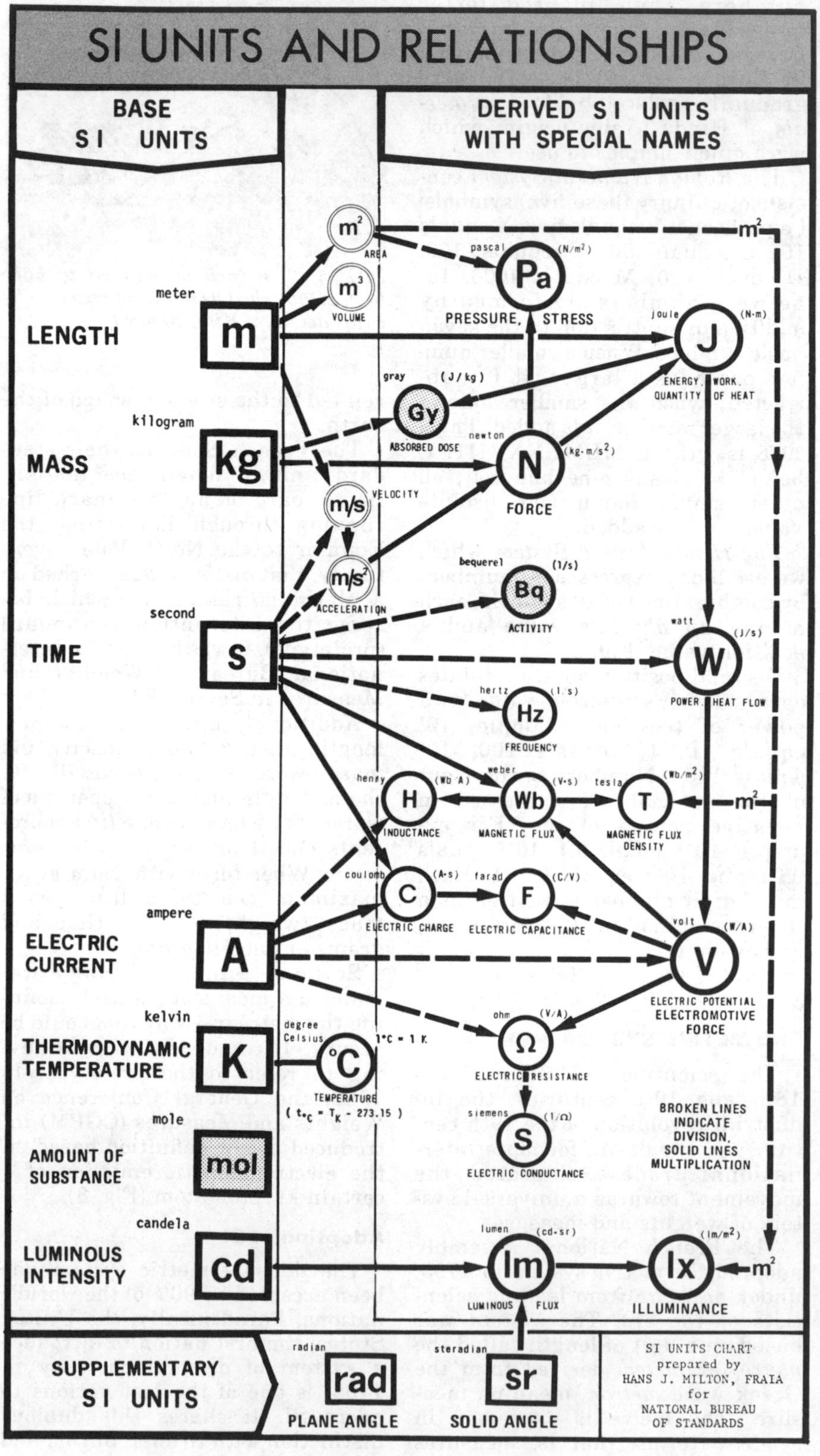

the two Yemens. Yet the U.S. was an original party of the 1875 Treaty and International Metric Convention which established the General Conference of Weights and Measures (CGPM).

The CGPM meets at least every six years. It controls the International Bureau of Standards which preserves the metric standards and compares national standards with them. The CGPM also conducts research to establish new standards. The U.S. National Bureau of Standards (NBS) represents our country in these activities.

Though the metric system was authorized by Congress in 1866, it was not until 1975 that the Metric Conversion Act established the 17 member Metric Board to plan, publicize and implement a voluntary change to the metric system based on SI.

Structure of SI

Originally the metric system provided a set of units to measure length, area, volume, capacity and mass, based on the metre and the kilogram. In 1881, the International Electrical Congress adopted a unit of time to produce the *centimetre-gram-second (cgs) system*. About 1900, metric measurements began to be based on the *metre-kilogram-second (MKS) system*. In 1935, the *ampere* became the fourth base unit, resulting in the *metre-kilogram-second-ampere (MKSA) system*. The 10th CGPM added the *kelvin* as a unit of temperature and *candela* as the unit of luminous intensity.

International System of Units (SI) In 1960, the CGPM formally introduced the *International System of Units* for which the abbreviation is *SI* in all languages. Many nations using the cgs, MKS or MKSA system are converting to the SI metric system (Fig. 6).

The growth of thermal dynamics, nuclear physics and electronic science caused further refinement of SI. In 1971, a seventh base unit was added, the *mole*, which expresses the amount of a substance (Fig. 8).

The Base Units shown in Figure 7 are the building blocks of SI. Except for the *kilogram* they are reproduceable under controlled conditions

FIG. 7 BASE UNITS IN SI

Quantity	Unit	Symbol
length	metre	m
mass	kilogram	kg
time	second	s
electric current	ampere	A
thermodynamic temperature	kelvin	K
amount of substance	mole	mol
luminous intensity	candela	cd

FIG. 8 SI TECHNICAL DEFINITIONS

Base Units:

Metre The metre is the length equal to 1 650 763.73 wavelengths in vacuum of the radiation corresponding to the transition between the levels $2p_{10}$ and $5d_5$ of the krypton-86 atom.

Kilogram This is the unit of mass; it is equal to the mass of the international prototype of the kilogram.

The kilogram is the only base unit whose name for historic reasons contains a prefix.

Second This is the measure of time with a duration of 9 192 631 770 periods of the radiation corresponding to the transition between the two hyperfine levels of the ground state of the cesium-133 atom.

Ampere This is the measure of current flow equal to that flow which, if maintained in two straight parallel conductors would produce between these conductors a force equal to 2×10^7 newton per metre of length. The conductors are assumed to be of infinite length, of negligible circular cross section, and placed one metre apart in vacuum.

Kelvin This is the unit of thermodynamic temperature equal to the fraction 1/273.16 of the thermodynamic temperature of the triple point of water.

The temperature interval of the degree Celsius and the kelvin are identical and the scales using these two units are related by a difference of exactly 273.15 kelvins.

Mole This is the amount of substance of a system which contains as many elementary entities as there are atoms in 0.012 kilogram of carbon-12. When the mole is used, the elementary entities must be specified and may be atoms, molecules, ions, electrons, other particles or specified groups of such particles.

Candela This is the luminous intensity in the perpendicular direction, of a surface of 1/600 000 square metre of blackbody at the temperature of freezing platinum under a pressure of 101 325 newtons per square metre.

Supplementary Units:

Radian This is the plane angle between two radii of a circle which cut off on the circumference an arc equal in length to the radius. The radian is equal to $360/2\pi$ and is an angle of approximately 57° 17′ 44.6″.

Steradian This is the solid angle which, having its vertex in the center of a sphere, cuts off an area of the surface of the sphere equal to that of a square with sides of length equal to the radius of the sphere. The total solid angle of a sphere is 4π steradians.

FIG. 9 DERIVED SI UNITS WITH SPECIAL NAMES

Physical Quantity	Symbol	Unit & Definition
Absorbed dose	(Gy)	The *gray* is the energy imparted by ionizing radiation to a mass of matter corresponding to one joule per kilogram.
Activity	(Bq)	The *becquerel* is the activity of a radionuclide having one spontaneous nuclear transition per second.
Electric capacitance	(F)	The *farad* is the capacitance of a capacitor between the plates of which there appears a difference of potential of one volt when it is charged by a means of electricity equal to one coulomb.
Electric conductance	(S)	The *siemens* is the electric conductance of a conductor in which a current of one ampere is produced by an electric potential difference of one volt.
Electric inductance	(H)	The *henry* is the inductance of a closed circuit in which an electromotive force of one volt is produced when the electric current in the circuit varies uniformly at a rate of one ampere per second.
Electric potential difference, electromotive force	(V)	The *volt* (unit of electric potential difference and electromotive force) is the difference of electric potential between two points of a conductor carrying a constant current of one ampere, when the power dissipated between these points is equal to one watt.
Electric resistance	(Ω)	The *ohm* is the electric resistance between two points of a conductor when a constant difference of potential of one volt, applied between these two points, produces in this conductor a current of one ampere, this conductor not being the source of any electromotive force.
Energy	(J)	The *joule* is the work done when the point of application of a force of one newton is displaced a distance of one metre in the direction of the force.
Force	(N)	The *newton* is that force which, when applied to a body having a mass of one kilogram, gives it an acceleration of one metre per second squared.
Frequency	(Hz)	The *hertz* is the frequency of a periodic phenomenon of which the period is one second.
Illuminance	(lx)	The *lux* is the illuminance produced by a luminous flux of one lumen uniformly distributed over a surface of one square metre.
Luminous flux	(lm)	The *lumen* is the luminous flux emitted in a solid angle of one steradian by a point source having a uniform intensity of one candela.
Magnetic flux	(Wb)	The *weber* is the magnetic flux which, linking a circuit of one turn, produces in it an electromotive force of one volt as it is reduced to zero at a uniform rate in one second.
Magnetic flux density	(T)	The *tesla* is the magnetic flux density given by a magnetic flux of one weber per square metre.
Power	(W)	The *watt* is the power which gives rise to the production of energy at the rate of one joule per second.
Pressure or stress	(Pa)	The *pascal* is the pressure or stress of one newton per square metre.
Quantity of electricity	(C)	The *coulomb* is the quantity of electricity transported in one second by a current of one ampere.

anywhere (Fig. 8). The kilogram is a platinum-iridium prototype kept by the International Bureau of Weights and Measures in Sevres, France.

Supplementary units, the *radian* (plane angle) and the *steradian* (solid angle) are also part of the basic SI structure (Fig. 8).

Derived Units are formed by combining base and supplementary units and other derived units according to algebraic relations linking corresponding quantities (Fig. 9).

Certain other units may also be used with SI.

Standard Prefixes allow a change in magnitude from the sub-atomic scale (10^{-18}) to the astronomic scale (10^{18}) (Fig. 10).

Non-SI Units for Use with SI To preserve the advantages of SI as a coherent system, the use of units from other systems should be minimized. There are acceptable but noncoherent traditional units retained for use with SI because of practical significance in general application (Fig. 11 & 12).

FIG. 10a **PREFERRED MULTIPLES, SUBMULTIPLES AND THEIR PREFIXES**

MULTIPLICATION FACTOR	PREFIX NAME	PREFIX SYMBOL	PRONUNCIATION
10^{12} or 1 000 000 000 000	tera	T	as in *terrace*
10^{9} or 1 000 000 000	giga	G	*jig'a*
10^{6} or 1 000 000	mega	M	as in *megaphone*
10^{3} or 1 000	kilo	k	as in *kilowatt*
10^{-3} or 0.001	milli	m	as in *military*
10^{-6} 0.000 001	micro	μ	as in *microphone*
10^{-9} or 0.000 000 001	nano	n	*nan'oh*
10^{-12} or 0.000 000 000 001	pico	p	*peek'oh*

FIG. 10b **NON-PREFERRED MULTIPLES AND THEIR PREFIXES**

MULTIPLICATION FACTOR	PREFIX NAME	PREFIX SYMBOL	PRONUNCIATION
10^{2} or 100	hecto	h	*heck'toe*
10^{1} or 10	deka	da	*deck'a*
10^{-1} or 0.1	deci	d	as in *decimal*
10^{-2} or 0.01	centi	c	as in *sentiment*

FIG. 11 **UNITS IN USE WITH SI**

Quantity	Unit	Symbol	Definition Symbol
time	minute	min	1 min = 60 s
	hour	h	1h = 60 min = 3600 s
	day	d	1 d = 24 h = 86 400 s
plane angle	degree	°	1° = (π/180) rad
volume	litre	L	1 L = 1 dm³ = 10^{-3} m³
mass	metric ton	t	1 t = 10^3 kg

FIG. 12 **UNITS ACCEPTED FOR LIMITED USE**

Quantity	Unit	Symbol	Definition
plane angle	minute	′	1′ = (1/60)° = (π/10 800) rad
	second	″	1″ = (1/60) = (π/648 000) rad
energy	kilowatthour	kWh	1 kWh = 3.6 MJ
area	barn	b	1 b = 10^{-28} m²
	hectare	ha	1 ha = 1 hm² = 10^4 m²
pressure	bar	bar	1 bar = 10^5 Pa

Time is expressed by the *second* (s); where it relates to life customs or calendar cycles, the *minute* (min), *hour* (h), *day* (d) and other calendar units may be necessary.

The Plane Angle unit is the *radian* (rad). Use of the degree (°) and its decimal submultiples is permissible when the radian is not convenient; use of the minute (′) and second (″) is discouraged.

Volume is expressed by the *cubic metre* (m³).

The *litre* (L) is acceptable for the measurement of liquids and gases.

Temperature is expressed by the *kelvin* (K). The degree *Celius* (°C), formerly called centigrade, may also be used with SI.

Mass is expressed by the *kilogram* (kg) or one of its multiples formed by attaching an SI pefix to *gram*, such as milligram and megagram. The correct SI unit, megagram (Mg), equal to 1 000 000 grams, should be used for measuring large masses formerly expressed in tons. In the interim, the metric ton (t), which equals one megagram, will also be used.

Area is expressed by the *square metre* (m²). The *hectare* (ha) an alternative name for square hectometre (hm²), equal to 10 000 square metres, should only be used for land or water measurement.

Energy should be expressed by the joule, together with its multiples. The *kilowatthour* (kWh) will continue to be used in electrical applications but should be replaced by the megajoule (MJ) whenever possible.

Pressure and Stress are expressed by the pascal (Pa)—which is equal to the newton per square metre (N/m²)—and prefixed units such as the kilopascal (kPa), megapascal (MPa), and gigapascal (GPa).

Old metric gravitational units for pressure and stress such as kilogram-force per square centimetre (kgf/cm²) should not be used; neither should units such as *bar* and *torr*. Use of the *millibar* will continue in order to permit meteorologists to communicate within their profession; however meteorologists should use the kilopascal (kPa) to present data to the public.

Metric Units to be Abandoned Many metric units in use before SI was adopted have been abandoned. SI supersedes all previous metric systems.

Centigram-Gram-Second (cgs) Units such as *abampere* and *statvolt* have been superseded by SI and are to be avoided.

Decimal Multiples of SI Units that cannot be handled by using the SI prefixes should not be used. For example, the *angstrom* (0.1 nm) should not be used.

Special Names Special names for multiples and submultiples of SI units should be avoided (except for *litre, metric ton* and *hectare* discussed under *Non-SI* Units for Use with SI). Figure 13 shows examples of units to be avoided, together with their SI equivalents.

Recommendations for the Use of SI

SI symbols are internationally agreed upon and many rules exist for their proper use. See Metrication Work File, Figures WF6 and WF7, for more detail.

Writing Unit Symbols In general, unit symbols (1) should be printed in *upright type* regardless of type style in surrounding text; (2) should be unaltered in the plural; (3)

should not be followed by a period except when used at the end of a sentence; (4) should be written in lowercase unless unit is derived from a proper name; (5) should have a space between the unit symbol and the numerical value.

Because the comma is used in some foreign countries as a decimal marker, it should not be used to separate digits into groups of three. Instead, digits should be divided into groups of three by using a space at least equal to the width of the letter "i" to separate these groups. In four digit numbers a space is not necessary. For example: 2.141 596; 73 722; 7372.

The word 'billion' means a thousand million (prefix *giga*) in the U.S., but a million million (prefix *tera*) in most other countries. Terms such as 'billion' and 'trillion' should not be used in technical writing.

Preferred Range of Values Whole numbers between 1 and 1000 should be used with the proper prefix wherever possible. For example, 725 m is preferred over 0.725 km or 725 000 mm.

The prefix that adequately covers the whole quantity range should be used. For example, if 725 m is part of a group of numbers shown in kilometres, it should be expressed as 0.725 km.

Exponential notation, instead of prefixes, is preferred in calculations. For example, 900 mm² equals 0.9 times 10^{-3} m².

One measurement unit should be used throughout a set of drawings so that numerical values can be represented by numbers only. In a drawing dimensioned in millimetres, 5 digit numbers are acceptable; for example, 32 845 mm.

FIG. 13 **SPECIAL NAMES TO BE AVOIDED**

Unit	Designation and Equivalent
fermi	1 fm = 10^{-15} m
micron	1 μm = 10^{-6} m
millimicron	1 nm = 10^{-9} m
are	1 dam² = 100 m²
gamma*	1 nT
γ (mass)	1 μg
λ (volume)	1 μl = 1 mm³
mho	1 S
candle	1 cd
candlepower	1 cd

*magnetic flux density

SI UNITS FOR DESIGN AND CONSTRUCTION

There are two principal factors in any measurement statement—a number, always written first, and a measurement unit. Correct selection of units for use in building construction is essential to minimize errors and optimize coordination (Fig. 14).

A list of recommended units for building design and construction and conversion charts for metric

FIG. 14 **SI UNITS FOR BUILDING CONSTRUCTION**

Activity	Quantity	Unit	Symbol
LAND SURVEYING	Linear measure	kilometre, metre	km, m
	Area	square kilometre	km²
		hectare (10 000 m²)	ha
		square metre	m²
EXCAVATING	Linear measure	metre, millimetre	m, mm
	Volume	cubic metre	m³
CONCRETING	Linear measure	metre, millimetre	m, mm
	Area	square metre	m²
	Volume	cubic metre	m³
	Temperature	degree Celsius	°C
	Water-capacity	litre	L
Constituents	Mass (weight)	kilogram, gram	kg, g
Reinforcement	Cross-section	square millimetre	mm²
TRUCKING	Distance	kilometre	km
	Mass (weight)	metric ton (1000 kg)	t
PAVING and PLASTERING	Linear measure	metre, millimetre	m, mm
	Area	square metre	m²
BRICKLAYING	Linear measure	metre, millimetre	m, mm
	Area	square metre	m²
	Mortar-volume	cubic metre	m³
CARPENTRY/JOINERY	Linear measure	metre, millimetre	m, mm
STEELWORKING	Linear measure	metre, millimetre	m, mm
	Mass (weight)	metric ton (1000 kg)	t
		kilogram, gram	kg, g
ROOFING	Linear measure	metre, millimetre	m, mm
	Area	square metre	m²
	Slope	millimetre/metre	mm/m
PAINTING	Linear measure	metre, millimetre	m, mm
	Area	square metre	m²
	Capacity	litre, millilitre	L, mL
GLAZING	Linear measure	metre, millimetre	m, mm
	Area	square metre	m²
PLUMBING	Linear measure	metre, millimetre	m, mm
	Mass (weight)	kilogram, gram	kg, g
	Capacity	litre	L
	Pressure	kilopascal	kPa
DRAINAGE	Linear measure	metre, millimetre	m, mm
	Area	hectare (10 000 m²)	ha
		square metre	m²
	Volume	cubic metre	m³
	Slope	millimetre/metre	mm/m
ELECTRICAL SERVICES	Linear measure	metre, millimetre	m, mm
	Frequency	hertz	Hz
	Power	watt, kilowatt	W, kW
	Energy	megajoule (1 kWh = 3.6 MJ)	MJ
	Electric current	ampere	A
	Electric potential	volt, kilovolt	V, kV
	Resistance	ohm	Ω
MECHANICAL SERVICES	Linear measure	metre, millimetre	m, mm
	Volume	cubic metre	m³
	Capacity	litre	L
	Airflow	metre/second	m/s
	Volume flow	cubic metre/second	m³/s
		litre/second	L/s
	Temperature	degree Celsius	°C
	Force	newton, kilonewton	N, kN
	Pressure	kilopascal	kPa
	Energy, Work	kilojoule, megajoule	kJ, MJ

FIG. 15 **COMPARISON OF CALCULATIONS IN CUSTOMARY UNITS AND IN METRIC (SI) UNITS**

DIVISION OF A LINEAR DIMENSION INTO EQUAL PARTS

Customary units	Metric (SI) units
A detailed drawing shows a height of 8'-10" between two floors connected by a stair with 14 risers, but not the height of each riser. What is the target height of risers so that none of them varies by more than 1/16"?	A detail drawing shows a height of 2690 mm between two floors connected by a stair with 14 risers, but not the height of each riser. What is the target height of risers so that none of them varies by more than 1 mm?
8'-10" = 106" 106 ÷ 14 = 7.57" 7.57" ≅ 7 9/16" (Check) 7 9/16" × 14 = 105 7/8" (-1/8") Target height of riser: 7 9/16" with 2 risers at 7 5/8" (+1/16" each)	2690 ÷ 14 = 192.1 mm (Check) 192 × 14 = 2688 mm (+2 mm) Target height of riser: 192 mm with 2 risers at 193 mm each.

Commentary: Division of measurement in customary units requires decimalization and change to and from decimal fractions.

AREA AND VOLUME CALCULATIONS TO DETERMINE ORDER QUANTITY

Customary units	Metric (SI) units
A flat concrete slab for an industrial structure has the following dimensions: Length: 25'-0"; width: 26'-6"; and thickness: 5". Determine the order quantity of concrete to the nearest half of a cubic yard.	A flat concrete slab for an industrial structure has the following dimensions: Length: 7.5 m; width: 8.4 m; and thickness: 125 mm (0.125 m). Determine the order quantity of concrete to the nearest half of a cubic meter.
$\frac{25 \times 26.5 \times 5}{12} = 276$ ft³ 276 ÷ 27 = 10.22 yd³ Order: 10 1/2 yd³	7.5 × 8.4 × 0.125 = 7.85 m³ Order: 8 m³

Commentary: Customary units need to be decimalized, and factors are required with inches and conversion of cubic feet to cubic yards. Metric volume is derived directly by multiplication of dimensions.

and customary values appear in 603 Metrication Work File.

Linear Measurement

The preferred SI units for building design and construction are the millimetre (mm) and the metre (m).

The *metre* and the *millimetre* have an inherently greater precision than the customary system. The customary system uses decimalized fractions in site work, but uses feet, inches and fractions in construction work. Using the millimetre as a principal unit of construction measurement avoids the use of fractions. One millimetre provides an accuracy of 1/25"; five digits can express any length up to 328 ft. The use of whole numbers greatly speeds all calculations and helps to avoid errors; calculations can be done twice as fast with only one-fifth the rate of error (Fig. 15). In the customary system fractions have to be reconciled to a common denominator and then added; inches have to be added and then converted to feet and inches; finally the feet are added.

The *centimetre* should be avoided for these reasons: (1) the order of magnitude between the millimetre and the centimetre is only ten and using both units would cause confusion; (2) the millimetre allows the use of whole numbers rather than fractions within appropriate tolerances for all building and product dimensions.

Unit symbols may be deleted if the drawing is designated "all dimensions shown in millimetres (or metres) unless otherwise noted."

Whole numbers always indicate millimetres. A five-digit number can indicate a length up to 328'; 327'10-11/16" = 99 941 mm. A length up to 32'9" can be shown by a four-digit number; a length of 3'3-5/16" by a three-digit number; 1/25" by a one millimetre.

Numbers carried to three decimal places always indicate metres. For example: 3.600; 0.300; 0.025.

The recommended use of millimetres and metres saves time and space in drawing, typing and computer applications. It also improves clarity in drawings with many dimensions.

The drawing scales used by architects and engineers have traditionally differed. Architects have generally used scales such as 1/4" equals 1' which resulted in non-decimal scale ratios; even when engineers used seemingly metric scales such as 1" = 100', the ratio is non-decimal (Fig. 16). Metrication will tend to unify these differences.

The consistent use of fewer scales has definite benefits: (1) Drawings prepared on transparent media drawn to the same scale may be easily superimposed, which makes drawing faster, easier and reduces the possibility of errors and omissions. (2) Drawings lend themselves to computer aided reduction and enlargement techniques. (3) Consistently sized drawings are easier to understand and speed construction in the field.

Area

The *square metre* (m^2) is the preferred unit for area measurement. Large areas can be expressed in square kilometres (km^2); or hectares (10 000 m^2); small areas in square millimetres (mm^2). The hectare (ha) should only be used for land and water measurement. The square centimetre (cm^2) should be avoided to minimize confusion. Sometimes the area of buildings or products is indicated by linear dimensions (example: 40 by 90 mm); generally the width is indicated first and height second.

Volume and Fluid Capacity

The cubic metre (m^3) is the preferred measurement unit for volume in construction and for large storage tank capacities.

The *litre* (L) and the millilitre (mL) are the measurement units of fluid capacity (liquid volume). The litre is equal to one thousandth of a cubic metre or one cubic decimetre (dm^3) and can be expressed as:
$1\ L = 10^{-3}\ m^3$ or $1\ L = 1\ dm^3$ or $1\ m^3 = 1000\ L$.

The cubic metre contains one billion (10^9) cubic millimetres (Fig. 11). Cubic centimetres and decimetres should be converted to the preferred cubic metres or millimetres whenever possible (Fig. 17).

Geometric Cross-sectional Properties

The expression of geometric cross-sectional properties of structural sections involves raising the unit of length to the third, fourth or sixth power with exponential notation (Fig. 10), as follows:

Modulus of section (S) is written in units of mm^3 or m^3. For example, ($1\ mm^3 = 0.000\ 000\ 001 = 10^{-9}\ m^3$).

Second moment of area, (I) and *torsional constant* (J) are written in units of mm^4 or m^4. For example, $1\ mm^4 = 10^{-12}\ m^4$.

Warping constant (C) is written in units of mm^6 or m^6 ($1\ mm^6 = 10^{-18}\ m^6$).

Figure 18 shows the cross-sectional properties of a wide flange beam 460 mm deep weighing 82 kg/m per unit of length.

Plane Angle

Wherever practicable, the radian (rad) should be used in engineering calculations. The customary units-degree (°), minute (′), and second (″) may continue to be used in cartography and surveying; the degree (°), with parts denoted by decimals (as in 26.25°) may continue to be used in engineering and construction.

Time Interval

The *second* (s) is the SI base unit. The *day* (d), *hour* (h), and *minute* (min) are non-SI alternatives. The minute, however, should be avoided as much as possible to minimize the variety of units in which time is a dimension.

Flow rates should be expressed in cubic metres per second, litres per second or cubic metres per hour:

• $1\ m^3/s = 1000\ L/s$ • (Do not use $60\ m^3/min$).

• $1\ L/s = 3.6\ m^3/h$ • (Do not use 60 L/min).

• $1\ m^3/h = 1000\ L/h$ • (Do not use 1667 L/min).

The month, which varies in length, should not be used, unless a specific calendar month is indicated.

The *calendar year* (symbol "a" for annum) represents 365 days equal to 31 536 000 seconds.

Temperature

The kelvin (K) is the base unit of temperature. The degree Celsius (°C), however, is acceptable and will replace the customary unit, Fahren-

FIG. 16 **METRIC AND NON-METRIC DRAWING SCALES**

TRADITIONAL SCALES [Expressed as Ratio]		METRIC SCALES	
		PREFERRED	OTHER
Full Size	[1:1]	1:1	
Half Full Size	[1:2]		1:2
4" = 1'-0"	[1:3]		
3" = 1'-0"	[1:4]		
		1:5	
2" = 1'-0"	[1:6]		
1 1/2" = 1'-0"	[1:8]		
		1:10	
1" = 1'-0"	[1:12]		
3/4" = 1'-0"	[1:16]		
		1:20	
1/2" = 1'-0"	[1:24]		
			1:25*
3/8" = 1'-0"	[1:32]		
1/4" = 1'-0"	[1:48]		
		1:50	
1" = 5'-0"	[1:60]		
3/16" = 1'-0"	[1:64]		
1/8" = 1'-0"	[1:96]		
		1:100	
1" = 10'-0"	[1:120]		
3/32" = 1'-0"	[1:128]		
1/16" = 1'-0"	[1:196]		
		1:200	

*Limited use

FIG. 17 **UNITS FOR VOLUME AND FLUID CAPACITY**

All Volumes	Fluid Volume Only	Limited Application	Relationships
m^3			$1\ m^3 = 1000\ L = 1000\ dm^3$
	L	dm^3	$1\ L = 1\ dm^3 = 10^{-3}\ m^3 = 1000\ mL$ $10^6\ mm^3$
	mL	cm^3	$1\ mL = 1\ cm^3 = 10^{-6}\ m^3$ $10^3\ mm^3$
mm^3			$1\ mm^3 = 10^{-9}\ m^3$

FIG. 18 **PROPERTIES OF A WIDE FLANGE BEAM 460 mm DEEP WEIGHING 82 kg/m**

Property	Designation and Units
Plastic Modulus	$Z_x = 1.835 \times 10^6\ mm^3\ (1.835 \times 10^{-3}\ m^3)$
Second Moment of Area	$I_{x-x} = 0.371 \times 10^9\ mm^4\ (0.371 \times 10^3\ m^4)$
Torsional Constant	$J = 0.691 \times 10^6\ mm^4\ (0.691 \times 10^{-3}\ m^4)$
Warping Constant	$C_w = 0.924 \times 10^{12}\ mm^6\ (0.924 \times 10^{-6}\ m^6)$

heit (°F), in everyday use (Fig. 19).

The temperature interval of one degree Celsius exactly equals one kelvin. A temperature expressed in degrees Celsius is equal to the temperature in kelvins less 273.15:

0 K = −273.15° C;

0° C = 273.15 K.

The zero point of the kelvin scale is absolute zero, which is the point at which all molecular movement stops (for further discussion see 105 Heat Control). There are no negative (minus) temperatures in the kelvin scale. The lack of negative numbers make kelvins useful for scientific applications; but the large numeric values make it cumbersome for everyday use. For example: room temperature may be expressed 68° F, 20° C or 293.15 K; body temperature as 98.6° F, 37° C or 310.15 K; boiling water as 212° F, 100° C or 373.15 K.

Mass, Weight, Force

SI differs not only from the customary system, but also from other forms of metric measurement, in the use of explicit and distinctly separate units for *mass* and *force*.

Considerable confusion exists in the use of the term *weight* to mean either *force* or *mass*.

FIG. 19 TEMPERATURE EQUIVALENTS

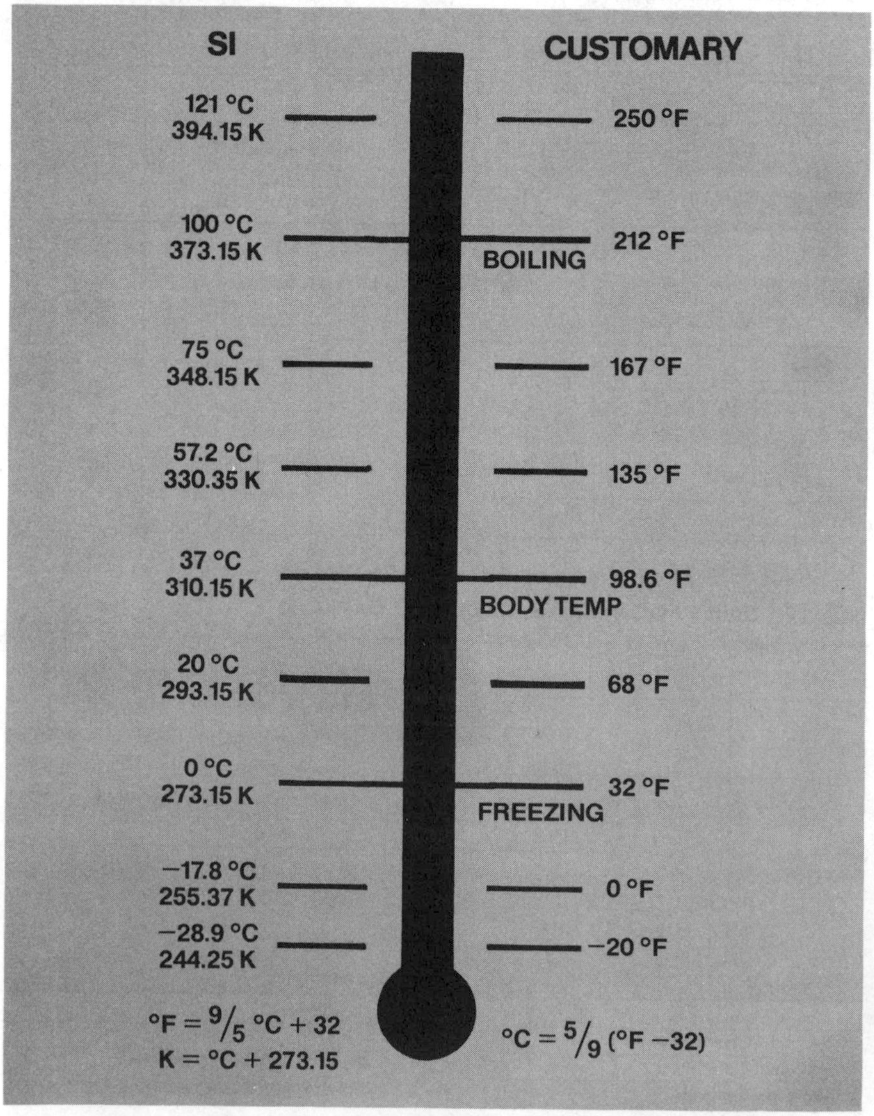

Mass In everyday use weight generally means *mass*. When one speaks of a person's weight, the quantity referred to is really mass.

Mass is the amount of matter there is in a body. An iron ball and a baseball of exactly the same size, take up the same amount of space but there is more matter (greater density or mass) per unit volume in the iron ball; therefore, the iron ball has more mass. *Density* is the ratio of mass to volume.

The *kilogram* (kg) is the SI unit restricted to *mass*, the quantity of matter in an object which is constant and independent of gravitational attraction. Multiples of the kilogram are formed by attaching an SI prefix to gram. The megagram (Mg) should be used to measure large masses formerly expressed in tons.

The name "ton" has been given to several large mass units that are widely used in commerce and technology: the long ton of 2240 lbs; the short ton of 2000 lbs; and metric ton of 1000 kg, also called "tonne." None of these terms are SI. The term "metric ton" should be restricted to commercial usage, and never used with prefixes. Use of the term "tonne" is discouraged.

Force In technical papers, weight has often been used to mean *force of gravity*, which is a particular force related to gravitational acceleration. The force of gravity varies from place to place on the earth and on other planets. A body of the same mass will weigh more or have more force on the earth than on the moon (Fig. 20).

To alleviate confusion, the term "weight" should be avoided in technical practice. Force of gravity should denote force. The *newton* (N) is the derived unit of force which equals mass times acceleration: $F = kg \cdot m/s^2$.

Previous metric systems have used the term "kilogram force" instead of newton, often dropping the suffix "force", which caused confusion. The use of kilogram-force should be avoided. The *newton* also supersedes the term kilopond (kp). The newton can also be used to designate pressure, stress, energy, work and quantity of heat, power and many electrical units.

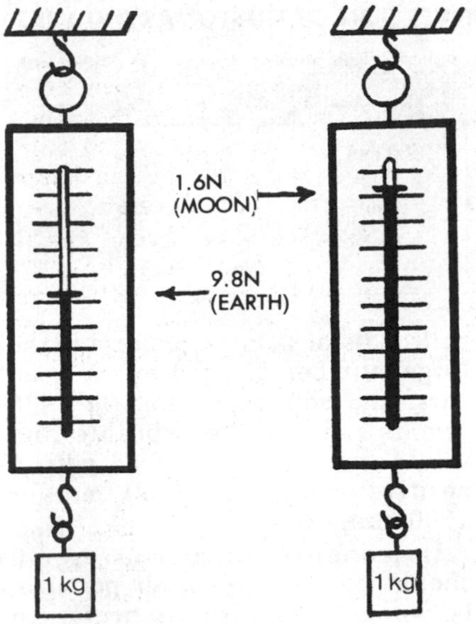

FIG. 20 One kg of matter will weigh over 6 times more newtons on earth than on the moon, although the mass (1 kg) is the same in both locations.

Pressure, Stress and Modulus of Elasticity The pascal (Pa), or newton per square metre, is the unit for both *pressure* and *stress*.

Pressure is a force applied uniformly over a surface and measured as a force per unit of area.

Stress is the effect of a force on a body, such as squeezing (compression), pulling (tension), bending (compression plus tension). Stress generally is accompanied by *strain*, a physical change in the body's shape or dimension.

The pascal and its prefixes megapascal (MPa) and kilopascal (kPa), replace a large number of customary units such as pounds per square inch (lbs./in.^{2} or psi).

It also supersedes a few non-SI metric units such as the bar (equal to 100 kPa) and the *millibar* (equal to 100 Pa).

Though it may sometimes be useful to report test results in N/mm^2 (identical to MN/m^2), calculations and results should be shown in MPa or kPa. Conversion factors may be found in 603 Metrication Work File.

Energy, Work and Quantity of Heat The *joule* (J) is the SI unit of energy, work and quantity of heat. It is equal to a newton meter (N•m) and a watt second (W•s). The joule supersedes many traditional units such as: Btu, therm, calorie, kilocalorie, foot pound-force.

The *kilowatthour* (kWh) has been the customary unit of energy consumption; the preferred unit is the megajoule (MJ). Since it would be needlessly costly to recalibrate electric meters, the kWh will continue to be used in electrical applications, but should not be introduced into new areas.

The joule should never be used for torque; for this property the designated unit is *newton metre* (N•m).

For dimensional consistency in rotational dynamics, torque should be expressed as newton metre per radian (N•m/rad) and moment of inertia as kilogram square metre per radian squared (kg•m^2/rad^2).

Power and Heat Flow Rate The *watt* (W), the unit for power and heat flow rate, is already in worldwide use as a unit for electrical power. It will replace some traditional units of power and heat flow rate: (1) horsepower (electric, boiler); (2) foot pound-force per hour (or minute or second); (3) Btu per hour, the calorie (or kilo calorie) per minute (or second); (4) the ton of refrigeration.

Electrical Engineering The only changes in electrical engineering will be the renaming of the unit of conductance to *siemens* (S) instead of mho and the use of hertz (Hz) for frequency instead of cycles per second (cps).

Lighting Units

The *candela* (cd) is the accepted unit for luminous intensity; it is already in common use and replaces candle and candlepower.

Luminance will be expressed in *candela per square metre* (cd/m^2), which replaces candela per square foot, foot lambert and lambert.

The SI unit, the *lumen* (lm) is also in common use already. Illuminance is expressed in *lux* (lx) which is equal to the lumen per square metre (lm/m^2) and replaces lumen per square foot and footcandle.

Dimensionless Quantities

Ratios such as relative humidity, (RH) specific gravity (SG), decibel (dB), remain unchanged when converting to SI.

CONVERSION OF NUMERICAL VALUES

Metrication is the process of change to a new set of reference units. It involves a change in numerical values by the ratio between customary and metric SI reference units; this ratio is known as the *conversion factor.*

Types of conversion are (1) *exact conversion,* (2) *soft conversion,* (3) *hard conversion* and (4) *rationalization.*

FIG. 21 **FRACTION, DECIMAL AND METRIC EQUIVALENTS***

Fractional Inch	Dec. Equiv.	Millimetres	Fractional Inch	Dec. Equiv.	Millimetres
1/16	0.0625	1.588	9/16	0.5625	14.288
1/12	0.0833	2.117	7/12	0.5833	14.817
1/10	0.1000	2.540	3/5	0.6000	15.240
1/8	0.1250	3.175	5/8	0.6250	15.875
1/6	0.1667	4.233	2/3	0.6667	16.933
3/16	0.1875	4.700	11/16	0.6875	17.463
1/5	0.2000	5.080	7/10	0.7000	17.780
1/4	0.2500	6.350	3/4	0.7500	19.050
3/10	0.3000	7.620	4/5	0.8000	20.320
5/16	0.3125	7.937	13/16	0.8125	20.638
1/3	0.3333	8.467	5/6	0.8333	21.167
3/8	0.3750	9.525	7/8	0.8750	22.225
2/5	0.4000	10.160	9/10	0.9000	22.860
5/12	0.4167	10.583	11/12	0.9167	23.283
7/16	0.4375	11.112	15/16	0.9375	23.813
1/2	0.5000	12.700	1"	1.0000	25.400

*To convert a decimal to percentage, carry the decimal point two places to the right. Thus, 63/64 or .9844, equals 98.44%.

Exact Conversion

An exact or a direct conversion is a precise calculation from a customary value to its exact SI equivalent, generally expressed to a number of decimal places.

Exact conversion may be needed in the early transitional period to SI units but only as the first step to the other types of conversion. Exact conversions will rarely, if ever, be needed in design and construction, but form an essential step before rounding or rationalization of traditional values can begin. Figure 21 shows exact conversion of fractions and decimals to three places. Full conversion charts appear in 603 Metrication Work File.

Rules For Exact Conversion Conversion of quantities should be handled carefully so that accuracy is neither sacrificed nor exaggerated. Conversion factors and quantities should not be rounded prior to multiplication or division.

Calculated values should be rounded to the minimum number of digits that will maintain the appropriate accuracy. For example, if a metric tape measure shows values to the nearest millimetre showing lineal values more precisely would be unnecessary if the final dimensions are to be applied using a tape measure.

For calculations using numbers of differing precision, the final answer should be *no more precise* than the least precise number (Fig. 22).

Soft Conversion

Soft conversion represents only a change in the way a product's size is described, but no significant change in the product's actual size. For example, a piece of gypsum lath which is 24" wide could be described as 609.6 mm if exact conversion is desired; if soft conversion is acceptable, the number could be rounded off to 610 mm. Such practice would give the appearance of metric conversion but would not result in the most ideal product width—600 mm—attained through hard conversion (Fig. 23).

As mentioned, soft conversion, is only a metric veneer; it implies a change made only on paper. Soft conversions should be limited to specifications, standards, codes and other technical data. Soft conversions should not be used in product sizing because it perpetuates customary practices and diminishes the advantages of SI.

FIG. 22 **CALCULATING EXACT CONVERSIONS**

addition or subtraction:	
original numbers	153 787 400
	647 200 000
	21 000 000
	821 987 400
rounded to the least precise and added:	
	154 000 000
	647 000 000
	21 000 000
	822 000 000
multiplication or division:	
175.5 × 3.475 = 609.8625	
Round the answer to 609.9 because the least precise number 175.5 has the decimal carried to only one place.	

FIG. 23 **EXACT, SOFT AND HARD CONVERSION OF CUSTOMARY VALUES**

Customary Value	Exact Conversion	Soft Conversion	Hard Conversion	% Change For Hard Conversion
24 inches [2 feet]	609.6 mm	610 mm	600 mm	−1.6
1 square [100 sq. ft]	9.2903 m²	9.3 m²	10 m²	+7.6
4 cu. yd.	3.058 m³	3.1 m³	3 m³	−1.9
1 quart	0.946 L	0.95 L	1 L	+5.7

Hard Conversion

Hard conversion represents a definite change from existing physical size or magnitude to a different, new *preferred* SI value. Such conversion generally leads to incompatibility between customary and metric properties or product characteristics. For example, a one quart container may be replaced by a one litre container, a 5.7% increase; or a 12 ft. length by a preferred length of 3600 mm, a 1.6% decrease.

Though more difficult to effect, hard conversion is a simpler and better form of change to SI. Figure 23 shows differences among customary, exact, soft and hard conversions.

However, hard conversion would be unrealistic and unacceptable in some cases, such as changing the standard railroad gage. The U.S. railroad gage currently has a nonpreferred dimension of 4′8-1/2″; the soft conversion is 1435 mm. Since railroad tracks must conform to the large number of existing railroad cars, the soft conversion to 1435 mm is perfectly acceptable and would cost little compared with a positive change to 1450 mm or 1500 mm.

In general, hard conversion should be used whenever possible, because it results in preferred values which are easier to work with. However, the decision to make a hard conversion must always be weighed against the cost of changing the many products or components which are intended to work together.

Rationalization

Rationalization is the development of standards for greater product compatibility and cost-effectiveness. These changes in product sizing are often the result of SI conversion. For instance, due to metrication, the number of reinforcing bar size has been reduced from 11 to 10 in Australia from 11 to 9 in Britain, and from 11 to 8 in Canada.

The objectives of rationalization are (1) to reduce the variety of products by eliminating unnecessary or uneconomical sizes; (2) to gain optimum steps in a new product range; (3) to harmonize various technical data, standards and codes by using preferred values; and (4) to simplify practices and procedures through less complicated numerical description.

PREFERRED DIMENSIONS AND COORDINATION

During the change to SI, it is important to select preferred dimensions for building products which will insure that they will work together when combined into walls, floors or other building assemblies. The use of preferred values which are multiples of a certain dimension in design or manufacture is termed dimensional (modular) coordination.

For example, 4″ is the most common building module in the customary system. The width of a wood stud is nominally 4″; the width of a brick and the height of a block are 8″; the spacing of studs and the length of a block are 16″; the width of gypsum lath and insulating sheathing is 24″; and so on.

In dimensional coordination, the dimensions of all building components are multiples of the basic building module.

The internationally recommended basic building module is 100 mm. It is small enough to provide flexibility in design, large enough to promote simplification of the number of component sizes and acceptable worldwide.

Advantages of Preferred Dimensions

The selection of preferred building dimensions, such as floor-to-floor height, will provide an incentive to produce standardized height components for use in high-rise buildings, such as precast stairs, ducts, shafts and cladding panels.

Preferred building dimensions will also (1) simplify distributors' inventory by reducing product variety, (2) make it easier to combine interrelated components into a building and (3) reduce the time and waste normally associated with assembling uncoordinated products.

In addition to reducing construction cost and time, the application of uniform preferred dimensions will simplify communications and give additional impetus to the use of computers for design and detailing.

The 100 mm Module

The 100 mm module has many compelling advantages:

(1) It is 1.6% smaller than the customary 4″ module so existing equipment can usually be used to make the new metric products.

(2) The use of the millimetre as the working unit in building design and construction will make all multimodular dimensions visible immediately. For example, 6400 mm is exactly 64 basic 100 mm modules; it is also 16 modules of 400 mm each, comparable to sixteen 16″ stud spaces. In customary units, the use of feet and inches hides the multiplier in modular dimensions. For example, it is not obvious that 21′4″ represents 64 whole 4″ modules, or that 26′8″ is a whole multiple of 16″ stud spacing.

(3) The modular square, 100 mm × 100 mm, has an area of 0.01 m^2. This relationship is useful in cost comparisons because the unit cost per modular square, in cents, has the same number as the unit cost per square metre in dollars. For example, carpet costing 15¢ a modular square would cost $15.00 a square metre.

(4) The modular cube, 100 mm × 100 mm × 100 mm, represents a volume of exactly 1 litre, equals also to 0.001 m^3. One litre of water, with a specific gravity of 1.0, has a mass of exactly 1 kilogram and this relationship is useful in many calculations. For example, a modular cube of concrete, with a specific gravity of 2.4, has a mass of 2.4 kilograms. The calculation of the mass of a 100 mm wide concrete wall, 2800 mm high and 7200 mm long is: $28 \times 72 \times 2.4 = 4\,834.4$ kg. The mass per unit length is 28×10 (1 metre) $\times 2.4 = 672$ kg/m.

(5) One dimension, the millimetre, expresses all building dimensions, including tolerances and limitations, which will enable much greater accuracy in construction.

INDUSTRY STANDARDS

Figure 24 shows some of the standards and recommendations for implementing SI published by the National Bureau of Standards (NBS), the American National Standard Institute (ANSI) and the American Society for Testing and Materials (ASTM).

FIG. 24 **INDUSTRY STANDARDS FOR SI PRACTICE**

Number	Title
NBS Special Publication 504	Metric Dimensional Coordination—The Issues and Precedent
NBS Special Publication 530	Metrication in Building Design, Production and Construction
NBS Technical Note 990	The Selection of Preferred Metric Values for Design and Construction
NBS Technical Note 938	Recommended Practice for the Use of Metric (SI) Units in Building Design and Construction
ASTM E 380-79	Standard for Metric Practice
ANSI/ASTM E 621-78*	Standard Practice for the Use of Metric (SI) Units in Building Design and Construction
ASTM E 713-80	A Standard for Scales for Metric Building Drawings

*Based on NBS Technical Note 938.

We gratefully acknowledge the U.S. Department of Commerce, National Bureau of Standards for the use of their publications as references and for permission to use photographs, charts and illustrations.

METRICATION 603

CONTENTS

METRICATION

Metrication will not only force a complete revision in the use of measurement and calculation, but will also offer a logical opportunity to revise standard product sizes, performance criteria and methods of testing. (An explanation of terminology important to the use of metric appears in 603 Metrication Main Text.)

THE METRIC SYSTEM AND SI

The International System of Units (SI) is a universal, rational, coherent, international and preferred measurement system which is derived from earlier decimal metric systems but supersedes them.

Adoption of SI

In 1960, the General Conference of Weights and Measures (CGPM) adopted SI, which has been accepted by more than 90% of the world's nations.

The 1975 Metric Conversion Act established the 17 member Metric Board to plan, publicize and implement a voluntary change to SI in the U.S.

Structure of SI

SI has fewer units than the U.S. customary system. Many SI units are already in use. The target date for complete SI conversion in the building industry is January 1, 1985.

It is recommended that the SI units and symbols shown in Fig. WF1 be gradually adopted for the specified activities in building construction (Fig. WF1).

International System of Units (SI) *Regardless of surrounding language or script, SI symbols have an agreed form and the same meaning worldwide. There is only one recognized SI unit for each physical quantity. All units are derived from 7 base and 2 supplementary units. (See Metrication Main Text, Figure 8 for technical definitions.)*

Algebraic relationships link the base, supplementary and derived units. Figure WF4 shows the SI units and Figure WF5 shows non-SI units which may be used with SI. SI supersedes all previous metric systems.

Standard Prefixes allow a change in magnitude from the sub-atomic scale (10^{-18}) to the astronomic scale (10^{18}) (Fig. WF2).

The Decimal Multiples of SI Units that are recommended appear in Figure WF2. Use of those multiples of SI units that cannot be handled by using the SI prefixes is discouraged.

FIG. WF1 **SI UNITS FOR BUILDING CONSTRUCTION**

Activity	Quantity	Unit	Symbol
LAND SURVEYING	Linear measure	kilometre, metre	km, m
	Area	square kilometre	km^2
		hectare (10 000 m^2)	ha
		square metre	m^2
EXCAVATING	Linear measure	metre, millimetre	m, mm
	Volume	cubic metre	m^3
CONCRETING	Linear measure	metre, millimetre	m, mm
	Area	square metre	m^2
	Volume	cubic metre	m^3
	Temperature	degree Celsius	°C
	Water-capacity	litre	L
Constituents	Mass (weight)	kilogram, gram	kg, g
Reinforcement	Cross-section	square millimetre	mm^2
TRUCKING	Distance	kilometre	km
	Mass (weight)	metric ton (1000 kg)	t
PAVING and PLASTERING	Linear measure	metre, millimetre	m, mm
	Area	square metre	m^2
BRICKLAYING	Linear measure	metre, millimetre	m, mm
	Area	square metre	m^2
	Mortar-volume	cubic metre	m^3
CARPENTRY/JOINERY	Linear measure	metre, millimetre	m, mm
STEELWORKING	Linear measure	metre, millimetre	m, mm
	Mass (weight)	metric ton (1000 kg)	t
		kilogram, gram	kg, g
ROOFING	Linear measure	metre, millimetre	m, mm
	Area	square metre	m^2
	Slope	millimetre/metre	mm/m
PAINTING	Linear measure	metre, millimetre	m, mm
	Area	square metre	m^2
	Capacity	litre, millilitre	L, mL
GLAZING	Linear measure	metre, millimetre	m, mm
	Area	square metre	m^2
PLUMBING	Linear measure	metre, millimetre	m, mm
	Mass (weight)	kilogram, gram	kg, g
	Capacity	litre	L
	Pressure	kilopascal	kPa
DRAINAGE	Linear measure	metre, millimetre	m, mm
	Area	hectare (10 000 m^2)	ha
		square metre	m^2
	Volume	cubic metre	m^3
	Slope	millimetre/metre	mm/m
ELECTRICAL SERVICES	Linear measure	metre, millimetre	m, mm
	Frequency	hertz	Hz
	Power	watt, kilowatt	W, kW
	Energy	megajoule (1 kWh = 3.6 MJ)	MJ
	Electric current	ampere	A
	Electric potential	volt, kilovolt	V, kV
	Resistance	ohm	Ω
MECHANICAL SERVICES	Linear measure	metre, millimetre	m, mm
	Volume	cubic metre	m^3
	Capacity	litre	L
	Airflow	metre/second	m/s
	Volume flow	cubic metre/second	m^3/s
		litre/second	L/s
	Temperature	degree Celsius	°C
	Force	newton, kilonewton	N, kN
	Pressure	kilopascal	kPa
	Energy, Work	kilojoule, megajoule	kJ, MJ

FIG. WF2 **PREFIX NAMES OF SOME POWERS OF TEN**

Multiplication Factor	Prefix	Symbol
1.000 000 000 000 000 000 = 10^{18}	exa	E
1 000 000 000 000 000 = 10^{15}	peta	P
1 000 000 000 000 = 10^{12}	tera	T
1 000 000 000 = 10^{9}	giga	G
1 000 000 = 10^{6}	mega	M
1 000 = 10^{3}	kilo	k
100 = 10^{2}	hecto	h
10 = 10^{1}	deka	da
0.1 = 10^{-1}	deci	d
0.01 = 10^{-2}	centi	c
0.001 = 10^{-3}	milli	m
0.000 001 = 10^{-6}	micro	μ
0.000 000 001 = 10^{-9}	nano	n
0.000 000 000 001 = 10^{-12}	pico	p
0.000 000 000 000 001 = 10^{-15}	femto	f
0.000 000 000 000 000 001 = 10^{-18}	atto	a

Special Names Special names for multiple and submultiples of SI units should be avoided except for *litre, metric ton* and *hectare* (Fig. WF3).

FIG. WF3 **SPECIAL NAMES TO BE AVOIDED**

Unit	Designation & Equivalent
fermi	1 fm = 10^{-15} m
micron	1 μ m = 10^{-6} m
millimicron	1 nm = 10^{-9} m
arc	1 dam^2 = 100 m^2
gamma*	1 nT
γ (mass)	1 μ g
λ (volume)	1μ l = 1 mm^3
mho	1 S
candle	1 cd
candlepower	1 cd

*Magnetic flux density.

Recommendations for the Use of SI

When SI units are used, they should be written following the style conventions shown in Figs. WF6 and WF7.

Writing Unit Symbols Unit symbols (1) should be printed or written in *upright type* regardless of type style in surrounding text; (2) should be unaltered in the plural; (3) should not be followed by a period except when used at the end of a sentence; (4) should not be capitalized, unless unit is derived from a proper name; (5) should have a space between the unit symbol and the numerical value.

The comma should not be used to separate digits into groups of three. Instead, digits should be divided into

FIG. WF4 **UNITS IN THE INTERNATIONAL SYSTEM—SI**

UNIT GROUP QUANTITY	UNIT NAME	SYMBOL	FORMULA	UNIT DERIVATION	REMARKS
Base Units:					
Length	metre	m			An alternative spelling is meter
Mass	kilogram	kg			
Time	second	s			Already in common use
Electric current	ampere	A			Already in common use
Thermodynamic temperature	kelvin	K			The customary unit for temperature is the degree Celsius (°C).
Amount of substance	mole	mol			The "mol" has no application in construction.
Luminous intensity	candela	cd			Already in common use
Supplementary Units:					
Plane angle	radian	rad			Already in common use
Solid angle	steradian	sr			Already in common use
Derived Units with Special Names:					
Frequency (of a periodic phenomenon)	hertz	Hz	1/s	s^{-1}	The hertz replaces "cycle per second."
Force	newton	N	$kg \cdot m/s^2$	$m \cdot kg \cdot s^{-2}$	
Pressure, stress, elastic modulus	pascal	Pa	N/m^2	$m^{-1} \cdot kg \cdot s^{-2}$	
Energy, work, quantity of heat	joule	J	$N \cdot m$	$m^2 \cdot kg \cdot s^{-2}$	
Power, radiant flux	watt	W	J/s	$m^2 \cdot kg \cdot s^{-3}$	Already in common use
Quantity of electricity, electric charge	coulomb	C	$A \cdot s$	$s \cdot A$	Already in common use
Electric potential, potential difference, electromotive force	volt	V	J/C or W/A	$m^2 \cdot kg \cdot s^{-3} \cdot A^{-1}$	Already in common use
Electric capacitance	farad	F	C/V	$m^{-2} \cdot kg^{-1} \cdot s^4 \cdot A^2$	Already in common use
Electric resistance	ohm	Ω	V/A	$m^2 \cdot kg \cdot s^{-3} \cdot A^{-2}$	Already in common use
Electric conductance	siemens	S	A/V or 1/Ω	$m^{-2} \cdot kg^{-1} \cdot s^3 \cdot A^2$	The "siemens" was formerly referred to as "mho."
Magnetic flux	weber	Wb	$V \cdot s$	$m^2 \cdot kg \cdot s^{-2} \cdot A^{-1}$	Already in common use
Magnetic flux density	tesla	T	Wb/m^2	$kg \cdot s^{-2} \cdot A^{-1}$	Already in common use
Electric inductance	henry	H	Wb/A	$m^2 \cdot kg \cdot s^{-2} \cdot A^{-2}$	Already in common use
Luminous flux	lumen	lm	$cd \cdot sr$	$cd \cdot sr$	Already in common use
Illuminance	lux	lx	lm/m^2	$m^{-2} \cdot cd \cdot sr$	
Absorbed dose	gray	Gy	J/kg	$m^2 \cdot s^{-2}$(*)	(*) kg is canceled out. No application in construction.
Derived Units with Generic Names:					
a. Units Expressed in Terms of One Base Unit:					
Area	square metre	m^2		m^2	
Volume, capacity	cubic metre	m^3		m^3	(1 m^3 = 1000 L)
Section modulus	metre to third power		m^3	m^3	
Second moment of area	metre to fourth power		m^4	m^4	
Curvature	reciprocal (of) metre		1/m	m^{-1}	
Rotational frequency	reciprocal (of) second		1/s	s^{-1}	Revolution per second (r/s) is used in specifications for rotating machinery.
Coefficient of linear thermal expansion	reciprocal (of) kelvin		1/K	K^{-1}	

FIG. WF4 **UNITS IN THE INTERNATIONAL SYSTEM—SI *(Continued)***

UNIT GROUP QUANTITY	UNIT NAME	SYMBOL	FORMULA	UNIT DERIVATION	REMARKS
b. Units Expressed in Terms of Two or More Base Units:					
Linear velocity	metre per second		m/s	$m \cdot s^{-1}$	
Linear acceleration	metre per second squared		m/s^2	$m \cdot s^{-2}$	
Kinematic viscosity	square metre per second		m^2/s	$m^2 \cdot s^{-1}$	
Volume rate of flow	cubic metre per second		m^3/s	$m^3 \cdot s^{-1}$	
Specific volume	cubic metre per kilogram		m^3/kg	$m^3 \cdot kg^{-1}$	
Mass per unit length	kilogram per metre		kg/m	$m^{-1} \cdot kg$	
Mass per unit area	kilogram per square metre		kg/m^2	$m^{-2} \cdot kg$	
Density (mass per unit volume)	kilogram per cubic metre		kg/m^3	$m^{-3} \cdot kg$	In this SI form, mass density is conveniently 1000 times specific gravity.
Moment of inertia	kilogram metre squared		$kg \cdot m^2$	$m^2 \cdot kg$	
Mass flow rate	kilogram per second		kg/s	$kg \cdot s^{-1}$	
Momentum	kilogram metre per second		$kg \cdot m/s$	$m \cdot kg \cdot s^{-1}$	
Angular momentum	kilogram metre squared per second		$kg \cdot m^2/s$	$m^2 \cdot kg \cdot s^{-1}$	
Magnetic field strength	ampere per metre		A/m	$m^{-1} \cdot A$	
Current density	ampere per square metre		A/m^2	$m^{-2} \cdot A$	
Luminance	candela per square metre		cd/m^2	$m^{-2} \cdot cd$	
c. Units Expressed in Terms of Base Units and/or Derived Units with Special Names:					
Moment of force, torque	newton metre		$N \cdot m$	$m^2 \cdot kg \cdot s^{-2}$	
Flexural rigidity	newton square metre		$N \cdot m^2$	$m^3 \cdot kg \cdot s^{-2}$	
Force per unit length, surface tension	newton per metre		N/m	$kg \cdot s^{-2}$ (1)	(1) m is canceled out
Dynamic viscosity	pascal second		$Pa \cdot s$	$m^{-1} \cdot kg \cdot s^{-1}$	
Impact ductility	joule per square metre		J/m^2	$kg \cdot s^{-2}$ (2)	(2) m^2 is canceled out
Combustion heat (per unit volume)	joule per cubic metre		J/m^3	$m^{-1} \cdot kg \cdot s^{-2}$	
Combustion heat (per unit mass), specific energy, specific latent heat	joule per kilogram		J/kg	$m^2 \cdot s^{-2}$(3)	(3) kg is canceled out
Heat capacity, entropy	joule per kelvin		J/K	$m^2 \cdot kg \cdot s^{-2} \cdot K^{-1}$	
Specific heat capacity, specific entropy	joule per kilogram kelvin		$J/(kg \cdot K)$	$m^2 \cdot s^{-2} \cdot K^{-1}$ (4)	(4) kg is canceled out
Heat flux density, irradiance, sound intensity	watt per square metre		W/m^2	$kg \cdot s^{-3}$ (5)	(5) m^2 is canceled out
Thermal conductivity	watt per metre kelvin		$W/(m \cdot K)$	$m \cdot kg \cdot s^{-3} \cdot K^{-1}$	
Coefficient of heat transfer	watt per square metre kelvin		$W/(m^2 \cdot K)$	$kg \cdot s^{-3} \cdot K^{-1}$ (6)	(6) m^2 is canceled out
Thermal resistance, thermal insulance	square metre kelvin per watt		$m^2 \cdot K/W$	$kg^{-1} \cdot s^3 \cdot K$ (7)	(7) m^2 is canceled out
Electric field strength	volt per metre		V/m	$m \cdot kg \cdot s^{-3} \cdot A^{-1}$	
Electric flux density	coulomb per square metre		C/m^2	$m^{-2} \cdot s \cdot A$	
Electric charge density	coulomb per cubic metre		C/m^3	$m^{-3} \cdot s \cdot A$	
Electric permittivity	farad per metre		F/m	$m^{-3} \cdot kg^{-1} \cdot s^4 \cdot A^2$	
Electric permeability	henry per metre		H/m	$m \cdot kg \cdot s^{-2} \cdot A^{-2}$	
Electric resistivity	ohm metre		$\Omega \cdot m$	$m^3 \cdot kg \cdot s^{-3}; \cdot A^{-2}$	
Electric conductivity	siemens per metre		S/m	$m^{-3} \cdot kg^{-1} \cdot s^3 \cdot A^2$	
Light exposure	lux second		$lx \cdot s$	$m^{-2} \cdot s \cdot cd \cdot sr$	
Luminous efficacy	lumen per watt		lm/W	$m^{-2} \cdot kg^{-1} \cdot s^3 \cdot cd \cdot sr$	
d. Units Expressed in Terms of Supplementary Units and Base and/or Derived Units:					
Angular velocity	radian per second		rad/s	$s^{-1} \cdot rad$	
Angular acceleration	radian per second squared		rad/s^2	$m^{-2} \cdot rad$	
Radiant intensity	watt per steradian		W/sr	$m^2 \cdot kg \cdot s^{-3} \cdot sr^{-1}$	
Radiance	watt per square metre steradian		$W/(m^2\ sr)$	$kg \cdot s^{-3} \cdot sr^{-1}$ (8)	(8) m^2 is canceled out

FIG. WF5 **OTHER UNITS WHOSE USE IS PERMITTED WITH SI**

QUANTITY	UNIT NAME	SYMBOL	RELATIONSHIP TO SI UNIT	REMARKS
Units for General Use:				
Volume	litre[1]	L[1]	1 L = 0.001 m^3 = 10^6 mm^3	An alternative spelling is liter. The litre may only be used with the SI prefix "milli."
Mass	metric ton[2]	t[2]	1 t = 1000 kg = (1 Mg)	
Time	minute	min	1 min = 60 s	
	hour	h	1 h = 3600 s = (60 min)	
	day (mean solar)	d	1 d = 86 400 s = (24 h)	
	year (calendar)	a	1a = 31 536 000 s = (365 d)	
Temperature interval	degree Celsius	°C	1°C = 1 K	The Celius temperature 0° C corresponds to 273.15 K exactly. ($t_{°C}$ = T_K—273.15)
Plane angle	degree (of arc)	°	1° = 0.017 453 rad = 17.453 mrad	1° = (π/180) rad
Velocity	kilometre per hour	km/h	1 km/h = 0.278 m/s	1 m/s = 3.6 km/h
Units Accepted for Limited Use:				
Area	hectare	ha	1 ha = 10 000 m^2	For use in land measurement.
Energy	kilowatthour	kWh	1 kWh = 3.6 MJ	For measurement of electrical energy consumption only.
Speed of rotation	revolution per minute	r/min	1 r/min = $\frac{1}{60}$ r/s = $\frac{2\pi}{60}$ rad/s	To measure rotational speed in slow-moving equipment only.

[1]The international symbol for "litre" is the lowercase "l", which can be easily confused with the numeral "1". Several English-speaking countries have adopted the script "ℓ" as symbol for "litre" in order to avoid any misinterpretation. The symbol "L" (capital ell) is recommended for United States use to prevent confusion.
[2]The international name for "metric ton" is "tonne." The metric ton is equal to the "megagram" (Mg).

groups of three by using a small space the width of the letter "i" to separate these groups. In four digit numbers a space is not necessary. For example:
2.141 596   73 722   7372   0.133 47

Terms such as billion and trillion should not be used in technical writing. These terms have different meanings in different countries.

SELECTION OF SI UNITS FOR DESIGN AND CONSTRUCTION

Correct selection of units for use in building design and construction is essential to minimize errors and optimize coordination.

Figures WF8 through WF13 list recommended units for building design and construction.

CONVERSION OF NUMERICAL VALUES

Metrication is the process of change to a new set of reference units. It involves a change in numerical values by the ratio between customary and metric SI reference units known as the conversion factor.

Types of conversion are: (1) exact conversion, (2) soft conversion, (3) hard conversion and (4) rationalization.

Exact Conversion

Exact conversion is a precise calculation from a customary value to its exact SI equivalent, generally expressed to a number of decimal places.

Exact conversion is needed in the early transitional period to SI units before other types of conversion become common. Figures WF14 and WF16 are exact conversion tables.

Rules for Exact Conversion Conversion factors and quantities should not be rounded prior to multiplication or division. Values resulting from calculations should be rounded to the minimum number of digits that will maintain the appropriate accuracy.

For example, a metric tape measure shows values to the nearest millimetre; giving results of calculated values more precisely than to the nearest millimetre is unnecessary.
(Text continues on page WF603-21.)

FIG. WF6 RULES AND RECOMMENDATIONS FOR THE PRESENTATION OF SI UNITS AND SYMBOLS

A. GENERAL	TYPICAL EXAMPLES	REMARKS
1. All unit names should either be denoted by correct symbols or be written in full. In the interest of simplification and to reduce the amount of writing use unit symbols rather than fully written forms.		
2. DO NOT USE mixtures of names and symbols.	USE: J/kg or joule per kilogram	NOT: joule per kg NOT: J/kilogram
B. SYMBOLS FOR UNIT QUANTITIES AND PREFIXES		
1. SI symbols are internationally agreed and there is only one symbol for each unit quantity. Multiples and submultiples are formed by using the unit symbol and attaching a prefix symbol in front of it.	m, kg, s, A, cd, K	See also B5 - B7
2. All unit symbols are shown in upright letters, and can be produced by a normal typewriter keyboard - with the exceptions of the symbols for the SI unit ohm and the prefix micro which are represented by Greek letters Ω and μ respectively.		EXCEPTIONS: Ω, μ
3. Unit symbols are NEVER followed by a period (full stop) except at the end of a sentence.	60 kg/m	NOT: 60 kg./m.
4. Unit symbols are normally written in lowercase, except for unit names derived from a proper name, in which case the initial is capitalized. Some units have symbols consisting of two letters from a proper name, of which only the first letter is capitalized. (The symbol for the unit name "ohm" is the capital Greek letter Ω).	m, kg, s, mol, cd etc. A, K, N, J, W, V etc. Pa, Hz, Wb, etc.	EXCEPTION: L
5. Prefixes for magnitudes from 10^6 to 10^{18} have capital upright letter symbols.	M, G, T etc.	See also C1
6. Prefixes for magnitudes from 10^{-18} through to 10^3 have lowercase upright letter symbols. (The symbol for 10^{-6} or micro is the lowercase Greek letter μ).	p, n, μ, m, k etc.	See also C1
7. Prefix symbols are directly attached to the unit symbol, without a space between them.	mm, kW, MN, etc.	NOT: m m, k W, M N
8. DO NOT USE compound prefixes to form a multiple or submultiple of a unit (e.g. USE nanometer, DO NOT USE micromillimeter or millimicrometer).	nm	NOT: μmm or mμm
9. In the case of the base unit kilogram, prefixes are attached to the 'gram' (e.g. milligram NOT microkilogram).	mg	NOT: μkg
10. USE ONLY ONE PREFIX when forming a multiple or a submultiple of a compound unit. Normally, the prefix should be attached to a unit in the numerator. An exception from this rule is made for the base unit kilogram.	km/s; mV/m	NOT: mm/μs; μV/mm EXCEPTION: MJ/kg NOT kJ/g
C. AREAS OF POSSIBLE CONFUSION REQUIRING SPECIAL CARE		
1. The symbols for SI units and the conventions that govern their use should be STRICTLY followed. A number of prefix and unit symbols use the same letter, but in different form. EXERCISE CARE to present the correct symbol for each quantity.	g (gram); G (giga) k (kilo); K (kelvin) m (milli); M (mega) m (meter); n (nano); N (newton)	OTHERS: c (centi); C (coulomb) °C (degree Celsius) s (second); S (siemens) t (metric ton); T (tesla) T (tera)
2. All prefix and unit symbols retain their prescribed form regardless of the surrounding typography. In printouts from limited character sets (telex, computer printers) special considerations apply to symbols for mega, micro, ohm and siemens. Where confusion is likely to arise, WRITE UNITS IN FULL.		

FIG. WF6 **RULES AND RECOMMENDATIONS FOR THE PRESENTATION OF SI UNITS AND SYMBOLS *(Continued)***

D. UNIT NAMES WRITTEN OUT IN FULL	TYPICAL EXAMPLES	REMARKS
1. Unit names, including prefixes are treated as common names and are not capitalized, except at the beginning of sentences or in titles. (The only exception is "Celsius" in "degree Celsius," where degree is considered as the unit name and is shown in lowercase, while Celsius represents an adjective and is capitalized).	meter, newton, etc.	NOT: Meter, Newton EXCEPTION: degree Celsius
2. Where a prefix is attached to an SI unit to form a multiple or submultiple, the combination is written as one word. (There are three cases where the final vowel of the prefix is omitted in the combination: megohm, kilohm, and hectare.)	millimeter; kilowatt	NOT: milli-meter NOT: kilo watt
3. Where a compound unit is formed by multiplication of two units, the use of a space between units is preferred, but a hyphen is acceptable and in some situations more appropriate, to avoid any risk of misinterpretation.	newton meter or newton-meter	NOT: newtonmeter
4. Where a compound unit is formed by division of two units, this is expressed by inserting per between the numerator and the denominator.	meter per second joule per kelvin	NOT: meter/second NOT: joule/kelvin
5. Where the numerical value of a unit is written in full, the unit should also be written in full.	seven meters	NOT: seven m
E. PLURALS		
1. Units written in full are subject to the normal rules of grammar. For any unit with a numerical value greater than one (1), an "s" is added to the written unit to denote the plural.	1.2 meters; 2.3 newtons; 33.2 kilograms	BUT: 0.8 meter
2. The following units have the same plural as singular when written out in full: hertz, lux, siemens.	350 kilohertz 12.5 lux	
3. Symbols NEVER change in the plural.	2.3 N; 33.2 kg	NOT: 2.3 Ns; 33.2 kgs
F. COMPOUND UNIT SYMBOLS - PRODUCTS AND QUOTIENTS		
1. The product of two units is indicated by a dot placed at mid-height between the unit symbols.	kN·m; Pa·s	NOT: kNm; Pas NOT: kN m; Pa s
2. To express a derived unit formed by division, any one of the following methods may be used:		See also F3 and F5
a. a solidus (slash, /)	kg/m^3; W/(m·K)	See also F3 and F5
b. a horizontal line between numerator and denominator	$\frac{kg}{m^3}$; $\frac{W}{m \cdot K}$	
c. a negative index (or negative power)	$kg \cdot m^{-3}$; $W \cdot m^{-1} \cdot K^{-1}$	
3. Only one solidus may be used in any combination.	m/s^2; $m \cdot kg/(s^3 \cdot A)$	NOT: m/s/s NOT: $m \cdot kg/s^3/A$
4. DO NOT USE the abbreviation "p" for per in the expression of a division.	km/h	NOT: kph or k.p.h.
5. Where the denominator is a product, this should be shown in parentheses.	$W/(m^2 \cdot K)$	

FIG. WF7 PRESENTATION OF NUMERICAL VALUES IN CONJUNCTION WITH SI*

A. DECIMAL MARKER	TYPICAL EXAMPLES	REMARKS
1. Whereas most European countries use the comma on the line as the decimal marker and this practice is advocated by ISO, a special exception is made for documents in the English language which have traditionally used the point (dot) on the line, or period, as decimal marker.		See also under G.
2. The recommended decimal marker for use in the United States is the point on the line (period), and the comma should not be used.	9.9 ; 15.375	NOT: 9,9 ; 15,375
In handwritten documents the decimal marker may be shown slightly above the line for clear identification.	9·9 ; 15·375	
3. Always show a zero before the decimal point for all numbers smaller than 1.0 (one).	0.1 ; 0.725	NOT: .1 ; .725
B. SPACING		
1. Always leave a gap between the numerical value associated with a symbol and the symbol, of at least half a space in width.	900 MHz ; 200 mg ; 10^6 mm^2 or 10^6 mm^2	NOT: 900MHz ; 200mg NOT: 10^6mm^2
In the case of the symbol for the "degree Celsius" this space is optional, but the degree symbol must always be attached to Celsius.	20°C or 20 °C	NOT: 20° C
2. In non-SI expressions of plane angle (°, ', "), DO NOT LEAVE A SPACE between the numerical value and the symbol.	27°30' (of arc)	NOT: 27 ° 30 '
3. Always leave a space on each side of signs for multiplication, division, addition, and subtraction	100 mm x 100 mm ; 36 MPa + 8 MPa	NOT: 100 mmx100 mm NOT: 36 MPa+ 8 MPa
C. FRACTIONS		
1. Avoid common fractions in connection with SI units.	WRITE: 0.5 kPa	NOT: 1/2 kPa
2. Always use decimal notation to express fractions of any number larger than 1.0 (one).	1.5 ; 16.375	NOT: 1-1/2 ; 16-3/8
3. While the most common fractions such as half, third, quarter, and fifth will remain in speech, always show decimal notation in written, typed, or printed material.	0.5 ; 0.33 ; 0.25 ; 0.2	NOT: 1/2 ; 1/3 ; 1/4 ; 1/5
D. POWERS OF UNITS AND EXPONENTIAL NOTATION		
1. When writing unit names with a modifier 'squared' or 'cubed', the following rules should be applied:		
a. In the case of area and volume, the modifier is written before the unit name as "square" and "cubic".	cubic meter ; square millimeter	NOT: meter cubed ; millimeter squared
b. In all other cases, the modifier is shown after the unit name as "squared", "cubed", "to the fourth power", etc.	meter per second squared	NOT: meter per square second; (or 'meter per second per second')
c. The abbreviations "sq." for square, and "cu." for cubic should not be used.		NOT: sq. millimeter NOT: cu. meter
2. For unit symbols with modifiers (such as square, cubic, fourth power etc.) always show the superscript immediately after the symbol.	m^2 ; mm^3 ; s^4	NOT: m ^{2}; mm ^{3}; s ^{4}

*Metre is the preferred international spelling applied to this base unit and all of its multiples and submultiples.

FIG. WF7 PRESENTATION OF NUMERICAL VALUES IN CONJUNCTION WITH SI* *(Continued)*

	TYPICAL EXAMPLES	REMARKS
3. Show the superscript as a reduced size numeral raised half a line space. Where a typewriter without superscript numerals is used, the full size numeral should be raised half a line space, provided that this does not encroach on print in the line above.	mm^3, m/s^2	PERMITTED: mm^3, m/s^2
4. Where an exponent is attached to a prefixed symbol, it indicates that that multiple (or submultiple) is raised to the power expressed by the exponent.	$1\ mm^3 = (10^{-3}\ m)^3 = 10^{-9}\ m^3$ $1\ km^2 = (10^3\ m)^2 = 10^6\ m^2$	NOT: $1\ mm^3 = 10^{-3}\ m^3$
E. RATIOS		
1. Do not mix units in expressing a ratio of like unit quantities.	0.01 m/m 0.03 m^2/m^2	NOT: 10 mm/m NOT: 30 000 mm^2/m^2
2. Wherever possible, use a nonquantitative expression (ratio or percentage) to indicate measurement of slopes, deflections, etc.		PREFERRED: 1:100; 0.01; 1% 1:33; 0.03; 3%
F. RANGE		
1. The choice of the appropriate prefix to indicate a multiple or submultiple of an SI unit is governed by convenience to obtain numerical values within a practical range and to eliminate non-significant digits.		
2. In preference, use prefixes representing ternary powers of 10 (10 raised to a power which is a multiple of 3).	milli, kilo, mega	AVOID: centi, deci, deka, hecto
3. Select prefixes so that the numerical value or values occur in a common range between 0.1 and 1000.	120 kN 3.94 mm 14.5 MPa	IN LIEU OF: 120 000 N 0.003 94 m 14 500 kPa
4. Compatibility with the general range must be a consideration; e. g., if all dimensions on a drawing are shown in millimeters (mm), a range from 1 to 99 999 (a maximum of five numerals) would be acceptable to avoid mixing of units.		NOTE: Drawings should show "All dimensions in millimeters".
G. PRESENTATION AND TABULATION OF NUMBERS		
1. In numbers with many digits it has been common practice in the United State to separate digits into groups of three by means of commas. This practice must not be used with SI, to avoid confusion. It is recommended international practice to arrange digits in long numbers in groups of three from the decimal marker, with a gap of not less than half a space, and not more than a full space, separating each group.	54 375.260 55 54 375.260 55	NOT: 54,375.260,55 NOT: 53475.26055
2. For individual numbers with four digits before (or after) the decimal marker this space is not necessary.	4500; 0.0355	
3. In all tabulations of numbers with five or more digits before and/or after the decimal marker, group digits into groups of three: e.g., 12.5255; 5735; 98 300 ; 0.425 75	12.525 5 5 735 98 300 0.425 75 104 047.951 25	
H. USE OF UNPREFIXED UNITS IN CALCULATIONS		
Errors in calculations involving compound units can be minimized if all prefixed units are reverted back to coherent base or derived units, with numerical values expressed in powers-of-ten notation.	PREFERRED: $136\ kJ = 136 \times 10^3\ J$ $20\ MPa = 20 \times 10^6\ Pa$ $1.5\ t\ (Mg) = 1.5 \times 10^3\ kg$	ALSO ACCEPTABLE: (or 1.36×10^5 J) (or 2×10^7 Pa)

*Metre is the preferred international spelling applied to this base unit and all of its multiples and submultiples.

FIG. WF8 SPACE AND TIME: GEOMETRY, KINEMATICS AND PERIODIC PHENOMENA*

QUANTITY AND SI UNIT SYMBOL	PREFERRED UNITS (SYMBOLS)	OTHER ACCEPTABLE UNITS	UNIT NAME	TYPICAL APPLICATIONS	REMARKS
LENGTH (m)	m		meter	*ARCHITECTURE AND GENERAL ENGINEERING* Levels, overall dimensions, spans, column heights, etc., in engineering computations. *ESTIMATING AND SPECIFICATION* Trenches, curbs, fences, lumber lengths, pipes and conduits; lengths of building materials generally. *LAND SURVEYING* Boundary and cadastral surveys; survey plans; heights, geodetic surveys, contours. *HYDRAULIC ENGINEERING* Pipe and channel lengths, depth of storage tanks or reservoirs, height of potentiometric head, hydraulic head, piezometric head.	USE meters on all drawings with scale ratios between 1:200 and 1:2000. Where required for purposes of accuracy, show dimensions to three decimal places. An alternative spelling is "metre".
	mm		millimeter	*ARCHITECTURE AND GENERAL ENGINEERING* Spans, dimensions in buildings, dimensions of building products; depth and width of sections; displacement, settlement, deflection, elongation; slump of concrete, size of aggregate; radius of gyration, eccentricity; detailed dimensions generally; rainfall. *ESTIMATING AND SPECIFICATION* Lumber cross sections; thicknesses, diameters, sheet metal gages, fasteners; all other building product dimensions. *HYDRAULIC ENGINEERING* Pipe diameters; radii of ground water wells; height of capillary rise; precipitation, evaporation.	USE millimeters on drawings with scale ratios between 1:1 and 1:200. AVOID the use of centimeters (cm). Where 'cm' is shown in documents, such as for snow depth, body dimensions, or carpet sizes, etc. CONVERT to 'mm' or 'm'.
	km		kilometer	Distances for transportation purposes geographical or statistical applications in surveying; long pipes and channels.	
	μm		micrometer	Thickness of coatings (paint, galvanizing etc.), thin sheet materials, size of fine aggregate.	
AREA (m^2)	m^2		square meter	*GENERAL APPLICATIONS* Small land areas; area of cross-section of earthworks, channels and larger pipes; surface area of tanks and small reservoirs; areas in general. *ESTIMATING AND SPECIFICATION* Site clearing; floor areas; paving, masonry construction, roofing, wall and floor finishes, plastering, paintwork, glass areas, membranes, lining materials, insulation, reinforcing mesh, formwork; areas of all building components.	($1\ m^2 = 10^6\ mm^2$) Replaces sq.ft.; sq.yd. and square. SPECIFY masonry construction by wall area x wall thickness.
	mm^2		square millimeter	Area of cross section for structural and other sections, bars, pipes, rolled and pressed shapes, etc.	AVOID the use of cm^2 (square centimeter) by conversion to mm^2. ($1\ cm^2 = 10^2\ mm^2 = 100\ mm^2$)
	km^2		square kilometer	Large catchment areas or land areas.	
		ha	hectare	Land areas; irrigation areas; areas on boundary and other survey plans.	($1\ ha = (10^2\ m)^2 = 10^4\ m^2 = 10\,000\ m^2$)

*Metre and litre are preferred international spellings applied to these base units and all of their multiples and submultiples, such as kilometre and millilitre.

FIG. WF8 **SPACE AND TIME: GEOMETRY, KINEMATICS AND PERIODIC PHENOMENA*** ***(Continued)***

QUANTITY AND SI UNIT SYMBOL	PREFERRED UNITS (SYMBOLS)	OTHER ACCEPTABLE UNITS	UNIT NAME	TYPICAL APPLICATIONS	REMARKS
VOLUME, CAPACITY (m^3)	m^3		cubic meter	*GENERAL APPLICATIONS* Volume, capacity (large quantities); volume of earthworks, excavations, filling, waste removal; concrete, sand, all bulk materials supplied by volume, and large quantities of lumber. *HYDRAULIC ENGINEERING* Water distribution, irrigation, diversions, sewage, storage capacity, underground basins.	$1\ m^3 = 1000\ L$ As far as possible, USE the cubic meter as the preferred unit of volume for all engineering purposes.
	mm^3		cubic millimeter	Volume, capacity (small quantities)	
		L	liter	Volume of fluids and containers for fluids; liquid materials, domestic water supply, consumption; volume/capacity of fuel tanks	The liter and its multiples or submultiples may be used for domestic and industrial supplies of liquids $1\ L = 1\ dm^3 = 1000\ cm^3$ $1\ mL = 1\ cm^3$ See also Section 9.3
		mL	milliliter	Volume of fluids and containers for fluids (limited application only)	
		cm^3	cubic centimeter	Limited application only (small quantities)	$1\ cm^3 = 1000\ mm^3 = 10^{-6}\ m^3$
MODULUS OF SECTION (m^3)	mm^3		millimeter to third power	Geometric properties of structural sections, such as plastic section modulus, elastic section modulus, etc.	See also Section 9.4
	m^3		meter to third power		
SECOND MOMENT OF AREA (m^4)	mm^4		millimeter to fourth power	Geometric properties of structural sections, such as moment of inertia of a section, torsional constant of a cross section.	See also Section 9.4
	m^4		meter to fourth power		
PLANE ANGLE (rad)	rad		radian	Generally used in calculations only to preserve coherence.	Slopes and gradients may be expressed as a ratio or as a percentage: 26.57° = 1 : 2 = 50 % = 0.4637 rad (1 rad = 57.2958°) See also Section 9.5
	mrad		milliradian		
		(__°)	degree (of arc)	*GENERAL APPLICATIONS* Angular measurement in construction (generally using decimalized degrees); angle of rotation, torsion, shear resistance, friction, internal friction, etc. *LAND SURVEYING* Bearings shown on boundary and cadastral survey plans; geodetic surveying	
TIME, TIME INTERVAL (s)	s		second	Time used in methods of test; all calculations involving derived units with a time component, in order to preserve coherence.	AVOID the use of minute (min) as far as possible
		h	hour	Time used in methods of test; all calculations involving labor time, plant hire, maintenance periods, etc.	(1 h = 3600 s) (1 d = 86 400 s = 86.4 ks)
		d	day		
		a	annum (year)		
FREQUENCY (Hz)	Hz		hertz	Frequency of sound, vibration, shock; frequency of electro-magnetic waves	($1\ Hz = 1/s = s^{-1}$) Replaces cycle(s) per second (c/s or cps)
	kHz		kilohertz		
	MHz		megahertz		

*Metre and litre are preferred international spellings applied to these base units and all of their multiples and submultiples, such as kilometre and millilitre.

FIG. WF8 **SPACE AND TIME: GEOMETRY, KINEMATICS AND PERIODIC PHENOMENA*** ***(Continued)***

QUANTITY AND SI UNIT SYMBOL	PREFERRED UNITS (SYMBOLS)	OTHER ACCEPTABLE UNITS	UNIT NAME	TYPICAL APPLICATIONS	REMARKS
ROTATIONAL FREQUENCY, SPEED OF ROTATION (s^{-1})		r/s	revolution per second	Widely used in the specification of rotational speed of machinery; Use r/min (revolutions per minute) only for slow moving machinery	(1 r/s = 2 π rad/s) = 60 r/min)
VELOCITY, SPEED (m/s)	m/s		meter per second	Calculations involving rectilinear motion, velocity and speed in general; wind velocity; velocity of fluids; pipe flow velocity	(1 m/s = 3.6 km/h)
		km/h	kilometer per hour	Wind speed; speed used in transportation; speed limits	
		mm/h	millimeter per hour	Rainfall intensity	
ANGULAR VELOCITY (rad/s)	rad/s		radian per second	Calculations involving rotational motion	
LINEAR ACCELERATION (m/s^2)	m/s^2		meter per second squared	Kinematics, and calculation of dynamic forces	Recommended value of acceleration of gravity for use in U.S: $g_{US} = 9.8\ m/s^2$ See also page 26
VOLUME RATE OF FLOW (m^3/s)	m^3/s		cubic meter per second	Volumetric flow in general; flow in pipes, ducts, channels, rivers; irrigation spray demand	(1 m^3/s = 1000 L/s) See also Section 9.6.3
		m^3/h	cubic meter per hour		
		m^3/d	cubic meter per day		
		L/s	liter per second	Volumetric flow of fluids only	
		L/d	liter per day		

*Metre and litre are preferred international spellings applied to these base units and all of their multiples and submultiples, such as kilometre and millilitre.

FIG. WF9 MECHANICS: STATICS AND DYNAMICS*

QUANTITY AND SI UNIT SYMBOL	PREFERRED UNITS (SYMBOLS)	OTHER ACCEPTABLE UNITS	UNIT NAME	TYPICAL APPLICATIONS	REMARKS
MASS (kg)	kg		kilogram	Mass of materials in general, mass of structural elements and machinery	USE kilograms (kg) in calculations and specifications
	g		gram	Mass of samples of material for testing	Masses greater than 10^4 kg (10 000 kg) may be conveniently expressed in metric tons (t): 1 t = 10^3 kg = 1 Mg = 1000 kg
		t	metric ton	Mass of large quantities of materials, such as structural steel, reinforcement, aggregates, concrete, etc.; ratings of lifting equipment.	
MASS PER UNIT LENGTH (kg/m)	kg/m		kilogram per meter	Mass per unit length of sections, bars, and similar items of uniform cross section	Also known as "Linear Density"
		g/m	gram per meter	Mass per unit length of wire and similar material of uniform cross section	
MASS PER UNIT AREA (kg/m^2)	kg/m^2		kilogram per square meter	Mass per unit area of slabs, plates, and similar items of uniform thickness or depth; rating for load-carrying capacities on floors (display on notices only)*	*DO NOT USE in stress calculations
		g/m^2	gram per square meter	Mass per unit area of thin sheet materials, coatings, etc.	
MASS DENSITY, CONCENTRATION (kg/m^3)	kg/m^3		kilogram per cubic meter	Density of materials in general; mass per unit volume of materials in a concrete mix; evaluation of masses of structures and materials	Also known as "Mass per Unit Volume" (1 kg/m^3 = 1 g/L) (1 g/m^3 = 1 mg/L)
		g/m^3	gram per cubic meter	Mass per unit volume (concentration) in pollution control	
		$\mu g/m^3$	microgram per cubic meter		
MOMENTUM (kg·m/s)	kg·m/s		kilogram meter per second	Used in applied mechanics; evaluation of impact and dynamic forces	
MOMENT OF INERTIA ($kg \cdot m^2$)	$kg \cdot m^2$		kilogram square meter	Rotational dynamics. Evaluation of the retraining forces required for propellers, windmills, etc.	See also Section 9.11
MASS PER UNIT TIME (kg/s)	kg/s		kilogram per second	Rate of transport of material on conveyors and other materials handling equipment	1 kg/s = 3.6 t/h
		t/h	metric ton per hour		
FORCE (N)	N		newton	Unit of force for use in calculations	1 N = 1 $kg \cdot m/s^2$ See also Section 9.8
	kN		kilonewton	Forces in structural elements, such as columns, piles, ties, pre-stressing tendons, etc.; concentrated forces; axial forces; reactions; shear force; gravitational force	
FORCE PER UNIT LENGTH (N/m)	N/m		newton per meter	Unit for use in calculations	
	kN/m		kilonewton per meter	Transverse force per unit length on a beam, column, etc.; force distribution in a linear direction	
MOMENT OF FORCE TORSIONAL OR BENDING MOMENT, (N·m)	N·m		newton meter	Bending moments (in structural sections), torsional moment; overturning moment; tightening tension for high strength bolts; torque in engine drive shafts, axles, etc.	See also Sections 9.10.4 and 9.11
	kN·m		kilonewton meter		
	MN·m		meganewton meter		

*Metre is the preferred international spelling applied to this base unit and all of its multiples and submultiples.

FIG. WF9 MECHANICS: STATICS AND DYNAMICS* *(Continued)*

QUANTITY AND SI UNIT SYMBOL	PREFERRED UNITS (SYMBOLS)	OTHER ACCEPTABLE UNITS	UNIT NAME	TYPICAL APPLICATIONS	REMARKS
PRESSURE, STRESS, MODULUS OF ELASTICITY (Pa)	Pa		pascal	Unit for use in calculations; low differential pressure in fluids	($1 \text{ Pa} = 1 \text{ N/m}^2$)
	kPa		kilopascal	Uniformly distributed pressure (loads) on floors; soil bearing pressure; wind pressure (loads), snow loads, dead and live loads; pressure in fluids; differential pressure (e.g., in ventilating systems)	Where wind pressure, snow loads, dead and live loads are shown in kN/m^2 CHANGE units to kPa
	MPa		megapascal	Modulus of elasticity; stress (ultimate, proof, yield, permissible, calculated, etc.) in structural materials; concrete and steel strength grades	$1 \text{ MPa} = 1 \text{ MN/m}^2 = 1 \text{ N/mm}^2$
	GPa		gigapascal	Modulus of elasticity in high strength materials	
	µPa		micropascal	Sound pressure (20 µPa is the reference quantity for sound pressure level)	
COMPRESSIBILITY (Pa^{-1})	1/Pa		reciprocal (of) pascal	Settlement analysis, (coefficient of compressibility), bulk compressibility	($1/\text{Pa} = 1 \text{ m}^2/\text{N}$)
	1/kPa		reciprocal (of) kilopascal		
DYNAMIC VISCOSITY (Pa·s)	Pa·s		pascal second	Shear stresses in fluids	($1 \text{ Pa·s} = 1 \text{ Ns/m}^2$) The centipoise (cP) $= 10^{-3}$ Pa·s WILL NOT BE USED
	mPa·s		millipascal second		
KINEMATIC VISCOSITY (m^2/s)	m^2/s		square meter per second		The centistokes (cSt) $= 10^{-6} \text{m}^2/\text{s}$ WILL NOT BE USED $1 \text{ cSt} = 1 \text{ mm}^2/\text{s}$
	mm^2/s		square millimeter per second	Computation of Reynold's number, settlement analysis (coeff. of consolidation)	
WORK, ENERGY (J)	J		joule	Energy absorbed in impact testing of materials; energy in general; calculations involving mechanical and electrical energy	
	kJ		kilojoule		
	MJ		megajoule		
		kWh	kilowatthour	Electrical energy applications only	1 kWh = 3.6 MJ
IMPACT STRENGTH (J/m^2)	J/m^2		joule per square meter	Impact strength; impact ductility	
	kJ/m^2		kilojoule per square meter		
POWER (W)	W		watt	Power in general (mechanical, electrical, thermal); input/output rating, etc. of motors, engines, heating and ventilating plant and other equipment in general	
	kW		kilowatt		
	MW		megawatt	Power input/output rating etc. of heavy power plant	
	pW		picowatt	Sound power level (1 pW is the reference quantity for sound power level)	

*Metre is the preferred international spelling applied to this base unit and all of its multiples and submultiples.

FIG. WF10 **THERMAL EFFECTS, HEAT TRANSFER***

QUANTITY AND SI UNIT SYMBOL	PREFERRED UNITS (SYMBOLS)	OTHER ACCEPTABLE UNITS	UNIT NAME	TYPICAL APPLICATIONS	REMARKS
TEMPERATURE VALUE (K)	K		kelvin	Expression of thermodynamic temperature; calculations involving units of temperature	($t_{°C} = T_K - 273.15$)
		°C	degree Celsius	Common temperature scale for use in meteorology and general applications; ambient temperature values	Temperature values will normally be measured in °C (degrees Celsius)
TEMPERATURE INTERVAL (K)	K		kelvin	Heat transfer calculations; temperature intervals in test methods, etc.	(1 K = 1°C) The use of K (kelvin) in compound units is recommended
		°C	degree Celsius		
COEFFICIENT OF LINEAR THERMAL EXPANSION (1/K)	1/K		reciprocal (of) kelvin	Expansion of materials subject to a change in temperature (generally expressed as a ratio per kelvin or degree Celsius)	
		1/°C	reciprocal (of) degree Celsius		
HEAT, QUANTITY OF HEAT (J)	J		joule	Thermal energy calculations. Enthalpy, latent heat, sensible heat	
	kJ		kilojoule		
	MJ		megajoule		
SPECIFIC ENERGY, SPECIFIC LATENT HEAT; COMBUSTION HEAT (mass basis) (J/kg)	J/kg		joule per kilogram	Heat of transition; heat and energy contained in materials; combustion heat per unit mass; calorific value of fuels (mass basis); specific sensible heat, specific latent heat in psychrometric calculations	
	kJ/kg		kilojoule per kilogram		
	MJ/kg		megajoule per kilogram		
ENERGY DENSITY, COMBUSTION HEAT (Volume basis) (J/m^3)	J/m^3		joule per cubic meter	Combustion heat per unit volume	
	kJ/m^3		kilojoule per cubic meter		($1\ kJ/m^3 = 1\ J/L$)
	MJ/m^3		megajoule per cubic meter	Calorific value of fuels (volume basis)	($1\ MJ/m^3 = 1\ kJ/L$)
HEAT CAPACITY, ENTROPY (J/K)	J/K		joule per kelvin	Thermal behavior of materials, heat transmission calculations, entropy	
	kJ/K		kilojoule per kelvin		
SPECIFIC HEAT CAPACITY, SPECIFIC ENTROPY (J/(kg·K))	J/(kg·K)		joule per kilogram kelvin	Thermal behavior of materials, heat transmission calculations	
	kJ/(kg·K)		kilojoule per kilogram kelvin		
HEAT FLOW RATE (W)	W		watt	Heat flow rate through walls, windows, etc.; heat demand	(1 W = 1 J/s)
	kW		kilowatt		
POWER DENSITY, HEAT FLUX DENSITY, IRRADIANCE (W/m^2)	W/m^2		watt per square meter	Density of power or heat flow through building walls and other heat transfer surfaces; heat transmission calculations	
	kW/m^2		kilowatt per square meter		

*Metre and litre are the preferred international spellings applied to these units and all of their multiples and submultiples.

FIG. WF10 **THERMAL EFFECTS, HEAT TRANSFER*** ***(Continued)***

QUANTITY AND SI UNIT SYMBOL	PREFERRED UNITS (SYMBOLS)	OTHER ACCEPTABLE UNITS	UNIT NAME	TYPICAL APPLICATIONS	REMARKS
HEAT RELEASE RATE	W/m^3		watt per cubic meter	Rate of heat release per unit volume over time (for gases and liquids)	($W/m^3 = J/(m^3 \cdot s)$)
(W/m^3)	kW/m^3		kilowatt per cubic meter		
THERMAL CONDUCTIVITY, ($W/(m \cdot K)$)	$W/(m \cdot K)$		watt per meter kelvin	Estimation of thermal behavior of materials and systems; heat transmission calculations Thermal conductivity of structural and building materials in fire-resistance testing, insulation, etc.	$1\ W/(m \cdot K) = 1\ W/(m \cdot °C)$ (k value)
COEFFICIENT OF HEAT TRANSFER, (THERMAL CONDUCTANCE)	$W/(m^2 \cdot K)$		watt per square meter kelvin	Heat transfer calculations for buildings, building components and equipment. Transmittance of construction elements	(U value)
($W/(m^2 \cdot K)$)	$kW/(m^2 \cdot K)$		kilowatt per square meter kelvin		
THERMAL RESISTIVITY $(m \cdot K)/W$	$(m \cdot K)/W$		meter kelvin per watt	Heat transmission calculations (reciprocal of thermal conductivity)	
THERMAL INSULANCE, (THERMAL RESISTANCE) $(m^2 \cdot K)/W$	$(m^2 \cdot K)/W$		square meter kelvin per watt	Heat transmission calcuations (reciprocal of thermal conductance)	(R value)

*Metre and litre are the preferred international spellings applied to these units and all of their multiples and submultiples.

FIG. WF11 **ELECTRICITY AND MAGNETISM***

QUANTITY AND SI UNIT SYMBOL	PREFERRED UNITS (SYMBOLS)	OTHER ACCEPTABLE UNITS	UNIT NAME	TYPICAL APPLICATIONS	REMARKS
ELECTRIC CURRENT (A)	A		ampere	Maintenance rating of an electrical installation. Leakage current	
	kA mA μA		kiloampere milliampere microampere		
MAGNETOMOTIVE FORCE, MAGNETIC POTENTIAL DIFFERENCE (A)				Used in the calculations involved in magnetic circuits	
MAGNETIC FIELD STRENGTH, MAGNETIZATION (A/m)	A/m		ampere per meter	Magnetic field strength used in calculation of magnetic circuitry such as transformers, magnetic amplifiers, and general cores	(1 kA/m = 1 A/mm)
	kA/m		kiloampere per meter		
CURRENT DENSITY (A/m^2)	A/m^2		ampere per square meter	Design of cross-sectional area of electrical conductor	
	kA/m^2		kiloampere per square meter		
		A/mm^2	ampere per square millimeter		($1\ A/mm^2 = 1\ MA/m^2$)
ELECTRIC CHARGE, QUANTITY OF ELECTRICITY (C)	C		coulomb	The voltage on a unit with capacitive type characteristics may be related to the amount of charge present (e.g. electrostatic precipitators). Storage battery capacities	1 C = 1 A·s DO NOT USE ampere hour: 1 A·h = 3.6 kC
	kC μC nC pC		kilocoulomb microcoulomb nanocoulomb picocoulomb		
ELECTRIC POTENTIAL, POTENTIAL DIFFERENCE, ELECTROMOTIVE FORCE (V)	V		volt		1 V = 1 W/A
	MV kV mV μV		megavolt kilovolt millivolt microvolt		
ELECTRIC FIELD STRENGTH (V/m)	V/m		volt per meter	The electric field strength gives the potential gradient at points in space. This may be used to calculate or test electrical parameters such as dielectric strength.	
	MV/m kV/m mV/m μV/m		megavolt per meter kilovolt per meter millivolt per meter microvolt per meter		
ACTIVE POWER (W)	W		watt	The useful power of an electrical circuit is expressed in 'watts' (W). (The apparent power in an electrical circuit is expressed in 'volt-amperes', (V·A)).	1 W = 1 V·A
	GW MW kW mW μW		gigawatt megawatt kilowatt milliwatt microwatt		
CAPACITANCE (F)	F		farad	Electronic components. Electrical design and performance calculators.	1 F = 1 C/V
	mF μF nF pF		millifarad microfarad nanofarad picofarad		

*Metre is the preferred international spelling applied to this base unit and all of its multiples and submultiples.

FIG. WF11 **ELECTRICITY AND MAGNETISM*** ***(Continued)***

QUANTITY AND SI UNIT SYMBOL	PREFERRED UNITS (SYMBOLS)	OTHER ACCEPTABLE UNITS	UNIT NAME	TYPICAL APPLICATIONS	REMARKS
RESISTANCE (Ω)	Ω GΩ MΩ kΩ mΩ		ohm gigaohm megohm kilohm milliohm	The design of electrical devices with resistance, such as motors, generators, heaters, electrical distribution systems, etc.	1 Ω = 1 V/A
CONDUCTANCE, ADMITTANCE SUSCEPTANCE (S)	S MS kS mS μS		siemens megasiemens kilosiemens millisiemens microsiemens		The siemens (S) was formerly known as mho
RESISTIVITY (Ω·m)	Ω·m GΩ·m MΩ·m kΩ·m mΩ·m μΩ·m nΩ·m		ohm meter gigaohm meter megohm meter kilohm meter milliohm meter microohm meter nanoohm meter		
(ELECTRICAL) CONDUCTIVITY (S/m)	S/m MS/m kS/m μS/m		siemens per meter megasiemens per meter kilosiemens per meter microsiemens per meter	A parameter for measuring water quality.	
MAGNETIC FLUX, FLUX OF MAGNETIC INDUCTION (Wb)	mWb		milliweber	Used in the calculations involved in magnetic circuits.	1 Wb = 1 V·s
MAGNETIC FLUX DENSITY, MAGNETIC INDUCTION (T)	T mT μT nT		tesla millitesla microtesla nanotesla	Used in the calculations involved in magnetic circuits.	1 T = 1 Wb/m^2
MAGNETIC VECTOR POTENTIAL (Wb/m^2)	kWb/m^2		kiloweber per square meter	Used in the calculations involved in magnetic circuits.	
SELF INDUCTANCE, MUTUAL INDUCTANCE, PERMEANCE (H)	H mH μH nH pH		henry millihenry microhenry nanohenry picohenry	Used in analysis and calculations involving transformers.	1 H = 1 Wb/A
RELUCTANCE (1/H)	1/H		reciprocal of henry	Design of motors and generators	
PERMEABILITY (H/m)	H/m μH/m nH/m		henry per meter microhenry per meter nanohenry per meter	Permeability gives the relationship between the magnetic flux density and the magnetic field strength.	

*Metre is the preferred international spelling applied to this base unit and all of its multiples and submultiples.

FIG. WF12 **LIGHTING***

QUANTITY AND SI UNIT SYMBOL	PREFERRED UNITS (SYMBOLS)	OTHER ACCEPTABLE UNITS	UNIT NAME	TYPICAL APPLICATIONS	REMARKS
LUMINOUS INTENSITY (cd)	cd		candela		
SOLID ANGLE (sr)	sr		steradian		
LUMINOUS FLUX (lm)	lm klm		lumen kilolumen	Luminous flux of light sources, lamps and light bulbs	1 lm = 1 cd·sr Already in general use
QUANTITY OF LIGHT (lm·s)	lm·s	lm·h	lumen second lumen hour		1 lm·h = 3600 lm/s
LUMINANCE (cd/m^2)	cd/m^2 kcd/m^2	cd/mm^2	candela per square meter kilocandela per square meter candela per square millimeter	Assessment of surface brightness; luminance of light sources, lamps and light bulbs; calculation of glare in lighting layouts	Replaces stilb (1 sb = 10^4 cd/m^2) and apostilb (1 apostilb = $cd/\pi m^2$)
ILLUMINANCE (lx)	lx klx		lux kilolux	Luminous flux per unit area used in determination of illumination levels and design/evaluation of interior lighting layouts. (Outdoor daylight illumination on a horizontal plane ranges up to 100 klx)	a) Formerly referred to as illumination 1 lx = 1 lm/m^2 b) Replaces (1 ph = 10^4 lx) c) Luminous exitance is described in lm/m^2
LIGHT EXPOSURE (lx·s)	lx·s klx·s		lux second kilolux second		
LUMINOUS EFFICACY (lm/W)	lm/W		lumen per watt	Rating of luminous efficacy of artificial light sources	

*Metre is the preferred international spelling applied to this base unit and all of its multiples and submultiples.

FIG. WF13 ACOUSTICS*

QUANTITY AND SI UNIT SYMBOL	PREFERRED UNITS (SYMBOLS)	OTHER ACCEPTABLE UNITS	UNIT NAME	TYPICAL APPLICATIONS	REMARKS
WAVE LENGTH (m)	m mm		meter millimeter	Definition of sound wave pitch	
AREA OF ABSORPTIVE SURFACE (m^2)	m^2		square meter	Calculations of room absorption	
PERIOD, PERIODIC TIME (s)	s ms		second millisecond	Measurement of time and reverberation time	
FREQUENCY (Hz)	Hz kHz		hertz kilohertz	Frequency ranges in sound absorption calculations and sound pressure measurement	1 Hz = 1 cycle per second (cps)
(INSTANTANEOUS) SOUND PRESSURE (Pa)	Pa mPa μPa		pascal millipascal micropascal	Measurement of sound pressure; reference level for sound pressure is 20 μPa, but sound pressure is shown in decibels (dB) based upon a logarithmic scale Sound pressure level L_p = $= 20 \log_{10} \frac{\text{actual pressure (Pa)}}{20 \times 10^{-6} \text{ (Pa)}}$	Do NOT USE dyne (1 dyn = 10 μPa)
SOUND POWER, SOUND ENERGY FLUX (W)	W mW μW pW		watt milliwatt microwatt picowatt	Measurement of sound power; reference level for sound power is 1 pW Sound power level, L = $= 10 \log_{10} \frac{\text{actual power (W)}}{10^{-12} \text{ (W)}}$	
SOUND INTENSITY (W/m^2)	W/m^2 pW/m^2		watt per square meter picowatt per square meter	Measurement of sound intensity; reference level for sound intensity is 1 pW/m^2 Sound intensity level L_I = $= 10 \log_{10} \frac{\text{actual intensity } (W/m^2)}{10^{-12} \ (W/m^2)}$ dB	
SPECIFIC ACOUSTIC IMPEDANCE (Pa·s/m)	Pa·s/m		pascal second per meter	Sound impedance measurement	(1 Pa·s/m = 1 N·s/m^3)
ACOUSTIC IMPEDANCE, RESISTANCE (Pa·s/m^3)	Pa·s/m^3		pascal second per cubic meter	Sound impedance measurement	

*Metre is the preferred international spelling applied to this base unit and all of its multiples and submultiples.

(continued from WF603-5).

For calculations using numbers of differing precision, the final answer should be no more precise than the least precise number.

FIG. WF14 FRACTION, DECIMAL AND METRIC EQUIVALENTS*

Fractional Inch	Dec. Equiv.	Millimetres	Fractional Inch	Dec. Equiv.	Millimetres
1/64	0.0156	0.397	17/32	0.5313	13.494
1/32	0.0313	0.794	35/64	0.5469	13.891
3/64	0.0469	1.191	9/16	0.5625	14.288
1/16	0.0625	1.588	37/64	0.5781	14.684
5/64	0.0781	1.984	7/12	0.5833	14.817
1/12	0.0833	2.117	19/32	0.5938	15.081
3/32	0.0938	2.381	3/5	0.6000	15.240
1/10	0.1000	2.540	39/64	0.6094	15.478
7/64	0.1094	2.778	5/8	0.6250	15.875
1/8	0.1250	3.175	41/64	0.6406	16.272
9/64	0.1406	3.572	21/32	0.6563	16.669
1/6	0.1667	4.233	2/3	0.6667	16.933
11/64	0.1719	4.366	43/64	0.6719	17.066
3/16	0.1875	4.700	11/16	0.6875	17.463
1/5	0.2000	5.080	7/10	0.7000	17.780
13/64	0.2031	5.159	45/64	0.7031	17.859
7/32	0.2188	5.556	23/32	0.7188	18.256
15/64	0.2344	5.953	47/64	0.7344	18.653
1/4	0.2500	6.350	3/4	0.7500	19.050
17/64	0.2656	6.747	49/64	0.7656	19.447
9/32	0.2813	7.144	25/32	0.7813	19.844
19/64	0.2969	7.541	51/64	0.7969	20.241
3/10	0.3000	7.620	4/5	0.8000	20.320
5/16	0.3125	7.937	13/16	0.8125	20.638
1/3	0.3333	8.467	53/64	0.8281	21.034
11/32	0.3438	8.731	5/6	0.8333	21.167
23/64	0.3594	9.128	27/32	0.8438	21.431
3/8	0.3750	9.525	55/64	0.8594	21.828
25/64	0.3906	9.922	7/8	0.8750	22.225
2/5	0.4000	10.160	57/64	0.8906	22.622
13/32	0.4063	10.319	9/10	0.9000	22.860
5/12	0.4167	10.583	29/32	0.9063	23.019
27/64	0.4219	10.716	11/12	0.9167	23.283
7/16	0.4375	11.112	59/64	0.9219	23.416
29/64	0.4531	11.509	15/16	0.9375	23.813
15/32	0.4688	11.906	61/64	0.9531	24.209
31/64	0.4844	12.303	31/32	0.9688	24.606
1/2	0.5000	12.700	63/64	0.9844	25.003
33/64	0.5156	13.097	1"	1.0000	25.400

*To convert a decimal to percentage, carry the decimal point two places to the right. Thus, 63/64, or .9844, equals 98.44%.

Soft Conversion and Hard Conversion

Soft conversion represents only a change in the way a product's size is described but no change in the product's actual size. This means that soft converted products are in production now. Hard conversion represents a definite change to a new preferred value (Fig. WF15).

For example, a piece of gypsum lath which is 24" wide could be described as 609.6 mm if exact conversion is required or the number could be rounded to 610 mm which is well within existing tolerances. Such practice would give the appearance of a metric conversion but would not result in the most ideal product width—600 mm—attained through hard conversion.

Rationalization

Rationalization is the development of an optimum assortment of product sizes and is a byproduct of metrication. It results in changes in products and standards.

An example of rationalization due to metrication is the change to fewer metric concrete reinforcing bar sizes: from 11 to 10 in Australia, 11 to 9 in Britain and 11 to 8 in Canada. (See 603 Metrication Main Text for further discussion).

PREFERRED DIMENSIONS AND DIMENSIONAL COORDINATION

During the change to SI, it is important to select preferred dimensions for building products which will insure that they will work together when combined into building assemblies. The use of preferred values in design or manufacture is termed dimensional (modular) coordination.

For example, 4" is the most common building module in the customary system. The width of a wood stud is *(Text continues on page WF603-25.)*

FIG. WF15 EXACT CONVERSION, SOFT CONVERSION, AND HARD CONVERSION OF CUSTOMARY VALUES

Customary Value	Exact Conversion	Soft Conversion	Hard Conversion	% Change
24 inches [2 feet]	609.6 mm	610 mm	600 mm	−1.6
1 square [100 sq. ft.]	9.2903 m^2	9.3 m^2	10 m^2	+7.6
4 cu. yd.	3.058 m^3	3.1 m^3	3 m^3	−1.9
1 quart	0.946 L	0.95 L	1 L	+5.7

FIG. WF16 CONVERSION FACTORS FOR THE MOST COMMON UNITS USED IN BUILDING DESIGN AND CONSTRUCTION*

METRIC TO CUSTOMARY			CUSTOMARY TO METRIC		
LENGTH					
1 km	= 0.621 371	mile (international)	1 mile (international)	= 1.609 344	km
	= 49.7096	chain	1 chain	= 20.1168	m
1 m	= 1.093 61	yd	1 yd	= 0.9144	m
	= 3.280 84	ft	1 ft	= 0.3048	m
1 mm	= 0.039 370 1	in		= 304.8	mm
			1 in	= 25.4	mm
			(1 U.S. survey foot	= 0.304 800 6	m)*
AREA					
$1\ km^2$	= 0.386 101	$mile^2$ (U.S. survey)	$1\ mile^2$ (U.S. survey)	= 2.590 00	km^2
1 ha	= 2.471 04	acre (U.S. survey)	1 acre (U.S. survey)	= 0.404 687	ha
$1\ m^2$	= 1.195 99	yd^2		= 4046.87	m^2
	= 10.7639	ft^2	$1\ yd^2$	= 0.836 127	m^2
$1\ mm^2$	= 0.001 550	in^2	$1\ ft^2$	= 0.092 903	m^2
			$1\ in^2$	= 645.16	mm^2
VOLUME, MODULUS OF SECTION					
$1\ m^3$	= $0.810\ 709 \times 10^{-3}$	acre feet	1 acre ft	= 1233.49	m^3
	= 1.307 95	yd^3	$1\ yd^3$	= 0.764 555	m^3
	= 35.3147	ft^3	100 board ft	= 0.235 974	m^3
	= 423.776	board ft	$1\ ft^3$	= 0.028 316 8	m^3
$1\ mm^3$	= 61.0237×10^{-6}	in^3		= 28.3168	L (dm^3)
			$1\ in^3$	= 16 387.1	mm^3
				= 16.3871	mL (cm^3)
(FLUID) CAPACITY					
1 L	= 0.035 314 7	ft^3	1 gal (U.S. liquid)**	= 3.785 41	L
	= 0.264 172	gal (U.S.)	1 qt (U.S. liquid)	= 946.353	mL
	= 1.056 69	qt (U.S.)	1 pt (U.S. liquid)	= 473.177	mL
1 mL	= 0.061 023 7	in^3	1 fl oz (U.S.)	= 29.5735	mL
	= 0.033 814	fl oz (U.S.)			
			** 1 gal(UK) approx. 1.2 gal(U.S.)		
SECOND MOMENT OF AREA					
$1\ mm^4$	= $2.402\ 51 \times 10^{-6}$	in^4	$1\ in^4$	= 416 231	mm^4
				= $0.416\ 231 \times 10^{-6}$	m^4
PLANE ANGLE					
1 rad	= 57° 17' 45"	(degree)	1° (degree)	= 0.017 453 3	rad
	= 57.2958°	(degree)		= 17.4533	mrad
	= 3437.75'	(minute)	1' (minute)	= 290.888	µrad
	= 206 265"	(second)	1" (second)	= 4.848 14	µrad
VELOCITY, SPEED					
1 m/s	= 3.280 84	ft/s	1 ft/s	= 0.3048	m/s
	= 2.236 94	mile/h	1 mile/h	= 1.609 344	km/h
1 km/h	= 0.621 371	mile/h		= 0.447 04	m/s
ACCELERATION					
$1\ m/s^2$	= 3.280 84	ft/s^2	$1\ ft/s^2$	= 0.3048	m/s^2

*Conversion factors are taken to six significant figures, where appropriate. Underlined values denote exact conversions.

FIG. WF16 **CONVERSION FACTORS FOR THE MOST COMMON UNITS USED IN BUILDING DESIGN AND CONSTRUCTION* *(Continued)***

METRIC TO CUSTOMARY			CUSTOMARY TO METRIC		
VOLUME RATE OF FLOW					
1 m³/s	= 35.3147	ft³/s	1 ft³/s	= 0.028 316 8	m³/s
	= 22.8245	million gal/d	1 ft³/min	= 0.471 947	L/s
	= 0.810 709 x 10⁻³	acre ft/s	1 gal/min	= 0.063 090 2	L/s
1 L/s	= 2.118 88	ft³/min	1 gal/h	= 1.051 50	mL/s
	= 15.850 3	gal/min	1 million gal/d	= 43.8126	L/s
	= 951.022	gal/h	1 acre ft/s	= 1233.49	m³/s
TEMPERATURE INTERVAL					
1 °C	= <u>1 K</u>	= <u>1.8</u> °F	1 °F	= 0.555 556	°C or K
				= 5/9 °C = 5/9 K	
EQUIVALENT TEMPERATURE VALUE ($t_{°C} = T_K - 273.15$)					
$t_{°C}$	= 5/9 (t_F − 32)		t_F	= 9/5 $t_{°C}$ + 32	
MASS					
1 kg	= 2.204 62	lb (avoirdupois)	1 ton (short)***	= 0.907 185	metric ton
	= 35.2740	oz (avoirdupois)		= 907.185	kg
1 metric ton	= 1.102 31	ton (short, 2000 lb)	1 lb	= 0.453 592	kg
	= 2204.62	lb	1 oz	= 28.3495	g
1 g	= 0.035 274	oz	1 pennyweight	= 1.555 17	g
	= 0.643 015	pennyweight			
			***(1 long ton (2240 lb)	= 1016.05	kg)
MASS PER UNIT LENGTH					
1 kg/m	= 0.671 969	lb/ft	1 lb/ft	= 1.488 16	kg/m
1 g/m	= 3.547 99	lb/mile	1 lb/mile	= 0.281 849	g/m
MASS PER UNIT AREA					
1 kg/m²	= 0.204 816	lb/ft²	1 lb/ft²	= 4.882 43	kg/m²
1 g/m²	= 0.029 494	oz/yd²	1 oz/yd²	= 33.9057	g/m²
	= 3.277 06 x 10⁻³	oz/ft²	1 oz/ft²	= 305.152	g/m²
DENSITY (MASS PER UNIT VOLUME)					
1 kg/m³	= 0.062 428	lb/ft³	1 lb/ft³	= 16.0185	kg/m³
	= 1.685 56	lb/yd³	1 lb/yd³	= 0.593 276	kg/m³
1 t/m³	= 0.842 778	ton/yd³	1 ton/yd³	= 1.186 55	t/m³
MOMENT OF INERTIA					
1 kg·m²	= 23.7304	lb·ft²	1 lb·ft²	= 0.042 140 1	kg·m²
	= 3417.17	lb·in²	1 lb·in²	= 292.640	kg·mm²
MASS PER UNIT TIME					
1 kg/s	= 2.204 62	lb/s	1 lb/s	= 0.453 592	kg/s
1 t/h	= 0.984 207	ton/h	1 ton/h	= 1.016 05	t/h

*Conversion factors are taken to six significant figures, where appropriate. Underlined values denote exact conversions.

FIG. WF16 **CONVERSION FACTORS FOR THE MOST COMMON UNITS USED IN BUILDING DESIGN AND CONSTRUCTION*** ***(Continued)***

METRIC TO CUSTOMARY			CUSTOMARY TO METRIC		
FORCE					
1 MN	= 112.404	tonf (ton-force)	1 tonf (ton-force)	= 8.896 44	kN
1 kN	= 0.112 404	tonf	1 kip (1000 lbf)	= 4.448 22	kN
	= 224.809	lbf (pound-force)	1 lbf (pound-force)	= 4.448 22	N
1 N	= 0.224 809	lbf			
MOMENT OF FORCE, TORQUE					
1 N·m	= 0.737 562	lbf·ft	1 lbf·ft	= 1.355 82	N·m
	= 8.850 75	lbf·in	1 lbf·in	= 0.112 985	N·m
1 kN·m	= 0.368 781	tonf·ft	1 tonf·ft	= 2.711 64	kN·m
	= 0.737 562	kip·ft	1 kip·ft	= 1.355 82	kN·m
FORCE PER UNIT LENGTH					
1 N/m	= 0.068 521 8	lbf/ft	1 lbf/ft	= 14.5939	N/m
1 kN/m	= 0.034 260 9	tonf/ft	1 lbf/in	= 175.127	N/m
			1 tonf/ft	= 29.187 8	kN/m
PRESSURE, STRESS, MODULUS OF ELASTICITY (FORCE PER UNIT AREA) (1 Pa = 1 N/m²)					
1 MPa	= 0.072 518 8	tonf/in²	1 tonf/in²	= 13.7895	MPa
	= 10.4427	tonf/ft²	1 tonf/ft²	= 95.7605	kPa
	= 145.038	lbf/in²	1 kip/in²	= 6.894 76	MPa
1 kPa	= 20.8854	lbf/ft²	1 lbf/in²	= 6.894 76	kPa
			1 lbf/ft²	= 47.8803	Pa
WORK, ENERGY, HEAT (1 J = 1 N·m = 1 W·s)					
1 MJ	= 0.277 778	kWh	1 kWh (550 ft·lbf/s)	= 3.6	MJ
1 kJ	= 0.947 817	Btu	1 Btu (Int. Table)	= 1.055 06	kJ
1 J	= 0.737 562	ft·lbf		= 1055.06	J
			1 ft·lbf	= 1.355 82	J
POWER, HEAT FLOW RATE					
1 kW	= 1.341 02	hp (horsepower)	1 hp	= 0.745 700	kW
1 W	= 3.412 14	Btu/h		= 745.700	W
	= 0.737 562	ft·lbf/s	1 Btu/h	= 0.293 071	W
			1 ft·lbf/s	= 1.355 82	W
HEAT FLUX DENSITY					
1 W/m²	= 0.316 998	Btu/(ft²·h)	1 Btu/(ft²·h)	= 3.154 59	W/m²
COEFFICIENT OF HEAT TRANSFER					
1 W/(m²·K)	= 0.176 110	Btu/(ft²·h·°F)	1 Btu/(ft²·h·°F)	= 5.678 26	W/(m²·K)
THERMAL CONDUCTIVITY					
1 W/(m·K)	= 0.577 789	Btu/(ft·h·°F)	1 Btu/(ft·h·°F)	= 1.730 73	W/(m·K)

*Conversion factors are taken to six significant figures, where appropriate. Underlined values denote exact conversions.

FIG. WF16 CONVERSION FACTORS FOR THE MOST COMMON UNITS USED IN BUILDING DESIGN AND CONSTRUCTION* *(Continued)*

METRIC TO CUSTOMARY			CUSTOMARY TO METRIC		
CALORIFIC VALUE (MASS AND VOLUME BASIS)					
1 kJ/kg (1 J/g)	= 0.429 923	Btu/lb	1 Btu/lb	= 2.326 (= 2.326	kJ/kg (J/g)
1 kJ/m³	= 0.026 839 2	Btu/ft³	1 Btu/ft³	= 37.2589	kJ/m³
THERMAL CAPACITY (MASS AND VOLUME BASIS)					
1 kJ/(kg·K)	= 0.238 846	Btu/(lb·°F)	1 Btu/(lb·°F)	= 4.1868	kJ/(kg·K)
1 kJ/(m³·K)	= 0.014 910 7	Btu/(ft³·°F)	1 Btu/(ft³·°F)	= 67.0661	kJ/(m³·K)
ILLUMINANCE					
1 lx (lux)	= 0.092 903	lm/ft² (footcandle)	1 lm/ft² (footcandle)	= 10.7639	lx (lux)
LUMINANCE					
1 cd/m²	= 0.092 903	cd/ft²	1 cd/ft²	= 10.7639	cd/m²
	= 0.291 864	footlambert	1 footlambert	= 3.426 26	cd/m²
1 kcd/m²	= 0.314 159	lambert	1 lambert	= 3.183 01	kcd/m²

*Conversion factors are taken to six significant figures, where appropriate. Underlined values denote exact conversions.

(continued from WF603-21)
nominally 4"; the width of a brick and height of a block are 8"; the spacing of studs and the length of a block are 16"; the width of gypsum lath and insulating sheathing is 24" and so on.

The internationally recommended basic building module is 100 mm. It is small enough to provide flexibility in design, large enough to promote simplification of the number of component sizes and acceptable worldwide.

Advantages of Preferred Dimensions

The selection of preferred building dimensions, such as floor-to-floor height, will provide an incentive to produce standardized story height components for use in high-rise buildings.

Application of an industry-wide system of preferred dimensions will reduce construction cost and time; simplify communications, and give additional impetus to the use of computers for design and detailing.

The 100 mm Module

The 100 mm module has many advantages:

(1) It is 1.6% smaller than the customary 4" module.

(2) The millimetre as the working unit in building design and construction will make all multimodular dimensions visible immediately. For example, 6400 mm is exactly 64 modules.

(3) The modular square, 100 mm × 100 mm, has an area of 0.01 m². This relationship is useful in price-cost comparisons because the unit cost per modular square, in cents, has the same number as the unit cost per square metre in dollars. As an example, for carpet costing $15.00 per square metre, the cost per modular square would be 15¢.

FIG. WF17 SI STANDARDS

Number	Title
NBS Special Publication 504	Metric Dimensional Coordination—The Issues and Precedent
NBS Special Publication 530	Metrication in Building Design, Production and Construction
NBS Technical Note 990	The Selection of Preferred Metric Values for Design and Construction
NBS Technical Note 938	Recommended Practice for the Use of Metric (SI) Units in Building Design and Construction
ASTM E 380-79	Standard for Metric Practice
ANSI/ASTM E 621-78*	Standard Practice for the Use of Metric (SI) Units in Building Design and Construction
ASTM E 713-80	A Standard for Scales for Metric Building Drawings

*Based on NBS Technical Note 938.

(4) The modular cube, 100 mm × 100 mm × 100 mm represents a volume of exactly 1 litre (0.001 m^3); one litre of water, with a specific gravity of 1.0, has a mass of exactly 1 kilogram. This relationship is useful in many calculations. For example, a modular cube of concrete, with a specific gravity of 2.4, has a mass of 2.4 kilograms.

(5) One dimension, the millimetre, expresses all building dimensions, including tolerances and limitations enabling much greater accuracy.

INDUSTRY STANDARDS

Figure WF17 shows some of the standards and recommendations published by the National Bureau of Standards (NBS), the American National Standard Institute (ANSI) and the American Society for Testing and Materials (ASTM).

CONSTRUCTION VOLUME INDEX 690

a

b

c

d

e

f

g

p

q

t

u

v

y

z